S0-AXP-123

Work It Out
You *can* and *will* succeed in this course if you learn to solve problems and then practice frequently. The following features are there to help you along the way.

FINDING THE SOLUTION: A PROCEDURE

EXAMPLE 2   Solving Exponential Equations Using the Logarithms

OBJECTIVE
Solve exponential equations when both sides are not expressed with the same base.

EXAMPLE
Solve for x: $5 \cdot 2^{x-3} = 17$.

Step 1 Isolate the exponential expression on one side of the equation.

1. $2^{x-3} = \dfrac{17}{5} = 3.4$

Step 2 Take the common or natural logarithm of both sides.

2. $\ln 2^{x-3} = \ln(3.4)$

Step 3 Use the power rule, $\log_a M^r = r \log_a M$.

3. $(x - 3)\ln 2 = \ln(3.4)$

Step 4 Solve for the variable.

4. $x - 3 = \dfrac{\ln(3.4)}{\ln 2}$

$x = \dfrac{\ln(3.4)}{\ln 2} + 3 \approx 4.766$

Practice Problem 2   Solve for x: $7 \cdot 3^{x+1} = 11$.

Finding the Solution
Finding the Solution boxes present important multistep procedures in a two-column format, illustrating the numbered steps of a procedure with a worked example.

End-of-Section Exercises
Each section ends with three levels of exercises for you to practice the math and apply your understanding:

- **Basic Skills and Concepts**
- **Applying the Concepts**
- **Beyond the Basics**

A EXERCISES Basic Skills and Concepts

1. The domain of the function $y = \log_a x$ is _____, and its range is _____.
2. The logarithmic form $y = \log_a x$ is equivalent to the exponential form _____.
3. The logarithm with base 10 is called the _____ logarithm, and the logarithm with base e is called the _____ logarithm.
4. $a^{\log_a x} =$ _____, and $\log_a a^x =$ _____.
5. *True or False* The graph of $\log_a x$, $a > 0, a \neq 1$, is an increasing function.
6. *True or False* The graph of $y = \log_a x$, $a > 0$, and $a \neq 1$, has no horizontal asymptote.

In Exercises 31–40, evaluate each expression without using a calculator.

31. $\log_5 125$ 32. $\log_9 81$
33. $\log 10,000$ 34. $\log_3 \dfrac{1}{3}$
35. $\log_2 \dfrac{1}{8}$ 36. $\log_4 \dfrac{1}{64}$
37. $\log_3 \sqrt{27}$ 38. $\log_{27} 3$
39. $\log_{16} 2$ 40. $\log_5 \sqrt{125}$

In Exercises 41–52, solve each equation.

41. $\log_5 x = 2$
42. $\log_5 x = -2$

B EXERCISES Applying the Concepts

In Exercises 95–102, use the model $A = A_0 e^{kt}$.

95. **Doubling your money.** How long would it take to double your money if you invested P dollars at the rate of 8% compounded continuously?
96. **Investment goal.** How long would it take to grow your investment from $10,000 to $120,000 at the rate of 10% compounded continuously?
97. **Rate for doubling your money.** At what annual rate of return, compounded continuously, would your investment double in six years?
98. **Rate-of-investment goal.** At what annual rate of return, compounded continuously, would your investment grow from $8000 to $50,000 in 25 years?
99. **Population of Canada.** The population of Canada was 31.3 million in 2000 and 33.4 million in 2007. Determine the time from 2000 until Canada's population (a) doubles and (b) triples.
100. **Population of Canada.** In Exercise 99, find the time from 2000 until Canada's population reaches 50 million.
101. **Water contamination.** A chemical is spilled into a reservoir of pure water. The concentration of chemical in the contaminated water is 4%. In one month, 20% of the water in the reservoir is replaced with clean water.
   a. What will be the concentration of the contaminant one year from now?
   b. For water to be safe for drinking, the concentration of this contaminant cannot exceed 0.01%. How long will it be before the water is safe for drinking?
102. **Toxic chemicals in a lake.** In a lake, one-fourth of the water is replaced by clean water every year. Sixteen thousand cubic meters of soluble toxic chemical spill takes place in the lake. Let $T(n)$ represent the amount of toxin left after n years.
   a. Find a formula for $T(n)$.
   b. How much toxin will be left after 12 years?
   c. When will 80% of the toxin be eliminated?

C EXERCISES Beyond the Basics

107. Sketch the graph of $f(x) = \begin{cases} \ln x & \text{if } x > 0 \\ \ln(-x) & \text{if } x < 0 \end{cases}$.
108. Sketch the graph of $f(x) = \begin{cases} \ln(1-x) & \text{if } x < 1 \\ e^{-x} & \text{if } x \geq 1 \end{cases}$.
109. Find the domain of $f(x) = \log_3 \left(\dfrac{x-2}{x+1} \right)$.
110. Find the domain of $g(x) = \log_2 \left(\dfrac{x+3}{x-2} \right)$.
111. **Finding the domain.** Find the domain of each function.
   a. $f(x) = \log_2(\log_3 x)$
   b. $f(x) = \log(\ln(x-1))$
   c. $f(x) = \ln(\log(x-1))$
   d. $f(x) = \log(\log(\log(x-1)))$
112. **Finding the inverse.** Find the inverse of each function in Exercise 111.
113. **Present value of an investment.** Recall that if P dollars is invested in an account at an interest rate r compounded continuously, then the amount A (called the *future value of P*) in the account t years from now will be $A = P e^{rt}$. Solving the equation for P, we get $P = A e^{-rt}$. In this formulation, P is called the present value of the investment.
   a. Find the present value of $100,000 at 7% compounded continuously for 20 years.
   b. Find the interest rate r compounded continuously that is needed to have $50,000 be the present value of $75,000 in ten years.
114. Your uncle is 40 years old, and he wants to have an annual pension of $50,000 each year at age 65. What is the present value of his pension if the money can be invested at all times at a continuously compounded interest rate of
   a. 5%?
   b. 8%?
   c. 10%?

PRACTICE TEST A

1. Solve the equation $5^{-x} = 125$.
2. Solve the equation $\log_2 x = 5$.
3. State the range of $y = -e^x + 1$ and find the asymptote of its graph.
4. Evaluate $\log_2 \dfrac{1}{8}$.
5. Solve the exponential equation $\left(\dfrac{1}{4}\right)^{2-x} = 4$.
6. Evaluate $\log 0.001$ without using a calculator.
7. Rewrite the expression $\ln 3 + 5 \ln x$ in condensed form.
8. Solve the equation $5x^2 e^x - x^4 e^x = 0$.
9. Solve the equation $e^{2x} + e^x - 6 = 0$.
10. Rewrite the expression $\ln \dfrac{2x^3}{(x+1)^5}$ in expanded logarithmic form.
11. Evaluate $\ln e^{-5}$.
12. Give the equation for the graph obtained by shifting the graph of $y = \ln x$ three units up and one unit right.
13. Sketch the graph of $y = 3^{x-1} + 2$.
14. State the domain of the function $f(x) =$
15. Write $3\ln x + \ln(x^3 + 2) - \dfrac{1}{2}\ln(3x^2$ condensed form.
16. Solve the equation $\log x = \log 6 - \log$
17. Find x if $\log_x 9 = 2$.
18. Rewrite the expression $\ln \sqrt{2x^3 y^2}$ in ex rithmic form.
19. Suppose $15,000 is invested in a savings 7% interest per year. Write the formula in the account after t years if the interes quarterly.
20. Suppose the number of Hispanic people living in the United States is approximated by $H = 15,000 e^{0.02t}$, where $t = 0$ represents 1960. According to this model, about how many Hispanic people were living in the United States in 1980?

PRACTICE TEST B

1. Solve the equation $3^{-x} = 9$.
   a. $\{2\}$ b. $\{-2\}$ c. $\left\{\dfrac{1}{2}\right\}$ d. $\left\{-\dfrac{1}{2}\right\}$ e. $\{\ln 2\}$
2. Solve the equation $\log_5 x = 2$.
   a. $\{10\}$ b. $\{25\}$ c. $\left\{\dfrac{5}{2}\right\}$ d. $\left\{\dfrac{2}{5}\right\}$ e. $\{2\}$
3. State the range and asymptote of $y = e^{-x} - 1$.
   a. $(0, \infty); y = -1$ b. $(-1, \infty); y = -1$
   c. $(-\infty, \infty); y = -1$ d. $(-1, \infty); y = 1$
   e. $(\infty, 1); y = 1$
4. Evaluate $\log_4 64$.
   a. 16   b. 8   c. 2   d. 3   e. 4

5. Find the solution of the exponential equation $\left(\dfrac{1}{3}\right)^{1-x} = 3$.
   a. $\left\{-\dfrac{1}{3}\right\}$ b. $\left\{\dfrac{1}{3}\right\}$ c. $\{1\}$ d. $\{2\}$
6. Evaluate $\log 0.01$.
   a. -1.99999999 b. -2 c. 2   d. 100
7. Which of the following expressions is equivalent to $\ln 7 + 2 \ln x$?
   a. $\ln(7 + 2x)$ b. $\ln(7x^2)$ c. $\ln(9x)$ d. $\ln(14x)$
8. Solve the equation $3x^2 e^x + x^3 e^x = 0$.
   a. $\{-3\}$ b. $\{0, -3\}$ c. $\{0\}$ d. $\varnothing$

End-of-Chapter Material
Each chapter concludes with a **Summary of Definitions, Concepts and Formulas, Review Exercises,** two **Practice Tests,** and a **Cumulative Review** to prove your mastery of the concepts presented in the chapter.

Annotated Instructor's Edition

Precalculus

A UNIT CIRCLE APPROACH

Annotated Instructor's Edition

Precalculus
A UNIT CIRCLE APPROACH

J. S. Ratti
University of South Florida

Marcus McWaters
University of South Florida

Addison-Wesley

Boston San Francisco New York
London Toronto Sydney Tokyo Singapore Madrid
Mexico City Munich Paris Cape Town Hong Kong Montreal

Publisher:	Greg Tobin
Executive Editor:	Anne Kelly
Sr. Project Editor:	Joanne Dill
Associate Project Editor:	Leah Goldberg
Editorial Assistant:	Sarah Gibbons
Senior Managing Editor:	Karen Wernholm
Senior Production Supervisor:	Peggy McMahon
Text Designer:	Carolyn Deacy Design
Cover Designer:	Barbara T. Atkinson
Cover Photo:	Alberto Paredes/ age fotostock
Photo Researcher:	Beth Anderson
Senior Media Producer:	Ceci Fleming
Math XL Project Supervisor:	Edward Chappell
QA Manager, Assessment Content:	Marty Wright
Marketing Manager:	Katherine Greig
Marketing Coordinator:	Katherine Minton
Senior Media Buyer:	Ginny Michaud
Digital Assets Manager:	Marianne Groth
Senior Author Support/ Technology Specialist:	Joe Vetere
Rights and Permissions Advisor:	Shannon Barbe
Sr. Manufacturing Manager:	Carol Melville
Production Coordination, Technical Illustrations, and Composition:	Pre-Press PMG
Situational Art:	Scientific Illustrators

Note to Instructors: The Annotated Instructor's Edition contains answers to all exercises. Where space permits, answers appear on the same page as their corresponding exercises—either adjacent to the exercise or at the top or bottom of the exercise set—or answers appear in available space on the adjacent page. Where space does not permit answers to appear in the exercise set, those answers can he found in the back of the book.

NOTICE: This work is protected by U.S. copyright laws and is provided solely for the use of college instructors in reviewing course materials for classroom use. Dissemination or sale of this work, or any part (including on the World Wide Web), will destroy the integrity of the work and is not permitted. The work and materials from it should never be made available to students except by instructors using the accompanying text in their classes. All recipients of this work are expected to abide by these restrictions and to honor the intended pedagogical purposes and the needs of other instructors who rely on these materials.

The Library of Congress has already cataloged the Student Edition as follows:

Library of Congress Cataloging-in-Publication Data

Ratti, J. S.
    Precalculus : a unit circle approach / J. S. Ratti, Marcus McWaters.
        p. cm.
    ISBN 0-321-53709-2 (student ed.)—ISBN 0-321-56507-X (annotated instructor's ed.)
    1. Functions—Textbooks.   2. Trigonometry—Textbooks.   3. Algebra—Textbooks.
    I. McWaters, Marcus M.   II. Title.
QA331.R358 2009
515—dc22   2009005441

For permission to use copyrighted material, grateful acknowledgment has been made to the copyright holders on page C-1 in the back of the book, which is hereby made part of this copyright page.

Many of the designations used by manufacturers and sellers to distinguish their products are claimed as trademarks. Where those designations appear in this book, and Pearson Education was aware of a trademark claim, the designations have been printed in initial caps or all caps.

Copyright © 2010 Pearson Education, Inc. All rights reserved. No part of this publication may be reproduced, stored in a retrieval system, or transmitted, in any form or by any means, electronic, mechanical, photocopying, recording, or otherwise, without the prior written permission of the publisher. Printed in the United States of America. For information on obtaining permission for use of material in this work, please submit a written request to Pearson Education, Inc., Rights and Contracts Department, 501 Boylston Street, Suite 900, Boston, MA 02116, fax your request to 617-671-3447, or e-mail at http://www.pearsoned.com/legal/permissions.htm.

1 2 3 4 5 6 7 8 9 10—QWT—11 10 09

Addison-Wesley
is an imprint of

www.pearsonhighered.com

ISBN-13 978-0-321-56507-5
ISBN-10 0-321-56507-X

To Our Children

Shammi, Raju, Sumeet
and
Sharon, Renee

Foreword

Many challenges face today's precalculus students and instructors. Students arrive in this course with varying levels of comprehension from their previous courses. Students often resort to memorization in order to pass the course instead of truly learning the concepts presented. As a result, a textbook needs to get students to a common starting point and engage them in becoming active learners, without sacrificing the solid mathematics that is necessary for conceptual understanding. Instructors are faced with the task of producing students who understand precalculus, are prepared for the next step, and find mathematics useful and interesting. Our goal is to try to help students and instructors achieve all of this and more.

In this text, there is a strong emphasis on both concept development and real-life applications. The clearly explained, well-developed and in-depth coverage of topics such as functions, graphing, the difference quotient, and limiting processes provides thorough preparation for the study of calculus and will substantially improve all students' comprehension of precalculus. Just-in-time review throughout the text ensures that all students are beginning with the same foundation of algebra skills. Numerous applications are used to motivate students to apply the concepts and skills they learn in precalculus to other courses, including the physical and biological sciences, engineering, and economics, and to on-the-job and everyday problem solving. Students are given ample opportunities throughout this book to think about important mathematical ideas and to practice and apply algebraic skills.

Another goal was to create a book that clearly shows the relevance of the material that students are learning. Throughout the text, we emphasize the reason the material being covered is important and how it can be applied. By thoroughly developing mathematical concepts with clearly defined terminology, students see the *why* behind those concepts, paving the way for deeper understanding, better retention, less reliance on rote memorization, and (ultimately) more success.

This focus on students does not mean that we neglected instructors. As instructors ourselves, we know how essential it is to use a book in which you believe and that helps you teach mathematics. To that end, the level of the book was carefully selected so that the material would be accessible to students and provide them with an opportunity to grow. Our hope is that once you have looked through this textbook, you will see that we fulfilled our initial goals of writing for today's students as well as for you, the instructor.

Contents

CHAPTER 1

Graphs and Functions 1

CHAPTER 2

Polynomial and Rational Functions 107

CHAPTER 3

Exponential and Logarithmic Functions 188

CHAPTER 4

Trigonometric Functions 253

CHAPTER 5

CHAPTER 6

Systems of Equations and Inequalities 480

CHAPTER 10

Further Topics in Algebra 680

APPENDIX A

Review 751

Preface

Students begin precalculus classes with widely varying backgrounds. Some haven't taken a math course in several years and may need to spend time reviewing prerequisite topics, while others are ready to jump right in to new and challenging material. We have provided review material in the Appendix and in some of the early sections of other chapters in such a way that it can be used or omitted as is appropriate for your course. In addition, students may follow several different paths after completing a precalculus course. Many will continue their study of mathematics in courses such as finite mathematics, statistics, and calculus. For others, this course may be their last mathematics course. Responding to the current and future needs of all of these students was essential in creating this text. Overall, we present our content in a systematic way that illustrates how to study and what to review. We believe that if students use this textbook well, they will succeed in this course.

Features

CHAPTER OPENER Each chapter opener includes a description of applications relevant to the content of the chapter, a series of images related to these and other applications, and the list of topics that will be covered in the chapter. In one page, students see what they are going to learn and why they are learning it.

REVIEW On the first page of each section is a list of topics that students should review prior to starting the chapter. Section references with page numbers accompany the suggested review material so that students can readily find the material. The **Objectives** of the section are then clearly stated and numbered. Each numbered objective is paired with a similarly numbered subsection so that students can quickly find the section material for an objective.

APPLICATION The discussion in each section begins with a motivating anecdote or piece of information that is tied to an application problem. This problem is solved in an example later in the section, using the mathematics covered in the section.

SECTION OPENERS These section openers lend continuity to the section and its content, utilizing material from a variety of fields: the physical and biological sciences (including health sciences), economics, art and architecture, the history of mathematics, and more. Of special interest are contemporary topics such as the greenhouse effect and global warming, CAT scans, and computer graphics.

DEFINITIONS AND THEOREMS All are boxed for emphasis and titled for ease of reference, as are lists of rules and properties.

FIGURES All figures are titled to make it easy to identify what is being illustrated.

EXAMPLES The examples include a wide range of computational, conceptual, and modern applied problems carefully selected to build confidence, competency, and understanding. Every example has a title that indicates its purpose. For every example, clarifying side comments are provided for each step in the detailed solution.

PRACTICE PROBLEM All examples are followed by a Practice Problem for students to try so that they can check their understanding of the concept covered.

PROCEDURE BOXES These boxes, interspersed throughout the text, present important procedures in numbered steps. Special **Finding the Solution** boxes present important multistep procedures, such as the steps for doing synthetic division, in a two-column format. The steps of the procedure are given in the left column, and an example is worked, following these steps, in the right column. This approach provides students with a clear model with which they can compare when encountering difficulty in their work. These boxes are a part of the numbered examples.

MAIN FACTS These boxes summarize information related to equations and their graphs, such as those of the conic sections.

HISTORICAL NOTES When appropriate, historical notes appear in the margin, giving students information on key people or ideas in the history and development of mathematics. This information is included to add flavor to the subject matter.

TECHNOLOGY CONNECTIONS Although the use of graphing calculators is optional in this book, Technology Connections give students tips on using calculators to solve problems, check answers, and reinforce concepts.

WARNINGS appear as appropriate throughout the text to let students know of common errors and pitfalls that can trip them up in their thinking or calculations.

RECALL Periodically, students are reminded in a margin Recall note of a key idea they learned earlier in the text that will help them work through a current problem.

STUDY TIPS Students are given hints for handling newly introduced concepts.

BY THE WAY These marginal notes provide students with additional interesting information on nonessential topics to keep them engaged in the mathematics presented.

EXERCISES The heart of any textbook are its exercises. Knowing this, we made sure that the quantity, quality, and variety of exercises meet the needs of all students. The problems in each exercise set are carefully graded to strengthen the skills developed in the associated section. Exercises are divided into three categories: **A: Basic Skills and Concepts** (developing fundamental skills), **B: Applying the Concepts** (using the section's material to solve real-world problems), and **C: Beyond the Basics** (providing more challenging exercises that give students an opportunity to reach beyond the material covered in the section). Exercises are paired so that the even-numbered A Exercises closely follow the preceding odd-numbered exercises. All application exercises are titled and relevant to the topics of the section. The A and B Exercises are intended for a typical student, while the C Exercises, generally more theoretical in nature, are suitable for honors students, special assignments, or extra credit. **Critical Thinking** exercises, appearing as appropriate, are designed to develop students' higher-level thinking skills, and calculator problems are included where needed. Finally, **Group Projects** are provided at the end of many exercise sets so that students can work together to reinforce and extend one another's comprehension of the material.

END-OF-CHAPTER The chapter-ending material includes a **Summary of Definitions, Concepts, and Formulas; Review Exercises**; and **two Practice Tests**. The chapter summary, a brief description of key topics indicating where the material occurs in the text, encourages students to reread sections rather than memorize definitions out of context. The Review Exercises provide students with an opportunity to practice what they have learned in the chapter. Then students can take Practice Test A in the usual open-ended format and Practice Test B, covering the same topics, in multiple-choice format. All tests are designed to increase student comprehension and verify that students have mastered the skills and concepts in the chapter. Mastery of these materials should indicate a true comprehension of the chapter and the likelihood of success on the associated in-class examination.

CUMULATIVE REVIEW EXERCISES These exercises appear at the end of every chapter, starting with Chapter 2, to remind students that mathematics is not modular and that what is learned in the first part of the book will be useful in later parts of the book and on the final examination.

Supplements

Student's Solutions Manual

- By Beverly Fusfield
- Provides detailed worked-out solutions to the odd-numbered end-of-section and Chapter Review exercises and solutions to all of the Practice Problems, Practice Tests and Cumulative Review problems
- ISBN-13: 978-0-321-62106-1; ISBN-10: 0-321-62106-9

Graphing Calculator Manual

- By Darryl Nester, *Bluffton University*
- Provides instructions and keystroke operations for the TI-83/83+, TI-84+, TI-86, and TI-89
- Keyed directly to text Examples and Technology Connections
- ISBN-13: 978-0-321-62102-3; ISBN-10: 0-321-62102-6

Video Lectures on DVD-Rom with Optional Subtitles

- Videos feature Section Summaries and Example Solutions. Section Summaries cover key definitions and procedures for most sections. Example Solutions walk students through the detailed solution process for many examples in the textbook.
- There are over 25 hours of video instruction specifically filmed for this book, making it ideal for distance learning or supplemental instruction on your home computer or in a campus computer lab.
- Videos include optional subtitles in English and Spanish.
- ISBN-13: 978-0-321-62103-0; ISBN-10: 0-321-62103-4

A Review of Algebra

- By Heidi Howard, *Florida Community College at Jacksonville*
- Provides additional support for those students needing further algebra review
- ISBN-13: 978-0-201-77347-7; ISBN-10: 0-201-77347-3

Annotated Instructor's Edition

- Answers included on the same page beside the text exercises where possible for quick reference
- ISBN-13: 978-0-321-56507-5; ISBN-10: 0-321-56507-X

Instructor's Solutions Manual

- By Beverly Fusfield
- Complete solutions provided for all end-of-section exercises, including the Critical Thinking and Group Projects, Practice Problems, Chapter Review exercises, Practice Tests, and Cumulative Review problems
- ISBN-13: 978-0-321-62098-9; ISBN-10: 0-321-62098-4

Instructor's Testing Manual

- By James Lapp
- Includes diagnostic pretests, chapter tests, and additional test items, grouped by section, with answers provided
- (Available online within MyMathLab or from the Instructor Resource Center at www.pearsonhighered.com/irc)

TestGen®

- Enables instructors to build, edit, print, and administer tests
- Features a computerized bank of questions developed to cover all text objectives
- (Available for download from Pearson Education's online catalog: www.pearsonhighered.com/testgen)

PowerPoint Lecture Slides

- Features presentations written and designed specifically for this text, including figures and examples from the text
- (Available online within MyMathLab or from the Instructor Resource Center at www.pearsonhighered.com/irc)

Pearson Math Adjunct Support Center

The **Pearson Math Adjunct Support Center** (www.pearsontutorservices.com/math-adjunct.html) is staffed by qualified instructors with more than 50 years of combined experience at both the community college and university level. Assistance is provided for faculty in the following areas:

- Suggested syllabus consultation
- Tips on using materials packed with your book
- Book-specific content assistance
- Teaching suggestions, including advice on classroom strategies

Media Resources

MathXL ## MathXL® Online Course (access code required)

MathXL® is an online homework, tutorial, and assessment system that accompanies Pearson's textbooks in mathematics or statistics.

- **Interactive homework exercises**, correlated to your textbook at the objective level, are algorithmically generated for unlimited practice and mastery. Most exercises are free-response and provide guided solutions, sample problems, and learning aids for extra help.
- **Personalized Study Plan**, generated when students complete a test or quiz, indicates which topics have been mastered and links to tutorial exercises for topics students have not mastered.
- **Multimedia learning aids**, such as video lectures and animations, help students independently improve their understanding and performance.
- **Gradebook**, designed specifically for mathematics and statistics, automatically tracks students' results and gives you control over how to calculate final grades.
- **MathXL Exercise Builder** allows you to create static and algorithmic exercises for your online assignments. You can use the library of sample exercises as an easy starting point.
- **Assessment Manager** lets you create online homework, quizzes, and tests that are automatically graded. Select just the right mix of questions from the MathXL exercise bank, instructor-created custom exercises, and/or TestGen test items.

MathXL is available to qualified adopters. For more information, visit our website at www.mathxl.com, or contact your Pearson sales representative.

MathXL® Tutorials on CD (ISBN-13: 978-0-321-62101-6; ISBN-10: 0-321-62101-8)

This interactive tutorial CD-ROM provides algorithmically generated practice exercises that are correlated at the objective level to the exercises in the textbook. Every practice exercise is accompanied by an example and a guided solution designed to involve students in the solution process. Selected exercises may also include a video clip to help students visualize concepts. The software provides helpful feedback for incorrect answers and can generate printed summaries of students' progress.

MyMathLab ## MyMathLab® Online Course (access code required)

MyMathLab® is a text-specific, easily customizable online course that integrates interactive multimedia instruction with textbook content. MyMathLab gives you the tools you need to deliver all or a portion of your course online, whether your students are in a lab setting or working from home.

- **Interactive homework exercises**, correlated to your textbook at the objective level, are algorithmically generated for unlimited practice and mastery. Most exercises are free-response and provide guided solutions, sample problems, and learning aids for extra help.
- **Personalized Study Plan**, generated when students complete a test or quiz, indicates which topics have been mastered and links to tutorial exercises for topics students have not mastered.

- **Multimedia learning aids**, such as video lectures, animations, and a complete multimedia textbook, help students independently improve their understanding and performance.
- **Assessment Manager** lets you create online homework, quizzes, and tests that are automatically graded. Select just the right mix of questions from the MyMathLab exercise bank, instructor-created custom exercises, and/or TestGen test items.
- **Gradebook**, designed specifically for mathematics and statistics, automatically tracks students' results and gives you control over how to calculate final grades. You can also add offline (paper-and-pencil) grades to the gradebook.
- **MathXL Exercise Builder** allows you to create static and algorithmic exercises for your online assignments. You can use the library of sample exercises as an easy starting point.
- **Pearson Tutor Center** (www.pearsontutorservices.com) access is automatically included with MyMathLab. The Tutor Center is staffed by qualified math instructors who provide textbook-specific tutoring for students via toll-free phone, fax, email, and interactive Web sessions.

MyMathLab is powered by CourseCompass™, Pearson Education's online teaching and learning environment, and by MathXL®, our online homework, tutorial, and assessment system. MyMathLab is available to qualified adopters. For more information, visit www.mymathlab.com or contact your Pearson sales representative.

InterAct Math Tutorial Web site: www.interactmath.com

Get practice and tutorial help online! This interactive tutorial Web site provides algorithmically generated practice exercises that correlate directly to the exercises in the textbook. Students can retry an exercise as many times as they like with new values each time for unlimited practice and mastery. Every exercise is accompanied by an interactive guided solution that provides helpful feedback for incorrect answers, and students can view a worked-out sample problem that takes them through an exercise similar to the one they're working on.

Video Lectures on DVD-Rom with Optional Subtitles (ISBN-13: 978-0-321-62103-0; ISBN-10: 0-321-62103-4)

The video lectures for this text are available on DVD-Rom, making it easy and convenient for students to watch the videos from a computer at home or on campus. The videos feature an engaging team of mathematics instructors who present Section Summaries and Example Solutions. Section Summaries cover key definitions and procedures from most sections. Example Solutions walk students through the detailed solution process for many examples in the textbook. The format provides distance-learning students with comprehensive video instruction for most sections in the book, but also allows students needing only small amounts of review to watch instruction on a specific skill or procedure. The videos have optional text subtitles, which can be easily turned off or on for individual student needs. Subtitles are available in English and Spanish.

Acknowledgments

We would like to express our gratitude to the reviewers of this first edition, who provided such invaluable insights and comments. Their contributions helped shape the development of the text and carry out the vision stated in the preface.

Reviewers

Alison Ahlgren, *University of Illinois at Urbana–Champaign*
Mohammed Aslam, *Georgia Perimeter College*
Ratan Barua, *Miami-Dade College*
Sam Bazzi, *Henry Ford Community College*
Diane Burleson, *Central Piedmont Community College*
Melissa Cass, *State University of New York–New Paltz*
Charles Conrad, *Volunteer State Community College*
Baiqiao Deng, *Columbus State University*
Gene Garza, *Samford University*
Bobbie Jo Hill, *Coastal Bend College*
Yvette Janecek, *Coastal Bend College*
Mohammed Kazemi, *University of North Carolina at Charlotte*
David Keller, *Kirkwood Community College*
Rebecca Leefers, *Michigan State University*
Paul Morgan, *College of Southern Idaho*
Kathy Nickell, *College of DuPage*
Catherine Pellish, *Front Range Community College*
Betty Peterson, *Mercon County*
Marshall Ransom, *Georgia Southern University*
Dr. Traci Reed, *St. Johns River Community College*
Linda Reist, *Macomb Community College*
Jeri Rogers, *Seminole Community College–Oviedo Campus*
Jason Rose, *College of Southern Idaho*
Delphy Shaulis, *University of Colorado–Boulder*
Cindy Shaw, *Moraine Valley Community College*
Cynthia Sikes, *Georgia Southern University*
James Smith, *Columbia State Community College*
Jacqueline Stone, *University of Maryland–College Park*
Kay Stroope, *Phillips County Community College*
Jo Tucker, *Tarrant County College–Southeast*
Tom Worthing, *Hutchinson Community College*
Vivian Zabrocki, *Montana State University–Billings*

Contributors

Jeremy Alm, *Iowa State University*
Ratan Barua, *Miami-Dade College*
Abby Baumgardner, *Blinn College–Bryan*
Edward Bender, *Century College*
Michael Flom, *Normandale Community College*
Tom Hayes, *Montana State University*
Anna Katsoulis, *East Carolina University*
Nicole Lang, *North Hennepin Community College*
Connie Novicoff, *Metropolitan State College of Denver*
Paul Nuñez, *Mesa Community College*

Rita Marie O'Brien, *Navarro College*
William Radulovich, *Florida Community College at Jacksonville*
Jason Ramirez, *Highline Community College*
Donna Saye, *Georgia Southern University*
Carol Schmidt, *Lincoln Land Community College*
Julie Turnbow, *Collin County Community College–Preston*
Chock Wong, *Chaminade University*

Class Testers

Shana Funderburk, *University of North Carolina at Charlotte*
Bobbie Jo Hill, *Coastal Bend College*
Anna Katsoulis, *East Carolina University*
Marcus McWaters, *University of South Florida*
Jeff Norris, *Paris Junior College*
Paul Nuñez, *Mesa Community College*
William Radulovich, *Florida Community College–Jacksonville*
Julie Turnbow, *Collin County Community College–Preston*

Market Development

Our sincerest thanks go to the legion of dedicated individuals who worked tirelessly to make this book possible. We express special thanks to Elka Block for the excellent work she did as the development editor on the text and Frank Purcell for his work on the art development. We would also like to express our gratitude to our typist, Beverly DeVine-Hoffmeyer, for her amazing patience and skill. We must also thank Dr. Praveen Rohatgi, Dr. Nalini Rohatgi, and Dr. Bhupinder Bedi for the consulting they provided on all material relating to medicine. We particularly want to thank Professor Mile Krajcevski for many helpful discussions and suggestions, particularly for improving the exercise sets. Further gratitude is due to Irena Andreevska, Gokarna Aryal, Ferene Tookos, and Christine Fitch for their assistance on the answers to the exercises in the text. In addition, we would like to thank Beverly Fusfield, Douglas Ewert, Carrie Green, Tom Wegleitner, and Elka Block and Frank Purcell of Twin Prime Editorial for their meticulous accuracy in checking the text. Thanks are due as well to Laura Hakala and Pre-Press PMG for their excellent production work. Finally, our thanks are extended to the professional and remarkable staff at Addison-Wesley. In particular, we would like to thank Greg Tobin, Publisher; Anne Kelly, Executive Editor; Joanne Dill, Senior Project Editor; Leah Goldberg, Associate Project Editor; Peggy McMahon, Senior Production Supervisor; Katherine Greig, Marketing Manager; Katherine Minton, Marketing Assistant; Barbara Atkinson, Designer; Cecilia Fleming, Media Producer; Karen Wernholm, Senior Managing Editor; and Joseph Vetere, Senior Author/Technical Art Support.

We invite all who use this book to send suggestions for improvements to Marcus McWaters at mmm@cas.usf.edu.

About the Authors

J. S. Ratti

EDUCATION

PhD Mathematics Wayne State University

TEACHING

Wayne State University (teaching assistant and instructor)

University of Nevada at Las Vegas (instructor)

Oakland University (assistant professor)

University of South Florida (professor and past chairman)

Undergraduate courses taught: college algebra, trigonometry, finite mathematics, calculus (all levels), set theory, differential equations, linear algebra

Graduate courses taught: number theory, abstract algebra, real analysis, complex analysis, graph theory

AWARDS

USF Research Council Grant

USF Teaching Incentive Program (TIP) Award

USF Outstanding Undergraduate Teaching Award

Academy of Applied Sciences grants

RESEARCH

Complex analysis, real analysis, graph theory, probability

PERSONAL INTERESTS

Fan of Tampa Bay Buccaneers

Marcus McWaters

EDUCATION

BS Major: mathematics Minor: physics
Louisiana State University New Orleans

PhD Major: mathematics Minor: philosophy
University of Florida

TEACHING

Louisiana State University (teaching assistant)

University of Florida (teaching assistant and graduate fellow)

University of South Florida (associate professor and department chair)

Undergraduate courses taught: college algebra, trigonometry, finite mathematics, calculus, business calculus, life science calculus, advanced calculus, vector calculus, set theory, differential equations, linear algebra

Graduate courses taught: graph theory, number theory, discrete mathematics, geometry, advanced linear algebra, topology, mathematical logic, modern algebra, real analysis

AWARDS

Graduate Fellow, University of Florida (Center of Excellence Grant)

USF Research Council grant

USF Teaching Incentive Program (TIP) Award

Provost's Award

RESEARCH

Topology, algebraic topology, topological algebra

Founding member, USF Center for Digital and Computational Video

PERSONAL INTERESTS

Traveling with my wife and two daughters, theater, waterskiing, racquetball

Graphs and Functions

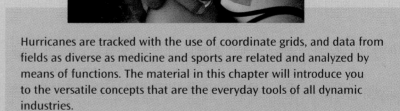

Hurricanes are tracked with the use of coordinate grids, and data from fields as diverse as medicine and sports are related and analyzed by means of functions. The material in this chapter will introduce you to the versatile concepts that are the everyday tools of all dynamic industries.

Graphs of Equations

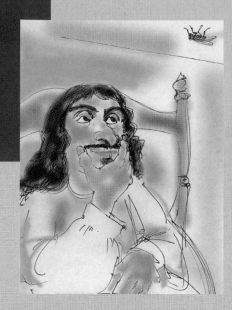

Before Starting this Section, Review

1. The number line (Appendix A, page 735)

2. The Pythagorean Theorem (Appendix A, page 790)

3. Equivalent equations (Appendix A, page 793)

4. Completing squares (Appendix A, page 811)

Objectives

1 Plot points in the Cartesian coordinate plane.

2 Find the distance between two points.

3 Find the midpoint of a line segment.

4 Sketch a graph by plotting points.

5 Find the intercepts of a graph.

6 Find the symmetries in a graph.

7 Find the equation of a circle.

A FLY ON THE CEILING

It is said that one day the French mathematician René Descartes noticed a fly buzzing around on a ceiling made of square tiles. He watched the fly for a long time and wondered how he could mathematically describe its location. Finally, he realized that he could describe the fly's position by its distance from the walls of the room. Descartes had just discovered the coordinate plane! In fact, the coordinate plane is sometimes called the Cartesian plane in his honor. The discovery led to the development of analytic geometry, the first mathematical blending of algebra and geometry.

Although the basic idea of graphing with coordinate axes dates all the way back to Apollonius in the second century B.C., Descartes, who lived in the 1600s, gets the credit for coming up with the two-axis system we use today. In Example 2, we will see how the Cartesian plane helps visualize data on smoking. ■

1 Plot points in the Cartesian coordinate plane.

The Coordinate Plane

A visually powerful device for exploring relationships between numbers is the Cartesian plane. A pair of real numbers in which the order is specified is called an **ordered pair** of real numbers. The ordered pair (a, b) has **first component** a and **second component** b. Two ordered pairs (x, y) and (a, b) are **equal**, and we write $(x, y) = (a, b)$, if and only if $x = a$ and $y = b$.

Just as the real numbers are identified with points on a line, called the *number line* or the *coordinate line*, the sets of ordered pairs of real numbers are identified with points on a plane called the **coordinate plane** or the **Cartesian plane**.

We begin with two coordinate lines, one horizontal and one vertical, that intersect at their zero points. The horizontal line (with positive numbers to the right) is usually called the **x-axis**, while the vertical line (with positive numbers up) is usually called the **y-axis**. Their point of intersection is called the **origin**. The x-axis and y-axis are called **coordinate axes**, and the plane they form is sometimes called the **xy-plane**. The axes divide the plane into four regions called **quadrants**, which are numbered as shown in Figure 1.1. The points on the axes themselves do not belong to any of the quadrants.

The notation $P(a, b)$, or $P = (a, b)$, designates the point P whose *first component* is a and whose *second component* is b. In an xy-plane, the first component, a, is called the **x-coordinate** of $P(a, b)$ and the second component, b, is called the

René Descartes

(1596–1650)

Descartes was born at La Haye, near Tours in southern France. He is often called the father of modern science. Descartes established a new, clear way of thinking about philosophy and science by accepting only those ideas that could be proved by or deduced from first principles. He took as his philosophical starting point the statement **Cogito ergo sum:** "I think; therefore, I am." Descartes made major contributions to modern mathematics, including the Cartesian coordinate system and the theory of equations.

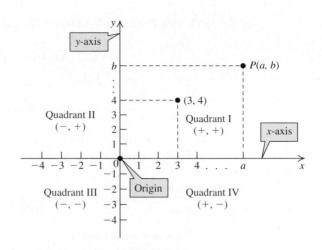

FIGURE 1.1 Quadrants in a plane

y-**coordinate** of $P(a, b)$. The signs of the x- and y-coordinates for each quadrant are shown in Figure 1.1. The point corresponding to the ordered pair (a, b) is called the **graph of the ordered pair (a, b)**. However, we frequently ignore the distinction between an ordered pair and its graph.

EXAMPLE 1   Graphing Points

Graph the following points in the xy-plane:

$A(3, 1)$, $B(-2, 4)$, $C(-3, -4)$, $D(2, -3)$, and $E(-3, 0)$.

SOLUTION

Figure 1.2 shows a coordinate plane, along with the graph of the given points. These points are located by moving left, right, up, or down starting from the origin $(0, 0)$.

$A(3, 1)$	3 units right, 1 unit up	$D(2, -3)$	2 units right, 3 units down
$B(-2, 4)$	2 units left, 4 units up	$E(-3, 0)$	3 units left, 0 units up
$C(-3, -4)$	3 units left, 4 units down		or down

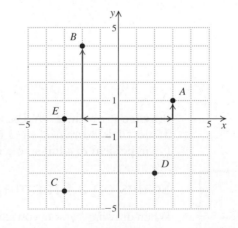

FIGURE 1.2 Graphing points

■ ■ ■

Practice Problem 1   Graph the points in the xy-plane:

$$P(-2, 2), Q(4, 0), R(5, -3), S(0, -3), \text{ and } T\left(-2, \frac{1}{2}\right). \qquad ■$$

EXAMPLE 2   **Graphing Data on Adult Smokers in the United States**

The data in Table 1.1 show the prevalence of smoking among adults aged 18 years and older in the United States over the years 1998–2004.

TABLE 1.1

Year	1998	1999	2000	2001	2002	2003	2004
Percent of adult smokers	24.1	23.5	23.2	22.7	22.4	21.6	20.9

Source: Center for Disease Control and Prevention, National Health Interview Survey.

   Graph the ordered pairs (year, percent of adult smokers), where the first coordinate represents a year and the second represents the percent of adult smokers in that year.

SOLUTION

We let x represent the years 1998 through 2004 and y represent the percent of adult smokers in each year. Since no data are given for the years 0 through 1997, we show a break in the x-axis. Alternately, we could declare a year—say, 1997—as 0. Similar comments apply to the y-axis. The graph of the points (1998, 24.1), (1999, 23.5), (2000, 23.2), and so on is shown in Figure 1.3. The graph shows that the percent of adult smokers has been declining every year since 1998.

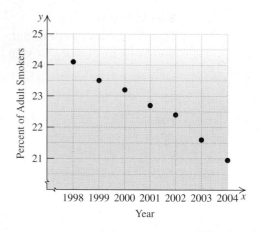

FIGURE 1.3 Declining percentage of smokers   ■ ■ ■

Practice Problem 2   In Example 2, suppose the percent of adult smokers shown in Table 1.1 is decreased by 3 in each year. Write the corresponding ordered pairs (year, percent of adult smokers).   ■

Scales on a Graphing Utility

When drawing a graph, you can use different scales for the x- and y-axes. Similarly, the scale can be set separately for each coordinate axis on a graphing utility. Once scales are set, you get a **viewing rectangle**, where your graphs are displayed. For example, in Figure 1.4, the scale for the x-axis is 1 (the distance between each tick mark), whereas the scale for the y-axis is 2. Read more about viewing rectangles in your graphing calculator manual.

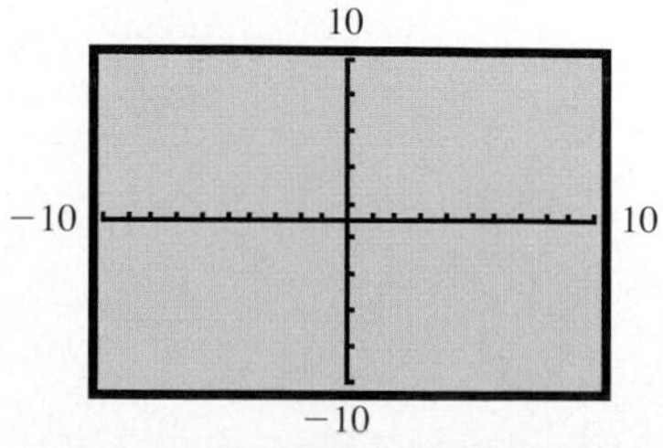

FIGURE 1.4 Viewing rectangle

2 Find the distance between two points.

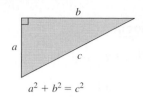

$$a^2 + b^2 = c^2$$

FIGURE 1.5 Pythagorean Theorem

The Distance Formula

If a Cartesian coordinate system has the same unit of measurement, such as inches or centimeters, on both axes, we can calculate the distance between any two points in that plane.

Recall that the Pythagorean Theorem states that in a **right triangle** with hypotenuse of length c and the legs of lengths a and b,

$$a^2 + b^2 = c^2, \qquad \text{Pythagorean Theorem}$$

as shown in Figure 1.5.

Suppose we want to compute the distance between the two points $P(x_1, y_1)$ and $Q(x_2, y_2)$. We draw a horizontal line through the point Q and a vertical line through the point P to form the right triangle PQS, as shown in Figure 1.6.

The length of the horizontal side of the triangle is $|x_2 - x_1|$, and the length of the vertical side is $|y_2 - y_1|$. The distance between P and Q, denoted $d(P, Q)$, is the length of the hypotenuse.

$$[d(P,Q)]^2 = |x_2 - x_1|^2 + |y_2 - y_1|^2 \qquad \text{Pythagorean Theorem}$$

$$d(P,Q) = \sqrt{|x_2 - x_1|^2 + |y_2 - y_1|^2} \qquad \text{Take the square root of both sides.}$$

$$d(P,Q) = \sqrt{(x_2 - x_1)^2 + (y_2 - y_1)^2} \qquad |a|^2 = a^2$$

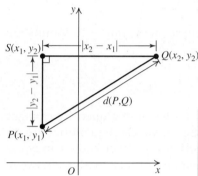

FIGURE 1.6 Visualizing the distance formula

THE DISTANCE FORMULA IN THE COORDINATE PLANE

Let $P = (x_1, y_1)$ and $Q = (x_2, y_2)$ be any two points in the coordinate plane. Then the distance between P and Q, denoted $d(P, Q)$, is given by the **distance formula**:

$$d(P,Q) = \sqrt{(x_2 - x_1)^2 + (y_2 - y_1)^2}.$$

EXAMPLE 3   **Finding the Distance Between Two Points**

Find the distance between the points $P = (-2, 5)$ and $Q = (3, -4)$.

SOLUTION

Let $(x_1, y_1) = (-2, 5)$ and $(x_2, y_2) = (3, -4)$. Then

$$x_1 = -2, y_1 = 5, x_2 = 3, \text{ and } y_2 = -4.$$

$$d(P,Q) = \sqrt{(x_2 - x_1)^2 + (y_2 - y_1)^2} \qquad \text{Distance formula}$$

$$= \sqrt{[3 - (-2)]^2 + (-4 - 5)^2} \qquad \text{Substitute the values for } x_1, x_2, y_1, y_2.$$

$$= \sqrt{5^2 + (-9)^2} \approx 10.3 \qquad \text{Use a calculator.} \qquad ■ ■ ■$$

Practice Problem 3 Find the distance between the points $(-5, 2)$ and $(-4, 1)$. ■

3 Find the midpoint of a line segment.

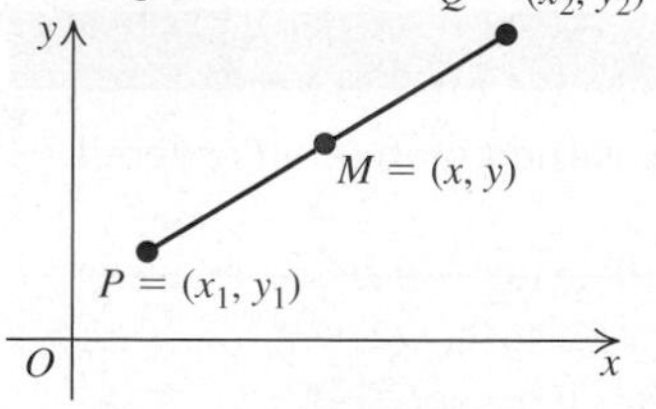

The Midpoint Formula

THE MIDPOINT FORMULA

The coordinates of the midpoint $M = (x, y)$ of the line segment joining $P = (x_1, y_1)$ and $Q = (x_2, y_2)$ are given by (see Figure 1.7):

$$M = (x, y) = \left(\frac{x_1 + x_2}{2}, \frac{y_1 + y_2}{2} \right).$$

FIGURE 1.7 Midpoint of a segment

The midpoint formula can be proved by showing that $d(P, M) = d(Q, M)$. See Exercise 89.

STUDY TIP

The coordinates of the midpoint of a line segment are found by taking the average of the x-coordinates and the average of the y-coordinates of the endpoints.

EXAMPLE 4   **Finding the Midpoint of a Line Segment**

Find the midpoint of the line segment joining the points $P(-3, 6)$ and $Q(1, 4)$.

SOLUTION

Let $(x_1, y_1) = (-3, 6)$ and $(x_2, y_2) = (1, 4)$. Then

$$x_1 = -3, y_1 = 6, x_2 = 1, \text{ and } y_2 = 4.$$

$$\text{Midpoint} = \left(\frac{x_1 + x_2}{2}, \frac{y_1 + y_2}{2}\right) = \left(\frac{-3 + 1}{2}, \frac{6 + 4}{2}\right) \qquad \text{Substitute values for } x_1, x_2, y_1, y_2.$$

$$= (-1, 5) \qquad \text{Simplify.}$$

The midpoint is $M = (-1, 5)$.   ■ ■ ■

Practice Problem 4   Find the midpoint of the line segment whose endpoints are $(5, -2)$ and $(6, -1)$.   ■

4 Sketch a graph by plotting points.

Graph of an Equation

We say that a *relation* exists between two quantities when one quantity changes with respect to the other quantity. The two changing (or varying) quantities are often represented by *variables*. A relation may sometimes be described by an equation or a formula. The following equations are examples of relationships between two variables:

$$y = 2x + 1; \quad x^2 + y^2 = 4; \quad y = x^2; \quad x = y^2; \quad F = \frac{9}{5}C + 32;$$

$$\text{and} \quad q = -3p^2 + 30.$$

An ordered pair (a, b) is said to **satisfy** an equation with variables x and y if, when a is substituted for x and b is substituted for y in the equation, the resulting statement is true. For example, the ordered pair $(2, 5)$ satisfies the equation $y = 2x + 1$ because replacing x with 2 and y with 5 yields $5 = 2(2) + 1$, which is a true statement. The ordered pair $(5, -2)$ does not satisfy this equation since replacing x with 5 and y with -2 yields $-2 = 2(5) + 1$, which is false. An ordered pair that satisfies an equation is called a **solution** of the equation.

In an equation involving x and y, if the value of y can be found given the value of x, then we say that y is the **dependent variable** and x is the **independent variable**. In the equation $y = 2x + 1$, for any real number x, there is a corresponding value of y. Hence, we have infinitely many solutions of the equation $y = 2x + 1$. When these solutions are graphed or plotted as points in the coordinate plane, they constitute the *graph of the equation*. Therefore, the graph of an equation is a geometric picture of its solutions set.

GRAPH OF AN EQUATION

The **graph of an equation** in two variables, such as x and y, is the graph of all ordered pairs (a, b) in the coordinate plane that satisfy the equation.

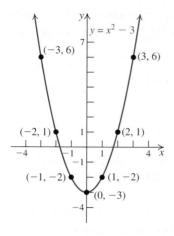

FIGURE 1.8 Graph of an equation

TECHNOLOGY CONNECTION

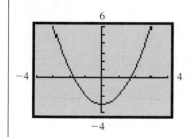

The calculator graph of $y = x^2 - 3$ reinforces the result of Example 5.

EXAMPLE 5   Sketching a Graph by Plotting Points

Sketch the graph of $y = x^2 - 3$.

SOLUTION

The equation has infinitely many solutions. To find a few, we choose integer values of x between -3 and 3. Then we find the corresponding values of y as shown in Table 1.2.

TABLE 1.2

x	$y = x^2 - 3$	(x, y)
-3	$y = (-3)^2 - 3 = 9 - 3 = 6$	$(-3, 6)$
-2	$y = (-2)^2 - 3 = 4 - 3 = 1$	$(-2, 1)$
-1	$y = (-1)^2 - 3 = 1 - 3 = -2$	$(-1, -2)$
0	$y = 0^2 - 3 = 0 - 3 = -3$	$(0, -3)$
1	$y = 1^2 - 3 = 1 - 3 = -2$	$(1, -2)$
2	$y = 2^2 - 3 = 4 - 3 = 1$	$(2, 1)$
3	$y = 3^2 - 3 = 9 - 3 = 6$	$(3, 6)$

We plot the solutions (x, y) and join them with a smooth curve to sketch the graph of $y = x^2 - 3$ shown in Figure 1.8.   ■ ■ ■

Practice Problem 5   Sketch the graph of $y = -x^2 + 1$.   ■

The bowl-shaped curve sketched in Figure 1.8 is called a *parabola*. You can find parabolic shapes in everyday settings, such as the path of a thrown ball or in the reflector behind a car's headlight.

Example 5 suggests how to sketch the graph of any equation by plotting points. We summarize the steps for this technique.

SKETCHING A GRAPH BY PLOTTING POINTS

Step 1 Make a representative table of solutions of the equation.

Step 2 Plot the solutions as ordered pairs in the Cartesian coordinate plane.

Step 3 Connect the solutions in Step 2 with a smooth curve.

Comment This technique has obvious pitfalls. For instance, many different curves pass through the four points in Figure 1.9. Assume that these points are solutions

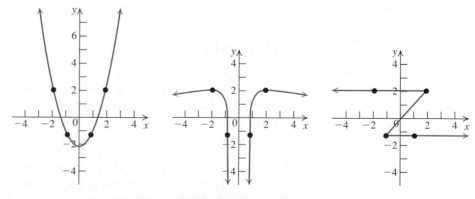

FIGURE 1.9 Several graphs through four points

of a given equation. We cannot guarantee that any curve through these points is the actual graph of the equation. However, in general, the more solutions plotted, the more accurate the graph.

5 Find the intercepts of a graph.

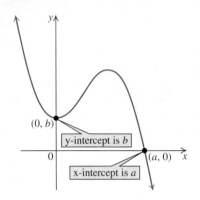

FIGURE 1.10 Intercepts of a graph

Intercepts

Let's examine the points where a graph intersects (crosses or touches) the coordinate axes. Since all points on the *x*-axis have a *y*-coordinate of 0, any point where a graph intersects the *x*-axis has the form $(a, 0)$. See Figure 1.10. The number *a* is called an **x-intercept** of the graph. Similarly, any point where a graph intersects the *y*-axis has the form $(0, b)$, and the number *b* is called a **y-intercept** of the graph.

FINDING THE INTERCEPTS OF THE GRAPH OF AN EQUATION

Step 1 To find the *x*-intercepts of the graph of an equation, set $y = 0$ in the equation and solve for *x*.

Step 2 To find the *y*-intercepts of the graph of an equation, set $x = 0$ in the equation and solve for *y*.

STUDY TIP

Do not try to calculate the *x*-intercept by setting $x = 0$. An *x*-intercept is the *x*-coordinate of a point where the graph touches or crosses the *x*-axis, so the *y*-coordinate must be 0.

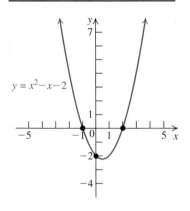

FIGURE 1.11 Intercepts of a graph

6 Find the symmetries in a graph.

EXAMPLE 6   Finding Intercepts

Find the *x*- and *y*-intercepts of the graph of the equation $y = x^2 - x - 2$.

SOLUTION

Step 1 Set $y = 0$ in the equation and solve for *x*.

$$
\begin{array}{lll}
0 = x^2 - x - 2 & \text{Set } y = 0. \\
0 = (x + 1)(x - 2) & \text{Factor.} \\
x + 1 = 0 \quad \text{or} \quad x - 2 = 0 & \text{Zero-product property} \\
x = -1 \quad \text{or} \qquad x = 2 & \text{Solve each equation for } x.
\end{array}
$$

The *x*-intercepts are -1 and 2.

Step 2 Set $x = 0$ in the equation and solve for *y*.

$$
\begin{array}{ll}
y = 0^2 - 0 - 2 & \text{Set } x = 0. \\
y = -2 & \text{Solve for } y.
\end{array}
$$

The *y*-intercept is -2.

The graph of the equation $y = x^2 - x - 2$ is shown in Figure 1.11. ■ ■ ■

Practice Problem 6   Find the intercepts of the graph of $y = 2x^2 + 3x - 2$. ■

Symmetry

The concept of **symmetry** helps us sketch graphs of equations. A graph has symmetry if one portion of the graph is a *mirror image* of another portion. As shown in Figure 1.12, if a line *l* is an **axis of symmetry**, or **line of symmetry**, we can construct the mirror image of any point *P* not on *l* by first drawing the perpendicular line segment from *P* to *l*. Then we extend this segment an equal distance on the other side to a point *P'* so that the line *l* perpendicularly bisects the line segment $\overline{PP'}$. In Figure 1.12, we say that the point *P'* is the *symmetric image* of the point *P* with respect to the line *l*. Symmetry lets us use information about part of the graph to draw the remainder of the graph.

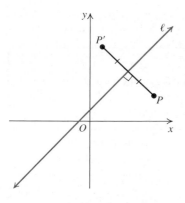

FIGURE 1.12 Symmetric points about a line

There are three basic types of symmetries.

TESTS FOR SYMMETRIES

1. A graph is **symmetric with respect to (or about) the y-axis** if for every point (x, y) on the graph the point $(-x, y)$ is also on the graph. See Figure 1.13(a).

2. A graph is **symmetric with respect to (or about) the x-axis** if for every point (x, y) on the graph the point $(x, -y)$ is also on the graph. See Figure 1.13(b).

3. A graph is **symmetric with respect to (or about) the origin** if for every point (x, y) on the graph the point $(-x, -y)$ is also on the graph. See Figure 1.13(c).

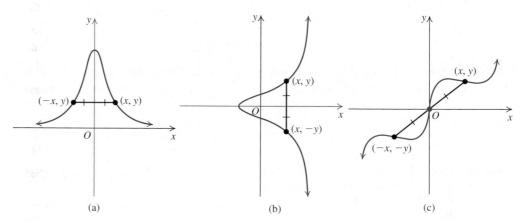

(a)   (b)   (c)

FIGURE 1.13 **Three types of symmetry**

TECHNOLOGY CONNECTION

The calculator graph of $y = \dfrac{1}{x^2 + 5}$ reinforces the result of Example 7.

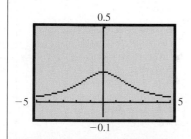

STUDY TIP

Note that if *only* even powers of x appear in an equation, then the graph is automatically symmetric with respect to the y-axis since for any integer n, $(-x)^{2n} = x^{2n}$. Consequently, we obtain the original equation when we replace x with $-x$.

EXAMPLE 7   **Checking for Symmetry**

Determine whether the graph of the equation $y = \dfrac{1}{x^2 + 5}$ is symmetric with respect to the y-axis.

SOLUTION

Replace x with $-x$ to see if $(-x, y)$ also satisfies the equation.

$$y = \frac{1}{x^2 + 5} \qquad \text{Original equation}$$

$$y = \frac{1}{(-x)^2 + 5} \qquad \text{Replace } x \text{ with } -x.$$

$$y = \frac{1}{x^2 + 5} \qquad \text{Simplify: } (-x)^2 = x^2.$$

Since replacing x with $-x$ gives us the original equation, the graph of $y = \dfrac{1}{x^2 + 5}$ is symmetric with respect to the y-axis.   ■ ■ ■

Practice Problem 7   Check the graph of $x^2 - y^2 = 1$ for symmetry with respect to the y-axis.   ■

FINDING THE SOLUTION: A PROCEDURE

EXAMPLE 8 **Sketching a Graph Using Symmetry**

OBJECTIVE

Use symmetry to sketch the graph of an equation.

EXAMPLE

Use symmetry to sketch the graph of $y = 4x - x^3$.

Step 1 Test for all three symmetries.
About the x-axis: Replace y with $-y$.
About the y-axis: Replace x with $-x$.
About the origins: Replace x with $-x$ and y with $-y$.

1.

x-axis	y-axis	origin
Replace y by $-y$	Replace x by $-x$	Replace x by $-x$ and y by $-y$
$-y = 4x - x^3$	$y = 4(-x) - (-x)^3$	$-y = 4(-x) - (-x)^3$
$y = -4x + x^3$	$y = -4x + x^3$	$-y = -4x + x^3$
No	No	$y = 4x - x^3$
		Yes

Step 2 Make a table of values using any symmetries found in Step 1.

2. Origin symmetry: If (x, y) is on the graph, so is $(-x, -y)$. Use only positive x values in the table.

x	0	0.5	1	1.5	2	2.5
$y = 4x - x^3$	0	1.875	3	2.625	0	-5.625

Step 3 Plot the points from the table and draw a smooth curve through them.

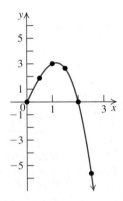

Graph of $y = 4x - x^3$, $x \geq 0$

Step 4 Extend the portion of the graph found in Step 3 using symmetry.

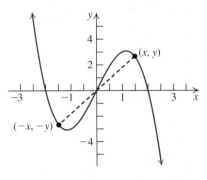

Graph of $y = 4x - x^3$ ■ ■ ■

Practice Problem 8 Use symmetry to sketch the graph of $y = x^4 - 4x^2$. ■

7 Find the equation of a circle.

Circles

Sometimes a curve that is described geometrically can also be described by an algebraic equation. We illustrate this situation in the case of a circle.

CIRCLE

A **circle** is a set of points in a Cartesian coordinate plane that are a fixed distance r from a specified point (h, k). The fixed distance r is called the **radius** of the circle, and the specified point (h, k) is called the **center** of the circle.

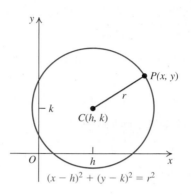

FIGURE 1.14

A point $P(x, y)$ is on the circle if and only if its distance from the center $C(h, k)$ is r. Using the notation for the distance between the points P and C, we have

$$d(P, C) = r$$
$$\sqrt{(x - h)^2 + (y - k)^2} = r \qquad \text{Distance formula}$$
$$(x - h)^2 + (y - k)^2 = r^2 \qquad \text{Square both sides.}$$

The equation $(x - h)^2 + (y - k)^2 = r^2$ is an equation of a circle with radius r and center (h, k). A point (x, y) is on the circle of radius r and center $C(h, k)$ if and only if it satisfies this equation. Figure 1.14 is the graph of a circle with center $C(h, k)$ and radius r.

THE STANDARD FORM FOR THE EQUATION OF A CIRCLE

The equation of a circle with center (h, k) and radius r is

(1) $$(x - h)^2 + (y - k)^2 = r^2$$

and is called the **standard form** of an equation of a circle.

TECHNOLOGY CONNECTION

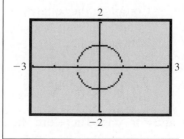

To graph the equation $x^2 + y^2 = 1$ on a graphing calculator, we first solve for y.

$$y^2 = 1 - x^2 \qquad \text{Subtract } x^2 \text{ from both sides.}$$
$$y = \pm\sqrt{1 - x^2} \qquad \text{Square root property}$$

We then graph the two equations

$$Y_1 = \sqrt{1 - x^2} \quad \text{and}$$
$$Y_2 = -\sqrt{1 - x^2}$$

in the same window. The graph of y_1 is the upper semicircle $(y \geq 0)$, and the graph of y_2 is the lower semicircle $(y \leq 0)$. The calculator graph does not quite look like a circle. We use the ZSquare option to make the display look like a circle.

EXAMPLE 9 Finding the Equation of a Circle

Find the standard form of the equation of the circle with center $(-3, 4)$ and radius 7.

SOLUTION

$$(x - h)^2 + (y - k)^2 = r^2 \qquad \text{Center–radius form}$$
$$[x - (-3)]^2 + (y - 4)^2 = 7^2 \qquad \text{Replace } h \text{ with } -3, k \text{ with } 4, \text{ and } r \text{ with } 7.$$
$$(x + 3)^2 + (y - 4)^2 = 49 \qquad \text{We usually eliminate double negatives.} \quad ■ ■ ■$$

Practice Problem 9 Find the standard form of the equation of the circle with center $(3, -6)$ and radius 10. ■

If an equation in two variables can be written in standard form, then its graph is a circle with center (h, k) and radius r.

EXAMPLE 10 Graphing a Circle

Specify the center and radius and graph each circle.

a. $x^2 + y^2 = 1$

b. $(x + 2)^2 + (y - 3)^2 = 25$

SOLUTION

a. The equation $x^2 + y^2 = 1$ can be rewritten as

$$(x - 0)^2 + (y - 0)^2 = 1^2.$$

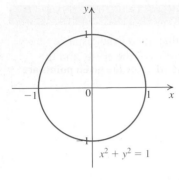

FIGURE 1.15 The unit circle

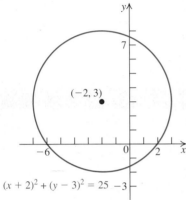

FIGURE 1.16 Circle with radius 5 and center at (−2, 3)

Comparing this equation with equation (1), we conclude that the given equation is an equation of a circle with center $(0, 0)$ and radius 1. The graph is shown in Figure 1.15. This circle is called the **unit circle**.

b. Rewriting the equation $(x + 2)^2 + (y − 3)^2 = 25$ as

$$[x − (−2)]^2 + (y − 3)^2 = 5^2, \qquad x + 2 = x − (−2)$$

we see that the graph of this equation is a circle with center $(−2, 3)$ and radius 5. The graph is shown in Figure 1.16. ■ ■ ■

Practice Problem 10 Graph the equation $(x − 2)^2 + (y + 1)^2 = 36$. ■

If we expand the squared expressions in the standard equation of a circle,

(1) $$(x − h)^2 + (y − k)^2 = r^2,$$

and then simplify, we obtain an equation of the form

(2) $$x^2 + y^2 + ax + by + c = 0.$$

GENERAL FORM OF THE EQUATION OF A CIRCLE

The **general form** of the equation of a circle is

$$x^2 + y^2 + ax + by + c = 0.$$

On the other hand, if we are given an equation in general form, we can convert it to standard form by completing the squares on the x- and y-terms. This gives

(3) $$(x − h)^2 + (y − k)^2 = d.$$

If $d > 0$, the graph of equation (3) is a circle with center (h, k) and radius $\sqrt{d}$. If $d = 0$, the graph of equation (3) is the point (h, k). If $d < 0$, there is no graph.

EXAMPLE 11 **Converting the General Form to Standard Form**

Find the center and radius of the circle with equation

$$x^2 + y^2 − 6x + 8y + 10 = 0.$$

SOLUTION
Complete the squares on both the x-terms and y-terms to get standard form.

$x^2 + y^2 − 6x + 8y + 10 = 0$	Original equation
$(x^2 − 6x) + (y^2 + 8y) = −10$	Group the x-terms and y-terms.
$(x^2 − 6x + 9) + (y^2 + 8y + 16) = −10 + 9 + 16$	Complete the squares by adding 9 and 16 to both sides.
$(x − 3)^2 + (y + 4)^2 = 15$	Factor and simplify.
$(x − 3)^2 + [y − (−4)]^2 = \left(\sqrt{15}\right)^2$	Rewrite in standard form.

The last equation tells us that we have $h = 3$, $k = −4$, and $r = \sqrt{15}$. Therefore, the circle has center $(3, −4)$ and radius $\sqrt{15} \approx 3.9$. ■ ■ ■

Practice Problem 11 Find the center and radius of the circle with equation $x^2 + y^2 + 4x − 6y − 12 = 0$. ■

RECALL

To make $x^2 + bx$ a perfect square, add $\left(\dfrac{1}{2}b\right)^2$ to get

$$x^2 + bx + \left(\frac{1}{2}b\right)^2 = \left(x + \frac{b}{2}\right)^2.$$

This completes the square.

SECTION 1.1 ■ Exercises

A EXERCISES Basic Skills and Concepts

1. A point with a negative first coordinate and a positive second coordinate lies in the _____second_____ quadrant.

2. If $(-2, 4)$ is a point on a graph that is symmetric with respect to the y-axis, then the point ____(2, 4)____ is also on the graph.

3. If $(0, -5)$ is a point of a graph, then -5 is a(n) _____y-_____ intercept of the graph.

4. An equation in standard form of a circle with center $(1, 0)$ and radius 2 is $(x - 1)^2 + y^2 = 4$.

5. *True or False.* The graph of the equation $3x^2 - 2x + y + 3 = 0$ is a circle. False

6. *True or False.* If a graph is symmetric about the x-axis, then it must have at least one x-intercept. False

7. Plot and label each of the given points in a Cartesian coordinate plane and state the quadrant, if any, in which each point is located. $(2, 2), (3, -1), (-1, 0),$ $(-2, -5), (0, 0), (-7, 4), (0, 3), (-4, 2)$ †

8. **a.** Write the coordinates of any five points on the x-axis. What do these points have in common? †
   b. Graph the points $(-2, 1), (0, 1), (0.5, 1), (1, 1),$ and $(2, 1)$. Describe the set of all points of the form $(x, 1)$, where x is a real number. †

9. **a.** If the x-coordinate of a point is 0, where does that point lie? On the y-axis
   b. Graph the points $(-1, 1), (-1, 1.5), (-1, 2),$ $(-1, 3),$ and $(-1, 4)$. Describe the set of all points of the form $(-1, y)$, where y is a real number. †

10. Let $P(x, y)$ be a point in a coordinate plane. In which quadrant does P lie
    a. if x and y are both negative? Quadrant III
    b. if x and y are both positive? Quadrant I
    c. if x is positive and y is negative? Quadrant IV
    d. if x is negative and y is positive? Quadrant II

In Exercises 11–16, find (a) the distance between P and Q and (b) the coordinates of the midpoint of the line segment PQ.

11. $P(2, 1), Q(2, 5)$ **a.** 4 **b.** $(2, 3)$

12. $P(3, 5), Q(-2, 5)$ **a.** 5 **b.** $(0.5, 5)$

13. $P(-1, -5), Q(2, -3)$ **a.** $\sqrt{13}$ **b.** $(0.5, -4)$

14. $P(-4, 1), Q(-7, -9)$ **a.** $\sqrt{109}$ **b.** $(-5.5, -4)$

15. $P(v - w, t), Q(v + w, t)$ **a.** $2|w|$ **b.** (v, t)

16. $P(t, k), Q(k, t)$ **a.** $\sqrt{2}|t - k|$ **b.** $\left(\dfrac{t + k}{2}, \dfrac{t + k}{2}\right)$

In Exercises 17–20, determine whether the given points are collinear. Points are *collinear* if they can be labeled P, Q, and R so that $d(P, Q) + d(Q, R) = d(P, R)$.

17. $(0, 0), (1, 2), (-1, -2)$ Yes

18. $(3, 4), (0, 0), (-3, -4)$ Yes

19. $(9, 6), (0, -3), (3, 1)$ No

20. $(-2, 3), (3, 1), (2, -1)$ No

In Exercises 21–26, identify the triangle PQR as an *isosceles* (two sides of equal length), *equilateral* (three sides of equal length), or *scalene* triangle (three sides of different lengths).

21. $P(-5, 5), Q(-1, 4), R(-4, 1)$ Isosceles

22. $P(3, 2), Q(6, 6), R(-1, 5)$ Isosceles

23. $P(-4, 8), Q(0, 7), R(-3, 5)$ Scalene

24. $P(6, 6), Q(-1, -1), R(-5, 3)$ Scalene

25. $P(1, -1), Q(-1, 1), R(-\sqrt{3}, -\sqrt{3})$ Equilateral

26. $P(-.5, -1), Q(-1.5, 1), R(\sqrt{3} - 1, \sqrt{3}/2)$ Equilateral

In Exercises 27–30, determine whether the given points are on the graph of the equation.

	Equation	Points
27.	$y = x - 1$	$(-3, -4,), (1, 0), (4, 3), (2, 3)$ †
28.	$y = \sqrt{x + 1}$	$(3, 2), (0, 1), (8, -3), (8, 3)$ †
29.	$y = \dfrac{1}{x}$	$\left(-3, \dfrac{1}{3}\right), (1, 1), (0, 0), \left(2, \dfrac{1}{2}\right)$ †
30.	$y^2 = x$	$(1, -1), (1, 1)\ (0, 0), (2, -\sqrt{2})$

Each of the points is on the graph.

In Exercises 31–42, graph each equation by plotting points. Let $x = -3, -2, -1, 0, 1, 2,$ and 3, where applicable.

31. $y = x + 1$ †

32. $y = x - 1$ †

33. $y = 2x$ †

34. $y = \dfrac{1}{2}x$ †

35. $y = |x|$ †

36. $y = |x + 1|$ †

37. $y = 4 - x^2$ †

38. $y = x^2 - 4$ †

39. $y = \sqrt{9 - x^2}$ †

40. $y = -\sqrt{9 - x^2}$ †

41. $y = x^3$ †

42. $y = -x^3$ †

43. Write the x- and y-intercepts of the graph.
    x-intercepts: $-3, 0, 3$; y-intercepts: $-2, 0, 2$

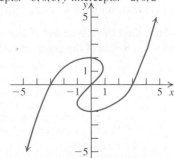

†Due to space constrictions, answers to these exercises may be found in the Answers beginning on page A–1 in the back of the book.

44. Write the x- and y-intercepts of the graph.
x-intercepts: $-2, 4$; y-intercept: -4

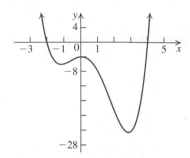

In Exercises 45–54, find the x- and y-intercepts of the graph of each equation.

45. $3x + 4y = 12$ †

46. $\dfrac{x}{5} + \dfrac{y}{3} = 1$ †

47. $2x + 3y = 5$ †

48. $\dfrac{x}{2} - \dfrac{y}{3} = 1$ †

49. $y = x^2 - 6x + 8$ †

50. $x = y^2 - 5y + 6$ †

51. $x^2 + y^2 = 4$ †

52. $y = \sqrt{9 - x^2}$ †

53. $y = \sqrt{x^2 - 1}$ †

54. $xy = 1$ No x- or y-intercepts

In Exercises 55–62, test each equation for symmetry with respect to the x-axis, the y-axis, and the origin.

55. $y = x^2 + 1$ †

56. $x = y^2 + 1$ †

57. $y = x^3 + x$ †

58. $y = 2x^3 - x$ †

59. $y = 5x^4 + 2x^2$ †

60. $y = -3x^6 + 2x^4 + x^2$ †

61. $y = -3x^5 + 2x^3$ †

62. $y = 2x^2 - |x|$ †

In Exercises 63 and 64, specify the center and the radius of each circle.

63. $(x + 1)^2 + (y - 3)^2 = 16$ Center: $(-1, 3)$; radius: 4

64. $(x + 2)^2 + (y + 3)^2 = 11$ Center: $(-2, -3)$; radius: $\sqrt{11}$

In Exercises 65–70, find the standard form of the equation of a circle that satisfies the given conditions. Graph each equation.

65. Center $(3, 1)$; radius 3 $(x - 3)^2 + (y - 1)^2 = 9$

66. Center $(-1, 2)$; radius 2 $(x + 1)^2 + (y - 2)^2 = 4$

67. Center $(-3, -2)$; and touches the y-axis
$(x + 3)^2 + (y + 2)^2 = 9$

68. Center $(3, -4)$ and passes through the point $(-1, 5)$
$(x - 3)^2 + (y + 4)^2 = 97$

69. Diameter with endpoints $(2, -3)$ and $(8, 5)$
$(x - 5)^2 + (y - 1)^2 = 25$

70. Diameter with endpoints $(7, 4)$ and $(-3, 6)$
$(x - 2)^2 + (y - 5)^2 = 26$

In Exercises 71–74, a. Find the center and radius of each circle. b. Find the x- and y-intercepts of the graph of each circle.

71. $x^2 + y^2 - 2x - 2y - 4 = 0$ **b.** x-intercepts: $1 \pm \sqrt{5}$; y-intercepts: $1 \pm \sqrt{5}$

72. $x^2 + y^2 - 4x - 2y - 15 = 0$ **b.** x-intercepts: $2 \pm \sqrt{19}$; y-intercepts: $-3, 5$

73. $2x^2 + 2y^2 + 4y = 0$ **b.** x-intercept: 0; y-intercepts: $-2, 0$

74. $3x^2 + 3y^2 + 6x = 0$ **b.** x-intercepts: $-2, 0$; y-intercepts: 0

71a. Circle; center: $(1, 1)$; radius: $\sqrt{6}$
72a. Circle; center: $(2, 1)$; radius: $2\sqrt{5}$
73a. Circle; center: $(0, -1)$; radius: 1
74a. Circle; center: $(-1, 0)$; radius: 1

B EXERCISES Applying the Concepts

75. **Population.** The table shows the total population of the United States rounded to the nearest million (181 represents 181,000,000). Graph the data in a Cartesian coordinate system. †

Year	Total Population (millions)
1960	181
1965	194
1970	205
1975	216
1980	228
1985	238
1990	250
1995	267
2000	282

Source: U.S. Census Bureau.

In Exercises 76–79, use the following *vital statistics* table:

Year	Rate per 1000 population			
	Births	Deaths	Marriages	Divorces
1950	24.1	9.6	11.1	2.6
1955	25.0	9.3	9.3	2.3
1960	23.7	9.5	8.5	2.2
1965	19.4	9.4	9.3	2.5
1970	18.4	9.5	10.6	3.5
1975	14.6	8.6	10.0	4.8
1980	15.9	8.8	10.6	5.2
1985	15.8	8.8	10.1	5.0
1990	16.7	8.6	9.4	4.7
1995	14.8	8.8	8.9	4.4
2000	14.4	8.7	8.5	4.2

Source: U.S. Census Bureau.

The table shows the rate per 1000 population of live births, deaths, marriages, and divorces from 1950 to 2000 at five-year intervals. Graph the data in a Cartesian coordinate system and connect the points with line segments.

76. Graph (year, births). †   **77.** Graph (year, deaths). †

78. Graph (year, marriages). †   **79.** Graph (year, divorces). †

In Exercises 80 and 81, use the midpoint formula to find the estimate.

80. **Murders in USA.** In the United States, 22,000 murders were committed in 1995 and 18,000 in 1997. Assuming that this decline is linear, estimate the number of murders committed in 1996. (*Source:* U.S. Census Bureau.) 20,000

81. Spending on prescription drugs. Americans spent 141 billion dollars on prescription drugs in 2001 and 252 billion dollars in 2005. Assuming that this trend continued, estimate the amount spent on prescription drugs during 2002, 2003, and 2004. (*Source: U.S. Census Bureau.*)
$168.75 billion in 2002, $196.5 billion in 2003, and $224.25 billion in 2004

82. Corporate profits. The equation $P = -0.5t^2 - 3t + 8$ describes the monthly profits (in millions of dollars) of ABCD Corp. for the year 2008, with $t = 0$ representing July 2008.

    a. How much profit did the corporation make in March 2008? $12 million

    b. How much profit did the corporation make in October 2008? −$5.5 million

    c. Sketch the graph of the equation. †

    d. Find the t-intercepts. What do they represent? †

    e. Find the P-intercept. What does it represent?
    8. It represents the profit in July 2008.

83. Female students in colleges. The equation

$$P = -0.002t^2 + 0.093t + 8.18$$

models the approximate number (in millions) of female college students in the United States for the academic years 1995–2001, with $t = 0$ representing 1995.

    a. Sketch the graph of the equation. †

    b. Find the positive t-intercept. What does it represent? No t-intercept

    c. Find the P-intercept. What does it represent? 8.18. It represents the
(*Source: Statistical Abstracts of the United States.*) number of female
students in 1995.

84. Motion. An object is thrown up from the top of a building that is 320 feet high. The equation $y = -16t^2 + 128t + 320$ gives the object's height (in feet) above the ground at any time t (in seconds) after the object is thrown.

    a. What is the height of the object after $0, 1, 2, 3, 4, 5,$ and 6 seconds? †

    b. Sketch the graph of the equation $y = -16t^2 + 128t + 320$. †

    c. What part of the graph represents the physical aspects of the problem? $0 \le t \le 10$

    d. What are the intercepts of this graph, and what do they mean? t-intercept: 10. It represents the time when the object hits the ground. y-intercept: 320. It shows the height of the building.

85. Diving for treasure. A treasure-hunting team of divers is placed in a computer-controlled diving cage. The equation $d = \dfrac{40}{3}t - \dfrac{2}{9}t^2$ describes the depth d (in feet) that the cage will descend in t minutes.

    a. Sketch the graph of the equation $d = \dfrac{40}{3}t - \dfrac{2}{9}t^2$. †

    b. What part of the graph represents the physical aspects of the problem? $0 \le t \le 60$

    c. What is the total time of the entire diving experiment? 1 hr

---

C EXERCISES Beyond the Basics

86. Use coordinates to prove that the diagonals of a parallelogram bisect each other.
[*Hint:* Choose $(0, 0), (a, 0), (b, c),$ and $(a + b, c)$ as the vertices of a parallelogram.]

87. Let $A(2, 3), B(5, 4),$ and $C(3, 8)$ be three points in a coordinate plane. Find the coordinates of the point D such that the points $A, B, C,$ and D form a parallelogram with

    a. AB as one of the diagonals. $(4, -1)$

    b. AC as one of the diagonals. $(0, 7)$

    c. BC as one of the diagonals. $(6, 9)$

[*Hint:* Use Exercise 86.]

88. Prove that if the diagonals of a quadrilateral bisect each other, the quadrilateral is a parallelogram. [*Hint:* Choose $(0, 0), (a, 0), (b, c),$ and (x, y) as the vertices of a quadrilateral and show that $x = a + b, y = c$.]

89. Use the notation in the accompanying figure to prove the midpoint formula.

$$M(x, y) = \left(\frac{x_1 + x_2}{2}, \frac{y_1 + y_2}{2} \right)$$

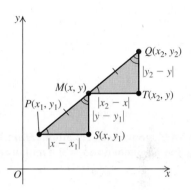

90. Trisecting a line segment. Let $A(x_1, y_1)$ and $B(x_2, y_2)$ be two points in a coordinate plane.

    a. Show that the point $C\left(\dfrac{2x_1 + x_2}{3}, \dfrac{2y_1 + y_2}{3} \right)$ is one-third of the way from A to B.
[*Hint:* Show that (i) $d(A, C) + d(C, B) = d(A, B)$ and (ii) $d(A, C) = \dfrac{1}{3}d(A, B)$.]

    b. Show that the point $D\left(\dfrac{x_1 + 2x_2}{3}, \dfrac{y_1 + 2y_2}{3} \right)$ is two-thirds of the way from A to B. *The points C and D are called the points of trisection of segment AB.*

    c. Find the points of trisection of the line segment joining $(-1, 2)$ and $(4, 1)$. $\left(\dfrac{2}{3}, \dfrac{5}{3} \right)$ and $\left(\dfrac{7}{3}, \dfrac{4}{3} \right)$

In Exercises 91–93, sketch the complete graph by using the given symmetry.

91. †

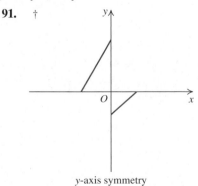

y-axis symmetry

92. †

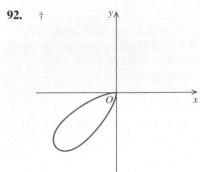

origin symmetry

93. †

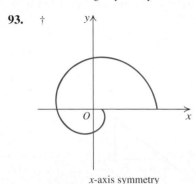

x-axis symmetry

In Exercises 94–97, graph each equation. [*Hint:* Use the zero-product property of numbers: $ab = 0$ if and only if $a = 0$ or $b = 0$.]

94. $(y - x)(y - 2x) = 0$ †

95. $y^2 = x^2$ †

96. $(y - x)(x^2 + y^2 - 4) = 0$ †

97. $(y - x + 1)(x^2 + y^2 - 6x + 8y + 24) = 0$ †

98. In the same coordinate system, sketch the graphs of the two circles with equations $x^2 + y^2 - 4x + 2y - 20 = 0$ and $x^2 + y^2 - 4x + 2y - 31 = 0$ and find the area of the region bounded by the two circles. †

Critical Thinking

In Exercises 99–101, describe the set of points $P(x, y)$ in the xy-plane that satisfy the given condition.

99. a. $x = 0$ y-axis   **b.** $y = 0$ x-axis

100. a. $xy = 0$ **b.** $xy \neq 0$

101. a. $xy > 0$ **b.** $xy < 0$

102. Sketch the graph of $y^2 = 2x$ and explain how this graph is related to the graphs of $y = \sqrt{2x}$ and $y = -\sqrt{2x}$. †

103. Show that a graph that is symmetric with respect to the x-axis and y-axis must be symmetric with respect to the origin. Give an example to show that the converse is not true. See the graph of $y = x$.

GROUP PROJECTS

1. a. Show that a circle with diameter having endpoints $A(0, 1)$ and $B(6, 8)$ intersects the x-axis at the roots of the equation $x^2 - 6x + 8 = 0$.

   b. Show that a circle with diameter having endpoints $A(0, 1)$ and $B(a, b)$ intersects the x-axis at the roots of the equation $x^2 - ax + b = 0$.

   c. Use graph paper, ruler, and compass to approximate the roots of the equation $x^2 - 3x + 1 = 0$.

2. a. Show that a circle with diameter having endpoints $A(1, 0)$ and $B(10, 7)$ intersects the y-axis at the roots of the equation $y^2 - 7y + 10 = 0$.

   b. Show that a circle with diameter having endpoints $A(1, 0)$ and $B(a, b)$ intersects the y-axis at the roots of the equation $y^2 - by + a = 0$.

Answers:
100. a. The union of the x-axis and y-axis
   b. The plane without the x-axis and y-axis
101. a. Quadrants I and III
   b. Quadrants II and IV

Lines

Before Starting this Section, Review

1. Evaluating algebraic expressions (Appendix A, page 756)
2. Graphs of equations (Section 1.1, page 6)

Objectives

1. Find the slope of a line.
2. Write the point–slope form of the equation of a line.
3. Write the slope–intercept form of the equation of a line.
4. Recognize the equations of horizontal and vertical lines.
5. Recognize the general form of the equation of a line.
6. Find equations of parallel and perpendicular lines.

A TEXAN'S TALL TALE

Gunslinger Wild Bill Longley's *first* burial took place on October 11, 1878, after he was hanged before a crowd of thousands in Giddings, Texas.

Rumors persisted that Longley's hanging had been a hoax and that he had somehow faked his death and escaped execution. In 2001, Longley's descendants had his grave opened to determine whether the remains matched Wild Bill's description: a tall white male, age 27. Both the skeleton and some personal effects suggested that this was indeed Wild Bill. Modern science lent a hand too: The DNA of Wild Bill's sister's descendant Helen Chapman was a perfect match.

Now the notorious gunman could be buried back in the Giddings cemetery—for the *second* time. How did the scientists conclude from the skeletal remains that Wild Bill was approximately 6 feet tall? (See Example 8.) ■

---

1 Find the slope of a line.

Slope of a Line

In this section, we study various forms of first-degree equations in two variables. Since the graphs of these equations are straight lines, they are called **linear equations**. Just as we measure weight or temperature by a number, we measure the "steepness" of a line by a number called its **slope**.

Consider two points $P(x_1, y_1)$ and $Q(x_2, y_2)$ on a nonvertical line, as shown in Figure 1.17. We say that the *rise* is the change in y-coordinates between the points (x_1, y_1) and (x_2, y_2) and that the *run* is the corresponding change in the x-coordinates. A positive run indicates change to the right, as shown in Figure 1.17; a negative run indicates change to the left. Similarly, a positive rise indicates upward change; a negative rise indicates downward change.

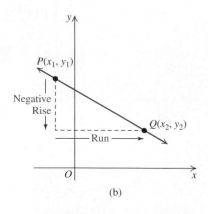

FIGURE 1.17 **Slope of a line**

SLOPE OF A LINE

The **slope** of a nonvertical line that passes through the points $P(x_1, y_1)$ and $Q(x_2, y_2)$ is denoted by m and is defined by

$$m = \frac{\text{rise}}{\text{run}} = \frac{y_2 - y_1}{x_2 - x_1}, \quad x_1 \neq x_2.$$

The slope of a vertical line is undefined.

Since $m = \dfrac{\text{rise}}{\text{run}} = \dfrac{\text{change in } y\text{-coordinates}}{\text{change in } x\text{-coordinates}}$, if the change in x is one unit to the right, then the change in y equals the slope of the line. The slope of a line is therefore the **change in y per unit change in x**. In other words, the slope of a line measures the rate of change of y with respect to x. Roofs, staircases, graded landscapes, and mountainous roads all have slopes. For example, if the pitch (slope) of a section of the roof is 0.4, then for every horizontal distance of 10 feet in that section, the roof ascends 4 feet.

EXAMPLE 1 Finding and Interpreting the Slope of a Line

Sketch the graph of the line that passes through $P(1, -1)$ and $Q(3, 3)$. Find and interpret the slope of the line.

SOLUTION

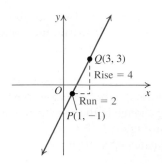

FIGURE 1.18 **Interpreting slope**

The graph of the line is sketched in Figure 1.18. The slope m of this line is given by

$$m = \frac{\text{rise}}{\text{run}} = \frac{\text{change in } y\text{-coordinates}}{\text{change in } x\text{-coordinates}}$$

$$= \frac{(y\text{-coordinate of } Q) - (y\text{-coordinate of } P)}{(x\text{-coordinate of } Q) - (x\text{-coordinate of } P)}$$

$$= \frac{(3) - (-1)}{(3) - (1)} = \frac{3 + 1}{3 - 1} = \frac{4}{2} = 2.$$

Interpretation

A slope of 2 means that the value of y increases two units for every one unit increase in the value of x. ■ ■ ■

Practice Problem 1 Find the slope of the line containing the points $(-7, 5)$ and $(6, -3)$. ■

Figure 1.19 shows several lines passing through the origin. The slope of each line is labeled.

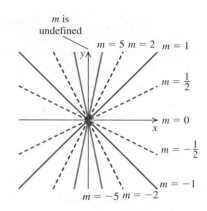

FIGURE 1.19 Slopes of lines

MAIN FACTS ABOUT SLOPES OF LINES

1. Scanning graphs from left to right, lines with positive slopes rise and lines with negative slopes fall.
2. The greater the absolute value of the slope, the steeper the line.
3. The slope of a vertical line is undefined.
4. The slope of a horizontal line is zero.

◆ **WARNING**

Be careful when using the formula for finding the slope of a line joining the points (x_1, y_1) and (x_2, y_2).

You can use either

$$m = \frac{y_2 - y_1}{x_2 - x_1} \quad \text{or} \quad m = \frac{y_1 - y_2}{x_1 - x_2},$$

but it is incorrect to use

$$m = \frac{y_1 - y_2}{x_2 - x_1} \quad \text{or} \quad m = \frac{y_2 - y_1}{x_1 - x_2}.$$

---

2 Write the point–slope form of the equation of a line.

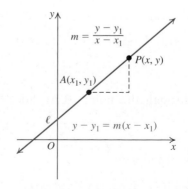

FIGURE 1.20 Point–slope form

Equation of a Line

We now find the equation of a line ℓ that passes through the point $A(x_1, y_1)$ and has slope m. Let $P(x, y)$, with $x \neq x_1$, be any point in the plane. Then $P(x, y)$ is on the line ℓ if and only if the slope of the line passing through $P(x, y)$ and $A(x_1, y_1)$ is m. This is true if and only if

$$\frac{y - y_1}{x - x_1} = m.$$

See Figure 1.20.

Multiplying both sides by $x - x_1$ gives $y - y_1 = m(x - x_1)$, which is also satisfied when $x = x_1$ and $y = y_1$.

THE POINT–SLOPE FORM OF THE EQUATION OF A LINE

If a line has slope m and passes through the point (x_1, y_1), then the **point–slope form** of an equation of the line is

$$y - y_1 = m(x - x_1).$$

■■■ **EXAMPLE 2** **Finding an Equation of a Line with Given Point and Slope**

Find the point–slope form of the equation of the line passing through the point $(1, -2)$ and with slope $m = 3$. Then solve for y.

SOLUTION

$$
\begin{aligned}
y - y_1 &= m(x - x_1) && \text{Point–slope form} \\
y - (-2) &= 3(x - 1) && \text{Substitute } x_1 = 1, y_1 = -2, \text{and } m = 3. \\
y + 2 &= 3x - 3 && \text{Simplify.} \\
y &= 3x - 5 && \text{Solve for } y.
\end{aligned}
$$
■ ■ ■

Practice Problem 2   Find the point–slope form of the equation of the line passing through the point $(-2, -3)$ and with slope $-\dfrac{2}{3}$. Then solve for y. ■

■■■ **EXAMPLE 3** **Finding an Equation of a Line Passing Through Two Given Points**

Find the point–slope form of the equation of the line ℓ passing through the points $(-2, 1)$ and $(3, 7)$. Then solve for y.

SOLUTION
We first find the slope m of the line ℓ.

$$ m = \frac{7 - 1}{3 - (-2)} = \frac{6}{3 + 2} = \frac{6}{5} \qquad m = \frac{y_2 - y_1}{x_2 - x_1} $$

$$
\begin{aligned}
y - y_1 &= m(x - x_1) && \text{Point–slope form} \\
y - 7 &= \frac{6}{5}(x - 3) && \text{Substitute } x_1 = 3, y_1 = 7, \text{and } m = \frac{6}{5}. \\
y &= \frac{6}{5}x + \frac{17}{5} && \text{Add 7 to both sides and simplify.}
\end{aligned}
$$
■ ■ ■

Practice Problem 3   Find the point–slope from of the equation of a line passing through the points $(-3, -4)$ and $(-1, 6)$. Then solve for y. ■

In Example 3, using the point $(-2, 1)$ for (x_1, y_1) in the point–slope form would give you the same answer.

3 Write the slope–intercept form of the equation of a line.

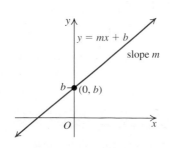

FIGURE 1.21 Slope–intercept form

■■■ **EXAMPLE 4** **Finding an Equation of a Line with a Given Slope and y-intercept**

Find the point–slope form of the equation of the line with slope m and y-intercept b. Then solve for y.

SOLUTION
Since the line has y-intercept b, the line passes through the point $(0, b)$. See Figure 1.21.

$$
\begin{aligned}
y - y_1 &= m(x - x_1) && \text{Point–slope form} \\
y - b &= m(x - 0) && \text{Substitute } x_1 = 0 \text{ and } y_1 = b. \\
y - b &= mx && \text{Simplify.} \\
y &= mx + b && \text{Solve for } y.
\end{aligned}
$$
■ ■ ■

Practice Problem 4   Find the point–slope form of the equation of the line with slope 2 and y-intercept -3. Then solve for y. ■

Historical Note

No one is sure why the letter m is used for slope. Many people suggest that m comes from the French **monter** (to climb), but Descartes, who was French, did not use m. Professor John Conway of Princeton University has suggested that m could stand for "modulus of slope."

SLOPE–INTERCEPT FORM OF THE EQUATION OF A LINE

The **slope–intercept form** of the equation of the line with slope m and y-intercept b is

$$y = mx + b.$$

The linear equation in the form $y = mx + b$ displays the slope m (the coefficient of x) and the y-intercept b (the constant term). The number m tells which way and how much the line is tilted; the number b tells where the line intersects the y-axis.

EXAMPLE 5   **Graph Using the Slope and y-intercept**

Graph the line whose equation is $y = \dfrac{2}{3}x + 2$.

SOLUTION

The equation

$$y = \frac{2}{3}x + 2$$

is in the slope-intercept form with slope $\dfrac{2}{3}$ and y-intercept 2. To sketch the graph, find two points on the line and draw a line through the two points. Use the y-intercept as one of the points, and then use the slope to locate a second point. Since $m = \dfrac{2}{3}$, let 2 be the rise and 3 be the run. From the point $(0, 2)$, move three units to the right (run) and two units up (rise). This gives $(0 + \text{run}, 2 + \text{rise}) = (0 + 3, 2 + 2) = (3, 4)$ as the second point.

The line we want joins the points $(0, 2)$ and $(3, 4)$ and is shown in Figure 1.22.   ■ ■ ■

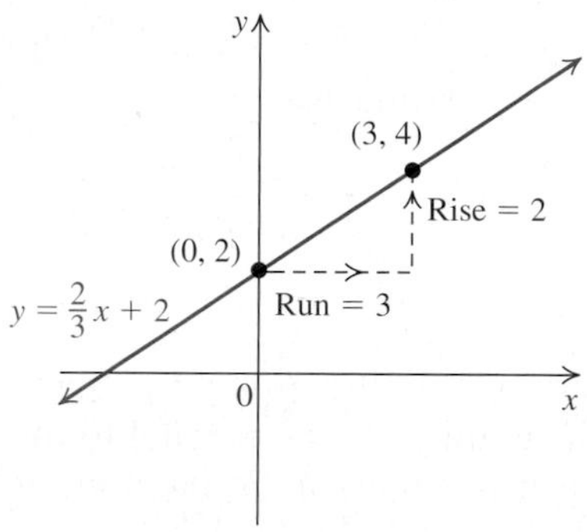

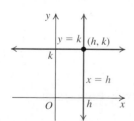

FIGURE 1.22 Locating a second point on a line

Practice Problem 5   Graph the line with slope $-\dfrac{2}{3}$ that contains the point $(0, 4)$.   ■

4 Recognize the equations of horizontal and vertical lines.

Equations of Horizontal and Vertical Lines

Consider the horizontal line through the point (h, k). The y-coordinate of every point on this line is k. So we can write its equation as $y = k$. This line has slope $m = 0$ and y-intercept k. Similarly, an equation of the vertical line through the point (h, k) is $x = h$. This line has undefined slope and x-intercept h. See Figure 1.23.

FIGURE 1.23 Vertical and horizontal lines

HORIZONTAL AND VERTICAL LINES

An equation of a horizontal line through (h, k) is $y = k$.

An equation of a vertical line through (h, k) is $x = h$.

EXAMPLE 6   **Recognizing Horizontal and Vertical Lines**

Discuss the graph of each equation in the xy-plane.

a. $y = 2$ **b.** $x = 4$

TECHNOLOGY CONNECTION

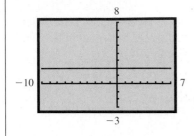

Calculator graph of $y = 2$

STUDY TIP

Students sometimes find it easy to remember that if a linear equation

(1) does not have y in it, the graph is parallel to the y-axis or
(2) does not have x in it, the graph is parallel to the x-axis.

TECHNOLOGY CONNECTION

You cannot graph $x = 4$ by using the Y= key on your calculator.

5 Recognize the general form of the equation of a line.

SOLUTION

a. The equation $y = 2$ may be considered as an equation in two variables x and y by writing

$$0 \cdot x + y = 2.$$

Any ordered pair of the form $(x, 2)$ is a solution of this equation. Some solutions are $(-1, 2), (0, 2), (2, 2)$, and $(7, 2)$. It follows that the graph is a line parallel to the x-axis and two units above it, as shown in Figure 1.24. Its slope is 0.

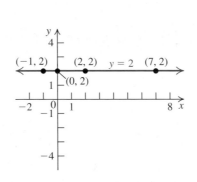

FIGURE 1.24 **Horizontal line**   FIGURE 1.25 **Vertical line**

b. The equation $x = 4$ may be written as

$$x + 0 \cdot y = 4.$$

Any ordered pair of the form $(4, y)$ is a solution of this equation. Some solutions are $(4, -5), (4, 0), (4, 2)$, and $(4, 6)$. The graph is a line parallel to the y-axis and four units to the right of it, as shown in Figure 1.25. Its slope is undefined. ■ ■ ■

Practice Problem 6   Sketch the graphs of the lines $x = -3$ and $y = 7$.   ■

General Form of the Equation of a Line

An equation of the form

$$ax + by + c = 0,$$

where a, b, and c are constants and not both a and b are zero, is called the *general form* of the linear equation. Consider two possible cases: $b \neq 0$ and $b = 0$.

Suppose $b \neq 0$: we can isolate y on one side and rewrite the equation in slope–intercept form.

$ax + by + c = 0$	Original equation
$by = -ax - c$	Subtract $ax + c$ from both sides.
$y = -\dfrac{a}{b}x - \dfrac{c}{b}.$	Divide by b.

The result is an equation of a line in slope–intercept form with slope $-\dfrac{a}{b}$ and y-intercept $-\dfrac{c}{b}$.

Suppose $b = 0$: we know that $a \neq 0$ since not both a and b are zero, and we can solve for x.

$$ax + 0 \cdot y + c = 0 \qquad \text{Replace } b \text{ with } 0.$$
$$ax + c = 0 \qquad \text{Simplify.}$$
$$x = -\frac{c}{a} \qquad \text{Solve for } x.$$

The graph of the equation $x = -\dfrac{c}{a}$ is a vertical line.

GENERAL FORM OF THE EQUATION OF A LINE

The graph of every linear equation

$$ax + by + c = 0,$$

where a, b, and c are constants and not both a and b are zero, is a line. The equation $ax + by + c = 0$ is called the **general form** of the equation of a line.

TECHNOLOGY CONNECTION

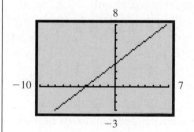 Calculator graph of
$y = \dfrac{3}{4}x + 3$

EXAMPLE 7 Graphing a Linear Equation

Find the slope, y-intercept, and x-intercept of the line with equation

$$3x - 4y + 12 = 0.$$

Then sketch the graph.

SOLUTION

First, solve for y to write the equation in slope–intercept form:

$$3x - 4y + 12 = 0 \qquad \text{Original equation}$$
$$4y = 3x + 12 \qquad \text{Subtract } 3x + 12; \text{ multiply by } -1.$$
$$y = \frac{3}{4}x + 3 \qquad \text{Divide by 4.}$$

This equation tells us that the slope m is $\dfrac{3}{4}$ and the y-intercept is 3. To find the x-intercept, set $y = 0$ in the original equation and obtain $3x + 12 = 0$, or $x = -4$. So the x-intercept is -4.

We can sketch the graph of the equation if we can find two points on the graph. So we use the intercepts and sketch the line joining the points $(-4, 0)$ and $(0, 3)$. The graph is shown in Figure 1.26. To help prevent errors, find and graph a third point on the line. ■ ■ ■

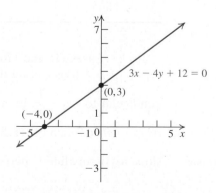

FIGURE 1.26

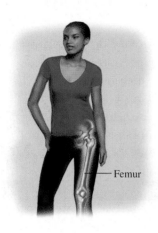

Femur

Practice Problem 7   Sketch the graph of the equation $3x + 4y = 24$.   ■

Forensic scientists use various clues to determine the identity of deceased individuals. Scientists know that certain bones serve as excellent sources of information about a person's height. The length of the *femur*, the long bone that stretches from the hip (pelvis) socket to the kneecap (patella), can be used to estimate a person's height. Knowing both the gender and race of the individual increases the accuracy of this relationship between the length of the bone and the height of the person.

EXAMPLE 8   **Inferring Height from the Femur**

The height H of a human male is related to the length x of his femur by the formula

$$H = 2.6x + 65,$$

where all measurements are in centimeters. The femur of Wild Bill Longley measured between 45 and 46 centimeters. Estimate the height of Wild Bill.

SOLUTION

Substituting $x = 45$ and $x = 46$ into the preceding formula, we have possible heights

$$H_1 = (2.6)(45) + 65 = 182$$
$$\text{and}\quad H_2 = (2.6)(46) + 65 = 184.6.$$

Therefore, Wild Bill's height was between 182 and 184.6 centimeters.

Converting to inches, we conclude that his height was between $\dfrac{182}{2.54} \approx$ 71.7 inches and $\dfrac{184.6}{2.54} \approx 72.7$ inches. He was approximately 6 feet tall.   ■ ■ ■

Practice Problem 8   Suppose the femur of a person measures between 43 and 44 centimeters. Estimate the height of the person in centimeters.   ■

> **RECALL**
>
> 1 inch = 2.54 centimeters

---

6 Find equations of parallel and perpendicular lines.

Parallel and Perpendicular Lines

We can use slope to decide whether two lines are parallel, perpendicular, or neither.

PARALLEL AND PERPENDICULAR LINES

Let ℓ_1 and ℓ_2 be two distinct lines with slopes m_1 and m_2, respectively. Then

ℓ_1 is parallel to ℓ_2 if and only if $m_1 = m_2$.

ℓ_1 is perpendicular to ℓ_2 if and only if $m_1 \cdot m_2 = -1$.

Any two vertical lines are parallel, and any horizontal line is perpendicular to any vertical line.

In Exercises 115 and 116, you are asked to prove these assertions. The statement $m_1 \cdot m_2 = -1$ tells us that if the slope of a line ℓ is m, then the slope of a line perpendicular to ℓ is $-\dfrac{1}{m}$. In other words, the slopes of perpendicular lines are *negative reciprocals* of each other. For example, if the slope m_1 of a line ℓ_1 is $-\dfrac{3}{4}$, then the slope m_2 of any line ℓ_2 perpendicular to ℓ_1 is $-\dfrac{1}{-\dfrac{3}{4}} = \dfrac{4}{3}$. See Figure 1.27.

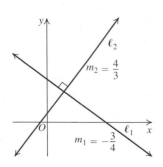

FIGURE 1.27 Perpendicular lines

FINDING THE SOLUTION: A PROCEDURE

EXAMPLE 9	Finding Equations of Parallel and Perpendicular Lines

OBJECTIVE

Let $\ell : ax + by + c = 0$. Find the equation of each line through the point (x_1, y_1):

(a) ℓ_1 parallel to ℓ

(b) ℓ_2 perpendicular to ℓ

Step 1   Find the slope m of ℓ.

Step 2   Write slope of ℓ_1 and ℓ_2.

The slope m_1 of ℓ_1 is m.

The slope m_2 of ℓ_2 is $-\dfrac{1}{m}$.

Step 3   Write equations of ℓ_1 and ℓ_2.

Use point–slope form to write equations of ℓ_1 and ℓ_2.

Simplify to write equations in the requested form.

EXAMPLE

Let $\ell : 2x - 3y + 6 = 0$. Find in general form the equation of each line through the point $(2, 8)$:

(a) ℓ_1 parallel to ℓ

(b) ℓ_2 perpendicular to ℓ

1.

$\ell : 2x - 3y + 6 = 0$	Original equation
$-3y = -2x - 6$	Subtract $2x + 6$.
$y = \dfrac{2}{3}x + 2$	Divide by -3.

The slope of ℓ is $m = \dfrac{2}{3}$.

2. $m_1 = \dfrac{2}{3}$ ℓ_1 parallel to ℓ

  $m_2 = -\dfrac{3}{2}$ ℓ_2 perpendicular to ℓ

3.

$\ell_1 : y - 8 = \dfrac{2}{3}(x - 2)$	Point–slope form
$2x - 3y + 20 = 0$	Simplify.
$\ell_2 : y - 8 = -\dfrac{3}{2}(x - 2)$	Point–slope form
$3x + 2y - 22 = 0$	Simplify.

■ ■ ■

Practice Problem 9   Find in general form the equation of the line

a. through $(-2, 5)$ and parallel to the line containing $(2, 3)$ and $(5, 7)$.

b. through $(3, -4)$ and perpendicular to the line $4x + 5y + 1 = 0$. ■

Summary

1.   We summarize different forms of equations whose graphs are lines.

Form	Equation	Slope	Other information
General	$ax + by + c = 0$ $a \neq 0$, or $b \neq 0$	$-\dfrac{a}{b}$ if $b \neq 0$	x-intercept is $-\dfrac{c}{a}$, $a \neq 0$.
		Undefined if $b = 0$	y-intercept is $-\dfrac{c}{b}$, $b \neq 0$.
Point–slope	$y - y_1 = m(x - x_1)$	m	Line passes through (x_1, y_1).
Slope–intercept	$y = mx + b$	m	x-intercept is $-\dfrac{b}{m}$ if $m \neq 0$; y-intercept is b.
Vertical line through (h, k)	$x = h$	undefined	x-intercept is h; no y-intercept if $h \neq 0$.
Horizontal line through (h, k)	$y = k$	0	When $k \neq 0$, no x-intercept; y-intercept is k. When $k = 0$, the graph is the x-axis.

2.   Parallel and perpendicular lines: Two lines with slopes m_1 and m_2 are
(a) parallel if $m_1 = m_2$ and (b) perpendicular if $m_1 \cdot m_2 = -1$.

SECTION 1.2 ■ Exercises

A EXERCISES Basic Skills and Concepts

1. The number that measures the "steepness" of a line is called the ____slope____.

2. In the slope formula $m = \dfrac{y_2 - y_1}{x_2 - x_1}$, the number $y_2 - y_1$ is called the ____rise____.

3. If two lines have the same slope, then they are ____parallel____.

4. The y-intercept of the line with equation $y = 3x - 5$ is ____-5____.

5. A line perpendicular to a line with slope $-\dfrac{1}{5}$ has slope ____5____.

6. *True or False.* A line with undefined slope is perpendicular to every line with slope 0. True

7. *True or False.* If $(1, 2)$ is a point on a line with slope 4, then so is the point $(2, 4)$. False

8. *True or False.* If the points $(1, 3)$ and $(2, y)$ are on a line with negative slope, then $y > 3$. False

In Exercises 9–16, find the slope of the line through the given pair of points. Without plotting any points, state whether the line is rising, falling, horizontal, or vertical.

9. $(1, 3), (4, 7)$ $\dfrac{4}{3}$, rising

10. $(0, 4), (2, 0)$ -2, falling

11. $(3, -2), (-2, 3)$ -1, falling

12. $(-3, 7), (-3, -4)$ Undefined, vertical

13. $(3, -2), (2, -3)$ 1, rising

14. $(0.5, 2), (3, -3.5)$ -2.2, falling

15. $(\sqrt{2}, 1), (1 + \sqrt{2}, 5)$ 4, rising

16. $(1 - \sqrt{3}, 0), (1 + \sqrt{3}, 3\sqrt{3})$ $\dfrac{3}{2}$, rising

17. In the figure, identify the line with the given slope m.
    a. $m = 1$ ℓ_3
    b. $m = -1$ ℓ_2
    c. $m = 0$ ℓ_4
    d. m is undefined. ℓ_1

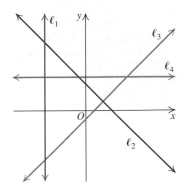

18. Find the slope of each line. (The scale is the same on both axes.) ℓ_1 has slope 1, ℓ_2 has slope 0, ℓ_3 has slope 2, and ℓ_4 has slope $-\dfrac{4}{3}$

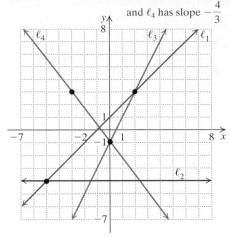

In Exercises 19 and 20, write an equation whose graph is described.

19. a. Graph is the x-axis. $y = 0$
    b. Graph is the y-axis. $x = 0$

20. a. Graph consists of all points with a y-coordinate of -4. $y = -4$
    b. Graph consists of all points with an x-coordinate of 5. $x = 5$

In Exercises 21–26, find an equation in slope–intercept form of the line that passes through the given point and has slope m. Also sketch the graph of the line by locating the second point with the rise-and-run method.

21. $(0, 4); m = \dfrac{1}{2}$ †

22. $(0, 4); m = -\dfrac{1}{2}$ †

23. $(2, 1); m = -\dfrac{3}{2}$ †

24. $(-1, 0); m = \dfrac{2}{5}$ †

25. $(5, -4); m = -\dfrac{3}{5}$ †

26. $(5, -4); m$ is undefined. †

In Exercises 27–48, use the given conditions to find an equation in slope–intercept form of each nonvertical line. Write vertical lines in the form $x = h$.

27. Passing through $(0, 1)$ and $(1, 0)$ $y = -x + 1$

28. Passing through $(0, 1)$ and $(1, 3)$ $y = 2x + 1$

29. Passing through $(-1, 3)$ and $(3, 3)$ $y = 3$

30. Passing through $(-5, 1)$ and $(2, 7)$ †

31. Passing through $(-2, -1)$ and $(1, 1)$ †

32. Passing through $(-1, -3)$ and $(6, -9)$ †

33. Passing through $\left(\dfrac{1}{2}, \dfrac{1}{4}\right)$ and $(0, 2)$ $y = -\dfrac{7}{2}x + 2$

34. Passing through $(4, -7)$ and $(4, 3)$ $x = 4$

35. A vertical line through $(5, 1.7)$ $x = 5$

†Due to space constrictions, answers to these exercises may be found in the Answers beginning on page A–1 in the back of the book.

36. A horizontal line through $(1.4, 1.5)$ $y = 1.5$

37. A horizontal line through $(0, 0)$ $y = 0$

38. A vertical line through $(0, 0)$ $x = 0$

39. $m = 0$; y-intercept $= 14$ $y = 14$

40. $m = 2$; y-intercept $= 5$ $y = 2x + 5$

41. $m = -\dfrac{2}{3}$; y-intercept $= -4$ $y = -\dfrac{2}{3}x - 4$

42. $m = -6$; y-intercept $= -3$ $y = -6x - 3$

43. x-intercept $= -3$; y-intercept $= 4$ †

44. x-intercept $= -5$; y-intercept $= -2$ †

45. Parallel to $y = 5$; passing through $(4, 7)$ $y = 7$

46. Parallel to $x = 5$; passing through $(4, 7)$ $x = 4$

47. Perpendicular to $x = -4$; passing through $(-3, -5)$ $y = -5$

48. Perpendicular to $y = -4$; passing through $(-3, -5)$ $x = -3$

49. Let ℓ_1 be a line with slope $m = -2$. Determine whether the given line ℓ_2 is parallel to ℓ_1, perpendicular to ℓ_1, or neither.
   a. ℓ_2 is the line through the points $(1, 5)$ and $(3, 1)$. Parallel
   b. ℓ_2 is the line through the points $(7, 3)$ and $(5, 4)$. Neither
   c. ℓ_2 is the line through the points $(2, 3)$ and $(4, 4)$. Perpendicular

50. A particle initially at $(-1, 3)$ moves along a line of slope $m = 4$ to a new position (x, y).
   a. Find y if $x = 2$. $y = 15$
   b. Find x if $y = 9$. $x = \dfrac{1}{2}$

In Exercises 51–58, find the slope and intercepts from the equation of the line. Sketch the graph of each equation.

51. $x + 2y - 4 = 0$ †   **52.** $x = 3y - 9$ †

53. $3x - 2y + 6 = 0$ †   **54.** $2x = -4y + 15$ †

55. $x - 5 = 0$ †   **56.** $2y + 5 = 0$ †

57. $x = 0$ †   **58.** $y = 0$ †

59. Show that an equation of a line through the points $(a, 0)$ and $(0, b)$, with a and b nonzero, can be written in the form

$$\frac{x}{a} + \frac{y}{b} = 1.$$

This form is called the **two-intercept form** of the equation of a line.

In Exercises 60–64, use the two-intercept form of the equation of a line given in Exercise 59.

60. Find an equation of the line whose x-intercept is 4 and y-intercept is 3. $\dfrac{x}{4} + \dfrac{y}{3} = 1$

61. Find the x- and y-intercepts of the graph of the equation $2x + 3y = 6$.

62. Repeat Exercise 61 for the equation $3x - 4y + 12 = 0$.

63. Find an equation of the line that has equal intercepts and passes through the point $(3, -5)$. $y = -x - 2$

64. Find an equation of the line making intercepts on the axes equal in magnitude but opposite in sign and passing through the point $(-5, -8)$. $y = x - 3$

61. $\dfrac{x}{3} + \dfrac{y}{2} = 1$; x-intercept $= 3$; y-intercept $= 2$ **62.** $-\dfrac{x}{4} + \dfrac{y}{3} = 1$; x-intercept $= -4$; y-intercept $= 3$

65. Find an equation of the line passing through the points $(2, 4)$ and $(7, 9)$. Use the equation to show that the three points $(2, 4)$ and $(7, 9)$, and $(-1, 1)$ are on the same line.

66. Use the technique of Exercise 65 to check whether the points $(7, 2)$, $(2, -3)$, and $(5, 1)$ are on the same line.

In Exercises 67–74, determine whether each pair of lines are parallel, perpendicular, or neither.

67. $x = -1$ and $2x + 7 = 9$ Parallel

68. $x = 0$ and $y = 0$ Perpendicular

69. $2x + 3y = 7$ and $y = 2$ Neither

70. $y = 3x + 1$ and $6y + 2x = 0$ Perpendicular

71. $10x + 2y = 3$ and $y + 1 = -5x$ Parallel

72. $4x + 3y = 1$ and $3 + y = 2x$ Neither

73. $3x + 8y = 7$ and $5x - 7y = 0$ Neither

74. $x = 4y + 8$ and $y = -4x + 1$ Perpendicular

In Exercises 75–80, find the equation of the line in slope–intercept form satisfying the given conditions.

75. Parallel to $x + y = 1$; passing through $(1, 1)$ $y = -x + 2$

76. Parallel to $y = 6x + 5$; y-intercept of -2 $y = 6x - 2$

77. Perpendicular to $3x - 9y = 18$; passing through $(-2, 4)$ $y = -3x - 2$

78. Perpendicular to $-2x + y = 14$; passing through $(0, 0)$

79. Perpendicular to $y = 6x + 5$; y-intercept of 4

80. Parallel to $-2x + 3y - 7 = 0$; passing through $(1, 0)$

78. $y = -\dfrac{1}{2}x$ **79.** $y = -\dfrac{1}{6}x + 4$ **80.** $y = \dfrac{2}{3}x - \dfrac{2}{3}$

B EXERCISES Applying the Concepts

81. **Jet takeoff.** Upon takeoff, a jet climbs to 4 miles as it passes over 40 miles of land below it. Find the slope of the jet's climb. $\dfrac{1}{10}$

82. **Road gradient.** Driving down the Smoky Mountains in Tennessee, Samantha descends 2000 feet in elevation while moving 4 miles horizontally away from a high point on the straight road. Find the slope (gradient) of the road (1 mile $= 5280$ feet). $|m| = \dfrac{25}{264}$

In Exercises 83–90, write an equation of a line in slope–intercept form. To describe this situation, interpret (a) the variables x and y and (b) the meaning of the slope and the y-intercept.

83. **Christmas savings account.** You currently have $130 in a Christmas savings account. You deposit $7 per week into this account. (Ignore interest.)

84. **Golf club charges.** Your golf club charges a yearly membership fee of $1000 and $35 per session of golf. †

85. **Wages.** Judy's job at a warehouse pays $11 per hour up to 40 hours per week. If she works over 40 hours in a week, she is paid 1.5 times her usual wage. [*Hint:* You will need two equations.] †

65. $y = x + 2$. Since the coordinates of the point $(-1, 1)$ satisfy this equation, $(-1, 1)$ also lies on this line.   **66.** They are not collinear.   **83. a.** x: time in weeks; y: amount of money in the account; $y = 7x + 130$ **b.** Slope: weekly deposit in the account; y-intercept: initial deposit

86. **Paying off a refrigerator.** You bought a new refrigerator for $700. You made a down payment of $100 and promised to pay $15 per month. (Ignore interest.) †

87. **Converting currency.** In 2007, according to the International Monetary Fund, 1 U.S. dollar equaled 40.5 Indian rupees. Convert currency from rupees to dollars. †

88. **Life expectancy.** In 2007, the life expectancy of a female born in the United States was 80.9 years and was increasing at a rate of 0.17 per year. (Assume that this rate of increase remains constant.) †

89. **Manufacturing.** A manufacturer produces 50 TVs at a cost of $17,500 and 75 TVs at a cost of $21,250. †

90. **Demand.** The demand for a product is zero units when the price per unit is $100 and 1000 units when the price per unit is zero dollars. †

Exercises 91 and 92 deal with the topic of *depreciation*. When filing income tax returns, taxpayers (businesses and individuals) can claim deductions for depreciation on items such as cars, computers, and buildings used for business purposes. The government allows these deductions because the value of such assets decreases (depreciates) over time. The simplest method for finding the depreciated value is called *straight-line depreciation* and assumes that the item's value decreases linearly with time.

91. **Depreciating a tractor.** The value V of a tractor purchased for $14,000 and depreciated linearly at the rate of $1400 per year is given by $v = -1400t + 14,000$, where t represents the number of years since the purchase. Find the value of the tractor after (a) two years and (b) six years. When will the tractor have no value? †

92. **Depreciating a computer.** A computer is purchased for $3000 and is to be depreciated linearly over four years. Assume that it has no value at the end of the four years.
    a. Find the amount depreciated per year. $750 per yr
    b. Find a linear equation relating the value V of the computer after t years of purchase. $V = -750t + 3000$
    c. Graph the equation in part (b). †

93. **Playing for a concert.** A famous band is considering playing for a concert and charging $1000, plus $1.50 per person attending the concert. Write an equation relating the income y of the band to the number x of people attending the concert. $y = 1.5x + 1000$

94. **Car rental.** The U Drive car rental agency charges $30.00 per day and $0.25 per mile. **a.** $C = 0.25x + 30$
    a. Find an equation relating the cost C of renting a car from U Drive for a one-day trip covering x miles.
    b. Draw the graph of the equation in part (a). †
    c. How much does it cost to rent the car for one day covering 60 miles? $45
    d. How many miles were driven if the one-day rental cost was $47.75? 71

95. **Female prisoners in Florida.** The number of females in Florida's prisons rose from 2425 in 2000 to 4026 in 2006. (*Source:* Florida Department of Corrections.)
    a. Find a linear equation relating the number y of women prisoners to the year t. (Take $t = 0$ to represent 2000.) $y = 266.8t + 2425$

b. Draw the graph of the equation from part (a). †
c. How many women prisoners were there in 2003? 3225
d. Predict the number of women prisoners in 2010. 5093

96. **Fahrenheit and Celsius.** In the Fahrenheit temperature scale, water boils at 212°F and freezes at 32°F. In the Celsius scale, water boils at 100°C and freezes at 0°C. Assume that the Fahrenheit temperature F and Celsius temperature C are linearly related. **a.** $F = \frac{9}{5}C + 32$
    a. Find the equation in the slope–intercept form relating F and C, with C as independent variable.
    b. What is the meaning of slope in part (a)? †
    c. Find the Fahrenheit temperatures, to the nearest degree, corresponding to 40°C, 25°C, −5°C, and −10°C. †
    d. Find the Celsius temperatures, to the nearest degree, corresponding to 100°F, 90°F, 75°F, −10°F, and −20°F. †
    e. The normal body temperature in humans ranges from 97.6°F to 99.6°F. Convert this temperature range to degrees Celsius. 36.44°C to 37.55°
    f. When is the Celsius temperature the same numerical value as the Fahrenheit temperature? at −40°

97. **Demand equation.** A demand equation expresses a relationship between the demand q (the number of items demanded) and the price p per unit. A linear demand equation has the form $q = mp + b$. The slope m (usually negative) measures the change in demand per unit change in price. The y-intercept b gives the demand if the items were given away. A merchant can sell 480 T-shirts per week at a price of $4 each, but can sell only 400 per week if the price per T-shirt is increased to $4.50.
    a. Find a linear demand equation for T-shirts.
    b. Predict the demand for T-shirts if the price per T-shirt is $5. 320
       a. $q = -160p + 1120$

98. **Supply equation.** A supply equation expresses a relationship between the supply q (the number of units a supplier is willing to make available) and the unit price p (the price per item). A linear supply equation has the form $q = mp + b$. The slope m is usually positive. A manufacturer of ceiling fans can supply 70 fans per day at $50 apiece and will supply 100 fans a day at a price of $60 per fan.
    a. Find a linear supply equation for the fans from this manufacturer. $q = 3p - 80$
    b. Predict the supply at a price of $62 per fan. 106

99. **Lake pollution.** In 2002, tests showed that each 1000 liters of water from Lake Heron contained 7 milligrams of a polluting mercury compound. Two years later tests showed that 8 milligrams of the compound were contained in each 1000 liters of the lake's water. Assuming that the increase in pollutants is linear, find a linear equation relating the year in which the test was given to the number of milligrams of pollutant per 1000 liters of the lake's water. What does your equation predict the pollutant content per 1000 liters of water in Lake Heron to be in 2010? $y = 0.5x + 7; 11$

100. **Selling shoes.** The Alpha shoe manufacturer has determined that the annual cost of making x pairs of its best-selling shoes, α1 (alpha one), is $25 per pair plus

$75,000 in fixed overhead costs. Each pair of shoes that is manufactured is sold wholesale for $50.

a. Find the equations that model the cost, the revenue, and the profit (Profit = Revenue − Cost). Verify that all three equations are linear. †

b. Graph the profit equation on the xy-coordinate system. †

c. What is the slope of the line in part (b)? What is the practical meaning of this slope? †

d. Find the intercepts of the line in part (b). What is the practical meaning of these intercepts?
y-intercept = −75,000; loss if 0 pairs of shoes are sold

101. Cost. The cost of producing four modems is $210.20, and the cost of producing ten modems is $348.80.

a. Write a linear equation in slope–intercept form relating cost to the number of modems produced. Sketch a graph of the equation. †

b. What is the practical meaning of the slope and the intercepts of the equation in part (a)? †

c. Predict the cost of producing 12 modems. 395

102. Viewers of a TV show. After five months, the number of viewers of *The Weekly Show* on TV was 5.73 million. After eight months, the number of viewers of the show rose to 6.27 million. Assume that the linear model applies.

a. Write an equation expressing the number of viewers V after x months. $V = 0.18x + 4.83$

b. Interpret the slope and the intercepts of the line in part (a). †

c. Predict the number of viewers of the show after 11 months. 6.81 million

103. Health insurance coverage. In Michigan, 11.7% of the population was not covered by health insurance in 2000. In 2005, the percentage of uninsured people rose to 12.7%. Use the linear model to predict the percentage of people in Michigan that will not be covered in 2010. 13.7% (*Source:* Michigan Department of Community Health.)

104. Oil consumption. The total world consumption of oil increased from 73.22 million barrels per day in 1999 to 82.59 million barrels per day in 2004. Create a linear model involving a relation between the year and the daily oil consumption in that year. Let $t = 0$ represent the year 1999. (*Source:* www.cia.gov) $c = 1.874t + 73.22$

C EXERCISES Beyond the Basics

105. The graph of a line has slope $m = 3$. If $P(-2, 3)$ and $Q(1, c)$ and are points on the graph, find c. $c = 12$

106. Find a real number c such that the line $3x - cy - 2 = 0$ has a y-intercept of -4. $c = \dfrac{1}{2}$

In Exercises 107–110, show that the three given points are collinear by using (a) slopes and (b) the distance formula.

107. $(0, 1), (1, 3),$ and $(-1, -1)$

108. $(1, 2), (-1, 4),$ and $(2, 1)$

109. $(1, 2), (0, -3),$ and $(-1, -8)$

110. $(1, .5), (2, 0),$ and $(0.5, 0.75)$

111. Show that the triangle with vertices $A(1, 1), B(-1, 4),$ and $C(5, 8)$ is a right triangle by using (a) slopes and (b) the converse of the Pythagorean Theorem. †

112. Show that the four points, $A(-4, -1), B(1, 2), C(3, 1),$ and, $D(-2, -2)$ are the vertices of a parallelogram. [*Hint:* Opposite sides of a parallelogram are parallel.] †

113. Show that the four points $A(-10, 9), B(-2, 24), C(5, 1),$ and $D(13, 16)$ are vertices of a square. †

114. a. Find an equation of the line that is the perpendicular bisector of the line segment with endpoints $A(2, 3)$ and $B(-1, 5)$. $y = \dfrac{3}{2}x + \dfrac{13}{4}$

b. Show that $y = x$ is the perpendicular bisector of the line segment with endpoints and $A(a, b)$ and $B(b, a)$.

115. Show that if two nonvertical lines are parallel, their slopes are equal. Start by letting the two lines be $\ell_1: y = m_1x + b_1$ and $\ell_2: y = m_2x + b_2 \ (b_1 > b_2)$. Show that if (x_1, y_1) and (x_2, y_2) are on ℓ_1, then $(x_1, y_1 - (b_1 - b_2))$ and $((x_2, y_2 - (b_1 - b_2)))$ must be on ℓ_2. Then by computing m_1 and m_2, show that $m_1 = m_2$. (See the accompanying figure.)

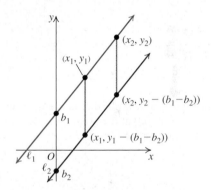

116. In the accompanying figure, note that lines ℓ_1 and ℓ_2 have slopes m_1 and m_2. Assume that ℓ_1 and ℓ_2 are perpendicular. Use the distance formula and the converse of the Pythagorean Theorem to show that $m_1 \cdot m_2 = -1$.

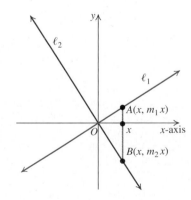

117. Geometry. Use a coordinate plane to prove that if two sides of a quadrilateral are congruent (equal in length) and parallel, then so are the other two sides.

118. Geometry. Use a coordinate plane to prove that the midpoints of the four sides of a quadrilateral are the vertices of a parallelogram.

119. Geometry. Find the coordinates of the point B of a line segment AB, given that $A = (2, 2)$, $d(A, B) = 12.5$, and the slope of the line segment AB is $\frac{4}{3}$. $(9.5, 12)$ or $(-5.5, -8)$

120. Geometry. Show that an equation of a circle with (x_1, y_1) and (x_2, y_2) as endpoints of a diameter can be written in the form

$$(x - x_1)(x - x_2) + (y - y_1)(y - y_2) = 0.$$

[*Hint:* Use the fact that when the two endpoints of a diameter and any other point on the circle are used as vertices for a triangle, the angle opposite the diameter is a right angle.]

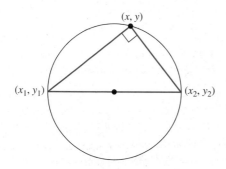

121. Geometry. Find an equation of the tangent line to the circle $x^2 + y^2 = 169$ at the point $(-5, 12)$. (See the accompanying figure.) $y = \frac{5}{12}x + \frac{169}{12}$

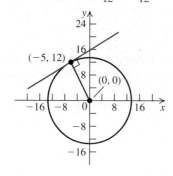

122. Geometry. Show that $xx_1 + yy_1 = a^2$ is an equation of the tangent line to the circle $x^2 + y^2 = a^2$ at the point (x_1, y_1) on the circle.

In Exercises 123–126, use a graphing device to graph the given lines on the same screen. Interpret your observations about the family of lines.

123. $y = 2x + b$ for $b = 0, \pm 2, \pm 4$

124. $y = mx + 2$ for $m = 0, \pm 2, \pm 4$

125. $y = m(x - 1)$ for $m = 0, \pm 3, \pm 5$

126. $y = m(x + 1) - 2$ for $m = 0, \pm 3, \pm 5$

Critical Thinking

127. By considering slopes, show that the points $(1, -1)$, $(-2, 5)$, and $(3, -5)$ lie on the same line.

128. By considering slopes, show that the points $(-9, 6)$, $(-2, 14)$, and $(-1, -1)$ are the vertices of a right triangle.

129. a. The equation $y + 2x + k = 0$ describes a family of lines for each value of k. Sketch the graphs of the members of this family for $k = 3, 0$, and -2. What is the common characteristic of the family? †

  b. Repeat part (a) for the family of lines
  $y - kx + 4 = 0$. †

Answers:
123. Lines with slope 2 and different y-intercepts
124. Lines with y-intercept 2 and different slopes
125. Lines passing through the point $(1, 0)$ with different slopes
126. Lines passing through the point $(-1, -2)$ with different slopes
127. The slope of the line that passes through $(1, -1)$ and $(-2, 5)$ is -2.
The slope of the line that passes through $(1, -1)$ and $(3, -5)$ is -2.
The slope of the line that passes through $(-2, 5)$ and $(3, -5)$ is -2.
128. The slope of the line that passes through $(-9, 6)$ and $(-2, 14)$ is $\frac{8}{7}$.
The slope of the line that passes through $(-9, 6)$ and $(-1, -1)$ is $-\frac{7}{8}$.
The slope of the line that passes through $(-2, 14)$ and $(-1, -1)$ is -15.

Functions

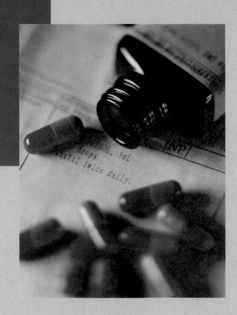

Before Starting this Section, Review

1. Graph of an equation (Section 1.1, page 6)

2. Equation of a circle (Section 1.1, page 11)

Objectives

1 Use functional notation and find function values.

2 Find the domain of a function.

3 Identify the graph of a function.

4 Find the average rate of change of a function.

5 Solve applied problems by using functions.

THE UPS AND DOWNS OF DRUG LEVELS

Medicines in the form of a pill have a much harder time getting to work than you do. First, they are swallowed and dumped into a pool of acid in your stomach. Then they dissolve; leave the stomach; and begin to be absorbed, mostly through the lining of the small intestine. Next, the blood around the intestine carries the medicines to the liver, an organ designed to metabolize (break down) and remove foreign material from the blood. Because of this property of the liver, most drugs have a tough time getting past it. The amount of drug that makes it into your bloodstream, compared with the amount that you put in your mouth, is called the drug's *bioavailability*. If a drug has low bioavailability, then much of the drug is destroyed before it reaches the bloodstream. The amount of drug you take in a pill is calculated to correct for this deficiency so that the amount you need actually ends up in your blood.

Once the drugs are past the liver, the bloodstream carries them throughout the rest of the body in about one minute. The study of the ways drugs are absorbed and move through the body is called *pharmacokinetics,* or PK for short. PK measures the ups and downs of drug levels in your body. In Example 9, we consider an application of *functions* to the levels of drugs in the bloodstream.

(Adapted from *The ups and downs of drug levels,* by Bob Munk, www.thebody.com/content/treat/art1062.html). ■

1 Use functional notation and find function values.

Functions

There are many ways of expressing a relationship between two quantities. For example, if you are paid $10 per hour, then the relation between the number of hours, x, that you work and the amount of money, y, that you earn may be expressed by the equation $y = 10x$. Replacing x with 40 yields $y = 400$ and indicates that 40 hours of work corresponds to a $400 paycheck. A special relationship such as $y = 10x$ in which to each element x in one set there corresponds a unique element y in another set is called a *function*. We say that your pay is a function of the number of hours you work. Because the value of y depends on the given value of x, y is called the **dependent variable** and x is called the **independent variable**. Any symbol may be used for the dependent or independent variables.

DEFINITION OF FUNCTION

A **function** from a set X to a set Y is a rule that assigns to each element of X one and only one corresponding element of Y. The set X is the **domain** of the function. The set of those elements of Y that correspond (are assigned) to the elements of X is the **range** of the function.

To decide whether a correspondence is a function, we must check whether *any* domain element is paired with more than one range element. If this happens, that domain element does not have a *unique* corresponding element of Y and the correspondence is not a function. See the correspondence diagrams in Figure 1.28.

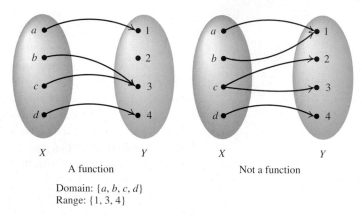

A function   Not a function

Domain: $\{a, b, c, d\}$
Range: $\{1, 3, 4\}$

FIGURE 1.28 Correspondence diagrams

BY THE WAY . . .

The functional notation
$y = f(x)$ was first used by the
great Swiss mathematician
Leonhard Euler in the
*Commentarii Academia
Petropolitanae,* published
in 1735.

We usually use single letters such as f, F, g, G, h, and H as the name of a function. If we choose f as the name of a function, then for each x in the domain of f, there corresponds a unique y in its range. The number y is denoted by $f(x)$, read as "f of x" or as "f at x." We call $f(x)$ the **value of f at the number x** and say that f *assigns* the value $f(x)$ to x.

Tables and graphs can be used to describe functions. The data in Table 1.3 (reproduced from Section 1.1) show the prevalence of smoking among adults aged 18 years and older in the United States over the years 1998–2004.

TABLE 1.3

Year	1998	1999	2000	2001	2002	2003	2004
Percent of adult smokers	24.1	23.5	23.2	22.7	22.4	21.6	20.9

Source: Centers for Disease Control, National Health Interview Survey.

The graph in Figure 1.29 is a **scatter diagram** that gives a visual representation of the same information. Scatter diagrams show the relationship between two variables by displaying data as points of a graph. The variable that might be considered the independent variable is plotted on the x-axis, and the dependent variable is plotted on the y-axis. Here the percent of adult smokers depends on the year and is plotted on the y-axis.

FIGURE 1.29 **Scatter diagram**

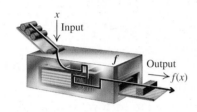

FIGURE 1.30 **The function f, as a machine**

The function described by Table 1.3 and Figure 1.29 has for its domain the set $\{1998, 1999, 2000, 2001, 2002, 2003, 2004\}$ and for its range the set $\{24.1, 23.5, 23.2, 22.7, 22.4, 21.6, 20.9\}$. The function assigns to each year in the domain the percent of adult smokers 18 years and older in the United States during that year.

Yet another nice way of picturing the concept of a function is as a "machine," as shown in Figure 1.30. If x is in the domain of a function, then the machine accepts it as an "input" and produces the value $f(x)$ as the "output."

◆ **WARNING**

In writing $y = f(x)$, do not confuse f, which is the name of the function, with $f(x)$, which is a number in the range of f that the function assigns to x. The symbol $f(x)$ does not mean "f times x."

STUDY TIP

The definition of a function is independent of the letters used to denote the function and the variables. For example,

$$s(x) = x^2, g(y) = y^2,$$
$$\text{and } h(t) = t^2$$

represent the same function.

Functions Defined by Equations When the relation defined by an equation in two variables is a function, we can often solve the equation for the dependent variable in terms of the independent variable. For example, the equation $y - x^2 = 0$ can be solved for y in terms of x:

$$y = x^2.$$

Now we can replace the dependent variable, in this case, y, with functional notation $f(x)$ and express the function as

$$f(x) = x^2 \quad \text{(read "} f \text{ of } x \text{ equals } x^2 \text{")}.$$

Here $f(x)$ plays the role of y and is the value of the function f at x. For example, if $x = 3$ is an element (input value) in the domain of f, the corresponding element (output value) in the range is found by replacing x with 3 in the equation

$$f(x) = x^2,$$

so that $\qquad f(3) = 3^2 = 9. \qquad$ Replace x with 3.

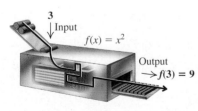

FIGURE 1.31 **The function f, as a machine**

We say that the value of the function f at 3 is 9. See Figure 1.31. In other words, the number 9 in the range corresponds to the number 3 in the domain and the ordered pair $(3, 9)$ is an ordered pair of the function f.

If a function g is defined by an equation such as $y = x^2 - 6x + 8$, the notation

$$y = x^2 - 6x + 8 \quad \text{and} \quad g(x) = x^2 - 6x + 8$$

define the same function.

EXAMPLE 1   **Determining Whether an Equation Defines a Function**

Determine whether y is a function of x for each equation.

a. $6x^2 - 3y = 12$ **b.** $y^2 - x^2 = 4$

SOLUTION

Solve each equation for y in terms of x. If more than one value of y corresponds to the same value of x, then y is not a function of x.

a.

$6x^2 - 3y = 12$	Original equation
$6x^2 - 3y + 3y - 12 = 12 + 3y - 12$	Add $3y - 12$ to both sides.
$6x^2 - 12 = 3y$	Simplify.
$2x^2 - 4 = y$	Divide by 3.

The last equation shows that only one value of y corresponds to each value of x. For example, if $x = 0$, then $y = 0 - 4 = -4$. So y is a function of x.

b.

$y^2 - x^2 = 4$	Original equation
$y^2 - x^2 + x^2 = 4 + x^2$	Add x^2 to both sides.
$y^2 = x^2 + 4$	Simplify.
$y = \pm\sqrt{x^2 + 4}$	Square root property

The last equation shows that two values of y correspond to each value of x. For example, if $x = 0$, then $y = \pm\sqrt{0^2 + 4} = \pm\sqrt{4} = \pm 2$. Both $y = 2$ and $y = -2$ correspond to $x = 0$. Therefore, y is not a function of x.   ■ ■ ■

Practice Problem 1   Determine whether y is a function of x for each equation.

a. $2x^2 - y^2 = 1$ **b.** $x - 2y = 5$ ■

TECHNOLOGY CONNECTION

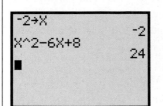 A graphing calculator allows you to store a number in a variable and then find the value of a function at that number. This screen shows the evaluation of the function $g(x) = x^2 - 6x + 8$ from Example 2b, for $x = -2$.

```
-2→X
              -2
X^2-6X+8
              24
■
```

EXAMPLE 2   **Evaluating a Function**

Let g be the function defined by the equation

$$y = x^2 - 6x + 8.$$

Evaluate each function value.

a. $g(3)$ **b.** $g(-2)$ **c.** $g\left(\dfrac{1}{2}\right)$ **d.** $g(a + 2)$ **e.** $g(x + h)$

SOLUTION

Since the function is named g, we replace y with $g(x)$ and write

$$g(x) = x^2 - 6x + 8 \qquad \text{Replace } y \text{ with } g(x).$$

In this notation, the independent variable x is a placeholder. We can write $g(x) = x^2 - 6x + 8$ as

$$g(\ ) = (\ )^2 - 6(\ ) + 8.$$

a.

$g(x) = x^2 - 6x + 8$	The given equation
$g(3) = 3^2 - 6(3) + 8$	Replace x with 3 at each occurrence of x.
$= 9 - 18 + 8 = -1$	Simplify.

The statement $g(3) = -1$ means that the value of the function g at 3 is -1.
    Just as in part **a**, we can evaluate g at any value x in its domain:

$$g(x) = x^2 - 6x + 8 \qquad \text{Original equation}$$

b. $g(-2) = (-2)^2 - 6(-2) + 8 = 24$ Replace x with -2 and simplify.

c. $g\left(\dfrac{1}{2}\right) = \left(\dfrac{1}{2}\right)^2 - 6\left(\dfrac{1}{2}\right) + 8 = \dfrac{21}{4}$ Replace x with $\dfrac{1}{2}$ and simplify.

d. $g(a + 2) = (a + 2)^2 - 6(a + 2) + 8$ Replace x with $(a + 2)$ in $g(x)$.

$\quad\quad\quad\quad = a^2 + 4a + 4 - 6a - 12 + 8$ Recall:
$\quad\quad\quad\quad\quad\quad\quad\quad\quad\quad\quad\quad\quad\quad\quad\quad (x + y)^2 = x^2 + 2xy + y^2.$

$\quad\quad\quad\quad = a^2 - 2a$ Simplify.

e. $g(x + h) = (x + h)^2 - 6(x + h) + 8$ Replace x with $(x + h)$ in $g(x)$.

$\quad\quad\quad\quad = x^2 + 2xh + h^2 - 6x - 6h + 8$ Simplify.   ■ ■ ■

Practice Problem 2   Let g be the function defined by the equation $y = -2x^2 + 5x$. Evaluate each function value.

a. $g(0)$ **b.** $g(-1)$ **c.** $g(x + h)$ ■

2 Find the domain of a function.

The Domain of a Function

Sometimes a function does not have a specified domain.

AGREEMENT ON DOMAIN

If the domain of a function that is defined by an equation is not specified, then we agree that the domain of the function is the largest set of real numbers that result in real numbers as outputs.

BY THE WAY . . .

The word *domain* comes from the Latin word *dominus,* or *master,* which in turn is related to *domus,* meaning "home." *Domus* is also the root for the words *domestic* and *domicile.*

When we use our agreement to find the domain of a function, first we usually find the values of the variable that do not result in real number outputs. Then we exclude those numbers from the domain. Remember that

1. division by zero is undefined.

2. the square root (or any even root) of a negative number is not a real number.

EXAMPLE 3   **Finding the Domain of a Function**

Find the domain of each function.

a. $f(x) = \dfrac{1}{1 - x^2}$ **b.** $g(x) = \sqrt{x}$ **c.** $h(x) = \dfrac{1}{\sqrt{x - 1}}$ **d.** $P(t) = 2t + 1$

SOLUTION

a. The function f is not defined when the denominator $1 - x^2$ is 0. Since $1 - x^2 = 0$ if $x = 1$ or $x = -1$, the domain of f is the set $\{x | x \neq -1$ and $x \neq 1\}$, which in interval notation is written $(-\infty, -1) \cup (-1, 1) \cup (1, \infty)$.

b. Since the square root of a negative number is not a real number, negative numbers are excluded from the domain of g. The domain of $g(x) = \sqrt{x}$ is $\{x | x \geq 0\}$, or $[0, \infty)$ in interval notation.

c. The function $h(x) = \dfrac{1}{\sqrt{x - 1}}$ has *two* restrictions. The square root of a negative number is not a real number, so $\sqrt{x - 1}$ is a real number only if $x - 1 \geq 0$. However, we cannot allow $x - 1 = 0$ because $\sqrt{x - 1}$ is in the denominator. Therefore, we must have $x - 1 > 0$, or $x > 1$. The domain of h must be $\{x | x > 1\}$, or in interval notation $(1, \infty)$.

d. When any real number is substituted for t in $y = 2t + 1$, a unique real number is determined. The domain is the set of all real numbers: $\{t \mid t \text{ is a real number}\}$, or in interval notation $(-\infty, \infty)$.   ■ ■ ■

Practice Problem 3   Find the domain of the function

$$f(x) = \frac{1}{\sqrt{1 - x}}.$$ ■

3 Identify the graph of a function.

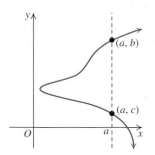

FIGURE 1.32 Graph does not represent a function

Graphs of Functions

The *graph of a function f* is the set of ordered pairs $(x, f(x))$ such that x is in the domain of f. That is, the graph of f is the graph of the equation $y = f(x)$. We sketch the graph of $y = f(x)$ by plotting points and joining them with a smooth curve. The graph of a function provides valuable visual information about the function.

Figure 1.32 shows that not every curve in the plane is the graph of a function. In Figure 1.32, a vertical line intersects the curve at two distinct points (a, b) and (a, c). This curve cannot be the graph of $y = f(x)$ for any function f because having $f(a) = b$ and $f(a) = c$ means that f assigns two different range values to the same domain element a. We express this statement in a slightly different way as follows:

VERTICAL-LINE TEST

If no vertical line intersects the graph of a curve (or scatterplot) at more than one point, then the curve (or scatterplot) is the graph of a function.

EXAMPLE 4   **Identifying the Graph of a Function**

Determine which graphs in Figure 1.33 are graphs of functions.

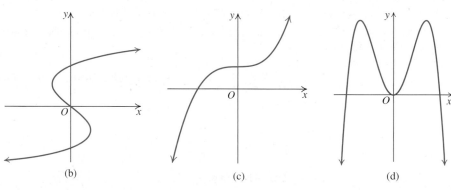

(a)   (b)   (c)   (d)

FIGURE 1.33 The vertical-line test

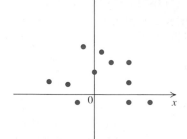

SOLUTION

The graphs in Figures 1.33(a) and 1.33(b) are not graphs of functions because a vertical line can be drawn through the two points farthest to the left in Figure 1.33(a) and the y-axis is one of many vertical lines that contain more than one point on the graph in Figure 1.33(b). The graphs in Figures 1.33(c) and 1.33(d) are the graphs of functions because no vertical line intersects either graph at more than one point.   ■ ■ ■

Practice Problem 4   Decide whether the graph in the margin is the graph of a function.   ■

EXAMPLE 5   **Examining the Graph of a Function**

Let $y = f(x) = x^2 - 2x - 3$.

a. Is the point $(1, -3)$ on the graph of f?

b. Find all values of x so that $(x, 5)$ is on the graph of f.

c. Find all y-intercepts of the graph of f.

d. Find all x-intercepts of the graph of f.

SOLUTION

a. We check whether $(1, -3)$ satisfies the equation $y = x^2 - 2x - 3$.

$$-3 \overset{?}{=} (1)^2 - 2(1) - 3 \qquad \text{Replace } x \text{ with 1 and } y \text{ with } -3.$$
$$-3 \overset{?}{=} -4 \, \text{No!}$$

So $(1, -3)$ is *not* on the graph of f. See Figure 1.34.

b. Substitute 5 for y in $y = x^2 - 2x - 3$ and solve for x.

$$5 = x^2 - 2x - 3$$
$$0 = x^2 - 2x - 8 \qquad \text{Subtract 5 from both sides.}$$
$$0 = (x - 4)(x + 2) \qquad \text{Factor.}$$
$$x - 4 = 0 \quad \text{or} \quad x + 2 = 0 \qquad \text{Zero-product property}$$
$$x = 4 \quad \text{or} \quad x = -2. \qquad \text{Solve for } x.$$

The point $(x, 5)$ is on the graph of f only when $x = -2$ and $x = 4$. Both $(-2, 5)$ and $(4, 5)$ are on the graph of f. See Figure 1.34.

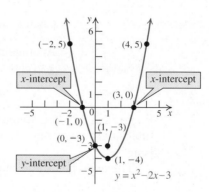

FIGURE 1.34

c. Find all points (x, y) with $x = 0$ in $y = x^2 - 2x - 3$.

$$y = 0^2 - 2(0) - 3 \qquad \text{Replace } x \text{ with 0 to find the } y\text{-intercept.}$$
$$y = -3 \qquad \text{Simplify.}$$

The only y-intercept is -3. (See Figure 1.34.)

d. Find all points (x, y) with $y = 0$ in $y = x^2 - 2x - 3$.

$$0 = x^2 - 2x - 3$$
$$0 = (x + 1)(x - 3) \qquad \text{Factor.}$$
$$x + 1 = 0 \quad \text{or} \quad x - 3 = 0 \qquad \text{Zero-product property}$$
$$x = -1 \quad \text{or} \quad x = 3. \qquad \text{Solve for } x.$$

The x-intercepts of the graph of f are -1 and 3. See Figure 1.34.   ■■■

Practice Problem 5 Let $y = f(x) = x^2 + 4x - 5$.

a. Is the point $(2, 7)$ on the graph of f?

b. Find all values of x so that $(x, -8)$ is on the graph of f.

c. Find the y-intercept of the graph of f.

d. Find any x-intercepts of the graph of f. ■

SUMMARY

A function is usually described in one or more of the following four ways:

- A correspondence diagram
- A table of values
- An equation or a formula
- A scatter diagram or a graph

Input	Output
x	y or $f(x)$
First coordinate	Second coordinate
Independent variable	Dependent variable
Domain is the set of all inputs.	Range is the set of all outputs.

---

4 Find the average rate of change of a function.

Average Rate of Change

Suppose your salary increases from $25,000 a year to $40,000 a year over a five-year period. You then have

Change in salary: $40,000 − $25,000 = $15,000

Average rate of change in salary:

$$\frac{\$40,000 - \$25,000}{5} = \frac{\$15,000}{5} = \$3000 \text{ per year.}$$

Regardless of when you actually received the raises during the five-year period, the final salary is the same as if you received your average annual increase, $3000, *each* year.

The average rate of change can be defined in a more general setting.

THE AVERAGE RATE OF CHANGE OF A FUNCTION

Let $(a, f(a))$ and $(b, f(b))$ be two points on the graph of a function f. Then the **average rate of change** of $f(x)$ as x changes from a to b is defined by

$$\frac{f(b) - f(a)}{b - a}, a \neq b.$$

FINDING THE SOLUTION: A PROCEDURE

| EXAMPLE 6 | Finding the Average Rate of Change |

OBJECTIVE

Find the average rate of change of a function f as x changes from a to b.

Step 1 Find $f(a)$ and $f(b)$.

Step 2 Use the values from Step 1 in the definition of average rate of change.

EXAMPLE

Find the average rate of change of $f(x) = 2 - 3x^2$ as x changes from $x = 1$ to $x = 3$.

1. $f(1) = 2 - 3(1)^2 = -1$ $a = 1$
    $f(3) = 2 - 3(3)^2 = -25$ $b = 3$

2. $\dfrac{f(b) - f(a)}{b - a} = \dfrac{-25 - (-1)}{3 - 1} = -\dfrac{24}{2}$
    $= -12$ ■ ■ ■

Practice Problem 6 Find the average rate of change of $f(x) = 1 - x^2$ as x changes from 2 to 4. ■

| EXAMPLE 7 | Finding the Average Rate of Change |

Find the average rate of change of $f(x) = 2x^2 - 3$ as x changes from $x = c$ to $x = c + h, h \neq 0$.

SOLUTION

Average rate of change $= \dfrac{f(c + h) - f(c)}{(c + h) - c}$ Definition

$= \dfrac{[2(c + h)^2 - 3] - [2(c)^2 - 3]}{c + h - c}$ Find $f(c + h)$ and $f(c)$.

$= \dfrac{[2(c^2 + 2ch + h^2) - 3] - (2c^2 - 3)}{h}$ Expand the binomial square.

$= \dfrac{4ch + 2h^2}{h} = \dfrac{\cancel{h}(4c + 2h)}{\cancel{h}}$ Simplify; $h \neq 0$.

$= 4c + 2h$ ■ ■ ■

Practice Problem 7 Find the average rate of change of $f(x) = 1 - x^2$ as x changes from $x = c$ to $x = c + h, h \neq 0$. ■

The average rate of change calculated in Example 7 is called a *difference quotient*. The difference quotient is an important concept in calculus.

DIFFERENCE QUOTIENT

For a function *f*, the **difference quotient** is

$$\frac{f(x + h) - f(x)}{h}, \quad h \neq 0.$$

| EXAMPLE 8 | Evaluating and Simplifying a Difference Quotient |

Let $f(x) = 2x^2 - 3x + 5$. Find and simplify $\dfrac{f(x + h) - f(x)}{h}, h \neq 0$.

SOLUTION

First, we find $f(x + h)$.

$$f(x) = 2x^2 - 3x + 5 \qquad \text{Original equation}$$
$$f(x + h) = 2(x + h)^2 - 3(x + h) + 5 \qquad \text{Replace } x \text{ with } (x + h).$$
$$= 2(x^2 + 2xh + h^2) - 3(x + h) + 5 \qquad (x + h)^2 = x^2 + 2xh + h^2$$
$$= 2x^2 + 4xh + 2h^2 - 3x - 3h + 5 \qquad \text{Use the distributive property.}$$

Then we substitute into the difference quotient.

$$\frac{f(x + h) - f(x)}{h} = \frac{\overbrace{(2x^2 + 4xh + 2h^2 - 3x - 3h + 5)}^{f(x+h)} - \overbrace{(2x^2 - 3x + 5)}^{f(x)}}{h}$$

$$= \frac{2x^2 + 4xh + 2h^2 - 3x - 3h + 5 - 2x^2 + 3x - 5}{h} \qquad \text{Subtract.}$$

$$= \frac{4xh - 3h + 2h^2}{h} \qquad \text{Simplify.}$$

$$= \frac{\cancel{h}(4x - 3 + 2h)}{\cancel{h}} \qquad \text{Factor out } h.$$

$$= 4x - 3 + 2h \qquad \text{Remove the common factor } h. \qquad ■ ■ ■$$

Practice Problem 8 Repeat Example 8 for $f(x) = -x^2 + x - 3$. ■

5 Solve applied problems by using functions.

Applications

Level of Drugs in Bloodstream When you take a dose of medication, the drug level in the blood goes up quickly and soon reaches its *peak*, called $C_{\max}$ (the maximum concentration). As your liver or kidneys remove the drug, your blood's drug levels drop until the next dose enters your bloodstream. The lowest drug level is called the *trough*, or $C_{\min}$ (the minimum concentration). The ideal dose should be strong enough to be effective without causing too many side effects. We start by setting upper and lower limits on the drug's blood levels, shown by the two horizontal lines on a pharmacokinetics (PK) graph. See Figure 1.35. The upper line represents the drug level at which people start to develop serious side effects. The lower line represents the minimum drug level that provides the desired effect.

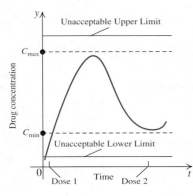

FIGURE 1.35 Drug levels

EXAMPLE 9 **Cholesterol-Reducing Drugs**

Many drugs used to lower high blood cholesterol levels are called *statins* and are very popular and widely prescribed. These drugs, along with proper diet and exercise, help prevent heart attacks and strokes. Recall from the introduction to this section that bioavailability is the amount of a drug you have ingested that makes it into your bloodstream. A statin with a bioavailability of 30% has been prescribed for Boris to treat his cholesterol levels. He is to take 20 milligrams daily. Every day his body filters out half the statin. Find the maximum concentration of the statin in the bloodstream on each of the first ten days of using the drug and graph the result.

SOLUTION

Since the statin has 30% bioavailability and Boris takes 20 milligrams per day, the maximum concentration in the bloodstream is 30% of 20 milligrams, or $20(0.3) =$ 6 milligrams from each day's prescription. Because half the statin is filtered out of the body each day, the daily maximum concentration is

$$\frac{1}{2}(\text{previous day's maximum concentration}) + 6.$$

Table 1.4 shows the maximum concentration of the drug for each of the first ten days. After the first day, each number in the second column is computed (to three decimal places) by adding 6 to one-half the number above it. The graph is shown in Figure 1.36.

TABLE 1.4

Day	Maximum Concentration
1	6.000
2	9.000
3	10.500
4	11.250
5	11.625
6	11.813
7	11.906
8	11.953
9	11.977
10	11.988

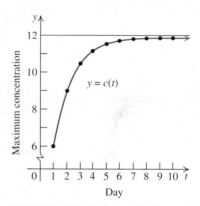

FIGURE 1.36 Maximum drug concentration

The graph shows that the maximum concentration of the statin in the bloodstream approaches 12 milligrams if Boris continues to take one pill every day. ■ ■ ■

Practice Problem 9 In Example 9, use Table 1.4 to find the domain and the range of the function. Also compute the function value at 6. ■

Functions in Economics

The important concepts of *breaking even, earning a profit,* or *taking a loss* in economics will be discussed throughout this book. Here we define some of the elementary functions in economics.

Let x represent the number of units of an item manufactured and sold at a price of p dollars. Then the cost $C(x)$ of producing x items includes the **fixed cost** (such as rent, insurance, product design, setup, promotion, and overhead) and the **variable cost** (which depends on the number of items produced at a certain cost per item).

Linear Cost Function A linear cost function $C(x)$ is given by

$$C(x) = \text{(variable cost)} + \text{(fixed cost)}$$
$$= ax + b,$$

where the fixed cost is b and the **marginal cost** (cost of producing each item) is a dollars per item.

Average cost, denoted by $\overline{C}(x)$, is defined by $\overline{C}(x) = \dfrac{C(x)}{x}$.

Linear Price–Demand Function Suppose x items can be sold (demanded) at a price of p dollars per item. Then a linear demand function usually has the form

$$p(x) = -mx + d \qquad \text{expressing } p \text{ as a function of } x,$$

or

$$x(p) = -np + k \qquad \text{expressing } x \text{ as a function of } p.$$

Both expressions indicate that a higher price will result in fewer items sold. The constants $m, d, n,$ and k depend on the given situation.

Revenue Function $R(x)$

Revenue = (Price per item) × (Number of items sold)

$$R(x) = p \cdot x = (-mx + d)x \qquad\qquad p = p(x) = -mx + d$$

Profit Function $P(x)$

$$\text{Profit} = \text{Revenue} - \text{Cost}$$
$$P(x) = R(x) - C(x)$$

A manufacturing company will

  (i) have a *profit* if $P(x) > 0$.

 (ii) *break even* if $P(x) = 0$.

(iii) suffer a *loss* if $P(x) < 0$.

EXAMPLE 10   **Breaking Even**

Metro Entertainment Co. spent $100,000 on production costs for its off-Broadway play *Bride and Prejudice*. Once the play runs, each performance costs $1000 per show and the revenue from each show is $2400. Using x to represent the number of shows,

a. write the cost function $C(x)$.

b. write the revenue function $R(x)$.

c. write the profit function $P(x)$.

d. determine how many showings of *Bride and Prejudice* must be held for Metro to break even.

SOLUTION

a. $C(x) = (\text{Variable cost}) + (\text{fixed cost})$
$$= 1000x + 100{,}000$$

b. $R(x) = (\text{Revenue per show})(\text{Number of shows})$
$$= 2400x$$

c. $P(x) = R(x) - C(x)$
$$= 2400x - (1000x + 100{,}000)$$
$$= 2400x - 1000x - 100{,}000 = 1400x - 100{,}000$$

d. Metro will break even when $P(x) = 0$.

$$1400x - 100{,}000 = 0$$
$$1400x = 100{,}000$$
$$x = \frac{100{,}000}{1400} \approx 71.429. \qquad \text{Divide both sides by 1400.}$$

Only a whole number of shows is possible, so Metro needs 72 shows to break even.   ■■■

Practice Problem 10   Suppose in Example 10 that once the play runs, each performance costs $1200 and the revenue from each show is $2500. How do the results in Example 10 change?   ■

SECTION 1.3 ■ Exercises

A EXERCISES Basic Skills and Concepts

1. In the functional notation $y = f(x)$, x is the __independent__ variable.

2. If $f(-2) = 7$, then -2 is in the __domain__ of the function f and 7 is in the __range__ of f.

3. If the point $(9, -14)$ is on the graph of a function f, then $f(9) = $ __-14__.

4. If $(3, 7)$ and $(3, 0)$ are points on a graph, then the graph cannot be the graph of a(n) __function__.

5. To find the x-intercepts of the graph of an equation in x and y, we solve the equation __$y = 0$__.

6. *True or False.* If $(3, 6)$ and $(5, 22)$ are points on the graph of f, then the average rate of change of f as x changes from 3 to 5 is 14. __False__

7. *True or False.* If $x = -7$, then $-x$ is in the domain of $f(x) = \sqrt{x}$. __True__

8. *True or False.* The domain of $f(x) = \dfrac{1}{\sqrt{x + 2}}$ is all real x, $x \neq -2$. __False__

In Exercises 9–14, determine the domain and range of each function. Explain why each non-function is not a function.

9.

Domain: $\{a, b, c\}$; range: $\{d, e\}$; function

10.
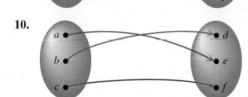
Domain: $\{a, b, c\}$; range: $\{d, e, f\}$; function

11.
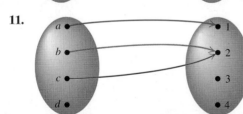
Domain; $\{a, b, c\}$; range: $\{1, 2\}$; function

12.
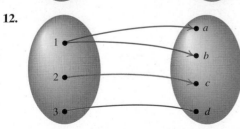
Domain: $\{1, 2, 3\}$; range: $\{a, b, c, d\}$; not a function

Domain: $\{-3, -1, 0, 1, 2, 3\}$; range: $\{-8, -3, 0, 1\}$; function

13.

x	-3	-1	0	1	2	3
y	-8	0	1	0	-3	-8

14.

x	0	3	8	0	3	8
y	-1	-2	-3	1	2	2

Domain: $\{0, 3, 8\}$; range: $\{-3, -2, -1, 1, 2\}$; not a function

In Exercises 15–28, determine whether each equation defines y as a function of x.

15. $x + y = 2$ Yes

16. $x = y - 1$ Yes

17. $y = \dfrac{1}{x}$ Yes

18. $xy = -1$ Yes

19. $y = |x - 1|$ Yes

20. $x = |y|$ No

21. $y = \dfrac{1}{\sqrt{2x - 5}}$ Yes

22. $y = \dfrac{1}{\sqrt{x^2 - 1}}$ Yes

23. $2 - y = 3x$ Yes

24. $3x - 5y = 15$ Yes

25. $x^2 + y = 8$ Yes

26. $x = y^2$ No

27. $x^2 + y^3 = 5$ Yes

28. $x + y^3 = 8$ Yes

In Exercises 29–32, let $f(x) = x^2 - 3x + 1$, $g(x) = \dfrac{2}{\sqrt{x}}$, and $h(x) = \sqrt{2 - x}$.

29. Find $f(0), g(0), h(0), f(a)$, and $f(-x)$.

30. Find $f(1), g(1), h(1), g(a)$, and $g(x^2)$. †

31. Find $f(-1), g(-1), h(-1), h(c)$, and $h(-x)$. †

32. Find $f(4), g(4), h(4), g(2 + k)$, and $f(a + k)$. †

33. Let $f(x) = \dfrac{2x}{\sqrt{4 - x^2}}$. Find each function value.

   a. $f(0)$ 0
   b. $f(1)$ $\dfrac{2\sqrt{3}}{3}$
   c. $f(2)$ Not defined
   d. $f(-2)$ Not defined
   e. $f(-x)$ $\dfrac{-2x}{\sqrt{4 - x^2}}$

34. Let $g(x) = 2x + \sqrt{x^2 - 4}$. Find each function value.

   a. $g(0)$ Not defined
   b. $g(1)$ Not defined
   c. $g(2)$ 4
   d. $g(-3)$ $-6 + \sqrt{5}$
   e. $g(-x)$ $-2x + \sqrt{x^2 - 4}$

29. $f(0) = 1$, $g(0)$ is not defined, $h(0) = \sqrt{2}$, $f(a) = a^2 - 3a + 1$, $f(-x) = x^2 + 3x + 1$

†Due to space constrictions, answers to these exercises may be found in the Answers beginning on page A–1 in the back of the book.

In Exercises 35–48, find the domain of each function.

35. $f(x) = -8x + 7$ **36.** $f(x) = 2x^2 - 11$ $(-\infty, \infty)$

$(-\infty, \infty)$

37. $f(x) = \dfrac{1}{x - 9}$ **38.** $f(x) = \dfrac{1}{x + 9}$

39. $h(x) = \dfrac{2x}{x^2 - 1}$ **40.** $h(x) = \dfrac{x - 3}{x^2 - 4}$

41. $G(x) = \dfrac{\sqrt{x - 3}}{x + 2}$ $[3, \infty)$ **42.** $G(x) = \dfrac{\sqrt{x + 3}}{1 - x}$

43. $f(x) = \dfrac{3}{\sqrt{4 - x}}$ $(-\infty, 4)$ **44.** $f(x) = \dfrac{x}{\sqrt{2 - x}}$ $(-\infty, 2)$

45. $F(x) = \dfrac{x + 4}{x^2 + 3x + 2}$ **46.** $F(x) = \dfrac{1 - x}{x^2 + 5x + 6}$

47. $g(x) = \dfrac{\sqrt{x^2 + 1}}{x}$ **48.** $g(x) = \dfrac{1}{x^2 + 1}$ $(-\infty, \infty)$

In Exercises 49–54, use the vertical-line test to determine whether the given graph represents a function.

49. Yes

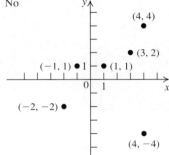

50. No

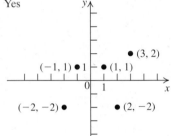

51. Yes

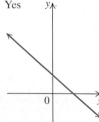

52. No

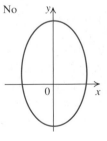

53.   Yes

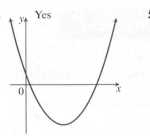

54. No

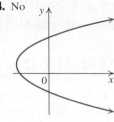

In Exercises 55–58, the graph of a function is given. Find the indicated function values.

55. $f(-4), f(-1), f(3), f(5)$, $f(-4) = -2, f(-1) = 1,$
$f(3) = 5, f(5) = 7$

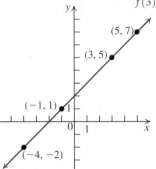

56. $g(-2), g(1), g(3), g(4)$ $g(-2) = 5, g(1) = -4,$
$g(3) = 0, g(4) = 5$

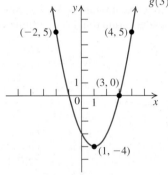

57. $h(-2), h(-1), h(0), h(1)$ $h(-2) = -5, h(-1) = 4,$
$h(0) = 3, h(1) = 4$

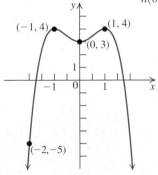

Answers:
37. $(-\infty, 9) \cup (9, \infty)$ **38.** $(-\infty, -9) \cup (-9, \infty)$
39. $(-\infty, -1) \cup (-1, 1) \cup (1, \infty)$ **40.** $(-\infty, -2) \cup (-2, 2) \cup (2, \infty)$
42. $[-3, 1) \cup (1, \infty)$ **45.** $(-\infty, -2) \cup (-2, -1) \cup (-1, \infty)$
46. $(-\infty, -3) \cup (-3, -2) \cup (-2, \infty)$ **47.** $(-\infty, 0) \cup (0, \infty)$

58. $f(-1), f(0), f(1)$ $f(-1) = 4, f(0) = 0, f(1) = -4$

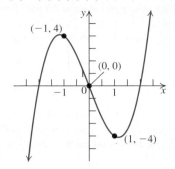

59. Let $h(x) = x^2 - x + 1$. Find x such that $(x, 7)$ is on the graph of h. $x = -2$ or 3

60. Let $H(x) = x^2 + x + 8$. Find x such that $(x, 7)$ is on the graph of H.   No solution

61. Let $f(x) = -2(x + 1)^2 + 7$.
  a. Is $(1, 1)$ a point of the graph of f?   No
  b. Find all x such that $(x, 1)$ is on the graph of f.   †
  c. Find all y-intercepts of the graph of f.   5
  d. Find all x-intercepts of the graph of f.   †

62. Let $g(x) = -3x^2 - 12x$.
  a. Is $(-2, 10)$ a point of the graph of f?   No
  b. Find x such that $(x, 12)$ is on the graph of g. $x = -2$
  c. Find all y-intercepts of the graph of f.   0
  d. Find all x-intercepts of the graph of f. $-4, 0$

In Exercises 63–74, find the average rate of change of the function as x changes from a to b.

63. $f(x) = -2x + 7; a = -1, b = 3$ -2

64. $f(x) = 4x - 9; a = -2, b = 2$ 4

65. $g(x) = 2x^2; a = 0, b = 5$ 10

66. $g(x) = -4x^2; a = -1, b = 4$ -12

67. $h(x) = x^2 - 1; a = -2, b = 0$ -2

68. $h(x) = 2 - x^2; a = 3, b = 4$ -7

69. $f(x) = (3 - x)^2; a = 1, b = 3$ -2

70. $f(x) = (x - 2)^2; a = -1, b = 5$ 0

71. $g(x) = x^3; a = -1, b = 3$ 7

72. $g(x) = -x^3; a = -1, b = 3$ -7

73. $h(x) = \dfrac{1}{x}; a = 2, b = 6$ $-\dfrac{1}{12}$

74. $h(x) = \dfrac{4}{x + 3}; a = -2, b = 4$ $-\dfrac{4}{7}$

For each function in Exercises 75–84, compute (for $h \neq 0$):

  a. $f(x + h)$ **b.** $f(x + h) - f(x)$ **c.** $\dfrac{f(x + h) - f(x)}{h}$

75. $f(x) = x$ †

76. $f(x) = 3x + 2$ †

77. $f(x) = x^2$ †

78. $f(x) = x^2 - x$ †

79. $f(x) = 2x^2 + 3x$ †

80. $f(x) = 3x^2 - 2x + 5$ †

81. $f(x) = 4$ †

82. $f(x) = -3$ †

83. $f(x) = \dfrac{1}{x}$ †

84. $f(x) = -\dfrac{1}{x}$ †

B EXERCISES Applying the Concepts

In Exercises 85–88, state whether the given relation is a function and explain why.

85. To each day of the year there corresponds the high temperature in your hometown on that day.
Yes because there is only one high temperature every day.

86. To each year since 1950 there corresponds the cost of a first-class postage stamp on January 1 of that year.

87. To each letter of the word *NUTS* there corresponds the states whose names start with that letter.
No because Nevada and North Carolina both start with *N*.

88. To each day of the week there corresponds the first names of people currently living in the United States born on that day of the week.   No because people with different first names may have the same birthday.

In Exercises 89 and 90, use the table listing the Super Bowl winners for the years 1995–2008.

Year	Super Bowl Winner
1995	San Francisco
1996	Dallas
1997	Green Bay
1998	Denver
1999	Denver
2000	St. Louis
2001	Baltimore
2002	New England
2003	Tampa Bay
2004	New England
2005	New England
2006	Pittsburgh
2007	Indianapolis
2008	New York Giants

Source: www.superbowlhistory.net

89. Consider the correspondence whose domain values are the years 1995–2008 and whose range values are the corresponding winning teams. Is this a function? Explain your answer.   Yes because there is only one winner in each year.

90. Consider the correspondence whose domain values are the names of the winning teams in the table and whose range values are the corresponding years. Is this a function? Explain your answer.
No because New England won more than once.

91. **Square tiles.** The area $A(x)$ of a square tile is a function of the length x of the square's side. Write a function rule for the area of a square tile. Find and interpret $A(4)$. $A(x) = x^2; A(4) = 16; A(4)$ represents the area of a square tile with side of length 4.

92. **Cube.** The volume $V(x)$ of a cube is a function of the length x of the cube's edge. Write a function rule for the volume of a cube. Evaluate the function for a cube with sides of length 3 inches. $V(x) = x^3; V(3) = 27$ in.^{3}

Answers:
86. Yes because the cost of first-class postage on January 1 of any year is a single value.

93. **Surface area.** Is the total surface area S of a cube a function of the edge x of the cube? If it is not a function, explain why not. If it is a function, write the function rule $S(x)$ and evaluate $S(3)$. It is a function. $S(x) = 6x^2$; $S(3) = 54$

94. **Measurement.** One meter equals about 39.37 inches. Write a function rule for converting inches to meters. Evaluate the function for 59 inches. $f(x) = \dfrac{x}{39.37}$; $f(59) \approx 1.5$

95. **Cost.** A computer notebook manufacturer has a daily fixed cost of $10,500 and a marginal cost of $210 per notebook.
    a. Find the daily cost $C(x)$ of manufacturing x notebooks per day. $C(x) = 210x + 10,500$
    b. Find the cost of producing 50 notebooks per day. $21,000
    c. Find the average cost of a computer notebook if 50 notebooks are manufactured each day. $420
    d. How many notebooks should be manufactured each day so that the average cost per notebook is $315? 100

96. **Cost.** The Just Juice Company has a monthly cost of $20,000 and a marginal cost of $4 per case of juice.
    a. Find the monthly cost $C(x)$ if the company produces x cases of juice per month. $C(x) = 4x + 20,000$
    b. Find the cost of producing 12,000 cases of juice per month. $68,000
    c. Find the average cost of a case of juice in part (b). $5.67
    d. How many cases of juice should be produced each month so that the average cost per case is $4.50? 40,000

97. **Price–demand.** Analysts for an electronics company determined that the price–demand function for its 30-inch LCD televisions is
    $$p(x) = 1275 - 25x, \quad 1 \le x \le 30,$$
    where p is the wholesale price per TV in dollars and x, in thousands, is the number of TVs that can be sold.
    a. Find and interpret $p(5)$, $p(15)$, and $p(30)$. †
    b. Sketch the graph of $y = p(x)$. †
    c. How many TVs can be sold at $650 per TV? 25

98. **Revenue.**
    a. Using the price–demand function $p(x) = 1275 - 25x, 1 \le x \le 30$, of Exercise 97, write the company's revenue function $R(x)$ and state its domain. $R(x) = 1275x - 25x^2$; domain: $[0, 30]$
    b. Find and interpret $R(1)$, $R(5)$, $R(10)$, $R(15)$, $R(20)$, $R(25)$, and $R(30)$. †
    c. Using the data from part (b), sketch the graph of $y = R(x)$. †
    d. Find the number of TVs that must be sold to generate a revenue of $4.7 million. [*Hint:* Solve $4700 = 1275x - 25x^2$] 4000

99. **Breaking even.** Peerless Publishing Company intends to publish Harriet Henrita's next novel. The estimated cost is $75,000 plus $5.50 for each copy printed. The selling price per copy is $15. The bookstores retain 40% of the selling price as commission. Let x represent the number of copies printed and sold.
    a. Find the cost function $C(x)$. $C(x) = 5.5x + 75,000$
    b. Find the revenue function $R(x)$. $R(x) = 9x$

    c. Find the profit function $P(x)$. $P(x) = 3.5x - 75,000$
    d. How many copies of the novel must be sold for Peerless to break even? 21,429
    e. What is the company profit if 46,000 copies are sold? $86,000

100. **Breaking even.** Capital Records Company plans to produce a CD by the popular rapper Rapit. The fixed cost is $500,000 and the variable cost is $0.50 per CD. The company sells each CD to record stores for $5. Let x represent the number of CDs produced and sold.
     a. How many disks must be sold for the company to break even? 111,111
     b. How many CDs must be sold for the company to make a profit of $750,000? 277,778

101. **Motion of a projectile.** A stone thrown upward with an initial velocity of 128 feet per second will attain a height of h feet in t seconds, where
     $$h(t) = 128t - 16t^2, \quad 0 \le t \le 8.$$
     a. What is the domain of h? $[0, 8]$
     b. Find $h(2)$, $h(4)$, and $h(6)$.
     c. How long will it take the stone to hit the ground?
     d. Sketch a graph of $y = h(t)$. †

102. **Housing affordability in the United States.** The following table lists the median sale price of existing homes during 1991–2006.

Year	Price in thousands of dollars
1991	99.7
1993	103.1
1995	111.1
1997	121.8
1999	133.3
2001	147.8
2003	182.1
2005	219.6
2006	221.9

Source: National Association of Realtors.

Find the average rate of change of a median-priced home during each period.
     a. 1995–1997 5.35
     b. 1991–2006 8.15
     c. Use your results in (a) and (b) to estimate the median sale price of a home in 1996.
        Based on (a): $116,450; based on (b): $140,450

103. **Drug prescription.** A certain drug has been prescribed to treat an infection. The patient receives an injection of 16 milliliters of the drug every four hours. During this same four-hour period, the kidneys filter out one-fourth of the drug. Find the concentration of the drug

Answers:
101. b. $h(2) = 192, h(4) = 256, h(6) = 192$ **c.** 8 sec

after 4 hours, 8 hours, 12 hours, 16 hours, and 20 hours. Sketch the graph of the concentration of the drug in the bloodstream as a function of time. †

104. Drug prescription. Every day Jane takes one 500-milligram aspirin tablet that has 80% bioavailability. During the same day, three-fourths of the aspirin is metabolized. Find the maximum concentration of the aspirin in the bloodstream on each of the first ten days of using the drug and graph the result. †

C EXERCISES Beyond the Basics

105. Give an example (if possible) of a function matching each description. Some have many different correct answers.
  a. Its graph is symmetric with respect to the y-axis.
  b. Its graph is symmetric with respect to the x-axis.
  c. Its graph is symmetric with respect to the origin.
  d. Its graph consists of a single point. $f(x) = \sqrt{-x^2}$
  e. Its graph is a horizontal line. $f(x) = 1$
  f. Its graph is a vertical line. Not possible

106. Explain whether $f(x)$ and $g(x)$ are equal.
  a. $f(x) = 2(\sqrt{x})^2 + 5$, $g(x) = 2x + 5$
  b. $f(x) = 2(\sqrt[3]{x})^3 + 5$, $g(x) = 2x + 5$

In Exercises 107–112, find an expression for $f(x)$ by solving for y and replacing y with $f(x)$. Then find the domain of f and compute $f(4)$.

107. $3x - 5y = 15$

108. $x = \dfrac{y}{y - 1}$

109. $x = \dfrac{2}{y - 4}$

110. $xy - 3 = 2y$

111. $(x^2 + 1)y + x = 2$

112. $yx^2 - \sqrt{x} = -2y$

In Exercises 113–118, state whether f and g represent the same function. Explain your reasons.

113. $f(x) = 3x - 4, 0 \le x \le 8$ $g(x) = 3x - 4$

114. $f(x) = 3x^2$ $g(x) = 3x^2, -5 \le x \le 5$

115. $f(x) = x - 1$ $g(x) = \dfrac{x^2 - 1}{x + 1}$

116. $f(x) = \dfrac{x + 2}{x^2 - x - 6}$ $g(x) = \dfrac{1}{x - 3}$

Answers:
105. a. $f(x) = |x|$ **b.** $f(x) = 0$ **c.** $f(x) = x$
106. a. No because the domain of f is $[0, \infty)$ while the domain of g is $(-\infty, \infty)$ **b.** Yes because $(\sqrt[3]{x})^3 = x$ for every real number x
107. $f(x) = \dfrac{3}{5}x - 3$; domain: $(-\infty, \infty)$; $f(4) = -\dfrac{3}{5}$
108. $f(x) = \dfrac{x}{x - 1}$; domain: $(-\infty, 1) \cup (1, \infty)$; $f(4) = \dfrac{4}{3}$
109. $f(x) = \dfrac{4x + 2}{x}$; domain: $(-\infty, 0) \cup (0, \infty)$; $f(4) = \dfrac{9}{2}$
110. $f(x) = \dfrac{3}{x - 2}$; domain: $(-\infty, 2) \cup (2, \infty)$; $f(4) = \dfrac{3}{2}$
111. $f(x) = \dfrac{2 - x}{x^2 + 1}$; domain: $(-\infty, \infty)$; $f(4) = -\dfrac{2}{17}$
112. $f(x) = \dfrac{\sqrt{x}}{x^2 + 2}$; domain: $[0, \infty)$; $f(4) = \dfrac{1}{9}$

117. $f(x) = \dfrac{x^2 - 4}{x - 2}, \ 2 < x < \infty$
   $g(x) = x + 2, \ 2 < x < \infty$

118. $f(x) = \dfrac{x - 2}{x^2 - 6x + 8}, \ 2 < x < \infty$
   $g(x) = \dfrac{1}{x - 4}, \ 2 < x < \infty$

119. Let $f(x) = ax^2 + ax - 3$. If $f(2) = 15$, find a. 3

120. Let $g(x) = x^2 + bx + b^2$. If $g(6) = 28$, find b.

121. Let $h(x) = \dfrac{3x + 2a}{2x - b}$. If $h(6) = 0$ and $h(3)$ are undefined, find a and b. $a = -9, b = 6$

122. Let $f(x) = 2x - 3$. Find $f(x^2)$ and $[f(x)]^2$.

123. If $g(x) = x^2 - \dfrac{1}{x^2}$, show that $g(x) + g\left(\dfrac{1}{x}\right) = 0$.

124. If $f(x) = \dfrac{x - 1}{x + 1}$, show that $f\left(\dfrac{x - 1}{x + 1}\right) = -\dfrac{1}{x}$.

125. If $f(x) = \dfrac{x + 3}{4x - 5}$ and $t = \dfrac{3 + 5x}{4x - 1}$, show that $f(t) = x$.

Critical Thinking

126. Write an equation of a function with each domain. Answers will vary.
  a. $[2, \infty)$
  b. $(2, \infty)$
  c. $(-\infty, 2]$
  d. $(-\infty, 2)$

127. Consider the graph of the function
$$y = f(x) = ax^2 + bx + c, a \ne 0.$$
  a. Write an equation whose solution yields the x-intercepts. $ax^2 + bx + c = 0$
  b. Write an equation whose solution is the y-intercept. $y = c$
  c. Write (if possible) a condition under which the graph of $y = f(x)$ has no x-intercepts. $b^2 - 4ac < 0$
  d. Write (if possible) a condition under which the graph of $y = f(x)$ has no y-intercepts. Not possible

113. No because they have different domains
114. No because they have different domains
115. No because g is not defined at -1
116. No because f is not defined at -2
117. Yes because $f(x) = \dfrac{x^2 - 4}{x - 2} = \dfrac{(x + 2)(x - 2)}{x - 2} = x + 2 = g(x)$, since $x = 2$ is outside the given domain
118. Yes because $f(x) = \dfrac{x - 2}{x^2 - 6x + 8} = \dfrac{x - 2}{(x - 4)(x - 2)}$
$= \dfrac{1}{x - 4} = g(x)$, since $x = 2$ is outside the given domain
120. $b = -4$ or $b = -2$
122. $f(x^2) = 2x^2 - 3, [f(x)]^2 = 4x^2 - 12x + 9$
126. a. $y = \sqrt{x - 2}$ **b.** $y = \dfrac{1}{\sqrt{x - 2}}$ **c.** $y = \sqrt{2 - x}$
  d. $y = \dfrac{1}{\sqrt{2 - x}}$

A Library of Functions

Before Starting this Section, Review

1. Equation of a line (Chapter 1, page 19)
2. Absolute value (Appendix A, page 755)
3. Graph of an equation (Section 1.1, page 6)

Objectives

1. Relate linear functions to linear equations.
2. Determine whether a function is increasing or decreasing on an interval.
3. Recognize relative maximum and minimum values.
4. Identify even and odd functions.
5. Graph some important functions.
6. Evaluate and graph piecewise functions.

THE MEGATOOTH SHARK

The giant "Megatooth" shark (*Carcharodon megalodon*), once estimated to be 100 to 120 feet in length, is the largest meat-eating fish that ever lived. Its actual length has been the subject of scientific controversy. Sharks first appeared in the oceans more than 400 million years ago, almost 200 million years before the dinosaurs. A shark's skeleton is made of cartilage that decomposes rather quickly, making complete shark fossils rare. Consequently, scientists rely on calcified vertebrae, fossilized teeth, and small skin scales to reconstruct the evolutionary record of sharks.

The great white shark is the closest living relative to the giant "Megatooth" shark and has been used as a model to reconstruct the "Megatooth." John Maisey, curator at the American Museum of Natural History in New York City, used a partial set of "Megatooth" teeth to make a comparison with the jaws of the great white shark. In Example 2, we will use a formula to calculate the length of the "Megatooth" on the basis of the height of the tooth of the largest known "Megatooth" specimen. ■

---

1 Relate linear functions to linear equations.

Linear Functions

We know from Section 1.2 that the graph of a linear equation $y = mx + b$ is a straight line with slope m and y-intercept b. A *linear function* has a similar definition.

LINEAR FUNCTIONS

Let m and b be real numbers. The function $f(x) = mx + b$ is called a **linear function**. If $m = 0$, the function $f(x) = b$ is called a **constant function**. If $m = 1$ and $b = 0$, the resulting function $f(x) = x$ is called the **identity function**. See Figure 1.37.

The domain of a linear function is the interval $(-\infty, \infty)$ because the expression $mx + b$ is defined for any real number x. Figure 1.37 shows that the graph extends indefinitely to the left and right. The range of a nonconstant linear function also is the interval $(-\infty, \infty)$ because the graph extends indefinitely upward and downward. The range of a constant function $f(x) = b$ is the single real number b. See Figure 1.37(c).

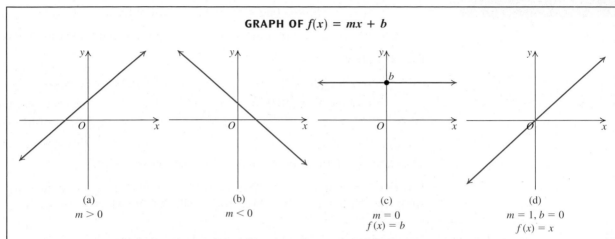

GRAPH OF $f(x) = mx + b$

(a)
$m > 0$

(b)
$m < 0$

(c)
$m = 0$
$f(x) = b$

(d)
$m = 1, b = 0$
$f(x) = x$

FIGURE 1.37 The graph of a linear function is a nonvertical line with slope *m* and *y*-intercept *b*.

EXAMPLE 1   **Writing a Linear Function**

Write a linear function g for which $g(1) = 4$ and $g(-3) = -2$.

SOLUTION

First, we need to find the equation of the line passing through the two points:

$$(x_1, y_1) = (1, 4) \text{ and } (x_2, y_2) = (-3, -2)$$

The slope of the line is given by

$$m = \frac{y_2 - y_1}{x_2 - x_1} = \frac{-2 - 4}{-3 - 1} = \frac{-6}{-4} = \frac{3}{2}.$$

Next, we use the point–slope form of a line.

$$y - y_1 = m(x - x_1) \quad \text{Point–slope form of a line}$$

$$y - 4 = \frac{3}{2}(x - 1) \quad \text{Substitute } x_1 = 1, y_1 = 4, \text{ and } m = \frac{3}{2}.$$

$$y - 4 = \frac{3}{2}x - \frac{3}{2} \quad \text{Distributive property}$$

$$y = \frac{3}{2}x + \frac{5}{2} \quad \text{Add } 4 = \frac{8}{2} \text{ to both sides and simplify.}$$

$$g(x) = \frac{3}{2}x + \frac{5}{2} \quad \text{Function notation}$$

The graph of the function $y = g(x)$ is shown in Figure 1.38. ■ ■ ■

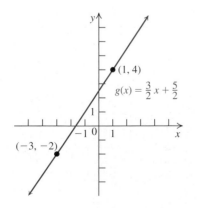

FIGURE 1.38 Point–slope form

Practice Problem 1   Write a linear function g for which $g(-2) = 2$ and $g(1) = 8$.

■

EXAMPLE 2   **Determining the Length of the "Megatooth" Shark**

The largest known "Megatooth" specimen is a tooth that has a total height of 15.6 centimeters. Calculate the length of the "Megatooth" shark from which it came by using the formula

$$\text{Shark length} = (0.96)(\text{height of tooth}) - 0.22,$$

where shark length is measured in meters and tooth height is measured in centimeters.

SOLUTION

Shark length $= (0.96)(\text{height of tooth}) - 0.22$	Formula
Shark length $= (0.96)(15.6) - 0.22$	Replace height of tooth with 15.6.
$= 14.756$	Simplify.

The shark's length is approximately 14.8 meters (48.4 feet).   ■ ■ ■

Practice Problem 2   If the "Megatooth" specimen was a tooth measuring 16.4 cm, what was the length of the "Megatooth" shark from which it came?   ■

2 Determine whether a function is increasing or decreasing on an interval.

Increasing and Decreasing Functions

Classifying functions according to how one variable changes with respect to the other variable can be very useful. Imagine a particle moving from left to right along the graph of a function. This means that the *x*-coordinate of the point on the graph is getting larger. If the corresponding *y*-coordinate of the point is getting larger, getting smaller, or staying the same, then the function is called *increasing, decreasing,* or *constant,* respectively.

INCREASING, DECREASING, AND CONSTANT FUNCTIONS

Let f be a function and let x_1 and x_2 be any two numbers in an open interval (a, b) contained in the domain of f. See Figure 1.39. The symbols a and b may represent real numbers, $-\infty$, or ∞. Then

(i) f is called an **increasing function** on (a, b) if $x_1 < x_2$ implies $f(x_1) < f(x_2)$.

(ii) f is called a **decreasing function** on (a, b) if $x_1 < x_2$ implies $f(x_1) > f(x_2)$.

(iii) f is **constant** on (a, b) if $x_1 < x_2$ implies $f(x_1) = f(x_2)$.

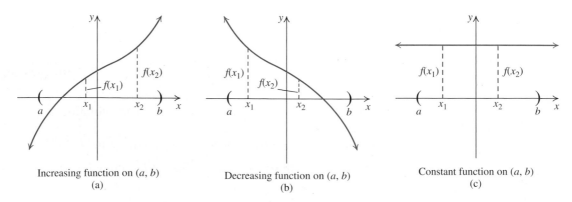

Increasing function on (a, b)	Decreasing function on (a, b)	Constant function on (a, b)
(a)	(b)	(c)

FIGURE 1.39 Increasing, decreasing, and constant functions

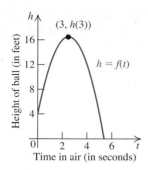

FIGURE 1.40 The function $f(t)$ increases on $(0, 3)$ and decreases on $(3, 6)$.

Geometrically, the graph of an increasing function on an open interval (a, b) rises as x-values increase on (a, b). This is because, by definition, as x-values increase from x_1 to x_2, the y-values also increase, from $f(x_1)$ to $f(x_2)$. Similarly, the graph of a decreasing function on (a, b) falls as x-values increase. The graph of a constant function is horizontal, or flat, over the interval (a, b).

Not every function always increases or decreases. The function that assigns the height of a ball tossed in the air to the length of time it is in the air is an example of a function that increases, but also decreases, over different intervals of its domain. See Figure 1.40.

EXAMPLE 3   Tracking the Behavior of a Function

From the graph of the function g in Figure 1.41(a), find the intervals over which g is increasing, is decreasing, or is constant.

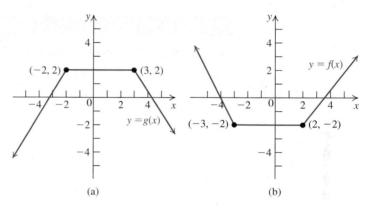

(a)   (b)

FIGURE 1.41

SOLUTION

The function g is increasing on the interval $(-\infty, -2)$. It is constant on the interval $(-2, 3)$. It is decreasing on the interval $(3, \infty)$. ■ ■ ■

Practice Problem 3   Using the graph in Figure 1.41(b), find the intervals over which f is increasing, is decreasing, or is constant. ■

Relative Maximum and Minimum Values

The y-coordinate of a point that is higher (lower) than any nearby point on a graph is called a *relative maximum* (*relative minimum*) value of the function. Relative maximum and minimum values are the y-coordinates of points corresponding to the peaks and troughs, respectively, of the graph.

STUDY TIP

When specifying the intervals over which a function f is increasing, decreasing, or constant, you need to use the intervals in the *domain* of f, *not* in the range of f.

3 Recognize relative maximum and minimum values.

DEFINITION OF RELATIVE MAXIMUM AND RELATIVE MINIMUM

If a is in the domain of a function f, we say that the value $f(a)$ is a **relative minimum of f** if there is an interval (x_1, x_2) containing a such that

$$f(a) \le f(x) \text{ for every } x \text{ in the interval } (x_1, x_2).$$

We say that the value $f(a)$ is a **relative maximum of f** if there is an interval (x_1, x_2) containing a such that

$$f(a) \ge f(x) \text{ for every } x \text{ in the interval } (x_1, x_2).$$

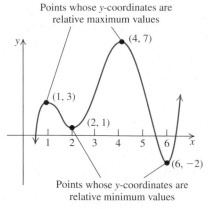

Points whose *y*-coordinates are
relative maximum values

Points whose *y*-coordinates are
relative minimum values

FIGURE 1.42 Relative maximum and minimum values

4 Identify even and odd functions.

The graph of a function can help you find or approximate maximum or minimum values (if any) of the function. For example, the graph of a function f in Figure 1.42 shows that

- f has two relative maxima:
  - **(i)** at $x = 1$ with relative maximum value $f(1) = 3$, and
  - **(ii)** at $x = 4$ with relative maximum value $f(4) = 7$
- f has two relative minima:
  - **(i)** at $x = 2$ with relative minimum value $f(2) = 1$, and
  - **(ii)** at $x = 6$ with relative minimum value $f(6) = -2$

Even-Odd Functions and Symmetry

This topic, *even-odd functions*, uses ideas of symmetry discussed in Section 1.1.

EVEN FUNCTION

A function f is called an **even function** if, for each x in the domain of f, $-x$ is also in the domain of f and

$$f(-x) = f(x).$$

The graph of an even function is symmetric with respect to the *y*-axis.

EXAMPLE 4   Graphing the Squaring Function

Show that the squaring function $f(x) = x^2$ is an even function and sketch its graph.

SOLUTION
The function $f(x) = x^2$ is even because

$$
\begin{aligned}
f(-x) &= (-x)^2 && \text{Replace } x \text{ with } -x. \\
&= x^2 && (-x)^2 = x^2 \\
&= f(x) && \text{Replace } x^2 \text{ with } f(x).
\end{aligned}
$$

To sketch the graph of $f(x) = x^2$, we make a table of values. See Table 1.5(a). We then plot the points that are in the first quadrant and join them with a smooth curve. Next, we use *y*-axis symmetry to find additional points on the graph of f in the second quadrant, which are also listed in Table 1.5(b). See Figure 1.43.

TABLE 1.5(a)

x	0	$\frac{1}{2}$	1	$\frac{3}{2}$	2	$\frac{5}{2}$	3
$f(x) = x^2$	0	$\frac{1}{4}$	1	$\frac{9}{4}$	4	$\frac{25}{4}$	9

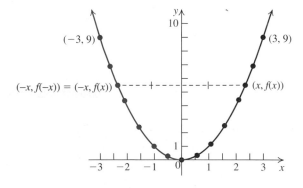

FIGURE 1.43 Graph of $f(x) = x^2$

TABLE 1.5(b)

x	0	$-\dfrac{1}{2}$	-1	$-\dfrac{3}{2}$	-2	$-\dfrac{5}{2}$	-3
$f(x) = x^2$	0	$\dfrac{1}{4}$	1	$\dfrac{9}{4}$	4	$\dfrac{25}{4}$	9

■ ■ ■

STUDY TIP

The graph of a non-zero function cannot be symmetric with respect to the x-axis because the two ordered pairs (x, y) and $(x, -y)$ would then be points of the graph on the same vertical line.

Practice Problem 4   Show that the function $f(x) = -x^2$ is an even function and sketch its graph.

■

ODD FUNCTION

A function f is an **odd function** if for each x in the domain of f, $-x$ is also in the domain of f and

$$f(-x) = -f(x).$$

The graph of an odd function is symmetric with respect to the origin.

EXAMPLE 5   **Graphing the Cubing Function**

Show that the cubing function defined by $g(x) = x^3$ is an odd function and sketch its graph.

SOLUTION

The function $g(x) = x^3$ is an odd function because

$$
\begin{aligned}
g(-x) &= (-x)^3 && \text{Replace } x \text{ with } -x. \\
&= -x^3 && (-x)^3 = -x^3 \\
&= -g(x) && \text{Replace } x^3 \text{ with } g(x).
\end{aligned}
$$

To sketch the graph of $g(x) = x^3$, we plot points in the first quadrant and then use symmetry about the origin to extend the graph to the third quadrant. The graph is shown in Figure 1.44.

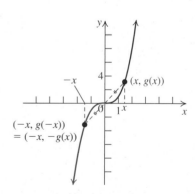

FIGURE 1.44 Graph of $g(x) = x^3$ ■ ■ ■

Practice Problem 5   Show that the function $f(x) = -x^3$ is an odd function and sketch its graph.

■

The function $h(x) = 2x + 3$, whose graph is sketched in Figure 1.45, is neither even nor odd, and its graph is neither symmetric with respect to the y-axis nor symmetric with respect to the origin. To see this algebraically, notice that $h(-x) = 2(-x) + 3 = -2x + 3$, which is not equivalent to $h(x)$ or $-h(x)$ (since $-h(x) = -2x - 3$).

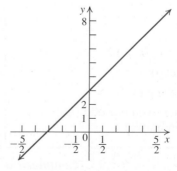

FIGURE 1.45 Graph of
$h(x) = 2x + 3$

5 Graph some important functions.

Basic Functions

As you progress through this course and future mathematics courses, you will repeatedly come across a small list of basic functions. The next box lists some of these common algebraic functions, along with their properties. You should try to produce these graphs by plotting points and using symmetries. The unit length in all of the graphs shown is the same on both axes.

A LIBRARY OF BASIC FUNCTIONS

Constant Function
$f(x) = c$

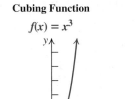

Domain: $(-\infty, \infty)$
Range: $\{c\}$
Constant on $(-\infty, \infty)$
Even function (y-axis symmetry)

Identity Function
$f(x) = x$

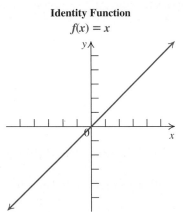

Domain: $(-\infty, \infty)$
Range: $(-\infty, \infty)$
Increasing on $(-\infty, \infty)$
Odd function (origin symmetry)

Squaring Function
$f(x) = x^2$

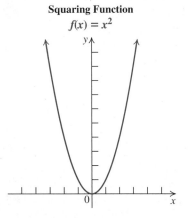

Domain: $(-\infty, \infty)$
Range: $[0, \infty)$
Decreasing on $(-\infty, 0)$
Increasing on $(0, \infty)$
Even function (y-axis symmetry)

Cubing Function
$f(x) = x^3$

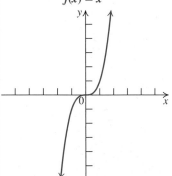

Domain: $(-\infty, \infty)$
Range: $(-\infty, \infty)$
Increasing on $(-\infty, \infty)$
Odd function (origin symmetry)

Absolute Value Function
$f(x) = |x|$

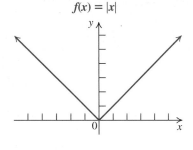

Domain: $(-\infty, \infty)$
Range: $[0, \infty)$
Decreasing on $(-\infty, 0)$
Increasing on $(0, \infty)$
Even function (y-axis symmetry)

Square Root Function
$f(x) = \sqrt{x} = x^{1/2}$

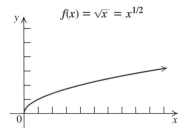

Domain: $[0, \infty)$
Range: $[0, \infty)$
Increasing on $(0, \infty)$
Neither even nor odd (no symmetry)

(Continued)

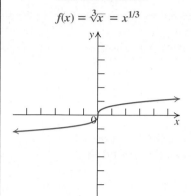

Cube Root Function

$$f(x) = \sqrt[3]{x} = x^{1/3}$$

Domain: $(-\infty, \infty)$

Range: $(-\infty, \infty)$

Increasing on $(-\infty, \infty)$

Odd function (origin symmetry)

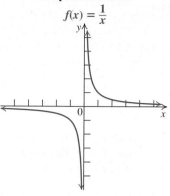

Reciprocal Function

$$f(x) = \frac{1}{x}$$

Domain: $(-\infty, 0) \cup (0, \infty)$

Range: $(-\infty, 0) \cup (0, \infty)$

Decreasing on $(-\infty, 0) \cup (0, \infty)$

Odd function (origin symmetry)

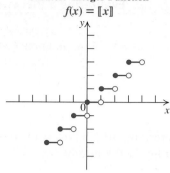

Greatest–Integer Function

$$f(x) = [\![x]\!]$$

Domain: $(-\infty, \infty)$

Range: $\{\dots -3, -2, -1, 0, 1, 2, 3, \dots\}$

Neither odd nor even (no symmetry)

6 Evaluate and graph piecewise functions.

Piecewise Functions

In the definition of some functions, different rules for assigning output values are used on different parts of the domain. Such functions are called **piecewise functions**. For example, in Peach County, Georgia, a section of the interstate highway has a speed limit of 55 miles per hour (mph). If you are caught speeding between 56 and 74 mph, your fine is $50 plus $3 for every mile per hour over 55 mph. For 75 mph and higher, your fine is $150 plus $5 for every mile per hour over 75 mph.

Let $f(x)$ be the piecewise function that represents your fine for speeding at x miles per hour. We express $f(x)$ as a piecewise function:

$$f(x) = \begin{cases} 50 + 3(x - 55), & 56 \le x < 75 \\ 150 + 5(x - 75), & x \ge 75 \end{cases}$$

The first line of the function means that if $56 \le x < 75$, your fine is $f(x) = 50 + 3(x - 55)$. Suppose you are caught driving 60 mph. Then we have

$$f(x) = 50 + 3(x - 55) \qquad \text{Expression used for } 56 \le x < 75$$
$$f(60) = 50 + 3(60 - 55) \qquad \text{Substitute 60 for } x.$$
$$= 65 \qquad \text{Simplify.}$$

Your fine for speeding at 60 mph is $65. The second line of the function means that if $x \ge 75$, your fine is $f(x) = 150 + 5(x - 75)$. Suppose you are caught driving 90 mph. Then we have

$$f(x) = 150 + 5(x - 75) \qquad \text{Expression used for } x \ge 75$$
$$f(90) = 150 + 5(90 - 75) \qquad \text{Substitute 90 for } x.$$
$$= 225 \qquad \text{Simplify.}$$

Your fine for speeding at 90 mph is $225.

FINDING THE SOLUTION: A PROCEDURE

EXAMPLE 6   **Evaluating a Piecewise Function**

OBJECTIVE	EXAMPLE
Evaluate F(a) for piecewise function F.	*Let*

$$F(x) = \begin{cases} x^2 & \text{if } x < 1 \\ 2x + 1 & \text{if } x \geq 1 \end{cases}$$

Find $F(0)$ and $F(2)$.

Step 1   **Determine which line of the function applies to the number a.**

1. Let $a = 0$. Because $0 < 1$, use the first line, $F(x) = x^2$. Let $a = 2$. Because $2 > 1$, use the second line, $F(x) = 2x + 1$.

Step 2   **Evaluate $F(a)$ using the line chosen in Step 1.**

2. $F(0) = (0)^2 = 0$
   $F(2) = 2(2) + 1 = 5$ ■ ■ ■

Practice Problem 6

Let $f(x) = \begin{cases} x^2 & \text{if } x \leq -1 \\ 2x & \text{if } x > -1 \end{cases}$. Find $f(-2)$ and $f(3)$.   ■

Graphing Piecewise Functions

Let's consider how to graph a piecewise function. The **absolute value function**, $f(x) = |x|$, can be expressed as a piecewise function by using the definition of absolute value:

$$f(x) = |x| = \begin{cases} -x & \text{if } x < 0 \\ x & \text{if } x \geq 0 \end{cases}$$

The first line in the function means that if $x < 0$, we use the equation $y = f(x) = -x$. So, if $x = -3$, then

$$y = f(-3) = -(-3) \qquad \text{Substitute } -3 \text{ for } x \text{ in } f(x) = -x$$
$$= 3 \qquad \text{Simplify.}$$

Thus, $(-3, 3)$ is a point on the graph of $y = |x|$. However, if $x \geq 0$, we use the second line in the function, which is $y = f(x) = x$. So, if $x = 2$, then

$$y = f(2) = 2 \qquad \text{Substitute } 2 \text{ for } x \text{ in } f(x) = x$$

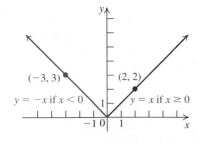

FIGURE 1.46 Graph of $y = |x|$

So $(2, 2)$ is a point on the graph of $y = |x|$. The two pieces $y = -x$ and $y = x$ are linear functions. We graph the appropriate parts of these lines ($y = -x$ for $x < 0$ and $y = x$ for $x \geq 0$) to form the graph of $y = |x|$. See Figure 1.46.

EXAMPLE 7   Graphing a Piecewise Function

Let

$$F(x) = \begin{cases} x^2 & \text{if } x < 1 \\ 3x + 1 & \text{if } x \geq 1 \end{cases}$$

Sketch the graph of $y = F(x)$.

SOLUTION

In the definition of F, the formula changes at $x = 1$. We call such numbers the **breakpoints** of the formula. For the function F, the only breakpoint is 1. Generally, to graph a piecewise function, we graph the function separately over the open intervals determined by the breakpoints and then graph the function at the breakpoints themselves. For the function $y = F(x)$, we graph the equation $y = x^2$ on the interval $(-\infty, 1)$. See Figure 1.47(a). Next, we graph the equation $y = 3x + 1$ on the interval $(1, \infty)$ and at the breakpoint 1, where $y = F(1) = 3(1) + 1 = 4$. See Figure 1.47(b).

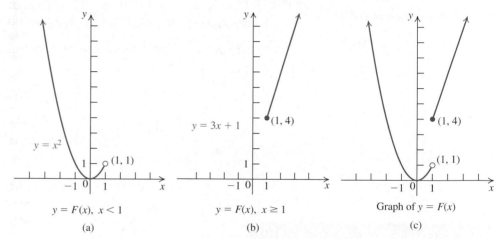

$$y = F(x), x < 1$$
(a)

$$y = F(x), x \geq 1$$
(b)

Graph of $y = F(x)$
(c)

FIGURE 1.47 A piecewise function

TECHNOLOGY CONNECTION

To graph the greatest integer function, enter $Y_1 = \text{int}(x)$. If you have your calculator set to graph in **connected** mode, you will get the following incorrect graph:

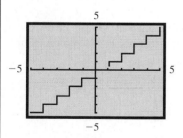

For the graph to appear correctly, change from the **connected** to the **dot** mode:

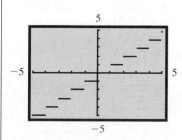

Note that the step from $0 \leq x \leq 1$ is obscured by the x-axis.

Combining these portions, we obtain the graph of $y = F(x)$, shown in Figure 1.47(c). When we come to the edge of the part of the graph we are working with, we draw

(i) a closed circle if that point is included.

(ii) an open circle if the point is excluded.

You may find it helpful to think of this procedure as following the graph of $y = x^2$ when x is less than 1 and then jumping to the graph of $y = 3x + 1$ when x is equal to or greater than 1.   ■ ■ ■

Practice Problem 7   Let $F(x) = \begin{cases} x^2 & \text{if } x \leq -1 \\ 2x & \text{if } x > -1 \end{cases}$. Sketch the graph of $y = F(x)$.

■

Some piecewise functions are called **step functions**. Their graphs look like the steps of a staircase.

The **greatest integer function** is denoted by $[\![x]\!]$, or $\text{int}(x)$, where $[\![x]\!] = $ the greatest integer less than or equal to x. For example,

$$[\![2]\!] = 2, \quad [\![2.3]\!] = 2, \quad [\![2.7]\!] = 2, \quad [\![2.99]\!] = 2$$

because 2 is the greatest integer less than or equal to 2, 2.3, 2.7, and 2.99. Similarly,

$$[\![-2]\!] = -2, \quad [\![-1.9]\!] = -2, \quad [\![-1.1]\!] = -2, \quad [\![-1.001]\!] = -2$$

because -2 is the greatest integer less than or equal to $-2, -1.9, -1.1$, and -1.001.

In general, if m is an integer such that $m \leq x < m + 1$, then $[\![x]\!] = m$. In other words, if x is between two consecutive integers m and $m + 1$, then $[\![x]\!]$ is assigned the smaller integer m.

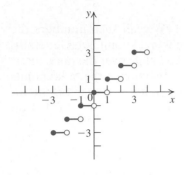

FIGURE 1.48 Graph of y = [x]

> ### EXAMPLE 8   Graphing a Step Function
>
> Graph the greatest integer function $f(x) = [\![x]\!]$.
>
> **SOLUTION**
>
> Choose a typical closed interval between two consecutive integers—say, the interval $[2, 3]$. We know that between 2 and 3, the greatest integer function's value is 2. In symbols, if $2 \le x < 3$, then $[\![x]\!] = 2$. Similarly, if $1 \le x < 2$, then $[\![x]\!] = 1$. Therefore, the values of $[\![x]\!]$ are constant between each pair of consecutive integers and jump by one unit at each integer. The graph of $f(x) = [\![x]\!]$ is shown in Figure 1.48. ■■■

Practice Problem 8 Find the values of $f(x) = [\![x]\!]$ for $x = -3.4$ and $x = 4.7$. ■

The greatest integer function f can be interpreted as a piecewise function:

$$f(x) = [\![x]\!] = \begin{cases} \vdots \\ -2 \text{ if } -2 \le x < -1 \\ -1 \text{ if } -1 \le x < 0 \\ 0 \text{ if } 0 \le x < 1 \\ 1 \text{ if } 1 \le x < 2 \\ \vdots \end{cases}$$

SECTION 1.4 ■ Exercises

A EXERCISES Basic Skills and Concepts

1. The graph of the linear function $f(x) = b$ is a(n) <u>horizontal</u> line.

2. The absolute value function can be expressed as a piecewise function by writing $f(x) = |x| = $ _____.

3. The piecewise function $f(x) = \begin{cases} x^2 + 2 \text{ if } x \le 1 \\ ax \text{ if } x > 1 \end{cases}$ will have a break at $x = 1$ unless $a = $ <u>3</u>.

4. If $(3, -2)$ is a point on the graph of f and $f(x) \le -2$ for every x in the interval $(2, 4)$, then -2 is a(n) <u>relative maximum</u> of f.

5. The function $f(x) = -x^4 - x^2 + 1$ is an example of an <u>even</u> function.

6. *True or False.* The function $f(x) = -x^{202} + x^{12} - x^2 + 3$ is an even function. True

7. *True or False.* The function $f(x) = x^3 + x + 1$ is an odd function. False

8. *True or False.* The function $f(x) = [\![x]\!]$ is increasing on the interval $(0, 3)$. False

In Exercises 9–18, write a linear function f that has the indicated values. Sketch the graph of f.

9. $f(0) = 1, f(-1) = 0$ †
10. $f(1) = 0, f(2) = 1$ †
11. $f(-1) = 1, f(2) = 7$ †
12. $f(-1) = -5, f(2) = 4$ †
13. $f(1) = 1, f(2) = -2$ †
14. $f(1) = -1, f(3) = 5$ †
15. $f(-2) = 2, f(2) = 4$ †
16. $f(2) = 2, f(4) = 5$ †
17. $f(0) = -1, f(3) = -3$ †
18. $f(1) = \frac{1}{4}, f(4) = -2$ †

2. $\begin{cases} -x \text{ if } x < 0 \\ x \text{ if } x \ge 0 \end{cases}$

In Exercises 19–28, sketch the graph and find the intervals over which the given function is increasing, is decreasing, or is constant.

19. $f(x) = 2$
20. $g(x) = -3$
21. $h(x) = 3x + 4$
22. $g(x) = -2x + 5$
23. $f(x) = 5x^2$
24. $h(x) = -x^3$
25. $g(x) = 2|x|$
26. $f(x) = -|x|$
27. $f(x) = -\sqrt[3]{x}$
28. $g(x) = -\sqrt{x}$

In Exercises 29–42, determine whether the given function is even, odd, or neither. The domain for each of the functions in Exercises 29–32 is $\{-3, -2, -1, 0, 1, 2, 3\}$.

	x	-3	-2	-1	0	1	2	3	
29.	$f(x)$	5	-2	1	0	1	-2	5	Even
30.	$g(x)$	4	-3	0	2	0	3	-4	Neither
31.	$h(x)$	3	2	1	0	5	2	-3	Neither
32.	$p(x)$	2	3	4	0	-4	-3	-2	Odd

33. $f(x) = 2x^4 + 4$ Even
34. $g(x) = 3x^4 - 5$ Even
35. $f(x) = 5x^3 - 3x$ Odd
36. $g(x) = 2x^3 + 4x$ Odd
37. $f(x) = \dfrac{1}{x^2 + 4}$ Even
38. $g(x) = \dfrac{x}{x^2 + 1}$ Odd

†Due to space constrictions, answers to these exercises may be found in the Answers beginning on page A–1 in the back of the book.

Answers: **19.** Constant on $(-\infty, \infty)$ **20.** Constant on $(-\infty, \infty)$ **21.** Increasing on $(-\infty, \infty)$ **22.** Decreasing on $(-\infty, \infty)$ **23.** Decreasing on $(-\infty, 0)$; increasing on $(0, \infty)$ **24.** Decreasing on $(-\infty, \infty)$

39. $f(x) = \dfrac{x^3}{x^2 + 1}$ Odd

40. $g(x) = \dfrac{x^4 + 3}{2x^3 - 3x}$ Odd

Neither

41. $f(x) = \dfrac{x^2 - 2x}{5x^4 + 4x^2 + 7}$

42. $g(x) = \dfrac{x^2 + 7}{3x^4 + 16x^2 + 9}$ Even

In Exercises 43–50, the graph of a function is given. Use the graph to find each of the following:

   a. The domain and the range of the function
   b. The intercepts, if any
   c. The intervals on which the function is increasing, is decreasing, or is constant
   d. Whether the function is even, odd, or neither

43. †   **44.** †

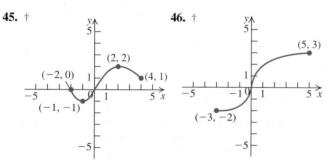

45. †   **46.** †

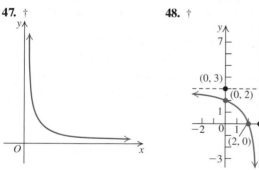

47. †   **48.** †

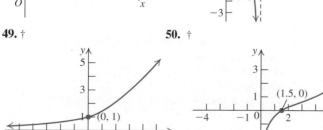

49. †   **50.** †

51. Let

$$f(x) = \begin{cases} x \text{ if } x \geq 2 \\ 2 \text{ if } x < 2 \end{cases}$$

   a. Find $f(1), f(2),$ and $f(3)$. $f(1) = 2, f(2) = 2, f(3) = 3$
   b. Sketch the graph of $y = f(x)$. †

52. Let

$$g(x) = \begin{cases} 2x \text{ if } x < 0 \\ x \text{ if } x \geq 0 \end{cases}$$

   $g(-1) = -2, g(0) = 0,$
   a. Find $g(-1), g(0),$ and $g(1)$. $g(1) = 1$
   b. Sketch the graph of $y = g(x)$. †

53. Let

$$f(x) = \begin{cases} 1 \text{ if } x > 0 \\ -1 \text{ if } x < 0 \end{cases}$$

   a. Find $f(-15)$ and $f(12)$. $f(-15) = -1, f(12) = 1$
   b. Sketch the graph of $y = f(x)$. †   Domain:
   c. Find the domain and the range of f. $(-\infty, 0) \cup (0, \infty)$;
   range: $\{-1, 1\}$

54. Let

$$g(x) = \begin{cases} 2x + 4 \text{ if } x > 1 \\ x + 2 \text{ if } x \leq 1 \end{cases}$$

   $g(-3) = -1, g(1) = 3,$
   a. Find $g(-3), g(1),$ and $g(3)$. $g(3) = 10$
   b. Sketch the graph of $y = g(x)$. †   Domain: $(-\infty, \infty)$;
   c. Find the domain and the range of g.   range: $(-\infty, 3] \cup (6, \infty)$

In Exercises 55–60, sketch the graph of each piecewise function. From the graphs find the range of each function.

55. $f(x) = \begin{cases} x^2 & \text{if } x \geq 2 \\ 3x - 2 \text{ if } x < 2 \end{cases}$ **56.** $g(x) = \begin{cases} 2x^2 \text{ if } x \geq 1 \\ |x| \text{ if } x < 1 \end{cases}$
†   †

57. $g(x) = \begin{cases} \sqrt[3]{x} \text{ if } x \geq 8 \\ -|x| \text{ if } x < 8 \end{cases}$ **58.** $h(x) = \begin{cases} \dfrac{1}{x} \text{ if } x \geq 2 \\ x \text{ if } x < 2 \end{cases}$
†   †

59. $f(x) = \begin{cases} 2 \text{ if } x \geq 3 \\ [\![x]\!] \text{ if } 1 \leq x < 3 \\ -|x| \text{ if } x < 1 \end{cases}$ †

60. $f(x) = \begin{cases} -x^2 \text{ if } x \geq 1 \\ |x| \text{ if } -2 \leq x < 1 \\ 2x \text{ if } x < -2 \end{cases}$ †

B EXERCISES   Applying the Concepts

61. **Converting volume.** To convert the volume of a liquid measured in ounces to a volume measured in liters, we use the fact that 1 liter ≈ 33.81 ounces. Let x denote the volume measured in ounces and y denote the volume measured in liters.

   a. Write an equation for the linear function $y = f(x)$. What are the domain and range of f? †
   b. Compute $f(3)$. What does it mean? †
   c. How many liters of liquid are in a typical soda can containing 12 ounces of liquid?   0.3549

62. Boiling point and elevation. The boiling point B of water (in degrees Fahrenheit) at elevation h (in thousands of feet) above sea level is given by the linear function $B(h) = -1.8h + 212$.
   a. Find and interpret the intercepts of the function $y = B(h)$. †
   b. Find the domain of this function. †
   c. Find the elevation at which water boils at 98.6°F. Why is this height dangerous to humans? 63,000 feet.
It is dangerous because 98.6°F is the temperature of human blood.
63. Pressure at sea depth. The pressure P in atmospheres (atm) at a depth d feet is given by the linear function

$$P(d) = \frac{1}{33}d + 1.$$

   a. Find and interpret the intercepts of $y = P(d)$. †
   b. Find $P(0), P(10), P(33)$, and $P(100)$. †
   c. Find the depth at which the pressure is 5 atmospheres. 132 ft

64. Speed of sound. The speed V of sound (in feet per second) in air at temperature T (in degrees Fahrenheit) is given by the linear function $V(T) = 1055 + 1.1T$.
   a. Find the speed of sound at 90°F. 1154 ft/s
   b. Find the temperature at which the speed of sound is 1100 feet per second. 40.91°F

65. Manufacturer's cost. A manufacturer of printers has a total cost per day consisting of a fixed overhead of $6000 plus product costs of $50 per printer.
   a. Express the total cost C as a function of the number x of printers produced. $C = 50x + 6000$
   b. Draw the graph of $y = C(x)$. Interpret the y-intercept. †
   c. How many printers were manufactured on a day when the total cost was $11,500? 110

66. Supply function. When the price p of a commodity is $10 per unit, 750 units of it are sold. For every $1 increase in the unit price, the supply q increases by 100 units.
   a. Write the linear function $q = f(p)$. $f(p) = 100p - 250$
   b. Find the supply when the price is $15 per unit. 1250
   c. What is the price at which 1750 units can be supplied? $20

67. Apartment rental. Suppose a two-bedroom apartment in a neighborhood near your school rents for $900 per month. If you move in after the first of the month, the rent is prorated; that is, it is reduced linearly.
   a. Suppose you move into the apartment x days after the first of the month. (Assume that the month has 30 days.) Express the rent R as a function of x. $R = 900 - 30x$
   b. Compute the rent if you move in six days after the first of the month. $720
   c. When did you move into the apartment if the landlord charged you a rent of $600? Ten days after the first of the month
68. College admissions. The average SAT scores (mathematics and critical reading) for incoming students at Central State College (CSC) have been rising linearly in recent years. In 2002, the average SAT score was 1120; in 2004, it was 1150. (The SAT testing format changed considerably in 2005.)

   a. Express the average SAT score at CSC as a function of time. †
   b. If the trend continues, what will be the average SAT score of incoming students at CSC in 2010? 1240
   c. If the trend continues, when will the average SAT score at CSC be 1300? In 2014

69. Breathing capacity. Suppose the average maximum breathing capacity for humans drops linearly from 100% at age 20 to 40% at age 80. What age corresponds to 50% capacity? 70

70. Drug dosage for children. If a is the adult dosage of a medicine and t is the child's age, then the *Friend's rule* for the child's dosage y is given by the formula

$$y = \frac{2}{25}ta.$$

   a. Suppose the adult dosage is 60 milligrams. Find the dosage for a five-year-old child. 24 mg
   b. How old would a child have to be in order to be prescribed an adult dosage? 12.5 yr

71. Air pollution. In Ballerenia, the average number y of deaths per month was observed to be linearly related to the concentration x of sulfur dioxide in the air. Suppose there are 30 deaths when $x = 150$ milligrams per cubic meter and 50 deaths when $x = 420$ milligrams per cubic meter.
   a. Write y as a function of x.
   b. Find the number of deaths when $x = 350$ milligrams per cubic meter. 45
   c. If the number of deaths per month is 45, what is the concentration of sulfur dioxide in the air? 352.5 mg/m³

72. Child shoe sizes. Children's shoe sizes start with size 0, having an insole length L of $3\frac{11}{12}$ inches, and each full size being $\frac{1}{3}$ inch longer.
   a. Write the equation $y = L(S)$, where S is the shoe size.
   b. Find the length of the insole of a child's shoe of size 4. 5.25 in
   c. What size (to the nearest half-size) shoe will fit a child whose insole length is 6.1 inches? 6.5

73. State income tax. Suppose a state's income tax code states that the tax liability T on x dollars of taxable income is as follows:

$$T(x) = \begin{cases} 0.04x & \text{if } 0 \le x < 20{,}000 \\ 800 + 0.06x & \text{if } x \ge 20{,}000 \end{cases}$$

   a. Graph the function $y = T(x)$. †
   b. Find the tax liability on each taxable income.
      (i) $12,000 $480
      (ii) $20,000 $2000
      (iii) $50,000 $3800
   c. Find your taxable income if you had each tax liability.
      (i) $600 $15,000
      (ii) $1200 This value is not possible as a liability.
      (iii) $2300 $25,000

Answers:
71. a. $y = \frac{2}{27}(x - 150) + 30$

72. a. $y = \frac{1}{3}S + \frac{47}{12}$, where S is the size and y is the length

74. Federal income tax. The tax table for the 2007 U.S. Income Tax for a single taxpayer is as follows:

If taxable income is over	But not over	The tax is
$0	$7,825	10% of the amount over $0.
$7,825	$31,850	$782.50 plus 15% of the amount over $7,825.
$31,850	$77,100	$4386.25 plus 25% of the amount over $31,850.
$77,100	$160,850	$15,698.75 plus 28% of the amount over $77,100.
$160,850	$349,700	$39,148.75 plus 33% of the amount over $160,850.
$349,700	No limit	$101,469.25 plus 35% of the amount over $349,700.

Source: Internal Revenue Service.

a. Express the information given in the table as a six-part piecewise function f, using x as a variable representing taxable income. †
b. Find the tax liability on each taxable income.
  (i) $35,000 $5173.75
  (ii) $100,000 $22,110.75
  (iii) $500,000 $154,074.25
c. Find your taxable income if you had each tax liability.
  (i) $3500 $25,941.67
  (ii) $12,700 $65,105.00
  (iii) $35,000 $146,003.04

C EXERCISES Beyond the Basics

75.
$$\text{Let } f(x) = \begin{cases} 3x + 5, & -3 \le x < -1 \\ 2x + 1, & -1 \le x < 2 \\ 2 - x, & 2 \le x \le 4 \end{cases}$$

a. Find the following:
  (i) $f(-2)$ -1
  (ii) $f(-1)$ -1
  (iii) $f(3)$ -1
b. Find x when $f(x) = 2$. $x = \frac{1}{2}$
c. Sketch a graph of $y = f(x)$. †

76.
$$\text{Let } g(x) = \begin{cases} 3, & -3 \le x < -1 \\ -6x - 3, & -1 \le x < 0 \\ 3x - 3, & 0 \le x \le 1 \end{cases}$$

a. Find the following:
  (i) $g(-2)$ 3
  (ii) $g\left(\frac{1}{2}\right)$ $-\frac{3}{2}$

  (iii) $g(0)$ -3
  (iv) $g(-1)$ 3
b. Find x when **(i)** $g(x) = 0$ and **(ii)** $2g(x) + 3 = 0$. †
c. Sketch a graph of $y = g(x)$. †

In Exercises 77 and 78, a function is given. For each function,
a. find the domain and range.
b. find the intervals over which the function is increasing, decreasing, or constant.
c. state whether the function is odd, even, or neither.

77. $f(x) = x - [\![x]\!]$ †   **78.** $f(x) = \dfrac{1}{[\![x]\!]}$ †

79. Let $f(x) = \dfrac{|x|}{x}, x \ne 0$. Find $|f(x) - f(-x)|$. 2

80. The Windchill Index (WCI). Suppose the outside air temperature is T degrees Fahrenheit and the wind speed is v miles per hour. A formula for the WCI based on observations is as follows:

$$WCI = \begin{cases} T, & 0 \le v \le 4 \\ 91.4 + (91.4 - T)(0.0203v - 0.304\sqrt{v} - 0.474), & 4 < v < 45 \\ 1.6T - 55, & v \ge 45 \end{cases}$$

a. Find the WCI to the nearest degree if the outside air temperature is 40°F and
  (i) $v = 2$ miles per hour. 40
  (ii) $v = 16$ miles per hour. 21
  (iii) $v = 50$ miles per hour. 9
b. Find the air temperature to the nearest degree if
  (i) the WCI is -58°F and $v = 36$ miles per hour. -4°F
  (ii) the WCI is -10°F and $v = 49$ miles per hour. 28°F

Critical Thinking

81. Postage-rate function. The U.S. Postal Service uses the function $f(x) = -[\![-x]\!]$, called the **ceiling integer function**, to determine postal charges. For instance, if you have an envelope that weighs 3.2 ounces, you will be charged the rate for $f(3.2) = -[\![-3.2]\!] = -(-4) = 4$ ounces. In 2007, it cost 41¢ for the first ounce (or a fraction of it) and 17¢ for each additional ounce (or a fraction of it) to mail a first-class letter in the United States.
a. Express the cost C (in cents) of mailing a letter weighing x ounces. $C(x) = 17(f(x) - 1) + 41$
b. Graph the function $y = C(x)$. †
c. What are the domain and range of $y = C(x)$? †

82. The cost C of parking a car at the metropolitan airport is $4 for the first hour and $2 for each additional hour or fraction thereof. Write the cost function $C(x)$, where x is the number of parking hours, in terms of the greatest integer function. $C(x) = 2[\![x]\!] + 4$

83. Car rentals. The weekly cost of renting a compact car from U-Rent is $150. There is no charge for driving the first 100 miles, but there is a 20¢ charge for each mile (or fraction thereof) driven over 100 miles.
a. Express the cost C of renting a car for a week and driving it for x miles. †
b. Sketch the graph of $y = C(x)$. †
c. How many miles were driven if the weekly rental cost was $190? 300

Transformations of Functions

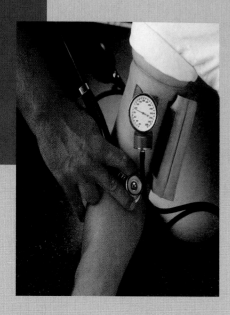

| Before Starting this Section, Review | Objectives |

Before Starting this Section, Review

1. Graphs of basic functions (Section 1.2, page 36)
2. Symmetry (Section 1.1, page 8)
3. Completing the square (Appendix A, page 812)

Objectives

1. Learn the meaning of transformations.
2. Use vertical or horizontal shifts to graph functions.
3. Use reflections to graph functions.
4. Use stretching or compressing to graph functions.

MEASURING BLOOD PRESSURE

Blood pressure is the force of blood per unit area against the walls of the arteries. Blood pressure is recorded as two numbers: the systolic pressure (as the heart beats) and the diastolic pressure (as the heart relaxes between beats). If your blood pressure is "120 over 80," it is rising to a maximum of 120 millimeters of mercury as the heart beats and is falling to a minimum of 80 millimeters of mercury as the heart relaxes.

The most precise way to measure blood pressure is to place a small glass tube in an artery and let the blood flow out and up as high as the heart can pump it. That was the way Stephen Hales (1677–1761), the first investigator of blood pressure, learned that a horse's heart could pump blood 8 feet, 3 inches, up a tall tube. Such a test, however, consumed a great deal of blood. Jean Louis Marie Poiseuille (1797–1869) greatly improved the process by using a mercury-filled manometer, which allowed a smaller, narrower tube. Even so, a blood pressure check in the 1840s was a very different experience than what people routinely undergo today.

In Example 8, we investigate Poiseuille's Law for arterial blood flow. ■

1 Learn the meaning of transformations.

Transformations

If a new function is formed by performing certain operations on a given function f, then the graph of the new function is called a **transformation** of the graph of f. For example, the graphs of $y = |x| + 2$ and $y = |x| - 3$ are transformations of the graph of $y = |x|$; that is, the graph of each is a special modification of the graph of $y = |x|$.

2 Use vertical or horizontal shifts to graph functions.

Vertical and Horizontal Shifts

EXAMPLE 1 **Graphing Vertical Shifts**

Let $f(x) = |x|, g(x) = |x| + 2$, and $h(x) = |x| - 3$. Sketch the graphs of these functions on the same coordinate plane. Describe how the graphs of g and h relate to the graph of f.

SOLUTION

Make a table of values and graph the equations $y = f(x)$, $y = g(x)$, and $y = h(x)$.

TABLE 1.6

TABLE 1.6

x	$y = \|x\|$	$y = \|x\| + 2$	$y = \|x\| - 3$
-5	5	7	2
-3	3	5	0
-1	1	3	-2
0	0	2	-3
1	1	3	-2
3	3	5	0
5	5	7	2

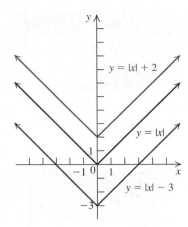

FIGURE 1.49 Vertical shifts of $y = \|x\|$

TECHNOLOGY CONNECTION

Many graphing calculators use abs(x) for $|x|$. The graphs of $y = |x|$, $y = |x| - 3$, and $y = |x| + 2$ illustrate that these graphs are identical in shape. They differ only in vertical placement.

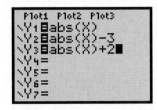

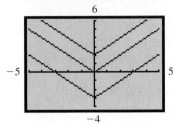

Notice that in Table 1.6 and Figure 1.49, for each value of x, the value of $y = |x| + 2$ is 2 more than the value of $y = |x|$. So the graph of $y = |x| + 2$ is the graph of $y = |x|$ shifted two units up.

We also note that for each value of x, the value of $y = |x| - 3$ is 3 less than the value of $y = |x|$. So the graph of $y = |x| - 3$ is the graph of $y = |x|$ shifted three units down. ■ ■ ■

Practice Problem 1 Let

$$f(x) = x^3, g(x) = x^3 + 1, \text{ and } h(x) = x^3 - 2.$$

Sketch the graphs of all three functions on the same coordinate plane. Describe how the graphs of g and h relate to the graph of f. ■

Example 1 illustrates the concept of the vertical (up or down) shift of a graph.

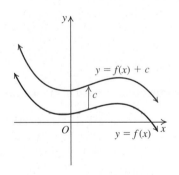

VERTICAL SHIFT

Let $c > 0$. The graph of $y = f(x) + c$ is the graph of $y = f(x)$ shifted c units *up*, and the graph of $y = f(x) - c$ is the graph of $y = f(x)$ shifted c units *down*. (See Figure 1.50.)

Next, we consider the operation that shifts a graph horizontally.

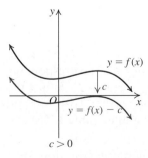

FIGURE 1.50 Vertical shifts

EXAMPLE 2 Writing Functions for Horizontal Shifts

Let $f(x) = x^2, g(x) = (x - 2)^2$, and $h(x) = (x + 3)^2$. A table of values for f, g, and h is given in Table 1.7. The three functions f, g, and h are graphed on the same coordinate plane in Figure 1.51. Describe how the graphs of g and h relate to the graph of f.

TECHNOLOGY CONNECTION

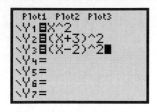

The graphs of $y = x^2$, $y = (x + 3)^2$, and $y = (x - 2)^2$ illustrate that these graphs are identical in shape. They differ only in horizontal placement.

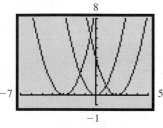

TABLE 1.7

x	$y = x^2$	$y = (x - 2)^2$	x	$y = x^2$	$y = (x + 3)^2$
-4	**16**	36	-4	16	1
-3	**9**	25	-3	9	0
-2	**4**	16	-2	4	1
-1	**1**	9	-1	**1**	4
0	**0**	4	0	**0**	9
1	**1**	1	1	**1**	16
2	4	**0**	2	**4**	25
3	9	1	3	**9**	36
4	16	4	4	**16**	49
	(a)			**(b)**	

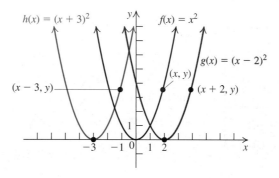

FIGURE 1.51 Horizontal shifts

SOLUTION

First, notice that all three functions are squaring functions.

a. Replacing x with $x - 2$ on the right side of the defining equation for f gives the equation for g.

$$f(x) = \quad x^2 \qquad \text{Defining equation for } f$$
$$\downarrow$$
$$g(x) = (x - 2)^2 \qquad \text{Replace } x \text{ with } x - 2 \text{ to obtain the defining equation for } g.$$

The x-intercept of f is 0. We can find the x-intercept of g by solving $g(x) = (x - 2)^2 = 0$ for x. We get $x - 2 = 0$, or $x = 2$.

This means that $(2, 0)$ is a point on the graph of g, whereas $(0, 0)$ is a point on the graph of f. In general, each point (x, y) on the graph of f has a corresponding point $(x + 2, y)$ on the graph of g. So the graph of $g(x) = (x - 2)^2$ is just the graph of $f(x) = x^2$ shifted two units to the *right*. Noticing that $f(0) = 0$ and $g(2) = 0$ will help you remember which way to shift the graph. Table 1.7(a) and Figure 1.51 illustrate these ideas.

b. Replacing x with $x + 3$ on the right-hand side of the defining equation for f gives the equation for h.

$$f(x) = \quad x^2 \qquad \text{Defining equation for } f$$
$$\downarrow$$
$$h(x) = (x + 3)^2 \qquad \text{Replace } x \text{ with } x + 3 \text{ to obtain the defining equation for } h.$$

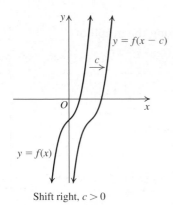

Shift right, $c > 0$

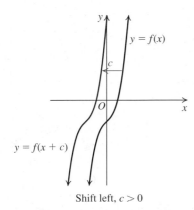

Shift left, $c > 0$

FIGURE 1.52 Horizontal shifts

To find the x-intercept of h, we solve $h(x) = (x + 3)^2 = 0$. We get $x + 3 = 0$ and find that $x = -3$. This means that $(-3, 0)$ is a point on the graph of h, whereas $(0, 0)$ is a point on the graph of f. In general, each point (x, y) on the graph of f has a corresponding point $(x - 3, y)$ on the graph of h. So the graph of $h(x) = (x + 3)^2$ is just the graph of $f(x) = x^2$ shifted three units to the *left*. Noticing that $f(0) = 0$ and $h(-3) = 0$ will help you remember which way to shift the graph.

Table 1.7(b) and Figure 1.51 confirm these considerations.   ■ ■ ■

Practice Problem 2   Let

$$f(x) = x^3, g(x) = (x - 1)^3, \text{ and } h(x) = (x + 2)^3.$$

Sketch the graphs of all three functions on the same coordinate plane. Describe how the graphs of g and h relate to the graph of f.   ■

HORIZONTAL SHIFTS

Let $c > 0$. The graph of $y = f(x - c)$ is the graph of $y = f(x)$ shifted c units to the right. The graph of $y = f(x + c)$ is the graph of $y = f(x)$ shifted c units to the left. See Figure 1.52.

Replacing x with $x - c$ in the equation $y = f(x)$ results in a function whose graph is the graph of f shifted c units to the right $(c > 0)$. Similarly, replacing x with $x + c$ in the equation $y = f(x)$ results in a function whose graph is the graph of f shifted c units to the left $(c > 0)$.

FINDING THE SOLUTION: A PROCEDURE

EXAMPLE 3	Graphing Combined Vertical and Horizontal Shifts

OBJECTIVE
Sketch the graph of $g(x) = f(x - c) + b$, where f is a function whose graph is known.

EXAMPLE
Sketch the graph of $g(x) = \sqrt{x + 2} - 3$.

Step 1   Identify and graph the known function f.

1. Choose $f(x) = \sqrt{x}$.

    The graph of $y = \sqrt{x}$ is shown in Step 4.

Step 2   Identify the constants b and c.

2. $g(x) = \sqrt{x - (-2)} + (-3)$, so $c = -2$ and $b = -3$.

Step 3   Graph $y = f(x - c)$ by shifting the graph of f horizontally to the

    (i) right if $c > 0$.
    (ii) left if $c < 0$.

3. Since $c = -2 < 0$, the graph of $y = \sqrt{x + 2}$ is the graph of f shifted horizontally two units to the left. See the blue graph in Step 4.

continued on next page

Step 4 Graph $y = f(x - c) + b$ by shifting the graph of $y = f(x - c)$ vertically

   (i) up if $b > 0$.
   (ii) down if $b < 0$.

4. Graph $y = \sqrt{x + 2} - 3$ by shifting the graph of $y = \sqrt{x + 2}$ three units down.

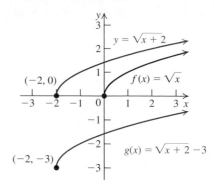

Horizontal and vertical shifts   ■ ■ ■

Practice Problem 3 Sketch the graph of

$$f(x) = \sqrt{x - 2} + 3.$$
■

3 Use reflections to graph functions.

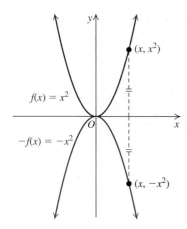

FIGURE 1.53 Reflection in the x-axis

Reflections

Comparing the Graphs of $y = f(x)$ and $y = -f(x)$ Consider the graph of $f(x) = x^2$, shown in Figure 1.53. Table 1.8 gives some values for $f(x)$ and $g(x) = -f(x)$. Note that the y-coordinate of each point in the graph of $g(x) = -f(x)$ is the opposite of the y-coordinate of the corresponding point on the graph of $f(x)$. So the graph of $y = -x^2$ is the reflection of the graph of $y = x^2$ in the x-axis. This means that the points (x, x^2) and $(x, -x^2)$ are the same distance from, but on opposite sides of, the x-axis. See Figure 1.53.

TABLE 1.8

x	$f(x) = x^2$	$g(x) = -f(x) = -x^2$
-3	9	-9
-2	4	-4
-1	1	-1
0	0	0
1	1	-1
2	4	-4
3	9	-9

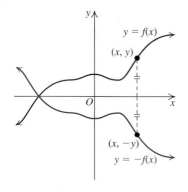

FIGURE 1.54 Reflection in the x-axis

REFLECTION IN THE x-AXIS

The graph of $y = -f(x)$ is a reflection of the graph of $y = f(x)$ in the x-axis. If a point (x, y) is on the graph of $y = f(x)$, then the point $(x, -y)$ is on the graph of $y = -f(x)$. See Figure 1.54.

Comparing the Graphs of $y = f(x)$ and $y = f(-x)$ To compare the graphs of $f(x)$ and $g(x) = f(-x)$, consider the graph of $f(x) = \sqrt{x}$. Then $f(-x) = \sqrt{-x}$. See Figure 1.55. Table 1.9 gives some values for $f(x)$ and $g(x) = f(-x)$. The domain of f is $[0, \infty)$, and the domain of g is $(-\infty, 0]$. Each point (x, y) on the graph of f has a corresponding point $(-x, y)$ on the graph of g. So the graph of $y = f(-x)$ is the reflection of the graph of $y = f(x)$ in the y-axis. This means that the points $(x, \sqrt{x})$ and $(-x, \sqrt{x})$ are the same distance from, but on opposite sides of, the y-axis. See Figure 1.55.

TABLE 1.9

x	$f(x) = \sqrt{x}$	$-x$	$g(x) = \sqrt{-x}$
-4	Undefined	4	2
-1	Undefined	1	1
0	0	0	0
1	1	-1	Undefined
4	2	-4	Undefined

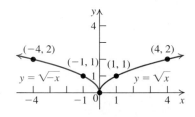

FIGURE 1.55 Graphing $y = \sqrt{-x}$

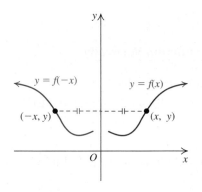

FIGURE 1.56 Reflection in the y-axis

REFLECTION IN THE y-AXIS

The graph of $y = f(-x)$ is a reflection of the graph of $y = f(x)$ in the y-axis. If a point (x, y) is on the graph of $y = f(x)$, then the point $(-x, y)$ is on the graph of $y = f(-x)$. See Figure 1.56.

EXAMPLE 4   Combining Transformations

Explain how the graph of $y = -|x - 2| + 3$ can be obtained from the graph of $y = |x|$.

SOLUTION

Start with the graph of $y = |x|$. Follow the point $(0, 0)$ on $y = |x|$. See Figure 1.57(a).

Step 1   Shift the graph of $y = |x|$ two units to the right to obtain the graph of $y = |x - 2|$. The point $(0, 0)$ moves to $(2, 0)$. See Figure 1.57(b).

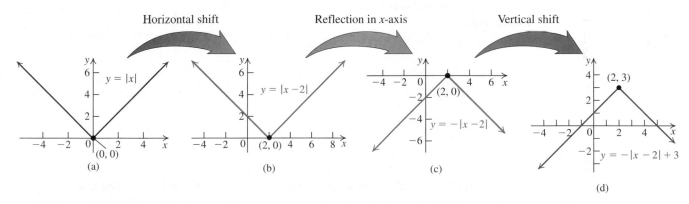

FIGURE 1.57 Transformations of $y = |x|$

Step 2   Reflect the graph of $y = |x - 2|$ in the x-axis to obtain the graph of $y = -|x - 2|$. The point $(2, 0)$ remains $(2, 0)$. See Figure 1.57(c).

Step 3   Finally, shift the graph of $y = -|x - 2|$ three units up to obtain the graph of $y = -|x - 2| + 3$. The point $(2, 0)$ moves to $(2, 3)$. See Figure 1.57(d).

■ ■ ■

Practice Problem 4   Explain how the graph of $y = -(x - 1)^2 + 2$ can be obtained from the graph of $y = x^2$.

■

4 Use stretching or compressing to graph functions.

Stretching or Compressing

The transformations that shift or reflect a graph change only the graph's position, not its shape. Such transformations are called **rigid transformations**. We now look at transformations that distort the shape of a graph, called **nonrigid transformations**. We consider the relationship of the graphs of $y = af(x)$ and $y = f(bx)$ to the graph of $y = f(x)$.

Comparing the Graphs of $y = f(x)$ and $y = af(x)$

EXAMPLE 5 **Stretching or Compressing a Function Vertically**

Let $f(x) = |x|$, $g(x) = 2|x|$, and $h(x) = \frac{1}{2}|x|$. Sketch the graphs of f, g, and h on the same coordinate plane and describe how the graphs of g and h are related to the graph of f.

SOLUTION

The graphs of $y = |x|$, $y = 2|x|$, and $y = \frac{1}{2}|x|$ are sketched in Figure 1.58. Table 1.10 gives some typical function values.

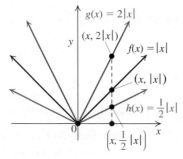

FIGURE 1.58 Vertical stretch and compression

TABLE 1.10

x	$f(x) = |x|$	$g(x) = 2|x|$	$h(x) = \frac{1}{2}|x|$
-2	2	4	1
-1	1	2	$\frac{1}{2}$
0	0	0	0
1	1	2	$\frac{1}{2}$
2	2	4	1

The graph of $y = 2|x|$ is the graph of $y = |x|$ vertically stretched (expanded) by multiplying each of its y-coordinates by 2. It is twice as high as the graph of $|x|$ at every real number x. The result is a taller V-shaped curve. See Figure 1.58.

The graph $y = \frac{1}{2}|x|$ is the graph of $y = |x|$ vertically compressed (shrunk) by multiplying each of its y-coordinates by $\frac{1}{2}$. It is half as high as the graph of $|x|$ at every real number x. The result is a flatter V-shaped curve. See Figure 1.58. ■ ■ ■

Practice Problem 5 Let $f(x) = \sqrt{x}$ and $g(x) = 2\sqrt{x}$. Sketch the graphs of f and g on the same coordinate plane and describe how the graph of g is related to the graph of f. ■

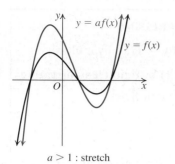

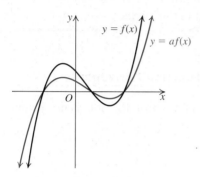

FIGURE 1.59 **Vertical stretch or compression**

VERTICAL STRETCHING OR COMPRESSING

The graph of $y = af(x)$ is obtained from the graph of $y = f(x)$ by multiplying the y-coordinate of each point on the graph of $y = f(x)$ by a and leaving the x-coordinate unchanged. The result (see Figure 1.59) is as follows:

1. A **vertical stretch** away from the x-axis if $a > 1$
2. A **vertical compression** toward the x-axis if $0 < a < 1$

If $a < 0$, first graph $y = |a|f(x)$ by stretching or compressing the graph of $y = f(x)$ vertically. Then reflect the resulting graph in the x-axis.

Comparing the Graphs of $y = f(x)$ and $y = f(bx)$ Given a function $f(x)$, let's explore the effect of the constant b in graphing the function $y = f(bx)$. Consider the graphs of $y = f(x)$ and $y = f(2x)$ in Figure 1.60(a). Multiplying the *independent* variable x by 2 compresses the graph $y = f(x)$ horizontally toward the y-axis. Because the value $2x$ is twice the value of x, a point on the x-axis will be only half as far from the origin when $y = f(2x)$ has the same y value as $y = f(x)$. See Figure 1.60(a).

STUDY TIP

Notice that replacing the variable x with $x \pm c$, or ax, in $y = f(x)$ results in a horizontal change in the graphs. However, replacing the y value $f(x)$ by $f(x) \pm c$, or $bf(x)$, results in a vertical change in the graph of $y = f(x)$.

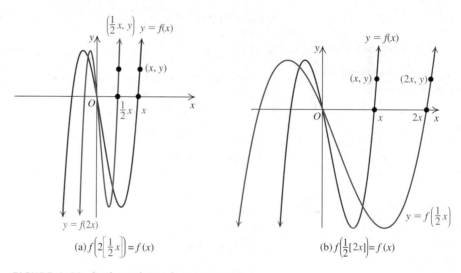

FIGURE 1.60 **Horizontal stretch or compression**

Now consider the graphs of $y = f(x)$ and $y = f\left(\dfrac{1}{2}x\right)$ in Figure 1.60(b). Multiplying the *independent* variable x by $\dfrac{1}{2}$ stretches the graph of $y = f(x)$ horizontally away from the y-axis. The value $\dfrac{1}{2}x$ is half the value of x so that a point on the x-axis will be twice as far from the origin for $y = f\left(\dfrac{1}{2}x\right)$ to have the same y value as $y = f(x)$. See Figure 1.60(b).

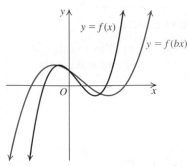

0 < b < 1 : stretch horizontally

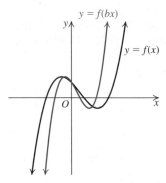

b > 1 : compress horizontally

FIGURE 1.61 Horizontal stretch and compression

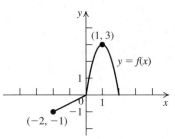

FIGURE 1.62

HORIZONTAL STRETCHING OR COMPRESSING

The graph of $y = f(bx)$ is obtained from the graph of $y = f(x)$ by multiplying the x-coordinate of each point on the graph of $y = f(x)$ by $\dfrac{1}{b}$ and leaving the y-coordinate unchanged. The result (see Figure 1.61) is as follows:

1. A **horizontal stretch** away from the y-axis if $0 < b < 1$
2. A **horizontal compression** toward the y-axis if $b > 1$

If $b < 0$, first graph $f(|b|x)$ by stretching or compressing the graph of $y = f(x)$ horizontally. Then reflect the graph of $y = f(|b|x)$ in the y-axis.

EXAMPLE 6   **Stretching or Compressing a Function Horizontally**

Using the graph of the function $y = f(x)$ in Figure 1.62, whose formula is not given, sketch the following graphs.

a. $f\left(\dfrac{1}{2}x\right)$ **b.** $f(2x)$ **c.** $f(-2x)$

SOLUTION

a. To graph $y = f\left(\dfrac{1}{2}x\right)$, we stretch the graph of $y = f(x)$ horizontally by a factor of 2. In other words, we transform each point (x, y) in Figure 1.62 to the point $(2x, y)$ in Figure 1.63(a).

b. To graph $y = f(2x)$, we compress the graph of $y = f(x)$ horizontally by a factor of $\dfrac{1}{2}$. Therefore, we transform each point (x, y) in Figure 1.62 to $\left(\dfrac{1}{2}x, y\right)$ in Figure 1.63(b).

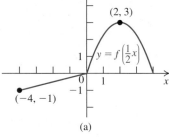

(a)

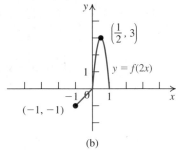

(b)

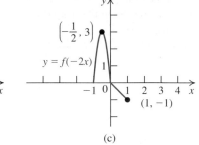

(c)

FIGURE 1.63

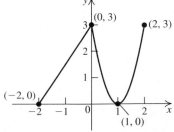

c. To graph $y = f(-2x)$, we reflect the graph of $y = f(2x)$ in Figure 1.63(b) in the y-axis. So we transform each point (x, y) in Figure 1.63(b) to the point $(-x, y)$ in Figure 1.63(c). ■ ■ ■

Practice Problem 6   The graph of a function $y = f(x)$ is given in the margin. Sketch the graphs of the following functions.

a. $f\left(\dfrac{1}{2}x\right)$ **b.** $f(2x)$ ■

Multiple Transformations in Sequence

When graphing requires more than one transformation of a basic function, it is helpful to perform transformations in the following order:

1. Horizontal shifts   **2.** Stretch or compress   **3.** Reflections   **4.** Vertical shifts

EXAMPLE 7   **Combining Transformations**

Sketch the graph of the function $f(x) = 3 - 2(x - 1)^2$.

SOLUTION

Begin with the basic function $y = x^2$. Then apply the necessary transformations in a sequence of steps. The result of each step is shown in Figure 1.64.

Step 1 $y = x^2$ Identify a related function whose graph is familiar. In this case, use $y = x^2$. See Figure 1.64(a).

Step 2 $y = (x - 1)^2$ Replace x with $x - 1$; shift the graph of $y = x^2$ one unit to the right. See Figure 1.64(b).

Step 3 $y = 2(x - 1)^2$ Multiply by 2; stretch the graph of $y = (x - 1)^2$ vertically by a factor of 2. See Figure 1.64(c).

Step 4 $y = -2(x - 1)^2$ Multiply by -1. Reflect the graph of $y = 2(x - 1)^2$ in the x-axis. See Figure 1.64(d).

Step 5 $y = 3 - 2(x - 1)^2$ Add 3. Shift the graph of $y = -2(x - 1)^2$ three units up. See Figure 1.64(e).

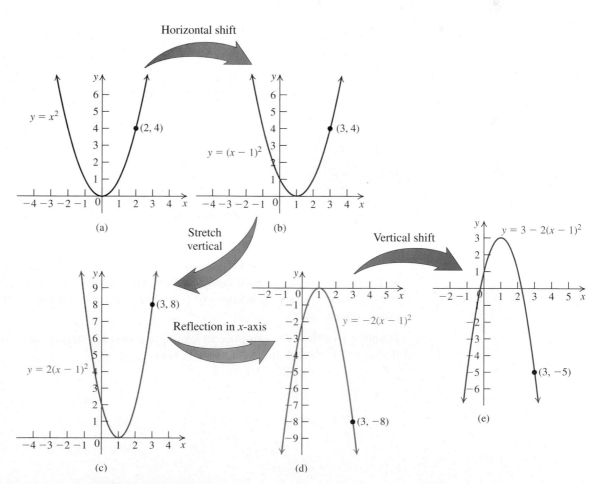

FIGURE 1.64 Multiple transformations

Practice Problem 7   Sketch the graph of the function

$$f(x) = 3\sqrt{x + 1} - 2.$$ ■

FIGURE 1.65

EXAMPLE 8	**Using Poiseuille's Law for Arterial Blood Flow**

For an artery with radius R, the velocity v of the blood flow at a distance r from the center of the artery is given by

$$v = c(R^2 - r^2),$$

where c is a constant that is determined for a particular artery. See Figure 1.65.

To emphasize that the velocity v depends on the distance r from the center of the artery, we write $v = v(r)$, or $v(r) = c(R^2 - r^2)$. Starting with the graph of $y = r^2$, sketch the graph of $y = v(r)$, where the artery has radius $R = 3$ and $c = 10^4$.

SOLUTION

$$v(r) = c(R^2 - r^2) \qquad \text{Original equation}$$
$$v(r) = 10^4(9 - r^2) \qquad \text{Substitute } c = 10^4 \text{ and } R = 3.$$
$$v(r) = 9 \cdot 10^4 - 10^4 r^2 \qquad \text{Distributive property}$$

Sketch the graph of $y = v(r)$ in Figure 1.66 through a sequence of transformations: stretch the graph of $y = r^2$ vertically by a factor of 10^4, reflect the resulting graph in the x-axis, and shift this graph $9 \cdot 10^4$ units up. Figure 1.66 shows the graph of $y = v(r) = 10^4(9 - r^2)$.

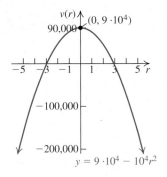

FIGURE 1.66 **Velocity of flow, r units from the artery's center**

The velocity of the blood decreases from the center of the artery ($r = 0$) to the artery wall ($r = 3$), where it ceases to flow. The portions of the graph in Figure 1.66 corresponding to negative values of r, or values of r greater than 3, do not have any physical significance. ■ ■ ■

Practice Problem 8   Suppose in Example 8 that the artery has radius $R = 3$ and that $c = 10^3$. Sketch the graph of $y = v(r)$. ■

	Summary of Transformations of $y = f(x)$		
To graph	**Draw the graph of f and**	**Make these changes to the equation $y = f(x)$**	**Change the graph point (x, y) to**
Vertical shifts for $c > 0$,			
$y = f(x) + c$	Shift the graph of f up c units.	Add c to $f(x)$.	$(x, y + c)$
$y = f(x) - c$	Shift the graph of f down c units.	Subtract c from $f(x)$.	$(x, y - c)$
Horizontal shifts for $c > 0$,			
$y = f(x + c)$	Shift the graph of f to the left c units.	Replace x with $x + c$.	$(x - c, y)$
$y = f(x - c)$	Shift the graph of f to the right c units.	Replace x with $x - c$.	$(x + c, y)$
Reflection in the x-axis			
$y = -f(x)$	Reflect the graph of f in the x-axis.	Multiply $f(x)$ by -1.	$(x, -y)$
Reflection in the y-axis			
$y = f(-x)$	Reflect the graph of f in the y-axis.	Replace x with $-x$.	$(-x, y)$
Vertical stretching or compressing:			
$y = af(x)$	Multiply each y-coordinate of $y = f(x)$ by $\|a\|$. The graph of $y = f(x)$ is stretched vertically away from the x-axis if $a > 1$ and is compressed vertically toward the x-axis if $0 < a < 1$. If $a < 0$, the graph is first reflected in the x-axis and then vertically stretched or compressed.	Multiply $f(x)$ by a.	(x, ay)
Horizontal stretching or compressing:			
$y = f(bx)$	Multiply each x-coordinate of $y = f(x)$ by $\dfrac{1}{\|b\|}$. The graph of $y = f(x)$ is stretched away from the y-axis if $0 < b < 1$ and is compressed toward the y-axis if $b > 1$. If $b < 0$, sketch the graph of $y = f(\|b\|x)$; then reflect it in the y-axis.	Replace x with bx.	$\left(\dfrac{x}{b}, y\right)$

SECTION 1.5 ■ Exercises

A EXERCISES Basic Skills and Concepts

1. The graph of $y = f(x) - 3$ is found by vertically shifting the graph of $y = f(x)$ three units ___down___.

2. The graph of $y = f(x + 5)$ is found by horizontally shifting the graph of $y = f(x)$ five units to the ___left___.

3. The graph of $y = f(bx)$ is a horizontal compression of the graph of $y = f(x)$ if b ___>1___.

4. The graph of $y = f(-x)$ is found by reflecting the graph of $y = f(x)$ in the ___y-axis___.

5. *True or False.* The graph of $y = f(x)$ and $y = f(-x)$ cannot be the same. False

6. *True or False.* You get the same graph by shifting the graph of $y = x^2$ two units up, reflecting the shifted graph in the x-axis or by reflecting the graph of $y = x^2$ in the x-axis, and then shifting the reflected graph up two units. False

In Exercises 7–20, describe the transformations that produce the graphs of g and h from the graph of f.

7. $f(x) = \sqrt{x}$
   a. $g(x) = \sqrt{x} + 2$ shift two units up   b. $h(x) = \sqrt{x} - 1$ shift one unit down

8. $f(x) = |x|$
   a. $g(x) = |x| + 1$ shift one unit up   b. $h(x) = |x| - 2$ shift two units down

9. $f(x) = x^2$
   a. $g(x) = (x + 1)^2$ shift one unit left   b. $h(x) = (x - 2)^2$ shift two units right

10. $f(x) = \dfrac{1}{x}$
    a. $g(x) = \dfrac{1}{x + 2}$ shift two units left   b. $h(x) = \dfrac{1}{x - 3}$ shift three units right

11. $f(x) = \sqrt{x}$
    a. $g(x) = \sqrt{x + 1} - 2$ shift one unit left and two units down
    b. $h(x) = \sqrt{x - 1} + 3$ shift one unit right and three units up

12. $f(x) = x^2$
    a. $g(x) = -x^2$ reflect in x-axis   b. $h(x) = (-x)^2$ reflect in y-axis

13. $f(x) = |x|$
    a. $g(x) = -|x|$ reflect in x-axis   b. $h(x) = |-x|$ reflect in y-axis

14. $f(x) = \sqrt{x}$
    a. $g(x) = 2\sqrt{x}$ stretch vertically by a factor of 2   b. $h(x) = \sqrt{2x}$ compress horizontally by a factor of 2

15. $f(x) = \dfrac{1}{x}$
    a. $g(x) = \dfrac{2}{x}$ stretch vertically by a factor of 2   b. $h(x) = \dfrac{1}{2x}$ compress horizontally by a factor of 2

16. $f(x) = x^3$
    a. $g(x) = (x - 2)^3 + 1$ shift two units right and one unit up
    b. $h(x) = -(x + 1)^3 + 2$ shift one unit left, reflect in x-axis, and shift two units up

17. $f(x) = \sqrt{x}$
    a. $g(x) = -\sqrt{x} + 1$ reflect in x-axis, shift one unit up   b. $h(x) = \sqrt{-x} + 1$ reflect in y-axis, shift one unit up

18. $f(x) = [\![x]\!]$
    a. $g(x) = [\![x - 1]\!] + 2$ shift one unit right and two units up   b. $h(x) = 3[\![x]\!] - 1$ stretch vertically by a factor of 3, shift one unit down

19. $f(x) = \sqrt[3]{x}$
    a. $g(x) = \sqrt[3]{x} + 1$ shift one unit up   b. $h(x) = \sqrt[3]{x + 1}$ shift one unit left

20. $f(x) = \sqrt[3]{x}$
    a. $g(x) = 2\sqrt[3]{1 - x} + 4$
    b. $h(x) = -\sqrt[3]{x - 1} + 3$

In Exercises 21–32, match each function with its graph (a)–(l).

21. $y = -|x| + 1$ e
22. $y = -\sqrt{-x}$ c
23. $y = \sqrt{x^2}$ g
24. $y = \dfrac{1}{2}|x|$ h
25. $y = \sqrt{x + 1}$ i
26. $y = 2|x| - 3$ a
27. $y = 1 - 2\sqrt{x}$ b
28. $y = -|x - 1| + 1$ k
29. $y = (x - 1)^2$ l
30. $y = -x^2 + 3$ f
31. $y = -2(x - 3)^2 - 1$ d
32. $y = 3 - \sqrt{1 - x}$ j

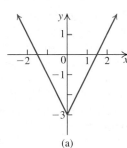

(a)

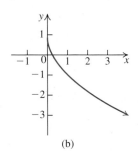

(b)

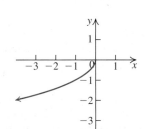

(c)

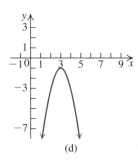

(d)

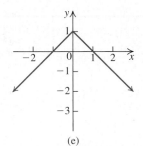

(e)

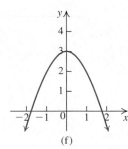

(f)

Answers:
20. a. shift one unit left, stretch vertically by a factor of 2, reflect in y-axis, and shift four units up.
    b. shift one unit right, reflect in x-axis, and shift three units up

†Due to space constrictions, answers to these exercises may be found in the Answers beginning on page A–1 in the back of the book.

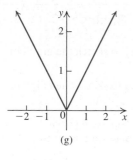

(g)

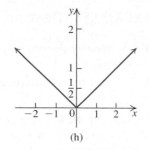

(h)

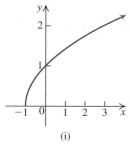

(i)

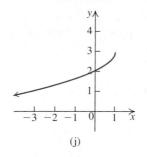

(j)

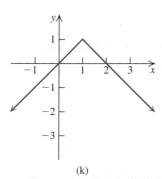

(k)

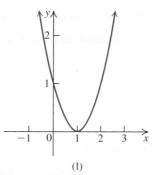

(l)

In Exercises 33–62, graph each function by starting with a function from the library of functions and then using the techniques of shifting, compressing, stretching, and/or reflecting.

33. $f(x) = x^2 - 2$ †
34. $f(x) = x^2 + 3$ †

35. $g(x) = \sqrt{x} + 1$ †
36. $g(x) = \sqrt{x} - 4$ †

37. $h(x) = |x + 1|$ †
38. $h(x) = |x - 2|$ †

39. $f(x) = (x - 3)^3$ †
40. $f(x) = (x + 2)^3$ †

41. $g(x) = (x - 2)^2 + 1$ †
42. $g(x) = (x + 3)^2 - 5$ †

43. $h(x) = -\sqrt{x}$ †
44. $h(x) = \sqrt{-x}$ †

45. $f(x) = -\dfrac{1}{x}$ †
46. $f(x) = -\dfrac{1}{2x}$ †

47. $g(x) = \dfrac{1}{2}|x|$ †
48. $g(x) = 4|x|$ †

49. $h(x) = -x^3 + 1$ †
50. $h(x) = -(x + 1)^3$ †

51. $f(x) = 2(x + 1)^2 - 1$ †
52. $f(x) = -(x - 1)^2$ †

53. $g(x) = 5 - x^2$ †
54. $g(x) = 2 - (x + 3)^2$ †

55. $h(x) = |1 - x|$ †
56. $h(x) = -2\sqrt{x - 1}$ †

57. $f(x) = -|x + 3| + 1$ †
58. $f(x) = 2 - \sqrt{x}$ †

59. $g(x) = -\sqrt{-x} + 2$ †
60. $g(x) = 3\sqrt{2 - x}$ †

61. $h(x) = 2[\![x + 1]\!]$ †
62. $h(x) = [\![-x]\!] + 1$ †

In Exercises 63–70, write an equation for a function whose graph fits the given description.

63. The graph of $f(x) = x^3$ is shifted two units up. $y = x^3 + 2$

64. The graph of $f(x) = \sqrt{x}$ is shifted three units left. $y = \sqrt{x + 3}$

65. The graph of $f(x) = |x|$ is reflected in the *x*-axis. $y = -|x|$

66. $f(x) = \sqrt{x}$ is reflected in the *y*-axis. $y = \sqrt{-x}$

67. The graph of $f(x) = x^2$ is shifted three units right and two units up. $y = (x - 3)^2 + 2$

68. The graph of $f(x) = \sqrt{x}$ is shifted three units left, reflected in the *x*-axis, and shifted two units down. $y = -\sqrt{x + 3} - 2$

69. The graph of $f(x) = x^3$ is shifted four units left, stretched vertically by a factor of 3, reflected in the *y*-axis, and shifted two units up. $y = 3(4 - x)^3 + 2$

70. The graph of $f(x) = |x|$ is shifted four units right, stretched vertically by a factor of 2, reflected in the *x*-axis, and shifted three units down. $y = -2|x - 4| - 3$

In Exercises 71–78, graph the function $y = g(x)$, given the following graph of $y = f(x)$.

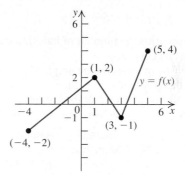

71. $g(x) = f(x) + 1$ †
72. $g(x) = -2f(x)$ †

73. $g(x) = f\left(\dfrac{1}{2}x\right)$ †
74. $g(x) = f(-2x)$ †

75. $g(x) = f(x - 1)$ †
76. $g(x) = f(2 - x)$ †

77. $g(x) = -2f(x + 1) + 3$ †
78. $g(x) = -f(-x + 1) - 2$ †

B EXERCISES Applying the Concepts

In Exercises 79–82, let *f* be the function that associates the employee number *x* of each employee of the ABC Corporation with his or her annual salary $f(x)$ in dollars.

79. Across-the-board raise. Each employee was awarded an across-the-board raise of $800 per year. Write a function $g(x)$ to describe the new salary. $g(x) = f(x) + 800$

80. Percentage raise. Suppose each employee was awarded a 5% raise. Write a function $h(x)$ to describe the new salary. $h(x) = 1.05f(x)$

81. Across-the-board and percentage raise. Suppose each employee was awarded a $500 across-the-board raise and an additional 2% of his or her increased salary. Write a function $p(x)$ to describe these new salaries. $p(x) = 1.02(x + 500)$

82. Across-the-board percentage raise. Suppose the employees making $30,000 or more received a 2% raise, while those making less than $30,000 received a 10% raise. Write a piecewise function to describe these new salaries.

83. Health plan. The ABC Corporation pays for its employees' health insurance at an annual cost (in dollars) given by

$$C(x) = 5,000 + 10\sqrt{x - 1},$$

where x is the number of employees covered.
a. Use transformations on the graph of $y = \sqrt{x}$ to sketch the graph of $y = C(x)$. †
b. Assuming the company has 400 employees, find its annual outlay for the health coverage. $5199.75

84. In Exercise 83, assume that the insurance company increases the annual cost of health insurance for the ABC Corporation by 10%. Write a function that reflects the new annual cost of health coverage. $y = 5500 + 11\sqrt{x - 1}$

85. Demand. The weekly demand for paper hats produced by Mythical Manufacturers is given by

$$x(p) = 109,561 - (p + 1)^2,$$

where x represents the number of hats that can be sold at a price of p cents each.
a. Use transformations on the graph of $y = p^2$ to sketch the graph of $y = x(p)$. †
b. Find the price at which 69,160 hats can be sold. $2
c. Find the price at which no hats can be sold. $3.30

86. Revenue. The weekly demand for cashmere sweaters produced by the Wool Shop, Inc., is $x(p) = -3p + 600$. The revenue is given by $R(p) = -3p^2 + 600p$. Describe how to sketch the graph of $y = R(p)$ by applying transformations to the graph of $y = p^2$. [*Hint:* First, write $R(p)$ in the form $-3(p - h)^2 + k$.]

87. Daylight. At 60° north latitude, the graph of $y = f(t)$ gives the number of hours of daylight. (On the t-axis, note that 1 = January; 12 = December.) †

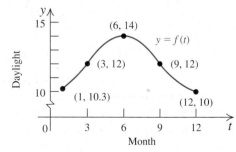

Sketch the graph of $y = f(t) - 12$.

88. Use the graph of $y = f(t)$ of Exercise 87 to sketch the graph of $y = 24 - f(t)$. Interpret the result. †

C EXERCISES Beyond the Basics

89. Use transformations on $y = [\![x]\!]$ to sketch the graph of $y = -2[\![x - 1]\!] + 3$. †

90. Use transformations on $y = [\![x]\!]$ to sketch the graph of $y = 3[\![x + 2]\!] - 1$. †

In Exercises 91–98, by completing the square on each quadratic expression, use transformations on $y = x^2$ to sketch the graph of $y = f(x)$.

91. $f(x) = x^2 + 4x$
[*Hint:* $x^2 + 4x = (x^2 + 4x + 4) - 4 = (x + 2)^2 - 4$.] †

92. $f(x) = x^2 - 6x$ †

93. $f(x) = -x^2 + 2x$ †

94. $f(x) = -x^2 - 2x$ †

95. $f(x) = 2x^2 - 4x$ †

96. $f(x) = 2x^2 + 6x + 3.5$ †

97. $f(x) = -2x^2 - 8x + 3$ †

98. $f(x) = -2x^2 + 2x - 1$ †

In Exercises 99–104, sketch the graph of each function.

99. $y = |2x + 3|$ †   **100.** $y = |[\![x]\!]|$ †

101. $y = |4 - x^2|$ †   **102.** $y = \left|\dfrac{1}{x}\right|$ †

103. $y = \sqrt{|x|}$ †   **104.** $y = [\![|x|]\!]$ †

Critical Thinking

105. If f is a function with x-intercept -2 and y-intercept 3, find the corresponding x-intercept and y-intercept, if possible, for
a. $y = f(x - 3)$ **b.** $y = 5f(x)$
c. $y = -f(x)$ **d.** $y = f(-x)$

106. If f is a function with x-intercept 4 and y-intercept -1, find the corresponding x-intercept and y-intercept, if possible, for
a. $y = f(x + 2)$ **b.** $y = 2f(x)$
c. $y = -f(x)$ **d.** $y = f(-x)$

107. Suppose the graph of $y = f(x)$ is given. Using transformations, explain how the graph of $y = g(x)$ differs from the graph of $y = h(x)$.
a. $g(x) = f(x) + 3$, $h(x) = f(x + 3)$
b. $g(x) = f(x) - 1$, $h(x) = f(x - 1)$
c. $g(x) = 2f(x)$, $h(x) = f(2x)$
d. $g(x) = -3f(x)$, $h(x) = f(-3x)$

108. Suppose the graph of $y = f(-4x)$ is given. Explain how to obtain the graph of $y = f(x)$.

105. a. x-intercept 1   **b.** x-intercept -2; y-intercept 15   **c.** x-intercept -2; y-intercept $-$
d. x-intercept 2; y-intercept 3   **106. a.** x-intercept 2   **b.** x-intercept 4; y-intercept -2
c. x-intercept 4; y-intercept 1   **d.** x-intercept -4; y-intercept -1
107. a. The graph of g is the graph of h shifted three units to the right and three units up.
b. The graph of g is the graph of h shifted one unit to the left and one unit down.   **c.** The graph of g is the graph of h stretched horizontally and vertically by a factor of 2.   **d.** The graph of g is the graph of h stretched horizontally by a factor of 3, reflected in the y-axis, stretched vertically by a factor of 3, and reflected in the x-axis.   **108.** Stretch the graph $y = f(-4x)$ horizontally by a factor of 4 and reflect the resulting graph in the y-axis.

Answers:

82. $j(x) = \begin{cases} 1.1(x) & \text{if } (x) < 30{,}000 \\ 1.02(x), & \text{if } (x) \geq 30{,}000 \end{cases}$

86. $R(p) = -3(p - 100)^2 + 30{,}000$; shifting 100 units to the right, stretching by a factor of 3, reflecting in the x-axis, and shifting 30,000 units up

Combining Functions; Composite Functions

Before Starting this Section, Review

1. Domain of function (Section 1.3, page 35)
2. Area of a circle (Appendix A, page 791)
3. Rational inequalities (Appendix A, page 833)

Objectives

1. Learn basic operations on functions.
2. Form composite functions.
3. Find the domain of a composite function.
4. Decompose a function.
5. Apply composition to practical problems.

EXXON VALDEZ OIL SPILL DISASTER

The oil tanker *Exxon Valdez* departed from the Trans-Alaska Pipeline terminal at 9:12 P.M., March 23, 1989. After passing through the Valdez Narrows, the tanker encountered icebergs in the shipping lanes, and Captain Hazelwood ordered the vessel to be taken out of the shipping lanes to go around the icebergs. For reasons that remain unclear, the ship failed to make the turn back into the shipping lanes, and the tanker ran aground on Bligh Reef at 12:04 A.M., March 24, 1989.

At the time of the accident, the *Exxon Valdez* was carrying 53 million gallons of oil, and approximately 11 million gallons were spilled. The amount of spilled oil is roughly equivalent to 125 Olympic-sized swimming pools. This spill was one of the largest ever to occur in the United States and is widely considered the worst worldwide in terms of damage to the environment. The spill covered a large area, stretching for 460 miles from Bligh Reef to the tiny village of Chigmik on the Alaska Peninsula. In Example 6, we calculate the area covered by an oil spill. ■

---

1 Learn basic operations on functions.

Combining Functions

Just as numbers can be added, subtracted, multiplied, and divided to produce new numbers, so functions can be added, subtracted, multiplied, and divided to produce other functions. To add, subtract, multiply, or divide functions, we simply add, subtract, multiply, or divide their range, or output values.

For example, if $f(x) = x^2 - 2x + 4$ and $g(x) = x + 2$, then

$$f(x) + g(x) = (x^2 - 2x + 4) + (x + 2) = x^2 - x + 6.$$

This new function $y = x^2 - x + 6$ is called the sum function $f + g$.

SUM, DIFFERENCE, PRODUCT, AND QUOTIENT OF FUNCTIONS

Let f and g be two functions. The **sum** $f + g$, the **difference** $f - g$, the **product** fg, and the **quotient** $\dfrac{f}{g}$ are functions whose domains consist of those values of x that are common to the domains of f and g, except that for $\dfrac{f}{g}$, all x for which $g(x) = 0$ must also be excluded. These functions are defined as follows:

- Sum $\qquad\qquad\qquad (f + g)(x) = f(x) + g(x)$
- Difference $\qquad\qquad (f - g)(x) = f(x) - g(x)$
- Product $\qquad\qquad\ (fg)(x) = f(x) \cdot g(x)$
- Quotient $\qquad\qquad \left(\dfrac{f}{g}\right)(x) = \dfrac{f(x)}{g(x)},\quad$ provided that $g(x) \neq 0$.

EXAMPLE 1   Combining Functions

Let $f(x) = x^2 - 6x + 8$ and $g(x) = x - 2$. Find each of the following functions.

a. $(f - g)(4)$ **b.** $\left(\dfrac{f}{g}\right)(3)$ **c.** $(f + g)(x)$ **d.** $(f - g)(x)$

e. $(fg)(x)$ **f.** $\left(\dfrac{f}{g}\right)(x)$

SOLUTION

a. $\qquad\quad f(4) = 4^2 - 6(4) + 8 = 0$

$\qquad\qquad g(4) = 4 - 2 = 2$

$\qquad (f - g)(4) = f(4) - g(4)$

$\qquad\qquad\qquad = 0 - 2 = -2$

b. $\qquad f(3) = 3^2 - 6(3) + 8 = -1$

$\qquad\quad g(3) = 3 - 2 = 1$

$\qquad \left(\dfrac{f}{g}\right)(3) = \dfrac{f(3)}{g(3)} = \dfrac{-1}{1} = -1$

c. $(f + g)(x) = f(x) + g(x)$ $\qquad\qquad$ Defintion of sum

$\qquad\qquad\quad = (x^2 - 6x + 8) + (x - 2)$ $\qquad$ Add $f(x)$ and $g(x)$.

$\qquad\qquad\quad = x^2 - 5x + 6$ $\qquad\qquad\qquad$ Combine terms.

d. $(f - g)(x) = f(x) - g(x)$ $\qquad\qquad$ Defintion of difference

$\qquad\qquad\quad = (x^2 - 6x + 8) - (x - 2)$ $\qquad$ Subtract $g(x)$ from $f(x)$.

$\qquad\qquad\quad = x^2 - 7x + 10$ $\qquad\qquad\qquad$ Combine terms.

e. $(fg)(x) = f(x) \cdot g(x)$ $\qquad\qquad\qquad$ Defintion of product

$\qquad\qquad = (x^2 - 6x + 8)(x - 2)$ $\qquad\quad$ Multiply $f(x)$ and $g(x)$.

$\qquad\qquad = x^2(x - 2) - 6x(x - 2) + 8(x - 2)$ $\quad$ Distributive property

$\qquad\qquad = x^3 - 2x^2 - 6x^2 + 12x + 8x - 16$ $\quad$ Distributive property

$\qquad\qquad = x^3 - 8x^2 + 20x - 16$ $\qquad\qquad$ Combine terms.

f. $\left(\dfrac{f}{g}\right)(x) = \dfrac{f(x)}{g(x)},\quad g(x) \neq 0$ $\qquad$ Definition of quotient

$\qquad\qquad = \dfrac{x^2 - 6x + 8}{x - 2},\quad x - 2 \neq 0$ $\qquad$ Divide $f(x)$ by $g(x)$.

$\qquad\qquad = \dfrac{(x - 2)(x - 4)}{x - 2},\quad x \neq 2$ $\qquad$ Factor the numerator.

$\qquad\qquad = x - 4,\quad x \neq 2$

Since f and g are polynomials, the domain of f and g is the set of all real numbers, or in interval notation, $(-\infty, \infty)$. So the domain for $f + g$, $f - g$, and fg is $(-\infty, \infty)$. However, for $\dfrac{f}{g}$, we must exclude 2, since $g(2) = 0$. The domain for $\dfrac{f}{g}$ is the set of all real numbers x, $x \neq 2$; in interval notation, $(-\infty, 2) \cup (2, \infty)$. ■ ■ ■

Practice Problem 1   Let $f(x) = 3x - 1$ and $g(x) = x^2 + 2$. Find $(f + g)(x)$, $(f - g)(x)$, $(fg)(x)$, and $\left(\dfrac{f}{g}\right)(x)$. ■

◆ **WARNING**

When rewriting a function such as

$$\left(\frac{f}{g}\right)(x) = \frac{(x - 2)(x - 4)}{x - 2}, \quad x \neq 2,$$

by removing the common factor, remember to specify "$x \neq 2$."

We identify the domain before we simplify, and write

$$\left(\frac{f}{g}\right)(x) = x - 4, \quad x \neq 2.$$

---

2 Form composite functions.

Composition of Functions

There is yet another way to construct a new function from two functions. Figure 1.67 shows what happens when you apply one function $g(x) = x^2 - 1$ to an input (domain) value x and then apply a second function $f(x) = \sqrt{x}$ to the output value, $g(x)$, from the first function. In this case, the output from the first function becomes the input for the second function.

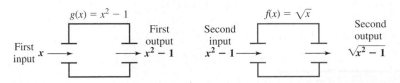

FIGURE 1.67 Composition of functions

COMPOSITION OF FUNCTIONS

If f and g are two functions, the composition of the function f with the function g is written as $f \circ g$ and is defined by the equation

$$(f \circ g)(x) = f(g(x)) \quad \text{(read "f of g of x")},$$

where the domain of $f \circ g$ consists of those values x in the domain of g for which $g(x)$ is in the domain of f.

Figure 1.68 may help clarify the definition of $f \circ g$.

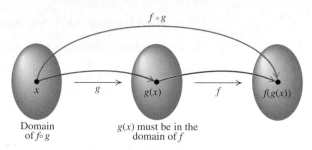

FIGURE 1.68 Composition of function f with g

Evaluating $(f \circ g)(x)$ By definition,

$$(f \circ g)(x) = f(g(x))$$

inner function

outer function

Thus, in order to evaluate $(f \circ g)(x)$, we must

(i) evaluate the inner function g at x.

(ii) use the result $g(x)$ as the input for the outer function f.

TECHNOLOGY CONNECTION

By defining Y_1 as $f(x)$ and Y_2 as $g(x)$, you can use a graphing calculator to verify the results of Example 2. The screens show the results for **a** and **b**.

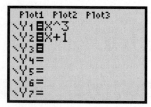

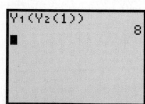

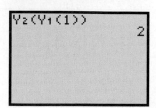

EXAMPLE 2 **Evaluating a Composite Function**

Let $f(x) = x^3$ and $g(x) = x + 1$. Find each of the following.

a. $(f \circ g)(1)$ **b.** $(g \circ f)(1)$ **c.** $(f \circ f)(-1)$

SOLUTION

a. $(f \circ g)(1) = f(g(1))$ Definition of $f \circ g$
           $= f(2)$ $g(1) = 1 + 1 = 2$
           $= 2^3 = 8$ Evaluate $f(2)$ and simplify.

b. $(g \circ f)(1) = g(f(1))$ Definition of $g \circ f$
           $= g(1)$ $f(1) = 1^3 = 1$
           $= 1 + 1 = 2$ Evaluate $g(1)$ and simplify.

c. $(f \circ f)(-1) = f(f(-1))$ Definition of $f \circ f$
            $= f(-1)$ $f(-1) = (-1)^3 = -1$
            $= (-1^3) = -1$ Evaluate $f(-1)$ and simplify.   ■ ■ ■

Practice Problem 2 Let $f(x) = -5x$ and $g(x) = x^2 + 1$. Find each of the following.

a. $(f \circ g)(0)$ **b.** $(g \circ f)(0)$ ■

EXAMPLE 3 **Finding Composite Functions**

Let $f(x) = 2x + 1$ and $g(x) = x^2 - 3$. Find each composite function.

a. $(f \circ g)(x)$ **b.** $(g \circ f)(x)$ **c.** $(f \circ f)(x)$

SOLUTION

a. $(f \circ g)(x) = f(g(x))$ Definition of $f \circ g$
           $= f(x^2 - 3)$ Replace $g(x)$ with $x^2 - 3$.
           $= 2(x^2 - 3) + 1$ Replace x with $x^2 - 3$ in $f(x) = 2x - 1$.
           $= 2x^2 - 5$ Simplify.

b. $(g \circ f)(x) = g(f(x))$ Definition of $g \circ f$

$\qquad\qquad = g(2x + 1)$ Replace $f(x)$ with $2x + 1$.

$\qquad\qquad = (2x + 1)^2 - 3$ Replace x with $2x + 1$ in $g(x) = x^2 - 3$.

$\qquad\qquad = 4x^2 + 4x + 1 - 3$ $(A + B)^2 = A^2 + 2AB + B^2$

$\qquad\qquad = 4x^2 + 4x - 2$ Simplify.

c. $(f \circ f)(x) = f(f(x))$ Defintion of $f \circ f$

$\qquad\qquad = f(2x + 1)$ Replace $f(x)$ with $2x + 1$.

$\qquad\qquad = 2(2x + 1) + 1$ Replace x with $2x + 1$ in $f(x) = 2x + 1$.

$\qquad\qquad = 4x + 3$ Simplify.   ■ ■ ■

Practice Problem 3   Let $f(x) = 2 - x$ and $g(x) = 2x^2 + 1$. Find $(g \circ f)(x)$ and $(f \circ g)(x)$.   ■

Parts **a** and **b** of Example 3 illustrate that, in general, $f \circ g \neq g \circ f$. In other words, the composition of functions is not commutative.

◆ **WARNING**   Note that $f(g(x))$ is not $f(x) \cdot g(x)$. In $f(g(x))$, the output of g is *used as an input* for f; in contrast, in $f(x) \cdot g(x)$, the output $f(x)$ is *multiplied* by the output $g(x)$.

---

3 Find the domain of a composite function.

Domain of Composite Functions

Let f and g be two functions. The domain of the composite function $f \circ g$ consists of those values of x in the domain of g for which $g(x)$ is in the domain of f. So to find the domain of $f \circ g$, we first find the domain of g. Then from the domain of g, we exclude those values of x such that $g(x)$ is not in the domain of f.

EXAMPLE 4   **Finding the Domain of a Composite Function**

Let $f(x) = x + 1$ and $g(x) = \dfrac{1}{x}$.

a. Find $(f \circ g)(x)$ and its domain.   **b.** Find $(g \circ f)(x)$ and its domain.

SOLUTION

a. $(f \circ g)(x) = f(g(x))$ Definition of $f \circ g$

$\qquad\qquad = f\left(\dfrac{1}{x}\right)$ Replace $g(x)$ with $\dfrac{1}{x}$.

$\qquad\qquad = \dfrac{1}{x} + 1$ Replace x with $\dfrac{1}{x}$ in $f(x) = x + 1$.

Because 0 is not in the domain of $g(x) = \dfrac{1}{x}$, we must exclude 0 from the domain of $f \circ g$. This is the only restriction on the domain of $f \circ g$ since the domain of f is $(-\infty, \infty)$. So the domain of $f \circ g$ is all real $x, x \neq 0$. In interval notation, the domain of $f \circ g$ is $(-\infty, 0) \cup (0, \infty)$.

b. $(g \circ f)(x) = g(f(x))$ Definition of $g \circ f$

$\qquad\qquad = g(x + 1)$ Replace $f(x)$ with $x + 1$.

$\qquad\qquad = \dfrac{1}{x + 1}$ Replace x with $x + 1$ in $g(x) = \dfrac{1}{x}$.

The domain of f is $(-\infty, \infty)$, so we need to exclude from the domain of $g \circ f$ only values of x for which $f(x)$ is not in the domain of g. That is, values of x such

---

STUDY TIP

If the function $f \circ g$ can be simplified, determine the domain *before* simplifying. For example, if $f(x) = x^2$ and $g(x) = \sqrt{x}$,

$(f \circ g)(x) = f(g(x)) = f(\sqrt{x})$.

Before simplifying

$(f \circ g)(x) = (\sqrt{x})^2, x \geq 0$

After simplifying

$(f \circ g)(x) = x, x \geq 0$

that $f(x) = 0$. If $f(x) = 0$, then $x + 1 = 0$ and $x = -1$. Therefore, the domain of $g \circ f$ is all real numbers x, $x \neq -1$, or $(-\infty, -1) \cup (-1, \infty)$.   ■ ■ ■

Practice Problem 4   Let $f(x) = \sqrt{x}$ and $g(x) = 2x + 5$. Find $(g \circ f)(x)$ and its domain.   ■

4 Decompose a function.

Decomposition of a Function

In the composition of two functions, we combine two functions and create a new function. However, sometimes it is more useful to use the concept of composition to *decompose* a function into simpler functions. For example, consider the function $H(x) = \sqrt{x^2 - 2}$. A natural way to write $H(x)$ as the composition of two functions is to let $f(x) = \sqrt{x}$ and $g(x) = x^2 - 2$ so that $f(g(x)) = f(x^2 - 2) = \sqrt{x^2 - 2} = H(x)$.

A function may be decomposed into simpler functions in several different ways by choosing various "inner" functions.

FINDING THE SOLUTION: A PROCEDURE

EXAMPLE 5   **Decomposing a Function**

OBJECTIVE
Write a function H as a composite of simpler functions f and g so that $H = f \circ g$.

EXAMPLE
Write

$$H(x) = \frac{1}{\sqrt{2x^2 + 1}}$$

as $f \circ g$.

Step 1   Define $g(x)$ as any expression in the defining equation for H.

1. Let $g(x) = 2x^2 + 1$.

Step 2   To get $f(x)$ from the defining equation for H: (1) replace the letter H with f (2) replace the expression chosen for $g(x)$ with x.

2. $H(x) = \dfrac{1}{\sqrt{2x^2 + 1}}$

becomes

$f(x) = \dfrac{1}{\sqrt{x}}$ Replace the expression $2x^2 + 1$ from Step 1 with x and replace H with f.

Step 3   Now we have

$$H(x) = f(g(x))$$
$$= f \circ g(x).$$

3. $f(g(x)) = f(2x^2 + 1)$

$\qquad = \dfrac{1}{\sqrt{2x^2 + 1}}$

$\qquad = H(x)$ ■ ■ ■

Practice Problem 5   Repeat Example 5 letting $g(x) = \sqrt{2x^2 + 1}$. Find $f(x)$ and write $H(x) = f(g(x))$.   ■

5 Apply composition to practical problems.

Applications

EXAMPLE 6   **Calculating the Area of an Oil Spill from a Tanker**

Suppose oil spills from a tanker into the Pacific Ocean and the area of the spill is a perfect circle. (Winds, tides, and coastlines prevent real spills from forming circles.) The radius of this oil slick increases (because oil continues to spill) at the rate of 2 miles per hour.

a. Express the area of the oil slick as a function of time.

b. Calculate the area covered by the oil slick in six hours.

SOLUTION

The area A of the oil slick is a function of its radius r:

$$A = f(r) = \pi r^2$$

The radius is also a function of time. We know that r increases at the rate of 2 miles per hour. So in t hours, the radius r of the oil slick is given by the function

$$r = g(t) = 2t \qquad \text{Distance} = \text{Rate} \times \text{Time}$$

a. Therefore, the area A of the oil slick is a function of the radius, which is itself a function of time. It is a composite function given by

$$A = f(g(t)) = f(2t) = \pi(2t)^2$$
$$= 4\pi t^2.$$

b. Substitute $t = 6$ into A to find the area covered by the oil in six hours:

$$A = 4\pi(6)^2 = 4\pi(36)$$
$$= 144\pi \approx 452 \text{ square miles} \qquad ■ ■ ■$$

Practice Problem 6   What would the result of Example 6 be if the radius of the oil slick increased at a rate of 3 miles per hour? ■

EXAMPLE 7   **Applying Composition to Sales**

A car dealer offers an 8% discount off the manufacturer's suggested retail price (MSRP) of x dollars for any new car on his lot. At the same time, the manufacturer offers a \$4000 rebate for each purchase of a car.

a. Write a function $r(x)$ that represents the price after only the rebate.
b. Write a function $d(x)$ that represents the price after only the dealer's discount.
c. Write the functions $(r \circ d)(x)$ and $(d \circ r)(x)$. What do they represent?
d. Calculate $(d \circ r)(x) - (r \circ d)(x)$. Interpret this expression.

SOLUTION

a. The MSRP is x dollars, and the manufacturer's rebate is \$4000 for each car; so

$$r(x) = x - 4000$$

represents the price of a car after the rebate.

b. The dealer's discount is 8% of x, or $0.08x$; therefore,

$$d(x) = x - 0.08x = 0.92x$$

represents the price of the car after the dealer's discount.

c. **(i)** $(r \circ d)(x) = r(d(x))$ Apply the dealer's discount first.
$$= r(0.92x) \qquad d(x) = 0.92x$$
$$= 0.92x - 4000 \quad \text{Evaluate } r(0.92x) \text{ using } r(x) = x - 4000.$$

Then $(r \circ d)(x) = 0.92x - 4000$ represents the price when the dealer's discount is applied first.

(ii) $(d \circ r)(x) = d(r(x))$ Apply the manufacturer's rebate first.
$$= d(x - 4000) \qquad r(x) = x - 4000$$
$$= 0.92(x - 4000) \quad \text{Evaluate } d(x - 4000) \text{ using } d(x) = 0.92x.$$
$$= 0.92x - 3680$$

Thus, $(d \circ r)(x) = 0.92x - 3680$ represents the price when the manufacturer's rebate is applied first.

d. $(d \circ r)(x) - (r \circ d)(x) = (0.92x - 3680) - (0.92x - 4000)$ Subtract function values.

$$= 320 \text{ dollars}$$

This equation shows that any car, regardless of its price, will cost $320 more if you apply the rebate first and then the discount.   ■ ■ ■

Practice Problem 7   How will the result of Example 7 change if the dealer offers a 6% discount and the manufacturer offers a $4500 rebate?   ■

SECTION 1.6 ■ Exercises

A EXERCISES Basic Skills and Concepts

1. If $f(2) = 12$ and $g(2) = 4$, then $(g - f)(2) =$ ___−8___.

2. If $f(x) = 2x - 1$ and $f(x) + g(x) = 0$, then $g(x) =$ ___1 − 2x___.

3. If $f(x) = x$ and $g(x) = 1$, then $(f \circ g)(x) =$ ___1___.

4. If $f(1) = 4$ and $g(4) = 7$, then $(g \circ f)(1) =$ ___7___.

5. *True or False.* If $f(x) = 2x$ and $g(x) = \dfrac{1}{2x}$, then $(f \circ g)(x) = x$. False

6. *True or False.* It cannot happen that $f \circ g = g \circ f$. False

In Exercises 7–10, functions f and g are given. Find each of the given values.

a. $(f + g)(-1)$ **b.** $(f - g)(0)$

c. $(f \cdot g)(2)$ **d.** $\left(\dfrac{f}{g}\right)(1)$

7. $f(x) = 2x; g(x) = -x$ †

8. $f(x) = 1 - x^2; g(x) = x + 1$ †

9. $f(x) = \dfrac{1}{\sqrt{x + 2}}; g(x) = 2x + 1$ †

10. $f(x) = \dfrac{x}{x^2 - 6x + 8}; g(x) = 3 - x$ †

In Exercises 11–18, functions f and g are given. Find each of the following functions and state its domain.

a. $f + g$ **b.** $f - g$ **c.** $f \cdot g$

d. $\dfrac{f}{g}$ **e.** $\dfrac{g}{f}$

11. $f(x) = x - 3; g(x) = x^2$ †

12. $f(x) = 2x - 1; g(x) = x^2$ †

13. $f(x) = x^3 - 1; g(x) = 2x^2 + 5$ †

14. $f(x) = x^2 - 4; g(x) = x^2 - 6x + 8$ †

15. $f(x) = 2x - 1; g(x) = \sqrt{x}$ †

16. $f(x) = 1 - \dfrac{1}{x}; g(x) = \dfrac{1}{x}$ †

17. $f(x) = \dfrac{2}{x + 1}, g(x) = \dfrac{x}{x + 1}$ †

18. $f(x) = \dfrac{x^2 + 3x + 2}{x^3 + 4x}$ †

$g(x) = \dfrac{2x^3 + 8x}{x^2 + x - 2}$

In Exercises 19 and 20, use each diagram to evaluate $(g \circ f)(x)$. Then evaluate $(g \circ f)(2)$ and $(g \circ f)(-3)$.

19. $x \rightarrow \boxed{f(x) = x^2 - 1} \rightarrow \rightarrow \boxed{g(x) = 2x + 3} \rightarrow$

$g(f(x)) = 2x^2 + 1; g(f(2)) = 9; g(f(-3)) = 19$

20. $x \rightarrow \boxed{f(x) = |x + 1|} \rightarrow \rightarrow \boxed{g(x) = 3x^2 - 1} \rightarrow$

$g(f(x)) = 3|x^2 + 2x + 1| - 1; g(f(2)) = 26; g(f(-3)) = 11$

In Exercises 21–32, let $f(x) = 2x + 1$ and $g(x) = 2x^2 - 3$. Evaluate each expression.

21. $(f \circ g)(2)$ 11   **22.** $(g \circ f)(2)$ 47

23. $(f \circ g)(-3)$ 31   **24.** $(g \circ f)(-5)$ 159

25. $(f \circ g)(0)$ −5   **26.** $(g \circ f)\left(\dfrac{1}{2}\right)$ 5

27. $(f \circ g)(-c)$ $4c^2 - 5$ **28.** $(f \circ g)(c)$ $4c^2 - 5$

29. $(g \circ f)(a)$ $8a^2 + 8a - 1$

30. $(g \circ f)(-a)$ $8a^2 - 8a - 1$

31. $(f \circ f)(1)$ 7   **32.** $(g \circ g)(-1)$ −1

In Exercises 33–38, the functions f and g are given. Find $f \circ g$ and its domain.

33. $f(x) = \dfrac{2}{x + 1}; g(x) = \dfrac{1}{x} \dfrac{2x}{x + 1}; x \neq 0, x \neq -1$

34. $f(x) = \dfrac{1}{x - 1}; g(x) = \dfrac{2}{x + 3} \dfrac{x + 3}{x + 1}; x \neq -1, x \neq -3$

35. $f(x) = \sqrt{x - 3}; g(x) = 2 - 3x \sqrt{-3x - 1}; \left(-\infty, -\dfrac{1}{3}\right]$

36. $f(x) = \dfrac{x}{x - 1}; g(x) = 2 + 5x \dfrac{2 + 5x}{1 + 5x}; x \neq -\dfrac{1}{5}$

37. $f(x) = |x|; g(x) = x^2 - 1 |x^2 - 1|; (-\infty, \infty)$

38. $f(x) = 3x - 2; g(x) = |x - 1| 3|x - 1| - 2; (-\infty, \infty)$

†Due to space constrictions, answers to these exercises may be found in the Answers beginning on page A-1 in the back of the book.

In Exercises 39–52, the functions *f* and *g* are given. Find each composite function and describe its domain.

a. $f \circ g$ **b.** $g \circ f$
c. $f \circ f$ **d.** $g \circ g$

39. $f(x) = 2x - 3; g(x) = x + 4$ †

40. $f(x) = x - 3; g(x) = 3x - 5$ †

41. $f(x) = 1 - 2x; g(x) = 1 + x^2$ †

42. $f(x) = 2x - 3; g(x) = 2x^2$ †

43. $f(x) = 2x^2 + 3x; g(x) = 2x - 1$ †

44. $f(x) = x^2 + 3x; g(x) = 2x$ †

45. $f(x) = x^2; g(x) = \sqrt{x}$ †

46. $f(x) = x^2 + 2x; g(x) = \sqrt{x + 2}$ †

47. $f(x) = \dfrac{1}{2x - 1}; g(x) = \dfrac{1}{x^2}$ †

48. $f(x) = x - 1; g(x) = \dfrac{x}{x + 1}$ †

49. $f(x) = |x|; g(x) = -2$ †

50. $f(x) = 3; g(x) = 5$ †

51. $f(x) = 1 + \dfrac{1}{x}; g(x) = \dfrac{1 + x}{1 - x}$ †

52. $f(x) = \sqrt[3]{x + 1}; g(x) = x^3 + 1$ †

In Exercises 53–62, express the given function *H* as a composition of two functions *f* and *g* such that $H(x) = (f \circ g)(x)$.

53. $H(x) = \sqrt{x + 2}$ $f(x) = \sqrt{x}; g(x) = x + 2$

54. $H(x) = |3x + 2|$ $f(x) = |x|; g(x) = 3x + 2$

55. $H(x) = (x^2 - 3)^{10}$ $f(x) = x^{10}; g(x) = x^2 - 3$

56. $H(x) = \sqrt{3x^2 + 5}$ $f(x) = \sqrt{x} + 5; g(x) = 3x^2$

57. $H(x) = \dfrac{1}{3x - 5}$ $f(x) = \dfrac{1}{x}; g(x) = 3x - 5$

58. $H(x) = \dfrac{5}{2x + 3}$ $f(x) = \dfrac{5}{x}; g(x) = 2x + 3$

59. $H(x) = \sqrt[3]{x^2 - 7}$ $f(x) = \sqrt[3]{x}; g(x) = x^2 - 7$

60. $H(x) = \sqrt[4]{x^2 + x + 1}$ $f(x) = \sqrt[4]{x}; g(x) = x^2 + x + 1$

61. $H(x) = \dfrac{1}{|x^3 - 1|}$ $f(x) = \dfrac{1}{|x|}; g(x) = x^3 - 1$

62. $H(x) = \sqrt[3]{1 + \sqrt{x}}$ $f(x) = \sqrt[3]{x}; g(x) = 1 + \sqrt{x}$

B EXERCISES Applying the Concepts

63. **Cost, revenue, and profit.** A retailer purchases *x* shirts from a wholesaler at a price of $12 per shirt. Her selling price for each shirt is $22. The state has 7% sales tax. Interpret each of the following functions.

a. $f(x) = 12x$ Cost function

b. $g(x) = 22x$ Revenue function
c. $h(x) = g(x) + 0.07g(x)$
d. $P(x) = g(x) - f(x)$ Profit function

64. **Cost, revenue, and profit.** The demand equation for a product is given by $x = 5000 - 5p$. where *x* is the number of units produced and sold at price *p* (in dollars) per unit. The cost (in dollars) of producing *x* units is given by $C(x) = 4x + 12,000$. Express each of the following as a function of price.

a. Cost $C(p) = 32,000 - 20p$
b. Revenue $R(p) = 5000p - 5p^2$
c. Profit $P(p) = -5p^2 + 5020p - 32,000$

65. **Cost and revenue.** A manufacturer of radios estimates that his daily cost of producing *x* radios is given by the equation $C = 350 + 5x$. The equation $R = 25x$ represents the revenue in dollars from selling *x* radios.

a. Write and simplify the profit function

$$P(x) = R(x) - C(x). \quad P(x) = 20x - 350$$

b. Find $P(20)$. What does the number $P(20)$ represent?
c. How many radios should the manufacturer produce and sell to have a daily profit of $500? 43
d. Find the composite function $(R \circ x)(C)$. What does this function represent?
  [*Hint:* To find $x(C)$, solve $C = 350 + 5x$ for *x*.]

66. **Mail order.** You order merchandise worth *x* dollars from D-Bay Manufacturers. The company charges you sales tax of 4% of the purchase price plus a shipping-and-handling fee of $3 plus 2% of the after-tax purchase price.

a. Write a function $g(x)$ that represents the sales tax.
b. What does the function $h(x) = x + g(x)$ represent?
c. Write a function $f(x)$ that represents the shipping-and-handling fee.
d. What does the function $T(x) = h(x) + f(x)$ represent?

67. **Consumer issues.** You work in a department store in which employees are entitled to a 30% discount on their purchases. You also have a coupon worth $5 off any item.

a. Write a function $f(x)$ that models discounting an item by 30%. $f(x) = 0.7x$
b. Write a function $g(x)$ that models applying the coupon. $g(x) = x - 5$
c. Use a composition of your two functions from (a) and (b) to model your cost for an item if the clerk applies the discount first and then the coupon. $(g \circ f)(x) = 0.7x - 5$
d. Use a composition of your two functions from (a) and (b) to model your cost for an item if the clerk applies the coupon first and then the discount.
e. Use the composite functions from (c) and (d) to find how much more an item costs if the clerk applies the coupon first. $1.50

Answers:
65. b. $P(20) = 50$; profit made after 20 radios were sold
d. $(R \circ x)(C) = 5C - 1,750$. The function represents the revenue in terms of the cost *C*.

63. c. Selling price of *x* shirts after sales tax
66. a. $g(x) = 0.04x$ **b.** The after-tax selling price of merchandise worth x-dollars **c.** $f(x) = 0.02h(x) + 3$; the after-tax shipping-and-handling fee for merchandise for a price of *x* dollars **d.** The total price of merchandise worth *x* dollars, including the shipping-and-handling fee **67. d.** $(f \circ g)(x) = 0.7(x - 5)$

68. Consumer issues. You work in a department store in which employees are entitled to a 20% discount on their purchases. In addition, the store is offering a 10% discount on all items.
   a. Write a function $f(x)$ that models discounting an item by 20%. $f(x) = 0.8x$
   b. Write a function $g(x)$ that models discounting an item by 10%. $g(x) = 0.9x$
   c. Use a composition of your two functions from (a) and (b) to model taking the 20% discount first. †
   d. Use a composition of your two functions from (a) and (b) to model taking the 10% discount first. †
   e. Use the composite functions from (c) and (d) to decide which discount you would prefer to take first.
   The discounts are the same.

69. Test grades. Professor Harsh gave a test to his college algebra class, and nobody got more than 80 points (out of 100) on the test. One problem worth 8 points had insufficient data, so nobody could solve that problem. The professor adjusts the grades for the class by (1) increasing everyone's score by 10% and (2) giving everyone 8 bonus points. Let x represent the original score of a student.
   a. Write statements (1) and (2) as functions $f(x)$ and $g(x)$, respectively. $f(x) = 1.1x; g(x) = x + 8$
   b. Find $(f \circ g)(x)$ and explain what it means. †
   c. Find $(g \circ f)(x)$ and explain what it means. †
   d. Evaluate $(f \circ g)(70)$ and $(g \circ f)(70)$. 85.8 and 85.0
   e. Does $(f \circ g)(x) = (g \circ f)(x)$? No
   f. Suppose a score of 90 or better results in an A. What is the lowest original score that will result in an A if the professor uses
   (i) $(f \circ g)(x)$; **(ii)** $(g \circ f)(x)$? **(i)** 73.82 **(ii)** 74.55

70. Sales commissions. Henrita works as a salesperson in a department store. Her weekly salary is $200 plus a 3% bonus on weekly sales over $8000. Suppose her sales in a week are x dollars.
   a. Let $f(x) = 0.03x$. What does this mean?
   b. Let $g(x) = x - 8000$. What does this mean?
   c. Which composite function, $(f \circ g)(x)$ or $(g \circ f)(x)$, represents Henrita's bonus? $(f \circ g)(x)$
   d. What was Henrita's salary for the week in which her sales were $17,500? $485
   e. What were Henrita's sales for the week in which her salary was $521? $18,700

71. Enhancing a fountain. A circular fountain has a radius of x feet. A circular fence is installed around the fountain at a distance of 30 feet from its edge.
   a. Write the function $f(x)$ that represents the area of the fountain. $f(x) = \pi x^2$
   b. Write the function $g(x)$ that represents the entire area enclosed by the fence. $g(x) = \pi(x + 30)^2$
   c. What area does the function $g(x) - f(x)$ represent?
   d. The cost of the fence was $4200, installed at the rate of $10.50 per running foot (per perimeter foot). You are to prepare an estimate for paving the area between the

fence and the fountain at $1.75 per square foot. To the nearest dollar, what should your estimate be? $16,052

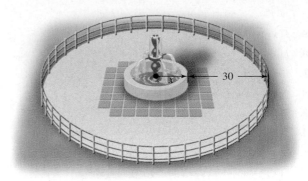

72. Running track design. An outdoor running track with semicircular ends and parallel sides is constructed. The length of each straight portion of the sides is 180 meters. The track has a uniform width of 4 meters throughout. The inner radius of each semicircular end is x meters.
   a. Write the function $f(x)$ that represents the area enclosed within the *outer* edge of the running track.
   b. Write the function $g(x)$ that represents the area of the field enclosed within the *inner* edge of the running track. $g(x) = 360x + \pi x^2$
   c. What does the function $f(x) - g(x)$ represent?
   d. Suppose the inner perimeter of the track is 900 meters.
   (i) Find the area of the track. 3650.26
   (ii) Find the outer perimeter of the track. 925.13

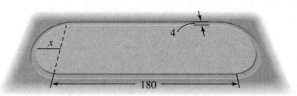

73. Area of a disk. The area A of a circular disk of radius r units is given by $A = f(r) = \pi r^2$. Suppose a metal disk is being heated and its radius r is increasing according to the equation $r = g(t) = 2t + 1$, where t is time in hours.
   a. Find $(f \circ g)(t)$. $(f \circ g)(t) = \pi(2t + 1)^2$
   b. Determine A as a function of time. $A(t) = \pi(2t + 1)^2$
   c. Compare parts (a) and (b). They are the same.

74. Volume of a balloon. The volume V of a spherical balloon of radius r inches is given by the formula $V = f(r) = \dfrac{4}{3}\pi r^3$. Suppose the balloon is being inflated and its radius is increasing at the rate of 2 inches per second. That is, $r = g(t) = 2t$, where t is time in seconds.
   a. Find $(f \circ g)(t)$.
   b. Determine V as a function of time.
   c. Compare parts (a) and (b). They are the same.

Answers:
70. a. A function that models 3% of an amount x **b.** The amount of money that qualified for a 3% bonus **71. c.** The area between the fountain and the fence

72. a. $f(x) = 1440 + 360x + \pi(x + 4)^2$ **c.** The area of the track
74. a. $(f \circ g)(t) = \dfrac{32}{3}\pi t^3$ **b.** $V(t) = \dfrac{32}{3}\pi t^3$

75. Currency exchange. On July 13, 2007, each British pound (symbol £) was worth \$2.0337 and each Mexican peso was worth £0.04567 (*Source:* www.xe.com/ucc)
   a. Write a function f that converts pounds to dollars.
   b. Write a function, g that converts pesos to pounds.
   c. Which function $f \circ g$ or $g \circ f$, converts pesos to dollars? $(f \circ g)(x)$
   d. How much was 1000 pesos worth in dollars on July 13, 2007? \$92.88

C EXERCISES Beyond the Basics

76. Odd and even functions. State whether each of the following functions is an odd function, an even function, or neither. Justify your statement.
   a. The sum of two even functions Even function
   b. The sum of two odd functions Odd function
   c. The sum of an even function and an odd function Neither function
   d. The product of two even functions Even function
   e. The product of two odd functions Even function
   f. The product of an even function and an odd function Odd function

77. Odd and even functions. State whether the composite function $f \circ g$ is an odd function, an even function, or neither in the following situations.
   a. f and g are odd functions. Odd function
   b. f and g are even functions. Even function
   c. f is odd and g is even. Even function
   d. f is even and g is odd. Even function

78. Let $h(x)$ be any function whose domain contains $-x$ whenever it contains x. Show that
   a. $f(x) = h(x) + h(-x)$ is an even function.
   b. $g(x) = h(x) - h(-x)$ is an odd function.
   c. $h(x)$ can always be expressed as the sum of an even function and an odd function.

79. Use Exercise 78 to write each of the following functions as the sum of an even function and an odd function.
   a. $h(x) = x^2 - 2x + 3$
   b. $h(x) = [\![x]\!] + x$

Critical Thinking

80. Let $f(x) = \dfrac{1}{[\![x]\!]}$ and $g(x) = \sqrt{x(2 - x)}$. Find the domain of each of the following functions.
   a. $f(x)$ **b.** $g(x)$ Domain: $[0, 2]$
   c. $f(x) + g(x)$ **d.** $\dfrac{f(x)}{g(x)}$ Domain: $[1, 2)$

81. For each of the following functions, find the domain of $f \circ f$.
   a. $f(x) = \dfrac{1}{\sqrt{-x}}$ Domain: $\varnothing$ **b.** $f(x) = \dfrac{1}{\sqrt{1 - x}}$

Answers:
75. a. $f(x) = 2.0337x$ **b.** $g(x) = 0.04567x$
78. c. $h(x) = \dfrac{h(x) + h(-x)}{2} + \dfrac{h(x) - h(-x)}{2}$
79. a. $h(x) = o(x) + e(x)$, where $e(x) = x^2 + 3$ and $o(x) = -2x$
   b. $h(x) = o(x) + e(x)$, where
   $e(x) = \dfrac{[\![x]\!] + [\![-x]\!]}{2}$ and $o(x) = x + \dfrac{[\![x]\!] - [\![-x]\!]}{2}$

80. a. Domain: $(-\infty, 0) \cup [1, \infty)$ **c.** Domain: $[1, 2]$
81. b. Domain: $(-\infty, 0)$

Inverse Functions

Before Starting this Section, Review	Objectives

Before Starting this Section, Review

1. Vertical-line test (Section 1.3, page 36)
2. Domain and range of a function (Section 1.3, page 35)
3. Composition of functions (Section 1.6, page 79)
4. Line of symmetry (Section 1.1, page 8)
5. Solving linear and quadratic equations (Appendix A, page 814)

Objectives

1. Define an inverse function.
2. Find the inverse function.
3. Use inverse functions to find the range of a function.
4. Apply inverse functions in the real world.

WATER PRESSURE ON UNDERWATER DEVICES

The pressure at any depth in the ocean is due to the weight of the overlying water: the deeper you go, the greater the pressure. Pressure is a vital consideration in the design of all tools and vehicles that work beneath the water's surface. Every 33 feet of depth increases the pressure by approximately 15 pounds per square inch (1 atmosphere). Computing the pressure precisely requires a somewhat complicated equation that takes into account the water's salinity, temperature, and slight compressibility. However, a basic formula is given by "pressure equals depth times 15, divided by 33." Thus, we have pressure (psi) = depth (ft) × 15(psi per atm)/33(ft per atm), where *psi* means "pounds per square inch" and *atm* abbreviates "atmosphere."

For example, at 99 feet below the surface, the pressure is 45 psi. At 1 mile (5280 feet) below the surface, the pressure is $(5280 \times 15)/33 = 2400$ psi. Suppose the pressure gauge on a diving bell breaks and shows a reading of 1800 psi. We can determine the bell's depth when the gauge failed by using an inverse function. We will return to this problem in Example 9. ∎

---

1 Define an inverse function.

Inverses

DEFINITION OF A ONE-TO-ONE FUNCTION

A function is a **one-to-one function** if each y-value in its range corresponds to only one x-value in its domain.

Let f be a one-to-one function. Then the preceding definition says that for any two numbers x_1 and x_2 in the domain of f, if $x_1 \neq x_2$, then $f(x_1) \neq f(x_2)$. That is, f *is a one-to-one function if different x-values correspond to different y-values.*

Figure 1.69(a) represents a one-to-one function, and Figure 1.69(b) represents a function that is not one-to-one. The relation shown in Figure 1.69(c) is not a function.

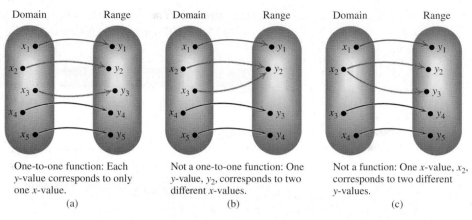

FIGURE 1.69 Deciding whether a diagram represents a function, a one-to-one function, or neither

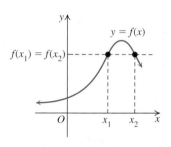

FIGURE 1.70 The function f is *not* one-to-one

With a one-to-one function, different x-values correspond to different y-values. Consequently, if a function f is *not* one-to-one, then there are at least two numbers x_1 and x_2 in the domain of f such that $x_1 \neq x_2$ and $f(x_1) = f(x_2)$. Geometrically, this means that the horizontal line passing through the two points $(x_1, f(x_1))$ and $(x_2, f(x_2))$ contains these *two* points on the graph of f. See Figure 1.70. The geometric interpretation of a one-to-one function is called the *horizontal-line test*.

HORIZONTAL-LINE TEST

A function f is one-to-one if no horizontal line intersects the graph of f in more than one point.

EXAMPLE 1   Using the Horizontal-Line Test

Use the horizontal-line test to determine which of the following functions are one-to-one.

a. $f(x) = 2x + 5$ **b.** $g(x) = x^2 - 1$ **c.** $h(x) = 2\sqrt{x}$

SOLUTION

We see that no horizontal line intersects the graphs of f (Figure 1.71(a)) or h (Figure 1.71(c)) in more than one point; therefore, the functions f and h are one-to-one. The function g is not one-to-one since the horizontal line $y = 3$ (among others) in Figure 1.71(b) intersects the graph of g at more than one point.

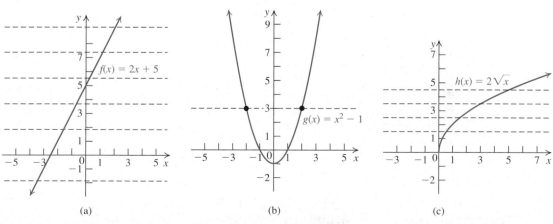

FIGURE 1.71 Horizontal-line test

■ ■ ■

Practice Problem 1 Use the horizontal-line test to determine whether $f(x) = (x - 1)^2$ is a one-to-one function.   ■

DEFINITION OF f^{-1} FOR A ONE-TO-ONE FUNCTION f

Let f represent a one-to-one function. Then if y is in the range of f, there is only one value of x in the domain of f such that $f(x) = y$. We define the inverse of f, called the **inverse function of f**, denoted f^{-1}, by $f^{-1}(y) = x$ if and only if $y = f(x)$.

From this definition we have the following:

> Domain of f = Range of f^{-1} and   Range of f = Domain of f^{-1}

◆ **WARNING** The notation $f^{-1}(x)$ does not mean $\dfrac{1}{f(x)}$. The expression $\dfrac{1}{f(x)}$ represents the reciprocal of $f(x)$ and is sometimes written as $[f(x)]^{-1}$.

EXAMPLE 2   Relating the Values of a Function and Its Inverse

Assume that f is a one-to-one function.

a. If $f(3) = 5$, find $f^{-1}(5)$.   **b.** If $f^{-1}(-1) = 7$, find $f(7)$.

SOLUTION

By definition, $f^{-1}(y) = x$ if and only if $y = f(x)$.

a. Let $x = 3$ and $y = 5$. Now reading the definition from right to left, $5 = f(3)$ if and only if $f^{-1}(5) = 3$. Thus, $f^{-1}(5) = 3$.

b. Let $y = -1$ and $x = 7$. Now $f^{-1}(-1) = 7$ if and only if $f(7) = -1$. Thus, $f(7) = -1$.   ■ ■ ■

Practice Problem 2 Assume that f is a one-to-one function.

a. If $f(-3) = 12$, find $f^{-1}(12)$.   **b.** If $f^{-1}(4) = 9$, find $f(9)$.   ■

Consider the following input–output diagram for $f^{-1} \circ f$:

STUDY TIP

The reason that only a one-to-one function can have an inverse is that if $x_1 \neq x_2$ but $f(x_1) = y$ and $f(x_2) = y$ then $f^{-1}(y)$ must be x_1 as well as x_2. This is not possible since $x_1 \neq x_2$.

The preceding diagram suggests the following:

INVERSE FUNCTION PROPERTY

Let f denote a one-to-one function. Then

1. $(f \circ f^{-1})(x) = f(f^{-1}(x)) = x$ for every x in the domain of f^{-1}.
2. $(f^{-1} \circ f)(x) = f^{-1}(f(x)) = x$ for every x in the domain of f.

Further, if g is any function such that (for the values of x in these equations)
$$f(g(x)) = x \quad \text{and} \quad g(f(x)) = x, \quad \text{then} \quad g = f^{-1}.$$

One interpretation of the equation $(f^{-1} \circ f)(x) = x$ is that f^{-1} undoes anything that f does to x. For example, let

$$f(x) = x + 2 \qquad f \text{ adds 2 to any input } x.$$

To undo what f does to x, we should subtract 2 from x. That is, the inverse of f should be

$$g(x) = x - 2. \qquad g \text{ subtracts 2 from } x.$$

Let's verify that $g(x) = x - 2$ is indeed the inverse function of x.

$$
\begin{aligned}
(f \circ g)(x) &= f(g(x)) & &\text{Definition of } f \circ g \\
&= f(x - 2) & &\text{Replace } g(x) \text{ with } x - 2. \\
&= (x - 2) + 2 & &\text{Replace } x \text{ with } x - 2 \text{ in } f(x) = x + 2. \\
&= x & &\text{Simplify.}
\end{aligned}
$$

We leave it for you to check that $(f \circ g)(x) = x$.

EXAMPLE 3　**Verifying Inverse Functions**

Verify that the following pairs of functions are inverses of each other:

$$f(x) = 2x + 3 \quad \text{and} \quad g(x) = \frac{x - 3}{2}$$

SOLUTION

$$
\begin{aligned}
(f \circ g)(x) &= f(g(x)) & &\text{Definition of } f \circ g \\
&= f\!\left(\frac{x - 3}{2}\right) & &\text{Replace } g(x) \text{ with } \frac{x - 3}{2}. \\
&= 2\!\left(\frac{x - 3}{2}\right) + 3 & &\text{Replace } x \text{ with } \frac{x - 3}{2} \text{ in } f(x) = 2x + 3. \\
&= x & &\text{Simplify.}
\end{aligned}
$$

Thus, $f(g(x)) = x$.

$$
\begin{aligned}
g(x) &= \frac{x - 3}{2} & &\text{Given function } g \\
(g \circ f)(x) &= g(f(x)) & &\text{Definition of } g \text{ of } f \\
&= g(2x + 3) & &\text{Replace } f(x) \text{ with } 2x + 3. \\
&= \frac{(2x + 3) - 3}{2} & &\text{Replace } x \text{ with } 2x + 3 \text{ in } g(x) = \frac{x - 3}{2}. \\
&= x & &\text{Simplify.}
\end{aligned}
$$

Thus, $g(f(x)) = x$. Since $f(g(x)) = g(f(x)) = x$, f and g are inverses of each other.　■ ■ ■

Practice Problem 3 Verify that $f(x) = 3x - 1$ and $g(x) = \dfrac{x + 1}{3}$ are inverses of each other. ∎

In Example 3, notice how the functions f and g neutralize (undo) the effect of each other. The function f takes an input x, *multiplies* it by 2, and *adds* 3; g neutralizes (or undoes) this effect by *subtracting* 3 and *dividing* by 2. This process is illustrated in Figure 1.72. Notice that g reverses the operations performed by f *and* the order in which they are done.

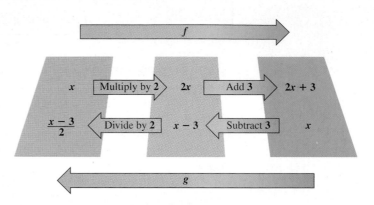

FIGURE 1.72 Function g undoes f

---

2 Find the inverse function.

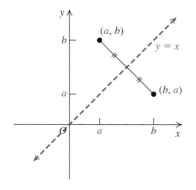

FIGURE 1.73 Relationship between points (a, b) and (b, a)

Finding the Inverse Function

Let $y = f(x)$ be a one-to-one function; then f has an inverse function. Suppose (a, b) is a point on the graph of f. Then $b = f(a)$. This means that $a = f^{-1}(b)$, so (b, a) is a point on the graph of f^{-1}. The points (a, b) and (b, a) are symmetric with respect to the line $y = x$, as shown in Figure 1.73. (See Exercises 85 and 86.) That is, if the graph paper is folded along the line $y = x$, the points (a, b) and (b, a) will coincide. Therefore, we have the following property:

SYMMETRY PROPERTY OF THE GRAPHS OF f AND f^{-1}

The graph of a function f and the graph of f^{-1} are symmetric with respect to the line $y = x$.

> **EXAMPLE 4**   **Finding the Graph of f^{-1} from the Graph of f**

The graph of a function f is shown in Figure 1.74. Sketch the graph of f^{-1}.

SOLUTION

By the horizontal-line test, f is a one-to-one function, so its inverse will also be a function.

The graph of f consists of two line segments: one joining the points $(-3, -5)$ and $(-1, 2)$ and the other joining the points $(-1, 2)$ and $(4, 3)$.

The graph of f^{-1} is the reflection of the graph of f in the line $y = x$. The reflections of the points $(-3, -5), (-1, 2)$, and $(4, 3)$ in the line $y = x$ are $(-5, -3)$, $(2, -1)$, and $(3, 4)$, respectively.

STUDY TIP

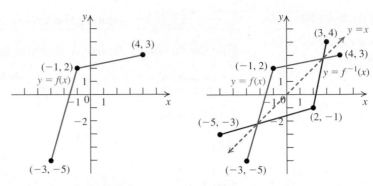

STUDY TIP

It is helpful to notice that points on the *x*-axis, $(a, 0)$, reflect to points on the *y*-axis, $(0, a)$, and conversely. Also notice that points on the line $y = x$ are unaffected by reflection in the line $y = x$.

FIGURE 1.74 Graph of f   **FIGURE 1.75 Graph of f^{-1}**

The graph of f^{-1} consists of two line segments: one joining the points $(-5, -3)$ and $(2, -1)$ and the other joining the points $(2, -1)$ and $(3, 4)$. See Figure 1.75.   ■ ■ ■

Practice Problem 4   Use the graph of a function f in Figure 1.76 to sketch the graph of f^{-1}.   ■

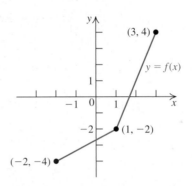

FIGURE 1.76 Graphing f^{-1} from the graph of f

The symmetry between the graphs of f and f^{-1} tells us that we can find an equation for the inverse function $y = f^{-1}(x)$ from the equation of a one-to-one function $y = f(x)$ by interchanging the roles of x and y in the equation $y = f(x)$. This results in the equation $x = f(y)$. Then we solve the equation $x = f(y)$ for y in terms of x to get $y = f^{-1}(x)$.

FINDING THE SOLUTION: A PROCEDURE

| **EXAMPLE 5** | **Finding an equation for f^{-1}** |

OBJECTIVE	**EXAMPLE**
Find the inverse of a function.	*Find the inverse of $f(x) = 3x - 4$.*
Step 1 Replace $f(x)$ with y in the equation defining $f(x)$.	$y = 3x - 4$
Step 2 Interchange x and y.	$x = 3y - 4$
Step 3 Solve the equation in Step 2 for y.	$x + 4 = 3y$ Add 4 to both sides.
	$\dfrac{x + 4}{3} = y$ Divide both sides by 3.
Step 4 Replace y with $f^{-1}(x)$.	$f^{-1}(x) = \dfrac{x + 4}{3}$

■ ■ ■

Practice Problem 5   Find the inverse of $f(x) = -2x + 3$.   ■

STUDY TIP

Interchanging x and y in Step 2 is done so we can write f^{-1} in the usual format with x as the independent variable and y as the dependent variable.

EXAMPLE 6   Finding the Inverse Function

Find the inverse of the one-to-one function

$$f(x) = \frac{x+1}{x-2}, \quad x \neq 2.$$

SOLUTION

Step 1 $\quad y = \dfrac{x+1}{x-2}$ Replace $f(x)$ with y.

Step 2 $\quad x = \dfrac{y+1}{y-2}$ Interchange x and y.

Step 3   Solve $x = \dfrac{y+1}{y-2}$ for y.   This is the most challenging step.

$\quad x(y-2) = y+1$ Multiply both sides by $y-2$.

$\quad xy - 2x = y+1$ Distributive property

$\quad xy - 2x + 2x - y = y+1+2x-y$ Add $2x-y$ to both sides.

$\quad xy - y = 2x+1$ Simplify.

$\quad y(x-1) = 2x+1$ Factor out y.

$\quad y = \dfrac{2x+1}{x-1}$ Divide both sides by $x-1$, assuming that $x \neq 1$.

Step 4 $f^{-1}(x) = \dfrac{2x+1}{x-1}, \ x \neq 1$ Replace y with $f^{-1}(x)$.

To see if our calculations are accurate, we compute $f(f^{-1}(x))$ and $f^{-1}(f(x))$.

$$f^{-1}(f(x)) = f^{-1}\left(\frac{x+1}{x-2}\right) = \frac{2\left(\dfrac{x+1}{x-2}\right)+1}{\dfrac{x+1}{x-2}-1} = \frac{2x+2+x-2}{x+1-x+2} = \frac{3x}{3} = x$$

You should also check that $f(f^{-1}(x)) = x$.   ■ ■ ■

Practice Problem 6   Find the inverse of the one-to-one function $f(x) = \dfrac{x}{x+3}$, $x \neq -3$.   ■

3 Use inverse functions to find the range of a function.

Finding the Range of a Function

In Section 1.4, we mentioned that it is not always easy to determine the range of a function that is defined by an equation. However, suppose a function f has inverse f^{-1}. Then the range of f is the domain of f^{-1}.

EXAMPLE 7   Finding the Domain and Range

RECALL

Remember that the domain of a function given by a formula is the largest set of real numbers for which the outputs are real numbers.

Find the domain and the range of the function $f(x) = \dfrac{x+1}{x-2}$ of Example 6.

SOLUTION

The domain of $f(x) = \dfrac{x+1}{x-2}$ is the set of all real numbers x such that $x \neq 2$.

In interval notation, the domain of f is $(-\infty, 2) \cup (2, \infty)$.

From Example 6, $f^{-1}(x) = \dfrac{2x + 1}{x - 1}, x \neq 1$; therefore,

$$\text{Range of } f = \text{Domain of } f^{-1} = \{x | x \neq 1\}.$$

In interval notation, the range of f is $(-\infty, 1) \cup (1, \infty)$. ■ ■ ■

Practice Problem 7 Find the domain and the range of the function $f(x) = \dfrac{x}{x + 3}$. ■

If a function f is not one-to-one, then it does not have an inverse function. Sometimes by changing its domain, we can produce an interesting function that does have an inverse. (This technique is frequently used in trigonometry.) We saw in Example 1(b) that $g(x) = x^2 - 1$ is not a one-to-one function; so g does not have an inverse function. However, the horizontal-line test shows that the function

$$G(x) = x^2 - 1, \quad x \geq 0$$

with domain $[0, \infty)$ is one-to-one. See Figure 1.77. Therefore, G has an inverse function G^{-1}.

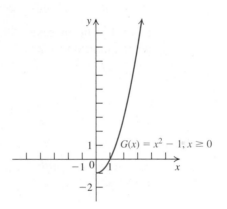

FIGURE 1.77 **The function G has an inverse.**

TECHNOLOGY CONNECTION

 You can also see that the graphs of $G(x) = x^2 - 1, x \geq 0$ and $G^{-1}(x) = \sqrt{x + 1}$ are symmetric in the line $y = x$ with a graphing calculator.

Enter $Y_1, Y_2,$ and Y_3 as $G(x), G^{-1}(x),$ and $f(x) = x$, respectively.

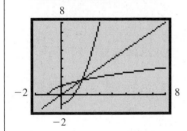

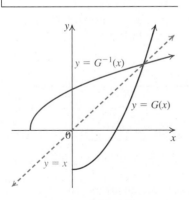

FIGURE 1.78 **Graphs of G and G^{-1}**

EXAMPLE 8   **Finding an Inverse Function**

Find the inverse of $G(x) = x^2 - 1, x \geq 0$.

SOLUTION

Step 1 $\quad y = x^2 - 1, \quad x \geq 0$ Replace $G(x)$ with y.

Step 2 $\quad x = y^2 - 1, \quad y \geq 0$ Interchange x and y.

Step 3 $\quad x + 1 = y^2, \quad y \geq 0$ Add 1 to both sides.

$$y = \pm\sqrt{x + 1}, \quad y \geq 0 \quad \text{Solve for } y.$$

Since $y \geq 0$ from Step 2, we reject $y = -\sqrt{x + 1}$. So we choose

$$y = \sqrt{x + 1}.$$

Step 4 $G^{-1}(x) = \sqrt{x + 1}$ Replace y with $G^{-1}(x)$.

The graphs of G and G^{-1} are shown in Figure 1.78. ■ ■ ■

Practice Problem 8 Find the inverse of $G(x) = x^2 - 1, x \leq 0$. ■

4 Apply inverse functions in the real world.

Applications

Functions that model real-life situations are frequently expressed as formulas with letters that remind you of the variable they represent. In finding the inverse of a function expressed by a formula, interchanging the letters could be very confusing. Accordingly, we omit Step 2, keep the letters the same, and just solve the formula for the other variable.

EXAMPLE 9   Water Pressure on Underwater Devices

At the beginning of this section, we wrote the formula for finding the water pressure p (in pounds per square inch) at a depth d (in feet) below the surface. This formula can be written as $p = \dfrac{15d}{33}$. Suppose the pressure gauge on a diving bell breaks and shows a reading of 1800 psi. How far below the surface was the bell when the gauge failed?

SOLUTION

We want to find the unknown depth in terms of the known pressure. This depth is given by the inverse of the function $p = \dfrac{15d}{33}$. To find the inverse, we solve the given equation for d.

$$p = \frac{15d}{33} \qquad \text{Original equation}$$

$$33p = 15d \qquad \text{Multiply both sides by 33.}$$

$$d = \frac{33p}{15} \qquad \text{Solve for } d.$$

Now we can use this formula to find the depth when the gauge read 1800 psi. We let $p = 1800$.

$$d = \frac{33p}{15} \qquad \text{Depth from pressure equation}$$

$$d = \frac{33(1800)}{15} \qquad \text{Replace } p \text{ with 1800.}$$

$$d = 3960$$

The device was 3960 feet below the surface when the gauge failed.   ■ ■ ■

Practice Problem 9   In Example 9, suppose the pressure gauge showed a reading of 1650 psi. Determine the depth of the bell when the gauge failed.   ■

SECTION 1.7 ■ Exercises

A EXERCISES Basic Skills and Concepts

1. If no horizontal line intersects the graph of a function f in more than one point, then f is a(n) __one-to-one__ function.

2. A function f is one-to-one when different x-values correspond to __different y-values__.

3. If $f(x) = 3x$, then $f^{-1}(x) =$ __$\frac{1}{3}x$__.

4. The graphs of a function f and its inverse f^{-1} are symmetric in the line __$y = x$__.

5. *True or False*. If a function f has an inverse, then the domain of the inverse function is the range of f. True

6. *True or False*. It is possible for a function to be its own inverse, that is, for $f = f^{-1}$. True

†Due to space constrictions, answers to these exercises may be found in the Answers beginning on page A–1 in the back of the book.

In Exercises 7–14, the graph of a function is given. Use the horizontal-line test to determine whether the function is one-to-one.

7.

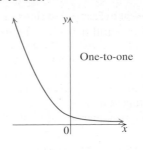

One-to-one

8.

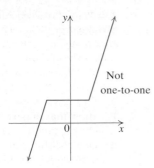

Not one-to-one

9.

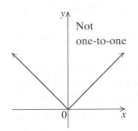

Not one-to-one

10.

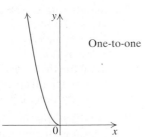

One-to-one

11.

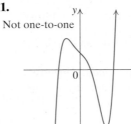

Not one-to-one

12.

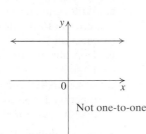

Not one-to-one

13.

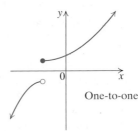

One-to-one

14.

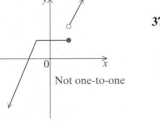

Not one-to-one

In Exercises 15–24, assume that the function f is one-to-one.

15. If $f(2) = 7$,   find $f^{-1}(7)$. 2

16. If $f^{-1}(4) = -7$,   find $f(-7)$. 4

17. If $f(-1) = 2$,   find $f^{-1}(2)$. −1

18. If $f^{-1}(-3) = 5$,   find $f(5)$. −3

19. If $f(a) = b$,   find $f^{-1}(b)$. a

20. If $f^{-1}(c) = d$,   find $f(d)$. c

21. Find $(f^{-1} \circ f)(337)$. 337

22. Find $(f \circ f^{-1})(25\pi)$. 25π

23. Find $(f \circ f^{-1})(-1580)$. −1580

24. Find $(f^{-1} \circ f)(9728)$. 9728

25. For $f(x) = 2x - 3$, find each of the following.
 a. $f(3)$ 3   **b.** $f^{-1}(3)$ 3
 c. $(f \circ f^{-1})(19)$ 19   **d.** $(f \circ f^{-1})(5)$ 5

26. For $f(x) = x^3$, find each of the following.
 a. $f(2)$ 8   **b.** $f^{-1}(8)$ 2
 c. $(f \circ f^{-1})(15)$ 15   **d.** $(f^{-1} \circ f)(27)$ 27

27. For $f(x) = x^3 + 1$, find each of the following.
 a. $f(1)$ 2   **b.** $f^{-1}(2)$ 1   **c.** $(f \circ f^{-1})(269)$ 269

28. For $g(x) = \sqrt[3]{2x^3 - 1}$, find each of the following.
 a. $g(1)$ 1   **b.** $g^{-1}(1)$ 1   **c.** $(g^{-1} \circ g)(135)$ 135

In Exercises 29–34, show that f and g are inverses of each other by verifying that $f(g(x)) = x = g(f(x))$.

29. $f(x) = 3x + 1$; $g(x) = \dfrac{x - 1}{3}$

30. $f(x) = 2 - 3x$; $g(x) = \dfrac{2 - x}{3}$

31. $f(x) = x^3$; $g(x) = \sqrt[3]{x}$

32. $f(x) = \dfrac{1}{x}$; $g(x) = \dfrac{1}{x}$

33. $f(x) = \dfrac{x - 1}{x + 2}$; $g(x) = \dfrac{1 + 2x}{1 - x}$

34. $f(x) = \dfrac{3x + 2}{x - 1}$; $g(x) = \dfrac{x + 2}{x - 3}$

In Exercises 35–40, the graph of a function f is given. Sketch the graph of f^{-1}.

35. †

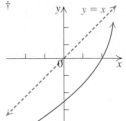

36. †

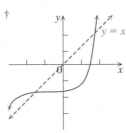

37. †

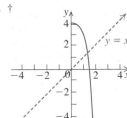

38. †

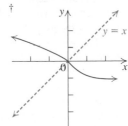

39. †

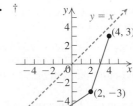

40. †

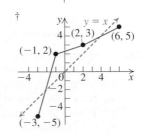

In Exercises 41–52,

a. Determine whether the given function is a one-to-one function.

b. If the function is one-to-one, find its inverse.

c. Sketch the graph of the function and its inverse on the same coordinate axes.

d. Give the domain and intercepts of each one-to-one function and its inverse function.

41. $f(x) = 15 - 3x$ †

42. $g(x) = 2x + 5$ †

43. $f(x) = \sqrt{4 - x^2}$ Not one-to-one

44. $f(x) = -\sqrt{9 - x^2}$ Not one-to-one

45. $f(x) = \sqrt{x} + 3$ †

46. $f(x) = 4 - \sqrt{x}$ †

47. $g(x) = \sqrt[3]{x + 1}$ †

48. $h(x) = \sqrt[3]{1 - x}$ †

49. $f(x) = \dfrac{1}{x - 1}, x \neq 1$ †

50. $g(x) = 1 - \dfrac{1}{x}, x \neq 0$ †

51. $f(x) = 2 + \sqrt{x + 1}$ †

52. $f(x) = -1 + \sqrt{x + 2}$ †

53. Find the domain and range of the function f of Exercise 33. Domain: $(-\infty, -2) \cup (-2, \infty)$; range: $(-\infty, 1) \cup (1, \infty)$

54. Find the domain and range of the function f of Exercise 34. Domain: $(-\infty, 1) \cup (1, \infty)$; range: $(-\infty, 3) \cup (3, \infty)$

In Exercises 55–58, assume that the given function is one-to-one. Find the inverse of the function. Also find the domain and the range of the given function.

55. $f(x) = \dfrac{x + 1}{x - 2}, x \neq 2$ †

56. $g(x) = \dfrac{x + 2}{x + 1}, x \neq -1$ †

57. $f(x) = \dfrac{1 - 2x}{1 + x}, x \neq -1$ †

58. $h(x) = \dfrac{x - 1}{x - 3}, x \neq 3$ †

In Exercises 59–66, find the inverse of each function and sketch the graph of the function and its inverse on the same coordinate axes.

59. $f(x) = -x^2, x \geq 0$ †

60. $g(x) = -x^2, x \leq 0$ †

61. $f(x) = |x|, x \geq 0$ †

62. $g(x) = |x|, x \leq 0$ †

63. $f(x) = x^2 + 1, x \leq 0$ †

64. $g(x) = x^2 + 5, x \geq 0$ †

65. $f(x) = -x^2 + 2, x \leq 0$ †

66. $g(x) = -x^2 - 1, x \geq 0$ †

B EXERCISES Applying the Concepts

67. Temperature scales. Scientists use the **Kelvin** temperature scale, in which the lowest possible temperature (called absolute zero) is 0 K. (K denotes degrees Kelvin.)
The function $K(C) = C + 273$ gives the relationship between the Kelvin temperature (K) and Celsius temperature (C).
a. Find the inverse function of $K(C) = C + 273$. What does it represent? †
b. Use the inverse function from (a) to find the Celsius temperature corresponding to 300 K. 27°C
c. A comfortable room temperature is 22°C. What is the corresponding Kelvin temperature? 295 K

68. Temperature scales. The boiling point of water is 373 K, or 212°F; the freezing point of water is 273 K, or 32°F. The relationship between Kelvin and Fahrenheit temperatures is linear.
a. Write a linear function expressing $K(F)$ in terms of F. †
b. Find the inverse of the function in part (a). What does it mean? †
c. A normal human body temperature is 98.6°F. What is the corresponding Kelvin temperature? 310 K

69. Composition of functions. Use Exercises 67 and 68 and the composition of functions to
a. write a function that expresses F in terms of C.
b. write a function that expresses C in terms of F.
a. $F = \dfrac{9}{5}C + 32$ **b.** $C = \dfrac{5}{9}F - \dfrac{160}{9}$

70. Celsius and Fahrenheit temperatures. Show that the functions in (a) and (b) of Exercise 69 are inverse functions.

71. Currency exchange. Alisha went to Europe last summer. She discovered that when she exchanged her U.S. dollars for euros, she received 25% fewer euros than the number of dollars she exchanged. (She got 75 euros for every 100 U.S. dollars.) When she returned to the United States, she got 25% more dollars than the number of euros she exchanged.
a. Write each conversion function. †
b. Show that in part (a), the two functions are not inverse functions. †
c. Does Alisha gain or lose money after converting both ways? She loses money.

72. Hourly wages. Anwar is a short-order cook in a diner. He is paid $4 per hour plus 5% of all food sales per hour. His average hourly wage w in terms of the food sales of x dollars is $w = 4 + 0.05x$.
a. Write the inverse function. What does it mean? †
b. Use the inverse function to estimate the hourly sales at the diner if Anwar averages $12 per hour. $160

73. Hourly wages. In Exercise 72, suppose, in addition, that Anwar is guaranteed a minimum wage of $7 per hour.
a. Write a function expressing his hourly wage w in terms of food sales per hour. [*Hint:* Use a piecewise function.] †
b. Does the function in part (a) have an inverse? Explain. †
c. If the answer in part (b) is yes, find the inverse function. If the answer is no, restrict the domain so that the new function has an inverse. Restriction of the domain: $[60, \infty)$

74. Simple pendulum. If a pendulum is released at a certain point, the **period** is the time it takes for the pendulum to swing along its path and return to the point from which it was released. The period T (in seconds) of a simple pendulum is a function of its length l (in feet) and is given by $T = (1.11)l$.
a. Find the inverse function. What does it mean? †
b. Use the inverse function to calculate the length of the pendulum if its period is two seconds. 1.8 ft
c. The convention center in Portland, Oregon, has the longest pendulum in the United States. The pendulum's length is 90 feet. Find the period. 99.9 sec

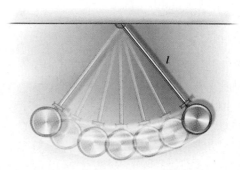

75. Water supply. Suppose x is the height of the water above the opening at the base of a water tank. The velocity V of water that flows from the opening at the base is a function of x and is given by $V(x) = 8\sqrt{x}$.

a. Find the inverse function. What does it mean? †
b. Use the inverse function to calculate the height of the water in the tank when the flow is **(i)** 30 feet per second and **(ii)** 20 feet per second. **(i)** 14.0625 ft **(ii)** 6.25 ft

76. Physics. A projectile is fired from the origin over horizontal ground. Its altitude y (in feet) is a function of its horizontal distance x (in feet) and is given by $y = 64x - 2x^2$.
a. Find the inverse function. †
b. Use the inverse function to compute the horizontal distance when the altitude of the projectile is **(i)** 32 feet, **(ii)** 256 feet, and **(iii)** 512 feet. **(i)** 0.51 ft **(ii)** 4.69 ft **(iii)** 16 ft.

77. Loan repayment. Chris purchased a car at 0% interest for five years. After making a down payment, she agreed to pay the remaining $36,000 in monthly payments of $600 per month for 60 months.
a. What does the function $f(x) = 36,000 - 600x$ represent? The balance due after x months
b. Find the inverse of the function in part (a). What does the inverse function represent? †
c. Use the inverse function to find the number of months remaining to make payments if the balance due is $22,000. 24

78. Demand function. A marketing survey finds that the number x (in millions) of computer chips the market will purchase is a function of its price p (in dollars) and is estimated by $x = 8p^2 - 32p + 1200, 0 < p \le 2$.
a. Find the inverse function. What does it mean? †
b. Use the inverse function to estimate the price of a chip if the demand is 1180.5 million chips. $0.75

C EXERCISES Beyond the Basics

In Exercises 79 and 80, show that f and g are inverses of each other by verifying that $f(g(x)) = x = g(f(x))$.

79.

x	1	3	4
$f(x)$	3	5	2

x	3	5	2
$g(x)$	1	3	4

80.

x	1	2	3	4
$f(x)$	-2	0	-3	1

x	-2	0	-3	1
$g(x)$	1	2	3	4

81. Let $f(x) = \sqrt{4 - x^2}$.
a. Sketch the graph of $y = f(x)$. †
b. Is f one-to-one? No
c. Find the domain and the range of f. Domain: $[-2, 2]$; range: $[0, 2]$

82. Let $f(x) = \dfrac{[x] + 2}{[x] - 2}$, where $[x]$ is the greatest integer function.
a. Find the domain of f. Domain: $(-\infty, 2) \cup [3, \infty)$
b. Is f one-to-one? No

83. Let $f(x) = \dfrac{x}{x + 1} = 1 - \dfrac{1}{x + 1}$.
a. Show that f is one-to-one. †
b. Find the inverse of f. †
c. Find the domain and the range of f. Domain: $(-\infty, -1) \cup (-1, \infty)$; range: $(-\infty, 1) \cup (1, \infty)$

84. Let $g(x) = \sqrt{1 - x^2}, 0 \le x \le 1$.
a. Show that g is one-to-one. †
b. Find the inverse of g. $g^{-1}(x) = g(x) = \sqrt{1 - x^2}$
c. Find the domain and the range of g. Domain = Range = $[0, 1]$

85. Let $P(3, 7)$ and $Q(7, 3)$ be two points in the plane.
a. Show that the midpoint M of the line segment PQ lies on the line $y = x$. †
b. Show that the line $y = x$ is the perpendicular bisector of the line segment $\overline{PQ}$; that is, show that the line $y = x$ contains the midpoint found in (a) and makes a right angle with $\overline{PQ}$. You will have shown that P and Q are symmetric about the line $y = x$. †

86. Use the procedure outlined in Exercise 85 to show that the points (a, b) and (b, a) are symmetric about the line $y = x$. (See Figure 1.73.)

87. The graph of $f(x) = x^3$ is given in Section 1.4.
a. Sketch the graph of $g(x) = (x - 1)^3 + 2$ by using transformations of the graph of f. †
b. Find $g^{-1}(x)$. $g^{-1}(x) = \sqrt[3]{x - 2} + 1$
c. Sketch the graph of $g^{-1}(x)$ by reflecting the graph of g in part (a) about the line $y = x$. †

88. Inverse of composition of functions. We show that if f and g have inverses, then

$$(f \circ g)^{-1} = g^{-1} \circ f^{-1}. \text{ (Note the order.)}$$

Let $f(x) = 2x - 1$ and $g(x) = 3x + 4$.
a. Find the following:
 (i) $f^{-1}(x)$ † **(ii)** $g^{-1}(x)$ †
 (iii) $(f \circ g)(x)$ † **(iv)** $(g \circ f)(x)$ †
 (v) $(f \circ g)^{-1}(x)$ † **(vi)** $(g \circ f)^{-1}(x)$ †
 (vii) $(f^{-1} \circ g^{-1})(x)$ † **(viii)** $(g^{-1} \circ f^{-1})(x)$ †
b. From part (a), conclude that
 (i) $(f \circ g)^{-1} = g^{-1} \circ f^{-1}$. †
 (ii) $(g \circ f)^{-1} = f^{-1} \circ g^{-1}$. †

89. Repeat Exercise 88 for $f(x) = 2x + 3$ and $g(x) = x^3 - 1$. †

Critical Thinking

90. Does every odd function have an inverse? Explain. †

91. Is there an even function that has an inverse? Explain. †

92. Does every increasing or decreasing function have an inverse? Explain. Yes because increasing and decreasing functions are one-to-one

93. A relation R is a set of ordered pairs (x, y). The inverse of R is the set of ordered pairs (y, x).
a. Give an example of a function whose inverse relation is not a function. $R = \{(-1, 1), (0, 0), (1, 1)\}$
b. Give an example of a relation R whose inverse is a function. $R = \{(-1, 1), (0, 0), (1, 2)\}$

SUMMARY ■ Definitions, Concepts, and Formulas

1.1 Graphs of Equations

i. Ordered Pair. A pair of numbers in which the order is specified is called an ordered pair of numbers.

ii. The Distance Formula. The distance between two points $P(x_1, y_1)$ and $Q(x_2, y_2)$, denoted by $d(P, Q)$, is given by

$$d(P, Q) = \sqrt{(x_2 - x_1)^2 + (y_2 - y_1)^2}.$$

iii. The Midpoint Formula. The coordinates of the midpoint $M(x, y)$ of the line segment joining $P(x_1, y_1)$ and $Q(x_2, y_2)$ are given by

$$M = (x, y) = \left(\frac{x_1 + x_2}{2}, \frac{y_1 + y_2}{2} \right).$$

iv. The graph of an equation in two variables, say, x and y, is the set of all ordered pairs (a, b) in the coordinate plane that satisfy the given equation. A graph of an equation, then, is a picture of its solution set.

v. Sketching the graph of an equation by plotting points

Step 1 Make a representative table of solutions of the equation.

Step 2 Plot the solutions in Step 1 as ordered pairs in the coordinate plane.

Step 3 Connect the representative solutions in Step 2 by a smooth curve.

vi. Intercepts

1. The x-intercept is the x-coordinate of a point on the graph where the graph touches or crosses the x-axis. To find the x-intercepts, set $y = 0$ in the equation and solve for x.

2. The y-intercept is the y-coordinate of a point on the graph where the graph touches or crosses the y-axis. To find the y-intercepts, set $x = 0$ in the equation and solve for y.

vii. Symmetry

A graph is symmetric with respect to

a. the y-axis if for every point (x, y) on the graph the point $(-x, y)$ is also on the graph. That is, replacing x with $-x$ in the equation produces an equivalent equation.

b. the x-axis if for every point (x, y) on the graph the point $(x, -y)$ is also on the graph. That is, replacing y with $-y$ in the equation produces an equivalent equation.

c. the *origin* if for every point (x, y) on the graph the point $(-x, -y)$ is also on the graph. That is, replacing x with $-x$ and y with $-y$ in the equation produces an equivalent equation.

viii. Circle

A circle is a set of points in the coordinate plane that is at a fixed distance r from a fixed point (h, k). The fixed distance r is called the radius of the circle, and the fixed point (h, k) is called the center of the circle. The equation

$$(x - h)^2 + (y - k)^2 = r^2$$

is called the *standard equation* of a circle.

1.2 Lines

i. The slope m of a nonvertical line through the points $P(x_1, y_1)$ and $Q(x_2, y_2)$ is

$$m = \frac{\text{rise}}{\text{run}} = \frac{(y\text{-coordinate of } Q) - (y\text{-coordinate of } P)}{(x\text{-coordinate of } Q) - (x\text{-coordinate of } P)}$$
$$= \frac{y_2 - y_1}{x_2 - x_1}.$$

The slope of a vertical line is undefined. The slope of a horizontal line is 0.

ii. Equation of a line

$y - y_1 = m(x - x_1)$	Point–slope form
$y = mx + b$	Slope–intercept form
$y = k$	Horizontal line
$x = k$	Vertical line
$ax + by + c = 0$	General form

iii. Parallel and Perpendicular Lines

Two distinct lines with respective slopes m_1 and m_2 are

1. parallel if $m_1 = m_2$.
2. perpendicular if $m_1 m_2 = -1$.

1.3 Functions

i. A **function** from a set X to a set Y is a rule that assigns to each element of X one and only one corresponding element of Y. The set X is the **domain** of the function. The set of those elements of Y that correspond (are assigned) to the elements of X is the **range** of the function.

ii. If a function f is defined by an equation, then its domain is the largest set of real numbers for which $f(x)$ is a real number.

iii. Vertical-line test. If each vertical line intersects a graph at no more than one point, the graph is the graph of a function.

iv. The **average rate of change** of $f(x)$ as x changes from a to b is defined by $\dfrac{f(b) - f(a)}{b - a}, b \neq a.$

v. The quantity $\dfrac{f(x+h)-f(x)}{h}$, $h \neq 0$, is called the **difference quotient**.

1.4 A Library of Functions

i. For **piecewise functions**, different rules are used in different parts of the domain.

ii. A function f is **increasing**, **decreasing**, or **constant** on an interval depending on whether its graph is respectively rising, falling, or staying the same as you move from left to right on the graph.

iii. A function f is **even** if $f(-x) = f(x)$. The graph of an even function is symmetric with respect to the y-axis. A function f is **odd** if $f(-x) = -f(x)$. The graph of an odd function is symmetric with respect to the origin.

iv. Basic Functions

Identity function	$f(x) = x$
Constant function	$f(x) = c$
Squaring function	$f(x) = x^2$
Cubing function	$f(x) = x^3$
Absolute value function	$f(x) = |x|$
Square root function	$f(x) = \sqrt{x}$
Cube root function	$f(x) = \sqrt[3]{x}$
Reciprocal function	$f(x) = \dfrac{1}{x}$
Greatest integer function	$f(x) = [\![x]\!]$

1.5 Transformations of Functions

i. Vertical and horizontal shifts: Let f be a function and c be a positive number.

To Graph	Shift the graph of f by c units
$f(x) + c$	up
$f(x) - c$	down
$f(x - c)$	right
$f(x + c)$	left

ii. Reflections:

To Graph	Reflect the graph of f in the
$-f(x)$	x-axis
$f(-x)$	y-axis

iii. Stretching and compressing: The graph of $g(x) = af(x)$ for $a > 0$ has the same shape as the graph of $f(x)$ and is taller if $a > 1$ and flatter if $0 < a < 1$. If a is negative, the graph of $y = |a|f(x)$ is reflected in the x-axis.

The graph of $g(x) = f(bx)$ for $b > 0$ is obtained from the graph of f by stretching away from the y-axis if $0 < b < 1$ and compressing horizontally toward the y-axis if $b > 1$. If b is negative, the graph of $y = f(|b|x)$ is reflected in the y-axis.

1.6 Combining Functions; Composite Functions

i. Given two functions f and g, for all values of x for which both $f(x)$ and $g(x)$ are defined, the functions can be combined to form **sum**, **difference**, **product**, and **quotient** functions.

ii. The **composition** of f and g is defined by $(f \circ g)(x) = f(g(x))$.

The input of f is the output of g. The domain of $f \circ g$ is the set of all x's in the domain of g such that $g(x)$ is in the domain of f.

In general $f \circ g \neq g \circ f$.

1.7 Inverse Functions

i. A function f is **one-to-one** if for any x_1 and x_2 in the domain of f, $f(x_1) = f(x_2)$ implies $x_1 = x_2$.

ii. Horizontal-line test. If each horizontal line intersects the graph of a function f in at most one point, then f is a one-to-one function.

iii. Inverse function. Let f be a one-to-one function. Then g is the inverse of f, and we write $g = f^{-1}$ if $(f \circ g)(x) = x$ for every x in the domain of g and $(g \circ f)(x) = x$ for every x in the domain of f. The graph of f^{-1} is the reflection of the graph of f in the line $y = x$.

REVIEW EXERCISES

Basic Skills and Concepts

In Exercises 1–8, state whether the given statement is true or false.

1. $(0, 5)$ is the midpoint of the line segment joining $(-3, 1)$ and $(3, 11)$. False

2. The equation $(x + 2)^2 + (y + 3)^2 = 5$ is the equation of a circle with center $(2, 3)$ and radius 5. False

3. If a graph is symmetric with respect to the x-axis and the y-axis, then it must be symmetric with respect to the origin. True

4. If a graph is symmetric with respect to the origin, then it must be symmetric with respect to the x-axis and the y-axis. False

†Due to space constrictions, answers to these exercises may be found in the Answers beginning on page A-1 in the back of the book.

5. In the graph of the equation of the line $3y = 4x + 9$, the slope is 4 and the y-intercept is 9. False

6. If the slope of a line is 2, then the slope of any line perpendicular to it is $\dfrac{1}{2}$. False

7. The slope of a vertical line is undefined. True

8. The equation $(x - 2)^2 + (y + 3)^2 = -25$ is the equation of a circle with center $(2, -3)$ and radius 5. False

In Exercises 9–14, find
a. the distance between the point P and Q.
b. the coordinates of the midpoint of the line segment PQ.
c. the slope of the line containing the points P and Q.

9. $P(3, 5), Q(-1, 3)$ †   **10.** $P(-3, 5), Q(3, -1)$ †

11. $P(4, -3), Q(9, -8)$ †   **12.** $P(2, 3), Q(-7, -8)$ †

13. $P(2, -7), Q(5, -2)$ †   **14.** $P(-5, 4), Q(10, -3)$ †

15. Show that the points $A(0, 5), B(-2, -3)$, and $C(3, 0)$ are the vertices of a right triangle.

16. Show that the points $A(1, 2), B(4, 8), C(7, -1)$, and $D(10, 5)$ are the vertices of a rhombus. †

17. Which of the points $(-6, 3)$ and $(4, 5)$ is closer to the origin? $(4, 5)$

18. Which of the points $(-6, 4)$ and $(5, 10)$ is closer to the point $(2, 3)$? $(5, 10)$

19. Find a point on the x-axis that is equidistant from the points $(-5, 3)$ and $(4, 7)$. $\left(\dfrac{31}{18}, 0\right)$

20. Find a point on the y-axis that is equidistant from the points $(-3, -2)$ and $(2, -1)$. $(0, -4)$

In Exercises 21–24, specify whether the given graph has axis symmetry or origin symmetry (or neither).

21. †   **22.** †

 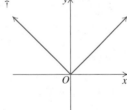   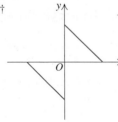

23. †   **24.** †

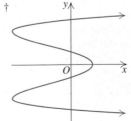   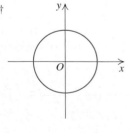

In Exercises 25–34, sketch the graph of the given equation. List all intercepts and describe any symmetry of the graph.

25. $x + 2y = 4$ †   **26.** $3x - 4y = 12$ †

27. $y = -2x^2$ †   **28.** $x = -y^2$ †

29. $y = x^3$ †   **30.** $x = -y^3$ †

31. $y = x^2 + 2$ †   **32.** $y = 1 - x^2$ †

33. $x^2 + y^2 = 16$ †   **34.** $y = x^4 - 4$ †

In Exercises 35–37, find the standard form of the equation of the circle that satisfies the given conditions.

35. Center $(2, -3)$, radius 5 $(x - 2)^2 + (y + 3)^2 = 25$

36. Diameter with endpoints $(5, 2)$ and $(-5, 4)$
$x^2 + (y - 3)^2 = 26$

37. Center $(-2, -5)$, touching the y-axis
$(x + 2)^2 + (y + 5)^2 = 4$

In Exercises 38–42, describe and sketch the graph of the given equation and give the x- and y- intercepts.

38. $2x - 5y = 10$ †   **39.** $\dfrac{x}{2} - \dfrac{y}{5} = 1$ †

40. $(x + 1)^2 + (y - 3)^2 = 16$ †

41. $x^2 + y^2 - 2x + 4y - 4 = 0$ †

42. $3x^2 + 3y^2 - 6x - 6 = 0$ †

In Exercises 43–47, find the slope–intercept form of the equation of the line that satisfies the given conditions.

43. Passing through $(1, 2)$ with slope -2 $y = -2x + 4$

44. x-intercept 2, y-intercept 5 $y = -\frac{5}{2}x + 5$

45. Passing through $(1, 3)$ and $(-1, 7)$ $y = -2x + 5$

46. Passing through $(1, 2)$, parallel to the x-axis $y = 2$

47. Passing through $(1, 3)$, perpendicular to the x-axis $x = 1$

48. Determine whether the lines in each pair are parallel, perpendicular, or neither.
a. $y = 3x - 2$ and $y = 3x + 2$ Parallel
b. $3x - 5y + 7 = 0$ and $5x - 3y + 2 = 0$ Neither
c. $ax + by + c = 0$ and $bx - ay + d = 0$ Perpendicular
d. $y + 2 = \dfrac{1}{3}(x - 3)$ and $y - 5 = 3(x - 3)$ Neither

In Exercises 49–58, graph each equation. State which equations determine y as a function of x.

49. $x = y^2$ †

50. $y = x^2$ †

51. $x - y = 1$ †   **52.** $y = \sqrt{x - 2}$ †

53. $x^2 + y^2 = 0.04$ †   **54.** $x = -\sqrt{y}$ †

55. $x = 1$ †   **56.** $y = 2$ †

57. $y = |x + 1|$ †   **58.** $x = y^2 + 1$ †

In Exercises 59–76, let $f(x) = 3x + 1$ and $g(x) = x^2 - 2$. Find each of the following.

59. $f(-2)$ -5 **60.** $g(-2)$ 2

61. x if $f(x) = 4$ $x = 1$ **62.** x if $g(x) = 2$ $x = \pm 2$

63. $(f + g)(1)$ 3 **64.** $(f - g)(-1)$ -1

65. $(f \cdot g)(-2)$ -10 **66.** $(g \cdot f)(0)$ -2

67. $(f \circ g)(3)$ 22 **68.** $(g \circ f)(-2)$ 23

69. $(f \circ g)(x)$ $3x^2 - 5$ **70.** $(g \circ f)(x)$ $9x^2 + 6x - 1$

71. $(f \circ f)(x)$ $9x + 4$ **72.** $(g \circ g)(x)$ $x^4 - 4x^2 + 2$

73. $f(a + h)$ $3a + 3h + 1$ **74.** $g(a - h)$ $a^2 - 2ah + h^2 - 2$

75. $\dfrac{f(x + h) - f(x)}{h}$ 3 **76.** $\dfrac{g(x + h) - g(x)}{h}$ $2x + h$

In Exercises 77–82, graph each function and state its domain and range. Determine the intervals over which the function is increasing, decreasing, or constant.

77. $f(x) = -3$ †   **78.** $f(x) = x^2 - 2$ †

79. $g(x) = \sqrt{3x - 2}$ †   **80.** $h(x) = \sqrt{36 - x^2}$ †

81. $f(x) = \begin{cases} x + 1 & \text{if } x \geq 0 \\ -x + 1 & \text{if } x < 0 \end{cases}$ †

82. $g(x) = \begin{cases} x & \text{if } x \geq 0 \\ x^2 & \text{if } x < 0 \end{cases}$ †

In Exercises 83–86, use transformations to graph each pair of functions on the same coordinate axes.

83. $f(x) = \sqrt{x}, g(x) = \sqrt{x + 1}$ †

84. $f(x) = |x|, g(x) = 2|x - 1| + 3$ †

85. $f(x) = \dfrac{1}{x}, g(x) = -\dfrac{1}{x - 2}$ †

86. $f(x) = x^2, g(x) = (x + 1)^2 - 2$ †

In Exercises 87–92, state whether each function is odd, even, or neither. Discuss the symmetry of each graph.

87. $f(x) = x^2 - x^4$ †   **88.** $f(x) = x^3 + x$ †

89. $f(x) = |x| + 3$ †   **90.** $f(x) = 3x + 5$ †

91. $f(x) = \sqrt{x}$ †   **92.** $f(x) = \dfrac{2}{x}$ †

In Exercises 93–96, express each function as a composition of two functions.

93. $f(x) = \sqrt{x^2 - 4}$ †   **94.** $g(x) = (x^2 - x + 2)^{50}$ †

95. $h(x) = \sqrt{\dfrac{x - 3}{2x + 5}}$ †   **96.** $H(x) = (2x - 1)^3 + 5$ †

In Exercises 97–100, determine whether the given function is one-to-one. If the function is one-to-one, find its inverse and sketch the graph of the function and its inverse on the same coordinate axes.

97. $f(x) = x + 2$ †   **98.** $f(x) = -2x + 3$ †

99. $f(x) = \sqrt[3]{x - 2}$ †   **100.** $f(x) = 8x^3 - 1$ †

In Exercises 101 and 102, assume that f is a one-to-one function. Find f^{-1} and find the domain and range of f.

101. $f(x) = 4 + \sqrt{x - 1}$ †

102. $f(x) = -3 + \sqrt{x + 2}$ †

103. For the following graph of a function f
  a. write a formula for f as a piecewise function. †
  b. find the domain and range of f. Domain: $[-3, 3]$; range: $[-3, 4]$
  c. find the intercepts of the graph of f. x-intercept: -2; y-intercept: 1
  d. draw the graph of $y = f(-x)$. †
  e. draw the graph of $y = -f(x)$. †
  f. draw the graph of $y = f(x) + 1$. †

  g. draw the graph of $y = f(x + 1)$. †
  h. draw the graph of $y = 2f(x)$. †
  i. draw the graph of $y = f(2x)$. †
  j. draw the graph of $y = f\left(\dfrac{1}{2}x\right)$. †

  k. explain why f is one-to-one.
  l. draw the graph of $y = f^{-1}(x)$. †

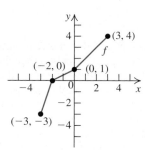

103. **k.** It satisfies the horizontal-line test.

Applying the Concepts

104. **Scuba diving.** The pressure P on the body of a scuba diver increases linearly as she descends to greater depths d in seawater. The pressure at a depth of 10 feet is 19.2 pounds per square inch, and the pressure at a depth of 25 feet is 25.95 pounds per square inch.
  a. Write the equation relating P and d (with d as independent variable) in slope–intercept form. $P = 0.45d + 14.7$
  b. What is the meaning of the slope and the y-intercept in part (a)? †
  c. Determine the pressure on a scuba diver who is at a depth of 160 feet. 86.7 lb/in.^2
  d. Find the depth at which the pressure on the body of the scuba diver is 104.7 pounds per square inch. 200 ft

105. **Waste disposal.** The Sioux Falls, South Dakota, council considered the following data regarding the cost of disposing of waste material:

Year	1996	2000
Waste (in pounds)	87,000	223,000
Cost	$54,000	$173,000

Source: Sioux Falls Health Department.

Assume that the cost C (in dollars) is linearly related to the waste w (in pounds).
  a. Write the linear equation relating C and w in slope–intercept form. $C = 0.875w - 22{,}125$
  b. Explain the meaning of the slope and the intercepts of the equation in part (a). †
  c. Suppose the city has projected 609,000 pounds of waste for the year 2012. Calculate the projected cost of disposing of the waste for the year 2012.
  d. Suppose the city's projected budget for waste collection in 2012 is 1 million dollars. How many pounds of waste can the city handle?

105. **c.** $510,750
  d. 1,168,142.86 pounds

106. Checking your speedometer. Zoe checks the accuracy of her speedometer by driving a few measured (not odometer) miles at a constant 60 miles per hour while her friend checks his watch at the beginning and end of the measured distance.

   a. How does a watch tell Zoe's friend whether the speedometer is accurate? What mathematics is involved?

   b. Why shouldn't Zoe use her odometer to measure the number of miles?

107. Playing blackjack. Chloe, a bright mathematician, read a book about counting cards in blackjack and went to Las Vegas to try her luck. She started with $100 and discovered that the amount of money she had at time t hours after the start of the game could be expressed by the function $f(t) = 100 + 55t - 3t^2$.

   a. What amount of money had Chloe won or lost in the first two hours? She won $98.

   b. What was the average rate at which Chloe was winning or losing money during the first two hours?

   c. Did she lose all of her money? If so, when?

   d. If she played until she lost all of her money, what was the average rate at which she was losing her money? $5/hr

106. a. 60 mph = 1 mi/min, so if the speedometer is correct, the number of minutes elapsed must be equal to the distance in miles.

   b. Because the odometer reading is based on the speedometer

107. b. She was winning at a rate of $49/h.

   c. Yes. After 20 hours

108. Volume discounting. Major cola distributors sell a case (containing 24 cans) of cola to the retailer at a price of $4. They offer a discount of 20% for purchases over 100 cases and a discount of 25% for purchases over 500 cases. Write a piecewise function that describes this pricing scheme. Take x as the number of cases purchased and $f(x)$ as the price paid.

109. Air pollution. The daily level L of carbon monoxide in a city is a function of the number of automobiles in the city. Suppose $L(x) = 0.5\sqrt{x^2 + 4}$, where x is the number of automobiles (in hundred thousands). Suppose further that the number of automobiles in a given city is growing according to the formula $x(t) = 1 + 0.002t^2$, where t is time in years measured from now.

   a. Form a composite function describing the daily pollution level as a function of time.

   b. What pollution level is expected in five years? 1.13

110. Toy manufacturing. After being in business for t years, a toy manufacturer is making $x = 5000 + 50t + 10t^2$ GI Jimmy toys per year. The sale price p in dollars per toy has risen according to the formula $p = 10 + 0.5t$. Write a formula for the manufacturer's yearly revenue as

   a. a function of time t. $5t^3 + 125t^2 + 3000t + 50,000$

   b. a function of price p. $40p^3 - 700p^2 + 8000p$

108. $f(x) = \begin{cases} 4x, & \text{if } 0 \le x \le 100 \\ 3.2x + 80, & \text{if } 100 < x \le 500 \\ 3x + 180, & \text{if } x > 500 \end{cases}$

109. a. $0.5\sqrt{0.000004t^4 + 0.004t^2 + 5}$

PRACTICE TEST A

1. Determine which symmetries the graph of the equation $3x + 2xy^2 = 1$ has. Symmetric about the x-axis

2. Find the x- and y-intercepts of the graph of $y = x^2(x - 3)(x + 1)$. x-intercepts: $-1, 0, 3$; y-intercept: 0

3. Graph the equation $x^2 + y^2 - 2x - 2y = 2$ and give the x- and y- intercepts. †

4. Write the slope–intercept form of the equation of the line with slope -1 and passing through the point $(2, 7)$. $y = -x + 9$

5. Write an equation of the line parallel to the line $8x - 2y = 7$ and passing through $(2, -1)$. $y = 4x - 9$

6. Use $f(x) = -2x + 1$ and $g(x) = x^2 + 3x + 2$ to find $(fg)(2)$. -36

7. Use $f(x) = 2x - 3$ and $g(x) = 1 - 2x^2$ to evaluate $g(f(2))$. -1

8. Use $f(x) = x^2 - 2x$ to find $(f \circ f)(x)$. $x^4 - 4x^3 + 2x^2 + 4x$

9. If $f(x) = \begin{cases} x^3 - 2 & \text{if } x \le 0 \\ 1 - 2x^2 & \text{if } x > 0 \end{cases}$, find (a) $f(-1)$, (b) $f(0)$, and (c) $f(1)$.

10. Find the domain of the function $f(x) = \dfrac{\sqrt{x}}{\sqrt{1 - x}}$. $[0, 1)$

11. Find the domain of the function $f(x) = \sqrt{x^2 + x - 6}$. $(-\infty, -3] \cup [2, \infty)$

12. Determine the average rate of change of the function $f(x) = 2x + 7$ between $x = 1$ and $x = 4$. 2

13. Determine whether the function $f(x) = 2x^4 - \dfrac{3}{x^2}$ is even, odd, or neither. Even

14. Find the intervals where the function shown is increasing or decreasing. Increasing on $(-\infty, 0) \cup (2, \infty)$; decreasing on $(0, 2)$

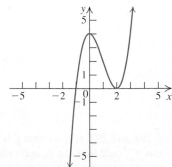

15. Suppose the graph of f is given. Describe how the graph of $y = f(x - 3)$ can be obtained from the graph of f. Shift the graph of f three units to the right.

16. A ball is thrown upward from the ground. After t seconds, the height h (in feet) above the ground is given by $h = 25 - (2t - 5)^2$. How many seconds does it take for the ball to reach a height of 25 feet? 2.5 sec

17. If f is a one-to-one function and $f(2) = 7$, find $f^{-1}(7)$. 2

Answers:

9. a. $f(-1) = -3$ **b.** $f(0) = -2$ **c.** $f(1) = -1$

18. Find the inverse function $f^{-1}(x)$ of the one-to-one function $f(x) = 1 + \dfrac{1}{x}$. $f^{-1}(x) = \dfrac{1}{x-1}$

19. Now that Jo has saved $1000 for a car, her dad takes over and deposits $100 into her account each month. Write an equation that relates the total amount of money, A, in Jo's account to the number of months, x, that Jo's dad had been putting money into her account. $A(x) = 100x + 1000$

20. The cost C in dollars for renting a car for one day is a function of the number of miles traveled, m. For a car renting for $30.00 per day and $0.25 per mile, this function is given by

$$C(m) = 0.25m + 30.$$

   a. Find the cost of renting the car for one day and driving 230 miles. $87.50

   b. If the charge for renting the car for one day is $57.50, how many miles were driven? 110 mi

PRACTICE TEST B

1. The graph of the equation $|x| + 2|y| = 2$ is symmetric with respect to which of the following? d

   I. the origin   **II.** the x-axis   **III.** the y-axis
   a. I only   **b.** II only
   c. III only   **d.** I, II, and III

2. Which are the x- and y-intercepts of the graph of $y = x^2 - 9$? b

   a. x-intercept 3, y-intercept -9
   b. x-intercepts ± 3, y-intercept -9
   c. x-intercept 3, y-intercept 9
   d. x-intercepts ± 3, y-intercepts ± 9

3. The slope of a line is undefined if the line is parallel to d

   a. the x-axis   **b.** the line $x - y = 0$
   c. the line $x + y = 0$ **d.** the y-axis

4. Which is an equation of the circle with center $(0, -5)$ and radius 7? d

   a. $x^2 + y^2 - 10y = 0$ **b.** $x^2 + (y - 5)^2 = 7$
   c. $x^2 + (y - 5)^2 = 49$ **d.** $x^2 + y^2 + 10y = 24$

5. Which is an equation of the line passing through $(2, -3)$ and having slope -1? c

   a. $y + 3 = -(x + 2)$ **b.** $y - 3 = -(x - 2)$
   c. $y + 3 = -(x - 2)$ **d.** $y + 3 = -(x + 2)$

6. What is a second point on the line through $(3, 2)$ and having slope $-\dfrac{1}{2}$? b

   a. $(2, 4)$ **b.** $(5, 1)$
   c. $(7, 2)$ **d.** $(4, 4)$

7. Which is an equation of the line parallel to the line $6x - 3y = 5$ and passing through $(-1, 2)$? d

   a. $y - 2 = -2(x + 1)$ **b.** $y - 2 = 6(x + 1)$
   c. $y - 2 = -6(x + 1)$ **d.** $y - 2 = 2(x + 1)$

8. Use $f(x) = 3x - 5$ and $g(x) = 2 - x^2$ to find $(f \circ g)(x)$. b

   a. $-9x^2 + 30x - 23$ **b.** $-3x^2 + 1$
   c. $3x^2 + 1$ **d.** $9x^2 - 30x + 23$

9. Use $f(x) = 2x^2 - x$ to find $(f \circ f)(x)$. a

   a. $8x^4 - 8x^3 + x$ **b.** $8x^4 + x$
   c. $4x^4 - 4x^3 + x^2 - x$ **d.** $4x^4 - x^3 + x^2$
   e. $-x^4 + 6x^2 - x$

10. If $g(t) = \dfrac{1 - t}{1 + t}$, find $g(a - 1)$. c

   a. $\dfrac{2 - a}{2 + a}$ **b.** $\dfrac{1 - a}{1 + a}$
   c. $\dfrac{2 - a}{a}$ **d.** -1

11. Which of the following is the domain of the function $f(x) = \sqrt{x} + \sqrt{1 - x}$? a

   a. $[0, 1]$ **b.** $(-\infty, -1] \cup [0, +\infty)$
   c. $(-\infty, 1]$ **d.** $[0, +\infty)$

12. What is the domain of the function $f(x) = \sqrt{x^2 + 6x - 7}$? b

   a. $(-\infty, -7) \cup (1, \infty)$ **b.** $(-\infty, -7] \cup [1, \infty)$
   c. $[-7, 1]$ **d.** $(1, \infty)$

13. Which description of the behavior of the graph shown here is correct? a

   a. increasing on $(0, 3)$; decreasing on $(-3, 0) \cup (3, 4)$
   b. increasing on $(-1, 3)$; decreasing on $(2, -1) \cup (3, 1)$
   c. increasing on $(-3, 3)$; decreasing on $(2, -1) \cup (3, 1)$
   d. increasing on $(1.2, 4)$; decreasing on $(-3, -1.4)$

14. Suppose the graph of f is given. Describe how to obtain the graph of $y = f(x + 4)$ from the graph of f. a

   a. Shift the graph of f four units to the left.
   b. Shift the graph of f four units down.
   c. Shift the graph of f four units up.
   d. Shift the graph of f four units to the right.
   e. Reflect the graph of f in the x-axis.

15. Suppose the graph of f is given. Describe how to obtain the graph of $y = -2f(x - 4)$ from the graph of f. b
   a. Shift the graph of f four units to the left, shrink it vertically by a factor of $\frac{1}{2}$, and reflect it in the x-axis.
   b. Shift the graph of f four units to the right, stretch it vertically by a factor of 2, and reflect it in the x-axis.
   c. Shift the graph of f four units to the left, shrink it vertically by a factor of $\frac{1}{2}$, and reflect it in the x-axis.
   d. Shift the graph of f four units to the right, then stretch it vertically by a factor of 2, and reflect it in the y-axis.

16. Which of the following is a one-to-one function? d
   a. $f(x) = |x + 3|$ **b.** $f(x) = x^2$
   c. $f(x) = \sqrt{x^2 + 9}$ **d.** $f(x) = x^3 + 2$

17. If f is a one-to-one function and $f(3) = 5$, then $f^{-1}(5) =$ c
   a. $\dfrac{1}{5}$ **b.** $\dfrac{1}{3}$
   c. 3   **d.** -5

18. Find the inverse function $f^{-1}(x)$ of $f(x) = \dfrac{1 - 3x}{5 + 2x}$. c
   a. $\dfrac{1 + 3x}{5 - 2x}$ **b.** $\dfrac{5 + 2x}{1 - 3x}$
   c. $\dfrac{1 - 5x}{3 + 2x}$ **d.** $\dfrac{3 + 2x}{1 - 5x}$

19. The ideal weight w (in pounds) for a man of height x inches is found by subtracting 190 from 5 times his height. Write a linear equation that gives the ideal weight of a man of height x inches. Then find the ideal weight of a man whose height is 70 inches. b
   a. $w = \dfrac{x + 190}{5}$, 160 lb   **b.** $w = 5x - 190$, 160 lb.
   c. $w = 190 - \dfrac{x}{5}$, 176 lb   **d.** $w = \dfrac{x + 190}{5}$, 52 lb

20. Cassie rented an intermediate sedan at \$25 per day $+$\$0.20 per mile. How many miles can Cassie travel for \$50? a
   a. 125   **b.** 1025
   c. 250   **d.** 175

Polynomial and Rational Functions

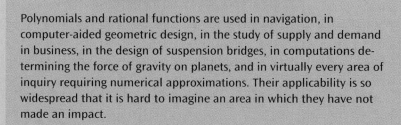

Polynomials and rational functions are used in navigation, in computer-aided geometric design, in the study of supply and demand in business, in the design of suspension bridges, in computations determining the force of gravity on planets, and in virtually every area of inquiry requiring numerical approximations. Their applicability is so widespread that it is hard to imagine an area in which they have not made an impact.

Quadratic Functions

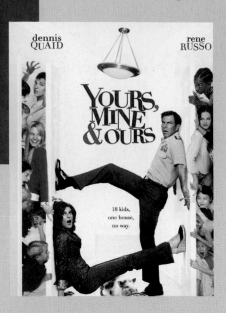

dennis QUAID   rene RUSSO

Before Starting this Section, Review

1. Linear functions (Section 1.4, page 48)
2. Completing the square (Appendix A, page 811)
3. Quadratic formula (Appendix A, page 813)
4. Transformations (Section 1.5, page 62)

Objectives

1 Graph a quadratic function in standard form.

2 Graph a quadratic function.

3 Solve problems modeled by quadratic functions.

YOURS, MINE, AND OURS

The movie *Yours, Mine, and Ours,* released in 1968, is a comedy about the ultimate blended family. The movie was based on the true story of Helen North, a widow with eight children, who married Frank Beardsley, a widower with ten children. The May 1964 issue of *Time* magazine had the following announcement:

May. 1, 1964

Born. To Francis Beardsley, 48, chief warrant officer at the Navy postgraduate school at Monterey, Calif., and Helen North Beardsley, 34: their second child, second daughter; in Carmel, Calif. The couple's 19 other children (he had ten, she eight from previous marriages) all voted on a name for the new Beardsley, came up "unanimously," with Helen Monica.

The film *Yours, Mine, and Ours* details the nuances of everyday life in a house full of children. The poster shown in the margin is from a remake of this popular story. In Example 5, we discuss the maximum possible number of fights among the children in such a family. ■

1 Graph a quadratic function in standard form.

Quadratic Function

A function whose defining equation is a polynomial is called a *polynomial function.* In this section, you will study **quadratic functions**, which are polynomial functions of degree 2.

QUADRATIC FUNCTION

A function of the form

$$f(x) = ax^2 + bx + c,$$

where $a, b,$ and c, are real numbers with $a \neq 0$, is called a **quadratic function.**

Note that the squaring function $s(x) = x^2$ (whose graph is a parabola) is the quadratic function $f(x) = ax^2 + bx + c$ with $a = 1$, $b = 0$, and $c = 0$. We will show that we can graph any quadratic function by transforming the graph of $s(x) = x^2$. Therefore, since the graph of $y = x^2$ is a parabola, the graph of any quadratic function is also a parabola.

First, we compare the graphs of $f(x) = ax^2$ for $a > 0$ and $g(x) = ax^2$ for $a < 0$. These are quadratic functions with $b = 0$ and $c = 0$.

$f(x) = ax^2, a > 0$	$g(x) = ax^2, a < 0$
The graph of $f(x) = ax^2$ is obtained from the graph of the squaring function $s(x) = x^2$ by vertically stretching or compressing it. Here are three such graphs on the same axes for $a = 2, a = 1,$ and $a = \dfrac{1}{2}.$	The graph of $g(x) = ax^2$ is the reflection of the graph of $f(x) = \lvert a \rvert x^2$ in the x-axis. Here are three graphs, for $a = -2, a = -1,$ and $a = -\dfrac{1}{2}.$

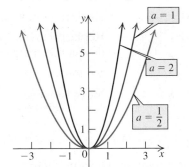

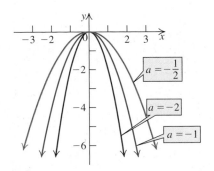

Standard Form of a Quadratic Function

Any quadratic function $f(x) = ax^2 + bx + c$ with $a \neq 0$ can be rewritten in the form $f(x) = a(x - h)^2 + k$, called the **standard form** of the quadratic function f. By using transformations (see Section 1.5), we can obtain the graph of $f(x) = a(x - h)^2 + k$ from the graph of the squaring function $s(x) = x^2$ as follows:

$$s(x) = x^2 \qquad \text{The squaring function}$$

$$g(x) = (x - h)^2 \qquad \text{Horizontal translation to the right if } h > 0 \text{ and to the left if } h < 0$$

$$G(x) = a(x - h)^2 \qquad \text{Vertical stretching or compressing. If } a < 0, \text{ the graph is also reflected in the } x\text{-axis.}$$

$$f(x) = a(x - h)^2 + k \qquad \text{Vertical translation, up if } k > 0 \text{ and down if } k < 0$$

These transformations show that the graph of a quadratic function $f(x) = a(x - h)^2 + k$ is a parabola that opens up if $a > 0$ and down if $a < 0$. Its domain is $(-\infty, \infty)$. The parabola is symmetric with respect to the vertical line $x = h$. The line of symmetry is called the **axis** (or **axis of symmetry**) of the parabola. The point (h, k) where the axis meets the parabola is called the **vertex** of the parabola. The vertex of any parabola that opens up is the *lowest point* of the graph, and the vertex of any parabola that opens down is the *highest point* of the graph. Therefore, k is the **minimum value** for any parabola that opens up or k is the **maximum value** for any parabola that opens down. The standard form is particularly useful because it immediately identifies the vertex (h, k).

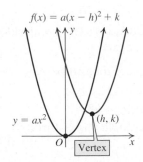

$f(x) = a(x - h)^2 + k$

$y = ax^2$

(h, k)

Vertex

FIGURE 2.1 Quadratic functions

THE STANDARD FORM OF A QUADRATIC FUNCTION

The quadratic function

$$f(x) = a(x - h)^2 + k, \ a \neq 0$$

is in **standard form.** The graph of f is a parabola with **vertex** (h, k). (See Figure 2.1.) The parabola is symmetric with respect to the line $x = h$, called the **axis** of the parabola. If $a > 0$, the parabola opens up, and if $a < 0$, the parabola opens down. If $a > 0$, k is the **minimum value** of f, and if $a < 0$, k is the **maximum value** of f.

EXAMPLE 1	Finding a Quadratic Function

Find the standard form of the quadratic function f whose graph has vertex $(-3, 4)$ and passes through the point $(-4, 7)$. Does f have a maximum or a minimum value?

SOLUTION

$f(x) = a(x - h)^2 + k$	Standard form of quadratic function
$f(x) = a[x - (-3)]^2 + 4$	Replace h with -3 and k with 4.
$y = a(x + 3)^2 + 4$	Replace $f(x)$ with y, and simplify.
$7 = a(-4 + 3)^2 + 4$	Replace x with -4 and y with 7 because $(-4, 7)$ lies on the graph of $y = f(x)$.
$7 = a + 4$	Simplify.
$a = 3.$	Solve for a.

The standard form is

$$f(x) = 3(x + 3)^2 + 4. \qquad a = 3, h = -3, k = 4$$

Since $a = 3 > 0$, f has a minimum value of 4 at $x = -3$.   ■ ■ ■

Practice Problem 1   Find the standard form of the quadratic function f whose graph has vertex $(1, -5)$ and passes through the point $(3, 7)$. Does f have a maximum or a minimum value?   ■

FINDING THE SOLUTION: A PROCEDURE

EXAMPLE 2	Graphing a Quadratic Function in Standard Form

OBJECTIVE
Sketch the graph of $f(x) = a(x - h)^2 + k$.

EXAMPLE
Sketch the graph of $f(x) = -3(x + 2)^2 + 12$.

Step 1   **The graph is a parabola** because it has the form $f(x) = a(x - h)^2 + k$. Identify a, h, and k.

The graph of $f(x) = -3(x + 2)^2 + 12$

$$= (-3)[x - (-2)]^2 + 12$$
$$\qquad\quad \uparrow \qquad\quad \uparrow \qquad\quad \uparrow$$
$$\qquad\quad a \qquad\quad h \qquad\quad k$$

is a parabola; $a = -3$, $h = -2$, and $k = 12$.

Step 2   **Determine how the parabola opens.** If $a > 0$, the parabola opens *up*. If $a < 0$, it opens *down*.

Since $a = -3 < 0$, the parabola opens down.

Step 3   **Find the vertex (h, k).** If $a > 0$ (or $a < 0$), the function f has a minimum (or a maximum) value k at $x = h$.

The vertex $(h, k) = (-2, 12)$. Since the parabola opens down, the function f has a maximum value of 12 at $x = -2$.

Step 4   **Find the x-intercepts (if any).** Set $f(x) = 0$ and solve the equation $a(x - h)^2 + k = 0$ for x. If the solutions are real numbers, they are the x-intercepts. If not, the parabola lies above the x-axis (when $a > 0$) or below the x-axis (when $a < 0$).

$0 = -3(x + 2)^2 + 12$	Set $f(x) = 0$.
$3(x + 2)^2 = 12$	Add $3(x + 2)^2$ to both sides.
$(x + 2)^2 = 4$	Divide by 3.
$x + 2 = \pm 2$	Square root property
$x = -2 \pm 2$	Subtract 2.
$x = 0$ or $x = -4$	Solve for x.

The x-intercepts are 0 and -4. The parabola passes through the points $(0, 0)$ and $(-4, 0)$.

Step 5 Find the y-intercept. Replace x with 0. Then $f(0) = ah^2 + k$ is the y-intercept.

$f(0) = -3(0 + 2)^2 + 12 = -3(4) + 12 = 0$. The y-intercept is 0.

Step 6 Sketch the graph. Plot the points found in Steps 3–5 and join them forming a parabola. Show the axis $x = h$ of the parabola by drawing a dashed line.

The axis of the parabola is the vertical line $x = -2$. The graph of the parabola is shown in the figure.

If there are no x-intercepts, draw the half of the parabola that passes through the vertex and the y-intercept. Then use the axis of symmetry to draw the other half.

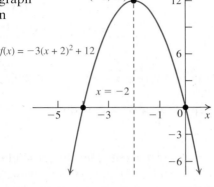

$f(x) = -3(x + 2)^2 + 12$

Practice Problem 2   Graph the quadratic function $f(x) = -2(x + 1)^2 + 3$.   ■

2 Graph a quadratic function.

Graphing a Quadratic Function $f(x) = ax^2 + bx + c$

A quadratic function $f(x) = ax^2 + bx + c$ can be changed to the standard form $f(x) = a(x - h)^2 + k$ by *completing the square*.

Converting $f(x) = ax^2 + bx + c$ to Standard Form

$f(x) = ax^2 + bx + c$	Original function
$= a\left(x^2 + \dfrac{b}{a}x\right) + c$	Factor out a from $ax^2 + bx$.
$= a\left(x^2 + \dfrac{b}{a}x + \dfrac{b^2}{4a^2} - \dfrac{b^2}{4a^2}\right) + c$	To complete the square, add and subtract $\left(\dfrac{1}{2} \cdot \dfrac{b}{a}\right)^2 = \dfrac{b^2}{4a^2}$ within the parentheses.
$= a\left(x^2 + \dfrac{b}{a}x + \dfrac{b^2}{4a^2}\right) - a \cdot \dfrac{b^2}{4a^2} + c$	Distributive property
$= a\left(x + \dfrac{b}{2a}\right)^2 - \dfrac{b^2}{4a} + c$	$\left(x + \dfrac{b}{2a}\right)^2 = x^2 + \dfrac{b}{a}x + \dfrac{b^2}{4a^2}$

Comparing this form with the standard form $f(x) = a(x - h)^2 + k$, we have the following:

$$h = -\frac{b}{2a} \qquad x\text{-coordinate of the vertex}$$

$$\text{and} \quad k = c - \frac{b^2}{4a} \qquad y\text{-coordinate of the vertex}$$

Notice that $f\left(-\dfrac{b}{2a}\right) = a\left(-\dfrac{b}{2a} + \dfrac{b}{2a}\right)^2 - \dfrac{b^2}{4a} + c$ Replace x with $-\dfrac{b}{2a}$ in the last equation for $f(x)$.

$$= -\frac{b^2}{4a} + c = k.$$

So to find the vertex $(h, k) = \left(-\dfrac{b}{2a},\ f\left(-\dfrac{b}{2a}\right)\right)$, you find the x-coordinate $h = -\dfrac{b}{2a}$ of the vertex and then calculate $k = f\left(-\dfrac{b}{2a}\right)$ to find its y-coordinate.

FINDING THE SOLUTION: A PROCEDURE

EXAMPLE 3   Graphing a Quadratic Function

OBJECTIVE

Graph $f(x) = ax^2 + bx + c, a \neq 0$.

EXAMPLE

Sketch the graph of $f(x) = 2x^2 + 8x - 10$.

Step 1. Identify a, b, and c.

In the equation $y = f(x) = 2x^2 + 8x - 10$, $a = 2, b = 8$, and $c = -10$.

Step 2. Determine how the parabola opens. If $a > 0$, the parabola opens *up*; if $a < 0$, the parabola opens *down*.

Since $a = 2 > 0$, the parabola opens up.

Step 3. Find the vertex (h, k). Use the formula

$$(h, k) = \left(-\frac{b}{2a},\ f\left(-\frac{b}{2a}\right)\right).$$

$$h = -\frac{b}{2a} = -\frac{8}{2(2)} = -2$$
$$k = f(h) = f(-2)$$
$$= 2(-2)^2 + 8(-2) - 10 \qquad \text{Replace } h \text{ with } -2.$$
$$= -18 \qquad \text{Simplify.}$$

The function f has a minimum value of -18 at $x = -2$.

Step 4   Find the x-intercepts (if any). Let $f(x) = 0$ and solve $ax^2 + bx + c = 0$. If the solutions are real numbers, they are the x-intercepts. If not, the parabola lies above the x-axis (when $a > 0$) or below the x-axis (when $a < 0$).

$$2x^2 + 8x - 10 = 0 \qquad \text{Set } f(x) = 0.$$
$$2(x^2 + 4x - 5) = 0 \qquad \text{Factor out 2.}$$
$$2(x + 5)(x - 1) = 0 \qquad \text{Factor.}$$
$$x + 5 = 0 \quad \text{or} \quad x - 1 = 0 \qquad \text{Zero-product property}$$
$$x = -5 \quad \text{or} \quad x = 1 \qquad \text{Solve for } x.$$

The x-intercepts are at the points $(-5, 0)$ and $(1, 0)$.

Step 5   Find the y-intercept. Let $x = 0$. The result, $f(0) = c$, is the y-intercept.

Set $x = 0$ to obtain $f(0) = 2(0)^2 + 8(0) - 10 = -10$. The y-intercept is at the point $(0, -10)$.

Step 6   The parabola is symmetric with respect to its axis, $x = -\dfrac{b}{2a}$. Use this symmetry to find additional points.

The axis of symmetry is $x = -2$. The symmetric image of $(0, -10)$ with respect to the axis $x = -2$ is $(-4, -10)$.

$f(x) = 2x^2 + 8x - 10$

Step 7   Draw a parabola through the points found in Steps 3–6.

The parabola passing through the points in Steps 3–6 is sketched in the figure.

If there are no x-intercepts, draw the half of the parabola that passes through the vertex and the y-intercept. Then use the axis of symmetry to draw the other half.

Practice Problem 3   Graph the quadratic function $f(x) = 3x^2 - 3x - 6$. ■

> **EXAMPLE 4**   **Identifying the Characteristics of a Quadratic Function from Its Graph**

The graph of the function $f(x) = -2x^2 + 8x - 5$ is shown in Figure 2.2.

a. Find the domain and range of f.

b. Solve the inequality $-2x^2 + 8x - 5 > 0$.

SOLUTION

The graph in Figure 2.2 is a parabola that opens down and whose vertex is $(2, 3)$.

a. The domain of f is $(-\infty, \infty)$. Since the parabola opens down, the maximum value of f is the y-coordinate of its vertex $(2, 3)$. So the range of f is $(-\infty, 3]$.

b. The x-intercepts of f are $\dfrac{4 - \sqrt{6}}{2} \approx 0.775$ and $\dfrac{4 + \sqrt{6}}{2} \approx 3.225$. The graph of f is above the x-axis between these intercepts. So $y = -2x^2 + 8x - 5 > 0$ for the x-values in the interval $\left(\dfrac{4 - \sqrt{6}}{2}, \dfrac{4 + \sqrt{6}}{2}\right)$, and the solution set of the inequality $-2x^2 + 8x - 5 > 0$ is the interval $\left(\dfrac{4 - \sqrt{6}}{2}, \dfrac{4 + \sqrt{6}}{2}\right)$.   ■ ■ ■

Practice Problem 4   Graph the function $f(x) = 3x^2 - 6x - 1$ and solve the inequality $3x^2 - 6x - 1 \le 0$.   ■

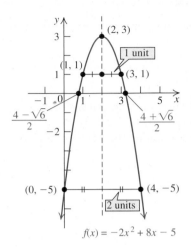

FIGURE 2.2

3 Solve problems modeled by quadratic functions.

Applications

Many applications of quadratic functions involve finding the maximum or minimum value of the function.

> **RECALL**

For a quadratic function $f(x) = ax^2 + bx + c$, the vertex is $(h, k) = \left(-\dfrac{b}{2a}, f\left(-\dfrac{b}{2a}\right)\right)$ and k is the function's minimum value if $a > 0$ or its maximum value if $a < 0$.

> **EXAMPLE 5**   *Yours, Mine, and Ours*

A widower with ten children marries a widow who also has children. After their marriage, they have their own children. If the total number of children is 24 and we assume that the children of the same parents do not fight, find the maximum possible number of fights among the children. (In this example, a fight between Sean and Misty, no matter how many times they fight, is considered one fight.)

SOLUTION

Suppose the widow had x number of children from her previous marriages. Then the couple has $24 - 10 - x = 14 - x$ additional children after their marriage. Since the children of the same parents do not fight, there are no fights among the ten children the widower brought into the marriage, among the x children the widow brought into the marriage, or among the $14 - x$ children of the couple (widower and widow).

The possible number of fights among the children of

(i) the widower (10 children) and the widow (x children) is $10x$.

(ii) the widower (10 children) and the couple ($14 - x$ children) is $10(14 - x)$.

(iii) the widow (x children) and the couple ($14 - x$ children) is $x(14 - x)$.

The possible number y of all fights is given by

$$y = 10x + 10(14 - x) + x(14 - x)$$
$$= 10x + 140 - 10x + 14x - x^2 \qquad \text{Distributive property}$$
$$= 140 + 14x - x^2 \qquad \text{Simplify.}$$

> **BY THE WAY . . .**

This example illustrates a *counting principle* that we will study in Section 10.6.

In the quadratic function $y = f(x) = -x^2 + 14x + 140$, we have $a = -1, b = 14$, and $c = 140$.

The vertex (h, k) is given by

$$h = -\frac{b}{2a} = -\frac{14}{2(-1)} = 7,$$
$$k = f(7) = -(7)^2 + 14(7) + 140 = 189.$$

Since $a < 0$, the function f has a maximum value. So the maximum possible number of fights among the children is 189. ■ ■ ■

Practice Problem 5 Repeat Example 5 if the widower had eight children from his previous marriage. ■

Note: The domain of a function that models a real-life problem may consist of only integer values. However, in solving these problems we frequently find it convenient to extend the domain to a suitable interval of real numbers containing the relevant integers.

SECTION 2.1 ■ Exercises

A EXERCISES Basic Skills and Concepts

1. A point where the axis of the parabola meets the parabola is called the _____vertex_____.

2. The vertex of the graph of $f(x) = -2(x + 3)^2 - 5$ is _____$(-3, -5)$_____.

3. *True or False* The graph of the function in Exercise 2 opens down. True

4. *True or False* The graph of $f(x) = -2 - x + x^2$ opens down. False

5. The x-coordinate of the vertex of the parabola of Exercise 4 is _____. $\frac{1}{2}$

6. *True or False* The y-coordinate of the vertex of the parabola $f(x) = x^2 - 2x + 5$ is $f(1)$. True

7. *True or False* If $a > 0$, then the minimum value of $f(x) = ax^2 + bx + c$ is $f\left(-\dfrac{b}{2a}\right)$. True

8. *True or False* If $a < 0$, then $f(x) = ax^2 + bx + c$ has no minimum value. True

In Exercises 9–16, match each quadratic function with its graph.

9. $y = -\dfrac{1}{3}x^2$ f

10. $y = -3x^2$ d

11. $y = -3(x + 1)^2$ a

12. $y = 2(x + 1)^2$ e

13. $y = (x - 1)^2 + 2$ h

14. $y = (x - 1)^2 - 3$ b

15. $y = 2(x + 1)^2 - 3$ g

16. $y = -3(x + 1)^2 + 2$ c

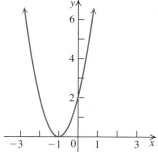

(c)

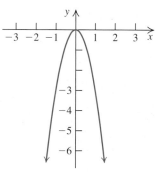

(d)

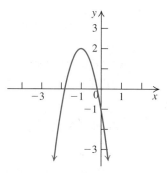

(e)

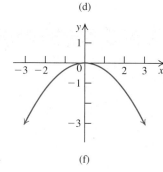

(f)

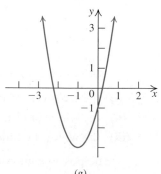

(g)

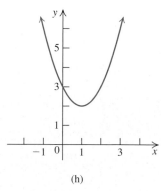

(h)

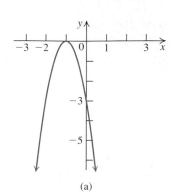

(a)

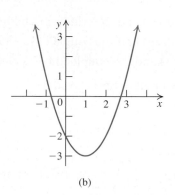

(b)

In Exercises 17–20, find a quadratic function of the form $y = ax^2$ that passes through the given point.

17. $(2, -8)$ $y = -2x^2$

18. $(-3, 3)$ $y = \dfrac{1}{3}x^2$

19. $(2, 20)$ $y = 5x^2$

20. $(-3, -6)$ $y = -\dfrac{2}{3}x^2$

†Due to space constrictions, answers to these exercises may be found in the Answers beginning on page A–1 in the back of the book.

In Exercises 21–30, find the quadratic function $y = f(x)$ that has the given vertex and whose graph passes through the given point. Write the function in standard form.

21. Vertex $(0, 0)$; passing through $(-2, 8)$ $y = 2x^2$

22. Vertex $(2, 0)$; passing through $(1, 3)$ $y = 3(x - 2)^2$

23. Vertex $(-3, 0)$; passing through $(-5, -4)$ $y = -(x + 3)^2$

24. Vertex $(0, 1)$; passing through $(-1, 0)$ $y = -x^2 + 1$

25. Vertex $(2, 5)$; passing through $(3, 7)$ $y = 2(x - 2)^2 + 5$

26. Vertex $(-3, 4)$; passing through $(0, 0)$ †

27. Vertex $(2, -3)$; passing through $(-5, 8)$ †

28. Vertex $(-3, -2)$; passing through $(0, -8)$ †

29. Vertex $\left(\dfrac{1}{2}, \dfrac{1}{2}\right)$; passing through $\left(\dfrac{3}{4}, -\dfrac{1}{4}\right)$ †

30. Vertex $\left(-\dfrac{3}{2}, -\dfrac{5}{2}\right)$; passing through $\left(1, \dfrac{55}{8}\right)$ †

In Exercises 31–34, the graph of a quadratic function $y = f(x)$ is given. Find the standard form of the function.

31.

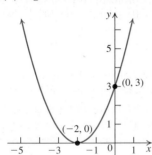

$y = \dfrac{3}{4}(x + 2)^2$

32.

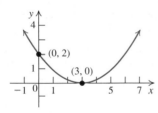

$y = \dfrac{2}{9}(x - 3)^2$

33.

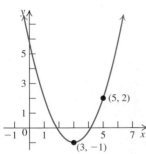

$y = \dfrac{3}{4}(x - 3)^2 - 1$

34.

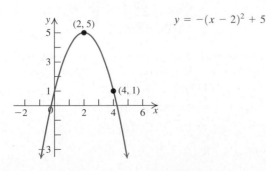

$y = -(x - 2)^2 + 5$

In Exercises 35–42, graph the given function by writing it in the standard form $y = a(x - h)^2 + k$ and then using transformations on $y = x^2$. Find the vertex, the axis, and the intercepts of the parabola.

35. $y = x^2 + 4x$ †   36. $y = x^2 - 2x + 2$ †

37. $y = 6x - 10 - x^2$ †   38. $y = 8 + 3x - x^2$ †

39. $y = 2x^2 - 8x + 9$ †   40. $y = 3x^2 + 12x - 7$ †

41. $y = -3x^2 + 18x - 11$ †   42. $y = -5x^2 - 20x + 13$ †

In Exercises 43–50, (a) determine whether the graph of the given quadratic function opens up or down, (b) find the vertex $\left(-\dfrac{b}{2a},\, f\left(-\dfrac{b}{2a}\right)\right)$, (c) find the axis of symmetry, (d) find the x- and y-intercepts, and (e) sketch the graph of the function.

43. $y = x^2 - 8x + 15$ †   44. $y = x^2 + 8x + 13$ †

45. $y = x^2 - x - 6$ †   46. $y = x^2 + x - 2$ †

47. $y = x^2 - 2x + 4$ †   48. $y = x^2 - 4x + 5$ †

49. $y = 6 - 2x - x^2$ †   50. $y = 2 + 5x - 3x^2$ †

In Exercises 51–58, a quadratic function f is given. (a) Determine whether the given quadratic function has a maximum value or a minimum value. Then find this value. (b) Find the range of f.

51. $f(x) = x^2 - 4x + 3$ †   52. $f(x) = -x^2 + 6x - 8$ †

53. $f(x) = -4 + 4x - x^2$ †   54. $f(x) = x^2 - 6x + 9$ †

55. $f(x) = 2x^2 - 8x + 3$ †   56. $f(x) = 3x^2 + 12x - 5$ †

57. $f(x) = -4x^2 + 12x + 7$ † 58. $f(x) = 8x - 5 - 2x^2$ †

In Exercises 59–64, solve the given quadratic inequality by sketching the graph of the corresponding quadratic function.

59. $x^2 - 4 \le 0$ †   60. $x^2 + 4x + 5 < 0$ †

61. $x^2 - 4x + 3 > 0$ †   62. $x^2 + x - 2 > 0$ †

63. $-6x^2 + x + 7 > 0$ †   64. $-5x^2 + 9x - 4 \le 0$ †

B EXERCISES Applying the Concepts

65. **Maximizing revenue.** A revenue function is given by $R(x) = 15 + 114x - 3x^2$, where x is the number of units produced and sold. Find the value of x for which the revenue is maximum.   19

66. **Maximizing revenue.** A demand function $p = 200 - 4x$, where p is the price per unit and x is the number of units produced and sold (the demand). Find x for which the revenue is maximum. [Recall that $R(x) = x \cdot p$.]   25

67. **Minimizing cost.** The total cost of a product is $C = 200 - 50x + x^2$, where x is the number of units produced. Find the value of x for which the total cost is minimum.   25

68. **Maximizing profit.** A manufacturer produces x items of a product at a total cost of $(50 + 2x)$ dollars. The demand function for the product is $p = 100 - x$. Find the value of x for which the profit is maximum. How much is the maximum profit? [Recall that Profit = Revenue − Cost.] $x = 49$; maximum profit: \$2351

69. Geometry. Find the dimensions of a rectangle of maximum area if the perimeter of the rectangle is 80 units. What is the maximum area?

70. Enclosing area. A rancher with 120 meters of fence intends to enclose a rectangular region along a river (which serves as a natural boundary requiring no fence). Find the maximum area that can be enclosed. 1800 m²

71. Enclosing playing fields. You have 600 meters of fencing, and you plan to lay out two identical playing fields. The fields can be side-by-side and separated by a fence, as shown in the figure. Find the dimensions and the maximum area of each field. Dimensions: 75 m × 100 m; area: 7500 m²

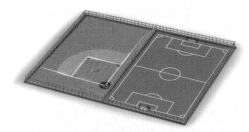

72. Enclosing area on a budget. You budget $2400 for constructing a rectangular enclosure that consists of a high surrounding fence and a lower inside fence that divides the enclosure in half. The high fence costs $8 per foot, and the low fence costs $4 per foot. Find the dimensions and the maximum area of each half of the enclosure. Dimensions: 60 ft × 37.5 ft; area: 2250 ft²

73. Maximizing apple yield. From past surveys and records, an orchard owner in northern Michigan has determined that if 26 apple trees per acre are planted, each tree yields 500 apples, on average. The yield decreases by ten apples per tree for each additional tree planted. How many trees should be planted per acre to achieve the maximum total yield? 38

74. Buying and selling beef. A steer weighing 300 pounds gains 8 pounds per day and costs $1 a day to keep. You bought this steer today at a market price of $1.50 per pound. However, the market price is falling 2¢ per pound per day. When should you sell the steer to maximize your profit? After 16 days

75. Going on a tour. A group of students plans a tour. The charge per student is $72 if 20 students go on the trip. If more than 20 students participate, the charge per student is reduced by $2 times the number of students over 20. Find the number of students that will furnish the maximum revenue. What is the maximum revenue?
28 students; maximum revenue: $1568

Answers:
69. Square with 20-unit sides; maximum area: 400 square units

76. Computer disks. A 5-inch-radius computer disk contains information in units called *bytes* that are arranged in concentric tracks on the disk. Assume that *m* bytes per inch can be put on each track. The tracks are uniformly spread (that is, there is a fixed number, say, *p*, of tracks per inch, measured radially across the disk). If the number of bytes on each track must be the same, where should the innermost track be located to get the maximum number of bytes on the disk? 2.5 inches from the center

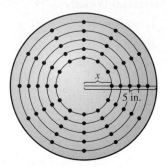

77. Motion of a projectile. A projectile is fired straight up with a velocity of 64 ft/s. Its altitude (height) *h* after *t* seconds is given by

$$h(t) = -16t^2 + 64t.$$

a. What is the maximum height of the projectile? 64 ft
b. When does the projectile hit the ground? After 4 sec

78. Motion of a projectile. From the top of a building, a projectile is shot 400 feet high with a velocity of 64 feet per second. Its altitude (height) *h* after *t* seconds is given by

$$h(t) = -16t^2 + 64t + 400.$$

a. What is the maximum height of the projectile? 464 ft
b. When does the projectile hit the ground? After 7.39 sec

79. Architecture. A window is to be constructed in the shape of a rectangle surmounted by a semicircle. If the perimeter of the window is 18 feet, find its dimensions for the maximum area. †

80. Architecture. The shape of the Gateway Arch in St. Louis, Missouri, is a *catenary* curve, which closely resembles a parabola. The function

$$y = f(x) = -0.00635x^2 + 4x$$

models the shape of the arch, where y is the height in feet and x is the horizontal distance from the base of the left side of the arch in feet.

a. Graph the function $y = f(x)$. †
b. What is the width of the arch at the base? 629.92 ft
c. What is the maximum height of the arch? 629.92 ft

81. Football. The altitude or height h (in feet) of a punted football can be modeled by the quadratic function

$$h = -0.01x^2 + 1.18x + 2,$$

where x (in feet) is the horizontal distance from the point of impact with the punter's foot. (See accompanying figure.)

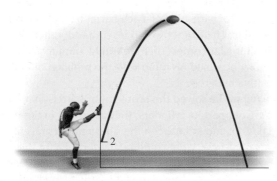

a. Find the vertex of the parabola. $(59, 36.81)$
b. If the opposing team fails to catch the ball, how far away from the punter does the ball hit the ground (to the nearest foot)? 120 ft
c. What is the maximum height of the punted ball (to the nearest foot)? 37 ft
d. The nearest defensive player is 6 feet from the point of impact. How high must the player reach to block the punt (to the nearest foot)? 9 ft
e. How far downfield has the ball traveled when it reaches a height of 7 feet for the second time? 113.6 ft

82. Field hockey. A *scoop* in field hockey is a pass that propels the ball from the ground into the air. Suppose a player makes a scoop with an upward velocity of 30 feet per second. The function

$$h = -16t^2 + 30t$$

models the altitude or height h in feet at time t in seconds. Will the ball ever reach a height of 16 feet? No

C EXERCISES Beyond the Basics

In Exercises 83–88, find a quadratic function of the form $f(x) = ax^2 + bx + c$ that satisfies the given conditions.

83. The graph of f is obtained by shifting the graph of $y = 3x^2$ two units horizontally to the left and three units vertically up. $f(x) = 3x^2 + 12x + 15$

84. The graph of f passes through the point $(1, 5)$ and has vertex $(-3, -2)$. $f(x) = \dfrac{7}{16}x^2 + \dfrac{21}{8}x + \dfrac{31}{16}$

85. The graph of f has vertex $(1, -2)$ and y-intercept 4.
$f(x) = 6x^2 - 12x + 4$

86. The graph of f has x-intercepts 2 and 6 and y-intercept 24. $f(x) = 2x^2 - 16x + 24$

87. The graph of f has x-intercept 7 and y-intercept 14, and $x = 3$ is the axis of symmetry. $f(x) = -2x^2 + 12x + 14$

88. The graph of f is obtained by shifting the graph of $y = x^2 + 2x + 2$ three units horizontally to the right and two units vertically down. $f(x) = x^2 - 4x + 3$

In Exercises 89–92, find two quadratic functions, one opening up and the other down, whose graphs have the given x-intercepts.

89. $-2, 6$ Opening up: $x^2 - 4x - 12$; opening down: $-x^2 + 4x + 12$

90. $-3, 5$ Opening up: $x^2 - 2x - 15$; opening down: $-x^2 + 2x + 15$

91. $-7, -1$ Opening up: $x^2 + 8x + 7$; opening down: $-x^2 - 8x - 7$

92. $2, 10$ Opening up: $x^2 - 12x + 20$; opening down: $-x^2 + 12x - 20$

93. Let $f(x) = 4x - x^2$. Solve $f(a + 1) - f(a - 1) = 0$ for a. $a = 2$

94. Let $f(x) = x^2 + 1$. Find $(f \circ f)(x)$.
$(f \circ f)(x) = x^4 + 2x^2 + 2$

Critical Thinking

95. Symmetry about the line $x = h$. The graph of a polynomial function $y = f(x)$ is symmetric in the line $x = h$ if $f(h + p) = f(h - p)$ for every number p. For $f(x) = ax^2 + bx + c, a \neq 0$, set $f(h + p) = f(h - p)$ and show that $h = -\dfrac{b}{2a}$. This exercise proves that the graph of a quadratic function is symmetric in its axis $x = -\dfrac{b}{2a}$.

96. In the graph of $y = 2x^2 - 8x + 9$, find the coordinates of the point symmetric to the point $(-1, 19)$ in the axis of symmetry. $(5, 19)$

Polynomial Functions

MAGYAR POSTA

1 Ft

Johannes
Kepler
1571-1630

Johannes Kepler (1571–1630)

Kepler was born in Weil in southern Germany and studied at the University of Tübingen. He studied with Michael Mastlin (professor of mathematics), who privately taught him the Copernican theory of the sun-centered universe while hardly daring to recognize it openly in his lectures. Kepler knew that to work out a detailed version of Copernicus's theory, he had to have access to the observations of the Danish astronomer Tycho Brahe (1546–1601). Kepler's correspondence with Brahe resulted in his appointment as Brahe's assistant. Brahe died about 18 months after Kepler's arrival. During those 18 months, Kepler learned enough about Brahe's work to use the material in working out his own major project. He was appointed as Imperial Mathematician by Emperor Rudolf to succeed Tycho Brahe and spent the next 11 years in Prague. He produced the famous three laws of planetary motion. His investigations were published in 1609 in his book *Nova Astronomia*.

Before Starting this Section, Review

1. Graphs of equations (Section 1.1, page 6)
2. Solving linear equations (Appendix A, page 793)
3. Solving quadratic equations (Appendix A, page 813)
4. Transformations (Section 1.5, page 62)
5. Relative maximum and minimum values (Section 1.4, page 51)

Objectives

1. Learn properties of the graphs of polynomial functions.
2. Determine the end behavior of polynomial functions.
3. Find the zeros of a polynomial function by factoring.
4. Identify the relationship between degrees, real zeros, and turning points.
5. Graph polynomial functions.

KEPLER'S WEDDING RECEPTION

Kepler is best known as an astronomer who discovered the three laws of planetary motion. However, Kepler's primary field was not astronomy, but mathematics. He served as a mathematician in Austrian Emperor Mathias's court. After the death of his wife Barbara, Kepler remarried in 1613. His new wife, Susanna, had a crash course in Kepler's character. Legend has it that at his own wedding reception, Kepler observed that the Austrian vintners could quickly and mysteriously compute the capacities (volumes) of a variety of wine barrels. Each barrel had a hole, called a *bunghole*, in the middle of its side. The vintner would insert a rod, called the *bungrod*, into the hole until it hit the far corner. Then he would announce the volume. During the reception, Kepler occupied his mind with the problem of finding the volume of a wine barrel with a bungrod.

In Example 10, we show you how he solved the problem for barrels that are perfect cylinders. But barrels are not perfect cylinders: they are wider in the middle and narrower at the top and bottom and in between the sides make a graceful curve that appears to be an arc of a circle. What could be the formula for the volume of such a shape? Kepler tackled this problem in his important book *Nova Stereometria Doliorum Vinariorum* (1615) and developed a complete mathematical theory in relation to it. ■

---

1 Learn properties of the graphs of polynomial functions.

Polynomial Functions

We begin with some terminology. A **polynomial function of degree n** is a function of the form

$$f(x) = a_n x^n + a_{n-1} x^{n-1} + \cdots + a_2 x^2 + a_1 x + a_0,$$

where n is a nonnegative integer and the *coefficients* $a_n, a_{n-1}, \ldots, a_2, a_1, a_0$ are real numbers with $a_n \neq 0$. The term $a_n x^n$ is called the **leading term**, the number a_n (the coefficient of x^n) is called the **leading coefficient**, and a_0 is the **constant term**. A constant function $f(x) = a(a \neq 0)$, which may be written as $f(x) = ax^0$, is a polynomial of degree 0. Because the zero function $f(x) = 0$ can be written in several ways, such as $0 = 0x^6 + 0x = 0x^5 + 0 = 0x^4 + 0x^3 + 0$, no degree is assigned to it.

In Section 1.4, we discussed polynomial functions of degree 0 and 1. In Section 2.1, we studied polynomial functions of degree 2. Polynomials of degree 3, 4, and 5 are also called **cubic**, **quartic**, and **quintic polynomials**, respectively. In this section, we concentrate on graphing polynomial functions of degree 3 or more.

Here are some common properties shared by all polynomial functions.

COMMON PROPERTIES OF POLYNOMIAL FUNCTIONS

1. The domain of a polynomial function is the set of all real numbers.
2. The graph of a polynomial function is a **continuous curve**. This means that the graph has no holes or gaps and can be drawn on a sheet of paper without lifting the pencil. See Figure 2.3.

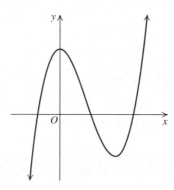

A continuous curve

(a)

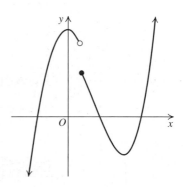

A discontinuous curve;
cannot be the graph of
a polynomial function

(b)

FIGURE 2.3

3. The graph of a polynomial function is a **smooth curve**. This means that the graph of a polynomial function does not contain any sharp corners. See Figure 2.4.

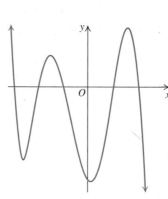

A smooth and continuous curve

(a)

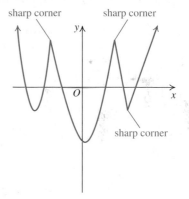

A continuous but not a smooth
curve; cannot be the graph of a
polynomial function

(b)

FIGURE 2.4

EXAMPLE 1   **Polynomial Functions**

State which functions are polynomial functions. For each polynomial function, find its degree, the leading term, and the leading coefficient.

a. $f(x) = 5x^4 - 2x + 7$

b. $g(x) = 7x^2 - x + 1,\ 1 \le x \le 5$

SOLUTION

a. $f(x) = 5x^4 - 2x + 7$ is a polynomial function. Its degree is 4, the leading term is $5x^4$, and the leading coefficient is 5.

b. $g(x) = 7x^2 - x + 1, 1 \leq x \leq 5$ is not a polynomial function because its domain is not $(-\infty, \infty)$.   ■ ■ ■

Practice Problem 1   Repeat Example 1 for each function.

a. $f(x) = \dfrac{x^2 + 1}{x - 1}$ **b.** $g(x) = 2x^7 + 5x^2 - 17$ ■

Power Functions

The simplest nth-degree polynomial function is a function of the form $f(x) = ax^n$ and is called a power function.

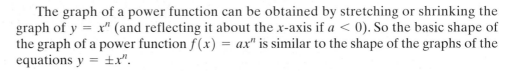

POWER FUNCTION

A function of the form

$$f(x) = ax^n$$

is called a **power function of degree n,** where a is a nonzero real number and n is a positive integer.

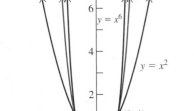

FIGURE 2.5 Power functions of even degree

The graph of a power function can be obtained by stretching or shrinking the graph of $y = x^n$ (and reflecting it about the x-axis if $a < 0$). So the basic shape of the graph of a power function $f(x) = ax^n$ is similar to the shape of the graphs of the equations $y = \pm x^n$.

Power Functions of Even Degree   Let $f(x) = ax^n$. If n is even, then $(-x)^n = x^n$. So in this case, $f(-x) = a(-x)^n = ax^n = f(x)$. Therefore, a power function is an even function when n is even, so its graph is symmetric with respect to the y-axis.

The graph of $y = x^n$ (when n is even) is similar to the graph of $y = x^2$; the graphs of $y = x^2, y = x^4$, and $y = x^6$ are shown in Figure 2.5.

Notice that each of these graphs passes through the points $(-1, 1)$, $(0, 0)$, and $(1, 1)$. On the interval $-1 < x < 1$, the larger the exponent, the flatter the graph. (See Exercise 95.) However, on the interval $(-\infty, -1) \cup (1, \infty)$, the larger the exponent is, the more rapidly the graph rises. (See Exercise 96.)

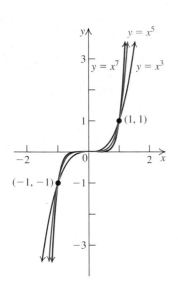

FIGURE 2.6 Power functions of odd degree

Power Functions of Odd Degree   Again, let $f(x) = ax^n$. If n is odd, then $(-x)^n = -x^n$. In this case, we have $f(-x) = a(-x)^n = -ax^n = -f(x)$. So the power function $f(x) = ax^n$ is an odd function when n is odd; its graph is symmetric with respect to the origin. When n is odd, the graph of $f(x) = x^n$ is similar to the graph of $y = x^3$. The graphs of $y = x^3, y = x^5$, and $y = x^7$ are shown in Figure 2.6. If n is an odd positive integer, then the graph of $y = x^n$ passes through the points $(-1, -1), (0, 0)$, and $(1, 1)$. The larger the value of n is, the flatter the graph is near the origin (in the interval $-1 < x < 1$) and the more rapidly it rises (or falls) when $|x| > 1$.

2   Determine the end behavior of polynomial functions.

End Behavior of Polynomial Functions

Let's study the behavior of the function $y = f(x)$ when the independent variable x is large in absolute value. First, we consider the expression x *approaches infinity*. The notation is $x \to \infty$ and means that x gets larger and larger without bound: it can

STUDY TIP

Here is a scheme for remembering the axis directions associated with the symbols $-\infty$ and ∞.

$x \rightarrow -\infty$	Left
$x \rightarrow \infty$	Right
$y \rightarrow -\infty$	Down
$y \rightarrow \infty$	Up

assume values greater than 1, 10, 100, 1000, ..., without end. Similarly, $x \rightarrow -\infty$ (read "x approaches negative infinity") means that x can assume values less than $-1, -10, -100, -1000, \ldots$, without end. Of course, x may be replaced by any other independent variable.

The behavior of a function $y = f(x)$ as $x \rightarrow \infty$ or $x \rightarrow -\infty$ is called the **end behavior** of the function. In the accompanying box, we describe the behavior of the power function $y = f(x) = ax^n$ as $x \rightarrow \infty$ or as $x \rightarrow -\infty$. Every polynomial function exhibits end behavior similar to one of the four functions $y = x^2$, $y = -x^2$, $y = x^3$, or $y = -x^3$.

End Behavior for the Graph of $y = f(x) = ax^n$

n **Even**	n **Odd**
Case 1 $a > 0$	**Case 3** $a > 0$

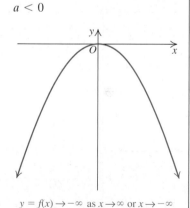

$y = f(x) \rightarrow \infty$ as $x \rightarrow \infty$ or $x \rightarrow -\infty$

(a)

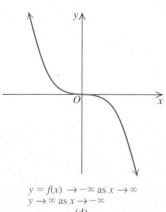

$y = f(x) \rightarrow \infty$ as $x \rightarrow \infty$
$y \rightarrow -\infty$ as $x \rightarrow -\infty$

(b)

The graph rises to the left and right, similar to $y = x^2$.	The graph rises to the right and falls to the left, similar to $y = x^3$.
Case 2 $a < 0$	**Case 4** $a < 0$

$y = f(x) \rightarrow -\infty$ as $x \rightarrow \infty$ or $x \rightarrow -\infty$

(c)

$y = f(x) \rightarrow -\infty$ as $x \rightarrow \infty$
$y \rightarrow \infty$ as $x \rightarrow -\infty$

(d)

The graph falls to the left and right, similar to $y = -x^2$.	The graph rises to the left and falls to the right, similar to $y = -x^3$.

In the next example, we show that the end behavior of the given polynomial function is determined by its leading term. If a polynomial $P(x)$ has approximately the same values as $f(x) = ax^n$ when $|x|$ is very large, we write $P(x) \approx ax^n$ when $|x|$ is very large.

TECHNOLOGY CONNECTION

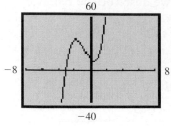

Calculator graph of
$P(x) = 2x^3 + 5x^2 - 7x + 11$

TABLE 2.1

x	$\dfrac{5}{x}$
10	0.5
10^2	0.05
10^3	0.005
10^4	0.0005

EXAMPLE 2   **Understanding the End Behavior of a Polynomial Function**

Let $P(x) = 2x^3 + 5x^2 - 7x + 11$ be a polynomial function of degree 3. Show that $P(x) \approx 2x^3$ when $|x|$ is very large.

SOLUTION

$$P(x) = 2x^3 + 5x^2 - 7x + 11 \qquad \text{Given polynomial}$$

$$= x^3\left(2 + \frac{5}{x} - \frac{7}{x^2} + \frac{11}{x^3}\right) \qquad \text{Distributive property}$$

When $|x|$ is very large, the terms $\frac{5}{x}$, $-\frac{7}{x^2}$, and $\frac{11}{x^3}$ are close to 0. Table 2.1 shows some values of $\frac{5}{x}$ as x increases. (You should compute the values of $-\frac{7}{x^2}$ and $\frac{11}{x^3}$ for $x = 10, 10^2, 10^3$, and 10^4.) When $|x|$ is very large, we then have

$$P(x) = x^3\left(2 + \frac{5}{x} - \frac{7}{x^2} + \frac{11}{x^3}\right) \approx x^3(2 + 0 - 0 + 0)$$

$$= 2x^3.$$

So the polynomial $P(x) \approx 2x^3$ when $|x|$ is very large.   ■ ■ ■

Practice Problem 2   Let $P(x) = 4x^3 + 2x^2 + 5x - 17$. Show that $P(x) \approx 4x^3$ when $|x|$ is very large.   ■

We can use the technique of Example 2 to show that the end behavior of any polynomial function is determined by its leading term.

The end behaviors of polynomial functions are shown in the following graphs.

THE LEADING-TERM TEST

Let $f(x) = a_n x^n + a_{n-1} x^{n-1} + \cdots + a_1 x + a_0 \, (a_n \neq 0)$ be a polynomial function. Its leading term is $a_n x^n$. The behavior of the graph of f as $x \to \infty$ or $x \to -\infty$ is similar to one of the following four graphs and is described as shown in each case.

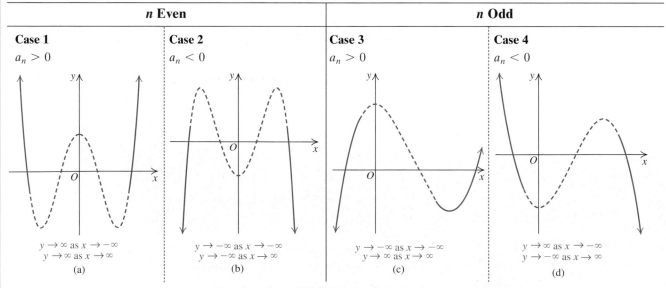

This test does not describe the middle portion of each graph, shown by the dashed lines.

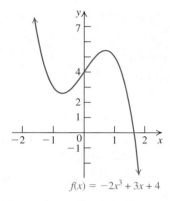

$f(x) = -2x^3 + 3x + 4$

FIGURE 2.7 End behavior of
$f(x) = -2x^3 + 3x + 4$

3 Find the zeros of a
polynomial function
by factoring.

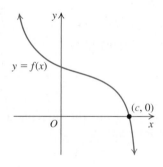

FIGURE 2.8 A zero of f

EXAMPLE 3   Using the Leading-Term Test

Use the leading-term test to determine the end behavior of the graph of

$$y = f(x) = -2x^3 + 3x + 4.$$

SOLUTION
Here $n = 3$ (odd) and $a_n = -2 < 0$. Case 4 applies. The graph of $f(x)$ rises to the left and falls to the right. See Figure 2.7. This behavior is described as $y \rightarrow \infty$ as $x \rightarrow -\infty$ and $y \rightarrow -\infty$ as $x \rightarrow \infty$.   ■ ■ ■

Practice Problem 3   What is the end behavior of $f(x) = -2x^4 + 5x^2 + 3$?   ■

Zeros of a Function

Let f be a function. An input c in the domain of f that produces output 0 is called a *zero* of the function f. For example, if $f(x) = (x - 3)^2$, then 3 is a zero of f because $f(3) = (3 - 3)^2 = 0$. In other words, a number c is a **zero** of f if $f(c) = 0$. The zeros of a function f can be obtained by solving the equation $f(x) = 0$. One way to solve a polynomial equation $f(x) = 0$ is to factor $f(x)$ and then use the zero-product property.

If c is a real number and $f(c) = 0$, then c is called a **real zero** of f. Geometrically, this means that the graph of f has an x-intercept at $x = c$.

REAL ZEROS OF POLYNOMIAL FUNCTIONS

If f is a polynomial function and c is a real number, then the following statements are equivalent:

1. c is a **zero** of f.

2. c is a **solution** (or **root**) of the equation $f(x) = 0$.

3. c is an **x-intercept** of the graph of f. The point $(c, 0)$ is on the graph of f. See Figure 2.8.

EXAMPLE 4   Finding the Zeros of a Polynomial Function

Find all real zeros of each polynomial function.

a. $f(x) = x^3 + 2x^2 - x - 2$ **b.** $g(x) = x^3 - 2x^2 + x - 2$

SOLUTION
a. We factor $f(x)$ and then solve the equation $f(x) = 0$.

$$
\begin{aligned}
f(x) &= x^3 + 2x^2 - x - 2 && \text{Given function} \\
&= (x^3 + 2x^2) - (x + 2) && \text{Group terms.} \\
&= x^2(x + 2) - 1(x + 2) && \text{Distributive property} \\
&= (x + 2)(x^2 - 1) && \text{Distributive property} \\
&= (x + 2)(x + 1)(x - 1) && \text{Factor } x^2 - 1.
\end{aligned}
$$

$(x + 2)(x + 1)(x - 1) = 0$ Set $f(x) = 0$.

$x + 2 = 0, \quad x + 1 = 0, \quad$ or $\quad x - 1 = 0$ Zero-product property

$x = -2, \quad x = -1, \quad$ or $\quad x = 1$ Solve each equation.

The zeros of $f(x)$ are $-2, -1$, and 1. (See the calculator graph of f.)

TECHNOLOGY CONNECTION

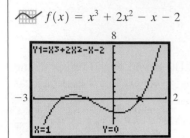

$f(x) = x^3 + 2x^2 - x - 2$

We can verify that the zeros of f are -2, -1, and 1 using the TRACE function.

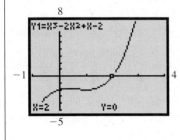

$g(x) = x^3 - 2x^2 + x - 2$

b. By grouping terms, we factor $g(x)$ to obtain

$$
\begin{aligned}
g(x) &= x^3 - 2x^2 + x - 2 \\
&= x^2(x - 2) + 1 \cdot (x - 2) && \text{Group terms.} \\
&= (x - 2)(x^2 + 1) && \text{Factor out } x - 2. \\
0 &= (x - 2)(x^2 + 1) && \text{Set } g(x) = 0. \\
x - 2 &= 0, \quad \text{or} \quad x^2 + 1 = 0 && \text{Zero-product property}
\end{aligned}
$$

Solving for x, we obtain 2 as the only real zero of $g(x)$ since $x^2 + 1 > 0$ for all real numbers x. The calculator graph of g supports our conclusion. ■ ■ ■

Practice Problem 4 Find all real zeros of $f(x) = 2x^3 - 3x^2 + 4x - 6$. ■

Finding the zeros of an unfactorable polynomial of degree 3 or more is one of the most important problems of mathematics. You will learn more about this topic in Sections 2.3 and 2.4. For now, we will approximate the zeros by using the *Intermediate Value Theorem*. Recall that the graph of a polynomial function $f(x)$ is a continuous curve. Suppose you locate two numbers a and b such that the function values $f(a)$ and $f(b)$ have opposite signs (one positive and the other negative). Then the continuous graph of f connecting the points $(a, f(a))$ and $(b, f(b))$ must cross the x-axis (at least once) somewhere between a and b. In other words, the function f has a zero between a and b. See Figure 2.9.

THE INTERMEDIATE VALUE THEOREM

Let a and b be two numbers with $a < b$. If f is a polynomial function such that $f(a)$ and $f(b)$ have opposite signs, then there is at least one number c, with $a < c < b$, for which $f(c) = 0$.

Notice that Figure 2.9(c) shows more than one zero of f. That is why the Intermediate Value Theorem says that f has *at least one* zero between a and b.

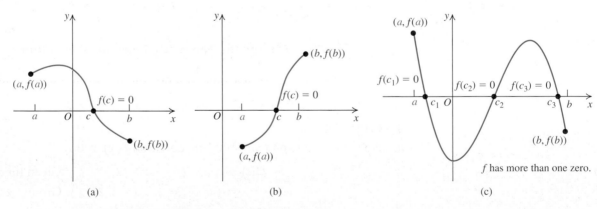

FIGURE 2.9 Each function has at least one zero between a and b

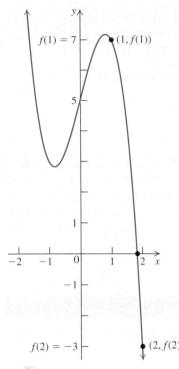

FIGURE 2.10 A zero lies between 1 and 2

4 Identify the relationship between degrees, real zeros, and turning points.

EXAMPLE 5   **Using the Intermediate Value Theorem**

Show that the function $f(x) = -2x^3 + 4x + 5$ has a real zero between 1 and 2.

SOLUTION

$$
\begin{aligned}
f(1) &= -2(1)^3 + 4(1) + 5 &&\text{Replace } x \text{ with 1 in } f(x).\\
&= -2 + 4 + 5 = 7 &&\text{Simplify.}\\
f(2) &= -2(2)^3 + 4(2) + 5 &&\text{Replace } x \text{ with 2.}\\
&= -2(8) + 8 + 5 = -3 &&\text{Simplify.}
\end{aligned}
$$

Since $f(1)$ is positive and $f(2)$ is negative, they have opposite signs. So by the Intermediate Value Theorem, the polynomial function $f(x) = -2x^3 + 4x + 5$ has a real zero between 1 and 2. See Figure 2.10.   ■ ■ ■

Practice Problem 5   Show that $f(x) = 2x^3 - 3x - 6$ has a real zero between 1 and 2.   ■

In Example 5, if you want to find the zero between 1 and 2 more accurately, you can divide the interval $[1, 2]$ into tenths and find the value of the function at each of the points $1, 1.1, 1.2, \ldots, 1.9,$ and 2. You will find that

$$
f(1.8) = 0.536 \quad \text{and} \quad f(1.9) = -1.118.
$$

So by the Intermediate Value Theorem, f has a zero between 1.8 and 1.9. By continuing this process, you can find the zero of f between 1.8 and 1.9 to any degree of accuracy.

Zeros and Turning Points

In Section 2.4, you will study the Fundamental Theorem of Algebra. One of its consequences is the following fact:

REAL ZEROS OF A POLYNOMIAL FUNCTION

A polynomial function of degree n with real coefficients has, *at most*, n real zeros.

EXAMPLE 6   **Finding the Number of Real Zeros**

Find the number of distinct real zeros of the following polynomial functions of degree 3.

a. $f(x) = (x - 1)(x + 2)(x - 3)$ **b.** $g(x) = (x + 1)(x^2 + 1)$

c. $h(x) = (x - 3)^2(x + 1)$

SOLUTION

a.
$$
\begin{aligned}
f(x) = (x - 1)(x + 2)(x - 3) &= 0 &&\text{Set } f(x) = 0.\\
x - 1 = 0, \quad \text{or} \quad x + 2 = 0, \quad \text{or} \quad x - 3 &= 0 &&\text{Zero-product property}\\
x = 1 \quad \text{or} \quad x = -2 \quad \text{or} \quad x &= 3 &&\text{Solve for } x.
\end{aligned}
$$

So $f(x)$ has three real zeros: 1, -2, and 3.

b.
$$
\begin{aligned}
g(x) = (x + 1)(x^2 + 1) &= 0 &&\text{Set } g(x) = 0.\\
x + 1 = 0 \quad \text{or} \quad x^2 + 1 &= 0 &&\text{Zero-product property}\\
x &= -1 &&\text{Solve for } x.
\end{aligned}
$$

Because $x^2 + 1 > 0$ for all real x, the equation $x^2 + 1 = 0$ has no real solution. So -1 is the only real zero of $g(x)$.

c.
$$h(x) = (x - 3)^2(x + 1) = 0 \quad \text{Set } h(x) = 0.$$
$$(x - 3)^2 = 0, \quad \text{or} \quad x + 1 = 0 \quad \text{Zero-product property}$$
$$x = 3, \quad \text{or} \quad x = -1 \quad \text{Solve for } x.$$

The real zeros of $h(x)$ are 3 and -1. Thus, $h(x)$ has two distinct real zeros.

■ ■ ■

Practice Problem 6 Find the distinct real zeros of

$$f(x) = (x + 1)^2(x - 3)(x + 5). \qquad ■$$

In Example 6c, the factor $(x - 3)$ appears twice in $h(x)$. We say that 3 is a zero repeated *twice*, or that 3 is a *zero of multiplicity* 2. Notice that the graph of $y = h(x)$ in Figure 2.11 *touches* the x-axis at $x = 3$, where h has a zero of multiplicity 2 (an even number). In contrast, the graph of h *crosses* the x-axis at $x = -1$, where it has a zero of multiplicity 1 (an odd number). We next state the geometric significance of the multiplicity of a zero.

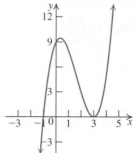

FIGURE 2.11 Graph of
$h(x) = (x - 3)^2(x + 1)$

MULTIPLICITY OF A ZERO

If c is a zero of a polynomial function $f(x)$ and the corresponding factor $(x - c)$ occurs exactly m times when $f(x)$ is factored completely, then c is called a **zero of multiplicity m**.

1. If m is odd, the graph of f crosses the x-axis at $x = c$.

2. If m is even, the graph of f touches but does not cross the x-axis at $x = c$.

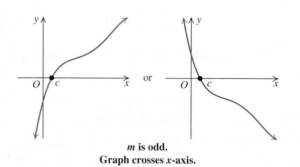

m is odd.
Graph crosses x-axis.

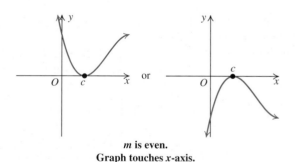

m is even.
Graph touches x-axis.

EXAMPLE 7 **Finding the Zeros and Their Multiplicity**

Find the zeros of the polynomial function $f(x) = x^2(x + 1)(x - 2)$ and give the multiplicity of each zero.

SOLUTION

The polynomial $f(x)$ is already in factored form.

$$f(x) = x^2(x + 1)(x - 2) = 0 \qquad \text{Set } f(x) = 0.$$
$$x^2 = 0, \quad \text{or} \quad x + 1 = 0, \quad \text{or} \quad x - 2 = 0 \qquad \text{Zero-product property}$$
$$x = 0 \quad \text{or} \quad x = -1 \quad \text{or} \quad x = 2 \qquad \text{Solve for } x.$$

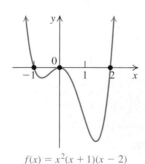

$f(x) = x^2(x + 1)(x - 2)$

FIGURE 2.12

The function has three distinct zeros: 0, -1, and 2. The zero $x = 0$ has multiplicity 2, and each of the zeros -1 and 2 has multiplicity 1. The graph of f is given in Figure 2.12.

Notice that the graph of f in Figure 2.12 crosses the x-axis at -1 and 2 (the zeros of odd multiplicity), while it touches but does not cross the x-axis at $x = 0$ (the zero of even multiplicity). ■ ■ ■

Practice Problem 7   Write the zeros of $f(x) = (x - 1)^2(x + 3)(x + 5)$ and give the multiplicity of each. ■

Turning Points   The graph of $f(x) = 2x^3 - 3x^2 - 12x + 5$ is shown in Figure 2.13. The value $f(-1) = 12$ is a **local** (or **relative**) **maximum** value of f. Similarly, the value $f(2) = -15$ is a **local** (or **relative**) **minimum** value of f. (See section 1.4 for definitions.) The graph points corresponding to the local maximum and local minimum values of a polynomial function are the **turning points**. At each turning point, the graph changes direction from increasing to decreasing or from decreasing to increasing.

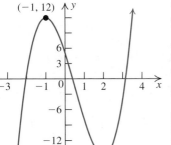

The graph of $f(x) = 2x^3 - 3x^2 - 12x + 5$ has turning points at $(-1, 12)$ and $(2, -15)$.

FIGURE 2.13 Turning points

NUMBER OF TURNING POINTS

If $f(x)$ is a polynomial of degree n, then the graph of f has, *at most*, $(n - 1)$ turning points.

EXAMPLE 8   **Finding the Number of Turning Points**

Use a graphing calculator and the window $-10 \le x \le 10$; $-30 \le y \le 30$ to find the number of turning points of the graph of each polynomial function.

a. $f(x) = x^4 - 7x^2 - 18$

b. $g(x) = x^3 + x^2 - 12x$

c. $h(x) = x^3 - 3x^2 + 3x - 1$

SOLUTION

The graphs of f, g, and h are shown in Figure 2.14.

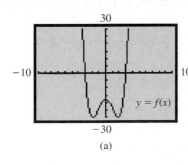
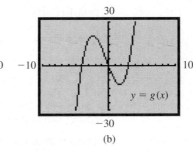
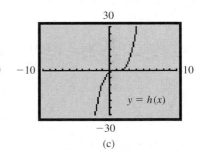

(a)   (b)   (c)

FIGURE 2.14 Turning points on a calculator graph

a. Function f has three turning points: two local minimum and one local maximum.

b. Function g has two turning points: one local maximum and one local minimum.

c. Function h has no turning points: it is increasing on the interval $(-\infty, \infty)$. ■ ■ ■

Practice Problem 8   Find the number of turning points of the graph of $f(x) = -x^4 + 3x^2 - 2$. ■

5 Graph polynomial functions.

Graphing a Polynomial Function

You can graph a polynomial function by constructing a table of values for a large number of points, plotting the points, and connecting the points with a smooth curve. (In fact, a graphing calculator uses this method.) Generally, however, this process is a poor way to graph a function by hand because it is tedious, is error-prone, and may give misleading results. Similarly, a graphing calculator may give misleading results if the viewing window is not chosen appropriately. Next, we discuss a better strategy for graphing a polynomial function.

FINDING THE SOLUTION: A PROCEDURE

EXAMPLE 9   Graphing a Polynomial Function

OBJECTIVE	**EXAMPLE**
Sketch the graph of a polynomial function.	*Sketch the graph of $f(x) = -x^3 - 4x^2 + 4x + 16$.*

Step 1   Determine the end behavior. Apply the leading-term test.

Since the degree, 3, of f is odd and the leading coefficient, -1, is negative, the end behavior (Case 4, page 122) is as shown in this sketch.

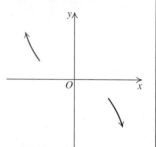

Step 2   Find the real zeros of the polynomial function. Set $f(x) = 0$. Solve the equation $f(x) = 0$ by factoring or by using a graphing calculator. The real zeros will give you the x-intercepts of the graph.

$$-x^3 - 4x^2 + 4x + 16 = 0$$
$$-(x + 4)(x + 2)(x - 2) = 0$$
$$x = -4, x = -2, \text{ or } x = 2$$

There are three zeros, each of multiplicity 1, so the graph crosses the x-axis at each zero.

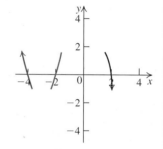

Step 3   Find the y-intercept by computing $f(0)$.

The y-intercept is $f(0) = 16$. The graph passes through the point $(0, 16)$.

Step 4   Use symmetry (if any) Check whether the function is odd, even, or neither.

$$f(-x) = -(-x)^3 - 4(-x)^2 + 4(-x) + 16$$
$$= x^3 - 4x^2 - 4x + 16$$

Since $f(-x) \neq f(x)$, there is no symmetry in the y-axis. Also, since $f(-x) \neq -f(x)$, there is no symmetry with respect to the origin.

Step 5   Determine the sign of $f(x)$ by using arbitrarily chosen "test numbers" in the intervals defined by the x-intercepts. Find the intervals on which the graph lies above or below the x-axis.

The three zeros $-4, -2$, and 2 divide the x-axis into four intervals: $(-\infty, -4), (-4, -2), (-2, 2)$, and $(2, \infty)$.

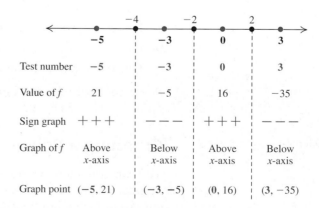

Test number	-5	-3	0	3
Value of f	21	-5	16	-35
Sign graph	$+++$	$---$	$+++$	$---$
Graph of f	Above x-axis	Below x-axis	Above x-axis	Below x-axis
Graph point	$(-5, 21)$	$(-3, -5)$	$(0, 16)$	$(3, -35)$

Step 6 Draw the graph. To check whether the graph is drawn correctly, use the fact that the number of turning points is less than the degree of the polynomial.

The graph connects all of the points we have found with a smooth curve.

The number of turning points is 2, which is less than 3, the degree of f.

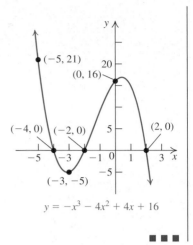

$y = -x^3 - 4x^2 + 4x + 16$

■ ■ ■

Practice Problem 9 Graph $f(x) = -x^4 + 5x^2 - 4$.   ■

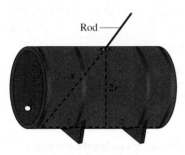

Rod

FIGURE 2.15 A wine barrel

Volume of a Wine Barrel Kepler first analyzed the problem of finding the volume of a wine barrel for a cylindrical barrel. (Review the information about this problem given in the introduction to this section and see Figure 2.15.) The volume V of a cylinder with radius r and height h is given by the formula $V = \pi r^2 h$. In Figure 2.15, $h = 2z$, so $V = 2\pi r^2 z$. Let x represent the value measured by the rod. Then by the Pythagorean Theorem,

$$x^2 = 4r^2 + z^2.$$

Kepler's challenge was to calculate V, given only x—in other words, to express V as a function of x. To solve this problem, Kepler observed the key fact that Austrian wine barrels were made with the same height-to-diameter ratio. Kepler assumed that the winemakers would have made a smart choice for this ratio, one that would maximize the volume for a fixed value of r. He calculated the ratio to be $\sqrt{2}$.

EXAMPLE 10   Volume of a Wine Barrel

Express the volume V of the wine barrel in Figure 2.15 as a function of x. Assume that $\dfrac{\text{height}}{\text{diameter}} = \dfrac{2z}{2r} = \dfrac{z}{r} = \sqrt{2}$.

SOLUTION

We have the following relationships:

$$V = 2\pi r^2 z \qquad \text{Volume with radius } r \text{ and height } 2z$$

$$\frac{z}{r} = \sqrt{2} \qquad \text{Given ratio}$$

$$x^2 = 4r^2 + z^2 \qquad \text{Pythagorean Theorem}$$

$$\frac{x^2}{r^2} = 4 + \frac{z^2}{r^2} \qquad \text{Divide both sides by } r^2.$$

$$\frac{x^2}{r^2} = 4 + 2 = 6 \qquad \text{Since } \frac{z}{r} = \sqrt{2}, \frac{z^2}{r^2} = \left(\frac{z}{r}\right)^2 = (\sqrt{2})^2 = 2.$$

From $\dfrac{x^2}{r^2} = 6$, $r^2 = \dfrac{x^2}{6}$, so $r = \dfrac{x}{\sqrt{6}}$. Using $\dfrac{z}{r} = \sqrt{2}$, we obtain

$$z = \sqrt{2}\, r = \sqrt{2}\,\frac{x}{\sqrt{6}} \qquad \text{Replace } r \text{ with } \frac{x}{\sqrt{6}}.$$

$$= \frac{x}{\sqrt{3}} \qquad \frac{\sqrt{2}}{\sqrt{6}} = \frac{\sqrt{2}}{\sqrt{2}\cdot\sqrt{3}} = \frac{1}{\sqrt{3}}$$

Substituting $z = \dfrac{x}{\sqrt{3}}$ and $r^2 = \dfrac{x^2}{6}$ in the expression for V, we have

$$V = 2\pi r^2 z$$

$$= 2\pi \left(\frac{x^2}{6}\right)\left(\frac{x}{\sqrt{3}}\right) \qquad \text{Replace } z \text{ with } \frac{x}{\sqrt{3}} \text{ and } r^2 \text{ with } \frac{x^2}{6}.$$

$$= \frac{\pi}{3\sqrt{3}} x^3 \qquad\qquad \text{Simplify.}$$

$$\approx 0.6046 x^3 \qquad\qquad \text{Use a calculator.}$$

Consequently, Austrian vintners could easily calculate the volume of a wine barrel by using the formula $V = 0.6046 x^3$ where x was the length measured by the rod.

■ ■ ■

Practice Problem 10 Use Kepler's formula to compute the volume of wine (in liters) in a barrel with $x = 70$ cm. Note: 1 decimeter (dm) = 10 cm and 1 liter = $(1\text{ dm})^3$. ■

SECTION 2.2 ■ Exercises

A EXERCISES Basic Skills and Concepts

1. Consider the polynomial $2x^5 - 3x^4 + x - 6$. The degree of this polynomial is _____5_____, its leading term is _____$2x^5$_____, its leading coefficient is _____2_____, and its constant term is _____-6_____ .

2. The domain of a polynomial function is the _set of all real numbers_.

3. The behavior of the function $y = f(x)$ as $x \to \infty$ or $x \to -\infty$ is called the _end behavior_ of the function.

4. The end behavior of any polynomial function is determined by its _leading_ term.

5. A number c for which $f(c) = 0$ is called a(n) _____zero_____ of the function f.

6. If c is a zero of even multiplicity for a function f, then the graph of f _____touches_____ the x-axis at c.

7. The graph of a polynomial function of degree n has, at most, _____$n-1$_____ turning points.

8. The zeros of a polynomial function $y = f(x)$ are determined by solving the equation _____$f(x) = 0$_____ .

In Exercises 9–14, for each polynomial function, find the degree, the leading term, and the leading coefficient.

9. $f(x) = 2x^5 - 5x^2$ †

10. $f(x) = 3 - 5x - 7x^4$ †

11. $f(x) = \dfrac{2x^3 + 7x}{3}$ †

12. $f(x) = -7x + 11 + \sqrt{2}x^3$ †

13. $f(x) = \pi x^4 + 1 - x^2$ †

14. $f(x) = 5$ †

In Exercises 15–22, explain why the given function is not a polynomial function.

15. $f(x) = x^2 + 3|x| - 7$ Presence of $|x|$

16. $f(x) = 2x^3 - 5x, \quad 0 \le x \le 2$ Domain not $(-\infty, \infty)$

17. $f(x) = \dfrac{x^2 - 1}{x + 5}$ Division by a polynomial is not allowed

18. $f(x) = \begin{cases} 2x + 3, & x \ne 0 \\ 1, & x = 0 \end{cases}$ Graph is not continuous

19. $f(x) = 5x^2 - 7\sqrt{x}$ Presence of $\sqrt{x}$

20. $f(x) = x^{4.5} - 3x^2 + 7$ Noninteger exponent

21. $f(x) = \dfrac{x^2 - 1}{x - 1}, \quad x \ne 1$ Division by a polynomial is not allowed

22. $f(x) = 3x^2 - 4x + 7x^{-2}$ Negative exponent

In Exercises 23–28, explain why the given graph cannot be the graph of a polynomial function.

23.

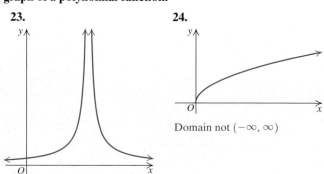

Graph not continuous

24.

Domain not $(-\infty, \infty)$

---

†Due to space constrictions, answers to these exercises may be found in the Answers beginning on page A–1 in the back of the book.

25.

Graph not continuous

26.

Has a sharp corner

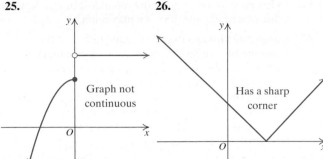

27.

Not a function

28.

Has sharp corners

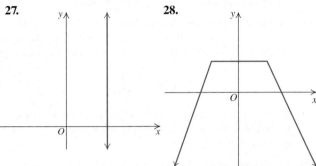

In Exercises 29–34, match the polynomial function with its graph. Use the leading-term test and the y-intercept.

29. $f(x) = -x^4 + 3x^2 + 4$ c

30. $f(x) = x^6 - 7x^4 + 7x^2 + 15$ f

31. $f(x) = x^4 - 10x^2 + 9$ a

32. $f(x) = x^3 + x^2 - 17x + 15$ e

33. $f(x) = x^3 + 6x^2 + 12x + 8$ d

34. $f(x) = -x^3 - 2x^2 + 11x + 12$ b

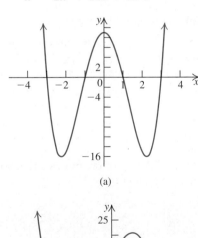

(a)

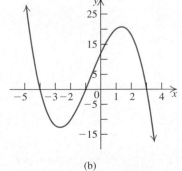

(b)

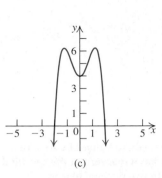

(c)

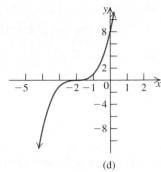

(d)

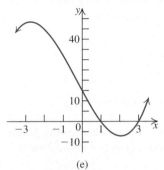

(e)

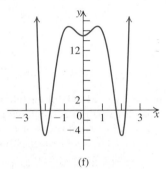

(f)

In Exercises 35–38, the graph of a polynomial function f is given. Find the unique equation for $f(x)$ from its graph.

35.

36.

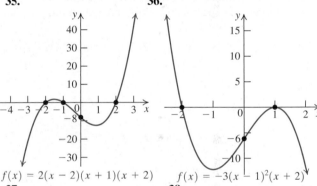

$f(x) = 2(x - 2)(x + 1)(x + 2)$ $f(x) = -3(x - 1)^2(x + 2)$

37.

38.

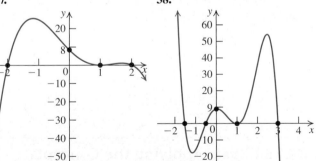

$f(x) = -2(x - 1)^2(x - 2)(x + 2)$

In Exercises 39–46, (a) find the real zeros of each polynomial function and state the multiplicity for each zero and (b) state whether the graph crosses or touches but does not cross the x-axis at each x-intercept. (c) What is the maximum possible number of turning points?

39. $f(x) = 5(x - 1)(x + 5)$ †

40. $f(x) = 3(x + 2)^2(x - 3)$ †

38. $f(x) = -4(x - 3)(x - 1)^2\left(x + \dfrac{1}{2}\right)\left(x + \dfrac{3}{2}\right)$

41. $f(x) = (x + 1)^2(x - 1)^3$ †

42. $f(x) = 2(x + 3)^2(x + 1)(x - 6)$ †

43. $f(x) = -5\left(x + \dfrac{2}{3}\right)\left(x - \dfrac{1}{2}\right)^2$ †

44. $f(x) = x^3 - 6x^2 + 8x$ †

45. $f(x) = -x^4 + 6x^3 - 9x^2$ †

46. $f(x) = x^4 - 5x^2 - 36$ †

In Exercises 47–50, use the Intermediate Value Theorem to show that each polynomial $P(x)$ has a real zero in the specified interval. Approximate this zero to two decimal places.

47. $P(x) = x^4 - x^3 - 10; [2, 3]$ 2.09

48. $P(x) = x^4 - x^2 - 2x - 5; [1, 2]$ 1.87

49. $P(x) = x^5 - 9x^2 - 15; [2, 3]$ 2.28

50. $P(x) = x^5 + 5x^4 + 8x^3 + 4x^2 - x - 5; [0, 1]$ 0.68

In Exercises 51–64, for each polynomial function f,

a. describe the end behavior of f.

b. find the real zeros of f. Determine whether the graph of f crosses or touches but does not cross the x-axis at each x-intercept.

c. use the zeros of f and test numbers to find the intervals over which the graph of f is above or below the x-axis.

d. determine the y-intercept.

e. find any symmetries of the graph of the function.

f. determine the maximum possible number of turning points.

g. sketch the graph of f.

51. $f(x) = x^2 + 3$ †   **52.** $f(x) = x^2 - 4x + 4$ †

53. $f(x) = x^2 + 4x - 21$ †

54. $f(x) = -x^2 + 4x + 12$ †

55. $f(x) = -2x^2(x + 1)$ †

56. $f(x) = x - x^3$ †

57. $f(x) = x^2(x - 1)^2$ †

58. $f(x) = x^2(x + 1)(x - 2)$ †

59. $f(x) = (x - 1)^2(x + 3)(x - 4)$ †

60. $f(x) = (x + 1)^2(x^2 + 1)$ †

61. $f(x) = -x^2(x^2 - 1)(x + 1)$ †

62. $f(x) = -x^2(x^2 - 4)(x + 2)$ †

63. $f(x) = x(x + 1)(x - 1)(x + 2)$ †

64. $f(x) = x^2(x^2 + 1)(x - 2)$ †

B EXERCISES Applying the Concepts

65. Drug reaction. Let $f(x) = 3x^2(4 - x)$.

a. Find the zeros of f and their multiplicities. †

b. Sketch the graph of $y = f(x)$. †

c. How many turning points are in the graph of f? 2

d. A patient's reaction $R(x)$ ($R(x) \geq 0$) to a certain drug, the size of whose dose is x, is observed to be $R(x) = 3x^2(4 - x)$. What is the domain of the function $R(x)$? What portion of the graph of $f(x)$ constitutes the graph of $R(x)$? †

 66. In Exercise 65, use a graphing calculator to estimate the value of x for which $R(x)$ is a maximum. $x = 2.67$

67. Cough and air velocity. The normal radius of the windpipe of a human adult at rest is approximately 1 cm. When you cough, your windpipe contracts to, say, a radius r. The velocity v at which the air is expelled depends on the radius r and is given by the formula $v = a(1 - r)r^2$, where a is a positive constant.

a. What is the domain of v? $[0, 1]$

b. Sketch the graph of $v(r)$. †

 68. In Exercise 67, use a graphing calculator to estimate the value of r for which v is a maximum. 0.67

69. Maximizing revenue. A manufacturer of slacks has discovered that the number x of khakis sold to retailers per week at a price p dollars per pair is given by the formula

$$p = 27 - \left(\dfrac{x}{300}\right)^2.$$

a. Write the weekly revenue $R(x)$ as a function of x.

b. What is the domain of $R(x)$? $[0, 900\sqrt{3}]$

c. Graph $y = R(x)$. †

 70. a. In Exercise 69, use a graphing calculator to estimate the number of slacks sold to maximize the revenue. 900

b. What is the maximum revenue from part (a)? $16,200

71. Maximizing production. An orange grower in California hires migrant workers to pick oranges during the season. He has 12 employees, and each can pick 400 oranges per hour. He has discovered that if he adds more workers, the production per worker decreases due to lack of supervision. When x new workers (above the 12) are hired, each worker picks $400 - 2x^2$ oranges per hour.

a. Write the number $N(x)$ of oranges picked per hour as a function of x. $N(x) = (x + 12)(400 - 2x^2)$

b. Find the domain of $N(x)$. $[0, 10\sqrt{2}]$

c. Graph the function $y = N(x)$. †

72. a. In Exercise 71, estimate the maximum number of oranges that can be picked per hour. 5950

b. Find the number of employees involved in part (a). 17

73. Making a box. From a rectangular 8×15 piece of cardboard, four congruent squares with sides of length x are cut out, one at each corner. (See the accompanying figure.) The resulting crosslike piece is then folded to form a box.

a. Find the volume V of the box as a function of x.

b. Sketch the graph of $y = V(x)$. †

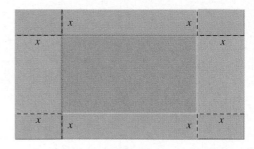

Answers:

69. a. $R(x) = 27x - \dfrac{1}{90,000}x^3$ **73. a.** $V(x) = x(8 - 2x)(15 - 2x)$

74. In Exercise 73, use a graphing calculator to estimate the largest possible value of the volume of the box. 90.74

75. Mailing a box. According to Postal Service regulations, the girth plus the length of a parcel sent by Priority Mail may not exceed 108 inches. $V(x) = x^2(108 - 4x)$
  a. Express, as a function of x, the volume V of a rectangular parcel with two square faces (with sides of length x inches) that can be sent by mail.
  b. Sketch a graph of $V(x)$. †

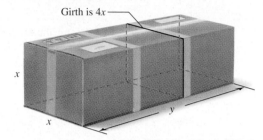

Girth is 4x

x

y

x

76. In Exercise 75, use a graphing calculator to estimate the largest possible volume of the box that can be sent by Priority Mail. 11,664 in.^{3}

77. Luggage design. AAA Airlines requires that the total outside dimensions (length + width + height) of a piece of checked-in luggage not exceed 62 inches. You have designed a suitcase in the shape of a box whose height (x inches) equals its width. Your suitcase meets AAA's requirements.
  a. Write the volume V of your suitcase as a function of x. $V(x) = x^2(62 - 2x)$
  b. Graph the function $y = V(x)$. †

78. In Exercise 77, use a graphing calculator to find the approximate dimensions of the piece of luggage with the largest volume that you can check in on this airline. 20.67 in. × 20.67 in. × 20.67 in.

79. Luggage dimensions. American Airlines requires that the total dimensions (length + width + height) of a carry-on bag not exceed 45 inches. You have designed a rectangular boxlike bag whose length is twice its width x. Your bag meets American's requirements.
  a. Write the volume V of your bag as a function of x, where x is measured in inches. $V(x) = 2x^2(45 - 3x)$
  b. Graph the function $y = V(x)$. †

80. In Exercise 79, use a graphing calculator to estimate the dimensions of the bag with the largest volume that you can carry on an American flight. (*Source:* www.aa.com) 10 in. × 15 in. × 20 in.

81. Worker productivity. An efficiency study of the morning shift at an assembly plant indicates that the number y of units produced by an average worker t hours after 6 A.M. may be modeled by the formula $y = -t^3 + 6t^2 + 16t$. Sketch the graph of $y = f(t)$. †

82. In Exercise 81, estimate the time in the morning when a worker is most efficient. 11 A.M.

83. Death rates from AIDS. According to the *Mortality and Morbidity* report of the U.S. Centers for Disease Control, the following table gives the death rates per 100,000 population from AIDS in the United States for the years 1993–2000.

Year	Death Rate	Year	Death Rate
1993	14.5	1997	6.1
1994	16.2	1998	4.9
1995	16.3	1999	5.4
1996	11.7	2000	5.2

To model the death rate, a data analysis program produced the cubic polynomial

$$P(t) = 0.21594t^3 - 2.226t^2 + 3.7583t + 14.6788,$$

where $t = 0$ represents 1993.

Sketch the graph of $y = P(t)$. †

84. In Exercise 83, when does the highest point occur for $0 \le t \le 20$? At $t = 20$

C EXERCISES Beyond the Basics

In Exercises 85–94, use transformations of the graph of $y = x^4$ or $y = x^5$ to graph each function $f(x)$. Find the zeros of $f(x)$ and their multiplicity.

85. $f(x) = (x - 1)^4$ †

86. $f(x) = -(x + 1)^4$ †

87. $f(x) = x^4 + 2$ †

88. $f(x) = x^4 + 81$ †

89. $f(x) = -(x - 1)^4$ †

90. $f(x) = \dfrac{1}{2}(x - 1)^4 - 8$ †

91. $f(x) = x^5 + 1$ †

92. $f(x) = (x - 1)^5$ †

93. $f(x) = 8 - \dfrac{(x + 1)^5}{4}$ †

94. $f(x) = 81 + \dfrac{(x + 1)^5}{3}$ †

95. Prove that if $0 < x < 1$, then $0 < x^4 < x^2$. This shows that the graph of $y = x^4$ is flatter than the graph of $y = x^2$ on the interval $(0, 1)$.

96. Prove that if $x > 1$, then $x^4 > x^2$. This shows that the graph of $y = x^4$ is higher than the graph of $y = x^2$ on the interval $(1, \infty)$.

97. Sketch the graph of $y = x^2 - x^4$. Use a graphing calculator to estimate the maximum vertical distance between the graphs $y = x^2$ and $y = x^4$ on the interval $(0, 1)$. Also estimate the value of x at which this maximum distance occurs. †

98. Repeat Exercise 97 by sketching the graph of
$y = x^3 - x^5$. †

In Exercises 99–102, write a polynomial with the given characteristics.

99. A polynomial of degree 3 that has exactly two
x-intercepts $f(x) = x^2(x + 1)$

100. A polynomial of degree 3 that has three distinct
x-intercepts and whose graph rises to the left and falls
to the right

101. A polynomial of degree 4 that has exactly two distinct
x-intercepts and whose graph falls to the left and
right $f(x) = 1 - x^4$

102. A polynomial of degree 4 that has exactly three distinct
x-intercepts and whose graph rises to the left and
right $f(x) = (x + 1)x^2(x - 1)$

In Exercises 103–106, find the smallest possible degree of a polynomial with the given graph. Explain your reasoning.

103.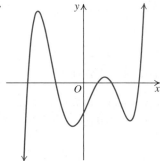
5 because the graph has five x-intercepts

104.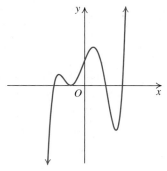
5 because the graph has four turning points

Answers:
100. $f(x) = -x(x + 1)(x - 1)$

105.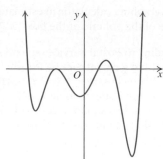
6 because the graph has five turning points

106.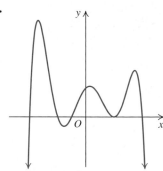
6 because the graph has five turning points

Critical Thinking

107. Is it possible for the graph of a polynomial function to
have no y-intercepts? Explain.

108. Is it possible for the graph of a polynomial function to
have no x-intercepts? Explain.

109. Is it possible for the graph of a polynomial function of
degree 3 to have exactly one local maximum and no
local minimum? Explain. Not possible

110. Is it possible for the graph of a polynomial function of
degree 4 to have exactly one local maximum and
exactly one local minimum? Explain. Not possible

107. No because the domain of any polynomial function is
$(-\infty, \infty)$, which includes the point $x = 0$.
108. Yes. The graph of $f(x) = x^2 + 1$ has no x-intercept

Dividing Polynomials and the Rational Zeros Test

Before Starting this Section, Review

1. Multiplying and dividing polynomials (Appendix A, page 764)

2. Synthetic division (Appendix A, page 767)

3. Evaluating a function (Section 1.3, page 34)

4. Factoring trinomials (Appendix A, page 770)

Objectives

1. Learn the Division Algorithm.

2. Use the Remainder and Factor Theorems.

3. Use the Rational Zeros Test.

GREENHOUSE EFFECT AND GLOBAL WARMING

The **greenhouse effect** refers to the fact that Earth is surrounded by an atmosphere that acts like the transparent cover of a greenhouse, allowing sunlight to filter through while trapping heat. It is becoming increasingly clear that we are experiencing *global warming*. This means that Earth's atmosphere is becoming warmer near its surface. Scientists who study climate agree that global warming is due largely to the emission of "greenhouse gases," such as carbon dioxide (CO_2), chlorofluorocarbons (CFCs), methane (CH_4), ozone (O_3), and nitrous oxide (N_2O). The temperature of Earth's surface increased by $1°F$ over the twentieth century. The late 1990s and early 2000s were some of the hottest years ever recorded. Projections of future warming suggest a global temperature increase of between $2.5°F$ and $10.4°F$ by 2100.

Global warming will result in long-term changes in climate, including a rise in average temperatures, unusual rainfalls, and more storms and floods. The sea level may rise so much that people will have to move away from coastal areas. Some regions of the world may become too dry for farming.

The production and consumption of energy is one of the major causes of greenhouse gas emissions. In Example 5, we look at the pattern for the consumption of petroleum in the United States. ■

---

1 Learn the division algorithm.

The Division Algorithm

Dividing polynomials is similar to dividing integers. The fact that 5 divides 10 evenly is expressed as $\frac{10}{5} = 2$ or as $10 = 5 \cdot 2$. By comparison, in equation

RECALL

A polynomial $P(x)$ is an expression of the form

$$a_n x^n + \cdots + a_1 x + a_0.$$

(1) $x^3 - 1 = (x - 1)(x^2 + x + 1)$ Factor, using difference of cubes.

dividing both sides by $x - 1$ we obtain

$$\frac{x^3 - 1}{x - 1} = x^2 + x + 1, \quad x \neq 1.$$

STUDY TIP

It is helpful to compare the division process for integers with that of polynomials. Notice that

$$3\overline{)7}$$

with quotient 2, 6, 1

can also be written as

$$7 = 3 \cdot 2 + 1$$

or

$$\frac{7}{3} = 2 + \frac{1}{3}$$

BY THE WAY . . .

An *algorithm* is a finite sequence of precise steps that lead to a solution of a problem

POLYNOMIAL FACTOR

A polynomial $D(x)$ is a **factor** of a polynomial $F(x)$ if there is a polynomial $Q(x)$ such that $F(x) = D(x) \cdot Q(x)$.

Next, consider the equation

$$(2) \qquad x^3 - 2 = (x - 1)(x^2 + x + 1) - 1. \qquad \text{Subtract 1 from both sides of equation (1).}$$

Dividing both sides by $(x - 1)$, we can rewrite equation (2) in the form

$$\frac{x^3 - 2}{x - 1} = x^2 + x + 1 - \frac{1}{x - 1}, \quad x \neq 1.$$

Here we say that when the **dividend** $x^3 - 2$ is divided by the **divisor** $x - 1$, the **quotient** is $x^2 + x + 1$ and the **remainder** is -1. This result is an example of the *division algorithm*:

THE DIVISION ALGORITHM

If a polynomial $F(x)$ is divided by a polynomial $D(x)$, and $D(x)$ is not the zero polynomial, then there are unique polynomials $Q(x)$ and $R(x)$ such that

$$\frac{F(x)}{D(x)} = Q(x) + \frac{R(x)}{D(x)} \qquad \text{or}$$

$$\underset{\uparrow}{F(x)} = \underset{\uparrow}{D(x)} \cdot \underset{\uparrow}{Q(x)} + \underset{\uparrow}{R(x)}$$

$$\text{Dividend} \qquad \text{Divisor} \qquad \text{Quotient} \qquad \text{Remainder}$$

Either $R(x)$ is the *zero polynomial* or the degree of $R(x)$ is less than the degree of $D(x)$.

The division algorithm says that the dividend is equal to the divisor times the quotient plus the remainder. The expression $\dfrac{F(x)}{D(x)}$ is **improper** if degree of $F(x) \geq$ degree of $D(x)$. However, the expression $\dfrac{R(x)}{D(x)}$ is always **proper** because by the division algorithm, degree of $R(x) <$ degree of $D(x)$.

For an improper expression $\dfrac{F(x)}{D(x)}$, you may recall the **long division** and the **synthetic division** processes shown in the next example.

EXAMPLE 1   **Using Long Division and Synthetic Division**

Use long division and synthetic division to find the quotient and remainder when $2x^4 + x^3 - 16x^2 + 18$ is divided by $x + 2$.

SOLUTION

Long Division

$$
\begin{array}{r}
2x^3 - 3x^2 - 10x + 20 \\
x + 2\overline{)2x^4 + x^3 - 16x^2 + 0x + 18} \\
\underline{2x^4 + 4x^3} \\
-3x^3 - 16x^2 + 0x + 18 \\
\underline{-3x^3 - 6x^2} \\
-10x^2 + 0x + 18 \\
\underline{-10x^2 - 20x} \\
20x + 18 \\
\underline{20x + 40} \\
-22
\end{array}
$$

Synthetic Division

$$
\begin{array}{r|rrrrr}
-2 & 2 & 1 & -16 & 0 & 18 \\
   &   & -4 & 6 & 20 & -40 \\
\hline
   & 2 & -3 & -10 & 20 & \underline{|-22}
\end{array}
$$

The quotient is $2x^3 - 3x^2 - 10x + 20$, and the remainder is -22. So the result is

$$
\frac{2x^4 + x^3 - 16x^2 + 18}{x + 2} = 2x^3 - 3x^2 - 10x + 20 + \frac{-22}{x + 2}
$$

$$
= 2x^3 - 3x^2 - 10x + 20 - \frac{22}{x + 2}. \qquad \blacksquare\blacksquare\blacksquare
$$

Practice Problem 1   Use synthetic division to divide $2x^3 + x^2 - 18x - 7$ by $x - 3$. ■

2 Use the Remainder and Factor Theorems.

The Remainder and Factor Theorems

In Example 1, we discovered that when the polynomial

$$
F(x) = 2x^4 + x^3 - 16x^2 + 18 \text{ is divided by } x + 2,
$$

the remainder is -22, a constant.

Let's evaluate F at -2. We have:

$$
F(-2) = 2(-2)^4 + (-2)^3 - 16(-2)^2 + 18 \qquad \text{Replace } x \text{ with } -2 \text{ in } F(x).
$$
$$
= 2(16) - 8 - 16(4) + 18 = -22 \qquad \text{Simplify.}
$$

We see that $F(-2) = -22$, the remainder we found when we divided $F(x)$ by $x + 2$. Is this result a coincidence? No, the following theorem asserts that such a result is always true.

THE REMAINDER THEOREM

If a polynomial $F(x)$ is divided by $x - a$, then the remainder R is given by

$$
R = F(a).
$$

Note that in the statement of the Remainder Theorem, $F(x)$ plays the dual role of representing a polynomial and the function defined by the same polynomial.

The Remainder Theorem has a very simple proof: the division algorithm says that if a polynomial $F(x)$ is divided by a polynomial $D(x) \neq 0$, then

$$
F(x) = D(x) \cdot Q(x) + R(x),
$$

where $Q(x)$ is the quotient and $R(x)$ is the remainder. Suppose that $D(x)$ is a linear polynomial of the form $x - a$. Since the degree of $R(x)$ is less than the degree of $D(x)$, $R(x)$ is a constant. Call this constant R. In this case, by the division algorithm, we have

$$F(x) = (x - a)Q(x) + R.$$

Replacing x with a in $F(x)$, we obtain

$$F(a) = (a - a)Q(a) + R = 0 + R = R.$$

This proves the Remainder Theorem.

EXAMPLE 2 Using the Remainder Theorem

Find the remainder when the polynomial

$$F(x) = 2x^5 - 4x^3 + 5x^2 - 7x + 2$$

is divided by $x - 1$.

SOLUTION

We could find the remainder by using long division or synthetic division, but a quicker way is to evaluate $F(x)$ when $x = 1$. By the Remainder Theorem, $F(1)$ is the remainder.

$$F(1) = 2(1)^5 - 4(1)^3 + 5(1)^2 - 7(1) + 2 \qquad \text{Replace } x \text{ with 1 in } F(x).$$
$$= 2 - 4 + 5 - 7 + 2 = -2$$

The remainder is -2. ■ ■ ■

Practice Problem 2 Use the Remainder Theorem to find the remainder when

$$F(x) = x^{110} - 2x^{57} + 5 \text{ is divided by } x - 1. \qquad ■$$

Sometimes the Remainder Theorem is used in the other direction: to evaluate a polynomial at a specific value of x. This use is illustrated in the next example.

EXAMPLE 3 Using the Remainder Theorem

Let $f(x) = x^4 + 3x^3 - 5x^2 + 8x + 75$. Find $f(-3)$.

SOLUTION

One way of solving this problem is to evaluate $f(x)$ when $x = -3$.

$$f(-3) = (-3)^4 + 3(-3)^3 - 5(-3)^2 + 8(-3) + 75 \qquad \text{Replace } x \text{ with } -3.$$
$$= 6 \qquad\qquad\qquad\qquad\qquad\qquad\qquad\qquad\quad \text{Simplify.}$$

Another way is to use synthetic division to find the remainder when the polynomial $f(x)$ is divided by $(x + 3)$.

$$
\begin{array}{r|rrrrr}
-3 & 1 & 3 & -5 & 8 & 75 \\
   &   & -3 & 0 & 15 & -69 \\
\hline
   & 1 & 0 & -5 & 23 & \,|\,6
\end{array}
$$

The remainder is 6. By the Remainder Theorem, $f(-3) = 6$. ■ ■ ■

Practice Problem 3 Use synthetic division to find $f(-2)$, where

$$f(x) = x^4 + 10x^2 + 2x - 20. \qquad ■$$

Note that if we divide any polynomial $F(x)$ by $x - a$, we have

$$F(x) = (x - a)Q(x) + R \qquad \text{Division algorithm}$$
$$F(x) = (x - a)Q(x) + F(a) \qquad R = F(a), \text{ by the Remainder Theorem}$$

The last equation says that if $F(a) = 0$, then

$$F(x) = (x - a)Q(x)$$

and $(x - a)$ is a factor of $F(x)$. Conversely, if $(x - a)$ is a factor of $F(x)$, then dividing $F(x)$ by $(x - a)$ gives a remainder of 0. Therefore, by the Remainder Theorem, $F(a) = 0$. This argument proves the Factor Theorem.

THE FACTOR THEOREM

A polynomial $F(x)$ has $(x - a)$ as a factor if and only if $F(a) = 0$.

The Factor Theorem shows that if $F(x)$ is a polynomial, then the following problems are equivalent:

1. Factoring $F(x)$

2. Finding the zeros of the function $F(x)$ defined by the polynomial expression

3. Solving (or finding the roots of) the polynomial equation $F(x) = 0$

Note that if a is a zero of the polynomial function $F(x)$, then $F(x) = (x - a)Q(x)$. Any solution of the equation $Q(x) = 0$ is also a zero of $F(x)$. Since the degree of $Q(x)$ is less than the degree of $F(x)$, it is simpler to solve the equation $Q(x) = 0$ to find other possible zeros of $F(x)$. The equation $Q(x) = 0$ is called the **depressed equation** of $F(x)$.

The next example illustrates how to use the Factor Theorem and the depressed equation to solve a polynomial equation.

EXAMPLE 4   **Using the Factor Theorem**

Given that 2 is a zero of the function $f(x) = 3x^3 + 2x^2 - 19x + 6$, solve the polynomial equation $3x^3 + 2x^2 - 19x + 6 = 0$.

SOLUTION

Since 2 is a zero of $f(x)$, we have $f(2) = 0$. The Factor Theorem tells us that $(x - 2)$ is a factor of $f(x)$. Next, we use synthetic division to divide $f(x)$ by $(x - 2)$.

$$
\begin{array}{r|rrrr}
2 & 3 & 2 & -19 & 6 \\
  &   & 6 & 16 & -6 \\
\hline
  & 3 & 8 & -3 & 0 \quad \longleftarrow \text{Remainder}
\end{array}
$$

$$\underbrace{}_{\text{coefficients of the quotient}}$$

The coefficients of the quotient gives the depressed equation, and because the remainder is 0,

$$f(x) = 3x^3 + 2x^2 - 19x + 6 = (x - 2)(\underbrace{3x^2 + 8x - 3}_{\text{quotient}}).$$

Any solution of the depressed equation $3x^2 + 8x - 3 = 0$ is a zero of f. Because this equation is of degree 2, any method of solving a quadratic equation may be used to solve it.

$$(3x^2 + 8x - 3) = 0 \qquad \text{Depressed equation}$$
$$(3x - 1)(x + 3) = 0 \qquad \text{Factor}$$
$$3x - 1 = 0 \quad \text{or} \quad x + 3 = 0 \qquad \text{Zero-product property}$$
$$x = \frac{1}{3} \quad \text{or} \quad x = -3 \qquad \text{Solve each equation.}$$

The solution set is $\left\{ -3, \dfrac{1}{3}, 2 \right\}$. ■ ■ ■

Practice Problem 4   Solve the equation $3x^3 - x^2 - 20x - 12 = 0$, given that one solution is -2. ■

Application Energy is measured in British thermal units (BTU). One BTU is the amount of energy required to raise the temperature of 1 pound of water 1°F at or near 39.2°F; 1 million BTU is equivalent to 8 gallons of gasoline. Fuels such as coal, petroleum, and natural gas are measured in quadrillion BTU (1 quadrillion = 10^{15}), abbreviated "quads."

EXAMPLE 5   **Petroleum Consumption**

The petroleum consumption C (in quads) in the United States from 1970–1995 can be modeled by the function

$$C(x) = 0.003x^3 - 0.139x^2 + 1.621x + 28.995,$$

where $x = 0$ represents 1970, $x = 1$ represents 1971, and so on.

The model indicates that $C(5) = 34$ quads of petroleum were consumed in 1975. Find another year between 1975 and 1995 when the model indicates that 34 quads of petroleum were consumed. (The model was slightly modified from the data obtained from the *Statistical Abstracts of the United States.*)

SOLUTION

We are given that

$$C(5) = 34$$
$$C(x) = (x - 5)Q(x) + 34 \qquad \text{Division algorithm and Remainder Theorem}$$
$$C(x) - 34 = (x - 5)Q(x) \qquad \text{Subtract 34 from both sides.}$$

Therefore, 5 is a zero of $F(x) = C(x) - 34$

$$= 0.003x^3 - 0.139x^2 + 1.621x - 5.005.$$

Finding another year between 1975 and 1995 when the petroleum consumption was 34 quads requires us to find another zero of $F(x) = C(x) - 34$ that is between 5 and 25. Because 5 is a zero of $F(x)$, we use synthetic division to find the quotient $Q(x)$.

```
5| 0.003   -0.139    1.621   -5.005
            0.015    -0.62    5.005
   ─────────────────────────────────
   0.003   -0.124    1.001  |   0
```

We solve the depressed equation $Q(x) = 0$.

$$0.003x^2 - 0.124x + 1.001 = 0 \qquad Q(x) = 0.003x^2 - 0.124x + 1.001$$

$$x = \frac{0.124 \pm \sqrt{(-0.124)^2 - 4(0.003)(1.001)}}{2(0.003)} \qquad \text{Quadratic formula}$$

$$x = 11 \text{ or } x = 30.\overline{3} \qquad \text{Use a calculator.}$$

Because x needs to be between 5 and 25, this model tells us that in 1981 ($x = 11$), the petroleum consumption was also 34 quads.   ■ ■ ■

Practice Problem 5   The value of $C(x) = 0.23x^3 - 4.255x^2 + 0.345x + 41.05$ is 10 when $x = 3$. Find another positive number x for which $C(x) = 10$.   ■

3 Use the Rational Zeros Test.

The Rational Zeros Test

We use the Factor Theorem to determine whether a number a is a zero of a polynomial function $F(x)$. We use the following test (see Exercise 81 for a proof) to find *possible* rational zeros of a polynomial function with integer coefficients.

THE RATIONAL ZEROS TEST

If $F(x) = a_n x^n + a_{n-1} x^{n-1} + \cdots + a_2 x^2 + a_1 x + a_0$ is a polynomial function with integer coefficients $(a_n \neq 0, a_0 \neq 0)$ and $\dfrac{p}{q}$ is a rational number in lowest terms that is a zero of $F(x)$, then

1. p is a factor of the constant term a_0.
2. q is a factor of the leading coefficient a_n.

EXAMPLE 6   **Using the Rational Zeros Test**

Find all rational zeros of $F(x) = 2x^3 + 5x^2 - 4x - 3$.

SOLUTION

First, we list all *possible* rational zeros of $F(x)$.

$$\text{Possible rational zeros } = \frac{p}{q} = \frac{\text{Factors of the constant term, } -3}{\text{Factors of the leading coefficient, } 2}$$

Factors of -3: $\pm 1, \pm 3$

Factors of 2: $\pm 1, \pm 2$

All combinations can be found by using just the *positive* factors of the leading coefficient. The possible rational zeros are

$$\frac{\pm 1}{1}, \frac{\pm 1}{2}, \frac{\pm 3}{1}, \frac{\pm 3}{2}$$

or

$$\pm 1, \pm \frac{1}{2}, \pm \frac{3}{2}, \pm 3.$$

We use synthetic division to determine whether a rational zero exists among these eight candidates. We start by testing 1. If 1 is not a rational zero, we will test other possibilities.

$$
\begin{array}{r|rrrr}
1 & 2 & 5 & -4 & -3 \\
  &   & 2 & 7 & 3 \\
\hline
  & 2 & 7 & 3 & 0
\end{array}
$$

The zero remainder tells us that $(x - 1)$ is a factor of $F(x)$, with the other factor being $2x^2 + 7x + 3$. So

$$F(x) = (x - 1)(2x^2 + 7x + 3)$$
$$= (x - 1)(2x + 1)(x + 3) \quad \text{Factor } 2x^2 + 7x + 3.$$

We find the zeros of $F(x)$ by solving the equation $(x - 1)(2x + 1)(x + 3) = 0$.

$$x - 1 = 0 \quad \text{or} \quad 2x + 1 = 0 \quad \text{or} \quad x + 3 = 0 \quad \text{Zero-product property}$$
$$x = 1 \quad \text{or} \quad x = -\frac{1}{2} \quad \text{or} \quad x = -3 \quad \text{Solve each equation.}$$

The solution set is $\left\{1, -\frac{1}{2}, -3\right\}$. The rational zeros of F are $-3, -\frac{1}{2}$, and 1. ■ ■ ■

Practice Problem 6   Find all rational zeros of $F(x) = 2x^3 + 3x^2 - 6x - 8$. ■

SECTION 2.3 ■ Exercises

A EXERCISES Basic Skills and Concepts

1. In the division $\dfrac{x^4 - 2x^3 + 5x^2 - 2x + 1}{x^2 - 2x + 3} = x^2 + 2 +$

   $\dfrac{2x - 5}{x^2 - 2x + 3}$, the dividend is _____, the divisor

   is _____, the quotient is _____, and the remainder is _____.

2. If $P(x), Q(x)$, and $F(x)$ are polynomials and $F(x) = P(x) \cdot Q(x)$, then the factors of $F(x)$ are $\underset{P(x)}{\underline{\quad\quad}}$ and $\underset{Q(x)}{\underline{\quad\quad}}$.

3. The Remainder Theorem sates that if a polynomial $F(x)$ is divided by $(x - a)$, then the remainder $R = \underline{\quad F(a) \quad}$.

4. The Factor Theorem states that $(x - a)$ is a factor of a polynomial $F(x)$ if and only if $\underline{\quad F(a) \quad} = 0$.

5. *True* or *False* The only possible rational zeros of $P(x) = 3x^4 - 5x^2 + 2x + 1$ are $\pm1, \pm3$. False

6. *True* or *False* The only possible rational zeros of $P(x) = x^3 - 5x^2 + 3x - 7$ are ±7. False

In Exercises 7–16, use synthetic division to find the quotient and the remainder when the first polynomial is divided by the second polynomial.

7. $x^3 - x^2 - 7x + 2; x - 1$

8. $2x^3 - 3x^2 - x + 2; x + 2$

9. $x^3 + 4x^2 - 7x - 10; x + 2$

10. $x^3 + x^2 - 13x + 2; x - 3$

11. $x^4 - 3x^3 + 2x^2 + 4x + 5; x - 2$

12. $x^4 - 5x^3 - 3x^2 + 10; x - 1$

13. $2x^3 + 4x^2 - 3x + 1; x - \dfrac{1}{2}$

14. $3x^3 + 8x^2 + x + 1; x - \dfrac{1}{3}$

15. $2x^3 - 5x^2 + 3x + 2; x + \dfrac{1}{2}$

16. $3x^3 - 2x^2 + 8x + 2; x + \dfrac{1}{3}$ The quotient is $3x^2 - 3x + 9$, and the remainder is -1.

In Exercises 17–20, use synthetic division and the Remainder Theorem to find each function value. Check your answer by evaluating the function at the given x-value.

17. $f(x) = x^3 + 3x^2 + 1$
   a. $f(1)$ 5
   b. $f(-1)$ 3
   c. $f\left(\dfrac{1}{2}\right)$ $\dfrac{15}{8}$
   d. $f(10)$ 1301

18. $g(x) = 2x^3 - 3x^2 + 1$
   a. $g(-2)$ −27
   b. $g(-1)$ −4
   c. $g\left(-\dfrac{1}{2}\right)$ 0
   d. $g(7)$ 540

19. $h(x) = x^4 + 5x^3 - 3x^2 - 20$
   a. $h(1)$ −17
   b. $h(-1)$ −27
   c. $h(-2)$ −56
   d. $h(2)$ 24

20. $f(x) = x^4 + 0.5x^3 - 0.3x^2 - 20$
   a. $f(0.1)$ −20.0024
   b. $f(0.5)$ −19.95
   c. $f(1.7)$ −10.0584
   d. $f(-2.3)$ 0.3136

†Due to space constrictions, answers to these exercises may be found in the Answers beginning on page A–1 in the back of the book.
Answers:
1. $x^4 - 2x^3 + 5x^2 - 2x + 1$; $x^2 - 2x + 3$; $x^2 + 2$; $2x - 5$ 7. The quotient is $x^2 - 7$, and the remainder is -5.   8. The quotient is $2x^2 - 7x + 13$, and the remainder is -24.   9. The quotient is $x^2 + 2x - 11$, and the remainder is 12.   10. The quotient is $x^2 + 4x - 1$, and the remainder is -1.
11. The quotient is $x^3 - x^2 + 4$, and the remainder is 13.   12. The quotient is $x^3 - 4x^2 - 7x - 7$, and the remainder is 3.   13. The quotient is $2x^2 + 5x - \frac{1}{2}$, and the remainder is $\frac{3}{4}$.   14. The quotient is $3x^2 + 9x + 4$, and the remainder is $\frac{7}{3}$.   15. The quotient is $2x^2 - 6x + 6$, and the remainder is -1.

In Exercises 21–28, use the Factor Theorem to show that the linear polynomial is a factor of the second polynomial. Check your answer by synthetic division.

21. $x - 1; 2x^3 + 3x^2 - 6x + 1$

22. $x - 3; 3x^3 - 9x^2 - 4x + 12$

23. $x + 1; 5x^4 + 8x^3 + x^2 + 2x + 4$

24. $x + 3; 3x^4 + 9x^3 - 4x^2 - 9x + 9$

25. $x - 2; x^4 + x^3 - x^2 - x - 18$

26. $x + 3; x^5 + 3x^4 + x^2 + 8x + 15$

27. $x + 2; x^6 - x^5 - 7x^4 + x^3 + 8x^2 + 5x + 2$

28. $x - 2; 2x^6 - 5x^5 + 4x^4 + x^3 - 7x^2 - 7x + 2$

In Exercises 29–32, find the value of k for which the first polynomial is a factor of the second polynomial.

29. $x + 1; x^3 + 3x^2 + x + k$ [*Hint:* Let $f(x) = x^3 + 3x^2 + x + k$. Then $f(-1) = 0$.] $k = -1$

30. $x - 1; -x^3 + 4x^2 + kx - 2$ $k = -1$

31. $x - 2; 2x^3 + kx^2 - kx - 2$ $k = -7$

32. $x - 1; k^2 - 3kx^2 - 2kx + 6$ $k = 2$ or 3

In Exercises 33–36, use the Factor Theorem to show that the first polynomial *is not* a factor of the second polynomial. [*Hint:* Show that the remainder is not zero.]

33. $x - 2; -2x^3 + 4x^2 - 4x + 9$ Remainder: 1

34. $x + 3; -3x^3 - 9x^2 + 5x + 12$ Remainder: -3

35. $x + 2; 4x^4 + 9x^3 + 3x^2 + x + 4$ Remainder: 6

36. $x - 3; 3x^4 - 8x^3 + 5x^2 + 7x - 3$ Remainder: 90

In Exercises 37–40, find the set of possible rational zeros of the given function.

37. $f(x) = 3x^3 - 4x^2 + 5$ $\{\pm\frac{1}{3}, \pm 1, \pm\frac{5}{3}, \pm 5\}$

38. $g(x) = 2x^4 - 5x^2 - 2x + 1$ $\{\pm\frac{1}{2}, \pm 1\}$

39. $h(x) = 4x^4 - 9x^2 + x + 6$ $\{\pm\frac{1}{4}, \pm\frac{1}{2}, \pm\frac{3}{4}, \pm 1, \pm\frac{3}{2}, \pm 2, \pm 3, \pm 6\}$

40. $F(x) = 6x^6 + 5x^5 + x - 35$

In Exercises 41–56, find all rational zeros of the given polynomial function.

41. $f(x) = x^3 - x^2 - 4x + 4$ $\{-2, 1, 2\}$

42. $f(x) = x^3 + x^2 + 2x + 2$ $\{-1\}$

43. $f(x) = x^3 - 4x^2 + x + 6$ $\{-1, 2, 3\}$

44. $f(x) = x^3 + 3x^2 + 2x + 6$ $\{-3\}$

45. $g(x) = 2x^3 + x^2 - 13x + 6$ $\{-3, \frac{1}{2}, 2\}$

46. $g(x) = 3x^3 - 2x^2 + 3x - 2$ $\{\frac{2}{3}\}$

47. $g(x) = 6x^3 + 13x^2 + x - 2$ $\{-2, -\frac{1}{2}, \frac{1}{3}\}$

48. $g(x) = 2x^3 + 3x^2 + 4x + 6$ $\{-\frac{3}{2}\}$

49. $h(x) = 3x^3 + 7x^2 + 8x + 2$ $\{-\frac{1}{3}\}$

50. $h(x) = 2x^3 + x^2 + 8x + 4$ $\{-\frac{1}{2}\}$

51. $h(x) = x^4 - x^3 - x^2 - x - 2$ $\{-1, 2\}$

52. $h(x) = 2x^4 + 3x^3 + 8x^2 + 3x - 4$ $\{-1, \frac{1}{2}\}$

53. $F(x) = x^4 - x^3 - 13x^2 + x + 12$ $\{-3, -1, 1, 4\}$

Answers:

40. $\left\{ \begin{matrix} \pm\frac{1}{6}, \pm\frac{1}{2}, \pm 1, \pm\frac{5}{6}, \pm\frac{5}{2}, \pm 5, \pm 7, \pm\frac{35}{2}, \\ \pm\frac{1}{3}, \pm\frac{5}{6}, \pm\frac{7}{6}, \pm\frac{7}{3}, \pm\frac{7}{2}, \pm\frac{35}{6}, \pm\frac{35}{3}, \pm 35 \end{matrix} \right\}$

54. $G(x) = 3x^4 + 5x^3 + x^2 + 5x - 2$ $\{-2, \frac{1}{3}\}$

55. $F(x) = x^4 - 2x^3 + 10x^2 - x + 1$ No rational zeros

56. $G(x) = x^6 + 2x^4 + x^2 + 2$ No rational zeros

In Exercises 57–60, first find all rational zeros of f, then use the depressed equation (see Example 4) to find all roots of the equation $f(x) = 0$.

57. $f(x) = x^3 + 5x^2 - 8x + 2$ $\{1, -3 \pm \sqrt{11}\}$

58. $f(x) = x^3 - 7x^2 - 5x + 3$ $\{-1, 4 \pm \sqrt{13}\}$

59. $f(x) = x^4 - 3x^3 + 3x - 1$

60. $f(x) = x^4 - 6x^3 - 7x^2 + 54x - 18$ $\{3, -3, 3 \pm \sqrt{7}\}$

B EXERCISES Applying the Concepts

61. Geometry. The area of a rectangle is $(2x^4 - 2x^3 + 5x^2 - x + 2)$ square centimeters. Its length is $(x^2 - x + 2)$ cm. Find its width. $2x^2 + 1$

62. Geometry. The volume of a rectangular solid is $(x^4 + 3x^3 + x + 3)$ cubic inches. Its length and width are $(x + 3)$ and $(x + 1)$ inches, respectively. Find its height. $x^2 - x + 1$

63. Geometry. Find a point on the parabola $y = x^2$ whose distance from $(18, 0)$ is $4\sqrt{17}$. $(2, 4)$

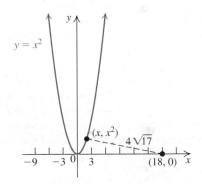

64. Making a box. A square piece of tin 18 inches on each side is to be made into a box, without a top, by cutting a square from each corner and folding up the flaps to form the sides. What size corners should be cut so that the volume of the box is 432 cubic inches? $x = 3$ in.

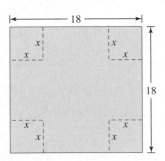

65. Cost function. The total cost of a product (in dollars) is given by $C(x) = 3x^3 - 6x^2 + 108x + 100$, where x is the demand for the product. Find the value of x that gives a total cost of $\$628$. $x = 4$

59. $\left\{ 1, -1, \dfrac{3 \pm \sqrt{5}}{2} \right\}$

66. Profit. For the product in Exercise 65, the demand function is $p(x) = 330 + 10x - x^2$. Find the value of x that gives the profit of $910. $x = 5$

67. Production cost. The total cost per month, in thousands of dollars, of producing x printers (with x measured in hundreds) is given by $C(x) = x^3 - 15x^2 + 5x + 50$. How many units must be produced so that the total monthly cost is $125,000? 1500

68. Break even. If the demand function for the printers in Exercise 67 is $p(x) = -3x + 3 + \dfrac{74}{x}$, find the number of printers that must be produced and sold to break even. 1200

69. Total energy consumption. The total energy consumption (in quads) in the United States from 1970 to 1995 is approximated by the function

$$C(x) = 0.0439x^3 - 1.4911x^2 + 14.224x + 41.76,$$

where $x = 0$ represents 1970. The model indicates that 81 quads of energy were consumed in 1979. Find another year after 1979 when the model indicates that 81 quads of energy were consumed. (The model was modified from the data obtained from the *Statistical Abstracts of the United States*.) 1990

70. Lottery sales. In the lottery game called *Lotto*, players typically select six digits from a large field of numbers. Varying prizes are offered for matching three through six numbers drawn by the lottery. Lotto sales (in billions of dollars) in the United States from 1983 through 2003 are approximated by the function

$$s(x) = 0.0039x^3 - 0.2095x^2 + 3.2201x - 2.7505,$$

where $x = 0$ represents 1980. The model indicates that $8.6 billion in Lotto sales occurred in 1985. Find another year after 1985 when the model indicates that 8.6 billion in Lotto sales occurred. (The model was modified from the data obtained from the *Statistical Abstracts of the United States*.) 2001

C EXERCISES Beyond the Basics

71. Show that $\dfrac{1}{2}$ is a root of multiplicity 2 of the equation

$$4x^3 + 8x^2 - 11x + 3 = 0.$$

72. Show that $-\dfrac{2}{3}$ is a root of multiplicity 2 of the equation

$$9x^3 + 3x^2 - 8x - 4 = 0.$$

73. If n is a positive integer, find the values of n for which each of the following is true.
 a. $x + a$ is a factor of $x^n + a^n$. *n is an odd integer.*
 b. $x + a$ is a factor of $x^n - a^n$. *n is an even integer.*
 c. $x - a$ is a factor of $x^n + a^n$. *There is no such n.*
 d. $x - a$ is a factor of $x^n - a^n$. *n is any positive integer.*

74. Use Exercise 73 to prove the following:
 a. $7^{11} - 2^{11}$ is exactly divisible by 5.
 b. $(19)^{20} - (10)^{20}$ is exactly divisible by 261.

In Exercises 75–78, show that the given number is irrational. [*Hint:* To show that $x = 1 + \sqrt{3}$ is irrational, form the equivalent equations $x - 1 = \sqrt{3}$, $(x - 1)^2 = 3$, and $x^2 - 2x - 2 = 0$ and show that the last equation has no rational roots.]

75. $\sqrt{3}$ **76.** $\sqrt[3]{4}$

77. $3 - \sqrt{2}$ **78.** $(9)^{\frac{2}{3}}$

In Exercises 79 and 80, find a polynomial equation of least possible degree with integer coefficients whose roots include the given numbers.

79. $1, -1, 0$ $x^3 - x = 0$ **80.** $2, 1, -2$
                                          $x^3 - x^2 - 4x + 4 = 0$

81. To prove the rational zeros test, we assume the **Fundamental Theorem of Arithmetic**, which states that "Every integer has a unique prime factorization." Supply reasons for each step in the following proof:

  (i) $F\left(\dfrac{p}{q}\right) = 0$ and $\dfrac{p}{q}$ is in lowest terms.

  (ii) $a_n\left(\dfrac{p}{q}\right)^n + a_{n-1}\left(\dfrac{p}{q}\right)^{n-1} + \cdots + a_1\left(\dfrac{p}{q}\right) + a_0 = 0$

 (iii) $a_n p^n + a_{n-1} p^{n-1} q + \cdots + a_1 p q^{n-1} + a_0 q^n = 0$

 (iv) $a_n p^n + a_{n-1} p^{n-1} q + \cdots + a_1 p q^{n-1} = -a_0 q^n$

  (v) p is a factor of the left side of (iv).

 (vi) p is a factor of $-a_0 q^n$.

 (vii) p is a factor of a_0.

(viii) $a_{n-1} p^{n-1} q + \cdots + a_1 p q^{n-1} + a_0 q^n = -a_n p^n$

 (ix) q is a factor of the left side of the equation in (viii).

  (x) q is a factor of $-a_n p^n$.

 (xi) q is a factor of a_n.

82. Find all rational roots of the equation

$$2x^4 + 2x^3 + \frac{1}{2}x^2 + 2x - \frac{3}{2} = 0. \quad \left\{-\frac{3}{2}, \frac{1}{2}\right\}$$

[*Hint:* Multiply both sides of the equation by the lowest common denominator.]

83. Synthetic division is a process of dividing a polynomial $F(x)$ by $x - a$. Modify the process to use synthetic division so that it finds the remainder when the first polynomial is divided by the second.
 a. $2x^3 + 3x^2 + 6x - 2$; $2x - 1$
    $\left[Hint: 2x - 1 = 2\left(x - \dfrac{1}{2}\right).\right]$ Remainder: 2
 b. $2x^3 - x^2 - 4x + 1$; $2x + 3$ Remainder: -2

84. Let $g(x) = 3x^3 - 5cx^2 - 3c^2x + 3c^3$. Use synthetic division to find each function value.
 a. $g(-c)$ $-2c^3$ **b.** $g(2c)$ c^3

Critical Thinking

85. True or False.
 a. $\dfrac{1}{3}$ is a possible zero of $P(x) = x^5 - 3x^2 + x + 3$.
                                        False
 b. $\dfrac{2}{5}$ is a possible zero of $P(x) = 2x^7 - 5x^4 - 17x + 25$.
                                        False

Zeros of a Polynomial Function

Before Starting this Section, Review

1. Long division (Appendix A, page 765)

2. Synthetic division (Appendix A, page 767)

3. Rational zeros test (Section 2.3, page 141)

4. Complex numbers (Appendix A, page 803)

5. Conjugate of a complex number (Appendix A, page 805)

Objectives

1. Find the possible number of positive and negative zeros of polynomials.

2. Find the bounds on the real zeros of polynomials.

3. Learn basic facts about the complex zeros of polynomials.

4. Use the Conjugate Pairs Theorem to find zeros of polynomials.

Gerolamo Cardano (1501–1576)

Cardano was trained as a physician, but his range of interests included philosophy, music, gambling, mathematics, astrology, and the occult sciences. Cardano was a colorful man, full of contradictions. His lifestyle was deplorable even by the standards of the times, and his personal life was filled with tragedies. His wife died in 1546. He saw one son executed for wife poisoning. He cropped the ears of a second son who attempted the same crime. In 1570, he was imprisoned for six months for heresy when he published the horoscope of Jesus. Cardano spent the last few years in Rome, where he wrote his autobiography, *De Propria Vita,* in which he did not spare the most shameful revelations.

CARDANO ACCUSED OF STEALING FORMULA

In sixteenth-century Italy, getting and keeping a teaching job at a university was difficult. University appointments were mostly temporary. One way a professor could convince the administration that he was worthy of continuing in his position was by winning public challenges. Two contenders for a position would present each other with a list of problems, and later in a public forum, each person would present his solutions to the other's problems. As a result, scholars often kept secret their new methods of solving problems.

In 1535, in a public contest primarily about solving the cubic equation, Nicolo Tartaglia (1499–1557) won over his opponent Antonio Maria Fiore. Gerolamo Cardano (1501–1576), in the hope of learning the secret of solving a cubic equation, invited Tartaglia to visit him. After many promises and much flattery, Tartaglia revealed his method on the promise, probably given under oath, that Cardano would keep it confidential. When Cardano's book *Ars Magna* (The Great Art) appeared in 1545, the formula and the method of proof were fully disclosed. Angered at this apparent breach of a solemn oath and feeling cheated out of the rewards of his monumental work, Tartaglia accused Cardano of being a liar and a thief for stealing his formula. Thus began the bitterest feud in the history of mathematics, carried on with name-calling and mudslinging of the lowest order. As Tartaglia feared, the formula (see Group Project 1) has forever since been known as Cardano's formula for the solution of the cubic equation. ▨

---

1 Find the possible number of positive and negative zeros of polynomials.

Descartes's Rule of Signs

In Section 2.3, we used the Rational Zeros Test to find the possible rational zeros of a polynomial function with integer coefficients. We now investigate briefly some other useful tests for the real zeros of polynomial functions. One such test is called *Descartes's Rule of Signs.*

If the terms of a polynomial are written in descending order, we say that a **variation of sign** occurs when the signs of two consecutive terms differ. For example, in the polynomial $2x^5 - 5x^3 - 6x^2 + 7x + 3$, the signs of the terms are $+\ -\ -\ +\ +$. Therefore, there are two variations of sign, as follows:

$$+\quad -\quad -\quad +\quad +$$

There are three variations of sign in

$$x^5 - 3x^3 + 2x^2 + 5x - 7.$$

DESCARTES'S RULE OF SIGNS

Let $F(x)$ be a polynomial function with real coefficients and with terms written in descending order.

1. The number of positive zeros of F is equal to the number of variations of sign of $F(x)$ or is less than that number by an even integer.

2. The number of negative zeros of F is equal to the number of variations of sign of $F(-x)$ or is less than that number by an even integer.

When using Descartes's Rule, a zero of multiplicity m should be counted as m zeros.

EXAMPLE 1 **Using Descartes's Rule of Signs**

Find the possible number of positive and negative zeros of

$$f(x) = x^5 - 3x^3 - 5x^2 + 9x - 7.$$

SOLUTION

There are three variations of sign in $f(x)$:

$$f(x) = x^5 - 3x^3 - 5x^2 + 9x - 7$$

The number of positive zeros of $f(x)$ must be either 3 or $3 - 2 = 1$. Furthermore,

$$f(-x) = (-x)^5 - 3(-x)^3 - 5(-x)^2 + 9(-x) - 7$$
$$= -x^5 + 3x^3 - 5x^2 - 9x - 7.$$

The polynomial $f(-x)$ has two variations of sign; therefore, the number of negative zeros of $f(x)$ is either 2 or $2 - 2 = 0$. ■ ■ ■

Practice Problem 1 Find the possible number of positive and negative zeros of

$$f(x) = 2x^5 + 3x^2 + 5x - 1.$$ ■

2 Find the bounds on the real zeros of polynomials.

Bounds on the Real Zeros

If a polynomial function $F(x)$ has no zeros greater than a number k, then k is called an **upper bound** on the zeros of $F(x)$. Similarly, if $F(x)$ has no zeros less than a number k, then k is called a **lower bound** on the zeros of $F(x)$. These bounds have the following rules.

RULES FOR BOUNDS

Let $F(x)$ be a polynomial function with real coefficients and a positive leading coefficient. Suppose $F(x)$ is synthetically divided by $x - k$.

1. If $k > 0$ and each number in the last row is zero or positive, then k is an upper bound on the zeros of $F(x)$.

2. If $k < 0$ and numbers in the last row alternate in sign, then k is a lower bound on the zeros of $F(x)$.

Zeros in the last row can be regarded as positive or negative.

| EXAMPLE 2 | Finding the Bounds on the Zeros |

Find upper and lower bounds on the zeros of

$$F(x) = x^4 - x^3 - 5x^2 - x - 6.$$

SOLUTION

By the Rational Zeros Test, the possible rational zeros of $F(x)$ are $\pm 1, \pm 2, \pm 3$, and ± 6. There is only one variation of sign in $F(x) = x^4 - x^3 - 5x^2 - x - 6$, so $F(x)$ has only one positive zero. First, we try synthetic division by $x - k$ with $k = 1, 2, 3, \ldots$. The first integer that makes all numbers in the last row 0 or positive is an upper bound on the zeros of $F(x)$.

$$
\begin{array}{r|rrrrr}
1 & 1 & -1 & -5 & -1 & -6 \\
  &   & 1 & 0 & -5 & -6 \\
\hline
  & 1 & 0 & -5 & -6 & -12
\end{array}
$$ Synthetic division with $k = 1$

$$
\begin{array}{r|rrrrr}
2 & 1 & -1 & -5 & -1 & -6 \\
  &   & 2 & 2 & -6 & -14 \\
\hline
  & 1 & 1 & -3 & -7 & -20
\end{array}
$$ Synthetic division with $k = 2$

$$
\begin{array}{r|rrrrr}
3 & 1 & -1 & -5 & -1 & -6 \\
  &   & 3 & 6 & 3 & 6 \\
\hline
  & 1 & 2 & 1 & 2 & 0
\end{array}
$$ Synthetic division with $k = 3$

←—— All numbers are 0 or positive.

By the rule for bounds, 3 is an upper bound on the zeros of $F(x)$.

We now try synthetic division by $x - k$ with $k = -1, -2, -3, \ldots$. The first negative integer for which the numbers in the last row alternate in sign is a lower bound on the zeros of $F(x)$.

$$
\begin{array}{r|rrrrr}
-1 & 1 & -1 & -5 & -1 & -6 \\
   &   & -1 & 2 & 3 & -2 \\
\hline
   & 1 & -2 & -3 & 2 & -8
\end{array}
$$ Synthetic division with $k = -1$

$$
\begin{array}{r|rrrrr}
-2 & 1 & -1 & -5 & -1 & -6 \\
   &   & -2 & 6 & -2 & 6 \\
\hline
   & 1 & -3 & 1 & -3 & 0
\end{array}
$$ Synthetic division with $k = -2$

These numbers alternate in sign ←—— if we count 0 as positive.

By the rule of bounds, -2 is a lower bound on the zeros of $F(x)$.   ■ ■ ■

Practice Problem 2   Find upper and lower bounds on the zeros of

$$f(x) = 2x^3 + 5x^2 + x - 2.$$ ■

3 Learn basic facts about the complex zeros of polynomials.

The Factor Theorem connects the concept of *factors* and *zeros* of a polynomial, and we continue to investigate this connection.

Because the equation $x^2 + 1 = 0$ has no real roots, the polynomial function $P(x) = x^2 + 1$ has no real zeros. However, if we replace x with the complex number i, we have

$$P(i) = i^2 + 1 = (-1) + 1 = 0.$$

Consequently, i is a complex number for which $P(i) = 0$; that is, i is a *complex zero* of $P(x)$. In Section 2.2, we stated that a polynomial of degree n can have, at most, n real zeros. We now extend our number system to allow the coefficients of polynomials and variables to represent complex numbers. When we want to emphasize that complex numbers may be used in this way, we call the polynomial $P(x)$ a **complex polynomial**. If $P(z) = 0$ for a complex number z, we say that z is a **zero** or a **complex zero** of $P(x)$. In the complex number system, every nth-degree polynomial equation

RECALL

A number z of the form $z = a + bi$ is a **complex number,** where a and b are real numbers and $i = \sqrt{-1}$.

has *exactly* n roots and every nth-degree polynomial can be factored into exactly n linear factors. This fact follows from the Fundamental Theorem of Algebra.

FUNDAMENTAL THEOREM OF ALGEBRA

Every polynomial

$$P(x) = a_n x^n + a_{n-1} x^{n-1} + \cdots + a_1 x + a_0 \quad (n \geq 1, a_n \neq 0)$$

with complex coefficients $a_n, a_{n-1}, \ldots, a_1, a_0$ has at least one complex zero.

RECALL

If $b = 0$, the complex number $a + bi$ is a real number. So the set of real numbers is a subset of the set of complex numbers.

This theorem was proved by the mathematician Karl Friedrich Gauss at age 20. The proof is beyond the scope of this book, but if we are allowed to use complex numbers, we can use the theorem to prove that every polynomial has a complete factorization.

FACTORIZATION THEOREM FOR POLYNOMIALS

If $P(x)$ is a complex polynomial of degree $n \geq 1$, it can be factored into n (not necessarily distinct) linear factors of the form

$$P(x) = a(x - r_1)(x - r_2) \cdots (x - r_n),$$

where $a, r_1, r_2, \ldots, r_n$ are complex numbers.

To prove the Factorization Theorem for Polynomials, we notice that if $P(x)$ is a polynomial of degree 1 or higher with leading coefficient a, then the Fundamental Theorem of Algebra states that there is a zero r_1 for which $P(r_1) = 0$. By the Factor Theorem, $(x - r_1)$ is a factor of $P(x)$ and

$$P(x) = (x - r_1)q_1(x),$$

where the degree of $q_1(x)$ is $n - 1$ and its leading coefficient is a. If $n - 1$ is positive, then we apply the Fundamental Theorem to $q_1(x)$; this gives us a zero, say, r_2, of $q_1(x)$. By the Factor Theorem, $(x - r_2)$ is a factor of $q_1(x)$ and

$$q_1(x) = (x - r_2)q_2(x),$$

where the degree of $q_1(x)$ is $n - 2$ and its leading coefficient is a. Then

$$P(x) = (x - r_1)(x - r_2)q_2(x).$$

This process can be continued until $P(x)$ is completely factored as

$$P(x) = a(x - r_1)(x - r_2) \cdots (x - r_n),$$

where a is the leading coefficient and $r_1, r_2, \ldots, r_n$ are zeros of $P(x)$. The proof is now complete.

In general, the numbers $r_1, r_2, \ldots, r_n$ may not be distinct. As with the polynomials we encountered previously, if the factor $(x - r)$ appears m times in the complete factorization of $P(x)$, then we say that r is a zero of *multiplicity* m. The polynomial $P(x) = (x - 2)^3(x - i)^2$ has zeros

2 (of multiplicity 3) and i (of multiplicity 2).

The polynomial $P(x) = (x - 2)^3(x - i)^2$ has only 2 and i as its zeros, although it is of degree 5.

EXAMPLE 3   **Constructing a Polynomial Whose Zeros Are Given**

Find a polynomial $P(x)$ of degree 4 with a leading coefficient of 2 and zeros $-1, 3, i$, and $-i$. Write $P(x)$

a. in completely factored form.   **b.** by expanding the product found in part **a.**

SOLUTION

a. Since $P(x)$ has degree 4, we write:

$$P(x) = a(x - r_1)(x - r_2)(x - r_3)(x - r_4). \quad \text{Complete factored form}$$

Now replace the leading coefficient a with 2 and the zeros r_1, r_2, r_3, and r_4 with $-1, 3, i$, and $-i$ (in any order). Then

$$P(x) = 2[x - (-1)](x - 3)(x - i)[x - (-i)]$$
$$= 2(x + 1)(x - 3)(x - i)(x + i). \quad \text{Simplify.}$$

b. $P(x) = 2(x + 1)(x - 3)(x - i)(x + i)$

$\quad\quad\quad = 2(x + 1)(x - 3)(x^2 + 1) \quad\quad \text{Expand } (x - i)(x + i).$

$\quad\quad\quad = 2(x + 1)(x^3 - 3x^2 + x - 3) \quad\quad \text{Expand } (x - 3)(x^2 + 1).$

$\quad\quad\quad = 2(x^4 - 2x^3 - 2x^2 - 2x - 3) \quad\quad \text{Expand}$
$\quad\quad\quad\quad\quad\quad\quad\quad\quad\quad\quad\quad\quad\quad\quad\quad (x + 1)(x^3 - 3x^2 + x - 3).$

$\quad\quad\quad = 2x^4 - 4x^3 - 4x^2 - 4x - 6 \quad\quad \text{Distributive property}$ ■ ■ ■

Practice Problem 3   Find a polynomial $P(x)$ of degree 4 with a leading coefficient of 3 and zeros $-2, 1, 1 + i, 1 - i$. Write $P(x)$

a. in completely factored form.   **b.** by expanding the product found in part **a.**

---

4 Use the Conjugate Pairs Theorem to find zeros of polynomials.

Conjugate Pairs Theorem

In Example 3, we constructed a polynomial that had both i and its conjugate, $-i$, among its zeros. Our next result says that, for all polynomials with real coefficients, nonreal zeros occur in conjugate pairs.

RECALL

The conjugate of a complex number $z = a + bi$ is $\bar{z} = a - bi$. For two complex numbers z_1 and z_2, we have (Appendix A)

$$\overline{z_1 + z_2} = \overline{z_1} + \overline{z_2}$$
$$\overline{z_1 z_2} = \overline{z_1}\,\overline{z_2}.$$

CONJUGATE PAIRS THEOREM

If $P(x)$ is a polynomial function whose coefficients are real numbers and if $z = a + bi$ is a zero of P, then its conjugate, $\bar{z} = a - bi$, is also a zero of P.

To prove the Conjugate Pairs Theorem, let

$$P(x) = a_n x^n + a_{n-1} x^{n-1} + \cdots + a_1 x + a_0,$$

where $a_n, a_{n-1}, \ldots, a_1, a_0$ are real numbers. Now suppose that $P(z) = 0$. Then we must show that $P(\bar{z}) = 0$ also. We will use the fact that the conjugate of a real number $a = a + 0i$ is identical to the real number a (since $\bar{a} = a - 0i = a$) and then use the conjugate sum and product properties $\overline{z_1 + z_2} = \overline{z_1} + \overline{z_2}$, and $\overline{z_1 z_2} = \overline{z_1}\,\overline{z_2}$.

$$P(z) = a_n z^n + a_{n-1} z^{n-1} + \cdots + a_1 z + a_0 \quad\quad \text{Replace } x \text{ with } z.$$

$$P(\bar{z}) = a_n (\bar{z})^n + a_{n-1}(\bar{z})^{n-1} + \cdots + a_1 \bar{z} + a_0 \quad\quad \text{Replace } z \text{ with } \bar{z}.$$

$$= \bar{a}_n \overline{z^n} + \bar{a}_{n-1} \overline{z^{n-1}} + \cdots + \bar{a}_1 \bar{z} + \bar{a}_0 \quad\quad (\bar{z})^k = \overline{z^k}, \overline{a_k} = a_k$$

$$= \overline{a_n z^n} + \overline{a_{n-1} z^{n-1}} + \cdots + \overline{a_1 z} + \overline{a_0} \quad\quad \overline{z_1}\,\overline{z_2} = \overline{z_1 z_2}$$

$$= \overline{a_n z^n + a_{n-1} z^{n-1} + \cdots + a_1 z + a_0} \quad\quad \overline{z_1} + \overline{z_2} = \overline{z_1 + z_2}$$

$$= \overline{P(z)} = \overline{0} = 0 \quad\quad P(z) = 0$$

We see that if $P(z) = 0$, then $P(\bar{z}) = 0$, and the theorem is proved.

This theorem has several uses. If, for example, we know that $2 - 5i$ is a zero of a polynomial with real coefficients, then we know that $2 + 5i$ is also a zero. Because nonreal zeros occur in conjugate pairs, we know that there will always be an *even* number of nonreal zeros. Therefore, any polynomial of odd degree with real coefficients has at least one zero that is a real number.

ODD-DEGREE POLYNOMIALS WITH REAL ZEROS

Any polynomial $P(x)$ of odd degree with real coefficients must have at least one real zero.

For example, the polynomial $P(x) = 5x^7 + 2x^4 + 9x - 12$ has at least one zero that is a real number because $P(x)$ has degree 7 (odd degree) and has real numbers as coefficients.

STUDY TIP

In problems like Example 4, once you determine all the zeros, you can use the Factorization Theorem to write a polynomial with those zeros.

EXAMPLE 4   Using the Conjugate Pairs Theorem

A polynomial $P(x)$ of degree 9 with real coefficients has the following zeros: 2, of multiplicity 3; $4 + 5i$, of multiplicity 2; and $3 - 7i$. Write all nine zeros of $P(x)$.

SOLUTION

Since complex zeros occur in conjugate pairs, the conjugate $4 - 5i$ of $4 + 5i$ is a zero of multiplicity 2 and the conjugate $3 + 7i$ of $3 - 7i$ is a zero of $P(x)$. The nine zeros of $P(x)$ are

$$2, 2, 2, 4 + 5i, 4 + 5i, 4 - 5i, 4 - 5i, 3 + 7i, \text{ and } 3 - 7i. \qquad \blacksquare \blacksquare \blacksquare$$

Practice Problem 4   A polynomial $P(x)$ of degree 8 with real coefficients has the following zeros: -3 and $2 - 3i$, each of multiplicity 2, and i. Write all eight zeros of $P(x)$. $\qquad \blacksquare$

Every polynomial of degree n has exactly n zeros and can be factored into a product of n linear factors. If the polynomial has real coefficients, then, by the Conjugate Pairs Theorem, its nonreal zeros occur as conjugate pairs. Consequently, if $r = a + bi$ is a zero, then so is $\bar{r} = a - bi$, and both $x - r$ and $x - \bar{r}$ are linear factors of the polynomial. Multiplying these factors, we have:

$$
\begin{aligned}
(x - r)(x - \bar{r}) &= x^2 - (r + \bar{r})x + r\bar{r} && \text{Use FOIL.} \\
&= x^2 - 2ax + (a^2 + b^2) && r + \bar{r} = 2a; r\bar{r} = a^2 + b^2
\end{aligned}
$$

The quadratic polynomial $x^2 - 2ax + (a^2 + b^2)$ has real coefficients and is *irreducible* (cannot be factored any further) *over the real numbers*. This result shows that each pair of nonreal conjugate zeros can be combined into one *irreducible* quadratic factor.

**FACTORIZATION THEOREM FOR A POLYNOMIAL
WITH REAL COEFFICIENTS**

Every polynomial with real coefficients can be uniquely factored over the real numbers as a product of linear factors and/or irreducible quadratic factors.

EXAMPLE 5   Finding the Complex Zeros of a Polynomial

Given that $2 - i$ is a zero of $P(x) = x^4 - 6x^3 + 14x^2 - 14x + 5$, find the remaining zeros.

SOLUTION

Since $P(x)$ has real coefficients, the conjugate $\overline{2 - i} = 2 + i$, is also a zero. By the Factorization Theorem, the linear factors $[x - (2 - i)]$ and $[x - (2 + i)]$ appear in the factorization of $P(x)$. Consequently, their product

$$
\begin{aligned}
[x - (2 - i)][x - (2 + i)] &= (x - 2 + i)(x - 2 - i) \\
&= [(x - 2) + i][(x - 2) - i] \quad \text{Regroup.} \\
&= (x - 2)^2 - i^2 \quad (A + B)(A - B) = A^2 - B^2 \\
&= x^2 - 4x + 4 + 1 \quad \text{Expand } (x - 2)^2, i^2 = -1. \\
&= x^2 - 4x + 5 \quad \text{Simplify.}
\end{aligned}
$$

is also a factor of $P(x)$. We divide $P(x)$ by $x^2 - 4x + 5$ to find the other factor.

RECALL

Dividend =
Quotient · Divisor + Remainder

$$
\begin{array}{r}
x^2 - 2x + 1 \quad \leftarrow \text{Quotient} \\
\text{Divisor} \rightarrow x^2 - 4x + 5 \overline{)\, x^4 - 6x^3 + 14x^2 - 14x + 5} \quad \leftarrow \text{Dividend} \\
\underline{x^4 - 4x^3 + 5x^2} \\
-2x^3 + 9x^2 - 14x \\
\underline{-2x^3 + 8x^2 - 10x} \\
x^2 - 4x + 5 \\
\underline{x^2 - 4x + 5} \\
0 \quad \leftarrow \text{Remainder}
\end{array}
$$

$$
\begin{aligned}
P(x) &= (x^2 - 2x + 1)(x^2 - 4x + 5) \quad P(x) = \text{Quotient} \cdot \text{Divisor} \\
&= (x - 1)(x - 1)(x^2 - 4x + 5) \quad \text{Factor } x^2 - 2x + 1. \\
&= (x - 1)(x - 1)[x - (2 - i)][x - (2 + i)] \quad \text{Factor } x^2 - 4x + 5.
\end{aligned}
$$

The zeros of $P(x)$ are 1 (of multiplicity 2), $2 - i$, and $2 + i$. ■ ■ ■

Practice Problem 5 Given that $2i$ is a zero of $P(x) = x^4 - 3x^3 + 6x^2 - 12x + 8$, find the remaining zeros. ■

EXAMPLE 6 **Finding the Zeros of a Polynomial**

Find all zeros of the polynomial $P(x) = x^4 - x^3 + 7x^2 - 9x - 18$.

SOLUTION

Since the degree of $P(x)$ is 4, $P(x)$ has four zeros. The Rational Zeros Test tells us that the possible rational zeros are

$$\pm 1, \pm 2, \pm 3, \pm 6, \pm 9, \pm 18.$$

Testing these possible zeros by synthetic division, we find that 2 is a zero.

$$
\begin{array}{r}
2 | \;\; 1 \quad -1 \quad 7 \quad -9 \quad -18 \\
\underline{\quad\quad 2 \quad\; 2 \quad 18 \quad\;\; 18} \\
1 \quad\;\; 1 \quad 9 \quad\;\; 9 \quad\;\;\; 0
\end{array}
$$

$P(2) = 0$, so $(x - 2)$ is a factor of $P(x)$ and the quotient is $x^3 + x^2 + 9x + 9$.

We can solve the depressed equation $x^3 + x^2 + 9x + 9 = 0$ by factoring by grouping.

$$
\begin{aligned}
x^2(x + 1) + 9(x + 1) &= 0 \quad \text{Group terms, distributive property} \\
(x^2 + 9)(x + 1) &= 0 \quad \text{Distributive property} \\
x^2 + 9 = 0 \quad \text{or} \quad x + 1 &= 0 \quad \text{Zero-product property} \\
x^2 = -9 \quad \text{or} \quad x &= -1 \quad \text{Solve each equation.} \\
x = \pm\sqrt{-9} = \pm 3i \quad & \quad \text{Solve } x^2 = -9 \text{ for } x.
\end{aligned}
$$

The four zeros of $P(x)$ are $-1, 2, 3i$, and $-3i$. The complete factorization of $P(x)$ is

$$P(x) = x^4 - x^2 + 7x - 9x - 18 = (x + 1)(x - 2)(x - 3i)(x + 3i).$$

■ ■ ■

Practice Problem 6   Find all zeros of the polynomial

$$P(x) = x^4 - 8x^3 + 22x^2 - 28x + 16.$$

■

SECTION 2.4 ■ Exercises

A EXERCISES Basic Skills and Concepts

1. If the terms of a polynomial are written in descending order, then a variation of sign occurs when the signs of two ___consecutive___ terms differ.

2. The number of positive zeros of a polynomial function $P(x)$ is equal to the number of ___changes___ of sign of $P(x)$ or is less than that number by ___an even integer___.

3. If a polynomial $P(x)$ has no zero greater than a number k, then k is called a(n) ___upper bound___ of the zeros of $P(x)$. If $P(x)$ has no zero less than a number m, then m is called a(n) ___lower bound___ of the zeros of $P(x)$.

4. If $z = a + bi$ is a zero of a polynomial $P(x)$ with real coefficients, then ___$\bar{z} = a - bi$___ is also a zero of $P(x)$.

5. *True or False* Every polynomial of odd degree with real coefficients has at least one real zero. True

6. *True or False* There is a polynomial with real coefficients that has $2 + 3i$ as its only nonreal zero. False

In Exercises 7–14, determine the possible number of positive and negative zeros of the given function by using Descartes's Rule of Signs.

7. $f(x) = 5x^3 - 2x^2 - 3x + 4$ †

8. $g(x) = 3x^3 + x^2 - 9x - 3$ †

9. $f(x) = 2x^3 + 5x^2 - x + 2$ †

10. $g(x) = 3x^4 + 8x^3 - 5x^2 + 2x - 3$ †

11. $h(x) = 2x^5 - 5x^3 + 3x^2 + 2x - 1$ †

12. $F(x) = 5x^6 - 7x^4 + 2x^3 - 1$ †

13. $G(x) = -3x^4 - 4x^3 + 5x^2 - 3x + 7$ †

14. $H(x) = -5x^5 + 3x^3 - 2x^2 - 7x + 4$ †

In Exercises 15–22, determine upper and lower bounds on the zeros of the given function.

15. $f(x) = 3x^3 - x^2 + 9x - 3$ Upper bound: $\frac{1}{3}$; lower bound: -1

16. $g(x) = 2x^3 - 3x^2 - 14x + 21$ Upper bound: $\frac{7}{2}$; lower bound: -3

17. $F(x) = 3x^3 + 2x^2 + 5x + 7$ Upper bound: 1; lower bound: $-\frac{7}{3}$

18. $G(x) = x^3 + 3x^2 + x - 4$ Upper bound: 1; lower bound: -4

19. $h(x) = x^4 + 3x^3 - 15x^2 - 9x + 31$ Upper bound: 31; lower bound: -31

20. $H(x) = 3x^4 - 20x^3 + 28x^2 + 19x - 13$

21. $f(x) = 6x^4 + x^3 - 43x^2 - 7x + 7$

22. $g(x) = 6x^4 + 23x^3 + 25x^2 - 9x - 5$

In Exercises 23–32, find all solutions of the equation in the complex number system.

23. $x^2 + 25 = 0$ $x = \pm 5i$

24. $9x^2 + 16 = 0$ $x = \pm\frac{4}{3}i$

25. $(x - 2)^2 + 9 = 0$
$x = 2 \pm 3i$

26. $(x - 1)^2 + 27 = 0$
$x = 1 \pm 3i\sqrt{3}$

27. $x^2 + 4x + 4 = -9$
$x = -2 \pm 3i$

28. $x^2 + 2x + 1 = -16$
$x = -1 \pm 4i$

29. $x^3 - 8 = 0$
$\{2, -1 + i\sqrt{3}, -1 - i\sqrt{3}\}$

30. $x^4 - 1 = 0$

31. $(x - 2)(x - 3i)(x + 3i) = 0$ $\{2, 3i, -3i\}$

32. $(x - 1)(x - 2i)(2x - 6i)(2x + 6i) = 0$ $\{1, 2i, 3i, -3i\}$

In Exercises 33–38, find the remaining zeros of a polynomial $P(x)$ with real coefficients and with the specified degree and zeros.

33. Degree 3; zeros: $2, 3 + i$ $3 - i$

34. Degree 3; zeros: $2, 2 - i$ $2 + i$

35. Degree 4; zeros: $0, 1, 2 - i$ $2 + i$

36. Degree 4; zeros: $-i, 1 - i$ $i, 1 + i$

37. Degree 6; zeros: $0, 5, i, 3i$ $-i, -3i$

38. Degree 6; zeros: $2i, 4 + i, i - 1$ $-1 - i, -2i, 4 - i$

In Exercises 39–42, find the polynomial $P(x)$ with real coefficients having the specified degree, leading coefficient, and zeros.

39. Degree 4; leading coefficient 2; zeros: $5 - i, 3i$ †

40. Degree 4; leading coefficient -3; zeros: $2 + 3i, 1 - 4i$ †

41. Degree 5; leading coefficient 7; zeros: 5 (multiplicity 2), 1, $3 - i$ $P(x) = 7x^5 - 119x^4 + 777x^3 - 2415x^2 + 3500x - 1750$

42. Degree 6; leading coefficient 4; zeros: 3, 0 (multiplicity 3), $2 - 3i$ $P(x) = 4x^6 - 28x^5 + 100x^4 - 156x^3$

In Exercises 43–46, use the given zero to find all the zeros for each function.

43. $P(x) = x^4 + x^3 + 9x^2 + 9x$, zero: $3i$ $-1, 0, 3i, -3i$

44. $P(x) = x^4 - 2x^3 + x^2 + 2x - 2$, zero: $1 - i$
$-1, 1, 1 + i, 1 - i$

45. $P(x) = x^5 - 5x^4 + 2x^3 + 22x^2 - 20x$, zero: $3 - i$
$-2, 0, 1, 3 + i, 3 - i$

46. $P(x) = 2x^5 - 11x^4 + 19x^3 - 17x^2 + 17x - 6$, zero: i
$\frac{1}{2}, 2, 3, i, -i$

In Exercises 47–56, find all zeros of each polynomial function.

47. $P(x) = x^3 - 9x^2 + 25x - 17$ $1, 4 + i, 4 - i$

48. $P(x) = x^3 - 5x^2 + 7x + 13$ $-1, 3 + 2i, 3 - 2i$

49. $P(x) = 3x^3 - 2x^2 + 22x + 40$ $-\frac{4}{3}, 1 + 3i, 1 - 3i$

50. $P(x) = 3x^3 - x^2 + 12x - 4$ $\frac{1}{3}, 2i, -2i$

51. $P(x) = 2x^4 - 10x^3 + 23x^2 - 24x + 9$ $1, \frac{3}{2}, \frac{3}{2} + \frac{3}{2}i, \frac{3}{2} - \frac{3}{2}i$

†Due to space constrictions, answers to these exercises may be found in the Answers beginning on page A–1 in the back of the book.

Answers: **20.** Upper bound: 13; lower bound: -1 **21.** Upper bound: $\frac{7}{2}$; lower bound: $-\frac{7}{2}$ **22.** Upper bound: $\frac{1}{2}$; lower bound: -5

52. $P(x) = 9x^4 + 30x^3 + 14x^2 - 16x + 8$

53. $P(x) = x^4 - 4x^3 - 5x^2 + 38x - 30$ †

54. $P(x) = x^4 + x^3 + 7x^2 + 9x - 18$ $-2, 1, 3i, -3i$

55. $P(x) = 2x^5 - 11x^4 + 19x^3 - 17x^2 + 17x - 6$ †

56. $P(x) = x^5 - 2x^4 - x^3 + 8x^2 - 10x + 4$ †

In Exercises 57–60, find an equation of a polynomial function of least degree having the given complex zeros, intercepts, and graph.

57. f has complex zero $3i$

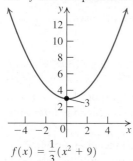

$f(x) = \dfrac{1}{3}(x^2 + 9)$

58. f has complex zero $-i$

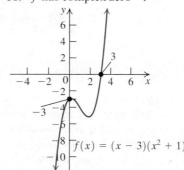

$f(x) = (x - 3)(x^2 + 1)$

59. f has complex zeros i and $2i$

60. f has complex zero $-2i$

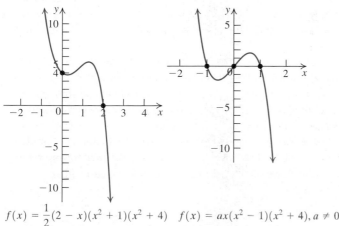

$f(x) = \dfrac{1}{2}(2 - x)(x^2 + 1)(x^2 + 4)$ $f(x) = ax(x^2 - 1)(x^2 + 4), a \neq 0$

C EXERCISES Beyond the Basics

61. The solutions -1 and 1 of the equation $x^2 = 1$ are called the square roots of 1. The solutions of the equation $x^3 = 1$ are called the cube roots of 1. Find the cube roots of 1. How many are there? †

62. The solutions of the equation $x^n = 1$, where n is a positive integer, are called the nth roots of 1 or "nth roots of unity." Explain the relationship between the solutions of the equation $x^n = 1$ and the zeros of the complex polynomial $P(x) = x^n - 1$. How many nth roots of 1 are there? †

63. Show that the polynomial function $P(x) = x^6 + 2x^4 + 3x^2 + 4$ has six nonreal zeros.

64. Show that the polynomial function $F(x) = x^5 + x^3 + 2x + 1$ has four nonreal zeros.

65. Show that the polynomial function $f(x) = x^7 + 5x + 3$ has six complex zeros.

66. Show that the polynomial function $g(x) = x^3 - x^2 - 1$ has two complex zeros.

Answers: **52.** $-2, \dfrac{1}{3} + \dfrac{1}{3}i, \dfrac{1}{3} - \dfrac{1}{3}i$

In Exercises 67–70, find an equation with real coefficients of a polynomial function f that has the given characteristics. Then write the end behavior of the graph of $y = f(x)$.

67. Degree: 3; Zeros: $2, 1 + 2i$; y-intercept: 40 †

68. Degree: 3; Zeros: $1, 2 - 3i$; y-intercept: -26 †

69. Degree: 4; Zeros: $1, -1, 3 + i$; y-intercept: 20 †

70. Degree: 4; Zeros: $1 - 2i, 3 - 2i$; y-intercept: 130 †

Critical Thinking

71. Show that if $r_1, r_2, \ldots, r_n$ are the roots of the equation

$$a_n x^n + a_{n-1} x^{n-1} + \cdots + a_1 x + a_0 = 0 \quad (a_n \neq 0),$$

then the sum of the roots satisfies

$$r_1 + r_2 + \cdots + r_n = -\frac{a_{n-1}}{a_n}$$

and the product of the roots satisfies

$$r_1 \cdot r_2 \cdot \cdots \cdot r_n = (-1)^n \frac{a_0}{a_n}.$$

[*Hint:* Factor the polynomial; then multiply it out, using $r_1, r_2, \ldots, r_n$, and compare coefficients with those of the original polynomial.]

GROUP PROJECTS

72. In his book *Ars Magna*, Cardano explained how to solve cubic equations. He considered the following example:

$$x^3 + 6x = 20.$$

  a. Explain why this equation has exactly one real solution.

  b. Cardano explained the method as follows: "I take two cubes v^3 and u^3 whose difference is 20 and whose product is 2, that is, a third of the coefficient of x. Then, I say that $x = v - u$ is a solution of the equation." Show that if $v^3 - u^3 = 20$ and $vu = 2$, then $x = v - u$ is indeed the solution of the equation $x^3 + 6x = 20$. †

  c. Solve the system

$$v^3 - u^3 = 20$$
$$vu = 2$$

$u = \sqrt[3]{-10 \pm 6\sqrt{3}}$,

$v = \dfrac{2}{\sqrt[3]{-10 \pm 6\sqrt{3}}}, v = \sqrt[3]{10 \pm 6\sqrt{3}}$

to find u and v.

  d. Consider the equation $x^3 + px = q$, where p is a positive number. Using your work in parts (a), (b), and (c) as a guide, show that the unique solution of this equation is

$$x = \sqrt[3]{\frac{q}{2} + \sqrt{\left(\frac{q}{2}\right)^2 + \left(\frac{p}{3}\right)^3}} - \sqrt[3]{-\frac{q}{2} + \sqrt{\left(\frac{q}{2}\right)^2 + \left(\frac{p}{3}\right)^3}}.$$

  e. Consider an arbitrary cubic equation

$$x^3 + ax^2 + bx + c = 0.$$

Show that the substitution $x = y - \dfrac{a}{3}$ allows you to write the cubic equation as

$$y^3 + py = q.$$

73. Use Cardano's method to solve the equation $x^3 + 6x^2 + 10x + 8 = 0$. $\{-4, -1 + i, -1 - i\}$

Rational Functions

Before Starting this Section, Review

1. Domain of a function (Section 1.3, page 35)
2. Zeros of a function (Section 2.2, page 123)
3. Symmetry (Section 1.1, page 8)
4. Quadratic formula (Appendix A, page 813)

Objectives

1. Define a rational function.
2. Find vertical asymptotes (if any).
3. Find horizontal asymptotes (if any).
4. Graph rational functions.
5. Graph rational functions with oblique asymptotes.
6. Graph a revenue curve.

FEDERAL TAXES AND REVENUES

Federal governments levy income taxes on their citizens and corporations to pay for defense, infrastructure, and social services. The relationship between tax rates and tax revenues shown in Figure 2.16 is a "revenue curve."

All economists agree that a zero tax rate produces no tax revenues, and that no one would bother to work with a 100% tax rate, so tax revenue would be zero. It then follows that in a given economy, there is some optimal tax rate T between zero and 100% at which people are willing to maximize their output and still pay their taxes. This optimal rate brings in the most revenue for the government. However, since no one knows the actual shape of the revenue curve, it is impossible to find the exact value of T.

Notice in Figure 2.16 that between the two extreme rates are two rates (such as a and b in the figure) that will collect the same amount of revenue. In a democracy, politicians can argue that taxes are currently too high (at some point b in Figure 2.16) and should therefore be reduced to encourage incentives and harder work (this is *supply-side economics*); at the same time, the government will generate more revenue. Others can argue that we are well to the left of T (at some point a in Figure 2.16), so the tax rate should be raised (for the rich) to generate more revenue. In Example 9, we sketch the graph of a revenue curve for a hypothetical economy. ■

FIGURE 2.16 A revenue curve

---

1. Define a rational function.

Rational Functions

Recall that a sum, difference, or product of two polynomial functions is also a polynomial function. However, the quotient of two polynomial functions is generally *not* a polynomial function. We call the quotient of two polynomial functions a *rational function*.

RATIONAL FUNCTION

A function f that can be expressed in the form

$$f(x) = \frac{N(x)}{D(x)},$$

where the numerator $N(x)$ and the denominator $D(x)$ are polynomials and $D(x)$ is not the zero polynomial, is called a **rational function.** The domain of f consists of all real numbers for which $D(x) \neq 0$.

Examples of rational functions are

$$f(x) = \frac{x - 1}{2x + 1}, \quad g(y) = \frac{5y}{y^2 + 1}, \quad h(t) = \frac{3t^2 - 4t + 1}{t - 2}, \quad \text{and} \quad F(z) = 3z^4 - 5z^2 + 6z + 1.$$

($F(z)$ has the constant polynomial function $D(z) = 1$ as its denominator.)
By contrast, $S(x) = \dfrac{\sqrt{1 - x^2}}{x + 1}$ and $G(x) = \dfrac{|x|}{x}$ are not rational functions.

EXAMPLE 1   Finding the Domain of a Rational Function

Find the domain of each rational function.

a. $f(x) = \dfrac{3x^2 - 12}{x - 1}$ **b.** $g(x) = \dfrac{x}{x^2 - 6x + 8}$ **c.** $h(x) = \dfrac{x^2 - 4}{x - 2}$

SOLUTION

a. The domain of $f(x) = \dfrac{3x^2 - 12}{x - 1}$ is the set of all real numbers x for which
$x - 1 \neq 0$ (that is, $x \neq 1$), or in interval notation, $(-\infty, 1) \cup (1, \infty)$.

b. The domain of g is the set of all real numbers x for which the denominator $D(x)$
is nonzero.

$$
\begin{array}{lll}
D(x) = x^2 - 6x + 8 = 0 & & \text{Set } D(x) = 0. \\
(x - 2)(x - 4) = 0 & & \text{Factor } D(x). \\
x - 2 = 0 \quad \text{or} \quad x - 4 = 0 & & \text{Zero-product property} \\
x = 2 \quad \text{or} \quad x = 4 & & \text{Solve for } x.
\end{array}
$$

The domain of g is the set of all real numbers x except 2 and 4, or in interval
notation, $(-\infty, 2) \cup (2, 4) \cup (4, \infty)$.

c. The domain of $h(x) = \dfrac{x^2 - 4}{x - 2}$ is the set of all real numbers x except 2, or in
interval notation, $(-\infty, 2) \cup (2, \infty)$.   ■ ■ ■

Practice Problem 1   Find the domain of the rational function $f(x) = \dfrac{x - 3}{x^2 - 4x - 5}$. ■

RECALL

The graph of a polynomial
function has no holes, gaps, or
sharp corners.

In Example 1, part (**c**), note that the functions $h(x) = \dfrac{x^2 - 4}{x - 2} = \dfrac{(x - 2)(x + 2)}{x - 2}$
and $H(x) = x + 2$ are not equal: the domain of $H(x)$ is the set of all real numbers,
but the domain of $h(x)$ is the set of all real numbers x except 2. The graph of
$y = H(x)$ is a line with slope 1 and y-intercept 2. The graph of $y = h(x)$ is the same
line as $y = H(x)$, except that the point $(2, 4)$ is missing. See Figure 2.17.

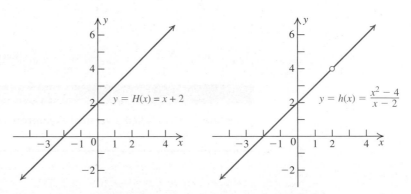

FIGURE 2.17 Functions $H(x)$ and $h(x)$ have different graphs.

As with the quotient of integers, if the polynomials $N(x)$ and $D(x)$ have no common factors, then the rational function $f(x) = \dfrac{N(x)}{D(x)}$ is said to be in **lowest terms**. The zeros of $N(x)$ and $D(x)$ will play an important role in graphing the function f.

2 Find vertical asymptotes (if any).

Vertical Asymptotes

Consider the function $f(x) = \dfrac{1}{x}$. We know that $f(0)$ is undefined, meaning that f is not assigned any value at $x = 0$. How does $f(x)$ behave "near" the excluded value $x = 0$? Consider the following table.

x	1	0.1	0.01	0.001	0.0002	10^{-9}	10^{-50}
$f(x) = \dfrac{1}{x}$	1	10	100	1000	5000	10^{9}	10^{50}

We see that as x approaches 0 (with $x > 0$), $f(x)$ increases without bound. In symbols, we write "as $x \to 0^{+}, f(x) \to \infty$." This statement is read "As x approaches 0 *from the right, $f(x)$ approaches infinity.*"

The symbol $x \to a^{+}$, read "x approaches a from the right," refers only to values of x that are greater than a. The symbol $x \to a^{-}$ is read "x approaches a from the left" and refers only to values of x that are less than a.

For the function $f(x) = \dfrac{1}{x}$, we can make another table of values that includes negative values of x near zero, such as $x = -1, -0.1, -0.001, \dots, -10^{-9}$, and so on. We can conclude that as $x \to 0^{-}, f(x) \to -\infty$ and say that as x approaches 0 *from the left, $f(x)$ approaches negative infinity.*

The graph of $f(x) = \dfrac{1}{x}$ is the *reciprocal function* from Chapter 1 (see Figure 2.18).

The vertical line $x = 0$ (the y-axis) is called a *vertical asymptote.*

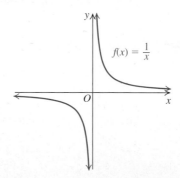

FIGURE 2.18 Reciprocal function

VERTICAL ASYMPTOTE

The line $x = a$ is called a **vertical asymptote** of the graph of a function f if $|f(x)| \to \infty$ as $x \to a^{+}$ or as $x \to a^{-}$.

This definition says that if the line $x = a$ is a vertical asymptote of the graph of a rational function f, then the graph of f *near* $x = a$ behaves like one of the graphs in Figure 2.19.

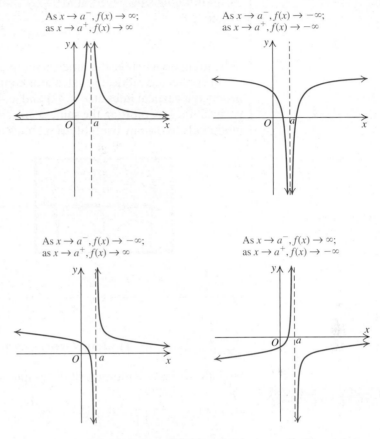

As $x \to a^-$, $f(x) \to \infty$;
as $x \to a^+$, $f(x) \to \infty$

As $x \to a^-$, $f(x) \to -\infty$;
as $x \to a^+$, $f(x) \to -\infty$

As $x \to a^-$, $f(x) \to -\infty$;
as $x \to a^+$, $f(x) \to \infty$

As $x \to a^-$, $f(x) \to \infty$;
as $x \to a^+$, $f(x) \to -\infty$

FIGURE 2.19 Behavior of a rational function near a vertical asymptote

If $x = a$ is a vertical asymptote for a rational function f, then a is not in the domain; so the graph of f *does not cross* the line $x = a$.

LOCATING VERTICAL ASYMPTOTES OF RATIONAL FUNCTIONS

If $f(x) = \dfrac{N(x)}{D(x)}$ is a rational function, where $N(x)$ and $D(x)$ do not have a common factor and a is a real zero of $D(x)$, then the line with equation $x = a$ is a vertical asymptote of the graph of f.

This means that the vertical asymptotes (if any) are found by locating the real zeros of the denominator.

TECHNOLOGY CONNECTION

Graphing calculators give very different graphs for

$$g(x) = \frac{1}{x^2 - 9}$$

depending on whether the mode is set to *connected* or *dot*.

In connected mode, the calculator connects the dots between plotted points, producing vertical lines at $x = -3$ and $x = 3$. Dot mode avoids graphing these vertical lines by plotting the same points, but not connecting them. However, dot mode fails to display continuous sections of the graph.

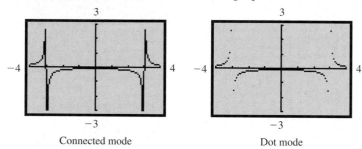

Connected mode   Dot mode

EXAMPLE 2   **Finding Vertical Asymptotes**

Find all vertical asymptotes of the graph of each rational function.

a. $f(x) = \dfrac{1}{x - 1}$ **b.** $g(x) = \dfrac{1}{x^2 - 9}$ **c.** $h(x) = \dfrac{1}{x^2 + 1}$

SOLUTION

a. There are no common factors in the numerator and denominator of $f(x) = \dfrac{1}{x - 1}$, and the only *zero* of the denominator is 1. Therefore, $x = 1$ is a vertical asymptote of $f(x)$.

b. The rational function $g(x)$ is in lowest terms. Factoring $x^2 - 9 = (x + 3)(x - 3)$, we see that the *zeros* of the denominator are -3 and 3. Therefore, the lines $x = -3$ and $x = 3$ are the two vertical asymptotes of $g(x)$.

c. Because the denominator $x^2 + 1$ has no real zeros, the graph of $h(x)$ has *no* vertical asymptotes. ■ ■ ■

Practice Problem 2   Find the vertical asymptotes of the graph of

$$f(x) = \frac{x + 1}{x^2 + 3x - 10}.$$ ■

The next example illustrates that the graph of a rational function may have gaps (missing points) with or without vertical asymptotes.

EXAMPLE 3   **Rational Function Whose Graph Has a Hole**

Find all vertical asymptotes of the graph of each rational function.

a. $h(x) = \dfrac{x^2 - 9}{x - 3}$ **b.** $g(x) = \dfrac{x + 2}{x^2 - 4}$

TECHNOLOGY CONNECTION

Calculator graph for

$$h(x) = \frac{x^2 - 9}{x - 3}$$

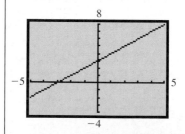

A calculator graph does not show the hole in the graph, but a calculator table explains the situation.

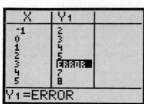

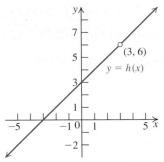

FIGURE 2.20 Graph with a hole

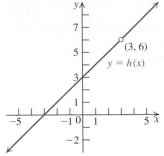

FIGURE 2.21 Graph with a hole and an asymptote

3 Find horizontal asymptotes (if any).

SOLUTION

a. $h(x) = \dfrac{x^2 - 9}{x - 3}$ Given function

$ = \dfrac{(x + 3)(x - 3)}{x - 3}$ Factor $x^2 - 9 = (x + 3)(x - 3)$ in the numerator.

$ = x + 3, \quad \text{if } x \neq 3$ Simplify.

The graph of $h(x)$ is the line $y = x + 3$ with a gap (or hole) at $x = 3$. See Figure 2.20.

The graph of $h(x)$ has no vertical asymptote at $x = 3$, the zero of the denominator, because the numerator and the denominator have the factor $(x - 3)$ in common.

b. $g(x) = \dfrac{x + 2}{x^2 - 4}$ Given function

$ = \dfrac{x + 2}{(x + 2)(x - 2)}$ Factor the denominator.

$ = \dfrac{1}{x - 2}, \quad x \neq -2$ Simplify.

The graph of $g(x)$ has a hole at $x = -2$ and a vertical asymptote at $x = 2$. See Figure 2.21.   ■ ■ ■

Practice Problem 3 Find all vertical asymptotes of the graph of $f(x) = \dfrac{3 - x}{x^2 - 9}$. ■

Horizontal Asymptotes

A vertical asymptote identifies the behavior of the graph of a function near a specific real number but says nothing about the function's end behavior. Consider again the function $f(x) = \dfrac{1}{x}$. The larger we make x, the closer $\dfrac{1}{x}$ is to 0: as $x \to \infty, f(x) = \dfrac{1}{x} \to 0$. Similarly, as $x \to -\infty, f(x) = \dfrac{1}{x} \to 0$. The x-axis, with equation $y = 0$, is a *horizontal asymptote* of the graph of $f(x) = \dfrac{1}{x}$. In fact, for any real number a and any positive integer n, as $|x| \to \infty, g(x) = \dfrac{a}{x^n} \to 0$. So the x-axis is a horizontal asymptote of the graph of $g(x) = \dfrac{a}{x^n}$.

HORIZONTAL ASYMPTOTE

The line $y = k$ is a **horizontal asymptote** of the graph of a function f if $f(x) \to k$ as $x \to \infty$ or as $x \to -\infty$.

In general, a line $y = k$ is a horizontal asymptote if either $f(x) \to k$ as $x \to \infty$ or $f(x) \to k$ as $x \to -\infty$. For example, in Figure 2.22, the line with equation $y = 2$ is a horizontal asymptote of the graph shown because $f(x) \to 2$ as $x \to \infty$. However, it can be shown that this situation is not possible for a rational function.

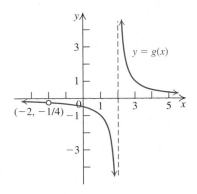

FIGURE 2.22 Horizontal asymptote

The graph of a function $y = f(x)$ can cross its horizontal asymptote (in contrast to a vertical asymptote). The graph of $g(x) = \dfrac{x + 2}{x^2 - 4}$, shown in Figure 2.21, has a horizontal asymptote: $y = 0$. This can be verified algebraically. Divide the numerator and the denominator of g by x^2.

$$g(x) = \frac{x + 2}{x^2 - 4}$$

$$= \frac{\dfrac{x + 2}{x^2}}{\dfrac{x^4 - 4}{x^2}} \qquad \text{Divide numerator and denominator by } x^2.$$

$$= \frac{\dfrac{x}{x^2} + \dfrac{2}{x^2}}{\dfrac{x^2}{x^2} - \dfrac{4}{x^2}} \qquad \frac{a \pm b}{c} = \frac{a}{c} \pm \frac{b}{c}$$

$$= \frac{\dfrac{1}{x} + \dfrac{2}{x^2}}{1 - \dfrac{4}{x^2}} \qquad \text{Simplify each expression.}$$

As $|x| \to \infty$, the expressions $\dfrac{1}{x}, \dfrac{2}{x^2}$, and $\dfrac{4}{x^2}$ all approach 0, so

$$g(x) \to \frac{0 + 0}{1 - 0} = \frac{0}{1} = 0.$$

Since $g(x) \to 0$ as $|x| \to \infty$, the line $y = 0$ (the x-axis) is a horizontal asymptote.

Note that although a rational function can have more than one vertical asymptote (see Example 2(b)), it can have, at most, one horizontal asymptote. We can find the horizontal asymptote (if any) of a rational function by dividing the numerator and denominator by the highest power of x that appears in the denominator and investigating the resulting expression as $|x| \to \infty$. Alternatively, one may use the following rules.

RULES FOR LOCATING HORIZONTAL ASYMPTOTES

Let f be a rational function given by

$$f(x) = \frac{N(x)}{D(x)} = \frac{a_n x^n + a_{n-1} x^{n-1} + \cdots + a_2 x^2 + a_1 x + a_0}{b_m x^m + b_{m-1} x^{m-1} + \cdots + b_2 x^2 + b_1 x + b_0}, \quad a_n \neq 0, b_m \neq 0,$$

in lowest terms. To find whether the graph of f has one horizontal asymptote or no horizontal asymptote, we compare the degree of the numerator, n, with that of the denominator, m:

1. If $n < m$, then the x-axis ($y = 0$) is the horizontal asymptote.

2. If $n = m$, then the line with equation $y = \dfrac{a_n}{b_m}$ is the horizontal asymptote.

3. If $n > m$, then the graph of f has *no* horizontal asymptote.

EXAMPLE 4   **Finding the Horizontal Asymptote**

Find the horizontal asymptote (if any) of the graph of each rational function.

a. $f(x) = \dfrac{5x + 2}{1 - 3x}$ **b.** $g(x) = \dfrac{2x}{x^2 + 1}$ **c.** $h(x) = \dfrac{3x^2 - 1}{x + 2}$

SOLUTION

a. The numerator and denominator of $f(x) = \dfrac{5x + 2}{1 - 3x}$ are both of degree 1. The leading coefficient of the numerator is 5 and that of the denominator is -3. By Rule 2, the line $y = \dfrac{5}{-3} = -\dfrac{5}{3}$ is the horizontal asymptote of the graph of f.

b. For the function $g(x) = \dfrac{2x}{x^2 + 1}$, the degree of the numerator is 1 and that of the denominator is 2. By Rule 1, the line $y = 0$ (the x-axis) is the horizontal asymptote.

c. The degree of the numerator, 2, of $h(x)$ is greater than the degree of its denominator, 1. By Rule 3, the graph of h has no horizontal asymptote.   ■ ■ ■

Practice Problem 4   Find the horizontal asymptote (if any) of the graph of each function.

a. $f(x) = \dfrac{2x - 5}{3x + 4}$ **b.** $g(x) = \dfrac{x^2 + 3}{x - 1}$ ■

---

4 Graph rational functions.

Graphing Rational Functions

FINDING THE SOLUTION: A PROCEDURE

EXAMPLE 5   **Graphing a Rational Function**

OBJECTIVE	EXAMPLE
Graph $f(x) = \dfrac{N(x)}{D(x)}$, where $f(x)$ is in lowest terms.	Sketch the graph of $f(x) = \dfrac{2x^2 - 2}{x^2 - 9}$.
Step 1 Find the intercepts. The x-intercepts are found by solving the equation $N(x) = 0$. The y-intercept is $f(0)$.	$\begin{aligned} 2x^2 - 2 &= 0 &&\text{Set } N(x) = 0. \\ 2(x - 1)(x + 1) &= 0 &&\text{Factor.} \\ x - 1 = 0 \text{ or } x + 1 &= 0 &&\text{Zero-product property} \\ x = 1 \text{ or } x &= -1 &&\text{Solve for } x. \end{aligned}$ The x-intercepts are -1 and 1, so the graph passes through the points $(-1, 0)$ and $(1, 0)$. $$f(0) = \frac{2(0)^2 - 2}{(0)^2 - 9} \qquad \text{Replace } x \text{ with 0 in } f(x).$$ $$= \frac{2}{9} \qquad\qquad \text{Simplify.}$$ The y-intercept is $\dfrac{2}{9}$, so the graph of f passes through the point $\left(0, \dfrac{2}{9}\right)$.

continued on the next page

Step 2 **Find the vertical asymptotes (if any).** Solve $D(x) = 0$ to find the vertical asymptotes of the graph. Sketch the vertical asymptotes.

$$x^2 - 9 = 0 \qquad \text{Set } D(x) = 0.$$
$$(x - 3)(x + 3) = 0 \qquad \text{Factor.}$$
$$x - 3 = 0 \quad \text{or} \quad x + 3 = 0 \qquad \text{Zero-product property}$$
$$x = 3 \quad \text{or} \qquad x = -3 \qquad \text{Solve for } x.$$

The vertical asymptotes of the graph of f are the lines $x = -3$ and $x = 3$.

Step 3 **Find the horizontal asymptote (if any).** Use the rules on page 160 for finding the horizontal asymptote of a rational function.

Since $n = m = 2$, by Rule 2 the horizontal asymptote is

$$y = \frac{\text{Leading coefficient of } N(x)}{\text{Leading coefficient of } D(x)} = \frac{2}{1} = 2.$$

Step 4 **Test for symmetry.** If $f(-x) = f(x)$, then f is symmetric with respect to the y-axis. If $f(-x) = -f(x)$, then f is symmetric with respect to the origin.

$$f(-x) = \frac{2(-x)^2 - 2}{(-x)^2 - 9} = \frac{2x^2 - 2}{x^2 - 9} = f(x).$$ The graph of f is symmetric in the y-axis. This is the only symmetry.

Step 5 **Find the sign of $f(x)$.** Use sign graphs and test numbers associated with the zeros of $N(x)$ and $D(x)$ to determine where the graph of f is above the x-axis and where it is below the x-axis.

The zeros of $N(x)$ and $D(x)$ are $-3, -1, 1,$ and 3. These zeros divide the x-axis into five intervals. (See the figure.)

	-3		-1		1		3	
Test number	-4		-2		0		2		4
Value of f at the test number	$\dfrac{30}{7}$		$-\dfrac{6}{5}$		$\dfrac{2}{9}$		$-\dfrac{6}{5}$		$\dfrac{30}{7}$
Sign graph	$+++$		$---$		$+++$		$---$		$+++$
Graph point	$\left(-4, \dfrac{30}{7}\right)$		$\left(-2, -\dfrac{6}{5}\right)$		$\left(0, \dfrac{2}{9}\right)$		$\left(2, -\dfrac{6}{5}\right)$		$\left(4, \dfrac{30}{7}\right)$
Graph of f	Above x-axis		Below x-axis		Above x-axis		Below x-axis		Above x-axis

Step 6 **Sketch the graph.** Plot the points and asymptotes found in Steps 1–5 and use symmetry to sketch the graph of f.

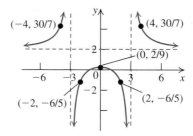

■ ■ ■

Practice Problem 5 Sketch the graph of $f(x) = \dfrac{2x}{x^2 - 1}$.

■

EXAMPLE 6   Graphing a Rational Function

Sketch the graph of $f(x) = \dfrac{x^2 + 2}{(x + 2)(x - 1)}$.

SOLUTION

Step 1 Since $x^2 + 2 > 0$, the graph has no x-intercepts.

$$f(0) = \frac{0^2 + 2}{(0 + 2)(0 - 1)} \qquad \text{Replace } x \text{ with 0 in } f(x).$$

$$= -1 \qquad\qquad\qquad \text{Simplify.}$$

The y-intercept is -1.

Step 2 Set $(x + 2)(x - 1) = 0$. Solving for x, we have $x = -2$ or $x = 1$. The vertical asymptotes are the lines $x = -2$ and $x = 1$.

Step 3 $f(x) = \dfrac{x^2 + 2}{x^2 + x - 2}$. By Rule 2, page 160, the horizontal asymptote is

$y = \dfrac{1}{1} = 1$, because the leading coefficient for both the numerator and the denominator is 1.

Step 4 **Symmetry.** None

Step 5 **Sign intervals.** The zeros of the denominator, -2 and 1, yield the following table:

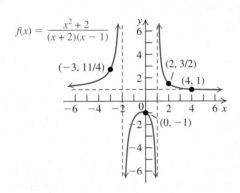

$$f(x) = \frac{x^2 + 2}{(x + 2)(x - 1)}$$

FIGURE 2.23 Graph crossing horizontal asymptote

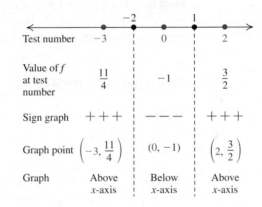

Step 6 The graph of f is shown in Figure 2.23.   ■ ■ ■

Practice Problem 6 Sketch the graph of $f(x) = \dfrac{2x^2 - 1}{2x^2 + x - 3}$.   ■

Notice in Figure 2.23 that the graph of f crosses the horizontal asymptote $y = 1$. To find the point of intersection, set $f(x) = 1$ and solve for x.

$$\frac{x^2 + 2}{x^2 + x - 2} = 1 \qquad\qquad \text{Set } f(x) = 1.$$

$$x^2 + 2 = x^2 + x - 2 \qquad \text{Multiply both sides by } x^2 + x - 2.$$

$$x = 4 \qquad\qquad\qquad \text{Solve for } x.$$

The graph of f crosses the horizontal asymptote at the point $(4, 1)$.

We know that the graphs of polynomial functions are continuous (all in one piece). The next example shows that the graph of a rational function (other than a polynomial function) can also be continuous.

EXAMPLE 7   **Graphing a Rational Function**

Sketch a graph of $f(x) = \dfrac{x^2}{x^2 + 1}$.

SOLUTION

Step 1   Since $f(0) = 0$ and setting $f(x) = 0$, we have $x = 0$. Therefore, 0 is both the x-intercept and the y-intercept for the graph of f.

Step 2   Because $x^2 + 1 > 0$ for all x (the domain is the set of all real numbers because there are no real zeros for the denominator), there are no vertical asymptotes.

Step 3   By Rule 2 for locating horizontal asymptotes, the horizontal asymptote is $y = 1$.

Step 4 $f(-x) = \dfrac{(-x)^2}{(-x)^2 + 1} = \dfrac{x^2}{x^2 + 1} = f(x)$. The graph is symmetric with respect to the y-axis.

Step 5   The graph of $y = f(x)$ is always above the x-axis except at $x = 0$.

Step 6   The graph of $y = f(x)$ is shown in Figure 2.24.   ■ ■ ■

Practice Problem 7   Sketch the graph of $f(x) = \dfrac{x^2 + 1}{x^2 + 2}$.   ■

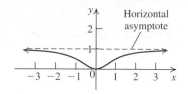

FIGURE 2.24 A continuous rational function

5 Graph rational functions with oblique asymptotes.

Oblique Asymptotes

Suppose

$$f(x) = \frac{N(x)}{D(x)}$$

and the degree of $N(x)$ is greater than the degree of $D(x)$. We know from Rule 3 on page 160 that the graph of f has no horizontal asymptote. Use either long division or synthetic division to obtain

$$f(x) = \frac{N(x)}{D(x)} = Q(x) + \frac{R(x)}{D(x)},$$

where the degree of $R(x)$ is less than the degree of $D(x)$. Rule 1 on page 160 tells us that the x-axis is a horizontal asymptote for $\dfrac{R(x)}{D(x)}$; that is, as $x \to \infty$ or as $x \to -\infty$, the expression $\dfrac{R(x)}{D(x)} \to 0$.

Therefore, as $x \to \pm\infty$,

$$f(x) \to Q(x) + 0 = Q(x).$$

This means that as $|x|$ gets very large, the graph of $f(x)$ behaves like the graph of the polynomial $Q(x)$. If the degree of $N(x)$ is exactly one more than the degree of $D(x)$, then $Q(x)$ will have the linear form $mx + b$. In this case, $f(x)$ is said to have an **oblique** (or **slanted**) asymptote. Consider, for example, the function

$$f(x) = \frac{x^2 + x}{x - 1} = x + 2 + \frac{2}{x - 1}.$$

Now as $x \to \pm\infty$, the expression $\dfrac{2}{x - 1} \to 0$ so that the graph of f approaches the graph of the oblique asymptote—the line $y = x + 2$, as shown in Figure 2.25.

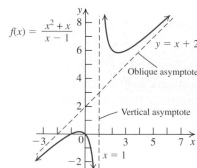

$f(x) = \dfrac{x^2 + x}{x - 1}$

FIGURE 2.25 Graph with vertical and oblique asymptotes

EXAMPLE 8   **Graphing a Rational Function with an Oblique Asymptote**

Sketch the graph of $f(x) = \dfrac{x^2 - 4}{x + 1}$.

SOLUTION

Step 1   **Intercepts:** $f(0) = \dfrac{0 - 4}{0 + 1} = -4$, so the y-intercept is -4. Set $x^2 - 4 = 0$.
We have $x = -2$ and $x = 2$, so the x-intercepts are -2 and 2.

Step 2   **Vertical asymptotes:** Set $x + 1 = 0$. We get $x = -1$, so the line $x = -1$ is a vertical asymptote.

Step 3   **Asymptotes:** Since the degree of the numerator is greater than the degree of the denominator, $f(x)$ has no horizontal asymptote. However, by long division,

$$f(x) = \frac{x^2 - 4}{x + 1} = x - 1 + \frac{-3}{x + 1} = x - 1 - \frac{3}{x + 1}.$$

As $x \to \pm\infty$, the expression $\dfrac{3}{x + 1} \to 0$, so the line $y = x - 1$ is an oblique asymptote: the graph gets close to the line $y = x - 1$ as $x \to \infty$ and as $x \to -\infty$.

Step 4   **Symmetry:** You should check that there is no y-axis or origin symmetry.

Step 5   **Sign of f:** Use the intervals determined by the zeros of the numerator and denominator:

$$f(x) = \frac{x^2 - 4}{x + 1} = \frac{(x + 2)(x - 2)}{x + 1}$$

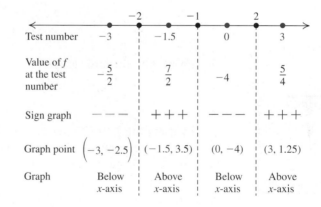

	-2	-1	2	
Test number	-3	-1.5	0	3
Value of f at the test number	$-\dfrac{5}{2}$	$\dfrac{7}{2}$	-4	$\dfrac{5}{4}$
Sign graph	$-\,-\,-$	$+\,+\,+$	$-\,-\,-$	$+\,+\,+$
Graph point	$\left(-3, -2.5\right)$	$(-1.5, 3.5)$	$(0, -4)$	$(3, 1.25)$
Graph	Below x-axis	Above x-axis	Below x-axis	Above x-axis

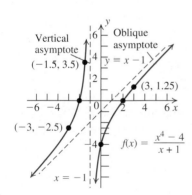

FIGURE 2.26

Step 6   The graph of f is shown in Figure 2.26.   ■ ■ ■

Practice Problem 8   Sketch the graph of $f(x) = \dfrac{x^2 + 2}{x - 1}$.   ■

6   Graph a revenue curve.

Graph of a Revenue Curve

EXAMPLE 9   **Graphing a Revenue Curve**

The revenue curve for an economy of a country is given by

$$R(x) = \frac{x(100 - x)}{x + 10},$$

where x is the tax rate in percent and $R(x)$ is the tax revenue in billions of dollars.

a. Find and interpret $R(10)$, $R(20)$, $R(30)$, $R(40)$, $R(50)$, and $R(60)$.

b. Sketch the graph of $y = R(x)$ for $0 \le x \le 100$.

c. Use a graphing calculator to estimate the tax rate that yields the maximum revenue.

SOLUTION

a. $R(10) = \dfrac{10(100 - 10)}{10 + 10} = \$45$ billion. This means that if the income is taxed at the rate of 10%, the total revenue for the government will be $45 billion.

$$\text{Similarly, } R(20) \approx \$53.3 \text{ billion,}$$
$$R(30) = \$52.5 \text{ billion,}$$
$$R(40) = \$48 \quad \text{billion,}$$
$$R(50) \approx \$41.7 \text{ billion, and}$$
$$R(60) \approx \$34.3 \text{ billion.}$$

b. The graph of the function $y = R(x)$ for $0 \le x \le 100$ is shown in Figure 2.27.

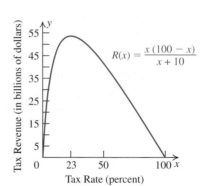

$$R(x) = \frac{x(100 - x)}{x + 10}$$

FIGURE 2.27

c. From the calculator graph of

$$Y = \frac{100x - x^2}{x + 10},$$

by using the TRACE feature, you can see that the tax rate of about 23% produces the maximum tax revenue of about $53.7 billion for the government. (See the Group Project for an algebraic proof.)   ■ ■ ■

Practice Problem 9   Repeat Example 9 for $R(x) = \dfrac{x(100 - x)}{x + 20}$.   ■

TECHNOLOGY CONNECTION

▨ TRACE

$$y = \frac{100x - x^2}{x + 10}$$

to approximate the maximum revenue.

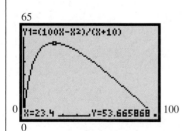

SECTION 2.5 ■ Exercises

A EXERCISES Basic Skills and Concepts

1. A rational function can be expressed in the form

   _____ .

2. The line $x = a$ is a vertical asymptote of f if $|f(x)| \to \infty$ as $x \to$ ___a^+___ or as $x \to$ ___a^-___ .

3. The line $y = k$ is a horizontal asymptote of f if $f(x) \to k$ as $x \to$ ___∞___ or as $x \to$ ___$-\infty$___ .

4. If an asymptote is neither horizontal nor vertical, then it is called a(n) _____ . oblique asymptote

5. *True or False* Every rational function has at least one vertical asymptote. False

†Due to space constrictions, answers to these exercises may be found in the Answers beginning on page A–1 in the back of the book.

Answer: **1.** $\dfrac{N(x)}{D(x)}$, where $N(x)$ and $D(x)$ are polynomials and $D(x)$ is not the zero polynomial

6. *True or False* Every rational function has, at most, one horizontal asymptote.　True

In Exercises 7–14, find the domain of each rational function.

7. $f(x) = \dfrac{x-3}{x+4}$

8. $f(x) = \dfrac{x+1}{x-1}$

9. $g(x) = \dfrac{x-1}{x^2+1}$　$(-\infty, \infty)$

10. $g(x) = \dfrac{x+2}{x^2+4}$　$(-\infty, \infty)$

11. $h(x) = \dfrac{x-3}{x^2-x-6}$

12. $h(x) = \dfrac{x-7}{x^2-6x-7}$

13. $F(x) = \dfrac{2x+3}{x^2-6x+8}$

14. $F(x) = \dfrac{3x-2}{x^2-3x+2}$

In Exercises 15–24, use the graph of the rational function $f(x)$ to complete each statement.

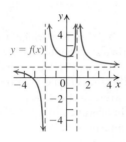

15. As $x \to 1^+$, $f(x) \to$ _____ . ∞

16. As $x \to 1^-$, $f(x) \to$ _____ . ∞

17. As $x \to -2^+$, $f(x) \to$ _____ . ∞

18. As $x \to -2^-$, $f(x) \to$ _____ . $-\infty$

19. As $x \to \infty$, $f(x) \to$ _____ . 1

20. As $x \to -\infty$, $f(x) \to$ _____ . 1

21. The domain of $f(x)$ is _____ . $(-\infty, -2) \cup (-2, 1) \cup (1, \infty)$

22. There are ____two____ vertical asymptotes.

23. The equations of the vertical asymptotes of the graph are ___$x = -2$___ and ___$x = 1$___.

24. The equation of the horizontal asymptote of the graph is ___$y = 1$___.

In Exercises 25–34, find the vertical asymptotes, if any, of the graph of each rational function.

25. $f(x) = \dfrac{x}{x-1}$　$x = 1$

26. $f(x) = \dfrac{x+3}{x-2}$　$x = 2$

27. $g(x) = \dfrac{(x+1)(2x-2)}{(x-3)(x+4)}$　$x = -4$ and $x = 3$

28. $g(x) = \dfrac{(2x-1)(x+2)}{(2x+3)(3x-4)}$　$x = -\dfrac{3}{2}$ and $x = \dfrac{4}{3}$

29. $h(x) = \dfrac{x^2-1}{x^2+x-6}$

30. $h(x) = \dfrac{x^2-4}{3x^2+x-4}$

31. $f(x) = \dfrac{x^2-6x+8}{x^2-x-12}$

32. $f(x) = \dfrac{x^2-9}{x^3-4x}$

33. $g(x) = \dfrac{2x+1}{x^2+x+1}$

34. $g(x) = \dfrac{x^2-36}{x^2+5x+9}$

In Exercises 35–42, find the horizontal asymptote, if any, of the graph of each rational function.

35. $f(x) = \dfrac{x+1}{x^2+5}$　$y = 0$

36. $f(x) = \dfrac{2x-1}{x^2-4}$　$y = 0$

37. $g(x) = \dfrac{2x-3}{3x+5}$　$y = \dfrac{2}{3}$

38. $g(x) = \dfrac{3x+4}{-4x+5}$　$y = -\dfrac{3}{4}$

39. $h(x) = \dfrac{x^2-49}{x+7}$　No horizontal asymptote

40. $h(x) = \dfrac{x+3}{x^2-9}$　$y = 0$

41. $f(x) = \dfrac{2x^2-3x+7}{3x^3+5x+11}$　$y = 0$

42. $f(x) = \dfrac{3x^3+2}{x^2+5x+11}$　No horizontal asymptote

In Exercises 43–48, match the rational function with its graph.

43. $f(x) = \dfrac{2}{x-3}$　d

44. $f(x) = \dfrac{x-2}{x+3}$　f

45. $f(x) = \dfrac{1}{x^2-2x}$　e

46. $f(x) = \dfrac{x}{x^2+1}$　b

47. $f(x) = \dfrac{x^2+2x}{x-3}$　a

48. $f(x) = \dfrac{x^2}{x^2-4}$　c

a.

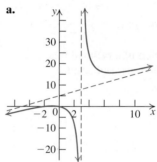

b.

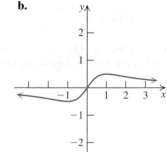

c.

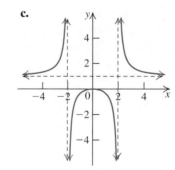

d.

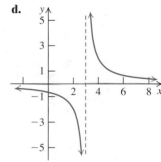

e.

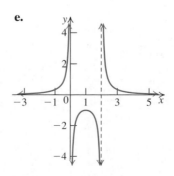

f.

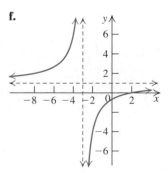

29. $x = -3$ and $x = 2$　**30.** $x = -\frac{4}{3}$ and $x = 1$　**31.** $x = -3$
32. $x = -2$, $x = 0$, and $x = 2$　**33.** No vertical asymptote
34. No vertical asymptote

Answers:
7. $(-\infty, -4) \cup (-4, \infty)$　**8.** $(-\infty, 1) \cup (1, \infty)$　**11.** $(-\infty, -2) \cup (-2, 3) \cup (3, \infty)$
12. $(-\infty, -1) \cup (-1, 7) \cup (7, \infty)$　**13.** $(-\infty, 2) \cup (2, 4) \cup (4, \infty)$　**14.** $(-\infty, 1) \cup (1, 2) \cup (2, \infty)$

In Exercises 49–64, use the six-step procedure on pages 161–162 to graph each rational function.

49. $f(x) = \dfrac{2x}{x - 3}$ †

50. $f(x) = \dfrac{-x}{x - 1}$ †

51. $f(x) = \dfrac{x}{x^2 - 4}$ †

52. $f(x) = \dfrac{x}{1 - x^2}$ †

53. $h(x) = \dfrac{-2x^2}{x^2 - 9}$ †

54. $h(x) = \dfrac{4 - x^2}{x^2}$ †

55. $f(x) = \dfrac{2}{x^2 - 2}$ †

56. $f(x) = \dfrac{-2}{x^2 - 3}$ †

57. $g(x) = \dfrac{x + 1}{(x - 2)(x + 3)}$ †

58. $g(x) = \dfrac{x - 1}{(x + 1)(x - 2)}$ †

59. $h(x) = \dfrac{x^2}{x^2 + 1}$ †

60. $h(x) = \dfrac{2x^2}{x^2 + 4}$ †

61. $f(x) = \dfrac{x^2 - 4}{x^3 - 9x}$ †

62. $f(x) = \dfrac{x^2 - 1}{x^2 + 2x - 8}$ †

63. $g(x) = \dfrac{(x - 2)^2}{x - 2}$ †

64. $g(x) = \dfrac{(x - 1)^2}{x - 1}$ †

In Exercises 65–68, find an equation of a rational function having the given asymptotes, intercepts, and graph.

65.

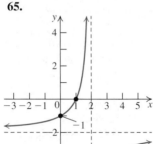

66.

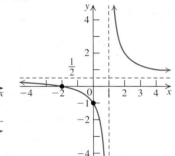

$f(x) = \dfrac{-2(x - 1)}{x - 2}$

$f(x) = \dfrac{(x + 2)}{2(x - 1)}$

67.

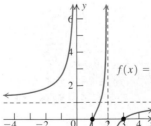

$f(x) = \dfrac{(x - 1)(x - 3)}{x(x - 2)}$

68.

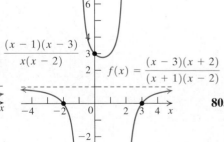

$f(x) = \dfrac{(x - 3)(x + 2)}{(x + 1)(x - 2)}$

In Exercises 69–76, find the oblique asymptote and sketch the graph of each rational function.

69. $f(x) = \dfrac{2x^2 + 1}{x}$ †

70. $f(x) = \dfrac{x^2 - 1}{x}$ †

71. $g(x) = \dfrac{x^3 - 1}{x^2}$ †

72. $g(x) = \dfrac{2x^3 + x^2 + 1}{x^2}$ †

73. $h(x) = \dfrac{x^2 - x + 1}{x + 1}$ †

74. $h(x) = \dfrac{2x^2 - 3x + 2}{x - 1}$ †

75. $f(x) = \dfrac{x^3 - 2x^2 + 1}{x^2 - 1}$ †

76. $h(x) = \dfrac{x^3 - 1}{x^2 - 4}$ †

B EXERCISES Applying the Concepts

For Exercises 77 and 78, use the following definition: Given a cost function C, the **average cost** of the first x items is found from the formula

$$\overline{C}(x) = \frac{C(x)}{x}, x > 0.$$

a. $C(x) = 0.5x + 2000$

b. $\overline{C}(x) = 0.5 + \dfrac{2000}{x}$

77. Average cost. The Genuine Trinket Co. manufactures "authentic" trinkets for gullible tourists. Fixed daily costs are $2000, and it costs $0.50 to produce each trinket.
a. Write the cost function C for producing x trinkets.
b. Write the average cost $\overline{C}$ of producing x trinkets.
c. Find and interpret $\overline{C}(100), \overline{C}(500)$, and $\overline{C}(1000)$. †
d. Find and interpret the horizontal asymptote of the rational function $\overline{C}(x)$. †

78. Average cost. The monthly cost C of producing x portable CD players is given by

$$C(x) = -0.002x^2 + 6x + 7000.$$

a. Write the average cost function $\overline{C}(x)$. †
b. Find and interpret $\overline{C}(100), \overline{C}(500)$, and $\overline{C}(1000)$. †
c. Find and interpret the oblique asymptote of $\overline{C}(x)$. †

79. Biology: birds collecting seeds. A bird is collecting seed from a field that contains 100 grams of seed. The bird collects x grams of seed in t minutes, where

$$t = f(x) = \frac{4x + 1}{100 - x}, 0 < x < 100.$$

a. Sketch the graph of $t = f(x)$. †
b. How long does it take the bird to collect
  (i) 50 grams? about 4 min
  (ii) 75 grams? about 12 min
  (iii) 95 grams? about 76 min
  (iv) 99 grams? 397 min
c. Complete the following statements if applicable:
  (i) As $x \to 100^-, f(x) \to$ _____ . ∞
  (ii) As $x \to 100^+, f(x) \to$ _____ . †
d. Does the bird ever collect all of the seed from the field? No

80. Criminology. Suppose the city of Las Vegas decides to be crime-free. The estimated cost of catching and convicting $x\%$ of the criminals is given by

$$C(x) = \frac{1000}{100 - x} \text{ million dollars.}$$

a. Find and interpret $C(50), C(75), C(90)$, and $C(99)$. †
b. Sketch a graph of $C(x), 0 \le x < 100$. †
c. What happens to $C(x)$ as $x \to 100^-$? Approaches ∞
d. What percentage of the criminals can be caught and convicted for $30 million? 66.67%

81. Environment. Suppose an environmental agency decides to get rid of the impurities from the water of a polluted river. The estimated cost of removing $x\%$ of the impurities is given by

$$C(x) = \frac{3x^2 + 50}{x(100 - x)} \text{ billion dollars.}$$

a. How much will it cost to remove 50% of the impurities? $3.02 billion
b. Sketch the graph of $y = C(x)$. †
c. Estimate the percentage of the impurities that can be removed at a cost of $30 billion. 90.89%

82. Biology. The growth function $g(x)$ describes the growth rate of organisms as a function of some nutrient concentration x. Suppose

$$g(x) = a\frac{x}{k + x}, \quad x \ge 0, †$$

where a and k are positive constants.
a. Find the horizontal asymptote of the graph of $g(x)$. Use the asymptote to explain why a is called the *saturation level* of the nutrient. †
b. Show that k is the half-saturation constant; that is, show that if $x = k$, then $g(x) = \frac{a}{2}$. †

83. Population of bacteria. The population P (in thousands) of a colony of bacteria at time t (in hours) is given by

$$P(t) = \frac{8t + 16}{2t + 1}, \quad t \ge 0.$$

a. Find the initial population of the colony. (Find the population at $t = 0$ hours.) 16,000
b. What is the long-term behavior of the population? (What happens when $t \to \infty$?) The population will stabilize at 4000.

84. Drug concentration. The concentration c (in milligrams per liter) of a drug in the bloodstream of a patient at time $t \ge 0$ (in hours since the drug was injected) is given by

$$c(t) = \frac{5t}{t^2 + 1}, \quad t \ge 0.$$

a. By plotting points or by using a graphing calculator, sketch the graph of $y = c(t)$. †
b. Find and interpret the horizontal asymptote of the graph of $y = c(t)$. †
c. Find the approximate time when the concentration of drug in the bloodstream is maximal. 1 hr after the injection
d. At what time is the concentration level equal to 2 milligrams per liter? Half an hour and 2 hr after the injection

85. Book publishing. The printing and binding cost for a college algebra book is $10. The editorial cost is $200,000. The first 2500 books are samples and are given free to professors. Let x be the number of college algebra books produced.
a. Write a function f describing the average cost of saleable books. †
b. Find the average cost of a saleable book if 10,000 books are produced. $40
c. How many books must be produced to bring the average cost of a saleable book under $20? More than 25,000

d. Find the vertical and horizontal asymptotes of the graph of $y = f(x)$. How are they related to this situation? †

86. A "phony" sale. A jewelry merchant at a local mall wants to have a " p percent off" sale. She marks up the stock by q percent to break even with respect to the prices before the sale.
a. Show that $q = \frac{p}{1 - p}$, $0 < p < 1$.
b. Sketch the graph of q in part **a**. †
c. How much should she mark up the stock to break even if she wants to have a "25% off" sale? 33.33%

C EXERCISES Beyond the Basics

In Exercises 87–96, use transformations of the graph of $y = \frac{1}{x}$ or $y = \frac{1}{x^2}$ to graph each rational function $f(x)$.

87. $f(x) = -\frac{2}{x}$ †

88. $f(x) = \frac{1}{2 - x}$ †

89. $f(x) = \frac{1}{(x - 2)^2}$ †

90. $f(x) = \frac{1}{x^2 + 2x + 1}$ †

91. $f(x) = \frac{1}{(x - 1)^2} - 2$ †

92. $f(x) = \frac{1}{(x + 2)^2} + 3$ †

93. $f(x) = \frac{1}{x^2 + 12x + 36}$ †

94. $f(x) = \frac{3x^2 + 18x + 28}{x^2 + 6x + 9}$ †

95. $f(x) = \frac{x^2 - 2x + 2}{x^2 - 2x + 1}$ †

96. $f(x) = \frac{2x^2 + 4x - 3}{x^2 + 2x + 1}$ †

97. Sketch and discuss the graphs of the rational functions of the form $y = \frac{a}{x^n}$, where a is a nonzero real number and n is a positive integer, in the following cases:
a. $a > 0$ and n is odd. †
b. $a < 0$ and n is odd. †
c. $a > 0$ and n is even. †
d. $a < 0$ and n is even. †

98. Graphing the reciprocal of a polynomial function. Let $f(x)$ be a polynomial function and $g(x) = \frac{1}{f(x)}$. Justify the following statements about the graphs of $f(x)$ and $g(x)$:
a. If c is a zero of $f(x)$, then $x = c$ is a vertical asymptote of the graph of $g(x)$. †
b. If $f(x) > 0$, then $g(x) > 0$, and if $f(x) < 0$, then $g(x) < 0$. †
c. The graphs of f and g intersect for those values (and only those values) of x for which $f(x) = g(x) = \pm 1$. †
d. When f is increasing (decreasing, constant) on an interval of its domain, g is decreasing (increasing, constant) on that interval. †

99. Use Exercise 98 to sketch the graphs of $f(x) = x^2 - 4$ and $g(x) = \frac{1}{x^2 - 4}$ on the same coordinate axes. †

100. Use Exercise 98 to sketch the graphs of $f(x) = x^2 + 1$ and $g(x) = \frac{1}{x^2 + 1}$ on the same coordinate axes. †

101. Recall that the inverse of a function $f(x)$ is written as $f^{-1}(x)$, while the reciprocal of $f(x)$ can be written as either $\dfrac{1}{f(x)}$ or $[f(x)]^{-1}$. The point of this exercise is to make clear that the reciprocal of a function has nothing to do with the inverse of a function. Let $f(x) = 2x + 3$. Find both $[f(x)]^{-1}$ and $f^{-1}(x)$. Compare the two functions. Graph all three functions on the same coordinate axes. †

102. Let $f(x) = \dfrac{x-1}{x+2}$. Find $[f(x)]^{-1}$ and $f^{-1}(x)$. Graph all three functions on the same coordinate axes. †

103. Sketch the graph of $g(x) = \dfrac{2x^3 + 3x^2 + 2x - 4}{x^2 - 1}$. Discuss the end behavior of $g(x)$. †

104. Sketch the graph of $f(x) = \dfrac{x^3 - 2x^2 + 1}{x - 2}$. Discuss the end behavior of $f(x)$. †

In Exercises 105 and 106, the graph of the rational function $f(x)$ crosses the horizontal asymptote. Graph $f(x)$ and find the point of intersection of the curve with its horizontal asymptote.

105. $f(x) = \dfrac{x^2 + x - 2}{x^2 - 2x - 3}$ † **106.** $f(x) = \dfrac{4 - x^2}{2x^2 - 5x - 3}$ †

In Exercises 107–110, find an equation of a rational function f satisfying the given conditions.

107. Vertical asymptote: $x = 3$
Horizontal asymptote: $y = -1$ $f(x) = \dfrac{2-x}{x-3}$
x-intercept: 2

108. Vertical asymptote: $x = -1, x = 1$
Horizontal asymptote: $y = 1$ $f(x) = \dfrac{x^2}{x^2 - 1}$
x-intercept: 0

109. Vertical asymptote: $x = 0$
Slant asymptote: $y = -x$ $f(x) = \dfrac{-x^2 + 1}{x}$
x-intercepts: $-1, 1$

110. Vertical asymptote: $x = 3$
Slant asymptote: $y = x + 4$ $f(x) = \dfrac{x^2 + x - 6}{x - 3}$
y-intercept: 2; $f(4) = 14$

Critical Thinking

In Exercises 111–116, make up a rational function $f(x)$ that has all of the characteristics given in the exercise.

111. Has a vertical asymptote at $x = 2$, a horizontal asymptote at $y = 1$, and a y-intercept at $(0, -2)$ †

112. Has vertical asymptotes at $x = 2$ and $x = -1$, a horizontal asymptote at $y = 0$, a y-intercept at $(0, 2)$, and an x-intercept at 4 †

113. Has $f\left(\dfrac{1}{2}\right) = 0$; $f(x) \to 4$ as $x \to \pm\infty$, $f(x) \to \infty$ as $x \to 1^-$, and $f(x) \to \infty$ as $x \to 1^+$. †

114. Has $f(0) = 0$; $f(x) \to 2$ as $x \to \pm\infty$; has no vertical asymptotes and is symmetric about the y-axis †

115. Has $y = 3x + 2$ as an oblique asymptote; has a vertical asymptote at $x = 1$ †

116. Is it possible for the graph of a rational function to have both a horizontal and an oblique asymptote? Explain. No because in that case, there would be values of x with two different corresponding function values; so it would not be a fuction.

GROUP PROJECTS Finding the Range and Turning Points

Find the range of the rational function $y = \dfrac{1}{x^2 + x - 2}$. First, solve this equation for x as follows:

$$y = \dfrac{1}{x^2 + x - 2} \qquad \text{Original equation}$$

$$(x^2 + x - 2)y = 1 \qquad \begin{array}{l}\text{Multiply both sides}\\\text{by } (x^2 + x - 2).\end{array}$$

$$x^2 y + xy - 2y = 1$$

$$yx^2 + yx - (2y + 1) = 0 \qquad \begin{array}{l}\text{This is a quadratic}\\\text{equation in } x.\end{array}$$

Use the quadratic formula with $a = y, b = y$, and $c = -(2y + 1)$:

$$x = \dfrac{-y \pm \sqrt{y^2 + 4y(2y + 1)}}{2y}$$

(1) $\qquad x = \dfrac{-y \pm \sqrt{9y^2 + 4y}}{2y}$.

Finally, for x to be a real number, the **discriminant** must be nonnegative. So $9y^2 + 4y \geq 0$, or $y(9y + 4) \geq 0$.

Sign graph / Test numbers:

$$+ \qquad - \qquad +$$

Test numbers: $-1 \qquad -\dfrac{4}{9} \quad -\dfrac{1}{4} \ 0 \qquad 1$

Using the sign graph, we have $y \leq \dfrac{-4}{9}$ or $y \geq 0$. However, Equation (1) is not defined for $y = 0$. Thus, the range of $f(x) = \dfrac{1}{x^2 + x - 2}$ is $\left(-\infty, \dfrac{-4}{9}\right] \cup (0, \infty)$. The value $y = -\dfrac{4}{9}$ is a local maximum value. Substitute $y = -\dfrac{4}{9}$ in (1) to obtain $x = -\dfrac{1}{2}$. Hence, $\left(-\dfrac{1}{2}, -\dfrac{4}{9}\right)$ is a local maximum point on the graph of f.

Use the preceding technique to find the range and turning points (if any) of the graph of each function.

1. $f(x) = \dfrac{x}{x^2 - 4}$

2. $f(x) = \dfrac{x^2 - 4}{x + 1}$ (See Example 8)

3. $R(x) = \dfrac{x(100 - x)}{x + 10}$ (See Example 9)

4. $f(x) = \dfrac{x^2 + 2}{(x + 2)(x - 1)}$ (See Example 6) †

1. Range: $(-\infty, \infty)$, no turning points
2. Range: $(-\infty, \infty)$, no turning points
3. Range: $(-\infty, 120 - 20\sqrt{11}) \cup (120 + 20\sqrt{11}, \infty)$; turning points: $(-10 + 10\sqrt{11}, 120 - 20\sqrt{11})$ is a local maximum, and $(-10 - 10\sqrt{11}, 120 + 20\sqrt{11})$ is a local minimum.

Variation

Sir Isaac Newton 1642–1727

Issac Newton was a mathematician and physicist, one of the foremost scientific intellects of all time. He entered Cambridge University in 1661 and was elected a Fellow of Trinity College in 1667 and Lucasian Professor of Mathematics in 1669. He remained at the university, lecturing most years, until 1696.

As a firm opponent of the attempt by King James II to make the universities into Catholic institutions, Newton was elected Member of Parliament for the University of Cambridge to the Convention Parliament of 1689, and he sat again in 1701–1702.

Meanwhile, in 1696, he moved to London as warden of the Royal Mint. He became master of the Mint in 1699, an office he retained until his death. He was elected a Fellow of the Royal Society of London in 1671, and in 1703, he became president, being annually reelected for the rest of his life. One of his major works, *Opticks*, appeared the next year; he was knighted in Cambridge in 1705.

Before Starting this Section, Review

1. Equation of a line (Section 1.2, page 19)
2. Circumference and area of a circle (Appendix A, page 791)

Objectives

1. Solve direct variation problems.
2. Solve inverse variation problems.
3. Solve joint and combined variation problems.

NEWTON AND THE APPLE

According to legend, Isaac Newton was sitting under an apple tree, and when an apple fell on his head, he suddenly thought of the Law of Universal Gravitation. As in many such legends, the story may not be true in detail, but it contains elements of the truth.

Before we tell you a more realistic version of this story, let's recall some terminology associated with the motion of objects.

$$\text{Recall that speed} = \frac{\text{distance}}{\text{time}}.$$

(i) The **velocity** of an object is the speed of an object in a specific direction.

(ii) **Acceleration:** If the velocity of an object is changing, we say that the object is accelerating. Acceleration is the rate of change of velocity.

(iii) **Force** = Mass × Acceleration

What really happened with the apple? Perhaps the correct version of the story is that upon observing an apple falling from a tree, Newton concluded that the apple accelerated because the velocity of the apple changed from when it began hanging on the tree (zero velocity) and then moved to the ground. He decided that a force must act on the apple to cause this acceleration. He called this force "gravity" and the associated acceleration (Force = Mass × Acceleration) the "acceleration due to gravity." He conjectured further that if the force of gravity reaches the top of the apple tree, it might reach even farther; might it reach all the way to the moon? In that case, to keep the moon moving in a circular orbit, rather than wandering into outer space, Earth must exert a gravitational force on the moon. Newton realized that the force that brought the apple to the ground and the force that holds the moon in orbit were the same. He concluded that any two objects in the universe exert gravitational attraction on each other, whereupon he gave us the *Law of Universal Gravitation*. (See Example 6.)

Some scientists believe that it was Kepler's Third Law of Planetary Motion (see Exercise 53), not an apple, that led Newton to his Law of Universal Gravitation. ∎

---

1 Solve direct variation problems.

Direct Variation

Two types of relationships between quantities (variables) occur so frequently in mathematics that they are given special names: *direct variation* and *inverse variation*.

Direct variation describes any relationship between two quantities in which any increase (or decrease) in one causes a proportional increase (or decrease) in the other. For example, the sales tax in Tampa, Florida, in 2008 was 7%, so that the sales tax on a

STUDY TIP

The statements

"*y* varies directly as *x*,"
"*y* varies as *x*,"
"*y* is directly proportional to *x*,"
"*y* is proportional to *x*,"
and "$y = kx, k \neq 0$"

have the same meaning.

$1000 purchase was $70. If we *double* the purchase to $2000, the sales tax also *doubles*, to $140. If we *reduce* the purchase by *half* to $500, the sales tax is also *reduced* by half, to $35. We say that the sales tax *varies directly* as the purchase price. The equation

$$\text{Sales tax} = (0.07)(\text{Purchase price})$$

expresses the fact that the sales tax was a constant (0.07) multiple of the purchase price.

DIRECT VARIATION

A quantity *y* is said to **vary directly** as the quantity *x*, or *y* is **directly proportional** to *x*, if there is a constant *k* such that $y = kx$. This constant *k* is called the **constant of variation** or the **constant of proportionality**.

FINDING THE SOLUTION: A PROCEDURE

EXAMPLE 1   Solving the Variation Problems

OBJECTIVE
Solve variation problems.

EXAMPLE
Suppose y varies as x and $x = \dfrac{4}{3}$ when y = 20. Find y when $x = 8$.

Step 1   Write the equation with the constant of variation, *k*.

$$y = kx \qquad \text{\textit{y} varies as \textit{x}.}$$

Step 2   Substitute the given values of the variables into the equation in Step 1 to find the value of the constant *k*.

$$20 = k\left(\frac{4}{3}\right) \qquad \text{Replace \textit{y} with 20 and \textit{x} with } \frac{4}{3}.$$

$$20\left(\frac{3}{4}\right) = k\left(\frac{4}{3}\right)\left(\frac{3}{4}\right) \qquad \text{Multiply both sides by } \frac{3}{4}.$$

$$15 = k \qquad \text{Simplify to solve for \textit{k}.}$$

Step 3   Rewrite the equation in Step 1 with the value of *k* from Step 2.

$$y = 15x \qquad \text{Replace \textit{k} with 15.}$$

Step 4   Use the equation from Step 3 to answer the question posed in the problem.

$$y = 15(8) \qquad \text{Replace \textit{x} with 8.}$$
$$y = 120 \qquad \text{Simplify.}$$

So $y = 120$ when $x = 8$.   ■ ■ ■

Practice Problem 1   Suppose *y* varies directly as *x*. If *y* is 6 when *x* is 30, find *y* when $x = 120$.   ■

EXAMPLE 2   Direct Variation in Electrical Circuits

The current in a circuit connected to a 220-volt battery is 50 amperes. If the current is directly proportional to the voltage of the attached battery, what voltage battery is needed to produce a current of 75 amperes?

SOLUTION
Let I = current in amperes and V = voltage in volts of the battery.

Step 1 $I = kV$ *I* varies directly as *V*.

Step 2 $50 = k(220)$ Substitute $I = 50, V = 220$.

$$\frac{50}{220} = k \qquad \text{Solve for \textit{k}.}$$

$$\frac{5}{22} = k \qquad \text{Simplify.}$$

Step 3 $I = \dfrac{5}{22}V$ Replace k with $\dfrac{5}{22}$ in $I = kV$.

Step 4 $75 = \dfrac{5}{22}V$ Substitute $I = 75$ in Step 3.

$\dfrac{22}{5} \cdot 75 = V$ Solve for V.

$330 = V$ Simplify.

A battery of 330 volts is needed to produce 75 amperes of current.   ■ ■ ■

Practice Problem 2 In Example 2, if the current in the circuit is 60 amperes, what voltage battery is needed to produce a current of 75 amperes?   ■

We can generalize the concept of direct variation to variation with powers.

DIRECT VARIATION WITH POWERS

"A quantity y varies directly as the nth power of x" means

$$y = kx^n,$$

where k is a nonzero constant and $n > 0$.

Note that when $n = 1$, y varies directly as x. In the formula for the area of a circle, $A = \pi r^2$, A varies directly as the square (or second power) of the radius r, with π as the constant of variation.

EXAMPLE 3   Solving a Problem Involving Direct Variation with Powers

Suppose you had forgotten the formula for the volume of a sphere but were told that the volume V of a sphere varies directly as the cube of its radius r. In addition, you are given that $V = 972\pi$ when $r = 9$. Find V when $r = 6$.

SOLUTION

Step 1 $V = kr^3$ V varies directly as r^3.

Step 2 $972\pi = k(9)^3$ Replace V with 972π and r with 9.

$972\pi = k(729)$ $9^3 = 729$

$k = \dfrac{972\pi}{729}$ Solve for k.

$= \dfrac{4}{3}\pi$ Simplify.

Step 3 $V = \dfrac{4}{3}\pi r^3$ Substitute $k = \dfrac{4}{3}\pi$ in Step 1.

Step 4 $V = \dfrac{4}{3}\pi(6)^3$ Replace r with 6.

$= 288\pi$ cubic units   ■ ■ ■

Practice Problem 3 If y varies directly as the square of x and $y = 48$ when $x = 2$, find y when $x = 5$.   ■

2 Solve inverse variation problems.

Inverse Variation

INVERSE VARIATION

A quantity y **varies inversely** as the quantity x, or y is **inversely proportional** to x, if

$$y = \frac{k}{x}$$

where k is a nonzero constant.

Suppose a plane takes four hours to fly from Atlanta to Denver at an average speed of 300 miles per hour. If we *increase* the average speed to 600 miles per hour (twice the previous speed), the flight time *decreases* to two hours (half the previous time). For the Atlanta–Denver trip, the time of flight varies inversely as the speed of the plane:

$$\text{time} = \frac{\text{distance}}{\text{speed}}$$

$$= \frac{k}{\text{speed}}, \text{where } k = \text{distance between Atlanta and Denver}$$

We see that the distance is the constant of variation when time and speed are the variables.

EXAMPLE 4 Solving an Inverse Variation Problem

Suppose y varies inversely as x and $y = 35$ when $x = 11$. Find y if $x = 55$.

SOLUTION

Step 1 $y = \dfrac{k}{x}$ y varies inversely as x.

Step 2 $35 = \dfrac{k}{11}$ Substitute $y = 35, x = 11$.

          $35 \cdot 11 = k$ Solve for k.

            $385 = k$ Simplify.

Step 3 $y = \dfrac{385}{x}$ Replace k with 385 in $y = \dfrac{k}{x}$.

Step 4 $y = \dfrac{385}{55}$ Replace x with 55.

          $y = 7$ Simplify.   ■ ■ ■

Practice Problem 4 A varies inversely as B, and $A = 12$ when $B = 5$. Find A when $B = 3$.   ■

INVERSE VARIATION WITH POWERS

A quantity y varies inversely as the nth power of x means that $y = \dfrac{k}{x^n}$, where k is a nonzero constant and $n > 0$.

BY THE WAY . . .

Light travels at a speed of 299,792,458 meters per second. The distance that light travels in a year is so large that it is a useful unit of distance in astronomy.

1. One light year is approximately 9.46×10^{15} m.
2. The nearest star (other than the sun) is 4.3 light years away.
3. The Milky Way (our galaxy) is about 100,000 light years in diameter.
4. The most distant objects that astronomers can see are about 18 billion light years away. Thus, the light that we presently see from these objects began its journey to us about 18 billion years ago. Since this is close to the estimated age of the universe, that light is a kind of fossil record of the universe not long after its birth.

3 Solve joint and combined variation problems.

EXAMPLE 5   Solving a Problem Involving Inverse Variation with Powers

The intensity of light varies inversely as the square of the distance from the light source. If Rita doubles her distance from a lamp, what happens to the intensity of light at her new location?

SOLUTION

Let I be the intensity of light at a distance d from the light source. Since I is inversely proportional to the square of d, we have

$$I = \frac{k}{d^2}.$$

If we replace d with $2d$, the intensity I_1 at the new location is given by

$$I_1 = \frac{k}{(2d)^2} \qquad \text{Replace } d \text{ with } 2d \text{ in } I = \frac{k}{d^2}.$$

$$I_1 = \frac{k}{4d^2} \qquad (2d)^2 = 2^2 d^2 = 4d^2$$

Therefore,

$$I_1 = \frac{I}{4} \qquad \text{Replace } \frac{k}{d^2} \text{ with } I.$$

This equation tells us that if Rita doubles her distance from the light source, the intensity of light at the new location will be one-fourth the intensity at the original location.   ■ ■ ■

Practice Problem 5   If y varies inversely as the square root of x and $y = \frac{3}{4}$ when $x = 16$, find x when $y = 2$.   ■

Joint and Combined Variation

Sometimes different types of variations occur in a problem that has more than one independent variable.

JOINT AND COMBINED VARIATION

The expression *z varies jointly as x and y* means that $z = kxy$ for some nonzero constant k. Similarly, if n and m are positive numbers, then the expression *z varies jointly as the nth power of x and mth power of y* means that $z = kx^n y^m$ for some nonzero constant k.

For example, the volume V of a right circular cylinder with radius r and height h is given by

$$V = \pi r^2 h.$$

Using the vocabulary of variation, we say, "The volume V of a right circular cylinder varies jointly as its height h and the square of its radius r." The constant of variation is π.

We can combine the joint variation and inverse variation. For example, the relationship

$$w = \frac{kx^2 y^3}{\sqrt{z}}$$

can be described as "w varies jointly as the square of x and cube of y and inversely as the square root of z." The same relationship also can be stated as "w varies directly as $x^2 y^3$ and inversely as $\sqrt{z}$."

EXAMPLE 6   **Newton's Law of Universal Gravitation**

Newton's Law of Universal Gravitation says that every object in the universe attracts every other object with a force acting along the line of the centers of the two objects. Also, this attracting force is directly proportional to the product of the two masses and inversely proportional to the square of the distance between the two objects.

a. Write the law symbolically.

b. Estimate the value of g (the acceleration due to gravity) near the surface of Earth. Use the following estimates: radius of Earth $R_E = 6.38 \times 10^6$ meters, and mass of Earth $M_E = 5.98 \times 10^{24}$ kilograms.

SOLUTION

a. Let m_1 and m_2 be the masses of the two objects and r be the distance between their centers. See Figure 2.28. Let F denote the gravitational force betwen the objects.

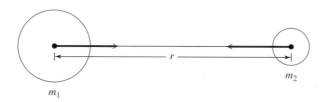

FIGURE 2.28   **Gravitational attraction**

Then according to Newton's Law of Universal Gravitation,

$$F = G \cdot \frac{m_1 m_2}{r^2}.$$

The constant of proportionality G is called the *universal gravitational constant*. It is "universal" because it is thought to be the same at all places and all times. If the masses m_1 and m_2 are measured in kilograms, r is measured in meters, and the force F is measured in newtons, then the value of G is approximately 6.67×10^{-11} cubic meters per kilogram per second squared. This value was experimentally verified by Henry Cavendish in 1795.

b. *Estimating the value of* g (acceleration due to gravity)

We use Newton's Laws: Force = Mass × Acceleration = $m \cdot g$

$$\text{Force} = G \cdot \frac{m_1 m_2}{r^2}$$

For an object of mass m near the surface of Earth (see Figure 2.29):

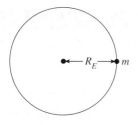

FIGURE 2.29   **Finding g on Earth**

$$\text{Force} = G \cdot \frac{m_1 m_2}{r^2} \qquad \text{Law of Universal Gravitation}$$

$$\cancel{m} \cdot g = G \cdot \frac{\cancel{m} M_E}{R_E^2} \qquad \text{Force} = m \cdot g; \text{ replace } m_1 \text{ with } m \text{ and } m_2 \text{ with } M_E.$$

$$g = G \cdot \frac{M_E}{R_E^2} \qquad \text{Divide both sides by } m.$$

$$= \frac{(6.67 \times 10^{-11} \text{ m}^2/\text{kg/sec}^2) \cdot (5.98 \times 10^{24} \text{ kg})}{(6.38 \times 10^6)^2 \text{ m}} \qquad \text{Substitute appropriate values for } G, M_E, \text{ and } R_E.$$

$$\approx 9.8 \text{ m/sec}^2 \qquad \text{Use a calculator.} \qquad ■ ■ ■$$

Practice Problem 6   The mass of Mars is about 6.42×10^{23} kilograms, and its radius is about 3397 kilometers. What is the acceleration due to gravity near the surface of Mars?   ■

We summarize the steps involved in solving most variation problems.

SOLVING VARIATION PROBLEMS

Step 1 Write the equation that models the problem.

a. y varies directly with x.	$y = kx$
b. y varies with the nth power of x.	$y = kx^n$
c. y varies inversely with x.	$y = \dfrac{k}{x}$
d. y varies inversely with the nth power of x.	$y = \dfrac{k}{x^n}$
e. z varies jointly with the nth power of x and the mth power of y.	$z = kx^n y^m$
f. z varies directly with the nth power of x and inversely with the mth power of y.	$z = \dfrac{kx^n}{y^m}$

Step 2 Substitute the given values in the equation in Step 1 and solve for k.

Step 3 Rewrite the equation in Step 1 with the value of k from Step 2.

Step 4 Use the equation from Step 3 to answer the question posed in the problem.

If a variation problem uses a more complicated relationship than those in Step 1, simply write the appropriate equation involving k and the variables and then proceed with Steps 2–4.

SECTION 2.6 ▪ Exercises

A EXERCISES Basic Skills and Concepts

In Exercises 1–8, write each statement as an equation.

1. P varies directly as T. $P = kT$

2. P varies inversely as V. $P = \dfrac{k}{V}$

3. y varies as the square root of x. $y = k\sqrt{x}$

4. F varies jointly as m_1 and m_2. $F = km_1m_2$

5. V varies jointly as the cube of x and fourth power of y.

6. w varies directly as the square root of x and inversely as y.

7. z varies jointly as x and u and inversely as square of v.

8. S varies jointly as the square root of x, the square of y, and the cube of z and inversely as the square of u.

In Exercises 9–24, use the four-step procedure to solve for the variable requested.

9. x varies directly as y, and $x = 15$ when $y = 30$. Find x if $y = 28$. $k = \dfrac{1}{2}, x = 14$

10. y varies directly as x, and $y = 3$ when $x = 2$. Find y when $x = 7$. $k = \dfrac{3}{2}, y = \dfrac{21}{2}$

11. s varies directly as the square of t, and $s = 64$ when $t = 2$. Find s when $t = 5$. $k = 16, s = 400$

12. y varies directly as the cube of x, and $y = 270$ when $x = 3$. Find x when $y = 80$. $k = 10, x = 2$

13. r varies inversely as u, and $r = 3$ when $u = 11$. Find r if $u = \dfrac{1}{3}$. $k = 33, r = 99$

14. y varies inversely as z, and $y = 24$ when $z = \dfrac{1}{6}$. Find y if $z = 1$. $k = 4, y = 4$

15. B varies inversely with the cube of A. If $B = 1$ when $A = 2$, find B when $A = 4$. $k = 8, B = \dfrac{1}{8}$

16. y varies inversely with the cube root of x, and $y = 10$ when $x = 2$. Find x if $y = 40$. $k = 10\sqrt[3]{2}, x = \dfrac{1}{32}$

17. z varies jointly as x and y, and $z = 42$ when $x = 2$ and $y = 3$. Find y if $z = 56$ and $x = 2$. $k = 7, y = 4$

18. m varies directly as q and inversely as p, and $m = \dfrac{1}{2}$ when $p = 26$ and $q = 13$. Find m when $p = 14$ and $q = 7$. $k = 1, m = \dfrac{1}{2}$

†Due to space constrictions, answers to these exercises may be found in the Answers beginning on page A–1 in the back of the book.

Answers:

5. $V = kx^3y^4$ 6. $w = \dfrac{k\sqrt{x}}{y}$ 7. $z = \dfrac{kxu}{v^2}$ 8. $S = \dfrac{k\sqrt{x}y^2z^3}{u^2}$

19. z varies directly as the square of x, and $z = 32$ when $x = 4$. Find z if $x = 5$. $k = 2, z = 50$

20. u varies inversely as the cube of t, and $u = 9$ when $t = 2$. Find u if $t = 6$. $k = 72, u = \frac{1}{3}$

21. P varies jointly as T and the square of Q, and $P = 36$ when $T = 17$ and $Q = 6$. Find P when $T = 4$ and $Q = 9$. †

22. a varies jointly as b and the square root of c, and $a = 9$ when $b = 13$ and $c = 81$. Find a when $b = 5$ and $c = 9$. †

23. z varies directly as the square root of x and inversely as the square of y, and $z = 24$ when $x = 16$ and $y = 3$. Find x when $z = 27$ and $y = 2$. $k = 54, x = 4$

24. z varies jointly as u and the cube of v and inversely as the square of w; $z = 9$ when $u = 4$, $v = 3$, and $w = 2$. Find w when $u = 27$, $v = 2$, and $z = 8$. $k = \frac{1}{3}, w = \pm 3$

In Exercises 25–28, solve for the variable requested without determining the constant of variation, k. Use the fact that if

$$x_1 = ky_1 \text{ and } x_2 = ky_2, \text{ then } \frac{x_1}{y_1} = k = \frac{x_2}{y_2} \text{ so that } \frac{x_1}{y_1} = \frac{x_2}{y_2}.$$

25. If y is proportional to x and if $y = 12$ when $x = 16$, find y when $x = 8$. $y = 6$

26. If z varies directly as w and if $z = 17$ when $w = 22$, find z when $w = 110$. $z = 85$

27. If y is directly proportional to x and if $y = 100$ for the value x_0 of x, find y when x_0 is doubled—that is, when $x = 2x_0$. $y = 200$

28. If x varies directly as the square root of y and if $x = 2$ when $y = 9$, find y when $x = 3$. $y = \frac{81}{4}$

B EXERCISES Applying the Concepts

29. **Hubble constant.** The American astronomer Edwin Powell Hubble (1889–1953) is renowned for having determined that there are other galaxies in the universe beyond the Milky Way. In 1929, he stated that the galaxies observed at a particular time recede from each other at a speed that is directly proportional to the distance between them. The constant of proportionality is denoted by the letter H. Write Hubble's statement in the form of an equation.

30. **Malthusian doctrine.** The British scientist Thomas Robert Malthus (1766–1834) was a pioneer of population science and economics. In 1798, he stated that populations grow faster than the means that can sustain them. One of his arguments was that the rate of change, R, of a given population is directly proportional to the size P of the population. Write the equation that describes Malthus's argument. $R = kP$, where k is a constant.

31. **Converting units of length.** Use the fact that 1 foot ≈ 30.5 centimeters.
    a. Write an equation that expresses the fact that a length measured in centimeters is directly proportional to the length measured in feet. †
    b. Convert the following measurements into centimeters.
        (i) 8 feet †   (ii) 5 feet 4 inches †

    c. Convert the following measurements into feet.
        (i) 57 centimeters †   (ii) 1 meter 24 centimeters †

32. **Converting weight.** Use the fact that 1 kilogram ≈ 2.20 pounds.
    a. Write an equation that expresses the fact that a weight measured in pounds is directly proportional to the weight measured in kilograms. †
    b. Convert the following measurements to pounds.
        (i) 125 grams   0.275 lb
        (ii) 4 kilograms   8.8 lb
        (iii) 2.4 kilograms   5.28 lb
    c. Convert the following measurements to kilograms.
        (i) 27 pounds   12.27 kg
        (ii) 160 pounds   72.73 kg

33. **Protein from soybeans.** The quantity P of protein obtained from dried soybeans is directly proportional to the quantity Q of dried soybeans used. If 20 grams of dried soybeans produces 7 grams of protein, how much protein can be produced from 100 grams of soybeans?   35 g

34. **Wages.** A person's weekly wages W are directly proportional to the number of hours h worked per week. If Samantha earned $600 for a 40-hour week, how much would she earn if she worked only 25 hours in a particular week? What does the constant of proportionality mean here? $375. The constant of proportionality is the hourly wage.

35. **Physics.** The distance d that an object falls varies directly as the square of the time t during which it is falling. If an object falls 64 feet in 2 seconds, how long will it take the object to fall 9 feet? $\frac{3}{4}$ sec

36. **Physics. Hooke's Law** states that the force F required to stretch a spring by x units is directly proportional to x. If a force of 10 pounds stretches a spring by 4 inches, find the force required to stretch a spring by 6 inches.   15 lb

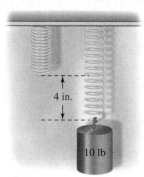

4 in.

10 lb

37. **Chemistry. Boyle's Law** states that at a constant temperature, the pressure P of a compressed gas is inversely proportional to its volume V. If the pressure is 20 pounds per square inch when the volume of the gas is 300 cubic inches, what is the pressure when the gas is compressed to 100 cubic inches?   60 lb/in.^{2}

38. **Chemistry.** In the Kelvin temperature scale, the lowest possible temperature (called absolute zero) is 0 K, where K denotes degrees Kelvin. The relationship between Kelvin temperature (T_K) and Celsius temperature (T_C) is given by $T_K = T_C + 273$.
    The pressure P exerted by a gas varies directly as its temperature T_K and inversely as its volume V. Assume that at a temperature of 260 K, a gas occupies 13 cubic inches at a pressure of 36 pounds per square inch.

Answers:
29. $y = Hx$, where y is the speed of the galaxies and x is the distance between them.

a. Find the volume of the gas when the temperature is 300 K and the pressure is 40 pounds per square inch. 13.5 in.³
b. Find the pressure when the temperature is 280 K and the volume is 39 cubic inches. 12.92 lb/in.²

39. Weight. The weight of an object varies inversely as the square of the object's distance from the center of Earth. The radius of Earth is 3960 miles.
  a. If an astronaut weighs 120 pounds on the surface of Earth, how much does she weigh 6000 miles above the surface of Earth? 18.97 lb
  b. If a miner weighs 200 pounds on the surface of Earth, how much does he weigh 10 miles below the surface of Earth? 201.01 lb

40. Making a profit. Suppose you wanted to make a profit by buying gold by weight at one altitude and selling at another altitude for the same price per unit weight. Should you buy or sell at the higher altitude? (Use the information from Exercise 39.) Buy

In Exercises 41 and 42, use Newton's Law of Universal Gravitation. (See Example 6.)

41. Gravity on the moon. The mass of the moon is about 7.4×10^{22} kilograms, and its radius is about 1740 kilometers. How much is the acceleration due to gravity on the surface of the moon? 1.63 m/s²

42. Gravity on the sun. The mass of the sun is about 2×10^{30} kilograms, and its radius is 696,000 kilometers. How much is the acceleration due to gravity on the surface of the sun? 2.75×10^2 m/s²

43. Illumination. The intensity I of illumination from a light source is inversely proportional to the square of the distance d from the source. Suppose the intensity is 320 candlepower at a distance of 10 feet from a light source.
  a. What is the intensity at 5 feet from the source? †
  b. How far away from the source will the intensity be 400 candlepower? 8.94 ft from the source

44. Speed and skid marks. Police estimate that the speed s of a car in miles per hour varies directly as the square root of d, where d in feet is the length of the skid marks left by a car traveling on a dry concrete pavement. A car traveling 48 miles per hour leaves skid marks of 96 feet.
  a. Write an equation relating s and d. $s = 2\sqrt{6d}$
  b. Use the equation in part (**a**) to estimate the speed of a car whose skid marks stretched (i) 60 feet, (ii) 150 feet, and (iii) 200 feet. †
  c. Suppose you are driving 70 miles per hour and slam on your brakes. How long will your skid marks be? 204.17 ft

45. Simple pendulum. *Periodic motion* is motion that repeats itself over successive equal intervals of time. The time

required for one complete repetition of the motion is called the *period*. The period of a simple pendulum varies directly as the square root of its length. What is the effect on the length if the period is doubled?
  (*Note:* The amplitude of a pendulum has no effect on its period. This is what makes pendulums such good timekeepers. Because they invariably lose energy due to friction, their amplitude decreases but their period remains constant.) The length is multiplied by 4.

46. Biology. The volume V of a lung is directly proportional to its internal surface area A. A lung with volume 400 cubic centimeters from a certain species has an average internal surface area of 100 square centimeters. Find the volume of a lung of a member of this species if the lung's internal surface area is 120 square centimeters. 480 cm³

47. Horsepower. The horsepower H of an automobile engine varies directly as the square of the piston radius R and the number N of pistons.
  a. Write the given information in equation form. †
  b. What is the effect on the horsepower if the piston radius is doubled? The horsepower is multiplied by 4.
  c. What is the effect on the horsepower if the number of pistons is doubled? The horsepower is doubled.
  d. What is the effect on the horsepower if the radius of the pistons is cut in half and the number of pistons is doubled? The horsepower is cut in half.

48. Safe load. The safe load that a rectangular beam can support varies jointly as the width and square of the depth of the beam and inversely as its length. A beam 4 inches wide, 6 inches deep, and 25 feet long can support a safe load of 576 pounds. Find the safe load for a beam that is of the same material but is 6 inches wide, 10 inches deep, and 20 feet long. 3000 lb

C EXERCISES Beyond the Basics

49. Energy from a windmill. The energy E from a windmill varies jointly as the square of the length l of the blades and the cube of the wind velocity v.
  a. Express the given information as an equation with k as the constant of variation. $E = kl^2v^3$
  b. If blades of length 10 feet and wind velocity 8 miles per hour generate 1920 watts of electric power, find k. †
  c. How much electric power would be generated if the blades were 8 feet long and the wind velocity was 25 miles per hour? 37,500 W
  d. If the velocity of the wind doubles, what happens to E? †
  e. If the length of blades doubles, what happens to E? †
  f. What happens to E if both the length of the blades and the wind velocity are doubled? E is multiplied by 32.

50. Intensity of light. The intensity of light, I_d, at a distance d from the source of light varies directly as the intensity I of the source and inversely as d^2. It is known that at a distance of 2 meters, a 100-watt bulb produces an intensity of approximately 2 watts per square meter.
  a. Find the constant of variation, k. $k = 0.08$

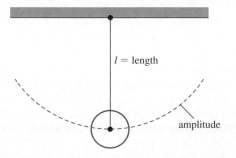
l = length

amplitude

b. Suppose a 200-watt bulb is located on a wall 2 meters above the floor. What is the intensity of light at a point A that is 3 meters from the wall? 1.23 W/m^2

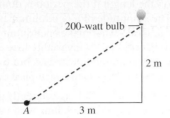

200-watt bulb
2 m
A 3 m

c. What is the intensity of illumination at the point A in the figure if the bulb is raised by 1 meter? 0.89 W/m^2

51. Metabolic rate. Metabolism is the sum total of all physical and chemical changes that take place within an organism. According to the laws of thermodynamics, all of these changes will ultimately release heat, so the metabolic rate is a measure of heat production by an animal. Biologists have found that the normal resting metabolic rate of a mammal is directly proportional to the $\frac{3}{4}$ power of its body weight.

The resting metabolic rate of a person weighing 75 kilograms is 75 watts.
a. Find the constant of proportionality, k. $k = 2.94$
b. Estimate the resting metabolic rate of a brown bear weighing 450 kilograms. 287.25 W
c. What is the effect on metabolic rate if the body weight is multiplied by 4? †
d. What is the approximate weight of an animal with a metabolic rate of 250 watts? 373.93 kg

52. Comparing gravitational forces. The masses of the sun, Earth, and the moon are 2×10^{30} kilograms, 6×10^{24} kilograms, and 7.4×10^{22} kilograms, respectively. The Earth–sun distance is about 400 times the Earth–moon distance. Use Newton's Law of Universal Gravitation to compare the gravitational attraction between the sun and Earth with that between Earth and the moon. †

53. Kepler's Third Law. Suppose an object of mass M_1 orbits around an object of mass M_2. Let r be the average distance in meters between the centers of the two objects and let T be the *orbital period* (the time in seconds the object completes one orbit). Kepler's Third Law states that T^2 is directly proportional to r^3 and inversely proportional to $M_1 + M_2$. The constant of proportionality is $\frac{4\pi^2}{G}$.
a. Write the equation that expresses Kepler's Third Law. †
b. Earth orbits the sun once each year at a distance of about 1.5×10^8 kilometers. Find the mass of the sun. Use these estimates: $2.01 \times 10^{30} \text{ kg}$
Mass of Earth + Mass of sun ≈ Mass of sun,
$G = 6.67 \times 10^{-11}$, 1 year $= 3.15 \times 10^7$ seconds

54. Use Kepler's Third Law. The moon orbits Earth in about 27.3 days at an average distance of about 384,000 kilometers. Find the mass of Earth. [*Hint:* Convert the orbital period, defined in Exercise 53, to seconds; also use Mass of Earth + Mass of moon ≈ Mass of Earth.]
$6.02 \times 10^{24} \text{ kg}$

55. Spreading a rumor. Let P be the population of a community. The rate R (per day) at which a rumor spreads in the community is jointly proportional to the number N of people who have already heard the rumor and the number $(P - N)$ of people who have not yet heard the rumor.

A rumor about the president of a college is spread in his college community of 10,000 people. Five days after the rumor started, 1000 people had heard it, and it was spreading at the rate of 45 additional people per day.
a. Write an equation relating R, P, and N. †
b. Find the constant k of variation. †
c. Find the rate at which the rumor was spreading when one-half the college community had heard it. †
d. How many people had heard the rumor when it was spreading at the rate of 100 people per day?
  a. $R = kN(P - N)$, where k is the constant of variation.
  b. $k = \dfrac{5}{10^6}$ **c.** 125 people per day   **d.** 2764 or 7236

Critical Thinking

56. Electricity. The current I in an electric circuit varies directly as the voltage V and inversely as the resistance R. If the resistance is increased by 20%, what percent increase must occur in the voltage to increase the current by 30%? 56%

57. Precious stones. The value of a precious stone is proportional to the square of its weight.
a. Calculate the loss incurred by cutting a diamond worth $1000 into two pieces whose weights are in the ratio 2:3. $480
b. A precious stone worth $25,000 is accidentally dropped and broken into three pieces, the weights of which are in the ratio 5:9:11. Calculate the loss incurred due to breakage. $15,920
c. A diamond breaks into five pieces, the weights of which are in the ratio 1:2:3:4:5. If the resulting loss is $85,000, find the value of the original diamond. Also calculate the value of a diamond whose weight is twice that of the original diamond.

58. Bus service. The profit earned in running a bus service is jointly proportional to the distance and the number of passengers over a certain fixed number. The profit is $80 when 30 passengers are carried a distance of 40 km and is $180 when 35 passengers are carried 60 km. What is the minimum number of passengers that results in no loss? 20

59. Weight of a sphere. The weight of a sphere is directly proportional to the cube of its radius. A metal sphere has a hollow space about its center in the form of a concentric sphere, and its weight is $\frac{7}{8}$ times the weight of a solid sphere of the same radius and material. Find the ratio of the inner to the outer radius of the hollow sphere. 1:2

60. Train speed. A locomotive engine can go 24 miles per hour, and its speed is reduced by a quantity that varies directly as the square root of the number of cars it pulls. Pulling four cars, its speed is 20 miles per hour. Find the greatest number of cars the engine can pull. 144

Answer:
57. c. The value of the original diamond is $112,500. The value of a diamond whose weight is twice that of the original is $450,000.

SUMMARY ▪ Definitions, Concepts, and Formulas

2.1 Quadratic Functions

i. A **quadratic function** f is a function of the form

$$f(x) = ax^2 + bx + c, a \neq 0.$$

ii. The **standard form** of a quadratic function is

$$f(x) = a(x - h)^2 + k, a \neq 0.$$

iii. The graph of a quadratic function is a transformation of the graph of $y = x^2$.

iv. The graph of a quadratic function is a parabola with vertex

$$(h, k) = \left(-\frac{b}{2a}, f\left(-\frac{b}{2a} \right) \right).$$

v. The maximum (if $a < 0$) or minimum (if $a > 0$) value of a quadratic function $f(x) = ax^2 + bx + c$ occurs at the vertex of the parabola.

2.2 Polynomial Functions

i. A function f of the form
$$f(x) = a_n x^n + a_{n-1} x^{n-1} + \cdots + a_1 x + a_0, a_n \neq 0$$
is a polynomial function of degree n.

ii. The graph of a polynomial function is smooth and continuous.

iii. The end behavior of the graph of a polynomial function depends upon the sign of the leading coefficient and the degree of the polynomial.

iv. A real number c is a **zero** of a function f if $f(c) = 0$. Geometrically, c is an x-intercept of the graph of $y = f(x)$.

v. If in the factorization of a polynomial function $f(x)$ the factor $(x - a)$ occurs exactly m times, then a is a zero of **multiplicity** m. If m is odd, the graph of $y = f(x)$ crosses the x-axis at a; if m is even, the graph touches but does not cross the x-axis at a.

vi. If the degree of a polynomial function $f(x)$ is n, then $f(x)$ has, at most, n real zeros and the graph of $f(x)$ has at most $(n - 1)$ turning points.

vii. **The Intermediate Value Theorem:** Let $f(x)$ be a polynomial function and a and b be two numbers such that $a < b$. If $f(a)$ and $f(b)$ have opposite signs, then there is at least one number c, with $a < c < b$, for which $f(c) = 0$.

viii. See page 128 for graphing a polynomial function.

2.3 Dividing Polynomials and the Rational Zeros Test

i. Division Algorithm: If a polynomial $F(x)$ is divided by a polynomial $D(x) \neq 0$, there are unique polynomials $Q(x)$ and $R(x)$ such that $F(x) = D(x)Q(x) + R(x)$, where either $R(x) = 0$ or deg $R(x) <$ deg $D(x)$. In words, "The dividend equals the product of the divisor and the quotient plus the remainder."

ii. Remainder Theorem: If a polynomial $F(x)$ is divided by $(x - a)$, the remainder is $F(a)$.

iii. Factor Theorem: A polynomial function $F(x)$ has $(x - a)$ as a factor if and only if $F(a) = 0$.

iv. Rational Zeros Test: If $\dfrac{p}{q}$ is a rational zero in lowest terms for a polynomial function with integer coefficients, then p is a factor of the constant term and q is a factor of the leading coefficient.

2.4 Zeros of a Polynomial Function

i. Descartes's Rule of Signs. Let $F(x)$ be a polynomial function with real coefficients.
  a. The number of positive zeros of F is equal to the number of variations of signs of $F(x)$ or is less than that number by an even integer.
  b. The number of negative zeros of F is equal to the number of variations of sign of $F(-x)$ or is less than that number by an even integer.

ii. Rules for Bounds on the Zeros. Suppose a polynomial $F(x)$ is synthetically divided by $x - k$.
  a. If $k > 0$ and each number in the last row is zero or positive, then k is an upper bound on the zeros of $F(x)$.
  b. If $k < 0$ and the numbers in the last row alternate in sign, then k is a lower bound on the zeros of $F(x)$.

iii. The Fundamental Theorem of Algebra. An nth-degree polynomial equation has at least one complex zero.

iv. Factorization Theorem for Polynomials. If $P(x)$ is a polynomial of degree $n \geq 1$, it can be factored into n (not necessarily distinct) linear factors of the form $P(x) = a(x - r_1)(x - r_2) \cdots (x - r_n)$, where a, $r_1, r_2, \ldots, r_n$ are complex numbers.

v. Number of Zeros Theorem. A polynomial of degree n has exactly n complex zeros provided that a zero of multiplicity k is counted k times.

vi. Conjugate Pairs Theorem. If $a + bi$ is a zero of the polynomial function P (with real coefficients), then $a - bi$ is also a zero of P.

iv. The line $y = k$ is a horizontal asymptote of the graph of f if $f(x) \to k$ as $x \to \infty$ or as $x \to -\infty$.

v. A procedure for graphing rational functions is given on page 161.

2.6 Variation

k is a nonzero constant called the constant of variation.

2.5 Rational Functions

i. A function $F(x) = \dfrac{N(x)}{D(x)}$, where $N(x)$ and $D(x)$ are polynomials and $D(x) \neq 0$, is called a rational function. The domain of $F(x)$ is the set of all real numbers except the real zeros of $D(x)$.

ii. The line $x = a$ is a vertical asymptote of the graph of f if $|f(x)| \to \infty$ as $x \to a^+$ or as $x \to a^-$.

iii. If $\dfrac{N(x)}{D(x)}$ is in lowest terms, then the graph of $F(x)$ has vertical asymptotes at the real zeros of $D(x)$.

Variation	Equation
y varies directly with x.	$y = kx$
y varies with the nth power of x.	$y = kx^n$
y varies inversely with x.	$y = \dfrac{k}{x}$
y varies inversely with the nth power of x.	$y = \dfrac{k}{x^n}$
z varies jointly with the nth power of x and the mth power of y.	$z = kx^n y^m$
z varies directly with the nth power of x and inversely with the mth power of y.	$z = \dfrac{kx^n}{y^m}$

REVIEW EXERCISES

Basic Skills and Concepts

In Exercises 1–10, graph each quadratic function by finding (i) whether the parabola opens up or down, (ii) its vertex, (iii) its axis, (iv) its x-intercepts, (v) its y-intercept, and (vi) the intervals over which the function is increasing and decreasing.

1. $y = (x - 1)^2 + 2$ †
2. $y = (x + 2)^2 - 3$ †
3. $y = -2(x - 3)^2 + 4$ †
4. $y = -\dfrac{1}{2}(x + 1)^2 + 2$ †
5. $y = -2x^2 + 3$ †
6. $y = 2x^2 + 4x - 1$ †
7. $y = 2x^2 - 4x + 3$ †
8. $y = -2x^2 - x + 3$ †
9. $y = 3x^2 - 2x + 1$ †
10. $y = 3x^2 - 5x + 4$ †

In Exercises 11–14, determine whether the given quadratic function has a maximum or a minimum value and then find that value.

11. $f(x) = 3 - 4x + x^2$
12. $f(x) = 8x - 4x^2 - 3$
13. $f(x) = -2x^2 - 3x + 2$
14. $f(x) = \dfrac{1}{2}x^2 - \dfrac{3}{4}x + 2$

In Exercises 15–18, graph each polynomial function by using transformations on the appropriate function $y = x^n$.

15. $f(x) = (x + 1)^3 - 2$ †
16. $f(x) = (x + 1)^4 + 2$ †
17. $f(x) = (1 - x)^3 + 1$ †
18. $f(x) = x^4 + 3$ †

In Exercises 19–24, for each polynomial function f,

(i) determine the end behavior of f.
(ii) determine the zeros of f. State the multiplicity of each zero. Determine whether the graph of f crosses or only touches the axis at each x-intercept.
(iii) find the x- and y-intercepts of the graph of f.
(iv) use test numbers to find the intervals over which the graph of f is above or below the x-axis.
(v) find any symmetry.
(vi) sketch the graph of $y = f(x)$.

19. $f(x) = x(x - 1)(x + 2)$ †
20. $f(x) = x^3 - x$ †
21. $f(x) = -x^2(x - 1)^2$ †
22. $f(x) = -x^3(x - 2)^2$ †
23. $f(x) = -x^2(x^2 - 1)$ †
24. $f(x) = -(x - 1)^2(x^2 + 1)$ †

In Exercises 25–28, divide by using long division.

25. $\dfrac{6x^2 + 5x - 13}{3x - 2}$
26. $\dfrac{8x^2 - 14x + 15}{2x - 3}$
27. $\dfrac{8x^4 - 4x^3 + 2x^2 - 7x + 165}{x + 1}$
28. $\dfrac{x^3 - 3x^2 + 4x + 7}{x^2 - 2x + 6}$ Quotient: $x - 1$; remainder: $13 - 4x$

In Exercises 29–32, divide by using synthetic division.

29. $\dfrac{x^3 - 12x + 3}{x - 3}$ †
30. $\dfrac{-4x^3 + 3x^2 - 5x}{x - 6}$ †

†Due to space constrictions, answers to these exercises may be found in the Answers beginning on page A–1 in the back of the book.

Answers:
11. Minimum: $(2, -1)$ **12.** Maximum: $(1, 1)$ **13.** Maximum: $\left(-\dfrac{3}{4}, \dfrac{25}{8}\right)$ **14.** Minimum: $\left(\dfrac{3}{4}, \dfrac{55}{32}\right)$ **25.** Quotient: $2x + 3$; remainder: -7
26. Quotient: $4x - 1$; remainder: 12 **27.** Quotient: $8x^3 - 12x^2 + 14x - 21$; remainder: 186

31. $\dfrac{2x^4 - 3x^3 + 5x^2 - 7x + 165}{x + 1}$ Quotient: $2x^3 - 5x^2 + 10x - 17$; remainder: 182

32. $\dfrac{3x^5 - 2x^4 + x^2 - 16x - 132}{x + 2}$ Quotient: $3x^4 - 8x^3 + 16x^2 - 31x + 46$; remainder: -224

In Exercises 33–36, a polynomial function $f(x)$ and a constant c are given. Find $f(c)$ by (i) evaluating the function and (ii) using synthetic division and the Remainder Theorem.

33. $f(x) = x^3 - 3x^2 + 11x - 29; c = 2$ -11

34. $f(x) = 2x^3 + x^2 - 15x - 2; c = -2$ 16

35. $f(x) = x^4 - 2x^2 - 5x + 10; c = -3$ 88

36. $f(x) = x^5 + 2; c = 1$ 3

In Exercises 37–40, a polynomial function $f(x)$ and a constant c are given. Use synthetic division to show that c is a zero of $f(x)$. Use the result to final all zeros of $f(x)$.

37. $f(x) = x^3 - 7x^2 + 14x - 8; c = 2$ 1, 2, 4

38. $f(x) = 2x^3 - 3x^2 - 12x + 4; c = -2$ †

39. $f(x) = 3x^3 + 14x^2 + 13x - 6; c = \dfrac{1}{3}$ $-3, -2, \dfrac{1}{3}$

40. $f(x) = 4x^3 + 19x^2 - 13x + 2; c = \dfrac{1}{4}$ †

In Exercises 41 and 42, use the Rational Zeros Test to list all possible rational zeros of $f(x)$.

41. $f(x) = x^4 + 3x^3 - x^2 - 9x - 6$ $-6, -3, -2, -1, 1, 2, 3, 6$

42. $f(x) = 9x^3 - 36x^2 - 4x + 16$ †

In Exercises 43–46, use Descartes's Rule of Signs and the Rational Zeros Test to find all real zeros of each polynomial function.

43. $f(x) = 5x^3 + 11x^2 + 2x$ $-2, -\dfrac{1}{5}, 0$

44. $f(x) = x^3 + 2x^2 - 5x - 6$ $-3, -1, 2$

45. $f(x) = x^3 + 3x^2 - 4x - 12$ $-3, -2, 2$

46. $f(x) = 2x^3 - 9x^2 + 12x - 5$ $1, \dfrac{5}{2}$

In Exercises 47–52, find all of the zeros of $f(x)$, real and nonreal, that are not given.

47. $f(x) = x^3 - 7x + 6$; one zero is 2. $-3, 1, 2$

48. $f(x) = x^4 + x^3 - 3x^2 - x + 2$; 1 is a zero of multiplicity 2. $-2, -1, 1$

49. $f(x) = x^4 - 2x^3 + 6x^2 - 18x - 27$; two zeros are -1 and 3. $-1, 3, 3i, -3i$

50. $f(x) = 4x^3 - 19x^2 + 32x - 15$; one zero is $2 - i$.

51. $f(x) = x^4 + 2x^3 + 9x^2 + 8x + 20$; one zero is $-1 + 2i$.

52. $f(x) = x^5 - 7x^4 + 24x^3 - 32x^2 + 64$; $2 + 2i$ is a zero of multiplicity 2. $-1, 2 + 2i, 2 - 2i$

In Exercises 53–58, solve each equation in the complex number system.

53. $x^3 - x^2 - 4x + 4 = 0$ $\{-2, 1, 2\}$

54. $2x^3 + x^2 - 12x - 6 = 0$

55. $4x^3 - 7x - 3 = 0$

56. $x^3 + x^2 - 8x - 6 = 0$ $\{-3, 1 - \sqrt{3}, 1 + \sqrt{3}\}$

57. $x^4 - x^3 - x^2 - x - 2 = 0$ $\{-1, 2, i, -i\}$

58. $x^4 - x^3 - 13x^2 + x + 12 = 0$ $\{-3, -1, 1, 4\}$

In Exercises 59 and 60, show that the given equation has no rational roots.

59. $x^3 + 13x^2 - 6x - 2 = 0$ †

60. $3x^4 - 9x^3 - 2x^2 - 15x - 5 = 0$ †

In Exercises 61 and 62, use the Intermediate Value Theorem to find the value of the real root between 1 and 2 of each equation to two decimal places.

61. $x^3 + 6x^2 - 28 = 0$ 1.88

62. $x^3 + 3x^2 - 3x - 7 = 0$ 1.60

In Exercises 63–70, graph each rational function by following the six-step procedure outlined in Section 2.5.

63. $f(x) = 1 + \dfrac{1}{x}$ †

64. $f(x) = \dfrac{2 - x}{x}$ †

65. $f(x) = \dfrac{x}{x^2 - 1}$ †

66. $f(x) = \dfrac{x^2 - 9}{x^2 - 4}$ †

67. $f(x) = \dfrac{x^3}{x^2 - 9}$ †

68. $f(x) = \dfrac{x + 1}{x^2 - 2x - 8}$ †

69. $f(x) = \dfrac{x^4}{x^2 - 4}$ †

70. $f(x) = \dfrac{x^2 + x - 6}{x^2 - x - 12}$ †

Applying the Concepts

71. Variation. If y varies directly as x and if $y = 12$ when $x = 4$, find y when $x = 5$. $y = 15$

72. Variation. If p varies inversely as q and if $p = 4$ when $q = 3$, find p when $q = 4$. $p = 3$

73. Variation. If s varies directly as the square of t and if $s = 20$ when $t = 2$, find s when $t = 3$. $s = 45$

74. Variation. If y varies inversely as x^2 and if $y = 3$ when $x = 8$, find x when $y = 12$. $x = \pm 4$

75. Missile path. A missile fired from the origin of a coordinate system follows a path described by the equation $y = -\dfrac{1}{10}x^2 + 20x$, where the x-axis is at ground level. Sketch the missile's path and determine what its maximum altitude is and where it hits the ground. †

76. Minimizing area. Suppose a wire 20 centimeters long is to be cut into two pieces, each of which will be formed into a square. Find the size of each piece that minimizes the total area. Both pieces must be 10 cm.

77. A farmer wants to fence off three identical adjoining rectangular pens, each 400 square feet in area. (See the figure.) What should be the width and length of each pen so that the least amount of fence is used?

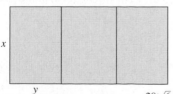

$x = 10\sqrt{6}$ ft, $y = \dfrac{20\sqrt{6}}{3}$ ft

Answers:

50. $\dfrac{3}{4}, 2 + i, 2 - i$ **51.** $2i, -2i, -1 + 2i, -1 - 2i$ **54.** $\left\{-\dfrac{1}{2}, -\sqrt{6}, \sqrt{6}\right\}$ **55.** $\left\{-1, -\dfrac{1}{2}, \dfrac{3}{2}\right\}$

78. Suppose the outer boundary of the pens in Exercise 77 requires heavy fence that costs $5 per foot and two internal partitions cost $3 per foot. What dimensions x and y will minimize the cost?

79. Electric circuit. In the circuit shown in the figure, the voltage $V = 100$ volts and the resistance $R = 50$ ohms. We want to determine the size of the remaining resistor (x ohms). The power absorbed by the circuit is given by

$$p(x) = \frac{V^2 x}{(R + x)^2}.$$

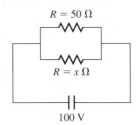

$R = 50\ \Omega$

$R = x\ \Omega$

$100\ V$

a. Graph the function $y = p(x)$. †
b. Use a graphing calculator to find the value of x that maximizes the power absorbed. $x = 50$

80. Maximizing area. A sheet of paper for a poster is 18 square feet in area. The margins at the top and bottom are 9 inches each, and the margin on each side is 6 inches. What should the dimensions of the paper be if the printed area is to be a maximum? $36\sqrt{3}$ in. × $24\sqrt{3}$ in.

81. Maximizing profit. A manufacturer makes and sells printers to retailers at $24 per unit. The total daily cost C in dollars of producing x printers is given by

$$C(x) = 150 + 3.9x + \frac{3}{1000}x^2.$$

$P(x) = -\dfrac{3}{1000}x^2 + 20.1x - 150$

a. Write the profit P as a function of x.
b. Find the number of printers the manufacturer should produce and sell to achieve maximum profit. 3350
c. Find the average cost $\overline{C}(x) = \dfrac{C(x)}{x}$. Graph $y = \overline{C}(x)$. †

82. Meteorology. The function $p = \dfrac{69.1}{a + 2.3}$ relates the atmospheric pressure p in inches of mercury to the altitude a in miles from the surface of Earth.

a. Find the pressure on Mount Kilimanjaro at an altitude of 19,340 feet. 11.59 in.
b. Is there an altitude at which the pressure is 0? No

83. Wages. An employee's wages are directly proportional to the time he or she has worked. Sam earned $280 for 40 hours. How much would Sam earn if he worked for 35 hours? $245

84. Car's stopping distance. The distance required for a car to come to a stop after its brakes are applied is directly proportional to the square of its speed. If the stopping distance for a car traveling 30 miles per hour is 25 feet,

what is the stopping distance for a car traveling 66 miles per hour? 121 ft

85. Illumination. The amount of illumination from a source of light varies directly as the intensity of the source and inversely as the square of the distance from the source. At what distance from a light source of intensity 300 candlepower will the illumination be one-half the illumination 6 inches away from the source? $6\sqrt{2}$ in.

86. Chemistry. *Charles's Law* states that at a constant pressure, the volume V of a gas is directly proportional to its temperature T (in Kelvin degrees). If a bicycle tube is filled with 1.2 cubic feet of air at a temperature of 295 K, what will be the volume of the air in the tube if the temperature rises to 310 K while the pressure stays the same? 1.26 ft³

87. Electric circuits. The current I (measured in amperes) in an electrical circuit varies inversely as the resistance R (measured in ohms) when the voltage is held constant. The current in a certain circuit is 30 amperes when the resistance is 300 ohms.

a. Find the current in the circuit if the resistance is decreased to 250 ohms. 36 amp
b. What resistance will yield a current of 60 amperes? 150 Ω

88. Safe load. The safe load that a circular column can support varies directly as the square root of its radius and inversely as the square of its length. A pillar with radius 4 inches and length 12 feet can safely support a 20-ton load. Find the load that a pillar of the same material with diameter 6 inches and length 10 feet can safely support. 24.94 tons

89. Spread of disease. An infectious cold virus spreads in a community at a rate R (per day) that is jointly proportional to the number of people who are infected with the virus and the number of people in the community who are not infected yet. After the tenth day of the start of a certain infection, 15% of the total population of 20,000 people of Pollutville had been infected and the virus was spreading at the rate of 255 people per day.

a. Find the constant of proportionality, k.
b. At what rate is the disease spreading when one-half the population is infected? 500 people per day
c. Find the number of people infected when the rate of infection reached 95 people per day. 1000 and 19,000

90. Coulomb's law. The electric charge is measured in coulombs. (The charges on an electron and a proton, which are equal and opposite, are approximately 1.602×10^{-19} coulomb.) Coulomb's Law states that the force F between two particles is jointly proportional to their charges q_1 and q_2 and inversely proportional to the square of the distance between the two particles. Two charges are acted upon by a repulsive force of 96 units. What is the force if the distance between the particles is quadrupled? 6 units

89. a. $k = \dfrac{1}{200,000}$

Answers:

78. $x = 5\sqrt{30}$ ft, $y = \dfrac{8\sqrt{30}}{3}$ ft

PRACTICE TEST A

1. Find the x-intercepts of the graph of
   $f(x) = x^2 - 6x + 2$. x-intercepts: $3 \pm \sqrt{7}$

2. Graph $f(x) = 3 - (x + 2)^2$. †

3. Find the vertex of the parabola described by
   $y = -7x^2 + 14x + 3$. $(1, 10)$

4. Find the domain of the function
   $$f(x) = \frac{x^2 - 1}{x^2 + 3x - 4}.\quad (-\infty, -4) \cup (-4, 1) \cup (1, \infty).$$

5. Find the quotient and remainder of $\dfrac{x^3 - 2x^2 - 5x + 6}{x + 2}$.
   Quotient: $x^2 - 4x + 3$; remainder: 0

6. Graph the polynomial function $P(x) = x^5 - 4x^3$. †

7. Find all of the zeros of $f(x) = 2x^3 - 2x^2 - 8x + 8$,
   given that 2 is one of the zeros. $-2, 1, 2$

8. Find the quotient of $\dfrac{-6x^3 + x^2 + 17x + 3}{2x + 3}$.
   Quotient: $-3x^2 + 5x + 1$

9. Use the Remainder Theorem to find the value $P(-2)$ of
   the polynomial $P(x) = x^4 + 5x^3 - 7x^2 + 9x + 17$.

10. Find all rational roots of the equation $x^3 - 5x^2 - 4x + 20 = 0$ and then find the irrational roots if there are any. $-2, 2, 5$

11. Find the zeros of the polynomial function
    $f(x) = x^4 + x^3 - 15x^2$.

12. For $P(x) = 2x^{18} - 5x^{13} + 6x^3 - 5x + 9$, list all possible
    rational zeros found by the Rational Zeros Test, but do
    not check to see which values are actually zeros.

Answers:

9. $P(-2) = -53$ 11. $0, -\dfrac{1}{2} + \dfrac{\sqrt{61}}{2}, -\dfrac{1}{2} - \dfrac{\sqrt{61}}{2}$

12. $-9, -\dfrac{9}{2}, -3, -\dfrac{3}{2}, -1, -\dfrac{1}{2}, \dfrac{1}{2}, 1, \dfrac{3}{2}, 3, \dfrac{9}{2}, 9$

13. Describe the end behavior of $f(x) = (x + 3)^3(x - 5)^2$.

14. Find the zeros and the multiplicity of each zero for
    $f(x) = (x^2 - 4)(x + 2)^2$.

15. Determine how many positive and how many negative real
    zeros the polynomial function $P(x) = 3x^6 + 2x^3 - 7x^2 + 8x$ can have. Number of positive zeros: 0 or 2; number
    of negetive zeros: 1

16. Find the horizontal and the vertical asymptotes of the
    graph of Horizontal asymptote: $y = 2$; vertical asymptotes:
    $x = -4$ and $x = 5$
    $$f(x) = \frac{2x^2 + 3}{x^2 - x - 20}.$$

17. Write an equation that expresses the statement "y is
    directly proportional to x and inversely proportional to
    the square of t."

18. In Problem 17, suppose $y = 6$ when $x = 8$ and $t = 2$.
    Find y if $x = 12$ and $t = 3$. $y = 4$

19. The cost C of producing x thousand units of a product is
    given by 15 thousand units
    $$C = x^2 - 30x + 335 \text{ (dollars)}.$$
    Find the value of x for which the cost is minimum.

20. From a rectangular 8×17 piece of cardboard, four
    congruent squares with sides of length x are cut out, one
    at each corner. The sides can then be folded to form a
    box. Find the volume V of the box as a function of x.

13. $f(x) \to -\infty$ as $x \to -\infty$ and $f(x) \to \infty$ as $x \to \infty$.

14. $x = -2$, multiplicity: 3; $x = 2$, multiplicity: 1

17. $y = \dfrac{kx}{t^2}$, where k is a constant   20. $V(x) = x(8 - 2x)(17 - 2x)$

PRACTICE TEST B

1. Find the x-intercepts of the graph of $f(x) = x^2 + 5x + 3$. b

   a. $\dfrac{-5 \pm i\sqrt{13}}{2}$ b. $\dfrac{-5 \pm \sqrt{13}}{2}$

   c. $\dfrac{-5 \pm i\sqrt{37}}{2}$ d. $\dfrac{-5 \pm \sqrt{37}}{2}$

2. Which is the graph of $f(x) = 4 - (x - 2)^2$? d

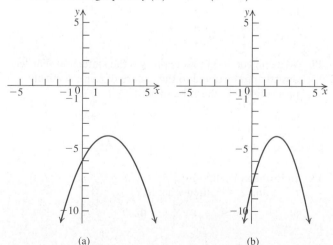

(a)   (b)

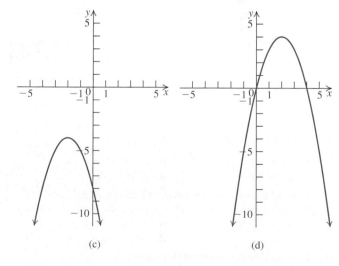

(c)   (d)

3. Find the vertex of the parabola described by
   $y = 6x^2 + 12x - 5$. a
   a. $(-1, -11)$ b. $(1, -5)$
   c. $(-1, 13)$ d. $(1, 13)$

4. Which of the following is *not* in the domain of the function $f(x) = \dfrac{x^2}{x^2 + x - 6}$? b

I. -3
II. 0
III. 2

a. I and II
b. I and III
c. II and III
d. I only

5. Find the quotient and remainder when $x^3 - 8x + 6$ is divided by $x + 3$. d

a. $x^2 - 8; 2$
b. $x^2 - 8; 0$
c. $x^2 - 3x + 1; x + 3$
d. $x^2 - 3x + 1; 3$

6. Which is the graph of the polynomial $P(x) = x^4 + 2x^3$? c

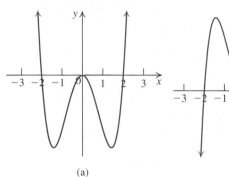

(a)

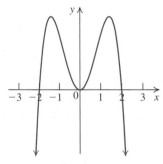

(b)

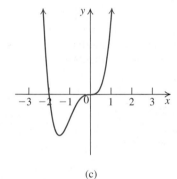

(c)

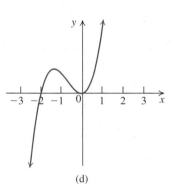

(d)

7. Find all of the zeros of $f(x) = 3x^3 - 26x^2 + 61x - 30$ given that 3 is one of the zeros (that is, $f(3) = 0$). c

a. $3, -5, -\dfrac{2}{3}$
b. $3, 2, \dfrac{5}{3}$
c. $3, 5, \dfrac{2}{3}$
d. $3, -2, \dfrac{5}{3}$

8. Find the quotient $\dfrac{-10x^3 + 21x^2 - 17x + 12}{2x - 3}$. b

a. $x^2 - 3x + 4$
b. $-5x^2 + 3x - 4$
c. $x^2 + 3x - 4$
d. $-5x^2 - 4$

9. Use the Remainder Theorem to find the value $P(-3)$ of the polynomial $P(x) = x^4 + 4x^3 + 7x^2 + 10x + 15$. c

a. 13
b. 15
c. 21
d. 6

10. Find all rational roots of the equation $-x^3 + x^2 + 8x - 12 = 0$ and then find the irrational roots if there are any. a

a. -3 and 2
b. -3 and $\sqrt{2}$
c. $-1, -2$, and 3
d. -3 and $\sqrt{3}$

11. Find the zeros of the polynomial function $f(x) = x^3 + x^2 - 30x$. a

a. $x = -6, x = 5, x = 0$
b. $x = 0, x = -6$
c. $x = 4, x = 5$
d. $x = 0, x = 4, x = 5$

12. For $P(x) = x^{30} - 4x^{25} + 6x^2 + 60$, list all possible rational zeros found by the Rational Zeros Test (but do not check to see which values are actually zeros). d

a. $\pm 2, \pm 3, \pm 4, \pm 5, \pm 6, \pm 8, \pm 10, \pm 12, \pm 15, \pm 20, \pm 30$, and ± 60
b. $\pm 1, \pm 2, \pm 3, \pm 4, \pm 5, \pm 6, \pm 10, \pm 12, \pm 15, \pm 18, \pm 20, \pm 24, \pm 30$, and ± 60
c. $\pm 1, \pm 3, \pm 4, \pm 5, \pm 6, \pm 12, \pm 15, \pm 20, \pm 30$, and ± 60
d. $\pm 1, \pm 2, \pm 3, \pm 4, \pm 5, \pm 6, \pm 10, \pm 12, \pm 15, \pm 20, \pm 30$, and ± 60

13. Which of the following correctly describes the end behavior of $f(x) = (x + 1)^2(x - 2)^2$? c

a. $\begin{cases} y \to \infty \text{ as } x \to -\infty \\ y \to -\infty \text{ as } x \to \infty \end{cases}$
b. $\begin{cases} y \to -\infty \text{ as } x \to -\infty \\ y \to -\infty \text{ as } x \to \infty \end{cases}$
c. $\begin{cases} y \to \infty \text{ as } x \to -\infty \\ y \to \infty \text{ as } x \to \infty \end{cases}$
d. $\begin{cases} y \to -\infty \text{ as } x \to -\infty \\ y \to \infty \text{ as } x \to \infty \end{cases}$

14. Find the zeros and the multiplicity of each zero for $f(x) = (x^2 - 1)(x + 1)^2$. b

a. $\begin{cases} \text{zero } 1, & \text{multiplicity 1} \\ \text{zero } -1, & \text{multiplicity 2} \end{cases}$
b. $\begin{cases} \text{zero } 1, & \text{multiplicity 1} \\ \text{zero } -1, & \text{multiplicity 3} \end{cases}$
c. $\begin{cases} \text{zero } 1, & \text{multiplicity 2} \\ \text{zero } -1, & \text{multiplicity 2} \end{cases}$
d. $\begin{cases} \text{zero } i, & \text{multiplicity 2} \\ \text{zero } -1, & \text{multiplicity 2} \end{cases}$

15. Determine how many positive and how many negative real zeros the polynomial $P(x) = x^5 - 4x^3 - x^2 + 6x - 3$ can have. c

a. positive 3, negative 2
b. positive 2 or 0, negative 2 or 0
c. positive 3 or 1, negative 2 or 0
d. positive 2 or 0, negative 3 or 1

16. The horizontal and the vertical asymptotes of the graph of $f(x) = \dfrac{x^2 + x - 2}{x^2 + x - 12}$ are c

a. $x = -2, y = 3, y = -4$.
b. $x = 1, y = 3, y = -4$.
c. $y = 1, x = 3, x = -4$.
d. $y = -2, x = 3, x = -4$.

17. Write an equation that expresses the statement that S is directly proportional to the square of t and inversely proportional to the cube of x. d

a. $S = \dfrac{kt^3}{\sqrt[3]{x}}$
b. $S = kt^2x^3$
c. $S = \dfrac{kt^3}{x^2}$
d. $S = \dfrac{kt^2}{x^3}$

18. In Problem 17, suppose $S = 27$ if $t = 3$ and $x = 1$. If $t = 6$ and $x = 3$, then what is S? b

a. 54
b. 4
c. $\dfrac{4}{3}$
d. 9

19. The cost C of producing x thousand units of a product is given by b

$$C = x^2 - 24x + 319 \text{ (dollars)}.$$

Find the value of x for which the cost is minimum.
   a. 319   **b.** 12   **c.** 175   **d.** 24

20. From a rectangular 10×12 piece of cardboard, four congruent squares with sides of length x are cut out, one at each corner. The sides can then be folded to form a box. Find the volume V of the box as a function of x. d
   a. $V = 2x(10 - x)(12 - x)$
   b. $V = 2x(10 - 2x)(12 - 2x)$
   c. $V = x(10 - x)(12 - x)$
   d. $V = x(10 - 2x)(12 - 2x)$

CUMULATIVE REVIEW EXERCISES ▪ Chapters 1 and 2

1. Find the distance between the points $P(-1,3)$ and $Q(2,5)$. $\sqrt{13}$

2. Find the midpoint of the line segment joining $P(2,-5)$ and $Q(-8,-3)$. $M(-3,-4)$

3. Find the x- and y-intercepts of the graph of the equation $y = x^2 - 2x - 8$ and sketch the graph. †

In Exercises 4 and 5, find the slope and intercepts of the line and sketch the graph.

4. $x + 3y - 6 = 0$ †   5. $x = 2y - 6$ †

6. Find an equation in standard form of the circle with center $(2, -3)$ and radius 4. $(x - 2)^2 + (y + 3)^2 = 16$

7. Find the center and radius of the circle with equation
$$x^2 + y^2 + 2x - 4y - 4 = 0.$$
Center: $(-1, 2)$; radius: 3

In Exercises 8 and 9, find the slope–intercept form of the line satisfying the given condition.

8. The line has slope 3 and passes through $(1, -2)$. $y = 3x - 5$

9. The line is parallel to $2x + 3y = 5$ and passes through $(1, 3)$. $y = -\dfrac{2}{3}x + \dfrac{11}{3}$

In Exercises 10 and 11, find the domain of each function.

10. $f(x) = \dfrac{1}{2x + 3}$ 11. $p(x) = \dfrac{1}{\sqrt{4 - 2x}}$

12. Let $f(x) = x^2 - 2x + 3$. Find $f(-2), f(3), f(x + h)$, and $\dfrac{f(x + h) - f(x)}{h}$.

13. Let $f(x) = \sqrt{x}$ and $g(x) = x^2 + 1$. Find each of the following.
   a. $f(g(x))$ **b.** $g(f(x))$
   c. $f(f(x))$ **d.** $g(g(x))$

14. Let $f(x) = \begin{cases} 3x + 2 & \text{if } x \le 2 \\ 4x - 1 & \text{if } 2 < x \le 3 \\ 6 & \text{if } x > 3 \end{cases}$.
   a. Find $f(1), f(3)$, and $f(4)$. $f(1) = 5, f(3) = 11, f(4) = 6$
   b. Sketch the graph of $y = f(x)$. †

15. Let $f(x) = 2x - 3$. Find $f^{-1}(x)$. $f^{-1}(x) = \dfrac{1}{2}x + \dfrac{3}{2}$

16. Use transformations on $y = \sqrt{x}$ to sketch the graph of each function.
   a. $f(x) = \sqrt{x + 2}$ †   **b.** $g(x) = -2\sqrt{x + 1} + 3$ †

17. Use the Rational Zeros Test to list all possible rational zeros of $f(x) = 2x^4 - 3x^2 + 5x - 6$.

In Exercises 18–21, graph each function.

18. $f(x) = 2x^2 - 4x + 1$ †

19. $f(x) = -x^2 + 2x + 3$ †

20. $f(x) = (x - 1)^2(x + 2)$ †

21. $f(x) = \dfrac{x^2 - 1}{x^2 - 4}$ †

22. Let $f(x) = x^4 - 3x^3 + 2x^2 + 2x - 4$. Given that $1 + i$ is a zero of f, find all zeros of f. $-1, 2, 1 + i, 1 - i$

23. Suppose that y varies as the square root of x and that $y = 6$ when $x = 4$. Find y if $x = 9$. 9

24. A drug manufactured by a pharmaceutical company is sold in bulk at a price of \$150 per unit. The total production cost (in dollars) for x units in one week is
$$C(x) = 0.02x^2 + 100x + 3000.$$
How many units of the drug must be manufactured and sold in a week to maximize the profit? What is the maximum profit? 1250, \$28,250

25. The profit (in dollars) for a product is given by
$$P(x) = 0.02x^3 + 48.8x^2 - 2990x + 25,000,$$
where x is the number of units produced and sold. One break-even point occurs when $x = 10$. Use synthetic division to find another break-even point for the product. $x = 50$

Answers:

10. Domain: $\left(-\infty, -\dfrac{3}{2}\right) \cup \left(-\dfrac{3}{2}, \infty\right)$ 11. Domain $(-\infty, 2)$.   12. $f(-2) = 11, f(3) = 6, f(x + h) = x^2 + 2(h - 1)x + h^2 - 2h + 3$,

$\dfrac{f(x + h) - f(x)}{h} = 2x + h - 2$ 13. **a.** $f(g(x)) = \sqrt{x^2 + 1}$ **b.** $g(f(x)) = x + 1$ **c.** $f(f(x)) = \sqrt[4]{x}$ **d.** $g(g(x)) = x^4 + 2x^2 + 2$

17. $-6, -3, -2, -\dfrac{3}{2}, -1, -\dfrac{1}{2}, \dfrac{1}{2}, 1, \dfrac{3}{2}, 2, 3, 6$

3 Exponential and Logarithmic Functions

The world of finance, the growth of many populations, the molecular activity that results in an atomic explosion, and countless everyday events are modeled with exponential and logarithmic functions.

Exponential Functions

Before Starting this Section, Review

1. Integer exponents (Appendix A, page 758)

2. Rational exponents (Appendix A, page 785)

3. Graphing and transformations (Section 1.5, page 62)

4. One-to-one functions (Section 1.7, page 88)

5. Increasing and decreasing functions (Section 1.4, page 50)

Objectives

1. Define an exponential function.
2. Graph exponential functions.
3. Solve exponential equations.
4. Use transformations on exponential functions.

FOOLING A KING

One night in northwest India, a wise man named Shashi invented a wonderful new game called "Shatranj" (chess). The next morning he took it to the king Rai Bhalit, who marveled at it and said to Shashi, "Name your reward." Shashi merely requested that 1 grain of wheat be placed on the first square of the chessboard, 2 grains on the second, 4 on the third, 8 on the fourth, and so on, for all 64 squares. The king agreed to his request, thinking the man was an eccentric fool for asking for only a few grains of wheat when he could have had gold, jewels, or even his daughter's hand in marriage.

You may know the end of this story. Much more wheat was needed to satisfy Shashi's request. The king could never fulfill his promise to Shashi, so instead he had him beheaded.

The details of this rapidly growing wheat phenomenon are given in the margin. Table 3.1 shows the number of grains of wheat that need to be stacked on each square of the chessboard.

These data can be modeled by the function

$$g(n) = 2^{n-1},$$

where $g(n)$ is the number of grains of wheat and n is the number of the square (or the nth square) on the chessboard. Each square has twice the number of grains as the previous square. The function g is an example of an *exponential function* with base 2, with n restricted to $1, 2, 3, \ldots, 64$—the number of the chessboard squares. The name *exponential function* comes from the fact that the variable n occurs in the exponent. The base 2 is the *growth factor*.

When the growth rate of a quantity is directly proportional to the existing amount, the growth can be modeled by an exponential function. For example, exponential functions are often used to model the growth of investments and populations, the cell division of living organisms, and the decay of radioactive material. ■

TABLE 3.1 Grains of Wheat on a Chessboard

Square number	Grains placed on this square
1	$1 \, (= 2^0)$
2	$2 \, (= 2^1)$
3	$4 \, (= 2^2)$
4	$8 \, (= 2^3)$
5	$16 \, (= 2^4)$
6	$32 \, (= 2^5)$
.	.
.	.
.	.
63	2^{62}
64	2^{63}

Exponential Functions

EXPONENTIAL FUNCTION

A function f of the form

$$f(x) = a^x, a > 0 \text{ and } a \neq 1,$$

is called an **exponential function with base a.** Its domain is $(-\infty, \infty)$.

In the definition of the exponential function, we rule out the base $a = 1$ because in this case, the function is simply the constant function $f(x) = 1^x$, or $f(x) = 1$. We exclude negative bases so that the domain can include all real numbers. For example, a cannot be -2 because $f\left(\dfrac{1}{2}\right) = (-2)^{1/2} = \sqrt{-2}$ is not a real number. Other functions that have variables in the exponent, such as $f(x) = 4 \cdot 3^x$, $g(x) = -5 \cdot 4^{3x-2}$, and $h(x) = c \cdot a^x$ ($a > 0$ and $a \neq 1$), are also called exponential functions.

Evaluate Exponential Functions

We can evaluate exponential functions by using the laws of exponents and/or calculators, as in the next example.

EXAMPLE 1   **Evaluating Exponential Functions**

---

a. Let $f(x) = 3^{x-2}$. Find $f(4)$.

b. Let $g(x) = -2 \cdot 10^x$. Find $g(-2)$.

c. Let $h(x) = \left(\dfrac{1}{9}\right)^x$. Find $h\left(-\dfrac{3}{2}\right)$.

d. Let $F(x) = 4^x$. Find $F(3.2)$.

SOLUTION

a. $f(4) = 3^{4-2} = 3^2 = 9$

b. $g(-2) = -2 \cdot 10^{-2} = -2 \cdot \dfrac{1}{10^2} = -2 \cdot \dfrac{1}{100} = -0.02$

c. $h\left(-\dfrac{3}{2}\right) = \left(\dfrac{1}{9}\right)^{-\frac{3}{2}} = (9^{-1})^{-\frac{3}{2}} = 9^{\frac{3}{2}} = (\sqrt{9})^3 = 27$

d. $F(3.2) = 4^{3.2} \approx 84.44850629$ Use a calculator.   ■ ■ ■

Practice Problem 1 Let $f(x) = \left(\dfrac{1}{4}\right)^x$. Find $f(2), f(0), f(-1), f\left(\dfrac{5}{2}\right)$, and $f\left(-\dfrac{3}{2}\right)$.

■

TECHNOLOGY CONNECTION

You can use either the $\boxed{\wedge}$ or $\boxed{x^y}$ key on your calculator to evaluate exponential functions. For example, to evaluate $4^{3.2}$, press

4 $\boxed{\wedge}$ 3.2 $\boxed{\text{ENTER}}$

or

4 $\boxed{x^y}$ 3.2 $\boxed{=}$. You will see the following display:

```
4^(3.2)
        84.44850629
```

Here the displayed number is an approximate value, so we should write $4^{3.2} \approx 84.44850629$.

Because the domain of an exponential function is $(-\infty, \infty)$, in Example 1, we evaluated exponential functions at some rational numbers. But what is the meaning of a^x when x is an irrational number? For example, what do expressions such as $3^{\sqrt{2}}$ and 2^π mean? It turns out that the definition of a^x (with $a > 0$ and x irrational) requires methods discussed in calculus. However, we can understand the basis for the definition from the following discussion. Suppose we want to define the number 2^π. We use several numbers that approximate π. A calculator shows that $\pi \approx 3.14159265 \ldots$. We successively approximate 2^π by using the rational powers shown in Table 3.2.

TABLE 3.2

Values of $f(x) = 2^x$ for rational values of x that approach π					
x	3	3.1	3.14	3.141	3.1415
2^x	$2^3 = 8$	$2^{3.1} = 8.5\ldots$	$2^{3.14} = 8.81\ldots$	$2^{3.141} = 8.821\ldots$	$2^{3.1415} = 8.8244\ldots$

It can be shown that the powers $2^3, 2^{3.1}, 2^{3.14}, 2^{3.141}, 2^{3.1415}, \ldots$ approach exactly one number. We define 2^π as that number. Table 3.2 shows that $2^\pi \approx 8.82$ (correct to two decimal places).

For our work with exponential functions, we need the following fundamental facts:

1. Exponential functions $f(x) = a^x$ are defined for all real numbers x.

2. The graph of an exponential function is a continuous (unbroken) curve.

3. The rules of exponents hold for all real numbers.

RULES OF EXPONENTS

Let a, b, x, and y be real numbers with $a > 0$ and $b > 0$.

$$a^x \cdot a^y = a^{x+y} \qquad (a^x)^y = a^{xy}$$

$$\frac{a^x}{a^y} = a^{x-y} \qquad a^0 = 1$$

$$(ab)^x = a^x b^x \qquad a^{-x} = \frac{1}{a^x} = \left(\frac{1}{a}\right)^x$$

---

2 Graph exponential functions.

Graphing Exponential Functions

Let's see how to sketch the graph of an exponential function. Although the domain is the set of all real numbers, we usually evaluate the functions only for the integer values of x (for ease of computation). To evaluate a^x for noninteger values of x, use your calculator. Recall that the graph of a function $f(x)$ is the graph of the equation $y = f(x)$. We can use either $f(x) = a^x$ or $y = a^x$ to represent a given function.

EXAMPLE 2 **Graphing an Exponential Function with Base $a > 1$**

Graph the exponential function $f(x) = 3^x$.

SOLUTION

First, make a table of a few values of x and the corresponding values of y.

x	-3	-2	-1	0	1	2	3
$y = 3^x$	$\frac{1}{27}$	$\frac{1}{9}$	$\frac{1}{3}$	1	3	9	27

Next, plot the points (see Figure 3.1(a)) and draw a smooth curve through them. See Figure 3.1(b).

TECHNOLOGY CONNECTION

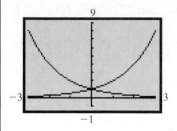

Graph $y_1 = 2^x$ and $y_2 = \left(\dfrac{1}{2}\right)^x$ on the same screen.

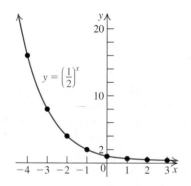

Notice that the graphs of y_1 and y_2 are reflections of each other in the y-axis.

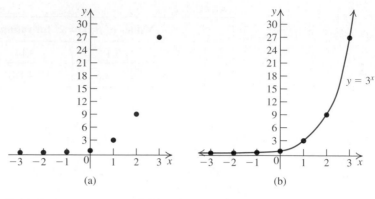

(a)   (b)

FIGURE 3.1 **An exponential function $y = a^x$ with $a > 1$**

The graph in Figure 3.1(b) is typical of the graphs of exponential functions $f(x) = a^x$ when $a > 1$. Note that the x-axis is the horizontal asymptote of the graph of $y = a^x$.   ■ ■ ■

Practice Problem 2   Sketch the graph of

$$f(x) = 2^x.$$ ■

Let's now sketch the graph of an exponential function $f(x) = a^x$ when $0 < a < 1$.

EXAMPLE 3   **Graphing an Exponential Function $f(x) = a^x$, with $0 < a < 1$**

Sketch the graph of $y = \left(\dfrac{1}{2}\right)^x$.

SOLUTION
Make a table similar to the one in Example 2.

x	-3	-2	-1	0	1	2	3
$y = \left(\dfrac{1}{2}\right)^x$	8	4	2	1	$\dfrac{1}{2}$	$\dfrac{1}{4}$	$\dfrac{1}{8}$

FIGURE 3.2 An exponential function $y = a^x$ with $0 < a < 1$

Plotting these points and drawing a smooth curve through them, we get the graph of $y = \left(\dfrac{1}{2}\right)^x$, shown in Figure 3.2. In this graph, as x increases in the positive direction, $y = \left(\dfrac{1}{2}\right)^x$ decreases toward 0. This graph is typical of the graph of an exponential function $y = a^x$ with $0 < a < 1$.   ■ ■ ■

Practice Problem 3   Sketch the graph of

$$f(x) = \left(\dfrac{1}{3}\right)^x.$$ ■

RECALL

The graph of $y = f(-x)$ is the reflection in the y-axis of the graph of $y = f(x)$.

Since $y = \left(\dfrac{1}{2}\right)^x = (2^{-1})^x = 2^{-x}$, the graph of $y = \left(\dfrac{1}{2}\right)^x$ can also be obtained by reflecting the graph of $y = 2^x$ in the y-axis. The calculator graph in the margin supports this observation.

In Figure 3.3, we sketch the graphs of four exponential functions on the same set of axes. These graphs illustrate some general properties of exponential functions.

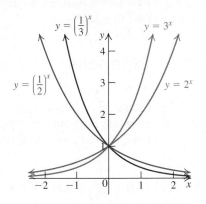

**FIGURE 3.3 Graphs of some
exponential functions**

Properties of Exponential Functions Let $y = f(x) = a^x, a > 0, a \neq 1$.

1. The domain of $f(x) = a^x$ is $(-\infty, \infty)$.

2. The range of $f(x) = a^x$ is $(0, \infty)$: the entire graph lies above the x-axis.

3. For $a > 1$,
    (i) f is an increasing function, so the graph rises to the right.
    (ii) as $x \to \infty$, $y \to \infty$.
    (iii) as $x \to -\infty$, $y \to 0$.

4. For $0 < a < 1$,
    (i) f is a decreasing function, so the graph falls to the right.
    (ii) as $x \to -\infty$, $y \to \infty$.
    (iii) as $x \to \infty$, $y \to 0$.

5. Each exponential function f is one-to-one. So,
    (i) if $a^m = a^n$, then $m = n$.
    (ii) f has an inverse.

6. The graph of $f(x) = a^x$ has no x-intercepts, so it never crosses the x-axis. No value of x will cause $f(x) = a^x$ to equal 0.

7. The graph of $f(x) = a^x$ has y-intercept 1, so $y = f(0) = a^0 = 1$.

8. From 3(iii) and 4(iii), we conclude that the x-axis is a horizontal asymptote for every exponential function of the form $f(x) = a^x$.

> **RECALL**
>
> A function f is one-to-one if $f(x_1) = f(x_2)$ implies that $x_1 = x_2$.

> **REMINDER**
>
> Recall that a line $y = b$ is called a horizontal asymptote if $f(x) \to b$ as $x \to \infty$ or as $x \to -\infty$.

EXAMPLE 4   **Finding a Base Value for an Exponential Function**

Find a if the graph of the exponential function $f(x) = a^x$ contains the point $(2, 49)$.

SOLUTION
Write the function in the form $y = f(x)$.

$$y = a^x$$
$$49 = a^2 \quad \text{The point } (2, 49) \text{ is on the graph.}$$
$$\pm 7 = a \quad \text{Solve for } a.$$

We discard the -7 because the base for an exponential function must be positive. Therefore, $a = 7$ and the exponential function is $f(x) = 7^x$. ■ ■ ■

Practice Problem 4 Find a if the exponential function $f(x) = a^x$ contains the point $(-1, 5)$. ■

3 Solve exponential equations.

Exponential Equations

We can use the one-to-one property of the exponential function (if $a^u = a^v$, then $u = v$) to solve an exponential equation. This technique works for equations in which both sides can be written as powers of the same base. Later in the chapter you will learn how to solve more general exponential equations.

FINDING THE SOLUTION: A PROCEDURE

EXAMPLE 5 **Solving Exponential Equations**

OBJECTIVE
Solve an exponential equation when both sides can be written as powers of the same base.

EXAMPLE
Solve for x: $5^{x(x-3)} = \dfrac{1}{25}$.

Step 1 Write (if necessary) both sides of the exponential equation as powers of the same base.

1. $5^{x(x-3)} = \dfrac{1}{25}$ Given equation

    $5^{x(x-3)} = 5^{-2}$ $\dfrac{1}{25} = 5^{-2}$

Step 2 Use the one-to-one property of exponential functions to equate exponents.

2. $x(x - 3) = -2$ If $a^u = a^v$, then $u = v$.

Step 3 Solve for the variable.

3. $x^2 - 3x = -2$ Distributive property

     $x^2 - 3x + 2 = 0$ Write in standard form.

     $(x - 1)(x - 2) = 0$ Factor.

     $x - 1 = 0$ or $x - 2 = 0$ Zero-product property

     $x = 1$ or $x = 2$ Solve for x.

Step 4 Check your solutions.

4. We leave the check that $x = 1$ and $x = 2$ satisfy the equation to you.

The solution set is $\{1, 2\}$. ■ ■ ■

Practice Problem 5 Solve for x: $3^{2x+1} = 243$. ■

5 Use transformations on exponential functions.

Transformations on Exponential Functions

We can apply the transformations discussed in Section 1.5 to graph functions related to the basic exponential function $f(x) = a^x$.

REMINDER

The graph of the function $f(x) = a^x$ is the graph of the equation $y = a^x$.

Transformations on Exponential Function $f(x) = a^x$

Transformation	Equation	Effect on Graph
Horizontal Shift	$y = a^{x+b}$ $= f(x + b)$	Shift the graph of $y = a^x$ (i) b units left if $b > 0$. (ii) $\lvert b \rvert$ units right if $b < 0$.
Vertical Shift	$y = a^x + b$ $= f(x) + b$	Shift the graph of $y = a^x$ (i) b units up if $b > 0$. (ii) $\lvert b \rvert$ units down if $b < 0$.
Stretching or Compressing (vertically)	$y = ca^x = cf(x)$	Multiply the y coordinates by c. The graph of $y = a^x$ is vertically (i) stretched if $c > 1$. (ii) compressed if $0 < c < 1$.
Reflection	$y = -a^x = -f(x)$ $y = a^{-x} = f(-x)$	Reflect the graph of $y = a^x$ in the x-axis. Reflect the graph of $y = a^x$ in the y-axis.

EXAMPLE 6   **Sketching Graphs**

Use transformations to sketch the graph of each function.

a. $f(x) = 3^x - 4$ **b.** $f(x) = 3^{x+1}$

c. $f(x) = -3^x$ **d.** $f(x) = -3^x + 2$

State the domain and the range of each function and the horizontal asymptote of its graph.

SOLUTION

Figure 3.4 describes the transformations of the exponential function $f(x) = 3^x$.

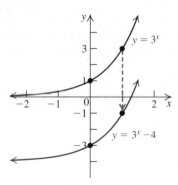

a. Subtract 4 from 3^x: Shift the graph of $y = 3^x$ four units down.

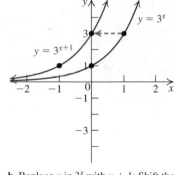

b. Replace x in 3^x with $x + 1$: Shift the graph of $y = 3^x$ one unit left.

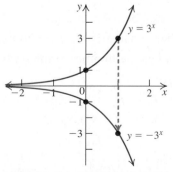

c. Multiply 3^x by -1: Reflect the graph of $y = 3^x$ in the x-axis.

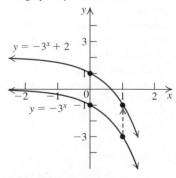

d. Add 2 to -3^x: Shift the graph of $y = -3^x$ from part **c** two units up.

FIGURE 3.4

a. Domain: $(-\infty, \infty)$; range: $(-4, \infty)$; horizontal asymptote: $y = -4$

b. Domain: $(-\infty, \infty)$; range: $(0, \infty)$; horizontal asymptote: x-axis $(y = 0)$

c. Domain: $(-\infty, \infty)$; range: $(-\infty, 0)$; horizontal asymptote: x-axis $(y = 0)$

d. Domain: $(-\infty, \infty)$; range: $(-\infty, 2)$; horizontal asymptote: $y = 2$ ■ ■ ■

Practice Problem 6   Sketch the graph of $y = 3^{x-1} + 4$.   ■

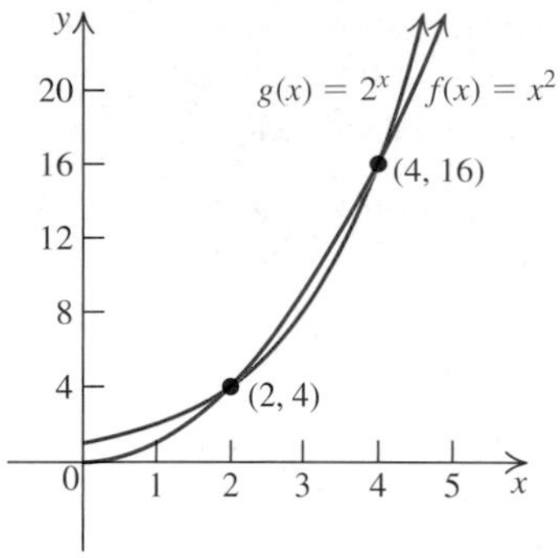

EXAMPLE 7   **Comparing Exponential and Power Functions**

Compare the graphs of $f(x) = x^2$ and $g(x) = 2^x$ for $x \geq 0$.

SOLUTION

Figure 3.5 shows the graphs of f and g over the interval $[0, 5]$. We notice the following:

(i) The graph of $g(x) = 2^x$ is higher than the graph of $f(x) = x^2$ in the interval $[0, 2)$; so $2^x > x^2$ for $0 \leq x < 2$.

(ii) The graph of f intersects the graph of g at $x = 2$; so $2^x = x^2$ at $x = 2$. Both function values are 4, and $(2, 4)$ is a point on both graphs.

(iii) In the interval $(2, 4)$, the graph of f is higher than the graph of g; that is, $x^2 > 2^x$ for $2 < x < 4$.

(iv) The graph of f intersects the graph of g again at $x = 4$; that is, $2^x = x^2$ at $x = 4$. Both function values are 16, and $(4, 16)$ is a point on both graphs.

(v) For $x > 4$, the graph of g is higher than the graph of f; that is, $2^x > x^2$ for $x > 4$.

■ ■ ■

Practice Problem 7   Compare the graphs of $f(x) = x^4$ and $g(x) = 4^x$ for $x \geq 0$.   ■

The result of Example 7 can be generalized. That is, for large values of x, the exponential function with base $a > 1$ is greater than any power function. In other words, for large values of x, we have

$$a^x > x^n \qquad (a > 1 \text{ and } n \text{ any positive integer}).$$

In fact, for large values of x, any exponential function $y = a^x (a > 1)$ eventually becomes greater than any polynomial function.

EXAMPLE 8   **Bacterial Growth**

A technician to the French microbiologist Louis Pasteur noticed that a certain culture of bacteria in milk doubles every hour. If the bacteria count $B(t)$ is modeled by the equation

$$B(t) = 2000 \cdot 2^t,$$

with t in hours, find

a. the initial number of bacteria.

b. the number of bacteria after ten hours.

c. the time when the number of the bacteria will be 32,000.

SOLUTION

a. Initial size $B_0 = B(0) = 2000 \cdot 2^0 = 2000 \cdot 1 = 2000$ Substitute $t = 0$.

b. $B(10) = 2000 \cdot 2^{10} = 2,048,000$ Substitute $t = 10$.

FIGURE 3.5 **Graphs of $f(x) = x^2$ and $g(x) = 2^x$**

Louis Pasteur

(1822–1895)

The French microbiologist Louis Pasteur's discovery that most infectious diseases are caused by germs is one of the most important in medical history. His contributions to microbiology and medicine may be summarized as follows:

1. He championed changes in hospital practices to minimize the spread of disease by microbes.

2. He discovered that weakened forms of a microbe could be used as an immunization against the disease.

3. He discovered that rabies was transmitted by agents so small that they could not be seen under a microscope. Thus, he revealed the world of viruses.

4. He developed *pasteurization*, a process by which harmful microbes in perishable food products are destroyed with heat without destroying the food.

c. We have to find t when $B(t) = 32,000$ in the equation $B(t) = 2000 \cdot 2^t$.

$$32,000 = 2000 \cdot 2^t$$

$16 = 2^t$	Divide both sides by 2000.
$2^4 = 2^t$	Rewrite 16 as 2^4.
$4 = t$	If $a^x = a^y$, then $x = y$.

After four hours, the number of bacteria will be 32,000. ■ ■ ■

Practice Problem 8   In Example 8, find the time when the bacteria will be 128,000. ■

SECTION 3.1 ■ Exercises

A EXERCISES Basic Skills and Concepts

1. For the exponential function $f(x) = a^x, a > 0, a \neq 1$, the domain is __(−∞, ∞)__ and the range is __(0, ∞)__.

2. The graph of $f(x) = 3^x$ has y-intercept __1__ and has __no__ x-intercept.

3. The horizontal asymptote of the graph of $y = \left(\frac{1}{3}\right)^x$ is the __x-axis__.

4. The exponential function $f(x) = a^x$ is increasing if a is _____ and is decreasing if a is _____.

5. *True or False* The graphs of $y = 2^x$ and $y = \left(\frac{1}{2}\right)^x$ are symmetric with respect to the x-axis. False

6. *True or False* For the equation $y = a^x \, (a > 0, a \neq 1)$, $y \to \infty$ as $x \to \infty$. False

In Exercises 7–14, explain whether the given equation defines an exponential function. Give reasons for your answers. Write the base for each exponential function.

7. $y = x^3$ No, the base is not a constant.

8. $y = 4^x$ Yes, the base is 4.

9. $y = 2^{-x}$ Yes, the base is $\frac{1}{2}$.

10. $y = (-5)^x$ No, the base is not a positive constant.

11. $y = x^x$ No, the base is not a constant.

12. $y = 4^2$ No, the exponent is not a variable.

13. $y = 0^x$ No, the base is not a positive constant.

14. $y = (1.8)^x$ Yes, the base is 1.8.

In Exercises 15–24, evaluate each exponential function for the given value. (Use a calculator if necessary.)

15. $f(x) = 5^{x-1}, f(3)$ 25

16. $f(x) = -2^{x+1}, f(3)$ −16

17. $f(x) = 3^{1-x}, f(3.2)$ ≈ 0.089

18. $f(x) = 6^{x-1}, f(2.8)$ ≈ 25.158

19. $g(x) = -\left(\frac{1}{2}\right)^{x+1}, g(-3.5)$ ≈ −5.657

Answer:
4. greater than 1; greater than 0 and less than 1

20. $g(x) = -\left(\frac{2}{3}\right)^{2x-1}, g(2.5)$ $-\frac{16}{81} \approx -0.198$

21. $g(x) = \left(\frac{1}{4}\right)^{1-2x}, g\left(\frac{3}{2}\right)$ 16

22. $g(x) = -\left(\frac{2}{5}\right)^{3x-1}, g\left(\frac{4}{3}\right)$ −0.064

23. $h(x) = 1 - 3^{-x}, h(1)$ $\frac{2}{3}$

24. $h(x) = 3 - 5^{-x}, h(2)$ $\frac{74}{25}$

In Exercises 25–30, find the exponential function of the form $f(x) = a^x$ that contains the given graph point.

25. $(2, 16)$ $a = 4$

26. $\left(-2, \frac{1}{9}\right)$ $a = 3$

27. $(3, 216)$ $a = 6$

28. $\left(3, \frac{1}{125}\right)$ $a = \frac{1}{5}$

29. $(-2, 16)$ $a = \frac{1}{4}$

30. $\left(\frac{1}{3}, \frac{1}{2}\right)$ $a = \frac{1}{8}$

In Exercises 31 and 32, find the function of the form $f(x) = c \cdot a^x$ that contains the two given points.

31. $(0, 3)$ and $(2, 12)$ $y = 3 \cdot 2^x$

32. $(0, 5)$ and $(1, 15)$ $y = 5 \cdot 3^x$

In Exercises 33–40, sketch the graph of the given function by making a table of values. (Use a calculator if necessary.)

33. $f(x) = 4^x$ †

34. $g(x) = 10^x$ †

35. $g(x) = \left(\frac{3}{2}\right)^{-x}$ †

36. $h(x) = 7^{-x}$ †

37. $h(x) = \left(\frac{1}{4}\right)^x$ †

38. $f(x) = \left(\frac{1}{10}\right)^x$ †

39. $f(x) = (1.3)^{-x}$ †

40. $g(x) = (0.7)^{-x}$ †

41. How are the graphs in Exercises 33 and 37 related? Can we obtain the graph of Exercise 37 from that of Exercise 33? If so, how? Yes, reflection in the y-axis

42. Repeat Exercise 41 for the graphs in Exercises 34 and 38. Yes, reflection in the y-axis

†Due to space constrictions, answers to these exercises may be found in the Answers beginning on page A–1 in the back of the book.

Match each exponential function given in Exercises 43–46 with one of the graphs labeled (a), (b), (c), and (d).

43. $f(x) = 5^x$ c

44. $f(x) = -5^x$ b

45. $f(x) = 5^{-x}$ a

46. $f(x) = 5^{-x} + 1$ d

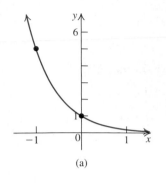

(a)

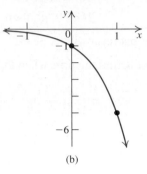

(b)

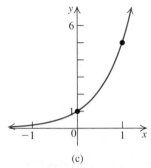

(c)

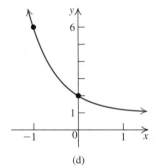

(d)

In Exercises 47–56, graph the equation by using transformations to the graph of the appropriate function in Exercises 33–40.

47. $y = 4^{x-1}$ †

48. $y = -10^x$ †

49. $y = 4^{-x}$ †

50. $y = -\left(\dfrac{1}{4}\right)^x$ †

51. $y = 7^x + 2$ †

52. $y = 4 - 7^x$ †

53. $y = 4 + \left(\dfrac{3}{2}\right)^x$ †

54. $y = \left(\dfrac{1}{10}\right)^x - 1$ †

55. $y = -(1.3)^{-x}$ †

56. $y = 2 - (0.7)^{-x}$ †

In Exercises 57–64, start with the basic exponential function $f(x) = 3^x$ and use transformations to sketch the graph of the function $g(x)$. State the domain and range of $g(x)$ and the horizontal asymptote of its graph.

57. $g(x) = 3^{x-1}$ †

58. $g(x) = 3^{x-1} - 2$ †

59. $g(x) = \dfrac{1}{2}3^{x+1}$ †

60. $g(x) = \dfrac{1}{2}3^{x+1} - 2$ †

61. $g(x) = -2 \cdot 3^{x+1} + 1$ † **62.** $g(x) = 2 \cdot 3^{x+1} - 1$ †

63. $g(x) = -2 \cdot 3^{x-1} + 4$ † **64.** $g(x) = \dfrac{1}{2} \cdot 3^{-x+1} - 2$ †

In Exercises 65–68, write an equation of each graph in the final position.

65. The graph of $y = 2^x$ is shifted two units left and five units up. $y = 2^{x+2} + 5$

66. The graph of $y = 3^x$ is reflected in the y-axis and then shifted three units right. $y = 3^{-x+3}$

67. The graph of $y = \left(\dfrac{1}{2}\right)^x$ is stretched by a factor of 2 and then shifted five units down. $y = 2\left(\dfrac{1}{2}\right)^x - 5$

68. The graph of $y = 2^{-x}$ is reflected in the x-axis and then shifted three units up. $y = -2^{-x} + 3$

In Exercises 69–84, solve each equation for x by first rewriting both sides as powers of the same base.

69. $3^x = 81$ $x = 4$

70. $9^x = 243$ $x = \dfrac{5}{2}$

71. $4^{|x|} = 128$ $x = \pm\dfrac{7}{2}$

72. $3^{-x} = 27$ $x = -3$

73. $2^{-x+1} = 16$ $x = -3$

74. $4^{x-1} = 1$ $x = 1$

75. $2^x = 4^{2x+1}$ $x = -\dfrac{2}{3}$

76. $(\sqrt{2})^x = \dfrac{1}{16}$ $x = -8$

77. $\left(\dfrac{3}{5}\right)^x = \dfrac{125}{27}$ $x = -3$

78. $\left(\dfrac{2}{3}\right)^x = \left(\dfrac{9}{4}\right)^3$ $x = -6$

79. $2^{|x|} = 16$ $x = \pm 4$

80. $4^{-|x|} = \dfrac{1}{128}$ $x = \pm\dfrac{7}{2}$

81. $3^x = (9)^{x-1} \cdot (27)^{1-3x}$ **82.** $5^{x^2} = \left(\dfrac{1}{25}\right)^{-8}$ $x = \pm 4$

83. $9^x - 3^x = 0$ $x = 0$ **84.** $2^{x^2} = (8)^{2x} \cdot \dfrac{1}{256}$. $x = 4, 2$

81. $x = \dfrac{1}{8}$

B EXERCISES Applying the Concepts

85. Suppose a metal block is cooling so that its temperature T (in °C) is given by $T = 200 \cdot 4^{-0.1t}$, where t is in hours.
 a. Find the temperature after
  (i) 2 hours 151.572 **(ii)** 3.5 hours 123.114
 b. How long has the cooling been taking place if the block now has a temperature of 100°C? $t = 5$ hours
 c. Find the eventual temperature $(t \rightarrow \infty)$. 0°

86. Suppose the worldwide sales (in millions) of a computer chip called pentium p are approximated by

$$s(t) = 50 - 30 \cdot 2^{-t},$$

where t represents the number of years the pentium p has been on the market.
 a. Find each function value.
  (i) $s(0)$ †   **(ii)** $s(1)$ †
  (iii) $s(5)$ †   **(iv)** $s(10)$ †
 b. Graph $y = s(x)$. Find the horizontal asymptote. What is the meaning of this horizontal asymptote? †

87. The population (in thousands) of people of East Indian origin in the United States is approximated by the function

$$p(t) = 1600(2)^{0.1047t},$$

where t is the number of years since 2000.
 a. Find the population of this group in
  (i) 2000. 1600   **(ii)** 2008. 2859
 b. Predict the population in 2015. 4752 just after 2019
 c. When will the population reach 6.4 million?
 (The function $p(t)$ is modeled by the authors from the data from the weekly magazine *India Abroad*.)

88. Suppose we have a large sheet of paper 0.015 centimeter thick and we tear the paper in half and put the pieces on top of each other. We keep tearing and stacking in this manner, always tearing each piece in half. How high will the resulting pile of paper be if we continue the process of tearing and stacking
   a. 30 times? $0.015 \times 2^{30} \approx 16,106,127 \text{ cm}$
   b. 40 times? $0.015 \times 2^{40} \approx 16,492,674,420 \text{ cm}$
   c. 50 times? [*Hint:* Use a calculator.]
           $0.015 \times 2^{50} \approx 1.689 \times 10^{13} \text{ cm}$

89. A jar with a volume of 1000 cubic centimeters contains bacteria that doubles in number every minute. If the jar is full in 60 minutes, how long did it take for the container to be half full? 59 minutes

90. A box contains radioactive plutonium. The number of kilograms $p(t)$ at time t years is given by the formula

$$p(t) = 2^{-0.0001754\,t}.$$

Find the amount of plutonium at the following times:
   a. $t = 0$ 1 kg   **b.** $t = 800$ 0.9073 kg
   c. $t = 1000$ 0.8855 kg   **d.** $t = 10,000$ 0.2965 kg

91. In Exercise 90, how long will it take until only one-half kilogram of plutonium is left? 5701.25 yr

92. The number of bacteria present after t minutes is given by the formula $n(t) = 1000(3)^{0.1t}$.
   a. Find the number of bacteria after
      (i) 5 minutes. 1732
      (ii) 10 minutes. 3000
   b. How long will it take for the bacteria population to reach 9000? 20 minutes

---

C EXERCISES Beyond the Basics

93. Sketch the graph of $f(x) = 2^{|x|} = \begin{cases} 2^x & \text{if } x \geq 0 \\ 2^{-x} & \text{if } x < 0 \end{cases}$
   Check for symmetry. Find the range of this function. †

94. Sketch the graph of $g(x) = 2^{-|x|}$. †

95. Sketch the graph of $g(x) = 2^{x^2}$. Check for symmetry. Find the domain and range of this function. †

96. a. Sketch the graph of $h(x) = 2^{-x^2}$. †
   b. Find the domain and range of $h(x)$.
   c. Why is it not possible to obtain the graph of $h(x)$ by reflecting the graph of $g(x)$ of Exercise 95 in the y-axis? Because $g(-x) = 2^{(-x)^2} = 2^{x^2} \neq h(x)$

97. Let $f(x) = 3^x + 3^{-x}$ and $g(x) = 3^x - 3^{-x}$. Find each of the following:
   a. $f(x) + g(x)$ $f(x) + g(x) = 2 \cdot 3^x$
   b. $f(x) - g(x)$ $f(x) - g(x) = 2 \cdot 3^{-x}$
   c. $[f(x)]^2 - [g(x)]^2$ $[f(x)]^2 - [g(x)]^2 = 4$
   d. $[f(x)]^2 + [g(x)]^2$ $[f(x)]^2 + [g(x)]^2 = 2(3^{2x} + 3^{-2x})$

98. Repeat Exercise 97 for the functions $f(x) = a^x + a^{-x}$ and $g(x) = a^x - a^{-x}$.

In Exercises 99–102, solve each equation for x.

99. $\left(\dfrac{3}{5}\right)^{2x-1} = \left(\dfrac{25}{9}\right)^{3x+1}$ $-\dfrac{1}{8}$ **100.** $\left(\dfrac{3}{4}\right)^{x^2+8} = \left(\dfrac{64}{27}\right)^{2x}$ $-2, -4$

101. $\left(\dfrac{1}{2}\right)^{|x-3|} = \left(\dfrac{1}{8}\right)^5$ 18, −12   **102.** $\left(\dfrac{2}{3}\right)^{|x+2|} = \left(\dfrac{81}{16}\right)^{-3}$ 10, −14

In Exercises 103–106, solve the given equation for x. [*Hint:* These equations are quadratic in form.]

103. $2^{2x} - 12 \cdot 2^x + 32 = 0$ $x = 2, 3$

104. $3^{2x} - 4 \cdot 3^{x+1} + 27 = 0$ $x = 1, 2$

105. $4(2^x + 2^{-x}) = 17$ $x = 2, -2$

106. $4^{1+x} + 4^{1-x} = 10$ $x = \dfrac{1}{2}, -\dfrac{1}{2}$

107. Growth of bacteria. The bacterium *Escherichia coli*, commonly known as *E. coli*, is found in the human digestive tract. Suppose a colony of *E. coli* started with a single cell and each *E. coli* cell divides into two cells one-third hour after its birth. Suppose also that no *E. coli* cells die.
   a. Write a function that models the number of *E. coli* cells after t hours. $y(t) = 2^{3t}$
   b. Given that the mass of a single *E. coli* cell is approximately $5 \cdot 2^{-43}$ grams, find the mass of the colony after ten hours. 5×2^{-13} gm
   c. When will the mass of the colony be 20 grams? 15 hr

108. Internet users. The number of Internet users worldwide was estimated to be 0.4 billion in 2000 and 1.1 billion in 2005. The growth can be approximated by an exponential function.
   a. Find an exponential function of the form $y = c \cdot 2^{kx}$ that models the number of Internet users, where $x = 0$ corresponds to 2000 and y corresponds to the number of users in billions. $y = 0.4 \cdot 2^{0.292x}$
   b. Graph the function and estimate the number of users in 2008. †
   c. Use the graph to estimate the year when there are 2.5 billion users. In 2009

109. Spread of AIDS. In 1987, the number of AIDS cases in the United States was estimated at 50,000. In 2005, this number was estimated at 1 million. Assume that the trend continued.
   a. Set up a function of the form

$$f(t) = c \cdot 2^{kt}$$
$$f(t) = 50,000 \cdot 2^{0.2401t}$$

   to describe the number of cases of AIDS, where t is time measured in years from 1987.
   b. According to the model in (a), how many cases of AIDS occurred in 2000? 435,000
   c. Use the model to predict the number of cases of AIDS in 2012. 3,206,000
   d. According to the model, when was the number of cases of AIDS 1.6 million? Some time in 2008 ($t = 21$).

Critical Thinking

110. Give a convincing argument to show that the equation $2^x = k$ has exactly one solution for every $k > 0$. Support your argument with graphs.

111. Find all solutions of the equation $2^x = 2x$. How do you know you have found all solutions?

98. a. $f(x) + g(x) = 2 \cdot a^x$ **b.** $f(x) + g(x) = 2 \cdot a^{-x}$
   c. $[f(x)]^2 - [g(x)]^2 = 4$ **d.** $[f(x)]^2 + [g(x)]^2 = 2(a^{2x} + a^{-2x})$
110. The function $y = 2^x$ is an increasing function and is thus one-to-one. The horizontal line $y = k$ for $k > 0$ intersects the graph of $y = 2^x$ at exactly one point.
111. Sketch the graphs of $y = 2^x$ and $y = 2x$. These graphs intersect at $x = 2$ and $x = 1$.

Answers:
96. b. Domain $= (-\infty, \infty)$
   Range $= (0, 1]$

The Natural Exponential Function

Before Starting this Section, Review

1. Exponential function (Section 3.1, page 190)
2. Graphing and transformations (Section 1.5, page 62)

Objectives

1. Develop a compound interest formula.
2. Understand the number *e*.
3. Graph exponential functions.
4. Evaluate exponential functions.

DESCENDANT SEEKS LOAN REPAYMENT

George Washington could not tell a lie, but apparently he could leave behind a whopping debt.

In 1777, Jacob DeHaven, a wealthy Pennsylvania trader, responded to a desperate plea from Washington when it seemed that the Revolutionary War was about to be lost. Mr. DeHaven loaned $450,000 worth of gold and supplies to the Continental Congress to rescue the troops at Valley Forge. His loan provided badly needed supplies, provisions, and salaries for the Continental Army. George Washington went on to defeat the British in the war. After the war, the government offered to repay DeHaven's loan in continental dollars, the then worthless currency of the fledging country. But he refused and died penniless in 1812.

In 1990, DeHaven's descendants sued the U.S. government for repayment of the loan. "If I owed the government this money, I would be expected to pay the principal, interest, and any penalties that might have been incurred," said Carolyn Cokerham, one of nearly 2000 living descendants of DeHaven. Many of his descendants have tried unsuccessfully to get the money repaid through the years. How much money did DeHaven's descendants claim the government owed them on the 1990 anniversary of the loan at the then prevailing interest rate of 6%? (See Example 5.) ▪

To answer the DeHaven and related financial questions, let's first review some terminology concerning interest.

SIMPLE INTEREST

A fee charged for borrowing a lender's money is the **interest**, denoted by *I*.

The original, or initial, amount of money borrowed is the **principal**, denoted by *P*.

The period of time during which the borrower pays back the principal plus the interest is the **time**, denoted by *t*.

The **interest rate** is the percent charged for the use of the principal for the given period. The interest rate, denoted by *r*, is expressed as a decimal. Unless stated otherwise, the period is assumed to be one year; that is, *r* is an *annual* rate.

The amount of interest computed only on the principal is called **simple interest**.

RECALL

$8\% = 8$ percent

$\quad = 8$ per hundred

$\quad = \dfrac{8}{100}$

$\quad = 0.08$

When money is deposited with a bank, the bank becomes the borrower. For example, depositing $1000 in an account at 8% interest means that the principal P is $1000 and the interest rate r is 0.08 for the bank as the borrower.

SIMPLE INTEREST FORMULA

The simple interest I on a principal P at a rate r (expressed as a decimal) per year for t years is

$$I = Prt. \quad (1)$$

EXAMPLE 1   **Calculating Simple Interest**

Juanita has deposited $8000 in a bank for five years at a simple interest rate of 6%.

a. How much interest will she receive?

b. How much money will be in her account at the end of five years?

SOLUTION

a. $\quad P = \$8000, r = 0.06, \text{ and } t = 5.$

$\qquad I = Prt$ $\hspace{4.5cm}$ Equation (1)

$\qquad\quad = \$8000\,(0.06)(5)$ $\hspace{2.7cm}$ Substitute values.

$\qquad\quad = \$2400$ $\hspace{4.6cm}$ Simplify.

b. In five years, the amount A she will receive is the original principal plus the interest earned:

$$A = P + I$$
$$= \$8000 + \$2400$$
$$= \$10{,}400$$

■ ■ ■

Practice Problem 1   Find the amount that will be in a bank account if $10,000 is deposited at a simple interest rate of 7.5% for two years.   ■

Given P and r, the amount $A(t)$ calculated at simple interest and due in t years is found by using the formula

$$A(t) = P + Prt. \quad (2)$$

Equation (2) is a linear function of t. Simple interest problems are examples of **linear growth**, which take place when the growth of a quantity occurs at a constant rate and so can be modeled by a linear function.

1 Develop a compound interest formula.

Compound Interest

In the real world, simple interest is rarely used for periods of more than one year. Instead, we use **compound interest**—the interest paid on both the principal and the accrued (previously earned) interest.

To illustrate compound interest, suppose $1000 is deposited in a bank account paying 4% annual interest. At the end of one year, the account will contain the original $1000 plus the 4% interest earned on the $1000:

$$\$1000 + (0.04)(\$1000) = \$1040$$

Similarly, if P represents the initial amount deposited at an interest rate r (expressed as a decimal) per year, then the amount A_1 in the account after one year is,

$$A_1 = P + rP \qquad \text{Principal } P \text{ plus interest earned}$$
$$= P(1 + r) \qquad \text{Factor out } P.$$

During the second year, the account earns interest on the new principal A_1. The amount A_2 in the account after the second year will be equal to A_1 plus the interest on A_1.

$$A_2 = A_1 + rA_1 \qquad A_1 \text{ plus interest earned on } A_1$$
$$= A_1(1 + r) \qquad \text{Factor out } A_1.$$
$$= P(1 + r)(1 + r) \qquad A_1 = P(1 + r)$$
$$= P(1 + r)^2$$

The amount A_3 in the account after the third year is

$$A_3 = A_2 + rA_2 \qquad A_2 \text{ plus interest earned on } A_2$$
$$= A_2(1 + r) \qquad \text{Factor out } A_2.$$
$$= P(1 + r)^2(1 + r) \qquad A_2 = P(1 + r)^2$$
$$= P(1 + r)^3$$

In general, the amount A in the account after t years is given by

$$A = P(1 + r)^t. \qquad (3)$$

We say that this type of interest is **compounded annually** because it is paid once a year.

EXAMPLE 2   Calculating Compound Interest

Juanita deposits $8000 in a bank at the interest rate of 6% compounded annually for five years.

a. How much money will she have in her account after five years?

b. How much interest will she receive?

SOLUTION

a. Here $P = \$8000, r = 0.06$, and $t = 5$; so

$$A = P(1 + r)^t$$
$$= \$8000(1 + 0.06)^5 \qquad P = \$8000, r = 0.06, t = 5$$
$$= \$8000(1.06)^5$$
$$= \$10{,}705.80 \qquad \text{Use a calculator.}$$

b. Interest $= A - P = \$10{,}705.80 - \$8000 = \$2705.80.$ ▪▪▪

Practice Problem 2   Repeat Example 2 assuming that the bank pays 7.5% interest compounded annually. ▪

Comparing Examples 1 and 2, we see that compounding Juanita's interest made her money grow faster. We expect this because the function $A(t) = P(1 + r)^t$ is an exponential function with base $(1 + r)$.

Banks or financial institutions usually pay savings account interest more than once a year. They pay a smaller amount of interest more frequently. Suppose the annual interest rate r quoted (also called the **nominal rate**) is compounded n times per year (at equal intervals) instead of annually. Then for each period, the interest rate is $\dfrac{r}{n}$, and there are $n \cdot t$ periods in t years. Accordingly, we can restate the formula $A(t) = P(1 + r)^t$ as follows:

COMPOUND INTEREST FORMULA

$$A = P\left(1 + \frac{r}{n}\right)^{nt} \quad (4)$$

A = amount after t years

P = principal

r = annual interest rate (expressed as a decimal number)

n = number of times interest is compounded each year

t = number of years

The total amount accumulated after t years, denoted by A, is also called the **future value** of the investment.

EXAMPLE 3   **Using Different Compounding Periods to Compare Future Values**

BY THE WAY . . .

Compounding that occurs 1, 2, 4, 12, and 365 times a year is known as compounding annually, semiannually, quarterly, monthly, and daily, respectively.

If \$100 is deposited in a bank that pays 5% annual interest, find the future value A after one year if the interest is compounded

(i) annually.

(ii) semiannually.

(iii) quarterly.

(iv) monthly.

(v) daily.

SOLUTION

In the following computations, $P = 100$, $r = 0.05$, and $t = 1$. Only n, the number of times interest is compounded each year, changes. Since $t = 1$, $nt = n(1) = n$.

(i) Annual Compounding: $A = P\left(1 + \dfrac{r}{n}\right)^{nt}$ $n = 1; t = 1$

$A = 100(1 + 0.05) = \$105.00$

(ii) Semiannual Compounding: $A = P\left(1 + \dfrac{r}{2}\right)^2$ $n = 2; t = 1$

$A = 100\left(1 + \dfrac{0.05}{2}\right)^2$

$\approx \$105.06$ Use a calculator.

(iii) Quarterly Compounding: $A = P\left(1 + \dfrac{r}{4}\right)^4$ $n = 4; t = 1$

$A = 100\left(1 + \dfrac{0.05}{4}\right)^4$

$\approx \$105.09$ Use a calculator.

(iv) Monthly Compounding: $A = P\left(1 + \dfrac{r}{12}\right)^{12}$ $n = 12; t = 1$

$A = 100\left(1 + \dfrac{0.05}{12}\right)^{12}$

$\approx \$105.12$ Use a calculator.

2 Understand the number *e*.

Leonhard Euler

(1707–1783)

Leonhard Euler was the son of a Calvinist minister from the vicinity of Basel, Switzerland. At 13, Euler entered the University of Basel, pursuing a career in theology, as his father wanted. At the university, Euler was tutored by Johann Bernoulli, of the famous Bernoulli family of mathematicians. Euler's interest and skills led him to abandon his theological studies and take up mathematics. He obtained his master's degree in philosophy at the age of 16. In 1727, Peter the Great invited him to join the Academy at St. Petersburg. In 1741, Euler moved to the Berlin Academy, where he stayed until 1766. He then returned to St. Petersburg, where he remained for the rest of his life.

Euler was incredibly prolific, contributing to many areas of mathematics, including number theory, combinatorics, and analysis, as well as its applications to areas such as music and naval architecture. He wrote over 1100 books and papers and left so much unpublished work that it took 47 years after he died for all of his work to be published. During Euler's life, his papers accumulated so quickly that he kept a large pile of articles awaiting publication. The Berlin Academy published the papers on top of this pile, so later results often were published before results they depended on or superseded. Euler had 13 children and was able to continue his work while a child or two bounced on his knees. He was blind for the last 17 years of his life, but because of his fantastic memory, this affliction did not diminish his mathematical output. The project of publishing his collected works, undertaken by the Swiss Society of Natural Science, is still going on and will require more than 75 volumes.

(v) Daily Compounding:

$$A = P\left(1 + \frac{r}{365}\right)^{365} \qquad n = 365; t = 1$$

$$A = 100\left(1 + \frac{0.05}{365}\right)^{365}$$

$$\approx \$105.13 \qquad \text{Use a calculator.}$$

■ ■ ■

Practice Problem 3   Repeat Example 3 assuming that $5000 is deposited at a 6.5% annual rate.   ■

Compound Interest Formula

Notice in Example 3 that the future value A increases with n, the number of compounding periods. (Of course P, r, and t are fixed.) The question is, If n increases indefinitely (100, 1000, 10,000 times, and so on), does the amount A also increase indefinitely? Let's see why the answer is no. Let $h = \dfrac{n}{r}$. Then we have

$$A = P\left(1 + \frac{r}{n}\right)^{nt} \qquad \text{Equation (4)}$$

$$= P\left[\left(1 + \frac{r}{n}\right)^{\frac{n}{r}}\right]^{rt} \qquad nt = \frac{n}{r}\cdot rt$$

$$= P\left[\left(1 + \frac{1}{h}\right)^{h}\right]^{rt} \qquad h = \frac{n}{r}, \text{so } \frac{1}{h} = \frac{r}{n} \qquad (5)$$

Table 3.3 shows the expression $\left(1 + \dfrac{1}{h}\right)^{h}$ as h takes on increasingly larger values.

TABLE 3.3

h	$\left(1 + \dfrac{1}{h}\right)^{h}$
1	2
2	2.25
10	2.59374
100	2.70481
1000	2.71692
10,000	2.71815
100,000	2.71827
1,000,000	2.71828

Table 3.3 suggests that as h gets larger and larger, $\left(1 + \dfrac{1}{h}\right)^{h}$ gets closer and closer to a fixed number. This observation can be proven, and the fixed number is denoted by e in honor of the famous mathematician Leonhard Euler (pronounced "oiler"). The number e, an irrational number, is sometimes called the **Euler number**.

The value of e to 15 places is

$$e = 2.718281828459045.$$

TECHNOLOGY CONNECTION

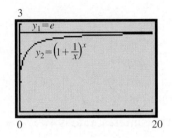

Graphs of $y_1 = e$ and $y_2 = \left(1 + \dfrac{1}{x}\right)^x$ in the window $0 < x < 20$, $0 < y < 3$

As $x \to \infty$, $\left(1 + \dfrac{1}{x}\right)^x \to e$.

We sometimes write

$$e = \lim_{h \to \infty} \left(1 + \frac{1}{h}\right)^h,$$

which means that when h is very large, $\left(1 + \dfrac{1}{h}\right)^h$ has a value very close to e. Note that in equation (5), as n gets very large, the quantity $h = \dfrac{n}{r}$ also gets very large because r is fixed. Therefore, the expression inside the brackets approaches e; so the compounded amount $A = P\left(1 + \dfrac{r}{n}\right)^{nt}$ approaches $A = Pe^{rt}$.

Continuous Compounding When interest is compounded continuously, the amount A after one year, with principal P and interest rate r (written as a decimal number), is Pe^r; the amount A after t years is given by the following formula:

CONTINUOUS COMPOUND INTEREST FORMULA

$$A = Pe^{rt} \qquad (6)$$

A = amount after t years

P = principal

r = annual interest rate (expressed as a decimal number)

t = number of years

TECHNOLOGY CONNECTION

Most calculators have the $\boxed{e^x}$ or $\boxed{\text{EXP}}$ key for calculating e^x for a given value of x.

On a TI-83 calculator, to evaluate a number such as $\boxed{e^5}$, you press $\boxed{e^x}$ 5 $\boxed{\text{ENTER}}$ and read the display 148.4131591.

EXAMPLE 4   **Calculating Continuous Compound Interest**

Find the amount when a principal of \$8300 is invested at a 7.5% annual rate of interest compounded continuously for eight years and three months.

SOLUTION

We use formula (6), with $P = \$8300$ and $r = 0.075$. We convert eight years and three months to 8.25 years.

$$A = \$8300e^{(0.075)(8.25)} \qquad \text{Use the formula } A = Pe^{rt}.$$
$$\approx \$15{,}409.83 \qquad \text{Use a calculator.} \qquad \blacksquare\ \blacksquare\ \blacksquare$$

Practice Problem 4   Repeat Example 4 assuming that \$9000 is invested at a 6% annual rate. $\blacksquare$

EXAMPLE 5   **Calculating the Amount of Repaying a Loan**

How much money did the government owe DeHaven's descendants for 213 years on the \$450,000 loan at the interest rate of 6%?

SOLUTION

a. With simple interest,

$$A = P + Prt = P(1 + rt)$$
$$= \$450{,}000\,[1 + (0.06)(213)] = \$6.201 \text{ million.}$$

b. With interest compounded yearly,

$$A = P(1 + r)^t = \$450,000 \, (1 + 0.06)^{213}$$
$$\approx \$1.105 \times 10^{11} = \$110.500 \text{ billion.}$$

c. With interest compounded quarterly,

$$A = P\left(1 + \frac{r}{4}\right)^{4t} = \$450,000 \left(1 + \frac{0.06}{4}\right)^{4(213)}$$
$$\approx \$1.45305 \times 10^{11} = \$145.305 \text{ billion.}$$

d. With interest compounded continuously,

$$A = Pe^{rt} = \$450,000e^{0.06 \, (213)}$$
$$\approx \$1.5977 \times 10^{11} = \$159.770 \text{ billion.}$$

Notice the dramatic difference between quarterly and continuous compounding and the dramatic difference between simple and compound interest. ■ ■ ■

Practice Problem 5 Repeat Example 5 assuming that the interest rate is 4%. ■

3 Graph exponential functions.

The Natural Exponential Function

The exponential function

$$f(x) = e^x$$

with base e is so prevalent in the sciences that it is often referred to as *the* exponential function or the natural exponential function. We use a calculator to find e^x to two decimal places for $x = -2, -1, 0, 1,$ and 2 in Table 3.4.

TABLE 3.4

x	e^x
-2	0.14
-1	0.37
0	1
1	2.72
2	7.39

The graph of $f(x) = e^x$ is sketched in Figure 3.6 by using the ordered pairs in Table 3.4.

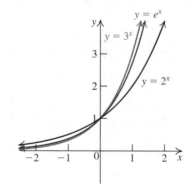

FIGURE 3.6 The graph of $y = e^x$ is between the graphs $y = 2^x$ and $y = 3^x$.

Since $2 < e < 3$, the graph of $y = e^x$ lies between the graphs $y = 2^x$ and $y = 3^x$. The function $f(x) = e^x$ has all the properties of exponential functions with base $a > 1$ listed in Section 3.1.

We can apply the transformations from Section 1.5 to the natural exponential function.

BY THE WAY . . .

Transcendental numbers and transcendental functions.

Numbers that are solutions of polynomial equations with rational coefficients are called **algebraic:** -2 is algebraic because it satisfies the equation $x + 2 = 0$, and $\sqrt{3}$ is algebraic because it satisfies the equation $x^2 - 3 = 0$. Numbers that are not algebraic are called **transcendental,** a term coined by Euler to describe numbers such as e and π that appear to "transcend the power of algebraic methods."

There is a somewhat similar distinction between functions. The exponential and logarithmic functions, which are the main subject of this chapter, are transcendental functions.

EXAMPLE 6   Sketching a Graph

Use transformations to sketch the graph of

$$g(x) = e^{x-1} + 2.$$

SOLUTION

We start with the natural exponential function $f(x) = e^x$ and shift its graph one unit right to obtain the graph of $y = e^{x-1}$.

We then shift the graph of $y = e^{x-1}$ up two units to obtain the graph of $g(x) = e^{x-1} + 2$. See Figure 3.7.

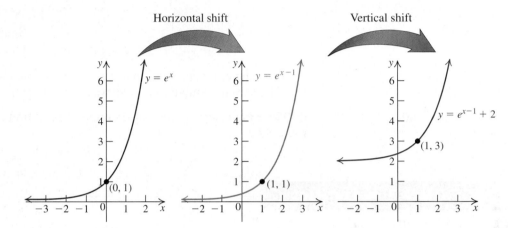

FIGURE 3.7 Graphing $g(x) = e^{x-1} + 2$ ■ ■ ■

Practice Problem 6   Sketch the graph of $g(x) = -e^{x-1} - 2$.   ■

Numerous applications of the exponential function are based on the fact that many different quantities have a growth (or decay) rate proportional to their size. Using calculus, we can show that in such cases, we obtain the following mathematical model:

MODEL FOR EXPONENTIAL GROWTH OR DECAY

$$A(t) = A_0\, e^{kt} \quad (7)$$

$A(t) = $ the amount at time t

$A_0 = A(0)$, the initial amount

$k = $ relative rate of growth $(k > 0)$ or decay $(k < 0)$

$t = $ time

---

4 Evaluate exponential functions.

EXAMPLE 7

In the year 2000, the human population of the world was approximately 6 billion and the annual rate of growth was about 2.1%. Using the model given in equation (7), estimate the population of the world in the following years.

a. 2030   **b.** 1990

SOLUTION

a. The year 2000 corresponds to $t = 0$. So $A_0 = 6$ (billion), $k = 0.021$, and 2030 corresponds to $t = 30$.

$$A(30) = 6e^{(0.021)(30)} \qquad A(t) = A_0 e^{kt}$$

$$A(30) \approx 11.265663 \qquad \text{Use a calculator.}$$

Thus, the model predicts that if the rate of growth is 2.1% per year, over 11.26 billion people will be in the world in 2030.

b. The year 1990 corresponds to $t = -10$ (because 1990 is ten years prior to 2000). We have

$$A(-10) = 6e^{(0.021)(-10)} \qquad A(t) = A_0 e^{kt}$$

$$= 6e^{(-0.21)} \qquad \text{Simplify.}$$

$$\approx 4.8635055 \qquad \text{Use a calculator.}$$

Thus, the model predicts that the world had over 4.86 billion people in 1990. (The actual population in 1990 was 5.28 billion.) ■ ■ ■

Practice Problem 7 Repeat Example 7 assuming that the annual rate of growth was 2.3%. ■

SECTION 3.2 ■ Exercises

A EXERCISES Basic Skills and Concepts

1. The formula for simple interest is $I =$ _____Prt_____.

2. The value of e to two decimal places is _____2.72_____.

3. The formula for continuous compound interest is $A =$ _____Pe^{rt}_____.

4. As x gets larger and larger, the value of $\left(1 + \dfrac{1}{x}\right)^x$ gets closer to _____e_____.

5. *True or False* Let $y = e^{-x}$. As $x \to \infty$, $y \to 0$. True

6. *True or False* The graph of $y = e^x + 1$ is obtained by shifting the graph of $y = e^x$ horizontally one unit to the right. False

In Exercises 7–12, find the simple interest for each value of principal P, rate r per year, and time t.

7. $P = \$5000, r = 10\%, t = 5$ years $2500

8. $P = \$10,000, r = 5\%, t = 10$ years $5000

9. $P = \$6750, r = 4.6\%, t = 4$ years and 6 months $1397.25

10. $P = \$8620, r = 5.7\%, t = 6$ years and 3 months $3070.88

11. $P = \$7800, r = 6\dfrac{7}{8}\%, t = 10$ years and 9 months $5764.69

12. $P = \$8670, r = 4\dfrac{1}{8}\%, t = 6$ years and 8 months $2384.25

In Exercises 13–18, find (a) the future value of the given principal P and (b) the interest earned in the given period.

13. $P = \$5250$ at 8% compounded quarterly for 10 years
    a. $11,592.21 **b.** $6342.21

14. $P = \$3500$ at 6.5% compounded annually for 13 years
    a. $7936.21 **b.** $4436.21

15. $P = \$6240$ at 7.5% compounded monthly for 12 years
    a. $15,305.00 **b.** $9065.00

16. $P = \$8000$ at 6.5% compounded daily for 15 years
    a. $21,207.47 **b.** $13,207.47

17. $P = \$7500$ at 5% compounded continuously for 10 years
    a. $12,365.41 **b.** $4865.41

18. $P = \$8500$ at $4\dfrac{3}{4}\%$ compounded continuously for 8 years
    a. $12,429.42 **b.** $3929.42

In Exercises 19–22, find the principal P that will generate the given future value A.

19. $A = \$10,000$ at 8% compounded annually for 10 years
    $4631.93

20. $A = \$10,000$ at 8% compounded quarterly for 10 years
    $4528.90

21. $A = \$10,000$ at 8% compounded daily for 10 years
    $4493.68

22. $A = \$10,000$ at 8% compounded continuously for 10 years $4493.29

In Exercises 23–30, starting with the graph of $y = e^x$, use transformations to sketch the graph of each function and state its horizontal asymptote.

23. $f(x) = e^{-x}$ †

24. $f(x) = -e^x$ †

25. $f(x) = e^{x-2}$ †

26. $f(x) = e^{2-x}$ †

27. $f(x) = 1 + e^x$ †

28. $f(x) = 2 - e^{-x}$ †

29. $f(x) = -e^{x-2} + 3$ †

30. $g(x) = 3 + e^{2-x}$ †

†Due to space constrictions, answers to these exercises may be found in the Answers beginning on page A–1 in the back of the book.

In Exercises 31 and 32, sketch the graph of each function.

31. $f(x) = e^{|x|}$ †

32. $g(x) = e^{-|x|}$ †

B EXERCISES Applying the Concepts

33. Price appreciation. In 2008, the median price of a house in Miami was $190,000. Assuming a rate of increase of 3% per year, what can we expect the price of such a house to be in 2013? $220,262

34. Price appreciation. In 2008, the median price of a compact car was $16,800. If the rate of increase is assumed to be 4% per year, what can we expect the price of such a car to be in 2010? $18,171

35. Investment. How much should a mother invest at the time her son is born to provide him with $80,000 at age 21? Assume that the interest is 7% compounded quarterly. $18,629.40

36. Investment. Sabrina is planning to retire in 20 years. She calculates that she will need $100,000 in addition to the money she will get from her retirement plans (including Social Security). How much should she invest today at 10% compounded daily to accomplish her objective? $13,537.24

37. Depreciation. Trans Trucking Co. purchased a truck for $80,000. The company depreciated the truck at the end of each year at the rate of 15% of its current value. What is the value of the truck at the end of the fifth year? $35,496.43

38. Depreciation. Suppose you bought a car for $21,600. After using the car for three years, you wreck it. The insurance company will pay you the depreciated value of the car at the rate of 20% of its previous year's value. How much would you receive from the insurance company if your car was a total loss? $11,059.20

39. Manhattan Island purchase. In 1626, Peter Du Minuit purchased Manhattan Island from the Native Americans for 60 Dutch guilders (about $24). Suppose the $24 was invested in 1626 at a 6% rate. How much money would that investment be worth in 2006 if the interest was
a. simple interest. $571
b. compounded annually. $99,183,639,920
c. compounded monthly. $180,905,950,100
d. compounded continuously. $191,480,886,300

40. Investment. Ms. Ann Scheiber retired from government service in 1941 with a monthly pension of $83 and $5000 in savings. At the time of her death in January 1995 at the age of 101, Ms. Scheiber had turned the $5000 into $22 million through shrewd investments in the stock market. She bequeathed all of it to Yeshiva University in New York City. What annual rate of return compounded annually would turn $5000 to a whopping $22 million in 54 years? 16.81%

41. Population. The population of Sometown, USA, was 12,000 in 1990 and grew to 15,000 in 2000. Assume that the population will continue to grow exponentially at the same constant rate. What will be the population of Sometown in 2010? $\left[Hint: \text{Show that } e^k = \left(\dfrac{5}{4}\right)^{1/10}. \right]$ 18,750

42. Population. The population of the United States was about 280 million in 2000 and 300 million in 2006. Assume that the population will continue to grow exponentially at the same rate. What is the predicted population of the United States in 2020? 352.4 million

43. Medicine. Tests show that a new ointment X helps heal wounds. If A_0 square millimeters is the area of the original wound, then the area A of the wound after n days of application of the ointment X is given by $A = A_0 e^{-0.43n}$. If the area of the original wound was 10 square millimeters, find the area of the wound after ten days of application of the new ointment. 0.1357 mm²

44. Medicine. In the model of Exercise 43, suppose the area of the wound after 20 days of application of the ointment is 0.1 square millimeter. What was the area of the original wound? 543.2 mm²

45. Cooling. The temperature T (in °C) of coffee at time t minutes after its removal from the microwave is given by the equation

$$T = 25 + 73e^{-0.28t}.$$

Find the temperature of the coffee at each time listed.
a. $t = 0$ 98°C
b. $t = 10$ 29.44°C
c. $t = 20$ 25.27°C
d. after a long time 25°C

46. Medicine. The radioactive substance iodine-131 is used in measuring heart, liver, and thyroid activity. The quantity Q (in grams) remaining t days after the element is purchased is given by the equation

$$Q = 40e^{-0.086643397t}.$$

Calculate the amount at each time listed.
a. initially purchased 40 g
b. 8 days after purchase 20 g
c. 16 days after purchase 10 g
d. 24 days after purchase 5 g

C EXERCISES Beyond the Basics

In Exercises 47–52, sketch the graph of the given function.

47. $f(x) = \begin{cases} e^{-x} & \text{if } x > 0 \\ e^{x} & \text{if } x \le 0 \end{cases}$ †

48. $f(x) = \begin{cases} e^{x} & \text{if } x > 0 \\ -e^{x} & \text{if } x \le 0 \end{cases}$ †

49. $f(x) = \begin{cases} 1 + x & \text{if } x > 0 \\ e^{x} & \text{if } x \le 0 \end{cases}$ †

50. $g(x) = \begin{cases} 1 + x^2 & \text{if } x > 0 \\ e^{-x} & \text{if } x \le 0 \end{cases}$ †

51. $f(x) = \begin{cases} e^{-x} & \text{if } x > 0 \\ 1 & \text{if } x \le 0 \end{cases}$ †

52. $g(x) = \begin{cases} e^{-x} & \text{if } x \le 0 \\ 1 & \text{if } 0 < x < 1 \\ x^2 & \text{if } x \ge 1 \end{cases}$ †

53. Let $f(x) = e^x$. Show that

a. $\dfrac{f(x + h) - f(x)}{h} = e^x\left(\dfrac{e^h - 1}{h}\right)$.

b. $f(x + y) = f(x)f(y)$.

c. $f(-x) = \dfrac{1}{f(x)}$.

54. You need the following definition for this exercise: Let $0! = 1; 1! = 1; 2! = 2 \cdot 1$, and, in general, $n! = n(n - 1)(n - 2) \cdots 3 \cdot 2 \cdot 1$. The symbol $n!$ is read "n factorial." Now let

$$S_n = 2 + \frac{1}{2!} + \frac{1}{3!} + \frac{1}{4!} + \cdots + \frac{1}{n!}.$$

Compute S_n for $n = 5, 10$, and 15. Compare the resulting values with the value of e on page 204. †

55. The hyperbolic cosine function (abbreviated as cosh x, often pronounced as "kosh x or coshine x") is defined by cosh $x = \dfrac{e^x + e^{-x}}{2}$.

a. What is the domain of cosh x? $(-\infty, \infty)$
b. Is cosh x odd, even, or neither? even
c. Sketch the graph of $y = $ cosh x by plotting points. †

56. The hyperbolic sine function (abbreviated as sinh x, often pronounced as "shine x") is defined by

$$\sinh x = \frac{e^x - e^{-x}}{2}.$$

a. What is the domain of sinh x? $(-\infty, \infty)$
b. Is sinh x odd, even, or neither? odd
c. Sketch the graph of $y = $ sinh x by plotting points. †

In Exercises 57 and 58, prove that the given equations are true for all real numbers.

57. cosh x + sinh $x = e^x$

58. $(\cosh x)^2 - (\sinh x)^2 = 1$

59. Define the **tangent hyperbolic function** (tanh x) by tanh $x = \dfrac{e^x - e^{-x}}{e^x + e^{-x}}$ and the **secant hyperbolic function** (sech x) by sech $x = \dfrac{1}{\cosh x}$. Then show that $1 - (\tanh x)^2 = (\text{sech } x)^2$.

60. If an investment of P dollars returns A dollars after one year, the **effective annual interest rate** or **annual yield** y is defined by the equation

$$A = P(1 + y).$$

Show that if the sum of P dollars is invested at a nominal rate r per year, compounded m times per year, the effective annual yield y is given by the equation

$$y = \left(1 + \frac{r}{m}\right)^m - 1.$$

61. Calculate the effective annual yield if money is invested at 8% annual rate compounded
a. quarterly. 8.2432%
b. monthly. 8.2999%
c. daily. 8.3277%
d. continuously. 8.3287%
[*Hint:* Use the formula from Exercise 60.]

62. Sumeet's investment in a mutual fund increased from $30,000 to $45,000 in five years. What was his effective annual yield? 8.447%
[*Hint:* Use $A = P(1 + y)^t$, where t is the number of years and y is the effective annual yield.]

63. Fidelity Federal offers three types of investments paying, respectively,

9.5 percent continuous interest, yields 9.9658%
10 percent simple interest, and yields 10%
9.6 percent interest compounded monthly yields 10.03%

Which investment will yield the greatest return? An interest rate of 9.6% compounded monthly will yield the greatest return.

64. The relative rate of growth of $A(t) = Pe^{kt}$ in the interval $[t, t + h]$ is defined by $\dfrac{A(t + h) - A(t)}{A(t)}$.

a. Show that $\dfrac{A(t + 1) - A(t)}{A(t)} = e^k - 1$.

b. Show that $\dfrac{A(t + h) - A(t)}{A(t)} = e^{kh} - 1$.

Critical Thinking

65. Let $f(x) = e^{-x^2}$.
a. Calculate $f(0), f(1), f(2)$, and $f(3)$. (Use a calculator.)
b. Find the axis of symmetry of f. y-axis
c. Find $\lim f(x)$ as $x \to \infty$ and as $x \to -\infty$. The limit is 0.
d. Sketch the graph of f. †

66. An important function in statistics is the probability density function, defined by

$$f(x) = \frac{1}{\sigma\sqrt{2\pi}} e^{-\frac{(x - \mu)^2}{2\sigma^2}},$$

where σ (the Greek letter sigma) and μ (the Greek letter mu) are constants with $\sigma > 0$ and μ a real number.
a. Find the axis of symmetry of f. $x = \mu$
b. Find $\lim f(x)$ as $x \to \infty$ and as $x \to -\infty$. The limit is 0.
c. Describe how to sketch the graph of f from the graph of $y = e^{-x^2}$. Answers will vary.

Answers:

65. **a.** $f(0) = 1, f(1) = \dfrac{1}{e} \approx 0.3679,$

$f(2) \approx 0.0183, f(3) \approx 0.0001234$

Logarithmic Functions

Before Starting this Section, Review

1. Finding inverse of a one-to-one function (Section 1.7, page 90)
2. Polynomial and rational inequalities (Appendix A, page 830)
3. Graphing transformations (Section 1.5, page 62)

Objectives

1. Define logarithmic functions.
2. Evaluate logarithms.
3. Find the domains of logarithmic functions.
4. Graph logarithmic functions.
5. Use logarithms to evaluate exponential equations.

THE MCDONALD'S COFFEE CASE

Stella Liebeck of Albuquerque, New Mexico, was in the passenger seat of her grandson's car when she was severely burned by McDonald's coffee in February 1992. Liebeck ordered coffee that was served in a foam cup at the McDonald's drive-through window. While attempting to add cream and sugar to her coffee, Liebeck spilled the entire cup into her lap. A vascular surgeon determined that Liebeck suffered full-thickness burns (third-degree burns) over 6% of her body. She sued McDonald's for damages.

During the preliminary phase of the trial, McDonald's said that on the basis of a consultant's advice, the company brewed its coffee at 195° to 205°F and held it at 180°F to 190°F to maintain optimal taste.

Coffee served at home is generally between 135°F and 140°F. Prior to the Liebeck case, the prestigious Shriner's Burn Institute of Cincinnati had published warnings to the franchise food industry that beverages above 130°F were causing serious scald burns. Liebeck's lawsuit was settled out of court, and both parties agreed to keep the award for damages confidential.

Avoiding Further Lawsuits Corporate lawyers informed McDonald's management that further lawsuits from customers who spill their coffee can be avoided if the temperature of the coffee is 125°F or less when it is delivered to the customers. In Example 10, we will learn how long to wait after brewing the coffee before delivering it to customers. ■

---

1 Define logarithmic functions.

Logarithmic Functions

Recall from Section 3.1 that every exponential function

$$f(x) = a^x, a > 0, a \neq 1,$$

is a *one-to-one* function and therefore has an inverse function. Let's find the inverse of an exponential function. First, we consider the exponential function $f(x) = 3^x$. If we think of an exponential function f as "putting an exponent on" a particular base, then the inverse function f^{-1} must "lift the exponent off" to undo the effect of f, as illustrated below.

$$x \xrightarrow{\quad f \quad} 3^x \xrightarrow{\quad f^{-1} \quad} x$$

Let's try to find the inverse of $f(x) = 3^x$ by the procedure outlined in Section 1.7.

Step 1 Replace $f(x)$ in the equation $f(x) = 3^x$ with y: $y = 3^x$.

Step 2 Interchange x and y in the equation in Step 1: $x = 3^y$.

Step 3 Solve the equation for y in Step 2.

In Step 3, we need to solve the equation $x = 3^y$ for y. Finding y is easy when $3 = 3^y$ or $27 = 3^y$, but what if $5 = 3^y$? There actually is a real number y that solves the equation, $5 = 3^y$, but we have not introduced the name for this number yet. (We were in a similar position concerning the solution of $y^3 = 7$ before we introduced the name for the solution, namely, $\sqrt[3]{7}$). We can say that y is "the exponent on 3 that gives 5." Instead, we use the word *logarithm* (or *log* for short) to describe this exponent. We use the notation $\log_3 5$ (read as "log of 5 with base 3") to express the solution of the equation $5 = 3^y$. This log notation is a new way to write information about exponents: $y = \log_3 x$ means $x = 3^y$. In other words, $\log_3 x$ *is the exponent to which 3 must be raised to obtain* x. Therefore, the inverse function of the exponential function $f(x) = 3^x$ is written

$$f^{-1}(x) = \log_3 x \quad \text{or} \quad y = \log_3 x, \text{where } y = f^{-1}(x).$$

The previous discussion could be repeated for any base a instead of 3 provided that $a > 0$ and $a \neq 1$.

DEFINITION OF THE LOGARITHMIC FUNCTION

For $x > 0$, $a > 0$, and $a \neq 1$,

$$y = \log_a x \quad \text{if and only if} \quad x = a^y.$$

The function $f(x) = \log_a x$ is called the **logarithmic function with base a**.

The definition of the logarithm says that the two equations

$$y = \log_a x \text{ (\textbf{logarithmic form})}$$
$$x = a^y \text{ (\textbf{exponential form})}$$

are equivalent. For instance, $\log_3 81 = 4$ because 3 must be raised to the fourth power to obtain 81.

EXAMPLE 1   **Converting from Exponential to Logarithmic Form**

Write each exponential equation in logarithmic form.

a. $4^3 = 64$ **b.** $\left(\dfrac{1}{2}\right)^4 = \dfrac{1}{16}$ **c.** $a^{-2} = 7$

SOLUTION
a. $4^3 = 64$ is equivalent to $\log_4 64 = 3$.
b. $\left(\dfrac{1}{2}\right)^4 = \dfrac{1}{16}$ is equivalent to $\log_{1/2} \dfrac{1}{16} = 4$.
c. $a^{-2} = 7$ is equivalent to $\log_a 7 = -2$.   ■ ■ ■

Practice Problem 1   Write each exponential equation in logarithmic form.

a. $2^{10} = 1024$ **b.** $9^{-1/2} = \dfrac{1}{3}$ **c.** $p = a^q$ ■

EXAMPLE 2   **Converting from Logarithmic Form to Exponential Form**

Write each logarithmic equation in exponential form.

a. $\log_3 243 = 5$ **b.** $\log_2 5 = x$ **c.** $\log_a N = x$

SOLUTION

a. $\log_3 243 = 5$ is equivalent to $243 = 3^5$.

b. $\log_2 5 = x$ is equivalent to $5 = 2^x$.

c. $\log_a N = x$ is equivalent to $N = a^x$.   ■ ■ ■

Practice Problem 2   Write each logarithmic equation in exponential form.

a. $\log_2 64 = 6$ **b.** $\log_v u = w$ ■

2 Evaluate logarithms.

Evaluating Logarithms

The technique of converting from logarithmic form to exponential form can be used to evaluate some logarithms by inspection.

| EXAMPLE 3 | Evaluating Logarithms |

Find the value of each of the following logarithms.

a. $\log_5 25$ **b.** $\log_2 16$ **c.** $\log_{1/3} 9$

d. $\log_7 7$ **e.** $\log_6 1$ **f.** $\log_4 \dfrac{1}{2}$

SOLUTION

Logarithmic Form	Exponential Form	Value
a. $\log_5 25 = y$	$25 = 5^y$ or $5^2 = 5^y$	$y = 2$
b. $\log_2 16 = y$	$16 = 2^y$ or $2^4 = 2^y$	$y = 4$
c. $\log_{1/3} 9 = y$	$9 = \left(\dfrac{1}{3}\right)^y$ or $3^2 = 3^{-y}$	$y = -2$
d. $\log_7 7 = y$	$7 = 7^y$ or $7^1 = 7^y$	$y = 1$
e. $\log_6 1 = y$	$1 = 6^y$ or $6^0 = 6^y$	$y = 0$
f. $\log_4 \dfrac{1}{2} = y$	$\dfrac{1}{2} = 4^y$ or $2^{-1} = 2^{2y}$	$y = -\dfrac{1}{2}$

■ ■ ■

Practice Problem 3   Evaluate.

a. $\log_3 9$ **b.** $\log_9 \dfrac{1}{3}$ **c.** $\log_{1/2} 32$ ■

| EXAMPLE 4 | Using the Definition of Logarithm |

Solve each equation.

a. $\log_5 x = -3$ **b.** $\log_3 \dfrac{1}{27} = y$

c. $\log_z 1000 = 3$ **d.** $\log_2 (x^2 - 6x + 10) = 1$

SOLUTION

a. $\log_5 x = -3$

$\qquad x = 5^{-3}$ Exponential form

$\qquad x = \dfrac{1}{5^3} = \dfrac{1}{125}$ $a^{-n} = \dfrac{1}{a^n}$

b. $\log_3 \dfrac{1}{27} = y$

$\dfrac{1}{27} = 3^y$ Exponential form

$3^{-3} = 3^y$ $\dfrac{1}{27} = \dfrac{1}{3^3} = 3^{-3}$

$-3 = y$ If $a^x = a^y$, then $x = y$.

c. $\log_z 1000 = 3$

$1000 = z^3$ Exponential form

$10^3 = z^3$ $1000 = 10^3$

$10 = z$ Take the cube root of both sides.

d. $\log_2 (x^2 - 6x + 10) = 1$

$x^2 - 6x + 10 = 2^1$ Exponential form

$x^2 - 6x + 8 = 0$ $2^1 = 2$; subtract 2 from both sides.

$(x - 2)(x - 4) = 0$ Factor.

$x - 2 = 0$ or $x - 4 = 0$ Zero-product property

$x = 2$ or $x = 4$ Solve for x.   ■ ■ ■

Practice Problem 4   Solve each equation.

a. $\log_2 x = 3$ **b.** $\log_z 125 = 3$ **c.** $\log_3(x^2 - x - 5) = 0$ ■

---

3   Find the domains of logarithmic functions.

Domain of Logarithmic Functions

Since the exponential function $f(x) = a^x$ has domain $(-\infty, \infty)$ and range $(0, \infty)$, its inverse function $y = \log_a x$ **has domain $(0, \infty)$ and range $(-\infty, \infty)$**. Therefore, the logarithms of 0 and of negative numbers are not defined; so expressions such as $\log_a(-2)$ and $\log_a(0)$ are meaningless.

EXAMPLE 5   **Finding the Domain**

Find the domain of $f(x) = \log_3(2 - x)$.

SOLUTION

Since the domain of the logarithmic function is $(0, \infty)$, the expression $2 - x$ must be positive. The domain of f is the set of all real numbers x, where

$$2 - x > 0$$
$$2 > x \qquad \text{Solve for } x.$$

The domain of f is $(-\infty, 2)$.   ■ ■ ■

Practice Problem 5   Find the domain of $f(x) = \log_{10} \sqrt{1 - x}$.   ■

Since $a^1 = a$ and $a^0 = 1$ for any base a, expressing these equations in logarithmic notation gives

$$\log_a a = 1 \quad \text{and} \quad \log_a 1 = 0.$$

Also, letting $f(x) = a^x$ and $f^{-1}(x) = \log_a x$, the equation

$$f^{-1}(f(x)) = x \text{ can be written as } \log_a a^x = x$$

and the equation

$$f(f^{-1}(x)) = x \text{ can be written as } a^{\log_a x} = x.$$

We summarize these relationships as basic properties of logarithms.

> **BASIC PROPERTIES OF LOGARITHMS**
>
> For any base $a > 0$, with $a \neq 1$,
> 1. $\log_a a = 1$.
> 2. $\log_a 1 = 0$.
> 3. $\log_a a^x = x$ for any real number x.
> 4. $a^{\log_a x} = x$ for any $x > 0$.

4   Graph logarithmic functions.

Graphs of Logarithmic Functions

We now look at other properties of logarithmic functions and begin with an example showing how to graph a logarithmic function.

EXAMPLE 6   **Sketching a Graph**

Sketch the graph of $y = \log_3 x$.

Plotting points (Method 1) To find selected ordered pairs on the graph of $y = \log_3 x$, we choose the x-values to be powers of 3. We can easily compute the logarithms of these values by using Property 3, namely, $\log_3 3^x = x$. We compute the y values in Table 3.5.

RECALL

Remember that "$\log_3 x$" means the exponent on 3 that gives x.

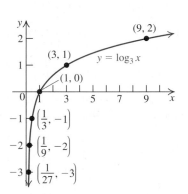

FIGURE 3.8 Graph by plotting points

TABLE 3.5

x	$y = \log_3 x$	(x, y)
$\dfrac{1}{27}$	Since $3^{-3} = \dfrac{1}{27}$, $y = \log_3 \dfrac{1}{27} = -3$	$\left(\dfrac{1}{27}, -3\right)$
$\dfrac{1}{9}$	Since $3^{-2} = \dfrac{1}{9}$, $y = \log_3 \dfrac{1}{9} = -2$	$\left(\dfrac{1}{9}, -2\right)$
$\dfrac{1}{3}$	Since $3^{-1} = \dfrac{1}{3}$, $y = \log_3 \dfrac{1}{3} = -1$	$\left(\dfrac{1}{3}, -1\right)$
1	Since $3^0 = 1$, $y = \log_3 1 = 0$	$(1, 0)$
3	Since $3^1 = 3$, $y = \log_3 3 = 1$	$(3, 1)$
9	Since $3^2 = 9$, $y = \log_3 9 = 2$	$(9, 2)$

    Plotting these ordered pairs and connecting them with a smooth curve gives us the graph of $y = \log_3 x$, shown in Figure 3.8.

Using the inverse function (Method 2) Since $y = \log_3 x$ is the inverse of the function $y = 3^x$, we first graph the exponential function $y = 3^x$. Then reflect the graph of $y = 3^x$ in the line $y = x$. Both graphs are shown in Figure 3.9.   ■ ■ ■

Practice Problem 6   Sketch the graph of

$$y = \log_2 x.$$

    ■

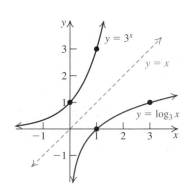

FIGURE 3.9 Graph by using the inverse function

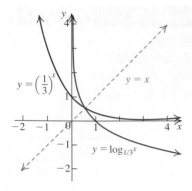

FIGURE 3.10

We can graph $y = \log_{1/3} x$ either by plotting points or by using the graph of its inverse. See Figure 3.10. The graph of $y = \log_3 x$ shown in Figure 3.9 is increasing. The graph of $y = \log_{1/3} x$ shown in Figure 3.10, on the other hand, is decreasing.

In general, logarithmic functions have two basic shapes, determined by the base a. If $a > 1$, the graph of $y = \log_a x$ is increasing, as in Figure 3.11(a). If $0 < a < 1$, the graph is decreasing, as in Figure 3.11(b).

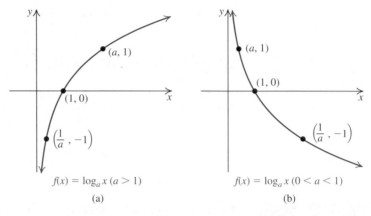

FIGURE 3.11 Basic shapes of $y = \log_a x$

STUDY TIP

To get the correct shape for a logarithmic function $y = \log_a x$, it is helpful to graph the three points

$$(a, 1), (1, 0), \text{ and } \left(\frac{1}{a}, -1\right).$$

PROPERTIES OF EXPONENTIAL AND LOGARITHMIC FUNCTIONS	
Exponential Function $y = a^x$	**Logarithmic Function $y = \log_a x$**
1. The domain is $(-\infty, \infty)$, and the range is $(0, \infty)$.	The domain is $(0, \infty)$, and the range is $(-\infty, \infty)$.
2. The y-intercept is 1, and there is no x-intercept.	The x-intercept is 1, and there is no y-intercept.
3. The x-axis $(y = 0)$ is the horizontal asymptote.	The y-axis $(x = 0)$ is the vertical asymptote.
4. The function is one-to-one; that is, $a^u = a^v$ if and only if $u = v$.	The function is one-to-one; that is, $\log_a u = \log_a v$ if and only if $u = v$.
5. The function is increasing if $a > 1$ and decreasing if $0 < a < 1$.	The function is increasing if $a > 1$ and decreasing if $0 < a < 1$.

Once we know the shape of a logarithmic graph, we can shift it vertically or horizontally, stretch it, compress it, check answers with it, and interpret the graph.

EXAMPLE 7   Using Transformations

Start with the graph of $f(x) = \log_3 x$ and use transformations to sketch the graph of each function.

a. $f(x) = \log_3 x + 2$ **b.** $f(x) = \log_3 (x - 1)$
c. $f(x) = -\log_3 x$ **d.** $f(x) = \log_3 (-x)$

State the domain and range and the vertical asymptote for the graph of each function.

SOLUTION

We start with the graph of $f(x) = \log_3 x$ and use the transformations shown in Figure 3.12.

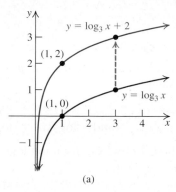

(a)

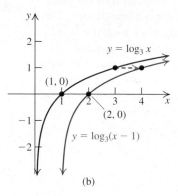

(b)

a. $f(x) = \log_3 x + 2$

Adding 2 to $\log_3 x$ shifts the graph two units up. Domain: $(0, \infty)$; range: $(-\infty, \infty)$; vertical asymptote: $x = 0$ (y-axis)

b. $f(x) = \log_3 (x - 1)$

Replacing x with $x - 1$ in $\log_3 x$ shifts the graph one unit right. Domain: $(1, \infty)$; range $(-\infty, \infty)$; vertical asymptote: $x = 1$

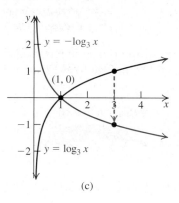

(c)

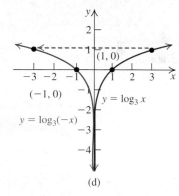

(d)

c. $f(x) = -\log_3 (x)$

Multiplying $\log_3 x$ by -1 reflects the graph in the x-axis. Domain: $(0, \infty)$; range: $(-\infty, \infty)$; vertical asymptote: $x = 0$

d. $f(x) = \log_3 (-x)$

Replacing x with $-x$ in $\log_3 x$ reflects the graph in the y-axis. Domain: $(-\infty, 0)$; range: $(-\infty, \infty)$; vertical asymptote: $x = 0$

FIGURE 3.12 Transformations on $y = \log_3 x$ ■ ■ ■

Practice Problem 7   Use transformations to sketch the graph of

$$y = -\log_2 (x - 3).$$ ■

Common Logarithms

The logarithm with base 10 is called the **common logarithm** and is denoted by omitting the base, so

$$\log x = \log_{10} x.$$

$$y = \log x \,(x > 0) \quad \text{if and only if} \quad x = 10^y.$$

TECHNOLOGY CONNECTION

You can find the common logarithms of any positive number with the LOG key on your calculator. The exact key sequence will vary with different calculators. On most calculators, if you want to find log 3.7, press

3.7 LOG ENTER

or

LOG 3.7 ENTER

and see the display
0.5682017.

Although the display reads 0.5682017, this number really is an approximate value; so you should write $\log_{10} 3.7 \approx 0.5682017$.

Applying the basic properties of logarithms (see page 215) to common logarithms, we have the following:

1. $\log 10 = 1$
2. $\log 1 = 0$
3. $\log 10^x = x$
4. $10^{\log x} = x$

We use property (3) to evaluate the common logarithms of numbers that are powers of 10. For example,

$$\log 1000 = \log 10^3 = 3$$
$$\text{and}\quad \log 0.01 = \log 10^{-2} = -2.$$

We use a calculator to find the common logarithms of numbers that are not powers of 10 by pressing the LOG key.

The graph of $y = \log x$ is similar to the graph in Figure 3.11(a) because the base for $\log x$ is understood to be 10.

EXAMPLE 8   Using Transformations to Sketch a Graph

Sketch the graph of

$$y = 2 - \log (x - 2).$$

SOLUTION

We start with the graph of the function $f(x) = \log x$ and use the order of transformations (see page 71) indicated in Figure 3.13 to sketch the graph.

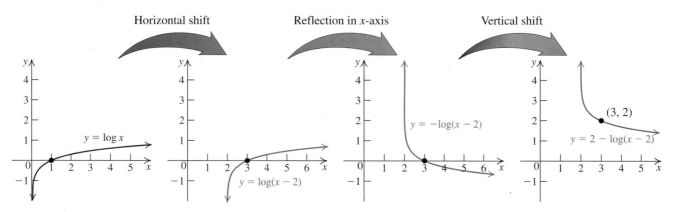

Step 1: Replacing x with $x - 2$ in $\log x$ shifts the graph of $y = \log x$ two units right.

Step 2: Multiplying $\log (x - 2)$ by -1 reflects the graph of $y = \log (x - 2)$ in the x-axis.

Step 3: Adding 2 to $-\log (x - 2)$ shifts the graph of $y = -\log (x - 2)$ two units up.

FIGURE 3.13 Transformations on $y = \log x$

The domain of $f(x) = 2 - \log (x - 2)$ is $(2, \infty)$, the range is $(-\infty, \infty)$, and the vertical asymptote is the line $x = 2$.   ■ ■ ■

Practice Problem 8   Sketch the graph of

$$y = \log (x - 3) - 2.$$ ■

TECHNOLOGY CONNECTION

To evaluate natural logarithms, we use the $\boxed{\text{LN}}$ key on a calculator. To find $\ln(2.3)$ on most calculators, press

2.3 $\boxed{\text{LN}}$ $\boxed{\text{ENTER}}$
or
$\boxed{\text{LN}}$ 2.3 $\boxed{\text{ENTER}}$

and see the display 0.832909123.

Natural Logarithms

In most applications in calculus and the sciences, the convenient base for logarithms is the number e. The logarithms with base e are called **natural logarithms** and are denoted by $\ln x$ (read "*ell en x*" or "*lawn x*") so that

$$\ln x = \log_e x.$$

$$y = \ln x \,(x > 0) \quad \text{if and only if} \quad x = e^y.$$

Applying the basic properties of logarithms (see page 215) to natural logarithms, we have the following:

1. $\ln e = 1$
2. $\ln 1 = 0$
3. $\ln e^x = x$
4. $e^{\ln x} = x$

We can use property (3) to evaluate the natural logarithms of powers of e.

EXAMPLE 9   **Evaluating the Natural Logarithm Function**

Evaluate each expression.

a. $\ln e^4$ **b.** $\ln \dfrac{1}{e^{2.5}}$ **c.** $\ln 3$

SOLUTION

a. $\ln e^4 = 4$ Property (3)

b. $\ln \dfrac{1}{e^{2.5}} = \ln e^{-2.5} = -2.5$ Property (3)

c. $\ln 3 \approx 1.0986123$ Use a calculator.   ■ ■ ■

Practice Problem 9   Evaluate each expression.

a. $\ln \dfrac{1}{e}$ **b.** $\ln 2$ ■

5 Use logarithms to solve exponential equations.

Investments

EXAMPLE 10   **Doubling Your Money**

a. How long will it take to double your money if it earns 6.5% compounded continuously?

b. At what rate of return, compounded continuously, would your money double in 5 years?

SOLUTION

If P dollars is invested and you want to double it, then the final amount $A = 2P$.

a. $A = Pe^{rt}$ Continuous compounding formula
    $2P = Pe^{0.065t}$ $A = 2P, r = 0.065$
    $2 = e^{0.065t}$ Divide both sides by P.
    $\ln 2 = 0.065t$ Logarithmic form of $x = e^y$.

$$\frac{\ln 2}{0.065} = t \qquad \text{Divide both sides by 0.065.}$$

$$t \approx 10.66 \qquad \text{Use a calculator.}$$

It will take 11 years to double your money.

b. $\quad A = Pe^{rt} \qquad\qquad$ Continuous compounding formula

$\qquad 2P = Pe^{5r} \qquad\qquad A = 2P, t = 5$

$\qquad\; 2 = e^{5r} \qquad\qquad$ Divide both sides by P.

$\quad \ln 2 = 5r \qquad\qquad$ Logarithmic form of $x = e^y$.

$$\frac{\ln 2}{5} = r \qquad\qquad \text{Divide both sides by 5.}$$

$\qquad\; r \approx 0.1386 \qquad$ Use a calculator.

Your investment will double in 5 years at the rate of 13.86%. $\qquad$ ■ ■ ■

Practice Problem 10 $\quad$ Repeat Example 10 for tripling $(A = 3P)$ your money. $\quad$ ■

Newton's Law of Cooling

When a cool drink is removed from a refrigerator and placed in a room at normal temperature, the drink warms to the temperature of the room. When removed from the oven, a pizza baked at a high temperature cools to the temperature of the room.

In situations such as these, the rate at which an object's temperature changes at any given time is proportional to the difference between its temperature and the temperature of the surrounding medium. This observation is called *Newton's Law of Cooling*, although, as in the case of a cool drink, it applies to warming as well.

NEWTON'S LAW OF COOLING

Newton's Law of Cooling states that

$$T = T_S + (T_0 - T_S)e^{-kt},$$

where T is the temperature of the object at time t, T_S is the surrounding temperature, and T_0 is the value of T at $t = 0$.

EXAMPLE 11 $\quad$ **McDonald's Hot Coffee**

The local McDonald's franchise has discovered that when coffee is poured from a coffeemaker whose contents are 180°F into a noninsulated pot, after 1 minute, the coffee cools to 165°F if the room temperature is 72°F. How long should the employees wait before pouring the coffee from this noninsulated pot into cups to deliver it to customers at 125°F?

SOLUTION

We use Newton's Law of Cooling with $T_0 = 180$ and $T_S = 72$.

$$T = 72 + (180 - 72)e^{-kt}$$

$$T = 72 + 108e^{-kt} \qquad\qquad (1) \qquad \text{Simplify.}$$

From the given data, we have $T = 165$ when $t = 1$.

$$165 = 72 + 108e^{-k} \qquad \text{Substitute data into equation (1).}$$

$$93 = 108e^{-k} \qquad \text{Subtract 72 from both sides.}$$

$$\frac{93}{108} = e^{-k} \qquad \text{Solve for } e^{-k}.$$

$$\ln\left(\frac{93}{108}\right) = -k \qquad \text{Write in logarithmic form.}$$

$$-0.1495317 = -k \qquad \text{Use a calculator.}$$

$$k = 0.1495317$$

With this value of k, equation (1) becomes

$$T = 72 + 108e^{-0.1495317t} \qquad (2)$$

$$125 = 72 + 108e^{-0.1495317t} \qquad \text{Replace } T \text{ with 125 in equation (2).}$$

$$\frac{125 - 72}{108} = e^{-0.1495317t} \qquad \text{Solve for } e^{-0.1495317t}.$$

$$\frac{53}{108} = e^{-0.1495317t} \qquad \text{Simplify.}$$

$$\ln\left(\frac{53}{108}\right) = -0.1495317t \qquad \text{Write in logarithmic form.}$$

$$t = \frac{-1}{0.1495317} \ln\left(\frac{53}{108}\right) \qquad \text{Solve for } t.$$

$$\approx 4.76 \qquad \text{Use a calculator.}$$

The employees should wait approximately 4.76 minutes (realistically, about 5 minutes) to deliver the coffee at $125°F$ to the customers. ▪ ▪ ▪

Practice Problem 11 Repeat Example 11 assuming that the coffee is to be delivered to the customers at $120°F$. ▪

In the next example, we use the mathematical model

$$A = A_0 e^{rt} \qquad (3)$$

where A is the quantity after t units of time, A_0 is the initial (original) quantity when the experiment started ($t = 0$), and r is the growth or decay rate per period.

EXAMPLE 12 **Chemical Toxins in a Lake**

In a large lake, one-fifth of the water is replaced by clean water each year. A chemical spill deposits 60,000 cubic meters of soluble toxic waste into the lake.

a. How much of this toxin will be left in the lake after four years?

b. When will the toxic chemical be reduced to 6000 cubic meters?

SOLUTION

Since one-fifth of the water in the lake is replaced by clean water every year, the decay rate for the toxin is $\frac{1}{5}$; so $r = -\frac{1}{5}$. With $A_0 = 60{,}000$ and $r = -\frac{1}{5}$, equation (3) becomes

$$A = 60{,}000e^{(-1/5)t} \qquad (4)$$

where A is the amount of toxin (in cubic meters) after t years.

a. $A = 60{,}000e^{(-1/5)(4)}$ Substitute $t = 4$ in equation (4).

$\approx 26{,}959.74 \text{ m}^3$ Use a calculator.

b. $6000 = 60{,}000e^{(-1/5)t}$ Replace A with 6000 in equation (4).

$0.1 = e^{(-1/5)t}$ Divide both sides by 60,000.

$\ln(0.1) = -\dfrac{1}{5}t$ Logarithmic form

$t = -5\ln(0.1)$ Solve for t.

$\approx 11.51 \text{ years}$ Use a calculator.   ■ ■ ■

Practice Problem 12 Repeat Example 12 assuming that one-sixth of the water in the lake is replaced by clean water each year.   ■

SECTION 3.3 ■ Exercises

A EXERCISES Basic Skills and Concepts

1. The domain of the function $y = \log_a x$ is $\underline{(0, \infty)}$, and its range is $\underline{(-\infty, \infty)}$.

2. The logarithmic form $y = \log_a x$ is equivalent to the exponential form $\underline{a^y = x}$.

3. The logarithm with base 10 is called the $\underline{\text{common}}$ logarithm, and the logarithm with base e is called the $\underline{\text{natural}}$ logarithm.

4. $a^{\log_a x} = \underline{\quad x \quad}$, and $\log_a a^x = \underline{\quad x \quad}$.

5. *True or False* The graph of $\log_a x, a > 0, a \neq 1$, is an increasing function. False

6. *True or False* The graph of $y = \log_a x, a > 0$, and $a \neq 1$, has no horizontal asymptote. True

In Exercises 7–18, write each exponential equation in logarithmic form.

7. $5^2 = 25$ $\log_5 25 = 2$ 8. $81^{1/2} = 9$ $\log_{81} 9 = \dfrac{1}{2}$

9. $(49)^{-\frac{1}{2}} = \dfrac{1}{7}$ 10. $\left(\dfrac{1}{16}\right)^{1/2} = \dfrac{1}{4}$ $\log_{\frac{1}{16}}\left(\dfrac{1}{4}\right) = \dfrac{1}{2}$
   $\log_{49}\left(\dfrac{1}{7}\right) = -\dfrac{1}{2}$

11. $\left(\dfrac{1}{16}\right)^{-\frac{1}{2}} = 4$ $\log_{\frac{1}{16}} 4 = -\dfrac{1}{2}$ 12. $(a^2)^2 = a^4$ $\log_{a^2} a^4 = 2$

13. $10^0 = 1$ $\log_{10} 1 = 0$ 14. $10^4 = 10{,}000$ $\log_{10} 10{,}000 = 4$

15. $(10)^{-1} = 0.1$ 16. $3^x = 5$ $\log_3 5 = x$
   $\log_{10} 0.1 = -1$

17. $a^2 = 5$ 18. $a^e = \pi$ $\log_a \pi = e$
   $\log_a 5 = 2$

In Exercises 19–30, write each logarithmic equation in exponential form.

19. $\log_2 32 = 5$ $2^5 = 32$ 20. $\log_7 49 = 2$ $7^2 = 49$

21. $\log_{10} 100 = 2$ $10^2 = 100$ 22. $\log_{10} 10 = 1$ $10^1 = 10$

23. $\log_{10} 1 = 0$ $10^0 = 1$ 24. $\log_a 1 = 0$ $a^0 = 1$

25. $\log_{10} 0.01 = -2$ 26. $\log_{1/5} 5 = -1$ $\left(\dfrac{1}{5}\right)^{-1} = 5$
   $10^{-2} = 0.01$

27. $\log_8 2 = \dfrac{1}{3}$ $8^{\frac{1}{3}} = 2$ 28. $\log 1000 = 3$ $10^3 = 1000$

29. $\ln 2 = x$ $e^x = 2$ 30. $\ln \pi = a$ $e^a = \pi$

In Exercises 31–40, evaluate each expression without using a calculator.

31. $\log_5 125$ 3   32. $\log_9 81$ 2

33. $\log 10{,}000$ 4   34. $\log_3 \dfrac{1}{3}$ -1

35. $\log_2 \dfrac{1}{8}$ -3 36. $\log_4 \dfrac{1}{64}$ -3

37. $\log_3 \sqrt{27}$ $\dfrac{3}{2}$ 38. $\log_{27} 3$ $\dfrac{1}{3}$

39. $\log_{16} 2$ $\dfrac{1}{4}$ 40. $\log_5 \sqrt{125}$ $\dfrac{3}{2}$

In Exercises 41–52, solve each equation.

41. $\log_5 x = 2$ $x = 25$

42. $\log_5 x = -2$ $x = \dfrac{1}{25}$

43. $\log_5 (x - 2) = 3$ $x = 127$

44. $\log_5 (x + 1) = -2$ $x = -\dfrac{24}{25}$

45. $\log_2 \left(\dfrac{1}{16}\right) = x$ $x = -4$

46. $\log_x \left(\dfrac{1}{5}\right) = -1$ $x = 5$

47. $\log_{16} \sqrt{x - 1} = \dfrac{1}{4}$ $x = 5$

48. $\log_{27} \sqrt[3]{1 - x} = \dfrac{1}{3}$ $x = -26$

49. $\log_a (x - 1) = 0$ $x = 2$

50. $\log_a (x^2 + 5x + 7) = 0$ $x = -3, -2$

51. $\log_2 (x^2 - 7x + 14) = 1$ $x = 3, 4$

52. $\log (x^2 + 5x + 16) = 1$ $x = -3, -2$

In Exercises 53–60, find the domain of each function.

53. $f(x) = \log_2 (x + 1)$ 54. $f(x) = \log_2 (x + 3)$

55. $g(x) = \log_3 (x - 5)$ 56. $g(x) = \log_3 (x - 8)$

Answers:

53. $(-1, \infty)$ 54. $(-3, \infty)$ 55. $(5, \infty)$ 56. $(8, \infty)$

†Due to space constrictions, answers to these exercises may be found in the Answers beginning on page A–1 in the back of the book.

57. $f(x) = \log_2 \sqrt{x - 1}$ **58.** $f(x) = \log_2 \sqrt{x - 3}$

59. $g(x) = \log_4(-x)$ **60.** $g(x) = \log_4 \sqrt{-x}$

61. Match each logarithmic function with one of the graphs labeled (a)–(f).
- **a.** $f(x) = \log x$ f
- **b.** $f(x) = -\log|x|$ a
- **c.** $f(x) = -\log(-x)$ d
- **d.** $f(x) = \log(x - 1)$ b
- **e.** $y = (\log x) - 1$ e
- **f.** $f(x) = \log(-1 - x)$ c

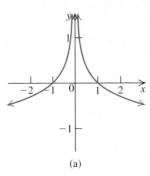

(a)

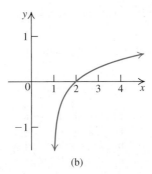

(b)

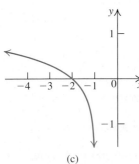

(c)

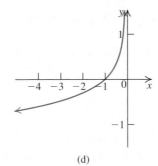

(d)

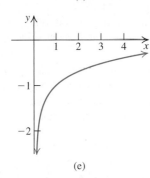

(e)

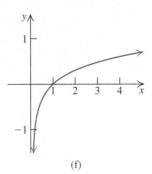

(f)

62. For each given graph, find a function of the form $f(x) = \log_a x$ that represents the graph.

a. $f(x) = \log_2 x$ **b.** $f(x) = \log_3 x$

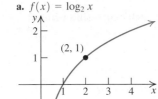

(a)

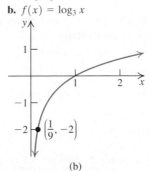

(b)

Answers:
57. $(1, \infty)$ **58.** $(3, \infty)$ **59.** $(-\infty, 0)$ **60.** $(-\infty, 0)$

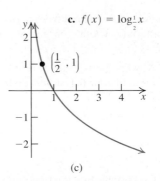

c. $f(x) = \log_{\frac{1}{2}} x$ **d.** $f(x) = \log_4 x$

(c)   (d)

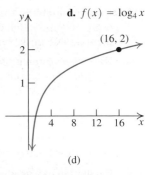

In Exercises 63–70, graph the given function by using transformations on the appropriate basic graph of the form $y = \log_a x$. State the domain and range of the function and the vertical asymptote of the graph.

63. $f(x) = \log_4(x + 3)$ †   **64.** $g(x) = \log_{1/2}(x - 1)$ †

65. $y = -\log_5 x$ †   **66.** $y = 2\log_7(x)$ †

67. $y = \log_{1/5}(-x)$ †   **68.** $y = 1 + \log_{1/5}(-x)$ †

69. $y = |\log_3 x|$ †   **70.** $y = \log_3|x|$ †

In Exercises 71–78, begin with the graph of $f(x) = \log_2 x$ and use transformations to sketch the graph of each function. Find the domain and range of the function and the vertical asymptote of the graph.

71. $y = \log_2(x - 1)$ †   **72.** $y = \log_2(-x)$ †

73. $y = \log_2(3 - x)$ †   **74.** $y = \log_2 x$ †

75. $y = 2 + \log_2(3 - x)$ †   **76.** $y = 4 - \log_2(3 - x)$ †

77. $y = \log_2|x|$ †   **78.** $y = \log_2 x^2$ †

In Exercises 79–84, evaluate each expression.

79. $\log_4(\log_3 81)$ 1   **80.** $\log_4[\log_3(\log_2 8)]$ 0

81. $5^{\log_5 7}$ 7   **82.** $10^{\log 13}$ 13

83. $e^{\ln 5}$ 5   **84.** $e^{\ln e^2}$ e^2

In Exercises 85–88, solve each equation.

85. $\log x = 2$ $x = 100$ **86.** $\log(x - 1) = 1$ $x = 11$

87. $\ln x = 1$ $x = e$ **88.** $\ln x = 0$ $x = 1$

In Exercises 89–94, begin with the graph of $y = \ln x$ and use transformations to sketch the graph of each of the given functions.

89. $y = \ln(x + 2)$ †   **90.** $y = \ln(2 - x)$ †

91. $y = -\ln(2 - x)$ †   **92.** $y = 2\ln x$ †

93. $y = 3 - 2\ln x$ †   **94.** $y = 1 - \ln(1 - x)$ †

B EXERCISES Applying the Concepts

In Exercises 95–102, use the model $A = A_0 e^{kt}$.

95. **Doubling your money.** How long would it take to double your money if you invested P dollars at the rate of 8% compounded continuously? 8.66 yr

96. **Investment goal.** How long would it take to grow your investment from $10,000 to $120,000 at the rate of 10% compounded continuously? 24.85 yr

97. Rate for doubling your money. At what annual rate of return, compounded continuously, would your investment double in six years? 11.55%

98. Rate-of-investment goal. At what annual rate of return, compounded continuously, would your investment grow from $8000 to $50,000 in 25 years? 7.33%

99. Population of Canada. The population of Canada was 31.3 million in 2000 and 33.4 million in 2007. Determine the time from 2000 until Canada's population (a) doubles and (b) triples. **a.** 74.72 yr **b.** 118.43 yr

100. Population of Canada. In Exercise 99, find the time from 2000 until Canada's population reaches 50 million. 50.49 yr

101. Water contamination. A chemical is spilled into a reservoir of pure water. The concentration of chemical in the contaminated water is 4%. In one month, 20% of the water in the reservoir is replaced with clean water.
a. What will be the concentration of the contaminant one year from now? 0.36%
b. For water to be safe for drinking, the concentration of this contaminant cannot exceed 0.01%. How long will it be before the water is safe for drinking? 29.96 months

102. Toxic chemicals in a lake. In a lake, one-fourth of the water is replaced by clean water every year. Sixteen thousand cubic meters of soluble toxic chemical spill takes place in the lake. Let $T(n)$ represent the amount of toxin left after n years.
a. Find a formula for $T(n)$. $T(n) = 16{,}000e^{-0.25t}$
b. How much toxin will be left after 12 years? 796.59 m^3
c. When will 80% of the toxin be eliminated? 6.4 yr

103. Newton's Law of Cooling. A thermometer is taken from a room at 75°F to the outdoors, where the temperature is 20°F. The reading on the thermometer drops to 50°F after one minute.
a. Find the reading on the thermometer after
  (i) Five minutes. 22.66°F
  (ii) Ten minutes. 20.13°F
  (iii) One hour. 20°F
b. How long will it take for the reading to drop to 22°F? 5.5 min

104. Newton's Law of Cooling. The last bit of ice in a picnic cooler has melted ($T_0 = 32°F$). The temperature in the park is 85°F. After 30 minutes, the temperature in the cooler is 40°F. How long will it take for the temperature inside the cooler to reach 50°F? 76 min

105. Cooking salmon. A salmon filet initially at 50°F is cooked in an oven at a constant temperature of 400°F. After ten minutes, the temperature of the filet rises to 160°F. How long does it take until the salmon is medium rare at 220°F? 17.6 min

106. The time of murder. A forensic specialist took the temperature of a victim's body lying in a street at 2:10 A.M. and found it to be 85.7°F. At 2:40 A.M., the temperature of the body was 84.8°F. When was the murder committed if the air temperature during the night was 55°F? [Remember, normal body temperature is 98.6°F.] At 8:17 P.M. (353 min earlier than 2:10 A.M., assuming that the temperature of the live body was 98.6°F)

C EXERCISES Beyond the Basics

107. Sketch the graph of $f(x) = \begin{cases} \ln x & \text{if } x > 0 \\ \ln(-x) & \text{if } x < 0 \end{cases}$. †

108. Sketch the graph of $f(x) = \begin{cases} \ln(1 - x) & \text{if } x < 1 \\ e^{-x} & \text{if } x \geq 1 \end{cases}$. †

109. Find the domain of $f(x) = \log_3\left(\dfrac{x - 2}{x + 1}\right)$.
$(-\infty, -1) \cup (2, \infty)$

110. Find the domain of $g(x) = \log_2\left(\dfrac{x + 3}{x - 2}\right)$.
$(-\infty, -3) \cup (2, \infty)$

111. Finding the domain. Find the domain of each function.
a. $f(x) = \log_2(\log_3 x)$ $(1, \infty)$
b. $f(x) = \log(\ln(x - 1))$ $(2, \infty)$
c. $f(x) = \ln(\log(x - 1))$ $(2, \infty)$
d. $f(x) = \log(\log(\log(x - 1)))$ $(11, \infty)$

112. Finding the inverse. Find the inverse of each function in Exercise 111.

113. Present value of an investment. Recall that if P dollars is invested in an account at an interest rate r compounded continuously, then the amount A (called the *future value of P*) in the account t years from now will be $A = Pe^{rt}$. Solving the equation for P, we get $P = Ae^{-rt}$. In this formulation, P is called the present value of the investment.
a. Find the present value of $100,000 at 7% compounded continuously for 20 years. $24,659.69
b. Find the interest rate r compounded continuously that is needed to have $50,000 be the present value of $75,000 in ten years. 4.05%

114. Your uncle is 40 years old, and he wants to have an annual pension of $50,000 each year at age 65. What is the present value of his pension if the money can be invested at all times at a continuously compounded interest rate of
a. 5%? $286,504.80
b. 8%? $84,584.55
c. 10%? $41,042.50

Critical Thinking

In Exercises 115 and 116 evaluate each expression without using a calculator

115. $2^{\log_2 3} - 3^{\log_3 2}$ 1

116. $(\log_3 4 + \log_2 9)^2 - (\log_3 4 - \log_2 9)^2$. 16

117. Solve for x: $\log_3[\log_4(\log_2 x)] = 0$. 16

Answers:
112. a. $y = 3^{2^x}$
  b. $y = e^{10^x} + 1$
  c. $y = 10^{e^x} + 1$
  d. $y = 1 + 10^{10^{10^x}}$

Rules of Logarithms

Before Starting this Section, Review

1. Definition of logarithm (Section 3.3, page 212)
2. Basic properties of logarithms (Section 3.3, page 215)
3. Rules of exponents (Section 3.1, page 191)

Objectives

1 Learn the rules of logarithms.
2 Change the base of a logarithm.
3 Apply logarithms in growth and decay.

King Tut's Golden Mask

THE BOY KING TUT

Today the most famous pharaoh of Ancient Egypt is King Nebkheperure Tutankhamun, popularly called "King Tut." He was only 9 years old when he became a pharaoh. In 2005, a team of Egyptian scientists headed by Dr. Zahi Hawass determined that he was 19 years old when he died (around 1346 B.C.).

Tutankhamun was a short-lived boy king who, unlike the great Egyptian Kings Khufu (builder of the Great Pyramid), Amenhotep III (builder of temples throughout Egypt), and Ramesses II (prolific builder and usurper), accomplished nothing significant during his reign. In fact, little was known about him prior to Howard Carter's discovery of his tomb (and the amazing treasures it held) in the Valley of the Kings on November 4, 1922. Carter, with his benefactor Lord Carnarvon at his side, entered the tomb's burial chamber on November 26, 1922. Lord Carnarvon died seven weeks after entering the burial chamber, giving rise to the theory of the "curse" of King Tut.

The work on the tomb continued until 1933. The tomb contained a pristine mummy of an Egyptian king, lying intact in his original burial furniture. He was accompanied by a small slice of the royal world of the pharaohs: golden chariots, statues of gold and ebony, a fleet of miniature ships to accommodate his trip to the hereafter, his throne of gold, bottles of perfume, precious jewelry, and more. The "Treasures of Tutankhamun" exhibition, first shown at the British Museum in London in 1972, traveled to many countries, including the United States. In 2005, a new exhibition, "Tutankhamun and the Golden Age of the Pharaohs," visited the United States.

In Example 6, we discuss the age of a work of art found in King Tut's tomb. ■

1 Learn the rules of logarithms.

Rules of Logarithms

Recall the basic properties of logarithms from Section 3.3 for $a > 0, a \neq 1$:

$$\log_a a = 1 \qquad (1)$$
$$\log_a 1 = 0 \qquad (2)$$
$$\log_a a^x = x, x \text{ real} \qquad (3)$$
$$a^{\log_a x} = x, x > 0 \qquad (4)$$

In this section, we will discuss some important rules of logarithms that are helpful in calculations, simplifications, and applications.

Rules of Logarithms

Let M, N, and a be positive real numbers with $a \neq 1$ and let r be any real number.

Rule	Description	Examples
1. *Product Rule:* $\log_a(MN) = \log_a M + \log_a N$	The logarithm of the product of two (or more) numbers is the sum of the logarithms of the numbers.	$\ln(5 \cdot 7) = \ln 5 + \ln 7$ $\log(3x) = \log 3 + \log x$ $\log_2(5 \cdot 17) = \log_2 5 + \log_2 17$
2. *Quotient Rule:* $\log_a\left(\dfrac{M}{N}\right) = \log_a M - \log_a N$	The logarithm of the quotient of two numbers is the difference of the logarithms of the numbers.	$\ln \dfrac{5}{7} = \ln 5 - \ln 7$ $\log\left(\dfrac{5}{x}\right) = \log 5 - \log x$
3. *Power Rule:* $\log_a M^r = r \log_a M$	The logarithm of a number to the power r is r times the logarithm of the number.	$\ln 5^7 = 7 \ln 5$ $\log(5)^{3/2} = \dfrac{3}{2}\log 5$ $\log_2(7)^{-3} = -3\log_2 7$

Rules 1, 2, and 3 follow from the corresponding rules of the exponents:

$$a^u \cdot a^v = a^{u+v} \qquad \text{Product rule}$$

$$\frac{a^u}{a^v} = a^{u-v} \qquad \text{Quotient rule}$$

$$(a^u)^r = a^{ur} \qquad \text{Power rule}$$

For example, to prove the product rule for logarithms, we let

$$\log_a M = u \qquad \text{and} \qquad \log_a N = v.$$

The corresponding exponential forms of these equations are

$$M = a^u \qquad \text{and} \qquad N = a^v.$$

Then

$$
\begin{aligned}
MN &= a^u \cdot a^v \\
MN &= a^{u+v} \qquad && \text{Product rule of exponents} \\
\log_a MN &= u + v \qquad && \text{Logarithmic form} \\
\log_a MN &= \log_a M + \log_a N \qquad && \text{Replace } u \text{ with } \log_a M \text{ and } v \text{ with } \log_a N.
\end{aligned}
$$

This proves the product rule of logarithms. In Exercise 92, you are asked to prove the quotient rule and the power rule similarly by using the corresponding rules for exponents.

REMINDER

Radicals can also be written as exponents. Recall that

$$\sqrt{x} = x^{1/2}$$

and $\qquad \sqrt[3]{x} = x^{1/3}.$

In general,

$$\sqrt[m]{x} = x^{1/m}.$$

EXAMPLE 1 **Using Rules of Logarithms to Evaluate Expressions**

Given that $\log_5 z = 3$ and $\log_5 y = 2$, evaluate each expression.

a. $\log_5(yz)$ **b.** $\log_5(125y^7)$ **c.** $\log_5\sqrt{\dfrac{z}{y}}$ **d.** $\log_5(z^{1/30}y^5)$

SOLUTION

a. $\log_5(yz) = \log_5 y + \log_5 z \qquad$ Product rule

$ = 2 + 3 = 5 \qquad$ Use the given values and simplify.

b. $\log_5(125y^7) = \log_5 125 + \log_5 y^7$ Product rule

$\quad\quad\quad\quad\quad = \log_5 5^3 + \log_5 y^7$ $125 = 5^3$

$\quad\quad\quad\quad\quad = 3 + 7\log_5 y$ Power rule and $\log_a a = 1$

$\quad\quad\quad\quad\quad = 3 + 7(2) = 17$ Use the given values and simplify.

c. $\log_5\sqrt{\dfrac{z}{y}} = \log_5\left(\dfrac{z}{y}\right)^{1/2}$ Rewrite radical as exponent.

$\quad\quad\quad = \dfrac{1}{2}\log_5\dfrac{z}{y}$ Power rule

$\quad\quad\quad = \dfrac{1}{2}(\log_5 z - \log_5 y)$ Quotient rule

$\quad\quad\quad = \dfrac{1}{2}(3 - 2) = \dfrac{1}{2}$ Use the given values and simplify.

d. $\log_5(z^{1/30}y^5) = \log_5 z^{1/30} + \log_5 y^5$ Product rule

$\quad\quad\quad = \dfrac{1}{30}\log_5 z + 5\log_5 y$ Power rule

$\quad\quad\quad = \dfrac{1}{30}(3) + 5(2)$ Use the given values.

$\quad\quad\quad = 0.1 + 10 = 10.1$ Simplify.   ■ ■ ■

Practice Problem 1   Evaluate each expression for the given values in Example 1.

a. $\log_5(y/z)$ **b.** $\log_5(y^2z^3)$ ■

In many applications in more advanced mathematics courses, the rules of logarithms are used in both directions; that is, the rules are read from left to right and from right to left. For example, by the product rule of logarithms, we have

$$\log_2 3x = \log_2 3 + \log_2 x.$$

The expression $\log_2 3 + \log_2 x$ is the *expanded form* of $\log_2 3x$, while $\log_2 3x$ is the *condensed form,* or the *single logarithmic form,* of $\log_2 3 + \log_2 x$.

EXAMPLE 2   **Writing Expressions in Expanded Form**

Write each expression in expanded form.

a. $\log_2\dfrac{x^2(x-1)^3}{(2x+1)^4}$ **b.** $\log_c\sqrt{x^3y^2z^5}$

SOLUTION

a. $\log_2\dfrac{x^2(x-1)^3}{(2x+1)^4} = \log_2 x^2(x-1)^3 - \log_2(2x+1)^4$ Quotient rule

$\quad\quad\quad = \log_2 x^2 + \log_2(x-1)^3 - \log_2(2x+1)^4$ Product rule

$\quad\quad\quad = 2\log_2 x + 3\log_2(x-1) - 4\log_2(2x+1)$ Power rule

b. $\log_c \sqrt{x^3 y^2 z^5} = \log_c (x^3 y^2 z^5)^{1/2}$ $\qquad$ $\sqrt{a} = a^{1/2}$

$\qquad\qquad\qquad = \dfrac{1}{2} \log_c x^3 y^2 z^5$ $\qquad$ Power rule

$\qquad\qquad\qquad = \dfrac{1}{2} [\log_c x^3 + \log_c y^2 + \log_c z^5]$ $\qquad$ Product rule

$\qquad\qquad\qquad = \dfrac{1}{2} [3 \log_c x + 2 \log_c y + 5 \log_c z]$ $\qquad$ Power rule

$\qquad\qquad\qquad = \dfrac{3}{2} \log_c x + \log_c y + \dfrac{5}{2} \log_c z$ $\qquad$ Distributive property ■ ■ ■

Practice Problem 2 Write each expression in expanded form.

a. $\ln \dfrac{2x - 1}{x + 4}$ $\qquad$ **b.** $\log \sqrt{\dfrac{4xy}{z}}$ $\qquad\qquad\qquad\qquad\qquad$ ■

EXAMPLE 3 **Writing Expressions in Condensed Form**

Write each expression in condensed form.

a. $\log 3x - \log 4y$ $\qquad\qquad\qquad$ **b.** $2 \ln x + \dfrac{1}{2} \ln (x^2 + 1)$

c. $2 \log_2 5 + \log_2 9 - \log_2 75$ $\qquad$ **d.** $\dfrac{1}{3} [\ln x + \ln (x + 1) - \ln (x^2 + 1)]$

SOLUTION

a. $\log 3x - \log 4y = \log \left(\dfrac{3x}{4y} \right)$ $\qquad$ Quotient rule

b. $2 \ln x + \dfrac{1}{2} \ln (x^2 + 1) = \ln x^2 + \ln (x^2 + 1)^{1/2}$ $\qquad$ Power rule

$\qquad\qquad\qquad\qquad = \ln (x^2 \sqrt{x^2 + 1})$ $\qquad$ Product rule; $\sqrt{a} = a^{1/2}$

c. $2 \log_2 5 + \log_2 9 - \log_2 75 = \log_2 5^2 + \log_2 9 - \log_2 75$ $\qquad$ Power rule

$\qquad\qquad\qquad\qquad = \log_2 (25 \cdot 9) - \log_2 75$ $\qquad$ $5^2 = 25$; Product rule

$\qquad\qquad\qquad\qquad = \log_2 \dfrac{25 \cdot 9}{75}$ $\qquad$ Quotient rule

$\qquad\qquad\qquad\qquad = \log_2 3$ $\qquad$ $\dfrac{25 \cdot 9}{75} = \dfrac{9}{3} = 3$

d. $\dfrac{1}{3} [\ln x + \ln (x + 1) - \ln (x^2 + 1)] = \dfrac{1}{3} [\ln x(x + 1) - \ln (x^2 + 1)]$ $\qquad$ Product rule

$\qquad\qquad\qquad\qquad\qquad = \dfrac{1}{3} \ln \left[\dfrac{x(x + 1)}{x^2 + 1} \right]$ $\qquad$ Quotient rule

$\qquad\qquad\qquad\qquad\qquad = \ln \sqrt[3]{\dfrac{x(x + 1)}{x^2 + 1}}$ $\qquad$ Power rule; $a^{1/3} = \sqrt[3]{a}$

$\qquad\qquad\qquad\qquad\qquad\qquad\qquad\qquad\qquad$ ■ ■ ■

Practice Problem 3 Write in condensed form: $\dfrac{1}{2} [\log (x + 1) + \log (x - 1)]$. ■

Be careful when using the rules of logarithms to simplify expressions. For example, there is no property that allows you to rewrite $\log_a (x + y)$.

In general,

$$\log_a (x + y) \neq \log_a x + \log_a y. \quad (5)$$

To illustrate this inequality, let $x = 100$, $y = 10$, and $a = 10$. Then the value of the left side of (5) is

$$\log(100 + 10) = \log(110) \approx 2.0414.$$

The value of the right side of (5) is

$$\begin{aligned} \log 100 + \log 10 &= \log 10^2 + \log 10 \\ &= 2\log 10 + \log 10 \qquad \text{Power rule} \\ &= 2(1) + 1 = 3. \end{aligned}$$

Therefore, $\log(100 + 10) \neq \log 100 + \log 10$.

◆ **WARNING**

To avoid common errors, be aware that in general,

$$\log_a(M + N) \neq \log_a M + \log_a N.$$
$$\log_a(M - N) \neq \log_a M - \log_a N.$$
$$(\log_a M)(\log_a N) \neq \log_a MN.$$
$$\log_a\!\left(\frac{M}{N}\right) \neq \frac{\log_a M}{\log_a N}.$$
$$\frac{\log_a M}{\log_a N} \neq \log_a M - \log_a N.$$
$$(\log_a M)^r \neq r\log_a M.$$

---

2 Change the base of a logarithm.

Change of Base

Calculators usually come with two types of log keys: a key for natural logarithms (base e, $\boxed{\text{LN}}$) and a key for common logarithms (base 10, $\boxed{\text{LOG}}$). These two types of logarithms are frequently used in applications. Sometimes, however, we need to evaluate logarithms for bases other than e and 10. The *change-of-base formula* helps us evaluate these logarithms.

Suppose we are given $\log_b x$ and we want to find an equivalent expression in terms of logarithms with base a. We let

$$u = \log_b x. \quad (6)$$

In exponential form, we have

$$x = b^u. \quad (7)$$

Then

$$\begin{aligned} \log_a x &= \log_a b^u && \text{Take } \log_a \text{ of each side of (7).} \\ \log_a x &= u\log_a b && \text{Power rule} \\ u &= \frac{\log_a x}{\log_a b} && \text{Solve for } u. \\ \log_b x &= \frac{\log_a x}{\log_a b} \quad (8) && \text{Substitute for } u \text{ from (6).} \end{aligned}$$

TECHNOLOGY CONNECTION

To graph a logarithmic function whose base is not e or 10, use the change-of-base formula. For example, graph $y = \log_5 x$ by entering $y_1 = \text{LN}(X)/\text{LN}(5)$.

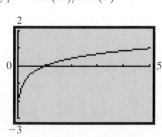

Equation (8) is the change-of-base formula. By choosing $a = 10$ and $a = e$, we can state the change-of-base formula as follows:

CHANGE-OF-BASE FORMULA

Let a, b, and x be positive real numbers with $a \neq 1$ and $b \neq 1$. Then $\log_b x$ can be converted to a different base as follows:

$$\log_b x = \frac{\log_a x}{\log_a b} = \frac{\log x}{\log b} = \frac{\ln x}{\ln b}$$

$$\text{(base } a\text{)} \quad \text{(base } 10\text{)} \quad \text{(base } e\text{)}$$

EXAMPLE 4   Using a Change of Base to Compute Logarithms

Compute $\log_5 13$ by changing to **a.** common logarithms and **b.** natural logarithms.

SOLUTION

a. $\log_5 13 = \dfrac{\log 13}{\log 5}$ Change to base 10.

≈ 1.59369 Use a calculator.

b. $\log_5 13 = \dfrac{\ln 13}{\ln 5}$ Change to base e.

≈ 1.59369 Use a calculator.   ■ ■ ■

Practice Problem 4   Find $\log_3 15$.   ■

EXAMPLE 5   Matching Data to an Exponential Curve

Find the exponential function of the form $f(x) = ae^{bx}$ that passes through the points $(0, 2)$ and $(3, 8)$.

SOLUTION

We need to find the values of a and b. Since the graph passes through the point $(0, 2)$, we have

$$2 = f(0) = ae^{b(0)} = ae^0 = a \cdot 1 = a;$$

so $a = 2$. Next, for the graph to pass through the point $(3, 8)$, we must have

$$8 = f(3) = ae^{b(3)} = 2e^{3b} \qquad \text{Replace } a \text{ with 2.}$$

Now solve $8 = 2e^{3b}$ for b.

$$4 = e^{3b} \qquad \text{Divide both sides by 2.}$$
$$\ln 4 = 3b \qquad \text{Logarithmic form}$$
$$b = \frac{1}{3}\ln 4 \qquad \text{Divide both sides by 3.}$$

Therefore, the desired exponential function is

$$f(x) = 2e^{(\frac{1}{3}\ln 4)x}.$$ ■ ■ ■

Practice Problem 5   Find an exponential function $f(x) = ae^{bx}$ that passes through the points $(0, 5)$ and $(1, 20)$.   ■

3 Apply logarithms in growth and decay.

Growth and Decay

Exponential growth (or decay) occurs when a quantity grows (or decreases) at a rate proportional to its size. The standard growth formula is

$$A(t) = A_0 e^{kt} \qquad (9)$$

where $A(t)$ = the amount of substance (or population) at time t,

$A_0 = A(0)$ is the initial amount,

k = relative rate of growth (if $k > 0$) or decay (if $k < 0$), and

t = time.

Radiocarbon Dating

Ordinary carbon, called carbon-12 (^{12}C), is stable and does not decay. However, carbon-14 (^{14}C) is a form of carbon that decays radioactively with a **half-life** (*time required for half of any given mass to decay*) of 5700 years. Sunlight constantly produces carbon-14 in Earth's atmosphere. When a living organism breathes or eats, it absorbs carbon-12 and carbon-14. After the organism dies, no more carbon-14 is absorbed; so the age of its remains can be calculated by determining how much carbon-14 has decayed. The method of radiocarbon dating was developed by the American scientist W. F. Libby.

EXAMPLE 6   **King Tut's Treasure**

In 1960, a group of specialists from the British Museum in London investigated whether a piece of art containing organic material found in Tutankhamun's tomb had been made during his reign or (as some historians claimed) whether it belonged to an earlier period. We know that King Tut died in 1346 B.C. and ruled Egypt for ten years. What percent of the amount of carbon-14 originally contained in the object should be present in 1960 if the object was made during Tutankhamun's reign?

SOLUTION

Because the half-life of carbon-14 is approximately 5700 years, we can rewrite the equation $A(t) = A_0 e^{kt}$ as

$$A(t) = A_0 e^{5700k}.$$

Now we can find the value of k.

$$\frac{1}{2} A_0 = A_0 e^{5700k} \qquad \text{Replace } A(t) \text{ with } \frac{1}{2} A_0.$$

$$\frac{1}{2} = e^{5700k} \qquad \text{Divide both sides by } A_0.$$

$$\ln\left(\frac{1}{2}\right) = 5700k \qquad \text{Logarithmic form}$$

$$k = \frac{\ln\left(\frac{1}{2}\right)}{5700} \qquad \text{Solve for } k.$$

$$\approx -0.0001216 \qquad \text{Use a calculator.}$$

Substituting this value of k into equation (9), we obtain

$$A(t) = A_0 e^{-0.0001216t} \qquad (10)$$

Let x represent the percent of the original amount of carbon-14 remaining in the object after t years. Then xA_0 is the amount of carbon-14 in the object t years after the object was made.

$$xA_0 = A_0 e^{-0.0001216t} \qquad \text{Replace } A(t) \text{ with } xA_0 \text{ in equation (10).}$$
$$x = e^{-0.0001216t} \qquad (11) \qquad \text{Divide both sides by } A_0.$$

The time t that elapsed between King Tut's death and 1960 is

$$t = 1960 + 1346 \qquad \text{1346 B.C. to 1960 A.D.}$$
$$= 3306.$$

Then x_1, the percent of the original amount of carbon-14 remaining in the object (after 3306 years), is

$$x_1 = e^{-0.0001216(3306)} \qquad \text{Replace } t \text{ with 3306 in (11).}$$
$$\approx 0.66897 = 66.897\%.$$

Since King Tut began ruling Egypt ten years before he died, the time t_1 that elapsed from the beginning of his reign to 1960 is 3316 years. So x_2, the percent of the original amount of carbon-14 remaining in the object after 3316 years is

$$x_2 = e^{-0.0001216(3316)} \qquad \text{Replace } t \text{ with 3316 in (11).}$$
$$\approx 0.66816 = 66.816\%.$$

Therefore, a piece of art made during King Tut's reign would have between 66.816% and 66.897% of carbon-14 remaining in 1960. ■ ■ ■

Practice Problem 6   Repeat Example 6 assuming that the art object was made 194 years before King Tut's death. ■

SECTION 3.4 ■ Exercises

A EXERCISES Basic Skills and Concepts

1. $\log_a MN = \underline{\quad \log_a M \quad} + \underline{\quad \log_a N \quad}$.

2. $\underline{\quad \log_a \frac{M}{N} \quad} = \log_a M - \log_a N$.

3. $\log_a M^r = \underline{\quad r \log_a M \quad}$.

4. The change-of-base formula for base e is $\log_a M = \underline{\dfrac{\ln M}{\ln a}}$.

5. *True or False* $\log_a (u + v) = \log_a u + \log_a v$. False

6. *True or False* $\log \dfrac{x}{10} = \log x - 1$. True

In Exercises 7–18, given that $\log x = 2$, $\log y = 3$, $\log 2 \approx 0.3$, and $\log 3 \approx 0.48$, evaluate each expression without using a calculator.

7. $\log 6$ 0.78

8. $\log 4$ 0.6

9. $\log 5 \left[\text{Hint: } 5 = \dfrac{10}{2} \right]$ 0.7

10. $\log (3x)$ 2.48

11. $\log \left(\dfrac{2}{x} \right)$ −1.7

12. $\log x^2$ 4

13. $\log (2x^2 y)$ 7.3

14. $\log xy^3$ 11

15. $\log \sqrt[3]{x^2 y^4}$ $\dfrac{16}{3}$

16. $\log (\log x^2)$ 0.6

17. $\log \sqrt[3]{48}$ 0.56

18. $\log_2 3$ 1.6

In Exercises 19–30, write each expression in expanded form.

19. $\ln [x(x - 1)]$ †

20. $\log_2 \left(\dfrac{x - 1}{x + 3} \right)$ †

21. $\log \dfrac{\sqrt{x^2 + 1}}{x + 3}$ †

22. $\ln \dfrac{x(x + 1)}{(x - 1)^2}$ †

23. $\log_3 \dfrac{(x - 1)^2}{(x + 1)^5}$ †

24. $\log_4 \left(\dfrac{x^2 - 9}{x^2 - 6x + 8} \right)^{\frac{2}{3}}$ †

25. $\log_b xyz$ †

26. $\log_b \sqrt{xyz}$ †

27. $\log_b x^2 y^3 z$ †

28. $\log_b \dfrac{y^2}{xz^3}$ †

29. $\log \sqrt{x \sqrt{yz}}$ †

30. $\log_b \left(x \log_b b^{\sqrt{yz}} \right)$ †

In Exercises 31–42, write each expression in condensed form.

31. $\log_2 x + \log_2 7$ $\log_2 (7x)$

32. $\log_2 x - \log_2 3$ $\log_2 \left(\dfrac{x}{3} \right)$

33. $\log \dfrac{5}{2} - \log \dfrac{7}{9}$ $\log \left(\dfrac{45}{14} \right)$

34. $2 \log (x - 1)$ $\log (x - 1)^2$

†Due to space constrictions, answers to these exercises may be found in the Answers beginning on page A–1 in the back of the book.

35. $\frac{1}{2} \log x - \log y + \log z \quad \log\left(\frac{z\sqrt{x}}{y}\right)$

36. $\frac{1}{2}(\log x + \log y) \quad \log(xy)^{1/2}$

37. $\frac{1}{5}(\log_2 z + 2\log_2 y) \quad \log_2(zy^2)^{1/5}$

38. $\frac{1}{3}(\log x - 2\log y + 3\log z) \quad \log\left(\frac{xz^3}{y^2}\right)^{1/3}$

39. $\ln x + 2\ln y + 3\ln z \quad \ln(xy^2z^3)$

40. $2\ln x - 3\ln y + 4\ln z \quad \ln\left(\frac{x^2z^4}{y^3}\right)$

41. $2\ln x - \frac{1}{2}\ln(x^2+1) \quad \ln\left(\frac{x^2}{\sqrt{x^2+1}}\right)$

42. $2\ln x + \frac{1}{2}\ln(x^2-1) - \frac{1}{2}\ln(x^2+1) \quad \ln\left(\frac{x^2\sqrt{x^2-1}}{\sqrt{x^2+1}}\right)$

In Exercises 43–50, use the change-of-base formula and a calculator to evaluate each logarithm.

43. $\log_2 5 \quad 2.322$

44. $\log_4 11 \quad 1.730$

45. $\log_{1/2} 3 \quad -1.585$

46. $\log_{\sqrt{3}} 12.5 \quad 4.598$

47. $\log_{\sqrt{5}} \sqrt{17} \quad 1.760$

48. $\log_{15} 123 \quad 1.777$

49. $\log_2 7 + \log_4 3 \quad 3.6$

50. $\log_2 9 - \log_{\sqrt{2}} 5 \quad -1.474$

In Exercises 51–62, find the value of each expression without using a calculator.

51. $\log_3\sqrt{3} \quad \frac{1}{2}$

52. $\log_{1/4} 4 \quad -1$

53. $\log_3(\log_2 8) \quad 1$

54. $2^{\log_2 2} \quad 2$

55. $5^{2\log_5 3 + \log_5 2} \quad 18$

56. $e^{3\ln 2 - 2\ln 3} \quad \frac{8}{9}$

57. $10^{3\log 2 - \log 4} \quad 2$

58. $10^{\log_{1/10} 3} \quad \frac{1}{3}$

59. $10^{\log_{100} 3} \quad \sqrt{3}$

60. $\log 4 + 2\log 5 \quad 2$

61. $\log_2 160 - \log_2 5 \quad 5$

62. $e^{\log_e 25} \quad 5$

In Exercises 63–66, find the exponential function of the form $f(x) = ae^{bx}$ that passes through the given points.

63. $(0,2)$ and $(2,8)$ $f(x) = 2e^{(\ln 2)x}$

64. $(0,8)$ and $(2,5)$ $f(x) = 8e^{\left(\ln\frac{5}{8}\right)\frac{x}{2}}$

65. $(0,100)$ and $(5,1000)$ $f(x) = 100e^{(\ln 10)\frac{x}{5}}$

66. $(0,10)$ and $(10,2)$ $f(x) = 10e^{-(\ln 5)\frac{x}{10}}$

B EXERCISES Applying the Concepts

In Exercises 67–70, assume the exponential growth model $A(t) = A_0e^{kt}$ and a world population of 6.5 billion in 2006.

67. Rate of population growth. If the world adds about 90 million people in 2007, what is the rate of growth? Rate of growth $\approx 1.375\%$.

68. Population. Use the rate of growth from Exercise 67 to estimate the year in which the world population will be
  a. 12 billion. 2051
  b. 20 billion. 2088

69. Population growth. If the population must stay below 20 billion during the next 100 years, what is the maximum acceptable annual rate of growth? 1.12%

70. Population decline. If world population must decline below 5 billion during the next 25 years, what must be the annual rate of decline? 1.05%

71. Continuous compounding. If $1000 is deposited in a bank at 10% interest compounded continuously, how many years will it take for the money to grow to $3500? 12.53 yr

72. Continuous compounding. At what rate of interest compounded continuously will an investment double in six years? 11.55%

73. Half-life. Iodine-131, a radioactive substance that is effective in locating brain tumors, has a half-life of only eight days. A hospital purchased 20 grams of the substance but had to wait five days before it could be used. How much of the substance was left after five days? 12.969 g

74. Half-life. Plutonium-241 has a half-life of 13 years. A laboratory purchased 10 grams of the substance but did not use it for two years. How much of the substance was left after two years? 8.9885 g

75. Half-life. Tritium is used in nuclear weapons to increase their power. It decays at the rate of 5.5% per year. Calculate the half-life of tritium. 12.60 yr

76. Half-life. Sixty grams of magnesium-28 decayed into 7.5 grams after 63 hours. What is the half-life of magnesium-28? 21 hr

Effective medicine dosage. Use the following model in Exercises 77–80: The concentration $C(t)$ of a drug administered to a patient intravenously jumps to its highest level almost immediately. The concentration subsequently decays exponentially according to the law

$$C(t) = C_0 e^{-kt},$$

where $C(0) = C_0$. The physician administering the drug would like to have

$$m < C(t) < M,$$

where m is the concentration below which the drug is ineffective and M is the concentration above which the drug is dangerous.

77. Drug concentration. Suppose that immediately after a certain drug is administered, the concentration is 5 milligrams per milliliter. Ten hours later the concentration drops to 1.5 milligrams per milliliter. Determine the value of k for this drug. $k = 0.1204$

78. Drug concentration. For the drug of Exercise 77, suppose the maximum safe level is $M = 12$ milligrams per milliliter and the minimum effective level is $m = 3$ milligrams per milliliter. If the initial concentration is M, find the maximum possible time between doses of this drug. 11.5 hr

79. Half-life. The half-life of Valium is 36 hours. Suppose a patient receives 16 milligrams of the drug at 8 A.M. How much Valium is in the patient's blood at 4 P.M. the same day? 13.7 g

80. **Half-life.** The half-life of aspirin in the blood is 12 hours. Estimate the time required for the aspirin to decay to 90% of the original amount ingested. 1.82 hr

C EXERCISES Beyond the Basics

81. Show that
$$\log_b(\sqrt{x^2+1}-x) + \log_b(\sqrt{x^2+1}+x) = 0.$$

82. Show that
$$\log_b(\sqrt{x+1}+\sqrt{x}) = -\log_b(\sqrt{x+1}-\sqrt{x}).$$

83. Show that $(\log_b a)(\log_a b) = 1$.

84. Show that $\log \dfrac{a}{b} + \log \dfrac{b}{a} = 0$.

85. Write $\log_2 x = \dfrac{\ln x}{\ln 2}$ by the change-of-base formula. Then use a graphing calculator to sketch the graph of $y = \log_2 x$. †

86. Sketch the graph of $y = \log_5(x+3)$. †

87. By the power rule for logarithms, $\log x^2 = 2 \log x$. However, the graphs of $f(x) = \log x^2$ and $g(x) = 2 \log x$ are not identical. Explain why.

88. Simplify each expression.
   a. $\log\left(\dfrac{a}{b}\right) + \log\left(\dfrac{b}{a}\right) + \log\left(\dfrac{c}{a}\right) + \log\left(\dfrac{a}{c}\right)$ 0
   b. $\log\left(\dfrac{a^2}{bc}\right) + \log\left(\dfrac{b^2}{ca}\right) + \log\left(\dfrac{c^2}{ab}\right)$ 0
   c. $\log_2 3 \cdot \log_3 4$ 2
   d. $\log_a b \cdot \log_b c \cdot \log_c a$ 1

89. Find the number of digits in N if $\log_2(\log_2 N) = 4$. 5

90. Let $f(x) = \log_a x$. Show that
   a. $f\left(\dfrac{1}{x}\right) = -f(x) = \log_{1/a} x$.
   b. $\dfrac{f(x+h) - f(x)}{h} = \log_a\left(1 + \dfrac{h}{x}\right)^{\frac{1}{h}}, h \neq 0$

91. State whether each of the following is true or false for all permissible values of the variable.
   a. $\log(x+2) = \log x + \log 2$ False
   b. $\log 2x = \log x + \log 2$ True
   c. $\log_2 3x = 3 \log_2 x$ False

   d. $(\log_3 2x)^4 = 4 \log_3 2x$ False
   e. $\log_5 x^2 = 2 \log_5 x$ True
   f. $\log \dfrac{x}{3} = \log x - \log 3$ True
   g. $\log \dfrac{x}{4} = \dfrac{\log x}{\log 4}$ False
   h. $\ln\left(\dfrac{1}{x}\right) = -\ln x$ True
   i. $\log_2 x^2 = (\log_2 x)(\log_2 x)$ False
   j. $\log |10x| = 1 + \log |x|$ True

92. a. Prove the quotient rule for logarithms:
      $$\log_a \dfrac{M}{N} = \log_a M - \log_a N$$
   b. Prove the power rule for logarithms:
      $$\log_a M^r = r \log_a M$$

93. If $\log\left(\dfrac{a+b}{3}\right) = \dfrac{1}{2}(\log a + \log b)$, show that $a^2 + b^2 - 7ab = 0$.

94. Evaluate $\dfrac{1}{1 + \log_c ab} + \dfrac{1}{1 + \log_a bc} + \dfrac{1}{1 + \log_b ca}$.

   [*Hint:* Convert to common logarithms and then simplify.] 1

Critical Thinking

95. **What went wrong?** Find the error in the following argument:

$$3 < 4$$
$$3 \log \dfrac{1}{2} < 4 \log \dfrac{1}{2}$$
$$\log\left(\dfrac{1}{2}\right)^3 < \log\left(\dfrac{1}{2}\right)^4$$
$$\log \dfrac{1}{8} < \log \dfrac{1}{16}$$
$$\dfrac{1}{8} < \dfrac{1}{16}$$
$$2 < 1$$

Since $\log \dfrac{1}{2}$ is a negative number, $3 < 4$ implies that $3 \log \dfrac{1}{2} > 4 \log \dfrac{1}{2}$.

Answer:
87. The domain of $2 \log x$ is $(0, \infty)$, while the domain of $\log(x^2)$ is $(-\infty, 0) \cup (0, \infty)$.

Exponential and Logarithmic Equations

Before Starting this Section, Review

1. One-to-one property of exponential functions (Section 3.1, page 193).
2. One-to-one property of logarithmic functions (Section 3.3, page 216).
3. Rules of logarithms (Section 3.4, page 226).

Objectives

1 Solve exponential equations.

2 Solve applied problems involving exponential equations.

3 Solve logarithmic equations.

4 Use the logistic growth model.

LOGISTIC GROWTH MODEL

In Section 3.4, we discussed the population growth equation

$$A(t) = A_0 e^{kt}. \qquad (1)$$

Equation (1) is a realistic model for a biological population that has plenty of food, enough space to grow, and no threat from predators.

However, most populations are constrained by limitations on resources such as the plants that populations eat. In the 1830s, the Belgian scientist P. F. Verhulst added another constraint to the population growth model: *There is a limited sustainable maximum population M of a species that the habitat can support.* In population biology, *M* is called the *carrying capacity*.

Verhulst replaced equation (1) with

$$P(t) = \frac{M}{1 + ae^{-kt}} \qquad (2)$$

where *k* is the rate of growth and *a* is a positive constant. In 1840, using the data from the first five U.S. censuses, he made a prediction of the U.S. population for the year 1940. As things turned out, it was off by less than 1%.

The mathematical model represented by equation (2) is called the *logistic growth model* or the *Verhulst model*. Verhulst introduced the term *logistique* to refer to the "loglike" qualities of the S-shaped graph of equation (2). (See Figure 3.14.)

To sketch the graph of equation (2), we notice that the *y*-intercept is

$$P(0) = \frac{M}{1 + ae^0} = \frac{M}{1 + a}.$$

As $t \to \infty$, the quantity $ae^{-kt} \to 0$ (since *k* is positive). Consequently, the denominator $1 + ae^{-kt} \to 1$; therefore, $P(t) \to M$ and the line $y = M$ is a horizontal asymptote. Similarly, as $t \to -\infty$, $e^{-kt} \to \infty$. In this case, the denominator $1 + ae^{-kt} \to \infty$ and $P(t) \to 0$. The *x*-axis, or $y = 0$, is another horizontal asymptote. See the graph of $y = \dfrac{6}{1 + 2e^{-1.5t}}$ in Figure 3.14.

In Example 9, we will use the logistic function defined by equation (2). ■

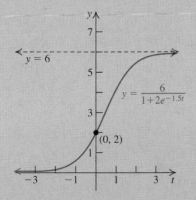

FIGURE 3.14 A logistic curve

1 Solve exponential equations.

Solving Exponential Equations

An equation containing terms of the form $a^x (a > 0, a \neq 1)$ is called an **exponential equation**. Here are some examples of simple exponential equations:

$$2^x = 15$$
$$9^x = 3^{x+1}$$
$$7^x = 13^{2x}$$
$$5 \cdot 2^{x-3} = 17$$

If both sides of an equation can be expressed as a power of the same base, then we can use the one-to-one property of exponential functions to solve the equation. We used this strategy in Section 3.1.

RECALL

The one-to-one property of exponential and logarithmic functions states that

- if $a^x = a^y$, then $x = y$.
- if $\log_a x = \log_a y$, then $x = y$.

| **EXAMPLE 1** | **Solving an Exponential Equation** |

Solve each equation.

a. $25^x = 125$ **b.** $9^x = 3^{x+1}$

SOLUTION

a.

$$(5^2)^x = 5^3 \qquad \text{Rewrite 25 as } 5^2 \text{ and 125 as } 5^3.$$
$$5^{2x} = 5^3 \qquad \text{Power rule for exponents}$$
$$2x = 3 \qquad \text{One-to-one property}$$
$$x = \frac{3}{2} \qquad \text{Solve for } x.$$

b.

$$(3^2)^x = 3^{x+1} \qquad \text{Rewrite 9 as } 3^2.$$
$$3^{2x} = 3^{x+1} \qquad \text{Power rule of exponents}$$
$$2x = x + 1 \qquad \text{One-to-one property}$$
$$x = 1 \qquad \text{Solve for } x. \qquad \blacksquare\blacksquare\blacksquare$$

Practice Problem 1   Solve each equation.

a. $3^x = 243$ **b.** $8^x = 4$ $\qquad\qquad\qquad\qquad\qquad\qquad\qquad$ ■

We can use logarithms to solve those equations for which both sides cannot be readily expressed with the same base. For example, the fact that $\log 2^x = x \log 2$ allows us to solve the equation $2^x = 9$ by first taking the log of both sides:

$$2^x = 9$$
$$\log 2^x = \log 9$$
$$x \log 2 = \log 9$$
$$x = \frac{\log 9}{\log 2}$$

The general procedure for solving exponential equations using logarithms is given next.

FINDING THE SOLUTION: A PROCEDURE

EXAMPLE 2   Solving Exponential Equations Using the Logarithms

OBJECTIVE	**EXAMPLE**
Solve exponential equations when both sides are not expressed with the same base.	*Solve for x:* $5 \cdot 2^{x-3} = 17$.
Step 1 Isolate the exponential expression on one side of the equation.	**1.** $2^{x-3} = \dfrac{17}{5} = 3.4$
Step 2 Take the common or natural logarithm of both sides.	**2.** $\ln 2^{x-3} = \ln (3.4)$
Step 3 Use the power rule, $\log_a M^r = r \log_a M$.	**3.** $(x - 3) \ln 2 = \ln (3.4)$
Step 4 Solve for the variable.	**4.** $x - 3 = \dfrac{\ln (3.4)}{\ln 2}$
	$\ x = \dfrac{\ln (3.4)}{\ln 2} + 3 \approx 4.766$

■ ■ ■

Practice Problem 2   Solve for x: $7 \cdot 3^{x+1} = 11$.   ■

EXAMPLE 3   Solving an Exponential Equation with Different Bases

Solve the equation $5^{2x-3} = 3^{x+1}$ and approximate the answer to three decimal places.

SOLUTION

When different bases are involved, we begin with Step 2 of the procedure above.

$\ln 5^{2x-3} = \ln 3^{x+1}$	Take the natural log of both sides.
$(2x - 3) \ln 5 = (x + 1) \ln 3$	Power rule of logarithms
$2x \ln 5 - 3 \ln 5 = x \ln 3 + \ln 3$	Distributive property
$2x \ln 5 - x \ln 3 = \ln 3 + 3 \ln 5$	Collect terms containing x on one side.
$x(2 \ln 5 - \ln 3) = \ln 3 + 3 \ln 5$	Factor out x on the left side.
$x = \dfrac{\ln 3 + 3 \ln 5}{2 \ln 5 - \ln 3}$	Solve for x to obtain an *exact* solution.
≈ 2.795	Use a calculator.

■ ■ ■

Practice Problem 3   Solve the equation $3^{x+1} = 2^{2x}$.   ■

EXAMPLE 4   An Exponential Equation of Quadratic Form

Solve the equation $3^x - 8 \cdot 3^{-x} = 2$.

SOLUTION

$3^x(3^x - 8 \cdot 3^{-x}) = 2(3^x)$	Multiply both sides by 3^x.
$3^{2x} - 8 \cdot 3^0 = 2 \cdot 3^x$	Distributive property; $3^x \cdot 3^{-x} = 3^0$
$3^{2x} - 8 = 2 \cdot 3^x$	$3^0 = 1$
$3^{2x} - 2 \cdot 3^x - 8 = 0$	Subtract $2 \cdot 3^x$ from both sides.

The last equation is quadratic in form. To see this, we let $y = 3^x$; then $y^2 = (3^x)^2 = 3^{2x}$. So,

$$3^{2x} - 2 \cdot 3^x - 8 = 0$$

$$y^2 - 2y - 8 = 0 \qquad \text{Substitute: } y = 3^x, y^2 = 3^{2x}.$$

$$(y + 2)(y - 4) = 0 \qquad \text{Factor.}$$

$$y + 2 = 0 \quad \text{or} \quad y - 4 = 0 \qquad \text{Zero-product property}$$

$$y = -2 \quad \text{or} \qquad y = 4 \qquad \text{Solve for } y.$$

$$3^x = -2 \quad \text{or} \qquad 3^x = 4 \qquad \text{Replace } y \text{ with } 3^x.$$

But $3^x = -2$ is not possible because $3^x > 0$ for all real numbers x. Solve $3^x = 4$.

$$\ln 3^x = \ln 4 \qquad \text{Take the natural log of both sides.}$$

$$x \ln 3 = \ln 4 \qquad \text{Power rule of logarithms}$$

$$x = \frac{\ln 4}{\ln 3} \qquad \text{An exact solution}$$

$$\approx 1.262 \qquad \text{Use a calculator.} \qquad\qquad ■ ■ ■$$

Practice Problem 4   Solve the equation $e^{2x} - 4e^x - 5 = 0$. ■

---

2 Solve applied problems involving exponential equations.

Applications of Exponential Equations

EXAMPLE 5   **Solving a Population Growth Problem**

The following table shows the approximate population and annual growth rate of the United States and Pakistan in 2005.

Country	Population	Annual Population Growth Rate
United States	295 million	1.0%
Pakistan	162 million	3.1%

Source: www.cia.gov

Use the alternate population model $P(t) = P_0(1 + r)^t$, where P_0 is the initial population and t is the time in years since 2005. Assume that the growth rate for each country stays the same.

a. Estimate the population of each country in 2015.

b. In what year will the population of the United States be 350 million?

c. In what year will the population of Pakistan be the same as the population of the United States?

SOLUTION
The given model is

$$P(t) = P_0(1 + r)^t.$$

a. In 2005, the initial year, the U.S. population was $P_0 = 295$. In ten years, 2015, the population of the United States will be

$$295(1 + 0.01)^{10} \approx 325.86 \text{ million.} \qquad P_0 = 295, r = 0.01, t = 10$$

In the same way, the population of Pakistan in 2015 will be

$$162(1 + 0.031)^{10} \approx 219.84 \text{ million.} \qquad P_0 = 162, r = 0.031, t = 10$$

b. To find the year when the population of the United States will be 350 million, we solve for t in the equation

$$350 = 295(1 + 0.01)^t. \qquad P(t) = 350, P_0 = 295, r = 0.01$$

$$\frac{350}{295} = (1.01)^t \qquad \text{Divide both sides by 295.}$$

$$\ln\left(\frac{350}{295}\right) = \ln(1.01)^t \qquad \text{Take natural logs of both sides.}$$

$$\ln\left(\frac{350}{295}\right) = t\ln(1.01) \qquad \text{Power rule for logarithms}$$

$$t = \frac{\ln\left(\frac{350}{295}\right)}{\ln(1.01)} \approx 17.18 \qquad \text{Solve for } t \text{ and use a calculator.}$$

The population of the United States will be 350 million approximately 17.18 years after 2005—that is, sometime in 2022.

c. To find when the population of the two countries will be same, we solve for t in the equation

$$295(1 + 0.01)^t = 162(1 + 0.031)^t \qquad \text{Equate populations for both countries.}$$

$$295(1.01)^t = 162(1.031)^t$$

$$\frac{295}{162} = \left(\frac{1.031}{1.01}\right)^t \qquad \text{Divide both sides by } 162(1.01)^t \text{ and rewrite.}$$

$$\ln\left(\frac{295}{162}\right) = \ln\left(\frac{1.031}{1.01}\right)^t \qquad \text{Take natural logs of both sides.}$$

$$\ln\left(\frac{295}{162}\right) = t\ln\left(\frac{1.031}{1.01}\right) \qquad \text{Power rule for logarithms}$$

$$t = \frac{\ln\left(\frac{295}{162}\right)}{\ln\left(\frac{1.031}{1.01}\right)} \qquad \text{Solve for } t.$$

$$\approx 29.13 \text{ years} \qquad \text{Use a calculator.}$$

Therefore, the two populations will be equal in about 29.13 years, that is, during 2034. ■ ■ ■

Practice Problem 5 Repeat Example 5 assuming that the rates of growth for the United States and Pakistan are 1.1% and 3.3%, respectively. ■

3 Solve logarithmic equations.

Solving Logarithmic Equations

Equations that contain terms of the form $\log_a x$ are called **logarithmic equations**. Here are some examples of logarithmic equations:

$$\log_2 x = 4$$
$$\log_3(2x - 1) = \log_3(x + 2)$$
$$\log_2(x - 3) + \log_2(x - 4) = 1$$

To solve an equation such as $\log_2 x = 4$, we rewrite it in the equivalent exponential form:

$$\log_2 x = 4 \text{ means } x = 2^4 = 16$$

Thus, $x = 16$ is a solution of the equation $\log_2 x = 4$. Because the domain of logarithmic functions consists of positive numbers, we must check apparent solutions of logarithmic equations in the original equation.

EXAMPLE 6   Solving a Logarithmic Equation

Solve $4 + 3 \log_2 x = 1$.

SOLUTION

$$
\begin{array}{ll}
4 + 3 \log_2 x = 1 & \text{Original equation} \\
3 \log_2 x = 1 - 4 & \text{Subtract 4 from both sides.} \\
3 \log_2 x = -3 & \text{Simplify.} \\
\log_2 x = -1 & \text{Divide both sides by 3.} \\
x = 2^{-1} & \text{Exponential form} \\
x = \dfrac{1}{2} & a^{-n} = \dfrac{1}{a^n}
\end{array}
$$

Check   Substitute $x = \dfrac{1}{2} = 2^{-1}$ into $4 + 3 \log_2 x = 1$.

$$
\begin{array}{ll}
4 + 3 \log_2 2^{-1} \overset{?}{=} 1 & \\
4 + 3(-1) \log_2 2 \overset{?}{=} 1 & \text{Power rule for logarithms} \\
4 - 3 \overset{?}{=} 1 & \log_2 2 = 1 \\
1 \overset{?}{=} 1 & \text{Yes}
\end{array}
$$

This check shows that the solution set is $\left\{ \dfrac{1}{2} \right\}$.   ■ ■ ■

Practice Problem 6   Solve $1 + 2 \ln x = 4$.   ■

If each side of an equation can be expressed as a single logarithm with the same base, then we can use the one-to-one property of logarithms to solve the equation.

REMINDER

The Product rule says that
$\log_a M + \log_a N = \log_a MN$.

EXAMPLE 7   Using the One-to-One Property of Logarithms

Solve $\log_4 x + \log_4 (x + 1) = \log_4 (x - 1) + \log_4 6$.

SOLUTION

$$
\begin{array}{ll}
\log_4 x + \log_4 (x + 1) = \log_4 (x - 1) + \log_4 6 & \text{Original equation} \\
\log_4 [x(x + 1)] = \log_4 [6(x - 1)] & \text{Product rule for logarithms} \\
x (x + 1) = 6(x - 1) & \text{One-to-one property} \\
x^2 + x = 6x - 6 & \text{Distributive property} \\
x^2 - 5x + 6 = 0 & \text{Add } -6x + 6 \text{ to both sides.} \\
(x - 2)(x - 3) = 0 & \text{Factor.} \\
x - 2 = 0 \text{ or } x - 3 = 0 & \text{Zero-product property} \\
x = 2 \text{ or } x = 3 & \text{Solve for } x.
\end{array}
$$

Now we check these possible solutions in the original equation.

Check $x = 2$	**Check $x = 3$**
$\log_4 2 + \log_4 (2+1) \overset{?}{=} \log_4 (2-1) + \log_4 6$	$\log_4 3 + \log_4 (3+1) \overset{?}{=} \log_4 (3-1) + \log_4 6$
$\log_4 2 + \log_4 3 \overset{?}{=} \log_4 1 + \log_4 6$	$\log_4 3 + \log_4 4 \overset{?}{=} \log_4 2 + \log_4 6$
$\log_4 (2 \cdot 3) \overset{?}{=} \log_4 (1 \cdot 6)$ Yes	$\log_4 (3 \cdot 4) \overset{?}{=} \log_4 (2 \cdot 6)$ Yes

The solution set is $\{2, 3\}$.   ■ ■ ■

Practice Problem 7   Solve $\ln (x + 5) + \ln (x + 1) = \ln (x - 1)$.   ■

EXAMPLE 8   **Using the Product and Quotient Rules**

Solve the following equations.

a. $\log_2 (x - 3) + \log_2 (x - 4) = 1$ **b.** $\log_2 (x + 4) + \log_2 (x + 3) = 1$

SOLUTION

a.

$\log_2 (x - 3) + \log_2 (x - 4) = 1$	Original equation
$\log_2 [(x - 3)(x - 4)] = 1$	Product rule for logarithms
$(x - 3)(x - 4) = 2^1$	Exponential form
$x^2 - 7x + 10 = 0$	Write in standard form.
$(x - 2)(x - 5) = 0$	Factor.
$x - 2 = 0$ or $x - 5 = 0$	Zero-product property
$x = 2$ or $x = 5$	Solve for x.

We check the possible solutions in the original equation.

Check $x = 2$	**Check $x = 5$**
$\log_2 (2 - 3) + \log_2 (2 - 4) \overset{?}{=} 1$	$\log_2 (5 - 3) + \log_2 (5 - 4) \overset{?}{=} 1$
$\log_2 (-1) + \log_2 (-2) \overset{?}{=} 1$ No	$\log_2 2 + \log_2 1 \overset{?}{=} 1$
Since logarithms are not defined for negative numbers, the number 2 is an extraneous solution of the original equation.	$1 + 0 \overset{?}{=} 1$
	$1 \overset{?}{=} 1$ Yes

The solution set is $\{5\}$.

b.

$\log_2 (x + 4) + \log_2 (x + 3) = 1$	Original equation
$\log_2 [(x + 4)(x + 3)] = 1$	Product rule
$(x + 4)(x + 3) = 2^1$	Exponential form
$x^2 + 7x + 10 = 0$	Write in standard form.
$(x + 2)(x + 5) = 0$	Factor.
$x = -2$ or $x = -5$	Solve for x.

We check the possible solutions in the original equation.

Check $x = -2$	**Check $x = -5$**
$\log_2 (-2 + 4) + \log_2 (-2 + 3) \overset{?}{=} 1$	$\log_2 (-5 + 4) + \log_2 (-5 + 3) \overset{?}{=} 1$
$\log_2 2 + \log_2 1 \overset{?}{=} 1$	$\log_2 (-1) + \log_2 (-2) \overset{?}{=} 1$
$1 + 0 \overset{?}{=} 1$ Yes	$\log_2 (-1)$ and $\log_2 (-2)$ are undefined. No

The solution set is $\{-2\}$.   ■ ■ ■

Practice Problem 8   Solve $\log_3 (x - 8) + \log_3 x = 2$.   ■

STUDY TIP

Remember that "$\log_3 x$" means the exponent on 3 that gives x.

4 Use the logistic growth model.

Pierre François Verhulst

(1804–1849)

Pierre Verhulst was born and educated in Brussels, Belgium. He received his PhD from the University of Ghent in 1825. Influenced by Lambert Quetelet, he became interested in social statistics. Verhulst's research on the law of population growth is important. Before Quetelet and Verhulst, scientists believed that an increasing population as a function of time was given by

$$P(t) = P_0(1 + k)^t \quad \text{or}$$
$$P(t) = P_0 e^{kt}.$$

Verhulst showed that there are forces that tend to prevent the population growth according to these laws. He discovered the model given by equation (2).

EXAMPLE 9   **Using the Logistic Growth Model**

Suppose the carrying capacity M of the human population on Earth is 35 billion. In 1987, the world population was about 5 billion. Use the logistic growth model of P. F. Verhulst to calculate the average rate, k, of growth of the population, given that the population was about 6 billion in 2003.

$$P(t) = \frac{M}{1 + ae^{-kt}} \quad (2) \qquad \text{From the section introduction}$$

SOLUTION

If 1987 represents $t = 0$, then $P(0) = 5$ and $M = 35$; so equation (2) becomes

$$5 = \frac{35}{1 + ae^{-k(0)}}$$

$$5 = \frac{35}{1 + a}$$

$$5(1 + a) = 35 \qquad \text{Multiply both sides by } 1 + a.$$

$$a = 6 \qquad \text{Solve for } a.$$

Equation (2), with $M = 35$ and $a = 6$, becomes

$$P(t) = \frac{35}{1 + 6e^{-kt}}. \quad (3)$$

We now solve equation (3) for k, given that $t = 16$ (for 2003) and $P(16) = 6$.

$$6 = \frac{35}{1 + 6e^{-16k}}$$

$$6 + 36e^{-16k} = 35 \qquad \text{Multiply by } 1 + 6e^{-16k} \text{ and distribute.}$$

$$e^{-16k} = \frac{29}{36} \qquad \text{Isolate } e^{-16k}.$$

$$-16k = \ln\left(\frac{29}{36}\right) \qquad \text{Logarithmic form}$$

$$k = -\frac{1}{16} \ln\left(\frac{29}{36}\right) \qquad \text{Solve for } k.$$

$$k \approx 0.0135 = 1.35\% \qquad \text{Use a calculator.}$$

The average growth rate of the world population was approximately 1.35%. ■ ■ ■

Practice Problem 9   Repeat Example 9 assuming that the estimated world population in 2005 was about 6.5 billion. ■

The magnitude M of an earthquake is defined in the Richter scale by:

$$M = \log\left(\frac{I}{I_0}\right) \quad (4)$$

where I is the intensity of the earthquake and I_0 is a zero-level intensity earthquake below which we would not be aware of Earth's movement.

EXAMPLE 10   **Comparing Two Earthquakes**

Compare the intensity of the Mexico City earthquake of 1985 which registered 8.1 on the Richter scale to that of the San Francisco earthquake of 1989 which measured 7.1 on the Richter scale.

SOLUTION

Let I_M and I_S denote the intensities of the Mexico City and San Francisco earthquakes, respectively. By equation (4), we have

$$8.1 = \log\left(\frac{I_M}{I_0}\right) \qquad 7.1 = \log\left(\frac{I_S}{I_0}\right) \qquad \text{Use given values in equation (4)}$$

$$\frac{I_M}{I_0} = 10^{8.1} \qquad \frac{I_S}{I_0} = 10^{7.1} \qquad \text{Exponential form of } y = \log x$$

$$I_M = 10^{8.1}I_0 \qquad I_S = 10^{7.1}I_0 \qquad \text{Solve for } I_M \text{ and } I_S$$

Dividing I_M by I_S and simplifying, we get

$$\frac{I_M}{I_S} = \frac{10^{8.1}I_0}{10^{7.1}I_0} = \frac{10^{8.1}}{10^{7.1}} = 10^{8.1-7.1} = 10^1 = 10$$

$$I_M = 10I_S$$

This equation shows that the intensity of the Mexico City earthquake was ten times that of the San Francisco earthquake.   ■ ■ ■

Practice Problem 10   Compare the intensity of the Assam (India) earthquake of 1950, I_A, that registered 8.4 on the Richter scale, to that of the Tangshan (China) earthquake of 1976, I_T, that registered 7.9 on the Richter scale.   ■

SECTION 3.5 ■ Exercises

A EXERCISES Basic Skills and Concepts

1. An equation that contains terms of the form a^x is called a(n) __exponential__ equation.

2. An equation that contains terms of the form $\log_a x$ is called a(n) __logarithmic__ equation.

3. The equation $y = \dfrac{m}{1 + ae^{-bx}}$ represents a(n) __logistic__ model.

4. *True or False* A logistic curve always has two horizontal asymptotes. True

5. *True or False* Since the domain of $f(x) = \log_a x$ is the interval $(0, \infty)$, a logarithmic equation cannot have a negative solution. False

6. *True or False* The equation $8^{2x} = 4^{3x}$ is true for all real values of x. True

In Exercises 7–22, solve each equation.

7. $2^x = 16$ $x = 4$
8. $3^x = 243$ $x = 5$
9. $8^x = 32$ $x = \frac{5}{3}$
10. $5^{x-1} = 1$ $x = 1$
11. $4^{|x|} = 128$ $x = \pm\frac{7}{2}$
12. $9^{|x|} = 243$ $x = \pm\frac{5}{2}$
13. $5^{-|x|} = 625$ $\varnothing$
14. $3^{-|x|} = 81$ $\varnothing$
15. $\ln x = 0$ $x = 1$
16. $\ln(x - 1) = 1$ $x = 1 + e$
17. $\log_2 x = -1$ $x = \frac{1}{2}$
18. $\log_2(x + 1) = 3$ $x = 7$
19. $\log_3 |x| = 2$ $x = \pm 9$
20. $\log_2 |x + 1| = 3$ $x = 7, -9$

21. $\frac{1}{2}\log x - 2 = 0$ $x = 10{,}000$

22. $\frac{1}{3}\log(x + 1) - 1 = 0$ $x = 999$

In Exercises 23–50, solve each exponential equation and approximate the result correct to three decimal places.

23. $2^x = 3$ $x = 1.585$
24. $3^x = 5$ $x = 1.465$
25. $2^{2x+3} = 15$ $x = 0.453$
26. $3^{2x+5} = 17$ $x = -1.21$
27. $5 \cdot 2^x - 7 = 10$ $x = 1.766$
28. $3 \cdot 5^x + 4 = 11$ $x = 0.526$
29. $3 \cdot 4^{2x-1} + 4 = 14$ $x = 0.934$
30. $2 \cdot 3^{4x-5} - 7 = 10$ $x = 1.737$
31. $5^{1-x} = 2^x$ $x = 0.699$
32. $3^{2x-1} = 2^{x+1}$ $x = 1.191$
33. $2^{1-x} = 3^{4x+6}$ $x = -1.160$
34. $5^{2x+1} = 3^{x-1}$ $x = -1.277$
35. $2 \cdot 3^{x-1} = 5^{x+1}$ $x = -3.944$
36. $5 \cdot 2^{2x+1} = 7 \cdot 3^{x-1}$ $x = -5.059$
37. $\dfrac{5}{2 + 3^x} = 4$ $\varnothing$
38. $\dfrac{7}{3 + 5 \cdot 2^x} = 4$ $\varnothing$
39. $(1.065)^t = 2$ $t = 11.007$
40. $(1.0725)^t = 2$ $t = 9.903$
41. $2^{2x} - 4 \cdot 2^x = 21$ $x = 2.807$
42. $4^x - 4^{-x} = 2$ $x = 0.636$
43. $9^x - 6 \cdot 3^x + 8 = 0$ $x = 0.631, 1.262$
44. $\dfrac{3^x + 5 \cdot 3^{-x}}{3} = 2$ $x = 0, 1.465$

---

†Due to space constrictions, answers to these exercises may be found in the Answers beginning on page A–1 in the back of the book.

45. $\dfrac{3^x - 3^{-x}}{3^x + 3^{-x}} = \dfrac{1}{4}$ $x = 0.232$

46. $\dfrac{e^x - e^{-x}}{e^x + e^{-x}} = \dfrac{1}{3}$ $x = 0.347$

47. $\dfrac{4}{2 + 3^x} = 1$ $x = \dfrac{\ln 2}{\ln 3}$

48. $\dfrac{7}{2^x - 1} = 3$ $x = \dfrac{\ln 10 - \ln 3}{\ln 2}$

49. $\dfrac{17}{5 - 3^x} = 7$ $x = \dfrac{\ln 18 - \ln 7}{\ln 3}$

50. $\dfrac{15}{3 + 2 \cdot 5^x} = 4$ $x = \dfrac{\ln 3 - \ln 8}{\ln 5}$

In Exercises 51–68, solve each logarithmic equation.

51. $3 + \log (2x + 5) = 2$ $x = -\frac{49}{20}$

52. $1 + \log (3x - 4) = 0$ $x = \frac{41}{30}$

53. $\log (x^2 - x - 5) = 0$ $x = 3, -2$

54. $\log (x^2 - 6x + 9) = 0$ $x = 4, 2$

55. $\log_4 (x^2 - 7x + 14) = 1$ $x = 5, 2$

56. $\log_4 (x^2 + 5x + 10) = 1$ $x = -3, -2$

57. $\ln (2x - 3) - \ln (x + 5) = 0$ $x = 8$

58. $\log (x + 8) + \log (x - 1) = 1$ $x = 2$

59. $\log x + \log (x + 9) = 1$ $x = 1$

60. $\log_5 (3x - 1) - \log_5 (2x + 7) = 0$ $x = 8$

61. $\log_a (5x - 2) - \log_a (3x + 4) = 0$ $x = 3$

62. $\log (x - 1) + \log (x + 2) = 1$ $x = 3$

63. $\log_6 (x + 2) + \log_6 (x - 3) = 1$ $x = 4$

64. $\log_2 (3x - 2) - \log_2 (5x + 1) = 3$ $\varnothing$

65. $\log_3 (2x - 7) - \log_3 (4x - 1) = 2$ $\varnothing$

66. $\log_4 \sqrt{x + 3} - \log_4 \sqrt{2x - 1} = \dfrac{1}{4}$ $x = \frac{5}{3}$

67. $\log_7 3x + \log_7 (2x - 1) = \log_7 (16x - 10)$ $x = \frac{2}{3}, \frac{5}{2}$

68. $\log_3 (x + 1) + \log_3 2x = \log_3 (3x + 1)$ $x = 1$

B EXERCISES Applying the Concepts

69. Investment. Find the time required for an investment of $10,000 to grow to $18,000 at an annual interest rate of 6% if the interest is compounded
  a. yearly. 10.087 yr
  b. quarterly. 9.870 yr
  c. monthly. 9.821 yr
  d. daily. 9.797 yr
  e. continuously. 9.796 yr

70. Investment. How long will it take for an investment of $100 to double in value if the annual rate of interest is 7.2% compounded
  a. yearly. 9.970 yr
  b. quarterly. 9.713 yr
  c. monthly. 9.656 yr
  d. continuously. 9.627 yr

In Exercises 71–76, use the following table, which shows various statistics for Canada, Mexico, and the United Kingdom in 2000 and assume the same annual growth rate to continue for each country. Use the continuous compounding model.

Item	Canada	Mexico	United Kingdom
Population	33 million	108.7 million	60.8 million
Annual growth rate	0.9%	1.2%	0.3%
Price/liter of milk	$1.97	$0.92	$0.70
Price/kilogram of bread	$2.05	$1.48	$1.20
Annual inflation rate	2%	3.4%	3%

71. Annual growth rate. Find all of the data for each country for each year.
  a. 2012 †
  b. 2018 †

72. Annual growth rate. When will the population of Mexico reach 250 million people? In 2069 (after 69.4 yr)

73. Annual growth rate. In what year will the population of Canada be 60 million people? In 2066 (after 66.6 yr)

74. Annual growth rate. In Mexico in 2000, the average person spent approximately $500 on food. Use the 2000 inflation rate to calculate the cost of this food in the year
  a. 2012. $751.90
  b. 2018. $922.06

75. Annual growth rate. In each country, when will the price of a kilogram of bread be $10?

76. Annual growth rate. In what year will the price of a liter of milk be the same in Mexico and Canada?

77. Inflation. In the fictional land of Sardonia, the price of a car costing $20,000 in U.S. currency will double in eight years. What will the car cost in five years? What is the annual rate of inflation? $r \approx 8.66\%$; $30,837.52

78. Doubling time. An amount P dollars is deposited in a regular bank account. How long does it take to double the initial investment (called the **doubling time**) if
  a. the interest rate is r, compounded annually. $\dfrac{\ln 2}{\ln(1 + r)}$
  b. the interest rate is r (per year), compounded continuously. $t = \dfrac{\ln 2}{r}$ yr

79. Population growth. It is estimated that in t years from 2000, the population of Sardonia will be

$$P(t) = \dfrac{32}{5 + 3e^{-kt}} \text{ million.}$$

  a. Find the average rate of growth if the population of Sardonia in 2004 was 4.2 million. $k \approx 3.4\%$
  b. What was the population of Sardonia in 2002? ≈ 4.1 million
  c. Sketch the graph of $y = P(t)$. †
  d. What will the population of Sardonia be in 2020?
  e. What will happen to the population in the long run?

Answers:
75. Canada: in 2079 (after 79.2 yr)
  Mexico: in 2056 (after 56.2 yr)
  United Kingdom: in 2070 (after 70.7 yr)
76. In 2054 (after 54.4 yr) **79. d.** ≈ 4.9 million **e.** 6.4 million

80. Epidemic outbreak. The number of people in a community who became infected in an epidemic t weeks after its outbreak is given by the function

$$f(t) = \frac{20,000}{1 + ae^{-kt}},$$

where 20,000 people of the community are susceptible to the disease. If 1000 people were infected initially and 8999 had been infected by the end of the fourth week,
  a. Find the number of people infected after 8 weeks. 18,542
  b. After how many weeks will 12,400 people be infected? 5 weeks

81. Spreading a rumor. A jealous student started a malicious rumor on an isolated college campus of 5000 students. The number of people who heard the rumor within t days after it was started is given by the function

$$R(t) = \frac{5000}{1 + ae^{-kt}}.$$

Half the students had heard the rumor within ten days.
  a. Find a and k and graph the function $y = R(t)$. †
  b. How many students would have heard the rumor within 15 days after it started? 4930
  c. How many students altogether will hear the rumor? 5000

82. Biological growth. In his laboratory experiment in 1934, G. F. Gause placed paramecia (unicellular microorganisms) in 5 cubic centimeters of a saline (salt) solution with a constant amount of food and measured their growth on a daily basis. He found that the population $P(t)$ after t days was approximated by

$$P(t) = \frac{4490}{1 + e^{5.4094 - 1.0255t}}, \quad t \ge 0.$$

  a. What was the initial size of the paramecia? 20
  b. What was the carrying capacity of the medium? 4490
  c. Graph the equation $y = P(t)$. †

83. Marriage rate. The marriage rate in the United States in 2006 was 0.73%, and there were about 2,164,000 marriages that year. (*Source:* Department of Health and Human Services.)
    Use the model $M(t) = M_0(1 + r)^t$, with a constant marriage rate and $t = 0$ corresponding to the year 2006 to estimate
  a. the number of marriages in 2026. 2.50 million
  b. the year in which the number of marriages will be 4 million. In 2084

84. Divorce rate. The divorce rate in the United States in 2006 was 0.36%, and there were 1,125,000 divorces that year. Use the model $D(t) = D_0(1 + r)^t$, with a constant divorce rate and $t = 0$ corresponding to the year 2006 to estimate
  a. the number of divorces in 2026. 1.21 million
  b. the year in which the number of divorces will be 2 million. In 2160

85. Light absorption in physics. The light intensity I (in lumens) at a depth of x feet in Lake Elizabeth is given by

$$\log\left(\frac{I}{12}\right) = -0.025x.$$

  a. Find the light intensity at a depth of 30 feet. 2.134 lm
  b. At what depth is the light intensity 4 lumens? 19.08 ft

86. Rule of 70. Bankers use the **rule of 70** to estimate the doubling time for money invested at different rates. The rule of 70 states that

$$\text{Doubling time} \approx \frac{70}{100r} \text{ years,}$$

where r is the annual interest rate (expressed as a decimal number). Explain why this formula works. $\ln 2 \approx 0.69$

Richter scale. In Exercises 87–90, use the following information. The energy E (measured in joules) released by an earthquake of magnitude M on the Richter scale is given by the equation

$$\log E = 4.4 + 1.5M$$

87. The Great China Earthquake of 1920 registered 8.6 on the Richter scale. Let $I_0 = 1$. $\quad I = 10^{8.6}\,\text{W/m}^2$
  a. What was the intensity of this earthquake?
  b. How many joules of energy were released?
$\qquad\qquad\qquad\qquad\qquad\qquad E = 10^{17.3}$ joules
88. Suppose one earthquake registers one more point on the Richter scale than another. $\quad I_2 = 10I$
  a. How are their corresponding intensities related?
  b. How are their released energies related? $E_2 = 10^{1.5}E_1$

89. If one earthquake is 150 times as intense as another, what is the difference in the Richter scale readings of the two earthquakes? $\log 150 \approx 2.2$

90. If the energy released by one earthquake is 150 times that of another, what is the difference in the Richter scale readings of the two earthquakes? $\frac{1}{1.5}\log 150 \approx 1.45$

Sound. In Exercises 91–94, use the following information. The *loudness* (level of sound) of a sound is measured in *decibels* (dB). It is one-tenth of a *bel* (named after Alexander Graham Bell). The loudness L of a sound is related to its intensity I by the equation

$$L = 10\log\left(\frac{I}{I_0}\right)$$

where $I_0 = 10^{-12}$ watts per square meter (W/m^2) is the intensity of the faintest sound that the human ear can detect.
91. Find the dB level L of a TV at average volume from 12 ft away if it has an intensity of $250 \times 10^{-7}\,(\text{W/m}^2)$. ≈ 74 dB

92. Compare the intensity of a 65 dB sound to that of a 42 dB sound. $\approx 200I_1$

93. Find the intensity of a 73 dB sound. $10^{-4.7}\,\text{W/m}^2$

94. The intensity of one sound is 5000 times that of another sound. Find the difference in the dB levels of the two sounds. ≈ 37 dB

C EXERCISES Beyond the Basics

95. Solve for t: $P = \dfrac{M}{1 + e^{-kt}}$. $t = \dfrac{1}{k}\ln\left(\dfrac{P}{M - P}\right)$

96. Find k so that
  a. $2^x = e^{kx}$.   **b.** $e^x = 2^{kx}$. $k = \dfrac{1}{\ln 2} \approx 1.443$
  $k = \ln 2 \approx 0.693$

97. If $x = \dfrac{e^y - e^{-y}}{e^y + e^{-y}}$, show that $y = \dfrac{1}{2}\ln\left(\dfrac{1 + x}{1 - x}\right)$.

98. If $x = \dfrac{a^y + a^{-y}}{a^y - a^{-y}}$, show that $y = \dfrac{1}{2}\log_a\left(\dfrac{x + 1}{x - 1}\right)$.

99. If $\dfrac{\log x}{2} = \dfrac{\log y}{3} = \dfrac{\log z}{5}(= k)$, show that
  a. $xy = z$.   **b.** $y^2 z^2 = x^8$.
  [*Hint:* First, express x, y, and z in terms of k.]

100. The pressure P and the volume V of a gas are related by the formula

$$PV^\alpha = k,$$

where α and k are constants that depend on the gas. Find α for a given gas with $P_1 = 2$ atmospheres (atm), $V_1 = 3.2$ cubic feet, $P_2 = 3$ atm, and $V_2 = 2.3$ cubic feet. $\alpha = 1.228$

101. **Annuity.** The future value A of an ordinary annuity with payment size R, periodic rate i, and a term of n payments is given by the formula $A = R\dfrac{(1 + i)^n - 1}{i}$.
Solve this equation for n.

102. The present value P of an annuity with payment size R, periodic rate i, and a term of n payments is given by the formula

$$P(1 + i)^n = R\dfrac{(1 + i)^n - 1}{i}.$$

Solve this equation for n.

Answers:

101. $n = \dfrac{\ln\left(\dfrac{R + iA}{R}\right)}{\ln(1 + i)}$ **102.** $n = \dfrac{\ln\left(\dfrac{R}{R - iP}\right)}{\ln(1 + i)}$

In Exercises 103–106, solve each equation for x.

103. $(\log x)^2 = \log x$ $x = 1, 10$

104. $\log_2 x + \log_4 x = 6$ $x = 16$

105. $\log_4 x^2(x - 1)^2 - \log_2(x - 1) = 1$ $x = 2$

106. $\log(\log x^{10}) = 1$ $x = 10$

107. Find the value of $\log_2 \log_2 \log_3 \log_3 27^3$. 0

108. If $f(x) = \log_a\left(\dfrac{1 + x}{1 - x}\right)$, show that $f\left(\dfrac{2x}{1 + x^2}\right) = 2f(x)$.

109. If $\log 3 = 0.477$ and $(1000)^x = 3$, then the value of x is a
  a. 0.159.   **b.** 10.   **c.** 0.0477.   **d.** 0.0159.

110. The simplified value of $\log(\log x^2) - \log(\log x)$ is c
  a. 2.   **b.** $\dfrac{1}{2}$.   **c.** $\log 2$.   **d.** $2\log x$.

Critical Thinking

In Exercises 111 and 112, x, y, and z represent positive numbers.

111. Evaluate $\dfrac{x^{\log y}}{x^{\log z}} \cdot \dfrac{y^{\log z}}{y^{\log x}} \cdot \dfrac{z^{\log x}}{z^{\log y}}$ 1

112. Evaluate $\dfrac{1}{\log_{xy} xyz} + \dfrac{1}{\log_{yz} xyz} + \dfrac{1}{\log_{zx} xyz}$ 2

113. **Logistic function.** For $f(t) = \dfrac{P}{1 + ae^{-kt}}$ it can be shown that the maximum rate of growth occurs when $f(t) = \dfrac{P}{2}$.
Find the time when the rate of growth is maximum.
$$t = \dfrac{\ln a}{k}$$

SUMMARY ■ Definitions, Concepts, and Formulas

3.1 Exponential Functions

i. A function $f(x) = a^x$, with $a > 0$ and $a \neq 1$, is called an exponential function with base a.

Rules of exponents: $a^x a^y = a^{x+y}$, $\dfrac{a^x}{a^y} = a^{x-y}$, $(a^x)^y = a^{xy}$,

$$a^0 = 1, a^{-x} = \dfrac{1}{a^x}$$

ii. Exponential functions are one-to-one: If $a^x = a^y$, then $x = y$.

iii. If $a > 1$, then $f(x) = a^x$ is an increasing function; $f(x) \to \infty$ as $x \to \infty$ and $f(x) \to 0$ as $x \to -\infty$.

iv. If $0 < a < 1$, then $f(x) = a^x$ is a decreasing function; $f(x) \to 0$ as $x \to \infty$ and $f(x) \to \infty$ as $x \to -\infty$.

v. The graph of $f(x) = a^x$ has y-intercept 1, and the x-axis is a horizontal asymptote.

3.2 The Natural Exponential Function

i. *Simple-interest formula.* If P dollars is invested at an interest rate r per year for t years, then the simple interest is given by the formula $I = Prt$.

ii. *Compound interest.* P dollars invested at annual rate r compounded n times per year for t years amounts to
$$A_n = P\left(1 + \frac{r}{n}\right)^{nt}.$$

iii. The Euler constant $e = \lim_{h \to \infty}\left(1 + \frac{1}{h}\right)^h \approx 2.718$.

iv. *Continuous compounding.* P dollars invested at an annual rate r compounded continuously for t years amounts to $A = Pe^{rt}$.

v. The function $f(x) = e^x$ is the natural exponential function.

3.3 Logarithmic Functions

i. For $a > 0$ and $a \neq 1$, $y = \log_a x$ if and only if $x = a^y$.

ii. *Basic properties:* $\log_a a = 1$, $\log_a 1 = 0$,
$$\log_a a^x = x, a^{\log_a x} = x \qquad \text{Inverse properties}$$

iii. The domain of $\log_a x$ is $(0, \infty)$, the range is $(-\infty, \infty)$, and the y-axis is a vertical asymptote. The x-intercept is 1.

iv. Logarithmic functions are one-to-one: If $\log_a x = \log_a y$, then $x = y$.

v. If $a > 1$, then $f(x) = \log_a x$ is an increasing function; $f(x) \to \infty$ as $x \to \infty$, and $f(x) \to -\infty$ as $x \to 0^+$.

vi. If $0 < a < 1$, then $f(x) = \log_a x$ is a decreasing function; $f(x) \to -\infty$ as $x \to \infty$, and $f(x) \to \infty$ as $x \to 0^+$.

vii. The common logarithmic function is $y = \log x$ (base 10); the natural logarithmic function is $y = \ln x$ (base e).

3.4 Rules of Logarithms

i. *Rules of logarithms:*
$$\log_a MN = \log_a M + \log_a N \qquad \text{Product rule}$$
$$\log_a \frac{M}{N} = \log_a M - \log_a N \qquad \text{Quotient rule}$$
$$\log_a M^r = r \log_a M \qquad \text{Power rule}$$

ii. *Change-of-base formula:*
$$\log_b x = \frac{\log_a x}{\log_a b} = \frac{\log x}{\log b} = \frac{\ln x}{\ln b}$$
$$\text{(base } a\text{)} \quad \text{(base 10)} \quad \text{(base } e\text{)}$$

3.5 Exponential and Logarithmic Equations

An *exponential equation* is an equation in which a variable occurs in one or more exponents.

A *logarithmic equation* is an equation that involves the logarithm of a function of the variable.

Exponential and logarithmic equations are solved by using some or all of the following techniques:

i. Using the one-to-one property of exponential and logarithmic functions

ii. Converting from exponential to logarithmic form or vice versa

iii. Using the product, quotient, and power rules for exponents and logarithms

REVIEW EXERCISES

Basic Skills

In Exercises 1–10, state whether the given statement is true or false.

1. The function $f(x) = a^x$ is exponential if $a > 0$. False

2. The graph of $f(x) = 4^x$ approaches the x-axis as $x \to -\infty$. True

3. The domain of $f(x) = \log(2 - x)$ is $(-\infty, 2]$. False

4. The equation $u = 10^v$ means $\log u = v$. True

5. The inverse of $f(x) = \ln x$ is $g(x) = e^x$. True

6. The graph of $y = a^x (a > 0, a \neq 1)$ always contains the points $(0, 1)$ and $(1, a)$. True

7. $\ln(M + N) = \ln M + \ln N$ False

8. $\log\sqrt{300} = 1 + \frac{1}{2}\log 3$ True

9. The functions $f(x) = 2^{-x}$ and $g(x) = \left(\frac{1}{2}\right)^x$ have the same graph. True

10. $\ln u = \frac{\log u}{\log e}$ True

In Exercises 11–18, match the function with its graph in (a)–(h).

11. $f_1(x) = \log_2 x$ h

12. $f_2(x) = 2^x$ c

13. $f_3(x) = \log_2(3 - x)$ f

14. $f_4(x) = \log_{1/2} x$ d or e

15. $f_5(x) = -\log_2 x$ d or e

†Due to space constrictions, answers to these exercises may be found in the Answers beginning on page A–1 in the back of the book.

16. $f_6(x) = \left(\dfrac{1}{2}\right)^x$ b

17. $f_7(x) = 3 - 2^{-x}$ a

18. $f_8(x) = \dfrac{6}{1 + 2e^{-x}}$ g

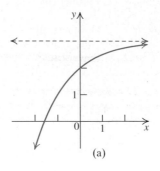

(a)

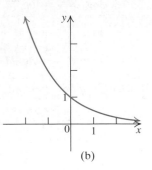

(b)

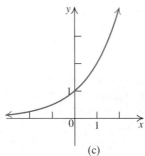

(c)

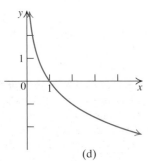

(d)

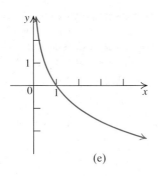

(e)

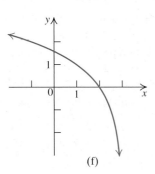

(f)

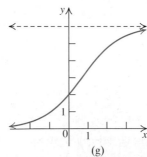

(g)

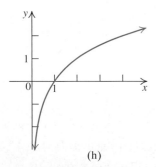

(h)

In Exercises 19–30, graph each function using transformations on an appropriate graph. Determine the domain, range, and asymptotes (if any).

19. $f(x) = 2^{-x}$ †

20. $g(x) = 2^{-0.5x}$ †

21. $h(x) = 3 + 2^{-x}$ †

22. $f(x) = e^{|x|}$ †

23. $g(x) = e^{-|x|}$ †

24. $h(x) = e^{-x+1}$ †

25. $f(x) = \ln(-x)$ †

26. $g(x) = 2\ln|x|$ †

27. $h(x) = 2\ln(x - 1)$ †

28. $f(x) = 2 - \left(\dfrac{1}{2}\right)^x$ †

29. $g(x) = 2 - \ln(-x)$ †

30. $h(x) = 3 + 2\ln(5 + x)$ †

In Exercises 31–34, sketch the graph of the given function using two steps:

a. find the y-intercept. **b.** find $\lim\limits_{x \to \infty} f(x)$ and $\lim\limits_{x \to -\infty} f(x)$.

31. $f(x) = 3 - 2e^{-x}$ †

32. $f(x) = \dfrac{5}{2 + 3e^{-x}}$ †

33. $f(x) = e^{-x^2}$ †

34. $f(x) = 3 - \dfrac{6}{1 + 2e^{-x}}$ †

In Exercises 35–38, find a and k and then evaluate the function.

35. Let $f(x) = a(2^{kx})$, with $f(0) = 10$ and $f(3) = 640$. Find $f(2)$. $a = 10, k = 2, f(2) = 160$

36. Let $f(x) = 50 - a(5^{kx})$, with $f(0) = 10$ and $f(2) = 0$. Find $f(1)$. $a = 40, k = 0.06932, f(1) = 5.28$

37. Let $f(x) = \dfrac{4}{1 + ae^{-kx}}$, with $f(0) = 1$ and $f(3) = \dfrac{1}{2}$. Find $f(4)$. $a = 3, k = -0.2824, f(4) = 0.38898$

38. Let $f(x) = 6 - \dfrac{3}{1 + ae^{-kx}}$, with $f(0) = 5$ and $f(4) = 4$. Find $f(10)$. $a = 2, k = 0.3466, f(10) = 3.1764$

In Exercises 39–42, write each logarithm in expanded form.

39. $\ln(xy^2z^3)$ $\ln x + 2\ln y + 3\ln z$

40. $\log(x^3\sqrt{y - 1})$ $3\log x + \dfrac{1}{2}\log(y - 1)$

41. $\ln\left[\dfrac{x\sqrt{x^2 + 1}}{(x^2 + 3)^2}\right]$ $\ln x + \dfrac{1}{2}\ln(x^2 + 1) - 2\ln(x^2 + 3)$

42. $\ln\sqrt{\dfrac{x^3 + 5}{x^3 - 7}}$ $\dfrac{1}{2}[\ln(x^3 + 5) - \ln(x^3 - 7)]$

In Exercises 43–48, write y as a function of x.

43. $\ln y = \ln x + \ln 3$ $y = 3x$

44. $\ln y = \ln(C) + kx$; C and k are constants. $y = Ce^{kx}$

45. $\ln y = \ln(x - 3) - \ln(y + 2)$ $y = -1 \pm \sqrt{x - 2}$

46. $\ln y = \ln x - \ln(x^2y) - 2\ln y$ $y = \left(\dfrac{1}{x}\right)^{1/4}$

47. $\ln y = \dfrac{1}{2}\ln(x - 1) + \dfrac{1}{2}\ln(x + 1) - \ln(x^2 + 1)$

$y = \dfrac{\sqrt{x^2 - 1}}{x^2 + 1}$

48. $\ln(y - 1) = \dfrac{1}{x} + \ln(y)$ $y = \dfrac{1}{1 - e^{1/x}}$

In Exercises 49–66, solve each equation.

49. $3^x = 81$ 4

50. $5^{x-1} = 625$ 5

51. $2^{x^2+2x} = 16$ $-1 \pm \sqrt{5}$

52. $3^{x^2-6x+8} = 1$ $\{2, 4\}$

53. $3^x = 23$

54. $2^{x-1} = 5.2$

55. $273^x = 19$

56. $27 = 9^x \cdot 3^{x^2}$ $\{1, -3\}$

57. $3^k = e^{kx}$ $\ln 3$

58. $a^{2x-1} = b$

59. $x^{\ln x} = e$

60. $2^{2^x} = 8$

61. $\log_3(x + 2) - \log_3(x - 1) = 1$ 2.5

62. $\log_5(3x + 7) + \log_5(x - 5) = 2$ 6

63. $2\ln 3x = 3\ln x$ 9

64. $2\log x = \ln e$ $\sqrt{10}$

65. $2^x - 8 \cdot 2^{-x} - 7 = 0$ 3

66. $3^x - 24 \cdot 3^{-x} = 10$

Answers:

53. $\dfrac{\ln 23}{\ln 3} \approx 2.854$ **54.** $\dfrac{\ln 5.2}{\ln 2} + 1 \approx 3.379$ **55.** $\dfrac{\ln 19}{\ln 273} \approx 0.525$

58. $\dfrac{1}{2}\left(\dfrac{\ln b}{\ln a} + 1\right)$ **59.** $\left\{e, \dfrac{1}{e}\right\}$ **60.** $\dfrac{\ln 3}{\ln 2} \approx 1.585$ **66.** $\dfrac{\ln 12}{\ln 3} \approx 2.262$

Applying the Concepts

67. You have $7000 to invest for seven years. Which investment will provide the greater return, 5% compounded yearly or 4.75% compounded monthly? †

68. How long will it take $8000 to grow to $20,000 if the rate of interest is 7% compounded continuously? 13.09 yr

69. The formula $P(t) = 33e^{0.003t}$ models the population of Canada, in millions, t years after 2007.
   a. Estimate the population of Canada in 2017. 34 million
   b. According to this model, when will the population of Canada be 60 million? In 2206 (after 199.3 yr)

70. The formula $C(t) = 100 + 25e^{.03t}$ models the average cost of a house in Sometown, U.S.A., t years after 2000. The cost is expressed in thousands of dollars.
   a. Sketch the graph of $y = C(t)$. †
   b. Estimate the average cost in 2010. $133.75 thousand
   c. According to this model, when will the average cost of a house in Sometown be $250,000? 59.73 yr

71. An experimental drug was injected into the bloodstream of a rat. The concentration $C(t)$ of the drug (in micrograms per milliliter of blood) after t hours was modeled by the function

$$C(t) = 0.3e^{-0.47t}, 0.5 \le t \le 10.$$

   a. Graph the function $y = C(t)$ for $0.5 \le t \le 10$. †
   b. When will the concentration of the drug be 0.029 micrograms per milliliter? (1 microgram $= 10^{-6}$ gram) ≈ 5 hr

72. X-ray technicians are shielded by a wall insulated with lead. The equation $x = \dfrac{1}{152} \log\left(\dfrac{I_0}{I}\right)$ measures the thickness x (in centimeters) of the lead insulation required to reduce the initial intensity I_0 of X-rays to the desired intensity I.
   a. What thickness of lead is required to reduce the intensity of X-rays to one-tenth their initial intensity? thickness = 1/152 cm
   b. What thickness of lead is required to reduce the intensity of X-rays to $\dfrac{1}{40}$ their initial intensity? thickness = 0.01053987 cm
   c. How much is the intensity reduced if the lead insulation is 10 centimeters thick? $I = \dfrac{1}{10^{1520}} I_0$

73. Chai (a certain kind of tea) is made by adding boiling water (212° F) to the chai mix. Suppose you make chai in a room with the air temperature at 75°F. According to Newton's Law of Cooling, the temperature of the chai t minutes after it is boiled is given by a function of the form $f(t) = 75 + ae^{-kt}$. After one minute, the temperature of the chai falls to 200°F. How long will it take for the chai to be drinkable at 150°F? $t = 6.57$ min

74. Approximately $P(t) = \dfrac{6}{1 + 5e^{-0.7t}}$ thousand people caught a new form of influenza within t weeks of its outbreak.
   a. Sketch the graph of $y = P(t)$. †
   b. How many people had the disease initially? 1000

   c. How many people contracted the disease within four weeks? 4601
   d. If the trend continues, how many people in all will contract the disease? 6000

75. The population density x miles from the center of a town called Greenville is approximated by the equation $D(x) = 5e^{0.08x}$, in thousands of people per square mile.
   a. What is the population density at the center of Greenville? 5000 people/square mile
   b. What is the population density 5 miles from the center of Greenville? 7459 people/square mile
   c. Approximately how far from the center of Greenville would the density be 15,000 people per square mile? ≈ 13.73 miles

76. In 2000, the population of the United States was 280 million and the number of vehicles was 200 million. If the population of the United States is growing at the rate of 1% per year while the number of vehicles is growing at the rate of 3%, in what year will there be an average of one vehicle per person? In 2017 (after 16.82 yr)

77. **Light intensity.** The *Bouguer–Lambert Law* states that the intensity I of sunlight filtering down through water at a depth x (in meters) decreases according to the exponential decay function $I = I_0 e^{-kx}$, where I_0 is the intensity at the surface and $k > 0$ is an absorption constant that depends on the murkiness of the water. Suppose the absorption constant of Carrollwood Lake was experimentally determined to be $k = 0.73$. How much light has been absorbed by the water at a depth of 2 meters? $I = 0.2322I_0$

78. The strength of a TV signal usually fades due to a damping effect of cable lines. If I_0 is the initial strength of the signal, then its strength I at a distance x miles is measured by the formula $I = I_0 e^{-kx}$, where $k > 0$ is a damping constant that depends on the type of wire used. Suppose the damping constant has been measured experimentally to be $k = 0.003$. What fraction of the signal is lost at a distance of 10 miles? 20 miles? $\approx 3\%$; $\approx 6\%$

79. **Walking speed in a city.** In 1976, Marc and Helen Bernstein discovered that in a city with population p, the average speed s (in feet per second) that a person walks on main streets can be approximated by the formula

$$s(p) = 0.04 + 0.86 \ln p.$$

   a. What is the average walking speed of pedestrians in Tampa (population 470,000)? 11.27 ft/sec
   b. What is the average walking speed of pedestrians in Bowman, Georgia (population 450)? 5.29 ft/sec
   c. What is the estimated population of a town in which the estimated average walking speed is 4.6 feet per second? 200.8

80. **Drinking and driving.** Just after Eric had his last drink, the alcohol level in his bloodstream was 0.26 (milligram of alcohol per milliliter of blood). After one-half hour, his alcohol level was 0.18. The alcohol level $A(t)$ in a person follows the exponential decay law:

$$A(t) = A_0 e^{-kt},$$

where $k > 0$ depends on the individual.
   a. What is the value of A_0 for Eric? 0.26
   b. What is the value of k for Eric? $k = 0.74$

c. If the legal driving limit for alcohol level is 0.08, how long should Eric wait (after his last drink) before he will legally be able to drive? $t = 1.59\,\text{hr} \approx 1\,\text{hr}\,34\,\text{min}$

81. Bacteria culture. The mass $m(t)$, in grams, of a bacteria culture grows according to the logistic growth model

$$m(t) = \frac{6}{1 + 5e^{-0.7t}},$$

where t is time measured in days.
a. What is the initial mass of the culture? 1 g
b. What happens to the mass in the long run?
c. When will the mass reach 5 grams? 4.6 days
b. Remains stationary at 6 g

82. A learning model. The number of units $n(t)$ produced per day after t days of training is given by

$$n(t) = 60(1 - e^{-kt}),$$

where $k > 0$ is a constant that depends on the individual.
a. Estimate k for Rita, who produced 20 units after one day of training. $k = 0.41$
b. How many units will Rita produce per day after ten days of training? $58.90 \approx 59$ units
c. How many days should Rita be trained if she is expected to produce 40 units per day? $2.7 \approx 3$ days

PRACTICE TEST A

1. Solve the equation $5^{-x} = 125$. $x = -3$

2. Solve the equation $\log_2 x = 5$. $x = 32$

3. State the range of $y = -e^x + 1$ and find the asymptote of its graph. Range $= (-\infty, 1)$; horizontal asymptote : $y = 1$

4. Evaluate $\log_2 \frac{1}{8}$. -3

5. Solve the exponential equation $\left(\dfrac{1}{4}\right)^{2-x} = 4$. $x = 3$

6. Evaluate $\log 0.001$ without using a calculator. -3

7. Rewrite the expression $\ln 3 + 5 \ln x$ in condensed form. $\ln 3x^5$

8. Solve the equation $5x^3 e^x - x^4 e^x = 0$. $\{0, 5\}$

9. Solve the equation $e^{2x} + e^x - 6 = 0$. $\ln 2$

10. Rewrite the expression $\ln \dfrac{2x^3}{(x+1)^5}$ in expanded logarithmic form. $\ln 2 + 3 \ln x - 5 \ln (x+1)$

11. Evaluate $\ln e^{-5}$. -5

12. Give the equation for the graph obtained by shifting the graph of $y = \ln x$ three units up and one unit right. $y = \ln (x-1) + 3$

13. Sketch the graph of $y = 3^{x-1} + 2$. †

14. State the domain of the function $f(x) = \ln (-x) + 4$. $(-\infty, 0)$

15. Write $3 \ln x + \ln (x^3 + 2) - \dfrac{1}{2} \ln (3x^2 + 2)$ in condensed form. $\ln \dfrac{x^3(x^3 + 2)}{\sqrt{3x^2 + 2}}$

16. Solve the equation $\log x = \log 6 - \log (x - 1)$. $x = 3$

17. Find x if $\log_x 9 = 2$. $x = 3$

18. Rewrite the expression $\ln \sqrt{2x^3 y^2}$ in expanded logarithmic form. $\frac{1}{2}[\ln 2 + 3 \ln x + 2 \ln y]$

19. Suppose $15,000 is invested in a savings account paying 7% interest per year. Write the formula for the amount in the account after t years if the interest is compounded quarterly. $A(t) = 15,000(1.0175)^{4t}$

20. Suppose the number of Hispanic people living in the United States is approximated by $H = 15,000e^{0.02t}$, where $t = 0$ represents 1960. According to this model, about how many Hispanic people were living in the United States in 1980? 22,377

PRACTICE TEST B

1. Solve the equation $3^{-x} = 9$. b
    a. $\{2\}$ **b.** $\{-2\}$ **c.** $\left\{\dfrac{1}{2}\right\}$ **d.** $\left\{-\dfrac{1}{2}\right\}$ **e.** $\{\ln 2\}$

2. Solve the equation $\log_5 x = 2$. b
    a. $\{10\}$ **b.** $\{25\}$ **c.** $\left\{\dfrac{5}{2}\right\}$ **d.** $\left\{\dfrac{2}{5}\right\}$ **e.** $\{2\}$

3. State the range and asymptote of $y = e^{-x} - 1$. b
    a. $(0, \infty); y = -1$ **b.** $(-1, \infty); y = -1$
    c. $(-\infty, \infty); y = -1$ **d.** $(-1, \infty); y = 1$
    e. $(\infty, 1); y = 1$

4. Evaluate $\log_4 64$. d
    a. 16 **b.** 8 **c.** 2 **d.** 3 **e.** 4

5. Find the solution of the exponential equation $\left(\dfrac{1}{3}\right)^{1-x} = 3$. d
    a. $\left\{-\dfrac{1}{3}\right\}$ **b.** $\left\{\dfrac{1}{3}\right\}$ **c.** $\{1\}$ **d.** $\{2\}$

6. Evaluate $\log 0.01$. b
    a. -1.99999999 **b.** -2 **c.** 2 **d.** 100

7. Which of the following expressions is equivalent to $\ln 7 + 2 \ln x$? b
    a. $\ln (7 + 2x)$ **b.** $\ln (7x^2)$ **c.** $\ln (9x)$ **d.** $\ln (14x)$

8. Solve the equation $3x^2 e^x + x^3 e^x = 0$. b
    a. $\{-3\}$ **b.** $\{0, -3\}$ **c.** $\{0\}$ **d.** $\varnothing$

9. Solve the equation $e^{2x} - e^x - 6 = 0$. d

　　a. $\{-\ln 2\}$　**b.** $\{-\ln 3\}$　**c.** $\{\ln 6\}$　**d.** $\{\ln 3\}$

10. Rewrite the expression $\ln \dfrac{3x^2}{(x+1)^{10}}$ in expanded logarithmic form. d

　　a. $\ln 6x - 10 \ln x + 1$
　　b. $2 \ln 3x - 10 \ln (x+1)$
　　c. $2 \ln 3 + 2 \ln x - 10 \ln (x+1)$
　　d. $\ln 3 + 2 \ln x - 10 \ln (x+1)$

11. Find $\ln e^{3x}$. b

　　a. 3　**b.** $3x$　**c.** $3+x$　**d.** x

12. The equation for the graph obtained by shifting the graph of $y = \log_3 x$ two units up and three units left is d

　　a. $y = \log_3 (x-3) + 2$　**b.** $y = \log_3 (x+3) - 2$.
　　c. $y = \log_3 (x-3) - 2$　**d.** $y = \log_3 (x+3) + 2$.

13. Which of the following graphs is the graph of $y = 5^{x+1} - 4$? b

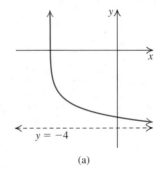

(a)

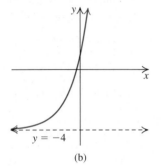

(b)

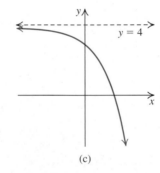

(c)

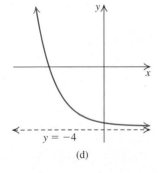

(d)

14. Find the domain of the function $f(x) = \ln (1-x) + 3$. a

　　a. $(-\infty, 1)$　**b.** $(1, \infty)$　**c.** $(-\infty, -1)$
　　d. $(-\infty, 3)$　**e.** $(3, \infty)$

15. Write $\ln x - 2 \ln (x^2 + 1) + \dfrac{1}{2} \ln (x^4 + 1)$ in condensed form. a

　　a. $\ln \dfrac{x\sqrt{x^4+1}}{(x^2+1)^2}$　　**b.** $\ln \dfrac{x}{x^2+1}$
　　c. $\ln (x - (x^2+1)^2 + (x^4+1)^{1/2})$
　　d. $2 \ln \dfrac{x(1+x^4)}{x^2+1}$

16. Solve the equation $\log x = \log 12 - \log (x+1)$. d

　　a. $\left(\dfrac{11}{2}\right)$　**b.** $\left(\dfrac{13}{2}\right)$　**c.** $\{3, -4\}$　**d.** $\{3\}$　**e.** $\left\{\dfrac{2}{12}\right\}$

17. Find x if $\log_x 16 = 4$. b

　　a. 4　**b.** 2　**c.** 64　**d.** $\dfrac{1}{4}$　**e.** 16

18. Rewrite the expression $\ln \sqrt[3]{5x^2y^3}$ in expanded logarithmic form. c

　　a. $\dfrac{1}{3}(\ln 5x^2 + \ln y^3)$　　**b.** $\dfrac{1}{3}(2 \ln 5x + 3 \ln y)$
　　c. $\dfrac{1}{3}(\ln 5 + 2 \ln x + 3 \ln y)$　**d.** $\dfrac{1}{3} \ln 5 + 2 \ln x + 3 \ln y$
　　e. $\dfrac{1}{3} \ln 5 + 2 + \ln x + 3 + \ln y$

19. Suppose $12,000 is invested in a savings account paying 10.5% interest per year. Write the formula for the amount in the account after t years if the interest is compounded monthly. d

　　a. $A = 12{,}000(1.105)^t$　**b.** $A = 12{,}000(1.525)^{2t}$
　　c. $A = 12{,}000(1.2625)^{4t}$　**d.** $A = 12{,}000(1.00875)^{12t}$

20. The population of a certain city is growing according to the model $P = 10{,}000 \log_5 (t+5)$, where t is time in years. If $t = 0$ corresponds to the year 2000, what will the population of the city be in 2020? b

　　a. 30,000　　　　**b.** 20,000
　　c. 50,000　　　　**d.** 10,000

CUMULATIVE REVIEW EXERCISES ▪ Chapters 1–3

1. Find the x- and y-intercepts of the graph of the equation $x^2 + (y-1)^2 = 1$. Check for symmetry with respect to both axes and the origin. x-intercept: 0; y-intercepts: 0, 2; symmetric in y-axis

2. Graph the equation $y = x^3$. †

3. Find the center and radius of the circle $x^2 + 2x + y^2 - 4y - 20 = 0$. Graph the circle. †

4. Write the slope–intercept form of the equation of the line that passes through $(-2, 4)$ and is perpendicular to the line $2x + 3y = 17$. $y = \dfrac{3}{2}x + 7$

5. Find the domain of $f(x) = \log \dfrac{x-2}{x+3}$. $(-\infty, -3) \cup (2, \infty)$

6. Evaluate the function $f(x) = \begin{cases} 2x - 3, & x < 1 \\ 2x^2 + 1, & x \geq 1 \end{cases}$ at each value specified.

　　a. $f(-2)$　−7　**b.** $f(0)$　−3　**c.** $f(3)$. 19

7. Identify the basic function and use transformations to sketch the graph of the function $f(x) = 3\sqrt{x+1} - 2$. †

8. Let $f(x) = 3\sqrt{x}$ and $g(x) = -\dfrac{1}{x}$.

　　Find and write the domain of $(f \circ g)(x)$ in interval notation. $(f \circ g)(x) = 3\sqrt{-\dfrac{1}{x}}$. Domain of $f \circ g$ is $(-\infty, 0)$.

9. Determine whether the function has an inverse function; if so, find the inverse function.
   a. $f(x) = 2x^3 + 1$
   b. $f(x) = |x|$ No
   c. $f(x) = \ln x$. Yes, $f^{-1}(x) = e^x$

10. Divide
    a. $\dfrac{2x^3 + 3x + 1}{x^2 + 1}$ using long division
    b. $\dfrac{2x^4 + 7x^3 - 2x + 3}{x + 3}$ using synthetic division

11. Find a polynomial function of least degree with integer coefficients that has the given zeros.
    a. $1, 2, -3$ $(x - 1)(x - 2)(x + 3) = x^3 - 7x + 6$
    b. $1, -1, i, -i$ $x^4 - 1$

12. Use rules of logarithms to rewrite the following expressions in expanded form.
    a. $\log_2 5x^3$ $\log_2 5 + 3\log_2 x$
    b. $\log_a \sqrt[3]{\dfrac{xy^2}{z}}$ $\frac{1}{3}[\log_a x + 2\log_a y - \log_a z]$
    c. $\ln \dfrac{3\sqrt{x}}{5y}$ $\ln 3 + \frac{1}{2}\ln x - \ln 5 - \ln y$

13. Use rules of logarithms to rewrite the following expressions in condensed form.
    a. $\dfrac{1}{2}(\log x + \log y)$ $\log \sqrt{xy}$
    b. $3\ln x - 2\ln y$ $\ln \dfrac{x^3}{y^2}$

14. Use a calculator to evaluate the expression. Round your answer to three decimal places.
    a. $\log_9 17$ 1.289
    b. $\log_3 25$ 2.930
    c. $\log_{1/2} 0.3$ 1.737

15. Solve each equation.
    a. $\log_6 (x + 3) + \log_6 (x - 2) = 1$ {3}
    b. $5^{x^2 - 4x + 5} = 25$ {1, 3}
    c. $3.1^{x-1} = 23$ 3.77

16. Find all possible rational zeros of
    $$f(x) = x^4 - x^3 - 2x^2 - 2x + 4.$$
    Also find upper and lower bounds for the zeros of f.

17. Let $f(x) = (x - 1)^3(x + 2)^2(x - 3)$.
    a. Find the zeros of f along with their multiplicities.
    b. What is the behavior of f near each zero?
    c. What is the end behavior of f?
    d. Sketch the graph of f. †

18. Let $f(x) = \dfrac{(x - 1)(x + 2)}{(x - 3)(x + 4)}$.
    a. Find the vertical asymptotes of f (if any).
    b. Find the horizontal asymptotes of f (if any).
    c. Find the intervals over which $f(x) > 0$ or $f(x) < 0$.
    d. Sketch the graph of f. †

16. Possible rational zeros $\{\pm 1, \pm 2, \pm 4\}$. Rational zeros: $\{1, 2\}$; upper bound: 3; lower bound: -1

17. a. $1, 1, 1, -2, -2, 3$
b. At $x = 1$ and $x = 3$, the graph crosses the x-axis. At $x = -2$, the graph touches but does not cross the x-axis.
c. $f(x) \rightarrow \infty$ as $x \rightarrow \infty$; $f(x) \rightarrow \infty$ as $x \rightarrow -\infty$

18. a. Vertical asymptotes: $x = 3, x = -4$
b. Horizontal asymptotes: $y = 1$
c. $f(x) > 0$ on $(-\infty, -4) \cup (-2, 1) \cup (3, \infty)$
   $f(x) < 0$ on $(-4, -2) \cup (1, 3)$

Answers:

9. a. Yes, $f^{-1}(x) = \left(\dfrac{x - 1}{2}\right)^{1/3}$

10. a. $Q = 2x, R = x + 1, D = x^2 + 1$;
   $2x + \dfrac{x + 1}{x^2 + 1}$
b. $Q = 2x^3 + x^2 - 3x + 7, R = -18, D = x + 3$;
   $2x^3 + x^2 - 3x + 7 - \dfrac{18}{x + 3}$

Trigonometric Functions

Angles are vital for activities ranging from taking photographs to designing layouts for city streets. Scientists use angles and basic trigonometry to measure remote distances such as the distance between mountain peaks and the distances between planets. Angles are universally used to specify locations on Earth via longitude and latitude. In this chapter, we begin the study of trigonometry and its many uses.

Angles and Their Measure

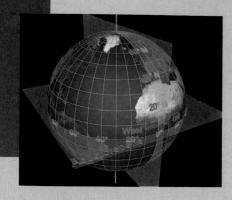

Before Starting this Section, Review

1. Circumference and area of a circle (Appendix A, page 791)

Objectives

1. Learn the vocabulary associated with angles.
2. Use degree and radian measure.
3. Convert between degree and radian measure.
4. Find complements and supplements.
5. Find the length of an arc of a circle.
6. Compute linear and angular speed.
7. Find the area of a sector.

LATITUDE AND LONGITUDE

Any location on Earth can be described by two numbers, its *latitude* and its *longitude*. To understand how these two location numbers are assigned, we think of Earth as being a perfect sphere.

Lines of latitude are parallel circles of different size around the sphere representing Earth. The longest circle is the equator, with latitude 0, while at the poles, the circles shrink to a point. Lines of longitude (also called *meridians*) are circles of identical size that pass through the North Pole and the South Pole as they circle the globe. Each of these circles crosses the equator. The equator is divided into 360 degrees, and the longitude of a location is the number of degrees where the meridian through that location meets the equator. For historical reasons, the meridian near the old Royal Astronomical Observatory in Greenwich, England, is the one chosen as 0 longitude.

Today this prime meridian is marked with a band of brass that stretches across the yard of the Observatory. Longitude is measured from the prime meridian, with positive values going east (0 to 180) and negative values going west (0 to −180). Both ±180-degree longitudes share the same line, in the middle of the Pacific Ocean. At the equator, the distance on Earth's surface for each one degree of latitude or longitude is just over 69 miles.

Every location on Earth is then identified by the meridian and the latitude lines that pass through it. In Example 7, we explain how latitude values are assigned and how they can be used to compute distances between cities having the same longitude. ■

---

1. Learn the vocabulary associated with angles.

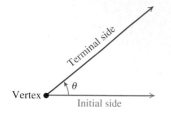

FIGURE 4.1 An angle

Angles

A **ray** is a part of a line made up of a point, called the **endpoint**, and all of the points on one side of the endpoint. An **angle** is formed by rotating a ray about its endpoint. The angle's **initial side** is the ray's original position, while the angle's **terminal side** is the ray's position after the rotation. The endpoint is called the **vertex** of the angle. A curved arrow drawn near the angle's vertex indicates both the direction and amount of rotation from the initial side to the terminal side. See Figure 4.1.

The Greek letters α (alpha), β (beta), γ (gamma), and θ (theta) are often used to name angles. If the rotation is counterclockwise, the result is a **positive angle**; if the rotation is clockwise, the result is a **negative angle**. See Figure 4.2. Rotation in both directions is unrestricted. Angles that have the same initial and terminal sides are called **coterminal** angles. In Figure 4.2(c), α and β are different angles but are coterminal.

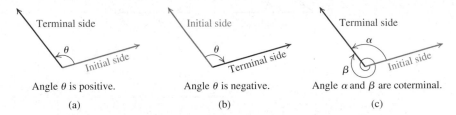

FIGURE 4.2 **Positive, negative, and coterminal angles**

An angle in a rectangular coordinate system is in **standard position** if its vertex is at the origin and its initial side lies on the positive x-axis. All of the angles in Figure 4.3 are in standard position. An angle in standard position is **quadrantal** if its terminal side lies on a coordinate axis; it is said to **lie in a quadrant** if its terminal side lies in that quadrant. See Figure 4.3.

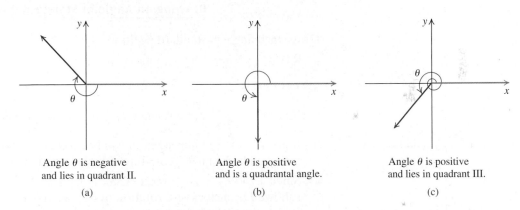

FIGURE 4.3 **Angles in standard position**

We measure angles by determining the amount of rotation from the initial side to the terminal side, indicated by a curved arrow that also shows direction. Two units of measurement for angles are *degrees* and *radians*.

2 Use degree and radian measure.

Measuring Angles by Using Degrees

A measure of **one degree** (denoted by 1°) is assigned to an angle resulting from a rotation $\dfrac{1}{360}$ of a complete revolution counterclockwise about the vertex. An angle formed by rotating the initial side counterclockwise one full rotation so that the terminal and initial sides coincide has measure 360 degrees, written as 360°.

Angles are classified according to their measures. As illustrated in Figure 4.4, an **acute angle** has measure between 0° and 90°; a **right angle** has measure 90°, or one-fourth of a revolution; an **obtuse angle** has measure between 90° and 180°; and a **straight angle** has measure 180°, or half a revolution. Figure 4.4 also shows several angles measured in degrees.

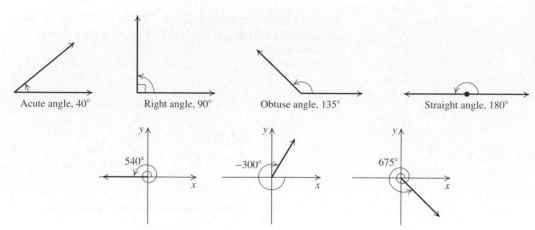

FIGURE 4.4 **Degree measure of various angles**

The measure of an angle has no numerical limit because the terminal side can be rotated without limitation.

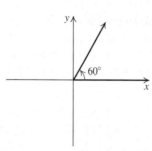

FIGURE 4.5

EXAMPLE 1   **Drawing an Angle in Standard Position**

Draw each angle in standard position.

a. 60°   **b.** 135°   **c.** −240°   **d.** 405°

SOLUTION

a. Since $60 = \frac{2}{3}(90)$, a 60° angle is $\frac{2}{3}$ of a 90° angle. See Figure 4.5.

FIGURE 4.6

b. Since $135 = 90 + 45$, a 135° angle is a counterclockwise rotation of 90°, followed by half a 90° counterclockwise rotation. See Figure 4.6.

c. Since $-240 = -180 - 60$, a −240° angle is a clockwise rotation of 180°, followed by a clockwise rotation of 60°. See Figure 4.7.

d. Since $405 = 360 + 45$, a 405° angle is one complete counterclockwise rotation, followed by half a 90° counterclockwise rotation. See Figure 4.8.   ■ ■ ■

Practice Problem 1   Draw a 225° angle in standard position.   ■

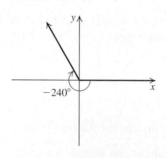

FIGURE 4.7

Today it is common to divide degrees into fractional parts using *decimal degree notation* such as 30.5°. Traditionally, however, degrees were expressed in terms of *minutes* and *seconds*. One **minute**, denoted by 1′, is defined as $\frac{1}{60}(1°)$, and one **second**, denoted 1″, is defined as $\frac{1}{3600}(1°)$. An angle measuring 27 degrees, 14 minutes, and 39 seconds is written as 27°14′39″. Since $\frac{1}{3600} = \frac{1}{60} \cdot \frac{1}{60}$, we have

$$1'' = \frac{1}{3600}(1°) = \frac{1}{60}\left[\frac{1}{60}(1°)\right] = \frac{1}{60}(1').$$

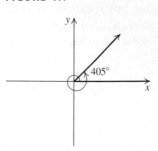

FIGURE 4.8

RELATIONSHIP BETWEEN DEGREES, MINUTES, AND SECONDS		
$1' = \frac{1}{60}(1°) = \left(\frac{1}{60}\right)°$	$1° = 60'$	$1° = (3600)''$
$1'' = \frac{1}{3600}(1°) = \left(\frac{1}{3600}\right)°$	$1'' = \frac{1}{60}(1')$	$1' = 60''$

TECHNOLOGY CONNECTION

Many calculators can perform conversions between angle measurements in decimal degree form and in degrees, minutes, and seconds (DMS) form. From the ANGLE menu you get the following:

```
24°8'15"
        24.1375
67.526°▸DMS
      67°31'33.6"
```

Let's use these relationships to convert between degree, minute, second (DMS) notation, and decimal degree notation. For example,

$$30' = 30 \cdot 1' = 30 \cdot \frac{1}{60}(1°) = 0.5(1°) = 0.5°.$$

So

$$18°30' = 18° + 30' = 18° + 0.5° = 18.5°.$$

Since

$$0.2° = \frac{2}{10}(1°) = \frac{2}{10}(60') = 12',$$

then

$$39.2° = 39° + 0.2° = 39° + 12' = 39°12'.$$

EXAMPLE 2   Converting Between DMS and Decimal Notation

a. Convert $24°8'15''$ to decimal degree notation, rounded to two decimal places.

b. Convert $67.526°$ to DMS notation, rounded to the nearest second.

SOLUTION

a. $24°8'15'' = 24° + 8 \cdot 1' + 15 \cdot 1''$ Rewrite.

$= 24° + 8\left(\frac{1}{60}\right)° + 15\left(\frac{1}{3600}\right)°$ $1' = \left(\frac{1}{60}\right)°, 1'' = \left(\frac{1}{3600}\right)°$

$\approx 24.14°$ Use a calculator.

b. $67.526° = 67° + 0.526 \cdot 1°$ Rewrite.

$= 67° + 0.526(60')$ $1° = 60'$

$= 67° + 31.56'$ Use a calculator.

$= 67° + 31' + 0.56 \cdot 1'$ Rewrite.

$= 67° + 31' + 0.56(60'')$ $1' = 60''$

$= 67° + 31' + 33.6'' \approx 67°31'34''$ Use a calculator and round to nearest second. ■ ■ ■

Practice Problem 2

a. Convert $13°9'22''$ to decimal degree notation, rounded to two decimal places.

b. Convert $41.275°$ to DMS notation, rounded to the nearest second. ■

Radian Measure

An angle whose vertex is at the center of a circle is called a **central angle**. A central angle intercepts the arc of the circle from the initial side to the terminal side. A positive central angle that intercepts an arc of the circle of length equal to the radius of the circle is said to have measure **1 radian**. See Figure 4.9.

Because the circumference of a circle of radius r is $2\pi r \approx 6.28r$, six arcs of length r can be consecutively marked off on a circle of radius r, leaving only a small portion of the circumference uncovered. See Figure 4.10.

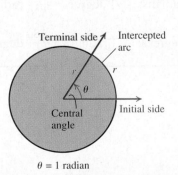

$\theta = 1$ radian

FIGURE 4.9 One radian

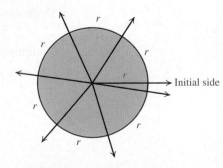

Circumference $= 2\pi r \approx 6.28r$

FIGURE 4.10

Each of the positive central angles in Figure 4.10 that intercepts an arc of length r has measure 1 radian; so the measure of one complete revolution is a little more than 6 radians. In this book, we assume (unless otherwise specified) that all central angles are positive. From geometry, we know that the measure of a central angle is proportional to the length of the arc it intercepts. This fact leads to the following relation.

RADIAN MEASURE OF A CENTRAL ANGLE

The radian measure θ of a central angle that intercepts an arc of length s on a circle of radius r is given by

$$\theta = \frac{s}{r}\text{ radians.}$$

Notice that when $s = r$, we have $\theta = \dfrac{r}{r} = 1$ radian.

3 Convert between degree and radian measure.

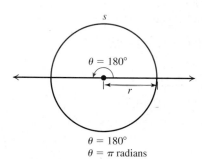

$\theta = 180°$

$\theta = \pi$ radians

FIGURE 4.11

Relationship Between Degrees and Radians

To see the relationship between degrees and radians, consider the central angle in Figure 4.11 with measure $\theta = 180°$ in a circle of radius r. The length of the arc, s, that is intercepted by this angle is half the circumference of the circle. So

$$s = \frac{1}{2}(\text{circumference})$$

$$s = \frac{1}{2}(2\pi r) = \pi r \qquad \text{Circumference} = 2\pi r$$

We can now find the radian measure for $\theta = 180°$.

$$180° = \theta = \frac{s}{r}\text{ radians} \qquad \text{Radian measure of a central angle}$$

$$= \frac{\pi r}{r}\text{ radians} \qquad \text{Replace } s \text{ with } \pi r.$$

$$= \pi\text{ radians} \qquad \text{Simplify.}$$

We have the relationship $\boxed{180° = \pi \text{ radians}}$. Dividing both sides of this equation by 180 or π gives the conversion formulas between degrees and radians.

CONVERTING BETWEEN DEGREES AND RADIANS

Degrees to radians: 1 degree $= \dfrac{\pi}{180}$ radian $\theta° = \theta\left(\dfrac{\pi}{180}\right)$ radians

Radians to degrees: 1 radian $= \dfrac{180}{\pi}$ degrees θ radians $= \theta\left(\dfrac{180}{\pi}\right)$ degrees

---

TECHNOLOGY CONNECTION

A graphing calculator can convert degrees to radians from the ANGLE menu. The calculator should be set in Radian mode.

```
180°
30°      3.141592654
55°      .5235987756
         .9599310886
```

EXAMPLE 3   Converting from Degrees to Radians

Convert each angle from degrees to radians.

a. 30°   **b.** 90°   **c.** −225°   **d.** 55°

SOLUTION

To convert degrees to radians, multiply degrees by $\dfrac{\pi}{180°}$.

a. $30° = 30° \cdot \dfrac{\pi}{180°} = \dfrac{30\pi}{180} = \dfrac{\pi}{6}$ radian

b. $90° = 90° \cdot \dfrac{\pi}{180°} = \dfrac{90\pi}{180} = \dfrac{\pi}{2}$ radians

c. $-225° = -225° \cdot \dfrac{\pi}{180°} = \dfrac{-225\pi}{180} = -\dfrac{5\pi}{4}$ radians

d. $55° = 55° \cdot \dfrac{\pi}{180°} = \dfrac{55\pi}{180} \approx 0.96$ radian   ■ ■ ■

Practice Problem 3   Convert −45° to radians.   ■

If an angle is a fraction of a complete rotation, we usually write it in radian measure as a fractional multiple of π, rather than as a decimal. For example, we write 30° as $\dfrac{\pi}{6}$ radian, which is the exact value, instead of writing the approximation 30° ≈ 0.52 radian.

EXAMPLE 4   Converting from Radians to Degrees

Convert each angle in radians to degrees.

a. $\dfrac{\pi}{3}$ radians   **b.** $-\dfrac{3\pi}{4}$ radians   **c.** 1 radian

SOLUTION

To convert radians to degrees, multiply radians by $\dfrac{180°}{\pi}$.

a. $\dfrac{\pi}{3}$ radians $= \dfrac{\pi}{3} \cdot \dfrac{180°}{\pi} = \left(\dfrac{180}{3}\right)° = 60°$

b. $-\dfrac{3\pi}{4}$ radians $= -\dfrac{3\pi}{4} \cdot \dfrac{180°}{\pi} = \left(-\dfrac{3}{4}\right)180° = -135°$

c. 1 radian $= 1 \cdot \dfrac{180°}{\pi} \approx 57.3°$ ■ ■ ■

Practice Problem 4   Convert $\dfrac{3\pi}{2}$ radians to degrees.   ■

TECHNOLOGY CONNECTION

A graphing calculator can convert radians to degrees from the ANGLE menu. The calculator should be set in degree mode.

```
(π/3)ʳ
              60
(-3π/4)ʳ
             -135
1ʳ
       57.29577951
```

Figure 4.12 includes the degree and radian measures of some commonly used angles. With practice, you should be able to make these conversions without using the figure.

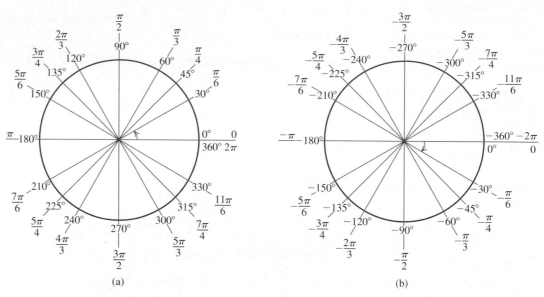

(a)   (b)

FIGURE 4.12 Frequently used angles

4 Find complements and supplements.

Complements and Supplements

Two positive angles are **complements** (or **complementary angles**) if the sum of their measures is 90°. So, two angles with measures 50° and 40° are complements because 50° + 40° = 90°. Each angle is the complement of the other. Two positive angles are **supplements** (or **supplementary angles**) if the sum of their measures is 180°. So, two angles with measures 120° and 60° are supplements because 120° + 60° = 180°. Each angle is the supplement of the other.

STUDY TIP

A positive angle with measure ≥ 90° does not have a complement. A positive angle with measure ≥ 180° does not have a supplement.

EXAMPLE 5   **Finding Complements and Supplements**

Find the complement and the supplement of the given angle or explain why the angle has no complement or supplement.

a. 73°   **b.** 110°

SOLUTION

a. If θ represents the complement of 73°, then $\theta + 73° = 90°$; so $\theta = 90° - 73° = 17°$. The complement of 73° is 17°.
If α represents the supplement for 73°, then $\alpha + 73° = 180°$; so $\alpha = 180° - 73° = 107°$. The supplement of 73° is 107°.

b. There is no complement of a 110° angle because 110° > 90°.
If β represents the supplement of 110°, then $\beta = 180° - 110° = 70°$. The supplement of 110° is 70°. ■ ■ ■

Practice Problem 5   Find the complement and the supplement for 67° or explain why 67° does not have a complement or a supplement. ■

With radian measure, two positive angles are complements when the sum of their measures is $\dfrac{\pi}{2}$ radians and are supplements when that sum is π radians.

5 Find the length of an arc of a circle.

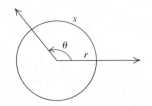

FIGURE 4.13

Length of an Arc of a Circle

Recall that the formula for the radian measure θ of a central angle that intercepts an arc of length s is $\theta = \dfrac{s}{r}$. See Figure 4.13. Multiplying both sides of this equation by r yields $s = r\theta$, a formula for the length s of the arc intercepted by a central angle with radian measure θ in a circle of radius r.

> **ARC LENGTH FORMULA**
>
> $$s = r\theta$$

EXAMPLE 6   Finding Arc Length of a Circle

A circle has a radius of 18 inches. Find the length of the arc intercepted by a central angle with measure $210°$.

SOLUTION

We must first convert the central angle measure from degrees to radians.

$$s = r\theta \qquad \text{Arc length formula}$$

$$s = 18\left(\frac{7\pi}{6}\right) \qquad\qquad \theta = 210° = 210°\left(\frac{\pi}{180°}\right) = \frac{7\pi}{6} \text{ radians}$$

$$= 21\pi \approx 65.97 \text{ inches} \qquad \text{Simplify and use a calculator.} \qquad ■■■$$

Practice Problem 6   A circle has a radius of 2 meters. Find the length of the arc intercepted by a central angle with measure $225°$.   ■

EXAMPLE 7   Finding the Distance Between Cities

We discussed longitude and latitude in the introduction to this section. We determine the latitude of a location L by first finding the point of intersection, P, between the meridian through L and the equator. The latitude of L is the angle formed by rays drawn from the center of Earth to points L and P, with the ray through P being the initial ray. See Figure 4.14(a).

Billings, Montana, is due north of Grand Junction, Colorado. Find the distance between Billings (latitude $45°48'$ N) and Grand Junction (latitude $39°7'$ N). Use 3960 miles as the radius of Earth. The N in $45°48'$ N means that the location is north of the equator.

SOLUTION

Because Billings is due north of Grand Junction, the same meridian passes through both cities. The distance between the cities is the length of the arc, s, on this meridian intercepted by the central angle, θ, that is the difference in their latitudes. See Figure 4.14(b).

The measure of angle θ is $45°48' - 39°7' = 6°41'$. To use the arc length formula, we must convert this angle to radians.

$$\theta = 6°41' \approx 6.6833° = 6.6833\left(\frac{\pi}{180}\right) \text{ radian} \approx 0.117 \text{ radian}$$

We use this value of θ and $r = 3960$ miles in the arc length formula:

$$s \approx (3960)(0.117) \text{ miles} \approx 463 \text{ miles} \qquad s = r\theta$$

The distance between Billings and Grand Junction is about 463 miles.   ■■■

The latitude of L

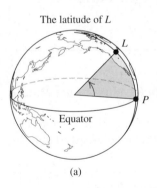

(a)

Distance on a meridian

(b)

FIGURE 4.14 Distance between cities

Practice Problem 7   Chicago, Illinois, is due north of Pensacola, Florida. Find the distance between Chicago (latitude $41°51'$ N) and Pensacola (latitude $30°25'$ N). Use $r = 3960$ miles.   ■

6 Compute linear and angular speed.

Linear and Angular Speed

If a skater skates 100 feet in 5 seconds, then her *average speed, v,* is given by

$$v = \frac{\text{distance}}{\text{time}} = \frac{100 \text{ feet}}{5 \text{ seconds}} = 20 \text{ feet per second.}$$

Suppose now that the skater is skating around the circumference of a circular skating rink with radius 40 feet. See Figure 4.15. In five seconds, the skater travels an arc of length 100 feet and moves through an angle θ, where

$$\theta = \frac{s}{r} = \frac{100}{40} = 2.5 \text{ radians.}$$

We say that the skater is traveling at an (average) *angular speed* of ω (omega), given by $\omega = \frac{\theta}{t} = \frac{2.5}{5} = 0.5$ radian per second. We call the average speed, 20 feet per second, the (average) *linear speed* to distinguish it from the (average) *angular speed*, 0.5 radian per second. Notice that $v = 20 = 40(0.5) = r\omega$. We define these ideas more generally as follows.

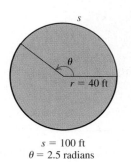

$s = 100 \text{ ft}$
$\theta = 2.5 \text{ radians}$

FIGURE 4.15

LINEAR AND ANGULAR SPEED

Suppose an object travels around a circle of radius r. If the object travels through a central angle of θ radians and an arc of length s, in time t, then

1. $v = \dfrac{s}{t}$ is the (average) **linear speed** of the object.

2. $\omega = \dfrac{\theta}{t}$ is the (average) **angular speed** of the object.

Further, since $s = r\theta$, by replacing s with $r\theta$ in **1** and then using **2** to replace $\dfrac{\theta}{t}$ with ω, we have

3. $v = r\omega$.

EXAMPLE 8   **Finding Angular and Linear Speed from Revolutions per Minute**

A model plane is attached to a swivel so that it flies in a circular path at the end of a 12-foot wire at the rate of 15 revolutions per minute. Find the angular speed and the linear speed of the plane.

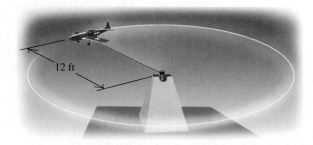

12 ft

SOLUTION

Because angular speed is measured in radians per unit of time, we must convert revolutions per minute into radians per minute. Recall that

$$1 \text{ revolution} = 2\pi \text{ radians.}$$

Then 15 revolutions $= 15 \cdot 2\pi = 30\pi$ radians. So, the angular speed ω is 30π radians per minute. Linear speed is given by

$$v = r\omega$$

$\qquad = 12 \cdot 30\pi$ feet per minute   Replace r with 12 and ω with 30π radians per minute.

$\qquad \approx 1131$ feet per minute   Use a calculator. ■ ■ ■

Practice Problem 8   Find the angular speed and the linear speed of the plane in Example 8 assuming that the wire is 10 feet long and the plane flies at a rate of 18 revolutions per minute. ■

7 Find the area of a sector.

Area of a Sector

A **sector** of a circle is a region bounded by the two sides of a central angle and the intercepted arc. Suppose θ is the radian measure of a central angle of a circle of radius r. See Figure 4.16. To find a formula for the area A of the sector formed by θ, we use the proportion

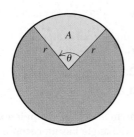

FIGURE 4.16 A sector of a circle

$$\frac{\text{area of a sector}}{\text{area of a circle}} = \frac{\text{length of the intercepted arc}}{\text{circumference of the circle}}$$ Fact from geometry

$$\frac{A}{\pi r^2} = \frac{r\theta}{2\pi r}$$ Area of a circle $= \pi r^2$, circumference $= 2\pi r$ Arc length formula $s = r\theta$

$$A = \pi r^2 \left(\frac{r\theta}{2\pi r} \right)$$ Multiply both sides by πr^2.

$$= \frac{1}{2} r^2 \theta$$ Simplify.

AREA OF A SECTOR

The area A of a sector of a circle of radius r formed by a central angle with radian measure θ is

$$A = \frac{1}{2} r^2 \theta.$$

EXAMPLE 9   **Finding the Area of a Sector of a Circle**

How many square inches of pizza have you eaten (rounded to the nearest square inch) if you eat a sector of an 18-inch diameter pizza whose edges form a 30° angle? See Figure 4.17.

SOLUTION

We must convert 30° to radians to use the sector area formula. The pizza's radius is

FIGURE 4.17 A slice of pizza

half its diameter, or 9 inches, and $30° = \dfrac{\pi}{6}$ radian.

$$A = \frac{1}{2} r^2 \theta$$ Area of a sector formula

$$= \frac{1}{2} 9^2 \left(\frac{\pi}{6} \right)$$ Replace r with 9 and θ with $\dfrac{\pi}{6}$.

$$\approx 21 \text{ square inches}$$ Use a calculator.

You have eaten about 21 square inches of pizza. ■ ■ ■

Practice Problem 9   Find the area of a sector of the circle of radius 10 inches formed by an angle of 60°. Round the answer to two decimal places. ■

SECTION 4.1 ■ Exercises

A EXERCISES Basic Skills and Concepts

1. A negative angle is formed by rotating the initial side in the __clockwise__ direction.

2. An angle is in standard position if its vertex is at the origin of a coordinate system and its initial side lies on the __positive x-axis__.

3. One second (1″) is __one-sixtieth__ of a minute.

4. *True or False* One radian is smaller than one degree. False

5. *True or False* The circumference of a circle with radius 1 meter is approximately 6.28 meters. True

6. *True of False* Every acute angle has a complement and a supplement. True

In Exercises 7–14, draw each angle in standard position.

7. $30°$ †
8. $150°$ †
9. $-120°$ †
10. $-330°$ †
11. $\dfrac{5\pi}{3}$ †
12. $\dfrac{11\pi}{6}$ †
13. $-\dfrac{4\pi}{3}$ †
14. $-\dfrac{13\pi}{4}$ †

In Exercises 15–20, convert each angle to decimal degree notation. Round your answers to two decimal places.

15. $70°45'$ 70.75°
16. $38°38'$ 38.63°
17. $23°42'30''$ 23.71°
18. $45°50'50''$ 45.85°
19. $-15°42'57''$ −15.72°
20. $-70°18'13''$ −70.30°

In Exercises 21–26, convert each angle to DMS notation. Round your answers to the nearest second.

21. $27.32°$ 27°19'12″
22. $120.64°$ 120°38'24″
23. $13.347°$ 13°20'49″
24. $110.433°$ 110°25'59″
25. $19.0511°$ 19°3'4″
26. $82.7272°$ 82°43'38″

In Exercises 27–36, convert each angle from degrees to radians. Express each answer as a multiple of π.

27. $45°$ $\dfrac{\pi}{4}$
28. $60°$ $\dfrac{\pi}{3}$
29. $-180°$ $-\pi$
30. $-210°$ $-\dfrac{7\pi}{6}$
31. $315°$ $\dfrac{7\pi}{4}$
32. $330°$ $\dfrac{11\pi}{6}$
33. $480°$
34. $450°$
35. $-510°$ $-\dfrac{17\pi}{6}$
36. $-420°$ $-\dfrac{7\pi}{3}$

In Exercises 37–46, convert each angle from radians to degrees.

37. $\dfrac{\pi}{2}$ 90°
38. $\dfrac{3\pi}{4}$ 135°
39. $-\dfrac{5\pi}{4}$ −225°
40. $-\dfrac{3\pi}{2}$ −270°
41. $\dfrac{5\pi}{3}$ 300°
42. $\dfrac{11\pi}{6}$ 330°
43. $\dfrac{5\pi}{2}$ 450°
44. $\dfrac{17\pi}{6}$ 510°
45. $-\dfrac{11\pi}{4}$ −495°
46. $-\dfrac{7\pi}{3}$ −420°

In Exercises 47–50, convert each angle from degrees to radians. Round your answers to two decimal places.

47. $12°$ 0.21 radian
48. $127°$ 2.22 radians
49. $-84°$ −1.47 radians
50. $-175°$ −3.05 radians

In Exercises 51–54, convert each angle from radians to degrees. Round your answers to two decimal places.

51. 0.94 53.86°
52. 5 286.48°
53. −8.21 −470.40°
54. −6.28 −359.82°

In Exercises 55–60, find the complement and the supplement of the given angle or explain why the angle has no complement or supplement.

55. $47°$
56. $75°$
57. $120°$
58. $160°$ †
59. $210°$ †
60. $-50°$ †

In Exercises 61–76, use the following notations:
θ = central angle of a circle, r = radius of a circle,
s = length of the intercepted arc, v = linear velocity,
ω = angular velocity, A = area of the sector of a circle, and
t = time.

In each case, find the missing quantity. Round your answers to three decimal places.

61. $r = 25$ inches, $s = 7$ inches, $\theta = ?$ 0.28 radian
62. $r = 5$ feet, $s = 6$ feet, $\theta = ?$ 1.2 radians
63. $r = 10.5$ centimeters, $s = 22$ centimeters, $\theta = ?$ 2.095 radians
64. $r = 60$ meters, $s = 120$ meters, $\theta = ?$ 2 radians
65. $r = 3$ m, $\theta = 25°$, $s = ?$ 1.309 m
66. $r = 0.7$ m, $\theta = 357°$, $s = ?$ 4.362 m
67. $r = 6.5$ m, $\theta = 12$ radians, $s = ?$ 78 m
68. $r = 6$ m, $\theta = \dfrac{\pi}{6}$ radians, $s = ?$ 3.142 m
69. $r = 6$ m, $\omega = 10$ rad/min, $v = ?$ 60 m/min
70. $r = 3.2$ feet, $\omega = 5$ rad/sec, $v = ?$ 16 ft/sec
71. $v = 20$ feet/sec, $r = 10$ feet, $\omega = ?$ 2 rad/sec
72. $v = 10$ m/min, $r = 6$ m, $\omega = ?$ $\dfrac{5}{3}$ rad/min
73. $r = 10$ in., $\theta = 4$ rad, $A = ?$ 200 in²
74. $r = 1.5$ feet, $\theta = 60°$, $A = ?$ 1.178 ft²

†Due to space constrictions, answers to these exercises may be found in the Answers beginning on page A–1 in the back of the book.

Answers:

33. $\dfrac{8\pi}{3}$ 34. $\dfrac{5\pi}{2}$

55. Complement: 43°; supplement: 133°
56. Complement: 15°; supplement: 105°
57. Complement: none because the measure of the angle is greater than 90°; supplement: 60°

75. $A = 20 \text{ ft}^2, \theta = 60°, r = ?$ 6.180 ft

76. $A = 60 \text{ m}^2, \theta = 2 \text{ rad}, r = ?$ 7.746 m

---

B EXERCISES Applying the Concepts

77. Wheel ratios. An automobile is pushed so that its wheels turn three-quarters of a revolution. If the tires have a radius of 15 inches, how many inches does the car move? 70.69 inches

78. Nautical miles. A nautical mile is the length of an arc of the equator intercepted by a central angle of $1'$. Using 3960 miles for the value of the radius of the Earth, express 1 nautical mile in terms of standard (**statute**) miles. 1.15 miles

79. Angles on a clock. What is the radian measure of the smaller central angle made by the hands of a clock at 4:00? Express your answer as a rational multiple of π. $\dfrac{2\pi}{3}$ radians

80. Angles on a clock. What is the radian measure of the larger central angle made by the hands of a clock at 7:00? Express your answer as a rational multiple of π. $\dfrac{7\pi}{6}$ radians

81. Diameter of a pizza. You are told that a slice of pizza whose edges form a 25° angle with an outer crust edge 4 inches long was found in a gym locker. What was the diameter of the original pizza? Round your answer to the nearest inch. 18 inches

82. Arc of a pendulum. How far does the tip of a 5-foot pendulum travel as it swings through an angle of 30°? Round your answer to two decimal places. 2.62 feet

83. Lifting a shark. A shark is suspended by a wire wound around a pulley of radius 4 inches. The shark must be raised 6 inches more to be completely off the ground. Through how many degrees must the pulley be rotated counterclockwise to accomplish the 6-inch lift? Round your answer to the nearest degree. 86°

84. Security cameras. A security camera rotates through an angle of 120°. To the nearest foot, what is the arc width of the field of view 40 feet from the camera? 84 feet

85. Pulley wheels. Two pulleys are connected by a belt so that when one pulley rotates, the linear speeds of the belt and both pulleys are the same. (See the figure.) The radius of the smaller pulley is 2 inches, and the radius of the larger pulley is 5 inches. A point on the belt travels at a rate of 600 inches per minute.
  a. Find the angular speed of the larger pulley.
  b. Find the angular speed of the smaller pulley.
  a. 120 radians per minute
  b. 300 radians per minute

86. Ferris wheel speed. A Ferris wheel in Vienna, Austria, has a diameter of approximately 61 meters. Assume that it takes 90 seconds for the Ferris wheel to make one complete revolution. $\omega = 0.07$ radian per second
  a. Find the angular speed of the Ferris wheel. Round your answer to two decimal places.
  b. Find the linear speed of the Ferris wheel in meters per second. Round your answer to two decimal places.
  $v = 2.13$ meters per second

87. Bicycle speed. A bicycle's wheels are 24 inches in diameter. If the bike is traveling at a rate of 25 miles per hour, find the angular speed of the wheels. Round your answer to two decimal places. 132,000 radians per hour

88. Bicycle speed. A bicycle's wheels are 30 inches in diameter. If the angular speed of the wheels is 11 radians per second, find the speed of the bicycle in inches per second. 165 inches per second

89. Hard disk speed. Hard disks are circular disks that store data in your computer. Suppose a circular hard disk is 3.75 inches in diameter. If the disk rotates 7200 revolutions per minute, what is the linear speed of a point on the edge of the disk (in inches per minute)? 84,823 inches per minute

In Exercises 90–93, the latitude of any location on Earth is the angle formed by the two rays drawn from the center of Earth to the location and to the equator. The ray through the location is the initial ray. Use 3960 miles as the radius of Earth.

90. Distance between cities. Indianapolis, Indiana, is due north of Montgomery, Alabama. Find the distance between Indianapolis (north latitude 39°44′ N) and Montgomery (latitude 32°23′ N). ≈ 508 miles

91. Distance between cities. Pittsburgh, Pennsylvania, is due north of Charleston, South Carolina. Find the distance between Pittsburgh (north latitude 40°30′ N) and Charleston (latitude 32°54′ N). ≈ 525 miles

92. Distance between cities. Amsterdam, Netherlands, is due north of Lyon, France. Find the distance between Amsterdam (north latitude 52°23′ N) and Lyon (latitude 45°42′ N). ≈ 462 miles

93. Distance between cities. Adana, Turkey (north latitude 36°59′ N), is due north of Jerusalem, Israel (latitude 31°47′ N). Find the distance between Adana and Jerusalem. ≈ 359 miles

94. Difference in latitudes. Miles City, Montana, is 440 miles due north of Boulder, Colorado. Find the difference in the latitudes of these two cities. ≈ 6.366°

95. Difference in latitudes. Lincoln, Nebraska, is 554 miles due north of Dallas, Texas. Find the difference in the latitudes of these two cities. ≈ 8.016°

96. Linear speed at the equator. Earth rotates on an axis that goes through both the North and South Poles. It makes one complete revolution in 24 hours. Find the linear speed (in miles per hour) of a location on the equator. ≈ 1037 miles per hour

C EXERCISES Beyond the Basics

97. Use the fact that $\theta + \pi$ and $\theta - \pi$ are coterminal angles to prove that $\dfrac{5\pi}{3}$ and $-\dfrac{\pi}{3}$ are coterminal.

98. Show that if α is the complement of θ, then $\alpha + \dfrac{\theta}{2}$ is the complement of $\dfrac{\theta}{2}$.

99. Find the smallest positive integer n so that θ and $\theta - \dfrac{n\pi}{2}$ are coterminal. 4

100. Find the smallest positive integer n so that θ and $\theta + \dfrac{n\pi}{3}$ are coterminal. 6

101. Show that if the linear speed of a point on a rotating circular disk of radius r is v, then the linear speed of the midpoint on a radius is $\dfrac{1}{2}v$.

102. Show that if A is the area of a sector of a circle of radius r formed by a central angle with radian measure θ, then $4A$ is the area of a sector of a circle of radius $2r$ formed by a central angle with radian measure θ.

103. Show that if A is the area of a sector of a circle of radius r formed by a central angle with radian measure θ, then $\dfrac{1}{2}A$ is the area of a circle of radius r formed by a central angle of radian measure $\dfrac{1}{2}\theta$.

104. Pulley wheels. Two pulleys are connected by a belt (see figure) so that when one pulley rotates, the linear speeds of the belt and both pulleys are the same. Suppose the radius of the smaller pulley is r and the

radius of the larger pulley is r_1. If the angular speed of the smaller pulley is ω and the angular speed of the larger pulley is ω_1, show that $\dfrac{r}{\omega_1} = \dfrac{r_1}{\omega}$.

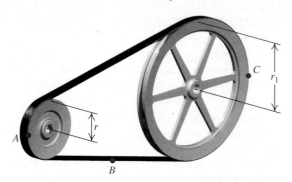

105. Pulley wheels. In Exercise 104, let $r = 6$ inches and $r_1 = 8$ inches. If the point B in the figure of Exercise 104 has a velocity of 88 feet per second, find
  a. the angular and the linear velocity of point A.
  b. the angular and the linear velocity of point C.

105. a. $\omega_A = 176$ radians per second; $v_A = 1056$ inches per second
  b. $\omega_C = 132$ radians per second; $v_C = 1056$ inches per second

Critical Thinking

106. Arrange the following radian measures from smallest to largest: $\pi, \dfrac{3\pi}{2}, 3, 4$. $3, \pi, 4, \dfrac{3\pi}{2}$

107. Which line of latitude has a smaller radius, the line of latitude through a city of latitude 30°40′13″ or the line of latitude through a city of latitude 40°30′13″? The line of latitude through 40°30′13″

108. Assume that θ is a central angle in a circle of radius r that intercepts an arc of length s. Show that if θ has radian measure less than $\dfrac{\pi}{2}$, then the central angle that intercepts an arc of length $\dfrac{\pi r}{2} - s$ and θ are complementary angles.

GROUP PROJECT

In about 250 B.C., the philosopher Eratosthenes estimated the radius of Earth based on what might appear to be scant information. He knew that the city of Syene (modern Aswan) is directly south of the city of Alexandria and that Syene and Alexandria are (in modern units) about 500 miles apart. He also knew that in Syene at noon on a midsummer day, an upright rod casts no shadow but makes a 7°12′ angle with a vertical rod at Alexandria and that then the sun's rays at Syene are effectively parallel to the sun's rays at Alexandria. By measuring the length of the shadow in Alexandria at noon on the summer solstice when there was no shadow in Syene, explain how Eratosthenes could measure the circumference of Earth.

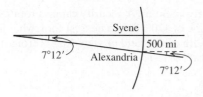

The Unit Circle; Trigonometric Functions of an Angle

AN EGYPTIAN ASTRONOMER; OR, HIPPARCHUS AT ALEXANDRIA.

Before Starting this Section, Review

1. Rationalizing a denominator (Appendix A, page 782)

2. Distance between two points (Section 1.1, page 5)

3. Unit circle (Section 1.1, page 12)

Objectives

1. Define the trigonometric functions using the unit circle.

2. Find exact trigonometric function values using a point on the unit circle.

3. Find trigonometric function values of quadrantal angles.

4. Find trigonometric function values of any angle.

5. Approximate trigonometric function values using a calculator.

MEASURING THE HEAVENS

Early astronomers used the lengths of chords intercepting the arcs of circles to compute distances that arose in astronomical problems. A major problem was to find the length of the chord intercepted by a given angle. For a circle of radius 1 unit, the length of the chord intercepted by an angle x (radians) is $2 \sin\left(\dfrac{x}{2}\right)$, where $\sin\left(\dfrac{x}{2}\right)$ refers to a function introduced in this section. (See Example 5.) In about 140 B.C., the first known table of chord lengths was produced by the Greek astronomer and mathematician Hipparchus. Although these tables are lost, it is claimed that Hipparchus wrote 12 books on tables of chords. For this reason, Hipparchus is sometimes called the founder of trigonometry. ■

---

1 Define the trigonometric functions using the unit circle.

RECALL

A line is *tangent* to a circle if it intersects the circle in exactly one point.

Trigonometric Functions and the Unit Circle

Recall from Section 1.1 that a circle with radius 1 centered at the origin of a rectangular coordinate system is a **unit circle**. Its equation is $x^2 + y^2 = 1$. With the aid of a unit circle, we can interpret length on a number line as the length of an arc and as the radian measure of the angle that intercepts the arc.

Recall from Section 4.1 that a positive central angle with measure θ radians in a circle of radius r intercepts an arc of length $s = r\theta$. In a unit circle, $r = 1$; so the length, s, of the intercepted arc is $s = 1 \cdot \theta$, or $s = \theta$. That is, the *radian measure* and the *arc length* of an arc intercepted by a central angle in a unit circle are identical. See Figure 4.18(a). Now see Figure 4.18(b), where the real number line is drawn tangent to the unit circle at the point $(1, 0)$. The origin for the number line is placed at $(1, 0)$. The unit length on the number line is the same as on the coordinate axes, and the positive numbers are above the x-axis.

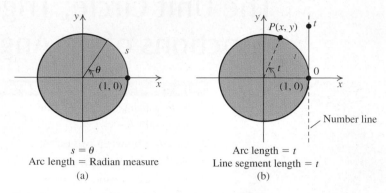

FIGURE 4.18 Relationships on a unit circle

For a real number t ($t \geq 0$) on the tangent number line, we "wrap" the segment from 0 to t *counterclockwise* around the unit circle, arriving at a point $P(x, y)$ that marks the endpoint of an arc of length t from $(1, 0)$ to P. In this way, each real number t ($t \geq 0$) corresponds to a unique point P on the unit circle. Because the circumference of the unit circle is 2π, if $t > 2\pi$, you must go around the circle more than once to arrive at the point P.

For $t < 0$, we "wrap" the segment from 0 to t *clockwise* around the circle, this time arriving at a point $P(x, y)$ that marks the endpoint of an arc of length $|t|$. (See Figure 4.19.) The negative angle having the x-axis for its initial side and the ray from the origin through the point $P(x, y)$ as its terminal side has radian measure t.

The correspondence between real numbers and endpoints of arcs on the unit circle is used to define six important functions, called the **trigonometric functions**. The full names of these functions are **sine**, **cosine**, **tangent**, **cosecant**, **secant**, and **cotangent**. The customary abbreviations for the values of these functions of a real number t are **sin t**, **cos t**, **tan t**, **csc t**, **sec t**, and **cot t**, respectively. Because the definitions of trigonometric functions are based on the unit circle, they are often called **circular functions**.

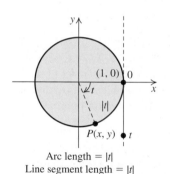

Arc length $= |t|$
Line segment length $= |t|$

FIGURE 4.19 Wrapping t, with $t < 0$

UNIT CIRCLE DEFINITIONS OF THE TRIGONOMETRIC FUNCTIONS OF REAL NUMBERS

Let t be any real number and let $P(x, y)$ be the point on the unit circle associated with t. Then

$$\sin t = y \qquad\qquad \csc t = \frac{1}{y} \ (y \neq 0)$$

$$\cos t = x \qquad\qquad \sec t = \frac{1}{x} \ (x \neq 0)$$

$$\tan t = \frac{y}{x} \ (x \neq 0) \qquad \cot t = \frac{x}{y} \ (y \neq 0)$$

Note that each function in the second column is the *reciprocal* of the corresponding function in the first column. That is,

$$\csc t = \frac{1}{y} = \frac{1}{\sin t} \qquad \sec t = \frac{1}{x} = \frac{1}{\cos t} \qquad \cot t = \frac{x}{y} = \frac{1}{\dfrac{y}{x}} = \frac{1}{\tan t}$$

From the definitions of the tangent and secant functions, we see that these functions are not defined when $x = 0$. Similarly, the cotangent and cosecant functions are not defined when $y = 0$.

A point P on the unit circle associated with a real number t has coordinates $(\cos t, \sin t)$ because $x = \cos t$ and $y = \sin t$. See Figure 4.20.

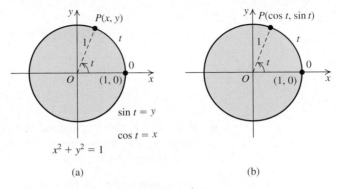

sin $t = y$

cos $t = x$

$x^2 + y^2 = 1$

(a)   (b)

FIGURE 4.20 Unit circle definitions of sin t and cos t

2 Find exact trigonometric function values using a point on the unit circle.

Finding Exact Trigonometric Function Values

EXAMPLE 1 Evaluating Trigonometric Functions

Find the values (if any) of the six trigonometric functions of each value of t.

a. $t = 0$ **b.** $t = \dfrac{\pi}{2}$ **c.** $t = \pi$ **d.** $t = \dfrac{3\pi}{2}$ **e.** $t = -3\pi$

SOLUTION

For each value of t, find the corresponding point $P(x, y)$ on the unit circle. (See Figure 4.21.) Then use the definitions of the trigonometric functions.

a. $t = 0$ corresponds to the point $(x, y) = (1, 0)$.

$$\sin 0 = y = 0 \qquad \csc 0 = \frac{1}{y} \text{ is undefined}$$

$$\cos 0 = x = 1 \qquad \sec 0 = \frac{1}{x} = \frac{1}{1} = 1$$

$$\tan 0 = \frac{y}{x} = \frac{0}{1} = 0 \qquad \cot 0 = \frac{x}{y} \text{ is undefined}$$

b. $t = \dfrac{\pi}{2}$ corresponds to the point $(x, y) = (0, 1)$.

$$\sin \frac{\pi}{2} = y = 1 \qquad \csc \frac{\pi}{2} = \frac{1}{y} = \frac{1}{1} = 1$$

$$\cos \frac{\pi}{2} = x = 0 \qquad \sec \frac{\pi}{2} = \frac{1}{x} \text{ is undefined}$$

$$\tan \frac{\pi}{2} = \frac{y}{x} \text{ is undefined} \qquad \cot \frac{\pi}{2} = \frac{y}{x} = \frac{0}{1} = 0$$

c. $t = \pi$ corresponds to the point $(x, y) = (-1, 0)$.

$$\sin \pi = y = 0 \qquad \csc \pi = \frac{1}{y} \text{ is undefined}$$

$$\cos \pi = x = -1 \qquad \sec \pi = \frac{1}{x} = \frac{1}{-1} = -1$$

$$\tan \pi = \frac{y}{x} = \frac{0}{-1} = 0 \qquad \cot \pi = \frac{x}{y} \text{ is undefined}$$

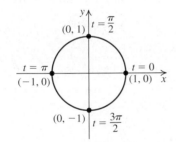

FIGURE 4.21 Points corresponding to various values of t

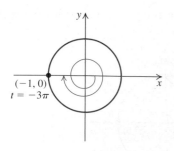

FIGURE 4.22

d. $t = \dfrac{3\pi}{2}$ corresponds to the point $(x, y) = (0, -1)$.

$$\sin\frac{3\pi}{2} = y = -1 \qquad\qquad \csc\frac{3\pi}{2} = \frac{1}{y} = \frac{1}{-1} = -1$$

$$\cos\frac{3\pi}{2} = x = 0 \qquad\qquad \sec\frac{3\pi}{2} = \frac{1}{x} \text{ is undefined}$$

$$\tan\frac{3\pi}{2} = \frac{x}{y} \text{ is undefined} \qquad \cos\frac{3\pi}{2} = \frac{x}{y} = \frac{0}{-1} = 0$$

e. If $|t| > 2\pi$, you go around the circle more than once to find the point corresponding to t. From Figure 4.22, we see that $t = -3\pi$ corresponds to exactly the same point, $(-1, 0)$, as does $t = \pi$.

$$\sin(-3\pi) = \sin\pi = 0 \qquad \cos(-3\pi) = \cos\pi = -1$$
$$\tan(-3\pi) = \tan\pi = 0 \qquad \csc(-3\pi) = \csc\pi \text{ is undefined}$$
$$\sec(-3\pi) = \sec\pi = -1 \qquad \cot(-3\pi) = \cot\pi \text{ is undefined} \qquad ■■■$$

Practice Problem 1 Repeat Example 1 for $t = \dfrac{5\pi}{2}$.   ■

RECALL

If a triangle has two equal angles, it is an isosceles triangle.

$\theta = \dfrac{\pi}{4}; x = \dfrac{\sqrt{2}}{2}; y = \dfrac{\sqrt{2}}{2}$

FIGURE 4.23 $P(x, y)$ for $\dfrac{\pi}{4}$

Trigonometric Function Values for $t = \dfrac{\pi}{4}$ Because arc length equals radian measure for (positive) central angles in the unit circle, the angle θ in Figure 4.23 has radian measure $\dfrac{\pi}{4}$, or 45°. Therefore, in the right triangle with sides of length x and y, both acute angles are 45° angles. So the triangle is isosceles; that is, $x = y$. We have:

$$x^2 + y^2 = 1 \qquad\qquad \text{Equation for the unit circle}$$
$$x^2 + x^2 = 1 \qquad\qquad y = x \text{ when } t = \frac{\pi}{4}$$
$$2x^2 = 1 \qquad\qquad \text{Simplify.}$$
$$x^2 = \frac{1}{2} \qquad\qquad \text{Divide both side by 2.}$$
$$x = \sqrt{\frac{1}{2}} = \frac{1}{\sqrt{2}} \cdot \frac{\sqrt{2}}{\sqrt{2}} = \frac{\sqrt{2}}{2} \qquad x \text{ is positive because } P(x, y) \text{ is in quadrant I.}$$

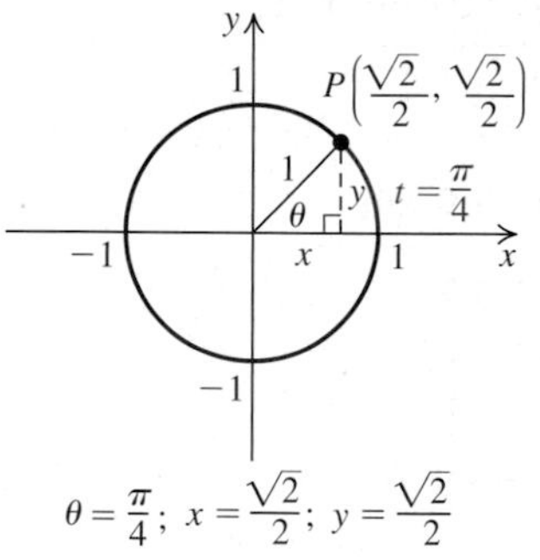

So, $y = x = \dfrac{\sqrt{2}}{2}$, when $t = \dfrac{\pi}{4}$.

Therefore, $(x, y) = \left(\dfrac{\sqrt{2}}{2}, \dfrac{\sqrt{2}}{2}\right)$ is the point on the unit circle associated with $t = \dfrac{\pi}{4}$. We have

$$\sin\frac{\pi}{4} = y = \frac{\sqrt{2}}{2} \quad \cos\frac{\pi}{4} = x = \frac{\sqrt{2}}{2} \quad \tan\frac{\pi}{4} = \frac{y}{x} = 1$$

$$\csc\frac{\pi}{4} = \frac{1}{y} = \sqrt{2} \quad \sec\frac{\pi}{4} = \frac{1}{x} = \sqrt{2} \quad \cot\frac{\pi}{4} = \frac{x}{y} = 1$$

3 Find trigonometric function values of quadrantal angles.

Trigonometric Functions of an Angle

Given an angle θ in standard position, let $P(x, y)$ be the point where the terminal ray of θ intersects the unit circle. See Figure 4.24.

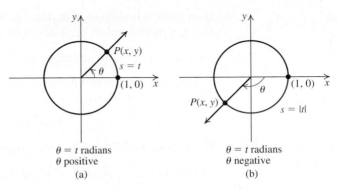

FIGURE 4.24

Because $r = 1$ on the unit circle, the arc length formula $s = r\theta$ becomes $s = t$, where θ is a positive angle with radian measure t. For negative angles, $s = |t|$. Previously, we defined the trigonometric functions for a real number t using the same point $P(x, y)$ that corresponds to an angle θ with radian measure t. We now define the trigonometric functions of an *angle*.

DEFINITION OF THE TRIGONOMETRIC FUNCTION OF AN ANGLE

If θ is an angle with radian measure t, then

$$\sin \theta = \sin t \qquad \cos \theta = \cos t \qquad \tan \theta = \tan t$$

$$\csc \theta = \csc t \qquad \sec \theta = \sec t \qquad \cot \theta = \cot t$$

If θ is given in degrees, convert θ to radians before using these equations.

EXAMPLE 2   Finding the Trigonometric Function Values of a Quadrantal Angle

Find the trigonometric function values of $90°$.

SOLUTION

Since $90° = \dfrac{\pi}{2}$ radians, we can use the results of Example 1 for $t = \dfrac{\pi}{2}$ and $\theta = 90°$.

$$\sin 90° = \sin\left(\frac{\pi}{2} \text{ radians}\right) = \sin\left(\frac{\pi}{2}\right) = 1$$

$$\cos 90° = \cos\left(\frac{\pi}{2} \text{ radians}\right) = \cos\left(\frac{\pi}{2}\right) = 0$$

$$\tan 90° = \tan\left(\frac{\pi}{2} \text{ radians}\right) = \tan\left(\frac{\pi}{2}\right) \text{ is undefined}$$

$$\csc 90° = \csc\left(\frac{\pi}{2} \text{ radians}\right) = \csc\left(\frac{\pi}{2}\right) = 1$$

$$\sec 90° = \sec\left(\frac{\pi}{2} \text{ radians}\right) = \sec\left(\frac{\pi}{2}\right) \text{ is undefined}$$

$$\cot 90° = \cot\left(\frac{\pi}{2} \text{ radians}\right) = \cot\left(\frac{\pi}{2}\right) = 0 \qquad\qquad ■■■$$

Practice Problem 2   Find the trigonometric function values for $-270°$.   ■

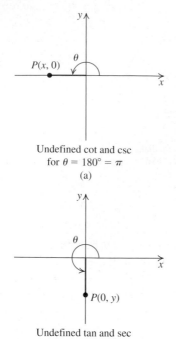

Undefined cot and csc
for $\theta = 180° = \pi$
(a)

Undefined tan and sec
for $\theta = 270° = \dfrac{3\pi}{2}$
(b)

FIGURE 4.25 Undefined
trigonometric function values

If an angle θ is measured in degrees, we use the degree symbol to indicate a trigono-metric function value, such as cos 60° and tan 28°. If θ is measured in radians, we do not use a symbol (for example, $\sin \dfrac{\pi}{4}$, sec π, and cos 16). Notice that the 0 coordinate of a point on the terminal side of a quadrantal angle cannot appear in a denominator. (See Figure 4.25(a), where the terminal side is on the x-axis.) The y-coordinate is 0; so the cotangent $\left(\cot \theta = \dfrac{x}{y} \right)$ and the cosecant $\left(\csc \theta = \dfrac{r}{y} \right)$ functions are undefined. If the terminal side is on the y-axis, as in Figure 4.25(b), then the x-coordinate is 0 and the tangent $\left(\tan \theta = \dfrac{y}{x} \right)$ and the secant $\left(\sec \theta = \dfrac{r}{x} \right)$ functions are undefined.

Table 4.1 summarizes some facts about quadrantal angles, all of which can be obtained as was done in Example 2. (See Figure 4.26.)

TABLE 4.1 Trigonometric Function Values of Quadrantal Angles

θ (degrees)	θ (radians)	$\sin \theta$	$\cos \theta$	$\tan \theta$	$\cot \theta$	$\sec \theta$	$\csc \theta$
0°	0	0	1	0	undefined	1	undefined
90°	$\dfrac{\pi}{2}$	1	0	undefined	0	undefined	1
180°	π	0	−1	0	undefined	−1	undefined
270°	$\dfrac{3\pi}{2}$	−1	0	undefined	0	undefined	−1
360°	2π	0	1	0	undefined	1	undefined

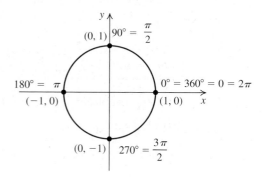

FIGURE 4.26 Quadrantal angles

4 Find trigonometric function values of any angle.

Trigonometric Function Values of an Angle θ

Suppose the terminal ray of an angle θ in standard position intersects the unit circle at the point $P_1(x_1, y_1)$, as illustrated for an acute angle θ in Figure 4.27(a). Let $P(x, y)$ be any point on the terminal side of θ and $r = \sqrt{x^2 + y^2}$, as shown in Figure 4.27(b). The triangles POQ and P_1OQ_1 are similar because they have equal angles; so the ratios of their corresponding sides are equal. We have

$$\sin \theta = y_1 = \frac{y_1}{1} = \frac{y}{r}, \quad \cos \theta = x_1 = \frac{x_1}{1} = \frac{x}{r}, \quad \tan \theta = \frac{y_1}{x_1} = \frac{y}{x},$$ and so on. We can now

evaluate the trigonometric functions for any angle θ using any point on its terminal side.

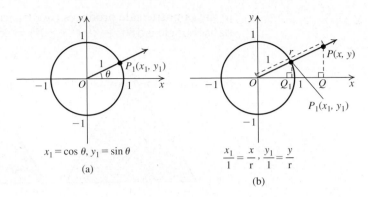

FIGURE 4.27

VALUES OF THE TRIGONOMETRIC FUNCTIONS OF AN ANGLE θ

Let $P(x, y)$ be any point on the terminal ray of an angle θ in standard position (other than the origin) and let $r = \sqrt{x^2 + y^2}$. Then $r > 0$, and

$$\sin \theta = \frac{y}{r} \qquad\qquad \csc \theta = \frac{r}{y} \, (y \neq 0)$$

$$\cos \theta = \frac{x}{r} \qquad\qquad \sec \theta = \frac{r}{x} \, (x \neq 0)$$

$$\tan \theta = \frac{y}{x} \, (x \neq 0) \qquad \cot \theta = \frac{x}{y} \, (y \neq 0)$$

EXAMPLE 3   **Finding Trigonometric Function Values**

Suppose θ is an angle whose terminal side contains the point $P(-1, 3)$. Find the exact values of the six trigonometric functions of θ.

SOLUTION

Because $x = -1$ and $y = 3$ (see Figure 4.28), we have

$$r = \sqrt{x^2 + y^2} \qquad\qquad \text{Definition of } r$$
$$= \sqrt{(-1)^2 + 3^2} = \sqrt{10} \qquad \text{Substitute and simplify.}$$

Replacing x with -1, y with 3, and r with $\sqrt{10}$ in the definition of the trigonometric functions, we have

$$\sin \theta = \frac{y}{r} = \frac{3}{\sqrt{10}} = \frac{3\sqrt{10}}{10} \qquad \csc \theta = \frac{r}{y} = \frac{\sqrt{10}}{3}$$

$$\cos \theta = \frac{x}{r} = \frac{-1}{\sqrt{10}} = -\frac{\sqrt{10}}{10} \qquad \sec \theta = \frac{r}{x} = \frac{\sqrt{10}}{-1} = -\sqrt{10}$$

$$\tan \theta = \frac{y}{x} = \frac{3}{-1} = -3 \qquad\qquad \cot \theta = \frac{x}{y} = \frac{-1}{3} = -\frac{1}{3} \qquad ■ ■ ■$$

Practice Problem 3   Suppose θ is an angle whose terminal side contains the point $P(2, -5)$. Find the exact values of the six trigonometric functions of θ. ■

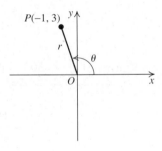

$P(-1, 3)$

FIGURE 4.28

Trigonometric Function Values for $\dfrac{\pi}{6} = 30°$ and $\dfrac{\pi}{3} = 60°$

Figure 4.29(a) shows an equilateral triangle with sides of length 1. Each angle, then, is $60° = \dfrac{\pi}{3}$. A segment from the vertex of one of the angles drawn perpendicular to the opposite side produces two congruent right triangles. The third angle in each right triangle is $180° - (90° + 60°) = 30°$. See Figure. 4.29(b).

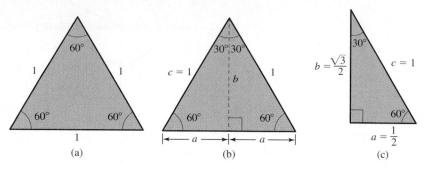

FIGURE 4.29 A 30°–60°–90° triangle

Because the corresponding legs of the right triangles in Figure 4.29(b) have the same length, a, the base length is $2a$. So $2a = 1$ and $a = \dfrac{1}{2}$. Then

$$a^2 + b^2 = c^2 \qquad \text{The Pythagorean Theorem}$$

$$\frac{1}{4} + b^2 = 1 \qquad a = \frac{1}{2}, c = 1$$

$$b^2 = 1 - \frac{1}{4} = \frac{3}{4}$$

$$b = \sqrt{\frac{3}{4}} = \frac{\sqrt{3}}{2} \qquad b > 0 \text{ because } b \text{ represents length.}$$

Figure 4.29(c) shows the resulting 30°–60°–90° triangle.

EXAMPLE 4 **Finding Exact Trigonometric Function Values of $\dfrac{\pi}{6} = 30°$**

Find the exact trigonometric function values of $\dfrac{\pi}{6} = 30°$.

SOLUTION

Arrange the 30°–60°–90° triangle in Figure 4.29(c) with the 30° angle in standard position. (See Figure 4.30.) The point $(x, y) = \left(\dfrac{\sqrt{3}}{2}, \dfrac{1}{2}\right)$ is on the terminal side of $\theta = \dfrac{\pi}{6} = 30°$. Then

$$\sin \frac{\pi}{6} = \sin 30° = \frac{y}{r} = \frac{1}{2} \qquad\qquad \csc \frac{\pi}{6} = \csc 30° = \frac{1}{\frac{1}{2}} = 2$$

$$\cos \frac{\pi}{6} = \cos 30° = \frac{x}{r} = \frac{\sqrt{3}}{2} \qquad \sec \frac{\pi}{6} = \sec 30° = \frac{1}{\frac{\sqrt{3}}{2}} = \frac{2}{\sqrt{3}} = \frac{2\sqrt{3}}{3}$$

$$\tan \frac{\pi}{6} = \tan 30° = \frac{y}{x} = \frac{\frac{1}{2}}{\frac{\sqrt{3}}{2}} = \frac{1}{\sqrt{3}} = \frac{\sqrt{3}}{3} \qquad \cot \frac{\pi}{6} = \cot 30° = \frac{\frac{\sqrt{3}}{2}}{\frac{1}{2}} = \sqrt{3} \quad ■■■$$

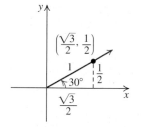

FIGURE 4.30 The sine and cosine of 30°

Practice Problem 4   Position the 30°–60°–90° triangle with the 60° angle in standard position to find the exact trigonometric function values of $\dfrac{\pi}{3} = 60°$.   ■

We summarize the trigonometric function values for $\dfrac{\pi}{6}, \dfrac{\pi}{4}$, and $\dfrac{\pi}{3}$.

Exact Trigonometric Function Values for $\dfrac{\pi}{6}, \dfrac{\pi}{4}$, and $\dfrac{\pi}{3}$							
θ (degrees)	θ (radians)	$\sin\theta$	$\cos\theta$	$\tan\theta$	$\csc\theta$	$\sec\theta$	$\cot\theta$
30°	$\dfrac{\pi}{6}$	$\dfrac{1}{2}$	$\dfrac{\sqrt{3}}{2}$	$\dfrac{\sqrt{3}}{3}$	2	$\dfrac{2\sqrt{3}}{3}$	$\sqrt{3}$
45°	$\dfrac{\pi}{4}$	$\dfrac{\sqrt{2}}{2}$	$\dfrac{\sqrt{2}}{2}$	1	$\sqrt{2}$	$\sqrt{2}$	1
60°	$\dfrac{\pi}{3}$	$\dfrac{\sqrt{3}}{2}$	$\dfrac{1}{2}$	$\sqrt{3}$	$\dfrac{2\sqrt{3}}{3}$	2	$\dfrac{\sqrt{3}}{3}$

EXAMPLE 5   **Finding Chord Length on the Unit Circle**

Find the length of the chord of the unit circle intercepted by an angle of θ radians.

SOLUTION

Let $P(x, y)$ be the point where the terminal ray of the angle $\dfrac{\theta}{2}$ intersects the unit circle. See Figure 4.31. Then $y = \sin\left(\dfrac{\theta}{2}\right)$ and $y = $ half of the length of the specified chord. Therefore, $2\sin\left(\dfrac{\theta}{2}\right) = 2y = $ length of the specified chord.   ■ ■ ■

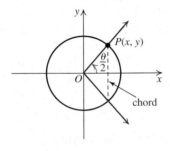

FIGURE 4.31

Practice Problem 5   Find the length of the chord of unit circle intercepted by an angle of 60°.   ■

Since the value of each trigonometric function of an angle in standard position is completely determined by the position of the terminal side, the following statements are true.

1. Coterminal angles are assigned identical values by the six trigonometric functions.

2. The signs of the values of the trigonometric functions are determined by the quadrant containing the terminal side.

3. For any integer n, θ and $\theta + n360°$ (in degree measure) are coterminal angles and θ and $\theta + 2\pi n$ (in radian measure) are coterminal angles.

Some frequently used consequences of these observations are given next.

TRIGONOMETRIC FUNCTION VALUES OF COTERMINAL ANGLES	
θ in degrees	**θ in radians**
$\sin\theta = \sin(\theta + n360°)$	$\sin\theta = \sin(\theta + 2\pi n)$
$\cos\theta = \cos(\theta + n360°)$	$\cos\theta = \cos(\theta + 2\pi n)$
These equations hold for any integer n.	

EXAMPLE 6 **Trigonometric Function Values of Coterminal Angles**

Find the exact values for

a. $\sin 2580°$ **b.** $\cos \dfrac{17\pi}{4}$.

SOLUTION

a. $2580° = 60° + 2520° = 60° + 7(360°)$; so $\sin 2580° = \sin 60° = \dfrac{\sqrt{3}}{2}$

b. $\dfrac{17\pi}{4} = \dfrac{\pi}{4} + \dfrac{16\pi}{4} = \dfrac{\pi}{4} + 2(2\pi)$; so $\cos \dfrac{17\pi}{4} = \cos \dfrac{\pi}{4} = \dfrac{\sqrt{2}}{2}$ ■ ■ ■

Practice Problem 6 Find the exact values for **a.** $\cos 1830°$ **b.** $\sin \dfrac{31\pi}{3}$ ■

5 Approximate trigonometric function values using a calculator.

Evaluating Trigonometric Functions Using a Calculator

When you are using a calculator to find the values of the trigonometric functions, the first step is to set the *mode* of measurement. If you are working with an angle given in degrees, set the mode to *degrees*; otherwise, set the mode to *radians*.

Your calculator has keys labeled $\boxed{\text{SIN}}$, $\boxed{\text{COS}}$, and $\boxed{\text{TAN}}$ but no keys for directly evaluating the cosecant, secant, or cotangent functions. However, these functions are the reciprocal functions of sine, cosine, and tangent functions, respectively. Therefore, to evaluate these functions, use the $\boxed{x^{-1}}$ key with the appropriate function.

EXAMPLE 7 **Approximating Trigonometric Function Values Using a Calculator**

Use a calculator to find the approximate value of each expression. Round your answers to two decimal places.

a. $\sin 71°$ **b.** $\tan \dfrac{5\pi}{7}$ **c.** $\sec 1.3$

SOLUTION

a. Set the MODE to degrees. $\sin 71° \approx 0.9455185756 \approx 0.95$

b. Set the MODE to radians. $\tan \dfrac{5\pi}{7} \approx -1.253960338 \approx -1.25$

c. Set the MODE to radians. $\sec 1.3 = \dfrac{1}{\cos 1.3} \approx 3.738334127 \approx 3.74$ ■ ■ ■

Practice Problem 7 Repeat Example 7 for each expression.

a. $\cos 114°$ **b.** $\cot 3.6$ ■

TECHNOLOGY CONNECTION

The first screen shows how $\sin 71°$ is displayed on a graphing calculator in degree mode.

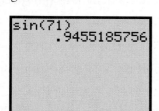

This screen shows how $\tan \dfrac{5\pi}{7}$ and $\sec 1.3$ are displayed on a graphing calculator in radian mode.

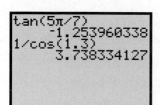

◆ **WARNING**   The calculator keys $\boxed{\text{SIN}^{-1}}$, $\boxed{\text{COS}^{-1}}$, and $\boxed{\text{TAN}^{-1}}$ do not represent the reciprocal functions for the sine, cosine, and tangent functions, respectively. The functions $\sin^{-1}$, $\cos^{-1}$, and $\tan^{-1}$ are discussed in Section 4.6.

A EXERCISES Basic Skills and Concepts

1. If $P(x, y) = \left(\dfrac{1}{3}, \dfrac{2\sqrt{2}}{3}\right)$ is the point on the unit circle corresponding to t, then $\cos t = $ ____$\dfrac{1}{3}$____.

2. The central angle with radian measure 3.8 intercepts an arc of length ____3.8____ feet on a circle with radius 1 foot.

3. If $P(x, y)$ is a point on the unit circle corresponding to a quadrantal angle, then either x or y equals ____0____.

4. *True or False* The trigonometric function values for $t = \pi$ are identical to those for $t = -\pi$. True

5. *True or False* If $(4, 3)$ is a point on the terminal side of θ, then $\cos \theta = 4$. False

6. *True or False* The arc of a circle with radius 1 inch that is intercepted by an angle with radian measure -3.2 has length 3.2 inches. True

In Exercises 7–14, $P(x, y)$ is the point on the unit circle that corresponds to the real number t. Find the exact values of the six trigonometric functions of t.

7. $\left(\dfrac{2\sqrt{2}}{3}, \dfrac{1}{3}\right)$ †

8. $\left(\dfrac{1}{2}, \dfrac{\sqrt{3}}{2}\right)$ †

9. $\left(-\dfrac{1}{3}, \dfrac{2\sqrt{2}}{3}\right)$ †

10. $\left(-\dfrac{\sqrt{3}}{2}, \dfrac{1}{2}\right)$ †

11. $\left(\dfrac{1}{5}, -\dfrac{2\sqrt{6}}{5}\right)$ †

12. $\left(\dfrac{1}{2}, -\dfrac{\sqrt{3}}{2}\right)$ †

13. $\left(-\dfrac{2\sqrt{2}}{3}, -\dfrac{1}{3}\right)$ †

14. $\left(-\dfrac{2\sqrt{6}}{5}, -\dfrac{1}{5}\right)$ †

In Exercises 15–24, find each exact value. Do not use a calculator.

15. $\cos(5\pi)$ -1

16. $\sin(3\pi)$ 0

17. $\tan(4\pi)$ 0

18. $\cot\left(\dfrac{7\pi}{2}\right)$ 0

19. $\csc\dfrac{5\pi}{2}$ 1

20. $\sec(7\pi)$ -1

21. $\sin(-2\pi)$ 0

22. $\cos(-5\pi)$ -1

23. $\cos\left(-\dfrac{3\pi}{2}\right)$ 0

24. $\sin\left(-\dfrac{\pi}{2}\right)$ -1

In Exercises 25–32, a point on the terminal side of an angle θ is given. Find the exact values of the six trigonometric functions of θ.

25. $(-4, 3)$ †

26. $(-3, 5)$ †

27. $(-\sqrt{3}, -1)$ †

28. $(-1, -2)$ †

29. $(3, 3)$ †

30. $(-2, -2)$ †

31. $(12, -5)$ †

32. $(7, -2)$ †

In Exercises 33–48, use a calculator to find the approximate value of each expression. Round your answers to two decimal places.

33. $\cos 14°$ 0.97

34. $\sin 20°$ 0.34

35. $\tan 32°$ 0.62

36. $\cot 67°$ 0.42

37. $\sec 34°$ 1.21

38. $\csc 72°$ 1.05

39. $\cos\dfrac{\pi}{5}$ 0.81

40. $\sin\dfrac{\pi}{7}$ 0.43

41. $\tan\dfrac{2\pi}{9}$ 0.84

42. $\cot\dfrac{3\pi}{7}$ 0.23

43. $\sec\dfrac{\pi}{6}$ 1.15

44. $\csc\dfrac{\pi}{8}$ 2.61

45. $\sin(-41°)$ -0.66

46. $\cos(-54°)$ 0.59

47. $\sin(-17.3)$ 1.00

48. $\cos(-38.6)$ 0.62

In Exercises 49–60, find the exact value of each expression. Do not use a calculator.

49. $\sin 45° + \cos 30°$ †

50. $\cos 60° + \sin 30°$ 1

51. $\sin 180° - \cos 90°$ 0

52. $\cos 180° - \sin 90°$ -2

53. $\sin 30° \sec 60°$ 1

54. $\sin 270° \csc 45°$ $-\sqrt{2}$

55. $\sin\dfrac{\pi}{4} - \cos \pi$ $1 + \dfrac{\sqrt{2}}{2}$

56. $\sec \pi + \sin\dfrac{\pi}{6}$ $-\dfrac{1}{2}$

57. $\tan\dfrac{\pi}{4} - \cot\dfrac{\pi}{3}$ $1 - \dfrac{\sqrt{3}}{3}$

58. $\csc\dfrac{\pi}{2} + \cos\dfrac{\pi}{3}$ $\dfrac{3}{2}$

59. $\sin\dfrac{3\pi}{2}\tan\dfrac{\pi}{4}$ -1

60. $\cos\dfrac{\pi}{2}\sec \pi$ 0

In Exercises 61–68, use the figure and a straightedge to approximate each trigonometric function value to the nearest tenth. (The grid lines are 0.1 unit apart.) Then use a calculator to find each value to the nearest hundredth.

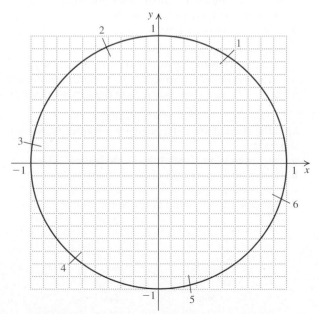

†Due to space constrictions, answers to these exercises may be found in the Answers beginning on page A–1 in the back of the book.

61. sin 0.5 0.48

62. cos 1 0.54

63. cos 2 −0.42

64. sin 2.5 0.60

65. sin 3.5 −0.35

66. cos 4 −0.65

67. cos 5 0.28

68. cos 5.5 0.71

74. Height of a kite. A kite is flying at the end of 100 feet of string that is taut. The height, h, of the kite is given by $h = 100 \sec \theta$, where θ is the angle the string makes with the ground. Find the kite's height if

a. $\theta = 30°$. 115.47 ft   **b.** $\theta = 60°$. 200 ft

B EXERCISES Applying the Concepts

69. Tide patterns. The depth of water, d feet, in a channel t hours after midnights is $d = 3 \cos\left(\dfrac{\pi}{6} t\right) + 10$. Find the channel depth at

a. 6 P.M. (low tide). 7   **b.** noon (high tide). 13

70. Light refraction. A fish swimming d feet below the surface is viewed from a line of sight that makes an angle of θ degrees with the vertical. Because the light is refracted by the water, the fish appears to be A feet below the surface where

$$A = \frac{3d \cos \theta}{\sqrt{7 + 9 \cos^2 \theta}}.$$

What is the apparent depth of a fish swimming 2 feet below the surface when viewed at an angle of 60° from the vertical? 0.99 foot

71. Biorhythms. Biorhythm patterns are based on the belief that an individual's emotional, intellectual, and physical states vary rhythmically over a 28-day cycle. The most positive score is 100. If your intellectual level t days after birth is given by $I(t) = 100 \sin\left(\dfrac{2\pi}{28} t\right)$, what is your intellectual level on your twenty-first birthday? (Count one year as 365 days.) −100

72. Pollution levels. On a typical day, a particular rural city's air pollution level (suspended particulate levels in mg/m³) is

$$P(t) = 0.5 + 17t - 0.7t^2 + 0.9 \sin(3t + 1),$$

where t is the number of hours after midnight.

a. Find the pollution level at noon. 103.12
b. Find the pollution level at 10 P.M. 34.93

73. Sound levels. The decibel level, D, of a sound directed eastward is measured 3 yards from the source. If the ray from the sound source through a point 3 yards away makes an angle θ with an eastward ray from the source point, then

$$D = 25 + 15 \cos \theta.$$

Find the decibel level at a point 3 yards from the sound source if

a. the point is due east of the sound source. 40 db
b. the point is due north of the sound source. 25 db
c. the point is due west of the sound source. 10 db
d. the point is due south of the sound source. 25 db

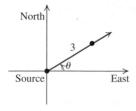

C EXERCISES Beyond the Basics

In Exercises 75–84, let $\alpha = 30°$, $\beta = 60°$, $f(\theta) = \tan \theta$, and $g(\theta) = \cot \theta$. Find the exact value of each expression. Do not use a calculator.

75. $f(\alpha)$ $\dfrac{\sqrt{3}}{3}$

76. $g(\alpha)$ $\sqrt{3}$

77. $[f(\alpha)]^2$ $\dfrac{1}{3}$

78. $g(3\alpha)$ 0

79. $3g(\beta)$ $\sqrt{3}$

80. $(fg)(\alpha)$ 1

81. $\left(\dfrac{f}{g}\right)(\beta)$ 3

82. $3f(\alpha) + g(\alpha)$ $2\sqrt{3}$

83. $f(\alpha) - g\left(\dfrac{\beta}{2}\right)$ $-\dfrac{2\sqrt{3}}{3}$

84. $f(2\alpha) - g(\beta)$ $\dfrac{2\sqrt{3}}{3}$

In Exercises 85–87, use the figure to find a triangle congruent to triangle *POM* that justifies the statements in each exercise.

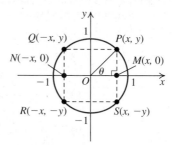

85. $\sin \theta = \sin(180° - \theta)$, and $\cos \theta = -\cos(180° - \theta)$.

86. $\sin \theta = -\sin(180° + \theta)$, and $\cos \theta = -\cos(180° + \theta)$.

87. $\sin \theta = -\sin(360° - \theta)$, and $\cos \theta = \cos(360° - \theta)$.

In Exercises 88–90, use the results of Exercises 85–87.

88. Find sin 135° and cos 135°. $\sin 135° = \dfrac{\sqrt{2}}{2}$; $\cos 135° = -\dfrac{\sqrt{2}}{2}$

89. Find sin 225° and cos 225°. $\sin 225° = -\dfrac{\sqrt{2}}{2}$; $\cos 225° = -\dfrac{\sqrt{2}}{2}$

90. Find sin 315° and cos 315°.

In Exercises 91–96, find all possible values of t that correspond to the given point $P(x, y)$ on the unit circle.

91. $P(-1, 0)$ †

92. $P(0, -1)$ †

93. $P\left(\dfrac{\sqrt{2}}{2}, \dfrac{-\sqrt{2}}{2}\right)$ †

94. $P\left(-\dfrac{\sqrt{2}}{2}, -\dfrac{\sqrt{2}}{2}\right)$ †

95. $P\left(-\dfrac{\sqrt{3}}{2}, \dfrac{1}{2}\right)$ †

96. $P\left(\dfrac{1}{2}, \dfrac{-\sqrt{3}}{2}\right)$ †

Critical Thinking

97. What is the smallest positive value for $\sec \theta$? 1

98. What is the largest negative value for $\csc \theta$? −1

99. Which number is larger, $\tan^2 \theta$ or $\sec^2 \theta$? $\sec^2 \theta$

Answers:
85. $\triangle QON$ **86.** $\triangle RON$ **87.** $\triangle SOM$

90. $\sin 315° = -\dfrac{\sqrt{2}}{2}$; $\cos 315° = \dfrac{\sqrt{2}}{2}$

Some Properties of the Trigonometric Functions

Before Starting this Section, Review

1. Equation of a circle (Section 1.1, page 11)
2. Angles (Section 4.1, page 254)
3. Similarity (Appendix A, page 789)

Objectives

1 Determine the signs of the trigonometric functions in each quadrant.

2 Find a reference angle.

3 Use basic trigonometric identities.

GOLF AND THE SINE FUNCTION

The sine and cosine functions show up in surprising places. The angle that a golf ball makes with the ground on take off (and its initial velocity) determines the rest of its flight. Moreover, the time it takes for the ball to hit the ground after reaching its maximum height is the same as if it had been dropped straight down from that height. The horizontal motion has no effect on the vertical motion. Suppose we ignore air resistance and measure time in seconds and distance in feet. Then the equation $h = v_0 t \sin\theta - 16t^2$ gives the height, h, of the golf ball after t seconds, where θ is the initial angle the ball makes with the ground and v_0 is the ball's initial velocity. The horizontal distance, d, that the ball travels in t seconds is $d = tv_0 \cos\theta$. In Example 6, we use these equations to investigate the flight of a golf ball. ■

1 Determine the signs of the trigonometric functions in each quadrant.

```
         y
   II          I
 (−, +)       (+, +)
 sin θ > 0,
 csc θ > 0   All positive
 Others < 0
─────────────────── x
 tan θ > 0,   cos θ > 0,
 cot θ > 0    sec θ > 0
 Others < 0   Others < 0
 (−, −)       (+, −)
   III          IV
```

FIGURE 4.32 Signs of the trigonometric functions

Signs of the Trigonometric Functions

Suppose angle θ is not quadrantal and its terminal side contains the point (x, y). We know that the denominator used in defining the trigonometric functions, $r = \sqrt{x^2 + y^2}$, is always *positive*. Therefore, the signs of x and y determine the signs of the trigonometric functions.

If θ lies in quadrant I, then both x and y are positive; so all six trigonometric function values are positive. However, if θ lies in quadrant II, then x is negative and y is positive; so only $\sin\theta = \dfrac{y}{r}$ and $\csc\theta = \dfrac{r}{y}$ are positive. If θ lies in quadrant III, then x and y are both negative; so only $\tan\theta = \dfrac{y}{x}$ and $\cot\theta = \dfrac{x}{y}$ are positive. If θ lies in quadrant IV, then x is positive and y is negative; so only $\cos\theta = \dfrac{x}{r}$ and $\sec\theta = \dfrac{r}{x}$ are positive. Figure 4.32 summarizes the signs of the trigonometric functions.

EXAMPLE 1 **Determining the Quadrant in Which an Angle Lies**

If $\tan\theta > 0$ and $\cos\theta < 0$, in which quadrant does θ lie?

SOLUTION

Because $\tan\theta > 0$, θ lies either in quadrant I or in quadrant III. However, $\cos\theta > 0$ for θ in quadrant I; so θ must lie in quadrant III. ■■■

Practice Problem 1 If $\sin\theta > 0$ and $\cos\theta < 0$, in which quadrant does θ lie? ■

STUDY TIP

The sentence *All Students Take Calculus* can be useful in remembering which trigonometric functions are positive in quadrants I–IV. Use the first letter of each word in the phrase

> A ll functions (I)
>
> S ine (II)
>
> T angent (III)
>
> C osine (IV)

and recall that a and $\dfrac{1}{a}$ have the same sign.

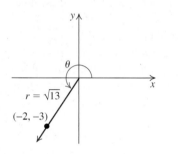

FIGURE 4.33

2 Find a reference angle.

STUDY TIP

When a reference angle is found, a common error is to use the y-axis as one of the sides. Remember that the two sides of a reference angle are always the terminal side of the original angle and the positive or the negative x-axis.

EXAMPLE 2   **Evaluating Trigonometric Functions**

Given that $\tan \theta = \dfrac{3}{2}$ and $\cos \theta < 0$, find the exact values of $\sin \theta$ and $\sec \theta$.

SOLUTION

Because $\tan \theta > 0$ and $\cos \theta < 0$, θ lies in quadrant III. We want to identify a point (x, y) in quadrant III that is on the terminal side of θ. We have

$$\tan \theta = \frac{y}{x} = \frac{3}{2}$$

and because the point (x, y) is in quadrant III, both x and y must be negative. (See Figure 4.33.) If we choose $x = -2$ and $y = -3$, then

$$\tan \theta = \frac{y}{x} = \frac{-3}{-2} = \frac{3}{2}$$

and $r = \sqrt{x^2 + y^2} = \sqrt{(-2)^2 + (-3)^2} = \sqrt{4 + 9} = \sqrt{13}$.

Using $x = -2$, $y = -3$, and $r = \sqrt{13}$, we can find $\sin \theta$ and $\sec \theta$.

$$\sin \theta = \frac{y}{r} = \frac{-3}{\sqrt{13}} = -\frac{3\sqrt{13}}{13} \qquad \sec \theta = \frac{r}{x} = \frac{\sqrt{13}}{-2} = -\frac{\sqrt{13}}{2} \quad \blacksquare\blacksquare\blacksquare$$

Practice Problem 2   Given that $\tan \theta = -\dfrac{4}{5}$ and $\cos \theta > 0$, find the exact values of $\sin \theta$ and $\sec \theta$.   ■

Reference Angle

For any angle θ, there is a corresponding acute angle called its *reference angle*, whose trigonometric function values are identical to those of θ, except possibly for the sign.

DEFINITION OF A REFERENCE ANGLE

Let θ be an angle in standard position that is not a quadrantal angle. The **reference angle** for θ is the positive acute angle θ' ("theta prime") formed by the terminal side of θ and the x-axis.

If θ lies in quadrant I, then $\theta = \theta'$. The reference angle for a positive angle θ (in degrees) in each of the four quadrants is shown in Figure 4.34.

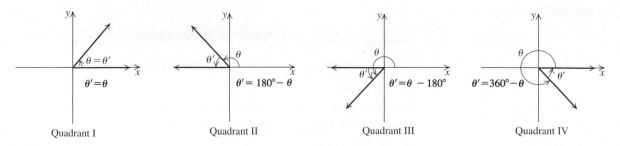

FIGURE 4.34 Reference angles

If θ is measured in radians and $0 < \theta < 2\pi$, we have the following:

$$\theta' = \theta \qquad \text{in quadrant I}$$
$$\theta' = \pi - \theta \qquad \text{in quadrant II}$$
$$\theta' = \theta - \pi \qquad \text{in quadrant III}$$
$$\theta' = 2\pi - \theta \qquad \text{in quadrant IV}$$

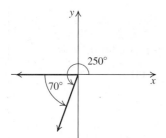

FIGURE 4.35

EXAMPLE 3   Identifying Reference Angles

Find the reference angle θ' for each angle θ.

a. $\theta = 250°$ **b.** $\theta = \dfrac{3\pi}{5}$ **c.** $\theta = 5.75$

SOLUTION

a. Because $250°$ lies in quadrant III, the reference angle is $\theta' = \theta - 180°$. So, $\theta' = 250° - 180° = 70°$. See Figure 4.35.

b. Because $\dfrac{3\pi}{5}$ lies in quadrant II, the reference angle is $\theta' = \pi - \theta$. So,

$$\theta' = \pi - \frac{3\pi}{5} = \frac{5\pi}{5} - \frac{3\pi}{5} = \frac{2\pi}{5}. \text{ See Figure 4.36.}$$

c. Since no degree symbol appears in $\theta = 5.75$, θ has radian measure. Now $\dfrac{3\pi}{2} \approx 4.71$ and $2\pi \approx 6.28$. So θ lies in quadrant IV and $\theta' = 2\pi - \theta$. So, $\theta' = 2\pi - 5.75 \approx 6.28 - 5.75 = 0.53$. See Figure 4.37.   ■ ■ ■

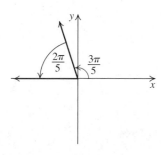

FIGURE 4.36

Practice Problem 3   Find the reference angle θ' for each angle θ.

a. $\theta = 175°$ **b.** $\theta = \dfrac{5\pi}{3}$ **c.** $\theta = 8.22$ ■

You will learn later, there are convenient methods for finding the values of the trigonometric functions for $-\theta$ from the values of the function for θ. Consequently, we do not need to find reference angles for negative angles.

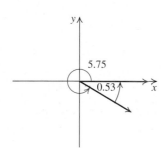

FIGURE 4.37

Using Reference Angles   A reference angle is used to find the values of trigonometric functions of any angle θ. For example, consider the reference angle θ' of the angle θ in Figure 4.38. Let $P(x, y)$ be a point on the terminal side of θ in quadrant III. Then $Q(|x|, |y|)$ is in quadrant I. From the definition of trigonometric functions, we have

$$\cos \theta = \frac{x}{r} = \frac{-|x|}{r} \qquad x \text{ is negative.}$$

The right triangles POM and QON in Figure 4.38 are congruent by SAS congruence. Since θ' is an acute angle in triangle QON, we have

$$\cos \theta' = \frac{|x|}{r}.$$

Here, $\cos \theta = -\cos \theta'$; in general, $\cos \theta$ and $\cos \theta'$ have the same value except possibly for the sign ($\cos \theta = \pm \cos \theta'$). The same is true for the other five trigonometric functions.

We next give a three-step procedure for using reference angles to find trigonometric function values.

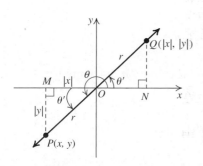

FIGURE 4.38 Reference angle

FINDING THE SOLUTION: A PROCEDURE

EXAMPLE 4   **Finding Trigonometric Function Values Using the Reference Angles**

OBJECTIVE	EXAMPLE
Find the value of any trigonometric function of a positive angle θ.	*Find* sin 1320°.
Step 1 If $\theta > 360°$, find a coterminal angle between 0° and 360°. Otherwise go to Step 2.	Because $1320° = 3(360°) + 240°$, 240° is coterminal with 1320°.
Step 2 Find the reference angle θ' for the angle resulting from Step 1. Write the trigonometric function of θ'.	Because 240° is in quadrant III, its reference angle θ' is $$\theta' = 240° - 180° = 60°$$ and $$\sin \theta' = \sin 60° = \frac{\sqrt{3}}{2} \quad \text{See page 272.}$$
Step 3 Choose the correct sign for θ based on the quadrant in which it lies.	Angle θ and its coterminal angle, 240°, lie in quadrant II, where the sine is *negative*. So, $$\sin 1320° = \sin 240° = -\sin 60° = -\frac{\sqrt{3}}{2} \quad \blacksquare\blacksquare\blacksquare$$

Practice Problem 4   Find the exact value of cos 1035°.

EXAMPLE 5   **Using the Reference Angle to Find Values of Trigonometric Functions**

Find the exact value of each expression.

a. $\tan 330°$ **b.** $\sec \dfrac{59\pi}{6}$

SOLUTION

a. Step 1   Because 330° is between 0° and 360°, we find its reference angle.

   Step 2   Because 330° is in quadrant IV, its reference angle θ' is as follows:
$$\theta' = 360° - 330° = 30°$$
$$\tan \theta' = \tan 30° = \frac{\sqrt{3}}{3} \qquad \text{See page 272.}$$

   Step 3   In quadrant IV, $\tan \theta$ is negative; so
$$\tan 330° = -\tan 30° = -\frac{\sqrt{3}}{3}.$$

b. Step 1   Since the radian measure of $\dfrac{59\pi}{6}$ is greater than 2π, its degree measure is greater than 360°.
$$\frac{59\pi}{6} = \frac{11\pi + 48\pi}{6} = \frac{11\pi}{6} + 8\pi;$$
so $\dfrac{11\pi}{6}$ is an angle between 0 and 2π that is coterminal with $\dfrac{59\pi}{6}$.

Step 2   Because $\dfrac{11\pi}{6}$ is in quadrant IV, its reference angle θ' is as follows:

$$\theta' = 2\pi - \frac{11\pi}{6} = \frac{\pi}{6}$$

$$\sec\theta' = \sec\frac{\pi}{6} = \frac{2\sqrt{3}}{3} \qquad \text{See page 272.}$$

Step 3   In quadrant IV, $\sec\theta > 0$; so

$$\sec\frac{59\pi}{6} = \sec\frac{11\pi}{6} = \sec\frac{\pi}{6} = \frac{2\sqrt{3}}{3}. \qquad\qquad ■ ■ ■$$

Practice Problem 5   Find the exact value of each expression.

a. $\csc 1035°$ **b.** $\cot\dfrac{17\pi}{6}$ ■

EXAMPLE 6   **The Flight of a Golf Ball**

A golf ball is hit on a level fairway with an initial velocity of 128 ft/sec and an initial angle of flight of 30°. Find its range (the horizontal distance it traveled before hitting the ground) and its maximum height.

SOLUTION

We use the height equation from the section introduction.

$$h = v_0 t \sin\theta - 16t^2$$
$$h = 128t \sin 30° - 16t^2 \qquad \text{Given } v_0 = 128 \text{ and } \theta = 30°$$
$$= 64t - 16t^2 = 16t(t - 4) \qquad \text{Replace } \sin 30° \text{ with } \frac{1}{2} \text{ and simplify.}$$

The graph of $h = 64t - 16t^2$ is a parabola; the portion of the graph with $h(t) \geq 0$ represents the flight path of the ball. (See Figure 4.39.) The vertex is $(2, 64)$ because $t = \dfrac{-64}{2(-16)} = 2$ and $h(2) = 64(2) - 16(2)^2 = 64$. So the maximum height of the ball is 64 feet. Because $h(4) = 0$, the ball remains in flight for four seconds. The distance, d, traveled after four seconds is the range, where:

$$d = tv_0 \cos\theta \qquad\qquad \text{Horizontal distance equation}$$
$$d = 4(128) \cos\theta = 4(128) \cos 30° \qquad \text{Given } v_0 = 128 \text{ and } \theta = 30°$$
$$= 4(128)\frac{\sqrt{3}}{2} \approx 443 \qquad\qquad \text{Use a calculator.}$$

The ball reaches a maximum height of 64 feet and has a range of 443 feet.   ■ ■ ■

Practice Problem 6   Repeat Example 6 for a ball with initial velocity 140 ft/sec and initial angle of flight of 45°.   ■

Basic Trigonometric Identities

Because the equation for the unit circle is

$$x^2 + y^2 = 1, \qquad \text{The center is } (0,0); \text{ the radius is } 1.$$

we have the following identity:

$$(\cos t)^2 + (\sin t)^2 = 1. \qquad \text{Replace } x \text{ with } \cos t \text{ and } y \text{ with } \sin t.$$

By agreement, we write $\cos^2 t$ instead of $(\cos t)^2$ and $\sin^2 t$ instead of $(\sin t)^2$. A similar agreement holds for powers of all of the trigonometric functions.

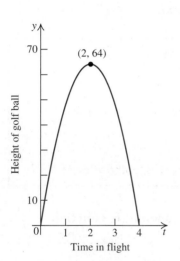

FIGURE 4.39

RECALL

The vertex of the graph of a parabola $f(x) = ax^2 + bx + c$ is

$$\left(-\frac{b}{2a}, f\left(-\frac{b}{2a}\right)\right).$$

---

3 Use basic trigonometric identities.

The equation $x^2 + y^2 = 1$ yields two other useful identities.

$$\frac{x^2}{x^2} + \frac{y^2}{x^2} = \frac{1}{x^2}, x \neq 0 \qquad \text{Divide both sides by } x^2.$$

$$1 + \left(\frac{y}{x}\right)^2 = \left(\frac{1}{x}\right)^2 \qquad \text{Simplify.}$$

$$1 + \left(\frac{\sin t}{\cos t}\right)^2 = \left(\frac{1}{\cos t}\right)^2 \qquad \text{Replace } x \text{ with } \cos t \text{ and } y \text{ with } \sin t.$$

$$1 + \tan^2 t = \sec^2 t \qquad \frac{\sin t}{\cos t} = \tan t, \frac{1}{\cos t} = \sec t$$

Dividing both sides of the equation $x^2 + y^2 = 1$ by y^2, and then replacing x with $\cos t$ and y with $\sin t$, leads to the identity

$$1 + \cot^2 t = \csc^2 t.$$

The three identities resulting from the equation $x^2 + y^2 = 1$ are called the **Pythagorean identities** because the Pythagorean Theorem is the basis for this equation. We list these and several other useful identities next.

BASIC TRIGONOMETRIC IDENTITIES

Quotient and Reciprocal Identities

$$\tan t = \frac{\sin t}{\cos t} \qquad \cot t = \frac{\cos t}{\sin t} \qquad \csc t = \frac{1}{\sin t} \qquad \sec t = \frac{1}{\cos t} \qquad \cot t = \frac{1}{\tan t}$$

Pythagorean Identities

$$\cos^2 t + \sin^2 t = 1 \qquad 1 + \tan^2 t = \sec^2 t \qquad 1 + \cot^2 t = \csc^2 t$$

EXAMPLE 7   **Finding the Exact Value of a Trigonometric Function Using a Pythagorean Identity**

a. Given $\sin t = \dfrac{1}{3}$ and $\cos t < 0$, find $\cos t$ and $\tan t$.

b. Given $\sec t = -2$ and $\tan t > 0$, find $\tan t$.

SOLUTION

a. Use the Pythagorean identity involving $\sin t$.

$$\cos^2 t + \sin^2 t = 1 \qquad \qquad \text{Pythagorean identity}$$

$$\cos^2 t + \left(\frac{1}{3}\right)^2 = 1 \qquad \qquad \text{Replace } \sin t \text{ with } \frac{1}{3}.$$

$$\cos^2 t = 1 - \frac{1}{9} = \frac{8}{9} \qquad \text{Isolate the } t \text{ term and simplify.}$$

$$\cos t = \pm\sqrt{\frac{8}{9}} = \pm\frac{2\sqrt{2}}{3} \qquad \text{Square root property}$$

$$\cos t = -\frac{2\sqrt{2}}{3} \qquad \qquad \cos t < 0 \text{ is given.}$$

$$\tan t = \frac{\sin t}{\cos t} = \frac{\dfrac{1}{3}}{-\dfrac{2\sqrt{2}}{3}} \qquad \text{Replace } \sin t \text{ with } \frac{1}{3} \text{ and } \cos t \text{ with } -\frac{2\sqrt{2}}{3}.$$

$$= -\frac{1}{2\sqrt{2}} = -\frac{\sqrt{2}}{4} \qquad \text{Simplify and rationalize the denominator.}$$

b. Use the Pythagorean identity involving $\sec t$.

$$1 + \tan^2 t = \sec^2 t \qquad \text{Pythagorean identity}$$
$$1 + \tan^2 t = (-2)^2 \qquad \text{Replace } \sec t \text{ with } -2.$$
$$\tan^2 t = 3 \qquad \text{Subtract 1 from both sides and simplify.}$$
$$\tan t = \pm\sqrt{3} \qquad \text{Square root property}$$
$$\tan t = \sqrt{3} \qquad \tan t > 0 \qquad\qquad\blacksquare\ \blacksquare\ \blacksquare$$

Practice Problem 7 Given $\cos t = -\dfrac{2}{3}$ and $\sin t > 0$, find $\sin t$ and $\tan t$. ■

We will discuss and work with trigonometric identities in detail in Chapter 5.

SECTION 4.3 ■ Exercises

A EXERCISES Basic Skills and Concepts

1. If $\sin \theta < 0$, then θ lies in either quadrant ____III____ or quadrant ____IV____.

2. Two angles (with radian measure between 0 and 2π) whose sine value is $\dfrac{1}{2}$ are ____$\dfrac{\pi}{6}$____ and ____$\dfrac{5\pi}{6}$____.

3. The only two trigonometric functions that have positive values in quadrant IV are the ____cosine____ and ____secant____ functions.

4. If $\tan \theta = \dfrac{1}{2}$, then $\cot \theta = $ ____2____.

5. *True or False* If $\tan \theta$ and $\cot \theta$ are both defined, then $\tan \theta \cot \theta = 1$. True

6. *True or False* The reference angle for $\theta = \dfrac{7\pi}{6}$ is $\dfrac{\pi}{3}$. False

7. *True or False* The exact value of $\sin 330°$ is $\dfrac{1}{2}$. False

8. *True or False* A basic identity is $1 + \tan^2 t = \sec^2 t$. True

In Exercises 9–16, use the given information to find the quadrant in which each angle θ lies.

9. $\sin \theta < 0$ and $\cos \theta < 0$
Quadrant III
10. $\sin \theta < 0$ and $\tan \theta > 0$
Quadrant III
11. $\sin \theta > 0$ and $\cos \theta < 0$
Quadrant II
12. $\tan \theta > 0$ and $\csc \theta < 0$
Quadrant III
13. $\cos \theta > 0$ and $\csc \theta < 0$
Quadrant IV
14. $\cos \theta < 0$ and $\cot \theta > 0$
Quadrant III
15. $\sec \theta < 0$ and $\csc \theta > 0$
Quadrant II
16. $\sec \theta < 0$ and $\tan \theta > 0$
Quadrant III

In Exercises 17–24, find the exact values of the remaining trigonometric functions of θ from the given information.

17. $\cos \theta = -\dfrac{5}{13}, \theta$ in quadrant III †

18. $\tan \theta = -\dfrac{3}{4}, \theta$ in quadrant IV †

19. $\cot \theta = -\dfrac{3}{4}, \theta$ in quadrant II †

20. $\sec \theta = \dfrac{4}{\sqrt{7}}, \theta$ in quadrant IV †

21. $\sin \theta = \dfrac{3}{5}, \tan \theta < 0$ †   **22.** $\cot \theta = \dfrac{3}{2}, \sec \theta > 0$ †

23. $\sec \theta = 3, \sin \theta < 0$ †   **24.** $\tan \theta = -2, \sin \theta > 0$ †

In Exercises 25–42, find the reference angle for each angle.

25. $120°$ $60°$ **26.** $275°$ $85°$

27. $500°$ $40°$ **28.** $420°$ $60°$

29. $240°$ $60°$ **30.** $120°$ $60°$

31. $315°$ $45°$ **32.** $510°$ $30°$

33. $225°$ $45°$ **34.** $\dfrac{17\pi}{3}$ $\dfrac{\pi}{3}$

35. $\dfrac{19\pi}{4}$ $\dfrac{\pi}{4}$ **36.** $\dfrac{28\pi}{6}$ $\dfrac{\pi}{3}$

37. $\dfrac{31\pi}{6}$ $\dfrac{\pi}{6}$ **38.** $\dfrac{5\pi}{6}$ $\dfrac{\pi}{6}$

39. $\dfrac{7\pi}{6}$ $\dfrac{\pi}{6}$ **40.** $\dfrac{13\pi}{4}$ $\dfrac{\pi}{4}$

41. $\dfrac{9\pi}{4}$ $\dfrac{\pi}{4}$ **42.** $\dfrac{10\pi}{3}$ $\dfrac{\pi}{3}$

In Exercises 43–50, insert $+$ or $-$ in the blank so that the resulting statement is true for the angle θ and its reference angle θ'.

43. If θ is in quadrant I, $\sin \theta = $ ____$+$____ $\sin \theta'$.

44. If θ is in quadrant I, $\cos \theta = $ ____$+$____ $\cos \theta'$.

45. If θ is in quadrant II, $\sin \theta = $ ____$+$____ $\sin \theta'$.

46. If θ is in quadrant II, $\tan \theta = $ ____$-$____ $\tan \theta'$.

47. If θ is in quadrant III, $\cos \theta = $ ____$-$____ $\cos \theta'$.

48. If θ is in quadrant III, $\sec \theta = $ ____$-$____ $\sec \theta'$.

49. If θ is in quadrant IV, $\cot \theta = $ ____$-$____ $\cot \theta'$.

50. If θ is in quadrant IV, $\csc \theta = $ ____$-$____ $\csc \theta'$.

†Due to space constrictions, answers to these exercises may be found in the Answers beginning on page A–1 in the back of the book.

In Exercises 51–62, find the exact value of each expression.

51. $\cos 120°$

52. $\sin 315°$

53. $\tan 510°$

54. $\cot 750°$ $\sqrt{3}$

55. $\sec 210°$

56. $\csc 300°$

57. $\sin \dfrac{7\pi}{6}$ $-\dfrac{1}{2}$

58. $\cos \dfrac{4\pi}{3}$ $-\dfrac{1}{2}$

59. $\sec \dfrac{15\pi}{4}$ $\sqrt{2}$

60. $\cot \dfrac{19\pi}{4}$ -1

61. $\sin \dfrac{29\pi}{3}$ $-\dfrac{\sqrt{3}}{2}$

62. $\tan \dfrac{55\pi}{3}$ $\sqrt{3}$

In Exercises 63–70, use the Pythagorean identities to find the exact value of each expression.

63. Given $\sin t = \dfrac{2}{5}$ and $\cos t < 0$, find $\cos t$. $-\dfrac{\sqrt{21}}{5}$

64. Given $\cos t = \dfrac{1}{6}$ and $\sin t < 0$, find $\sin t$. $-\dfrac{\sqrt{35}}{6}$

65. Given $\sec t = -5$ and $\tan t < 0$, find $\tan t$. $-2\sqrt{6}$

66. Given $\cot t = 4$ and $\csc t < 0$, find $\csc t$. $-\sqrt{17}$

67. Given $\tan t = -3$ and $\sec t > 0$, find $\sec t$. $\sqrt{10}$

68. Given $\csc t = -6$ and $\cos t > 0$, find $\cos t$. $\dfrac{\sqrt{35}}{6}$

69. Given $\sin t = -\dfrac{3}{4}$ and $\cos t < 0$, find $\sec t$. $-\dfrac{4\sqrt{7}}{7}$

70. Given $\cos t = \dfrac{3}{5}$ and $\sin t > 0$, find $\csc t$. $\dfrac{5}{4}$

B EXERCISES Applying the Concepts

71. Polarization. One lens is removed from a pair of polarized sunglasses and put on top of the other lens, both aligned identically. If the top lens is rotated through an angle θ, the fraction of light passing through the aligned lenses is $\cos^2 \theta$. Find θ in degrees so that a rotation through θ allows only $\dfrac{1}{4}$ as much light through the lenses as when they are aligned identically. $\theta = 60°$ or $\theta = 300°$

72. Cable cost. Cable must be laid from a point on the shore of a lake to an island. The locations and distances are shown in the figure. It costs $700(5 - \cot \theta) + 1200 \csc \theta$ dollars to lay the cable, where θ is the angle the cable to the island makes with the shore. Find the cost if

a. $\theta = 30°$. **b.** $\theta = 45°$. **c.** $\theta = 60°$. **d.** $\theta = 90°$.
a. \$4687.56 **b.** \$4497.06 **c.** \$4481.50 **d.** \$4700

73. Household voltage. The voltage in a common household electrical outlet is given by $V = 166 \cos (120\pi t)$, where V is measured in volts and t in seconds. Find the voltage when t equals

a. $\dfrac{1}{120}$ sec. -166 volts **b.** $\dfrac{1}{60}$ sec. 166 volts

74. Volume of a cone. The volume, V in feet^3, of the cone in the figure is $9\pi \cos^2 \theta \sin \theta$. Find V if

a. $\theta = 30°$.

b. $\theta = 60°$.

a. $\dfrac{27\pi}{8}$ ft^3 **b.** $\dfrac{9\pi\sqrt{3}}{8}$ ft^3

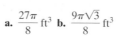

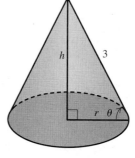

75. Piston action. A rod OA of length r rotates counterclockwise about a fixed point O. A second rod AB with length $L > 2r$ is connected to a piston at point B. If all lengths are given in centimeters, the distance x between the points O and B is $x = r \cos \theta + \sqrt{L^2 - r^2 \sin^2 \theta}$. Find x if

a. $\theta = 0°$. **b.** $\theta = 90°$. **c.** $\theta = 180°$.

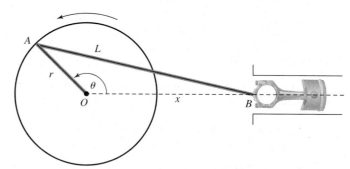

76. Observing London. The Millennium Wheel in London is a large, continuously turning observational wheel with diameter 135 m (approximately). The height, h, of the passenger capsules above the base is $68 + 67.5 \sin (\theta - 90°)$, where θ is the angle the capsule arm makes with a vertical ray downward from the wheel's center. (See the figure.) How high is the capsule when

a. $\theta = 0°$? **b.** $\theta = 90°$? **c.** $\theta = 180°$?

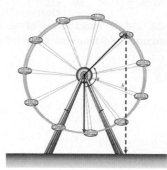

Answers:

51. $-\dfrac{1}{2}$ **52.** $-\dfrac{\sqrt{2}}{2}$ **53.** $-\dfrac{\sqrt{3}}{3}$ **55.** $-\dfrac{2\sqrt{3}}{3}$ **56.** $-\dfrac{2\sqrt{3}}{3}$

75. a. $L + r$ cm **b.** $\sqrt{L^2 - r^2}$ cm **c.** $L - r$ **76. a.** 0.5 m **b.** 68 m **c.** 135.5 m

In Exercises 77 and 78, use the following information.

Assuming level ground and neglecting air resistance, the formulas governing the flight of a golf ball in the introduction and Example 5 apply to any object projected upward from ground level at an angle θ. That is, if the initial velocity, v_0, is given in ft/sec and time is measured in seconds, then t seconds into flight gives the following:

$$h = v_0 t \sin \theta - 16t^2 \quad \text{the object's height}$$

$$d = t v_0 \cos \theta \quad \text{the horizontal distance traveled} \\ \text{(the range)}$$

77. If $\theta = 30°$ and the initial velocity is 80 ft/sec, find
   a. the object's maximum height. 25 ft
   b. the object's range. 173 ft

78. If $\theta = 30°$ and the initial velocity is 80 ft/sec, find the time the object was in flight. 2.5 sec

C EXERCISES Beyond the Basics

Use the information given for Exercises 77 and 78 for Exercises 79–81.

79. For an angle θ and initial velocity v_0, find an expression for the object's maximum height. [*Hint:* Find the vertex of $h = v_0 t \sin \theta - 16t^2$.]

80. For an angle θ and initial velocity v_0, find an expression for the object's range. [*Hint:* First find the time the object is in flight.]

81. Show that height is a (quadratic) function of distance, that is $h = ad^2 + bd + c$. [*Hint:* First, solve $d = tv_0 \cos \theta$ for t.]

82. Use the values $\dfrac{\sqrt{0}}{2}, \dfrac{\sqrt{1}}{2}, \dfrac{\sqrt{2}}{2}, \dfrac{\sqrt{3}}{2}$, and $\dfrac{\sqrt{4}}{2}$ and the figure to write the values for $\sin \theta$ and $\cos \theta$ for $\theta = 0°$, $\theta = 30°$, $\theta = 45°$, $\theta = 60°$, and $\theta = 90°$. [*Hint:* Note that $\dfrac{\sqrt{4}}{2} = 1, \dfrac{\sqrt{1}}{2} = \dfrac{1}{2}$, and $\dfrac{\sqrt{0}}{2} = 0$.]

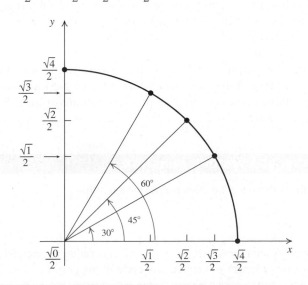

83. Use the coordinates of a point (x, y) on the terminal side of an angle θ to show that $\cos(\theta + \pi) = -\cos \theta$.

84. Use the coordinates of a point (x, y) on the terminal side of an angle θ to show that $\sin(\theta + \pi) = -\sin \theta$.

85. Use the coordinates of a point (x, y) on the terminal side of an angle θ to show that $\tan(\theta + \pi) = \tan \theta$.

86. Find a formula for the reference angle θ', of a negative angle θ, with $-180° < \theta < -90°$. $\theta + 180°$

87. Find a formula for the reference angle θ', of a negative angle θ, with $-270° < \theta < -180°$. $-(\theta + 180°)$

88. Find a formula for the reference angle θ', of a negative angle θ, with $-360° < \theta < -270°$. $\theta + 360°$

89. If $\sin \theta = -\dfrac{7}{16}$ and θ is in quadrant IV, find $\sin(-\theta)$. $\dfrac{7}{16}$

90. If $\cos \theta = 0.03$ and θ is in quadrant IV, find $\cos(-\theta)$. 0.03

91. If $\tan \theta = 5$ and θ is in quadrant I, find $\tan(-\theta)$. -5

92. If $\sec \theta = -2$ and θ is in quadrant II, find $\sec(-\theta)$. -2

93. If $\csc \theta = -3$ and θ is in quadrant IV, find $\csc(-\theta)$. 3

94. Find all values of θ, with $-360° < \theta < 0°$, measured in degrees, for which $\tan \theta = 1$. $-135°, -315°$

Critical Thinking

95. Show that $\sin \theta \leq 1$ and $\cos \theta \leq 1$ for any acute angle θ. [*Hint:* For any real numbers x and y, $x^2 + y^2 \geq x^2$.]

96. Show that $-1 \leq \sin \theta \leq 1$ and $-1 \leq \cos \theta \leq 1$ for any angle θ.

97. Show that if $0 \leq \theta \leq \dfrac{\pi}{2}$, then $\cos \theta = \sqrt{1 - \sin^2 \theta}$.

98. Show that if $\dfrac{\pi}{2} \leq \theta \leq \pi$, then $\cos \theta = -\sqrt{1 - \sin^2 \theta}$.

Answers:

79. $\dfrac{(v_0 \sin \theta)^2}{64}$ **80.** $\dfrac{(v_0)^2 \sin \theta \cos \theta}{16}$ **81.** $h = (\tan \theta)d - (16 \sec^2 \theta)\dfrac{d^2}{v_0^2}$

82. $\sin 0° = \dfrac{\sqrt{0}}{2} = 0 \quad \cos 0° = \dfrac{\sqrt{4}}{2} = 1$

$\sin 30° = \dfrac{\sqrt{1}}{2} = \dfrac{1}{2} \quad \cos 30° = \dfrac{\sqrt{3}}{2}$

$\sin 45° = \dfrac{\sqrt{2}}{2} \quad \cos 45° = \dfrac{\sqrt{2}}{2}$

$\sin 60° = \dfrac{\sqrt{3}}{2} \quad \cos 60° = \dfrac{\sqrt{1}}{2} = \dfrac{1}{2}$

$\sin 90° = \dfrac{\sqrt{4}}{2} = 1 \quad \cos 90° = \dfrac{\sqrt{0}}{2} = 0$

83. $\cos(\theta + \pi) = \dfrac{-x}{r} = -\dfrac{x}{r} = -\cos \theta$

84. $\sin(\theta + \pi) = \dfrac{-y}{r} = -\dfrac{y}{r} = -\sin \theta$

85. $\tan(\theta + \pi) = \dfrac{-y}{-x} = \dfrac{y}{x} = \tan \theta$

Graphs of the Sine and Cosine Functions

Before Starting this Section, Review

1. Equation of a circle (Section 1.1, page 11)
2. Transformations of functions (Section 1.5, page 62)
3. Unit circle definitions of the sine and cosine (Section 4.2, page 268)

Objectives

1. Define periodic functions.
2. Graph the sine and cosine functions.
3. Find the amplitude and period of sinusoidal functions.
4. Find the phase shifts and graph sinusoidal functions of the forms $y = a \sin b(x - c)$ and $y = a \cos b(x - c)$.

LENGTH OF DAYS

During winter, because the Northern Hemisphere tilts away from the sun, it receives less direct solar radiation than does the Southern Hemisphere, which tilts toward the sun. In the Northern Hemisphere, the summer solstice (the longest day of the year) occurs on or around June 21 and the winter solstice (the shortest day of the year) occurs on or about December 21. In the Southern Hemisphere, the seasons are reversed. The number of hours of daylight varies at different latitudes. The length of a day is one of the many phenomena that can be described by the trigonometric functions we study in this section. In Example 11, we see how to model the number of daylight hours that the city of Paris enjoys throughout the year. ■

---

1 Define periodic functions.

Periodic Functions

Many phenomena and behaviors repeat a particular pattern over and over. For example, no matter where you live, the sun appears in a dependable pattern throughout the year. Naturalists find that certain animal populations vary in a regular pattern over time. Lighthouse beams sweep out regular patterns. Blood pressure for a typical individual fluctuates rhythmically. *Periodic functions* can often be used to model this type of repetitious activity.

In this section, we focus on the graphs of the sine and cosine functions and transformations of these graphs. To graph these functions on an xy-coordinate system, we use x as the independent variable and y as the dependent variable. So we write $y = \sin x$ rather than $y = \sin t$.

DEFINITION OF A PERIODIC FUNCTION

A function f is said to be **periodic** if there is a positive number p such that

$$f(x + p) = f(x)$$

for every x in the domain of f.

The smallest value of p (if there is one) for which $f(x + p) = f(x)$ is called the **period** of f. The graph of f over any interval of length p is called one **cycle** of the graph.

Because $\sin(x + 2\pi) = \sin x$ and $\cos(x + 2\pi) = \cos x$, both the sine and cosine functions are periodic functions. The period for each of the sine and cosine functions is 2π. See Exercise 80.

2 Graph the sine and cosine functions.

Graph of the Sine Function

To sketch the graph of $y = \sin x$ for $0 \le x \le 2\pi$, we can use Table 4.2. This table of approximate values for $y = \sin x$ uses common values of x.

TABLE 4.2

x	0	$\dfrac{\pi}{6}$	$\dfrac{\pi}{4}$	$\dfrac{\pi}{3}$	$\dfrac{\pi}{2}$	$\dfrac{2\pi}{3}$	$\dfrac{3\pi}{4}$	$\dfrac{5\pi}{6}$	π	$\dfrac{7\pi}{6}$	$\dfrac{5\pi}{6}$	$\dfrac{4\pi}{3}$	$\dfrac{3\pi}{2}$	$\dfrac{5\pi}{3}$	$\dfrac{7\pi}{4}$	$\dfrac{11\pi}{6}$	2π
$\sin x$	0	0.5	0.71	0.87	1	0.87	0.71	0.5	0	−0.5	−0.71	−0.87	−1	−0.87	−0.71	−0.5	0

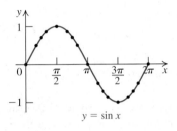

FIGURE 4.40 Graph of the sine function for $0 \le x \le 2\pi$

Connecting these points with a smooth curve gives us the graph in Figure 4.40. Only $0, \dfrac{\pi}{2}, \pi, \dfrac{3\pi}{2}$, and 2π are marked on the x-axis, but all of the ordered pairs $(x, \sin x)$ given in Table 4.2 are plotted on the curve.

Recall from Section 4.2 that $(\cos x, \sin x)$ is the point on the unit circle associated with an arc of length $|x|$. In this situation, the horizontal axis is labeled the u-axis and x corresponds to a circular arc. See Figure 4.41. Consequently, $-1 \le \sin x \le 1$; that is, the range of the sine function is $[-1, 1]$.

The input, x, for $y = \sin x$ can be any real number; so the domain of the sine function is $(-\infty, \infty)$. Because the sine function is periodic with period 2π, we can sketch the complete graph of $y = \sin x$ by extending the graph in Figure 4.40 indefinitely to the left and to the right in every successive interval of length 2π. See Figure 4.42.

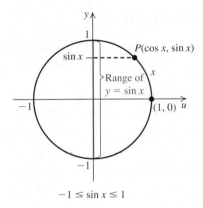

FIGURE 4.41 Range of $y = \sin x$

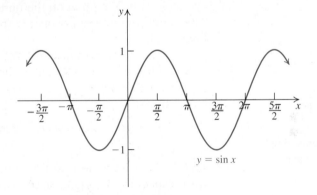

FIGURE 4.42 The sine curve

Graph of the Cosine Function

We can use the same method for graphing $y = \cos x$ that we used for $y = \sin x$. Table 4.3 gives the values for $y = \cos x$, where x has common values with $0 \le x \le 2\pi$.

TABLE 4.3

x	0	$\dfrac{\pi}{6}$	$\dfrac{\pi}{4}$	$\dfrac{\pi}{3}$	$\dfrac{\pi}{2}$	$\dfrac{2\pi}{3}$	$\dfrac{3\pi}{4}$	$\dfrac{5\pi}{6}$	π	$\dfrac{7\pi}{6}$	$\dfrac{5\pi}{6}$	$\dfrac{4\pi}{3}$	$\dfrac{3\pi}{2}$	$\dfrac{5\pi}{3}$	$\dfrac{7\pi}{4}$	$\dfrac{11\pi}{6}$	2π
$\cos x$	1	0.87	0.71	0.5	0	−0.5	−0.71	−0.87	−1	−0.87	−0.71	−0.5	0	0.5	0.71	0.87	1

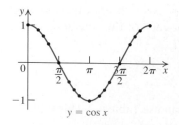

FIGURE 4.43 Graph of the cosine function for 0 ≤ x ≤ 2π

Connecting these points with a smooth curve gives the graph in Figure 4.43. Again, only $0, \dfrac{\pi}{2}, \pi, \dfrac{3\pi}{2}$, and 2π are marked on the x-axis, but all of the ordered pairs $(x, \cos x)$ given in Table 4.3 are plotted on the curve. Because $(\cos x, \sin x)$ is a point on the unit circle, we have $-1 \le \cos x \le 1$; so the range of the cosine function is $[-1, 1]$. See Figure 4.44.

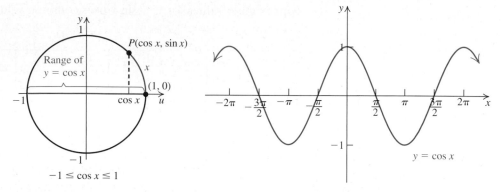

FIGURE 4.44 Range of y = cos x   **FIGURE 4.45 The cosine curve**

As with the sine function, the values of x for $y = \cos x$ are not restricted; so the domain of the cosine function is $(-\infty, \infty)$. We can draw the complete graph of $y = \cos x$ by extending the graph in Figure 4.43 indefinitely to the left and right in every successive interval of length 2π. See Figure 4.45.

Properties of the Sine and Cosine Functions

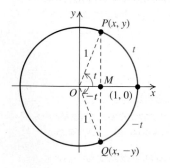

FIGURE 4.46 Comparing the sine and cosine of t and −t

Let $P(x, y)$ be the point on the unit circle associated with a positive real number t. The angle that intercepts the arc from $(1, 0)$ to $P(x, y)$ has measure t radians. See Figure 4.46.

The two right triangles OMP and OMQ shown in Figure 4.46 are congruent by SSS congruence. Consequently, the negative angle in standard position with the terminal side passing through the point $Q(x, -y)$ has measure $-t$. Since the circle has radius 1,

$$P(x, y) = (\cos t, \sin t) \qquad \text{and} \qquad Q(x, -y) = (\cos(-t), \sin(-t)).$$

So we have $\cos(-t) = x = \cos t$ and $\sin(-t) = -y = -\sin t$. Since $\sin(-t) = -\sin t$, the sine function is an *odd* function; so its graph is symmetric with respect to the origin. Similarly, since $\cos(-t) = \cos t$, the cosine function is an *even* function; so its graph is symmetric with respect to the y-axis.

PROPERTIES OF THE SINE AND COSINE FUNCTIONS	
Sine Function	**Cosine Function**
1. Period: 2π	**1.** Period: 2π
2. Domain: $(-\infty, \infty)$	**2.** Domain: $(-\infty, \infty)$
3. Range: $[-1, 1]$	**3.** Range: $[-1, 1]$
4. Odd: $\sin(-t) = -\sin t$	**4.** Even: $\cos(-t) = \cos t$

There are five **key points** in every cycle of the graphs of the sine and cosine functions that are very helpful for graphing these functions and their transformations. The key points are the x-intercepts, high points, and low points. See Figure 4.47.

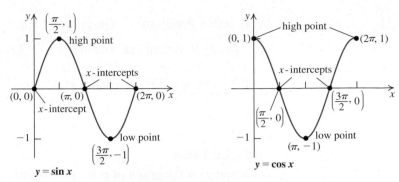

FIGURE 4.47 **Key points of the sine and cosine graphs**

3 Find the amplitude and
period of sinusoidal functions.

Amplitude and Period

The graphs of the sine and cosine functions and their transformations are called **sinusoidal** graphs or curves. In this section we study sinusoidal curves of the form $y = af[b(x - c)] + d$, where f is a sine or cosine function. We first consider transformations of the form $y = a \sin bx$ and $y = a \cos bx$. The number $|a|$ in these equations is called the **amplitude**, which is the function's largest value. We begin by letting $b = 1$.

EXAMPLE 1   **Graphing $y = a \sin x$**

Graph $y = 3 \sin x$, $y = \dfrac{1}{3} \sin x$, and $y = \sin x$ on the same coordinate system over the interval $[-2\pi, 2\pi]$.

SOLUTION

We begin with the graph of $y = \sin x$ and multiply the y-coordinate of each point (including the key points) on this graph by 3 to get the graph of $y = 3 \sin x$. This results in stretching the graph of $y = \sin x$ vertically by a factor of 3. Since $3(0) = 0$, the x-intercepts are unchanged. However, the maximum y value is $3(1) = 3$ and the minimum y value is $3(-1) = -3$. See Figure 4.48. Similarly, we multiply the y-coordinate on the graph by $\dfrac{1}{3}$ to get the graph of $y = \dfrac{1}{3} \sin x$. This results in compressing the graph of $y = \sin x$ vertically by a factor of $\dfrac{1}{3}$. As before, the x-intercepts are unchanged; but here the maximum y value is $\dfrac{1}{3}$ and the minimum y value is $-\dfrac{1}{3}$. See Figure 4.48.

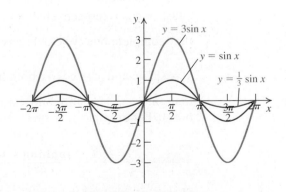

FIGURE 4.48 **Varying a in $y = a \sin x$**   ■ ■ ■

Practice Problem 1 Graph $y = 5 \sin x$, $y = \frac{1}{5} \sin x$, and $y = \sin x$ on the same coordinate system over the interval $[-2\pi, 2\pi]$. ■

EXAMPLE 2 **Graphing $y = a \cos x$**

Graph $y = -2 \cos x$ over the interval $[-2\pi, 2\pi]$. Find the amplitude and range of the function.

SOLUTION

Begin with the graph of $y = \cos x$. Multiply the y-coordinate of each point on that graph by 2 to stretch the graph vertically by a factor of 2 and get the graph of $y = 2 \cos x$. Pay particular attention to the key points. Notice that the x-intercepts are unchanged. Next, reflect the graph of $y = 2 \cos x$ in the x-axis to produce the graph of $y = -2 \cos x$. See Figure 4.49. The amplitude is $|-2| = 2$, the largest value attained by $y = -2 \cos x$; the smallest function value attained is -2. Therefore, the range of $y = -2 \cos x$ is $[-2, 2]$.

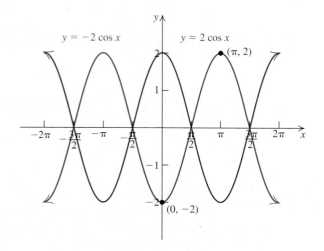

FIGURE 4.49 Graph of $y = -2 \cos x$ ■ ■ ■

Practice Problem 2 Graph $y = 5 \cos x$ over the interval $[-2\pi, 2\pi]$. Find the amplitude and range of the function. ■

Both the sine and cosine functions have period 2π. For $b > 0, 0 \le bx \le 2\pi$ is equivalent to $0 \le x \le \frac{2\pi}{b}$. Therefore, as x takes on the values from 0 to $\frac{2\pi}{b}$, the expression bx takes on the values from 0 to 2π.

The graphs of $y = \sin bx$ and $y = \cos bx$ have similar shapes to those of $y = \sin x$ and $y = \cos x$, respectively, but they are compressed or expanded horizontally so that they complete one cycle over any interval of length $\frac{2\pi}{b}$. Consequently, the functions $y = \sin bx$ and $y = \cos bx$ both have period $\frac{2\pi}{b}$. When $b < 0$, we can use the facts that $\sin(-x) = -\sin x$ and $\cos(-x) = \cos x$ to rewrite the given function with $b > 0$. For example, $y = \sin(-3x) = -\sin 3x$.

EXAMPLE 3 **Graphing $y = \sin bx$**

Sketch one cycle of the graphs of $y = \sin 3x$, $y = \sin \frac{1}{3} x$, and $y = \sin x$ on the same coordinate system.

SOLUTION

We begin with the graph of $y = \sin x$ and adjust its period, 2π, by dividing 2π by 3 to find the period $\dfrac{2\pi}{3}$ of $y = \sin 3x$. We then compress the graph of $y = \sin x$ horizontally so that one cycle of $y = \sin 3x$ is completed on the interval $\left[0, \dfrac{2\pi}{3}\right]$. Divide the interval $0 \le x \le \dfrac{2\pi}{3}$ into four equal parts to find the x-coordinates for the key points: $0, \dfrac{\pi}{6}, \dfrac{\pi}{3}, \dfrac{\pi}{2}$ and $\dfrac{2\pi}{3}$. Since the amplitude is 1, the y values of the key points are unchanged. See Figure 4.50.

   Similarly, to find the period of $y = \sin \dfrac{1}{3}x$, we divide 2π by $\dfrac{1}{3}$, resulting in $\dfrac{2\pi}{\frac{1}{3}} = 6\pi$. Therefore, one cycle for $y = \sin \dfrac{1}{3}x$ is completed on the interval $[0, 6\pi]$. Divide the interval $0 \le x \le 6\pi$ into four equal parts to get the x-coordinates for the key points: $0, \dfrac{3\pi}{2}, 3\pi, \dfrac{9\pi}{2}$, and 6π. Again, the y values of the key points are unchanged. See Figure 4.50.

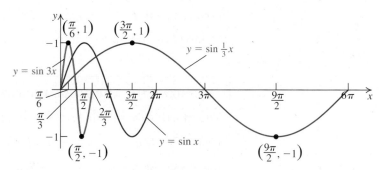

FIGURE 4.50 Horizontally compressing or expanding the graph of $y = \sin x$

■ ■ ■

Practice Problem 3   Sketch one cycle of the graphs of $y = \sin \dfrac{1}{2}x$ and $y = \sin x$ on the same coordinate system.

■

EXAMPLE 4   Graphing $y = a \cos bx$

Graph $y = 3 \cos \dfrac{1}{2}x$ over a one-period interval.

SOLUTION

We begin with the graph of $y = \cos x$ and find the period of $y = \cos \dfrac{1}{2}x$ by dividing 2π by $\dfrac{1}{2}$ to get

$$\frac{2\pi}{\frac{1}{2}} = 4\pi \qquad \text{Replace } b \text{ with } \frac{1}{2} \text{ in } \frac{2\pi}{b}.$$

The period of $y = \cos \dfrac{1}{2}x$ is 4π. So, the graph of $y = \cos \dfrac{1}{2}x$ is obtained by horizontally stretching the graph of $y = \cos x$ by a factor of 2.

TECHNOLOGY CONNECTION

To graph $y = 3 \cos \dfrac{1}{2}x$

with a graphing calculator, choose Radian mode and these WINDOW settings:

$$X \text{ min} = 0$$
$$X \text{ max} = 4\pi$$
$$X \text{ scl} = \pi/2$$
$$Y \text{ min} = -4$$
$$Y \text{ max} = 4$$
$$Y \text{ scl} = 1$$

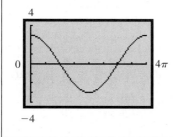

Next, we multiply the y-coordinate of each point on the graph of $y = \cos \dfrac{1}{2}x$ by 3 to stretch this graph vertically by a factor of 3. To help sketch the graphs, divide the period, 4π, into four equal parts, starting at 0, to find the x-coordinates for the key points. These x-coordinates: $0, \pi, 2\pi, 3\pi$, and 4π, correspond to the highest points, the lowest point, and the x-intercepts of the graph. The key points on the graph are $(0, 3), (\pi, 0), (2\pi, -3), (3\pi, 0)$ and $(4\pi, 3)$. See Figure 4.51. The graph can then be extended indefinitely to the left and right. The amplitude is 3; so the largest value of y is 3. The range is $[-3, 3]$.

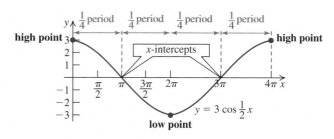

FIGURE 4.51 One cycle of $y = 3 \cos \dfrac{1}{2}x$ ■ ■ ■

Practice Problem 4   Graph $y = 2 \cos 2x$ over a one-period interval.   ■

The next box summarizes information used to graph $y = a \sin bx$ and $y = a \cos bx$.

CHANGING THE AMPLITUDE AND PERIOD OF THE SINE AND COSINE FUNCTIONS

The functions $y = a \sin bx$ and $y = a \cos bx$ $(b > 0)$ have amplitude $|a|$ and period $\dfrac{2\pi}{b}$. If $a > 0$, the graphs of $y = a \sin bx$ and $y = a \cos bx$ are similar to the graphs of $y = \sin x$ and $y = \cos x$, respectively, with three changes.

1. The range is $[-a, a]$.

2. One cycle is completed over the interval $\left[0, \dfrac{2\pi}{b}\right]$.

3. The x-coordinates for the key points are $0, \dfrac{\pi}{2b}, \dfrac{\pi}{b}, \dfrac{3\pi}{2b}$, and $\dfrac{2\pi}{b}$. The y-values at each of these numbers is one of: $-a, 0$, or a.

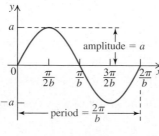
$$y = a \sin bx, (a > 0)$$

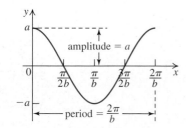
$$y = a \cos bx, (a > 0)$$

If $a < 0$, the graphs are the reflections of the graph of $y = |a| \sin bx$ and $y = |a| \cos bx$, respectively, in the x-axis.

4 Find the phase shifts and graph sinusoidal functions of the forms $y = a \sin b(x - c)$ and $y = a \cos b(x - c)$.

Phase Shift

We now consider sinusoidal graphs with horizontal shifts.

EXAMPLE 5   **Graphing $y = a \sin (x - c)$**

Graph $y = \sin\left(x - \dfrac{\pi}{2}\right)$ over a one-period interval.

SOLUTION

Recall that for any function f, the graph of $y = f(x - c)$ is the graph of $y = f(x)$ shifted right c units if $c > 0$ and left $|c|$ units if $c < 0$. Comparing the equation

$$y = \sin\left(x - \frac{\pi}{2}\right) \text{ with } y = \sin x, \text{ we see that } y = \sin\left(x - \frac{\pi}{2}\right) \text{ has the form}$$

$$y = \sin(x - c), \text{where } c = \frac{\pi}{2}. \text{Therefore, the graph of } y = \sin\left(x - \frac{\pi}{2}\right) \text{ is the graph}$$

of $y = \sin x$ shifted right $\dfrac{\pi}{2}$ units. See Figure 4.52.   ■ ■ ■

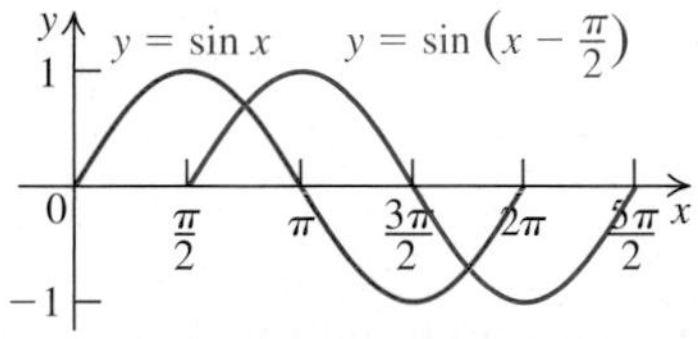

FIGURE 4.52 Graph of

$y = \sin\left(x - \dfrac{\pi}{2}\right)$

Practice Problem 5   Graph $y = \cos\left(x - \dfrac{\pi}{4}\right)$ over a one-period interval.   ■

Before describing the general procedure for graphing the sinusoidal functions $y = a \sin b(x - c)$ and $y = a \cos b(x - c)$, with $b > 0$, we find the value of x for which $b(x - c) = 0$. This number is called the **phase shift**. Let

$$b(x - c) = 0$$
$$x - c = 0 \qquad \text{Divide both sides by } b.$$
$$x = c \qquad \text{Add } c \text{ to both sides.}$$

The phase shift is the amount by which the graphs of $y = a \sin bx$ or $y = a \cos bx$ are shifted horizontally. These graphs are shifted right if $c > 0$ and left if $c < 0$. To graph a cycle of either function, we start the cycle at $x = c$ and then divide the period into four equal parts. We describe this procedure next.

FINDING THE SOLUTION: A PROCEDURE

EXAMPLE 6   **Graphing $y = a \sin [b(x - c)]$ and $y = a \cos [b(x - c)]$**

OBJECTIVE
Graph a function of the form $y = a \sin [b(x - c)]$ or $y = a \cos [b(x - c)]$, with $b > 0$, by finding the amplitude, period, and phase shift.

EXAMPLE
Find the amplitude, period, and phase shift, and sketch the graph of $y = 3 \sin\left[2\left(x - \dfrac{\pi}{4}\right)\right]$ over a one-period interval.

Step 1   Find the amplitude, period, and phase shift.

amplitude $= |a|$

period $= \dfrac{2\pi}{b}$

phase shift $= c$

If $c > 0$, shift to the right.

If $c < 0$, shift to the left.

1. $y = 3 \sin\left[2\left(x - \dfrac{\pi}{4}\right)\right]$

$\qquad\qquad \uparrow \qquad \uparrow \qquad \uparrow$

$a = 3, b = 2, c = \dfrac{\pi}{4}$

amplitude $= |3| = 3$

period $= \dfrac{2\pi}{2} = \pi$

phase shift $= \dfrac{\pi}{4}$ Shift right because $\dfrac{\pi}{4} > 0$.

continued on the next page

Step 2 The cycle begins at $x = c$. One complete cycle occurs over the interval $\left[c, c + \dfrac{2\pi}{b} \right]$.

Step 3 Divide the interval $\left[c, c + \dfrac{2\pi}{b} \right]$ into four equal parts, each of length $\dfrac{1}{4}(\text{period}) = \dfrac{1}{4}\left(\dfrac{2\pi}{b} \right)$. This gives the x-coordinates for the five key points:

$$c, c + \frac{1}{4}\left(\frac{2\pi}{b} \right), c + \frac{1}{2}\left(\frac{2\pi}{b} \right),$$
$$c + \frac{3}{4}\left(\frac{2\pi}{b} \right), \text{ and } c + \frac{2\pi}{b}.$$

Step 4 If $a > 0$, for $y = a \sin[b(x - c)]$, sketch one cycle of the sine curve through the key points $(c, 0)$,

$$\left(c + \frac{\pi}{2b}, a \right), \left(c + \frac{\pi}{b}, 0 \right),$$
$$\left(c + \frac{3\pi}{2b}, -a \right), \text{ and } \left(c + \frac{2\pi}{b}, 0 \right).$$

For $y = a \cos[b(x - c)]$, sketch one cycle of the cosine curve through the key points (c, a), $\left(c + \dfrac{\pi}{2b}, 0 \right), \left(c + \dfrac{\pi}{b}, -a \right)$,

$$\left(c + \frac{3\pi}{2b}, 0 \right), \text{ and } \left(c + \frac{2\pi}{b}, a \right).$$

If $a < 0$, reflect the graph of $y = |a| \sin b(x - c)$ or $y = |a| \cos b(x - c)$ in the x-axis.

2. Begin the cycle at $x = \dfrac{\pi}{4}$ and graph one cycle over

$$\left[\frac{\pi}{4}, \frac{\pi}{4} + \pi \right] = \left[\frac{\pi}{4}, \frac{5\pi}{4} \right]$$

3. Divide the interval $\left[\dfrac{\pi}{4}, \dfrac{5\pi}{4} \right]$ into four equal parts each having length $\dfrac{1}{4}(\text{period}) = \dfrac{1}{4}(\pi) = \dfrac{\pi}{4}$.

The x-coordinates of the five key points are:
$$\frac{\pi}{4}, \frac{\pi}{4} + \frac{\pi}{4} = \frac{\pi}{2}, \frac{\pi}{4} + \frac{\pi}{2} = \frac{3\pi}{4}, \frac{\pi}{4} + \frac{3\pi}{4} = \pi, \text{ and}$$
$$\frac{\pi}{4} + \pi = \frac{5\pi}{4}$$

4. Sketch one cycle of the sine curve through the five key points: $\left(\dfrac{\pi}{4}, 0 \right), \left(\dfrac{\pi}{2}, 3 \right), \left(\dfrac{3\pi}{4}, 0 \right), (\pi, -3)$, and $\left(\dfrac{5\pi}{4}, 0 \right)$.

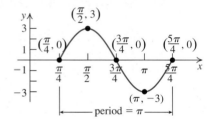

■ ■ ■

Practice Problem 6 Find the amplitude, period, and phase shift and sketch the graph of the function

$$y = \frac{1}{3} \cos\left[\frac{2}{5}\left(x + \frac{\pi}{6} \right) \right].$$

■

The equations $\sin(-x) = -\sin x$ and $\cos(-x) = \cos x$ allow us to graph $y = a \sin b(x - c)$ and $y = a \cos b(x - c)$ when $b < 0$. For example, to graph $y = 3 \sin\left[(-2)\left(x - \dfrac{\pi}{2} \right) \right]$ we rewrite the equation as $y = -3 \sin\left[2\left(x - \dfrac{\pi}{2} \right) \right]$.

EXAMPLE 7 **Graphing $y = a \cos b(x - c)$**

Graph $y = -\cos\left[2\left(x + \dfrac{\pi}{2} \right) \right]$ over a one-period interval.

SOLUTION

Use the steps in Example 6 for the procedure for graphing $y = a \cos b(x - c)$.

TECHNOLOGY CONNECTION

To graph
$$y = -\cos 2\left(x + \frac{\pi}{2}\right) \text{ with a}$$
graphing calculator, choose radian mode and enter these WINDOW settings:

$$X \min = -\pi/2$$
$$X \max = \pi/2$$
$$X \text{ scl} = \pi/4$$
$$Y \min = -2$$
$$Y \max = 2$$
$$Y \text{ scl} = 1$$

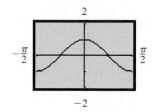

Step 1 Here $y = -\cos\left[2\left(x + \frac{\pi}{2}\right)\right] = (-1)\cos 2\left[x - \left(-\frac{\pi}{2}\right)\right]$

$$a = -1 \quad b = 2 \quad c = -\frac{\pi}{2}$$

amplitude $= |a| = |-1| = 1$ period $= \dfrac{2\pi}{b} = \dfrac{2\pi}{2} = \pi$

phase shift $= c = -\dfrac{\pi}{2}$ Shift left because $c < 0$.

Step 2 The cycle begins at $x = -\dfrac{\pi}{2}$.

One cycle is graphed over the interval $\left[-\dfrac{\pi}{2}, -\dfrac{\pi}{2} + \pi\right] = \left[-\dfrac{\pi}{2}, \dfrac{\pi}{2}\right]$.

Step 3 $\dfrac{1}{4}(\text{period}) = \dfrac{1}{4}(\pi) = \dfrac{\pi}{4}$

The x-coordinates of the five key points are:

starting point $= -\dfrac{\pi}{2}, -\dfrac{\pi}{2} + \dfrac{\pi}{4} = -\dfrac{\pi}{4}, -\dfrac{\pi}{2} + \dfrac{\pi}{2} = 0, -\dfrac{\pi}{2} + \dfrac{3\pi}{4} = \dfrac{\pi}{4}$,

and $-\dfrac{\pi}{2} + \dfrac{2\pi}{2} = \dfrac{\pi}{2}$

Step 4 Because $a = -1 < 0$, we graph $y = |-1|\cos\left[2\left(x + \dfrac{\pi}{2}\right)\right]$ by sketching

one cycle, beginning at $\left(-\dfrac{\pi}{2}, 1\right)$ and going through the points $\left(-\dfrac{\pi}{4}, 0\right)$,

$(0, -1), \left(\dfrac{\pi}{4}, 0\right)$, and $\left(\dfrac{\pi}{2}, 1\right)$. We then reflect this graph in the x-axis to

obtain the graph of $y = -\cos\left[2\left(x + \dfrac{\pi}{2}\right)\right]$.

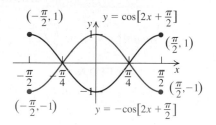

Practice Problem 7 Graph $y = -2\cos\left[3(x + \pi)\right]$ over a one-period interval. ■

To graph the function of the form $y = a\sin(bx - k)$ or $y = a\cos(bx - k)$, with $b > 0$, we first note that

$$bx - k = b\left(x - \frac{k}{b}\right) \quad \text{Factor out } b.$$

Consequently, to graph these functions, we rewrite their equations.

$$y = a\sin(bx - k) = a\sin\left[b\left(x - \frac{k}{b}\right)\right]$$
$$y = a\cos(bx - k) = a\cos\left[b\left(x - \frac{k}{b}\right)\right]$$

We can then use the procedures for graphing the functions of the form $y = a\sin b(x - c)$ and $y = a\cos b(x - c)$ by letting $\dfrac{k}{b} = c$.

EXAMPLE 8 **Finding the Period and Phase Shift**

Find the period and the phase shift of each function.

a. $y = 2\cos\left(3x - \dfrac{\pi}{2}\right)$ **b.** $y = 3\sin(\pi x + 2)$

SOLUTION

a.
$$y = 2\cos\left(3x - \frac{\pi}{2}\right) \quad \text{The given function}$$

$$= 2\cos\left[3\left(x - \frac{\pi}{6}\right)\right] \quad \text{Distributive property}$$

$$\underset{a}{\uparrow} \quad \underset{b}{\uparrow} \quad \underset{c}{\uparrow} \qquad \text{Compare with } y = a\cos b(x - c).$$

$$\text{Period} = \frac{2\pi}{b} = \frac{2\pi}{3} \qquad \text{Phase shift} = c = \frac{\pi}{6}$$

b.
$$y = 3\sin(\pi x + 2) \qquad \text{The given function}$$

$$y = 3\sin\pi\left[x - \left(-\frac{2}{\pi}\right)\right] \qquad \text{Distributive property}$$

$$\underset{a}{\uparrow} \quad \underset{b}{\uparrow} \qquad \underset{c}{\uparrow}$$

$$\text{Period} = \frac{2\pi}{b} = \frac{2\pi}{\pi} = 2 \qquad \text{Phase shift} = c = -\frac{2}{\pi} \qquad \blacksquare\blacksquare\blacksquare$$

Practice Problem 8 Find the period and the phase shift of each function.

a. $y = 3\sin\left(2x + \dfrac{\pi}{4}\right)$ **b.** $y = \cos(\pi x - 1)$ ■

EXAMPLE 9 **Graphing $y = a\sin(bx - k)$**

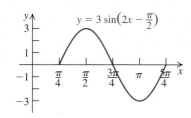

FIGURE 4.53

Graph $y = 3\sin\left(2x - \dfrac{\pi}{2}\right)$ over a one-period interval.

SOLUTION

Because $2x - \dfrac{\pi}{2} = 2\left(x - \dfrac{\pi}{4}\right)$, we can write $y = 3\sin\left(2x - \dfrac{\pi}{2}\right) =$

$3\sin\left[2\left(x - \dfrac{\pi}{4}\right)\right]$. This is the sinusoidal curve we graphed in Example 6 to show

the procedure for graphing the function $y = a\sin b(x - c)$ and $y = a\cos b(x - c)$.

The graph is shown again in Figure 4.53. ■■■

Practice Problem 9 Graph $y = 2\cos(3x - \pi)$ over a one-period interval. ■

Vertical Shifts

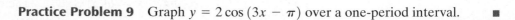

Recall that the graph of $y = f(x) + d$ results from shifting the graph of $y = f(x)$ vertically d units up if $d > 0$ and $|d|$ units down if $d < 0$. The next example shows this effect on a sinusoidal graph.

EXAMPLE 10   **Graphing** $y = a \cos b(x - c) + d$

Graph $y = \cos\left[2\left(x + \dfrac{\pi}{2}\right)\right] + 3$ over a one-period interval.

SOLUTION

In Example 7, we graphed the function $y = \cos\left[2\left(x + \dfrac{\pi}{2}\right)\right]$ in order to graph $y = -\cos\left[2\left(x + \dfrac{\pi}{2}\right)\right]$. To graph $y = \cos\left[2\left(x + \dfrac{\pi}{2}\right)\right] + 3$, shift the graph of $y = \cos\left[2\left(x + \dfrac{\pi}{2}\right)\right]$, up three units. See Figure 4.54.

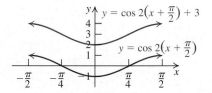

FIGURE 4.54 Graph of $y = \cos\left[2\left(x + \dfrac{\pi}{2}\right)\right] + 3$ ■ ■ ■

Practice Problem 10   Graph $y = \left[4 \sin\left(3x - \dfrac{\pi}{3}\right)\right] - 2$ over a one-period interval.   ■

EXAMPLE 11   **Modeling the Number of Daylight Hours in Paris**

Table 4.4 gives the average number of daylight hours in Paris each month. Let y represent the number of daylight hours in Paris in month x to find a function of the form $y = a \sin\left[b(x - c)\right] + d$ that models the hours of daylight throughout the year.

SOLUTION

Plot the values given in Table 4.4, representing the months by the integers 1 through 12 (Jan. $= 1, \ldots,$ Dec. $= 12$). Then sketch a function of the form $y = a \sin\left[b(x - c)\right] + d$ that models the points just graphed. See Figure 4.55.

TABLE 4.4

Daylight hours in Paris	
January	8.8
February	10.2
March	11.9
April	13.7
May	15.3
June	16.1
July	15.7
August	14.3
September	12.6
October	10.8
November	9.2
December	8.3

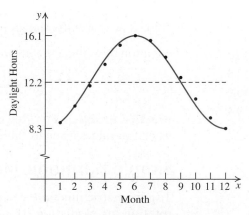

FIGURE 4.55 Hours of daylight in Paris each month

We want to find the values for the constants a, b, c, and d that will produce the graph shown in Figure 4.55. From Table 4.4, the range of the function is $[8.3, 16.1]$; so

$$\text{Amplitude} = a = \frac{\text{highest value} - \text{lowest value}}{2} = \frac{(16.1 - 8.3)}{2} = 3.9.$$

The sine graph has been shifted vertically by $\frac{1}{2}$ (highest value + lowest value), the average of the highest and lowest values. Thus, this average value gives

$$d = \frac{1}{2}(16.1 + 8.3) = 12.2.$$

The weather repeats every 12 months; so the period is 12. Because the period $= \frac{2\pi}{b}$, we can write

$$12 = \frac{2\pi}{b} \qquad \text{Replace period with 12.}$$

$$b = \frac{2\pi}{12} = \frac{\pi}{6} \qquad \text{Solve for } b.$$

So far, we can write $y = 3.9 \sin\left[\left(\frac{\pi}{6}\right)(x - c)\right] + 12.2$. The horizontally "unshifted" graph of $y = 3.9 \sin\left(\frac{\pi}{6}x\right) + 12.2$ is superimposed on Figure 4.55 in Figure 4.55A. This graph must be shifted to the right to fit the plotted data. To determine the phase shift, find the value of c when y has the greatest value. This value is $y = 16.1$ when $x = 6$ (in June) so we now have

$$16.1 = 3.9 \sin\left[\left(\frac{\pi}{6}\right)(6 - c)\right] + 12.2 \qquad \text{Replace } x \text{ with 6 and } y \text{ with 16.1.}$$

$$1 = \sin\left[\left(\frac{\pi}{6}\right)(6 - c)\right] \qquad \begin{array}{l}\text{Subtract 12.2 from both sides and} \\ \text{divide by 3.9.}\end{array}$$

Now $\sin\left[\left(\frac{\pi}{6}\right)(6 - c)\right]$ will first equal 1 at $\frac{\pi}{2}$. So

$$\frac{\pi}{6}(6 - c) = \frac{\pi}{2} \qquad \sin t = 1 \text{ when } t = \frac{\pi}{2}.$$

$$c = 3 \qquad \text{Solve for } c.$$

Note that this shifts the graph of $y = 3.9 \sin\left(\frac{\pi}{6}x\right) + 12.2$ three units right. The equation describing the hours of daylight in Paris *in month x* is

$$y = 3.9 \sin\left[\frac{\pi}{6}(x - 3)\right] + 12.2. \qquad \blacksquare\blacksquare\blacksquare$$

Practice Problem 11 Rework Example 11 for the city of Fargo, North Dakota, using the values given in Table 4.5. ∎

Simple Harmonic Motion

Trigonometric functions can often describe or model motion caused by vibration, rotation, or oscillation. The motions associated with sound waves, radio waves, alternating electric current, a vibrating guitar string, or the swing of a pendulum are examples of such motion. Another example is the movement of a ball suspended by a spring. If the ball is pulled downward and then released (ignoring the effects of

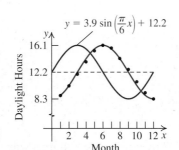

FIGURE 4.55A

TABLE 4.5

Daylight hours in Fargo	
January	9
February	10.3
March	11.9
April	13.6
May	15.1
June	15.8
July	15.4
August	14.2
September	12.5
October	10.9
November	9.4
December	8.6

friction and air resistance), the ball will repeatedly move up and down past its rest, or equilibrium, position. See Figure 4.56.

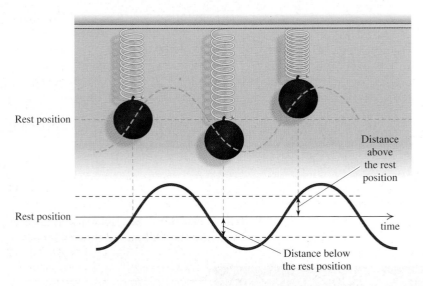

FIGURE 4.56 Simple harmonic motion

SIMPLE HARMONIC MOTION

An object whose position relative to an equilibrium position at time *t* can be described by either

$$y = a \sin \omega t \qquad \text{or} \qquad y = a \cos \omega t \quad (\omega > 0)$$

is in **simple harmonic motion**.

The *amplitude,* $|a|$, is the maximum distance the object reaches from its equilibrium position. The *period* of the motion, $\dfrac{2\pi}{\omega}$, is the time it takes for the object to complete one full cycle. The **frequency** of the motion is $\dfrac{\omega}{2\pi}$ and gives the number of cycles completed per unit of time.

If the distance above and below the rest position of the ball is graphed as a function of time, a sine or a cosine curve results. The amplitude of the curve is the maximum distance above the rest position the ball travels, and one cycle is completed as the ball travels from its highest position to its lowest position and back to its highest position.

EXAMPLE 12   Simple Harmonic Motion of a Ball Attached to a Spring

Suppose a ball attached to a spring is pulled down 6 inches and released and the resulting simple harmonic motion has a period of eight seconds. Write an equation for the ball's simple harmonic motion.

SOLUTION

We must first choose between an equation of the form $y = a \sin \omega t$ or $y = a \cos \omega t$. We start tracking the motion of the ball when $t = 0$. Note that for $t = 0$,

$$y = a \sin (\omega \cdot 0) = a \cdot \sin 0 = a \cdot 0 = 0 \text{ and } y = a \cos (\omega \cdot 0) = a \cdot \cos 0 = a \cdot 1 = a.$$

If we choose to start tracking the ball's motion when we release it after pulling it down 6 inches, we should choose $a = -6$ and $y = -6 \cos \omega t$. Because we pulled the ball *down* in order to start, a is negative. We now have the form of the equation of motion.

$$y = -6 \cos \omega t \qquad \text{When } t = 0, y = -6.$$

$$\text{period} = \frac{2\pi}{\omega} = 8 \qquad \text{The period is given as eight seconds.}$$

$$\omega = \frac{2\pi}{8} = \frac{\pi}{4} \qquad \text{Solve fo } \omega.$$

Replacing ω with $\dfrac{\pi}{4}$ in $y = -6 \cos \omega t$ yields the equation of the ball's simple harmonic motion:

$$y = -6 \cos \frac{\pi}{4} t$$ ■ ■ ■

Practice Problem 12 Rework Example 12 with the motion period of three seconds and the ball pulled down 4 inches. ■

SECTION 4.4 ■ Exercises

A EXERCISES Basic Skills and Concepts

1. The lowest point on the graph of $y = \cos x$, $0 \le x \le 2\pi$, occurs when $x =$ _____ π.

2. The highest point on the graph of $y = \sin x, 0 \le x \le 2\pi$, occurs when $x =$ _____ $\frac{\pi}{2}$.

3. The range of the cosine function is ___$[-1, 1]$___.

4. The maximum value of $y = -2 \sin x$ is ___2___.

5. *True or False* The amplitude of $y = -3 \cos x$ is -3. False

6. *True or False* The range of $y = \sin 3x$ is $[-3, 3]$. False

7. *True or False* The period of $y = 7 \cos 2x$ is π. True

8. *True or False* The x-coordinate of the first key point for $y = \sin\left(x + \dfrac{\pi}{2}\right)$ is $-\dfrac{\pi}{2}$. True

In Exercises 9–28, sketch the graph of each given equation over the interval $[-2\pi, 2\pi]$.

9. $y = 2 \sin x$ †

10. $y = 4 \cos x$ †

11. $y = -\dfrac{1}{2} \sin x$ †

12. $y = -2 \sin x$ †

13. $y = \dfrac{3}{2} \cos x$ †

14. $y = \dfrac{5}{4} \sin x$ †

15. $y = \cos 2x$ †

16. $y = \sin 4x$ †

17. $y = \cos \dfrac{2}{3} x$ †

18. $y = \sin \dfrac{4}{3} x$ †

19. $y = \cos\left(x + \dfrac{\pi}{2}\right)$ †

20. $y = \sin\left(x + \dfrac{\pi}{4}\right)$ †

21. $y = \cos\left(x - \dfrac{\pi}{3}\right)$ †

22. $y = \sin(x - \pi)$ †

23. $y = 2 \cos\left(x - \dfrac{\pi}{2}\right)$ †

24. $y = 2 \sin\left(x + \dfrac{\pi}{3}\right)$ †

25. $y = \sin x + 1$ †

26. $y = \cos x - 2$ †

27. $y = -\cos x + 1$ †

28. $y = \sin x - 3$ †

In Exercises 29–36, find the amplitude, period, and phase shift of each given function.

29. $y = 5 \cos(x - \pi)$ †

30. $y = 3 \sin\left(x - \dfrac{\pi}{8}\right)$ †

31. $y = 7 \cos\left[9\left(x + \dfrac{\pi}{6}\right)\right]$ †

32. $y = 11 \sin\left[8\left(x + \dfrac{\pi}{3}\right)\right]$ †

33. $y = -6 \cos\left[\dfrac{1}{2}(x + 2)\right]$ †

34. $y = -8 \sin\left[\dfrac{1}{5}(x + 9)\right]$ †

35. $y = 0.9 \sin\left[0.25\left(x - \dfrac{\pi}{4}\right)\right]$ †

36. $y = \sqrt{5} \cos[\pi(x + 1)]$ †

In Exercises 37–44, graph each function over a one-period interval.

37. $y = -4 \cos\left(x + \dfrac{\pi}{6}\right)$ †

38. $y = -3 \sin\left(x - \dfrac{\pi}{6}\right)$ †

39. $y = \dfrac{5}{2} \sin\left[2\left(x - \dfrac{\pi}{4}\right)\right]$ †

40. $y = \dfrac{3}{2} \cos\left[2\left(x + \dfrac{\pi}{3}\right)\right]$ †

41. $y = -5 \cos\left[4\left(x - \dfrac{\pi}{6}\right)\right]$ †

42. $y = -3 \sin\left[4\left(x + \dfrac{\pi}{6}\right)\right]$ †

43. $y = \dfrac{1}{2} \sin\left[4\left(x + \dfrac{\pi}{4}\right)\right] + 2$ †

44. $y = -\dfrac{1}{2} \cos\left[2\left(x - \dfrac{\pi}{2}\right)\right] - 3$ †

†Due to space constrictions, answers to these exercises may be found in the Answers beginning on page A–1 in the back of the book.

In Exercises 45–54, write each function in the form
$y = a \sin b(x - c)$ **or** $y = a \cos b(x - c)$. **Find each period and phase shift.**

45. $y = 4 \cos \left(2x + \dfrac{\pi}{3}\right)$ †   **46.** $y = 5 \sin \left(3x + \dfrac{\pi}{2}\right)$ †

47. $y = -\dfrac{3}{2} \sin (2x - \pi)$ †   **48.** $y = -2 \cos \left(5x - \dfrac{\pi}{4}\right)$ †

49. $y = 3 \cos \pi x$ †   **50.** $y = \sin \dfrac{\pi x}{3}$ †

51. $y = \dfrac{1}{2} \cos \left(\dfrac{\pi x}{4} + \dfrac{\pi}{4}\right)$ †   **52.** $y = -\sin \left(\dfrac{\pi x}{6} + \dfrac{\pi}{6}\right)$ †

53. $y = 2 \sin (\pi x + 3)$ †   **54.** $y = -\cos \left(\pi x - \dfrac{1}{4}\right)$ †

B EXERCISES Applying the Concepts

55. Pulse and blood pressure. Blood pressure is given by two numbers written as a fraction: $\dfrac{\text{systolic}}{\text{diastolic}}$. The *systolic* reading is the maximum pressure in an artery, and the *diastolic* reading is the minimum pressure in an artery. As the heart beats, the systolic measurement is taken; when the heart rests, the diastolic measurement is taken. A reading of $\dfrac{120}{80}$ is considered normal. Your pulse is the number of heartbeats per minute. Suppose Desmond's blood pressure after t minutes is given by

$$p(t) = 20 \sin (140\pi t) + 122,$$

where $p(t)$ is the pressure in millimeters of mercury.
  a. Find the period and explain how it relates to pulse. †
  b. Graph the function p over one period. †
  c. What is Desmond's blood pressure? †

56. Electrical circuit voltage. The voltage $V(t)$ in an electrical circuit is given by $V(t) = 4.5 \cos 160\pi t$, where t is measured in seconds.
  a. Find the amplitude and the period for $V(t)$. †
  b. Find the frequency of $V(t)$, that is, the number of cycles completed in one second. Frequency = 80
  c. Graph $V(t)$ over the interval $\left[0, \dfrac{1}{40}\right]$. †
  d. Use a graphing calculator to verify your graph.

57. Voltage of an electric outlet. The voltage across the terminals of a typical electrical outlet is approximated by $V(t) = 156 \sin (120\pi t)$, where t is measured in seconds.
  a. Find the amplitude and the period for $V(t)$. †
  b. Find the frequency of $V(t)$, that is, the number of cycles completed in one second. Frequency = 60
  c. Graph $V(t)$ over the interval $\left[0, \dfrac{1}{30}\right]$. †
  d. Use a graphing calculator to verify your graph.

58. Oscillating ball. Suppose a ball attached to a spring is pulled down 5 inches and released and that the resulting simple harmonic motion has a period of ten seconds. Write an equation of the ball's simple harmonic motion.

$$y = -5 \cos \left(\dfrac{\pi}{5} t\right)$$

59. Kangaroo population. The kangaroo population in a certain region is given by the function

$$N(t) = 650 + 150 \sin 2t,$$

where the time t is measured in years.
  a. What is the largest number of kangaroos present in the region at any time? 800 kangaroos
  b. What is the smallest number of kangaroos present in the region at any time? 500 kangaroos
  c. How much time elapses between occurrences of the largest and the smallest kangaroo population?

60. Lion and deer populations. The number of deer in a region is given by $D(t) = 450 \sin \left(\dfrac{3t}{5}\right) + 1200$, where t is in years and the number of lions in that same region is given by $L(t) = 225 \sin \left(\dfrac{2}{5}(t - 2)\right) + 500$, where t is in years.
  a. Graph both functions on the same coordinate plane for $0 \le t \le 10$. †
  b. What can you say about the relationship between these populations based on the graphs obtained in (a)? †
  c. Use a graphing calculator to verify your graph.

61. Daylight hours in London. The table gives the average number of daylight hours in London each month, where $x = 1$ represents January. Let y represent the number of daylight hours in London in month x. Find a function of the form $y = a \sin b(x - c) + d$ that models the hours of daylight throughout the year. $y = 4.35 \sin \dfrac{\pi}{6}(x - 3) + 12.25$

Jan	Feb	Mar	Apr	May	June
8.4	10	11.9	13.9	15.6	16.6

July	Aug	Sept	Oct	Nov	Dec
16.1	14.6	12.6	10.7	8.9	7.9

62. Daylight hours in Washington, D.C. The table gives the average number of daylight hours in Washington, D.C., each month, where $x = 1$ represents January. Let y represent the number of daylight hours in Washington, D.C., in month x. Find a function of the form $y = a \sin b(x - c) + d$ that models the hours of daylight throughout the year. $y = 2.65 \sin \dfrac{\pi}{6}(x - 3) + 12.15$

Jan	Feb	Mar	Apr	May	June
9.8	10.8	12	13.2	14.3	14.8

July	Aug	Sept	Oct	Nov	Dec
14.6	13.6	12.4	11.2	10.1	9.5

In Exercises 63 and 64, use the table of average monthly temperatures (where Jan = 1, . . . , Dec = 12) to find a function of the form $y = a \sin b(x - c) + d$ **that models the temperature in the given city. The four seasons of the year match remarkably well with the quarter periods of the sine function. Since the temperatures repeat each year, the period**

Answers:
59. c. $\dfrac{\pi}{2}$ years ≈ 1.57 years

is 12 months; so $b = \dfrac{2\pi}{12} = \dfrac{\pi}{6}$. The amplitude $a = \dfrac{1}{2}$ (high

temperature − low temperature), and the vertical shift

$d = \dfrac{1}{2}$ (high temperature + low temperature). Since the

maximum value a of an unshifted sine curve with period 12

occurs at $\dfrac{1}{4}(12) = 3$ units from the start of the cycle, you can

find c by solving the equation $c =$ hottest month $- 3$.

63. **Average temperature.** The monthly average temperatures for Augusta, Maine, are given in the table.

Month	1	2	3	4	5	6	7	8	9	10	11	12
°F	19	22	31	43	54	63	70	67	59	48	37	24

Source: www.weatherbase.com/

$$y = 25.5 \sin \dfrac{\pi}{6}(x - 4) + 44.5$$

   a. Find a function of the form $y = a \sin b(x - c) + d$ that models these temperatures.

   b. Graph the ordered pairs and the function on the same coordinate system.

   c. Compute the function values for January, April, July, and October and compare them with the table values for these months. †

 d. Use a graphing calculator to verify your graph.

64. **Average temperatures.** The monthly average temperatures for Nashville, Tennessee, are given in the table.

Month	1	2	3	4	5	6	7	8	9	10	11	12
°F	40	45	53	63	71	79	83	81	74	63	52	44

Source: www.weatherbase.com/

$$y = 21.5 \sin \dfrac{\pi}{6}(x - 4) + 61.5$$

   a. Find a function of the form $y = a \sin b(x - c) + d$ that models these temperatures.

   b. Graph the ordered pairs and the function on the same coordinate system. †

   c. Compute the function values for January, April, July, and October and compare them to the table values for these months.

 d. Use a graphing calculator to verify your graph.
   c. January $= f(1) = 40$; April $= f(4) = 61.5$;
   July $= f(7) = 83$; October $= f(10) = 61.5$

C EXERCISES Beyond the Basics

In Exercises 65–78, graph each equation over the interval $[0, 2\pi]$.

65. $y = 2 \sin (3x + \pi)$ †   **66.** $y = 4 \cos \left(2x + \dfrac{\pi}{2} \right)$ †

67. $y = -\dfrac{1}{2} \cos \left(\dfrac{2}{3}x - \dfrac{\pi}{9} \right)$ † **68.** $y = -2 \sin (\pi x - 2)$ †

69. $y = 3 \sin \left(-x + \dfrac{\pi}{4} \right)$ †   **70.** $y = -2 \sin \left(-2x + \dfrac{\pi}{3} \right)$ †

71. $y = 2 \cos \left(-3x + \dfrac{\pi}{2} \right) + 2$ †

72. $y = -3 \cos \left(-\dfrac{x}{\pi} + \dfrac{1}{2} \right) + 3$ †

73. $y = |\sin x|$ †   **74.** $y = |\cos x|$ †

75. $y = x \sin x$ †   **76.** $y = x \cos x$ †

77. $y = \sqrt{x} \sin x$ †   **78.** $y = |x \sin x|$ †

79. Graph the equations $y = x^2$, $y = -x^2$, and $y = x^2 \sin x$ on the same coordinate plane. †

80. We know that $\sin (x + 2\pi) = \sin x$ for all real numbers x. Consequently, if 2π is not the period for the sine function, then $\sin (x + p) = \sin x$ for some real number p, with $0 < p < 2\pi$, and all real numbers x.

   a. By evaluating $\sin (x + p) = \sin x$ for $x = 0$, show that if the period is not 2π, then $p = \pi$.

   b. Find a value of x for which $\sin (x + \pi) \neq \sin x$ and conclude that the period of the sine function is 2π.

In Exercises 81–84, write an equation for each graph.

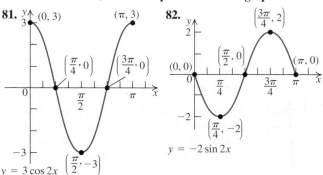

81. $y = 3 \cos 2x$

82. $y = -2 \sin 2x$

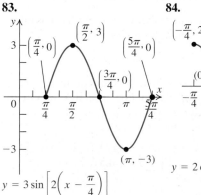
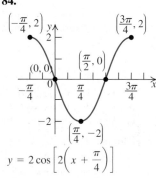

83. $y = 3 \sin \left[2\left(x - \dfrac{\pi}{4} \right) \right]$

84. $y = 2 \cos \left[2\left(x + \dfrac{\pi}{4} \right) \right]$

85. Let $f(x) = a \sin [b(x - c)] + d$. If $f(k) = a + d$, write an expression for c in terms of k and b. $c = k - \dfrac{\pi}{2b}$

Critical Thinking

86. Suppose a function of the form $y = a \sin b(x - c) + d$, with $a > 0$, has period $= 20$.

   a. In a one-period interval, how many units are between the value of x at which the maximum value is attained and the value of x at which the minimum value is attained? 10 units

   b. In a one-period interval, how many units are between the value of x at which y has value d and the value of x at which the maximum value is attained? 5 units

87. Suppose the minimum value of a function of the form $y = a \cos b(x - c) + d$, with $a > 0$, occurs at a value of x that is five units from the value of x at which the function has the maximum value. What is the period of the function? 10

Graphs of the Other Trigonometric Functions

Before Starting this Section, Review

1. Vertical asymptotes (Section 2.5, page 156)

2. Transformations of functions (Section 1.5, page 62)

Objectives

1 Graph the tangent and cotangent functions.

2 Graph the cosecant and secant functions.

THE U.S.S. ENTERPRISE AND MACH NUMBERS

Star Trek fans probably know quite a bit about the warp speeds attainable by the Starfleet's U.S.S. Enterprise. Near Earth, Mach numbers provide a more relevant measure of speed.

The Mach number (named after the Austrian physicist Ernst Mach) is the ratio of the speed of the aircraft to the speed of sound. For example, an aircraft traveling at Mach 2 is traveling at twice the speed of sound. An airplane flying at less than the speed of sound is traveling at *subsonic* speed; at the speed of sound (Mach 1), the plane is traveling at *transonic* speed; at speeds greater than the speed of sound, it is traveling at *supersonic* speeds; and at speeds greater than 5 times the speed of sound, it is traveling at *hypersonic* speeds. In Example 4, we see how at supersonic and hypersonic speeds Mach numbers are given as values of the cosecant function and we graph a range of Mach numbers that are attainable by various military aircraft. ∎

In this section, you learn to sketch the graphs of the tangent, cotangent, cosecant, and secant functions.

1 Graph the tangent and cotangent functions.

Graphs of the Tangent and Cotangent Functions

The graph of $y = \tan x$ The tangent function differs from the sine and cosine functions in three significant ways.

1. The tangent function has period π. Figure 4.57 illustrates that $\tan(t + \pi) = \tan t$. Therefore, values of the tangent function repeat every π units.

2. Because $\tan x = \dfrac{\sin x}{\cos x}$, we have $\tan x = 0$ when $\sin x = 0$, and $\tan x$ is undefined when $\cos x = 0$. In particular, the tangent function is undefined at $\pm\dfrac{\pi}{2}, \pm\dfrac{3\pi}{2}, \pm\dfrac{5\pi}{2}, \dots$.

3. The tangent function has no amplitude; that is, there are no minimum or maximum y-values. Another way to say this is that the range of $y = \tan x$ is $(-\infty, \infty)$.

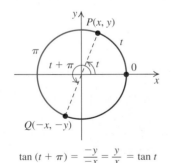

$\tan(t + \pi) = \dfrac{-y}{-x} = \dfrac{y}{x} = \tan t$

FIGURE 4.57 $\tan(t + \pi) = \tan t$

Note that $\tan(-x) = \dfrac{\sin(-x)}{\cos(-x)} = \dfrac{-\sin x}{\cos x} = -\tan x$. Therefore, the tangent is an odd function; so its graph is symmetric with respect to the origin. To graph the tangent function, we use the approximate values in Table 4.6 for $y = \tan x$, where x is in the interval $\left[0, \dfrac{\pi}{2}\right)$.

305

TECHNOLOGY CONNECTION

To graph the tangent, cotangent, secant, or cosecant functions on a graphing calculator, use Dot mode and appropriate window settings.

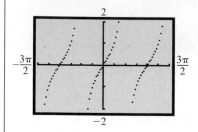

$Y_1 = \tan x$, Dot mode

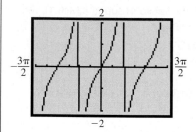

$Y_1 = \tan x$, Connected mode

TABLE 4.6

x	0	$\dfrac{\pi}{6}$	$\dfrac{\pi}{4}$	$\dfrac{\pi}{3}$	$\dfrac{7\pi}{18}$	$\dfrac{4\pi}{9}$	$\dfrac{17\pi}{36}$
$y = \tan x$	0	0.6	1	1.7	2.7	5.7	11.4

As x approaches $\dfrac{\pi}{2}$, the values of $\tan x$ continue to increase; in fact, they increase indefinitely. The decimal value of $\dfrac{\pi}{2}$ is approximately 1.5707963. When $x = \dfrac{89\pi}{180} \approx 1.55$, $\tan x > 57$; when $x = 1.57$, $\tan x > 1255$; and when $x = 1.5707$, $\tan x > 10{,}000$. The graph of $y = \tan x$ has a vertical asymptote at $x = \dfrac{\pi}{2}$. See Figure 4.58(a).

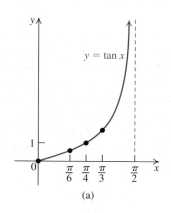
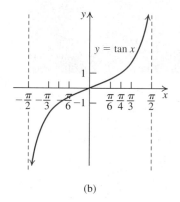

(a)   (b)

FIGURE 4.58 **Graphing $y = \tan x$**

We can extend the graph of $y = \tan x$ to the interval $\left(-\dfrac{\pi}{2}, \dfrac{\pi}{2}\right)$ by using the fact that the graph is symmetric with respect to the origin. See Figure 4.58(b).

Because the period of the tangent function is π and we have graphed $y = \tan x$ over an interval of length π, we can draw the complete graph of $y = \tan x$ by repeating the graph in Figure 4.58(b) indefinitely to the left and right, over intervals of length π. Figure 4.59 shows three cycles of the graph.

The graph of $y = \cot x$ The cotangent function also has period π. Since $\cot x = \dfrac{\cos x}{\sin x}$, we have $\cot x = 0$ when $\cos x = 0$, and $\cot x$ is undefined when $\sin x = 0$. In particular, the cotangent function is undefined at $0, \pm\pi, \pm2\pi, \pm3\pi, \ldots$

The cotangent is an odd function because

$$\cot(-x) = \frac{\cos(-x)}{\sin(-x)} = \frac{\cos x}{-\sin x} = -\cot x.$$

The fact that $\cot x = \dfrac{1}{\tan x}$ tells us that $\tan x$ and $\cot x$ have the same sign everywhere they are both defined and that when $|\tan x|$ is large, $|\cot x|$ is small (and conversely). Figure 4.60 shows three cycles of the graph of the cotangent function.

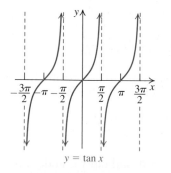

$y = \tan x$

FIGURE 4.59

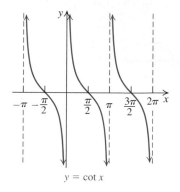

$y = \cot x$

FIGURE 4.60

Main Facts About $y = \tan x$ and $y = \cot x$

	$y = \tan x$	$y = \cot x$
Period	π	π
Domain	All real numbers except odd multiples of $\dfrac{\pi}{2}$	All real numbers except integer multiples of π
Range	$(-\infty, \infty)$	$(-\infty, \infty)$
Vertical Asymptotes	At $x =$ odd multiples of $\dfrac{\pi}{2}$	At $x =$ integer multiples of π
x-intercepts	At integer multiples of π	At odd multiples of $\dfrac{\pi}{2}$
Symmetry	$\tan(-x) = -\tan x$, odd function, so symmetric with respect to the origin	$\cot(-x) = -\cot x$, odd function, so symmetric with respect to the origin

Graphing $y = a \tan b(x - c)$ The procedure for graphing $y = a \tan b(x - c)$ is based on the essential features of the graph of $y = \tan x$.

FINDING THE SOLUTION: A PROCEDURE

EXAMPLE 1   **Graphing $y = a \tan b(x - c)$ or $y = a \cot b(x - c)$**

OBJECTIVE

Graph a function of the form
$y = a \tan [b(x - c)]$ *or* $y = a \cot [b(x - c)]$, *where $b > 0$, by finding the period and phase shift.*

Step 1   Find the following:

vertical stretch factor $= |a|$

period $= \dfrac{\pi}{b}$

phase shift $= c$

Step 2   Locate two adjacent vertical asymptotes. For $y = a \tan [b(x - c)]$, solve

$b(x - c) = -\dfrac{\pi}{2}$ and $b(x - c) = \dfrac{\pi}{2}$.

For $y = a \cot [b(x - c)]$, solve
$b(x - c) = 0$ and $b(x - c) = \pi$.

Step 3   Divide the interval on the x-axis between the two vertical asymptotes into four equal parts, each of length $\dfrac{1}{4}\left(\dfrac{\pi}{b}\right)$.

EXAMPLE

Graph $y = 3 \tan \left[2\left(x - \dfrac{\pi}{4} \right) \right]$.

1. $y = 3 \tan \left[2\left(x - \dfrac{\pi}{4} \right) \right]$

$a = 3 \qquad b = 2 \qquad c = \dfrac{\pi}{4}$

vertical stretch factor $= |a| = |3| = 3$

period $= \dfrac{\pi}{b} = \dfrac{\pi}{2}$ phase shift $= c = \dfrac{\pi}{4}$

2. $2\left(x - \dfrac{\pi}{4} \right) = -\dfrac{\pi}{2}$ $\qquad\qquad$ $2\left(x - \dfrac{\pi}{4} \right) = \dfrac{\pi}{2}$

$x - \dfrac{\pi}{4} = -\dfrac{\pi}{4}$ $\qquad\qquad$ $x - \dfrac{\pi}{4} = \dfrac{\pi}{4}$

$x = -\dfrac{\pi}{4} + \dfrac{\pi}{4}$ $\qquad\qquad$ $x = \dfrac{\pi}{4} + \dfrac{\pi}{4}$

$x = 0$ $\qquad\qquad\qquad$ $x = \dfrac{\pi}{2}$

3. The interval $\left(0, \dfrac{\pi}{2} \right)$ has length $\dfrac{\pi}{2}$, and $\dfrac{1}{4}\left(\dfrac{\pi}{2}\right) = \dfrac{\pi}{8}$.

The division points of the interval $\left(0, \dfrac{\pi}{2} \right)$ are

$0 + \dfrac{\pi}{8} = \dfrac{\pi}{8}, 0 + 2\left(\dfrac{\pi}{8}\right) = \dfrac{\pi}{4},$ and $0 + 3\left(\dfrac{\pi}{8}\right) = \dfrac{3\pi}{8}$

continued on the next page

Step 4 Evaluate the function at the three x values found in Step 3 that are the division points of the interval.

4.

x	$y = 3\tan\left[2\left(x - \dfrac{\pi}{4}\right)\right]$
$\dfrac{\pi}{8}$	-3
$\dfrac{\pi}{4}$	0
$\dfrac{3\pi}{8}$	3

Step 5 Sketch the vertical asymptotes using the values found in Step 2. Connect the points in Step 4 with a smooth curve in the standard shape of a cycle for the given function. Repeat the graph to the left and right over intervals of length $\dfrac{\pi}{b}$.

5.

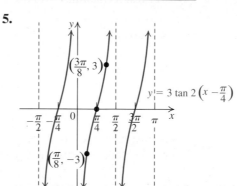

■ ■ ■

Practice Problem 1 Graph $y = -3\tan\left(x + \dfrac{\pi}{4}\right)$. ■

The fact that both the tangent and cotangent functions are odd functions allows us to graph $y = a\tan[b(x - c)]$ and $y = a\cot[b(x - c)]$ when $b < 0$. For example, to graph $y = 3\tan\left[(-2)\left(x - \dfrac{\pi}{2}\right)\right]$ we rewrite the equation as $y = -3\tan\left[2\left(x - \dfrac{\pi}{2}\right)\right]$.

EXAMPLE 2 **Graphing $y = a\cot[b(x - c)]$**

Graph $y = -4\cot\left(x - \dfrac{\pi}{2}\right)$ over the interval $[-\pi, 2\pi]$.

SOLUTION

Step 1 For $y = -4\cot\left(x - \dfrac{\pi}{2}\right)$, we have $a = -4$, $b = 1$, and $c = \dfrac{\pi}{2}$. Therefore,

vertical stretch factor $= |-4| = 4$ period $= \dfrac{\pi}{1} = \pi$ phase shift $= \dfrac{\pi}{2}$

Step 2 Locate two adjacent asymptotes for a cotangent function, solve the equations:

$$x - \dfrac{\pi}{2} = 0 \quad \text{and} \quad x - \dfrac{\pi}{2} = \pi.$$

$$x = \dfrac{\pi}{2} \qquad\qquad x = \dfrac{3\pi}{2}$$

Step 3 The interval $\left(\dfrac{\pi}{2}, \dfrac{3\pi}{2}\right)$ has length π, the period of the cotangent function.

The division points of $\left(\dfrac{\pi}{2}, \dfrac{3\pi}{2}\right)$ are $\dfrac{\pi}{2} + \dfrac{1}{4}(\pi) = \dfrac{3\pi}{4}$,

$\dfrac{\pi}{2} + \dfrac{2}{4}(\pi) = \pi$, and $\dfrac{\pi}{2} + \dfrac{3}{4}(\pi) = \dfrac{5\pi}{4}$.

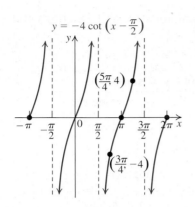

FIGURE 4.61 Graph of

$$y = -4\cot\left(x - \frac{\pi}{2}\right)$$

Step 4   Evaluate the function at $\dfrac{3\pi}{4}$, π, and $\dfrac{5\pi}{4}$.

x	$y = -4\cot\left(x - \dfrac{\pi}{2}\right)$
$\dfrac{3\pi}{4}$	-4
π	0
$\dfrac{5\pi}{4}$	4

Step 5   Sketch the vertical asymptotes at $x = \dfrac{\pi}{2}$ and $x = \dfrac{3\pi}{2}$. Draw the cycle for the cotangent function through the three graph points found in Step 4. Repeat the graph to the left or right over intervals of length π. See Figure 4.61.   ■ ■ ■

Practice Problem 2   Graph $y = -3\cot\left(x + \dfrac{\pi}{4}\right)$ over the interval $\left(-\dfrac{3\pi}{4}, \dfrac{5\pi}{4}\right)$. ▨

2   Graph the cosecant and secant functions.

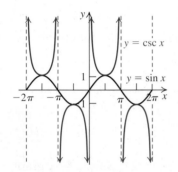

FIGURE 4.62 Graph of $y = \csc x$

Graphs of the Secant and Cosecant Functions

We can use the same technique to graph the secant and cosecant functions that we used to graph the cotangent function. Both functions are reciprocal functions; that is,

$$\csc x = \frac{1}{\sin x} \quad \text{and} \quad \sec x = \frac{1}{\cos x}$$

Consider the first equation. For every x in the domain of the cosecant function, $\sin x$ and $\csc x$ have the same sign, and when $|\sin x|$ is small, $|\csc x|$ is large (and conversely).

   Now $\csc x$ is undefined when $\sin x = 0$. Consequently, $\csc x$ is undefined at $0, \pm\pi, \pm2\pi, \pm3\pi, \ldots$ In fact, the graph of the cosecant function has a vertical asymptote at the integer multiples of π. Because $\csc x = 1$ whenever $\sin x = 1$ and $\csc x = -1$ when $\sin x = -1$, the graphs of $y = \csc x$ and $y = \sin x$ coincide if $\sin x = \pm 1$. See Figure 4.62.

   Because $\sec x = \dfrac{1}{\cos x}$, the relationship between the graphs of $y = \sec x$ and $y = \cos x$ is similar to the relationship between the graphs of $y = \sin x$ and $y = \csc x$. See Figure 4.63.

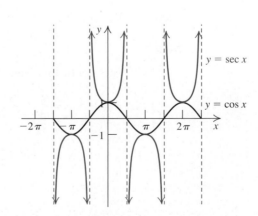

FIGURE 4.63 Graph of $y = \sec x$

We summarize the essential facts about the secant and cosecant functions and their graphs.

Main Facts About $y = \csc x$ and $y = \sec x$		
	$y = \csc x$	$y = \sec x$
Period	2π	2π
Domain	All real numbers except integer multiples of π	All real numbers except odd multiples of $\dfrac{\pi}{2}$
Range	$(-\infty, -1] \cup [1, \infty)$	$(-\infty, -1] \cup [1, \infty)$
Vertical Asymptotes	$x =$ integer multiples of π	$x =$ odd multiples of $\dfrac{\pi}{2}$
x-intercepts	No x-intercepts	No x-intercepts
Symmetry	$\csc(-x) = -\csc x$, odd function, origin symmetry	$\sec(-x) = \sec x$, even function with y-axis symmetry

The procedures for graphing the functions

$$y = a \csc [b(x - c)] \text{ and } y = a \sec [b(x - c)]$$

rely on those previously described for the related sine and cosine functions.

EXAMPLE 3 **Graphing $y = a \csc [b(x - c)]$**

Graph $y = 3 \csc 2x$ over a two-period interval.

SOLUTION

Because $\csc 2x = \dfrac{1}{\sin 2x}$, $3 \csc 2x = 3 \sin 2x$ when $\sin 2x = \pm 1$. So we first graph $y = 3 \sin 2x$.

We follow the steps for graphing a function of this type given in Section 4.4.

Step 1 $y = 3 \sin 2x$

amplitude $= 3$ period $= \dfrac{2\pi}{2} = \pi$ no phase shift, because $c = 0$

Step 2   The starting point for the cycle is $x = 0$. One cycle is graphed over $[0, \pi]$.

Step 3 $\dfrac{1}{4}(\text{period}) = \dfrac{1}{4}\pi = \dfrac{\pi}{4}$

The x-coordinates of the five key points are:

$$x = 0, \quad x = \frac{\pi}{4}, \quad x = 2\left(\frac{\pi}{4}\right) = \frac{\pi}{2}, \quad x = 3\left(\frac{\pi}{4}\right) = \frac{3\pi}{4}, \quad x = 4\left(\frac{\pi}{4}\right) = \pi.$$

Step 4   Sketch the graph of $y = 3 \sin 2x$ through the key points $(0, 0)$, $\left(\dfrac{\pi}{4}, 3\right)$, $\left(\dfrac{\pi}{2}, 0\right)$, $\left(\dfrac{3\pi}{4}, -3\right)$, and $(\pi, 0)$. See Figure 4.64.

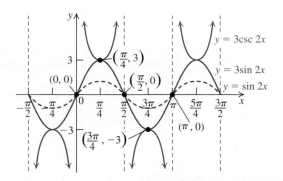

FIGURE 4.64 Graph of $y = 3 \csc 2x$

Step 5   Use the graph from Step 4 to extend the graph to the interval $\left(-\dfrac{\pi}{2}, \dfrac{3\pi}{2}\right)$.

Now use the technique that generated the graph of $y = \csc x$ from that of $y = \sin x$ to graph $y = 3 \csc 2x$. Start at the common graph points $\left(\dfrac{\pi}{4}, 3\right)$ and $\left(\dfrac{3\pi}{4}, -3\right)$; then use the reciprocal relationship $y = 3 \csc 2x = 3\left(\dfrac{1}{\sin 2x}\right)$. See Figure 4.64.   ■■■

Practice Problem 3   Graph $y = 2 \sec 3x$ over a two-period interval.   ■

EXAMPLE 4   Graphing a Range of Mach Numbers

When a plane travels at supersonic and hypersonic speeds, small disturbances in the atmosphere are transmitted downstream within a cone. The cone intersects the ground; Figure 4.65 shows the edge of the cone's intersection with the ground. The sound waves strike the edge of the cone at a right angle. The speed of the sound wave is represented by leg s of the right triangle in Figure 4.65. The plane is moving at speed v, which is represented by the hypotenuse of the right triangle in this figure.

The Mach number, M, is given by

$$M = M(x) = \frac{\text{speed of the aircraft}}{\text{speed of sound}} = \frac{v}{s} = \csc\left(\frac{x}{2}\right),$$

where x is the angle at the vertex of the cone. Graph the Mach number function, $M(x)$, as the angle at the vertex of the cone varies. What is the range of Mach numbers associated with the interval $\left[\dfrac{\pi}{4}, \pi\right)$?

SOLUTION

Because $\csc\dfrac{x}{2} = \dfrac{1}{\sin\dfrac{x}{2}}$, first graph $y = \sin\dfrac{x}{2}$. For convenience, we have sketched the graph of $y = \sin\dfrac{x}{2}$ over the interval $[0, 2\pi]$ in Figure 4.66. The graph of $y = \csc\dfrac{x}{2}$ is

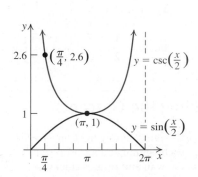

FIGURE 4.65 Sonic cone

FIGURE 4.66 Mach numbers

sketched over the interval $(0, 2\pi)$ using the reciprocal relationship between the sine graph and the cosecant graph.

$$\text{For } x = \frac{\pi}{4}, \quad y = \csc\frac{x}{2} = \csc\frac{\pi}{8} \approx 2.6.$$

$$\text{For } x = \pi, \quad y = \csc\frac{x}{2} = \csc\frac{\pi}{2} = 1.$$

The range of Mach numbers associated with the interval $\left[\frac{\pi}{4}, \pi\right)$ is $(1, 2.6]$. ■ ■ ■

Practice Problem 4 In Example 4, what is the range of Mach numbers associated with the interval $\left[\frac{\pi}{8}, \frac{\pi}{4}\right]$? ■

SECTION 4.5 ■ Exercises

A EXERCISES Basic Skills and Concepts

1. The tangent function has period ____π____.

2. The value of x with $0 \le x \le \pi$ for which the tangent function is undefined is ____$\frac{\pi}{2}$____.

3. If $0 \le x \le \pi$ and $\cot x = 0$, then $x =$ ____$\frac{\pi}{2}$____.

4. The maximum value of $y = \csc x$ for $\pi \le x \le 2\pi$ is ____-1____.

5. *True or False* The range of $y = \cot x$ is all real numbers. **True**

6. *True or False* The vertical stretch factor for
$y = -3\tan\left(x - \frac{\pi}{4}\right)$ is -3. **False**

7. *True or False* The phase shift for $y = \tan 3\left(x - \frac{\pi}{2}\right)$ is $\frac{\pi}{6}$. **False**

8. *True or False* The secant function is the inverse of the sine function. **False**

In Exercises 9–26, graph each function over a one-period interval.

9. $y = \tan\left(x - \frac{\pi}{4}\right)$ †

10. $y = \tan\left(x + \frac{\pi}{4}\right)$ †

11. $y = \cot\left(x + \frac{\pi}{4}\right)$ †

12. $y = \cot\left(x - \frac{\pi}{4}\right)$ †

13. $y = \tan 2x$ †

14. $y = \tan\frac{x}{2}$ †

15. $y = \cot\frac{x}{2}$ †

16. $y = \cot 2x$ †

17. $y = -\tan x$ †

18. $y = -\cot x$ †

19. $y = 3\tan x$ †

20. $y = 3\cot x$ †

21. $y = \sec 2x$ †

22. $y = \sec\frac{x}{2}$ †

23. $y = \csc 3x$ †

24. $y = \csc\frac{x}{3}$ †

25. $y = \sec(x - \pi)$ †

26. $y = \csc(x - \pi)$ †

In Exercises 27–48, graph each function over a two-period interval.

27. $y = \tan 2\left(x + \frac{\pi}{2}\right)$ †

28. $y = \tan 2\left(x - \frac{\pi}{2}\right)$ †

29. $y = \cot 2\left(x - \frac{\pi}{2}\right)$ †

30. $y = \cot 2\left(x + \frac{\pi}{2}\right)$ †

31. $y = \tan\frac{1}{2}(x + 2\pi)$ †

32. $y = \tan\frac{1}{2}(x - 2\pi)$ †

33. $y = \cot\frac{1}{2}(x - 2\pi)$ †

34. $y = \cot\frac{1}{2}(x + 2\pi)$ †

35. $y = -\frac{1}{2}\tan\frac{x}{2}$ †

36. $y = -\frac{1}{2}\cot\frac{x}{2}$ †

37. $y = \sec 4\left(x - \frac{\pi}{4}\right)$ †

38. $y = \sec\frac{1}{2}\left(x - \frac{\pi}{2}\right)$ †

39. $y = -2\sec 3x$ †

40. $y = -3\csc 4x$ †

41. $y = 3\csc\left(x + \frac{\pi}{2}\right)$ †

42. $y = 3\sec\left[2\left(x - \frac{\pi}{6}\right)\right]$ †

43. $y = \tan\left[\frac{2}{3}\left(x - \frac{\pi}{2}\right)\right]$ †

44. $y = 2\cot\left[2\left(x - \frac{\pi}{6}\right)\right]$ †

45. $y = -5\tan\left[2\left(x + \frac{\pi}{3}\right)\right]$ †

46. $y = -3\cot\left[\frac{1}{2}\left(x - \frac{\pi}{3}\right)\right]$ †

47. $y = \frac{1}{3}\cot[2(x - \pi)]$ †

48. $y = \frac{1}{2}\tan\left[4\left(x - \frac{\pi}{6}\right)\right]$ †

B EXERCISES Applying the Concepts

49. **Prison searchlight.** A dual-beam rotating light on a movie set is positioned as a spotlight shining on a prison wall. The light is 20 feet from the wall and rotates clockwise. The light shines on point P on the wall when first turned on ($t = 0$). After t seconds, the distance (in feet)

---

†Due to space constrictions, answers to these exercises may be found in the Answers beginning on page A–1 in the back of the book.

from the beam on the wall to the point P is given by the function

$$d(t) = 20 \tan \frac{\pi t}{5}.$$

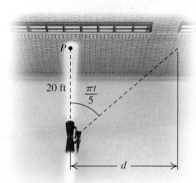

When the light beam is to the right of P, the value of d is positive, and when the beam is to the left of P, the value of d is negative.

a. Graph d over the interval $0 \le t \le 5$. †

b. Since $d(t)$ is undefined for $t = 2.5$, where is the rotating light pointing when $t = 2.5$?
The light is pointing parallel to the wall.

50. Prison searchlight. In Exercise 49, find the value for b assuming that the light beam is to sweep the entire wall in 10 seconds and $d(t) = 20 \tan bt$. $\frac{\pi}{10}$

C EXERCISES Beyond the Basics

51. Show that if f is a periodic function with period p, then the reciprocal function $\frac{1}{f}$ is also a periodic function with period p.

52. Show that if f is an odd function, then the reciprocal function $\frac{1}{f}$ is also an odd function.

53. Show that if f is an even function, then the reciprocal function $\frac{1}{f}$ is also an even function.

In Exercises 54–63, graph each function over a two-period interval.

54. $y = 3 \tan(-2x)$ †

55. $y = 2 \cot(-3x)$ †

56. $y = -2 \sec\left(-\dfrac{x}{2}\right)$ †

57. $y = -\dfrac{1}{2} \csc\left(-\dfrac{x}{5}\right)$ †

58. $y = \cot(\pi - x)$ †

59. $y = -\tan\left(\dfrac{\pi}{2} - x\right)$ †

60. $y = -\sec(\pi - 4x)$ †

61. $y = 2 \sec(\pi - 2x)$ †

62. $y = 2 \tan\left(\dfrac{\pi}{2} - 2x\right) + 1$ †

63. $y = 4 \cot(\pi - 4x) + 3$ †

In Exercises 64–67, write an equation for each graph.

64.

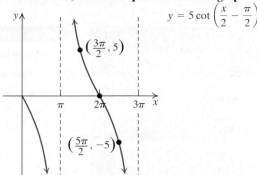

$y = 5 \cot\left(\dfrac{x}{2} - \dfrac{\pi}{2}\right)$

65.

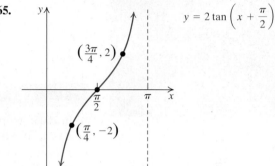

$y = 2 \tan\left(x + \dfrac{\pi}{2}\right)$

66.

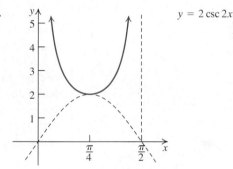

$y = 2 \csc 2x$

67.

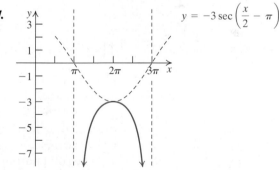

$y = -3 \sec\left(\dfrac{x}{2} - \pi\right)$

Critical Thinking

68. For what number k, with $-2\pi < k < 0$, is $x = k$ a vertical asymptote for $y = \cot\dfrac{1}{2}\left(x - \dfrac{\pi}{4}\right)$? $k = -\dfrac{7\pi}{4}$

69. For what number b does $y = \sec bx$ have period $\dfrac{\pi}{3}$? 6

70. Write an equation in the form $y = a \csc b(x - c)$ that has the same graph as $y = -2 \sec x$. $y = -2 \csc\left(x + \dfrac{\pi}{2}\right)$

Inverse Trigonometric Functions

Before Starting this Section, Review

1. Inverse functions (Section 1.7, page 91)
2. Composition of functions (Section 1.6, page 79)
3. Exact values of the trigonometric functions (Section 4.2, page 269)

Objectives

1 Graph and apply the inverse sine function.

2 Graph and apply the inverse cosine function.

3 Graph and apply the inverse tangent function.

4 Evaluate inverse trigonometric functions using a calculator.

5 Find exact values of composite functions involving the inverse trigonometric functions.

RETAIL THEFT

In 2005, security cameras at Filene's Basement store in Boston filmed a theft coordinated by a thief and an accomplice. The accomplice distracted the salesperson so the thief could steal a $16,000 necklace. Retail theft is a major concern for all retail outlets, from Tiffany & Co. to Walmart.

The National Retail Federation, the industry's largest trade group, and the Retail Industry Leaders Association have instituted password-protected national crime databases online. These databases allow retailers to share information about thefts and determine whether they have been a target of individual shoplifters who steal for themselves or a target of organized crime. In addition to participating in the shared databases, many large retailers have their own organized anti-crime squads. One estimate puts loss to organized theft at over $30 billion annually. In Example 9, we see how methods in this section can be used in an attempt to prevent loss from theft. ■

1 Graph and apply the inverse sine function.

The Inverse Sine Function

Recall that a function f has an inverse that is also a function if no horizontal line intersects the graph of f in more than one point. Because every horizontal line $y = b$, where $-1 \le b \le 1$, intersects the graph of $y = \sin x$ at more than one point, the sine function fails the horizontal line test, so is not one-to-one, and consequently has no inverse.

The solid portion of the sine graph shown in Figure 4.67 is the graph of $y = \sin x$ for $-\dfrac{\pi}{2} \le x \le \dfrac{\pi}{2}$. If we restrict the domain of $y = \sin x$ to the interval $\left[-\dfrac{\pi}{2}, \dfrac{\pi}{2}\right]$, the resulting function

$$y = \sin x, -\frac{\pi}{2} \le x \le \frac{\pi}{2}$$

is one-to-one (it passes the horizontal line test); so its inverse is also a function. Notice that the restricted function takes on all values in the range of $y = \sin x$, which is $[-1, 1]$, and that each of these y-values corresponds to exactly one x-value in the restricted domain $\left[-\dfrac{\pi}{2}, \dfrac{\pi}{2}\right]$. The inverse function for $y = \sin x, -\dfrac{\pi}{2} \le x \le \dfrac{\pi}{2}$, is called the **inverse sine**, or **arcsine**, function and is denoted by $\sin^{-1} x$ or by arcsin x.

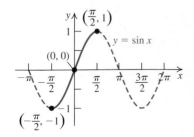

FIGURE 4.67 $y = \sin x,$ $-\dfrac{\pi}{2} \le x \le \dfrac{\pi}{2}$

314

RECALL

If two functions are inverses, their graphs are symmetric with respect to the line $y = x$.

INVERSE SINE FUNCTION

The equation $y = \sin^{-1} x$ means $\sin y = x$, where $-1 \le x \le 1$ and $-\dfrac{\pi}{2} \le y \le \dfrac{\pi}{2}$.

Read $y = \sin^{-1} x$ as "y equals inverse sine at x."

The range of $y = \sin x$ is $[-1, 1]$; so the domain of $y = \sin^{-1} x$ is $[-1, 1]$. The domain of the restricted sine function is $\left[-\dfrac{\pi}{2}, \dfrac{\pi}{2}\right]$; so the range of $y = \sin^{-1} x$ is $\left[-\dfrac{\pi}{2}, \dfrac{\pi}{2}\right]$. We can graph $y = \sin^{-1} x$ by reflecting the graph of $y = \sin x$, for $-\dfrac{\pi}{2} \le x \le \dfrac{\pi}{2}$, in the line $y = x$. See Figure 4.68.

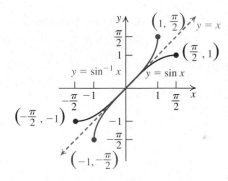

FIGURE 4.68 Graph of $y = \sin^{-1} x$

EXAMPLE 1   **Finding the Exact Values for $y = \sin^{-1} x$**

Find the exact values of y.

a. $y = \sin^{-1}\dfrac{\sqrt{3}}{2}$ **b.** $y = \sin^{-1}\left(-\dfrac{1}{2}\right)$ **c.** $y = \sin^{-1} 3$

SOLUTION

a. The equation $y = \sin^{-1}\dfrac{\sqrt{3}}{2}$ means $\sin y = \dfrac{\sqrt{3}}{2}$ and $-\dfrac{\pi}{2} \le y \le \dfrac{\pi}{2}$.

Because $\sin\dfrac{\pi}{3} = \dfrac{\sqrt{3}}{2}$ and $-\dfrac{\pi}{2} \le \dfrac{\pi}{3} \le \dfrac{\pi}{2}$, we have

$y = \sin^{-1}\dfrac{\sqrt{3}}{2} = \dfrac{\pi}{3}$.

STUDY TIP

You can also read $y = \sin^{-1}(x)$ as "y is the number in the interval $\left[-\dfrac{\pi}{2}, \dfrac{\pi}{2}\right]$ whose sine is x."

b. The equation $y = \sin^{-1}\left(-\dfrac{1}{2}\right)$ means $\sin y = -\dfrac{1}{2}$ and $-\dfrac{\pi}{2} \le y \le \dfrac{\pi}{2}$.

Because $\sin\left(-\dfrac{\pi}{6}\right) = -\dfrac{1}{2}$ and $-\dfrac{\pi}{2} \le \dfrac{\pi}{6} \le \dfrac{\pi}{2}$, we have

$y = -\dfrac{\pi}{6}$.

c. Since 3 is not in the domain of the inverse sine function, $[-1, 1]$, $\sin^{-1} 3$ does not exist.   ■ ■ ■

Practice Problem 1   Find the exact values of y.

a. $y = \sin^{-1}\left(-\dfrac{\sqrt{3}}{2}\right)$ **b.** $y = \sin^{-1}(-1)$ ■

2 Graph and apply the inverse cosine function.

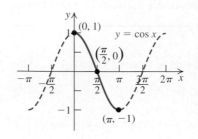

FIGURE 4.69 $y = \cos x, 0 \leq x \leq \dfrac{\pi}{2}$

STUDY TIP

You can also read $y = \cos^{-1} x$ as "y is the number in the interval $[0, \pi]$ whose cosine is x."

The Inverse Cosine Function

When we restrict the domain of $y = \cos x$ to the interval $[0, \pi]$, the resulting function, $y = \cos x$ (with $0 \leq x \leq \pi$), is one-to-one. No horizontal line intersects the graph of $y = \cos x$, with $0 \leq x \leq \pi$, in more than one point. See Figure 4.69. Consequently, the restricted cosine function has an inverse function.

The inverse function for $y = \cos x, 0 \leq x \leq \pi$, is called the **inverse cosine**, or **arccosine**, function and is denoted by $\cos^{-1} x$, or by arccos x.

INVERSE COSINE FUNCTION

The equation $y = \cos^{-1} x$ means $\cos y = x$ where $-1 \leq x \leq 1$ and $0 \leq y \leq \pi$. Read $y = \cos^{-1} x$ as "y equals inverse cosine at x."

Reflecting the graph of $y = \cos x$, for $0 \leq x \leq \pi$, in the line, $y = x$ produces the graph of $y = \cos^{-1} x$, shown in Figure 4.70.

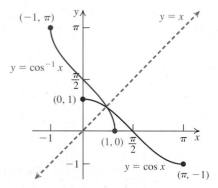

FIGURE 4.70 Graph of $y = \cos^{-1} x$

EXAMPLE 2 **Finding an Exact Value for $\cos^{-1} x$**

Find the exact value of y.

a. $y = \cos^{-1} \dfrac{\sqrt{2}}{2}$ **b.** $y = \cos^{-1}\left(-\dfrac{1}{2}\right)$

SOLUTION

a. The equation $y = \cos^{-1} \dfrac{\sqrt{2}}{2}$ means $\cos y = \dfrac{\sqrt{2}}{2}$ and $0 \leq y \leq \pi$.

Since $\cos \dfrac{\pi}{4} = \dfrac{\sqrt{2}}{2}$ and $0 \leq \dfrac{\pi}{4} \leq \pi$, we have $y = \dfrac{\pi}{4}$.

b. The equation $y = \cos^{-1}\left(-\dfrac{1}{2}\right)$ means $\cos y = -\dfrac{1}{2}$ and $0 \leq y \leq \pi$.

Since $\cos \dfrac{2\pi}{3} = -\dfrac{1}{2}$ and $0 \leq \dfrac{2\pi}{3} \leq \pi$, we have $y = \dfrac{2\pi}{3}$.   ■■■

Practice Problem 2 Find the exact value of y.

a. $y = \cos^{-1}\left(-\dfrac{\sqrt{2}}{2}\right)$ **b.** $y = \cos^{-1} \dfrac{1}{2}$ ■

3 Graph and apply the inverse tangent function.

The Inverse Tangent Function

The *inverse tangent function* results from restricting the domain of the tangent function to the interval $\left(-\dfrac{\pi}{2}, \dfrac{\pi}{2}\right)$ to obtain a one-to-one function. The inverse of this restricted tangent function is the **inverse tangent**, or **arctangent**, function.

STUDY TIP

You can also read $y = \tan^{-1} x$ as "y is the number in the interval $\left(-\dfrac{\pi}{2}, \dfrac{\pi}{2}\right)$ whose tangent is x."

INVERSE TANGENT FUNCTION

The equation $y = \tan^{-1} x$ means $\tan y = x$, where $-\infty < x < \infty$ and $-\dfrac{\pi}{2} < y < \dfrac{\pi}{2}$.

Read $y = \tan^{-1} x$ as "y equals the inverse tangent at x."

The graph of $y = \tan^{-1} x$ is obtained by reflecting the graph of $y = \tan x$, with $-\dfrac{\pi}{2} < x < \dfrac{\pi}{2}$, in the line $y = x$. Figure 4.71 shows the graph of the restricted tangent function. Figure 4.72 shows the graph of $y = \tan^{-1} x$.

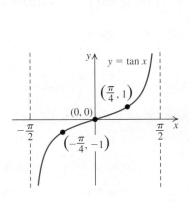

FIGURE 4.71 $y = \tan x$, $-\dfrac{\pi}{2} < x < \dfrac{\pi}{2}$

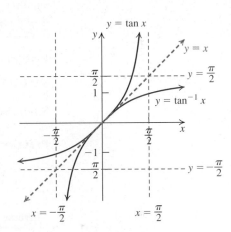

FIGURE 4.72 Graph of $y = \tan^{-1} x$

EXAMPLE 3   Finding Exact Values for $\tan^{-1} x$

Find the exact value of y.

a. $y = \tan^{-1} 0$ **b.** $y = \tan^{-1}(-\sqrt{3})$

SOLUTION

a. Since $\tan 0 = 0$ and $-\dfrac{\pi}{2} < 0 < \dfrac{\pi}{2}$, we have $y = 0$.

b. Since $\tan\left(-\dfrac{\pi}{3}\right) = -\sqrt{3}$ and $-\dfrac{\pi}{2} < -\dfrac{\pi}{3} < \dfrac{\pi}{2}$, we have $y = -\dfrac{\pi}{3}$.   ■ ■ ■

Practice Problem 3   Find the exact value of $y = \tan^{-1} \dfrac{\sqrt{3}}{3}$.   ■

Other Inverse Trigonometric Functions

Sometimes the ranges of the *inverse secant* and *inverse cosecant* functions differ from those used in this text. Always check the definitions of the domains of these two functions when they are used outside this course.

Inverse cotangent $y = \cot^{-1} x$ means $\cot y = x$,

where $-\infty < x < \infty$ and $0 < y < \pi$.

Inverse cosecant $y = \csc^{-1} x$ means $\csc y = x$,

where $|x| \geq 1$ and $-\dfrac{\pi}{2} \leq y \leq \dfrac{\pi}{2}, y \neq 0$.

Inverse secant $y = \sec^{-1} x$ means $\sec y = x$,

where $|x| \geq 1$ and $0 \leq y \leq \pi, y \neq \dfrac{\pi}{2}$.

EXAMPLE 4   **Finding the Exact Value for $\csc^{-1} x$**

Find the exact value for $y = \csc^{-1} 2$.

SOLUTION

Since $\csc \dfrac{\pi}{6} = 2$ and $-\dfrac{\pi}{2} \leq \dfrac{\pi}{6} \leq \dfrac{\pi}{2}$, we have $y = \csc^{-1} 2 = \dfrac{\pi}{6}$. ■ ■ ■

Practice Problem 4   Find the exact value of $y = \sec^{-1} 2$. ■

Inverse Trigonometric Functions			
Inverse Function	**Equivalent to**	**Domain**	**Range**
$y = \sin^{-1} x$	$\sin y = x$	$[-1, 1]$	$\left[-\dfrac{\pi}{2}, \dfrac{\pi}{2}\right]$
$y = \cos^{-1} x$	$\cos y = x$	$[-1, 1]$	$[0, \pi]$
$y = \tan^{-1} x$	$\tan y = x$	$(-\infty, \infty)$	$\left(-\dfrac{\pi}{2}, \dfrac{\pi}{2}\right)$
$y = \cot^{-1} x$	$\cot y = x$	$(-\infty, \infty)$	$(0, \pi)$
$y = \csc^{-1} x$	$\csc y = x$	$(-\infty, -1] \cup [1, \infty)$	$\left[-\dfrac{\pi}{2}, 0\right) \cup \left(0, \dfrac{\pi}{2}\right]$
$y = \sec^{-1} x$	$\sec y = x$	$(-\infty, -1] \cup [1, \infty)$	$\left[0, \dfrac{\pi}{2}\right) \cup \left(\dfrac{\pi}{2}, \pi\right]$

4 Evaluate inverse trigonometric functions using a calculator.

Using a Calculator with Inverse Functions

In Section 4.3, we defined the six trigonometric functions of *real numbers*, and in this section, we have defined the corresponding six inverse trigonometric functions of real numbers. For example,

$$\sin \frac{\pi}{4} = \frac{\sqrt{2}}{2}$$

TECHNOLOGY CONNECTION

The secondary functions on your calculator, labeled SIN^{-1}, COS^{-1}, and TAN^{-1}, are associated with the keys labeled $\boxed{\text{SIN}}$, $\boxed{\text{COS}}$, and $\boxed{\text{TAN}}$, respectively. Consult your manual to learn how to access these secondary functions. The screen shows several values for the trigonometric inverse functions on a calculator set to Radian mode.

```
cos-1(3/4)
            .723
sin-1( -0.86)
          -1.035
tan-1( -6.25)
          -1.412
```

and

$$\sin^{-1}\frac{\sqrt{2}}{2} = \frac{\pi}{4}.$$

However, because we also defined the trigonometric functions of *angles* in Section 4.3, it is meaningful when working with angles in degree measure to write a statement such as

$$\sin^{-1}\frac{\sqrt{2}}{2} = 45°.$$

When using the inverse trigonometric functions on a calculator to find a real number (or equivalently, an angle measured in radians), make sure you set your calculator to Radian mode.

When using a calculator to find $\csc^{-1} x$ or $\sec^{-1} x$, find $\sin^{-1}\frac{1}{x}$ and $\cos^{-1}\frac{1}{x}$, respectively. For example, if $\csc^{-1} 5 = \theta$, then $\csc\theta = 5$, or $\frac{1}{\sin\theta} = 5$. So $\sin\theta = \frac{1}{5}$ and $\theta = \sin^{-1}\left(\frac{1}{5}\right)$. However, to find $\cot^{-1} x$, begin by finding $\tan^{-1}\frac{1}{x}$; this gives you a value in the interval $\left(-\frac{\pi}{2}, \frac{\pi}{2}\right)$. If $x \geq 0$, this is the correct value, but **for $x < 0$, $\cot^{-1}(x) = \pi + \tan^{-1}\frac{1}{x}$**; so that $\cot^{-1}(x)$ is in the interval $\left(\frac{\pi}{2}, \pi\right)$.

When using a calculator to find an unknown angle measure in degrees, make sure you set your calculator to degree measure.

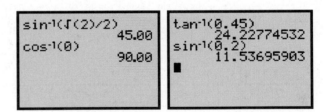

EXAMPLE 5   **Using a Calculator to Find the Values of Inverse Functions**

Use a calculator to find the value of y in radians rounded to four decimal places.

a. $y = \sin^{-1} 0.75$ **b.** $y = \cot^{-1} 2.8$ **c.** $y = \cot^{-1}(-2.3)$

SOLUTION

Set your calculator to Radian mode.

a. $y = \sin^{-1} 0.75 \approx 0.8481$

b. $y = \cot^{-1} 2.8 = \tan^{-1}\left(\frac{1}{2.8}\right) \approx 0.3430$

c. $y = \cot^{-1}(-2.3) = \pi + \tan^{-1}\left(-\frac{1}{2.3}\right) \approx 2.7315$ ■ ■ ■

Practice Problem 5   Use a calculator to find the value of y in radians rounded to four decimal places.

a. $y = \cos^{-1} 0.22$ **b.** $y = \csc^{-1} 3.5$ **c.** $y = \cot^{-1}(-4.7)$ ■

EXAMPLE 6 **Using a Calculator to Find the Values of Inverse Functions**

Use a calculator to find the value of y in degrees rounded to four decimal places.

a. $y = \tan^{-1} 0.99$ **b.** $y = \sec^{-1} 25$ **c.** $y = \cot^{-1}(-1.3)$

SOLUTION

Set your calculator to Degree mode.

a. $y = \tan^{-1} 0.99 \approx 44.7121°$

b. $y = \sec^{-1} 25 = \cos^{-1} \dfrac{1}{25} \approx 87.7076°$

c. $y = \cot^{-1}(-1.3) = 180° + \tan^{-1}\left(-\dfrac{1}{1.3}\right) \approx 142.4314°$ ■ ■ ■

Practice Problem 6 Repeat Example 6 for each expression.

a. $y = \cot^{-1} 0.75$ **b.** $y = \csc^{-1} 13$ **c.** $y = \tan^{-1}(-12)$ ■

---

5 Find exact values of composite functions involving the inverse trigonometric functions.

Composition of Trigonometric and Inverse Trigonometric Functions

Recall that if f is a one-to-one function with inverse f^{-1}, then $f^{-1}[f(x)] = x$ for every x in the domain of f and $f[f^{-1}(x)] = x$ for every x in the domain of f^{-1}. This leads to the following formulas for the inverse sine, cosine, and tangent functions.

Inverse Sine	Inverse Cosine	Inverse Tangent
$\sin^{-1}(\sin x) = x$, $-\dfrac{\pi}{2} \le x \le \dfrac{\pi}{2}$	$\cos^{-1}(\cos x) = x$, $0 \le x \le \pi$	$\tan^{-1}(\tan x) = x$, $-\dfrac{\pi}{2} < x < \dfrac{\pi}{2}$
$\sin(\sin^{-1} x) = x$, $-1 \le x \le 1$	$\cos(\cos^{-1} x) = x$, $-1 \le x \le 1$	$\tan(\tan^{-1} x) = x$, $-\infty < x < \infty$

EXAMPLE 7 **Finding the Exact Value of $\sin^{-1}(\sin x)$ and $\cos^{-1}(\cos x)$**

Find the exact value of

a. $\sin^{-1}\left[\sin\left(-\dfrac{\pi}{8}\right)\right]$ **b.** $\cos^{-1}\left(\cos\dfrac{5\pi}{4}\right)$

SOLUTION

a. Because $-\dfrac{\pi}{2} \le -\dfrac{\pi}{8} \le \dfrac{\pi}{2}$, we have $\sin^{-1}\left[\sin\left(-\dfrac{\pi}{8}\right)\right] = -\dfrac{\pi}{8}$.

b. We cannot use the formula $\cos^{-1}(\cos x) = x$ for $x = \dfrac{5\pi}{4}$ because $\dfrac{5\pi}{4}$ is not in the interval $[0, \pi]$. However, $\cos\dfrac{5\pi}{4} = \cos\left(2\pi - \dfrac{5\pi}{4}\right) = \cos\dfrac{3\pi}{4}$ and $\dfrac{3\pi}{4}$ is in the interval $[0, \pi]$. Therefore,

$$\cos^{-1}\left(\cos\dfrac{5\pi}{4}\right) = \cos^{-1}\left(\cos\dfrac{3\pi}{4}\right) = \dfrac{3\pi}{4}.$$ ■ ■ ■

Practice Problem 7 Find the exact value of $\sin^{-1}\left(\sin\dfrac{3\pi}{2}\right)$. ■

To find the exact values of expressions involving the composition of a trigonometric function and the inverse of a *different* trigonometric function, we use points on the terminal side of an angle in standard position.

EXAMPLE 8 **Finding the Exact Value of a Composite Trigonometric Expression**

Find the exact value of

a. $\cos\left(\tan^{-1}\dfrac{2}{3}\right)$.   **b.** $\sin\left[\cos^{-1}\left(-\dfrac{1}{4}\right)\right]$.

SOLUTION

a. Let θ represent the radian measure of the angle in the interval $\left(-\dfrac{\pi}{2},\dfrac{\pi}{2}\right)$,

with $\tan\theta = \dfrac{2}{3}$. Then since $\tan\theta$ is positive, θ must be positive. We have

$$\theta = \tan^{-1}\frac{2}{3} \quad \text{and} \quad 0 < \theta < \frac{\pi}{2}.$$

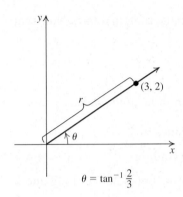

$\theta = \tan^{-1}\dfrac{2}{3}$

FIGURE 4.73

Figure 4.73 shows θ in standard position. If (x, y) is a point on the terminal side of θ, then $\tan\theta = \dfrac{y}{x}$.

Consequently, we can choose the point with coordinates $(3, 2)$ to determine the terminal side of θ. Then $x = 3$, $y = 2$ and we have

$$\tan\theta = \frac{2}{3} \quad \text{and} \quad \cos\theta = \frac{3}{r}, \text{where}$$

$r = \sqrt{x^2 + y^2} = \sqrt{3^2 + 2^2} = \sqrt{9 + 4} = \sqrt{13}$. So

$$\cos\left(\tan^{-1}\frac{2}{3}\right) = \cos\theta = \frac{3}{r} = \frac{3}{\sqrt{13}} = \frac{3\sqrt{13}}{13}.$$

b. Let θ represent the radian measure of the angle in $[0, \pi]$, with $\cos\theta = -\dfrac{1}{4}$. Then since $\cos\theta$ is negative, θ is in quadrant II, so

$$\theta = \cos^{-1}\left(-\frac{1}{4}\right) \quad \text{and} \quad \frac{\pi}{2} < \theta < \pi.$$

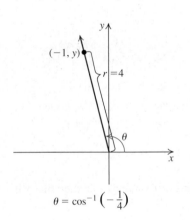

$\theta = \cos^{-1}\left(-\dfrac{1}{4}\right)$

FIGURE 4.74

Figure 4.74 shows θ in standard position. If (x, y) is a point on the terminal side of θ, and r is the distance between (x, y) and the origin, then $\sin\theta = \dfrac{y}{r}$. We choose the point with coordinates $(-1, y)$, a distance of 4 units from the origin, on the terminal side of θ. Then,

$$\cos\theta = -\frac{1}{4} \quad \text{and} \quad \sin\theta = \frac{y}{4}, \text{where}$$

$$r = \sqrt{x^2 + y^2} = \sqrt{(-1)^2 + y^2} \quad \text{or} \quad r^2 = 1 + y^2$$

$$4^2 = 1 + y^2 \qquad \text{Replace } r \text{ with 4.}$$

$$15 = y^2 \qquad \text{Simplify.}$$

$$\sqrt{15} = y \qquad y \text{ is positive.}$$

Thus,

$$\sin\left[\cos^{-1}\left(-\frac{1}{4}\right)\right] = \sin\theta = \frac{y}{r} = \frac{\sqrt{15}}{4}. \qquad \blacksquare\ \blacksquare\ \blacksquare$$

Practice Problem 8 Find the exact value of $\cos\left[\sin^{-1}\left(-\dfrac{1}{3}\right)\right]$.   ■

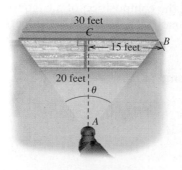

FIGURE 4.75

| EXAMPLE 9 | **Finding the Rotation Angle for a Security Camera** |

A security camera is to be installed 20 feet away from the center of a jewelry counter. The counter is 30 feet long. What angle, to the nearest degree, should the camera rotate through so that it scans the entire counter? See Figure 4.75.

SOLUTION

The counter center C, the camera A, and a counter end B form a right triangle. The angle at vertex A is $\dfrac{\theta}{2}$, where θ is the angle through which the camera rotates. Note that

$$\tan\frac{\theta}{2} = \frac{15}{20} = \frac{3}{4}$$

$$\frac{\theta}{2} = \tan^{-1}\frac{3}{4} \approx 36.87° \qquad \text{Use a calculator in Degree mode.}$$

$$\theta \approx 73.74° \qquad\qquad \text{Multiply both sides by 2.}$$

Set the camera to rotate through 74° to scan the entire counter. ■ ■ ■

Practice Problem 9   Rework Example 9 for a counter that is 20 feet long and a camera set 12 feet from the center of the counter. ■

SECTION 4.6 ■ Exercises

A EXERCISES Basic Skills and Concepts

1. The domain of $f(x) = \sin^{-1} x$ is ___[-1, 1]___.
2. The range of $f(x) = \tan^{-1} x$ is _____. $\left(-\dfrac{\pi}{2}, \dfrac{\pi}{2}\right)$
3. The exact value of $y = \cos^{-1}\dfrac{1}{2}$ is ___$\dfrac{\pi}{3}$___.
4. $\sin^{-1}(\sin \pi) = $ ___0___.
5. *True or False* If $-1 \le x \le 0$, then $\sin^{-1} x \le 0$. True
6. *True or False* If $-1 \le x \le 0$, then $\cos^{-1} x \le 0$. False
7. *True or False* The domain of $f(x) = \cos^{-1} x$ is $0 \le x \le \pi$. False
8. *True or False* The value of $\tan^{-1}\left(\dfrac{1}{\sqrt{3}}\right)$ is $\dfrac{\pi}{3}$. False

In Exercises 9–32, find each exact value of y or state that y is undefined.

9. $y = \sin^{-1} 0$ 0
10. $y = \cos^{-1} 0$ $\dfrac{\pi}{2}$
11. $y = \sin^{-1}\left(-\dfrac{1}{2}\right)$ $-\dfrac{\pi}{6}$
12. $y = \cos^{-1}\left(-\dfrac{\sqrt{3}}{2}\right)$ $\dfrac{5\pi}{6}$
13. $y = \cos^{-1}(-1)$ π
14. $y = \sin^{-1}\dfrac{1}{2}$ $\dfrac{\pi}{6}$
15. $y = \cos^{-1}\dfrac{\pi}{2}$ Undefined
16. $y = \sin^{-1}\pi$ Undefined
17. $y = \tan^{-1}\sqrt{3}$ $\dfrac{\pi}{3}$
18. $y = \tan^{-1} 1$ $\dfrac{\pi}{4}$
19. $y = \tan^{-1}(-1)$ $-\dfrac{\pi}{4}$
20. $y = \tan^{-1}\left(-\dfrac{\sqrt{3}}{3}\right)$ $-\dfrac{\pi}{6}$
21. $y = \cot^{-1}(-1)$ $-\dfrac{\pi}{4}$
22. $y = \cot^{-1} 1$ $\dfrac{\pi}{4}$
23. $y = \sin^{-1}\left(-\dfrac{\sqrt{2}}{2}\right)$ $-\dfrac{\pi}{4}$
24. $y = \cos^{-1}\left(\dfrac{\sqrt{3}}{2}\right)$ $\dfrac{\pi}{6}$
25. $y = \cot^{-1}\sqrt{3}$ $\dfrac{\pi}{6}$
26. $y = \cot^{-1}(-\sqrt{3})$ $-\dfrac{\pi}{6}$
27. $y = \cos^{-1}(-2)$ Undefined
28. $y = \sin^{-1}\sqrt{3}$ Undefined
29. $y = \sec^{-1}(-2)$ $\dfrac{2\pi}{3}$
30. $y = \sec^{-1}\sqrt{2}$ $\dfrac{\pi}{4}$
31. $y = \csc^{-1}\dfrac{2\sqrt{3}}{3}$ $\dfrac{\pi}{3}$
32. $y = \csc^{-1}(2)$ $\dfrac{\pi}{6}$

In Exercises 33–44, find each exact value of y or state that y is undefined.

33. $y = \sin\left(\sin^{-1}\dfrac{1}{8}\right)$ $\dfrac{1}{8}$
34. $y = \cos\left(\cos^{-1}\dfrac{1}{5}\right)$ $\dfrac{1}{5}$
35. $y = \cos(\cos^{-1} 0.6)$ 0.6
36. $y = \sin(\sin^{-1} 0.8)$ 0.8
37. $y = \tan^{-1}\left(\tan\dfrac{\pi}{7}\right)$ $\dfrac{\pi}{7}$
38. $y = \tan^{-1}\left(\tan\dfrac{\pi}{4}\right)$ $\dfrac{\pi}{4}$
39. $y = \tan(\tan^{-1} 247)$ 247
40. $y = \tan(\tan^{-1} 7)$ 7
41. $y = \sin^{-1}\left(\sin\dfrac{4\pi}{3}\right)$ $-\dfrac{\pi}{3}$
42. $y = \cos^{-1}\left(\cos\dfrac{5\pi}{3}\right)$ $\dfrac{\pi}{3}$
43. $y = \tan^{-1}\left(\tan\dfrac{2\pi}{3}\right)$ $-\dfrac{\pi}{3}$
44. $y = \tan\left(\tan^{-1}\dfrac{2\pi}{3}\right)$ $\dfrac{2\pi}{3}$

In Exercises 45–54, use a calculator to find each value of y in degrees rounded to two decimal places.

45. $y = \cos^{-1} 0.6$ 53.13°
46. $y = \sin^{-1} 0.23$ 13.30°
47. $y = \sin^{-1}(-0.69)$ −43.63°
48. $y = \cos^{-1}(-0.57)$ 124.75°
49. $y = \sec^{-1}(3.5)$ 73.40°
50. $y = \csc^{-1}(6.8)$ 8.46°

†Due to space constrictions, answers to these exercises may be found in the Answers beginning on page A–1 in the back of the book.

51. $y = \tan^{-1} 14$ 85.91°

52. $y = \tan^{-1} 50$ 88.85°

53. $y = \tan^{-1}(-42.147)$
−88.64°

54. $y = \tan^{-1}(-0.3863)$
−21.12°

In Exercises 55–66, use a sketch to find each exact value of y.

55. $y = \cos\left(\sin^{-1}\dfrac{2}{3}\right)$ $\dfrac{\sqrt{5}}{3}$

56. $y = \sin\left(\cos^{-1}\dfrac{3}{4}\right)$ $\dfrac{\sqrt{7}}{4}$

57. $y = \sin\left[\cos^{-1}\left(-\dfrac{4}{5}\right)\right]$ $\dfrac{3}{5}$

58. $y = \cos\left(\sin^{-1}\dfrac{3}{5}\right)$ $\dfrac{4}{5}$

59. $y = \cos\left(\tan^{-1}\dfrac{5}{2}\right)$ $\dfrac{2\sqrt{29}}{29}$

60. $y = \sin\left(\tan^{-1}\dfrac{13}{5}\right)$ $\dfrac{13\sqrt{194}}{194}$

61. $y = \tan\left(\cos^{-1}\dfrac{4}{5}\right)$ $\dfrac{3}{4}$

62. $y = \tan\left[\sin^{-1}\left(-\dfrac{3}{4}\right)\right]$ $-\dfrac{3\sqrt{7}}{7}$

63. $y = \sin(\tan^{-1} 4)$ $\dfrac{4\sqrt{17}}{17}$

64. $y = \cos(\tan^{-1} 3)$ $\dfrac{\sqrt{10}}{10}$

65. $y = \tan(\sec^{-1} 2)$ $\sqrt{3}$

66. $y = \tan[\csc^{-1}(-2)] - \dfrac{\sqrt{3}}{3}$

B EXERCISES Applying the Concepts

67. Sprinkler rotation. A sprinkler rotates back and forth through an angle θ, as shown in the figure. At a distance of 5 feet from the sprinkler, the rays that form the sides of angle θ are 6 feet apart. Find θ. 62°

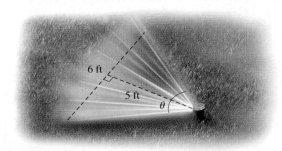

68. Irradiating flowers. A tray of flowers is being irradiated by a beam from a rotating lamp, as shown in the figure. If the tray is 8 feet long and the lamp is 2 feet from the center of the tray, through what angle should the lamp rotate to irradiate the full length of the tray? 127°

69. Motorcycle racing. A video camera is set up 110 feet at a right angle to a straight quarter-mile racetrack, as shown in the figure. The starting line is to the left, and the finish

is to the right. Through what angle must the camera rotate to film the entire race? 159°

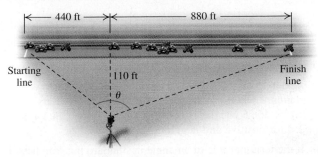

70. Camera's viewing angle. The viewing angle for the 35-millimeter camera is given (in degrees) by

$$\theta = 2\tan^{-1}\frac{18}{x},$$ where x is the focal length of the lens.

The focal length on most adjustable cameras is marked in millimeters on the lens mount.

a. Find the viewing angle, in degrees, if the focal length is 50 millimeters. 39.6°

b. Find the viewing angle, in degrees, if the focal length is 200 millimeters. 10.3°

C EXERCISES Beyond the Basics

In Exercises 71–73, determine whether each function is increasing or decreasing on its domain.

71. $y = \sin^{-1} x$ Increasing

72. $y = \cos^{-1} x$ Decreasing

73. $y = \tan^{-1} x$ Increasing

74. Show that

a. $\sec^{-1} x \neq \dfrac{1}{\cos^{-1} x}$.

b. $\sec^{-1} x = \cos^{-1}\dfrac{1}{x}$.

75. Graph the function $y = \cot^{-1} x$. †

76. Graph the function $y = \csc^{-1} x$. †

77. Graph the function $y = \sec^{-1} x$. †

78. For what values of x is $\cot^{-1}(\cot x) = x$ true? $(0, \pi)$

79. For what values of x is $\sec^{-1}(\sec x) = x$ true?

80. For what values of x is $\csc^{-1}(\csc x) = x$ true?

Critical Thinking

In Exercises 81–88, evaluate each expression in terms of x.

81. $\sin(\cos^{-1} x), |x| \leq 1$ $\dfrac{}{\sqrt{1-x^2}}$

82. $\tan(\sin^{-1} x), |x| < 1$

83. $\cos(\tan^{-1} x)$

84. $\sin(\cot^{-1} x)$

85. $\cos(\sin^{-1} x), |x| \leq 1$ $\sqrt{1-x^2}$

86. $\tan(\cos^{-1} x), |x| \leq 1, x \neq 0$

87. $\sin(\tan^{-1} x)$ $\dfrac{x}{\sqrt{x^2+1}}$

88. $\cos(\cot^{-1} x)$ $\dfrac{x}{\sqrt{x^2+1}}$

Answers:

79. $\left[0, \dfrac{\pi}{2}\right) \cup \left(\dfrac{\pi}{2}, \pi\right]$

80. $\left[-\dfrac{\pi}{2}, 0\right) \cup \left(0, \dfrac{\pi}{2}\right]$

82. $\dfrac{x}{\sqrt{1-x^2}}$

83. $\dfrac{1}{\sqrt{1+x^2}}$

84. $\dfrac{1}{\sqrt{1+x^2}}$

86. $\dfrac{\sqrt{1-x^2}}{x}$

SUMMARY ■ Definitions, Concepts, and Formulas

4.1 Angles and Their Measure

i. An **angle** is formed by rotating a ray around its endpoint.

ii. An angle in a rectangular coordinate system is in **standard position** if its vertex is at the origin and its initial side coincides with the positive x-axis. (See page 254 for positive and negative angles.)

iii. If the terminal side of an angle in standard position lies on the x-axis or the y-axis, the angle is a **quadrantal angle**.

iv. Angles can be measured in **degrees**, where $1° = \dfrac{1}{360}$ of a complete revolution.

v. An **acute angle** is an angle with measure between $0°$ and $90°$, a **right angle** is an angle with measure $90°$, an **obtuse angle** is an angle with measure between $90°$ and $180°$, and a **straight angle** is an angle with measure $180°$.

vi. Angles that have the same initial and terminal sides are **coterminal** angles.

vii. Angles can also be measured in **radians**. The radian measure of a central angle θ that intercepts an arc of length s on a circle of radius r is $\theta = \dfrac{s}{r}$ radians.

viii. To convert from degrees to radians, multiply degrees by $\dfrac{\pi}{180°}$. To convert from radians to degrees, multiply radians by $\dfrac{180°}{\pi}$.

ix. Two positive angles are **complements** if their sum is $90°$. Two positive angles are **supplements** if their sum is $180°$.

x. The formula for the length s of the arc intercepted by a central angle θ in a circle of radius r is $s = r\theta$.

xi. If an object travels at a constant speed v on a circle of radius r through an angle of θ radians and an arc of length s, in time t, then the (average) **linear speed** of the object is $v = \dfrac{s}{t}$ and the (average) **angular speed** ω of the object is $\omega = \dfrac{\theta}{t}$. Further, $v = r\omega$.

xii. The area of a sector $= \dfrac{1}{2}r^2\theta$, where r is the radius of the circle and θ is in radians.

4.2 The Unit Circle; Trigonometric Functions of an Angle

i. Unit circle If the terminal side of an angle t intersects the unit circle at the point (x, y), then $\sin t = y$, $\cos t = x$, $\tan t = \dfrac{y}{x}$, $\cot t = \dfrac{x}{y}$, $\sec t = \dfrac{1}{x}$, and $\csc t = \dfrac{1}{y}$. The domain of the sine and cosine functions is $(-\infty, \infty)$, and their range is $[-1, 1]$.

Quotient Identities

$$\tan x = \frac{\sin x}{\cos x} \qquad \cot x = \frac{\cos x}{\sin x}$$

ii. Trigonometric function values of coterminal angles

θ in degrees	θ in radians
$\sin \theta = \sin (\theta + n360°)$	$\sin \theta = \sin (\theta + 2\pi n)$
$\cos \theta = \cos (\theta + n360°)$	$\cos \theta = \cos (\theta + 2\pi n)$

These equations hold for any integer n.

TABLE 4.1

$\theta°$	θ radians	$\sin \theta$	$\cos \theta$	$\tan \theta$	$\csc \theta$	$\sec \theta$	$\cot \theta$
$30°$	$\dfrac{\pi}{6}$	$\dfrac{1}{2}$	$\dfrac{\sqrt{3}}{2}$	$\dfrac{\sqrt{3}}{3}$	2	$\dfrac{2\sqrt{3}}{3}$	$\sqrt{3}$
$45°$	$\dfrac{\pi}{4}$	$\dfrac{\sqrt{2}}{2}$	$\dfrac{\sqrt{2}}{2}$	1	$\sqrt{2}$	$\sqrt{2}$	1
$60°$	$\dfrac{\pi}{3}$	$\dfrac{\sqrt{3}}{2}$	$\dfrac{1}{2}$	$\sqrt{3}$	$\dfrac{2\sqrt{3}}{3}$	2	$\dfrac{\sqrt{3}}{3}$

4.3 Some Properties of the Trigonometric Functions

i. *Signs of the trigonometric functions*

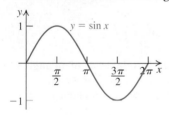

ii. The *reference angle* for θ in standard position is the positive acute angle θ' formed by the terminal side of θ and the *x*-axis. Reference angles are used to find the values of trigonometric functions for *any* angle θ.

iii. *Basic trigonometric identities*

Quotient and reciprocal identities

$$\tan t = \frac{\sin t}{\cos t}, \cot t = \frac{\cos t}{\sin t},$$

$$\csc t = \frac{1}{\sin t}, \sec t = \frac{1}{\cos t}, \cot t = \frac{1}{\tan t}$$

Pythagorean identities

$$\cos^2 t + \sin^2 t = 1, 1 + \tan^2 t = \sec^2 t, 1 + \cot^2 t = \csc^2 t$$

4.4 Graphs of the Sine and Cosine Functions

i. Sine and cosine function graphs and properties

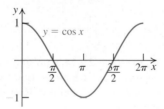

Sine Function
1. Period: 2π
2. Domain: $(-\infty, \infty)$
3. Range: $[-1, 1]$
4. Odd: $\sin(-t) = -\sin t$

Cosine Function
1. Period: 2π
2. Domain: $(-\infty, \infty)$
3. Range: $[-1, 1]$
4. Even: $\cos(t) = \cos t$

ii. The functions $y = a \sin b(x - c) + d$ and $y = a \cos b(x - c) + d$ $(b > 0)$ have amplitude $|a|$, period $\frac{2\pi}{b}$, phase shift c, and vertical shift d.

4.5 Graphs of the Other Trigonometric Functions

i. Tangent and cotangent function graphs and properties

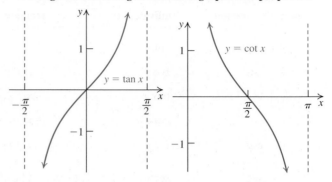

Tangent Function
1. Period: π
2. Domain: All real numbers except for odd multiples of $\frac{\pi}{2}$
3. Range: $(-\infty, \infty)$
4. Odd: $\tan(-x) = -\tan x$

Cotangent Function
1. Period: π
2. Domain: All real numbers except for integer multiples of π
3. Range: $(-\infty, \infty)$
4. Odd: $\cot(-x) = -\cot x$

ii. Secant and cosecant function graphs and properties

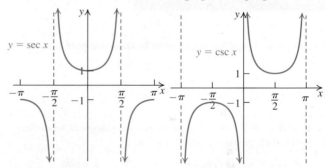

Secant Function
1. Period: 2π
2. Domain: All real numbers except for odd multiples of $\frac{\pi}{2}$
3. Range: $(-\infty, -1] \cup [1, \infty)$
4. Even: $\sec(-x) = \sec x$

Cosecant Function
1. Period: 2π
2. Domain: All real numbers except for integer multiples of π
3. Range: $(-\infty, -1] \cup [1, \infty)$
4. Odd: $\csc(-x) = -\csc x$

4.6 Inverse Trigonometric Functions

i.

Inverse Trigonometric Functions

Inverse Function	Equivalent to	Domain	Range
$y = \sin^{-1} x$	$\sin y = x$	$[-1, 1]$	$\left[-\dfrac{\pi}{2}, \dfrac{\pi}{2}\right]$
$y = \cos^{-1} x$	$\cos y = x$	$[-1, 1]$	$[0, \pi]$
$y = \tan^{-1} x$	$\tan y = x$	$(-\infty, \infty)$	$\left(-\dfrac{\pi}{2}, \dfrac{\pi}{2}\right)$
$y = \cot^{-1} x$	$\cot y = x$	$(-\infty, \infty)$	$(0, \pi)$
$y = \csc^{-1} x$	$\csc y = x$	$(-\infty, -1] \cup [1, \infty)$	$\left[-\dfrac{\pi}{2}, 0\right) \cup \left(0, \dfrac{\pi}{2}\right]$
$y = \sec^{-1} x$	$\sec y = x$	$(-\infty, -1] \cup [1, \infty)$	$\left[0, \dfrac{\pi}{2}\right) \cup \left(\dfrac{\pi}{2}, \pi\right]$

ii. Points on the terminal sides of angles in the standard position are used to find exact values of the composition of a trigonometric function and a different inverse trigonometric function.

iii. (a) $\sin^{-1}(\sin x) = x$ for $-\dfrac{\pi}{2} \le x \le \dfrac{\pi}{2}$ and
$\sin(\sin^{-1} x) = x$ for $-1 \le x \le 1$

(b) $\cos^{-1}(\cos x) = x$ for $0 \le x \le \pi$ and
$\cos(\cos^{-1} x) = x$ for $-1 \le x \le 1$

(c) $\tan^{-1}(\tan x) = x$ for $-\dfrac{\pi}{2} < x < \dfrac{\pi}{2}$ and
$\tan(\tan^{-1} x) = x$ for $-\infty < x < \infty$

REVIEW EXERCISES

In Exercises 1–4, draw each angle in standard position.

1. $240°$ †

2. $-150°$ †

3. $-\dfrac{2\pi}{3}$ †

4. $\dfrac{7\pi}{3}$ †

In Exercises 5–7, convert each angle from degrees to radians. Express each answer as a multiple of π.

5. $20°$ $\dfrac{\pi}{9}$

6. $36°$ $\dfrac{\pi}{5}$

7. $-60°$ $-\dfrac{\pi}{3}$

In Exercises 8–10, convert each angle from radians to degrees.

8. $\dfrac{7\pi}{10}$ $126°$

9. $\dfrac{5\pi}{18}$ $50°$

10. $-\dfrac{4\pi}{9}$ $-80°$

In Exercises 11 and 12, find the radian measure of a central angle of a circle of radius r that intercepts each arc of length s. Round your answers to three decimal places.

11. 15 inches, $s = 40$ inches 2.667

12. 9 inches, $s = 17$ inches 1.889

In Exercises 13 and 14, find the length of the arc on each circle of radius r intercepted by a central angle θ. Round your answers to three decimal places.

13. $r = 2$ meters, $\theta = 36°$ 1.257 m

14. $r = 0.9$ meter, $\theta = 12°$ 0.188 m

In Exercises 15 and 16, find the area of the sector of a circle of radius r formed by the central angle θ. Round your answers to three decimal places.

15. $r = 3$ feet, $\theta = 32°$ 2.513 ft²

16. $r = 4$ meters, $\theta = 47°$ 6.562 m²

In Exercises 17–20, use each given trigonometric function value of θ to find the five other trigonometric function values of an acute angle θ. Rationalize the denominators where necessary.

17. $\cos\theta = \dfrac{2}{9}$ †

18. $\sin\theta = \dfrac{1}{5}$ †

19. $\tan\theta = \dfrac{5}{3}$ †

20. $\cot\theta = \dfrac{5}{4}$ †

In Exercises 21–24, a point on the terminal side of an angle θ is given. For each angle, find the exact values of the six trigonometric functions. Rationalize the denominators where necessary.

21. $(2, 8)$ †

22. $(-3, 7)$ †

23. $(-\sqrt{5}, 2)$ †

24. $(-\sqrt{3}, -\sqrt{6})$ †

In Exercises 25–28, use the given information to find the quadrant in which θ lies.

25. $\sin\theta < 0$ and $\cos\theta > 0$ IV

26. $\sin\theta > 0$ and $\tan\theta > 0$ I

27. $\sin\theta < 0$ and $\cot\theta > 0$ III

28. $\cos\theta < 0$ and $\csc\theta > 0$ II

In Exercises 29–32, find the exact values of the remaining trigonometric functions of θ from the given information.

29. $\cos\theta = -\dfrac{4}{5}$, θ in quadrant III $\sin\theta = -\dfrac{3}{5}$, $\tan\theta = \dfrac{3}{4}$, $\cot\theta = \dfrac{4}{3}$, $\sec\theta = -\dfrac{5}{4}$, $\csc\theta = -\dfrac{5}{3}$

30. $\tan\theta = -\dfrac{5}{12}$, θ in quadrant IV $\sin\theta = -\dfrac{5}{13}$, $\cos\theta = \dfrac{12}{13}$, $\cot\theta = -\dfrac{12}{5}$, $\sec\theta = \dfrac{13}{12}$, $\csc\theta = -\dfrac{13}{5}$

†Due to space constrictions, answers to these exercises may be found in the Answers beginning on page A–1 in the back of the book.

31. $\sin\theta = \dfrac{3}{5}$, θ in quadrant II $\cos\theta = -\dfrac{4}{5}$, $\tan\theta = -\dfrac{3}{4}$,
$\cot\theta = -\dfrac{4}{3}$, $\sec\theta = -\dfrac{5}{4}$, $\csc\theta = \dfrac{5}{3}$

32. $\csc\theta = -\dfrac{5}{4}$, θ in quadrant III $\sin\theta = -\dfrac{4}{5}$, $\cos\theta = -\dfrac{3}{5}$,
$\tan\theta = \dfrac{4}{3}$, $\cot\theta = \dfrac{3}{4}$, $\sec\theta = -\dfrac{5}{3}$

In Exercises 33–36, find the exact value of each trigonometric function by using reference angles. Do not use a calculator.

33. $\cos 150°$ $-\dfrac{\sqrt{3}}{2}$ **34.** $\sin(-300°)$ $\dfrac{\sqrt{3}}{2}$

35. $\tan 390°$ $\dfrac{\sqrt{3}}{3}$ **36.** $\cot -405°$ -1

In Exercises 37–40, sketch the graph of each given equation over the interval $[-2\pi, 2\pi]$.

37. $y = -\dfrac{3}{2}\cos x$ † **38.** $y = \dfrac{5}{2}\sin x$ †

39. $y = 3\sin\left(x - \dfrac{\pi}{3}\right)$ † **40.** $y = 4\cos\left(x + \dfrac{3\pi}{2}\right)$ †

In Exercises 41–44, find the amplitude or vertical stretch factor, period, and phase shift of each given function.

41. $y = 14\sin\left(2x + \dfrac{\pi}{7}\right)$ † **42.** $y = 21\cos\left(8x + \dfrac{\pi}{9}\right)$ †

43. $y = 6\tan\left(2x + \dfrac{\pi}{5}\right)$ † **44.** $y = -11\cot\left(6x + \dfrac{\pi}{12}\right)$ †

In Exercises 45–50, graph each function over a two-period interval.

45. $y = -2\cos\left(x + \dfrac{2\pi}{3}\right)$ † **46.** $y = 3\cos(x - \pi)$ †

47. $y = 5\tan\left[2\left(x - \dfrac{\pi}{4}\right)\right]$ † **48.** $y = -\cot\left[2\left(x + \dfrac{\pi}{4}\right)\right]$ †

49. $y = \dfrac{1}{2}\sec\dfrac{x}{2}$ † **50.** $y = -3\csc\dfrac{x}{2}$ †

In Exercises 51–58, find the exact value of y or state that y is undefined.

51. $y = \cos^{-1}\left(\cos\dfrac{5\pi}{8}\right)$ $\dfrac{5\pi}{8}$ **52.** $y = \sin^{-1}\left(\sin\dfrac{7\pi}{6}\right)$ $-\dfrac{\pi}{6}$

53. $y = \tan^{-1}\left(\tan -\dfrac{2\pi}{3}\right)$ $\dfrac{\pi}{3}$ **54.** $y = \sin\left(\cos^{-1}\dfrac{1}{2}\right)$ $\dfrac{\sqrt{3}}{2}$

55. $y = \cos\left(\sin^{-1}\dfrac{\sqrt{2}}{2}\right)$ $\dfrac{\sqrt{2}}{2}$ **56.** $y = \cos\left(\tan^{-1}\dfrac{3}{4}\right)$ $\dfrac{4}{5}$

57. $y = \tan\left[\cos^{-1}\left(-\dfrac{1}{2}\right)\right]$ $-\sqrt{3}$ **58.** $y = \tan\left[\sin^{-1}\left(-\dfrac{\sqrt{3}}{2}\right)\right]$ $-\sqrt{3}$

59. A tractor rolls backward so that its wheels turn a quarter of a revolution. If the tires have a radius of 28 inches, how many inches has the tractor moved? ≈ 44 inches

60. What is the radian measure of the smaller central angle made by the hands of a clock at 5:00? Express your answer as a rational multiple of π. $\dfrac{5\pi}{6}$

61. New Orleans, Louisiana, is due south of Dubuque, Iowa. Find the distance between New Orleans (north latitude 29°59′ N) and Dubuque (north latitude 42°31′ N). Use 3960 miles as the value of the radius of Earth. ≈ 866 miles

62. A horse on the outside row of a merry-go-round is 20 feet from the center, and its linear speed is 314 feet per minute. Find its angular speed. 15.7 radians/min

63. A point on the rim of a pottery wheel with a 28 inch diameter has a linear speed of 21 inches per second. Find its angular speed. $\dfrac{3}{2}$ radian sec

64. A NASA spacecraft has a circular orbit around Mars about 400 km above its surface. If it takes two hours to complete one orbit and the radius of Mars is 6780 km, find the linear speed of the spacecraft. 22,557 km/hr

65. A geostationary satellite has a circular orbit 36,000 km above Earth's surface. It takes 24 hours for the satellite to complete its orbit, appearing to be stationary above a fixed location on Earth. If the radius of Earth is 6400 km, find the linear speed of the satellite. 11,100 km/hr

66. The table gives the average number of daylight hours in Santa Fe, New Mexico, each month. Let y represent the number of daylight hours in Santa Fe in month x and find a function of the form $y = a\sin b(x - c) + d$ that models the hours of daylight throughout the year, where $x = 1$ represents January. $y = 2.25\sin\dfrac{\pi}{6}(x - 4) + 12.05$

Jan	Feb	Mar	Apr	May	June
10.1	9.9	12.0	12.7	14.1	14.1

July	Aug	Sept	Oct	Nov	Dec
14.3	13.5	12.0	11.3	10.0	9.8

67. Suppose a ball attached to a spring is pulled down 11 inches and released and the resulting simple harmonic motion has a period of five seconds. Write an equation of the ball's simple harmonic motion. $y = -11\cos\dfrac{2\pi}{5}t$

PRACTICE TEST A

1. Convert 140° to radians. 2.4435

2. Convert $\dfrac{7\pi}{5}$ to degrees. 252°

3. If $(2, -5)$ is a point on the terminal side of an angle θ, find $\cos\theta$. $\dfrac{2\sqrt{29}}{29}$

4. Find the radian measure of a central angle of a circle of radius 10 centimeters that intercepts an arc of length 15 centimeters. 1.5

5. Find the area of a sector of a circle of radius 15 centimeters that intersects an arc of length 10 centimeters. 75 cm²

6. If $\cos \theta = \dfrac{2}{7}$ for an acute angle θ, find $\sin \theta$. $\dfrac{3\sqrt{5}}{7}$

7. If $\tan \theta < 0$ and $\csc \theta > 0$, find the quadrant in which θ lies. Quadrant II

8. If $\cot \theta = -\dfrac{5}{12}$ and θ is in quadrant IV, find $\sec \theta$. $\dfrac{13}{5}$

9. What is the reference angle for $217°$? $37°$

10. If $\cos \dfrac{3\pi}{7} = 0.223$, find $\sin \dfrac{\pi}{14}$. 0.223

11. From a seat 65 meters high on a Ferris wheel, the angle of depression to a hot dog stand is $42°$. How many meters, to the nearest meter, is the hot dog stand from a point directly below the seat? 72 m

12. Give the amplitude and range for $y = -7 \sin x$.
    Amplitude $= 7$; range $= [-7, 7]$

13. Find the amplitude, period, and phase shift for
    $y = 19 \sin 12(x + 3\pi)$.   Amplitude $= 19$; period $= \dfrac{\pi}{6}$; phase shift $= -3\pi$

14. Suppose a ball attached to a spring is pulled down 7 inches and released and the resulting simple harmonic motion has a period of four seconds. Write an equation for the ball's simple harmonic motion. $y = -7 \cos\left(\dfrac{\pi}{2}t\right)$

15. Graph $y = -\cos \dfrac{x}{2}$ over a one-period interval. †

16. Graph $y = \sin 2(x - \pi)$ over a one-period interval. †

17. Graph $y = \tan\left(x + \dfrac{\pi}{4}\right)$ over a one-period interval. †

18. Find the exact value of $y = \sin^{-1}\left(-\dfrac{\sqrt{3}}{2}\right)$. $-\dfrac{\pi}{3}$

19. Find the exact value of $y = \cos^{-1}\left[\cos \dfrac{4\pi}{3}\right]$. $\dfrac{2\pi}{3}$

20. Find the exact value of $y = \tan\left[\sin^{-1}\left(-\dfrac{1}{3}\right)\right]$. $-\dfrac{\sqrt{2}}{4}$

PRACTICE TEST B

Multiple choice—select the correct answer.

1. Convert $160°$ to radians. b
   a. $\dfrac{7\pi}{9}$ b. $\dfrac{8\pi}{9}$ c. $\dfrac{4\pi}{3}$ d. $\dfrac{7\pi}{3}$

2. Convert $\dfrac{13\pi}{5}$ to degrees. d
   a. $36°$ b. $72°$ c. $108°$ d. $468°$

3. If $(-3, 1)$ is a point on the terminal side of an angle θ, find $\cos \theta$. a
   a. $-\dfrac{3\sqrt{10}}{10}$ b. $-\dfrac{\sqrt{10}}{10}$ c. $\dfrac{\sqrt{10}}{10}$ d. $\dfrac{3\sqrt{10}}{10}$

4. Find the radian measure of a central angle of a circle of radius 8 that intercepts an arc of length 4. a
   a. $\dfrac{1}{2}$ b. 2   c. 32   d. 128

5. Find the area of a sector of a circle of radius 6 centimeters that intercepts an arc of length 3 centimeters. d
   a. $\dfrac{9}{2}$ cm^2 b. $\dfrac{27}{2}$ cm^2 c. 6 cm^2 d. 9 cm^2

6. If $\cos \theta = \dfrac{3}{4}$ for an acute angle θ, find $\sin \theta$. b
   a. $\dfrac{3}{\sqrt{7}}$ b. $\dfrac{\sqrt{7}}{4}$ c. $\dfrac{\sqrt{7}}{3}$ d. $\dfrac{4}{3}$

7. If $\sin \theta < 0$ and $\sec \theta > 0$, find the quadrant in which θ lies. d
   a. I   b. II   c. III   d. IV

8. If $\tan \theta = \dfrac{12}{5}$ and θ is in quadrant III, find $\csc \theta$. b
   a. $-\dfrac{5}{12}$ b. $-\dfrac{13}{12}$ c. $\dfrac{5}{13}$ d. $\dfrac{5}{12}$

9. What is the reference angle for $640°$? c
   a. $10°$ b. $60°$ c. $80°$ d. $280°$

10. In which quadrant is it true that both $\cos \theta > 0$ and $\csc \theta > 0$? a
    a. I   b. II   c. III   d. IV

11. A point on the rim of a wheel with a radius of 50 centimeters has an angular speed of 0.8 radian per second. What is the linear speed of the point? c
    a. 4 centimeters per second
    b. 6.25 centimeters per second
    c. 40 centimeters per second
    d. 62.5 centimeters per second

12. If $y = a \cos bx$ has an amplitude of $\dfrac{1}{2}$ and a period of 4π, what are the values of a and b? a
    a. $a = \dfrac{1}{2}, b = \dfrac{1}{2}$ b. $a = 2, b = \dfrac{1}{2}$
    c. $a = \dfrac{1}{2}, b = 2$ d. $a = 2, b = 2$

13. Which of the following statements is *false*? b
    a. $\sin(-x) = -\sin x$ b. $\cos(-x) = -\sin x$
    c. $\tan(-x) = -\tan x$ d. $\cot(-x) = -\cot x$

14. The equation $y = -5 \cos \dfrac{\pi}{4}t$ describes the harmonic of a ball attached to a spring. What is the period of the motion? d
    a. $\dfrac{\pi}{4}$ b. -5 c. 5   d. 8

15. Find the amplitude a, period b, and phase shift c for
    $y = -7 \cos\left[\dfrac{1}{6}\left(x - \dfrac{\pi}{12}\right)\right]$. d
    a. $a = 7, b = \dfrac{1}{6}\pi, c = \dfrac{\pi}{12}$
    b. $a = 7, b = \dfrac{1}{6}\pi, c = -\dfrac{\pi}{12}$

c. $a = 7, b = 12\pi, c = -\dfrac{\pi}{12}$

d. $a = 7, b = 12\pi, c = \dfrac{\pi}{12}$

16. This is the graph of which function? c

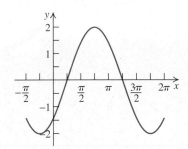

a. $y = -2\sin\left(x + \dfrac{\pi}{2}\right)$ **b.** $y = 2\sin\left(x + \dfrac{\pi}{2}\right)$

c. $y = -2\cos\left(x + \dfrac{\pi}{4}\right)$ **d.** $y = 2\cos\left(x + \dfrac{\pi}{4}\right)$

17. Find the exact value of $y = \cos^{-1}\left(\cos\dfrac{11\pi}{3}\right)$. c

a. $\dfrac{11\pi}{3}$ **b.** $\dfrac{5\pi}{3}$ **c.** $\dfrac{\pi}{3}$ **d.** $-\dfrac{\pi}{3}$

18. Find the exact value of $y = \cos\left[\sin^{-1}\left(-\dfrac{4}{5}\right)\right]$. a

a. $\dfrac{3}{5}$ **b.** $-\dfrac{3}{5}$ **c.** $\dfrac{3}{4}$ **d.** $-\dfrac{4}{3}$

19. Find the exact value of $y = \cos^{-1}\left(\cos\dfrac{7\pi}{4}\right)$. d

a. $\dfrac{2\pi}{3}$ **b.** $\dfrac{-2\pi}{3}$ **c.** $\dfrac{7\pi}{4}$ **d.** $\dfrac{\pi}{4}$

20. Find the exact value of $y = \cos\left[\tan^{-1}\left(\dfrac{5}{3}\right)\right]$. d

a. $\dfrac{3}{5}$ **b.** $\dfrac{5}{34}$ **c.** $\dfrac{5\sqrt{34}}{34}$ **d.** $\dfrac{3\sqrt{34}}{34}$

CUMULATIVE REVIEW EXERCISES ▪ Chapters 1–4

1. Solve the equation $3x^2 - 30x + 50 = -24$. $5 \pm \dfrac{\sqrt{3}}{3}$

2. Find all real number solutions of the equation $(\sqrt{t} + 1)^2 + 2(\sqrt{t} + 1) = 3$. 0

3. Solve the inequality $\dfrac{4 - x}{2x - 4} > 0$ and write the answer in interval notation. $(2, 4)$

4. Write the slope–intercept form of the equation of the line containing the points $(3, -1)$ and $(-6, 5)$. $y = -\dfrac{2}{3}x + 1$

5. Find the domain of the function $f(x) = \ln\left(\dfrac{x}{2 - x}\right)$. $(0, 2)$

6. Identify the basic function and then use transformations to sketch the graph of the function $f(x) = 2(x - 3)^2 - 5$. †

7. Use transformations to sketch the graph of the function $f(x) = 4e^x + 1$. †

8. Find and write the domain of $(f \circ g)(x)$ in interval notation if $f(x) = \ln x$ and $g(x) = -\dfrac{1}{x}$. $(-\infty, 0)$

9. Determine whether each function has an inverse function. If so, find an equation for $f^{-1}(x)$.
   a. $f(x) = 3x + 7$ †   **b.** $f(x) = |x|$ †

10. Use synthetic division to find the quotient and remainder.
   $\dfrac{2x^4 + 4x^3 + x^2 + x - 2}{x + 2}$ $2x^3 + x - 1$, remainder 0

11. Find the horizontal and vertical asymptotes of the graph of $f(x) = \dfrac{10 - 3x^2}{x^2 + 4x - 5}$. $y = -3, x = -5, x = 1$

12. Use rules of logarithms to rewrite each expression in the expanded form.
   a. $\log\sqrt[3]{\dfrac{x^2 y}{z}}$ †   **b.** $\ln\left(\dfrac{7x^4}{5\sqrt{y}}\right)$ †

13. Sketch the graph of $f(x) = \begin{cases} -\ln x & \text{if } x > 0 \\ \ln|x| & \text{if } x < 0 \end{cases}$ †

14. Solve the equation $\log(x^2 + 9x) = 1$. $\{-10, 1\}$

15. Find all possible rational zeros of the function $f(x) = 2x^4 + 5x^3 + x^2 + 10x - 6$. †

16. Use the given trigonometric function value of θ to find the five other trigonometric function values of an acute angle θ. Rationalize the denominators where necessary.
   a. $\cos\theta = \dfrac{3}{7}$ †   **b.** $\sin\theta = \dfrac{2}{11}$ †

17. Find the exact values of the remaining trigonometric functions of θ from the given information.
   a. $\cos\theta = -\dfrac{4}{5}, \theta$ in quadrant II †
   b. $\cot\theta = \dfrac{5}{12}, \theta$ in quadrant III †

18. Sketch the graph of each equation over the interval $[-2\pi, 2\pi]$.
   a. $y = -\dfrac{1}{2}\cos(x + \pi)$ †
   b. $y = -5\sin\left(x + \dfrac{5\pi}{3}\right)$ †

19. Find the exact value of y or state that y is undefined.
   a. $y = \cos\left[\sin^{-1}\left(-\dfrac{1}{5}\right)\right]$ $\dfrac{2\sqrt{6}}{5}$
   b. $y = \sin\left(\tan^{-1}\dfrac{7}{2}\right)$ $\dfrac{7\sqrt{53}}{53}$

20. Find the exact value of $y = \cos^{-1}\left(\cos\dfrac{5\pi}{3}\right)$. $\dfrac{\pi}{3}$

Analytic Trigonometry

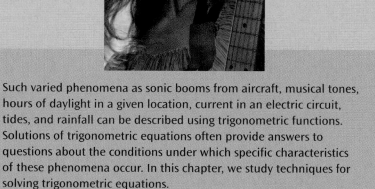

Such varied phenomena as sonic booms from aircraft, musical tones, hours of daylight in a given location, current in an electric circuit, tides, and rainfall can be described using trigonometric functions. Solutions of trigonometric equations often provide answers to questions about the conditions under which specific characteristics of these phenomena occur. In this chapter, we study techniques for solving trigonometric equations.

Trigonometric Identities and Equations

Before Starting this Section, Review

1. Signs of the trigonometric functions (Section 4.3, page 279)

2. LCD and adding fractions (Appendix A)

3. Factoring and special products (Appendix A)

Objectives

1 Use fundamental trigonometric identities to evaluate trigonometric functions.

2 Simplify a complicated trigonometric expression.

3 Prove that a given equation is not an identity.

4 Verify a trigonometric identity.

Hipparchus of Rhodes

(190–120 B.C.)

Hipparchus was born in Nicacea (now called Iznik) in Bithynia (northwest Turkey) but spent much of his life in Rhodes (Greece). Many historians consider him the founder of trigonometry. He introduced trigonometric functions in the form of a chord table—the ancestor of the sine table found in ancient Indian astronomical works. With his chord table, Hipparchus could solve the height–distance problems of plane trigonometry. In his time, there was no Greek term for trigonometry because it was not counted as a branch of mathematics. Rather, trigonometry was just an aid to solve problems in astronomy.

VISUALIZING TRIGONOMETRIC FUNCTIONS

Trigonometric functions are commonly defined as the ratios of two sides corresponding to an angle in a right triangle. The Greek mathematician Hipparchus was the first person to relate trigonometric functions to a circle. His work allows us to visualize the trigonometric functions of an acute angle θ as the length of various segments associated with a unit circle. See Figure 5.1. The figure shows several similar triangles. For example, triangles PCT and OCP are similar, so

$$\frac{CT}{PC} = \frac{PC}{OC} \qquad \text{Sides are proportional.}$$

From Figure 5.1, we use the definitions of trigonometric functions to get

$$PC = \sin\theta, \quad OC = \cos\theta, \quad \text{and} \quad CT = OT - OC = \sec\theta - \cos\theta.$$

Substituting these values into the equation $\dfrac{CT}{PC} = \dfrac{PC}{OC}$, we obtain

$$\frac{\sec\theta - \cos\theta}{\sin\theta} = \frac{\sin\theta}{\cos\theta}.$$

In Example 6, we verify that this equation is an identity. Recall that an *identity* is an equation that is true for all numbers for which both sides are defined.

In Section 4.2, the six trigonometric functions were defined in terms of the coordinates of a point $P(x, y)$ on a unit circle. Consequently, these functions are related to each other and any expression containing these functions can be written in several equivalent ways.

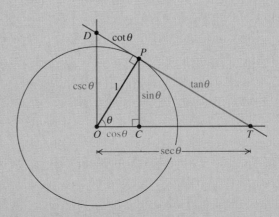

FIGURE 5.1 The unit circle and trigonometric functions

Fundamental Trigonometric Identities

In Chapter 4, we used the definitions of the trigonometric functions to establish reciprocal, quotient, Pythagorean, and even-odd identities. These are called **fundamental trigonometric identities**.

FUNDAMENTAL TRIGONOMETRIC IDENTITIES

1. **Reciprocal Identites**

$$\csc x = \frac{1}{\sin x} \qquad \sec x = \frac{1}{\cos x} \qquad \cot x = \frac{1}{\tan x}$$

$$\sin x = \frac{1}{\csc x} \qquad \cos x = \frac{1}{\sec x} \qquad \tan x = \frac{1}{\cot x}$$

2. **Quotient Identities**

$$\tan x = \frac{\sin x}{\cos x} \qquad \cot x = \frac{\cos x}{\sin x}$$

3. **Pythagorean Identities**

$$\sin^2 x + \cos^2 x = 1 \qquad 1 + \tan^2 x = \sec^2 x \qquad 1 + \cot^2 x = \csc^2 x$$

4. **Even–Odd Identities**

$$\sin(-x) = -\sin x \qquad \cos(-x) = \cos x \qquad \tan(-x) = -\tan x$$

$$\csc(-x) = -\csc x \qquad \sec(-x) = \sec x \qquad \cot(-x) = -\cot x$$

---

1 Use fundamental trigonometric identities to evaluate trigonometric functions.

Evaluating Trigonometric Functions

The next example illustrates the use of a Pythagorean identity to evaluate a related function.

STUDY TIP

Remember that the terminal side of an angle determines the quadrant in which the angle lies. It does *not* matter whether the angle is measured in radians or in degrees.

EXAMPLE 1 **Using a Pythagorean Identity**

If $\cos \theta = -\dfrac{5}{13}$ and $\pi < \theta < \dfrac{3\pi}{2}$, find $\tan \theta$.

SOLUTION

Since $\cos \theta = -\dfrac{5}{13}$, the reciprocal identity says that $\sec \theta = -\dfrac{13}{5}$.

$$\tan^2 \theta = \sec^2 \theta - 1 \qquad\qquad 1 + \tan^2 \theta = \sec^2 \theta$$
$$\tan \theta = \sqrt{\sec^2 \theta - 1} \qquad\quad \text{Because } \theta \text{ is in quadrant III, } \tan \theta \text{ is positive.}$$
$$= \sqrt{\left(-\frac{13}{5}\right)^2 - 1} \qquad\quad \text{Replace } \sec \theta \text{ with } -\frac{13}{5}.$$
$$= \sqrt{\frac{169}{25} - 1} = \sqrt{\frac{144}{25}} = \frac{12}{5}$$

■ ■ ■

Practice Problem 1 If $\sin \theta = \dfrac{4}{5}$ and $\dfrac{\pi}{2} < \theta < \pi$, find $\cot \theta$. ■

The next example shows how fundamental trigonometric identities are used to evaluate the remaining trigonometric functions from one specified trigonometric function value.

> **EXAMPLE 2**   **Using the Fundamental Trigonometric Identities**

If $\cot x = \dfrac{3}{4}$ and $\pi < x < \dfrac{3\pi}{2}$, find the values of the remaining trigonometric functions.

SOLUTION

When given $\cot x$, we can find $\csc x$ by using the identity:

$$\csc^2 x = 1 + \cot^2 x \qquad \text{Pythagorean identity}$$

$$= 1 + \left(\frac{3}{4}\right)^2 \qquad \text{Replace } \cot x \text{ with } \frac{3}{4}.$$

$$\csc^2 x = 1 + \frac{9}{16} = \frac{25}{16} \qquad \text{Simplify.}$$

$$\csc x = -\frac{5}{4} \qquad \begin{array}{l}\text{Square root property: because } x \text{ is} \\ \text{in quadrant III, } \csc x \text{ is negative.}\end{array}$$

$$\sin x = -\frac{4}{5} \qquad \sin x = \frac{1}{\csc x}$$

Multiplying both sides of the quotient identity $\dfrac{\cos x}{\sin x} = \cot x$ by $\sin x$, we have

$$\cos x = \cot x \sin x$$

$$= \left(\frac{3}{4}\right)\left(-\frac{4}{5}\right) \qquad \text{Substitute values of } \cot x \text{ and } \sin x.$$

$$= -\frac{3}{5} \qquad \text{Simplify.}$$

We have the following values of the trigonometric functions:

$$\sin x = -\frac{4}{5} \qquad \cos x = -\frac{3}{5} \qquad \tan x = \frac{4}{3}$$

$$\csc x = -\frac{5}{4} \qquad \sec x = -\frac{5}{3} \qquad \cot x = \frac{3}{4} \qquad\qquad ∎ ∎ ∎$$

Practice Problem 2   If $\tan x = -1$ and $\dfrac{\pi}{2} < x < \pi$, find the values of the remaining trigonometric functions. ∎

2 Simplify a complicated trigonometric expression.

Simplifying a Trigonometric Expression

Recall that when simplifying algebraic expressions, we can use properties of real numbers, factoring techniques, and special product formulas. We can also rationalize the denominator or find the least common denominator. To simplify trigonometric expressions, we use these same techniques together with the fundamental trigonometric identities.

EXAMPLE 3   **Expressing All Trigonometric Functions in Terms of Sines and Cosines**

Rewrite the given expression in terms of sines and cosines and then simplify the resulting expression.

$$\cot x + \tan x + \csc x \sec x$$

SOLUTION

$$\cot x + \tan x + \csc x \sec x$$

$$= \frac{\cos x}{\sin x} + \frac{\sin x}{\cos x} + \frac{1}{\sin x} \cdot \frac{1}{\cos x} \qquad \text{Quotient and Reciprocal identities}$$

$$= \frac{\cos^2 x}{\sin x \cos x} + \frac{\sin^2 x}{\sin x \cos x} + \frac{1}{\sin x \cos x} \qquad \text{Write each fraction with common denominator } \sin x \cos x.$$

$$= \frac{\cos^2 x + \sin^2 x + 1}{\sin x \cos x} \qquad \text{Add numerators.}$$

$$= \frac{1 + 1}{\sin x \cos x} = \frac{2}{\sin x \cos x} \qquad \sin^2 x + \cos^2 x = 1 \qquad ■■■$$

Practice Problem 3   Rewrite the expression in terms of sines and cosines and then simplify the resulting expression.

$$\frac{\tan x}{\sec x + 1} + \frac{\tan x}{\sec x - 1} \qquad ■$$

---

3 Prove that a given equation is not an identity.

Trigonometric Equations and Identities

To verify that an equation is a trigonometric identity, you must *prove* that both sides of the equation are equal for all values of the variable for which both sides are defined.

Consider the following equation:

$$\cos x = \sqrt{1 - \sin^2 x} \qquad (1)$$

Equation (1) is true for all values of x in the interval $\left[-\frac{\pi}{2}, \frac{\pi}{2}\right]$. See Figure 5.2(a).

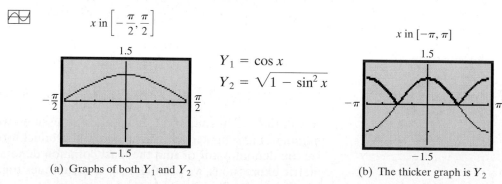

$$Y_1 = \cos x$$
$$Y_2 = \sqrt{1 - \sin^2 x}$$

(a) Graphs of both Y_1 and Y_2 (b) The thicker graph is Y_2

FIGURE 5.2

Since the domain of the sine and cosine functions is the interval $(-\infty, \infty)$, both sides of equation (1) are defined for all real numbers. But for any value of x in the interval $\left(\frac{\pi}{2}, \pi\right]$, the left side of equation (1) has a negative value (because $\cos x$ is negative in quadrant II), while the right side of equation (1) has a positive value. Therefore, equation (1) is *not* an identity. Figure 5.2(b) illustrates that a graphing calculator can help verify that a given equation is *not* an identity. Figure 5.2(a), however, shows that a graphing calculator cannot prove that a given equation *is* an identity because you might happen to use a viewing window where the graphs coincide.

EXAMPLE 4   **Proving That an Equation Is Not an Identity**

Prove that the following equation is not an identity.

$$(\sin x - \cos x)^2 = \sin^2 x - \cos^2 x$$

SOLUTION

To prove that the equation is not an identity we find at least one value of x that results in both sides being defined but not equal.

Let $x = 0$. We know that $\sin 0 = 0$ and $\cos 0 = 1$.

$$\text{The left side} = (\sin x - \cos x)^2 = (0 - 1)^2 = 1 \qquad \text{Replace } x \text{ with } 0.$$
$$\text{The right side} = \sin^2 x - \cos^2 x = (0)^2 - (1)^2 = -1 \qquad \text{Replace } x \text{ with } 0.$$

For $x = 0$, the equation's two sides are not equal; so it is not an identity.   ■ ■ ■

Practice Problem 4   Prove that the equation $\cos x = 1 - \sin x$ is not an identity. ■

4 Verify a trigonometric identity.

Process of Verifying Trigonometric Identities

Verifying a trigonometric identity differs from solving an equation. In verifying an identity, we are given an equation and want to show that it is true for *all* values of the variable for which both sides of the equation are defined. We will use the following method.

VERIFYING TRIGONOMETRIC IDENTITIES

To verify that an equation is an identity, transform one side of the equation into the other side by a sequence of steps, each of which produces an identity. The steps involved can be algebraic manipulations or can use known identities. Note that in verifying an identity, we *do not* just perform the same operation on both sides of the equation.

Methods of Verifying Trigonometric Identities

Verifying trigonometric identities requires practice and experience; there is no set procedure. In the following examples, we suggest five guidelines for verifying trigonometric identities.

1. Start with the more complicated side and transform it to the simpler side.

EXAMPLE 5   **Verifying an Identity**

Verify the identity:

$$\frac{\csc^2 x - 1}{\cot x} = \cot x$$

SOLUTION

We start with the left side of the equation because it appears more complicated than the right side.

$$\frac{\csc^2 x - 1}{\cot x} = \frac{\cot^2 x}{\cot x} \qquad \csc^2 x = 1 + \cot^2 x$$

$$= \cot x \qquad \text{Remove the common factor, } \cot x.$$

We have shown that the left side of the equation is equal to the right side, which verifies the identity.   ■ ■ ■

TECHNOLOGY CONNECTION

In Example 5, we use a graphing calculator to graph $Y_1 = \dfrac{\csc^2 x - 1}{\cot x}$ and $Y_2 = \cot x$. Although not a definitive method of proof, the graphs appear identical; so the equation in Example 5 *appears* to be an identity.

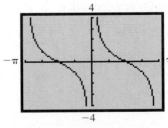

Graph of Y_1

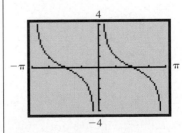

Graph of Y_2

Practice Problem 5   Verify the identity:

$$\frac{1 - \sin^2 x}{\cos x} = \cos x$$ ■

2. Stay focused on the final expression.

While working on one side of the equation, stay focused on your goal of converting it to the form on the other side. This often helps in deciding what your next step should be.

EXAMPLE 6   **Verifying an Identity**

Verify the following identity proposed in the introduction to this section:

$$\frac{\sec \theta - \cos \theta}{\sin \theta} = \frac{\sin \theta}{\cos \theta}$$

SOLUTION

We start with the more complicated left side.

$$\frac{\sec \theta - \cos \theta}{\sin \theta} = \frac{\dfrac{1}{\cos \theta} - \cos \theta}{\sin \theta} \qquad \sec \theta = \frac{1}{\cos \theta}$$

$$= \frac{\left(\dfrac{1}{\cos \theta} - \cos \theta\right)\cos \theta}{\sin \theta \cos \theta} \qquad \text{Multiply numerator and denominator by } \cos \theta.$$

$$= \frac{\dfrac{1}{\cos \theta} \cdot \cos \theta - \cos^2 \theta}{\sin \theta \cos \theta} \qquad \text{Distributive property}$$

$$= \frac{1 - \cos^2 \theta}{\sin \theta \cos \theta} \qquad \text{Simplify.}$$

$$= \frac{\sin^2 \theta}{\sin \theta \cos \theta} \qquad \sin^2 \theta + \cos^2 \theta = 1$$

$$= \frac{\sin \theta}{\cos \theta} \qquad \text{Remove the common factor } \sin \theta.$$

The left side is identical to the right side; the given equation is an identity.   ■ ■ ■

Practice Problem 6   In Figure 5.1 on page 331, triangles OPD and TCP are similar. Thus, $\dfrac{OD}{OP} = \dfrac{PT}{CT}$ leads to

$$\frac{\csc \theta}{1} = \frac{\tan \theta}{\sec \theta - \cos \theta}.$$

Verify that this equation is an identity.   ■

3. Convert to sines and cosines.

It may be helpful to rewrite all trigonometric functions in the equation in terms of sines and cosines and then simplify.

EXAMPLE 7   **Verifying by Rewriting with Sines and Cosines**

Verify the identity $\cot^4 x + \cot^2 x = \cot^2 x \csc^2 x$.

SOLUTION

We start with the more complicated left side.

$$\cot^4 x + \cot^2 x = \frac{\cos^4 x}{\sin^4 x} + \frac{\cos^2 x}{\sin^2 x} \qquad \cot x = \frac{\cos x}{\sin x}$$

$$= \frac{\cos^4 x}{\sin^4 x} + \frac{\cos^2 x}{\sin^2 x} \cdot \frac{\sin^2 x}{\sin^2 x} \qquad \text{The LCD is } \sin^4 x.$$

$$= \frac{\cos^4 x + \cos^2 x \sin^2 x}{\sin^4 x} \qquad \text{Add rational expressions.}$$

$$= \frac{\cos^2 x(\cos^2 x + \sin^2 x)}{\sin^4 x} \qquad \text{Factor out } \cos^2 x.$$

$$= \frac{\cos^2 x(1)}{\sin^4 x} \qquad \cos^2 x + \sin^2 x = 1$$

$$= \frac{\cos^2 x}{\sin^2 x} \cdot \frac{1}{\sin^2 x} \qquad \begin{array}{l}\text{Factor to obtain the form on the} \\ \text{right side.}\end{array}$$

$$= \cot^2 x \csc^2 x \qquad \cot x = \frac{\cos x}{\sin x}, \csc x = \frac{1}{\sin x}$$

Because the left side is identical to the right side, the given equation is an identity.

■ ■ ■

Practice Problem 7   Verify the identity $\tan^4 x + \tan^2 x = \tan^2 x \sec^2 x$.   ■

You could also verify the identity in Example 7 as follows:

$$\cot^4 x + \cot^2 x = \cot^2 x(\cot^2 x + 1) \qquad \text{Distributive property}$$
$$= \cot^2 x \csc^2 x \qquad \text{Pythagorean identity}$$

This approach shows that rewriting the expression using only sines and cosines is not always the quickest way to verify an identity. It is a useful approach when you are stuck, however.

4. Work on both sides.

Sometimes it is helpful to work separately on both sides of the equation. To verify the identity $P(x) = Q(x)$, for example, we transform the left side $P(x)$ into $R(x)$ by using algebraic manipulations and known identities. Then $P(x) = R(x)$ is an identity. Next, we transform the right side, $Q(x)$, into $R(x)$ so that the equation $R(x) = Q(x)$ is an identity. It then follows that $P(x) = Q(x)$ is an identity. See the next example.

TECHNOLOGY CONNECTION

In Example 8, we use the TABLE feature of a graphing calculator in Radian mode to create a table of values for

$$Y_1 = \frac{1}{1 - \sin x} - \frac{1}{1 + \sin x}$$

and

$$Y_2 = \frac{\tan^2 x + \sec^2 x + 1}{\csc x},$$

for different values of x.

X	Y₁	Y₂
-3	-.288	-.288
-2.5	-1.865	-1.865
-2	-10.5	-10.5
-1.5	-398.7	-398.7
-1	-5.765	-5.765
-.5	-1.245	-1.245
0	0	ERROR

X=0

X	Y₁	Y₂
0	0	ERROR
.5	1.245	1.245
1	5.7649	5.7649
1.5	398.7	398.7
2	10.501	10.501
2.5	1.8649	1.8649
3	.28797	.28797

X=0

Note that Y_1 equals 0 when $x = 0$ but that Y_2 is undefined when $x = 0$. While not a definitive method of proof, the table shows that the equation in Example 8 *appears* to be an identity because the table values are identical for values of x for which both Y_1 and Y_2 are defined.

EXAMPLE 8 — Verifying an Identity by Transforming Both Sides Separately

Verify the identity:

$$\frac{1}{1 - \sin x} - \frac{1}{1 + \sin x} = \frac{\tan^2 x + \sec^2 x + 1}{\csc x}$$

SOLUTION

We start with the left side of the equation.

$$\frac{1}{1 - \sin x} - \frac{1}{1 + \sin x} \qquad \text{Begin with the left side.}$$

$$= \frac{1}{1 - \sin x} \cdot \frac{1 + \sin x}{1 + \sin x} - \frac{1}{1 + \sin x} \cdot \frac{1 - \sin x}{1 - \sin x} \qquad \begin{array}{l}\text{Rewrite using the LCD,}\\ (1 - \sin x)(1 + \sin x).\end{array}$$

$$= \frac{(1 + \sin x) - (1 - \sin x)}{(1 - \sin x)(1 + \sin x)} \qquad \text{Subtract fractions.}$$

$$= \frac{1 + \sin x - 1 + \sin x}{1 - \sin^2 x} \qquad \begin{array}{l}\text{Rewrite the numerator;}\\ \text{multiply in the denominator.}\end{array}$$

$$= \frac{2 \sin x}{\cos^2 x} \qquad \text{Simplify}; 1 - \sin^2 x = \cos^2 x$$

We now try to convert the right side to the form $\dfrac{2 \sin x}{\cos^2 x}$.

$$\frac{\tan^2 x + \sec^2 x + 1}{\csc x}$$

$$= \frac{(\tan^2 x + 1) + \sec^2 x}{\csc x} \qquad \text{Regroup terms.}$$

$$= \frac{\sec^2 x + \sec^2 x}{\csc x} \qquad \tan^2 x + 1 = \sec^2 x$$

$$= \frac{2 \sec^2 x}{\csc x} \qquad \text{Simplify.}$$

$$= 2 \left(\frac{1}{\csc x} \right) \sec^2 x \qquad \text{Rewrite.}$$

$$= 2 \sin x \left(\frac{1}{\cos^2 x} \right) \qquad \text{Reciprocal identities}$$

$$= \frac{2 \sin x}{\cos^2 x} \qquad \text{Multiply.}$$

Because both sides of the original equation are equal to $\dfrac{2 \sin x}{\cos^2 x}$, the identity is verified. ■ ■ ■

Practice Problem 8 Verify the identity:

$$\tan \theta + \sec \theta = \frac{\csc \theta + 1}{\cot \theta}$$

■

5. Use conjugates.

It may be helpful to multiply both the numerator and the denominator of a fraction by the same factor. Recall that the sum $a + b$ and the difference $a - b$ are the *conjugates* of each other.

EXAMPLE 9   **Verifying an Identity by Using a Conjugate**

Verify the identity:

$$\frac{\cos x}{1 + \sin x} = \frac{1 - \sin x}{\cos x}$$

SOLUTION

We start with the left side of the equation.

$$\frac{\cos x}{1 + \sin x} = \frac{\cos x(1 - \sin x)}{(1 + \sin x)(1 - \sin x)}$$ Multiply the numerator and the denominator by $1 - \sin x$, the conjugate of $1 + \sin x$.

$$= \frac{\cos x(1 - \sin x)}{1 - \sin^2 x}$$ $(a + b)(a - b) = a^2 - b^2$

$$= \frac{\cos x(1 - \sin x)}{\cos^2 x}$$ $1 - \sin^2 x = \cos^2 x$

$$= \frac{1 - \sin x}{\cos x}$$ Remove the common factor $\cos x$.   ■ ■ ■

Practice Problem 9   Verify the identity $\dfrac{\tan x}{\sec x + 1} = \dfrac{\sec x - 1}{\tan x}$.   ■

We summarize guidelines and hints to help you verify trigonometric identities.

GUIDELINES FOR VERIFYING TRIGONOMETRIC IDENTITIES

Algebra Operations

Review the procedure for combining fractions by finding the least common denominator.

Fundamental Trigonometric Identities

Review the fundamental trigonometric identities summarized on page 332. Look for an opportunity to apply the fundamental trigonometric identities when working on either side of the identity to be verified. Become thoroughly familiar with alternative forms of fundamental identities. For example, $\sin^2 x = 1 - \cos^2 x$ and $\sec^2 x - \tan^2 x = 1$ are alternative forms of the fundamental identities $\sin^2 x + \cos^2 x = 1$ and $\sec^2 x = 1 + \tan^2 x$, respectively.

1. **Start with the more complicated side.**
   It is generally helpful to start with the more complicated side of an identity and simplify it until it becomes identical to the other side. See Example 5.

2. **Stay focused on the final expression.**
   While working on one side of the identity, stay focused on your goal of converting it to the form on the other side. See Example 6.

   The following three techniques are *sometimes* helpful.

3. **Option: Convert to sines and cosines.**
   Rewrite one side of the identity in terms of sines and cosines. See Example 7.

4. **Option: Work on both sides.**
   Transform each side separately to the same equivalent expression.
   See Example 8.

5. **Option: Use conjugates.**
   In expressions containing $1 + \sin x, 1 - \sin x, 1 + \cos x, 1 - \cos x, \sec x + \tan x$, and so on, multiply the numerator and the denominator by its conjugate and then use one of the Pythagorean identities. See Example 9.

SECTION 5.1 ■ Exercises

A EXERCISES Basic Skills and Concepts

1. An equation that is true for all values of the variable in its domain is called a(n) ___identity___.

2. To show that an equation is not an identity, we must find at least ___one___ value(s) of the variable that results in both sides being defined but ___not equal___.

3. $\sin^2 x +$ ___$\cos^2 x$___ $= 1; 1 +$ ___$\tan^2 x$___ $= \sec^2 x; \csc^2 x - \cot^2 x =$ ___1___.

4. *True or False* $\tan^2 x \cot x = \tan x$ is not an identity because $\tan^2 x \cot x$ is not defined for $x = 0$ but $\tan x$ is defined for $x = 0$. False

5. *True or False* The value of $x = \dfrac{3\pi}{2}$ can be used to show that $\sin x = \sqrt{1 - \cos^2 x}$ is not an identity. True

6. *True or False* If $\cos \theta < 0$ and $\pi < \theta < \dfrac{3\pi}{2}$, then $\tan \theta < 0$. False

7. If $\sin \theta = -\dfrac{12}{13}$ and $\pi < \theta < \dfrac{3\pi}{2}$, find $\cos \theta$. $-\dfrac{5}{13}$

8. If $\sec \theta = \dfrac{5}{4}$ and $\dfrac{3\pi}{2} < \theta < 2\pi$, find $\tan \theta$. $-\dfrac{3}{4}$

9. If $\cot \theta = \dfrac{1}{2}$ and $\pi < \theta < \dfrac{3\pi}{2}$, find $\csc \theta$. $-\dfrac{\sqrt{5}}{2}$

10. If $\cos \theta = -\dfrac{1}{3}$ and $\dfrac{\pi}{2} < \theta < \pi$, find $\tan \theta$. $-2\sqrt{2}$

11. If $\sin x = -\dfrac{2}{3}$ and $\pi < x < \dfrac{3\pi}{2}$, find $\cot x$. $\dfrac{\sqrt{5}}{2}$

12. If $\tan x = \dfrac{1}{2}$ and $\pi < x < \dfrac{3\pi}{2}$, find $\cos x$. $-\dfrac{2\sqrt{5}}{2}$

In Exercises 13–24, use the fundamental identities and the given information to find the exact values of the remaining trigonometric functions of x.

13. $\cos x = -\dfrac{3}{5}$ and $\sin x = \dfrac{4}{5}$ †

14. $\cos x = \dfrac{3}{5}$ and $\sin x = -\dfrac{4}{5}$ †

15. $\sin x = \dfrac{1}{\sqrt{3}}$ and $\cos x = \sqrt{\dfrac{2}{3}}$ †

16. $\sin x = \dfrac{1}{\sqrt{3}}$ and $\cos x = -\sqrt{\dfrac{2}{3}}$ †

17. $\tan x = \dfrac{1}{2}$ and $\sec x = -\dfrac{\sqrt{5}}{2}$ †

18. $\tan x = \dfrac{1}{2}$ and $\sec x = \dfrac{\sqrt{5}}{2}$ †

19. $\csc x = 3$ and $\cot x = 2\sqrt{2}$ †

20. $\csc x = 3$ and $\cot x = -2\sqrt{2}$ †

21. $\cot x = \dfrac{12}{5}$ and $\sin x = -\dfrac{5}{13}$ †

22. $\cot x = -\dfrac{12}{5}$ and $\sin x = -\dfrac{5}{13}$ †

23. $\sec x = 3$ and $\dfrac{3\pi}{2} < x < 2\pi$ †

24. $\tan x = -2$ and $\dfrac{\pi}{2} < x < \pi$ †

In Exercises 25–34, use the fundamental identities and appropriate algebraic operations to simplify each expression.

25. $(1 + \tan x)(1 - \tan x) + \sec^2 x$ 2

26. $(\sec x - 1)(\sec x + 1) - \tan^2 x$ 0

27. $(\sec x + \tan x)(\sec x - \tan x)$ 1

28. $\dfrac{\sec^2 x - 4}{\sec x - 2}$ $\sec x + 2$

29. $\csc^4 x - \cot^4 x$ $\csc^2 x + \cot^2 x$

30. $\sin x \cos x(\tan x + \cot x)$ 1

31. $\dfrac{\sec x \csc x(\sin x + \cos x)}{\sec x + \csc x}$ 1

32. $\dfrac{1}{\csc x + 1} - \dfrac{1}{\csc x - 1}$ $-2 \tan^2 x$

33. $\dfrac{\tan^2 x - 2 \tan x - 3}{\tan x + 1}$ $\tan x - 3$

34. $\dfrac{\tan^2 x + \sec x - 1}{\sec x - 1}$ $\sec x + 2$

In Exercises 35–40, use a graphing calculator to determine whether the equation could be an identity by graphing each side of the equation on the same screen. Then prove or disprove the statement algebraically.

35. $\cos^2 x - \sin^2 x = 2 \cos^2 x - 1$ †

36. $(\sin x + \cos x)^2 - 1 = 2 \sin x \cos x$ †

37. $\sin x + \cos x = 1$ †

38. $\dfrac{\sin 2x}{2} = \sin x$ †

39. $\dfrac{\sin^2 x}{1 + \cos x} = 1 - \cos x$ †

40. $\sin x = \cos x \tan x$ †

In Exercises 41–46, use the TABLE feature of a graphing calculator in Radian mode to calculate each side of the equation for different values of x. Determine whether the equation could be an identity. Then prove or disprove the statement algebraically.

41. $\sin x \cot x = \cos x$ †

42. $\sec x \cot x = \csc x$ †

†Due to space constrictions, answers to these exercises may be found in the Answers beginning on page A–1 in the back of the book.

43. $\dfrac{\tan x - 1}{\tan x + 1} = \dfrac{1 - \cot x}{1 + \cot x}$ †

44. $\sin^2 x \sec^2 x + 1 = \sec^2 x$ †

45. $\tan 2x = 2 \tan x$ †

46. $(1 - \sin x)^2 = \cos x$ †

In Exercises 47–82, verify each identity.

47. $\sin x \tan x + \cos x = \sec x$

48. $\cos x \cot x + \sin x = \csc x$

49. $\dfrac{1 - 4 \cos^2 x}{1 - 2 \cos x} = 1 + 2 \cos x$

50. $\dfrac{9 - 16 \sin^2 x}{3 + 4 \sin x} = 3 - 4 \sin x$

51. $(\cos x - \sin x)(\cos x + \sin x) = 1 - 2 \sin^2 x$

52. $(\sin x - \cos x)(\sin x + \cos x) = 1 - 2 \cos^2 x$

53. $\sin^2 x \cot^2 x + \sin^2 x = 1$

54. $\tan^2 x - \sin^2 x = \sin^4 x \sec^2 x$

55. $\sin^3 x - \cos^3 x = (\sin x - \cos x)(1 + \sin x \cos x)$

56. $\sin^3 x + \cos^3 x = (\sin x + \cos x)(1 - \sin x \cos x)$

57. $\cos^4 x - \sin^4 x = 1 - 2 \sin^2 x$

58. $\cos^4 x - \sin^4 x = 2 \cos^2 x - 1$

59. $\dfrac{1}{1 - \sin x} + \dfrac{1}{1 + \sin x} = 2 \sec^2 x$

60. $\dfrac{1}{1 - \cos x} + \dfrac{1}{1 + \cos x} = 2 \csc^2 x$

61. $\dfrac{1}{\csc x - 1} - \dfrac{1}{\csc x + 1} = 2 \tan^2 x$

62. $\dfrac{1}{\sec x - 1} + \dfrac{1}{\sec x + 1} = 2 \sec x \cot^2 x$

63. $\sec^2 x + \csc^2 x = \sec^2 x \csc^2 x$

64. $\cot^2 x + \tan^2 x = \sec^2 x \csc^2 x - 2$

65. $\dfrac{1}{\sec x - \tan x} + \dfrac{1}{\sec x + \tan x} = \dfrac{2}{\cos x}$

66. $\dfrac{1}{\csc x + \cot x} + \dfrac{1}{\csc x - \cot x} = \dfrac{2}{\sin x}$

67. $\dfrac{\sin x}{1 + \cos x} = \csc x - \cot x$

68. $\dfrac{\sin x}{1 - \cos x} = \csc x + \cot x$

69. $(\sin x + \cos x)^2 = 1 + 2 \sin x \cos x$

70. $(\sin x - \cos x)^2 = 1 - 2 \sin x \cos x$

71. $(1 + \tan x)^2 = \sec^2 x + 2 \tan x$

72. $(1 - \cot x)^2 = \csc^2 x - 2 \cot x$

73. $\dfrac{\tan x \sin x}{\tan x + \sin x} = \dfrac{\tan x - \sin x}{\tan x \sin x}$

74. $\dfrac{\cot x \cos x}{\cot x + \cos x} = \dfrac{\cot x - \cos x}{\cot x \cos x}$

75. $(\tan x + \cot x)^2 = \sec^2 x + \csc^2 x$

76. $(1 + \cot^2 x)(1 + \tan^2 x) = \dfrac{1}{\sin^2 x \cos^2 x}$

77. $\dfrac{\sin^2 x - \cos^2 x}{\sec^2 x - \csc^2 x} = \sin^2 x \cos^2 x$

78. $\left(\tan x + \dfrac{1}{\cot x}\right)\left(\cot x + \dfrac{1}{\tan x}\right) = 4$

79. $\dfrac{\tan x}{1 + \sec x} + \dfrac{1 + \sec x}{\tan x} = 2 \csc x$

80. $\dfrac{\cot x}{1 + \csc x} + \dfrac{1 + \csc x}{\cot x} = 2 \sec x$

81. $\dfrac{\sin x + \tan x}{\cos x + 1} = \tan x$

82. $\dfrac{\sin x}{1 + \tan x} = \dfrac{\cos x}{1 + \cot x}$

B EXERCISES Applying the Concepts

83. **Length of a ladder.** A ladder x feet long makes an angle θ with the horizontal and reaches a height of 20 feet. Then $x = \dfrac{20}{\sin \theta}$. Use a reciprocal identity to rewrite this formula. $20 \csc \theta$

84. **Distance from a building.** From a distance of x feet to a 60-foot-high building, the angle of elevation is θ degrees. Then $x = \dfrac{60}{\tan \theta}$. Use a reciprocal identity to rewrite this formula. $60 \cot \theta$

85. If two nonvertical lines with slopes m_1 and m_2 $(m_1 > m_2)$ intersect, then the acute angle θ between the lines satisfies the equation $m_1 \cos \theta - \sin \theta = m_2 \cos \theta + m_1 m_2 \sin \theta$. Rewrite this equation in terms of a single trigonometric function. $m_1 - \tan \theta = m_2 + m_1 m_2 \tan \theta$

86. The angles to the top of a tower measured from two locations d units apart are α and β as shown in the figure. The height h of the tower satisfies the equation $h = \dfrac{d \sin \alpha \sin \beta}{\cos \alpha \sin \beta - \sin \alpha \cos \beta}$. Rewrite this equation in terms of cotangents. $\dfrac{d}{\cot \alpha - \cot \beta}$

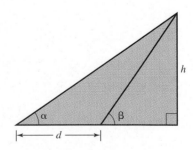

C EXERCISES Beyond the Basics

In Exercises 87–92, verify each identity.

87. $\dfrac{1 - \sin x}{1 - \sec x} - \dfrac{1 + \sin x}{1 + \sec x} = 2 \cot x (\cos x - \csc x)$

88. $\dfrac{\sec x + \tan x}{\csc x + \cot x} - \dfrac{\sec x - \tan x}{\csc x - \cot x} = 2(\sec x - \csc x)$

89. $(1 - \tan x)^2 + (1 - \cot x)^2 = (\sec x - \csc x)^2$

90. $\sec^6 x - \tan^6 x = 1 + 3 \tan^2 x + 3 \tan^4 x$

91. $\dfrac{\cot x + \csc x - 1}{\cot x + \csc x + 1} = \dfrac{1 - \sin x}{\cos x}$

92. $\dfrac{\tan x + \sec x - 1}{\tan x - \sec x + 1} = \dfrac{1 + \sin x}{\cos x}$

In Exercises 93–102, simplify each expression.

93. $(\sin u \cos v + \cos u \sin v)^2 + (\cos u \cos v - \sin u \sin v)^2$ 1

94. $(\sin u \cos v - \cos u \sin v)^2 + (\cos u \cos v + \sin u \sin v)^2$ 1

95. $3(\sin^4 x + \cos^4 x) - 2(\sin^6 x + \cos^6 x)$ 1

96. $\sin^6 x + \cos^6 x + 3 \cos^2 x - 3 \cos^4 x$ 1

97. $\dfrac{1 - \cos\theta}{1 + \cos\theta} + \dfrac{1 + \cos\theta}{1 - \cos\theta} - 4 \cot^2\theta$ 2

98. $\dfrac{\cos^3\theta + \sin^3\theta}{\cos\theta + \sin\theta} + \dfrac{\cos^3\theta - \sin^3\theta}{\cos\theta - \sin\theta}$ 2

99. $\dfrac{\sec^2\theta + 2 \tan^2\theta}{1 + 3 \tan^2\theta}$ 1

100. $\dfrac{\csc^2\alpha + \sec^2\alpha}{\csc^2\alpha - \sec^2\alpha} - \dfrac{1 + \tan^2\alpha}{1 - \tan^2\alpha}$ 0

101. $\cos^2 x (3 - 4 \cos^2 x)^2 + \sin^2 x (3 - 4 \sin^2 x)^2$ 1

102. $(\sin\theta + \csc\theta)^2 + (\cos\theta + \sec\theta)^2 - \tan^2\theta - \cot^2\theta$ 7

103. If $x = r \cos u \cos v$, $y = r \cos u \sin v$, and $z = r \sin u$, verify that $x^2 + y^2 + z^2 = r^2$.

104. If $\tan x + \cot x = 2$, show that
   a. $\tan^2 x + \cot^2 x = 2$.
   b. $\tan^3 x + \cot^3 x = 2$.

105. If $\sec x + \cos x = 2$, show that
   a. $\sec^2 x + \cos^2 x = \sec^4 x + \cos^4 x = 2$
   b. $\sec^3 x + \cos^3 x = 2$.

106. If $\sin x + \sin^2 x = 1$, show that $\cos^2 x + \cos^4 x = 1$.

Critical Thinking

107. If x and y are real numbers, explain why $\sec\theta$ cannot be equal to $\dfrac{xy}{x^2 + y^2}$.

108. Find values of t for which $\sin\theta = \dfrac{1 + t^2}{1 - t^2}$ is possible. $t = 0$

109. Find values of a and b for which $\cos^2\theta = \dfrac{a^2 + b^2}{2ab}$ is possible. $a = b$, for $a, b \neq 0$

110. Find values of a and b for which $\csc^2\theta = \dfrac{2ab}{a^2 + b^2}$ is possible. $a = b$, for $a, b \neq 0$

GROUP PROJECT

111. Prove that the following inequalities are true for $0 < \theta < \dfrac{\pi}{2}$.
   a. $\sin^2\theta + \csc^2\theta \geq 2$
   b. $\cos^2\theta + \sec^2\theta \geq 2$
   c. $\sec^2\theta + \csc^2\theta \geq 4$

Trigonometric Equations

Before Starting this Section, Review

1. Reference angle (Section 4.3, page 280)

2. Periods of trigonometric functions (Sections 4.4 and 4.5, page 288)

3. Factoring techniques (Appendix A, page 769)

4. Solving linear and quadratic equations (Appendix A, page 814)

Objectives

1 Solve trigonometric equations of the form $a \sin(x - c) = k$, $a \cos(x - c) = k$, and $a \tan(x - c) = k$.

2 Solve trigonometric equations involving multiple angles.

3 Solve trigonometric equations by using the zero-product property.

4 Solve trigonometric equations that contain more than one trigonometric function.

5 Solve trigonometric equations by squaring both sides.

BY THE WAY ...

The expression *once in a blue moon* means "*not often*," but what exactly is a blue moon? The popular definition says that a second full moon in a single calendar month is called a *blue moon*. On average, a blue moon occurs once every two and one-half years.

THE PHASES OF THE MOON

The moon orbits Earth and completes one revolution in about 29.5 days (a lunar month). During this period, the angle between the sun, Earth, and the moon changes. See Figure 5.3. This causes different amounts of the illuminated moon to face Earth; consequently, the shape of the moon appears to change. The shape varies from a **new moon**—the phase of the moon when the moon is not visible from Earth (this happens when the moon is between the sun and Earth)—to a **half moon**—the phase of the moon when the moon looks like half a circular disk—to a **full moon**—the phase of the moon when the moon looks like a full circular disk in the sky—back to a half moon and then to a new moon. The portion of the moon visible from Earth on any given day is called the **phase of the moon**. In Example 10, we express the phases of the moon by a sinusoidal equation.

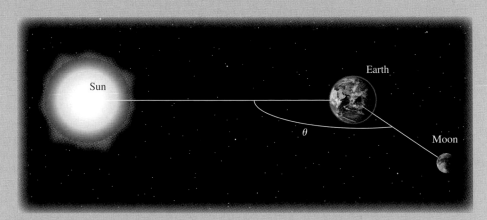

FIGURE 5.3   Phases of the moon

Trigonometric Equations

Recall that a *trigonometric equation* is an equation that contains a trigonometric function with a variable. An identity is an equation that is true for all values in the domain of the variable. We verified trigonometric identities in Section 5.1. Now we will work with trigonometric equations that are satisfied by some but not all values of the variable. For example, $\sin x = \dfrac{1}{2}$ is satisfied by $x = \dfrac{\pi}{6}$ but not by $x = 0$.

Solving a trigonometric equation means finding its *solution set*. Some trigonometric equations do not have a solution. For example, the equation $\sin x = 2$ has no solution because $-1 \le \sin x \le 1$ for all real numbers x. However, if a trigonometric equation has a solution then it has infinitely many solutions, because the trigonometric functions are periodic. Also note that we cannot find an exact solution for every equation. In such cases, we can approximate the solution by using a calculator.

In this section, we study three types of equations:

1. Equations of the form

$$a \sin (x - c) = k, \qquad a \cos (x - c) = k, \qquad a \tan (x - c) = k,$$

where a, c, and k are constants

2. Equations that can be solved by factoring by using the zero-product property

3. Equations equivalent to a polynomial equation in one trigonometric function

---

<table>
<tr><td>**1**</td><td>Solve trigonometric equations of the form</td></tr>
</table>

$a \sin (x - c) = k,$

$a \cos (x - c) = k,$ and

$a \tan (x - c) = k.$

Trigonometric Equations of the Form $a \sin (x - c) = k$, $a \cos (x - c) = k$, and $a \tan (x - c) = k$

We begin with some simple equations.

EXAMPLE 1 Solving a Trigonometric Equation

Find all solutions of each equation. Express all solutions in radians.

a. $\sin x = \dfrac{\sqrt{2}}{2}$ **b.** $\cos \theta = -\dfrac{\sqrt{3}}{2}$ **c.** $\tan x = -\sqrt{3}$

SOLUTION

a. We first find the solutions of $\sin x = \dfrac{\sqrt{2}}{2}$ in the interval $[0, 2\pi)$. We know that $\sin \dfrac{\pi}{4} = \dfrac{\sqrt{2}}{2}$ and that $\sin x$ is positive only in quadrants I and II. See Figure 5.4. The first and second quadrant angles with reference angle $\dfrac{\pi}{4}$ are

$$x = \frac{\pi}{4} \quad \text{and} \quad x = \pi - \frac{\pi}{4} = \frac{3\pi}{4}.$$

So $x = \dfrac{\pi}{4}$ and $x = \dfrac{3\pi}{4}$ are the only solutions in the interval $[0, 2\pi)$. Since the sine function is periodic with period 2π, adding any integer multiple of 2π to either

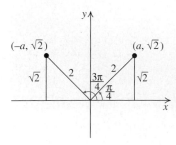

FIGURE 5.4 $\sin \dfrac{\pi}{4} = \dfrac{\sqrt{2}}{2}$

of these values makes the sine of the resulting value still $\frac{\sqrt{2}}{2}$. All solutions of the equation $\sin x = \frac{\sqrt{2}}{2}$ are therefore given by

$$x = \frac{\pi}{4} + 2n\pi \quad \text{or} \quad x = \frac{3\pi}{4} + 2n\pi, \text{for any integer } n.$$

b. As in part **a**, we first find all solutions in the interval $[0, 2\pi)$. We know that $\cos\frac{\pi}{6} = \frac{\sqrt{3}}{2}$ and that $\cos\theta$ is negative only in quadrants II and III. See Figure 5.5. The second and third quadrant angles having reference angle $\frac{\pi}{6}$ are

$$\theta = \pi - \frac{\pi}{6} = \frac{5\pi}{6} \quad \text{and} \quad \theta = \pi + \frac{\pi}{6} = \frac{7\pi}{6}.$$

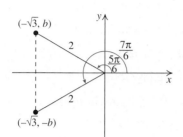

FIGURE 5.5 $\cos\dfrac{5\pi}{6} = -\dfrac{\sqrt{3}}{2}$

$\qquad\qquad = \cos\dfrac{7\pi}{6}$

So $\theta = \frac{5\pi}{6}$ and $\theta = \frac{7\pi}{6}$ are the only solutions in the interval $[0, 2\pi)$.

The period of cosine function is 2π, so all solutions are given by

$$\theta = \frac{5\pi}{6} + 2n\pi \quad \text{or} \quad \theta = \frac{7\pi}{6} + 2n\pi, \text{for any integer } n.$$

c. Because $\tan x$ is a periodic function with period π, we first find all solutions of $\tan x = -\sqrt{3}$ in the interval $[0, \pi)$. We know that $\tan\frac{\pi}{3} = \sqrt{3}$ and that $\tan x$ is negative in quadrant II. See Figure 5.6. The second quadrant angle with reference angle $\frac{\pi}{3}$ is

$$x = \pi - \frac{\pi}{3} = \frac{2\pi}{3}.$$

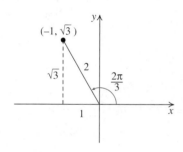

FIGURE 5.6 $\tan\dfrac{2\pi}{3} = -\sqrt{3}$

So $x = \frac{2\pi}{3}$ is the only solution in the interval $[0, \pi)$. Because the period of $\tan x$ is π, all solutions of $\tan x = -\sqrt{3}$ are given by

$$x = \frac{2\pi}{3} + n\pi, \text{for any integer } n. \qquad ■■■$$

Practice Problem 1 Find all solutions of each equation. Express solutions in radians.

a. $\sin x = 1$ **b.** $\cos x = 1$ **c.** $\tan x = 1$ ■

In Example 1, we were able to find the exact solutions of trigonometric equations because the angles involved were common angles. In the next example, we find approximate solutions of trigonometric equations involving other angles.

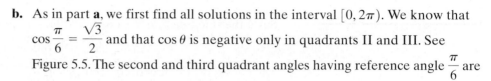

EXAMPLE 2   **Finding Approximate Solutions**

a. Find all solutions of $\sin x = 0.3$. Round solutions to the nearest tenth of a degree.

b. Find all solutions of $\cot x = -3.5$ in the interval $[0, 2\pi)$. Round solutions to four decimal places.

RECALL

If x' denotes the reference angle for x, then $\sin x = \pm \sin x'$, $\cos x = \pm \cos x'$, and $\tan x = \pm \tan x'$. Similar identities hold for $\sec x$, $\csc x$, and $\cot x$.

SOLUTION

a. We use the function $\sin^{-1}$ to find the reference angle x' for any angle x with $\sin x = 0.3$.

$$\sin x' = |\sin x| = 0.3 \qquad \text{Given equation}$$
$$x' = \sin^{-1}(0.3) \qquad 0° < x' < 90°$$
$$x' \approx 17.5° \qquad \text{Use a calculator in Degree mode.}$$

Since the sine function has positive values in quadrants I and II, we have

$$x \approx 17.5° \quad \text{or} \quad x \approx 180° - 17.5° = 162.5°.$$

The period of sine function is $360°$; so all solutions are given by

$$x \approx 17.5° + n \cdot 360° \quad \text{or} \quad x \approx 162.5° + n \cdot 360°, \text{for any integer } n.$$

b. Since $\cot x = -3.5 = -\dfrac{7}{2}$, we have

$$\tan x = -\frac{2}{7} \qquad \text{Reciprocal identity}$$

We first find the reference angle x' for any angle x with $\tan x = -\dfrac{2}{7}$.

$$\tan x' = |\tan x| = \left| -\frac{2}{7} \right| = \frac{2}{7}$$
$$x' = \tan^{-1}\left(\frac{2}{7}\right) \qquad 0 < x' < \frac{\pi}{2}$$
$$x' \approx 0.2783 \qquad \text{Use a calculator in Radian mode.}$$

Because the cotangent function is negative in quadrants II and IV, we have

$$x \approx \pi - 0.2783 \approx 2.8633 \quad \text{and} \quad x \approx 2\pi - 0.2783 \approx 6.0049.$$

These are the only solutions in the interval $[0, 2\pi)$. ■ ■ ■

Practice Problem 2

a. Find all solutions of $\sec x = 1.5$. Round solutions to the nearest tenth of a degree.

b. Find all solutions of $\tan x = -2$ in the interval $[0, 2\pi)$. Round solutions to four decimal places. ■

RECALL

The period of $y = \tan x$ and $y = \cot x$ is π. The period of the other four trigonometric functions is 2π.

In most of our examples, we will find the solutions in the interval $[0, 2\pi)$ for equations involving sine, cosine, secant, and cosecant functions. To find all solutions of such equations, add $2n\pi$ to these solutions. Similarly, to find all solutions of an equation involving tangent or cotangent function, add $n\pi$ to each solution found in $[0, \pi)$. When an equation has only a finite number of solutions, we write the solutions using set notation.

EXAMPLE 3 **Solving a Linear Trigonometric Equation**

Find all solutions in the interval $[0, 2\pi)$ of the equation

$$2 \sin\left(x - \frac{\pi}{4}\right) + 1 = 2.$$

SOLUTION

Replace $x - \dfrac{\pi}{4}$ with θ in the given equation.

$$2 \sin \theta + 1 = 2$$

$$\sin \theta = \frac{1}{2} \qquad \text{Solve for } \sin \theta.$$

The reference angle is $\dfrac{\pi}{6}$ because $\sin \dfrac{\pi}{6} = \dfrac{1}{2}$. (See page 275 or recall from memory.)
Since $\sin \theta$ is positive only in quadrants I and II, the solutions of $\sin \theta = \dfrac{1}{2}$ in $[0, 2\pi)$ are

$$\theta = \frac{\pi}{6} \quad \text{or} \quad \theta = \pi - \frac{\pi}{6} = \frac{5\pi}{6}.$$

Now $\theta = x - \dfrac{\pi}{4}$, so $\theta = \dfrac{\pi}{6}$ $\qquad$ or $\qquad$ $\theta = \dfrac{5\pi}{6}$

$$
\begin{array}{c|c}
x - \dfrac{\pi}{4} = \dfrac{\pi}{6} & x - \dfrac{\pi}{4} = \dfrac{5\pi}{6} \\[2mm]
x = \dfrac{\pi}{6} + \dfrac{\pi}{4} & x = \dfrac{5\pi}{6} + \dfrac{\pi}{4} \\[2mm]
x = \dfrac{2\pi}{12} + \dfrac{3\pi}{12} = \dfrac{5\pi}{12} & x = \dfrac{10\pi}{12} + \dfrac{3\pi}{12} = \dfrac{13\pi}{12}
\end{array}
$$

Replace θ with $x - \dfrac{\pi}{4}$.

Add $\dfrac{\pi}{4}$ to both sides.

Simplify.

The solution set in the interval $[0, 2\pi)$ is $\left\{ \dfrac{5\pi}{12}, \dfrac{13\pi}{12} \right\}$. ■ ■ ■

Practice Problem 3 Find all solutions in the interval $[0°, 360°)$ of the equation
$$2 \sec (x - 30)° - 1 = \sec (x - 30)° + 3. \qquad ■$$

2 Solve trigonometric equations involving multiple angles.

Equations Involving Multiple Angles

If x is the measure of an angle, then for any real number k, the number kx is a **multiple angle** of x. Equations such as

$$\sin 3x = \frac{1}{2}, \quad \cos \frac{1}{2}x = 0.7, \quad \text{and} \quad \sin 2x + \sin 4x = 0$$

involve multiple angles. Be careful when finding all solutions of such equations in the interval $[0, 2\pi)$, as you will see in Example 4.

EXAMPLE 4 Solving a Trigonometric Equation Containing Multiple Angles

Find all solutions of the equation $\cos 3x = \dfrac{1}{2}$ in the interval $[0, 2\pi)$.

SOLUTION

Recall that $\cos \theta = \dfrac{1}{2}$ for the acute angle $\theta = \dfrac{\pi}{3}$. Because $\cos \theta$ is also positive in quadrant IV, we have $\theta = 2\pi - \dfrac{\pi}{3} = \dfrac{5\pi}{3}$. The period of the cosine function is 2π, so replacing θ with $3x$, all solutions of the equation $\cos 3x = \dfrac{1}{2}$ are given by

$$3x = \frac{\pi}{3} + 2n\pi \quad \text{or} \quad 3x = \frac{5\pi}{3} + 2n\pi \qquad \text{For any integer } n$$

$$x = \frac{\pi}{9} + \frac{2n\pi}{3} \quad \text{or} \quad x = \frac{5\pi}{9} + \frac{2n\pi}{3} \qquad \text{Divide both sides by 3.}$$

TECHNOLOGY CONNECTION

On a graphing calculator, you can see that the graphs of $Y_1 = \cos 3x$ and $Y_2 = \dfrac{1}{2}$ intersect in six points, which correspond to the six solutions in the interval $[0, 2\pi)$.

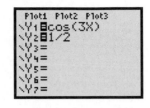

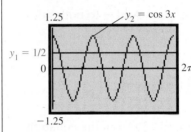

To find solutions in the interval $[0, 2\pi)$, try the following:

$n = -1$	$x = \dfrac{\pi}{9} - \dfrac{2\pi}{3} = -\dfrac{5\pi}{9}$	$x = \dfrac{5\pi}{9} - \dfrac{2\pi}{3} = -\dfrac{\pi}{9}$	
$n = 0$	$x = \dfrac{\pi}{9}$	$x = \dfrac{5\pi}{9}$	
$n = 1$	$x = \dfrac{\pi}{9} + \dfrac{2\pi}{3} = \dfrac{7\pi}{9}$	$x = \dfrac{5\pi}{9} + \dfrac{2\pi}{3} = \dfrac{11\pi}{9}$	
$n = 2$	$x = \dfrac{\pi}{9} + \dfrac{4\pi}{3} = \dfrac{13\pi}{9}$	$x = \dfrac{5\pi}{9} + \dfrac{4\pi}{3} = \dfrac{17\pi}{9}$	
$n = 3$	$x = \dfrac{\pi}{9} + 2\pi = \dfrac{19\pi}{9}$	$x = \dfrac{5\pi}{9} + 2\pi = \dfrac{23\pi}{9}$	

The values resulting from $n = -1$ are too small and those resulting from $n = 3$ are too large to be in the interval $[0, 2\pi)$. The solutions correspond to $n = 0, 1,$ and 2:

$$\left\{ \frac{\pi}{9}, \frac{5\pi}{9}, \frac{7\pi}{9}, \frac{11\pi}{9}, \frac{13\pi}{9}, \frac{17\pi}{9} \right\}$$ ■ ■ ■

Practice Problem 4 Find all solutions of the equation $\sin 2x = \dfrac{1}{2}$ in the interval $[0, 2\pi)$. ■

EXAMPLE 5 **Solving a Trigonometric Equation Containing a Multiple Angle**

Find all solutions of the equation $\sin \dfrac{x}{2} = -\dfrac{\sqrt{3}}{2}$ in the interval $[0, 2\pi)$.

SOLUTION

We know that $\sin \dfrac{\pi}{3} = \dfrac{\sqrt{3}}{2}$. Since $\sin\theta$ is negative only in quadrants III and IV, the solutions for $\sin\theta = -\dfrac{\sqrt{3}}{2}$ are

$$\theta = \pi + \frac{\pi}{3} = \frac{4\pi}{3} \quad \text{and} \quad \theta = 2\pi - \frac{\pi}{3} = \frac{5\pi}{3}.$$

TECHNOLOGY CONNECTION

The graphs of $Y_1 = \sin \dfrac{x}{2}$ and $Y_2 = \dfrac{-\sqrt{3}}{2}$ confirm that the equation in Example 5 has no solution.

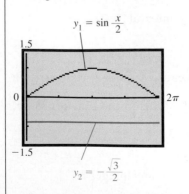

Since the sine function has period 2π, all solutions of $\sin \dfrac{x}{2} = -\dfrac{\sqrt{3}}{2}$ are given by

$$\frac{x}{2} = \frac{4\pi}{3} + 2n\pi \quad \text{or} \quad \frac{x}{2} = \frac{5\pi}{3} + 2n\pi \qquad \text{For any integer } n$$

$$x = \frac{8\pi}{3} + 4n\pi \quad \text{or} \quad x = \frac{10\pi}{3} + 4n\pi \qquad \text{Multiply both sides by 2.}$$

For any integer n, the values of x in both expressions lie outside the interval $[0, 2\pi)$. For example, for $n = 0, x = \dfrac{8\pi}{3} > 2\pi$ and for $n = -1, x = \dfrac{8\pi}{3} - 4\pi = -\dfrac{2\pi}{3} < 0$. The equation $\sin \dfrac{x}{2} = -\dfrac{\sqrt{3}}{2}$, therefore, has no solution in the interval $[0, 2\pi)$, so the solution set is $\varnothing$. ■ ■ ■

Practice Problem 5 Find all solutions of the equation $\tan \dfrac{x}{2} = \dfrac{1}{\sqrt{3}}$ in the interval $[0, 2\pi)$. ■

3 Solve trigonometric equations by using the zero-product property.

Trigonometric Equations and the Zero-Product Property

EXAMPLE 6   **Solving an Equation by Using the Zero-Product Property**

Find all solutions of the equation $(2 \sin x - 1)(7 \tan x + 2) = 0$ in the interval $[0, 2\pi)$. Round solutions to multiples of π or four decimal places.

SOLUTION

$$(2 \sin x - 1)(7 \tan x + 2) = 0 \qquad \text{Given equation}$$

$2 \sin x - 1 = 0$ or $7 \tan x + 2 = 0$	Zero-product property

$$\sin x = \frac{1}{2} \qquad\qquad \tan x = -\frac{2}{7} \qquad \begin{array}{l}\text{Solve for } \sin x \text{ and} \\ \tan x.\end{array}$$

$$x = \frac{\pi}{6} \quad \text{or} \quad x = \frac{5\pi}{6} \quad\Big|\quad x \approx \pi - 0.2783 \quad \text{or} \quad x \approx 2\pi - 0.2783$$

$$x \approx 2.8633 \quad \text{or} \quad x \approx 6.0049$$

So the solution set of the given equation is

$$\left\{ \frac{\pi}{6}, \frac{5\pi}{6}, 2.8633, 6.0049 \right\}.$$ ■ ■ ■

Practice Problem 6   Find all solutions of the equation

$$(\sin x - 1)(\sqrt{3} \tan x + 1) = 0$$

in the interval $[0, 2\pi)$.   ■

TECHNOLOGY CONNECTION

You can confirm that the equation in Example 7 has the two solutions we found in $[0, 2\pi)$ by graphing $Y_1 = 2 \sin^2 x - 5 \sin x + 2$ and inspecting the x-intercepts using TRACE .

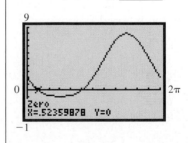

EXAMPLE 7   **Solving a Quadratic Trigonometric Equation**

Find all solutions of the equation $2 \sin^2 \theta - 5 \sin \theta + 2 = 0$. Express the solutions in radians.

SOLUTION

The equation is quadratic in $\sin \theta$. We will use the zero-product property to solve for $\sin \theta$ and then solve the resulting equation for θ.

$2 \sin^2 \theta - 5 \sin \theta + 2 = 0$	$2x^2 - 5x + 2 = 0$, if $x = \sin \theta$
$(2 \sin \theta - 1)(\sin \theta - 2) = 0$	$(2x - 1)(x - 2) = 0$; factor.
$2 \sin \theta - 1 = 0$ or $\sin \theta - 2 = 0$	Zero-product property

$$\sin \theta = \frac{1}{2} \qquad\qquad \sin \theta = 2 \qquad \text{Solve for } \sin \theta.$$

$$\theta = \frac{\pi}{6} \quad \text{or} \quad \theta = \frac{5\pi}{6} \quad\Big|\quad \text{No solution} \qquad -1 \le \sin \theta \le 1$$

So $\theta = \dfrac{\pi}{6}$ and $\theta = \dfrac{5\pi}{6}$ are the only solutions in the interval $[0, 2\pi)$. Because $\sin \theta$ has period 2π, all possible solutions are

$$\theta = \frac{\pi}{6} + 2n\pi \quad \text{or} \quad \theta = \frac{5\pi}{6} + 2n\pi, \text{ for any integer } n.$$ ■ ■ ■

Practice Problem 7   Find all solutions of the equation $2 \cos^2 \theta - \cos \theta - 1 = 0$. Express the solutions in radians.   ■

4 Solve trigonometric equations that contain more than one trigonometric function.

Equations with More Than One Trigonometric Function

When a trigonometric equation contains more than one trigonometric function, we can sometimes use fundamental trigonometric identities to convert the equation into an equation containing only one trigonometric function.

EXAMPLE 8   **Solving a Trigonometric Equation Using Identities**

Find all solutions of the equation $2 \sin^2 \theta + \sqrt{3} \cos \theta + 1 = 0$ in the interval $[0, 2\pi)$.

SOLUTION

Because the equation contains sines and cosines, we will use the Pythagorean identity $\sin^2 \theta + \cos^2 \theta = 1$ to convert the equation into one containing only cosines.

$$2 \sin^2 \theta + \sqrt{3} \cos \theta + 1 = 0 \qquad \text{Given equation}$$
$$2(1 - \cos^2 \theta) + \sqrt{3} \cos \theta + 1 = 0 \qquad \text{Replace } \sin^2 \theta \text{ with } 1 - \cos^2 \theta.$$
$$2 - 2 \cos^2 \theta + \sqrt{3} \cos \theta + 1 = 0 \qquad \text{Distributive property}$$
$$3 - 2 \cos^2 \theta + \sqrt{3} \cos \theta = 0 \qquad \text{Combine terms.}$$
$$2 \cos^2 \theta - \sqrt{3} \cos \theta - 3 = 0 \qquad \text{Multiply both sides by } -1.$$

Solve the last equation for $\cos \theta$.

$$\cos \theta = \frac{-(-\sqrt{3}) \pm \sqrt{(-\sqrt{3})^2 - 4(2)(-3)}}{2(2)} \qquad \begin{array}{l} \text{Use } a = 2, b = -\sqrt{3}, \text{ and} \\ c = -3 \text{ in the quadratic formula.} \end{array}$$

$$= \frac{\sqrt{3} \pm \sqrt{3 + 24}}{4} \qquad \text{Simplify.}$$

$$= \frac{\sqrt{3} \pm \sqrt{27}}{4} = \frac{\sqrt{3} \pm 3\sqrt{3}}{4} \qquad \sqrt{27} = \sqrt{9 \cdot 3} = \sqrt{9}\sqrt{3} = 3\sqrt{3}$$

So

$$\cos \theta = \frac{\sqrt{3} + 3\sqrt{3}}{4} \qquad \text{or} \qquad \cos \theta = \frac{\sqrt{3} - 3\sqrt{3}}{4}$$

$$= \frac{4\sqrt{3}}{4} = \sqrt{3} \qquad\qquad = \frac{(-2)\sqrt{3}}{4} = -\frac{\sqrt{3}}{2}$$

$$\cos \theta = \sqrt{3} > 1$$

Because $-1 \le \cos \theta \le 1$, the equation $\cos \theta = \sqrt{3}$ has no solution.

The equation $\cos \theta = -\dfrac{\sqrt{3}}{2}$ has two solutions in the interval $[0, 2\pi)$:

$$\theta = \pi - \frac{\pi}{6} = \frac{5\pi}{6} \text{ and}$$

$$\theta = \pi + \frac{\pi}{6} = \frac{7\pi}{6}$$

The solution set for the given equation in the interval $[0, 2\pi)$ is $\left\{\dfrac{5\pi}{6}, \dfrac{7\pi}{6}\right\}$. ■ ■ ■

Practice Problem 8   Find all solutions of the equation $2 \cos^2 \theta + 3 \sin \theta - 3 = 0$ in the interval $[0, 2\pi)$. ■

5 Solve trigonometric equations by squaring both sides.

Extraneous Solutions

Next, we solve a trigonometric equation by squaring both sides of the equation and then using an identity. Recall that squaring both sides of an equation may result in extraneous solutions. Therefore, you must check all possible solutions.

STUDY TIP

The Pythagorean identities

$$\sin^2 x + \cos^2 x = 1$$
$$\tan^2 x + 1 = \sec^2 x$$
$$\cot^2 x + 1 = \csc^2 x$$

can sometimes be used to rewrite a trigonometric equation after both sides are squared.

EXAMPLE 9   **Solving a Trigonometric Equation by Squaring**

Find all solutions of the equation $\sqrt{3}\cos x = \sin x + 1$ in the interval $[0, 2\pi)$.

SOLUTION

$(\sqrt{3}\cos x)^2 = (\sin x + 1)^2$	Square both sides of the equation.
$3\cos^2 x = \sin^2 x + 2\sin x + 1$	Expand the binomial.
$3(1 - \sin^2 x) = \sin^2 x + 2\sin x + 1$	Pythagorean identity
$3 - 3\sin^2 x = \sin^2 x + 2\sin x + 1$	Distributive property
$-4\sin^2 x - 2\sin x + 2 = 0$	Collect like terms.
$2\sin^2 x + \sin x - 1 = 0$	Divide both sides by -2.
$(2\sin x - 1)(\sin x + 1) = 0$	Factor.
$2\sin x - 1 = 0 \quad \text{or} \quad \sin x + 1 = 0$	Zero-product property
$\sin x = \dfrac{1}{2} \qquad\qquad \sin x = -1$	Solve for $\sin x$.
$x = \dfrac{\pi}{6} \text{ or } x = \pi - \dfrac{\pi}{6} = \dfrac{5\pi}{6} \qquad x = \dfrac{3\pi}{2}$	Solve for x.

The possible solutions are $\dfrac{\pi}{6}, \dfrac{5\pi}{6}$, and $\dfrac{3\pi}{2}$.

Check:

$x = \dfrac{\pi}{6}$	$x = \dfrac{5\pi}{6}$	$x = \dfrac{3\pi}{2}$
$\sqrt{3}\cos\dfrac{\pi}{6} \stackrel{?}{=} \sin\dfrac{\pi}{6} + 1$	$\sqrt{3}\cos\dfrac{5\pi}{6} \stackrel{?}{=} \sin\dfrac{5\pi}{6} + 1$	$\sqrt{3}\cos\dfrac{3\pi}{2} \stackrel{?}{=} \sin\dfrac{3\pi}{2} + 1$
$\sqrt{3}\left(\dfrac{\sqrt{3}}{2}\right) \stackrel{?}{=} \dfrac{1}{2} + 1$	$\sqrt{3}\left(-\dfrac{\sqrt{3}}{2}\right) \stackrel{?}{=} \dfrac{1}{2} + 1$	$\sqrt{3}(0) \stackrel{?}{=} -1 + 1$
$\dfrac{3}{2} \stackrel{?}{=} \dfrac{3}{2}$ ✓	$-\dfrac{3}{2} \stackrel{?}{=} \dfrac{3}{2}$	$0 \stackrel{?}{=} 0$ ✓
Yes	No	Yes

The solution set of the equation in the interval $[0, 2\pi)$ is $\left\{\dfrac{\pi}{6}, \dfrac{3\pi}{2}\right\}$. ■ ■ ■

Practice Problem 9   Find all solutions of the equation $\sqrt{3}\cot\theta + 1 = \sqrt{3}\csc\theta$ in the interval $[0, 2\pi)$. ■

TECHNOLOGY CONNECTION

〰 You can confirm that the equation in Example 9 has the two solutions in $[0, 2\pi)$ by graphing $Y_1 = \sqrt{3}\cos x$ and the right side $Y_2 = \sin x + 1$. Find the points of intersection using INTERSECT.

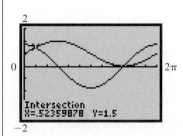

EXAMPLE 10   **The Phases of the Moon**

The moon orbits Earth in 29.5 days. The portion F of the moon visible from Earth x days after the new moon is given by the equation

$$F = 0.5\sin\left[\frac{\pi}{14.75}\left(x - \frac{14.75}{2}\right)\right] + 0.5.$$

Find x when 75% of the moon is visible from Earth.

SOLUTION

We are asked to find x when $F = 75\% = 0.75$. We solve the equation:

$$0.75 = 0.5 \sin\left[\frac{\pi}{14.75}\left(x - \frac{14.75}{2}\right)\right] + 0.5 \qquad \text{Replace } F \text{ with } 0.75.$$

$$0.25 = 0.5 \sin\left[\frac{\pi}{14.75}\left(x - \frac{14.75}{2}\right)\right] \qquad \text{Subtract } 0.5 \text{ from both sides.}$$

$$\frac{1}{2} = \sin\left[\frac{\pi}{14.75}\left(x - \frac{14.75}{2}\right)\right] \qquad \text{Divide both sides by } 0.5.$$

Thus, $\theta = \dfrac{\pi}{14.75}\left(x - \dfrac{14.75}{2}\right)$ is a solution of the equation $\sin \theta = \dfrac{1}{2}$.

We know that $\sin \dfrac{\pi}{6} = \dfrac{1}{2}$. Because $\sin \theta$ is positive only in quadrants I and II, we have

$$\frac{\pi}{14.75}\left(x - \frac{14.75}{2}\right) = \frac{\pi}{6} \quad \text{or} \quad \frac{\pi}{14.75}\left(x - \frac{14.75}{2}\right) = \frac{5\pi}{6}$$

$$x - \frac{14.75}{2} = \frac{14.75}{6} \qquad\qquad x - \frac{14.75}{2} = \frac{5(14.75)}{6} \qquad \begin{array}{l}\text{Multiply both} \\ \text{sides by } 14.75/\pi.\end{array}$$

$$x = \frac{14.75}{2} + \frac{14.75}{6} \qquad\qquad x = \frac{14.75}{2} + \frac{5(14.75)}{6} \qquad \text{Solve for } x.$$

$$x \approx 10 \qquad\qquad\qquad\qquad x \approx 20 \qquad \begin{array}{l}\text{Use a calculator and} \\ \text{round to the nearest day.}\end{array}$$

So 75% of the moon is visible 10 days and 20 days after the new moon. ■ ■ ■

Practice Problem 10 In Example 10, find x when 30% of the moon is visible from Earth. ■

SECTION 5.2 ■ Exercises

A EXERCISES Basic Skills and Concepts

1. The equation $\sin x = \dfrac{1}{2}$ has ___two___ solution(s) in $[0, 2\pi)$.

2. All solutions of $\sin x = \dfrac{1}{2}$ are given by _____ and _____, for any integer n.

3. The equation $\cos x = 1$ has ___one___ solutions(s) in $[0, 2\pi)$.

4. All solutions of $\tan x = -1$ are given by $\dfrac{-\dfrac{\pi}{4} + n\pi}{\quad}$, for any integer n.

5. *True or False* The equation $\sec x = \dfrac{1}{2}$ has two solutions in $[0, 2\pi)$. False

6. *True or False* All solutions of $\sin x = 0$ are given by $x = n\pi$, for any integer n. True

In Exercises 7–16, find all solutions of each equation. Express the solutions in radians.

7. $\cos x = 0$

8. $\sin x = 0$

9. $\tan x = -1$

10. $\cot x = -1$

11. $\cos x = \dfrac{\sqrt{2}}{2}$

12. $\sin x = \dfrac{\sqrt{3}}{2}$

13. $\cot x = \sqrt{3}$ $\dfrac{\pi}{6} + n\pi$

14. $\tan x = -\dfrac{\sqrt{3}}{3}$ $\dfrac{5\pi}{6} + n\pi$

15. $\cos x = -\dfrac{1}{2}$

16. $\sin x = -\dfrac{\sqrt{3}}{2}$

In Exercises 17–26, find all solutions of each equation. Express the solutions in degrees.

17. $\tan x = \dfrac{\sqrt{3}}{3}$ $30° + 180°n$

18. $\cot x = 1$ $45° + 180°n$

19. $\sin x = -\dfrac{1}{2}$

20. $\cos x = \dfrac{1}{2}$

†Due to space constrictions, answers to these exercises may be found in the Answers beginning on page A–1 in the back of the book.

Answers:

2. $\dfrac{\pi}{6} + 2n\pi; \dfrac{5\pi}{6} + 2n\pi$ 7. $\dfrac{\pi}{2} + 2n\pi, \dfrac{3\pi}{2} + 2n\pi$

8. $0 + 2n\pi, \pi + 2n\pi$ 9. $\dfrac{3\pi}{4} + n\pi$ 10. $\dfrac{3\pi}{4} + n\pi$

11. $\dfrac{\pi}{4} + 2n\pi, \dfrac{7\pi}{4} + 2n\pi$ 12. $\dfrac{\pi}{3} + 2n\pi, \dfrac{2\pi}{3} + 2n\pi$

15. $\dfrac{2\pi}{3} + 2n\pi, \dfrac{4\pi}{3} + 2n\pi$ 16. $\dfrac{4\pi}{3} + 2n\pi, \dfrac{5\pi}{3} + 2n\pi$

19. $210° + 360°n, 330° + 360°n$ 20. $60° + 360°n, 300° + 360°n$

21. $\csc x = 1$ $90° + 360°n$ **22.** $\sec x = -1$ $180° + 360°n$

23. $\sqrt{3} \csc x - 2 = 0$ †   **24.** $\sqrt{3} \sec x + 2 = 0$ †

25. $2 \sec x - 4 = 0$ †   **26.** $2 \csc x + 4 = 0$ †

In Exercises 27–32, find all solutions of each equation in the interval $[0°, 360°)$. Round the answers to the nearest tenth of a degree.

27. $\sin \theta = 0.4$ $\{23.6°, 156.4°\}$ **28.** $\cos \theta = 0.6$ $\{53.1°, 306.9°\}$

29. $\sec \theta = 7.2$ $\{82.0°, 278.0°\}$ **30.** $\csc \theta = -4.5$ $\{192.8°, 347.2°\}$

31. $\tan(\theta - 30°) = -5$ $\{131.3°, 311.3°\}$ **32.** $\cot(\theta + 30°) = 6$ $\{159.5°, 339.5°\}$

In Exercises 33–38, find all solutions of each equation in the interval $[0, 2\pi)$. Round the solutions to four decimal places.

33. $\csc x = -2$ $\{3.6652, 5.7596\}$ **34.** $3 \sin x - 1 = 0$ $\{0.3398, 2.8018\}$

35. $3 \tan x + 4 = 0$ $\{2.2143, 5.3559\}$ **36.** $2 \sec x - 7 = 0$ $\{1.2810, 5.0021\}$

37. $2 \csc x + 5 = 0$ $\{3.5531, 5.8717\}$ **38.** $\cos x = 0.1106$ $\{1.4600, 4.8232\}$

In Exercises 39–46, find all solutions of each equation in the interval $[0, 2\pi)$.

39. $\sin\left(x + \dfrac{\pi}{4}\right) = \dfrac{1}{2}$ $\left\{\dfrac{7\pi}{12}, \dfrac{23\pi}{12}\right\}$

40. $2 \cos\left(x - \dfrac{\pi}{4}\right) + 1 = 0$ $\left\{\dfrac{11\pi}{12}, \dfrac{19\pi}{12}\right\}$

41. $\sec\left(x - \dfrac{\pi}{8}\right) + 2 = 0$ †   **42.** $\csc\left(x + \dfrac{\pi}{8}\right) - 2 = 0$ †

43. $\sqrt{3} \tan\left(x - \dfrac{\pi}{6}\right) - 1 = 0$ $\left\{\dfrac{\pi}{3}, \dfrac{4\pi}{3}\right\}$

44. $\cot\left(x + \dfrac{\pi}{6}\right) + 1 = 0$ $\left\{\dfrac{7\pi}{12}, \dfrac{19\pi}{12}\right\}$

45. $2 \sin\left(x - \dfrac{\pi}{3}\right) + 1 = 0$ $\left\{\dfrac{\pi}{6}, \dfrac{3\pi}{2}\right\}$

46. $2 \cos\left(x + \dfrac{\pi}{3}\right) + \sqrt{2} = 0$ $\left\{\dfrac{5\pi}{12}, \dfrac{11\pi}{12}\right\}$

In Exercises 47–56, find all solutions of each equation in the interval $[0, 2\pi)$.

47. $\cos 2x = \dfrac{1}{2}$ †   **48.** $\csc 2x = \dfrac{1}{2}$ $\varnothing$

49. $\tan 2x = \dfrac{\sqrt{3}}{3}$ †   **50.** $\cot 2x = \dfrac{\sqrt{3}}{3}$ †

51. $\sin 3x = \dfrac{1}{2}$ †   **52.** $\cos 3x = \dfrac{\sqrt{3}}{2}$ †

53. $\cos \dfrac{x}{2} = \dfrac{1}{2}$ $\left\{\dfrac{2\pi}{3}\right\}$ **54.** $\csc \dfrac{x}{2} = 2$ $\left\{\dfrac{\pi}{3}, \dfrac{5\pi}{3}\right\}$

55. $\tan \dfrac{x}{3} = 1$ $\left\{\dfrac{3\pi}{4}\right\}$ **56.** $\cot \dfrac{x}{3} = \sqrt{3}$ $\left\{\dfrac{\pi}{2}\right\}$

In Exercises 57–64, find all solutions of each equation in the interval $[0, 2\pi)$.

57. $(\sin x + 1)(\tan x - 1) = 0$ $\left\{\dfrac{\pi}{4}, \dfrac{3\pi}{2}, \dfrac{5\pi}{4}\right\}$

58. $(2 \cos x + 1)(\sqrt{3} \tan x - 1) = 0$ †

59. $(\csc x - 2)(\cot x + 1) = 0$ †

60. $(\sqrt{3} \sec x - 2)(\sqrt{3} \cot x + 1) = 0$ †

61. $(\tan x + 1)(2 \sin x - 1) = 0$ †

62. $(2 \sin x - \sqrt{3})(2 \cos x - 1) = 0$ †

63. $(\sqrt{2} \sec x - 2)(2 \sin x + 1) = 0$ †

64. $(\cot x - 1)(\sqrt{2} \csc x + 2) = 0$ †

In Exercises 65–72, find all solutions of each equation in the interval $[0, 2\pi)$.

65. $4 \sin^2 x = 1$ †   **66.** $4 \cos^2 x = 1$ †

67. $\tan^2 x = 1$ †   **68.** $\sec^2 x = 2$ †

69. $3 \csc^2 x = 4$ †   **70.** $3 \cot^2 x = 1$ †

71. $2 \sin^2 \theta - \sin \theta - 1 = 0$ †   **72.** $2 \cos^2 \theta - 5 \cos \theta + 2 = 0$ †

In Exercises 73–82, use trigonometric identities to solve each equation in the interval $[0, 2\pi)$.

73. $\sin x = \cos x$ †   **74.** $\sqrt{3} \sin x + \cos x = 0$ †

75. $3 \sin^2 x = \cos^2 x$ †   **76.** $3 \cos^2 x = \sin^2 x$ †

77. $\cos^2 x - \sin^2 x = 1$ $0, \pi$ **78.** $2 \sin^2 x + \cos x - 1 = 0$ †

79. $2 \cos^2 x - 3 \sin x - 3 = 0$ †

80. $2 \sin^2 x - \cos x - 1 = 0$ †

81. $\sqrt{3} \sec^2 x - 2 \tan x - 2\sqrt{3} = 0$ †

82. $\csc^2 x - (\sqrt{3} + 1) \cot x + (\sqrt{3} - 1) = 0$ †

In Exercises 83–86, solve each trigonometric equation in the interval $[0, 2\pi)$ by first squaring both sides.

83. $\sqrt{3} \sin x = 1 + \cos x$ †   **84.** $\tan x + 1 = \sec x$ $\{0\}$

85. $\sqrt{3} \tan \theta + 1 = \sqrt{3} \sec \theta$

86. $\sqrt{3} \cot \theta + 1 = \sqrt{3} \csc \theta$

85. $\left\{\dfrac{\pi}{6}\right\}$ **86.** $\left\{\dfrac{\pi}{3}\right\}$

B EXERCISES Applying the Concepts

87. **Angle of elevation.** Find the angle of elevation of the sun if the shadow of a tree 24 feet high is $8\sqrt{3}$ feet in length.   60°

88. **Leaning Tower of Pisa.** The Leaning Tower of Pisa is 179 feet high, and at noon, its shadow measures 16.5 feet. At what angle is the tower inclined from the vertical?   ≈ 5.3°

89. **Returning spacecraft.** When a spacecraft reenters the atmosphere, the angle of reentry off the horizontal must be between 5.1° and 7.1°. Under 5.1°, the craft would "skip" back into space, and over 7.1°, the acceleration forces would be too high and the craft would crash. On reentry, a spacecraft descends 60 kilometers vertically while traveling 605 kilometers. Find the angle of reentry off the horizontal to the nearest tenth of a degree.   ≈ 5.7°

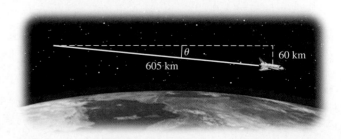

90. Electric current. The electric current I (in amperes) produced by an alternator in time t (in seconds) is given by $I = 60 \sin (120\pi t)$. Find the smallest possible positive value of t (rounded to four decimal places) such that
   a. $I = 30$ amperes. ≈ 0.0014 sec
   b. $I = -20$ amperes. ≈ 0.0092 sec

91. Simple harmonic motion. A simple harmonic motion is described by the equation $x = 6 \sin \left(\dfrac{\pi}{2}t\right)$, where x is the displacement in feet and t is time in seconds. Find positive values of t for which the displacement is 3 feet. $\dfrac{1}{3} + 4n \sec, \dfrac{5}{3} + 4n \sec, n$ an integer

92. Simple harmonic motion. Repeat Exercise 91 assuming that the simple harmonic motion is described by the equation $x = 6 \cos \left(\dfrac{\pi}{2}t\right)$. $\dfrac{2}{3} + 4n \sec, \dfrac{10}{3} + 4n \sec, n$ an integer

93. Overcoat sales. The monthly sales of overcoats of the menswear store Suit Yourself in Winterland are approximated by the equation

$$S(x) = 500 + 500 \sin \left[\frac{\pi}{4}(x - 2)\right],$$

where x is the month, with $x = 1$ for January. Find the months (after appropriate rounding) in which the number of overcoats sold is
   a. 0. August
   b. 1000. April, December
   c. 500. February, June, October

94. Number of tourists. The weekly number of tourists visiting a tropical island is approximated by the equation

$$y = 20 + 10 \sin \left[\frac{\pi}{26}(x - 14)\right],$$

where y is the number of visitors (in thousands) in the xth week of the year, starting with $x = 1$ for the first week in January. Find the week (after appropriate rounding) in which the number of tourists to the island is
   a. 30,000. 27
   b. 25,000. 19, 36
   c. 15,000. 44, 10

95. Average monthly temperature. The average monthly sea surface temperature T (in degrees Fahrenheit) on Paradise Island is approximated by the equation

$$T = 10.5 \sin \left[\frac{\pi}{6}(x - 5)\right] + 73.5,$$

where x is the month, with $x = 1$ representing January. Find the months (after appropriate rounding) when the average sea surface temperature is
   a. 78.75° F. June, October
   b. 84° F. August

96. Average monthly snowfall. The average monthly snowfall y (in inches) in Rochester, New York, in the xth month can be approximated by

$$y = 12 + 12 \sin \left[\frac{\pi}{4}(x - 2)\right], 1 \le x \le 8,$$

where October = 1, November = 2,..., May = 8. Find the months (after appropriate rounding) when the average snowfall in Rochester is
   a. 24 inches. January
   b. 18 inches. February, December

C EXERCISES Beyond the Basics

In Exercises 97–108, find all solutions of each equation in the interval $[0, 2\pi)$.

97. $\sin^4 2x = 1$

98. $2 \cos 6x = 1$

99. $\tan^2 2\theta = 1$

100. $\sin x \cos x = \sin x$ $\{0, \pi\}$

101. $\sin x \cos x - \dfrac{\sqrt{3}}{2} \sin x + \dfrac{1}{2} \cos x - \dfrac{\sqrt{3}}{4} = 0$
   [*Hint:* Factor by grouping.]

102. $2 \sin x \tan x + 2 \sin x + \tan x + 1 = 0$
   [*Hint:* Factor by grouping.]

103. $(\tan 2x - 1)(\sin x + 1) = 0$

104. $(\sqrt{3} \cot 2x - 1)(2 \cos x - 1) = 0$

105. $\tan^2 x = 4$

106. $3 \cos^2 x + \cos x = 0$

107. $\cos x - \sec x + 1 = 0$

108. $3 \csc x + 2 \cot^2 x = 5$

Critical Thinking

109. Solve the following equation: $\sin x \cos x = \dfrac{1}{2}$.
   [*Hint:* Square both sides and replace $\cos^2 x$ with $1 - \sin^2 x$.]

110. Follow the hint given for Exercise 109 to show that the equation $\sin x \cos x = 1$ has no real-number solutions.

Answers:

97. $\left\{\dfrac{\pi}{4}, \dfrac{3\pi}{4}, \dfrac{5\pi}{4}, \dfrac{7\pi}{4}\right\}$

98. $\{0.1745, 0.8727, 1.2217, 1.9199, 2.2689, 2.9671, 3.3161, 4.0143, 4.3633, 5.0615, 5.4105, 6.1087\}$ **99.** $\left\{\dfrac{\pi}{8}, \dfrac{3\pi}{8}, \dfrac{5\pi}{8}, \dfrac{7\pi}{8}, \dfrac{9\pi}{8}, \dfrac{11\pi}{8}, \dfrac{13\pi}{8}, \dfrac{15\pi}{8}\right\}$

101. $\left\{\dfrac{\pi}{6}, \dfrac{7\pi}{6}, \dfrac{11\pi}{6}\right\}$ **102.** $\left\{\dfrac{3\pi}{4}, \dfrac{7\pi}{6}, \dfrac{7\pi}{4}, \dfrac{11\pi}{6}\right\}$

103. $\left\{\dfrac{\pi}{8}, \dfrac{5\pi}{8}, \dfrac{9\pi}{8}, \dfrac{3\pi}{2}, \dfrac{13\pi}{8}\right\}$ **104.** $\left\{\dfrac{\pi}{6}, \dfrac{\pi}{3}, \dfrac{2\pi}{3}, \dfrac{7\pi}{6}, \dfrac{5\pi}{3}\right\}$

105. $\{1.107, 2.034, 4.249, 5.176\}$ **106.** $\{1.5708, 1.911, 4.373, 4.7124\}$

107. $\{0.905, 5.379\}$ **108.** $\{0.911, 2.231, 3.512, 5.913\}$ **109.** $\left\{\dfrac{\pi}{4}, \dfrac{5\pi}{4}\right\}$

Sum and Difference Formulas

Before Starting this Section, Review

1. Distance formula (Section 1.1, page 5)
2. Fundamental trigonometric identities (Section 4.3, page 283)
3. Trigonometric functions of common angles (Section 4.2, page 272)

Objectives

1 Use the sum and difference formulas for cosine.

2 Use cofunction identities.

3 Use the sum and difference formulas for sine.

4 Use the sum and difference formulas for tangent.

PURE TONES IN MUSIC

A vibrating part of an instrument transmits its vibrations to the air, and they travel away as sound waves. The number of complete cycles per second traveled by a wave is called its *frequency* and is denoted by *f*. The unit of frequency, cycles per second, is also called hertz (abbreviated Hz) after Gustav Hertz, the discoverer of radio waves. The note A in the octave above middle C has a frequency of 440 Hz, or 440 cycles per second. Therefore, the time between waves is $\frac{1}{440}$ second. We define this time between waves as the period $T: T = \frac{1}{f}$. You hear the sound produced by note A when the vibration from the source travels to your ear and puts pressure on your eardrum.

A *pure tone* is a tone in which the vibration is a simple harmonic of just one frequency. The simple harmonic of a pure tone with frequency *f* can be described by the equation

$$y = a \sin (2\pi ft),$$

where $|a|$ is the amplitude (indicating loudness), *t* is the time in seconds, and *y* is the pressure of a pure tone on an eardrum (in pounds per square foot). In Example 10, we consider the pressure due to two pure tones of the same frequency. ■

---

1 Use the sum and difference formulas for cosine.

Sum and Difference Formulas for Cosine

In this section, we develop identities involving the sum and difference of two variables, *u* and *v*. The variables represent any two real numbers, or angles, in radian or degree measure.

SUM AND DIFFERENCE FORMULAS FOR COSINE

$$\cos (u + v) = \cos u \cos v - \sin u \sin v$$
$$\cos (u - v) = \cos u \cos v + \sin u \sin v$$

To prove the second formula, we assume that $0 < v < u < 2\pi$, although this identity is true for all real numbers *u* and *v*. Figure 5.7(a) shows points *P* and *Q* on the unit circle on the terminal sides of angles *u* and *v*. The definition of the circular functions tells us that $P = (\cos u,\ \sin u)$ and $Q = (\cos v,\ \sin v)$. In Figure 5.7(b), we rotated angle $(u - v)$ to standard position and labeled point B on the unit circle and the terminal side of angle $(u - v)$.

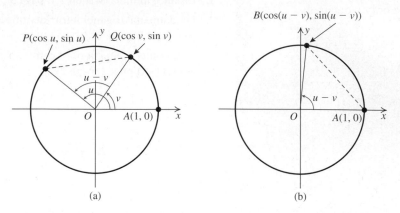

FIGURE 5.7

RECALL

The distance $d(P, Q)$ between two points $P(x_1, y_1)$ and $Q(x_2, y_2)$ is given by

$$d(P, Q) = \sqrt{(x_2 - x_1)^2 + (y_2 - y_1)^2}.$$

We know that triangle POQ in Figure 5.7(a) is congruent to triangle BOA in Figure 5.7(b) by SAS (side-angle-side) congruence; so

$$d(P, Q) = d(A, B)$$
$$[d(P, Q)]^2 = [d(A, B)]^2 \qquad (1) \qquad \text{Square both sides.}$$

Now use the distance formula to simplify the left side of equation (1).

$$
\begin{aligned}
[d(P, Q)]^2 &= (\cos u - \cos v)^2 + (\sin u - \sin v)^2 &&\text{Distance formula} \\
&= \cos^2 u - 2\cos u \cos v + \cos^2 v && (a - b)^2 = a^2 - 2ab + b^2 \\
&\quad + \sin^2 u - 2\sin u \sin v + \sin^2 v && \\
&= (\cos^2 u + \sin^2 u) + (\cos^2 v + \sin^2 v) &&\text{Combine terms.} \\
&\quad - 2(\cos u \cos v + \sin u \sin v) && \\
&= 1 + 1 - 2(\cos u \cos v + \sin u \sin v) &&\text{Pythagorean identity} \\
[d(P, Q)]^2 &= 2 - 2(\cos u \cos v + \sin u \sin v) &&\text{Simplify.}
\end{aligned}
$$

Now simplify the right side of equation (1).

$$
\begin{aligned}
[d(A, B)]^2 &= [\cos(u - v) - 1]^2 + [\sin(u - v) - 0]^2 &&\text{Distance formula} \\
&= \cos^2(u - v) - 2\cos(u - v) + 1 + \sin^2(u - v) &&\text{Expand binomial.} \\
&= [\cos^2(u - v) + \sin^2(u - v)] + 1 - 2\cos(u - v) &&\text{Combine terms.} \\
&= 1 + 1 - 2\cos(u - v) &&\text{Pythagorean identity} \\
[d(A, B)]^2 &= 2 - 2\cos(u - v) &&\text{Simplify.}
\end{aligned}
$$

Because $[d(A, B)]^2 = [d(P, Q)]^2$,

$$2 - 2\cos(u - v) = 2 - 2(\cos u \cos v + \sin u \sin v) \qquad \text{Substitute in each side.}$$
$$\mathbf{\cos(u - v) = \cos u \cos v + \sin u \sin v} \qquad \text{Simplify.}$$

We have proved the formula for the cosine of the difference of two angles.

TECHNOLOGY CONNECTION

Although a calculator will give only a decimal approximation for irrational numbers, you can use one to check your work. Remember to use Radian mode.

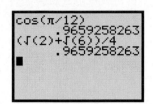

EXAMPLE 1 **Using the Difference Formula for Cosine**

Find the exact value of $\cos \dfrac{\pi}{12}$ by using $\dfrac{\pi}{12} = \dfrac{\pi}{3} - \dfrac{\pi}{4}$.

SOLUTION

We use the exact values of the trigonometric functions of $\dfrac{\pi}{3}$ and $\dfrac{\pi}{4}$ as well as the difference formula for cosine.

$$\cos \frac{\pi}{12} = \cos \left(\frac{\pi}{3} - \frac{\pi}{4} \right) \qquad\qquad \frac{\pi}{12} = \frac{\pi}{3} - \frac{\pi}{4}$$

$$= \cos \frac{\pi}{3} \cos \frac{\pi}{4} + \sin \frac{\pi}{3} \sin \frac{\pi}{4} \qquad \text{Formula for } \cos(u - v)$$

$$= \frac{1}{2} \cdot \frac{\sqrt{2}}{2} + \frac{\sqrt{3}}{2} \cdot \frac{\sqrt{2}}{2} \qquad \text{Use exact values. (See page 275.)}$$

$$= \frac{\sqrt{2} + \sqrt{6}}{4} \qquad\qquad \text{Multiply and add.} \qquad ■■■$$

Practice Problem 1 Find the exact value of $\cos 15°$ by using $15° = 45° - 30°$. ■

The difference formula for $\cos(u - v)$ is true for all real numbers and angles u and v. We can use this formula and the odd-even identities to prove the formula for $\cos(u + v)$.

$$\cos(u + v) = \cos[u - (-v)] \qquad\qquad u + v = u - (-v)$$

$$= \cos u \cos(-v) + \sin u \sin(-v) \quad \text{Difference formula for cosine}$$

$$= \cos u \cos v + \sin u (-\sin v) \qquad \cos(-v) = \cos v; \sin(-v) = -\sin v$$

$$\mathbf{\cos(u + v) = \cos u \cos v - \sin u \sin v} \qquad \text{Simplify.}$$

We have proved the formula for the cosine of the sum of two angles or two real numbers.

RECALL

The cosine function is even: $\cos(-x) = \cos(x)$.
The sine function is odd: $\sin(-x) = -\sin x$.

EXAMPLE 2 **Using the Sum Formula for Cosine**

Find the exact value of $\cos 75°$ by using $75° = 45° + 30°$.

SOLUTION

$$\cos 75° = \cos(45° + 30°) \qquad\qquad 75° = 45° + 30°$$

$$= \cos 45° \cos 30° - \sin 45° \sin 30° \qquad \text{Sum formula for cosine}$$

$$= \frac{\sqrt{2}}{2} \cdot \frac{\sqrt{3}}{2} - \frac{\sqrt{2}}{2} \cdot \frac{1}{2} \qquad \text{Use exact values.}$$

$$= \frac{\sqrt{6}}{4} - \frac{\sqrt{2}}{4} \qquad\qquad \text{Multiply.}$$

$$= \frac{\sqrt{6} - \sqrt{2}}{4} \qquad\qquad \text{Simplify.} \qquad ■■■$$

Practice Problem 2 Find the exact value of $\cos \dfrac{7\pi}{12}$ by using $\dfrac{7\pi}{12} = \dfrac{\pi}{3} + \dfrac{\pi}{4}$. ■

TECHNOLOGY CONNECTION

It will give only a decimal approximation for irrational numbers.
You can use a calculator in Degree mode to check your work.

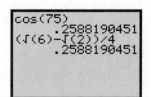

2 Use cofunction identities.

Cofunction Identities

Two trigonometric functions f and g are called **cofunctions** if

$$f\left(\frac{\pi}{2} - x \right) = g(x) \quad \text{and} \quad g\left(\frac{\pi}{2} - x \right) = f(x).$$

We derive two *cofunction identities*.

$$\cos(u - v) = \cos u \cos v + \sin u \sin v \qquad \text{Difference formula for cosine}$$

$$\cos\left(\frac{\pi}{2} - v\right) = \cos\frac{\pi}{2}\cos v + \sin\frac{\pi}{2}\sin v \qquad \text{Replace } u \text{ with } \frac{\pi}{2}.$$

$$= 0 \cdot \cos v + 1 \cdot \sin v = \sin v \qquad \cos\frac{\pi}{2} = 0; \sin\frac{\pi}{2} = 1$$

The result is a cofuntion identity that holds for any real number v or angle v in radian measure:

$$\boldsymbol{\cos\left(\frac{\pi}{2} - v\right) = \sin v}$$

If we replace v with $\frac{\pi}{2} - v$ in the identity $\cos\left(\frac{\pi}{2} - v\right) = \sin v$, we have

$$\cos\left[\frac{\pi}{2} - \left(\frac{\pi}{2} - v\right)\right] = \sin\left(\frac{\pi}{2} - v\right) \qquad \text{Replace } v \text{ with } \frac{\pi}{2} - v.$$

$$\cos v = \sin\left(\frac{\pi}{2} - v\right) \qquad \text{Simplify.}$$

So $\qquad\qquad \boldsymbol{\sin\left(\frac{\pi}{2} - v\right) = \cos v.}$

BASIC COFUNCTION IDENTITIES

If v is any real number or angle measured in radians, then

$$\cos\left(\frac{\pi}{2} - v\right) = \sin v$$

$$\sin\left(\frac{\pi}{2} - v\right) = \cos v$$

If angle v is measured in degrees, then replace $\frac{\pi}{2}$ with $90°$ in these identities.

EXAMPLE 3　Using Cofunction Identities

Prove that for any real number x, $\tan\left(\frac{\pi}{2} - x\right) = \cot x$.

SOLUTION

$$\tan\left(\frac{\pi}{2} - x\right) = \frac{\sin\left(\frac{\pi}{2} - x\right)}{\cos\left(\frac{\pi}{2} - x\right)} \qquad \text{Quotient identity}$$

$$= \frac{\cos x}{\sin x} \qquad \text{Use cofunction identities.}$$

$$= \cot x \qquad \text{Quotient identity} \qquad ■ ■ ■$$

Practice Problem 3　Prove that for any real number x, $\sec\left(\frac{\pi}{2} - x\right) = \csc x$. ■

3 Use the sum and difference formulas for sine.

Sum and Difference Formulas for Sine

To prove the difference formula for the sine function, we start with a cofunction identity.

$$\sin(u - v) = \cos\left[\frac{\pi}{2} - (u - v)\right] \qquad \text{Cofunction identity}$$

$$= \cos\left[\left(\frac{\pi}{2} - u\right) + v\right] \qquad \frac{\pi}{2} - (u - v) = \left(\frac{\pi}{2} - u\right) + v$$

$$= \cos\left(\frac{\pi}{2} - u\right)\cos v - \sin\left(\frac{\pi}{2} - u\right)\sin v \qquad \text{Sum formula for cosine}$$

$$= \sin u \cos v - \cos u \sin v \qquad \text{Cofunction identities}$$

We have the difference formula for sine:

$$\sin(u - v) = \sin u \cos v - \cos u \sin v$$

which holds for all real numbers u and v. If we replace v with $-v$, we derive the sum formula for sine.

$$\sin(u + v) = \sin[u - (-v)] \qquad u + v = u - (-v)$$

$$= \sin u \cos(-v) - \cos u \sin(-v) \qquad \text{Difference formula for sine}$$

$$= \sin u \cos v - \cos u (-\sin v) \qquad \cos(-v) = \cos v; \\ \sin(-v) = -\sin v$$

$$= \sin u \cos v + \cos u \sin v \qquad \text{Simplify.}$$

This proves the sum formula for sine:

$$\sin(u + v) = \sin u \cos v + \cos u \sin v.$$

EXAMPLE 4   **Using the Difference Formula for Sine**

Prove the identity $\sin(\pi - x) = \sin x$.

SOLUTION

$$\sin(\pi - x) = \sin\pi\cos x - \cos\pi\sin x \qquad \text{Difference formula for sine}$$

$$= (0)\cos x - (-1)\sin x \qquad \sin\pi = 0; \cos\pi = -1$$

$$= \sin x \qquad \text{Simplify.} \qquad ■■■$$

Practice Problem 4   Prove the identity $\sin(\pi + x) = -\sin x$.   ■

EXAMPLE 5   **Using the Sum Formula for Sine**

Find the exact value of $\sin 63° \cos 27° + \cos 63° \sin 27°$ without using a calculator.

SOLUTION

The given expression is the right side of the sum formula for sine:

$$\sin(u + v) = \sin u \cos v + \cos u \sin v$$

$$\sin 63° \cos 27° + \cos 63° \sin 27° = \sin(63° + 27°) \qquad u = 63°, v = 27°$$

$$= \sin 90° = 1 \qquad ■■■$$

Practice Problem 5   Find the exact value of $\sin 43° \cos 13° - \cos 43° \sin 13°$ without using a calculator.   ■

EXAMPLE 6 **Finding the Exact Value of a Sum**

Let $\sin u = -\dfrac{3}{5}$ and $\cos v = \dfrac{12}{13}$, with $\pi < u < \dfrac{3\pi}{2}$ and $\dfrac{3\pi}{2} < v < 2\pi$. Find the exact value of $\sin(u + v)$.

SOLUTION

Given $\sin u = -\dfrac{3}{5}$ with u in quadrant III, we find the exact value of $\cos u$.

$$\cos^2 u = 1 - \sin^2 u \qquad \text{Pythagorean identity}$$
$$\cos u = -\sqrt{1 - \sin^2 u} \qquad \text{In quadrant III, } \cos u \text{ is negative.}$$
$$= -\sqrt{1 - \left(-\dfrac{3}{5}\right)^2} \qquad \text{Replace } \sin u \text{ with } -\dfrac{3}{5}.$$
$$= -\sqrt{1 - \dfrac{9}{25}} \qquad \left(-\dfrac{3}{5}\right)^2 = \dfrac{9}{25}$$
$$= -\sqrt{\dfrac{16}{25}} \qquad 1 - \dfrac{9}{25} = \dfrac{25}{25} - \dfrac{9}{25} = \dfrac{16}{25}$$
$$\cos u = -\dfrac{4}{5}$$

Similarly, given $\cos v = \dfrac{12}{13}$ with v in quadrant IV, we find the exact value of $\sin v$.

$$\sin v = -\sqrt{1 - \cos^2 v} \qquad \text{In quadrant IV, } \sin v \text{ is negative.}$$
$$= -\sqrt{1 - \left(\dfrac{12}{13}\right)^2} \qquad \text{Replace } \cos v \text{ with } \dfrac{12}{13}.$$
$$= -\sqrt{\dfrac{25}{169}} \qquad 1 - \left(\dfrac{12}{13}\right)^2 = 1 - \dfrac{144}{169} = \dfrac{169 - 144}{169}$$
$$\sin v = -\dfrac{5}{13}$$

$$\sin(u + v) = \sin u \cos v + \cos u \sin v \qquad \text{Sum formula for sine}$$
$$= \left(-\dfrac{3}{5}\right)\left(\dfrac{12}{13}\right) + \left(-\dfrac{4}{5}\right)\left(-\dfrac{5}{13}\right) \qquad \text{Replace values of } \sin u, \sin v, \cos u, \text{ and } \cos v.$$
$$= -\dfrac{36}{65} + \dfrac{20}{65} = -\dfrac{16}{65} \qquad \text{Multiply and add.}$$

The exact value of $\sin(u + v)$ is $-\dfrac{16}{65}$. ■ ■ ■

Practice Problem 6 For the angles u and v in Example 6, find the exact value of $\cos(u + v)$. ■

EXAMPLE 7 **Verifying an Identity by Using a Sum or Difference Formula**

Verify the identity $\dfrac{\sin(x - y)}{\cos x \cos y} = \tan x - \tan y$.

SOLUTION

We start with the more complicated left side.

$$\dfrac{\sin(x - y)}{\cos x \cos y} = \dfrac{\sin x \cos y - \cos x \sin y}{\cos x \cos y} \qquad \text{Formula for } \sin(u - v)$$
$$= \dfrac{\sin x \cos y}{\cos x \cos y} - \dfrac{\cos x \sin y}{\cos x \cos y} \qquad \dfrac{a - b}{c} = \dfrac{a}{c} - \dfrac{b}{c}$$

$$= \frac{\sin x}{\cos x} - \frac{\sin y}{\cos y} \qquad \text{Remove common factors.}$$

$$= \tan x - \tan y \qquad \text{Quotient identity}$$

Since the left side is identical to the right side, the given equation is an identity.   ■ ■ ■

Practice Problem 7   Verify the following identity: $\dfrac{\cos (x + y)}{\sin x \sin y} = \cot x \cot y - 1$ ■

Reduction Formula

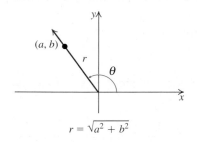

y

(a, b)

r

θ

x

$r = \sqrt{a^2 + b^2}$

FIGURE 5.8 $a = r \cos \theta$
$b = r \sin \theta$

Figure 5.8 suggests that a point (a, b) is on the terminal side of the angle θ if and only if

$$\cos \theta = \frac{a}{\sqrt{a^2 + b^2}} \quad \text{and} \quad \sin \theta = \frac{b}{\sqrt{a^2 + b^2}}.$$

Using these values for $\cos \theta$ and $\sin \theta$ in the sum formula for the sine, we obtain

$$\sin (x + \theta) = \sin x \cos \theta + \cos x \sin \theta \qquad \text{Sum formula for sine}$$

$$\sin (x + \theta) = \sin x \frac{a}{\sqrt{a^2 + b^2}} + \cos x \frac{b}{\sqrt{a^2 + b^2}}$$

Multiplying both sides of this equation by $\sqrt{a^2 + b^2}$, we have the reduction formula:

$$\sqrt{a^2 + b^2} \sin (x + \theta) = a \sin x + b \cos x$$

REDUCTION FORMULA

If (a, b) is any point on the terminal side of an angle θ (radians) in standard position, then

$$a \sin x + b \cos x = \sqrt{a^2 + b^2} \sin (x + \theta)$$

for any real number x.

EXAMPLE 8   **Using the Reduction Formula**

Find an angle θ, in radians, and a real number A such that

$$\sin x - \sqrt{3} \cos x = A \sin (x + \theta).$$

SOLUTION
Note that by the reduction formula

$$\sin x - \sqrt{3} \cos x = a \sin x + b \cos x \qquad a = 1 \text{ and } b = -\sqrt{3}$$

$$= A \sin (x + \theta),$$

where $A = \sqrt{a^2 + b^2} = \sqrt{1^2 + (-\sqrt{3})^2} = \sqrt{4} = 2$ and θ is any angle in standard position that has the point $(a, b) = (1, -\sqrt{3})$ on its terminal side. One such angle is $\theta = \tan^{-1}\left(\dfrac{b}{a}\right) = \tan^{-1}(-\sqrt{3}) = -\dfrac{\pi}{3}$. Then with $A = 2$ and $\theta = -\dfrac{\pi}{3}$, we have

$$\sin x - \sqrt{3} \cos x = 2 \sin \left[x + \left(-\frac{\pi}{3} \right) \right] = 2 \sin \left(x - \frac{\pi}{3} \right). \qquad ■ ■ ■$$

Practice Problem 8   Find an angle θ, in radians, and a real number A such that

$$\sin x + \sqrt{3} \cos x = A \sin (x + \theta). \qquad ■$$

EXAMPLE 9 Using the Reduction Formula to Sketch a Graph

Sketch the graph of the equation $y = \sin x - \sqrt{3} \cos x$.

SOLUTION

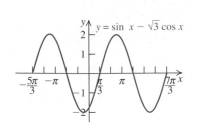

FIGURE 5.9

Example 8 shows that $\sin x - \sqrt{3} \cos x = 2 \sin \left(x - \dfrac{\pi}{3} \right)$. Therefore, the graphs

of $y = \sin x - \sqrt{3} \cos x$ and $y = 2 \sin \left(x - \dfrac{\pi}{3} \right)$ are the same. The graph of

$y = 2 \sin \left(x - \dfrac{\pi}{3} \right)$ is a sine wave with amplitude 2, period 2π, and phase shift

$\dfrac{\pi}{3}$ units to the right. Two cycles of the graph of $y = 2 \sin \left(x - \dfrac{\pi}{3} \right) =$

$\sin x - \sqrt{3} \cos x$ are shown in Figure 5.9. ■ ■ ■

Practice Problem 9 Sketch two cycles of the graph of $y = \sin x + \sqrt{3} \cos x$. ■

EXAMPLE 10 Combining Two Pure Tones

Suppose the pressure exerted by two pure tones (in pounds per square foot) after t seconds is given by

$$y_1 = 0.3 \sin (800\pi t) \quad \text{and} \quad y_2 = 0.4 \cos (800\pi t).$$

Find the amplitude, period, frequency, and phase shift for the total pressure $y = y_1 + y_2$.

SOLUTION

We have $y = 0.3 \sin (800\pi t) + 0.4 \cos (800\pi t)$. We use the reduction formula with $a = 0.3$ and $b = 0.4$ to rewrite this equation in the form

$$y = A \sin (800\pi t + \theta) = A \sin \left[800\pi \left(t + \dfrac{\theta}{800\pi} \right) \right],$$

where

$$A = \sqrt{a^2 + b^2} = \sqrt{(0.3)^2 + (0.4)^2} = 0.5$$

and $(0.3, 0.4)$ is a point on the terminal side of θ. One such angle is $\theta = \tan^{-1} \left(\dfrac{0.4}{0.3} \right)$.

For the total pressure,

$$y = A \sin \left[800\pi \left(t + \dfrac{\theta}{800\pi} \right) \right]$$

$$= 0.5 \sin \left[800\pi \left(t + \dfrac{\theta}{800\pi} \right) \right] \qquad A = 0.5$$

amplitude $= 0.5$

$$\text{period} = \dfrac{2\pi}{800\pi} = \dfrac{1}{400}$$

frequency $= 400$ Frequency $= \dfrac{1}{\text{Period}}$

phase shift $= -\dfrac{\theta}{800\pi} \approx -0.00037$ Use a calculator in Radian mode with

$$\theta = \tan^{-1} \left(\dfrac{0.4}{0.3} \right) \qquad ■ ■ ■$$

Practice Problem 10 Find the amplitude, period, frequency, and phase shift of the total pressure due to the pressure from the two pure tones

$$y_1 = 0.1 \sin (400t) \quad \text{and} \quad y_2 = 0.2 \cos (400t). \qquad ■$$

4 Use the sum and difference formulas for tangent.

Sum and Difference Formulas for Tangent

We use the quotient identity $\tan x = \dfrac{\sin x}{\cos x}$ and the formulas for $\sin (u - v)$ and $\cos (u - v)$ to derive a difference formula for tangent.

$$\tan (u - v) = \frac{\sin (u - v)}{\cos (u - v)} \qquad \text{Quotient identity}$$

$$= \frac{\sin u \cos v - \cos u \sin v}{\cos u \cos v + \sin u \sin v} \qquad \begin{array}{l}\text{Formulas for } \sin (u - v) \text{ and} \\ \cos (u - v)\end{array}$$

$$= \frac{\dfrac{\sin u \cos v}{\cos u \cos v} - \dfrac{\cos u \sin v}{\cos u \cos v}}{\dfrac{\cos u \cos v}{\cos u \cos v} + \dfrac{\sin u \sin v}{\cos u \cos v}} \qquad \begin{array}{l}\text{Divide numerator and denominator} \\ \text{by } \cos u \cos v.\end{array}$$

$$= \frac{\dfrac{\sin u}{\cos u} - \dfrac{\sin v}{\cos v}}{1 + \dfrac{\sin u \sin v}{\cos u \cos v}} \qquad \text{Simplify.}$$

$$= \frac{\tan u - \tan v}{1 + \tan u \tan v} \qquad \text{Quotient identity}$$

We have derived the difference formula for the tangent:

$$\boxed{\tan (u - v) = \frac{\tan u - \tan v}{1 + \tan u \tan v}}$$

Replacing v with $-v$ in the difference formula, we have

$$\tan (u + v) = \tan [u - (-v)] \qquad u + v = u - (-v)$$

$$= \frac{\tan u - \tan (-v)}{1 + \tan u \tan (-v)} \qquad \text{Formula for } \tan (u - v)$$

$$= \frac{\tan u - (-\tan v)}{1 + \tan u (-\tan v)} \qquad \tan (-v) = -\tan v$$

$$= \frac{\tan u + \tan v}{1 - \tan u \tan v} \qquad \text{Simplify.}$$

We now have the sum formula for tangent:

$$\boxed{\tan (u + v) = \frac{\tan u + \tan v}{1 - \tan u \tan v}}$$

EXAMPLE 11 **Verifying an Identity**

Verify the identity $\tan (\pi - x) = -\tan x$.

SOLUTION

Apply the difference formula for the tangent to $\tan (\pi - x)$.

$$\tan (\pi - x) = \frac{\tan \pi - \tan x}{1 + \tan \pi \tan x} \qquad \begin{array}{l}\text{Replace } u \text{ with } \pi \text{ and } v \text{ with } x \text{ in} \\ \text{the formula for } \tan (u - v).\end{array}$$

$$= \frac{0 - \tan x}{1 + 0 \cdot \tan x} \qquad \tan \pi = 0$$

$$= -\tan x \qquad \text{Simplify.}$$

Therefore, the given equation is an identity. ■ ■ ■

Practice Problem 11 Verify the identity $\tan (\pi + x) = \tan x$. ■

Summary

The important identities involving the sum, difference, and cofunctions of two numbers or angles are summarized next.

SUM AND DIFFERENCE FORMULA

$$\cos(u - v) = \cos u \cos v + \sin u \sin v \qquad \cos(u + v) = \cos u \cos v - \sin u \sin v$$

$$\sin(u - v) = \sin u \cos v - \cos u \sin v \qquad \sin(u + v) = \sin u \cos v + \cos u \sin v$$

$$\tan(u - v) = \frac{\tan u - \tan v}{1 + \tan u \tan v} \qquad \tan(u + v) = \frac{\tan u + \tan v}{1 - \tan u \tan v}$$

COFUNCTION IDENTITIES

$$\sin\left(\frac{\pi}{2} - x\right) = \cos x \qquad \cos\left(\frac{\pi}{2} - x\right) = \sin x \qquad \tan\left(\frac{\pi}{2} - x\right) = \cot x$$

$$\csc\left(\frac{\pi}{2} - x\right) = \sec x \qquad \sec\left(\frac{\pi}{2} - x\right) = \csc x \qquad \cot\left(\frac{\pi}{2} - x\right) = \tan x$$

REDUCTION FORMULA

If (a, b) is any point on the terminal side of an angle θ (radians) in standard position, then for any real number x,

$$a \sin x + b \cos x = \sqrt{a^2 + b^2} \sin(x + \theta).$$

SECTION 5.3 ■ Exercises

A EXERCISES Basic Skills and Concepts

1. $\sin(A + B) = \underline{\sin A \cos B + \cos A \sin B}$

2. $\cos A \cos B - \sin A \sin B = \underline{\cos(A + B)}$.

3. $\tan(A + B) = \underline{\hspace{2cm}}$. $\dfrac{\tan A + \tan B}{1 - \tan A \tan B}$

4. *True or False* $\cos\left[\dfrac{\pi}{2} + x\right] = \sin x$. False

5. *True or False* $\sin\left[\dfrac{\pi}{2} + x\right] = \cos x$. True

6. *True or False* $\tan(x + y) = \tan x + \tan y$. False

In Exercises 7–26, find the exact value of each expression.

7. $\sin(45° + 30°)$ †

8. $\sin(45° - 30°)$ †

9. $\sin(60° - 45°)$ †

10. $\sin(60° + 45°)$ †

11. $\sin(-105°)$ †

12. $\cos 285°$ †

13. $\tan 225°$ 1

14. $\tan(-165°)$ $2 - \sqrt{3}$

15. $\sin\left(\dfrac{\pi}{6} + \dfrac{\pi}{4}\right)$ $\dfrac{\sqrt{6} + \sqrt{2}}{4}$

16. $\cos\left(\dfrac{\pi}{3} - \dfrac{\pi}{4}\right)$ $\dfrac{\sqrt{6} + \sqrt{2}}{4}$

17. $\tan\left(\dfrac{\pi}{4} - \dfrac{\pi}{6}\right)$ $2 - \sqrt{3}$

18. $\cot\left(\dfrac{\pi}{3} - \dfrac{\pi}{4}\right)$ $2 + \sqrt{3}$

19. $\sec\left(\dfrac{\pi}{3} + \dfrac{\pi}{4}\right)$ $-\sqrt{6} - \sqrt{2}$

20. $\csc\left(\dfrac{\pi}{4} - \dfrac{\pi}{3}\right)$ $-\sqrt{6} - \sqrt{2}$

21. $\cos\dfrac{-5\pi}{12}$ $\dfrac{-\sqrt{6} - \sqrt{2}}{4}$

22. $\sin\dfrac{7\pi}{12}$ $\dfrac{\sqrt{6} + \sqrt{2}}{4}$

23. $\tan\dfrac{19\pi}{12}$ $-2 - \sqrt{3}$

24. $\sec\dfrac{\pi}{12}$ $\sqrt{6} - \sqrt{2}$

25. $\tan\dfrac{17\pi}{12}$ $2 + \sqrt{3}$

26. $\csc\dfrac{11\pi}{12}$ $\sqrt{6} + \sqrt{2}$

In Exercises 27–40, verify each identity.

27. $\sin\left(x + \dfrac{\pi}{2}\right) = \cos x$

28. $\cos\left(x + \dfrac{\pi}{2}\right) = -\sin x$

29. $\sin\left(x - \dfrac{\pi}{2}\right) = -\cos x$

30. $\cos\left(x - \dfrac{\pi}{2}\right) = \sin x$

31. $\tan\left(x + \dfrac{\pi}{2}\right) = -\cot x$

32. $\tan\left(x - \dfrac{\pi}{2}\right) = -\cot x$

---

†Due to space constrictions, answers to these exercises may be found in the Answers beginning on page A–1 in the back of the book.

33. $\csc(x + \pi) = -\csc x$

34. $\sec(x + \pi) = -\sec x$

35. $\cos\left(x + \dfrac{3\pi}{2}\right) = \sin x$

36. $\cos\left(x - \dfrac{3\pi}{2}\right) = -\sin x$

37. $\tan\left(x - \dfrac{3\pi}{2}\right) = -\cot x$

38. $\tan\left(x + \dfrac{3\pi}{2}\right) = -\cot x$

39. $\cot(3\pi - x) = -\cot x$

40. $\csc\left(\dfrac{5\pi}{2} - x\right) = \sec x$

In Exercises 41–50, find the exact value of each expression without using a calculator.

41. $\sin 56° \cos 34° + \cos 56° \sin 34°$ 1

42. $\cos 57° \cos 33° - \sin 57° \sin 33°$ 0

43. $\cos 331° \cos 61° + \sin 331° \sin 61°$ 0

44. $\cos 110° \sin 70° + \sin 110° \cos 70°$ 0

45. $\dfrac{\tan 129° - \tan 84°}{1 + \tan 129° \tan 84°}$ 1   **46.** $\dfrac{\tan 28° + \tan 17°}{1 - \tan 28° \tan 17°}$ 1

47. $\sin\dfrac{7\pi}{12}\cos\dfrac{3\pi}{12} - \cos\dfrac{7\pi}{12}\sin\dfrac{3\pi}{12}$ $\dfrac{\sqrt{3}}{2}$

48. $\cos\dfrac{5\pi}{12}\cos\dfrac{\pi}{12} - \sin\dfrac{5\pi}{12}\sin\dfrac{\pi}{12}$ 0

49. $\dfrac{\tan\dfrac{5\pi}{12} - \tan\dfrac{2\pi}{12}}{1 + \tan\dfrac{5\pi}{12}\tan\dfrac{2\pi}{12}}$ 1   **50.** $\dfrac{\tan\dfrac{5\pi}{12} + \tan\dfrac{7\pi}{12}}{1 - \tan\dfrac{5\pi}{12}\tan\dfrac{7\pi}{12}}$ 0

In Exercises 51–56, find the exact value of each expression, given that $\tan u = \dfrac{3}{4}$, with u in quadrant III, and $\sin v = \dfrac{5}{13}$, with v in quadrant II.

51. $\sin(u - v)$ $\dfrac{56}{65}$ **52.** $\sin(u + v)$ $\dfrac{16}{65}$

53. $\cos(u + v)$ $\dfrac{63}{65}$ **54.** $\cos(u - v)$ $\dfrac{33}{65}$

55. $\tan(u + v)$ $\dfrac{16}{63}$ **56.** $\tan(u - v)$ $\dfrac{56}{33}$

In Exercises 57–62, find the exact value of each expression, given that $\cos\alpha = -\dfrac{2}{5}$, with α in quadrant II, and $\sin\beta = -\dfrac{3}{7}$, with β in quadrant IV.

57. $\sin(\alpha - \beta)$ †   **58.** $\cos(\alpha - \beta)$ †

59. $\csc(\alpha + \beta)$ †   **60.** $\sec(\alpha + \beta)$ †

61. $\cot(\alpha - \beta)$ †   **62.** $\cot(\alpha + \beta)$ †

In Exercises 63–70, verify each identity.

63. $\dfrac{\sin(x + y)}{\cos x \cos y} = \tan x + \tan y$

64. $\dfrac{\sin(x + y)}{\sin x \sin y} = \cot x + \cot y$

65. $\dfrac{\cos(x + y)}{\cos x \cos y} = 1 - \tan x \tan y$

66. $\dfrac{\cos(x - y)}{\sin x \sin y} = \cot x \cot y + 1$

67. $\dfrac{\cos(x - y)}{\sin(x + y)} = \dfrac{1 + \cot x \cot y}{\cot x + \cot y}$

68. $\dfrac{\cos(x + y)}{\sin(x - y)} = \dfrac{1 - \cot x \cot y}{\cot x - \cot y}$

69. $\dfrac{\sin(x - y)}{\sin(x + y)} = \dfrac{\cot y - \cot x}{\cot y + \cot x}$

70. $\dfrac{\cos(x + y)}{\cos(x - y)} = \dfrac{\cot x \cot y - 1}{\cot x \cot y + 1}$

In Exercises 71–78, rewrite each equation in the form $y = A \sin(x + \theta)$; graph the resulting equation.

71. $y = \sin x + \cos x$ †   **72.** $y = \sin x - \cos x$ †

73. $y = 3\sin x + 4\cos x$ †   **74.** $y = 4\sin x - 3\cos x$ †

75. $y = 5\sin x - 12\cos x$ † **76.** $y = 12\sin x + 5\cos x$ †

77. $y = \cos x - 3\sin x$ †   **78.** $y = 3\cos x - 4\sin x$ †

B EXERCISES Applying the Concepts

79. **Analytic geometry.** Suppose two nonvertical lines l_1 and l_2 intersect in a plane. The slope of l_1 is m_1, and the slope of l_2 is m_2. Note that $m_1 = \tan\theta_1$ and $m_2 = \tan\theta_2$. Show that the tangent of the acute angle α between the lines l_1 and l_2 is given by $\tan\alpha = \left|\dfrac{m_2 - m_1}{1 + m_1 m_2}\right|$.

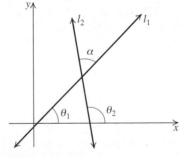

In Exercises 80–82, use Exercise 79 to find the acute angle between the lines l_1 and l_2. $\tan^{-1}(3/5) \approx 0.5404$

80. **Angle between lines.** l_1: $y = x + 5$, l_2: $y = 4x + 2$.

81. **Angle between lines.** l_1: $y = 2x + 5$, l_2: $6x - 3y = 21$. 0

82. **Angle between lines.** l_1: $y = 2x - 4$, l_2: $x + 2y = 7$. $\dfrac{\pi}{2}$

83. **Combining two pure tones.** The pressure exerted by two pure tones in pounds per square foot is given by the equations $y_1 = 0.4\sin(400t)$ and $y_2 = 0.3\cos(400t)$. Find the amplitude and the phase shift for the total pressure $y = y_1 + y_2$. $A = 0.5; \theta \approx -0.00161$

84. Combining two pure tones. Repeat Exercise 83 for
$y_1 = 0.05 \sin{(600t)}$ and $y_2 = 0.12 \cos{(600t)}$.
$A \approx 0.13; \theta \approx -0.00196$

85. Simple harmonic motion. A simple harmonic motion is
described by the equation $x = 0.12 \sin 2t + 0.5 \cos 2t$,
where x is the distance from the equilibrium position and
t is the time in seconds.
a. Find the amplitude of the motion. $A \approx 0.5142$
b. Find the frequency and the phase shift.

86. Simple harmonic motion. Repeat Exercise 85 assuming
that the motion is described by the equation

$$x = \frac{1}{2} \sin 3t + \frac{1}{3} \cos 3t.$$

C EXERCISES Beyond the Basics

87. Difference quotient. Let $f(x) = \sin x$. Show that
$$\frac{f(x + h) - f(x)}{h} = \sin x \left(\frac{\cos h - 1}{h} \right) + \cos x \left(\frac{\sin h}{h} \right).$$

88. Difference quotient. Let $f(x) = \cos x$. Show that
$$\frac{f(x + h) - f(x)}{h} = \cos x \left(\frac{\cos h - 1}{h} \right) - \sin x \left(\frac{\sin h}{h} \right).$$

In Exercises 89–100, verify each identity.

89. $\cos u \cos{(u + v)} + \sin u \sin{(u + v)} = \cos v$

90. $\sin{(x + y)} \cos y - \cos{(x + y)} \sin y = \sin x$

91. $\sin 5x \cos 3x - \cos 5x \sin 3x = \sin 2x$

92. $\sin{(2x - y)} \cos y + \cos{(2x - y)} \sin y = \sin 2x$

93. $\sin \left(\dfrac{\pi}{2} + x - y \right) = \cos x \cos y + \sin x \sin y$

94. $\cos \left(\dfrac{\pi}{2} + x - y \right) = \cos x \sin y - \sin x \cos y$

95. $\sin{(x + y)} \sin{(x - y)} = \sin^2 x - \sin^2 y$

96. $\cos{(\alpha + \beta)} \cos{(\alpha - \beta)} = \cos^2 \alpha - \sin^2 \beta$

97. $\sin^2 \left(\dfrac{\pi}{4} + \dfrac{x}{2} \right) - \sin^2 \left(\dfrac{\pi}{4} - \dfrac{x}{2} \right) = \sin x$

98. $\cos^2 \left(\dfrac{\pi}{4} + \dfrac{x}{2} \right) - \sin^2 \left(\dfrac{\pi}{4} - \dfrac{x}{2} \right) = 0$

99. $\dfrac{\sin{(\alpha - \beta)}}{\sin \alpha \sin \beta} + \dfrac{\sin{(\beta - \gamma)}}{\sin \beta \sin \gamma} + \dfrac{\sin{(\gamma - \alpha)}}{\sin \gamma \sin \alpha} = 0$

100. $\dfrac{\sin{(\alpha - \beta)}}{\cos \alpha \cos \beta} + \dfrac{\sin{(\beta - \gamma)}}{\cos \beta \cos \gamma} + \dfrac{\sin{(\gamma - \alpha)}}{\cos \gamma \cos \alpha} = 0$

Answers:

85. b. Frequency $= \dfrac{1}{\pi}$; phase shift ≈ -0.6676

86. a. $A = \dfrac{\sqrt{13}}{6} \approx 0.6009$

b. Frequency $= \dfrac{3}{2\pi}$; phase shift ≈ -0.196

In Exercises 101–110, find the exact value of each expression.

101. $\cos^2 15° - \cos^2 30° + \cos^2 45° - \cos^2 60° + \cos^2 75°$ $\dfrac{1}{2}$

102. $\cos^2 \left(\dfrac{\pi}{8} \right) + \cos^2 \left(\dfrac{3\pi}{8} \right) + \cos^2 \left(\dfrac{5\pi}{8} \right) + \cos^2 \left(\dfrac{7\pi}{8} \right)$ 2

103. $\cos 60° + \cos 80° + \cos 100°$ $\dfrac{1}{2}$

104. $\sin 30° - \sin 70° + \sin 110°$ $\dfrac{1}{2}$

105. $\sin \left[\tan^{-1} \left(-\dfrac{3}{4} \right) + \cos^{-1} \left(\dfrac{4}{5} \right) \right]$ 0

106. $\cos \left[\sin^{-1} \left(-\dfrac{3}{5} \right) + \cos^{-1} \left(\dfrac{3}{5} \right) \right]$ $\dfrac{24}{25}$

107. $\sin \left[\sin^{-1} \left(\dfrac{3}{5} \right) - \cos^{-1} \left(\dfrac{4}{5} \right) \right]$ 0

108. $\cos \left[\sin^{-1} \left(\dfrac{3}{5} \right) + \tan^{-1} \left(-\dfrac{4}{3} \right) \right]$ $\dfrac{24}{25}$

109. $\tan \left[\cos^{-1} \left(\dfrac{4}{5} \right) + \tan^{-1} \left(\dfrac{2}{3} \right) \right]$ $\dfrac{17}{6}$

110. $\tan \left[\sin^{-1} \left(-\dfrac{3}{5} \right) + \cos^{-1} \left(\dfrac{4}{5} \right) \right]$ 0

111. Show that $\tan 70° - \tan 20° = 2 \tan 50°$.

112. Show that $\sin{(\theta + n\pi)} = (-1)^n \sin \theta$.

113. Find the exact value of
$\tan 1° \tan 2° \tan 3° \cdots \tan 88° \tan 89°$. 1

**In Exercises 114–116, assume that x and y are positive
numbers in the domain of the inverse function considered.
Prove each identity.**

114. $\sin^{-1} x \pm \sin^{-1} y = \sin^{-1}{(x\sqrt{1 - y^2} \pm y\sqrt{1 - x^2})}$
[*Hint:* Let $u = \sin^{-1} x$ and $v = \sin^{-1} y$ and consider
$\sin{(u \pm v)}$.]

115. $\cos^{-1} x + \cos^{-1} y = \cos^{-1}{(xy - \sqrt{1 - x^2}\sqrt{1 - y^2})}$

116. $\sin^{-1} x + \cos^{-1} x = \dfrac{\pi}{2}$

Critical Thinking

117. Explain why we cannot use the formula for $\tan{(u - v)}$
to verify that $\tan \left(\dfrac{\pi}{2} - x \right) = \cot x$.

118. Find a formula for $\cot{(x - y)}$ in terms of $\cot x$ and
$\cot y$.
$\dfrac{1 + \cot x \cot y}{\cot y - \cot x}$

Double-Angle and Half-Angle Formulas

Before Starting this Section, Review

1. Sum formulas for sine, cosine, and tangent functions (Section 5.3, page 355)

2. Pythagorean identities (Section 5.1, page 352)

3. Trigonometric functions of common angles (Section 4.2, page 275)

4. Trigonometric equations (Section 5.2, page 343)

Objectives

1. Use double-angle formulas.
2. Use power-reducing formulas.
3. Use half-angle formulas.
4. Solve trigonometric equations involving multiple angles and half-angles.

COST OF USING AN ELECTRIC BLANKET

Three basic properties of electricity—*voltage, current*, and *power*—together with the rate charged by your electric company, are used to find the cost of using an electric device.

Voltage (*V*) is measured in *volts*. It is the force that pushes electricity through a wire. *Current* (*I*), measured in *amperes* (amps), measures how much electricity is moving through the device per second. *Power* (*P*), measured in *watts*, gives the energy consumed per second by an electric device and is defined by the equation

$$P = VI.$$

Your electric company bills you by the kilowatt-hour. When you turn on an electric device that consumes 1000 watts for one hour, it consumes 1 kilowatt-hour. Suppose your electric company charges 8¢ per kilowatt-hour. An electric blanket might use 250 watts (depending on the setting). If you turn it on for ten hours, it will consume $250 \times 10 = 2500 = 2.5$ kilowatt-hours. This will cost you $(2.5)(8) = 20$¢.

In Example 5, we use trigonometric identities to compute the *wattage rating* of an electric blanket. ∎

---

1 Use double-angle formulas.

Double-Angle Formulas

Suppose an angle measures x (radians or degrees); then $2x$ is double the measure of x. *Double-angle formulas* express trigonometric functions of $2x$ in terms of functions of x. We begin with the sum formulas for the sine, cosine, and tangent functions.

$$\sin (u + v) = \sin u \cos v + \cos u \sin v$$
$$\cos (u + v) = \cos u \cos v - \sin u \sin v$$
$$\tan (u + v) = \frac{\tan u + \tan v}{1 - \tan u \tan v}$$

To find the double-angle formula for the sine, replace u and v with x in the sum formula for sine.

$\sin (u + v) = \sin u \cos v + \cos u \sin v$	Sum formula for sine
$\sin (x + x) = \sin x \cos x + \cos x \sin x$	Replace both u and v with x.
$\sin 2x = 2 \sin x \cos x$	Simplify.

The identity $$\sin 2x = 2 \sin x \cos x$$

is the double-angle formula for the sine function. We derive double-angle formulas for the cosine and tangent functions by replacing both u and v with x in the sum formulas.

$$\cos 2x = \cos^2 x - \sin^2 x \qquad \tan 2x = \frac{2 \tan x}{1 - \tan^2 x}$$

To derive two other useful forms for $\cos 2x$, first replace $\cos^2 x$ with $1 - \sin^2 x$ and then replace $\sin^2 x$ with $1 - \cos^2 x$ in the formula for $\cos 2x$.

$$
\begin{aligned}
\cos 2x &= \cos^2 x - \sin^2 x \\
&= (1 - \sin^2 x) - \sin^2 x \\
\cos 2x &= 1 - 2 \sin^2 x
\end{aligned}
\qquad \bigg| \qquad
\begin{aligned}
\cos 2x &= \cos^2 x - \sin^2 x \\
&= \cos^2 x - (1 - \cos^2 x) \\
\cos 2x &= 2 \cos^2 x - 1
\end{aligned}
$$

DOUBLE-ANGLE FORMULAS

$$\sin 2x = 2 \sin x \cos x \qquad \cos 2x = \cos^2 x - \sin^2 x$$

$$\tan 2x = \frac{2 \tan x}{1 - \tan^2 x} \qquad \cos 2x = 1 - 2 \sin^2 x$$

$$\cos 2x = 2 \cos^2 x - 1$$

EXAMPLE 1   **Using Double-Angle Formulas**

If $\cos \theta = -\dfrac{3}{5}$ and θ is in quadrant II, find the exact value of each expression.

a. $\sin 2\theta$ **b.** $\cos 2\theta$ **c.** $\tan 2\theta$

SOLUTION

We use identities to find $\sin \theta$ and $\tan \theta$.

$$
\begin{aligned}
\sin \theta &= \sqrt{1 - \cos^2 \theta} & & \theta \text{ is in quadrant II.} \\
&= \sqrt{1 - \left(-\frac{3}{5}\right)^2} & & \cos \theta = -\frac{3}{5} \\
&= \frac{4}{5} & & \text{Simplify.}
\end{aligned}
$$

$$
\tan \theta = \frac{\sin \theta}{\cos \theta} = \frac{4/5}{-3/5} = -\frac{4}{3} \qquad \cos \theta = -\frac{3}{5} \text{ is given}
$$

a.
$$
\begin{aligned}
\sin 2\theta &= 2 \sin \theta \cos \theta & & \text{Double-angle formula for sine} \\
&= 2\left(\frac{4}{5}\right)\left(-\frac{3}{5}\right) & & \text{Replace } \sin \theta \text{ with } \frac{4}{5} \text{ and } \cos \theta \text{ with } -\frac{3}{5}. \\
&= -\frac{24}{25} & & \text{Simplify.}
\end{aligned}
$$

b. $\cos 2\theta = \cos^2 \theta - \sin^2 \theta$ Double-angle formula for cosines

$$= \left(-\frac{3}{5}\right)^2 - \left(\frac{4}{5}\right)^2 \quad \text{Replace } \cos \theta \text{ with } -\frac{3}{5} \text{ and } \sin \theta \text{ with } \frac{4}{5}.$$

$$= \frac{9}{25} - \frac{16}{25} = -\frac{7}{25} \quad \text{Simplify.}$$

c. $\tan 2\theta = \dfrac{2 \tan \theta}{1 - \tan^2 \theta}$ Double-angle formula for tangent

$$= \frac{2\left(-\dfrac{4}{3}\right)}{1 - \left(-\dfrac{4}{3}\right)^2} \quad \text{Replace } \tan \theta \text{ with } -\frac{4}{3}.$$

$$= \frac{-\dfrac{8}{3}}{1 - \dfrac{16}{9}} = \frac{-\dfrac{8}{3}}{-\dfrac{7}{9}}$$

$$= \left(-\frac{8}{3}\right)\left(-\frac{9}{7}\right) = \frac{24}{7} \quad \text{Simplify.}$$

You can also find $\tan 2\theta$ by using parts **a** and **b**:

$$\tan 2\theta = \frac{\sin 2\theta}{\cos 2\theta} = \frac{-\dfrac{24}{25}}{-\dfrac{7}{25}} = \frac{24}{7} \qquad \blacksquare\ \blacksquare\ \blacksquare$$

Practice Problem 1 If $\sin x = \dfrac{12}{13}$ and $\dfrac{\pi}{2} < x < \pi$, find the exact value of each expression.

a. $\sin 2x$ **b.** $\cos 2x$ **c.** $\tan 2x$ ■

EXAMPLE 2 **Using Double-Angle Formulas**

Find the exact value of each expression.

a. $1 - 2 \sin^2 \left(\dfrac{\pi}{12}\right)$ **b.** $\dfrac{2 \tan 22.5°}{1 - \tan^2 22.5°}$

SOLUTION

a. The given expression is the right side of the following formula, where $\theta = \dfrac{\pi}{12}$.

$$\cos 2\theta = 1 - 2 \sin^2 \theta \qquad \text{A double-angle formula for cosine}$$

$$1 - 2 \sin^2 \left(\frac{\pi}{12}\right) = \cos 2\left(\frac{\pi}{12}\right) \qquad \text{Replace } \theta \text{ with } \frac{\pi}{2}; \text{ interchange sides.}$$

$$= \cos \frac{\pi}{6} = \frac{\sqrt{3}}{2}$$

b. The given expression is the right side of the following formula, where $\theta = 22.5°$.

$$\tan 2\theta = \frac{2 \tan \theta}{1 - \tan^2 \theta} \qquad \text{Double-angle formula for tangent}$$

$$\frac{2 \tan (22.5°)}{1 - \tan^2 2(22.5°)} = \tan 2(22.5°) \qquad \begin{array}{l}\text{Replace } \theta \text{ with } 22.5°; \\ \text{interchange sides.}\end{array}$$

$$= \tan 45° = 1 \qquad \blacksquare \blacksquare \blacksquare$$

Practice Problem 2 Find the exact value of each expression.

a. $2 \cos^2 \left(\dfrac{\pi}{12}\right) - 1$ **b.** $\cos^2 22.5° - \sin^2 22.5°$ ■

EXAMPLE 3   **Finding a Triple-Angle Formula for Sine**
_____

Verify the identity $\sin 3x = 3 \sin x - 4 \sin^3 x$.

SOLUTION

We begin by writing $3x$ as $2x + x$ and use the sum formula for the sine and the double-angle formulas to verify this identity.

$\sin 3x = \sin (2x + x)$

$\qquad = \sin 2x \cos x + \cos 2x \sin x$ $\qquad$ Sum formula for sine

$\qquad = (2 \sin x \cos x) \cos x + (1 - 2 \sin^2 x) \sin x$ $\qquad$ Replace $\sin 2x$ with $2 \sin x \cos x$ and $\cos 2x$ with $1 - 2 \sin^2 x$.

$\qquad = 2 \sin x \cos^2 x + \sin x - 2 \sin^3 x$ $\qquad$ Multiply; distributive property

$\qquad = 2 \sin x(1 - \sin^2 x) + \sin x - 2 \sin^3 x$ $\qquad$ $\cos^2 x = 1 - \sin^2 x$

$\qquad = 2 \sin x - 2 \sin^3 x + \sin x - 2 \sin^3 x$ $\qquad$ Distributive property

$\qquad = 3 \sin x - 4 \sin^3 x$ $\qquad$ Simplify.

We have verified the identity by showing that the left and right sides of the equation are equal. $\qquad \blacksquare \blacksquare \blacksquare$

Practice Problem 3 Verify the identity $\cos 3x = 4 \cos^3 x - 3 \cos x$. ■

We can use the double-angle formulas to express trigonometric functions of $4\theta, 6\theta$, and 8θ in terms of $2\theta, 3\theta$, and 4θ, respectively. For example, the identities

$$\cos 4\theta = 2 \cos^2 2\theta - 1, \quad \sin 6\theta = 2 \sin 3\theta \cos 3\theta, \quad \text{and} \quad \tan 8\theta = \frac{2 \tan 4\theta}{1 - \tan^2 4\theta}$$

can be directly derived from the appropriate double-angle formulas.

2 Use power-reducing formulas.

Power-Reducing Formulas

The purpose of *power-reducing formulas* is to express $\sin^2 x$, $\cos^2 x$, and $\tan^2 x$ in terms of trigonometric functions with powers not greater than 1. These formulas are useful in calculus.

POWER-REDUCING FORMULAS

$$\sin^2 x = \frac{1 - \cos 2x}{2} \qquad \cos^2 x = \frac{1 + \cos 2x}{2} \qquad \tan^2 x = \frac{1 - \cos 2x}{1 + \cos 2x}$$

We can derive the first power-reducing formula by using the appropriate formula for $\cos 2x$.

$$\cos 2x = 1 - 2 \sin^2 x \qquad \text{Double-angle formula for } \cos 2x \text{ in terms of sine}$$

$$2 \sin^2 x = 1 - \cos 2x \qquad \text{Add } 2 \sin^2 x - \cos 2x \text{ to both sides and simplify.}$$

$$\sin^2 x = \frac{1 - \cos 2x}{2} \qquad \text{Divide both sides by 2.}$$

Similarly, we can derive the second formula.

$$\cos 2x = 2 \cos^2 x - 1 \qquad \text{Double-angle formula for } \cos 2x \text{ in terms of cosine}$$

$$1 + \cos 2x = 2 \cos^2 x \qquad \text{Add 1 to both sides.}$$

$$\frac{1 + \cos 2x}{2} = \cos^2 x \qquad \text{Divide both sides by 2.}$$

We begin with the quotient identity to prove the third formula.

$$\tan^2 x = \frac{\sin^2 x}{\cos^2 x} \qquad \text{Quotient identity}$$

$$= \frac{\dfrac{1 - \cos 2x}{2}}{\dfrac{1 + \cos 2x}{2}} \qquad \text{Power-reducing formulas for } \sin^2 x \text{ and } \cos^2 x$$

$$= \frac{1 - \cos 2x}{1 + \cos x} \qquad \text{Multiply numerator and denominator by 2. Simplify.}$$

EXAMPLE 4   Using Power-Reducing Formulas

Write an equivalent expression for $\cos^4 x$ that contains only first powers of cosines of multiple angles.

SOLUTION

We use the power-reducing formulas repeatedly.

$$\cos^4 x = (\cos^2 x)^2 \qquad\qquad\qquad a^4 = (a^2)^2$$

$$= \left(\frac{1 + \cos 2x}{2} \right)^2 \qquad\qquad \text{Power-reducing formula}$$

$$= \frac{1}{4} (1 + 2 \cos 2x + \cos^2 2x) \qquad \text{Expand the binomial.}$$

$$= \frac{1}{4} \left(1 + 2 \cos 2x + \frac{1 + \cos 4x}{2} \right) \qquad \begin{array}{l}\text{Power-reducing formula for } \cos^2 x; \\ \text{replace } x \text{ with } 2x.\end{array}$$

$$= \frac{1}{4} \left(1 + 2 \cos 2x + \frac{1}{2} + \frac{1}{2} \cos 4x \right) \qquad \frac{a + b}{c} = \frac{a}{c} + \frac{b}{c}$$

$$= \frac{1}{4} + \frac{2}{4} \cos 2x + \frac{1}{8} + \frac{1}{8} \cos 4x \qquad \text{Distributive property}$$

$$= \frac{3}{8} + \frac{1}{2} \cos 2x + \frac{1}{8} \cos 4x \qquad \text{Simplify.} \qquad\qquad ■ ■ ■$$

Practice Problem 4 Write an equivalent expression for $\sin^4 x$ that contains only first powers of cosines of multiple angles. ■

Alternating Current

Alternating current (AC) is the electric current that reverses direction, usually many times per second. Most electrical generators produce alternating current. The *wattage rating* of an electric device is $\dfrac{1}{\sqrt{2}} \approx 0.7071$ times the maximum wattage of the device.

EXAMPLE 5 **Finding the Wattage Rating of an Electric Blanket**

The voltage V of household current is given by $V = 170 \sin(120\pi t)$ volts, where t is in seconds. Suppose the amount of current passing through an electric blanket is $I = 0.1 \sin(120\pi t)$ amps, where t is in seconds. Find the wattage rating of the blanket.

SOLUTION

We have $P = VI$

Power = voltage × current
watts = volts × amps

$$P = [170 \sin(120\pi t)][0.1 \sin(120\pi t)] \qquad \text{Substitute the given values of } V \text{ and } I.$$

$$= 17 \sin^2(120\pi t) \qquad \text{Multiply.}$$

$$= 17\left[\frac{1 - \cos(240\pi t)}{2}\right] \qquad \text{Power-reducing formula}$$

$$P = 8.5 - 8.5 \cos(240\pi t) \qquad \text{Simplify.}$$

The maximum value of P is 17 watts when $\cos(240\pi t) = -1$; so the wattage rating for this blanket is $\dfrac{17}{\sqrt{2}} \approx 12$ watts. ■ ■ ■

Practice Problem 5 In Example 5, find the wattage rating of a lightbulb for which $I = 0.83 \sin(120\pi t)$ amps. ■

3 Use half-angle formulas.

Half-Angle Formulas

If an angle measures θ, then $\dfrac{\theta}{2}$ is half the measure of θ. Half-angle formulas express trigonometric functions of $\dfrac{\theta}{2}$ in terms of functions of θ. To derive the half-angle formulas, we replace x with $\dfrac{\theta}{2}$ in the power-reducing formulas and then take the square root of both sides. For example,

$$\cos^2 x = \frac{1 + \cos 2x}{2} \qquad \text{Power-reducing formula for cosines}$$

$$\cos^2 \frac{\theta}{2} = \frac{1 + \cos 2\left(\frac{\theta}{2}\right)}{2} \qquad \text{Replace } x \text{ with } \frac{\theta}{2}.$$

$$\cos^2 \frac{\theta}{2} = \frac{1 + \cos \theta}{2} \qquad \text{Simplify.}$$

$$\cos \frac{\theta}{2} = \pm\sqrt{\frac{1 + \cos \theta}{2}} \qquad \text{Square root property}$$

We call the last equation a *half-angle formula* for cosine. The sign $+$ or $-$ depends on the quadrant in which $\dfrac{\theta}{2}$ lies. We can derive half-angle formulas for sine and tangent in a similar manner.

> **HALF-ANGLE FORMULAS**
>
> $$\sin\frac{\theta}{2} = \pm\sqrt{\frac{1 - \cos\theta}{2}} \qquad \cos\frac{\theta}{2} = \pm\sqrt{\frac{1 + \cos\theta}{2}} \qquad \tan\frac{\theta}{2} = \pm\sqrt{\frac{1 - \cos\theta}{1 + \cos\theta}}$$
>
> The sign $+$ or $-$ depends on the quadrant in which $\dfrac{\theta}{2}$ lies.

EXAMPLE 6   **Using Half-Angle Formulas**

Use a half-angle formula to find the exact value of $\cos 157.5°$.

SOLUTION

Because $157.5° = \dfrac{315°}{2}$, we use the half-angle formula for $\cos\dfrac{\theta}{2}$ with $\theta = 315°$.

Because $\dfrac{\theta}{2} = 157.5°$ lies in quadrant II, $\cos\dfrac{\theta}{2}$ is negative.

$$\cos 157.5° = \cos\frac{315°}{2}$$

$$= -\sqrt{\frac{1 + \cos 315°}{2}} \qquad\qquad \text{Half-angle formula for cosine}$$

$$= -\sqrt{\frac{1 + \cos 45°}{2}} \qquad\qquad \begin{aligned}&\cos 315° = \cos(360° - 45°)\\&= \cos 45°\end{aligned}$$

$$= -\sqrt{\frac{\left(1 + \dfrac{\sqrt{2}}{2}\right)\cdot 2}{2\cdot 2}} \qquad\qquad \text{Replace } \cos 45° \text{ with } \dfrac{\sqrt{2}}{2}. \text{ Multiply}$$
$$\text{numerator and denominator by 2.}$$

$$= -\sqrt{\frac{2 + \sqrt{2}}{2\cdot 2}} = -\frac{\sqrt{2 + \sqrt{2}}}{2} \qquad\qquad \text{Simplify.} \qquad ■ ■ ■$$

Practice Problem 6   Find the exact value of $\sin 112.5°$. ▪

EXAMPLE 7   **Finding the Exact Value**

Given that $\sin\theta = -\dfrac{5}{13}, \pi < \theta < \dfrac{3\pi}{2}$, find the exact value of each expression.

a. $\sin\dfrac{\theta}{2}$ **b.** $\tan\dfrac{\theta}{2}$

SOLUTION

Because θ lies in quadrant III, $\cos\theta$ is negative; so

$$\cos\theta = -\sqrt{1 - \sin^2\theta} \qquad\qquad \text{Pythagorean identity}$$

$$\cos\theta = -\sqrt{1 - \left(-\frac{5}{13}\right)^2} \qquad\qquad \text{Replace } \sin\theta \text{ with } -\frac{5}{13}.$$

$$= -\sqrt{1 - \frac{25}{169}} = -\sqrt{\frac{169 - 25}{169}} = -\frac{12}{13} \qquad \text{Simplify.}$$

a. Because θ lies in quadrant III, $\dfrac{\theta}{2}$ lies in quadrant II $\left(\text{if } \pi < \theta < \dfrac{3\pi}{2}, \text{ then}\right.$ $\left.\dfrac{\pi}{2} < \dfrac{\theta}{2} < \dfrac{3\pi}{4}\right)$. The half-angle formula gives

$$\sin\frac{\theta}{2} = \sqrt{\frac{1 - \cos\theta}{2}} \qquad\qquad \text{In quadrant II, } \sin\frac{\theta}{2} \text{ is positive.}$$

$$= \sqrt{\frac{1 - \left(-\dfrac{12}{13}\right)}{2}} \qquad\qquad \text{Replace } \cos\theta \text{ with } -\frac{12}{13}.$$

$$= \sqrt{\frac{\dfrac{25}{13}}{2}} = \frac{5}{\sqrt{26}} = \frac{5\sqrt{26}}{26} \qquad \text{Simplify.}$$

b. From the half-angle formula, we have

$$\tan\frac{\theta}{2} = -\sqrt{\frac{1 - \cos\theta}{1 + \cos\theta}} \qquad\qquad \text{In quadrant II, } \tan\frac{\theta}{2} \text{ is negative.}$$

$$= -\sqrt{\frac{1 - \left(-\dfrac{12}{13}\right)}{1 + \left(-\dfrac{12}{13}\right)}} \qquad\qquad \cos\theta = -\frac{12}{13}.$$

$$= -\sqrt{\frac{\dfrac{25}{13}}{\dfrac{1}{13}}} = -5 \qquad\qquad \text{Simplify.} \qquad\qquad ■\ ■\ ■$$

Practice Problem 7 For θ in Example 7, find $\cos\dfrac{\theta}{2}$. ■

EXAMPLE 8 **Verifying an Identity Containing Half-Angles**

Verify the identity $\quad \sin x \cos\dfrac{x}{2} = \sin\dfrac{x}{2}(1 + \cos x)$.

SOLUTION

The left side contains $\sin x$, and the right side contains $\sin\dfrac{x}{2}$. We replace x with $\dfrac{x}{2}$ in the double-angle identity for sine.

$$\sin 2x = 2\sin x \cos x \qquad\qquad \text{Double-angle identity for sine}$$

$$\sin x = 2\sin\frac{x}{2}\cos\frac{x}{2} \qquad\qquad \text{Replace } x \text{ with } \frac{x}{2}; \sin 2\left(\frac{x}{2}\right) = \sin x.$$

Begin with the left side of the original equation and use this expression for $\sin x$.

$$\sin x \cos\frac{x}{2} = 2\sin\frac{x}{2}\cos\frac{x}{2}\cos\frac{x}{2} \qquad\qquad \text{Replace } \sin x \text{ with } 2\sin\frac{x}{2}\cos\frac{x}{2}.$$

$$= 2\sin\frac{x}{2}\cos^2\frac{x}{2}$$

$$= 2\sin\frac{x}{2}\left(\frac{1 + \cos 2\left(\dfrac{x}{2}\right)}{2}\right) \qquad\qquad \text{Power-reducing formula for cosine}$$

$$= \sin\frac{x}{2}(1 + \cos x) \qquad\qquad \cos 2\left(\frac{x}{2}\right) = \cos x; \text{ simplify.}$$

Because the left side is identical to the right side of the equation, the identity is verified.   ■ ■ ■

Practice Problem 8   Verify the identity $\sin \dfrac{x}{2} \sin x = \cos \dfrac{x}{2}(1 - \cos x)$.   ■

4 Solve trigonometric equations involving multiple angles and half-angles.

Trigonometric Equations Involving Multiple Angles and Half-Angles

EXAMPLE 9   **Solving a Trigonometric Equation Involving Multiple Angles**

Find all solutions of the equation $2 \sin \theta - \sin 3\theta = 0$ in the interval $[0, 2\pi)$.

SOLUTION

$$
\begin{array}{ll}
2 \sin \theta - \sin 3\theta = 0 & \text{Given equation.} \\[4pt]
2 \sin \theta - (3 \sin \theta - 4 \sin^3 \theta) = 0 & \text{Use } \sin 3\theta = 3 \sin \theta - 4 \sin^3 \theta \text{ from} \\
& \text{Example 3.} \\[4pt]
2 \sin \theta - 3 \sin \theta + 4 \sin^3 \theta = 0 & \text{Distributive property} \\[4pt]
4 \sin^3 \theta - \sin \theta = 0 & \text{Simplify.} \\[4pt]
\sin \theta (4 \sin^2 \theta - 1) = 0 & \text{Factor.} \\[4pt]
\sin \theta (2 \sin \theta - 1)(2 \sin \theta + 1) = 0 & \text{Factor the difference of squares.}
\end{array}
$$

$$
\begin{array}{lll}
\sin \theta = 0 \quad \text{or} & 2 \sin \theta - 1 = 0 \quad \text{or} & 2 \sin \theta + 1 = 0 & \text{Zero-product} \\
& & & \text{property} \\[6pt]
& \sin \theta = \dfrac{1}{2} & \sin \theta = -\dfrac{1}{2} & \text{Solve for } \sin \theta. \\[10pt]
\theta = 0 \quad \text{or} & \theta = \dfrac{\pi}{6} \quad \text{or} & \theta = \pi + \dfrac{\pi}{6} = \dfrac{7\pi}{6} \quad \text{or} & \text{Solve for } \theta \text{ in} \\
& & & [0, 2\pi). \\[10pt]
\theta = \pi & \theta = \pi - \dfrac{\pi}{6} = \dfrac{5\pi}{6} & \theta = 2\pi - \dfrac{\pi}{6} = \dfrac{11\pi}{6}
\end{array}
$$

The solution set is $\left\{ 0, \dfrac{\pi}{6}, \dfrac{5\pi}{6}, \pi, \dfrac{7\pi}{6}, \dfrac{11\pi}{6} \right\}$.   ■ ■ ■

Practice Problem 9   Find all solutions of the equation $\sin \theta - \cos 2\theta = 0$ in the interval $[0, 2\pi)$.   ■

EXAMPLE 10   **Solving a Trigonometric Equation Involving Half-Angles**

Find all solutions of the equation $\sin x - \cos \dfrac{x}{2} = 0$ in the interval $[0, 2\pi)$.

SOLUTION

Note that for x to be a solution in the interval $[0, 2\pi)$, $\dfrac{x}{2}$ must be in the interval $[0, \pi)$.

$$
\begin{array}{ll}
\sin x - \cos \dfrac{x}{2} = 0 & \text{Given equation} \\[10pt]
2 \sin \dfrac{x}{2} \cos \dfrac{x}{2} - \cos \dfrac{x}{2} = 0 \qquad 2 \sin \dfrac{x}{2} \cos \dfrac{x}{2} = \sin 2\left(\dfrac{x}{2}\right) = \sin x \\[10pt]
\cos \dfrac{x}{2}\left(2 \sin \dfrac{x}{2} - 1\right) = 0 & \text{Factor.}
\end{array}
$$

continued on the next page

TECHNOLOGY CONNECTION

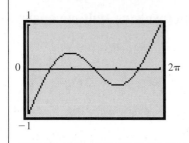

You can check your solutions graphically. Sketch the graph of

$$Y_1 = \sin x - \cos \frac{x}{2}$$ and see

where it crosses the x-axis.

$$\cos \frac{x}{2} = 0 \quad \text{or} \quad 2\sin \frac{x}{2} - 1 = 0 \qquad \text{Zero-product property}$$

$$\frac{x}{2} = \frac{\pi}{2} \qquad \qquad \sin \frac{x}{2} = \frac{1}{2} \quad \text{Solve for } \frac{x}{2} \text{ in } [0, \pi), \text{ then } x \text{ is in } [0, 2\pi).$$

$$x = \pi \qquad \qquad \frac{x}{2} = \frac{\pi}{6} \quad \text{or} \quad \frac{x}{2} = \frac{5\pi}{6}$$

$$x = \frac{\pi}{3} \quad \text{or} \quad x = \frac{5\pi}{3}$$

The solution set is $\left\{ \dfrac{\pi}{3}, \pi, \dfrac{5\pi}{3} \right\}$. ■ ■ ■

Practice Problem 10 Find all solutions of the equation $\sin x + \cos \dfrac{x}{2} = 0$ in the interval $[0, 2\pi)$. ■

SECTION 5.4 ■ Exercises

A EXERCISES Basic Skills and Concepts

1. The double-angle formula for $\sin 2x$ is $\sin 2x = \underline{2\sin x \cos x}$.

2. In the double-angle formula $\cos 2x = \cos^2 x - \sin^2 x$, replace $\cos^2 x$ with $1 - \sin^2 x$ to obtain a double-angle formula $\cos 2x = \underline{1 - 2\sin^2 x}$ in terms of $\sin^2 x$. Solve this formula for $\sin^2 x$ to obtain the power-reducing formula $\sin^2 x = \underline{\qquad}^\dagger$.

3. The formula for $\cos 2x$ in terms of $\cos^2 x$ is $\cos 2x = \underline{2\cos^2 x - 1}$. Solve this formula for $\cos^2 x$ to obtain the power-reducing formula $\cos^2 x = \underline{\qquad}^\dagger$.

4. *True or False* $\tan 2x = \dfrac{1 - \cos 2x}{1 + \cos 2x}$. False

5. *True or False* $\dfrac{1}{2}\tan 2x = \tan x$. False

6. *True or False* $\cos \dfrac{\theta}{2} = -\sqrt{\dfrac{1 + \cos \theta}{2}}$, $\pi < \theta < 2\pi$. True

In Exercises 7–12, use the given information about the angle θ to find the exact value of

a. $\sin 2\theta$ b. $\cos 2\theta$ c. $\tan 2\theta$.

7. $\sin \theta = \dfrac{3}{5}$, θ in quadrant II   a. $-\dfrac{24}{25}$ b. $\dfrac{7}{25}$ c. $-\dfrac{24}{7}$

8. $\cos \theta = -\dfrac{5}{13}$, θ in quadrant III   a. $\dfrac{120}{169}$ b. $-\dfrac{119}{169}$ c. $-\dfrac{120}{119}$

9. $\tan \theta = 4$, $\sin \theta < 0$ a. $\dfrac{8}{17}$ b. $-\dfrac{15}{17}$ c. $-\dfrac{8}{15}$

10. $\sec \theta = -\sqrt{3}$, $\sin \theta > 0$ a. $-\dfrac{2\sqrt{2}}{3}$ b. $-\dfrac{1}{3}$ c. $2\sqrt{2}$

11. $\tan \theta = -2$, $\dfrac{\pi}{2} < \theta < \pi$ a. $-\dfrac{4}{5}$ b. $-\dfrac{3}{5}$ c. $\dfrac{4}{3}$

12. $\cot \theta = -7$, $\dfrac{3\pi}{2} < \theta < 2\pi$ a. $-\dfrac{7}{25}$ b. $\dfrac{24}{25}$ c. $-\dfrac{7}{24}$

In Exercises 13–22, use a double-angle formula to find the exact value of each expression.

13. $1 - 2\sin^2 75° \quad -\dfrac{\sqrt{3}}{2}$

14. $\dfrac{2\tan 75°}{1 - \tan^2 75°} \quad -\dfrac{\sqrt{3}}{3}$

15. $2\cos^2 105° - 1 \quad -\dfrac{\sqrt{3}}{2}$

16. $1 - 2\sin^2 165° \quad \dfrac{\sqrt{3}}{2}$

17. $\dfrac{2\tan 165°}{1 - \tan^2 165°} \quad -\dfrac{\sqrt{3}}{3}$

18. $2\cos^2 165° - 1 \quad \dfrac{\sqrt{3}}{2}$

19. $1 - 2\sin^2 \dfrac{\pi}{8} \quad \dfrac{\sqrt{2}}{2}$

20. $2\cos^2\left(-\dfrac{\pi}{8}\right) - 1 \quad \dfrac{\sqrt{2}}{2}$

21. $\dfrac{2\tan\left(-\dfrac{5\pi}{12}\right)}{1 - \tan^2\left(-\dfrac{5\pi}{12}\right)} \quad \dfrac{\sqrt{3}}{3}$

22. $1 - 2\sin^2\left(-\dfrac{7\pi}{12}\right) \quad -\dfrac{\sqrt{3}}{2}$

In Exercises 23 and 24, verify each "quadruple-angle" formula.

23. $\sin 4\theta = \cos \theta(4\sin \theta - 8\sin^3 \theta)$

24. $\cos 4\theta = 8\cos^4 \theta - 8\cos^2 \theta + 1$

In Exercises 25–32, verify each identity.

25. $\cos^4 x - \sin^4 x = \cos 2x$

26. $1 + \cos 2x + 2\sin^2 x = 2$

27. $\dfrac{1 + \sin 2x}{\cos 2x} = \dfrac{\cos x + \sin x}{\cos x - \sin x}$

†Due to space constrictions, answers to these exercises may be found in the Answers beginning on page A–1 in the back of the book.

28. $\dfrac{\cos 2x}{\sin 2x} + \dfrac{\sin x}{\cos x} = \csc 2x$ **29.** $\dfrac{\sin 2x}{\sin x} - \dfrac{\cos 2x}{\cos x} = \sec x$

30. $\dfrac{\cos 3x}{\sin 3x} + \dfrac{\sin x}{\cos x} = \dfrac{\cos 2x}{\sin 3x \cos x}$

31. $\tan 2x + \tan x = \dfrac{\sin 3x}{\cos 2x \cos x}$

32. $\tan 2x - \tan x = \tan x \sec 2x$

In Exercises 33–42, use the power-reducing formulas to rewrite each expression that does not contain trigonometric functions of power greater than 1.

33. $4\sin^2 x \cos^2 x$ $\dfrac{1 - \cos 4x}{2}$ **34.** $\sin^2 x \cos^2 x$ $\dfrac{1 - \cos 4x}{8}$

35. $4\sin x \cos x(1 - 2\sin^2 x)$ $\sin 4x$

36. $4\sin x \cos x(2\cos^2 x - 1)$ $\sin 4x$

37. $2\sin 3x \cos 3x(2\cos^2 3x - 1)$ $\dfrac{\sin 12x}{2}$

38. $\sin 8x(1 - 2\sin^2 4x)$ $\sin 8x \cos 8x$

39. $\sin \dfrac{x}{2} \cos \dfrac{x}{2}\left(1 - 2\sin^2 \dfrac{x}{2}\right)$ $\dfrac{\sin x \cos x}{2}$

40. $\sin x\left(2\cos^2 \dfrac{x}{2} - 1\right)$ $\sin x \cos x$

41. $8\sin^4 \dfrac{x}{2}$ $\cos 2x - 4\cos x + 3$

42. $8\cos^4 \dfrac{x}{2}$ $\cos 2x + 4\cos x + 3$

In Exercises 43–54, use half-angle formulas to find the exact value of each expression.

43. $\sin \dfrac{\pi}{12}$ $\dfrac{\sqrt{2 - \sqrt{3}}}{2}$

44. $\sin \dfrac{\pi}{8}$ $\dfrac{\sqrt{2 - \sqrt{2}}}{2}$

45. $\cos \dfrac{\pi}{8}$ $\dfrac{\sqrt{2 + \sqrt{2}}}{2}$

46. $\tan \dfrac{\pi}{8}$ $\dfrac{\sqrt{2 - \sqrt{2}}}{\sqrt{2 + \sqrt{2}}}$

47. $\sin\left(-\dfrac{3\pi}{8}\right)$ $-\dfrac{\sqrt{2 + \sqrt{2}}}{2}$

48. $\cos\left(-\dfrac{3\pi}{8}\right)$ $\dfrac{\sqrt{2 - \sqrt{2}}}{2}$

49. $\tan\left(\dfrac{7\pi}{8}\right)$ $-\dfrac{\sqrt{2 - \sqrt{2}}}{\sqrt{2 + \sqrt{2}}}$

50. $\sec\left(-\dfrac{7\pi}{8}\right)$ $-\dfrac{\sqrt{2}\sqrt{2 - \sqrt{2}}}{\sqrt{2 - \sqrt{2}}}$

51. $\tan 112.5°$ $-\sqrt{2} - 1$

52. $\cos 112.5°$ $-\dfrac{\sqrt{2 - \sqrt{2}}}{2}$

53. $\sin(-75°)$ $-\dfrac{\sqrt{2 + \sqrt{3}}}{2}$

54. $\tan(-105°)$ $\dfrac{\sqrt{2 + \sqrt{3}}}{\sqrt{2 - \sqrt{3}}}$

In Exercises 55–62, use the information about the angle θ to find the exact value of

a. $\sin \dfrac{\theta}{2}$. **b.** $\cos \dfrac{\theta}{2}$. **c.** $\tan \dfrac{\theta}{2}$.

55. $\sin \theta = \dfrac{4}{5}, \dfrac{\pi}{2} < \theta < \pi$ **a.** $\dfrac{2\sqrt{5}}{5}$ **b.** $\dfrac{\sqrt{5}}{5}$ **c.** 2

56. $\cos \theta = -\dfrac{12}{13}, \pi < \theta < \dfrac{3\pi}{2}$ **a.** $\dfrac{5\sqrt{26}}{26}$ **b.** $-\dfrac{\sqrt{26}}{26}$ **c.** -5

57. $\tan \theta = -\dfrac{2}{3}, \dfrac{\pi}{2} < \theta < \pi$ †

58. $\cot \theta = \dfrac{3}{4}, \pi < \theta < \dfrac{3\pi}{2}$ †

59. $\sin \theta = \dfrac{1}{5}, \cos \theta < 0$ † **60.** $\cos \theta = \dfrac{2}{3}, \sin \theta < 0$ †

61. $\sec \theta = \sqrt{5}, \sin \theta > 0$ † **62.** $\csc \theta = \sqrt{7}, \tan \theta < 0$ †

In Exercises 63–72, verify each identity.

63. $\left(\sin \dfrac{t}{2} + \cos \dfrac{t}{2}\right)^2 = 1 + \sin t$

64. $\left(\sin \dfrac{t}{2} - \cos \dfrac{t}{2}\right)^2 = 1 - \sin t$

65. $2\cos^2 \dfrac{x}{2} = \dfrac{\sin^2 x}{1 - \cos x}$ **66.** $2\sin^2 \dfrac{x}{2} = \dfrac{\sin^2 x}{1 + \cos x}$

67. $\tan \dfrac{x}{2} = \dfrac{\sin x}{1 + \cos x}$ **68.** $\tan \dfrac{x}{2} = \dfrac{1 - \cos x}{\sin x}$

69. $\sin^2 \dfrac{x}{2} + \cos x = \cos^2 \dfrac{x}{2}$

70. $\cos^2 x + \cos x = 2\cos^2 \dfrac{x}{2} - \sin^2 x$

71. $\sin x = \dfrac{2\tan \dfrac{x}{2}}{1 + \tan^2 \dfrac{x}{2}}$ **72.** $\cos x = \dfrac{1 - \tan^2 \dfrac{x}{2}}{1 + \tan^2 \dfrac{x}{2}}$

In Exercises 73–80, find all solutions of each equation in the interval $[0, 2\pi)$. Use a graphing calculator to verify your solutions.

73. $\sin x - \sin 2x = 0$ **74.** $\sin x - \cos 2x = 0$

75. $\cos 2t - 2\sin^2 t = 0$ † **76.** $\cos 2t + 2\cos^2 t = 0$ †

77. $\sin \dfrac{x}{2} + \sin x = 0$ $\left\{0, \dfrac{4\pi}{3}\right\}$

78. $\cos \dfrac{x}{2} + \cos x = 0$ $\left\{\dfrac{2\pi}{3}\right\}$

79. $\tan \dfrac{x}{2} + \tan x = 0$ $\left\{0, \dfrac{2\pi}{3}, \dfrac{4\pi}{3}\right\}$

80. $\cos x - 2\sin \dfrac{x}{2} \cos \dfrac{x}{2} = 0$ $\left\{\dfrac{\pi}{4}, \dfrac{5\pi}{4}\right\}$

73. $\left\{0, \dfrac{\pi}{3}, \pi, \dfrac{5\pi}{3}\right\}$ **74.** $\left\{\dfrac{\pi}{6}, \dfrac{3\pi}{2}, \dfrac{5\pi}{6}\right\}$

B EXERCISES Applying the Concepts

In Exercises 81–86, assume that the voltage of the current is given by $V = 170 \sin(120\pi t)$, where t is in seconds. Find the wattage rating $\left(\dfrac{\text{maximum wattage}}{\sqrt{2}}\right)$ of each electric device if the current flowing through the device is I amps.

81. Light bulb. $I = 0.832 \sin(120\pi t)$. 100 watts

82. Microwave oven. $I = 7.487 \sin(120\pi t)$. 900 watts

83. Toaster. $I = 9.983 \sin(120\pi t)$. 1200 watts

84. Refrigerator. $I = 6.655 \sin(120\pi t)$. 800 watts

85. Vacuum cleaner. $I = 4.991 \sin(120\pi t)$. 600 watts

86. Television. $I = 2.917 \sin(120\pi t)$. 351 watts

87. Voltage of current. Suppose the voltage of a current is given by $V = 170 \sin(120\pi t)$. Find the first two positive times when the voltage is 85 volts. $\dfrac{1}{720}, \dfrac{1}{144}$

88. Voltage of current. Suppose the voltage of a current is given by $V = 160 \sin(120\pi t)$. Find the first two positive times when the voltage is $80\sqrt{3}$ volts. $\dfrac{1}{360}, \dfrac{1}{180}$

89. **Throwing a football.** A quarterback throws a ball with an initial velocity of v_0 feet per second at an angle θ with the horizontal. The horizontal distance x in feet that the ball is thrown is modeled by the equation $x = \dfrac{v_0^2}{16} \sin\theta \cos\theta$. For a fixed v_0, find the angle that produces the maximum distance x. $\left\{\dfrac{\pi}{4}\right\}$

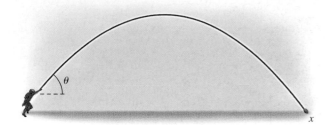

90. **Angle of elevation.** A, B, and C are three points on the same horizontal line. The segment CT represents a vertical tower. The angle of elevation of T from the point B is twice that from the point A. The lengths of the segments AB and BC are 80 meters and 50 meters, respectively. Show that the angle of elevation θ from A in the figure is given by $\sin^{-1}\left(\dfrac{\sqrt{3}}{4}\right)$.

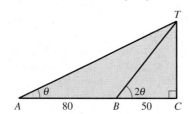

C EXERCISES Beyond the Basics

In Exercises 91–101, verify each identity.

91. $\tan 3x = \dfrac{3\tan x - \tan^3 x}{1 - 3\tan^2 x}$

92. $\dfrac{\sin 3x + \cos 3x}{\cos x - \sin x} = 1 + 2\sin 2x$

93. $\dfrac{\sin x - \cos x}{\sin x + \cos x} - \dfrac{\sin x + \cos x}{\sin x - \cos x} = 2\tan 2x$

94. $\dfrac{2}{\tan x + \cot x} = \sin 2x$

95. $\dfrac{2\sin x}{\cos 3x} = \tan 3x - \tan x$

96. $\cot x - \tan x = 2\cot 2x$

97. $\dfrac{1 - \tan^2\left(\dfrac{\pi}{4} - x\right)}{1 + \tan^2\left(\dfrac{\pi}{4} - x\right)} = \sin 2x$

98. $\dfrac{1 + \sin 2x - \cos 2x}{1 + \sin 2x + \cos 2x} = \tan x$

99. $\sin^2\dfrac{\pi}{8} + \sin^2\dfrac{3\pi}{8} + \sin^2\dfrac{5\pi}{8} + \sin^2\dfrac{7\pi}{8} = 2$

100. $\sin^2\left(\dfrac{\pi}{8} + \dfrac{x}{2}\right) - \sin^2\left(\dfrac{\pi}{8} - \dfrac{x}{2}\right) = \dfrac{1}{\sqrt{2}}\sin x$

101. $\sqrt{2 + \sqrt{2 + 2\cos 4x}} = 2\cos x, 0 < x < \dfrac{\pi}{4}$

102. Solve $\sin 2x + \dfrac{1}{\sin 2x} = \dfrac{5}{2}, 0 \le x < 2\pi$

103. Show that (a) $\sin\dfrac{\pi}{10} = \dfrac{\sqrt{5} - 1}{4}$.

(b) $\cos\dfrac{\pi}{10} = \dfrac{\sqrt{10 + 2\sqrt{5}}}{4}$.

[*Hint:* Let $x = \dfrac{\pi}{10}$. Then $5x = \dfrac{\pi}{2}$ or $3x = \dfrac{\pi}{2} - 2x$.

$\cos 3x = \cos\left(\dfrac{\pi}{2} - 2x\right) = \sin 2x$. So,

$4\cos^3 x - 3\cos x = 2\sin x \cos x$ or
$4\cos^2 x - 3 = 2\sin x$ or $4(1 - \sin^2 x) - 3 = 2\sin x$.
Solve this quadratic equation for $\sin x$ to obtain (a).]

104. Use Exercise 103 to find the exact value of each expression.

 a. $\sin\dfrac{\pi}{5}$ b. $\cos\dfrac{\pi}{5}$ c. $\sin\dfrac{3\pi}{10}$ d. $\cos\dfrac{7\pi}{10}$.

105. a. Use the identities $\sin 2\theta = 2\sin\theta \cos\theta$ and the cofunction identities to show that

$$\sin\theta = 4\sin\dfrac{\theta}{4}\sin\left(\dfrac{\pi}{2} - \dfrac{\theta}{4}\right)\sin\left(\dfrac{\pi}{2} - \dfrac{\theta}{2}\right)$$

is an identity.

 b. Solve the equation $\theta = \dfrac{\pi}{2} - \dfrac{\theta}{2}$; use the result in part **a** to show that $\sin\dfrac{\pi}{12}\sin\dfrac{5\pi}{12} = \dfrac{1}{4}$.

 c. Solve the equation $\theta = \dfrac{\pi}{2} - \dfrac{\theta}{4}$; use the result in part **a** to show that $\sin\dfrac{\pi}{10}\sin\dfrac{3\pi}{10} = \dfrac{1}{4}$.

Critical Thinking

106. Use the figure to find the exact value of each expression.

 a. $\sin\theta$ b. $\cos\theta$ c. $\sin 2\theta$

 d. $\cos 2\theta$ e. $\tan 2\theta$ f. $\sin\dfrac{\theta}{2}$

 g. $\cos\dfrac{\theta}{2}$ h. $\tan\dfrac{\theta}{2}$

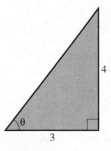

Answers:

102. $\left\{\dfrac{\pi}{12}, \dfrac{5\pi}{12}, \dfrac{13\pi}{12}, \dfrac{17\pi}{12}\right\}$ 104. a. $\dfrac{(\sqrt{5} - 1)(\sqrt{10 + 2\sqrt{5}})}{8}$ b. $\dfrac{\sqrt{5} + 1}{4}$ c. $\dfrac{\sqrt{5} + 1}{4}$

d. $-\dfrac{(\sqrt{5} - 1)(\sqrt{10 + 2\sqrt{5}})}{8}$ 106. a. $\dfrac{4}{5}$ b. $\dfrac{3}{5}$ c. $\dfrac{24}{25}$ d. $-\dfrac{7}{25}$ e. $-\dfrac{24}{7}$ f. $\dfrac{\sqrt{5}}{5}$ g. $\dfrac{2\sqrt{5}}{5}$ h. $\dfrac{1}{2}$

Product-to-Sum and Sum-to-Product Formulas

Before Starting this Section, Review

1. Sum and difference formulas for sines and cosines (Section 5.3, page 355)
2. Fundamental identities (Section 5.1, page 332)

Objectives

1 Derive product-to-sum formulas.

2 Derive sum-to-product formulas.

3 Verify trigonometric identities involving multiple angles.

TOUCH-TONE PHONES

Dual-tone multi-frequency (DTMF), also known as Touch-Tone, tone dialing, or push-button dialing, is a method for instructing a telephone switching network to dial a telephone number. The system was developed by Bell Labs in the late 1950s, and Touch-Tone phones were introduced to the public at the 1964 New York World's Fair. The dual-tone keypad on a Touch-Tone phone has four rows and three columns. Each row represents a *low* frequency, and each column represents a *high* frequency, as shown in the figure. Pressing a button on a Touch-Tone phone produces a unique sound. The sound is produced by the combination of two tones (one of low frequency and the other of high frequency) associated with that button, as shown in the figure. In other words, the sound produced is given by $y = y_1 + y_2$, with $y_1 = \sin(2\pi l t)$ and $y_2 = \sin(2\pi h t)$, where l and h are the button's low and high frequencies (in hertz = Hz), respectively.

For example, if you press a single button, such as "1," it will send a sinusoidal tone of two frequencies 697 Hz and 1209 Hz. The sound produced by pressing "1" is represented by the equation

$$y = \sin[2\pi(697)t] + \sin[2\pi(1209)t]$$

or

$$y = \sin(1394\pi t) + \sin(2418\pi t).$$

The two tones are the reason for calling such phones *dual-tone multi-frequency phones*. In Example 6, we analyze the sound produced by the sum of two sine waves. (*Source:* http://en.wikipedia.org.) ■

1 Derive product-to-sum formulas.

Product-to-Sum Formulas

The following formulas allow us to rewrite products of sines or cosines as sums or differences.

PRODUCT-TO-SUM FORMULAS

$$\cos x \cos y = \frac{1}{2}[\cos(x - y) + \cos(x + y)]$$

$$\sin x \sin y = \frac{1}{2}[\cos(x - y) - \cos(x + y)]$$

$$\sin x \cos y = \frac{1}{2}[\sin(x + y) + \sin(x - y)]$$

$$\cos x \sin y = \frac{1}{2}[\sin(x + y) - \sin(x - y)]$$

These formulas are fairly easy to prove. For example, to prove the first formula

$$\cos x \cos y = \frac{1}{2}[\cos (x - y) + \cos (x + y)],$$

we write the difference and sum formulas for cosine and add.

$$\cos (x - y) = \cos x \cos y + \sin x \sin y$$
$$\underline{\cos (x + y) = \cos x \cos y - \sin x \sin y}$$

$\cos (x - y) + \cos (x + y) = 2 \cos x \cos y + 0$		Add.
$2 \cos x \cos y = \cos (x - y) + \cos (x + y)$		Interchange sides.
$\mathbf{\cos x \cos y = \dfrac{1}{2}[\cos (x - y) + \cos (x + y)]}$		Multiply both sides by $\dfrac{1}{2}$.

Similarly, subtracting $\cos (x + y)$ from $\cos (x - y)$ gives the second formula:

$$\mathbf{\sin x \sin y = \frac{1}{2}[\cos (x - y) - \cos (x + y)]}$$

We can prove the third and fourth formulas in the box in the same way.

EXAMPLE 1   Using a Product-to-Sum Formula

Write $\sin 5\theta \cos 3\theta$ as the sum or difference of two trigonometric functions.

SOLUTION

Use the formula $\sin x \cos y = \dfrac{1}{2}[\sin (x + y) + \sin (x - y)]$.

$$\sin 5\theta \cos 3\theta = \frac{1}{2}[\sin (5\theta + 3\theta) + \sin (5\theta - 3\theta)] \qquad \text{Replace } x \text{ with } 5\theta \text{ and } y \text{ with } 3\theta.$$

$$= \frac{1}{2}[\sin 8\theta + \sin 2\theta] \qquad \text{Simplify.}$$

$$= \frac{1}{2}\sin 8\theta + \frac{1}{2}\sin 2\theta \qquad \text{Distributive property}$$

■ ■ ■

Practice Problem 1   Write $\cos 3x \cos x$ as the sum or difference of two trigonometric functions. ■

EXAMPLE 2   Using a Product-to-Sum Formula

Find the exact value of $\sin 75° \sin 15°$.

SOLUTION

Use the formula $\sin x \sin y = \dfrac{1}{2}[\cos (x - y) - \cos (x + y)]$.

$$\sin 75° \sin 15° = \frac{1}{2}[\cos (75° - 15°) - \cos (75° + 15°)] \qquad \text{Replace } x \text{ with } 75° \text{ and } y \text{ with } 15°.$$

$$= \frac{1}{2}[\cos 60° - \cos 90°] \qquad \text{Simplify.}$$

$$= \frac{1}{2}\left[\frac{1}{2} - 0\right] = \frac{1}{4} \qquad \begin{aligned} &\cos 60° = \frac{1}{2}; \\ &\cos 90° = 0 \end{aligned} \quad ■ ■ ■$$

Practice Problem 2   Find the exact value of $\sin 15° \cos 75°$. ■

2 Derive sum-to-product formulas.

Sum-to-Product Formula

The following formulas allow us to do just the opposite—rewrite sums or differences of sines or cosines as products.

SUM-TO-PRODUCT FORMULAS

$$\cos x + \cos y = 2 \cos\left(\frac{x+y}{2}\right)\cos\left(\frac{x-y}{2}\right)$$

$$\cos x - \cos y = -2 \sin\left(\frac{x+y}{2}\right)\sin\left(\frac{x-y}{2}\right)$$

$$\sin x + \sin y = 2 \sin\left(\frac{x+y}{2}\right)\cos\left(\frac{x-y}{2}\right)$$

$$\sin x - \sin y = 2 \sin\left(\frac{x-y}{2}\right)\cos\left(\frac{x+y}{2}\right)$$

We prove these formulas by using the product-to-sum formulas. For example, to prove the first formula, we start with the right side of the equation and use the product-to-sum formulas.

$$2 \cos\left(\frac{x+y}{2}\right)\cos\left(\frac{x-y}{2}\right)$$

$$= 2 \cdot \frac{1}{2}\left[\cos\left(\frac{x+y}{2} - \frac{x-y}{2}\right) + \cos\left(\frac{x+y}{2} + \frac{x-y}{2}\right)\right] \qquad \text{Use the formula for the product of cosines.}$$

$$= \cos\left(\frac{x+y-x+y}{2}\right) + \cos\left(\frac{x+y+x-y}{2}\right) \qquad \text{Simplify.}$$

$$= \cos\left(\frac{2y}{2}\right) + \cos\left(\frac{2x}{2}\right) \qquad \text{Simplify.}$$

$$= \cos y + \cos x \qquad \text{Simplify.}$$

You can verify the other three formulas in the box in a similar way. (See Exercise 87.)

EXAMPLE 3   Using Sum-to-Product Formulas

Write each expression as a product of two trigonometric functions and simplify where possible.

a. $\sin 4\theta - \sin 6\theta$ **b.** $\cos 65° + \cos 55°$

SOLUTION

a. Use the formula $\sin x - \sin y = 2 \sin\left(\frac{x-y}{2}\right)\cos\left(\frac{x+y}{2}\right)$.

$$\sin 4\theta - \sin 6\theta = 2 \sin\left(\frac{4\theta - 6\theta}{2}\right)\cos\left(\frac{4\theta + 6\theta}{2}\right) \qquad \text{Replace } x \text{ with } 4\theta \text{ and } y \text{ with } 6\theta.$$

$$= 2 \sin(-\theta)\cos(5\theta) \qquad \text{Simplify.}$$

$$= -2 \sin \theta \cos 5\theta \qquad \sin(-\theta) = -\sin \theta$$

b. Use the formula $\cos x + \cos y = 2 \cos \left(\dfrac{x + y}{2} \right) \cos \left(\dfrac{x - y}{2} \right)$.

$$\cos 65° + \cos 55° = 2 \cos \left(\frac{65° + 55°}{2} \right) \cos \left(\frac{65° - 55°}{2} \right) \qquad \text{Replace } x \text{ with } 65° \text{ and } y \text{ with } 55°.$$

$$= 2 \cos 60° \cos 5° \qquad \text{Simplify.}$$

$$= 2 \left(\frac{1}{2} \right) \cos 5° = \cos 5° \qquad \cos 60° = \frac{1}{2}$$

■ ■ ■

Practice Problem 3 Write each expression as a product of two trigonometric functions and simplify where possible.

a. $\cos 2x - \cos 4x$ **b.** $\sin 43° + \sin 17°$ ■

In the next example, we show how to express a sum of a sine and a cosine as a product by first using a cofunction identity.

EXAMPLE 4   **Expressing a Sum of a Sine and a Cosine as a Product**

Write $\sin 5\theta + \cos 3\theta$ as a product of two trigonometric functions.

SOLUTION
Use the formula

$$\cos x + \cos y = 2 \cos \left(\frac{x + y}{2} \right) \cos \left(\frac{x - y}{2} \right).$$

$$\sin 5\theta + \cos 3\theta = \cos \left(\frac{\pi}{2} - 5\theta \right) + \cos 3\theta \qquad \text{Cofunction identity}$$

$$= 2 \cos \left(\frac{\frac{\pi}{2} - 5\theta + 3\theta}{2} \right) \cos \left(\frac{\frac{\pi}{2} - 5\theta - 3\theta}{2} \right) \qquad \begin{array}{l} \text{Sum-to-product} \\ \text{formula for} \\ \text{cosines} \end{array}$$

$$= 2 \cos \left(\frac{\pi}{4} - \theta \right) \cos \left(\frac{\pi}{4} - 4\theta \right) \qquad \text{Simplify.}$$

■ ■ ■

STUDY TIP

In Example 4, you could also write $\sin 5\theta + \cos 3\theta =$ $\sin 5\theta + \sin \left(\dfrac{\pi}{2} - 3\theta \right)$ and then use the sum-to-product formula.

Practice Problem 4 Write $\sin x - \cos x$ as a product of two trigonometric functions.   ■

3 Verify trigonometric identities involving multiple angles.

Verify Trigonometric Identities

The next example illustrates the use of sum-to-product formulas to verify identities.

EXAMPLE 5   **Verifying an Identity**

Verify the identity $\dfrac{\sin 5\theta + \sin 9\theta}{\cos 5\theta - \cos 9\theta} = \cot 2\theta$.

SOLUTION

$$\frac{\sin 5\theta + \sin 9\theta}{\cos 5\theta - \cos 9\theta} = \frac{2 \sin \dfrac{5\theta + 9\theta}{2} \cos \dfrac{5\theta - 9\theta}{2}}{-2 \sin \dfrac{5\theta + 9\theta}{2} \sin \dfrac{5\theta - 9\theta}{2}}$$ Use the formulas for $\sin x + \sin y$ and $\cos x - \cos y$.

$$= \frac{2 \cancel{\sin 7\theta} \cos (-2\theta)}{-2 \cancel{\sin 7\theta} \sin (-2\theta)}$$ Simplify.

$$= \frac{\cos (-2\theta)}{-\sin (-2\theta)}$$ Remove the common factors.

$$= \frac{\cos 2\theta}{\sin 2\theta}$$ $\cos (-x) = \cos x$, and $\sin (-x) = -\sin x$

$$= \cot 2\theta$$ Quotient identity   ■ ■ ■

Practice Problem 5   Verify the identity $\dfrac{\cos 5x - \cos x}{\sin x - \sin 5x} = \tan 3x$.   ■

Analyzing Touch-Tone Phones

When a sound of frequency f_1 is combined with a sound of frequency f_2, the resulting sound has frequency $\dfrac{f_1 + f_2}{2}$. In a Touch-Tone phone, when you press a button with a low frequency l and a high frequency h, the sound produced is modeled by

$$y = \sin (2\pi l t) + \sin (2\pi h t)$$

$$y = 2 \sin \left[2\pi \left(\frac{l + h}{2} \right) t \right] \cos \left[2\pi \left(\frac{l - h}{2} \right) t \right]$$ Sum-to-product formula

$$y = 2 \cos \left[2\pi \left(\frac{h - l}{2} \right) t \right] \sin \left[2\pi \left(\frac{l + h}{2} \right) t \right]$$ $\cos (-\theta) = \cos \theta$

The last equation suggests that the resulting sound may be thought of as a sine wave with frequency of $\dfrac{l + h}{2}$ and variable amplitude of $2 \cos \left[2\pi \left(\dfrac{h - l}{2} \right) t \right]$.

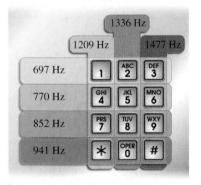

FIGURE 5.10

TECHNOLOGY CONNECTION

graph of
$y = 2 \cos (484\pi t) \sin [2\pi (1094) t]$

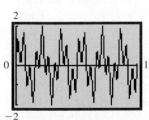

EXAMPLE 6   **Touch-Tone Phones**

a. Write an expression that models the tone of the sound produced by pressing the "8" button on your Touch-Tone phone. See Figure 5.10.
b. Rewrite the expression from part **a** as a product of two trigonometric functions.
c. Write the frequency and the variable amplitude of the sound from part **a**.

SOLUTION

a. Pressing the "8" key produces the sound given by

$$y = \sin [2\pi (852) t] + \sin [2\pi (1336) t] \qquad l = 852, h = 1336$$

b. $y = 2 \sin \left[2\pi \left(\dfrac{852 + 1336}{2} \right) t \right] \cos \left[2\pi \left(\dfrac{852 - 1336}{2} \right) t \right]$ Sum-to-product formula

$$= 2 \cos [2\pi (242) t] \sin [2\pi (1094) t] \qquad\qquad \cos (-\theta) = \cos \theta$$

$$= 2 \cos (484\pi t) \sin [2\pi (1094) t]$$

c. The frequency is 1094 Hz, and the variable amplitude is $2 \cos (484\pi t)$.   ■ ■ ■

Practice Problem 6   Repeat Example 6, assuming that you press the "1" button on your Touch-Tone phone.   ■

SECTION 5.5 ■ Exercises

A EXERCISES Basic Skills and Concepts

1. We can rewrite the product of two sines as a difference of two cosines by using the formula $\sin x \sin y =$ _____ . $\frac{1}{2}[\cos(x-y) - \cos(x+y)]$

2. We can rewrite the product of a sine and a cosine as the sum of two sines by using the formula $\sin x \cos y =$ _____ . $\frac{1}{2}[\sin(x+y) + \sin(x-y)]$

3. We can rewrite the sum of two cosines as a product of two cosines by using the formula $\cos x + \cos y =$ _____ . $2\cos\left(\frac{x+y}{2}\right)\cos\left(\frac{x-y}{2}\right)$

4. *True or False* $\cos x - \cos y = 2\sin\left(\frac{x+y}{2}\right)\sin\left(\frac{y-x}{2}\right)$.   True

5. *True or False* $\sin x + \sin y = \sin(x+y)$.   False

6. *True or False* $\sin(2x+1) - \sin(2x-1) = 2\sin(1)\cos 2x$.   True

In Exercises 7–22, use the product-to-sum formulas to rewrite each expression as the sum or difference of two functions. Simplify where possible.

7. $\sin x \cos x$ $\frac{1}{2}\sin 2x$

8. $\cos x \cos x$ $\frac{1}{2} + \frac{1}{2}\cos 2x$

9. $\sin x \sin x$ †

10. $\cos x \sin x$ †

11. $\sin 25° \cos 5°$ †

12. $\sin 40° \sin 20°$ †

13. $\cos 140° \cos 20°$ †

14. $\cos 70° \sin 20°$ †

15. $\sin\frac{7\pi}{12}\sin\frac{\pi}{12}$ $\frac{1}{4}$

16. $\sin\frac{3\pi}{8}\cos\frac{\pi}{8}$ $\frac{1}{2} + \frac{\sqrt{2}}{4}$

17. $\cos\frac{5\pi}{8}\sin\frac{\pi}{8}$ $\frac{\sqrt{2}}{4} - \frac{1}{2}$

18. $\cos\frac{5\pi}{3}\cos\frac{\pi}{3}$ $\frac{1}{4}$

19. $\sin 5\theta \cos\theta$ †

20. $\cos 3\theta \sin 2\theta$ †

21. $\cos 4x \cos 3x$ †

22. $\sin 5x \sin 2x$ †

In Exercises 23–30, find the exact value of each expression.

23. $\sin 37.5° \sin 7.5°$ †

24. $\cos 52.5° \cos 7.5°$ †

25. $\sin 67.5° \cos 22.5°$ †

26. $\cos 105° \sin 75°$ †

27. $\sin\frac{5\pi}{24}\cos\frac{\pi}{24}$ $\frac{1}{4} + \frac{\sqrt{2}}{4}$

28. $\sin\frac{7\pi}{12}\sin\frac{\pi}{12}$ $\frac{1}{4}$

29. $\cos\frac{13\pi}{24}\cos\frac{5\pi}{24}$ $\frac{1}{4} - \frac{\sqrt{2}}{4}$

30. $\cos\frac{7\pi}{24}\sin\frac{\pi}{24}$ $\frac{\sqrt{3}}{4} - \frac{\sqrt{2}}{4}$

In Exercises 31–50, use sum-to-product formulas to rewrite each expression as a product. Simplify where possible.

31. $\cos 40° - \cos 20°$ $-\sin 10°$

32. $\sin 22° + \sin 8°$ $2\sin 15° \cos 7°$

33. $\sin 32° - \sin 16°$ $2\sin 8° \cos 24°$

34. $\cos 47° + \cos 13°$ $\sqrt{3}\cos 17°$

35. $\sin\frac{\pi}{5} + \sin\frac{2\pi}{5}$ $2\sin\frac{3\pi}{10}\cos\frac{\pi}{10}$

36. $\cos\frac{\pi}{12} + \cos\frac{\pi}{3}$ $2\cos\frac{5\pi}{24}\cos\frac{\pi}{8}$

37. $\cos\frac{1}{2} + \cos\frac{1}{3}$ †

38. $\sin\frac{2}{3} - \sin\frac{1}{4}$ $2\sin\frac{5}{24}\cos\frac{11}{24}$

39. $\cos 3x + \cos 5x$ $2\cos 4x \cos x$

40. $\sin 5x - \sin 3x$ $2\sin x \cos 4x$

41. $\sin 7x + \sin(-x)$ $2\sin 3x \cos 4x$

42. $\cos 7x - \cos 3x$ $-2\sin 5x \sin 2x$

43. $\sin x + \cos x$ †

44. $\cos x - \sin x$ †

45. $\sin 2x - \cos 2x$ †

46. $\cos 3x + \sin 3x$ †

47. $\sin 3x + \cos 5x$ †

48. $\sin 5x - \cos x$ †

49. $a(\sin x + \cos x)$ †

50. $a(\sin bx + \cos bx)$ †

In Exercises 51–60, verify each identity.

51. $\dfrac{\sin x + \sin 3x}{\cos x + \cos 3x} = \tan 2x$

52. $\dfrac{\sin 2x + \sin 4x}{\cos 2x + \cos 4x} = \tan 3x$

53. $\dfrac{\cos 3x - \cos 7x}{\sin 7x + \sin 3x} = \tan 2x$

54. $\dfrac{\cos 12x - \cos 4x}{\sin 4x - \sin 12x} = \tan 8x$

55. $\dfrac{\cos 2x + \cos 2y}{\cos 2x - \cos 2y} = \cot(y-x)\cot(y+x)$

56. $\dfrac{\sin 2x + \sin 2y}{\sin 2x - \sin 2y} = \dfrac{\tan(x+y)}{\tan(x-y)}$

57. $\sin x + \sin 2x + \sin 3x = \sin 2x(1 + 2\cos x)$

58. $\cos x + \cos 2x + \cos 3x = \cos 2x(1 + 2\cos x)$

59. $\sin 2x + \sin 4x + \sin 6x = 4\cos x \cos 2x \sin 3x$

60. $\cos x + \cos 3x + \cos 5x + \cos 7x$
    $= 4\cos x \cos 2x \cos 4x$

B EXERCISES Applying the Concepts

61. **Touch-Tone phone.** $y = \sin[2\pi(852)t] + \sin[2\pi(1209)t]$
    a. Write an expression that models the tone by pressing the "7" button on your Touch-Tone phone. (See Figure 5.11, page 383.)
    b. Rewrite the expression from part (a) as a product of two functions. $y = 2\cos(357\pi t)\sin[2\pi(1030.5)t]$
    c. Write the frequency and the variable amplitude of the sound from part (a). $f = 1030.5$ Hz, $A = 2\cos(357\pi t)$

62. **Touch-Tone phone.** Repeat Exercise 61 assuming that you press the "#" button on your Touch-Tone phone.   †

63. **Beats in music.** When two musical tones of slightly different frequencies are played, the sound you hear will fluctuate in volume according to the difference in their frequencies. These are called *beat frequencies*. (Piano tuners use the beat frequency to adjust piano wire until it is at the same frequency as the tuning fork.) Let $y_1 = 0.05\cos(112\pi t)$ and $y_2 = 0.05\cos(120\pi t)$ model two tones.
    a. Write $y = y_1 + y_2$ as the product of two functions. $y = 0.1\cos(116\pi t)\cos(4\pi t)$
    b. How many beats are there? $f = 4$ beats per unit time

†Due to space constrictions, answers to these exercises may be found in the Answers beginning on page A–1 in the back of the book.

64. **Beats in music.** Repeat Exercise 63 for the two tones represented by $y_1 = 0.04 \sin(110\pi t)$ and $y_2 = 0.04 \sin(114\pi t)$. **a.** $y = 0.08 \cos(2\pi t) \sin(112\pi t)$ **b.** $f = 2$ beats per unit time

C EXERCISES Beyond the Basics

In Exercises 65 and 66, sketch the graph of f. What is the amplitude and period?

65. $f(x) = \cos(2x + 1) + \cos(2x - 1)$

66. $f(x) = \sin(3x + 1) + \sin(3x - 1)$

In Exercises 67–76, verify each identity.

67. $\cos 40° + \cos 50° + \cos 70° + \cos 80°$
$= \cos 10° + \cos 20°$

68. $\sin 10° + \sin 20° + \sin 40° + \sin 50° = 2 \cos 5° \cos 15°$

69. $\cos 4\theta \cos \theta - \cos 6\theta \cos 9\theta = \sin 5\theta \sin 10\theta$

70. $\cos 38° \cos 46° - \sin 14° \sin 22° = \dfrac{1}{2} \cos 24°$

71. $\sin 25° \sin 35° - \sin 25° \sin 85° - \sin 35° \sin 85° = -\dfrac{3}{4}$

72. $\cos 20° \cos 40° \cos 60° \cos 80° = \dfrac{1}{16}$

73. $\dfrac{\sin 3x \cos 5x - \sin x \cos 7x}{\sin x \sin 7x + \cos 3x \cos 5x} = \tan 2x$

74. $\dfrac{\sin 11x \sin x + \sin 7x \sin 3x}{\cos 11x \sin x + \cos 7x \sin 3x} = \tan 8x$

75. $\dfrac{\cos x + \cos 3x + \cos 5x + \cos 7x}{\sin x + \sin 3x + \sin 5x + \sin 7x} = \cot 4x$

76. $\dfrac{\sin 3x + \sin 5x + \sin 7x + \sin 9x}{\cos 3x + \cos 5x + \cos 7x + \cos 9x} = \tan 6x$

In Exercises 77–82, find all solutions of each equation in the interval $[0, 2\pi)$. Use a graphing calculator to verify your solutions.

77. $\cos \theta + \cos 5\theta = 0$ **78.** $\sin \theta + \sin 3\theta = 0$

79. $\cos 3\theta - \cos 7\theta = 0$ **80.** $\sin 5\theta - \sin 3\theta = 0$

81. $(\sin \theta + \sin 3\theta) - (\cos \theta + \cos 3\theta) = 0$

82. $(\sin \theta + \sin 2\theta + \sin 3\theta) - (\cos \theta + \cos 2\theta + \cos 3\theta) = 0$

83. Verify the identity $\cos 7x = 2 \cos 6x \cos x - 2 \cos 4x \cos x + 2 \cos 2x \cos x - \cos x$.

84. **a.** Verify the identity
$$4 \sin \theta \sin\left(\dfrac{\pi}{3} + \theta\right) \sin\left(\dfrac{\pi}{3} - \theta\right) = \sin 3\theta.$$
[*Hint:* Recall that $\sin 3\theta = 3 \sin \theta - 4 \sin^3 \theta$.]
b. Use part (a) to show that
$$\sin 20° \sin 40° \sin 60° \sin 80° = \dfrac{3}{16}.$$

85. Verify the identity
$$\dfrac{\sin(x + 3y) + \sin(3x + y)}{\sin 2x + \sin 2y} = 2 \cos(x + y).$$

86. Verify the identity $\sin^3 x \sin 3x + \cos^3 x \cos 3x = \cos^3 2x$.

87. Verify the following sum-to-product formulas.

a. $\cos x - \cos y = -2 \sin\left(\dfrac{x + y}{2}\right) \sin\left(\dfrac{x - y}{2}\right)$

b. $\sin x + \sin y = 2 \sin\left(\dfrac{x + y}{2}\right) \cos\left(\dfrac{x - y}{2}\right)$

c. $\sin x - \sin y = 2 \sin\left(\dfrac{x - y}{2}\right) \cos\left(\dfrac{x + y}{2}\right)$

Critical Thinking

88. Supply the reason for each step in verifying the identity
$$\sin \dfrac{\pi}{14} \sin \dfrac{3\pi}{14} \sin \dfrac{5\pi}{14} = \dfrac{1}{8}.$$

$$\sin \dfrac{\pi}{14} \sin \dfrac{3\pi}{14} \sin \dfrac{5\pi}{14} = \cos \dfrac{3\pi}{7} \cos \dfrac{2\pi}{7} \cos \dfrac{\pi}{7}$$

$$= \dfrac{8 \sin \dfrac{\pi}{7} \cos \dfrac{\pi}{7} \cos \dfrac{2\pi}{7} \cos \dfrac{3\pi}{7}}{8 \sin \dfrac{\pi}{7}}$$

$$= \dfrac{2 \sin \dfrac{4\pi}{7} \cos \dfrac{3\pi}{7}}{8 \sin \dfrac{\pi}{7}}$$

$$= \dfrac{\sin \pi + \sin \dfrac{\pi}{7}}{8 \sin \dfrac{\pi}{7}}$$

$$= \dfrac{1}{8}$$

GROUP PROJECTS

Assume that $A + B + C = 180°$. Verify each identity.

89. $\sin 2A + \sin 2B + \sin 2C = 4 \sin A \sin B \sin C$

90. $\cos 2A + \cos 2B + \cos 2C = -1 - 4 \cos A \cos B \cos C$

Answers:

65. The amplitude is $2 \cos 1 \approx 1.081$. The period is π.

66. The amplitude is $2 \cos 1 \approx 1.081$. The period is $\dfrac{2\pi}{3}$.

77. $\left\{\dfrac{\pi}{6}, \dfrac{\pi}{4}, \dfrac{\pi}{2}, \dfrac{3\pi}{4}, \dfrac{5\pi}{6}, \dfrac{7\pi}{6}, \dfrac{5\pi}{4}, \dfrac{3\pi}{2}, \dfrac{7\pi}{4}, \dfrac{11\pi}{6}\right\}$ **78.** $\left\{0, \dfrac{\pi}{2}, \pi, \dfrac{3\pi}{2}\right\}$

79. $\left\{0, \dfrac{\pi}{5}, \dfrac{2\pi}{5}, \dfrac{\pi}{2}, \dfrac{3\pi}{5}, \dfrac{4\pi}{5}, \pi, \dfrac{6\pi}{5}, \dfrac{7\pi}{5}, \dfrac{3\pi}{2}, \dfrac{8\pi}{5}, \dfrac{9\pi}{5}\right\}$

80. $\left\{0, \dfrac{\pi}{8}, \dfrac{3\pi}{8}, \dfrac{5\pi}{8}, \dfrac{7\pi}{8}, \pi, \dfrac{9\pi}{8}, \dfrac{11\pi}{8}, \dfrac{13\pi}{8}, \dfrac{15\pi}{8}\right\}$

81. $\left\{\dfrac{\pi}{8}, \dfrac{\pi}{2}, \dfrac{5\pi}{8}, \dfrac{9\pi}{8}, \dfrac{3\pi}{2}, \dfrac{13\pi}{8}\right\}$ **82.** $\left\{\dfrac{\pi}{8}, \dfrac{5\pi}{8}, \dfrac{2\pi}{3}, \dfrac{9\pi}{8}, \dfrac{4\pi}{3}, \dfrac{13\pi}{8}\right\}$

SUMMARY ■ **Definitions, Concepts, and Formulas**

5.1 Trigonometric Identities and Equations

Reciprocal Identities

$$\csc x = \frac{1}{\sin x} \qquad \sec x = \frac{1}{\cos x} \qquad \cot x = \frac{1}{\tan x}$$

Quotient Identities

$$\tan x = \frac{\sin x}{\cos x} \qquad \cot x = \frac{\cos x}{\sin x}$$

Pythagorean Identities

$$\sin^2 x + \cos^2 x = 1$$
$$1 + \tan^2 x = \sec^2 x$$
$$1 + \cot^2 x = \csc^2 x$$

Even-Odd Identities

$$\sin(-x) = -\sin x \quad \cos(-x) = \cos x \quad \tan(-x) = -\tan x$$
$$\csc(-x) = -\csc x \quad \sec(-x) = \sec x \quad \cot(-x) = -\cot x$$

See page 335 for *guidelines for verifying trigonometric identities*.

5.2 Trigonometric Equations

A trigonometric equation is an equation that contains a trigonometric function with a variable. Values of the variable that satisfy a trigonometric equation are its *solutions*. Solving a trigonometric equation means to find its *solution set*.

Algebraic techniques are used to solve equations that are in linear, quadratic, or factorable form. Sometimes trigonometric identities are used to solve trigonometric equations.

5.3 Sum and Difference Formulas

Sum and Difference Formulas

$$\cos(u - v) = \cos u \cos v + \sin u \sin v$$
$$\cos(u + v) = \cos u \cos v - \sin u \sin v$$
$$\sin(u - v) = \sin u \cos v - \cos u \sin v$$
$$\sin(u + v) = \sin u \cos v + \cos u \sin v$$
$$\tan(u - v) = \frac{\tan u - \tan v}{1 + \tan u \tan v}$$
$$\tan(u + v) = \frac{\tan u + \tan v}{1 - \tan u \tan v}$$

Cofunction Identities

$$\sin\left(\frac{\pi}{2} - x\right) = \cos x \qquad \cos\left(\frac{\pi}{2} - x\right) = \sin x$$
$$\tan\left(\frac{\pi}{2} - x\right) = \cot x \qquad \csc\left(\frac{\pi}{2} - x\right) = \sec x$$
$$\sec\left(\frac{\pi}{2} - x\right) = \csc x \qquad \cot\left(\frac{\pi}{2} - x\right) = \tan x$$

Reduction Formula

If (a, b) is any point on the terminal side of an angle θ in standard position, then $a \sin x + b \cos x = \sqrt{a^2 + b^2} \sin(x + \theta)$ for any real number x.

5.4 Double-Angle and Half-Angle Formulas

Double-Angle Formulas

$$\sin 2x = 2 \sin x \cos x$$
$$\cos 2x = \cos^2 x - \sin^2 x$$
$$\cos 2x = 1 - 2 \sin^2 x$$
$$\cos 2x = 2 \cos^2 x - 1$$
$$\tan 2x = \frac{2 \tan x}{1 - \tan^2 x}$$

Power-Reducing Formulas

$$\sin^2 x = \frac{1 - \cos 2x}{2}$$
$$\cos^2 x = \frac{1 + \cos 2x}{2}$$
$$\tan^2 x = \frac{1 - \cos 2x}{1 + \cos 2x}$$

Half-Angle Formulas

$$\sin\frac{\theta}{2} = \pm\sqrt{\frac{1 - \cos\theta}{2}}$$
$$\cos\frac{\theta}{2} = \pm\sqrt{\frac{1 + \cos\theta}{2}}$$
$$\tan\frac{\theta}{2} = \pm\sqrt{\frac{1 - \cos\theta}{1 + \cos\theta}}$$

where the sign $+$ or $-$ depends on the quadrant in which $\frac{\theta}{2}$ lies.

5.5 Product-to-Sum and Sum-to-Product Formulas

Product-to-Sum Formulas

$$\cos x \cos y = \frac{1}{2}\left[\cos(x-y) + \cos(x+y)\right]$$

$$\sin x \sin y = \frac{1}{2}\left[\cos(x-y) - \cos(x+y)\right]$$

$$\sin x \cos y = \frac{1}{2}\left[\sin(x+y) + \sin(x-y)\right]$$

$$\cos x \sin y = \frac{1}{2}\left[\sin(x+y) - \sin(x-y)\right]$$

Sum-to-Product Formulas

$$\cos x + \cos y = 2\cos\left(\frac{x+y}{2}\right)\cos\left(\frac{x-y}{2}\right)$$

$$\cos x - \cos y = -2\sin\left(\frac{x+y}{2}\right)\sin\left(\frac{x-y}{2}\right)$$

$$\sin x + \sin y = 2\sin\left(\frac{x+y}{2}\right)\cos\left(\frac{x-y}{2}\right)$$

$$\sin x - \sin y = 2\sin\left(\frac{x-y}{2}\right)\cos\left(\frac{x+y}{2}\right)$$

REVIEW EXERCISES

Basis Skills and Concepts

In Exercises 1–4, use the information given to find the exact value of the remaining trigonometric functions of θ.

1. $\sin\theta = -\dfrac{2}{3}$ and $\cos\theta < 0$ †

2. $\tan\theta = -\dfrac{1}{2}$ and $\csc\theta > 0$ †

3. $\sec\theta = 3$ and $\tan\theta < 0$ †

4. $\csc\theta = 5$ and $\cot\theta < 0$ †

In Exercises 5–20, verify each identity.

5. $(\sin x + \cos x)^2 + (\sin x - \cos x)^2 = 2$

6. $(1 - \tan x)^2 + (1 + \tan x)^2 = 2\sec^2 x$

7. $\dfrac{1 - \tan^2\theta}{1 + \tan^2\theta} = \cos^2\theta - \sin^2\theta$

8. $\dfrac{\sin x + \tan x}{\csc x + \cot x} = \sin^2 x \sec x$

9. $\dfrac{\sin\theta}{1 + \cos\theta} + \dfrac{\sin\theta}{1 - \cos\theta} = 2\csc\theta$

10. $\tan^2 x \sin^2 x = \tan^2 x - \sin^2 x$

11. $\dfrac{\sin\theta}{1 - \cot\theta} + \dfrac{\cos\theta}{1 - \tan\theta} = \cos\theta + \sin\theta$

12. $\dfrac{\tan\theta - \sin\theta}{\sin^3\theta} = \dfrac{\sec\theta}{1 + \cos\theta}$

13. $\dfrac{\tan x}{\sec x - 1} + \dfrac{\tan x}{\sec x + 1} = 2\csc x$

14. $\dfrac{1}{\csc x - \cot x} - \dfrac{1}{\cot x + \csc x} = 2\cot x$

15. $\dfrac{1 + \sin\theta}{1 - \sin\theta} = (\sec\theta + \tan\theta)^2$

16. $\dfrac{1 - \cos\theta}{1 + \cos\theta} = (\csc\theta - \cot\theta)^2$

17. $\dfrac{\sec x - \tan x}{\sec x + \tan x} = (\sec x - \tan x)^2$

18. $\dfrac{\csc x + \cot x}{\csc x - \cot x} = (\csc x + \cot x)^2$

19. $\dfrac{\sin x - \cos x + 1}{\sin x + \cos x - 1} = \dfrac{\sin x + 1}{\cos x}$

[*Hint:* Multiply the numerator and the denominator of the left side by $\sin x + (1 - \cos x)$.]

20. $\sin^2 x \cot x = \dfrac{2\cos x}{\sin x + \csc x + \cos^2 x \csc x}$

In Exercises 21–26, find all solutions of each equation in the interval $[0, 2\pi)$. Use a graphing calculator to verify your solutions.

21. $2\cos^2 x - 1 = 0$

22. $3\tan^2 x - 1 = 0$

23. $2\cos^2 x - \cos x - 1 = 0$ †

24. $2\sin^2 x - 5\sin x - 3 = 0$ †

25. $2\sin 3x - 1 = 0$ †

26. $2\cos(2x - 1) - 1 = 0$ †

In Exercises 27–32, solve each equation for θ in the interval $[0°, 360°)$. If necessary, write the answers to the nearest tenth of a degree. Use a graphing calculator to verify your solutions.

27. $3\sin\theta - 1 = 0$ $\{19.5°, 160.5°\}$

28. $4\cos\theta + 1 = 0$ $\{104.5°, 255.5°\}$

29. $2\sqrt{3}\cos 2\theta - 3 = 0$ $\{15°, 165°, 195°, 345°\}$

30. $(1 - 2\sin 2\theta)(\sqrt{3} + 2\cos 2\theta) = 0$ $\{15°, 75°, 105°, 195°, 255°, 285°\}$

31. $(\sqrt{3}\tan\theta + 1)(2\cos\theta + 1) = 0$ $\{120°, 150°, 240°, 330°\}$

32. $\sqrt{3}\cos\theta = \sin\theta + 1$ $\{30°, 270°\}$

In Exercises 33–40, find the exact value of each expression.

33. $\cos 15°$ †

34. $\sin 105°$ †

35. $\csc 75°$ $\sqrt{6} - \sqrt{2}$

36. $\tan 75°$ †

Answers:

21. $\left\{\dfrac{\pi}{4}, \dfrac{3\pi}{4}, \dfrac{5\pi}{4}, \dfrac{7\pi}{4}\right\}$

22. $\left\{\dfrac{\pi}{6}, \dfrac{5\pi}{6}, \dfrac{7\pi}{6}, \dfrac{11\pi}{6}\right\}$

†Due to space constrictions, answers to these exercises may be found in the Answers beginning on page A–1 in the back of the book.

37. $\sin 41° \cos 49° + \cos 41° \sin 49°$ 1

38. $\cos 50° \cos 10° - \sin 50° \sin 10°$ $\frac{1}{2}$

39. $\dfrac{\tan 69° + \tan 66°}{1 - \tan 69° \tan 66°}$ -1

40. $2 \cos 75° \cos 15°$ $\frac{1}{2}$

In Exercises 41–44, let $\sin u = \dfrac{4}{5}$, $\cos v = \dfrac{5}{13}$, **for** $0 < u \le \dfrac{\pi}{2}$, $0 < v \le \dfrac{\pi}{2}$. **Find the value of each expression.**

41. $\sin (u - v)$ $-\dfrac{16}{65}$
42. $\cos (u + v)$ $-\dfrac{33}{65}$

43. $\cos (u - v)$ $\dfrac{63}{65}$
44. $\tan (u - v)$ $-\dfrac{16}{63}$

In Exercises 45–62, verify each identity.

45. $\sin (x - y) \cos y + \cos (x - y) \sin y = \sin x$

46. $\cos (x - y) \cos y - \sin (x - y) \sin y = \cos x$

47. $\dfrac{\sin (u + v)}{\sin (u - v)} = \dfrac{\tan u + \tan v}{\tan u - \tan v}$

48. $\dfrac{\tan (x + y)}{\cot (x - y)} = \dfrac{\tan^2 x - \tan^2 y}{1 - \tan^2 x \tan^2 y}$

49. $\dfrac{\sin 4x}{\sin 2x} - \dfrac{\cos 4x}{\cos 2x} = \sec 2x$

50. $\dfrac{\sin 3x}{\sin x} - \dfrac{\cos 3x}{\cos x} = 2$

51. $\sin \left(x - \dfrac{\pi}{6} \right) + \cos \left(x + \dfrac{\pi}{3} \right) = 0$

52. $\cos 2x = \cos^4 x - \sin^4 x$

53. $1 + \tan \theta \tan 2\theta = \sec 2\theta$

54. $\dfrac{\sin \theta + \sin 2\theta}{1 + \cos \theta + \cos 2\theta} = \tan \theta$

55. $\dfrac{\sin \theta + \sin 2\theta}{\cos \theta + \cos 2\theta} = \tan \dfrac{3\theta}{2}$

56. $\dfrac{\sin \theta - \sin 2\theta}{\cos \theta - \cos 2\theta} = -\cot \dfrac{3\theta}{2}$

57. $\dfrac{\sin 5x - \sin 3x}{\sin 5x + \sin 3x} = \dfrac{\tan x}{\tan 4x}$

58. $\dfrac{\cos 5x - \cos 3x}{\cos 5x + \cos 3x} = -\tan x \tan 4x$

59. $\dfrac{\tan 3x + \tan x}{\tan 3x - \tan x} = 2 \cos 2x$

60. $\dfrac{\sin 4x}{2(1 + \cos 4x)} = \dfrac{\tan x}{1 - \tan^2 x}$

61. $\dfrac{\sin 3x + \sin 5x + \sin 7x + \sin 9x}{\cos 3x + \cos 5x + \cos 7x + \cos 9x} = \tan 6x$

62. $\dfrac{\cos x + \cos 3x + \cos 5x + \cos 7x}{\sin x + \sin 3x + \sin 5x + \sin 7x} = \cot 4x$

In Exercises 63 and 64, sketch the graph of each equation by first converting it to the form $y = A \sin (x + \theta)$.

63. $y = \sqrt{3} \sin x + \cos x$ †   **64.** $y = \sin x + \sqrt{3} \cos x$ †

In Exercises 65 and 66, assume that $A + B + C = 180°$. **Verify each identity.**

65. $\sin 2A + \sin 2B - \sin 2C = 4 \cos A \cos B \sin C$

66. $\tan A + \tan B + \tan C = \tan A \tan B \tan C$

Applying the Concepts

67. A searchlight beam from a Coast Guard boat shines on a lake at an angle θ to the surface of the water. Viewed by a swimmer under water, the beam appears to make an angle α with the surface of the water, as shown in the figure. Under these circumstances, $\cos \theta = \dfrac{4}{3} \cos \alpha$. Find α if $\cos \theta = \dfrac{2}{3}$. 60°

68. The rabbit population of an area is described by $N(t) = 2400 + 500 \sin \pi t$, where $N(t)$ gives the number of rabbits in the area after t years. When will the rabbit population first reach 1900? 1.5 years

69. The voltage from an alternating current generator is given by $V(t) = 160 \cos 120\pi t$, where t is measured in seconds. Find the time t when the voltage first equals 80. $\dfrac{1}{360}$ sec

70. A jeweler wants to make a gold pendant in the shape of an isosceles triangle. Express the area of the pendant in terms of $\dfrac{\theta}{2}$ and x, where θ is the angle included by two sides of equal length x. $x^2 \sin \dfrac{\theta}{2} \cos \dfrac{\theta}{2}$

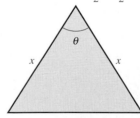

71. If the area of the pendant in Exercise 70 is 1 square inch, find the angle included between the two sides of equal length when each measures 2 inches. 30°

PRACTICE TEST A

1. If $\sin \theta = \dfrac{3}{5}$ and $\cos \theta < 0$, find $\tan \theta$. $-\dfrac{3}{4}$

2. If $\tan x = \dfrac{2}{3}$ and $\csc x < 0$, find $\cos x$. $-\dfrac{3\sqrt{13}}{13}$

In Problems 3–8, verify each identity.

3. $\dfrac{1 - \sin^2 x}{\sin^2 x} = \cot^2 x$

4. $2 \sin x \cos x = (\sin x + \cos x + 1)(\sin x + \cos x - 1)$

5. $\sin x \sin\left(\dfrac{\pi}{2} - x\right) = \dfrac{\sin 2x}{2}$

6. $\dfrac{\sin 2x}{2(\cos x + \sin x)} = \dfrac{\sin x}{1 + \tan x}$

7. $\dfrac{\sin 2x + \sin 4x}{\cos 2x + \cos 4x} = \tan 3x$

8. $\cos(x + y)\cos(x - y) = \cos^2 x + \cos^2 y - 1$

In Problems 9 and 10, find all solutions of each equation.

9. $\tan(-x) = 1$ $-\dfrac{\pi}{4} + n\pi$ 10. $\sin 4x = \dfrac{1}{2}$
                                     $\dfrac{\pi}{24} + n\dfrac{\pi}{2}, \dfrac{5\pi}{24} + n\dfrac{\pi}{2}$

In Problems 11 and 12, find all solutions of each equation in the interval $[0, 2\pi)$.

11. $\cos\dfrac{x}{3} = \dfrac{\sqrt{2}}{2}$ $\left\{\dfrac{3\pi}{4}\right\}$ 12. $\sin 2x + \cos x = 0$
                                                $\left\{\dfrac{\pi}{2}, \dfrac{7\pi}{6}, \dfrac{3\pi}{2}, \dfrac{11\pi}{6}\right\}$

In Problems 13–15, find the exact value of each expression.

13. $\sin 56° + \cos 146°$ 0

14. $\cos 48° \cos 12° - \sin 48° \sin 12°$ $\dfrac{1}{2}$

15. $\sin\dfrac{5\pi}{12}$ $\dfrac{\sqrt{6} + \sqrt{2}}{4}$

16. If $\sin \theta = \dfrac{4}{5}$, find the exact value of $\cos 2\theta$. $-\dfrac{7}{25}$

17. If $\tan \theta = \dfrac{3}{4}$, find the exact value of $\sin 2\theta$. $\dfrac{24}{25}$

18. Determine whether the function $y = \cos\left(\dfrac{\pi}{2} - x\right) + \tan x$ is odd, even, or neither.   Odd

19. Determine whether $\sin x + \sin y = \sin(x + y)$ is an identity.   No; let $x = y = \dfrac{\pi}{2}$.

20. A golf ball is hit on a level fairway with an initial velocity of $v_0 = 128$ ft/sec on an initial angle of flight θ (in degrees). The range (in feet) of the ball is given by the expression $\dfrac{v_0^2 \sin 2\theta}{32}$. Find the value(s) of θ if the range is 512 feet.

PRACTICE TEST B

1. If $\sin \theta = -\dfrac{12}{13}$ and $\pi < \theta < \dfrac{3\pi}{2}$, find $\sec \theta$.   a
   - **a.** $-\dfrac{13}{5}$
   - **b.** $-\dfrac{12}{5}$
   - **c.** $-\dfrac{5}{12}$
   - **d.** $\dfrac{13}{5}$

2. Which expression is equal to $(\sin x + \cos x)^2 + (\sin x - \cos x)^2$?   b
   - **a.** 1
   - **b.** 2
   - **c.** $\sin x \cos x$
   - **d.** $4 \sin x \cos x$

3. Which expression is equal to $\dfrac{\sin x}{1 + \cos x} + \dfrac{\sin x}{1 - \cos x}$?   d
   - **a.** $2 \sin x$
   - **b.** $2 \cos x$
   - **c.** $2 \sec x$
   - **d.** $2 \csc x$

4. Which expression is equal to $\dfrac{\tan x + \tan y}{\cot x + \cot y}$?   b
   - **a.** $\cot x \cot y$
   - **b.** $\tan x \tan y$
   - **c.** $\sec x \csc y$
   - **d.** $\sin x \cos y$

5. Which expression is equal to $\dfrac{\sin x + \sin y}{\cos x + \cos y} + \dfrac{\cos x - \cos y}{\sin x - \sin y}$?   a
   - **a.** 0
   - **b.** 1
   - **c.** $\tan x \tan y$
   - **d.** $\tan x + \tan y$

6. Which expression is equal to $\dfrac{\sin 2x}{1 + \cos 2x}$?   b
   - **a.** $\cot x$
   - **b.** $\tan x$
   - **c.** $\tan 2x$
   - **d.** $\cot 2x$

7. Which expression is equal to $\dfrac{1 - \cos 2x}{1 + \cos 2x}$?   c
   - **a.** $\sin^2 x$
   - **b.** $\cos^2 x$
   - **c.** $\tan^2 x$
   - **d.** $\cot^2 x$

8. Which expression is equal to $\dfrac{\sin 3x - \sin x}{\cos x - \cos 3x}$?   d
   - **a.** $\tan x$
   - **b.** $\tan 2x$
   - **c.** $\cot x$
   - **d.** $\cot 2x$

In Problems 9–11, find all solutions of each equation.

9. $\cot(-x) = -1$ c

    a. $x = \dfrac{\pi}{4} + 2\pi n$, for any integer n

    b. $x = \dfrac{3\pi}{4} + 2\pi n$, for any integer n

    c. $x = \dfrac{\pi}{4} + \pi n$, for any integer n

    d. $x = \dfrac{3\pi}{4} + \pi n$, for any integer n

10. $2\cos 4x = -1$ c

    a. $x = \dfrac{2\pi}{3} + 2\pi n$ and $x = \dfrac{4\pi}{3} + 2\pi n$, for any integer n

    b. $x = \dfrac{\pi}{3} + \dfrac{n\pi}{2}$ and $x = \dfrac{5\pi}{3} + \dfrac{n\pi}{2}$, for any integer n

    c. $x = \dfrac{\pi}{6} + \dfrac{n\pi}{2}$ and $x = \dfrac{\pi}{3} + \dfrac{n\pi}{2}$, for any integer n

    d. $x = \dfrac{\pi}{12} + 2\pi n$ and $x = \dfrac{5\pi}{12} + 2\pi n$, for any integer n

11. $\sin \dfrac{x}{3} = \dfrac{\sqrt{2}}{2}$ c

    a. $x = \dfrac{\pi}{4} + 2\pi n$ and $x = \dfrac{3\pi}{4} + 2\pi n$, for any integer n

    b. $x = \dfrac{\pi}{4} + 6n\pi$ and $x = \dfrac{\pi}{12} + 6n\pi$, for any integer n

    c. $x = \dfrac{3\pi}{4} + 6\pi n$ and $x = \dfrac{9\pi}{4} + 6\pi n$, for any integer n

    d. $x = \dfrac{3\pi}{4} + 2\pi n$ and $x = \dfrac{5\pi}{4} + 2\pi n$, for any integer n

12. Find all solutions of $\cos x - \sin 2x = 0$ for the interval $[0, 2\pi)$. a

    a. $\left\{\dfrac{\pi}{6}, \dfrac{\pi}{2}, \dfrac{5\pi}{6}, \dfrac{3\pi}{2}\right\}$ b. $\left\{\dfrac{\pi}{6}, \dfrac{\pi}{3}, \dfrac{5\pi}{6}, \dfrac{7\pi}{6}\right\}$

    c. $\left\{\dfrac{\pi}{2}, \dfrac{5\pi}{6}, \pi, \dfrac{3\pi}{2}\right\}$ d. $\left\{\dfrac{\pi}{6}, \dfrac{\pi}{2}, \dfrac{5\pi}{6}, \dfrac{11\pi}{6}\right\}$

In Problems 13 and 14, find the exact value of each expression.

13. $\sin 71° + \cos 161°$ d

    a. 1.892   b. -1.892

    c. 1   d. 0

14. $\sin 53° \cos 37° + \cos 53° \sin 37°$ d

    a. $\dfrac{1}{2}$ b. $\dfrac{\sqrt{3}}{2}$

    c. 0   d. 1

15. If $\sin \theta = \dfrac{2}{3}$, find the exact value of $\cos 2\theta$. b

    a. $\dfrac{1}{3}$ b. $\dfrac{1}{9}$

    c. $\dfrac{14}{9}$ d. $\dfrac{\sqrt{5}}{3}$

16. If $\tan \theta = \dfrac{4}{3}$, find the exact value of $\sin 2\theta$. a

    a. $\dfrac{24}{25}$ b. $\dfrac{4}{5}$

    c. $\dfrac{3}{5}$ d. $\dfrac{8}{5}$

17. Which one of the following best describes the symmetry of the graph of $y = \sin\left(\dfrac{\pi}{2} - x\right) + \cot x$? d

    a. Symmetric with respect to the origin
    b. Symmetric with respect to the x-axis
    c. Symmetric with respect to the y-axis
    d. Not symmetric to either axis or the origin

18. Which pair of values results in a false statement for $\tan(x + y) = \tan x + \tan y$? c

    a. $x = 0, y = \pi$ b. $x = \dfrac{\pi}{4}, y = -\dfrac{\pi}{4}$

    c. $x = \dfrac{\pi}{4}, y = \dfrac{5\pi}{4}$ d. $x = \dfrac{\pi}{4}, y = \dfrac{3\pi}{4}$

19. Identify all solutions of $2\cos^2 x = \dfrac{3}{2}$ in the interval $[0, 2\pi)$. d

    a. $\left\{\dfrac{\pi}{6}\right\}$ b. $\left\{\dfrac{\pi}{6}, \dfrac{5\pi}{6}\right\}$

    c. $\left\{\dfrac{\pi}{6}, \dfrac{11\pi}{6}\right\}$ d. $\left\{\dfrac{\pi}{6}, \dfrac{5\pi}{6}, \dfrac{7\pi}{6}, \dfrac{11\pi}{6}\right\}$

20. A golf ball is hit on a level fairway with an initial velocity of $v_0 = 128$ ft/sec and an initial angle of flight θ (in degrees). The range (in feet) of the ball is given by the expression $\dfrac{v_0^2 \sin 2\theta}{32}$. Find the value(s) of θ if the range is $256\sqrt{3}$ feet. b

    a. $45°$ b. $30°, 60°$

    c. $45°, 135°$ d. $150°, 175°$

CUMULATIVE REVIEW EXERCISES ■ Chapters 1–5

In Exercises 1–6, solve each equation or inequality.

1. $x^2 + x - 1 = 0$ $x = -\dfrac{1}{2} \pm \dfrac{\sqrt{5}}{2}$

2. $\log_2(x + 1) + \log_2(x - 1) = 1$ $x = \sqrt{3}$

3. $5^{-x} = 9$ $x = -\dfrac{\log 9}{\log 5}$

4. $\sin 2x - \cos x = 0$ $(0 \le x < 2\pi)$ $\left\{\dfrac{\pi}{6}, \dfrac{\pi}{2}, \dfrac{5\pi}{6}, \dfrac{3\pi}{2}\right\}$

5. $x^3 - 4x > 0$ $(-2, 0) \cup (2, \infty)$

6. $\dfrac{2x - 3}{x + 2} - 1 < 0$ $(-2, 5)$

7. If $f(x) = 2x^2 - x + 3$, find $\dfrac{f(x+h) - f(x)}{h}$. $4x - 1 + 2h$

8. If $f(x) = \dfrac{x+2}{2x-1}$, find $f^{-1}(x)$. $f^{-1}(x) = \dfrac{x+2}{2x-1}$

In Exercises 9–16, sketch the graph of each function $f(x)$ by using transformations on the given basic function.

9. $f(x) = x^2 - 4x - 7$;   use $y = x^2$. †

10. $f(x) = 2\sqrt{x-3} + 1$;   use $y = \sqrt{x}$. †

11. $f(x) = 2^{x-1} - 2$;   use $y = 2^x$. †

12. $f(x) = 2\ln(x+1)$;   use $y = \ln x$. †

13. $f(x) = \dfrac{1}{2}\sin 2x$;   use $y = \sin x$. †

14. $f(x) = -2\cos(3x - 1)$; use $y = \cos x$. †

15. $f(x) = -3\csc x$;   use $y = 3\sin x$. †

16. $f(x) = 2\tan 4x$;   use $y = \tan x$. †

17. Simplify $\dfrac{2\csc^2 x - 5\cot x - 5}{2\cot x + 1}$. $\cot x - 3$

18. Use an identity to find the exact value of $\sec 15°$. $2\sqrt{2 - \sqrt{3}}$

19. Verify the identity $\sin\left(x - \dfrac{3\pi}{2}\right) = \cos x$.

20. The angles to the top of a tower from two points on the ground at distances 800 and 3200 feet are complementary angles. Find the height of the tower.

6 Applications of Trigonometric Functions

Applications of trigonometry permeate virtually every field, from architecture and bionics to medicine and physics. In this chapter, we introduce vectors and polar coordinates and investigate the many applications of trigonometry to real-world problems.

Right-Triangle Trigonometry

Before Starting this Section, Review

1. Rationalizing the denominator (Appendix A, page 782)
2. Acute angle definition (Section 4.1, page 255)
3. Pythagorean Theorem (Appendix A, page 790)
4. Similar triangles (Appendix A, page 789)
5. Functions (Section 1.3, page 31)

Objectives

1 Express the trigonometric functions using a right triangle.

2 Evaluate trigonometric functions of angles in a right triangle.

3 Solve right triangles.

4 Use right-triangle trigonometry in applications.

Mount Kilimanjaro

MEASURING MOUNT KILIMANJARO

Trigonometry developed as a result of attempts to solve practical problems in astronomy, navigation, and land measurement. The word *trigonometry*, coined by Pitiscus in 1594, is derived from two Greek words, *trigonon* (triangle) and *metron* (measure), and means "triangle measurement."

   Measuring objects that are generally inaccessible is one of the many successes of trigonometry. In Example 7, we use trigonometry to approximate the height of Mount Kilimanjaro, Tanzania. Mount Kilimanjaro is the highest mountain in Africa. ■

---

1 Express the trigonometric functions using a right triangle.

RECALL

In a right triangle, the side opposite the right angle is called the hypotenuse and the other two sides are called the legs.

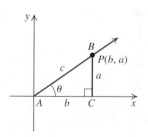

FIGURE 6.2

Trigonometric Ratios and Functions

Consider a right triangle with one of its acute angles labeled θ. In Figure 6.1, the capital letters A, B, and C designate the vertices of the triangle and the lowercase letters a, b, and c, respectively, represent the lengths of the sides opposite these angles.

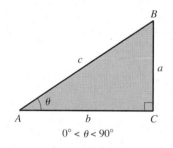

a = length of the side opposite θ
b = length of the side adjacent to θ
c = length of the hypotenuse

$0° < \theta < 90°$

FIGURE 6.1 Right triangle

Now place this triangle in standard position, as shown in Figure 6.2. The point corresponding to the vertex B is $P(b, a)$. Because P is on the terminal side of θ, we can use it to express the trigonometric functions of θ as ratios of the sides of a right triangle. We use the words *opposite* for the length of the leg opposite θ, *adjacent* for the length of the leg adjacent θ, and *hypotenuse* for the length of the hypotenuse.

TRIGONOMETRIC FUNCTIONS OF AN ANGLE θ IN A RIGHT TRIANGLE

$$\sin \theta = \frac{\text{opposite}}{\text{hypotenuse}} = \frac{a}{c} \qquad\qquad \csc \theta = \frac{\text{hypotenuse}}{\text{opposite}} = \frac{c}{a}$$

$$\cos \theta = \frac{\text{adjacent}}{\text{hypotenuse}} = \frac{b}{c} \qquad\qquad \sec \theta = \frac{\text{hypotenuse}}{\text{adjacent}} = \frac{c}{b}$$

$$\tan \theta = \frac{\text{opposite}}{\text{adjacent}} = \frac{a}{b} \qquad\qquad \cot \theta = \frac{\text{adjacent}}{\text{opposite}} = \frac{b}{a}$$

RECALL

Two triangles are *similar* if their corresponding angles are congruent and their corresponding sides are proportional. Two similar triangles have the same shape but not necessarily the same size.

Because any two right triangles having acute angle θ are similar, the ratios of corresponding sides depend only on the angle θ, not on the size of the triangle. Three right triangles with acute angle θ are shown in Figure 6.3.

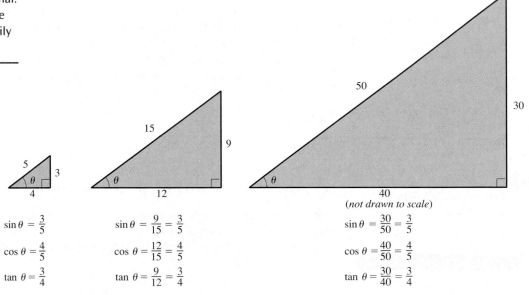

FIGURE 6.3 Three similar triangles

Notice that the value of each of the six trigonometric functions is independent of the triangle used to find it. Recall that $\csc \theta$, $\sec \theta$, and $\cot \theta$ are the reciprocals of $\sin \theta$, $\cos \theta$, and $\tan \theta$, respectively. Consequently, for any of the triangles in Figure 6.3, we have $\csc \theta = \dfrac{5}{3}$, $\sec \theta = \dfrac{5}{4}$, and $\cot \theta = \dfrac{4}{3}$.

2 Evaluate trigonometric functions of angles in a right triangle.

Evaluating Trigonometric Functions

EXAMPLE 1 **Finding the Values of Trigonometric Functions**

Find the exact values for the six trigonometric functions of the angle θ in Figure 6.4.

SOLUTION

To find the values for the six trigonometric functions of θ, we must first find the value of c, the length of the hypotenuse.

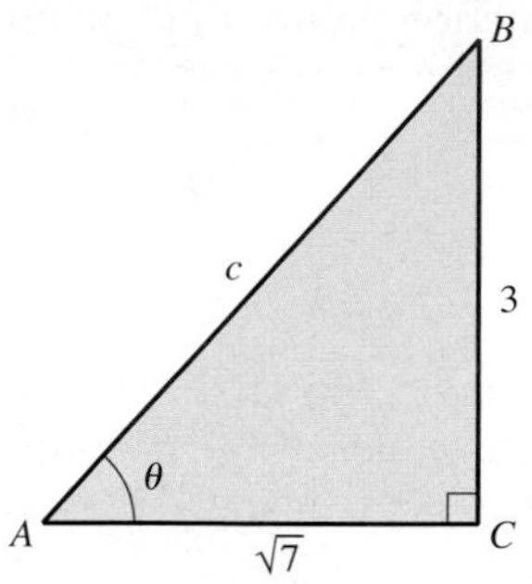

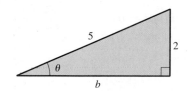

FIGURE 6.4

$$\begin{aligned} a^2 + b^2 &= c^2 && \text{Pythagorean Theorem} \\ (3)^2 + (\sqrt{7})^2 &= c^2 && \text{Replace } a \text{ with 3 and } b \text{ with } \sqrt{7}. \\ 9 + 7 &= c^2 \\ 16 &= c^2 \\ 4 &= c && c \text{ is a positive number.} \end{aligned}$$

Now, with $c = 4$, $a = 3$, and $b = \sqrt{7}$, we have:

$$\sin\theta = \frac{\text{opposite}}{\text{hypotenuse}} = \frac{3}{4} \qquad \csc\theta = \frac{\text{hypotenuse}}{\text{opposite}} = \frac{4}{3}$$

$$\cos\theta = \frac{\text{adjacent}}{\text{hypotenuse}} = \frac{\sqrt{7}}{4} \qquad \sec\theta = \frac{\text{hypotenuse}}{\text{adjacent}} = \frac{4}{\sqrt{7}} = \frac{4\sqrt{7}}{7}$$

$$\tan\theta = \frac{\text{opposite}}{\text{adjacent}} = \frac{3}{\sqrt{7}} = \frac{3\sqrt{7}}{7} \qquad \cot\theta = \frac{\text{adjacent}}{\text{opposite}} = \frac{\sqrt{7}}{3}$$

These are exact values; using a calculator would yield approximate values. ■ ■ ■

Practice Problem 1 Find the exact values for the six trigonometric functions of the acute angle θ in a right triangle if the length of the side opposite θ is 4 and the length of the hypotenuse is 5. ■

EXAMPLE 2 Finding the Remaining Trigonometric Function Values from a Given Value

Find the other five trigonometric function values of θ, given that θ is an acute angle of a right triangle with $\sin\theta = \frac{2}{5}$.

SOLUTION

Because

$$\sin\theta = \frac{2}{5} = \frac{\text{opposite}}{\text{hypotenuse}},$$

FIGURE 6.5

we draw a right triangle with hypotenuse of length 5 and the side opposite θ of length 2. See Figure 6.5. Then

$$\begin{aligned} 5^2 &= 2^2 + b^2 && \text{Pythagorean Theorem} \\ b^2 &= 5^2 - 2^2 && \text{Subtract } 2^2 \text{ from both sides.} \\ b^2 &= 21 && \text{Simplify.} \\ b &= \sqrt{21}. && b \text{ is a positive number.} \end{aligned}$$

For this triangle we have:

$$\begin{aligned} c &= \text{length of the hypotenuse} = 5 \\ a &= \text{length of the opposite side} = 2 \\ b &= \text{length of the adjacent side} = \sqrt{21} \end{aligned}$$

So,

$$\sin\theta = \frac{\text{opposite}}{\text{hypotenuse}} = \frac{2}{5} \qquad \csc\theta = \frac{\text{hypotenuse}}{\text{opposite}} = \frac{5}{2}$$

$$\cos\theta = \frac{\text{adjacent}}{\text{hypotenuse}} = \frac{\sqrt{21}}{5} \qquad \sec\theta = \frac{\text{hypotenuse}}{\text{adjacent}} = \frac{5}{\sqrt{21}} = \frac{5\sqrt{21}}{21}$$

$$\tan\theta = \frac{\text{opposite}}{\text{adjacent}} = \frac{2}{\sqrt{21}} = \frac{2\sqrt{21}}{21} \qquad \cot\theta = \frac{\text{adjacent}}{\text{opposite}} = \frac{\sqrt{21}}{2}$$

■ ■ ■

Practice Problem 2 Find the other five trigonometric function values of θ, given that θ is an acute angle of a right triangle with $\cos \theta = \dfrac{1}{3}$. ▪

Complements

Because θ and $90° - \theta$ are acute angles in the same right triangle, the same six possible ratios of the lengths of the sides are used to compute the values of the trigonometric functions for both angles. Notice that the leg opposite θ is the leg adjacent to $90° - \theta$; so

$$\sin \theta = \frac{\text{opposite to } \theta}{\text{hypotenuse}} = \frac{\text{adjacent to } (90° - \theta)}{\text{hypotenuse}} = \cos(90° - \theta).$$

Similarly, $\cos \theta = \sin(90° - \theta)$ and $\tan \theta = \cot(90° - \theta)$.

Taking reciprocals of these values extends this relationship to the remaining three trigonometric functions.

The prefix *co-* (in *cosine, cotangent,* and *cosecant*) stands for *complement.* In general, for any acute angle θ in a right triangle, the other acute angle is its complement, $90° - \theta$. See Figure 6.6. The pairs of functions—sine and cosine, tangent and cotangent, secant and cosecant—are cofunctions of each other.

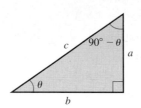

FIGURE 6.6 *a* **is opposite** θ **and adjacent to** $90° - \theta$**.**
b **is adjacent to** θ **and opposite** $90° - \theta$**.**

STUDY TIP

Here is a scheme for remembering the cofunction identities for acute angles A and B with $A + B = 90°$, or $A + B = \dfrac{\pi}{2}$,

complementary angles
↓   ↓
$\sin A = \cos B$
↑   ↑
cofunctions

The sine and cosine may be replaced with any other cofunction pairs.

COMPLEMENTARY RELATIONSHIPS

The value of any trigonometric function of an acute angle θ is equal to the cofunction of the complement of θ. This is true whether θ is measured in degrees or in radians.

θ in degrees

$$\sin \theta = \cos(90° - \theta) \qquad \cos \theta = \sin(90° - \theta)$$
$$\tan \theta = \cot(90° - \theta) \qquad \cot \theta = \tan(90° - \theta)$$
$$\sec \theta = \csc(90° - \theta) \qquad \csc \theta = \sec(90° - \theta)$$

If θ is measured in radians, replace $90°$ with $\dfrac{\pi}{2}$.

EXAMPLE 3   **Finding Trigonometric Function Values of a Complementary Angle**

a. Given that $\cot 68° \approx 0.4040$, find $\tan 22°$.

b. Given that $\cos 72° \approx 0.3090$, find $\sin 18°$.

SOLUTION

a. Note that $68° = 90° - 22°$. So,

$$\tan 22° = \cot(90° - 22°) = \cot 68° \approx 0.4040.$$

b. Here $72° = 90° - 18°$. So,

$$\sin 18° = \cos(90° - 18°) = \cos 72° \approx 0.3090. \qquad ▪▪▪$$

Practice Problem 3

a. Given that $\csc 21° \approx 2.7904$, find $\sec 69°$.

b. Given that $\tan 75° \approx 3.7321$, find $\cot 15°$. ▪

3 Solve right triangles.

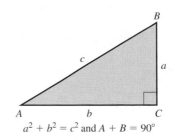

FIGURE 6.7 Labeling a right triangle

$a^2 + b^2 = c^2$ and $A + B = 90°$

Solving Right Triangles

We use right-triangle trigonometry to find unknown lengths in triangles and unknown distances in applications. Using inverse trigonometric functions, we also find unknown angle measures in right triangles as well as unknown angle measures in applied problems. To **solve a triangle** means to find all unknown side lengths and angle measures. In this section, we discuss how to solve right triangles. In the next two sections, we will discuss how to solve triangles that do not contain right angles, which are called *oblique* triangles. When solving right triangles, we solve each triangle ABC, where $\angle C$ is always the right angle and $\angle A$ and $\angle B$ are the acute angles. (We frequently use A as a shorthand way to write measure of $\angle A$.) The side lengths are a, b, and c, where a is the length of the side opposite $\angle A$, and so on. Because $\angle C$ is always the right angle, the length of the hypotenuse is always c and the lengths of the legs are a and b. See Figure 6.7. With this notation, $a^2 + b^2 = c^2$ and $A + B = 90°$.

In applications, degree measure is traditionally used for the angles. Here are some useful facts about solving right triangles:

1. The measure of one angle (the right angle) is automatically given: $C = 90°$.

2. Using the Pythagorean Theorem, the length of the third side of any right triangle can be found if two of the side lengths are known, or if all three side lengths are known, the Pythagorean Theorem can be used as a check.

3. Knowing the measure of one acute angle allows us to find the other because the two acute angles of any right triangle are complementary.

4. We can solve a right triangle if we are given *either* of the following combinations of measurements:
   a. The lengths of any two sides
   b. The length of any one side and the measure of either of the acute angles.

EXAMPLE 4   **Solving a Right Triangle, Given One Acute Angle and One Side**

Solve right triangle ABC if $A = 23°$ and $c = 5.8$.

SOLUTION

Sketch the triangle. See Figure 6.8.

To find a, find a trigonometric function that involves a and the given parts A and c.

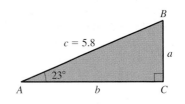

FIGURE 6.8

$$\sin A = \frac{a}{c} \qquad \text{Definition of sine}$$

$$c \sin A = a \qquad \text{Multiply both sides by } c.$$

$$5.8 \sin 23° = a \qquad \text{Substitute 5.8 for } c \text{ and } 23° \text{ for } A.$$

$$a \approx 2.3 \qquad \text{Use a calculator; round to the nearest tenth.}$$

To find b, find a trigonometric function that involves b and the given parts A and c.

$$\cos A = \frac{b}{c} \qquad \text{Definition of cosine}$$

$$c \cos A = b \qquad \text{Multiply both sides by } c.$$

$$5.8 \cos 23° = b \qquad \text{Substitute 5.8 for } c \text{ and } 23° \text{ for } A.$$

$$b \approx 5.3 \qquad \text{Use a calculator; round to the nearest tenth.}$$

Once a and b have been found, the Pythagorean Theorem can be used as a check.
Finally, because $\angle A$ and $\angle B$ are complementary, $B = 90° - 23° = 67°$.   ■ ■ ■

Practice Problem 4   Solve right triangle ABC if $c = 25.8$ and $A = 56°$.   ■

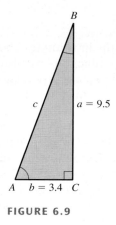

FIGURE 6.9

EXAMPLE 5 **Solving a Right Triangle, Given Two Sides**

Solve right triangle ABC if $a = 9.5$ and $b = 3.4$.

SOLUTION

Sketch the triangle. See Figure 6.9.

To find A, use a trigonometric function that involves A and the given parts a and b.

$$\tan A = \frac{a}{b} \qquad \text{Definition of tangent}$$

$$\tan A = \frac{9.5}{3.4} \qquad \text{Substitute 9.5 for } a \text{ and 3.4 for } b.$$

$$A = \tan^{-1}\left(\frac{9.5}{3.4}\right) \approx 70.3° \qquad \text{Use a calculator; round to one decimal place.}$$

To find c, find a trigonometric function that involves c, A, and a.

$$\sin A = \frac{a}{c} \qquad \text{Definition of sine}$$

$$c = \frac{a}{\sin A} \qquad \text{Multiply both sides by } \frac{c}{\sin A}.$$

$$c \approx \frac{9.5}{\sin 70.3°} \qquad \text{Substitute 9.5 for } a \text{ and 70.3° for } A.$$

$$c \approx 10.1 \qquad \text{Use a calculator; round to the nearest tenth.}$$

Now that all three sides have been found, the Pythagorean Theorem can be used as a check.

Finally, $B = 90° - A$; so $B \approx 90° - 70.3°$, or $B \approx 19.7°$. ■ ■ ■

Practice Problem 5 Solve right triangle ABC if $b = 24.5$ and $c = 36.9$. ■

---

4 Use right-triangle trigonometry in applications.

Applications

Angles that are measured between a line of sight and a horizontal line occur in many applications. If the line of sight is *above* the horizontal line, the angle between these two lines is called the **angle of elevation**. If the line of sight is *below* the horizontal line, the angle between the two lines is called the **angle of depression**.

For example, suppose you are taking a picture of a friend who has climbed to the top of the Arc de Triomphe in Paris. See Figure 6.10. The angle your eyes rotate through as your gaze changes from looking straight ahead to looking at your friend is the angle of elevation.

The angle your friend's eyes rotate through as she changes from looking straight ahead to looking down at you (no pun intended) is the angle of depression.

STUDY TIP

Remember that an angle of elevation or an angle of depression is the angle between the line of sight and a *horizontal* line.

 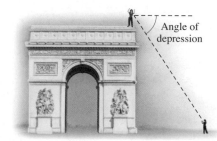

FIGURE 6.10 Angles of elevation and depression

FIGURE 6.11 Height of a cloud

STUDY TIP

When solving a triangle, to find:
Length, use the $\boxed{\text{SIN}}$, $\boxed{\text{COS}}$, and $\boxed{\text{TAN}}$ keys; Angles, use the $\boxed{\text{SIN}^{-1}}$, $\boxed{\text{COS}^{-1}}$, and $\boxed{\text{TAN}^{-1}}$ keys.

EXAMPLE 6   Finding the Height of a Cloud

To determine the height of a cloud at night, a farmer shines a spotlight straight upward to a spot on the cloud. The angle of elevation to this same spot on the cloud from a point located 150 feet horizontally from the spotlight is 68.2°. Find the height of the cloud.

SOLUTION

In Figure 6.11, h represents the height of the cloud, in feet. We have:

$$\tan 68.2° = \frac{h}{150} \qquad \text{Definition of tangent}$$

$$h = 150 \tan 68.2° \qquad \text{Multiply both sides by 150.}$$

$$h \approx 375 \qquad \text{Use a calculator.}$$

The height of the cloud is approximately 375 feet.   ■ ■ ■

Practice Problem 6   From an observation deck 425 feet high on the top of a lighthouse, the angle of depression to a ship at sea is 4.2°. How many miles is the ship from a point at sea level directly below the observation deck?   ■

EXAMPLE 7   Measuring the Height of Mount Kilimanjaro

A surveyor wants to measure the height of Mount Kilimanjaro by using the known height of a nearby mountain. The nearby location is at an altitude of 8720 feet, the distance between that location and Mount Kilimanjaro's peak is 4.9941 miles, and the angle of elevation from the lower location is 23.75°. See Figure 6.12. Use this information to find the approximate height of Mount Kilimanjaro.

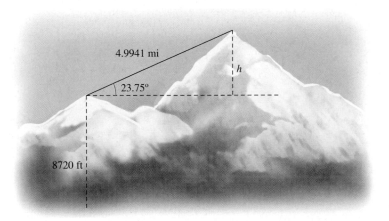

FIGURE 6.12 Height of Mount Kilimanjaro

SOLUTION

The sum of the side length h and the location height of 8720 feet gives the approximate height of Mount Kilimanjaro. Let h be measured in miles. Use the definition of $\sin\theta$, for $\theta = 23.75°$.

$$\sin\theta = \frac{\text{opposite}}{\text{hypotenuse}} = \frac{h}{4.9941}$$

$$h = (4.9941)\sin\theta \qquad \text{Multiply both sides by 4.9941.}$$

$$= (4.9941)\sin 23.75° \qquad \text{Replace } \theta \text{ with 23.75°.}$$

$$h \approx 2.0114 \qquad \text{Use a calculator.}$$

Because 1 mile = 5280 feet,

$$2.0114 \text{ miles} = (2.0114)(5280)$$
$$\approx 10{,}620 \text{ feet}.$$

Thus, the height of Mount Kilimanjaro

$$\approx 10{,}620 + 8720 = 19{,}340 \text{ feet}. \qquad \blacksquare\,\blacksquare\,\blacksquare$$

Practice Problem 7   The height of the nearby location to Mount McKinley, Alaska, is 12,870 feet, its distance to Mount McKinley's peak is 3.3387 miles, and the angle of elevation θ is 25°. What is the approximate height of Mount McKinley?   ■

EXAMPLE 8   Finding the Width of a River

To find the width of a river, a surveyor sights straight across the river from a point A on her side to a point B on the opposite side. See Figure 6.13. She then walks 200 feet upstream to a point C. The angle θ that the line of sight from point C to point B makes with the riverbank is 58°. How wide is the river?

SOLUTION

The points A, B, and C are the vertices of a right triangle with acute angle $\theta = 58°$.
Let w represent the width of the river. From Figure 6.13, we have:

$$\frac{w}{200} = \tan 58° \qquad \tan \theta = \frac{\text{opposite}}{\text{adjacent}}$$

$$w = 200 \tan 58° \qquad \text{Multiply both sides by 200.}$$

$$w \approx 320.07 \text{ feet} \qquad \text{Use a calculator.}$$

The river is about 320 feet wide at the point A.

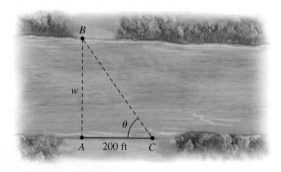

FIGURE 6.13 Width of a river   ■ ■ ■

Practice Problem 8   From a point on the edge of one of two parallel rooftop sides, a point directly opposite is identified on the other rooftop. From a second point 25 feet from the first point along the rooftop edge, a second sighting to the point on the opposite rooftop is made, and the angle θ that this line of sight makes with the rooftop is 60°. How far apart are the two buildings? Round your answer to the nearest foot.

■

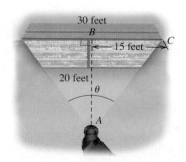

FIGURE 6.14 Security camera

EXAMPLE 9   Finding the Rotation Angle for a Security Camera

A security camera is to be installed 20 feet away from the center of a jewelry counter. The counter is 30 feet long. Through what angle, to the nearest degree, should the camera rotate so that it scans the entire counter? See Figure 6.14.

SOLUTION

The counter center B, the camera A, and one end of the counter C form a right triangle. The angle at vertex A is $\dfrac{\theta}{2}$, where θ is the angle through which the camera rotates. See Figure 6.14. We have:

$$\tan \frac{\theta}{2} = \frac{15}{20} = \frac{3}{4} \qquad \text{Definition of tangent}$$

$$\frac{\theta}{2} = \tan^{-1}\left(\frac{3}{4}\right) \approx 36.87° \qquad \text{Use a calculator in Degree mode.}$$

$$\theta \approx 73.74° \qquad \text{Multiply both sides by 2.}$$

Set the camera to rotate through 74° to scan the entire counter.   ■ ■ ■

Practice Problem 9 Rework Example 9 for a counter that is 20 feet long and a camera set 12 feet from the center of the counter.   ■

SECTION 6.1 ■ Exercises

A EXERCISES Basic Skills and Concepts

1. If θ is an acute angle and $\sin \theta = \dfrac{2\sqrt{2}}{3}$, then $\cos \theta = \underline{\dfrac{1}{3}}$.

2. In a right triangle with sides of lengths 1, 5, and $\sqrt{26}$, the tangent of the angle opposite the side of length 5 is $\underline{\dfrac{5}{}}$.

3. If θ is an acute angle (measured in degrees) in a right triangle and $\cos \theta = 0.7$, then $\sin(90° - \theta) = \underline{0.7}$.

4. *True or False* If θ is an acute angle in a right triangle and $\tan \theta = \dfrac{2}{5}$, then the length of the leg opposite θ is 2.   False

5. *True or False* The leg adjacent to angle θ in a right triangle is the leg opposite the angle $90° - \theta$. True

6. *True or False* The length of the hypotenuse of a right triangle having a leg of length 1 must be greater than 1.   True

In Exercises 7–12, find the exact values for the six trigonometric functions of the angle θ in each figure. Rationalize the denominator where necessary.

7.

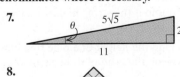

$$\sin\theta = \frac{2\sqrt{5}}{25} \qquad \csc\theta = \frac{5\sqrt{5}}{2}$$
$$\cos\theta = \frac{11\sqrt{5}}{25} \qquad \sec\theta = \frac{5\sqrt{5}}{11}$$
$$\tan\theta = \frac{2}{11} \qquad \cot\theta = \frac{11}{2}$$

8.

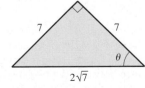

$$\sin\theta = \frac{\sqrt{7}}{2} \qquad \csc\theta = \frac{2\sqrt{7}}{7}$$
$$\cos\theta = \frac{\sqrt{7}}{2} \qquad \sec\theta = \frac{2\sqrt{7}}{7}$$
$$\tan\theta = 1 \qquad \cot\theta = 1$$

9.

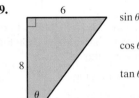

$$\sin\theta = \frac{3}{5} \qquad \csc\theta = \frac{5}{3}$$
$$\cos\theta = \frac{4}{5} \qquad \sec\theta = \frac{5}{4}$$
$$\tan\theta = \frac{3}{4} \qquad \cot\theta = \frac{4}{3}$$

10.

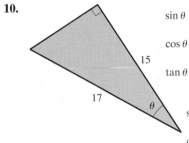

$$\sin\theta = \frac{8}{17} \qquad \csc\theta = \frac{17}{8}$$
$$\cos\theta = \frac{15}{17} \qquad \sec\theta = \frac{17}{15}$$
$$\tan\theta = \frac{8}{15} \qquad \cot\theta = \frac{15}{8}$$

11.

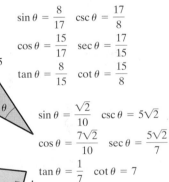

$$\sin\theta = \frac{\sqrt{2}}{10} \qquad \csc\theta = 5\sqrt{2}$$
$$\cos\theta = \frac{7\sqrt{2}}{10} \qquad \sec\theta = \frac{5\sqrt{2}}{7}$$
$$\tan\theta = \frac{1}{7} \qquad \cot\theta = 7$$

12.

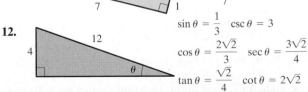

$$\sin\theta = \frac{1}{3} \qquad \csc\theta = 3$$
$$\cos\theta = \frac{2\sqrt{2}}{3} \qquad \sec\theta = \frac{3\sqrt{2}}{4}$$
$$\tan\theta = \frac{\sqrt{2}}{4} \qquad \cot\theta = 2\sqrt{2}$$

In Exercises 13–18, use each given trigonometric function value of θ to find the five other trigonometric function values of the acute angle θ. Rationalize the denominators where necessary.

13. $\cos\theta = \dfrac{2}{3}$ †   **14.** $\sin\theta = \dfrac{3}{4}$ †   **15.** $\tan\theta = \dfrac{5}{3}$ †

16. $\cot\theta = \dfrac{6}{11}$ †   **17.** $\sec\theta = \dfrac{13}{12}$ †   **18.** $\csc\theta = 4$ †

†Due to space constrictions, answers to these exercises may be found in the Answers beginning on page A–1 in the back of the book.

In Exercises 19–24, find the trigonometric function values of each corresponding complementary angle.

19. Given that $\sin 58° \approx 0.8480$, find $\cos 32°$.   0.8480

20. Given that $\cos 37° \approx 0.7986$, find $\sin 53°$.   0.7986

21. Given that $\tan 27° \approx 0.5095$, find $\cot 63°$.   0.5095

22. Given that $\cot 49° \approx 0.8693$, find $\tan 41°$.   0.8693

23. Given that $\sec 65° \approx 2.3662$, find $\csc 25°$.   2.3662

24. Given that $\csc 78° \approx 1.0223$, find $\sec 12°$.   1.0223

In Exercises 25–32, solve each right triangle; angle C is the right angle. Round to the nearest tenth where necessary.

25. $A = 50°, c = 9.2$ $B = 40°, a \approx 7.0, b \approx 5.9$

26. $B = 28°, c = 10.5$ $A = 62°, a \approx 9.3, b \approx 4.9$

27. $A = 36°, a = 12.0$ $B = 54°, c \approx 20.4, b \approx 16.5$

28. $B = 74°, b = 6.3$ $A = 16°, a \approx 1.8, c \approx 6.6$

29. $a = 12.5, b = 6.2$ $c \approx 14.0, A \approx 63.6°, B \approx 26.4°$

30. $a = 4.3, b = 8.1$ $c \approx 9.2, A \approx 28°, B \approx 62°$

31. $b = 9.4, c = 14.5$ $a \approx 11, A \approx 49.6°, B \approx 40.4°$

32. $a = 6.2, c = 18.7$ $b \approx 17.6, A \approx 19.4°, B \approx 70.6°$

In Exercises 33–39, use the figure and the given values to find each specified side length, angle, and trigonometric function value. Round your answer to three decimal places.

33. $a = 8, b = 10$ Find $c, \sin\theta$, and $\tan\theta$.

34. $a = 18, b = 3$ Find $c, \sin\theta$, and $\cos\theta$.

35. $a = 23, b = 7$ Find $\cos\theta$ and $\tan\theta$.

36. $a = 19, b = 27$ Find $c, \cos\theta$, and $\tan\theta$.

37. $\theta = 30°, a = 9$ Find $\sin\theta, b$, and c.

38. $\theta = 30°, a = 5$ Find b and c.

39. $\theta = 60°, a = 7$ Find α, b, and c.

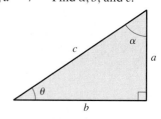

B EXERCISES Applying the Concepts

40. Positioning a ladder. An 18-foot ladder is resting against a wall of a building in such a way that the top of the ladder is 14 feet above the ground. How far is the foot of the ladder from the base of the building? ≈ 11 feet

41. Positioning a ladder. The ladder from Exercise 40 is placed against a wall so that the foot of the ladder is 7 feet from the base of the wall. How high up the wall will the ladder reach? ≈ 17 feet

In Exercises 42 and 43, use the following definitions: The *vertical rise* of a ski lift is the vertical distance traveled when a lift car goes from the bottom terminal to the top terminal. The *slope length* is the linear distance a lift car

travels as it moves from the bottom terminal to the top terminal. We treat the slope length as the hypotenuse and the vertical rise as the side opposite the angle formed by the horizontal line at the top terminal and the lift cables.

42. Ski lift dimensions. Find the horizontal distance a lift car passes over if the vertical rise is 295 meters and the slope length is 1070 meters. ≈ 1029 meters

43. Ski lift dimensions. Find the vertical rise (to the nearest foot) for a ski lift with slope length 2050 feet if the angle formed by a horizontal line at the top terminal and the lift cable is 16°. ≈ 565 feet

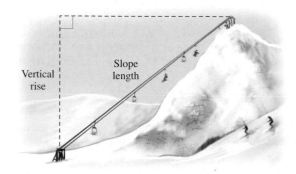

44. Ski lift height. If a skier travels 5000 feet up a ski lift whose angle of inclination is 30°, how high above his starting level is he?   2500 feet

45. Shadow length. At a time when the sun's rays make a 50° angle with the horizontal, how long is the shadow cast by a 6-foot man standing on flat ground? ≈ 5 feet

46. Height of a tree. The angle relative to the horizontal from the top of a tree to a point 110 feet from its base (on flat ground) is 15°. Find the height of the tree to the nearest foot. ≈ 29 feet

47. Height of a cliff. The angle of elevation from a rowboat moored 75 feet from a cliff is 73.8°. Find the height of the cliff to the nearest foot. ≈ 258 feet

48. Height of a flagpole. A flagpole is supported by a 30-foot tension wire attached to the flagpole and to a stake in the ground. If the ground is flat and the angle the wire makes with the ground is 40°, how high up the flagpole is the wire attached? Round your answer to the nearest foot. ≈ 19 feet

49. Flying a kite. A girl 5 feet tall is flying a kite. How long must the string be in order for her to raise the kite 200 feet above the ground if the string makes a $28°4'$ angle with the ground? ≈ 414 feet

50. Height of a shadow. Find the height of a pine tree that casts a 93-foot shadow on the ground if the angle of elevation of the sun is $24°35'$. ≈ 43 feet

51. Width of a river. Sal and his friend are on opposite sides of a river. To find the width of the river, each of them hammered a stake on his bank of the river in such a way that the distance between the two stakes approximates the width of the river. Sal walks 10 feet away from his stake along the river and finds that the

Answers:

33. $c \approx 12.806, \sin\theta \approx 0.625, \tan\theta \approx 0.781$
34. $c \approx 18.248, \sin\theta \approx 0.986, \cos\theta \approx 0.164$

35. $\cos\theta \approx 0.291, \tan\theta \approx 3.286$
36. $c \approx 33.015, \cos\theta \approx 0.818, \tan\theta \approx 0.704$
37. $\sin\theta = 0.5, b \approx 15.588, c = 18$ **38.** $b \approx 8.660, c = 10$
39. $\alpha = 30°, b \approx 4.041, c \approx 8.083$

line of sight from his new position to his friend's stake across the river and the line of sight to his own stake form a 73°34′ angle. How wide is the river? ≈ 34 feet

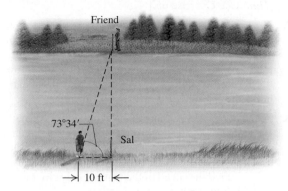

Friend

73°34′

Sal

10 ft

52. **Measuring the Empire State Building.** From a neighboring building's 13th-floor window that is 128 meters above the ground, the angle of elevation of the top of the Empire State Building is 59°20′, while the angle of depression of the base is 40°29′. How far away from the neighboring building and how tall is the Empire State Building? †

53. **Sighting a mortar station.** The angle of depression from a helicopter 3000 feet directly above an allied mortar station to an enemy supply depot is 13°. How far is the depot from the mortar station? ≈ 12,994 feet

54. **Distance between Earth and the moon.** If the radius of Earth is 3960 miles and angle C (the latitude of C) is 89°3′, find the distance between the center of Earth and the center of the moon. ≈ 238,844 miles

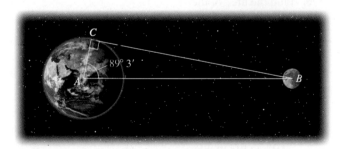

C

89° 3′

A

B

55. **Sprinkler rotation.** A sprinkler rotates back and forth through an angle θ, as shown in the figure. At a distance of 5 feet from the sprinkler, the rays that form the sides of angle θ are 6 feet apart. Find θ. 62°

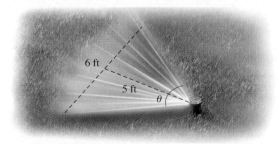

6 ft

5 ft

θ

56. **Irradiating flowers.** A tray of flowers is being irradiated by a beam from a rotating lamp, as shown in the figure. If the tray is 8 feet long and the lamp is 2 feet from the center of the tray, through what angle

should the lamp rotate to irradiate the full length of the tray? 127°

2 ft   θ

8 ft

In Exercises 57 and 58, use the figure to find the radius r for the given latitudes (values of θ). Use 3960 miles for the radius of Earth.

57. **Latitude.** Find the radius of the 50th parallel (of latitude). ≈ 2545 miles

58. **Latitude.** Find the radius of the 53rd parallel (of latitude). ≈ 2383 miles

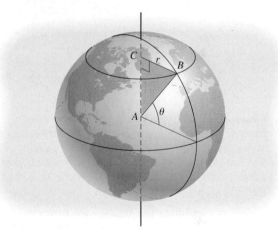

C r B

A θ

59. **Motorcycle racing.** A video camera is set up 110 feet at a right angle to a straight quarter-mile racetrack, as shown in the figure. The starting line is to the left, and the finish is to the right. Through what angle must the camera rotate to film the entire race? 159°

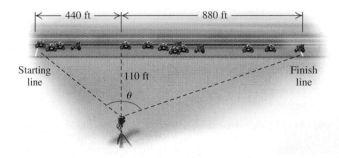

440 ft   880 ft

Starting line

110 ft

θ

Finish line

60. **Camera's viewing angle.** The viewing angle for the 35-millimeter camera is given (in degrees) by

$\theta = 2\tan^{-1}\dfrac{18}{x}$, where x is the focal length of the lens. The focal length on most adjustable cameras is marked in millimeters on the lens mount.

a. Find the viewing angle, in degrees, if the focal length is 50 millimeters. 39.6°

b. Find the viewing angle, in degrees, if the focal length is 200 millimeters. 10.3°

C EXERCISES Beyond the Basics

61. **Height of an airplane.** From a tower 150 feet high with the sun directly overhead, an airplane and its shadow have an angle of elevation of 67.4° and an angle of depression of 8.1°, respectively. Find the height of the airplane.
    ≈ 2682 feet

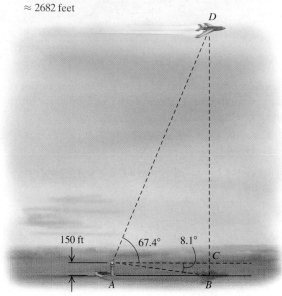

62. **Camera angle.** Raphael stands 12 feet from a large painting on a wall. He sets his camera 5 feet 6 inches above the floor. If the bottom edge of the painting is 4 feet above the floor and the angle of elevation from the camera to the top of the painting is 50°, find the height of the painting to the nearest tenth of a foot. ≈ 15.8 feet

63. **Height of a building.** The WCNC-TV Tower in Dallas, North Carolina, is 600 meters high. Suppose a building is erected such that the base of the building is on the same plane as the base of the tower, the angle of elevation from the top of the building to the top of the tower is 75.24°, and the angle of depression from the top of the building to the foot of the tower is 60.05°. How high would the building have to be? (*Source:* http://en.wikipedia.org) ≈ 188.27 meters

64. **Height of a building.** Suppose from the top of the Empire State Building (which is about 1250 feet high) the angle of depression to the top of a second building is 55.8° and the angle of depression to the bottom of the second building is 67.7°. Find the height of the second building. ≈ 496 feet

65. In the figure below, show that $h = \dfrac{d}{\cot \alpha - \cot \beta}$.

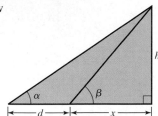

66. **Height of a plane.** A plane flies directly over two tracking stations that are 500 meters apart on a level plain. The plane flies in the direction from Station A to Station B. After the plane has passed Station B, an observer at Station B records the angle of elevation as 38.9°. At the same time, an observer at station A records the angle of elevation as 36.9°. What is the plane's altitude? [*Hint:* Use Exercise 65.] ≈ 5402 meters

67. **Measuring the Statue of Liberty.** While sailing toward the Statue of Liberty, a sailor in a boat observed that at a certain point, the angle of elevation of the tip of the torch was 25°. After sailing another 100 meters toward the statue, the angle of elevation became 41°50′. How tall is the Statue of Liberty? ≈ 97 meters

68. **Speed of a plane.** A plane flying at an altitude of 12,000 feet finds the angle of depression for a building ahead of the plane to be 10.4°. With the building still straight ahead, two minutes later the angle of depression is 25.6°. Find the speed of the plane relative to the ground (in feet per minute). ≈ 20,168.5 feet per minute

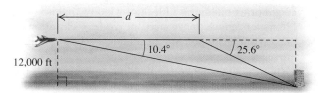

69. In the figure below, show that $h = \dfrac{d}{\cot \alpha + \cot \beta}$.

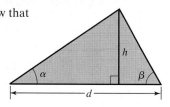

70. Find the area of the triangle in the figure. ≈ 20 square units

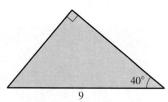

[*Hint:* Use Exercise 69.]

71. Find the area of triangle ABC in the figure. ≈ 48 square units

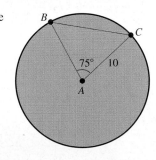

Critical Thinking

72. Draw a right triangle with acute angle θ. Then use the definitions of $\sin \theta$ and $\cos \theta$ together with the Pythagorean Theorem to show that $\sin^2 \theta + \cos^2 \theta = 1$.

73. In a right triangle with acute angle θ, what value must the adjacent side have for the length of the opposite side to represent $\tan \theta$? 1

74. Use the result of Exercise 73 to compare $\tan \alpha$ and $\tan \beta$ if α and β are acute angles and the measure of α is less than the measure of β. $\tan \beta > \tan \alpha$

The Law of Sines

Before Starting this Section, Review

1. Trigonometric functions (Section 4.2, page 268)
2. Inverse functions (Section 1.7, page 90)
3. Geometric properties of a triangle (Appendix A, page 789)

Objectives

1 Learn vocabulary and conventions for solving triangles.

2 Derive the Law of Sines.

3 Solve AAS and ASA triangles by using the Law of Sines.

4 Solve for possible triangles in the ambiguous SSA case.

5 Find the area of an SAS triangle.

Mount Everest

In 1852, an unsung hero Radhanath Sickdhar managed to *calculate* the height of Peak XV, an icy peak in the Himalayas. The highest mountain in the world—later named Mount Everest—stood at 29,002 feet. Sickdhar's calculations used triangulations as well as the phenomenon called refraction—the bending of light rays by the density of Earth's atmosphere. Like George Everest, Sickdhar may have never seen Mount Everest.

THE GREAT TRIGONOMETRIC SURVEY OF INDIA

Triangulation is the process of finding the coordinates and the distance to a point by using the Law of Sines. Triangulation was used in surveying the Indian subcontinent. The survey, which was called "one of the most stupendous works in the whole history of science," was begun by William Lambton, a British army officer, in 1802. It started in the south of India and extended north to Nepal, a distance of approximately 1600 miles. Lasting several decades and employing thousands of workers, the survey was named the Great Trigonometric Survey (GTS). In 1818, George Everest (1790–1866), a Welsh surveyor and geographer, was appointed assistant to Lambton. Everest, who succeeded Lambton in 1823 and was later promoted as surveyor general of India, completed the GTS. Mount Everest was surveyed and named after Everest by his successor Andrew Waugh. In Example 2, we use the Law of Sines to estimate the height of a mountain. ■

1 Learn vocabulary and conventions for solving triangles.

Solving Oblique Triangles

The process of finding the unknown side lengths and angle measures in a triangle is called *solving a triangle*. In Section 6.1, you learned techniques of solving right triangles. We now discuss triangles that are not necessarily right triangles. A triangle with no right angle is called an **oblique triangle**. From geometry, we know that two triangles with any of the following sets of equal parts will always be **congruent**. In other words, the following sets of given parts determine a unique triangle:

ASA: Two angles and the included side

AAS: Two angles and a nonincluded side

SAS: Two sides and the included angle

SSS: All three sides

Although **SSA** (two sides and a nonincluded angle) does not guarantee a unique triangle, there can be, at most, two triangles with the given parts. On the other hand, **AAA** (or **AA**) guarantees only *similar* triangles, not congruent ones, so there are infinitely many triangles with the same three angle measures.

For any triangle ABC, we let each letter such as A stand for both the vertex A and the measure of the angle at A. As usual, the angles of a triangle ABC are labeled A, B, and C and the lengths of their opposite sides are labeled, a, b, and c, respectively. See Figure 6.15.

RECALL

Two triangles are *congruent* if corresponding sides are congruent (equal in length) and corresponding angles are congruent (equal in measure). An *included side* is the side between two angles, and an *included angle* is the angle between two sides.

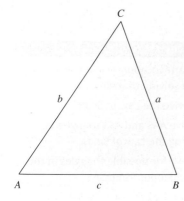

FIGURE 6.15 Labeling a triangle

2 Derive the Law of Sines.

RECALL

An *altitude* of a triangle is a segment drawn from any vertex of the triangle perpendicular to the opposite side, or to an extension of the opposite side.

To solve an oblique triangle, given at least one side and two other measures, we consider four possible cases:

Case 1. Two angles and a side are known (**AAS** and **ASA** triangles).

Case 2. Two sides and an angle opposite one of the given sides are known (**SSA** triangles).

Case 3. Two sides and their included angle are known (**SAS** triangles).

Case 4. All three sides are known (**SSS** triangles).

We use the Law of Sines to analyze triangles in Cases 1 and 2. In the next section, we use the *Law of Cosines* to solve triangles in Cases 3 and 4.

The Law of Sines

The Law of Sines states that in a triangle ABC with sides of length a, b, and c,

$$\frac{\sin A}{a} = \frac{\sin B}{b} = \frac{\sin C}{c}.$$

We can derive the Law of Sines by using the right triangle definition of the sine of an acute angle θ:

$$\sin \theta = \frac{\text{opposite}}{\text{hypotenuse}}$$

We begin with an oblique triangle ABC. See Figures 6.16 and 6.17. Draw altitude CD from the vertex C to the side AB (or its extension, as in Figure 6.17). Let h be the length of segment CD.

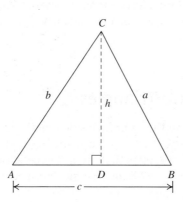

FIGURE 6.16 Altitude *CD* with acute ∠*B*

In right triangle CDA,

$$\frac{h}{b} = \sin A \quad \text{or}$$

$$h = b \sin A.$$

In right triangle CDB,

$$\frac{h}{a} = \sin B \quad \text{or}$$

$$h = a \sin B.$$

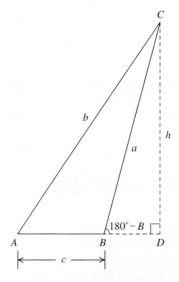

FIGURE 6.17 Altitude *CD* with obtuse ∠*B*

In right triangle CDA,

$$\frac{h}{b} = \sin A \quad \text{or}$$

$$h = b \sin A.$$

In right triangle CDB,

$$\frac{h}{a} = \sin(180° - B) = \sin B \quad \text{or}$$

$$h = a \sin B.$$

Since in each figure $h = b \sin A$ and $h = a \sin B$, we have

$$b \sin A = a \sin B$$

$$\frac{b \sin A}{ab} = \frac{a \sin B}{ab} \qquad \text{Divide both sides by } ab.$$

$$\frac{\sin A}{a} = \frac{\sin B}{b} \qquad \text{Simplify.}$$

Similarly, by drawing altitudes from the other two vertices, we can show that

$$\frac{\sin B}{b} = \frac{\sin C}{c} \quad \text{and} \quad \frac{\sin C}{c} = \frac{\sin A}{a}.$$

We have proved the Law of Sines.

THE LAW OF SINES

In any triangle ABC, with sides of length a, b, and c,

$$\frac{\sin A}{a} = \frac{\sin B}{b}, \quad \frac{\sin B}{b} = \frac{\sin C}{c}, \quad \text{and} \quad \frac{\sin C}{c} = \frac{\sin A}{a}.$$

We can rewrite these relations in compact notation:

$$\frac{\sin A}{a} = \frac{\sin B}{b} = \frac{\sin C}{c}, \quad \text{or equivalently,} \quad \frac{a}{\sin A} = \frac{b}{\sin B} = \frac{c}{\sin C}.$$

---

3 Solve AAS and ASA triangles by using the Law of Sines.

Solving AAS and ASA Triangles

We next learn to solve AAS and ASA triangles of Case 1, where two angles and a side are given. Note that if two angles of a triangle are given, then you also know the third angle (since the sum of the angles of a triangle is 180°). So in Case 1, all three angles and one side of a triangle are known.

FINDING THE SOLUTION: A PROCEDURE

| EXAMPLE 1 | Solving the AAS and ASA Triangles |

OBJECTIVE
Solve a triangle given two angles and a side.

EXAMPLE
Solve triangle ABC with $A = 62°$, $c = 14$ feet, $B = 74°$.
Round side lengths to the nearest tenth.

Step 1 **Find the third angle.** Find the measure of the third angle by subtracting the measures of the known angles from 180°.

1. $C = 180° - A - B$
   $= 180° - 62° - 74°$
   $= 44°$

Step 2 **Make a chart.** Make a chart of the six parts of the triangle, including the known and the unknown parts. Sketch the triangle.

2.

$A = 62°$	$a = ?$
$B = 74°$	$b = ?$
$C = 44°$	$c = 14$

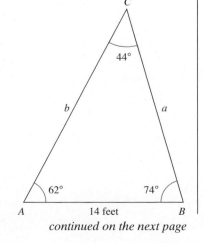

continued on the next page

Step 3   Apply the Law of Sines. Select two ratios from the Law of Sines in which three of the four quantities are known. Solve for the fourth quantity. Use the form of the Law of Sines in which the unknown quantity is in the numerator.

3.

$$\frac{b}{\sin B} = \frac{c}{\sin C} \qquad\qquad \frac{a}{\sin A} = \frac{c}{\sin C}$$

$$\frac{b}{\sin 74°} = \frac{14}{\sin 44°} \qquad\qquad \frac{a}{\sin 62°} = \frac{14}{\sin 44°}$$

$$b = \frac{14 \sin 74°}{\sin 44°} \qquad\qquad a = \frac{14 \sin 62°}{\sin 44°}$$

$$b \approx 19.4 \text{ feet} \qquad\qquad a \approx 17.8 \text{ feet}$$

Step 4   Show the solution. Show the solution by completing the chart.

4.

$A = 62°$	$a \approx 17.8$ feet
$B = 74°$	$b \approx 19.4$ feet
$C = 44°$	$c = 14$

Practice Problem 1   Solve triangle ABC with $A = 70°$, $B = 65°$, and $a = 16$ inches. Round side lengths to the nearest tenth.   ■

EXAMPLE 2   Height of a Mountain

From a point on a level plain at the foot of a mountain, a surveyor finds the angle of elevation of the peak of the mountain to be 20°. She walks 3465 meters closer (on a direct line between the first point and the base of the mountain) and finds the angle of elevation to be 23°. Estimate the height of the mountain to the nearest meter.

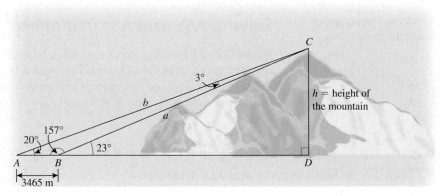

FIGURE 6.18 Height of a mountain

SOLUTION

In Figure 6.18, consider triangle ABC.

$$\angle ABC = 180° - 23° = 157°$$

Find the third angle in triangle ABC.

$$C = 180° - 20° - 157° = 3°$$

Apply the Law of Sines.

$$\frac{a}{\sin 20°} = \frac{3465}{\sin 3°}$$

$$a = \frac{3465 \sin 20°}{\sin 3°} \approx 22{,}644 \text{ meters}$$

Now in triangle BCD, $\sin 23° = \dfrac{h}{a}$. So, $h = a \sin 23° \approx 22{,}644 \sin 23° \approx 8848$ meters.

The mountain is approximately 8848 meters high.   ■ ■ ■

Practice Problem 2 From a point on a level plain at the foot of a mountain, the angle of elevation of the peak is 40°. If you move 2500 feet farther (on a direct line extending from the first point and the base of the mountain), the angle of elevation of the peak is 35°. How high above the plain is the peak? Round your answer to the nearest foot.   ■

4 Solve for possible triangles in the ambiguous SSA case.

Solving SSA Triangles—the Ambiguous Case

In Case 1, where we know two angles and a side, we obtain a unique triangle. However, if we know the lengths of two sides and the measure of the angle opposite one of these sides, then we could have a result that is (1) not a triangle, (2) exactly one triangle, or (3) two different triangles. For this reason, Case 2 is called the **ambiguous case**.

Suppose we want to draw triangle ABC with given measures of A, a, and b. We draw angle A with a horizontal initial side, and we place the point C on the terminal side of A so that the length of segment AC is b. We need to locate point B on the initial side of A so that the measure of the segment CB is a. See Figure 6.19.

The number of possible triangles depends upon the length h of the altitude from C to the initial side of angle A. Since $\sin A = \dfrac{h}{b}$, we have $h = b \sin A$. Various possibilities for forming triangle ACB are illustrated in Figure 6.20 (where A is an acute angle) and Figure 6.21 (where A is an obtuse angle).

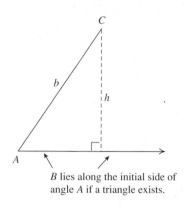

B lies along the initial side of angle A if a triangle exists.

FIGURE 6.19 Need to locate B

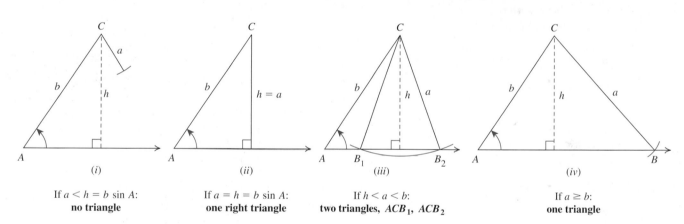

If $a < h = b \sin A$:
no triangle

If $a = h = b \sin A$:
one right triangle

If $h < a < b$:
two triangles, ACB_1, ACB_2

If $a \geq b$:
one triangle

FIGURE 6.20 Acute angle A

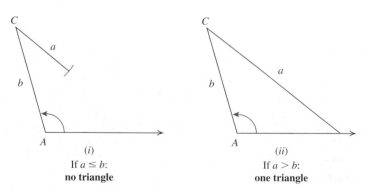

If $a \leq b$:
no triangle

If $a > b$:
one triangle

FIGURE 6.21 Obtuse angle A

Rather than memorize all of the cases involved in solving SSA triangles, use the following procedure.

FINDING THE SOLUTION: A PROCEDURE

EXAMPLE 3   Solving the SSA Triangles

OBJECTIVE

Solve a triangle if two sides and an angle opposite one of them is given.

Step 1 **Make a chart** of the six parts, the known and the unknown parts.

Step 2 **Apply the Law of Sines** to the two ratios in which three of the four quantities are known. Solve for the sine of the fourth quantity. Use the form of the Law of Sines in which the unknown quantity is in the numerator.

Step 3 If the sine of the angle, say θ, in Step 2 is greater than 1, there is no triangle with the given measurements. If $\sin\theta$ is between 0 and 1, go to Step 4.

Step 4 Let $\sin\theta$ be x, with $0 < x \leq 1$. If $x \neq 1$, then θ has two possible values:
  (i) $\theta_1 = \sin^{-1}x$, so $0 < \theta_1 < 90°$
  (ii) $\theta_2 = 180° - \sin^{-1}x$.

Step 5 If $x \neq 1$, with (known angle) $+ \theta_1 < 180°$ and (known angle) $+ \theta_2 < 180°$, then there are two triangles. Otherwise, there is only one triangle, and if $x = 1$, it is a right triangle.

Step 6 Find the third angle of the triangle(s).

Step 7 Use the Law of Sines to find the remaining side(s).

Step 8 Show the solution(s).

EXAMPLE

Solve triangle ABC with $B = 32°$, $b = 100$ feet, and $c = 150$ feet. Round each answer to the nearest tenth.

1.

$A = ?$	$a = ?$
$B = 32°$	$b = 100$
$C = ?$	$c = 150$

Three quantities are known.

2. $\dfrac{\sin C}{c} = \dfrac{\sin B}{b}$ The Law of Sines

$\sin C = \dfrac{c \sin B}{b}$ Multiply both sides by c.

$\qquad = \dfrac{(150)\sin 32°}{100}$ Substitute values.

$\sin C \approx 0.7949$ Use a calculator.

4. Two possible values of C:

$$C_1 \approx \sin^{-1}(0.7949) \approx 52.6°$$
$$C_2 \approx 180° - 52.6° = 127.4°$$

5. $B + C_1 = 32° + 52.6° = 84.6° < 180°$, and
$B + C_2 = 32° + 127.4° = 159.4° < 180°$.
We have two triangles with the given measurements.

6. $\angle BAC_1 \approx 180° - 32° - 52.6° = 95.4°$
$\angle BAC_2 \approx 180° - 32° - 127.4° = 20.6°$

7. $\dfrac{a_1}{\sin\angle BAC_1} = \dfrac{b}{\sin B}$ $\qquad$ $\dfrac{a_2}{\sin\angle BAC_2} = \dfrac{b}{\sin B}$

$a_1 = \dfrac{b\sin\angle BAC_1}{\sin B}$ $\qquad$ $a_2 = \dfrac{b\sin\angle BAC_2}{\sin B}$

$a_1 = \dfrac{100\sin 95.4°}{\sin 32°}$ $\qquad$ $a_2 = \dfrac{(100)\sin 20.6°}{\sin 32°}$

$a_1 \approx 187.9$ feet $\qquad$ $a_2 \approx 66.4$ feet

8. $\triangle BAC_1$ $\qquad\qquad\qquad$ $\triangle BAC_2$

$\angle BAC_1 \approx 95.4°$	$a_1 \approx 187.9$
$B = 32°$	$b = 100$
$C_1 \approx 52.6°$	$c = 150$

$\angle BAC_2 \approx 20.6°$	$a_2 \approx 66.4$
$B = 32°$	$b = 100$
$C_2 \approx 127.4°$	$c = 150$

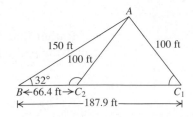

■ ■ ■

Practice Problem 3 Solve triangle ABC with $C = 35°$, $b = 15$ feet, and $c = 12$ feet. Round each answer to the nearest tenth. ■

EXAMPLE 4 **Solving an SSA Triangle (No Solution)**

Solve triangle ABC with $A = 50°$, $a = 8$ inches, and $b = 15$ inches.

SOLUTION

Step 1 **Make a chart.**

$A = 50°$	$a = 8$
$B = ?$	$b = 15$
$C = ?$	$c = ?$

←— Three quantities are known.

Step 2 **Apply the Law of Sines.**

$$\frac{\sin B}{b} = \frac{\sin A}{a} \qquad \text{The Law of Sines}$$

$$\sin B = \frac{b \sin A}{a} \qquad \text{Multiply both sides by } b.$$

$$\sin B = \frac{(15) \sin 50°}{8} \qquad \text{Substitute values.}$$

$$\sin B \approx 1.44 \qquad \text{Use a calculator.}$$

Step 3 Since $\sin B \approx 1.44 > 1$, we conclude that no triangle has the given measures. ■ ■ ■

Practice Problem 4 Solve triangle ABC with $A = 65°$, $a = 16$ meters, and $b = 30$ meters. ■

EXAMPLE 5 **Solving an SSA Triangle (One Solution)**

Solve triangle ABC with $C = 40°$, $c = 20$ meters, and $a = 15$ meters. Round each answer to the nearest tenth.

SOLUTION

Step 1 **Make a chart.**

$A = ?$	$a = 15$
$B = ?$	$b = ?$
$C = 40°$	$c = 20$

Three quantities are known.

Step 2 **Apply the Law of Sines.**

$$\frac{\sin A}{a} = \frac{\sin C}{c} \qquad \text{The Law of Sines}$$

$$\sin A = \frac{a \sin C}{c} \qquad \text{Multiply both sides by } a.$$

$$= \frac{(15) \sin 40°}{20} \qquad \text{Substitute values.}$$

$$\sin A \approx 0.4821 \qquad \text{Use a calculator.}$$

Step 4 $A = \sin^{-1}(0.4821) \approx 28.8°$. Two possible values of A are $A_1 \approx 28.8°$ and $A_2 \approx 180° - 28.8° = 151.2°$.

Step 5   Since $C + A_2 = 40° + 151.2° = 191.2° > 180°$, there is no triangle with vertex A_2. Thus, only one triangle has measure of angle $A_1 \approx 28.8°$.

Step 6   The third angle at B has measure $\approx 180° - 40° - 28.8° = 111.2°$.

Step 7   **Find the remaining side length.**

$$\frac{b}{\sin B} = \frac{c}{\sin C} \qquad \text{The Law of Sines}$$

$$b = \frac{c \sin B}{\sin C} \qquad \text{Multiply both sides by } \sin B.$$

$$= \frac{(20) \sin (111.2°)}{\sin 40°} \qquad \text{Substitute values.}$$

$$b \approx 29.0 \text{ m} \qquad \text{Use a calculator.}$$

Step 8   **Show the solution.** See Figure 6.22.

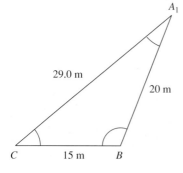

$A_1 \approx 28.8°$	$a = 15$ meters
$B \approx 111.2°$	$b \approx 29.0$ meters
$C = 40°$	$c = 20$ meters

FIGURE 6.22   ■ ■ ■

Practice Problem 5   Solve triangle ABC with $C = 60°$, $c = 50$ feet, and $a = 30$ feet.
                                                              ■

Bearings

In navigation and surveying, directions are usually given by using *bearings*. A **bearing** is the measure of an acute angle from due north or due south. The bearing N 40° E means 40° to the east of due north, whereas S 30° W means 30° to the west of due south. See Figure 6.23.

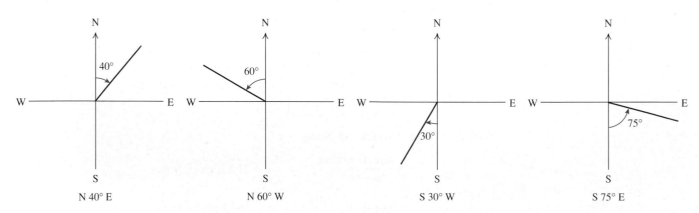

FIGURE 6.23 Bearings

| **EXAMPLE 6** | **Navigation Using Bearings** |

A ship sailing due west at 20 miles per hour records the bearing of an oil rig at N 55.4° W. An hour and a half later the bearing of the same rig is N 66.8° E.

a. How far is the ship from the oil rig the second time?

b. How close did the ship pass to the oil rig?

SOLUTION

a. In an hour and a half, the ship travels $(1.5)(20) = 30$ miles due west. The oil rig (C), the starting point for the ship (A), and the position of the ship after an hour and a half (B) form the vertices of the triangle ABC shown in Figure 6.24.

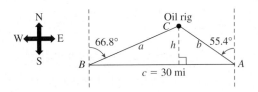

FIGURE 6.24 Navigation

Then $A = 90° - 55.4° = 34.6°$ and $B = 90° - 66.8° = 23.2°$. Therefore, $C = 180° - 34.6° - 23.2° = 122.2°$.

$$\frac{a}{\sin A} = \frac{c}{\sin C} \qquad \text{The Law of Sines}$$

$$a = \frac{c \sin A}{\sin C} \qquad \text{Multiply both sides by } \sin A.$$

$$= \frac{30 \sin 34.6°}{\sin 122.2°} \approx 20 \text{ miles} \qquad \begin{array}{l}\text{Substitute values and use a}\\\text{calculator.}\end{array}$$

The ship is approximately 20 miles from the oil rig when the second bearing is taken.

b. The shortest distance between the ship and the oil rig is the length of the segment h in triangle ABC.

$$\sin B = \frac{h}{a} \qquad \text{Right-triangle definition of the sine}$$

$$h = a \sin B \qquad \text{Multiply both sides by } a.$$

$$h \approx 20 \sin 23.2° \qquad \text{Substitute values.}$$

$$\approx 7.9 \qquad \text{Use a calculator.}$$

The ship passes within 7.9 miles of the oil rig.   ▪ ▪ ▪

Practice Problem 6   Rework Example 6 assuming that the second bearing is taken after two hours and is N 60.4° E and the ship is traveling due west at 25 miles per hour.   ▪

5 Find the area of an SAS triangle.

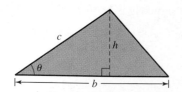

FIGURE 6.25 **Acute angle θ**

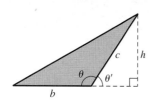

FIGURE 6.26 **Obtuse angle θ**

Area of a Triangle

Recall that the area K of a triangle is given by the formula

$$\text{Area} = \frac{1}{2}(\text{base})(\text{height}),$$

or

$$K = \frac{1}{2}bh.$$

If θ is an acute angle in a triangle (see Figure 6.25), then $\sin\theta = \dfrac{h}{c}$; so $h = c\sin\theta$. The area of this triangle is given by

$$K = \frac{1}{2}bh = \frac{1}{2}bc\sin\theta \qquad \text{Replace } h \text{ with } c\sin\theta.$$

The angle θ shown in Figure 6.26 is an obtuse angle. Here $\theta' = 180° - \theta$ is the reference angle for θ and $\sin\theta = \sin\theta' = \dfrac{h}{c}$. So,

$$h = c\sin\theta' = c\sin\theta$$
$$K = \frac{1}{2}bh = \frac{1}{2}bc\sin\theta \qquad \text{Replace } h \text{ with } c\sin\theta.$$

AREA OF A TRIANGLE

In any triangle, if θ is the included angle between sides of lengths b and c, the area K of the triangle is given by the formula:

$$K = \frac{1}{2}bc\sin\theta$$

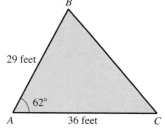

FIGURE 6.27

EXAMPLE 7 **Finding the Area of a Triangle**

Find the area of the triangle ABC in Figure 6.27.

SOLUTION

We are given the angle of measure 62° between the sides of lengths 36 feet and 29 feet. The area K of the triangle ABC is given by

$$K = \frac{1}{2}bc\sin\theta \qquad\qquad \text{Area formula}$$

$$= \frac{1}{2}(36)(29)\sin 62° \qquad b = 36, c = 29, \theta = 62°$$

$$\approx 460.9 \text{ square feet} \qquad \text{Use a calculator.} \qquad ■ ■ ■$$

Practice Problem 7 Find the area of triangle ABC assuming that the lengths of the sides AC and BC are 27 feet and 38 feet, respectively, and the measure of the angle between these sides is 47°. ■

SECTION 6.2 ■ Exercises

A EXERCISES Basic Skills and Concepts

1. If you know two angles of a triangle, then you can determine the third angle because the sum of all three angles is _____180_____ degrees.

2. The Law of Sines states that if a, b, and c are the sides opposite angles A, B, and C, then $\dfrac{a}{\sin A}$ = $\dfrac{b}{\sin B}$ = $\dfrac{c}{\sin C}$.

3. If you are given any two angles and one side, then there is exactly _____one_____ triangle(s) possible.

4. For given data a, A, and b, there can be no triangle, exactly _____one_____ triangle, or _____two_____ triangles.

5. *True or False* The Law of Sines allows you to solve a triangle if you know two sides and the angle opposite one of them. True

6. *True or False* A triangle is uniquely determined if any two sides and one angle are given. False

In Exercises 7–14, solve each triangle. Round each answer to the nearest tenth.

7.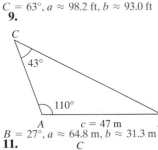
$C = 63°$, $a \approx 98.2$ ft, $b \approx 93.0$ ft

8.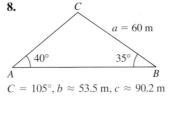
$C = 105°$, $b \approx 53.5$ m, $c \approx 90.2$ m

9.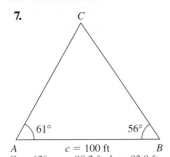
$B = 27°$, $a \approx 64.8$ m, $b \approx 31.3$ m

10.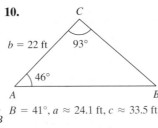
$B = 41°$, $a \approx 24.1$ ft, $c \approx 33.5$ ft

11.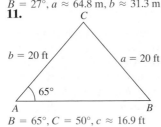
$B = 65°$, $C = 50°$, $c \approx 16.9$ ft

12.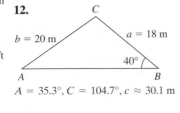
$A = 35.3°$, $C = 104.7°$, $c \approx 30.1$ m

13.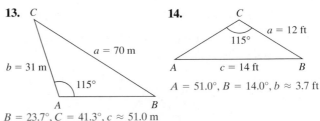
$B = 23.7°$, $C = 41.3°$, $c \approx 51.0$ m

14.
$A = 51.0°$, $B = 14.0°$, $b \approx 3.7$ ft

In Exercises 15–26, solve triangle ABC. Round each answer to the nearest tenth.

15. $A = 40°$, $B = 35°$, $a = 100$ meters
$C = 105°$, $b \approx 89.2$ m, $c \approx 150.3$ m

16. $A = 80°$, $B = 20°$, $a = 100$ meters
$C = 80°$, $b \approx 34.7$ m, $c \approx 100$ m

17. $A = 46°$, $C = 55°$, $a = 75$ centimeters
$B = 79°$, $b \approx 102.3$ cm, $c \approx 85.4$ cm

18. $A = 35°$, $C = 98°$, $a = 75$ centimeters
$B = 47°$, $b \approx 95.6$ cm, $c \approx 129.5$ cm

19. $A = 35°$, $C = 47°$, $c = 60$ feet
$B = 98°$, $a \approx 47.1$ ft, $b \approx 81.2$ ft

20. $A = 44°$, $C = 76°$, $c = 40$ feet
$B = 60°$, $a \approx 28.6$ ft, $b \approx 35.7$ ft

21. $B = 43°$, $C = 67°$, $b = 40$ inches
$A = 70°$, $a \approx 55.1$ in., $c \approx 54.0$ in.

22. $B = 95°$, $C = 35°$, $b = 100$ inches
$A = 50°$, $a \approx 76.9$ in., $c \approx 57.6$ in.

23. $B = 110°$, $C = 46°$, $c = 23.5$ feet
$A = 24°$, $a \approx 13.3$ ft, $b \approx 30.7$ ft

24. $B = 67°$, $C = 63°$, $c = 16.8$ feet
$A = 50°$, $a \approx 14.4$ ft, $b \approx 17.4$ ft

25. $A = 35.7°$, $B = 45.8°$, $c = 30$ meters
$C = 98.5°$, $a \approx 17.7$ m, $b \approx 21.7$ m

26. $A = 64.5°$, $B = 54.3°$, $c = 40$ meters
$C = 61.2°$, $a \approx 41.2$ m, $b \approx 37.1$ m

In Exercises 27–46, solve each SSA triangle. Indicate whether the given measurements result in no triangle, one triangle, or two triangles. Solve each resulting triangle. Round each answer to the nearest tenth.

27. $A = 40°$, $a = 23$, $b = 20$ $B \approx 34°$, $C \approx 106°$, $c \approx 34.4$

28. $A = 36°$, $a = 30$, $b = 24$ $B \approx 28°$, $C \approx 116°$, $c \approx 45.9$

29. $A = 30°$, $a = 25$, $b = 50$ $C = 60°$, $B = 90°$, $c = 25\sqrt{3}$

30. $A = 60°$, $a = 20\sqrt{3}$, $b = 40$ $B = 90°$, $C = 30°$, $c = 20$

31. $A = 40°$, $a = 10$, $b = 20$ No triangle exists.

32. $A = 62°$, $a = 30$, $b = 40$ No triangle exists.

33. $A = 95°$, $a = 18$, $b = 21$ No triangle exists.

34. $A = 110°$, $a = 37$, $b = 41$ No triangle exists.

35. $A = 100°$, $a = 40$, $b = 34$ $B \approx 56.8°$, $C \approx 23.2°$, $c \approx 16.0$

36. $A = 105°$, $a = 70$, $b = 30$ $B \approx 24.5°$, $C \approx 50.5°$, $c \approx 56.0°$

37. $B = 50°$, $b = 22$, $c = 40$ No triangle exists.

38. $B = 64°$, $b = 45$, $c = 60$ No triangle exists.

39. $B = 46°$, $b = 35$, $c = 40$

40. $B = 32°$, $b = 50$, $c = 60$

41. $B = 97°$, $b = 27$, $c = 30$ No triangle exists.

42. $B = 110°$, $b = 19$, $c = 21$ No triangle exists.

Answers:

39. Solution 1: $C \approx 55.3°$, $A \approx 78.7°$, $a \approx 47.7$
Solution 2: $C \approx 124.7°$, $A \approx 9.3°$, $a \approx 7.9$

40. Solution 1: $C \approx 39.5°$, $A \approx 108.5°$, $a \approx 89.5$
Solution 2: $C \approx 140.5°$, $A \approx 7.5°$, $a \approx 12.3$

†Due to space constrictions, answers to these exercises may be found in the Answers beginning on page A–1 in the back of the book.

43. $A = 42°, a = 55, c = 62$

44. $A = 34°, a = 6, c = 8$

45. $C = 40°, a = 3.3, c = 2.1$ No triangle exists.

46. $C = 62°, a = 50, c = 100$ $A ≈ 26.2°, B ≈ 91.8°, b ≈ 113.2$

In Exercises 47–54, find the area of a triangle *ABC* (see the figure) with the given information.

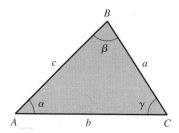

47. $α = 57°, b = 30$ inches, $c = 52$ inches 654.2 sq in.

48. $β = 110°, a = 20$ centimeters, $c = 27$ centimeters 253.7 sq cm

49. $γ = 46°, a = 15$ kilometers, $b = 22$ kilometers 118.7 sq km

50. $α = 146.7°, b = 16.7$ feet, $c = 18.6$ feet 85.3 sq ft

51. $β = 38.6°, a = 12$ millimeters, $c = 16.7$ millimeters 62.5 sq mm

52. $β = 112.5°, a = 151.6$ feet, $c = 221.8$ feet 15,532.7 sq ft

53. $γ = 107.3°, a = 271$ feet, $b = 194.3$ feet 25,136.6 sq ft

54. $α = 131.8°, b = 15.7$ millimeters, $c = 18.2$ millimeters 106.5 sq mm

B EXERCISES Applying the Concepts

55. Finding distance. Angela wants to find the distance from point *A* to her friend Carmen's house at point *C* on the other side of the river. She knows the distance from *A* to Betty's house at *B* is 540 feet. (See the figure.) The measurement of angles *A* and *B* are 57° and 46°, respectively. Calculate the distance from *A* to *C*. Round to the nearest foot. ≈ 399 ft

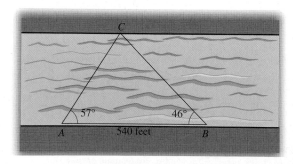

56. Finding distance. In Exercise 55, find the width of the river assuming that the houses are on the (very straight) banks of the river. ≈ 334 ft

57. Target A laser beam with an angle of elevation of 42° is reflected by a target and is received 1200 yards from the point of origin. Assume that the trajectory of the beam forms (approximately) an isosceles triangle.
a. Find the total distance the beam travels. Round to the nearest yard. 1615 yd

b. What is the height of the target? Round to the nearest yard. 540 yd

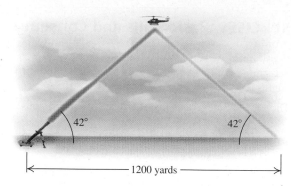

58. Height of a flagpole. Two surveyors stand 200 feet apart with a flagpole between them. Suppose the "transit" at each location is 5 feet high. The transit measures the angles of elevation of the top of the flagpole at the two locations to be 30° and 25°, respectively. Find the height of the flagpole. Round to the nearest foot. 57 ft

59. Radio beacon. A ship sailing due east at the rate of 16 miles per hour records the bearing of a radio beacon at N 36.5° E. Two hours later the bearing of the same beacon is N 55.7° W. Round each answer to the nearest tenth of a mile. 25.7 mi
a. How far is the ship from the beacon the second time?
b. How close to the beacon did the ship pass? 14.5 mi

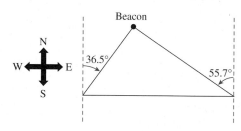

60. Lighthouse. A boat sailing due north at the rate of 14 miles per hour records the bearing of a lighthouse as N 8.4° E. Two hours later the bearing of the same lighthouse is N 30.9° E. Round the answers to the nearest tenth of a mile.
a. How far is the ship from the lighthouse the second time? 10.7 mi
b. If the boat follows the same course and speed, how close to the lighthouse will it approach? 5.5 mi

61. Geostationary satellite. A camel rider was traveling due west at night in the desert. At midnight, his angle of elevation to a satellite was 89°. After traveling west for 20 miles, his angle of elevation to the satellite was 89.05°. The satellite is at a constant height above the ground. How high is the satellite? Round the answer to the nearest mile. 22,912 mi

62. Height of a tower. A flagpole 20 feet tall is placed on top of a tower. At a point on the ground, a surveyor measures the angle of elevation of the bottom of the flag to be 69.6° and that of the top of the flag to be 70.9°. What is the height of the tower to the nearest foot? 270 ft

Answers:
43. Solution 1: $C ≈ 49.0°, B ≈ 89.0°, b ≈ 82.2$ **44.** Solution 1: $C ≈ 48.2°, B ≈ 97.8°, b ≈ 10.6$
   Solution 2: $C ≈ 131.0°, B ≈ 7.0°, b ≈ 10.0$ Solution 2: $C ≈ 131.8°, B ≈ 14.2°, b ≈ 2.6$

63. **Area of the Bermuda Triangle.** The angle formed by the lines of sight from Ft. Lauderdale, Florida, to Bermuda and to San Juan, Puerto Rico, is approximately 67°. The distance from Ft. Lauderdale to Bermuda is 1026 miles, and the distance from Ft. Lauderdale to San Juan is 1046 miles. Find the area of the Bermuda Triangle, which is formed with these cities as vertices. 493,941 square miles

64. **Landscaping.** A triangular region between the three streets shown in the figure will be landscaped for $30 a square foot. How much will the landscaping cost? $5250

C EXERCISES Beyond the Basics

65. Suppose CD bisects angle C of triangle ABC. Show that $\dfrac{AD}{DB} = \dfrac{AC}{CB}$. (See the figure.)

    [*Hint:* Apply the Law of Sines to triangles ACD and DCB.]

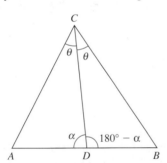

66. A flagpole 15 feet high stands on a building 75 feet high. To an observer at a height of 90 feet, the building and the flagpole intercept equal angles. Find the distance of the observer from the top of the flagpole. [*Hint:* Use Exercise 65.] 18.4 ft

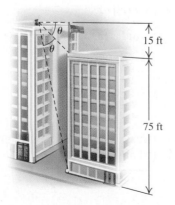

67. A point on an island is located 24 miles southwest of a dock. A ship leaves the dock at 1 P.M., traveling west at 12 miles per hour. At what time(s) to the nearest minute is the ship 20 miles from the point? 1:32 P.M., 3:18 P.M.

68. A side of a parallelogram is 60 inches long and makes angles of 28° and 42° with the two diagonals. What are the lengths of the diagonals? Round each answer to the nearest tenth of an inch. 60.0 in.; 85.4 in.

69. **Mollweide's formula.** In a triangle ABC, prove that

$$\frac{b - c}{a} = \frac{\sin\dfrac{B - C}{2}}{\cos\dfrac{A}{2}}.$$

[*Hint:* Let $\dfrac{a}{\sin A} = \dfrac{b}{\sin B} = \dfrac{c}{\sin C} = k$. Write an expression for $\dfrac{b - c}{a}$ in terms of sines and then use Sum-to-Product and Half-Angle formulas. Note that because the formula contains all six parts of a triangle, it can be used as a check for the solutions of a triangle.]

70. **Mollweide's formula.** In a triangle ABC, prove that

$$\frac{b + c}{a} = \frac{\cos\dfrac{B - C}{2}}{\sin\dfrac{A}{2}}.$$

71. **Law of Tangents.** In a triangle ABC, prove that

$$\frac{b - c}{b + c} = \frac{\tan\left(\dfrac{B - C}{2}\right)}{\tan\left(\dfrac{B + C}{2}\right)}.$$

72. Two sides of a triangle are $\sqrt{3} + 1$ feet and $\sqrt{3} - 1$ feet, and the measure of the angle between these sides is 60°. Use the Law of Tangents (Exercise 71) to find the measure of the difference of the remaining angles. 90°

Critical Thinking

In Exercises 73 and 74, give the most specific description you can for a triangle satisfying the specified conditions. Explain your reasoning.

73. In a triangle ABC, $a \sin A = b \sin B$. An isosceles triangle with $m\angle A = m\angle B$

74. In a triangle ABC, $a \cos A = b \cos B$. An isosceles right triangle

75. Use the Law of Sines to show that any isosceles triangle has two equal angles.

76. Use the Law of Sines to show that any triangle with two equal angles is isosceles.

The Law of Cosines

Before Starting this Section, Review

1. Distance formula (Section 1.1, page 5)
2. Inverse functions (Section 1.7, page 90)
3. Area of a triangle (Section 6.2, page 414)

Objectives

1. Derive the Law of Cosines.
2. Use the Law of Cosines to solve SAS triangles.
3. Use the Law of Cosines to solve SSS triangles.
4. Use Heron's formula to find the area of a triangle.

GENERALIZING THE PYTHAGOREAN THEOREM

Suppose a Boeing 747 jumbo jet is flying over Disney World in Orlando, Florida, at 552 miles per hour and is heading due south to Brazil. Twenty minutes later an F-16 fighter jet heading due east passes over Disney World at a speed of 1250 miles per hour. By using the Pythagorean Theorem, we can easily calculate the distance d between the two planes t hours after the F-16 passes over Disney World. We have the following:

$$d = \sqrt{(1250t)^2 + \left[\frac{1}{3}(552) + 552t\right]^2} \qquad 20 \text{ minutes} = \frac{1}{3} \text{ hour}$$

Suppose all the other facts in the problem are unchanged except that the F-16 has a bearing of N 37°E. Now we can no longer apply the Pythagorean Theorem. In Example 2, we use the Law of Cosines, described in this section, to solve this problem. ■

1 Derive the Law of Cosines.

The Law of Cosines

The Pythagorean Theorem states that the relationship $c^2 = a^2 + b^2$ holds in a right triangle ABC, where c represents the length of the hypotenuse. This relationship is not true if the triangle is not a right triangle. The **Law of Cosines** is a generalization of the Pythagorean Theorem that is true in any triangle. This law will be used to solve triangles in which two sides and the included angle are known (SAS triangles), as well as those triangles in which all three sides are known (SSS triangles).

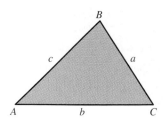

FIGURE 6.28

THE LAW OF COSINES

In triangle ABC with sides of lengths a, b, and c (as in Figure 6.28),

$$a^2 = b^2 + c^2 - 2bc \cos A,$$
$$b^2 = c^2 + a^2 - 2ca \cos B,$$
$$c^2 = a^2 + b^2 - 2ab \cos C.$$

In words, the square of any side of a triangle is equal to the sum of the squares of the length of the other two sides, less twice the product of the lengths of the other sides and the cosine of their included angle.

Derivation of the Law of Cosines

To derive the Law of Cosines, place triangle ABC in a rectangular coordinate system with the vertex A at the origin and the side c along the positive x-axis. See Figure 6.29.

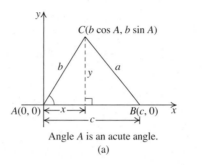
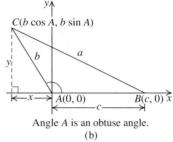

Angle A is an acute angle.
(a)

Angle A is an obtuse angle.
(b)

FIGURE 6.29 Two cases for the Law of Cosines

Label the coordinates of the vertices as shown in Figure 6.29. In Figures 6.29(a) and 6.29(b), the point $C(x, y)$ on the terminal side of angle A has coordinates $(b \cos A, b \sin A)$, which are found using the cosine and sine definitions.

$$\cos A = \frac{x}{b} \qquad \sin A = \frac{y}{b}$$

$$x = b \cos A \qquad y = b \sin A$$

The point B has coordinates $(c, 0)$. Applying the distance formula to the line segment joining $B(c, 0)$ and $C(b \cos A, b \sin A)$, we have

$$a = d(C, B)$$
$$a^2 = [d(C, B)]^2 \qquad \text{Square both sides.}$$
$$a^2 = (b \cos A - c)^2 + (b \sin A - 0)^2 \qquad \text{Distance formula}$$
$$a^2 = b^2 \cos^2 A - 2bc \cos A + c^2 + b^2 \sin^2 A \qquad \text{Expand binomial.}$$
$$a^2 = b^2(\sin^2 A + \cos^2 A) + c^2 - 2bc \cos A \qquad \text{Regroup terms.}$$
$$a^2 = b^2 + c^2 - 2bc \cos A \qquad \sin^2 A + \cos^2 A = 1$$

The last equation is one of the forms of the Law of Cosines. Similarly, by placing the vertex B and then the vertex C at the origin, we obtain the other two forms. Notice that the Law of Cosines becomes the Pythagorean Theorem if the included angle is $90°$ because $\cos 90° = 0$.

2 Use the Law of Cosines to solve SAS Triangles.

Solving SAS Triangles

Let's solve SAS triangles (Case 3 of Section 6.2).

FINDING THE SOLUTION: A PROCEDURE

EXAMPLE 1 Solving the SAS Triangles

OBJECTIVE	EXAMPLE

OBJECTIVE

Solve triangles in which the measures of two sides and the included angle are known.

EXAMPLE

Solve triangle ABC with $a = 15$ inches, $b = 10$ inches, and $C = 60°$. Round each answer to the nearest tenth.

Step 1 Use the appropriate form of the Law of Cosines to find the side opposite the given angle.

1. Find side c opposite angle C.

$$c^2 = a^2 + b^2 - 2ab \cos C \qquad \text{The Law of Cosines}$$

$$c^2 = (15)^2 + (10)^2 - 2(15)(10) \cos 60° \qquad \text{Substitute values.}$$

$$c^2 = 225 + 100 - 2(15)(10)\left(\frac{1}{2}\right) \qquad \cos 60° = \frac{1}{2}$$

$$c^2 = 175 \qquad \text{Simplify.}$$
$$c = \sqrt{175} \approx 13.2 \qquad \text{Use a calculator.}$$

Step 2 Use the **Law of Sines** to find the angle opposite the shorter of the two given sides. Note that this angle is always an acute angle.

2. Find angle B.

$$\frac{\sin B}{b} = \frac{\sin C}{c} \qquad \text{The Law of Sines}$$

$$\sin B = \frac{b \sin C}{c} \qquad \text{Multiply both sides by } b.$$

$$\sin B = \frac{10 \sin 60°}{\sqrt{175}} \qquad \text{Substitute values.}$$

$$B = \sin^{-1}\left(\frac{10 \sin 60°}{\sqrt{175}}\right) \approx 40.9° \qquad 0° < B \le 90°$$

Step 3 Use the angle sum formula to find the third angle.

3. $A \approx 180° - 60° - 40.9° \approx 79.1° \qquad A + B + C = 180°$

Step 4 Write the solution.

4.

$A \approx 79.1°$	$a = 15$ inches
$B \approx 40.9°$	$b = 10$ inches
$C = 60°$	$c \approx 13.2$ inches

■ ■ ■

Practice Problem 1 Solve triangle ABC with $c = 25$, $a = 15$, and $B = 60°$. Round each answer to the nearest tenth. ■

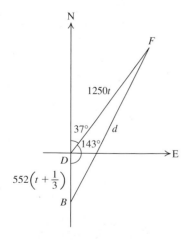

FIGURE 6.30

EXAMPLE 2 Using the Law of Cosines

Suppose a Boeing 747 is flying over Disney World headed due south at 552 miles per hour. Twenty minutes later an F-16 passes over Disney World with a bearing of N 37°E at a speed of 1250 miles per hour. Find the distance between the two planes three hours after the F-16 passes over Disney World. Round the answer to the nearest tenth.

SOLUTION

Suppose the F-16 has been traveling for t hours after passing over Disney World. Then because the Boeing 747 had a head start of 20 minutes $= \frac{1}{3}$ hour, the Boeing 747 has been traveling $\left(t + \frac{1}{3}\right)$ hours due south. The distance d between the two planes is shown in Figure 6.30. Using the Law of Cosines in triangle FDB, we have

$$d^2 = (1250t)^2 + \left[552\left(t + \frac{1}{3}\right)\right]^2 - 2(1250t) \cdot 552\left(t + \frac{1}{3}\right)\cos 143°$$

$$d^2 \approx 28{,}469{,}270.04 \qquad \text{Substitute } t = 3; \text{ use a calculator.}$$

$$d \approx 5335.7 \text{ miles} \qquad \text{Use a calculator.} \qquad ■■■$$

Practice Problem 2   Repeat Example 2 assuming that the F-16 is traveling at a speed of 1375 miles per hour due N 75°E and the Boeing 747 is traveling with a bearing of S 12°W at 550 miles per hour.   ■

3 Use the Law of Cosines to solve SSS triangles.

Solving SSS Triangles

Let's solve SSS triangles (Case 4 of Section 6.2).

FINDING THE SOLUTION: A PROCEDURE

EXAMPLE 3   **Solving the SSS Triangles**

OBJECTIVE	EXAMPLE
Solve triangles in which the measures of the three sides are known.	*Solve triangle ABC with a = 3.1 feet, b = 5.4 feet, and c = 7.2 feet. Round answers to the nearest tenths.*

OBJECTIVE

Solve triangles in which the measures of the three sides are known.

Step 1   Use the Law of Cosines to find the angle opposite the longest side.

Step 2   Use the Law of Sines to find either of the two remaining acute angles.

Step 3   Use the angle sum formula to find the third angle.

Step 4   Write the solution.

EXAMPLE

Solve triangle ABC with a = 3.1 feet, b = 5.4 feet, and c = 7.2 feet. Round answers to the nearest tenths.

1. Because c is the longest side, we first find angle C.

$$c^2 = a^2 + b^2 - 2ab\cos C \qquad \text{The Law of Cosines}$$

$$2ab\cos C = a^2 + b^2 - c^2 \qquad \begin{array}{l}\text{Add } 2ab\cos C - c^2 \\ \text{to both sides.}\end{array}$$

$$\cos C = \frac{a^2 + b^2 - c^2}{2ab} \qquad \text{Solve for } \cos C.$$

$$\cos C = \frac{(3.1)^2 + (5.4)^2 - (7.2)^2}{2(3.1)(5.4)} \qquad \text{Substitute values.}$$

$$\cos C \approx -0.39 \qquad \text{Use a calculator.}$$

$$C \approx \cos^{-1}(-0.39) \approx 113° \qquad 0° < C < 180°$$

2. Find angle B.

$$\frac{\sin B}{b} = \frac{\sin C}{c} \qquad \text{The Law of Sines}$$

$$\sin B = \frac{b\sin C}{c} \qquad \text{Multiply both sides by } b.$$

$$B = \sin^{-1}\left(\frac{b\sin C}{c}\right) \qquad 0° < B < 90°$$

$$B = \sin^{-1}\left(\frac{5.4\sin 113°}{7.2}\right) \qquad \text{Substitute values.}$$

$$B \approx 43.7° \qquad \text{Use a calculator.}$$

3. $A \approx 180° - 43.7° - 113° \qquad A + B + C = 180°$

$A \approx 23.3° \qquad \text{Simplify.}$

4.

$A \approx 23.3°$	$a = 3.1$ feet
$B = 43.7°$	$b = 5.4$ feet
$C \approx 113°$	$c = 7.2$ feet

■■■

Practice Problem 3 Solve triangle ABC with $a = 4.5$, $b = 6.7$, and $c = 5.3$. Round each answer to the nearest tenth. ∎

EXAMPLE 4 | **Solving an SSS Triangle**

Solve triangle ABC with $a = 2$ meters, $b = 9$ meters, and $c = 5$ meters. Round each answer to the nearest tenth.

SOLUTION
Step 1 We first find B, the angle opposite the longest side.

$$b^2 = c^2 + a^2 - 2ca \cos B \qquad \text{The Law of Cosines}$$

$$\cos B = \frac{c^2 + a^2 - b^2}{2ca} \qquad \text{Solve for } \cos B.$$

$$\cos B = \frac{5^2 + 2^2 - 9^2}{2(5)(2)} \qquad \text{Substitute values.}$$

$$\cos B = -2.6 \qquad \text{Simplify.} \qquad ∎∎∎$$

Since the range of the cosine function is $[-1, 1]$, there is no angle B for which $\cos B = -2.6$. This means that a triangle with the given information cannot exist. Since $2 + 5 < 9$, you can also use the Triangle Inequality from geometry to see that there is no such triangle.

Practice Problem 4 Solve triangle ABC with $a = 2$ inches, $b = 3$ inches, and $c = 6$ inches. ∎

> **RECALL**
>
> The Triangle Inequality states that in any triangle, the sum of the lengths of any two sides of a triangle is greater than the length of the third side.

4 Use Heron's formula to find the area of a triangle.

Heron's Area Formula

The Law of Cosines can be used to derive a formula for the area of a triangle if the length of the *three* sides is known. The formula is called **Heron's formula**.

> **RECALL**
>
> In Section 6.2, you learned that if two sides and their included angle are known, the area K of a triangle ABC is given by
>
> $$K = \frac{1}{2}ab \sin C$$
>
> $$= \frac{1}{2}bc \sin A$$
>
> $$= \frac{1}{2}ca \sin B.$$

HERON'S FORMULA FOR SSS TRIANGLES

The area K of a triangle with sides of lengths, a, b, and c is given by

$$K = \sqrt{s(s - a)(s - b)(s - c)},$$

where $s = \frac{1}{2}(a + b + c)$ is the **semiperimeter**.

In Exercise 60, we ask you to derive Heron's formula.

EXAMPLE 5 | **Using Heron's Formula**

Find the area of triangle ABC with $a = 29$ inches, $b = 25$ inches, and $c = 40$ inches. Round the answer to the nearest tenth.

SOLUTION
We first find s:

$$s = \frac{a + b + c}{2}$$

$$= \frac{29 + 25 + 40}{2} = 47$$

$$\text{Area of the triangle} = \sqrt{s(s-a)(s-b)(s-c)} \qquad \text{Heron's formula}$$
$$= \sqrt{47(47-29)(47-25)(47-40)} \qquad \text{Substitute values.}$$
$$\approx 360.9 \text{ square inches.} \qquad \text{Use a calculator.}$$

■ ■ ■

Practice Problem 5 Find the area of triangle ABC with $a = 11$ meters, $b = 17$ meters, and $c = 20$ meters.   ■

EXAMPLE 6   **Using Heron's Formula**

A triangular swimming pool has side lengths 23 feet, 17 feet, and 26 feet. How many gallons of water will fill the pool to a depth of 5 feet? Round the answer to the nearest whole number.

SOLUTION

To calculate the volume of water, we first calculate the area of the triangular surface.

We have $a = 23$, $b = 17$, and $c = 26$. So $s = \dfrac{1}{2}(a + b + c) = 33$.

By Heron's formula, the area K of the triangular surface is

$$K = \sqrt{s(s-a)(s-b)(s-c)}$$
$$= \sqrt{33(33-23)(33-17)(33-26)} \qquad \text{Substitute values for } a, b, c, \text{ and } s.$$
$$\approx 192.2498 \text{ square feet.}$$

$$\text{The volume of water} = \text{surface area} \times \text{depth}$$
$$\approx 192.2498 \times 5 \approx 961.25 \text{ cubic feet.}$$

One cubic foot contains approximately 7.5 gallons of water. So $961.25 \times 7.5 \approx 7209$ gallons of water will fill the pool.   ■ ■ ■

Practice Problem 6 Repeat Example 6 assuming that the swimming pool has side lengths 25 feet, 30 feet, and 33 feet and the depth of the pool is 5.5 feet.   ■

SECTION 6.3 ■ Exercises

A EXERCISES Basic Skills and Concepts

1. One form of the Law of Cosines is
   $c^2 = a^2 + b^2 - 2ab \cos (\underline{\quad C \quad})$.

2. If we take the angle in the Law of Cosines to be $90°$, then we get the $\underline{\text{Pythagorean}}$ Theorem.

3. Triangles with SAS given are solved by the Law of Cosines, as are triangles with $\underline{\quad\text{three}\quad}$ sides given.

4. When one angle is found by the Law of Cosines, the other can be found with the Law of $\underline{\quad\text{Sines}\quad}$.

5. *True or False* The Law of Cosines is used to solve triangles when two angles and a side are given. False

6. *True or False* For a triangle with SSS given, we can find angle A by the formula
   $$A = \cos^{-1}\left(\frac{b^2 + c^2 - a^2}{2bc}\right). \text{ True}$$

In Exercises 7–10, solve each triangle. Round each answer to the nearest tenth.

7.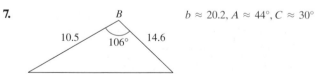
   $b \approx 20.2, A \approx 44°, C \approx 30°$

8.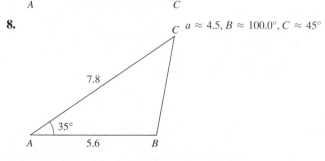
   $a \approx 4.5, B \approx 100.0°, C \approx 45°$

†Due to space constrictions, answers to these exercises may be found in the Answers beginning on page A–1 in the back of the book.

9.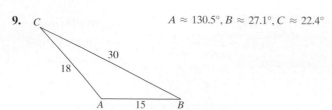

$A \approx 130.5°, B \approx 27.1°, C \approx 22.4°$

10.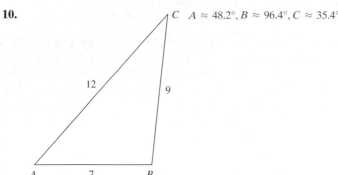

C $A \approx 48.2°, B \approx 96.4°, C \approx 35.4°$

In Exercises 11–26, solve each triangle *ABC*. Round each answer to the nearest tenth. All sides are measured in feet.

11. $a = 15, b = 9, C = 120°$ $c \approx 21, A \approx 38.2°, B \approx 21.8°$

12. $a = 14, b = 10, C = 75°$ $c \approx 15.0, A \approx 64.8°, B \approx 40.2°$

13. $b = 10, c = 12, A = 62°$ $a \approx 11.5, B \approx 50.4°, C \approx 67.6°$

14. $b = 11, c = 16, A = 110°$ $a \approx 22.3, B \approx 27.6°, C \approx 42.4°$

15. $c = 12, a = 15, b = 11$ $A \approx 81.3°, B \approx 46.4°, C \approx 52.3°$

16. $c = 16, a = 11, b = 13$ $A \approx 43.0°, B \approx 53.8°, C \approx 83.2°$

17. $a = 9, b = 13, c = 18$ $A \approx 28.3°, B \approx 43.3°, C \approx 108.4°$

18. $a = 14, b = 6, c = 10$ $A = 120°, B \approx 21.8°, C \approx 38.2°$

19. $a = 2.5, b = 3.7, c = 5.4$ $A \approx 23.7°, B \approx 36.4°, C \approx 119.9°$

20. $a = 4.2, b = 2.9, c = 3.6$ $A \approx 79.7°, B \approx 42.8°, C \approx 57.5°$

21. $b = 3.2, c = 4.3, A = 97.7°$ $a \approx 5.7, B \approx 33.8°, C \approx 48.5°$

22. $b = 5.4, c = 3.6, A = 79.2°$ $a \approx 5.9, B \approx 64.0°, C \approx 36.8°$

23. $c = 4.9, a = 3.9, B = 68.3°$ $b \approx 5.0, A \approx 46.3°, C \approx 65.4°$

24. $c = 7.8, a = 9.8, B = 95.6°$ $b \approx 13.1, A \approx 48.1°, C \approx 36.3°$

25. $a = 2.3, b = 2.8, c = 3.7$ $A \approx 38.4°, B \approx 49.1°, C \approx 92.5°$

26. $a = 5.3, b = 2.9, c = 4.6$ $A \approx 86.8°, B \approx 33.1°, C \approx 60.1°$

In Exercises 27–34, find the area of each triangle by using Heron's formula. Round each answer to the nearest tenth. All sides are measured in inches.

27. $a = 2, b = 3, c = 4$ 2.9 square inches

28. $a = 50, b = 100, c = 130$ 2245.0 square inches

29. $a = 50, b = 50, c = 75$ 1240.2 square inches

30. $a = 100, b = 100, c = 125$ 4878.9 square inches

31. $a = 7.5, b = 4.5, c = 6.0$ 13.5 square inches

32. $a = 8.5, b = 9.0, c = 4.5$ 18.9 square inches

33. $a = 3.7, b = 5.1, c = 4.2$ 7.7 square inches

34. $a = 9.8, b = 5.7, c = 6.5$ 17.7 square inches

B EXERCISES   Applying the Concepts

In Exercises 35–48, round each answer to the nearest tenth.

35. Chord length. Find the length of the chord intercepted by a central angle of 42° in a circle of radius 8 feet. 5.7 feet

36. Central angle. Find the measure of the central angle of a circle of radius 6 feet that intercepted a chord of length 3.5 feet. 33.9°

37. Tunnel length. Engineers must bore a straight tunnel through the base of a mountain. They select a reference point *C* on the plain surrounding the mountain. The distance from *C* to portal *A* (the starting tunnel point) and portal *B* (the ending tunnel point) is 2352 yards and 1763 yards, respectively. The measure of $\angle ACB$ is 41°. How long (to the nearest yard) is the tunnel? 1543.1 yards

38. Pond length. A surveyor needs to find the length of a pond (see figure), but does so indirectly. She selects a point *A* on one side of the pond and measures distances to the points *B* and *C* at opposite ends of the pond to be 537 yards and 823 yards, respectively. The measure of $\angle BAC$ is 130°. (See figure.) How long is the pond? 1238.5 yards

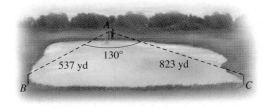

39. Roof truss. A roof truss is made in the shape of an inverted V. The lengths of the two edges are 11 feet and 23 feet. The edges meet at the peak, making a 65° angle. Find
a. the width of the truss. 20.9 feet
b. the height of the peak. 11.0 feet

40. Solar panels. A roof of a house addition is being built to accept solar energy panels as shown in the figure. Find
a. the length of the edge *AC* of the truss. 20.8 feet
b. the measure of $\angle BAC$. 79.9°

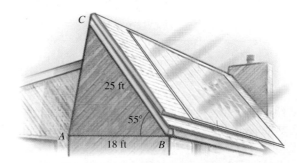

41. Real estate. You have inherited a commercial triangular-shaped lot of side lengths 400 feet, 250 feet, and 274 feet. The neighboring property is selling for $1 million per acre. How much is your lot worth? (*Hint:* 1 acre = 43,560 square feet.) $780,000

42. Real estate. Find the area of the quadrangular lot shown in the figure. All measurements are in feet. 1210.5 square feet

43. Swimming pool. Find the number of gallons of water in a triangular swimming pool with sides 11 feet, 16 feet, and 19 feet and a depth of 5 feet. Round your answer to the nearest whole number. [Recall: Approximately 7.5 gallons of water are in 1 cubic foot.] 3297 gallons

44. Hikers. Two hikers, Sonia and Tony, leave the same point at the same time. Sonia walks due east at the rate of 3 miles per hour, and Tony walks 45° northeast at the rate of 4.3 miles per hour. How far apart are the hikers after three hours? 9.1 miles

45. Distance between ships. Two ships leave the same port— Ship A at 1.00 P.M. and ship B at 3:30 P.M. Ship A sails on a bearing of S 37° E at 18 miles per hour, and B sails on a bearing of N 28° E at 20 miles per hour. How far apart are the ships at 8.00 P.M.? 183.2 miles

46. Navigation. A ship is traveling due north. At two different points A and B, the navigator of the ship sites a lighthouse at the point C, as shown in the accompanying figure.
a. Determine the distance from B to C. 3689 meters
b. How much farther due north must the ship travel to reach the point closest to the lighthouse? 1634.5 meters

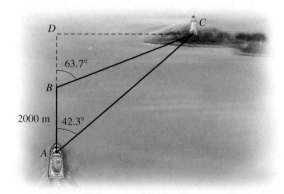

47. Tangential circles. Three circles of radii 1.2 inches, 2.2 inches, and 3.1 inches are tangent to each other externally.

Find the angles of the triangle formed by joining the centers of the circles. $A \approx 86.2°, B \approx 54.0°, C \approx 39.8°$

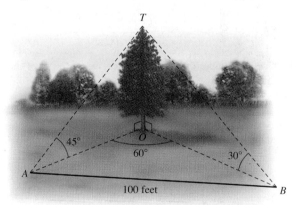

48. Height of a tree. A tree is planted at a point O on horizontal ground. Two points A and B on the ground are 100 feet apart. The angles of elevation of the top of the tree T from the points A and B are 45° and 30°, respectively. The measure of $\angle AOB$ is 60°. Find the height of the tree. 66.4 feet

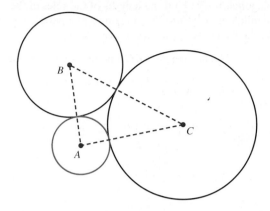

C EXERCISES Beyond the Basics

In Exercises 49–52, round each answer to the nearest tenth.

49. A parallelogram has adjacent sides 8 cm and 13 cm. If the shorter diagonal is 11 cm long, find the length of the longer diagonal. 18.6 cm

50. Find the area of the parallelogram of Exercise 49. 87.6 cm^2

51. A parallelogram has adjacent sides 10 cm and 15 cm long, and the angle between them is 40°. Find the length of the two diagonals. 9.8 cm, 23.6 cm

52. The length of the two diagonals of a parallelogram are 16 cm and 22 cm. The acute angle between these diagonals is 63°. Find the lengths of the sides of the parallelogram. [*Hint:* The diagonals of a parallelogram bisect each other.] 10.3 cm, 16.3 cm

In Exercises 53–60, triangle ABC has sides a, b, and c and $s = \frac{1}{2}(a + b + c)$. Use the Law of Cosines to prove each identity.

53. $1 - \cos A = \dfrac{(a - b + c)(a + b - c)}{2bc}$

54. $1 - \cos A = \dfrac{2(s - b)(s - c)}{bc}$

55. $1 + \cos A = \dfrac{(b + c + a)(b + c - a)}{2bc}$

56. $1 + \cos A = \dfrac{2s(s - a)}{bc}$

57. Use half-angle formula and Exercise 54 to prove the following:

$$\sin \frac{A}{2} = \sqrt{\frac{(s - b)(s - c)}{bc}}$$

58. Use half-angle formula and Exercise 56 to prove the following:

$$\cos \frac{A}{2} = \sqrt{\frac{s(s - a)}{bc}}$$

59. Use Exercises 57 and 58 to prove that

$$\sin A = \frac{2}{bc}\sqrt{s(s - a)(s - b)(s - c)}.$$

60. Use Exercise 59 to prove Heron's formula.

61. Use the Law of Cosines to show that for a triangle with legs of lengths a, b, and c (with $a \le b$),

$$b - a < c < b + a.$$

Critical Thinking

62. Solve triangle ABC with vertices $A(-2, 1)$, $B(5, 3)$, and $C(3, 6)$.

63. Solve triangle ABC with vertices $A(-3, -5)$, $B(6, 10)$, and $C(3, -2)$.

GROUP PROJECTS

You can use the Law of Cosines to solve triangle ABC in the ambiguous case of Section 6.2. For example, to solve triangle ABC with $B = 150°$, $b = 10$, and $c = 6$, use the Law of Cosines to write $b^2 = a^2 + c^2 - 2ac \cos B$.

Substitute values of b, c, and B in this equation to obtain a quadratic equation in a. Find the roots of this equation and interpret your results. Again use the Law of Cosines to find A. Then find $C = 180° - A - B$.

Use this technique to solve each triangle ABC.

64. $B = 150°$, $b = 10$, and $c = 6$ $a \approx 4.3$, $A \approx 12.5°$, $C \approx 17.5°$

65. $A = 30°$, $a = 6$, and $b = 10$

66. $A = 60°$, $a = 12$, and $c = 15$ No triangle can be formed.

Answers:
62. $a = \sqrt{13}$, $b = 5\sqrt{2}$, $c = \sqrt{53}$, $A \approx 29.1°$, $B \approx 72.2°$, $C \approx 78.7°$
63. $a = 3\sqrt{17}$, $b = 3\sqrt{5}$, $c = 3\sqrt{34}$, $A \approx 32.5°$, $B \approx 16.9°$, $C \approx 130.6°$
65. $c \approx 11.98$, $C \approx 93.6°$, $B \approx 56.4°$, $c \approx 5.34$, $C \approx 26.4°$, $B \approx 123.6°$

Vectors

Before Starting this Section, Review	**Objectives**
1. Trigonometric functions (Section 4.2, page 268)	**1** Represent vectors geometrically.
	2 Represent vectors algebraically.
2. Inverse trigonometric functions (Section 4.6, page 314)	**3** Find a unit vector in the direction of **v**.
	4 Write a vector in terms of its magnitude and direction.
	5 Use vectors in applications.

THE MEANING OF FORCE

A **force** is a push or pull resulting from an object's *interaction* with another object. When the interacting objects are physically in contact with each other, the resulting force is called a *contact force*. Examples of contact forces include friction, tension, air resistance, and applied forces. A noncontact force is called an *action-at-a-distance* force. Examples of such forces include gravitational, electrical, and magnetic forces. The standard units of measurement for the magnitude of a force are pounds (lb) in the English system and newtons (N) in the metric system. The conversion factor between the systems is

1 pound of force = 4.45 newtons.

In Examples 8–10, we discuss the *resultant* of two or more forces acting on an object. ■

Vectors

STUDY TIP

The *magnitude* of a vector is similar to the *absolute value* of a real number. Like absolute value, the magnitude of a vector cannot be negative.

Many physical quantities such as length, area, volume, mass, and temperature are completely described by their magnitudes in appropriate units. Such quantities are called **scalar quantities**. Other physical quantities such as velocity, acceleration, and force are completely described only if *both* a magnitude (size) and a direction are specified. For example, the movement of wind is usually described by its speed (magnitude) and its direction, say, 15 miles per hour southwest. The wind speed and wind direction together form a **vector quantity** called the *wind velocity*.

Geometric Vectors

1 Represent vectors geometrically.

We can represent a **vector** geometrically by a directed line segment with an arrowhead. The arrow specifies the direction of the vector, and its length describes the magnitude. The tail of the arrow is the vector's **initial point**, and the tip of the arrow is its **terminal point**. We denote vectors by lowercase boldface type, such as **a**, **b**, **i**, **j**, **u**, **v**, and **w**. When discussing vectors, we refer to real numbers as **scalars**. Scalars will be denoted by lowercase italic type, such as *a*, *b*, *x*, *y*, and *z*.

As in Figure 6.31, if the initial point of a vector **v** is P and the terminal point is Q, we write

$$\mathbf{v} = \overrightarrow{PQ}.$$

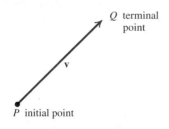

FIGURE 6.31 A vector

The **magnitude** (or **norm**) of a vector $\mathbf{v} = \overrightarrow{PQ}$, denoted by $\|\mathbf{v}\|$, or $\|\overrightarrow{PQ}\|$, is the length of the vector **v** and is a scalar quantity.

Equivalent Vectors

Two vectors having the same length and same direction are called **equivalent vectors**. Since a vector is determined by its length and direction only, equivalent vectors are regarded as **equal** even though they may be located in different positions. If **v** and **w** are equivalent, we write **v** = **w**. See Figure 6.32.

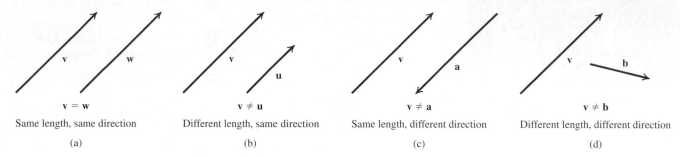

v = **w**	**v** ≠ **u**	**v** ≠ **a**	**v** ≠ **b**
Same length, same direction	Different length, same direction	Same length, different direction	Different length, different direction
(a)	(b)	(c)	(d)

FIGURE 6.32 Equivalent vectors have the same length and direction.

The vector of length zero is called the **zero vector** and is denoted by **0**. The zero vector has zero magnitude and arbitrary direction. If vectors **v** and **a**, as in Figure 6.32(c), have the same length and opposite direction, then **a** is the **opposite vector** of **v** and we write **a** = −**v**.

Adding Vectors

We add two vectors **v** and **w**, as shown in Figure 6.33.

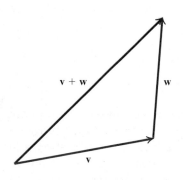

FIGURE 6.33 Adding v and w.

GEOMETRIC VECTOR ADDITION

Let **v** and **w** be any two vectors. Position the terminal point of **v** so that it coincides with the initial point of **w**. The *sum* **v** + **w** is the **resultant vector** whose initial point coincides with the initial point of **v**, and its terminal point coincides with the terminal point of **w**.

In Figure 6.34, we have constructed two sums, **v** + **w** and **w** + **v**. It is clear that

$$\boxed{\mathbf{v} + \mathbf{w} = \mathbf{w} + \mathbf{v}}$$

and that the sum coincides with the diagonal of the parallelogram determined by **v** and **w** when **v** and **w** have the same initial point.

Vector subtraction is defined just like the subtraction of real numbers. For any two vectors **v** and **w**, **v** − **w** = **v** + (−**w**), where −**w** is the opposite of **w**. See Figure 6.35.

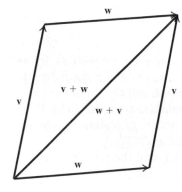

FIGURE 6.34 v + w = w + v

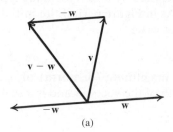
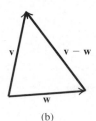

(a)	(b)

FIGURE 6.35 Vector v − w

Figure 6.35(b) shows that you can construct the difference vector **v** − **w** without first drawing −**w**: place the vectors **v** and **w** so that their initial points coincide. Then the vector from the terminal point of **w** to the terminal point of **v** is the vector **v** − **w**.

A second basic arithmetic operation for vectors is multiplying vectors by real numbers.

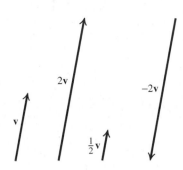

FIGURE 6.36 Scalar multiples of v

Let **v** be a vector and c a scalar (a real number). The vector c**v** is called the **scalar multiple** of **v**. See Figure 6.36.

If $c > 0$, c**v** has the same direction as **v** and magnitude $c\|\mathbf{v}\|$.

If $c < 0$, c**v** has the opposite direction of **v** and magnitude $|c|\|\mathbf{v}\|$.

If $c = 0$, c**v** = 0**v** = **0**.

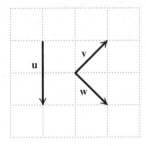

FIGURE 6.37

EXAMPLE 1 **Geometric Vectors**

Use the vectors **u**, **v**, and **w** in Figure 6.37 to graph each vector.

a. **u** − 2**w**   **b.** 2**v** − **u** + **w**

SOLUTION
The graphs are shown in Figure 6.38.

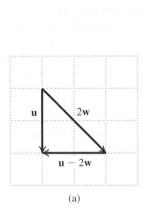

(a)

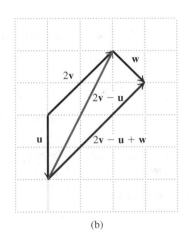

(b)

FIGURE 6.38   ■ ■ ■

Practice Problem 1 Use the vectors **u**, **v**, and **w** of Figure 6.37 to graph each vector.

a. 2**w** + **v**   **b.** 2**w** + **v** − **u**   ■

2 Represent vectors algebraically.

Algebraic Vectors

We now consider vectors in the Cartesian coordinate plane. Since the location of the initial point of a vector is not relevant, we typically draw vectors with their initial point at the origin. Such a vector is called a **position vector**. Note that the terminal point of a position vector will completely determine the vector, so specifying the terminal point will specify the vector. For the position vector **v** (see Figure 6.39) with initial point at the origin O and terminal point at $P(v_1, v_2)$, we denote the vector by

$$\mathbf{v} = \overrightarrow{OP} = \langle v_1, v_2 \rangle.$$

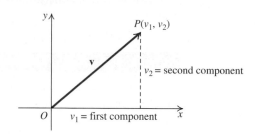

FIGURE 6.39 A position vector

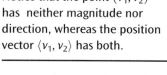

STUDY TIP

Notice that the point (v_1, v_2) has neither magnitude nor direction, whereas the position vector $\langle v_1, v_2 \rangle$ has both.

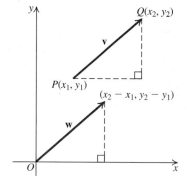

FIGURE 6.40 Two equivalent vectors

We call v_1 and v_2 the **components** of the vector **v**; v_1 is the **first component**, and v_2 is the **second component**. Note the difference between the notations for the *point* (v_1, v_2) and the *position vector* $\langle v_1, v_2 \rangle$. The magnitude of the position vector $\mathbf{v} = \langle v_1, v_2 \rangle$ follows directly from the Pythagorean Theorem. We have $\|\mathbf{v}\| = \sqrt{v_1^2 + v_2^2}$.

If equivalent vectors **v** and **w** are located so that their initial points are at the origin, then their terminal points must coincide (because the equivalent vectors have the same length and same direction). Thus, for the vectors

$$\mathbf{v} = \langle v_1, v_2 \rangle \quad \text{and} \quad \mathbf{w} = \langle w_1, w_2 \rangle,$$
$$\mathbf{v} = \mathbf{w} \quad \text{if and only if} \quad v_1 = w_1 \quad \text{and} \quad v_2 = w_2.$$

We can use congruent triangles to show that any vector **v** with initial point $P(x_1, y_1)$ and terminal point $Q(x_2, y_2)$ is equivalent to the position vector $\mathbf{w} = \langle x_2 - x_1, y_2 - y_1 \rangle$. Any vector in the Cartesian plane can therefore be represented by a position vector. See Figure 6.40.

REPRESENTING A VECTOR AS A POSITION VECTOR

The vector $\overrightarrow{PQ}$ with initial point $P(x_1, y_1)$ and terminal point $Q(x_2, y_2)$ is equal to the position vector

$$\mathbf{w} = \langle x_2 - x_1, y_2 - y_1 \rangle.$$

EXAMPLE 2   **Representing a Vector in the Cartesian Plane**

Let **v** be the vector with initial point $P(4, -2)$ and terminal point $Q(-1, 3)$. Write **v** as a position vector.

SOLUTION

Because **v** has initial point $P(4, -2)$, $x_1 = 4$ and $y_1 = -2$.

Because **v** has terminal point $Q(-1, 3)$, $x_2 = -1$ and $y_2 = 3$.

So

$$\mathbf{v} = \langle x_2 - x_1, y_2 - y_1 \rangle \qquad \text{Position vector representation of } \overrightarrow{PQ}$$
$$\mathbf{v} = \langle -1 - 4, 3 - (-2) \rangle \qquad \text{Substitute values.}$$
$$\mathbf{v} = \langle -5, 5 \rangle \qquad \text{Simplify.} \qquad \blacksquare\blacksquare\blacksquare$$

Practice Problem 2   Let **w** be the vector with initial point $P(-2, 7)$ and terminal point $Q(1, -3)$. Write **w** as a position vector. ■

The zero vector is $\mathbf{0} = \langle 0, 0 \rangle$, and the opposite of the vector $\mathbf{v} = \langle v_1, v_2 \rangle$ is $-\mathbf{v} = \langle -v_1, -v_2 \rangle$. The following properties describe the arithmetic operations on vectors, using components.

ARITHMETIC OPERATIONS ON VECTORS

If $\mathbf{v} = \langle v_1, v_2 \rangle$ and $\mathbf{w} = \langle w_1, w_2 \rangle$ are vectors and c is any scalar, then

$$\mathbf{v} + \mathbf{w} = \langle v_1 + w_1, v_2 + w_2 \rangle \qquad \text{Vector addition}$$
$$\mathbf{v} - \mathbf{w} = \langle v_1 - w_1, v_2 - w_2 \rangle \qquad \text{Vector subtraction}$$
$$c\mathbf{v} = \langle cv_1, cv_2 \rangle \qquad \text{Scalar multiplication}$$

EXAMPLE 3   **Operations on Vectors**

Let $\mathbf{v} = \langle 2, 3 \rangle$ and $\mathbf{w} = \langle -4, 1 \rangle$. Find each expression.

a. $\mathbf{v} + \mathbf{w}$ **b.** $-2\mathbf{v}$ **c.** $2\mathbf{v} - \mathbf{w}$ **d.** $\|2\mathbf{v} - \mathbf{w}\|$

SOLUTION

a. $\mathbf{v} + \mathbf{w} = \langle 2, 3 \rangle + \langle -4, 1 \rangle = \langle 2 - 4, 3 + 1 \rangle = \langle -2, 4 \rangle$

b. $-2\mathbf{v} = -2\langle 2, 3 \rangle = \langle -2 \cdot 2, -2 \cdot 3 \rangle = \langle -4, -6 \rangle$

c. $2\mathbf{v} - \mathbf{w} = 2\langle 2, 3 \rangle - \langle -4, 1 \rangle$

$\qquad\qquad = \langle 4, 6 \rangle - \langle -4, 1 \rangle \qquad$ Scalar multiplication

$\qquad\qquad = \langle 4 - (-4), 6 - 1 \rangle \qquad$ Vector subtraction

$\qquad\qquad = \langle 8, 5 \rangle \qquad$ Simplify.

d. $\|2\mathbf{v} - \mathbf{w}\| = \|\langle 8, 5 \rangle\| \qquad\qquad$ From part **c**

$\qquad\qquad = \sqrt{8^2 + 5^2} \qquad\qquad \mathbf{v} = \langle v_1, v_2 \rangle; \|\mathbf{v}\| = \sqrt{v_1^2 + v_2^2}$

$\qquad\qquad = \sqrt{64 + 25} = \sqrt{89} \qquad$ Simplify.   ■ ■ ■

Practice Problem 3   Let $\mathbf{v} = \langle -1, 2 \rangle$ and $\mathbf{w} = \langle 2, -3 \rangle$. Find each expression.

a. $\mathbf{v} + \mathbf{w}$ **b.** $3\mathbf{w}$ **c.** $3\mathbf{w} - 2\mathbf{v}$ **d.** $\|3\mathbf{w} - 2\mathbf{v}\|$ ■

3 Find a unit vector in the direction of **v**.

Unit Vectors

Recall that if $c > 0$ and $\mathbf{v}$ is any vector, then the vector $c\mathbf{v}$ has the same direction as $\mathbf{v}$ and its length is given by

$$\|c\mathbf{v}\| = c\|\mathbf{v}\|.$$

If $\mathbf{v}$ is a nonzero vector and $c = \dfrac{1}{\|\mathbf{v}\|}$, then $\left\|\dfrac{1}{\|\mathbf{v}\|}\mathbf{v}\right\| = \dfrac{1}{\|\mathbf{v}\|}\|\mathbf{v}\| = 1$. Consequently, $\dfrac{1}{\|\mathbf{v}\|}\mathbf{v}$ is a vector of length 1 in the same direction as $\mathbf{v}$. A vector of length 1 is called a **unit vector**.

EXAMPLE 4   **Finding a Unit Vector**

Find a unit vector $\mathbf{u}$ in the direction of $\mathbf{v} = \langle 3, -4 \rangle$.

SOLUTION

First, find magnitude of $\mathbf{v} = \langle 3, -4 \rangle$.

$$\|\mathbf{v}\| = \sqrt{(3)^2 + (-4)^2} \qquad \text{Formula for magnitude}$$
$$\|\mathbf{v}\| = \sqrt{25} = 5 \qquad \text{Simplify.}$$

Now let

$$\mathbf{u} = \frac{1}{\|\mathbf{v}\|}\mathbf{v} \qquad \text{Multiply } \mathbf{v} \text{ by the scalar } \frac{1}{\|\mathbf{v}\|}.$$

$$\mathbf{u} = \frac{1}{5}\langle 3, -4 \rangle \qquad \text{Substitute values.}$$

$$\mathbf{u} = \left\langle \frac{3}{5}, -\frac{4}{5} \right\rangle \qquad \text{Scalar multiplication}$$

Check that $\|\mathbf{u}\| = 1$:

$$\|\mathbf{u}\| = \sqrt{\left(\frac{3}{5}\right)^2 + \left(-\frac{4}{5}\right)^2} = \sqrt{\frac{9}{25} + \frac{16}{25}} = \sqrt{\frac{25}{25}} = 1$$

Therefore, $\mathbf{u} = \left\langle \frac{3}{5}, -\frac{4}{5} \right\rangle$ is a unit vector in the same direction as $\mathbf{v}$. (Note that $\mathbf{u}$ is a scalar multiple of $\mathbf{v}$.) ■ ■ ■

Practice Problem 4 Find a unit vector in the direction of $\mathbf{v} = \langle -12, 5 \rangle$. ■

Vectors in **i**, **j** Form

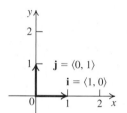

FIGURE 6.41 Standard unit vectors

In a Cartesian coordinate plane, two important unit vectors lie along the positive coordinate axes:

$$\mathbf{i} = \langle 1, 0 \rangle \quad \text{and} \quad \mathbf{j} = \langle 0, 1 \rangle$$

The unit vectors $\mathbf{i}$ and $\mathbf{j}$ are called **standard unit vectors**. See Figure 6.41.

Every vector $\mathbf{v} = \langle v_1, v_2 \rangle$ can be expressed in terms of $\mathbf{i}$ and $\mathbf{j}$ as follows:

$$\begin{aligned}\mathbf{v} = \langle v_1, v_2 \rangle &= \langle v_1, 0 \rangle + \langle 0, v_2 \rangle \qquad && \text{Vector addition} \\ &= v_1\langle 1, 0 \rangle + v_2\langle 0, 1 \rangle && \text{Scalar multiplication} \\ &= v_1\mathbf{i} + v_2\mathbf{j} && \text{Replace } \langle 1, 0 \rangle \text{ with } \mathbf{i} \text{ and } \langle 0, 1 \rangle \text{ with } \mathbf{j}.\end{aligned}$$

The scalars v_1 and v_2 are called the *horizontal* and *vertical components of* $\mathbf{v}$, respectively. A vector $\mathbf{v}$ from $(0, 0)$ to (v_1, v_2) can therefore be represented in the form

$$\mathbf{v} = v_1\mathbf{i} + v_2\mathbf{j}$$

with

$$\|\mathbf{v}\| = \sqrt{v_1^2 + v_2^2}.$$

EXAMPLE 5 **Vectors Involving i and j**

Find each expression for $\mathbf{u} = 4\mathbf{i} + 7\mathbf{j}$ and $\mathbf{v} = 2\mathbf{i} + 5\mathbf{j}$.

a. $\mathbf{u} - 3\mathbf{v}$ **b.** $\|\mathbf{u} - 3\mathbf{v}\|$

SOLUTION

a.
$$\begin{aligned}\mathbf{u} - 3\mathbf{v} &= (4\mathbf{i} + 7\mathbf{j}) - 3(2\mathbf{i} + 5\mathbf{j}) \\ &= 4\mathbf{i} + 7\mathbf{j} - 6\mathbf{i} - 15\mathbf{j} \qquad && \text{Scalar multiplication} \\ &= (4 - 6)\mathbf{i} + (7 - 15)\mathbf{j} && \text{Group terms.} \\ &= -2\mathbf{i} - 8\mathbf{j} && \text{Simplify.}\end{aligned}$$

b.
$$\begin{aligned}\|\mathbf{u} - 3\mathbf{v}\| &= \|-2\mathbf{i} - 8\mathbf{j}\| \\ &= \sqrt{(-2)^2 + (-8)^2} \qquad && \|v_1\mathbf{i} + v_2\mathbf{j}\| = \sqrt{v_1^2 + v_2^2} \\ &= \sqrt{68} = 2\sqrt{17} && \text{Simplify.}\end{aligned}$$
■ ■ ■

Practice Problem 5 Find each expression for $\mathbf{u} = -3\mathbf{i} + 2\mathbf{j}$ and $\mathbf{v} = \mathbf{i} + 4\mathbf{j}$.

a. $3\mathbf{u} + 2\mathbf{v}$ **b.** $\|3\mathbf{u} + 2\mathbf{v}\|$ ■

4 Write a vector in terms of its magnitude and direction.

Vector in Terms of Magnitude and Direction

Let $\mathbf{v} = \langle v_1, v_2 \rangle$ be a position vector and suppose θ is the smallest positive angle that $\mathbf{v}$ makes with the positive x-axis. The angle θ is called the **direction angle** of $\mathbf{v}$. See Figure 6.42.

The length (or magnitude) of the vector $\overrightarrow{OP}$ is $\|\mathbf{v}\|$; so,

$$\frac{v_1}{\|\mathbf{v}\|} = \cos\theta \quad \text{and} \quad \frac{v_2}{\|\mathbf{v}\|} = \sin\theta \qquad \text{Definitions of cosine and sine}$$

$$v_1 = \|\mathbf{v}\| \cos\theta \qquad v_2 = \|\mathbf{v}\| \sin\theta \qquad \text{Multiply both sides by } \|\mathbf{v}\|.$$

Hence,

$$\mathbf{v} = v_1\mathbf{i} + v_2\mathbf{j} \qquad\qquad \text{Write } \mathbf{v} \text{ in terms of } \mathbf{i} \text{ and } \mathbf{j}.$$

$$\mathbf{v} = \|\mathbf{v}\| \cos\theta\,\mathbf{i} + \|\mathbf{v}\| \sin\theta\,\mathbf{j} \qquad \text{Replace values of } v_1 \text{ and } v_2.$$

$$\mathbf{v} = \|\mathbf{v}\|(\cos\theta\,\mathbf{i} + \sin\theta\,\mathbf{j}) \qquad \text{Factor out } \|\mathbf{v}\|.$$

FIGURE 6.42 $\mathbf{v} = \|\mathbf{v}\|(\cos\theta\,\mathbf{i} + \sin\theta\,\mathbf{j})$

VECTOR IN TERMS OF MAGNITUDE AND DIRECTION

The formula

$$\mathbf{v} = \|\mathbf{v}\|(\cos\theta\,\mathbf{i} + \sin\theta\,\mathbf{j})$$

expresses a vector $\mathbf{v}$ in terms of its magnitude $\|\mathbf{v}\|$ and its direction angle θ.

Note that any angle coterminal with the direction angle θ of a vector $\mathbf{v}$ can be used in place of θ in the formula $\mathbf{v} = \|\mathbf{v}\|(\cos\theta\,\mathbf{i} + \sin\theta\,\mathbf{j})$.

EXAMPLE 6 **Writing a Vector with Given Length and Direction Angle**

Write the vector of magnitude 3 that makes an angle of $\dfrac{\pi}{3}$ with the positive x-axis.

SOLUTION

If $\mathbf{v}$ is the required vector, then

$$\mathbf{v} = \|\mathbf{v}\|(\cos\theta\,\mathbf{i} + \sin\theta\,\mathbf{j}) \qquad \text{Write } \mathbf{v} \text{ in terms of its magnitude and direction.}$$

$$\mathbf{v} = 3\left(\cos\frac{\pi}{3}\,\mathbf{i} + \sin\frac{\pi}{3}\,\mathbf{j}\right) \qquad \text{Substitute } \|\mathbf{v}\| = 3 \text{ and } \theta = \frac{\pi}{3}.$$

$$\mathbf{v} = 3\left(\frac{1}{2}\,\mathbf{i} + \frac{\sqrt{3}}{2}\,\mathbf{j}\right) \qquad \cos\frac{\pi}{3} = \frac{1}{2}, \sin\frac{\pi}{3} = \frac{\sqrt{3}}{2}$$

$$\mathbf{v} = \frac{3}{2}\,\mathbf{i} + \frac{3\sqrt{3}}{2}\,\mathbf{j} \qquad \text{Distributive property} \qquad\qquad ■ ■ ■$$

Practice Problem 6 Write the vector of magnitude 2 that makes an angle of $\dfrac{11\pi}{6}$ with the positive x-axis. ■

We can find the direction angle θ of a position vector $\mathbf{v} = \langle v_1, v_2 \rangle$ by using the equation $\tan\theta = \dfrac{v_2}{v_1}$ and the quadrant in which the point (v_1, v_2) lies.

EXAMPLE 7   **Finding the Direction Angle of a Vector**

Find the direction angle of the vector $\mathbf{v} = -4\mathbf{i} + 3\mathbf{j}$.

SOLUTION

$$\mathbf{v} = -4\mathbf{i} + 3\mathbf{j}$$
$$= \langle -4, 3 \rangle \qquad \text{Write } \mathbf{v} \text{ as a position vector.}$$
$$\tan \theta = \frac{3}{-4} = -\frac{3}{4}$$

The reference angle θ' is given by

$$\theta' = \left| \tan^{-1}\left(-\frac{3}{4}\right) \right| \approx |-36.87°| = 36.87°.$$

Since the point $(-4, 3)$ lies in quadrant II, we have

$$\theta = 180° - \theta' \approx 180° - 36.87° = 143.13°.$$

The direction angle of $\mathbf{v}$ is 143.13°. See Figure 6.43.   ■ ■ ■

Practice Problem 7   Find the direction angle of $\mathbf{v} = 2\mathbf{i} - 3\mathbf{j}$.   ■

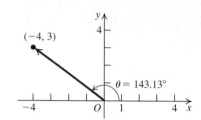

FIGURE 6.43 The direction angle of $\mathbf{v} = -4\mathbf{i} + 3\mathbf{j}$

5 Use vectors in applications.

Applications of Vectors

It is an experimental fact that forces behave like vectors. This means that if a system of forces acts on a particle, that particle will move as though it were acted on by a single force equal to the vector sum of the forces. This single force is called the **resultant** of the system of forces. If the particle does not move, the resultant is zero. We say that the particle is in **equilibrium.**

EXAMPLE 8   **Finding the Resultant**

Find the magnitude and bearing of the resultant $\mathbf{R}$ of two forces $\mathbf{F}_1$ and $\mathbf{F}_2$, where $\mathbf{F}_1$ is a 50 lb force acting northward and $\mathbf{F}_2$ is a 40 lb force acting eastward.

SOLUTION

We set up the coordinate system with the y-axis pointing northward. Then $\mathbf{F}_1 = 50\mathbf{j}$ and $\mathbf{F}_2 = 40\mathbf{i}$. See Figure 6.44.

$$\mathbf{R} = \mathbf{F}_1 + \mathbf{F}_2 = \mathbf{F}_2 + \mathbf{F}_1 = 40\mathbf{i} + 50\mathbf{j}$$
$$\|\mathbf{R}\| = \sqrt{(40)^2 + (50)^2} \approx 64.0$$
$$\tan \theta = \frac{50}{40} \qquad\qquad \text{Direction angle}$$
$$\theta = \tan^{-1}\left(\frac{50}{40}\right) \approx 51.3° \qquad \text{Use a calculator.}$$

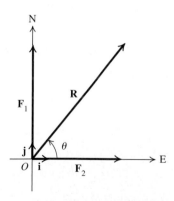

FIGURE 6.44 The resultant R

The angle between $\mathbf{R}$ and the y-axis (north direction) is $90° - 51.3° = 38.7°$. Therefore, $\mathbf{R}$ is a force of 64.0 lb in the direction N 38.7° E.   ■ ■ ■

Practice Problem 8   Repeat Example 8 assuming that $\mathbf{F}_1$ is a 40 lb force acting northward and $\mathbf{F}_2$ is a 30 lb force acting westward.   ■

EXAMPLE 9   Using Vectors in Air Navigation

An F-15 fighter jet is flying over Mount Rushmore at an airspeed (speed in still air) of 800 miles per hour on a bearing of N 30° E. The velocity of wind is 40 miles per hour in the direction of S 45° E. Find the actual speed and direction (relative to the ground) of the plane. Round each answer to the nearest tenth.

SOLUTION

Set up a coordinate system with north along the positive y-axis. See Figure 6.45.
    Let **v** be the air velocity of the plane,
        w be the wind velocity, and
        r be the resultant ground velocity of the plane.
Writing each vector in terms of its magnitude and the direction angle θ that the vector makes with the positive x-axis, we have

$$\mathbf{v} = 800\,(\cos 60°\mathbf{i} \ + \ \sin 60°\mathbf{j}) \qquad 90° - 30° = 60°$$
$$\mathbf{w} = 40[\cos 315°\mathbf{i} + \sin 315°\mathbf{j}] \qquad 270° + 45° = 315°$$
$$= 40[\cos 45°\mathbf{i} - \sin 45°\mathbf{j}] \qquad \cos 315° = \cos 45°; \sin 315° = -\sin 45°$$

The resultant **r** is given by

$$\mathbf{r} = \mathbf{v} + \mathbf{w}$$

$$\mathbf{r} = 800(\cos 60°\mathbf{i} \ + \ \sin 60°\mathbf{j}) + 40(\cos 45°\mathbf{i} - \sin 45°\mathbf{j}) \qquad$$ Substitute values of **v** and **w**.

$$\mathbf{r} = (800 \cos 60° \ + \ 40 \cos 45°)\mathbf{i} + (800 \sin 60° - 40 \sin 45°)\mathbf{j} \qquad$$ Add vectors.

$$\|\mathbf{r}\| = \sqrt{(800 \cos 60° \ + \ 40 \cos 45°)^2 + (800 \sin 60° - 40 \sin 45°)^2} \qquad \|v_1\mathbf{i} + v_2\mathbf{j}\| = \sqrt{v_1^2 + v_2^2}$$

$$\|\mathbf{r}\| \approx 790.6 \qquad$$ Use a calculator.

The ground speed of the F-15 is approximately 790.6 miles per hour. To find the actual direction (bearing) of the plane, we first find the direction angle θ of **r**.

$$\theta = \tan^{-1}\left(\frac{800 \sin 60° \ - \ 40 \sin 45°}{800 \cos 60° \ + \ 40 \cos 45°}\right) \qquad \text{For } \mathbf{r} = r_1\mathbf{i} + r_2\mathbf{j}, \theta = \tan^{-1}\left(\frac{r_2}{r_1}\right)$$

$$\approx 57.2° \qquad \text{Use a calculator.}$$

The angle between **r** and the y-axis (direction north) is

$$90° - 57.2° = 32.8°.$$

The bearing of the F-15 is approximately N 32.8° E.   ■ ■ ■

Practice Problem 9   Repeat Example 9 assuming that the bearing of the plane is N 60° W and the direction of the wind is S 30°W.   ■

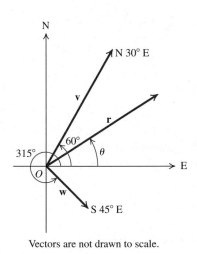

N

N 30° E

v

r

315°

60°

θ

O

E

w

S 45° E

Vectors are not drawn to scale.

FIGURE 6.45 Velocity vectors

EXAMPLE 10   Obtaining an Equilibrium

What single force must be added to the system of forces **A**, **B**, **C**, and **D** shown in Figure 6.46 to obtain a system that is in equilibrium?

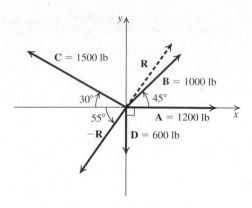

FIGURE 6.46 A system in equilibrium

SOLUTION

We first find the resultant **R** of the given system. The simplest procedure is to find and add the corresponding horizontal components and the vertical components of each force.

The horizontal component R_1 of **R** is given by

$$R_1 = A_1 \qquad + B_1 \qquad + C_1 \qquad + D_1$$
$$= 1200 \cos 0° + 1000 \cos 45° + 1500 \cos 150° + 600 \cos (270°)$$
$$\approx 608.07$$

Similarly, the vertical component R_2 of **R** is given by

$$R_2 = A_2 \qquad + B_2 \qquad + C_2 \qquad + D_2$$
$$= 1200 \sin 0° + 1000 \sin 45° + 1500 \sin 150° + 600 \sin (270°)$$
$$\approx 857.11$$
$$\mathbf{R} = R_1\mathbf{i} \qquad + R_2\mathbf{j}$$
$$= 608.07\mathbf{i} + 857.11\mathbf{j}$$
$$\|\mathbf{R}\| = \sqrt{(608.07)^2 + (857.11)^2} \approx 1051 \text{ lb}$$
$$\theta = \tan^{-1}\left(\frac{R_2}{R_1}\right) = \tan^{-1}\left(\frac{857.11}{608.07}\right)$$
$$\approx 55.0° \qquad \text{Use a calculator}$$

The force that must be added to obtain equilibrium is the negative of **R**. This is a force of 1051 lb, and −**R** makes an angle of 180° + 55° = 235° with the positive x-axis. See Figure 6.46. ■ ■ ■

Practice Problem 10 Find the resultant of the following system of forces acting simultaneously at a point: **u** = 200 lb in the direction N 40° E, **v** = 300 lb in the direction N 70° W, and **w** = 400 lb in the direction S 20° E. ■

> **RECALL**
>
> For a vector **v** with direction angle θ, the horizontal component $v_1 = \|\mathbf{v}\| \cos \theta$ and vertical component $v_2 = \|\mathbf{v}\| \sin \theta$.

SECTION 6.4 ■ Exercises

A EXERCISES Basic Skills and Concepts

1. A vector is a quantity that is characterized by a magnitude and a(n) ___direction___.

2. The resultant of **v** and **w** is the vector sum **v** + **w** and is represented by the diagonal of the ___parallelogram___ with adjacent sides **v** and **w**.

3. If **v** = $\langle a, b \rangle$, then $\|\mathbf{v}\| =$ ___$\sqrt{a^2 + b^2}$___, and its direction angle $\theta = \tan^{-1}($ ___$\frac{b}{a}$___).

4. If **v** is a nonzero vector, then the unit vector **u** in the direction of **v** is given by $\mathbf{u} = \dfrac{1}{\|\mathbf{v}\|}($ ___**v**___).

†Due to space constrictions, answers to these exercises may be found in the Answers beginning on page A–1 in the back of the book.

5. *True or False* The zero vector **0** is the only vector with no direction specified. True

6. *True or False* The vector −**v** has the same magnitude as **v** but the opposite direction. True

In Exercises 7–14, use the vectors u, v, and w in the accompanying figure to graph each vector.

7. **u** + **v** †

8. 3**v** †

9. 2**u** − **w** †

10. **w** − 2**v** †

11. **u** + **v** + **w** †

12. 2**u** − **w** + **v** †

13. **w** − 2**v** + **u** †

14. 2**v** + 3**w** − **u** †

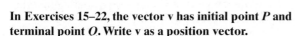

In Exercises 15–22, the vector v has initial point P and terminal point Q. Write v as a position vector.

15. $P(3, 6), Q(2, 9)$ $\langle -1, 3 \rangle$ **16.** $P(6, -4), Q(1, 1)$ $\langle -5, 5 \rangle$

17. $P(-5, -2), Q(-3, -4)$ † **18.** $P(0, 0), Q(-3, -6)$ $\langle -3, -6 \rangle$

19. $P(-1, 4), Q(2, -3)$ $\langle 3, -7 \rangle$ **20.** $P(3.5, 2.7), Q(-1.5, 1.3)$ †

21. $P\left(\frac{1}{2}, \frac{3}{4}\right), Q\left(-\frac{1}{2}, -\frac{7}{4}\right)$ † **22.** $P\left(-\frac{2}{3}, \frac{4}{9}\right), Q\left(\frac{1}{3}, -\frac{2}{3}\right)$ †

In Exercises 23–26, determine whether the vectors $\overrightarrow{AB}$ and $\overrightarrow{CD}$ are equivalent. [Hint: Write $\overrightarrow{AB}$ and $\overrightarrow{CD}$ as position vectors.]

23. $A(1, 0), B(3, 4), C(-1, 2), D(1, 6)$ Equivalent

24. $A(-1, 2), B(3, -2), C(2, 5), D(6, 1)$ Equivalent

25. $A(2, -1), B(3, 5), C(-1, 3), D(-2, -3)$ Not equivalent

26. $A(5, 7), B(6, 3), C(-2, 1), D(-3, 5)$ Not equivalent

In Exercises 27–34, let $v = \langle -1, 2 \rangle$ and $w = \langle 3, -2 \rangle$. Find each expression.

27. $\|v\|$ $\sqrt{5}$ **28.** $\|w\|$ $\sqrt{13}$

29. $v - w$ $\langle -4, 4 \rangle$ **30.** $v + w$ $\langle 2, 0 \rangle$

31. $2v - 3w$ $\langle -11, 10 \rangle$ **32.** $2w - 3v$ $\langle 9, -10 \rangle$

33. $\|2v - 3w\|$ $\sqrt{221}$ **34.** $\|2w - 3v\|$ $\sqrt{181}$

In Exercises 35–40, find a unit vector u in the direction of the given vector.

35. $\langle 1, -1 \rangle$ † **36.** $\langle 1, 3 \rangle$ †

37. $\langle -4, 3 \rangle$ † **38.** $\langle 5, -12 \rangle$ †

39. $\langle \sqrt{2}, \sqrt{2} \rangle$ † **40.** $\langle \sqrt{5}, -2 \rangle$ †

In Exercises 41–46, find each expression for $u = 2i − 5j$ and $v = −3i − 2j$.

41. $u + v$ $-i - 7j$ **42.** $u - v$ $5i - 3j$

43. $2u - 3v$ $13i - 4j$ **44.** $2v + 3u$ $-19j$

45. $\|2u - 3v\|$ $\sqrt{185}$ **46.** $\|2v + 3u\|$ 19

In Exercises 47–54, write the vector v in the form $v_1 i + v_2 j$, given $\|v\|$ and the angle θ that v makes with the positive x-axis.

47. $\|v\| = 2, \theta = 30°$ $\sqrt{3}i + j$ **48.** $\|v\| = 5, \theta = 45°$ †

49. $\|v\| = 4, \theta = 120°$ † **50.** $\|v\| = 3, \theta = 150°$ †

51. $\|v\| = 3, \theta = \frac{5\pi}{3}$ † **52.** $\|v\| = 4, \theta = \frac{11\pi}{6}$ †

53. $\|v\| = 7, \theta = -\frac{\pi}{3}$ † **54.** $\|v\| = 8, \theta = \frac{3\pi}{4}$ †

In Exercises 55–62, find the magnitude and the direction angle of the vector v.

55. $v = 10(\cos 60°i + \sin 60°j)$ $10, 60°$

56. $v = -4(\cos 30°i + \sin 30°j)$ $4, 210°$

57. $v = -3(\cos 30°i - \sin 30°j)$ $3, 150°$

58. $v = 2(\cos 300°i - \sin 300°j)$ $2, 60°$

59. $v = 5i + 12j$ $13, 67.38°$ **60.** $v = 12i - 5j$ $13, 337.38°$

61. $v = -4i - 3j$ $5, 216.87°$ **62.** $v = -5i + 12j$ $13, 112.62°$

B EXERCISES Applying the Concepts

In Exercises 63–72 write each answer to the nearest tenth of a unit.

63. Wind and vectors. A wind is blowing 25 miles per hour in the direction N 67° E. Express the wind velocity **v** in the form $ai + bj$. $23.0i + 9.8j$

64. Wind and vectors. The velocity **v** of a wind is given by $v = 5i − 12j$, where the vectors **i** and **j** represent 1-mile-per-hour winds blowing east and north, respectively. Find the speed (magnitude of **v**) and the direction of the wind. 13 mi/hr, S 22.6° E

65. Resultant force. Find the magnitude and bearing of the resultant **R** of two forces **F₁** and **F₂**, where **F₁** is a force of 25 lb acting due south and **F₂** is a force of 32 lb acting due west. $40.61, 218.0°$

66. Repeat Exercise 65 assuming that **F₁** is doubled. $59.36, 237.38°$

67. Finding components. A force of 80 lb acts in the direction N 65° W. Find its east and north components. $-72.50, 33.81$

68. Repeat Exercise 67 assuming that a force of 60 lb acts in the direction S 32° W. $-31.80, -50.88$

69. Resultant force. Two forces of 300 lb and 400 lb act at the origin and make angles of 30° and 60°, respectively, with the positive x-axis. Find the resultant. †

70. Repeat Exercise 69 assuming that the forces have the same magnitude but that the angle each force makes with the positive x-axis is doubled. †

71. Air navigation. A plane flies at an airspeed (speed in still air) of 500 miles per hour on a bearing of N 35° E. An east wind (wind from east to west) is blowing at 30 miles per hour. Find the plane's ground speed and direction. 483.4 mi/hr, N 32.1° E

72. Air navigation. A plane flies at an airspeed of 550 miles per hour on a straight course from an airfield *F*. A 30-mile-per-hour wind is blowing from the west. What must be the plane's bearing so that after one hour of flying time the plane is due north of *F*? N 3.1° W

73. River navigation. A river flowing southward has a current of 4 miles per hour. A motorboat in still water maintains a speed of 15 miles per hour. The motorboat starts at the west shore on a bearing of N 40° E. What is the actual speed and direction of the boat? 12.2 mi/hr, N 52.2° E

74. River crossing. A fisherwoman on the west bank of the river of Exercise 73 sees several people fishing at a spot P directly east of her. If she owns the boat of Exercise 73, what direction should she take to land on the spot P? N 74.5° E

In Exercises 75 and 76, find a single force that must be added to the system of forces shown in the figure to obtain a system that is in equilibrium.

75.

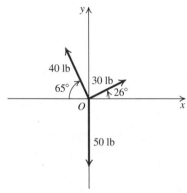

76.

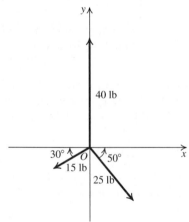

Answers:
75. A force of magnitude 10.08 pounds making an angle of 176.61° with the positive x-axis **76.** A force of magnitude 13.70 pounds making an angle of 257.01° with the positive x-axis

C EXERCISES Beyond the Basics

In Exercises 77 and 78 calculate the magnitude of the forces F_1 and F_2 assuming that the system is in equilibrium.

77.

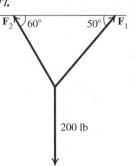

78.

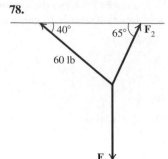

79. Find the terminal point of $\mathbf{v} = 4\mathbf{i} - 3\mathbf{j}$ assuming that the initial point is $(-2, 1)$. $(2, -2)$

80. Find the initial point of $\mathbf{v} = \langle -3, 5 \rangle$ assuming that the terminal point is $(4, 0)$. $(7, -5)$

81. Let $\mathbf{u} = \langle -1, 2 \rangle$ and $\mathbf{v} = \langle 3, 5 \rangle$. Find a vector $\mathbf{x} = \langle x_1, x_2 \rangle$ that satisfies the equation $2\mathbf{u} - \mathbf{x} = 2\mathbf{x} + 3\mathbf{v}$.

82. Let $P(3, 5)$ and $Q(7, -4)$ be two points in a coordinate plane. Use vectors to find the coordinates of the point on the line segment joining P and Q that is $\dfrac{3}{4}$ of the way from P to Q. $\left(6, -\dfrac{7}{4} \right)$

83. Use vectors to find the lengths of the diagonals of the parallelogram determined by $\mathbf{i} + 2\mathbf{j}$ and $\mathbf{i} - \mathbf{j}$. $3, \sqrt{5}$

84. Use vectors to find the fourth vertex of a parallelogram, with three vertices at $(0, 0), (2, 3)$ and $(6, 4)$. [*Hint:* There is more than one answer.] $(4, 1), (-4, -1), (8, 7)$

Critical Thinking

Let $A = (1, 2), B = (4, 3)$, and $C = (6, 1)$ be points in the plane. Find $P = (x, y)$ so that each of the following is true.

85. $\overrightarrow{AB} = \overrightarrow{CP}$ $(9, 2)$

86. $\overrightarrow{PA} = \overrightarrow{CB}$ $(3, 0)$

87. $\overrightarrow{AP} = \overrightarrow{CB}$ $(-1, 4)$

88. $\overrightarrow{PA} = \overrightarrow{BC}$ $(-1, 4)$

77. $\|\mathbf{F}_1\| = 106.42$ lb, $\|\mathbf{F}_2\| = 136.81$ lb **78.** $\|\mathbf{F}_1\| = 137.13$ lb, $\|\mathbf{F}_2\| = 108.76$ lb **81.** $\left\langle -\dfrac{11}{3}, -\dfrac{11}{3} \right\rangle$

The Dot Product

Before Starting this Section, Review

1. Pythagorean Theorem (Appendix A, page 790)

2. The Law of Sines (Section 6.2, page 405)

3. The Law of Cosines (Section 6.3, page 418)

Objectives

1 Define the dot product of two vectors.

2 Find the angle between two vectors.

3 Define orthogonal vectors.

4 Find the projection of a vector onto another vector.

5 Decompose a vector into two orthogonal vectors.

6 Use the definition of work.

HOW TO MEASURE WORK

If you move an object a distance of d units by applying a force F in the direction of motion, then physicists define the work W done on the object by the force F to be

$$W = Fd$$

$$\text{Work} = (\text{force})(\text{distance}).$$

If the force is measured in pounds and the distance is measured in feet, then the unit of work is *foot-pounds*. Unfortunately, it is not always possible to exert a force in the direction you would like. For example, if you are pulling a child's wagon, then you exert a force in a direction dictated by the position of the handle. In this section, we give a precise definition of work done by a constant force, and in Example 10, we compute the work done in one such instance. ∎

---

1 Define the dot product of two vectors.

The Dot Product

We know that scalar multiplication and vector addition and subtraction produce vectors. A new operation, the *dot product* of two vectors, produces a *scalar*.

THE DOT PRODUCT

For two vectors $\mathbf{v} = \langle v_1, v_2 \rangle$ and $\mathbf{w} = \langle w_1, w_2 \rangle$, the dot product of $\mathbf{v}$ and $\mathbf{w}$, denoted $\mathbf{v} \cdot \mathbf{w}$, is defined as follows:

$$\mathbf{v} \cdot \mathbf{w} = \langle v_1, v_2 \rangle \cdot \langle w_1, w_2 \rangle = v_1 w_1 + v_2 w_2$$

EXAMPLE 1 **Finding the Dot Product**

Find the dot product $\mathbf{v} \cdot \mathbf{w}$.

a. $\mathbf{v} = \langle 2, -3 \rangle$ and $\mathbf{w} = \langle 3, 4 \rangle$ **b.** $\mathbf{v} = -3\mathbf{i} + 5\mathbf{j}$ and $\mathbf{w} = 2\mathbf{i} + 3\mathbf{j}$

SOLUTION

a. $\mathbf{v} \cdot \mathbf{w} = \langle 2, -3 \rangle \cdot \langle 3, 4 \rangle$

$\qquad = (2)(3) + (-3)(4) = -6$ Definition of dot product

b. $\mathbf{v} \cdot \mathbf{w} = (-3\mathbf{i} + 5\mathbf{j}) \cdot (2\mathbf{i} + 3\mathbf{j})$

$= \langle -3, 5 \rangle \cdot \langle 2, 3 \rangle$ Rewrite as position vectors.

$= (-3)(2) + (5)(3) = 9$ Definition of dot product   ■ ■ ■

Practice Problem 1   Find the dot product $\mathbf{v} \cdot \mathbf{w}$.

a. $\mathbf{v} = \langle 1, 2 \rangle$ and $\mathbf{w} = \langle -2, 5 \rangle$ **b.** $\mathbf{v} = 4\mathbf{i} - 3\mathbf{j}$ and $\mathbf{w} = -3\mathbf{i} - 4\mathbf{j}$ ■

2 Find the angle between two vectors.

The Angle Between Two Vectors

We use the Law of Cosines to find another formula for the dot product. The new formula will be used to find the angle between any two vectors.

Let θ be the angle between the two vectors $\mathbf{v} = \langle v_1, v_2 \rangle$ and $\mathbf{w} = \langle w_1, w_2 \rangle$. See Figure 6.47. Apply the Law of Cosines to the triangle OPQ in Figure 6.47.

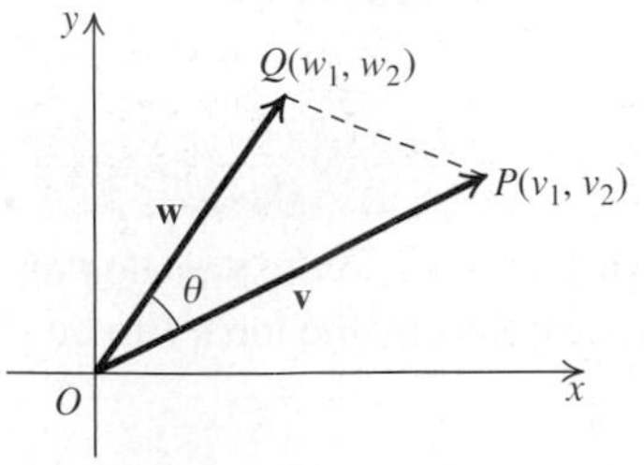

FIGURE 6.47 Angle between two vectors

$[d(P, Q)]^2 = \|\mathbf{v}\|^2 + \|\mathbf{w}\|^2 - 2\|\mathbf{v}\| \|\mathbf{w}\| \cos \theta$ The Law of Cosines

$(v_1 - w_1)^2 + (v_2 - w_2)^2$
$= (v_1^2 + v_2^2) + (w_1^2 + w_2^2) - 2\|\mathbf{v}\| \|\mathbf{w}\| \cos \theta$ Use the distance formula and square of the magnitudes of $\mathbf{v}$ and $\mathbf{w}$.

$v_1^2 - 2v_1 w_1 + w_1^2 + v_2^2 - 2v_2 w_2 + w_2^2$
$= v_1^2 + v_2^2 + w_1^2 + w_2^2 - 2\|\mathbf{v}\| \|\mathbf{w}\| \cos \theta$ $(A - B)^2 = A^2 - 2AB + B^2$

$-2v_1 w_1 - 2v_2 w_2 = -2\|\mathbf{v}\| \|\mathbf{w}\| \cos \theta$ Subtract $v_1^2 + w_1^2 + v_2^2 + w_2^2$ from both sides.

$v_1 w_1 + v_2 w_2 = \|\mathbf{v}\| \|\mathbf{w}\| \cos \theta$ Divide both sides by -2.

$\mathbf{v} \cdot \mathbf{w} = \|\mathbf{v}\| \|\mathbf{w}\| \cos \theta$ Definition of $\mathbf{v} \cdot \mathbf{w}$

The last equation provides us with another formula for $\mathbf{v} \cdot \mathbf{w}$. Solving the last equation for $\cos \theta$ gives us a formula for finding the angle between two vectors.

THE DOT PRODUCT AND THE ANGLE BETWEEN TWO VECTORS

If θ ($0° \leq \theta \leq 180°$) is the angle between two nonzero vectors $\mathbf{v}$ and $\mathbf{w}$, then

$$\mathbf{v} \cdot \mathbf{w} = \|\mathbf{v}\| \|\mathbf{w}\| \cos \theta \quad \text{or} \quad \cos \theta = \frac{\mathbf{v} \cdot \mathbf{w}}{\|\mathbf{v}\| \|\mathbf{w}\|}.$$

EXAMPLE 2   **Finding the Dot Product**

If $\mathbf{v}$ and $\mathbf{w}$ are two vectors of magnitudes 5 and 7, respectively, and the angle between them is 75°, find $\mathbf{v} \cdot \mathbf{w}$. Round the answer to the nearest tenth.

SOLUTION

$\mathbf{v} \cdot \mathbf{w} = \|\mathbf{v}\| \|\mathbf{w}\| \cos \theta$ Formula for $\mathbf{v} \cdot \mathbf{w}$

$= (5)(7) \cos 75°$ Substitute given values.

≈ 9.1 Use a calculator.   ■ ■ ■

Practice Problem 2   Repeat Example 2 assuming that the angle between $\mathbf{v}$ and $\mathbf{w}$ is 60°.   ■

Two vectors $\mathbf{v}$ and $\mathbf{w}$ are **parallel** if there is a nonzero scalar, c, so that $\mathbf{v} = c\mathbf{w}$. The angle θ between parallel vectors is either 0° or 180°. So, $\mathbf{v}$ and $\mathbf{w}$ are parallel if $\mathbf{v} \cdot \mathbf{w} = \pm \|\mathbf{v}\| \|\mathbf{w}\|$.

| **EXAMPLE 3** | **Deciding Whether Two Vectors Are Parallel** |

Determine whether the vectors $\mathbf{v} = 2\mathbf{i} - 3\mathbf{j}$ and $\mathbf{w} = -4\mathbf{i} + 6\mathbf{j}$ are parallel.

SOLUTION

Since $\mathbf{w} = -2(2\mathbf{i} - 3\mathbf{j}) = -2\mathbf{v}$, the vectors $\mathbf{v}$ and $\mathbf{w}$ are parallel. Note that

$$\cos\theta = \frac{\mathbf{v} \cdot \mathbf{w}}{\|\mathbf{v}\|\|\mathbf{w}\|} = \frac{-8 - 18}{\sqrt{13}\sqrt{52}} = \frac{-26}{\sqrt{13}(2\sqrt{13})} = \frac{-26}{26} = -1,$$

so that $\theta = \pi$ is the angle between $\mathbf{v}$ and $\mathbf{w}$.   ■ ■ ■

Practice Problem 3   Determine whether $\mathbf{v} = -\mathbf{i} - 2\mathbf{j}$ and $\mathbf{w} = 2\mathbf{i} + \mathbf{j}$ are parallel. ■

Because the angle θ between two nonzero vectors $\mathbf{v}$ and $\mathbf{w}$ satisfies $\cos\theta = \dfrac{\mathbf{v} \cdot \mathbf{w}}{\|\mathbf{v}\|\|\mathbf{w}\|}$ and $0° \le \theta \le 180°$, we have $\theta = \cos^{-1}\left(\dfrac{\mathbf{v} \cdot \mathbf{w}}{\|\mathbf{v}\|\|\mathbf{w}\|}\right)$.

| **EXAMPLE 4** | **Finding the Angle Between Two Vectors** |

Find the angle θ (in degrees) between the vectors $\mathbf{v} = 2\mathbf{i} + 3\mathbf{j}$ and $\mathbf{w} = -3\mathbf{i} + 4\mathbf{j}$. Round the answer to the nearest tenth of a degree.

SOLUTION

$$\cos\theta = \frac{\mathbf{v} \cdot \mathbf{w}}{\|\mathbf{v}\|\|\mathbf{w}\|}$$ Formula for the cosine of the angle between $\mathbf{v}$ and $\mathbf{w}$

$$\cos\theta = \frac{(2\mathbf{i} + 3\mathbf{j}) \cdot (-3\mathbf{i} + 4\mathbf{j})}{\|2\mathbf{i} + 3\mathbf{j}\|\|-3\mathbf{i} + 4\mathbf{j}\|}$$ Substitute expressions for $\mathbf{v}$ and $\mathbf{w}$.

$$\cos\theta = \frac{(2)(-3) + (3)(4)}{\sqrt{2^2 + 3^2}\sqrt{(-3)^2 + 4^2}}$$ Use formulas for the dot product and the magnitude.

$$\cos\theta = \frac{6}{5\sqrt{13}}$$ Simplify.

$$\theta = \cos^{-1}\left(\frac{6}{5\sqrt{13}}\right)$$ $0° \le \theta \le 180°$

$$\theta \approx 70.6°$$ Use a calculator.   ■ ■ ■

Practice Problem 4   Find the angle (in degrees) between the vectors $\mathbf{v} = 2\mathbf{i} - 3\mathbf{j}$ and $\mathbf{w} = 3\mathbf{i} + 5\mathbf{j}$. Round the answer to the nearest tenth of a degree.   ■

We can verify the following properties of the dot product by writing the vectors as position vectors and using the definitions of vector addition, scalar multiplication, and the dot product.

PROPERTIES OF THE DOT PRODUCT

If $\mathbf{u}, \mathbf{v}$, and $\mathbf{w}$ are vectors and c is a scalar, then

1. $\mathbf{u} \cdot \mathbf{v} = \mathbf{v} \cdot \mathbf{u}$
2. $\mathbf{u} \cdot (\mathbf{v} + \mathbf{w}) = \mathbf{u} \cdot \mathbf{v} + \mathbf{u} \cdot \mathbf{w}$
3. $\mathbf{0} \cdot \mathbf{v} = 0$
4. $\mathbf{v} \cdot \mathbf{v} = \|\mathbf{v}\|^2$
5. $(c\mathbf{u}) \cdot \mathbf{v} = c(\mathbf{u} \cdot \mathbf{v}) = \mathbf{u} \cdot (c\mathbf{v})$

> ### EXAMPLE 5   Finding the Magnitude of the Sum of Two Vectors

Let **v** and **w** be two vectors in the plane of magnitudes 12 and 7, respectively. The angle between **v** and **w** is 65°. Find the magnitude of **v** + 2**w** to the nearest tenth.

SOLUTION

$$
\begin{aligned}
\|\mathbf{v} + 2\mathbf{w}\|^2 &= (\mathbf{v} + 2\mathbf{w}) \cdot (\mathbf{v} + 2\mathbf{w}) && \text{Property 4} \\
&= (\mathbf{v} + 2\mathbf{w}) \cdot \mathbf{v} + (\mathbf{v} + 2\mathbf{w}) \cdot (2\mathbf{w}) && \text{Property 2} \\
&= \mathbf{v} \cdot \mathbf{v} + 2(\mathbf{w} \cdot \mathbf{v}) + 2(\mathbf{v} \cdot \mathbf{w}) + 4(\mathbf{w} \cdot \mathbf{w}) && \text{Properties 1 and 5} \\
&= \|\mathbf{v}\|^2 + 4(\mathbf{v} \cdot \mathbf{w}) + 4\|\mathbf{w}\|^2 && \text{Properties 1 and 4} \\
&= \|\mathbf{v}\|^2 + 4\|\mathbf{v}\|\|\mathbf{w}\| \cos\theta + 4\|\mathbf{w}\|^2 && \mathbf{v} \cdot \mathbf{w} = \|\mathbf{v}\|\|\mathbf{w}\| \cos\theta \\
&= (12)^2 + 4(12)(7) \cos 65° + 4(7)^2 && \text{Substitute values.} \\
\|\mathbf{v} + 2\mathbf{w}\|^2 &\approx 481.9997 && \text{Use a calculator.} \\
\|\mathbf{v} + 2\mathbf{w}\| &\approx 21.95 \approx 22.0 && \text{Take the positive square root.}
\end{aligned}
$$

■ ■ ■

Practice Problem 5   Repeat Example 4 assuming that $\|\mathbf{v}\| = 20$ and $\|\mathbf{w}\| = 12$ and the angle between **v** and **w** is 54°.   ■

3   Define orthogonal vectors.

Orthogonal Vectors

We know that if **v** and **w** are two nonzero vectors and $\theta \, (0° \le \theta \le 180°)$ is the measure of the angle between them, then

$$\mathbf{v} \cdot \mathbf{w} = \|\mathbf{v}\|\|\mathbf{w}\| \cos\theta \qquad \text{Form of the dot product}$$

From this equation, it follows that

$$\cos\theta = 0 \text{ if and only if } \mathbf{v} \cdot \mathbf{w} = 0.$$

Because $0° \le \theta \le 180°$, we conclude that

$$\theta = 90° \text{ if and only if } \mathbf{v} \cdot \mathbf{w} = 0.$$

DEFINITION OF ORTHOGONAL VECTORS

Two vectors **v** and **w** are **orthogonal** (perpendicular) if and only if $\mathbf{v} \cdot \mathbf{w} = 0$.

Since $\mathbf{0} \cdot \mathbf{w} = 0$ for any vector **w**, it follows from the definition that the zero vector is orthogonal to every vector.

> ### EXAMPLE 6   Deciding Whether Vectors Are Orthogonal

Determine whether $\mathbf{v} = 2\mathbf{i} - 3\mathbf{j}$ and $\mathbf{w} = 3\mathbf{i} + 2\mathbf{j}$ are orthogonal.

SOLUTION

Since $\mathbf{v} \cdot \mathbf{w} = (2\mathbf{i} - 3\mathbf{j}) \cdot (3\mathbf{i} + 2\mathbf{j}) = \langle 2, -3 \rangle \cdot \langle 3, 2 \rangle = 6 - 6 = 0$, **v** and **w** are orthogonal.   ■ ■ ■

Practice Problem 6   Determine whether $\mathbf{v} = 4\mathbf{i} + 8\mathbf{j}$ and $\mathbf{w} = 2\mathbf{i} - \mathbf{j}$ are orthogonal.   ■

EXAMPLE 7   **Orthogonal Vectors**

Find the scalar c so that the vectors $\mathbf{v} = 3\mathbf{i} + 2\mathbf{j}$ and $\mathbf{w} = 4\mathbf{i} + c\mathbf{j}$ are orthogonal.

SOLUTION

The vectors $\mathbf{v} = 3\mathbf{i} + 2\mathbf{j} = \langle 3, 2 \rangle$ and $\mathbf{w} = 4\mathbf{i} + c\mathbf{j} = \langle 4, c \rangle$ are orthogonal if and only if

$$
\begin{aligned}
\mathbf{v} \cdot \mathbf{w} &= 0 && \text{Definition of orthogonality} \\
\langle 3, 2 \rangle \cdot \langle 4, c \rangle &= 0 \\
(3)(4) + (2)(c) &= 0 && \text{Definition of dot product} \\
12 + 2c &= 0 && \text{Simplify.} \\
c &= -6 && \text{Solve for } c.
\end{aligned}
$$
■ ■ ■

Practice Problem 7   Find the scalar k so that the vectors $\mathbf{v} = \langle k, -2 \rangle$ and $\mathbf{w} = \langle 4, 6 \rangle$ are orthogonal. ■

4 Find the projection of a vector onto another vector.

Projection of a Vector

A geometric interpretation of the dot product is obtained by considering the *projection* of a vector onto another vector.

Let $\overrightarrow{OP}$ and $\overrightarrow{OQ}$ be representations of the nonzero vectors $\mathbf{v}$ and $\mathbf{w}$, respectively. The projections of $\overrightarrow{OP}$ in the direction of $\overrightarrow{OQ}$ is the directed line segment $\overrightarrow{OR}$, where R is the foot of the perpendicular from P to the line containing $\overrightarrow{OQ}$. See Figure 6.48. Then the vector represented by $\overrightarrow{OR}$ is called the **vector projection of v onto w** and is denoted by $\text{proj}_{\mathbf{w}}\mathbf{v}$.

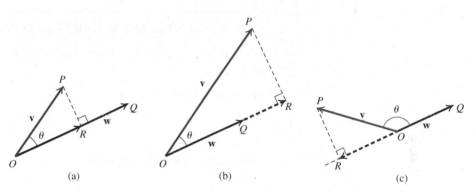

FIGURE 6.48 Vector projection of v onto w

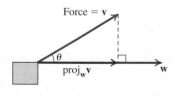

Force = v

$\text{proj}_{\mathbf{w}}\mathbf{v}$

FIGURE 6.49

In Figure 6.49, if a force $\mathbf{v}$ is used to pull a box, then the effective force that moves the box in the direction of $\mathbf{w}$ is the vector projection of $\mathbf{v}$ onto $\mathbf{w}$.

Let $\theta\,(0° \le \theta \le 180°)$ be the measure of the angle between nonzero vectors $\mathbf{v}$ and $\mathbf{w}$. The number $\|\mathbf{v}\| \cos \theta$ is called the **scalar projection of v onto w**. We can express this number in terms of the dot product.

$$
\begin{aligned}
\text{Scalar projection } \mathbf{v} \text{ onto } \mathbf{w} &= \|\mathbf{v}\| \cos \theta \\
&= \frac{\|\mathbf{v}\|\|\mathbf{w}\| \cos \theta}{\|\mathbf{w}\|} && \text{Multiply numerator and denominator by } \|\mathbf{w}\|. \\
&= \frac{\mathbf{v} \cdot \mathbf{w}}{\|\mathbf{w}\|} && \mathbf{v} \cdot \mathbf{w} = \|\mathbf{v}\|\|\mathbf{w}\| \cos \theta
\end{aligned}
$$

If θ is acute, then $\text{proj}_{\mathbf{w}}\mathbf{v}$ has magnitude $\|\mathbf{v}\| \cos \theta$ and direction $\dfrac{\mathbf{w}}{\|\mathbf{w}\|}$.

If θ is obtuse, $\cos\theta < 0$ and $\text{proj}_{\mathbf{w}}\mathbf{v}$ has magnitude $-\|\mathbf{v}\|\cos\theta$ and direction $-\dfrac{\mathbf{w}}{\|\mathbf{w}\|}$. See Figure 6.50.

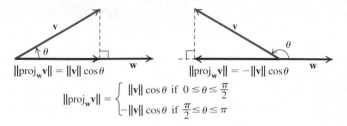

$$\|\text{proj}_{\mathbf{w}}\mathbf{v}\| = \|\mathbf{v}\|\cos\theta \qquad \|\text{proj}_{\mathbf{w}}\mathbf{v}\| = -\|\mathbf{v}\|\cos\theta$$

$$\|\text{proj}_{\mathbf{w}}\mathbf{v}\| = \begin{cases} \|\mathbf{v}\|\cos\theta & \text{if } 0 \le \theta \le \dfrac{\pi}{2} \\ -\|\mathbf{v}\|\cos\theta & \text{if } \dfrac{\pi}{2} \le \theta \le \pi \end{cases}$$

FIGURE 6.50 Magnitude and direction of $\text{proj}_{\mathbf{w}}\mathbf{v}$

Because $(-\|\mathbf{v}\|\cos\theta)\left(-\dfrac{\mathbf{w}}{\|\mathbf{w}\|}\right) = (\|\mathbf{v}\|\cos\theta)\left(\dfrac{\mathbf{w}}{\|\mathbf{w}\|}\right)$, in both cases ($\theta$ acute or obtuse), we have

$$\text{proj}_{\mathbf{w}}\mathbf{v} = (\|\mathbf{v}\|\cos\theta)\left(\dfrac{\mathbf{w}}{\|\mathbf{w}\|}\right)$$

$$= \left(\dfrac{\mathbf{v}\cdot\mathbf{w}}{\|\mathbf{w}\|}\right)\dfrac{\mathbf{w}}{\|\mathbf{w}\|} \qquad \|\mathbf{v}\|\cos\theta = \dfrac{\mathbf{v}\cdot\mathbf{w}}{\|\mathbf{w}\|}$$

$$= \dfrac{\mathbf{v}\cdot\mathbf{w}}{\|\mathbf{w}\|^2}\mathbf{w}$$

VECTOR AND SCALAR PROJECTIONS OF v ONTO w

Let $\mathbf{v}$ and $\mathbf{w}$ be two nonzero vectors and let $\theta(0° \le \theta \le 180°)$ be the angle between them.

The **vector projection of v onto w** is:

$$\text{proj}_{\mathbf{w}}\mathbf{v} = \left(\dfrac{\mathbf{v}\cdot\mathbf{w}}{\|\mathbf{w}\|^2}\right)\mathbf{w}$$

The **scalar projection of v onto w** is:

$$\|\mathbf{v}\|\cos\theta = \dfrac{\mathbf{v}\cdot\mathbf{w}}{\|\mathbf{w}\|}$$

EXAMPLE 8 **Finding the Vector Projection of v onto w**

Given $\mathbf{v} = \langle -2, 3\rangle$ and $\mathbf{w} = \langle 3, -4\rangle$, find the vector and scalar projections of $\mathbf{v}$ onto $\mathbf{w}$.

SOLUTION

$$\|\mathbf{v}\| = \sqrt{(-2)^2 + 3^2} = \sqrt{13} \text{ and } \|\mathbf{w}\| = \sqrt{3^2 + (-4)^2} = \sqrt{25} = 5$$

$$\mathbf{v}\cdot\mathbf{w} = \mathbf{w}\cdot\mathbf{v} = \langle -2, 3\rangle \cdot \langle 3, -4\rangle$$

$$= (-2)(3) + 3(-4) = -6 - 12 = -18$$

a. $\text{proj}_{\mathbf{w}}\mathbf{v} = \dfrac{\mathbf{v} \cdot \mathbf{w}}{\|\mathbf{w}\|^2}\mathbf{w}$ Vector projection of **v** onto **w**

$\quad\quad\quad = -\dfrac{18}{25}\langle 3, -4 \rangle$ Substitute values.

$\quad\quad\quad = \left\langle -\dfrac{54}{25}, \dfrac{72}{25} \right\rangle$ Scalar multiplication

b. $\dfrac{\mathbf{v} \cdot \mathbf{w}}{\|\mathbf{w}\|}$ Scalar projection of **v** onto **w**

$\quad\quad = -\dfrac{18}{5}$ $\mathbf{v} \cdot \mathbf{w} = -18, \|\mathbf{w}_1\| = 5$

Practice Problem 8   Given $\mathbf{v} = \langle 1, -2 \rangle$ and $\mathbf{w} = \langle 2, 5 \rangle$, find the vector and scalar projections of **v** onto **w**.   ■

5 Decompose a vector into two orthogonal vectors.

Decomposition of a Vector

Given two vectors **v** and **w**, adding these vectors produces the resultant vector **v** + **w**. In many applications, it is necessary to *decompose* a vector into two orthogonal vectors. We use the vector $\text{proj}_{\mathbf{w}}\mathbf{v}$ to write **v** as the sum of two orthogonal vectors.

DECOMPOSITION OF A VECTOR

Let **v** and **w** be two nonzero vectors. The vector **v** can be written in the form

$$\mathbf{v} = \mathbf{v}_1 + \mathbf{v}_2,$$

where $\mathbf{v}_1$ is parallel to **w** and $\mathbf{v}_2$ is orthogonal to **w**. We have

$$\mathbf{v}_1 = \text{proj}_{\mathbf{w}}\mathbf{v} = \left(\frac{\mathbf{v} \cdot \mathbf{w}}{\|\mathbf{w}\|^2} \right)\mathbf{w} \text{ and } \mathbf{v}_2 = \mathbf{v} - \mathbf{v}_1.$$

The expression $\mathbf{v} = \mathbf{v}_1 + \mathbf{v}_2$ is called the **decomposition of v with respect to w**. The vectors $\mathbf{v}_1$ and $\mathbf{v}_2$ are called **components of v**.

Because $\mathbf{v}_1$ is a scalar multiple of **w**, $\mathbf{v}_1$ is parallel to **w**. In Exercise 84, we ask you to prove that $\mathbf{v}_2$ is orthogonal to **w**.

FINDING THE SOLUTION: A PROCEDURE

EXAMPLE 9   **Decomposing a Vector into Components**

OBJECTIVE	EXAMPLE
Express $\mathbf{v} = \mathbf{v}_1 + \mathbf{v}_2$, *where* $\mathbf{v}_1$ *is parallel to* **w** *and* $\mathbf{v}_2$ *is orthogonal to* **w**.	*Find the decomposition* $\mathbf{v} = \mathbf{v}_1 + \mathbf{v}_2$ *of* $\mathbf{v} = \langle -2, 10 \rangle$ *with respect to* $\mathbf{w} = \langle 3, -2 \rangle$.
Step 1 Compute. $\mathbf{v} \cdot \mathbf{w}$ and $\|\mathbf{w}\|^2$	**1.** $\mathbf{v} \cdot \mathbf{w} = \langle -2, 10 \rangle \cdot \langle 3, -2 \rangle$ $\quad\quad = (-2)(3) + (10)(-2) = -26$ $\|\mathbf{w}\|^2 = (3)^2 + (-2)^2 = 9 + 4 = 13$
Step 2 Compute. $\mathbf{v}_1 = \text{proj}_{\mathbf{w}}\mathbf{v} = \left(\dfrac{\mathbf{v} \cdot \mathbf{w}}{\|\mathbf{w}\|^2} \right)\mathbf{w}$	**2.** $\mathbf{v}_1 = \left(\dfrac{\mathbf{v} \cdot \mathbf{w}}{\|\mathbf{w}\|^2} \right)\mathbf{w} = \dfrac{-26}{13}\langle 3, -2 \rangle$ From Step 1 $\quad\quad = -2\langle 3, -2 \rangle = \langle -6, 4 \rangle$ Simplify.

continued on the next page

Step 3 Compute.

$$\mathbf{v}_2 = \mathbf{v} - \mathbf{v}_1$$

Check that $\mathbf{v}_2$ is orthogonal to $\mathbf{w}$ by showing $\mathbf{v}_2 \cdot \mathbf{w} = 0$.

Step 4 Write the decomposition.

$$\mathbf{v} = \mathbf{v}_1 + \mathbf{v}_2$$

3. $\mathbf{v}_2 = \mathbf{v} - \mathbf{v}_1 = \langle -2, 10 \rangle - \langle -6, 4 \rangle$
$$= \langle -2 - (-6), 10 - 4 \rangle$$
$$= \langle 4, 6 \rangle$$

Check:

$$\mathbf{v}_2 \cdot \mathbf{w} = \langle 4, 6 \rangle \cdot \langle 3, -2 \rangle = (4)(3) + (6)(-2) = 0$$

4. $\mathbf{v} = \langle -2, 10 \rangle = \quad \mathbf{v}_1 \quad + \quad \mathbf{v}_2$
$$= \underbrace{\langle -6, 4 \rangle}_{\text{Parallel to } \mathbf{w}} + \underbrace{\langle 4, 6 \rangle}_{\text{Orthogonal to } \mathbf{w}}$$

Practice Problem 9 Repeat Example 9 with $\mathbf{v} = \langle 5, 1 \rangle$ and $\mathbf{w} = \langle 4, 4 \rangle$.

7 Use the definition of work.

Work

DEFINITION OF WORK

The work W done by a constant force $\mathbf{F}$ in moving an object from a point P to a point Q is defined by

$$W = \mathbf{F} \cdot \overrightarrow{PQ} = \|\mathbf{F}\|\|\overrightarrow{PQ}\| \cos \theta,$$

where θ is the angle between $\mathbf{F}$ and $\overrightarrow{PQ}$.

EXAMPLE 10 Computing Work

A child pulls a wagon along level ground, with a force of 40 pounds along the handle on the wagon that makes an angle of 42° with the horizontal. How much work has she done by pulling the wagon 150 feet?

SOLUTION

The work done is as follows:

$$\begin{aligned} W &= \mathbf{F} \cdot \overrightarrow{PQ} & \text{Definition of work} \\ &= \|\mathbf{F}\|\|\overrightarrow{PQ}\| \cos \theta & \mathbf{u} \cdot \mathbf{v} = \|\mathbf{u}\|\|\mathbf{v}\| \cos \theta \\ &= (40)(150) \cos 42° & \text{Substitute given values.} \\ &\approx 4458.87 \text{ foot-pounds} & \text{Use a calculator.} \end{aligned}$$

Practice Problem 10 In Example 10, compute the work done if the handle on the wagon makes an angle of 60° with the horizontal.

SECTION 6.5 ■ Exercises

A EXERCISES Basic Skills and Concepts

1. The dot product of $\mathbf{v} = \langle a_1, a_2 \rangle$ and $\mathbf{w} = \langle b_1, b_2 \rangle$ is defined as $\mathbf{v} \cdot \mathbf{w} = \underline{a_1 b_1 + a_2 b_2}$.

2. If θ is the angle between the vectors $\mathbf{v}$ and $\mathbf{w}$, then $\mathbf{v} \cdot \mathbf{w} = \underline{\|\mathbf{v}\|\|\mathbf{w}\| \cos \theta}$.

3. If $\mathbf{v}$ and $\mathbf{w}$ are orthogonal, then $\mathbf{v} \cdot \mathbf{w} = \underline{\quad 0 \quad}$.

4. If $\mathbf{v}$ and $\mathbf{w}$ are nonzero vectors with $\mathbf{v} \cdot \mathbf{w} = 0$, then $\mathbf{v}$ and $\mathbf{w}$ are $\underline{\text{orthogonal}}$.

5. *True or False* If $\mathbf{v} \cdot \mathbf{w} < 0$, then the angle between $\mathbf{v}$ and $\mathbf{w}$ is an obtuse angle. True

6. *True or False* To find a vector perpendicular to a given vector, switch components and change the sign of one of them. True

†Due to space constrictions, answers to these exercises may be found in the Answers beginning on page A–1 in the back of the book.

In Exercises 7–14, find u · v.

7. $\mathbf{u} = \langle 1, -2 \rangle, \mathbf{v} = \langle 3, 5 \rangle$ −7
8. $\mathbf{u} = \langle 1, 3 \rangle, \mathbf{v} = \langle -3, 1 \rangle$ 0
9. $\mathbf{u} = \langle 2, -6 \rangle, \mathbf{v} = \langle 3, 1 \rangle$ 0
10. $\mathbf{u} = \langle -1, -2 \rangle, \mathbf{v} = \langle -3, -4 \rangle$ 11
11. $\mathbf{u} = \mathbf{i} - 3\mathbf{j}, \mathbf{v} = 8\mathbf{i} - 2\mathbf{j}$ 14
12. $\mathbf{u} = 6\mathbf{i} - \mathbf{j}, \mathbf{v} = 2\mathbf{i} + 7\mathbf{j}$ 5
13. $\mathbf{u} = 4\mathbf{i} - 2\mathbf{j}, \mathbf{v} = 3\mathbf{j}$ −6
14. $\mathbf{u} = -2\mathbf{i}, \mathbf{v} = 5\mathbf{j}$ 0

In Exercises 15–20, find u · v, where θ is the angle between the vectors u and v. Round answers to the nearest tenths.

15. $\|\mathbf{u}\| = 2, \|\mathbf{v}\| = 5, \theta = \dfrac{\pi}{6}$ $5\sqrt{3}$

16. $\|\mathbf{u}\| = 3, \|\mathbf{v}\| = 4, \theta = \dfrac{\pi}{3}$ 6

17. $\|\mathbf{u}\| = 5, \|\mathbf{v}\| = 4, \theta = 65°$ 8.5

18. $\|\mathbf{u}\| = 5, \|\mathbf{v}\| = 3, \theta = 78°$ 3.1

19. $\|\mathbf{u}\| = 3, \|\mathbf{v}\| = 7, \theta = 120°$ −10.5

20. $\|\mathbf{u}\| = 6, \|\mathbf{v}\| = 4, \theta = 150°$ −20.8

In Exercises 21–30, find the angle between the vectors v and w. Round each answer to the nearest tenth of a degree.

21. $\mathbf{v} = \langle 2, 3 \rangle, \mathbf{w} = \langle 3, 4 \rangle$ 3.2°
22. $\mathbf{v} = \langle -1, 1 \rangle, \mathbf{w} = \langle 5, 12 \rangle$ 67.6°
23. $\mathbf{v} = -2\mathbf{i} - 3\mathbf{j}, \mathbf{w} = \mathbf{i} + \mathbf{j}$ 168.7°
24. $\mathbf{v} = \mathbf{i} - \mathbf{j}, \mathbf{w} = 3\mathbf{i} - 4\mathbf{j}$ 8.1°
25. $\mathbf{v} = 2\mathbf{i} + 5\mathbf{j}, \mathbf{w} = -5\mathbf{i} + 2\mathbf{j}$ 90°
26. $\mathbf{v} = 3\mathbf{i} - 7\mathbf{j}, \mathbf{w} = 7\mathbf{i} + 3\mathbf{j}$ 90°
27. $\mathbf{v} = \mathbf{i} + \mathbf{j}, \mathbf{w} = 3\mathbf{i} + 3\mathbf{j}$ 0°
28. $\mathbf{v} = -2\mathbf{i} + 3\mathbf{j}, \mathbf{w} = -4\mathbf{i} + 6\mathbf{j}$ 0°
29. $\mathbf{v} = -2\mathbf{i} - 3\mathbf{j}, \mathbf{w} = -4\mathbf{i} + 6\mathbf{j}$ 112.6°
30. $\mathbf{v} = -3\mathbf{i} + 4\mathbf{j}, \mathbf{w} = 6\mathbf{i} - 8\mathbf{j}$ 180°

In Exercises 31–38, let v and w be two vectors in the plane of magnitudes 12 and 16, respectively. The angle between v and w is 50°. Find each expression. Round answers to the nearest tenths.

31. $\|\mathbf{v} + \mathbf{w}\|$ 25.4
32. $\|\mathbf{v} - \mathbf{w}\|$ 12.4
33. $\|2\mathbf{v} + \mathbf{w}\|$ 36.4
34. $\|2\mathbf{v} - \mathbf{w}\|$ 18.4
35. Angle between $\mathbf{v} + \mathbf{w}$ and $\mathbf{w}$ 21.2°
36. Angle between $\mathbf{v} - \mathbf{w}$ and $\mathbf{v}$ 82.0°
37. Angle between $2\mathbf{v} + \mathbf{w}$ and $\mathbf{v}$ 19.7°
38. Angle between $2\mathbf{v} - \mathbf{w}$ and $\mathbf{w}$ 91.8°

In Exercises 39–44, determine whether the vectors v and w are orthogonal.

39. $\mathbf{v} = 2\mathbf{i} + 3\mathbf{j}, \mathbf{w} = -3\mathbf{i} + 2\mathbf{j}$ Orthogonal
40. $\mathbf{v} = -4\mathbf{i} - 5\mathbf{j}, \mathbf{w} = 5\mathbf{i} - 4\mathbf{j}$ Orthogonal
41. $\mathbf{v} = \langle 2, 7 \rangle, \mathbf{w} = \langle -7, -2 \rangle$ Not orthogonal
42. $\mathbf{v} = \langle -3, 4 \rangle, \mathbf{w} = \langle 4, -3 \rangle$ Not orthogonal
43. $\mathbf{v} = \langle 12, 6 \rangle, \mathbf{w} = \langle 2, -4 \rangle$ Orthogonal
44. $\mathbf{v} = \langle -9, 6 \rangle, \mathbf{w} = \langle 4, 6 \rangle$ Orthogonal

In Exercises 45–50, determine whether the vectors v and w are parallel.

45. $\mathbf{v} = 2\mathbf{i} + 3\mathbf{j}, \mathbf{w} = \dfrac{2}{3}\mathbf{i} + \mathbf{j}$ Parallel
46. $\mathbf{v} = 5\mathbf{i} - 2\mathbf{j}, \mathbf{w} = -10\mathbf{i} + 4\mathbf{j}$ Parallel
47. $\mathbf{v} = \langle 3, 5 \rangle, \mathbf{w} = \langle -6, 10 \rangle$ Not parallel
48. $\mathbf{v} = \langle -4, -2 \rangle, \mathbf{w} = \left\langle -1, \dfrac{1}{2} \right\rangle$ Not parallel
49. $\mathbf{v} = \langle 6, 3 \rangle, \mathbf{w} = \langle 2, 1 \rangle$ Parallel
50. $\mathbf{v} = \langle 2, 5 \rangle, \mathbf{w} = \langle -4, -10 \rangle$ Parallel

In Exercises 51–54, determine the scalar c so that the vectors v and w are (a) orthogonal and (b) parallel.

51. $\mathbf{v} = \langle 2, 3 \rangle, \mathbf{w} = \langle 3, c \rangle$ $c = -2, c = \dfrac{9}{2}$
52. $\mathbf{v} = \langle 1, -2 \rangle, \mathbf{w} = \langle c, 4 \rangle$ $c = 8, c = -2$
53. $\mathbf{v} = \langle -4, 3 \rangle, \mathbf{w} = \langle c, -6 \rangle$ $c = -\dfrac{9}{2}, c = 8$
54. $\mathbf{v} = \langle -1, -3 \rangle, \mathbf{w} = \langle -2, c \rangle$ $c = \dfrac{2}{3}, c = -6$

In Exercises 55–62, find $\text{proj}_{\mathbf{w}}\mathbf{v}$.

55. $\mathbf{v} = \langle 3, 5 \rangle, \mathbf{w} = \langle 4, 1 \rangle$ $\langle 4, 1 \rangle$
56. $\mathbf{v} = \langle 3, 5 \rangle, \mathbf{w} = \langle -4, 2 \rangle$ $\left\langle \dfrac{2}{5}, -\dfrac{1}{5} \right\rangle$
57. $\mathbf{v} = \langle 0, 2 \rangle, \mathbf{w} = \langle 3, 4 \rangle$ $\left\langle \dfrac{24}{25}, \dfrac{32}{25} \right\rangle$
58. $\mathbf{v} = \langle 2, 0 \rangle, \mathbf{w} = \langle 4, -3 \rangle$ $\left\langle \dfrac{32}{25}, -\dfrac{24}{25} \right\rangle$
59. $\mathbf{v} = 5\mathbf{i} - 3\mathbf{j}, \mathbf{w} = \mathbf{i}$ $\langle 5, 0 \rangle$
60. $\mathbf{v} = 2\mathbf{i} + 7\mathbf{j}, \mathbf{w} = \mathbf{j}$ $\langle 0, 7 \rangle$
61. $\mathbf{v} = \langle a, b \rangle, \mathbf{w} = \mathbf{i} + \mathbf{j}$ $\left\langle \dfrac{a}{2} + \dfrac{b}{2}, \dfrac{a}{2} + \dfrac{b}{2} \right\rangle$
62. $\mathbf{v} = \langle a, b \rangle, \mathbf{w} = \mathbf{i} - \mathbf{j}$ $\left\langle \dfrac{a}{2} - \dfrac{b}{2}, -\dfrac{a}{2} + \dfrac{b}{2} \right\rangle$

In Exercises 63–70, decompose v with respect to w.

63. $\mathbf{v} = \langle 2, 3 \rangle, \mathbf{w} = \mathbf{i}$ $\langle 2, 0 \rangle + \langle 0, 3 \rangle$
64. $\mathbf{v} = \langle 4, -3 \rangle, \mathbf{w} = \mathbf{j}$ $\langle 0, -3 \rangle + \langle 4, 0 \rangle$
65. $\mathbf{v} = \mathbf{i}, \mathbf{w} = \langle 1, 1 \rangle$ $\left\langle \dfrac{1}{2}, \dfrac{1}{2} \right\rangle + \left\langle \dfrac{1}{2}, -\dfrac{1}{2} \right\rangle$
66. $\mathbf{v} = \mathbf{j}, \mathbf{w} = \langle 4, -3 \rangle$ $\left\langle -\dfrac{12}{25}, \dfrac{9}{25} \right\rangle + \left\langle \dfrac{12}{25}, \dfrac{16}{25} \right\rangle$
67. $\mathbf{v} = \langle 4, 2 \rangle, \mathbf{w} = \langle -2, 4 \rangle$ $\langle 0, 0 \rangle + \langle 4, 2 \rangle$
68. $\mathbf{v} = \langle 5, 2 \rangle, \mathbf{w} = \langle 3, 5 \rangle$ $\left\langle \dfrac{75}{34}, \dfrac{125}{34} \right\rangle + \left\langle \dfrac{95}{34}, -\dfrac{57}{34} \right\rangle$
69. $\mathbf{v} = \langle 4, -1 \rangle, \mathbf{w} = \langle 3, 4 \rangle$ $\left\langle \dfrac{24}{25}, \dfrac{32}{25} \right\rangle + \left\langle \dfrac{76}{25}, -\dfrac{57}{25} \right\rangle$
70. $\mathbf{v} = \langle -3, 5 \rangle, \mathbf{w} = \langle 3, -4 \rangle$ $\left\langle -\dfrac{87}{25}, \dfrac{116}{25} \right\rangle + \left\langle \dfrac{12}{25}, \dfrac{9}{25} \right\rangle$

B EXERCISES Applying the Concepts

71. **Finding resultant force.** Two forces of 6 and 8 pounds act at an angle of 30° to each other. Find the resultant force and the angle of the resultant force to each of the two forces. 13.51 lb, 12.8°, 17.2°

72. Repeat Exercise 71 assuming that the two forces act at an angle of 60° to each other. 12.2 lb, 25.3°, 34.7°

73. **Work.** A force $\mathbf{F} = 64\mathbf{i} - 20\mathbf{j}$ (in pounds) is applied to an object. The resulting movement of the object is represented by the vector $\mathbf{d} = 10\mathbf{i} + 4\mathbf{j}$ (in feet). Find the work done by the force. 560 foot-pounds

74. Work. A car is pushed 150 feet on a level street by a force of 60 pounds at an angle of 18° with the horizontal. Find the work done. 8559.5 foot-pounds.

75. Inclined plane. A car is being towed up an inclined plane making an angle of 18° with the horizontal as shown in the figure. Find the force **F** needed to make the component of **F** parallel to the ground equal to 5250 pounds.
5520.2 pounds.

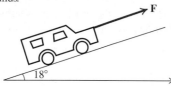

76. Inclined plane. In Exercise 75, find the work done by the force **F** to tow the car up 10 feet along the inclined plane. 55,202 foot-pounds

77. Work. A vector **F** represents a force that has a magnitude of 10 pounds, and $\dfrac{2\pi}{3}$ is the angle for its direction. Find the work done by the force in moving an object from the origin to the point $(-6, 3)$. Distance is measured in feet. 56 foot-pounds

78. Work. Two forces $\mathbf{F}_1 = 3\mathbf{i} - 2\mathbf{j}$ and $\mathbf{F}_2 = -4\mathbf{i} + 3\mathbf{j}$ act on an object and move it along a straight line from the point $(8, 4)$ to the point $(3, 5)$. Find the work done by the two forces acting together. The magnitudes of $\mathbf{F}_1$ and $\mathbf{F}_2$ are measured in pounds, and distance is measured in feet. 6 foot-pounds

C EXERCISES Beyond the Basics

79. Orthogonality on a circle. In the figure, $\overrightarrow{AB}$ is the diameter of a circle with center $O(0,0)$ and radius a. Show that $\overrightarrow{CA}$ and $\overrightarrow{CB}$ are perpendicular. [*Hint:* $x^2 + y^2 = a^2$.]

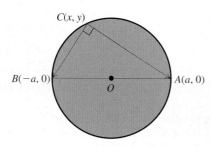

80. Cauchy-Schwarz inequality. If **v** and **w** are two vectors, then $|\mathbf{v} \cdot \mathbf{w}| \leq \|\mathbf{v}\|\|\mathbf{w}\|$. When is $|\mathbf{v} \cdot \mathbf{w}| = \|\mathbf{v}\|\|\mathbf{w}\|$? [*Hint:* $|\mathbf{v} \cdot \mathbf{w}| = \|\mathbf{v}\|\|\mathbf{w}\| \cos \theta$.] $\theta = 0$

81. Verify that for any vectors **u**, **v**, and **w**,
$\mathbf{u} \cdot (\mathbf{v} + \mathbf{w}) = \mathbf{u} \cdot \mathbf{v} + \mathbf{u} \cdot \mathbf{w}$.

82. Prove that $(\mathbf{v} + \mathbf{w}) \cdot (\mathbf{v} + \mathbf{w}) = \|\mathbf{v}\|^2 + 2(\mathbf{v} \cdot \mathbf{w}) + \|\mathbf{w}\|^2$.

83. Triangle inequality. Use Exercises 80 and 82 to prove that $\|\mathbf{v} + \mathbf{w}\| \leq \|\mathbf{v}\| + \|\mathbf{w}\|$.

84. In the decomposition $\mathbf{v} = \mathbf{v}_1 + \mathbf{v}_2$ (see page 445) with respect to **w**, show that $\mathbf{v}_2 = \mathbf{v} - \text{proj}_\mathbf{w}\mathbf{v}$ is orthogonal to **w**.

85. Show that the vectors $\mathbf{v} = \langle 2, -5 \rangle$ and $\mathbf{w} = \langle 5, 2 \rangle$ are perpendicular to each other.

86. Let $\mathbf{v} = 3\mathbf{i} + 7\mathbf{j}$. Find a vector perpendicular to **v**. $7\mathbf{i} - 3\mathbf{j}$

87. Orthogonal unit vectors. Find the unit vector orthogonal to $\mathbf{v} = 3\mathbf{i} - 4\mathbf{j}$. $\left\langle \dfrac{4}{5}, \dfrac{3}{5} \right\rangle$

88. Orthogonal unit vectors. Let $\mathbf{u}_1$ and $\mathbf{u}_2$ be orthogonal unit vectors and $\mathbf{v} = x\mathbf{u}_1 + y\mathbf{u}_2$. Find
  a. $\mathbf{v} \cdot \mathbf{u}_1$ and x
  b. $\mathbf{v} \cdot \mathbf{u}_2$. y

89. Right triangle. Use the dot product to show that ABC is a right triangle and identify the right angle assuming that $A = (-2, 1)$, $B = (3, 1)$, and $C = (-1, -1)$.

90. Parallelogram law. Show that for any two vectors **u** and **v**,

$$\|\mathbf{u} + \mathbf{v}\|^2 + \|\mathbf{u} - \mathbf{v}\|^2 = 2\|\mathbf{u}\|^2 + 2\|\mathbf{v}\|^2.$$

[*Hint:* $\|\mathbf{a}\|^2 = \mathbf{a} \cdot \mathbf{a}$.]

91. Polarization identity. Show that for any two vectors **u** and **v**,

$$\|\mathbf{u} + \mathbf{v}\|^2 - \|\mathbf{u} - \mathbf{v}\|^2 = 4(\mathbf{u} \cdot \mathbf{v}).$$

92. Show that **u** and **v** are orthogonal if and only if $\|\mathbf{u} + \mathbf{v}\| = \|\mathbf{u} - \mathbf{v}\|$.

93. Show that the distance d from a point $P(x_0, y_0)$ to the line $l: ax + by + c = 0$ is

$$d = \frac{|ax_0 + by_0 + c|}{\sqrt{a^2 + b^2}}.$$

[*Hint:* **i.** Choose two points $Q(x_1, y_1)$ and $R(x_2, y_2)$ on l.
ii. Show $\mathbf{n} = \langle a, b \rangle$ is perpendicular to l, by showing that $\mathbf{n} \cdot \overrightarrow{QR} = 0$.
iii. Show that $\|\text{proj}_\mathbf{n}\overrightarrow{QP}\| = d$.]

Critical Thinking

94. *True or False* If $\mathbf{u} \cdot \mathbf{v} = \mathbf{u} \cdot \mathbf{w}$ and $\mathbf{u} \neq 0$, then $\mathbf{v} = \mathbf{w}$. Give reasons for your answer.
False; consider $\mathbf{u} = \langle 1, 0 \rangle$, $\mathbf{v} = \langle 0, 2 \rangle$, and $\mathbf{w} = \langle 0, 3 \rangle$.

Polar Coordinates

Before Starting this Section, Review

1. Coordinate plane (Section 1.1, page 2)
2. Degree and radian measure of angles (Section 4.1, page 255)
3. Trigonometric functions (Section 4.2, page 268)

Objectives

1. Plot points using polar coordinates.
2. Convert points between polar and rectangular forms.
3. Convert equations between rectangular and polar forms.
4. Graph polar equations.

BIONICS

On large construction sites and maintenance projects, it is common to see mechanical systems that lift, dig, and scoop just like biological systems having arms, legs, and hands. The technique of solving mechanical problems with our knowledge of biological systems is called **bionics**. Bionics is used to make robotic devices for the remote-control handling of radioactive materials and the machines that are sent to the moon or to Mars to scoop soil samples, bring them into the space vehicle, perform analysis on them, and radio the results back to Earth. In Example 4, we use polar coordinates to find the reach of a robotic arm.

Up to now, we have used a rectangular coordinate system to identify points in the plane. In this section, we introduce another coordinate system, called a **polar coordinate system**, that allows us to describe many curves with rather simple equations. The polar coordinate system is believed to have been introduced by Jakob Bernoulli (1654–1705). ■

---

1 Plot points using polar coordinates.

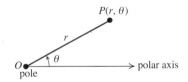

FIGURE 6.51 Polar coordinates (r, θ) of a point

Polar Coordinates

In a rectangular coordinate system, a point in the plane is described in terms of the horizontal and vertical directed distances from the origin. In a polar coordinate system, we draw a horizontal ray, called the **polar axis**, in the plane; its endpoint is called the **pole**. A point P in the plane is described by an ordered pair of numbers (r, θ), the **polar coordinates** of P.

Figure 6.51 shows the point $P(r, \theta)$ in the polar coordinate system, where r is the "directed distance" (see Figures 6.51 and 6.52) from the pole O to the point P, and θ is a directed angle from the polar axis to the line segment OP. The angle θ may be measured in degrees or radians. As usual, positive angles are measured counterclockwise from the polar axis and negative angles are measured clockwise from the polar axis.

To locate a point with polar coordinates $(5, 40°)$, we draw an angle $\theta = 40°$ counterclockwise from the polar axis. We then measure five units along the terminal side of θ. In Figure 6.52, we have plotted the points with polar coordinates $(5, 40°)$, $(3, 220°)$, $(4, -30°)$, and $(2, 110°)$.

STUDY TIP

When plotting a point in polar coordinates, measure the angle θ first. Then measure a distance of r units from the pole along the terminal ray of θ. Notice that in working with polar coordinates, you use the second coordinate first.

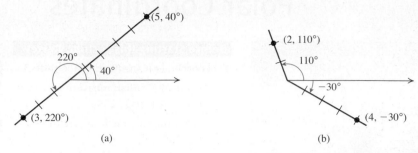

(a)   (b)

FIGURE 6.52 Plotting points using polar coordinates

Unlike rectangular coordinates, the polar coordinates of a point are not unique. For example, the polar coordinates $(3, 60°)$, $(3, 420°)$, and $(3, -300°)$ represent the same point. See Figure 6.53. In general, if a point P has polar coordinates (r, θ), then for any integer n,

$$(r, \underbrace{\theta + n \cdot 360°}_{\theta \text{ in degrees}}) \qquad \text{or} \qquad (r, \underbrace{\theta + 2n\pi}_{\theta \text{ in radians}})$$

are also polar coordinates of P.

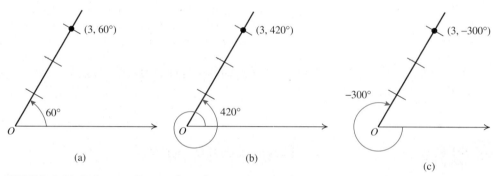

(a)   (b)   (c)

FIGURE 6.53 Polar coordinates of a point are not unique.

When curves are graphed in polar coordinates, it is convenient to allow r to be negative. That's why we defined r as *a directed distance*. If r is negative, then instead of the point being on the terminal side of the angle, it is $|r|$ units on the extension of the terminal side, which is the ray from the pole that extends in the direction opposite the terminal side. For example, the point $P(-4, -30°)$ shown in Figure 6.54 is the same point as $(4, 150°)$. Some other polar coordinates for the same point P are $(-4, 330°)$, $(4, -210°)$, and $(4, 510°)$.

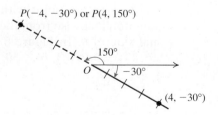

FIGURE 6.54 Negative r

EXAMPLE 1   **Finding Different Polar Coordinates**

a. Plot the point P with polar coordinates $(3, 225°)$.

Find another pair of polar coordinates of P with the following properties.

b. $r < 0$ and $0° < \theta < 360°$

c. $r < 0$ and $-360° < \theta < 0°$

d. $r > 0$ and $-360° < \theta < 0°$

SOLUTION

The point $P(3, 225°)$ is plotted by first drawing $\theta = 225°$ counterclockwise from the polar axis. Since $r = 3 > 0$, the point P is on the terminal side of the angle, three units from the pole. See Figure 6.55(a).

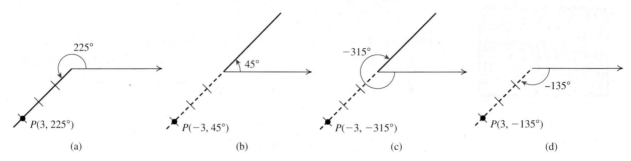

FIGURE 6.55 **Plotting points**

The answer to parts, **b, c,** and **d** are, respectively, $(-3, 45°)$, $(-3, -315°)$, and $(3, -135°)$. See Figure 6.55(b)–(d). ■ ■ ■

Practice Problem 1

a. Plot the point P with polar coordinates $(2, -150°)$.

Find another pair of polar coordinates of P with the following properties.

b. $r < 0$ and $0° < \theta < 360°$

c. $r > 0$ and $0° < \theta < 360°$

d. $r < 0$ and $-360° < \theta < 0°$ ■

2 Convert points between polar and rectangular forms.

Converting Between Polar and Rectangular Forms

Frequently, we need to use polar and rectangular coordinates in the same problem. To do this, we let the positive x-axis of the rectangular coordinate system serve as the polar axis and the origin serve as the pole. Each point P then has polar coordinates (r, θ) and rectangular coordinates (x, y), as shown in Figure 6.56.

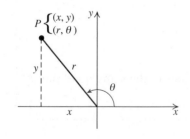

FIGURE 6.56 $P = (x, y)$ or $P = (r, \theta)$

From Figure 6.56, we see the following relationships: $x^2 + y^2 = r^2$, $\sin \theta = \dfrac{y}{r}$, $\cos \theta = \dfrac{x}{r}$, and $\tan \theta = \dfrac{y}{x}$. These relationships hold whether $r > 0$ or $r < 0$, and they are used to convert a point's coordinates between rectangular and polar forms.

CONVERTING FROM POLAR TO RECTANGULAR COORDINATES

To convert the polar coordinates (r, θ) of a point to rectangular coordinates, use the equations

$$x = r \cos \theta \quad \text{and} \quad y = r \sin \theta.$$

TECHNOLOGY CONNECTION

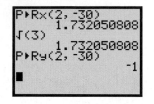 A graphing calculator can be used to check conversions from polar to rectangular coordinates. Choose the appropriate options from the ANGLE menu. The x- and y-coordinates must be calculated separately. The screen shown here is from a calculator set in Degree mode.

```
P▸Rx(2,-30)
        1.732050808
√(3)
        1.732050808
P▸Ry(2,-30)
                  -1
■
```

EXAMPLE 2   Converting from Polar to Rectangular Coordinates

Convert the polar coordinates of each point to rectangular coordinates.

a. $(2, -30°)$ **b.** $\left(-4, \dfrac{\pi}{3}\right)$

SOLUTION

a.

$x = r\cos\theta$	$y = r\sin\theta$	Conversion equations
$x = 2\cos(-30°)$	$y = 2\sin(-30°)$	Replace r with 2 and θ with $-30°$.
$x = 2\cos 30°$	$y = -2\sin 30°$	$\cos(-\theta) = \cos\theta$, $\sin(-\theta) = -\sin\theta$
$x = 2\left(\dfrac{\sqrt{3}}{2}\right) = \sqrt{3}$	$y = -2\left(\dfrac{1}{2}\right) = -1$	$\cos 30° = \dfrac{\sqrt{3}}{2}, \sin 30° = \dfrac{1}{2}$

The rectangular coordinates of $(2, -30°)$ are $(\sqrt{3}, -1)$.

b.

$x = r\cos\theta$	$y = r\sin\theta$	Conversion equations
$x = -4\cos\dfrac{\pi}{3}$	$y = -4\sin\dfrac{\pi}{3}$	Replace r with -4 and θ with $\dfrac{\pi}{3}$.
$x = -4\left(\dfrac{1}{2}\right) = -2$	$y = -4\left(\dfrac{\sqrt{3}}{2}\right) = -2\sqrt{3}$	$\cos\dfrac{\pi}{3} = \dfrac{1}{2}, \sin\dfrac{\pi}{3} = \dfrac{\sqrt{3}}{2}$

The rectangular coordinates of $\left(-4, \dfrac{\pi}{3}\right)$ are $(-2, -2\sqrt{3})$.   ■ ■ ■

Practice Problem 2   Convert the polar coordinates of each point to rectangular coordinates.

a. $(-3, 60°)$ **b.** $\left(2, -\dfrac{\pi}{4}\right)$ ■

When converting from rectangular coordinates (x, y) of a point to polar coordinates (r, θ), remember that infinitely many representations are possible. We will find the polar coordinates (r, θ) of the point (x, y) that have $r > 0$ and $0° \leq \theta < 360°$ (or $0 \leq \theta < 2\pi$).

CONVERTING FROM RECTANGULAR TO POLAR COORDINATES

To convert the rectangular coordinates (x, y) of a point to polar coordinates:

1. Find the quadrant in which the given point (x, y) lies.
2. Use $r = \sqrt{x^2 + y^2}$ to find r.
3. Find θ by using $\tan\theta = \dfrac{y}{x}$ and choose θ so that it lies in the same quadrant as the point (x, y).

EXAMPLE 3   Converting from Rectangular to Polar Coordinates

Find polar coordinates (r, θ) of the point P with $r > 0$ and $0 \leq \theta < 2\pi$, whose rectangular coordinates are $(x, y) = (-2, 2\sqrt{3})$.

TECHNOLOGY CONNECTION

A graphing calculator can also be used to check conversions from rectangular to polar coordinates. Choose the appropriate options on the ANGLE menu. The r and θ coordinates must be calculated separately. The screen shown here is from a calculator set in Radian mode.

```
R▶Pr(-2,2√(3))
                4
R▶Pθ(-2,2√(3))
        2.094395102
2π/3
        2.094395102
```

SOLUTION

1. The point $P(-2, 2\sqrt{3})$ lies in quadrant II with $x = -2$ and $y = 2\sqrt{3}$.

2.
$$
\begin{aligned}
r &= \sqrt{x^2 + y^2} && \text{Formula for computing } r \\
&= \sqrt{(-2)^2 + (2\sqrt{3})^2} && \text{Substitute values of } x \text{ and } y. \\
&= \sqrt{4 + 12} = \sqrt{16} = 4 && \text{Simplify.}
\end{aligned}
$$

3.
$$
\tan \theta = \frac{y}{x} \qquad \text{Definition of } \tan \theta
$$

$$
\tan \theta = \frac{2\sqrt{3}}{-2} = -\sqrt{3} \qquad \text{Substitute values and simplify.}
$$

Since the tangent is negative in Quadrants II and IV, the equation $\tan \theta = -\sqrt{3}$ gives $\theta = \pi - \dfrac{\pi}{3} = \dfrac{2\pi}{3}$ or $\theta = 2\pi - \dfrac{\pi}{3} = \dfrac{5\pi}{3}$. Because P lies in quadrant II, we choose $\theta = \dfrac{2\pi}{3}$. So the required polar coordinates are $\left(4, \dfrac{2\pi}{3}\right)$.   ■ ■ ■

Practice Problem 3   Find the polar coordinates of the point P with $r > 0$ and $0 \le \theta < 2\pi$, whose rectangular coordinates are $(x, y) = (-1, -1)$.   ■

EXAMPLE 4   **Positioning a Robotic Hand**

For the robotic arm of Figure 6.57, find the polar coordinates of the hand relative to the shoulder. Round all answers to the nearest tenth.

SOLUTION

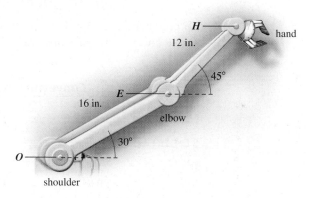

FIGURE 6.57 Robotic arm

We need to find the polar coordinates (r, θ) of the hand H, with the shoulder O as the pole. Since segment $\overline{OE}$ measures 16 inches and makes an angle of $30°$ with the horizontal, we can write the vector $\overrightarrow{OE}$ in terms of its components:

$$
\overrightarrow{OE} = \langle 16 \cos 30°, 16 \sin 30° \rangle
$$

Similarly,

$$
\overrightarrow{EH} = \langle 12 \cos 45°, 12 \sin 45° \rangle
$$

We know that

$$
\begin{aligned}
\overrightarrow{OH} &= \overrightarrow{OE} + \overrightarrow{EH} && \text{Resultant vector} \\
&= \langle 16 \cos 30°, 16 \sin 30° \rangle + \langle 12 \cos 45°, 12 \sin 45° \rangle \\
&= \langle 16 \cos 30° + 12 \cos 45°, 16 \sin 30° + 12 \sin 45° \rangle && \text{Vector addition}
\end{aligned}
$$

So the rectangular coordinates (x, y) of H are

$$x = 16\cos 30° + 12\cos 45°, \quad y = 16\sin 30° + 12\sin 45°.$$

We now convert the coordinates (x, y) of H to polar coordinates (r, θ).

$r = \sqrt{x^2 + y^2}$	Formula for r
$= \sqrt{(16\cos 30° + 12\cos 45°)^2 + (16\sin 30° + 12\sin 45°)^2}$	Substitute values for x and y.
≈ 27.8 inches	Use a calculator.
$\theta = \tan^{-1}\left(\dfrac{y}{x}\right)$	Definition of inverse tangent
$= \tan^{-1}\left(\dfrac{16\sin 30° + 12\sin 45°}{16\cos 30° + 12\cos 45°}\right) \approx 36.4°$	Substitute values and use a calculator.

The polar coordinates of H relative to the pole O are $(r, \theta) = (27.8, 36.4°)$. ■ ■ ■

Practice Problem 4   Repeat Example 4 with $\overline{OE} = 15$ inches and $\overline{EH} = 10$ inches. ■

---

3 Convert equations between rectangular and polar forms.

Converting Equations Between Rectangular and Polar Forms

An equation that has the rectangular coordinates x and y as variables is called a **rectangular** (or **Cartesian**) **equation**. Similarly, an equation where the polar coordinates r and θ are the variables is called a **polar equation**. Some examples of polar equations are

$$r = \sin\theta, \quad r = 1 + \cos\theta, \quad \text{and} \quad r = \theta.$$

To convert a rectangular equation to a polar equation, we simply replace x with $r\cos\theta$ and y with $r\sin\theta$ and then simplify where possible.

EXAMPLE 5   **Converting an Equation from Rectangular to Polar Form**

Convert the equation $x^2 + y^2 - 3x + 4 = 0$ to polar form.

SOLUTION

$x^2 + y^2 - 3x + 4 = 0$	Given equation
$(r\cos\theta)^2 + (r\sin\theta)^2 - 3(r\cos\theta) + 4 = 0$	$x = r\cos\theta, y = r\sin\theta$
$r^2\cos^2\theta + r^2\sin^2\theta - 3r\cos\theta + 4 = 0$	$(ab)^2 = a^2b^2$
$r^2(\cos^2\theta + \sin^2\theta) - 3r\cos\theta + 4 = 0$	Factor out r^2 from the first two terms.
$r^2 - 3r\cos\theta + 4 = 0$	$\sin^2\theta + \cos^2\theta = 1$

The equation $r^2 - 3r\cos\theta + 4 = 0$ is a polar form of the given rectangular equation. ■ ■ ■

Practice Problem 5   Convert the equation $x^2 + y^2 - 2y + 3 = 0$ to polar form. ■

Converting an equation from polar to rectangular form will frequently require some ingenuity in order to use the substitutions:

$$r^2 = x^2 + y^2, \quad r\cos\theta = x, \quad r\sin\theta = y, \quad \text{and} \quad \tan\theta = \frac{y}{x}.$$

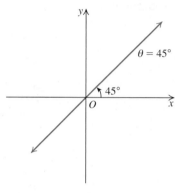

FIGURE 6.58 $x^2 + y^2 = 3^2$ or $r = 3$

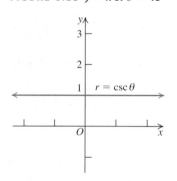

FIGURE 6.59 $y = x$ or $\theta = 45°$

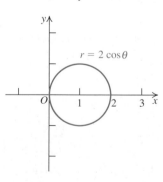

FIGURE 6.60 $y = 1$ or $r\sin\theta = 1$

EXAMPLE 6 **Converting an Equation from Polar to Rectangular Form**

Convert each polar equation to a rectangular equation and identify its graph.

a. $r = 3$　**b.** $\theta = 45°$　**c.** $r = \csc\theta$　**d.** $r = 2\cos\theta$

SOLUTION

a.

$\quad r = 3$　　Given polar equation

$\quad r^2 = 9$　　Square both sides.

$\quad x^2 + y^2 = 9$　　Replace r^2 with $x^2 + y^2$.

The graph of $r = 3$ (or $x^2 + y^2 = 3^2$) is a circle with center at the pole and radius = 3 units. See Figure 6.58.

b.

$\quad \theta = 45°$　　Given polar equation

$\quad \tan\theta = \tan 45°$　　If $x = y$, then $\tan x = \tan y$.

$\quad \dfrac{y}{x} = 1$　　$\tan\theta = \dfrac{y}{x}$ and $\tan 45° = 1$

$\quad y = x$　　Multiply both sides by x.

The graph of $\theta = 45°$ (or $y = x$) is a line through the pole, making an angle of 45° with the polar axis. See Figure 6.59.

c.

$\quad r = \csc\theta$　　Given polar equation

$\quad r = \dfrac{1}{\sin\theta}$　　Reciprocal identity

$\quad r\sin\theta = 1$　　Multiply both sides by $\sin\theta$.

$\quad y = 1$　　Replace $r\sin\theta$ with y.

The graph of $r = \csc\theta$ (or $y = 1$) is a horizontal line. See Figure 6.60.

d.

$\quad r = 2\cos\theta$　　Given polar equation

$\quad r^2 = 2r\cos\theta$　　Multiply both sides by r.

$\quad x^2 + y^2 = 2x$　　Replace r^2 with $x^2 + y^2$ and $r\cos\theta$ with x.

$\quad x^2 - 2x + y^2 = 0$　　Subtract $2x$ from both sides and simplify.

$\quad (x^2 - 2x + 1) + y^2 = 1$　　Add 1 to both sides to complete the square.

$\quad (x - 1)^2 + y^2 = 1$　　Factor perfect square trinomial.

The graph of $r = 2\cos\theta$ (or $(x - 1)^2 + y^2 = 1$) is a circle with center at $(1, 0)$ and radius 1 unit. See Figure 6.61.　■ ■ ■

Practice Problem 6 Convert each polar equation to a rectangular equation and identify its graph.

a. $r = 5$　**b.** $\theta = -\dfrac{\pi}{3}$　**c.** $r = \sec\theta$　**d.** $r = 2\sin\theta$　■

FIGURE 6.61 $(x - 1)^2 + y^2 = 1$ or $r = 2\cos\theta$

4 Graph polar equations.

The Graph of a Polar Equation

Many important polar equations do not convert "nicely" to rectangular equations. To graph these equations, we plot points in polar coordinates. The graph of a polar equation

$$r = f(\theta)$$

is the set of all points $P(r, \theta)$ that have at least one polar coordinate representation that satisfies the equation. We make a table of several ordered pair solutions (r, θ) of the equation, plot these points, and join then with a smooth curve.

EXAMPLE 7 **Sketching a Graph by Plotting Points**

Sketch the graph of the polar equation $r = \theta \ (\theta \geq 0)$ by plotting points.

SOLUTION

We choose values of θ that are integer multiples of $\dfrac{\pi}{2}$. Since $r = \theta$, the polar coordinates of points on the curve are as follows:

$$(0,0),\left(\frac{\pi}{2},\frac{\pi}{2}\right),(\pi,\pi),\left(\frac{3\pi}{2},\frac{3\pi}{2}\right),(2\pi,2\pi),\left(\frac{5\pi}{2},\frac{5\pi}{2}\right),\dots$$

Plot these points and join them with a smooth curve to obtain the graph of $r = \theta$ shown in Figure 6.62. This curve is called the **spiral of Archimedes**.

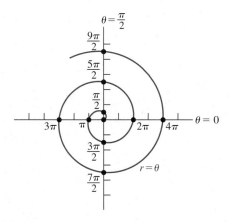

FIGURE 6.62 Spiral of Archimedes ■ ■ ■

Practice Problem 7 Sketch the graph of the polar equation $r = 2\theta(\theta \leq 0)$ by plotting points. ■

As with Cartesian equations, it is helpful to determine symmetry in the graph of a polar equation. If the graph of a polar equation has symmetry, then we can sketch its graph by plotting fewer points.

TESTS FOR SYMMETRY IN POLAR COORDINATES

The graph of a polar equation has the symmetry described below if an equivalent equation results when making the replacements shown.

Symmetry with respect to the polar axis	**Symmetry with respect to the line $\theta = \dfrac{\pi}{2}$**	**Symmetry with respect to the pole**
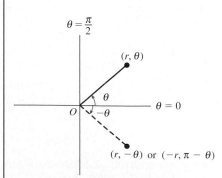	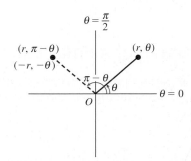	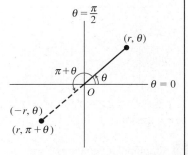
Replace (r,θ) with $(r,-\theta)$ or $(-r,\pi-\theta)$.	Replace (r,θ) with $(r,\pi-\theta)$ or $(-r,-\theta)$.	Replace (r,θ) with $(r,\pi+\theta)$ or $(-r,\theta)$.

Note that if a polar equation has two of these symmetries, then it must have the third symmetry.

TECHNOLOGY CONNECTION

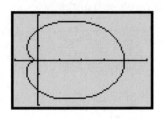 A graphing calculator can be used to graph polar equations. The calculator must be set in Polar (Pol) mode. Either Radian or Degree mode may be used, but make sure that your WINDOW settings are appropriate for your choice. See your owner's manual for details on polar graphing. The screen shows the graph of $r = 2(1 + \cos \theta)$.

EXAMPLE 8 **Sketching the Graph of a Polar Equation**

Sketch the graph of the polar equation $r = 2(1 + \cos \theta)$.

SOLUTION

Since $\cos \theta$ is an even function (that is, $\cos(-\theta) = \cos \theta$), the graph of $r = 2(1 + \cos \theta)$ is symmetric in the polar axis. Thus, to graph this equation, it is sufficient to compute values for $0 \le \theta \le \pi$. We construct Table 6.1 by substituting values for θ at increments of $\frac{\pi}{6}$, calculating the corresponding values of r. Also, we use the fact that $\cos(-\theta) = \cos \theta$.

TABLE 6.1

θ	0	$\frac{\pi}{6}$	$\frac{\pi}{3}$	$\frac{\pi}{2}$	$\frac{2\pi}{3}$	$\frac{5\pi}{6}$	π
$r = 2(1 + \cos \theta)$	4	$2 + \sqrt{3} \approx 3.73$	3	2	1	$2 - \sqrt{3} \approx 0.27$	0

We plot the points (r, θ), draw a smooth curve through them, and reflect the curve in the polar axis.

The graph of the equation $r = 2(1 + \cos \theta)$ is shown in Figure 6.63. This type of curve is called a *cardioid* because it resembles a heart.

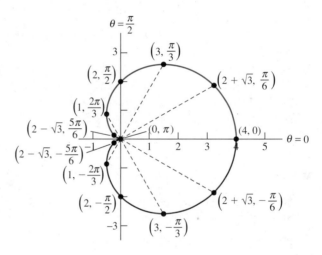

FIGURE 6.63 $r = 2(1 + \cos \theta)$ ■ ■ ■

Practice Problem 8 Sketch the graph of the polar equation $r = 2(1 + \sin \theta)$. ■

FINDING THE SOLUTION: A PROCEDURE

EXAMPLE 9 **Graphing a Polar Equation**

OBJECTIVE
Sketch the graph of a polar equation $r = f(\theta)$, where f is a periodic function.

Step 1 Test for symmetry.

EXAMPLE
Sketch the graph of $r = \cos 2\theta$.

1. Replace θ with $-\theta$,
$$\cos(-2\theta) = \cos 2\theta \qquad (1)$$
Replace θ with $\pi - \theta$,
$$\cos 2(\pi - \theta) = \cos(2\pi - 2\theta) = \cos 2\theta \qquad (2)$$
From (1) and (2), we conclude that the graph is symmetric about the polar axis and about the line $\theta = \frac{\pi}{2}$.

continued on the next page

Step 2   Analyze the behavior of r, symmetric from Step 1, and then find selected points.

2. To get a sense of how r behaves for different values of θ, we make the following table:

As θ goes from:	r varies from:
0 to $\dfrac{\pi}{4}$	1 to 0
$\dfrac{\pi}{4}$ to $\dfrac{\pi}{2}$	0 to -1
$\dfrac{\pi}{2}$ to $\dfrac{3\pi}{4}$	-1 to 0
$\dfrac{3\pi}{4}$ to π	0 to 1

Because the curve is symmetric about $\theta = 0$ and $\theta = \dfrac{\pi}{2}$, we find selected points for $0 \le \theta \le \dfrac{\pi}{2}$:

θ	0	$\dfrac{\pi}{6}$	$\dfrac{\pi}{4}$	$\dfrac{\pi}{3}$	$\dfrac{\pi}{2}$
$r = \cos 2\theta$	1	$\dfrac{1}{2}$	0	$-\dfrac{1}{2}$	-1

Step 3   Sketch the graph of $r = f(\theta)$ using the points (r, θ) found in Step 2.

3.

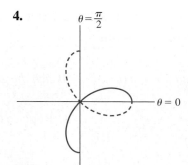

Step 4   Use symmetries to complete the graph.

4.

Using symmetry about the polar axis   Using symmetry about the line $\theta = \dfrac{\pi}{2}$

The curve of $r = \cos 2\theta$ is called a *four-leafed rose*.   ■ ■ ■

Practice Problem 9   Sketch the graph of $r = \sin 2\theta$.   ■

STUDY TIP

If a curve $r = f(\theta)$ is symmetric about the line $\theta = \dfrac{\pi}{2}$ (y-axis), you can first sketch the portion of the graph for $-\dfrac{\pi}{2} \le \theta \le \dfrac{\pi}{2}$ or

for $0 \le \theta \le \dfrac{\pi}{2}$ and $\pi \le \theta \le \dfrac{3\pi}{2}$, then complete the graph by symmetry.

EXAMPLE 10   **Graphing a Polar Equation**

Sketch a graph of $r = 1 + 2\sin\theta$.

SOLUTION

Step 1   **Symmetry.** Replacing (r, θ) with $(r, \pi - \theta)$, we obtain an equivalent equation. The graph is symmetric with respect to the line $\theta = \dfrac{\pi}{2}$.

Step 2   **Analyze and make a table of values.** We first analyze the behavior of r:

As θ goes from $-\dfrac{\pi}{2}$ to $-\dfrac{\pi}{6}$, r varies from -1 to 0.

As θ goes from $-\dfrac{\pi}{6}$ to 0, r varies from 0 to 1.

As θ goes from 0 to $\dfrac{\pi}{2}$, r varies from 1 to 3.

As θ goes from $\dfrac{\pi}{2}$ to π, r varies from 3 to -1.

So it is sufficient to first sketch the graph for $-\dfrac{\pi}{2} \le \theta \le \dfrac{\pi}{2}$. Table 6.2 gives the coordinates of some points on the graph.

TABLE 6.2

θ	$-\dfrac{\pi}{2}$	$-\dfrac{\pi}{3}$	$-\dfrac{\pi}{4}$	$-\dfrac{\pi}{6}$	0	$\dfrac{\pi}{6}$	$\dfrac{\pi}{4}$	$\dfrac{\pi}{3}$	$\dfrac{\pi}{2}$
r	-1	$1 - \sqrt{3}$	$1 - \sqrt{2}$	0	1	2	$1 + \sqrt{2}$	$1 + \sqrt{3}$	3

Step 3   Sketch the graph for $-\dfrac{\pi}{2} \le \theta \le \dfrac{\pi}{2}$ using the points (r, θ) found in Step 2.

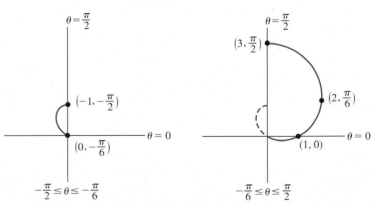

FIGURE 6.64 Graph of $r = 1 + 2\sin\theta$ for $-\dfrac{\pi}{2} \le \theta \le \dfrac{\pi}{2}$

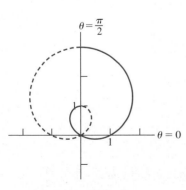

FIGURE 6.65 Graph of
$r = 1 + 2\sin\theta$

Step 4   Complete the graph by symmetry with respect to the line $\theta = \dfrac{\pi}{2}$. The sketch is shown in Figure 6.65. This curve is called a *limaçon*.   ■ ■ ■

Practice Problem 10   Sketch a graph of $r = 1 + 2\cos\theta$.   ■

The graph of an equation of the form

$$r = a \pm b\cos\theta \qquad \text{or} \qquad r = a \pm b\sin\theta$$

is a **limaçon**. If $b > a$, as in Example 10, the limaçon has a loop. If $b = a$, the limaçon is a **cardioid**, a heart-shaped curve. The curve in Example 8 is a cardioid. If $b < a$, the limaçon has a shape similar to the one shown in Example 11.

STUDY TIP

The symmetry for the curve

$r = 3 + 2\sin\theta$ about $\theta = \dfrac{\pi}{2}$

cannot be decided from
the failure of the test:
replace (r, θ) with $(-r, -\theta)$.

EXAMPLE 11   **Graphing a Limaçon**

Sketch a graph of $r = 3 + 2\sin\theta$.

SOLUTION

Step 1   The graph is symmetric with respect to the line $\theta = \dfrac{\pi}{2}$ because if (r, θ) is
replaced with $(r, \pi - \theta)$, an equivalent equation is obtained.

Step 2   Analyze the behavior of $r = 3 + 2\sin\theta$. As θ goes from 0 to $\dfrac{\pi}{2}$, r increases
from 3 to 5; as θ goes from $\dfrac{\pi}{2}$ to π, r decreases from 5 to 3; as θ goes from
π to $\dfrac{3\pi}{2}$, r decreases from 3 to 1; and as θ goes from $\dfrac{3\pi}{2}$ to 2π, r increases
from 1 to 3. Because of symmetry about the line $\theta = \dfrac{\pi}{2}$, we find selected
points for $0 \le \theta \le \dfrac{\pi}{2}$ and $\pi \le \theta \le \dfrac{3\pi}{2}$. Table 6.3 gives polar coordinates
of some points in these intervals on the graph.

TABLE 6.3

θ	0	$\dfrac{\pi}{6}$	$\dfrac{\pi}{3}$	$\dfrac{\pi}{2}$	π	$\dfrac{7\pi}{6}$	$\dfrac{4\pi}{3}$	$\dfrac{3\pi}{2}$
r	3	4	$3 + \sqrt{3}$	5	3	2	$3 - \sqrt{3}$	1

Step 3   Sketch the portion of the graph for $0 \le \theta \le \dfrac{\pi}{2}$ and $\pi \le \theta \le \dfrac{3\pi}{2}$. See
Figure 6.66.

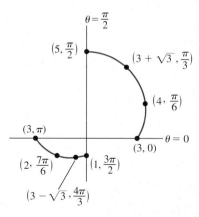

FIGURE 6.66 $r = 3 + 2\sin\theta$; $0 \le \theta \le \dfrac{\pi}{2}$ and $\pi \le \theta \le \dfrac{3\pi}{2}$

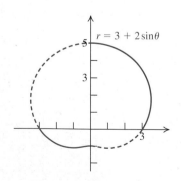

FIGURE 6.67 Graph of
$r = a + b\cos\theta$ with $a > b$

Step 4   Complete the graph by symmetry with respect to the line $\theta = \dfrac{\pi}{2}$. See
Figure 6.67.   ■ ■ ■

Practice Problem 11   Sketch a graph of $r = 2 - \cos\theta$.   ■

Note that even if a polar equation fails a test for a given symmetry, its graph nevertheless may exhibit that symmetry. See the margin Study Tip. In other words, the tests
for polar coordinate symmetry are *sufficient* but not *necessary* for showing symmetry.

The following graphs have simple polar equations.

LIMAÇONS				
The graphs of $r = a \pm b \cos\theta$, $r = a \pm b \sin\theta$ $(a > 0, b > 0)$ are called **limaçons** (French for "snail").	Inner loop if $\dfrac{a}{b} < 1$	Cardioid if $\dfrac{a}{b} = 1$	Dimpled if $1 < \dfrac{a}{b} < 2$	Not Dimpled if $\dfrac{a}{b} \geq 2$

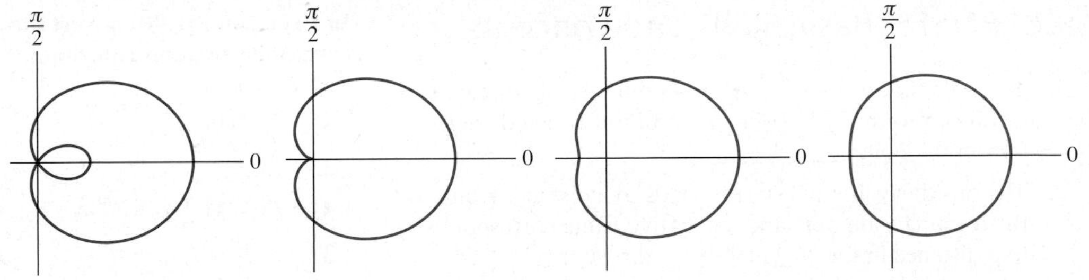

ROSE CURVES				
The graphs of $r = a \cos n\theta$, $r = a \sin n\theta$ $(a > 0)$ are called **rose curves**. If n is odd, the rose has n petals. If n is even, the rose has $2n$ petals.	$r = a \cos 3\theta$	$r = a \sin 5\theta$	$r = a \sin 2\theta$	$r = a \cos 4\theta$

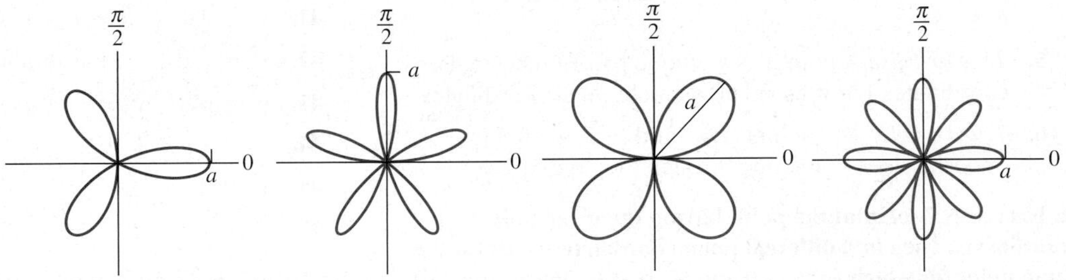

CIRCLES AND LEMNISCATES				
	$r = a \cos\theta$ Circle	$r = a \sin\theta$ Circle	$r^2 = a^2 \sin 2\theta$ Lemniscate	$r^2 = a^2 \cos 2\theta$ Lemniscate

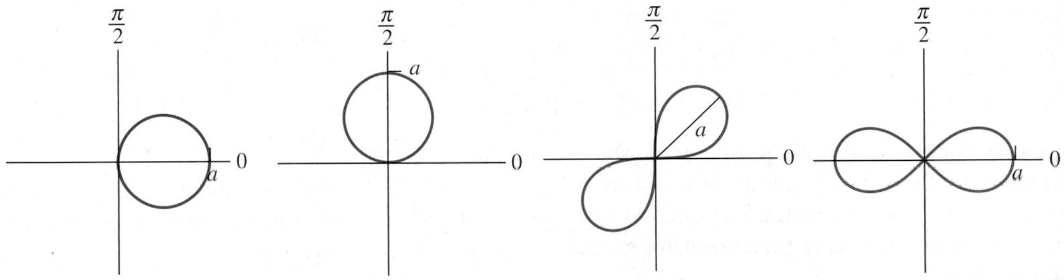

SPIRALS $(a > 0, k > 0)$			
	$r = a\theta$	$r = \dfrac{a}{\theta}$	$r = a^{k\theta}$

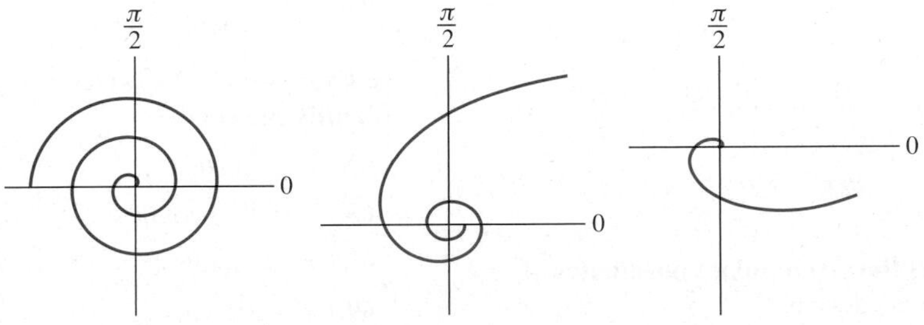

SECTION 6.6 ■ Exercises

A EXERCISES Basic Skills and Concepts

1. The polar coordinates (r, θ) of a point are the directed distance r to the ___origin___ and the directed angle θ from the polar ___axis___.

2. The positive value of r corresponds to a distance r on the terminal side of θ, and a negative value corresponds to a distance in the ___opposite___ direction.

3. The substitutions $x = $ ___$r \cos \theta$___ and $y = $ ___$r \sin \theta$___ transform a rectangular equation into a polar equation for the same curve.

4. Polar coordinates are found from the rectangular coordinates by the formulas $r = $ ___$\sqrt{x^2 + y^2}$___ and $\theta = $ _____ $\cdot \tan^{-1}\left(\dfrac{y}{x}\right)$

5. *True or False* A point has a unique pair of rectangular coordinates, but it has many pairs of polar coordinates. True

6. *True or False* The points $(2, \theta)$ and $(-2, \theta)$ are on opposite sides of the line through the pole. True

In Exercises 7–14, plot the point having the given polar coordinates. Then find different polar coordinates (r, θ) for the same point for which (a) $r < 0$ and $0° \le \theta < 360°$; (b) $r < 0$ and $-360° < \theta < 0°$; (c) $r > 0$ and $-360° < \theta < 0°$.

7. $(4, 45°)$ †

8. $(5, 150°)$ †

9. $(3, 90°)$ †

10. $(2, 240°)$ †

11. $(3, 60°)$ †

12. $(4, 210°)$ †

13. $(6, 300°)$ †

14. $(2, 270°)$ †

In Exercises 15–22, plot the point having the given polar coordinates. Then give two different pairs of polar coordinates of the same point, (a) one with the given value of r and (b) one with r having the opposite sign of the given value of r.

15. $\left(2, -\dfrac{\pi}{3}\right)$ †

16. $\left(\sqrt{2}, -\dfrac{\pi}{4}\right)$ †

17. $\left(-3, \dfrac{\pi}{6}\right)$ †

18. $\left(-2, \dfrac{3\pi}{4}\right)$ †

19. $\left(-2, -\dfrac{\pi}{6}\right)$ †

20. $(-2, -\pi)$ †

21. $\left(4, \dfrac{7\pi}{6}\right)$ †

22. $(2, 3)$ †

In Exercises 23–30, convert the given polar coordinates of each point to rectangular coordinates.

23. $(3, 60°)$ $\left(\dfrac{3}{2}, \dfrac{3\sqrt{3}}{2}\right)$

24. $(-2, -30°)$ $(-\sqrt{3}, 1)$

25. $(5, -60°)$ $\left(\dfrac{5}{2}, -\dfrac{5\sqrt{3}}{2}\right)$

26. $(-3, 90°)$ $(0, -3)$

27. $(3, \pi)$ $(-3, 0)$

28. $\left(\sqrt{2}, -\dfrac{\pi}{4}\right)$ $(1, -1)$

29. $\left(-2, -\dfrac{5\pi}{6}\right)$ $(\sqrt{3}, 1)$

30. $\left(-1, \dfrac{7\pi}{6}\right)$ $\left(\dfrac{\sqrt{3}}{2}, \dfrac{1}{2}\right)$

In Exercises 31–38, convert the rectangular coordinates of each point to polar coordinates (r, θ) with $r > 0$ and $0 \le \theta < 2\pi$.

31. $(1, -1)$ $\left(\sqrt{2}, \dfrac{7\pi}{4}\right)$

32. $(-\sqrt{3}, 1)$ $\left(2, \dfrac{5\pi}{6}\right)$

33. $(3, 3)$

34. $(-4, 0)$ $(4, \pi)$

35. $(3, -3)$ $\left(3\sqrt{2}, \dfrac{7\pi}{4}\right)$

36. $(2\sqrt{3}, 2)$ $\left(4, \dfrac{\pi}{6}\right)$

37. $(-1, \sqrt{3})$ $\left(2, \dfrac{2\pi}{3}\right)$

38. $(-2, -2\sqrt{3})$ $\left(4, \dfrac{4\pi}{3}\right)$

In Exercises 39–46, convert each rectangular equation to polar form.

39. $x^2 + y^2 = 16$ $r = 4$

40. $x + y = 1$ $r = \dfrac{1}{\cos \theta + \sin \theta}$

41. $y^2 = 4x$ $r = 4 \cot \theta \csc \theta$

42. $x^3 = 3y^2$ $r = 3 \tan^2 \theta \sec \theta$

43. $y^2 = 6y - x^2$ $r = 6 \sin \theta$

44. $x^2 - y^2 = 1$ $r^2 = \dfrac{1}{\cos^2 \theta - \sin^2 \theta}$

45. $xy = 1$ $r^2 = \dfrac{1}{\sin \theta \cos \theta}$

46. $x^2 + y^2 - 4x + 6y = 12$ $r^2 - 4r \cos \theta + 6r \sin \theta = 12$

In Exercises 47–54, convert each polar equation to rectangular form. Identify each curve.

47. $r = 2$ †

48. $r = -3$ †

49. $\theta = \dfrac{3\pi}{4}$ †

50. $\theta = -\dfrac{\pi}{6}$ †

51. $r = 4 \cos \theta$ †

52. $r = 4 \sin \theta$ †

53. $r = -2 \sin \theta$ †

54. $r = -3 \cos \theta$ †

In Exercises 55–62, sketch the graph of each polar equation by transforming it to rectangular coordinates.

55. $r = -2$ †

56. $r \cos \theta = -1$ †

57. $r \sin \theta = 2$ †

58. $r = \sin \theta$ †

59. $r = 2 \sec \theta$ †

60. $r + 3 \cos \theta = 0$ †

61. $r = \dfrac{1}{\cos \theta + \sin \theta}$ †

62. $r = \dfrac{6}{2 \cos \theta + 3 \sin \theta}$ †

In Exercises 63–78, sketch the graph of each polar equation. Identify the curve.

63. $r = \sin \theta$ †

64. $r = \cos \theta$ †

65. $r = 1 - \cos \theta$ †

66. $r = 1 + \sin \theta$ †

67. $r = \cos 3\theta$ †

68. $r = \sin 3\theta$ †

69. $r = \sin 4\theta$ †

70. $r = \cos 4\theta$ †

71. $r = 1 - 2 \sin \theta$ †

72. $r = 2 - 4 \sin \theta$ †

73. $r = 2 + 4 \cos \theta$ †

74. $r = 2 - 4 \cos \theta$ †

75. $r = 3 - 2 \sin \theta$ †

76. $r = 5 + 3 \sin \theta$ †

77. $r = 4 + 3 \cos \theta$ †

78. $r = 5 + 2 \cos \theta$ †

Answers:

33. $\left(3\sqrt{2}, \dfrac{\pi}{4}\right)$

†Due to space constrictions, answers to these exercises may be found in the Answers beginning on page A–1 in the back of the book.

B EXERCISES Applying the Concepts

In Exercises 79–82, determine the position of the hand relative to the shoulder by considering the robotic arm illustrated in the figure. Round all answers to the nearest tenth.

79. $\alpha = 45°, \beta = 30°$ $(21.8, 39.6°)$

80. $\alpha = -30°, \beta = 60°$ $(16.1, -0.3°)$

81. $\alpha = -70°, \beta = 0°$ $(18.3, -45.8°)$

82. $\alpha = 47°, \beta = 17°$ $(21.3, 36.2°)$

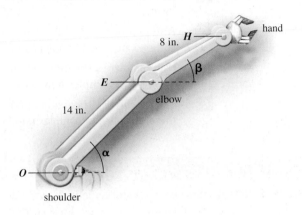

C EXERCISES Beyond the Basics

In Exercises 83 and 84, convert each rectangular equation to polar form. Assume that $x \geq 0, y \geq 0$ and that $\sqrt{x^2 + y^2} < \dfrac{\pi}{2}$.

83. $y = x \tan\left(\sqrt{x^2 + y^2}\right)$ $r = \theta$

84. $y = x \tan\left(\ln\sqrt{x^2 + y^2}\right)$ $r = e^\theta$

In Exercises 85–90, convert each polar equation to rectangular form.

85. $r(1 - \sin\theta) = 3$ $y = \dfrac{x^2}{6} - \dfrac{3}{2}$

86. $r(1 + \cos\theta) = 2$ $y^2 = 4 - 4x$

87. $r\left(1 + \dfrac{1}{2}\cos\theta\right) = 1$ $y^2 = 1 - x - \dfrac{3x^2}{4}$

88. $r\left(1 + \dfrac{3}{4}\sin\theta\right) = 3$ $\dfrac{16x^2}{7} + \left(y + \dfrac{36}{7}\right)^2 = \dfrac{2304}{49}$

89. $r(1 - 3\cos\theta) = 5$ $y^2 - 8x^2 - 30x - 25 = 0$

90. $r(1 - 2\sin\theta) = 4$ $x^2 - 3y^2 - 16y - 16 = 0$

91. Prove that the distance between the points $P(r_1, \theta_1)$ and $Q(r_2, \theta_2)$ is given by

$$d(P, Q) = \sqrt{r_1^2 + r_2^2 - 2r_1 r_2 \cos(\theta_1 - \theta_2)}.$$

92. Prove that the equation $r = a \sin\theta + b \cos\theta$ represents a circle. Find its center and radius.

93. Prove that the area K of the triangle whose polar coordinates are $(0, 0)$, (r_1, θ_1), and (r_2, θ_2) is given by

$$K = \dfrac{1}{2}r_1 r_2 \sin(\theta_2 - \theta_1). \text{ [Assume that } 0 \leq \theta_1 <$$

$\theta_2 \leq \pi$ and $r_1 > 0, r_2 > 0$.]

94. Show that the polar equation of a circle with center (a, θ_0) and radius a is given by $r = 2a \cos(\theta - \theta_0)$.

Discuss the cases $\theta_0 = 0$ and $\theta_0 = \dfrac{\pi}{2}$.

 In Exercises 95–104, use a graphing calculator to graph the polar equation. Identify any horizontal or vertical asymptotes that appear on the screen but are not part of the graph.

95. $r = e^\theta$ (logarithmic spiral) †

96. $r = e^{\frac{\theta}{3}}$ (logarithmic spiral) †

97. $r^2 = 4 \sin 2\theta$ (lemniscate) †

98. $r^2 = 9 \sin 2\theta$ (lemniscate) †

99. $r^2 = 16 \cos 2\theta$ (lemniscate) †

100. $r^2 = 4 \cos 2\theta$ (lemniscate) †

101. $r = 2 \sin\theta \tan\theta$ (cissoid) †

102. $(r - 2)^2 = 8\theta$ (parabolic spiral) †

103. $r = 2 \csc\theta + 3$ (conchoid) †

104. $r = 2 \sec\theta - 1$ (conchoid) †

Critical Thinking

Determine whether each statement is true or false. Explain your reasoning.

105. The points (r, θ) and $(-r, \theta)$ are symmetric with respect to the y-axis. False

106. The points (r, θ) and $(-r, -\theta)$ are symmetric with respect to the y-axis. True

107. The points (r, θ) and $(r, -\theta)$ are symmetric with respect to the x-axis. True

108. The polar coordinates $(r, -\theta)$ and $(-r, \pi - \theta)$ represent the same point in the plane. True

Answer:

92. Center $\left(\dfrac{b}{2}, \dfrac{a}{2}\right)$; radius $\sqrt{\dfrac{b^2}{4} + \dfrac{a^2}{4}}$

Polar Form of Complex Numbers; DeMoivre's Theorem

Before Starting this Section, Review

1. Addition of complex numbers (Appendix A, page 804)
2. Conversion between polar and rectangular coordinates (Section 6.6, page 451)

Objectives

1 Represent complex numbers geometrically.

2 Find the absolute value of a complex number.

3 Write a complex number in polar form.

4 Find products and quotients of complex numbers in polar form.

5 Use DeMoivre's Theorem to find powers of a complex number.

6 Use DeMoivre's Theorem to find the nth roots of a complex number.

ALTERNATING CURRENT CIRCUITS

In the early days of the study of alternate current (AC) circuits, scientists concluded that AC circuits were somehow different from the battery-powered direct current (DC) circuits. However, both types of circuits obey exactly the same physical and mathematical laws. The breakthrough in understanding the AC circuits came in 1893 by Charles Steinmetz. He explained that in AC circuits, the voltage, current, and resistance (called *impedance* in AC circuits) do not act like scalars, but alternate in direction, and possess frequency and phase shift. He advocated the use of the polar form of complex numbers to represent the magnitude, frequency, and phase shift for the AC circuit quantities: voltage, current, and impedance. Steinmetz eventually became known as "the wizard who generated electricity from the square root of minus one." In Example 6, we use complex numbers to compute the total impedance in an AC circuit. ▪

---

1 Represent complex numbers geometrically.

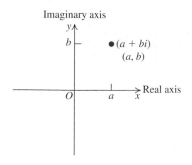

FIGURE 6.68 Complex plane

Geometric Representation of Complex Numbers

Because each complex number $a + bi$ determines a unique ordered pair (a, b) of real numbers, we can represent the set of complex numbers geometrically by points in a rectangular coordinate system. Specifically, we can represent the complex number $a + bi$ by the point (a, b) in a rectangular coordinate system. We call the plane in this system the **complex plane**. The x-axis is also called the **real axis** because the real part of a complex number is plotted along the x-axis. Similarly, the y-axis is also called the **imaginary axis**. See Figure 6.68. We can think of a complex number $a + bi$ as a position vector with initial point $(0, 0)$ and terminal point (a, b).

EXAMPLE 1 Plotting Complex Numbers

Plot each complex number in the complex plane.

$$1 + 3i, \quad -2 + 2i, \quad -3, \quad -2i, \quad \text{and} \quad 3 - i$$

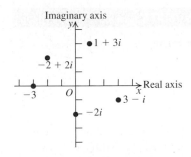

FIGURE 6.69 Plotting complex numbers

2 Find the absolute value of a complex number.

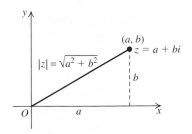

FIGURE 6.70 Absolute value of z

SOLUTION

Figure 6.69 shows the points representing the complex numbers $1 + 3i$, $-2 + 2i$, -3, $-2i$ and $3 - i$.   ■ ■ ■

Practice Problem 1   Plot each complex number in the complex plane.

$$2 + 3i, \quad -3 + 2i, \quad 4, \quad -2 - 3i, \quad \text{and} \quad 3 - 2i$$ ■

The Absolute Value of a Complex Number

Let (a, b) represent the complex number $z = a + bi$ in a rectangular coordinate system. Then the distance from the origin O to the point (a, b) is $\sqrt{a^2 + b^2}$. See Figure 6.70. The number $\sqrt{a^2 + b^2}$ is called the *absolute value* (or **magnitude** or **modulus**) of the complex number $z = a + bi$ and is denoted by $|z|$.

ABSOLUTE VALUE OF A COMPLEX NUMBER

The **absolute value** of a complex number $z = a + bi$:

$$|z| = |a + bi| = \sqrt{a^2 + b^2}$$

EXAMPLE 2   **Finding the Absolute Value of a Complex Number**

Find the absolute value of each complex number.

a. $4 + 3i$ **b.** $2 - 3i$ **c.** $-4 + i$ **d.** $-2 - 2i$ **e.** $-3i$

SOLUTION

In each case, we use the formula $|a + bi| = \sqrt{a^2 + b^2}$ and simplify.

a. $|4 + 3i| = \sqrt{4^2 + 3^2} = \sqrt{25} = 5$
b. $|2 - 3i| = \sqrt{2^2 + (-3)^2} = \sqrt{4 + 9} = \sqrt{13}$
c. $|-4 + i| = |-4 + 1 \cdot i| = \sqrt{(-4)^2 + (1)^2} = \sqrt{16 + 1} = \sqrt{17}$
d. $|-2 - 2i| = \sqrt{(-2)^2 + (-2)^2} = \sqrt{4 + 4} = \sqrt{4 \cdot 2} = 2\sqrt{2}$
e. $|-3i| = |0 - 3i| = \sqrt{0^2 + (-3)^2} = \sqrt{0 + 9} = 3$ ■ ■ ■

Practice Problem 2   Find the absolute value of each complex number.

a. $-5 + 12i$ **b.** -7 **c.** i **d.** $a - bi$ ■

3 Write a complex number in polar form.

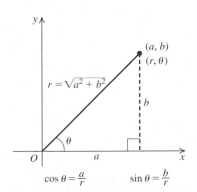

FIGURE 6.71 Polar form of
$z = a + bi$

Polar Form of a Complex Number

A complex number z written as $z = a + bi$ is said to be in **rectangular form**. The point (a, b) has polar coordinates (r, θ), where $r = \sqrt{a^2 + b^2}$, $a = r \cos \theta$, and $b = r \sin \theta$. See Figure 6.71. The complex number $z = a + bi$ can therefore be written in the form

$$z = r \cos \theta + (r \sin \theta)i = r(\cos \theta + i \sin \theta).$$

We call $r(\cos \theta + i \sin \theta)$ the **polar form** or the **trigonometric form** of z.

POLAR FORM OF A COMPLEX NUMBER

The complex number $z = a + bi$ can be written in **polar form**

$$z = r(\cos\theta + i\sin\theta),$$

where $a = r\cos\theta$, $b = r\sin\theta$, $r = \sqrt{a^2 + b^2}$, and $\tan\theta = \dfrac{b}{a}$.

When a nonzero complex number is written in polar form, the positive number r is the **modulus** or **absolute value** of z; the angle θ is called the **argument** of z (written $\theta = \arg z$).

Note that the angle θ in the polar representation of z is not unique because for any integer n,

$$r(\cos\theta + i\sin\theta) = r[\cos(\theta + n\cdot 360°) + i\sin(\theta + n\cdot 360°)].$$

Two complex numbers in polar form are therefore equal if and only if their *moduli* (plural of *modulus*) are equal and their arguments differ by a multiple of 360° (or a multiple of 2π).

FINDING THE SOLUTION: A PROCEDURE

EXAMPLE 3   Writing a Complex Number in the Polar Form

OBJECTIVE

Write a complex number $z = a + bi$ in polar form.

Step 1   Find r. Identify a and b. Use the formula $r = \sqrt{a^2 + b^2}$.

Step 2   Find θ. Use $\tan\theta = \dfrac{b}{a}$ to find the possible values of θ. Choose θ in the quadrant in which (a, b) lies. In general, using a calculator, a value of θ is given by:

$$\theta = \begin{cases} \tan^{-1}\dfrac{b}{a} & \text{if } a > 0 \\[2mm] 180° + \tan^{-1}\dfrac{b}{a} & \text{if } a < 0 \end{cases}$$

Step 3   Write in polar form. From Steps 1 and 2, write the polar form $z = r(\cos\theta + i\sin\theta)$.

EXAMPLE

Write $z = \sqrt{3} - i$ in polar form. Express the argument θ in degrees, $0° \le \theta < 360°$.

1. $z = \sqrt{3} - i$, $a = \sqrt{3}$, $b = -1$
$$r = \sqrt{(\sqrt{3})^2 + (-1)^2} = \sqrt{3 + 1} = 2.$$

2. $\tan\theta = \dfrac{b}{a} = \dfrac{-1}{\sqrt{3}} = -\dfrac{1}{\sqrt{3}}$. We know that $\tan 30° = \dfrac{1}{\sqrt{3}}$
and the tangent is negative in quadrants II and IV; so either $\theta = 180° - 30° = 150°$ or $\theta = 360° - 30° = 330°$. Because $(a, b) = (\sqrt{3}, -1)$ lies in quadrant IV, we have $\theta = 330°$. See the figure.

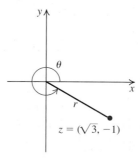

$z = (\sqrt{3}, -1)$

3. $z = 2(\cos 330° + i\sin 330°)$ $r = 2, \theta = 330°$

■ ■ ■

Practice Problem 3   Write $z = -1 - i$ in polar form, letting $0 \le \theta < 2\pi$.   ■

EXAMPLE 4   **Writing a Complex Number in Rectangular Form**

Write the complex number $z = 2\left(\cos\dfrac{\pi}{6} + i\sin\dfrac{\pi}{6}\right)$ in rectangular form.

SOLUTION

$$z = 2\left(\cos\frac{\pi}{6} + i\sin\frac{\pi}{6}\right) \qquad \text{Given complex number}$$

$$= 2\cos\frac{\pi}{6} + 2i\sin\frac{\pi}{6} \qquad \text{Distributive property}$$

$$= 2\left(\frac{\sqrt{3}}{2}\right) + 2i\left(\frac{1}{2}\right) \qquad \cos\frac{\pi}{6} = \frac{\sqrt{3}}{2}, \sin\frac{\pi}{6} = \frac{1}{2}$$

$$= \sqrt{3} + i \qquad \text{Simplify.}$$

The rectangular form of $z = 2\left(\cos\dfrac{\pi}{6} + i\sin\dfrac{\pi}{6}\right)$ is $\sqrt{3} + i$.   ■ ■ ■

Practice Problem 4   Write $z = 4\left(\cos\dfrac{5\pi}{3} + i\sin\dfrac{5\pi}{3}\right)$ in rectangular form.   ■

4 Find products and quotients of complex numbers in polar form.

Product and Quotient in Polar Form

The polar representation of complex numbers leads to an interesting interpretation of the product and quotient of two complex numbers.

PRODUCT AND QUOTIENT RULES FOR TWO COMPLEX NUMBERS IN POLAR FORM

Let $z_1 = r_1(\cos\theta_1 + i\sin\theta_1)$ and $z_2 = r_2(\cos\theta_2 + i\sin\theta_2)$ be two complex numbers in polar form. Then

$$z_1z_2 = r_1r_2[\cos(\theta_1 + \theta_2) + i\sin(\theta_1 + \theta_2)] \qquad \text{Product rule}$$

and

$$\frac{z_1}{z_2} = \frac{r_1}{r_2}[\cos(\theta_1 - \theta_2) + i\sin(\theta_1 - \theta_2)], z_2 \neq 0 \qquad \text{Quotient rule}$$

In words: to multiply two complex numbers in polar form, we multiply their moduli and add their arguments; to divide two complex numbers, we divide their moduli and subtract their arguments.

We ask you to prove these rules in Exercises 95 and 96.

EXAMPLE 5   **Finding the Product and Quotient of Two Complex Numbers**

Let $z_1 = 3(\cos 65° + i\sin 65°)$ and $z_2 = 4(\cos 15° + i\sin 15°)$.

Find z_1z_2 and $\dfrac{z_1}{z_2}$. Leave the answers in polar form.

SOLUTION

$$z_1z_2 = 3(\cos 65° + i\sin 65°)\cdot 4(\cos 15° + i\sin 15°) \qquad \text{Product of given numbers}$$

$$= 3\cdot 4[\cos(65° + 15°) + i\sin(65° + 15°)] \qquad \text{Multiply moduli and add arguments.}$$

$$= 12(\cos 80° + i\sin 80°) \qquad \text{Simplify.}$$

continued on the next page

$$\frac{z_1}{z_2} = \frac{3(\cos 65° + i \sin 65°)}{4(\cos 15° + i \sin 15°)} \qquad \text{Quotient of given numbers}$$

$$= \frac{3}{4}[\cos(65° - 15°) + i\sin(65° - 15°)] \qquad \begin{array}{l}\text{Divide moduli and subtract}\\\text{arguments.}\end{array}$$

$$= \frac{3}{4}(\cos 50° + i\sin 50°) \qquad \text{Simplify.} \qquad \blacksquare\blacksquare\blacksquare$$

Practice Problem 5   Let $z_1 = 5(\cos 75° + i \sin 75°)$ and $z_2 = 2(\cos 60° + i\sin 60°)$. Find $z_1 z_2$ and $\dfrac{z_1}{z_2}$. Leave the answers in polar form.   ■

EXAMPLE 6   **Using Complex Numbers in AC Circuits**

In a parallel circuit, the total impedance Z_t is given by

$$Z_t = \frac{Z_1 Z_2}{Z_1 + Z_2}.$$

Find Z_t, if

$$Z_1 = 9(\cos 90° + i \sin 90°) \text{ and } Z_2 = 4[\cos(-60°) + i \sin(-60°)].$$

SOLUTION

We first calculate $Z_1 Z_2$ and $Z_1 + Z_2$.

$$Z_1 Z_2 = 9(\cos 90° + i \sin 90°) \cdot 4[\cos(-60°) + i \sin(-60°)] \qquad \begin{array}{l}\text{Substitute values for}\\ Z_1 \text{ and } Z_2.\end{array}$$

$$= 9 \cdot 4[\cos(90° - 60°) + i \sin(90° - 60°)] \qquad \begin{array}{l}\text{Multiply moduli and}\\\text{add arguments.}\end{array}$$

$$= 36(\cos 30° + i \sin 30°) \qquad \text{Simplify.}$$

To add complex numbers in polar form, convert to rectangular form, perform addition, and then convert back to polar form.

$$Z_1 = 9 \cos 90° + 9i \sin 90° = 9i \qquad \cos 90° = 0, \sin 90° = 1$$

$$Z_2 = 4[\cos(-60°) + i\sin(-60°)]$$

$$\quad = 4\cos 60° - 4i \sin 60° \qquad \cos(-\theta) = \cos\theta, \sin(-\theta) = -\sin\theta$$

$$\quad = 4\left(\frac{1}{2}\right) - 4i\left(\frac{\sqrt{3}}{2}\right) \qquad \cos 60° = \frac{1}{2}, \sin 60° = \frac{\sqrt{3}}{2}$$

$$\quad = 2 - 2\sqrt{3}i \qquad \text{Simplify.}$$

$$Z_1 + Z_2 = 9i + 2 - 2\sqrt{3}i$$

$$Z_1 + Z_2 = 2 + (9 - 2\sqrt{3})i \qquad \text{Regroup.}$$

To write $Z_1 + Z_2$ in polar form, find the values of r and θ.

$$r = \sqrt{a^2 + b^2}$$

$$\quad = \sqrt{2^2 + (9 - 2\sqrt{3})^2} \qquad \text{Replace } a \text{ with 2 and } b \text{ with } 9 - 2\sqrt{3}.$$

$$\quad \approx 5.89 \qquad \text{Use a calculator.}$$

$$\tan\theta = \frac{b}{a} = \frac{9 - 2\sqrt{3}}{2} \approx 2.768$$

$$\theta \approx \tan^{-1}(2.768) \approx 70° \qquad \text{Use a calculator.}$$

The polar form of $Z_1 + Z_2$ is approximately $5.89(\cos 70° + i \sin 70°)$. Thus,

$$Z_t = \frac{Z_1 Z_2}{Z_1 + Z_2} \approx \frac{36(\cos 30° + i \sin 30°)}{5.89(\cos 70° + i \sin 70°)}$$

$$= \frac{36}{5.89}[\cos (30° - 70°) + i \sin (30° - 70°)] \qquad \text{Substitute values for } Z_1 Z_2 \text{ and } Z_1 + Z_2.$$

$$= \frac{36}{5.89}[\cos (-40°) + i \sin (-40°)] \qquad \text{Divide moduli and subtract.}$$

$$= \frac{36}{5.89}(\cos 40° - i \sin 40°) \qquad \begin{array}{l}\cos (-\theta) = \cos \theta, \\ \sin (-\theta) = -\sin \theta\end{array}$$

$$\approx 6.11(\cos 40° - i \sin 40°). \qquad \frac{36}{5.89} \approx 6.11 \qquad ■■■$$

Practice Problem 6 Repeat Example 6 with

$$Z_1 = 4(\cos 45° + i \sin 45°) \quad \text{and} \quad Z_2 = 6(\cos 0° + i \sin 0°). \qquad ■$$

5 Use DeMoivre's Theorem to find powers of a complex number.

Powers of Complex Numbers in Polar Form

Let $z = r(\cos \theta + i \sin \theta)$ be a complex number in polar form. Then

$$z^2 = z \cdot z$$
$$= r(\cos \theta + i \sin \theta) \cdot r(\cos \theta + i \sin \theta) \qquad \text{Form product } z \cdot z.$$
$$= (r \cdot r)[\cos (\theta + \theta) + i \sin (\theta + \theta)] \qquad \text{Multiply moduli and add arguments.}$$
$$= r^2(\cos 2\theta + i \sin 2\theta) \qquad \text{Simplify.}$$

$$z^3 = z^2 \cdot z$$
$$= r^2(\cos 2\theta + i \sin 2\theta) \cdot r(\cos 2\theta + i \sin 2\theta) \qquad \text{Polar form of } z^2$$
$$= (r^2 \cdot r)[\cos (2\theta + \theta) + i \sin (2\theta + \theta)] \qquad \text{Multiply moduli and add arguments.}$$
$$= r^3(\cos 3\theta + i \sin 3\theta). \qquad \text{Simplify.}$$

In a similar way, you can show that

$$z^4 = r^4(\cos 4\theta + i \sin 4\theta).$$

DeMoivre's Theorem (whose proof is omitted) follows this pattern.

DEMOIVRE'S THEOREM

Let $z = r(\cos \theta + i \sin \theta)$ be a complex number in polar form. Then for any integer n,

$$z^n = r^n(\cos n\theta + i \sin n\theta).$$

EXAMPLE 7   **Finding the Power of a Complex Number**

Let $z = 1 + i$. Use DeMoivre's Theorem to find each power of z. Write answers in rectangular form.

a. z^{16} **b.** z^{-10}

SOLUTION

We first convert $z = 1 + i$ to polar form. We find r and θ.

$$r = \sqrt{a^2 + b^2} = \sqrt{1^2 + 1^2} = \sqrt{2}$$

TECHNOLOGY CONNECTION

A graphing calculator can be used to find powers of complex numbers or to check your work. The calculator must be set in $a + bi$ mode.

```
(1+i)^(-10)
          -.03125i
Ans►Frac
          -1/32i
■
```

and

$$\tan \theta = \frac{b}{a} = \frac{1}{1} = 1, \text{so } \theta = \frac{\pi}{4}$$

a. $z = \sqrt{2}\left(\cos \frac{\pi}{4} + i \sin \frac{\pi}{4}\right)$ Polar form of $z = 1 + i$

$z^{16} = \left[\sqrt{2}\left(\cos \frac{\pi}{4} + i \sin \frac{\pi}{4}\right)\right]^{16}$ Raise both sides to the 16th power.

$z^{16} = (\sqrt{2})^{16}\left[\cos\left(16 \cdot \frac{\pi}{4}\right) + i \sin\left(16 \cdot \frac{\pi}{4}\right)\right]$ DeMoivre's Theorem

$\quad\quad = 2^8[\cos(4\pi) + i \sin(4\pi)]$ $(\sqrt{2})^{16} = (2^{\frac{1}{2}})^{16} = 2^{\frac{1}{2} \cdot 16} = 2^8$

$\quad\quad = 256(1 + i \cdot 0) = 256$ $\cos 4\pi = 1, \sin 4\pi = 0$

b. $z = \sqrt{2}\left(\cos \frac{\pi}{4} + i \sin \frac{\pi}{4}\right)$ Polar form of $z = 1 + i$

$z^{-10} = \left[\sqrt{2}\left(\cos \frac{\pi}{4} + i \sin \frac{\pi}{4}\right)\right]^{-10}$ Raise both sides to the -10th power.

$z^{-10} = (\sqrt{2})^{-10}\left[\cos\left(-10 \cdot \frac{\pi}{4}\right) + i \sin\left(-10 \cdot \frac{\pi}{4}\right)\right]$ DeMoivre's Theorem

$\quad\quad = \frac{1}{32}\left[\cos\left(-\frac{5\pi}{2}\right) + i \sin\left(-\frac{5\pi}{2}\right)\right]$ Simplify; $(\sqrt{2})^{-10} = (2^{1/2})^{-10} = 2^{-5} = \frac{1}{32}$

$\quad\quad = \frac{1}{32}[0 + i(-1)] = -\frac{1}{32}i$ $\cos\left(-\frac{5\pi}{2}\right) = 0, \sin\left(-\frac{5\pi}{2}\right) = -1$

■ ■ ■

Practice Problem 7 Let $z = -1 + i$. Use DeMoivre's Theorem to find each power of z. Write the answers in rectangular form.

a. z^8 **b.** z^{-12} ■

6 Use DeMoivre's Theorem to find the nth roots of a complex number.

Roots of Complex Numbers

Let z and w be two complex numbers and let n be a positive integer. The complex number z is called an **nth root of w** if

$$z^n = w.$$

We can use DeMoivre's Theorem to find the nth roots of a complex number.

DEMOIVRE'S nth ROOTS THEOREM

The nth roots of a complex number $w = r(\cos \theta + i \sin \theta)$, where $r > 0$ and θ is in degrees, are given by

$$z_k = r^{1/n}\left[\cos\left(\frac{\theta + 360°k}{n}\right) + i \sin\left(\frac{\theta + 360°k}{n}\right)\right], \text{ for } k = 0, 1, 2, \ldots, n - 1.$$

If θ is in radians, replace $360°$ with 2π in z_k.

The result says that there are exactly n nth roots of a nonzero complex number.

EXAMPLE 8   **Finding the Roots of a Complex Number**

Find the three cube roots of $1 + i$ in polar form, with the argument in degrees.

SOLUTION
In Example 7, we showed that

$$1 + i = \sqrt{2}\left(\cos\frac{\pi}{4} + i\sin\frac{\pi}{4}\right) \qquad \text{Polar form of } 1 + i$$
$$1 + i = \sqrt{2}(\cos 45° + i\sin 45°). \qquad \text{Replace } \frac{\pi}{4} \text{ with } 45°.$$

RECALL

$\sqrt{2} = 2^{1/2}$, so
$(\sqrt{2})^{1/3} = (2^{1/2})^{1/3} = 2^{1/6}$.

From DeMoivre's Theorem for finding complex roots, with $n = 3$, we have

$$z_k = (\sqrt{2})^{1/3}\left[\cos\left(\frac{45° + 360°k}{3}\right) + i\sin\left(\frac{45° + 360°k}{3}\right)\right], \quad k = 0, 1, 2.$$

By substituting $k = 0, 1$, and 2 in the expression for z_k and simplifying, we find the three cube roots.

$$z_0 = 2^{1/6}\left[\cos\left(\frac{45° + 360°\cdot 0}{3}\right) + i\sin\left(\frac{45° + 360°\cdot 0}{3}\right)\right] \qquad k = 0$$
$$= 2^{1/6}(\cos 15° + i\sin 15°) \qquad\qquad\qquad \text{Simplify.}$$
$$z_1 = 2^{1/6}\left[\cos\left(\frac{45° + 360°\cdot 1}{3}\right) + i\sin\left(\frac{45° + 360°\cdot 1}{3}\right)\right] \qquad k = 1$$
$$= 2^{1/6}(\cos 135° + i\sin 135°) \qquad\qquad\qquad \text{Simplify.}$$
$$z_2 = 2^{1/6}\left[\cos\left(\frac{45° + 360°\cdot 2}{3}\right) + i\sin\left(\frac{45° + 360°\cdot 2}{3}\right)\right] \qquad k = 2$$
$$= 2^{1/6}(\cos 255° + i\sin 255°) \qquad\qquad\qquad \text{Simplify.}$$

TECHNOLOGY CONNECTION

A graphing calculator can be used to check complex roots. The calculator must be set in $a + bi$ mode. The screen shown here is from a calculator also set in Degree mode.

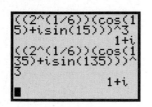

The complex numbers z_0, z_1, and z_2 are the three cube roots of the complex number $1 + i$.   ■ ■ ■

Practice Problem 8   Find the three cube roots of $-1 + i$ in polar form, with the argument in degrees.   ■

The solutions of the so-called *cyclotomic* equation $z^n = 1$ are called the **nth roots of unity**. The name refers to the close association of this equation with the regular n-gons inscribed in a circle.

EXAMPLE 9   **Finding nth Roots of Unity**

Find the complex sixth roots of unity. Write the roots in polar form and represent them geometrically.

SOLUTION
The polar form for $1 = 1 + 0i$ is

$$1 = 1(\cos 0° + i\sin 0°).$$

From DeMoivre's Theorem for finding complex roots with $n = 6$, we have

$$z_k = (1)^{1/6}\left[\cos\left(\frac{0° + 360°\cdot k}{6}\right) + i\sin\left(\frac{0° + 360°\cdot k}{6}\right)\right] \qquad k = 0, 1, 2, 3, 4, 5$$

$z_0 = 1(\cos 0° + i\sin 0°)$ Let $k = 0$ in z_k and simplify.
$z_1 = 1(\cos 60° + i\sin 60°)$ Let $k = 1$ in z_k and simplify.
$z_2 = 1(\cos 120° + i\sin 120°)$ Let $k = 2$ in z_k and simplify.

continued on the next page

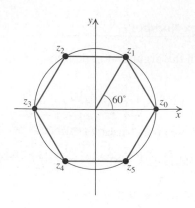

FIGURE 6.72 Sixth roots of unity

$$z_3 = 1(\cos 180° + i \sin 180°) \quad \text{Let } k = 3 \text{ in } z_k \text{ and simplify.}$$
$$z_4 = 1(\cos 240° + i \sin 240°) \quad \text{Let } k = 4 \text{ in } z_k \text{ and simplify.}$$
$$z_5 = 1(\cos 300° + i \sin 300°) \quad \text{Let } k = 5 \text{ in } z_k \text{ and simplify.}$$

Geometrically, the sixth roots of unity form a regular hexagon and are equally spaced at 60° intervals on a unit circle. See Figure 6.72. ▪ ▪ ▪

Practice Problem 9 Find the complex fourth roots of unity. Write the roots in polar form. ▪

SECTION 6.7 ▪ Exercises

A EXERCISES Basic Skills and Concepts

1. If $z = a + bi$, then $|z| = \underline{\sqrt{a^2 + b^2}}$ and arg $z = \theta = \underline{\qquad}$. $\tan^{-1}\left(\dfrac{b}{a}\right)$

2. If $|z| = r$ and arg $z = \theta$, then the polar form of z is $z = \underline{r(\cos\theta + i\sin\theta)}$.

3. To multiply two complex numbers in polar form, multiply their $\underline{\text{moduli}}$ and $\underline{\text{add}}$ their arguments.

4. DeMoivre's Theorem states that $[r(\cos\theta + i\sin\theta)]^n = \underline{r^n(\cos n\theta + i\sin n\theta)}$.

5. *True or False* If $z = r(\cos\theta + i\sin\theta)$, then $\dfrac{1}{z} = z^{-1} = \dfrac{1}{r}(\cos\theta - i\sin\theta)$. True

6. *True or False* The n complex nth roots of 1 are equally spaced points on the unit circle. True

In Exercises 7–14, plot each complex number and find its absolute value.

7. $z = 2$ 2 †
8. $z = -3$ 3 †
9. $z = -4i$ 4 †
10. $z = 2i$ 2 †
11. $z = 3 + 4i$ 5 †
12. $z = -1 + 2i$ $\sqrt{5}$ †
13. $z = -2 - 3i$ $\sqrt{13}$ †
14. $z = 2 - 5i$ $\sqrt{29}$ †

In Exercises 15–26, write each complex number in polar form. Express the argument θ in degrees, with $0 \le \theta < 360°$.

15. $1 + \sqrt{3}i$
$2(\cos 60° + i \sin 60°)$
16. $-1 + \sqrt{2}i$
$\sqrt{3}(\cos 125.3° + i \sin 125.3°)$
17. $-1 + i$
$\sqrt{2}(\cos 135° + i \sin 135°)$
18. $1 - i$
$\sqrt{2}(\cos 315° + i \sin 315°)$
19. i $\cos 90° + i \sin 90°$
20. $-i$ $\cos 270° + i \sin 270°$
21. 1 $\cos 0° + i \sin 0°$
22. -1 $\cos 180° + i \sin 180°$
23. $3 - 3i$ †
24. $4\sqrt{3} + 4i$
$8(\cos 30° + i \sin 30°)$
25. $2 - 2\sqrt{3}i$
$4(\cos 300° + i \sin 300°)$
26. $2 + 3i$
$\sqrt{13}(\cos 56.3° + i \sin 56.3°)$

In Exercises 27–38, write each complex number in rectangular form.

27. $2(\cos 60° + i \sin 60°)$ $1 + \sqrt{3}i$
28. $4(\cos 120° + i \sin 120°)$ $-2 + 2\sqrt{3}i$
29. $3(\cos\pi + i \sin\pi)$ -3
30. $5\left(\cos\dfrac{\pi}{2} + i \sin\dfrac{\pi}{2}\right)$ $5i$
31. $5(\cos 240° + i \sin 240°)$ $-\dfrac{5}{2} - \dfrac{5\sqrt{3}}{2}i$
32. $2(\cos 300° + i \sin 300°)$ $1 - \sqrt{3}i$
33. $8(\cos 0° + i \sin 0°)$ 8
34. $2(\cos(-90°)° + i \sin(-90°))$ $-2i$
35. $6\left(\cos\dfrac{5\pi}{6} + i \sin\dfrac{5\pi}{6}\right)$ $-3\sqrt{3} + 3i$
36. $4\left(\cos\dfrac{3\pi}{4} + i \sin\dfrac{3\pi}{4}\right)$ $-2\sqrt{2} + 2\sqrt{2}i$
37. $3\left(\cos\left(-\dfrac{\pi}{3}\right) + i \sin\left(-\dfrac{\pi}{3}\right)\right)$ $\dfrac{3}{2} - \dfrac{3\sqrt{3}}{2}i$
38. $5\left(\cos\left(-\dfrac{7\pi}{6}\right) + i \sin\left(-\dfrac{7\pi}{6}\right)\right)$ $-\dfrac{5\sqrt{3}}{2} + \dfrac{5}{2}i$

In Exercises 39–50, find $z_1 z_2$ and $\dfrac{z_1}{z_2}$. Write each answer in polar form.

39. $z_1 = 4(\cos 75° + i \sin 75°)$, $z_2 = 2(\cos 15° + i \sin 15°)$ †
40. $z_1 = 6(\cos 90° + i \sin 90°)$, $z_2 = 2(\cos 45° + i \sin 45°)$ †
41. $z_1 = 5(\cos 240° + i \sin 240°)$, $z_2 = 2(\cos 60° + i \sin 60°)$
42. $z_1 = 10(\cos 135° + i \sin 135°)$,
$z_2 = 4(\cos 225° + i \sin 225°)$ †
43. $z_1 = 3(\cos 40° + i \sin 40°)$, $z_2 = 5(\cos 20° + i \sin 20°)$ †
44. $z_1 = 5(\cos 65° + i \sin 65°)$, $z_2 = 2(\cos 25° + i \sin 25°)$ †
45. $z_1 = 1 + i$, $z_2 = 1 - i$ †
46. $z_1 = 1 + \sqrt{3}i$, $z_2 = 1 + i$ †
47. $z_1 = -\sqrt{3} + i$, $z_2 = 2 + 2i$ †
48. $z_1 = 3 + 3i$, $z_2 = 2 - 2\sqrt{3}i$ †
49. $z_1 = 2\sqrt{2} + 2i$, $z_2 = \sqrt{3} - i$ †
50. $z_1 = 4\sqrt{3} + 4i$, $z_2 = 3 - 3i$ †

In Exercises 51–64, use DeMoivre's Theorem to find the indicated power. Write the answers in rectangular form.

51. $\left[2\left(\cos\dfrac{\pi}{3} + i \sin\dfrac{\pi}{3}\right)\right]^{12}$ 2^{12}
52. $\left[\sqrt{2}\left(\cos\dfrac{\pi}{4} + i \sin\dfrac{\pi}{4}\right)\right]^{20}$ -2^{10}

†Due to space constrictions, answers to these exercises may be found in the Answers beginning on page A–1 in the back of the book.

53. $\left[2\left(\cos\left(-\dfrac{3\pi}{4}\right)+i\sin\left(-\dfrac{3\pi}{4}\right)\right)\right]^6$ $-64i$

54. $\left(\cos\dfrac{\pi}{27}+i\sin\dfrac{\pi}{27}\right)^{-9}$ $\dfrac{1}{2}-\dfrac{\sqrt{3}}{2}i$

55. $\left[2\left(\cos\left(\dfrac{3\pi}{4}\right)+i\sin\left(\dfrac{3\pi}{4}\right)\right)\right]^{-6}$ $-\dfrac{1}{64}i$

56. $\left[3\left(\cos\left(-\dfrac{5\pi}{6}\right)+i\sin\left(-\dfrac{5\pi}{6}\right)\right)\right]^{16}$ $3^{16}\left(-\dfrac{1}{2}+\dfrac{\sqrt{3}}{2}i\right)$

57. $\left[2\left(\cos\left(-\dfrac{\pi}{4}\right)+i\sin\left(-\dfrac{\pi}{4}\right)\right)\right]^{-10}$ $\dfrac{1}{2^{10}}i$

58. $\left[\dfrac{1}{2}\left(\cos\left(-\dfrac{3\pi}{4}\right)+i\sin\left(-\dfrac{3\pi}{4}\right)\right)\right]^{-6}$ $64i$

59. $(1-i)^{12}$ -64 **60.** $(1-\sqrt{3}i)^{10}$ $-512+512\sqrt{3}i$

61. i^{25} i **62.** $(3+3i)^8$ $104{,}976$

63. $\left(\dfrac{1}{2}-\dfrac{\sqrt{3}}{2}i\right)^{-8}$ $-\dfrac{1}{2}+\dfrac{\sqrt{3}}{2}i$ **64.** $\left(\dfrac{\sqrt{3}}{2}+\dfrac{1}{2}i\right)^{10}$ $\dfrac{1}{2}-\dfrac{\sqrt{3}}{2}i$

In Exercises 65–72, find all complex roots. Write your answers in polar form, with the argument θ in degrees with $0 \le \theta < 360°$.

65. Cube roots of 64 †   **66.** Cube roots of -64 †

67. Square roots of i †   **68.** Fourth roots of $-i$ †

69. Sixth roots of -1 †   **70.** Eighth roots of unity †

71. Square roots of $1-\sqrt{3}i$ **72.** Cube roots of $4+4\sqrt{3}i$
        †   †

B EXERCISES Applying the Concepts

73. Trigonometry. Use DeMoivre's Theorem to verify each identity.
  a. $\sin 3\theta = 3\sin\theta - 4\sin^3\theta$
  b. $\cos 3\theta = 4\cos^3\theta - 3\cos\theta$

74. Geometry. Show that the area K of a triangle with vertices at the origin O, z_1, and z_2 is given by
$$K = \dfrac{1}{2}|z_1||z_2|\sin\left(\left|\arg\dfrac{z_1}{z_2}\right|\right).$$

75. AC circuits. Use Ohm's Law, $I = \dfrac{V}{Z}$, to find the current I, given voltage $V = 120(\cos 60° + i\sin 60°)$ and impedance $Z = 8\cos(30° + i\sin 30°)$. $15(\cos 30° + i\sin 30°)$

76. AC circuits. Use Ohm's Law from Exercise 75 to find the voltage V, given $I = 6(\cos 40° + i\sin 40°)$ and $Z = 16\cos(110° + i\sin 110°)$. $96(\cos 150° + i\sin 150°)$

In Exercises 77 and 78, use the formula $Z_1 = \dfrac{Z_1 Z_2}{Z_1 + Z_2}$ to find the total impedance in a parallel circuit for the given values of Z_1 and Z_2.

77. $Z_1 = 16(\cos 180° + i\sin 180°)$ and
$Z_2 = 2(\cos 150° + i\sin 150°)$ $1.8(\cos 153.2° + i\sin 153.2°)$

78. $Z_1 = 12(\cos 270° + i\sin 270°)$ and
$Z_2 = 3(\cos 60° + i\sin 60°)$ $3.8(\cos 50.9° + i\sin 50.9°)$

102. $\cos(72k)° + i\sin(72k)°, k = 1, 2, 3,$ and 4

C EXERCISES Beyond the Basics

In Exercises 79–84, use DeMoivre's Theorem to compute each power in polar form, with θ in degrees, with $0 \le \theta < 360°$.

79. $(1-\sqrt{3}i)^{10}(-2+2i)^{-6}$ $2(\cos 30° + i\sin 30°)$

80. $\left(\dfrac{1}{2}+\dfrac{\sqrt{3}}{2}i\right)^8 (2-2i)^{-6}$ $\dfrac{1}{512}(\cos 30° + i\sin 30°)$

81. $\left(\sin\dfrac{\pi}{6}+i\cos\dfrac{\pi}{6}\right)^{10}$ $\cos 240° + i\sin 240°$

82. $\left(\sin\dfrac{2\pi}{3}+i\cos\dfrac{2\pi}{3}\right)^6$ $\cos 180° + i\sin 180°$

83. $\left(\sin\dfrac{5\pi}{3}+i\cos\dfrac{5\pi}{3}\right)^{-8}$ $\cos 240° + i\sin 240°$

84. $\left(\sin\dfrac{7\pi}{6}+i\cos\dfrac{7\pi}{6}\right)^{-10}$ $\cos 120° + i\sin 120°$

In Exercises 85–88, find all complex solutions of each equation.

85. $z^4 = 1$ †   **86.** $z^8 = -1$ †

87. $z^3 = 1+i$ †   **88.** $z^7 = (1-i)^2$ †

89. Let $z = \cos\theta + i\sin\theta, z \ne 0$. Prove that
$$\dfrac{1}{z} = \cos\theta - i\sin\theta.$$

In Exercises 90–93, let $z = \cos\theta + i\sin\theta, z \ne 0$. Use Exercise 89 and DeMoivre's Theorem to prove each identity.

90. $z + \dfrac{1}{z} = 2\cos\theta$ **91.** $z - \dfrac{1}{z} = 2i\sin\theta$

92. $z^n + \dfrac{1}{z^n} = 2\cos n\theta$ **93.** $z^n - \dfrac{1}{z^n} = 2i\sin n\theta$

94. Prove that
$$(1 + \cos\theta + i\sin\theta)^n = \left(2\cos\dfrac{\theta}{2}\right)^n\left(\cos\dfrac{n\theta}{2} + i\sin\dfrac{n\theta}{2}\right).$$

95. Prove the product rule for two complex numbers in polar form.

96. Prove the quotient rule for two complex numbers in polar form.

97. Find the product $(3 + 2i)(5 + i)$ using FOIL. Then deduce the identity: $\tan^{-1}\left(\dfrac{2}{3}\right) + \tan^{-1}\left(\dfrac{1}{5}\right) = \dfrac{\pi}{4}$.
[*Hint:* $\arg(z_1 z_2) = \arg z_1 + \arg z_2$.]

98. Find the product $(p + q + i)(p^2 + pq + 1 + iq)$, then deduce the identity: $\tan^{-1}\left(\dfrac{1}{p+q}\right) + \tan^{-1}\left(\dfrac{q}{p^2 + pq + 1}\right) = \tan^{-1}\left(\dfrac{1}{p}\right)$.

99. Expand $(2 + 3i)^4$. Then deduce the identity:
$$4\tan^{-1}\left(\dfrac{3}{2}\right) - \tan^{-1}\left(\dfrac{120}{119}\right) = \pi.$$

100. Find the product $(5 + i)^4(-239 + i)$. Then deduce the identity: $4\tan^{-1}\left(\dfrac{1}{5}\right) - \tan^{-1}\left(\dfrac{1}{239}\right) = \dfrac{\pi}{4}$.

101. Find the three solutions of the equation $x^3 + x^2 + x + 1 = 0$. [*Hint:* Multiply both sides by $(x - 1)$, solve the new equation, and reject the root $x = 1$.] $\{1, i, -i\}$

102. Find the four solutions of the equation $x^4 + x^3 + x^2 + x + 1 = 0$.

Critical Thinking

103. Let n be a positive integer. State whether each of the following is true or false.

a. $\left(\cos\dfrac{\pi}{6} + i\sin\dfrac{\pi}{6}\right)^n = \cos\dfrac{n\pi}{6} + i\sin\dfrac{n\pi}{6}$ True

b. $\left(\sin\dfrac{\pi}{3} + i\cos\dfrac{\pi}{3}\right)^n = \sin\dfrac{n\pi}{3} + i\cos\dfrac{n\pi}{3}$ False

c. $\left[\cos\left(-\dfrac{\pi}{3}\right) + i\sin\left(-\dfrac{\pi}{3}\right)\right]^n = \cos\dfrac{n\pi}{3} - i\sin\dfrac{n\pi}{3}$ True

d. $\left(\cos\dfrac{\pi}{6} + i\sin\dfrac{\pi}{3}\right)^n = \cos\dfrac{n\pi}{6} + i\sin\dfrac{n\pi}{3}$ False

e. $(\cos\theta + i\sin\theta)^n = \cos n\theta + i\sin n\theta$ True

f. $(\cos\theta - i\sin\theta)^n = \cos n\theta - i\sin n\theta$ True

g. $(\cos\theta + i\sin\theta)^{-n} = \cos n\theta - i\sin n\theta$ True

h. $(\cos\theta - i\sin\theta)^{-n} = \cos n\theta + i\sin n\theta$ True

i. $(\sin\theta + i\cos\theta)^n = \sin n\theta + i\cos n\theta$ False

j. $(\sin\theta + i\cos\theta)^n = \cos n\left(\dfrac{\pi}{2} - \theta\right) + i\sin n\left(\dfrac{\pi}{2} - \theta\right)$ True

---

SUMMARY ■ Definitions, Concepts, and Formulas

6.1 Right-Triangle Trigonometry

i. The right-triangle definitions for trigonometric functions of an acute angle θ are given by

$$\sin\theta = \frac{\text{opposite side}}{\text{hypotenuse}} = \frac{a}{c} \qquad \csc\theta = \frac{\text{hypotenuse}}{\text{opposite side}} = \frac{c}{a}$$

$$\cos\theta = \frac{\text{adjacent side}}{\text{hypotenuse}} = \frac{b}{c} \qquad \sec\theta = \frac{\text{hypotenuse}}{\text{adjacent side}} = \frac{c}{b}$$

$$\tan\theta = \frac{\text{opposite side}}{\text{adjacent side}} = \frac{a}{b} \qquad \cot\theta = \frac{\text{adjacent side}}{\text{opposite side}} = \frac{b}{a}$$

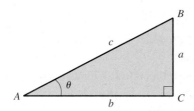

ii. Two acute angles are complements if their sum is 90° or $\dfrac{\pi}{2}$. The value of any trigonometric function of an acute angle θ is equal to the cofunction of the complement of θ.

iii. To solve a right triangle means to find the measurements of missing angles and the sides.

6.2 The Law of Sines

i. In a triangle ABC with sides of length $a, b,$ and c,
$$\frac{\sin A}{a} = \frac{\sin B}{b} = \frac{\sin C}{c}, \text{ or equivalently, } \frac{a}{\sin A} = \frac{b}{\sin B} = \frac{c}{\sin C}.$$

ii. The Law of Sines is used to solve AAS, ASA, and SSA triangles. The SSA case is called the *ambiguous case*

because the information provided may lead to two triangles, one triangle, or no triangle.

6.3 The Law of Cosines

i. In a triangle ABC with sides of lengths $a, b,$ and c,

$$a^2 = b^2 + c^2 - 2bc\cos A;$$
$$b^2 = c^2 + a^2 - 2ca\cos B;$$
$$c^2 = a^2 + b^2 - 2ab\cos C.$$

ii. The Law of Cosines is used to solve SAS and SSS triangles. This law is also used to prove Heron's formula.

iii. Heron's formula. The area K of a triangle with sides of lengths $a, b,$ and c is given by

$$K = \sqrt{s(s - a)(s - b)(s - c)},$$

where $s = \dfrac{1}{2}(a + b + c)$ is the *semiperimeter*.

6.4 Vectors

i. A vector **v** is a quantity that has both magnitude and direction. A vector is geometrically represented by an arrow.

ii. The magnitude of the scalar product $c\mathbf{v}$ is $|c|$ times the magnitude of **v**. If $c > 0$, $c\mathbf{v}$ has the direction of **v**. If $c < 0$, $c\mathbf{v}$ has the opposite direction of **v**. The zero vector **0** has magnitude 0 but arbitrary direction.

iii. To add the vectors **v** and **w** geometrically, place **w** so that its initial point coincides with the terminal point of **v** and draw a vector from the initial point of **v** to the terminal point of **w**.

iv. Algebraic vectors. A vector $\mathbf{v}$ with initial point at the origin and terminal point (v_1, v_2) is written as $\mathbf{v} = \langle v_1, v_2 \rangle$; v_1 and v_2 are the horizontal and vertical components of $\mathbf{v}$, respectively.

- The magnitude of $\mathbf{v} = \|\mathbf{v}\| = \sqrt{v_1^2 + v_2^2}$. The direction angle of $\mathbf{v}$ is the angle θ that $\mathbf{v}$ makes with the positive x-axis. We have $\cos\theta = \dfrac{v_1}{\|\mathbf{v}\|}$ and $\sin\theta = \dfrac{v_2}{\|\mathbf{v}\|}$, $0° \le \theta < 360°$.

- The horizontal and vertical components of vector $\mathbf{v}$ with magnitude r and direction angle θ are given by

$$v_1 = r\cos\theta, \ \ v_2 = r\sin\theta.$$

v. Vector operations. For vectors $\mathbf{v} = \langle v_1, v_2 \rangle = v_1\mathbf{i} + v_2\mathbf{j}$ and $\mathbf{w} = \langle w_1, w_2 \rangle = w_1\mathbf{i} + w_2\mathbf{j}$, the following operations apply:

- **Vector addition:** $\mathbf{v} + \mathbf{w} = \langle v_1 + w_1, v_2 + w_2 \rangle$
- **Scalar multiplication:** $c\mathbf{v} = \langle cv_1, cv_2 \rangle$

6.5 The Dot Product

i. For two vectors $\mathbf{v} = \langle v_1 v_2 \rangle$ and $\mathbf{w} = \langle w_1 w_2 \rangle$, the dot product

$$\mathbf{v} \cdot \mathbf{w} = v_1 w_1 + v_2 w_2 = \|\mathbf{v}\|\|\mathbf{w}\|\cos\theta,$$

where θ is the angle between $\mathbf{v}$ and $\mathbf{w}$.

ii. Vectors $\mathbf{v}$ and $\mathbf{w}$ are

1. orthogonal (perpendicular) if $\mathbf{v} \cdot \mathbf{w} = 0$.
2. parallel if $\mathbf{v} \cdot \mathbf{w} = \pm\|\mathbf{v}\|\|\mathbf{w}\|$.

iii. Vector projection of $\mathbf{v}$ onto $\mathbf{w}$ is given by

$$\text{proj}_\mathbf{w}\mathbf{v} = \frac{\mathbf{v} \cdot \mathbf{w}}{\|\mathbf{w}\|^2}\mathbf{w}.$$

iv. A nonzero vector $\mathbf{v}$ can be written as $\mathbf{v} = \mathbf{v}_1 + \mathbf{v}_2$, where $\mathbf{v}_1 = \text{proj}_\mathbf{w}\mathbf{v}$ is parallel to $\mathbf{w}$ and $\mathbf{v}_2 = \mathbf{v} - \mathbf{v}_1$ is orthogonal to $\mathbf{w}$. The scalar projection of $\mathbf{v}$ onto $\mathbf{w}$ is

$$\frac{\mathbf{v} \cdot \mathbf{w}}{\|\mathbf{w}\|}.$$

v. The **work** W done by a constant force $\mathbf{F}$ in moving an object from a point P to a point Q is $W = \mathbf{F} \cdot \overline{PQ}$.

6.6 Polar Coordinates

i. The ordered pair (r, θ) represents a point P in the plane that is a directed distance of r units from the origin (pole) O, where θ is the directed angle between the polar axis (positive x-axis) and the line segment OP. The ordered pair (r, θ) gives the polar coordinates of P.

ii. The polar coordinates of a point $P(r, \theta)$ are not unique. In fact, if n is any integer, then

$$(r, \theta), (r, \theta + 2n\pi), \text{ and } (-r, \theta + \pi + 2n\pi)$$

are all polar coordinates of the same point.

iii. To convert a point from polar coordinates (r, θ) to rectangular coordinates, use the equations $x = r\cos\theta, \ y = \sin\theta$.

iv. To convert a point from rectangular coordinates (x, y) to polar coordinates (r, θ), use the procedure on page 452.

v. A polar equation is an equation whose variables are r and θ.

6.7 Polar Form of Complex Numbers; DeMoivre's Theorem

i. Geometric representation. A complex number $z = a + bi$ can be represented as an ordered pair (a, b) in the complex plane.

ii. Modulus. $|z| = |a + bi| = \sqrt{a^2 + b^2}$, $|z|$ is called the **modulus** of z.

iii. Polar form. The polar form of $z = a + bi$ is $z = r(\cos\theta + i\sin\theta)$, where $r = \sqrt{a^2 + b^2}$, $a = r\cos\theta$, $b = r\sin\theta$, and $\tan\theta = \dfrac{b}{a}$. The number r is the modulus of z, and θ is called the argument of z (written arg z).

iv. Multiplication and division. Let $z_1 = r_1(\cos\theta_1 + i\sin\theta_1)$ and $z_2 = r_2(\cos\theta_2 + i\sin\theta_2)$. Then

$$z_1 z_2 = r_1 r_2 [\cos(\theta_1 + \theta_2) + i\sin(\theta_1 + \theta_2)]$$

and

$$\frac{z_1}{z_2} = \frac{r_1}{r_2}[\cos(\theta_1 - \theta_2) + i\sin(\theta_1 - \theta_2)], \ \ z_2 \ne 0.$$

v. DeMoivre's Theorem. Let $z = r(\cos\theta + i\sin\theta)$ be a complex number in polar form. Then for any integer n,

$$[r(\cos\theta + i\sin\theta)]^n = r^n(\cos n\theta + i\sin n\theta).$$

vi. Roots of complex numbers. The n nth roots of a complex number $r(\cos\theta + i\sin\theta)$ are given by

$$z_k = \sqrt[n]{r}\left[\cos\left(\frac{\theta + 2\pi k}{n}\right) + i\sin\left(\frac{\theta + 2\pi k}{n}\right)\right] \text{ or }$$

$$z_k = \sqrt[n]{r}\left[\cos\left(\frac{\theta + 360°k}{n}\right) + i\sin\left(\frac{\theta + 360°k}{n}\right)\right],$$

where $k = 0, 1, 2, \ldots, n - 1$.

REVIEW EXERCISES

Basic Skills and Concepts

In Exercises 1–10, solve right triangle ABC with right angle at C. Round each answer to the nearest tenth.

1. $A = 30°$, $a = 6$
   $B = 60°, b ≈ 10.4, c = 12$
2. $B = 35°$, $b = 5$
   $A = 55°, a ≈ 7.1, c ≈ 8.7$
3. $A = 37°$, $b = 4$
   $B = 53°, a ≈ 3.0, c ≈ 5.0$
4. $B = 43°$, $a = 10$
   $A = 47°, b ≈ 9.3, c ≈ 13.7$
5. $A = 40°$, $c = 12$
   $B = 50°, a ≈ 7.7, b ≈ 9.2$
6. $B = 50°$, $c = 8$
   $A = 40°, a ≈ 5.1, b ≈ 6.1$
7. $a = 3$, $b = 5$
   $A ≈ 31.0°, B ≈ 59.0°, c ≈ 5.8$
8. $a = 5$, $b = 10$
   $A ≈ 26.6°, B ≈ 63.4°, c ≈ 11.2$
9. $a = 4$, $c = 6$
   $A ≈ 41.8°, B ≈ 48.2°, b ≈ 4.5$
10. $b = 3$, $c = 7$
    $A ≈ 64.6°, B ≈ 25.4°, a ≈ 6.3$

In Exercises 11–18, use the Law of Sines to solve each triangle ABC. Round each answer to the nearest tenth.

11. $A = 40°$, $B = 35°$, $c = 100$ $C = 105°, a ≈ 66.5, c ≈ 59.4$

12. $B = 30°$, $C = 80°$, $a = 100$ $A = 70°, b ≈ 53.2, c ≈ 104.8$

13. $A = 45°$, $a = 25$, $b = 75$ No triangle exists.

14. $B = 36°$, $a = 12.5$, $b = 8.7$ $A ≈ 57.6°, C ≈ 86.4°, c ≈ 14.8;$ $A ≈ 122.4°, C ≈ 21.6°, c ≈ 5.4$

15. $A = 48.5°$, $C = 57.3°$, $b = 47.3$ $B = 74.2°, a ≈ 36.8, c ≈ 41.4$

16. $A = 67°$, $a = 100$, $c = 125$ No triangle exists.

17. $A = 65.2°$, $a = 21.3$, $b = 19$ $B ≈ 54.1°, C ≈ 60.7°, c ≈ 20.5$

18. $C = 53°$, $a = 140$, $c = 115$ $A ≈ 76.5°, B ≈ 50.5°, b ≈ 111.2;$ $A ≈ 103.5°, B ≈ 23.5°, b ≈ 57.4$

19. In triangle ABC, let $c = 20$ and $B = 60°$. Find a value of b such that C has **(i)** two possible values, **(ii)** exactly one value, **(iii)** no value. †

20. Repeat Exercise 19 for $c = 20$ and $B = 150°$. †

In Exercises 21–28, use the Law of Cosines to solve each triangle ABC. Round each answer to the nearest tenth.

21. $a = 60$, $b = 90$, $c = 125$ $A ≈ 26.6°, B ≈ 42.1°, C ≈ 111.3°$

22. $a = 15$, $b = 9$, $C = 120°$ $A ≈ 38.2°, B ≈ 21.8°, c = 21$

23. $a = 40$, $c = 38$, $B = 80°$ $A ≈ 51.7°, C ≈ 48.3°, b ≈ 50.2$

24. $a = 10$, $b = 20$, $c = 22$ $A ≈ 27.0°, B ≈ 65.3°, C ≈ 87.7°$

25. $a = 2.6$, $b = 3.7$, $c = 4.8$ $A ≈ 32.5°, B ≈ 49.8°, C ≈ 97.7°$

26. $a = 15$, $c = 26$, $B = 115°$ $A ≈ 22.8°, C ≈ 42.2°, b ≈ 35.1$

27. $a = 12$, $b = 7$, $C = 130°$ $A ≈ 32.0°, B ≈ 18.0°, c ≈ 17.3$

28. $b = 75$, $c = 100$, $A = 80°$ $B ≈ 40.3°, C ≈ 59.7°, a ≈ 114.1$

In Exercises 29–32, find the area of each triangle ABC with the given information. Round each answer to the nearest square unit.

29. $a = 5$ meters, $b = 7$ meters, $c = 10$ meters 16 square meters

30. $a = 2.4$ meters, $b = 3.4$ meters, $c = 4.4$ meters 4 square meters

31. $A = 65°$, $b = 6$ feet, $c = 4$ feet 11 square feet

32. $A = 115°$, $b = 20$ inches, $a = 30$ inches 140 square inches

In Exercises 33 and 34, let v be the vector with initial point P and terminal point Q. Write v as a position vector in terms of i and j.

33. $P(3, 5), Q(2, 7)$ $-\mathbf{i} + 2\mathbf{j}$
34. $P(-1, 3), Q(5, -4)$ $6\mathbf{i} - 7\mathbf{j}$

In Exercises 35–38, let $\mathbf{v} = \langle -2, 3 \rangle$ and $\mathbf{w} = \langle 5, -6 \rangle$. Find each expression.

35. $5\mathbf{v}$ $\langle -10, 15 \rangle$
36. $2\mathbf{v} + \mathbf{w}$ $\langle 1, 0 \rangle$
37. $3\mathbf{v} - 2\mathbf{w}$ $\langle -16, 21 \rangle$
38. $\|\mathbf{v} + \mathbf{w}\|$ $3\sqrt{2}$

In Exercises 39–42, find the unit vector u in the direction of the vector v.

39. $\mathbf{v} = \mathbf{i} + \mathbf{j}$ $\frac{\sqrt{2}}{2}\mathbf{i} + \frac{\sqrt{2}}{2}\mathbf{j}$
40. $\mathbf{v} = 2\mathbf{i} - 7\mathbf{j}$ †
41. $\mathbf{v} = \langle 3, -5 \rangle$ †
42. $\mathbf{v} = \langle -5, -2 \rangle$ †

In Exercises 43–46, write v in terms of i and j for the given magnitude $\|\mathbf{v}\|$ and direction angle θ.

43. $\|\mathbf{v}\| = 6$, $\theta = 30°$ $3\sqrt{3}\mathbf{i} + 3\mathbf{j}$
44. $\|\mathbf{v}\| = 20$, $\theta = 120°$ $-10\mathbf{i} + 10\sqrt{3}\mathbf{j}$
45. $\|\mathbf{v}\| = 12$, $\theta = 225°$ $-6\sqrt{2}\mathbf{i} - 6\sqrt{2}\mathbf{j}$
46. $\|\mathbf{v}\| = 10$, $\theta = -30°$ $5\sqrt{3}\mathbf{i} - 5\mathbf{j}$

In Exercises 47–50, find the dot product $\mathbf{v} \cdot \mathbf{w}$ and $\text{proj}_\mathbf{w} \mathbf{v}$.

47. $\mathbf{v} = \langle 2, -3 \rangle$, $\mathbf{w} = \langle 3, 4 \rangle$ $-6; \text{proj}_\mathbf{w} \mathbf{v} = \left\langle -\frac{18}{25}, -\frac{24}{25} \right\rangle$

48. $\mathbf{v} = \langle -1, -2 \rangle$, $\mathbf{w} = \langle 4, -1 \rangle$ $-2; \text{proj}_\mathbf{w} \mathbf{v} = \left\langle -\frac{8}{17}, \frac{2}{17} \right\rangle$

49. $\mathbf{v} = 2\mathbf{i} - 5\mathbf{j}$, $\mathbf{w} = 5\mathbf{i} + 2\mathbf{j}$ $0; \text{proj}_\mathbf{w} \mathbf{v} = \langle 0, 0 \rangle$

50. $\mathbf{v} = 2\mathbf{i} + \mathbf{j}$, $\mathbf{w} = 2\mathbf{i} - \mathbf{j}$ $3; \text{proj}_\mathbf{w} \mathbf{v} = \left\langle \frac{6}{5}, -\frac{3}{5} \right\rangle$

In Exercises 51–54, use the dot product to find the angle θ $(0 \le \theta \le \pi)$ between the vectors v and w. Round your answers to the nearest tenth of a degree.

51. $\mathbf{v} = 2\mathbf{i} + 3\mathbf{j}$, $\mathbf{w} = -\mathbf{i} + 2\mathbf{j}$ $60.3°$

52. $\mathbf{v} = \mathbf{i} + 4\mathbf{j}$, $\mathbf{w} = -4\mathbf{i} - \mathbf{j}$ $118.1°$

53. $\mathbf{v} = \langle 1, 1 \rangle$, $\mathbf{w} = \langle -3, 2 \rangle$ $101.3°$

54. $\mathbf{v} = \langle 1, 5 \rangle$, $\mathbf{w} = \langle 3, -1 \rangle$ $97.1°$

In Exercises 55–58, plot each point in polar coordinates and find its rectangular coordinates.

55. $(24, 30°)$ †
56. $\left(12, -\frac{\pi}{3} \right)$ †
57. $\left(-2, -\frac{\pi}{4} \right)$ †
58. $\left(-3, -\frac{3\pi}{4} \right)$ †

In Exercises 59–62, convert the rectangular coordinates of each point to polar coordinates (r, θ), with $r > 0$ and $0 \le \theta < 2\pi$.

59. $(-2, 2)$ $\left(2\sqrt{2}, \frac{3\pi}{4} \right)$
60. $(\sqrt{3}, 1)$ $\left(2, \frac{\pi}{6} \right)$
61. $(2\sqrt{3}, -2)$ $\left(4, \frac{11\pi}{6} \right)$
62. $(-2, -2\sqrt{3})$ $\left(4, \frac{4\pi}{3} \right)$

In Exercises 63–66, convert each rectangular equation to a polar equation. $r = \frac{12}{3\cos\theta + 2\sin\theta}$

63. $3x + 2y = 12$
64. $x^2 + y^2 = 36$ $r = 6$
65. $x^2 + y^2 = 8x$ $r = 8\cos\theta$
66. $x^2 + y^2 = 6y$ $r = 6\sin\theta$

In Exercises 67–72, convert each polar equation to a rectangular equation.

67. $r = -3$ $x^2 + y^2 = 9$
68. $\theta = \frac{5\pi}{6}$ $y = -\frac{\sqrt{3}}{3}x$

†Due to space constrictions, answers to these exercises may be found in the Answers beginning on page A–1 in the back of the book.

69. $r = 3 \csc \theta$ $y = 3$
70. $r = 2 \sec \theta$ $x = 2$
71. $r = 1 - 2 \sin \theta$
$(x^2 + y^2 + 2y)^2 = x^2 + y^2$
72. $r = 3 \cos \theta$ $x^2 + y^2 = 3x$

In Exercises 73–76, write each complex number in polar form. Express the arguments in radians.

73. $-3i$ †
74. $-1 + i$ †
75. $5\sqrt{3} - 5i$ †
76. $-2 - 2\sqrt{3}i$ †

In Exercises 77–80, write each complex number in rectangular form.

77. $2(\cos 45° + i \sin 45°)$ †
78. $3(\cos 240° + i \sin 240°)$ †
79. $6\left(\cos \dfrac{3\pi}{4} + i \sin \dfrac{3\pi}{4}\right)$ †
80. $4\left(\cos \dfrac{7\pi}{6} + i \sin \dfrac{7\pi}{6}\right)$ †

In Exercises 81–84, find z_1z_2 and $\dfrac{z_1}{z_2}$. Leave your answers in polar form.

81. $z_1 = 3(\cos 25° + \sin 25°)$, $z_2 = 2(\cos 10° + i \sin 10°)$ †
82. $z_1 = 4(\cos 300° + i \sin 300°)$, $z_2 = 2(\cos 20° + i \sin 20°)$
83. $z_1 = 2\left(\cos \dfrac{5\pi}{6} + i \sin \dfrac{5\pi}{6}\right)$, $z_2 = 3\left(\cos \dfrac{\pi}{3} + i \sin \dfrac{\pi}{3}\right)$ †
84. $z_1 = 5\left(\cos \dfrac{4\pi}{3} + i \sin \dfrac{4\pi}{3}\right)$, $z_2 = 15\left(\cos \dfrac{\pi}{3} + i \sin \dfrac{\pi}{3}\right)$ †

In Exercises 85–88, use DeMoivre's Theorem to find the indicated power. Write your answers in polar form.

85. $[3(\cos 40° + i \sin 40°)]^3$ †
86. $\left[4\left(\cos \dfrac{\pi}{6} + i \sin \dfrac{\pi}{6}\right)\right]^6$ †
87. $(2 - 2\sqrt{3}i)^6$ †
88. $(2 - 2i)^7$ †

In Exercises 89–92, find all complex roots. Write your answers in polar form, with the arguments in degrees.

89. Cube roots of -125 †
90. Fourth roots of $-16i$ †
91. Fifth roots of $-1 + \sqrt{3}i$ †
92. Sixth roots of $1 - i$ †

Applying the Concepts

93. **Width of a river.** A surveyor wants to measure the width of a river. She stands at a point A, with point B opposite her on the other side of the river. From A, she walks 320 feet along the bank to a point C. The line CA makes an angle of 40° with the line CB. What is the width of the river? 268.5 feet

94. **Angle of elevation of a hill.** A tower 120 feet tall is located at the top of a hill. At a point 460 feet down the hill, the angle between the surface of the hill and the line of sight to the top of the tower is 12°. Find the angle of elevation of the hill to a horizontal plane. 25.2°

95. **Distance between cars.** Two cars start from a point A along two straight roads. The angle between the two roads is 72°. The speeds of the two cars are 55 miles per hour and 65 miles per hour. How far apart are the two cars after 80 minutes? Round to the nearest tenth of a mile. 94.7 miles

96. **Triangular plot.** A triangular plot of land has sides of lengths 310 feet, 415 feet, and 175 feet. Find the largest angle between the sides. 114.8°

97. **Area.** Find the area of the triangular plot in Exercise 96. 24,625 square feet

98. **Resultant force.** Two forces of magnitudes 16 and 20 pounds are acting on an object. The bearings of the forces are N 75° E and S 20° E, respectively. Find the magnitude and the direction of the resultant force, to the nearest tenth. 26.7 pounds, S 56.7° E

99. **Work.** A force of magnitude 30 pounds at an angle of 48° is used to pull a wagon 60 feet. Find the work done. 1204.4 foot-pounds

100. **AC current.** The total impedance in an AC circuit is given by $Z = \dfrac{Z_1Z_2}{Z_1 + Z_2}$. Find Z if $Z_1 = 20\left(\cos \dfrac{\pi}{6} + i \sin \dfrac{\pi}{6}\right)$ and $Z_2 = 10\left(\cos \dfrac{2\pi}{3} + i \sin \dfrac{2\pi}{3}\right)$. $4\sqrt{5}(\cos 93.5° + i \sin 93.5°)$

PRACTICE TEST A

1. In problems 1–4, solve triangle ABC. Round each answer to the nearest tenth. $B = 30°, a \approx 23.7, b \approx 13.1$

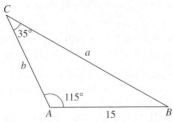

2. $A = 42°, B = 37°, a = 50$ meters. $C = 101°, b \approx 45.0$ m, $c \approx 73.4$ m

3.

$A \approx 28.2°$
$C \approx 45.8°$
$b \approx 71.1$

4. $a = 30$ feet, $b = 20$ feet, $c = 25$ feet. $A \approx 82.8°, B \approx 41.4°, C \approx 55.8°$
5. Solve right triangle ABC if $a = 5.6$ and $b = 4.1$. $c \approx 6.9, A \approx 53.8°, B \approx 36.2°$
6. Find the measure of the central angle of a circle of radius 8 feet that intercepts a chord of length 4.5 feet. Round to the nearest tenth of a degree. 32.7°

7. A surveyor, starting from point A, walks 580 feet in the direction N 70.0° E. From that point, she walks 725 feet in the direction N 35.0° W. How far, to the nearest foot, is she from her starting point? 803 feet

8. The vector **v** has initial point $P(3, 5)$ and terminal point $Q(2, -7)$. Write **v** as a position vector. $\langle -1, -12 \rangle$

9. If $\mathbf{v} = \langle -2, 3 \rangle$ and $\mathbf{w} = \langle 1, 5 \rangle$, find $\mathbf{v} - \mathbf{w}$. $\langle -3, -2 \rangle$

10. If $\mathbf{v} = -2\mathbf{i} + 3\mathbf{j}$ and $\mathbf{w} = \mathbf{i} + 5\mathbf{j}$, find $2\mathbf{v} - 3\mathbf{w}$. $-7\mathbf{i} - 9\mathbf{j}$

11. Let $\|\mathbf{v}\| = 3$ and suppose **v** makes an angle of $\theta = -30°$ with the positive x-axis. Write the vector **v** in the form $v_1\mathbf{i} + v_2\mathbf{j}$. $\frac{3\sqrt{3}}{2}\mathbf{i} - \frac{3}{2}\mathbf{j}$

12. Find the dot product $\mathbf{v} \cdot \mathbf{w}$ if $\mathbf{v} = 4\mathbf{i} + 3\mathbf{j}$ and $\mathbf{w} = -\mathbf{i} + 7\mathbf{j}$. 17

13. Find the angle $\theta(0 \le \theta \le 180°)$ between the vectors $\mathbf{v} = 3\mathbf{i} - 4\mathbf{j}$ and $\mathbf{w} = -2\mathbf{i} + 5\mathbf{j}$. Round the answer to the nearest tenth of a degree. 164.9°

14. Convert $(-2, -45°)$ to rectangular coordinates $(-\sqrt{2}, \sqrt{2})$

15. Convert $(-\sqrt{3}, -1)$ to polar coordinates with $r > 0$ and $0 \le \theta < 2\pi$. $\left(2, \frac{7\pi}{6}\right)$

16. Convert the polar equation $r = -3$ to rectangular form. Identify the curve having this equation.

17. Write $3\left(\cos\left(\frac{2\pi}{3}\right) + i\sin\left(\frac{2\pi}{3}\right)\right)$ in rectangular form. $-\frac{3}{2} + \frac{3\sqrt{3}}{2}i$

18. Write $\frac{z_1}{z_2}$ in polar form, if $z_1 = 1 - \sqrt{3}i$, $z_2 = -1 + i$, with the argument given in degrees. $\sqrt{2}(\cos 165° + i\sin 165°)$

19. Use DeMoivre's Theorem to find $\left[\sqrt{5}\left(\cos\left(\frac{\pi}{4}\right) + i\sin\left(\frac{\pi}{4}\right)\right)\right]^{-6}$. Write the answer in rectangular form. $\frac{1}{125}i$

20. Find all fourth roots of $1 + i$. Express the argument in degrees.
$2^{\frac{1}{8}}(\cos 11.25° + i\sin 11.25°)$, $2^{\frac{1}{8}}(\cos 101.25° + i\sin 101.25°)$,
$2^{\frac{1}{8}}(\cos 191.25° + i\sin 191.25°)$, $2^{\frac{1}{8}}(\cos 281.25° + i\sin 281.25°)$

Answers:

16. $x^2 + y^2 = 9$; circle with center $(0, 0)$ and radius 3.

PRACTICE TEST B

1. In triangle ABC, which is closest to the correct value for b? a
   a. $b = 13.4$
   b. $b = 20.8$
   c. $b = 20.2$
   d. $b = 0.05$

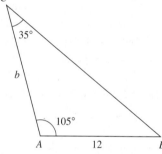

2. In triangle ABC, if $A = 27°$, $B = 50°$, and $a = 25$ meters, which is closest to the correct value for c? c
   a. $c = 0.03$ meter   **b.** $c = 42.2$ meters
   c. $c = 53.7$ meters   **d.** $c = 0.02$ meter

3. Which statement is true for the measurements $A = 25°$, $a = 25$ feet, $b = 30$ feet? c
   (The angle measurements in choices (c) and (d) are given to the nearest tenth.)
   a. No triangle has these measurements.
   b. One triangle has these measurements.
   c. Two triangles, with $c_1 = 48.7$ and $c_2 = 5.6$, have these measurements (in feet).
   d. Two triangles, with $c_1 = 46$ and $c_2 = 6.4$, have these measurements (in feet).

4. In triangle ABC, which is closest to the correct value for b?
   a. $b = 35.8$
   b. $b = 57.2$
   c. $b = 18.7$
   d. $b = 48.4$
   d

5. A 3-foot shrub is 4.5 feet away from a 12-foot streetlight. What length shadow is cast by the shrub? a
   a. 1.5 feet   **b.** 2 feet   **c.** 2.67 feet   **d.** 4 feet

6. In triangle ABC, with $a = 12$ feet, $b = 13$ feet, and $c = 7$ feet, which is closest to the correct value for A? b
   a. 24°   **b.** 66°   **c.** 81.8°   **d.** 32.2°

7. Find the measure of the central angle of a circle of radius 9 feet that intercepts a chord of length 3 feet. Round to the nearest tenth of a degree. d
   a. 70.8°   **b.** 63.6°   **c.** 26.4°   **d.** 19.2°

8. A hunter leaves a straight east–west road at a point A, walking 2 miles at a bearing N 40° W to a point B. From point B, he walks at a bearing of S 20° E and arrives back on the road at a point C. How far, to the nearest tenth of a mile, is point B from point C? c
   a. 4.5 miles   **b.** 2.9 miles
   c. 1.6 miles   **d.** 5.3 miles

9. The vector **v** has initial point $P(-2, 5)$ and terminal point $Q(3, -1)$. Write **v** as a position vector. a
   a. $\langle 5, -6 \rangle$ **b.** $\langle 1, -6 \rangle$ **c.** $\langle 1, -4 \rangle$ **d.** $\langle 5, -4 \rangle$

10. If $\mathbf{v} = \langle 1, -4 \rangle$ and $\mathbf{w} = \langle 3, -2 \rangle$, find $\mathbf{v} - \mathbf{w}$. c
   a. $\langle 4, -2 \rangle$ **b.** $\langle 2, 2 \rangle$ **c.** $\langle -2, -2 \rangle$ **d.** $\langle 4, -6 \rangle$

11. Let $\mathbf{v} = 7\mathbf{i} + 3\mathbf{j}$ and $\mathbf{w} = -2\mathbf{i} + \mathbf{j}$. Find $2\mathbf{v} - 3\mathbf{w}$. a
   a. $20\mathbf{i} + 3\mathbf{j}$ **b.** $20\mathbf{i} - 3\mathbf{j}$ **c.** $8\mathbf{i} - 6\mathbf{j}$ **d.** $8\mathbf{i} + 6\mathbf{j}$

12. Let $\|\mathbf{v}\| = 6$ and suppose **v** makes an angle of $\theta = -60°$ with the positive x-axis. Write the vector **v** in the form $v_1\mathbf{i} + v_2\mathbf{j}$. d
   a. $3\sqrt{3}\,\mathbf{i} - 3\mathbf{j}$ **b.** $3\sqrt{3}\,\mathbf{i} + 3\mathbf{j}$
   c. $3\mathbf{i} + 3\sqrt{3}\mathbf{j}$ **d.** $3\mathbf{i} - 3\sqrt{3}\mathbf{j}$

13. Find the dot product $\mathbf{v} \cdot \mathbf{w}$ if $\mathbf{v} = -2\mathbf{i} - 5\mathbf{j}$ and $\mathbf{w} = 9\mathbf{i} - 4\mathbf{j}$. d
   a. -38 **b.** 38 **c.** $2\mathbf{i} + 20\mathbf{j}$ **d.** 2

14. Find the angle between the vectors $\mathbf{v} = 7\mathbf{i} - \mathbf{j}$ and $\mathbf{w} = \mathbf{i} + \sqrt{2}\mathbf{j}$. Round the answer to the nearest degree. b
   a. $84°$ **b.** $63°$ **c.** $32°$ **d.** $17°$

15. Convert $(-4, -30°)$ to rectangular coordinates. c
   a. $(-2\sqrt{3}, -2)$ **b.** $(2, -2\sqrt{3})$
   c. $(-2\sqrt{3}, 2)$ **d.** $(2\sqrt{3}, -2)$

16. Convert $(3, -\sqrt{3})$ to polar coordinates, with $r > 0$ and $0 \le \theta < 360°$. b
   a. $(12, 330°)$ **b.** $(2\sqrt{3}, 330°)$
   c. $(2\sqrt{3}, 150°)$ **d.** $(\sqrt{3}, 150°)$

17. Which polar equation has a circle for its graph? b
   a. $r = -4\cos\theta + 1$ **b.** $r = -2$
   c. $r\sin\theta = 2$ **d.** $r\cos\theta = 1$

18. Write $5\left(\cos\left(\dfrac{7\pi}{3}\right) + i\sin\left(\dfrac{7\pi}{3}\right)\right)$ in rectangular form. c
   a. $5 + 5\sqrt{3}\,\mathbf{i}$ **b.** $5 - 5\sqrt{3}\,\mathbf{i}$
   c. $\dfrac{5}{2} + \dfrac{5\sqrt{3}}{2}\mathbf{i}$ **d.** $\dfrac{5}{2} - \dfrac{5\sqrt{3}}{2}\mathbf{i}$

19. Write $\dfrac{z_1}{z_2}$ in polar form, if $z_1 = i$ and $z_2 = 1 - i$. c
   a. $1(\cos 135° + i\sin 135°)$
   b. $\sqrt{2}\,(\cos 135° + i\sin 135°)$
   c. $\dfrac{\sqrt{2}}{2}(\cos 135° + i\sin 135°)$
   d. $\dfrac{\sqrt{2}}{2}(\cos 315° + i\sin 315°)$

20. Use DeMoivre's Theorem to find
   $\left[\sqrt{2}\left(\cos\left(\dfrac{\pi}{12}\right) + i\sin\left(\dfrac{\pi}{12}\right)\right)\right]^4$. Write the answer in rectangular form. c
   a. $-2 + 2\sqrt{3}i$ **b.** $\dfrac{1}{2} + \dfrac{\sqrt{3}}{2}i$
   c. $2 + 2\sqrt{3}i$ **d.** $\dfrac{1}{2} - \dfrac{\sqrt{3}}{2}i$

CUMULATIVE REVIEW EXERCISES ▪ Chapters 1–6

1. If $f(x) = \sqrt{x}$ and $g(x) = \dfrac{1}{x - 2}$, find the domain of $\dfrac{g}{f}$.

2. If $f(x) = 2x + 7$, find $f^{-1}(x)$.

3. Sketch the graph of $y = -2x^2 + 8x + 10$. †

4. Solve the inequality $x^2 - 6x + 8 > 0$. $(-\infty, 2) \cup (4, \infty)$

In Exercises 5 and 6, solve each equation for all solutions in the interval $[0, 2\pi)$. Give your answers in radians.

5. $\sin\theta\cos\theta = -\dfrac{\sqrt{3}}{4}$ **6.** $4\sin^2\theta - 1 = 0$

In Exercises 7 and 8, find all complex roots. Write your answers in polar form, with the argument in degrees.

7. Fourth roots of $\sqrt{3} + i$ **8.** Sixth roots of -64

9. Find the equation of the line in slope-intercept form that passes through the point $(-1, 5)$ and is perpendicular to the line $2x + 3y + 6 = 0$. $y = \dfrac{3}{2}x + \dfrac{13}{2}$

10. Write the following in condensed form:
   $2\log x + \dfrac{1}{2}\log(y + 1) - \log(3x + 1)$. $\log\dfrac{x^2\sqrt{y + 1}}{3x + 1}$

In Exercises 11–14, use transformations to sketch the graph of each function.

11. $f(x) = 2\sqrt{x - 1} - 3$ †   **12.** $f(x) = -|x + 1| + 2$ †

13. $f(x) = -2\cos(3x + \pi)$ †   **14.** $f(x) = 2\sec 3x$ †

15. If $\cos\theta = -\dfrac{5}{13}$ and $\sin\theta > 0$, find $\tan\theta$. $-\dfrac{12}{5}$

16. Use an identity to find the exact value of $\cos 65°\cos 35° + \sin 65°\sin 35°$. $\dfrac{\sqrt{3}}{2}$

17. Verify the identity
   $\dfrac{1}{1 - \cos x} + \dfrac{1}{1 + \cos x} = 2(1 + \cot^2 x)$.

18. Verify the identity $\dfrac{1 + \cos 2x}{\sin 2x} = \cot x$.

19. Solve triangle ABC with $A = 65°$, $B = 46°$, $c = 60$ meters. $C = 69°, a \approx 58.2$ m, $b \approx 46.2$ m

20. A ladder leaning against a vertical wall makes an angle of $47°$ with the ground. The foot of the ladder is 2.3 meters from the wall. Find the length of the ladder. Round to the nearest tenth. 3.4 m

Answers:

1. $(0, 2) \cup (2, \infty)$ **2.** $f^{-1}(x) = \dfrac{x - 7}{2}$

5. $\left\{\dfrac{2\pi}{3}, \dfrac{5\pi}{6}, \dfrac{5\pi}{3}, \dfrac{11\pi}{6}\right\}$ **6.** $\left\{\dfrac{\pi}{6}, \dfrac{5\pi}{6}, \dfrac{7\pi}{6}, \dfrac{11\pi}{6}\right\}$

7. $2^{\frac{1}{4}}(\cos 7.5° + i\sin 7.5°)$, $2^{\frac{1}{4}}(\cos 97.5° + i\sin 97.5°)$, $2^{\frac{1}{4}}(\cos 187.5° + i\sin 187.5°)$, $2^{\frac{1}{4}}(\cos 277.5° + i\sin 227.5°)$

8. $2(\cos 30° + i\sin 30°)$, $2(\cos 90° + i\sin 90°)$, $2(\cos 150° + i\sin 150°)$, $2(\cos 210° + i\sin 210°)$, $2(\cos 270° + i\sin 270°)$, $2(\cos 330° + i\sin 330°)$

7 Systems of Equations and Inequalities

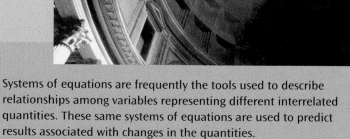

Systems of equations are frequently the tools used to describe relationships among variables representing different interrelated quantities. These same systems of equations are used to predict results associated with changes in the quantities.

The quantities involved can come from the study of medicine, taxes, elections, business, agriculture, city planning, or any discipline in which multiple quantities interact.

Systems of Linear Equations in Two Variables

Al-Khwarizmi (780–850)

Al-Khwarizmi came to Baghdad from the town of Khwarizm, which is now called *Khivga* and is part of Uzbekistan. The term *algorithm* is a corruption of the name al-khwarizmi. Originally, the word *algorism* was used for the rules for performing arithmetic by using decimal notation. *Algorism* evolved into the word *algorithm* by the eighteenth century. The word *algorithm* now means a finite set of precise instructions for performing a computation or for solving a problem. (The picture shown is a Soviet postage stamp from 1983 depicting Al-Khwarizmi.)

Before Starting this Section, Review

1. Graphs of equations (Section 1.1, page 6)

2. Graphs of linear equations (Section 1.2, page 23)

Objectives

1 Verify a solution to a system of equations.

2 Solve a system of equations by the graphical method.

3 Solve a system of equations by the substitution method.

4 Solve a system of equations by the elimination method.

5 Solve applied problems by solving systems of equations.

ALGEBRA–IRAQ CONNECTION

The most important contributions of medieval Islamic mathematicians lie in the area of algebra. One of the greatest Islamic scholars was Muhammad ibn Musa al-Khwarizmi (A.D. 780–850). Al-Khwarizmi was one of the first scholars in the House of Wisdom, an academy of scientists, established by caliph al-Mamun in the city of Baghdad in what is now Iraq. Jews, Christians, and Muslims worked together in scholarly pursuits in the academy during this period. The eminent scholar al-Khwarizmi wrote several books on astronomy and mathematics. Western Europeans first learned about algebra from his books. The word *algebra* comes from the Arabic *al-jabr*, part of the title of his book *Kitab al-jabr wal-muqabala*.

This book was translated into Latin and was a widely used text. The Arabic word *al-jabr* means "restoration," as in restoring broken parts. (At one time, it was not unusual to see the sign "Algebrista y Sangrador," meaning "bone setter and blood letter," at the entrance of a Spanish barber's shop. The sign informed customers of the barber's side business.) Al-Khwarizmi used the term in the mathematical sense of removing a negative quantity on one side of an equation and restoring it as a positive quantity on the other side. ∎

---

1 Verify a solution to a system of equations.

System of Equations

A set of equations with common variables is called a **system of equations**. If each equation in a system of equations is linear, then the set of equations is called a **system of linear equations** or a **linear system of equations**. However, if at least one equation in a system of equations is nonlinear, then the set of equations is called a **nonlinear system of equations**. In this section, we will solve systems of two linear equations in two variables, such as:

$$\begin{cases} 2x - y = 5 \\ x + 2y = 5 \end{cases}$$

Notice that we designate a system of equations by using a left brace. A system of equations is sometimes referred to as a set of *simultaneous equations*. A **solution of a system of equations in two variables** x and y is an ordered pair of numbers (a, b) such that when x is replaced with a and y is replaced with b, the resulting equations are true. The **solution set of a system of equations** is the set of all solutions of the system.

EXAMPLE 1 ── Verifying a Solution

Verify that the ordered pair $(3, 1)$ is a solution of the system of linear equations

$$\begin{cases} 2x - y = 5 & (1) \\ x + 2y = 5 & (2) \end{cases}$$

SOLUTION

To verify that $(3, 1)$ is a solution, replace x with 3 and y with 1 in both equations.

Equation (1)	Equation (2)
$2x - y = 5$	$x + 2y = 5$
$2(3) - 1 \overset{?}{=} 5$	$3 + 2(1) \overset{?}{=} 5$
$5 = 5 \checkmark$	$5 = 5 \checkmark$

Since the ordered pair $(3, 1)$ satisfies both equations, it is a solution of the system.

■ ■ ■

Practice Problem 1 Verify that $(1, 3)$ is a solution of the system

$$\begin{cases} x + y = 4 \\ 3x - y = 0 \end{cases}$$

■

In this section, you will learn three methods of solving a system of two equations in two variables: the *graphical method,* the *substitution method,* and the *elimination method.*

2 Solve a system of equations by the graphical method.

Graphical Method

Recall that the graph of a linear equation

$$ax + by = c \qquad (a \text{ and } b \text{ not both zero})$$

is a line. Consider a system of linear equations:

$$\begin{cases} a_1 x + b_1 y = c_1 \\ a_2 x + b_2 y = c_2 \end{cases}$$

A solution of this system is a point (an ordered pair) that satisfies both equations. To find or estimate the solution set of this system, therefore, we graph both equations on the same coordinate axes and find the coordinates of any points of intersection. This procedure is called the **graphical method**.

EXAMPLE 2 ── Solving a System by the Graphical Method

Use the graphical method to solve the system of equations.

$$\begin{cases} 2x - y = 4 & (1) \\ 2x + 3y = 12 & (2) \end{cases}$$

SOLUTION

Step 1 **Graph both equations on the same coordinate axes.**

(i) We first graph equation (1) by finding its intercepts.

a. To find the y-intercept, set $x = 0$ and solve for y.
We have $2(0) - y = 4$, or $y = -4$; so the y-intercept is -4.

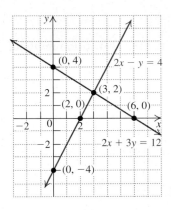

FIGURE 7.1 Graphical solution

TECHNOLOGY CONNECTION

In Example 2, solve each equation for y. Graph

$$Y_1 = 2x - 4 \quad \text{and}$$

$$Y_2 = -\frac{2}{3}x + 4$$

on the same screen.

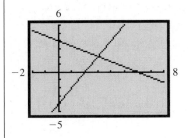

Use the **intersect** feature on your calculator to find the point of intersection.

b. To find the x-intercept, set $y = 0$ and solve for x.
We have $2x - 0 = 4$, or $x = 2$; so the x-intercept is 2.
The points $(0, -4)$ and $(2, 0)$ are on the graph of equation (1), which is sketched in red in Figure 7.1.

(ii) We now graph equation (2) using intercepts.
Here the x-intercept is 6 and the y-intercept is 4; so joining the points $(0, 4)$ and $(6, 0)$ produces the line sketched in blue in Figure 7.1.

Step 2 **Find the point(s) of intersection of the two graphs.** We observe in Figure 7.1 that the point of intersection of the two graphs is $(3, 2)$.

Step 3 **Check your solution(s).** Replace x with 3 and y with 2 in equations (1) and (2).

Equation (1)	Equation (2)
$2x - y = 4$	$2x + 3y = 12$
$2(3) - 2 \stackrel{?}{=} 4$	$2(3) + 3(2) \stackrel{?}{=} 12$
$4 = 4 \; \checkmark$	$12 = 12 \; \checkmark$

Step 4 **Write the solution set for the system.**
The solution set is $\{(3, 2)\}$.   ■ ■ ■

Practice Problem 2 Solve the system of equations graphically.

$$\begin{cases} x + y = 2 \\ 4x + y = -1 \end{cases}$$ ■

If a system of equations has at least one solution (as in Example 2), the system is **consistent**. A system of equations with no solution is **inconsistent**, and its solution set is the empty set, $\varnothing$.

A system of two linear equations in two variables must have one of the following types of solution sets.

1. One solution (the lines intersect; see Figure 7.2(a)): the system is consistent, and the equations in the system are said to be **independent**.

2. No solution (the lines are parallel; see Figure 7.2(b)): the system is inconsistent.

3. Infinitely many solutions (the lines coincide; see Figure 7.2(c)): the system is consistent, and the equations in the system are said to be **dependent**.

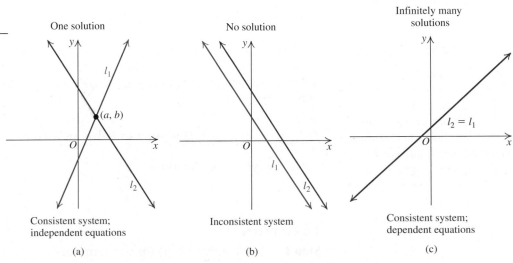

FIGURE 7.2 Possible solution sets of a system of two linear equations

3 Solve a system of equations by the substitution method.

Substitution Method

Another way to solve a linear system of equations is with the **substitution method**.

FINDING THE SOLUTION: A PROCEDURE

EXAMPLE 3   **The Substitution Method**

OBJECTIVE	**EXAMPLE**
Reduce the solution of the system to the solution of one equation in one variable by substitution.	*Solve the system.*

OBJECTIVE
Reduce the solution of the system to the solution of one equation in one variable by substitution.

EXAMPLE
Solve the system.

$$\begin{cases} 2x - 5y = 3 & (1) \\ y - 2x = 9 & (2) \end{cases}$$

We choose equation (2) and express y in terms of x.

Step 1 Solve for one variable. Choose one of the equations and express one of its variables in terms of the other variable.

$y = 2x + 9 \quad$ (3)   Solve Equation (2) for y.

Step 2 Substitute. Substitute the expression found in Step 1 into the other equation to obtain an equation in one variable.

$2x - 5y = 3$ Equation (1)
$2x - 5(2x + 9) = 3$ Substitute $2x + 9$ for y.
$2x - 10x - 45 = 3$ Distributive property

Step 3 Solve the equation obtained in Step 2.

$-8x - 45 = 3$ Combine like terms.
$-8x = 3 + 45$ Add 45 to both sides.
$x = -6$ Solve for x.

Step 4 Back-substitute. Substitute the value(s) you found in Step 3 back into the expression you found in Step 1. The result is the solution(s).

$y = 2x + 9$ Equation (3) from Step 1
$y = 2(-6) + 9 = -3$ Substitute $x = -6$.
Because $x = -6$ and $y = -3$, the solution set is $\{(-6, -3)\}$.

Step 5 Check. Check your answer(s) in the original equations.

Check: $x = -6$ and $y = -3$.

Equation (1)	Equation (2)
$2(-6) - 5(-3) \overset{?}{=} 3$	$-3 - 2(-6) \overset{?}{=} 9$
$-12 + 15 \quad = 3 ✓$	$-3 + 12 \quad = 9 ✓$ ■ ■ ■

Practice Problem 3   Solve $\begin{cases} x - y = 5 \\ 2x + y = 7 \end{cases}$ ■

STUDY TIP

It is easy to make arithmetic errors in the many steps involved in the process of solving a system of equations. You should check your solution by substituting it into each equation in the original system.

It is possible that in the process of solving the equation in Step 3, you obtain an equation of the form $0 = k$ (which can be written as $0x = k$ or $0y = k$), where k is a *nonzero* constant. In such cases, the false statement $0 = k$ indicates that the system is *inconsistent*.

EXAMPLE 4   **Attempting to Solve an Inconsistent System of Equations**

Solve the system of equations.

$$\begin{cases} x + y = 3 & (1) \\ 2x + 2y = 9 & (2) \end{cases}$$

SOLUTION

Step 1   Solve equation (1) for y in terms of x.

$$y = 3 - x \quad \text{Add } -x \text{ to both sides.}$$

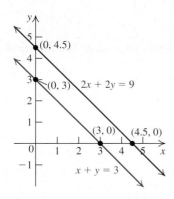

FIGURE 7.3 Inconsistent system

Step 2   Substitute this expression into equation (2).

$$2x + 2y = 9 \qquad \text{Equation (2)}$$
$$2x + 2(3 - x) = 9 \qquad \text{Replace } y \text{ with } 3 - x \text{ (from Step 1).}$$

Step 3   Solve for x.

$$2x + 6 - 2x = 9 \qquad \text{Distributive property}$$
$$\underbrace{0 = 3}_{\text{False}} \qquad \text{Subtract 6 from both sides and simplify.}$$

Since the equation $0 = 3$ is false, the system is inconsistent. Figure 7.3 shows that the graphs of the two equations in this system are parallel lines. Because the lines do not intersect, the system has no solution. The solution set is $\varnothing$.   ■ ■ ■

Practice Problem 4   Solve the system of equations.

$$\begin{cases} x - 3y = 1 \\ -2x + 6y = 3 \end{cases}$$ ■

In Step 3, it is also possible to end up with an equation of the form $0 = 0$. In such cases, the equations are *dependent* and the system has infinitely many solutions.

EXAMPLE 5   **Solving a Dependent System**

Solve the system of equations.

$$\begin{cases} 4x + 2y = 12 & (1) \\ -2x - y = -6 & (2) \end{cases}$$

SOLUTION

Step 1   Solve equation (2) for y in terms of x.

$$-2x - y = -6 \qquad \text{Equation (2)}$$
$$-y = -6 + 2x \qquad \text{Add } 2x \text{ to both sides.}$$
$$y = 6 - 2x \qquad \text{Multiply both sides by } -1.$$

Step 2   Substitute $(6 - 2x)$ for y in equation (1).

$$4x + 2y = 12 \qquad \text{Equation (1)}$$
$$4x + 2(6 - 2x) = 12 \qquad \text{Replace } y \text{ with } 6 - 2x \text{ (from Step 1).}$$

Step 3   Solve for x.

$$4x + 12 - 4x = 12 \qquad \text{Distributive property}$$
$$\underbrace{0 = 0}_{\text{True}} \qquad \text{Subtract 12 from both sides and simplify.}$$

The equation $0 = 0$ is true for *every* value of x. Thus, *any* value of x can be used in the equation $y = 6 - 2x$ for back-substitution.

The solutions of the system are of the form $(x, 6 - 2x)$, and the solution set is

$$\{(x, 6 - 2x)\}.$$

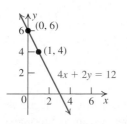

FIGURE 7.4 Dependent equations

In other words, the solution set consists of the infinite number of ordered pairs (x, y) lying on the line with equation $4x + 2y = 12$, as shown in Figure 7.4. You can find particular solutions by replacing x with any real number. For example, if we let $x = 0$ in $(x, 6 - 2x)$, we find that $(0, 6 - 2(0)) = (0, 6)$ is a solution of the system. Similarly, letting $x = 1$, we find that $(1, 4)$ is a solution.   ■ ■ ■

Practice Problem 5   Solve the system of equations.

$$\begin{cases} -2x + y = -3 \\ 4x - 2y = 6 \end{cases}$$ ■

4 Solve a system of equations by the elimination method.

Elimination Method

The **elimination method** of solving a system of equations is also called the **addition method**.

FINDING THE SOLUTION: A PROCEDURE

EXAMPLE 6 The Elimination Method

OBJECTIVE
Solve a system of two linear equations by first eliminating one variable.

EXAMPLE
Solve the system.

$$\begin{cases} 2x + 3y = 21 & (1) \\ 3x - 4y = 23 & (2) \end{cases}$$

We arbitrarily select y as the variable to be eliminated.

Step 1 **Adjust the coefficients.** If necessary, multiply both equations by appropriate numbers to get two new equations in which the coefficients of the variable to be eliminated are opposites.

$$\begin{cases} 8x + 12y = 84 \\ 9x - 12y = 69 \end{cases}$$
Multiply equation (1) by 4.
Multiply equation (2) by 3.

Coefficients are opposites of each other.

Step 2 **Add the equations.** Add the resulting equations to get an equation in one variable.

$$\begin{array}{r} 8x + 12y = 84 \\ 9x - 12y = 69 \\ \hline 17x \quad\quad = 153 \end{array}$$

Adding the two equations from Step 1 eliminates the y-terms.

Step 3 **Solve** the resulting equation.

$17x = 153$ Equation from Step 2

$x = \dfrac{153}{17}$ Divide both sides by 17.

$x = 9$ Simplify.

Step 4 **Back-substitute** the value you found into one of the original equations to solve for the other variable.

$2x + 3y = 21$ Equation (1)

$2(9) + 3y = 21$ Replace x with 9.

$18 + 3y = 21$ Simplify.

$3y = 3$ Subtract 18 from both sides.

$y = 1$ Solve for y.

Step 5 **Write** the solution set from Steps 3 and 4.

The solution set is $\{(9, 1)\}$.

Step 6 **Check** your solution(s) in the original equations (1) and (2).

Check $x = 9$ and $y = 1$.

Equation (1)	Equation (2)
$2(9) + 3(1) \overset{?}{=} 21$	$3(9) - 4(1) = 23$
$18 + 3 = 21$ ✓	$27 - 4 = 23$ ✓

■ ■ ■

Practice Problem 6 Solve the system.

$$\begin{cases} 3x + 2y = 3 \\ 9x - 4y = 4 \end{cases}$$

■

Just as in the substitution method, if you add the equations in Step 2 and the resulting equation becomes $0 = k$, where $k \neq 0$, then the system is inconsistent: it has no solutions. As an illustration, if you solve the system of equations in Example 4 by the elimination method, you will obtain the equation $0 = 3$. You now conclude that the system is inconsistent. Similarly, in Step 2, if you obtain $0x + 0y = 0$ (or $0 = 0$), then the equations are dependent.

STUDY TIP

You should use the elimination method when the terms involving one of the variables can easily be eliminated by adding multiples of the equations. Otherwise, use substitution to solve the system. The graphing method is usually used to confirm or visually interpret the result obtained from the other two methods.

The next example illustrates how some nonlinear systems can be solved by substituting variables to create a linear system and then solving the new system by the elimination method.

EXAMPLE 7   **Using the Elimination Method**

Solve the system.

$$\begin{cases} \dfrac{2}{x} + \dfrac{5}{y} = -5 & (1) \\[2mm] \dfrac{3}{x} - \dfrac{2}{y} = -17 & (2) \end{cases}$$

SOLUTION

Replace $\dfrac{1}{x}$ with u and $\dfrac{1}{y}$ with v. (Note that $\dfrac{2}{x} = 2\left(\dfrac{1}{x}\right) = 2u$, and so on.) Equations (1) and (2) become

$$\begin{cases} 2u + 5v = -5 & (3) \\ 3u - 2v = -17 & (4) \end{cases}$$

Now we solve the new system for u and v using the elimination method.

Step 1   We choose to eliminate the variable u.

$6u + 15v = -15$	(5)	Multiply equation (3) by 3.
$-6u + 4v = 34$	(6)	Multiply equation (4) by -2.

Step 2 $19v = 19$ (7)   Add equations (5) and (6).

Step 3 $v = 1$ Solve equation (7) for v.

Step 4 $2u + 5v = -5$ Equation (3)

$2u + 5(1) = -5$ Back-substitute $\dot{v} = 1$.

$u = -5$ Solve for u.

Step 5   Now solve for x and y, the variables in the original system.

$u = \dfrac{1}{x}$ and $v = \dfrac{1}{y}$ Original substitution

$-5 = \dfrac{1}{x}$ $\Big|$ $1 = \dfrac{1}{y}$ Replace u with -5 and v with 1.

$x = -\dfrac{1}{5}$ $\Big|$ $y = 1$ Solve for x and y.

The solution set of the original system is $\left\{\left(-\dfrac{1}{5}, 1\right)\right\}$.

Step 6   Verify that the ordered pair $\left(-\dfrac{1}{5}, 1\right)$ is the solution of the original system of equations (1) and (2).   ■ ■ ■

Practice Problem 7   Solve the system.

$$\begin{cases} \dfrac{4}{x} + \dfrac{3}{y} = 1 \\[2mm] \dfrac{2}{x} - \dfrac{6}{y} = 3 \end{cases}$$ ■

5 Solve applied problems by solving systems of equations.

Applications

We can deduce from Section 1.3 that as the price of a product increases, demand for it decreases, and as the price increases, the supply of the product also increases. The **equilibrium point** is the ordered pair (x, p) such that the number of units, x, and the price per unit, p, satisfy *both* the demand and supply equations.

STUDY TIP

The graph of a system of equations allows you to use geometric intuition to understand algebra. If the geometric and algebraic representations of a solution do not agree, you must go back and find the error(s).

EXAMPLE 8   **Finding the Equilibrium Point**

Find the equilibrium point if the supply and demand functions for a new brand of digital video recorder (DVR) are given by the system

$$p = 60 + 0.0012x \quad (1) \quad \text{Supply equation}$$
$$p = 80 - 0.0008x \quad (2) \quad \text{Demand equation}$$

where p is the price in dollars and x is the number of units.

SOLUTION

We substitute the value of p from equation (1) into equation (2) and solve the resulting equation.

$$p = 80 - 0.0008x \qquad \text{Demand equation}$$
$$60 + 0.0012x = 80 - 0.0008x \qquad \text{Replace } p \text{ with } 60 + 0.0012x.$$
$$0.002x = 20 \qquad \text{Collect like terms and simplify.}$$
$$x = \frac{20}{0.002} = 10{,}000 \qquad \text{Solve for } x.$$

The equilibrium point occurs when the supply and demand for DVRs is 10,000 units. To find the price p, we back-substitute $x = 10{,}000$ into either of the original equations (1) or (2).

$$p = 60 + 0.0012x \qquad \text{Equation (1)}$$
$$= 60 + 0.0012(10{,}000) \qquad \text{Replace } x \text{ with } 10{,}000.$$
$$= 72 \qquad \text{Simplify.}$$

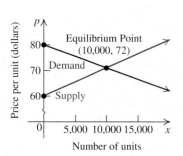

FIGURE 7.5 Price of DVRs

The equilibrium point is $(10{,}000, 72)$. See Figure 7.5.

Check. Verify that the ordered pair $(10{,}000, 72)$ satisfies both equations (1) and (2). ■ ■ ■

Practice Problem 8 Find the equilibrium point.

$$\begin{cases} p = 20 + 0.002x & \text{Supply equation} \\ p = 77 - 0.008x & \text{Demand equation} \end{cases}$$ ■

In general, to solve an applied problem involving two unknowns, we need to set up two equations using two variables to represent the two unknowns. The general rule is that *the number of equations formed must be equal to the number of unknown quantities involved.*

SECTION 7.1 ■ Exercises

A EXERCISES Basic Skills and Concepts

1. The ordered pair (a, b) is a(n) ____solution____ of a system of equations in x and y providing that when x is replaced with a and y is replaced with b, the resulting equations are true.

2. The two nongraphical methods for solving a system of equations are the ____substitution____ and ____elimination____ methods.

3. If in the process of solving a system of equations you get an equation of the form $0 = k$, where k is not zero, then the system is ____inconsistent____.

4. If in the process of solving a system of equations you get an equation of the form $0 = 0$, then the system has ____dependent equations____

5. *True or False* A system consisting of two identical equations has no solution. False

6. *True or False* If in the process of solving a system of two equations in x and y you get the equation $x = 5$, then the system has exactly one solution. False

In Exercises 7–12, determine which ordered pairs are solutions of each system of equations.

7. $\begin{cases} 2x + 3y = 3 \\ 3x - 4y = 13 \end{cases}$ $(1, -3), (3, -1), (6, 3), \left(5, \dfrac{1}{2}\right)$ $(3, -1)$

8. $\begin{cases} x + 2y = 6 \\ 3x + 6y = 18 \end{cases}$ $(2, 2), (-2, 4), (0, 3), (1, 2)$
   $(2, 2), (-2, 4), (0, 3)$

9. $\begin{cases} 5x - 2y = 7 \\ -10x + 4y = 11 \end{cases}$ $\left(\dfrac{5}{4}, 1\right), \left(0, \dfrac{11}{4}\right), (1, -1), (3, 4)$
   None of them

10. $\begin{cases} x - 2y = -5 \\ 3x - y = 5 \end{cases}$ $(1, 3), (-5, 0), (3, 4), (3, -4)$ $(3, 4)$

11. $\begin{cases} x + y = 1 \\ \dfrac{1}{2}x + \dfrac{1}{3}y = 2 \end{cases}$ $(0, 1), (1, 0), \left(\dfrac{2}{3}, \dfrac{3}{2}\right), (10, -9)$ $(10, -9)$

12. $\begin{cases} \dfrac{2}{x} + \dfrac{3}{y} = 2 \\ \dfrac{6}{x} + \dfrac{18}{y} = 9 \end{cases}$ $(3, 2), (2, 3), (4, 3), (3, 4)$ $(2, 3)$

In Exercises 13–22, estimate the solution(s) (if any) of each system by the graphical method. Check your solution(s). For any dependent equations, write your answer with x being arbitrary.

13. $\begin{cases} x + y = 3 \\ x - y = 1 \end{cases}$ †

14. $\begin{cases} x + y = 10 \\ x - y = 2 \end{cases}$ †

15. $\begin{cases} x + 2y = 6 \\ 2x + y = 6 \end{cases}$ †

16. $\begin{cases} 2x - y = 4 \\ x - y = 3 \end{cases}$ †

17. $\begin{cases} 3x - y = -9 \\ y = 3x + 6 \end{cases}$ †

18. $\begin{cases} 5x + 2y = 10 \\ y = -\dfrac{5}{2}x - 5 \end{cases}$ †

19. $\begin{cases} x + y = 7 \\ y = 2x \end{cases}$ †

20. $\begin{cases} y - x = 2 \\ y + x = 9 \end{cases}$ †

21. $\begin{cases} 3x + y = 12 \\ y = -3x + 12 \end{cases}$ †

22. $\begin{cases} 2x + 3y = 6 \\ 6y = -4x + 12 \end{cases}$ †

In Exercises 23–36, determine whether each system is *consistent* or *inconsistent*. If the system is consistent, determine whether the equations are dependent or independent. Do not solve the system.

23. $\begin{cases} y = -2x + 3 \\ y = 3x + 5 \end{cases}$
    Independent

24. $\begin{cases} 3x + y = 5 \\ 2x + y = 4 \end{cases}$ Independent

25. $\begin{cases} 2x + 3y = 5 \\ 3x + 2y = 7 \end{cases}$
    Independent

26. $\begin{cases} 2x - 4y = 5 \\ 3x + 5y = -6 \end{cases}$
    Independent

27. $\begin{cases} 3x + 5y = 7 \\ 6x + 10y = 14 \end{cases}$ Dependent

28. $\begin{cases} 3x - y = 2 \\ 9x - 3y = 6 \end{cases}$ Dependent

29. $\begin{cases} x + 2y = -5 \\ 2x - y = 4 \end{cases}$ Independent

30. $\begin{cases} x + 2y = -2 \\ 2x - 3y = 5 \end{cases}$ Independent

31. $\begin{cases} 2x - 3y = 5 \\ 6x - 9y = 10 \end{cases}$ Inconsistent

32. $\begin{cases} 3x + y = 2 \\ 15x + 5y = 15 \end{cases}$ Inconsistent

33. $\begin{cases} -3x + 4y = 5 \\ \dfrac{9}{2}x - 6y = \dfrac{15}{2} \end{cases}$ Inconsistent

34. $\begin{cases} 6x + 5y = 11 \\ 9x + \dfrac{15}{2}y = 21 \end{cases}$ † Inconsistent

35. $\begin{cases} 7x - 2y = 3 \\ 11x - \dfrac{3}{2}y = 8 \end{cases}$ Independent

36. $\begin{cases} 4x + 7y = 10 \\ 10x + \dfrac{35}{2}y = 25 \end{cases}$ † Dependent

In Exercises 37–46, solve each system of equations by the substitution method. Check your solutions. For any dependent equations, write your answer in the ordered pair form given in Example 5.

37. $\begin{cases} y = 2x + 1 \\ 5x + 2y = 9 \end{cases}$ $\left\{\left(\dfrac{7}{9}, \dfrac{23}{9}\right)\right\}$

38. $\begin{cases} x = 3y - 1 \\ 2x - 3y = 7 \end{cases}$ $\{(8, 3)\}$

39. $\begin{cases} 3x - y = 5 \\ x + y = 7 \end{cases}$ $\{(3, 4)\}$

40. $\begin{cases} 2x + y = 2 \\ 3x - y = -7 \end{cases}$ $\{(-1, 4)\}$

41. $\begin{cases} 2x - y = 5 \\ -4x + 2y = 7 \end{cases}$ $\varnothing$

42. $\begin{cases} 3x + 2y = 5 \\ -9x - 6y = 15 \end{cases}$ $\varnothing$

43. $\begin{cases} \dfrac{2}{3}x + y = 3 \\ 3x + 2y = 1 \end{cases}$ $\{(-3, 5)\}$

44. $\begin{cases} x - 2y = 3 \\ 4x + 6y = 3 \end{cases}$ $\left\{\left(\dfrac{12}{7}, -\dfrac{9}{14}\right)\right\}$

45. $\begin{cases} x - 2y = 5 \\ -3x + 6y = -15 \end{cases}$ $\left\{\left(x, \dfrac{1}{2}(x - 5)\right)\right\}$

46. $\begin{cases} x + y = 3 \\ 2x + 2y = 6 \end{cases}$ $\{(x, 3 - x)\}$

In Exercises 47–56, solve each system of equations by the elimination method. Check your solutions. For any dependent equations, write your answer as in Example 5.

47. $\begin{cases} x - y = 1 \\ x + y = 5 \end{cases}$ $\{(3, 2)\}$

48. $\begin{cases} 2x - 3y = 5 \\ 3x + 2y = 14 \end{cases}$ $\{(4, 1)\}$

49. $\begin{cases} x + y = 0 \\ 2x + 3y = 3 \end{cases}$ $\{(-3, 3)\}$

50. $\begin{cases} x + y = 3 \\ 3x + y = 1 \end{cases}$ $\{(-1, 4)\}$

51. $\begin{cases} 5x - y = 5 \\ 3x + 2y = -10 \end{cases}$ $\{(0, -5)\}$ **52.** $\begin{cases} 3x - 2y = 1 \\ -7x + 3y = 1 \end{cases}$ $\{(-1, -2)\}$

53. $\begin{cases} x - y = 2 \\ -2x + 2y = 5 \end{cases}$ $\varnothing$ **54.** $\begin{cases} 4x + 7y = -3 \\ -8x - 14y = 6 \end{cases}$

55. $\begin{cases} 4x + 6y = 12 \\ 2x + 3y = 6 \end{cases}$ **56.** $\begin{cases} x + y = 5 \\ 2x + 2y = -10 \end{cases}$ $\varnothing$

In Exercises 57–76, use any method to solve each system of equations. For any dependent equations, write your answer as in Example 5.

57. $\begin{cases} 2x + y = 9 \\ 2x - 3y = 5 \end{cases}$ $\{(4, 1)\}$ **58.** $\begin{cases} x + 2y = 10 \\ x - 2y = -6 \end{cases}$ $\{(2, 4)\}$

59. $\begin{cases} 2x + 5y = 2 \\ x + 3y = 2 \end{cases}$ $\{(-4, 2)\}$ **60.** $\begin{cases} 4x - y = 6 \\ 3x - 4y = 11 \end{cases}$ $\{(1, -2)\}$

61. $\begin{cases} 2x + 3y = 7 \\ 3x + y = 7 \end{cases}$ $\{(2, 1)\}$ **62.** $\begin{cases} x = 3y + 4 \\ x = 5y + 10 \end{cases}$ $\{(-5, -3)\}$

63. $\begin{cases} 2x + 3y = 9 \\ 3x + 2y = 11 \end{cases}$ $\{(3, 1)\}$ **64.** $\begin{cases} 3x - 4y = 0 \\ y = \dfrac{2x + 1}{3} \end{cases}$ $\{(4, 3)\}$

65. $\begin{cases} \dfrac{x}{4} + \dfrac{y}{6} = 1 \\ x + 2(x - y) = 7 \end{cases}$ $\left\{\left(\dfrac{19}{6}, \dfrac{5}{4}\right)\right\}$ **66.** $\begin{cases} \dfrac{x}{3} + \dfrac{y}{5} = 12 \\ x - y = 4 \end{cases}$ $\{(24, 20)\}$

67. $\begin{cases} 3x = 2(x + y) \\ 3x - 5y = 2 \end{cases}$ $\{(4, 2)\}$ **68.** $\begin{cases} y + x + 2 = 0 \\ y + 2x + 1 = 0 \end{cases}$ $\{(1, -3)\}$

69. $\begin{cases} 0.2x + 0.7y = 1.5 \\ 0.4x - 0.3y = 1.3 \end{cases}$ $\{(4, 1)\}$ **70.** $\begin{cases} 0.6x + y = -1 \\ x - 0.5y = 7 \end{cases}$ $\{(5, -4)\}$

71. $\begin{cases} \dfrac{2}{x} + \dfrac{5}{y} = -5 \\ \dfrac{3}{x} - \dfrac{2}{y} = -17 \end{cases}$ **72.** $\begin{cases} \dfrac{2}{x} + \dfrac{1}{y} = 3 \\ \dfrac{4}{x} - \dfrac{2}{y} = 0 \end{cases}$ $\left\{\left(\dfrac{4}{3}, \dfrac{2}{3}\right)\right\}$

73. $\begin{cases} \dfrac{3}{x} + \dfrac{1}{y} = 4 \\ \dfrac{6}{x} - \dfrac{1}{y} = 2 \end{cases}$ $\left\{\left(-\dfrac{1}{5}, 1\right)\right\}$ $\left\{\left(\dfrac{3}{2}, \dfrac{1}{2}\right)\right\}$ **74.** $\begin{cases} \dfrac{6}{x} + \dfrac{3}{y} = 0 \\ \dfrac{4}{x} + \dfrac{9}{y} = -1 \end{cases}$ $\{(14, -7)\}$

75. $\begin{cases} \dfrac{5}{x} + \dfrac{10}{y} = 3 \\ \dfrac{2}{x} - \dfrac{12}{y} = -2 \end{cases}$ $\{(5, 5)\}$ **76.** $\begin{cases} \dfrac{3}{x} + \dfrac{4}{y} = 1 \\ \dfrac{6}{x} + \dfrac{4}{y} = 3 \end{cases}$ $\left\{\left(\dfrac{3}{2}, -4\right)\right\}$

B EXERCISES Applying the Concepts

In Exercises 77–80, the demand and supply functions of a product are given. In each case, _p_ represents the price in dollars per unit and _x_ represents the number of units in hundreds. Find the equilibrium point.

77. $\begin{cases} 2p + x = 140 & \text{Demand equation} \\ 12p - x = 280 & \text{Supply equation} \end{cases}$ $(80, 30)$

78. $\begin{cases} 7p + x = 150 & \text{Demand equation} \\ 10p - x = 20 & \text{Supply equation} \end{cases}$ $(80, 10)$

79. $\begin{cases} 2p + x = 25 & \text{Demand equation} \\ x - p = 13 & \text{Supply equation} \end{cases}$ $(17, 4)$

80. $\begin{cases} p + 2x = 96 & \text{Demand equation} \\ p - x = 39 & \text{Supply equation} \end{cases}$ $(19, 58)$

In Exercises 81–96, use a system of equations to solve each problem.

81. Diameter of a pizza. The sum of the diameters of the largest and the smallest pizza sold at the Monster Pizza Shop is 29 inches. The difference in their diameters is 13 inches. Find the diameters of the largest and the smallest pizza. †

82. Calories in hamburgers. The sum of the number of calories in a hamburger from Boston Burger and a hamburger from Carmen's Broiler is 1130. The difference in the number of calories in the hamburgers is 40. If the Boston Burger has the larger number of calories, how many calories are in each restaurant's hamburger? †

83. Trash composition. Paper and plastic together account for 48% (by weight) of the total trash collected. If the weight of paper trash collected is five times the weight of plastic trash, what percent of the total trash collected is paper and what percent is plastic? 8% of the total trash collected is plastic, and 40% is paper.

84. Cost of food and clothing. The average monthly combined cost of food and clothing for the Martínez family is $1000. If they spend four times as much for food as they do for clothes, what is the average monthly cost of each? †

85. Mardi Gras parade "throws." Levon paid 40¢ for each string of beads and 30¢ for each doubloon (a special coin) he bought to throw from his Mardi Gras float. He paid a total of $265 for the two items. His doubloons and beads combined to a total of 770 items. How many doubloons and how many strings of beads did Levon buy? 340 beads and 430 doubloons

86. Halloween candy. Janet bought 135 pieces of candy to give away on Halloween. She bought two kinds of candy, paying 24¢ apiece for one kind and 18¢ apiece for the other. If she spent $26.70 for the candy, how many pieces of each kind did she buy? 40 at 24¢ each and 95 at 18¢ each

In Exercises 87 and 88, use the information in the following table.

Food Description McDonald's	Fat (gms)	Carb (gms)	Protein (gms)
Breakfast Burrito	20	21	13
Egg McMuffin	12	27	17

Source: www.shapefit.com/mcdonalds.html

Suppose you ate breakfast at McDonald's during one workweek (Monday–Friday).

87. McDonald's breakfast. If you consumed 123 grams of carbohydrates and 77 grams of protein, how many of each breakfast item did you eat during the workweek? 3 Egg McMuffins and 2 Breakfast Burritos

88. McDonald's breakfast. If you consumed 68 grams of fat and 129 grams of carbohydrates, how many of each breakfast item did you eat during the workweek? 4 Egg McMuffins and 1 Breakfast Burrito

Answers:

54. $\left\{\left(x, \dfrac{-3 - 4x}{7}\right)\right\}$ **55.** $\left\{\left(x, 2 - \dfrac{2}{3}x\right)\right\}$

89. Investment. Last year Mrs. García invested $50,000. She put part of the money in a real estate venture that paid 7.5% for the year and the rest in a small-business venture that returned 12% for the year. The combined income from the two investments for the year totaled $5190. How much did she invest at each rate? $18,000 at 7.5% and $32,000 at 12%

90. Investment. Mr. Sharma invested a total of $30,000 in two ventures for a year. The annual return from one of them was 8%, and the other paid 10.5% for the year. He received a total income of $2550 from both investments. How much did he invest at each rate? $24,000 at 8% and $6000 at 10.5%

91. Tutoring other students. A student earns twice as much per hour for tutoring as she does working at McDougal's. If her average wage is $11.25 per hour, how much does she earn per hour at each job? $7.50 per hour at McDougal's and $15 per hour for tutoring

92. Plumber's earnings. A plumber earns $15 per hour more than her apprentice. For a 40-hour week, their combined earnings were $2200. What is the hourly rate for each? The plumber earns $35 per hour, and her apprentice earns $20 per hour.

93. Mixture problem. An herb that sells for $5.50 per pound is mixed with tea that sells for $3.20 per pound to produce a 100-pound mix that is worth $3.66 per pound. How many pounds of each ingredient does the mix contain? 20 pounds of herb and 80 pounds of tea

94. Mixture problem. A chemist had a solution of 60% acid. She added some distilled water, reducing the acid concentration to 40%. She then added 1 more liter of water to further reduce the acid concentration to 30%. How much of the 30% solution did she then have? 4 liters

95. Airplane speed. With the help of a tail wind, a plane travels 3000 kilometers in 5 hours. The return trip against the wind requires 6 hours. Assume that the direction and the wind speed are constant. Find both the speed of the plane in still air and the wind speed. [*Hint:* Remember that the wind helps the plane in one direction and hinders it in the other.] The speed of the plane is 550 km/hr, and the wind speed is 50 km/hr.

96. Motorboat speed. A motorboat travels up a stream a distance of 12 miles in 2 hours. If the current had been twice as strong, the trip would have taken 3 hours. How long should it take for the return trip down the stream? 1.2 hours

97. Break-even analysis. A local publishing company publishes *City Magazine*. The production and setup costs are $30,000, and the cost of producing each magazine is $2. Each magazine sells for $3.50. Assume that x magazines are published and sold.
   a. Write the total cost function $y = C(x)$ and the revenue function $y = R(x)$. $C(x) = 30,000 + 2x; R(x) = 3.5x$
   b. Graph both functions from part (a) on the same coordinate plane. †
   c. How many magazines must be sold to break even? 20,000 magazines

98. Break-even analysis. An electronic manufacturer plans to make digital portable music players (MP4s). Fixed costs will be $750,000, and it will cost $50 to manufacture each MP4, which will be sold for $125 to the retailers. Assume that x MP4s are manufactured and sold. $C(x) = 750,000 + 50x; R(x) = 125x$
   a. Write the total cost function $y = C(x)$ and the revenue function $y = R(x)$.
   b. Graph both functions from part (a) on the same coordinate plane. †
   c. How many MP4s must be sold to break even? 10,000 MP4s

99. Making a job decision. Shanaysha has job offers from department stores A and B to sell major appliances. Store A offers her a fixed salary of $400 per week. Store B offers her $150 per week plus a commission of 4% of her weekly sales. How much should Shanaysha's weekly sales be for the offer from Store B to be better than that from Store A? Her weekly sales should be more than $6250.

100. Making a job decision. Sheena has job offers from companies A and B to sell insurance. Company A offers her $25,000 per year plus 2% of her yearly sales premiums. Company B offers her $30,000 per year plus 1% of her yearly sales premiums. How much must Sheena earn in yearly sales premiums for the offer from Company B to be the better offer? She must earn less than $500,000 in sales premiums.

C EXERCISES Beyond the Basics

101. Solve for x and y.

$$\begin{cases} \dfrac{40}{x+y} + \dfrac{2}{x-y} = 5 \\ \dfrac{25}{x+y} - \dfrac{2}{x-y} = 1 \end{cases} \quad \left\{ \left(\dfrac{1261}{204}, \dfrac{949}{204} \right) \right\}$$

[*Hint:* Let $u = \dfrac{1}{x+y}$ and $v = \dfrac{1}{x-y}$, solve for u and v, and then solve for x and y.]

102. Solve for x and y. If $xy = 0$, then $\{(0,0)\}$; If $xy \neq 0$, then $\left\{ \left(1, \dfrac{3}{2} \right) \right\}$

$$\begin{cases} 6x + 3y = 7xy \\ 3x + 9y = 11xy \end{cases}$$

[*Hint:* Case (*i*), $xy = 0$; case (*ii*), $xy \neq 0$: divide by xy.]

103. For what value of the constant c does the following system have only one solution?

$$\begin{cases} x + 2y = 7 \\ 3x + 5y = 11 \quad c = 2 \\ cx + 3y = 4 \end{cases}$$

[*Hint:* Solve the first two equations for x and y and then substitute the solution into the third equation.]

104. Repeat Exercise 103 for the following system of equations.

$$\begin{cases} 5x + 7y = 11 \\ 13x + 17y = 19 \quad c = 4 \\ 3x + cy = 5 \end{cases}$$

105. Find the value of c for which the following system of equations represents the same line.

$$\begin{cases} 4x + 7y = 10 \\ 10x + cy = 25 \end{cases} \quad c = \frac{35}{2}$$

106. Find the value of c for which the following system of equations is inconsistent.

$$\begin{cases} 2x + cy = 11 \\ 5x - 7y = 5 \end{cases} \quad c = -\frac{14}{5}$$

107. If $(x - 2)$ is a factor of both $f(x) = x^3 - 4x^2 + ax + b$ and $g(x) = x^3 - ax^2 + bx + 8$, find the values of a and b. $a = 4, b = 0$

108. Let $l_1: a_1x + b_1y + c_1 = 0$ and $l_2: a_2x + b_2y + c_2 = 0$ be two non-parallel lines. Then for any constant k, the line $(a_1x + b_1y + c_1) + k(a_2x + b_2y + c_2) = 0$ is an equation of the line passing through the point of intersection of l_1 and l_2.

In Exercises 109–111, write your answer in slope–intercept form. [*Hint:* Use Exercise 108.]

109. Find an equation of the line that passes through the point of intersection of the lines with equations $2x + y = 3$ and $x - 3y = 12$ and that is parallel to the line with equation $3x + 2y = 8$. $y = -\frac{3}{2}x + \frac{3}{2}$

110. Find an equation of the line that passes through the point of intersection of the lines with equations $5x + 2y = 7$ and $6x - 5y = 38$ and that passes through the point $(1, 3)$. $y = -\frac{7}{2}x + \frac{13}{2}$

111. Find an equation of the line that passes through the point of intersection of the lines with equations $2x + 5y + 7 = 0$ and $13x - 10y + 3 = 0$ and that is perpendicular to the line with equation $7x + 13y = 8$. $y = \frac{13}{7}x + \frac{6}{7}$

Critical Thinking

112. Consider the system of equations

$$\begin{cases} a_1x + b_1y = c_1 \\ a_2x + b_2y = c_2 \end{cases}$$

where $a_1, b_1, c_1, a_2, b_2,$ and c_2 are constants. Find a relationship (an equation) among these constants such that the system of equations has
  a. only one solution. Find the solution in terms of the constants.
  b. no solution.
  c. infinitely many solutions.

113. Let $3x + 4y - 12 = 0$ be the equation of a line l_1 and let $P(5, 8)$ be a point.
  a. Find an equation of a line l_2 passing through the point P and perpendicular to l_1. $3y - 4x = 4$
  b. Find the point Q of the intersection of the two lines l_1 and l_2.
  c. Find the distance $d(P, Q)$. This is the distance from the point P to the line l_1. 7

114. Use the procedure outlined in Exercise 113 to show that the (perpendicular) distance from a point $P(x_1, y_1)$ to the line $ax + by + c = 0$ is given by $\dfrac{|ax_1 + by_1 + c|}{\sqrt{a^2 + b^2}}$.

115. Use the formula from Exercise 114 to find the distance from the given point P to the given line l.
  a. $P(2, 3), l: x + y - 7 = 0$ $\sqrt{2}$
  b. $P(-2, 5), l: 2x - y + 3 = 0$ $\dfrac{6\sqrt{5}}{5}$
  c. $P(3, 4), l: 5x - 2y - 7 = 0$ 0
  d. $P(0, 0), l: ax + by + c = 0$ $\dfrac{|c|}{\sqrt{a^2 + b^2}}$ if $a \neq 0$ or $b \neq 0$

Answers:

112. a. $x = \dfrac{b_2c_1 - b_1c_2}{a_1b_2 - a_2b_1}, y = \dfrac{a_1c_2 - a_2c_1}{a_1b_2 - a_2b_1}$, where $a_1b_2 - a_2b_1 \neq 0$

  b. $\dfrac{c_1}{b_1} \neq \dfrac{c_2}{b_2}, \dfrac{a_1}{a_2} = \dfrac{b_1}{b_2}$ **c.** $\dfrac{a_1}{a_2} = \dfrac{b_1}{b_2} = \dfrac{c_1}{c_2}$

113. b. $\left(\dfrac{4}{5}, \dfrac{12}{5}\right)$

114. a. $b(x - x_1) - a(y - y_1) = 0$

  b. $\left(\dfrac{b^2x_1 - aby_1 - ac}{a^2 + b^2}, \dfrac{-abx_1 + a^2y_1 - bc}{a^2 + b^2}\right)$

Systems of Linear Equations in Three Variables

Before Starting this Section, Review

1. Solving systems of linear equations (Section 7.1, page 481)

2. Solving a system of equations in two variables (Section 7.1, page 483)

Objectives

1. Solve a system of linear equations in three variables.

2. Classify systems as consistent and inconsistent.

3. Solve nonsquare systems.

4. Interpret linear systems in three variables geometrically.

5. Use linear systems in applications.

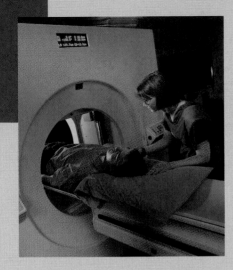

Cross section with tumor

FIGURE 7.6 Cross section with tumor

CAT SCANS

Computerized axial tomography imaging is more commonly known as a "CAT scan" or "CT scan." Tomography is from the Greek word *tomos*, meaning "slice or section," and *graphia*, meaning "describing." The CAT scan machine was invented in 1972 by British engineer Godfrey Hounsfield and, independently, by physicist Allan Cormack of Tufts University in Massachusetts. Hounsfield never attended a university, but he started experimenting with electrical and mechanical devices as a boy. He shared a Nobel Prize in 1979 with Cormack for inventing the scanner technology and was honored with knighthood in England for his contributions to medicine and science. CAT scanners use X-rays measured outside the patient's body, together with algebraic computations, to construct pictures of the patient's internal body organs, including the brain. The scanner does not "take" any pictures; instead, it *computes* pictures. The idea behind the CAT scanner is simple. When an X-ray passes through an object, it loses intensity. The denser the object, the more intensity the X-ray loses. Since tumors are denser than healthy tissues, an unexpected loss of intensity may indicate the presence of a tumor. To obtain sufficient information, a large number of properly arranged X-rays is needed. Imagine a cross section of body tissue with a superimposed mathematical grid. Each region bounded by the grid lines is called a *grid cell*. In an actual CAT scan, each grid cell is 2 square millimeters and contains 10,000 or more biological cells. The grid shown in Figure 7.6 would require many X-rays. Actual CAT scanners work with grids that have 512 $\times$ 512 grid cells and require hundreds of thousands of X-ray beams. In Example 6, we discuss a simple situation in which X-rays pass through 3 grid cells. ■

Adapted with permission from *Mathematics Teacher,* May 1996. © 1996 by the National Council of Teachers of Mathematics.

1. Solve a system of linear equations in three variables.

Systems of Linear Equations

A **linear equation in the variables** $x_1, x_2, \ldots, x_n$ is an equation that can be written in the form

$$a_1x_1 + a_2x_2 + \cdots + a_nx_n = b,$$

where b and the *coefficients* $a_1, a_2, \ldots, a_n$ are real numbers. The subscript n can be any positive integer. When n is small—say, between 2 and 5 (as in most textbook examples and exercises)—we normally use $x, y, z, u,$ and v instead of $x_1, x_2, x_3, x_4,$ and x_5. However, in real-life applications, n might be 100 or 1000 or even greater.

A **system of linear equations** (or a **linear system**) in three variables is a collection of two or more linear equations involving the same variables. For example,

$$\begin{cases} x + 3y + z = 0 \\ 2x - y + z = 5 \\ 3x - 3y + 2z = 10 \end{cases}$$

is a system of three linear equations in the three variables $x, y,$ and z. An ordered triple (a, b, c) is a **solution** of a system of three linear equations in three variables $x, y,$ and z if each equation in the system is a true statement when $a, b,$ and c are substituted for $x, y,$ and z, respectively.

EXAMPLE 1   Verifying a Solution

Determine whether the ordered triple $(2, -1, 3)$ is a solution of the following linear system.

$$\begin{cases} 5x + 3y - 2z = 1 & (1) \\ x - y + z = 6 & (2) \\ 2x + 2y - z = -1 & (3) \end{cases}$$

SOLUTION

We replace x with 2, y with -1, and z with 3 in all three equations.

Equation (1)	Equation (2)	Equation (3)
$5x + 3y - 2z = 1$	$x - y + z = 6$	$2x + 2y - z = -1$
$5(2) + 3(-1) - 2(3) \stackrel{?}{=} 1$	$2 - (-1) + 3 \stackrel{?}{=} 6$	$2(2) + 2(-1) - (3) \stackrel{?}{=} -1$
$10 - 3 - 6 \stackrel{?}{=} 1$	$2 + 1 + 3 \stackrel{?}{=} 6$	$4 - 2 - 3 \stackrel{?}{=} -1$
$1 = 1$ ✓	$6 = 6$ ✓	$-1 = -1$ ✓

Since the ordered triple $(2, -1, 3)$ satisfies all three equations, it is a solution of the system. ■ ■ ■

Practice Problem 1 Determine whether the ordered triple $(2, -3, 2)$ is a solution of the following linear system.

$$\begin{cases} x + y + z = 1 \\ 3x + 4y + z = -4 \\ 2x + y + 2z = 5 \end{cases}$$

■

We will now use a procedure called *Gaussian elimination* to convert a system of three equations in the variables $x, y,$ and z into an equivalent system in *triangular form* with leading coefficient 1, as shown below.

$$\begin{cases} x + ay + bz = p & (1) \\ y + cz = q & (2) \\ z = z_0 & (3) \end{cases}$$

We can then solve the system by back-substitution. To get a system in triangular form, we use the following three operations to obtain equivalent systems (systems with the same solution set as the original system).

OPERATIONS THAT PRODUCE EQUIVALENT SYSTEMS

1. Interchange the position of any two equations.
2. Multiply any equation by a nonzero constant.
3. Add a nonzero multiple of one equation to another.

FINDING THE SOLUTION: A PROCEDURE

EXAMPLE 2   **The Gaussian Elimination Method**

OBJECTIVE

Solve a system of three equations in three variables by first converting the system to the form

$$\begin{cases} x + ay + bz = p \\ \quad\ y + cz = q \\ \qquad\quad z = z_0 \end{cases}$$

EXAMPLE

Solve the system of equations.

$$\begin{cases} \qquad\ 3y - z = 5 & (1) \\ 2x + y + z = 9 & (2) \\ x + 2y + 2z = 3 & (3) \end{cases}$$

Step 1 Rearrange the equations, if necessary, to obtain an x-term with a nonzero coefficient in the first equation. Then multiply the first equation by the reciprocal of the coefficient of the x-term to get 1 as a leading coefficient.

Interchange equations (1) and (3).

$$\begin{cases} x + 2y + 2z = 3 & (3) \\ 2x + y + z = 9 & (2) \\ \qquad\ 3y - z = 5 & (1) \end{cases}$$

The first equation (equation (3)) already has a leading coefficient of 1.

Step 2 By adding appropriate multiples of the first equation, eliminate any x-terms from the second and third equations. Then multiply the resulting second equation by the reciprocal of the coefficient of the y-term to get 1 as a leading coefficient.

To eliminate x from equation (2), add -2 times equation (3) to equation (2).

$$\begin{array}{llll} -2x - 4y - 4z = -6 & & -2(x + 2y + 3z = 3) \\ \underline{\ 2x + y + z = \quad 9} & (2) \\ \qquad -3y - 3z = \quad 3 & (4) & \text{Add.} \end{array}$$

$$\begin{cases} x + 2y + 2z = 3 & (3) \\ \qquad\ y + z = -1 & (5) \quad \text{Multiply equation (4) by } -\dfrac{1}{3}. \\ \qquad\ 3y - z = 5 & (1) \end{cases}$$

Step 3 Add an appropriate multiple of the second equation from Step 2 to eliminate any y-term from the third equation. Then solve the resulting equation for z.

To eliminate y from equation (1), add -3 times equation (5) to equation (1).

$$\begin{array}{lll} -3y - 3z = \quad 3 & & -3(y + z = -1) \\ \underline{\ 3y - z = \quad 5} & (1) \\ \qquad -4z = \quad 8 & & \text{Add.} \\ \qquad\quad z = -2 & & \text{Solve for } z. \end{array}$$

Step 4 Back-substitute the value of z from Step 3 into one of the equations in Step 3 that contains only y and z and solve for y.

$$\begin{array}{ll} 3y - z = 5 & \text{Equation (1) from Step 3} \\ 3y - (-2) = 5 & \text{Substitute } z = -2. \\ 3y + 2 = 5 & \text{Simplify} \\ y = 1 & \text{Solve for } y. \end{array}$$

continued on the next page

Step 5 Back-substitute the values of y and z from Steps 3 and 4 in any equation containing x, y, and z and solve for x.

$$x + 2y + 2z = 3 \qquad \text{Equation (3) from Step 1}$$
$$x + 2(1) + 2(-2) = 3 \qquad \text{Substitute } y = 1, z = -2.$$
$$x = 5 \qquad \text{Solve for } x.$$

Step 6 Write the solution set.

The solution set is $\{(5, 1, -2)\}$.

Step 7 Check your answer in the original equations.

Check: Verify the solution by substituting $x = 5$, $y = 1$, and $z = -2$ in each of the original equations.

$$3(1) - (-2) = 5 \; \checkmark \quad (1)$$
$$2(5) + (1) + (-2) = 9 \; \checkmark \quad (2)$$
$$5 + 2(1) + 2(-2) = 3 \; \checkmark \quad (3) \qquad ■ ■ ■$$

Practice Problem 2 Solve the system.

$$\begin{cases} 2x + 5y \quad\;\; = 1 \\ x - 3y + 2z = 1 \\ -x + 2y + \;\; z = 7 \end{cases}$$

■

2 Classify systems as consistent and inconsistent.

Number of Solutions of a Linear System

As in the case of two variables, a system of linear equations in three variables may be **consistent** (having at least one solution) or **inconsistent** (having no solution). The equations in a consistent system may be **dependent** (the system has infinitely many solutions) or **independent** (the system has one solution).

INCONSISTENT SYSTEM

If in the process of converting a linear system to triangular form an equation of the form $0 = a$ occurs, where $a \neq 0$, then the system has no solution and is *inconsistent*.

EXAMPLE 3   **Attempting to Solve a Linear System with No Solution**

Solve the system of equations.

$$\begin{cases} x - \;\; y + 2z = \;\; 5 & (1) \\ 2x + \;\; y + \;\; z = \;\; 7 & (2) \\ 3x - 2y + 5z = 20 & (3) \end{cases}$$

SOLUTION

Steps 1–2 To eliminate x from equation (2), add -2 times equation (1) to equation (2).

$$\begin{array}{ll} -2x + 2y - 4z = -10 & -2(x - y + 2z = 5) \\ \underline{2x + \;\; y + \;\; z = \quad 7} & (2) \\ 3y - 3z = \;\; -3 & (4) \quad \text{Add.} \end{array}$$

Next, we eliminate x using equations (1) and (3).

$$\begin{array}{ll} -3x + 3y - 6z = -15 & -3(x - y + 2z = 5) \\ \underline{3x - 2y + 5z = \quad 20} & (3) \\ y - \;\; z = \quad 5 & (5) \quad \text{Add.} \end{array}$$

We now have the system:

$$\begin{cases} x - y + 2z = 5 & (1) \\ 3y - 3z = -3 & (4) \\ y - z = 5 & (5) \end{cases}$$

STUDY TIP

An alternate method for Step 3 is to solve the 2 × 2 system of equations (4) and (5), then back substitute into equation (1) to solve for the third variable.

Step 3   Multiply equation (4) by $\dfrac{1}{3}$ to obtain

$$y - z = -1 \quad (6)$$

We can eliminate y using equations (5) and (6).

$$\begin{array}{ll} -y + z = 1 & \qquad -1(y - z = 1) \\ \underline{y - z = 5} & (5) \\ 0 = 6 & (7) \qquad \text{Add.} \end{array}$$

We now have the system in triangular form:

$$\begin{cases} x - y + 2z = 5 & (1) \\ y - z = -1 & (4) \\ \underbrace{0 = 6}_{\text{False}} & (7) \end{cases}$$

This system is equivalent to the original system. Because equation (7) is false, we conclude that the solution set of the system is $\varnothing$ and the system is inconsistent. ■ ■ ■

Practice Problem 3   Solve the system of equations.

$$\begin{cases} 2x + 2y + 2z = 12 \\ -3x + y - 11z = -6 \\ 2x + y + 4z = -8 \end{cases}$$ ▨

DEPENDENT EQUATIONS

If in the process of converting a linear system to triangular form,

  (i) an equation of the form $0 = a(a \neq 0)$ *does not* occur but

  (ii) an equation of the form $0 = 0$ *does* occur, then the system of equations has infinitely many solutions and the equations are *dependent*.

EXAMPLE 4   **Solving a System with Infinitely Many Solutions**

Solve the system of equations.

$$\begin{cases} x - y + z = 7 & (1) \\ 3x + 2y - 12z = 11 & (2) \\ 4x + y - 11z = 18 & (3) \end{cases}$$

SOLUTION

Steps 1–2   Eliminate x using equations (1) and (2).

$$\begin{array}{ll} -3x + 3y - 3z = -21 & \qquad -3(x - y + z = 7) \\ \underline{3x + 2y - 12z = 11} & (2) \\ 5y - 15z = -10 & (4) \qquad \text{Add.} \end{array}$$

Eliminate x using equations (1) and (3).

$$
\begin{array}{rll}
-4x + 4y - 4z = -28 & \qquad & -4(x - y + z = 7) \\
\underline{4x + y - 11z = 18} & \quad (3) & \\
5y - 15z = -10 & \quad (5) & \text{Add.}
\end{array}
$$

We now have the following equivalent system:

$$
\begin{cases}
x - y + z = 7 & (1) \\
5y - 15z = -10 & (4) \\
5y - 15z = -10 & (5)
\end{cases}
$$

Step 3 We eliminate y using equations (4) and (5).

$$
\begin{array}{rll}
-5y + 15z = 10 & \qquad & -1(5y - 15z = -10) \\
\underline{5y - 15z = -10} & \quad (5) & \\
0 = 0 & \quad (6) & \text{Add.}
\end{array}
$$

We finally have the system in triangular form:

$$
\begin{cases}
x - y + z = 7 & (1) \\
5y - 15z = -10 & (4) \\
\underbrace{0 = 0}_{\text{True}} & (6)
\end{cases}
$$

The equation $0 = 0$ is equivalent to $0z = 0$, which is true for every value of z. Solving equation (4) for y, we have $y = 3z - 2$. Substituting $y = 3z - 2$ into equation (1) and solving for x, we obtain

$$
\begin{array}{ll}
x = y - z + 7 & \text{Solve equation (1) for } x. \\
x = (3z - 2) - z + 7 & \text{Substitute } y = 3z - 2. \\
x = 2z + 5 & \text{Simplify.}
\end{array}
$$

Thus, for every real number z, the triple $(x, y, z) = (2z + 5, 3z - 2, z)$ is a solution of the system. For example, when $z = 0$, the triple $(5, -2, 0)$ is a solution. When $z = 1$, the triple $(7, 1, 1)$ is a solution. Therefore, the given system has infinitely many solutions, one for each real number z, and the equations are dependent. The solution set for the system is $\{(2z + 5, 3z - 2, z)\}$. ■■■

Practice Problem 4 Solve the system of equations.

$$
\begin{cases}
x + y + z = 5 \\
-4x - y - 8z = -29 \\
2x + 5y - 2z = 1
\end{cases}
$$
■

3 Solve nonsquare systems.

Nonsquare Systems

Sometimes in a linear system, the number of equations is not the same as the number of variables. Such systems are called **nonsquare linear systems**.

EXAMPLE 5 **Solving a Nonsquare Linear System**

Solve the system of linear equations.

$$
\begin{cases}
x + 5y + 3z = 7 & (1) \\
2x + 11y - 4z = 6 & (2)
\end{cases}
$$

SOLUTION

Steps 1–2   We eliminate x using equations (1) and (2).

$$
\begin{array}{rl}
-2x - 10y - 6z = -14 & \qquad -2(x + 5y + 3z = 7) \\
\underline{2x + 11y - 4z = 6} & \qquad (2) \\
y - 10z = -8 & \qquad (3) \qquad \text{Add.}
\end{array}
$$

We obtain the equivalent system:

$$
\begin{cases}
x + 5y + 3z = 7 & (1) \\
y - 10z = -8 & (3)
\end{cases}
$$

Steps 3–4   Since there is no third equation, Steps 3 and 4 are not needed.

Step 5   Solve equation (3) for y in terms of z to obtain

$$y = 10z - 8.$$

Step 6   Back-substitute $y = 10z - 8$ into equation (1) and solve for x.

$$
\begin{array}{ll}
x + 5y + 3z = 7 & \text{Equation (1)} \\
x + 5(10z - 8) + 3z = 7 & \text{Replace } y \text{ with } 10z - 8. \\
x = -53z + 47 & \text{Solve for } x.
\end{array}
$$

Steps 5 and 6 mean that z determines both the x and y values. The system has infinitely many solutions, one for each real number z. The solution set is $\{(-53z + 47, 10z - 8, z)\}$.   ■ ■ ■

Practice Problem 5   Solve the system of linear equations.

$$
\begin{cases}
x + 3y + 2z = 4 \\
2x + 7y - z = 5
\end{cases}
$$
    ■

4 Interpret linear systems in three variables geometrically.

Geometric Interpretation

The graph of a linear equation in three variables, such as $ax + by + cz = d$ (where a, b, and c are not all zero), is a plane in three-dimensional space. Figure 7.7 shows possible situations for a system of three linear equations in three variables.

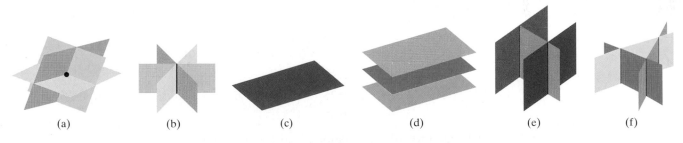

(a)   (b)   (c)   (d)   (e)   (f)

FIGURE 7.7

a. Three planes intersect in a single point. The system has only one solution.

b. Three planes intersect in one line. The system has infinitely many solutions.

c. Three planes coincide with each other. The system has infinitely many solutions.

d. There are three parallel planes. The system has no solution.

e. Two parallel planes are intersected by a third plane. The system has no solution.

f. Three planes have no point in common. The system has no solution.

5 Use linear systems in applications.

An Application to CAT Scans

CAT scanners report X-ray data in units called *linear attenuation units* (LAUs), also called *Hounsfield numbers*. The data are reported in such a way that if an X-ray passes through consecutive grid cells, the total weakening of the beam is the sum of the individual decreases. Table 7.1 gives the range of LAUs that determine whether the grid cell contains healthy tissue, tumorous tissue, a bone, or a metal.

TABLE 7.1 CAT Scanner Ranges

Type of Tissue	LAU values
Healthy tissue	0.1625–0.2977
Tumorous tissue	0.2679–0.3930
Bone	0.38757–0.5108
Metal	1.54–∞

Reprinted with permission from *Mathematics Teacher*, May 1996. © 1996 by the National Council of Teachers of Mathematics.

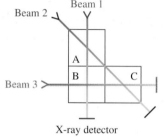

Beam 2   Beam 1

Beam 3

A B C

X-ray detector

Three grid cells, three X-rays

FIGURE 7.8 Reprinted with permission from *Mathematics Teacher*, May 1996. © 1996 by the National Council of Teachers of Mathematics.

EXAMPLE 6   **A CAT Scan with Three Grid Cells**

Let A, B, and C be three grid cells as shown in Figure 7.8. A CAT scanner reports the following data for a patient named Monica.

(i) Beam 1 is weakened by 0.80 unit as it passes through grid cells A and B.

(ii) Beam 2 is weakened by 0.55 unit as it passes through grid cells A and C.

(iii) Beam 3 is weakened by 0.65 unit as it passes through grid cells B and C.

Use Table 7.1 to determine which grid cells contain each type of tissue listed.

SOLUTION

Suppose grid cell A weakens the beam by x units, grid cell B weakens it by y units, and grid cell C weakens it by z units. Then we have the system:

$$\begin{cases} x + y \quad\quad = 0.80 & (1) \quad (\text{Beam 1}) \\ x \quad\quad + z = 0.55 & (2) \quad (\text{Beam 2}) \\ \quad\quad y + z = 0.65 & (3) \quad (\text{Beam 3}) \end{cases}$$

To solve this system of equations, we use the elimination procedure. Adding -1 times equation (1) to equation (2), we obtain the equivalent system:

$$\begin{cases} x + \quad y \quad\quad = \quad 0.80 & (1) \\ \quad -y + z = -0.25 & (4) \quad -1 \text{ times equation (1) plus equation (2)} \\ \quad\quad y + z = \quad 0.65 & (3) \end{cases}$$

Add equation (4) to equation (3) to get the system

$$\begin{cases} x + \quad y \quad\quad = \quad 0.80 & (1) \\ \quad -y + \quad z = -0.25 & (4) \\ \quad\quad 2z = \quad 0.40 & (5) \quad \text{Equation (4) plus equation (3)} \end{cases}$$

Multiply equation (5) by $\frac{1}{2}$ to obtain $z = 0.20$. Then back-substitute $z = 0.20$ in equation (4) to get $-y + 0.20 = -0.25$, or $y = 0.45$. Next, back-substitute $y = 0.45$ in equation (1) to get $x + 0.45 = 0.80$. Solving for x gives $x = 0.35$. Now referring to Table 7.1, we conclude that cell A contains tumorous tissue (because $x = 0.35$), cell B contains a bone (because $y = 0.45$), and cell C contains healthy tissue (because $z = 0.20$). ■ ■ ■

Practice Problem 6 In Example 6, suppose **(i)** Beam 1 is weakened by 0.65 unit as it passes through grid cells A and B, **(ii)** Beam 2 is weakened by 0.70 unit as it passes through grid cells A and C, and **(iii)** Beam 3 is weakened by 0.55 unit as it passes through grid cells B and C. Use Table 7.1 to determine which grid cells contain each type of tissue listed. ■

SECTION 7.2 ■ Exercises

A EXERCISES Basic Skills and Concepts

1. Systems of equations that have the same solution sets are called ___equivalent___ systems.

2. Adding a(n) ___constant___ multiple of one equation to another equation in the system produces an equivalent system.

3. If any of the equations in a system has no solution, then the system is ___inconsistent___.

4. If the number of equations is different from the number of variables in a linear system, then the system is called a(n) ___nonsquare___ system.

5. *True or False* The graph of the equation $x - 2y + z = 1$ is a plane in three-dimensional space. True

6. *True or False* If $\{(2z, 1 - z, z + 3)\}$ is the solution set of a linear system in three variables, then $(0, 1, 3)$ is a solution of this system. True

In Exercises 7–10, determine whether the given ordered triple (x, y, z) is a solution of the given linear system.

7. $\begin{cases} 2x - 2y - 3z = 1 \\ \quad\;\; 3y + 2z = -1 \\ \quad\;\; y + z = 0 \end{cases}$ $(1, -1, 1)$ Yes

8. $\begin{cases} 3x - 2y + z = 2 \\ \quad\;\; y + z = 5 \\ x \quad\quad + 3z = 8 \end{cases}$ $(2, 3, 2)$ Yes

9. $\begin{cases} x + 3y - 2z = 0 \\ 2x - y + 4z = 5 \\ x - 11y + 14z = 0 \end{cases}$ $(-10, 8, 7)$ No

10. $\begin{cases} x - 4y + 7z = 14 \\ 3x + 8y - 2z = 13 \\ 7x - 8y + 26z = 5 \end{cases}$ $(4, 1, 2)$ No

In Exercises 11–14, solve each triangular linear system.

11. $\begin{aligned} x + y + z &= 4 \\ y - 2z &= 4 \\ z &= -1 \end{aligned}$ $\{(3, 2, -1)\}$

12. $\begin{aligned} x + 3y + z &= 3 \\ y + 2z &= 8 \\ z &= 5 \end{aligned}$ $\{(4, -2, 5)\}$

13. $\begin{aligned} x - 5y + 3z &= -1 \\ y - 2z &= -6 \\ z &= 4 \end{aligned}$ $\{(-3, 2, 4)\}$

14. $\begin{aligned} x + 3y + 5z &= 0 \\ y + 7z &= 2 \\ z &= \frac{1}{2} \end{aligned}$ $\left\{\left(2, -\frac{3}{2}, \frac{1}{2}\right)\right\}$

15. In the system in Exercise 7, interchange equations (2) and (3). Write the new equivalent system. In the new system, eliminate y from the last equation. Convert the system to triangular form. †

16. In Exercise 8, interchange equations (1) and (3). In the new equivalent system, eliminate x from the last equation. Convert the system to triangular form. †

17. Convert the system of Exercise 9 to triangular form. †

18. Convert the system of Exercise 10 to triangular form. †

In Exercises 19–22, convert the system to triangular form and then use back-substitution to solve the system.

19. $\begin{cases} x - y + z = 6 \\ \quad\;\; 2y + 3z = 5 \\ \quad\quad\;\; 2z = 6 \end{cases}$ $\{(1, -2, 3)\}$

20. $\begin{cases} 4x + 5y + 2z = -3 \\ \quad\;\; 3y - z = 14 \\ \quad\quad\; -3z = 15 \end{cases}$ $\{(-2, 3, -5)\}$

21. $\begin{cases} 4x + 4y + 4z = 7 \\ 3x - 8y \quad\quad = 14 \\ \quad\quad\quad\;\; 4z = -1 \end{cases}$ $\left\{\left(\frac{30}{11}, -\frac{8}{11}, -\frac{1}{4}\right)\right\}$

22. $\begin{cases} 5x + 10y + 10z = 0 \\ \quad\;\; 2y + 3z = -0.6 \\ \quad\quad\quad\; 4z = 1.6 \end{cases}$ $\{(1, -0.9, 0.4)\}$

In Exercises 23–44, find the solution set of each linear system. Identify inconsistent systems and dependent equations. For dependent equations, write your answer as in Example 4.

23. $\begin{cases} x + y + z = 6 \\ x - y + z = 2 \\ 2x + y - z = 1 \end{cases}$ $\{(1, 2, 3)\}$

24. $\begin{cases} x + y + z = 6 \\ 2x + 3y - z = 5 \\ 3x - 2y + 3z = 8 \end{cases}$ $\{(1, 2, 3)\}$

25. $\begin{cases} 2x + 3y + z = 9 \\ x + 2y + 3z = 6 \\ 3x + y + 2z = 8 \end{cases}$ †

26. $\begin{cases} 4x + 2y + 3z = 6 \\ x + 2y + 2z = 1 \\ 2x - y + z = -1 \end{cases}$ $\{(3, 3, -4)\}$

27. $\begin{cases} x - 4y + 7z = 14 \\ 3x + 8y - 2z = 13 \\ 7x - 8y + 26z = 5 \end{cases}$ Inconsistent

28. $\begin{cases} x + y + 4z = 6 \\ x + 2y - 2z = 8 \\ 7x + 10y + 10z = 60 \end{cases}$ Inconsistent

29. $\begin{cases} 2x + 3y + 2z = 7 \\ x + 3y - z = -2 \\ x - y + 2z = 8 \end{cases}$ $\{(3, -1, 2)\}$

30. $\begin{cases} x - y + 2z = 3 \\ 2x + 2y + z = 3 \\ x + y + 3z = 4 \end{cases}$ $\{(1, 0, 1)\}$

†Due to space constrictions, answers to these exercises may be found in the Answers beginning on page A–1 in the back of the book.

31. $\begin{cases} 4x - 2y + z = 5 \\ 2x + y - 2z = 4 \\ x + 3y - 2z = 6 \end{cases}$ $\{(2,2,1)\}$

32. $\begin{cases} x - 3y + 2z = 9 \\ 2x + 4y - 3z = -9 \\ 3x - 2y + 5z = 12 \end{cases}$ $\{(1,-2,1)\}$

33. $\begin{cases} 2x + y + z = 6 \\ x + y - z = 1 \\ x + y + 2z = 4 \end{cases}$ $\{(3,-1,1)\}$

34. $\begin{cases} 2x + y - 3z = 7 \\ x - y - 2z = 4 \\ 3x + 3y + 2z = 4 \end{cases}$ $\{(2,0,-1)\}$

35. $\begin{cases} x - y - z = 3 \\ x + 9y + z = 3 \\ 2x + 3y - z = 6 \end{cases}$ †

36. $\begin{cases} x + y - z = 2 \\ 3x - y - z = 10 \\ 3x + y - 2z = 8 \end{cases}$ †

37. $\begin{cases} x + y = 0 \\ y + 2z = -4 \\ y + z = 4 \end{cases}$ $\{(12,-12,4)\}$

38. $\begin{cases} 2x + 4y + 3z = 6 \\ x + 2z = -1 \\ x - 2y + z = -5 \end{cases}$ $\{(-1,2,0)\}$

39. $\begin{cases} 2y - z = -4 \\ x + z = 3 \\ 2x + 3y = -1 \end{cases}$ $\{(1,-1,2)\}$

40. $\begin{cases} x + y = 9 \\ 2y + 3z = 7 \\ x - 2z = 4 \end{cases}$ $\{(10,-1,3)\}$

41. $\begin{cases} 3x - 2z = 11 \\ 2x + y = 8 \\ 2y + 3z = 1 \end{cases}$ $\{(3,2,-1)\}$

42. $\begin{cases} 2x + y = 4 \\ x + 2z = 3 \\ 3y - z = 5 \end{cases}$ $\{(1,2,1)\}$

43. $\begin{cases} 2x + 6y + 11 = 0 \\ 6y - 18z + 1 = 0 \end{cases}$ †

44. $\begin{cases} 3x + 5y - 15 = 0 \\ 6x + 20y - 6z = 11 \end{cases}$ Dependent; $\left\{ \left(\dfrac{49}{6} - z, \dfrac{-19}{10} + \dfrac{3}{5}z, z \right) \right\}$

B EXERCISES Applying the Concepts

In Exercises 45–52, use a system of equations to solve each problem.

45. Investment. Miguel invested $20,000 in three different funds that paid 4%, 5%, and 6% for the year. The total income for the year from the three funds was $1060. The income from the 6% fund was twice the income from the 5% fund. What amount was invested in each fund?

46. Number problem. The sum of the digits in a three-digit number is 14. The sum of the hundreds digit and the units digit is equal to the tens digit. If the hundreds digit and the units digit are interchanged, the number is increased by 297. What is the original number? 275

47. Election campaign. Alex, Becky, and Courtney volunteered to stuff 741 envelopes with newsletters for Senator Douglas's reelection campaign. Alex could assemble 124 per hour; Becky, 118 per hour; and Courtney, 132 per hour. They worked a total of 6 hours. The sum of the number of hours that Becky and Courtney spent was twice what Alex spent. How long did each of them work? †

48. Age of students. A college algebra class of 38 students at Central State College was made up of people who were 18, 19, and 20 years of age. The average of their ages was 18.5 years. How many of each age were in the class if the number of 18-year-olds was eight more than the combined number of 19- and 20-year-olds? †

49. Coins in a machine. A vending machine's coin box contains nickels, dimes, and quarters. The total number of coins in the box is 300. The number of dimes is three times the number of nickels and quarters together. If the box contains $30.05, find the number of nickels, dimes, and quarters that it contains. 56 nickels, 225 dimes, and 19 quarters

50. Sports. The Wildcats scored 46 points in a football game. Twice the number of points resulting from the sum of field goals and extra points equals two more than the number of points from touchdowns. Five times the number of points scored by field goals equals twice the number of points from touchdowns. Find the number of points resulting from touchdowns, field goals, and extra points. [A touchdown = 6 points, a field goal = 3 points, and an extra point = 1 point.] 5 touchdowns, 4 field goals, and 4 extra points

51. Weekly wage. Amy worked 53 hours one week and was paid at three different rates. She earned $7.40 per hour for her normal daytime work, $9.20 per hour for night work, and $11.75 per hour for holiday work. If her total gross wages for the week were $452.20 and the number of regular daytime hours she worked exceeded the combined hours of night and holiday work by 9 hours, how many hours of each category of work did Amy perform? 31 hours for normal daytime work, 14 hours at night, and 8 hours on a holiday

52. Components of a product. A manufacturer buys three components— A, B, and C —for use in making a toaster. She used as many units of A as she did of B and C combined. The cost of A, B, and C is $4, $5, and $6 per unit, respectively. If she purchased 100 units of these components for a total of $480, how many units of each did she purchase? 50 units at $4, 20 units at $5, and 30 units at $6

In Exercises 53–56, use Table 7.1 and Figure 7.8 to determine which grid cells of the patients contain healthy tissue, tumorous tissue, bone, or metal.

Beam decrease, in LAUs

	Patient	Beam 1	Beam 2	Beam 3
53.	Vicky	0.54	0.40	0.52 †
54.	Yolanda	0.65	0.80	0.75 †
55.	Nina	0.51	0.49	0.44 †
56.	Srinivasan	0.44	2.21	2.23 †

C EXERCISES Beyond the Basics

In Exercises 57–60, write a linear equation of the form $x + by + cz = d$ that is satisfied by all three of the given ordered triples.

57. $(1,0,0), (0,1,0), (0,0,1)$ $x + y + z = 1$

58. $\left(\dfrac{1}{3}, 0, 0 \right), (0,4,3), (1,2,2)$ $x + \dfrac{4}{3}y - \dfrac{5}{3}z = \dfrac{1}{3}$

59. $(3,-4,0), \left(0, \dfrac{1}{4}, \dfrac{1}{2} \right), (1,1,-4)$ $x + \dfrac{2}{3}y + \dfrac{1}{3}z = \dfrac{1}{3}$

60. $(0,1,-10), \left(\dfrac{1}{8}, 0, \dfrac{1}{4} \right), \left(1, \dfrac{1}{3}, -2 \right)$ $x - \dfrac{111}{14}y - \dfrac{11}{14}z = -\dfrac{1}{14}$

Answers:
45. $4000 invested at 4%, $6000 invested at 5%, and $10,000 invested at 6%

In Exercises 61–64, find an equation of the parabola of the form $y = ax^2 + bx + c$ **that passes through the three given points.**

61. $(0, 1), (-1, 0), (1, 4)$ $y = x^2 + 2x + 1$

62. $(0, 2), (-1, 30), (2, 6)$ $y = 10x^2 - 18x + 2$

63. $(1, 2), (-1, 4), (2, 4)$ $y = x^2 - x + 2$

64. $(0, 3), (-1, 4), (1, 6)$ $y = 2x^2 + x + 3$

In Exercises 65–68, find an equation of the circle of the form $x^2 + y^2 + ax + by + c = 0$ **that passes through the three given points.**

65. $(0, 4), (2\sqrt{2}, 2\sqrt{2}), (-4, 0)$ $x^2 + y^2 - 16 = 0$

66. $(0, 3), (0, -1), (\sqrt{3}, 2)$ $x^2 + y^2 - 2y - 3 = 0$

67. $(1, 2), (6, -3), (4, 1)$ $x^2 + y^2 - 2x + 6y - 15 = 0$

68. $(5, 6), (-1, 6), (3, 2)$ $x^2 + y^2 - 4x - 10y + 19 = 0$

In Exercises 69 and 70, solve the system of equations.

69.
$$\begin{cases} \dfrac{1}{x} + \dfrac{3}{y} - \dfrac{1}{z} = 5 \\ \dfrac{2}{x} + \dfrac{4}{y} + \dfrac{6}{z} = 4 \\ \dfrac{2}{x} + \dfrac{3}{y} + \dfrac{1}{z} = 3 \end{cases} \left\{ \left(-\dfrac{9}{14}, \dfrac{9}{19}, -\dfrac{9}{2} \right) \right\}$$

70.
$$\begin{cases} \dfrac{1}{x} + \dfrac{2}{y} + \dfrac{3}{z} = 8 \\ \dfrac{2}{x} + \dfrac{5}{y} + \dfrac{9}{z} = 16 \\ \dfrac{3}{x} - \dfrac{4}{y} - \dfrac{5}{z} = 32 \end{cases} \left\{ \left(\dfrac{2}{19}, -\dfrac{2}{3}, 2 \right) \right\}$$

$$\left[\textit{Hint: Let } \dfrac{1}{x} = u, \dfrac{1}{y} = v, \text{ and } \dfrac{1}{z} = w. \text{ Solve for } u, v, \text{ and } w. \right]$$

In Exercises 71 and 72, find the value of c **for which the given system has a unique solution.**

71.
$$\begin{cases} x + 2y - 5z - 9 = 0 \\ 3x - y + 2z - 14 = 0 \\ 2x + 3y - z - 3 = 0 \\ cx - 5y + z + 3 = 0 \end{cases} c = -\dfrac{16}{5}$$

72.
$$\begin{cases} x + 5y + z = 42 \\ 3x + y - 3z = c \\ x - y + 3z = 4 \\ x + y - z = 0 \end{cases} c = -16$$

73. Find a quadratic function of the form $y = ax^2 + bx + c$ whose graph passes through the points $(-1, -1), (0, 5)$, and $(2, 5)$. $y = -2x^2 + 4x + 5$

74. Use Table 7.1 to investigate four grid cells, arranged in a square, as shown in the given figure. Figures (a) and (b) reprinted with permission from *Mathematics Teacher,* May 1996. © 1996 by National Council of Teachers of Mathematics.

a. For Figure (a), the following data were observed.

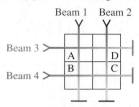

Four grid cells and four X-rays
(a)

1. Beam 1 decreased by 0.60 unit.
2. Beam 2 decreased by 0.75 unit.
3. Beam 3 decreased by 0.65 unit.
4. Beam 4 decreased by 0.70 unit.

Is there sufficient information in Figure (a) to determine which grid cells may contain tumorous tissue?
    No. The system has infinitely many solutions.

b. From two additional X-rays, as shown in Figure (b), the following data were observed.

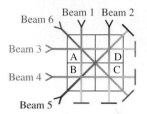

Four grid cells and six X-rays
(b)

5. Beam 5 decreased by 0.85 unit.
6. Beam 6 decreased by 0.50 unit.

Show that the six X-rays represented in the two figures are sufficient to locate tumors in this situation but that, in fact, it is not necessary to use all six rays.

Critical Thinking

75. Write a linear system of three equations in the variables x, y, and z that has $\{(1, -1, 2)\}$ as its solution set.

76. Write a linear system of three equations in the variables x, y, and z such that
a. the system has no solution.
b. the system has infinitely many solutions.

Answers:
74. b. The system has a unique solution. To see that cell C contains tumorous tissue, use the data for beams 2, 4, and 5 to find that the cell C weakens the beam by 0.3 unit.

75. $\begin{cases} x + y - z = -2 \\ 2x - y + 3z = 9 \\ x + y + z = 2 \end{cases}$ Answers may vary.

76. a. $\begin{cases} x + 3y + 3z = 15 \\ x + 2y + z = 1 \\ 2y + 4z = 11 \end{cases}$ Answers may vary.

b. $\begin{cases} x + 2y - z = 4 \\ x + 3y + 2z = 5 \\ y + 3z = 1 \end{cases}$ Answers may vary.

Systems of Linear Inequalities; Linear Programming

Before Starting this Section, Review

1. Intersection of sets (Appendix A, page 754)

2. Graphing a line (Section 1.2, page 21)

3. Solving systems of equations (Section 7.1, page 483)

Objectives

1 Graph a linear inequality in two variables.

2 Graph linear systems of inequalities in two variables.

3 Apply systems of inequalities to linear programming.

George Dantzig 1914–2005

George Dantzig received numerous honors and awards for his work. Dantzig attributed his success in mathematics to his father, who encouraged him to develop his analytic power by making him solve thousands of geometry problems while George was still in high school.

During his first year of graduate studies, Dantzig arrived late in one of Professor Neyman's classes at the University of California at Berkeley. On the blackboard were two problems that Dantzig assumed had been assigned for homework. He submitted his homework a little late. He was surprised to learn that these were the two famous unsolved problems in statistics. The solutions of these problems became part of Dantzig's dissertation for his Ph.D. degree from Berkeley.

THE SUCCESS OF LINEAR PROGRAMMING

From 1941 to 1946, George Dantzig was head of the Combat Analysis Branch, U.S. Army Air Corps Headquarters Statistical Control. During this period, he became an expert on planning methods involving troop supply problems that arose during World War II. With a war on several fronts, the size of the problem of coordinating supplies was daunting. This problem was known as *programming*, a military term which at that time referred to plans or schedules for training, logistical supply, or the deployment of forces. Dantzig mechanized the planning process by introducing *linear programming*, where *programming* has the military meaning.

In recent years, linear programming has been applied to problems in almost all areas of human life. A quick check in the library at a university may uncover entire books on linear programming applied to business, agriculture, city planning, and many other areas. Even feedlot operators use linear programming to decide how much and what kinds of food they should give their livestock each day. In Example 7 and in the exercises, we consider several linear programming problems involving two variables. ■

1 Graph a linear inequality in two variables.

Graph of a Linear Inequality in Two Variables

The statements $x + y > 4, 2x + 3y < 7, y \geq x,$ and $x + y \leq 9$ are examples of linear inequalities in the variables x and y. As with equations, a **solution of an inequality in two variables** x and y is an ordered pair (a, b) that results in a true statement when x is replaced with a and y is replaced with b in the inequality. For example, the ordered pair $(3, 2)$ is a solution of the inequality $4x + 5y > 11$ because $4(3) + 5(2) = 22 > 11$ is a true statement. The ordered pair $(1, 1)$ is *not* a solution of the inequality $4x + 5y > 11$ because $4(1) + 5(1) = 9 > 11$ is a false statement. The set of all solutions of an inequality is called the **solution set of the inequality**. The **graph of an inequality in two variables** is the graph of the solution set of the inequality.

EXAMPLE 1   Graphing a Linear Inequality

Graph the inequality $2x + y > 6$.

SOLUTION

First, graph the *equation* $2x + y = 6$. See Figure 7.9. Notice that if we solve this equation for y, we have $y = -2x + 6$. Any point $P(a, b)$ whose y-coordinate, b, is *greater* than $6 - 2a$ must be above this line. Therefore, the graph of $y > -2x + 6$, which is equivalent to $2x + y > 6$, consists of all points (x, y) in the plane that lie above the line $2x + y = 6$. The graph of the inequality $2x + y > 6$ is shaded in Figure 7.10.

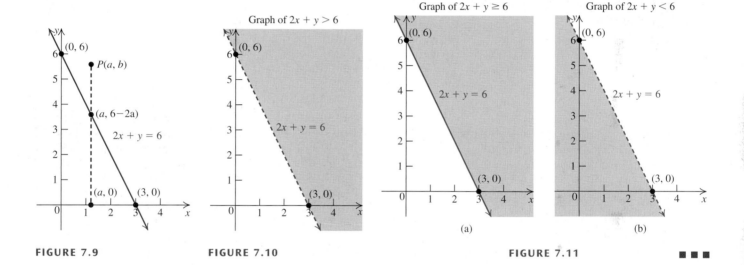

FIGURE 7.9   **FIGURE 7.10**   **FIGURE 7.11**   ▪ ▪ ▪

Practice Problem 1   Graph the inequality $3x + y < 6$.   ▪

In Figure 7.10, the graph of the equation $2x + y = 6$ is drawn as a dashed line to indicate that points on the line are *not* included in the solution set of the inequality $2x + y > 6$. In Figure 7.11, however, the graph of the inequality $2x + y \geq 6$ shows $2x + y = 6$ as a solid line: the solid line indicates that the points on the line *are* included in the solution set. Note that the region below the line $2x + y = 6$ in Figure 7.11(b) is the graph of the inequality $2x + y < 6$.

In general, the graph of an inequality's *corresponding equation* (found by changing the inequality symbol to an equal sign) is a line that separates the plane into two regions, each called a *half plane*. The line itself is called the **boundary** of each region. *All* of the points in one region satisfy the inequality, and *none* of the points in the other region are solutions. You can therefore test one point, called a *test point*, to determine which region represents the solution set.

FINDING THE SOLUTION: A PROCEDURE

EXAMPLE 2 Graphing a Linear Inequality in Two Variables

OBJECTIVE	EXAMPLE

OBJECTIVE

Graph an inequality in two variables.

Step 1 Replace the inequality symbol with an equal ($=$) sign.

Step 2 Sketch the graph of the *corresponding equation* in Step 1. Use a dashed line for the boundary if the given inequality sign is $<$ or $>$ and a solid line if the inequality symbol is $\leq$ or $\geq$.

Step 3 The graph in Step 2 divides the plane into two regions. Select a *test point* in either region, but not on the graph of the equation in Step 1.

Step 4 **(i)** If the coordinates of the test point satisfy the inequality, then so do all of the points in that region. Shade that region.

(ii) If the test point's coordinates do not satisfy the inequality, shade the region that does not contain the test point.

The shaded region (including the boundary if it is solid) is the graph of the inequality.

EXAMPLE

Graph the linear inequality $y \geq -2x + 4$.

$$y = -2x + 4$$

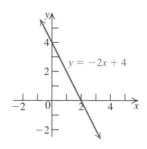

The graph of $y = -2x + 4$.

The line drawn is solid, because the given inequality sign is $\geq$.

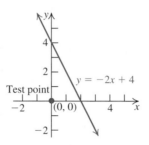

Select $(0, 0)$ as a test point.

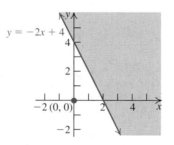

Since $0 \geq -2(0) + 4$ is false, the coordinates of the test point $(0, 0)$ do *not* satisfy the inequality $y \geq -2x + 4$.

The graph of the solution set of $y \geq -2x + 4$ is on the side of the line that does not contain $(0, 0)$. It is shaded in the figure.

■ ■ ■

Practice Problem 2 Graph the linear inequality $y \leq -2x + 4$. ■

EXAMPLE 3 Graphing Inequalities

Sketch the graph of each of the following inequalities.

a. $x \geq 2$ **b.** $y < 3$ **c.** $x + y < 4$

SOLUTION

Steps	Inequality		
1. Change the inequality to an equality.	**a.** $x \geq 2$ $x = 2$	**b.** $y < 3$ $y = 3$	**c.** $x + y < 4$ $x + y = 4$
2. Graph the equation in Step 1 with a dashed line ($<$ or $>$) or a solid line ($\leq$ or $\geq$).			
3. Select a test point.	Test $(0, 0)$ in $x \geq 2$; $0 \geq 2$ is a false statement. The region not containing $(0, 0)$, together with the vertical line, is the solution set.	Test $(0, 0)$ in $y < 3$; $0 < 3$ is a true statement. The region containing $(0, 0)$ is the solution set.	Test $(0, 0)$ in $x + y < 4$; $0 + 0 < 4$ is a true statement. The region containing $(0, 0)$ is the solution set.
4. Shade the solution set.			

Practice Problem 3   Graph the inequality $x + y > 9$.

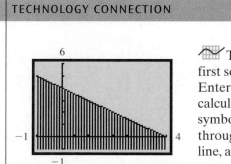

> **TECHNOLOGY CONNECTION**
>
> To shade a graph on a graphing calculator, first solve the corresponding equation for y. Enter the equation using the $\boxed{Y=}$ key. On some calculators, you can move the cursor to the symbol to the left of Y_1. Press $\boxed{\text{ENTER}}$ to cycle through the shading options: ◣ shades below the line, and ◤ shades above it.

2 Graph linear systems of inequalities in two variables.

Systems of Linear Inequalities in Two Variables

An ordered pair (a, b) is a **solution of a system of inequalities** involving two variables if it is a solution of each of its inequalities. For example, $(0, 1)$ is a solution of the system

$$\begin{cases} 2x + y \le 1 \\ x - 3y < 0 \end{cases}$$

because $2(0) + 1 \le 1$ and $0 - 3(1) < 0$ are both true statements. The **solution set of a system of inequalities** is the *intersection* of the solution sets of all inequalities in the system. To find the graph of this solution set, graph the solution set of each inequality in the system (with different-color shadings) on the same coordinate system. The region common to all of the solution sets, which is where all of the shaded parts overlap, is the solution set.

TECHNOLOGY CONNECTION

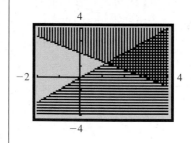 A graphing calculator uses overlapping crosshatch shading to display the solution set of a system of inequalities. The solution set for the system

$$\begin{cases} 2x + 3y > 6 & (1) \\ y - x \le 0 & (2) \end{cases}$$

is shown here.

EXAMPLE 4 **Graphing a System of Two Inequalities**

Graph the solution set of the system of inequalities.

$$\begin{cases} 2x + 3y > 6 & (1) \\ y - x \le 0 & (2) \end{cases}$$

SOLUTION
Graph each inequality separately in the same coordinate plane.

Inequality 1	**Inequality 2**
$2x + 3y > 6$	$y - x \le 0$

Step 1 $2x + 3y = 6$ | **Step 1** $y - x = 0$

Step 2 Sketch $2x + 3y = 6$ as a dashed line by joining the points $(0, 2)$ and $(3, 0)$. | **Step 2** Sketch $y - x = 0$ as a solid line by joining the points $(0, 0)$ and $(1, 1)$.

Step 3 Test $(0, 0)$ in | **Step 3** Test $(1, 0)$ in

$$\begin{array}{c} 2x + 3y > 6 \\ 2 \cdot 0 + 3 \cdot 0 \overset{?}{>} 6 \\ 0 \overset{?}{>} 6 \quad \text{False} \end{array}$$

$$\begin{array}{c} y - x \le 0 \\ 0 - 1 \overset{?}{\le} 0 \\ -1 \overset{?}{\le} 0 \quad \text{True} \end{array}$$

Step 4 Shade as in Figure 7.12(a). | **Step 4** Shade as in Figure 7.12(b).

The graph of the solution set of the system of inequalities (1) and (2) (the region where the shadings overlap) is shown shaded in purple in Figure 7.12(c). ■ ■ ■

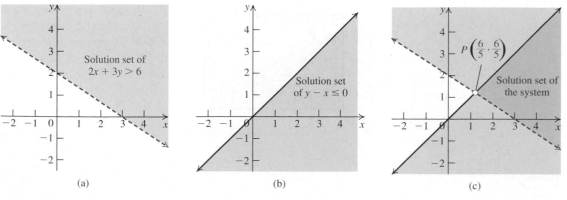

(a)   (b)   (c)

FIGURE 7.12

Practice Problem 4   Graph the solution set of the system of inequalities.

$$\begin{cases} 3x - y > 3 \\ x - y \geq 1 \end{cases}$$

■

The point $P\left(\dfrac{6}{5}, \dfrac{6}{5}\right)$ in Figure 7.12(c) is the point of intersection of the two lines $2x + 3y = 6$ and $y - x = 0$. Such a point is called a **corner point**, or **vertex**, of the solution set. It is found by solving the system of equations $\begin{cases} 2x + 3y = 6 \\ y - x = 0 \end{cases}$. In this case, however, the point P is not in the solution set of the given system because the ordered pair $\left(\dfrac{6}{5}, \dfrac{6}{5}\right)$ satisfies only inequality (2), but not inequality (1) of Example 4. We show P as an open circle in the figure.

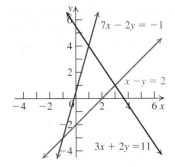

FIGURE 7.13

STUDY TIP

In working with linear inequalities, you must use the test point to graph the solution set. *Do not* assume that just because the symbol $\leq$ or $<$ appears in the inequality, the graph of the inequality is below the line.

EXAMPLE 5   Solving a System of Three Linear Inequalities

Sketch the graph and label the vertices of the solution set of the system of linear inequalities.

$$\begin{cases} 3x + 2y \leq 11 & (1) \\ x - y \leq 2 & (2) \\ 7x - 2y \geq -1 & (3) \end{cases}$$

SOLUTION

First, on the same coordinate plane, sketch the graphs of the three linear equations that correspond to the three inequalities. Since either $\leq$ or $\geq$ occurs in all three inequalities, all of the equations are graphed as *solid* lines. See Figure 7.13.

Notice that $(0, 0)$ satisfies each of the inequalities but none of the equations. You can use $(0, 0)$ as the test point for each inequality.

Make the following conclusions:

(i) Because $(0, 0)$ lies below the line $3x + 2y = 11$, the region on or below the line $3x + 2y = 11$ is in the solution set of inequality (1).

(ii) Because $(0, 0)$ is above the line $x - y = 2$, the region on or above the line $x - y = 2$ is in the solution set of inequality (2).

(iii) Finally, because $(0, 0)$ lies below the line $7x - 2y = -1$, the region on or below the line $7x - 2y = -1$ belongs to the solution set of inequality (3).

The region common to the regions in **(i)**, **(ii)**, and **(iii)** (including the boundaries) is the solution set of the given system of inequalities. The solution set is the shaded region in Figure 7.14, including the sides of the triangle.

Figure 7.14 also shows the vertices of the solution set. These vertices are obtained by solving each pair of equations in the system. Because all vertices are solutions of the given system, they are shown as (filled-in) points.

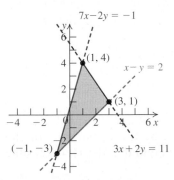

FIGURE 7.14 Graph of the solution set

a. To find the vertex $(3, 1)$, solve the system $\begin{cases} 3x + 2y = 11 & (1) \\ x - y = 2 & (2) \end{cases}$

Solve equation (2) for x to obtain $x = 2 + y$. Substitute $x = 2 + y$ in equation (1).

$$3(2 + y) + 2y = 11$$
$$6 + 3y + 2y = 11 \qquad \text{Distributive property}$$
$$y = 1 \qquad \text{Solve for } y.$$

Back-substitute $y = 1$ in equation (2) to find $x = 3$. The vertex is the point $(3, 1)$.

b. Solve the system of equations $\begin{cases} x - y = 2 \\ 7x - 2y = -1 \end{cases}$ by the substitution method to find the vertex $(-1, -3)$.

c. Solve the system of equations $\begin{cases} 3x + 2y = 11 \\ 7x - 2y = -1 \end{cases}$ by the elimination method to find the vertex $(1, 4)$.

■ ■ ■

Practice Problem 5 Sketch the graph (and label the corner points) of the solution set of the system of inequalities.

$$\begin{cases} 2x + 3y < 16 \\ 4x + 2y \geq 16 \\ 2x - y \leq 8 \end{cases}$$

■

3 Apply systems of inequalities to linear programming.

Applications: Linear Programming

Recall that if a function f with domain $[a, b]$ has a largest and a smallest value, the largest value is called the *maximum value* and the smallest value is called the *minimum value*. The process of finding the maximum or minimum value of a quanity is called **optimization**. For example, for the function

$$f(x) = x^2, \quad -2 \leq x \leq 3,$$

the maximum value of f occurs at $x = 3$ and is given by $f(3) = 3^2 = 9$, while the minimum value occurs at $x = 0$ and is given by $f(0) = 0^2 = 0$. See Figure 7.15. This is an example of an optimization problem involving one variable.

Now we will investigate optimization problems involving two variables, say, x and y. Assume the following conditions:

1. The quantity f to be maximized or minimized can be written as a linear expression in x and y; that is,

$$f = ax + by, \text{where } a \neq 0, b \neq 0 \text{ are constants.}$$

2. The domain of the variables x and y is restricted to a region S that is determined by (is a solution set of) a system of linear inequalities.

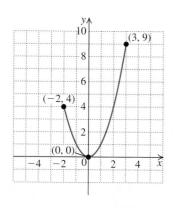

FIGURE 7.15 Optimization

A problem that satisfies conditions (1) and (2) is called a **linear programming problem**.

The inequalities that determine the region S are called **constraints**, the region S is called the **set of feasible solutions**, and $f = ax + by$ is called the **objective function**. A point in S at which f reaches its maximum (or minimum) value, together with the value of f at that point, is called an **optimal solution**.

Now consider the following linear programming problem.

Find values of x and y that will make $f = 2x + 3y$ as large as possible (in other words, that will result in a maximum value for f), where x and y are restricted by the following constraints:

$$\begin{cases} 3x + 5y \leq 15 \\ x \geq 0 \\ y \geq 0 \end{cases}$$

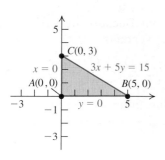

FIGURE 7.16 Feasible solutions

To solve this problem, first sketch the graph of the permissible values of x and y; that is, find the set of feasible solutions determined by the constraints of the problem. Using the method explained in Example 5, you find that the set of feasible solutions is the shaded region shown in Figure 7.16.

Now you have to find the values of $f = 2x + 3y$ at each of the points in the shaded region of Figure 7.16 and then see where f takes on the maximum value.

The shaded region, however, has infinitely many points. Since you cannot evaluate f at each point, you need another method to solve the problem.

Fortunately, it can be shown that if there is an optimal solution, it must occur at one of the vertices (corner points) of the feasible solution set. This means that when there is an optimal solution, *we can find the maximum (or minimum) value of f by first computing the value of f at each vertex and then picking the largest (or smallest) value that results.*

FINDING THE SOLUTION: A PROCEDURE

| EXAMPLE 6 | Solving a Linear Programming Problem |

OBJECTIVE
Solve a linear programming problem.

EXAMPLE
Maximize $f = 2x + 3y$ subject to the constraints given below.

Step 1 Write an expression for the quantity to be maximized or minimized. This expression is the objective function.

Maximize the objective function

$$f = 2x + 3y$$

Step 2 Write all constraints as linear inequalities.

subject to the constraints

$$\begin{cases} 3x + 5y \le 15 \\ \quad\quad x \ge 0 \\ \quad\quad y \ge 0 \end{cases}$$

Step 3 Graph the solution set of the constraint inequalities. This set is the set of feasible solutions.

See Figure 7.16 for the solution set of the constraints.

Step 4 Find all vertices of the solution set in Step 3 by solving all pairs of equations corresponding to the constraint inequalities.

Find

$A(0, 0)$ by solving $\begin{cases} x = 0 \\ y = 0 \end{cases}$

$B(5, 0)$ by solving $\begin{cases} y = 0 \\ 3x + 5y = 15 \end{cases}$

and

$C(0, 3)$ by solving $\begin{cases} x = 0 \\ 3x + 5y = 15 \end{cases}$

Step 5 Find the values of the objective function at each of the vertices of Step 4.

At $A(0, 0)$, $f = 2(0) + 3(0) = 0$.
At $B(5, 0)$, $f = 2(5) + 3(0) = 10$.
At $C(0, 3)$, $f = 2(0) + 3(3) = 9$.

Step 6 The largest of the values (if any) in Step 5 is the maximum value of the objective function, and the smallest of the values (if any) in Step 5 is the minimum value of the objective function.

At $B(5, 0)$, the objective function f has a maximum value of 10. Although you are not asked for the minimum value, it is 0 at $A(0, 0)$.

■ ■ ■

Practice Problem 6   Maximize $f = 4x + 5y$ subject to the constraints

$$\begin{cases} 5x + 7y \le 35 \\ \quad\quad x \ge 0 \\ \quad\quad y \ge 0 \end{cases}$$

■

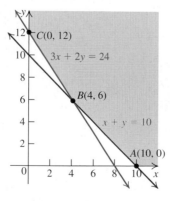

EXAMPLE 7   **Nutrition; Minimizing Calories**

Fat Albert wants to go on a crash diet and needs your help in designing a lunch menu. The menu is to include two items: soup and salad. The vitamin units (milligrams) and calorie counts in each ounce of soup and salad are given in Table 7.2.

TABLE 7.2

Item	Vitamin A	Vitamin C	Calories
Soup	1	3	50
Salad	1	2	40

The menu must provide at least:

10 units of vitamin A.

24 units of vitamin C.

Find the number of ounces of each item in the menu needed to provide the required vitamins with the fewest number of calories.

SOLUTION

a. State the problem mathematically.

Step 1 **Write the objective function.** Let the lunch menu contain x ounces of soup and y ounces of salad. You need to minimize the total number of calories in the menu. Since the soup has 50 calories per ounce (see Table 7.2), x ounces of soup has $50x$ calories. Likewise, the salad has $40y$ calories. The number f of calories in the two items is therefore given by the objective function:

$$f = 50x + 40y$$

Step 2 **Write the constraints.** Since each ounce of soup provides 1 unit of vitamin A, and each ounce of salad provides 1 unit of vitamin A, the two items together provide $1 \cdot x + 1 \cdot y = x + y$ units of vitamin A. The lunch must have at least 10 units of vitamin A; this means that

$$x + y \geq 10.$$

Similarly, the two items provide $3x + 2y$ units of vitamin C. The menu must provide at least 24 units of vitamin C, so

$$3x + 2y \geq 24.$$

The fact that x and y cannot be negative means that $x \geq 0$ and $y \geq 0$. You now have four constraints.

Summarize your information. Find x and y such that the value of

$$f = 50x + 40y \qquad \text{Objective function from Step 1}$$

is a *minimum*, with the restrictions:

$$\begin{cases} x + y \geq 10 \\ 3x + 2y \geq 24 \\ x \geq 0 \\ y \geq 0 \end{cases} \qquad \text{Constraints from Step 2}$$

b. Solve the linear programming problem.

Step 3 **Graph the set of feasible solutions.** The set of feasible solutions is shaded in Figure 7.17. This set is bounded by the lines whose equations are

$$x + y = 10, \, 3x + 2y = 24, \, x = 0, \text{ and } y = 0.$$

FIGURE 7.17

Step 4 **Find the vertices.** The vertices of the set of feasible solutions are

$$A(10, 0), \text{found by solving} \begin{cases} y = 0 \\ x + y = 10 \end{cases}$$

$$B(4, 6), \text{found by solving} \begin{cases} x + y = 10 \\ 3x + 2y = 24 \end{cases}$$

$$C(0, 12), \text{found by solving} \begin{cases} x = 0 \\ 3x + 2y = 24 \end{cases}$$

Step 5 **Find the value of f at the vertices.** The value of f at each of the vertices is given in Table 7.3.

TABLE 7.3

Vertex (x, y)	Value of $f = 50x + 40y$
$(10, 0)$	$50(10) + 40(0) = 500$
$(4, 6)$	$50(4) + 40(6) = 440$
$(0, 12)$	$50(0) + 40(12) = 480$

Step 6 **Find the maximum or minimum value of f.** From Table 7.3, the smallest value of f is 440, which occurs when $x = 4$ and $y = 6$.

c. State the conclusion.

The lunch menu for Fat Albert should contain 4 ounces of soup and 6 ounces of salad. His intake of 440 calories will be as small as possible under the given constraints. ■ ■ ■

Practice Problem 7 How does the answer to Example 7 change if 1 ounce of soup contains 30 calories and 1 ounce of salad contains 60 calories? ■

Notice in Example 7 that the value of the objective function $f = 50x + 40y$ can be made as large as you want by choosing x or y (or both) to be very large. So f has a minimum value of 440 but no maximum value.

SECTION 7.3 ■ Exercises

A EXERCISES Basic Skills and Concepts

1. In the graph of $x - 3y > 1$, the corresponding equation $x - 3y = 1$ is graphed as a(n) _____dashed_____ line.

2. If a test point from one of the two regions determined by an inequality's corresponding equation satisfies the inequality, then _____all points_____ in that region satisfy the inequality.

3. In a system of inequalities containing both $2x + y > 5$ and $2x - y \le 3$, the point of intersection of the lines $2x + y = 5$ and $2x - y = 3$ is _not a solution_ of the system

4. The process of finding the maximum or minimum value of a quantity is called _optimization_.

5. *True or False* There are always infinitely many solutions of an inequality $ax + by < c$, where $a, b,$ and c are real numbers and $a \ne 0$. True

6. *True or False* The ordered pair $(1, 1)$ is a solution of the inequality $x + y < 2$. False

In Exercises 7–22, graph each inequality.

7. $x \ge 0$ †   **8.** $y \ge 0$ †   **9.** $x > -1$ †

10. $y > 2$ †   **11.** $x \ge 2$ †   **12.** $y \le 3$ †

13. $y - x < 0$ †   **14.** $y + 2x \le 0$ †

15. $x + 2y < 6$ †   **16.** $3x + 2y < 12$ †

17. $2x + 3y \ge 12$ †   **18.** $2x + 5y \ge 10$ †

19. $x - 2y \le 4$ †   **20.** $3x - 4y \le 12$ †

21. $3x + 5y < 15$ †   **22.** $5x + 7y < 35$ †

In Exercises 23–28, determine which ordered pairs are solutions of each system of inequalities.

23. $\begin{cases} x + y < 2 \\ 2x + y \ge 6 \end{cases}$

  a. $(0, 0)$ **b.** $(-4, -1)$ **c.** $(3, 0)$ **d.** $(0, 3)$

  None

24. $\begin{cases} x - y < 2 \\ 3x + 4y \ge 12 \end{cases}$

  a. $(0, 0)$ **b.** $\left(\dfrac{16}{5}, \dfrac{3}{5}\right)$ **c.** $(4, 0)$ **d.** $\left(\dfrac{1}{2}, \dfrac{1}{2}\right)$

  None

†Due to space constrictions, answers to these exercises may be found in the Answers beginning on page A–1 in the back of the book.

25. $\begin{cases} y < x + 2 \\ x + y \le 4 \end{cases}$ $(0,0), (1,0), (0,1), (1,1)$

   a. $(0,0)$ **b.** $(1,0)$ **c.** $(0,1)$ **d.** $(1,1)$

26. $\begin{cases} y \le 2 - x \\ x + y \le 1 \end{cases}$ $(0,0), (0,1), (-1,-1)$

   a. $(0,0)$ **b.** $(0,1)$ **c.** $(1,2)$ **d.** $(-1,-1)$

27. $\begin{cases} 3x - 4y \le 12 \\ x + y \le 4 \\ 5x - 2y \ge 6 \end{cases}$

   a. $(0,0)$ **b.** $(2,0)$ **c.** $(3,1)$ **d.** $(2,2)$
   $(2,0), (3,1), (2,2)$

28. $\begin{cases} 2 + y \le 3 \\ x - 2y \le 3 \\ 5x + 2y \ge 3 \end{cases}$

   a. $(0,0)$ **b.** $(1,0)$ **c.** $(1,2)$ **d.** $(2,1)$
   $(1,0), (2,1)$

In Exercises 29–44, graph the solution set of each system of linear inequalities and label the vertices (if any) of the solution set.

29. $\begin{cases} x + y \le 1 \\ x - y \le -1 \end{cases}$ †

30. $\begin{cases} x + y \ge 1 \\ y - x \ge 1 \end{cases}$ †

31. $\begin{cases} 3x + 5y \le 15 \\ 2x + 2y \le 8 \end{cases}$ †

32. $\begin{cases} -2x + y \le 2 \\ 3x + 2y \le 4 \end{cases}$ †

33. $\begin{cases} 2x + 3y \le 6 \\ 4x + 6y \ge 24 \end{cases}$ †

34. $\begin{cases} x + 2y \ge 6 \\ 2x + 4y \le 4 \end{cases}$ †

35. $\begin{cases} 3x + y \le 8 \\ 4x + 2y \ge 4 \end{cases}$ †

36. $\begin{cases} x + 2y \le 12 \\ 2x + 4y \ge 8 \end{cases}$ †

37. $\begin{cases} x \ge 0 \\ y \ge 0 \\ x + y \le 1 \end{cases}$ †

38. $\begin{cases} x \ge 0 \\ y \ge 0 \\ x + y > 2 \end{cases}$ †

39. $\begin{cases} x \ge 0 \\ y \ge 0 \\ 2x + 3y \ge 6 \end{cases}$ †

40. $\begin{cases} x \ge 0 \\ y \ge 0 \\ 3x + 4y \le 12 \end{cases}$ †

41. $\begin{cases} x + y \le 1 \\ x - y \ge 1 \\ 2x + y \le 1 \end{cases}$ †

42. $\begin{cases} x + y \le 1 \\ x - y \ge 1 \\ 2x + y \ge 1 \end{cases}$ †

43. $\begin{cases} x + y \le 1 \\ x - y < 2 \\ -x + y \le 3 \\ x + y \ge -4 \end{cases}$ †

44. $\begin{cases} x + y \le -1 \\ 2x - y < -2 \\ -2x + y \le -3 \\ 2x + 2y \le -4 \end{cases}$ †

In Exercises 45–50, use the following figure to indicate the regions (A–G) that correspond to the graph of the given system of inequalities.

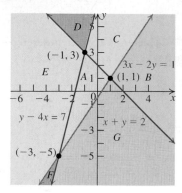

45. $\begin{cases} x + y \le 2 \\ 3x - 2y \le 1 \\ y - 4x \le 7 \end{cases}$ A

46. $\begin{cases} x + y \ge 2 \\ 3x - 2y \le 1 \\ y - 4x \le 7 \end{cases}$ C

47. $\begin{cases} x + y \ge 2 \\ 3x - 2y \ge 1 \\ y - 4x \le 7 \end{cases}$ B

48. $\begin{cases} x + y \ge 2 \\ 3x - 2y \le 1 \\ y - 4x \ge 7 \end{cases}$ D

49. $\begin{cases} x + y \le 2 \\ 3x - 2y \le 1 \\ y - 4x \le 7 \end{cases}$ E

50. $\begin{cases} x + y \le 2 \\ 3x - 2y \ge 1 \\ y - 4x \ge 7 \end{cases}$ F

In Exercises 51–56, find the maximum and the minimum value of each objective function f. The graph of the set of feasible solutions is as follows:

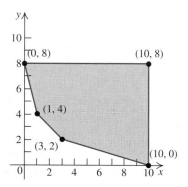

51. $f = x + y$ Max 18, Min 5   **52.** $f = 2x + y$ Max 28, Min 6

53. $f = x + 2y$ Max 26, Min 7   **54.** $f = 2x + 5y$ Max 60, Min 16

55. $f = 5x + 2y$ Max 66, Min 13   **56.** $f = 8x + 5y$ Max 120, Min 28

In Exercises 57–64, you are given an objective function f and a set of constraints. Find the set of feasible solutions determined by the given constraints and then maximize or minimize the objective function as directed.

57. Maximize $f = 9x + 13y$, subject to the constraints $x \ge 0, y \ge 0, 5x + 8y \le 40, 3x + y \le 12$. †

58. Maximize $f = 7x + 6y$, subject to the constraints $x \ge 0$, $y \ge 0, 2x + 3y \le 13, x + y \le 5$.
Maximum value is 35 at $(5,0)$.

59. Maximize $f = 5x + 7y$, subject to the constraints $x \ge 0$, $y \ge 0, x + y \le 70, x + 2y \le 100, 2x + y \le 120$.
Maximum value is 410 at $(40, 30)$.

60. Maximize $f = 2x + y$, subject to the constraints $x \ge 0$, $y \ge 0, x + y \ge 5, 2x + 3y \le 21, 4x + 3y \le 24$.
Maximum value is 12 at $(6, 0)$.

61. Minimize $f = x + 4y$, subject to the constraints $x \ge 0$, $y \ge 0, x + 3y \ge 3, 2x + y \ge 2$.
Minimum value is 3 at $(3, 0)$.

62. Minimize $f = 5x + 2y$, subject to the constraints $x \ge 0, y \ge 0, 5x + y \ge 10, x + y \ge 6$.
Minimum value is 15 at $(1, 5)$.

63. Minimize $f = 13x + 15y$, subject to the constraints $x \ge 0, y \ge 0, 3x + 4y \ge 360, 2x + y \ge 100$.
Minimum value is 1364 at $(8, 84)$.

64. Minimize $f = 40x + 37y$, subject to the constraints $x \ge 0, y \ge 0, 10x + 3y \ge 180, 2x + 3y \ge 60$.
Minimum value is 970 at $(15, 10)$.

B EXERCISES Applying the Concepts

65. Agriculture: crop planning. A farmer owns 240 acres of cropland. He can use his land to produce corn and soybeans. A total of 320 labor hours is available to him during the production period of both of these commodities. Each acre for corn production requires 2 hours of labor, and each acre for soybean production requires 1 hour of labor. If the profit per acre in corn production is $50 and that in soybean production is $40, find the number of acres that he should allocate to the production of corn and soybeans so as to maximize his profit. †

66. Repeat Exercise 65 assuming that the profit per acre in corn production is $30 and that in soybean production is $40. †

67. **Manufacturing: production scheduling.** Two machines (I and II) each produce two grades of plywood: Grade A and Grade B. In 1 hour of operation, machine I produces 20 units of Grade A and 10 units of Grade B plywood. Also in 1 hour of operation, machine II produces 30 units of Grade A and 40 units of Grade B plywood. The machines are required to meet a production schedule of at least 1400 units of Grade A and 1200 units of Grade B plywood. If the cost of operating machine I is $50 per hour and the cost of operating machine II is $80 per hour, find the number of hours each machine should be operated so as to minimize the cost. †

68. Repeat Exercise 67 assuming that the cost of operating machine I is $70 per hour and the cost of operating machine II is $90 per hour. †

69. **Advertising.** The Sundial Cheese Company wants to allocate its $6000 advertising budget to two local media—television and newspaper—so that the exposure (number of viewers) to its advertisements is maximized. There are 60,000 viewers exposed to each minute of television time and 20,000 readers exposed to each page of newspaper advertising. Each minute of television time costs $1000 and each page of newspaper advertising costs $500. How should the company allocate its budget if it has to buy at least 1 minute of television time and at least two pages of newspaper advertising? †

70. **Advertising.** The Fantastic Bread Company allocates a maximum of $24,000 to advertise its product to television and newspaper. Each hour of television costs $4000, and each page of newspaper advertising costs $3000. The exposure from each hour of television time is assumed to be 120,000, and the exposure from each page of newspaper advertising is 80,000. Furthermore, the board of directors requires at least 2 hours of television time and one page of newspaper advertising. How should the advertising budget be divided to maximize exposure to the advertisements? †

71. **Agriculture.** A Florida citrus company has 480 acres of land for growing oranges and grapefruit. Profits per acre are $40 for oranges and $30 for grapefruit. The total labor hours available during the production are 800. Each acre for oranges uses 2 hours of labor, and each acre for grapefruit uses 1 hour of labor. If the fixed costs are $3000, what is the maximum profit? *The maximum profit is $17,600 at (320,160).*

72. **Agriculture.** The Florida Juice Company makes two types of fruit punch—Fruity and Tangy—by blending orange juice and apple juice into a mixture. The fruit punch is sold in 5-gallon bottles. A bottle of Fruity earns a profit of $3, and a bottle of Tangy earns a $2 profit. A bottle of Fruity requires 3 gallons of orange juice and 2 gallons of apple juice, while a bottle of Tangy requires 4 gallons of orange juice and 1 gallon of apple juice. If 200 gallons of orange juice and 120 gallons of apple juice are available, find the maximum profit for the company. *The maximum profit is $184 at (56, 8).*

73. **Manufacturing.** The Fantastic Furniture Company manufactures rectangular and circular tables. Each rectangular table requires 1 hour of assembling and 1 hour of finishing, and each circular table requires 1 hour of assembling and 2 hours of finishing. The company's 20 assemblers and 30 finishers each work 40 hours per week. Each rectangular table brings a profit of $3, and each circular table brings a profit of $4. If all of the tables manufactured are sold, how many of each type should be made to maximize profits? †

74. **Manufacturing.** A company produces a portable global positioning system (GPS) and a portable digital video disc (DVD) player, each of which requires three stages of processing. The length of time for processing each unit is given in the following table:

Stage	GPS (hr/unit)	DVD (hr/unit)	Maximum process capacity (hr/day)
I	12	12	840
II	3	6	300
III	8	4	480
Profit per unit	$6	$4	

How many of each product should the company produce per day in order to maximize profit? †

75. **Building houses.** A contractor has 100 units of concrete, 160 units of wood, and 400 units of glass. He builds terraced houses and cottages. A terraced house requires 1 unit of concrete, 2 units of wood, and 2 units of glass; a cottage requires 1 unit of concrete, 1 unit of wood, and 5 units of glass. A terraced house sells for $40,000, and a cottage sells for $45,000. How many houses of each type should the contractor build to obtain the maximum amount of money? †

76. **Hospitality.** Mrs. Adams owns a motel consisting of 300 single rooms and an attached restaurant that serves breakfast. She knows that 30% of the male guests and 50% of the female guests will eat in the restaurant. Suppose she makes a profit of $18.50 per day from every guest who eats in her restaurant and $15 per day from every guest who does not eat in her restaurant. If Mrs. Adams never has more than 125 female guests and never more than 220 male guests, find the number of male and female guests that would provide the maximum profit. †

77. **Hospitality.** In Exercise 76, find Mrs. Adams's maximum profit during a season in which she always has at least 100 guests if she makes a profit of $15 per day from every guest who eats in her restaurant while she suffers a loss of $2 per day from every guest who does not eat in her restaurant. *Mrs. Adams's maximum profit is $1355 when she has 175 male guests and 125 female guests.*

78. Transportation. Major Motors, Inc., must produce at least 5000 luxury cars and 12,000 medium-priced cars. It must produce at most 30,000 units of compact cars. The company owns one factory in Michigan and one in North Carolina. The Michigan factory produces 20, 40, and 60 units of luxury, medium-priced, and compact cars, respectively, per day, while the North Carolina factory produces 10, 30, and 20 luxury, medium-priced, and compact cars, respectively, per day. If the Michigan factory costs $60,000 per day to operate and the North Carolina factory costs $40,000 per day to operate, find the number of days each factory should operate to minimize the cost and meet the requirements. †

79. Nutrition. Elisa Epstein is buying two types of frozen meals: a meal of enchiladas with vegetables and a vegetable loaf meal. She wants to guarantee that she gets a minimum of 60 grams of carbohydrates, 40 grams of proteins, and 35 grams of fats. The enchilada meal contains 5, 3, and 5 grams of carbohydrates, proteins, and fats, respectively, per kilogram, while the vegetable loaf contains 2, 2, and 1 gram of carbohydrates, proteins, and fats, respectively, per kilogram. If the enchilada meal costs $3.50 per kilogram and the vegetable loaf costs $2.25 per kilogram, how many kilograms of each should Elisa buy to minimize the cost and still meet the minimum requirement? †

80. Temporary help. The personnel director of the main post office plans to hire extra helpers during the holiday season. Since office space is limited, the number of temporary workers cannot exceed 10. She hires workers sent by Super Temps and by Ready Aid. From her past experience, she knows that a typical worker sent by Super Temps can handle 300 letters and 80 packages per day and a typical worker from Ready Aid can handle 600 letters and 40 packages per day. The post office expects that the additional daily volume of letters and packages will be at least 3900 and 560, respectively. The daily wages for a typical worker from Super Temps and Ready Aid are $96 and $86, respectively. How many workers from each agency should be hired to keep the daily wages to a minimum? †

C EXERCISES Beyond the Basics

In Exercises 81–88, write a system of linear inequalities that has the given graph.

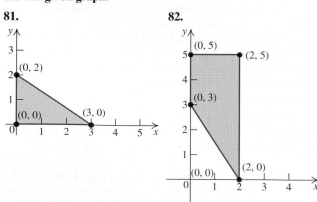

83.   **84.**

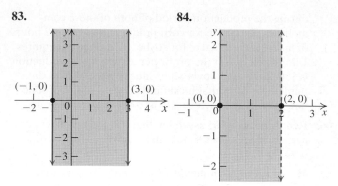

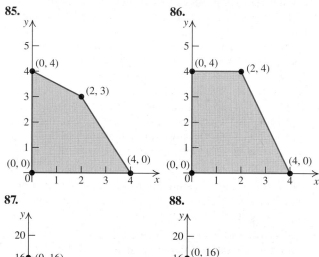

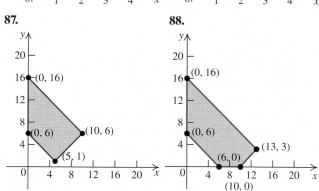

Critical Thinking

In Exercises 89–90, graph the solution set of each inequality.

89. $6 \le 2x + 3y \le 12$ †   **90.** $|x + 2y| < 4$ †

Answers:

81. $\begin{cases} x \ge 0 \\ y \ge 0 \\ y \le \dfrac{-2}{3}x + 2 \end{cases}$ **82.** $\begin{cases} x \ge 0 \\ x \le 2 \\ y \ge 0 \\ y \ge \dfrac{-3}{2}x + 3 \\ y \le 5 \end{cases}$ **83.** $\begin{cases} x \ge -1 \\ x \le 3 \end{cases}$ **84.** $\begin{cases} x \ge 0 \\ x < 2 \end{cases}$

85. $\begin{cases} x \ge 0 \\ y \ge 0 \\ y \le \dfrac{-1}{2}x + 4 \\ y \le \dfrac{-3}{2}x + 6 \end{cases}$ **86.** $\begin{cases} x \ge 0 \\ y \ge 0 \\ y \le 4 \\ y \le -2x + 8 \end{cases}$ **87.** $\begin{cases} x \ge 0 \\ y \ge x - 4 \\ y \ge -x + 6 \\ y \le -x + 16 \end{cases}$

88. $\begin{cases} x \ge 0 \\ y \ge 0 \\ y \le -x + 16 \\ y \ge x - 10 \\ y \ge -x + 6 \end{cases}$

Nonlinear Systems of Equations and Inequalities

| **Before Starting this Section, Review** | **Objectives** |

Before Starting this Section, Review

1. Solve systems of linear equations (Section 7.1, page 482)

2. Solve systems of linear inequalities (Section 7.3, page 508)

Objectives

1 Solve nonlinear systems of equations.

2 Solve a nonlinear system of inequalities.

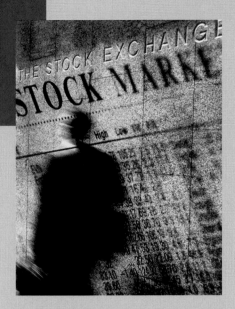

New York Stock Exchange

INVESTING IN THE STOCK MARKET

A *stock* is a certificate that shows that you own a small fraction of a corporation. When you own stock in a company, you are referred to as a *shareholder*. When you buy stock, you are paying for a small percentage of everything the company owns, including a slice of its profits (or losses).

Suppose you inherit some money and want to invest $1000 in the stock market. Where do you invest your money? (This is a serious question. You may have to do a great deal of research on the companies that interest you, or you may have to get help from a broker.)

Suppose you get a tip from a friend who heard it from his brother-in-law (a taxi driver who overheard his customer's conversation) that Microsoft is coming out with a new version of Microsoft Windows that is going to double the company's business. Where do you buy Microsoft stock? You go to a broker. A brokerage house (such as Merrill Lynch) is a supermarket of stocks. You can tell any broker that you have $1000 and want to buy as much Microsoft stock as you can. The broker might reply, "A share of Microsoft at this time [the price may fluctuate every second] costs $31 [the actual closing price on October 17, 2007, was $31], and I am going to charge you $55 for my services. So you can buy $\dfrac{1000 - 55}{31} \approx 30.48$, or 30 shares of Microsoft." You then give the broker the money, and you become part owner of Microsoft. (Usually, the broker does not give you the paper stock certificate, but rather transfers ownership to you.)

You can make a profit on your investment if you sell your stock for more than the total amount you paid for it. You can also make money while you own certain stocks by receiving dividends. A *dividend* on a share of stock is a return on your investment and represents the portion of the company's profits allocated to that share. The dividend can be in the form of cash or additional shares of the company. In Example 3, we discuss such a scenario. ■

1 Solve nonlinear systems of equations.

Nonlinear Systems of Equations

In Sections 7.1, 7.2, and 7.3, we discussed systems of *linear* equations and inequalities. We now consider *nonlinear* systems of equations and inequalities involving two variables: in these systems, at least one equation or inequality is nonlinear.

To determine whether an ordered pair is a solution of a nonlinear system, you check whether it is a solution of each equation or inequality in the system. Thus, $(2, 0)$ is a solution of the system

$$\begin{cases} x^2 + 3y > 3 & (1) \\ 9x^2 - 4y^2 \le 36 & (2) \end{cases}$$

because $2^2 + 3(0) > 3$ and $9(2)^2 - 4(0)^2 \le 36$ are both true statements.

517

EXAMPLE 1   **Using Substitution to Solve a Nonlinear System**

Solve the system of equations by the substitution method.

$$\begin{cases} 4x + y = -3 & (1) \\ -x^2 + y = 1 & (2) \end{cases}$$

SOLUTION

We follow the same steps that were used in the substitution method for solving a system of linear equations in Section 7.1.

Step 1   **Solve for one variable.** In equation (2), you can express y in terms of x:

$$y = x^2 + 1 \quad (3) \quad \text{Solve equation (2) for } y.$$

Step 2   **Substitute.** Substitute $x^2 + 1$ for y in equation (1).

$$\begin{aligned} 4x + y &= -3 & &\text{Equation (1)} \\ 4x + (x^2 + 1) &= -3 & &\text{Replace } y \text{ with } (x^2 + 1). \\ 4x + x^2 + 4 &= 0 & &\text{Add 3 to both sides.} \\ x^2 + 4x + 4 &= 0 & &\text{Rewrite in descending powers of } x. \end{aligned}$$

Step 3   **Solve** the equation resulting from Step 2.

$$\begin{aligned} (x + 2)(x + 2) &= 0 & &\text{Factor.} \\ x + 2 &= 0 & &\text{Zero-product property} \\ x &= -2 & &\text{Solve for } x. \end{aligned}$$

Step 4   **Back-substitution.** Substitute $x = -2$ in equation (3) to find the corresponding y-value.

$$\begin{aligned} y &= x^2 + 1 & &\text{Equation (3)} \\ y &= (-2)^2 + 1 & &\text{Replace } x \text{ with } -2. \\ y &= 5 & &\text{Simplify.} \end{aligned}$$

Since $x = -2$ and $y = 5$, the apparent solution set of the system is $\{(-2, 5)\}$.

Step 5   **Check.** Replace x with -2 and y with 5 in equation (1) and equation (2).

$$\begin{array}{ll} 4x + y = -3 & -x^2 + y = 1 \\ 4(-2) + 5 \stackrel{?}{=} -3 & -(-2)^2 + 5 \stackrel{?}{=} 1 \\ -8 + 5 = -3 \;\checkmark & -4 + 5 = 1 \;\checkmark \end{array}$$

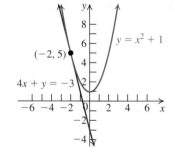

FIGURE 7.18

The graphs of the line $4x + y = -3$ and the parabola $y = x^2 + 1$ confirm that the solution set is $\{(-2, 5)\}$. See Figure 7.18.   ■ ■ ■

Practice Problem 1   Solve the system of equations by the substitution method.

$$\begin{cases} x^2 + y = 2 & (1) \\ 2x + y = 3 & (2) \end{cases} \qquad \blacksquare$$

EXAMPLE 2   **Using Elimination to Solve a Nonlinear System**

Solve the system of equations by the elimination method.

$$\begin{cases} x^2 + y^2 = 25 & (1) \\ x^2 - y = 5 & (2) \end{cases}$$

TECHNOLOGY CONNECTION

To use a graphing calculator to graph the system in Example 2, you must solve $x^2 + y^2 = 25$ for y and enter the two resulting equations.

$$Y_1 = \sqrt{25 - x^2}$$
$$Y_2 = -\sqrt{25 - x^2}$$

Then enter

$$Y_3 = x^2 - 5$$

and use the ZSquare feature.

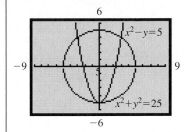

SOLUTION

Step 1 **Adjust the coefficients.** Because x has the same power in both equations, we choose the variable x for elimination. We multiply equation (2) by -1 and obtain the equivalent system.

$$\begin{cases} x^2 + y^2 = 25 & (1) \\ -x^2 + y = -5 & (3) \end{cases} \quad \text{Multiply equation (2) by } -1.$$

Step 2 $y^2 + y = 20$ (4)   Add equations (1) and (3).

Step 3 Solve the equation found in Step 2: $y^2 + y = 20$.

$$\begin{aligned} y^2 + y - 20 &= 0 & \text{Subtract 20 from both sides.} \\ (y + 5)(y - 4) &= 0 & \text{Factor.} \\ y + 5 = 0 \quad \text{or} \quad y - 4 &= 0 & \text{Zero-product property} \\ y = -5 \quad \text{or} \qquad y &= 4 & \text{Solve for } y. \end{aligned}$$

Step 4 **Back-substitute** the values in one of the original equations to solve for the other variable.

(i) Substitute $y = -5$ in equation (2) and solve for x.

$$\begin{aligned} x^2 - y &= 5 & \text{Equation (2)} \\ x^2 - (-5) &= 5 & \text{Replace } y \text{ with } -5. \\ x^2 &= 0 & \text{Simplify.} \\ x &= 0 & \text{Solve for } x. \end{aligned}$$

So, $(0, -5)$ is a solution of the system.

(ii) Substitute $y = 4$ in equation (2) and solve for x.

$$\begin{aligned} x^2 - y &= 5 & \text{Equation (2)} \\ x^2 - 4 &= 5 & \text{Replace } y \text{ with } 4. \\ x^2 &= 9 & \text{Add 4 to both sides.} \\ x &= \pm 3 & \text{Square root property} \end{aligned}$$

So, $(3, 4)$ and $(-3, 4)$ are solutions of the system.
From **(i)** and **(ii)**, the apparent solution set for the system is $\{(0, -5), (3, 4), (-3, 4)\}$.

Step 5 **Check.**

	$(0, -5)$	$(3, 4)$	$(-3, 4)$
$x^2 + y^2 = 25$	$0^2 + (-5)^2 \overset{?}{=} 25$	$3^2 + 4^2 \overset{?}{=} 25$	$(-3)^2 + 4^2 \overset{?}{=} 25$
	$25 = 25$ ✓	$9 + 16 = 25$	$9 + 16 \overset{?}{=} 25$
		$25 = 25$ ✓	$25 = 25$ ✓
$x^2 - y = 5$	$0^2 - (-5) \overset{?}{=} 5$	$3^2 - 4 \overset{?}{=} 5$	$(-3)^2 - 4 \overset{?}{=} 5$
	$5 = 5$ ✓	$9 - 4 = 5$ ✓	$9 - 4 = 5$ ✓

The graphs of the circle $x^2 + y^2 = 25$ and the parabola $y = x^2 - 5$ sketched in Figure 7.19 confirm the three solutions. ■ ■ ■

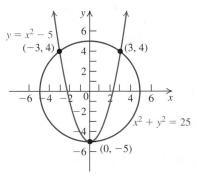

FIGURE 7.19

Practice Problem 2 Solve the system of equations.

$$\begin{cases} x^2 + 2y^2 = 34 \\ x^2 - y^2 = 7 \end{cases}$$

EXAMPLE 3   **Investing in Stocks**

Danielle bought 240 shares of Alpha Airlines stock at \$40 per share and paid \$100 in broker fees. She kept these shares for three years, during which time she received a number of additional shares of the stock as dividends. At the end of three years, her dividends were worth $2400 and she sold all of her stock and made a profit of $7100 (after paying $100 in broker's commission). How many shares of the stock did she receive as dividends? What was the selling price of each share of the stock?

SOLUTION

Let x = number of shares she received as dividends
p = selling price per share

Then

$$xp = 2400 \quad (1) \quad \text{The total dividend is worth \$2400.}$$

The number of shares she sold at p dollars per share is $240 + x$.

$$\text{Revenue} = (240 + x)p$$
$$\text{Cost} = (240)(40) + 100 = 9700$$

number of shares bought   commission

price she paid per share

and

$$\text{Revenue} - \text{Cost} = \text{Profit}$$
$$(240 + x)p - 9700 = 7100$$
$$(240 + x)p = 16{,}800 \qquad \text{Add 9700 to both sides.}$$
$$240p + xp = 16{,}800 \quad (2) \quad \text{Distributive property}$$

We solve the following system of equations:

$$\begin{cases} xp = 2400 & (1) \quad \text{Revenue from the dividends} \\ 240p + xp = 16{,}800 & (2) \quad \text{Revenue} - \text{Cost} = \text{Profit} \end{cases}$$

Substitute $xp = 2400$ from equation (1) into equation (2):

$$240p + \quad xp = 16{,}800 \quad (2)$$
$$240p + 2400 = 16{,}800 \quad (3) \quad \text{Substitute 2400 for } xp \text{ in (2).}$$
$$240p = 14{,}400 \qquad \text{Subtract 2400 from both sides.}$$
$$p = 60 \qquad \text{Solve for } p.$$

Back-substitute $p = 60$ in equation (1).

$$xp = 2400$$
$$x(60) = 2400 \qquad \text{Replace } p \text{ with 60.}$$
$$x = \frac{2400}{60} = 40 \qquad \text{Solve for } x.$$

Danielle received $x = 40$ shares as dividends and sold her stock at $p = \$60$ per share. ■ ■ ■

Practice Problem 3   Rework Example 3 assuming that Danielle's dividends were worth $1950 after three years and her profit was $7850. ■

2 Solve a nonlinear system of inequalities.

Nonlinear Systems of Inequalities

We first describe a procedure for graphing a nonlinear inequality. Recall from Section 7.3 that the graph of an inequality in two variables is the graph of the solution set of the inequality. This statement applies to both linear and nonlinear inequalities. We use the same *test-point* procedure that we discussed for linear inequalities in Section 7.3.

FINDING THE SOLUTION: A PROCEDURE

EXAMPLE 4   **Graphing a Nonlinear Inequality in Two Variables**

OBJECTIVE

Graph a nonlinear inequality in two variables.

Step 1   Replace the inequality symbol with an equal ($=$) sign.

Step 2   Sketch the graph of the *corresponding equation* in Step 1. Use a dashed curve if the given inequality sign is $<$ or $>$ and a solid curve if the inequality symbol is $\leq$ or $\geq$.

Step 3   The graph in Step 2 divides the plane into two regions. Select a *test point* in either region, but not on the graph.

Step 4   **(i)** If the coordinates of the test point satisfy the inequality, then so do all of the points in that region. Shade that region.

(ii) If the test point's coordinates do not satisfy the inequality, shade the region that does not contain the test point.

The shaded region (including the boundary if it is a solid curve) is the graph of the inequality.

EXAMPLE

Graph $y > x^2 - 2$.

$$y = x^2 - 2$$

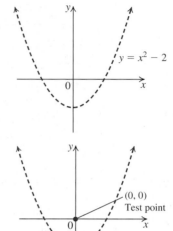

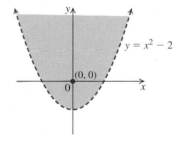

Because $0 > 0^2 - 2 = -2$, the coordinates of the test point, $(0, 0)$, satisfy the inequality $y > x^2 - 2$.

The graph of the solution set of $y > x^2 - 2$ is shown shaded.

■ ■ ■

Practice Problem 4   Graph $y \leq x^2 - 2$.   ■

The procedure for solving a nonlinear system of inequalities is identical to the procedure for solving a linear system of inequalities.

EXAMPLE 5   **Solving a Nonlinear System of Inequalities**

Graph the solution set of the system of inequalities.

$$\begin{cases} y \leq 4 - x^2 & (1) \\ y \geq \dfrac{3}{2}x - 3 & (2) \\ y \geq -6x - 3 & (3) \end{cases}$$

SOLUTION

Graph each inequality separately in the same coordinate plane. Since $(0,0)$ is not a solution of any of the corresponding equations, use $(0,0)$ as a test point for each inequality.

	Inequality (1)	Inequality (2)	Inequality (3)
	$y \leq 4 - x^2$	$y \geq \dfrac{3}{2}x - 3$	$y \geq -6x - 3$
Step 1	$y = 4 - x^2$	$y = \dfrac{3}{2}x - 3$	$y = -6x - 3$
Step 2	The graph of $y = 4 - x^2$ is a parabola opening down with vertex $(0,4)$. Sketch it as a solid curve.	The graph of $y = \dfrac{3}{2}x - 3$ is a line through the points $(0, -3)$ and $(2, 0)$. Sketch a solid line.	The graph of $y = -6x - 3$ is a line through the points $(0, -3)$ and $\left(-\dfrac{1}{2}, 0\right)$. Sketch a solid line.
Step 3	Testing $(0,0)$ in $y \leq 4 - x^2$ gives $0 \leq 4$, a true statement. The solution set is below the parabola.	Testing $(0,0)$ in $y \geq \dfrac{3}{2}x - 3$ gives $0 \geq -3$, a true statement. The solution set is above the line.	Testing $(0,0)$ in $y \geq -6x - 3$ gives $0 \geq -3$, a true statement. The solution set is above the line.
Step 4	Shade the region as in Figure 7.20(a).	Shade the region as in Figure 7.20(b).	Shade the region as in Figure 7.20(c).

The region common to all three graphs is the graph of the solution set of the given system of inequalities. See Figure 7.20(d).

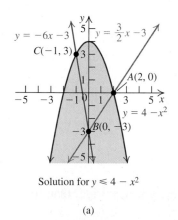

Solution for $y \leq 4 - x^2$

(a)

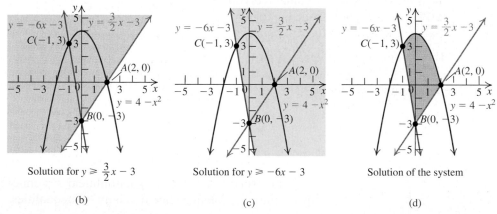

Solution for $y \geq \dfrac{3}{2}x - 3$ Solution for $y \geq -6x - 3$ Solution of the system

(b)   (c)   (d)

FIGURE 7.20

Use the substitution method to locate the points of intersection A, B, and C of the corresponding equation. See Figure 7.20(d). Solve all pairs of corresponding equations.

The point $A(2, 0)$ is the solution of the system:

$$\begin{cases} y = 4 - x^2 & \text{Boundary of inequality (1)} \\ y = \dfrac{3}{2}x - 3 & \text{Boundary of inequality (2)} \end{cases}$$

The point $B(0, -3)$ is the solution of the system:

$$\begin{cases} y = \dfrac{3}{2}x - 3 & \text{Boundary of inequality (2)} \\ y = -6x - 3 & \text{Boundary of inequality (3)} \end{cases}$$

Finally, the point $C(-1, 3)$ is the solution of the system:

$$\begin{cases} y = -6x - 3 & \text{Boundary of inequality (1)} \\ y = 4 - x^2 & \text{Boundary of inequality (2)} \end{cases}$$ ■ ■ ■

Practice Problem 5   Graph the solution set of the system of inequalities.

$$\begin{cases} y \geq x^2 + 1 \\ y \leq -x + 13 \\ y < 4x + 13 \end{cases}$$

SECTION 7.4 ■ Exercises

A EXERCISES Basic Skills and Concepts

1. In a nonlinear system of equations or inequalities, __at least one__ equation or inequality must be nonlinear.

2. Both the __substitution__ and __elimination__ methods can also be used to solve nonlinear systems of equations and inequalities.

3. When graphing an inequality's corresponding equation, use a(n) __solid__ curve if the inequality symbol is ≤ or ≥.

4. When graphing a system of inequalities, any point in the plane can serve as a test point provided that the point is not on __any__ of the graphs of the corresponding equations.

5. *True or False* If $4x + 3y = 8$ is one of the equations in a system of equations, then the system is linear. False

6. *True or False* If $x^2 + 2y < 5$ is one of the inequalities in a system of inequalities, then the point $(2, 1)$ cannot be used as a test point in graphing the system. False

In Exercises 7–14, determine which of the given ordered pairs are solutions of each system of equations.

7. $\begin{cases} 2x + 3y = 3 \\ x - y^2 = 2 \end{cases}$ $(1, -3), (3, -1), (6, 3), \left(5, \dfrac{1}{2}\right)$
$(3, -1)$

8. $\begin{cases} x + 2y = 6 \\ y = x^2 \end{cases}$ $(2, 2), (-2, 4), (0, 3), (1, 2)$
$(-2, 4)$

9. $\begin{cases} 5x - 2y = 7 \\ x^2 + y^2 = 2 \end{cases}$ $\left(\dfrac{5}{4}, 1\right), \left(0, \dfrac{11}{4}\right), (1, -1), (3, 4)$
$(1, -1)$

10. $\begin{cases} x - 2y = -5 \\ x^2 + y^2 = 25 \end{cases}$ $(1, 3), (-5, 0), (3, 4), (3, -4)$
$(-5, 0), (3, 4)$

11. $\begin{cases} 4x^2 + 5y^2 = 180 \\ x^2 - y^2 = 9 \end{cases}$ $(5, 4), (-5, 4), (3, 0), (-5, -4)$
$(5, 4), (-5, 4), (-5, -4)$

12. $\begin{cases} x^2 - y^2 = 3 \\ x^2 - 4x + y^2 = -3 \end{cases}$ $(2, 1), (2, -1), (-2, 1), (-2, -1)$
$(2, 1), (2, -1)$

13. $\begin{cases} y = e^{x-1} \\ y = 2x - 1 \end{cases}$ $(1, 1), (0, -1), (2, e), (-1, -3)$
$(1, 1)$

14. $\begin{cases} y = \ln(x + 1) \\ y = x \end{cases}$ $(1, 1), (0, 0), (-1, -1), (e - 1, 1)$
$(0, 0)$

In Exercises 15–30, solve each system of equations by the substitution method. Check your solutions.

15. $\begin{cases} y = x^2 \\ y = x + 2 \end{cases}$ $\{(-1, 1), (2, 4)\}$

16. $\begin{cases} y = x^2 \\ x + y = 6 \end{cases}$ $\{(2, 4), (-3, 9)\}$

17. $\begin{cases} x^2 - y = 6 \\ x - y = 0 \end{cases}$ $\{(3, 3), (-2, -2)\}$

18. $\begin{cases} x^2 - y = 6 \\ 5x - y = 0 \end{cases}$ $\{(-1, -5), (6, 30)\}$

19. $\begin{cases} x^2 + y^2 = 9 \\ x = 3 \end{cases}$ $\{(3, 0)\}$

20. $\begin{cases} x^2 + y^2 = 9 \\ y = 3 \end{cases}$ $\{(0, 3)\}$

21. $\begin{cases} x^2 + y^2 = \dfrac{5}{2} \\ x - y = -3 \end{cases}$ $\left\{\left(-\dfrac{5}{2}, 1\right), (-1, 2)\right\}$

22. $\begin{cases} x^2 + y^2 = 13 \\ 2x - 3y = 0 \end{cases}$ $\{(3, 2), (-3, -2)\}$

23. $\begin{cases} x^2 - 4x + y^2 = -2 \\ x - y = 2 \end{cases}$ $\{(1, -1), (3, 1)\}$

24. $\begin{cases} x^2 - 8y + y^2 = -6 \\ 2x - y = 1 \end{cases}$ $\{(1, 1), (3, 5)\}$

25. $\begin{cases} x - y = -2 \\ xy = 3 \end{cases}$ $\{(-3, -1), (1, 3)\}$

26. $\begin{cases} x - 2y = 4 \\ xy = 6 \end{cases}$ $\{(-2, -3), (6, 1)\}$

27. $\begin{cases} 4x^2 + y^2 = 25 \\ x + y = 5 \end{cases}$ $\{(2, 3), (0, 5)\}$

28. $\begin{cases} x^2 + 4y^2 = 16 \\ x + 2y = 4 \end{cases}$ $\{(4, 0), (0, 2)\}$

29. $\begin{cases} x^2 - y^2 = 24 \\ 5x - 7y = 0 \end{cases}$ $\{(7, 5), (-7, -5)\}$

30. $\begin{cases} x^2 - 7y^2 = 9 \\ x - y = 3 \end{cases}$ $\{(3, 0), (4, 1)\}$

In Exercises 31–40, solve each system of equations (only real solutions) by the elimination method. Check your solutions.

31. $\begin{cases} x^2 + y^2 = 20 \\ x^2 - y^2 = 12 \end{cases}$ †

32. $\begin{cases} x^2 + 8y^2 = 9 \\ 3x^2 + y^2 = 4 \end{cases}$ †

33. $\begin{cases} x^2 + 2y^2 = 12 \\ 7y^2 - 5x^2 = 8 \end{cases}$ $\{(2, 2), (-2, 2), (2, -2), (-2, -2)\}$

34. $\begin{cases} 3x^2 + 2y^2 = 77 \\ x^2 - 6y^2 = 19 \end{cases}$ $\{(-5, 1), (5, 1), (-5, -1), (5, -1)\}$

†Due to space constrictions, answers to these exercises may be found in the Answers beginning on page A–1 in the back of the book.

35. $\begin{cases} x^2 - y = 2 \\ 2x - y = 4 \end{cases}$ $\varnothing$

36. $\begin{cases} x^2 + 3y = 0 \\ x - y = -12 \end{cases}$ $\varnothing$

37. $\begin{cases} x^2 + y^2 = 5 \\ 3x^2 - 2y^2 = -5 \end{cases}$ †

38. $\begin{cases} x^2 + 4y^2 = 5 \\ 9x^2 - y^2 = 8 \end{cases}$ †

39. $\begin{cases} x^2 + y^2 + 2x = 9 \\ x^2 + 4y^2 + 3x = 14 \end{cases}$ †

40. $\begin{cases} x^2 + y^2 = 9 \\ x^2 + y^2 - 18x = 0 \end{cases}$ †

In Exercises 41–54, use any method to solve each system of equations.

$\{(5,3),(3,5)\}$

$\left\{(3,2),\left(\dfrac{23}{3}, \dfrac{-22}{3}\right)\right\}$

41. $\begin{cases} x + y = 8 \\ xy = 15 \end{cases}$

42. $\begin{cases} 2x + y = 8 \\ x^2 - y^2 = 5 \end{cases}$

43. $\begin{cases} x^2 + y^2 = 2 \\ 3x^2 + 3y^2 = 9 \end{cases}$ $\varnothing$

44. $\begin{cases} x^2 + y^2 = 5 \\ 3x^2 - y^2 = 11 \end{cases}$ †

45. $\begin{cases} y^2 = 4x + 4 \\ y = 2x - 2 \end{cases}$

46. $\begin{cases} xy = 125 \\ y = x^2 \end{cases}$ $\{(5,25)\}$

$\{(0,-2),(3,4)\}$

47. $\begin{cases} x^2 + 4y^2 = 25 \\ x - 2y + 1 = 0 \end{cases}$ †

48. $\begin{cases} y = x^2 - 5x + 4 \\ 3x + y = 3 \end{cases}$ $\{(1,0)\}$

49. $\begin{cases} x^2 - 3y^2 = 1 \\ x^2 + 4y^2 = 8 \end{cases}$ †

50. $\begin{cases} 4x^2 - y^2 = 12 \\ 4y^2 - x^2 = 12 \end{cases}$ †

51. $\begin{cases} x^2 - xy + 5x = 4 \\ 2x^2 - 3xy + 10x = -2 \end{cases}$ $\left\{(2,5),\left(-7,\dfrac{-10}{7}\right)\right\}$

52. $\begin{cases} x^2 - xy + x = -4 \\ 3x^2 - 2xy - 2x = 4 \end{cases}$ $\left\{(-2,-3),\left(6,\dfrac{23}{3}\right)\right\}$

53. $\begin{cases} x^2 + y^2 - 8x = -8 \\ x^2 - 4y^2 + 6x = 0 \end{cases}$ †

54. $\begin{cases} 4x^2 + y^2 - 9y = -4 \\ 4x^2 - y^2 - 3y = 0 \end{cases}$ †

In Exercises 55–68, use the procedure for graphing a nonlinear inequality in two variables to graph each inequality.

55. $y \le x^2 + 2$ †

56. $y < x^2 - 3$ †

57. $y > x^2 - 1$ †

58. $y \ge x^2 + 1$ †

59. $y \le (x - 1)^2 + 3$ †

60. $y > -(x + 1)^2 + 4$ †

61. $x^2 + y^2 > 4$ †

62. $x^2 + y^2 \le 9$ †

63. $(x - 1)^2 + (y - 2)^2 \le 9$ †

64. $(x + 2)^2 + (y - 1)^2 > 4$ †

65. $y < 2^x$ †

66. $y \ge e^x$ †

67. $y < \log x$ †

68. $y \ge \ln x$ †

For Exercises 69–74, use the following figure to indicate the region or regions that correspond to the graph of each system of inequalities.

69. $\begin{cases} y \ge x^2 - 5 \\ x^2 + y^2 \le 25 \\ x \ge 0 \end{cases}$ D, K

70. $\begin{cases} y \ge x^2 - 5 \\ x^2 + y^2 \ge 25 \\ x \ge 0 \end{cases}$ H

71. $\begin{cases} y \le x^2 - 5 \\ x^2 + y^2 \le 25 \\ x \ge 0 \end{cases}$ E, L

72. $\begin{cases} y \le x^2 - 5 \\ x^2 + y^2 \ge 25 \\ y \ge 0 \end{cases}$ A, F

73. $\begin{cases} y \le x^2 - 5 \\ x^2 + y^2 \le 25 \\ y \ge 0 \end{cases}$ E, B

74. $\begin{cases} y \ge x^2 - 5 \\ x^2 + y^2 \le 25 \\ y \ge 0 \end{cases}$ C, D

In Exercises 75–82, graph the region determined by each system of inequalities and label all points of intersection.

75. $\begin{cases} x + y \le 6 \\ y \ge x^2 \end{cases}$ †

76. $\begin{cases} x + y \ge 6 \\ y \ge x^2 \end{cases}$ †

77. $\begin{cases} y \ge 2x + 1 \\ y \le -x^2 + 2 \end{cases}$ †

78. $\begin{cases} y \le 2x + 1 \\ y \le -x^2 + 2 \end{cases}$ †

79. $\begin{cases} y \ge x \\ x^2 + y^2 \le 1 \end{cases}$ †

80. $\begin{cases} y \le x \\ x^2 + y^2 \le 1 \end{cases}$ †

81. $\begin{cases} x^2 + y^2 \le 25 \\ x^2 + y \le 5 \end{cases}$ †

82. $\begin{cases} x^2 + y^2 \le 25 \\ x^2 + y \ge 5 \end{cases}$ †

B EXERCISES Applying the Concepts

Exercises 83 and 84 show the demand and supply functions of a product. In each case, p represents the price per unit and x represents the number of units, in hundreds. Determine the price that gives market equilibrium and the number of units that are demanded and supplied at that price.

83. $\begin{cases} p + 2x^2 = 96 \\ p - 13x = 39 \end{cases}$ $(3, 78)$

84. $\begin{cases} p^2 + 6p + 3x = 75 \\ -p + x = 13 \end{cases}$ $(16, 3)$

In Exercises 85–92, use systems of equations to solve the problem.

85. Numbers. Find two positive numbers whose sum is 24 and whose product is 143. 11 and 13

86. Numbers. Find two positive numbers whose difference is 13 and whose product is 114. 6 and 19

87. Commercial land. A commercial parcel of land is in the form of a trapezoid with two right angles, as shown in the figure. The oblique side is 100 meters in length, and two sides that meet at the vertex of one of the right angles are equal. The perimeter of the land is 360 meters. Find

a. the lengths of the remaining sides. $x = 60$ m, $y = 140$ m

b. the area of the land. 6000 m²

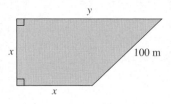

88. Dividing a pasture. The area of a rectangular pasture is 8750 square feet. The pasture is divided into three smaller pastures by two fences parallel to the shorter sides. The widths of two of the larger pastures are the same, and the width of the third is one-half that of the others. Find the dimensions of the original pasture if the perimeter of the smaller of the subdivisions is 190 feet. 25 ft by 350 ft

89. Chartering a fishing boat. A group of students of a fraternity chartered a fishing boat that would cost $960. Before the date of the fishing trip, eight more students joined the group and agreed to pay their share of the cost of the charter. The cost per student was reduced by $6. Find the original number of students in the group and the original cost per student. 32 people and $30

90. Carpet sale. A salesperson sold a square carpet and a rectangular carpet whose length was 6 feet more than its width. The combined area of the two carpets was 540 square feet. The price of the square carpet was $12 per square yard, and the price of the rectangular carpet was $10 per square yard. If the rectangular piece was $50 more than the square piece, find the dimensions of each carpet. 15 ft by 15 ft and 15 ft by 21 ft

91. Buying and selling stocks. Flat Fee Brokerage, Inc., charges $100 commission for every transaction (purchase or sale) you make. Anju bought a block of stock containing over 400 shares that cost her $10,000, including the commission. After one year, she received a stock dividend of 30 shares. She then sold all of her stock for $3 more per share than it cost and made a profit of $1900 (after paying the commission). How many shares did she buy, and what was the original price of each share? 450 original shares, purchased at $22 per share

92. Trading stocks. Repeat Exercise 91 with the following details: Anju does not receive dividends and decides to keep 100 shares of the stock. She sells the remaining stock for $10 more per share than it cost and makes a profit of $3900 after paying the commission. 660 original shares, purchased at $15 per share

C EXERCISES Beyond the Basics

In Exercises 93–102, graph each inequality.

93. $xy \geq 1$ † **94.** $xy < 3$ †

95. $y < 2^x$ † **96.** $y \leq 2^{x-1}$ †

97. $y \geq 3^{x-2}$ † **98.** $y > -3^x$ †

99. $y \geq \ln x$ † **100.** $y < -\ln x$ †

101. $y > \ln(x-1)$ † **102.** $y \leq -\ln(x+1)$ †

In Exercises 103–104, graph the solution set of each inequality.

103. $1 \leq x^2 + y^2 \leq 9$ † **104.** $x^2 \leq y \leq 4x^2$ †

In Exercises 105–110, solve each system of equations and verify the solution graphically.

105. $\begin{cases} y = 3^x + 4 \\ y = 3^{2x} - 2 \end{cases}$ † **106.** $\begin{cases} y = \ln x + 1 \\ y = \ln x - 2 \end{cases}$ †

107. $\begin{cases} y = 2^x + 3 \\ y = 2^{2x} + 1 \end{cases}$ † **108.** $\begin{cases} y = 2^x + 3 \\ y = 4^x + 1 \end{cases}$ †

109. $\begin{cases} x = 3^y \\ 3^{2y} = 3x - 2 \end{cases}$ † **110.** $\begin{cases} x = 3^y \\ x^2 = 3^{y+1} - 2 \end{cases}$ †

In Exercises 111–114, the given circle intersects the given line at the points A and B. Find the coordinates of A and B and compute the distance $d(A, B)$.

111. Circle: $x^2 + y^2 - 7x + 5y + 6 = 0$; line: x-axis

112. The circle of Exercise 111 and the y-axis

113. Circle: $x^2 + y^2 + 2x - 4y - 5 = 0$; line: $x - y + 1 = 0$

114. Circle: $x^2 + y^2 - 6x - 8y - 50 = 0$; line: $2x + y - 5 = 0$

115. Show that the line with equation $x + 2y + 6 = 0$ is a tangent line to the circle with equation $x^2 + y^2 - 2x + 2y - 3 = 0$. (A line that intersects the circle at exactly one point is a **tangent line** to the circle.)

116. Repeat Exercise 115 for the line with equation $3x - 4y = 0$ and the circle with equation $x^2 + y^2 - 2x - 4y + 4 = 0$.

Critical Thinking

117. Consider a system of two equations in two variables.
$$\begin{cases} y = mx + b \\ y = Ax^3 + Bx^2 + Cx + D, A \neq 0 \end{cases} †$$
Explain whether it is possible for this system to have
a. no real solutions.
b. one real solution.
c. two real solutions.
d. three real solutions.
e. four real solutions.
f. more than four real solutions.

118. Repeat Exercise 117 for the system.
$$\begin{cases} y = mx + b \\ y = Ax^4 + Bx^3 + Cx^2 + Dx + E, A \neq 0 \end{cases} †$$

Answers:
111. $A(1, 0), B(6, 0), d(A, B) = 5$
112. $A(0, -3), B(0, -2), d(A, B) = 1$
113. $A(-2, -1), B(2, 3), d(A, B) = 4\sqrt{2}$
114. $A(1 + \sqrt{14}, 3 - 2\sqrt{14}), B(1 - \sqrt{14}, 3 + 2\sqrt{14}), d(A, B) = 2\sqrt{70}$
115. $(0, -3)$ **116.** $\left(\dfrac{8}{5}, \dfrac{6}{5}\right)$

Partial-Fraction Decomposition

Before Starting this Section, Review

1. Division of polynomials (Appendix A, page 766)
2. Factoring polynomials (Appendix A, page 769)
3. Irreducible polynomials (Appendix A, page 769)
4. Rational expressions (Appendix A, page 773)

Objectives

1. Become familiar with partial-fraction decomposition.

2. Decompose $\dfrac{P(x)}{Q(x)}$, when $Q(x)$ has only distinct linear factors.

3. Decompose $\dfrac{P(x)}{Q(x)}$, when $Q(x)$ has repeated linear factors.

4. Decompose $\dfrac{P(x)}{Q(x)}$, when $Q(x)$ has distinct irreducible quadratic factors.

5. Decompose $\dfrac{P(x)}{Q(x)}$, when $Q(x)$ has repeated irreducible quadratic factors.

Georg Simon Ohm (1787–1854)

Georg Ohm was born in Erlangen, Germany. His father, Johann, a mechanic and a self-educated man, was interested in philosophy and mathematics and gave his children an excellent education in physics, chemistry, mathematics, and philosophy through his own teachings. The Ohm's law named in Georg Ohm's honor appeared in his famous pamphlet *Die galvanische Kette, mathematisch bearbeitet* (1827). Ohm's work was not appreciated in Germany, where it was first published. Eventually, it was recognized, first by the British Royal Society, which awarded Ohm the Copley Medal in 1841. In 1849, Ohm became a curator of the Bavarian Academy's science museum and began to lecture at the University of Munich. Finally, in 1852 (two years before his death), Ohm achieved his lifelong ambition of becoming chair of physics at the University of Munich.

OHM'S LAW

If a voltage is applied across a resistor (such as a wire), a current will flow. Ohm's law states that in a circuit at constant temperature,

$$I = \frac{V}{R}$$

where V is the voltage (measured in volts), I is the current (measured in amperes), and R is the resistance (measured in ohms). A **parallel circuit** consists of two or more resistors connected as shown in Figure 7.21. In the figure, the current I from the source divides into two currents, I_1 and I_2, with I_1 going through one resistor and I_2 going through the other, so that $I = I_1 + I_2$. Suppose we want to find the total resistance R in the circuit due to the two resistors R_1 and R_2 connected in parallel. Ohm's law can be used to show that in the parallel circuit in Figure 7.21, $\dfrac{1}{R} = \dfrac{1}{R_1} + \dfrac{1}{R_2}$. This result is called the *parallel property of resistors*. In Example 7, we examine the parallel property of resistors by using partial fractions.

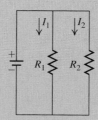

FIGURE 7.21 Parallel resistors

1 Become familiar with partial-fraction decomposition.

Partial Fractions

Recall that to add two rational expressions such as $\dfrac{2}{x+3}$ and $\dfrac{3}{x-1}$ required you to find the least common denominator, rewrite expressions with the same denominator, and then add the corresponding numerators. This procedure gives

$$\frac{2}{x+3} + \frac{3}{x-1} = \frac{2(x-1)}{(x+3)(x-1)} + \frac{3(x+3)}{(x-1)(x+3)} = \frac{5x+7}{(x+3)(x-1)}$$

In some applications of algebra to more advanced mathematics, you need a *reverse* procedure for *splitting* a fraction such as $\dfrac{5x+7}{(x-1)(x+3)}$ into the simpler fractions to obtain

$$\frac{5x+7}{(x+3)(x-1)} = \frac{2}{x+3} + \frac{3}{x-1}.$$

Each of the two fractions on the right is called a **partial fraction**. Their sum is called the **partial-fraction decomposition** of the rational expression on the left.

A rational expression $\dfrac{P(x)}{Q(x)}$ is called **improper** if the degree $P(x) \geq$ degree $Q(x)$ and is called **proper** if the degree $P(x) <$ degree $Q(x)$. Examples of improper rational expressions are

$$\frac{x^3}{x^2-1}, \quad \frac{(x+2)(x-1)}{(x-2)(x+3)}, \quad \text{and} \quad \frac{x^2+x+1}{2x+3}.$$

Using long division, we can express an improper rational expression $\dfrac{P(x)}{Q(x)}$ as the sum of a polynomial and a proper rational expression:

$$\frac{P(x)}{Q(x)} \quad = \quad S(x) \quad + \quad \frac{R(x)}{Q(x)}$$

$$\uparrow \qquad\qquad \uparrow \qquad\qquad \uparrow$$

| Improper Expression | Polynomial | Proper Expression |

RECALL

Any polynomial $Q(x)$ with real coefficients can be factored so that each factor is linear or is an irreducible quadratic factor.

We therefore restrict our discussion of the decomposition of $\dfrac{P(x)}{Q(x)}$ into partial fractions to cases involving proper rational expressions. We also assume that $P(x)$ and $Q(x)$ have no common factor.

The problem of decomposing a proper fraction $\dfrac{P(x)}{Q(x)}$ into partial fractions depends on the type of factors in the denominator $Q(x)$. In this section, we consider four cases for $Q(x)$.

2 Decompose $\dfrac{P(x)}{Q(x)}$, when $Q(x)$ has only distinct linear factors.

$Q(x)$ Has Only Distinct Linear Factors

CASE 1: THE DENOMINATOR IS THE PRODUCT OF DISTINCT (NONREPEATED) LINEAR FACTORS

Suppose $Q(x)$ can be factored as

$$Q(x) = (x - a_1)(x - a_2) \cdots (x - a_n),$$

with no factor repeated. The partial-fraction decomposition of $\dfrac{P(x)}{Q(x)}$ is of the form

$$\frac{P(x)}{Q(x)} = \frac{A_1}{x - a_1} + \frac{A_2}{x - a_2} + \cdots + \frac{A_n}{x - a_n}$$

where $A_1, A_2, \ldots, A_n$ are constants to be determined.

We can find the constants $A_1, A_2, \ldots, A_n$ by using the following procedure. Note that if the number of constants is small, we use the letters $A, B, C, \ldots$, instead of $A_1, A_2, A_3, \ldots$.

FINDING THE SOLUTION: A PROCEDURE

EXAMPLE 1 Partial-Fraction Decomposition

OBJECTIVE

Find the partial-fraction decomposition of a rational expression.

Step 1 Write the form of the partial-fraction decomposition with the unknown constants $A, B, C, \ldots$ in the numerators of the decomposition.

Step 2 Multiply both sides of the equation in Step 1 by the original denominator. Use the distributive property and eliminate common factors. Simplify.

Step 3 Write both sides of the equation in Step 2 in descending powers of x and equate the coefficients of like powers of x.

Step 4 Solve the linear system resulting from Step 3 for the constants $A, B, C, \ldots$.

Step 5 Substitute the values you found for $A, B, C, \ldots$ into the equation in Step 1 and write the partial-fraction decomposition.

EXAMPLE

Find the partial fraction decomposition of $\dfrac{3x + 26}{(x - 3)(x + 4)}$.

$$\frac{3x + 26}{(x - 3)(x + 4)} = \frac{A}{x - 3} + \frac{B}{x + 4}$$

Multiply both sides by $(x - 3)(x + 4)$ and simplify to obtain

$$3x + 26 = (x + 4)A + (x - 3)B$$
$$3x + 26 = (A + B)x + (4A - 3B)$$
$$\begin{cases} 3 = A + B & \text{Equate coefficients of } x. \\ 26 = 4A - 3B & \text{Equate constant coefficients.} \end{cases}$$

Solving the system of equations in Step 3, we obtain $A = 5$ and $B = -2$.

$$\frac{3x + 26}{(x - 3)(x + 4)} = \frac{5}{x - 3} + \frac{-2}{x + 4} \qquad \begin{array}{l} \text{Replace } A \text{ with 5 and} \\ B \text{ with } -2 \text{ in Step 1.} \end{array}$$
$$\frac{3x + 26}{(x - 3)(x + 4)} = \frac{5}{x - 3} - \frac{2}{x + 4}$$

■ ■ ■

Practice Problem 1 Find the partial fraction decomposition of $\dfrac{2x - 7}{(x + 1)(x - 2)}$. ■

| EXAMPLE 2 | Finding the Partial-Fraction Decomposition When the Denominator Has Only Distinct Linear Factors |

Find the partial-fraction decomposition of the expression

$$\frac{10x - 4}{x^3 - 4x}.$$

SOLUTION

First, factor the denominator.

$$x^3 - 4x = x(x^2 - 4) \qquad \text{Distributive property}$$
$$= x(x - 2)(x + 2) \qquad x^2 - 4 = (x - 2)(x + 2)$$

Each of the factors x, $x - 2$, and $x + 2$ becomes a denominator for a partial fraction.

Step 1 The partial-fraction decomposition is given by

$$\frac{10x - 4}{x(x - 2)(x + 2)} = \frac{A}{x} + \frac{B}{x - 2} + \frac{C}{x + 2} \qquad \begin{array}{l}\text{Use } A, B, \text{ and } C \text{ instead of} \\ A_1, A_2, \text{ and } A_3.\end{array}$$

Step 2 Multiply both sides by the common denominator $x(x - 2)(x + 2)$.

Distribute

$$\cancel{x}(\cancel{x - 2})(\cancel{x + 2})\left[\frac{10x - 4}{\cancel{x}(\cancel{x - 2})(\cancel{x + 2})}\right] = x(x - 2)(x + 2)\left[\frac{A}{x} + \frac{B}{x - 2} + \frac{C}{x + 2}\right]$$

$$10x - 4 = A(x - 2)(x + 2) + Bx(x + 2) + Cx(x - 2) \quad (1) \qquad \begin{array}{l}\text{Distributive property;} \\ \text{simplify.}\end{array}$$

$$= A(x^2 - 4) + B(x^2 + 2x) + C(x^2 - 2x) \qquad \text{Multiply.}$$
$$= Ax^2 - 4A + Bx^2 + 2Bx + Cx^2 - 2Cx \qquad \text{Distributive property}$$
$$= (A + B + C)x^2 + (2B - 2C)x - 4A. \qquad \text{Combine like terms.}$$

Step 3 Now use the fact that *two equal polynomials have equal corresponding coefficients*. Writing $10x - 4 = 0x^2 + 10x - 4$, we have

$$0x^2 + 10x - 4 = (A + B + C)x^2 + (2B - 2C)x - 4A.$$

Equating corresponding coefficients leads to the system of equations.

$$\begin{cases} A + B + C = 0 & \text{Equate coefficients of } x^2. \\ 2B - 2C = 10 & \text{Equate coefficients of } x. \\ -4A = -4 & \text{Equate constant coefficients.} \end{cases}$$

Step 4 Solve the system of equations in Step 3 to obtain $A = 1$, $B = 2$, and $C = -3$. (See Exercise 57.)

Step 5 The partial-fraction decomposition is

$$\frac{10x - 4}{x^3 - 4x} = \frac{1}{x} + \frac{2}{x - 2} + \frac{-3}{x + 2} = \frac{1}{x} + \frac{2}{x - 2} - \frac{3}{x + 2}.$$

Alternative Solution of Example 2

In Example 2, to find the constants A, B, and C, we *equated coefficients* of like powers of x and solved the resulting system. An alternative (and sometimes quicker) method is to *substitute* well-chosen values for x in equation (1) found in Step 2:

$$10x - 4 = A(x - 2)(x + 2) + Bx(x + 2) + Cx(x - 2) \qquad (1)$$

TECHNOLOGY CONNECTION

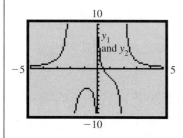

You can reinforce your partial-fraction connection graphically:

The graph of

$$Y_1 = \frac{10x - 4}{x^3 - 4x}$$

appears to be identical to the graph of

$$Y_2 = \frac{1}{x} + \frac{2}{x - 2} - \frac{3}{x + 2}.$$

The graphs of Y_1 and Y_2 are the same.

---

3 Decompose $\dfrac{P(x)}{Q(x)}$, when $Q(x)$ has repeated linear factors.

Substitute $x = 2$ in equation (1) to cause the terms containing A and C to be 0.

$$10(2) - 4 = A(2 - 2)(2 + 2) + B(2)(2 + 2) + C(2)(2 - 2)$$

$$16 = 8B \qquad \text{Simplify.}$$

$$2 = B \qquad \text{Solve for } B.$$

Now substitute $x = -2$ in equation (1) to get

$$10(-2) - 4 = A(-2 - 2)(-2 + 2) + B(-2)(-2 + 2) + C(-2)(-2 - 2)$$

$$-24 = 8C \qquad \text{Simplify.}$$

$$-3 = C \qquad \text{Solve for } C.$$

Finally, substitute $x = 0$ in equation (1) to get

$$10(0) - 4 = A(0 - 2)(0 + 2) + B(0)(0 + 2) + C(0)(0 - 2)$$

$$-4 = -4A \qquad \text{Simplify.}$$

$$1 = A \qquad \text{Solve for } A.$$

Because we found the same values for A, B, and C, the partial-fraction decomposition will also be the same. ■ ■ ■

Practice Problem 2 Find the partial-fraction decomposition of

$$\frac{3x^2 + 4x + 3}{x^3 - x}.$$ ■

$Q(x)$ Has Repeated Linear Factors

> **CASE 2: THE DENOMINATOR HAS A REPEATED LINEAR FACTOR**
>
> Let $(x - a)^m$ be the linear factor $(x - a)$ that is repeated m times in $Q(x)$.
>
> Then the portion of the partial-fraction decomposition of $\dfrac{P(x)}{Q(x)}$ that corresponds to the factor $(x - a)^m$ is
>
> $$\frac{A_1}{x - a} + \frac{A_2}{(x - a)^2} + \cdots + \frac{A_m}{(x - a)^m}$$
>
> where $A_1, A_2, \ldots, A_m$ are constants.

EXAMPLE 3 **Finding the Partial-Fraction Decomposition When the Denominator Has Repeated Linear Factors**

Find the partial-fraction decomposition of: $\dfrac{x + 4}{(x + 3)(x - 1)^2}$

SOLUTION

Step 1 The linear factor $(x - 1)$ is repeated twice, and the factor $(x + 3)$ is nonrepeating. So the partial-fraction decomposition has the form:

$$\frac{x + 4}{(x + 3)(x - 1)^2} = \frac{A}{x + 3} + \frac{B}{x - 1} + \frac{C}{(x - 1)^2}$$

Step 2–4 Multiply both sides of the equation in Step 1 by the original denominator, $(x + 3)(x - 1)^2$, then use the distributive property on the right side and simplify to get

$$x + 4 = A(x - 1)^2 + B(x + 3)(x - 1) + C(x + 3).$$

Substitute $x = 1$ in the last equation to obtain

$$1 + 4 = A(1 - 1)^2 + B(1 + 3)(1 - 1) + C(1 + 3).$$

This simplifies to $5 = 4C$, or $C = \dfrac{5}{4}$.

To find A, we substitute $x = -3$ to get

$$-3 + 4 = A(-3 - 1)^2 + B(-3 + 3)(-3 - 1) + C(-3 + 3).$$

This simplifies to $1 = 16A$, or $A = \dfrac{1}{16}$.

To find B, we replace x with any convenient number, say, 0. We have

$$0 + 4 = A(0 - 1)^2 + B(0 + 3)(0 - 1) + C(0 + 3)$$
$$4 = A - 3B + 3C \qquad\qquad\qquad \text{Simplify.}$$

Now replace A with $\dfrac{1}{16}$ and C with $\dfrac{5}{4}$ in the last equation.

$$4 = \frac{1}{16} - 3B + 3\left(\frac{5}{4}\right)$$

$$64 = 1 - 48B + 60 \qquad \text{Multiply both sides by 16.}$$

$$B = -\frac{1}{16} \qquad\qquad \text{Solve for } B.$$

We have $A = \dfrac{1}{16}$, $B = -\dfrac{1}{16}$, and $C = \dfrac{5}{4}$.

Step 5 Substitute $A = \dfrac{1}{16}$, $B = -\dfrac{1}{16}$, and $C = \dfrac{5}{4}$ in the decomposition in Step 1 to get

$$\frac{x + 4}{(x + 3)(x - 1)^2} = \frac{\frac{1}{16}}{x + 3} + \frac{-\frac{1}{16}}{x - 1} + \frac{\frac{5}{4}}{(x - 1)^2}$$

or

$$\frac{x + 4}{(x + 3)(x - 1)^2} = \frac{1}{16(x + 3)} - \frac{1}{16(x - 1)} + \frac{5}{4(x - 1)^2} \qquad ■ ■ ■$$

Practice Problem 3 Find the partial fraction decomposition of

$$\frac{x + 5}{x(x - 1)^2}. \qquad\qquad ■$$

4 Decompose $\dfrac{P(x)}{Q(x)}$, when $Q(x)$ has distinct irreducible quadratic factors.

$Q(x)$ Has Distinct Irreducible Quadratic Factors

> **CASE 3: THE DENOMINATOR HAS A DISTINCT (NONREPEATED) IRREDUCIBLE QUADRATIC FACTOR**
>
> Suppose $ax^2 + bx + c$ is an irreducible quadratic factor of $Q(x)$. Then the portion of the partial-fraction decomposition of $\dfrac{P(x)}{Q(x)}$ that corresponds to $ax^2 + bx + c$ has the form:
>
> $$\frac{Ax + B}{ax^2 + bx + c}$$

RECALL

The expression $ax^2 + bx + c$ is irreducible if $b^2 - 4ac < 0$. In that case, $ax^2 + bx + c = 0$ has no real solutions.

EXAMPLE 4 **Finding the Partial-Fraction Decomposition When the Denominator Has a Distinct Irreducible Quadratic Factor**

Find the partial-fraction decomposition of: $\dfrac{3x^2 - 8x + 1}{(x - 4)(x^2 + 1)}$

SOLUTION

Step 1 The factor $x - 4$ is linear and $x^2 + 1$ is irreducible; so the partial fraction decomposition has the form:

$$\frac{3x^2 - 8x + 1}{(x - 4)(x^2 + 1)} = \frac{A}{x - 4} + \frac{Bx + C}{x^2 + 1}$$

Step 2 Multiply both sides of the decomposition in Step 1 by the original denominator, $(x - 4)(x^2 + 1)$, and simplify to get

$$3x^2 - 8x + 1 = A(x^2 + 1) + (Bx + C)(x - 4).$$

Substitute $x = 4$ to find $17 = 17A$, or $A = 1$.

Step 3 Collect like terms and write both sides of the equation in Step 2 in descending powers of x.

$$3x^2 - 8x + 1 = (A + B)x^2 + (-4B + C)x + (A - 4C)$$

Equate corresponding coefficients to get

$$\begin{cases} A + B = 3 & (1) \quad \text{Equate the coefficients of } x^2. \\ -4B + C = -8 & (2) \quad \text{Equate the coefficients of } x. \\ A - 4C = 1 & (3) \quad \text{Equate the constant coefficients.} \end{cases}$$

Step 4 Substitute $A = 1$ from Step 2 in equation (1) to get $1 + B = 3$, or $B = 2$. Substitute $A = 1$ in equation (3) to get $1 - 4C = 1$, or $C = 0$.

Step 5 Substitute $A = 1, B = 2$, and $C = 0$ into the decomposition in Step 1 to get:

$$\frac{3x^2 - 8x + 1}{(x - 4)(x^2 + 1)} = \frac{1}{x - 4} + \frac{2x}{x^2 + 1}$$ ▪ ▪ ▪

Practice Problem 4 Find the partial-fraction decomposition of:

$$\frac{3x^2 + 5x - 2}{x(x^2 + 2)}$$ ▪

5 Decompose $\dfrac{P(x)}{Q(x)}$, when $Q(x)$ has repeated irreducible quadratic factors.

$Q(x)$ Has Repeated Irreducible Quadratic Factors

CASE 4: THE DENOMINATOR HAS A REPEATED IRREDUCIBLE QUADRATIC FACTOR

Suppose the denominator $Q(x)$ has a factor $(ax^2 + bx + c)^m$, where $m \geq 2$ is an integer and $ax^2 + bx + c$ is irreducible. Then the portion of the partial-fraction decomposition of $\dfrac{P(x)}{Q(x)}$ that corresponds to the factor $(ax^2 + bx + c)^m$ has the form:

$$\frac{A_1 x + B_1}{ax^2 + bx + c} + \frac{A_2 x + B_2}{(ax^2 + bx + c)^2} + \cdots + \frac{A_m x + B_m}{(ax^2 + bx + c)^m}$$

| EXAMPLE 5 | **Finding the Partial-Fraction Decomposition When the Denominator Has a Repeated Irreducible Quadratic Factor** |

Find the partial-fraction decomposition of: $\dfrac{2x^4 - x^3 + 13x^2 - 2x + 13}{(x - 1)(x^2 + 4)^2}$.

SOLUTION

Step 1 The denominator has a nonrepeating linear factor $(x - 1)$ and an irreducible quadratic factor $(x^2 + 4)$ that appears twice. The decomposition therefore has the form:

$$\frac{2x^4 - x^3 + 13x^2 - 2x + 13}{(x - 1)(x^2 + 4)^2} = \frac{A}{x - 1} + \frac{Bx + C}{x^2 + 4} + \frac{Dx + E}{(x^2 + 4)^2}$$

Step 2 Multiply both sides of the decomposition in Step 1 by the original denominator $(x - 1)(x^2 + 4)^2$, use the distributive property, and eliminate common factors:

$$2x^4 - x^3 + 13x^2 - 2x + 13 =$$
$$A(x^2 + 4)^2 + (Bx + C)(x - 1)(x^2 + 4) + (Dx + E)(x - 1).$$

Substitute $x = 1$ and simplify to obtain $25 = 25A$, or $A = 1$.

Steps 3–4 Multiply the factors on the right side of the equation in Step 2 and collect like terms to obtain

$$2x^4 - x^3 + 13x^2 - 2x + 13 =$$
$$(A + B)x^4 + (-B + C)x^3 + (8A + 4B - C + D)x^2 + (-4B + 4C - D + E)x + (16A - 4C - E).$$

Equate corresponding coefficients to get the system:

$$\begin{cases} A + \ B & = \ \ 2 \quad (1) \quad \text{Coefficient of } x^4 \\ \ \ - \ B + \ C & = -1 \quad (2) \quad \text{Coefficient of } x^3 \\ 8A + 4B - \ C + D & = 13 \quad (3) \quad \text{Coefficient of } x^2 \\ \ \ - 4B + 4C - D + E = -2 \quad (4) \quad \text{Coefficient of } x \\ 16A \qquad - 4C \quad - E = 13 \quad (5) \quad \text{Constant coefficient} \end{cases}$$

Now back-substitute $A = 1$ from Step 2 in equation (1) to get $B = 1$. Again, back-substitute $B = 1$ in equation (2) to get $C = 0$.

Back-substitute $A = 1$ and $C = 0$ in equation (5) to obtain $E = 3$. Again, back-substitute $A = 1$, $B = 1$, and $C = 0$ in equation (3) to get $D = 1$. Hence, $A = 1$, $B = 1$, $C = 0$, $D = 1$, and $E = 3$.

Step 5 Substitute $A = 1$, $B = 1$, $C = 0$, $D = 1$, and $E = 3$ in the equation in Step 1 to obtain the partial-fraction decomposition.

$$\frac{2x^4 - x^3 + 13x^2 - 2x + 13}{(x - 1)(x^2 + 4)^2} = \frac{1}{x - 1} + \frac{x}{x^2 + 4} + \frac{x + 3}{(x^2 + 4)^2} \qquad ■ ■ ■$$

Practice Problem 5 Find the partial-fraction decomposition of:

$$\frac{x^2 + 3x + 1}{(x^2 + 1)^2} \qquad ■$$

| EXAMPLE 6 | **Learning a Clever Way to Add** |

Express the sum

$$\frac{1}{2 \cdot 3} + \frac{1}{3 \cdot 4} + \frac{1}{4 \cdot 5} + \cdots + \frac{1}{2009 \cdot 2010}$$

as a fraction of whole numbers in lowest terms.

SOLUTION

Each term in the sum is of the form: $\dfrac{1}{k(k+1)}$. From the partial-fraction decomposition, we have:

$$\frac{1}{k(k+1)} = \frac{1}{k} - \frac{1}{k+1}$$

Therefore, by using this decomposition for each term, we get:

$$\frac{1}{2\cdot 3} + \frac{1}{3\cdot 4} + \frac{1}{4\cdot 5} + \cdots + \frac{1}{2009\cdot 2010}$$

$$= \left(\frac{1}{2} - \frac{1}{3}\right) + \left(\frac{1}{3} - \frac{1}{4}\right) + \left(\frac{1}{4} - \frac{1}{5}\right) + \cdots + \left(\frac{1}{2009} - \frac{1}{2010}\right).$$

After removing parentheses, we see that most terms cancel in pairs $\Big($ such as $-\dfrac{1}{3}$ and $\dfrac{1}{3}\Big)$, leaving only the first and the last term. The original sum is therefore referred to as a *telescoping sum* that "collapses" to

$$\frac{1}{2} - \frac{1}{2010} = \frac{1005 - 1}{2010} = \frac{1004}{2010} = \frac{502}{1005}. \qquad \blacksquare\blacksquare\blacksquare$$

Practice Problem 6   Express the sum $\dfrac{1}{4\cdot 5} + \dfrac{1}{5\cdot 6} + \dfrac{1}{6\cdot 7} + \cdots + \dfrac{1}{3111\cdot 3112}$ as a fraction of whole numbers in lowest terms. $\qquad\blacksquare$

EXAMPLE 7   **Calculating the Resistance in a Circuit**

The total resistance R due to two resistors connected in parallel is given by

$$R = \frac{x(x+5)}{3x+10}.$$

Use partial-fraction decomposition to write $\dfrac{1}{R}$ in terms of partial fractions. What does each term represent?

SOLUTION

From $R = \dfrac{x(x+5)}{3x+10}$, we get the following:

$$\frac{1}{R} = \frac{3x+10}{x(x+5)} \qquad \text{Take the reciprocal of both sides.}$$

$$\frac{3x+10}{x(x+5)} = \frac{A}{x} + \frac{B}{x+5} \qquad \text{Form of partial-fraction decomposition}$$

$$3x+10 = A(x+5) + Bx \qquad \text{Multiply both sides by } x(x+5) \text{ and simplify.}$$

Substitute $x = 0$ in the last equation to obtain $5A = 10$, or $A = 2$. Now substitute $x = -5$ in the same equation to get $-5 = -5B$, or $B = 1$. Thus, we have

$$\frac{1}{R} = \frac{3x+10}{x(x+5)} = \frac{2}{x} + \frac{1}{x+5}$$

$$\frac{1}{R} = \frac{1}{\dfrac{x}{2}} + \frac{1}{x+5} \qquad \text{Rewrite the right side.}$$

RECALL

The total resistance R in a circuit due to two resistors R_1 and R_2 connected in parallel satisfies the equation

$$\frac{1}{R} = \frac{1}{R_1} + \frac{1}{R_2}.$$

The last equation says that if two resistors with resistances $R_1 = \dfrac{x}{2}$ and $R_2 = x + 5$ are connected in parallel, they will produce a total resistance R, where $R = \dfrac{x(x + 5)}{3x + 10}$. ■■■

Practice Problem 7   Repeat Example 7 assuming that $R = \dfrac{(x + 1)(x + 2)}{4x + 7}$. ■

SECTION 7.5 ■ Exercises

A EXERCISES Basic Skills and Concepts

1. In a rational expression, if the degree of the numerator, $P(x)$, is less than the degree of the denominator, $Q(x)$, then the expression is ____proper____.

2. The numerators in the partial-fraction decomposition of a rational expression are constants if the denominator, $Q(x)$, can be factored into ___linear factors___.

3. In a rational expression, if $(x - 8)$ is a linear factor that is repeated three times in the denominator, then the portion of the expression's partial-fraction decomposition that corresponds to $(x - 8)^3$ has ____three____ terms.

4. *True or False* The form of the partial-fraction decomposition of a rational expression depends only on its denominator. True

5. *True or False* Both $(x - 1)^2$ and $(x + 2)^2$ are irreducible quadratic factors of the polynomial $Q(x) = (x - 1)^2(x + 2)^2$. False

6. *True or False* If the denominator of a rational expression is $Q(x) = x^3 - 9x$ then $x, x - 3$, and $x + 3$ become denominators in its partial-fraction decomposition. True

In Exercises 7–16, write the form of the partial-fraction decomposition of each rational expression. You do not need to solve for the constants.

7. $\dfrac{1}{(x - 1)(x + 2)}$ $\dfrac{A}{x - 1} + \dfrac{B}{x + 2}$

8. $\dfrac{x}{(x + 1)(x - 3)}$ $\dfrac{A}{x + 1} + \dfrac{B}{x - 3}$

9. $\dfrac{1}{x^2 + 7x + 6}$ $\dfrac{A}{x + 6} + \dfrac{B}{x + 1}$

10. $\dfrac{3}{x^2 - 6x + 8}$ $\dfrac{A}{x - 2} + \dfrac{B}{x - 4}$

11. $\dfrac{2}{x^3 - x^2}$ $\dfrac{A}{x^2} + \dfrac{B}{x} + \dfrac{C}{x - 1}$

12. $\dfrac{x - 1}{(x + 2)^2(x - 3)}$ †

13. $\dfrac{x^2 - 3x + 3}{(x + 1)(x^2 - x + 1)}$ †

14. $\dfrac{2x + 3}{(x - 1)^2(x^2 + x + 1)}$ †

15. $\dfrac{3x - 4}{(x^2 + 1)^2}$ †

16. $\dfrac{x - 2}{(2x + 3)^2(x^2 + 3)^2}$ †

In Exercises 17–48, find the partial-fraction decomposition of each rational expression.

17. $\dfrac{2x + 1}{(x + 1)(x + 2)}$ $\dfrac{3}{x + 2} - \dfrac{1}{x + 1}$

18. $\dfrac{7}{(x - 2)(x + 5)}$ $\dfrac{1}{x - 2} - \dfrac{1}{x + 5}$

19. $\dfrac{1}{x^2 + 4x + 3}$ †

20. $\dfrac{x}{x^2 + 5x + 6}$ $\dfrac{3}{x + 3} - \dfrac{2}{x + 2}$

21. $\dfrac{2}{x^2 + 2x}$ $\dfrac{-1}{x + 2} + \dfrac{1}{x}$

22. $\dfrac{x + 9}{x^2 - 9}$ $\dfrac{2}{x - 3} - \dfrac{1}{x + 3}$

23. $\dfrac{x}{(x + 1)(x + 2)(x + 3)}$ †

24. $\dfrac{x^2}{(x - 1)(x^2 + 5x + 4)}$ †

25. $\dfrac{x - 1}{(x + 1)^2}$ $\dfrac{-2}{(x + 1)^2} + \dfrac{1}{x + 1}$

26. $\dfrac{3x + 2}{x^2(x + 1)}$ $\dfrac{2}{x^2} + \dfrac{1}{x} - \dfrac{1}{x + 1}$

27. $\dfrac{2x^2 + x}{(x + 1)^3}$ †

28. $\dfrac{x}{(x - 1)(x + 1)}$ †

29. $\dfrac{-x^2 + 3x + 1}{x^3 + 2x^2 + x}$ †

30. $\dfrac{5x^2 - 8x + 2}{x^3 - 2x^2 + x}$ †

31. $\dfrac{1}{x^2(x + 1)^2}$ †

32. $\dfrac{2}{(x - 1)^2(x + 3)^2}$ †

33. $\dfrac{1}{(x^2 - 1)^2}$ †

34. $\dfrac{3}{(x^2 - 4)^2}$ †

35. $\dfrac{x - 1}{(2x - 3)^2}$ †

36. $\dfrac{6x + 1}{(3x + 1)^2}$ †

37. $\dfrac{6x + 7}{4x^2 + 12x + 9}$ †

38. $\dfrac{2x + 3}{9x^2 + 30x + 25}$ †

39. $\dfrac{x - 3}{x^3 + x^2}$ $\dfrac{4}{x} - \dfrac{3}{x^2} - \dfrac{4}{x + 1}$

40. $\dfrac{1}{x^3 + x}$ $\dfrac{-x}{x^2 + 1} + \dfrac{1}{x}$

41. $\dfrac{x^2 + 2x + 4}{x^3 + x^2}$ $\dfrac{-2}{x} + \dfrac{3}{x + 1} + \dfrac{4}{x^2}$

42. $\dfrac{x^2 + 2x - 1}{(x - 1)(x^2 + 1)}$ †

43. $\dfrac{x}{x^4 - 1}$ †

44. $\dfrac{3x^3 - 5x^2 + 12x + 4}{x^4 - 16}$ †

45. $\dfrac{1}{x(x^2 + 1)^2}$ †

46. $\dfrac{x^2}{(x^2 + 2)^2}$ †

47. $\dfrac{2x^2 + 3x}{(x^2 + 1)(x^2 + 2)}$ $\dfrac{4 - 3x}{x^2 + 2} + \dfrac{-2 + 3x}{x^2 + 1}$

48. $\dfrac{x^2 - 2x}{(x^2 + 9)(x^2 + x + 7)}$ $\dfrac{-x}{x^2 + x + 7} + \dfrac{x}{x^2 + 9}$

B EXERCISES Applying the Concepts

In Exercises 49–52, express each sum as a fraction of whole numbers in lowest terms.

49. $\dfrac{1}{1 \cdot 2} + \dfrac{1}{2 \cdot 3} + \dfrac{1}{3 \cdot 4} + \cdots + \dfrac{1}{n(n + 1)}$ $\dfrac{n}{n + 1}$

50. $\dfrac{2}{1 \cdot 3} + \dfrac{2}{2 \cdot 4} + \dfrac{2}{3 \cdot 5} + \cdots + \dfrac{2}{100 \cdot 102}$ $\dfrac{7625}{5151}$

†Due to space constrictions, answers to these exercises may be found in the Answers beginning on page A–1 in the back of the book.

51. $\dfrac{2}{1\cdot 3} + \dfrac{2}{3\cdot 5} + \dfrac{2}{5\cdot 7} + \cdots + \dfrac{2}{(2n-1)(2n+1)} \quad \dfrac{2n}{2n+1}$

52. $\dfrac{2}{1\cdot 2\cdot 3} + \dfrac{2}{2\cdot 3\cdot 4} + \dfrac{2}{3\cdot 4\cdot 5} + \cdots + \dfrac{2}{100\cdot 101\cdot 102} \quad \dfrac{2575}{5151}$

$\Bigg[$ *Hint*: Write

$$\frac{2}{k(k+1)(k+2)} = \frac{1}{k} - \frac{2}{k+1} + \frac{1}{k+2}$$
$$= \frac{1}{k} - \frac{1}{k+1} - \frac{1}{k+1} + \frac{1}{k+2} \Bigg]$$

In Exercises 53–56, the total resistance R in a circuit is given. Use partial-fraction decomposition to write $\dfrac{1}{R}$ in terms of partial fractions. Interpret each term in the partial fraction.

53. Circuit resistance. $R = \dfrac{(x+1)(x+3)}{2x+4} \quad \dfrac{1}{R} = \dfrac{1}{x+1} + \dfrac{1}{x+3}$

54. Circuit resistance. $R = \dfrac{(x+2)(x+4)}{7x+20}$

55. Circuit resistance. $R = \dfrac{R_1 R_2 R_3}{R_1 R_2 + R_2 R_3 + R_3 R_1}$

56. Circuit resistance. $R = \dfrac{x(x+2)(x+4)}{3x^2 + 12x + 8}$

C EXERCISES Beyond the Basics

57. Solve the system of equations in Example 2 to verify the values of the constants $A, B,$ and $C.$

Answers:

54. $\dfrac{1}{R} = \dfrac{3}{x+2} + \dfrac{4}{x+4} = \dfrac{1}{\left(\dfrac{x+2}{3}\right)} + \dfrac{1}{\left(\dfrac{x+4}{4}\right)}$

55. $\dfrac{1}{R} = \dfrac{1}{R_1} + \dfrac{1}{R_2} + \dfrac{1}{R_3}$ **56.** $\dfrac{1}{R} = \dfrac{1}{x} + \dfrac{1}{x+2} + \dfrac{1}{x+4}$

In Exercises 58–66, find the partial-fraction decomposition of each rational expression.

58. $\dfrac{1}{x^3+1}$ **59.** $\dfrac{4x}{(x^2-1)^2}$

60. $\dfrac{2x+3}{(x^2+1)(x+1)}$ **61.** $\dfrac{x+1}{(x^2+1)(x-1)^2}$

62. $\dfrac{x}{(x^2+1)^2(x-1)}$ **63.** $\dfrac{x^3}{(x+1)^2(x+2)^2}$

64. $\dfrac{2x^3+2x^2-1}{(x^2+x+1)^2}$ **65.** $\dfrac{15x}{x^3-27}$

66. $\dfrac{2x^2-9x+10}{x^3+8}$

67. Find the partial-fraction decomposition of:

$$\frac{4x}{x^4+4}$$

$[$ *Hint*: $x^4 + 4 = x^4 + 4 + 4x^2 - 4x^2 = (x^2+2)^2 - (2x)^2$
$= (x^2 - 2x + 2)(x^2 + 2x + 2).]$

68. Find the partial-fraction decomposition of:

$$\frac{x^2 - 8x + 18}{(x-5)^3}$$

In Exercises 69–72, find the partial-fraction decomposition by first using long division.

69. $\dfrac{x^2+4x+5}{x^2+3x+2}$ **70.** $\dfrac{(x-1)(x+2)}{(x+3)(x-4)}$

71. $\dfrac{2x^4 + x^3 + 2x^2 - 2x - 1}{(x-1)(x^2+1)}$

72. $\dfrac{x^4}{(x-1)(x-2)(x-3)}$

58. $\dfrac{-x+2}{3(x^2-x+1)} + \dfrac{1}{3(x+1)}$ **59.** $\dfrac{1}{(x-1)^2} - \dfrac{1}{(x+1)^2}$

60. $\dfrac{-x+5}{2(x^2+1)} + \dfrac{1}{2(x+1)}$ **61.** $\dfrac{-1}{2(x-1)} + \dfrac{x-1}{2(x^2+1)} + \dfrac{1}{(x-1)^2}$

62. $\dfrac{-x+1}{2(x^2+1)^2} + \dfrac{1}{4(x-1)} + \dfrac{-x-1}{4(x^2+1)}$

63. $\dfrac{-4}{x+2} - \dfrac{8}{(x+2)^2} - \dfrac{1}{(x+1)^2} + \dfrac{5}{x+1}$

64. $\dfrac{2x}{x^2+x+1} + \dfrac{-1-2x}{(x^2+x+1)^2}$ **65.** $\dfrac{5}{3(x-3)} + \dfrac{-5x+15}{3(x^2+3x+9)}$

66. $\dfrac{3}{x+2} + \dfrac{-x-1}{x^2-2x+4}$ **67.** $\dfrac{1}{x^2-2x+2} - \dfrac{1}{x^2+2x+2}$

68. $\dfrac{3}{(x-5)^3} + \dfrac{1}{x-5} + \dfrac{2}{(x-5)^2}$ **69.** $1 + \dfrac{2}{x+1} - \dfrac{1}{x+2}$

70. $1 - \dfrac{4}{7(x+3)} + \dfrac{18}{7(x-4)}$ **71.** $2x + 3 + \dfrac{2x-1}{x^2+1} + \dfrac{1}{x-1}$

72. $x + 6 - \dfrac{16}{x-2} + \dfrac{81}{2(x-3)} + \dfrac{1}{2(x-1)}$

SUMMARY ■ Definitions, Concepts, and Formulas

7.1 Systems of Linear Equations in Two Variables

A **system of equations** is a set of equations with common variables.

A **solution** of a system of equations in two variables x and y is an ordered pair of numbers (a, b) that satisfies all equations in the system. The solution set of a system of two linear equations of the form $ax + by = c$ is the point(s) of intersection of the graphs of the equations.

Systems with no solution are called **inconsistent**, and systems with at least one solution are **consistent**. A system of two linear equations in two variables may have one solution (**independent equations**), no solution (**inconsistent system**), or infinitely many solutions (**dependent equations**).

Three methods of solving a system of equations are (1) the graphical method, (2) the substitution method, and (3) the elimination or addition method.

7.2 Systems of Linear Equations in Three Variables

A **linear equation** in n variables $x_1, x_2, \ldots, x_n$ is an equation that can be written in the form

$$a_1 x_1 + a_2 x_2 + \cdots + a_n x_n = b,$$

where b and the coefficients $a_1, a_2, \ldots, a_n$ are real numbers.

A system of linear equations can be solved by transforming it into an *equivalent system* that is easier to solve.
The following operations produce equivalent systems:

1. Interchange the position of any two equations.

2. Multiply any equation by a nonzero constant.

3. Add a nonzero multiple of one equation to another.

The **Gaussian elimination method** is a procedure for converting a system of linear equations into an equivalent system in *triangular form*.

A linear system may have **(i)** only one solution, **(ii)** infinitely many solutions, or **(iii)** no solution.

7.3 Systems of Linear Inequalities; Linear Programming

A statement of the form $ax + by < c$ is a linear inequality in the variables x and y. The symbol $<$ may be replaced by $\leq$, $>$, or $\geq$. A procedure for graphing a linear inequality in two variables is described on page 506.

The graph of the solution set of a system of inequalities is obtained by **(i)** graphing each inequality of the system in the same coordinate plane and then **(ii)** finding the region that is common to every graph in the system.

Linear Programming

In a linear programming model, a quantity f (to be maximized or minimized) that can be expressed in the form $f = ax + by$ is called an **objective function** of the variables x and y. The **constraints** are the restrictions placed on the variables x and y that can be expressed as a system of linear inequalities.

A procedure for solving a linear programming problem is given on page 511.

7.4 Nonlinear Systems of Equations and Inequalities

A system of equations (inequalities) in which at least one equation (inequality) is nonlinear is called a **nonlinear system of equations (inequalities)**. The methods of substitution, elimination, or graphing is sometimes used to solve a nonlinear system.

7.5 Partial-Fraction Decomposition

In partial-fraction decomposition, we reverse the addition of rational expressions. A rational expression $\dfrac{P(x)}{Q(x)}$ is called a proper fraction if $\deg P(x) < \deg Q(x)$.

There are four cases to consider in decomposing $\dfrac{P(x)}{Q(x)}$ into partial fractions:

1. $Q(x)$ has only distinct linear factors. (See page 528.)
2. $Q(x)$ has repeated linear factors. (See page 530.)
3. $Q(x)$ has distinct irreducible quadratic factors. (See page 531.)
4. $Q(x)$ has repeated irreducible quadratic factors. (See page 532.)

REVIEW EXERCISES

Basic Skills and Concepts

In Exercises 1–18, solve each system of equations by using the method of your choice. Identify systems with no solution and systems with infinitely many solutions.

1. $\begin{cases} 3x - y = -5 \\ x + 2y = 3 \end{cases}$ $\{(-1,2)\}$

2. $\begin{cases} x + 3y + 6 = 0 \\ y = 4x - 2 \end{cases}$ $\{(0,-2)\}$

3. $\begin{cases} 2x + 4y = 3 \\ 3x + 6y = 10 \end{cases}$ $\varnothing$

4. $\begin{cases} x - y = 2 \\ 2x - 2y = 9 \end{cases}$ $\varnothing$

5. $\begin{cases} 3x - y = 3 \\ \dfrac{1}{2}x + \dfrac{1}{3}y = 2 \end{cases}$ $\{(2,3)\}$

6. $\begin{cases} 0.02y - 0.03x = -0.04 \\ 1.5x - y = 3 \end{cases}$ $\varnothing$

7. $\begin{cases} x + 3y + z = 0 \\ 2x - y + z = 5 \\ 3x - 3y + 2z = 10 \end{cases}$ †

8. $\begin{cases} 2x + y = 11 \\ 3y - z = 5 \\ x + 2z = 1 \end{cases}$ †

9. $\begin{cases} x + y = 1 \\ 3y + 2z = 0 \\ 2x - 3z = 7 \end{cases}$ †

10. $\begin{cases} 2x - 3y + z = 2 \\ x - 3y + 2z = -1 \\ 2x + 3y + 2z = 3 \end{cases}$ †

11. $\begin{cases} x + y + z = 1 \\ x + 5y + 5z = -1 \\ 3x - y - z = 4 \end{cases}$ $\varnothing$

12. $\begin{cases} x + 3y - 2z = -4 \\ 2x + 6y - 4z = 3 \\ x + y + z = 1 \end{cases}$ $\varnothing$

13. $\begin{cases} x + 4y + 3z = 1 \\ 2x + 5y + 4z = 4 \end{cases}$ †

14. $\begin{cases} 3x - y + 2z = 9 \\ x - 2y + 3z = 2 \end{cases}$ †

15. $\begin{cases} 3x - y = 2 \\ x + 2y = 9 \\ 3x + y = 10 \end{cases}$ $\varnothing$

16. $\begin{cases} x - 2y = 1 \\ 3x + 4y = 11 \\ 2x + 2y = 7 \end{cases}$ $\varnothing$

17. $\begin{cases} x + y + z = 3 \\ 2x - y + z = 4 \\ x + 4y + 2z = 5 \end{cases}$ †

18. $\begin{cases} x - y + z = 4 \\ x + 2y + 3z = 2 \\ 2x + y + 4z = 6 \end{cases}$ †

In Exercises 19–24, graph each system of inequalities.

19. $\begin{cases} x + y \leq 1 \\ x - y \leq 1 \\ x \geq 0 \end{cases}$ †

20. $\begin{cases} 2x + 3y \leq 6 \\ 4x - 3y \leq 12 \\ x \geq 0 \end{cases}$ †

21. $\begin{cases} 4x + y \leq 7 \\ 2x + 5y \geq -1 \\ x - 2y \geq -5 \end{cases}$ †

22. $\begin{cases} 7x - 2y + 6 \leq 0 \\ x + y + 14 \geq 0 \\ 2x - 3y + 9 \geq 0 \end{cases}$ †

23. $\begin{cases} x + y - 3 \leq 0 \\ 3x - y - 9 \leq 0 \\ y + 3 \geq 0 \\ 7x + 4y + 23 \geq 0 \end{cases}$ †

24. $\begin{cases} x + 5y - 11 \leq 0 \\ 5x + y - 7 \leq 0 \\ x - 5y - 17 \leq 0 \\ 7x + y + 25 \geq 0 \end{cases}$ †

In Exercises 25–28, solve each linear programming problem.

25. Maximize $z = 2x + 3y$, subject to the constraints $x \geq 0, y \geq 0, 3x + 7y \leq 21$. Max of 14 at $x = 7, y = 0$

26. Minimize $z = -5x + 2y$, subject to the constraints $x \geq -2, y \leq 1, x - 2y + 2 \leq 0$. Min of 2 at $x = 0, y = 1$

27. Minimize $z = x + 3y$, subject to the constraints $2x - 3y + 2 \geq 0, 4x + y - 10 \leq 0, x - 3y - 9 \leq 0, 3x + y + 3 \geq 0$. Min of -9 at $x = 0, y = -3$

28. Maximize $z = 2x + 5y$, subject to the constraints $x + 3y - 3 \leq 0, 3x - y - 9 \leq 0, x + 4y + 10 \geq 0, 3x - 2y + 2 \geq 0$. Max of 6 at $x = 3, y = 0$

In Exercises 29–34, solve each nonlinear system of equations.

29. $\begin{cases} x + 3y = 1 \\ x^2 - 3x = 7y + 3 \end{cases}$ †

30. $\begin{cases} 4x - y = 3 \\ y^2 - 2y = x - 2 \end{cases}$ †

31. $\begin{cases} x - y = 4 \\ 5x^2 + y^2 = 24 \end{cases}$ †

32. $\begin{cases} x - 2y = 7 \\ 2x^2 + 3y^2 = 29 \end{cases}$ †

33. $\begin{cases} xy = 2 \\ x^2 + 2y^2 = 9 \end{cases}$ †

34. $\begin{cases} xy = 3 \\ 2x^2 + 3y^2 = 21 \end{cases}$ †

In Exercises 35–40, graph each nonlinear system of inequalities.

35. $\begin{cases} x - y \leq 1 \\ x^2 + y^2 \leq 13 \end{cases}$ †

36. $\begin{cases} x - y \geq 1 \\ x^2 + y^2 \leq 13 \end{cases}$ †

37. $\begin{cases} x - y \leq 1 \\ x^2 + y^2 \leq 13 \\ x \geq 0 \\ y \geq 0 \end{cases}$ †

38. $\begin{cases} x - y \geq 1 \\ x^2 + y^2 \geq 13 \\ x \leq 4 \\ y \geq 0 \end{cases}$ †

39. $\begin{cases} y \leq 3x \\ x^2 + y^2 \leq 25 \\ x \geq 0 \\ y \geq 0 \end{cases}$ †

40. $\begin{cases} y \geq 3x \\ x^2 + y^2 \leq 25 \\ x \geq 0 \\ y \geq 0 \end{cases}$ †

In Exercises 41–46, find the partial-fraction decomposition of each rational expression.

41. $\dfrac{x + 4}{x^2 + 5x + 6}$ †

42. $\dfrac{x + 14}{x^2 + 3x - 4}$ †

43. $\dfrac{3x^2 + x + 1}{x(x - 1)^2}$ †

44. $\dfrac{3x^2 + 2x + 3}{(x^2 - 1)(x + 1)}$ †

45. $\dfrac{x^2 + 2x + 3}{(x^2 + 4)^2}$ †

46. $\dfrac{2x}{x^4 - 1}$ †

Applying the Concepts

47. **Investment.** A speculator invested part of $15,000 in a high-risk venture and received a return of 12% at the end of the year. The rest of the $15,000 was invested at 4% annual interest. The combined annual income from the two sources was $1300. How much was invested at each rate? $8750 was invested at 12%, and $6250 was invested at 4%.

48. **Agriculture.** A farmer earns a profit of $525 per acre of tomatoes and $475 per acre of soybeans. His soybean acreage is 5 acres more than twice his tomato acreage. If his total profit from the two crops is $24,500, how many acres of tomatoes and how many acres of soybeans does he have? 15 acres of tomatoes and 35 acres of soybeans

49. **Geometry.** The area of a rectangle is 63 square feet, and its perimeter is 33 feet. What are the dimensions of the rectangle? 10.5 ft by 6 ft

†Due to space constrictions, answers to these exercises may be found in the Answers beginning on page A–1 in the back of the book.

50. **Numbers.** Twice the sum of the reciprocals of two numbers is 13, and the product of the numbers is $\frac{1}{9}$. Find the numbers. $\frac{2}{9}$ and $\frac{1}{2}$

51. **Geometry.** The hypotenuse of a right triangle is 17. If one leg of the triangle is increased by 1 and the other leg is increased by 4, the hypotenuse becomes 20. Find the sides of the triangle. 15 and 8

52. **Paper route.** Chris covers her paper route, which is 21 miles long, by 7:30 A.M. each day. If her average rate of travel were 1 mile faster each hour, she would cover the route by 7 A.M. What time does she start in the morning? 4 A.M.

53. **Agriculture.** A rectangular pasture with an area of 6400 square meters is divided into three smaller pastures by two fences parallel to the shorter sides. The width of two of the smaller pastures is the same, and the width of the third is twice that of the others. Find the dimensions of the original pasture if the perimeter of the larger of the subdivisions is 240 m. The original width is 40 m, and the original length is 160 m.

54. **Leasing.** Budget Rentals leases its compact cars for $23.00 per day plus $0.17 per mile. Dollar Rentals leases the same car for $24 per day plus $0.22 per mile.
    a. Find the cost functions describing the daily cost of leasing from each company. †
    b. Graph the two functions in part (a) on the same coordinate axes. †
    c. From which company should you lease the car if you plan to drive **(i)** 50 miles per day; **(ii)** 60 miles per day; **(iii)** 70 miles per day? For (i), (ii), and (iii), you should lease from Budget Rentals.

55. **Break-even analysis.** Auto-Sprinkler Corp. manufactures seven-day, 24-hour variable timers for lawn sprinklers. The corporation has a monthly fixed cost of $60,000 and a production cost of $12 for each timer manufactured. Each timer sells for $20.
    a. Write the cost function and the revenue function for selling x timers per month. †
    b. Graph the two functions in part (a) on the same coordinate plane and hence find the break-even point graphically. †
    c. Find the break-even point algebraically. (7500, 150,000)
    d. How many timers should be sold in a month to realize a profit (before taxes) of 15% of the cost? 11,129 timers

56. **Equilibrium quantity and price.** The demand equation for a product is $p = \dfrac{4000}{x}$, where p is the price per unit and x is the number of units of the product. The supply equation for the product is $p = \dfrac{x}{20} + 10$.
    a. Find the equilibrium quantity. 200
    b. Find the equilibrium price. $20
    c. Graph the supply and demand functions on the same coordinate plane and label the equilibrium point. †

57. **Apartment lease.** Alisha and Sunita signed a lease on an apartment for nine months. At the end of six months, Alisha got married and moved out. She paid the landlady an amount equal to the difference between double-occupancy rental and single-occupancy rental for the remaining three months, and Sunita paid the single-occupancy rate for the same three months. If the nine-month rental cost Alisha $2250 and Sunita $3150, what were the single and double monthly rates? $330 for the single and $360 for the double

58. **Seating arrangement.** The 600 graduating seniors at Central State College are seated in rows, each of which contains the same number of chairs, and every chair is occupied. If five more chairs were in each row, everyone could be seated in four fewer rows. How many chairs are in each row? 25 chairs and 24 rows

59. **Passing a final exam.** Twenty-six students in a college algebra class took a final exam on which the passing score was 70. The mean score of those who passed was 78, and the mean score of those who failed was 26. The mean of all scores was 72. How many students failed the exam? 3

60. **Finding numbers.** The average of a and b is 2.5, the average of b and c is 3.8, and the average of a and c is 3.1. Find the numbers a, b, and c. $a = 1.8, b = 3.2, c = 4.4$

61. **May–December match.** Steve and his wife Janet have the same birthday. When Steve was as old as Janet is now, he was twice as old as Janet was then. When Janet becomes as old as Steve is now, the sum of their ages will be 119. How old are Steve and Janet now? Janet is 34, and Steve is 51.

62. **Percentage increase.** Let x and y be positive real numbers. Increasing x by y percent gives 46, while increasing y by x percent gives 21. Find x and y. $x = 40, y = 15$

63. **Criminals rob a bank.** Three criminals—Butch, Sundance, and Billy—robbed a bank and divided the loot in the following manner: Since Butch planned the job and drove the getaway car, he got 75% as much as Sundance and Billy put together. Since Sundance was an expert safecracker, he got $500 more than 50% of Billy's and Butch's cut put together. Billy got $1000 less than three times the difference of Butch and Sundance's cut. How much did they steal, and what was each criminal's take? They stole $35,000, and $15,000 went to Butch, $12,000 went to Sundance, and $8000 went to Billy.

64. **Curve fitting.** Find all of the curves of the form $y = ax^2 + bx + c$ that contain the points $(0, 1)$, $(1, 0)$, and $(-1, 6)$. $y = 2x^2 - 3x + 1$

65. **Using exponents.** Given that $2^x = 8^y$ and $9^y = 3^{x-2}$, find x and y. $x = 6, y = 2$

66. **Area of an octagon.** A square of side length 8 has its corners cut off to make it a regular octagon. Find the area of the octagon. (See the figure.) ≈ 53.02 square units

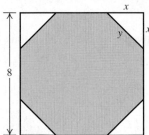

67. Building houses. A builder has 42 units of material and 32 units of labor available during a given period. She builds two-story houses or one-story houses or some of both. Suppose she makes a profit of $10,000 on each two-story house and $4000 on each one-story house. A two-story house requires seven units of material and one unit of labor, whereas a one-story house requires one unit of material and two units of labor. How many houses of each type should she build to make a maximum profit? What is the maximum profit?
She should build 4 two-story houses and 14 one-story houses to obtain a maximum profit of $96,000.

68. Minimizing cost. An animal food is to be a mixture of two products X and Y. The content and cost of 1 pound of each product is given in the following table.

Product	Protein grams	Fat grams	Carbohydrates grams	Cost
X	180	2	240	$0.75
Y	36	8	200	$0.56

How much of each product should be used to minimize the cost if each bag must contain at least 612 grams of protein, at least 22 grams of fat, and at most 1880 grams of carbohydrates? 3 of product X and 2 of product Y

PRACTICE TEST A

In Problems 1–7, solve each system of equations.

1. $\begin{cases} 2x - y = 4 \\ 2x + y = 4 \end{cases}$ $\{(2, 0)\}$

2. $\begin{cases} x + 2y = 8 \\ 3x + 6y = 24 \end{cases}$ $\{(8 - 2y, y)\}$

3. $\begin{cases} -2x + y = 4 \\ 4x - 2y = 4 \end{cases}$ $\varnothing$

4. $\begin{cases} 3x + 3y = -15 \\ 2x - 2y = -10 \end{cases}$ $\{(-5, 0)\}$

5. $\begin{cases} \dfrac{5}{3}x + \dfrac{y}{2} = 14 \\ \dfrac{2}{3}x - \dfrac{y}{8} = 3 \end{cases}$ $\{(6, 8)\}$

6. $\begin{cases} y = x^2 \\ 3x - y + 4 = 0 \end{cases}$ $\{(-1, 1), (4, 16)\}$

7. $\begin{cases} x - 3y = -4 \\ 2x^2 + 3x - 3y = 8 \end{cases}$ $\left\{ \left(-3, \dfrac{1}{3}\right), (2, 2) \right\}$

8. Two gold bars together weigh a total of 485 pounds. One bar weighs 15 pounds more than the other.
   (a) Write a system of equations that describes these relationships.
   (b) How much does each bar weigh? $x = 250, y = 235$

9. Convert the given system to triangular form.

$$\begin{cases} 2x + y + 2z = 4 \\ 4x + 6y + z = 15 \\ 2x + 2y + 7z = -1 \end{cases}$$

10. Use back-substitution to solve the given linear system.

$$\begin{cases} x - 6y + 3z = -2 \\ 9y - 5z = 2 \\ 2z = 10 \end{cases} \quad \{(1, 3, 5)\}$$

In Exercises 11–13, solve the system of equations.

11. $\begin{cases} x + y + z = 8 \\ 2x - 2y + 2z = 4 \\ x + y - z = 12 \end{cases}$ $(7, 3, -2)$

12. $\begin{cases} x + z = -1 \\ 3y + 2z = 5 \\ 3x - 3y + z = -8 \end{cases}$ $\left(x, \dfrac{2x}{3} + \dfrac{7}{3}, -x - 1 \right)$

Answers:

8. a. $\begin{cases} x + y = 485 \\ x - y = 15 \end{cases}$ **9.** $x + \dfrac{1}{2}y + z = 2$

$$y - \dfrac{3}{4}z = \dfrac{7}{4}$$

$$z = \dfrac{-27}{23}$$

13. $\begin{cases} 2x - y + z = 2 \\ x + y - z = -1 \\ x - 5y + 5z = 7 \end{cases}$ $\left(\dfrac{1}{3}, \dfrac{-4}{3} + z, z \right)$

14. A vending machine's coin box contains nickels, dimes, and quarters. The total number of coins in the box is 300. The number of dimes is three times the number of nickels and quarters together. If $30.65 is in the box, find the number of nickels, dimes, and quarters that it contains. 53 nickels, 225 dimes, and 22 quarters

For Problems 15 and 16, write the form of the partial-fraction decomposition of the given rational expression. You do not need to solve for the constants.

15. $\dfrac{2x}{(x - 5)(x + 1)}$

16. $\dfrac{-5x^2 + x - 8}{(x - 2)(x^2 + 1)^2}$

17. Find the partial-fraction decomposition of the rational expression

$$\dfrac{x + 3}{(x + 4)^2(x - 7)}$$

18. Graph the inequality $3x + y < 6$. †

19. Graph the solution set of the system of inequalities

$$\begin{cases} y - x^2 \le 3 \\ y - x > 0 \end{cases} †$$

20. Maximize $z = 2x + y$, subject to the constraints $x \ge 0, y \ge 0, x + 3y \le 3$. Maximum value is 6 at $(3, 0)$.

15. $\dfrac{A}{(x - 5)} + \dfrac{B}{(x + 1)}$ **16.** $\dfrac{A}{x - 2} + \dfrac{Bx + C}{x^2 + 1} + \dfrac{Dx + E}{(x^2 + 1)^2}$

17. $-\dfrac{10}{121(x + 4)} + \dfrac{1}{11(x + 4)^2} + \dfrac{10}{121(x - 7)}$

PRACTICE TEST B

In Problems 1–7, solve each system of equations.

1. $\begin{cases} 2x - 3y = 16 \\ x - y = 7 \end{cases}$ a

    a. $\{(5, -2)\}$ **b.** $\{(8, 0)\}$

    c. $\{(0, -7)\}$ **d.** $\{(4, -3)\}$

2. $\begin{cases} 6x - 9y = -2 \\ 3x - 5y = -6 \end{cases}$ c

    a. $\{(-3, 0)\}$ **b.** $\left\{\left(0, \dfrac{5}{6}\right)\right\}$

    c. $\left\{\left(\dfrac{44}{3}, 10\right)\right\}$ **d.** $\{(1, 1)\}$

3. $\begin{cases} 2x - 5y = 9 \\ 4x - 10y = 18 \end{cases}$ b

    a. $\left\{\left(0, \dfrac{9}{2}\right)\right\}$ **b.** $\left\{\left(\dfrac{5}{2}y + \dfrac{9}{2}, y\right)\right\}$

    c. $\left\{\left(x, \dfrac{5}{2}x + \dfrac{9}{2}\right)\right\}$ **d.** $\varnothing$

4. $\begin{cases} 3x + 5y = 1 \\ -6x - 10y = 2 \end{cases}$ c

    a. $\left\{\left(0, -\dfrac{1}{5}\right)\right\}$ **b.** $\{(2, -1)\}$

    c $\varnothing$ **d.** $\left\{\left(-\dfrac{5}{3}y + \dfrac{1}{3}, y\right)\right\}$

5. $\begin{cases} \dfrac{1}{5}x + \dfrac{2}{5}y = 1 \\ \dfrac{1}{4}x - \dfrac{1}{3}y = \dfrac{-5}{12} \end{cases}$ b

    a. $\{(-15, 10)\}$ **b.** $\{(1, 2)\}$

    c. $\{(-10, -15)\}$ **d.** $\{(1, 0)\}$

6. $\begin{cases} x - 3y = \dfrac{1}{2} \\ -2x + 6y = -1 \end{cases}$ d

    a. $\left\{\left(\dfrac{1}{2}, 0\right)\right\}$ **b.** $\left\{\left(2, \dfrac{1}{2}\right)\right\}$

    c. $\varnothing$ **d.** $\left\{\left(3y + \dfrac{1}{2}, y\right)\right\}$

7. $\begin{cases} 3x + 4y = 12 \\ 3x^2 + 16y^2 = 48 \end{cases}$ d

    a. $\{(0, 4), (3, 0)\}$ **b.** $\{(0, -4)\}$

    c. $\{(0, -4), (3, 0)\}$ **d.** $\left\{(4, 0), \left(2, \dfrac{3}{2}\right)\right\}$

8. Student tickets for a dance cost \$2, and nonstudent tickets cost \$5. Three hundred tickets were sold, and the total ticket receipts were \$975. How many of each type of ticket were sold? d

    a. 150 student tickets   **b.** 200 student tickets
         150 nonstudent tickets   100 nonstudent tickets

    c. 210 student tickets   **d.** 175 student tickets
         90 nonstudent tickets   125 nonstudent tickets

9. Convert the given system to triangular form. c

$$\begin{cases} x + 3y + 3z = 4 \\ 2x + 5y + 4z = 5 \\ x + 2y + 2z = 6 \end{cases}$$

    a. $\begin{aligned} x + 3y + 3z &= 4 \\ y + 2z &= 8 \\ z &= 2 \end{aligned}$ **b.** $\begin{aligned} x + 3y + 3z &= 4 \\ 2y + z &= 1 \\ z &= 5 \end{aligned}$

    c. $\begin{aligned} x + 3y + 3z &= 4 \\ y + 2z &= 3 \\ z &= 5 \end{aligned}$ **d.** $\begin{aligned} x + 3y + 3z &= 4 \\ 2y + z &= -2 \\ z &= 5 \end{aligned}$

10. Use back-substitution to solve the given linear system.

$$\begin{cases} 2x + y + z = 0 \\ 10y - 2z = 4 \\ 3z = 9 \end{cases} \text{ a}$$

    a. $\{(-2, 1, 3)\}$ **b.** $\{(0, -3, 3)\}$
    c. $\{(0, 2, 8)\}$ **d.** $\{(-1, 1, 1)\}$

11. Solve the given system of equations or state that the system is inconsistent.

$$\begin{cases} 2x + 13y + 6z = 1 \\ 3x + 10y + 11z = 15 \\ 2x + 10y + 8z = 8 \end{cases} \text{ b}$$

    a. $\{(3, -1, -4)\}$ **b.** $\{(1, -1, 2)\}$
    c. $\varnothing$ **d.** $(1, 6, 8)\}$

12. Solve the given system of equations or state that the system is inconsistent.

$$\begin{cases} 4x - y + 7z = -2 \\ 2x + y + 11z = 13 \\ 3x - y + 4z = -3 \end{cases} \text{ d}$$

    a. $\{(-3z + 1, -5z + 11, z)\}$
    b. $\{(-14, -14, 5)\}$
    c. $\left\{\left(-\dfrac{19}{5}, 3, \dfrac{8}{5}\right)\right\}$
    d. $\varnothing$

13. Solve the given system of equations or state that the system is inconsistent.

$$\begin{cases} x - 2y + 3z = 4 \\ 2x - y + z = 1 \\ x + y - 2z = -3 \end{cases} \text{ c}$$

    a. $\{(1, 6, 5)\}$ **b.** $\{(0, 1, 2)\}$
    c. $\{(x, 5x + 1, 3x + 2)\}$ **d.** $\varnothing$

14. Forty-six students will go to one of the countries France, Italy, and Spain for six weeks during the summer. The number of students going to France or Italy is four more than the number going to Spain. The number of students going to France is two less than the number going to Spain. How many students are going to each country? b

a. France: 23
Italy: 21
Spain: 2

b. France: 19
Italy: 6
Spain: 21

c. France: 21
Italy: 6
Spain: 19

d. France: 2
Italy: 21
Spain: 23

For Problems 15 and 16, write the form of the partial-fraction decomposition of the given rational functions. You do not need to solve for the constants.

15. $\dfrac{x}{(x + 2)(x - 7)}$ c

a. $\dfrac{A}{x + 2} + \dfrac{Bx}{x - 7}$

b. $\dfrac{Ax}{x + 2} + \dfrac{Bx}{x - 7}$

c. $\dfrac{A}{x + 2} + \dfrac{B}{x - 7}$

d. $\dfrac{A}{x + 2} + \dfrac{Bx + C}{x - 7}$

16. $\dfrac{7 - x}{(x - 3)(x + 5)^2}$ a

a. $\dfrac{A}{x - 3} + \dfrac{B}{x + 5} + \dfrac{C}{(x + 5)^2}$

b. $\dfrac{A}{x - 3} + \dfrac{B}{x + 5} + \dfrac{Cx + D}{(x + 5)^2}$

c. $\dfrac{A}{x + 3} + \dfrac{B}{(x + 5)^2}$

d. $\dfrac{A}{x + 3} + \dfrac{Bx + C}{(x + 5)^2}$

17. Find the partial-fraction decomposition of the rational expression

$$\frac{x^2 + 15x + 18}{x^3 - 9x}. \text{ c}$$

a. $\dfrac{-2}{x} + \dfrac{26}{x - 9}$

b. $\dfrac{-2}{x} + \dfrac{4}{x + 3} - \dfrac{1}{x - 3}$

c. $\dfrac{-2}{x} + \dfrac{4}{x - 3} - \dfrac{1}{x + 3}$

d. $\dfrac{-2}{x} + \dfrac{3x - 15}{x^2 - 9}$

18. Which of the graphs given at the right is the graph of $x + 3y > 3$? b

19. Which of the graphs given at the right is the graph of the solution set of the system of inequalities

$$\begin{cases} y - x^2 + 4 \geq 0 \\ 3x - y \leq 0 \end{cases}? \text{ a}$$

20. Maximize $z = 3x + 21y$, subject to the constraints

$$x \geq 0, y \geq 0, 2x + y \leq 8, 2x + 3y \leq 16. \text{ b}$$

a. 84

b. 112

c. $\dfrac{16}{3}$

d. 12

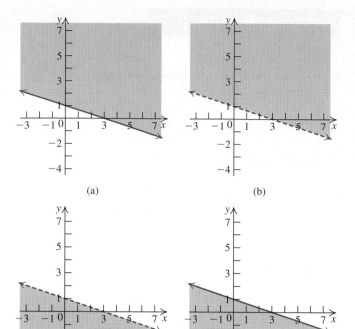

(a)   (b)

(c)   (d)

Figure for Exercise 18

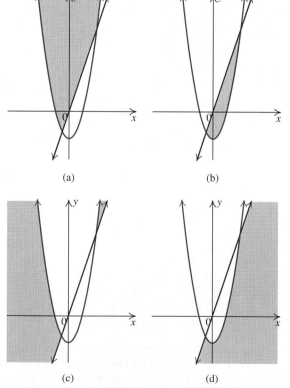

(a)   (b)

(c)   (d)

Figure for Exercise 19

CUMULATIVE REVIEW EXERCISES ▪ Chapters 1–7

In Exercises 1–8, solve each equation or inequality.

1. $\dfrac{1}{x-1} + \dfrac{4}{x-4} = \dfrac{5}{x-5}$ $\left\{\dfrac{5}{2}\right\}$

2. $2\left(x + \dfrac{1}{x}\right)^2 - 7\left(x + \dfrac{1}{x}\right) + 5 = 0$

3. $\sqrt{3x-5} = x - 3$ $\{7\}$ **4.** $\dfrac{x-1}{x+3} \le 0$ $(-3, 1]$

5. $2^{x-1} = 5$ $\dfrac{\ln 2 + \ln 5}{\ln 2}$ **6.** $\log_x 16 = 4$ $\{2\}$

7. $\log(x-3) + \log(x-1) = \log(2x-5)$ $\{4\}$

8. $\tan x = -1, 0 \le x \le 2\pi$

In Exercises 9–12, use transformations to graph each function.

9. $f(x) = |x+1| - 2$ †

10. $f(x) = -(x-2)^2 + 3$ †

11. $f(x) = 2^{x-1} - 3$ †

12. $f(x) = 2 + 3\sin x$ †

13. Let $f(x) = 2x - 2$.
  a. Find $f^{-1}(x)$.
  b. Graph f and f^{-1} on the same coordinate plane. †

14. Let $f(x) = x^3 - x^2 + x - 6$
  a. List all possible rational zeros of f. $\pm 1, \pm 2, \pm 3, \pm 6$
  b. Use synthetic division to test the possible rational zeros and find a real zero. 2
  c. Use the zero from part (b) to find all of the (real or complex) zeros of $f(x)$.
  $\left\{2, -\dfrac{1}{2} + i\dfrac{\sqrt{11}}{2}, -\dfrac{1}{2} - i\dfrac{\sqrt{11}}{2}\right\}$

15. Expand and simplify: $\log_3(9x^4)$. $2 + 4\log_3(x)$

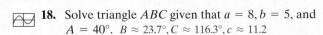 **16.** A radioactive substance decays so that the amount present in t years is given by $A = A_0 e^{-kt}$, where A_0 is the initial amount. Find the half-life of the substance if
  a. $k = 0.05$. 13.86
  b. $k = 0.0002$. 3465.74

17. Verify the identity: $\dfrac{1 + \sec x}{\tan^2 x} = \dfrac{\cos x}{1 - \cos x}$.

18. Solve triangle ABC given that $a = 8, b = 5$, and $A = 40°$. $B \approx 23.7°, C \approx 116.3°, c \approx 11.2$

In Exercises 19 and 20, solve the system of equations.

19. $\begin{cases} 5x - 2y + 25 = 0 \\ 4y - 3x - 29 = 0 \end{cases}$ $\{(-3, 5)\}$

20. $\begin{cases} 2x - y + z = 3 \\ x + 3y - 2z = 11 \\ 3x - 2y + z = 4 \end{cases}$ $\{(3, 2, -1)\}$

Answers:

2. $\left\{2, \dfrac{1}{2}, \dfrac{1}{2} + i\dfrac{\sqrt{3}}{2}, \dfrac{1}{2} - i\dfrac{\sqrt{3}}{2}\right\}$ **8.** $\left\{\dfrac{3\pi}{4}, \dfrac{7\pi}{4}\right\}$

13. a. $f^{-1}(x) = \dfrac{x+2}{2}$

8 Matrices and Determinants

The linear programming problems introduced in the previous chapter involving quantities from the study of such disciplines as medicine, traffic control, structural design, and city planning often involve an extremely large number of variables. To solve such problems successfully requires the use of powerful computers and new techinques. In this chapter, we study matrices and several methods of solving linear systems that are easily implemented on computers.

Matrices and Systems of Equations

Wassily Leontief 1906–1999

Leontief studied philosophy, sociology, and economics at the University of Leningrad and earned the degree of Learned Economist in 1925. He continued his studies at the University of Berlin, where he earned his PhD degree. In 1973, the Nobel Prize in Economic Sciences was awarded to Leontief for his work on input–output analysis of the U.S. economy.

Before Starting this Section, Review

1. Systems of equations (Section 7.1, page 481)
2. Equivalent systems (Section 7.2, page 495)
3. Gaussian elimination (Section 7.2, page 494)

Objectives

1. Define a matrix.
2. Use matrices to solve a system.
3. Use Gaussian elimination to solve a system.
4. Use Gauss–Jordan elimination to solve a system.

LEONTIEF INPUT–OUTPUT MODEL

In 1949, the Russian-born Harvard Professor Wassily Leontief opened the door to a new era in mathematical modeling in economics. He divided the U.S. economy into 500 "sectors," such as the coal, automotive, communications industries, and agriculture. For each sector, he wrote a linear equation that described how that sector distributed its output to other sectors of the economy. In 1949, the largest computer was the Mark II, and it could not handle the resulting system of 500 equations in 500 variables. Leontief then combined some of the sectors and reduced the system to 42 equations in 42 variables to solve the problem. As computers have become more powerful, scientists and engineers can now work on problems far more complex than they ever dreamed of a few decades ago. In Example 10, we explore Leontief's input–output model applied to a simple economy. ■

---

1 Define a matrix.

Definition of a Matrix

Suppose Great Builders, Incorporated (GBI), builds ranch, colonial, and modern houses. GBI wants to compare the cost of labor (in millions of dollars) and the cost of material (in millions of dollars) involved in building each type of house during one year. These data could be represented as in Table 8.1.

TABLE 8.1

Type of house	Ranch	Colonial	Modern
Cost of labor (in millions)	12	13	14
Cost of material (in millions)	9	11	10

Or it could be represented more compactly by the following array:

$$\begin{bmatrix} 12 & 13 & 14 \\ 9 & 11 & 10 \end{bmatrix}$$

This rectangular array of numbers is called a *matrix* (plural, *matrices*). The matrix has two rows (the costs) and three columns (the types of houses).

DEFINITION OF A MATRIX

A **matrix** is a rectangular array of numbers.

$$A = \begin{bmatrix} a_{11} & a_{12} & \cdots & a_{1n} \\ a_{21} & a_{22} & \cdots & a_{2n} \\ \vdots & \vdots & & \vdots \\ a_{m1} & a_{m2} & \cdots & a_{mn} \end{bmatrix} \begin{matrix} \leftarrow \text{Row 1} \\ \leftarrow \text{Row 2} \\ \\ \leftarrow \text{Row } m \end{matrix}$$

$$\underset{\text{Column 1}}{\uparrow} \quad \underset{\text{Column 2}}{\uparrow} \quad \underset{\text{Column } n}{\uparrow}$$

If a matrix A has m rows and n columns, then we say that A is of **order m by n** (written $m \times n$).

The **entry** or **element** in the ith row and jth column is a real number and is denoted by the *double-subscript* notation a_{ij}. We can call a_{ij} the (i, j)**th entry**. For example, a_{24} is the entry in the second row and fourth column of the matrix A. In general, we will use capital letters such as A, B, ... to denote matrices, and the corresponding lowercase letter $a_{ij}, b_{ij}, \ldots$ for their entries.

If A has n rows and n columns, then A is called a **square matrix of order n**. The entries $a_{11}, a_{22}, \ldots, a_{nn}$ form the **main diagonal** of A. A $1 \times n$ matrix is called a **row matrix**, and an $n \times 1$ matrix is called a **column matrix**.

EXAMPLE 1   Determining the Order of Matrices

Determine the order of each matrix. Identify square, row, and column matrices. Identify entries in the main diagonal of each square matrix.

a. $A = \begin{bmatrix} 3 \end{bmatrix}$ **b.** $B = \begin{bmatrix} 3 & 5 & -7 \end{bmatrix}$ **c.** $C = \begin{bmatrix} 0 & 1 \\ -3 & 4 \end{bmatrix}$ **d.** $D = \begin{bmatrix} 1 & 2 & 3 \\ 4 & 5 & 6 \\ 7 & 8 & 9 \end{bmatrix}$

SOLUTION

a. Matrix A with one row and one column is a 1×1 matrix. A is a square matrix of order 1. In A, $a_{11} = 3$. A is also a column and a row matrix.

b. Matrix B with one row and three columns is a 1×3 matrix. B is a row matrix.

c. Matrix C, a 2×2 matrix, is a square matrix of order 2. In C, the entries $c_{11} = 0$ and $c_{22} = 4$ form the main diagonal.

d. Matrix D, a 3×3 matrix, is a square matrix of order 3. In D, the entries $d_{11} = 1, d_{22} = 5$, and $d_{33} = 9$ form the main diagonal.   ■ ■ ■

Practice Problem 1   Determine the order of each matrix.

a. $\begin{bmatrix} -1 & 3 \\ 7 & 4 \\ 0 & 0 \end{bmatrix}$ **b.** $\begin{bmatrix} 3 & -8 \end{bmatrix}$

■

2 Use matrices to solve a system.

Using Matrices to Solve Linear Systems

To solve a system of linear equations by the elimination method, the particular symbols used for the variables do not matter; only the coefficients and the constants are important. We can display the constants and coefficients of a system in a matrix called the **augmented matrix** of the system. The matrix containing only the coefficients of the variables is the **coefficient matrix**. Consider the following system.

$$\begin{cases} x - y - z = 1 \\ 2x - 3y + z = 10 \\ x + y - 2z = 0 \end{cases}$$

Coefficients of x
Coefficients of y
Coefficients of z

Augmented Matrix:
$$\begin{bmatrix} 1 & -1 & -1 & 1 \\ 2 & -3 & 1 & 10 \\ 1 & 1 & -2 & 0 \end{bmatrix}$$

Constants

Coefficient Matrix:
$$\begin{bmatrix} 1 & -1 & -1 \\ 2 & -3 & 1 \\ 1 & 1 & -2 \end{bmatrix}$$

Notice that the numbers in the first column of the augmented matrix are the coefficients of x, those in the second column are the coefficients of y, and those in the third column are the coefficients of z. The constants on the right side of the system are found in the fourth column. The vertical line in the augmented matrix is to remind you of the equal signs in the equations. If a variable does not appear in one of the equations, represent it by a zero in the matrix.

EXAMPLE 2 **Writing the Augmented Matrix of a Linear System**

Write the augmented matrix of the linear system.

$$\begin{cases} 2x + 3z = 1 \\ 2z + y = 5 \\ -4x + 5y = 7 \end{cases}$$

SOLUTION

First, write the system with the variables lined up in columns and insert zeros as coefficients of any missing variable.

$$\begin{cases} 2x + 0y + 3z = 1 \\ 0x + 1y + 2z = 5 \\ -4x + 5y + 0z = 7 \end{cases}$$

The augmented matrix of the given system is:

$$\begin{bmatrix} 2 & 0 & 3 & 1 \\ 0 & 1 & 2 & 5 \\ -4 & 5 & 0 & 7 \end{bmatrix}$$

■ ■ ■

Practice Problem 2 Write the augmented matrix of the linear system.

$$\begin{cases} 3y - z = 8 \\ x + 4y = 14 \\ -2y + 9z = 0 \end{cases}$$

■

We can reverse this process and write a linear system from a given augmented matrix. For example, the augmented matrix

$$\begin{bmatrix} 9 & 2 & -1 & 6 \\ 3 & 4 & 7 & -8 \\ 0 & 1 & 5 & 11 \end{bmatrix}$$ corresponds to the system of equations $$\begin{cases} 9x + 2y - z = 6 \\ 3x + 4y + 7z = -8. \\ y + 5z = 11 \end{cases}$$

TECHNOLOGY CONNECTION

Many graphing calculators have you first name a matrix, specify the size, and then insert the entries for the matrix. No vertical line separates the column of constants, but this does not affect any of the matrix operations. The augmented matrix from Example 2 has been named A.

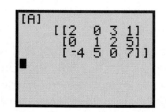

The basic strategy for solving a system of equations is to *replace the given system with an equivalent system* (one with the same solution set) *that is easier to solve*. In the previous chapter, we used three basic operations to solve a linear system:

1. Interchange two equations.
2. Multiply all of the terms in an equation by a nonzero constant.
3. Add a multiple of one equation to another equation. In other words, replace one equation with the sum of itself and a multiple of another equation.

In matrix terminology, the three correspond operations are called the **elementary row operations**. Two matrices are **row equivalent** if one can be obtained from the other by a sequence of elementary row operations, given next.

Elementary Row Operations		
Row Operation	**In Symbols**	**Description**
1. Interchange two rows.	$R_i \leftrightarrow R_j$	Interchange the ith and jth rows.
2. Multiply a row by a nonzero constant.	cR_j	Multiply the jth row by $c, c \neq 0$.
3. Add a multiple of one row to another row.	$cR_i + R_j \rightarrow R_j$	Replace the jth row by adding c times the ith row to it.

EXAMPLE 3 **Applying Elementary Row Operations**

Perform the indicated row operations (**a**), (**b**), and (**c**) in order on the following matrix:

$$A = \begin{bmatrix} 3 & -2 & -3 \\ 2 & 4 & 14 \end{bmatrix}$$

a. $R_1 \leftrightarrow R_2,$ **b.** $\dfrac{1}{2}R_1,$ **c.** $-3R_1 + R_2 \rightarrow R_2$

SOLUTION

a. $A = \begin{bmatrix} 3 & -2 & -3 \\ 2 & 4 & 14 \end{bmatrix} \xrightarrow{R_1 \leftrightarrow R_2} \begin{bmatrix} 2 & 4 & 14 \\ 3 & -2 & -3 \end{bmatrix} = B$ Interchange the 1st and 2nd rows.

b. $B = \begin{bmatrix} 2 & 4 & 14 \\ 3 & -2 & -3 \end{bmatrix} \xrightarrow{\frac{1}{2}R_1} \begin{bmatrix} 1 & 2 & 7 \\ 3 & -2 & -3 \end{bmatrix} = C$ Multiply the 1st row by $\frac{1}{2}$.

c. $C = \begin{bmatrix} 1 & 2 & 7 \\ 3 & -2 & -3 \end{bmatrix} \xrightarrow{-3R_1 + R_2 \rightarrow R_2} \begin{bmatrix} 1 & 2 & 7 \\ 0 & -8 & -24 \end{bmatrix}$ Add -3 times the 1st row to the 2nd row. ■ ■ ■

Practice Problem 3 Perform the row operations of Example 3 in order on the following matrix:

$$A = \begin{bmatrix} 3 & 4 & 5 \\ 2 & 4 & 6 \end{bmatrix}$$ ■

In the next example, we solve a system of equations both with and without matrix notation and place the results side by side for comparison. We solve the system by converting it into triangular form by first using elimination and then back-substitution.

TECHNOLOGY CONNECTION

Many graphing calculators provide all three row operations. Consider the matrix A.

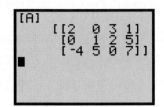

[A]
[[2 0 3 1]
 [0 1 2 5]
 [-4 5 0 7]]

Here are some examples of row operations on the matrix A.

Swap rows 1 and 3.

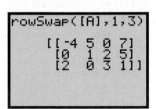

rowSwap([A],1,3)

[[-4 5 0 7]
 [0 1 2 5]
 [2 0 3 1]]

Multiply row 1 by 4.

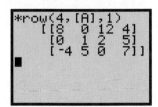

*row(4,[A],1)
[[8 0 12 4]
 [0 1 2 5]
 [-4 5 0 7]]

Add 3 times row 1 to row 2.

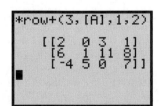

*row+(3,[A],1,2)

[[2 0 3 1]
 [6 1 11 8]
 [-4 5 0 7]]

EXAMPLE 4   **Comparing Linear Systems and Matrices**

Solve the system of linear equations.

$$\begin{cases} x - y - z = 1 & (1) \\ 2x - 3y + z = 10 & (2) \\ x + y - 2z = 0 & (3) \end{cases}$$

SOLUTION

Linear System

$$\begin{cases} x - y - z = 1 & (1) \\ 2x - 3y + z = 10 & (2) \\ x + y - 2z = 0 & (3) \end{cases}$$

Add -2 times equation (1) to equation (2).
Add -1 times equation (1) to equation (3).

$$\begin{cases} x - y - z = 1 & (1) \\ -y + 3z = 8 & (4) \\ 2y - z = -1 & (5) \end{cases}$$

Multiply equation (4) by -1.

$$\begin{cases} x - y - z = 1 & (1) \\ y - 3z = -8 & (6) \\ 2y - z = -1 & (5) \end{cases}$$

Add -2 times equation (6) to equation (5).

$$\begin{cases} x - y - z = 1 & (1) \\ y - 3z = -8 & (6) \\ 5z = 15 & (7) \end{cases}$$

Augmented Matrix

$$A = \begin{bmatrix} 1 & -1 & -1 & | & 1 \\ 2 & -3 & 1 & | & 10 \\ 1 & 1 & -2 & | & 0 \end{bmatrix}$$

Use the first row to produce zeros at the (2, 1) and (3, 1) positions.

$$\xrightarrow[{(-1)R_1+R_3 \to R_3}]{-2R_1+R_2 \to R_2} \begin{bmatrix} 1 & -1 & -1 & | & 1 \\ 0 & -1 & 3 & | & 8 \\ 0 & 2 & -1 & | & -1 \end{bmatrix}$$

Produce a 1 at the (2, 2) position.

$$\xrightarrow{(-1)R_2} \begin{bmatrix} 1 & -1 & -1 & | & 1 \\ 0 & 1 & -3 & | & -8 \\ 0 & 2 & -1 & | & -1 \end{bmatrix}$$

Use the second row to produce a zero at the (3, 2) position.

$$\xrightarrow{-2R_2+R_3 \to R_3} \begin{bmatrix} 1 & -1 & -1 & | & 1 \\ 0 & 1 & -3 & | & -8 \\ 0 & 0 & 5 & | & 15 \end{bmatrix}$$

The equation $5z = 15$, which corresponds to row 3 of the final matrix, gives the value $z = 3$. We find y by back-substitution.

$y - 3z = -8$	Equation (6); final matrix row 2
$y - 3(3) = -8$	Replace z with 3.
$y = 1$	Solve for y.

We find x by back-substitution.

$x - y - z = 1$	Equation (1); final matrix row 1
$x - 1 - 3 = 1$	Replace y with 1 and z with 3.
$x = 5$	Solve for x.

The solution of the system is $x = 5$, $y = 1$, and $z = 3$. You should check this solution by substituting it into the original system of equations. The solution set is $\{(5, 1, 3)\}$.   ■ ■ ■

Practice Problem 4 Solve the system of linear equations.

$$\begin{cases} x - 6y + 3z = -2 \\ 3x + 3y - 2z = -2 \\ 2x - 3y + z = -2 \end{cases}$$ ■

The last matrix in the solution of Example 4 is in *row-echelon form,* which is defined next. In the definition, a **nonzero row** means a row that contains at least one nonzero entry; a **leading entry** of a row is the leftmost nonzero entry in a nonzero row.

ROW-ECHELON FORM AND REDUCED ROW-ECHELON FORM

An $m \times n$ matrix is in **row-echelon form** if it has the following three properties:

1. The leading entry of each nonzero row is a 1.

2. The leading entry in a row is to the right of the leading entry in the row above it.

3. All nonzero rows are above the rows consisting entirely of zeros.

A matrix in row-echelon form having the following property is in **reduced row-echelon form:** Each leading 1 is the only nonzero entry in its column.

Property 2 says that the leading entries form an echelon (steplike) pattern that moves down and to the right.

The following matrices are in row-echelon form; the starred entries may be any value, including zero.

$$\begin{bmatrix} 1 & * & * \\ 0 & 1 & * \\ 0 & 0 & 1 \end{bmatrix} \qquad \begin{bmatrix} 1 & * & * & * \\ 0 & 1 & * & * \\ 0 & 0 & 1 & * \end{bmatrix} \qquad \begin{bmatrix} 1 & * & * & * \\ 0 & 0 & 1 & * \\ 0 & 0 & 0 & 1 \\ 0 & 0 & 0 & 0 \end{bmatrix}$$

The following matrices are in *reduced* row-echelon form because the entries above and below each leading 1 are zero. (Again, the starred entries may be any value, including zero.)

$$\begin{bmatrix} 0 & 1 & 0 & 0 \\ 0 & 0 & 1 & 0 \\ 0 & 0 & 0 & 1 \end{bmatrix} \qquad \begin{bmatrix} 1 & 0 & 0 & * \\ 0 & 1 & 0 & * \\ 0 & 0 & 1 & * \end{bmatrix} \qquad \begin{bmatrix} 1 & * & 0 & 0 \\ 0 & 0 & 1 & 0 \\ 0 & 0 & 0 & 1 \\ 0 & 0 & 0 & 0 \end{bmatrix}$$

An augmented matrix may be transformed into several different row-echelon form matrices by using different sequences of row operations. However, its *reduced* row-echelon form is *unique.* (See Exercise 91.)

3 Use Gaussian elimination to solve a system.

Gaussian Elimination

The method of solving a system of linear equations by transforming the augmented matrix of the system into row-echelon form and then using back-substitution to find the solution set is known as **Gaussian elimination.**

FINDING THE SOLUTION: A PROCEDURE

| EXAMPLE 5 | Solving Linear Systems by Using Gaussian Elimination |

OBJECTIVE

Solve a system of linear equations by Gaussian elimination.

Step 1 Write the augmented matrix of the system.

Step 2 Use elementary row operations to transform the augmented matrix into row-echelon form.

Step 3 Write the system of linear equations that corresponds to the last matrix in Step 2.

Step 4 Use the system of equations from Step 3, together with back-substitution, to find the system's solution set.

EXAMPLE

Solve the system of equations.

$$\begin{cases} x + 2y = -2 \\ 4x + 3y = 7 \end{cases}$$

$$A = \begin{bmatrix} 1 & 2 & | & -2 \\ 4 & 3 & | & 7 \end{bmatrix} \qquad \text{Augmented matrix}$$

$$\xrightarrow{-4R_1 + R_2 \to R_2} \begin{bmatrix} 1 & 2 & | & -2 \\ 0 & -5 & | & 15 \end{bmatrix} \qquad \begin{array}{l}\text{This operation produces a 0 in} \\ \text{the } (2, 1) \text{ position.}\end{array}$$

$$\xrightarrow{-\frac{1}{5}R_2} \begin{bmatrix} 1 & 2 & | & -2 \\ 0 & 1 & | & -3 \end{bmatrix} \qquad \begin{array}{l}\text{This operation produces a 1 in} \\ \text{the } (2, 2) \text{ position.}\end{array}$$

$$\begin{cases} x + 2y = -2 & (1) \\ \quad\quad y = -3 & (2) \end{cases}$$

$$\begin{array}{ll} x + 2y = -2 & \text{Equation (1)} \\ x + 2(-3) = -2 & \text{Replace } y \text{ with } -3. \\ \quad\quad x = 4 & \text{Solve for } x. \end{array}$$

You should check the solution set, $\{(4, -3)\}$, by substituting $x = 4$ and $y = -3$ into the original system of equations. ■■■

Practice Problem 5 Solve by Gaussian elimination.

$$\begin{cases} x - 2y = 1 \\ 2x + 3y = 16 \end{cases}$$ ■

| EXAMPLE 6 | Solving a System by Using Gaussian Elimination |

Solve by Gaussian elimination.

$$\begin{cases} 2x + y + z = 6 \\ -3x - 4y + 2z = 4 \\ x + y - z = -2 \end{cases}$$

STUDY TIP

Note that when we perform a row operation such as

$3R_1 + R_2 \to R_2$ on a matrix,

1. the entries in row 1 are *unchanged* and
2. we multiply each entry of row 1 by 3 and add this product to the corresponding entries of row 2 to obtain a *new row 2*.

SOLUTION

Step 1 $A = \begin{bmatrix} 2 & 1 & 1 & | & 6 \\ -3 & -4 & 2 & | & 4 \\ 1 & 1 & -1 & | & -2 \end{bmatrix}$ The augmented matrix of the system

Step 2 $\xrightarrow{R_1 \leftrightarrow R_3} \begin{bmatrix} 1 & 1 & -1 & | & -2 \\ -3 & -4 & 2 & | & 4 \\ 2 & 1 & 1 & | & 6 \end{bmatrix}$

$\xrightarrow[{-2R_1 + R_3 \to R_3}]{3R_1 + R_2 \to R_2} \begin{bmatrix} 1 & 1 & -1 & | & -2 \\ 0 & -1 & -1 & | & -2 \\ 0 & -1 & 3 & | & 10 \end{bmatrix}$

Continued on the next page

$$\xrightarrow{(-1)R_2} \begin{bmatrix} 1 & 1 & -1 & | & -2 \\ 0 & 1 & 1 & | & 2 \\ 0 & -1 & 3 & | & 10 \end{bmatrix}$$

$$\xrightarrow{R_2 + R_3 \rightarrow R_3} \begin{bmatrix} 1 & 1 & -1 & | & -2 \\ 0 & 1 & 1 & | & 2 \\ 0 & 0 & 4 & | & 12 \end{bmatrix}$$

$$\xrightarrow{\frac{1}{4}R_3} \begin{bmatrix} 1 & 1 & -1 & | & -2 \\ 0 & 1 & 1 & | & 2 \\ 0 & 0 & 1 & | & 3 \end{bmatrix}$$

The last matrix is in row-echelon form.

Step 3 The system of equations corresponding to the last matrix in Step 2 is

$$\begin{cases} x + y - z = -2 & (1) \\ y + z = 2 & (2) \\ z = 3 & (3) \end{cases}$$

Step 4 Equation (3) in Step 3 gives the value $z = 3$. Back-substitute $z = 3$ in equation (2).

$$y + z = 2 \qquad \text{Equation (2)}$$
$$y + 3 = 2 \qquad \text{Replace } z \text{ with 3.}$$
$$y = -1 \qquad \text{Solve for } y.$$

Now back-substitute $z = 3$ and $y = -1$ in equation (1).

$$x + y - z = -2 \qquad \text{Equation (1)}$$
$$x - 1 - 3 = -2$$
$$x = 2 \qquad \text{Solve for } x.$$

You should check the solution set, $\{(2, -1, 3)\}$, by substituting these values into the original system of equations. ■ ■ ■

Practice Problem 6 Solve the system of equations.

$$\begin{cases} 2x + y - z = 7 \\ x - 3y - 3z = 4 \\ 4x + y + z = 3 \end{cases}$$ ■

EXAMPLE 7 **Attempting to Solve a System with No Solution**

Solve the system of equations by Gaussian elimination.

$$\begin{cases} y + 5z = -4 \\ x + 4y + 3z = -2 \\ 2x + 7y + z = 8 \end{cases}$$

SOLUTION

Step 1
$$A = \begin{bmatrix} 0 & 1 & 5 & | & -4 \\ 1 & 4 & 3 & | & -2 \\ 2 & 7 & 1 & | & 8 \end{bmatrix}. \qquad \text{The augmented matrix of the system}$$

Step 2 $\xrightarrow{R_1 \leftrightarrow R_2} \begin{bmatrix} 1 & 4 & 3 & | & -2 \\ 0 & 1 & 5 & | & -4 \\ 2 & 7 & 1 & | & 8 \end{bmatrix}$

$$\xrightarrow{-2R_1+R_3\to R_3} \begin{bmatrix} 1 & 4 & 3 & | & -2 \\ 0 & 1 & 5 & | & -4 \\ 0 & -1 & -5 & | & 12 \end{bmatrix}$$ Row 2 already begins with a 0.

$$\xrightarrow{R_2+R_3\to R_3} \begin{bmatrix} 1 & 4 & 3 & | & -2 \\ 0 & 1 & 5 & | & -4 \\ 0 & 0 & 0 & | & 8 \end{bmatrix}$$

$$\xrightarrow{\frac{1}{8}R_3} \begin{bmatrix} 1 & 4 & 3 & | & -2 \\ 0 & 1 & 5 & | & -4 \\ 0 & 0 & 0 & | & 1 \end{bmatrix}$$ The matrix is now in row-echelon form.

Step 3   The system of equations corresponding to the last matrix in Step 2 is

$$\begin{cases} x + 4y + 3z = -2 \\ \quad\quad y + 5z = -4 \\ \quad\quad\quad\quad 0 = 1 \quad \text{A false statement} \end{cases}$$

The equation $0 = 1$ can be rewritten as $0x + 0y + 0z = 1$. Because this equation is never true, we conclude that this system is inconsistent. Because this system is equivalent to the original system, the original system is also inconsistent.

RECALL

A system is called *inconsistent* if it has no solution.

Step 4   The solution set for the system is $\varnothing$.   ■ ■ ■

Practice Problem 7   Solve the following system of equations by first transforming the augmented matrix into row-echelon form.

$$\begin{cases} 6x + 8y - 14z = 3 \\ 3x + 4y - 7z = 12 \\ 6x + 3y + z = 0 \end{cases}$$ ■

4   Use Gauss–Jordan elimination to solve a system.

TECHNOLOGY CONNECTION

Many graphing calculators can produce the unique reduced row-echelon form of an augmented matrix. Consider the augmented matrix A in Example 4.

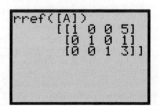

We use the "rref" (reduced row-echelon form) option to obtain the reduced row-echelon form.

Gauss–Jordan Elimination

If we continue the Gaussian elimination procedure until we have a *reduced* row-echelon form, the procedure is called **Gauss–Jordan elimination**.

EXAMPLE 8   **Solving a System of Equations by Gauss–Jordan Elimination**

Solve the system given in Example 4 by Gauss–Jordan elimination.

$$\begin{cases} x - y - z = 1 \\ 2x - 3y + z = 10 \\ x + y - 2z = 0 \end{cases} \quad \text{The given system}$$

SOLUTION

$$\begin{bmatrix} 1 & -1 & -1 & | & 1 \\ 0 & 1 & -3 & | & -8 \\ 0 & 0 & 5 & | & 15 \end{bmatrix}$$ The final augmented matrix of the system in Example 4

$$\xrightarrow{\frac{1}{5}R_3} \begin{bmatrix} 1 & -1 & -1 & | & 1 \\ 0 & 1 & -3 & | & -8 \\ 0 & 0 & 1 & | & 3 \end{bmatrix}$$ The matrix is now in row-echelon form.

$$\xrightarrow{R_2+R_1\to R_1} \begin{bmatrix} 1 & 0 & -4 & | & -7 \\ 0 & 1 & -3 & | & -8 \\ 0 & 0 & 1 & | & 3 \end{bmatrix}$$

$$\xrightarrow[3R_3+R_2\to R_2]{4R_3+R_1\to R_1} \begin{bmatrix} 1 & 0 & 0 & | & 5 \\ 0 & 1 & 0 & | & 1 \\ 0 & 0 & 1 & | & 3 \end{bmatrix}$$

We now have an equivalent matrix in reduced row-echelon form. The corresponding system of equations for the last augmented matrix is

$$\begin{cases} x = 5 \\ y = 1 \\ z = 3 \end{cases}$$

The solution set is therefore $\{(5, 1, 3)\}$, as in Example 4. ■ ■ ■

Practice Problem 8   Solve the system by using Gauss–Jordan elimination.

$$\begin{cases} 2x - 3y - 2z = 0 \\ x + y - 2z = 7 \\ 3x - 5y - 5z = 3 \end{cases}$$ ■

EXAMPLE 9   Solving a System with Infinitely Many Solutions

Solve the system of equations.

$$\begin{cases} x + 2y + 5z = 4 \\ y + 4z = 4 \\ 2x + 4y + 10z = 8 \end{cases}$$

SOLUTION

The augmented matrix A of the system is given below. We want to find the equivalent system in reduced row-echelon form.

Since $a_{11} = 1$ and $a_{21} = 0$, we need a zero at the $(3, 1)$ position.

$$A = \begin{bmatrix} 1 & 2 & 5 & | & 4 \\ 0 & 1 & 4 & | & 4 \\ 2 & 4 & 10 & | & 8 \end{bmatrix} \xrightarrow{-2R_1 + R_3 \rightarrow R_3} \begin{bmatrix} 1 & 2 & 5 & | & 4 \\ 0 & 1 & 4 & | & 4 \\ 0 & 0 & 0 & | & 0 \end{bmatrix}$$

Next, we need a zero at the $(1, 2)$ position.

$$\xrightarrow{-2R_2 + R_1 \rightarrow R_1} \begin{bmatrix} 1 & 0 & -3 & | & -4 \\ 0 & 1 & 4 & | & 4 \\ 0 & 0 & 0 & | & 0 \end{bmatrix}$$ The matrix is now in reduced row-echelon form.

The equivalent system is

$$x - 3z = -4$$
$$y + 4z = 4$$

Solving for x and y in terms of z, we obtain

$$x = 3z - 4$$
$$y = -4z + 4$$

Each real number z results in a solution with $y = -4z + 4$ and $x = 3z - 4$, giving infinitely many solutions of the form $(3z - 4, -4z + 4, z)$. The solution set is $\{(3z - 4, -4z + 4, z)\}$. ■ ■ ■

Practice Problem 9   Solve the system of equations.

$$\begin{cases} x + z = -1 \\ 3y + 2z = 5 \\ 3x - 3y + z = -8 \end{cases}$$ ■

EXAMPLE 10   Leontief Input–Output Model

Consider an economy that has steel, coal, and transportation industries. There are two types of demands (measured in dollars) on the production of each industry: interindustry demand and external consumer demand. The outputs and requirements of the three industries are shown in Figure 8.1. For example, $1.00 of transportation output requires $0.10 from steel and $0.01 from coal.

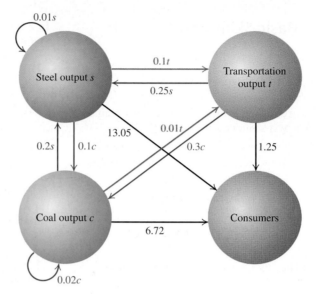

FIGURE 8.1 Output diagram of an economy

a. Write a system of equations that expresses the outputs of the three industries. Assume that all quantities are given in millions of dollars.

b. Verify that $s = 15, c = 10$, and $t = 8$ will meet both interindustry and consumer demand.

SOLUTION

a. To satisfy both consumer and interindustry demand, we obtain the following system of outputs. (s denotes total steel output; c, total coal output; t, total transportation output.)

$$\begin{cases} s = 0.01s + 0.1t + 0.1c + 13.05 & \text{Distribution of steel output} \\ c = 0.2s + 0.01t + 0.02c + 6.72 & \text{Distribution of coal output} \\ t = 0.25s + 0.3c + 1.25 & \text{Distribution of transportation output} \end{cases}$$

You can rewrite this system as

$$\begin{cases} 0.99s - 0.1c - 0.1t = 13.05 & (1) \\ -0.2s + 0.98c - 0.01t = 6.72 & (2) \\ -0.25s - 0.3c + t = 1.25 & (3) \end{cases}$$

b. To verify that the given numbers (obtained by solving the system above using matrices) satisfy these equations, we substitute $s = 15, c = 10$, and $t = 8$ into each equation.

$$(0.99)(15) - (0.1)(10) - (0.1)(8) = 13.05 \checkmark \quad (1)$$
$$(-0.20)(15) + (0.98)(10) - (0.01)(8) = 6.72 \checkmark \quad (2)$$
$$(-0.25)(15) - (0.30)(10) + 8 = 1.25 \checkmark \quad (3)$$

In other words, output levels of $s = 15, c = 10$, and $t = 8$ million dollars will meet interindustry and consumer demand. ■ ■ ■

Practice Problem 10 In Example 10, assume that consumer demand (in millions of dollars) is 12.46 for steel, 3 for coal, and 2.7 for transportation.

a. Write a system of equations to express the outputs for the three industries.

b. Verify that $s = 14, c = 6, t = 8$ will meet both interindustry and consumer demand. ■

SECTION 8.1 ■ Exercises

A EXERCISES Basic Skills and Concepts

1. A matrix is any rectangular array of ___numbers___.

2. The array of coefficients and constants in a linear system is called the ___augmented matrix___ of the system.

3. Two matrices are row-equivalent if one can be obtained from the other by a sequence of ___row operations___.

4. If a matrix is in row-echelon form and each leading entry 1 is the only nonzero entry in its column, then the matrix is in _____ form. reduced row-echelon

5. *True or False* If A is a 3×4 matrix, then each row of A has three entries. False

6. *True or False* Every $m \times n$ $(n \geq 2)$ matrix is the matrix of some linear system. True

7. *True or False* The augmented matrix for the system of equations False

$$\begin{cases} 2x + 3y \quad\;\; = 5 \\ \quad\; 3y - 4z = -1 \\ x + \quad\;\; 2z = 3 \end{cases} \text{ is } \begin{bmatrix} 2 & 3 & 5 \\ 3 & -4 & -1 \\ 1 & 2 & 3 \end{bmatrix}.$$

8. *True or False* The augmented matrix for the system of equations False

$$\begin{cases} 2x - y = 1 \\ 3x + y = 9 \\ 5x - 2y = 4 \end{cases} \text{ is } \begin{bmatrix} 2x & -y & 1 \\ 3x & y & 9 \\ 5x & -2y & 4 \end{bmatrix}.$$

In Exercises 9–14, determine the order of each matrix.

9. $[7]$ 1×1

10. $[1 \quad 3 \quad 5 \quad 9]$ 1×4

11. $\begin{bmatrix} 3 & 4.3 & 7.5 & 8 \\ 2 & -1 & -3 & 0 \end{bmatrix}$ 2×4

12. $\begin{bmatrix} -1 & 2 & 3 \\ 4 & -5 & 7 \\ 1 & 0 & 0 \end{bmatrix}$ 3×3

13. $\begin{bmatrix} 4 & 5 & 1 \\ 0 & 0 & 0 \end{bmatrix}$ 2×3

14. $\begin{bmatrix} -5 & 2 & 3 & 7 \\ 2.5 & e & -\pi & \dfrac{1}{2} \\ -e & \pi & 0 & 1 \end{bmatrix}$ 3×4

15. Let $A = \begin{bmatrix} 1 & 2 & 3 & 4 \\ 5 & 6 & 7 & 8 \\ 9 & 10 & 11 & 12 \end{bmatrix}$. Identify the entries $a_{13}, a_{31}, a_{33},$ and a_{34}. $a_{13} = 3, a_{31} = 9, a_{33} = 11, a_{34} = 12$

16. In matrix A from Exercise 15, identify the (i, j)th location of each entry.
    a. 7 a_{23}
    b. 10 a_{32}
    c. 4 a_{14}
    d. 12 a_{34}

17. Is the following array a matrix? Why or why not?

$$\begin{bmatrix} 2 & 0 & 4 \\ 1 & -3 & 5 \\ 0 & 0 & \end{bmatrix} \text{ No, it is not a rectangular array of numbers.}$$

18. Write the matrix A with the following entries:†

$$a_{13} = 5, a_{22} = 2, a_{11} = 6, a_{12} = -5, a_{23} = 4, \text{ and } a_{21} = 7$$

In Exercises 19–24, write the augmented matrix for each system of linear equations.

19. $\begin{cases} 2x + 4y = 2 \\ x - 3y = 1 \end{cases}$ †

20. $\begin{cases} x_1 + 2x_2 = 7 \\ 3x_1 + 5x_2 = 11 \end{cases}$ †

21. $\begin{cases} 5x_1 - 11 = 7x_2 \\ 17x_2 - 19 = 13x_1 \end{cases}$ †

22. $\begin{cases} 2v - 10 = 3u \\ 5u + 7 = v \end{cases}$ †

23. $\begin{cases} -x + 2y + 3z = 8 \\ 2x - 3y + 9z = 16 \\ 4x - 5y - 6z = 32 \end{cases}$ †

24. $\begin{cases} x - y \quad\;\;\; = 2 \\ 2x \quad\;\; + 3z = -5 \\ \quad\; y - 2z = 7 \end{cases}$ †

In Exercises 25–28, write the system of linear equations represented by each augmented matrix. Use $x, y,$ and z as the variables.

25. $\begin{bmatrix} 1 & 2 & -3 & 4 \\ -2 & -3 & 1 & 5 \\ 3 & -3 & 2 & 7 \end{bmatrix}$ †

26. $\begin{bmatrix} -1 & 2 & 3 & 6 \\ 2 & 3 & 1 & 2 \\ 4 & 3 & 2 & 1 \end{bmatrix}$ †

27. $\begin{bmatrix} 1 & -1 & 1 & 2 \\ 2 & 1 & -3 & 6 \end{bmatrix}$ †

28. $\begin{bmatrix} 1 & 1 & 1 & 2 \\ 1 & -1 & -1 & 4 \\ 2 & 3 & 1 & 6 \\ -1 & 1 & -1 & 8 \end{bmatrix}$ †

In Exercises 29–32, perform the indicated elementary row operations in the stated order.

29. $\begin{bmatrix} 2 & 3 & 5 \\ 1 & 2 & 3 \end{bmatrix}$; (i) $R_1 \leftrightarrow R_2$, (ii) $-2R_1 + R_2 \rightarrow R_2$, (iii) $-R_2$ †

30. $\begin{bmatrix} 2 & 4 & 2 \\ 1 & 5 & 7 \end{bmatrix}$; (i) $\frac{1}{2}R_1$, (ii) $(-1) R_1 + R_2 \rightarrow R_2$, (iii) $\frac{1}{3}R_2$ †

31. $\begin{bmatrix} 1 & 2 & 3 & 4 \\ 0 & 4 & -3 & 11 \\ 0 & 1 & 5 & -3 \end{bmatrix}$; (i) $R_2 \leftrightarrow R_3$, (ii) $-4R_2 + R_3 \rightarrow R_3$, (iii) $-\frac{1}{23}R_3$ †

32. $\begin{bmatrix} 1 & \dfrac{3}{2} & -2 & \dfrac{1}{2} \\ 0 & 2 & 3 & 4 \\ -2 & -1 & 7 & 3 \end{bmatrix}$; (i) $2R_1 + R_3 \rightarrow R_3$, (ii) $\frac{1}{2}R_2 + R_3 \rightarrow R_3$ †

In Exercises 33–36, identify the elementary row operation used and supply the missing entries in each row-equivalent matrix.

33. $\begin{bmatrix} 4 & 5 & -7 \\ 5 & 4 & -2 \end{bmatrix} \rightarrow \begin{bmatrix} 1 & ? & -\dfrac{7}{4} \\ 5 & 4 & -2 \end{bmatrix} \rightarrow \begin{bmatrix} 1 & ? & -\dfrac{7}{4} \\ 0 & -\dfrac{9}{4} & ? \end{bmatrix} \rightarrow$

$$\begin{bmatrix} 1 & ? & -\dfrac{7}{4} \\ 0 & 1 & ? \end{bmatrix} \text{†}$$

34. $\begin{bmatrix} 2 & 6 & -8 \\ 3 & -1 & 2 \end{bmatrix} \rightarrow \begin{bmatrix} 1 & 3 & -? \\ 3 & -1 & 2 \end{bmatrix} \rightarrow \begin{bmatrix} 1 & 3 & ? \\ 0 & -? & 14 \end{bmatrix} \rightarrow$

$$\begin{bmatrix} 1 & 3 & -? \\ 0 & 1 & ? \end{bmatrix} \text{†}$$

†Due to space constrictions, answers to these exercises may be found in the Answers beginning on page A–1 in the back of the book.

35. $\begin{bmatrix} 1 & 4 & 3 & 1 \\ 0 & -3 & -2 & 0 \\ 0 & 7 & 5 & -3 \end{bmatrix} \rightarrow \begin{bmatrix} 1 & 4 & 3 & 1 \\ 0 & 1 & ? & 0 \\ 0 & 7 & 5 & -3 \end{bmatrix} \rightarrow$

$\begin{bmatrix} 1 & 4 & 3 & 1 \\ 0 & 1 & ? & 0 \\ 0 & 0 & ? & -3 \end{bmatrix} \rightarrow \begin{bmatrix} 1 & 4 & 3 & 1 \\ 0 & 1 & ? & 0 \\ 0 & 0 & 1 & ? \end{bmatrix}$ †

36. $\begin{bmatrix} 1 & 1 & 1 & 3 \\ 2 & 3 & 3 & 8 \\ 1 & -3 & -2 & 5 \end{bmatrix} \rightarrow \begin{bmatrix} 1 & 1 & 1 & 3 \\ 0 & 1 & ? & 2 \\ 1 & -3 & -2 & 5 \end{bmatrix} \rightarrow$

$\begin{bmatrix} 1 & 1 & 1 & 3 \\ 0 & 1 & ? & 2 \\ 0 & -4 & ? & 2 \end{bmatrix} \rightarrow \begin{bmatrix} 1 & 1 & 1 & 3 \\ 0 & 1 & ? & 2 \\ 0 & 0 & ? & 10 \end{bmatrix}$ †

In Exercises 37–44, determine whether each matrix is in row-echelon form. If your answer is no, explain why. If your answer is yes, is the matrix in reduced row-echelon form?

37. $\begin{bmatrix} 0 & 1 & 2 \\ 1 & 0 & 3 \end{bmatrix}$ No, because Property 1 is not satisfied.

38. $\begin{bmatrix} 1 & 2 & 5 \\ 0 & 3 & 6 \end{bmatrix}$ No, because Property 1 is not satisfied.

39. $\begin{bmatrix} 1 & 0 & 0 & 2 \\ 0 & 1 & 0 & 3 \\ 0 & 0 & 1 & 4 \end{bmatrix}$

40. $\begin{bmatrix} 0 & 0 & 1 & 2 \\ 0 & 1 & 0 & 1 \\ 1 & 0 & 0 & 3 \end{bmatrix}$ No, Property 2 is not satisfied.

41. $\begin{bmatrix} 1 & 2 & 0 & 2 \\ 0 & 0 & 1 & 5 \end{bmatrix}$ The matrix is in reduced row-echelon form because Properties 1–4 are satisfied.

42. $\begin{bmatrix} 0 & 1 & -1 & -2 & 0 & 2 & 0 \\ 0 & 0 & 0 & 0 & 1 & 2 & 0 \end{bmatrix}$ The matrix is in reduced row-echelon form because Properties 1–4 are satisfied.

43. $\begin{bmatrix} 1 & 0 & 0 & -2 \\ 0 & 1 & 0 & 3 \\ 0 & 0 & 1 & 2 \\ 0 & 0 & 0 & 0 \end{bmatrix}$ The matrix is in reduced row-echelon form because Properties 1–4 are satisfied.

44. $\begin{bmatrix} 1 & 0 & 0 & 3 \\ 0 & 1 & 0 & 4 \\ 0 & 0 & 0 & 5 \\ 0 & 0 & 1 & 3 \end{bmatrix}$ No, Properties 1 and 2 are not satisfied.

In Exercises 45–54, the augmented matrix of a system of equations has been transformed to an equivalent matrix in row-echelon form or reduced row-echelon form. Using x, y, z, and w as variables, write the system of equations corresponding to the matrix. If the system is consistent, solve it.

45. $\left[\begin{array}{cc|c} 1 & 2 & 1 \\ 0 & 1 & -2 \end{array}\right]$ †

46. $\left[\begin{array}{cc|c} 1 & 0 & 2 \\ 0 & 1 & 3 \end{array}\right]$ †

47. $\left[\begin{array}{ccc|c} 1 & 4 & 2 & 2 \\ 0 & 0 & 1 & 3 \end{array}\right]$ †

48. $\left[\begin{array}{ccc|c} 1 & 2 & 3 & 4 \\ 0 & 0 & 0 & 1 \end{array}\right]$ †

49. $\left[\begin{array}{ccc|c} 1 & 2 & 3 & 2 \\ 0 & 1 & -2 & 4 \\ 0 & 0 & 1 & -1 \end{array}\right]$ †

50. $\left[\begin{array}{ccc|c} 1 & 0 & 2 & 12 \\ 0 & 1 & 3 & 12 \\ 0 & 0 & 1 & 5 \end{array}\right]$ †

51. $\left[\begin{array}{cccc|c} 1 & 0 & 0 & 0 & 2 \\ 0 & 1 & 0 & 0 & -5 \\ 0 & 0 & 1 & 2 & 3 \\ 0 & 0 & 0 & 0 & 0 \end{array}\right]$ †

52. $\left[\begin{array}{cccc|c} 1 & 0 & 0 & 0 & 4 \\ 0 & 1 & 0 & 0 & 3 \\ 0 & 0 & 0 & 1 & 2 \end{array}\right]$ †

53. $\left[\begin{array}{cccc|c} 1 & 0 & 0 & 0 & -5 \\ 0 & 1 & 0 & 0 & 4 \\ 0 & 0 & 1 & 2 & 3 \\ 0 & 0 & 0 & 1 & 0 \end{array}\right]$ †

54. $\left[\begin{array}{cccc|c} 1 & 0 & 0 & 0 & 3 \\ 0 & 1 & 0 & 0 & 2 \\ 0 & 0 & 1 & 0 & 0 \\ 0 & 0 & 0 & 0 & 1 \end{array}\right]$ †

In Exercises 55–68, solve each system of equations by Gaussian elimination.

55. $\begin{cases} x - 2y = 11 \\ 2x - y = 13 \end{cases}$ $\{(5, -3)\}$

56. $\begin{cases} 3x - 2y = 4 \\ 4x - 3y = 5 \end{cases}$ $\{(2, 1)\}$

57. $\begin{cases} 2x - 3y = 3 \\ 4x - y = 11 \end{cases}$ $\{(3, 1)\}$

58. $\begin{cases} 3x + 2y = 1 \\ 6x + 4y = 3 \end{cases}$ $\varnothing$

59. $\begin{cases} 3x - 5y = 4 \\ 4x - 15y = 13 \end{cases}$

60. $\begin{cases} -2x + 4y = 1 \\ 3x - 5y = -9 \end{cases}$

61. $\begin{cases} x - y = 1 \\ 2x + y = 5 \\ 3x - 4y = 2 \end{cases}$ $\{(2, 1)\}$

62. $\begin{cases} y = 2x + 1 \\ 3x + 2y + 1.5 = 0 \\ 4x - 2y + 2 = 0 \end{cases}$

63. $\begin{cases} x + y + z = 6 \\ x - y + z = 2 \\ 2x + y - z = 1 \end{cases}$ †

64. $\begin{cases} 2x + 4y + z = 5 \\ x + y + z = 6 \\ 2x + 3y + z = 6 \end{cases}$ †

65. $\begin{cases} 2x + 3y - z = 9 \\ x + y + z = 9 \\ 3x - y - z = -1 \end{cases}$ †

66. $\begin{cases} x + y + 2z = 4 \\ 2x - y + 3z = 9 \\ 3x - y - z = 2 \end{cases}$ †

67. $\begin{cases} 3x + 2y + 4z = 19 \\ 2x - y + z = 3 \\ 6x + 7y - z = 17 \end{cases}$ †

68. $\begin{cases} 4x + 3y + z = 8 \\ 2x + y + 4z = -4 \\ 3x + z = 1 \end{cases}$ †

In Exercises 69–74, solve each system of equations by Gauss–Jordan elimination.

69. $\begin{cases} x - y = 1 \\ x - z = -1 \\ 2x + y - z = 3 \end{cases}$ †

70. $\begin{cases} 4x + 5z = 7 \\ y - 6z = 8 \\ 3x + 4y = 9 \end{cases}$ †

71. $\begin{cases} x + y - z = 4 \\ x + 3y + 5z = 10 \\ 3x + 5y + 3z = 18 \end{cases}$ †

72. $\begin{cases} x + y + z = -5 \\ 2x - y - z = -4 \\ y + z = -2 \end{cases}$ †

73. $\begin{cases} x + 2y - z = 6 \\ 3x + y + 2z = 3 \\ 2x + 5y + 3z = 9 \end{cases}$ †

74. $\begin{cases} 2x + 4y - z = 9 \\ x + 3y - 3 = 4 \\ 3x + y + 2z = 7 \end{cases}$ †

75. The reduced row-echelon forms of the augmented matrices of three systems of equations are given. How many solutions does each system have?

a. $\left[\begin{array}{cc|c} 1 & 0 & 2 \\ 0 & 1 & 3 \end{array}\right]$ One solution

b. $\left[\begin{array}{ccc|c} 1 & 2 & 3 & 8 \\ 0 & 0 & 1 & 2 \\ 0 & 0 & 0 & 0 \end{array}\right]$ Infinitely many solutions

c. $\left[\begin{array}{cc|c} 1 & 0 & 2 \\ 0 & 0 & 3 \end{array}\right]$ No solution

In Exercises 76–78, determine whether the statement is true or false and justify your answer.

76. If A is a 5×7 matrix, then each column of A has seven entries. False. Each row of A has seven entries.

77. The matrix $E = \begin{bmatrix} 1 & 3 & 0 \\ 0 & 0 & 1 \\ 0 & 0 & 0 \end{bmatrix}$ is in reduced row-echelon form. True

78. If a matrix A is in reduced row-echelon form, then at least one of the entries in each column must be a 1. †

Answers:

39. The matrix is in reduced row-echelon form because Properties 1–4 are satisfied.

59. $\left\{\left(-\dfrac{1}{5}, -\dfrac{23}{25}\right)\right\}$ **60.** $\left\{\left(-\dfrac{31}{2}, -\dfrac{15}{2}\right)\right\}$ **62.** $\left\{\left(\dfrac{-1}{2}, 0\right)\right\}$

B EXERCISES Applying the Concepts

In Exercises 79–82, Leontief's input–output model of a simplified economy is described. (See Example 10.) For each exercise, do the following.

a. Set up (without solving) a linear system whose solution will represent the required production schedule.
b. Write an augmented matrix for the economy.
c. Solve the system of equations by using part (b) to find the production schedule that will meet interindustry and consumer demand.

79. **Two-sector economy.** Consider an economy with only two industries: A and B. Suppose industry B needs $0.10 worth of A's product for each $1.00 of output B produces and that industry A needs $0.20 worth of B's product for each $1.00 of output A produces. Consumer demand for A's product is $1000, and consumer demand for B's product is $780. †

80. **A company with two branches.** A company has two interacting branches: A and B. Branch A consumes $0.50 of its own output and $0.20 of B's output for every $1.00 it produces. Branch B consumes $0.60 of A's output and $0.30 of its own output for $1.00 of output. Consumer demand for A's product is $50,000, and consumer demand for B's product is $40,000. †

81. **Three-sector economy.** An economy has three sectors: labor, transportation, and food industries. Suppose the demand on $1.00 in labor is $0.40 for transportation and $0.20 for food; the demand on $1.00 in transportation is $0.50 for labor and $0.30 for transportation; and the demand on $1.00 in food production is $0.50 for labor, $0.05 for transportation, and $0.35 for food. Outside consumer demand for the current production period is $10,000 for labor, $20,000 for transportation, and $10,000 for food. †

82. **Three-sector economy.** Consider an economy with three industries A, B, and C, with outputs a, b, and c, respectively. Demand on the three industries is shown in the figure. †

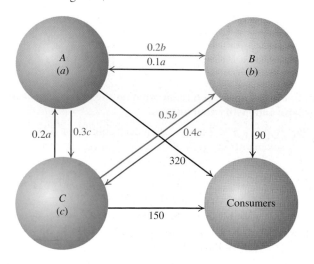

83. **Heat transfer.** In a study of heat transfer in a grid of wires, the temperature at an exterior node is maintained at a constant value (in °F) as shown in the accompanying figure. When the grid is in thermal equilibrium, the temperature at an interior node is the average of the temperatures at the four adjacent nodes. For instance

$$T_1 = \frac{0 + 0 + 300 + T_2}{4}, \text{ or } 4T_1 - T_2 = 300.$$ Find the temperatures T_1, T_2, and T_3 when the grid is in thermal equilibrium. $T_1 = 110, T_2 = 140, T_3 = 60$

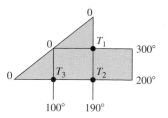

84. Repeat Exercise 83 for the following grid of wires.
$T_1 = 30, T_2 = \dfrac{75}{2}, T_3 = 40, T_4 = \dfrac{65}{2}$

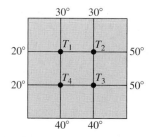

85. **Traffic flow.** The sketch shows the intersections of certain one-way streets. The traffic volume during one hour was observed. To keep the traffic moving, traffic controllers use the formula that the number of cars per hour entering an intersection equals the number of cars per hour exiting the intersection.
a. Set up a system of equations that keeps traffic moving. †
b. Solve the system of equations in part (a). †

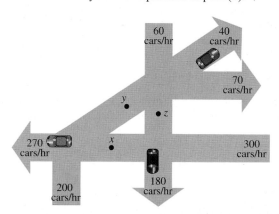

86. Repeat Exercise 85 for the following pattern of one-way streets. †

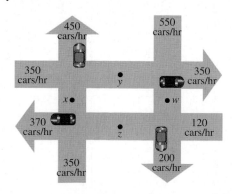

87. Modeling. A football was punted. Its height h above the ground at time t is given by the following table:

Time t (seconds)	Height h (feet)
$t = 0.5$	31
$t = 1$	51
$t = 2$	67

$h = -16t^2 + 64t + 3$

a. Use the system of equations to find the equation of the parabola of the form $h = at^2 + bt + c$.

b. What is the hang time (the time it takes for the ball to touch the ground) of the punt? approximately 4 sec

c. What is the maximum height of the ball? 67 ft

C EXERCISES Beyond the Basics

88. Partial fractions. Find the partial-fraction decomposition of the rational expression

$$\frac{2x^2 + 22x}{(x-1)^2(x+2)(x+3)}.$$

Use matrix methods to find the partial fractions.

In Exercises 89 and 90, solve each system of equations by Gauss–Jordan elimination.

89. $\begin{cases} x + y + z + w = 0 \\ x + 3y + 2z + 4w = 0 \\ 2x + z - w = 0 \end{cases}$ $\{(y + 2w, y, -2y - 3w, w)\}$

90. $\begin{cases} x - y + z - w = 4 \\ x + 2y + z + w = 2 \\ 2x + 3y + 4z + 5w = 5 \\ 3x + 4y + 2z - w = 8 \end{cases}$ $\{(1, 0, 2, -1)\}$

91. Let $A = \begin{bmatrix} 1 & 2 & -3 & 1 \\ 1 & 0 & -3 & 2 \\ 0 & 1 & 1 & 0 \\ 2 & 3 & 0 & -2 \end{bmatrix}$.

a. Find two different matrices B and C in row-echelon form that are row-equivalent to A. †

b. Show that the reduced row-echelon form of the matrices B and C of part (a) produce the same matrix. †

Answers:

88. $\dfrac{1}{x-1} + \dfrac{2}{(x-1)^2} - \dfrac{4}{x+2} + \dfrac{3}{x+3}$

92. Consider the following system of linear equations:

$$\begin{cases} x + y + z = 6 \\ x - y + z = 2 \\ 2x + y - z = 1 \end{cases}$$

a. Find two different matrices B and C in row-echelon form that yield the same solution set. †

b. Show that the reduced row-echelon form of the matrices B and C of part (a) produces the same matrix. †

93. a. Use the methods of this section to solve a general 2×2 system:

$$\begin{cases} ax + by = m \\ cx + dy = n \end{cases} †$$

b. Under what conditions on the coefficients does the system have **(i)** a unique solution, **(ii)** no solution, and **(iii)** infinitely many solutions? †

94. Construct three different augmented matrices for linear systems whose solution set is $x = 1, y = 0, z = -2$. There are many different correct answers. †

In Exercises 95 and 96, solve each system of equations by using row operations.

95. $\begin{cases} \log x + \log y + \log z = 6 \\ 3\log x - \log y + 3\log z = 10 \\ 5\log x + 5\log y - 4\log z = 3 \end{cases}$ $\{(10, 100, 1000)\}$

[*Hint:* Let $u = \log x$, $v = \log y$, and $w = \log z$.]

96. $\begin{cases} 4 \cdot 2^x - 3 \cdot 3^y + 5^z = 1 \\ 2^x + 4 \cdot 3^y - 2 \cdot 5^z = 10 \\ 2 \cdot 2^x - 2 \cdot 3^y + 3 \cdot 5^z = 4 \end{cases}$ $\left\{\left(1, 1, \dfrac{\ln 2}{\ln 5}\right)\right\}$

[*Hint:* Let $u = 2^x$, $v = 3^y$, and $w = 5^z$.]

97. Find the cubic function $y = ax^3 + bx^2 + cx + d$ whose graph passes through the points $(1, 5)$, $(-1, 1)$, $(2, 7)$, and $(-2, 11)$.

98. Find the cubic function $y = ax^3 + bx^2 + cx + d$ whose graph passes through the points $(1, 8)$, $(-1, 2)$, $(2, 8)$, and $(-2, 20)$.

Critical Thinking

$\begin{bmatrix} 0 \\ 0 \end{bmatrix}, \begin{bmatrix} 1 \\ 0 \end{bmatrix}$

99. Find the form of all 2×1 matrices in reduced row-echelon form.

100. Find the form of all 2×2 matrices in reduced row-echelon form.

101. Suppose a matrix A is transformed to a matrix B by an elementary row operation. Is there an elementary row operation that transforms B to A? Explain.
Yes, the inverse of the operation used to transform A to B

102. Are the following statements true or false? Justify your answers. Suppose a matrix A is in reduced row-echelon form.

a. If we delete a row of A, then the remaining matrix is in reduced row-echelon form. True

b. Repeat part (a), replacing *row* with *column*. False

97. $y = -x^3 + 2x^2 + 3x + 1$ **98.** $y = -2x^3 + 3x^2 + 5x + 2$

100. $\begin{bmatrix} 0 & 0 \\ 0 & 0 \end{bmatrix}, \begin{bmatrix} 1 & k \\ 0 & 0 \end{bmatrix}, \begin{bmatrix} 0 & 1 \\ 0 & 0 \end{bmatrix}, \begin{bmatrix} 1 & 0 \\ 0 & 1 \end{bmatrix}$, where k is some constant

Matrix Algebra

Before Starting this Section, Review

1. Properties of real numbers (Appendix A)
2. Solving systems of equations (Section 7.1, page 481)

Objectives

1 Define equality of two matrices.

2 Define matrix addition and scalar multiplication.

3 Define matrix multiplication.

4 Apply matrix multiplication to computer graphics.

COMPUTER GRAPHICS

The following fields illustrate the widespread and growing applications of computer graphics:

1. The *automobile industry*. Both computer-aided design (CAD) and computer-aided manufacturing (CAM) have revolutionized the automotive industry. Many months before a new car is built, engineers design and construct a *mathematical car*—a wire-frame model that exists only in computer memory and on graphics display terminals. The mathematical model organizes and influences each step of the design and manufacture of the car.

2. The *entertainment industry*. Computer graphics are used by the entertainment industry with spectacular success. They were used in *Star Wars* and other movies for special effects, and they are used in the design and creation of arcade games.

3. *Military*. From the beginning of the widespread use of computers (in the 1960s), computer graphics have been used by the military to prepare for or defend against war. Modern weaponry depends heavily on computer-generated images.

Engineers use matrix algebra techniques on graphic images to change their orientation of scale, to zoom in on them, or to switch between two- and three-dimensional views. In Example 10 and in the exercises, we examine how basic matrix algebra can be used to manipulate graphical images. ■

---

1 Define equality of two matrices.

Equality of Matrices

In the previous section, a matrix was defined as a rectangular array of numbers written within brackets. In this section and the next two, you will learn about the operations with and uses of matrices. A matrix may be represented in any of the following three ways:

1. A matrix may be denoted by capital letters, such as A, B, C, and so on.
2. A matrix may be denoted by its representative (i, j)th entry: $[a_{ij}]$, $[b_{ij}]$, $[c_{ij}]$, and so on.

James Joseph Sylvester

(1814–1897)

In 1850, Sylvester coined the term **matrix** to denote a rectangular array of numbers. Sylvester, who was born into a Jewish family in London and studied for several years at Cambridge, was not permitted to take his degree there for religious reasons. Therefore, he received his degree from Trinity College, Dublin. In 1871, he accepted the chair of mathematics at the newly opened Johns Hopkins University in Baltimore. While at Hopkins, Sylvester founded the *American Journal of Mathematics* and helped develop a tradition of graduate education in mathematics in the United States.

3. A matrix may be written as a rectangular array of numbers:

$$A = [a_{ij}] = \begin{bmatrix} a_{11} & a_{12} & \cdots & a_{1n} \\ a_{21} & a_{22} & \cdots & a_{2n} \\ \vdots & \vdots & \cdots & \vdots \\ a_{m1} & a_{m2} & \cdots & a_{mn} \end{bmatrix}$$

We now examine some *algebraic* properties of matrices.

EQUALITY OF MATRICES

Two matrices $A = [a_{ij}]$ and $B = [b_{ij}]$ are said to be **equal,** written $A = B$, if

1. A and B have the same order $m \times n$ (that is, A and B have the same number m of rows and same number n of columns).

2. $a_{ij} = b_{ij}$ for all i and j; that is, each (i, j)th entry of A is equal to the corresponding (i, j)th entry of B.

EXAMPLE 1   **Determining Equality of Matrices**

Find x and y such that

$$\begin{bmatrix} x + y & 2 \\ 3 & x - y \end{bmatrix} = \begin{bmatrix} 4 & 2 \\ 3 & 1 \end{bmatrix}$$

SOLUTION

Because equal matrices must have identical entries in corresponding positions, we have the following system of equations:

$$\begin{cases} x + y = 4 & (1) \\ x - y = 1 & (2) \end{cases}$$ Equate entries in the $(1, 1)$ position.
    Equate entries in the $(2, 2)$ position.

Solve this system: Add the two equations to obtain $x = \dfrac{5}{2}$. Substitute $x = \dfrac{5}{2}$ in equation $x - y = 1$ and solve for y to obtain $y = \dfrac{3}{2}$. So, $x = \dfrac{5}{2}$ and $y = \dfrac{3}{2}$.   ■ ■ ■

Practice Problem 1   Find x and y such that

$$\begin{bmatrix} 1 & 2x - y \\ x + y & 5 \end{bmatrix} = \begin{bmatrix} 1 & 1 \\ 2 & 5 \end{bmatrix}$$ ■

2 Define matrix addition and scalar multiplication.

Matrix Addition and Scalar Multiplication

Matrix addition differs somewhat from real number addition. You can add any two real numbers, but you can add two matrices only if they have the same order.

MATRIX ADDITION

If $A = [a_{ij}]$ and $B = [b_{ij}]$ are two $m \times n$ matrices, their **sum** $A + B$ is the $m \times n$ matrix defined by

$$A + B = [a_{ij} + b_{ij}],$$

for all i and j.

TECHNOLOGY CONNECTION

To find the sum of two matrices with a graphing calculator, name the matrices and then use the regular addition key with the names of the matrices.

Consider the matrices A and B in Example 2 as follows:

```
[A]
    [[-2  3 ]
     [4   -1]
     [0   2 ]]
```

```
[B]
    [[6   4 ]
     [-7  9 ]
     [8   -5]]
```

The sum is

```
[A]+[B]
    [[4   7 ]
     [-3  8 ]
     [8   -3]]
```

The definition of matrix addition says that

(i) we can add two matrices of the same order simply by adding their corresponding entries, but

(ii) the sum of two matrices of different orders is not defined.

EXAMPLE 2   **Adding Matrices**

Let $A = \begin{bmatrix} -2 & 3 \\ 4 & -1 \\ 0 & 2 \end{bmatrix}$, $B = \begin{bmatrix} 6 & 4 \\ -7 & 9 \\ -8 & -5 \end{bmatrix}$, and $C = \begin{bmatrix} 1 & 2 \\ 3 & 4 \end{bmatrix}$. Find each sum if possible.

a. $A + B$ **b.** $A + C$

SOLUTION

a. Since A and B have the same order, $A + B$ is defined. We have

$$A + B = \begin{bmatrix} -2 & 3 \\ 4 & -1 \\ 0 & 2 \end{bmatrix} + \begin{bmatrix} 6 & 4 \\ -7 & 9 \\ 8 & -5 \end{bmatrix} = \begin{bmatrix} -2+6 & 3+4 \\ 4-7 & -1+9 \\ 0+8 & 2-5 \end{bmatrix} = \begin{bmatrix} 4 & 7 \\ -3 & 8 \\ 8 & -3 \end{bmatrix}$$

b. $A + C$ is not defined because A and C do not have the same order. ■ ■ ■

Practice Problem 2   Let $A = \begin{bmatrix} 2 & -1 & 4 \\ 5 & 0 & 9 \end{bmatrix}$ and $B = \begin{bmatrix} -8 & 2 & 9 \\ 7 & 3 & 6 \end{bmatrix}$. Find $A + B$. ■

You can also multiply any matrix A by any real number c. The real number c is called a **scalar,** and the multiplication process is called *scalar multiplication*.

SCALAR MULTIPLICATION

Let $A = [a_{ij}]$ be an $m \times n$ matrix and let c be a real number. Then the **scalar product** of A and c is denoted by cA and is defined by

$$cA = [ca_{ij}].$$

In other words, to multiply a matrix A by a real number c, you multiply each entry of A by c. When $c = -1$, we frequently write $-A$ instead of $(-1)A$. Further, the *difference* of two matrices, $A - B$, is defined to be the sum $A + (-1)B$.

MATRIX SUBTRACTION

If A and B are two $m \times n$ matrices, then their **difference** is defined by

$$A - B = A + (-1)B.$$

Subtraction $A - B$ is performed by subtracting the corresponding entries of B from those of A.

EXAMPLE 3   **Performing Scalar Multiplication and Matrix Subtraction**

Let $A = \begin{bmatrix} 1 & 2 & 0 \\ -1 & 3 & 1 \\ 2 & -1 & 4 \end{bmatrix}$ and $B = \begin{bmatrix} 2 & 1 & 3 \\ 1 & 0 & -2 \\ -3 & 4 & 5 \end{bmatrix}$. Find the following.

a. $3A$ **b.** $2B$ **c.** $3A - 2B$

TECHNOLOGY CONNECTION

To find the difference of two matrices with a graphing calculator, name the matrices and then use the regular subtraction key with the names of the matrices. Similarly, to multiply a matrix by a scalar, use the name of the matrix and the regular multiplication key.

Consider the matrices A and B in Example 3 as follows:

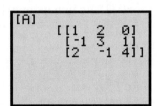

```
[A]
  [[1   2   0]
   [-1  3   1]
   [2  -1   4]]
```

```
[B]
  [[2   1   3]
   [1   0  -2]
   [-3  4   5]]
```

We find $3A - 2B$:

```
3*[A]-2*[B]
  [[-1   4   -6]
   [-5   9    7]
   [12 -11    2]]
■
```

SOLUTION

a. $3A = 3\begin{bmatrix} 1 & 2 & 0 \\ -1 & 3 & 1 \\ 2 & -1 & 4 \end{bmatrix} = \begin{bmatrix} 3(1) & 3(2) & 3(0) \\ 3(-1) & 3(3) & 3(1) \\ 3(2) & 3(-1) & 3(4) \end{bmatrix} = \begin{bmatrix} 3 & 6 & 0 \\ -3 & 9 & 3 \\ 6 & -3 & 12 \end{bmatrix}$

b. $2B = 2\begin{bmatrix} 2 & 1 & 3 \\ 1 & 0 & -2 \\ -3 & 4 & 5 \end{bmatrix} = \begin{bmatrix} 2(2) & 2(1) & 2(3) \\ 2(1) & 2(0) & 2(-2) \\ 2(-3) & 2(4) & 2(5) \end{bmatrix} = \begin{bmatrix} 4 & 2 & 6 \\ 2 & 0 & -4 \\ -6 & 8 & 10 \end{bmatrix}$

c. $3A - 2B = 3A + (-1)2B$

$= \begin{bmatrix} 3 & 6 & 0 \\ -3 & 9 & 3 \\ 6 & -3 & 12 \end{bmatrix} + \begin{bmatrix} -4 & -2 & -6 \\ -2 & 0 & 4 \\ 6 & -8 & -10 \end{bmatrix}$ Substitute from parts **a** and **b**, changing the sign of each entry in B.

$= \begin{bmatrix} 3 + (-4) & 6 + (-2) & 0 + (-6) \\ -3 + (-2) & 9 + 0 & 3 + 4 \\ 6 + 6 & -3 + (-8) & 12 + (-10) \end{bmatrix} = \begin{bmatrix} -1 & 4 & -6 \\ -5 & 9 & 7 \\ 12 & -11 & 2 \end{bmatrix}$ ■ ■ ■

Practice Problem 3 Let $A = \begin{bmatrix} 7 & -4 \\ 3 & 6 \\ 0 & -2 \end{bmatrix}$ and $B = \begin{bmatrix} 1 & -3 \\ 2 & 2 \\ 5 & 8 \end{bmatrix}$. Find $2A - 3B$. ■

An $m \times n$ matrix whose entries are all equal to 0 is called the $m \times n$ **zero matrix** and is denoted by $\mathbf{0}_{mn}$. When m and n are understood from the context, we simply write $\mathbf{0}$. The matrix $-A = (-1)A$ is called the **negative** of A. Many of the algebraic properties of matrix addition and scalar multiplication are similar to the familiar properties of real numbers.

MATRIX ADDITION AND SCALAR MULTIPLICATION PROPERTIES	
Let A, B, and C be $m \times n$ matrices and c and d be scalars.	
1. $A + B = B + A$	Commutative property of addition
2. $A + (B + C) = (A + B) + C$	Associative property of addition
3. $A + \mathbf{0} = \mathbf{0} + A = A$	Additive identity property
4. $A + (-A) = (-A) + A = \mathbf{0}$	Additive inverse property
5. $(cd)A = c(dA)$	Associative property of scalar multiplication
6. $1A = A$	Scalar identity property
7. $c(A + B) = cA + cB$	Distributive property
8. $(c + d)A = cA + dA$	Distributive property

In the next example, we use these properties to solve an equation involving matrices. Such an equation is called a **matrix equation**.

EXAMPLE 4   **Solving a Matrix Equation**

Solve the matrix equation $3A + 2X = 4B$ for X, where

$$A = \begin{bmatrix} 2 & 0 \\ 4 & 6 \end{bmatrix} \text{ and } B = \begin{bmatrix} 1 & 3 \\ 5 & 2 \end{bmatrix}$$

SOLUTION

$$3A + 2X = 4B \qquad \text{The given equation}$$

$$2X = 4B - 3A \qquad \text{Subtract the matrix } 3A \text{ from both sides.}$$

$$X = \frac{1}{2}(4B - 3A) \qquad \text{Multiply both sides by } \frac{1}{2}.$$

$$= \frac{1}{2}\left(4\begin{bmatrix} 1 & 3 \\ 5 & 2 \end{bmatrix} - 3\begin{bmatrix} 2 & 0 \\ 4 & 6 \end{bmatrix}\right) \qquad \text{Substitute for } A \text{ and } B.$$

$$= \frac{1}{2}\left(\begin{bmatrix} 4 & 12 \\ 20 & 8 \end{bmatrix} - \begin{bmatrix} 6 & 0 \\ 12 & 18 \end{bmatrix}\right) \qquad \text{Scalar multiplication}$$

$$= \frac{1}{2}\begin{bmatrix} -2 & 12 \\ 8 & -10 \end{bmatrix} \qquad \text{Subtract.}$$

$$= \begin{bmatrix} -1 & 6 \\ 4 & -5 \end{bmatrix} \qquad \text{Scalar multiplication}$$

You should check that the matrix $X = \begin{bmatrix} -1 & 6 \\ 4 & -5 \end{bmatrix}$ satisfies the given matrix equation. ■ ■ ■

Practice Problem 4 Solve the matrix equation $5A + 3X = 2B$ for X, where

$$A = \begin{bmatrix} 1 & -1 \\ 3 & 5 \end{bmatrix} \text{ and } B = \begin{bmatrix} 2 & 7 \\ -3 & -5 \end{bmatrix} \qquad ■$$

3 Define matrix multiplication.

Matrix Multiplication

You know that the sum of two matrices is defined only if both matrices are of the same order. In the case of multiplication, however, the general rule is that if A is an $m \times p$ matrix and B is a $p \times n$ matrix, then the *product AB* is defined; otherwise, it is not defined.

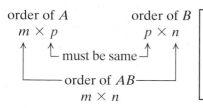

order of A order of B
$m \times p$ $p \times n$
└── must be same ──┘
└──── order of AB ────┘
$m \times n$

> **RULE FOR DEFINING THE PRODUCT AB**
>
> In order to define the product AB of two matrices A and B, the number of columns of A must be equal to the number of rows of B. If A is an $m \times p$ matrix and B is a $p \times n$ matrix, then the product AB is an $m \times n$ matrix.

EXAMPLE 5 **Determining the Order of the Product**

For each pair of matrices A and B, state whether the product matrix AB is defined. If it is, find the order of AB.

a. $A = \begin{bmatrix} 1 & 3 & 5 \end{bmatrix}, B = \begin{bmatrix} 2 \\ 3 \\ 7 \end{bmatrix}$ **b.** $A = \begin{bmatrix} 2 & 3 \\ -1 & 4 \\ 5 & 6 \end{bmatrix}, B = \begin{bmatrix} 2 & 1 & 3 & 5 \\ 4 & -1 & -2 & 6 \end{bmatrix}$

c. $A = \begin{bmatrix} 1 \\ -1 \\ 2 \end{bmatrix}, B = \begin{bmatrix} 2 & 1 & 4 \\ 0 & 2 & 2 \end{bmatrix}$

SOLUTION

a. Yes, the product AB is defined. A is a 1×3 matrix, and B is a 3×1 matrix; so the number of columns of A equals the number of rows of B. The product AB is a 1×1 matrix, which has a single entry.

b. Yes, AB is defined. A is a 3×2 matrix, and B is a 2×4 matrix; so the number of columns of A equals the number of rows of B. The product AB is a 3×4 matrix.

c. No, AB is not defined. A is a 3×1 matrix, and B is a 2×3 matrix; so the number of columns of A does not equal the number of rows of B. ■ ■ ■

Practice Problem 5 State whether the product matrix AB is defined. If AB is defined, find the order of the product matrix AB.

$$A = \begin{bmatrix} 1 & 4 & 7 \\ 0 & 2 & -2 \end{bmatrix}, \quad B = \begin{bmatrix} 4 \\ 3 \\ -1 \end{bmatrix}$$ ■

Before you learn to multiply matrices A and B, consider the next example.

EXAMPLE 6 Calculating the Total Revenue from Car Sales

A car dealership sold 10 sports cars (S), 30 compacts (C), and 45 mini-compacts (M). The average price per car of type S, C, and M was 42, 24, and 17 (in thousands of dollars), respectively. Find the total revenue received by the dealership from the sale of these cars.

SOLUTION
You can represent the number of cars of each type sold by a 1×3 matrix:

$$N = \begin{bmatrix} S & C & M \end{bmatrix} = \begin{bmatrix} 10 & 30 & 45 \end{bmatrix}$$

Using the price of each type of car, you can construct a 3×1 price matrix:

$$P = \begin{bmatrix} 42 \\ 24 \\ 17 \end{bmatrix}$$

Then the total revenue R received by the dealership from the sale of these cars last month is given by $R = 10(42) + 30(24) + 45(17) = 1905$ (or $1,905,000).

We define the right side of this equation to be the product NP of the two matrices N and P. In matrix notation,

$$NP = \begin{bmatrix} 10 & 30 & 45 \end{bmatrix} \begin{bmatrix} 42 \\ 24 \\ 17 \end{bmatrix} = 10(42) + 30(24) + 45(17) = 1905.$$ ■ ■ ■

Practice Problem 6 In Example 6, suppose the average price per car of types S, C, and M is 41, 26, and 19 (in thousands of dollars), respectively. Find the total revenue received by the dealership from the sale of these cars.   ■

Example 6 leads to the following definition.

PRODUCT OF $1 \times n$ AND $n \times 1$ MATRICES

Suppose A is a $1 \times n$ matrix and B is an $n \times 1$ matrix:

$$A = \begin{bmatrix} a_1 & a_2 & a_3 \cdots a_n \end{bmatrix} \text{ and } B = \begin{bmatrix} b_1 \\ b_2 \\ \vdots \\ b_n \end{bmatrix}$$

We define the product AB by

$$AB = a_1b_1 + a_2b_2 + \cdots + a_nb_n.$$

We make the following observations about this definition.

1. Since A is a $1 \times n$ matrix and B is an $n \times 1$ matrix, the product AB is defined and is a 1×1 matrix.

2. To find the product AB, multiply the leftmost entry of A and the top entry of B; then move to the right in A and down in B while multiplying corresponding (first, second, third, and so on) entries together; finally, add all of the resulting products.

EXAMPLE 7   Finding the Product AB

Find AB, where $A = [2 \quad 0 \quad 1 \quad -2]$ and $B = \begin{bmatrix} 3 \\ 1 \\ -1 \\ 4 \end{bmatrix}$.

SOLUTION

Since A has order 1×4 and B has order 4×1, AB is defined. The product rule gives

$$AB = 2(3) + 0(1) + 1(-1) + (-2)(4) = -3.$$

The product of the 1×4 matrix A and the 4×1 matrix B is therefore the 1×1 matrix $[-3]$, or the number -3. ■ ■ ■

Practice Problem 7   Find AB:

$$A = [3 \quad -1 \quad 2 \quad 7] \quad \text{and} \quad B = \begin{bmatrix} -2 \\ 0 \\ 1 \\ 5 \end{bmatrix}$$

■

In Example 7, you may think of the answer -3 as the 1×1 matrix $[-3]$ or simply as the number -3. We shall have occasion to make use of both points of view.

The rule for multiplying any two matrices can be reduced to the special case of multiplying a $1 \times p$ matrix by a $p \times 1$ matrix repeatedly.

MATRIX MULTIPLICATION

Let $A = [a_{ij}]$ be an $m \times p$ matrix and $B = [b_{ij}]$ be a $p \times n$ matrix. Then the **product** AB is the $m \times n$ matrix $C = [c_{ij}]$, where the entry c_{ij} of C is obtained by multiplying the ith row (matrix) of A by the jth column (matrix) of B.

The definition of the product AB says that

$$c_{ij} = a_{i1}b_{1j} + a_{i2}b_{2j} + \cdots + a_{ip}b_{pj}.$$

Written in full, the matrix product $AB = C$ in the definition is as follows:

$$\begin{bmatrix} a_{11} & a_{12} & \cdots & a_{1p} \\ a_{21} & a_{22} & \cdots & a_{2p} \\ \vdots & \vdots & \cdots & \vdots \\ a_{i1} & a_{i2} & \cdots & a_{ip} \\ \vdots & \vdots & \cdots & \vdots \\ a_{m1} & a_{m2} & \cdots & a_{mp} \end{bmatrix} \begin{bmatrix} b_{11} & b_{12} & \cdots & b_{1j} & \cdots & b_{1n} \\ b_{21} & b_{22} & \cdots & b_{2j} & \cdots & b_{2n} \\ \vdots & \vdots & \cdots & \vdots & \cdots & \vdots \\ b_{p1} & b_{p2} & \cdots & b_{pj} & \cdots & b_{pn} \end{bmatrix} = \begin{bmatrix} c_{11} & c_{12} & \cdots & c_{1j} & \cdots & c_{1n} \\ \vdots & \vdots & \cdots & \vdots & & \vdots \\ c_{i1} & c_{i2} & \cdots & c_{ij} & \cdots & c_{in} \\ \vdots & \vdots & \vdots & \vdots & & \vdots \\ c_{m1} & c_{m2} & \cdots & c_{mj} & \cdots & c_{mn} \end{bmatrix}$$

TECHNOLOGY CONNECTION

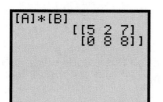 To find the product of two matrices using a graphing calculator, name the matrices and then use the regular multiplication key with the names of the matrices.

Consider the matrices A and B in Example 8.

We find the product AB.

```
[A]*[B]
        [[5 2 7]
         [0 8 8]]
```

EXAMPLE 8 **Finding the Product of Two Matrices**

Find the products AB and BA:

$$A = \begin{bmatrix} 1 & 2 \\ -1 & 3 \end{bmatrix} \quad \text{and} \quad B = \begin{bmatrix} 3 & -2 & 1 \\ 1 & 2 & 3 \end{bmatrix}$$

SOLUTION

Since A is of order 2×2 and the order of B is 2×3, the product AB is defined and has order 2×3.

Let $AB = C = [c_{ij}]$. By the definition of the product AB, each entry c_{ij} of C is found by multiplying the ith row of A by the jth column of B. For example c_{11} is found by multiplying the first row of A by the first column of B:

$$\begin{bmatrix} 1 & 2 \\ * & * \end{bmatrix} \begin{bmatrix} 3 & * & * \\ 1 & * & * \end{bmatrix} = \begin{bmatrix} 1(3) + 2(1) & * & * \\ * & * & * \end{bmatrix}$$

So

$$AB = \begin{bmatrix} 1 & 2 \\ -1 & 3 \end{bmatrix} \begin{bmatrix} 3 & -2 & 1 \\ 1 & 2 & 3 \end{bmatrix}$$

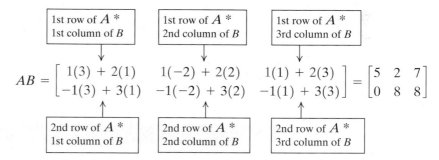

$$AB = \begin{bmatrix} 1(3) + 2(1) & 1(-2) + 2(2) & 1(1) + 2(3) \\ -1(3) + 3(1) & -1(-2) + 3(2) & -1(1) + 3(3) \end{bmatrix} = \begin{bmatrix} 5 & 2 & 7 \\ 0 & 8 & 8 \end{bmatrix}$$

The product BA is not defined because B is of order 2×3 and A is of order 2×2; that is, the number of columns of B is not the same as the number of rows of A. ■■■

Practice Problem 8 Find the products AB and BA:

$$A = \begin{bmatrix} 5 & 0 \\ 2 & -1 \end{bmatrix} \quad \text{and} \quad B = \begin{bmatrix} 8 & 1 \\ -2 & 6 \\ 0 & 4 \end{bmatrix}$$

■

In Example 8, it is obvious that $AB \neq BA$ because BA is undefined. Even if both AB and BA are defined, AB may not equal BA, as the next example illustrates.

EXAMPLE 9 **Finding the Product of Two Matrices**

Find the products AB and BA:

$$A = \begin{bmatrix} 2 & 3 \\ -1 & 0 \end{bmatrix} \quad \text{and} \quad B = \begin{bmatrix} 3 & 4 \\ 5 & 1 \end{bmatrix}$$

SOLUTION

$$AB = \begin{bmatrix} 2 & 3 \\ -1 & 0 \end{bmatrix} \begin{bmatrix} 3 & 4 \\ 5 & 1 \end{bmatrix} = \begin{bmatrix} 2(3) + 3(5) & 2(4) + 3(1) \\ -1(3) + 0(5) & -1(4) + 0(1) \end{bmatrix} = \begin{bmatrix} 21 & 11 \\ -3 & -4 \end{bmatrix}$$

$$BA = \begin{bmatrix} 3 & 4 \\ 5 & 1 \end{bmatrix} \begin{bmatrix} 2 & 3 \\ -1 & 0 \end{bmatrix} = \begin{bmatrix} 3(2) + 4(-1) & 3(3) + 4(0) \\ 5(2) + 1(-1) & 5(3) + 1(0) \end{bmatrix} = \begin{bmatrix} 2 & 9 \\ 9 & 15 \end{bmatrix}$$ ■■■

Practice Problem 9 Find the products AB and BA:

$$A = \begin{bmatrix} 7 & 1 \\ 0 & 3 \end{bmatrix} \quad \text{and} \quad B = \begin{bmatrix} 2 & -1 \\ 4 & 4 \end{bmatrix} \qquad ■$$

In Example 9, observe that $AB \neq BA$. Thus, in general, *matrix multiplication is not commutative*. Matrix multiplication, however, does share many of the properties of the multiplication of real numbers. Also, integral powers of square matrices are defined exactly as for real numbers. We write A^2 to mean AA, A^3 to mean AAA, and so on. We define A^0 by

$$A^0 = I_n = \begin{bmatrix} 1 & 0 & 0 & \cdots & 0 \\ 0 & 1 & 0 & \cdots & 0 \\ 0 & 0 & 1 & \cdots & 0 \\ \vdots & & & \vdots & \vdots \\ 0 & 0 & 0 & \cdots & 1 \end{bmatrix}.$$

The matrix I_n is called the **identity matrix**.

When the order of the matrix is clear from the context, we can drop the subscript n and write I for I_n. For any square matrix A, we have

$$\boxed{IA = A \text{ and } AI = A}$$

So, in matrix multiplication, the matrix I plays a role similar to that of the number 1 in the multiplication of real numbers.

PROPERTIES OF MATRIX MULTIPLICATION

Let A, B, and C be matrices and let c be a scalar. Assume that each product and sum is defined. Then

1. $(AB)C = A(BC)$ — Associative property of multiplication
2. (i) $A(B + C) = AB + AC$ — Distributive property
   (ii) $(A + B)C = AC + BC$ — Distributive property
3. $c(AB) = (cA)B = A(cB)$ — Associative property of scalar multiplication

---

In the exercises, you will be asked to verify these properties and compare them to similar properties of real numbers.

4 Apply matrix multiplication to computer graphics.

Computer Graphics

Letters used for labels on a computer screen are very simple two-dimensional graphics. The next example illustrates how a letter (or a figure) is transformed when the coordinates of the points on the figure are all multiplied by the same matrix.

FIGURE 8.2

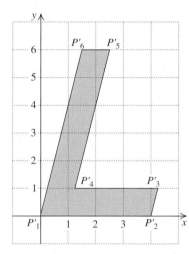

FIGURE 8.3

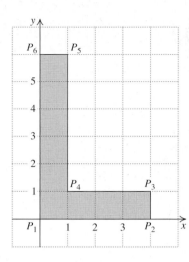

EXAMPLE 10 **Transforming a Letter**

The capital letter L in Figure 8.2 is determined by six points (or *vertices*) P_1 through P_6. The coordinates of the six points can be stored in a *data matrix D*. We draw line segments connecting these vertices in order.

$$
\begin{array}{c}
\qquad\quad P_1 \quad P_2 \quad P_3 \quad P_4 \quad P_5 \quad P_6 \\
\begin{array}{l} x\text{-coordinate} \\ y\text{-coordinate} \end{array}
\begin{bmatrix} 0 & 4 & 4 & 1 & 1 & 0 \\ 0 & 0 & 1 & 1 & 6 & 6 \end{bmatrix} = D
\end{array}
$$

If $A = \begin{bmatrix} 1 & 0.25 \\ 0 & 1 \end{bmatrix}$, compute AD and graph the figure it represents.

SOLUTION

The columns of the product matrix AD represent the transformed vertices of the letter L.

$$
AD = \begin{bmatrix} 1 & 0.25 \\ 0 & 1 \end{bmatrix}\begin{bmatrix} 0 & 4 & 4 & 1 & 1 & 0 \\ 0 & 0 & 1 & 1 & 6 & 6 \end{bmatrix}
$$

$$
= \begin{bmatrix} 1\cdot0 + 0.25\cdot0 & 1\cdot4 + 0.25\cdot0 & 1\cdot4 + 0.25\cdot1 & 1\cdot1 + 0.25\cdot1 & 1\cdot1 + 0.25\cdot6 & 1\cdot0 + 0.25\cdot6 \\ 0\cdot0 + 1\cdot0 & 0\cdot4 + 1\cdot0 & 0\cdot4 + 1\cdot1 & 0\cdot1 + 1\cdot1 & 0\cdot1 + 1\cdot6 & 0\cdot0 + 1\cdot6 \end{bmatrix}
$$

$$
\begin{array}{c}
P_1' \ \ P_2' \ \ P_3' \quad\ P_4' \quad\ P_5' \quad\ P_6' \\
= \begin{bmatrix} 0 & 4 & 4.25 & 1.25 & 2.5 & 1.5 \\ 0 & 0 & 1 & 1 & 6 & 6 \end{bmatrix}
\end{array}
$$

Figure 8.3 shows the transformed vertices and the transformed figure formed by these vertices. ▪ ▪ ▪

Practice Problem 10 In Example 10, use the matrix $A = \begin{bmatrix} 0 & 1 \\ 1 & 0.25 \end{bmatrix}$. Graph the figure represented by the matrix AD. ▪

SECTION 8.2 ▪ Exercises

A EXERCISES Basic Skills and Concepts

1. Two $m \times n$ matrices $A = [a_{ij}]$ and $B = [b_{ij}]$ are equal if __$a_{ij} = b_{ij}$__ for all i and j.

2. Let $A = [a_{ij}]$, $B = [b_{ij}]$, and $C = [c_{ij}]$. If $A + B = C$, then $c_{ij} =$ __$a_{ij} + b_{ij}$__ for all i and for all j.

3. The product of a $1 \times n$ matrix A and an $n \times 1$ matrix B is a __1×1__ matrix.

4. If A is an $m \times n$ matrix and B is an $n \times p$ matrix, then AB is defined and is a(n) __$m \times p$__ matrix.

5. *True or False* If AB and BA are defined, then $AB = BA$.
   False

6. *True or False* Any two square matrices can be multiplied.
   False

In Exercises 7–14, find the values of all variables.

7. $\begin{bmatrix} 2 \\ -3 \end{bmatrix} = \begin{bmatrix} x \\ u \end{bmatrix}$ $x = 2, u = -3$ 8. $\begin{bmatrix} -\dfrac{y}{2} \\ x \end{bmatrix} = \begin{bmatrix} 4 \\ 3 \end{bmatrix}$ $x = 3, y = -8$

9. $\begin{bmatrix} 2 & x \\ y & -3 \end{bmatrix} = \begin{bmatrix} 2 & -1 \\ 3 & -3 \end{bmatrix}$ $x = -1, y = 3$

10. $\begin{bmatrix} 2 - x & 1 \\ -2 & 3 + y \end{bmatrix} = \begin{bmatrix} 3 & 1 \\ -2 & -3 \end{bmatrix}$ $x = -1, y = -6$

11. $\begin{bmatrix} 2x - 3y & -4 \\ 5 & 3x + y \end{bmatrix} = \begin{bmatrix} 1 & -4 \\ 5 & 7 \end{bmatrix}$ $x = 2, y = 1$

12. $\begin{bmatrix} 3x + y & -17 \\ 19 & 2 \end{bmatrix} = \begin{bmatrix} -\dfrac{1}{2} & -17 \\ 19 & 2x + 3y \end{bmatrix}$ $x = -\dfrac{1}{2}, y = 1$

13. $\begin{bmatrix} x - y & 1 & 2 \\ 4 & 3x - 2y & 3 \\ 5 & 6 & 5x - 10y \end{bmatrix} = \begin{bmatrix} -1 & 1 & 2 \\ 4 & -1 & 3 \\ 5 & 6 & 6 \end{bmatrix}$ $\varnothing$

14. $\begin{bmatrix} x + y & 2 & 3 \\ 2 & x - y & 4 \\ 3 & 4 & 2x + 3y \end{bmatrix} = \begin{bmatrix} -1 & 2 & 3 \\ 2 & 5 & 4 \\ 3 & 4 & -5 \end{bmatrix}$ $x = 2, y = -3$

†Due to space constrictions, answers to these exercises may be found in the Answers beginning on page A–1 in the back of the book.

In Exercises 15–22, find each of the following if possible:

a. $A + B$ b. $A - B$
c. $-3A$ d. $3A - 2B$
e. $(A + B)^2$ f. $A^2 - B^2$

15. $A = \begin{bmatrix} 1 & 2 \\ 3 & 4 \end{bmatrix}, B = \begin{bmatrix} -1 & 0 \\ 2 & -3 \end{bmatrix}$ †

16. $A = \begin{bmatrix} \frac{1}{3} & 1 \\ -1 & 2 \end{bmatrix}, B = \begin{bmatrix} 1 & 0 \\ 2 & -\frac{1}{2} \end{bmatrix}$ †

17. $A = \begin{bmatrix} 2 & 3 \\ -4 & 5 \end{bmatrix}, B = \begin{bmatrix} 1 & 0 & 2 \\ 3 & 1 & 4 \end{bmatrix}$ †

18. $A = \begin{bmatrix} 1 & 2 & 3 \\ -1 & -3 & 4 \end{bmatrix}, B = \begin{bmatrix} 1 & 0 \\ 2 & 3 \\ 1 & 4 \end{bmatrix}$ †

19. $A = \begin{bmatrix} 4 & 0 & -1 \\ -2 & 5 & 2 \\ 0 & 0 & 1 \end{bmatrix}, B = \begin{bmatrix} 3 & 1 & 0 \\ 1 & -4 & 2 \\ 2 & 1 & 3 \end{bmatrix}$ †

20. $A = \begin{bmatrix} 1 & 0 & 2 \\ 2 & 1 & 0 \\ 0 & 1 & 3 \end{bmatrix}, B = \begin{bmatrix} 3 & -1 & 2 \\ 0 & 2 & 1 \\ 1 & 4 & -1 \end{bmatrix}$ †

21. $A = \begin{bmatrix} 1 & 2 & -3 \\ 3 & 4 & 5 \\ 2 & -1 & 0 \end{bmatrix}, B = \begin{bmatrix} 3 & 1 & 0 \\ 1 & -4 & 2 \\ 2 & 1 & 3 \end{bmatrix}$ †

22. $A = \begin{bmatrix} 1 & 0 & 2 \\ 2 & 1 & 0 \\ 0 & 1 & 3 \end{bmatrix}, B = \begin{bmatrix} 3 & -1 & 2 \\ 0 & 2 & 1 \\ 1 & 4 & -1 \end{bmatrix}$ †

In Exercises 23–30, solve each matrix equation for X, where

$$A = \begin{bmatrix} 2 & 3 & -1 \\ 1 & -2 & 4 \end{bmatrix} \text{ and } B = \begin{bmatrix} -2 & 1 & 0 \\ 2 & 3 & 4 \end{bmatrix}.$$

23. $A + X = B$ †   24. $B + X = A$ †
25. $2X - A = B$ †   26. $2X - B = A$ †
27. $2X + 3A = B$ †   28. $3X + 2A = B$ †
29. $2A + 3B + 4X = 0$ †   30. $3X - 2A + 5B = 0$ †

In Exercises 31–40, find each product if possible.

a. AB b. BA

31. $A = \begin{bmatrix} 1 & 2 \\ 3 & 4 \end{bmatrix}, B = \begin{bmatrix} -2 & 1 \\ 3 & 5 \end{bmatrix}$ †

32. $A = \begin{bmatrix} 3 & 2 \\ 1 & 5 \\ 0 & 1 \end{bmatrix}, B = \begin{bmatrix} 1 & 3 & -2 \\ 2 & 5 & 0 \end{bmatrix}$ †

33. $A = \begin{bmatrix} 2 & -1 & 0 \\ -3 & 1 & 2 \end{bmatrix}, B = \begin{bmatrix} 1 & 5 \\ -2 & 3 \\ 4 & 0 \end{bmatrix}$ †

34. $A = \begin{bmatrix} 1 & 2 & 3 \\ 4 & 5 & 6 \end{bmatrix}, B = \begin{bmatrix} -1 & -3 & -6 \\ 2 & 5 & 7 \end{bmatrix}$ The products AB and BA are not defined.

35. $A = \begin{bmatrix} 2 & 3 & 5 \end{bmatrix}, B = \begin{bmatrix} 1 \\ -2 \\ 4 \end{bmatrix}$ †

36. $A = \begin{bmatrix} 1 \\ 2 \\ -1 \end{bmatrix}, B = \begin{bmatrix} -3 & 0 & 2 \end{bmatrix}$ †

37. $A = \begin{bmatrix} 1 & 2 & 3 \end{bmatrix}, B = \begin{bmatrix} 1 & 2 & -1 \\ 0 & 3 & 1 \\ 2 & 0 & -3 \end{bmatrix}$ †

38. $A = \begin{bmatrix} 4 & -6 & 2 \\ 2 & 3 & 0 \\ 1 & 2 & -3 \end{bmatrix}, B = \begin{bmatrix} -2 & 0 & 1 \end{bmatrix}$ †

39. $A = \begin{bmatrix} 2 & 0 & 1 \\ 1 & 4 & 2 \\ 3 & -1 & 0 \end{bmatrix}, B = \begin{bmatrix} 3 & 1 & 0 \\ -1 & 2 & 0 \\ 4 & 5 & 2 \end{bmatrix}$ †

40. $A = \begin{bmatrix} 3 & -1 & 2 \\ 0 & 4 & -3 \\ 1 & -2 & 2 \end{bmatrix}, B = \begin{bmatrix} 2 & -5 & 0 \\ -1 & 2 & -1 \\ 3 & 0 & 2 \end{bmatrix}$ †

41. Let $A = \begin{bmatrix} 3 & 1 & 2 \\ 0 & 4 & 3 \\ 1 & -2 & 2 \end{bmatrix}$ and $B = \begin{bmatrix} 2 & 5 & 0 \\ 1 & 2 & -1 \\ 3 & 0 & 2 \end{bmatrix}$.

     Verify that $AB \neq BA$. †

42. Let $A = \begin{bmatrix} 5 & 2 & 4 \\ -6 & -3 & 10 \\ -2 & 6 & 5 \end{bmatrix}$ and $B = \begin{bmatrix} 4 & 1 & 2 \\ -3 & 0 & 5 \\ -1 & 3 & 4 \end{bmatrix}$.

     Verify that $AB = BA$. †

In Exercises 43–46, let

$$A = \begin{bmatrix} 1 & 2 \\ 3 & 4 \end{bmatrix}, B = \begin{bmatrix} 2 & -3 \\ 3 & 5 \end{bmatrix}, \text{ and } C = \begin{bmatrix} 0 & 1 \\ 2 & 4 \end{bmatrix}.$$

43. Verify the associative property for multiplication: $(AB)C = A(BC)$.

44. Verify the distributive property: $A(B + C) = AB + AC$.

45. Verify the distributive property: $(A + B)C = AC + BC$.

46. For any real number c, verify that $c(AB) = (cA)(B) = A(cB)$.

B EXERCISES Applying the Concepts

47. **Cost matrix.** The Build-Rite Co. is building an apartment complex. The cost of purchasing and transporting specific amounts of steel, glass, and wood (in appropriate units) from two different locations is given by the following matrices:

        Steel   Glass   Wood

$$A = \begin{bmatrix} 7 & 3 & 18 \\ 4 & 1 & 3 \end{bmatrix} \begin{matrix} \text{Cost of Material} \\ \text{Transportation Cost} \end{matrix}$$

$$B = \begin{bmatrix} 6 & 2 & 20 \\ 3 & 1 & 4 \end{bmatrix} \begin{matrix} \text{Cost of Material} \\ \text{Transportation Cost} \end{matrix}$$

Find the matrix representing the total cost of material and transportation for steel, glass, and wood from both locations. †

48. Repeat Exercise 47 if the order from location A is tripled and the order from location B is doubled. †

49. Stock purchase. Ms. Goodbyer plans to buy 100 shares of computer stock, 300 shares of oil stock, and 400 shares of automobile stock. The computer stock is selling for $60 a share, oil stock is selling for $38 a share, and automobile stock is selling for $17 a share. Use matrix multiplication to calculate the total cost of Ms. Goodbyer's purchases. $24,200

50. Product export. The International Export Corporation received an export order for three of its products, say, A, B, and C. The export order is (in thousands) for 50 of product A, 75 of product B, and 150 of product C. The material cost, labor, and profit (each in appropriate units) required to produce each unit of the product are given in the following table:

$$\begin{matrix} & \text{Materials} & \text{Labor} & \text{Profit} \\ & [5875 & 475 & 5125] \end{matrix}$$

$$\begin{matrix} & \text{Material} & \text{Labor} & \text{Profit} \\ \text{Product A} & \begin{bmatrix} 20 & 2 & 20 \\ \text{Product B} & 15 & 3 & 25 \\ \text{Product C} & 25 & 1 & 15 \end{bmatrix} \end{matrix}$$

Use matrix multiplication to compute the total material cost, total labor, and total profit if the entire export order is filled.

51. Executive compensation. The Multinational Oil Corporation pays its top executives a salary, a cash bonus, and shares of its stock annually. In 2005, the chairman of the board received $2.5 million in salary, a $1.5 million bonus, and 50,000 shares of stock; the president of the company received one-half the compensation of the chairman; and each of the four vice presidents was paid $100,000 in salary, a $150,000 bonus, and 5000 shares of stock.

a. Express payments to these executives in salary, bonus, and stock as a 3×3 matrix. †

b. Express the number of executives of each rank as a column matrix. †

c. Use matrix multiplication to compute the total amount the company paid to each executive in each category in 2005. †

52. Calories and protein. The Browns (B) and Newgards (N) are neighboring families. The Brown family has two men, three women, and one child, while the Newgard family has one man, one woman, and two children. Both families are weight watchers. They have the same dietician, who recommends the following daily allowance of calories and proteins:

Calories
male adults: 2400 calories
female adults: 1900 calories
children: 1800 calories
Protein (in grams)
male adults: 55
female adults: 45
children: 33

Represent the preceding information in matrix form. Use matrix multiplication to compute the total requirements of calories and proteins for each of the two families. †

In Exercise 53–56, a matrix A is given. Graph the figure represented by the matrix AD, where D is the matrix representation of the letter L of Example 10.

53. $A = \begin{bmatrix} 1 & 0 \\ 0 & -1 \end{bmatrix}$ †

54. $A = \begin{bmatrix} -1 & 0 \\ 0 & 1 \end{bmatrix}$ †

55. $A = \begin{bmatrix} 1 & 0 \\ 0.25 & 1 \end{bmatrix}$ †

56. $A = \begin{bmatrix} 1 & -0.25 \\ 0 & 1 \end{bmatrix}$ †

C EXERCISES Beyond the Basics

57. Let $A = \begin{bmatrix} 0 & 3 \\ 0 & 0 \end{bmatrix}$, $B = \begin{bmatrix} 2 & 1 \\ 3 & 0 \end{bmatrix}$, and $C = \begin{bmatrix} 5 & 4 \\ 3 & 0 \end{bmatrix}$.
Verify that $AB = AC$. This example shows that $AB = AC$ does not, in general, imply that $B = C$.

58. Let $A = \begin{bmatrix} 2 & -3 & -5 \\ -1 & 4 & 5 \\ 1 & -3 & -4 \end{bmatrix}$ and $B = \begin{bmatrix} -1 & 3 & 5 \\ 1 & -3 & -5 \\ -1 & 3 & 5 \end{bmatrix}$.
Verify that $AB = 0$. This example shows that $AB = 0$ does not imply that $A = 0$ or $B = 0$.

59. Select appropriate 2×2 matrices A and B to show that, in general, $(A + B)^2 \neq A^2 + 2AB + B^2$. †

60. Repeat Exercise 59 to show that, in general, $A^2 - B^2 \neq (A - B)(A + B)$. †

61. Let $A = \begin{bmatrix} 1 & -2 & 3 \\ 2 & -4 & 1 \\ 3 & -5 & 2 \end{bmatrix}$. Show that

$$3A^2 - 2A + I = \begin{bmatrix} 17 & -23 & 15 \\ -13 & 30 & 10 \\ -9 & 22 & 21 \end{bmatrix},$$

where $I = \begin{bmatrix} 1 & 0 & 0 \\ 0 & 1 & 0 \\ 0 & 0 & 1 \end{bmatrix}$.

62. Let $A = \begin{bmatrix} 1 & 3 & 2 \\ 2 & 0 & 3 \\ 1 & -1 & 1 \end{bmatrix}$. Show that

$$A^3 - 3A^2 + A - I = \begin{bmatrix} -3 & 14 & -3 \\ 5 & -2 & 8 \\ 5 & -7 & 6 \end{bmatrix},$$

where I is given in Exercise 61.

63. If possible, find a matrix B such that $\begin{bmatrix} 2 & 3 \\ 1 & 2 \end{bmatrix} B = \begin{bmatrix} 1 & 0 \\ 0 & 1 \end{bmatrix}$.

64. If possible find a matrix B such that $B = \begin{bmatrix} 2 & -3 \\ -1 & 2 \end{bmatrix}$
$\begin{bmatrix} 1 & -2 \\ -2 & 4 \end{bmatrix} B = \begin{bmatrix} 1 & 0 \\ 0 & 1 \end{bmatrix}$. Not possible

65. Find x and y if

$$\left(4\begin{bmatrix} 2 & -1 & 3 \\ 1 & 0 & 2 \end{bmatrix} - \begin{bmatrix} 1 & 2 & -1 \\ 2 & -3 & 4 \end{bmatrix} \right) \begin{bmatrix} 2 \\ -1 \\ 1 \end{bmatrix} = \begin{bmatrix} x \\ y \end{bmatrix}.$$

$x = 33, y = 5$

Answers:

57. $AB = \begin{bmatrix} 9 & 0 \\ 0 & 0 \end{bmatrix} = AC$

66. Find x, y, and z if

$$\begin{bmatrix} 3 & 2 & -1 \\ 4 & 9 & 2 \\ 5 & 0 & -2 \end{bmatrix} \begin{bmatrix} x \\ y \\ z \end{bmatrix} = \begin{bmatrix} 0 \\ 7 \\ 2 \end{bmatrix}. \quad x = 2, y = -1, z = 4$$

Critical Thinking

67. Let A be a $3 \times n$ matrix and B be a $5 \times m$ matrix.
  a. Under what conditions is AB defined? What is the order of AB when this product is defined?
  b. Repeat part (a) for BA.

68. Let $A = \begin{bmatrix} 3 & -1 & 0 \\ -2 & 3 & 1 \end{bmatrix}$, $B = \begin{bmatrix} 2 \\ 3 \\ 4 \end{bmatrix}$, and $C = \begin{bmatrix} -1 & 2 \end{bmatrix}$. In

which order should all three matrices be multiplied to produce a number? $(CA)B$

Answers:

67. a. AB is defined when $n = 5$, and the order of AB when this product is defined is $3 \times m$. **b.** BA is defined when $m = 3$, and the order of BA when this product is defined is $5 \times n$.

GROUP PROJECT

Market share. The Asma Corporation (A), Bronkial Brothers (B), and Coufmore Company (C) simultaneously decide to introduce a new cigarette at a time when each company has one-third of the market. During the year, the following occurs:
  (i) A retains 40% of its customers and loses 30% to B and 30% to C.
  (ii) B retains 30% of its customers and loses 60% to A and 10% to C.
  (iii) C retains 30% of its customers and loses 60% to A and 10% to B.

The foregoing information can be represented by the *transition matrix*

$$P = \begin{array}{c} A \\ B \\ C \end{array} \begin{bmatrix} 0.4 & 0.3 & 0.3 \\ 0.6 & 0.3 & 0.1 \\ 0.6 & 0.1 & 0.3 \end{bmatrix}$$

Write the initial market share as $X = \begin{bmatrix} \dfrac{1}{3} & \dfrac{1}{3} & \dfrac{1}{3} \end{bmatrix}$.

  a. Compute XP^2 and interpret your result.
  b. Compute XP^3, XP^4, and XP^5; observe the pattern for the long run. Describe the long-term market share for each company.

a. $XP^2 = \begin{bmatrix} \dfrac{37}{75} & \dfrac{19}{75} & \dfrac{19}{75} \end{bmatrix}$

b. $XP^3 = \begin{bmatrix} \dfrac{188}{375} & \dfrac{187}{750} & \dfrac{187}{750} \end{bmatrix}$

$XP^4 = \begin{bmatrix} \dfrac{937}{1875} & \dfrac{469}{1875} & \dfrac{469}{1875} \end{bmatrix}$

$XP^5 = \begin{bmatrix} \dfrac{4688}{9375} & \dfrac{4687}{18,750} & \dfrac{4687}{18,750} \end{bmatrix}$

$XP^n = \begin{bmatrix} 0.5 & 0.25 & 0.25 \end{bmatrix}$

The Matrix Inverse

Before Starting this Section, Review

1. Matrix multiplication (Section 8.2, page 566)

2. Equality of matrices (Section 8.2, page 561)

3. Gauss–Jordan elimination procedure (Section 8.1, page 553)

4. Leontieff input–output model (Section 8.1, page 554)

Objectives

1. Verify the multiplicative inverse of a matrix.

2. Find the inverse of a matrix.

3. Find the inverse of a 2×2 matrix.

4. Use matrix inverses to solve systems of linear equations.

5. Use matrix inverses in applied problems.

CODING AND DECODING MESSAGES

Imagine you are a secret agent who is looking for spies in different parts of the world. Periodically, you communicate with your headquarters via your laptop or by satellite. You expect that your communications will be intercepted. So before you broadcast any message, you must write it using a secret code. A message written with a secret code is called a **cryptogram**.

First, you might replace each letter in the message with a number and send the message as a sequence of numbers. The problem with this approach is that it is rather easy to crack such a code. For example, if letters such as *e* are always represented by the same symbol, then a code breaker (hacker) can guess which symbol represents each letter and eventually translate your message. In Example 8, we show you a better way to encode and decode a message using matrix multiplication. ∎

---

1 Verify the multiplicative inverse of a matrix.

The Multiplicative Inverse of a Matrix

In arithmetic, you know that if a is a nonzero real number, then a has a unique reciprocal, denoted by $\dfrac{1}{a}$, or a^{-1}. The number a^{-1} is called the **multiplicative inverse** of a, and

$$aa^{-1} = a^{-1}a = 1.$$

Similarly, certain square matrices have *multiplicative inverses*, with an identity matrix acting as "1."

THE INVERSE OF A MATRIX

Let A be an $n \times n$ matrix and let I be the $n \times n$ identity matrix that has 1s on the main diagonal and 0s elsewhere. If there is an $n \times n$ matrix B such that

$$AB = I \quad \text{and} \quad BA = I,$$

then B is called the **inverse** of A and we write $B = A^{-1}$ (read "A inverse").

Notice that if B is the inverse of A, then A is the inverse of B.

Arthur Cayley

(1821–1895)

English mathematicians of the first third of the nineteenth century attempted to extend the basic ideas of algebra and to determine exactly how much one can generalize the properties of integers to other types of quantities. Although Sylvester (see page 561) coined the term *matrix*, it was Arthur Cayley who developed the algebra of matrices. Cayley studied mathematics at Trinity College, Cambridge, but because there was no suitable teaching job available, he decided to become a lawyer. Although he became skilled in legal work, he regarded the law as just a way to support himself. During his legal career, he was able to write more than 300 mathematical papers. In 1863, Cambridge University established a new post in mathematics and offered it to Cayley. He accepted the job eagerly even though it paid less money than he made as a lawyer.

EXAMPLE 1   Verifying the Inverse of a Matrix

Show that B is the inverse of A:

$$A = \begin{bmatrix} 1 & 3 \\ 2 & 7 \end{bmatrix} \text{ and } B = \begin{bmatrix} 7 & -3 \\ -2 & 1 \end{bmatrix}$$

SOLUTION

You need to verify that $AB = I$ and $BA = I$.

$$AB = \begin{bmatrix} 1 & 3 \\ 2 & 7 \end{bmatrix}\begin{bmatrix} 7 & -3 \\ -2 & 1 \end{bmatrix} = \begin{bmatrix} 1(7) + 3(-2) & 1(-3) + 3(1) \\ 2(7) + 7(-2) & 2(-3) + 7(1) \end{bmatrix} = \begin{bmatrix} 1 & 0 \\ 0 & 1 \end{bmatrix} = I$$

$$BA = \begin{bmatrix} 7 & -3 \\ -2 & 1 \end{bmatrix}\begin{bmatrix} 1 & 3 \\ 2 & 7 \end{bmatrix} = \begin{bmatrix} 7(1) + (-3)(2) & 7(3) + (-3)(7) \\ (-2)(1) + 1(2) & (-2)(3) + 1(7) \end{bmatrix} = \begin{bmatrix} 1 & 0 \\ 0 & 1 \end{bmatrix} = I$$

Since $AB = I$ and $BA = I$, it follows that $B = A^{-1}$.   ■ ■ ■

Practice Problem 1   Show that B is the inverse of A:

$$A = \begin{bmatrix} 3 & 2 \\ 2 & 1 \end{bmatrix} \quad \text{and} \quad B = \begin{bmatrix} -1 & 2 \\ 2 & -3 \end{bmatrix}$$

■

Recall that, in general (even for square matrices), $AB \neq BA$. However, if A and B are square matrices with $AB = I$, then it can be shown that $BA = I$. Therefore, to determine whether A and B are inverses, you need only check that either $AB = I$ or $BA = I$.

In working with real numbers, you know that if $a = 0$, then a^{-1} does not exist. So it should not come as a surprise that if A is a square *zero* matrix, then A^{-1} does not exist. Surprisingly, there are also nonzero matrices that do not have inverses.

EXAMPLE 2   Proving That a Particular Nonzero Matrix Has No Inverse

Show that the matrix A does not have an inverse.

$$A = \begin{bmatrix} 2 & 0 \\ 0 & 0 \end{bmatrix}$$

SOLUTION

Suppose A has an inverse B, where

$$B = \begin{bmatrix} x & y \\ z & w \end{bmatrix}$$

Then you have

$$\begin{bmatrix} 2 & 0 \\ 0 & 0 \end{bmatrix}\begin{bmatrix} x & y \\ z & w \end{bmatrix} = \begin{bmatrix} 1 & 0 \\ 0 & 1 \end{bmatrix} \qquad AB = I$$

$$\begin{bmatrix} 2x & 2y \\ 0 & 0 \end{bmatrix} = \begin{bmatrix} 1 & 0 \\ 0 & 1 \end{bmatrix} \qquad \text{Multiply the two matrices on the left side.}$$

Since these two matrices are equal, you must have

$$0 = 1 \qquad \text{Equate the entries in the } (2, 2) \text{ position.}$$

Because $0 = 1$ is a false statement, the matrix A does not have an inverse.   ■ ■ ■

Practice Problem 2   Show that the matrix $A = \begin{bmatrix} 3 & 1 \\ 3 & 1 \end{bmatrix}$ does not have an inverse.

■

2 Find the inverse of a matrix.

Finding the Inverse of a Matrix

It can be shown that if a square matrix A has an inverse, then the inverse is unique. This means that a square matrix cannot have more than one inverse. If a matrix A has an inverse, then A is called **invertible** or **nonsingular**. A square matrix that does not have an inverse is called a **singular** matrix. The following discussion derives a technique for finding the inverse of an invertible matrix.

Suppose we want to find the inverse of the matrix $A = \begin{bmatrix} 1 & 2 \\ 3 & 5 \end{bmatrix}$. We first let $B = \begin{bmatrix} x & y \\ z & w \end{bmatrix}$ be the inverse of the matrix A. Then

$$AB = I \qquad \text{Definition of inverse}$$

$$\begin{bmatrix} 1 & 2 \\ 3 & 5 \end{bmatrix}\begin{bmatrix} x & y \\ z & w \end{bmatrix} = \begin{bmatrix} 1 & 0 \\ 0 & 1 \end{bmatrix} \qquad \text{Substitute for } A, B, \text{ and } I.$$

$$\begin{bmatrix} x + 2z & y + 2w \\ 3x + 5z & 3y + 5w \end{bmatrix} = \begin{bmatrix} 1 & 0 \\ 0 & 1 \end{bmatrix} \qquad \text{Multiply matrices } A \text{ and } B.$$

Equating entries in the last matrix yields two systems of linear equations.

$$\begin{cases} x + 2z = 1 \\ 3x + 5z = 0 \end{cases} \qquad \begin{matrix} \text{Equate the entries in the } (1, 1) \\ \text{and } (2, 1) \text{ positions.} \end{matrix}$$

$$\begin{cases} y + 2w = 0 \\ 3y + 5w = 1 \end{cases} \qquad \begin{matrix} \text{Equate the entries in the } (1, 2) \\ \text{and } (2, 2) \text{ positions.} \end{matrix}$$

Let's solve the two systems by Gauss–Jordan elimination.

System of equations	Matrix form
$\begin{cases} x + 2z = 1 \\ 3x + 5z = 0 \end{cases}$	$\begin{bmatrix} 1 & 2 & \vert & 1 \\ 3 & 5 & \vert & 0 \end{bmatrix}$
$\begin{cases} y + 2w = 0 \\ 3y + 5w = 1 \end{cases}$	$\begin{bmatrix} 1 & 2 & \vert & 0 \\ 3 & 5 & \vert & 1 \end{bmatrix}$

Because the coefficient matrix for both systems of equations is the original matrix

$$A = \begin{bmatrix} 1 & 2 \\ 3 & 5 \end{bmatrix},$$

the calculations involved in the Gauss–Jordan elimination on the coefficient matrix A will be the same for both systems. Therefore, you can solve both systems at once by considering the "doubly augmented" matrix

$$C = \begin{bmatrix} 1 & 2 & \vert & 1 & 0 \\ 3 & 5 & \vert & 0 & 1 \end{bmatrix}$$

Note that the matrix C is formed by adjoining the identity matrix I to matrix A:

$$C = [A \,\vert\, I]$$

Now apply Gauss–Jordan elimination to the matrix C.

$$C = \begin{bmatrix} 1 & 2 & \vert & 1 & 0 \\ 3 & 5 & \vert & 0 & 1 \end{bmatrix} \xrightarrow{-3R_1 + R_2 \to R_2} \begin{bmatrix} 1 & 2 & \vert & 1 & 0 \\ 0 & -1 & \vert & -3 & 1 \end{bmatrix} \xrightarrow{(-1)R_2} \begin{bmatrix} 1 & 2 & \vert & 1 & 0 \\ 0 & 1 & \vert & 3 & -1 \end{bmatrix}$$

$$\xrightarrow{-2R_2 + R_1 \to R_1} \begin{bmatrix} 1 & 0 & \vert & -5 & 2 \\ 0 & 1 & \vert & 3 & -1 \end{bmatrix}$$

TECHNOLOGY CONNECTION

To use a graphing calculator to find the inverse of an invertible matrix, name the matrix and then use the regular reciprocal key with the name of the matrix.

```
[A]
        [[1  2]
         [3  5]]
■
```

```
[A]⁻¹
        [[-5  2 ]
         [3  -1]]
■
```

Converting the last matrix back into two systems of equations, we have

$$\begin{bmatrix} 1 & 0 & | & -5 \\ 0 & 1 & | & 3 \end{bmatrix} \text{ represents } \begin{cases} 1 \cdot x + 0 \cdot z = -5 \\ 0 \cdot x + 1 \cdot z = 3 \end{cases} \text{ or } \begin{cases} x = -5 \\ z = 3 \end{cases}$$

Similarly,

$$\begin{bmatrix} 1 & 0 & | & 2 \\ 0 & 1 & | & -1 \end{bmatrix} \text{ represents } \begin{cases} 1 \cdot y + 0 \cdot w = 2 \\ 0 \cdot y + 1 \cdot w = -1 \end{cases} \text{ or } \begin{cases} y = 2 \\ w = -1 \end{cases}$$

We conclude that $x = -5$, $y = 2$, $z = 3$, and $w = -1$.

Since $B = \begin{bmatrix} x & y \\ z & w \end{bmatrix}$, it follows that

$$B = \begin{bmatrix} -5 & 2 \\ 3 & -1 \end{bmatrix} = A^{-1}$$

We have shown that to find the inverse of an invertible matrix A, we transform the matrix $[A \mid I]$ by a sequence of row operations into $[I \mid B]$, where $B = A^{-1}$.

PROCEDURE FOR FINDING THE INVERSE OF A MATRIX

Let A be an $n \times n$ matrix.

1. Form the $n \times 2n$ augmented matrix $[A \mid I]$, where I is the $n \times n$ identity matrix.

2. If there is a sequence of row operations that transforms A into I, then this same sequence of row operations will transform $[A \mid I]$ into $[I \mid B]$, where $B = A^{-1}$.

3. Check your work by showing that $AA^{-1} = I$.

If it is not possible to transform A into I by row operations, then A does not have an inverse. (This occurs if, at any step in the process, you obtain a matrix $[C \mid D]$ in which C has a row of zeros.)

EXAMPLE 3 **Finding the Inverse of a Matrix**

Find the inverse (if it exists) of the matrix $A = \begin{bmatrix} 1 & 2 & 3 \\ 2 & 5 & 7 \\ 2 & 4 & 6 \end{bmatrix}$.

SOLUTION

Step 1 Start by adjoining the identity matrix $I = \begin{bmatrix} 1 & 0 & 0 \\ 0 & 1 & 0 \\ 0 & 0 & 1 \end{bmatrix}$ to the matrix A to form the augmented matrix $[A \mid I]$.

$$[A \mid I] = \begin{bmatrix} 1 & 2 & 3 & | & 1 & 0 & 0 \\ 2 & 5 & 7 & | & 0 & 1 & 0 \\ 2 & 4 & 6 & | & 0 & 0 & 1 \end{bmatrix}$$

Step 2 Perform row operations on $[A \mid I]$ to transform it to a matrix of the form $[I \mid B]$.

$$\begin{bmatrix} 1 & 2 & 3 & | & 1 & 0 & 0 \\ 2 & 5 & 7 & | & 0 & 1 & 0 \\ 2 & 4 & 6 & | & 0 & 0 & 1 \end{bmatrix} \xrightarrow[\substack{-2R_1 + R_3 \to R_3}]{-2R_1 + R_2 \to R_2} \begin{bmatrix} 1 & 2 & 3 & | & 1 & 0 & 0 \\ 0 & 1 & 1 & | & -2 & 1 & 0 \\ 0 & 0 & 0 & | & -2 & 0 & 1 \end{bmatrix} = [C \mid D]$$

Because the third row of matrix C consists of zeros, it is impossible to transform A into I by row operations. Hence, A is *not* invertible. ■ ■ ■

Practice Problem 3   Find the inverse (if it exists) of the matrix $A = \begin{bmatrix} 1 & 4 & -2 \\ -1 & 1 & 2 \\ 3 & 7 & -6 \end{bmatrix}$.

■

EXAMPLE 4   **Finding the Inverse of a Matrix**

Find the inverse (if it exists) of the matrix $A = \begin{bmatrix} 1 & 1 & 0 \\ 0 & 3 & 1 \\ 2 & 3 & 3 \end{bmatrix}$.

SOLUTION

Step 1   Start with the matrix $[A \mid I]$:

$$[A \mid I] = \begin{bmatrix} 1 & 1 & 0 & | & 1 & 0 & 0 \\ 0 & 3 & 1 & | & 0 & 1 & 0 \\ 2 & 3 & 3 & | & 0 & 0 & 1 \end{bmatrix} \qquad I = I_3$$

Step 2   Use row operations on $[A \mid I]$. Then $[A \mid I] \rightarrow$

$$\xrightarrow[\substack{-2R_1 + R_3 \rightarrow R_3}]{\frac{1}{3}R_2 \rightarrow R_2} \begin{bmatrix} 1 & 1 & 0 & | & 1 & 0 & 0 \\ 0 & 1 & \frac{1}{3} & | & 0 & \frac{1}{3} & 0 \\ 0 & 1 & 3 & | & -2 & 0 & 1 \end{bmatrix} \xrightarrow{(-1)R_2 + R_3 \rightarrow R_3} \begin{bmatrix} 1 & 1 & 0 & | & 1 & 0 & 0 \\ 0 & 1 & \frac{1}{3} & | & 0 & \frac{1}{3} & 0 \\ 0 & 0 & \frac{8}{3} & | & -2 & -\frac{1}{3} & 1 \end{bmatrix}$$

$$\xrightarrow{\frac{3}{8}R_3} \begin{bmatrix} 1 & 1 & 0 & | & 1 & 0 & 0 \\ 0 & 1 & \frac{1}{3} & | & 0 & \frac{1}{3} & 0 \\ 0 & 0 & 1 & | & -\frac{6}{8} & -\frac{1}{8} & \frac{3}{8} \end{bmatrix} \xrightarrow{-\frac{1}{3}R_3 + R_2 \rightarrow R_2} \begin{bmatrix} 1 & 1 & 0 & | & 1 & 0 & 0 \\ 0 & 1 & 0 & | & \frac{2}{8} & \frac{3}{8} & -\frac{1}{8} \\ 0 & 0 & 1 & | & -\frac{6}{8} & -\frac{1}{8} & \frac{3}{8} \end{bmatrix}$$

$$\xrightarrow{(-1)R_2 + R_1 \rightarrow R_1} \begin{bmatrix} 1 & 0 & 0 & | & \frac{6}{8} & -\frac{3}{8} & \frac{1}{8} \\ 0 & 1 & 0 & | & \frac{2}{8} & \frac{3}{8} & -\frac{1}{8} \\ 0 & 0 & 1 & | & -\frac{6}{8} & -\frac{1}{8} & \frac{3}{8} \end{bmatrix}.$$

Hence, A is invertible and

$$A^{-1} = \begin{bmatrix} \frac{6}{8} & -\frac{3}{8} & \frac{1}{8} \\ \frac{2}{8} & \frac{3}{8} & -\frac{1}{8} \\ -\frac{6}{8} & -\frac{1}{8} & \frac{3}{8} \end{bmatrix} = \frac{1}{8}\begin{bmatrix} 6 & -3 & 1 \\ 2 & 3 & -1 \\ -6 & -1 & 3 \end{bmatrix}.$$

Step 3   You should verify that $AA^{-1} = I$. ■ ■ ■

Practice Problem 4   Find the inverse (if it exists) of the matrix $A = \begin{bmatrix} 1 & 2 & 3 \\ -2 & 3 & 1 \\ 4 & 5 & -2 \end{bmatrix}$.

■

3 Find the inverse of a 2 × 2 matrix.

A Rule for Finding the Inverse of a 2 × 2 Matrix

You can quickly determine whether any 2 × 2 matrix is invertible and, if so, what its inverse is.

A RULE FOR FINDING THE INVERSE OF A 2 × 2 MATRIX

The matrix

$$A = \begin{bmatrix} a & b \\ c & d \end{bmatrix}$$

is invertible if and only if $ad - bc \neq 0$. Moreover, if $ad - bc \neq 0$, then

$$A^{-1} = \frac{1}{ad - bc}\begin{bmatrix} d & -b \\ -c & a \end{bmatrix}.$$

If $ad - bc = 0$, the matrix A does not have an inverse.

EXAMPLE 5 **Finding the Inverse of a 2 × 2 Matrix**

Find the inverse (if it exists) of each matrix.

a. $A = \begin{bmatrix} 5 & 2 \\ 4 & 3 \end{bmatrix}$ **b.** $B = \begin{bmatrix} 4 & 6 \\ 2 & 3 \end{bmatrix}$

SOLUTION

a. For the matrix A, $a = 5, b = 2, c = 4$, and $d = 3$.
Here $ad - bc = (5)(3) - (2)(4) = 15 - 8 = 7 \neq 0$; so A is invertible and

$$A^{-1} = \frac{1}{7}\begin{bmatrix} 3 & -2 \\ -4 & 5 \end{bmatrix} \qquad \text{Substitute for } a, b, c, \text{ and } d \text{ in } \frac{1}{ad-bc}\begin{bmatrix} d & -b \\ -c & a \end{bmatrix}.$$

$$= \begin{bmatrix} \dfrac{3}{7} & -\dfrac{2}{7} \\ -\dfrac{4}{7} & \dfrac{5}{7} \end{bmatrix} \qquad \text{Scalar multiplication}$$

You should verify that $AA^{-1} = I$.

b. For the matrix B, $a = 4, b = 6, c = 2$, and $d = 3$; so $ad - bc = (4)(3) - (6)(2) = 12 - 12 = 0$. Therefore, the matrix B does not have an inverse. ■ ■ ■

Practice Problem 5 Find the inverse (if it exists) of each matrix.

a. $A = \begin{bmatrix} 8 & 2 \\ 4 & 1 \end{bmatrix}$ **b.** $B = \begin{bmatrix} 8 & -2 \\ 3 & 1 \end{bmatrix}$ ■

4 Use matrix inverses to solve systems of linear equations.

Solving Systems of Linear Equations by Using Matrix Inverses

Matrix multiplication can be used to write a system of linear equations in matrix form.

System of equations	Matrix form
$\begin{cases} 3x - 2y = 4 \\ 4x - 3y = 5 \end{cases}$	$\begin{bmatrix} 3 & -2 \\ 4 & -3 \end{bmatrix}\begin{bmatrix} x \\ y \end{bmatrix} = \begin{bmatrix} 4 \\ 5 \end{bmatrix}$
$\begin{cases} 2x_1 + 4x_2 - x_3 = 9 \\ 3x_1 + x_2 + 2x_3 = 7 \\ x_1 + 3x_2 - 3x_3 = 4 \end{cases}$	$\begin{bmatrix} 2 & 4 & -1 \\ 3 & 1 & 2 \\ 1 & 3 & -3 \end{bmatrix}\begin{bmatrix} x_1 \\ x_2 \\ x_3 \end{bmatrix} = \begin{bmatrix} 9 \\ 7 \\ 4 \end{bmatrix}$

Solving a system of linear equations amounts to solving a corresponding matrix equation of the form

$$AX = B,$$

where A is the coefficient matrix of the system, X is the column matrix containing the variables, and B is the column matrix of the system's constants.

If A is a square matrix *and* A is invertible, then the matrix equation $AX = B$ has a unique solution, $X = A^{-1}B$, as the following derivation shows.

$AX = B$	Given equation
$A^{-1}(AX) = A^{-1}B$	Be careful to multiply by A^{-1} on the left on both sides of the equation because multiplication of matrices is not commutative.
$(A^{-1}A)X = A^{-1}B$	Associative property
$IX = A^{-1}B$	$A^{-1}A = I$
$X = A^{-1}B$	I is the identity matrix.

EXAMPLE 6   Solving a Linear System by Using an Inverse Matrix

Use a matrix inverse to solve the linear system:
$$\begin{cases} x + y \phantom{{}+ z} = 4 \\ \phantom{x + {}} 3y + z = 7 \\ 2x + 3y + 3z = 21 \end{cases}$$

SOLUTION

Write the linear system in matrix form:

$$\begin{bmatrix} 1 & 1 & 0 \\ 0 & 3 & 1 \\ 2 & 3 & 3 \end{bmatrix} \begin{bmatrix} x \\ y \\ z \end{bmatrix} = \begin{bmatrix} 4 \\ 7 \\ 21 \end{bmatrix} \qquad \text{Use zeros for coefficients of missing variables.}$$

$$\underbrace{}_{A} \quad \underbrace{}_{X} \quad \underbrace{}_{B}$$

We need to solve the matrix equation $AX = B$ for X. Since the matrix A is invertible (see Example 4), the system has a unique solution $X = A^{-1}B$. Use the following:

$$A^{-1} = \frac{1}{8} \begin{bmatrix} 6 & -3 & 1 \\ 2 & 3 & -1 \\ -6 & -1 & 3 \end{bmatrix} \qquad \text{Computed in Example 4}$$

$$X = A^{-1}B$$

$$= \frac{1}{8} \begin{bmatrix} 6 & -3 & 1 \\ 2 & 3 & -1 \\ -6 & -1 & 3 \end{bmatrix} \begin{bmatrix} 4 \\ 7 \\ 21 \end{bmatrix} \qquad \text{Substitute for } A^{-1} \text{ and } B.$$

$$= \frac{1}{8} \begin{bmatrix} 6(4) - 3(7) + 1(21) \\ 2(4) + 3(7) - 1(21) \\ -6(4) - 1(7) + 3(21) \end{bmatrix} \qquad \text{Matrix multiplication}$$

$$= \frac{1}{8} \begin{bmatrix} 24 \\ 8 \\ 32 \end{bmatrix} = \begin{bmatrix} 3 \\ 1 \\ 4 \end{bmatrix} \qquad \text{Scalar multiplication}$$

The solution set is $\{(3, 1, 4)\}$, which you can check in the original system. ■ ■ ■

Practice Problem 6 Solve the linear system.

$$\begin{cases} 3x + 2y + 3z = 9 \\ 3x + y \quad\quad = 12 \\ x \quad\quad + z = 6 \end{cases}$$

5 Use matrix inverses in applied problems.

Applications of Matrix Inverses

The Leontief Input–Output Model We can use the inverse of a matrix to analyze input–output models that we studied in Section 8.1.

Suppose a simplified economy depends on two products: energy (E) and food (F). To produce one unit of E requires $\frac{1}{4}$ unit of E and $\frac{1}{2}$ unit of F. To produce one unit of F requires $\frac{1}{3}$ unit of E and $\frac{1}{4}$ unit of F. Then the interindustry consumption is given by the following matrix.

$$\begin{array}{cc} & \textbf{Input} \\ & E \quad F \end{array}$$

$$A = \begin{bmatrix} \dfrac{1}{4} & \dfrac{1}{3} \\ \dfrac{1}{2} & \dfrac{1}{4} \end{bmatrix} \begin{array}{l} E \\ F \end{array} \quad \textbf{Output}$$

The matrix A is the **interindustry technology input–output matrix**, or simply the **technology matrix**, of the system. If the system is producing x_1 units of energy and x_2 units of food, then the column matrix $X = \begin{bmatrix} x_1 \\ x_2 \end{bmatrix}$ is called the **gross production matrix**.

Consequently,

$$AX = \begin{bmatrix} \dfrac{1}{4} & \dfrac{1}{3} \\ \dfrac{1}{2} & \dfrac{1}{4} \end{bmatrix} \begin{bmatrix} x_1 \\ x_2 \end{bmatrix} = \begin{bmatrix} \dfrac{1}{4}x_1 + \dfrac{1}{3}x_2 \\ \dfrac{1}{2}x_1 + \dfrac{1}{4}x_2 \end{bmatrix} \begin{array}{l} \leftarrow \text{units consumed by } E \\ \leftarrow \text{units consumed by } F \end{array}$$

represents the interindustry consumptions. If the column matrix $D = \begin{bmatrix} d_1 \\ d_2 \end{bmatrix}$ represents consumer demand, then

$D = X - AX$	Gross production minus interindustry consumption
$= IX - AX$	I is the identity matrix.
$D = (I - A)X$	Distributive property
$(I - A)^{-1}D = (I - A)^{-1}(I - A)X$	Multiply both sides by $(I - A)^{-1}$ if $I - A$ is invertible.
$(I - A)^{-1}D = IX$	$(I - A)^{-1}(I - A) = I$
$= X$	I is the identity matrix.

So if $(I - A)$ is invertible, then the gross production matrix is $X = (I - A)^{-1}D$.

EXAMPLE 7 **Using the Leontief Input–Output Model**

In the preceding discussion, suppose the consumer demand for energy is 1000 units and for food is 3000 units. Find the level of production (X) that will meet interindustry and consumer demand.

SOLUTION

For the matrix A, we have

$$I - A = \begin{bmatrix} 1 & 0 \\ 0 & 1 \end{bmatrix} - \begin{bmatrix} \dfrac{1}{4} & \dfrac{1}{3} \\ \dfrac{1}{2} & \dfrac{1}{4} \end{bmatrix} = \begin{bmatrix} \dfrac{3}{4} & -\dfrac{1}{3} \\ -\dfrac{1}{2} & \dfrac{3}{4} \end{bmatrix}$$

RECALL

If $ad - bc \neq 0$, then for

$A = \begin{bmatrix} a & b \\ c & d \end{bmatrix}$, we have

$A^{-1} = \dfrac{1}{ad - bc} \begin{bmatrix} d & -b \\ -c & a \end{bmatrix}.$

Use the formula for the inverse of a 2×2 matrix.

$$(I - A)^{-1} = \cfrac{1}{\dfrac{3}{4} \cdot \dfrac{3}{4} - \left(-\dfrac{1}{3}\right)\left(-\dfrac{1}{2}\right)} \begin{bmatrix} \dfrac{3}{4} & \dfrac{1}{3} \\ \dfrac{1}{2} & \dfrac{3}{4} \end{bmatrix}$$

$$= \dfrac{48}{19} \begin{bmatrix} \dfrac{3}{4} & \dfrac{1}{3} \\ \dfrac{1}{2} & \dfrac{3}{4} \end{bmatrix} = \dfrac{1}{19} \begin{bmatrix} 36 & 16 \\ 24 & 36 \end{bmatrix} \qquad \text{Simplify.}$$

The matrix $D = \begin{bmatrix} 1000 \\ 3000 \end{bmatrix}$. Therefore, the gross production matrix X is:

$$X = (I - A)^{-1}D = \dfrac{1}{19} \begin{bmatrix} 36 & 16 \\ 24 & 36 \end{bmatrix} \begin{bmatrix} 1000 \\ 3000 \end{bmatrix} \qquad \begin{array}{l} \text{Substitute for} \\ (I - A)^{-1} \text{ and } D. \end{array}$$

$$= \dfrac{1}{19} \begin{bmatrix} 84{,}000 \\ 132{,}000 \end{bmatrix} = \begin{bmatrix} \dfrac{84{,}000}{19} \\ \dfrac{132{,}000}{19} \end{bmatrix}$$

To meet the consumer demand for 1000 units of energy and 3000 units of food, the energy produced must be $\dfrac{84{,}000}{19}$ units and food production must be $\dfrac{132{,}000}{19}$ units.

■ ■ ■

Practice Problem 7 In Example 7, find the level of production (X) that will meet both the interindustry and consumer demand if consumer demand for energy is 800 units and consumer demand for food is 3400 units.   ■

Cryptography

To encode or decode a message, first associate each letter in the message with a number, in the obvious manner. The blank corresponds to 0.

blank	A	B	C	D	E	F	G	H	I	J	K	L	M
0	1	2	3	4	5	6	7	8	9	10	11	12	13
	N	O	P	Q	R	S	T	U	V	W	X	Y	Z
	14	15	16	17	18	19	20	21	22	23	24	25	26

For example, the message

JOHNSON IS IN DANGER

becomes:

10	15	8	14	19	15	14	0	9	19	0	9	14	0	4	1	14	7	5	18
J	O	H	N	S	O	N		I	S		I	N		D	A	N	G	E	R

This message is easy to decode, so we will make the coding more complicated. First, partition these 20 numbers into groups of three, enclosed in brackets, and insert zeros at the end of the last bracket if necessary.

$$\begin{bmatrix} 10 & 15 & 8 \end{bmatrix} \begin{bmatrix} 14 & 19 & 15 \end{bmatrix} \begin{bmatrix} 14 & 0 & 9 \end{bmatrix} \begin{bmatrix} 19 & 0 & 9 \end{bmatrix} \begin{bmatrix} 14 & 0 & 4 \end{bmatrix} \begin{bmatrix} 1 & 14 & 7 \end{bmatrix} \begin{bmatrix} 5 & 18 & 0 \end{bmatrix}$$
$$\ \ \text{J}\ \ \ \text{O}\ \ \ \text{H}\ \ \ \text{N}\ \ \ \text{S}\ \ \ \text{O}\ \ \ \text{N}\quad\ \ \text{I}\ \ \ \text{S}\qquad\ \ \text{I}\ \ \ \text{N}\qquad\ \ \text{D}\ \ \ \text{A}\ \ \ \text{N}\ \ \ \text{G}\ \ \ \text{E}\ \ \ \text{R}$$

Let's write these seven groups of three numbers as the seven columns of a 3×7 matrix:

$$M = \begin{bmatrix} 10 & 14 & 14 & 19 & 14 & 1 & 5 \\ 15 & 19 & 0 & 0 & 0 & 14 & 18 \\ 8 & 15 & 9 & 9 & 4 & 7 & 0 \end{bmatrix}$$

Now choose any invertible 3×3 matrix A, say,

$$A = \begin{bmatrix} 1 & 2 & 3 \\ 1 & 3 & 3 \\ 1 & 2 & 4 \end{bmatrix}$$

We can use the matrix A to convert the message into code and then use A^{-1} to decode the message, as in the next example. The matrix A is called the **coding matrix**.

EXAMPLE 8   Encoding and Decoding a Message

Encode and decode the message

JOHNSON IS IN DANGER

by using the matrix A given previously.

SOLUTION

Step 1   Express the message numerically and partition the numbers into groups of three:

$$\begin{bmatrix} 10 & 15 & 8 \end{bmatrix} \begin{bmatrix} 14 & 19 & 15 \end{bmatrix} \begin{bmatrix} 14 & 0 & 9 \end{bmatrix} \begin{bmatrix} 19 & 0 & 9 \end{bmatrix} \begin{bmatrix} 14 & 0 & 4 \end{bmatrix} \begin{bmatrix} 1 & 14 & 7 \end{bmatrix} \begin{bmatrix} 5 & 18 & 0 \end{bmatrix}$$

Step 2   Write each group of three numbers in Step 1 as a column of a matrix M.

$$M = \begin{bmatrix} 10 & 14 & 14 & 19 & 14 & 1 & 5 \\ 15 & 19 & 0 & 0 & 0 & 14 & 18 \\ 8 & 15 & 9 & 9 & 4 & 7 & 0 \end{bmatrix}$$

Step 3   Select an invertible 3×3 matrix A, such as

$$A = \begin{bmatrix} 1 & 2 & 3 \\ 1 & 3 & 3 \\ 1 & 2 & 4 \end{bmatrix}$$

Step 4   Multiply the matrices in Steps 2 and 3 to form AM.

$$AM = \begin{bmatrix} 1 & 2 & 3 \\ 1 & 3 & 3 \\ 1 & 2 & 4 \end{bmatrix} \begin{bmatrix} 10 & 14 & 14 & 19 & 14 & 1 & 5 \\ 15 & 19 & 0 & 0 & 0 & 14 & 18 \\ 8 & 15 & 9 & 9 & 4 & 7 & 0 \end{bmatrix}$$

$$= \begin{bmatrix} 64 & 97 & 41 & 46 & 26 & 50 & 41 \\ 79 & 116 & 41 & 46 & 26 & 64 & 59 \\ 72 & 112 & 50 & 55 & 30 & 57 & 41 \end{bmatrix}$$

Matrix multiplication

The coded message sent will be the numbers from column 1 of AM, followed by the numbers from column 2, and so on.

64 79 72 97 116 112 41 41 50 46 46 55 26 26 30 50 64 57 41 59 41

Step 5   To decode the message in Step 4, write the numbers received in the message in groups of three and as columns of the matrix AM. Find the **decoding matrix** A^{-1} of the coding matrix A. You can verify that

$$A^{-1} = \begin{bmatrix} 6 & -2 & -3 \\ -1 & 1 & 0 \\ -1 & 0 & 1 \end{bmatrix}$$

Step 6   Compute the message matrix M.

$$M = A^{-1}(AM) = \begin{bmatrix} 6 & -2 & -3 \\ -1 & 1 & 0 \\ -1 & 0 & 1 \end{bmatrix}\begin{bmatrix} 64 & 97 & 41 & 46 & 26 & 50 & 41 \\ 79 & 116 & 41 & 46 & 26 & 64 & 59 \\ 72 & 112 & 50 & 55 & 30 & 57 & 41 \end{bmatrix}$$

$$= \begin{bmatrix} 10 & 14 & 14 & 19 & 14 & 1 & 5 \\ 15 & 19 & 0 & 0 & 0 & 14 & 18 \\ 8 & 15 & 9 & 9 & 4 & 7 & 0 \end{bmatrix} \quad \text{Matrix multiplication}$$

Step 7   Finally, convert the numerical message from the matrix M in Step 6 back into English.

JOHNSON IS IN DANGER.   ■ ■ ■

Practice Problem 8   Rework Example 8, replacing the message JOHNSON IS IN DANGER with the message JACK IS NOW SAFE.   ▨

SECTION 8.3 ■ Exercises

A EXERCISES Basic Skills and Concepts

1. For $n \times n$ matrices A and B, if $AB = I = BA$, then B is called the __inverse__ of A.

2. An $n \times n$ matrix A is invertible if there is a matrix B such that __$AB = BA$__ $= I$.

3. To find the inverse of an invertible matrix A, we transform $[A \mid I]$ by a sequence of row operations into $[I \mid B]$, where $B =$ __A^{-1}__.

4. $A = \begin{bmatrix} a & b \\ c & d \end{bmatrix}$ is invertible if and only if __$ad - bc \neq 0$__.

5. *True or False* Every square matrix is invertible. False

6. *True or False* $A = \begin{bmatrix} 8 & 16 \\ 4 & 8 \end{bmatrix}$ is invertible. False

In Exercises 7–16, determine whether B is the inverse of A by computing AB and BA.

7. $A = \begin{bmatrix} 1 & 2 \\ 1 & 3 \end{bmatrix}$, $B = \begin{bmatrix} 3 & -2 \\ -1 & 1 \end{bmatrix}$ Yes

8. $A = \begin{bmatrix} 3 & 2 \\ 4 & 3 \end{bmatrix}$, $B = \begin{bmatrix} 3 & -2 \\ -4 & 3 \end{bmatrix}$ Yes

9. $A = \begin{bmatrix} 3 & 2 \\ 1 & 4 \end{bmatrix}$, $B = \begin{bmatrix} \frac{2}{5} & -\frac{1}{5} \\ -\frac{1}{10} & \frac{3}{10} \end{bmatrix}$ Yes

10. $A = \begin{bmatrix} 2 & -3 \\ 4 & -3 \end{bmatrix}$, $B = \begin{bmatrix} -\frac{1}{2} & \frac{1}{2} \\ -\frac{2}{3} & \frac{1}{3} \end{bmatrix}$ Yes

11. $A = \begin{bmatrix} 1 & 0 & -1 \\ -1 & 1 & 1 \end{bmatrix}$, $B = \begin{bmatrix} 2 & 1 \\ 1 & 1 \\ 1 & 1 \end{bmatrix}$ No

12. $A = \begin{bmatrix} -2 & 1 & 3 \\ 0 & -1 & 1 \\ 1 & 2 & 0 \end{bmatrix}$, $B = \frac{1}{8}\begin{bmatrix} -2 & 6 & 4 \\ 1 & -3 & 2 \\ 1 & 5 & 2 \end{bmatrix}$ Yes

13. $A = \begin{bmatrix} 1 & -2 & 1 \\ -8 & 6 & -2 \\ 5 & -3 & 1 \end{bmatrix}$, $B = \frac{1}{2}\begin{bmatrix} 0 & 1 & 2 \\ 1 & 2 & 3 \\ 3 & 1 & 1 \end{bmatrix}$ No

14. $A = \begin{bmatrix} 2 & 3 & 1 \\ 1 & 2 & 3 \\ 3 & 1 & 2 \end{bmatrix}$, $B = \frac{1}{18}\begin{bmatrix} 1 & -5 & 7 \\ 7 & 1 & -5 \\ -5 & 7 & 1 \end{bmatrix}$ Yes

15. $A = \begin{bmatrix} 1 & 1 & 1 \\ 1 & 2 & 3 \\ -1 & 1 & -1 \end{bmatrix}$, $B = \frac{1}{4}\begin{bmatrix} 5 & -2 & -1 \\ 2 & 0 & 2 \\ -3 & 2 & -1 \end{bmatrix}$ Yes

16. $A = \begin{bmatrix} 1 & -1 & 0 \\ 0 & 1 & -1 \\ 1 & 0 & 1 \end{bmatrix}$, $B = \frac{1}{2}\begin{bmatrix} 1 & 1 & 1 \\ -1 & 1 & 1 \\ -1 & -1 & 1 \end{bmatrix}$ Yes

†Due to space constrictions, answers to these exercises may be found in the Answers beginning on page A–1 in the back of the book.

In Exercises 17–26, find A^{-1} (if it exists) by forming the matrix $[A \,|\, I]$ and using the row operation to obtain $[I \,|\, B]$, where $B = A^{-1}$.

17. $A = \begin{bmatrix} 2 & 0 \\ 1 & 3 \end{bmatrix}$ †

18. $A = \begin{bmatrix} 4 & 3 \\ 1 & 0 \end{bmatrix}$ †

19. $A = \begin{bmatrix} 2 & 4 \\ 3 & 6 \end{bmatrix}$

20. $A = \begin{bmatrix} 9 & 6 \\ 6 & 4 \end{bmatrix}$

21. $A = \begin{bmatrix} 1 & 6 & 4 \\ 0 & 2 & 3 \\ 0 & 1 & 2 \end{bmatrix}$ †

22. $A = \begin{bmatrix} -2 & 1 & 3 \\ 0 & -1 & 1 \\ 1 & 2 & 0 \end{bmatrix}$ †

23. $A = \begin{bmatrix} 2 & 4 & 3 \\ 0 & 1 & 1 \\ 2 & 2 & -1 \end{bmatrix}$ †

24. $A = \begin{bmatrix} 1 & 1 & 5 \\ 4 & 3 & -5 \\ 1 & 1 & 0 \end{bmatrix}$ †

25. $A = \begin{bmatrix} 1 & 2 & -1 \\ -1 & 1 & 2 \\ 2 & -1 & 1 \end{bmatrix}$ †

26. $A = \begin{bmatrix} 1 & 2 & 3 \\ 2 & 3 & 0 \\ 0 & 1 & 2 \end{bmatrix}$ †

In Exercises 27–32, find the inverse (if it exists) of $A = \begin{bmatrix} a & b \\ c & d \end{bmatrix}$ by using the formula $A^{-1} = \dfrac{1}{ad-bc}\begin{bmatrix} d & -b \\ -c & a \end{bmatrix}$. Check that $A^{-1}A = I$.

27. $A = \begin{bmatrix} 1 & 0 \\ 3 & 2 \end{bmatrix}$ †

28. $A = \begin{bmatrix} 3 & 4 \\ 5 & 6 \end{bmatrix}$ †

29. $A = \begin{bmatrix} 2 & -3 \\ -3 & 5 \end{bmatrix}$ †

30. $A = \begin{bmatrix} 3 & -4 \\ 2 & -2 \end{bmatrix}$ †

31. $A = \begin{bmatrix} a & -b \\ b & -a \end{bmatrix}, a^2 \neq b^2$ †

32. $A = \begin{bmatrix} 2 & -1 \\ 1 & -1 \end{bmatrix}$ †

In Exercises 33–36, write the given linear system of equations as a matrix equation $AX = B$.

33. $\begin{cases} 2x + 3y = -9 \\ x - 3y = 13 \end{cases}$ †

34. $\begin{cases} 5x - 4y = 7 \\ 4x - 3y = 5 \end{cases}$ †

35. $\begin{cases} 3x + 2y + z = 8 \\ 2x + y + 3z = 7 \\ x + 3y + 2z = 9 \end{cases}$ †

36. $\begin{cases} x + 3y + z = 4 \\ x - 5y + 2z = 7 \\ 3x + y - 4z = -9 \end{cases}$ †

In Exercises 37–40, write each matrix equation as a linear system of equations.

37. $\begin{bmatrix} 1 & -2 \\ 2 & 1 \end{bmatrix}\begin{bmatrix} x \\ y \end{bmatrix} = \begin{bmatrix} 0 \\ 5 \end{bmatrix}$ †

38. $\begin{bmatrix} 2 & 3 \\ 3 & -1 \end{bmatrix}\begin{bmatrix} x_1 \\ x_2 \end{bmatrix} = \begin{bmatrix} 0 \\ 11 \end{bmatrix}$ †

39. $\begin{bmatrix} 2 & 3 & 1 \\ 5 & 7 & -1 \\ 4 & 3 & 0 \end{bmatrix}\begin{bmatrix} x_1 \\ x_2 \\ x_3 \end{bmatrix} = \begin{bmatrix} -1 \\ 5 \\ 5 \end{bmatrix}$ †

40. $\begin{bmatrix} 3 & -2 & 3 \\ 5 & 0 & 4 \\ 2 & 7 & 0 \end{bmatrix}\begin{bmatrix} r \\ s \\ t \end{bmatrix} = \begin{bmatrix} 4 \\ 3 \\ -8 \end{bmatrix}$ †

41. Let $A = \begin{bmatrix} 1 & 2 & 5 \\ 2 & 3 & 8 \\ -1 & 1 & 2 \end{bmatrix}$. Show that $A^{-1} = \begin{bmatrix} 2 & -1 & -1 \\ 12 & -7 & -2 \\ -5 & 3 & 1 \end{bmatrix}$. †

Answers:
19. The inverse of matrix A does not exist. **20.** The inverse of matrix A does not exist.

In Exercises 42–44, use the result of Exercise 41 to solve the system of equations.

42. $\begin{cases} x + 2y + 5z = 4 \\ 2x + 3y + 8z = 6 \\ -x + y + 2z = 3 \end{cases}$ †

43. $\begin{cases} x_1 + 2x_2 + 5x_3 = 1 \\ 2x_1 + 3x_2 + 8x_3 = 3 \\ -x_1 + x_2 + 2x_3 = -3 \end{cases}$
$\{(2, -3, 1)\}$

44. $\begin{cases} x + 2y + 5z = -4 \\ 2x + 3y + 8z = -6 \\ -x + y + 2z = -\dfrac{5}{2} \end{cases}$ $\left\{\left(\dfrac{1}{2}, -1, -\dfrac{1}{2}\right)\right\}$

45. a. Find the inverse of the matrix $A = \begin{bmatrix} 1 & 1 & 1 \\ 1 & 2 & 3 \\ 1 & 4 & 9 \end{bmatrix}$. †

   b. Use the result of part (a) to solve the linear system:
   $\begin{cases} x + y + z = 6 \\ x + 2y + 3z = 14 \\ x + 4y + 9z = 36 \end{cases}$ $\{(1, 2, 3)\}$

46. a. Find the inverse of the matrix $A = \begin{bmatrix} 2 & 4 & -1 \\ 3 & 1 & 2 \\ 1 & 3 & -3 \end{bmatrix}$. †

   b. Use the result of part (a) to solve the linear system:
   $\begin{cases} 2x + 4y - z = 9 \\ 3x + y + 2z = 7 \\ x + 3y - 3z = 4 \end{cases}$ $\{(1, 2, 1)\}$

B EXERCISES Applying the Concepts

In Exercises 47–52, solve each linear system by using an inverse matrix.

47. $\begin{cases} 3x + 7y = 11 \\ -5x + 4y = 13 \end{cases}$ $\{(-1, 2)\}$

48. $\begin{cases} x - 7y = 3 \\ 2x + 3y = 23 \end{cases}$ $\{(10, 1)\}$

49. $\begin{cases} x + y + 2z = 7 \\ x - y - 3z = -6 \\ 2x + 3y + z = 4 \end{cases}$ $\{(2, -1, 3)\}$

50. $\begin{cases} x + y + z = 6 \\ 2x - 3y + 3z = 5 \\ 3x - 2y - z = -4 \end{cases}$ $\{(1, 2, 3)\}$

51. $\begin{cases} 2x + 2y + 3z = 7 \\ 5x + 3y + 5z = 3 \\ 3x + 5y + z = -5 \end{cases}$ $\{(-5, 1, 5)\}$

52. $\begin{cases} 9x + 7y + 4z = 12 \\ 6x + 5y + 4z = 5 \\ 4x + 3y + z = 7 \end{cases}$ $\{(3, -1, -2)\}$

In Exercises 53–56, technology matrix A, representing interindustry demand, and matrix D, representing consumer demand for the sectors in a two- or three-sector economy, are given. Use the matrix equation $X = (I - A)^{-1}D$ to find the production level X that will satisfy both demands.

53. Two-sector economy. In a two-sector economy,
$$A = \begin{bmatrix} 0.1 & 0.4 \\ 0.5 & 0.2 \end{bmatrix}, D = \begin{bmatrix} 50 \\ 30 \end{bmatrix} \quad X = \begin{bmatrix} 100 \\ 100 \end{bmatrix}$$

54. Two-sector economy. In a two-sector economy,
$$A = \begin{bmatrix} 0.2 & 0.5 \\ 0.1 & 0.6 \end{bmatrix}, D = \begin{bmatrix} 11 \\ 18 \end{bmatrix} \quad X = \begin{bmatrix} \dfrac{1340}{27} \\ \dfrac{1550}{27} \end{bmatrix}$$

55. Three-sector economy. In a three-sector economy,
$$A = \begin{bmatrix} 0.2 & 0.2 & 0 \\ 0.1 & 0.1 & 0.3 \\ 0.1 & 0 & 0.2 \end{bmatrix}, D = \begin{bmatrix} 400 \\ 600 \\ 800 \end{bmatrix}$$ †

56. Three-sector economy. In a three-sector economy,

$$A = \begin{bmatrix} 0.5 & 0.4 & 0.2 \\ 0.2 & 0.3 & 0.1 \\ 0.1 & 0.1 & 0.3 \end{bmatrix}, D = \begin{bmatrix} 500 \\ 300 \\ 200 \end{bmatrix}$$

In Exercises 57 and 58, use a 2 × 2 invertible coding matrix A of your choice. Remember to partition the numerical message into groups of two. (a) Write a cryptogram for each message. (b) Check by decoding.

57. Coding and decoding. CANNOT FIND COLOMBO.

58. Coding and decoding. FOUND HEAD OF TERRORIST NETWORK.

In Exercises 59 and 60, use a 3 × 3 invertible coding matrix A of your choice.

59. Code and then decode the message of Exercise 57.

60. Code and then decode the message of Exercise 58.

C EXERCISES Beyond the Basics

61. Suppose A is an invertible square matrix.
   a. Is A^{-1} invertible? Why or why not?
   b. If your answer to part (a) is yes, then what is $(A^{-1})^{-1}$?

62. Let $A^{-1} = \begin{bmatrix} 2 & 1 \\ 3 & -4 \end{bmatrix}$. Find A.

63. Let A be a square matrix. Show that if A^2 is invertible, then A must be invertible. [*Hint:* $A^2B = A(AB)$.]

64. Show that if A is invertible, then A^2 is invertible.

65. Show that if A and B are $n \times n$ invertible matrices, then AB is invertible and $(AB)^{-1} = B^{-1}A^{-1}$. (Note the order.)

66. Verify the result of Exercise 65 for

$$A = \begin{bmatrix} 1 & 2 \\ 3 & 4 \end{bmatrix}, B = \begin{bmatrix} -1 & 0 \\ 3 & -2 \end{bmatrix}$$

67. Let A, B, and C be $n \times n$ matrices such that $AB = AC$. Show that if A is invertible, then $B = C$.

68. Is the result of Exercise 67 true if A is not invertible?

69. Let A and B be $n \times n$ matrices such that $AB = A$. Show that if A is invertible, then $B = I$.

70. Give an example to show that the result of Exercise 69 is not true if A is not invertible.

71. Let $A = \begin{bmatrix} 3 & 4 \\ 2 & 3 \end{bmatrix}$.
   a. Show that the matrix A satisfies the equation

$$A^2 - 6A + I = 0.$$

   b. Use the equation in part (a) to show that $A^{-1} = 6I - A$.
   c. Use the equation in part (b) to find A^{-1}.

72. Let $A = \begin{bmatrix} 2 & -1 & 1 \\ -1 & 2 & -1 \\ 1 & -1 & 2 \end{bmatrix}$. Assume that A is invertible.
   a. Show that A satisfies the equation $A^3 - 6A^2 + 9A - 4I = 0$.
   b. Use the equation in part (a) to show that

$$A^{-1} = \frac{1}{4}[A^2 - 6A + 9I].$$

   [*Hint:* Consider the product $\frac{1}{4}[A^2 - 6A + 9I]A$.]
   c. Use part (b) to find A^{-1}. †

Critical Thinking

73. Suppose A and B are 2 × 2 matrices and $B^{-1} = \begin{bmatrix} 1 & 2 \\ 3 & 4 \end{bmatrix}$.

Find A
   a. if $AB = \begin{bmatrix} -1 & 3 \\ 0 & 2 \end{bmatrix}$ $A = \begin{bmatrix} 8 & 10 \\ 6 & 8 \end{bmatrix}$
   b. if $BA = \begin{bmatrix} -1 & 3 \\ 0 & 2 \end{bmatrix}$ $A = \begin{bmatrix} -1 & 7 \\ -3 & 17 \end{bmatrix}$
   c. if $B^{-1}AB = \begin{bmatrix} 1 & 1 \\ 1 & 2 \end{bmatrix}$
   d. if $BAB^{-1} = \begin{bmatrix} 1 & 1 \\ 1 & 2 \end{bmatrix}$

74. a. Is the matrix $\begin{bmatrix} x & -1 \\ 2 & x+2 \end{bmatrix}$ invertible for all real numbers x? Yes
   b. Is the matrix $\begin{bmatrix} x-1 & -1 \\ 3 & x-5 \end{bmatrix}$ invertible for all real numbers x? No

75. True or False? Justify your answers.
   a. If A and B are matrices for which $AB = BA$, then the formula $A^2B = BA^2$ must hold.
   b. If A and B are invertible matrices, then $A + B$ must be invertible. False. Let $A = I$ and $B = -I$.

76. Give an example of a 2 × 2 matrix A such that $A^2 = 0$. Show that the matrix $I + A$ is invertible and $(I + A)^{-1} = I - A$.

77. Is it true that every nonsingular matrix is a square matrix? Why or why not? True

Answers:

$$56.\ X = \begin{bmatrix} \dfrac{61,000}{27} \\ \dfrac{32,000}{27} \\ \dfrac{7000}{9} \end{bmatrix}$$

61. a. Yes. By the definition of *inverse*, if $AB = I$, then $BA = I$ and A and B are inverses. **b.** $(A^{-1})^{-1} = A$

$$62.\ A = \begin{bmatrix} \dfrac{4}{11} & \dfrac{1}{11} \\ \dfrac{3}{11} & -\dfrac{2}{11} \end{bmatrix}$$

68. Not necessarily

69. If A is invertible, there exists some matrix C such that $AC = I$ and $CA = I$. Then $AB = A \Leftrightarrow CAB = CA \Leftrightarrow IB = I \Leftrightarrow B = I$.

70. Let $A = \begin{bmatrix} 2 & 0 \\ 0 & 0 \end{bmatrix}$ and $B = \begin{bmatrix} 1 & 0 \\ 3 & 4 \end{bmatrix}$.

71. c. $A^{-1} = \begin{bmatrix} 3 & -4 \\ -2 & 3 \end{bmatrix}$

73. c. $A = \begin{bmatrix} -1 & -2 \\ \dfrac{5}{2} & 4 \end{bmatrix}$ **d.** $A = \begin{bmatrix} \dfrac{3}{2} & \dfrac{1}{2} \\ \dfrac{5}{2} & \dfrac{3}{2} \end{bmatrix}$

75. a. True. $A^2B = AAB = ABA = BAA = BA^2$ **76.** Let $A = \begin{bmatrix} 0 & -1 \\ 0 & 0 \end{bmatrix}$

Determinants and Cramer's Rule

Before Starting this Section, Review

1. Definition of a matrix (Section 8.1, page 546)

2. Systems of equations (Section 7.1, page 481)

3. Inverse of a 2 × 2 matrix (Section 8.3, page 578)

Objectives

1 Calculate the determinant of a 2 × 2 matrix.

2 Find minors and cofactors.

3 Evaluate the determinant of an $n \times n$ matrix.

4 Apply Cramer's Rule.

Seki Kōwa (1642–1708)

Seki Kōwa has been called "the Japanese Newton." He was born in the same year as Newton. Like Newton, he acquired much of his knowledge by self-study and was a brilliant problem solver. But the greatest similarity between the two men lies in the claim that Seki Kowa was the creator of *yenri* (calculus). Thus, Seki Kōwa probably invented the calculus in the East just as Newton and Leibniz invented the calculus in the West.

A LITTLE HISTORY OF DETERMINANTS

It appears that determinants were first investigated in 1683 by the outstanding Japanese mathematician Seki Kōwa in connection with systems of linear equations. Determinants were independently investigated in 1693 by the German mathematician Gottfried Leibniz (1646–1716). The French mathematician Alexandre-Théophile Vandermonde (1735–1796) was the first to give a coherent and systematic exposition of the theory of determinants. Determinants were used extensively in the 19th century by eminent mathematicians such as Cauchy, Jacobi, and Kronecker. Today determinants are of little numerical value in the large-scale matrix computations that occur so often. Nevertheless, a knowledge of determinants is useful in some applications of matrices. ■

---

1 Calculate the determinant of a 2 × 2 matrix.

The Determinant of a 2 × 2 Matrix

Associated with each square matrix A is a number called the *determinant* of A:

STUDY TIP

A determinant can be thought of as a function whose domain is the set of all square matrices and whose range is the set of real numbers.

DETERMINANT OF A 2 × 2 MATRIX

The **determinant** of the matrix

$$A = \begin{bmatrix} a & b \\ c & d \end{bmatrix},$$

denoted by $\det(A)$, $|A|$, or $\begin{vmatrix} a & b \\ c & d \end{vmatrix}$, is called a **2 by 2 determinant** and is defined by

$$\det(A) = ad - bc.$$

To remember this definition, note that you form products along diagonal elements as shown by the arrows.

$$\begin{bmatrix} a & b \\ c & d \end{bmatrix}$$
$$-bc \quad +ad$$

You attach a plus sign if you descend from left to right and a minus sign if you descend from right to left. Then you form the sum $ad - bc$.

EXAMPLE 1   **Calculating the Determinant of a 2 × 2 Matrix**

Evaluate each determinant.

a. $\begin{vmatrix} 3 & -4 \\ 1 & 5 \end{vmatrix}$ **b.** $\begin{vmatrix} -2 & 3 \\ 4 & -5 \end{vmatrix}$

SOLUTION

a. $\begin{vmatrix} 3 & -4 \\ 1 & 5 \end{vmatrix} = (3)(5) - (-4)(1) = 15 - (-4) = 15 + 4 = 19$

b. $\begin{vmatrix} -2 & 3 \\ 4 & -5 \end{vmatrix} = (-2)(-5) - (3)(4) = 10 - 12 = -2$ ■ ■ ■

Practice Problem 1   Evaluate each determinant.

a. $\begin{vmatrix} 5 & 1 \\ 3 & -7 \end{vmatrix}$ **b.** $\begin{vmatrix} 2 & -9 \\ -4 & 18 \end{vmatrix}$ ■

RECALL

A matrix of order n is an $n \times n$ square matrix.

Recall (page 578) that the matrix $A = \begin{bmatrix} a & b \\ c & d \end{bmatrix}$ is invertible if and only if $ad - bc \neq 0$. You can rephrase this property by saying that a *matrix of order 2 is invertible if and only if its determinant is nonzero*. If the matrix A is invertible, then its inverse can be expressed in terms of its determinant as:

$$A^{-1} = \frac{1}{ad - bc}\begin{bmatrix} d & -b \\ -c & a \end{bmatrix} = \frac{1}{\det(A)}\begin{bmatrix} d & -b \\ -c & a \end{bmatrix}$$

◆ **WARNING**

Note carefully the notations used for matrices and determinants. Brackets, [ ], denote a matrix, while vertical bars, | |, denote the determinant of a matrix. Remember that *the determinant of a matrix is a number.*

---

2 Find minors and cofactors.

Minors and Cofactors

To define the determinants of square matrices of order 3 or higher, we need some additional terminology.

STUDY TIP

Signs of cofactors

$$\begin{bmatrix} + & - & + & - & \cdots \\ - & + & - & + & \cdots \\ + & - & + & - & \cdots \\ - & + & - & + & \cdots \\ \vdots & \vdots & \vdots & \vdots & \end{bmatrix}$$

Note how the positions where $i + j$ is even alternate with those where $i + j$ is odd.

MINORS AND COFACTORS IN AN $n \times n$ MATRIX

Let A be an $n \times n$ square matrix. The **minor** M_{ij} of the element a_{ij} is the determinant of the $(n - 1) \times (n - 1)$ matrix obtained by deleting the ith row and the jth column of A. The **cofactor** A_{ij} of the entry a_{ij} is given by

$$A_{ij} = (-1)^{i+j}M_{ij}.$$

When $i + j$ is an even integer, $(-1)^{i+j} = 1$. When $i + j$ is an odd integer, $(-1)^{i+j} = -1$. Therefore,

$$A_{ij} = \begin{cases} M_{ij} \text{ if } i + j \text{ is an even integer} \\ -M_{ij} \text{ if } i + j \text{ is an odd integer} \end{cases}$$

Notice that to compute a minor in an $n \times n$ matrix, you must know how to compute the determinant of an $(n - 1) \times (n - 1)$ matrix. At this point in the text, because you have learned the definition of the determinant of a 2×2 matrix, you can find the minors and cofactors of entries in a 3×3 matrix.

EXAMPLE 2   **Finding Minors and Cofactors**

For the matrix $A = \begin{bmatrix} 2 & 3 & 4 \\ 5 & -3 & -6 \\ 0 & 1 & 7 \end{bmatrix}$, find:

a. the minors M_{11}, M_{23}, and M_{32}

b. the cofactors A_{11}, A_{23}, and A_{32}

SOLUTION

a.   **(i)** To find M_{11}, delete the first row and first column of the matrix A.

$$\begin{bmatrix} 2 & 3 & 4 \\ 5 & -3 & -6 \\ 0 & 1 & 7 \end{bmatrix}$$

Now find the determinant of the resulting matrix.

$$M_{11} = \begin{vmatrix} -3 & -6 \\ 1 & 7 \end{vmatrix} = (-3)(7) - (-6)(1) = -21 + 6 = -15$$

(ii) To find the minor M_{23}, delete the second row and third column of the matrix A.

$$\begin{bmatrix} 2 & 3 & 4 \\ 5 & -3 & -6 \\ 0 & 1 & 7 \end{bmatrix}$$

Now find the determinant of the resulting matrix.

$$M_{23} = \begin{vmatrix} 2 & 3 \\ 0 & 1 \end{vmatrix} = (2)(1) - (3)(0) = 2$$

(iii) To find M_{32}, delete the third row and second column of A.

$$\begin{bmatrix} 2 & 3 & 4 \\ 5 & -3 & -6 \\ 0 & 1 & 7 \end{bmatrix}$$

So $M_{32} = \begin{vmatrix} 2 & 4 \\ 5 & -6 \end{vmatrix} = (2)(-6) - (4)(5) = -12 - 20 = -32.$

b. To find the cofactors, use the formula $A_{ij} = (-1)^{i+j} M_{ij}$.

$A_{11} = (-1)^{1+1} M_{11} = (1)(-15) = -15$ $(-1)^{1+1} = (-1)^2 = 1; M_{11} = -15$

$A_{23} = (-1)^{2+3} M_{23} = (-1)(2) = -2$ $(-1)^{2+3} = (-1)^5 = -1; M_{23} = 2$

$A_{32} = (-1)^{3+2} M_{32} = (-1)(-32) = 32$ $(-1)^{3+2} = (-1)^5 = -1; M_{32} = -32$

■ ■ ■

Practice Problem 2   For the matrix $A = \begin{bmatrix} 3 & -1 & 2 \\ 4 & 5 & 6 \\ 7 & 1 & 2 \end{bmatrix}$, find

a. the minors M_{11}, M_{23}, and M_{32}

b. the cofactors A_{11}, A_{23}, and A_{32} ■

3 Evaluate the determinant of an $n \times n$ matrix.

The Determinant of an $n \times n$ Matrix

There are several ways of defining the determinant of a square matrix. Here we give an *inductive definition*. This means that the definition of the determinant of a matrix of order n uses the determinants of the matrices of order $(n - 1)$.

---

n BY n DETERMINANT

Let A be a square matrix of order $n \geq 3$. The **determinant** of A is the sum of the entries in any row of A (or column of A) multiplied by their respective cofactors.

Applying the preceding definition to find the determinant of a matrix is called **expanding by cofactors**. Expanding by the cofactors of the ith row gives

$$\det(A) = a_{i1}A_{i1} + a_{i2}A_{i2} + \cdots + a_{in}A_{in}$$

Similarly, expanding by the cofactors of the jth column yields

$$\det(A) = a_{1j}A_{1j} + a_{2j}A_{2j} + \cdots + a_{nj}A_{nj}$$

We usually say "expanding by the ith row (column)" instead of "expanding by the cofactors of the ith row (column)."

EXAMPLE 3   **Evaluating a 3 by 3 Determinant**

Find the determinant of $A = \begin{bmatrix} 2 & 3 & 4 \\ 1 & -2 & 2 \\ 3 & 4 & -1 \end{bmatrix}$.

SOLUTION

Expanding by the first row, we have $|A| = a_{11}A_{11} + a_{12}A_{12} + a_{13}A_{13}$ where

$$A_{11} = (-1)^{1+1}M_{11} = (1)\begin{vmatrix} -2 & 2 \\ 4 & -1 \end{vmatrix}$$

$$A_{12} = (-1)^{1+2}M_{12} = (-1)\begin{vmatrix} 1 & 2 \\ 3 & -1 \end{vmatrix}$$

$$A_{13} = (-1)^{1+3}M_{13} = (1)\begin{vmatrix} 1 & -2 \\ 3 & 4 \end{vmatrix}$$

Thus,

$$\begin{vmatrix} 2 & 3 & 4 \\ 1 & -2 & 2 \\ 3 & 4 & -1 \end{vmatrix} = 2(1)\begin{vmatrix} -2 & 2 \\ 4 & -1 \end{vmatrix} + 3(-1)\begin{vmatrix} 1 & 2 \\ 3 & -1 \end{vmatrix} + 4(1)\begin{vmatrix} 1 & -2 \\ 3 & 4 \end{vmatrix}$$

$$= 2(2 - 8) - 3(-1 - 6) + 4(4 + 6) \qquad \text{Evaluate determinants of order 2.}$$

$$= -12 + 21 + 40 = 49 \qquad \text{Simplify.} \qquad ■ ■ ■$$

Practice Problem 3   Find the determinant of $A = \begin{bmatrix} 2 & -3 & 7 \\ -2 & -1 & 9 \\ 0 & 2 & -9 \end{bmatrix}$.   ■

TECHNOLOGY CONNECTION

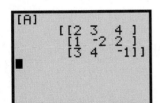 Use the Det option on a graphing calculator to find the determinant of a matrix.

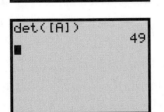

The definition says that we can expand the determinant by any row or column. If we expand by the third column in A in Example 3, we have

$$\begin{vmatrix} 2 & 3 & 4 \\ 1 & -2 & 2 \\ 3 & 4 & -1 \end{vmatrix} = 4(1) \begin{vmatrix} 1 & -2 \\ 3 & 4 \end{vmatrix} + 2(-1) \begin{vmatrix} 2 & 3 \\ 3 & 4 \end{vmatrix} + (-1)(1) \begin{vmatrix} 2 & 3 \\ 1 & -2 \end{vmatrix}$$

$$= 4(4 + 6) - 2(8 - 9) - 1(-4 - 3)$$

$$= 40 + 2 + 7 = 49$$

4 Apply Cramer's Rule.

Cramer's Rule

You have already learned several methods for solving a system of linear equations. Another method called **Cramer's Rule** uses determinants. To see a derivation of the rule, consider a simple system of two linear equations in two variables.

$$\begin{cases} a_1 x + b_1 y = c_1 & (1) \\ a_2 x + b_2 y = c_2 & (2) \end{cases}$$

We can solve this system by using the elimination method. Let's first eliminate y and solve for x.

$$\begin{array}{ll} a_1 b_2 x + b_1 b_2 y = c_1 b_2 & \text{Multiply equation (1) by } b_2. \\ \underline{-a_2 b_1 x - b_1 b_2 y = -c_2 b_1} & \text{Multiply equation (2) by } -b_1. \\ (a_1 b_2 - a_2 b_1)x = c_1 b_2 - c_2 b_1 & \text{Add to eliminate } y. \\ x = \dfrac{c_1 b_2 - c_2 b_1}{a_1 b_2 - a_2 b_1} & \text{Divide both sides by} \\ & a_1 b_2 - a_2 b_1 \text{ if } a_1 b_2 - a_2 b_1 \neq 0. \end{array}$$

Similarly, by eliminating x and solving for y, we have

$$y = \frac{a_1 c_2 - a_2 c_1}{a_1 b_2 - a_2 b_1} \qquad \text{if } a_1 b_2 - a_2 b_1 \neq 0$$

The solution of the system of equations (1) and (2) is therefore given by

$$x = \frac{c_1 b_2 - c_2 b_1}{a_1 b_2 - a_2 b_1} \qquad y = \frac{a_1 c_2 - a_2 c_1}{a_1 b_2 - a_2 b_1} \qquad \text{provided that } a_1 b_2 - a_2 b_1 \neq 0.$$

You can write the solution in the form of determinants

$$x = \frac{\begin{vmatrix} c_1 & b_1 \\ c_2 & b_2 \end{vmatrix}}{\begin{vmatrix} a_1 & b_1 \\ a_2 & b_2 \end{vmatrix}} \qquad y = \frac{\begin{vmatrix} a_1 & c_1 \\ a_2 & c_2 \end{vmatrix}}{\begin{vmatrix} a_1 & b_1 \\ a_2 & b_2 \end{vmatrix}} \qquad \text{provided that } D = \begin{vmatrix} a_1 & b_1 \\ a_2 & b_2 \end{vmatrix} \neq 0$$

Note that the denominator of both the x- and the y-values in the solution is the determinant D of the coefficient matrix of the linear system.

Suppose we write $D_x = \begin{vmatrix} c_1 & b_1 \\ c_2 & b_2 \end{vmatrix}$ and $D_y = \begin{vmatrix} a_1 & c_1 \\ a_2 & c_2 \end{vmatrix}$, which are obtained by replacing the column containing the coefficients of x in D and the column containing the coefficients of y in D, by the column of constants, respectively. Then the solution just found can be written as

$$x = \frac{D_x}{D} \qquad y = \frac{D_y}{D} \qquad \text{provided that } D \neq 0$$

Gabriel Cramer

(1704–1752)

Gabriel Cramer was one of three sons of a medical doctor in Geneva, Switzerland. Gabriel moved rapidly through his education in Geneva, and in 1722, while he was still 18 years old, he was awarded a PhD degree. Two years later he accepted the chair of mathematics at the Académie de Clavin in Geneva.

Cramer taught and traveled extensively and yet found time to produce articles and books covering a wide range of subjects. Although Cramer was not the first person to state the rule, it is universally known as "Cramer's Rule."

This representation of the solution is Cramer's Rule for solving a system of linear equations.

CRAMER'S RULE FOR SOLVING TWO EQUATIONS IN TWO VARIABLES

The system

$$\begin{cases} a_1x + b_1y = c_1 \\ a_2x + b_2y = c_2 \end{cases}$$

of two equations in two variables has a unique solution (x, y) given by

$$x = \frac{D_x}{D} \quad \text{and} \quad y = \frac{D_y}{D}$$

provided that $D \neq 0$, where

$$D = \begin{vmatrix} a_1 & b_1 \\ a_2 & b_2 \end{vmatrix}, \quad D_x = \begin{vmatrix} c_1 & b_1 \\ c_2 & b_2 \end{vmatrix}, \quad \text{and} \quad D_y = \begin{vmatrix} a_1 & c_1 \\ a_2 & c_2 \end{vmatrix}$$

EXAMPLE 4   **Using Cramer's Rule**

Use Cramer's Rule to solve the following system: $\begin{cases} 5x + 4y = 1 \\ 2x + 3y = 6 \end{cases}$

SOLUTION

Step 1   Form the determinant D of the coefficient matrix.

$$D = \begin{vmatrix} 5 & 4 \\ 2 & 3 \end{vmatrix} = 15 - 8 = 7$$

Since $D = 7 \neq 0$, the system has a unique solution.

Step 2   Replace the column of coefficients of x in D by the constant terms to get

$$D_x = \begin{vmatrix} 1 & 4 \\ 6 & 3 \end{vmatrix} = 3 - 24 = -21$$

Step 3   Replace the column of coefficients of y in D by the constant terms to get

$$D_y = \begin{vmatrix} 5 & 1 \\ 2 & 6 \end{vmatrix} = 30 - 2 = 28$$

Step 4   By Cramer's Rule,

$$x = \frac{D_x}{D} = \frac{-21}{7} = -3 \quad \text{and} \quad y = \frac{D_y}{D} = \frac{28}{7} = 4$$

The solution is $x = -3$ and $y = 4$, or $\{(-3, 4)\}$.

Check: You should check the solution in the original system.   ■ ■ ■

Practice Problem 4   Use Cramer's Rule to solve the following system:

$$\begin{cases} 2x + 3y = 7 \\ 5x + 9y = 4 \end{cases}$$

■

Cramer's Rule, as used in Example 4, can be generalized to a system of equations with more than two variables. We state the rule for three equations in three variables.

CRAMER'S RULE FOR SOLVING THREE EQUATIONS IN THREE VARIABLES

The system

$$\begin{cases} a_1x + b_1y + c_1z = k_1 \\ a_2x + b_2y + c_2z = k_2 \\ a_3x + b_3y + c_3z = k_3 \end{cases}$$

of three equations in three variables has a unique solution (x, y, z) given by

$$x = \frac{D_x}{D}, \quad y = \frac{D_y}{D}, \quad \text{and} \quad z = \frac{D_z}{D}$$

provided that $D \neq 0$, where

$$D = \begin{vmatrix} a_1 & b_1 & c_1 \\ a_2 & b_2 & c_2 \\ a_3 & b_3 & c_3 \end{vmatrix}, \quad D_x = \begin{vmatrix} k_1 & b_1 & c_1 \\ k_2 & b_2 & c_2 \\ k_3 & b_3 & c_3 \end{vmatrix}, \quad D_y = \begin{vmatrix} a_1 & k_1 & c_1 \\ a_2 & k_2 & c_2 \\ a_3 & k_3 & c_3 \end{vmatrix}, \quad \text{and} \quad D_z = \begin{vmatrix} a_1 & b_1 & k_1 \\ a_2 & b_2 & k_2 \\ a_3 & b_3 & k_3 \end{vmatrix}$$

STUDY TIP

When you apply Cramer's Rule, make sure that all linear equations are written in the form

$$ax + by + cz = k.$$

EXAMPLE 5 **Using Cramer's Rule**

Use Cramer's Rule to solve the system of equations.

$$\begin{cases} 7x + y + z - 1 = 0 \\ 5x + 3z - 2y = 4 \\ 4x - z + 3y = 0 \end{cases}$$

SOLUTION

First, rewrite the system of equations so that the terms on the left side of the equals sign are in the proper order and only the constant terms appear on the right side.

$$\begin{cases} 7x + y + z = 1 \\ 5x - 2y + 3z = 4 \\ 4x + 3y - z = 0 \end{cases}$$

Step 1 $D = \begin{vmatrix} 7 & 1 & 1 \\ 5 & -2 & 3 \\ 4 & 3 & -1 \end{vmatrix}$ D is the determinant of the coefficient matrix.

$$= 7\begin{vmatrix} -2 & 3 \\ 3 & -1 \end{vmatrix} - 5\begin{vmatrix} 1 & 1 \\ 3 & -1 \end{vmatrix} + 4\begin{vmatrix} 1 & 1 \\ -2 & 3 \end{vmatrix}$$ Expand by column 1.

$$= 7(2 - 9) - 5(-1 - 3) + 4(3 + 2)$$ Evaluate determinants of order 2.

$$= 7(-7) - 5(-4) + 4(5) = -9$$ Simplify.

Since $D \neq 0$, the system has a unique solution.

Step 2 $D_x = \begin{vmatrix} 1 & 1 & 1 \\ 4 & -2 & 3 \\ 0 & 3 & -1 \end{vmatrix}$ Replace the x-coefficients in the first column in D with the constants.

$= 1\begin{vmatrix} -2 & 3 \\ 3 & -1 \end{vmatrix} - 4\begin{vmatrix} 1 & 1 \\ 3 & -1 \end{vmatrix} + 0\begin{vmatrix} 1 & 1 \\ -2 & 3 \end{vmatrix}$ Expand by column 1.

$= 1(2 - 9) - 4(-1 - 3)$ Evaluate determinants of order 2.

$= 1(-7) - 4(-4) = 9$ Simplify.

Step 3 $D_y = \begin{vmatrix} 7 & 1 & 1 \\ 5 & 4 & 3 \\ 4 & 0 & -1 \end{vmatrix}$ Replace column 2 in D with the constants.

$= -1\begin{vmatrix} 5 & 3 \\ 4 & -1 \end{vmatrix} + 4\begin{vmatrix} 7 & 1 \\ 4 & -1 \end{vmatrix} - 0\begin{vmatrix} 7 & 1 \\ 5 & 3 \end{vmatrix}$ Expand by column 2 since it contains a zero.

$= -1(-5 - 12) + 4(-7 - 4)$ Evaluate the determinants.

$= 17 - 44 = -27$ Simplify.

Step 4 $D_z = \begin{vmatrix} 7 & 1 & 1 \\ 5 & -2 & 4 \\ 4 & 3 & 0 \end{vmatrix}$ Replace column 3 in D with the constants.

$= 1\begin{vmatrix} 5 & -2 \\ 4 & 3 \end{vmatrix} - 4\begin{vmatrix} 7 & 1 \\ 4 & 3 \end{vmatrix} + 0\begin{vmatrix} 7 & 1 \\ 5 & -2 \end{vmatrix}$ Expand by column 3.

$= (15 + 8) - 4(21 - 4)$ Evaluate the determinants.

$= 23 - 4(17) = -45$ Simplify.

Step 5   Cramer's Rule gives the following values:

$$x = \frac{D_x}{D} = \frac{9}{-9} = -1 \quad \text{From Steps 1 and 2}$$

$$y = \frac{D_y}{D} = \frac{-27}{-9} = 3 \quad \text{From Steps 1 and 3}$$

$$z = \frac{D_z}{D} = \frac{-45}{-9} = 5 \quad \text{From Steps 1 and 4}$$

The solution set is therefore $\{(-1, 3, 5)\}$.

Check: You should check the solution.   ■ ■ ■

Practice Problem 5   Use Cramer's Rule to solve the system.

$$\begin{cases} 3x + 2y + z = 4 \\ 4x + 3y + z = 5 \\ 5x + y + z = 9 \end{cases}$$

■

◆ **WARNING**   It is important to note that Cramer's Rule does not apply if $D = 0$. In this case, the system is either consistent and has dependent equations (has infinitely many solutions) or inconsistent (has no solution). When $D = 0$, use a different method for solving systems of equations to find the solution set.

SECTION 8.4 ■ Exercises

A EXERCISES Basic Skills and Concepts

1. The determinant of $\begin{bmatrix} a & b \\ c & d \end{bmatrix}$ is ___$ad - bc$___ .

2. The minor of an element is the determinant you get by deleting the ___row___ and the ___column___ containing that element.

3. To expand an $n \times n$ determinant, you multiply each element of some row (or column) by its ___cofactor___ and add the result.

4. A system of n linear equations in n variables has a unique solution provided the determinant of the _____ is not zero. coefficient matrix

5. *True or False* A 2 by 2 determinant is always positive. False

6. *True or False* To expand an $n \times n$ determinant, you multiply each element of some row by its minor and add the result. False

In Exercises 7–16, evaluate each determinant.

7. $\begin{vmatrix} 2 & 3 \\ 4 & 5 \end{vmatrix}$ -2

8. $\begin{vmatrix} 3 & -5 \\ 1 & 4 \end{vmatrix}$ 17

9. $\begin{vmatrix} 4 & -2 \\ 3 & -3 \end{vmatrix}$ -6

10. $\begin{vmatrix} -2 & \frac{1}{2} \\ 3 & 1 \end{vmatrix}$ $-\frac{7}{2}$

11. $\begin{vmatrix} -1 & -3 \\ -4 & -5 \end{vmatrix}$ -7

12. $\begin{vmatrix} \frac{1}{2} & \frac{1}{3} \\ \frac{1}{4} & \frac{1}{6} \end{vmatrix}$ 0

13. $\begin{vmatrix} \frac{3}{8} & \frac{1}{2} \\ -\frac{1}{9} & 5 \end{vmatrix}$ $\frac{139}{72}$

14. $\begin{vmatrix} -\frac{1}{2} & \frac{1}{3} \\ \frac{1}{4} & -\frac{1}{3} \end{vmatrix}$ $\frac{1}{12}$

15. $\begin{vmatrix} \sqrt{a} & \sqrt{b} \\ \sqrt{b} & \sqrt{a} \end{vmatrix}$ $a - b$

16. $\begin{vmatrix} \sqrt{3} & 1 \\ -1 & \sqrt{3} \end{vmatrix}$ 4

For Exercises 17–20, consider the matrix

$$A = \begin{bmatrix} 2 & -3 & 4 \\ 1 & -1 & 2 \\ 0 & 1 & 2 \end{bmatrix}$$

17. Find the minors.
   a. M_{21} **b.** M_{23} **c.** M_{32}

18. Find the cofactors.
   a. A_{21} **b.** A_{23} **c.** A_{32}

19. Find the minors.
   a. M_{11} **b.** M_{22} **c.** M_{31}

20. Find the cofactors.
   a. A_{11} **b.** A_{22} **c.** A_{31}

Answers:
17. **a.** -10 **b.** 2 **c.** 0 18. **a.** 10 **b.** -2 **c.** 0
19. **a.** -4 **b.** 4 **c.** -2 20. **a.** -4 **b.** 4 **c.** -2

In Exercises 21–36, evaluate each determinant.

21. $\begin{vmatrix} 1 & 0 & -1 \\ 0 & 2 & 2 \\ -1 & 0 & 0 \end{vmatrix}$ -2

22. $\begin{vmatrix} 2 & 0 & \frac{1}{2} \\ 1 & 0 & 2 \\ 4 & 0 & -5 \end{vmatrix}$ 0

23. $\begin{vmatrix} 1 & 2 & 3 \\ 0 & 3 & 4 \\ 0 & 0 & 4 \end{vmatrix}$ 12

24. $\begin{vmatrix} 2 & 3 & 4 \\ 0 & -4 & 6 \\ 0 & 0 & -5 \end{vmatrix}$ 40

25. $\begin{vmatrix} 1 & 0 & 0 \\ 2 & 0 & 0 \\ 3 & 4 & 5 \end{vmatrix}$ 0

26. $\begin{vmatrix} -1 & 0 & 0 \\ 3 & 2 & 0 \\ 4 & 5 & -3 \end{vmatrix}$ 6

27. $\begin{vmatrix} 1 & 6 & 0 \\ 2 & 5 & 3 \\ 3 & 4 & 0 \end{vmatrix}$ 42

28. $\begin{vmatrix} 1 & 0 & 2 \\ 0 & 5 & 7 \\ 3 & 1 & 0 \end{vmatrix}$ -37

29. $\begin{vmatrix} 3 & 4 & 1 \\ 1 & 4 & 3 \\ 4 & 3 & 1 \end{vmatrix}$ 16

30. $\begin{vmatrix} 3 & 1 & -2 \\ 4 & 0 & -4 \\ 2 & -1 & -3 \end{vmatrix}$ 0

31. $\begin{vmatrix} 0 & 1 & 6 \\ 1 & 0 & 4 \\ 8 & 3 & 1 \end{vmatrix}$ 49

32. $\begin{vmatrix} 0 & 2 & 3 \\ 1 & 0 & 1 \\ 3 & 2 & 0 \end{vmatrix}$ 12

33. $\begin{vmatrix} a & b & 0 \\ 0 & a & b \\ b & 0 & a \end{vmatrix}$ $a^3 + b^3$

34. $\begin{vmatrix} 0 & z & y \\ z & 0 & x \\ y & x & 0 \end{vmatrix}$ $2xyz$

35. $\begin{vmatrix} a & b & c \\ c & a & b \\ b & c & a \end{vmatrix}$ $a^3 + b^3 + c^3 - 3abc$

36. $\begin{vmatrix} u & v & w \\ w & v & u \\ u & v & w \end{vmatrix}$ 0

In Exercises 37–46, use Cramer's Rule (if applicable) to solve each system of equations.

37. $\begin{cases} x + y = 8 \\ x - y = -2 \end{cases}$ $\{(3, 5)\}$

38. $\begin{cases} 4x + 3y = -1 \\ 2x - 5y = 9 \end{cases}$ $\left\{ \left(\frac{11}{13}, -\frac{19}{13} \right) \right\}$

39. $\begin{cases} 5x + 3y = 11 \\ 2x + y = 4 \end{cases}$ $\{(1, 2)\}$

40. $\begin{cases} 2x - 7y = 13 \\ 5x + 6y = 9 \end{cases}$ $\{(3, -1)\}$

41. $\begin{cases} 2x + 9y = 4 \\ 3x - 2y = 6 \end{cases}$ $\{(2, 0)\}$

42. $\begin{cases} 5x + 3y = 1 \\ 2x - 5y = -12 \end{cases}$ $\{(-1, 2)\}$

43. $\begin{cases} 2x - 3y = 4 \\ 4x - 6y = 8 \end{cases}$ $\left\{ \left(\frac{3}{2}y + 2, y \right) \right\}$

44. $\begin{cases} 3x + y = 2 \\ 6x + 2y = 4 \end{cases}$ $\{(x, -3x + 2)\}$

45. $\begin{cases} \dfrac{2}{x} + \dfrac{3}{y} = 2 \\ \dfrac{5}{x} + \dfrac{8}{y} = \dfrac{31}{6} \end{cases}$ $\{(2, 3)\}$

46. $\begin{cases} \dfrac{3}{x} - \dfrac{6}{y} = 2 \\ \dfrac{4}{x} + \dfrac{7}{y} = -3 \end{cases}$ $\left\{ \left(-\frac{45}{4}, -\frac{45}{17} \right) \right\}$

$\left[\textit{Hint: For Exercises 45 and 46, let } u = \dfrac{1}{x} \textit{ and } v = \dfrac{1}{y}. \right]$

In Exercises 47–56, solve each system of equations by means of Cramer's Rule. Use technology to compute the determinants involved.

47. $\begin{cases} x - 2y + z = -1 \\ 3x + y - z = 4 \\ y + z = 1 \end{cases}$ $\{(1, 1, 0)\}$

48. $\begin{cases} x + 3y = 4 \\ y + 3z = 7 \\ z + 4x = 6 \end{cases}$ $\{(1, 1, 2)\}$

49. $\begin{cases} x + y - z = -3 \\ 2x + 3y + z = 2 \\ 2y + z = 1 \end{cases}$ **50.** $\begin{cases} 3x + y + z = 10 \\ x + y - z = 0 \\ 5x - 9y = 1 \end{cases}$

51. $\begin{cases} x + y + z = 3 \\ 2x - 3y + 5z = 4 \\ x + 2y - 4z = -1 \end{cases}$ **52.** $\begin{cases} x + 2y - z = 3 \\ 3x - y + z = 8 \\ x + y + z = 0 \end{cases}$

53. $\begin{cases} 2x - 3y + 5z = 11 \\ 3x + 5y - 2z = 7 \\ x + 2y - 3z = -4 \end{cases}$ **54.** $\begin{cases} x - 3y - 1 = 0 \\ 2x - y - 4z = 2 \\ y + 2z - 4 = 0 \end{cases}$

55. $\begin{cases} 5x + 2y + z = 12 \\ 2x + y + 3z = 13 \\ 3x + 2y + 4z = 19 \end{cases}$ **56.** $\begin{cases} 2x + y + z = 7 \\ 3x - y - z = -2 \\ x + 2y - 3z = -4 \end{cases}$

B EXERCISES Applying the Concepts

Assume that the *area of a triangle* with vertices (x_1, y_1), (x_2, y_2), and (x_3, y_3) is equal to the absolute value of D, where

$$D = \frac{1}{2} \begin{vmatrix} x_1 & y_1 & 1 \\ x_2 & y_2 & 1 \\ x_3 & y_3 & 1 \end{vmatrix}$$

In Exercises 57–60, use determinants to find the area of each triangle.

57. Area of a triangle. The triangle with vertices $(1, 2)$, $(-3, 4)$, and $(4, 6)$. 11

58. Area of a triangle. The triangle with vertices $(3, 1)$, $(4, 2)$, and $(5, 4)$. $\frac{1}{2}$

59. Area of a triangle. The triangle with vertices $(-2, 1)$, $(-3, -5)$, and $(2, 4)$. 10.5

60. Area of a triangle. The triangle with vertices $(-1, 2)$ and $(2, 4)$ and the origin. 4

Collinearity. Three points $A(x_1, y_1)$, $B(x_1, y_2)$, and $C(x_3, y_3)$ are *collinear* (lie on the same line) if and only if the area of the triangle ABC is 0. In terms of determinants, three points (x_1, y_1), (x_2, y_2), and (x_3, y_3) are collinear if and only if

$$\begin{vmatrix} x_1 & y_1 & 1 \\ x_2 & y_2 & 1 \\ x_3 & y_3 & 1 \end{vmatrix} = 0$$

In Exercises 61–64, use determinants to find whether the three given points are collinear.

61. $(0, 3)$, $(-1, 1)$, and $(2, 7)$ Yes

62. $(2, 0)$, $\left(1, \frac{1}{2}\right)$, and $(4, -1)$ Yes

63. $(0, -4)$, $(3, -2)$, and $(1, -4)$ No

64. $\left(0, -\frac{1}{4}\right)$, $(1, -1)$, and $(2, -2)$ No

Answers:
49. $\{(1, -1, 3)\}$ **50.** $\{(2, 1, 3)\}$ **51.** $\{(1, 1, 1)\}$ **52.** $\{(3, -1, -2)\}$
53. $\{(1, 2, 3)\}$ **54.** $\left\{\left(\frac{31}{7}, \frac{8}{7}, \frac{10}{7}\right)\right\}$ **55.** $\{(1, 2, 3)\}$ **56.** $\{(1, 2, 3)\}$

Equation of a line. As explained in the discussion of collinearity, three points (x, y), (x_1, y_1), and (x_2, y_2) all lie on the same line if and only if

$$D = \begin{vmatrix} x & y & 1 \\ x_1 & y_1 & 1 \\ x_2 & y_2 & 1 \end{vmatrix} = 0$$

In other words, a point $P(x, y)$ lies on the line passing through the points (x_1, y_1) and (x_2, y_2) if and only if the determinant $D = 0$. Therefore, the equation

$$\begin{vmatrix} x & y & 1 \\ x_1 & y_1 & 1 \\ x_2 & y_2 & 1 \end{vmatrix} = 0$$

is an equation of the line passing through the points (x_1, y_1) and (x_2, y_2).

In Exercises 65–68, use determinants to write an equation of the line passing through the given points. Then evaluate the determinant and write the equation of the line in slope–intercept form.

65. $(-1, -1)$, $(1, 3)$ $y = 2x + 1$ **66.** $(0, 4)$, $(1, 1)$ $y = -3x + 4$

67. $\left(0, \frac{1}{3}\right)$, $(1, 1)$ $y = \frac{2}{3}x + \frac{1}{3}$ **68.** $\left(1, -\frac{1}{3}\right)$, $(2, 0)$

$$y = \frac{1}{3}x - \frac{2}{3}$$

C EXERCISES Beyond the Basics

In Exercises 69–73, we list some properties of determinants. These properties are true for $n \times n$ matrices. Verify the properties for $n = 2$ or $n = 3$.

69. If each entry in any row or each entry in any column of a matrix A is zero, then $|A| = 0$. Verify each of the following.

a. $\begin{vmatrix} 0 & 0 \\ 2 & 5 \end{vmatrix} = 0$ b. $\begin{vmatrix} 1 & 0 \\ 3 & 0 \end{vmatrix} = 0$

c. $\begin{vmatrix} 1 & -2 & 3 \\ 0 & 0 & 0 \\ 4 & 5 & -7 \end{vmatrix} = 0$ d. $\begin{vmatrix} 1 & 0 \\ 4 & 5 & 0 \\ 6 & -7 & 0 \\ 8 & 15 & 0 \end{vmatrix} = 0$

70. If a matrix B is obtained from matrix A by interchanging two rows (or columns), then $|B| = -|A|$. Verify each of the following.

a. $\begin{vmatrix} 2 & 3 \\ 4 & 5 \end{vmatrix} = -\begin{vmatrix} 4 & 5 \\ 2 & 3 \end{vmatrix}$

b. $\begin{vmatrix} -5 & 3 \\ 2 & -4 \end{vmatrix} = -\begin{vmatrix} 3 & -5 \\ -4 & 2 \end{vmatrix}$

c. $\begin{vmatrix} 1 & 3 & 5 \\ 0 & 1 & 2 \\ 3 & -1 & 4 \end{vmatrix} = -\begin{vmatrix} 1 & 3 & 5 \\ 3 & -1 & 4 \\ 0 & 1 & 2 \end{vmatrix}$

71. If two rows (or columns) of a matrix A have corresponding entries that are equal, then $|A| = 0$. Verify the following.

a. $\begin{vmatrix} 2 & 3 & 5 \\ -1 & 4 & 8 \\ 2 & 3 & 5 \end{vmatrix} = 0$ b. $\begin{vmatrix} 3 & 4 & 3 \\ 1 & 2 & 1 \\ -1 & 6 & -1 \end{vmatrix} = 0$

72. If a matrix B is obtained from a matrix A by multiplying every entry of one row (or one column) by a real number c, then $|B| = c|A|$. Verify each of the following.

a. $\begin{vmatrix} -1 & 2 & 3 \\ 1 \cdot 5 & 4 \cdot 5 & 8 \cdot 5 \\ 2 & 3 & 6 \end{vmatrix} = 5 \begin{vmatrix} -1 & 2 & 3 \\ 1 & 4 & 8 \\ 2 & 3 & 6 \end{vmatrix}$

b. $\begin{vmatrix} 1 & 2 & 3 \\ -1 & 5 & 6 \\ 0 & 2 & 9 \end{vmatrix} = 3 \begin{vmatrix} 1 & 2 & 1 \\ -1 & 5 & 2 \\ 0 & 2 & 3 \end{vmatrix}$

73. If a matrix B is obtained from a matrix A by replacing any row (or column) of A by the sum of that row (or column) and c times another row (or column), then $|B| = |A|$.

Verify that the following two determinants are equal:

$$\begin{vmatrix} 2 & -3 & 4 \\ -4 & 7 & -8 \\ 5 & -1 & 3 \end{vmatrix} \xrightarrow{2R_1 + R_2 \to R_2} \begin{vmatrix} 2 & -3 & 4 \\ 0 & 1 & 0 \\ 5 & -1 & 3 \end{vmatrix}$$

74. Evaluate the determinant

$$D = \begin{vmatrix} 1 & 2 & 3 & 3 \\ 4 & 5 & 2 & 1 \\ 1 & 2 & 5 & 3 \\ 2 & 4 & 3 & 7 \end{vmatrix}$$

by following these steps:
(i) $(-1)R_1 + R_3 \to R_3$
(ii) Expand by the cofactors of row 3.
(iii) In **(ii)**, you will have a determinant of order 3. Evaluate it to obtain the value of D. -6

In Exercises 75–78, use the properties of determinants in Exercises 69–73 to obtain three zeros in one row (or column) and then evaluate the given determinant by evaluating the resulting determinant of order 3.

75. $\begin{vmatrix} 2 & 0 & -3 & 4 \\ 0 & 1 & 0 & 5 \\ 5 & 0 & -9 & 8 \\ 1 & 2 & 0 & 7 \end{vmatrix}$ 21

76. $\begin{vmatrix} -3 & 1 & 2 & 3 \\ 0 & -2 & 4 & 7 \\ -9 & 3 & 5 & 2 \\ 2 & -4 & 3 & -12 \end{vmatrix}$ 32

77. $\begin{vmatrix} 5 & 7 & 1 & 2 \\ 6 & 8 & 9 & 3 \\ 24 & 22 & 6 & 10 \\ 21 & 17 & 7 & 10 \end{vmatrix}$ 476

78. $\begin{vmatrix} 1 & 2 & 3 & 0 \\ 2 & -3 & 1 & 0 \\ -1 & -3 & 4 & 5 \\ 1 & 3 & 7 & 4 \end{vmatrix}$ −101

In Exercises 79–86, solve each equation for x.

79. $\begin{vmatrix} 3 & 2 \\ 6 & x \end{vmatrix} = 0$ 4

80. $\begin{vmatrix} 3 & 2 \\ x & 12 \end{vmatrix} = 0$ 18

81. $\begin{vmatrix} x & -2 \\ 1 & x-1 \end{vmatrix} = 0$

82. $\begin{vmatrix} 2x+7 & 4 \\ 1 & x \end{vmatrix} = 0$ $\frac{1}{2}, -4$

83. $\begin{vmatrix} 1 & -3 & 1 \\ 4 & 7 & x \\ 0 & 2 & 2 \end{vmatrix} = 0$ 23

84. $\begin{vmatrix} x & x+1 & x+2 \\ 2 & 3 & -1 \\ 3 & -2 & 4 \end{vmatrix} = 0$ $-\frac{37}{14}$

85. $\begin{vmatrix} x & 0 & 1 \\ 0 & x & 0 \\ 1 & 0 & x \end{vmatrix} = 0$ −1, 0, 1

86. $\begin{vmatrix} x & 2 & 3 \\ x & x & 1 \\ 2 & 0 & 1 \end{vmatrix} = -8$ 2, 6

In Exercises 87 and 88, use the formula on page 595 for the area of a triangle.

87. If the area of the triangle formed by the points $(-4, 2)$, $(0, k)$, and $(-2, k)$ is 28 square units, find k. 30

88. If the area of the triangle formed by the points $(2, 3)$, $(1, 2)$, and $(-2, k)$ is 3 square units, find k. −7

Critical Thinking

89. Solve the system by Cramer's Rule.

$$\begin{cases} \dfrac{1}{x-2} + \dfrac{3}{y+1} = 13 \\ \dfrac{4}{x-2} - \dfrac{5}{y+1} = 1 \end{cases} \quad \left\{ \left(\dfrac{9}{4}, -\dfrac{2}{3} \right) \right\}$$

[*Hint:* Do not try to simplify the equations by clearing the denominators because the resulting equations will not be linear. Instead, let $u = \dfrac{1}{x-2}$ and $v = \dfrac{1}{y+1}$. Solve the system for u and v and then use this solution to solve for x and y.]

In Exercises 90 and 91, use Cramer's Rule to solve each system of equations.

90. $\begin{cases} \dfrac{6}{x+1} + \dfrac{4}{y-1} = 7 \\ \dfrac{8}{x+1} + \dfrac{5}{y-1} = 9 \end{cases}$ $\{(1, 2)\}$

91. $\begin{cases} \dfrac{1}{3^x} + \dfrac{4}{4^y} = 25 \\ \dfrac{2}{3^x} - \dfrac{1}{4^y} = 14 \end{cases}$ $\{(-2, -1)\}$

Answers:
81. $\dfrac{1}{2} \pm i\dfrac{\sqrt{7}}{2}$

SUMMARY ▪ Definitions, Concepts, and Formulas

8.1 Matrices and Systems of Equations

i. Matrix. An $m \times n$ matrix is a rectangular array of numbers with m rows and n columns.

ii. Augmented matrix. The augmented matrix of a system of linear equations is a matrix whose entries are the coefficients of the variables in the system, together with the constants.

iii. Elementary row operations. For an augmented matrix of a system of linear equations, the following row operations will result in the matrix of an equivalent system:
1. Interchange any two rows.
2. Multiply any row by a nonzero constant.
3. Add a multiple of one row to another row.

iv. Row-echelon form and reduced row-echelon form. The elementary row operations are used to transform an augmented matrix to row-echelon form or reduced row-echelon form. (See page 550.)

v. Gaussian elimination is the method of transforming the augmented matrix to row-echelon form and then using back-substitution to find the solution set of a system of linear equations.

vi. Gauss–Jordan elimination. If you continue the Gaussian elimination procedure until a reduced row-echelon form is obtained, the procedure is called Gauss–Jordan elimination.

8.2 Matrix Algebra

i. Equality of matrices. Two matrices $A = [a_{ij}]$ and $B = [b_{ij}]$ are equal if **(i)** A and B have the same order $m \times n$ and **(ii)** all corresponding entries are equal.

ii. Matrix addition. The matrices A and B can be added or subtracted if they have the same order. Their sum or difference can be obtained by adding or subtracting the corresponding entries.

iii. Scalar multiplication. The product of a matrix by a real number (scalar) is obtained by multiplying every entry of the matrix by the real number.

iv. Matrix multiplication. The product AB of two matrices is defined only if the number of columns of A is equal to the number of rows of B. If A is an $m \times p$ matrix and B is a $p \times n$ matrix, then the product AB is an $m \times n$ matrix. The (i, j)th entry of AB is the sum of the products of the corresponding entries in the ith row of A and the jth column of B.

v. The identity matrix. The $n \times n$ identity matrix denoted by I_n or I is a matrix that has 1s on the main diagonal and 0s elsewhere. If A is an $n \times n$ matrix, then $AI = IA = A$.

8.3 The Inverse Matrix

i. Inverse of a matrix. Let A be an $n \times n$ matrix and I be the $n \times n$ identity matrix. If there is an $n \times n$ matrix B such that $AB = BA = I$, then A is **invertible** and B is called the **inverse** of A. We write $B = A^{-1}$.

ii. Finding the inverse. A procedure for finding A^{-1} (if it exists) is given on page 576.

iii. Inverse of a 2 × 2 matrix. The matrix $A = \begin{bmatrix} a & b \\ c & d \end{bmatrix}$ has an inverse if and only if $ad - bc \neq 0$. Moreover, if $ad - bc \neq 0$, then

$$A^{-1} = \frac{1}{ad - bc} \begin{bmatrix} d & -b \\ -c & a \end{bmatrix}.$$

iv. If A is invertible, then the solution of the matrix equation $AX = B$ is $X = A^{-1}B$.

8.4 Determinants and Cramer's Rule

i. Determinant. The determinant of a square matrix A is a real number denoted by $\det(A)$ or $|A|$.

ii. The determinant of a 2 × 2 matrix $A = \begin{bmatrix} a & b \\ c & d \end{bmatrix}$ is defined by $\det(A) = \begin{vmatrix} a & b \\ c & d \end{vmatrix} = ad - bc$.

iii. The determinant of a matrix of order $n \geq 3$ can be found by expanding by cofactors. (See page 589.)

iv. Cramer's Rule gives formulas for solving a square system of linear equations for which the determinant of the coefficient matrix is nonzero. (See page 591 for solving a system of two equations in two variables and page 592 for solving a system of three equations in three variables.)

REVIEW EXERCISES

Basic Skills and Concepts

In Exercises 1–4, determine the order of each matrix.

1. $\begin{bmatrix} -1 & 2 & 3 & 4 \end{bmatrix}$ 1×4 **2.** $\begin{bmatrix} 5 \end{bmatrix}$ 1×1

3. $\begin{bmatrix} 2 & 3 \\ -1 & 2 \\ 0 & 1 \end{bmatrix}$ 3×2 **4.** $\begin{bmatrix} 1 & -1 & 2 & -4 \\ 5 & 4 & 3 & 2 \end{bmatrix}$ 2×4

5. Let A be the matrix of Exercise 4. Identify the entries $a_{12}, a_{14}, a_{23},$ and a_{21}. $a_{12} = -1, a_{14} = -4, a_{23} = 3, a_{21} = 5$

6. Write the 2×3 matrix A with entries $a_{22} = 3, a_{21} = 4, a_{12} = 2, a_{13} = 5, a_{11} = -1,$ and $a_{23} = 1.$ †

In Exercises 7 and 8, write the augmented matrix for each system of linear equations.

7. $\begin{cases} 2x - 3y = 7 \\ 3x + y = 6 \end{cases}$ †   **8.** $\begin{cases} x + y - z = 6 \\ 2x - 3y - 2z = 2 \\ 5x - 3y + z = 8 \end{cases}$ †

In Exercises 9 and 10, convert each matrix to row-echelon form.

9. $\begin{bmatrix} 0 & 1 & 2 & 1 \\ 2 & 0 & 3 & 4 \\ 1 & -2 & 1 & 7 \end{bmatrix}$ †   **10.** $\begin{bmatrix} 3 & -1 & 2 & 12 \\ 1 & 1 & -1 & 2 \\ 1 & 2 & 1 & 7 \end{bmatrix}$ †

In Exercises 11 and 12, convert each matrix to reduced row-echelon form.

11. $\begin{bmatrix} 3 & 1 & 3 & 1 \\ 2 & 1 & 1 & 1 \\ 1 & -1 & -1 & 0 \end{bmatrix}$ †   **12.** $\begin{bmatrix} 3 & 4 & -4 & 2 \\ 2 & 1 & -2 & 1 \\ 1 & 1 & 2 & 1 \end{bmatrix}$ †

In Exercises 13–16, solve each system of equations by Gaussian elimination.

13. $\begin{cases} x + y - z = 0 \\ 2x + y - 2z = 3 \\ 3x - 2y + 3z = 9 \end{cases}$ †   **14.** $\begin{cases} x - y + 2z = 1 \\ x + 3y - z = 6 \\ 2x + y - 3z = 1 \end{cases}$ †

15. $\begin{cases} x - 2y + 3z = -2 \\ 2x - 3y + z = 9 \\ 3x - y + 2z = 5 \end{cases}$ **16.** $\begin{cases} 2x + y + z = 7 \\ x + 2y + z = 3 \\ x + 2y + 2z = 6 \end{cases}$ †

$\{(3, -2, -3)\}$

In Exercises 17–20, solve each system of equations by Gauss–Jordan elimination.

17. $\begin{cases} x - 2y - 2z = 11 \\ 3x + 4y - z = -2 \\ 4x + 5y + 7z = 7 \end{cases}$ **18.** $\begin{cases} x - y - 9z = 1 \\ 3x + 2y - z = 2 \\ 4x + 3y + 3z = 0 \end{cases}$

19. $\begin{cases} 2x - y + 3z = 4 \\ x + 3y + 3z = -2 \\ 3x + 2y - 6z = 6 \end{cases}$ †   **20.** $\begin{cases} x - y - 9z = 1 \\ 3x + 2y - z = 2 \\ 4x + 3y + 3z = 0 \end{cases}$

21. Find x and y if $\begin{bmatrix} x - y & 0 \\ 1 & x + y \end{bmatrix} = \begin{bmatrix} 1 & 0 \\ 1 & 3 \end{bmatrix}.$ $\{(2, 1)\}$

22. Find x and y if $\{(-1, 2)\}$

$\begin{bmatrix} 2x + 3y & -1 & -2 \\ 0 & 1 & x - y \\ 2 & 3x + 4y & 5 \end{bmatrix} = \begin{bmatrix} 4 & -1 & -2 \\ 0 & 1 & -3 \\ 2 & 5 & 5 \end{bmatrix}$

23. Let $A = \begin{bmatrix} 1 & 2 \\ -3 & 4 \end{bmatrix}$ and $B = \begin{bmatrix} 2 & -3 \\ -5 & 6 \end{bmatrix}.$ Find each of the following.

a. $A + B$ †   **b.** $A - B$ †   **c.** $2A$ †
d. $-3B$ †   **e.** $2A - 3B$ †

24. Let $A = \begin{bmatrix} 2 & 0 & -1 \\ 1 & 2 & 2 \\ -2 & 0 & 3 \end{bmatrix}$ and $B = \begin{bmatrix} 0 & 1 & -1 \\ 1 & -1 & 0 \\ -1 & 0 & 1 \end{bmatrix}.$

Find each of the following.

a. $A + B$ †   **b.** $A - B$ †   **c.** $2A + 3B$ †

25. For the matrices A and B of Exercise 23, solve the matrix equation $3A + 2B - 3X = \mathbf{0}$ for X. †

26. For the matrices A and B of Exercise 24, solve the matrix equation $A - 2X + 2B = \mathbf{0}$ for X. †

In Exercises 27–30, find the following products if possible.
a. AB **b.** BA

27. $A = \begin{bmatrix} 0 & 1 \\ 2 & 3 \end{bmatrix},$ $B = \begin{bmatrix} -1 & -1 \\ -3 & 4 \end{bmatrix}$ †

28. $A = \begin{bmatrix} 0 & 1 & 2 \\ -1 & 0 & 1 \end{bmatrix},$ $B = \begin{bmatrix} 1 & 2 \\ 3 & -1 \\ 0 & 4 \end{bmatrix}$ †

29. $A = \begin{bmatrix} 1 & 2 & -1 \end{bmatrix},$ $B = \begin{bmatrix} 2 \\ 3 \\ 1 \end{bmatrix}$ †

30. $A = \begin{bmatrix} 1 & 0 \\ 2 & -1 \\ 3 & -2 \end{bmatrix},$ $B = \begin{bmatrix} 1 & 2 & 3 & 4 \\ 2 & -1 & 2 & 3 \end{bmatrix}$ †

31. Find a matrix $A = \begin{bmatrix} x & y \\ z & w \end{bmatrix}$ such that

$$\begin{bmatrix} 1 & 2 \\ 3 & -1 \end{bmatrix} A = \begin{bmatrix} 5 & 6 \\ 1 & -3 \end{bmatrix} \begin{bmatrix} 1 & 0 \\ 2 & 3 \end{bmatrix}$$

32. Find a matrix $B = \begin{bmatrix} x & y \\ z & w \end{bmatrix}$ such that

$$B \begin{bmatrix} 1 & 2 \\ 3 & -1 \end{bmatrix} = \begin{bmatrix} 5 & 6 \\ 1 & -3 \end{bmatrix} \begin{bmatrix} \frac{23}{7} & \frac{4}{7} \\ -\frac{8}{7} & \frac{5}{7} \end{bmatrix}$$

Compare your answers for Exercises 31 and 32.

33. Find a matrix $A = \begin{bmatrix} x & y \\ z & w \end{bmatrix}$ such that

$$\begin{bmatrix} 5 & 3 \\ 4 & 2 \end{bmatrix} A = \begin{bmatrix} 1 & 0 \\ 0 & 1 \end{bmatrix} \begin{bmatrix} -1 & \frac{3}{2} \\ 2 & -\frac{5}{2} \end{bmatrix}$$

†Due to space constrictions, answers to these exercises may be found in the Answers beginning on page A–1 in the back of the book.

Answers:

17. $\{(5, -4, 1)\}$ **18.** $\{(-3, 5, -1)\}$ **20.** $\{(-3, 5, -1)\}$

34. Find a matrix $A = \begin{bmatrix} x & y \\ z & w \end{bmatrix}$ such that

$$A\begin{bmatrix} 5 & 3 \\ 4 & 2 \end{bmatrix} = \begin{bmatrix} 1 & 0 \\ 0 & 1 \end{bmatrix}\begin{bmatrix} -1 & \frac{3}{2} \\ 2 & -\frac{5}{2} \end{bmatrix}$$

Compare your answers for Exercises 33 and 34.

35. Let $A = \begin{bmatrix} 0 & 1 \\ -1 & 3 \end{bmatrix}$.
   a. Find A^{-1}. (Use the formula on page 578.) †
   b. Find $(A^2)^{-1}$. †
   c. Find $(A^{-1})^2$. †
   d. Use your answers in parts (b) and (c) to show that $(A^{-1})^2$ is the inverse of A^2.

36. Repeat Exercise 35 for $A = \begin{bmatrix} 3 & 4 \\ 2 & 3 \end{bmatrix}$. †

In Exercises 37–40, show that B is the inverse of A.

37. $A = \begin{bmatrix} 7 & 6 \\ 6 & 5 \end{bmatrix}$, $B = \begin{bmatrix} -5 & 6 \\ 6 & -7 \end{bmatrix}$

38. $A = \begin{bmatrix} -1 & 3 \\ 1 & -4 \end{bmatrix}$, $B = \begin{bmatrix} -4 & -3 \\ -1 & -1 \end{bmatrix}$

39. $A = \begin{bmatrix} 2 & -1 & 0 \\ 1 & 0 & 4 \\ 1 & -1 & 1 \end{bmatrix}$, $B = \frac{1}{5}\begin{bmatrix} 4 & 1 & -4 \\ 3 & 2 & -8 \\ -1 & 1 & 1 \end{bmatrix}$

40. $A = \begin{bmatrix} 2 & 0 & 1 \\ 0 & -1 & 2 \\ 1 & 0 & 1 \end{bmatrix}$, $B = \begin{bmatrix} 1 & 0 & -1 \\ -2 & -1 & 4 \\ -1 & 0 & 2 \end{bmatrix}$

In Exercises 41–44, find the inverse (if it exists) of each matrix.

41. $\begin{bmatrix} 3 & 1 \\ 2 & 4 \end{bmatrix}$ †

42. $\begin{bmatrix} 3 & -4 \\ 1 & 2 \end{bmatrix}$ †

43. $\begin{bmatrix} 1 & 2 & -2 \\ -1 & 3 & 0 \\ 0 & -2 & 1 \end{bmatrix}$ †

44. $\begin{bmatrix} 1 & 2 & 1 \\ 2 & 2 & 4 \\ 0 & 0 & 3 \end{bmatrix}$ †

In Exercises 45–50, use an inverse matrix (if possible) to solve each system of linear equations.

45. $\begin{cases} x + 3y = 7 \\ 2x + 5y = 4 \end{cases}$ $\{(-23, 10)\}$

46. $\begin{cases} 3x + 5y = 4 \\ 2x + 4y = 5 \end{cases}$ $\left\{\left(-\frac{9}{2}, \frac{7}{2}\right)\right\}$

47. $\begin{cases} x + 3y + 3z = 3 \\ x + 4y + 3z = 5 \\ x + 3y + 4z = 6 \end{cases}$ †

48. $\begin{cases} x + 2y + 3z = 6 \\ 2x + 4y + 5z = 8 \\ 3x + 5y + 6z = 10 \end{cases}$ †

49. $\begin{cases} x - y + z = 3 \\ 4x + 2y = 5 \\ 7x - y - z = 6 \end{cases}$ †

50. $\begin{cases} x + 2y + 4z = 7 \\ 4x + 3y - 2z = 6 \\ x - 3z = 4 \end{cases}$ $\{(91, -100, 29)\}$

In Exercises 51–54, find the determinant of each matrix.

51. $\begin{bmatrix} 2 & -5 \\ 3 & 4 \end{bmatrix}$ 23

52. $\begin{bmatrix} -1 & 4 \\ 11 & -3 \end{bmatrix}$ −41

53. $\begin{bmatrix} 12 & 21 \\ 4 & 7 \end{bmatrix}$ 0

54. $\begin{bmatrix} -7 & 9 \\ -4 & 5 \end{bmatrix}$ 1

In Exercises 55 and 56, find (a) the minors M_{12}, M_{23}, and M_{22} and (b) the cofactors A_{12}, A_{23}, and A_{22} of the given matrix A.

55. $A = \begin{bmatrix} 4 & -1 & 2 \\ -2 & -3 & 5 \\ 0 & 2 & -4 \end{bmatrix}$ † **56.** $A = \begin{bmatrix} 1 & 2 & 3 \\ 4 & 5 & 6 \\ 7 & 8 & 9 \end{bmatrix}$ †

In Exercises 57 and 58, find the determinant of each matrix A by expanding (a) by the second row and (b) by the third column.

57. $A = \begin{bmatrix} 1 & -2 & 3 \\ 4 & -1 & -2 \\ -2 & 1 & 5 \end{bmatrix}$ 35 **58.** $A = \begin{bmatrix} 1 & 4 & 3 \\ 6 & 8 & 10 \\ 2 & 5 & 4 \end{bmatrix}$ 8

In Exercises 59 and 60, find the determinant of each matrix A.

59. $A = \begin{bmatrix} 1 & 2 & 0 \\ 0 & 1 & 2 \\ 1 & 0 & 2 \end{bmatrix}$ 6 **60.** $A = \begin{bmatrix} 2 & 3 & 4 \\ 2 & 5 & 9 \\ 2 & 7 & 16 \end{bmatrix}$ 8

In Exercises 61–64, use Cramer's Rule to solve each system of equations.

61. $\begin{cases} 5x + 3y = 11 \\ 2x + y = 4 \end{cases}$ $\{(1,2)\}$ **62.** $\begin{cases} 2x - 7y - 13 = 0 \\ 5x + 6y - 9 = 0 \end{cases}$ †

63. $\begin{cases} 3x + y - z = 14 \\ x + 3y - z = 16 \\ x + y - 3z = -10 \end{cases}$ † **64.** $\begin{cases} 2x + 3y - 2z = 0 \\ 5y - 3x + 4z = 9 \\ 3x + 7y - 6z = 4 \end{cases}$ †

In Exercises 65 and 66, solve each equation for x.

65. $\begin{vmatrix} 1 & 2 & 4 \\ -3 & 5 & 7 \\ 1 & x & 4 \end{vmatrix} = 0$ 2 **66.** $\begin{vmatrix} 1 & -1 & 2 \\ 0 & x & 1 \\ 3 & 2 & x-1 \end{vmatrix} = 14$ †

67. Show that equation of the line passing through the points $(2, 3)$ and $(1, -4)$ is equivalent to the equation

$$\begin{vmatrix} x & y & 1 \\ 2 & 3 & 1 \\ 1 & -4 & 1 \end{vmatrix} = 0 \text{ (See page 595.) †}$$

68. Use a determinant to find the area of the triangle with vertices $A(1, 1)$, $B(4, 3)$, and $C(2, 7)$. (See page 595.) 8

Applying the Concepts

69. Maximizing profit. A company is considering which of three methods of production it should use in producing the three products A, B, and C. The number of units of each product produced by each method is shown in this matrix:

$$\begin{array}{c} \\ Method\ 1 \\ Method\ 2 \\ Method\ 3 \end{array} \begin{array}{ccc} A & B & C \\ \begin{bmatrix} 4 & 8 & 2 \\ 5 & 7 & 1 \\ 5 & 4 & 8 \end{bmatrix} \end{array}$$

The profit per unit of the products A, B, and C is \$10, \$4, and \$6, respectively. Use matrix multiplication to find which method maximizes total profit for the company.

70. Nutrition. Consider two families: Ardestanis (*A*) and Barkley (*B*). Family *A* consists of two adult males, three adult females, and one child. Family *B* consists of one adult male, one adult female, and two children. Their mutual dietician considers the age and weight of each member of the two families and recommends daily allowances for calories—adult males 2200, adult females 1700, and children 1500—and for protein—adult males 50 grams, adult females 40 grams, and children 30 grams.

Represent the information in the preceding paragraph by matrices. Use matrix multiplication to calculate the total requirements of calories and proteins for each of the two families. †

In Exercises 71–76, each problem produces a system of linear equations in two or three variables. Use the method of your choice from this chapter to solve the system of equations.

71. Airplane speed. An airplane traveled 1680 miles in 3 hours with the wind. It would have taken 3.5 hours to have made the same trip against the wind. Find the speed of the plane and the velocity of the wind, assuming that both are constant. The speed of the plane is 520 miles per hour, and the velocity of the wind is 40 miles per hour.

72. Walking speed. In 5 hours, Nertha walks 2 miles farther than Kristina walks in 4 hours; then in 12 hours, Kristina walks 10 miles farther than Nertha walks in 10 hours. How many miles per hour does each walk? Nertha walks 3.2 miles per hour, and Kristina walks 3.5 miles per hour.

73. Friends. Andrew, Bonnie, and Chauncie have $320 between them. Andrew has twice as much as Chauncie, and Bonnie and Chauncie together have $20 less than Andrew. How much money does each have? Andrew has $170, Bonnie has $65, and Chauncie has $85.

74. Ratio. Steve and Nicole entered a double-decker bus in London on which there were $\frac{5}{8}$ as many men as women.

After they boarded the bus, there were $\frac{7}{11}$ as many men as women on the bus. How many people were on the bus before they entered it? 20 men and 32 women

75. Hospital workers. Metropolitan Hospital employs three types of caregivers: registered nurses, licensed practical nurses, and nurse's aides. Each registered nurse is paid $75 per hour, each licensed practical nurse is paid $20 per hour, and each nurse's aide is paid $30 per hour. The hospital has budgeted $4850 per hour for the three categories of caregivers. The hospital requires that each caregiver spend some time in a class learning about advanced medical practices each week: registered nurses, 3 hours; licensed practical nurses, 5 hours; and nurse's aides, 4 hours. The advanced classes can accommodate a maximum of 530 person-hours each week. The hospital needs a total of 130 registered nurses, licensed practical nurses, and nurse's aides to meet the daily needs of the patients. Assume that the allowable payroll is met, the training classes are full, and the patients' needs are met. How many registered nurses, licensed practical nurses, and nurse's aides should the hospital employ? 30 registered nurses, 40 licensed, practical nurses, and 60 nurse's aides.

76. Sales tax. A drugstore sells three categories of items:
   (i) Medicine, which is not taxed
   (ii) Nonmedical items, which are taxed at the rate of 8%
   (iii) Beer and cigarettes, which are taxed at the "sin tax" rate of 20%

Last Friday the total sales (excluding taxes) of all three items were $18,500. The total tax collected was $1020. The sales of medicines exceeded the combined sales of the other items by $3500. Find the sales in each category. †

PRACTICE TEST A

1. Give the order of the matrix and the value of the designated entry of the matrix.

$$\begin{bmatrix} 3 & -1 & -3 & 0 \\ 2 & 4 & 1 & 7 \\ 5 & 0 & 3 & 2 \\ 8 & 10 & 9 & 6 \\ -4 & 0 & 0 & 3 \end{bmatrix} \quad a_{43} \quad 5 \times 4; 9$$

2. Write an augmented matrix for the system of equations.

$$\begin{cases} 7x - 3y + 9z = 5 \\ -2x + 4y + 3z = -12 \\ 8x - 5y + z = -9 \end{cases} †$$

3. Write the system of linear equations represented by the augmented matrix

$$\begin{bmatrix} 4 & 0 & -1 & | & -3 \\ 1 & 3 & 0 & | & 9 \\ 2 & 7 & 5 & | & 8 \end{bmatrix}$$

Use *x*, *y*, and *z* for the variables. †

4. Find values for the variables so that the matrices

$$\begin{bmatrix} x - 5 & y + 3 \\ z & 9 \end{bmatrix} = \begin{bmatrix} 5 & 13 \\ 0 & 9 \end{bmatrix}$$

are equal. $(10, 10, 0)$

In Problems 5 and 6, solve the system of equations by using matrices.

5. $\begin{cases} x + 2y + z = 6 \\ x + y - z = 7 \\ 2x - y + 2z = -3 \end{cases} \quad \{(2, 3, -2)\}$

6. $\begin{cases} 2x + y - 4z = 6 \\ -x + 3y - z = -2 \\ 2x - 6y + 2z = 4 \end{cases} \left\{ \left(\frac{20}{7} + \frac{11}{7}z, \frac{2}{7} + \frac{6}{7}z, z \right) \right\}$

7. Let $A = \begin{bmatrix} 5 & -2 \\ 4 & 0 \\ 7 & 6 \end{bmatrix}$ and $B = \begin{bmatrix} -3 & -1 \\ 0 & -8 \\ -4 & 6 \end{bmatrix}$. Find $A - B$. †

In Problems 8 and 9, find the product AB or state that AB is not defined.

8. $A = \begin{bmatrix} 3 & -7 & 2 \end{bmatrix}$, $B = \begin{bmatrix} 0 \\ 1 \\ 4 \end{bmatrix}$ $AB = \begin{bmatrix} 1 \end{bmatrix}$

9. $A = \begin{bmatrix} -4 & 1 & 2 \\ -9 & 0 & 3 \end{bmatrix}$, $B = \begin{bmatrix} 2 & 8 \\ 7 & -5 \end{bmatrix}$

The product AB is not defined.

For Problems 10–13, let

$A = \begin{bmatrix} -3 & 1 & 0 \\ 5 & 7 & 2 \end{bmatrix}$, $B = \begin{bmatrix} -1 & 4 \\ 8 & 2 \end{bmatrix}$, and $C = \begin{bmatrix} 1 & 5 \\ 0 & 4 \end{bmatrix}$.

10. Find $2A$. †   **11.** Find $A + BA$. †

12. Find C^2. †   **13.** Find C^{-1}. †

14. Find the inverse of the matrix

$$A = \begin{bmatrix} 2 & 1 & 3 \\ 1 & 2 & -1 \\ 3 & 1 & 5 \end{bmatrix} \dagger$$

15. Write the linear system

$$\begin{cases} 5x + 2y = 32 \\ 3x + y = 18 \end{cases}$$

as a matrix equation in the form $AX = B$, where A is the coefficient matrix and B is the constant matrix. †

16. Write the matrix equation

$$\begin{bmatrix} 12 & -3 \\ -2 & 7 \end{bmatrix} \begin{bmatrix} x \\ y \end{bmatrix} = \begin{bmatrix} 5 \\ -9 \end{bmatrix} \quad \begin{cases} 12x - 3y = 5 \\ -2x + 7y = -9 \end{cases}$$

as a system of linear equations without matrices.

17. Solve the matrix equation $A - 5X = 2B$ for X, where

$A = \begin{bmatrix} 1 & 5 & -2 \\ 4 & -2 & 7 \end{bmatrix}$ and $B = \begin{bmatrix} 2 & 5 & -11 \\ 18 & 8 & 11 \end{bmatrix}$.

In Problems 18 and 19, evaluate the determinant.

18. $\begin{vmatrix} \frac{1}{2} & -\frac{1}{4} \\ \frac{1}{2} & \frac{3}{4} \end{vmatrix} \quad \frac{1}{2}$ **19.** $\begin{vmatrix} 1 & 3 & 5 \\ 2 & 0 & 10 \\ -3 & 1 & -15 \end{vmatrix} \quad 0$

20. Use Cramer's Rule to write the solution of the system

$$\begin{cases} 2x - y + z = 3 \\ x + y + z = 6 \\ 4x + 3y - 2z = 4 \end{cases}$$

in the determinant form. (You do not need to find the solution.) †

Answers:

17. $X = \begin{bmatrix} -\frac{3}{5} & -1 & 4 \\ -\frac{32}{5} & -\frac{18}{5} & -3 \end{bmatrix}$

PRACTICE TEST B

1. Give the order of the matrix and the value of the designated entries of the matrix.

$$A = \begin{bmatrix} 0 & 3 & 0 & -10 & 6 \\ 14 & -5 & 7 & -7 & 8 \\ -15 & 9 & -6 & 13 & -2 \\ 0.4 & 11 & 4 & -3 & 0 \end{bmatrix} \quad a_{34} \; d$$

  a. Order: 5×4; $a_{34} = 6$
  b. Order: 4×5; $a_{34} = 11$
  c. Order: 20; $a_{34} = -4$
  d. Order: 4×5; $a_{34} = 13$

2. Write an augmented matrix for the following system of equations:

$$\begin{cases} 8x + 4y + 5z = 10 \\ -2x + 3y + 6z = -9 \; d \\ -2x + 8y + 6z = 4 \end{cases}$$

  a. $\begin{bmatrix} 8 & 4 & 5 \\ -2 & 3 & 6 \\ -2 & 8 & 6 \end{bmatrix}$ **b.** $\begin{bmatrix} 10 & 5 & 4 & 8 \\ -9 & 6 & 3 & -2 \\ 4 & 6 & 8 & -2 \end{bmatrix}$

  c. $\begin{bmatrix} 8 & -2 & -2 & 10 \\ 4 & 3 & 8 & -9 \\ 5 & 6 & 6 & 4 \end{bmatrix}$ **d.** $\begin{bmatrix} 8 & 4 & 5 & 10 \\ -2 & 3 & 6 & -9 \\ -2 & 8 & 6 & 4 \end{bmatrix}$

3. Write the system of linear equations represented by the following augmented matrix:

$$\begin{bmatrix} -1 & 7 & 0 & -6 \\ 2 & 0 & 6 & 5 \\ 0 & 8 & -8 & 4 \end{bmatrix}$$

Use x, y, and z for the variables. d

  a. $\begin{cases} x + 7y + z = -6 \\ 2x + 6z = 5 \\ 8y + 8z = 4 \end{cases}$ **b.** $\begin{cases} -x + 7y + z = -6 \\ 2x + 6y = 5 \\ 8x - 8y = 4 \end{cases}$

  c. $\begin{cases} -x + 7y + z = -6 \\ 2x + y + 6z = 5 \\ x + 8y - 8z = 4 \end{cases}$ **d.** $\begin{cases} -x + 7y = -6 \\ 2x + 6z = 5 \\ 8y - 8z = 4 \end{cases}$

4. Find values for the variables so that the matrices

$$\begin{bmatrix} x + 3 & y + 4 \\ 7 & 3 \end{bmatrix} \text{ and } \begin{bmatrix} 9 & 2 \\ 7 & z \end{bmatrix}$$

are equal. c
  a. $x = -6$; $y = 2$; $z = 3$ **b.** $x = 6$; $y = 9$; $z = 3$
  c. $x = 6$; $y = -2$; $z = 3$ **d.** $x = 9$; $y = 2$; $z = 3$

5. Find the value of z in the following system:

$$\begin{cases} 2x + y = 15 \\ 2y + z = 25 \; c \\ 2z + x = 26 \end{cases}$$

  a. 4 **b.** 9 **c.** 11 **d.** 12

6. Find the value of $4x + 3y + z$ in the following system:

$$\begin{cases} 2x + y = 17 \\ y + 2z = 15 \text{ d} \\ x + z = 9 \end{cases}$$

   a. 41   **b.** 43   **c.** 55   **d.** 45

7. Let $A = \begin{bmatrix} -1 & 4 \\ 0 & 4 \\ 8 & -4 \end{bmatrix}$ and $B = \begin{bmatrix} 7 & 2 \\ 17 & 4 \\ 2 & 2 \end{bmatrix}$. Find $A - B$. c

   a. $\begin{bmatrix} 1 & 2 \\ 7 & 0 \\ 6 & -2 \end{bmatrix}$ **b.** $\begin{bmatrix} 3 & -3 \\ 7 & 0 \\ -6 & 6 \end{bmatrix}$

   c. $\begin{bmatrix} -8 & 2 \\ -17 & 0 \\ 6 & -6 \end{bmatrix}$ **d.** $\begin{bmatrix} 1 & 5 \\ 7 & 8 \\ 10 & 0 \end{bmatrix}$

In Problems 8 and 9, find the product AB if possible.

8. $A = \begin{bmatrix} -8 & 2 & 9 \end{bmatrix}$, $B = \begin{bmatrix} 3 \\ 0 \\ -3 \end{bmatrix}$ b

   a. $\begin{bmatrix} -24 & 0 & -27 \end{bmatrix}$ **b.** $\begin{bmatrix} -51 \end{bmatrix}$

   c. $\begin{bmatrix} 210 \end{bmatrix}$ **d.** $\begin{bmatrix} -24 \\ 0 \\ -27 \end{bmatrix}$

9. $A = \begin{bmatrix} 3 & -2 & 1 \\ 0 & 4 & -2 \end{bmatrix}$, $B = \begin{bmatrix} 5 & 0 \\ -2 & 1 \end{bmatrix}$ a

   a. AB is not defined.   **b.** $\begin{bmatrix} 15 & -10 & 5 \\ -6 & 8 & -4 \end{bmatrix}$

   c. $\begin{bmatrix} 15 & -6 \\ -10 & 8 \\ 5 & -4 \end{bmatrix}$ **d.** $\begin{bmatrix} 15 & 0 \\ 0 & 4 \end{bmatrix}$

For Problems 10–13, let

$A = \begin{bmatrix} 2 & 1 & -3 \\ -5 & 2 & 1 \end{bmatrix}$, $B = \begin{bmatrix} -3 & 7 \\ 2 & 4 \end{bmatrix}$, and $C = \begin{bmatrix} 5 & 4 \\ 1 & 0 \end{bmatrix}$

10. Find $2A$. b

   a. $\begin{bmatrix} 1 & \dfrac{1}{2} & -\dfrac{3}{2} \\ -\dfrac{5}{2} & 1 & \dfrac{1}{2} \end{bmatrix}$ **b.** $\begin{bmatrix} 4 & 2 & -6 \\ -10 & 4 & 2 \end{bmatrix}$

   c. $\begin{bmatrix} 4 & 3 & -1 \\ -3 & 4 & 3 \end{bmatrix}$ **d.** $\begin{bmatrix} 0 & -1 & -5 \\ -7 & 0 & -1 \end{bmatrix}$

11. Find $A + BA$. c

   a. $\begin{bmatrix} 2 & -4 & 10 \\ -5 & 4 & 5 \end{bmatrix}$ **b.** $\begin{bmatrix} 4 & -2 & 6 \\ -10 & 4 & 2 \end{bmatrix}$

   c. $\begin{bmatrix} -39 & 12 & 13 \\ -21 & 12 & -1 \end{bmatrix}$ **d.** $\begin{bmatrix} -41 & 17 & -2 \\ -16 & 6 & 10 \end{bmatrix}$

12. Find C^2. c

   a. $\begin{bmatrix} 10 & 8 \\ 2 & 0 \end{bmatrix}$ **b.** $\begin{bmatrix} 41 & 5 \\ 5 & 1 \end{bmatrix}$

   c. $\begin{bmatrix} 29 & 20 \\ 5 & 4 \end{bmatrix}$ **d.** $\begin{bmatrix} 0 & 4 \\ 1 & 5 \end{bmatrix}$

13. Find C^{-1}. b

   a. $\begin{bmatrix} -\dfrac{5}{4} & 1 \\ \dfrac{1}{4} & 0 \end{bmatrix}$ **b.** $\begin{bmatrix} 0 & 1 \\ \dfrac{1}{4} & -\dfrac{5}{4} \end{bmatrix}$

   c. $\begin{bmatrix} \dfrac{1}{4} & -\dfrac{5}{4} \\ 0 & 1 \end{bmatrix}$ **d.** $\begin{bmatrix} 0 & -1 \\ -\dfrac{1}{4} & -\dfrac{5}{4} \end{bmatrix}$

14. Find the inverse of the following matrix:

$$A = \begin{bmatrix} 1 & 0 & 0 \\ 2 & 1 & 0 \\ 3 & -4 & 1 \end{bmatrix} \text{ d}$$

   a. $\begin{bmatrix} 1 & 0 & 0 \\ -2 & 1 & 0 \\ 3 & 4 & 1 \end{bmatrix}$ **b.** $\begin{bmatrix} 1 & 0 & 0 \\ -2 & 1 & 0 \\ -8 & 3 & 1 \end{bmatrix}$

   c. $\begin{bmatrix} 1 & -2 & -11 \\ 0 & 1 & 4 \\ 0 & 0 & 1 \end{bmatrix}$ **d.** $\begin{bmatrix} 1 & 0 & 0 \\ -2 & 1 & 0 \\ -11 & 4 & 1 \end{bmatrix}$

15. Write the linear system as a matrix equation

$$\begin{cases} 4x + 2y = 15 \\ -2x + 4y = 9 \end{cases}$$

in the form $AX = B$, where A is the coefficient matrix and B is the constant matrix. a

   a. $\begin{bmatrix} 4 & 2 \\ -2 & 4 \end{bmatrix}\begin{bmatrix} x \\ y \end{bmatrix} = \begin{bmatrix} 15 \\ 9 \end{bmatrix}$ **b.** $\begin{bmatrix} 4 & 2 \\ 4 & -2 \end{bmatrix}\begin{bmatrix} x \\ y \end{bmatrix} = \begin{bmatrix} 15 \\ 9 \end{bmatrix}$

   c. $\begin{bmatrix} 15 & 2 \\ 9 & -2 \end{bmatrix}\begin{bmatrix} x \\ y \end{bmatrix} = \begin{bmatrix} 4 \\ 4 \end{bmatrix}$ **d.** $\begin{bmatrix} 4 & -2 \\ 2 & 4 \end{bmatrix}\begin{bmatrix} x \\ y \end{bmatrix} = \begin{bmatrix} 15 \\ 9 \end{bmatrix}$

16. Write the matrix equation

$$\begin{bmatrix} -3 & 9 \\ -8 & -4 \end{bmatrix}\begin{bmatrix} x \\ y \end{bmatrix} = \begin{bmatrix} -4 \\ -5 \end{bmatrix}$$

as a system of linear equations. a

   a. $\begin{cases} -3x + 9y = -4 \\ -8x - 4y = -5 \end{cases}$ **b.** $\begin{cases} -3x + 9y = -4 \\ -4x - 8y = -5 \end{cases}$

   c. $\begin{cases} 9x - 3y = -4 \\ -8x - 4y = -5 \end{cases}$ **d.** $\begin{cases} -3x + 9y = 4 \\ -8x - 4y = 5 \end{cases}$

17. Let $A = \begin{bmatrix} -2 & -3 & 1 \\ -5 & 3 & -2 \end{bmatrix}$ and $B = \begin{bmatrix} -1 & -2 & -1 \\ 1 & 0 & 1 \end{bmatrix}$.

   Solve the matrix equation $A - 3X = -5B$ for X. b

   a. $X = \begin{bmatrix} -\dfrac{5}{2} & -\dfrac{9}{2} & -1 \\ -1 & \dfrac{3}{2} & \dfrac{1}{2} \end{bmatrix}$

   b. $X = \begin{bmatrix} -\dfrac{7}{3} & -\dfrac{13}{3} & -\dfrac{4}{3} \\ 0 & 1 & 1 \end{bmatrix}$

c. $X = \begin{bmatrix} -1 & \dfrac{3}{2} & -1 \\ -\dfrac{5}{2} & -\dfrac{9}{2} & -1 \end{bmatrix}$

d. $X = \begin{bmatrix} -\dfrac{1}{3} & \dfrac{1}{3} & \dfrac{8}{3} \\ -\dfrac{20}{3} & 3 & -\dfrac{11}{3} \end{bmatrix}$

In Problems 18 and 19, evaluate the determinant.

18. $\begin{vmatrix} -8 & 5 \\ -4 & -1 \end{vmatrix}$ d

a. -28 **b.** -44 **c.** -12 **d.** 28

19. $\begin{vmatrix} 2 & 3 & -2 \\ 3 & 0 & -3 \\ -3 & 0 & -5 \end{vmatrix}$ c

a. -72 **b.** 18 **c.** 72 **d.** -18

20. Solve the system of equations

$$\begin{cases} x + y + z = -6 \\ x - y + 3z = -22 \\ 2x + y + z = -10 \end{cases}$$

by using Cramer's Rule. b

a. $\varnothing$ **b.** $\{(-4, 3, -5)\}$
c. $\{(-5, -4, 3)\}$ **d.** $\{(-5, 3, -4)\}$

CUMULATIVE REVIEW EXERCISES ▪ Chapters 1–8

1. If the distance between the points $(2, -3)$ and $(-1, y)$ is five units, find y. $-7, 1$

In Problems 2–8, solve each equation or inequality.

2. $\dfrac{4}{x - 1} - \dfrac{3}{x + 2} = \dfrac{18}{(x + 2)(x - 1)}$ 7

3. $|2x - 5| = 3$ 4, 1

4. $4x^2 = 8x - 13$ $1 + \dfrac{3}{2}i, 1 - \dfrac{3}{2}i$

5. $\left(\dfrac{3x - 1}{x + 5}\right)^2 - 3\left(\dfrac{3x - 1}{x + 5}\right) - 28 = 0$ $-9, \dfrac{-19}{7}$

6. $\log_2|x| + \log_2|x + 6| = 4$ $2, -8$

7. $\dfrac{x + 2}{2x - 1} > 0$ $x < -2$ or $x > \dfrac{1}{2}$

8. $\cos x = -\dfrac{1}{2}, 0 \le x \le 2\pi$ $\left\{\dfrac{2\pi}{3}, \dfrac{4\pi}{3}\right\}$

9. Write all possible rational zeros of

$$f(x) = 4x^3 + 8x^2 - 11x + 3.$$
$\pm 1, \pm \dfrac{1}{2}, \pm \dfrac{1}{4}, \pm 3, \pm \dfrac{3}{2}, \pm \dfrac{3}{4}$

10. Show that $\dfrac{1}{2}$ is a zero of multiplicity 2 of the function f in Problem 9.

11. The horsepower required to propel a ship varies as the cube of the ship's speed. If the horsepower required for a speed of 15 miles per hour is 10,125, find the horsepower required for a speed of 20 miles per hour. 24,000 horsepower

12. A city covers an area of 400 square miles. At present, only 2 percent of its area is reserved for parks. In the unincorporated area outside the city limits, 20 percent of the area could be developed into parks. How many square miles of the incorporated area had to be annexed so that the city could develop 12 percent of its area as parks? 500

In Problems 13 and 14, solve the system of equations.

13. $\begin{cases} 3x + y = 2 \\ 4x + 5y = -1 \end{cases}$ $\{(1, -1)\}$

14. $\begin{cases} 2x + y = 5 \\ y^2 - 2y = -3x + 5 \end{cases}$ $\left\{(2,1), \left(\dfrac{5}{4}, \dfrac{5}{2}\right)\right\}$

15. Use transformations to sketch the graph of

$$f(x) = 1 + \cos\left(x - \dfrac{\pi}{4}\right). \ \dagger$$

16. Find the inverse of the matrix $\begin{bmatrix} 1 & 2 & -2 \\ -1 & 3 & 0 \\ 0 & -2 & 1 \end{bmatrix}$.

17. Use the result in Problem 16 to solve the following system of equations:

$$\begin{cases} x + 2y - 2z = 5 \\ -x + 3y \quad\;\; = 2 \\ -2y + z = -3 \end{cases} \{(1, 1, -1)\}$$

18. If $f(x) = x^2 + 3x - 1$ and $g(x) = x + 2$, find
(a) $F(x) = (f \circ g)(x)$ and **(b)** $F(4)$. **a.** $F(x) = x^2 + 7x + 9$

19. Let $f(x) = \dfrac{x}{x + 4}$. Find $f^{-1}(x)$. $f^{-1}(x) = \dfrac{-4x}{x - 1}$

20. Use the power-reducing identity to verify the identity:

$$\sin^4 x = \dfrac{1}{8}(3 - 4\cos 2x + \cos 4x)$$

Answers:

16. $A^{-1} = \begin{bmatrix} 3 & 2 & 6 \\ 1 & 1 & 2 \\ 2 & 2 & 5 \end{bmatrix}$ **18. b.** $F(4) = 53$

CHAPTER

9 Analytic Geometry

TOPICS

Greek mathematicians were fascinated by certain curves, called conic sections, now used to describe natural phenomena such as planetary orbits. Modern applications also include the construction of lenses, whispering galleries, navigation systems, and even medical devices. In this chapter, we investigate the conic sections and their many uses.

Conic Sections: Overview

Most of the sections in this chapter focus on the plane curves called **conics** or **conic sections**. As the name implies, these curves are the sections of a cone (similar to an ice cream cone) formed when a plane intersects the cone. We also discuss parametric equations of conics and other curves in the plane. ■

Apollonius of Perga

(c. 262 B.C.–190 B.C.)

The mathematician Apollonius was born in Perga, in southern Asia Minor (today known as Murtina in Turkey). He went to Alexandria to study with the successors of Euclid. He became famous in ancient times for his work on astronomy and his eight books on conics. He was able to discover and prove hundreds of beautiful and difficult theorems without modern algebraic symbolism. He is credited with introducing the terms *ellipse*, *hyperbola*, and *parabola*.

Euclid defined a cone as a surface generated by rotating a right triangle about one of its legs. However, the following description of a cone given by Apollonius is more appropriate.

RIGHT CIRCULAR CONE

Draw a circle on a plane. Draw a line *l*, called the **axis**, passing through the center of the circle and perpendicular to the plane. Choose a point *V* above the plane on this line. The surface consisting of all lines that simultaneously pass through the point *V* and the circle is called a **right circular cone** with **vertex** *V*. The vertex *V* separates the surface into two parts called *nappes of the cone*. See Figure 9.1.

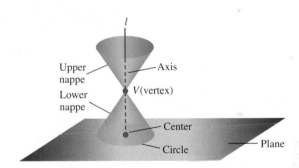

FIGURE 9.1 Parts of a cone

If we slice the cone with a plane, the intersections of the slicing plane and the cone are **conic sections**. If the slicing plane is horizontal (parallel to the first plane), the intersection is a **circle** (Figure 9.2(a)). If the slicing plane is inclined slightly from the horizontal, the intersection is an oval-shaped curve called an **ellipse** (Figure 9.2(b)). As the angle of the slicing plane increases, more elongated ellipses are formed (Figure 9.2(b)). If the angle of the slicing plane increases still further so that the slicing plane is parallel to one of the lines generating the cone, the intersection is called a **parabola** (Figure 9.2(c)). If the angle of the slicing plane increases yet further so that the slicing plane intersects both nappes of the cone, the resulting intersection is called a **hyperbola** (Figure 9.2(d)).

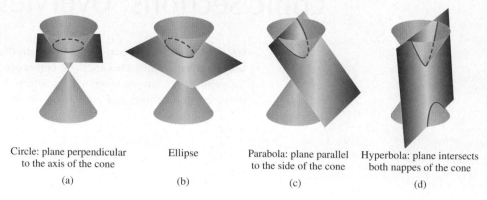

Circle: plane perpendicular
to the axis of the cone

(a)

Ellipse

(b)

Parabola: plane parallel
to the side of the cone

(c)

Hyperbola: plane intersects
both nappes of the cone

(d)

FIGURE 9.2 **Conic sections**

Intersections of the slicing plane through the vertex result in points and lines called **degenerate conic sections**. See Figure 9.3.

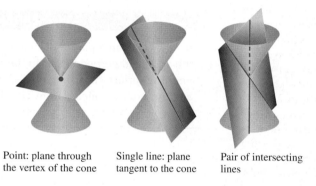

Point: plane through
the vertex of the cone

Single line: plane
tangent to the cone

Pair of intersecting
lines

FIGURE 9.3 **Degenerate conic sections**

Just as a circle (see page 11) is defined as the set of points in the plane at a fixed distance r (the radius) from a fixed point (the center), we can define each of the conic sections just described as a set of points in the plane that satisfy certain geometric conditions. It can be shown that these alternate definitions give the same family of curves described earlier. This enables us to keep our work in a two-dimensional setting. Using these definitions, we will derive the equations of the conic sections and show that an equation of the form $Ax^2 + Cy^2 + Dx + Ey + F = 0$ is (except in degenerate cases represented in Figure 9.3) the equation of a parabola, an ellipse, or a hyperbola. (A circle is a special case of an ellipse.)

We study these conic sections because of their practical applications. For example, a satellite dish, a flashlight lens, and a telescope lens have a parabolic shape, the planets travel in elliptical orbits, and comets travel in orbits that are either elliptical or hyperbolic. A comet with an elliptical orbit can be viewed from Earth more than once, while those with a hyperbolic orbit can be viewed only once!

The Parabola

Before Starting this Section, Review

1. Distance formula (Section 1.1, page 5)
2. Midpoint formula (Section 1.1, page 5)
3. Completing the square (Appendix A, page 811)
4. Line of symmetry (Section 1.1, page 8)
5. Slopes of perpendicular lines (Section 1.2, page 24)

Objectives

1 Define a parabola geometrically.
2 Find an equation of a parabola.
3 Translate a parabola.
4 Use the reflecting property of parabolas.

Hubble Telescope

Edwin Hubble (1889–1953)

Edwin Hubble is renowned for discovering that there are other galaxies in the universe beyond the Milky Way. Hubble then wanted to classify the galaxies according to their content, distance, shape, and brightness patterns. He made the momentous discovery that the galaxies were moving away from each other at a rate proportional to the distance between them (Hubble's Law). This led to the calculation of the point where the expansion began and to confirmation of the Big Bang theory of the origin of the universe. Recent estimates place the age of the universe at about 20 billion years.

THE HUBBLE SPACE TELESCOPE

In 1918, the 2.5-meter (100-inch) Hooker Telescope was installed at Mt. Wilson Observatory in Pasadena, California. With this telescope, astronomer Edwin Hubble measured the distances and velocities of the galaxies. His work led to today's concept of an expanding universe. In 1977, the National Aeronautics and Space Administration (NASA) named its largest, most complex, and most capable orbiting telescope in honor of Edwin Hubble.

On April 25, 1990, the Hubble Space Telescope was deployed into orbit from the space shuttle *Discovery*. However, due to a spherical aberration in the telescope's parabolic mirror, the Hubble had "blurred vision." (See Example 4.) In 1993, astronauts from the space shuttle *Endeavor* installed replacement instruments and supplemental optics to restore the telescope to full optimal performance.

Not since Galileo turned his telescope toward the heavens in 1610 has any event so changed our understanding of the universe as did the deployment of the Hubble Space Telescope. The first instrument of any kind that was specifically designed for routine servicing by spacewalking astronauts, the Hubble telescope brought stunning pictures from the end of the universe to the general public. ■

1 Define a parabola geometrically.

Geometric Definition of a Parabola

In Section 2.1, we named the graph of a quadratic function $y = ax^2 + bx + c$, $a \neq 0$, a parabola. In this section, we give a *geometric* definition of a parabola. Then we show that this definition leads to the graph of a quadratic function.

STUDY TIP

The distance from a point to a line is defined as the length of the perpendicular line segment from the point to the line.

PARABOLA

Let l be a line and F a point not on the line l. Let π be the plane determined by F and l. Then the set of all points P in the plane that are the same distance from F as they are from the line l is called a **parabola**. See Figure 9.4(b). Thus, a parabola is the set of points P for which $d(F, P) = d(P, l)$, where $d(P, l)$ denotes the distance between P and l.

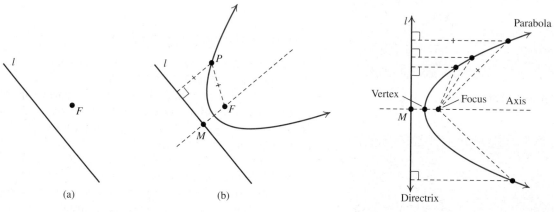

(a)   (b)

FIGURE 9.4 Defining a parabola

FIGURE 9.5

The line l is the **directrix** of the parabola, and the point F is the **focus**. The line through the focus *perpendicular* to the directrix is called the **axis**, or the **axis of symmetry**, of the parabola. The point at which the axis intersects the parabola is called the **vertex**. See Figure 9.5. Note that the vertex of a parabola is the midpoint of the segment $\overline{FM}$.

2 Find an equation of a parabola.

Equation of a Parabola

A coordinate system allows us to transform the geometric description of a parabola into an algebraic *equation of the parabola*. This equation is particularly simple if we place the vertex V at the origin and the focus on one of the coordinate axes. Suppose for a moment that the focus F is on the positive x-axis. Then F has coordinates of the form $(a, 0)$ with $a > 0$, as shown in Figure 9.6.

The vertex V lies on the parabola with focus F and directrix l; so we have the following:

$$d(V, l) = d(V, F) \qquad \text{Definition of parabola, where}$$
$$= a \qquad\qquad V \text{ is a point on the parabola}$$

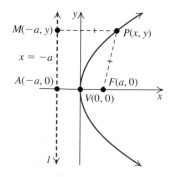

FIGURE 9.6 Deriving the equation of a parabola

The vertex V at the origin is located midway between the focus and the directrix; so the equation of the directrix is $x = -a$.

Suppose a point $P(x, y)$ lies on the parabola shown in Figure 9.6. Then the distance from the point P to the directrix l is the distance between the points $P(x, y)$ and $M(-a, y)$. So, a point $P(x, y)$ lies on this parabola if and only if

$$d(P, F) = d(P, M).$$

Using the distance formula on each side gives

$$\sqrt{(x-a)^2 + y^2} = \sqrt{(x+a)^2}$$

$$(x-a)^2 + y^2 = (x+a)^2 \qquad \text{Square both sides.}$$

$$x^2 - 2ax + a^2 + y^2 = x^2 + 2ax + a^2 \qquad \text{Expand squares.}$$

$$y^2 = 4ax \qquad \text{Simplify.}$$

The equation $y^2 = 4ax$ is called the **standard equation of a parabola with vertex (0, 0) and focus (a, 0)**. Similarly, if the focus of a parabola is placed on the negative x-axis, we obtain the equation $y^2 = -4ax$ as the standard equation of a parabola with vertex $(0, 0)$ and focus $(-a, 0)$.

By interchanging the roles of x and y, we find that $x^2 = 4ay$ is the standard equation of a parabola with vertex $(0, 0)$ and focus $(0, a)$. Similarly, the equation $x^2 = -4ay$ is the standard equation of a parabola with vertex $(0, 0)$ and focus $(0, -a)$.

Main facts about a parabola with $a > 0$

Standard Equation	$y^2 = 4ax$	$y^2 = -4ax$	$x^2 = 4ay$	$x^2 = -4ay$
Vertex	$(0,0)$	$(0,0)$	$(0,0)$	$(0,0)$
Description	Opens right	Opens left	Opens up	Opens down
Axis of Symmetry	$y = 0$ (x-axis)	$y = 0$ (x-axis)	$x = 0$ (y-axis)	$x = 0$ (y-axis)
Focus	$(a, 0)$	$(-a, 0)$	$(0, a)$	$(0, -a)$
Directrix	$x = -a$	$x = a$	$y = -a$	$y = a$

Graph

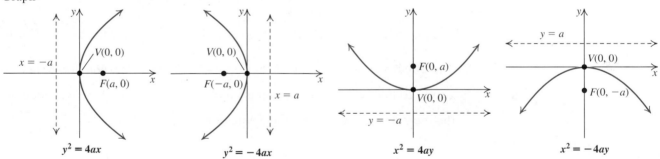

EXAMPLE 1   **Graphing a Parabola**

Graph each parabola and specify the vertex, focus, directrix, and axis.

a. $x^2 = -8y$ **b.** $y^2 = 5x$

SOLUTION

a. The equation $x^2 = -8y$ has the standard form $x^2 = -4ay$; so

$$-4a = -8 \qquad \text{Equate coefficients of } y.$$

$$a = 2 \qquad \text{Divide both sides by } -4.$$

The parabola opens down. The vertex is the origin, and the focus is $(0, -2)$. The directrix is the horizontal line $y = 2$; the axis of the parabola is the y-axis. You should plot some additional points to ensure the accuracy of your sketch.

TECHNOLOGY CONNECTION

You can use a graphing calculator to graph parabolas.

To graph $x^2 = 4ay$, just graph $Y_1 = \dfrac{1}{4a}x^2$.

For example, when $x^2 = -8y$, you have $4a = -8$; so you graph

$$Y_1 = -\dfrac{1}{8}x^2.$$

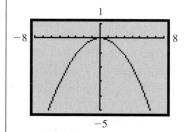

Since the focus is $(0, -2)$, substitute $y = -2$ in the equation $x^2 = -8y$ of the parabola to obtain

$$\begin{aligned}
x^2 &= (-8)(-2) &&\text{Replace } y \text{ with } -2 \text{ in } x^2 = -8y.\\
x^2 &= 16 &&\text{Simplify.}\\
x &= \pm 4 &&\text{Solve for } x.
\end{aligned}$$

Thus, the points $(4, -2)$ and $(-4, -2)$ are two symmetric points on the parabola to the right and the left of the focus. The graph of this parabola is sketched in Figure 9.7.

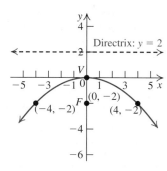

FIGURE 9.7 Graph of $x^2 = -8y$

b. The equation $y^2 = 5x$ has the standard form $y^2 = 4ax$.

$$\begin{aligned}
4a &= 5 &&\text{Equate coefficients of } x.\\
a &= \frac{5}{4} &&\text{Divide both sides by 4.}
\end{aligned}$$

The parabola opens to the right. The vertex is the origin, and the focus is $\left(\dfrac{5}{4}, 0\right)$.

The directrix is the vertical line $x = -a = -\dfrac{5}{4}$; the axis of the parabola is the x-axis. To find two symmetric points on the parabola that are above and below the focus, substitute $x = \dfrac{5}{4}$ in the equation of the parabola.

$$\begin{aligned}
y^2 &= 5x &&\text{Equation of the parabola}\\
y^2 &= 5\left(\frac{5}{4}\right) &&\text{Replace } x \text{ with } \frac{5}{4}.\\
y^2 &= \frac{25}{4} &&\text{Simplify.}\\
y &= \pm\frac{5}{2} &&\text{Solve for } y.
\end{aligned}$$

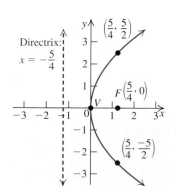

FIGURE 9.8 Graph of $y^2 = 5x$

Plot the two additional points $\left(\dfrac{5}{4}, \dfrac{5}{2}\right)$ and $\left(\dfrac{5}{4}, -\dfrac{5}{2}\right)$ on the parabola. The graph of $y^2 = 5x$ is sketched in Figure 9.8.　■ ■ ■

Practice Problem 1 Graph the parabola and specify the vertex, focus, directrix, and axis.

a. $x^2 = 12y$　　**b.** $y^2 = -6x$　　　　　　　　　■

The line segment joining the focus and two points symmetric with respect to the axis on a parabola is called the *latus rectum* of the parabola. Intuitively, the length of the latus rectum determines the "size of the opening" of the parabola.

TECHNOLOGY CONNECTION

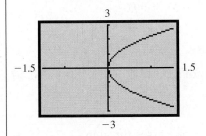

To graph $y^2 = 4ax$, solve the equation for y to obtain

$$y = \pm\sqrt{4ax}.$$

Then graph $Y_1 = \sqrt{4ax}$ and $Y_2 = -\sqrt{4ax}$. For example, to graph $y^2 = 5x$, you graph

$$Y_1 = \sqrt{5x},\ Y_2 = -\sqrt{5x}.$$

LATUS RECTUM

The line segment passing through the focus of a parabola, perpendicular to the axis, and having endpoints on the parabola is called the **latus rectum of the parabola**. Figure 9.9 shows that the length of the latus rectum for each of the graphs $y^2 = \pm4ax$ and $x^2 = \pm4ay$ for $a > 0$ is $4a$.

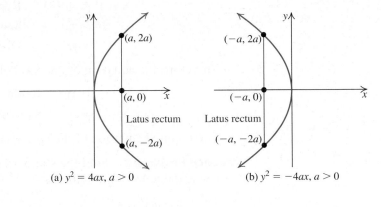

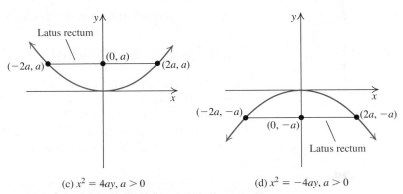

FIGURE 9.9 The latus rectum

EXAMPLE 2   **Finding the Equation of a Parabola**

Find the standard equation of a parabola with vertex $(0, 0)$ and satisfying the given description.

a. The focus is $(-3, 0)$.

b. The axis of the parabola is the y-axis, and the graph passes through the point $(-4, 2)$.

SOLUTION

a. The vertex $(0, 0)$ of the parabola and its focus $(-3, 0)$ are on the x-axis; so the parabola opens to the left. Its equation must be of the form $y^2 = -4ax$ with $a = 3$.

$$
\begin{aligned}
y^2 &= -4ax && \text{Form of the equation} \\
y^2 &= -4(3)x && \text{Replace } a \text{ with 3.} \\
y^2 &= -12x && \text{Simplify.}
\end{aligned}
$$

The equation of the parabola is $y^2 = -12x$.

b. Since the vertex of the parabola is $(0, 0)$, the axis is the y-axis, and the point $(-4, 2)$ is above the x-axis, the parabola opens up and the equation of the parabola is of the form

$$x^2 = 4ay.$$

This equation is satisfied by $x = -4$ and $y = 2$ because the parabola passes through the point $(-4, 2)$.

$$\begin{aligned}(-4)^2 &= 4a(2) &&\text{Replace } x \text{ with } -4 \text{ and } y \text{ with } 2. \\ 16 &= 8a &&\text{Simplify.} \\ 2 &= a &&\text{Divide both sides by 8.}\end{aligned}$$

Therefore, the equation of the parabola is

$$\begin{aligned}x^2 &= 4(2)y &&\text{Replace } a \text{ with } 2 \text{ in } x^2 = 4ay. \\ x^2 &= 8y &&\text{Simplify.}\end{aligned}$$

The required equation of the parabola is $x^2 = 8y$. ■ ■ ■

Practice Problem 2 Find the standard equation of a parabola with vertex $(0, 0)$ satisfying the following conditions.

a. The focus is $(0, 2)$.

b. The axis of the parabola is the x-axis, and the graph passes through $(1, 2)$. ■

3 Translate a parabola.

Translations of Parabolas

The equation of a parabola with vertex $(0, 0)$ and axis on a coordinate axis is of the form $x^2 = \pm 4ay$ or $y^2 = \pm 4ax$. You can use translations of the graphs of these equations to find the equation of a parabola with vertex (h, k) and axis of symmetry parallel to a coordinate axis. We will obtain these translations by replacing x with $x - h$ (a horizontal shift) and y with $y - k$ (a vertical shift).

Main facts about a parabola with vertex (h, k) and $a > 0$				
Standard Equation	$(y - k)^2 = 4a(x - h)$	$(y - k)^2 = -4a(x - h)$	$(x - h)^2 = 4a(y - k)$	$(x - h)^2 = -4a(y - k)$
Equation of axis	$y = k$	$y = k$	$x = h$	$x = h$
Description	Opens right	Opens left	Opens up	Opens down
Vertex	(h, k)	(h, k)	(h, k)	(h, k)
Focus	$(h + a, k)$	$(h - a, k)$	$(h, k + a)$	$(h, k - a)$
Directrix	$x = h - a$	$x = h + a$	$y = k - a$	$y = k + a$
Graph				

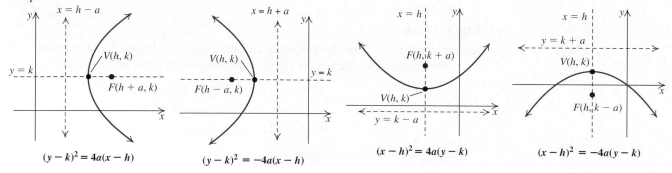

Expanding the standard equations of a parabola and collecting like terms results in an equation of the form

$$Ax^2 + Cy^2 + Dx + Ey + F = 0$$

with $A = 0$ or $C = 0$, but not both. Conversely, except in degenerate cases, any equation of this form can be put into one of the standard forms of a parabola by completing the squares on the x- or y- terms.

EXAMPLE 3   Graphing a Parabola

Find the vertex, focus, and directrix of the parabola $2y^2 - 8y - x + 7 = 0$. Sketch the graph of the parabola.

SOLUTION

First, complete the square on y. Then you can compare the resulting equation with one of the standard forms of the equation of a parabola.

$2y^2 - 8y - x + 7 = 0$	The given equation
$2y^2 - 8y = x - 7$	Isolate terms containing y.
$2(y^2 - 4y) = x - 7$	Factor out the common factor, 2.
$2(y^2 - 4y + 4) = x - 7 + 8$	Add $2\left(-\dfrac{4}{2}\right)^2 = 2(-2)^2 = 2 \cdot 4 = 8$ to both sides to complete the square.
$2(y - 2)^2 = x + 1$	$y^2 - 4y + 4 = (y - 2)^2$; simplify.
$(y - 2)^2 = \dfrac{1}{2}(x + 1)$	Divide both sides by 2.

The last equation is in one of the standard forms of the parabola. Comparing it with the form $(y - k)^2 = 4a(x - h)$, we have $h = -1$; $k = 2$; and $4a = \dfrac{1}{2}$, or $a = \dfrac{1}{8}$. The parabola opens to the right, the vertex is $(h, k) = (-1, 2)$, the focus is $(h + a, k) = \left(-1 + \dfrac{1}{8}, 2\right) = \left(-\dfrac{7}{8}, 2\right)$, and the directrix is the vertical line $x = h - a = -1 - \dfrac{1}{8} = -\dfrac{9}{8}$. The graph is shown in Figure 9.10.   ■ ■ ■

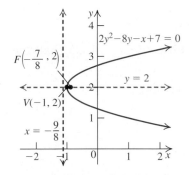

FIGURE 9.10 Converting $2y^2 - 8y - x + 7 = 0$ to standard form

Practice Problem 3   Find the vertex, focus, and directrix of the parabola

$$2x^2 - 8x - y + 7 = 0.$$

Sketch the graph of the parabola.   ■

4 Use the reflecting property of parabolas.

Reflecting Property of Parabolas

A property of parabolas that is useful in applications is the **reflecting property**: If a reflecting surface has parabolic cross sections with a common focus, then all light rays entering the surface parallel to the axis will be reflected through the focus. See Figure 9.11(a) on the next page. This property is used in reflecting telescopes and satellite antennas because the light rays or radio waves bouncing off a parabolic surface are reflected to the focus, where they are collected and amplified.

Conversely, if a light source is located at the focus of a parabolic reflector, the reflected rays will form a beam parallel to the axis. See Figure 9.11(b). This principle is used in flashlights, searchlights, and other such devices.

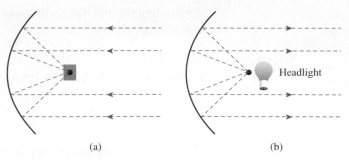

(a)   (b)

FIGURE 9.11

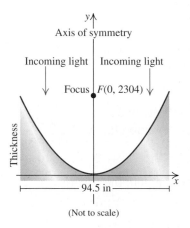

FIGURE 9.12 Hubble Space Telescope

<hr>

EXAMPLE 4 **Calculating Some Properties of the Hubble Space Telescope**

The parabolic mirror used in the Hubble Space Telescope has a diameter of 94.5 inches. See Figure 9.12. Find the equation of the parabola if its focus is 2304 inches from the vertex. What is the thickness of the mirror at the edges?

SOLUTION

Position the parabola so that its vertex is at the origin and its focus is on the positive y-axis. The equation of the parabola is of the form

$$x^2 = 4ay$$
$$x^2 = 4(2304)y \qquad a = 2304$$
$$x^2 = 9216y \qquad \text{Simplify.}$$

To find the thickness y of the mirror at the edge, substitute $x = 47.25$ (half the diameter) in the equation $x^2 = 9216y$ and solve for y.

$$(47.25)^2 = 9216y \qquad \text{Replace } x \text{ with 47.25.}$$
$$\frac{(47.25)^2}{9216} = y \qquad \text{Divide both sides by 9216.}$$
$$0.242248 \approx y \qquad \text{Use a calculator.}$$

Thus, the thickness of the mirror at the edges is approximately 0.242248 inch.

The Hubble telescope initially had "blurred vision" because the mirror was ground at the edges to a thickness of 0.24224 inch. In other words, it had been ground 0.000008 inch, or two-millionths of a meter, smaller than it should have been! The problem was fixed in 1993. ■ ■ ■

Practice Problem 4 A small imitation of the Hubble's parabolic mirror has a diameter of 3 inches. Find the equation of the parabola if its focus is 7.3 inches from the vertex. What is the thickness of the mirror at its edges, correct to four decimal places? ■

<hr>

SECTION 9.2 ■ Exercises

A EXERCISES Basic Skills and Concepts

1. A parabola is the set of all points P in the plane that are equidistant from a fixed line called the ___directrix___ and a fixed point not on the line called the ___focus___.

2. The line through the focus perpendicular to the directrix is called the ___latus rectum___ of the parabola.

3. The point at which the axis intersects the parabola is called the ___vertex___ of the parabola.

4. The graph of $y + 2 = 3(x - 4)^2$ is obtained by shifting the graph of $y = 3x^2$ to the right ___four___ units and shifting ___down___ 2 units.

5. *True or False* The vertex of a parabola is midway between its focus and its directrix. True

6. *True or False* A parabola is symmetric about its axis. True

In Exercises 7–14, find the focus and directrix of the parabola with the given equation. Then match each equation to one of the graphs labeled (a) through (h).

7. $x^2 = 2y$

8. $9x^2 = 4y$

9. $16x^2 = -9y$

10. $x^2 = -2y$

11. $y^2 = 2x$

12. $9y^2 = 16x$

13. $9y^2 = -16x$

14. $y^2 = -2x$ †

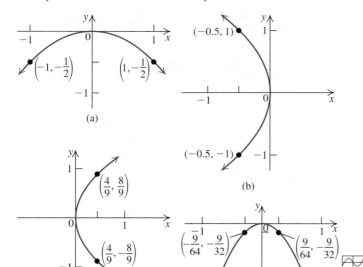

(a)

(b)

(c)

(d)

(e)

(f)

(g)

(h)

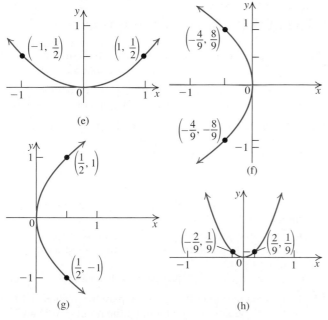

In Exercises 15–20, use a graphing device to graph each parabola.

15. $x^2 - 2y = 0$ †

16. $x^2 = 6y$ †

17. $y^2 = -3x$ †

18. $y^2 + 5x = 0$ †

19. $3x + 2y^2 = 0$ †

20. $y^2 - 5x = 0$ †

7. Focus: $\left(0, \dfrac{1}{2}\right)$; directrix: $y = -\dfrac{1}{2}$; graph: e

8. Focus: $\left(0, \dfrac{1}{9}\right)$; directrix: $y = -\dfrac{1}{9}$; graph: h

9. Focus: $\left(0, -\dfrac{9}{64}\right)$; directrix: $y = \dfrac{9}{64}$; graph: d

In Exercises 21–36, find the standard equation of the parabola that satisfies the given conditions. Also find the length of the latus rectum of each parabola.

21. Focus: $(0, 2)$; directrix: $y = 4$ †

22. Focus: $(0, 4)$; directrix: $y = -2$ †

23. Focus: $(-2, 0)$; directrix: $x = 3$ †

24. Focus: $(-1, 0)$; directrix: $x = -2$ †

25. Vertex: $(1, 1)$; directrix: $x = 3$ †

26. Vertex: $(1, 1)$ directrix: $y = 2$ †

27. Vertex: $(1, 1)$; directrix: $y = -3$ †

28. Vertex: $(1, 1)$; directrix: $x = -2$ †

29. Vertex: $(1, 0)$; focus: $(3, 0)$ †

30. Vertex: $(0, 1)$; focus: $(0, 2)$ †

31. Vertex: $(0, 1)$; focus: $(0, -2)$ †

32. Vertex: $(-1, 0)$; focus: $(-3, 0)$ †

33. Vertex: $(2, 3)$; directrix: $x = 4$ †

34. Vertex: $(2, 3)$; directrix: $y = 4$ †

35. Vertex: $(2, 3)$; directrix: $y = 1$ †

36. Vertex: $(2, 3)$; directrix: $x = 1$ †

In Exercises 37–40, use a graphing device to graph each parabola.

37. $(x + 1) = 4(y - 1)^2$ †

38. $(x - 2) = -4(y - 1)^2$ †

39. $(x - 1)^2 = -9(y + 1)$ †

40. $(x + 2)^2 = 6(y + 1)$ †

In Exercises 41–56, find the vertex, focus, and directrix of each parabola. Graph the equation.

41. $(y - 1)^2 = 2(x + 1)$ †

42. $(y + 2)^2 = 8(x - 3)$ †

43. $(x + 2)^2 = 3(y - 2)$ †

44. $(x - 3)^2 = 4(y + 1)$ †

45. $(y + 1)^2 = -6(x - 2)$ †

46. $(y - 2)^2 = -12(x + 3)$ †

47. $(x - 1)^2 = -10(y - 3)$ †

48. $(x + 2)^2 = -8(y + 3)$ †

49. $y = x^2 + 2x + 2$ †

50. $y = 3x^2 + 6x + 2$ †

51. $2y^2 + 4y - 2x + 1 = 0$ †

52. $3y^2 - 6y + x - 1 = 0$ †

53. $y + x^2 + x + \dfrac{5}{4} = 0$ †

54. $x = 8y^2 + 8y$ †

55. $x = 24y - 3y^2 - 2$ †

56. $y = 16x - 2x^2 + 3$ †

In Exercises 57–64, find two possible equations for the parabola with the given vertex V that passes through the given point P. The axis of symmetry of each parabola is parallel to a coordinate axis.

57. $V(0, 0), P(1, 2)$ †

58. $V(0, 0), P(-3, 2)$ †

59. $V(0, 1), P(2, 3)$ †

60. $V(1, 2), P(2, 1)$ †

61. $V(-2, 1), P(-3, 0)$ †

62. $V(1, -1), P(0, 0)$ †

63. $V(-1, 1), P(0, 2)$ †

64. $V(2, 3), P(3, -1)$ †

10. Focus: $\left(0, -\dfrac{1}{2}\right)$; directrix: $y = \dfrac{1}{2}$; graph: a

11. Focus: $\left(\dfrac{1}{2}, 0\right)$; directrix: $x = -\dfrac{1}{2}$; graph: g

12. Focus: $\left(\dfrac{4}{9}, 0\right)$; directrix: $x = -\dfrac{4}{9}$; graph: c

13. Focus: $\left(-\dfrac{4}{9}, 0\right)$; directrix: $x = \dfrac{4}{9}$; graph: f

B EXERCISES Applying the Concepts

65. **Satellite dish.** A parabolic satellite dish is 40 inches across and 20 inches deep (from the vertex to the plane of the rim). How far from the vertex must the signal-receiving (receptor) unit be located to ensure that it is at the focus of the parabolic cross sections? 5 in.

66. Repeat Exercise 65 assuming that the dish is 8 feet across and 3 feet deep. $\frac{4}{3}$ in.

67. **Solar heating.** A parabolic reflector mirror is to be used for the solar heating of water. The mirror is 18 feet across and 6 feet deep. Where should the heating element be placed to heat the water most rapidly? $3\frac{3}{8}$ ft from the vertex

68. **Flashlights.** A parabolic flashlight reflector is to be 4 inches across and 2 inches deep. Where should the lightbulb be placed? $\frac{1}{2}$ in. from the vertex

69. **Flashlight.** The shape of a parabolic flashlight reflector can be described by the equation $x = 4y^2$. Where should the lightbulb be placed? $\left(\frac{1}{16}, 0\right)$

70. Repeat Exercise 69 assuming that the parabolic flashlight reflector has the shape described by the equation $x = \frac{1}{2}y^2$. $\left(\frac{1}{2}, 0\right)$

71. **Satellite dish.** The shape of a parabolic satellite dish can be described by the equation $y = 4x^2$. Where should the microphone be placed? $\left(0, \frac{1}{16}\right)$

72. Repeat Exercise 71 assuming that the shape of the dish can be described by the equation $y = x^2$. $\left(0, \frac{1}{4}\right)$

73. **Suspension bridge.** The ends of a suspension bridge cable are fastened to two towers, which are 800 feet apart and 120 feet high. The cable hangs in the shape of a parabola and touches the roadbed midway between the towers. Find the length of the straight wire needed to suspend the roadbed from the cable at a distance of 250 feet from the tower. 16.875 ft

74. **Suspension bridge.** Repeat Exercise 73 assuming that the suspension bridge cable is at a height of 8 feet above the midway between the towers. 23.75 ft

75. **Kicking a football.** A football is kicked from the ground 70 yards along a parabolic path. The maximum height of the ball is 30 yards. Find the height of the ball 5 yards before it hits the ground. 7.96 yd

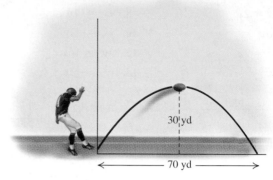

76. **Parabolic arch bridge.** A parabolic arch of a bridge has a 60-foot base and a height of 24 feet. Find the height of the arch at distances of 5, 10, and 20 feet from the center of the base. 5 ft: 23.33 ft; 10 ft: 21.33 ft; 20 ft: 13.33 ft

77. **Variable cost.** The average variable cost y, in dollars, of a monthly output of x tons of a company producing a metal is given by $y = \frac{1}{10}x^2 - 3x + 50$. Find the output and cost at the vertex of the parabola. Output: 15 tons; cost: $27.50

78. Repeat Exercise 77 assuming that the variable cost is given by $y = \frac{1}{8}x^2 - 2x + 60$. Output: 8 tons; cost: $52

C EXERCISES Beyond the Basics

79. Find the coordinates of the points at which the line $2x - 3y + 16 = 0$ intersects the parabola $y^2 = 16x$. (4, 8) and (16, 16)

80. Show that the line $y = 2x + 3$ intersects the parabola $y^2 = 24x$ at only one point. Find the point of intersection.

81. Find the equation of the directrix of a parabola with vertex $(-2, 2)$ and focus $(-6, 6)$. $y = x - 4$
 [*Hint:* Note that the axis of this parabola is *not* parallel to a coordinate axis. Recall that the vertex is midway between the focus and the directrix.]

82. A parabola has focus $(-4, 5)$, and the equation of its directrix is $3x - 4y = 18$.
 a. Find an equation for the axis of the parabola.
 b. Find the point of intersection of the axis and the directrix. $(2, -3)$
 c. Use part (b) to find the vertex of this parabola. $(-1, 1)$

83. Find the equation of two possible parabolas whose latus rectum is the line segment joining the points $(3, 5)$ and $(3, -3)$. $(y - 1)^2 = 8(x - 1)$ and $(y - 1)^2 = -8(x - 5)$

84. Repeat Exercise 83 assuming that the latus rectum is the line segment joining the points $(-5, 1)$ and $(3, 1)$.

85. Find an equation of the parabola with axis parallel to the y-axis, passing through the points $(0, 5)$, $(1, 4)$, and $(2, 7)$. †

86. Find an equation of the parabola with axis parallel to the x-axis, passing through the points $(-1, -1)$, $(3, 1)$, and $(4, 0)$. †

Answers:

80. The point of intersection is $\left(\frac{3}{2}, 6\right)$ 82. **a.** $y = -\frac{4}{3}x - \frac{1}{3}$

84. $(x + 1)^2 = 8(y + 1)$ and $(x + 1)^2 = -8(y - 3)$

87. Find the vertex, focus, directrix, and axis of the parabola

$$x^2 - 8x + 2y + 4 = 0. \ \dagger$$

88. Repeat Exercise 87 for the parabola

$$Ax^2 + Dx + Ey + F = 0, \quad A \neq 0, E \neq 0. \ \dagger$$

89. Find the vertex, focus, directrix, and axis of the parabola

$$y^2 + 3x - 6y + 15 = 0. \ \dagger$$

90. Repeat Exercise 89 for the parabola

$$Cy^2 + Dx + Ey + F = 0, \quad C \neq 0, D \neq 0. \ \dagger$$

In Exercises 91–94, use the following definition. A *tangent line* to a parabola is a line that is not parallel to the axis of the parabola and that intersects the parabola at exactly one point.

91. Find an equation of the tangent line at the point $(1, 3)$ on the parabola $y = 3x^2$ by using the following steps.

Step 1 Let m be the slope of the tangent line. Then the equation of the tangent line is
$y - 3 = m(x - 1)$. Solve this equation for y.

Step 2 Substitute the value of y from Step 1 in the equation $y = 3x^2$. You will obtain a quadratic equation in x.

Step 3 For the line in Step 1 to be the tangent line to the parabola, the roots of the quadratic equation in Step 2 must be equal. Set discriminant $= 0$ and solve for m.

Step 4 Use the value of m from Step 3 to write the equation of the tangent line to the parabola $y = 3x^2$.
$y = 6x - 3$

92. Show that the equation of the tangent line at a point (x_1, y_1) on the parabola $y = 4ax^2$ is $y = 8ax_1x - y_1$. [*Hint:* Follow the steps in the previous exercise and note that $y_1 = 4ax_1^2$.]

93. Show that the equation of the tangent line at point (x_1, y_1) on the parabola $x = 4ay^2$ is $y = \dfrac{1}{8ay_1}x + \dfrac{y_1}{2}$.

94. Reflection property of a parabola. Let $P(x_1 \ y_1)$ be a point on the parabola $y = 4ax^2$ with focus $F = (0, a)$ and the directrix $y = -a$. Let AP be the tangent line to the parabola. (See the figure.)

a. Use Exercise 92 to show that $d(F, A) = a + y_1$. Then use the definition of the parabola to show that $d(P, F) = a + y_1$. Conclude that $\angle\beta = \angle\gamma$.

This will suggest the following construction of the tangent line: Let B be a point on the axis of the parabola such that BP is perpendicular to the axis. Let V be the vertex of the parabola. Find the point A on the axis of the parabola such that $d(B, V) = d(A, V)$. Then the line AP is the tangent line to the parabola at the point P.

b. Use geometry to show that the angle between the y-axis and the line AP equals the angle between the lines AP and the line $x = x_1$ (which is parallel to the axis of parabola).

This demonstrates the reflection property of a parabola, which says that the tangent line to a parabola at a point P makes equal angles with

1. the line through P and the focus.
2. the axis of the parabola.

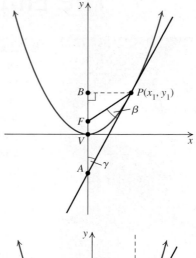

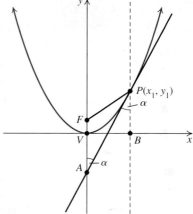

Critical Thinking

95. a. Is a parabola always the graph of a function? Why or why not? $\dagger$
  b. Find all possible values of the slope of the directrix of a parabola such that the parabola is always the graph of a function. 0

96. The equation $y^2 = 4kx$ describes a family of curves for each value of k. Sketch the graphs of the members of this family for $k = \pm 1, \pm 2, \pm 3$. What is the common characteristic of this family? $\dagger$

97. *True or False* The graph of $x = ay^2 + by + c \ (a \neq 0)$ is a parabola whose axis is parallel to the y-axis. False

GROUP PROJECTS

98. It can be proved (see page 492) that the distance d from a point (x_1, y_1) to a line $ax + by + c = 0$ is given by
$16x^2 + 9y^2 - 24xy - 158x - 194y + 976 = 0$
$$d = \frac{|ax_1 + by_1 + c|}{\sqrt{a^2 + b^2}}. \text{ Axis: } y = \frac{4}{3}x - \frac{1}{3}$$

Use this fact and the definition of a parabola to find an equation of the parabola whose focus is $(4, 5)$ and whose directrix is the line $3x + 4y = 7$. Find an equation for the axis of this parabola.

99. Given that the vertex of a parabola is $(6, -3)$ and the directrix is the line $3x - 5y + 1 = 0$, find the coordinates of the focus. $(9, -8)$

The Ellipse

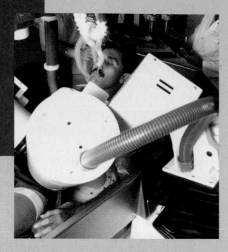

Lithotripter

| Before Starting this Section, Review | Objectives |

Before Starting this Section, Review

1. Distance formula (Section 1.1, page 5)
2. Completing the square (Appendix A, page 811)
3. Midpoint formula (Section 1.1, page 5)
4. Transformations of graphs (Section 1.5)
5. Symmetry (Section 1.1, page 8)

Objectives

1 Define an ellipse.
2 Find the equation of an ellipse.
3 Translate ellipses.
4 Use ellipses in applications.

WHAT IS LITHOTRIPSY?

Lithotripsy is a combination of two words: *litho* and *tripsy*. In Greek, the word *lith* denotes a stone and the word *tripsy* means "crushing." The complete medical term for lithotripsy is *extracorporeal shock-wave lithotripsy* (ESWL). Lithotripsy is a medical procedure in which a kidney stone is crushed into small sandlike pieces with the help of ultrasound high-energy shock waves, without breaking the skin.

The lithotripsy machine is called a *lithotripter*. The technology for the lithotripter was developed in Germany. The first successful treatment of a patient by a lithotripter was in February 1980 at Munich University. In the beginning, the patient had to lie in an elliptical tank filled with water and only small kidney stones were crushed. But as the medical research advanced, newer machines with better technology were manufactured. Now there is no need to keep the patient unconscious with an empty stomach, nor is the tub of water required.

In Example 5, you will see how the *reflecting property* of an ellipse is used in lithotripsy. ■

---

1 Define an ellipse.

Definition of Ellipse

ELLIPSE

An **ellipse** is the set of all points in the plane, the sum of whose distances from two fixed points is a constant. The fixed points are called the **foci** (the plural of *focus*) of the ellipse.

FIGURE 9.13 Drawing an ellipse

You can draw an ellipse with the use of thumbtacks, a fixed length of string, and a pencil. Stick a thumbtack at each of the two points (the foci) and tie one end of the string to each tack. Place a pencil inside the loop of the string and pull it taut. Move the pencil, keeping the string taut at all times. The pencil traces an ellipse, as shown in Figure 9.13.

The points of intersection of the ellipse with the line through the foci are called the **vertices** (plural of *vertex*). The line segment connecting the vertices is the **major axis** of the ellipse. The midpoint of the major axis is the **center** of the ellipse. The line segment that is perpendicular to the major axis at the center and with endpoints on the ellipse is called the **minor axis**. See Figure 9.14.

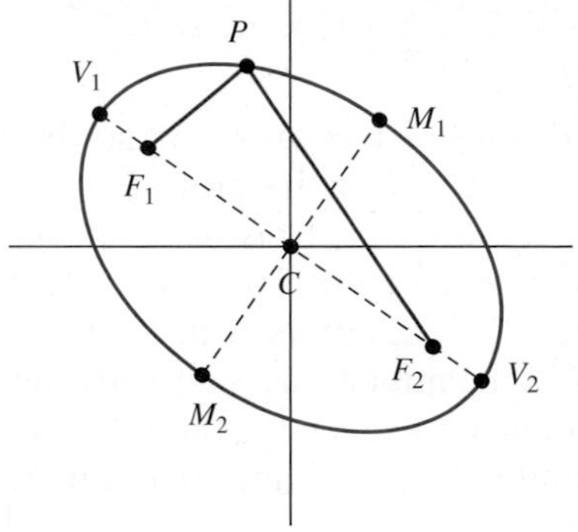

Points F_1 and F_2 are the foci.
Points V_1 and V_2 are the vertices.
Point C is the center.
Segment $\overline{V_1 V_2}$ is the major axis.
Segment $\overline{M_1 M_2}$ is the minor axis.
The sum of the distances
$PF_1 + PF_2$ from any point P on the
ellipse to the foci is constant.

FIGURE 9.14

2 Find the equation of an ellipse.

Equation of an Ellipse

To find a simple equation of an ellipse, we place the ellipse's major axis along one of the coordinate axes and the center of the ellipse at the origin. We will begin with the case in which the major axis lies along the x-axis. Let $F_1(-c, 0)$ and $F_2(c, 0)$ be the coordinates of the foci. See Figure 9.15. Note that the center at the origin is the midpoint of the line segment having the foci as endpoints. To simplify the equation, it is customary to let $2a$ be the constant distance referred to in the definition. Let $P(x, y)$ be a point on the ellipse. Then

$$d(P, F_1) + d(P, F_2) = 2a \qquad \text{Definition of ellipse}$$

$$\sqrt{(x + c)^2 + y^2} + \sqrt{(x - c)^2 + y^2} = 2a \qquad \text{Distance formula}$$

$$\sqrt{(x + c)^2 + y^2} = 2a - \sqrt{(x - c)^2 + y^2} \qquad \text{Isolate a radical.}$$

$$(x + c)^2 + y^2 = (2a - \sqrt{(x - c)^2 + y^2})^2 \qquad \text{Square both sides.}$$

$$x^2 + 2cx + c^2 + y^2 = 4a^2 - 4a\sqrt{(x - c)^2 + y^2} + (x - c)^2 + y^2 \qquad \text{Expand squares.}$$

$$x^2 + 2cx + c^2 + y^2 = 4a^2 - 4a\sqrt{(x - c)^2 + y^2} + x^2 - 2cx + c^2 + y^2 \qquad \text{Expand squares.}$$

$$4cx - 4a^2 = -4a\sqrt{(x - c)^2 + y^2} \qquad \text{Isolate the radical and simplify.}$$

$$cx - a^2 = -a\sqrt{(x - c)^2 + y^2} \qquad \text{Divide both sides by 4.}$$

$$(cx - a^2)^2 = a^2[(x - c)^2 + y^2] \qquad \text{Square both sides.}$$

$$c^2x^2 - 2a^2cx + a^4 = a^2(x^2 - 2cx + c^2 + y^2) \qquad \text{Expand squares.}$$

$$(c^2 - a^2)x^2 - a^2y^2 = a^2c^2 - a^4 \qquad \text{Simplify and rearrange.}$$

$$(a^2 - c^2)x^2 + a^2y^2 = a^2(a^2 - c^2). \quad (1) \qquad \text{Multiply both sides by } -1 \text{ and factor } a^4 - a^2c^2.$$

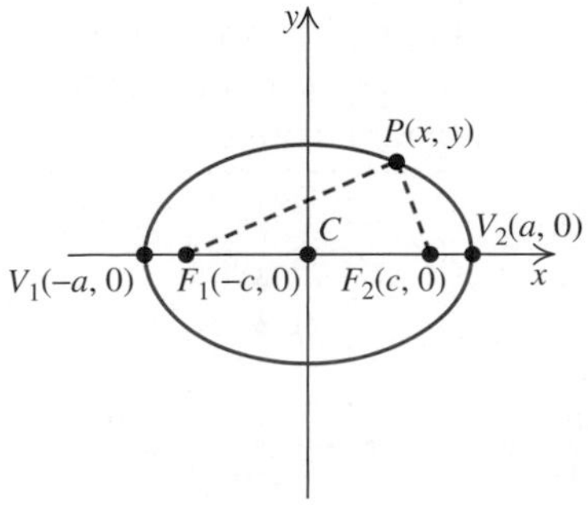

FIGURE 9.15

Now consider triangle F_1PF_2 in Figure 9.15. The *triangle inequality* from geometry states that the sum of the lengths of any two sides of a triangle is greater than the length of the third side; so for this triangle, we have

$$d(P, F_1) + d(P, F_2) > d(F_1, F_2)$$

$$2a > 2c \qquad 2a = d(P, F_1) + d(P, F_2); d(F_1, F_2) = 2c$$

$$a > c \qquad \text{Divide both sides by 2.}$$

$$a^2 > c^2. \qquad \text{Square both sides.}$$

Therefore, $a^2 - c^2 > 0$. Letting $b^2 = a^2 - c^2$, we obtain

$$(a^2 - c^2)x^2 + a^2y^2 = a^2(a^2 - c^2) \qquad \text{Equation (1)}$$

$$b^2x^2 + a^2y^2 = a^2b^2 \qquad \text{Replace } a^2 - c^2 \text{ with } b^2.$$

$$\frac{x^2}{a^2} + \frac{y^2}{b^2} = 1 \quad (2) \qquad \text{Divide both sides by } a^2b^2.$$

The last equation is the **standard form of the equation of an ellipse** with center $(0, 0)$ and foci $(-c, 0)$ and $(c, 0)$, where $b^2 = a^2 - c^2$.

To find the x-intercepts for the graph of this ellipse, set $y = 0$ in equation (2) to get $\dfrac{x^2}{a^2} = 1$.

The solutions are $x = \pm a$; so the x-intercepts of the graph are $-a$ and a. The ellipse's vertices are therefore $(-a, 0)$ and $(a, 0)$. The distance between the vertices $V_1(-a, 0)$ and $V_2(a, 0)$ is $2a$, so the length of the major axis is $2a$.

The y-intercepts of the ellipse (found by setting $x = 0$ in equation (2)) are $-b$ and b; so the endpoints of the minor axis are $(0, -b)$ and $(0, b)$. Consequently, the length of the minor axis is $2b$. (See the left-hand figure in the table below.)

Similarly, by reversing the roles of x and y, an equation of the ellipse with center $(0, 0)$ and foci $(0, -c)$ and $(0, c)$ on the y-axis is given by

$$\frac{x^2}{b^2} + \frac{y^2}{a^2} = 1, \text{where } b^2 = a^2 - c^2. \quad (3)$$

In equation (3), the major axis, of length $2a$, is along the y-axis and the minor axis, of length $2b$, is along the x-axis. (See the right-hand figure in the table below.)

If the major axis of an ellipse is along or parallel to the x-axis, the ellipse is called a **horizontal ellipse**, while an ellipse with major axis along or parallel to the y-axis is called a **vertical ellipse**.

We summarize the facts about horizontal and vertical ellipses with center $(0, 0)$. By the equation of a major or minor axis, we mean the equation of the line on which that axis lies.

Main facts about an ellipse with center $(0, 0)$

Standard Equation	$\dfrac{x^2}{a^2} + \dfrac{y^2}{b^2} = 1; \; a > b > 0$ (Horizontal ellipse)	$\dfrac{x^2}{b^2} + \dfrac{y^2}{a^2} = 1; \; a > b > 0$ (Vertical ellipse)
Relationship between a, b, and c	$b^2 = a^2 - c^2$	$b^2 = a^2 - c^2$
Major axis along	x-axis	y-axis
Length of major axis	$2a$	$2a$
Minor axis along	y-axis	x-axis
Length of minor axis	$2b$	$2b$
Vertices	$(\pm a, 0)$	$(0, \pm a)$
Foci	$(\pm c, 0)$	$(0, \pm c)$
Endpoints of minor axis	$(0, \pm b)$	$(\pm b, 0)$
Symmetry	The graph is symmetric with respect to the x-axis, y-axis, and origin.	The graph is symmetric with respect to the x-axis, y-axis, and origin.
Graph	$$\frac{x^2}{a^2} + \frac{y^2}{b^2} = 1$$ Horizontal ellipse	$$\frac{x^2}{b^2} + \frac{y^2}{a^2} = 1$$ Vertical ellipse

TECHNOLOGY CONNECTION

To graph the ellipse $\dfrac{x^2}{4} + \dfrac{y^2}{9} = 1$, first solve the equation for y

$$\frac{y^2}{9} = 1 - \frac{x^2}{4}$$

or $y^2 = 9\left(1 - \dfrac{x^2}{4}\right)$

so that $y = \pm 3\sqrt{1 - \dfrac{x^2}{4}}$.

Then graph both functions

$Y_1 = 3\sqrt{1 - \dfrac{x^2}{4}}$ and

$Y_2 = -3\sqrt{1 - \dfrac{x^2}{4}}$.

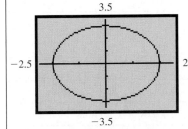

EXAMPLE 1   Finding an Equation of an Ellipse

Find the standard form of the equation of the ellipse that has vertex $(5, 0)$ and foci $(\pm 4, 0)$.

SOLUTION

Since the foci are $(-4, 0)$ and $(4, 0)$, the major axis is on the x-axis. You know that $c = 4$ and $a = 5$. Now find b^2:

$$b^2 = a^2 - c^2 \qquad \text{Relationship between } a, b, \text{ and } c$$
$$b^2 = (5)^2 - (4)^2 = 9 \qquad \text{Replace } c \text{ with 4 and } a \text{ with 5.}$$

Substituting 25 for a^2 and 9 for b^2 in the standard form for a horizontal ellipse, we get

$$\frac{x^2}{25} + \frac{y^2}{9} = 1. \qquad\qquad ■ ■ ■$$

Practice Problem 1   Find the standard form of the equation of the ellipse that has vertex $(0, 10)$ and foci $(0, -8)$ and $(0, 8)$.   ■

EXAMPLE 2   Graphing an Ellipse

Sketch a graph of the ellipse whose equation is $9x^2 + 4y^2 = 36$. Find the foci of the ellipse.

SOLUTION

First, write the equation in standard form:

$$9x^2 + 4y^2 = 36 \qquad \text{Given equation}$$
$$\frac{9x^2}{36} + \frac{4y^2}{36} = 1 \qquad \text{Divide both sides by 36.}$$
$$\frac{x^2}{4} + \frac{y^2}{9} = 1 \qquad \text{Simplify.}$$

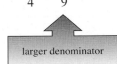

larger denominator

Because the denominator in the y^2-term is larger than the denominator in the x^2-term, the ellipse is a vertical ellipse. Here $a^2 = 9$ and $b^2 = 4$, so $c^2 = a^2 - b^2 = 9 - 4 = 5$. Thus, $a = 3$, $b = 2$, and $c = \sqrt{5}$. From the table on page 620, we know the following features of the ellipse:

Vertices:	$(0, \pm a) = (0, \pm 3)$
Foci:	$(0, \pm c) = (0, \pm\sqrt{5})$
Length of major axis:	$2a = 2(3) = 6$
Length of minor axis:	$2b = 2(2) = 4$

The graph of the ellipse is shown in Figure 9.16.   ■ ■ ■

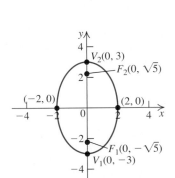

FIGURE 9.16 Ellipse with equation $9x^2 + 4y^2 = 36$

Practice Problem 2   Sketch a graph of the ellipse whose equation is $4x^2 + y^2 = 16$.   ■

3 Translate ellipses.

Translations of Ellipses

The graph of the equation

$$\frac{x^2}{a^2} + \frac{y^2}{b^2} = 1$$

is an ellipse with center $(0, 0)$ and major axis along a coordinate axis. As in the case of a parabola, horizontal and vertical shifts can be used to obtain the graph of an ellipse whose equation is

$$\frac{(x - h)^2}{a^2} + \frac{(y - k)^2}{b^2} = 1.$$

The center of such an ellipse is (h, k), and its major axis is parallel to a coordinate axis.

Main facts about horizontal and vertical ellipses with center (h, k)		
Standard Equation	$\dfrac{(x - h)^2}{a^2} + \dfrac{(y - k)^2}{b^2} = 1;$ $a > b > 0$ **(Horizontal ellipse)**	$\dfrac{(x - h)^2}{b^2} + \dfrac{(y - k)^2}{a^2} = 1;$ $a > b > 0$ **(Vertical ellipse)**
Center	(h, k)	(h, k)
Major axis along the line	$y = k$	$x = h$
Length of major axis	$2a$	$2a$
Minor axis along the line	$x = h$	$y = k$
Length of minor axis	$2b$	$2b$
Vertices	$(h + a, k), (h - a, k)$	$(h, k + a), (h, k - a)$
Endpoints of minor axis	$(h, k - b), (h, k + b)$	$(h - b, k), (h + b, k)$
Foci	$(h + c, k), (h - c, k)$	$(h, k + c), (h, k - c)$
Equation involving $a, b,$ and c	$c^2 = a^2 - b^2$	$c^2 = a^2 - b^2$
Symmetry	The graph is symmetric about the lines $x = h$ and $y = k$.	The graph is symmetric about the lines $x = h$ and $y = k$.
Graph	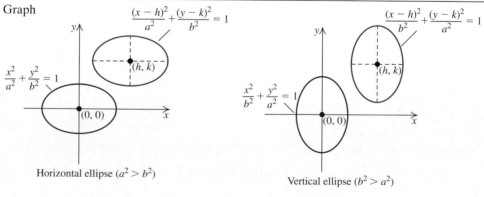	

Horizontal ellipse $(a^2 > b^2)$

Vertical ellipse $(b^2 > a^2)$

RECALL

The midpoint of the segment joining (x_1, y_1) and (x_2, y_2) is

$$\left(\frac{x_1 + x_2}{2}, \frac{y_1 + y_2}{2} \right).$$

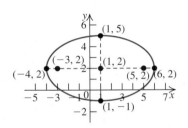

Center $(1, 2)$

$F_1(-3, 2)$ $F_2(5, 2)$

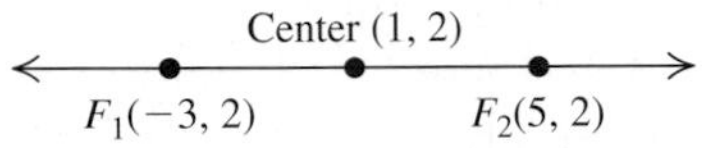

FIGURE 9.17 Finding the center given the foci

FIGURE 9.18 Ellipse with foci $(-3, 2)$ and $(5, 2)$

EXAMPLE 3   Finding the Equation of an Ellipse

Find an equation of the ellipse that has foci $(-3, 2)$ and $(5, 2)$ and that has a major axis of length 10.

SOLUTION

Since the foci $(-3, 2)$ and $(5, 2)$ lie on the horizontal line $y = 2$, the ellipse is a horizontal ellipse. The center of the ellipse is the midpoint of the line segment joining the foci. Using the midpoint formula yields

$$h = \frac{-3 + 5}{2} = 1 \quad \text{and} \quad k = \frac{2 + 2}{2} = 2.$$

The center of the ellipse is $(1, 2)$. See Figure 9.17. Since the length of the major axis is 10, the vertices must be at a distance $a = 5$ units from the center.

In Figure 9.17, the foci are four units from the center. Thus, $c = 4$. Now find b^2:

$$b^2 = a^2 - c^2 = (5)^2 - (4)^2 = 9$$

Since the major axis is horizontal, the standard form of the equation of the ellipse is

$$\frac{(x - h)^2}{a^2} + \frac{(y - k)^2}{b^2} = 1 \qquad a > b > 0$$

$$\frac{(x - 1)^2}{25} + \frac{(y - 2)^2}{9} = 1 \qquad \begin{array}{l}\text{Replace } h \text{ with } 1, k \text{ with } 2, \\ a^2 \text{ with } 25, \text{ and } b^2 \text{ with } 9.\end{array}$$

Now for this horizontal ellipse with center $(1, 2)$, we have $a = 5, b = 3$, and $c = 4$.

Vertices

$$(h \pm a, k) = (1 \pm 5, 2) = (-4, 2) \text{ and } (6, 2)$$

Endpoints of Minor Axis

$$(h, k \pm b) = (1, 2 \pm 3) = (1, -1) \text{ and } (1, 5)$$

Use the center and these four points to sketch the graph shown in Figure 9.18.   ■ ■ ■

Practice Problem 3   Find an equation of the ellipse that has foci at $(2, -3)$ and $(2, 5)$ and has a major axis of length 10.   ■

Both standard forms of the equations of an ellipse with center (h, k) can be put into the form

$$Ax^2 + Cy^2 + Dx + Ey + F = 0, \qquad (4)$$

where neither A nor C is zero and both have the same sign. For example, consider the ellipse with equation

$$\frac{(x - 1)^2}{4} + \frac{(y + 2)^2}{9} = 1.$$

$$9(x - 1)^2 + 4(y + 2)^2 = 36 \qquad \text{Multiply both sides by 36.}$$
$$9(x^2 - 2x + 1) + 4(y^2 + 4y + 4) = 36 \qquad \text{Expand squares.}$$
$$9x^2 - 18x + 9 + 4y^2 + 16y + 16 = 36 \qquad \text{Distributive property}$$
$$9x^2 + 4y^2 - 18x + 16y - 11 = 0 \qquad \text{Simplify.}$$

Comparing this equation with equation (4), we have

$$A = 9, C = 4, D = -18, E = 16, \text{ and } F = -11.$$

Conversely, equation (4) can be put into a standard form for an ellipse by completing the squares in both of the variables x and y.

EXAMPLE 4	Converting to Standard Form

Find the center, vertices, and foci of the ellipse with equation

$$3x^2 + 4y^2 + 12x - 8y - 32 = 0.$$

SOLUTION

Complete squares on x and y.

$3x^2 + 4y^2 + 12x - 8y - 32 = 0$	Original equation
$(3x^2 + 12x) + (4y^2 - 8y) = 32$	Group like terms.
$3(x^2 + 4x) + 4(y^2 - 2y) = 32$	Factor out 3 and 4.
$3(x^2 + 4x + 4) + 4(y^2 - 2y + 1) = 32 + 12 + 4$	Complete squares on x and y: add $3 \cdot 4$ and $4 \cdot 1$.
$3(x + 2)^2 + 4(y - 1)^2 = 48$	Simplify.
$\dfrac{(x + 2)^2}{16} + \dfrac{(y - 1)^2}{12} = 1$	Divide both sides by 48. (16 is the larger denominator.)

The last equation is the standard form for a horizontal ellipse with center $(-2, 1)$, $a^2 = 16$, $b^2 = 12$, and $c^2 = a^2 - b^2 = 16 - 12 = 4$. Thus, $a = 4$, $b = \sqrt{12} = 2\sqrt{3}$, and $c = 2$.

From the table on page 620, we have the following information:
The length of the major axis is $2a = 8$.
The length of the minor axis is $2b = 4\sqrt{3}$.

Center:	$(h, k) = (-2, 1)$
Foci:	$(h \pm c, k) = (-2 \pm 2, 1)$
	$= (-4, 1)$ and $(0, 1)$
Vertices:	$(h \pm a, k) = (-2 \pm 4, 1)$
	$= (-6, 1)$ and $(2, 1)$
Endpoints of minor axis:	$(h, k \pm b) = (-2, 1 \pm 2\sqrt{3})$
	$= (-2, 1 + 2\sqrt{3})$ and $(-2, 1 - 2\sqrt{3})$
	$\approx (-2, 4.46)$ and $(-2, -2.46)$

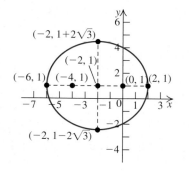

FIGURE 9.19 Ellipse with equation $3x^2 + 4y^2 + 12x - 8y - 32 = 0$

The graph of the ellipse is shown in Figure 9.19. ■ ■ ■

Practice Problem 4   Find the center, vertices, and foci of the ellipse with equation

$$x^2 + 4y^2 - 6x + 8y - 29 = 0.$$ ■

A *circle* is considered as an ellipse in which the two foci coincide at the center. Conversely, if the two foci of an ellipse coincide, then $2c = 0$ (remember, $2c$ is the distance between the two foci of an ellipse); so $c = 0$. Again, from the relationship $c^2 = a^2 - b^2 = 0$, we have $a = b$. Thus, the equation for the ellipse becomes the equation for a circle of radius $r = a$, having the same center as that of the ellipse.

4 Use ellipses in applications.

Applications

Ellipses have many applications. We mention just a few:

1. The orbits of the planets are ellipses with the sun at one focus. This fact alone is sufficient to explain the overwhelming importance of ellipses since the seventeenth century, when Kepler and Newton did their monumental work in astronomy.

2. Newton also reasoned that comets move in elliptical orbits about the sun. His friend Edmund Halley used this information to predict that a certain comet (now called Halley's comet) reappears about every 77 years. Halley saw the comet in 1682 and correctly predicted its return in 1759. The last sighting of the

comet was in 1986, and it is due to return in 2061. A recent study shows that Halley's comet has made approximately 2000 cycles so far and has about the same number to go before the sun erodes it away completely. (See Exercise 73.)

3. We can calculate the distance traveled by a planet in one orbit around the sun.

4. **The reflecting property of ellipses.** The *reflecting property* for an ellipse says that a ray of light originating at one focus will be reflected to the other focus. See Figure 9.20. Sound waves also follow such paths. This property is used in the construction of "whispering galleries," such as the gallery at St. Paul's Cathedral in London. Such rooms have ceilings whose cross sections are elliptical with common foci. As a result, sounds emanating from one focus are reflected by the ceiling to the other focus. Thus, a whisper at one focus may not be audible at all at a nearby place, but may nevertheless be clearly heard far off at the other focus.

Example 5 illustrates how the *reflecting property* of an ellipse is used in lithotripsy. Recall from the introduction to this section, a *lithotripter* is a machine that is designed to crush kidney stones into small sandlike pieces with the help of high-energy shock waves. To focus these shock waves accurately, a lithotripter uses the reflective property of ellipses. A patient is carefully positioned so that the kidney stone is at one focus of the ellipsoid (a three-dimensional ellipse) and the shock-producing source is placed at the other focus. The high-energy shock waves generated at this focus are concentrated on the kidney stone at the other focus, thus pulverizing it without harming the patient.

Edmund Halley

(1656–1742)

Edmund Halley was a friend of Isaac Newton. Although he is known primarily for calculating the orbit of Halley's comet, he was also a biologist, a geologist, a sea captain, a spy, and the author of the first actuarial mortality tables.

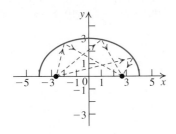

FIGURE 9.20

EXAMPLE 5   Lithotripsy

An elliptical water tank has a major axis of length 6 feet and a minor axis of length 4 feet. The source of high-energy shock waves from a lithotripter is placed at one focus of the tank. To smash the kidney stone of a patient, how far should the stone be positioned from the source?

SOLUTION

Since the length of the major axis of the ellipse is 6 feet, we have $2a = 6$; so $a = 3$. Similarly, the minor axis of 4 feet gives $2b = 4$, or $b = 2$. To find c, we use the equation $c^2 = a^2 - b^2$. We have

$$c^2 = (3)^2 - (2)^2 = 5. \text{ Therefore, } c = \pm\sqrt{5}.$$

If we position the center of ellipse at $(0, 0)$ and the major axis along the x-axis, then the foci of the ellipse are $(-\sqrt{5}, 0)$ and $(\sqrt{5}, 0)$. The distance between these foci is $2\sqrt{5} \approx 4.472$ feet. The kidney stone should be positioned 4.472 feet from the source of the shock waves. ■ ■ ■

Practice Problem 5   In Example 5, if the tank has a major axis of 8 feet and a minor axis of 4 feet, how far should the stone be positioned from the source? ■

SECTION 9.3 ■ Exercises

A EXERCISES Basic Skills and Concepts

1. An ellipse is the set of all points in the plane, the ___sum___ of whose distances from two fixed points is a constant.

2. The points of intersection of the ellipse with the line through the foci are called ___vertices___ of the ellipse.

3. The standard equation of an ellipse with center $(0,0)$, vertices $(\pm a, 0)$, and foci $(\pm c, 0)$ is $\frac{x^2}{a^2} + \frac{y^2}{b^2} = 1$, where $b^2 = $ ___$a^2 - c^2$___.

4. In the equation $\frac{x^2}{m^2} + \frac{y^2}{n^2} = 1$, if **(i)** $m^2 > n^2$, the graph is a(n) ___horizontal___ ellipse; **(ii)** if $m^2 < n^2$, the graph is a(n) ___vertical___ ellipse; **(iii)** if $m^2 = n^2$, the graph is a(n) ___circle___.

†Due to space constrictions, answers to these exercises may be found in the Answers beginning on page A–1 in the back of the book.

5. *True or False* The ellipse $\dfrac{x^2}{a^2} + \dfrac{y^2}{b^2} = 1$ lies inside the box formed by the four lines $x = \pm a, y = \pm b.$ True

6. *True or False* The graph of $Ax^2 + Cy^2 + Dx + Ey + F = 0$ is a degenerate conic or an ellipse if $AC > 0.$
   True (treating a circle as a special kind of ellipse)

In Exercises 7–26, find the vertices and foci for each ellipse. Graph each equation.

7. $\dfrac{x^2}{16} + \dfrac{y^2}{4} = 1$ †

8. $\dfrac{x^2}{4} + \dfrac{y^2}{16} = 1$ †

9. $\dfrac{x^2}{9} + y^2 = 1$ †

10. $x^2 + \dfrac{y^2}{4} = 1$ †

11. $\dfrac{x^2}{25} + \dfrac{y^2}{16} = 1$ †

12. $\dfrac{x^2}{25} + \dfrac{y^2}{9} = 1$ †

13. $\dfrac{x^2}{16} + \dfrac{y^2}{36} = 1$ †

14. $\dfrac{x^2}{9} + \dfrac{y^2}{16} = 1$ †

15. $x^2 + y^2 = 4$ †

16. $x^2 + y^2 = 16$ †

17. $x^2 + 4y^2 = 4$

18. $9x^2 + y^2 = 9$ †

19. $9x^2 + 4y^2 = 36$ †

20. $4x^2 + 25y^2 = 100$ †

21. $3x^2 + 4y^2 = 12$ †

22. $4x^2 + 5y^2 = 20$ †

23. $5x^2 - 10 = -2y^2$ †

24. $3y^2 = 21 - 7x^2$ †

25. $2x^2 + 3y^2 = 7$ †

26. $3x^2 + 4y^2 = 11$ †

In Exercises 27–40, find the standard form of the equation for the ellipse satisfying the given conditions. Graph the equation.

27. Foci: $(\pm 1, 0)$; vertex $(3, 0)$ †

28. Foci: $(\pm 3, 0)$; vertex $(5, 0)$ †

29. Foci: $(0, \pm 2)$; vertex $(0, 4)$ †

30. Foci: $(0, \pm 3)$; vertex $(0, -6)$ †

31. Foci: $(\pm 4, 0)$; y-intercepts: ± 3 †

32. Foci: $(\pm 3, 0)$; y-intercepts: ± 4 †

33. Foci: $(0, \pm 2)$; x-intercepts: ± 4 †

34. Foci: $(0, \pm 3)$; x-intercepts: ± 5 †

35. Center: $(0, 0)$, vertical major axis of length 10, and minor axis of length 6 †

36. Center: $(0, 0)$, vertical major axis of length 8, and minor axis of length 4 †

37. Center: $(0, 0)$; vertices: $(\pm 6, 0)$; $c = 3$ †

38. Center: $(0, 0)$; vertices: $(0, \pm 5)$; $c = 3$ †

39. Center: $(0, 0)$; foci: $(0, \pm 2)$; $b = 3$ †

40. Center: $(0, 0)$; foci: $(\pm 3, 0)$; $b = 2$ †

In Exercises 41–56, find the center, foci, and vertices of each ellipse. Graph each equation.

41. $\dfrac{(x - 1)^2}{4} + \dfrac{(y - 1)^2}{9} = 1$ †

42. $\dfrac{(x - 1)^2}{4} + \dfrac{(y - 1)^2}{16} = 1$ †

43. $\dfrac{x^2}{16} + \dfrac{(y + 3)^2}{4} = 1$ †

44. $\dfrac{(x + 2)^2}{4} + \dfrac{y^2}{9} = 1$ †

45. $3(x - 1)^2 + 4(y + 2)^2 = 12$ †

46. $9(x + 1)^2 + 4(y - 2)^2 = 36$ †

47. $4(x + 3)^2 + 5(y - 1)^2 = 20$ †

48. $25(x - 2)^2 + 4(y + 5)^2 = 100$ †

49. $5x^2 + 9y^2 + 10x - 36y - 4 = 0$ †

50. $9x^2 + 4y^2 + 72x - 24y + 144 = 0$ †

51. $9x^2 + 5y^2 + 36x - 40y + 71 = 0$ †

52. $3x^2 + 4y^2 + 12x - 16y - 32 = 0$ †

53. $x^2 + 2y^2 - 2x + 4y + 1 = 0$ †

54. $2x^2 + 2y^2 - 12x - 8y + 3 = 0$ †

55. $2x^2 + 9y^2 - 4x + 18y + 12 = 0$ No graph

56. $3x^2 + 2y^2 + 12x - 4y + 15 = 0$ No graph

In Exercises 57–60, use a graphing utility to sketch the graph of each ellipse.

57. $\dfrac{x^2}{3} + \dfrac{y^2}{5} = 1$ †

58. $\dfrac{x^2}{6} + \dfrac{y^2}{4} = 1$ †

59. $x^2 + 3y^2 = 21$ †

60. $5x^2 + y^2 = 10$ †

B EXERCISES Applying the Concepts

61. **Semielliptical arch.** A portion of an arch in the shape of a *semiellipse* (half of an ellipse) is 50 feet wide at the base and has a maximum height of 20 feet. Find the height of the arch at a distance of 10 feet from the center. (See the figure.) 18.3 ft

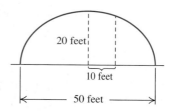

62. In Exercise 61, find the distance between the two points at the base where the height of the arch is 10 feet. 43.3 ft

63. **Semielliptical arch bridge.** A bridge is built in the shape of a semielliptical arch. The span of the bridge is 150 meters, and its maximum height is 45 meters. There are two vertical supports, each at a distance of 25 meters from the central position. Find their heights. 42.4 m

64. **Elliptical billiard table.** Some billiard tables (similar to pool tables, but without pockets) are manufactured in the shape of an ellipse. The foci of such tables are plainly marked for the convenience of the players. An elliptical billiard table is 6 feet long and 4 feet wide. Find the location of the foci. 2.2 ft from the center, along the major axis

65. "Ellipti-pool." In Exercise 64, suppose there is a pocket at one of the foci. A pool shark places the ball at the table and bets that he will be able to make the ball fall into the pocket 10 out of 10 times. Explain why you should or should not accept the bet. †

66. Whispering gallery. The vertical cross section of the dome of Statuary Hall in Washington, D.C., is semi-elliptical. The hall is 96 feet long and 23 feet high. It is said that John C. Calhoun used the whispering-gallery phenomenon to eavesdrop on his adversaries. Where should Calhoun and his foes be standing for the maximum whispering effect?

67. Whispering gallery. The vertical cross section of the dome of the Mormon Tabernacle in Salt Lake City, Utah, is semielliptical in shape. The cross section is 250 feet long with a maximum height of 80 feet. Where should two people stand to maximize the whispering effect?

In Exercises 68–71, use the fact (discovered by Johannes Kepler in 1609) that the planets move in elliptical orbits with the sun at one of the foci. Astronomers have measured the *perihelion* (the smallest distance from the planet to the sun) and *aphelion* (the largest distance from the planet to the sun) for each of the planets. The distances given in Table 9.1 are in millions of miles.

TABLE 9.1

Planet	Perihelion	Aphelion
Mercury	28.56	43.88
Venus	66.74	67.68
Earth	91.38	94.54
Mars	128.49	154.83
Jupiter	460.43	506.87
Saturn	837.05	936.37
Uranus	1699.45	1866.59
Neptune	2771.72	2816.42
Pluto	2749.57	4582.61

68. Mars. Find an equation of Mars's orbit about the sun.

Hint: Set up a coordinate system with the sun at one focus and the major axis lying on the *x*-axis. (See the figure.) Calculate a from the equation $2a =$ aphelion $+$ perihelion. Calculate c from the equation $c = a -$ perihelion. Calculate b^2 from the equation $b^2 = a^2 - c^2$. Write the equation $\dfrac{x^2}{a^2} + \dfrac{y^2}{b^2} = 1$. †

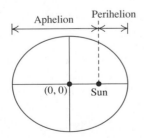

69. Earth. Write an equation for the orbit of Earth about the sun. †

Answers:

66. 5.9 ft from the wall on the opposite sides, along the major axis

67. 29 ft from the wall on the opposite sides, along the major axis.

70. Mercury. Write an equation for the orbit of the planet Mercury about the sun. †

71. Saturn. Write an equation for the orbit of the planet Saturn about the sun. †

72. Moon. The moon orbits Earth in an elliptical path with Earth at one focus. The major and minor axes of the orbit have lengths of 768,806 kilometers and 767,746 kilometers, respectively. Find the perihelion and aphelion of the orbit.
Perihelion: 364,224.1427 km; aphelion: 404,581.8573 km

73. Halley's comet. In 1682, Edmund Halley (1656–1742) identified the orbit of a certain comet (now called Halley's comet) as an elliptical orbit about the sun, with the sun at one focus. He calculated the length of its major axis to be approximately 5.39×10^9 kilometers and its minor axis to be approximately 1.36×10^9 kilometers. Find the perihelion and aphelion of the orbit of Halley's comet.

74. Orbit of a satellite. Assume that Earth is a sphere with a radius of 3960 miles. A satellite has an elliptical orbit around Earth with the center of Earth as one of its foci. The maximum and minimum heights of the satellite from the surface of Earth are 164 miles and 110 miles, respectively. Find an equation for the orbit of the satellite.
$$\frac{x^2}{16{,}785{,}409} + \frac{y^2}{16{,}784{,}680} = 1$$

C EXERCISES Beyond the Basics

75. Find an equation of the ellipse in standard form with center $(0, 0)$ and with major axis of length 10, passing through the point $\left(-3, \dfrac{16}{5}\right)$. †

76. Find an equation of the ellipse in standard form with center $(0, 0)$ and with major axis of length 6, passing through the point $\left(1, \dfrac{3\sqrt{5}}{2}\right)$. No solution

77. Find the lengths of the major and minor axes of the ellipse
$$\frac{x^2}{b^2} + \frac{y^2}{a^2} = 1,$$
given that the ellipse passes through the points $(2, 1)$ and $(1, -3)$.   Major axis: $\dfrac{2\sqrt{105}}{3}$; minor axis: $\dfrac{\sqrt{70}}{2}$

73. Perihelion: 8.72×10^7 km; aphelion: 5.3028×10^9 km

78. Find an equation of the ellipse in standard form with center $(2, 1)$, passing through the points $(10, 1)$ and $(6, 2)$.

79. The **eccentricity** of an ellipse, denoted by e, is defined as
$$e = \frac{\text{Distance between the foci}}{\text{Distance between the vertices}} = \frac{2c}{2a} = \frac{c}{a}.$$ What happens when $e = 0$?

In Exercises 80–84, use the definition of e given in Exercise 79 to determine the eccentricity of each ellipse. (This e has nothing to do with the number e, discussed in Chapter 3, that is the base for natural logarithms.)

80. $\dfrac{x^2}{16} + \dfrac{y^2}{9} = 1$ $\dfrac{\sqrt{7}}{4}$

81. $20x^2 + 36y^2 = 720$ $\dfrac{2}{3}$

82. $x^2 + 4y^2 = 1$ $\dfrac{\sqrt{3}}{2}$

83. $\dfrac{(x+1)^2}{25} + \dfrac{(y-2)^2}{9} = 1$

84. $x^2 + 2y^2 - 2x + 4y + 1 = 0$

85. An ellipse has its major axis along the x-axis and its minor axis along the y-axis. Its eccentricity is $\dfrac{1}{2}$, and the distance between the foci is 4. Find an equation of the ellipse.

86. Find an equation of the ellipse whose major axis has end points $(-3, 0)$ and $(3, 0)$ and that has eccentricity $\dfrac{1}{3}$.

87. A point $P(x, y)$ moves so that its distance from the point $(4, 0)$ is always one-half its distance from the line $x = 16$. Show that the equation of the path of P is an ellipse of eccentricity $\dfrac{1}{2}$.

88. A point, $P(x, y)$ moves so that its distance from the point $(0, 2)$ is always one-third its distance from the line $y = 10$. Show that the equation of the path of P is an ellipse of eccentricity $\dfrac{1}{3}$.

89. The **latus rectum of an ellipse** is the line segment that passes through a focus and that is perpendicular to the major axis and has endpoints on the ellipse. Show that the length of the latus rectum of an ellipse is $\dfrac{2b^2}{a}$, where b is half the length of the minor axis and a is half the length of the major axis.

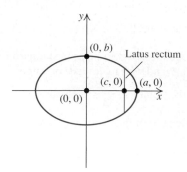

90. Find an equation of the ellipse with center $(0, 0)$, major axis along the x-axis, eccentricity $\dfrac{1}{\sqrt{2}}$, and latus rectum of length 3 units.

In Exercises 91–94, find all points of intersection of the given curves by solving systems of nonlinear equations. Sketch the graphs of the curves and show the points of intersection.

91. $\begin{cases} x + 3y = -2 \\ 4x^2 + 3y^2 = 7 \end{cases}$ †

92. $\begin{cases} x^2 = 2y \\ 2x^2 + y^2 = 12 \end{cases}$ †

93. $\begin{cases} x^2 + y^2 = 20 \\ 9x^2 + y^2 = 36 \end{cases}$ †

94. $\begin{cases} 9x^2 + 16y^2 = 36 \\ 18x^2 + 5y^2 = 45 \end{cases}$ †

In Exercises 95 and 96, use the following definition. A *tangent line* to an ellipse is a line that intersects the ellipse at exactly one point.

95. Find the equation of the tangent line to the ellipse with equation $\dfrac{x^2}{a^2} + \dfrac{y^2}{b^2} = 1$, with $a = \dfrac{5}{2}$ and $b = 5$ at the point $(2, 3)$. $y = -\dfrac{8}{3}x + \dfrac{25}{3}$
[*Hint:* See the steps for Exercise 91 in Section 9.2 and (in Step 3) write the discriminant as a perfect square.]

96. Show that the equation of the tangent line to the ellipse with equation $\dfrac{x^2}{a^2} + \dfrac{y^2}{b^2} = 1$, at the point (x_1, y_1), is
$$y = -\frac{b^2 x_1}{a^2 y_1} x + \frac{b^2 x_1^2}{b^2 y_1^2} + y_1.$$
[*Hint:* See the steps for Exercise 92 in Section 9.2 and show that $(a^2 y_1 m + b^2 x_1)^2 = 0$.]

Critical Thinking

97. Find the number of circles with a radius of three units that touch both of the coordinate axes. Write the equation of each circle.

98. Find the number of ellipses with a major axis and a minor axis of lengths six units and four units, respectively, that touch both axes. Write the equation of each ellipse.

97. 4 circles: $(x - 3)^2 + (y - 3)^2 = 9$
$(x + 3)^2 + (y - 3)^2 = 9$
$(x + 3)^2 + (y + 3)^2 = 9$
$(x - 3)^2 + (y + 3)^2 = 9$

98. Eight ellipses: $\dfrac{(x - 3)^2}{9} + \dfrac{(y - 2)^2}{4} = 1$,
$\dfrac{(x - 2)^2}{4} + \dfrac{(y - 3)^2}{9} = 1$,
$\dfrac{(x + 3)^2}{9} + \dfrac{(y - 2)^2}{4} = 1$, $\dfrac{(x + 2)^2}{4} + \dfrac{(y - 3)^2}{9} = 1$,
$\dfrac{(x + 3)^2}{9} + \dfrac{(y + 2)^2}{4} = 1$, $\dfrac{(x + 2)^2}{4} + \dfrac{(y + 3)^2}{9} = 1$,
$\dfrac{(x - 3)^2}{9} + \dfrac{(y + 2)^2}{4} = 1$, $\dfrac{(x - 2)^2}{4} + \dfrac{(y + 3)^2}{9} = 1$

Answers:
78. $(x - 2)^2 + 48(y - 1)^2 = 64$
79. If $e = 0$, the ellipse becomes a circle.
83. $\dfrac{4}{5}$ **84.** $\dfrac{\sqrt{2}}{2}$ **85.** $\dfrac{x^2}{16} + \dfrac{y^2}{12} = 1$ **86.** $\dfrac{x^2}{9} + \dfrac{y^2}{8} = 1$ **90.** $\dfrac{x^2}{9} + \dfrac{y^2}{\frac{9}{2}} = 1$

The Hyperbola

Before Starting this Section, Review

1. Distance formula (Section 1.1, page 5)

2. Midpoint formula (Section 1.1, page 5)

3. Completing the square (Appendix A, page 811)

4. Oblique asymptotes (Section 2.5, page 164)

5. Transformations of graphs (Section 1.5)

6. Symmetry (Section 1.1, page 8)

Objectives

1. Define a hyperbola.
2. Find the asymptotes of a hyperbola.
3. Graph a hyperbola.
4. Translate hyperbolas.
5. Use hyperbolas in applications.

Alfred Lee Loomis (1887–1975)

Alfred Loomis was an American lawyer, investment banker, physicist, and patron of scientific research. Besides inventing LORAN, he made significant contributions to the ground-based technology for landing airplanes by instruments. President Roosevelt recognized the value of Loomis's work and described him as second perhaps only to Churchill as the civilian most responsible for the Allied victory in World War II.

WHAT IS LORAN?

LORAN stands for LOng-RAnge Navigation. The federal government runs the LORAN system, which uses land-based radio navigation transmitters to provide users with information on position and timing. During World War II, Alfred Lee Loomis was selected to chair the National Defense Research Committee. Much of his work focused on the problem of creating a light system for plane-carried radar. As a result of this effort, he invented LORAN, the long-range navigation system whose offshoot LORAN-C remains in widespread use. The navigational method provided by LORAN is based on the principle of determining the points of intersection of two hyperbolas to fix the two-dimensional position of the receiver. In Example 8, we illustrate how this is done.

The LORAN-C system is now being supplemented by the global positioning system (GPS), which works on the same principle as LORAN-C. The GPS system consists of 24 satellites that orbit 11,000 miles above Earth. The GPS was created by the U.S. Department of Defense to provide precise navigation information for targeting weapons systems anywhere in the world. The GPS is available to civilian users worldwide in a degraded mode. It is now used in aviation, navigation, and automobiles. The system can provide your exact position on Earth anywhere, anytime.

The GPS and LORAN-C work together, giving a highly accurate, duplicate navigation system for the Defense Department. ▪

---

1 Define a hyperbola.

Definition of Hyperbola

HYPERBOLA

A **hyperbola** is the set of all points in the plane, the difference of whose distances from two fixed points is constant. The fixed points are called the **foci** of the hyperbola.

Figure 9.21 shows a hyperbola in *standard position*, with foci $F_1(-c,0)$ and $F_2(c,0)$ on the *x*-axis at equal distances from the origin. The two parts of the hyperbola are called **branches**.

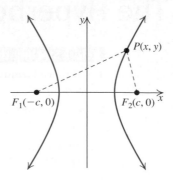

FIGURE 9.21 $PF_1 - PF_2$ **is a constant, for any P on the hyperbola**

As in the discussion of ellipses in Section 9.3 (see page 619), we take $2a$ as the positive constant referred to in the definition.

A point $P(x, y)$ lies on a hyperbola if and only if

$$|d(P, F_1) - d(P, F_2)| = 2a \qquad \text{Definition of hyperbola}$$

$$\sqrt{(x + c)^2 + y^2} - \sqrt{(x - c)^2 + y^2} = \pm 2a \qquad \text{Distance formula}$$

Just as in the case of an ellipse, we eliminate radicals and simplify (see Exercise 93) to obtain the equation:

$$(c^2 - a^2)x^2 - a^2y^2 = a^2(c^2 - a^2) \qquad (1)$$

$$b^2x^2 - a^2y^2 = a^2b^2 \qquad \text{Replace } (c^2 - a^2) \text{ with } b^2.$$

$$\frac{x^2}{a^2} - \frac{y^2}{b^2} = 1 \qquad (2) \qquad \text{Divide both sides by } a^2b^2.$$

Equation (2) is called the **standard form of the equation of a hyperbola** with center $(0, 0)$. The x-intercepts of the graph of equation (2) are $-a$ and a. The points corresponding to these x-intercepts are the **vertices** of the hyperbola. The distance between the vertices $V_1(-a, 0)$ and $V_2(a, 0)$ is $2a$. The line segment joining the two vertices is called the **transverse axis**. The midpoint of the transverse axis is the **center** of the hyperbola. The *center* is also the midpoint of the line segment joining the foci. See Figure 9.22. The line segment joining the points $(0, -b)$ and $(0, b)$ is called the **conjugate axis**.

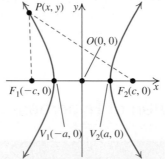

Points F_1 and F_2 are foci.
Points V_1 and V_2 are vertices.
Point O is the center.
Segment $\overline{V_1V_2}$ is the transverse axis.
The difference of the distances $\left|PF_1 - PF_2\right|$ from any point P on the hyperbola is constant.

FIGURE 9.22 Parts of a hyperbola

Similarly, an equation of a hyperbola with center $(0, 0)$ and foci $(0, -c)$ and $(0, c)$ on the y-axis is given by:

$$\frac{y^2}{a^2} - \frac{x^2}{b^2} = 1, \text{ where } b^2 = c^2 - a^2 \qquad (3)$$

Here the vertices are $(0, -a)$ and $(0, a)$. The transverse axis of length $2a$ of the graph of equation (3) lies on the y-axis, and its conjugate axis is the segment joining the points $(-b, 0)$ and $(0, b)$ of length $2b$ that lies on the x-axis.

Main facts about hyperbolas centered at $(0, 0)$

Standard Equation	$\dfrac{x^2}{a^2} - \dfrac{y^2}{b^2} = 1;\ a>0,\ b>0$	$\dfrac{y^2}{a^2} - \dfrac{x^2}{b^2} = 1;\ a>0,\ b>0$
Equation of transverse axis	$y = 0$ (x-axis)	$x = 0$ (y-axis)
Length of transverse axis	$2a$	$2a$
Equation of conjugate axis	$x = 0$ (y-axis)	$y = 0$ (x-axis)
Length of conjugate axis	$2b$	$2b$
Vertices	$(\pm a, 0)$	$(0, \pm a)$
Endpoints of conjugate axis	$(0, \pm b)$	$(\pm b, 0)$
Foci	$(\pm c, 0)$, where $c^2 = a^2 + b^2$	$(0, \pm c)$, where $c^2 = a^2 + b^2$
Description	Hyperbola has a *left branch* and a *right branch*. (Hyperbola opens left and right.)	Hyperbola has an *upper branch* and a *lower branch*. (Hyperbola opens up and down.)
Graph	Horizontal transverse axis	Vertical transverse axis

Graph (Horizontal transverse axis): $B_2(0, b)$, $V_1(-a, 0)$, $V_2(a, 0)$, $F_1(-c, 0)$, $F_2(c, 0)$, $B_1(0, -b)$

Graph (Vertical transverse axis): $F_2(0, c)$, $V_2(0, a)$, $B_1(-b, 0)$, $B_2(b, 0)$, $V_1(0, -a)$, $F_1(0, -c)$

EXAMPLE 1   Determining the Orientation of a Hyperbola

Does the hyperbola

$$\frac{x^2}{4} - \frac{y^2}{10} = 1$$

have its transverse axis on the x-axis or y-axis?

SOLUTION

The orientation (left–right branches or up–down branches) of a hyperbola is determined *by noting where the minus sign occurs in the standard equation.* From the table above, we see that the transverse axis is on the x-axis when the minus sign precedes the y^2-term and is on the y-axis when the minus sign precedes the x^2-term. In the equation in this example, the minus sign precedes the y^2-term, so the transverse axis is on the x-axis. See Figure 9.23. ■ ■ ■

Practice Problem 1   Does the hyperbola $\dfrac{y^2}{8} - \dfrac{x^2}{5} = 1$ have its transverse axis on the x-axis or the y-axis? ■

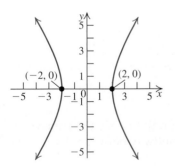

FIGURE 9.23 $\dfrac{x^2}{4} - \dfrac{y^2}{10} = 1$

> **EXAMPLE 2** **Finding the Vertices and Foci from the Equation of a Hyperbola**

Find the vertices and foci for the hyperbola

$$28y^2 - 36x^2 = 63.$$

SOLUTION

First, convert the equation to the standard form.

$$28y^2 - 36x^2 = 63 \qquad \text{Given equation}$$

$$\frac{28y^2}{63} - \frac{36x^2}{63} = 1 \qquad \text{Divide both sides by 63.}$$

$$\frac{4y^2}{9} - \frac{4x^2}{7} = 1 \qquad \text{Simplify.}$$

$$\frac{y^2}{\frac{9}{4}} - \frac{x^2}{\frac{7}{4}} = 1 \qquad \begin{array}{l}\text{Divide numerators and denominators}\\\text{of each term by 4.}\end{array}$$

The last equation is the equation of a hyperbola in standard form with center $(0, 0)$. Since the coefficient of x^2 is negative, the transverse axis lies on the y-axis.

Here $a^2 = \dfrac{9}{4}$ and $b^2 = \dfrac{7}{4}$. The equation $c^2 = a^2 + b^2$ gives the value of c^2.

$$c^2 = a^2 + b^2$$

$$= \frac{9}{4} + \frac{7}{4} \qquad \text{Replace } a^2 \text{ with } \frac{9}{4} \text{ and } b^2 \text{ with } \frac{7}{4}.$$

$$= \frac{16}{4} = 4 \qquad \text{Simplify.}$$

Now $a^2 = \dfrac{9}{4}$ gives $a = \dfrac{3}{2}$ and $c^2 = 4$ gives $c = 2$ because we require $a > 0$ and $c > 0$.

The vertices of the hyperbola are $\left(0, -\dfrac{3}{2}\right)$ and $\left(0, \dfrac{3}{2}\right)$. The foci of the hyperbola are $(0, -2)$ and $(0, 2)$. ■ ■ ■

Practice Problem 2 Find the vertices and foci for the hyperbola

$$x^2 - 4y^2 = 8. \qquad ■$$

> **EXAMPLE 3** **Finding the Equation of a Hyperbola**

Find the standard form of the equation of a hyperbola with vertices $(\pm 4, 0)$ and foci $(\pm 5, 0)$.

SOLUTION

Since the foci of the hyperbola, $(-5, 0)$ and $(5, 0)$ are on the x-axis, the transverse axis lies on the x-axis. The center of the hyperbola is midway between the foci, at $(0, 0)$. The standard form of such a hyperbola is

$$\frac{x^2}{a^2} - \frac{y^2}{b^2} = 1.$$

You need to find a^2 and b^2.

The distance a between the center $(0, 0)$ to either vertex, $(-4, 0)$ or $(4, 0)$, is 4; so $a = 4$ and $a^2 = 16$. The distance c between the center $(0, 0)$ to either focus, $(-5, 0)$ or $(5, 0)$, is 5; so $c = 5$ and $c^2 = 25$. Now use the equation $b^2 = c^2 - a^2$ to find $b^2 = c^2 - a^2 = 25 - 16 = 9$. Substitute $a^2 = 16$ and $b^2 = 9$ in $\dfrac{x^2}{a^2} - \dfrac{y^2}{b^2} = 1$ to get the standard form of the equation of the hyperbola

$$\frac{x^2}{16} - \frac{y^2}{9} = 1.$$

■ ■ ■

Practice Problem 3   Find the standard form of the equation of a hyperbola with vertices at $(0, \pm4)$ and foci at $(0, \pm6)$.

■

2 Find the asymptotes of a hyperbola.

The Asymptotes of a Hyperbola

Recall from Section 2.5 that a line $y = mx + b$ is a horizontal asymptote (if $m = 0$) or an oblique asymptote (if $m \ne 0$) of the graph of $y = f(x)$ if the distance from the line to the points on the graph of f approaches 0 as $x \to \infty$ or as $x \to -\infty$. When horizontal or oblique asymptotes exist, they provide us with information about the end behavior of the graph of f. This is especially helpful in graphing a hyperbola.

To find the asymptotes of the hyperbola with equation $\dfrac{x^2}{a^2} - \dfrac{y^2}{b^2} = 1$, we first solve this equation for y.

STUDY TIP

There is a convenient device that can be used for obtaining equations of the asymptotes of a hyperbola. Substitute 0 for the 1 on the right side of the equation of the hyperbola and then solve for y in terms of x. For example, for the hyperbola $\dfrac{x^2}{a^2} - \dfrac{y^2}{b^2} = 1$, replace the 1 on the right side by 0 to obtain $\dfrac{x^2}{a^2} - \dfrac{y^2}{b^2} = 0$. Solve this equation for y:

$$\frac{y^2}{b^2} = \frac{x^2}{a^2}$$

$$y^2 = \frac{b^2}{a^2}x^2$$

The lines $y = \pm\dfrac{b}{a}x$ are the asymptotes of the hyperbola.

$$\frac{x^2}{a^2} - \frac{y^2}{b^2} = 1 \qquad \text{Given equation of hyperbola}$$

$$-\frac{y^2}{b^2} = -\frac{x^2}{a^2} + 1 \qquad \text{Subtract } \frac{x^2}{a^2} \text{ from both sides.}$$

$$\frac{y^2}{b^2} = \frac{x^2}{a^2} - 1 \qquad \text{Multiply both sides by } -1.$$

$$y^2 = b^2\left(\frac{x^2}{a^2} - 1\right) \qquad \text{Multiply both sides by } b^2.$$

$$y^2 = b^2\left(\frac{x^2}{a^2} - \frac{x^2}{a^2}\cdot\frac{a^2}{x^2}\right) \qquad \text{Write } 1 = \frac{x^2}{a^2}\cdot\frac{a^2}{x^2}.$$

$$y^2 = \frac{b^2 x^2}{a^2}\left(1 - \frac{a^2}{x^2}\right) \qquad \text{Factor out } \frac{x^2}{a^2}.$$

$$y = \pm\frac{bx}{a}\sqrt{1 - \frac{a^2}{x^2}} \qquad \text{Square root property}$$

As $x \to \infty$ or as $x \to -\infty$, the quantity $\dfrac{a^2}{x^2}$ approaches 0. Thus, $\sqrt{1 - \dfrac{a^2}{x^2}}$ approaches 1 and the value of y approaches $\pm\dfrac{bx}{a}$. Therefore, the lines $y = \dfrac{b}{a}x$ and $y = -\dfrac{b}{a}x$ are the oblique asymptotes for the graph of the hyperbola $\dfrac{x^2}{a^2} - \dfrac{y^2}{b^2} = 1$.

Similarly, you can show that the lines $y = \dfrac{a}{b}x$ and $y = -\dfrac{a}{b}x$ are the oblique asymptotes for the graph of the hyperbola $\dfrac{y^2}{a^2} - \dfrac{x^2}{b^2} = 1$.

THE ASYMPTOTES OF A HYPERBOLA WITH CENTER (0, 0)

1. The graph of the hyperbola $\dfrac{x^2}{a^2} - \dfrac{y^2}{b^2} = 1$ has transverse axis along the x-axis and has the following two asymptotes:

$$y = \frac{b}{a}x \qquad \text{and} \qquad y = -\frac{b}{a}x$$

2. The graph of the hyperbola $\dfrac{y^2}{a^2} - \dfrac{x^2}{b^2} = 1$ has transverse axis along the y-axis and has the following two asymptotes:

$$y = \frac{a}{b}x \qquad \text{and} \qquad y = -\frac{a}{b}x$$

EXAMPLE 4   **Finding the Asymptotes of a Hyperbola**

Determine the asymptotes of each hyperbola.

a. $\dfrac{x^2}{4} - \dfrac{y^2}{9} = 1$ **b.** $\dfrac{y^2}{9} - \dfrac{x^2}{16} = 1$

SOLUTION

a. The hyperbola $\dfrac{x^2}{4} - \dfrac{y^2}{9} = 1$ is of the form $\dfrac{x^2}{a^2} - \dfrac{y^2}{b^2} = 1$; so $a = 2$ and $b = 3$.

Substituting these values into $y = \dfrac{b}{a}x$ and $y = -\dfrac{b}{a}x$, we get the asymptotes

$$y = \frac{3}{2}x \quad \text{and} \quad y = -\frac{3}{2}x.$$

b. The hyperbola $\dfrac{y^2}{9} - \dfrac{x^2}{16} = 1$ has the form $\dfrac{y^2}{a^2} - \dfrac{x^2}{b^2} = 1$; so $a = 3$ and $b = 4$.

Substituting these values into $y = \dfrac{a}{b}x$ and $y = -\dfrac{a}{b}x$, we get the asymptotes

$$y = \frac{3}{4}x \quad \text{and} \quad y = -\frac{3}{4}x.$$

The hyperbolas of **a** and **b** and their asymptotes are shown in Figures 9.24(a) and 9.24(b), respectively. ■ ■ ■

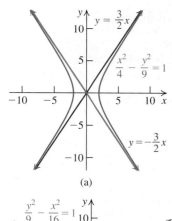

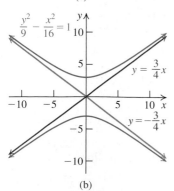

FIGURE 9.24 Asymptotes

Practice Problem 4   Determine the asymptotes of the hyperbola $\dfrac{y^2}{4} - \dfrac{x^2}{9} = 1$.

■

3 Graph a hyperbola.

Graphing a Hyperbola with Center $(0, 0)$

Consider the hyperbola with equation

$$\frac{x^2}{a^2} - \frac{y^2}{b^2} = 1.$$

The vertices of this hyperbola are $(-a, 0)$ and $(a, 0)$. The endpoints of the conjugate axis are $(0, -b)$ and $(0, b)$. The rectangle with vertices $(a, b), (-a, b), (-a, -b)$, and $(a, -b)$ is called the **fundamental rectangle** of the hyperbola. See Figure 9.25 on page 636. The diagonals of the fundamental rectangle have slopes $\frac{b}{a}$ and $-\frac{b}{a}$. Thus, the extensions of these diagonals are the asymptotes of the hyperbola.

FINDING THE SOLUTION: A PROCEDURE

EXAMPLE 5   **Graphing a Hyperbola Centered at $(0,0)$**

OBJECTIVE	EXAMPLE

OBJECTIVE

Sketch the graph of a hyperbola in the form

(i) $\dfrac{x^2}{a^2} - \dfrac{y^2}{b^2} = 1$ *or* **(ii)** $\dfrac{y^2}{a^2} - \dfrac{x^2}{b^2} = 1$

EXAMPLE

Sketch the graph of

a. $16x^2 - 9y^2 = 144$.

b. $25y^2 - 4x^2 = 100$.

Step 1 **Write the equation in standard form.** Determine transverse axis and the orientation of the hyperbola.

$16x^2 - 9y^2 = 144$

$\dfrac{16x^2}{144} - \dfrac{9y^2}{144} = \dfrac{144}{144}$

$\dfrac{x^2}{9} - \dfrac{y^2}{16} = 1$

Minus sign precedes y^2-term, the transverse axis is along the x-axis; the hyperbola opens left and right.

$25y^2 - 4x^2 = 100$

$\dfrac{25y^2}{100} - \dfrac{4x^2}{100} = \dfrac{100}{100}$

$\dfrac{y^2}{4} - \dfrac{x^2}{25} = 1$

Minus sign precedes x^2-term, the transverse axis is along the y-axis; the hyperbola opens up and down.

Step 2 **Locate vertices** and the endpoints of the conjugate axis.

Because $a^2 = 9$, $a = 3$. Also, $b^2 = 16$, so $b = 4$. The vertices are $(\pm 3, 0)$, and the endpoints of the conjugate axis are $(0, \pm 4)$.

Because $a^2 = 4$, $a = 2$. Also, $b^2 = 25$, so $b = 5$. The vertices are $(0, \pm 2)$, and the endpoints of the conjugate axis are $(\pm 5, 0)$.

Step 3 Lightly sketch the fundamental rectangle by drawing dashed lines parallel to the coordinate axes through the points in Step 2.

Draw dashed lines $x = 3$, $x = -3, y = 4$, and $y = -4$ to form the fundamental rectangle with vertices $(3, 4)$, $(-3, 4), (-3, -4)$, and $(3, -4)$.

Draw dashed lines $y = 2$, $y = -2, x = 5$, and $x = -5$ to form the fundamental rectangle with vertices $(5, 2)$, $(-5, 2), (-5, -2)$, and $(5, -2)$.

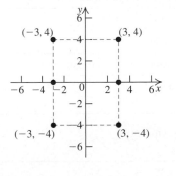

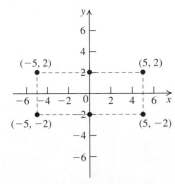

continued on the next page

Step 4 **Sketch the asymptotes.** Extend the diagonals of the fundamental rectangle. These are the asymptotes.

Step 5 **Sketch the graph.** Draw both branches of the hyperbola through the vertices, approaching the asymptotes. See Figures 9.25 and 9.26. The foci are located on the transverse axis, c units from the center, where $c^2 = a^2 + b^2$.

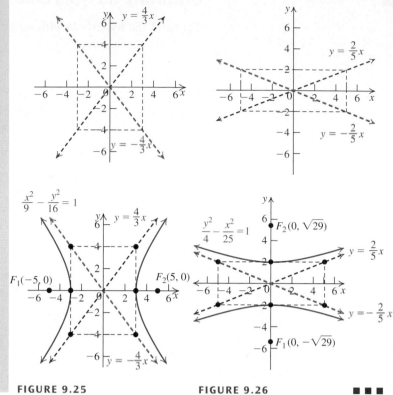

FIGURE 9.25   **FIGURE 9.26**   ■ ■ ■

Practice Problem 5 Sketch the graph of the equations.

a. $25x^2 - 4y^2 = 100$ **b.** $9y^2 - x^2 = 1$

4 Translate hyperbolas.

Translations of Hyperbolas

We can use horizontal and vertical shifts to find the standard form of the equations of hyperbolas centered at (h, k).

Main properties of hyperbolas centered at (h, k)		
Standard Equation	$\dfrac{(x-h)^2}{a^2} - \dfrac{(y-k)^2}{b^2} = 1$; $\,a>0, b>0$	$\dfrac{(y-k)^2}{a^2} - \dfrac{(x-h)^2}{b^2} = 1$; $\,a>0, b>0$
Equation of transverse axis	$y = k$	$x = h$
Length of transverse axis	$2a$	$2a$
Equation of conjugate axis	$x = h$	$y = k$
Length of conjugate axis	$2b$	$2b$
Center	(h, k)	(h, k)
Vertices	$(h - a, k)$ and $(h + a, k)$	$(h, k - a)$ and $(h, k + a)$
Endpoints of conjugate axis	$(h, k - b)$ and $(h, k + b)$	$(h - b, k)$ and $(h + b, k)$
Foci	$(h - c, k)$ and $(h + c, k)$; $c^2 = a^2 + b^2$	$(h, k - c)$ and $(h, k + c)$; $c^2 = a^2 + b^2$
Asymptotes	$y - k = \pm\dfrac{b}{a}(x - h)$	$y - k = \pm\dfrac{a}{b}(x - h)$

We now describe a procedure for sketching the graph of a hyperbola centered at (h, k).

FINDING THE SOLUTION: A PROCEDURE

| EXAMPLE 6 | Graphing a Hyperbola Centered at (h, k) |

OBJECTIVE

Sketch the graph of either

$$\frac{(x - h)^2}{a^2} - \frac{(y - k)^2}{b^2} = 1 \ or$$

$$\frac{(y - k)^2}{a^2} - \frac{(x - h)^2}{b^2} = 1$$

EXAMPLE

Sketch the graph of

a. $\dfrac{(x - 1)^2}{4} - \dfrac{(y + 2)^2}{9} = 1.$ **b.** $\dfrac{(y + 2)^2}{9} - \dfrac{(x - 1)^2}{4} = 1.$

Step 1 Plot the center (h, k) and draw horizontal and vertical dashed lines through the center.

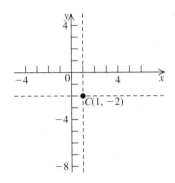
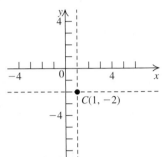

Step 2 Locate the vertices and the endpoints of the conjugate axis. Lightly sketch the fundamental rectangle, with sides parallel to the coordinate axes, through these points.

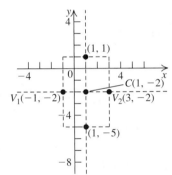
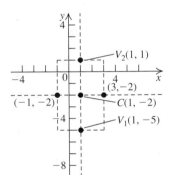

Step 3 Sketch dashed lines through opposite vertices of the fundamental rectangle. These are the asymptotes.

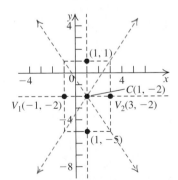
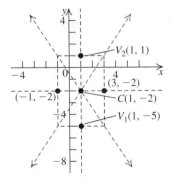

Step 4 Draw both branches of the hyperbola through the vertices and approaching the asymptotes.

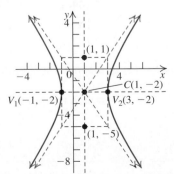
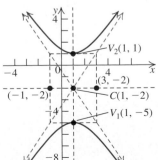

continued on the next page

Step 5   The foci are located on the transverse axis, c units from the center, where $c^2 = a^2 + b^2$.

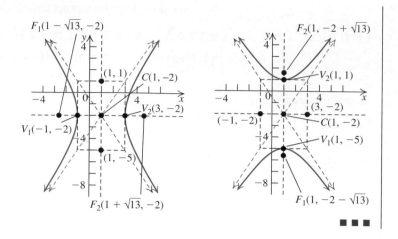

Practice Problem 6   Sketch the graph of the equations.

a. $\dfrac{(x+1)^2}{4} - \dfrac{(y-1)^2}{16} = 1$ **b.** $\dfrac{(y-1)^2}{4} - \dfrac{(x+1)^2}{9} = 1$ ■

Expanding the binomials in the equation of a hyperbola in standard form and collecting like terms results in an equation of the form

$$Ax^2 + Cy^2 + Dx + Ey + F = 0,$$

with A and C of opposite sign (that is, $AC < 0$.) Conversely, except in degenerate cases, any equation of this form can be put into one of the standard forms of a hyperbola by completing the squares on the x- and y-terms.

EXAMPLE 7   **Graphing a Hyperbola**

Show that $9x^2 - 16y^2 + 18x + 64y - 199 = 0$ is an equation of a hyperbola and then graph the hyperbola.

SOLUTION

Since we plan on completing squares, first group the x- and y-terms onto the left side and the constants onto the right side.

$9x^2 - 16y^2 + 18x + 64y - 199 = 0$	Given equation
$(9x^2 + 18x) + (-16y^2 + 64y) = 199$	Group terms.
$9(x^2 + 2x) - 16(y^2 - 4y) = 199$	Factor out 9 and -16.
$9(x^2 + 2x + 1) - 16(y^2 - 4y + 4) = 199 + 9 - 64$	Complete the squares: add $9 \cdot 1$ and $-16 \cdot 4$.
$9(x+1)^2 - 16(y-2)^2 = 144$	Factor and simplify.
$\dfrac{9(x+1)^2}{144} - \dfrac{16(y-2)^2}{144} = 1$	Divide both sides by 144 to obtain 1 on the right side.
$\dfrac{(x+1)^2}{16} - \dfrac{(y-2)^2}{9} = 1$	Simplify.

The last equation is the standard form of the equation of a hyperbola with center $(-1, 2)$.

Now we sketch the graph of this hyperbola.

Steps 1–2   **Locate the vertices.** For this hyperbola with center $(-1, 2)$, $a^2 = 16$ and $b^2 = 9$. Therefore, $a = 4$ and $b = 3$. From the table on page 636, with $h = -1$ and $k = 2$, the vertices are $(h - a, k) = (-1 - 4, 2) = (-5, 2)$ and $(h + a, k) = (-1 + 4, 2) = (3, 2)$.

Draw the fundamental rectangle. The vertices of the fundamental rectangle are $(3, -1)$, $(3, 5)$, $(-5, 5)$, and $(-5, -1)$.

Step 3   **Sketch the asymptotes.** Extend the diagonals of the fundamental rectangle to sketch the asymptotes:

$$y - 2 = \frac{3}{4}(x + 1) \quad \text{and} \quad y - 2 = -\frac{3}{4}(x + 1).$$

Step 4   **Sketch the graph.** Draw two branches of the hyperbola opening to the left and right, starting from the vertices $(-5, 2)$ and $(3, 2)$ and approaching the asymptotes. See Figure 9.27.   ■ ■ ■

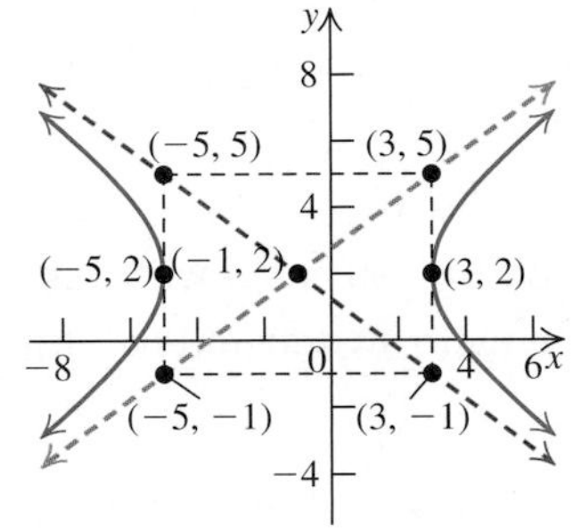

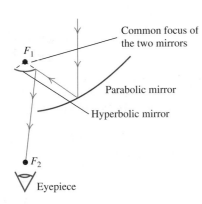

FIGURE 9.27 The hyperbola
$9x^2 - 16y^2 + 18x + 64y - 199 = 0$

Practice Problem 7   Show that $x^2 - 4y^2 - 2x + 16y - 20 = 0$ is an equation of a hyperbola and graph the hyperbola.   ■

5   Use hyperbolas in applications.

Applications

Hyperbolas have many applications. We list a few of them here:

1.   Comets that do not move in elliptical orbits around the sun almost always move in hyperbolic orbits. (In theory, they can also move in parabolic orbits.)

2.   Boyle's Law states that if a perfect gas is kept at a constant temperature, then its pressure P and volume V are related by the equation $PV = c$, where c is a constant. The graph of this equation is a hyperbola. In Exercise 94(b), you are asked to explore the case where $c = 8$. In this case, the transverse axis is not parallel to a coordinate axis.

3.   The hyperbola has the reflecting property that a ray of light from a source at one focus of a hyperbolic mirror (a mirror with hyperbolic cross sections) is reflected along the line through the other focus.

   The reflecting properties of the parabola and hyperbola are combined into one design for a reflecting telescope. See Figure 9.28. The parallel rays from a star are finally focused at the eyepiece at F_2.

4.   The definition of a hyperbola forms the basis of several important navigational systems.

FIGURE 9.28

How Does LORAN Work?

Suppose two stations A and B (several miles apart) transmit synchronized radio signals. The LORAN receiver in your ship measures the difference in reception times of these synchronized signals. The radio signals travel at a speed of 186,000 miles per second. Using this information, you can determine the difference $2a$ in the distance of your ship's receiver from the two transmitters. By the definition of a hyperbola, this information places your ship somewhere on a hyperbola with foci A and B. With two pairs of transmitters, the position of your ship can be determined at a point of intersection of the two hyperbolas. (See Exercises 87 and 88).

EXAMPLE 8   **Using LORAN**

LORAN navigational transmitters A and B are located at $(-130, 0)$ and $(130, 0)$, respectively. A receiver P on a fishing boat somewhere in the first quadrant listens to the pair (A, B) of transmissions and computes the difference of the distances from the boat to A and B as 240 miles. Find the equation of the hyperbola on which P is located.

SOLUTION

With reference to the transmitters A and B, P is located on a hyperbola with $2a = 240$, or $a = 120$, and the hyperbola has foci $(-130, 0)$ and $(130, 0)$.

$$\frac{x^2}{a^2} - \frac{y^2}{b^2} = 1 \qquad \text{Standard from of the equation of a hyperbola}$$

$$a^2 = (120)^2 = 14{,}400 \qquad \text{Replace } a \text{ with } 120.$$

$$b^2 = c^2 - a^2 = (130)^2 - (120)^2 = 2500 \qquad \text{Replace } c \text{ with } 130 \text{ and } a \text{ with } 120.$$

$$\frac{x^2}{14{,}400} - \frac{y^2}{2500} = 1 \qquad \text{Replace } a^2 \text{ with } 14{,}400 \text{ and } b^2 \text{ with } 2500.$$

The location of the ship therefore lies on a hyperbola (see Figure 9.29) with equation

$$\frac{x^2}{14{,}400} - \frac{y^2}{2500} = 1.$$

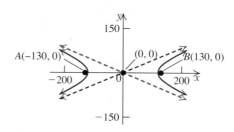

FIGURE 9.29   ■ ■ ■

Practice Problem 8   In Example 8, the transmitters A and B are located at $(-150, 0)$ and $(150, 0)$, respectively, and the difference of the distances from the boat to A and B is 260 miles. Find the equation of the hyperbola on which P is located.   ■

SECTION 9.4 ■ Exercises

A EXERCISES Basic Skills and Concepts

1. A hyperbola is a set of all points in the plane, the ___difference___ of whose distances from two fixed points is constant.

2. The line segment joining the two vertices of a hyperbola is called the ___transverse___ axis.

3. The standard equation of the hyperbola with center $(0, 0)$, vertices $(\pm a, 0)$, and foci $(\pm c, 0)$ is

$$\frac{x^2}{a^2} - \frac{y^2}{b^2} = 1, \text{ where } b^2 = \underline{} c^2 - a^2.$$

4. For the hyperbola $\dfrac{y^2}{a^2} - \dfrac{x^2}{b^2} = 1$, the vertices are ___$(0, a)$___ and ___$(0, -a)$___ and the foci are $(0, \pm c)$, where $c^2 = \underline{} a^2 + b^2$.

5. *True or False* The graph of $\dfrac{x^2}{a^2} - \dfrac{y^2}{b^2} = -1$ is a hyperbola. True

6. *True or False* The graph of $Ax^2 + Cy^2 + Dx + Ey + F = 0$ is a degenerate conic or a hyperbola if $AC < 0$. True

In Exercises 7–14, the equation of a hyperbola is given. Match each equation with its graph shown in (a)–(h).

7. $\dfrac{x^2}{9} - \dfrac{y^2}{4} = 1$ g

8. $\dfrac{x^2}{4} - \dfrac{y^2}{25} = 1$ c

9. $\dfrac{y^2}{25} - \dfrac{x^2}{4} = 1$ h

10. $\dfrac{y^2}{4} - \dfrac{x^2}{9} = 1$ e

11. $4x^2 - 25y^2 = 100$ d

12. $16x^2 - 49y^2 = 196$ a

13. $16y^2 - 9x^2 = 100$ b

14. $9y^2 - 16x^2 = 144$ f

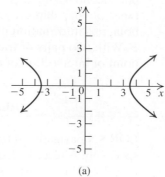

(a)

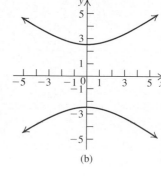

(b)

†Due to space constrictions, answers to these exercises may be found in the Answers beginning on page A–1 in the back of the book.

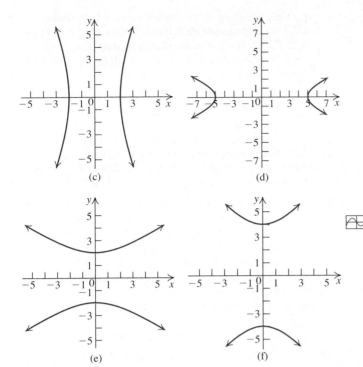

(c)   (d)

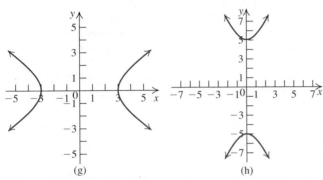

(e)   (f)

(g)   (h)

In Exercises 15–26, the equation of a hyperbola is given.
   a. Find and plot the vertices, the foci, and the transverse axis of the hyperbola.
   b. State how the hyperbola opens.
   c. Find and plot the vertices of the fundamental rectangle.
   d. Sketch the asymptotes and write their equations.
   e. Graph the hyperbola by using the vertices and the asymptotes.

15. $x^2 - \dfrac{y^2}{4} = 1$ †   **16.** $y^2 - \dfrac{x^2}{4} = 1$ †

17. $x^2 - y^2 = -1$ †   **18.** $y^2 - x^2 = -1$ †

19. $9y^2 - x^2 = 36$ †   **20.** $9x^2 - y^2 = 36$ †

21. $4x^2 - 9y^2 - 36 = 0$ †   **22.** $4x^2 - 9y^2 + 36 = 0$ †

23. $y = \pm\sqrt{4x^2 + 1}$ †   **24.** $y = \pm 2\sqrt{x^2 + 1}$ †

25. $y = \pm\sqrt{9x^2 - 1}$ †   **26.** $y = \pm 3\sqrt{x^2 - 4}$ †

In Exercises 27–40, find an equation of the hyperbola satisfying the given conditions. Graph the hyperbola.

27. Vertices: $(\pm 2, 0)$; foci: $(\pm 3, 0)$ †

28. Vertices: $(\pm 3, 0)$; foci: $(\pm 5, 0)$ †

29. Vertices: $(0, \pm 4)$; foci: $(0, \pm 6)$ †

30. Vertices: $(0, \pm 5)$; foci: $(0, \pm 8)$ †

31. Center: $(0, 0)$; vertex: $(0, 2)$; focus: $(0, 5)$ †

32. Center: $(0, 0)$; vertex: $(0, -1)$; focus: $(0, -4)$ †

33. Center: $(0, 0)$; vertex: $(1, 0)$; focus: $(-5, 0)$ †

34. Center: $(0, 0)$; vertex: $(-3, 0)$; focus: $(6, 0)$ †

35. Foci: $(0, \pm 5)$; the length of the transverse axis is 6. †

36. Foci: $(\pm 2, 0)$; the length of the transverse axis is 2. †

37. Foci: $(\pm\sqrt{5}, 0)$; asymptotes: $y = \pm 2x$ †

38. Foci: $(0, \pm 10)$; asymptotes: $y = \pm 3x$ †

39. Vertices: $(0, \pm 4)$; asymptotes: $y = \pm x$ †

40. Vertices: $(\pm 3, 0)$; asymptotes: $y = \pm 2x$ †

In Exercises 41–48, use a graphing device to graph each hyperbola.

41. $\dfrac{x^2}{2} - \dfrac{y^2}{4} = 1$ †   **42.** $\dfrac{x^2}{6} - \dfrac{y^2}{2} = 1$ †

43. $2y^2 - x^2 = 6$ †   **44.** $y^2 - 3x^2 = 12$ †

45. $(x + 1)^2 - 2y^2 = 4$ †   **46.** $3x^2 - (y - 2)^2 = 6$ †

47. $2x^2 - y^2 - 4x - 2y - 7 = 0$ †

48. $4y^2 - 3x^2 - 16y - 6x + 1 = 0$ †

In Exercises 49–68, the equation of a hyperbola is given.
   a. Find and plot the center, vertices, transverse axis, and asymptotes of the hyperbola.
   b. Use the vertices and the asymptotes to graph the hyperbola.

49. $\dfrac{(x - 1)^2}{9} - \dfrac{(y + 1)^2}{16} = 1$ †

50. $\dfrac{(y - 1)^2}{16} - \dfrac{(x + 1)^2}{4} = 1$ †

51. $\dfrac{(x + 2)^2}{25} - \dfrac{y^2}{49} = 1$ †

52. $\dfrac{(x + 1)^2}{9} - \dfrac{(y + 2)^2}{36} = 1$ †

53. $\dfrac{(x + 4)^2}{25} - \dfrac{(y + 3)^2}{49} = 1$ †

54. $\dfrac{(x - 3)^2}{9} - \dfrac{(y + 1)^2}{9} = 1$ †

55. $4x^2 - (y + 1)^2 = 25$ †   **56.** $9(x - 1)^2 - y^2 = 144$ †

57. $(y + 1)^2 - 9(x - 2)^2 = 25$ †

58. $6(x - 4)^2 - 3(y + 3)^2 = 4$ †

59. $x^2 - y^2 + 6x = 36$ †   **60.** $x^2 + 4x - 4y^2 = 12$ †

61. $x^2 - 4y^2 - 4x = 0$ †   **62.** $x^2 - 2y^2 - 8y = 12$ †

63. $2x^2 - y^2 + 12x - 8y + 3 = 0$ †

64. $y^2 - 9x^2 - 4y - 30x = 33$ †

65. $3x^2 - 18x - 2y^2 - 8y + 1 = 0$ †

66. $4y^2 - 9x^2 - 8y - 36x = 68$ †

67. $y^2 + 2\sqrt{2} - x^2 + 2\sqrt{2}x = 1$ †

68. $x^2 + x = \dfrac{y^2 - 1}{4}$ †

In Exercises 69–78, identify the conic section represented by each equation and sketch the graph.

69. $x^2 - 6x + 12y + 33 = 0$ †

70. $9y^2 = x^2 - 4x$ †

71. $y^2 - 9x^2 = -1$ †

72. $4x^2 + 9y^2 - 8x + 36y + 4 = 0$ †

73. $x^2 + y^2 - 4x + 8y = 16$ †

74. $2y^2 + 3x - 4y - 7 = 0$ †

75. $2x^2 - 4x + 3y + 8 = 0$ †

76. $y^2 - 9x^2 + 18x - 4y = 14$ †

77. $4x^2 + 9y^2 + 8x - 54y + 49 = 0$ †

78. $x^2 - 4y^2 + 6x + 12y = 0$ †

B EXERCISES Applying the Concepts

79. Sound of an explosion. Points A and B are 1000 meters apart, and it is determined from the sound of an explosion heard at these points at different times that the location of the explosion is 600 meters closer to A than to B. Show that the location of the explosion is restricted to points on a hyperbola and find the equation of the hyperbola. [*Hint:* Let the coordinates of A be $(-500, 0)$ and those of B be $(500, 0)$.] †

80. Sound of an explosion. Points A and B are 2 miles apart. The sound of an explosion at A was heard 3 seconds before it was heard at B. Assume that the sound travels at 1100 feet/second. Show that the location of the explosion is restricted to a hyperbola and find the equation of the hyperbola. [*Hint:* Recall that 1 mile = 5280 feet.] †

81. Thunder and lightning. Thunder is heard by Nicole and Juan, who are 8000 feet apart. Nicole hears the thunder 3 seconds before Juan does. Find the equation of the hyperbola whose points are possible locations where the lightning could have struck. Assume that the sound travels at 1100 feet/second. †

82. Locating source of thunder. In Exercise 81, Valerie is at a location midway between Nicole and Juan, and she hears the thunder 2 seconds after Juan does. Determine the location where the lightning strikes in relation to the three people involved. This is impossible. Since Valerie is at the origin, she must hear it.

83. LORAN. Two LORAN stations, A and B, are situated 300 kilometers apart along a straight coastline. Simultaneous radio signals are sent from each station to a ship. The ship receives the signal from A 0.0005 second before the signal from B. Assume that the radio signals travel 300,000 kilometers per second. Find the equation of the hyperbola on which the ship is located. †

84. LORAN. In Exercise 83, assuming that the ship were to follow the graph of the hyperbola, determine the location of the ship when it reaches the coastline. $(-75, 0)$

85. Target practice. A gun at G and a target at T are 1600 feet apart. The muzzle velocity of the bullet is 2000 feet per second. A person at P hears the crack of the gun and the thud of the bullet at the same time. Show that the

possible locations of the person are on a hyperbola. Find the equation of the hyperbola with G at $(-800, 0)$ and T at $(800, 0)$. Assume that the speed of sound is 1100 feet/second. [*Hint:* Show that $|d(P, G) - d(P, T)|$ is a constant.] †

86. LORAN. Two LORAN stations, A and B, lie on an east–west line with A 250 miles west of B. A plane is flying west on a line 50 miles north of the line AB. Radio signals are sent (traveling at 980 feet per microsecond) simultaneously from A and B, and the one sent from B arrives at the plane 500 microseconds before the one from A. Where is the plane? †

87. Using LORAN. LORAN navigational transmitters A, B, C, and D are located at $(-100, 0)$, $(100, 0)$, $(0, -150)$, and $(0, 150)$, respectively. A navigator P on a ship somewhere in the second quadrant listens to the pair (A, B) of transmitters and computes the difference of the distances from the ship to A and B as 120 miles. Similarly, by listening to the pair (C, D) of transmitters, navigator P computes the difference of the distances from the ship to C and D as 80 miles. Use a calculator to find the coordinates of the ship. $(-68.5748, 44.2719)$

88. Using LORAN. LORAN navigational transmitters A, B, and C are located at $(200, 0)$, $(-200, 0)$, and $(200, 1000)$, respectively. A navigator on a fishing boat listens to the pair (A, B) of transmitters and finds that the difference of the distances from the boat to A and B is 300 miles. She also finds that the difference of the distances from the boat to A and C is 400 miles. Use a calculator to find the possible locations of the boat. †

C EXERCISES Beyond the Basics

89. Write the standard form of the equation of the hyperbola for which the difference of the distances from any point on the hyperbola to $(-5, 0)$ and $(5, 0)$ is equal to (a) 2 units; (b) 4 units; (c) 8 units. †

90. Sketch the graphs of all three hyperbolas in Exercise 89 on the same coordinate plane. †

91. Write the standard form of the equation of the hyperbola for which the difference of the distances from any point on the hyperbola to the points F_1 and F_2 is two units, where
   a. F_1 is $(0, 6)$ and F_2 is $(0, -6)$; †
   b. F_1 is $(0, 4)$ and F_2 is $(0, -4)$; †
   c. F_1 is $(0, 3)$ and F_2 is $(0, -3)$. †

92. Sketch the graphs of all three hyperbolas in Exercise 91 on the same coordinate plane. †

93. Rewrite equation

$$\sqrt{(x + c)^2 + y^2} - \sqrt{(x - c)^2 + y^2} = \pm 2a$$

from page 630 as

$$\sqrt{(x + c)^2 + y^2} = \pm 2a + \sqrt{(x - c)^2 + y^2}.$$

Square both sides to obtain

$$(x + c)^2 + y^2 = 4a^2 \pm 4a\sqrt{(x - c)^2 + y^2} + (x - c)^2 + y^2.$$

Simplify and then isolate the radical; then square both sides again to show that the equation can be simplified to

$$(c^2 - a^2)x^2 - a^2y^2 = a^2(c^2 - a^2).$$

94. a. Let $p > 0$. Prove that the graph of the equation $xy = \dfrac{p^2}{2}$ is a hyperbola by showing that this equation is the equation of the hyperbola with foci (p, p) and $(-p, -p)$ and with difference $2p$ in distance.

$\left[\right.$*Hint:* Show that the equation

$$\sqrt{(x + p)^2 + (y + p)^2} - \sqrt{(x - p)^2 + (y - p)^2} = 2p$$

simplifies to $xy = \dfrac{p^2}{2}.$ $\left.\right]$

b. Use part (a) to sketch the graph of the hyperbola $xy = 8$. Find the coordinates of the foci.

95. A hyperbola for which the lengths of the transverse and conjugate axes are equal is called an **equilateral hyperbola.** Show that the asymptotes of an equilateral hyperbola are perpendicular to one another.

96. The **latus rectum of a hyperbola** is a line segment that passes through a focus, is perpendicular to the transverse axis, and has endpoints on the hyperbola. Show that the length of the latus rectum of the hyperbola

$$\frac{x^2}{a^2} - \frac{y^2}{b^2} = 1 \text{ is } \frac{2b^2}{a}.$$

97. The *eccentricity* **of a hyperbola**, denoted by e, is defined by

$$e = \frac{\text{Distance between the foci}}{\text{Distance between the vertices}} = \frac{2c}{2a} = \frac{c}{a}.$$

Show that for every hyperbola, $e > 1$. What happens when $e = 1$?

In Exercises 98–102, find the eccentricity and the length of the latus rectum of each hyperbola.

98. $\dfrac{x^2}{16} - \dfrac{y^2}{9} = 1$ †

99. $36x^2 - 25y^2 = 900$ †

100. $x^2 - y^2 = 49$ †

101. $8(x - 1)^2 - (y + 2)^2 = 2$ †

102. $5x^2 - 4y^2 - 10x - 8y - 19 = 0$ †

103. For the hyperbola $\dfrac{x^2}{a^2} - \dfrac{y^2}{b^2} = 1$ with eccentricity e, show that $b^2 = a^2(e^2 - 1)$.

104. Find an equation of a hyperbola if its foci are $(\pm 6, 0)$ and the length of its latus rectum is 10. †

105. A point $P(x, y)$ moves in the plane so that its distance from the point $(3, 0)$ is always twice its distance from the line $x = -1$. Show that the equation of the path of P is a hyperbola of eccentricity 2.

106. A point $P(x, y)$ moves in the plane so that its distance from the point $(0, -3)$ is always three times its distance from the line $y = 1$. Show that the equation of the path of P is a hyperbola of eccentricity 3.

In Exercises 107–112, find all points of intersection of the given curves. Make a sketch of the curves that shows the points of intersection.

107. $y - 2x - 20 = 0$ and $y^2 - 4x^2 = 36$ †

108. $x - 2y^2 = 0$ and $x^2 = 8y^2 + 5$ †

109. $x^2 + y^2 = 15$ and $x^2 - y^2 = 1$ †

110. $x^2 + 9y^2 = 9$ and $4x^2 - 25y^2 = 36$ †

111. $9x^2 + 4y^2 = 72$ and $x^2 - 9y^2 = -77$ †

112. $3y^2 - 7x^2 = 5$ and $9x^2 - 2y^2 = 1$ †

In Exercises 113 and 114, use the following definition. A *tangent line* **to a hyperbola is a line that is not parallel to either asymptote of the hyperbola and that intersects the hyperbola at only one point.**

113. Find the equation of the tangent line to the hyperbola with equation $\dfrac{x^2}{a^2} - \dfrac{y^2}{b^2} = 1$, with $a = \sqrt{\dfrac{5}{3}}$ and $b = \sqrt{\dfrac{5}{7}}$ at the point $(2, 1)$.
[*Hint:* See the steps for Exercise 91 in Section 9.2 and (in Step 3) write the discriminant as a perfect square.]

114. Show that the equation of the tangent line to the hyperbola with equation $\dfrac{x^2}{a^2} - \dfrac{y^2}{b^2} = 1$, at the point (x_1, y_1), is $y = \dfrac{b^2 x_1}{a^2 y_1}x - \dfrac{b^2 x_1^2}{a^2 y_1^2} + y_1$.
[*Hint:* See the steps for Exercise 92 in section 9.2 and show that $(a^2 y_1 m - b^2 x_1)^2 = 0.$]

Critical Thinking

115. Explain why each of the following represents a degenerate conic section. Where possible, sketch the graph.
a. $4x^2 - 9y^2 = 0$ †
b. $x^2 + 4y^2 + 6 = 0$ †
c. $8x^2 + 5y^2 = 0$ †
d. $x^2 + y^2 - 4x + 8y = -20$ †
e. $y^2 - 2x^2 = 0$ †

The tangent line equation near 113: $y = \dfrac{6}{7}x + \dfrac{5}{7}$

GROUP PROJECT

Discuss the graph of a quadratic equation in x and y of the form

$$Ax^2 + Cy^2 + Dx + Ey + F = 0, \quad AC \neq 0.$$

Rewrite this equation in the form

$$A\left(x^2 + \frac{D}{A}x + \frac{D^2}{4A^2}\right) + C\left(y^2 + \frac{E}{C}y + \frac{E^2}{4C^2}\right)$$
$$= \frac{D^2}{4A} + \frac{E^2}{4C} - F.$$

Include in your discussion the types of graphs you will obtain in each of the following cases:
i. $A > 0, C > 0$ **ii.** $A > 0, C < 0$
iii. $A < 0, C > 0$ **iv.** $A < 0, C < 0$.

In each case, discuss how the classification is affected by the sign of

$$\frac{D^2}{4A} + \frac{E^2}{4C} - F. \text{ †}$$

Rotation of Axes

Before Starting this Section, Review

1. Equation of a parabola (Section 9.2, page 608)
2. Equation of an ellipse (Section 9.3, page 619)
3. Equation of a hyperbola (Section 9.4, page 630)
4. Trigonometric functions (Chapters 4–6)

Objectives

1   Identify a conic.

2   Rotate coordinate axes.

3   Eliminate the xy-term in the equation of a conic.

4   Use the discriminant to identify a conic.

LAUNCHING A SPACE VEHICLE

Wind speed and wind direction are usually measured in east–west (EW) and north–south (NS) components. When describing wind characteristics, the standard or natural coordinate system uses the y-axis to represent the NS directions (with north in the direction of increasing positive numbers) and the x-axis to represent the EW directions (with east in the direction of increasing positive numbers). Sometimes information recorded relative to such a standard coordinate system must be interpreted in a coordinate system that is rotated with respect to the system. Consider, for example, space vehicle launches. The launch path may not be exactly due north, east, south, or west. So engineers need to convert the wind speed and direction, measured in EW and NS components, to coordinates in a plane where the axes are parallel and perpendicular to the launch direction. In Example 3, we show how to accomplish this conversion. ∎

---

1   Identify a conic.

Identifying a Conic

In Sections 9.2–9.4, we learned to identify the graph of an equation of the form

$$Ax^2 + Cy^2 + Dx + Ey + F = 0$$

as a parabola, an ellipse, or a hyperbola by the process of completing the squares. We summarize the results.

RECALL

The graph of equation (1) is a degenerate conic if it results in the following:

A pair of lines, a single line, a single point, or no points at all.

> **IDENTIFYING CONICS WITH AXES PARALLEL TO THE x- OR y-AXES**
>
> If we exclude the degenerate conics, the graph of the equation
>
> $$Ax^2 + Cy^2 + Dx + Ey + F = 0, \qquad (1)$$
>
> where A and C are not both zero, is
>
> 1. a parabola if either $A = 0$ or $C = 0$ ($AC = 0$). (See page 613.)
> 2. an ellipse (or a circle if $A = C$) if A and C have the same sign ($AC > 0$). (See page 623.)
> 3. a hyperbola if A and C have opposite signs ($AC < 0$). (See page 638.)

STUDY TIP

Only the *type* of conic that is the graph of

$Ax^2 + Cy^2 + Dx + Ey + F = 0$

can be determined without completing squares. To find specific characteristics or sketch the graph of the conic, you must complete the squares.

EXAMPLE 1   **Identifying a Conic**

Determine the type of conic that is the graph of each equation.

a. $5x^2 - 2y + 7 = 0$

b. $7x^2 - 4y^2 - 3x + 8y - 12 = 0$

c. $3x^2 + 2y^2 - 3x + 9 = 0$

SOLUTION

a. We compare each equation with equation (1). We have $A = 5$ and $C = 0$. Since $AC = 0$, we conclude that the graph is a parabola.

b. Here $A = 7$ and $C = -4$. Because A and C have opposite signs, the graph is a hyperbola.

c. Here $A = 3$ and $C = 2$. Because A and C have the same sign, the graph is an ellipse.   ■ ■ ■

Practice Problem 1   Determine the type of conic that is the graph of each equation.

a. $x^2 + 2y^2 - 11x + 3 = 0$ **b.** $8y^2 - x - y + 3 = 0$ ■

2 Rotate coordinate axes.

Rotation of Axes

The graph of the general second-degree equation

$$Ax^2 + Bxy + Cy^2 + Dx + Ey + F = 0 \qquad (2)$$

is also a conic (or a degenerate conic). When $B \neq 0$, the axes for the conic are not parallel to the x-axis or the y-axis. We can find the conic's axis (or axes) by rotating the x- and y-axes through a suitable angle θ. We choose θ so that the xy-term in equation (2) is eliminated. We then analyze the graph of the resulting equation.

In **a rotation of axes**, the x- and y-axes are rotated through an angle θ, keeping the original fixed. The new axes are labeled the x'-axis (read "x-prime axis") and y'-axis (read "y-prime axis"), respectively, as shown in Figure 9.30(a).

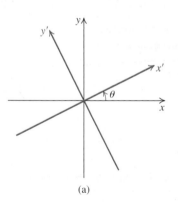

(a)   (b)

FIGURE 9.30 Rotation of axes

In Figure 9.30(b), the point P has two sets of coordinates: (x, y) in the old coordinate system and (x', y') in the new coordinate system. We let α denote the angle that the segment OP makes with the new x'-axis and let r denote the distance between P and the origin. In Figure 9.30(b), from the right triangle OPN, we have

$$\cos \alpha = \frac{x'}{r} \qquad \sin \alpha = \frac{y'}{r}.$$

So $x' = r \cos \alpha \qquad y' = r \sin \alpha.$

In the right triangle OPM in Figure 9.30(b), we have

$$\cos(\alpha + \theta) = \frac{x}{r} \qquad\qquad \sin(\alpha + \theta) = \frac{y}{r}.$$

So $\qquad\qquad x = r\cos(\alpha + \theta) \qquad\qquad y = r\sin(\alpha + \theta).$

Now $\quad x = r\cos(\alpha + \theta)$

$\qquad\quad = r(\cos\alpha\cos\theta - \sin\alpha\sin\theta) \qquad$ Sum formula for cosine

$\qquad\quad = (r\cos\alpha)\cos\theta - (r\sin\alpha)\sin\theta \qquad$ Distributive and associative properties

$\qquad\quad = x'\cos\theta - y'\sin\theta \qquad$ Replace $r\cos\alpha$ with x' and $r\sin\alpha$ with y'.

Similarly, $\quad y = r\sin(\alpha + \theta)$

$\qquad\quad = r(\sin\alpha\cos\theta + \cos\alpha\sin\theta) \qquad$ Sum formula for sine

$\qquad\quad = x'\sin\theta + y'\cos\theta \qquad$ Distribute, then replace $r\cos\alpha$ with x' and $r\sin\alpha$ with y'.

The equations $x = x'\cos\theta - y'\sin\theta$ and $y = x'\sin\theta - y'\cos\theta$ are two linear equations in x' and y'. This system of equations can be solved for x' and y' (see Exercise 70); the formulas relating the (x, y) and (x', y') coordinates of a point are given next.

ROTATION OF AXES FORMULAS

If the x-axis and y-axis are rotated through an angle θ, then the coordinates (x, y) of a point P determined by the xy-plane and the coordinates (x', y') of the point P determined by the $x'y'$-plane are related by the formulas:

$$x = x'\cos\theta - y'\sin\theta \qquad\qquad x' = x\cos\theta + y\sin\theta$$
$$y = x'\sin\theta + y'\cos\theta \qquad\qquad y' = -x\sin\theta + y\cos\theta$$

EXAMPLE 2   Rotating Axes to Eliminate *xy*-term

Rotate the coordinate axes through a $45°$ angle and express the equation $xy = 1$ in terms of the new $x'y'$ coordinates. Sketch the graph.

SOLUTION

Substitute $\theta = 45°$ in the rotation equations expressing x and y in terms of x' and y'. We have:

$$x = x'\cos 45° - y'\sin 45° = x'\frac{\sqrt{2}}{2} - y'\frac{\sqrt{2}}{2} = \frac{\sqrt{2}}{2}(x' - y')$$

$$y = x'\sin 45° + y'\cos 45° = x'\frac{\sqrt{2}}{2} + y'\frac{\sqrt{2}}{2} = \frac{\sqrt{2}}{2}(x' + y')$$

Then the equation $xy = 1$ becomes

$$\left[\frac{\sqrt{2}}{2}(x' - y')\right]\left[\frac{\sqrt{2}}{2}(x' + y')\right] = 1 \qquad \text{Replace } x \text{ with } \frac{\sqrt{2}}{2}(x' - y')$$

$$\text{and } y \text{ with } \frac{\sqrt{2}}{2}(x' + y'), \text{ in } xy = 1.$$

$$\left(\frac{\sqrt{2}}{2} \cdot \frac{\sqrt{2}}{2}\right)(x' - y')(x' + y') = 1 \qquad \text{Rearrange}$$

$$\frac{1}{2}(x'^2 - y'^2) = 1 \qquad \frac{\sqrt{2}}{2} \cdot \frac{\sqrt{2}}{2} = \frac{2}{4} = \frac{1}{2}; \text{ difference of squares}$$

$$\frac{x'^2}{2} - \frac{y'^2}{2} = 1 \qquad \text{Distributive property}$$

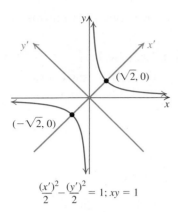

$$\frac{(x')^2}{2} - \frac{(y')^2}{2} = 1;\ xy = 1$$

FIGURE 9.31

The equation $\dfrac{x'^2}{2} - \dfrac{y'^2}{2} = 1$ is the standard form of the equation of a hyperbola with $a = b = \sqrt{2}$ in the $x'y'$-plane with center $(0,0)$, vertices $(\pm a, 0) = (\pm\sqrt{2}, 0)$, and asymptotes

$$y' = \pm\frac{b}{a}x' = \pm\frac{\sqrt{2}}{\sqrt{2}}x' = \pm x'.$$

Now $\qquad\qquad\qquad\qquad y' = x'$ $\qquad\qquad\qquad$ Equation of an asymptote

$\qquad -x\sin 45° + y\cos 45° = x\cos 45° + y\sin 45°$ $\qquad$ Write in xy-coordinates.

$$-x\frac{\sqrt{2}}{2} + y\frac{\sqrt{2}}{2} = x\frac{\sqrt{2}}{2} + y\frac{\sqrt{2}}{2}$$

$\qquad\qquad\qquad\qquad\quad x = 0$ $\qquad\qquad\qquad\qquad\quad$ Simplify.

Similarly, $y' = -x'$ leads to $y = 0$.

Therefore, the asymptotes $y' = x'$ and $y' = -x'$ are the y- and x-axes, respectively. The graph is shown in Figure 9.31. $\qquad$ ■ ■ ■

Practice Problem 2 Rotate the coordinate axes through a 45° angle and express the equation $xy = 2$ in terms of the new $x'y'$ coordinates. Sketch the graph. $\qquad$ ■

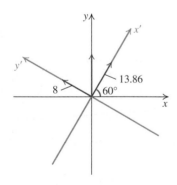

FIGURE 9.32 Launch direction

$\fbox{EXAMPLE 3}$ **Space Shuttles and Wind Speed**

Suppose that a space shuttle is being launched at a 30° angle to the east of due north as shown in Figure 9.32. Find the wind speed along and perpendicular to the initial launch path caused by a south-to-north wind whose speed is 16 knots.

SOLUTION

We represent the wind vector in the coordinate system in Figure 9.32. We want to find the coordinates of the points $(0, 16)$, which is the wind vector $\mathbf{v} = \langle 0, 16 \rangle$, in the $x'y'$ system determined by the shuttle.

We use the equations:

$x' = x\cos\theta + y\sin\theta \qquad\qquad y' = -x\sin\theta + y\cos\theta$

$x' = 0\cos 60° + 16\sin 60° \qquad y' = -0\sin 60° + 16\cos 60° \qquad$ Replace x with 0, y with 16, and θ with 60°.

$x' = 16\left(\dfrac{\sqrt{3}}{2}\right) \approx 13.86 \qquad y' = 16\left(\dfrac{1}{2}\right) = 8 \qquad$ Simplify.

Thus, the wind speed along the launch path is approximately 13.86 knots, and the wind speed perpendicular to the launch path is approximately 8 knots. See Figure 9.33. $\qquad$ ■ ■ ■

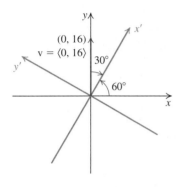

FIGURE 9.33 Wind speeds

Practice Problem 3 Rework Example 3 with a north-to-south wind speed of 12 knots. $\qquad$ ■

$\fbox{3}$ Eliminate the xy-term in the equation of a conic.

Eliminating the xy-Term

In Example 2, notice that we eliminated the Bxy term by rotating the axes through 45°.

In general, if we make the substitutions $x = x'\cos\theta - y'\sin\theta$ and $y = x'\sin\theta + y'\cos\theta$ in the equation

$$Ax^2 + Bxy + Cy^2 + Dx + Ey + F = 0$$

in which $B \neq 0$, we obtain an equation of the form

$$A'x'^2 + B'x'y' + C'y'^2 + D'x' + E'y' + F' = 0.$$

The coefficient B' in this equation (see Exercise 71) can be expressed in terms of the original coefficients as follows:

$$B' = B(\cos^2\theta - \sin^2\theta) - 2(A - C)\sin\theta\cos\theta \qquad \text{From Exercise 71}$$
$$= B\cos 2\theta - (A - C)\sin 2\theta \qquad \text{Replace } \cos^2\theta - \sin^2\theta \text{ with }$$
$$\cos 2\theta \text{ and replace } 2\sin\theta\cos\theta$$
$$\text{with } \sin 2\theta.$$

Then setting $B' = 0$, we have

$$B\cos 2\theta - (A - C)\sin 2\theta = 0$$

$$B\cos 2\theta = (A - C)\sin 2\theta \qquad \text{Add } (A - C)\sin 2\theta.$$

$$\frac{\cos 2\theta}{\sin 2\theta} = \frac{A - C}{B}, B \neq 0 \qquad \text{Divide by } B\sin 2\theta.$$

$$\cot 2\theta = \frac{A - C}{B}. \qquad \text{Quotient identity}$$

It is customary to choose an acute angle θ, that is, $0° \leq \theta < 90°$.

ELIMINATING AN *xy*-TERM BY ROTATION OF AXES

To transform the second-degree equation

$$Ax^2 + Bxy + Cy^2 + Dx + Ey + F = 0 \quad B \neq 0$$

to an equation of the form $A'x'^2 + C'y'^2 + D'x' + E'y' + F' = 0$ without $x'y'$ term, rotate the axes through an acute angle θ, where

$$\cot 2\theta = \frac{A - C}{B}.$$

FINDING THE SOLUTION: A PROCEDURE

EXAMPLE 4   **Rotating Axes to Eliminate the *xy*-term**

OBJECTIVE

Eliminate the xy-term in an equation of the form

$$Ax^2 + Bxy + Cy^2 + Dx + Ey + F = 0,$$
$$B \neq 0$$

by rotating the axes. Identify and graph the conic.

Step 1 Find the angle of rotation. Find the acute angle of rotation, θ, using the formula $\cot 2\theta = \dfrac{A - C}{B}$.

Step 2 Find $\cos\theta$ and $\sin\theta$. Find $\cos\theta$ and $\sin\theta$ by recognizing θ as a standard angle or by finding $\cos 2\theta$ and $\sin 2\theta$ from $\cot 2\theta$ and then using the half-angle formulas for $0 < \theta < 90°$.

$$\cos\theta = \sqrt{\frac{1 + \cos 2\theta}{2}}; \sin\theta = \sqrt{\frac{1 - \cos 2\theta}{2}}$$

EXAMPLE

Eliminate the xy-term in the equation

$$2x^2 + \sqrt{3}xy + y^2 - 5 = 0$$

by a suitable rotation of axes. Identify and graph the conic.

In $2x^2 + \sqrt{3}xy + y^2 - 5 = 0$, we have $A = 2$, $B = \sqrt{3}$, and $C = 1$. So $\cot 2\theta = \dfrac{2 - 1}{\sqrt{3}} = \dfrac{1}{\sqrt{3}} = \dfrac{\sqrt{3}}{3}$. So an appropriate choice for 2θ is $60°$. Since $2\theta = 60°$, we have $\theta = \dfrac{60°}{2} = 30°$.

$$\cos\theta = \cos 30° = \frac{\sqrt{3}}{2}; \quad \sin\theta = \sin 30° = \frac{1}{2}.$$

Step 3 **Convert from _xy_-coordinates to _x′y′_-coordinates.** Use Step 2 to replace _x_ with $x' \cos \theta - y' \sin \theta$ and _y_ with $x' \sin \theta + y' \cos \theta$ in the given equation. Collect like terms, and simplify.

$x = x' \cos 30° - y' \sin 30° = \dfrac{\sqrt{3}}{2}x' - \dfrac{1}{2}y'$ and

$y = x' \sin 30° + y' \cos 30° = \dfrac{1}{2}x' + \dfrac{\sqrt{3}}{2}y'.$

Then $2x^2 + \sqrt{3}xy + y^2 - 5 = 0$ becomes

$$2\left(\frac{\sqrt{3}}{2}x' - \frac{1}{2}y'\right)^2 + \sqrt{3}\left(\frac{\sqrt{3}}{2}x' - \frac{1}{2}y'\right)\left(\frac{1}{2}x' + \frac{\sqrt{3}}{2}y'\right)$$

$$+\left(\frac{1}{2}x' + \frac{\sqrt{3}}{2}y'\right)^2 - 5 = 0,$$ which (after collecting like

terms) simplifies to $\dfrac{x'^2}{2} + \dfrac{y'^2}{10} = 1.$

Step 4 **Identify the conic.** Identify the standard form of the conic in the _x′y′_-plane.

Using the standard form $\dfrac{(x')^2}{b^2} + \dfrac{(y')^2}{a^2} = 1$, we see that

$\dfrac{x'^2}{2} + \dfrac{y'^2}{10} = 1$ is the equation of an ellipse with center $(0, 0)$ in the _x′y′_-coordinate system.

The foci lie on the _y′_-axis because $10 > 2$.

$a^2 = 10, b^2 = 2,$ and $c^2 = a^2 - b^2 = 8$

Step 5 **Sketch the graph.** Sketch the graph of the conic in the _x′y′_-plane, using the procedures outlined in Sections 9.2–9.4.

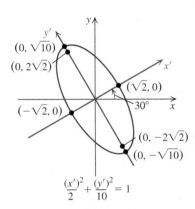

$$\frac{(x')^2}{2} + \frac{(y')^2}{10} = 1$$

■ ■ ■

Practice Problem 4 Eliminate the _xy_-term in the equation

$$5x^2 - 8xy + 5y^2 - 9 = 0$$

by a suitable rotation of axes. Identify and graph the conic.

■

EXAMPLE 5 **Identifying a Conic by Rotation of Axes**

Eliminate the _xy_-term in the equation $18x^2 - 48xy + 32y^2 - 20x - 15y - 25 = 0$ by a suitable rotation of axes. Identify and graph the conic.

SOLUTION

Step 1 In the given equation, $18x^2 - 48xy + 32y^2 - 20x - 15y - 25 = 0,$
$A = 18, B = -48,$ and $C = 32.$
So

$$\cot 2\theta = \frac{A - C}{B} = \frac{18 - 32}{-48} = \frac{-14}{-48} = \frac{7}{24}.$$

Step 2 Since $\cot 2\theta = \dfrac{7}{24} = \dfrac{x}{y}$ is positive, 2θ is in quadrant I.

From the sketch in Figure 9.34 (letting $x = 7$ and $y = 24$), we have

$$r = \sqrt{7^2 + 24^2} = \sqrt{625} = 25.$$

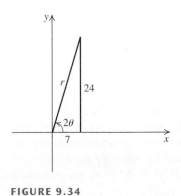

FIGURE 9.34

Then $\cos 2\theta = \dfrac{7}{25}$ and

$$\cos\theta = \sqrt{\frac{1 + \cos 2\theta}{2}} \qquad\qquad \sin\theta = \sqrt{\frac{1 - \cos 2\theta}{2}}$$

$$= \sqrt{\frac{1 + \dfrac{7}{25}}{2}} \qquad\qquad = \sqrt{\frac{1 - \dfrac{7}{25}}{2}}$$

$$= \sqrt{\frac{16}{25}} = \frac{4}{5} \qquad\qquad = \sqrt{\frac{9}{25}} = \frac{3}{5}$$

Step 3 $x = (\cos\theta)x' - (\sin\theta)y' = \dfrac{4}{5}x' - \dfrac{3}{5}y'$ Replace $\cos\theta$ with $\dfrac{4}{5}$ and $\sin\theta$

with $\dfrac{3}{5}$.

$$y = (\sin\theta)x' + (\cos\theta)y' = \frac{3}{5}x' + \frac{4}{5}y'$$

$18x^2 - 48xy + 32y^2 - 20x + 15y - 25 = 0$ becomes

$$18\left(\frac{4}{5}x' - \frac{3}{5}y'\right)^2 - 48\left(\frac{4}{5}x' - \frac{3}{5}y'\right)\left(\frac{3}{5}x' + \frac{4}{5}y'\right) + 32\left(\frac{3}{5}x' + \frac{4}{5}y'\right)^2$$

$$- 20\left(\frac{4}{5}x' - \frac{3}{5}y'\right) - 15\left(\frac{3}{5}x' + \frac{4}{5}y'\right) - 25 = 0.$$

Step 4 Collecting like terms and simplifying (see Exercise 69) yields

$50y'^2 - 25x' - 25 = 0$ or $y'^2 = \dfrac{1}{2}(x' + 1)$, which is an equation of a

parabola in the $x'y'$-plane.

Step 5 From $\cos\theta = \dfrac{4}{5}$, we find $\theta \approx 37°$.   Use a calculator.

Rotate the axes through approximately $37°$ as in Figure 9.35 and sketch the

graph of $y'^2 = \dfrac{1}{2}(x' + 1)$ in $x'y'$-plane. From the standard form $(y' - k)^2 =$

$4a(x' - h), a > 0$, we have $k = 0, h = -1$, and $4a = \dfrac{1}{2}\left(\text{or } a = \dfrac{1}{8}\right)$. So

the focus in the $x'y'$-plane is $(h + a, k) = \left(-1 + \dfrac{1}{8}, 0\right) = \left(-\dfrac{7}{8}, 0\right)$,

and the vertex is $(-1, 0)$. ■ ■ ■

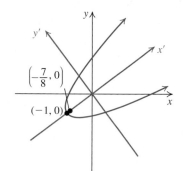

FIGURE 9.35

Practice Problem 5 Eliminate the xy-term in the equation $32x^2 - 48xy + 18y^2 -$
$15x - 20y = 0$ by a suitable rotation of axes. Identify and graph the conic. ■

4 Use the discriminant to identify a conic.

Identifying Conics by Using the Discriminant

By using the rotation of axes formulas, an equation of the form

$$Ax^2 + Bxy + Cy^2 + Dx + Ey + F = 0$$

can be transformed into an equation of the form

$$A'x'^2 + B'x'y' + C'y'^2 + D'x' + E'y' + F' = 0.$$

Each coefficient in the second equation can be expressed in terms of the coefficients in the first equation. (See Exercise 71.) It also can be shown that $B^2 - 4AC = B'^2 - 4A'C'$. (See Exercise 73.) If we can select a suitable rotation so that $B' = 0$, then $B'^2 - 4A'C' = -4A'C'$. From the test for identifying conics (see page 645) for the equation $A'x'^2 + C'y'^2 + D'x' + E'y' + F' = 0$, we conclude

TECHNOLOGY CONNECTION

To graph a second-degree equation of the form $Ax^2 + Bxy + Cy^2 + Dx + Ey + F = 0$ using a graphing calculator, you must first solve for y. Rewrite the equation as a quadratic equation in y, $Cy^2 + (Bx + E)y + (Ax^2 + Dx + F) = 0$, and use the quadratic formula to find two functions y_1, and y_2. Then graph Y_1 and Y_2 on your calculator in the same viewing window to produce the graph.

For example, to graph $2x^2 + 4xy + y^2 - x - y = 0$, rewrite the equation as $y^2 + (4x - 1)y + (2x^2 - x) = 0$. Solve for y to obtain

$$Y_1 = \frac{1 - 4x + \sqrt{8x^2 - 4x + 1}}{2}$$

and

$$Y_2 = \frac{1 - 4x - \sqrt{8x^2 - 4x + 1}}{2}.$$

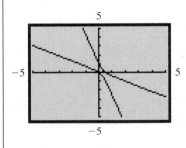

that the graph of this equation is a parabola if $A'C' = 0$, an ellipse if $A'C' > 0$, and a hyperbola if $A'C' < 0$. The quantity $B^2 - 4AC$ is called the **discriminant** of the equation $Ax^2 + Bxy + Cy^2 + Dx + Ey + F = 0$.

IDENTIFYING CONICS USING THE DISCRIMINANT

Except for degenerate cases, the graph of the equation

$$Ax^2 + Bxy + Cy^2 + Dx + Ey + F = 0,$$

where A and C are not both zero, is

1. a parabola if $B^2 - 4AC = 0$.
2. an ellipse (or a circle if $A = C$) if $B^2 - 4AC < 0$.
3. a hyperbola if $B^2 - 4AC > 0$.

EXAMPLE 6   **Identifying a Conic Using the Discriminant**

Use the discriminant to determine whether the graph of the non-degenerate conic with equation $3x^2 - 2\sqrt{5}xy + y^2 + 13x - 7y + 11 = 0$ is a parabola, an ellipse, or a hyperbola.

SOLUTION

We have $A = 3, B = -2\sqrt{5}$, and $C = 1$.

So, $B^2 - 4AC = (-2\sqrt{5})^2 - 4(3)(1) = 20 - 12 = 8$. Because $B^2 - 4AC > 0$, the graph of the given equation is a hyperbola.   ■ ■ ■

Practice Problem 6   Use the discriminant to determine whether the graph of the non-degenerate conic with equation $2x^2 + \sqrt{7}xy + 5y^2 - 3x + 1 = 0$ is a parabola, an ellipse, or a hyperbola.   ■

SECTION 9.5 ■ Exercises

A EXERCISES Basic Skills and Concepts

1. The general form for a second-degree equation is
   _____ . $Ax^2 + Bxy + Cy^2 + Dx + Ey + F = 0$

2. Rotation of axes is performed to eliminate the
   _____xy_____-term in a second-degree equation.

3. If the graph of an equation of the form
   $Ax^2 + Bxy + Cy^2 + Dx + Ey + F = 0$ is an ellipse,
   then _____ . $B^2 - 4AC < 0$

4. *True or False* The graph of a second-degree equation is a degenerate conic if it contains no points. True

5. *True or False* The graph of $x^2 + xy + y^2 - 3 = 0$ is a hyperbola. False

6. *True or False* If the coordinate axes are rotated through an acute angle θ and $\cot 2\theta < 0$, then $90° < 2\theta < 180°$.
   True

In Exercises 7–12, determine the type of conic that is the graph of the given equation.

7. $7y^2 - 13x + 2 = 0$
   parabola

8. $-12x^2 + 4x - 8y = 0$
   parabola

†Due to space constrictions, answers to these exercises may be found in the Answers beginning on page A–1 in the back of the book.

9. $3x^2 - 2y^2 + 4x - 9 = 0$ hyperbola

10. $-5x^2 + 3y^2 + 11x + 6y = 0$ hyperbola

11. $-2x^2 - y^2 + 5x + 2y + 1 = 0$ ellipse

12. $14x^2 + 3y^2 + 2y - 20 = 0$ ellipse

In Exercises 13–18, find the $x'y'$-coordinates of the given point if the coordinate axes are rotated through the given angle.

13. $(0, 1); \theta = 30°$ $\left(\dfrac{1}{2}, \dfrac{\sqrt{3}}{2}\right)$

14. $(-1, 0); \theta = 30°$ $\left(-\dfrac{\sqrt{3}}{2}, \dfrac{1}{2}\right)$

15. $(2, 1); \theta = 45°$

16. $(1, -1); \theta = 45°$ $(0, -\sqrt{2})$

17. $(3, -2); \theta = 60°$

18. $(-3, 4); \theta = 60°$ †

In Exercises 19–24, find the equation for the given conic in the $x'y'$-coordinates when the coordinate axes are rotated through an angle θ.

19. $x^2 + 3xy + y^2 + 5 = 0; \theta = 45°$ $\dfrac{y'^2}{10} - \dfrac{x'^2}{2} = 1$

20. $x^2 - 3xy + y^2 + 10 = 0; \theta = 45°$ $\dfrac{x'^2}{20} - \dfrac{y'^2}{4} = 1$

21. $2x^2 - 4\sqrt{3}xy + 6y^2 + 4\sqrt{3}x + 4y = 0; \theta = 30°$ $-x' = y'^2$

22. $2x^2 + 4\sqrt{3}xy + 6y^2 - \sqrt{3}x + y + 2 = 0; \theta = 60°$

23. $5x^2 + 4xy + 2y^2 - 6 = 0; \cos\theta = \dfrac{2\sqrt{5}}{5}, \sin\theta = \dfrac{\sqrt{5}}{5}$

24. $10x^2 + 6xy + 2y^2 - 11 = 0; \cos\theta = \dfrac{3\sqrt{10}}{10}, \sin\theta = \dfrac{\sqrt{10}}{10}$

In Exercises 25–36, find $\cos\theta$ and $\sin\theta$ such that rotation through the acute angle θ results in a new equation with no $x'y'$-term.

25. $4x^2 + 8xy + 4y^2 + \sqrt{2}x - \sqrt{2}y = 0$ $\cos\theta = \dfrac{\sqrt{2}}{2}, \sin\theta = \dfrac{\sqrt{2}}{2}$

26. $x^2 - 2xy + y^2 - \sqrt{2}x - \sqrt{2}y + 2 = 0$

27. $2x^2 + \sqrt{3}xy + y^2 - 10 = 0$

28. $2x^2 + 3xy + 2y^2 - 7 = 0$

29. $x^2 + 4xy + y^2 - 3 = 0$ $\cos\theta = \dfrac{\sqrt{2}}{2}, \sin\theta = \dfrac{\sqrt{2}}{2}$

30. $x^2 - 10xy + y^2 - 4 = 0$

31. $16x^2 - 24xy + 9y^2 - 10x + 70y + 25 = 0$ $\cos\theta = \dfrac{3}{5}, \sin\theta = \dfrac{4}{5}$

32. $18x^2 - 48xy + 32y^2 - 20x - 15y + 25 = 0$ †

33. $5x^2 - 6xy + 5y^2 - \sqrt{2}x + \sqrt{2}y - 6 = 0$ †

34. $3x^2 + 2xy + 3y^2 - 8 = 0$ †

35. $3x^2 - 2\sqrt{3}xy + y^2 - x - \sqrt{3}y + 2 = 0$ †

36. $-2x^2 - 12\sqrt{3}xy + 10y^2 - 4 = 0$ †

In Exercise 37–48, rotate the coordinate axes through an acute angle θ such that the new equation in x' and y' has no $x'y'$-term. Use the results from Exercises 25–36 for Exercises 37–48.

37. $4x^2 + 8xy + 4y^2 + \sqrt{2}x - \sqrt{2}y = 0$ $y' = 4x'^2$

38. $x^2 - 2xy + y^2 - \sqrt{2}x - \sqrt{2}y + 2 = 0$ $x' = y'^2 + 1$

39. $2x^2 + \sqrt{3}xy + y^2 - 10 = 0$ $\dfrac{x'^2}{4} + \dfrac{y'^2}{20} = 1$

40. $2x^2 + 3xy + 2y^2 - 7 = 0$ †

41. $x^2 + 4xy + y^2 - 3 = 0$ †

42. $x^2 - 10xy + y^2 - 4 = 0$ †

Answers:

15. $\left(\dfrac{3\sqrt{2}}{2}, \dfrac{\sqrt{2}}{-2}\right)$

17. $\left(\dfrac{3}{2} - \sqrt{3}, -\dfrac{3\sqrt{3}}{2} - 1\right)$

43. $16x^2 - 24xy + 9y^2 - 10x + 70y + 25 = 0$ †

44. $18x^2 - 48xy + 32y^2 - 20x - 15y + 25 = 0$ $x' = 2y'^2 + 1$

45. $5x^2 - 6xy + 5y^2 - \sqrt{2}x - \sqrt{2}y - 6 = 0$ $x'^2 - x' + 4y'^2 = 3$

46. $3x^2 + 2xy + 3y^2 - 8 = 0$ †

47. $3x^2 - 2\sqrt{3}xy + y^2 - x - \sqrt{3}y + 2 = 0$ $x' = 2y'^2 + 1$

48. $-2x^2 - 12\sqrt{3}xy + 10y^2 - 4 = 0$ †

In Exercises 49–52, sketch the graph of the degenerate conic or state that the graph contains no points.

49. $x^2 - 2xy + y^2 - 2 = 0$ †

50. $x^2 - 2xy + y^2 + x - y = 0$ †

51. $7x^2 + 6\sqrt{3}xy + 13y^2 + 4 = 0$ no points

52. $3x^2 - 2\sqrt{3}xy + y^2 - 1 = 0$ †

In Exercises 53–58, use the discriminant to identify the graph of the given non-degenerate conic as a parabola, an ellipse, or a hyperbola.

53. $3x^2 + 2xy + 3y^2 - 8 = 0$ ellipse

54. $7x^2 + 2xy + 7y^2 - 16 = 6$ ellipse

55. $9x^2 - 6xy + y^2 - 2x + 1 = 0$ parabola

56. $x^2 + 4xy - 2y^2 - 12 = 0$ hyperbola

57. $4x^2 + 24xy - 3y^2 - 60 = 0$ hyperbola

58. $4x^2 - 4xy + y^2 - 10y - 30 = 0$ parabola

In Exercises 59–64, sketch the graph using a graphing utility. These are the same equations given in Exercises 53–58.

59. $3x^2 + 2xy + 3y^2 - 8 = 0$ †

60. $7x^2 + 2xy + 7y^2 - 16 = 6$ †

61. $9x^2 - 6xy + y^2 - 2x + 1 = 0$ †

62. $x^2 + 4xy - 2y^2 - 12 = 0$ †

63. $4x^2 + 24xy - 3y^2 - 60 = 0$ †

64. $4x^2 - 4xy + y^2 - 10y - 30 = 0$ †

B EXERCISES Applying the Concepts

In Exercises 65–68, rework Example 3 with the given modifications.

65. Find the wind speed along and perpendicular to the initial launch path caused by an east-to-west wind of 15 knots. †

66. Find the wind speed along and perpendicular to the initial launch path caused by a west-to-east wind of 18 knots. †

67. Find the south-to-north wind speed caused by an 8-knot wind in the positive direction of the y'-axis of the coordinate system determined by the shuttle. 4 knots

68. Find the south-to-north wind speed caused by an 8-knot wind in the positive direction of the x'-axis of the coordinate system determined by the shuttle. 6.93 knots

22. $y' = -4x'^2 - 1$ 23. $\dfrac{x'^2}{1} + \dfrac{y'^2}{6} = 1$ 24. $\dfrac{x'^2}{1} + \dfrac{y'^2}{11} = 1$

26. $\cos\theta = \dfrac{\sqrt{2}}{2}, \sin\theta = \dfrac{\sqrt{2}}{2}$ 27. $\cos\theta = \dfrac{\sqrt{3}}{2}, \sin\theta = \dfrac{1}{2}$

28. $\cos\theta = \dfrac{\sqrt{2}}{2}, \sin\theta = \dfrac{\sqrt{2}}{2}$ 30. $\cos\theta = \dfrac{\sqrt{2}}{2}, \sin\theta = \dfrac{\sqrt{2}}{2}$

C EXERCISES Beyond the Basics

69. Show that the expression in Step 3 of Example 5 simplifies to $50y'^2 - 25x' - 25 = 0$.

70. Solve the system of equations

$$\begin{cases} x = x' \cos\theta - y' \sin\theta \\ y = x' \sin\theta + y' \cos\theta \end{cases}$$

for x' and y' in terms of x and y. [*Hint:* Multiply the first equation by $\cos\theta$ and the second equation by $\sin\theta$ and use elimination.]

71. Show that the rotation substitutions $x = x' \cos\theta - y' \sin\theta$ and $y = x' \sin\theta + y' \cos\theta$ in the equation $Ax^2 + Bxy + Cy^2 + Dx + Ey + F = 0$ transform the equation to the equation in the form.

$$A'x'^2 + B'x'y' + C'y'^2 + D'x' + E'y' + F' = 0$$

where

$A' = A \cos^2\theta + B \sin\theta \cos\theta + C \sin^2\theta$

$B' = B(\cos^2\theta - \sin^2\theta) - 2(A - C)\sin\theta \cos\theta$

$C' = A \sin^2\theta - B \sin\theta \cos\theta + C \cos^2\theta$

$D' = D \cos\theta + E \sin\theta$

$E' = -D \sin\theta + E \cos\theta$

$F' = F$

72. Shown that the quantity $A + C$ is invariant under rotation of axes, that is, $A + C = A' + C'$.

73. Show that the quantity $B^2 - 4AC$ is invariant under rotation of axes, that is, $B^2 - 4AC = B'^2 - 4A'C'$.

[*Hint:* Replace B', A', and C' with the formulas in Exercise 71 and use $\cos^4\theta + \sin^4\theta = 1 - 2\cos^2\theta \sin^2\theta$.]

74. Use the result of Exercise 73 to show that, excluding degenerate cases, the graph of the equation

$$Ax^2 + Bxy + Cy^2 + Dx + Ey + F = 0 \quad \text{is}$$

a. a parabola if $B^2 - 4AC = 0$.
b. an ellipse (or a circle) if $B^2 - 4AC < 0$.
c. a hyperbola if $B^2 - 4AC > 0$.

Answers:
70. $x' = x \cos\theta + y \sin\theta$
$y' = -x \sin\theta + y \cos\theta$

75. Explain why the equation $Ax^2 + Bxy + Cy^2 + Dx + Ey + F = 0$ cannot represent a circle if $B \neq 0$.

76. Show that the equation $x^2 + y^2 = r^2$ is transformed into the equation $x'^2 + y'^2 = r^2$ by rotation of axes through any angle θ.

77. It is possible to solve the equation

$$2(C - A)\sin\theta \cos\theta + B(\cos^2\theta - \sin^2\theta) = 0$$

for $\tan\theta$ rather than $\cot 2\theta$. Show that if θ is an acute angle, then $\tan\theta = \dfrac{(C - A) + \sqrt{(C - A)^2 + B^2}}{B}, B \neq 0$.

[*Hint:* If θ is an acute angle, then $\tan\theta > 0$.]

78. Show that the graph of the equation

$$\sqrt{x} + \sqrt{y} = \sqrt{a}, a > 0$$

is part of the graph of a parabola. [*Hint:* Eliminate radicals and use the discriminant.]

79. Show that the distance between two points $P(x, y)$ and $Q(x_1, y_1)$ in the xy-plane equals the distance $P'(x', y')$ and $Q'(x'_1, y'_1)$ in the $x'y'$-plane.

Critical Thinking

80. Use the result of Exercise 71 to find an angle of rotation such that the equation
$Ax^2 + Bxy + Cy^2 + Dx + Ey + F = 0$ becomes
$Cx^2 - Bxy + Ay^2 + Ex - Dy + F = 0$. $90°$

81. Find a value for F so that a rotation through an angle of $30°$ transforms the equation
$x^2 + \sqrt{3}xy + x - \sqrt{3}y + F = 0$ into the degenerate conic with equation $3x'^2 = (y' + 2)^2$. $F = -2$

Polar Equations of Conics

Before Starting this Section, Review

1. Definitions of conics (Sections 9.2–9.4)
2. Polar coordinates (Section 6.6)

Objectives

1 Define conics using the focus-directrix property.

2 Find the polar equation of a conic.

3 Graph the conic from its polar equation.

4 Apply polar equations to orbits of satellites.

SPUTNIK AND THE DAWNING OF THE SPACE AGE

On October 4, 1957, the Soviet Union successfully launched the world's first artificial satellite, *Sputnik,* into orbit around Earth. *Sputnik* was about the size of a basketball (diameter 23 inches, weight 184 pounds) and took about 98 minutes to orbit Earth. The *Sputnik* launch occurred during the Cold War, when the Americans and Russians regarded each other as enemies. The United States was shocked at the apparent Soviet superiority in rocket technology. Intense debate about America's standing led to the establishment of the National Aeronautics and Space Administration (NASA). The space race was on! Both the Americans and the Russians raced to develop spaceships to carry humans. On January 31, 1958, the United States successfully launched *Explorer I*, which discovered the radiation belts around Earth. In 1961, President Kennedy challenged the Russians to a race to the moon. In 1969, the United States got there first and the Russians refocused their efforts on a permanent presence in space with a space station.

In Example 7, we will find a polar equation of the elliptical path of *Sputnik*. ■

1 Define conics using the focus-directrix property.

Definition of a Conic

In Sections 9.2–9.4, we defined each conic as the parabola, the ellipse, and the hyperbola individually. We now explore a single property, the *eccentricity*, which allows us to use one definition for all of these curves.

BY THE WAY . . .

The symbol *e* is used both for the eccentricity of a conic and the Euler constant introduced in Chapter 3. The meaning should be clear from the context.

FOCUS-DIRECTRIX DEFINITION OF CONICS

Let *l* be a fixed line (called the **directrix**) in the plane and *F* a fixed point (called the **focus**) not on the line *l* and let $e > 0$ be a fixed number. The set of all points *P* in the plane that satisfy the ratio

$$\frac{d(P, F)}{d(P, l)} = e \quad \text{or} \quad d(P, F) = e \cdot d(P, l)$$

is called a **conic**. The number *e* in the equation is called the **eccentricity** of the conic. The conic (see Figure 9.36) is

an **ellipse** if $e < 1$.

a **parabola** if $e = 1$.

a **hyperbola** if $e > 1$.

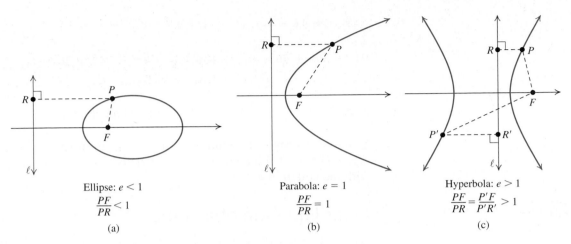

FIGURE 9.36 Identification of conics by the values of e

The line through the focus perpendicular to the directrix is called the **principal axis** (or simply the **axis**) of the conic. A point where the conic crosses the axis is called a **vertex**. The parabola has one vertex; the ellipse and hyperbola have two vertices. The point on the axis that is midway between the two vertices of an ellipse or a hyperbola is called the **center**. See Figure 9.37

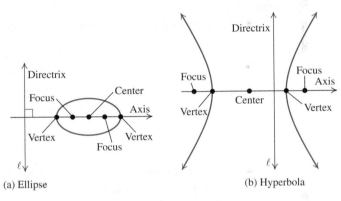

FIGURE 9.37

<hr/>

EXAMPLE 1   Finding the Equation of a Conic

Find an equation in the form $Ax^2 + Bxy + Cy^2 + Dx + Ey + F = 0$ of the conic with eccentricity $e = \dfrac{2}{3}$, focus $F(-2, 1)$, and directrix l with the equation $x = 3$.

SOLUTION

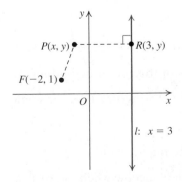

FIGURE 9.38

Figure 9.38 shows the focus F, the directrix l, and $P(x, y)$ representing any point on the conic. We define $\overline{PR}$ as the perpendicular from P to l, the directrix. The coordinates of R are $(3, y)$.

$$\frac{d(P, F)}{d(P, R)} = \frac{2}{3} \qquad\qquad \frac{d(P, F)}{d(P, l)} = e$$

$$3d(P, F) = 2d(P, R) \qquad\qquad \text{Cross multiply.}$$

$$3\sqrt{(x + 2)^2 + (y - 1)^2} = 2\sqrt{(x - 3)^2 + (y - y)^2} \qquad \text{Use distance formula.}$$

$$9[(x + 2)^2 + (y - 1)^2] = 4(x - 3)^2 \qquad\qquad \text{Square both sides.}$$

continued on the next page

$$9[x^2 + 4x + 4 + y^2 - 2y + 1] = 4(x^2 - 6x + 9) \qquad \text{Expand binomials.}$$

$$9x^2 + 36x + 36 + 9y^2 - 18y + 9 = 4x^2 - 24x + 36 \qquad \text{Distribute.}$$

$$5x^2 + 9y^2 + 60x - 18y + 9 = 0 \qquad \text{Simplify.}$$

Because $e = \dfrac{2}{3} < 1$, the equation is that of an ellipse. ■ ■ ■

Practice Problem 1 Find an equation in the form $Ax^2 + Bxy + Cy^2 + Dx + Ey + F = 0$ of the conic with eccentricity $e = 3$, focus $F(1, -2)$, and directrix l with the equation $y + 4 = 0$. ■

2 Find the polar equation of a conic.

Polar Equation of a Conic

We can find fairly simple polar equations of conics if we place the focus at the pole and let the directrix be vertical or horizontal. For simplicity, we use a mixture of the polar and Cartesian coordinate systems. For example, instead of writing the equation $r \sin \theta = 3$ or $r = 3 \csc \theta$ in polar coordinates, we use the simple form $y = 3$ in Cartesian coordinates.

Consider the case when the directrix is vertical and is p units to the left of the focus F, which is at the pole. The Cartesian equation of the directrix is $x = -p$.

In Figure 9.39, we see that

$$PF = r \text{ and } PR = p + r \cos \theta.$$

A point $P(r, \theta)$ lies on the conic if and only if

$$PF = e \cdot PR \qquad \text{Definition of a conic}$$
$$r = e(p + r \cos \theta) \qquad \text{Substitute for } PF \text{ and } PR.$$
$$r = ep + er \cos \theta \qquad \text{Distributive property}$$
$$r - er \cos \theta = ep \qquad \text{Rearrange}$$
$$r = \frac{ep}{1 - e \cos \theta} \qquad \text{Solve for } r.$$

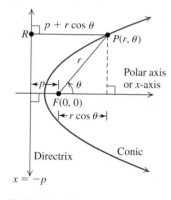

FIGURE 9.39

The polar equation of a conic with focus at the pole and vertical directrix p units to the left of the pole is

$$r = \frac{ep}{1 - e \cos \theta}.$$

(i) Since $\cos(-\theta) = \cos \theta$, the conic is symmetric about the x-axis.

(ii) If $e \neq 1$, we can find the polar coordinates of the vertices V_1 and V_2 by substituting $\theta = 0$ and $\theta = \pi$ in the polar equation:

$$V_1 = \left(\frac{ep}{1 - e \cos 0}, 0 \right) = \left(\frac{ep}{1 - e}, 0 \right) \text{ and } V_2 = \left(\frac{ep}{1 - e \cos \pi}, \pi \right) = \left(\frac{ep}{1 + e}, \pi \right)$$

Recall that in the polar coordinates system, $\left(-\dfrac{ep}{1 + e}, 0 \right)$ and $\left(\dfrac{ep}{1 + e}, \pi \right)$ represent the same point V_2. We can find the polar coordinates of the center by finding the midpoint of $V_1 = \left(\dfrac{ep}{1 - e}, 0 \right)$ and $V_2 = \left(-\dfrac{ep}{1 + e}, 0 \right)$. The center C of the conic

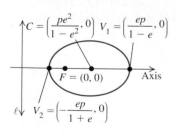

is $\quad C = \left(\dfrac{\dfrac{ep}{1 - e} - \dfrac{ep}{1 + e}}{2}, \dfrac{0 + 0}{2} \right) = \left(\dfrac{pe^2}{1 - e^2}, 0 \right)$. See Figure 9.40.

FIGURE 9.40 $r = \dfrac{ep}{1 - e \cos \theta}$,
$e < 1$

In a similar manner, we can derive a polar equation of a conic when the vertical directrix is p units to the right of the focus and when the directrix is horizontal. (See Exercise 51.) We summarize these results in the next box.

STANDARD FORM OF POLAR EQUATIONS OF CONICS

Polar equation of a conic with focus F at the pole (origin), with eccentricity e, having a distance of p units ($p > 0$) between the focus and the directrix l:

a. Vertical directrix left of the pole

$$r = \frac{ep}{1 - e \cos \theta}$$

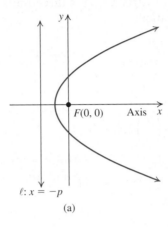

$\ell : x = -p$

(a)

b. Vertical directrix right of the pole

$$r = \frac{ep}{1 + e \cos \theta}$$

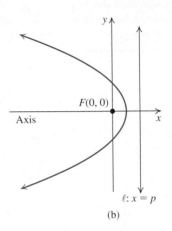

$\ell : x = p$

(b)

c. Horizontal directrix below the pole

$$r = \frac{ep}{1 - e \sin \theta}$$

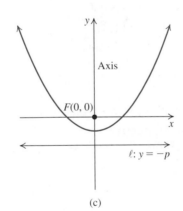

$\ell : y = -p$

(c)

d. Horizontal directrix above the pole

$$r = \frac{ep}{1 + e \sin \theta}$$

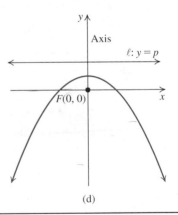

$\ell : y = p$

(d)

STUDY TIP

In the standard form for the polar equation of a conic, remember that in the denominator,

1. the cosine indicates a vertical directrix and

2. a minus sign indicates that the directrix is below, or to the left of, the pole.

EXAMPLE 2   **Writing Polar Equations of Conics**

Write a polar equation of each conic with focus at the pole eccentricity e and directrix l. Identify the conic.

a. $e = \dfrac{4}{5}$, l is $x = 3$ **b.** $e = 1$, l is $y = -2$ **c.** $e = \dfrac{3}{2}$, l is $y = -4$

SOLUTION

a. The vertical directrix $x = 3$ is three units to the right of the focus. We use the following equation:

$$r = \frac{ep}{1 + e\cos\theta} \qquad \text{Standard form}$$

$$= \frac{\left(\dfrac{4}{5}\right)(3)}{1 + \left(\dfrac{4}{5}\right)\cos\theta} \qquad \text{Substitute } e = \frac{4}{5} \text{ and } p = 3.$$

$$r = \frac{12}{5 + 4\cos\theta} \qquad \begin{array}{l}\text{Multiply numerator and denominator by}\\ \text{5 and simplify.}\end{array}$$

Because $e = \dfrac{4}{5} < 1$, the conic is an ellipse.

b. The horizontal directrix $y = -2$ is two units below the focus. We use the following equation:

$$r = \frac{ep}{1 - e\sin\theta} \qquad \text{Standard form}$$

$$r = \frac{(1)(2)}{1 - (1)\sin\theta} \qquad \text{Substitute } e = 1 \text{ and } p = 2.$$

$$r = \frac{2}{1 - \sin\theta} \qquad \text{Simplify.}$$

Because $e = 1$, the conic is a parabola.

c. The horizontal directrix $y = -4$ is four units below the focus. We use the following equation:

$$r = \frac{ep}{1 - e\sin\theta} \qquad \text{Standard form}$$

$$= \frac{\left(\dfrac{3}{2}\right)(4)}{1 - \left(\dfrac{3}{2}\right)\sin\theta} \qquad \text{Substitute } e = \frac{3}{2} \text{ and } p = 4.$$

$$r = \frac{12}{2 - 3\sin\theta} \qquad \begin{array}{l}\text{Multiply numerator and denominator by 2}\\ \text{and simplify.}\end{array}$$

Because $e = \dfrac{3}{2} > 1$, the conic is a hyperbola.   ■■■

Practice Problem 2   Write a polar equation for a conic with focus at the pole, eccentricity $e = \dfrac{3}{4}$, and directrix $y = 3$.   ■

3 Graph the conic from its polar equation.

Graphing a Conic

The graph of a polar equation of the form

$$r = \frac{ep}{1 \pm e \cos \theta} \qquad \text{Vertical directrix}$$

or

$$r = \frac{ep}{1 \pm e \sin \theta} \qquad \text{Horizontal directrix}$$

is a conic, where e is the eccentricity of the conic and p is the distance between the focus at the pole and the directrix. Remember that the conic is a parabola if $e = 1$, an ellipse if $0 < e < 1$, and a hyperbola if $e > 1$.

FINDING THE SOLUTION: A PROCEDURE

| EXAMPLE 3 | Graphing a Conic from Its Polar Equation |

OBJECTIVE

Sketch the graph of a conic given its polar equation.

EXAMPLE

Sketch the graph of the conic

$$r = \frac{5}{3 + 2 \cos \theta}.$$

Step 1 Use standard form. Write the equation of the conic in one of the standard forms.

The given equation is associated with the standard form

$$r = \frac{ep}{1 + e \cos \theta}.$$

$$r = \frac{5}{3 + 2 \cos \theta}$$

$$r = \frac{\dfrac{5}{3}}{1 + \dfrac{2}{3} \cos \theta} \qquad \begin{array}{l} \text{Divide numerator and} \\ \text{denominator by 3.} \end{array}$$

Step 2 Find e and p. Use the standard form in Step 1 to find the values of e and p.

Comparing $r = \dfrac{\dfrac{5}{3}}{1 + \dfrac{2}{3} \cos \theta}$ with $r = \dfrac{ep}{1 + e \cos \theta}$,

we have $e = \dfrac{2}{3}$ and $ep = \dfrac{5}{3}$.

$$\frac{2}{3} p = \frac{5}{3} \qquad \text{Substitute } e = \frac{2}{3}.$$

$$p = \frac{5}{2} \qquad \text{Solve for } p.$$

Step 3 Identify the conic. Use the value of e from Step 2 to identify the conic. Compare this equation with the standard form to find the directrix.

Because $e = \dfrac{2}{3} < 1$, the conic is an ellipse. By comparing our

equation with the standard form $r = \dfrac{ep}{1 + e \cos \theta}$, we see that

the focus of the ellipse is at the pole and the associated directrix

is the vertical line $x = \dfrac{5}{2} \left(\text{or } r \cos \theta = \dfrac{5}{2} \text{ in polar coordinates} \right).$

continued on the next page

Step 4 **Find vertices.** Locate the vertices by finding the points of intersection of the conic and the principal axis. The principal axis is perpendicular to the directrix. The principal axis is the axis of symmetry for a parabola. It is the major axis for an ellipse and the transverse axis for a hyperbola.

The principal axis is along the x-axis, so $\theta = 0$ and $\theta = \pi$.

When $\theta = 0$, $r = \dfrac{5}{3 + 2\cos 0} = \dfrac{5}{3 + 2} = 1$.

When $\theta = \pi$, $r = \dfrac{5}{3 + 2\cos \pi} = \dfrac{5}{3 - 2} = 5$.

The vertices are at $(1, 0)$ and $(5, \pi)$.

Step 5 **Identify symmetry.** The conic is symmetric about the principal axis.

The conic is symmetric about the x-axis.

Step 6 **Sketch the graph.** Plot points to sketch half of the conic and use symmetry to sketch the other half.

Sketch the upper half of the ellipse by plotting points from $\theta = 0$ to $\theta = \pi$. Then use symmetry to sketch the lower half of the ellipse.

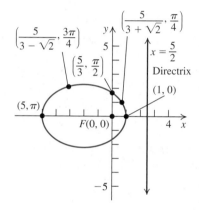

Practice Problem 3 Sketch the graph of the conic $r = \dfrac{5}{3 + 2\sin \theta}$.

EXAMPLE 4 Graphing a Parabola from Its Polar Form

Sketch the graph of the conic $r = \dfrac{12}{2 - 2\sin \theta}$.

SOLUTION

Step 1 The associated standard form is $r = \dfrac{ep}{1 - e\sin \theta}$, which has a horizontal directrix below the focus at the pole.

$$r = \frac{12}{2 - 2\sin \theta} \qquad \text{Given equation}$$

$$r = \frac{6}{1 - \sin \theta} \qquad \text{Divide numerator and denominator by 2.}$$

Step 2 $e = 1$ and $p = 6$

Step 3 **Identify the conic.** Because $e = 1$, the conic is a parabola with focus at the pole. The directrix is the horizontal line $y = -6$ (or $r\sin \theta = -6$ in polar coordinates).

Step 4 The principal axis is the y-axis. When $\theta = \dfrac{\pi}{2}$, $\sin \theta = 1$ and the denominator in $r = \dfrac{12}{2 - 2\sin \theta}$ is zero; so r is undefined. When $\theta = \dfrac{3\pi}{2}$, however, $r = 3$; so its vertex is at $\left(3, \dfrac{3\pi}{2}\right)$.

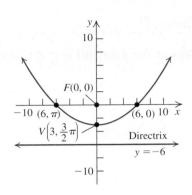

FIGURE 9.41

Step 5   The line $\theta = \dfrac{\pi}{2}$ (the y-axis) is the axis of symmetry.

Step 6   Sketch the left half of the parabola by plotting points; choose values of θ between $\dfrac{\pi}{2}$ and $\dfrac{3\pi}{2}$. Then use symmetry to sketch the right half of the parabola. See Figure 9.41.   ■ ■ ■

Practice Problem 4   Sketch the graph of the conic $r = \dfrac{6}{2 + 2\sin\theta}$.   ■

EXAMPLE 5   **Graphing a Hyperbola from Its Polar Form**

Sketch the graph of the conic $r = \dfrac{12}{2 - 4\cos\theta}$.

SOLUTION

Step 1   The associated standard form is $r = \dfrac{ep}{1 - e\cos\theta}$, which has a vertical directrix to the left of the focus at the pole.

$$r = \dfrac{12}{2 - 4\cos\theta}\qquad \text{Given equation}$$

$$r = \dfrac{6}{1 - 2\cos\theta}\qquad \text{Divide numerator and denominator by 2.}$$

Step 2 $e = 2$ and $ep = 6$

$$2p = 6\qquad \text{Substitute } e = 2.$$

$$p = 3\qquad \text{Solve for } p.$$

Step 3   Because $e = 2 > 1$, the conic is a hyperbola with focus at the pole; because $p = 3$, the directrix is the vertical line $x = -3$.

Step 4   The principal axis is the x-axis. The vertices are as follows:

$$V_1 = \left(\dfrac{ep}{1 + e}, \pi\right) = (2, \pi)\, \text{or}\, V_1 = (-2, 0)$$

$$V_2 = \left(\dfrac{ep}{1 - e}, 0\right) = (-6, 0)\, \text{or}\, V_2 = (6, \pi)$$

The center is at the midpoint of the line segment $\overline{V_2 V_1}$: $(-4, 0)$ or $(4, \pi)$. For the hyperbola $r = \dfrac{6}{1 - 2\cos\theta}$, the denominator $1 - 2\cos\theta$ equals 0 when $\cos\theta = \dfrac{1}{2}$, that is, when $\theta = \pm\dfrac{\pi}{3}$. If $0 \le \theta < \dfrac{\pi}{3}$, then the equation $r = \dfrac{6}{1 - 2\cos\theta}$ gives negative values of r and the points (r, θ) will lie on the left branch of the hyperbola. If $\dfrac{\pi}{3} < \theta \le \pi$, then $r > 0$ and the points will lie on the right branch of the hyperbola.

Step 5   The curve is symmetric about the x-axis.

Step 6   You can plot points for values of θ between 0 and π and sketch a portion of the hyperbola (shown in red in Figure 9.42.) Then you can use symmetry to sketch the remaining portion (shown in green in Figure 9.42) of the hyperbola.   ■ ■ ■

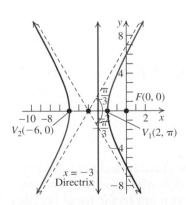

FIGURE 9.42 $r = \dfrac{12}{2 - 4\cos\theta}$

Practice Problem 5   Sketch the graph of the conic $r = \dfrac{8}{2 - 6\sin\theta}$.   ■

EXAMPLE 6   **Finding Polar Equations of a Conic**

Find a polar equation of the hyperbola having a focus at the pole and vertices at $\left(1, \dfrac{\pi}{2}\right)$ and $\left(3, \dfrac{\pi}{2}\right)$.

SOLUTION

Because the vertices of the hyperbola lie on the y-axis, the directrix is horizontal. Consider the polar form

$$r = \frac{ep}{1 + e \sin \theta}.$$

We can find its vertices when we substitute $\theta = \dfrac{\pi}{2}$ and $\theta = \dfrac{3\pi}{2}$.

When $\theta = \dfrac{\pi}{2}, r = \dfrac{ep}{1 + e}$; so

$$\frac{ep}{1 + e} = 1. \quad (1)$$

The point $\left(3, \dfrac{\pi}{2}\right)$ also has polar coordinates $\left(-3, \dfrac{3\pi}{2}\right)$.

When $\theta = \dfrac{3\pi}{2}, r = \dfrac{ep}{1 - e}$. Thus,

$$\frac{ep}{1 - e} = -3. \quad (2)$$

Solving equations (1) and (2) for ep and equating them gives

$$1 + e = -3(1 - e) \qquad \text{From (1), } ep = 1 + e; \text{ from (2),} \\ ep = -3(1 - e).$$

So

$$e = 2 \qquad \text{Solve for } e.$$

Since $ep = 1 + e$,

$$p = \frac{1 + e}{e} = \frac{1 + 2}{2} = \frac{3}{2} \qquad \text{Solve for } p, \text{ with } e = 2.$$

A polar equation of the hyperbola is therefore

$$r = \frac{ep}{1 + e \sin \theta} = \frac{2\left(\dfrac{3}{2}\right)}{1 + 2 \sin \theta} \qquad \text{Substitute values of } e \text{ and } p.$$

$$r = \frac{3}{1 + 2 \sin \theta} \qquad\qquad\qquad ■\ ■\ ■$$

Practice Problem 6   Find a polar equation of the hyperbola having a focus at the pole and vertices at $(2, \pi)$ and $(5, \pi)$.   ■

4 Apply polar equations to orbits of satellites.

Applications

According to Kepler's Law, if an object A moves in the gravitational field of a heavier object B, then the path of the object A must be a conic with a focus at the center of the mass of object B. In elliptical orbits, the closest point to the focus is called

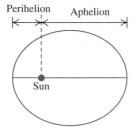

Perihelion   Aphelion

Sun

FIGURE 9.43

perigee (or **perihelion**) and the farthest point is called the **apogee** (or **aphelion**). See Figure 9.43.

EXAMPLE 7   Polar Equation of *Sputnik I*

Sputnik I had an elliptical orbit around Earth. It reached the maximum and minimum heights of 560 miles and 145 miles above Earth, respectively. Write a polar equation for the path of *Sputnik I* with the pole at the center of Earth. Assume that the radius of Earth is 3960 miles.

SOLUTION

Since the orbit of *Sputnik I* is elliptical, consider the form of its polar equation as

$$r = \frac{ep}{1 - e\cos\theta}, \text{ where } e < 1. \quad (1)$$

The maximum distance of *Sputnik I* from the center of Earth is

$$3960 + 560 = 4520 \text{ miles.}$$

The minimum distance of *Sputnik I* from the center of Earth is

$$3960 + 145 = 4105 \text{ miles.}$$

In equation (1), r has a maximum value when $\theta = 0$; so

$$\frac{ep}{1 - e} = 4520, \quad \text{or} \quad ep = 4520(1 - e)$$

Similarly, r has a minimum value when $\theta = \pi$; so

$$\frac{ep}{1 + e} = 4105, \quad \text{or} \quad ep = 4105(1 + e)$$

Equating the two expressions for ep, we have

$$4520(1 - e) = 4105(1 + e)$$

$$e = \frac{4520 - 4105}{4520 + 4105} \qquad \text{Solve for } e.$$

$$\approx 0.048$$

Because $\dfrac{ep}{1 - e} = 4520$, we have

$$p = \frac{4520(1 - e)}{e} \qquad \text{Solve for } p.$$

$$p = \frac{4520(1 - 0.048)}{0.048} \qquad \text{Replace } e \text{ with } 0.048.$$

$$p \approx 89{,}646.667$$

A polar equation for Sputnik *I* is

$$r = \frac{ep}{1 - e\cos\theta} = \frac{4303}{1 - 0.048\cos\theta} \qquad ep \approx 4303 \qquad ■ ■ ■$$

Practice Problem 7   The subplanet Pluto has an elliptical orbit around the sun. Its maximum and minimum distances from the center of the sun are 4582.61 million of miles and 2749.57 million of miles, respectively. Write a polar equation for the path of Pluto with the pole at the center of the sun.   ▨

SECTION 9.6 ■ Exercises

A EXERCISES Basic Skills and Concepts

1. Let p be a point on the graph of a conic with focus F and directrix l. The ratio $\dfrac{d(P, F)}{d(P, l)}$ is called __eccentricity__ of the conic. For eccentricity e, if $e < 1$, the conic is a(n) __ellipse__; if $e = 1$, the conic is a(n) __parabola__; and if $e > 1$, the conic is a(n) __hyperbola__.

In Exercises 2–5, write the polar form of a conic with focus at the pole, eccentricity e, and directrix p units from the focus.

2. Vertical directrix left of the pole $r = \dfrac{ep}{1 - e\cos\theta}$

3. Vertical directrix right of the pole $r = \dfrac{ep}{1 + e\cos\theta}$

4. Horizontal directrix above the pole $r = \dfrac{ep}{1 + e\sin\theta}$

5. Horizontal directrix below the pole $r = \dfrac{ep}{1 - e\sin\theta}$

6. The graph of a conic in the form $r = \dfrac{ep}{1 \pm e\cos\theta}$ is symmetric about the ___x___-axis. The graph of a conic in the form $r = \dfrac{ep}{1 \pm e\sin\theta}$ is symmetric about the ___y___-axis.

In Exercises 7–12, find a Cartesian equation of the conic having the given eccentricity e, focus F, and directrix l. Also identify the conic.

7. $e = 2; F(1, -4); l$ is $y = -3$ $\quad x^2 - 3y^2 - 2x - 16y - 19 = 0$; hyperbola

8. $e = 3; F(-3, 2); l$ is $x = 1$ $\quad -8x^2 + y^2 + 24x - 4y + 4 = 0$; hyperbola

9. $e = 1; F\left(-4, \dfrac{1}{2}\right); l$ is $y = -2$ $\quad x^2 + 8x - 5y + \dfrac{49}{4} = 0$; parabola

10. $e = 1; F(3, -2); l$ is $x = 0$ $\quad y^2 - 6x + 4y + 13 = 0$; parabola

11. $e = \dfrac{2}{3}; F(5, 3); l$ is $2x = 3$ $\quad 5x^2 + 9y^2 - 78x - 54y + 297 = 0$; ellipse

12. $e = \dfrac{3}{4}; F(0, -4); l$ is $y = -2$ $\quad 16x^2 + 7y^2 + 92y + 220 = 0$; ellipse

In Exercises 13–20, each polar equation represents a conic with a focus at the pole. Identify the conic.

13. $r = \dfrac{4}{1 - \cos\theta}$ parabola

14. $r = \dfrac{5}{4 + 6\cos\theta}$ hyperbola

15. $r = \dfrac{5}{4 - \sin\theta}$ ellipse

16. $r = \dfrac{1}{1 + \sin\theta}$ parabola

17. $r = \dfrac{2}{3 + \sin\theta}$ ellipse

18. $r = \dfrac{3}{2 + 3\sin\theta}$ hyperbola

19. $r = \dfrac{4}{1 + \sin\theta}$ parabola

20. $r = \dfrac{4}{1 - \cos\theta}$ parabola

In Exercises 21–32, identify the conic and sketch its graph.

21. $r = \dfrac{2}{1 - \cos\theta}$ †

22. $r = \dfrac{4}{1 + \cos\theta}$ †

23. $r = \dfrac{7}{3 + 4\cos\theta}$ †

24. $r = \dfrac{9}{4 + 5\cos\theta}$ †

25. $r = \dfrac{5}{3 + 2\cos\theta}$ †

26. $r = \dfrac{5}{3 - 2\cos\theta}$ †

27. $r = \dfrac{6}{2 + \sin\theta}$ †

28. $r = \dfrac{6}{2 - \sin\theta}$ †

29. $r = \dfrac{6}{1 - 2\sin\theta}$ †

30. $r = \dfrac{6}{1 + 2\sin\theta}$ †

31. $r = \dfrac{9}{7 + 2\sin\theta}$ †

32. $r = \dfrac{8}{5 - 3\sin\theta}$ †

In Exercises 33–44, find a polar equation of the conic that has a focus at the pole and that satisfies the given conditions.

33. Parabola; vertex $\left(6, \dfrac{3\pi}{2}\right)$ $\quad r = \dfrac{12}{1 - \sin\theta}$

34. Parabola; vertex $\left(4, \dfrac{\pi}{2}\right)$ $\quad r = \dfrac{8}{1 + \sin\theta}$

35. Parabola; vertex $(3, 0)$ $\quad r = \dfrac{6}{1 + \cos\theta}$

36. Parabola; vertex $(4, \pi)$

37. Ellipse; vertices $(4, 0)$ and $(6, \pi)$ $\quad r = \dfrac{24}{5 + \cos\theta}$

38. Ellipse; vertices $(2, 0)$ and $(10, \pi)$ $\quad r = \dfrac{10}{3 + 2\cos\theta}$

39. Ellipse; vertices $\left(3, \dfrac{\pi}{2}\right)$ and $\left(5, \dfrac{3\pi}{2}\right)$ $\quad r = \dfrac{15}{4 + \sin\theta}$

40. Ellipse; vertices $\left(2, \dfrac{\pi}{2}\right)$ and $\left(6, \dfrac{3\pi}{2}\right)$ $\quad r = \dfrac{6}{2 + \sin\theta}$

41. Hyperbola; vertices $(2, 0)$ and $(4, 0)$ $\quad r = \dfrac{8}{1 + 3\cos\theta}$

42. Hyperbola; vertices $(3, \pi)$ and $(7, \pi)$ $\quad r = \dfrac{21}{2 - 5\cos\theta}$

43. Hyperbola; vertices $\left(2, \dfrac{\pi}{2}\right)$ and $\left(6, \dfrac{\pi}{2}\right)$ $\quad r = \dfrac{6}{1 + 2\sin\theta}$

44. Hyperbola; vertices $\left(1, \dfrac{3\pi}{2}\right)$ and $\left(5, \dfrac{3\pi}{2}\right)$ $\quad r = \dfrac{5}{2 - 3\sin\theta}$

36. $r = \dfrac{8}{1 - \cos\theta}$

B EXERCISES Applying the Concepts

Planetary motion. **The planets travel in elliptical orbits with the sun at one focus. The table below lists the *perihelion* (the smallest distance from the planet to the sun) and the *aphelion* (the largest distance from the planet to the sun) for each of the planets. The distances given are in millions of miles. In Exercises 45–48, let the sun be the pole and the major axis be along the polar axis.**

Planet	Perihelion	Aphelion
Earth	91.38	94.54
Mars	128.49	154.83
Mercury	28.56	43.88
Saturn	837.05	936.37

†Due to space constrictions, answers to these exercises may be found in the Answers beginning on page A–1 in the back of the book.

45. Earth. Find the polar equation for the orbit of Earth.

46. Mars. Find the polar equation for the orbit of Mars.

47. Mercury. Find the polar equation for the orbit of Mercury.

48. Saturn. Find the polar equation for the orbit of Saturn.

49. Comet motion. A comet travels in a parabolic orbit around the sun at the focus. When the comet is 60 million kilometers from the sun, the line segment from the sun to the comet makes an angle of $\dfrac{\pi}{3}$ radians with the axis of the orbit. (a) Find a polar equation of the comet's orbit. (b) How close does the comet come to the sun?

50. Comet motion. A comet travels in a hyperbolic orbit around the sun at the focus. When the comet is 30 and 150 million miles from the sun, the line segment from the sun to the comet makes angles of $\dfrac{\pi}{3}$ and $\dfrac{4\pi}{3}$ radians, respectively, with the axis of the orbit.
 a. Find a polar equation of the comet's orbit.
 b. How close does the comet come to the sun?

C EXERCISES Beyond the Basics

51. Graph and interpret the conic with equation
$$r = \dfrac{6}{2 - \sin\left(\theta - \dfrac{\pi}{3}\right)}. \ ^\dagger$$

52. Show that the equation $r = 2\sec^2\dfrac{\theta}{2}$ is a polar equation of a parabola. Find an equation of the directrix.

53. Show that the equation $r = 2\csc^2\dfrac{\theta}{2}$ is a polar equation of a parabola. Find an equation of the directrix.

54. A **focal chord** of a conic is a chord that passes through one focus of the conic. (See the figure.) Let the focus F divide the chord into two segments of lengths r_1 and r_2. Show that for a fixed ellipse or a parabola, $\dfrac{1}{r_1} + \dfrac{1}{r_2}$ is a constant.

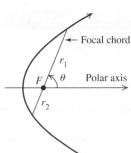

Focal chord

r_1

F θ Polar axis

r_2

Answers:

45. $r = \dfrac{92.933}{1 - 0.016997\cos\theta}$ **46.** $r = \dfrac{140.436}{1 - 0.092969\cos\theta}$

47. $r = \dfrac{34.600}{1 - 0.211485\cos\theta}$ **48.** $r = \dfrac{883.929}{1 - 0.056005\cos\theta}$

55. Show that for a hyperbola, the result in Exercise 54 is false unless we stay on one branch. $\Big[$*Hint:* Use the hyperbola $r = \dfrac{15}{1 - 4\cos\theta}$ and show what happens when $\theta = 0$ and when $\theta = \dfrac{\pi}{3}.\Big]$

56. Consider the polar equation $r = \dfrac{ep}{1 - e\cos\theta}$ of an ellipse $(e < 1)$.

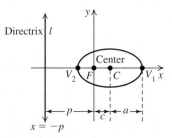

Show that in the figure, $c = ae$.

$\Big[$*Hint:* Write $V_1, V_2,$ and C in the form $(r_1, 0), (r_2, 0),$ and $(r_3, 0),$ respectively; then $r_3 = \dfrac{r_1 + r_2}{2}$ and $a = r_1 - r_3 = \dfrac{r_1 - r_2}{2}.\Big]$

57. Show that the polar equation of the ellipse in Exercise 56 can be expressed in the form $r = \dfrac{a(1 - e^2)}{1 - e\cos\theta}.$

58. Use Exercise 56 to write a Cartesian equation of the ellipse $r = \dfrac{3.2}{1 - 0.6\cos\theta}.$

$\Big[$*Hint:* Show $a = 5$ and $c = 3$. Find $b = \sqrt{a^2 - c^2}$. Then write the equation in the form $\dfrac{(x - c)^2}{a^2} + \dfrac{y^2}{b^2} = 1.\Big]$

Critical Thinking

59. *True or False* The graph of $r = \dfrac{ep}{1 - e\sin\theta}$ can be obtained by rotating the graph of $r = \dfrac{ep}{1 - e\cos\theta}$ counterclockwise through an angle of $\dfrac{\pi}{2}$ radians.

60. *True or False* The graph of $r = -\dfrac{4}{2 - \cos\theta}$ is the same as the graph of $r = \dfrac{4}{2 + \cos\theta}.$

49. $r = \dfrac{30}{1 - \cos\theta}$ with r in millions of miles; the closest approach to the sun is 15 million miles.

50. $r = \dfrac{150}{3 + 4\cos\theta}$ with r in millions of miles; the closest approach to the sun is approximately 21.43 million miles.

58. $\dfrac{(x - 3)^2}{25} + \dfrac{y^2}{16} = 1$ **59.** True **60.** True

Parametric Equations

| **Before Starting this Section, Review** | **Objectives** |

Before Starting this Section, Review

1. Equations of conics (Sections 9.2–9.4)
2. Fundamental trigonometric identities (Section 5.1)
3. Radian measure and arc length (Section 4.1)
4. Conversion of polar equations to Cartesian equations (Section 6.6)

Objectives

1. Graph plane curves described by parametric equations.
2. Eliminate the parameter.
3. Find parametric equations of a curve.

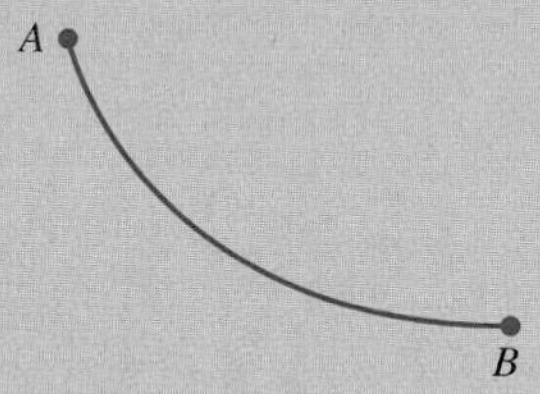

THE BRACHISTOCHRONE PROBLEM

Brachistochrone is a combination of two Greek words, *brachistos* meaning "shortest" and *chronos* meaning "time." In 1692, Johann Bernoulli posed the *brachistochrone problem* as a challenge to other mathematicians: Find the shape of a wire (or slide) along which a bead might slide in the least amount of time if it starts from a point *A* with zero speed and ends at a point *B* (not directly below *A*) under the influence of gravity and ignoring friction. The problem was solved by Isaac Newton, Jakob Bernoulli (Johann's brother), Gottfried Leibniz, Guillaume de L' Hôpital, and Johann Bernoulli.

You might think that the wire would form a straight line because a straight line is the shortest distance between the two points *A* and *B*. However, the correct shape is half of one arch of an inverted *cycloid*, shown in Figure 9.44. The brachistochrone curve (or the curve of fastest descent) is the cycloid investigated in Example 5. ∎

A

B

FIGURE 9.44

---

1 Graph plane curves described by parametric equations.

Parametric Equations

So far, we have used **Cartesian** or **rectangular equations** such as $2x^2 + 3y^2 = 6$ and $y^2 = 8x$ to represent curves in the plane. We have also represented curves using **polar equations** such as $r = 2 \sin \theta$ and $r = 1 - \cos \theta$. When describing the path of a particle moving in the plane, it is sometimes more helpful and informative to express each of the *x*- and *y*-coordinates of the path as a function of a third variable, called a **parameter**.

PARAMETRIC EQUATIONS OF A PLANE CURVE

Suppose $f(t)$ and $g(t)$ are functions defined on an interval *I*. A **plane curve** *C* is the set of points (x, y) where

$$x = f(t) \text{ and } y = g(t).$$

The equations $x = f(t)$ and $y = g(t)$ are called the **parametric equations** for the curve *C*, and *t* is called the **parameter** for *C*.

If *I* is the closed interval $[a, b]$, then $a \le t \le b$. The point $(f(a), g(a))$ is the **initial point** of the curve *C*, and the point $(f(b), g(b))$ is the **terminal point** of *C*. In applications involving the motion of a particle, the parameter *t* often represents time, but we also use other variables, such as θ (for an angle) and *s* (for distance) as parameters.

Graphing a Plane Curve

To sketch the graph of a plane curve C represented by the parametric equations $x = f(t)$ and $y = g(t)$, you plot points in the xy-plane corresponding to successive values of t. You then connect these points with a smooth curve. The path of a particle moving along the curve C with increasing values of t describes the **direction of increasing parameter** or the **orientation** of the curve C.

FINDING THE SOLUTION: A PROCEDURE

| EXAMPLE 1 | Graphing a Plane Curve |

OBJECTIVE

Sketch the graph of a plane curve represented by parametric equations.

Step 1 **Select values of t.** Select some values of t (in increasing order) in the given interval.

Step 2 **Make a table.** For each value of t in Step 1, use the given parametric equations to compute the x and y coordinates of the points (x, y) in the plane.

EXAMPLE

Sketch the graph represented by the parametric equations

$$x = t^2 - 3 \quad y = \frac{1}{2}t, \text{ where } -2 \le t \le 3.$$

Select integer values $t = -2, -1, 0, 1, 2,$ and 3 for t in the interval $[-2, 3]$.

t	$x = t^2 - 3$	$y = \dfrac{1}{2}t$	(x, y)
-2	$(-2)^2 - 3 = 1$	$\dfrac{1}{2}(-2) = -1$	$(1, -1)$
-1	$(-1)^2 - 3 = -2$	$\dfrac{1}{2}(-1) = -\dfrac{1}{2}$	$\left(-2, -\dfrac{1}{2}\right)$
0	$0^2 - 3 = -3$	$\dfrac{1}{2}(0) = 0$	$(-3, 0)$
1	$1^2 - 3 = -2$	$\dfrac{1}{2}(1) = \dfrac{1}{2}$	$\left(-2, \dfrac{1}{2}\right)$
2	$2^2 - 3 = 1$	$\dfrac{1}{2}(2) = 1$	$(1, 1)$
3	$3^2 - 3 = 6$	$\dfrac{1}{2}(3) = \dfrac{3}{2}$	$\left(6, \dfrac{3}{2}\right)$

Step 3 **Sketch the curve.** Plot the points (x, y) from Step 2 in an xy-plane. Connect the points by the order of increasing values of t with a smooth curve. Show the orientation by placing arrowheads on the curve.

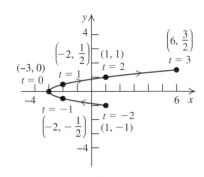

Practice Problem 1 Sketch the graph of $x = 2t, y = t^2 - 3, -2 \le t \le 2$.

2 Eliminate the parameter.

Eliminating the Parameter

In Example 1, we sketched the graph of a curve represented by parametric equations by plotting points. Sometimes we can identify or sketch the curve by first finding its Cartesian representation. We do this by eliminating the parameter in the given parametric equations. If the parametric equations involve trigonometric functions, we often can use trigonometric identities (see Example 3) to eliminate the parameter.

FINDING THE SOLUTION: A PROCEDURE

EXAMPLE 2   Eliminating the Parameter

OBJECTIVE

Convert parametric equations with parameter t to a Cartesian equation by eliminating the parameter. Sketch the graph and indicate its orientation.

Step 1   Solve for *t*. Solve one of the parametric equations for the parameter.

Step 2   Substitute. Substitute the expression for the parameter obtained in Step 1 into the other parametric equation.

Step 3   Simplify. Simplify the resulting Cartesian equation.

Step 4   Graph. Sketch the graph of the Cartesian equation in Step 3. Sketch the portion of the graph for the given values of the parameter and indicate its orientation.

EXAMPLE

Identify the parametric curve.

$$x = \sqrt{t+1} \quad y = \sqrt{t}, \text{where } t \geq 0,$$

by converting it to a Cartesian equation. Sketch its graph and show the orientation.

$$y = \sqrt{t} \qquad \text{A parametric equation}$$
$$y^2 = t \qquad \text{Square both sides.}$$

$$x = \sqrt{t+1} \qquad \text{The other parametric equation}$$
$$x = \sqrt{y^2+1} \qquad \text{Substitute } y^2 \text{ for } t \text{ from Step 1.}$$

$$x^2 = y^2 + 1 \qquad \text{Square both sides.}$$
$$x^2 - y^2 = 1 \qquad \text{Subtract } y^2 \text{ from both sides.}$$

The graph of $x^2 - y^2 = 1$ is a horizontal hyperbola. Because $t \geq 0$, we have $x = \sqrt{t+1} \geq 1$, $y = \sqrt{t} \geq 0$. The curve starts at the initial point $(1, 0)$ when $t = 0$ and rises along the hyperbola into the first quadrant as t increases. The parameter interval is $[0, \infty)$, so there is no terminal point. The graph, along with its orientation, is shown by a solid curve in the figure.

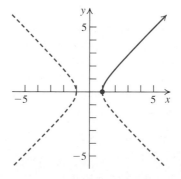

■ ■ ■

Practice Problem 2   Identify the parametric curve $x = t + 1$, $y = 3 + t^2$, $t \leq 0$ by eliminating the parameter and indicate the orientation of the graph.   ■

EXAMPLE 3   **Eliminating the Parameter**

Identify the parametric curve and indicate its orientation.

a. $x = \sin t, y = \cos^2 t, 0 \le t \le \dfrac{\pi}{2}$ **b.** $x = \sin t, y = \cos^2 t, t \ge 0$

SOLUTION

$$
\begin{aligned}
y &= \cos^2 t \\
&= 1 - \sin^2 t \quad \text{Pythagorean identity} \\
&= 1 - x^2 \quad\quad \sin t = x
\end{aligned}
$$

The graph of the Cartesian equation $y = 1 - x^2$ is shown in Figure 9.45(a).

a. When $t = 0$, $x = \sin 0 = 0$ and $y = \cos^2 0 = 1$. When $t = \dfrac{\pi}{2}$, $x = \sin\dfrac{\pi}{2} = 1$

and $y = \cos^2\dfrac{\pi}{2} = 0$. As t ranges from 0 to $\dfrac{\pi}{2}$, x values increase from 0 to 1, while y values decrease from 1 to 0. Thus, the parametric equations describe the parabolic arc $y = 1 - x^2$ with initial point $(0, 1)$ and terminal point $(1, 0)$. See Figure 9.45(b) for the oriented graph.

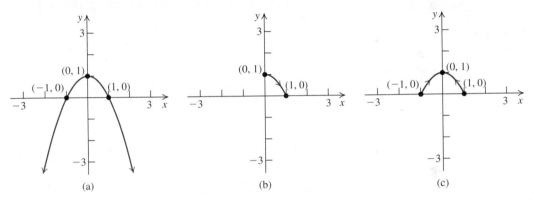

FIGURE 9.45

b. Note that for any value of t, we have $-1 \le x \le 1$ and $0 \le y \le 1$. Thus, as t ranges over the interval $[0, \infty)$, the parabolic arc $y = 1 - x^2$ shown in Figure 9.45(c) is traced back and forth an infinite number of times. The initial point is $(0, 1)$, and there is no terminal point.   ■ ■ ■

Practice Problem 3   Identify the parametric curve and indicate its orientation.

a. $x = \cos t, y = \sin t, 0 \le t \le \pi$ **b.** $x = \cos t, y = \sin t, t \ge 0$ ■

STUDY TIP

Be careful that the process of elimination does not include more (or less) points than those given by the parametric equations.

3 Find parametric equations of a curve.

Finding Parametric Equations

We now consider the reverse problem of finding parametric equations for a given curve. We call the parametric equations together with a parameter interval a **parameterization** of the curve. A given curve can be parameterized in infinitely many ways.

The easiest way to write the graph of any Cartesian equation $y = f(x)$ in parametric form is $x = t, y = f(t)$, with t in the domain of f. For example, a parameterization of the parabola $y = x^2 + 1$ is

$$x = t \qquad y = t^2 + 1$$

with $I = (-\infty, \infty)$.

You can convert any polar equation $r = g(\theta)$ to a parametric form as follows: First, convert the polar coordinates to Cartesian coordinates (see page 451):

$$x = r\cos\theta, \qquad y = r\sin\theta$$
$$x = g(\theta)\cos\theta \qquad y = g(\theta)\sin\theta \qquad \text{Replace } r \text{ with } g(\theta).$$

The equations $x = g(\theta)\cos\theta$, $y = g(\theta)\sin\theta$ give the parametric equations of $r = g(\theta)$ with parameter θ. For example, a parameterization of the polar equation $r = 1 + \cos\theta$ of cardioid is

$$x = (1 + \cos\theta)\cos\theta$$
$$y = (1 + \cos\theta)\sin\theta$$

with $I = [0, 2\pi]$.

The next example illustrates that the parametric equations of a curve are *not unique.*

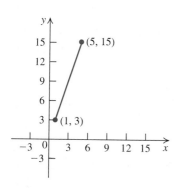

FIGURE 9.46

EXAMPLE 4 **Parameterizing a Line Segment**

The equation $y = 3x, 1 \le x \le 5$ represents the line segment joining the points $(1, 3)$ and $(5, 15)$. See Figure 9.46. Find a set of parametric equations and interpret the motion of a particle moving along this segment by using the following parameters.

a. $x = t$ **b.** $x = 1 + 4t$ **c.** $x = 5 - t$

SOLUTION

a. With $x = t$, we obtain the following parameterization of the line segment $y = 3x$, $1 \le x \le 5$:

$$x = t, y = 3t, 1 \le t \le 5$$

At time $t = 1$, the particle is at $(1, 3)$. It travels up the line segment and reaches the point $(5, 15)$ in $t = 5$ time units.

b. If $x = 1 + 4t$, then $y = 3x, 1 \le x \le 5$ becomes

$$y = 3(1 + 4t), 1 \le 1 + 4t \le 5 \qquad \text{Replace } x \text{ with } 1 + 4t.$$
$$y = 3 + 12t, 0 \le t \le 1 \qquad \text{Solve } 1 \le 1 + 4t \le 5 \text{ for } t.$$

The parametric equations are: $x = 1 + 4t$, $y = 3 + 12t$; $0 \le t \le 1$. At time $t = 0$, the particle is at $(1, 3)$. It travels the same line segment upward and reaches the point $(5, 15)$ in $t = 1$ time unit.

c. If $x = 5 - t$, then $y = 3x, 1 \le x \le 5$ becomes

$$y = 3(5 - t), 1 \le 5 - t \le 5 \qquad \text{Replace } x \text{ with } 5 - t.$$
$$= 15 - 3t, -4 \le -t \le 0 \qquad \text{Subtract 5 from all parts of } 1 \le 5 - t \le 5.$$
$$= 15 - 3t, 4 \ge t \ge 0 \qquad \text{Multiply the inequality } -4 \le -t \le 0 \text{ by } -1 \text{ and reverse the inequality signs.}$$
$$y = 15 - 3t, 0 \le t \le 4 \qquad \text{Rewrite.}$$

The parametric equations are: $x = 5 - t$, $y = 15 - 3t$; $0 \le t \le 4$. At time $t = 0$, the particle is at $(5, 15)$. It travels the same line segment (in the opposite direction) and reaches the point $(1, 3)$ in $t = 4$ time units. ■ ■ ■

Practice Problem 4 The graph of the equation $y = \sqrt{1 - x^2}, -1 \le x \le 1$ is the upper semicircle of radius 1 shown in Figure 9.47. Find a set of parametric equations and interpret the motion of a particle moving along this arc using the following parameters.

a. $x = \cos t$ **b.** $x = \cos 2t$ **c.** $x = -\cos t$ ■

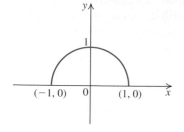

FIGURE 9.47

Remark. Example 4 shows the advantages of parametric equations over Cartesian equations in describing the motion of a particle. A Cartesian equation only indicates the path on which a particle is traveling. However, parametric equations of a curve give not only the path, but also information about the location of the particle at any specific time as well as the direction in which the particle is traveling.

EXAMPLE 5   Parametric Equations of a Cycloid

If a wheel (circle) of radius a rolls along a straight line, then a point P on the rim of the wheel traces a curve called a **cycloid**. Find parametric equations for a cycloid.

SOLUTION

Suppose the wheel rolls on the x-axis and the point P touches the x-axis at the origin. We let $t = 0$ when P is at the origin. Let $P(x, y)$ denote the location of the point P after the wheel rolls through an angle of t radians. In Figure 9.48, the measure of $\angle PCT$ is t radians.

When the wheel in Figure 9.48 rolls the distance OT, the radial line from the center C to the point P rotates through angle t. We treat t as a positive angle. So

$$OT = \overset{\frown}{PT} \quad \text{The distance } OT \text{ equals the arc length from } P \text{ to } T.$$
$$= at \quad \text{arc length} = (\text{radius})(\text{angle measured in radians})$$

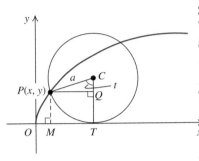

FIGURE 9.48 Graph of a cycloid

We now compute the coordinates of the point $P(x, y)$ in terms of the parameter t.

$$
\begin{aligned}
y &= PM \\
&= QT && \text{Opposite sides of a rectangle} \\
&= CT - CQ && \text{Figure 9.48} \\
&= a - a\cos t && \text{In triangle } PCQ, \cos t = \frac{CQ}{PC} = \frac{CQ}{a}. \\
&= a(1 - \cos t) && \text{Factor out } a. \\
x &= OM \\
&= OT - MT && \text{Figure 9.48} \\
&= at - PQ && OT = at, MT = PQ \\
&= at - a\sin t && \text{In triangle } PCQ, \sin t = \frac{PQ}{PC} = \frac{PQ}{a}. \\
&= a(t - \sin t)
\end{aligned}
$$

Thus, the parametric equations of a cycloid generated by a circle of radius a are given by

$$x = a(t - \sin t), \, y = a(1 - \cos t), \, -\infty < t < \infty \qquad ■ ■ ■$$

Practice Problem 5   A wheel of radius a rolls along a horizontal straight line; then a point P that is b units from the center with $b < a$ traces a curve called a **curtate cycloid**. See Figure 9.49. Find parametric equations for a curtate cycloid.   ■

TECHNOLOGY CONNECTION

Graphing calculators must be set in Radian and Parametric modes; the viewing window requires minimum and maximum values of x, y, and t.

The graph of

$$x = t - \sin(t)$$
$$y = 1 - \cos(t)$$

with $T_{\min} = 0, T_{\max} = 4\pi$ is shown.

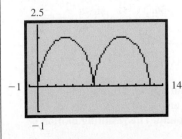

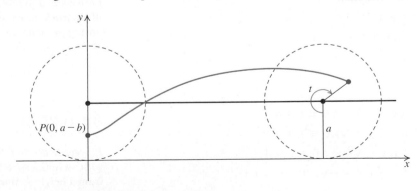

FIGURE 9.49 Graph of a curtate cycloid

SECTION 9.7 ■ Exercises

A EXERCISES Basic Concepts and Skills

In Exercises 1–8, let f and g be functions of t with $a \le t \le b$.

1. The set of points in the plane defined by the ordered pairs (x, y), where $x = f(t)$ and $y = g(t)$, is called a(n) __plane curve__.

2. The equations $x = f(t), y = g(t), a \le t \le b$ are called the _____ for the curve. parametric equations

3. The variable t is called the ___parameter___.

4. The point $(f(a), g(a))$ is called the ___initial point___.

5. The point $(f(b), g(b))$ is called the ___terminal point___.

6. The path along a parameterized curve C with increasing values of t is called the ___orientation___ of the curve.

7. *True or False* Every plane curve has a unique parameterization. False

8. *True or False* When you eliminate the parameter in a set of parametric equations, you obtain a Cartesian equation. True

In Exercises 9–12, sketch the graph of each curve by plotting points. Indicate the points on the curve that correspond to the integer values of t in the given interval.

9. $x = t - 1, y = 2t - 1, -2 \le t \le 2$ †

10. $x = t + 1, y = 3t + 2, -2 \le t \le 3$ †

11. $x = \dfrac{1}{2}t, y = t^2 + 1, -2 \le t \le 3$ †

12. $x = 2t^2 - 1, y = 2t, -2 \le t \le 2$ †

In Exercises 13–16, sketch the graph of each curve by plotting points. Indicate the points on the curve that correspond to $\theta = 0, \dfrac{\pi}{4}, \dfrac{\pi}{2}, \dfrac{3\pi}{4}, \pi$.

13. $x = \cos \theta, y = \sin \theta, 0 \le \theta \le \pi$ †

14. $x = \sin \theta, y = \cos \theta, 0 \le \theta \le \pi$ †

15. $x = 2 \cos \theta, y = \sin \theta, 0 \le \theta \le \pi$ †

16. $x = 2 \sin \theta, y = 3 \cos \theta, 0 \le \theta \le \pi$ †

In Exercises 17–28, eliminate the parameter to obtain a Cartesian equation of the curve. Graph the curve and indicate its orientation.

17. $x = t, y = 2t, -\infty < t < \infty$ †

18. $x = 3t, y = t, -\infty < t < \infty$ †

19. $x = 2t, y = 6t, 0 \le t \le 4$ †

20. $x = 2t + 1, y = -3t + \dfrac{3}{2}, -1 \le t \le \dfrac{1}{2}$ †

21. $x = \sqrt{t}, y = 3 - t, 0 \le t \le 4$ †

22. $x = t - 2, y = \sqrt{t}, 0 \le t \le 6$ †

23. $x = \sqrt{t}, y = \sqrt{4 - t}, 0 \le t \le 4$ †

24. $x = \sqrt{9 - t}, y = -\sqrt{t}, 0 \le t \le 9$ †

25. $x = 2\sqrt{2 - t}, y = 3\sqrt{t - 1}, 1 \le t \le 2$ †

26. $x = 3\sqrt{t - 2}, y = 4\sqrt{3 - t}, 2 \le t \le 3$ †

27. $x = t + 2, y = \sqrt{t^2 + 4t}, t \ge 0$ †

28. $x = 3\sqrt{t - 1}, y = 2\sqrt{t}, t \ge 1$ †

In Exercises 29–44, eliminate the parameter to obtain a Cartesian equation of the curve. Graph the curve and indicate its orientation.

29. $x = \sin t, y = \cos t, 0 \le t \le 2\pi$ †

30. $x = \cos t, y = \sin t, 0 \le t \le 2\pi$ †

31. $x = \cos^2 t, y = \sin^2 t, 0 \le t \le \dfrac{\pi}{2}$ †

32. $x = \sin^2 t, y = \cos^2 t, 0 \le t \le \pi$ †

33. $x = 3 \cos t, y = 2 \sin t, 0 \le t \le 2\pi$ †

34. $x = 3 \sin t, y = 2 \cos t, 0 \le t \le 4\pi$ †

35. $x = \cos \theta, y = \cos^2 2\theta, -\infty < \theta < \infty$ †

36. $x = \sin \theta, y = \cos^2 2\theta, -\infty < \theta < \infty$ †

37. $x = 1 - \cos t, y = 2 - \sin t, 0 \le t \le 2\pi$ †

38. $x = 3 - 2 \cos t, y = -1 + 5 \sin t, 0 \le t \le 2\pi$ †

39. $x = 2 \tan t - 1, y = 3 \sec t + 2, 0 \le t < \dfrac{\pi}{2}$ †

40. $x = 3 \csc t + 2, y = 5 \cot t - 1, 0 < t < \pi$ †

41. $x = e^t, y = e^{2t}, t \ge 0$ †   42. $x = 4e^{2t}, y = -e^t, t \ge 0$ †

43. $x = t^2, y = 2 \ln t, t \ge 1$ †   44. $x = 5 \ln t, y = t^5, t \ge 1$ †

In Exercises 45–48, consider the Cartesian equation $y = 3x + 2$ of a line. Find a set of parametric equations and interpret the motion of a particle moving along this line by using the parameter t with $-\infty < t < \infty$.

45. $x = t$ †   46. $x = \sin t$ †   47. $x = e^t$ †   48. $x = e^{-t}$ †

In Exercises 49–52, consider the Cartesian equation $y = x^2 - 4$ of a parabola. Find a set of parametric equations and interpret the motion of a particle moving along this parabola by using the parameter t with $-\infty < t < \infty$.

49. $x = -t$ †   50. $x = t^2$ †   51. $x = \cos t$ †   52. $x = e^t$ †

In Exercises 53–56, consider the Cartesian equation $x = \sqrt{1 - y^2}, -1 \le y \le 1$ of a right semicircle of radius 1. Find a set of parametric equations and interpret the motion of a particle moving along this semicircle by using the following parameters.

53. $x = \sin t$ †   54. $x = \sin 2t$ †

55. $x = -\sin t$ †   56. $x = -\sin 4t$ †

B EXERCISES Applying the Concepts

Projectile motion. Suppose an object is launched upward with an initial speed v_0 at an angle of θ with the horizontal from a height h above the ground. The resulting motion is

†Due to space constrictions, answers to these exercises may be found in the Answers beginning on page A–1 in the back of the book.

called *projectile motion.* **It can be shown that the parametric equations of the path of a projectile are**

$$x = (v_0 \cos\theta)t, \quad y = (v_0 \sin\theta)t - \frac{1}{2}gt^2 + h,$$

where *t* **is the time and** *g* **is the constant acceleration due to gravity** ($g \approx$ **32 ft/sec²** or **9.8 m/sec²**).

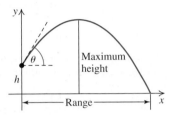

In Exercises 57–60, a golf ball is hit at an angle of 30° with an initial speed of 64 ft/sec. (Here $h = 0$**.)**

57. Write parametric equations that model the path of the golf ball. $x = 32\sqrt{3}t, y = 32t - 16t^2$

58. What is the location of the ball 2 seconds later?
On the ground approximately 110.9 feet from the starting point
59. When does the ball land on the ground? After 2 seconds

60. What is the range of the ball? Approximately 110.9 feet

In Exercises 61–64, a gun is fired from a tower 96 feet above the ground. The angle of elevation of the gun is 60°, and its muzzle speed is 1600 ft/sec.

61. Write parametric equations that model the path of the bullet. $x = 800t, y = 800\sqrt{3}t - 16t^2 + 96$

62. When does the bullet hit the ground?
After approximately 86.7 seconds
63. What is the range for the gun? Approximately 13.13 miles

64. Find a Cartesian equation for the path of the bullet. Identify the curve. $y = -\frac{1}{40,000}x^2 + \sqrt{3}x + 96$

C EXERCISES Beyond the Basics

65. a. Verify that the parametric equations

$$x = x_1 + (x_2 - x_1)t, y = y_1 + (y_2 - y_1)t, -\infty < t < \infty$$

represent the line passing through the points (x_1, y_1) and (x_2, y_2).
   b. Find parametric equations of the line segment from $(2, -3)$ to $(-1, 4)$. $x = 2 - 3t, y = -3 + 7t, 0 \le t \le 1$

66. Find parametric equations of the line segment from $(4, 1)$ to $(-2, -3)$. $x = 4 - 6t, y = 1 - 4t, 0 \le t \le 1$

67. Verify that the parametric equations

$$x = h + a\cos t, y = k + a\sin t, 0 \le t \le 2\pi$$

represent the curve that starts at the point $(h + a, k)$ and goes counterclockwise once around the circle with center (h, k) and radius a.

68. Find a parametric equation of the curve that starts at the point $(3, 0)$ and goes counterclockwise once around the circle with center at $(1, 0)$ and radius 2.
$x = 1 + 2\cos t, y = 2\sin t, 0 \le t \le 2\pi$
69. Find parametric equations of the curve that starts at $(-1, 5)$ and goes clockwise once around the circle with center $(-1, 2)$ and radius 3.
$x = -1 + 3\sin t, y = 2 + 3\cos t, 0 \le t \le 2\pi$
70. Verify that the parametric equations of the curve that starts at the point $(h + a, k)$ and goes counterclockwise

once around the ellipse $\dfrac{(x - h)^2}{a^2} + \dfrac{(y - k)^2}{b^2} = 1$ are

$x = h + a\cos t, y = k + b\sin t, 0 \le t \le 2\pi$.

71. Verify that the parametric equations of the hyperbola $\dfrac{(x - h)^2}{a^2} - \dfrac{(y - k)^2}{b^2} = 1$ are $x = h + a\sec t$, $y = k + b\tan t$. One branch is traced for $-\dfrac{\pi}{2} < t < \dfrac{\pi}{2}$; the other branch is traced for $\dfrac{\pi}{2} < t < \dfrac{2\pi}{2}$.

72. Verify that the parametric equation of the hyperbola $\dfrac{(y - k)^2}{a^2} - \dfrac{(x - h)^2}{b^2} = 1$ are $x = h + b\tan t$, $y = k + a\sec t$. One branch is traced for $-\dfrac{\pi}{2} < t < \dfrac{\pi}{2}$; the other branch is traced for $\dfrac{\pi}{2} < t < \dfrac{3\pi}{2}$.

In Exercises 73–77, describe the curve represented by the parametric equations.

73. $x = 2 + t^2, y = 2t + 1, t \ge 0$ †

74. $x = \dfrac{e^t + e^{-t}}{2}, y = \dfrac{e^t - e^{-t}}{2}, -\infty < t < \infty$ †

75. $x = \dfrac{1 - t^2}{1 + t^2}, y = \dfrac{2t}{1 + t^2}, t \ge 0$ †

76. $x = \dfrac{t^2 + 1}{t^2 - 1}, y = \dfrac{2t}{t^2 - 1}, t \ge 0; t \ne 1$ †

77. $x = 8(\cos t + \sin t), y = 15(\cos t - \sin t), -\infty < t < \infty$ †

78. a. Verify that the equations $x = at^2, y = 2at,$ $-\infty < t < \infty$ represent parametric equations of the parabola $y^2 = 4ax$.
   b. Suppose the points $t = t_1$ and $t = t_2$ of the parabola $y^2 = 4ax$ are the endpoints of a focal chord (a chord through the focus). Show that $t_1 t_2 = -1$.

Critical Thinking

79. Which of the following parametric equations represent the line $y = x + 1$? Explain.
   a. $x = 2t - 1, y = 2t$ **b.** $x = t^2 - 1, y = t^2$
   c. $x = t^3 + 1, y = t^3 + 3$ **d.** $x = e^t - 1, y = e^t$
   (a) represents the line; (b) and (d) represent rays included in the line but not the entire line, and (c) represents a different line.

GROUP PROJECT

80. Projectile motion. Eliminate the parameter t from the parametric equations $x = (v_0 \cos\theta)t, y = (v_0 \sin\theta)t - \dfrac{1}{2}gt^2$ to obtain a Cartesian equation.
   a. Show that the trajectory of a projectile is a parabola. †
   b. Find the vertex of the parabola. †
   c. Find the time of flight of the projectile. †
   d. Find the range. †
   e. Show that for a fixed initial speed v_0, the maximum range is obtained when $\theta = 45°$. Find a formula for this maximum range. †
   f. Show that doubling the initial speed of a projectile has the effect of multiplying both the range and the maximum height by a factor of four. †

SUMMARY ■ Definitions, Concepts, and Formulas

9.1 Conic Sections: Overview

Conic sections are the curves formed when a plane intersects the surface of a right circular cone. Except in some degenerate cases, the curves formed are the circle, the ellipse, the parabola, and the hyperbola.

9.2 The Parabola

i. **Definition.** A **parabola** is the set of all points in the plane that are the same distance from a fixed line and a fixed point not on the line. The fixed line is called the **directrix**, and the fixed point is called the **focus**.

ii. **Axis or axis of symmetry.** This is the line that passes through the focus and is perpendicular to the directrix.

iii. **Vertex.** This is the point at which the parabola intersects its axis. The vertex is halfway between the focus and the directrix.

iv. **Standard forms of equations of parabolas with vertex at (0, 0) and a > 0 are**

$$y^2 = \pm 4ax \text{ and } x^2 = \pm 4ay.$$

(See the table on page 609.)

v. The **latus rectum** of a parabola is the line segment that passes through the focus, is perpendicular to the axis of the parabola, and has endpoints on the parabola.

vi. **Standard forms of equations of parabolas with vertex (h, k) and a > 0** are $(y - k)^2 = \pm 4a(x - h)$ and $(x - h)^2 = \pm 4a(y - k)$. (See the table on page 612.)

9.3 The Ellipse

i. **Definition.** An **ellipse** is the set of all points in the plane, the sum of whose distances from two fixed points is a constant. The fixed points are called the **foci** of the ellipse.

ii. **Vertices.** These are the two points where the line through the foci intersects the ellipse.

iii. **Major axis.** This is the line segment joining the two vertices of the ellipse.

iv. **Center.** This is the midpoint of the line segment joining the foci. It is also the midpoint of the major axis.

v. **Minor axis.** This is the line segment that passes through the center of the ellipse, is perpendicular to the major axis, and has endpoints on the ellipse.

vi. **Standard forms of equations of ellipses with center (0, 0), a > b > 0, c < a, and b² = a² − c² are**

$$\frac{x^2}{a^2} + \frac{y^2}{b^2} = 1, \text{ with foci } (\pm c, 0) \text{ and vertices } (\pm a, 0),$$

and

$$\frac{x^2}{b^2} + \frac{y^2}{a^2} = 1, \text{ with foci } (0, \pm c) \text{ and vertices } (0, \pm a).$$

(See the table on page 620.)

vii. **Standard forms of equations of ellipses with center (h, k), a > b > 0, and b² = a² − c² are**

$$\frac{(x - h)^2}{a^2} + \frac{(y - k)^2}{b^2} = 1 \text{ and } \frac{(x - h)^2}{b^2} + \frac{(y - k)^2}{a^2} = 1.$$

(See the table on page 622.)

9.4 The Hyperbola

i. **Definition.** A **hyperbola** is the set of all points in the plane, the difference of whose distances from two fixed points is a constant. The fixed points are called the **foci** of the hyperbola.

ii. **Vertices.** These are the two points where the line through the foci intersects the hyperbola.

iii. **Transverse axis.** This is the line segment joining the two vertices of the hyperbola.

iv. **Center.** This is the midpoint of the line segment joining the foci. It is also the midpoint of the transverse axis.

v. **Standard forms of equations of hyperbolas with center (0, 0), c > a, and b² = c² − a² are**

$$\frac{x^2}{a^2} - \frac{y^2}{b^2} = 1, \text{ with foci } (\pm c, 0) \text{ and vertices } (\pm a, 0),$$

and

$$\frac{y^2}{a^2} - \frac{x^2}{b^2} = 1, \text{ with foci } (0, \pm c) \text{ and vertices } (0, \pm a).$$

(See the table on page 631.)

vi. **Conjugate axis.** This is a line segment of length $2b$ that passes through the center of the hyperbola and is perpendicularly bisected by the transverse axis.

vii. **Asymptotes.** The asymptotes of the hyperbola $\frac{x^2}{a^2} - \frac{y^2}{b^2} = 1$ are the two lines $y = \pm\frac{b}{a}x$; the asymptotes of the hyperbola $\frac{y^2}{a^2} - \frac{x^2}{b^2} = 1$ are the lines $y = \pm\frac{a}{b}x$.

viii. **Graphing.** A procedure for graphing a hyperbola centered at $(0, 0)$ is given on page 635.

ix. **Standard forms of equations of hyperbolas centered at (h, k) are**

$$\frac{(x - h)^2}{a^2} - \frac{(y - k)^2}{b^2} = 1 \text{ and } \frac{(y - k)^2}{a^2} - \frac{(x - h)^2}{b^2} = 1.$$

(See the table on page 636. Also see page 636 for a procedure for graphing such hyperbolas.)

9.5 Rotation of Axes

i. If we exclude the degenerate conics, the graph of the equation $Ax^2 + Cy^2 + Dx + Ey + F = 0$, where A and C are not both zero, is
(1) a parabola if $AC = 0$, (2) an ellipse (or a circle if $A = C$) if $AC > 0$, and (3) a hyperbola if $AC < 0$.

ii. Rotation of axes formulas. Suppose the x-axis is rotated through an angle θ and the new axes are labeled the x'-axis and y'-axis, respectively. If $P = (x, y)$ and $P = (x', y')$, then $x = x' \cos \theta - y' \sin \theta$ and $y = x' \sin \theta + y' \cos \theta$.

iii. Eliminating an xy-term by rotation of axes. To transform a second-degree equation $Ax^2 + Bxy + Cy^2 + Dx + Ey + F = 0$ to the form $A'x'^2 + C'y'^2 + D'x' + E'y' + F' = 0$ without the $x'y'$ term, rotate the axes through an acute angle θ, where

$$\cot 2\theta = \frac{A - C}{B}.$$

iv. Identify conics using the discriminant $B^2 - 4AC$. Except for degenerate cases, the graph of an equation $Ax^2 + Bxy + Cy^2 + Dx + Ey + F = 0$, with $AC \neq 0$, is
(1) a parabola if $B^2 - 4AC = 0$, (2) an ellipse if $B^2 - 4AC < 0$, and (3) a hyperbola if $B^2 - 4AC > 0$.

9.6 Polar Equations of Conics

i. Focus–Directrix definition of conics. The set of all points P in the plane for which the ratio of the distance from a fixed point (focus) and the fixed line (directrix) is a constant. The constant is denoted by e and is called the **eccentricity**. This conic is
(1) an ellipse if $e < 1$, (2) a parabola if $e = 1$, and (3) a hyperbola if $e > 1$.

ii. Standard forms of polar equations of conics. Polar equations of conics with focus at the pole, eccentricity e, and a distance of p units between the focus and directrix are as follows:

$$r = \frac{ep}{1 \pm e \cos \theta} \text{ and } r = \frac{ep}{1 \pm e \sin \theta}$$

(See page 658.)

iii. A procedure for graphing a conic given its polar equation is found on page 659.

9.7 Parametric Equations

i. Suppose $f(t)$ and $g(t)$ are functions defined on interval I. A **plane curve** C is a set of points (x, y), where $x = f(t)$ and $y = g(t)$. The equations $x = f(t)$ and $y = g(t)$ are the **parametric equations** for the curve C, and t is the **parameter**. The path of a particle moving along the curve C with increasing values of t describes the **orientation** (direction) of C.

ii. We can graph a plane curve represented by parametric equations by plotting points. (A procedure is described on page 667.)

iii. We can eliminate the parameter t in the parametric equations to convert to a Cartesian equation of the curve. It may be necessary to change the domain of the Cartesian equation to be consistent with the domain of the parameter t. (See page 669.)

iv. The easiest way to convert any Cartesian equation $y = f(x)$ in parametric form is $x = t$ and $y = f(t)$, with t in the domain of f. A polar equation $r = f(\theta)$ in parametric form is $x = f(\theta) \cos \theta$ and $y = f(\theta) \sin \theta$ with parameter θ.

REVIEW EXERCISES

In Exercises 1–8, sketch the graph of each parabola. Determine the vertex, focus, axis, and directrix of each parabola.

1. $y^2 = -6x$ †
2. $y^2 = 12x$ †
3. $x^2 = 7y$ †
4. $x^2 = -3y$ †
5. $(x - 2)^2 = -(y + 3)$ †
6. $(y + 1)^2 = 5(x + 2)$ †
7. $y^2 = -4y + 2x + 1$ †
8. $-x^2 + 2x + y = 0$ †

In Exercises 9–12, find an equation of the parabola satisfying the given conditions.

9. Vertex: $(0, 0)$; focus: $(-3, 0)$ $y^2 = -12x$
10. Vertex: $(0, 0)$; focus: $(0, 4)$ $x^2 = 16y$
11. Focus: $(0, 4)$; directrix: $y = -4$ $x^2 = 16y$
12. Focus: $(-3, 0)$; directrix: $x = 3$ $y^2 = -12x$

In Exercises 13–20, sketch the graph of each ellipse. Determine the foci, vertices, and endpoints of the minor axis of the ellipse.

13. $\dfrac{x^2}{25} + \dfrac{y^2}{4} = 1$ †
14. $\dfrac{x^2}{9} + \dfrac{y^2}{36} = 1$ †
15. $4x^2 + y^2 = 4$ †
16. $16x^2 + y^2 = 64$ †
17. $16(x + 1)^2 + 9(y + 4)^2 = 144$ †
18. $4(x - 1)^2 + 3(y + 2)^2 = 12$ †

†Due to space constrictions, answers to these exercises may be found in the Answers beginning on page A–1 in the back of the book.

19. $x^2 + 9y^2 + 2x - 18y + 1 = 0$ †

20. $4x^2 + y^2 + 8x - 10y + 13 = 0$ †

In Exercises 21–24, find an equation of the ellipse satisfying the given conditions.

21. Vertices: $(\pm 4, 0)$; endpoints of minor axis: $(0, \pm 2)$ †

22. Vertices: $(0, \pm 6)$; endpoints of minor axis: $(\pm 2, 0)$ †

23. Length of major axis 20; foci: $(\pm 5, 0)$ $\dfrac{x^2}{100} + \dfrac{y^2}{75} = 1$

24. Length of minor axis 16; foci: $(0, \pm 6)$ $\dfrac{x^2}{64} + \dfrac{y^2}{100} = 1$

In Exercises 25–32, sketch the graph of each hyperbola. Determine the vertices, the foci, and the asymptotes of each hyperbola.

25. $\dfrac{y^2}{16} - \dfrac{x^2}{4} = 1$ †

26. $\dfrac{x^2}{16} - \dfrac{y^2}{9} = 1$ †

27. $8x^2 - y^2 = 8$ †

28. $4y^2 - 4x^2 = 1$ †

29. $\dfrac{(x + 2)^2}{9} - \dfrac{(y - 3)^2}{4} = 1$ †

30. $\dfrac{(y + 1)^2}{6} - \dfrac{(x - 2)^2}{8} = 1$ †

31. $4y^2 - x^2 + 40y - 4x + 60 = 0$ †

32. $4x^2 - 9y^2 + 16x - 54y - 29 = 0$ †

In Exercises 33–36, find an equation of the hyperbola satisfying the given condition.

33. Vertices: $(\pm 1, 0)$; foci: $(\pm 2, 0)$ $x^2 - \dfrac{y^2}{3} = 1$

34. Vertices: $(0, \pm 2)$; foci: $(0, \pm 4)$ $\dfrac{y^2}{4} - \dfrac{x^2}{12} = 1$

35. Vertices: $(\pm 2, 0)$; asymptotes: $y = \pm 3x$ $\dfrac{x^2}{4} - \dfrac{y^2}{36} = 1$

36. Vertices: $(0, \pm 3)$; asymptotes: $y = \pm x$ $\dfrac{y^2}{9} - \dfrac{x^2}{9} = 1$

In Exercises 37–48, identify each equation as representing a circle, a parabola, an ellipse, or a hyperbola. Sketch the graph of the conic.

37. $5x^2 - 4y^2 = 20$ †

38. $x^2 - x + y = 1$ †

39. $3x^2 + 4y^2 + 8y - 12x - 6 = 0$ †

40. $2y - x^2 = 0$ †

41. $x^2 + y^2 + 2x - 3 = 0$ †

42. $2x + y^2 = 0$ †

43. $y^2 = x^2 + 3$ †

44. $x^2 = 10 - 3y^2$ †

45. $3x^2 + 3y^2 - 6x + 12y + 5 = 0$ †

46. $y + 2x - x^2 + 2y^2 = 0$ †

47. $9x^2 + 8y^2 = 36$ †

48. $2y^2 + 4y = 3x^2 - 6x + 9$ †

49. Find an equation of the hyperbola whose foci are the vertices of the ellipse $4x^2 + 9y^2 = 36$ and whose vertices are the foci of this ellipse. $\dfrac{x^2}{5} - \dfrac{y^2}{4} = 1$

50. Find an equation of the ellipse whose foci are the vertices of the hyperbola $9x^2 - 16y^2 = 144$ and whose vertices are the foci of this hyperbola. $\dfrac{x^2}{25} + \dfrac{y^2}{9} = 1$

In Exercises 51–54, find all points of intersection of the given curves and make a sketch.

51. $x^2 - 4y^2 = 36$ and $x - 2y - 20 = 0$ †

52. $y^2 - 8x^2 = 5$ and $y - 2x^2 = 0$ †

53. $3x^2 - 7y^2 = 5$ and $9y^2 - 2x^2 = 1$ †

54. $x^2 - y^2 = 1$ and $x^2 + y^2 = 7$ †

In Exercises 55–58, eliminate the xy-term in the equation by a suitable rotation. Identify and graph the conic.

55. $xy = 8$ †

56. $x^2 + xy + y^2 = 8$ †

57. $x^2 + 2xy + y^2 + x - y - 4 = 0$ †

58. $24xy + 7y^2 + 36 = 0$ †

In Exercises 59–66, the polar equation of a conic is given. Sketch the graph of the conic.

59. $r = \dfrac{3}{3 - \sin \theta}$ †

60. $r = \dfrac{6}{3 + 3 \sin \theta}$ †

61. $r = \dfrac{6}{3 + 2 \sin \theta}$ †

62. $r = \dfrac{6}{2 + 3 \sin \theta}$ †

63. $r = \dfrac{4}{2 - 3 \cos \theta}$ †

64. $r = \dfrac{4}{3 - 2 \cos \theta}$ †

65. $r = \dfrac{6}{2 + 3 \cos \theta}$ †

66. $r = \dfrac{6}{3 + 2 \cos \theta}$ †

In Exercises 67–74, eliminate the parameter to obtain a Cartesian equation of the curve. Graph the curve and indicate its orientation.

67. $x = t, y = 1 - t, 0 \le t \le 1$ †

68. $x = 3 \cos t, y = 3 \sin t, 0 \le t \le 2\pi$ †

69. $x = 3 \sin^2 t, y = 3 \cos^2 t, 0 \le t \le 2\pi$ †

70. $x = 3 \cos \theta, y = -3 \sin \theta, 0 \le \theta \le \pi$ †

71. $x = \sin t, y = \cos 2t, 0 \le t \le 2\pi$ †

72. $x = 3 \sec \theta, y = 3 \tan \theta, -\dfrac{\pi}{2} < \theta < \dfrac{\pi}{2}$ †

73. $x = t^3, y = t^2, -\infty < t < \infty$ †

74. $x = t^2 - 2t, y = t, -\infty < t < \infty$ †

75. **Parabolic curve.** Water flowing from the end of a horizontal pipe 20 feet above the ground describes a parabolic curve whose vertex is at the end of the pipe. If at a point 6 feet below the line of the pipe the flow of water has curved outward 8 feet beyond a vertical line through the end of the pipe, how far beyond this vertical line will the water strike the ground? 14.6 ft

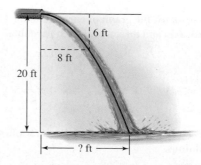

76. **Parabolic arch.** A parabolic arch has a height of 20 meters and a width of 36 meters at the base. If the vertex of the parabola is at the top of the arch, find the height of the arch at a distance of 9 meters from the center of the base. 15 m

77. Football. A football is 12 inches long, and a cross section containing a seam is an ellipse with a minor axis of length 7 inches. Suppose every cross section of the ball formed by a plane perpendicular to the major axis of the ellipse is a circle. Find the circumference of such a circular cross section located 2 inches from an end of the ball. 16.4 in

78. Production cost. A company assembles computers at two locations A and B that are 1000 miles apart. The per unit cost of production at location A is \$20 less than at location B. Assume that the route of delivery of the computers is along a straight line and that the delivery cost is 25¢ per unit per mile. Find the equation of the curve at any point of which the computers can be supplied from either location at the same cost.
[*Hint:* Take A at $(-500, 0)$ and B at $(500, 0)$.]
$$\frac{x}{40} = \sqrt{1 + \frac{y^2}{248{,}400}}$$

PRACTICE TEST A

1. Find the standard form of the parabola with focus $(0, 12)$ and directrix with equation $x = -12$. $(y - 12)^2 = 24(x + 6)$

2. Convert the equation $y^2 - 2y + 8x + 25 = 0$ to the standard form for a parabola by completing the square.
   $(y - 1)^2 = -8(x + 3)$

3. Find the vertex, focus, and directrix of the parabola with equation $(x + 2)^2 = -8(y - 1)$. Vertex: $(-2, 1)$;
   focus: $(-2, -1)$; directrix: $y = 3$

4. The base of a water slide is parabolic in shape and is 6 feet wide and 2 feet deep. Find the height of the slide 1 foot from its center. 2/9 ft

5. Find an equation of the parabola with vertex $(3, -1)$ and directrix $x = -3$. $(y + 1)^2 = 24(x - 3)$

In Problems 6 and 7, find the standard form of the equation of the ellipse satisfying the given conditions.

6. Foci: $(0, -2), (0, 2)$; vertices: $(0, -4), (0, 4)$ $\dfrac{x^2}{12} + \dfrac{y^2}{16} = 1$

7. Major axis horizontal with length 18; minor axis of length 4; center $(0, 0)$

8. Graph the ellipse with equation $9x^2 + y^2 = 9$. †

7. $\dfrac{x^2}{81} + \dfrac{y^2}{4} = 1$

9. Find the standard form of the equation of the hyperbola with foci $(0, -\sqrt{45}), (0, \sqrt{45})$ and vertices $(0, -6), (0, 6)$.

In Problems 10 and 11, convert the equation to the standard form for a hyperbola by completing the squares on x and y.

10. $y^2 - x^2 + 2x = 2$ $y^2 - (x - 1)^2 = 1$

11. $x^2 - 2x - 4y^2 - 16y = 19$ $\dfrac{(x - 1)^2}{4} - (y + 2)^2 = 1$

In Problems 12–20, identify the conic section and sketch its graph.

12. $x^2 - 9y^2 = -16$ †

13. $x^2 - 2y + 4y + 1 = 0$ †

14. $x^2 + y^2 + 2x - 6y = 3$ †

15. $25x^2 + 4y^2 = 100$ †

16. $xy = 1$ †

17. $2x^2 + 3xy + 2y^2 = 7$ †

18. $r = \dfrac{6}{4 - 3\cos\theta}$ †

19. $r = \dfrac{6}{3 + 4\cos\theta}$ †

20. $x = 4 - t, y = t^2$ †

9. $\dfrac{y^2}{36} - \dfrac{x^2}{9} = 1$

PRACTICE TEST B

1. Find the standard form of the equation of the parabola with focus $(-10, 0)$ and directrix $x = 10$. b
   a. $y^2 = -10x$ **b.** $y^2 = -40x$
   c. $x^2 = -40y$ **d.** $y^2 = 40x$

2. Convert the equation $y^2 - 4y - 5x + 24 = 0$ to the standard form for a parabola by completing the square. b
   a. $(y + 2)^2 = 5(x - 4)$ **b.** $(y - 2)^2 = 5(x - 4)$
   c. $(y + 2)^2 = -5(x - 4)$ **d.** $(y - 2)^2 = 5(x + 4)$

3. Find the vertex, focus, and directrix of the parabola with equation $(x - 3)^2 = 12(y - 1)$. a
   a. vertex: $(3, 1)$ focus: $(3, 4)$ directrix: $y = -2$
   b. vertex: $(-3, -1)$ focus: $(-3, 2)$ directrix: $y = -4$
   c. vertex: $(3, 1)$ focus: $(3, -2)$ directrix: $x = 4$
   d. vertex: $(1, 3)$ focus: $(1, 6)$ directrix: $y = 0$

4. The parabolic arch of a bridge has a 180-foot base and a height of 25 feet. Find the height of the arch 45 feet from the center of the base. d
  a. 12.5 ft   **b.** 6.25 ft
  c. 16.7 ft   **d.** 18.75 ft

5. Find an equation of the parabola with vertex $(-2, 1)$ and directrix $x = 2$. a
  a. $(y - 1)^2 = -16(x + 2)$
  b. $(y - 1)^2 = 16(x + 2)$
  c. $(y + 2)^2 = -16(x - 1)$
  d. $(y + 2)^2 = 16(x - 1)$

In Problems 6 and 7, find the standard form of the equation of the ellipse satisfying the given conditions.

6. Foci $(-3, 0), (3, 0)$; vertices $(-5, 0), (5, 0)$ b
  a. $\dfrac{x^2}{9} + \dfrac{y^2}{16} = 1$ **b.** $\dfrac{x^2}{25} + \dfrac{y^2}{16} = 1$
  c. $\dfrac{x^2}{9} + \dfrac{y^2}{25} = 1$ **d.** $\dfrac{x^2}{16} + \dfrac{y^2}{25} = 1$

7. Major axis vertical with length 16; length of minor axis $= 8$; center $(0, 0)$ a
  a. $\dfrac{x^2}{16} + \dfrac{y^2}{64} = 1$ **b.** $\dfrac{x^2}{64} + \dfrac{y^2}{256} = 1$
  c. $\dfrac{x^2}{8} + \dfrac{y^2}{64} = 1$ **d.** $\dfrac{x^2}{64} + \dfrac{y^2}{16} = 1$

8. Find the vertices and foci of the hyperbola with equation $\dfrac{x^2}{121} - \dfrac{y^2}{4} = 1$. c
  a. vertices: $(-11, 0), (11, 0)$ foci: $(-2, 0), (2, 0)$
  b. vertices: $(0, -1), (0, 11)$ foci: $(-5\sqrt{5}, 0), (5\sqrt{5}, 0)$
  c. vertices: $(-11, 0), (11, 0)$ foci: $(-5\sqrt{5}, 0), (5\sqrt{5}, 0)$
  d. vertices: $(-2, 0), (2, 0)$ foci: $(-5\sqrt{5}, 0), (5\sqrt{5}, 0)$

9. Find the standard form of the equation of the hyperbola with foci $(0, -10), (0, 10)$ and vertices $(0, -5), (0, 5)$. b
  a. $\dfrac{y^2}{25} - \dfrac{x^2}{100} = 1$ **b.** $\dfrac{y^2}{25} - \dfrac{x^2}{75} = 1$
  c. $\dfrac{x^2}{25} - \dfrac{y^2}{100} = 1$ **d.** $\dfrac{x^2}{25} - \dfrac{y^2}{75} = 1$

In Problems 10 and 11, convert the equation to the standard form for a hyperbola by completing the square on x and y.

10. $y^2 - 4x^2 - 2y - 16x - 19 = 0$ c
  a. $\dfrac{(y - 2)^2}{4} - (x + 4)^2 = 1$
  b. $\dfrac{(x - 1)^2}{4} - (y + 2)^2 = 1$

  c. $\dfrac{(y - 1)^2}{4} - (x + 2)^2 = 1$
  d. $(x + 2)^2 - \dfrac{(y - 1)^2}{4} = 1$

11. $4y^2 - 9x^2 - 16y - 36x - 56 = 0$ c
  a. $\dfrac{(y - 2)^2}{4} - \dfrac{(x + 2)^2}{9} = 1$
  b. $\dfrac{(x - 2)^2}{4} - \dfrac{(y + 2)^2}{9} = 1$
  c. $\dfrac{(y - 2)^2}{9} - \dfrac{(x + 2)^2}{4} = 1$
  d. $\dfrac{(y + 2)^2}{9} - \dfrac{(x - 2)^2}{4} = 1$

In Problems 12–20, identify the conic section.

12. $3x^2 + 2y^2 - 6x + 4y - 1 = 0$ c
  a. parabola   **b.** circle
  c. ellipse   **d.** hyperbola

13. $x^2 + 2x + 4y + 5 = 0$ a
  a. parabola   **b.** circle
  c. ellipse   **d.** hyperbola

14. $4x^2 - y^2 + 16x + 2y + 11 = 0$ d
  a. parabola   **b.** circle
  c. ellipse   **d.** hyperbola

15. $5y^2 - 3x^2 + 20y - 6x + 2 = 0$ d
  a. parabola   **b.** circle
  c. ellipse   **d.** hyperbola

16. $2x^2 - 5xy + y^2 = 7$ d
  a. parabola   **b.** circle
  c. ellipse   **d.** hyperbola

17. $x^2 + xy + y^2 = 10$ c
  a. parabola   **b.** circle
  c. ellipse   **d.** hyperbola

18. $r = \dfrac{2}{2 + \cos\theta}$ c
  a. parabola   **b.** circle
  c. ellipse   **d.** hyperbola

19. $r = \dfrac{8}{2 + 3\sin\theta}$ d
  a. parabola   **b.** circle
  c. ellipse   **d.** hyperbola

20. $x = 3\sec t, \quad y = 2\tan t$ d
  a. parabola   **b.** circle
  c. ellipse   **d.** hyperbola

CUMULATIVE REVIEW EXERCISES ▪ Chapters 1–9

1. Let $f(x) = x^2 - 3x + 2$. Find $\dfrac{f(x + h) - f(x)}{h}$. $\dfrac{f(x + h) - f(x)}{h} = 2x + h - 3$

2. Sketch the graph of $f(x) = \sqrt{x + 1} - 2$. †

3. Let $f(x) = 2x - 3$. Find the inverse function f^{-1}. Verify that $f(f^1(x)) = x$. †

4. Sketch the graph of $f(x) = \left(\dfrac{1}{2}\right)^{x+2}$. †

5. Solve the equation $\log_5(x - 1) + \log_5(x - 2) = 3\log_5\sqrt[3]{6}$. Solution: $x = 4$

6. Express

$$\log_a \sqrt[3]{x\sqrt{yz}}$$

in terms of logarithms of x, y, and z.

7. Solve the inequality

$$\frac{x}{x-2} \geq 1. \quad (2, \infty)$$

In Problems 8–12, solve each system of equations.

8. $\begin{cases} 1.4x - 0.5y = 1.3 \\ 0.4x + 1.1y = 4.1 \end{cases} \{(2, 3)\}$

9. $\begin{cases} 2x + y - 4z = 3 \\ x - 2y + 3z = 4 \\ -3x + 4y - z = -2 \end{cases} \{(4, 3, 2)\}$

10. $\begin{cases} y = x^2 - 1 \\ 3x^2 + 8y^2 = 8 \end{cases} \left\{(0, -1), \left(\pm\frac{\sqrt{26}}{4}, \frac{5}{8}\right)\right\}$

11. If $\cot\theta = \frac{5}{2}$ and $\sec\theta < 0$, find $\cos\theta$. $-\frac{5\sqrt{29}}{29}$

12. If $\cos\theta = -\frac{1}{3}$ and θ is in quadrant III, find $\sin 2\theta$. $\frac{4\sqrt{2}}{9}$

13. Convert the equation $x^2 + y^2 - 6x + 3 = 0$ to polar form. $r^2 - 6r\cos\theta + 3 = 0$

Answers:

6. $\frac{1}{3}\log_a x + \frac{1}{6}\log_a y + \frac{1}{6}\log_a z$

14. Find the determinant of the matrix $\begin{bmatrix} 1 & 4 & 7 \\ 2 & 5 & 8 \\ 3 & 6 & 9 \end{bmatrix}$. 0

15. Use Cramer's rule to solve the system of equations:

$$\begin{cases} 2x - 3y = -4 \\ 5x + 7y = 1 \end{cases} \left\{\left(-\frac{25}{29}, \frac{22}{29}\right)\right\}$$

16. Find the inverse of the matrix $A = \begin{bmatrix} 3 & -2 \\ -5 & 4 \end{bmatrix}$. $\begin{bmatrix} 2 & 1 \\ \frac{5}{2} & \frac{3}{2} \end{bmatrix}$

17. Solve the equation $2x^4 - 5x^2 + 3 = 0$. $\left\{1, -1, \frac{1}{2}\sqrt{6}, -\frac{1}{2}\sqrt{6}\right\}$

In Problems 18–20, an equation of a conic section is given. Identify the conic and sketch its graph.

18. $x^2 - y^2 = -4$ †

19. $9x^2 + 9y^2 = 144$ †

20. Sketch the graph of the rational function $f(x) = \frac{x}{x^2 - 16}$. †

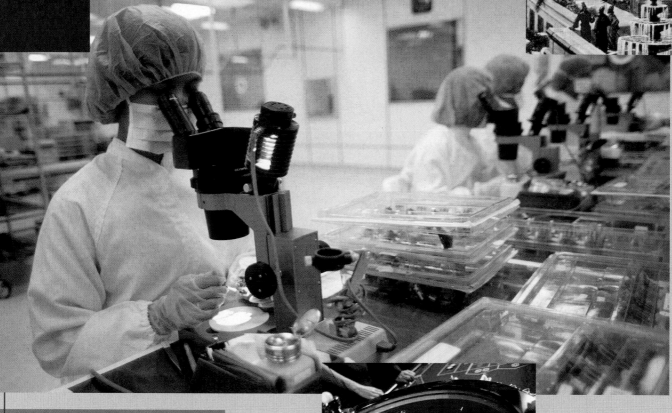

Probability was studied in ancient Eastern civilizations, including those in Babylonia and Egypt. Probability is most famously used in gambling, and Egyptians even used objects made from the heel bones of animals, called *tali*, as a type of dice. Today, probability is used not only in gambling, but also in polling, insurance, business, the life sciences, and many other areas. In this chapter, we learn to understand and compute the probability of an event.

Sequences and Series

Before Starting this Section, Review

1. Simplifying algebraic expressions (Appendix A, page 757)

2. Finding functional values (Section 1.3, page 31)

Objectives

1   Use sequence notation and find specific and general terms in a sequence.

2   Use factorial notation.

3   Use summation notation to write partial sums of a series.

THE FAMILY TREE OF THE HONEYBEE

The bee that most of us know best is the honeybee, an insect that lives in a colony and has an unusual family tree. Perhaps the most surprising fact about honeybees is that not all of them have two parents. This phenomenon begins with a special female in the colony called the *queen*. Many other female bees, called *worker bees*, live in the colony, but unlike the queen bee, they produce no eggs. Male bees do no work and are produced from the queen's unfertilized eggs; so male bees have no father, only a mother. The female bees are produced as a result of the queen bee mating with a male bee. Consequently, all female bees have two parents—a male and a female—whereas male bees have just one parent—a female. In this section, we study sequences, and in Example 5, we see how a famous sequence accurately counts a honeybee's ancestors. ∎

---

1   Use sequence notation and find specific and general terms in a sequence.

Sequences

The word *sequence* is used in mathematics in much the same way it is used in ordinary English. If someone saw a *sequence* of bad movies, you know that the person could list the first bad movie he or she saw, the second bad movie, and so on.

DEFINITION OF A SEQUENCE

An **infinite sequence** is a function whose domain is the set of positive integers. The function values, written as

$$a_1, a_2, a_3, a_4, \ldots, a_n, \ldots,$$

are called the **terms** of the sequence. The **nth term,** a_n, is called the **general term** of the sequence.

If the domain of a function consists of only the first n positive integers, the sequence is called a **finite sequence**. In computer science, finite sequences of the form $a_1, a_2, \ldots, a_n$ are called *strings*. Sometimes for convenience, we use 0 instead of 1 as the first subscript, and we write the sequence terms as $a_0, a_1, a_2, a_3, \ldots$. The nth term of this sequence is a_{n-1}.

BY THE WAY . . .

Notice in Example 1 that converting $a_n = 5n - 1$ to standard function notation gives us $f(n) = 5n - 1$, a linear function.

TECHNOLOGY CONNECTION

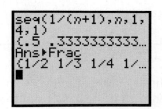 A graphing calculator can write the terms of a sequence and graph the sequence. You must first select Sequence mode.

The first four terms of

the sequence $a_n = \dfrac{1}{n + 1}$

and the graph of this sequence are shown here. Use the right arrow key ▶ to scroll to the right to display the sequence values corresponding to $n = 3$ and $n = 4$.

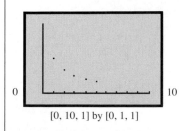

Now press GRAPH.

[0, 10, 1] by [0, 1, 1]

EXAMPLE 1   Writing the Terms of a Sequence from the General Term

Write the first four terms of each sequence.

a. $a_n = 5n - 1$ **b.** $a_n = \dfrac{1}{n + 1}$

SOLUTION

The first four terms of each sequence are found by replacing n with the integers 1, 2, 3, and 4 in the equation defining a_n.

a. $a_n = 5n - 1$ General term of the sequence

$a_1 = 5(1) - 1 = 4$ 1st term: Replace n with 1.

$a_2 = 5(2) - 1 = 9$ 2nd term: Replace n with 2.

$a_3 = 5(3) - 1 = 14$ 3rd term: Replace n with 3.

$a_4 = 5(4) - 1 = 19$ 4th term: Replace n with 4.

b. $a_n = \dfrac{1}{n + 1}$ General term of the sequence

$a_1 = \dfrac{1}{1 + 1} = \dfrac{1}{2}$ 1st term: Replace n with 1.

$a_2 = \dfrac{1}{2 + 1} = \dfrac{1}{3}$ 2nd term: Replace n with 2.

$a_3 = \dfrac{1}{3 + 1} = \dfrac{1}{4}$ 3rd term: Replace n with 3.

$a_4 = \dfrac{1}{4 + 1} = \dfrac{1}{5}$ 4th term: Replace n with 4.   ■■■

Practice Problem 1   Write the first four terms of the sequence with general term $a_n = -2^n$.   ■

We can use subscripts on variables other than a to represent the terms of a sequence. In Example 2, we use the variable b.

EXAMPLE 2   Writing the First Several Terms of a Sequence

Write the first six terms of the sequence defined by

$$b_n = (-1)^{n+1}\left(\frac{1}{n}\right).$$

SOLUTION

Recall that all even powers of -1 equal 1 and all odd powers of -1 equal -1.

$$b_1 = (-1)^{1+1}\left(\frac{1}{1}\right) = (-1)^2(1) = 1$$

$$b_2 = (-1)^{2+1}\left(\frac{1}{2}\right) = (-1)^3\left(\frac{1}{2}\right) = -\frac{1}{2}$$

$$b_3 = (-1)^{3+1}\left(\frac{1}{3}\right) = (-1)^4\left(\frac{1}{3}\right) = \frac{1}{3}$$

$$b_4 = (-1)^{4+1}\left(\frac{1}{4}\right) = (-1)^5\left(\frac{1}{4}\right) = -\frac{1}{4}$$

$$b_5 = (-1)^{5+1}\left(\frac{1}{5}\right) = (-1)^6\left(\frac{1}{5}\right) = \frac{1}{5}$$

$$b_6 = (-1)^{6+1}\left(\frac{1}{6}\right) = (-1)^7\left(\frac{1}{6}\right) = -\frac{1}{6}$$ ■■■

Practice Problem 2 Write the first six terms of the sequence defined by $a_n = \dfrac{(-1)^n}{n^2}$. ■

Frequently, the first few terms of a sequence exhibit a pattern that *suggests* a natural choice for the general term of the sequence.

EXAMPLE 3   Finding a General Term of a Sequence from a Pattern

Write the general term a_n for a sequence whose first five terms are given.

a. $1, 4, 9, 16, 25, \ldots$ **b.** $0, \dfrac{1}{2}, -\dfrac{2}{3}, \dfrac{3}{4}, -\dfrac{4}{5}, \ldots$

SOLUTION

Write the position number of the term above each term of the sequence and look for a pattern that connects the term to the position number of the term.

a.

n:	1	2	3	4	5	$\ldots$	n
term:	1,	4,	9,	16,	25,	$\ldots$,	a_n

Apparent pattern: Here $1 = 1^2, 4 = 2^2, 9 = 3^2, 16 = 4^2$, and $25 = 5^2$. Each term is the square of the position number of that term. This suggests $a_n = n^2$.

b.

n:	1	2	3	4	5	$\ldots$	n
term:	0,	$\dfrac{1}{2}$,	$-\dfrac{2}{3}$,	$\dfrac{3}{4}$,	$-\dfrac{4}{5}$,	$\ldots$,	a_n

Apparent pattern: When the terms alternate in sign and $n = 1$, we use factors such as $(-1)^n$ if we want to begin with the factor -1 or we use factors such as $(-1)^{n+1}$ if we want to begin with the factor 1. Notice that each term can be written as a quotient with denominator equal to the position number and numerator equal to one less than the position number, suggesting the general term $a_n = (-1)^n \left(\dfrac{n-1}{n} \right)$. ■ ■ ■

Practice Problem 3 Write a general term for a sequence whose first five terms are $0, -\dfrac{1}{2}, \dfrac{2}{3}, -\dfrac{3}{4}, \dfrac{4}{5}, \ldots$. ■

In Example 3, we found the general term of a sequence by finding a pattern in the first few terms of the sequence. However, when only a finite number of successive terms are given without a rule that defines the general term, a *unique* general term cannot be found.

To indicate why this is so, consider the two sequences $a_n = n^2$ and $b_n = n^2 + (n-1)(n-2)(n-3)(n-4)$. The first four terms of these two sequences are identical: $1, 4, 9, 16$. However, the fifth term is different: $a_5 = 25$, but $b_5 = 49$. Thus, either a_n or b_n could correctly describe a sequence whose first four terms are $1, 4, 9$, and 16. So you can *never* find a unique general term from a finite number of terms in a sequence.

Recursive Formulas

Up to this point, we have expressed the general term as a function of the position of the term in the sequence. A second approach is to define a sequence *recursively*. A **recursive formula** requires that one or more of the first few terms of the sequence be specified and all other terms be defined in relation to previously defined terms. An example will help clarify how this works.

Leonardo of Pisa (Fibonacci)

(1175–1250)

Leonardo, often known today by the name Fibonacci (son of Bonaccio) was born around 1175. He traveled extensively through North Africa and the Mediterranean, probably on business with his father. At each location, he met with Islamic scholars and absorbed the mathematical knowledge of the Islamic world. After returning to Pisa around 1200, he started writing books on mathematics. He was one of the earliest European writers on algebra. In his most famous book *Liber abaci*, or *Book of Calculations*, he introduced the Western world to the rules of computing with the new Hindu–Arabic numerals.

The **Fibonacci sequence** is a famous sequence that is defined recursively and shows up often in nature. In this sequence, we specify the first two terms as $a_0 = 1$ and $a_1 = 1$; each subsequent term is the sum of the two terms immediately preceding it. So we have

$$a_0 = 1, a_1 = 1, a_2 = a_0 + a_1 = 1 + 1 = 2 \qquad \text{Add the two given terms.}$$
$$a_1 = 1, a_2 = 2, a_3 = a_1 + a_2 = 1 + 2 = 3 \qquad \text{Add the two previous terms.}$$
$$a_2 = 2, a_3 = 3, a_4 = a_2 + a_3 = 2 + 3 = 5 \qquad \text{Add the two previous terms.}$$
$$a_3 = 3, a_4 = 5, a_5 = a_3 + a_4 = 3 + 5 = 8 \qquad \text{Add the two previous terms.}$$

The first six terms of the sequence are then

$$1, 1, 2, 3, 5, 8.$$

This sequence can also be defined in subscript notation:

$$a_0 = 1, a_1 = 1, a_k = a_{k-2} + a_{k-1}, \text{for all } k \geq 2$$

Replacing k with 2, 3, 4, and 5, respectively, in the equation $a_k = a_{k-2} + a_{k-1}$ will again produce the values $a_2 = 2$, $a_3 = 3$, $a_4 = 5$, and $a_5 = 8$.

EXAMPLE 4   **Finding Terms of a Recursively Defined Sequence**

Write the first five terms of the recursively defined sequence

$$a_1 = 4, \ a_{n+1} = 2a_n - 9.$$

SOLUTION

We are given the first term of the sequence: $a_1 = 4$.

$$a_2 = 2a_1 - 9 \qquad \text{Replace } n \text{ with 1 in } a_{n+1} = 2a_n - 9.$$
$$a_2 = 2(4) - 9 = -1 \qquad \text{Replace } a_1 \text{ with 4.}$$
$$a_3 = 2a_2 - 9 \qquad \text{Replace } n \text{ with 2 in } a_{n+1} = 2a_n - 9.$$
$$a_3 = 2(-1) - 9 = -11 \qquad \text{Replace } a_2 \text{ with } -1.$$
$$a_4 = 2a_3 - 9 \qquad \text{Replace } n \text{ with 3 in } a_{n+1} = 2a_n - 9.$$
$$a_4 = 2(-11) - 9 = -31 \qquad \text{Replace } a_3 \text{ with } -11.$$
$$a_5 = 2a_4 - 9 \qquad \text{Replace } n \text{ with 4 in } a_{n+1} = 2a_n - 9.$$
$$a_5 = 2(-31) - 9 = -71 \qquad \text{Replace } a_4 \text{ with } -31.$$

Thus, the first five terms of the sequence are

$$4, -1, -11, -31, -71. \qquad \blacksquare\ \blacksquare\ \blacksquare$$

Practice Problem 4   Write the first five terms of the recursively defined sequence $a_1 = -3, a_{n+1} = 2a_n + 5$. ■

EXAMPLE 5   **Diagramming the Family Tree of a Honeybee**

Find a sequence that accurately counts the ancestors of a male honeybee. The first term of the sequence should give the number of parents of a male honeybee, the second term the number of grandparents, the third term the number of great-grandparents, and so on.

SOLUTION

Recall from the introduction to this section that female honeybees have two parents—a male and a female—but that male honeybees have just one parent—a female.

First, we consider the family tree of a male honeybee. A male bee has one parent: a female. Since his mother had two parents, he has two grandparents. Because his grandmother had two parents and his grandfather had only one parent, he has three great-grandparents.

Now, to count the bee's ancestors, we count backward using Figure 10.1. You should recognize the first six terms of the Fibonacci sequence: 1, 2, 3, 5, 8. This sequence is defined by $a_0 = 1$, $a_1 = 1$, $a_2 = 2$, and $a_k = a_{k-2} + a_{k-1}$, for all $k \geq 3$. ■ ■ ■

Practice Problem 5   Find a sequence that accurately counts the ancestors of a female honeybee. ■

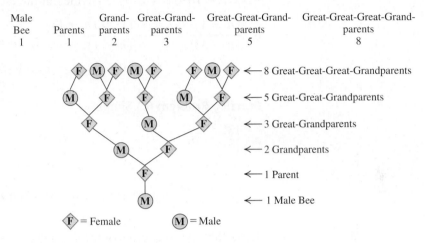

FIGURE 10.1

2 Use factorial notation.

Factorial Notation

BY THE WAY ...

The symbol $n!$ was first introduced by Christian Kremp of Strasbourg in 1808 as a convenience for the printer. Alternative symbols $\lfloor n$ and $\pi(n)$ were used until late in the nineteenth century.

   A typical calculator will compute approximate values of $n!$ up to $n = 69$.

$$69! \approx 1.7112 \times 10^{98}$$

Special types of products, called factorials, appear in some very important sequences.

DEFINITION OF FACTORIAL

For any positive integer n, **n factorial** (written $n!$) is defined as

$$n! = n \cdot (n-1) \cdots 4 \cdot 3 \cdot 2 \cdot 1.$$

As a special case, **zero factorial** (written $0!$) is defined as

$$0! = 1.$$

TECHNOLOGY CONNECTION

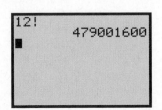

12!
         479001600

Here are the first eight values of $n!$

$0! = 1$	$4! = 4 \cdot 3 \cdot 2 \cdot 1 = 24$
$1! = 1$	$5! = 5 \cdot 4 \cdot 3 \cdot 2 \cdot 1 = 120$
$2! = 2 \cdot 1 = 2$	$6! = 6 \cdot 5 \cdot 4 \cdot 3 \cdot 2 \cdot 1 = 720$
$3! = 3 \cdot 2 \cdot 1 = 6$	$7! = 7 \cdot 6 \cdot 5 \cdot 4 \cdot 3 \cdot 2 \cdot 1 = 5040$

The values of $n!$ get large very quickly. For example, $10! = 3,628,800$ and $12! = 479,001,600$.

   In simplifying some expressions involving factorials, it is helpful to note that

$$\boxed{n! = n \cdot (n-1)!}$$

EXAMPLE 6   Simplifying a Factorial Expression

Simplify.

a. $\dfrac{16!}{14!}$ **b.** $\dfrac{(n+1)!}{(n-1)!}$

SOLUTION

a. $\dfrac{16!}{14!} = \dfrac{16 \cdot 15 \cdot 14!}{14!}$ $16! = 16 \cdot 15! = 16 \cdot 15 \cdot 14!$

$\qquad = 16 \cdot 15 = 240$ Divide out the factor $14!$

b. $\dfrac{(n+1)!}{(n-1)!} = \dfrac{(n+1) \cdot n \cdot (n-1)!}{(n-1)!}$ $(n+1)! = (n+1) \cdot n! = (n+1) \cdot n \cdot (n-1)!$

$\qquad = (n+1)n$ Divide out the common factor $(n-1)!$

■ ■ ■

Practice Problem 6   Simplify.

a. $\dfrac{13!}{12!}$ **b.** $\dfrac{n!}{(n-3)!}$

■

EXAMPLE 7   Writing Terms of a Sequence Involving Factorials

Write the first five terms of the sequence whose general term is $a_n = \dfrac{(-1)^{n+1}}{n!}$.

SOLUTION

Replace n in the formula for the general term with each positive integer from 1 through 5.

$$a_1 = \frac{(-1)^{1+1}}{1!} = \frac{(-1)^2}{1} = 1 \qquad a_4 = \frac{(-1)^{4+1}}{4!} = \frac{(-1)^5}{24} = -\frac{1}{24}$$

$$a_2 = \frac{(-1)^{2+1}}{2!} = \frac{(-1)^3}{2} = -\frac{1}{2} \qquad a_5 = \frac{(-1)^{5+1}}{5!} = \frac{(-1)^6}{120} = \frac{1}{120}$$

$$a_3 = \frac{(-1)^{3+1}}{3!} = \frac{(-1)^4}{6} = \frac{1}{6}$$

Using these five terms, we could write this sequence as

$$1, -\frac{1}{2}, \frac{1}{6}, -\frac{1}{24}, \frac{1}{120}, \ldots.$$

■ ■ ■

Practice Problem 7   Write the first five terms of the sequence whose general term is

$$a_n = \frac{(-1)^n 2^n}{n!}.$$

■

3 Use summation notation to write partial sums of a series.

Summation Notation

If we know the general term of a sequence, we can represent a *sum* of terms of the sequence by using *summation* (or *sigma*) *notation*. In this notation, the Greek letter Σ (capital sigma) indicates that we are to *add* the given terms. The letter Σ corresponds to the English letter *S*, the first letter of the word *sum*.

SUMMATION NOTATION

The sum of the first n terms of sequence $a_1, a_2, a_3, \ldots, a_n, \ldots$ is denoted by

$$\sum_{i=1}^{n} a_i = a_1 + a_2 + a_3 + \cdots + a_n.$$

The letter i in the summation notation is called the **index of summation**, n is called the **upper limit**, and 1 is called the **lower limit** of the summation.

In summation notation, the index need not start at $i = 1$. Also, any letter may be used in place of the index i.

TECHNOLOGY CONNECTION

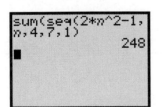 A graphing calculator in Sequence mode can sum terms of a series.

The viewing window shows $\sum_{j=4}^{7} (2j^2 - 1)$. Note that the calculator replaces the variable j with n.

```
sum(seq(2*n^2-1,
n,4,7,1)
              248
■
```

EXAMPLE 8   **Evaluating Sums Given in Summation Notation**

Find each sum.   **a.** $\displaystyle\sum_{i=1}^{9} i$ **b.** $\displaystyle\sum_{j=4}^{7} (2j^2 - 1)$ **c.** $\displaystyle\sum_{k=0}^{4} \frac{2^k}{k!}$

SOLUTION

a. Replace i with successive integers from 1 to 9 inclusive; then add.

$$\sum_{i=1}^{9} i = 1 + 2 + 3 + 4 + 5 + 6 + 7 + 8 + 9 = 45$$

b. The index of summation is j. Evaluate $(2j^2 - 1)$ for all consecutive integers from 4 through 7 inclusive; then add.

$$\sum_{j=4}^{7} (2j^2 - 1) = [2(4)^2 - 1] + [2(5)^2 - 1] + [2(6)^2 - 1] + [2(7)^2 - 1]$$

$$= 31 + 49 + 71 + 97 = 248$$

c. The index of summation is k. Evaluate $\dfrac{2^k}{k!}$ for consecutive integers 0 through 4, inclusive, then add.

$$\sum_{k=0}^{4} \frac{2^k}{k!} = \frac{2^0}{0!} + \frac{2^1}{1!} + \frac{2^2}{2!} + \frac{2^3}{3!} + \frac{2^4}{4!}$$

$$= \frac{1}{1} + \frac{2}{1} + \frac{4}{2} + \frac{8}{6} + \frac{16}{24} = 1 + 2 + 2 + \frac{4}{3} + \frac{2}{3} = 7 \quad \blacksquare\blacksquare\blacksquare$$

Practice Problem 8   Find the following sum: $\displaystyle\sum_{k=0}^{3} (-1)^k k!$ ■

The familiar properties of real numbers, such as the distributive, commutative, and associative properties, can be used to prove the summation properties listed next.

SUMMATION PROPERTIES

Let a_k and b_k represent the general terms of two sequences and let c represent any real number. Then

1. $\displaystyle\sum_{k=1}^{n} c = c \cdot n$ **2.** $\displaystyle\sum_{k=1}^{n} ca_k = c\sum_{k=1}^{n} a_k$

3. $\displaystyle\sum_{k=1}^{n} (a_k + b_k) = \sum_{k=1}^{n} a_k + \sum_{k=1}^{n} b_k$ **4.** $\displaystyle\sum_{k=1}^{n} (a_k - b_k) = \sum_{k=1}^{n} a_k - \sum_{k=1}^{n} b_k$

5. $\displaystyle\sum_{k=1}^{n} a_k = \sum_{k=1}^{j} a_k + \sum_{k=j+1}^{n} a_k$, for $1 \leq j < n$

Series

We all know how to add numbers, but do you think it is possible to add infinitely many numbers? It *is* sometimes possible to add infinitely many numbers, although a complete explanation of this kind of addition must be put off until a calculus course. For now, we need some definitions.

DEFINITION OF SERIES

Let $a_1, a_2, a_3, \ldots, a_k, \ldots$ be an infinite sequence. Then

1. The sum of the first n terms of the sequence is called the **nth partial sum** of the sequence and is denoted by

$$a_1 + a_2 + a_3 + \cdots + a_n = \sum_{i=1}^{n} a_i.$$

This sum is a **finite series**.

2. The sum of all terms of the infinite sequence is called an **infinite series** and is denoted by

$$a_1 + a_2 + a_3 + \cdots + a_n = \sum_{i=1}^{\infty} a_i.$$

EXAMPLE 9   Writing a Partial Sum in Summation Notation

Write each sum in summation notation.

a. $3 + 5 + 7 + \cdots + 21$ **b.** $\dfrac{1}{4} + \dfrac{1}{9} + \cdots + \dfrac{1}{49}$

SOLUTION

a. The finite series $3 + 5 + 7 + \cdots + 21$ is a sum of consecutive odd numbers from 3 to 21. Each of these numbers can be expressed in the form $2k + 1$, starting with $k = 1$ and ending with $k = 10$. With k as the index of summation, 1 as the lower limit, and 10 as the upper limit, we write

$$3 + 5 + 7 + \cdots + 21 = \sum_{k=1}^{10} (2k + 1).$$

b. The finite series $\dfrac{1}{4} + \dfrac{1}{9} + \cdots + \dfrac{1}{49}$ is a sum of fractions, each of which has numerator 1 and denominator k^2, starting with $k = 2$ and ending with $k = 7$. We write

$$\frac{1}{4} + \frac{1}{9} + \cdots + \frac{1}{49} = \sum_{k=2}^{7} \frac{1}{k^2}.$$

■ ■ ■

Practice Problem 9   Write the following in summation notation: $2 - 4 + 6 - 8 + 10 - 12 + 14$

■

SECTION 10.1 ■ Exercises

A EXERCISES Basic Skills and Concepts

1. An infinite sequence is a function whose domain is the set of _____. positive integers

2. If a sequence is defined by $a_n = 2n - \dfrac{10}{n}$, then $a_5 =$ _____. 8

3. By definition, $0! =$ _____. 1

4. Evaluate $\displaystyle\sum_{k=0}^{5} \frac{2k + 1}{2}$. 18

5. *True or False* If $a_1 = 3$ and $a_{n+1} = a_n^2$, then $a_3 = 27$. False

6. *True or False* $\dfrac{(2n)!}{n!} = 2$. False

In Exercises 7–26, write the first four terms of each sequence.

7. $a_n = 3n - 2$
$a_1 = 1, a_2 = 4, a_3 = 7, a_4 = 10$

8. $a_n = 2n + 1$
$a_1 = 3, a_2 = 5, a_3 = 7, a_4 = 9$

9. $a_n = 1 - \dfrac{1}{n}$ †

10. $a_n = 1 + \dfrac{1}{n}$ †

11. $a_n = -n^2$ †

12. $a_n = n^3$ †

13. $a_n = \dfrac{2n}{n + 1}$ †

14. $a_n = \dfrac{3n}{n^2 + 1}$ †

15. $a_n = (-1)^{n+1}$ †

16. $a_n = (-3)^{n-1}$ †

17. $a_n = 3 - \dfrac{1}{2^n}$ †

18. $a_n = \left(\dfrac{3}{2}\right)^n$ †

19. $a_n = 0.6$ †

20. $a_n = -0.4$ †

21. $a_n = \dfrac{(-1)^n}{n!}$ †

22. $a_n = \dfrac{n}{n!}$ †

23. $a_n = (-1)^n 3^{-n}$ †

24. $a_n = (-1)^n 3^{n-1}$ †

25. $a_n = \dfrac{e^n}{2n}$ †

26. $a_n = \dfrac{2^n}{e^n}$ †

In Exercises 27–40, write a general term a_n for each sequence. Assume that n begins with 1.

27. $1, 4, 7, 10, \ldots$
$a_n = 3n - 2$

28. $7, 9, 11, 13, \ldots$ $a_n = 2n + 5$

29. $\dfrac{1}{2}, \dfrac{1}{3}, \dfrac{1}{4}, \dfrac{1}{5}, \ldots$ $a_n = \dfrac{1}{n + 1}$

30. $\dfrac{2}{3}, \dfrac{3}{4}, \dfrac{4}{5}, \dfrac{5}{6}, \ldots$ $a_n = \dfrac{n + 1}{n + 2}$

31. $2, -2, 2, -2, \ldots$ $a_n = (-1)^{n+1}(2)$

32. $-3, 6, -9, 12, \ldots$ $a_n = (-1)^n(3n)$

33. $\dfrac{1}{2}, \dfrac{3}{4}, \dfrac{9}{8}, \dfrac{27}{16}, \ldots$ $a_n = \dfrac{3^{n-1}}{2^n}$

34. $\dfrac{-1}{2}, \dfrac{1}{4}, \dfrac{-1}{8}, \dfrac{1}{16}, \ldots$ $a_n = \left(-\dfrac{1}{2}\right)^n$

35. $1 \cdot 2, 2 \cdot 3, 3 \cdot 4, 4 \cdot 5, \ldots$ $a_n = n(n + 1)$

36. $\dfrac{-1}{1 \cdot 2}, \dfrac{1}{2 \cdot 3}, \dfrac{-1}{3 \cdot 4}, \dfrac{1}{4 \cdot 5}, \ldots$ $a_n = \dfrac{(-1)^n}{n(n + 1)}$

37. $2 + \dfrac{1}{2}, 2 - \dfrac{1}{3}, 2 + \dfrac{1}{4}, 2 - \dfrac{1}{5}, \ldots$ $a_n = 2 - \dfrac{(-1)^n}{n + 1}$

38. $1 + \dfrac{1}{2}, 1 + \dfrac{1}{3}, 1 + \dfrac{1}{4}, 1 + \dfrac{1}{5}, \ldots$ $a_n = 1 + \dfrac{1}{n + 1}$

39. $\dfrac{3^2}{2}, \dfrac{3^3}{3}, \dfrac{3^4}{4}, \dfrac{3^5}{5}, \ldots$ $a_n = \dfrac{3^{n+1}}{n + 1}$

40. $\dfrac{e}{2}, \dfrac{e^2}{4}, \dfrac{e^3}{8}, \dfrac{e^4}{16}, \ldots$ $a_n = \left(\dfrac{e}{2}\right)^n$

In Exercises 41–50, write the first five terms of each recursively defined sequence.

41. $a_1 = 2, a_{n+1} = a_n + 3$ †

42. $a_1 = 5, a_{n+1} = a_n - 1$ †

43. $a_1 = 3, a_{n+1} = 2a_n$ †

44. $a_1 = 1, a_{n+1} = \dfrac{1}{2}a_n$ †

45. $a_1 = 7, a_{n+1} = -2a_n + 3$ †

46. $a_1 = -4, a_{n+1} = -3a_n - 5$ †

47. $a_1 = 2, a_{n+1} = \dfrac{1}{a_n}$ †

48. $a_1 = -1, a_{n+1} = \dfrac{-1}{a_n}$ †

49. $a_1 = 25, a_{n+1} = \dfrac{(-1)^n}{5a_n}$ †

50. $a_1 = 12, a_{n+1} = \dfrac{(-1)^n}{3a_n}$ †

In Exercises 51–56, use a graphing calculator to (a) find the first ten terms of the sequence and (b) graph the first ten terms of the sequence.

51. $a_n = 3n^2 - 1$ †

52. $a_n = 4 - \dfrac{3}{n}$ †

53. $a_n = n\left(1 - \dfrac{1}{n}\right)$ †

54. $a_n = n^3 - n^2$ †

55. $a_n = 1 - \dfrac{1}{a_{n-1}}, a_1 = \dfrac{1}{2}$ †

56. $a_n = a_{n-1}^2, a_1 = 1$ †

In Exercises 57–64, simplify the factorial expression.

57. $\dfrac{3!}{5!}$ $\dfrac{1}{20}$

58. $\dfrac{8!}{10!}$ $\dfrac{1}{90}$

†Due to space constrictions, answers to these exercises may be found in the Answers beginning on page A–1 in the back of the book.

59. $\dfrac{12!}{11!}$ 12

60. $\dfrac{20!}{18!}$ 380

61. $\dfrac{n!}{(n+1)!}$ $\dfrac{1}{n+1}$

62. $\dfrac{(n-1)!}{(n-2)!}$ $n-1$

63. $\dfrac{(2n+1)!}{(2n)!}$ $2n+1$

64. $\dfrac{(2n+1)!}{(2n-1)!}$ $4n^2+2n$

In Exercises 65–76, find each sum.

65. $\displaystyle\sum_{k=1}^{7} 5$ 35

66. $\displaystyle\sum_{j=1}^{4} 12$ 48

67. $\displaystyle\sum_{j=0}^{5} j^2$ 55

68. $\displaystyle\sum_{k=0}^{4} k^3$ 100

69. $\displaystyle\sum_{i=1}^{5} (2i-1)$ 25

70. $\displaystyle\sum_{k=0}^{6} (1-3k)$ −56

71. $\displaystyle\sum_{j=1}^{7} \dfrac{j+1}{j}$ $\dfrac{1343}{140}$

72. $\displaystyle\sum_{k=0}^{5} \dfrac{1}{k+1}$ $\dfrac{49}{20}$

73. $\displaystyle\sum_{i=1}^{6} (-1)^i 3^{i-1}$ 182

74. $\displaystyle\sum_{k=0}^{4} (-1)^{k+1}k$ −2

75. $\displaystyle\sum_{k=1}^{3} (2-k^2)$ −8

76. $\displaystyle\sum_{j=0}^{4} (j^3+1)$ 105

In Exercises 77–84, write each sum in summation notation.

77. $1+3+5+7+\cdots+101$ $\displaystyle\sum_{k=1}^{51} (2k-1)$

78. $2+4+6+8+\cdots+102$ $\displaystyle\sum_{k=1}^{51} 2k$

79. $\dfrac{1}{5(1)}+\dfrac{1}{5(2)}+\dfrac{1}{5(3)}+\dfrac{1}{5(4)}+\cdots+\dfrac{1}{5(11)}$ $\displaystyle\sum_{k=1}^{11} \dfrac{1}{5k}$

80. $1+\dfrac{2}{2\cdot3}+\dfrac{2}{3\cdot4}+\dfrac{2}{4\cdot5}+\cdots+\dfrac{2}{9\cdot10}$ $\displaystyle\sum_{k=1}^{9} \dfrac{2}{k(k+1)}$

81. $1-\dfrac{1}{2}+\dfrac{1}{3}-\dfrac{1}{4}+\cdots-\dfrac{1}{50}$ $\displaystyle\sum_{k=1}^{50} \dfrac{(-1)^{k+1}}{k}$

82. $1-2+3-4+\cdots+(-50)$ $\displaystyle\sum_{k=1}^{50} (-1)^{k+1}k$

83. $\dfrac{1}{2}+\dfrac{2}{3}+\dfrac{3}{4}+\dfrac{4}{5}+\cdots+\dfrac{10}{11}$ $\displaystyle\sum_{k=1}^{10} \left(\dfrac{k}{k+1}\right)$

84. $2+\dfrac{2^2}{2}+\dfrac{2^3}{3}+\dfrac{2^4}{4}+\cdots+\dfrac{2^{10}}{10}$ $\displaystyle\sum_{k=1}^{10} \dfrac{2^k}{k}$

 **In Exercises 85–90, use a graphing calculator to find each sum.**

85. $\displaystyle\sum_{i=1}^{10} 12i^2$ 4620

86. $\displaystyle\sum_{i=1}^{50} (2i+7)$ 2900

87. $\displaystyle\sum_{k=5}^{30} \dfrac{7}{1-k^2}$ −1.345 (rounded to three decimal places)

88. $\displaystyle\sum_{k=10}^{25} \dfrac{(k+1)^2}{k}$ 312.987 (rounded to three decimal places)

89. $\displaystyle\sum_{j=1}^{100} (-1)^j j$ 50

90. $\displaystyle\sum_{j=8}^{50} \left(7+\dfrac{(-1)^j}{j}\right)$ 301.076 (rounded to three decimal places)

B EXERCISES Applying the Concepts

91. **Free fall.** In the absence of friction, a freely falling body will fall about 16 feet the first second, 48 feet the next second, 80 feet the third second, 112 feet the fourth second, and so on. How far has it fallen during
  a. the seventh second? 208 ft
  b. the *n*th second? $16+(n-1)32$ ft

92. **Bacterial growth.** A colony of 1000 bacteria doubles in size every hour. How many bacteria will there be in
  a. 2 hours? 4000
  b. 5 hours? 32,000
  c. *n* hours? $(1000)2^n$

93. **Workplace fines.** A contractor who quits work on a house was told she would be fined $50 if she did not resume work on Monday, $75 if she failed to resume work on Tuesday, $100 if she did not resume work on Wednesday, and so on (including weekends). How much will her fine be on the ninth day she fails to show up for work? $250

94. **Appreciating value.** A painting valued at $30,000 is expected to appreciate $1280 the first year, $1240 the second year, $1200 the third year, and so on. How much will the painting appreciate in
  a. the seventh year? $1040
  b. the tenth year? $920

95. **Cell phone use.** At the end of the first six months after a company began providing cell phones to its sales force, it averaged 600 cell minutes per month. For the next four years, its monthly cell phone use doubled every six months. How many cell minutes per month were being used at the end of three years? 19,200 min

96. **Motorcycle acceleration.** A motorcycle travels 10 yards the first second and then increases its speed by 20 yards per second in each succeeding second. At this rate, how far will the motorcycle travel during
  a. the eighth second? 150 yd
  b. the *n*th second? $20n-10$ yd

97. **Compound interest.** Suppose $10,000 is deposited into an account that earns 6% interest compounded semiannually. The balance in the account after *n* compounding periods is given by the sequence

$$A_n = 10,000\left(1+\dfrac{0.06}{2}\right)^n, n=1,2,3,\ldots.$$

  a. Find the first six terms of this sequence. †
  b. Find the balance in the account after 8 years. $16,047.10

98. **Compound interest.** Suppose $100 is deposited at the beginning of a year into an account that earns 8% interest compounded quarterly. The balance in the account after *n* compounding periods is

$$A_n = 100\left(1+\dfrac{0.08}{4}\right)^n, n=1,2,3,\ldots.$$

  a. Find the first six terms of this sequence. †
  b. Find the balance in the account after ten years. $220.8

99. **Real estate value.** Analysts estimate that a $100,000 condominium will increase 5% in value each year for the next seven years. Find the value of the condominium in each of those years. Write a formula for a sequence whose first seven terms give these values. †

100. **Diminishing savings.** Laura withdraws 10% of her $50,000 savings each year. Find the amount remaining in her account for each of the next ten years. Write a formula for a sequence whose first ten terms give these values. †

C EXERCISES Beyond the Basics

101. Find a formula for the nth term of the sequence defined recursively by $a_1 = \sqrt{2}$, $a_{n+1} = \sqrt{2a_n}$. [*Hint:* Write each of the first five terms as a power of 2.]

102. The triangular tiles used in the figures shown have white interiors and gold edges. A sequence of figures is obtained by adding one triangle to the previous figure.

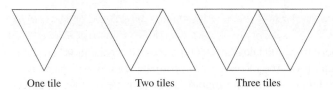

One tile   Two tiles   Three tiles

  a. Write a recursive sequence whose nth term, a_n, gives the number of gold edges in the nth figure.

  b. Write a recursive sequence whose nth term, b_n, gives the number of gold edges that lie on the perimeter of the nth figure. $b_1 = 3, b_{n+1} = b_n + 1$

103. The figure shows a sequence of successively smaller squares formed by connecting the midpoints of the sides of the preceding larger square. The area of the largest square is 1.

  a. Write the first five terms of the sequence whose general term, a_n, gives the area of the nth square.

  b. Write the general term, a_n, of the sequence.

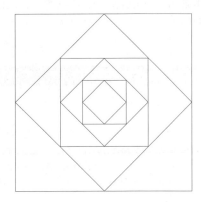

104. A sequence of concentric circles is designed so that the radius of the first circle is 1 and each successive circle has a radius that is twice the length of the radius of the preceding circle. Write a recursive sequence whose nth term, a_n, gives the area of the nth circle. $a_1 = \pi, a_{n+1} = 4a_n$

105. Find the first ten terms of the sequence defined recursively by $a_1 = a_2 = 1$, $a_n = a_{a_{n-1}} + a_{n-a_{n-1}}$. In 1989, a prize of \$1000 was offered to whoever first discovered (for this sequence) a value of n for which

$$\left| \frac{a_i}{i} \right| - \left| \frac{1}{2} \right| < \frac{1}{20} \text{ for all } i > n.$$

A mathematician named Colin L. Mallows of AT&T/Bell Laboratories found that $n = 1489$ and claimed the prize. (*Source:* http://el.media.mit.edu/logo-foundation/pubs/papers/easy_as_11223.html)

$a_1 = 1, a_2 = 1, a_3 = 2, a_4 = 2, a_5 = 3, a_6 = 4, a_7 = 4,$
$a_8 = 4, a_9 = 5, a_{10} = 6$

106. A sequence of numbers $a_0, a_1, a_2, a_3, \ldots$ satisfies the equation

$$a_n^2 = (-1)^n a_{n-1} + a_{n+1}.$$

If $a_0 = 1$ and $a_1 = 3$, find a_3. $a_3 = 97$

107. Find a sequence with general term a_m so that

$$\sum_{n=1}^{20} n^2 = \sum_{m=0}^{19} a_m.\quad a_m = (m+1)^2$$

108. Find a real number c such that

$$\sum_{n=1}^{50} (n-3)^2 = \sum_{n=1}^{50} (n^2 - 6n) + \sum_{n=1}^{50} c.\quad c = 9$$

109. Find a lower limit p and an upper limit q such that

$$\sum_{n=0}^{10} n^3 = \sum_{m=p}^{q} (m+2)^3.\quad p = -2, q = 8$$

110. Give an example of two sequences a_k and b_k such that

$$\sum_{k=1}^{20} a_k b_k \neq \left(\sum_{k=1}^{20} a_k \right)\left(\sum_{k=1}^{20} b_k \right).\quad \begin{array}{l} a_k = b_k = 1; \\ \text{answers may vary.} \end{array}$$

111. Show that $7! - 2(5!) = 40(5!)$.

112. Show that $\dfrac{10!}{6!\,4!} + \dfrac{10!}{5!\,5!} = \dfrac{11!}{6!\,5!}$.

Critical Thinking

113. Find the largest integer value for the upper limit k if

$$\sum_{n=1}^{k} (n^2 + n)\quad k = 5$$

$$= \sum_{n=1}^{k} [(n-1)(n-2)(n-3)(n-4)(n-5) + n^2 + n].$$

114. **The Ulam conjecture.** Define the sequence a_n recursively by

$$a_n = \begin{cases} \dfrac{a_{n-1}}{2} & \text{if } a_{n-1} \text{ is even.} \\[2mm] 3a_{n-1} + 1 & \text{if } a_{n-1} \text{ is odd} \end{cases}$$

The Ulam conjecture says that if your first term a_0 is any positive integer, the sequence a_n will eventually reach the integer 1. Verify the conjecture for $a_0 = 13$.

Answers:
101. $a_n = 2^{1-(1/2^n)}$ **102. a.** $a_1 = 3, a_{n+1} = a_n + 2$
103. a. $a_1 = 1, a_2 = \dfrac{1}{2}, a_3 = \dfrac{1}{4}, a_4 = \dfrac{1}{8}, a_5 = \dfrac{1}{16}$

b. $a_n = \dfrac{1}{2^{n-1}}$

114. The terms leading to 1 are 13, 40, 20, 10, 5, 16, 8, 4, 2, and 1.

Arithmetic Sequences; Partial Sums

Before Starting this Section, Review

1. The general term of a sequence (Section 10.1, page 681)

2. Partial sums (Section 10.1, page 688)

Objectives

1 Identify an arithmetic sequence and find its common difference.

2 Find the sum of the first n terms of an arithmetic sequence.

FALLING SPACE JUNK

On April 27, 2000, more than 300 kilograms (700 pounds) of "space junk" crashed to the ground in South Africa. The material was eventually identified as part of a Delta 2 rocket used to launch a global positioning satellite in 1996. Once an object begins a free fall toward Earth, it falls (in the absence of friction) 16 feet in the first second, 48 feet in the next second, 80 feet in the third second, and so on. The number of feet traveled in succeeding seconds is the sequence

$$16, 48, 80, 112, \ldots.$$

This is an example of an *arithmetic sequence*. We investigate such a sequence in Example 5. ■

1 Identify an arithmetic sequence and find its common difference.

When the difference between any two consecutive terms of a sequence is always the same number, the sequence is called an *arithmetic sequence*. In other words, the sequence $a_1, a_2, a_3, a_4, \ldots$ is an arithmetic sequence if $a_2 - a_1 = a_3 - a_2 = a_4 - a_3 = \ldots$.

DEFINITION OF ARITHMETIC SEQUENCE

The sequence

$$a_1, a_2, a_3, a_4, \ldots, a_n, \ldots$$

is an **arithmetic sequence**, or an **arithmetic progression**, if there is a number d such that each term in the sequence except the first is obtained from the preceding term by adding d to it. The number d is called the **common difference** of the arithmetic sequence. We have

$$d = a_{n+1} - a_n, \quad n \geq 1.$$

EXAMPLE 1 **Finding the Common Difference**

Find the common difference of each arithmetic sequence.

a. $3, 23, 43, 63, 83, \ldots$ **b.** $29, 19, 9, -1, -11, \ldots$

SOLUTION

a. The common difference is $23 - 3 = 20$ (or $43 - 23, 63 - 43,$ or $83 - 63$).

b. The common difference is $19 - 29 = -10$ (or $9 - 19, -1 - 9,$ or $-11 - (-1)$).

■ ■ ■

Practice Problem 1 Find the common difference of the arithmetic sequence

$$3, -2, -7, -12, -17, \ldots. \qquad ■$$

An arithmetic sequence can be completely specified by giving the first term a_1 and the common difference d. Let's see why. Suppose we are told that the sequence $a_1, a_2, a_3, a_4, \ldots$ is an arithmetic sequence and that $a_1 = 5$ and $d = 4$. Then

$$a_2 - a_1 = 4, \quad \text{so} \quad a_2 = a_1 + 4 = 5 + 4 = 9;$$
$$a_3 - a_2 = 4, \quad \text{so} \quad a_3 = a_2 + 4 = 9 + 4 = 13;$$
$$a_4 - a_3 = 4, \quad \text{so} \quad a_4 = a_3 + 4 = 13 + 4 = 17; \text{ and so on.}$$

Rewriting $d = a_{n+1} - a_n$ as $a_{n+1} = a_n + d$, we have a recursive definition of an arithmetic sequence.

RECURSIVE DEFINITION OF AN ARITHMETIC SEQUENCE

An arithmetic sequence $a_1, a_2, a_3, a_4, \ldots, a_n, \ldots$ can be defined recursively. The recursive formula

$$a_{n+1} = a_n + d \text{ for } n \geq 1$$

defines an arithmetic sequence with first term a_1 and common difference d.

Consider an arithmetic sequence with first term a_1 and common difference d. Using the recursive definition $a_{n+1} = a_n + d$, for $n \geq 1$, we write

$a_1 = a_1$ The first term is a_1.

$a_2 = a_1 + d$ Replace n with 1 in $a_{n+1} = a_n + d$.

$a_3 = a_2 + d = (a_1 + d) + d = a_1 + 2d$ Replace a_2 with $a_1 + d$; simplify.

$a_4 = a_3 + d = (a_1 + 2d) + d = a_1 + 3d$ Replace a_3 with $a_1 + 2d$; simplify.

$a_5 = a_4 + d = (a_1 + 3d) + d = a_1 + 4d$ Replace a_4 with $a_1 + 3d$; simplify.

$\vdots$

$a_n = a_{n-1} + d$

$\quad = [a_1 + (n - 2)d] + d$ Replace a_{n-1} with $a_1 + (n - 2)d$.

$\quad = a_1 + (n - 1)d.$ Simplify.

Thus, we can find the general term a_n from the values a_1 and d.

nTH TERM OF AN ARITHMETIC SEQUENCE

If a sequence $a_1, a_2, a_3, \ldots$ is an arithmetic sequence, then its nth term, a_n, is given by

$$a_n = a_1 + (n - 1)d,$$

where a_1 is the first term and d is the common difference.

EXAMPLE 2   Finding the nth Term of an Arithmetic Sequence

Find an expression for the nth term of the arithmetic sequence

$$3, 7, 11, 15, \ldots.$$

SOLUTION

Since $a_1 = 3$ and $d = 7 - 3 = 4$, we have

$$a_n = a_1 + (n - 1)d \qquad \text{Formula for the } n\text{th term}$$
$$= 3 + (n - 1)4 \qquad \text{Replace } a_1 \text{ with 3 and } d \text{ with 4.}$$
$$= 3 + 4n - 4 = 4n - 1 \quad \text{Simplify.}$$

Note that in the expression $a_n = 4n - 1$, we get the sequence $3, 7, 11, 15, \ldots$ by substituting $n = 1, 2, 3, 4, \ldots$.   ■ ■ ■

Practice Problem 2 Find an expression for the nth term of the arithmetic sequence

$$-3, 1, 5, 9, 13, 17, \ldots.$$ ■

EXAMPLE 3   **Finding the Common Difference of an Arithmetic Sequence**

Find the common difference d and the nth term a_n of the arithmetic sequence whose 5th term is 15 and whose 20th term is 45.

SOLUTION

$$a_n = a_1 + (n - 1)d \qquad \text{Formula for the } n\text{th term}$$
$$45 = a_1 + (20 - 1)d \qquad \text{Replace } n \text{ with 20; } a_n = a_{20} = 45.$$
$$(1) \quad 45 = a_1 + 19d. \qquad \text{Simplify.}$$

Also,

$$15 = a_1 + (5 - 1)d \qquad \text{Replace } n \text{ with 5; } a_n = a_5 = 15.$$
$$(2) \quad 15 = a_1 + 4d. \qquad \text{Simplify.}$$

Solving the system of equations

$$\begin{cases} a_1 + 19d = 45 & (1) \\ a_1 + 4d = 15 & (2) \end{cases}$$

gives $d = 2$ and $a_1 = 7$. We substitute $a_1 = 7$ and $d = 2$ in the formula for the nth term.

$$a_n = a_1 + (n - 1)d \qquad \text{Formula for the } n\text{th term}$$
$$a_n = 7 + (n - 1)2 \qquad \text{Replace } a_1 \text{ with 7 and } d \text{ with 2.}$$
$$= 7 + 2n - 2 = 2n + 5 \quad \text{Simplify.}$$

The nth term of this sequence is given by

$$a_n = 2n + 5, \quad n \geq 1.$$ ■ ■ ■

Practice Problem 3 Find the common difference d and the nth term a_n of the arithmetic sequence whose 4th term is 41 and whose 15th term is 8.   ■

2 Find the sum of the first n terms of an arithmetic sequence.

Sum of an Arithmetic Sequence

We can use the following formula in many applications involving the sum of an arithmetic sequence.

$$1 + 2 + 3 \cdots + n = \frac{n(n + 1)}{2}$$

Karl Friedrich Gauss

(1777–1855)

Karl Friedrich Gauss discovered this formula in his arithmetic class at the age of 8. One day, the story goes, the teacher became so incensed with the class that he assigned the students the task of adding up all of the numbers from 1 to 100. As Gauss's classmates dutifully began to add, Gauss walked up to the teacher and presented the answer, 5050. The story goes that the teacher was neither impressed nor amused. Here are Gauss's calculations:

$$S = 1 + 2 + 3 + \cdots + 100$$
$$\underline{S = 100 + 99 + 98 + \cdots + 1}$$
$$2S = \underbrace{101 + 101 + 101 + \cdots + 101}_{100 \text{ terms}}$$
$$2S = 100 \times 101$$
$$S = \frac{100 \times 101}{2} = 5050$$

It is said that Karl Friedrich Gauss (1777–1855) discovered this formula when he realized that any sum of numbers added in reverse order produces the same sum. Therefore, if S denotes the sum of the first n natural numbers, then

$$S = \quad 1 \quad + \quad 2 \quad + \quad 3 \quad + \cdots + \quad n$$
$$\underline{S = \quad n \quad + (n - 1) + (n - 2) + \cdots + \quad 1}$$
$$2S = \underbrace{(n + 1) + (n + 1) + (n + 1) + \cdots + (n + 1)}_{n \text{ terms}} \qquad \text{Add } S + S = 2S.$$

$$2S = n(n + 1)$$
$$S = \frac{n(n + 1)}{2} \qquad\qquad\qquad \text{Divide both sides by 2.}$$

For example, if $n = 100$, we have

$$S = 1 + 2 + 3 + \cdots + 100 = \frac{100(100 + 1)}{2}$$
$$= \frac{100(101)}{2} = 5050.$$

We can use this method to calculate the sum S_n of the first n terms of any arithmetic sequence.

First, we write the sum S_n of the first n terms:

$$S_n = a_1 + (a_1 + d) + (a_1 + 2d) + (a_1 + 3d) + \cdots + a_n$$

We can also write S_n (with the terms in reverse order) by starting with a_n and *subtracting* the common difference d:

$$S_n = a_n + (a_n - d) + (a_n - 2d) + (a_n - 3d) + \cdots + a_1$$

Adding the two equations for S_n, we find that the d's in the sums "drop out" (for example, $(a_1 + d) + (a_n - d) = a_1 + a_n$) and we have

$$2S_n = \underbrace{(a_1 + a_n) + (a_1 + a_n) + \cdots + (a_1 + a_n)}_{n \text{ terms}}$$

$$2S_n = n(a_1 + a_n)$$
$$S_n = n\left(\frac{a_1 + a_n}{2}\right) \qquad \text{Divide both sides by 2.}$$

SUM OF n TERMS OF AN ARITHMETIC SEQUENCE

Let $a_1, a_2, a_3, \ldots a_n$ be the first n terms of an arithmetic sequence with common difference d. The sum S_n of these n terms is given by

$$S_n = n\left(\frac{a_1 + a_n}{2}\right),$$

where $a_n = a_1 + (n - 1)d$.

EXAMPLE 4   **Finding the Sum of Terms of a Finite Arithmetic Sequence**

Find the sum of the following arithmetic sequence of numbers:

$$1 + 4 + 7 + \cdots + 25$$

SOLUTION

In this arithmetic sequence, $a_1 = 1$ and $d = 3$. Let's first find the number of terms.

$$a_n = a_1 + (n-1)d \qquad \text{Formula for } n\text{th term}$$
$$25 = 1 + (n-1)3 \qquad \text{Replace } a_n \text{ with 25, } a_1 \text{ with 1, and } d \text{ with 3.}$$
$$24 = (n-1)3 \qquad \text{Subtract 1 from both sides.}$$
$$8 = n-1 \qquad \text{Divide both sides by 3.}$$
$$n = 9 \qquad \text{Solve for } n.$$

$$S_n = n\left(\frac{a_1 + a_n}{2}\right) \qquad \text{Formula for } S_n$$

$$S_9 = 9\left(\frac{1+25}{2}\right) \qquad \text{Replace } n \text{ with 9, } a_1 \text{ with 1, and } a_n(=a_9) \text{ with 25.}$$

$$= 9(13) = 117 \qquad \text{Simplify.}$$

So $1 + 4 + 7 + \cdots + 25 = 117$. ■ ■ ■

Practice Problem 4 Find the sum $\dfrac{2}{3} + \dfrac{5}{6} + 1 + \dfrac{7}{6} + \dfrac{4}{3} + \dfrac{3}{2} + \dfrac{5}{3} + \dfrac{11}{6} + 2 + \dfrac{13}{6}$.

■

EXAMPLE 5 **Calculating the Distance Traveled by a Freely Falling Object**

In the introduction to this section, we described the arithmetic sequence 16, 48, 80, 112, ... whose terms gave the number of feet that freely falling space junk falls in successive seconds. For this sequence, find the following:

a. The common difference d

b. The nth term a_n

c. The distance the object travels in 10 seconds

SOLUTION

a. The common difference $d = a_2 - a_1 = 48 - 16 = 32$.

b. $a_n = a_1 + (n-1)d \qquad \text{Formula for the } n\text{th term}$
$\qquad = 16 + (n-1)32 \qquad \text{Replace } a_1 \text{ with 16 and } d \text{ with 32.}$
$\qquad = 32n - 16 \qquad \text{Simplify.}$

c. The sum of the first ten terms of this sequence gives the distance the object travels in ten seconds. We need to find a_n to use the formula.

$$a_n = 32n - 16 \qquad \text{From part b}$$
$$a_{10} = 32(10) - 16 \qquad \text{Replace } n \text{ with 10.}$$
$$= 304$$

$$S_n = n\left(\frac{a_1 + a_n}{2}\right) \qquad \text{Formula for } S_n$$

$$S_{10} = 10\left(\frac{16 + 304}{2}\right) \qquad \begin{array}{l}\text{Replace } n \text{ with 10, } a_1 \text{ with 16, and} \\ a_n(=a_{10}) \text{ with 304.}\end{array}$$

$$= 1600 \qquad \text{Simplify.}$$

The object falls 1600 feet in ten seconds. ■ ■ ■

Practice Problem 5 In Example 5, find the distance the object travels during the eleventh through fifteenth seconds. ■

SECTION 10.2 ■ Exercises

A EXERCISES Basic Skills and Concepts

1. If 5 is the common difference of an arithmetic sequence with general term a_n, then $a_{17} - a_{16} = $ _____5_____.

2. If 5 is the common difference of an arithmetic sequence with general terms a_n and $a_{21} = 7$, then $a_{22} = $ _____12_____.

3. If 14 is the term immediately following the sequence term 17 in an arithmetic sequence, then the common difference is _____-3_____.

4. If $a_1 = 2$ and $a_{11} = 22$ for an arithmetic sequence, then the common difference, d, is _____2_____.

5. *True or False* The common difference of an arithmetic sequence is always positive. False

6. *True or False* The sum of the first n terms of an arithmetic sequence always equals n times the average of its first and nth terms. True

In Exercises 7–20, determine whether each sequence is arithmetic. For those that are, find the first term a_1 and the common difference d.

7. $1, 2, 3, 4, 5, \ldots$ †   8. $1, 3, 5, 7, 9, \ldots$ †

9. $2, 5, 8, 11, 14, \ldots$ †   10. $10, 7, 4, 1, -2, \ldots$ †

11. $1, \dfrac{1}{2}, 0, -\dfrac{1}{2}, -\dfrac{1}{4}, \ldots$ †   12. $2, 4, 8, 16, 32, \ldots$ †

13. $1, -1, 2, -2, 3, \ldots$ †   14. $-\dfrac{1}{4}, \dfrac{1}{4}, \dfrac{3}{4}, \dfrac{5}{4}, \dfrac{7}{4}, \ldots$ †

15. $0.6, 0.2, -0.2, -0.6, -1, \ldots$ Arithmetic; $a_1 = 0.6, d = -0.4$

16. $2.3, 2.7, 3.1, 3.5, 3.9, \ldots$ Arithmetic; $a_1 = 2.3, d = 0.4$

17. $a_n = 2n + 6$ †   18. $a_n = 1 - 5n$ †

19. $a_n = 1 - n^2$ †   20. $a_n = 2n^2 - 3$ †

In Exercises 21–30, find an expression for the nth term of the arithmetic sequence.

21. $5, 8, 11, 14, 17, \ldots$ $a_n = 3n + 2$ 22. $4, 7, 10, 13, 16, \ldots$ $a_n = 3n + 1$

23. $11, 6, 1, -4, -9, \ldots$ $a_n = 16 - 5n$ 24. $9, 5, 1, -3, -7, \ldots$ $a_n = 13 - 4n$

25. $\dfrac{1}{2}, \dfrac{1}{4}, 0, -\dfrac{1}{4}, -\dfrac{1}{2}, \ldots$ †   26. $\dfrac{2}{3}, \dfrac{5}{6}, 1, \dfrac{7}{6}, \dfrac{4}{3}, \ldots$ $a_n = \dfrac{n + 3}{6}$

27. $-\dfrac{3}{5}, -1, -\dfrac{7}{5}, -\dfrac{9}{5}, -\dfrac{11}{5}, \ldots$ $a_n = -\dfrac{2n + 1}{5}$

28. $\dfrac{1}{2}, 2, \dfrac{7}{2}, 5, \dfrac{13}{2}, \ldots$ $a_n = \dfrac{3n - 2}{2}$

29. $e, 3 + e, 6, + e, 9 + e, 12 + e, \ldots$ $a_n = 3n + e - 3$

30. $2\pi, 2(\pi + 2), 2(\pi + 4), 2(\pi + 6), 2(\pi + 8) \ldots,$ $a_n = 2(2n + \pi - 2)$

In Exercises 31–36, find the common difference d and the nth term a_n of the arithmetic sequence with the specified terms.

31. 4th term 21; 10th term 60 $d = \dfrac{13}{2}, a_n = \dfrac{13}{2}n - 5$

32. 3rd term 15; 21st term 87 $d = 4, a_n = 4n + 3$

33. 7th term 8; 15th term -8 $d = -2, a_n = 22 - 2n$

34. 5th term 12; 18th term -1 $d = -1, a_n = 17 - n$

35. 3rd term 7; 23rd term 17 $d = \dfrac{1}{2}, a_n = \dfrac{n + 11}{2}$

36. 11th term -1; 31st term 5 $d = \dfrac{3}{10}, a_n = \dfrac{3n - 43}{10}$

In Exercises 37–46, find the sum of each arithmetic sequence.

37. $1 + 2 + 3 + \cdots + 50$ 1275

38. $2 + 4 + 6 + \cdots + 102$ 2652

39. $1 + 3 + 5 + \cdots + 99$ 2500

40. $5 + 10 + 15 + \cdots + 200$ 4100

41. $3 + 6 + 9 + \cdots + 300$ 15,150

42. $4 + 7 + 10 + \cdots + 301$ 15,250

43. $2 - 1 - 4 - \cdots - 34$ -208

44. $-3 - 8 - 13 - \cdots - 48$ -255

45. $\dfrac{1}{3} + 1 + \dfrac{5}{3} + \cdots + 7$ $\dfrac{121}{3}$

46. $\dfrac{3}{5} + 2 + \dfrac{17}{5} + \cdots + \dfrac{101}{5}$ 156

In Exercises 47–52, find the sum of the first n terms of the given arithmetic sequence.

47. $2, 7, 12, \ldots; n = 50$ 6225

48. $8, 10, 12, \ldots; n = 40$ 1880

49. $-15, -11, -7, \ldots; n = 20$ 460

50. $-20, -13, -6, \ldots; n = 25$ 1600

51. $3.5, 3.7, 3.9, \ldots; n = 100$ 1340

52. $-7, -6.5, -6, \ldots; n = 80$ 1020

In Exercises 53–58, find n for the given value of a_n in each arithmetic sequence.

53. $a_n = 75; 1, 3, 5, \ldots$ $n = 38$ 54. $a_n = 120; 2, 4, 6, \ldots$ $n = 60$

55. $a_n = 95; -1, 3, 7, \ldots$ $n = 25$ 56. $a_n = 83; -5, -3, -1, \ldots$ $n = 45$

57. $a_n = 50\sqrt{3}; 2\sqrt{3}, 4\sqrt{3}, 6\sqrt{3}, \ldots$ $n = 25$

58. $a_n = 73\pi; 3\pi, 5\pi, 7\pi, \ldots$ $n = 36$

B EXERCISES Applying the Concepts

In Exercises 59–65, assume that the indicated sequence is arithmetic.

59. **Orange picking.** When Eric started work as an orange picker, he picked 10 oranges in the first minute, 12 in the second minute, 14 in the third minute, and so on. How many oranges did Eric pick in the first half hour? 1170

60. **Exercise.** Walking up a steep hill, Jan walks 60 feet in the first minute, 57 feet in the second minute, 54 feet in the third minute, and so on.
   a. How far will Jan walk in the nth minute? $63 - 3n$ ft
   b. How far will Jan walk in the first 15 minutes? 585 ft

†Due to space constrictions, answers to these exercises may be found in the Answers beginning on page A–1 in the back of the book.

61. Contest winner. A contest winner will receive money each month for three years. The winner receives $50 the first month, $75 the second month, $100 the third month, and so on. How much money will the winner have collected after 30 months? $12,375

62. Salary. Darren took a 12-month temporary job with a monthly salary that increased a fixed amount each month. He can't recall the starting salary, but does remember that he was paid $820 at the end of the third month and $910 for his last month's work. How much was Darren's total pay for the entire 12 months? $10,260

63. Hourly wage. Antonio's new weekend job started at an hourly wage of $12.75. He is guaranteed a raise of 25¢ an hour every three months for the next four years. What will Antonio's hourly wage be at the end of four years? $16.75

64. Competing job offers. Denzel is considering offers from two companies. A marketing company pays $32,500 the first year and guarantees a raise of $1300 each year; an exporting company pays $36,000 the first year, with a guaranteed raise of $400 each year. Over a five-year period, which company will pay more? How much more?

65. Theater seating. A theater has 25 rows of seats. The first row has 20 seats, the second row has 22 seats, the third row has 24 seats, and so on. How many seats are in the theater? 1100

66. Bricks in a driveway. A brick driveway has 50 rows of bricks. The first row has 16 bricks, and the fiftieth row has 65 bricks. Assuming that the sequence of numbers giving the number of bricks in rows 1 through 50 is arithmetic, how many bricks does the driveway contain? 2025

67. Stacked logs. A stack of logs has 28 rows. The bottom row has 40 logs, and the top row has 8 logs. Assuming the sequence of numbers that gives the number of logs in succeeding rows is arithmetic, how many logs does the stack contain? 672

C EXERCISES Beyond the Basics

68. Elena wanted to teach her young daughter Sophia about saving. She gave Sophia 10¢ the first day and an additional 5¢ per day on each subsequent day. After a while, Sophia counted her money and found that she had saved a total of $3.25. How many days had she been saving? 10

Answers:
64. The exporting company will pay $8500 more.

69. Only 3 customers showed up for the opening day of a flea market. However, 9 came the second day, and an additional 6 customers came on each subsequent day. After several days, there was a total of 192 customers. How many days had the flea market been open? 8

70. Find the sum of all natural numbers between 45 and 100 that are divisible by 3. 1368

71. Find the sum of all natural numbers between 26 and 120 that are divisible by 7. 1029

72. A sequence is **harmonic** if the reciprocals of the terms of the sequence form an arithmetic sequence. Is the sequence $\frac{1}{2}, \frac{3}{5}, \frac{3}{4}, 1, \ldots$ harmonic? Explain your reasoning.

73. Consider the harmonic sequence $\frac{2}{5}, \frac{2}{7}, \frac{2}{9}, \ldots$
   a. Find the fourth term. $\frac{2}{11}$
   b. Find the nth term. $\frac{2}{2n+3}$

74. Show that if the numbers a, m, b are consecutive terms in an arithmetic sequence, then $m = \frac{a+b}{2}$. We call m the **arithmetic mean** of a and b.

75. If the numbers $a, m_1, m_2, \ldots, m_k, b$ form an arithmetic sequence, we say that the numbers $m_1, m_2, \ldots, m_k$ are k **arithmetic means** between a and b. Insert k arithmetic means between 1 and 30 so that the sum of the resulting series is 465. $k = 28, d = 1$

Critical Thinking

76. Find the nth term of an arithmetic sequence whose first term is 10 and whose 21st term is 0. $a_n = \frac{21-n}{2}$

77. If a_1 and d are the first term and common difference, respectively, of an arithmetic sequence, find the first term and difference of an arithmetic sequence whose terms are the negatives of the terms in the given sequence. First term $-a_1$; difference; $-d$

78. How many terms are in the series $\sum_{i=22}^{100} 17i^3$? 79

79. The sum of the first n counting numbers is 12,403. Find the value of n. $n = 157$

72. Yes, it is harmonic because the reciprocals form an arithmetic sequence.

Geometric Sequences and Series

Before Starting this Section, Review

1. The general term of a sequence (Section 10.1, page 681)
2. Partial sums (Section 10.1, page 688)
3. Compound interest (Section 3.2, page 201)

Objectives

1 Identify a geometric sequence and find its common ratio.

2 Find the sum of a finite geometric sequence.

3 Solve annuity problems.

4 Find the sum of an infinite geometric sequence.

SPREADING THE WEALTH

Cities across the nation compete to host a Super Bowl. Cities across the world compete to host the Olympic Games. What is the incentive? The *multiplier effect* is often listed among the benefits for hosting one of these events. The effect refers to the phenomonon that occurs when money spent directly on an event has a ripple effect on the economy that is many times the amount spent directly on the event.

Suppose, for example, that $10 million is spent directly by tourists for hotels, meals, tickets, cabs, and so on. Some portion of that amount will be respent by its recipients on groceries, clothes, gas, and so on.

This process is repeated over and over, affecting the economy much more than the initial $10 million spent. Under certain assumptions, the spending just described results in an *infinite geometric series*, whose sum is called the *multiplier*. Example 8 illustrates this effect. ■

---

1 Identify a geometric sequence and find its common ratio.

When the *ratio* of any two consecutive terms of a sequence is always the same number, the sequence is called a *geometric sequence*. In other words, the sequence $a_1, a_2, a_3, a_4, \ldots$ is a geometric sequence if

$$\frac{a_2}{a_1} = \frac{a_3}{a_2} = \frac{a_4}{a_3} = \cdots.$$

DEFINITION OF GEOMETRIC SEQUENCE

The sequence

$$a_1, a_2, a_3, a_4, \ldots, a_n, \ldots$$

is a **geometric sequence**, or a **geometric progression**, if there is a number r such that each term except the first in the sequence is obtained by multiplying the previous term by r. The number r is called the **common ratio** of the geometric sequence.

$$\frac{a_{n+1}}{a_n} = r, \quad n \geq 1$$

EXAMPLE 1 Finding the Common Ratio

Find the common ratio for each geometric sequence.

a. $2, 10, 50, 250, 1250, \ldots$ **b.** $-162, -54, -18, -6, -2, \ldots$

SOLUTION

a. The common ratio is $\dfrac{10}{2} = 5 \left(\text{or } \dfrac{50}{10}, \dfrac{250}{50}, \text{ or } \dfrac{1250}{250} \right)$.

b. The common ratio is $\dfrac{-54}{-162} = \dfrac{1}{3} \left(\text{or } \dfrac{-18}{-54}, \dfrac{-6}{-18}, \text{ or } \dfrac{-2}{-6} \right)$.   ■ ■ ■

Practice Problem 1 Find the common ratio for the geometric sequence 6, 18, 54, 162, 486, . . .   ■

A geometric sequence can be completely specified by giving the first term a_1 and the common ratio r. Suppose, for example, that we are told that the sequence $a_1, a_2, a_3, a_4, \ldots$ is geometric and that $a_1 = 3$ and $r = 4$. Then we can find a_2:

$$\frac{a_2}{a_1} = 4, \text{ so } a_2 = 4a_1 = 4 \cdot 3 = 12$$

Now we continue:

$$\frac{a_3}{a_2} = 4, \text{ so } a_3 = 4a_2 = 4 \cdot 12 = 48;$$

$$\frac{a_4}{a_3} = 4, \text{ so } a_4 = 4a_3 = 4 \cdot 48 = 192; \text{ and so on.}$$

By rewriting $\dfrac{a_{n+1}}{a_n} = r$ as $a_{n+1} = ra_n$, we have a recursive definition of a geometric sequence.

RECURSIVE DEFINITION OF A GEOMETRIC SEQUENCE

A geometric sequence $a_1, a_2, a_3, a_4, \ldots, a_n, \ldots$ can be defined recursively. The recursive formula

$$a_{n+1} = ra_n, \quad n \geq 1$$

defines a geometric sequence with the first term a_1 and the common ratio r.

EXAMPLE 2   Determining Whether a Sequence Is Geometric

Determine whether each sequence is geometric. If it is, find the first term and the common ratio.

a. $a_n = 4^n$ **b.** $b_n = \dfrac{3}{2^n}$ **c.** $c_n = 1 - 5^n$

SOLUTION

a. The sequence 4, 16, 64, 256, . . . , 4^n, . . . appears to be geometric, with $a_1 = 4$ and $r = 4$. To verify this, we check that $\dfrac{a_{n+1}}{a_n} = 4$ for all $n \geq 1$.

$$a_{n+1} = 4^{n+1} \qquad \text{Replace } n \text{ with } n + 1 \text{ in the general term } a_n = 4^n.$$

$$\frac{a_{n+1}}{a_n} = \frac{4^{n+1}}{4^n} = 4 \qquad \frac{4^{n+1}}{4^n} = \frac{4^n 4}{4^n} = 4$$

Consequently, $\dfrac{a_{n+1}}{a_n} = 4$ for all $n \geq 1$, and the sequence is geometric.

b. The sequence $\dfrac{3}{2}, \dfrac{3}{4}, \dfrac{3}{8}, \dfrac{3}{16}, \ldots$ appears to be geometric, with $b_1 = \dfrac{3}{2}$ and $r = \dfrac{1}{2}$.

To verify this, we check that $\dfrac{b_{n+1}}{b_n} = \dfrac{1}{2}$ for all $n \geq 1$.

$$b_{n+1} = \dfrac{3}{2^{n+1}} \qquad\qquad \text{Replace } n \text{ with } n + 1 \text{ in the general term } b_n = \dfrac{3}{2^n}.$$

$$\dfrac{b_{n+1}}{b_n} = \dfrac{3}{2^{n+1}} \div \dfrac{3}{2^n} \qquad\qquad \dfrac{b_{n+1}}{b_n} = b_{n+1} \div b_n$$

$$= \dfrac{3}{2^{n+1}} \cdot \dfrac{2^n}{3} = \dfrac{1}{2}$$

Consequently, $\dfrac{b_{n+1}}{b_n} = \dfrac{1}{2}$ for all $n \geq 1$, and the sequence is geometric.

c. The sequence $-4, -24, -124, -624, \ldots, 1 - 5^n, \ldots$ is *not* a geometric sequence because not all pairs of consecutive terms have identical quotients. The quotient of the first two terms is $\dfrac{-24}{-4} = 6$, but the quotient of the third and second terms is $\dfrac{-124}{-24} = \dfrac{31}{6}$. ■ ■ ■

Practice Problem 2 Determine whether the sequence $a_n = \left(\dfrac{3}{2}\right)^n$ is geometric. If so, find the first term and the common ratio. ■

From the equation $a_{n+1} = a_n r$ used in the recursive definition of a geometric sequence, we see that each term after the first is obtained by multiplying the preceding term by the common ratio r.

THE GENERAL TERM OF A GEOMETRIC SEQUENCE

Every geometric sequence can be written in the form

$$a_1, a_1 r, a_1 r^2, a_1 r^3, \ldots, a_1 r^{n-1}, \ldots,$$

where r is the common ratio. Since $a_1 = a_1(1) = a_1 r^0$, the **nth term of the geometric sequence** is

$$a_n = a_1 r^{n-1}, \text{ for } n \geq 1.$$

EXAMPLE 3	**Finding Terms in a Geometric Sequence**

For the geometric sequence $1, 3, 9, 27, \ldots$, find each of the following:

a. a_1 **b.** r **c.** a_n

SOLUTION

a. The first term of the sequence is given: $a_1 = 1$.

b. The sequence is geometric, so we can take the ratio of any two consecutive terms. The ratio of the first two terms gives us

$$r = \dfrac{3}{1} = 3.$$

c. $a_n = a_1 r^{n-1}$ Formula for the nth term

$\quad = (1)(3^{n-1})$ Replace a_1 with 1 and r with 3.

$\quad = 3^{n-1}$ ■ ■ ■

Practice Problem 3   For the geometric sequence $2, \dfrac{6}{5}, \dfrac{18}{25}, \dfrac{54}{125}, \dots$, find the following:

a. a_1 **b.** r **c.** a_n ■

Notice in Example 3 that converting $a_n = 3^{n-1}$ to standard function notation gives us $f(n) = 3^{n-1}$, an exponential function. A geometric sequence with first term a_1 and common ratio r can be viewed as an exponential function $f(n) = a_1 r^{n-1}$ with the set of natural numbers as its domain.

EXAMPLE 4   **Finding a Particular Term in a Geometric Sequence**

Find the 23rd term of a geometric sequence whose first term is 10 and whose common ratio is 1.2.

SOLUTION

$$a_n = a_1 r^{n-1} \qquad \text{Formula for the } n\text{th term}$$
$$a_{23} = 10(1.2)^{23-1} \qquad \text{Replace } n \text{ with } 23, a_1 \text{ with } 10, \text{ and } r \text{ with } 1.2.$$
$$= 10(1.2)^{22}$$
$$\approx 552.06 \qquad \text{Use a calculator.} \qquad \blacksquare \blacksquare \blacksquare$$

Practice Problem 4   Find the 18th term of a geometric sequence whose first term is 7 and whose common ratio is 1.5.   ■

2 Find the sum of a finite geometric sequence.

Finding the Sum of a Finite Geometric Sequence

We can find a formula for the sum S_n of the first n terms in a geometric sequence. (We have already found a similar formula for arithmetic sequences.)

$$S_n = a_1 + a_1 r + a_1 r^2 + a_1 r^3 + \cdots + a_1 r^{n-1} \qquad \text{Definition of } S_n$$
$$r S_n = a_1 r + a_1 r^2 + a_1 r^3 + a_1 r^4 + \cdots + a_1 r^n \qquad \text{Multiply both sides by } r.$$
$$S_n - r S_n = a_1 - a_1 r^n \qquad \text{Subtract the second equation from the first.}$$
$$(1 - r) S_n = a_1(1 - r^n) \qquad \text{Factor both sides.}$$
$$S_n = \frac{a_1(1 - r^n)}{1 - r}, r \neq 1 \qquad \text{Divide both sides by } 1 - r.$$

SUM OF THE TERMS OF A FINITE GEOMETRIC SEQUENCE

Let $a_1, a_2, a_3, \dots, a_n$ be the fist n terms of a geometric sequence with first term a_1 and common ratio r. The sum S_n of these terms is

$$S_n = \sum_{i=1}^{n} a_1 r^{i-1} = \frac{a_1(1 - r^n)}{1 - r}, r \neq 1.$$

A geometric sequence with $r = 1$ is a sequence in which all terms are identical; that is, $a_n = a_1(1)^{n-1} = a_1$. Consequently, the sum of the first n term, S_n, is

$$\underbrace{a_1 + a_1 + \cdots + a_1}_{n \text{ terms}} = na_1.$$

TECHNOLOGY CONNECTION

The following screen shows the result of Example 5a obtained using a graphing calculator.

```
sum(seq(5*0.7^(n
-1),n,1,15,1)
        16.58754064
■
```

EXAMPLE 5   **Finding the Sum of Terms of a Finite Geometric Sequence**

Find each sum.

a. $\displaystyle\sum_{i=1}^{15} 5(0.7)^{i-1}$ **b.** $\displaystyle\sum_{i=1}^{15} 5(0.7)^{i}$

SOLUTION

a. We can evaluate $\displaystyle\sum_{i=1}^{15} 5(0.7)^{i-1}$ by using the formula for $S_n = \displaystyle\sum_{i=1}^{n} a_1 r^{i-1}$, where $a_1 = 5, r = 0.7$, and $n = 15$.

$$S_{15} = \sum_{i=1}^{15} a_1 r^{i-1} = 5\left[\frac{1 - (0.7)^{15}}{1 - 0.7}\right] \qquad \text{Replace } a_1 \text{ with } 5, r = 0.7, \text{ and } n = 15.$$

$$\approx 16.588 \qquad\qquad\qquad \text{Use a calculator.}$$

b. $\displaystyle\sum_{i=1}^{15} 5(0.7)^{i} = (0.7)\sum_{i=1}^{15} 5(0.7)^{i-1}$ Factor out 0.7 from each term.

$$\approx (0.7)(16.588) \qquad \text{From part } \mathbf{a}$$

$$= 11.6116 \qquad\qquad\qquad\qquad\qquad \blacksquare\ \blacksquare\ \blacksquare$$

Practice Problem 5   Find the sum $\displaystyle\sum_{i=1}^{17} 3(0.4)^{i}$. ■

3 Solve annuity problems.

Annuities

One of the most important applications of finite geometric series is the computation of the value of an *annuity*. An **annuity** is a sequence of equal periodic payments. When a fixed rate of compound interest applies to all payments, the sum of all of the payments made plus all interest can be found by using the formula for the sum of a finite geometric sequence. The **value of an annuity** (also called the **future value of an annuity**) is the sum of all payments and interest. To develop a formula for the value of an annuity, we suppose P is deposited at the end of each year at an annual interest rate i compounded annually.

The compound interest formula $A = P(1 + i)^t$ gives the total value after t years when P earns an annual interest rate i (in decimal form) compounded once a year. At the end of the first year, the initial payment of P is made and the annuity's value is P. At the end of the second year, P is again deposited. At this time, the first deposit has earned interest during the second year. The value of the annuity after two years is

$$P \quad + \quad P(1 + i).$$

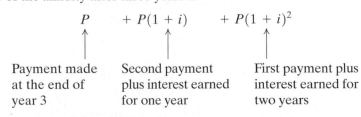

Payment made at the
end of year 2

First payment plus interest
earned for one year

The value of the annuity after three years is

$$P \quad + P(1 + i) \quad + P(1 + i)^2$$

Payment made
at the end of
year 3

Second payment
plus interest earned
for one year

First payment plus
interest earned for
two years

The value of the annuity after t years is

$$P + P(1 + i) + P(1 + i)^2 + P(1 + i)^3 + \cdots + P(1 + i)^{t-1}.$$

Payment made at
the end of year t

First payment plus interest
earned for $t - 1$ year

The value of the annuity after t years is the sum of a geometric sequence with first term $a_1 = P$ and common ratio $r = 1 + i$. We compute this sum A as follows:

RECALL

The sum S_n of the first n terms of a geometric sequence with first term a_1 and common ratio r is given by the formula

$$S_n = \frac{a_1(1 - r^n)}{1 - r}.$$

$$A = S_n = \frac{a_1(1 - r^n)}{1 - r} \qquad \text{Formula for } S_n$$

$$A = \frac{P[1 - (1 + i)^t]}{1 - (1 + i)} \qquad \text{Replace } n \text{ with } t, a_1 \text{ with } P, \text{ and } r \text{ with } 1 + i.$$

$$= \frac{P[1 - (1 + i)^t]}{-i} \qquad \text{Simplify.}$$

$$= P \frac{[(1 + i)^t - 1]}{i} \qquad \text{Multiply numerator and denominator by } -1.$$

If interest is compounded n times per year and equal payments are made at the end of each compounding period, we adjust the formula as we did in Section 3.2 to account for the more frequent compounding.

VALUE OF AN ANNUITY

Let P represent the payment in dollars made at the end of each of n compounding periods per year and let i be the annual interest rate. Then the value A of the annuity after t year is:

$$A = P\left[\frac{\left(1 + \dfrac{i}{n}\right)^{nt} - 1}{\dfrac{i}{n}} \right]$$

EXAMPLE 6   **Finding the Value of an Annuity**

An individual retirement account (IRA) is a common way to save money to provide funds after retirement. Suppose you make payments of $1200 into an IRA at the end of each year at an annual interest rate of 4.5% per year, compounded annually. What is the value of this annuity after 35 years?

SOLUTION

Each annuity payment is $P = \$1200$. The annual interest rate is 4.5% (so $i = 0.045$), and the number of years is $t = 35$. Because interest is compounded annually, $n = 1$. The value of the annuity is

$$A = 1200\left[\frac{\left(1 + \dfrac{0.045}{1}\right)^{(1)35} - 1}{\dfrac{0.045}{1}} \right]$$

$$= \$97{,}795.94 \qquad \text{Use a calculator.}$$

The value of the IRA after 35 years is $97,795.94.   ■ ■ ■

Practice Problem 6 If in Example 6 the end-of-year payments are $1500, the annual interest rate remains 4.5%, compounded annually, and the payments are made for 30 years, what is the value of the annuity? ■

4 Find the sum of an infinite geometric sequence.

Infinite Geometric Series

The sum S_n of the first n terms of a geometric series is given by the formula $S_n = \dfrac{a_1(1 - r^n)}{1 - r}$. If this finite sum S_n approaches a number S as $n \to \infty$ (that is, as n gets larger and larger), we say that S is the **sum of the infinite geometric series**, and we write

$$S = \sum_{i=1}^{\infty} a_1 r^{i-1}.$$

If r is any real number with $-1 < r < 1$ (equivalently, $|r| < 1$), then the value of r^n, like the value of $\left(\dfrac{1}{2}\right)^n$, gets closer and closer to 0 as n gets larger and larger. We indicate that the values of r^n approach 0 by writing

$$\lim_{n \to \infty} r^n = 0 \text{ if } |r| < 1.$$

The expression $\lim_{n \to \infty} r^n$ is read "the limit, as n approaches infinity, of r to the n" or "the limit, as n approaches infinity, of r to the nth power."

When $|r| < 1$,

$$S_n = \frac{a_1(1 - r^n)}{1 - r} \to \frac{a_1(1 - 0)}{1 - r} = \frac{a_1}{1 - r} \quad \text{as } n \to \infty.$$

SUM OF THE TERMS OF AN INFINITE GEOMETRIC SEQUENCE

If $|r| < 1$, the infinite sum

$$a_1 + a_1 r + a_1 r^2 + a_1 r^3 + \cdots + a_1 r^{n-1} + \cdots$$

is given by

$$S = \sum_{i=1}^{\infty} a_1 r^{i-1} = \frac{a_1}{1 - r}.$$

When $|r| \geq 1$, the infinite geometric series does not have a sum. This is because the value of r^n does not approach 0 as $n \to \infty$. For example, if $r = 2$, we have

$$2^1 = 2, 2^2 = 4, 2^3 = 8, 2^4 = 16, 2^5 = 32, 2^6 = 64, \ldots.$$

EXAMPLE 7 **Finding the Sum of an Infinite Geometric Series**

Find the sum $2 + \dfrac{3}{2} + \dfrac{9}{8} + \dfrac{27}{32} + \cdots$.

SOLUTION

The first term $a_1 = 2$, and the common ratio

$$r = \frac{\dfrac{3}{2}}{2} = \frac{3}{4}.$$

Because $|r| = \dfrac{3}{4} < 1$, we can use the formula for the sum of an infinite geometric series.

$$S = \frac{a_1}{1 - r} \qquad \text{Formula for } S$$

$$= \frac{2}{1 - \dfrac{3}{4}} \qquad \text{Replace } a_1 \text{ with 2 and } r \text{ with } \dfrac{3}{4}.$$

$$= 8 \qquad \text{Simplify.} \qquad\qquad ■■■$$

Practice Problem 7 Find the sum $3 + \dfrac{6}{3} + \dfrac{12}{9} + \dfrac{24}{27} + \cdots$. ■

EXAMPLE 8 **Calculating the Multiplier Effect**

The host city for the Super Bowl expects that tourists will spend $10,000,000. Assume that 80% of this money is spent again in the city, then 80% of this second round of spending is spent again, and so on. In the introduction to this section, we said that such a spending pattern results in a geometric series whose sum is called the *multiplier*. Find this series and its sum.

SOLUTION

We start our series with the $10,000,000 brought into the city and add the subsequent amounts spent.

$$10{,}000{,}000 + 10{,}000{,}000\,(0.80) + 10{,}000{,}000\,(0.80)^2 + 10{,}000{,}000\,(0.80)^3 + \cdots$$

| original 10,000,00 | 80% of 10,000,000 | 80% of previous amount | 80% of previous amount |

Using the formula $\displaystyle\sum_{i=1}^{\infty} ar^{i-1} = \frac{a}{1 - r}$ for the sum of an infinite geometric series, we have

$$\sum_{i=1}^{\infty} (10{,}000{,}000)(0.80)^{i-1} = \frac{\$10{,}000{,}000}{1 - 0.80} = \$50{,}000{,}000.$$

This amount should shed some light on why there is so much competition to serve as the host city for a major sports event. ■■■

Practice Problem 8 Find the series and the sum that results in Example 8 assuming that 85%, rather than 80%, occurs in the repeated spending. ■

SECTION 10.3 ■ Exercises

A EXERCISES Basic Skills and Concepts

1. If 5 is the common ratio of a geometric sequence with general term a_n, then $\dfrac{a_{63}}{a_{62}} = \underline{\quad 5 \quad}$.

2. If 5 is the common ratio of a geometric sequence with general term a_n and $a_{15} = -2$, then $a_{16} = \underline{\quad -10 \quad}$.

3. If 24 is the term immediately following the sequence term 8 in a geometric sequence, then the common ratio is $\underline{\quad 3 \quad}$.

4. An infinite geometric series $a_1 + a_1 r + a_1 r^2 + \cdots$ does not have a sum if $\underline{\quad |r| \quad} \geq 1$.

---

5. *True or False* If the sum of an infinite geometric series having the common ratio $r = \dfrac{1}{3}$ is $S = 15$, then $a_1 = 10$.
True

6. *True or False* The sum of an infinite geometric series can never be negative. False

In Exercises 7–26, determine whether each sequence is geometric. If it is, find the first term and the common ratio.

7. $3, 6, 12, 24, \ldots$ †   **8.** $2, 4, 8, 16, \ldots$ †

9. $1, 5, 10, 20, \ldots$ Not geometric   **10.** $1, 1, 3, 3, 9, 9, \ldots$ Not geometric

11. $1, -3, 9, -27, \ldots$ †   **12.** $-1, 2, -4, 8, \ldots$ †

13. $7, -7, 7, -7, \ldots$ †   **14.** $1, -2, 4, -8, \ldots$ †

15. $9, 3, 1, \dfrac{1}{3}, \ldots$ †   **16.** $5, 2, \dfrac{4}{5}, \dfrac{8}{25}, \ldots$ †

17. $a_n = \left(-\dfrac{1}{2}\right)^n$ †   **18.** $a_n = 5\left(\dfrac{2}{3}\right)^n$ †

19. $a_n = 2^{n-1}$ †   **20.** $a_n = -(1.06)^{n-1}$ †

21. $a_n = 7n^2 + 1$ Not geometric   **22.** $a_n = 1 - (2n)^2$ Not geometric

23. $a_n = 3^{-n}$ †   **24.** $a_n = 50(0.1)^{-n}$ †

25. $a_n = 5^{\frac{n}{2}}$ †   **26.** $a_n = 7^{\sqrt{n}}$ †

In Exercises 27–34, find the first term a_1, the common ratio r, and the nth term a_n for each geometric sequence.

27. $2, 10, 50, 250, \ldots$ $a_1 = 2, r = 5, a_n = (2)(5)^{n-1}$

28. $-3, -6, -12, -24, \ldots$ $a_1 = -3, r = 2, a_n = (-3)(2)^{n-1}$

29. $5, \dfrac{10}{3}, \dfrac{20}{9}, \dfrac{40}{27}, \ldots$ †   **30.** $1, \sqrt{3}, 3, 3\sqrt{3}, \ldots$ †

31. $0.2, -0.6, 1.8, -5.4, \ldots$ $a_1 = 0.2, r = -3, a_n = 0.2(-3)^{n-1}$

32. $1.3, -0.26, 0.052, -0.0104, \ldots$ $a_1 = 1.3, r = -0.2, a_n = \dfrac{1.3}{(-5)^{n-1}}$

33. $\pi^4, \pi^6, \pi^8, \pi^{10}, \ldots$ †   **34.** $e^2, 1, e^{-2}, e^{-4}, \ldots$ †
$a_1 = \pi^4, r = \pi^2, a_n = \pi^{2n+2}$

In Exercises 35–44, find the indicated term of each geometric sequence.

35. a_7 when $a_1 = 5$ and $r = 2$ $a_7 = 320$

36. a_7 when $a_1 = 8$ and $r = 3$ $a_7 = 5832$

37. a_{10} when $a_1 = 3$ and $r = -2$ $a_{10} = -1536$

38. a_{10} when $a_1 = 7$ and $r = -2$ $a_{10} = -3584$

39. a_6 when $a_1 = \dfrac{1}{16}$ and $r = 3$ $a_6 = \dfrac{243}{16}$

40. a_6 when $a_1 = \dfrac{1}{81}$ and $r = 3$ $a_6 = 3$

41. a_9 when $a_1 = -1$ and $r = \dfrac{5}{2}$ $a_9 = -\dfrac{390,625}{256}$

42. a_9 when $a_1 = -4$ and $r = \dfrac{3}{4}$ $a_9 = -\dfrac{6561}{16,384}$

43. a_{20} when $a_1 = 500$ and $r = -\dfrac{1}{2}$ $a_{20} = -\dfrac{125}{131,072}$

44. a_{20} when $a_1 = 1000$ and $r = -\dfrac{1}{10}$ $a_{20} = -\dfrac{1}{10^{16}}$

In Exercises 45–50, find the sum S_n of the first n terms of each geometric sequence.

45. $\dfrac{1}{10}, \dfrac{1}{2}, \dfrac{5}{2}, \dfrac{25}{2}, \ldots; n = 10$ **46.** $6, 2, \dfrac{2}{3}, \dfrac{2}{9}, \ldots; n = 10$

47. $\dfrac{1}{25}, -\dfrac{1}{5}, 1, -5, \ldots; n = 12$ $S_{12} = \dfrac{40,690,104}{25}$

48. $-10, \dfrac{1}{10}, \dfrac{-1}{1000}, \dfrac{1}{100,000}, \ldots; n = 12$ $S_{12} = -\dfrac{10^{24} - 1}{(101)(10)^{21}}$

49. $5, \dfrac{5}{4}, \dfrac{5}{4^2}, \dfrac{5}{4^3}, \ldots; n = 8$ $S_8 = \dfrac{109,225}{16,384}$

50. $2, \dfrac{2}{5}, \dfrac{2}{5^2}, \dfrac{2}{5^3}, \ldots; n = 8$ $S_8 = \dfrac{195,312}{78,125}$

In Exercises 51–58, find each sum.

51. $\displaystyle\sum_{i=1}^{5} \left(\dfrac{1}{2}\right)^{i-1}$ $\dfrac{31}{16}$ **52.** $\displaystyle\sum_{i=1}^{5} \left(\dfrac{1}{5}\right)^{i-1}$ $\dfrac{781}{625}$

53. $\displaystyle\sum_{i=1}^{8} 3\left(\dfrac{2}{3}\right)^{i-1}$ $\dfrac{6305}{729}$ **54.** $\displaystyle\sum_{i=1}^{8} 2\left(\dfrac{3}{5}\right)^{i-1}$ $\dfrac{384,064}{78,125}$

55. $\displaystyle\sum_{i=3}^{10} \dfrac{2^{i-1}}{4}$ 255 **56.** $\displaystyle\sum_{i=3}^{10} \dfrac{5^{i-1}}{2}$ $1,220,700$

57. $\displaystyle\sum_{i=1}^{20} \left(-\dfrac{3}{5}\right)\left(-\dfrac{5}{2}\right)^{i-1}$ $35(2)^{19}$ **58.** $\displaystyle\sum_{i=1}^{20} \left(-\dfrac{1}{4}\right)(3^{2-i})$ $-\dfrac{3^{20} - 1}{8(3)^{18}}$
$\dfrac{3(5^{20} - 2^{20})}{}$

In Exercises 59–64, use a graphing calculator to find each sum.

59. $\dfrac{1}{5} + \dfrac{1}{10} + \dfrac{1}{20} + \cdots + \dfrac{1}{320}$ $\dfrac{127}{320}$

60. $2 + 6 + 18 + \cdots + 1458$ 2186

61. $\displaystyle\sum_{n=1}^{10} 3^{n-2}$ 9841.333 (rounded to three decimal places)   **62.** $\displaystyle\sum_{n=1}^{15} \left(\dfrac{1}{7}\right)^{n-1}$ 1.167 (rounded to three decimal places)

63. $\displaystyle\sum_{n=1}^{10} (-1)^{n-1} 3^{2-n}$ **64.** $\displaystyle\sum_{n=1}^{20} (-4)^{3-n}$ 12.800 (rounded to three decimal places)
2.250 (rounded to three decimal places)

In Exercises 65–74, find each sum.

65. $\dfrac{1}{3} + \dfrac{1}{9} + \dfrac{1}{27} + \dfrac{1}{81} + \cdots$ $\dfrac{1}{2}$

66. $\dfrac{5}{2} + \dfrac{5}{4} + \dfrac{5}{8} + \dfrac{5}{16} + \cdots$ 5

67. $-\dfrac{1}{2} + \dfrac{1}{4} - \dfrac{1}{8} + \dfrac{1}{16} - \cdots$ $-\dfrac{1}{3}$

68. $-\dfrac{3}{2} + \dfrac{3}{4} - \dfrac{3}{8} + \dfrac{3}{16} + \cdots$ -1

69. $8 - 2 + \dfrac{1}{2} - \dfrac{1}{8} + \cdots$ $\dfrac{32}{5}$

70. $1 - \dfrac{3}{5} + \dfrac{9}{25} - \dfrac{27}{125} + \cdots$ $\dfrac{5}{8}$

71. $\displaystyle\sum_{n=0}^{\infty} 5\left(\dfrac{1}{3}\right)^n$ $\dfrac{15}{2}$ **72.** $\displaystyle\sum_{n=0}^{\infty} 3\left(\dfrac{1}{4}\right)^n$ 4

73. $\displaystyle\sum_{n=0}^{\infty} \left(-\dfrac{1}{4}\right)^n$ $\dfrac{4}{5}$ **74.** $\displaystyle\sum_{n=0}^{\infty} \left(-\dfrac{1}{3}\right)^n$ $\dfrac{3}{4}$

Answers:

45. $S_{10} = \dfrac{1,220,703}{5}$ **46.** $S_{10} = \dfrac{59,048}{6561}$

In Exercises 75–80, find *n* for the given value of a_n in each geometric sequence.

75. $a_n = 512$; $2, 4, 8, \ldots$. $n = 9$

76. $a_n = \dfrac{3}{1024}$; $3, \dfrac{3}{2}, \dfrac{3}{4}, \ldots$. $n = 11$

77. $a_n = -\dfrac{1}{4096}$; $1, -\dfrac{1}{2}, \dfrac{1}{4}, \ldots$. $n = 13$

78. $a_n = -\dfrac{1}{2187}$; $3, -1, \dfrac{1}{3}, \ldots$. $n = 9$

79. $a_n = -\dfrac{5}{5,764,801}$; $5, -\dfrac{5}{7}, \dfrac{5}{49}, \ldots$. $n = 9$

80. $a_n = \dfrac{1}{1,000,000}$; $1000, 100, 10, \ldots$. $n = 10$

B EXERCISES Applying the Concepts

81. **Population growth.** The population in a small town is increasing at the rate of 3% per year. If the present population is 20,000, what will the population be at the end of five years? 23,185

82. **Growth of a zoo collection.** The butterfly collection at a local zoo started with only 25 butterflies. If the number of butterflies doubled each month, how many butterflies were in the collection after 12 months? 102,400

83. **Savings growth.** Ramón deposits $100 on the last day of each month into a savings account that pays 6% annually, compounded monthly. What is the balance in the account after 36 compounding periods? $3933.61

84. **Savings growth.** Kat deposits $300 semiannually into a savings account for her son. If the account pays 8% annually, compounded semiannually, what is the balance in the account after 20 compounding periods? $8933.42

85. **Ancestors.** Every person has two parents, four grand-parents, eight great-grandparents, and so on. Find the number of ancestors a person has in the tenth generation back. 1024

86. **Bacterial growth.** A colony of bacteria doubles in number every day. If there are 1000 bacteria now, how many will there be
    a. in seven days? $1000(2^{(7-1)}) = 64,000$
    b. in *n* days? $1000(2^{n-1})$

87. **Investment.** An actuarial firm budgets a new position at $36,000 for the first year and a 5% raise each year thereafter. Find the total compensation a successful applicant for the job would receive during the first 20 years of employment. $1,190,374.35

88. **Investment.** A realtor offers two choices of payment for a small condominium:
    a. Pay $4000 per month for the next 25 months.
    b. Pay 1¢ the first month, 2¢ the second month, 4¢ the third month, and so on for 25 months.
    Which option, a or b, is the better choice for the buyer? Explain your reasoning. Option a is better.

89. **Pendulum motion.** The distance traveled by any point on a certain pendulum is 20% less than in the preceding swing. If the length traveled by the pendulum bob during the first swing is 56.25 centimeters, find the total distance the bob has traveled at the end of the fifth swing. 189.09 cm

90. **Depreciating a tractor.** Every year a tractor loses 20% of the value it had at the beginning of the year. If the tractor now has a value of $120,000, what will its value be in five years? $39,321.60

91. **Rebounding ball.** A particular ball always rebounds $\dfrac{3}{5}$ the distance it falls. If the ball is dropped from a height of 5 meters, how far will it travel before coming to a stop? 12.5 m

92. **Rebounding ball.** Repeat Exercise 91 assuming that the ball is dropped from a height of 9 meters. 22.5 m

C EXERCISES Beyond the Basics

93. The accompanying figure shows the first six squares in an infinite sequence of successively smaller squares formed by connecting the midpoints of the sides of the preceding larger square. The area of the largest square is 1. Find the total area of the indicated sections that are shaded in every other square as the number of squares increases to infinity. $\dfrac{2}{3}$

94. The accompanying figure shows the first three equi-lateral triangles in an infinite sequence of successively smaller equilateral triangles formed by connecting the midpoints of the sides of the preceding larger triangle. The largest triangle has sides of length 4. Find the total perimeter of all of the triangles. 24

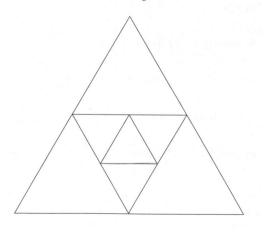

95. Prove that if $a_1, a_2, a_3, \ldots$ is a geometric sequence, then the sequence $\ln a_1, \ln a_2, \ln a_3, \ldots$ is an arithmetic sequence.

96. Prove that if $a_1, a_2, a_3, \ldots$ is a geometric sequence, then $a_1^2, a_2^2, a_3^2, \ldots$ is also a geometric sequence.

97. Show that if $a_1, a_2, a_3, \ldots$ is a geometric sequence with common ratio r, then $\dfrac{1}{a_1}, \dfrac{1}{a_2}, \dfrac{1}{a_3}, \ldots$ is also a geometric sequence. Also find its common ratio.

98. Show that if c is any nonzero real number, then $c^2, c, 1, \dfrac{1}{c}, \ldots$ is a geometric sequence.

99. Show that if x is any nonzero real number, then $x, 2, \dfrac{4}{x}, \dfrac{8}{x^2}, \ldots$ is a geometric sequence.

100. Show that if a and x are any two nonzero real numbers, then $\dfrac{x}{a}, -1, \dfrac{a}{x}, -\dfrac{a^2}{x^2}, \ldots$ is a geometric sequence.

101. Show that if a, x, and y are any three nonzero numbers, then $\dfrac{a}{x}, -\dfrac{a}{xy}, \dfrac{a}{xy^2}, \dfrac{-a}{xy^3}, \ldots$ is a geometric sequence.

102. Show that if $a_1, a_2, a_3, \ldots$ is an arithmetic sequence with common difference d, then $2^{a_1}, 2^{a_2}, 2^{a_3}, \ldots$ is a geometric sequence. Also find its common ratio.

Critical Thinking

103. If a_n denotes the nth term of an arithmetic sequence with $a_1 = 10$ and $d = 2.7$ and if b_n denotes the nth term of a geometric sequence with $b_1 = 10$ and $r = 2$, which number is larger, a_{1001} or b_{1001}? $\ b_{1001}$ is larger.

104. If the sum $\dfrac{a(1 - r^n)}{1 - r}$ of the first n terms of a geometric sequence is 1023, find a when $n = 10$ and $r = \dfrac{1}{2}$. $\ a = 512$

105. Find n and k so that the sum $5 + 5 \cdot 2 + 5 \cdot 2^2 + \cdots + 5 \cdot 2^{15}$ can be written as $\displaystyle\sum_{i=0}^{n} 5 \cdot 2^i$ or as $\displaystyle\sum_{i=1}^{k} 5 \cdot 2^{i-1}$. $\ n = 15, \ k = 16$

Group Projects

106. The sum of three consecutive terms of a geometric sequence is 35, and their product is 1000. Find the numbers. 5, 10, and 20

[*Hint:* Let $\dfrac{a}{r}$, a, and ar be the three terms of geometric sequence.]

107. The sum of three consecutive terms of an arithmetic sequence is 15. If 1, 4, and 19 are respectively added to the three terms, the resulting numbers form three consecutive terms of a geometric sequence. Find the numbers. 2, 5, and 8; or 26, 5, and -16

Mathematical Induction

Before Starting this Section, Review	Objective
1. Natural numbers (Appendix A, page 752)	**1** Prove statements by mathematical induction.
2. Inequalities	
3. Exponents	
4. Series (Section 10.1, page 688)	

JUMPING TO CONCLUSIONS; GOOD INDUCTION VERSUS BAD INDUCTION

A scientist had two large jars before him on the laboratory table. The jar on his left contained a hundred fleas; the jar on his right was empty. The scientist carefully lifted a flea from the jar on the left; placed the flea on the table between the two jars; stepped back; and said in a loud voice, "Jump!" The flea jumped and was put into the jar on the right. A second flea was carefully lifted from the jar on the left and placed on the table between the two jars. Again, the scientist stepped back and said in a loud voice, "Jump!" The flea jumped and was put into the jar on the right. In the same manner, the scientist treated each of the hundred fleas in the jar on the left, and each flea jumped as ordered. The two jars were then interchanged, and the experiment was continued with a slight difference. This time the scientist carefully lifted a flea from the jar on the left; *yanked off its hind legs;* placed the flea on the table between the jars; stepped back; and said in a loud voice, "Jump!" The flea did not jump and was put into the jar on the right. A second flea was carefully lifted from the jar on the left, its hind legs yanked off, and then placed on the table between the two jars. Again the scientist stepped back and said in a loud voice, "Jump!" The flea did not jump and was put into the jar on the right. In this manner, the scientist treated each of the hundred fleas in the jar on the left, and in no case did a flea jump when ordered. So the scientist recorded the following inductive statement in his notebook: "A flea, if its hind legs are yanked off, cannot hear." (*Source:* Modified from Howard Eves, *Mathematical Circles.*)

This, of course, is an example of truly bad reasoning. The principle of mathematical induction, demonstrated in Example 2, is the gold standard for reasoning in proving that statements about natural numbers are true. ■

1 Prove statements by mathematical induction.

Suppose your boss wants you to verify that each of a thousand bags contains two specific gold coins. There is only one way to be absolutely sure: You must check every bag. That would take a very long time, but you can eventually finish the job. However, consider the problem of verifying that a statement about the natural numbers such as

$$1^2 + 2^2 + 3^2 + \cdots + n^2 = \frac{n(n + 1)(2n + 1)}{6}$$

is true for all natural numbers n. Let's first test the statement for several values of n by comparing the value of $1^2 + 2^2 + 3^2 + \cdots + n^2$ in the second column of Table 10.1 with the value of $\dfrac{n(n + 1)(2n + 1)}{6}$ in the third column.

TABLE 10.1

n	$1^2 + 2^2 + 3^2 + \cdots + n^2$	$\dfrac{n(n+1)(2n+1)}{6}$	Equal?
1	$1^2 = 1$	$\dfrac{1(1+1)[2(1)+1]}{6} = 1$	Yes
2	$1^2 + 2^2 = 5$	$\dfrac{2(2+1)[2(2)+1]}{6} = 5$	Yes
3	$1^2 + 2^2 + 3^2 = 14$	$\dfrac{3(3+1)[2(3)+1]}{6} = 14$	Yes
4	$1^2 + 2^2 + 3^2 + 4^2 = 30$	$\dfrac{4(4+1)[2(4)+1]}{6} = 30$	Yes
5	$1^2 + 2^2 + 3^2 + 4^2 + 5^2 = 55$	$\dfrac{5(5+1)[2(5)+1]}{6} = 55$	Yes

We see that the statement is correct for the natural numbers 1, 2, 3, 4, and 5; furthermore, you can continue to verify that the statement is correct for any particular natural number n you try. But is it *always* correct? Clearly, we can't check every natural number!

The principle of *mathematical induction* provides a method for proving that statements about natural numbers are true for *all* natural numbers. The principle is based on the fact that after any natural number n, there is a next-larger natural number $n + 1$ and that any specified natural number can be reached by a finite number of such steps, starting from the natural number 1.

THE PRINCIPLE OF MATHEMATICAL INDUCTION

Let P_n be a statement that involves the natural number n with the following properties:

1. P_1 is true (the statement is true for the natural number 1).
2. And if P_k is a true statement, then P_{k+1} is a true statement.

Then the statement P_n is true for every natural number n.

A common physical interpretation is helpful in understanding why this principle works. Imagine an unending line of dominoes, as shown in Figure 10.2. Suppose the dominoes are arranged so that if *any* domino is knocked down, it will knock down the next domino behind it as well. What will happen if the first domino is knocked down? *All* of the dominoes will fall because

1. knocking down the first domino causes the second domino to be knocked down.
2. when the second domino is knocked down, it knocks down the third domino, and so on.

Determining the Statement P_{k+1} from the Statement P_k

To successfully use the Principle of Mathematical Induction, you must be able to determine the statement P_{k+1} from a given statement P_k. Suppose the given statement is

$$P_k: k \geq 1;$$

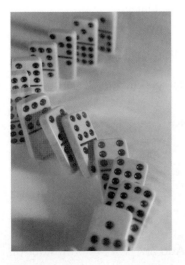

FIGURE 10.2

then we have

$$P_{k+1}: k + 1 \geq 1. \qquad \text{Replace } k \text{ with } k + 1 \text{ in } P_k.$$

It is important to recognize that since P_k says "k is greater than or equal to 1," the statement P_{k+1} says, "$k + 1$ is greater than or equal to 1." That is, P_{k+1} asserts the same property for $k + 1$ that P_k asserts for k.

EXAMPLE 1   Determining P_{k+1} from P_k

Find the statement P_{k+1} from the given statement P_k.

a. $P_k: k < 2^k$ **b.** $P_k: S_k = 3k^2 + 9$

c. $P_k: 1 + 2 + 2^2 + 2^3 + \cdots + 2^{k-1} = 2^k - 1$

d. $P_k: 3 + 6 + 9 + 12 + \cdots + 3k = \dfrac{3k(k + 1)}{2}$

SOLUTION

a. $P_k: k < 2^k, \quad P_{k+1}: k + 1 < 2^{k+1}$ Replace k with $k + 1$.

b. $P_k: S_k = 3k^2 + 9, \quad P_{k+1}: S_{k+1} = 3(k + 1)^2 + 9$ Replace k with $k + 1$.

c. $P_k: 1 + 2 + 2^2 + 2^3 + \cdots + 2^{k-1} = 2^k - 1$

$\quad\ P_{k+1}: 1 + 2 + 2^2 + 2^3 + \cdots + 2^{(k+1)-1} = 2^{k+1} - 1$ Replace k with $k + 1$.

d. $P_k: 3 + 6 + 9 + 12 + \cdots + 3k = \dfrac{3k(k + 1)}{2}$

$\quad\ P_{k+1}: 3 + 6 + 9 + 12 + \cdots + 3(k + 1) = \dfrac{3(k + 1)[(k + 1) + 1]}{2}$ Replace k with $k + 1$.

▪ ▪ ▪

Practice Problem 1   Find P_{k+1} for $P_k: (k + 3)^2 > k^2 + 9$.   ▪

EXAMPLE 2   Using Mathematical Induction

Use mathematical induction to prove that, for all natural numbers n,

$$2 + 4 + 6 + \cdots + 2n = n(n + 1).$$

SOLUTION

First, we verify that this statement is true for $n = 1$.

Check: $2(1) = 1(1 + 1)$ Replace n with 1 in the original statement.

          $2 = 2.$ ✓

The given statement is true for $n = 1$, so the first condition for the principle of mathematical induction holds.

    The second condition for the principle of mathematical induction requires two steps.

Step 1   Assume that the formula is true for some unspecified natural number k:

$$P_k: 2 + 4 + 6 + \cdots + 2k = k(k + 1) \qquad \text{Assumed true}$$

Step 2   On the basis of the assumption that P_k is true, show that P_{k+1} is true, which is the same as showing that

$$P_{k+1}: 2 + 4 + 6 + \cdots + 2(k + 1) = (k + 1)[(k + 1) + 1]. \qquad \text{Replace } k \text{ with } k + 1 \text{ in } P_k.$$

Begin by using P_k, the statement assumed to be true. Add $2(k + 1)$ to both sides of P_k in Step 1, which again results in a true statement.

Francesco Maurolico

(1494–1575)

The first known use of mathematical induction is in the work of the sixteenth-century mathematician Francesco Maurolico. Maurolico wrote extensively on the works of classical mathematics and made many contributions to geometry and optics. In his book *Arithmeticorum Libri Duo*, Maurolico presented a variety of properties of the integers, together with proofs of these properties. To prove some of the properties, he devised the method of mathematical induction. His first use of mathematical induction in this book was to prove that the sum of the first n odd positive integers equals n^2.

$$2 + 4 + 6 + \cdots + 2k = k(k + 1) \qquad \text{Assumed true}$$

$$2 + 4 + 6 + \cdots + 2k + 2(k + 1) = k(k + 1) + 2(k + 1) \qquad \text{Add } 2(k + 1) \text{ to both sides.}$$

$$2 + 4 + 6 + \cdots + 2k + 2(k + 1) = (k + 1)(k + 2) \qquad \text{Factor out the common factor } (k + 1) \text{ on the right.}$$

> It is not *necessary* to write $2k$ here because it is understood that $2k$ is the even integer that precedes $2(k + 1)$.

$$2 + 4 + 6 + \cdots + 2(k + 1) = (k + 1)[(k + 1) + 1] \qquad \text{Rewrite } (k + 2) \text{ as } (k + 1) + 1.$$

This last equation says that P_{k+1} is true if P_k is assumed to be true. Therefore, by the principle of mathematical induction, the statement

$$2 + 4 + 6 + \cdots + 2n = n(n + 1)$$

is true for every natural number n.   ■ ■ ■

Practice Problem 2   Use mathematical induction to prove that for all natural numbers n,

$$1 + 2 + 3 + \cdots + n = \frac{n(n + 1)}{2}.$$

■

EXAMPLE 3   Using Mathematical Induction

Use mathematical induction to prove that

$$2^n > n$$

for all natural numbers n.

SOLUTION

First, we show that the given statement is true for $n = 1$.

$$2^1 > 1 \qquad \text{Replace } n \text{ with 1 in the original statement.}$$

Thus, the inequality is true for $n = 1$.

Next, we assume that for some unspecified natural number k,

$$P_k: 2^k > k \quad \text{is true.}$$

Then we must use P_k to prove that P_{k+1} is true; that is,

$$P_{k+1}: 2^{k+1} > k + 1.$$

Now, by the product rule of exponents, $2^{k+1} = 2^k \cdot 2^1 = 2^k \cdot 2$; so we can get some information about 2^{k+1} by multiplying both sides of $P_k: 2^k > 2$ by 2.

$$2^k > k \qquad\qquad\qquad P_k \text{ is assumed true.}$$

$$2^k \cdot 2 > 2 \cdot k \qquad\qquad\quad \text{Multiply both sides by 2.}$$

$$2^{k+1} = 2^k \cdot 2 > 2k = k + k \geq k + 1 \qquad \text{Since } k \geq 1, k + k \geq k + 1.$$

Thus, $2^{k+1} > k + 1$ is true.

By the principle of mathematical induction, the statement

$$2^n > n$$

is true for every natural number n.   ■ ■ ■

Practice Problem 3   Use mathematical induction to prove that $3^n > n$ for all natural numbers n.   ■

SECTION 10.4 ■ Exercises

A EXERCISES Basic Skills and Concepts

1. Mathematical induction can only be used to prove statements about the <u>natural numbers</u>.

2. The first step in a mathematical induction proof that a statement P_n is true for all natural numbers n is that <u>P_1 is true</u>.

3. The second step in a mathematical induction proof is to assume that P_k is true for a natural number k and then show that <u>P_{k+1} is true</u>.

4. *True or False* If P_n is a statement about natural numbers and there is exactly one natural number, r for which P_r is false, then P_n is false. True

5. *True or False* The statement $e^n \geq n$ cannot be proved by mathematical induction because e is not a natural number.
   False

6. *True or False* The first step in proving that $2n \neq n$ for all natural numbers n is to note that $2 \neq 1$. True

In Exercises 7–10, find P_{k+1} from the given statement P_k.

7. $P_k: (k + 1)^2 - 2k = k^2 + 1$ $\begin{aligned}P_{k+1}: &(k + 2)^2 - 2(k + 1) \\ &= (k + 1)^2 + 1\end{aligned}$

8. $P_k: (1 + k)(1 - k) = 1 - k^2$ $P_{k+1}: (2 + k)(-k) = 1 - (k + 1)^2$

9. $P_k: 2^k > 5k$ $P_{k+1}: 2^{k+1} > 5(k + 1)$

10. $P_k: 1 + 3 + 5 + \cdots + (2k - 1) = k^2$
    $P_{k+1}: 1 + 3 + \cdots + (2k - 1)(2k + 1) = (k + 1)^2$

In Exercises 11–34, use mathematical induction to prove that each statement is true for all natural numbers n.

11. $2 + 4 + 6 + \cdots + 2n = n(n + 1)$

12. $1 + 3 + 5 + \cdots + (2n - 1) = n^2$

13. $4 + 8 + 12 + \cdots + 4n = 2n(n + 1)$

14. $3 + 6 + 9 + \cdots + 3n = \dfrac{3n(n + 1)}{2}$

15. $1 + 5 + 9 + \cdots + (4n - 3) = n(2n - 1)$

16. $3 + 8 + 13 + \cdots + (5n - 2) = \dfrac{n(5n + 1)}{2}$

17. $3 + 9 + 27 + \cdots + 3^n = \dfrac{3(3^n - 1)}{2}$

18. $5 + 25 + 125 + \cdots + 5^n = \dfrac{5(5^n - 1)}{4}$

19. $\dfrac{1}{1 \cdot 2} + \dfrac{1}{2 \cdot 3} + \dfrac{1}{3 \cdot 4} + \cdots + \dfrac{1}{n(n + 1)} = \dfrac{n}{n + 1}$

20. $\dfrac{1}{2 \cdot 4} + \dfrac{1}{4 \cdot 6} + \dfrac{1}{6 \cdot 8} + \cdots + \dfrac{1}{2n(2n + 2)} = \dfrac{n}{4(n + 1)}$

21. $2 \leq 2^n$ 22. $2n + 1 \leq 3^n$

23. $n(n + 2) < (n + 1)^2$ 24. $n \leq n^2$

25. $\dfrac{n!}{n} = (n - 1)!$ 26. $\dfrac{n!}{(n + 1)!} = \dfrac{1}{n + 1}$

27. $1^2 + 2^2 + 3^2 + \cdots + n^2 = \dfrac{n(n + 1)(2n + 1)}{6}$

28. $1^3 + 2^3 + 3^3 + \cdots + n^3 = \dfrac{n^2(n + 1)^2}{4}$

29. 2 is a factor of $n^2 + n$.

30. 2 is a factor of $n^3 + 5n$.

31. 6 is a factor of $n(n + 1)(n + 2)$.

32. 3 is a factor of $n(n + 1)(n - 1)$.

33. $(ab)^n = a^n b^n$ 34. $\left(\dfrac{a}{b}\right)^n = \dfrac{a^n}{b^n}$

B EXERCISES Applying the Concepts

35. **Counting hugs.** At a family reunion of n people ($n \geq 2$), each person hugs everyone else. Use mathematical induction to show that the number of hugs is $\dfrac{n^2 - n}{2}$.

36. **Winning prizes.** In a contest in which n prizes are possible, a contestant can win any number of the prizes (from 0 to n). Use mathematical induction to show that there are 2^n possible outcomes for the sets of prizes the contestant might win.
    [*Hint:* If $n = 1$, there are 2 ($= 2^1$) outcomes: winning 0 prizes or winning the one prize.]

37. **Koch's snowflake.** A geometric shape called the Koch snowflake can be formed by starting with an equilateral triangle, as in the figure. Assume that each side of the triangle has length 1. On the middle part of each side, construct an equilateral triangle with sides of length $\dfrac{1}{3}$ and erase the side of the smaller triangle that is on the side of the original triangle. Repeat this procedure for each of the smaller triangles. This process produces a figure that resembles a snowflake.
    a. Find a formula for the number of sides of the nth figure. Use mathematical induction to prove that this formula is correct. The number of sides of the nth figure is $3(4^{n-1})$.
    b. Find a formula for the perimeter of the nth figure. Use mathematical induction to prove that this formula is correct.
       The perimeter of the nth figure is $3\left(\dfrac{4}{3}\right)^{n-1}$.

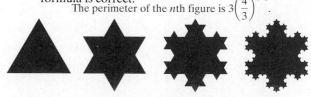

38. **Sierpinski's triangle.** A geometric figure known as Sierpinski's triangle is constructed by starting with an equilateral triangle as in the figure. Assume that each side of the triangle has length 1. Inside the original triangle, draw a second triangle by connecting the midpoints of the sides of the original triangle. This divides the original triangle into four identical

---

†Due to space constrictions, answers to these exercises may be found in the Answers beginning on page A–1 in the back of the book.

equilateral triangles. Repeat the process by constructing equilateral triangles inside each of the four smaller triangles, again using the midpoints of the sides of the triangle as vertices. Continuing this process with each group of smaller triangles produces Sierpinski's triangle. Coloring each set of smaller triangles helps in following the construction.

a. When a process is repeated over and over, each repetition is called an **iteration**. How many triangles will the fourth iteration have? 40

b. How many triangles will the fifth iteration have? 121

c. Find a formula for the number of triangles in the *n*th iteration. Use mathematical induction to prove that this formula is correct.

39. Towers of Hanoi. Three pegs are attached vertically to a horizontal board, as shown in the figure. One peg has *n* rings stacked on it, each ring smaller than the one below it. In a game known as the Tower of Hanoi puzzle, all of the rings must be moved to a different peg, with only one ring moved at a time, and no ring can be moved on top of a smaller ring. Determine the least number of moves that will accomplish this transfer. Use mathematical induction to prove that your answer is correct.

40. Exam answer sheets. An exam in which each question is answered with either "True" or "False" has *n* questions.

a. If every question is answered, how many answer sheets are possible
   (i) if $n = 1$?
   (ii) if $n = 2$?

b. Find a formula for the number of possible answer sheets for an exam with *n* questions (for any natural number *n*) and use mathematical induction to prove that the formula is correct.

Answers:

38. c. The number of triangles after the *n*th iteration is $\dfrac{3^n - 1}{2}$.

39. The smallest number of moves to accomplish the transfer is $2^n - 1$.
40. a. (i) 2   **(ii)** 4   **b.** The number of answer sheets is 2^n.

C EXERCISES Beyond the Basics

In Exercises 41–49, use mathematical induction to prove each statement for all natural numbers *n*.

41. 5 is a factor of $8^n - 3^n$.

42. 24 is a factor of $5^{2n} - 1$.

43. 64 is a factor of $3^{2n+2} - 8n - 9$.

44. 64 is a factor of $9^n - 8n - 1$.

45. 3 is a factor of $2^{2n+1} + 1$.

46. 5 is a factor of $2^{4n} - 1$.

47. $a - b$ is a factor of $a^n - b^n$.
   [*Hint*: $a^{k+1} - b^{k+1} = a(a^k - b^k) + b^k(a - b)$.]

48. If $a \neq 1$, then $1 + a + a^2 + \cdots + a^{n-1} = \dfrac{a^n - 1}{a - 1}$.

49. $\displaystyle\sum_{k=1}^{n+1} \frac{1}{n + k} \leq \frac{5}{6}$

Critical Thinking

50. Consider the statement "2 is a factor of $4n - 1$ for all natural numbers *n*." Then P_k: 2 is a factor of $4k - 1$.
   Assume that P_k is true so that $4k - 1 = 2m$ for some integer *m*. Then

$$\begin{aligned} P_{k+1} &= 4(k + 1) - 1 \\ &= 4k + 4 - 1 \\ &= (4k - 1) + 4 \\ &= 2m + 4 \\ &= 2(m + 2) \end{aligned}$$

   Thus, if P_k is true, then P_{k+1} is also true. However, 2 is never a factor of $4n - 1$ because $4n - 1$ is always an odd integer. Explain why this "proof" is not valid.

Extended principle of mathematical induction. **This principle states that if a statement P_n about natural numbers satisfies the two conditions**

(1) P_m is true for some natural number *m*

(2) P_k is true implies that P_{k+1} is also true, then P_n is true for all natural number *n*, with $n \geq m$.

In Exercises 51 and 52, use the extended principle of mathematical induction to prove the statement.

51. Prove that $2^n > n^2$ for all natural numbers $n, n > 4$.

52. Prove that $n! > n^3$ for all natural numbers $n, n \geq 6$.

The Binomial Theorem

Before Starting this Section, Review

1. Special products (Appendix A, page 764)
2. Factorials (Section 10.1, page 685)
3. Summation notation (Section 10.1, page 687)

Objectives

1. Use Pascal's Triangle to compute binomial coefficients.
2. Use Pascal's Triangle to expand a binomial power.
3. Use the Binomial Theorem to expand a binomial power.
4. Find the coefficient of a term in a binomial expansion.

Blaise Pascal (1623–1662)

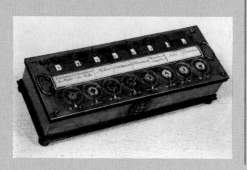

Pascaline

BLAISE PASCAL

Blaise Pascal was a French mathematician. An evident genius, he was 14 when he began to accompany his father to gatherings of mathematicians that were arranged by Father Marin Mersenne. At the age of 16, Pascal presented his own results at one of Mersenne's meetings.

Pascal's desire to help his father with his work collecting taxes led him to invent the first digital calculator, called the Pascaline. Pascal was also interested in atmospheric pressure, and in 1648, he observed that the pressure of the atmosphere decreased with height and that a vacuum existed above the atmosphere.

Pascal's intense interest in mathematics led him to produce important results related to conic sections and to engage in correspondence with Fermat in which he formulated the foundations for the *theory of probability*. Pascal died a painful death from cancer at the age of 39.

In this section, we study *Pascal's Triangle*. This triangle of numbers contains the important *binomial coefficients*. See Example 1. Pascal's work was influential in Newton's discovery of the general *Binomial Theorem* for fractional and negative powers. ■

Binomial Expansions Recall that a polynomial that has exactly two terms is called a *binomial*. In Appendix A, the formulas for expanding the binomial powers $(x + y)^2$ and $(x + y)^3$ are given. In the current section, we study a method for expanding $(x + y)^n$ for any positive integer n.

First consider these binomial expansions:

$$(x + y)^1 = x + y$$
$$(x + y)^2 = x^2 + 2xy + y^2$$
$$(x + y)^3 = x^3 + 3x^2y + 3xy^2 + y^3$$
$$(x + y)^4 = x^4 + 4x^3y + 6x^2y^2 + 4xy^3 + y^4$$
$$(x + y)^5 = x^5 + 5x^4y + 10x^3y^2 + 10x^2y^3 + 5xy^4 + y^5$$

The expansions of $(x + y)^2$ and $(x + y)^3$ should be familiar to you; the last three are left for you to verify. Each is a product of the previous binomial and $(x + y)$. For example, $(x + y)^4 = (x + y)^3(x + y)$.

The patterns for expansions of $(x + y)^n$ (with $n = 1, 2, 3, 4, 5$) suggest the following:

1. The expansion of $(x + y)^n$ has $n + 1$ terms.
2. The sum of the exponents on x and y in each term equals n.
3. The exponent on x starts at n ($x^n = x^n \cdot y^0$) in the first term and decreases by 1 for each term until it is 0 in the last term ($x^0 \cdot y^n = y^n$).
4. The exponent on y starts at 0 ($x^n = x^n \cdot y^0$) in the first term and increases by 1 for each term until it is n in the last term ($x^0 \cdot y^n = y^n$).
5. The variables x and y have symmetrical roles. That is, replacing x with y and y with x in the expansion of $(x + y)^n$ yields the same terms, just in a different order.

You may also have noticed that the coefficients of the first and last terms are both 1 and the coefficients of the second and the next-to-last terms are equal. In general, the coefficients of

$$x^{n-j}y^j \qquad \text{and} \qquad x^j y^{n-j}$$

are equal for $j = 0, 1, 2, \ldots, n$.

The coefficients in a binomial expansion of $(x + y)^n$ are called the **binomial coefficients**.

1 Use Pascal's Triangle to compute binomial coefficients.

Pascal's Triangle

As early as A.D. 1100, the Chinese scholar Chia Hsien had discovered the secret of the binomial coefficients that was later rediscovered by the French philosopher and mathematician Blaise Pascal (1623–1662). To understand Chia Hsien's and Pascal's construction of binomial coefficients, let's first look at the coefficients in these binomial expansions:

$$
\begin{aligned}
(x + y)^0 &= 1 \\
(x + y)^1 &= 1x + 1y \\
(x + y)^2 &= 1x^2 + 2xy + 1y^2 \\
(x + y)^3 &= 1x^3 + 3x^2y + 3xy^2 + 1y^3 \\
(x + y)^4 &= 1x^4 + 4x^3y + 6x^2y^2 + 4xy^3 + 1y^4 \\
(x + y)^5 &= 1x^5 + 5x^4y + 10x^3y^2 + 10x^2y^3 + 5xy^4 + 1y^5
\end{aligned}
$$

If we remove the variables and the plus signs and list only the coefficients, we get a triangle of numbers composed of the binomial coefficients. This triangle of numbers is known as *Pascal's Triangle*.

Note the symmetry in Pascal's Triangle. If the triangle were folded vertically down the middle, the numbers on each side of the crease would match. To create a new bottom row in the triangle, put the number 1 in the first and last places of the new row and add two neighboring entries in the previous row.

The top row is called the *zeroth row* because it corresponds to the binomial expansion of $(x + y)^0$. The next row is called the *first row* because it corresponds to the binomial expansion of $(x + y)^1$. All of the rows are named so that the *n*th *row* corresponds to the coefficients of $(x + y)^n$.

2 Use Pascal's Triangle to expand a binomial power.

EXAMPLE 1   **Using Pascal's Triangle to Expand a Binomial Power**

Expand $(4y - 2x)^5$.

SOLUTION

From the fifth row of Pascal's Triangle, we see that the binomial coefficients are

$$1, 5, 10, 10, 5, 1.$$

We must make some changes in the expansion

$$(x + y)^5 = x^5 + 5x^4y + 10x^3y^2 + 10x^2y^3 + 5xy^4 + y^5$$

to get the expansion for $(4y - 2x)^5$:

1. Replace x with $4y$.

2. Replace y with $-2x$.

$$\begin{aligned}(4y - 2x)^5 &= [4y + (-2x)]^5 = (4y)^5 + 5(4y)^4(-2x) \\ &\quad + 10(4y)^3(-2x)^2 + 10(4y)^2(-2x)^3 + 5(4y)(-2x)^4 + (-2x)^5 \\ &= 1024y^5 - 2560y^4x + 2560y^3x^2 - 1280y^2x^3 + 320yx^4 - 32x^5\end{aligned}$$

We note that expanding a *difference* results in alternating signs between terms.   ■ ■ ■

Practice Problem 1   Expand $(3y - x)^6$.   ■

3 Use the Binomial Theorem to expand a binomial power.

The coefficients in a binomial expansion can be computed by using ratios of certain factorials. We first introduce the symbol $\binom{n}{r}$.

DEFINITION OF $\binom{n}{r}$

If r and n are integers with $0 \le r \le n$, then we define

$$\binom{n}{r} = \frac{n!}{r!(n - r)!}$$

TECHNOLOGY CONNECTION

📉 Most graphing calculators can compute the binomial coefficients $\binom{n}{r}$. They frequently use the symbol *nCr*. Note that $\binom{5}{3}$ is shown as 5 nCr 3 on the viewing screen.

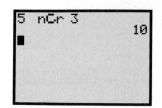

EXAMPLE 2   **Evaluating** $\binom{n}{r}$

Evaluate each binomial coefficient.

a. $\binom{4}{1}$ **b.** $\binom{5}{3}$ **c.** $\binom{9}{0}$ **d.** $\binom{35}{35}$

SOLUTION

a. $\binom{4}{1} = \dfrac{4!}{1!(4 - 1)!} = \dfrac{4!}{1!\,3!} = \dfrac{4 \cdot 3 \cdot 2 \cdot 1}{1(3 \cdot 2 \cdot 1)} = \dfrac{4}{1} = 4$

b. $\binom{5}{3} = \dfrac{5!}{3!(5 - 3)!} = \dfrac{5!}{3!\,2!} = \dfrac{5 \cdot 4 \cdot 3!}{3!\,2!} = \dfrac{5 \cdot 4}{2} = 5 \cdot 2 = 10$

c. $\dbinom{9}{0} = \dfrac{9!}{0!(9-0)!} = \dfrac{9!}{0!\,9!} = \dfrac{1}{1} = 1$ Recall that $0! = 1$.

d. $\dbinom{35}{35} = \dfrac{35!}{35!(35-35)!} = \dfrac{35!}{35!\,0!} = \dfrac{1}{1} = 1$ ■ ■ ■

Practice Problem 2   Evaluate each binomial coefficient.

a. $\dbinom{6}{2}$ **b.** $\dbinom{12}{9}$ ■

Examples 2(c) and 2(d) generalize for $n \geq 0$.

$$\dbinom{n}{0} = 1 \quad \text{and} \quad \dbinom{n}{n} = 1$$

The symbol $\dbinom{n}{r}$ is read "n choose r." It can be shown that $\dbinom{n}{r}$ is the number of ways of choosing a subset containing exactly r elements from a set with n elements.

The numbers $\dbinom{n}{r}$ show up as the coefficients in the expansion of a binomial power. For example, $(x + y)^4$ can be written as either

$$(x + y)^4 = x^4 + 4x^3y + 6x^2y^2 + 4xy^3 + y^4$$

or

$$(x + y)^4 = \dbinom{4}{0}x^4 + \dbinom{4}{1}x^3y + \dbinom{4}{2}x^2y^2 + \dbinom{4}{3}xy^3 + \dbinom{4}{4}y^4.$$

This result is the **Binomial Theorem** (for $n = 4$), which can be proved by mathematical induction. It provides an efficient method for expanding a binomial power. (See Exercise 64.) The Binomial Theorem can be used to expand a binomial power directly, without reference to Pascal's Triangle. This technique is particularly useful in expanding large powers of a binomial. For example, the expansion of $(x + y)^{20}$ would require that you produce 20 rows of the Pascal Triangle.

THE BINOMIAL THEOREM

If n is a natural number, then the binomial expansion of $(x + y)^n$ is given by

$$(x + y)^n = \dbinom{n}{0}x^n + \dbinom{n}{1}x^{n-1}y + \dbinom{n}{2}x^{n-2}y^2 + \cdots + \dbinom{n}{r}x^{n-r}y^r + \cdots + \dbinom{n}{n}y^n$$

$$= \sum_{r=0}^{n} \dbinom{n}{r}x^{n-r}y^r.$$

The coefficient $\dbinom{n}{r}$ of $x^{n-r}y^r$ is $\dfrac{n!}{r!(n-r)!}$.

EXAMPLE 3   Expanding a Binomial Power by Using the Binomial Theorem

Find the binomial expansion of $(x - 3y)^4$.

SOLUTION

$$(x - 3y)^4 = [x + (-3y)]^4$$

$$= \binom{4}{0}x^4 + \binom{4}{1}x^3(-3y) + \binom{4}{2}x^2(-3y)^2 + \binom{4}{3}x(-3y)^3 + \binom{4}{4}(-3y)^3$$

$$= \frac{4!}{0!\,4!}x^4 + \frac{4!}{1!\,3!}x^3(-3y) + \frac{4!}{2!\,2!}x^2(-3y)^2 + \frac{4!}{3!\,1!}x(-3y)^3 + \frac{4!}{4!\,0!}(-3y)^4$$

$$= x^4 + \frac{4\cdot 3!}{1!\,3!}x^3(-3y) + \frac{4\cdot 3\cdot 2!}{2!\,2!}x^2(-3y)^2 + \frac{4\cdot 3!}{3!\,1!}x(-3y)^3 + (-3y)^4$$

$$= x^4 + 4x^3(-3y) + \frac{4\cdot 3}{2!}x^2(9y)^2 + 4x(-27y^3) + (81y^4)$$

$$= x^4 - 12x^3y + 54x^2y^2 - 108xy^3 + 81y^4$$

■ ■ ■

Practice Problem 3   Find the binomial expansion of $(3x - y)^4$.   ■

④ Find the coefficient of a term in a binomial expansion.

The Binomial Theorem is also useful for finding a particular term and its coefficient in a binomial expansion.

EXAMPLE 4   Finding a Particular Coefficient in a Binomial Expansion

Find the coefficient of x^9y^3 in the expansion of $(x + y)^{12}$.

SOLUTION

The form of the expansion is.

$$(x + y)^{12} = \binom{12}{0}x^{12} + \binom{12}{1}x^{11}y + \binom{12}{2}x^{10}y^2 + \binom{12}{3}x^9y^3 + \cdots + \binom{12}{11}xy^{11} + \binom{12}{12}y^{12}.$$

From the fourth term, $\binom{12}{3}x^9y^3$, the coefficient of x^9y^3 is

$$\binom{12}{3} = \frac{12!}{3!(12 - 3!)} = \frac{12!}{3!\,9!} = \frac{12\cdot 11\cdot 10\cdot 9!}{3!\,9!} = 220.$$

■ ■ ■

Practice Problem 4   Find the coefficient of x^3y^9 in the expansion of $(x + y)^{12}$.   ■

The method illustrated in Example 4 allows us to find any particular term in a binomial expansion without writing out the complete expansion.

PARTICULAR TERM IN A BINOMIAL EXPRESSION

The term containing the factor x^r in the expansion of $(x + y)^n$ is

$$\binom{n}{n - r}x^ry^{n-r}.$$

EXAMPLE 5   **Finding a Particular Term in a Binomial Expansion**

Find the term containing x^{10} in the expansion of $(x + 2a)^{15}$.

SOLUTION

We begin with the formula for the term containing the factor x^r.

$$\binom{n}{n-r}x^r y^{n-r} = \binom{15}{15-10}x^{10}(2a)^{15-10}$$ Replace n with 15, r with 10, and y with $2a$.

$$= \binom{15}{5}x^{10}(2a)^5$$ Simplify.

$$= \frac{15!}{5!(15-5)!}x^{10}2^5 a^5$$ Recall that $\binom{n}{r} = \frac{n!}{r!(n-r)!}$.

$$= \frac{15!}{5!\,10!} \cdot 32x^{10}a^5$$ $2^5 = 32$

$$= \frac{15 \cdot 14 \cdot 13 \cdot 12 \cdot 11 \cdot \cancel{10!}}{5! \cdot \cancel{10!}} \cdot 32x^{10}a^5$$

$$= 96{,}096x^{10}a^5$$ Use a calculator. ■ ■ ■

Practice Problem 5   Find the term containing x^3 in the expansion of $(x + 2a)^{15}$. ■

SECTION 10.5 ■ Exercises

A EXERCISES Basic Skills and Concepts

1. The expansion of $(x + y)^n$ has ___ $n+1$ ___ terms.

2. In the expansion of $(x + y)^5$, the coefficient of $x^2 y^3$ is $\binom{n}{r}$ when $n = $ ___ 5 ___ and $r = $ ___ 3 ___.

3. Expanding a difference such as $(2x - y)^{10}$ results in ___ between terms. alternating signs

4. For any positive integer n, $\binom{n}{n} = $ ___ 1 ___.

5. *True or False* Every coefficient in the expansion of $(x + y)^n$, for $n > 1$, appears exactly twice. False

6. *True or False* If $n > 3$, then $2\binom{n}{2} = n$. False

In Exercises 7–16, evaluate each expression.

7. $\dfrac{6!}{3!}$ 120

8. $\dfrac{11!}{9!}$ 110

9. $\dfrac{12!}{11!}$ 12

10. $\dfrac{3!}{0!}$ 6

11. $\binom{6}{4}$ 15

12. $\binom{6}{3}$ 20

13. $\binom{9}{0}$ 1

14. $\binom{12}{0}$ 1

15. $\binom{7}{1}$ 7

16. $\binom{7}{3}$ 35

In Exercises 17–24, use Pascal's Triangle to expand each binomial.

17. $(x + 2)^4$ †

18. $(x + 3)^4$ †

19. $(x - 2)^5$ †

20. $(3 - x)^5$ †

21. $(2 - 3x)^3$ †

22. $(3 - 2x)^3$ †

23. $(2x + 3y)^4$ †

24. $(2x + 5y)^4$ †

In Exercises 25–46, use either the binomial theorem or Pascal's Triangle to expand each binomial.

25. $(x + 1)^4$ †

26. $(x + 2)^4$ †

27. $(x - 1)^5$ †

28. $(1 - x)^5$ †

29. $(y - 3)^3$ †

30. $(2 - y)^5$ †

31. $(x + y)^6$ †

32. $(x - y)^6$ †

33. $(1 + 3y)^5$ †

34. $(2x + 1)^5$ †

35. $(2x + 1)^4$ †

36. $(3x - 1)^4$ †

37. $(x - 2y)^3$ †

38. $(2x - y)^3$ †

39. $(2x + y)^4$ †

40. $(3x - 2y)^4$ †

41. $\left(\dfrac{x}{2} + 2\right)^7$ †

42. $\left(2 - \dfrac{x}{2}\right)^7$ †

43. $\left(a^2 - \dfrac{1}{3}\right)^4$ †

44. $\left(\dfrac{1}{2} - a^2\right)^4$ †

45. $\left(\dfrac{1}{x} + y\right)^3$ †

46. $\left(x + \dfrac{2}{y}\right)^3$ †

†Due to space constrictions, answers to these exercises may be found in the Answers beginning on page A–1 in the back of the book.

In Exercises 47–54, find the specified term.

47. $(x + y)^{10}$; term containing x^7 $120x^7y^3$

48. $(x + y)^{10}$; term containing y^7 $120x^3y^7$

49. $(x - 2)^{12}$; term containing x^3 $-112,640x^3$

50. $(2 - x)^{12}$; term containing x^3 $-112,640x^3$

51. $(2x + 3y)^8$; term containing x^6 $16,128x^6y^2$

52. $(2x + 3y)^8$; term containing y^6 $81,648x^2y^6$

53. $(5x - 2y)^{11}$; term containing y^9 $-704,000x^2y^9$

54. $(7x - y)^{15}$; term containing x^9 $201,969,803,035x^9y^6$

55. Find the value of $(1.2)^5$ by writing it in the form $(1 + 0.2)^5$ and applying the Binomial Theorem. 2.48832

56. Use the method in Exercise 55 to evaluate each expression.
   a. $(2.9)^4$ 70.7281   **b.** $(10.4)^3$ 1124.864

C EXERCISES Beyond the Basics

57. Find the middle term in the expansion of $\left(\sqrt{x} - \dfrac{2}{x^2} \right)^{10}$.

58. Find the middle term in the expansion of $\left(\sqrt{x} + \dfrac{1}{x^2} \right)^{10}$.

59. Find the middle term in the expansion of $(1 - x^2y^{-3})^{12}$.

60. Show that the middle term in the expansion of $(1 + x)^{2n}$ is

$$\frac{1 \cdot 3 \cdot 5 \cdots (2n - 1)}{n!} 2^n x^n.$$

61. Prove that $\dbinom{n}{0} + \dbinom{n}{1} + \dbinom{n}{2} + \cdots + \dbinom{n}{n} = 2^n$.

62. Prove that $\dbinom{n}{0} - \dbinom{n}{1} + \dbinom{n}{2} - \dbinom{n}{3} + \cdots$
   $+ (-1)^n \dbinom{n}{n} = 0$.

63. Prove that $\dbinom{k}{j} + \dbinom{k}{j - 1} = \dbinom{k + 1}{j}$.

64. Prove the Binomial Theorem by using the Principle of Mathematical Induction.

$\Bigg[$ *Hint:*

$(x + y)^{k+1} = (x + y)(x + y)^k = (x + y) \displaystyle\sum_{j=0}^{k} \binom{k}{j} x^{k-j} y^j$

$\quad = \displaystyle\sum_{j=0}^{k} \binom{k}{j} x^{k+1-j} y^j + \sum_{j=0}^{k} \binom{k}{j} x^{k-j} y^{j+1}. \Bigg]$

Use Exercise 63 to show that

$(x + y)^{k+1} = \displaystyle\sum_{j=0}^{k+1} \binom{k + 1}{j} x^{k+1-j} y^j. \Bigg]$

In Exercises 65–66, find the value of the expression without expanding any term.

65. $(2x - 1)^4 + 4(2x - 1)^3(3 - 2x) + 6(2x - 1)^2(3 - 2x)^2$
   $+ 4(2x - 1)(3 - 2x)^3 + (3 - 2x)^4$ 16

66. $(x + 1)^4 - 4(x + 1)^3(x - 1) + 6(x + 1)^2(x - 1)^2$
   $- 4(x + 1)(x - 1)^3 + (x - 1)^4$ 16

67. $(3x - 1)^5 + 5(3x - 1)^4(1 - 2x)$
   $+ 10(3x - 1)^3(1 - 2x)^2 + 10(3x - 1)^2(1 - 2x)^3$
   $+ 5(3x - 1)(1 - 2x)^4 + (1 - 2x)^5$ x^5

68. Find the constant term in each expansion.
   a. $\left(x^2 - \dfrac{1}{x} \right)^9$ 84   **b.** $\left(\sqrt{x} - \dfrac{2}{x^2} \right)^{10}$ 180

69. If the constant term in the expansion of $\left(kx - \dfrac{1}{x^2} \right)^6$ is 240, find k. $k = \pm 2$

70. If the constant term in the expansion of $\left(x^3 + \dfrac{k}{x^8} \right)^{11}$ is 1320, find k. $k = 2$

71. Show that there is no constant term in the expansion of $\left(2x^2 - \dfrac{1}{4x} \right)^{11}$.

Critical Thinking

72. Use the binomial expansion of $(x + y)^6$, with $x = 1$ and $y = 1$, to show that

$$2^6 = \binom{6}{0} + \binom{6}{1} + \binom{6}{2} + \binom{6}{3} + \binom{6}{4} + \binom{6}{5} + \binom{6}{6}.$$

73. Use the binomial expansion of $(x + y)^{10}$ to show that

$$\binom{10}{0} - \binom{10}{1} + \binom{10}{2} - \binom{10}{3} + \binom{10}{4} - \binom{10}{5}$$
$$+ \binom{10}{6} - \binom{10}{7} + \binom{10}{8} - \binom{10}{9} + \binom{10}{10} = 0.$$

74. Use the binomial expansion of $(x + y)^2$ to show that if $x > 0$ and $y > 0$, then $(x + y)^2 > x^2 + y^2$.

75. Use the binomial expansion of $(x + y)^n$ to show that if $x > 0$ and $y > 0$, then $(x + y)^n > x^n + y^n$.

76. Use the binomial expansion of $(1 + x)^n$ to show that if $x > 0$, then $(1 + x)^n > 1 + nx$.

Answers:

57. $-\dfrac{8064\sqrt{x}}{x^8}$ 58. $\dfrac{252\sqrt{x}}{x^8}$ 59. $924\dfrac{x^{12}}{y^{18}}$

Counting Principles

SOCIAL SECURITY
142-84-5194
THIS NUMBER HAS BEEN ESTABLISHED FOR
PATRICK MICHAEL GAFFNEY
SIGNATURE

Before Starting this Section, Review

1. Exponents (Appendix A, page 758)
2. Factorials (Section 10.1, page 685)

Objectives

1. Use the Fundamental Counting Principle.
2. Use the formula for permutations.
3. Use the formula for combinations.
4. Use the formula for distinguishable permutations.

THE MYSTERY OF SOCIAL SECURITY NUMBERS

Social Security numbers may seem to be assigned randomly, but they are not. Although the details are complicated, here is a brief introduction: First, the format for all Social Security numbers is NNN-NN-NNNN, where *N* must be one of the numbers 0, 1, 2, 3, 4, 5, 6, 7, 8, 9. U.S. citizens, permanent residents, and certain temporary residents are issued these numbers. For working people, the numbers help track tax contributions and Social Security benefits.

The first three digits in a Social Security number make up an *area number* assigned to a geographic location. The area number indicates the state in which the card was issued or the ZIP Code in the mailing address provided in the original application.

The middle two digits constitute a *group number* and are designed simply to break the Social Security number into convenient blocks for issuance. The exact procedure is not very enlightening.

The last four digits are *serial numbers* and are assigned in each group from 0001 through 9999.

Various restrictions limit the numbers that can be used. For example, numbers with all zeros in any of the digit blocks (000-xx-xxxx, xxx-00-xxxx, or xxx-xx-0000) are never issued. Even if there were no restrictions, at some point, there might be *more people than available numbers.* How many numbers can be issued in the Social Security format with no restrictions? The answer is 1,000,000,000 (1 billion). (The U.S. population in October 2006 was about 300,000,000 (300 million).)

In Example 3, you learn how to calculate the total number of possible Social Security numbers. ■

---

1. Use the Fundamental Counting Principle.

You may think that you already know all there is to know about counting. However, mathematicians have developed a variety of techniques that allow you to determine the number of items that are of interest to you without making the effort of listing and counting all of them. We start with an example that easily lists the items to be counted, but that also demonstrates a general counting procedure.

EXAMPLE 1   Counting Possible Car Selections

Suppose you are trying to decide between buying a sport-utility vehicle (SUV) and a four-door sedan. The SUV is available in black, red, and silver. The sedan is available in black, blue, and green. In how many ways can you choose a type of car and its color?

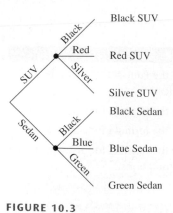

Black SUV
Red SUV
Silver SUV
Black Sedan
Blue Sedan
Green Sedan

FIGURE 10.3

SOLUTION

We can represent the possible solutions by using a *tree diagram*, as shown in Figure 10.3. There are two initial choices: a SUV and a sedan. For each of these two choices, there are three additional choices of color; so a total of $2 \cdot 3 = 6$ car selections is possible. ■■■

Practice Problem 1   Construct a tree diagram and determine the number of ways you can choose to take one of three courses—English, French, or math—in the morning, afternoon, or evening. ■

The car selection process in Example 1 involves making two choices. First, choose a type of car; second, choose a color. It is important in this process that the same *number* of choices be available for the second choice regardless of the result of the first choice. The exact rules for counting by this method are given next.

FUNDAMENTAL COUNTING PRINCIPLE

If a first choice can be made in p different ways, a second choice can be made in q different ways, a third choice can be made in r different ways, and so on, then the sequence of choices can be made in $p \cdot q \cdot r \cdots$ different ways.

EXAMPLE 2   **Counting Music Programs**

A disc jockey wants to open a program by picking a song from among 40 songs by one artist and close the program by picking a song from among 32 songs by a second artist. How many different choices are possible?

SOLUTION

Each choice of an opening and ending song sequence can be obtained as follows:
   First, choose an opening song in 40 ways.
   Second, choose the song to end the program in 32 ways.
   Since no matter which song is chosen to open the program, all 32 songs are available to end the program, the counting principle applies, and there are a total of

$$40 \cdot 32 = 1280$$

possible choices for the opening and ending sequences. ■■■

Practice Problem 2   Reba gets to choose both dinner and a movie for a Saturday night date. She can choose from any of seven restaurants and five movies. How many different choices are possible? ■

◆ **WARNING**   Be careful when using the Fundamental Counting Principle. Suppose you are trying to decide between buying a SUV and a convertible. The SUV is available only in black, red, or silver, and the convertible is available only in black or red. In how many ways can you choose a type of car and its color? The car and color can be selected by making two choices: type (SUV or convertible) followed by color. But in this situation, the number of ways you can make the second choice depends on the first choice. You can choose from *three* colors for a SUV, but only *two* colors for a convertible. The Fundamental Counting Principle *does not apply*. Fortunately, there are so few different choices here that you can easily list all five: black SUV, red SUV, silver SUV, black convertible, and red convertible.

EXAMPLE 3   **Counting Possible Social Security Numbers**

Social Security numbers have the format NNN-NN-NNNN, where each N must be one of the integers 0, 1, 2, 3, 4, 5, 6, 7, 8, and 9. Assuming that there are no other restrictions, how many such numbers are possible?

SOLUTION

There are nine positions in the format

NNN-NN-NNNN.

Since the number of choices available at each position is not affected by previous choices, the Fundamental Counting Principle can be used. Replace each of the letters N with a blank to be filled in with one of the ten digits 0, 1, 2, 3, 4, 5, 6, 7, 8, and 9.
   We have

$$\underline{10} \cdot \underline{10} \cdot \underline{10} \cdot \underline{10} \cdot \underline{10} \cdot \underline{10} \cdot \underline{10} \cdot \underline{10} \cdot \underline{10} = 10^9.$$

Thus, there are $10^9 = 1{,}000{,}000{,}000$ possible Social Security numbers.   ■ ■ ■

Practice Problem 3   In Example 3, suppose we fix the area number for the Social Security number to be 433. How many social security numbers can be assigned to this area—that is, numbers having the form 433-NN-NNNN?   ■

2 Use the formula for permutations.

Permutations

DEFINITION OF PERMUTATION

A **permutation** is an arrangement of n distinct objects in a fixed order in which no object is used more than once. The specific order is important: Each different ordering of the same objects is a different permutation.

EXAMPLE 4   **Counting Possible Arrangements for a Group Photograph**

How many different ways can five people be arranged in a row for a group photograph?

SOLUTION

To count the number of ways we can arrange five people in a row for a group photograph, we can assign numbers to each of the five photograph positions.

| 1 | 2 | 3 | 4 | 5 |

Then we proceed as follows:
   First, choosing any of the five people for Position 1 gives 5 ways.
   Second, choosing one of the four remaining people for Position 2 gives 4 ways.
   Third, choosing one of the three remaining people for Position 3 gives 3 ways.
   Fourth, choosing one of the remaining two people for Position 4 gives 2 ways.
   Fifth, choosing the last person remaining for Position 5 gives 1 way.
   Using the Fundamental Counting Principle, we see that there are

$$5 \cdot 4 \cdot 3 \cdot 2 \cdot 1 = 5!, \text{ or } 120,$$

possible arrangements.   ■ ■ ■

Practice Problem 4   How many different ways can seven books be arranged on a shelf?   ■

   This technique gives a general formula for counting permutations of n distinct objects.

NUMBER OF PERMUTATIONS OF *n* OBJECTS

The number of permutations of *n* distinct objects is

$$n! = n(n-1)\cdots 4\cdot 3\cdot 2\cdot 1.$$

That is, *n* distinct objects can be arranged in *n*! different ways.

Sometimes only some, but not all, of the available objects are to be arranged in a specific order, again with no object being used more than once. You may want to choose *r* objects from among *n* available objects. This type of ordering is called a **permutation of *n* objects taken *r* at a time**.

EXAMPLE 5 **Counting Prizewinners**

Three prizes (for first, second, and third place) are to be given out in a dog show having 27 contestants. In how many different ways can dogs come in first, second, and third?

SOLUTION

Consider how we might list the prize winners:

Any dog might win first prize; there are 27 possibilities.

Any remaining dog might win second prize; there are 26 possibilities.

Any remaining dog might win third prize; there are 25 possibilities. We can use the Fundamental Counting Principle: There are

$$27 \ \cdot \ 26 \ \cdot \ 25 = 17{,}550$$

$$\uparrow \qquad \uparrow \qquad \uparrow$$

1st   2nd   3rd
prize   prize   prize

distinct ways that the dogs can come in first, second, and third. This is an example of permutations of 27 objects taken 3 at a time. ■ ■ ■

Practice Problem 5 There are nine different rides at a state fair. A group has to decide which four rides they will go on and in what order. How many possibilities are there? ■

PERMUTATIONS OF *n* OBJECTS TAKEN *r* AT A TIME

The number of permutations of *n* distinct objects taken *r* at a time is denoted by $P(n, r)$, where

$$P(n, r) = \frac{n!}{(n-r)!}$$
$$= n(n-1)(n-2)\cdots(n-r+1).$$

EXAMPLE 6 **Using the Permutations Formula**

Use the formula for $P(n, r)$ to evaluate each expression.

a. $P(7, 3)$ **b.** $P(6, 0)$

TECHNOLOGY CONNECTION

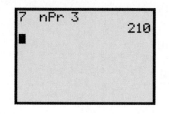 Most graphing calcula-
tors can compute $P(n, r)$.
They frequently use the symbol nPr. Note that $P(7, 3)$ is
shown as $7\ nPr\ 3$ on the
viewing screen.

```
7 nPr 3
              210
■
```

SOLUTION

a. $P(7, 3) = \dfrac{7!}{(7 - 3)!}$ Replace n with 7 and r with 3 in $P(n, r)$.

$= \dfrac{7!}{4!} = \dfrac{7 \cdot 6 \cdot 5 \cdot 4!}{4!} = 7 \cdot 6 \cdot 5 = 210$

b. $P(6, 0) = \dfrac{6!}{(6 - 0)!}$ Replace n with 6 and r with 0 in $P(n, r)$.

$= \dfrac{6!}{6!} = 1$ ■ ■ ■

Practice Problem 6   Evaluate.   **a.** $P(9, 2)$ **b.** $P(n, 0)$ ■

Let's apply the permutations formula to the problem of counting prizewinners in Example 5. The number of permutations of 27 dogs taken 3 at a time is

$$P(27, 3) = \frac{27!}{(27 - 3)!}$$

$$= \frac{27!}{24!} = \frac{27 \cdot 26 \cdot 25 \cdot 24!}{24!} = 27 \cdot 26 \cdot 25 = 17{,}550.$$

This is the same answer we found in Example 5.

3 Use the formula for combinations.

Combinations

Order is not always important in selecting objects. For example, if someone is bringing six movies on a vacation, the order in which the movies were chosen is unimportant.

DEFINITION OF A COMBINATION OF *n* DISTINCT OBJECTS TAKEN *r* AT A TIME

When *r* objects are chosen from *n* distinct objects without regard to order, we call the set of *r* objects a **combination of *n* objects taken *r* at a time.** The symbol $C(n, r)$ denotes the total number of combinations of *n* objects taken *r* at a time.

EXAMPLE 7   **Listing Combinations**

List all of the combinations of the four mascot names "bears," "bulls," "lions," and "tigers," taken two at a time. In other words, find $C(4, 2)$.

SOLUTION

If "bears" is one of the mascots chosen, the possibilities are {bears, bulls}, {bears, lions}, and {bears, tigers}.

If "bulls" is one of the mascots chosen, the remaining possibilities (excluding {bears, bulls}) are {bulls, lions} and {bulls, tigers}.

The final choice of two mascot names is {lions, tigers}.

The combinations of the four mascot names taken two at a time are {bears, bulls}, {bears, lions}, {bears, tigers}, {bulls, lions}, {bulls, tigers}, and {lions, tigers}.

Since there are six such combinations, we have $C(4, 2) = 6$.   ■ ■ ■

Practice Problem 7   List all of the combinations of the four mascot names "bears," "bulls," "lions," and "tigers," taken three at a time. In other words, find $C(4, 3)$.   ■

To find a formula for $C(n, r)$, we begin by comparing $C(n, r)$ with $P(n, r)$, the number of permutations of *n* objects taken *r* at a time. Suppose, for example, we have seven distinct objects—*A, B, C, D, E, F,* and *G*—and we pick three of them—*A, D,* and *E* (a combination of seven objects taken three at a time). This one combination

ADE gives 3! = 6 permutations of seven objects taken three at a time. Here are all of the possible arrangements of the letters *A, D,* and *E: ADE, AED, DAE, DEA, EAD,* and *EDA*. Since one combination generates six permutations, there are six (3!) times as many permutations of seven objects taken three at a time as there are combinations of seven objects taken three at a time. Thus, 3!(number of combinations) = (number of permutations).

$$3! \, C(7,3) = P(7,3)$$

$$C(7,3) = \frac{P(7,3)}{3!}$$

$$= \frac{\dfrac{7!}{(7-3)!}}{3!} \qquad \text{Replace } P(7,3) \text{ with } \frac{7!}{(7-3)!}.$$

$$= \frac{7!}{(7-3)!\,3!} = \frac{7!}{4!\,3!} = \frac{7 \cdot 6 \cdot 5 \cdot 4!}{4!\,3!} = \frac{7 \cdot 6 \cdot 5}{3 \cdot 2 \cdot 1} = 35$$

A similar process leads to a more general formula.

RECALL

Notice that the value of $C(n, r)$ is the same as that of $\dbinom{n}{r}$.

THE NUMBER OF COMBINATIONS OF *n* DISTINCT OBJECTS TAKEN *r* AT A TIME

The number of combinations of *n* distinct objects taken *r* at a time is

$$C(n, r) = \frac{n!}{(n-r)!\,r!}.$$

EXAMPLE 8 **Using the Combinations Formula**

Use the formula $C(n, r)$ to evaluate each expression.

a. $C(7, 5)$ **b.** $C(8, 0)$ **c.** $C(4, 4)$

SOLUTION

a. $C(7, 5) = \dfrac{7!}{(7-5)!\,5!}$ Replace *n* with 7 and *r* with 5 in $C(n, r)$.

$$= \frac{7!}{2!\,5!} = \frac{7 \cdot \overset{3}{6} \cdot 5!}{2 \cdot 1 \cdot 5!} = 21.$$

b. $C(8, 0) = \dfrac{8!}{(8-0)!\,0!}$ Substitute *n* = 8 and *r* = 0 in $C(n, r)$.

$$= \frac{8!}{8!\,0!} = 1.$$

TECHNOLOGY CONNECTION

Most graphing calculators can compute $C(n, r)$. Note that $C(4, 2)$ is shown as 4 nCr 2 on the viewing screen.

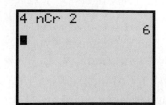
```
4 nCr 2
               6
■
```

c. $C(4, 4) = \dfrac{4!}{(4-4)!\,4!}$ Substitute *n* = 4 and *r* = 4 in $C(n, r)$.

$$= \frac{4!}{0!\,4!} = 1. \qquad\qquad ■■■$$

Practice Problem 8 Evaluate.

a. $C(9, 2)$ **b.** $C(n, 0)$ **c.** $C(n, n)$ ■

EXAMPLE 9 **Choosing Pizza Toppings**

How many different ways can five pizza toppings be chosen from the following choices: pepperoni, onions, mushrooms, green peppers, olives, tomatoes, mozzarella, and anchovies?

SOLUTION

Eight toppings are listed, so the question is, how many ways can we pick five things from eight things? That is, we need to find the number of combinations of eight objects taken five at a time, or $C(8, 5)$.

Use the formula $C(n, r) = \dfrac{n!}{(n - r)! \, r!}$.

$$C(8, 5) = \frac{8!}{(8 - 5)! \, 5!} \qquad \text{Replace } n \text{ with 8 and } r \text{ with 5.}$$

$$= \frac{8!}{3! \, 5!} = \frac{8 \cdot 7 \cdot 6}{3 \cdot 2 \cdot 1} = 56.$$

There are 56 ways that five of the eight pizza toppings can be chosen.   ■ ■ ■

Practice Problem 9   A box of assorted chocolates contains 12 different kinds of chocolates. In how many ways can three chocolates be chosen?   ■

4 Use the formula for distinguishable permutations.

Distinguishable Permutations

Suppose you were to stand in a line with identical twins. Although the three of you can arrange yourselves in 3! = 6 different ways, someone who could not tell the twins apart could distinguish only three different arrangements. These three arrangements are called *distinguishable permutations*. To better understand the idea of distinguishable permutations, suppose we have three tiles with the numbers 1, 2, and 3 printed on the fronts and the letters $M \, O \, M$ printed on the backs, respectively.

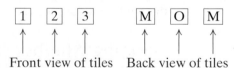

Front view of tiles   Back view of tiles

There are 3! = 6 permutations of the numbers 1, 2, 3.

Now consider the corresponding arrangements of the letters *M*, *O*, and *M* on the back sides of the tiles as the fronts of the tiles are arranged to show each of the six permutations of 1, 2, 3.

Front View of Tiles	Back View of Tiles
1 2 3	M O M
1 3 2	M M O
2 1 3	O M M
2 3 1	O M M
3 1 2	M M O
3 2 1	M O M

There are only three distinguishable arrangements that appear when the tiles are viewed from the back: MOM, MMO, and OMM.

The front view shows the permutations of three distinct objects, but the back view shows the permutations of three objects of which one is of one kind (the letter *O*) and two are of a second kind (the letter *M*).

DISTINGUISHABLE PERMUTATIONS

The number of distinguishable permutations of n objects of which n_1 are of one kind, n_2 are of a second kind, ..., and n_k are of a kth kind is

$$\frac{n!}{n_1! \cdot n_2! \cdots \cdot n_k!},$$

where $n_1 + n_2 + \cdots + n_k = n$.

EXAMPLE 10   **Distributing Gifts**

In how many ways can nine gifts be distributed among three children if each child receives three gifts?

SOLUTION

If we number the gifts 1 through 9 and use *A, B,* and *C* to represent the children, we can visualize any distribution of gifts as an arrangement of the letters *A, B,* and *C* (each letter being used three times) above the numbers 1 through 9. Here is one possible arrangement:

$$
\begin{array}{ccccccccc}
B & A & A & C & C & B & A & C & B \\
1 & 2 & 3 & 4 & 5 & 6 & 7 & 8 & 9
\end{array}
$$

In this arrangement, child *A* gets gifts 2, 3, and 7; child *B* gets gifts 1, 6, and 9; and child *C* gets gifts 4, 5, and 8.

To count all of the possible distributions of gifts, we count the number of permutations of nine objects, of which three are of one kind, three are of a second kind, and three are of a third kind:

$$
\frac{9!}{3!\,3!\,3!} = 1680
$$

There are 1680 ways the nine gifts can be distributed among the three children.

■ ■ ■

Practice Problem 10   In how many ways can six counselors be sent in pairs to three different locations?   ■

Deciding Whether to Use Permutations, Combinations, or the Fundamental Counting Principle

Suppose 23 students show up late for a private screening of a new movie. Suppose also that there is one seat available in each of the first, second, eighth, and tenth rows and that five students will be allowed to stand in the back of the theater.

Permutations If order is important, use permutations.	In how many ways can 4 of the 23 students be assigned the available seats in Rows 1, 2, 8, and 10? **Answer:** $\quad 23 \ \cdot \ 22 \ \cdot \ 21 \ \cdot \ 20 \ \cdot = 212{,}520$ ways $\quad$ Row 1 Row 2 Row 8 Row 10
Combinations If order is not important, use combinations.	In how many ways can 5 of the 19 students *not given seats* be chosen to stand in the back? **Answer:** $C(19, 5) = 11{,}628$ ways
Fundamental Counting Principle Whenever you are making consecutive choices and the number of choices at each stage is not affected by the way earlier choices were made, you can use the Fundamental Counting Principle.	In how many ways can students be assigned to the available seats and five students be chosen to stand in the back? **Answer:** $212{,}520 \times 11{,}628$ ways $\qquad\uparrow\qquad\qquad\qquad\uparrow$ Assign the four seats. Pick five students.

Notice that

- Since the Fundamental Counting Principle is used to count permutations, you can always use that principle instead of permutations.
- There are frequently equally good, although different, ways to use counting techniques to solve a problem. For example, we could first choose 5 students to stand in the back of the theater in $C(23, 5)$ ways and then assign the seats in Rows 1, 2, 8, and 10 to the 19 remaining students in $19 \cdot 18 \cdot 17 \cdot 16$ ways. The total number of ways students can be assigned to stand in the back and assigned to the available seats is then $C(23, 5) \cdot 19 \cdot 18 \cdot 17 \cdot 16$. You should verify that

$$23 \cdot 22 \cdot 21 \cdot 20 \cdot C(19, 5) = 212{,}520 \times 11{,}628.$$

SECTION 10.6 ■ Exercises

A EXERCISES Basic Skills and Concepts

1. Any arrangement of n distinct objects in a fixed order in which no object is used more than once is called a(n) __permutation__ .

2. The number of permutations of five distinct objects is __$5! = 120$__ .

3. When r objects are chosen from n distinct objects, the set of r objects is called a(n) __combination__ of n objects taken r at a time.

4. The number of distinguishable permutations of 10 objects of which 3 are of one kind and 7 are of a second kind is _____ . $\dfrac{10!}{3! \, 7!} = 120$

5. *True or False* When order is important in counting objects, we use permutations. True

6. *True or False* There are more permutations of n objects taken r at a time than there are combinations of n objects taken r at a time. True (as long as r is greater than 1)

In Exercises 7–14, use the formula for $P(n, r)$ to evaluate each expression.

7. $P(6, 1)$ 6
8. $P(7, 3)$ 210
9. $P(8, 2)$ 56
10. $P(10, 6)$ 151,200
11. $P(9, 9)$ 362,880
12. $P(5, 5)$ 120
13. $P(7, 0)$ 1
14. $P(4, 0)$ 1

In Exercises 15–22, use the formula for $C(n, r)$ to evaluate each expression.

15. $C(8, 3)$ 56
16. $C(7, 2)$ 21
17. $C(9, 4)$ 126
18. $C(10, 5)$ 252
19. $C(5, 5)$ 1
20. $C(6, 6)$ 1
21. $C(3, 0)$ 1
22. $C(5, 0)$ 1

In Exercises 23–26, use the Fundamental Counting Principle to solve each problem.

23. How many different 2-letter codes can be made up from the 26 capital letters of the alphabet if (a) repeated letters are allowed; (b) repeated letters are not allowed. **a.** 676 **b.** 650

24. How many possible answer sheets are there for a ten-question true–false exam if no answer is left blank? 1024

25. In how many ways can a president and vice president be chosen from an organization with 50 members? 2450

26. How many four-digit numbers can be formed with the digits 0, 1, 2, 3, 4, 5, 6, 7, 8, and 9 if the first digit cannot be 0? 9000

In Exercises 27–30, use the formula for counting permutations to solve each problem.

27. In how many different ways can the members of a family of four be seated in a row of four chairs? 24

28. Ten teams are entered in a bowling tournament. In how many ways can first, second, and third prizes be awarded? 720

29. Frederico has five shirts, three pairs of pants, and four ties that are appropriate for an interview. If he wears one of each type of apparel, how many different outfits can he wear? 60

30. A choreographer has to arrange eight pieces for a dance program. In how many ways can this be done? 40,320

In Exercises 31–34, use the formula for counting combinations to solve each problem.

31. A pizza menu offers pepperoni, peppers, mushrooms, sausage, and meatballs as extra toppings that can be added to your pizza. How many different pizzas can you order with two additional toppings? 10

32. In how many ways can 2 student representatives be chosen from a class of 15 students? 105

33. Students are allowed to choose 9 out of 11 problems to work for credit on an exam. How many ways can this be done? 55

34. You have to choose 3 out of 18 potential teammates to help you with a group assignment. How many ways can this be done? 816

†Due to space constrictions, answers to these exercises may be found in the Answers beginning on page A–1 in the back of the book.

In Exercises 35–40, solve the problem by any appropriate counting method.

35. How many ways can Ashley choose 4 out of 11 different canned goods to donate to charity? 330

36. How many ways can a computer store hire a salesclerk and a technician from seven applicants for the clerk position and four applicants for the technician position? 28

37. How many ways can six people be seated in a row with eight seats, leaving two seats vacant? 20,160

38. Seven students arrive to rent two canoes and three kayaks. How many ways can two students be selected to rent the canoes and three to rent the kayaks? 210

39. Dakota is trying to decide in which order to visit five prospective colleges. How many ways can this be done? 120

40. How many ways can the first 4 numbers be called from the 75 possible bingo numbers? 29,170,800

B EXERCISES Applying the Concepts

41. **Area codes.** How many three-digit area codes for telephones are possible if
    a. the first digit cannot be a 0 or 1 and the second digit must be a 0 or 1? 160
    b. there are no restrictions on the second digit? (Previous restrictions were abandoned in 1995.) 800

42. **Radio station call letters.** Radio stations in the United States have call letters that begin with either K or W (for example, WXYZ in Detroit). Some have a total of three letters, and others have four. How many different call-letter selections are possible? 36,504

43. **Fraternity letters.** The Greek alphabet contains 24 letters. How many 3-letter fraternity names can be made with Greek letters
    a. if no repetitions are allowed? 12,144
    b. if repetitions are allowed? 13,824

44. **Codes.** How many three-letter codes can be made from the first ten letters of the alphabet if
    a. no letters are repeated? 720
    b. letters may be repeated? 1000

45. **Supreme Court decisions.** In how many ways can the nine justices of the U.S. Supreme Court reach a majority decision? (The U.S. Supreme Court has nine justices who are appointed for life.) 256

46. **Committee selection.** A club consists of ten members. How many ways can a committee of four be selected? 210

47. **Wardrobe choices.** Al's wardrobe consists of eight pairs of slacks, eight shirts, and eight pairs of shoes. He does not want to wear the same outfit on any two days of the year. Will his current wardrobe allow him to do this? Yes

48. **Menu selection.** A French restaurant offers two choices of soup, three of salad, eight of an entree, and five of a dessert. How many different complete dinners can you order? 240

49. **Housing development.** A developer wants to build seven houses, each of a different design. How many ways can she arrange these homes on a street if
    a. four lots are on one side of the street and three are on the opposite side? 5040
    b. five lots are on one side of the street and two are on the opposite side? 5040

50. **Designing exercise sets.** The author of a mathematics textbook has seven problems for an exercise set, and she is trying to arrange them in increasing order of difficulty. How many ways can she select the problems if no two of them are equally difficult? How many ways can the author fail?

51. **Letter selection.** From the letters of the word *CHARITY*, groups of four letters are formed (without regard to order). How many of them will contain
    a. both A and R? 10
    b. A but not R? 10
    c. neither A nor R? 5

52. **Committee selection.** How many ways can a committee of 4 students and 2 professors be formed from 20 students and 8 professors? 135,660

53. **Inventory orders.** A manufacturer makes shirts in five different colors, seven different neck sizes, and three different sleeve lengths. How many shirts should a department store order to have one of each type? 105

54. **International book codes.** All recent books are identified by their International Standard Book Number (ISBN), a ten-digit code assigned by the publisher. A typical ISBN is 0-321-75526-2. The first digit, 0, represents the language of the book (English); the second block of digits, 321, represents the publishing company (Addison-Wesley); the third block of digits, 75526, is the number assigned by the publishing company to that book; and the final digit, 2, is the check digit, used to detect the most commonly made errors when ISBNs are copied. How many ISBNs are possible? (Remember that the previous nine digits determine the last digit.) 1 billion

55. **Political polling.** A senator sends out questionnaires asking his constituents to rank 10 issues of concern to them, such as crime, education, and taxes, in order of importance. Replies are filed in folders, and two replies are put in the same folder if and only if they give the same ranking to all 10 issues. Find the maximum number of folders that might be needed. 3,628,800

56. **Parcel delivery.** A UPS driver has to deliver 12 parcels to 12 different addresses. In how many orders can the driver make these deliveries? 479,001,600

Answers:
50. There is only one way to select the problems in increasing order of difficulty. The author can fail 5039 ways.

57. **Circus visit.** A father with seven children takes four of them at a time to a circus as often as he can, without taking the same four children more than once. What is the maximum number of times each child will go to the circus, and how many times will the father go?

58. **Arranging books.** From six history books, four biology books, and five economics books, how many ways can a person select three history books, two biology books, and four economics books and arrange them on a shelf?
217,728,000

59. **Bingo.** Each of four columns on a bingo card contains 5 different numbers chosen in any order from 15. Column "B" chooses from 1 to 15, column "I" from 16 to 30, . . . , and column "O" from 61 to 75. Only four numbers are under column "N" because of the free space. How many different cards are possible?   111,007,923,832,370,565

60. **Football conferences.** The Grid-Iron Football League in Sardonia is divided into two conferences—East and West—each having six members. Each team in the league must play each team in its conference twice and each team in the other conference once during a season. What is the total number of games played in the league?   96

61. **Married couples prohibited.** How many ways can a committee of four people be selected from five married couples if no committee is to include husband-and-wife pairs?   80

62. **Social club committees.** A social club contains seven women and four men. The club wants to select a committee of three to represent it at a state convention. How many of the possible committees contain at least one man?   130

63. **Coin gifts.** Among Corey's collection of early American coins, he has a half-dollar, a quarter, a nickel, and a penny. He wants to give his younger sister some or all of these four coins. How many different sets of coins can Cory give?   15

64. **Meal possibilities.** The university cafeteria offers two choices of soups, three of salad, five of main dishes, and six of desserts. How many different four-course meals can you choose?   180

In Exercises 65–68, how many distinguishable ways can the letters of each word be arranged?

65. TAIWAN 360

66. AMERICA 2520

67. SUCCESS 420

68. ARRANGE 1260

Answer:
57. The maximum number of times each child will go to the circus is 20. The father goes 35 times.

C EXERCISES Beyond the Basics

69. A company that supplies temporary help has a contract for 20 weeks to provide three workers per week to an insurance firm. The agreement specifies that in no two weeks can the same three workers be sent to work at the firm. How many different workers are necessary?   6

70. An FM radio station wants to start its evening programming with five songs every day for 21 days without using the same five songs on any two days. What is the smallest number of songs that can be used to accomplish this result?   7

71. In geometry, any two points determine a line. How many lines are determined by 30 given points, no 3 of which lie on the same line?   435

72. In geometry, any 3 noncollinear points determine a triangle. How many triangles can be determined by 21 given points, no 3 of which are collinear?   1330

In Exercises 73–80, solve each equation.

73. $\binom{m+1}{m-1} = 3!$ $m = 3$ 74. $6\binom{n-1}{2} = \binom{n+1}{4}$ $n = 8$

75. $2\binom{n-1}{2} = \binom{n}{3}$ $n = 6$ 76. $\frac{4}{3}\binom{k}{2} = \binom{k+1}{3}$ $k = 3$

77. $\binom{n}{0} + \binom{n}{1} + \binom{n}{2} + \cdots + \binom{n}{n} = 64$ $n = 6$

78. $\binom{k}{1} + \binom{k}{2} + \binom{k}{3} + \cdots + \binom{k}{k-1} = 126$ $k = 7$

79. $\binom{m}{4} = \binom{m}{5}$ $m = 9$ 80. $\binom{n}{7} = \binom{n}{3}$ $n = 10$

Critical Thinking

81. How many terms of the form $x^n y^m$ are possible if n and m can be any integer such that $1 \le n \le 5$ and $1 \le m \le 5$?   25

82. Use the Fundamental Counting Principle to determine the number of terms in the product

$$(a + b)(x^2 + 2xy + y^2).$$

Explain your reasoning.
[*Hint:* Each term in the product can be obtained by multiplying one term chosen from each factor.]   There are two terms in the first factor and three in the second, giving $(2)(3) = 6$ possibilities to multiply.

Probability

Before Starting this Section, Review

1. Set notation (Appendix A, page 758)
2. Fundamental Counting Principle (Section 10.6, page 724)
3. Permutations (Section 10.6, page 725)
4. Combinations (Section 10.6, page 727)

Objectives

1. Find the probability of an event.
2. Use the Additive Rule for finding probabilities.
3. Find the probability of mutually exclusive events.
4. Find the probability of the complement of an event.
5. Find experimental probabilities.

DOUBLE LOTTERY WINNER

Many people dream of winning a lottery, but only the most optimistic dream of winning it twice. When Evelyn Marie Adams won the New Jersey state lottery for the second time, *The New York Times* (February 14, 1986) claimed that the chances of one person winning the lottery twice were about 1 in 17 trillion. Two weeks later a letter from two statisticians appeared in the *Times* challenging this claim. Although they agreed that any particular person had very little chance of winning a lottery twice, they said that it was almost certain that someone in the United States would win twice. Further, they said that the odds were even for another double lottery win to occur within seven years. In less than two years, Robert Humphries won his second Pennsylvania lottery prize.

Evaluating how likely an event is to occur is the realm of probability, which we study in this section. In Example 4, we determine the probability of winning a lottery in which 6 numbers are randomly selected from the numbers 1 through 53.

The toss of a coin determines which team will kick the ball to the other team to start a football game. It seems everyone agrees that a coin toss gives both teams the same chance to choose whether to kick or receive the football. The ideas of "chance," "likelihood," and "probability" are all coupled with the desire to quantify expectations about the results of some process. ■

1 Find the probability of an event.

The Probability of an Event

Any process that terminates in one or more outcomes (results) is an **experiment**. For example, if we ask people what time of the day they typically eat their first meal or if we pull balls from a bag of colored balls, we have performed an experiment. The outcomes of the first experiment are the times of day given as responses to our question. The outcomes of the second experiment are the colors of the balls drawn from the bag.

The set of all possible outcomes of a given experiment is called the **sample space** of the experiment. Suppose the experiment is tossing a coin. Then we could denote the outcome "heads" by H; denote the outcome "tails" by T, and write the sample space, S, in set notation as

$$S = \{H, T\}.$$

A single die is a cube on which each face contains one, two, three, four, five, or six dots and no two faces contain the same number of dots. A roll of the die is an

experiment, the outcome of which is the number of dots showing on the upper face, and the sample space S can be written in set notation as

$$S = \{1, 2, 3, 4, 5, 6\}.$$

An event is any set of possible outcomes; that is, an **event** is any subset of a sample space. For the die-rolling experiment, one possible event is "The number showing is even." In set notation, we write this event as $E = \{2, 4, 6\}$. The event "The number showing is a 6" is written in set notation as $A = \{6\}$.

Outcomes of an experiment are said to be **equally likely** if no outcome should result more often than any other outcome when the experiment is run repeatedly. In both the coin-tossing and die-rolling experiments, all of the outcomes are equally likely.

PROBABILITY OF AN EVENT

If all of the outcomes in a sample space S are equally likely, then the **probability** of an event E, denoted $P(E)$, is the ratio of the number of outcomes in E, denoted $n(E)$, to the total number of outcomes in S, denoted $n(S)$. That is,

$$P(E) = \frac{n(E)}{n(S)}.$$

Because E is a subset of $S, 0 \le n(E) \le n(S)$. Dividing by $n(S)$ yields $0 \le \dfrac{n(E)}{n(S)} \le 1$. So the probability $P(E)$ of an event E is a number between 0 and 1 inclusive.

EXAMPLE 1   **Finding the Probability of an Event**

A single die is rolled. Write each event in set notation and find the probability of the event.

a. E_1: The number 2 is showing.

b. E_2: The number showing is odd.

c. E_3: The number showing is less than 5.

d. E_4: The number showing is greater than or equal to 1.

e. E_5: The number showing is 0.

SOLUTION

The sample space for the experiment of rolling a die is $S = \{1, 2, 3, 4, 5, 6\}$, so $n(S) = 6$. For any event E, $P(E) = \dfrac{n(E)}{n(S)}$.

a. $E_1 = \{2\}$, so $n(E_1) = 1$. $P(E_1) = \dfrac{n(E_1)}{n(S)} = \dfrac{1}{6}$

b. $E_2 = \{1, 3, 5\}$, so $n(E_2) = 3$. $P(E_2) = \dfrac{n(E_2)}{n(S)} = \dfrac{3}{6} = \dfrac{1}{2}$

c. $E_3 = \{1, 2, 3, 4\}$, so $n(E_3) = 4$. $P(E_3) = \dfrac{n(E_3)}{n(S)} = \dfrac{4}{6} = \dfrac{2}{3}$

d. $E_4 = \{1, 2, 3, 4, 5, 6\}$, so $n(E_4) = 6$. $P(E_4) = \dfrac{n(E_4)}{n(S)} = \dfrac{6}{6} = 1$

e. $E_5 = \varnothing$, so $n(E_5) = 0$. $P(E_5) = \dfrac{n(E_5)}{n(S)} = \dfrac{0}{6} = 0$ ■ ■ ■

Practice Problem 1 A single die is rolled. Write each event in set notation and give the probability of each event.

a. E_1: The number showing is even.

b. E_2: The number showing is greater than 4. ■

Notice in Example 1**d** that $E_4 = S$; so the event E_4 is certain to occur on every roll of the die. Every **certain event** has probability 1. On the other hand, in Example 1**e**, the event $E_5 = \varnothing$, so E_5 never occurs, and $P(E_5) = 0$. An **impossible event** always has probability 0.

EXAMPLE 2   Tossing a Coin

What is the probability of getting at least one tail when a coin is tossed twice?

SOLUTION

Tossing a coin once results in two equally likely outcomes: heads (H) and tails (T). Tossing the coin twice results in two equally likely outcomes for the first toss and, regardless of the first outcome, two equally likely outcomes for the second toss. By the Fundamental Counting Principle, there are $2 \cdot 2$, or 4, outcomes when we toss the coin twice. We can represent these outcomes as (first toss, second toss) pairs:

$$S = \{(H, H), (H, T), (T, H), (T, T)\}.$$

Because three of these outcomes, $E = \{(H, T), (T, H), (T, T)\}$, result in at least one tail we have

$$P(E) = \frac{n(E)}{n(S)} = \frac{3}{4}.$$ ■ ■ ■

Practice Problem 2 What is the probability of getting exactly one head when a coin is tossed twice?

Equally likely outcomes occur in most games of chance that involve tossing coins, rolling dice, or drawing cards. Lotteries and games involving spinning wheels also generate equally likely outcomes.

EXAMPLE 3   Rolling a Pair of Dice

What is the probability of getting a sum of 8 when a pair of dice is rolled?

SOLUTION

There are six equally likely outcomes for each die; so by the Fundamental Counting Principle, there are $6 \cdot 6$, or 36, equally likely outcomes when a pair of dice is rolled. It is useful to think of having two dice of different colors—say, red and green—in distinguishing the 36 outcomes. In this way, 2 on the red die and 6 on the green die is easily seen to be a different physical outcome from 6 on the red die and 2 on the green die, even though the same sum results.

We can represent the 36 outcomes as (red die, green die) pairs:

$$S = \{(1, 1), (1, 2), (1, 3), (1, 4), (1, 5), (1, 6),$$
$$(2, 1), (2, 2), (2, 3), (2, 4), (2, 5), (2, 6),$$
$$(3, 1), (3, 2), (3, 3), (3, 4), (3, 5), (3, 6),$$
$$(4, 1), (4, 2), (4, 3), (4, 4), (4, 5), (4, 6),$$
$$(5, 1), (5, 2), (5, 3), (5, 4), (5, 5), (5, 6),$$
$$(6, 1), (6, 2), (6, 3), (6, 4), (6, 5), (6, 6)\}$$

The highlighted ordered pairs in the sample space are the outcomes that have a sum of 8. In set notation, the event "The sum of the numbers on the dice is 8" is

$$E = \{(6, 2), (5, 3), (4, 4), (3, 5), (2, 6)\}.$$

Since $n(E) = 5$ and $n(S) = 36$,

$$P(E) = \frac{n(E)}{n(S)} = \frac{5}{36}.$$

So the probability of getting a sum of 8 is $\frac{5}{36}$.   ■ ■ ■

Practice Problem 3   What is the probability of getting a sum of 7 when a pair of dice is rolled?   ■

EXAMPLE 4   **Winning the Florida Lottery**

Table tennis balls numbered 1 through 53 are mixed together in a turning drum or basket. Six of the balls are chosen in a random way. To win the lottery, a player must have selected all six of the numbers chosen. The order in which the numbers are chosen does not matter. What is the probability of matching all six numbers? (*Source:* www.fllottery.com)

SOLUTION

Because the order in which the numbers are selected is not important, the sample space S consists of all sets of 6 numbers that can be selected from 53 numbers. Therefore, $n(S) = C(53, 6) = 22,957,480$. Because only 1 set of 6 numbers matches the 6 numbers drawn, the probability of the event "guessed all 6 numbers" is

$$P(\text{winning the lottery}) = P(\text{guessed all 6 numbers})$$

$$= \frac{1}{22,957,480} \approx 0.00000004.$$

Whether you say your chances of winning are 1 in 22,957,480 or about 0.00000004, you definitely know they are not very good!   ■ ■ ■

Practice Problem 4   A state lottery requires a winner to correctly select 6 out of 50 numbers. The order of the numbers does not matter. What is the probability of matching all six numbers?   ■

2 Use the Additive Rule for finding probabilities.

The Additive Rule

Consider the experiment of rolling one die. The sample space is $S = \{1, 2, 3, 4, 5, 6\}$, and $n(S) = 6$.

Let E be the event "The number rolled is greater than 4," and let F be the event "The number rolled is even." Then

$$E = \{5, 6\} \quad \text{and} \quad F = \{2, 4, 6\}.$$

The event "The number rolled is greater than 4 *and* even" is $E \cap F$.

$$E \cap F = \{6\}$$

Because $n(E) = 2, n(F) = 3, n(E \cap F) = 1$, and $n(S) = 6$, we have

$$P(E) = \frac{2}{6}, P(F) = \frac{3}{6}, P(E \cap F) = \frac{1}{6}.$$

The event "The number rolled is either greater than 4 *or* even" is $E \cup F$.

$$E \cup F = \{2, 4, 5, 6\}$$

Note that 6 is in *both* E and F, so it is counted once when $n(E)$ is counted and a second time when $n(F)$ is counted. However, 6 is counted only once when $n(E \cup F)$ is counted. We have $n(E \cup F) = n(E) + n(F) - n(E \cap F)$. That is, since 6 was counted twice when adding $n(E) + n(F)$, we must subtract $n(E \cap F)$ to get an accurate count for $E \cup F$.

$$n(E \cup F) = n(E) + n(F) - n(E \cap F)$$

$$\frac{n(E \cup F)}{n(S)} = \frac{n(E)}{n(S)} + \frac{n(F)}{n(S)} - \frac{n(E \cap F)}{n(S)} \qquad \text{Divide both sides by } n(S).$$

$$P(E \cup F) = P(E) + P(F) - P(E \cap F) \qquad \text{Definition of probability}$$

This example illustrates the *Additive Rule*.

THE ADDITIVE RULE

If E and F are events in a sample space, then

$$P(E \cup F) = P(E) + P(F) - P(E \cap F).$$

---

3 Find the probability of mutually exclusive events.

Mutually Exclusive Events

Two events are *mutually exclusive* if it is impossible for both to occur simultaneously. That is, E and F are **mutually exclusive** events if $E \cap F$ contains no outcomes. In this case, $P(E \cap F) = 0$ and the Additive Rule is somewhat simplified.

MUTUALLY EXCLUSIVE EVENTS

If E and F are any events in a sample space and $E \cap F = \varnothing$, then

$$P(E \cup F) = P(E) + P(F).$$

A standard deck of 52 playing cards uses four different designs called suits:

clubs, ♣; diamonds, ♦; hearts, ♥; and spades, ♠

Each suit has 13 cards: an ace, king, queen, and jack, and cards numbered 2 through 10. The diamond and heart symbols are red, and the club and spade symbols are black. For the purposes of this section, we assume that all decks of cards are standard and well shuffled, so that all cards drawn or dealt from a deck are equally likely.

EXAMPLE 5 **Drawing a Card from a Deck**

A card is drawn from a deck. What is the probability that the card is

a. a king? **b.** a spade?

c. a king or a spade? **d.** a heart or a spade?

SOLUTION

The sample space S is the 52-card deck; so $n(S) = 52$.

a. There are four kings, one in each suit; so

$$P(\text{king}) = \frac{4}{52} = \frac{1}{13}.$$

b. There are 13 spades in the deck; so

$$P(\text{spade}) = \frac{13}{52} = \frac{1}{4}.$$

c. By the Additive Rule, we have

$$P(\text{king or spade}) = P(\text{king}) + P(\text{spade}) - P(\text{king and spade}).$$

Because there is only one card that is both a king and a spade, $P(\text{king and spade}) = \frac{1}{52}$. From parts **a** and **b** in this example,

$$P(\text{king or spade}) = \frac{4}{52} + \frac{13}{52} - \frac{1}{52}$$

$$= \frac{16}{52} = \frac{4}{13}.$$

d. Because no card is both a heart and a spade, the events "Draw a heart" and "Draw a spade" are mutually exclusive. Thus,

$$P(\text{heart or spade}) = P(\text{heart}) + P(\text{spade})$$

$$= \frac{13}{52} + \frac{13}{52} = \frac{1}{2}.$$ ■ ■ ■

Practice Problem 5 What is the probability that a card pulled from a deck is a jack or a king?   ■

4 Find the probability of the complement of an event.

The Complement of an Event

The **complement of an event** E is the set of all outcomes in the sample space that are *not* in E. The complement of E is denoted by E'. Because every outcome in the sample space is in E or in E', the event $E \cup E'$ is a certain event and $P(E \cup E') = 1$. Also, E and E' are mutually exclusive events, so $P(E \cup E') = P(E) + P(E')$. Thus, $P(E) + P(E') = 1$ and $P(E') = 1 - P(E)$.

PROBABILITY OF THE COMPLEMENT

If E is any event and E' is its complement, then

$$P(E') = 1 - P(E).$$

EXAMPLE 6   **Finding the Probability of the Complement**

Brandy has applied to receive funding from her company to attend a conference. She knows that her name, along with the names of nine other deserving candidates, was put in a box from which two names will be drawn. What is the probability that Brandy will be one of the two selected?

SOLUTION

It is easier to determine the probability that Brandy will *not* be selected. Since there are nine candidates other than Brandy, there are $C(9, 2)$ ways of selecting two employees not including Brandy. There are $C(10, 2)$ ways of selecting two employees from all ten candidates. Hence, the probability that Brandy is not selected is

$$P(\text{Brandy is not selected}) = \frac{C(9, 2)}{C(10, 2)}$$

$$= \frac{9!}{2!\,7!} \cdot \frac{2!\,8!}{10!} = \frac{8}{10} = \frac{4}{5}.$$

The probability that Brandy is selected is

$$P(\text{Brandy is selected}) = 1 - P(\text{Brandy is not selected})$$

$$= 1 - \frac{4}{5} = \frac{1}{5}.$$ ■ ■ ■

Practice Problem 6   In Example 6, what is the probability of Brandy being selected if three names are drawn from the box?   ■

5 Find experimental probabilities.

Experimental Probabilities

The probabilities we have seen up to this point have been determined by some form of reasoning, not data. Data support the decision to assign the value $\frac{1}{2}$ as the probability of a head resulting from one toss of a coin, but the assignment is not *based* on data. This is also true for the probabilities associated with drawing a card, rolling a pair of dice, drawing a ball from an urn, and so on. Probability assigned this way is called **theoretical probability**.

In contrast, we can assign probability values based on observation and data. If an experiment is performed n times (where n is large) and a particular outcome occurs k times, then the number $\frac{k}{n}$ is the value for the probability of that outcome. Probability determined this way is called **experimental probability**. A statement such as "The probability that a 10-year-old child will live to be 95 years old is $\frac{3}{100,000}$" refers to experimental probability.

RELATIVE FREQUENCY PRINCIPLE

If an experiment is performed n times and an event E occurs k times, then we say that the experimental (or statistical) probability of the event, $P(E)$, is

$$P(E) = \frac{k}{n}.$$

EXAMPLE 7   **Using Data to Estimate Probabilities**

Table 10.2 shows the gender distribution of students at four California City colleges.

TABLE 10.2

College	Male	Female
Los Angeles	6,995	8,179
San Diego	12,251	14,914
San Francisco	16,857	2,000
Sacramento	9,040	11,838

Source: www.areaconnect.com

Find the probability that a student selected *at random* is:

a. a female.   **b.** attending Los Angeles City College.

c. a female attending Los Angeles City College.

Round your answers to the nearest hundredth.

SOLUTION

The term *randomly selected* means that every possible selection is equally likely. In our example, this means that each student is equally likely to be selected. With random selection, the percentage of students in a given category gives the probability that a student in that category will be chosen.

a. The total number of students is 82,074. The total number of female students is 36,931. Thus, $P(\text{female}) = \dfrac{36,931}{82,074} \approx 0.45$.

b. The total number of students is 82,074. The total number of students attending Los Angeles City College is 15,174. Hence,

$$P(\text{attends Los Angeles City College}) = \frac{15,174}{82,074} \approx 0.18.$$

c. There are 8179 female students attending Los Angeles City College. Consequently,

$$P(\text{female and attending Los Angeles City College}) = \frac{8179}{82,074} \approx 0.10. \qquad ■ ■ ■$$

Practice Problem 7   In Example 7, find the probability that a student selected at random is a male or attends Sacramento College.   ■

SECTION 10.7 ■ Exercises

A EXERCISES Basic Skills and Concepts

1. If no outcome of an experiment results more often than any other outcome, the outcomes are said to be _____. equally likely

2. A *certain* event has probability _____1_____, and an *impossible* event has probability _____0_____.

3. If it is impossible for two events to occur simultaneously, then the events are said to be mutually exclusive

4. If the probability that an event occurs is 0.7, then the probability that the event does not occur is _____0.3_____.

5. *True or False* The probability of any event is a number between 0 and 1. True (if "between 0 and 1" includes 0 and 1)

6. *True or False* If an experiment is performed 100 times and an event E occurs 65 times, then the *experimental probability* of the event is 0.65. True

In Exercises 7–12, write the sample space for each experiment.

7. Students are asked to rank Wendy's, McDonald's, and Burger King in order of preference. †

8. Students are asked to choose either a hamburger or a cheeseburger and then pick Wendy's, McDonalds, or Burger King as the restaurant from which they would prefer to order it. †

9. Two albums are selected from among *Thriller* (Michael Jackson), *The Wall* (Pink Floyd), *Eagles: Their Greatest Hits* (Eagles), and *Led Zeppelin IV* (Led Zeppelin). †

10. Three types of potato chips are selected from among regular salted, regular unsalted, barbecue, cheddar cheese, and sour cream and onion. †

11. Two blanks in a questionnaire are filled in, the first identifying race as white, African-American, Native American, Asian, other, or multiracial, and the second indicating gender as male or female. †

12. A six-sided die is chosen from a box containing a white, a red, and a green die. Then the die is tossed. †

In Exercises 13–16, find the probability of the given event if a year is selected at random.

13. Thanksgiving falls on a Sunday. 0

14. Thanksgiving falls on a Thursday. 1

15. Christmas falls on December 25. 1

16. Christmas falls on January 1. 0

In Exercises 17–26, classify each statement as an example of *theoretical probability* or *experimental probability*.

17. The probability that Ken will go to church next Sunday is 0.85. experimental

18. The probability of exactly two heads when a fair coin is tossed twice is $\dfrac{1}{4}$. theoretical

19. The probability of drawing a heart from a deck of cards is 0.25. theoretical

20. The probability that an American adult will watch a major network's evening newscast is 0.17. experimental

†Due to space constrictions, answers to these exercises may be found in the Answers beginning on page A–1 in the back of the book.

21. The probability that a person driving during the Memorial Day weekend will die in an automobile accident is $\frac{1}{58,000}$. experimental

22. The probability that a student will get a D or an F in a college statistics course is 0.21. experimental

23. The probability of drawing an ace from a deck of cards is $\frac{1}{13}$. theoretical

24. The probability that a person will be struck by lightning this year is $\frac{1}{727,000}$. experimental

25. The probability that a plane will take off on time from JFK International airport is 0.78. experimental

26. The probability that a college student cheats on an exam is 0.32. experimental

27. Match each given sentence with an appropriate probability. †

An event that	has probability
is very likely to happen	0
will surely happen	0.5
is a rare event	0.001
may or may not happen	1
will never happen	0.999

28. A coin is tossed three times. Rearrange the order of the following events from the most probable event (first) to the least probable event (last).
{exactly two heads}, {the sure event}, {at least one head}, {at least two heads}. †

In Exercises 29–34, a die is rolled. Write each event in set notation and give the probability of each event.

29. The number showing is a 1 or a 6. $\frac{1}{3}$

30. The number 5 is showing. $\frac{1}{6}$

31. The number showing is greater than 4. $\frac{1}{3}$

32. The number showing is a multiple of 3. $\frac{1}{3}$

33. The number showing is even. $\frac{1}{2}$

34. The number showing is less than 1. 0

In Exercises 35–40, a card is drawn randomly from a standard 52-card deck. Write each event in set notation and find its probability.

35. A queen is drawn. $\frac{1}{13}$

36. An ace is drawn. $\frac{1}{13}$

37. A heart is drawn. $\frac{1}{4}$

38. An ace of hearts or a king of hearts is drawn. $\frac{1}{26}$

39. A face card (jack, queen, or king) is drawn. $\frac{3}{13}$

40. A heart that is not a face card is drawn. $\frac{5}{26}$

In Exercises 41–46, a digit is chosen randomly from the digits 0 through 9; and then 3 is added to the digit and the sum recorded. Write each event in set notation and find the probability of the event.

41. The sum is 11. $\frac{1}{10}$

42. The sum is 0. 0

43. The sum is less than 4. $\frac{1}{10}$

44. The sum is greater than 8. $\frac{2}{5}$

45. The sum is even. $\frac{1}{2}$

46. The sum is odd. $\frac{1}{2}$

B EXERCISES Applying the Concepts

47. A blind date. The probability that Tony will like his blind date is 0.3. What is the probability that he will not like his blind date? 0.7

48. Drawing cards. The probability of drawing a spade from a standard 52-card deck is 0.25. What is the probability of drawing a card that is not a spade? 0.75

49. Gender and the Air Force. In 2006, about 20% of members of the U.S. Air Force were women. What is the probability that an Air Force member chosen at random is a woman? (*Source:* www.af.mil/news) $\frac{1}{5}$

50. Movie attendance. If 85% of people between the ages of 18 and 24 go to a movie at least once a month, what is the probability that a randomly chosen person in this age group will go to a movie this month? $\frac{17}{20}$

51. Gender and the Marines. In 2006, about 6% of members of the U.S. Marines were women. What is the probability that a Marine chosen at random is a man? (*Source:* DMDC-June 2005, TFDW-June 2005) $\frac{47}{50}$

52. Movie attendance. If 20% of people over 65 go to a movie at least once a month, what is the probability that a randomly chosen person over 65 will *not* go to a movie this month? $\frac{4}{5}$

53. Raffles. One hundred twenty-five tickets were sold at a raffle. If you have ten tickets, what is the probability of your winning the (only) prize? If there are two prizes, what is the probability of your winning both?

54. Committee selection. To select a committee with two members, two people will be selected at random from a group of four men and four women. What is the probability that a man and a woman will be selected? $\frac{4}{7}$

55. Gender of children. A couple has two children. Assuming that each child is equally likely to be a boy or a girl, find the probability of having
   a. two boys.
   b. two girls.
   c. one boy and one girl.

56. Gender of children. A couple has three children. Assuming that each child is equally likely to be a boy or a girl, find the probability that the couple will have
   a. all girls.
   b. at least two girls.
   c. at most two girls.
   d. exactly two girls.

Answers:

53. The probability of winning one prize is $\frac{2}{25}$. The probability of winning both is $\frac{9}{1550}$. **55. a.** $\frac{1}{4}$ **b.** $\frac{1}{4}$ **c.** $\frac{1}{2}$ **56. a.** $\frac{1}{8}$ **b.** $\frac{1}{2}$ **c.** $\frac{7}{8}$ **d.** $\frac{3}{8}$

57. **Car ownership.** The following table shows the number of cars owned by each of the 4340 families in a small town.

No. of families	37	1256	2370	526	131	120
No. of cars owned	0	1	2	3	4	5

A family is selected at random. Find the probability that this family owns
a. exactly two cars.   b. at most two cars.
c. exactly six cars.   d. at most six cars.
e. at least one car.

58. **Mammograms.** According to the American Cancer Society, 199 of 200 mammograms (for breast cancer screening) turn out to be normal. Find the probability that a mammogram of a woman chosen at random is not normal.

59. **Lactose intolerance.** Lactose intolerance (the inability to digest sugars found in dairy products) affects about 20% of non-Hispanic white Americans and about 75% of African, Asian, and Native Americans. What is the probability that a non-Hispanic white American has no lactose intolerance? Answer the same question for a person in the second group.

60. **Guessing on an exam.** A student has just taken a ten-question true–false test. If he must guess to answer each of the ten questions, what is the probability that
a. he scores 100% on the test.
b. he scores 90% on the test.
c. he scores at least 90% on the test.

61. **Committee selection.** A five-person committee is choosing a subcommittee of two of its members. Each member of the committee is a different age. The subcommittee members are chosen at random.
a. What is the probability that the subcommittee will consist of the two oldest members?
b. What is the probability that the subcommittee will consist of the oldest and the youngest members?

62. **How many children do mothers have?** The following data are based on statistics published by the U.S. Census Bureau on the eve of Mother's Day in 2000: Among the 35 million mothers in the United States in the age group from 15 to 44 in 1998, 10.7 million had one child, 14.3 million had two children, 6.7 million had three, and 3.3 million had four or more. If a mother from this age group is chosen at random, what is the probability that she has more than one child? 0.6943

63. **A test for depression.** Suppose that 10% of the U.S. population suffers from depression. A pharmaceutical company is believed to have developed a test with the following results: The test works for 90% of the people who are tested and are actually depressed (as diagnosed through other acceptable methods). In other words, if a patient is actually depressed, there is a probability of 0.9 that this patient will be diagnosed as a depressed patient by the pharmaceutical company. The test also works for 85% of people who are tested and not depressed.

Suppose the company tests 1000 people for depression. Complete the following table. †

Number of people who are	Depressed according to the test	Normal according to the test
actually depressed	90	10
actually normal	135	765

64. **Playing marbles.** Natasha and Deshawn are playing a game with marbles. Natasha has one marble (call it N), and Deshawn has two (call them M_1 and M_2). They roll their marbles toward a stake in the ground. The person whose marble is closest to the stake wins. Assume that measurements are accurate enough to eliminate ties. List all of the sample points. Use the notation M_1NM_2 to denote the sample point when M_1 is closest to the stake and M_2 is farthest from the stake. If the children are equally skillful (when the sample points are all equally likely), find the probability that Deshawn wins. †

65. **Gender among siblings.** If a family with four children is chosen at random, are the children more likely to be three of one gender and one of the other gender than two boys and two girls? (Assume that each child in a family is equally likely to be a boy or a girl.)
Three of one gender and one of the other is more likely.

66. **Hospital treatments.** Hospital records show that 11% of all patients admitted are admitted for surgical treatment, 15% are admitted for obstetrics, and 3% are admitted for both obstetrics and surgery. What is the probability that a new patient admitted is an obstetrics patient or a surgery patient or both? 0.23

67. **Pain relief medicine.** A pharmaceutical company testing a new over-the-counter pain relief medicine gave the new medicine, their old medicine, and placebos to different groups of patients. (A placebo is a sugar pill with no medicinal ingredients. However, it sometimes works, supposedly for psychological reasons.) The following table shows the results of the tests.

	Amount of Pain Relief		
Medicine	**Some**	**Complete**	**None**
Placebo	12	4	44
Old	25	25	10
New	18	40	12
Total	55	69	56

Find the probability that a randomly chosen person from the group of 180 people receiving one of the three types of medicines
a. received the placebo or had no pain relief.
b. received the new medicine or had complete pain relief.

59. The probability that a non-Hispanic white American has no lactose intolerance is 0.8. The probability for the second group is $\frac{1}{4}$.

Answers:
57. a. $\frac{237}{434}$ b. $\frac{3663}{4340}$ c. 0 d. 1 e. $\frac{4303}{4340}$ 58. $\frac{1}{200}$

60. a. $\frac{1}{1024}$ b. $\frac{5}{512}$ c. $\frac{11}{1024}$ 61. a. $\frac{1}{10}$ b. $\frac{1}{10}$ 67. a. $\frac{2}{5}$ b. $\frac{89}{180}$

C EXERCISES Beyond the Basics

68. A famous problem called the *birthday problem* asks for the probability that at least two people in a group of people have the same birthday (the same day and the same month but not necessarily the same year). The best way to approach this problem is to consider the complementary event that no two people have the same birthday. Assume that all birthdays are equally likely.
   a. Find the probability that in a group of two people, both have the same birthday.
   b. Find the probability that in a group of three people, at least two have the same birthday.
   c. Find the probability that in a group of 50 people, at least two have the same birthday.

69. What is the probability that a number between 1 and 1,000,000 contains the digit 3? [*Hint:* Consider the complementary event.]

70. A business executive types four letters to her clients and then types four envelopes addressed to the clients. What is the probability that if the four letters are randomly placed in envelopes, they are placed in correct envelopes?

71. Juan gave his telephone number to Maggie. She remembers that the first three digits are 407 and the remaining four digits consist of two 3s, one 6, and one 8. She is not certain about the order of the digits, however. Maggie dials 407-3638. What is the probability that Maggie dialed the correct number?

Critical Thinking

72. The March 2005 population survey by the U.S. Census Bureau found that among the 187 million people aged 25 or older, 59,840,000 reported high school graduation as the highest level of education attained. Find the probability that a person chosen at random from the 187 million in the survey would report his or her highest level of education as high school graduation.

73. Find the probability of randomly choosing a dark caramel from a box that contains only light and dark caramels and has twice as many light as dark caramels.

74. If a single die is rolled two times, what is the probability that the second roll will result in a number larger than the first roll?

Answers:

68. a. $\dfrac{1}{365}$ **b.** $\dfrac{1093}{133{,}225}$ **c.** $\dfrac{365^{49} - \dfrac{364!}{315!}}{365^{49}}$ **69.** $\dfrac{468{,}559}{1{,}000{,}000}$ **70.** $\dfrac{1}{24}$

71. $\dfrac{1}{12}$ **72.** $\dfrac{8}{25}$ **73.** $\dfrac{1}{3}$ **74.** $\dfrac{5}{12}$

10.1 Sequences and Series

i. An infinite sequence is a function whose domain is the set of positive integers. The function values $a_1, a_2, a_3, \ldots, a_n, \ldots$ are called the **terms** of the sequence. The term a_n is the **nth term**, or **general term**, of the sequence.

ii. *Recursive formula*

A sequence may be defined recursively, with the nth term of the sequence defined in relation to previous terms.

iii. *Factorial notation*

$$n! = n(n-1)\cdots 3 \cdot 2 \cdot 1 \text{ and } 0! = 1$$

iv. *Summation notation*

$$\sum_{i=1}^{n} a_i = a_1 + a_2 + a_3 + \cdots + a_n$$

v. *Series*

Given an infinite sequence $a_1, a_2, a_3, \ldots, a_n, \ldots,$

the sum $\sum_{i=1}^{n} a_i$ is called the **nth partial sum** and the sum

$$\sum_{i=1}^{\infty} a_i = a_1 + a_2 + a_3 + \cdots \text{ is called an } \textbf{infinite series}.$$

10.2 Arithmetic Sequences; Partial Sums

i. A sequence is an **arithmetic sequence** if each term after the first differs from the preceding term by a constant. The constant difference d between the consecutive terms is called the **common difference**. We have $d = a_n - a_{n-1}$ for all $n \geq 1$.

ii. The **nth term** a_n of the arithmetic sequence $a_1, a_2, a_3 \ldots, a_n, \ldots$ is given by $a_n = a_1 + (n-1)d, n \geq 1$, where d is the common difference.

iii. The **sum** S_n of the first n terms of the arithmetic sequence $a_1, a_2, \ldots, a_n, \ldots$ is given by the formula

$$S_n = n\left(\frac{a_1 + a_n}{2}\right).$$

10.3 Geometric Sequences and Series

i. A sequence is a **geometric sequence** if each term after the first is a constant multiple of the preceding term. The constant ratio r between the consecutive terms is called the **common ratio**. We have $\dfrac{a_{n+1}}{a_n} = r, n \geq 1$.

ii. The **nth term a_n of the geometric sequence** $a_1, a_2, a_3, \ldots, a_n, \ldots$ is given by $a_n = a_1 r^{n-1}, n \geq 1$, where r is the common ratio.

iii. The **sum** S_n of the first n terms of the geometric sequence $a_1, a_2, \ldots, a_n, \ldots$ with common ratio $r \neq 1$ is given by the formula

$$S_n = \frac{a_1(1 - r^n)}{1 - r}.$$

The **sum S of an infinite geometric sequence** is given by

$$S = \sum_{i=1}^{\infty} a_1 r^{i-1} = \frac{a_1}{1 - r} \text{ if } |r| < 1.$$

iv. An **annuity** is a sequence of equal payments made at equal times. Suppose P is the payment made at the end of each of n compounding periods per year and i is the annual interest rate. Then the value A of the annuity after t years is

$$A = P\frac{\left(1 + \dfrac{i}{n}\right)^{nt} - 1}{\dfrac{i}{n}}.$$

v. If $|r| < 1$, the infinite geometric series $a_1 + a_1 r + a_1 r^2 + \cdots + a_1 r^{n-1} + \cdots$ has the sum $S = \dfrac{a_1}{1 - r}$. When $|r| \geq 1$, the series does not have a sum.

10.4 Mathematical Induction

The statement P_n is true for all positive integers n if the following properties hold:

i. P_1 is true.

ii. If P_k is a true statement, then P_{k+1} is a true statement.

10.5 The Binomial Theorem

i. Binomial coefficient: $\dbinom{n}{j} = \dfrac{n!}{j!(n-j)!}, 0 \leq j < n$

$$\binom{n}{0} = 1 = \binom{n}{n}$$

ii. Binomial theorem:

$$(x + y)^n = \binom{n}{0}x^n + \binom{n}{1}x^{n-1}y + \binom{n}{2}x^{n-2}y^2 + \cdots$$
$$+ \binom{n}{j}x^{n-j}y^j + \cdots + \binom{n}{2}y^n$$

$$= \sum_{j=0}^{n} \binom{n}{j}x^{n-j}y^j$$

iii. The term containing the factor x^j in the expansion of $(x + y)^n$ is

$$\binom{n}{n-j}x^j y^{n-j}.$$

10.6 Counting Principles

i. **Fundamental Counting Principle:** If a first choice can be made in p different ways, a second choice in q different ways, a third choice in r different ways, and so on, then the sequence of choices can be made in $p \cdot q \cdot r \ldots$ different ways.

ii. A **permutation** is an arrangement of n distinct objects in a fixed order in which no object is used more than once. The order in an arrangement is important.

iii. **Permutation formula:** The number of permutations of n distinct objects taken r at a time is

$$P(n, r) = \frac{n!}{(n-r)!} = n(n-1)(n-2)\ldots(n-r+1).$$

iv. **Combination formula:** When r objects are chosen from n distinct objects without regard to order, we call the set of r objects a **combination** of n objects taken r at a time. The number of combinations of n distinct objects taken r at a time is denoted by $C(n, r)$, where

$$C(n, r) = \frac{n!}{(n-r)! \, r!}.$$

v. **Distinguishable permutations:** The number of permutations of n objects of which n_1 are of one kind, n_2 are of a second kind, $\ldots$, and n_k are of a kth kind is

$$\frac{n!}{n_1! \, n_2! \ldots n_k!},$$

where $n_1 + n_2 + \cdots + n_k = n$.

10.7 Probability

i. **Experiment:** Any process that terminates in one or more **outcomes** (results) is an experiment.

ii. **Sample space:** The set of all possible outcomes of an experiment is called the sample space of the experiment.

iii. **Event:** An event is any subset of a sample space.

iv. **Equally likely events:** Outcomes of an experiment are said to be equally likely if no outcome should result more often than any other outcome when the experiment is performed repeatedly.

v. **Probability:** If all of the outcomes in a sample space S are equally likely, the probability of an event E, denoted $P(E)$, is defined by

$$P(E) = \frac{n(E)}{n(S)},$$

where $n(E)$ is the number of outcomes in E and $n(S)$ is the total number of outcomes in S.

a. **The Additive Rule:** If E and F are events in a sample space S, then

$$P(E \text{ or } F) = P(E \cup F) = P(E) + P(F) - P(E \cap F).$$

b. If $E \cap F = \varnothing$, E and F are called **mutually exclusive** events. For mutually exclusive events E and F, $P(E \cup F) = P(E) + P(F)$.

c. $P(E') = P(\text{not } E) = 1 - P(E)$; the event E' is called the **complement** of the event E.

REVIEW EXERCISES

In Exercises 1–4, write the first five terms of each sequence.

1. $a_n = 2n - 3$ †

2. $a_n = \dfrac{n(n-2)}{2}$ †

3. $a_n = \dfrac{n}{2n+1}$ †

4. $a_n = (-2)^{n-1}$ †

In Exercises 5 and 6, write a general term a_n for each sequence.

5. $30, 28, 26, 24, \ldots$ $a_n = 32 - 2n$

6. $-1, 2, -4, 8, \ldots$ $a_n = -(-2)^{n-1}$

In Exercises 7–10, simplify each expression.

7. $\dfrac{9!}{8!}$ 9

8. $\dfrac{10!}{7!}$ 720

9. $\dfrac{(n+1)!}{n!}$ $n + 1$

10. $\dfrac{n!}{(n-2)!}$ $n^2 - n$

In Exercises 11–14, find each sum.

11. $\displaystyle\sum_{k=1}^{4} k^3$ 100

12. $\displaystyle\sum_{j=1}^{5} \dfrac{1}{2j}$ $\dfrac{137}{120}$

13. $\displaystyle\sum_{k=1}^{7} \dfrac{k+1}{k}$ $\dfrac{1343}{140}$

14. $\displaystyle\sum_{k=1}^{5} (-1)^k 3^{k+1}$ -549

In Exercises 15 and 16, write each sum in summation notation. $\displaystyle\sum_{k=1}^{50} \dfrac{1}{k}$ $\displaystyle\sum_{k=1}^{5} -(-2)^k$

15. $1 + \dfrac{1}{2} + \dfrac{1}{3} + \dfrac{1}{4} + \cdots + \dfrac{1}{50}$ 16. $2 - 4 + 8 - 16 + 32$

In Exercises 17–20, determine whether each sequence is arithmetic. If a sequence is arithmetic, find the first term a_1 and the common difference d.

17. $11, 6, 1, -4, \ldots$
Arithmetic; $a_1 = 11, d = -5$

18. $\dfrac{2}{3}, \dfrac{5}{6}, 1, \dfrac{7}{6}, \ldots$
Arithmetic; $a_1 = \dfrac{2}{3}, d = \dfrac{1}{6}$

19. $a_n = n^2 + 4$ Not arithmetic

20. $a_n = 3n - 5$
Arithmetic; $a_1 = -2, d = 3$

In Exercises 21–24, find an expression for the nth term of each arithmetic sequence.

21. $3, 6, 9, 12, 15, \ldots$ $a_n = 3n$

22. $5, 9, 13, 17, 21, \ldots$ $a_n = 4n + 1$

23. $x, x + 1, x + 2, x + 3, x + 4, \ldots$ $a_n = x + n - 1$

24. $3x, 5x, 7x, 9x, 11x, \ldots$ $a_n = (2n + 1)x$

In Exercises 25 and 26, find the common difference d and the nth term a_n for each arithmetic sequence.

25. 3rd term: 7; 8th term: 17 $d = 2, a_n = 2n + 1$

26. 5th term: -16; 20th term: -46 $d = -2, a_n = -2n - 6$

†Due to space constrictions, answers to these exercises may be found in the Answers beginning on page A–1 in the back of the book.

In Exercises 27 and 28, find each sum.

27. $7 + 9 + 11 + 13 + \cdots + 37$ 352

28. $\dfrac{1}{4} + \dfrac{1}{2} + \dfrac{3}{4} + 1 + \cdots + 15$ $\dfrac{915}{2}$

In Exercises 29 and 30, find the sum of the first n terms of each arithmetic sequence.

29. $3, 8, 13, \ldots; n = 40$ 4020

30. $-6, -5.5, -5, \ldots; n = 60$
    525

In Exercises 31–34, determine whether each sequence is geometric. If a sequence is geometric, find the first term a_1 and the common ratio r.

31. $4, -8, 16, -32, \ldots$
    Geometric: $a_1 = 4, r = -2$

32. $\dfrac{1}{5}, \dfrac{1}{10}, \dfrac{1}{20}, \dfrac{1}{40}, \ldots$
    Geometric: $a_1 = \dfrac{1}{5}, r = \dfrac{1}{2}$

33. $\dfrac{1}{3}, 1, \dfrac{5}{3}, 9, \ldots$ Not geometric

34. $a_n = 2^{-n}$
    Geometric: $a_1 = \dfrac{1}{2}, r = \dfrac{1}{2}$

In Exercises 35 and 36, for each geometric sequence, find the first term a_1, the common ratio r, and the nth term a_n.

35. $16, -4, 1, -\dfrac{1}{4}, \ldots$

36. $-\dfrac{5}{6}, -\dfrac{1}{3}, -\dfrac{2}{15}, -\dfrac{4}{75}, \ldots$

In Exercises 37 and 38, find the indicated term of each geometric sequence.

37. a_{10} when $a_1 = 2, r = 3$
    $a_{10} = 39{,}366$

38. a_{12} when $a_1 = -2, r = \dfrac{3}{2}$

In Exercises 39 and 40, find the sum S_n of the first n terms of each geometric sequence.

39. $\dfrac{1}{10}, \dfrac{1}{5}, \dfrac{2}{5}, \dfrac{4}{5}, \ldots; n = 12$
    $S_{12} = \dfrac{819}{2}$

40. $2, -1, \dfrac{1}{2}, -\dfrac{1}{4}, \ldots; n = 10$
    $S_n = \dfrac{341}{256}$

In Exercises 41–44, find each sum.

41. $\dfrac{1}{2} + \dfrac{1}{6} + \dfrac{1}{18} + \dfrac{1}{54} + \cdots$ $\dfrac{3}{4}$

42. $-5 - 2 - \dfrac{4}{5} - \dfrac{8}{25} - \cdots$ $-\dfrac{25}{3}$

43. $\displaystyle\sum_{i=1}^{\infty} \left(\dfrac{3}{5}\right)^i$ $\dfrac{3}{2}$

44. $\displaystyle\sum_{i=1}^{\infty} 7\left(-\dfrac{1}{4}\right)^i$ $-\dfrac{7}{5}$

In Exercises 45 and 46, use mathematical induction to prove that each statement is true for all natural numbers n.

45. $\displaystyle\sum_{k=1}^{n} 2^k = 2^{n+1} - 2$

46. $\displaystyle\sum_{k=1}^{n} k(k+1) = \dfrac{n(n+1)(n+2)}{3}$

In Exercises 47 and 48, evaluate each binomial coefficient.

47. $\dbinom{12}{7}$ $\dbinom{12}{7} = 792$

48. $\dbinom{11}{0}$ $\dbinom{11}{0} = 1$

In Exercises 49 and 50, find each binomial expansion.

49. $(x - 3)^4$
    $(x - 3)^4 = x^4 - 12x^3 + 54x^2 - 108x + 81$

50. $\left(\dfrac{x}{2} + 2\right)^6$

In Exercises 51 and 52, find the specified term.

51. $(x + 2)^{12}$; term containing x^5 $101{,}376x^5$

52. $(x + 2y)^{12}$; term containing x^7 $25{,}344x^7y^5$

Answers:

35. $a_1 = 16, r = -\dfrac{1}{4}, a_n = \dfrac{(-1)^{n-1}}{4^{n-3}}$

36. $a_1 = -\dfrac{5}{6}, r = \dfrac{2}{5}, a_n = -\dfrac{1}{3}\left(\dfrac{2}{5}\right)^{n-2}$

38. $a_{12} = -\dfrac{177{,}147}{1024}$

50. $\left(\dfrac{x}{2} + 2\right)^6 = \dfrac{1}{64}x^6 + \dfrac{3}{8}x^5 + \dfrac{15}{4}x^4 + 20x^3 + 60x^2 + 96x + 64$

In Exercises 53–64, use any method to solve the problem.

53. How many different numbers can be written with the digits 1, 2, 3, and 4 if no digit may be repeated? 24

54. How many ways can horses finish in first, second, and third place in a ten-horse race? 720

55. How many ways can seven people line up at a ticket window? 5040

56. In a building with seven entrances, how many ways can you enter the building and leave by a different entrance? 42

57. How many ways can five movies be listed on a display (vertically, one movie per line)? 120

58. How many different amounts of money can you make with a nickel, a dime, a quarter, and a half-dollar? 15

59. How many ways can three candies be taken from a box of ten candies? 120

60. How many ways can 2 face cards be drawn from a standard 52-card deck? 66

61. How many ways can 2 shirts and 3 pairs of pants be chosen from 12 shirts and 8 pairs of pants? 3696

62. How many doubles teams can be formed from eight tennis players? 28

63. In how many distinguishable ways can the letters of the word R E I T E R A T E be arranged? 15,120

64. A bookshelf has room for four books. How many different arrangements can be made on the shelf from six available books? 360

In Exercises 65–74, find the probability requested.

65. Four silver dollars, all with different dates, are randomly arranged on a horizontal display. What is the probability that the two with the most recent dates will be next to each other? $\dfrac{1}{2}$

66. A jar contains five white, three black, and two green stones. What is the probability of drawing a black stone on a single draw? $\dfrac{3}{10}$

67. The word R A N D O M has been spelled in a game of Scrabble. If two letters are chosen from this word at random, what is the probability that they are both consonants? $\dfrac{2}{5}$

68. The names of all 30 party guests are placed in a box, and one name is drawn to award a door prize. If you and your date are among the guests, what is the probability that one of you will win the prize? $\dfrac{1}{15}$

69. If two people are chosen at random from a group consisting of five men and three women, what is the probability that one man and one woman are chosen? $\dfrac{15}{28}$

70. A company determines that 32% of the e-mail it receives is junk mail. What is the probability that a randomly chosen e-mail at this company is not junk mail? $\dfrac{17}{25}$

71. If 2 cards are drawn from a standard 52-card deck, what is the probability that both are clubs? $\dfrac{1}{17}$

72. A battery manufacturer inspects 500 batteries and finds that 30 are defective. What is the probability that a randomly selected battery is not defective? $\dfrac{47}{50}$

73. A ball is taken at random from a pool table containing balls numbered 1 through 9. What is the probability of obtaining (a) ball number 7? (b) an even-numbered ball? (c) an odd-numbered ball? **a.** $\frac{1}{9}$ **b.** $\frac{4}{9}$ **c.** $\frac{5}{9}$

74. A clothes dryer load contains two shirts, four pairs of socks, and three nightgowns. You pull one item randomly from the dryer.
a. What is the probability that it is a sock? $\frac{8}{13}$
b. What is the probability that it is a nightgown? $\frac{3}{13}$

PRACTICE TEST A

In Problems 1 and 2, write the first five terms of each sequence. State whether the sequence is arithmetic or geometric.

1. $a_n = 3(5 - 4n)$ †

2. $a_n = -3(2^n)$ †

3. Write the first five terms of the sequence defined by $a_1 = -2$ and $a_n = 3a_{n-1} + 5$ for $n \geq 2$. †

4. Evaluate $\dfrac{(n-1)!}{n!}$. $\dfrac{1}{n}$

5. Find a_7 for the arithmetic sequence with $a_1 = -3$ and $d = 4$. $a_7 = 21$

6. Find a_8 for the geometric sequence with $a_1 = 13$ and $r = -\dfrac{1}{2}$. $a_8 = -\dfrac{13}{128}$

In Problems 7–9, find each sum.

7. $\displaystyle\sum_{k=1}^{20} (3k - 2)$ 590

8. $\displaystyle\sum_{k=1}^{5} \left(\frac{3}{4}\right)(2^{-k})$ $\dfrac{93}{128}$

9. $\displaystyle\sum_{k=1}^{\infty} 18\left(\frac{1}{100}\right)^k$; write the answer as a fraction in lowest terms. $\dfrac{2}{11}$

10. Evaluate $\dbinom{13}{0}$. 1

11. Expand $(1 - 2x)^4$. $16x^4 - 32x^3 + 24x^2 - 8x + 1$

12. Find the term containing x^1 in the expansion of $(2x + 1)^4$. $8x$

13. In how many ways can you answer every question on a true–false test containing ten questions? 1024

In Problems 14 and 15, evaluate each expression.

14. $P(9, 2)$ 72

15. $C(7, 5)$ 21

16. A firm needs to fill two positions—one in accounting and one in human resources. How many ways can these positions be filled if there are six applicants for the accounting position and ten applicants for the human resources position? 60

17. A state is considering using license plates consisting of two letters followed by a four-digit number. How many distinct license plate numbers can be formed? 6,760,000

18. A box contains five red balls and seven black balls. What is the probability that a ball drawn at random will be red? $\dfrac{5}{12}$

19. What is the probability that a 4 does *not* come up on one roll of a die? $\dfrac{5}{6}$

20. A box contains seven quarters, six dimes, and five nickels. If two coins are drawn at random, what is the probability that both are quarters? $\dfrac{7}{51}$

PRACTICE TEST B

In Problems 1 and 2, write the first five terms of each sequence. State whether the sequence is arithmetic or geometric.

1. $a_n = 4(2n - 3)$ c
  a. $-1, 1, 3, 5, 7$; arithmetic
  b. $-1, 1, 3, 5, 7$; geometric
  c. $-4, 4, 12, 20, 28$; arithmetic
  d. $-4, 4, 12, 20, 28$; geometric

2. $a_n = 2(4^n)$ d
  a. $2, 8, 32, 128, 512$; arithmetic
  b. $2, 8, 32, 128, 512$; geometric
  c. $8, 32, 128, 512, 2048$; arithmetic
  d. $8, 32, 128, 512, 2048$; geometric

3. Write the first five terms of the sequence defined by $a_1 = 5$ and $a_n = 2a_{n-1} + 4$, for $n \geq 2$. a
  a. $5, 14, 32, 68, 140$ **b.** $5, 14, 19, 24, 29$
  c. $5, 9, 13, 17, 21$ **d.** $5, 8, 10, 12, 14$

4. Evaluate $\dfrac{(n+2)!}{n+2}$. c
  a. $2!$ **b.** 1
  c. $(n+1)!$ **d.** $n+1$

5. Find a_8 for the arithmetic sequence with $a_1 = -6$ and $d = 3$. a
  a. 15   **b.** -271
  c. 30   **d.** 18

6. Find a_{10} for the geometric sequence with $a_1 = 247$ and $r = \dfrac{1}{3}$. b
  a. 250   **b.** $\dfrac{247}{19,683}$
  c. $\dfrac{247}{59,049}$ **d.** $\dfrac{247}{177,147}$

In Problems 7 and 8, find each sum.

7. $\displaystyle\sum_{k=1}^{45} (4k - 7)$ d
   a. 4185
   b. 3735
   c. 4027.5
   d. 3825

8. $\displaystyle\sum_{k=1}^{5} \left(\frac{4}{3}\right)(2^k)$ a
   a. $\dfrac{248}{3}$
   b. $\dfrac{251}{3}$
   c. $\dfrac{242}{3}$
   d. $\dfrac{287}{3}$

9. Evaluate $\displaystyle\sum_{k=1}^{\infty} 8(-0.3)^{k-1}$. b
   a. 6
   b. 6.15
   c. −6.15
   d. −4.15

10. Evaluate $\dbinom{12}{11}$. b
   a. $1.\overline{09}$
   b. 12
   c. 1
   d. 11!

11. Expand $(3x - 1)^4$. a
   a. $81x^4 - 108x^3 + 54x^2 - 12x + 1$
   b. $-81x^4 + 108x^3 - 54x^2 + 12x - 1$
   c. $-81x^4 + 108x^3 + 54x^2 + 12x + 1$
   d. $81x^4 - 108x^2 + 54x - 12$

12. Find the coefficient of x in the expansion of $(2x + 3)^3$. d
   a. 108
   b. 9
   c. 36
   d. 54

13. Karen has four necklaces, ten pairs of earrings, and three bracelets. How many ways can she choose a necklace, a pair of earrings, and a bracelet? c
   a. 40
   b. 17
   c. 120
   d. 240

In Problems 14 and 15, evaluate each expression.

14. $P(7, 3)$ a
   a. 210
   b. 840
   c. 1680
   d. 5040

15. $C(10, 7)$ d
   a. 3
   b. 720
   c. 620,400
   d. 120

16. How many two-digit numbers can be formed from the digits 0, 1, 2, 3, 4, 5, 6, 7, 8, and 9? No digit can be used more than once. b
   a. 45
   b. 81
   c. 3,628,880
   d. 100

17. You have selected eight CDs for your dad, but have only enough money to buy five of them. In how many ways can you select the five if you decide that a particular CD is a "must buy"? d
   a. 56
   b. 70
   c. 1680
   d. 35

18. What is the probability of rolling a 5 on one roll of a die? c
   a. $\dfrac{1}{5}$
   b. 4
   c. $\dfrac{1}{6}$
   d. 0.05

19. What is the probability of getting a sum greater than 9 when a pair of dice is rolled? a
   a. $\dfrac{1}{6}$
   b. $\dfrac{1}{4}$
   c. $\dfrac{1}{12}$
   d. $\dfrac{1}{18}$

20. What is the probability that a card drawn at random from a standard 52-card deck is *not* a spade? d
   a. $\dfrac{2}{5}$
   b. $\dfrac{1}{4}$
   c. $\dfrac{4}{13}$
   d. $\dfrac{3}{4}$

CUMULATIVE REVIEW EXERCISES ▪ Chapters 1–10

In Problems 1–9, solve each equation or inequality.

1. $|3x - 8| = |x|$ {2, 4}

2. $x^2(x^2 - 5) = -4$ {−2, −1, 1, 2}

3. $\log_2 (3x - 5) + \log_2 x = 1$ $x = 2$

4. $e^{2x} - e^x - 2 = 0$ $x = \ln 2$

5. $\dfrac{6}{x + 2} = \dfrac{4}{x}$ $x = 4$

6. $2(x - 8)^{-1} = (x - 2)^{-1}$ $x = -4$

7. $\sqrt{x - 3} = \sqrt{x} - 1$ $x = 4$

8. $x^2 + 4x \geq 0$ $(-\infty, -4] \cup [0, \infty)$

9. $18 \leq x^2 + 6x$ $(-\infty, -3 - 3\sqrt{3}] \cup [-3 + 3\sqrt{3}, \infty)$

In Problems 10–11, solve each system.

10. $\begin{cases} 5x + 4y = 6 \\ 4x - 3y = 11 \end{cases}$ $x = 2, y = -1$

11. $\begin{cases} x - 3y + 6z = -8 \\ 5x - 6y - 2z = 7 \\ 3x - 2y - 10z = 11 \end{cases}$ $x = 10, y = 7, z = \dfrac{1}{2}$

12. Graph the function $y = \cos\dfrac{2}{3}x$ over one period. †

13. Use identities to find the exact value of $\sin\left(-\dfrac{7\pi}{12}\right)$.

14. Write the complex number $2 + 2\sqrt{3}i$ in polar form.

15. Use transformations to sketch the graph of $f(x) = \sqrt{x - 3}$. †

16. Find the domain of $f(x) = \dfrac{x + 3}{x(x + 1)}$.

Answers: **13.** $\dfrac{-\sqrt{6} - \sqrt{2}}{4}$ **14.** $4\left(\cos\dfrac{\pi}{3} + i\sin\dfrac{\pi}{3}\right)$ **16.** $(-\infty, -1) \cup (-1, 0) \cup (0, \infty)$

17. Write $7 \ln x - \ln (x - 5)$ in condensed form.

18. If $A = \begin{bmatrix} 1 & 0 & -2 \\ 4 & 1 & 0 \\ 1 & 1 & 7 \end{bmatrix}$, find A^{-1}. $A^{-1} = \begin{bmatrix} 7 & -2 & 2 \\ -28 & 9 & -8 \\ 3 & -1 & 1 \end{bmatrix}$

19. Four students are pairing up to ride two tandem bicycles, each of which accommodates two riders, one in back and one in front. How many ways can this be done if we count arrangements as different if either the front or back rider is different? 12

Answers:

17. $\ln \left(\dfrac{x^7}{x - 5} \right)$

20. A room has 6 people who are 25 years old or older and 12 people whose age is between 17 and 24. Half the members of each group have gotten speeding tickets at some time. What is the probability that a person chosen at random from the room will be a person who is 25 years old or older or has gotten a speeding ticket at some time?

$\dfrac{2}{3}$

Review

The Real Numbers; Integer Exponents

Objectives

1 Classify sets of real numbers.

2 Specify sets of numbers in roster or set-builder notation.

3 Use interval notation.

4 Relate absolute value and distance on the real number line.

5 Evaluate algebraic expressions.

6 Use integer exponents.

7 Use the rules of exponents.

---

1 Classify sets of real numbers.

Classifying Sets of Numbers

The idea of a set is familiar to us all. We regularly refer to "a set of baseball cards," a "set of CDs," and "a set of dishes." In mathematics, as in everyday life, a **set** is a collection of objects. The objects in the set are called the **elements**, or **members**, of the set. In the study of algebra, we are interested primarily in sets of numbers. In listing the elements of a set, it is customary to enclose the listed elements in braces, { }, and separate them by commas.

We distinguish among various sets of numbers. The numbers we use to count with constitute the set of **natural numbers**: $\{1, 2, 3, 4, \ldots\}$. The three dots $\ldots$ can be read as "and so on" and indicate that the pattern continues indefinitely. The **whole numbers** are formed by adjoining the number 0 to the set of natural numbers to obtain $\{0, 1, 2, 3, 4, \ldots\}$. The **integers** consist of the set of natural numbers together with their opposites and 0: $\{\ldots, -4, -3, -2, -1, 0, 1, 2, 3, 4, \ldots\}$. The **rational numbers** consist of all numbers that *can* be expressed as the quotient, $\frac{a}{b}$, of two integers, where $b \neq 0$.

Examples of rational numbers are $\frac{1}{2}, \frac{5}{3}, -\frac{4}{17}$, and $\frac{7}{100}$. Any integer a can be expressed as the quotient of two integers by writing $a = \frac{a}{1}$. Consequently, every integer is also a rational number. The rational number $\frac{a}{b}$ can be written as a decimal by using long division. When you take two integers a and b and divide a by b, the result is either a **terminating decimal**, such as $\left(\frac{1}{2} = 0.5\right)$, or a **nonterminating repeating decimal**, such as $\left(\frac{2}{3} = 0.666\ldots\right)$.

There are decimals that neither terminate nor repeat. Decimals that neither terminate nor repeat represent **irrational numbers**. The rational numbers together with the irrational numbers form the **real numbers**.

Real Number Line

We associate the real numbers with points on a geometric line (imagined to be extended indefinitely in both directions) in such a way that each real number corresponds to exactly one point and each point corresponds to exactly one real number. The point is called the **graph** of the corresponding real number, and the real number is called the **coordinate** of the point. By agreement, **positive numbers** lie to the right of the point corresponding to 0 and **negative numbers** lie to the left of 0. See Figure A.1.

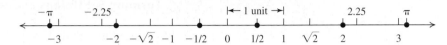

FIGURE A.1

Notice that $\frac{1}{2}$ and $-\frac{1}{2}$, and 2 and -2, and π and $-\pi$ correspond to pairs of points exactly the same distance from 0 but on opposite sides of 0.

When coordinates have been assigned to points on a line in the manner just described, the line is called a **real number line**, a **coordinate line**, a **real line**, or simply a **number line**. The point corresponding to 0 is called the **origin**.

Inequalities

The real numbers are **ordered** by their size. We say that *a* **is less than** *b* and write $a < b$, provided that $b = a + c$ for some *positive* number *c*. We also write $b > a$, meaning the same thing as $a < b$, and say that *b* **is greater than** *a*. On the real line, the numbers increase from left to right. Consequently, *a* **is to the left of** *b* **on the number line when** $a < b$. Similarly, *a* is to the right of *b* on the number line when $a > b$. We sometimes want to indicate that at least one of two conditions is correct: Either $a < b$ or $a = b$. In this case, we write $a \leq b$ or $b \geq a$. The four symbols, $<, >, \leq$, and $\geq$ are called **inequality symbols**.

The following order properties of inequalities are used throughout this text:

The *trichotomy property* says that if two real numbers are not equal, then one is larger than the other. The *transitive property* says that "less than" works like "smaller than" or "lighter than."

Frequently, we read $a > 0$ as "*a* is positive" instead of "*a* is greater than 0." We can also read $a < 0$ as "*a* is negative." If $a \geq 0$, then either $a > 0$ or $a = 0$ and we may say that "*a* is nonnegative."

2 Specify sets of numbers in roster or set-builder notation.

Sets

To specify a set, we do one of the following:

1. List the elements of the set (**roster method**)

2. Describe the elements of the set (often using **set-builder notation**)

Variables are helpful in describing sets when we use set-builder notation. The notation $\{x \mid x \text{ is a natural number less than } 6\}$ is in set-builder notation and describes the set $\{1, 2, 3, 4, 5\}$, using the roster method. We read $\{x \mid x \text{ is a natural number less than } 6\}$ as "the set of all *x* such that *x* is a natural number less than six." Generally, $\{x \mid x \text{ has property } P\}$ designates the set of all *x* such that (the vertical bar is read "such that") *x* has property *P*.

It may happen that a description fails to describe any number. For example, consider $\{x \mid x < 2 \text{ and } x > 7\}$. Of course, no number can be simultaneously less than 2 and greater than 7; so this set has no members. We refer to a set with no elements as the **empty set**, or **null set**, and use the special symbol $\varnothing$ to denote it.

We say that two sets *A* and *B* are **equal** if they have exactly the same elements. If each element of a set *A* is also an element of a set *B*, we say that *A* is a **subset** of *B*.

The symbol $\subseteq$ is used to denote the subset relation so that $A \subseteq B$ is read "A is a subset of B." For example, $\{0, 3\} \subseteq \{0, 1, 2, 3, 4\}$ because 0 and 3 are both elements of the set $\{0, 1, 2, 3, 4\}$.

The **union** of two sets A and B, denoted $A \cup B$, is the set consisting of all elements that are in A *or* B (or both). The **intersection** of A and B, denoted $A \cap B$, is the set consisting of all elements that are in both A *and* B. In other words, $A \cap B$ consists of the elements common to A and B.

EXAMPLE 1   Forming Set Unions and Intersections

Find $A \cap B$ and $A \cup B$ assuming that $A = \{-2, -1, 0, 1, 2\}$ and $B = \{-4, -2, 0, 2, 4\}$.

SOLUTION

$A \cap B = \{-2, 0, 2\}$, the set of elements common to both A and B.
$A \cup B = \{-4, -2, -1, 0, 1, 2, 4\}$, the set of elements that are in A or in B (or in both).

■ ■ ■

Practice Problem 1   Find $A \cap B$ and $A \cup B$ if $A = \{-3, -1, 0, 1, 3\}$ and $B = \{-4, -2, 0, 2, 4\}$.

■

When working with sets, we frequently designate a **universal set**, the set consisting of all elements under consideration. Then every set is a subset of the universal set, and the set of elements in the universal set that are not in a set A form the *complement of A*, denoted A'.

EXAMPLE 2   Finding the Complement of a Set

If $U = \{0, 1, 2, 3, 4, 5, 6, 7, 8, 9\}$ is the universal set and $A = \{0, 2, 4, 6, 8\}$, find A'.

SOLUTION

A' is the set of elements in U that are not in A; so $A' = \{1, 3, 5, 7, 9\}$.   ■ ■ ■

Practice Problem 2   If $U = \{0, 1, 2, 3, 4, 5, 6, 7, 8, 9\}$ is the universal set and $A = \{0, 3, 6, 9\}$, find A'.

■

3 | Use interval notation.

Intervals

We now turn our attention to graphing certain sets of numbers. That is, we graph each number in a given set. We are particularly interested in sets of real numbers, called **intervals**, whose graphs correspond to special sections of the number line.

If $a < b$, then the set of real numbers between a and b, but not including either a or b, is called the **open interval** from a to b and is denoted by (a, b). See Figure A.2. Using set-builder notation, we can write

$$(a, b) = \{x | a < x < b\}.$$

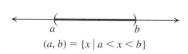

$(a, b) = \{x \mid a < x < b\}$

FIGURE A.2

We indicate graphically that the endpoints a and b are excluded from the open interval by drawing a left parenthesis at a and a right parenthesis at b. These parentheses enclose the numbers between a and b.

The **closed interval** from a to b is the set

$$[a, b] = \{x | a \le x \le b\}.$$

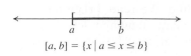

$[a, b] = \{x \mid a \le x \le b\}$

FIGURE A.3

The closed interval includes both endpoints a and b. We replace the parentheses by square brackets both in the interval notation and on the graph. See Figure A.3. Sometimes we want to include only one endpoint of an interval and exclude the other. Table A.1 on page 755 shows how this is done.

We also want to graph all numbers to the left or right of a given number, such as

$$\{x | x > 2\}.$$

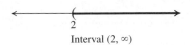

Interval $(2, \infty)$

FIGURE A.4

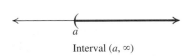

Interval (a, ∞)

FIGURE A.5

The number 2 is not included in this set, and we again indicate this fact graphically by drawing a left parenthesis at 2. We then graph all points to the right of 2, as shown in Figure A.4.

For any number a, we call $\{x \mid x > a\}$ an **unbounded** (or infinite) **interval** and denote this interval by (a, ∞). The symbol ∞ ("infinity") does not represent a number. The notation (a, ∞) is used to indicate the set of all real numbers that are greater than a, and the symbol ∞ is used to indicate that the interval extends indefinitely to the right of a. See Figure A.5.

The symbol $-\infty$ is another symbol that does not represent a number. The notation $(-\infty, a)$ is used to indicate the set of all real numbers that are less than a. The notation $(-\infty, \infty)$ represents the set of all real numbers.

Table A.1 lists various types of intervals that we use in this book. In the table, when two points a and b are involved, we assume that $a < b$.

TABLE A.1

Interval Notation	Set-Builder Notation	Graph
(a, b)	$\{x \mid a < x < b\}$	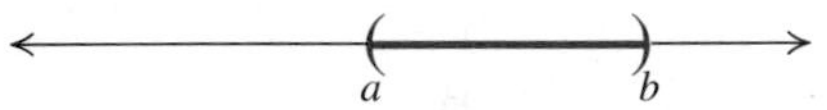
$[a, b]$	$\{x \mid a \le x \le b\}$	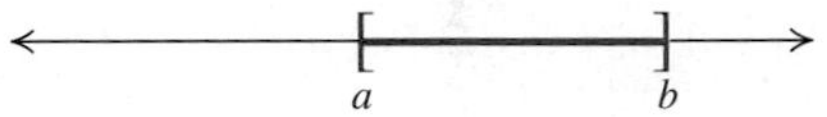
$(a, b]$	$\{x \mid a < x \le b\}$	
$[a, b)$	$\{x \mid a \le x < b\}$	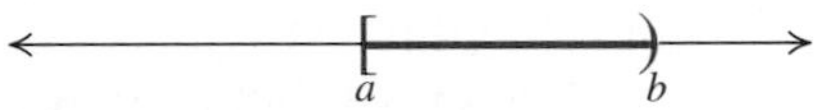
(a, ∞)	$\{x \mid x > a\}$	
$[a, \infty)$	$\{x \mid x \ge a\}$	
$(-\infty, b)$	$\{x \mid x < b\}$	
$(-\infty, b]$	$\{x \mid x \le b\}$	
$(-\infty, \infty)$	$(x \mid x \text{ is a real number})$	

STUDY TIP

The symbols ∞ and $-\infty$ are always used with parentheses, not square brackets. Also note that $<$ and $>$ correspond to parentheses and that $\le$ and $\ge$ correspond to square brackets.

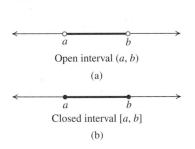

Open interval (a, b)

(a)

Closed interval $[a, b]$

(b)

FIGURE A.6

4 | Relate absolute value and distance on the real number line.

An alternative notation for indicating whether endpoints are included uses closed circles to show inclusion and open circles to show exclusion. See Figure A.6.

Absolute Value

The *absolute value* of a number a, denoted by $|a|$, is the distance between the origin and the point on the number line with coordinate a. The point with coordinate -3 is three units from the origin, so we write $|-3| = 3$ and say that the absolute value of -3 is 3. See Figure A.7.

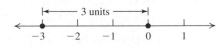

FIGURE A.7

ABSOLUTE VALUE

For any real number a, the **absolute value** of a, denoted $|a|$, is defined by

$$|a| = a \ \text{ if } a \ge 0 \quad \text{ and } \quad |a| = -a \ \text{ if } a < 0.$$

EXAMPLE 3   **Determining Absolute Value**

Find the absolute value of each of the following expressions.

a. $|4|$ **b.** $|-4|$ **c.** $|0|$ **d.** $|(-3) + 1|$

SOLUTION

a. $|4| = 4$ **b.** $|-4| = -(-4) = 4$

c. $|0| = 0$ **d.** $|(-3) + 1| = |-2| = -(-2) = 2$ ■ ■ ■

Practice Problem 3   Find the absolute value of each of the following.

a. $|-10|$ **b.** $|3 - 4|$ **c.** $|2(-3) + 7|$ ■

◆ **WARNING**   The absolute value of a number represents a distance. Because distance can never be negative, the absolute value of a number is never negative; it is always positive or zero. However, if a is not 0, $-|a|$ is always negative. Thus, $-|5.3| = -5.3$, $-|-4| = -4$, and $-|1.18| = -1.18$.

TECHNOLOGY CONNECTION

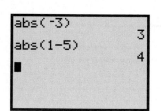 The absolute value function on your graphing calculator will find the value of the expression entered and then compute its absolute value.

```
abs(-3)
            3
abs(1-5)
            4
■
```

Distance Between Two Points on a Real Number Line

DISTANCE FORMULA ON A NUMBER LINE

If a and b are the coordinates of two points on a number line, then the distance between a and b, denoted by $d(a, b)$, is $|a - b|$. In symbols, $d(a, b) = |a - b|$.

EXAMPLE 4   **Finding the Distance Between Two Points**

Find the distance between -3 and 4 on the number line.

SOLUTION

Figure A.8 shows that the distance between -3 and 4 is seven units. The distance formula gives the same answer.

$$d(-3, 4) = |-3 - 4| = |-7| = -(-7) = 7.$$

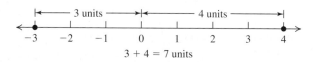

3 + 4 = 7 units

FIGURE A.8

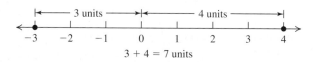

FIGURE A.9

Notice that reversing the order of -3 and 4 in this computation gives the same answer. That is, the distance between 4 and -3 is $|4 - (-3)| = |4 + 3| = |7| = 7$. It is always true that $|a - b| = |b - a|$. See Figure A.9.   ■ ■ ■

Practice Problem 4   Find the distance between -7 and 2 on the number line.   ■

5 Evaluate algebraic expressions.

Algebraic Expressions

In algebra, we use letters such as $a, b, x,$ and y to represent numbers. A letter that is used to represent one or more numbers is called a **variable**. A **constant** is either a specific number, such as 3 or $\frac{1}{2}$, or a letter that represents a fixed (but not necessarily specified) number. Any number, constant, variable, or parenthetical group (or any product of them) is called a **term**. A combination of terms using the ordinary operations of addition, subtraction, multiplication, and division (as well as exponentiation and roots, which are discussed later in this chapter) is called an **algebraic expression**, or simply an **expression**. Here are some examples of algebraic expressions:

$$x - 2 \qquad \frac{1}{x} + \sqrt{7} \qquad \frac{10}{y + 3} \qquad \sqrt{x} + |y| \div 5$$

If we replace each variable in an expression with a specific number and get a real number, the resulting number is called the **value of the algebraic expression**.

EXAMPLE 5   Evaluating an Algebraic Expression

Evaluate the following expressions for the given values of the variables.

a. $[(9 + x) \div 7] \cdot 3 - x$ for $x = 5$

b. $|x| - \dfrac{2}{y}$ for $x = 2$ and $y = -2$

SOLUTION

a. $\begin{aligned}[(9 + x) \div 7] \cdot 3 - x &= [(9 + 5) \div 7] \cdot 3 - 5 \qquad \text{Replace } x \text{ with 5.}\\ &= [14 \div 7] \cdot 3 - 5\\ &= 2 \cdot 3 - 5\\ &= 6 - 5\\ &= 1\end{aligned}$

b. $\begin{aligned}|x| - \frac{2}{y} &= |2| - \frac{2}{-2} \qquad \text{Replace } x \text{ with 2 and } y \text{ with } -2.\\ &= 2 - \frac{2}{-2}\\ &= 2 - (-1)\\ &= 3\end{aligned}$ ■ ■ ■

Practice Problem 5   Evaluate.

a. $(x - 2) \div 3 + x$ for $x = 3$ **b.** $7 - \dfrac{x}{|y|}$ for $x = -1, y = 3$ ■

The following two properties of real numbers are used frequently in algebra. The first property is used in simplifying expressions, and the second is used in solving equations.

Distributive property	$a \cdot (b + c) = a \cdot b + a \cdot c$
Zero-product property	If $a \cdot b = 0$, then $a = 0$ or $b = 0$.

The Domain of an Algebraic Expression

The **domain** of an algebraic expression having only one variable is the set of all possible numbers that may be used for the variable. For example, division by 0 is not defined, so no number can replace a variable if a denominator of 0 results. Thus, the domain of the expression $\dfrac{1}{x - 1}$ is the set of all real numbers $x, x \neq 1$.

EXAMPLE 6   **Finding the Domain of an Algebraic Expression**

Find the domain of the expression $\dfrac{5}{3-x}$.

SOLUTION

Because replacing x with 3 in $\dfrac{5}{3-x}$ results in a denominator of 0, the number 3 is not in its domain. In set notation, the domain is $\{x|x \neq 3\}$; in interval notation, it is $(-\infty, 3) \cup (3, \infty)$. ▪▪▪

Practice Problem 6   Find the domain of the variable x in the expression $\dfrac{2-x}{x+2}$. ▪

6 Use integer exponents.

Integer Exponents

The area of a square with side 5 feet is $5 \cdot 5 = 25$ square feet. The volume of a cube whose sides are each 5 feet is $5 \cdot 5 \cdot 5 = 125$ cubic feet. A shorter notation for $5 \cdot 5$ is 5^2 and for $5 \cdot 5 \cdot 5$ is 5^3. The number 5 is called the *base* for both 5^2 and 5^3. The number 2 is called the *exponent* in the expression 5^2 and indicates that the base 5 appears as a factor twice.

DEFINITION OF A POSITIVE INTEGER EXPONENT

If a is a real number and n is a positive integer, then

$$a^n = \underbrace{a \cdot a \cdot \cdots \cdot a}_{n \text{ factors}},$$

where a is the base and n is the exponent. We adopt the convention that $a^1 = a$.

We can evaluate exponents as follows:

$$5^3 = 5 \cdot 5 \cdot 5 = 125$$
$$(-3)^2 = (-3)(-3) = 9$$
$$-3^2 = -(3 \cdot 3) = -9$$
$$(-2)^3 = (-2)(-2)(-2) = -8$$

Pay careful attention to the fact that $(-3)^2 \neq -3^2$.

Negative exponents indicate the **reciprocal** of a number.

DEFINITION OF ZERO AND NEGATIVE INTEGER EXPONENTS

For any nonzero number a and any positive integer n,

$$a^0 = 1 \quad \text{and} \quad a^{-n} = \frac{1}{a^n}.$$

TECHNOLOGY CONNECTION

The notation for 5^3 on a graphing calculator is 5^3. Any expression on a calculator enclosed in parentheses and followed by ^n will be raised to the nth power. A common error is to forget parentheses when computing an expression such as $(-3)^2$, with the result being -3^2.

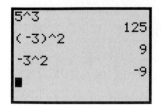

EXAMPLE 7   **Evaluating Zero or Negative Exponents**

Evaluate.

a. $(-5)^{-2}$ **b.** -5^{-2} **c.** 8^0 **d.** $\left(\dfrac{2}{3}\right)^{-3}$

SOLUTION

a. $(-5)^{-2} = \dfrac{1}{(-5)^2} = \dfrac{1}{25}$ **b.** $-5^{-2} = -\dfrac{1}{5^2} = -\dfrac{1}{25}$

c. $8^0 = 1$ **d.** $\left(\dfrac{2}{3}\right)^{-3} = \dfrac{1}{\left(\dfrac{2}{3}\right)^3} = \dfrac{1}{\dfrac{8}{27}} = \dfrac{27}{8}$ ■ ■ ■

Practice Problem 7   Evaluate.

a. 2^{-1} **b.** $\left(\dfrac{4}{5}\right)^0$ **c.** $\left(\dfrac{3}{2}\right)^{-2}$ ■

7 Use the rules of exponents.

Rules of Exponents

The following rules of exponents can be proved using the previous definitions. All expressions are assumed to be defined.

RULES OF EXPONENTS

$$a^m \cdot a^n = a^{m+n} \qquad \dfrac{a^m}{a^n} = a^{m-n} \qquad (a^m)^n = a^{mn}$$

$$(a \cdot b)^n = a^n \cdot b^n \qquad \left(\dfrac{a}{b}\right)^n = \dfrac{a^n}{b^n}$$

STUDY TIP

There are many different ways to simplify exponential expressions correctly. The order in which you apply the rules for exponents is a matter of personal preference.

EXAMPLE 8   **Using the Rules of Exponents**

Use the rules of exponents to write each expression without negative exponents.

a. $(-4x^2y^3)(7x^3y)$ **b.** $\left(\dfrac{x^5}{2y^{-3}}\right)^{-3}$

SOLUTION

a. $(-4x^2y^3)(7x^3y) = (-4)(7)x^2x^3y^3y$

$\qquad\qquad\qquad\quad = -28x^{2+3}y^{3+1}$

$\qquad\qquad\qquad\quad = -28x^5y^4$

b. $\left(\dfrac{x^5}{2y^{-3}}\right)^{-3} = \dfrac{(x^5)^{-3}}{(2y^{-3})^{-3}}$ $\left(\dfrac{a}{b}\right)^n = \dfrac{a^n}{b^n}$

$\qquad\qquad\quad = \dfrac{x^{5(-3)}}{2^{-3}(y^{-3})^{-3}}$ $(ab)^n = a^nb^n$

$\qquad\qquad\quad = \dfrac{x^{-15}}{2^{-3}y^{(-3)(-3)}}$ $(a^m)^n = a^{mn}$

$\qquad\qquad\quad = \dfrac{x^{-15}}{2^{-3}y^9}$ Simplify.

$\qquad\qquad\quad = \dfrac{x^{-15}x^{15}2^3}{x^{15}2^32^{-3}y^9}$ $\dfrac{a}{b} = \dfrac{ac}{bc}, c \neq 0$

$\qquad\qquad\quad = \dfrac{2^3}{x^{15}y^9}$ $a^0 = 1$

$\qquad\qquad\quad = \dfrac{8}{x^{15}y^9}$ ■ ■ ■

Practice Problem 8   Simplify each expression.

a. $(2x^4)^{-2}$ **b.** $\dfrac{x^2(-y)^3}{(xy^2)^3}$ ■

SECTION A.1 ■ Exercises

A EXERCISES Basic Skills and Concepts

1. The whole numbers are formed by adding the number ___0___ to the set of natural numbers.

2. The number -3 is an integer, but it is also a(n) ___a rational number, as well as a real number___.

3. If $a < b$, then a is to the ___left___ of b on the number line.

4. If a real number is not a rational number, it is a(n) ___irrational___ number.

In Exercises 5–8, fill in the blank with one of the symbols $=$, $<$, or $>$ to produce a true statement.

5. 3 ___$>$___ -2

6. Because x is less than $x + 1$, we write x ___$<$___ $x + 1$.

7. If $-x$ is positive, we write $-x$ ___$>$___ 0.

8. $\dfrac{-5}{2}$ ___$=$___ $-2\dfrac{1}{2}$

In Exercises 9–18, find each set. The universal set $U = \{-4, -3, -2, -1, 0, 1, 2, 3, 4\}$, $A = \{-4, -2, 0, 2, 4\}$, $B = \{-3, 0, 1, 2, 3, 4\}$, and $C = \{-4, -3, -2, -1, 0, 2\}$.

9. $A \cup B$ $\{-4, -3, -2, 0, 1, 2, 3, 4\}$
10. $A \cap B$ $\{0, 2, 4\}$
11. A' $\{-3, -1, 1, 3\}$
12. C' $\{1, 3, 4\}$
13. $A' \cup (B \cap C)$ $\{-3, -1, 0, 1, 2, 3\}$
14. $A \cap B'$ $\{-4, -2\}$
15. $(A \cup B)'$ $\{-1\}$
16. $(A \cap B)'$ $\{-4, -3, -2, -1, 1, 3\}$
17. $(A \cup B) \cap C$ $\{-4, -3, -2, 0, 2\}$
18. $(A \cup B) \cup C$ $\{-4, -3, -2, -1, 0, 1, 2, 3, 4\}$

In Exercises 19–24, use the roster method to write the given sets of numbers. Interpret "between a and b" to include both a and b.

19. The first three natural numbers $\{1, 2, 3\}$
20. The first five natural numbers $\{1, 2, 3, 4, 5\}$
21. The natural numbers between 3 and 7 $\{3, 4, 5, 6, 7\}$
22. The natural numbers between 8 and 11 $\{8, 9, 10, 11\}$
23. The negative integers between -3 and 5 $\{-3, -2, -1\}$
24. The whole numbers between -2 and 4 $\{0, 1, 2, 3, 4\}$

In Exercises 25–38, rewrite each expression without absolute value bars.

25. $|20|$ 20
26. $|12|$ 12
27. $-|-4|$ -4
28. $-|-17|$ -17
29. $\left|\dfrac{5}{7}\right|$ $\dfrac{5}{7}$
30. $\left|\dfrac{-3}{5}\right|$ $\dfrac{3}{5}$
31. $|5 - \sqrt{2}|$ $5 - \sqrt{2}$
32. $|\sqrt{2} - 5|$ $5 - \sqrt{2}$
33. $\dfrac{8}{|-8|}$ 1
34. $\dfrac{-8}{|8|}$ -1
35. $|5 + |-7||$ 12
36. $|5 - |-7||$ 2
37. $||7| - |4||$ 3
38. $||4| - |7||$ 3

In Exercises 39–46, use the absolute value to express the distance between the points with coordinates a and b on the number line. Then determine this distance by evaluating the absolute value expression.

39. $a = 3$ and $b = 8$ †
40. $a = 2$ and $b = 14$ †
41. $a = -6$ and $b = 9$ †
42. $a = -12$ and $b = 3$ †
43. $a = -20$ and $b = -6$ †
44. $a = -14$ and $b = -1$ †
45. $a = \dfrac{22}{7}$ and $b = -\dfrac{4}{7}$ †
46. $a = \dfrac{16}{5}$ and $b = -\dfrac{3}{5}$ †

In Exercises 47–58, graph each of the given intervals on a separate number line and write the inequality notation for each.

47. $[1, 4]$ †
48. $[-2, 2]$ †
49. $(14, 28)$ †
50. $\left(\dfrac{1}{2}, \dfrac{9}{2}\right)$ †
51. $(-3, 1]$ †
52. $[-6, -2)$ †
53. $[-3, \infty)$ †
54. $[0, \infty)$ †
55. $(-\infty, 5]$ †
56. $(-\infty, -1]$ †
57. $\left(-\dfrac{3}{4}, \dfrac{9}{4}\right)$ †
58. $\left(-3, -\dfrac{1}{2}\right)$ †

In Exercises 59–68, evaluate each expression for $x = 3$ and $y = -5$.

59. $2(x + y) - 3y$ 11
60. $-2(x + y) + 5y$ -21
61. $3|x| - 2|y|$ -1
62. $7|x - y|$ 56
63. $\dfrac{x - 3y}{2} + xy$ -6
64. $\dfrac{y + 3}{x} - xy$ $\dfrac{43}{3}$
65. $\dfrac{2(1 - 2x)}{y} - (-x)y$ -13
66. $\dfrac{3(2 - x)}{y} - (1 - xy)$ $-\dfrac{77}{5}$
67. $\dfrac{\dfrac{14}{x} + \dfrac{1}{2}}{\dfrac{-y}{4}}$ $\dfrac{62}{15}$
68. $\dfrac{\dfrac{4}{-y} + \dfrac{8}{x}}{\dfrac{y}{2}}$ $-\dfrac{104}{75}$

In Exercises 69–80, determine the domain of each expression. Write each domain in interval notation.

69. $\dfrac{3}{x - 1}$ $(-\infty, 1) \cup (1, \infty)$
70. $\dfrac{-5}{1 - x}$ $(-\infty, 1) \cup (1, \infty)$
71. $\dfrac{x + 2}{x}$ $(-\infty, 0) \cup (0, \infty)$
72. $\dfrac{x}{x + 7}$ $(-\infty, -7) \cup (-7, \infty)$
73. $\dfrac{1}{x} \div \dfrac{1}{x + 1}$ †
74. $\dfrac{2}{x} - \dfrac{3}{x - 5}$ †
75. $\dfrac{7}{(x - 1)(x + 2)}$ †
76. $\dfrac{x}{x(x + 3)}$ †
77. $\dfrac{x}{2 - x} + \dfrac{1}{x}$ †
78. $\dfrac{5}{x - 3} + \dfrac{4}{3 - x}$ †
79. $\dfrac{5}{x^2 + 1}$ $(-\infty, \infty)$
80. $\dfrac{x}{x^2 + 2}$ $(-\infty, \infty)$

†Due to space constrictions, answers to these exercises may be found in the Answers beginning on page A–1 in the back of the book.

In Exercises 81–90, name the exponent and the base.

81. 17^3 Exponent: 3; base: 17 **82.** 10^2 Exponent: 2; base: 10

83. 9^0 Exponent: 0; base: 9 **84.** $(-2)^0$ Exponent: 0; base: -2

85. $(-5)^5$ Exponent: 5; base: -5 **86.** $(-99)^2$ Exponent: 2; base: -99

87. -10^3 Exponent: 3; base: 10 **88.** $-(-3)^7$ Exponent: 7; base: -3

89. a^2 Exponent: 2; base: a **90.** $(-b)^3$ Exponent: 3; base: $-b$

In Exercise 91–116, evaluate each expression.

91. 6^1 6 **92.** 3^4 81

93. 7^0 1 **94.** $(-8)^0$ 1

95. $(2^3)^2$ 64 **96.** $(3^2)^3$ 729

97. $(3^2)^{-2}$ $\dfrac{1}{81}$ **98.** $(7^2)^{-1}$ $\dfrac{1}{49}$

99. $(5^{-2})^3$ $\dfrac{1}{15{,}625}$ **100.** $(5^{-1})^3$ $\dfrac{1}{125}$

101. $(4^{-3})\cdot(4^5)$ 16 **102.** $(7^{-2})\cdot(7^3)$ 7

103. $3^0 + 10^0$ 2 **104.** $5^0 - 9^0$ 0

105. $3^{-2} + \left(\dfrac{1}{3}\right)^2$ $\dfrac{2}{9}$ **106.** $5^{-2} + \left(\dfrac{1}{5}\right)^2$ $\dfrac{2}{25}$

107. $\dfrac{2^{11}}{2^{10}}$ 2 **108.** $\dfrac{3^6}{3^8}$ $\dfrac{1}{9}$

109. $\dfrac{(5^3)^4}{5^{12}}$ 1 **110.** $\dfrac{(9^5)^2}{9^8}$ 81

111. $\dfrac{2^5\cdot 3^{-2}}{2^4\cdot 3^{-3}}$ 6 **112.** $\dfrac{4^{-2}\cdot 5^3}{4^{-3}\cdot 5}$ 100

113. $\dfrac{-5^{-2}}{2^{-1}}$ $-\dfrac{2}{25}$ **114.** $\dfrac{-7^{-2}}{3^{-1}}$ $-\dfrac{3}{49}$

115. $\left(\dfrac{11}{7}\right)^{-2}$ $\dfrac{49}{121}$ **116.** $\left(\dfrac{13}{5}\right)^{-2}$ $\dfrac{25}{169}$

In Exercises 117–150, simplify each expression. Write your answers without negative exponents. Whenever an exponent is negative or zero, assume that the base is not zero.

117. x^4y^0 x^4 **118.** $x^{-1}y^0$ $\dfrac{1}{x}$

119. $x^{-1}y$ $\dfrac{y}{x}$ **120.** x^2y^{-2} $\dfrac{x^2}{y^2}$

121. $-8x^{-1}$ $-\dfrac{8}{x}$ **122.** $(-8x)^{-1}$ $-\dfrac{1}{8x}$

123. $x^{-1}(3y^0)$ $\dfrac{3}{x}$ **124.** $x^{-3}(3y^2)$ $\dfrac{3y^2}{x^3}$

125. $x^{-1}y^{-2}$ **126.** $x^{-3}y^{-2}$ $\dfrac{1}{x^3}$

127. $(x^{-3})^4$ **128.** $(x^{-5})^2$

129. $(x^{-11})^{-3}$ x^{33} **130.** $(x^{-4})^{-12}$ x^{48}

131. $-3(xy)^5$ $-3x^5y^5$ **132.** $-8(xy)^6$ $-8x^6y^6$

133. $4(xy^{-1})^2$ $\dfrac{4x^2}{y^2}$ **134.** $6(x^{-1}y)^3$ $\dfrac{6y^3}{x^3}$

135. $3(x^{-1}y)^{-5}$ $\dfrac{x^5}{y^5}$ **136.** $-5(xy^{-1})^{-6}$ $\dfrac{x^6}{y^6}$

137. $\dfrac{(x^3)^2}{(x^2)^5}$ $\dfrac{1}{x^4}$ **138.** $\dfrac{x^2}{(x^3)^4}$ $\dfrac{1}{x^{10}}$

139. $\left(\dfrac{2xy}{x}\right)^3$ $8y^3$ **140.** $\left(\dfrac{5xy}{x^3}\right)^4$ $\dfrac{625y^4}{x^8}$

141. $\left(\dfrac{-3x^2y}{x}\right)^5$ $-243x^5y^5$ **142.** $\left(\dfrac{-2xy^2}{y}\right)^3$ $-8x^3y^3$

143. $\left(\dfrac{-3x}{5}\right)^{-2}$ $\dfrac{25}{9x^2}$ **144.** $\left(\dfrac{-5y}{3}\right)^{-4}$ $\dfrac{81}{625y^4}$

145. $\left(\dfrac{4x^{-2}}{xy^5}\right)^3$ $\dfrac{64}{x^9y^{15}}$ **146.** $\left(\dfrac{3x^2y}{y^3}\right)^5$ $\dfrac{243x^{10}}{y^{10}}$

147. $\dfrac{x^3y^{-3}}{x^{-2}y}\cdot\dfrac{x^5}{y^4}$ **148.** $\dfrac{x^2y^{-2}}{x^{-1}y^2}\cdot\dfrac{x^3}{y^4}$

149. $\dfrac{27x^{-3}y^5}{9x^{-4}y^7}\cdot\dfrac{3x}{y^2}$ **150.** $\dfrac{15x^5y^{-2}}{3x^7y^{-3}}\cdot\dfrac{5y}{x^2}$

B EXERCISES Applying the Concepts

151. **Media players.** Let A = the set of people who own MP3 players and B = the set of people who own DVD players.
   a. Describe the set $A \cup B$. †
   b. Describe the set $A \cap B$. †
   c. Describe the set $A' \cap B$. †

152. **Standard car features.** The table indicates whether certain features are "standard" for each of three types of cars.

	Navigation system	Automatic Transmission	Leather seats
2008 Lexus SC 430	yes	yes	yes
2008 Lincoln Town Car	no	yes	yes
2008 Infiniti G37 coupe	no	no	yes

Source: www.autos.yahoo.com

Use the roster method to describe each of the following sets.   {Lexus SC 430}
   a. A = cars in which a navigation system is standard.
   b. B = cars in which an automatic transmission is standard.   {Lexus SC 430, Lincoln Town Car}
   c. C = cars in which leather seats are standard. †
   d. $A \cap B$ {Lexus SC 430}
   e. $B \cap C$ †   **f.** $A \cup B$ †
   g. $A \cup C$ †   **h.** $A \cap C'$ { } or $\varnothing$
   i. $(A \cap B)'$ {Lincoln Town Car, Infiniti G37}

153. **Blood pressure.** A group of college students had systolic blood pressure readings that ranged from 119.5 to 134.5 inclusive. Let x represent the value of the systolic blood pressure readings. Use inequalities to describe this range of values and graph the corresponding interval on a number line. †

154. **Population projections.** Population projections suggest that by the year 2050, the number of people 60 years old and older in the United States will be about 107 million. In 1950, the number of people 60 years old and older in the United States was 30 million. Let x represent the number (in millions) of people in the United States who are 60 years old and older. Use inequalities to describe this population range from 1950 to 2050 and graph the corresponding interval on a number line.
   Source: U.S. Census Bureau †

Answers:

125. $\dfrac{1}{xy^2}$ **126.** $\dfrac{1}{x^3y^2}$ **127.** $\dfrac{1}{x^{12}}$ **128.** $\dfrac{1}{x^{10}}$ **135.** $\dfrac{3x^5}{y^5}$ **136.** $-\dfrac{5y^6}{x^6}$

155. Heart rate. For exercise to be most beneficial, the optimum heart rate for a 20-year-old person is 120 beats per minute. Use absolute value notation to write an expression that describes the difference between the heart rate achieved by each of the following 20-year-old people and the ideal exercise heart rate. Then evaluate that expression.
   a. Latasha: 124 beats per minute $|124 - 120| = 4$
   b. Frances: 137 beats per minute $|137 - 120| = 17$
   c. Ignacio: 114 beats per minute $|114 - 120| = 6$

156. Downloading music. To download a 4 MB song with a 56 Kbs modem takes an average of 15 minutes. Use absolute value notation to write an expression that describes the difference between this average time and the actual time it took to download the following songs. Then evaluate that expression.
   a. *Believe* (Cher): 14 minutes $|15 - 14| = 1$
   b. *Caged Bird* (Alicia Keys): 17.5 minutes $|15 - 17.5| = 2.5$
   c. *Somewhere* (Barbra Streisand): 15 minutes
   $|15 - 15| = 0$

In Exercise 157–160, use the physical interpretation to determine the domain of the variable.

157. Depth and pressure. The pressure p, in pounds per square inch, is related to the depth d, in feet below the surface of an ocean, by $p = \dfrac{5}{11}d + 15$. What is the domain of the variable p? $[0, \infty)$

158. Volume. The volume V of a box that has a square base with a 2-foot side and a height of h feet is $V = 4h$. What is the domain of the variable h? $(0, \infty)$

159. Weight vs. height. The average weight w, in pounds, of a male between 5 feet and 5 feet 10 inches tall is related to his height h by $w = \left(\dfrac{11}{2}\right)h - 220$. What is the domain (in inches) of the variable h? $[60, 70]$

160. Pay per hour. The weekly pay A a worker receives for working a 40–hour week or less is given by $A = 10h$, where h represents the number of hours worked in that week. What is the domain of the variable h? $[0, 40]$

Negative calories. **In Exercises 161 and 162, use the fact that eating 100 grams of broccoli (a negative-calorie food) actually results in a net *loss* of 55 calories.**

161. If a cheeseburger has 522.5 calories, how many grams of broccoli would a person have to consume to have a net gain of zero calories? 950 g

162. Carmen ate 600 grams of broccoli and now wants to eat just enough French fries so that she has a net calorie intake of zero. If a small order of fries has 165 calories, how many orders does she have to eat? two orders

In Exercises 163 and 164, use the fact that when you uniformly stretch or shrink a three-dimensional object in every direction by a factor of a, the volume of the resulting figure is scaled by a factor of a^3. For example, when you uniformly scale (stretch or shrink) a three-dimensional object by a factor of 2, the volume of the resulting figure is scaled by a factor of 2^3, or 8.

163. A display in the shape of a baseball bat has a volume of 135 cubic feet. Find the volume of the display that results by uniformly scaling the figure by a factor of 2. 1080 ft^3

164. A display in the shape of a football has a volume of 675 cubic inches. Find the volume of the display that results by uniformly scaling the figure by a factor of 3.
$18,225 \text{ in}^3$

165. The area A of a square with side of length x is given by $A = x^2$. Use this relationship to
   a. verify that doubling the length of the side of a square floor increases the area of the floor by a factor of 2^2. $(2x)(2x) = 4x^2 = 2^2A$
   b. verify that tripling the length of the side of a square floor increases the area of the floor by a factor of 3^2. $(3x)(3x) = 3x^2 = 3^2A$

166. The area A of a circle with diameter d is given by $A = \pi\left(\dfrac{d}{2}\right)^2$. Use this relationship to
   a. verify that doubling the length of the diameter of a circular skating rink increases the area of the rink by a factor of 2^2.
   b. verify that tripling the length of the diameter of a circular skating rink increases the area of the rink by a factor of 3^2.

Answers:

166. a. $\pi\left(\dfrac{2d}{2}\right)^2 = 4\pi\left(\dfrac{d}{2}\right)^2 = 2^2A$ **b.** $\pi\left(\dfrac{3d}{2}\right)^2 = 9\pi\left(\dfrac{d}{2}\right)^2 = 3^2A$

Polynomials

Objectives

1 Use polynomial vocabulary.

2 Use special-product formulas.

3 Divide polynomials.

4 Use synthetic division.

5 Factor polynomials.

---

1 Use polynomial vocabulary.

Polynomial Vocabulary

We begin by reviewing the basic vocabulary of polynomials. A **monomial** is the simplest polynomial in the variable x; it contains one term and has the form ax^k, where a is a constant and k is either a positive integer or zero. The constant a is called the **coefficient** of the monomial. For $a \neq 0$, the integer k is called the **degree** of the monomial.

Consider the following monomials:

$2x^5$	The coefficient is 2, and the degree is 5.
$-3x^2$	The coefficient is -3, and the degree is 2.
-7	$-7 = -7(1) = -7x^0$. The coefficient is -7, and the degree is 0.
$8x$	The coefficient is 8, and the degree is $1(x = x^1)$.
x^{12}	The coefficient is $1(x^{12} = 1 \cdot x^{12})$, and the degree is 12.
$-x^4$	The coefficient is $-1(-x^4 = (-1) \cdot x^4)$, and the degree is 4.

The expression $5x^{-3}$ is not a monomial because the exponent of x is negative. Two monomials in the same variable with the same degree can be added or subtracted using the distributive property. For example, $-3x^5 - 9x^5 = (-3 - 9)x^5 = -12x^5$. Two monomials in the same variable with the same degree are called **like terms**.

POLYNOMIALS IN ONE VARIABLE

A **polynomial** in x is any sum of monomials in x. By combining like terms, we can write any polynomial in the form

$$a_n x^n + a_{n-1} x^{n-1} + \cdots + a_2 x^2 + a_1 x + a_0,$$

where n is either a positive integer or zero and $a_n, a_{n-1}, \ldots a_1, a_0$ are constants, called the **coefficients** of the polynomial. If $a_n \neq 0$, then n, the largest exponent on x, is called the **degree** and a_n is called the **leading coefficient** of the polynomial. The monomials $a_n x^n, a_{n-1} x^{n-1} \ldots, a_2 x^2, a_1 x,$ and a_0 are the **terms** of the polynomial. The monomial $a_n x^n$ is the **leading term** of the polynomial, and a_0 is the **constant term**.

Polynomials can be classified according to the number of terms they have. Polynomials with one term are called **monomials**, polynomials with two *unlike* terms are called **binomials**, and polynomials with three *unlike* terms are called **trinomials**. Polynomials with more than three unlike terms do not have special names.

By agreement, the only polynomial that has *no* degree is the **zero polynomial**, which results when all of the coefficients are 0. It is easiest to find the degree of a polynomial when it is written in **descending order**—that is, when the exponents decrease from left to right. In this case, the degree is the exponent of the leading term. A polynomial written in descending order is said to be in **standard form**.

The symbols $a_0, a_1, a_2, \ldots a_n$ in the general notation for a polynomial are just constants; the numbers to the lower right of a are called subscripts. We read the notation a_2 as "a sub 2." This type of notation is used when a large or indefinite number of constants are required.

2 Use special-product formulas.

Special Products

The basic multiplication rule for polynomials can be stated as follows:

MULTIPLYING POLYNOMIALS

To multiply two polynomials, multiply each term of one polynomial by every term of the other polynomial and combine like terms.

Particular polynomial products called *special products* occur frequently enough to deserve special attention. We introduce a very useful method, called F O I L (First, Outer, Inner, Last), for multiplying two binomials.

FOIL METHOD FOR $(A + B)(C + D)$

If A, B, C, and D represent any algebraic expressions, then

$$(A + B)(C + D) = \underset{F}{A \cdot C} + \underset{O}{A \cdot D} + \underset{I}{B \cdot C} + \underset{L}{B \cdot D}$$

EXAMPLE 1   Using the FOIL Method

Use the FOIL method to find the following products.

a. $(2x + 3)(x - 1) = \underset{F}{(2x)(x)} + \underset{O}{(2x)(-1)} + \underset{I}{(3)(x)} + \underset{L}{(3)(-1)}$

$\qquad\qquad\qquad = 2x^2 - 2x + 3x - 3$

$\qquad\qquad\qquad = 2x^2 + x - 3$

b. $(3x - 5)(4x - 6) = \underset{F}{(3x)(4x)} + \underset{O}{(3x)(-6)} + \underset{I}{(-5)(4x)} + \underset{L}{(-5)(-6)}$

$\qquad\qquad\qquad\qquad = 12x^2 - 18x - 20x + 30$

$\qquad\qquad\qquad\qquad = 12x^2 - 38x + 30$

c. $(x + a)(x + b) = \underset{F}{x \cdot x} + \underset{O}{x \cdot b} + \underset{I}{a \cdot x} + \underset{L}{a \cdot b}$

$\qquad\qquad\qquad = x^2 + bx + ax + a \cdot b$

$\qquad\qquad\qquad = x^2 + (b + a)x + ab$ ■ ■ ■

Practice Problem 1   Use FOIL to find each product.

a. $(4x - 1)(x + 7)$ **b.** $(3x - 2)(2x - 5)$ ■

The following special-product formulas, such as the formulas for squaring a binomial sum or difference, are used often and should be memorized. However, you should be able to derive them using FOIL if you forget them. We list several of these formulas next.

SPECIAL-PRODUCT FORMULAS

A and B represent any algebraic expression.

Formula	**Example**

Product giving a difference of squares

$$(A + B)(A - B) = A^2 - B^2 \qquad\qquad (5x + 2)(5x - 2) = (5x)^2 - 2^2$$
$$= 25x^2 - 4$$

Squaring a binomial sum or difference

$$(A + B)^2 = A^2 + 2AB + B^2 \qquad (3x + 2)^2 = (3x)^2 + 2 \cdot (3x)(2) + 2^2$$
$$= 9x^2 + 12x + 4$$

$$(A - B)^2 = A^2 - 2AB + B^2 \qquad (2x - 5)^2 = (2x)^2 - 2(2x)(5) + 5^2$$
$$= 4x^2 - 20x + 25$$

Cubing a binomial sum or difference

$$(A + B)^3 = A^3 + 3A^2B + 3AB^2 + B^3 \quad (x + 5)^3 = x^3 + 3x^2(5) + 3x(5)^2 + 5^3$$
$$= x^3 + 15x^2 + 75x + 125$$

$$(A - B)^3 = A^3 - 3A^2B + 3AB^2 - B^3 \quad (x - 4)^3 = x^3 - 3x^2(4) + 3x(4)^2 + 4^3$$
$$= x^3 - 12x^2 + 48x + 64$$

Products giving a sum or difference of cubes

$$(A + B)(A^2 - AB + B^2) = A^3 + B^3 \quad (x + 2)(x^2 - 2x + 4) = x^3 + 2^3 = x^3 + 8$$
$$(A - B)(A^2 + AB + B^2) = A^3 - B^3 \quad (x - 3)(x^2 + 3x + 9) = x^3 - 3^3 = x^3 - 27$$

---

3 Divide polynomials.

The Division Algorithm and Long Division

Dividing polynomials is similar to dividing integers. The fact that 5 divides 10 "evenly" is expressed as $\dfrac{10}{5} = 2$ or as $10 = 5 \cdot 2$. By comparison, the equation

$$(1) \qquad\qquad x^3 - 1 = (x - 1)(x^2 + x + 1) \qquad \text{Factor, using difference of cubes.}$$

suggests dividing both sides by $x - 1$ to obtain

$$\frac{x^3 - 1}{x - 1} = x^2 + x + 1, \quad x \neq 1.$$

We say that the polynomial $x - 1$ is a factor of $x^3 - 1$.

RECALL

A polynomial $P(x)$ is an expression of the form

$$a_x x^n + \cdots + a_0.$$

POLYNOMIAL FACTOR

A polynomial $D(x)$ is **a factor** of a polynomial $F(x)$ if there is a polynomial $Q(x)$ such that $F(x) = D(x) \cdot Q(x)$.

Next, consider the equation

(2) $x^3 - 2 = (x - 1)(x^2 + x + 1) - 1.$ Subtract 1 from both sides of equation (1).

Dividing both sides by $(x - 1)$, we can rewrite equation (2) in the form

$$\frac{x^3 - 2}{x - 1} = x^2 + x + 1 - \frac{1}{x - 1}, x \neq 1.$$

Here we say that when the **dividend** $x^3 - 2$ is divided by the **divisor** $x - 1$, the **quotient** is $x^2 + x + 1$ and the **remainder** is -1. In general, we state the following result, called the *division algorithm*.

THE DIVISION ALGORITHM

If a polynomial $F(x)$ is divided by a polynomial $D(x)$, with $D(x) \neq 0$, there are unique polynomials $Q(x)$ and $R(x)$ such that

$$F(x) \quad = \quad D(x) \quad \cdot \quad Q(x) \quad + \quad R(x).$$

| Dividend | Divisor | Quotient | Remainder |

Either $R(x)$ is the *zero polynomial* or the degree of $R(x)$ is less than the degree of $D(x)$.

The division algorithm says that the dividend is equal to the divisor times the quotient plus the remainder.

For a rational expression $\dfrac{F(x)}{D(x)}$, you may recall the **long-division** process shown below for finding the quotient and the remainder.

$$
\begin{array}{r}
3x + 4 \\
2x - 1 \overline{)\,6x^2 + 5x + 4\,} \\
\underline{6x^2 - 3x} \\
8x + 1 \\
\underline{8x - 4} \\
5
\end{array}
$$

Divisor → $2x - 1$
← Quotient
← Dividend
$(2x - 1)3x$
Subtract.
$(2x - 1)4$
Remainder →   Subtract.

FINDING THE SOLUTION: A PROCEDURE

EXAMPLE 2   **Using Long Division**

OBJECTIVE
Find the quotient and remainder when one polynomial is divided by another.

Step 1 Write the terms in the dividend and the divisor in descending powers of the variable.

EXAMPLE
Find the quotient and remainder when $2x^2 + x^5 + 7 + 4x^3$ is divided by $x^2 + 1 - x$.

Dividend: $x^5 + 4x^3 + 2x^2 + 7$ Divisor: $x^2 - x + 1$

Step 2 Insert terms with zero coefficients in the dividend for any missing powers of the variable.

Step 3 Divide the first term in the dividend by the first term in the divisor to obtain the first term in the quotient.

Step 4 Multiply the divisor by the first term in the quotient and subtract the product from the dividend.

Step 5 Treat the remainder obtained in Step 4 as a new dividend and repeat Steps 3 and 4. Continue this process until a remainder is obtained that is of lower degree than the divisor.

$$x^5 + 0x^4 + 4x^3 + 2x^2 + 0x + 7$$

$$\dfrac{x^5}{x^2} = x^3 \quad \text{Divide first terms.}$$

$$\begin{array}{r} x^3 \\ x^2 - x + 1 \overline{)\, x^5 + 0x^4 + 4x^3 + 2x^2 + 0x + 7} \end{array}$$

$$\begin{array}{r} x^3 \\ x^2 - x + 1 \overline{)\, x^5 + 0x^4 + 4x^3 + 2x^2 + 0x + 7} \\ (-)\underline{x^5 - x^4 + x^3} \qquad x^3(x^2 - x + 1) \\ x^4 + 3x^3 + 2x^2 + 0x + 7 \quad \text{Remainder} \end{array}$$

Subtract.

$$\dfrac{x^4}{x^2} = x^2 \qquad \dfrac{4x^3}{x^2} = 4x \qquad \dfrac{5x^2}{x^2} = 5$$

$$\begin{array}{r} x^3 + x^2 + 4x + 5 \\ x^2 - x + 1 \overline{)\, x^5 + 0x^4 + 4x^3 + 2x^2 + 0x + 7} \\ (-)\underline{x^5 - x^4 + x^3} \\ x^4 + 3x^3 + 2x^2 + 0x + 7 \leftarrow \text{New dividend} \\ (-)\underline{x^4 - x^3 + x^2} \qquad x^2(x^2 - x + 1) \\ 4x^3 + x^2 + 0x + 7 \quad \text{Remainder} \\ (-)\underline{4x^3 - 4x^2 + 4x} \qquad 4x(x^2 - x + 1) \\ 5x^2 - 4x + 7 \quad \text{Remainder} \\ (-)\underline{5x^2 - 5x + 5} \leftarrow 5(x^2 - x + 1) \\ x + 2 \quad \text{Remainder} \end{array}$$

Step 6 Write the quotient and the remainder.

Quotient $= x^3 + x^2 + 4x + 5$ Remainder $= x + 2$

■ ■ ■

Practice Problem 2 Divide $x^4 + 5x^2 + 2x + 6$ by $x^2 - x + 3$. ■

4 Use synthetic division.

Synthetic Division

You can decrease the work involved in finding the quotient and remainder when the divisor $D(x)$ is of the form $x - a$ by using the process known as *synthetic division*. The procedure is best explained by considering an example.

Long division	Synthetic division
$\begin{array}{r} 2x^2 + 1x + 6 \\ x - 3 \overline{)\, 2x^3 - 5x^2 + 3x - 14} \\ \underline{2x^3 - 6x^2} \\ 1x^2 + 3x - 14 \\ \underline{1x^2 - 3x} \\ 6x - 14 \\ \underline{6x - 18} \\ 4 \end{array}$	$\begin{array}{r} 3 \underline{\rvert\, 2 \quad -5 \quad 3 \quad -14} \\ 6 \quad 3 \quad 18 \\ \hline 2 \quad 1 \quad 6 \quad \rvert \quad 4 \leftarrow \text{Remainder} \end{array}$

Each circled term in the long division process is exactly the same as the term above it. Furthermore, the boxed terms are terms in the dividend, written in a new position. The synthetic division process eliminates these two sets of terms, resulting in a more efficient format.

The essential steps in the process can be carried out as follows:

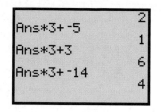

Synthetic division can be done on a graphing calculator. In this problem, enter the first coefficient, 2. Then repeatedly multiply the answer by 3 and add the next coefficient.

```
              2
Ans*3+ -5
              1
Ans*3+3
              6
Ans*3+ -14
              4
```

(i) Bring down 2 from the first line to the third line.

$$3 \rfloor\ \ 2\ \ -5\ \ \ 3\ \ -14$$
$$\downarrow$$
$$2$$

(ii) Multiply 2 in the third line by the 3 in the divisor position and place the product, 6, under −5; then add vertically to obtain 1.

$$3 \rfloor\ \ 2\ \ -5\ \ \ 3\ \ -14$$
$$6$$
$$2\ \ \ 1$$

(iii) Multiply 1 by 3 and place the product 3 under 3; then add vertically to obtain 6.

$$3 \rfloor\ \ 2\ \ -5\ \ \ 3\ \ -14$$
$$6\ \ \ 3$$
$$2\ \ \ 1\ \ \ 6$$

(iv) Finally, multiply 6 by 3 and place the product, 18, under −14; then add to obtain 4.

$$3 \rfloor\ \ 2\ \ -5\ \ \ 3\ \ -14$$
$$6\ \ \ 3\ \ \ 18$$
$$2\ \ \ 1\ \ \ 6\ \ \ |4$$

The first three numbers in the third line, namely, 2, 1, and 6, are the coefficients of the quotient: $2x^2 + x + 6$, and the last number, 4, is the remainder. The procedure for synthetic division is given next.

FINDING THE SOLUTION: A PROCEDURE

EXAMPLE 3   Using Synthetic Division

OBJECTIVE
Divide a polynomial $F(x)$ by $x - a$.

Step 1 Arrange the coefficients of $F(x)$ in order of *descending powers of x*, supplying zero as the coefficient of each missing power.

Step 2 Replace the divisor $x - a$ with a.

Step 3 Bring the first (leftmost) coefficient down below the line. Multiply it by a and write the resulting product one column to the right and above the line.

Step 4 Add the product obtained in Step 3 to the coefficient directly above it and write the resulting sum directly below it and below the line. This sum is the "newest" number below the line.

Step 5 Multiply the newest number below the line by a, write the resulting product one column to the right and above the line, and repeat Step 4.

EXAMPLE
Divide $2x^4 - 3x^2 + 5x - 63$ by $x + 3$.

$$\rfloor\ \ 2\ \ \ 0\ \ -3\ \ \ 5\ \ -63$$

Rewrite $x + 3 = x - (-3); a = -3$

$$-3 \rfloor\ \ 2\ \ \ 0\ \ -3\ \ \ 5\ \ -63$$

$$-3 \rfloor\ \ 2\ \ \ 0\ \ -3\ \ \ 5\ \ -63$$
$$-6$$
$$\otimes$$
$$2$$

$\otimes$ means "multiply by −3."

$$-3 \rfloor\ \ 2\ \ \ 0\ \ -3\ \ \ 5\ \ -63$$
$$-6$$
$$+\downarrow$$
$$2\ \ -6$$

$$-3 \rfloor\ \ 2\ \ \ 0\ \ -3\ \ \ 5\ \ -63$$
$$-6\ \ \ 18$$
$$\otimes\ \ +\downarrow$$
$$2\ \ -6\ \ \ 15$$

continued on the next page

Step 6 Repeat Step 5 until there is a product added to the constant term. Separate the last number below the line using a short vertical line.

$$
\begin{array}{r|rrrrr}
-3 & 2 & 0 & -3 & 5 & -63 \\
   &   & -6 & 18 & -45 & 120 \\
\hline
   & 2 & -6 & 15 & -40 & 57
\end{array}
$$

Step 7 The last number below the line is the remainder, and the other numbers, reading from left to right, are the coefficients of the quotient, which has degree one less than the dividend $F(x)$.

$$
\begin{array}{r|rrrr|r}
-3 & 2 & 0 & -3 & 5 & -63 \\
   &   & -6 & 18 & -45 & 120 \\
\hline
   & 2 & -6 & 15 & -40 & 57
\end{array}
$$

Dividend: degree 4

Quotient: degree 3

$$\underbrace{2x^3 - 6x^2 + 15x - 40}_{\text{Quotient}} \qquad \text{Remainder}$$

■ ■ ■

Practice Problem 3   Use synthetic division to divide $2x^3 + x^2 - 18x - 7$ by $x - 3$.

■

5 Factor polynomials.

Factoring Polynomials

In this section, we discuss only polynomials having integer coefficients. This restriction is called **factoring over the integers**.

To factor a polynomial means to write it as a *product of factors*. Consider the product

$$(x + 3)(x - 3) = x^2 - 9.$$

The polynomials $(x + 3)$ and $(x - 3)$ are called **factors** of the polynomial $x^2 - 9$.

When factoring a polynomial, the first thing to look for is a factor that is common to every term. This **common factor** can be "factored out" by the distributive property, $a(b + c) = ab + ac$.

Polynomial	Common Factor	Factored Form
$9 + 18y$	9	$9(1 + 2y)$
$2y^2 + 6y$	$2y$	$2y(y + 3)$
$7x^4 + 3x^3 + x^2$	x^2	$x^2(7x^2 + 3x + 1)$
$5x + 3$	No common factor	$5x + 3$

You can verify that any factorization is correct by multiplying the factors. We factored these polynomials by finding a *common factor* of their terms.

We frequently factor by reversing a product, such as $(x + a)(x + b) = x^2 + (a + b)x + ab$, to get the factoring form

$$x^2 + (a + b)x + ab = (x + a)(x + b).$$

For example, to factor $x^2 - 6x - 16$, we must find two integers a and b with $ab = -16$ and $a + b = -6$. Because 2 and -8 are factors of -16 and give a sum of -6,

$$x^2 - 6x - 16 = (x + 2)[x + (-8)] = (x + 2)(x - 8).$$

Polynomials that cannot be factored as a product of two polynomials (excluding the constant polynomials 1 and -1) are said to be **irreducible**. A polynomial is said to be **factored completely** when it is written as a product consisting of only irreducible factors.

EXAMPLE 4   Factoring Polynomials

Factor.

a. $16x^2 - 8x + 1$ **b.** $25x^2 - 49$ **c.** $x^4 - 16$ **d.** $x^3 - 64$

SOLUTION

a. $16x^2 - 8x + 1 = (4x)^2 - 2(4x)(1) + 1^2 = (4x - 1)^2$ (perfect square)

b. $25x^2 - 49 = (5x)^2 - 7^2 = (5x + 7)(5x - 7)$ (difference of squares)

c. $x^4 - 16 = (x^2 + 4)(x^2 - 4) = (x^2 + 4)(x + 2)(x - 2)$ (difference of squares)

d. $x^3 - 64 = x^3 - 4^3 = (x - 4)(x^2 + 4x + 16)$ (difference of cubes)   ■ ■ ■

You should use FOIL to check that $x^2 + 4$ is irreducible in part **c** and that $x^2 + 4x + 16$ is irreducible in part **d**.

Practice Problem 4   Factor.

a. $x^2 + 4x + 4$ **b.** $4x^2 - 25$ **c.** $x^4 - 81$ **d.** $x^3 - 125$ ■

> **STUDY TIP**
>
> You should always check your answer to a factoring problem by multiplying the factors to verify that the result is the original expression.

To factor the trinomial $Ax^2 + Bx + C$ as $(ax + b)(cx + d)$, we use FOIL and combine like terms to get $Ax^2 + Bx + C = acx^2 + (ad + bc)x + bd$. We can factor $Ax^2 + Bx + C$ if we find integers a, b, c, and d such that $A = ac, C = bd$, and $B = ad + bc$.

EXAMPLE 5   Factoring Using FOIL and by Grouping

Factor.   **a.** $6x^2 + 17x + 7$ **b.** $x^3 + 2x^2 + 3x + 6$

SOLUTION

a. $6x^2 + 17x + 7 = (2x + 1)(3x + 7)$ $6 = 2 \cdot 3, \ 7 = 1 \cdot 7, \ 17 = 2 \cdot 7 + 1 \cdot 3$

b. $x^3 + 2x^2 + 3x + 6 = (x^3 + 2x^2) + (3x + 6)$ Group terms with a common factor.

$\qquad = x^2(x + 2) + 3(x + 2)$ Factor out the common factor.

$\qquad = (x^2 + 3)(x + 2)$ Distributive property   ■ ■ ■

Practice Problem 5   Factor.   **a.** $5x^2 + 11x + 2$ **b.** $x^3 + 3x^2 + x + 3$ ■

SECTION A.2 ■ Exercises

A EXERCISES Basic Skills and Concepts

1. For a polynomial in x, the largest exponent on x is called the ___degree___ of the polynomial.

2. A polynomial written in descending order is said to be in ___standard form___.

3. A polynomial $D(x)$ is a factor of a polynomial $F(x)$ if there is a polynomial $Q(x)$ such that $F(x) = $ ___$D(x) \cdot Q(x)$___.

4. If a polynomial $F(x)$ is divided by a polynomial $D(x)$, $D(x) \neq 0$, then the remainder $R(x)$ is zero or the degree of $R(x)$ is ___less than___ the degree of $D(x)$.

In Exercises 5–8, determine whether the given expression is a polynomial. If it is, write it in standard from.

5. $1 + x^2 + 2x$
   Yes; $x^2 + 2x + 1$

6. $x - \dfrac{1}{x}$ No

7. $x^{-2} + 3x + 5$ No

8. $3x^4 + x^7 + 3x^5 - 2x + 1$
   Yes; $x^7 + 3x^5 + 3x^4 - 2x + 1$

In Exercises 9–12, find the degree and list the terms of the polynomial.

9. $7x + 3$
   Degree: 1; terms: $7x, 3$

10. $-3x^2 + 7$
    Degree 2; terms: $-3x^2, 7$

11. $x^2 - x^4 + 2x - 9$
    Degree 4; terms: $-x^4, x^2, 2x - 9$

12. $x + 2x^3 + 9x^7 - 21$
    Degree 7; terms: $9x^7, 2x^3, x, -21$

In Exercises 13–22, perform the indicated operations. Write the resulting polynomial in standard form.

13. $(x^3 + 2x^2 - 5x + 3) + (-x^3 + 2x - 4)$ $2x^2 - 3x - 1$

†Due to space constrictions, answers to these exercises may be found in the Answers beginning on page A–1 in the back of the book.

14. $(x^3 - 3x + 1) + (x^3 - x^2 + x - 3)$ $2x^3 - x^2 - 2x - 2$

15. $(2x^3 - x^2 + x - 5) - (x^3 - 4x + 3)$ $x^3 - x^2 + 5x - 8$

16. $(-x^3 + 2x - 4) - (x^3 + 3x^2 - 7x + 2)$ $-2x^3 - 3x^2 + 9x - 6$

17. $(-2x^4 + 3x^2 - 7x) - (8x^4 + 6x^3 - 9x^2 - 17)$
$-10x^4 - 6x^3 + 12x^2 - 7x + 17$

18. $3(x^2 - 2x + 2) + 2(5x^2 - x + 4)$ $13x^2 - 8x + 14$

19. $-2(3x^2 + x + 1) + 6(-3x^2 - 2x - 2)$ $-24x^2 - 14x - 14$

20. $2(5x^2 - x + 3) - 4(3x^2 + 7x + 1)$ $-2x^2 - 30x + 2$

21. $(3y^3 - 4y + 2) + (2y + 1) - (y^3 - y^2 + 4)$
$2y^3 + y^2 - 2y - 1$

22. $(5y^2 + 3y - 1) - (y^2 - 2y + 3) + (2y^2 + y + 5)$
$6y^2 + 6y + 1$

In Exercises 23–56, perform the indicated operations.

23. $6x(2x + 3)$ $12x^2 + 18x$ **24.** $7x(3x - 4)$ $21x^2 - 28x$

25. $(x + 1)(x^2 + 2x + 2)$ † **26.** $(x - 5)(2x^2 - 3x + 1)$ †

27. $(3x - 2)(x^2 - x + 1)$ † **28.** $(2x + 1)(x^2 - 3x + 4)$ †

29. $(x + 1)(x + 2)$ †   **30.** $(x + 2)(x + 3)$ †

31. $(3x + 2)(3x + 1)$ †   **32.** $(x + 3)(2x + 5)$ †

33. $(-4x + 5)(x + 3)$ †   **34.** $(-2x + 1)(x - 5)$ †

35. $(3x - 2)(2x - 1)$ †   **36.** $(x - 1)(5x - 3)$ †

37. $(2x - 3a)(2x + 5a)$ † **38.** $(5x - 2a)(x + 5a)$ †

39. $(x + 2)^2 - x^2$ $4x + 4$ **40.** $(x - 3)^2 - x^2$ $-6x + 9$

41. $(x + 3)^3 - x^3$ †   **42.** $(x - 2)^3 - x^3$
$-6x^2 + 12x - 8$

43. $(4x + 1)^2$ $16x^2 + 8x + 1$ **44.** $(3x + 2)^2$ $9x^2 + 12x + 4$

45. $(3x + 1)^3$ †   **46.** $(2x + 3)^3$ †

47. $(5 - 2x)(5 + 2x)$ † **48.** $(3 - 4x)(3 + 4x)$
$-16x^2 + 9$

49. $\left(x + \dfrac{3}{4}\right)^2$ $x^2 + \dfrac{3}{2}x + \dfrac{9}{16}$ **50.** $\left(x + \dfrac{2}{5}\right)^2$ $x^2 + \dfrac{4}{5}x + \dfrac{4}{25}$

51. $(2x - 3)(x^2 - 3x + 5)$ $2x^3 - 9x^2 + 19x - 15$

52. $(x - 2)(x^2 - 4x - 3)$ $x^3 - 6x^2 + 5x + 6$

53. $(1 + y)(1 - y + y^2)$ $y^3 + 1$

54. $(y + 4)(y^2 - 4y + 16)$ $y^3 + 64$

55. $(x - 6)(x^2 + 6x + 36)$ $x^3 - 216$

56. $(x - 1)(x^2 + x + 1)$ $x^3 - 1$

In Exercises 57–66, perform the indicated operations.

57. $(x + 2y)(3x + 5y)$ † **58.** $(2x + y)(7x + 2y)$ †

59. $(2x - y)(3x + 7y)$ † **60.** $(x - 3y)(2x + 5y)$ †

61. $(x - y)^2(x + y)^2$ † **62.** $(2x + y)^2(2x - y)^2$ †

63. $(x + y)(x - 2y)^2$ †   **64.** $(x - y)(x + 2y)^2$ †

65. $(x - 2y)^3(x + 2y)$ † **66.** $(2x + y)^3(2x - y)$ †

In Exercises 67–82, use long division to find the quotient and the remainder.

67. $\dfrac{2x^2 - 3x + 4}{x - 1}$ Quotient: $2x - 1$; remainder: 3

68. $\dfrac{b^2 + b - 6}{b + 3}$ Quotient: $b - 2$; remainder: 0

69. $\dfrac{6x^2 - x - 2}{2x + 1}$ Quotient: $3x - 2$; remainder: 0

70. $\dfrac{12m^2 + 7m - 12}{4m - 3}$ Quotient: $3m + 4$; remainder: 0

71. $\dfrac{4x^3 - 2x^2 + x - 3}{2x - 3}$ Quotient: $2x^2 + 2x + \dfrac{7}{2}$; remainder: $\dfrac{15}{2}$

72. $\dfrac{6x^3 + 6x^2 - 4x + 1}{3x - 2}$ Quotient: $2x^2 + \dfrac{10}{3}x + \dfrac{8}{9}$; remainder: $\dfrac{25}{9}$

73. $\dfrac{6x^3 + x + 4}{2x + 1}$ Quotient: $3x^2 - \dfrac{3}{2}x + \dfrac{5}{4}$; remainder: $\dfrac{11}{4}$

74. $\dfrac{3x^4 - 6x^2 + 3x - 7}{x + 1}$ Quotient: $3x^3 - 3x^2 - 3x + 6$; remainder: -13

75. $\dfrac{x^5 - 1}{x - 1}$ Quotient: $x^4 + x^3 + x^2 + x + 1$; remainder: 0

76. $\dfrac{6x^6 + 3x^2 + 10}{2x + 3}$ †

77. $\dfrac{x^6 + 5x^3 + 7x + 3}{x^2 + 2}$ Quotient: $x^4 - 2x^2 + 5x + 4$; remainder: $-3x - 5$

78. $\dfrac{4x^3 - 4x^2 - 9x + 5}{2x^2 - x - 5}$ Quotient: $2x - 1$: remainder: 0

79. $\dfrac{y^5 + 3y^4 - 6y^2 + 2y - 7}{y^2 + 2y - 3}$ Quotient: $y^3 + y^2 + y - 5$; remainder: $15y - 22$

80. $\dfrac{z^4 - 2z^2 + 1}{z^2 - 2z + 1}$ Quotient: $z^2 + 2z + 1$: remainder: 0

81. $\dfrac{6x^4 + 13x - 11x^3 - 10 - x^2}{3x^2 - 5 - x}$ Quotient: $2x^2 - 3x + 2$; remainder: 0

82. $\dfrac{4x^4 - 9x^2 - 3 - 8x^3 - 16x}{x + 2x^2 - 3}$ †

In Exercises 83–102, use synthetic division to find the quotient and the remainder.

83. $\dfrac{x^3 + 3x^2 - x - 12}{x - 2}$ Quotient: $x^2 + 5x + 9$; remainder: 6

84. $\dfrac{2x^3 + 4x^2 + x - 15}{x - 3}$ Quotient: $2x^2 + 10x + 31$; remainder: 78

85. $\dfrac{9x^3 - x^2 - 13x + 6}{x + 5}$ Quotient: $9x^2 - 46x + 217$; remainder: -1079

86. $\dfrac{x^3 - 7x^2 - 5x + 1}{x + 1}$ Quotient: $x^2 - 8x + 3$; remainder: -2

87. $\dfrac{x^4 + 2x^2 - x - 7}{x - 1}$ Quotient: $x^3 + x^2 + 3x + 2$; remainder: -5

88. $\dfrac{3x^4 - 6x^2 + 3x - 7}{x - 2}$ Quotient: $3x^3 + 6x^2 + 6x + 15$; remainder: 23

89. $\dfrac{-2x^3 + 5x^2 + 7x - 8}{x - 3}$ Quotient: $-2x^2 - x + 4$; remainder: 4

90. $\dfrac{-2x^3 - 5x^2 + 3x - 2}{x + 3}$ Quotient: $-2x^2 + x$; remainder: -2

91. $\dfrac{x^5 + x^4 - 7x^3 + 2x^2 + x - 1}{x - 1}$ Quotient: $x^4 + 2x^3 - 5x^2 - 3x - 2$; remainder: -3

92. $\dfrac{2x^5 + 4x^4 - 3x^3 - 7x^2 + 3x - 2}{x + 2}$
Quotient: $2x^4 - 3x^2 - x + 5$; remainder: -12

93. $\dfrac{x^5 + 1}{x + 1}$

94. $\dfrac{x^5 + 1}{x - 1}$

95. $\dfrac{x^6 + 2x^4 - x^3 + 5}{x + 1}$

96. $\dfrac{x^7 + 1}{x + 1}$

97. $\dfrac{2x^5 + 3x^2 + 7}{x - \dfrac{1}{2}}$

98. $\dfrac{6x^4 + 3x^3 + 1}{x - \dfrac{1}{3}}$

99. $\dfrac{2x^4 - 3x^2 + 5x + 1}{x + \dfrac{3}{2}}$

100. $\dfrac{6x^6 - 5x^4 + 3x^2 + 4}{x + \dfrac{2}{3}}$

101. $\dfrac{-5x^4}{x + 2}$

102. $\dfrac{3x^5}{x - 1}$

In Exercises 103–150, factor each polynomial completely. If a polynomial cannot be factored, state that it is irreducible.

103. $3x^3 - x^2$ $x^2(3x - 1)$ **104.** $2x^3 + 2x^2$ $2x^2(x + 1)$

105. $x^2 + 7x + 12$ **106.** $x^2 + 8x + 15$

107. $x^2 - 6x + 8$ **108.** $x^2 - 9x + 14$

109. $6x^2 + 17x + 12$ **110.** $8x^2 - 10x - 3$ $(2x - 3)(4x + 1)$

111. $x^2 + 6x + 9$ $(x + 3)^2$ **112.** $x^2 + 8x + 16$ $(x + 4)^2$

113. $9x^2 + 6x + 1$ $(3x + 1)^2$ **114.** $36x^2 + 12x + 1$ $(6x + 1)^2$

115. $x^2 - 64$ $(x - 8)(x + 8)$ **116.** $x^2 - 121$ $(x - 11)(x + 11)$

117. $16x^2 - 9$ **118.** $25x^2 - 49$

119. $x^3 - 27$ **120.** $x^3 - 216$

121. $8 - x^3$ **122.** $27 - x^3$

123. $x^3 - 3x^2 + x - 3$ **124.** $x^3 + 5x^2 + x + 5$

125. $x^3 - 5x^2 + x - 5$ **126.** $x^3 - 7x^2 + x - 7$

127. $x^4 - 1$ **128.** $x^4 - 81$

129. $20x^4 - 5$ **130.** $12x^4 - 75$

131. $1 - 16x^2$ $(1 - 4x)(1 + 4x)$ **132.** $4 - 25x^2$ $(2 - 5x)(2 + 5x)$

133. $x^2 - 6x + 9$ $(x - 3)^2$ **134.** $x^2 - 8x + 16$ $(x - 4)^2$

135. $4x^2 + 4x + 1$ $(2x + 1)^2$ **136.** $16x^2 + 8x + 1$ $(4x + 1)^2$

137. $2x^2 - 8x - 10$ **138.** $5x^2 - 10x - 40$

139. $2x^2 + 3x - 20$ **140.** $2x^2 - 7x - 30$

141. $x^2 - 12x + 36$ $(x - 6)^2$ **142.** $x^2 - 20x + 25$ Irreducible

143. $3x^5 + 12x^4 + 12x^3$ **144.** $2x^5 + 16x^4 + 32x^3$

145. $9x^2 - 1$ **146.** $16x^2 - 25$

147. $16x^2 + 24x + 9$ **148.** $4x^2 + 20x + 25$ $(2x + 5)^2$

149. $x^2 + 15$ Irreducible   **150.** $x^2 + 24$ Irreducible

Answers.
93. Quotient: $x^4 - x^3 + x^2 - x + 1$; remainder: 0
94. Quotient: $x^4 + x^3 + x^2 + x + 1$; remainder: 2
95. Quotient: $x^5 - x^4 + 3x^3 - 4x^2 + 4x - 4$; remainder: 9
96. Quotient: $x^6 - x^5 + x^4 - x^3 + x^2 - x + 1$; remainder: 0
97. Quotient: $2x^4 + x^3 + \dfrac{1}{2}x^2 + \dfrac{13}{4}x + \dfrac{13}{8}$; remainder: $\dfrac{125}{16}$

98. Quotient: $6x^3 + 5x^2 + \dfrac{5}{3}x + \dfrac{5}{9}$; remainder: $\dfrac{32}{27}$

99. Quotient: $2x^3 - 3x^2 + \dfrac{3}{2}x - \dfrac{11}{4}$; remainder: $-\dfrac{25}{8}$

100. Quotient: $6x^5 - 4x^4 - \dfrac{7}{3}x^3 + \dfrac{14}{9}x^2 + \dfrac{53}{27}x - \dfrac{106}{81}$; remainder:

$\dfrac{1184}{243}$ **101.** Quotient: $-5x^3 + 10x^2 - 20x + 40$; remainder: -80

102. Quotient: $3x^4 + 3x^3 + 3x^2 + 3x + 3$; remainder: 3
105. $(x + 3)(x + 4)$ **106.** $(x + 3)(x + 5)$
107. $(x - 2)(x - 4)$ **108.** $(x - 2)(x - 7)$ **109.** $(2x + 3)(3x + 4)$

117. $(4x - 3)(4x + 3)$ **118.** $(5x - 7)(5x + 7)$
119. $(x - 3)(x^2 + 3x + 9)$ **120.** $(x - 6)(x^2 + 6x + 36)$
121. $(2 - x)(4 + 2x + x^2)$ **122.** $(3 - x)(9 + 3x + x^2)$
123. $(x - 3)(x^2 + 1)$ **124.** $(x + 5)(x^2 + 1)$ **125.** $(x - 5)(x^2 + 1)$
126. $(x - 7)(x^2 + 1)$ **127.** $(x - 1)(x + 1)(x^2 + 1)$
128. $(x - 3)(x + 3)(x^2 + 9)$ **129.** $5(2x^2 - 1)(2x^2 + 1)$
130. $3(2x^2 - 5)(2x^2 + 5)$
137. $2(x + 1)(x - 5)$ **138.** $5(x + 2)(x - 4)$
139. $(2x - 5)(x + 4)$ **140.** $(2x + 5)(x - 6)$
143. $3x^3(x + 2)^2$ **144.** $2x^3(x + 4)^2$ **145.** $(3x + 1)(3x - 1)$
146. $(4x + 5)(4x - 5)$ **147.** $(4x + 3)^2$

Rational Expressions

Objectives

1 Reduce a rational expression to lowest terms.

2 Multiply and divide rational expressions.

3 Add and subtract rational expressions.

4 Identify and simplify complex fractions.

Rational Expressions

Recall that the quotient of two integers, $\frac{a}{b}$ $(b \neq 0)$, is a rational number. When we form the quotient of two polynomials, the result is called a **rational expression**. Following are some examples of rational expressions.

$$\frac{10}{17} \qquad \frac{x+2}{3} \qquad \frac{x+6}{(x-3)(x+4)} \qquad \frac{7}{x^2+1} \qquad \frac{x^2+2x-3}{x^2+5x}$$

We use the same language to describe rational expressions that we use to describe rational numbers. For $\frac{x^2+2x-3}{x^2+5x+6}$, we call x^2+2x-3 the **numerator** and x^2+5x+6 the **denominator**. As with rational numbers, we do not allow the denominator to be 0. We write $\frac{x+6}{(x-3)(x+4)}$, $x \neq 3$, $x \neq -4$ to indicate that 3 and -4 are not in the domain of this expression; the domain in interval notation is $(-\infty, -4) \cup (-4, 3) \cup (3, \infty)$.

1 Reduce a rational expression to lowest terms.

Lowest Terms for a Rational Expression

EXAMPLE 1 **Reducing a Rational Expression to Lowest Terms**

Simplify each expression.

a. $\dfrac{x^4+2x^3}{x+2}$, $x \neq -2$ **b.** $\dfrac{x^6-x-6}{x^3-3x^2}$, $x \neq 0$, $x \neq 3$.

SOLUTION

Factor each numerator and denominator and remove the common factors.

a. $\dfrac{x^4+2x^3}{x+2} = \dfrac{x^3(x+2)}{x+2} = \dfrac{x^3\cancel{(x+2)}}{\cancel{(x+2)}} = x^3$

b. $\dfrac{x^2-x-6}{x^3-3x^2} = \dfrac{(x+2)(x-3)}{x^2(x-3)} = \dfrac{(x+2)\cancel{(x-3)}}{x^2\cancel{(x-3)}} = \dfrac{x+2}{x^2}$ ∎

Practice Problem 1 Simplify each expression.

a. $\dfrac{2x^3+8x^2}{3x^2+12x}$ **b.** $\dfrac{x^2-4}{x^2+4x+4}$

◆ **WARNING**

Because x^2 is a *term*, not a *factor*, in both the numerator and denominator of $\dfrac{x^2 - 1}{x^2 + 2x + 1}$, you cannot remove x^2. You can remove only *factors* common to both numerator and denominator.

---

2 Multiply and divide rational expressions.

Multiplication and Division of Rational Expressions

We use the same rules for multiplying and dividing rational expressions as we do for rational numbers.

MULTIPLICATION AND DIVISION

For rational expressions $\dfrac{A}{B}$ and $\dfrac{C}{D}$,

$$\frac{A}{B} \cdot \frac{C}{D} = \frac{A \cdot C}{B \cdot D} \quad \text{and} \quad \frac{\dfrac{A}{B}}{\dfrac{C}{D}} = \frac{A}{B} \div \frac{C}{D} = \frac{A}{B} \cdot \frac{D}{C} = \frac{AD}{BC}$$

$$\text{if } B \neq 0, D \neq 0. \qquad\qquad \text{if } B \neq 0, C \neq 0, D \neq 0.$$

STUDY TIP

The best strategy to use when multiplying and dividing rational expressions is to factor each numerator and denominator completely and then remove the common factors.

EXAMPLE 2 **Multiplying and Dividing Rational Expressions**

Multiply or divide as indicated. Simplify your answer and leave it in factored form.

a. $\dfrac{x^2 + 3x + 2}{x^3 + 3x} \cdot \dfrac{2x^3 + 6x}{x^2 + x - 2}, \qquad x \neq 0, x \neq 1, x \neq -2$

b. $\dfrac{\dfrac{3x^2 + 11x - 4}{8x^3 - 40x^2}}{\dfrac{3x + 12}{4x^4 - 20x^3}}, \qquad x \neq 0, x \neq 5, x \neq -4$

SOLUTION

Factor each numerator and denominator and remove the common factors.

a. $\dfrac{x^2 + 3x + 2}{x^3 + 3x} \cdot \dfrac{2x^3 + 6x}{x^2 + x - 2} = \dfrac{(x + 2)(x + 1)}{x(x^2 + 3)} \cdot \dfrac{2x(x^2 + 3)}{(x - 1)(x + 2)}$

$$= \frac{\cancel{(x + 2)}(x + 1)(2\cancel{x})\cancel{(x^2 + 3)}}{\cancel{x}\cancel{(x^2 + 3)}(x - 1)\cancel{(x + 2)}} = \frac{2(x + 1)}{x - 1}$$

STUDY TIP

In the quotient of two rational expressions $\dfrac{\dfrac{A}{B}}{\dfrac{C}{D}} = \dfrac{A}{B} \cdot \dfrac{D}{C}$, note that three polynomials appear as denominators: B and D from $\dfrac{A}{B}$ and $\dfrac{C}{D}$, and C from $\dfrac{A}{B} \cdot \dfrac{D}{C}$.

b. $\dfrac{\dfrac{3x^2 + 11x - 4}{8x^3 - 40x^2}}{\dfrac{3x + 12}{4x^4 - 20x^3}} = \dfrac{3x^2 + 11x - 4}{8x^3 - 40x^2} \cdot \dfrac{4x^4 - 20x^3}{3x + 12} = \dfrac{(x + 4)(3x - 1)}{8x^2(x - 5)} \cdot \dfrac{4x^3(x - 5)}{3(x + 4)}$

$$= \frac{\cancel{(x + 4)}(3x - 1)\cancel{(4x^3)}\cancel{(x - 5)}}{\underset{2}{\cancel{8x^2}}\cancel{(x - 5)}(3)\cancel{(x + 4)}} \overset{x}{} = \frac{(3x - 1)x}{6} = \frac{x(3x - 1)}{6}$$

■ ■ ■

Practice Problem 2   Multiply or divide as indicated. Simplify your answer.

$$\dfrac{\dfrac{x^2 - 2x - 3}{7x^3 + 28x^2}}{\dfrac{4x + 4}{2x^4 + 8x^3}} \qquad x \ne 0, x \ne -4, x \ne -1$$ ■

3 Add and subtract rational expressions.

Addition and Subtraction of Rational Expressions

To add and subtract rational expressions, we use the same rules as for adding and subtracting rational numbers.

ADDITION AND SUBTRACTION OF RATIONAL EXPRESSIONS

For rational expressions $\dfrac{A}{D}$ and $\dfrac{C}{D}$ (same denominator $D \ne 0$),

$$\dfrac{A}{D} + \dfrac{C}{D} = \dfrac{A + C}{D} \qquad \text{and} \qquad \dfrac{A}{D} - \dfrac{C}{D} = \dfrac{A - C}{D}.$$

For rational expression $\dfrac{A}{B}$ and $\dfrac{C}{D}$ $(B \ne 0, D \ne 0)$,

$$\dfrac{A}{B} + \dfrac{C}{D} = \dfrac{AD + BC}{BD} \qquad \text{and} \qquad \dfrac{A}{B} - \dfrac{C}{D} = \dfrac{AD - BC}{BD}.$$

> **EXAMPLE 3**   **Adding and Subtracting Rational Expressions with the Same Denominator**

Add or subtract as indicated. Simplify your answer and leave both numerator and denominator in factored form.

a. $\dfrac{x - 6}{(x + 1)^2} + \dfrac{x + 8}{(x + 1)^2}, x \ne -1$

b. $\dfrac{3x - 2}{x^2 - 5x + 6} - \dfrac{2x + 1}{x^2 - 5x + 6}, x \ne 2, x \ne 3$

SOLUTION

a. $\dfrac{x - 6}{(x + 1)^2} + \dfrac{x + 8}{(x + 1)^2} = \dfrac{x - 6 + x + 8}{(x + 1)^2} = \dfrac{2x + 2}{(x + 1)^2} = \dfrac{2\cancel{(x + 1)}}{(x + 1)\cancel{(x + 1)}} = \dfrac{2}{x + 1}$

b. $\dfrac{3x - 2}{x^2 - 5x + 6} - \dfrac{2x + 1}{x^2 - 5x + 6} = \dfrac{(3x - 2) - (2x + 1)}{x^2 - 5x + 6} = \dfrac{3x - 2 - 2x - 1}{x^2 - 5x + 6}$

$$= \dfrac{x - 3}{(x - 2)(x - 3)} = \dfrac{\cancel{x - 3}}{(x - 2)\cancel{(x - 3)}} = \dfrac{1}{x - 2}$$

■ ■ ■

Practice Problem 3   Add or subtract as indicated. Simplify your answers.

a. $\dfrac{5x + 22}{x^2 - 36} + \dfrac{2(x + 10)}{x^2 - 36} \qquad x \ne 6, x \ne -6$

b. $\dfrac{4x + 1}{x^2 + x - 12} - \dfrac{3x + 4}{x^2 + x - 12} \qquad x \ne -4, x \ne 3$ ■

When adding or subtracting rational expressions with different denominators, we proceed (as with fractions) by finding a common denominator. The one most convenient to use is called the **least common denominator (LCD)**. It is the polynomial of least degree that contains each denominator as a factor. In the simplest case, the LCD is just the product of the denominators.

EXAMPLE 4 Adding and Subtracting When Denominators Have No Common Factor

Add or subtract as indicated. Simplify your answer and leave it in factored form.

a. $\dfrac{x}{x+1} + \dfrac{2x-1}{x+3}, x \neq 1, x \neq -3$ **b.** $\dfrac{2x}{x+1} - \dfrac{x}{x+2}, x \neq -1, x \neq -2$

SOLUTION

a. $\dfrac{x}{x+1} + \dfrac{2x-1}{x+3} = \dfrac{x(x+3)}{(x+1)(x+3)}$ LCD $= (x+1)(x+3)$

$$+ \dfrac{(2x-1)(x+1)}{(x+1)(x+3)}$$

$$= \dfrac{x(x+3) + (2x-1)(x+1)}{(x+1)(x+3)}$$

$$= \dfrac{x^2 + 3x + 2x^2 + 2x - x - 1}{(x+1)(x+3)}$$

$$= \dfrac{3x^2 + 4x - 1}{(x+1)(x+3)}$$

b. $\dfrac{2x}{x+1} - \dfrac{x}{x+2} = \dfrac{2x(x+2)}{(x+1)(x+2)}$ LCD $= (x+1)(x+2)$

$$- \dfrac{x(x+1)}{(x+1)(x+2)}$$

$$= \dfrac{2x(x+2) - x(x+1)}{(x+1)(x+2)} = \dfrac{2x^2 + 4x - x^2 - x}{(x+1)(x+2)}$$

$$= \dfrac{x^2 + 3x}{(x+1)(x+2)} = \dfrac{x(x+3)}{(x+1)(x+2)}$$ ■ ■ ■

Practice Problem 4   Add or subtract as indicated. Simplify your answers.

a. $\dfrac{2x}{x+2} + \dfrac{3x}{x-5}$, $x \neq -2, x \neq 5$ **b.** $\dfrac{5x}{x-4} - \dfrac{2x}{x+3}$, $x \neq 4, x \neq -3$.   ■

TO FIND THE LCD FOR RATIONAL EXPRESSIONS

1. Factor each denominator polynomial completely.

2. Form a product of the different irreducible factors of each polynomial. (Each distinct factor is used exactly once.)

3. Attach to each factor in this product the largest exponent that appears on this factor in *any* of the factored denominators.

EXAMPLE 5 Finding the LCD for Rational Expressions

Find the LCD for each pair of rational expressions.

a. $\dfrac{x+2}{x(x-1)^2(x+2)}, \dfrac{3x+7}{4x^2(x+2)^3}$ **b.** $\dfrac{x+1}{x^2-x-6}, \dfrac{2x-13}{x^2-9}$

SOLUTION

a. The denominators are already completely factored.

$4x(x - 1)(x + 2)$ Product of the different factors

$\text{LCD} = 4x^2(x - 1)^2(x + 2)^3$ The largest exponents are 2, 2, and 3.

b. $x^2 - x - 6 = (x + 2)(x - 3)$

      $x^2 - 9 = (x + 3)(x - 3)$

     $(x + 2)(x - 3)(x + 3)$ Product of the different factors

     $\text{LCD} = (x + 2)(x - 3)(x + 3)$ The largest exponent on each is 1.   ■ ■ ■

Practice Problem 5   Find the LCD for each pair of rational expressions.

a. $\dfrac{x^2 + 3x}{x^2(x + 2)^2(x - 2)}, \dfrac{4x^2 + 1}{3x(x - 2)^2}$ **b.** $\dfrac{2x - 1}{x^2 - 25}, \dfrac{3 - 7x^2}{(x^2 + 4x - 5)}$ ■

In adding or subtracting rational expressions with different denominators, the first step is to find the LCD. Here is the general procedure.

PROCEDURE FOR ADDING OR SUBTRACTING RATIONAL EXPRESSIONS WITH DIFFERENT DENOMINATORS.

Step 1   Find the LCD.

Step 2   Using $\dfrac{A}{B} = \dfrac{AC}{BC}$, write each rational expression as a rational expression with the LCD as the denominator.

Step 3   Following the order of operations, add or subtract numerators, keeping the LCD as the denominator.

Step 4   Simplify.

EXAMPLE 6   **Using the LCD to Add and Subtract Rational Expressions**

Add or subtract as indicated. Simplify your answer and leave it in factored form.

a. $\dfrac{3}{x^2 - 1} + \dfrac{x}{x^2 + 2x + 1}, x \neq 1, x \neq -1$ **b.** $\dfrac{x + 2}{x^2 - x} - \dfrac{3x}{4(x - 1)^2}, x \neq 0, x \neq 1$

SOLUTION

a. Note that $x^2 - 1 = (x - 1)(x + 1)$ and $x^2 + 2x + 1 = (x + 1)^2$; the LCD is $(x + 1)^2(x - 1)$.

$\dfrac{3}{x^2 - 1} = \dfrac{3}{(x - 1)(x + 1)} = \dfrac{3(x + 1)}{(x - 1)(x + 1)^2}$ Multiply numerator and denominator by $x + 1$.

$\dfrac{x}{x^2 + 2x + 1} = \dfrac{x}{(x + 1)^2} = \dfrac{x(x - 1)}{(x + 1)^2(x - 1)}$ Multiply numerator and denominator by $x - 1$.

Now add.

$$\dfrac{3}{x^2 - 1} + \dfrac{x}{x^2 + 2x + 1} = \dfrac{3(x + 1)}{(x - 1)(x + 1)^2} + \dfrac{x(x - 1)}{(x - 1)(x + 1)^2}$$

$$= \dfrac{3(x + 1) + x(x - 1)}{(x - 1)(x + 1)^2} = \dfrac{3x + 3 + x^2 - x}{(x - 1)(x + 1)^2}$$

$$= \dfrac{x^2 + 2x + 3}{(x - 1)(x + 1)^2}$$

b. The LCD is $4x(x - 1)^2$.

$$x^2 - x = x(x - 1)$$

$$\frac{x + 2}{x^2 - x} = \frac{x + 2}{x(x - 1)} = \frac{(x + 2) \cdot 4(x - 1)}{x(x - 1) \cdot 4(x - 1)} = \frac{4(x + 2)(x - 1)}{4x(x - 1)^2}$$ Multiply numerator and denominator by $4(x - 1)$.

$$\frac{3x}{4(x - 1)^2} = \frac{(3x)(x)}{4(x - 1)^2(x)} = \frac{3x^2}{4x(x - 1)^2}$$ Multiply numerator and denominator by x.

Now subtract.

$$\frac{x + 2}{x^2 - x} - \frac{3x}{4x(x - 1)^2} = \frac{4(x + 2)(x - 1)}{4x(x - 1)^2} - \frac{3x^2}{4x(x - 1)^2} = \frac{4(x + 2)(x - 1) - 3x^2}{4x(x - 1)^2}$$

$$= \frac{4x^2 + 4x - 8 - 3x^2}{4x(x - 1)^2} = \frac{x^2 + 4x - 8}{4x(x - 1)^2}$$ ■ ■ ■

Practice Problem 6   Add or subtract as indicated. Simplify your answers.

a. $\dfrac{4}{x^2 - 4x + 4} + \dfrac{x}{x^2 - 4}$, $x \neq 2, x = -2$

b. $\dfrac{2x}{3(x - 5)^2} - \dfrac{6x}{2(x^2 - 5x)}$, $x \neq 0, x \neq 5$ ■

4 Identify and simplify complex fractions.

Complex Fractions

A rational expression that contains another rational expression in its numerator or denominator (or both) is called a **complex rational expression** or **complex fraction**. To simplify a complex fraction, we write it as a rational expression in lowest terms.
    There are two effective methods of simplifying complex fractions.

PROCEDURES FOR SIMPLIFYING COMPLEX FRACTIONS

Method 1   Perform the operations indicated in both the numerator and the denominator of the complex fraction. Then multiply the resulting numerator by the reciprocal of the denominator.

Method 2   Multiply the numerator and the denominator of the complex fraction by the LCD of all rational expressions that appear in either the numerator or denominator. Simplify the result.

EXAMPLE 7   **Simplifying a Complex Fraction**

Simplify $\dfrac{\dfrac{1}{2} + \dfrac{1}{x}}{\dfrac{x^2 - 4}{2x}}$, $x \neq 0, x \neq 2, x \neq -2$, using each of the two methods.

SOLUTION

Method 1 $\dfrac{\dfrac{1}{2} + \dfrac{1}{x}}{\dfrac{x^2 - 4}{2x}} = \dfrac{\dfrac{x + 2}{2x}}{\dfrac{x^2 - 4}{2x}} = \dfrac{x + 2}{2x} \cdot \dfrac{2x}{x^2 - 4} = \dfrac{x + 2}{2x} \cdot \dfrac{2x}{(x + 2)(x - 2)}$

$$= \frac{\cancel{(x + 2)}\cancel{(2x)}}{\cancel{(2x)}\cancel{(x + 2)}(x - 2)} = \frac{1}{x - 2}$$

Method 2   The LCD of $\dfrac{1}{2}, \dfrac{1}{x}$, and $\dfrac{x^2 - 4}{2x}$ is $2x$.

$$\dfrac{\dfrac{1}{2} + \dfrac{1}{x}}{\dfrac{x^2 - 4}{2x}} = \dfrac{\left(\dfrac{1}{2} + \dfrac{1}{x}\right)(2x)}{\left(\dfrac{x^2 - 4}{2x}\right)(2x)} \qquad \text{Multiply numerator and denominator by the LCD.}$$

$$= \dfrac{\dfrac{1}{2}(2x) + \dfrac{1}{x}(2x)}{\left[\dfrac{(x^2 - 4)}{2x}\right](2x)} = \dfrac{x + 2}{x^2 - 4} = \dfrac{\cancel{x + 2}}{(x - 2)\cancel{(x + 2)}} = \dfrac{1}{x - 2} \qquad ■ ■ ■$$

Practice Problem 7   Simplify $\dfrac{\dfrac{5}{3x} + \dfrac{1}{3}}{\dfrac{x^2 - 25}{3x}}, x \neq 0, x \neq 5, x \neq -5.$ ■

EXAMPLE 8   **Simplifying a Complex Fraction**

Simplify $\dfrac{x}{2x - \dfrac{7}{3 - \dfrac{1}{2}}}, x \neq \dfrac{7}{5}.$

SOLUTION

We will use Method 1.

Begin with $\dfrac{7}{3 - \dfrac{1}{2}} = \dfrac{7}{\dfrac{5}{2}} = 7 \cdot \dfrac{2}{5} = \dfrac{14}{5}.$ Replace $3 - \dfrac{1}{2}$ with $\dfrac{5}{2}.$

Then $\dfrac{x}{2x - \dfrac{7}{3 - \dfrac{1}{2}}} = \dfrac{x}{2x - \dfrac{14}{5}}$ Replace $\dfrac{7}{3 - \dfrac{1}{2}}$ with $\dfrac{14}{5}.$

$$= \dfrac{x}{\dfrac{10x - 14}{5}} = x \cdot \dfrac{5}{10x - 14} = \dfrac{5x}{10x - 14} = \dfrac{5x}{2(5x - 7)}$$

■ ■ ■

Practice Problem 8   Simplify $\dfrac{5x}{3x - \dfrac{4}{2 - \dfrac{1}{3}}}, \quad x \neq \dfrac{4}{5}.$ ■

SECTION A.3 ■ Exercises

A EXERCISES Basic Skills and Concepts

1. The least common denominator for two rational expressions is the polynomial of least degree that contains _____ as a factor. both denominators

2. The first step in finding the LCD for two rational expressions is to _____ factor _____ the denominators completely.

3. If the denominators for two rational expressions are $x^2 + 2x$ and $x^2 - x - 2$, then the LCD is _____†_____ .

4. A rational expression that contains another rational expression in its numerator or denominator is called a(n) _____ . complex fraction

In Exercises 5–20, reduce each rational expression to lowest terms. Identify all numbers that must be excluded from the domain of the given rational expression and write the domain in interval notation.

5. $\dfrac{2x + 2}{x^2 + 2x + 1} \dfrac{2}{x + 1}, x \neq -1$ 6. $\dfrac{3x - 6}{x^2 - 4x + 4} \dfrac{3}{x - 2}, x \neq 2$

7. $\dfrac{3x + 3}{x^2 - 1} \dfrac{3}{x - 1}, x \neq -1, x \neq 1$ 8. $\dfrac{10 - 5x}{4 - x^2} \dfrac{5}{2 + x}, x \neq -2, x \neq 2$

9. $\dfrac{2x - 6}{9 - x^2} -\dfrac{2}{x + 3}, x \neq -3, x \neq 3$ 10. $\dfrac{15 + 3x}{x^2 - 25} \dfrac{3}{x - 5}, x \neq -5, x \neq 5$

11. $\dfrac{2x - 1}{1 - 2x} -1, x \neq \dfrac{1}{2}$ 12. $\dfrac{2 - 5x}{5x - 2} -1, x \neq \dfrac{2}{5}$

13. $\dfrac{x^2 - 6x + 9}{4x - 12} \dfrac{x - 3}{4}, x \neq 3$ 14. $\dfrac{x^2 - 10x + 25}{3x - 15} \dfrac{x - 5}{3}, x \neq 5$

15. $\dfrac{7x^2 + 7x}{x^2 + 2x + 1} \dfrac{7x}{x + 1}, x \neq -1$ 16. $\dfrac{4x^2 + 12x}{x^2 + 6x + 9} \dfrac{4x}{x + 3}, x \neq -3$

17. $\dfrac{x^2 - 11x + 10}{x^3 + 6x - 7}$ † 18. $\dfrac{x^2 + 2x - 15}{x^2 - 7x + 12}$ †

19. $\dfrac{6x^4 + 14x^3 + 4x^2}{6x^4 - 10x^3 - 4x^2}$ † 20. $\dfrac{3x^3 + x^2}{3x^4 - 11x^3 - 4x^2}$ †

In Exercises 21–38, multiply or divide as indicated. Simplify and leave the numerator and denominator in your answer in factored form.

21. $\dfrac{x - 3}{2x + 4} \cdot \dfrac{10x + 20}{5x - 15}$ 1 22. $\dfrac{6x + 4}{2x - 8} \cdot \dfrac{x - 4}{9x + 6} \dfrac{1}{3}$

23. $\dfrac{2x + 6}{4x - 8} \cdot \dfrac{x^2 + x - 6}{x^2 - 9}$ † 24. $\dfrac{25x^2 - 9}{4 - 2x} \cdot \dfrac{4 - x^2}{10x - 6}$ †

25. $\dfrac{x^2 - 7x}{x^2 - 6x - 7} \cdot \dfrac{x^2 - 1}{x^2}$ † 26. $\dfrac{x^2 - 9}{x^2 - 6x + 9} \cdot \dfrac{5x - 15}{x + 3}$ †

27. $\dfrac{x^2 - x - 6}{x^2 + 3x + 2} \cdot \dfrac{x^2 - 1}{x^2 - 9}$ † 28. $\dfrac{x^2 + 2x - 8}{x^2 + x - 20} \cdot \dfrac{x^2 - 16}{x^2 + 5x + 4}$ †

29. $\dfrac{2 - x}{x + 1} \cdot \dfrac{x^2 + 3x + 2}{x^2 - 4}$ -1 30. $\dfrac{3 - x}{x + 5} \cdot \dfrac{x^2 + 8x + 15}{x^2 - 9}$ -1

31. $\dfrac{x + 2}{6} \div \dfrac{4x + 8}{9} \dfrac{3}{8}$ 32. $\dfrac{x + 3}{20} \div \dfrac{4x + 12}{9} \dfrac{9}{80}$

33. $\dfrac{x^2 - 9}{x} \div \dfrac{2x + 6}{5x^2}$ † 34. $\dfrac{x^2 - 1}{3x} \div \dfrac{7x - 7}{x^2 + x} \dfrac{(x + 1)^2}{21}$

35. $\dfrac{x^2 + 2x - 3}{x^2 + 8x + 16} \div \dfrac{x - 1}{3x + 12} \dfrac{3(x + 3)}{x + 4}$

36. $\dfrac{x^2 + 5x + 6}{x^2 + 6x + 9} \div \dfrac{x^2 + 3x + 2}{x^2 + 7x + 12} \dfrac{x + 4}{x + 1}$

37. $\left(\dfrac{x^2 - 9}{x^3 + 8} \div \dfrac{x + 3}{x^3 + 2x^2 - x - 2} \right) \dfrac{1}{x^2 - 1} \dfrac{x - 3}{x^3 - 2x + 4}$

38. $\left(\dfrac{x^2 - 25}{x^2 - 3x - 4} \div \dfrac{x^2 + 3x - 10}{x^2 - 1} \right) \dfrac{x - 2}{x - 5} \dfrac{x - 1}{x - 4}$

In Exercises 39–56, add and subtract as indicated. Simplify and leave the numerator and denominator in your answer in factored form.

39. $\dfrac{x}{5} + \dfrac{3}{5} \dfrac{x + 3}{5}$ 40. $\dfrac{7}{4} - \dfrac{x}{4} \dfrac{7 - x}{4}$

41. $\dfrac{x}{2x + 1} + \dfrac{4}{2x + 1} \dfrac{x + 4}{2x + 1}$ 42. $\dfrac{2x}{7x - 3} + \dfrac{x}{7x - 3} \dfrac{3x}{7x - 3}$

43. $\dfrac{x^2}{x + 1} - \dfrac{x^2 - 1}{x + 1} \dfrac{1}{x + 1}$ 44. $\dfrac{2x + 7}{3x + 2} - \dfrac{x - 2}{3x + 2} \dfrac{x + 9}{3x + 2}$

45. $\dfrac{4}{3 - x} + \dfrac{2x}{x - 3} \dfrac{2(x - 2)}{x - 3}$ 46. $\dfrac{-2}{1 - x} + \dfrac{2 - x}{x - 1} - \dfrac{x - 4}{x - 1}$

47. $\dfrac{5x}{x^2 + 1} + \dfrac{2x}{x^2 + 1} \dfrac{7x}{x^2 + 1}$ 48. $\dfrac{x}{2(x - 1)^2} + \dfrac{3x}{2(x - 1)^2}$ †

49. $\dfrac{7x}{2(x - 3)} + \dfrac{x}{2(x - 3)}$ † 50. $\dfrac{4x}{4(x + 5)^2} + \dfrac{8x}{4(x + 5)^2}$ †

51. $\dfrac{x}{x^2 - 4} - \dfrac{2}{x^2 - 4} \dfrac{1}{x + 2}$ 52. $\dfrac{5x}{x^2 - 1} - \dfrac{5}{x^2 - 1} \dfrac{5}{x + 1}$

53. $\dfrac{x - 2}{2x + 1} - \dfrac{x}{2x - 1}$ † 54. $\dfrac{2x - 1}{4x + 1} - \dfrac{2x}{4x - 1}$ †

55. $\dfrac{-x}{x + 2} + \dfrac{x - 2}{x} - \dfrac{x}{x - 2} - \dfrac{x^3 + 2x^2 + 4x - 8}{x(x - 2)(x + 2)}$

56. $\dfrac{3x}{x - 1} + \dfrac{x + 1}{x} - \dfrac{2x}{x + 1} \dfrac{2x^3 + 6x^2 - x - 1}{x(x - 1)(x + 1)}$

In Exercises 57–64, find the LCD for each pair of national fractions.

57. $\dfrac{5}{3x - 6}, \dfrac{2x}{4x - 8} 12(x - 2)$ 58. $\dfrac{5x + 1}{7 + 21x}, \dfrac{1 - x}{3 + 9x}$ †

59. $\dfrac{3 - x}{4x^2 - 1}, \dfrac{7x}{(2x + 1)^2}$ † 60. $\dfrac{14x}{(3x - 1)^2}, \dfrac{2x + 7}{9x^2 - 1}$ †

61. $\dfrac{1 - x}{x^2 + 3x + 2}, \dfrac{3x + 12}{x^2 - 1}$ † 62. $\dfrac{5x + 9}{x^2 - x - 6}, \dfrac{x + 5}{x^2 - 9}$ †

63. $\dfrac{7 - 4x}{x^2 - 5x + 4}, \dfrac{x^2 - x}{x^2 + x - 2} (x - 1)(x - 4)(x + 2)$

64. $\dfrac{13x}{x^2 - 2x - 3}, \dfrac{2x^2 - 4}{x^2 + 3x + 2} (x - 3)(x + 1)(x + 2)$

†Due to space constrictions, answers to these exercises may be found in the Answers beginning on page A–1 in the back of the book.

In Exercises 65–80, perform the indicated operations and simplify the result. Leave the numerator and denominator in your answer in factored form.

65. $\dfrac{5}{x-3} + \dfrac{2x}{x^2-9}$ †

66. $\dfrac{3x}{x-1} + \dfrac{x}{x^2-1}$ †

67. $\dfrac{2x}{x^2-4} - \dfrac{x}{x+2}$ †

68. $\dfrac{3x-1}{x^2-16} - \dfrac{2x+1}{x-4}$ †

69. $\dfrac{x-2}{x^2+3x-10} + \dfrac{x+3}{x^2+x-6}$ $\dfrac{2x+3}{(x-2)(x+5)}$

70. $\dfrac{x+3}{x^2-x-2} + \dfrac{x-1}{x^2+2x+1}$ $\dfrac{2x^2+x+5}{(x-2)(x+1)^2}$

71. $\dfrac{2x-3}{9x^2-1} + \dfrac{4x-1}{(3x-1)^2}$ †   **72.** $\dfrac{3x+1}{(2x+1)^2} + \dfrac{x+3}{4x^2-1}$ †

73. $\dfrac{x-3}{x^2-25} - \dfrac{x-3}{x^2+9x+20}$ $\dfrac{9(x-3)}{(x+4)(x+5)(x-5)}$

74. $\dfrac{2x}{x^2-16} - \dfrac{2x-7}{x^2-7x+12}$ $\dfrac{7}{(x+4)(x-3)}$

75. $\dfrac{3}{x^2-4} + \dfrac{1}{2-x} - \dfrac{1}{2+x}$ $\dfrac{2x-3}{(2+x)(2-x)}$

76. $\dfrac{2}{5+x} + \dfrac{5}{x^2-25} + \dfrac{7}{5-x}$ $\dfrac{5(x+8)}{(5-x)(5+x)}$

77. $\dfrac{x+3a}{x-5a} - \dfrac{x+5a}{x-3a}$ †   **78.** $\dfrac{3x-a}{2x-a} - \dfrac{2x+a}{3x+a}$ †

79. $\dfrac{1}{x+h} - \dfrac{1}{x} - \dfrac{h}{x(x+h)}$ **80.** $\dfrac{1}{(x+h)^2} - \dfrac{1}{x^2}$ †

In Exercises 81–94, perform the indicated operations and simplify the result. Leave the numerator and denominator in your answer in factored form.

81. $\dfrac{\dfrac{2}{x} - \dfrac{2x}{3}}{\dfrac{3}{x^2}}$

82. $\dfrac{\dfrac{-6}{x} - \dfrac{2}{x^3}}{}$ $-3x^2$

83. $\dfrac{\dfrac{1}{x}}{1 - \dfrac{1}{x}}$ $\dfrac{1}{x-1}$

84. $\dfrac{\dfrac{1}{x^2}}{\dfrac{1}{x^2} - 1}$ $\dfrac{1}{(1-x)(1+x)}$

85. $\dfrac{\dfrac{1}{x} - 1}{\dfrac{1}{x} + 1}$ $\dfrac{1-x}{1+x}$

86. $\dfrac{2 - \dfrac{2}{x}}{1 + \dfrac{2}{x}}$ $\dfrac{2(x-1)}{x+2}$

87. $\dfrac{\dfrac{1}{x} - x}{1 - \dfrac{1}{x^2}}$ $-x$

88. $\dfrac{x - \dfrac{1}{x}}{\dfrac{1}{x^2} - 1}$ $-x$

89. $x - \dfrac{x}{x + \dfrac{1}{2}}$ $\dfrac{x(2x-1)}{2x+1}$

90. $x + \dfrac{x}{x + \dfrac{1}{2}}$ $\dfrac{x(2x+3)}{2x+1}$

91. $\dfrac{\dfrac{1}{x+h} - \dfrac{1}{x}}{h}$

$-\dfrac{1}{x(x+h)}, x \neq -h, x \neq 0$

92. $\dfrac{\dfrac{1}{(x+h)^2} - \dfrac{1}{x^2}}{h}$

$-\dfrac{2x+h}{x^2(x+h)^2}, x \neq -h, x \neq 0$

93. $\dfrac{\dfrac{1}{x-a} + \dfrac{1}{x+a}}{\dfrac{1}{x-a} - \dfrac{1}{x+a}}$

$\dfrac{x}{a}, x \neq -a, x \neq a$

94. $\dfrac{\dfrac{1}{x-a} + \dfrac{1}{x+a}}{\dfrac{x}{x-a} - \dfrac{a}{x+a}}$ $\dfrac{2x}{x^2+a^2}$

B EXERCISES Applying the Concepts

95. Bearing length. The length (in centimeters) of a bearing in the shape of a cylinder is given by $\dfrac{125.6}{\pi r^2}$, where r is the radius of the base of the cylinder. Find the length of a bearing with a diameter of 4 centimeters, using 3.14 as an approximation of π. 10 cm

96. Toy box height. The height (in feet) of an open toy box that is twice as long as it is wide is given by $\dfrac{13.5 - 2x^2}{6x}$, where x is the width of the box. Find the height of a toy box that is 1.5 feet wide. 1 ft

97. Diluting a mixture. A 100-gallon mixture of citrus extract and water is 3% citrus extract.
   a. Write a rational expression in x whose values give the percentage (in decimal form) of the mixture that is citrus extract when x gallons of water are added to the mixture. †
   b. Find the percentage of citrus extract in the mixture if 50 gallons of water are added to it. 2%

98. Acidity in a reservoir. A half-full 400,000-gallon reservoir is found to be 0.75% acid.
   a. Write a rational expression in x whose values give the percentage (in decimal form) of acid in the reservoir when x gallons of water are added to it. †
   b. Find the percentage of acid in the reservoir if 100,000 gallons of water are added to it. 0.5%

99. Diet lemonade. A company packages its powdered diet lemonade mix in containers in the shape of a cylinder. The top and bottom are made of a tin product that costs 5 cents per square inch. The side of the container is made of a cheaper material that costs 1 cent per square inch. The height of any cylinder can be found by dividing its volume by the area of its base.
   a. If x is the radius of the base (in inches), write a rational expression in x whose values give the cost of each container, assuming that the capacity of the container is 120 cubic inches. †
   b. Find the cost of a container whose base is 4 inches in diameter, using 3.14 as an approximation of π. $2.46

100. Storage containers. A company makes storage containers in the shape of an open box with a square base. The sides and base of the box are made of a plastic that costs 40 cents per square foot. The height of any box can be found by dividing its volume by the area of its base.
   a. If x is the length of the base (in feet), write a rational expression in x whose values give the cost of each container, assuming that the capacity of the container is 2.25 cubic feet.
   b. Find the cost of a container whose base is 1.5 feet long. $3.30

100. a. $\dfrac{40x^3 + 360}{x}$

Rational Exponents and Radicals

Objectives

1. Define and evaluate square roots.
2. Rationalize denominators and numerators.
3. Define and evaluate nth roots.
4. Combine like radicals.
5. Define and evaluate rational exponents.

---

1 Define and evaluate square roots.

Square Roots

When we raise the number 5 to the power 2, we write $5^2 = 25$. The reverse of this squaring process is called finding a **square root**. We know that 5 is a square root of 25 and write $\sqrt{25} = 5$. It is also true that $(-5)^2 = 25$; so -5 is *another* square root of 25.

The symbol $\sqrt{}$, called a **radical sign**, is used to distinguish between 5 and -5, the two square roots of 25. We write $5 = \sqrt{25}$ and $-5 = -\sqrt{25}$ and call $\sqrt{25}$ the *principal square root* of 25.

TECHNOLOGY CONNECTION

The square root symbol on a graphing calculator does not have a horizontal bar that goes completely over the number or expression whose square root is to be evaluated. You must enclose a rational number in parentheses before taking its square root and then use the "Frac" feature to have the answer appear as a fraction.

```
√(121)
                11
√(25/81)
        .5555555556
Ans▶Frac
              5/9
```

DEFINITION OF PRINCIPAL SQUARE ROOT
$\sqrt{a} = b$ means   (1) $b^2 = a$ and   (2) $b \geq 0$

If $\sqrt{a} = b$, then $a = b^2 \geq 0$. Consequently, the symbol $\sqrt{a}$ denotes a real number only when $a \geq 0$. The domain of the expression $\sqrt{x}$ is then $x \geq 0$, or in interval notation, $[0, \infty)$.

For any real number x,	$\sqrt{x^2} = \lvert x \rvert$
	and $(\sqrt{x})^2 = x$ if $x \geq 0$.

For example, $\sqrt{121} = \sqrt{11^2} = 11$ and $\sqrt{\dfrac{25}{81}} = \sqrt{\dfrac{5^2}{9^2}} = \sqrt{\left(\dfrac{5}{9}\right)^2} = \dfrac{5}{9}$.

2 Rationalize denominators and numerators.

Rationalizing Denominators

It is customary to rewrite a quotient involving square roots so that no radical appears in the denominator. The method for eliminating square roots in a denominator is called **rationalizing the denominator**. The special product $(A - B)(A + B) = A^2 - B^2$ is helpful when at least one term of a denominator binomial $A + B$ or $A - B$ is a square root. Here are some examples.

Denominator Factor	Multiply by	New Denominator Contains the Factor
$2 + \sqrt{3}$	$2 - \sqrt{3}$	$(2 + \sqrt{3})(2 - \sqrt{3}) = 2^2 - (\sqrt{3})^2 = 4 - 3 = 1$
$\sqrt{5} - \sqrt{2}$	$\sqrt{5} + \sqrt{2}$	$(\sqrt{5} - \sqrt{2})(\sqrt{5} + \sqrt{2}) = (\sqrt{5})^2 - (\sqrt{2})^2 = 5 - 2 = 3$
$1 + \sqrt{x}$	$1 - \sqrt{x}$	$(1 + \sqrt{x})(1 - \sqrt{x}) = 1^2 - (\sqrt{x})^2 = 1 - x, x \geq 0$

EXAMPLE 1   Rationalizing Denominators

Rationalize each denominator.

a. $\dfrac{1}{\sqrt{2}}$ **b.** $\dfrac{2}{7 - \sqrt{5}}$

SOLUTION

a. If we multiply numerator and denominator by $\sqrt{2}$, the denominator becomes $\sqrt{2} \cdot \sqrt{2} = (\sqrt{2})^2 = 2$. Consequently,

$$\frac{1}{\sqrt{2}} = \frac{1 \cdot \sqrt{2}}{\sqrt{2} \cdot \sqrt{2}} = \frac{\sqrt{2}}{2}.$$

b. We multiply numerator and denominator by $7 + \sqrt{5}$.

$$\frac{3}{7 - \sqrt{5}} = \frac{3 \cdot (7 + \sqrt{5})}{(7 - \sqrt{5})(7 + \sqrt{5})} = \frac{21 + 3\sqrt{5}}{7^2 - (\sqrt{5})^2} = \frac{21 + 3\sqrt{5}}{49 - 5} = \frac{21 + 3\sqrt{5}}{44}$$

■ ■ ■

Practice Problem 1   Rationalize each denominator.

a. $\dfrac{7}{\sqrt{8}}$ **b.** $\dfrac{3}{1 - \sqrt{7}}$ **c.** $\dfrac{x}{\sqrt{x} - 3}\ (x \geq 0)$ ■

3 Define and evaluate *n*th roots.

Other Roots

If n is a positive integer, we say that b is an *n*th root of a if $b^n = a$. We define the **principal *n*th root** of a real number a, denoted by the symbol $\sqrt[n]{a}$, as follows.

THE PRINCIPAL *n*th ROOT OF A REAL NUMBER

1. If a is positive $(a > 0)$, then $\sqrt[n]{a} = b$ provided that $b^n = a$ and $b > 0$.

2. If a is negative $(a < 0)$ and n is odd, then $\sqrt[n]{a} = b$ provided that $b^n = a$.

3. If a is negative $(a < 0)$ and n is even, then $\sqrt[n]{a}$ is not a real number.

We use the same vocabulary for *n*th roots that we used for square roots. That is, $\sqrt[n]{a}$ is called a **radical expression**, $\sqrt{}$ is the **radical sign**, and a is the **radicand**. The positive integer n is called the **index**. The index 2 is not written; we write $\sqrt{a}$ rather than $\sqrt[2]{a}$ for the principal square root of a. It is also common practice to call $\sqrt[3]{a}$ the **cube root** of a.

EXAMPLE 2   Finding Principal *n*th Roots

Find each root.

a. $\sqrt[3]{27}$ **b.** $\sqrt[3]{-64}$ **c.** $\sqrt[4]{16}$ **d.** $\sqrt[4]{(-3)^4}$ **e.** $\sqrt[8]{-46}$

SOLUTION

a. The radicand, 27, is *positive;* so $\sqrt[3]{27} = 3$ because $3^3 = 27$ and $3 > 0$.

b. The radicand, -64, is *negative;* so $\sqrt[3]{-64} = -4$ because $(-4)^3 = -64$.

c. The radicand, 16, is *positive;* so $\sqrt[4]{16} = 2$ because $2^4 = 16$ and $2 > 0$.

d. The radicand, $(-3)^4$, is *positive;* so $\sqrt[4]{(-3)^4} = 3$ because $3^4 = (-3)^4$ and $3 > 0$.

e. The radicand, -46, is *negative;* and the index, 8, is even; so $\sqrt[8]{-46}$ is not a real number. ■ ■ ■

Practice Problem 2   Find each root.

a. $\sqrt[3]{-8}$ b. $\sqrt[5]{32}$ c. $\sqrt[4]{81}$ d. $\sqrt[6]{-4}$ ■

The fact that the definition of the principal *n*th root depends on whether *n* is even or odd results in two rules to replace the square root rule $\sqrt{x^2} = |x|$.

If *n* is *odd*, then $\sqrt[n]{a^n} = a$.   If *n* is *even*, then $\sqrt[n]{a^n} = |a|$.

If *a* and *b* are real numbers and all of the indicated roots are defined, then

$$\sqrt[n]{ab} = \sqrt[n]{a} \cdot \sqrt[n]{b}, \quad \sqrt[n]{\frac{a}{b}} = \frac{\sqrt[n]{a}}{\sqrt[n]{b}}, \quad \text{and} \quad \sqrt[n]{a^m} = (\sqrt[n]{a})^m.$$

EXAMPLE 3   **Simplifying *n*th Roots**

Simplify.   a. $\sqrt[3]{135}$ b. $\sqrt[4]{162a^4}$

SOLUTION

a. Because $135 = 27 \cdot 5$ and 27 is a perfect cube ($3^3 = 27$), we have
$$\sqrt[3]{135} = \sqrt[3]{27 \cdot 5} = \sqrt[3]{27} \cdot \sqrt[3]{5} = 3\sqrt[3]{5}.$$

b. Because $162 = 81 \cdot 2$ and 81 is a perfect fourth power ($3^4 = 81$), we have
$$\sqrt[4]{162a^4} = \sqrt[4]{162}\sqrt[4]{a^4} = \sqrt[4]{81 \cdot 2}|a| = \sqrt[4]{81}\sqrt[4]{2}|a| = 3\sqrt[4]{2}|a|.$$ ■ ■ ■

Practice Problem 3   Simplify.

a. $\sqrt[3]{72}$ b. $\sqrt[4]{48a^2}$ ■

4   Combine like radicals.

Like Radicals

Radicals that have the same index and the same radicand are called **like radicals**. Like radicals can be combined with the use of the distributive property.

EXAMPLE 4   **Adding and Subtracting Radical Expressions**

Simplify.   a. $\sqrt{45} + 7\sqrt{20}$ b. $5\sqrt[3]{80x} - 3\sqrt[3]{270x}$

SOLUTION

a. We find perfect-square factors for both 45 and 20.
$$\sqrt{45} + 7\sqrt{20} = \sqrt{9 \cdot 5} + 7\sqrt{4 \cdot 5}$$
$$= 3\sqrt{5} + 14\sqrt{5} = (3 + 14)\sqrt{5} = 17\sqrt{5}$$

b. This time we find perfect-cube factors for both 80 and 270.

$$5\sqrt[3]{80x} - 3\sqrt[3]{270x} = 5\sqrt[3]{8 \cdot 10x} - 3\sqrt[3]{27 \cdot 10x}$$
$$= 10\sqrt[3]{10x} - 9\sqrt[3]{10x} = (10 - 9)\sqrt[3]{10x} = \sqrt[3]{10x} \quad ▪▪▪$$

Practice Problem 4   Simplify.

a. $3\sqrt{12} + 7\sqrt{3}$ **b.** $2\sqrt[3]{135x} - 3\sqrt[3]{40x}$ ▪

5 Define and evaluate rational exponents.

Rational Exponents

We know what 2^3 means, but what about $2^{1/3}$?

THE EXPONENT $\dfrac{1}{n}$

For any real number a and any integer $n > 1$,

$$a^{1/n} = \sqrt[n]{a}.$$

When n is even and $a < 0$, $\sqrt[n]{a}$ and $a^{1/n}$ are not real numbers.

We use this definition to evaluate some expressions with rational exponents.

$$16^{1/2} = \sqrt{16} = 4$$

$$(-27)^{1/3} = \sqrt[3]{-27} = \sqrt[3]{(-3)^3} = -3$$

$$\left(\frac{1}{16}\right)^{1/4} = \sqrt[4]{\frac{1}{16}} = \sqrt[4]{\left(\frac{1}{2}\right)^4} = \frac{1}{2}$$

$$32^{1/5} = \sqrt[5]{32} = \sqrt[5]{2^5} = 2$$

$(-5)^{1/4}$ is not a real number because the base is negative and 4 is even.

We next define $a^{m/n}$, where m and n are integers, when $n > 1$ and $\sqrt[n]{a}$ is a real number.

RATIONAL EXPONENTS

$$a^{m/n} = (\sqrt[n]{a})^m = \sqrt[n]{a^m}$$

provided that m and n are integers with no common factors, $n > 1$, and $\sqrt[n]{a}$ is a real number.

Note that the numerator m of the exponent $\dfrac{m}{n}$ is the *exponent of the radical expression* and the denominator n is the *index of the radical*. When n is even and $a < 0$, the symbol $a^{m/n}$ is not a real number.

We change rational exponents with negative denominators such as $\dfrac{3}{-5}$ and $\dfrac{-2}{-7}$ to $-\dfrac{3}{5}$ and $\dfrac{2}{7}$, respectively, before applying the definition. Note that $(-8)^{2/6} \neq [(-8)^{1/6}]^2$ because $(-8)^{1/6} = \sqrt[6]{-8}$ is not a real number but $(-8)^{2/6}$ is a real number. Writing $\dfrac{2}{6} = \dfrac{1}{3}$, we have

$$(-8)^{2/6} = (-8)^{1/3} = \sqrt[3]{-8} = -2.$$

We can now evaluate expressions having rational exponents.

$$8^{2/3} = (\sqrt[3]{8})^2 = 2^2 = 4$$

$$-16^{5/2} = -(\sqrt{16})^5 = -4^5 = -1024$$

$$100^{-3/2} = 100^{-3/2} = (\sqrt{100})^{-3} = \frac{1}{(\sqrt{100})^3} = \frac{1}{10^3} = \frac{1}{1000}$$

$$(-25)^{7/2} = [(-25)^{1/2}]^7 = (\sqrt{-25})^7, \text{ which is not a real number.}$$

EXAMPLE 5   Simplifying Expressions Having Rational Exponents

Express your answer with only positive exponents. Assume that x represents a positive real number. Simplify.

a. $2x^{1/3} \cdot 5x^{1/4}$ **b.** $\dfrac{21x^{-2/3}}{7x^{1/5}}$ **c.** $(x^{3/5})^{-1/6}$

SOLUTION

a. $2x^{1/3} \cdot 5x^{1/4} = 2 \cdot 5x^{1/3} \cdot x^{1/4} = 10x^{1/3+1/4} = 10x^{4/12+3/12} = 10x^{7/12}$

b. $\dfrac{21x^{-2/3}}{7x^{1/5}} = \left(\dfrac{21}{7}\right)\left(\dfrac{x^{-2/3}}{x^{1/5}}\right) = 3x^{-2/3-1/5} = 3x^{-10/15-3/15} = 3x^{-13/15} = 3 \cdot \left(\dfrac{1}{x^{13/15}}\right)$

$$= \dfrac{3}{x^{13/15}}$$

c. $(x^{3/5})^{-1/6} = x^{(3/5)(-1/6)} = x^{-1/10} = \dfrac{1}{x^{1/10}}$ ■ ■ ■

Practice Problem 5   Simplify. Express the answer with only positive exponents.

a. $4x^{1/2} \cdot 3x^{1/5}$ **b.** $\dfrac{25x^{-1/4}}{5x^{1/3}}, x > 0$ **c.** $(x^{2/3})^{-1/5}, x > 0$ ■

EXAMPLE 6   Simplifying an Expression Involving Negative Rational Exponents

Simplify $x(x + 1)^{-1/2} + 2(x + 1)^{1/2}$. Express your answer with only positive exponents and then rationalize the denominator. Assume that $(x + 1)^{-1/2}$ is defined.

SOLUTION

$$x(x + 1)^{-1/2} + 2(x + 1)^{1/2} = \frac{x}{(x + 1)^{1/2}} + \frac{2(x + 1)^{1/2}}{1} \qquad (x + 1)^{-1/2} = \frac{1}{(x + 1)^{1/2}}$$

$$= \frac{x}{(x + 1)^{1/2}} + \frac{2(x + 1)^{1/2}(x + 1)^{1/2}}{(x + 1)^{1/2}}$$

$$= \frac{x + 2(x + 1)}{(x + 1)^{1/2}} = \frac{3x + 2}{(x + 1)^{1/2}} \qquad (x + 1)^{1/2}(x + 1)^{1/2} = x + 1$$

To rationalize the denominator $(x + 1)^{1/2} = \sqrt{x + 1}$, multiply both numerator and denominator by $(x + 1)^{1/2}$.

$$\frac{3x + 2}{(x + 1)^{1/2}} = \frac{(3x + 2)(x + 1)^{1/2}}{(x + 1)^{1/2}(x + 1)^{1/2}} = \frac{(3x + 2)(x + 1)^{1/2}}{x + 1}$$ ■ ■ ■

Practice Problem 6   Simplify $x(x + 3)^{-1/2} + (x + 3)^{1/2}$.   ■

EXAMPLE 7   **Factoring an Expression Involving Rational Exponents**

Factor $\frac{4}{3}x^{1/3}(3x - 1) + 3x^{4/3}$.

SOLUTION

Begin by recognizing that $x^{4/3} = x \cdot x^{1/3}$; then factor out $\frac{1}{3}x^{1/3}$.

$$\frac{4}{3}x^{1/3}(3x - 1) + 3x^{4/3} = \frac{4}{3}x^{1/3}(3x - 1) + \frac{9}{3}x \cdot x^{1/3}$$

$$= \frac{1}{3}x^{1/3}[4(3x - 1) + 9x] = \frac{1}{3}x^{1/3}(21x - 4) \quad ■■■$$

Practice Problem 7   Factor $\frac{4}{3}x^{1/3}(x + 5) + x^{4/3}$.   ■

SECTION A.4 ■ Exercises

A EXERCISES Basic Skills and Concepts

1. Any positive number has ___two___ square roots.

2. To rationalize the denominator of an expression with denominator $2\sqrt{5}$, multiply the ___numerator___ and ___denominator___ by ___$\sqrt{5}$___.

3. Radicals that have the same index and the same radicand are called ___like radicals___.

4. The radical notation for $7^{1/3}$ is ___$\sqrt[3]{7}$___.

In Exercises 5–20, evaluate each root or state that the root is not a real number.

5. $\sqrt{64}$ 8
6. $\sqrt{100}$ 10
7. $\sqrt[3]{64}$ 4
8. $\sqrt[3]{125}$ 5
9. $\sqrt[3]{-27}$ −3
10. $\sqrt[3]{-216}$ −6
11. $\sqrt[3]{-\frac{1}{8}}$ $-\frac{1}{2}$
12. $\sqrt[3]{\frac{27}{64}}$ $\frac{3}{4}$
13. $\sqrt{(-3)^2}$ 3
14. $-\sqrt{4(-3)^2}$ −6
15. $\sqrt[4]{-16}$ Not a real number
16. $\sqrt[6]{-64}$ Not a real number
17. $-\sqrt[5]{-1}$ 1
18. $\sqrt[7]{-1}$ −1
19. $\sqrt[5]{(-7)^5}$ −7
20. $-\sqrt[5]{(-4)^5}$ 4

In Exercises 21–34, simplify each expression. Assume that all variables represent nonnegative real numbers.

21. $2\sqrt{3} + 5\sqrt{3}$ $7\sqrt{3}$
22. $7\sqrt{2} - \sqrt{2}$ $6\sqrt{2}$
23. $6\sqrt{5} - \sqrt{5} + 4\sqrt{5}$ $9\sqrt{5}$
24. $5\sqrt{7} + 3\sqrt{7} - 2\sqrt{7}$ $6\sqrt{7}$
25. $\sqrt{98x} - \sqrt{32x}$ $3\sqrt{2x}$
26. $\sqrt{45x} + \sqrt{20x}$ $5\sqrt{5x}$
27. $\sqrt[3]{24} - \sqrt[3]{81}$ $-\sqrt[3]{3}$
28. $\sqrt[3]{54} + \sqrt[3]{16}$ $5\sqrt[3]{2}$
29. $\sqrt[3]{3x} - 2\sqrt[3]{24x} + \sqrt[3]{375x}$ $2\sqrt[3]{3x}$
30. $\sqrt[3]{16x} + 3\sqrt[3]{54x} - \sqrt[3]{2x}$ $10\sqrt[3]{2x}$

31. $\sqrt{2x^5} - 5\sqrt{32x} + \sqrt{18x^3}$ $\sqrt{2x}(x^2 + 3x - 20)$
32. $2\sqrt{50x^5} + 7\sqrt{2x^3} - 3\sqrt{72x}$ $\sqrt{2x}(10x^2 + 7x - 18)$
33. $\sqrt{48x^5y} - 4y\sqrt{3x^3y} + y\sqrt{3xy^3}$ $(2x - y)^2\sqrt{3xy}$
34. $x\sqrt{8xy} + 4\sqrt{2xy^3} - \sqrt{18x^5y}$ $\sqrt{2xy}(2x + 4y - 3x^2)$

In Exercises 35–48, rationalize the denominator of each expression.

35. $\frac{2}{\sqrt{3}}$ $\frac{2\sqrt{3}}{3}$
36. $\frac{10}{\sqrt{5}}$ $2\sqrt{5}$
37. $\frac{7}{\sqrt{15}}$ $\frac{7\sqrt{15}}{15}$
38. $\frac{-3}{\sqrt{8}}$ $-\frac{3\sqrt{2}}{4}$
39. $\frac{1}{\sqrt{2} + x}$ $\frac{\sqrt{2} - x}{2 - x^2}$
40. $\frac{1}{\sqrt{5} + 2x}$ $\frac{\sqrt{5} - 2x}{5 - 4x^2}$
41. $\frac{3}{2 - \sqrt{3}}$ $3(2 + \sqrt{3})$
42. $\frac{-5}{1 - \sqrt{2}}$ $5(1 + \sqrt{2})$
43. $\frac{1}{\sqrt{3} + \sqrt{2}}$ $\sqrt{3} - \sqrt{2}$
44. $\frac{6}{\sqrt{5} - \sqrt{3}}$ $3(\sqrt{5} + \sqrt{3})$
45. $\frac{\sqrt{5} - \sqrt{2}}{\sqrt{5} + \sqrt{2}}$ $\frac{7 - 2\sqrt{10}}{3}$
46. $\frac{\sqrt{7} + \sqrt{3}}{\sqrt{7} - \sqrt{3}}$ $\frac{5 + \sqrt{21}}{2}$
47. $\frac{\sqrt{x + h} - \sqrt{x}}{\sqrt{x + h} + \sqrt{x}}$ †
48. $\frac{a\sqrt{x} + 3}{a\sqrt{x} - 3}$ $\frac{a^2x + 6a\sqrt{x} + 9}{a^2x - 9}$

In Exercises 49–58, evaluate each expression without using a calculator.

49. $25^{1/2}$ 5
50. $144^{1/2}$ 12
51. $(-8)^{1/3}$ −2
52. $(-27)^{1/3}$ −3
53. $8^{2/3}$ 4
54. $16^{3/2}$ 64
55. $-25^{-3/2}$ $-\frac{1}{125}$
56. $-9^{-3/2}$ $-\frac{1}{27}$
57. $\left(\frac{9}{25}\right)^{-3/2}$ $\frac{125}{27}$
58. $\left(\frac{1}{27}\right)^{-2/3}$ 9

†Due to space constrictions, answers to these exercises may be found in the Answers beginning on page A–1 in the back of the book.

In Exercises 59–70, simplify each expression, leaving your answer with only positive exponents. Assume that all variables represent positive numbers.

59. $x^{1/2} \cdot x^{2/5}$ $x^{9/10}$

60. $x^{3/5} \cdot 5x^{2/3}$ $5x^{19/15}$

61. $x^{3/5} \cdot x^{-1/2}$ $x^{1/10}$

62. $x^{5/3} \cdot x^{-3/4}$ $x^{11/12}$

63. $(8x^6)^{2/3}$ $4x^4$

64. $(16x^3)^{1/2}$ $4x^{3/2}$

65. $(27x^6y^3)^{-2/3}$ $\dfrac{1}{9x^4y^2}$

66. $(16x^4y^6)^{-3/2}$ $\dfrac{1}{64x^6y^9}$

67. $\dfrac{15x^{3/2}}{3x^{1/4}}$ $5x^{5/4}$

68. $\dfrac{20x^{5/2}}{4x^{2/3}}$ $5x^{11/6}$

69. $\left(\dfrac{x^{-1/4}}{y^{-2/3}}\right)^{-12}$ $\dfrac{x^3}{y^8}$

70. $\left(\dfrac{27x^{-5/2}}{y^{-3}}\right)^{-1/3}$ $\dfrac{x^{5/6}}{3y}$

In Exercises 71–78, convert each radical expression to its rational exponent form and then simplify. Assume that all variables represent positive numbers.

71. $\sqrt[4]{3^2}$ $3^{1/2}$

72. $\sqrt[4]{5^2}$ $5^{1/2}$

73. $\sqrt[3]{x^9}$ x^3

74. $\sqrt[3]{x^{12}}$ x^4

75. $\sqrt[3]{x^6y^9}$ x^2y^3

76. $\sqrt[3]{8x^3y^{12}}$ $2xy^4$

77. $\sqrt[4]{9}\sqrt{3}$ 3

78. $\sqrt[4]{49}\sqrt{7}$ 7

In Exercises 79–84, factor each expression. Use only positive exponents in your answer.

79. $\dfrac{4}{3}x^{1/3}(2x - 3) + 2x^{4/3}$ **80.** $\dfrac{4}{3}x^{1/3}(7x + 1) + 7x^{4/3}$

81. $4(3x + 1)^{1/3}(2x - 1) + 2(3x + 1)^{4/3}$

82. $4(3x - 1)^{1/3}(x + 2) + (3x - 1)^{4/3}$

83. $3(x^2 + 1)^{1/2}(2x^2 - x) + 2(x^2 + 1)^{3/2}$

84. $3(x^2 + 2)^{1/2}(x^2 - 2x) + (x^2 + 2)^{3/2}$

Answers:

79. $\dfrac{2}{3}x^{1/3}(7x - 6)$ **80.** $\dfrac{1}{3}x^{1/3}(49x + 4)$

81. $2(3x + 1)^{1/3}(7x - 1)$ **82.** $7(3x - 1)^{1/3}(x + 1)$

83. $(x^2 + 1)^{1/2}(8x^2 - 3x + 2)$ **84.** $2(x^2 + 2)^{1/2}(2x^2 - 3x + 1)$

B EXERCISES Applying the Concepts

85. Sign dimensions. A sign in the shape of an equilateral triangle of area A has sides of length $\sqrt{\dfrac{4A}{\sqrt{3}}}$ centimeters. Find the length of a side if the sign has an area of 692 square centimeters. (Use the approximation $\sqrt{3} \approx 1.73$.)   40 cm

86. Return on investment. For an initial investment of P dollars to mature to S dollars after two years when the interest is compounded annually, an annual interest rate of $r = \sqrt{\dfrac{S}{P}} - 1$ is required. Find r if $P = \$1,210,000$ and $S = \$1,411,344$.   8%

87. Terminal speed. The terminal speed V of a steel ball that weighs w grams and is falling in a cylinder of oil is given by the equation $V = \sqrt{\dfrac{w}{1.5}}$ centimeters per second. Find the terminal speed of a steel ball weighing 6 grams.   2 cm/sec

88. Electronic game current. The current I (in amperes) in a circuit for an electronic game using W watts and having a resistance of R ohms is given by $I = \sqrt{\dfrac{W}{R}}$. Find the current in a game circuit using 1058 watts and having a resistance of 2 ohms.   23 amp

Topics in Geometry

Triangles

The intuitive idea that two triangles have exactly the same shape (but not necessarily the same size) is captured in the concept of *similar triangles:*

Similar Triangles Two triangles are **similar** if and only if they have equal corresponding angles and their corresponding sides are proportional. See Figure A.10.

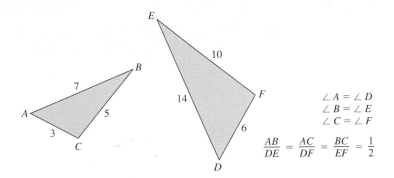

$$\angle A = \angle D$$
$$\angle B = \angle E$$
$$\angle C = \angle F$$

$$\frac{AB}{DE} = \frac{AC}{DF} = \frac{BC}{EF} = \frac{1}{2}$$

FIGURE A.10

Congruent Triangles Two triangles are *congruent* if and only if they have equal corresponding angles and sides.

Included Side A side of a triangle that lies between two angles is called the **included side** of the angles.

Included Angle The angle formed by two sides is the **included angle** of the sides.

Fortunately, not every part of the definition for congruent and similar triangles must be verified to decide whether two triangles are congruent, or similar.

CONGRUENT-TRIANGLE THEOREMS

SAS (Side–Angle–Side). If two sides and the included angle of one triangle are equal to the corresponding sides and the included angle of a second triangle, the two triangles are congruent.

ASA (Angle–Side–Angle). If two angles and the included side of one triangle are equal to the corresponding two angles and the included side of a second triangle, the two triangles are congruent.

SSS (Side–Side–Side). If the three sides of one triangle are equal to the corresponding sides of a second triangle, the two triangles are congruent.

SIMILAR-TRIANGLE THEOREMS

SAS (Side–Angle–Side). If two corresponding sides are proportional and the angles between the two sides are equal, the two triangles are similar.
AA (Angle–Angle). If two corresponding angles are equal, the two triangles are similar.
SSS (Side–Side–Side). If the lengths of all three sides of two triangles are proportional, the triangles are similar.

---

1 Use the Pythagorean Theorem and its converse.

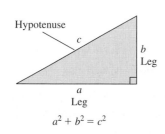

$a^2 + b^2 = c^2$

FIGURE A.11

Pythagorean Theorem and Its Converse

A **right triangle** is a triangle containing a right angle—that is, an angle of 90°. The side of the triangle opposite the 90° angle is called the **hypotenuse**; the other two sides are the **legs** of the triangle. Figure A.11 shows a right triangle with legs of length a and b and hypotenuse of length c. The symbol ⌐ is used to identify the right angle.

The *Pythagorean Theorem* says that the square of the length of a right triangle's hypotenuse is equal to the sum of the squares of the lengths of its legs.

PYTHAGOREAN THEOREM

If a, b, and c represent the lengths of the two legs and the hypotenuse, respectively, of a right triangle, then

$$a^2 + b^2 = c^2.$$

EXAMPLE 1   **The Pythagorean Theorem**

A right triangle has one leg of length 15 and a hypotenuse of length 39. Find the length of the remaining leg.

SOLUTION

Let c represent the length of the hypotenuse and a and b represent the legs of the right triangle. Then $c = 39$, $a = 15$, and $a^2 + b^2 = c^2$ can be written as follows:

$$15^2 + b^2 = 39^2$$
$$b^2 = 39^2 - 15^2 = 1296$$
$$b = 36$$

The length of the remaining leg is 36.   ▪ ▪ ▪

Practice Problem 1   A right triangle has legs of lengths 8 and 6. Find the length of the hypotenuse.   ▪

EXAMPLE 2   **The Pythagorean Theorem**

You want to change some prices advertised on a letter display fastened 15 feet high on the front wall of your business building. A garden along the front wall forces you to place a ladder 8 feet from the wall. What length ladder is required?

SOLUTION

If a ladder is placed 8 feet from the wall and rests 15 feet high on the wall, the ladder can be viewed as the hypotenuse of a right triangle with legs having lengths 8 feet and 15 feet. See Figure A.12.

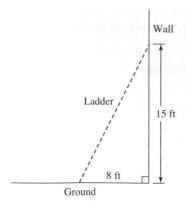

FIGURE A.12

Let c represent the length of the ladder.

$$c^2 = a^2 + b^2 \quad \text{Pythagorean Theorem}$$
$$c^2 = 8^2 + 15^2 \quad \text{Replace } a \text{ with 8 and } b \text{ with 15.}$$
$$c^2 = 289$$
$$c = 17$$

A 17-foot ladder will allow you to change the display.   ■ ■ ■

Practice Problem 2 In Example 2, suppose the display is fastened 8 feet high and the ladder must be placed 6 feet from the wall. What length ladder is required?   ■

CONVERSE OF THE PYTHAGOREAN THEOREM

Suppose a triangle has sides of length a, b, and c. If

$$c^2 = a^2 + b^2,$$

then the triangle is a right triangle, with a right angle opposite the longest side.

EXAMPLE 3   **Converse of the Pythagorean Theorem**

A triangle has sides of lengths 20, 21, and 29. Is it a right triangle?

SOLUTION

To determine whether the triangle is a right triangle, we first square the length of each side.

$$20^2 = 400, \quad 21^2 = 441, \quad \text{and} \quad 29^2 = 841$$
$$841 = 400 + 441$$

Because the square of the length of the largest side equals the sum of the squares of the lengths of the other two sides, the triangle is a right triangle.   ■ ■ ■

Practice Problem 3 A triangle has sides of lengths 2, 5, and 6. Is it a right triangle?   ■

2 Use geometry formulas.

Geometry Formulas

Figure A.13 lists some useful formulas from geometry.

Formulas for area (A), circumference (C), perimeter (P), and volume (V)

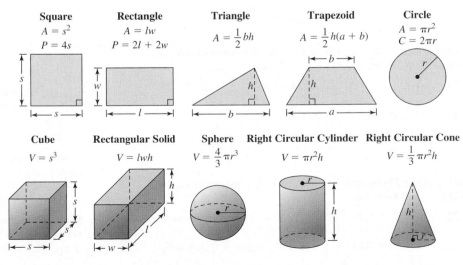

Square
$A = s^2$
$P = 4s$

Rectangle
$A = lw$
$P = 2l + 2w$

Triangle
$A = \frac{1}{2}bh$

Trapezoid
$A = \frac{1}{2}h(a + b)$

Circle
$A = \pi r^2$
$C = 2\pi r$

Cube
$V = s^3$

Rectangular Solid
$V = lwh$

Sphere
$V = \frac{4}{3}\pi r^3$

Right Circular Cylinder
$V = \pi r^2 h$

Right Circular Cone
$V = \frac{1}{3}\pi r^2 h$

FIGURE A.13

SECTION A.5 ■ Exercises

A EXERCISES Basic Skills and Concepts

1. The Pythagorean Theorem applies only to ___right___ triangles.

2. If the lengths of three sides of one triangle are equal to the lengths of the corresponding sides of a second triangle, then the two triangles are ___congruent___.

3. If two pairs of corresponding angles are equal, then the two triangles are ___similar___.

4. Because $6^2 + 8^2 = 10^2$, a triangle with sides of lengths 6, 8, and 10 is a(n) ___right triangle___.

In Exercises 5–10, the lengths of the legs of a right triangle are given. Find the length of the hypotenuse.

5. $a = 7, b = 24$ 25   6. $a = 5, b = 12$ 13

7. $a = 9, b = 12$ 15   8. $a = 10, b = 24$ 26

9. $a = 3, b = 3$ $3\sqrt{2}$ 10. $a = 2, b = 4$ $2\sqrt{5}$

In Exercises 11–14, the lengths of one leg, a, and the hypotenuse, c, of a right triangle are given. Find the length of the remaining leg.

11. $a = 15, c = 17$ 8   12. $a = 35, c = 37$ 12

13. $a = 14, c = 50$ 48   14. $a = 10, c = 26$ 24

In Exercises 15–18, find the area A and the perimeter P of a rectangle with length l and width w.

15. $l = 3, w = 5$ $A = 15, P = 16$

16. $l = 4, w = 3$ $A = 12, P = 14$

17. $l = 10, w = \dfrac{1}{2}$ $A = 5, P = 21$

18. $l = 16, w = \dfrac{1}{4}$ $A = 4, P = \dfrac{65}{2}$

In Exercises 19–22, find the area A of a triangle with base b and altitude h.

19. $b = 3, h = 4$ 6   20. $b = 6, h = 1$ 3

21. $b = \dfrac{1}{2}, h = 12$ 3   22. $b = \dfrac{1}{4}, h = \dfrac{2}{3}$ $\dfrac{1}{12}$

In Exercises 23–26, find the area A and the circumference C of a circle with radius r. (Use $\pi \approx 3.14$.)

23. $r = 1$ $A = 3.14; C = 6.28$ 24. $r = 3$ $A = 28.26; C = 18.84$

25. $r = \dfrac{1}{2}$ $A = 0.785; C = 3.14$

26. $r = \dfrac{3}{4}$ $A = 1.76625; C = 4.71$

In Exercises 27–30, find the volume V of a box with length l, width w, and height h.

27. $l = 1, w = 1, h = 5$ 5   28. $l = 2, w = 7, h = 4$ 56

29. $l = \dfrac{1}{2}, w = 10, h = 3$ 15 30. $l = \dfrac{1}{3}, w = \dfrac{3}{7}, h = 14$ 2

B EXERCISES Applying the Concepts

31. **Garden area.** Find the area of a rectangular garden if the width is 12 feet and the diagonal measures 20 feet. 192 ft^2

32. **Perimeter of a mural.** A mural is 6 feet wide, and its diagonal measures 10 feet. Find the perimeter of the mural. 28 ft

33. **Dimensions of a computer monitor.** The monitor on a laptop computer has a rectangular shape with a 15-inch diagonal. If the width of the monitor is 12 inches, what is its height? 9 in.

34. **Dimensions of a baseball diamond.** A baseball diamond is actually a square with 90-foot sides. How far does the catcher have to throw the ball to reach second base from home plate? $90\sqrt{2} \approx 127$ ft

35. **Repair problem.** A repairperson has to work on an air conditioner's overflow vent located 25 feet above the ground on the side of a house. Shrubs along the house require that a ladder be placed 10 feet from the house. How long must the ladder be for the repairperson to reach the overflow vent? $5\sqrt{29} \approx 27$ ft

36. **Building a ramp.** You must build a ramp that allows you to roll a cart from a storeroom into the back of a truck. The rear of the truck must remain 8 feet from the storeroom because of a curb. If the back of the truck is 4 feet above the storeroom floor, how long (in feet) must the ramp be? $4\sqrt{5} \approx 8.94$ ft

37. **Football field.** The playing surface of a football field (including the end zones) is 120 yards long and 53 yards wide. How far will a player jogging around the perimeter travel? 346 yd

38. **Reflecting pool.** The reflecting pool of the Lincoln Memorial has an approximate perimeter of 4400 feet. The length is approximately 1860 feet more than the width. Find the area of the reflecting pool. 345,100 ft^2

39. **Crop circle.** A crop circle with a diameter of 787 feet occurred at Milk Hill in Wiltshire, UK, August 2001. If you walk around the entire edge of the circle, how far will you have walked? (Use $\pi \approx 3.14$.) 2471 ft

40. **Bicycle tires.** The diameter of a bicycle tire is 26 inches. How far will the bicycle travel when the wheel makes one complete turn? (Use $\pi \approx 3.14$.) 81.64 in.

41. **Fishpond border.** A rectangular border a foot and a half wide is built around a rectangular fishpond. If the pond is 6 feet long and 4 feet wide, what is the area of the border? 39 ft^2

42. **Garden fence.** Casper wants to put a decorative fence around his rectangular garden. The garden is 5 feet wide, and its diagonal measures 13 feet. How many feet of fencing is required? 34 ft

43. **Oven dimensions.** An oven in the shape of a rectangular box is 2 feet wide, 2.5 feet high, and 3 feet deep. What is the volume of the oven? 15 ft^3

† Due to space constrictions answers to these exercise may be found in the Answer beginning on page A–1 in the back of the book.

Linear Equations; Applications

Objectives

1 Solve linear equations in one variable.

2 Learn a procedure for problem solving.

3 Solve simple interest problems.

4 Solve uniform-motion problems.

5 Solve work-rate problems.

6 Solve mixture problems.

---

1 Solve linear equations in one variable.

Definitions

An **equation in one variable** is a statement that two expressions, with at least one containing the variable, are equal. For example, $2x - 3 = 7$ is an equation in the variable x. The expressions $2x - 3$ and 7 are the *sides* of the equation. The **domain** of the variable in an equation is the set of all real numbers for which both sides of the equation are defined.

When the variable in an equation is replaced by a specific value from its domain, the resulting statement may be true or false. For example, in the equation $2x - 3 = 7$, if we let $x = 1$, the equation is false. However, if we replace x with 5, the equation is true. Those values (if any) of the variable that result in a true statement are called **solutions** or **roots** of the equation. Thus, 5 is a solution (or root) of the equation $2x - 3 = 7$. We also say that 5 **satisfies** the equation $2x - 3 = 7$. To **solve** an equation means to find all solutions of the equation; the set of all solutions of an equation is called its **solution set**.

An equation that is satisfied by every real number in the domain of the variable is called an **identity**. The equations

$$2(x + 3) = 2x + 6$$

$$\text{and} \quad x^2 - 9 = (x + 3)(x - 3)$$

are examples of identities. Some equations, such as $x = x + 5$, have no solution.

STUDY TIP

When finding the domain of a variable, remember that

1. Division by 0 is undefined.
2. The square root (or any even root) of a negative number is not a real number.

Solving an Equation

Equations that have the same solution set are called **equivalent equations**. For example, equations $x = 4$ and $3x = 12$ are equivalent equations with solution set $\{4\}$.

In general, to solve an equation in one variable, we replace the given equation with a sequence of equivalent equations until we obtain an equation whose solution is obvious, such as $x = 4$. The following operations yield equivalent equations.

Generating Equivalent Equations

Operations	Given Equation	Equivalent Equation
Simplify expressions on either side by eliminating parentheses, combining like terms, and so on.	$(2x - 1) - (x + 1) = 4$	$2x - 1 - x - 1 = 4$ or $x - 2 = 4$
Add (or subtract) the same expression on *both* sides of the equation.	$x - 2 = 4$	$x - 2 + 2 = 4 + 2$ or $x = 6$
Multiply (or divide) *both* sides of the equation by the same *nonzero* expression.	$2x = 8$	$\frac{1}{2} \cdot 2x = \frac{1}{2} \cdot 8$ or $x = 4$
Interchange the two sides of the equation.	$-3 = x$	$x = -3$

EXAMPLE 1   Solving a Linear Equation

Solve: $6x - [3x - 2(x - 2)] = 11$

SOLUTION

$6x - [3x - 2(x - 2)] = 11$	Original equation
$6x - [3x - 2x + 4] = 11$	Remove innermost parentheses by distributing $-2(x - 2) = -2x + 4$.
$6x - 3x + 2x - 4 = 11$	Remove square brackets and change the sign of each enclosed term.
$5x - 4 = 11$	Combine like terms.
$5x - 4 + 4 = 11 + 4$	Add 4 to both sides.
$5x = 15$	Simplify both sides in Step 3.
$\dfrac{5x}{5} = \dfrac{15}{5}$	Divide both sides by 5.
$x = 3$	Simplify.

The apparent solution is 3.

Check: Substitute $x = 3$ into the original equation. You should obtain $11 = 11$, Thus, 3 is the only solution of the original equation; so $\{3\}$ is its solution set. ■ ■ ■

Practice Problem 1   Solve $3x - [2x - 6(x + 1)] = -1$.   ■

2 Learn a procedure for problem solving.

Solving Applied Problems

Practical problems are often introduced through verbal descriptions that do not contain explicit requests for algebraic activity, such as "Solve the equation. . . . " Frequently, they describe a situation, perhaps in a physical or financial setting, and ask for a value for some specific quantity that will favorably resolve the situation. The process of translating a practical problem into a mathematical one is called **mathematical modeling**. Generally, we use variables to represent the quantities identified in the verbal description and represent relationships among these quantities with algebraic expressions or equations. Ideally, we reduce the problem to an equation that, when solved, also solves the practical problem. Then we solve the equation and check our result against the practical situation presented in the problem.

PROCEDURE FOR SOLVING APPLIED PROBLEMS

Step 1 Read the problem as many times as needed to understand it thoroughly. Pay close attention to the question asked to help identify the quantity the variable should represent.

Step 2 Assign a variable to represent the quantity you are looking for and when necessary, express all other unknown quantities in terms of this variable. Frequently, it is helpful to draw a diagram to illustrate the problem or to set up a table to organize the information.

Step 3 Write an equation that describes the situation.

Step 4 Solve the equation.

Step 5 Answer the question asked in the problem.

Step 6 Check the answer against the description of the original problem (not just the equation solved in Step 4).

3 Solve simple interest problems.

Interest is money paid for the use of borrowed money. For example, while your money is on deposit at a bank, the bank pays you for the right to use the money in your account. The money you are paid by the bank is interest on your deposit. On any loan, the money borrowed for use (and on which interest is paid) is called the **principal**. The **interest rate** is the percent of the principal charged for its use for a specific time period. The most straightforward type of interest computation is described next.

SIMPLE INTEREST

If a principal of P dollars is borrowed for a period of t years with interest rate r (expressed as a decimal) computed yearly, then the total interest paid at the end of t years is:

$$I = Prt$$

Interest computed with this formula is called **simple interest**. When interest is computed yearly, the rate r is called an *annual* interest rate (or *per annum* interest rate).

EXAMPLE 2   Solving Problems Involving Simple Interest

Ms. Sharpy invested a total of $10,000 in blue-chip and technology stocks. At the end of a year, the blue chips returned 12% and the tech stocks returned 8% on the original investments. How much was invested in each type of stock if the total interest earned (return on investment) was $1060?

SOLUTION

We need to find the amount invested in blue-chip stocks and the amount invested in technology stocks. The sum of these two investments is $10,000.

If $x =$ amount invested in blue-chip stocks, then $10,000 - x =$ amount invested in technology stocks.

We organize our information in the following table.

Investment	Principal P	Rate r	Time t	Interest $I = Prt$
Blue Chip	x	0.12	1	$0.12x$
Technology	$10{,}000 - x$	0.08	1	$0.08(10{,}000 - x)$

Interest from blue-chip	+	Interest from technology	=	Total interest

$$0.12x + 0.08(10{,}000 - x) = 1060 \qquad \text{Replace descriptions with algebraic expressions.}$$

$$12x + 8(10{,}000 - x) = 106{,}000 \qquad \text{Eliminate decimals.}$$

$$12x + 80{,}000 - 8x = 106{,}000$$

$$4x = 26{,}000 \qquad \text{Solve for } x.$$

$$x = 6500 \qquad \text{Dollars in blue-chip stocks}$$

$$10{,}000 - 6500 = 3500 \qquad \text{Dollars in technology stocks}$$

Ms. Sharpy invests $3500 in technology stocks and $6500 in blue-chip stocks.

Check in the original problem. Now $3500 + $6500 = $10,000, 12% of $6500 is $780, and 8% of $3500 is $280.

Thus, the total interest earned is $780 + $280 = $1060. ■ ■ ■

Practice Problem 2 One-fifth of my capital is invested at 5%, one-sixth of my capital at 8%, and the rest at 10% per year. If the annual interest on my capital is $130, what is my capital? ■

4 Solve uniform-motion problems.

Uniform Motion

If you drive 60 miles in two hours, your average speed is $\dfrac{60}{2} \, (= 30)$ miles per hour. If you multiply your average speed for the trip (30 miles per hour) by the time elapsed (two hours), you get the distance driven, 60 miles $(60 = 2 \cdot 30)$, When solving problems where "rate" refers to the average speed of an object, use the following formula.

UNIFORM MOTION FORMULA

If an object moves at a rate (average speed) r, then the distance d traveled in time t is:

$$d = rt$$

◆ **WARNING** Make sure you use consistent units of measurement. If, for example, the rate is given in miles per hour, the time should be given in hours. In such a problem, if the time is given as 15 minutes, change 15 minutes to $\dfrac{1}{4}$ hour before using the uniform-motion formula.

EXAMPLE 3   **Solving a Uniform-Motion Problem**

A motorcycle police officer is chasing a car that is speeding at 70 miles per hour. The police officer is 3 miles behind the car and is traveling 80 miles per hour. How long will it be before the officer overtakes the car?

SOLUTION

We are asked to find the amount of *time* before the policeman overtakes the car. Draw a sketch to help visualize the problem.

FIGURE A.14

Let x = distance in miles the car travels before being overtaken. Then $x + 3$ = distance in miles the motorcycle travels before overtaking the car. We organize our information in a table.

Object	d (in miles)	r (miles per hour)	$t = \left(\dfrac{d}{r}\right)$ (hours)
Car	x	70	$\dfrac{x}{70}$
Motorcycle	$x + 3$	80	$\dfrac{x + 3}{80}$

Because the time from the start of the chase to the interception point is the same for both the car and the motorcycle, we have

$$\frac{x}{70} = \frac{x + 3}{80} \qquad \text{The time is the same for both the car and the motorcycle.}$$

$$8x = 7(x + 3) \qquad \text{Multiply both sides by LCD, 560.}$$

$$8x = 7x + 21 \qquad \text{Distributive property}$$

$$x = 21 \qquad \text{Subtract } 7x \text{ from both sides.}$$

The time required to overtake the car is

$$t = \frac{x}{70} = \frac{21}{70} = \frac{3}{10} \qquad \text{Replace } x \text{ with 21.}$$

In $\dfrac{3}{10}$ of an hour, or 18 minutes, the police officer overtakes the car.

Check in the original problem. If the officer travels for $\dfrac{3}{10}$ of an hour at 80 miles per hour, he travels $d = (80)\left(\dfrac{3}{10}\right) = 24$ miles. The car traveling this same $\dfrac{3}{10}$ of an hour at 70 miles per hour travels $(70)\left(\dfrac{3}{10}\right) = 21$ miles. Because the police officer travels 3 more miles during that time than the car does (because he started 3 miles behind the car), he does indeed overtake the car.   ■ ■ ■

Practice Problem 3   A train 130 meters long crosses a bridge in 21 seconds. The train is moving at a speed of 25 meters per second. What is the length of the bridge?

■

EXAMPLE 4   Flying Time to the *Queen Elizabeth II*

The *Queen Elizabeth II* was 1000 miles from Britain and traveling toward Britain at 32 miles per hour while a Hercules aircraft was flying from Britain directly toward the ship averaging 300 miles per hour. How long would it take for the aircraft to meet the ship?

SOLUTION

The initial separation between the ship and the aircraft is 1000 miles. Let $t = $ time elapsed when ship and aircraft meet.
    Then $32t = $ distance the ship traveled and $300t = $ distance the aircraft traveled.

$$\boxed{\text{Distance the ship traveled}} \; + \; \boxed{\text{Distance the aircraft traveled}} \; = \; \boxed{\text{1000 miles}}$$

$32t + 300t = 1000$	Replace descriptions with algebraic expressions.
$332t = 1000$	Combine like terms.
$t = \dfrac{1000}{332} \approx 3.01$	Solve for t.

The aircraft and ship meet after about three hours.   ■ ■ ■

Practice Problem 4   How long would it take for the aircraft and ship in Example 4 to meet if the aircraft averaged 350 miles per hour and was 955 miles away?   ■

5 Solve work-rate problems.

Work Rate

Work-rate problems use an idea much like that of uniform motion. In problems involving rates of work, we assume that the work is done at a uniform rate.

WORK RATE

The portion of a job completed per unit of time is called the **rate of work**.

If a job can be completed in x units of time (seconds, hours, days, and so on), the portion of the job completed in *one unit of time* (1 second, 1 hour, 1 day, and so on) is $\dfrac{1}{x}$. The portion of the job completed in t *units of time* (t seconds, t hours, t days, and so on) is $t \cdot \dfrac{1}{x}$. When the portion of the job completed is 1, the job is done.

EXAMPLE 5   Solving a Work-Rate Problem

One copy machine copies twice as fast as another. If both copiers work together, they can finish a particular job in 2 hours. How long would it take each copier, working alone, to do the job?

SOLUTION

Because the speed of one copier is twice the speed of the other, if we find how long it takes the faster copier to do the job working alone, we will know the slower copier will take twice that time.

Let x = the number of hours for the faster copier to complete the job alone.

$2x$ = the number of hours for the slower copier to complete the job alone.

$\dfrac{1}{x}$ = portion of the job the faster copier does in 1 hour.

$\dfrac{1}{2x}$ = portion of the job the slower copier does in one hour.

The time it takes both copiers, working together, to do the job is 2 hours. Multiplying the portion of the job each copier does in 1 hour by 2 will give the portion of the completed job done by each copier. We organize our information in a table.

	Portion done in one hour	**Time to complete job, working together**	**Portion done by each copier**
Faster copier	$\dfrac{1}{x}$	2	$2\left(\dfrac{1}{x}\right) = \dfrac{2}{x}$
Slower copier	$\dfrac{1}{2x}$	2	$2\left(\dfrac{1}{2x}\right) = \dfrac{1}{x}$

Because the job is *completed* in 2 hours, the sum of the portion of the job completed by the faster copier and the portion of the job completed by the slower copier in 2 hours is 1. The model equation can now be written and solved.

$\dfrac{2}{x} + \dfrac{1}{x} = 1$ Faster copier portion + slower copier portion = 1.

$2 + 1 = x$ Multiply both sides by x.

$3 = x$

The faster copier could complete the job, working alone, in 3 hours. The slower copier would require $2(3) = 6$ hours to complete the job.

Check in the original problem. The faster copier completes $\dfrac{1}{3}$ of the job in 1 hour.

Similarly, the slower copier completes $\dfrac{1}{6}$ of the job in 1 hour.

In 2 hours, the faster copier completes $2\left(\dfrac{1}{3}\right) = \dfrac{2}{3}$ of the job and the slower copier completes $2\left(\dfrac{1}{6}\right) = \dfrac{2}{6} = \dfrac{1}{3}$ of the job. Because $\dfrac{2}{3} + \dfrac{1}{3} = 1$, the copiers complete the job in 2 hours.   ■ ■ ■

Practice Problem 5 A couple took turns washing their sports car on weekends. Jim could usually wash the car in 45 minutes, whereas Anita took 30 minutes for the same job. One weekend they were in a hurry to go to a party, so they worked together. How long did it take them?   ■

6 Solve mixture problems.

Mixtures

Mixture problems require combining various ingredients, such as water and antifreeze, to produce a blend with specific characteristics.

> **EXAMPLE 6**   **Solving a Mixture Problem**

A full 6-quart radiator contains 75% water and 25% pure antifreeze. How much of this mixture should be drained and replaced by pure antifreeze so that the resulting 6-quart mixture is 50% pure antifreeze?

SOLUTION

We are asked to find the quantity of the radiator mixture that should be drained and replaced by pure antifreeze. The mixture we drain is only 25% pure antifreeze, but is replaced with 100% pure antifreeze.

Because we want to find the quantity of mixture drained, let x = number of quart of mixture drained. Then x = number of quarts of pure antifreeze added and $0.25x$ = number of quarts of pure antifreeze drained.

Now we track the pure antifreeze.

$$
\boxed{\begin{array}{c}\text{Pure antifreeze}\\\text{in final mixture}\end{array}} = \boxed{\begin{array}{c}\text{Pure antifreeze in}\\\text{original mixture}\end{array}} - \boxed{\begin{array}{c}\text{Pure antifreeze}\\\text{drained from}\\\text{original mixture}\end{array}} + \boxed{\begin{array}{c}\text{Pure}\\\text{antifreeze}\\\text{added}\end{array}}
$$

$$
\begin{array}{ccccccc}
(50\% \text{ of } 6) & = & (25\% \text{ of } 6) & - & (25\% \text{ of } x) & + & x
\end{array}
$$

$(0.5)(6) = (0.25)(6) - 0.25x + x$ Replace descriptions with algebraic expressions.

$3 = 1.5 + 0.75x$ Solve for x.

$1.5 = 0.75x$

$2 = x$

Drain 2 quarts of mixture from the radiator.

Check in the original problem. Draining 2 quarts from the original 6 quarts leaves 4 quarts of mixture in the radiator. This mixture consists of 1 quart of pure antifreeze (25%) and 3 quarts of water (75%). Adding 2 quarts of antifreeze to the radiator produces 3 quarts of antifreeze mixed with 3 quarts of water. This is the desired 50% pure antifreeze solution. ■ ■ ■

Practice Problem 6   How many gallons of 40% sulfuric acid solution should be mixed with 20% sulfuric acid solution to obtain 50 gallons of 25% sulfuric acid solution? ■

SECTION A.6 ■ Exercises

A EXERCISES Basic Skills and Concepts

1. Equations that have the same solution are called <u>equivalent</u>.

2. Adding the same expression to or subtracting it from both sides of an equation results in a(n) <u>equivalent equation</u>

3. The process of translating a practical problem into a mathematical one is called <u>mathematical modeling</u>

4. Money paid for the use of borrowed money is called <u>interest</u>.

In Exercises 5–8, determine whether the given equation is an identity. If the equation is not an identity, give a value that demonstrates that fact.

5. $2x + 3 = 5x + 1$ No; $x = 0$

6. $3x + 4 = 6x + 2 - (3x - 2)$ Yes

7. $\dfrac{1}{x + 3} = \dfrac{1}{x} + \dfrac{1}{3}$ No; $x = 1$

8. $\dfrac{1}{x} + \dfrac{1}{2} = \dfrac{2 + x}{2x}$ Yes

†Due to space constrictions, answers to these exercises may be found in the Answers beginning on page A–1 in the back of the book.

In Exercises 9–34 solve each equation.

9. $3x + 5 = 14$ {3} **10.** $2x - 17 = 7$ {12}

11. $-10x + 12 = 32$ {−2} **12.** $-2x + 5 = 6$ $\left\{-\frac{1}{2}\right\}$

13. $3 - y = -4$ {7} **14.** $2 - 7y = 23$ {−3}

15. $7x + 7 = 2(x + 1)$ {−1} **16.** $3(x + 2) = 4 - x$ $\left\{-\frac{1}{2}\right\}$

17. $3(2 - y) + 5y = 3y$ {6} **18.** $9y - 3(y - 1) = 6 + y$

19. $4y - 3y + 7 - y = 2 - (7 - y)$ {12}

20. $3(y - 1) = 6y - 4 + 2y - 4y$ {1}

21. $2x + 3(x - 4) = 7x + 10$ {−11}

22. $3(2 - 3x) - 4x = 3x - 10$ {1}

23. $3x + \frac{x}{5} - 2 = \frac{1}{10} + 2x$ **24.** $\frac{1}{4} + 5x - \frac{x}{7} = \frac{5x}{14} + \frac{x}{2}$

25. $4[x + 2(3 - x)] = 2x + 1$

26. $3 - [x - 3(x + 2)] = 4$

27. $3(4y - 3) = 4[y - (4y - 3)]$

28. $5 - (6y + 9) + 2y = 2(y + 1)$ {−1}

29. $2x - 3(2 - x) = (x - 3) + 2x + 1$ {2}

30. $5(x - 3) - 6(x - 4) = -5$ {14}

31. $\frac{2x + 1}{9} - \frac{x + 4}{6} = 1$ {28}

32. $\frac{2 - 3x}{7} + \frac{x - 1}{3} = \frac{3x}{7}$ $\left\{-\frac{1}{11}\right\}$

33. $\frac{1 - x}{4} + \frac{5x + 1}{2} = 3 - \frac{2(x + 1)}{8}$ $\left\{\frac{4}{5}\right\}$

34. $\frac{x + 4}{3} + 2x - \frac{1}{2} = \frac{3x + 2}{6}$ $\left\{\frac{-3}{11}\right\}$

B EXERCISES Applying the Concepts

35. Real estate investment. A real estate agent invested a total of $4200. Part was invested in a high-risk real estate venture, and the rest was invested in a savings and loan. At the end of a year, the high-risk venture returned 15% on the women's investment and the savings and loan returned 8%. Find the amount invested in each if the agent's total income from her investments was $448. $1600 invested in high-risk venture and $2600 in savings and loan

36. Tax shelter. Mr. Mostafa received an inheritance of $7000. He put part of it in a tax shelter paying 9% and part in a bank paying 6%. If his annual interest totals $540, how much was invested at each rate? $4000 in a tax shelter and $3000 in a bank

37. Jogging. Angelina jogs 100 meters in the same time that Harry bicycles 150 meters. If Harry bicycles 15 meters per minute faster than Angelina jogs, find the rate of each. Angelina: 30 m/min; Harry: 45 m/min

38. Overtaking a lead. A car leaves New Orleans traveling 50 kilometers per hour. An hour later a second car leaves New Orleans following the first car and traveling 70 kilometers per hour. How long will it take the second car to overtake the first? 2.5 hr

39. Butterfat in milk. Two gallons of milk contain 2% butterfat. How many quarts of milk should be drained and replaced by pure butterfat to produce a solution of 2 gallons of milk containing 5% butterfat? (There are 4 quarts in a gallon.) $\frac{12}{49}$ qt

40. Gold alloy. A goldsmith has 120 grams of a gold alloy (a mixture of gold and one or more other metals) containing 75% pure gold. How much pure gold must be added to this alloy to obtain an alloy that is 80% pure gold? 30 g

41. Interest. Ms. Jordan invested $4900, part at 6% and the rest at 8%. If the yearly interest on each investment is the same, how much interest does she receive at the end of the year? $336

42. Interest. If $5000 is invested in a bank that pays 5% interest, how much more must be invested in bonds at 8% to earn 6% on the total investment? $2500

43. Separating planes. Two planes leave an airport traveling in opposite directions. One plane travels at 470 kilometers per hour, the other travels at 430 kilometers per hour. How long will it take them to be 2250 kilometers apart? 2.5 hr

44. Overtaking a bike. Lucas is $\frac{1}{3}$ of a mile from home, bicycling at 20 miles per hour, when his brother takes off on his bike to catch up. How fast must Lucas's brother go to catch Lucas a mile from home? 30 mph

45. Coffee blends. A coffee wholesaler wants to blend a coffee containing 35% chicory with a coffee containing 15% chicory to produce a 500-kilogram blend of coffee containing 18% chicory. How much of each type is required? 75 kg of the 35% chicory and 425 kg of the 15% chicory

46. Mixture of nuts. A mixture of nuts contains almonds worth $0.50 a pound, cashews worth $1.00 per pound, and pecans worth $0.75 per pound. If there are three times as many pounds of almonds as cashews in a 100-pound mixture worth $0.70 a pound, how many pounds of each nut does the mixture contain? 30 lb almonds, 10 lb cashews, 60 lb pecans

47. Draining a pool. An old pump can drain a pool in 6 hours working alone. A new pump can drain the pool in 4 hours working alone. How long will it take both pumps working together to empty the pool? 2 h and 24 min

48. Filling a blimp. One type of air blower can fill a blimp (or dirigible) in 6 hours working alone, while a second blower fills the same blimp in 9 hours working alone. How long will it take both blowers working together? 3 h and 36 min

49. Overtaking a lead. Karen notices that her husband left for the airport without his briefcase. Her husband drives 40 miles per hour and has a 15-minute head start. If Karen (with the briefcase) drives 60 miles per hour and the airport is 45 miles away, will she catch him before he arrives at the airport? Yes

50. Candy machine coin box. The coin box in a candy machine contained $17.90 in nickels, dimes, and quarters. It contained the same number of nickels as dimes and contained 136 coins in all. How many of each coin did it hold? nickels and dimes: 46; quarters: 44

Answers:

18. $\left\{\frac{3}{5}\right\}$ **23.** $\left\{\frac{7}{4}\right\}$ **24.** $\left\{-\frac{1}{16}\right\}$ **25.** $\left\{\frac{23}{6}\right\}$ **26.** $\left\{-\frac{5}{2}\right\}$ **27.** $\left\{\frac{7}{8}\right\}$

51. Increasing distances. Two cars start out together from the same place. They travel in opposite directions, one of them traveling 7 miles per hour faster than the other. After 3 hours, they are 621 miles apart. How fast is each car traveling?   100 mph and 107 mph

52. Shredding hay. A shredder distributes hay across 5/9 of a field in 10 hours working alone. By adding a second shredder, the entire field is finished in another 3 hours. How long would it take the second shredder to do the job working alone?   10 h and 48 min

53. Draining a pool. A swimming pool has two drainpipes. One can empty the pool in 3 hours, and the other can empty the pool in 7 hours. If both drains are open, how long will it take the pool to drain?   2 h and 6 min

54. Sorting letters. A new letter-sorting machine works twice as fast as the old one. Both sorters working together complete a job in 8 hours. How long would it have taken the new letter sorter to do the job alone?   12 hr

55. Travel time. Aya rode her bicycle from her house to a friend's house, averaging 16 kilometers per hour. It became dark before Aya was ready to return home, so her friend drove her home at a rate of 80 kilometers per hour. If Aya's total traveling time was 3 hours, how far away did her friend live?   40 km

56. Interest rates. Mr. Kaplan invests one sum of money at a certain rate of interest and half that sum at twice the first rate of interest. This arrangement turns out to yield 8% on the total investment. What are the two rates of interest?   6%, 12%

57. Plowing a field. A farmer can plow his field by himself in 15 days. If his son helps, they can do it in 6 days. How long would it take his son to plow the field by himself?   10 days

Many problems can be solved by modeling procedures introduced in this section other than those presented in the examples. Here are some types for you to try.

58. Box construction. An open box is to be constructed from a rectangular sheet of tin 3 meters wide by cutting out a 1-meter square from each corner and folding up the sides. The volume of the box is to be 2 cubic meters. What is the length of the tin rectangle?   4 m

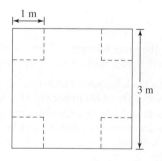

59. Age computation. Eric's grandfather is 57 years older than Eric. Five years from now, his grandfather will be four times as old as Eric is at that time. How old is the grandfather?   71 yr

60. Vending machine coins. A vending machine coin box contains nickels, dimes, and quarters. It contains three times as many nickels and four times as many quarters as it does dimes. If the box contains $96.25 in all, how many coins of each kind does it contain?   231 nickels, 77 dimes, 308 quarters

61. Theater ticket sales. A small theater company with 22 seats sold advance-sale tickets for $3.75 and tickets at the door for $4.50. If the company collected $84 and sold out completely, how many of each type of ticket did it sell?   20 tickets for $3.75 and 2 tickets for $4.50

62. Gumdrops. In a jar of red and green gumdrops, 12 more than half the gumdrops are red and the number of green gumdrops is 19 more than half the number of red gumdrops. How many of each color are in the jar?   86 red, 62 green

63. Real estate investment. A real estate investor bought acreage for $7200. After reselling three-fourths of the acreage at a profit of $30 per acre, she recovered the $7200. How many acres were sold?   60 acres

64. Suppose you average 75 miles per hour over the first half of your drive from Denver to Las Vegas, but your average speed for the entire trip is 60 miles per hour. What was your average speed for the second half of your drive?   50 mph

65. A mixture contains alcohol and water in a 5:1 ratio. Upon adding 5 liters of water, the ratio of alcohol to water becomes 5:2. Find the quantity of alcohol in the original mixture.   25 l

66. An alloy contains zinc and copper in a 5:8 ratio, and another alloy contains zinc and copper in a 5:3 ratio. Equal amounts of both alloys are melted together. Find the ratio of zinc to copper in the new alloy.   105 : 103

67. A man and a woman have the same birthday. When he was as old as she is now, the man was twice as old as the woman. When she becomes as old as he is now, the sum of their ages will be 119. How old is the man now?   51 yr

68. How many minutes is it before 6 P.M. if, 50 minutes ago, four times this number was the number of minutes past 3 P.M.?   26 min

69. Two pipes A and B can fill a tank in 24 minutes and 32 minutes, respectively. Initially, both pipes are opened simultaneously. After how many minutes should pipe B be turned off so that the tank is full in 18 minutes?   8 min

Complex Numbers

Objectives

1   Define a complex number.

2   Add and subtract complex numbers.

3   Multiply and divide complex numbers.

1   Define a complex number.

Complex Numbers

In the next section, you will learn methods of solving a quadratic equation $ax^2 + bx + c = 0$. Because the squares of real numbers are nonnegative (that is, $x^2 \geq 0$ for any real number x), the quadratic equation $x^2 = -1$ has no solution in the set of real numbers. To correct this shortcoming, we extend the real number system to a larger system called the complex number system. We define a new number, i, with the following properties:

$$i = \sqrt{-1}, \quad i^2 = -1$$

DEFINITION OF COMPLEX NUMBERS

A **complex number** is a number of the form

$$a + bi,$$

where a and b are real numbers and $i^2 = -1$.

The **real part** of the complex number $a + bi$ is a.

The **imaginary part** of the complex number $a + bi$ is b.

Two complex numbers are **equal** if and only if their real parts are equal and their imaginary parts are equal.

A complex number written in the form $a + bi$, where a and b are real numbers, is said to be in **standard form**. A complex number with real part 0, written as just bi, is called a **pure imaginary number**. Real numbers form a subset of complex numbers with imaginary part 0. For example, $-3 = -3 + 0i$.

We can express the **principal square root** of any negative number as the product of a real number and i.

> If $a > 0$ is any real number, then $\sqrt{-a} = (\sqrt{a})i$.

TECHNOLOGY CONNECTION

Most graphing calculators allow you to work with complex numbers by changing from "real" to $a + bi$ mode. Once the $a + bi$ mode is set, the $\sqrt{-1}$ is recognized as i. The i key is used to enter complex numbers such as $2 + 3i$.

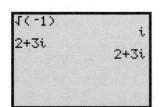

The real and imaginary parts of a complex number can then be found.

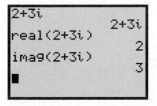

We identify the real and the imaginary parts of each complex number.

	Real part	Complex part
$2 + 5i$	2	5
$7 - 8i$	7	-8
$3i$	0	3
-9	-9	0
$3 + \sqrt{-25}$	3	5 (since $\sqrt{-25} = i\sqrt{25} = 5i$)

2 Add and subtract complex numbers.

Addition and Subtraction

ADDITION AND SUBTRACTION OF COMPLEX NUMBERS

For all real numbers $a, b, c,$ and $d,$

$$(a + bi) + (c + di) = (a + c) + (b + d)i \quad \text{and}$$
$$(a + bi) - (c + di) = (a - c) + (b - d)i.$$

TECHNOLOGY CONNECTION

In $a + bi$ mode, most graphing calculators perform addition and subtraction of complex numbers with the ordinary addition and subtraction keys.

```
(3+7i)+(2-4i)
            5+3i
(5+9i)-(6-8i)
          -1+17i
```

EXAMPLE 1 **Adding and Subtracting Complex Numbers**

Write the sum or difference of two complex numbers in standard form.

a. $(3 + 7i) + (2 - 4i)$ **b.** $(5 + 9i) - (6 - 8i)$

SOLUTION

a. $(3 + 7i) + (2 - 4i) = (3 + 2) + [7 + (-4)]i = 5 + 3i$
b. $(5 + 9i) - (6 - 8i) = (5 - 6) + [9 - (-8)]i = -1 + 17i$ ■ ■ ■

Practice Problem 1 Write the following complex numbers in standard form.

a. $(1 - 4i) + (3 + 2i)$ **b.** $(4 + 3i) - (5 - i)$ ■

◆ **WARNING** Recall that if a and b are nonnegative real numbers, then

$$\sqrt{a}\sqrt{b} = \sqrt{ab}.$$

However, this property is not true for nonreal numbers. For example,

$$\sqrt{-9}\sqrt{-9} = (3i)(3i) = 9i^2 = 9(-1) = -9,$$

but $$\sqrt{(-9)(-9)} = \sqrt{81} = 9.$$

Thus, $$\sqrt{-9}\sqrt{-9} \neq \sqrt{(-9)(-9)}.$$

3 Multiply and divide complex numbers.

Multiplying and Dividing Complex Numbers

We multiply complex numbers by first using FOIL (as we did with binomials) and then replacing i^2 with -1. The result, given next, need not be memorized.

MULTIPLYING COMPLEX NUMBERS

For all real numbers $a, b, c,$ and $d,$

$$(a + bi)(c + di) = (ac - bd) + (ad + bc)i.$$

EXAMPLE 2   **Multiplying Complex Numbers**

Write the following products in standard form.

a. $(3 - 5i)(2 + 7i)$ **b.** $-2i(5 - 9i)$

SOLUTION

a. $(3 - 5i)(2 + 7i) = \overset{F}{6} + \overset{O}{21i} - \overset{I}{10i} - \overset{L}{35i^2}$

$= 6 + 11i + 35$ Because $i^2 = -1, -35i^2 = 35$.

$= 41 + 11i$ Combine terms.

b. $-2i(5 - 9i) = -10i + 18i^2$ Distributive property

$= -10i - 18$ Because $i^2 = -1, 18i^2 = -18$.

$= -18 - 10i$ Standard form   ■ ■ ■

Practice Problem 2   Write the following products in standard form.

a. $(2 - 6i)(1 + 4i)$ **b.** $-3i(7 - 5i)$ ■

TECHNOLOGY CONNECTION

In $a + bi$ mode, most graphing calculators can compute the complex conjugate of a complex number.

```
conj(2+7i)
            2-7i
conj(5-3i)
            5+3i
■
```

THE CONJUGATE OF A COMPLEX NUMBER

If $z = a + bi$, then the conjugate (or complex conjugate) of z is denoted by $\bar{z}$ and defined by $\bar{z} = \overline{a + bi} = a - bi$.

For example, $\overline{2 + 7i} = 2 - 7i$ and $\overline{5 - 3i} = \overline{5 + (-3)i} = 5 - (-3)i = 5 + 3i$.

EXAMPLE 3   **Multiplying a Complex Number by Its Conjugate**

Find the product $z\bar{z}$ for the complex number $z = 2 + 5i$.

SOLUTION

If $z = 2 + 5i$, then $\bar{z} = 2 - 5i$.

$z\bar{z} = (2 + 5i)(2 - 5i) = 2^2 - (5i)^2$ Difference of squares

$= 4 - 25i^2$ $(5i)^2 = 5^2i^2 = 25i^2$

$= 4 - (-25)$ Because $i^2 = -1, 25i^2 = -25$.

$= 29$ Simplify.   ■ ■ ■

Practice Problem 3   Find the product $z\bar{z}$ for the complex number $z = 1 + 6i$.   ■

Notice that the product $z\bar{z}$ in Example 3 is a real number. To write the reciprocal of a nonzero complex number, or the quotient of two complex numbers, in the form $a + bi$, multiply the numerator and denominator by the conjugate of the denominator. The resulting denominator is a real number, as the next theorem states.

COMPLEX CONJUGATE PRODUCT THEOREM

If $z = a + bi$, then

$$z\bar{z} = a^2 + b^2.$$

TECHNOLOGY CONNECTION

In $a + bi$ mode, most graphing calculators perform division of complex numbers with the ordinary division key. The Frac option will display results, using rational numbers.

```
1/(2+i)
            .4-.2i
Ans▶Frac
          2/5-1/5i
■
```

```
(4+√(-25))/(2-√(
-9))
-.5384615385+1.…
Ans▶Frac
        -7/13+22/13i
```

The reciprocal of a complex number can also be found with the reciprocal key.

```
(2+i)⁻¹
            .4-.2i
Ans▶Frac
          2/5-1/5i
■
```

The complex conjugate product theorem provides a procedure for writing the quotient of two complex numbers in standard from. You do not need to memorize the final result.

DIVIDING COMPLEX NUMBERS

Let $z = a + bi$ $(z \neq 0)$ and $w = c + di$ be two complex numbers in standard form. Then

$$\frac{w}{z} = \frac{w\bar{z}}{z\bar{z}} \quad \text{Multiply numerator and denominator by } \bar{z}.$$

$$= \frac{(c + di)(a - bi)}{a^2 + b^2} = \frac{ac + bd + (ad - bc)i}{a^2 + b^2}$$

$$= \left(\frac{ac + bd}{a^2 + b^2}\right) + \left(\frac{ad - bc}{a^2 + b^2}\right)i$$

EXAMPLE 4   Dividing Complex Numbers

Write the following quotients in standard form.

a. $\dfrac{1}{2 + i}$ **b.** $\dfrac{4 + \sqrt{-25}}{2 - \sqrt{-9}}$

SOLUTION

a. The denominator is $2 + i$, so its conjugate is $2 - i$.

$$\frac{1}{2 + i} = \frac{1(2 - i)}{(2 + i)(2 - i)} \qquad \begin{array}{l}\text{Multiply numerator}\\\text{and denominator}\\\text{by } 2 - i.\end{array}$$

$$= \frac{2 - i}{2^2 + 1^2} = \frac{2 - i}{5} = \frac{2}{5} - \frac{1}{5}i$$

b. Write $\dfrac{4 + \sqrt{-25}}{2 - \sqrt{-9}}$ as $\dfrac{4 + 5i}{2 - 3i}$.

$$\frac{4 + 5i}{2 - 3i} = \frac{(4 + 5i)(2 + 3i)}{(2 - 3i)(2 + 3i)} \qquad \begin{array}{l}\text{Multiply numerator and}\\\text{denominator by } 2 + 3i.\end{array}$$

$$= \frac{8 + 12i + 10i + 15i^2}{2^2 + 3^2} \qquad \begin{array}{l}\text{Use FOIL in the numerator;}\\(2 - 3i)(2 + 3i) = 2^2 + 3^2.\end{array}$$

$$= \frac{-7 + 22i}{13} = -\frac{7}{13} + \frac{22}{13}i \qquad ■ ■ ■$$

Practice Problem 4   Write the following quotients in standard form.

a. $\dfrac{2}{1 - i}$ **b.** $\dfrac{-3i}{4 + \sqrt{-25}}$ ■

Powers of i

We already know the first two powers of i: $i^1 = i$ and $i^2 = -1$. For the following powers of i, notice the pattern.

$$i^1 = i \qquad\qquad\qquad i^5 = i^4 \cdot i = (1)i = i$$

$$i^2 = -1 \qquad\qquad\qquad i^6 = i^4 \cdot i^2 = (1) \cdot i^2 = i^2 = -1$$

$$i^3 = i^2 \cdot i = (-1)i = -i \qquad i^7 = i^4 \cdot i^3 = (1) \cdot i^3 = i^3 = -i$$

$$i^4 = i^2 \cdot i^2 = (-1)(-1) = 1 \qquad i^8 = i^4 \cdot i^4 = (1)(1) = 1$$

After i^4, the powers of i cycle indefinitely through the values i, -1, $-i$, and 1. If n is a positive integer, a quick way to evaluate i^n is to divide n by 4; then $i^n = i^r$, where r is the remainder: 0, 1, 2, or 3.

To see how this works, we evaluate i^{1003}. Dividing 1003 by 4, we get $1003 = 4(250) + 3$. Then $i^{1003} = i^{4(250)+3} = i^{4(250)} \cdot i^3 = (i^4)^{250} \cdot i^3 = (1)^{250} \cdot i^3 = 1 \cdot i^3 = i^3 = -i$. Similarly, dividing 26 by 4 gives the remainder 2; so $i^{26} = i^2 = -1$.

SECTION A.7 ■ Exercises

A EXERCISES Basic Skills and Concepts

1. The real part of the complex number $a + bi$ is ___a___ .

2. The imaginary part of the complex number $a + bi$ is ___b___ .

3. The complex conjugate of the complex number $a + bi$ is ___$a - bi$___ .

4. If $a > 0$, the principal square root of $-a$ is ___$(\sqrt{a})i$___ .

In Exercises 5–8, use the definition of equality of complex numbers to find the real numbers x and y such that the equation is true.

5. $2 + xi = y + 3i$ $x = 3, y = 2$

6. $x - 2i = 7 + yi$ $x = 7, y = -2$

7. $x - \sqrt{-16} = 2 + yi$ $x = 2, y = -4$

8. $3 + yi = x - \sqrt{-25}$ $x = 3, y = -5$

In Exercises 9–32, perform each operation and write the result in the standard form $a + bi$.

9. $(5 + 2i) + (3 + i)$ $8 + 3i$
10. $(6 + i) + (1 + 2i)$ $7 + 3i$
11. $(4 - 3i) - (5 + 3i)$ $-1 - 6i$
12. $(3 - 5i) - (3 + 2i)$ $-7i$
13. $(-2 - 3i) + (-3 - 2i)$ $-5 - 5i$
14. $(-5 - 3i) + (2 - i)$ $-3 - 4i$
15. $3(5 + 2i)$ $15 + 6i$
16. $4(3 + 5i)$ $12 + 20i$
17. $-4(2 - 3i)$ $-8 + 12i$
18. $-7(3 - 4i)$ $-21 + 28i$
19. $3i(5 + i)$ $-3 + 15i$
20. $2i(4 + 3i)$ $-6 + 8i$
21. $4i(2 - 5i)$ $20 + 8i$
22. $-3i(5 - 2i)$ $-6 - 15i$
23. $(3 + i)(2 + 3i)$ $3 + 11i$
24. $(4 + 3i)(2 + 5i)$ $-7 + 26i$
25. $(2 - 3i)(2 + 3i)$ 13
26. $(4 - 3i)(4 + 3i)$ 25
27. $(3 + 4i)(4 - 3i)$ $24 + 7i$
28. $(-2 + 3i)(-3 + 10i)$ $-24 - 29i$
29. $(\sqrt{3} - 12i)^2$ $-141 - 24\sqrt{3}i$
30. $(-\sqrt{5} - 13i)^2$ $-164 + 26\sqrt{5}i$
31. $(1 + 3i)^3$ $-26 - 18i$
32. $(1 - 2i)^3$ $-11 + 2i$

In Exercises 33–38, write the conjugate $\bar{z}$ of each complex number z. Then find $z\bar{z}$.

33. $z = 2 - 3i$ †
34. $z = 4 + 5i$ †
35. $z = \dfrac{1}{2} - 2i$ †
36. $z = \dfrac{2}{3} + \dfrac{1}{2}i$ †
37. $z = \sqrt{2} - 3i$ †
38. $z = \sqrt{5} + \sqrt{3}i$ †

In Exercises 39–52, write each quotient in the standard form $a + bi$.

39. $\dfrac{5}{-i}$ $5i$
40. $\dfrac{2}{-3i}$ $\dfrac{2}{3}i$
41. $\dfrac{-1}{1 + i}$ $-\dfrac{1}{2} + \dfrac{1}{2}i$
42. $\dfrac{1}{2 - i}$ $\dfrac{2}{5} + \dfrac{1}{5}i$
43. $\dfrac{5i}{2 + i}$ $1 + 2i$
44. $\dfrac{3i}{2 - i}$ $-\dfrac{3}{5} + \dfrac{6}{5}i$
45. $\dfrac{2 + 3i}{1 + i}$ $\dfrac{5}{2} + \dfrac{1}{2}i$
46. $\dfrac{3 + 5i}{4 + i}$ $1 + i$
47. $\dfrac{2 - 5i}{4 - 7i}$ $\dfrac{43}{65} - \dfrac{6}{65}i$
48. $\dfrac{3 + 5i}{1 + 3i}$ $\dfrac{9}{5} - \dfrac{2}{5}i$
49. $\dfrac{2 + \sqrt{-4}}{1 + i}$ 2
50. $\dfrac{5 - \sqrt{-9}}{3 + 2i}$ $\dfrac{9}{13} - \dfrac{19}{13}i$
51. $\dfrac{-2 + \sqrt{-25}}{2 - 3i}$ $-\dfrac{19}{13} + \dfrac{4}{13}i$
52. $\dfrac{-5 - \sqrt{-4}}{5 - \sqrt{-9}}$ $-\dfrac{19}{34} - \dfrac{25}{34}i$

In Exercises 53–56, find each power of i and simplify the expression.

53. i^{17} i
54. i^{125} i
55. i^{-7} i
56. i^{-24} 1

In Exercises 57–60, let $z_1 = a + bi$ and $z_2 = c + di$.

57. Prove that $\overline{\overline{z_1}} = z_1$.

58. Prove that $\overline{z_1 + z_2} = \overline{z_1} + \overline{z_2}$.

59. Prove that $\overline{z_1 z_2} = \overline{z_1}\,\overline{z_2}$. Use this fact to prove that $\overline{z^2} = (\overline{z})^2$.

60. Prove that $z_1 + \overline{z_1} = 2a$ and that $z_1 - \overline{z_1} = 2bi$.

In Exercises 61–64, let $z = 2 + 4i$ and $w = 3 - 2i$.

61. $\overline{w}z$ $14 - 8i$
62. $\overline{w}\overline{z}$ $-2 + 16i$
63. $\overline{z}w$ $-2 - 16i$
64. $\dfrac{w}{z}$ $-\dfrac{1}{10} - \dfrac{4}{5}i$

B EXERCISES Applying the Concepts

65. **Series circuits.** If the impedance of a resistor in a circuit is $Z_1 = 4 + 3i$ ohms and the impedance of a second resistor is $Z_2 = 5 - 2i$ ohms, find the total impedance of

†Due to space constrictions, answers to these exercises may be found in the Answers beginning on page A–1 in the back of the book.

the two resistors when they are placed in series (the sum of the two impedances). $9 + i$

Series circuit

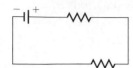

66. **Parallel circuits.** If the two resistors in Exercise 65 are connected in parallel, the total impedance is given by

$$\frac{Z_1 Z_2}{Z_1 + Z_2}.$$

Find the total impedance if the resistors in Exercise 65 are connected in parallel.

Parallel circuit

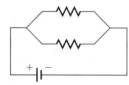

As with impedance, the current I and voltage V in a circuit can be represented by complex numbers. The three quantities (voltage V, impedance Z, and current I) are related by the equation $Z = \dfrac{V}{I}$. Thus, if two of these values are given, the value of the third can be found from this equation. In Exercises 67–72, find the value that is not specified.

66. $\dfrac{241}{82} + \dfrac{37}{82}i$

67. **Finding impedance.** $I = 7 + 5i$ $V = 35 + 70i$

68. **Finding impedance.** $I = 7 + 4i$ $V = 45 + 88i$

69. **Finding voltage.** $Z = 5 - 7i$ $I = 2 + 5i$

70. **Finding voltage.** $Z = 7 - 8i$ $I = \dfrac{1}{3} + \dfrac{1}{6}i$

71. **Finding current.** $V = 12 + 10i$ $Z = 12 + 6i$

72. **Finding current.** $V = 29 + 18i$ $Z = 25 + 6i$

67. $Z = \dfrac{595}{74} + \dfrac{315}{74}i$ 68. $Z = \dfrac{667}{65} + \dfrac{436}{65}i$ 69. $V = 45 + 11i$

70. $V = \dfrac{11}{3} - \dfrac{3}{2}i$ 71. $I = \dfrac{17}{15} + \dfrac{4}{15}i$ 72. $I = \dfrac{833}{661} + \dfrac{276}{661}i$

Quadratic Equations

Objectives

1 Solve a quadratic equation by factoring.
2 Solve a quadratic equation by the square root method.
3 Solve a quadratic equation by completing the square.
4 Solve a quadratic equation by using the quadratic formula.
5 Solve quadratic equations with complex solutions.

QUADRATIC EQUATION

A quadratic equation in the variable x is an equation equivalent to the equation

$$ax^2 + bx + c = 0,$$

where a, b, and c are real numbers and $a \neq 0$.

BY THE WAY ...

The word *quadratic* comes from the Latin *quadratus*, meaning square.

A quadratic equation written in the form $ax^2 + bx + c = 0$ is said to be in **standard form**. Quadratic equations are also called **second-degree equations**.

1 Solve a quadratic equation by factoring.

Factoring Method

Some quadratic equations written in standard form can be solved by factoring and using the **zero-product property**.

THE ZERO-PRODUCT PROPERTY

Let A and B be two algebraic expressions. Then $AB = 0$ if and only if $A = 0$ or $B = 0$.

STUDY TIP

The factoring method of solving an equation works *only when one of the sides of the equation is* 0.

EXAMPLE 1 **Solving a Quadratic Equation by Factoring**

Solve by factoring: $2x^2 + 5x = 3$

SOLUTION

$$
\begin{array}{ll}
2x^2 + 5x = 3 & \text{Original equation} \\
2x^2 + 5x - 3 = 0 & \text{Write in standard form.} \\
(2x - 1)(x + 3) = 0 & \text{Factor the left side.}
\end{array}
$$

We now set each factor equal to 0 and solve the resulting linear equations. The vertical line separates the computations leading to the two solutions.

$$
\begin{array}{l|ll}
2x - 1 = 0 & x + 3 = 0 & \text{Zero-product property} \\
2x = 1 & x = -3 & \text{Isolate the } x \text{ term on one side.} \\
x = \dfrac{1}{2} & x = -3 & \text{Solve for } x.
\end{array}
$$

Check: Check the two solutions in the original equation. The solution set is $\left\{\dfrac{1}{2}, -3\right\}$.

■ ■ ■

Practice Problem 1   Solve by factoring: $x^2 + 25x = -84$ ■

EXAMPLE 2   Solving a Quadratic Equation by Factoring

Solve by factoring: $x^2 + 16 = 8x$

SOLUTION

Write the equation in standard form.

$$x^2 - 8x + 16 = 0 \qquad \text{Subtract } 8x \text{ from both sides.}$$

Factor the left side of the equation.

$$(x - 4)(x - 4) = 0 \qquad \text{Perfect-square trinomial}$$

Set each factor equal to 0 and solve each resulting equation.

$$
\begin{array}{c|c}
x - 4 = 0 & x - 4 = 0 \\
x = 4 & x = 4
\end{array}
$$

Check the solution, 4, in the original equation. The solution set is $\{4\}$.   ■ ■ ■

Practice Problem 2   Solve by factoring: $x^2 - 6x = -9$ ■

In Example 2, because the factor $(x - 4)$ appears twice in the solution, the number 4 is called a **double root**, or a **root of multiplicity 2**, of the given equation.

2 Solve a quadratic equation by the square root method.

The Square Root Method

We can solve equations of the type $x^2 = d$ by using the square root method. Suppose we want to solve the equation $x^2 = 3$, which is equivalent to the equation $x^2 - 3 = 0$. Applying the factoring method, but *removing the restriction that coefficients and constants represent only integers,* we have

$$
\begin{aligned}
x^2 - 3 &= 0 \\
x^2 - (\sqrt{3})^2 &= 0 \qquad \text{Because } 3 = (\sqrt{3})^2. \\
(x + \sqrt{3})(x - \sqrt{3}) &= 0 \qquad \text{Factor (using real numbers).} \\
x + \sqrt{3} = \text{ or } x - \sqrt{3} &= 0 \qquad \text{Set each factor equal to zero.} \\
x = -\sqrt{3} \text{ or } x &= \sqrt{3} \qquad \text{Solve for } x.
\end{aligned}
$$

So the solutions of the equation $x^2 = 3$ are $-\sqrt{3}$ and $\sqrt{3}$, which we can write as $\pm\sqrt{3}$. Similarly, for any nonnegative real number d, the solutions of the quadratic equation $x^2 = d$ are $\pm\sqrt{d}$.

THE SQUARE ROOT PROPERTY

Suppose u is any algebraic expression and $d \geq 0$. If $u^2 = d$, then $u = \pm\sqrt{d}$.

EXAMPLE 3   Solving an Equation by the Square Root Method

Solve $(x - 3)^2 = 5$.

SOLUTION

$$
\begin{aligned}
(x - 3)^2 &= 5 \qquad \text{Original equation} \\
x - 3 &= \pm\sqrt{5} \qquad \text{Square root property} \\
x &= 3 \pm \sqrt{5} \qquad \text{Add 3 to both sides.}
\end{aligned}
$$

The solution set is $\{3 + \sqrt{5}, 3 - \sqrt{5}\}$.   ■ ■ ■

Practice Problem 3   Solve $(x + 2)^2 = 5$.   ■

3 Solve a quadratic equation by completing the square.

Completing the Square

We can use a method called **completing the square** to solve quadratic equations that cannot be solved by factoring and are not in the right form to use the square root method. In this method, we first write a given quadratic equation in the form $(x + k)^2 = d$. Then we solve the equation $(x + k)^2 = d$ by the square root method.

Recall that

$$(x + k)^2 = x^2 + 2kx + k^2.$$

Notice that the coefficient of x on the right side is $2k$, and half of this coefficient is k. We see that the constant term, k^2, in the trinomial $x^2 + 2kx + k^2$ is the *square of one-half the coefficient of* x.

PERFECT-SQUARE TRINOMIAL

A quadratic trinomial in x with coefficient of x^2 equal to 1 is a **perfect-square trinomial** if the constant term is the square of one-half the coefficient of x.

When the constant term is *not* the square of half the coefficient of x, we subtract the constant term from both sides and add a new constant term that will result in a perfect-square trinomial.

Let's look at some examples to learn what number to add to $x^2 + bx$ to create a perfect-square trinomial.

$x^2 + bx$	b	$\dfrac{b}{2}$	Add $\left(\dfrac{b}{2}\right)^2$	Perfect Square $\left(x + \dfrac{b}{2}\right)^2$
$x^2 + 6x$	6	3	$3^2 = 9$	$x^2 + 6x + 9 = (x + 3)^2$
$x^2 - 4x$	-4	-2	$(-2)^2 = 4$	$x^2 - 4x + 4 = (x - 2)^2$
$x^2 + 3x$	3	$\dfrac{3}{2}$	$\left(\dfrac{3}{2}\right)^2 = \dfrac{9}{4}$	$x^2 + 3x + \dfrac{9}{4} = \left(x + \dfrac{3}{2}\right)^2$
$x^2 - x$	-1	$-\dfrac{1}{2}$	$\left(-\dfrac{1}{2}\right)^2 = \dfrac{1}{4}$	$x^2 - x + \dfrac{1}{4} = \left(x - \dfrac{1}{2}\right)^2$

Therefore, to make $x^2 + bx$ a perfect square, we need to add

$$\left[\frac{1}{2}(\text{coefficient of } x)\right]^2 = \left[\frac{1}{2}(b)\right]^2 = \frac{b^2}{4}$$

so that

$$x^2 + bx + \frac{b^2}{4} = \left(x + \frac{b}{2}\right)^2.$$

We say that $\dfrac{b^2}{4}$ is added to $x^2 + bx$ to *complete the square*. See Figure A.15.

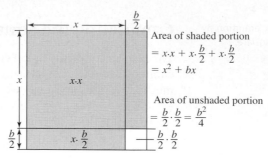

FIGURE A.15 **Area of a square** $= \left(x + \dfrac{b}{2} \right)^2$

METHOD OF COMPLETING THE SQUARE

Step 1 Rearrange the quadratic equation so that the terms in x^2 and x are on the left side of the equation and the constant term is on the right side.

Step 2 Make the coefficient of x^2 equal to 1 by dividing both sides of the equation by the original coefficient. (Steps **1** and **2** are interchangeable.)

Step 3 Add the square of one-half the coefficient of x to both sides of the equation.

Step 4 Write the equation in the form $(x + k)^2 = d$, using the fact that the left side is a perfect square.

Step 5 Take the square root of each side, prefixing $\pm$ to the right side.

Step 6 Solve the two equations from Step **5**.

EXAMPLE 4 **Solving a Quadratic Equation by Completing the Square**

Solve by completing the square: $3x^2 - 4x - 1 = 0$

SOLUTION

$$3x^2 - 4x - 1 = 0 \qquad \text{Note that the coefficient of } x^2 \text{ is 3.}$$

$$3x^2 - 4x = 1 \qquad \text{Add 1 to both sides.}$$

$$x^2 - \frac{4}{3}x = \frac{1}{3} \qquad \text{Divide both sides by 3.}$$

$$x^2 - \frac{4}{3}x + \left(-\frac{2}{3}\right)^2 = \frac{1}{3} + \left(-\frac{2}{3}\right)^2 \qquad \text{Add } \left[\frac{1}{2}\left(-\frac{4}{3}\right)\right]^2 = \left(-\frac{2}{3}\right)^2 \text{ to both sides.}$$

$$\left(x - \frac{2}{3}\right)^2 = \frac{7}{9} \qquad x^2 - \frac{4}{3}x + \left(-\frac{2}{3}\right)^2 = \left(x - \frac{2}{3}\right)^2$$

$$x - \frac{2}{3} = \pm\sqrt{\frac{7}{9}} \qquad \text{Take the square root of both sides.}$$

$$x - \frac{2}{3} = \pm\frac{\sqrt{7}}{3} \qquad \sqrt{\frac{7}{9}} = \frac{\sqrt{7}}{\sqrt{9}} = \frac{\sqrt{7}}{3}$$

$$x = \frac{2}{3} \pm \frac{\sqrt{7}}{3} = \frac{2 \pm \sqrt{7}}{3} \qquad \text{Add } \frac{2}{3} \text{ to both sides and combine.}$$

The solution set is $\left\{ \dfrac{2 - \sqrt{7}}{3}, \dfrac{2 + \sqrt{7}}{3} \right\}$. ■ ■ ■

Practice Problem 4 Solve by completing the square: $4x^2 - 24x + 25 = 0$ ■

4 Solve a quadratic equation by using the quadratic formula.

The Quadratic Formula

We can generalize the method of completing the square to derive a formula that gives a solution of *any* quadratic equation.

We solve the standard form of the quadratic equation by completing the square.

$$ax^2 + bx + c = 0, a \neq 0$$

$ax^2 + bx = -c$	Isolate the constant on the right side.
$x^2 + \dfrac{b}{a}x = -\dfrac{c}{a}$	Divide both sides by a.
$x^2 + \dfrac{b}{a}x + \left(\dfrac{b}{2a}\right)^2 = \left(\dfrac{b}{2a}\right)^2 - \dfrac{c}{a}$	Add the square of one-half the coefficient of x to both sides.
$\left(x + \dfrac{b}{2a}\right)^2 = \dfrac{b^2}{4a^2} - \dfrac{c}{a}$	The left side in Step 3 is a perfect square.
$\left(x + \dfrac{b}{2a}\right)^2 = \dfrac{b^2 - 4ac}{4a^2}$	Combine fractions on the right side.
$x + \dfrac{b}{2a} = \pm\sqrt{\dfrac{b^2 - 4ac}{4a^2}}$	Take the square roots of both sides.
$x = -\dfrac{b}{2a} \pm \dfrac{\sqrt{b^2 - 4ac}}{2|a|}$	Add $-\dfrac{b}{2a}$ to both sides; $\sqrt{4a^2} = \sqrt{4}\sqrt{a^2} = 2|a|$.
$x = -\dfrac{b}{2a} \pm \dfrac{\sqrt{b^2 - 4ac}}{2a}$	$\pm2|a| = \pm2a$ because $|a| = a$ or $|a| = -a$.
$x = \dfrac{-b \pm \sqrt{b^2 - 4ac}}{2a}$	Combine fractions.

This formula is called the **quadratic formula**.

THE QUADRATIC FORMULA

The solutions of the quadratic equation in the standard form $ax^2 + bx + c = 0$ with $a \neq 0$ are given by the formula:

$$x = \frac{-b \pm \sqrt{b^2 - 4ac}}{2a}$$

◆ **WARNING**   To use the quadratic formula, write the given quadratic equation in standard form. Then determine the values of a (coefficient of x^2), b (coefficient of x), and c (constant term).

EXAMPLE 5   **Solving a Quadratic Equation by Using the Quadratic Formula**

Solve $3x^2 = 5x + 2$ by using the quadratic formula.

SOLUTION

We first rewrite the equation in standard form.

$3x^2 - 5x - 2 = 0$	Subtract $5x + 2$ from both sides.
$3x^2 + (-5)x + (-2) = 0$	Identify values of $a, b,$ and c to be used in the quadratic formula.

$a \qquad b \qquad c$

continued on the next page

$$x = \frac{-b \pm \sqrt{b^2 - 4ac}}{2a} \qquad \text{Quadratic formula}$$

$$x = \frac{-(-5) \pm \sqrt{(-5)^2 - 4(3)(-2)}}{2(3)} \qquad \begin{array}{l}\text{Substitute 3 for } a, -5 \text{ for} \\ b, \text{ and } -2 \text{ for } c.\end{array}$$

$$= \frac{5 \pm \sqrt{25 + 24}}{6} \qquad \text{Simplify.}$$

$$= \frac{5 \pm \sqrt{49}}{6} = \frac{5 \pm 7}{6} \qquad \text{Simplify.}$$

Then

$$x = \frac{5 + 7}{6} = \frac{12}{6} = 2 \text{ or } x = \frac{5 - 7}{6} = \frac{-2}{6} = -\frac{1}{3}; \text{ the solution set is } \left\{-\frac{1}{3}, 2\right\}.$$

■ ■ ■

Practice Problem 5   Solve by using the quadratic formula: $6x^2 - x - 2 = 0$ ■

---

5 Solve quadratic equations with complex solutions.

Quadratic Equations with Complex Solutions

The methods of solving quadratic equations are also applicable to solving quadratic equations with complex solutions.

> **EXAMPLE 6**   **Solving Quadratic Equations with Complex Solutions**

Solve the equation $x^2 + 2x + 2 = 0$.

SOLUTION

$x^2 + 2x + 2 = 1 \cdot x^2 + 2x + 2 = 0$

$$x = \frac{-2 \pm \sqrt{2^2 - 4(1)(2)}}{2(1)} \qquad \begin{array}{l}\text{Use the quadratic formula} \\ \text{with } a = 1, b = 2, \text{ and } c = 2.\end{array}$$

$$= \frac{-2 \pm \sqrt{-4}}{2} = \frac{-2 \pm 2i}{2} = \frac{2(-1 \pm i)}{2} = -1 \pm i$$

The solution set is $\{-1 - i, -1 + i\}$. You should check these solutions.   ■ ■ ■

Practice Problem 6   Solve $x^2 = 4x - 13$.   ■

The Discriminant

If $a, b,$ and c are real numbers $(a \neq 0)$, we can predict the number and type of solutions that the equation $ax^2 + bx + c = 0$ will have without solving the equation. In the quadratic formula

$$x = \frac{-b \pm \sqrt{b^2 - 4ac}}{2a},$$

the quantity $b^2 - 4ac$ under the radical sign is called the **discriminant** of the equation. The discriminant reveals the type of solutions of the equation, as shown in the following table.

Discriminant	Number and Type of Solutions
$b^2 - 4ac > 0$	Two unequal real solutions
$b^2 - 4ac = 0$	One real solution (a root of multiplicity 2)
$b^2 - 4ac < 0$	Two nonreal complex solutions (complex conjugates)

EXAMPLE 7   **Using the Discriminant**

Use the discriminant to determine the number and type of solutions of each quadratic equation.

a. $x^2 - 4x + 2 = 0$ **b.** $2t^2 + 2t + 19 = 0$ **c.** $4x^2 + 4x + 1 = 0$

SOLUTION

Equation	$b^2 - 4ac$	Conclusion
a. $x^2 - 4x + 2 = 0$	$(-4)^2 - 4(1)(2) = 8 > 0$	Two unequal real solutions
b. $2t^2 + 2t + 19 = 0$	$(2)^2 - 4(2)(19) = -148 < 0$	Two nonreal complex conjugate solutions
c. $4x^2 + 4x + 1 = 0$	$(4)^2 - 4(4)(1) = 0$	One real repeated solution

■ ■ ■

Practice Problem 7   Determine the number and type of solutions.

a. $9x^2 - 6x + 1 = 0$ **b.** $x^2 - 5x + 3 = 0$ **c.** $2x^2 - 3x + 4 = 0$ ■

SECTION A.8 ■ Exercises

A EXERCISES Basic Skills and Concepts

1. Standard form for the quadratic equation $5x - 2x^2 = 8$ is $2x^2 - 5x + 8 = 0$.

2. If u is an algebraic expression, $d \geq 0$, and $u^2 = d$, then $u =$ $\pm\sqrt{d}$.

3. To make $x^2 + 10x$ a perfect square, we have to add 25 to $x^2 + 10x$.

4. If $ax^2 + bx + c$ does not factor readily, we can use the quadratic formula to solve the equation $ax^2 + bx + c = 0$.

In Exercises 5–24, solve each equation by factoring.

5. $x^2 - 5x = 0$ $\{0, 5\}$

6. $x^2 - 5x + 4 = 0$ $\{1, 4\}$

7. $x^2 + 5x = 14$ $\{-7, 2\}$

8. $x^2 - 11x = 12$ $\{-1, 12\}$

9. $x^2 = 5x + 6$ $\{-1, 6\}$

10. $x = x^2 - 12$ $\{-3, 4\}$

11. $2x^2 + 5x - 3 = 0$ †

12. $2x^2 - 9x + 10 = 0$ †

13. $3y^2 + 5y + 2 = 0$ †

14. $6x^2 + 11x + 4 = 0$ †

15. $5x^2 + 12x + 4 = 0$ †

16. $3x^2 - 2x - 5 = 0$ †

17. $2x^2 + x = 15$ †

18. $6x^2 = 1 - x$ †

19. $12x^2 - 10x = 12$ †

20. $-x^2 + 10x + 1200 = 0$ †

21. $18x^2 - 45x = -7$ †

22. $18x^2 + 57x + 45 = 0$ †

23. $4x^2 - 10x - 750 = 0$ †

24. $12x^2 + 43x + 36 = 0$ †

In Exercises 25–34, solve each equation by the square root method.

25. $3x^2 = 48$ $\{-4, 4\}$

26. $2x^2 = 50$ $\{-5, 5\}$

27. $x^2 + 1 = 5$ $\{-2, 2\}$

28. $2x^2 - 1 = 17$ $\{-3, 3\}$

29. $x^2 + 5 = 1$ $\{-2i, 2i\}$

30. $4x^2 + 9 = 0$ †

31. $(x - 1)^2 = 16$ $\{-3, 5\}$

32. $(2x - 3)^2 = 25$ $\{-1, 4\}$

33. $(3x - 2)^2 + 16 = 0$ †

34. $(2x + 3)^2 + 25 = 0$ †

In Exercises 35–44, add a constant term to the expression to make it a perfect square.

35. $x^2 + 4x$ 4

36. $y^2 + 10y$ 25

37. $x^2 + 6x$ 9

38. $y^2 - 8y$ 16

39. $x^2 - 7x$ $\dfrac{49}{4}$

40. $x^2 - 3x$ $\dfrac{9}{4}$

41. $x^2 + \dfrac{1}{3}x$ $\dfrac{1}{36}$

42. $x^2 - \dfrac{3}{2}x$ $\dfrac{9}{16}$

43. $x^2 + ax$ $\dfrac{a^2}{4}$

44. $x^2 - \dfrac{2a}{3}x$ $\dfrac{a^2}{9}$

In Exercises 45–58, solve each equation by completing the square.

45. $x^2 + 2x - 5 = 0$ †

46. $x^2 + 6x = -7$ †

47. $x^2 - 3x - 1 = 0$ †

48. $x^2 - x - 3 = 0$ †

49. $2r^2 + 3r = 9$ †

50. $3k^2 - 5k + 1 = 0$ †

51. $z^2 - 2z + 2 = 0$ †

52. $x^2 - 6x + 11 = 0$ †

53. $2x^2 - 20x + 49 = -7$ †

54. $4y^2 + 4y + 5 = 0$ †

55. $5x^2 - 6x = 4x^2 + 6x - 3$ †

56. $x^2 + 7x - 5 = x - x^2$ †

57. $5y^2 + 10y + 4 = 2y^2 + 3y + 1$ †

58. $3x^2 - 1 = 5x^2 - 3x - 5$ †

†Due to space constrictions, answers to these exercises may be found in the Answers beginning on page A–1 in the back of the book.

In Exercises 59–74, solve each equation by using the quadratic formula.

59. $x^2 + 2x - 4 = 0$ †
60. $m^2 + 3m + 2 = 0$ †
61. $6x^2 = 7x + 5$ †
62. $t^2 + 7 = 4t$ †
63. $3z^2 - 2z = 7$
64. $6y^2 + 11y = 10$
65. $3p^2 + 8p + 4 = 0$
66. $8(x^2 - x) = x^2 - 3$
67. $x^2 = 5(x - 1)$
68. $(x + 13)(x + 5) = -2$
69. $t(t + 1) = 3t^2 + 1$
70. $3(x^2 + 1) = 2x^2 + 4x + 1$
71. $2t^2 - 5 = 0$
72. $3k^2 - 48 = 0$ $\{-4, 4\}$
73. $9k^2 + 25 = 0$
74. $3k^2 + 4 = 0$

In Exercises 75–82, find the discriminant and determine the number and type of roots of each equation.

75. $4x^2 - 12x + 9 = 0$
76. $3x^2 - 4x + 3 = 0$ †
77. $2y^2 = 6 - y$ †
78. $9y^2 + 24y = -16$ †
79. $17x - 12 = 6x^2$ †
80. $5x^2 = 7x - 3$ †
81. $-3x^2 + 21 = 0$ †
82. $x^2 + 2x + \dfrac{1}{2} = 0$ †

In Exercises 83–88, find the values of k for which the given equation has equal roots.

83. $x^2 - kx + 3 = 0$ †
84. $x^2 + 3kx + 8 = 0$ †
85. $2x^2 + kx + k = 0$ †
86. $kx^2 + 2x + 6 = 0$ †
87. $x^2 + k^2 = 2(k + 1)x$ †
88. $kx^2 + (k + 3)x + 4 = 0$ †

89. If r and s are the roots of the quadratic equation $ax^2 + bx + c = 0$, show that

$$r + s = -\frac{b}{a} \quad \text{and} \quad r \cdot s = \frac{c}{a}.$$

90. Find the sum and the product of the roots of the following equations without solving the equations.
a. $3x^2 + 5x - 5 = 0$ †
b. $3x^2 - 7x = 1$ †
c. $\sqrt{3}x^2 = 3x + 4$ †
d. $(1 + \sqrt{2})x^2 - \sqrt{2}x + 5 = 0$ †

In Exercises 91 and 92, determine k so that the sum and the product of the roots are equal.

91. $2x^2 + (k - 3)x + 3k - 5 = 0$ 2
92. $5x^2 + (2k - 3)x - 2k + 3 = 0$ $(-\infty, \infty)$

93. Suppose r and s are the roots of the quadratic equation $ax^2 + bx + c = 0$. Show that you can write $ax^2 + bx + c = a(x - r)(x - s)$. Thus, the quadratic trinomial can be written in factored form.
[*Hint:* From Exercise 89, $b = -a(r + s)$ and $c = ars$. Substitute in $ax^2 + bx + c$ and factor.]

Answers:
63. $\left\{\dfrac{1 - \sqrt{22}}{3}, \dfrac{1 + \sqrt{22}}{3}\right\}$ **64.** $\left\{-\dfrac{5}{2}, \dfrac{2}{3}\right\}$ **65.** $\left\{-2, -\dfrac{2}{3}\right\}$
66. $\left\{\dfrac{4}{7} - \dfrac{\sqrt{5}}{7}i, \dfrac{4}{7} + \dfrac{\sqrt{5}}{7}i\right\}$ **67.** $\left\{\dfrac{5 - \sqrt{5}}{2}, \dfrac{5 + \sqrt{5}}{2}\right\}$
68. $\{-9 - \sqrt{14}, -9 + \sqrt{14}\}$ **69.** $\left\{\dfrac{1}{4} - \dfrac{\sqrt{7}}{4}i, \dfrac{1}{4} + \dfrac{\sqrt{7}}{4}i\right\}$
70. $\{2 - \sqrt{2}, 2 + \sqrt{2}\}$ **71.** $\left\{-\dfrac{\sqrt{10}}{2}, \dfrac{\sqrt{10}}{2}\right\}$ **73.** $\left\{-\dfrac{5}{3}i, \dfrac{5}{3}i\right\}$
74. $\left\{-\dfrac{2\sqrt{3}}{3}i, \dfrac{2\sqrt{3}}{3}i\right\}$ **75.** $D = 0$; real equal roots

94. Use Exercise 93 to factor each trinomial.
a. $4x^2 + 4x - 5$ $(2x + 1 + \sqrt{6})(2x + 1 - \sqrt{6})$
b. $25x^2 + 40x + 11$ $(5x + 4 + \sqrt{5})(5x + 4 - \sqrt{5})$
c. $25x^2 - 30x + 34$ $(5x - 3 + 5i)(5x + 3 - 5i)$
d. $72x^2 + 95x - 1000$ $(9x + 40)(8x - 25)$

B EXERCISES Applying the Concepts

95. Geometry. The length of a rectangle is 5 centimeters greater than its width. The area of the rectangle is 500 square centimeters. Find the dimensions of the rectangle. 20 cm by 25 cm

96. Geometry. The sides of a rectangle are in the ratio $3:2$. The area of the rectangle is 216 square centimeters. Find the dimensions of the rectangle. 12 cm by 18 cm

97. Cutting a wire. A length of wire 16 inches is to be cut into two pieces. Then each piece will be bent to form a square. Find the length of the two pieces if the sum of the areas of the two squares is 10 square inches. 4 in. and 12 in.

98. Cutting a wire. A piece of wire is 38 inches long. The wire is cut into two pieces. Then each piece is bent to form a square. Find the length of each piece if three times the area of the larger square exceeds the area of the smaller square by 95.75 square inches. 14 in. and 24 in.

99. Manufacturing. The surface area A of a cylinder with height h and radius r is given by the equation $A = 2\pi rh + 2\pi r^2$. A company makes soup cans by using 32π square inches of aluminum sheet for each can. If the height of the can is 6 inches, find the radius of the can. 2 in.

100. Making a box. A topless box is to be made from a square sheet of tin by cutting 5-inch squares from each of the four corners and then shaping the tin into an open box by turning up the sides. If the box is to hold 480 cubic inches, find the size of the piece of tin to be used. 19.8 in. by 19.8 in.

101. Making a box. A 2-inch square is cut from each corner of a rectangular piece of cardboard whose length exceeds the width by 4 inches. The sides are then turned up to form an open box. If the volume of the box is 64 cubic inches, find the dimensions of the box. 8 in. by 4 in. by 2 in.

102. Bus travel. Two buses leave Atlanta at the same time, one traveling west at 40 miles per hour and the other traveling north at 30 miles per hour. When will the buses be 400 miles apart? after 8 hours

103. Skydivers. Skydivers are in free fall from the time they jump out of a plane until they open their parachutes. A skydiver jumps from 5000 feet. The diver's height h above the ground t seconds after the jump is described by the equation $h = -16t^2 + 5000$. Find the time during which the diver is in free fall if the parachute opens at 1000 feet. 15.8 seconds

Solving Other Types of Equations

Objectives

1. Solve equations involving absolute value.
2. Solve equations by factoring.
3. Solve rational equations.
4. Solve equations involving radicals.
5. Solve equations that are quadratic in form.

---

1 Solve equations involving absolute value.

Equations Involving Absolute Value

Recall that geometrically, $|a|$ is the distance between the origin and a.

DEFINITION OF ABSOLUTE VALUE

$$|a| = a \text{ if } a \geq 0 \qquad \text{and} \qquad |a| = -a \text{ if } a < 0.$$

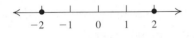

FIGURE A.16

Because the only two numbers on the number line that are exactly two units from the origin are 2 and -2, they are the only solutions of the equation $|x| = 2$. See Figure A.16.

EXAMPLE 1   Solving an Equation Involving Absolute Value

Solve the equation $|2x - 3| - 5 = 8$.

SOLUTION

First, isolate the absolute value expression to one side of the equation:

$$|2x - 3| = 13 \qquad \text{Add 5 to both sides.}$$

$2x - 3 = 13$	or	$2x - 3 = -13.$	Definition of absolute value
$2x = 16$		$2x = -10$	Add 3 to both sides of each equation.
$x = 8$		$x = -5$	Divide both sides by 2.

We leave it to you to check these solutions. The solution set is $\{-5, 8\}$. ∎ ∎ ∎

Practice Problem 1   Solve the equation $|6x - 3| - 8 = 1$. ∎

---

2 Solve equations by factoring.

Solving Equations by Factoring

The method of factoring used in solving quadratic equations can also be used to solve certain nonquadratic equations. Specifically, we use the zero-product property for solving those equations that can be expressed in factored form with 0 *on one side*.

EXAMPLE 2   Solving an Equation by Factoring

Solve by factoring: **a.** $x^4 = 9x^2$ **b.** $x^3 - 2x^2 = -x + 2$

SOLUTION

a. We move all terms to the left side, factor, and set each factor equal to zero.

$$x^4 = 9x^2 \qquad \text{Original equation}$$
$$x^4 - 9x^2 = 0 \qquad \text{Subtract } 9x^2 \text{ from both sides.}$$
$$x^2(x^2 - 9) = 0 \qquad x^4 - 9x^2 = x^2(x^2 - 9)$$
$$x^2(x + 3)(x - 3) = 0 \qquad x^2 - 9 = (x + 3)(x - 3)$$

$$x^2 = 0 \quad \text{or} \quad x + 3 = 0 \quad \text{or} \quad x - 3 = 0 \qquad \text{Zero-product property}$$
$$x = 0 \quad \text{or} \quad x = -3 \quad \text{or} \quad x = 3 \qquad \text{Solve each equation for } x.$$

> **RECALL**
>
> $A^2 - B^2 = (A + B)(A - B)$
>
> Difference of two squares

Check each possible solution.

Let $x = 0$	Let $x = -3$	Let $x = 3$
$(0)^4 \stackrel{?}{=} 9(0)^2$	$(-3)^4 \stackrel{?}{=} 9(-3)^2$	$(3)^4 \stackrel{?}{=} 9(3)^2$
$0 = 0 \checkmark$	$81 = 81 \checkmark$	$81 = 81 \checkmark$

b. Move all terms to the left side and factor by grouping.

$$x^3 - 2x^2 = -x + 2 \qquad \text{Original equation}$$
$$x^3 - 2x^2 + x - 2 = 0 \qquad \text{Add } x - 2 \text{ to both sides and simplify.}$$
$$(x^3 - 2x^2) + (x - 2) = 0 \qquad \text{Group terms.}$$
$$x^2(x - 2) + 1 \cdot (x - 2) = 0 \qquad \begin{array}{l}\text{Factor } x^2 \text{ from the first two terms and}\\ \text{rewrite } x - 2 \text{ as } 1 \cdot (x - 2).\end{array}$$
$$(x - 2)(x^2 + 1) = 0 \qquad \text{Factor out the common factor } (x - 2).$$

$$x - 2 = 0 \quad \bigg| \quad x^2 + 1 = 0 \qquad \text{Set each factor equal to zero.}$$
$$x = 2 \quad \bigg| \quad x^2 = -1$$
$$\bigg| \quad x = \pm i$$

Check each possible solution.

Let $x = 2$.	Let $x = i$.	Let $x = -i$.
$(2)^3 - 2(2)^2 \stackrel{?}{=} -2 + 2$	$(i)^3 - 2(i)^2 \stackrel{?}{=} -(i) + 2$	$(-i)^3 - 2(-i)^2 \stackrel{?}{=} -(-i) + 2$
$8 - 8 \stackrel{?}{=} 0$	$-i + 2 = -i + 2 \checkmark$	$i + 2 = i + 2 \checkmark$
$0 = 0 \checkmark$		

Thus, the solution set of the equation is $\{2, i, -i\}$.   ■ ■ ■

Practice Problem 2 Solve by factoring:

a. $x^4 = 4x^2$ **b.** $x^3 - 5x^2 = 4x - 20$ ■

3 Solve rational equations.

Rational Equations

If at least one algebraic expression with the variable in the denominator appears in an equation, the equation is a **rational equation**. When we multiply a rational equation by an expression containing the variable, we may introduce a solution that satisfies the new equation but does not satisfy the original equation. Such a solution is called an **extraneous solution** or **extraneous root**. See Exercises 47 and 48.

EXAMPLE 3	**Solving a Rational Equation**

Solve $\dfrac{1}{6} + \dfrac{1}{x + 1} = \dfrac{1}{x}$.

SOLUTION

The LCD from the denominators 6, $x + 1$, and x is $6x(x + 1)$. Multiply both sides of the equation by the LCD and make the right side 0.

$$6x(x + 1)\left[\frac{1}{6} + \frac{1}{1 + x}\right] = 6x(x + 1)\left[\frac{1}{x}\right] \qquad \text{Multiply both sides by the LCD.}$$

$$\frac{\cancel{6}x(x + 1)}{\cancel{6}} + \frac{6x\cancel{(x + 1)}}{\cancel{x + 1}} = \frac{6\cancel{x}(x + 1)}{\cancel{x}} \qquad \text{Distributive property}$$

$$x(x + 1) + 6x = 6(x + 1) \qquad \text{Simplify.}$$

$$x^2 + x + 6x = 6x + 6 \qquad \text{Distributive property}$$

$$x^2 + x - 6 = 0 \qquad \text{Subtract } 6x + 6 \text{ from both sides.}$$

$$(x + 3)(x - 2) = 0 \qquad \text{Factor.}$$

$$x + 3 = 0 \quad \text{or} \quad x - 2 = 0 \qquad \text{Set each factor equal to zero.}$$

$$x = -3 \quad \text{or} \quad x = 2 \qquad \text{Solve for } x.$$

Check the solutions, -3 and 2, in the original equation. The solution set is $\{-3, 2\}$.

■ ■ ■

Practice Problem 3 Solve $\dfrac{1}{x} - \dfrac{12}{5x + 10} = \dfrac{1}{5}$. ■

4 Solve equations involving radicals.

Equations Involving Radicals

If an equation involves radicals or rational exponents, the method of raising both sides to a positive integer power is often used to remove them. When this is done, the solutions of the new equation always contain the solutions of the original equation. However, in some cases, the new equation has *more* solutions than the original equation. For instance, consider the equation $x = 3$. If we square both sides, we get $x^2 = 9$. Notice that the given equation, $x = 3$, has only one solution, namely, 3, whereas the new equation, $x^2 = 9$, has two solutions: 3 and -3. Extraneous solutions may be introduced whenever we raise both sides of an equation to an *even* power. So *we* must *check all* solutions after doing so.

EXAMPLE 4 **Solving an Equation Involving a Radical**

Solve $\sqrt{2x + 1} + 1 = x$.

SOLUTION

First, we isolate the radical on one side of the equation.

$$\sqrt{2x + 1} = x - 1. \qquad \text{Subtract 1 from both sides.}$$

Next, we square both sides and simplify.

$$(\sqrt{2x + 1})^2 = (x - 1)^2$$

$$2x + 1 = x^2 - 2x + 1 \qquad (\sqrt{a})^2 = a$$

$$2x + 1 - 2x - 1 = x^2 - 2x + 1 - 2x - 1 \qquad \text{Subtract } 2x + 1 \text{ from both sides.}$$

$$0 = x^2 - 4x \qquad \text{Simplify.}$$

$$0 = x(x - 4) \qquad \text{Factor.}$$

$$x = 0 \quad \text{or} \quad x - 4 = 0 \qquad \text{Set each factor equal to zero.}$$

$$x = 0 \quad \text{or} \quad x = 4 \qquad \text{Solve for } x.$$

Check: $\sqrt{2(0) + 1} + 1 \stackrel{?}{=} 0$ | $\sqrt{2(4) + 1} + 1 \stackrel{?}{=} 4$
$\sqrt{1} + 1 \stackrel{?}{=} 0$ | $\sqrt{9} + 1 \stackrel{?}{=} 4$
$2 \neq 0$ | $4 = 4 \checkmark$

Because 0 is an extraneous root, the only solution is 4; the solution set is $\{4\}$.   ■ ■ ■

Practice Problem 4   Solve $\sqrt{6x + 4} + 2 = x$.   ■

EXAMPLE 5   **Solving an Equation Involving a Radical**

Solve $\sqrt[3]{2x + 1} + 4 = 3$.

SOLUTION

$$\sqrt[3]{2x + 1} = -1 \qquad \text{Subtract 4 from both sides.}$$
$$\left(\sqrt[3]{2x + 1}\right)^3 = (-1)^3 \qquad \text{Cube both sides.}$$
$$2x + 1 = -1 \qquad \left(\sqrt[3]{a}\right)^3 = a$$
$$2x = -2 \qquad \text{Isolate the } x \text{ term.}$$
$$x = -1 \qquad \text{Solve for } x.$$

You should always check your answers to avoid errors.   ■ ■ ■

Practice Problem 5   Solve $\sqrt[5]{3x - 7} = 2$.   ■

5 Solve equations that are quadratic in form.

Equations That Are Quadratic in Form

An equation in a variable x is **quadratic in form** if it can be written as

$$au^2 + bu + c = 0 \qquad (a \neq 0),$$

where u is an expression in the variable x. We solve the equation $au^2 + bu + c = 0$ for u. Then the solutions of the original equation can be obtained by replacing u with the expression in x that u represents.

EXAMPLE 6   **Solving an Equation That Is Quadratic in Form by Substitution**

Solve $(x^2 - 1)^2 - 6(x^2 - 1) - 16 = 0$.

SOLUTION

The equation becomes a quadratic equation if we replace $x^2 - 1$ with u.

$$(x^2 - 1)^2 - 6(x^2 - 1) - 16 = 0 \qquad \text{Original equation}$$
$$u^2 - 6u - 16 = 0 \qquad \text{Replace } x^2 - 1 \text{ with } u.$$
$$(u + 2)(u - 8) = 0 \qquad \text{Factor.}$$
$$u + 2 = 0 \quad \text{or} \quad u - 8 = 0 \qquad \text{Zero-product property}$$
$$u = -2 \quad \text{or} \quad u = 8 \qquad \text{Solve for } u.$$

Because $u = x^2 - 1$, we now find x from $u = -2$ and $u = 8$.

$$x^2 - 1 = -2 \quad | \quad x^2 - 1 = 8 \qquad \text{Replace } u \text{ with } x^2 - 1.$$
$$x^2 = -1 \quad | \quad x^2 = 9 \qquad \text{Add 1 to both sides of each equation.}$$
$$x = \pm i \quad | \quad x = \pm 3 \qquad \text{Solve for } x.$$

The equation has four apparent solutions: i, $-i$, 3, and -3.

Check: You should check that i, $-i$, 3, and -3 are all solutions of the equation.

Thus, $\{i, -i, 3, -3\}$ is the solution set of the original equation.   ■ ■ ■

Practice Problem 6   Solve $(x + 1)^2 - 3(x + 1) = 40$.   ■

SECTION A.9 ■ Exercises

A EXERCISES Basic Skills and Concepts

1. Both -7 and 7 are solutions of the equation $|x| = $ ____7____ .

2. We use the _____ to solve equations that can be expressed in factored form with 0 on one side. *zero-product property*

3. To solve an equation containing one radical expression, we first ____*isolate*____ the expression on ____*one*____ sides(s) of the equation.

4. The equation $3(x + 2)^2 - 9(x + 2) + 4 = 0$ becomes a quadratic equation in u if we replace ____$x + 2$____ with u in the equation.

In Exercises 5–28, solve each equation.

5. $|3x| = 9$ $\{-3, 3\}$
6. $|4x| = 24$ $\{-6, 6\}$

7. $|-2x| = 6$ $\{-3, 3\}$
8. $|-x| = 3$ $\{-3, 3\}$

9. $|x + 3| = 2$ $\{-5, -1\}$
10. $|x - 4| = 1$ $\{3, 5\}$

11. $|2x - 6| = 8$ $\{-1, 7\}$
12. $|3x - 6| = 9$ $\{-1, 5\}$

13. $|6x - 2| = 9$ †
14. $|6x - 3| = 9$ $\{-1, 2\}$

15. $|2x + 3| - 1 = 0$ $\{-2, -1\}$
16. $|2x - 3| - 1 = 0$ $\{1, 2\}$

17. $\frac{1}{2}|x| = 3$ $\{-6, 6\}$
18. $\frac{3}{5}|x| = 6$ $\{-10, 10\}$

19. $\left|\frac{1}{4}x + 2\right| = 3$ $\{-20, 4\}$
20. $\left|\frac{3}{2}x - 1\right| = 3$ $\left\{-\frac{4}{3}, \frac{8}{3}\right\}$

21. $6|2x - 1| - 8 = 10$ $\{-1, 2\}$
22. $5|4x - 1| + 10 = 15$ $\left\{0, \frac{1}{2}\right\}$

23. $2|3x - 4| - 7 = 7$ $\left\{-1, \frac{11}{3}\right\}$
24. $9|2x - 3| + 2 = -7$ $\varnothing$

25. $2|2x + 1| = -1$ $\varnothing$
26. $|3x + 7| = -2$ $\varnothing$

27. $|x^2 - 4| = 0$ $\{-2, 2\}$
28. $|9 - x^2| = 0$ $\{-3, 3\}$

In Exercises 29–38, find the real solutions of each equation by factoring.

29. $x^3 = 2x^2$ $\{0, 2\}$
30. $3x^4 - 27x^2 = 0$ $\{-3, 0.3\}$

31. $(\sqrt{x})^3 = \sqrt{x}$ $\{0, 1\}$
32. $(\sqrt{x})^5 = 16\sqrt{x}$ $\{0, 4\}$

33. $x + \sqrt[3]{x} = 0$ $\{0\}$
34. $x - \sqrt[3]{x} = 0$ $\{-1, 0, 1\}$

35. $x^4 - x^3 = x^2 - x$ $\{-1, 0, 1\}$
36. $x^3 - 36x = 16(x - 6)$ $\{-8, 2, 6\}$

37. $x^4 = 27x$ $\{0, 3\}$
38. $3x^4 = 24x$ $\{0, 2\}$

In Exercises 39–48, solve each equation by multiplying both sides by the LCD.

39. $\frac{x + 1}{3x - 2} = \frac{5x - 4}{3x + 2}$ $\left\{\frac{1}{4}, 2\right\}$
40. $\frac{x}{2x + 1} = \frac{3x + 2}{4x + 3}$ $\{-1\}$

41. $\frac{1}{x} + \frac{2}{x + 1} = 1$ †
42. $\frac{x}{x + 1} + \frac{x + 1}{x + 2} = 1$ †

43. $\frac{6x - 7}{x} - \frac{1}{x^2} = 5$ †
44. $\frac{1}{x} + \frac{2}{x + 1} + \frac{3}{x + 2} = 0$ †

45. $\frac{1}{x} + \frac{1}{x - 3} = \frac{7}{3x - 5}$ $\{-3, 5\}$
46. $\frac{x + 3}{x - 1} + \frac{x + 4}{x + 1} = \frac{8x + 5}{x^2 - 1}$ $\left\{-\frac{3}{2}, 2\right\}$

47. $\frac{1}{x - 4} - \frac{5}{x + 4} = \frac{8}{x^2 - 16}$ $\varnothing$

48. $\frac{x}{x - 3} - \frac{x - 4}{x + 2} = \frac{4x + 3}{x^2 - x - 6}$ $\varnothing$

In Exercises 49–58, solve each equation.

49. $x = \sqrt{6x - x^2}$ $\{0, 3\}$
50. $x + \sqrt{x + 6} = 0$ $\{-2\}$

51. $x - \sqrt{6x + 7} = 0$ $\{7\}$
52. $\sqrt{y + 6} = y$ $\{3\}$

53. $r + 11 = 6\sqrt{r + 3}$ $\{1, 13\}$
54. $\sqrt{6y - 11} = 2y - 7$ $\{6\}$

55. $\sqrt{3y + 1} = y - 1$ $\{5\}$
56. $t - \sqrt{3t + 6} = -2$ $\{-2, 1\}$

57. $\sqrt{5x^2 - 10x + 9} = 2x - 1$ $\{2, 4\}$

58. $x + \sqrt{x + 1} = 5$ $\{3\}$

In Exercises 59–68, solve each equation by using an appropriate substitution. Check your answers.

59. $x^{2/3} - 6x^{1/3} - 7 = 0$ †
60. $x^{2/3} - 3x^{1/3} + 2 = 0$ †

61. $(\sqrt{y} + 5)^2 - 9(\sqrt{y} + 5) + 20 = 0$ $\{0\}$

62. $(2\sqrt{t} + 1)^2 - 2(2\sqrt{t} + 1) - 3 = 0$ $\{1\}$

63. $(x^2 - 4)^2 - 3(x^2 - 4) - 4 = 0$ $\{-2\sqrt{2}, -\sqrt{3}, \sqrt{3}, 2\sqrt{2}\}$

64. $(x^2 + 2)^2 - 5(x^2 + 2) - 6 = 0$ $\{-2, 2, -i\sqrt{3}, i\sqrt{3}\}$

65. $(x^2 + 3)^2 + (x^2 + 3) - 6 = 0$ $\{-i\sqrt{6}, -i, i, i\sqrt{6}\}$

66. $(x^2 - 2)^2 + 7(x^2 - 2) + 10 = 0$ $\{-i\sqrt{3}, 0, i\sqrt{3}\}$

67. $(x^2 - 3x)^2 - 2(x^2 - 3x) - 8 = 0$ $\{-1, 1, 2, 4\}$

68. $(x^2 - 4x)^2 + 7(x^2 - 4x) + 12 = 0$ $\{1, 2, 3\}$

B EXERCISES Applying the Concepts

69. **Fractions.** The numerator of a fraction is two less than the denominator. The sum of the fraction and its reciprocal is $\frac{25}{12}$. Find the numerator and denominator if each is a positive integer. $\frac{6}{8}$

70. **Fractions.** The numerator of a fraction is five less than the denominator. The sum of the fraction and six times its reciprocal is $\frac{25}{2}$. Find the numerator and denominator if each is an integer. $\frac{5}{10}$

71. **Investment.** Latasha bought some stock for $1800. If the price of each share of the stock had been $18 less, she could have bought five more shares for the same $1800.
   a. How many shares of the stock did she buy? 20 shares
   b. How much did she pay for each share? $90

72. **Bus charter.** A civic club charters a bus for one day at a cost of $575. When two more people join the group, each person's cost decreases by $2. How many people were in the original group? 23 people

†Due to space constrictions, answers to these exercises may be found in the Answers beginning on page A–1 in the back of the book.

73. **Depth of a well.** A stone is dropped into a well, and the time it takes to hear the splash is 4 seconds. Find the depth of the well (to the nearest foot). Assume that the speed of sound is 1100 feet per second and that $d = 16t^2$ is the distance d an object falls in t seconds. 230 ft

74. **Rate of current.** A motorboat is capable of traveling at a speed of 10 miles per hour in still water. On a particular day, it took 30 minutes longer to travel a distance of 12 miles upstream than it took to travel the same distance downstream. What was the rate of the current in the stream on that day? 2 mph

75. **Train speed.** A freight train requires $2\frac{1}{2}$ hours longer to make a 300-mile journey than an express train does. If the express averages 20 miles per hour faster than the freight train, how long does it take the express train to make the trip? 5 hr

76. **Washing the family car.** A couple washed the family car in 24 minutes. Previously, when they each had washed the car alone, it took the husband 20 minutes longer to wash the car than it took the wife. How long did it take the wife to wash the car? 40 min

77. **Filling a swimming pool.** A small swimming pool can be filled by two hoses together in 1 hour and 12 minutes. The larger hose alone will fill the pool in 1 hour less than the smaller one. How long does it take the smaller hose to fill the pool? 3 hr

78. **The area of a rectangular plot.** The diagonal and the longer side of a rectangular plot together are three times the length of the shorter side. If the longer side exceeds the shorter side by 100 feet, what is the area of the plot? 120,000 ft²

Linear Inequalities

Objectives

1. Learn the vocabulary for discussing inequalities.
2. Solve and graph linear inequalities.
3. Solve and graph a combined inequality.
4. Solve and graph an inequality involving the reciprocal of a linear expression.
5. Solve inequalities involving absolute value.

---

1 Learn the vocabulary for discussing inequalities.

Inequalities

In general, an **inequality** is a statement that one algebraic expression is less than or is less than or equal to another algebraic expression. Some examples of inequalities are

$$4 < 5, \quad 3x + 2 \leq 14, \quad 5x + 7 > 3x + 23, \quad \text{and} \quad x^2 \geq 0.$$

The **domain** of a variable in an inequality is the set of all real numbers for which both sides of the inequality are defined. The real numbers that result in a true statement when those numbers are substituted for the variable in the inequality are called **solutions** of the inequality. Because replacing x with 1 in the inequality $3x + 2 \leq 14$ results in the true statement $3(1) + 2 \leq 14$, the number 1 is a solution of the inequality $3x + 2 \leq 14$. We also say that 1 *satisfies* the inequality $3x + 2 \leq 14$. To **solve** an inequality means to find all solutions of the inequality—that is, its solution set.

The fact that x^2 is never negative, or is **nonnegative**, is called the nonnegative identity.

THE NONNEGATIVE IDENTITY

$$x^2 \geq 0$$

for any real number x.

Two inequalities that have exactly the same solution set are called **equivalent inequalities**. The basic method of solving inequalities is similar to the method for solving equations: We replace a given inequality by a series of equivalent inequalities until we arrive at an equivalent inequality, such as $x < 5$, whose solution set we already know.

The following operations produce equivalent inequalities.

1. Simplifying one or both sides of an inequality by combining like terms and eliminating parentheses.
2. Adding or subtracting the same expression on both sides of the inequality.

Caution is required in multiplying or dividing both sides of an inequality by a real number or an expression representing a real number. Notice what happens when we multiply an inequality by -1.

We know that $2 < 3$, but how does $-2 = (-1)(2)$ compare with $-3 = (-1)(3)$? We have $-2 > -3$.

If we multiply (or divide) both sides of the inequality $2 < 3$ by -1, to get a correct result, we have to exchange the $<$ symbol for the $>$ symbol. This exchange of symbols is called reversing the **sense** or the **direction** of the inequality.

The following chart describes the way multiplication and division affect inequalities.

If C is a nonzero real number, then the following inequalities are all equivalent.

Sign of C	Inequality	Sense	Example
	$A < B$	$<$	$3x < 12$
C positive	$A \cdot C < B \cdot C$	Unchanged	$\frac{1}{3}(3x) < \frac{1}{3}(12)$
C positive	$\dfrac{A}{C} < \dfrac{B}{C}$	Unchanged	$\dfrac{3x}{3} < \dfrac{12}{3}$
C negative	$A \cdot C > B \cdot C$	Reversed	$-\frac{1}{3}(3x) > -\frac{1}{3}(12)$
C negative	$\dfrac{A}{C} > \dfrac{B}{C}$	Reversed	$\dfrac{3x}{-3} > \dfrac{12}{-3}$

Similar results apply when $<$ is replaced throughout with $\leq$, $>$, or $\geq$.

---

2 Solve and graph linear inequalities.

Linear Inequalities

A **linear inequality in one variable** is an inequality that is equivalent to one of the forms

$$ax + b < 0 \quad \text{or} \quad ax + b \leq 0,$$

where a and b represent real numbers and $a \neq 0$.

Inequalities such as $2x - 1 > 0$ and $2x - 1 \geq 0$ are linear inequalities because they are equivalent to $-2x + 1 < 0$ and $-2x + 1 \leq 0$, respectively.

STUDY TIP

A linear inequality becomes a linear equation when the inequality symbol is replaced with the equal sign ($=$).

EXAMPLE 1   **Solving and Graphing Linear Inequalities**

Solve the inequality $7x - 11 \leq 2(x - 3)$ and graph its solution set.

SOLUTION

$7x - 11 \leq 2(x - 3)$	Original inequality
$7x - 11 \leq 2x - 6$	Distributive property
$7x - 11 + 11 \leq 2x - 6 + 11$	Add 11 to both sides.
$7x \leq 2x + 5$	Simplify.
$7x - 2x \leq 2x + 5 - 2x$	Subtract $2x$ from both sides.
$5x \leq 5$	Simplify.
$\dfrac{5x}{5} \leq \dfrac{5}{5}$	Divide both sides by 5.
$x \leq 1$	Simplify.

The solution set is $\{x \mid x \leq 1\}$, or in interval notation, $(-\infty, 1]$. The graph is shown in Figure A.17. ■ ■ ■

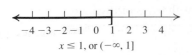

$x \leq 1$, or $(-\infty, 1]$

FIGURE A.17

Practice Problem 1   Solve the inequality $4x + 9 > 2(x + 6) + 1$ and graph its solution set. ■

3 Solve and graph a combined inequality.

Combining Two Inequalities

An inequality such as $-5 < 2x + 3 \leq 9$ is shorthand for $-5 < 2x + 3$ *and* $2x + 3 \leq 9$ and is called a **compound inequality**. Solving such inequalities requires no new principles.

EXAMPLE 2   **Solving and Graphing a Compound Inequality**

Solve the inequality $-5 < 2x + 3 \leq 9$ and graph its solution set.

SOLUTION

We must find all real numbers that are solutions of *both* inequalities

$$-5 < 2x + 3 \text{ and } 2x + 3 \leq 9.$$

Let's see what happens if we solve these inequalities separately.

$-5 < 2x + 3$	$2x + 3 \leq 9$	Original inequalities
$-5 - 3 < 2x + 3 - 3$	$2x + 3 - 3 \leq 9 - 3$	Subtract 3 from both sides.
$-8 < 2x$	$2x \leq 6$	Simplify.
$\dfrac{-8}{2} < \dfrac{2x}{2}$	$\dfrac{2x}{2} \leq \dfrac{6}{2}$	Divide both sides by 2.
$-4 < x$	$x \leq 3$	Simplify.

The solution of the original pair of inequalities consists of all real numbers x such that $-4 < x$ *and* $x \leq 3$. We can also write this solution as $\{x | -4 < x \leq 3\}$ or as $(-4, 3]$. The graph is shown in Figure A.18.

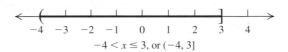

$-4 < x \leq 3$, or $(-4, 3]$

FIGURE A.18

Notice that we did the same thing to both inequalities in each step of the solution process. We can solve the compound inequality more efficiently by working on both inequalities simultaneously as follows:

$-5 < 2x + 3 \leq 9$	Original inequality
$-5 - 3 < 2x + 3 - 3 \leq 9 - 3$	Subtract 3 from each part.
$-8 < 2x \leq 6$	Simplify each part.
$\dfrac{-8}{2} < \dfrac{2x}{2} \leq \dfrac{6}{2}$	Divide each part by 2.
$-4 < x \leq 3.$	Simplify each part.

The solution set is $\{x | -4 < x \leq 3\}$, exactly the solution set we obtained previously.

■ ■ ■

Practice Problem 2   Solve and graph $-6 \leq 4x - 2 < 4$.   ■

4 Solve and graph an inequality involving the reciprocal of a linear expression.

Inequalities Involving the Reciprocal of a Linear Expression

To solve nonlinear inequalities, we need to be able to exchange certain nonlinear inequalities for equivalent linear inequalities. The **Reciprocal Sign Property** says that any number and its reciprocal have the same sign.

THE RECIPROCAL SIGN PROPERTY

If $x \neq 0$, then x and $\dfrac{1}{x}$ are both positive or are both negative. In symbols,

if $x > 0$, then $\dfrac{1}{x} > 0$ and if $x < 0$, then $\dfrac{1}{x} < 0$.

EXAMPLE 3   **Solving and Graphing an Inequality by Using the Reciprocal Sign Property**

Solve and graph $(3x - 12)^{-1} > 0$.

SOLUTION

$$(3x - 12)^{-1} > 0 \qquad \text{Original inequality}$$

$$\frac{1}{3x - 12} > 0 \qquad a^{-1} = \frac{1}{a}; \text{ here } a = 3x - 12.$$

$$3x - 12 > 0 \qquad \text{Reciprocal Sign Property}$$

$$3x > 12 \qquad \text{Add 12 to both sides.}$$

$$x > 4. \qquad \text{Divide both sides by 3.}$$

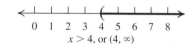

$x > 4$, or $(4, \infty)$

FIGURE A.19

The solution set is $\{x \mid x > 4\}$, or in interval notation, $(4, \infty)$. The graph is shown in Figure A.19.   ■ ■ ■

Practice Problem 3   Solve and graph $(2x - 8)^{-1} \geq 0$.   ■

5 Solve inequalities involving absolute value.

Inequalities Involving Absolute Value

If $|x| < 2$, then x is closer than two units to the origin. See Figure A.20(a). If $|x| > 2$, then x is farther than two units from the origin and is in either the interval $(-\infty, -2)$ or the interval $(2, \infty)$, that is, either $x < -2$ or $x > 2$. See Figure A.20(b).

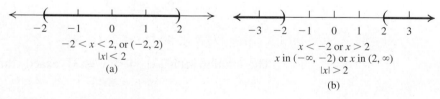

$-2 < x < 2$, or $(-2, 2)$
$|x| < 2$
(a)

$x < -2$ or $x > 2$
x in $(-\infty, -2)$ or x in $(2, \infty)$
$|x| > 2$
(b)

FIGURE A.20

This discussion suggests the following rules.

RULES FOR SOLVING ABSOLUTE VALUE INEQUALITIES

If $a > 0$ and u is an algebraic expression, then

1. $|u| < a$ is equivalent to $-a < u < a$.
2. $|u| \le a$ is equivalent to $-a \le u \le a$.
3. $|u| > a$ is equivalent to $u < -a$ or $u > a$.
4. $|u| \ge a$ is equivalent to $u \le -a$ or $u \ge a$.

EXAMPLE 4 **Solving an Inequality Involving Absolute Value**

Solve the inequality $|4x - 1| \le 9$ and graph the solution set.

SOLUTION

Rule 2 applies here, with $u = 4x - 1$ and $a = 9$.

$$|4x - 1| \le 9 \text{ is equivalent to}$$

$-9 \le 4x - 1 \le 9$	Rule 2, $-a \le u \le a$
$1 - 9 \le 4x - 1 + 1 \le 9 + 1$	Add 1 to each part.
$-8 \le 4x \le 10$	Simplify.
$-\dfrac{8}{4} \le \dfrac{4x}{4} \le \dfrac{10}{4}$	Divide by 4: the sense of the inequality is unchanged.
$-2 \le x \le \dfrac{5}{2}$	Simplify.

The solution set is $\left\{ x \middle| -2 \le x \le \dfrac{5}{2} \right\}$; that is, the solution set is the closed interval $\left[-2, \dfrac{5}{2} \right]$. See Figure A.21.   ■ ■ ■

Practice Problem 4 Solve $|3x + 3| \le 6$ and graph the solution set.   ■

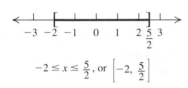

$-2 \le x \le \dfrac{5}{2}$, or $\left[-2, \dfrac{5}{2} \right]$

FIGURE A.21

EXAMPLE 5 **Solving an Inequality Involving Absolute Value**

Solve the inequality $|2x - 8| \ge 4$ and graph the solution set.

SOLUTION

$$|2x - 8| \ge 4 \text{ is equivalent to}$$

$2x - 8 \le -4$	or	$2x - 8 \ge 4.$	Rule 4, with $u = 2x - 8$ and $a = 4$.
$2x - 8 + 8 \le -4 + 8$		$2x - 8 + 8 \ge 4 + 8$	Add 8 to each part.
$2x \le 4$		$2x \ge 12$	Simplify.
$\dfrac{2x}{2} \le \dfrac{4}{2}$		$\dfrac{2x}{2} \ge \dfrac{12}{2}$	Divide both sides of each part by 2.
$x \le 2$		$x \ge 6$	Simplify.

The solution set is $\{x | x \le 2 \text{ or } x \ge 6\}$; that is, the solution set is the set of all real numbers x in either of the intervals $(-\infty, 2]$ or $[6, \infty)$. This set can also be written as $(-\infty, 2] \cup [6, \infty)$. See Figure A.22.   ■ ■ ■

Practice Problem 5 Solve $|2x + 3| \ge 6$ and graph the solution set.   ■

$x \le 2 \text{ or } x \ge 6$
$(-\infty, 2] \cup [6, \infty)$

FIGURE A.22

SECTION A.10 ■ Exercises

A EXERCISES Basic Skills and Concepts

1. The identity $x^2 \geq 0$ is called the <u>nonnegative identity</u>

2. Two inequalities that have exactly the same solution set are called <u>equivalent</u>.

3. When an inequality is multiplied or divided by a negative number, the sense of the resulting inequality is <u>reversed</u>.

4. If $x \neq 0$, then both x and $\dfrac{1}{x}$ are either <u>positive</u> or <u>negative</u>.

In Exercises 5–14, fill in the blank with the correct inequality symbol, using the rules for producing equivalent inequalities.

5. If $x < 8$, then $x - 8$ <u>$<$</u> 0.

6. If $x \leq -3$, then $x + 3$ <u>$\leq$</u> 0.

7. If $x \geq 5$, then $x - 5$ <u>$\geq$</u> 0.

8. If $x > 7$, then $x - 7$ <u>$>$</u> 0.

9. If $\dfrac{x}{2} < 6$, then x <u>$<$</u> 12.

10. If $\dfrac{x}{3} < 2$, then x <u>$<$</u> 6.

11. If $-2x \leq 4$, then x <u>$\geq$</u> -2.

12. If $-3x > 12$, then x <u>$<$</u> -4.

13. If $x + 3 < 2$, then x <u>$<$</u> -1.

14. If $x - 5 \leq 3$, then x <u>$\leq$</u> 8.

In Exercises 15–22, graph the solution set of each inequality and write it in interval notation.

15. $-2 < x < 5$ †

16. $-5 \leq x \leq 0$ †

17. $0 < x \leq 4$ †

18. $1 \leq x < 7$ †

19. $x \geq -1$ †

20. $x > 2$ †

21. $-5x \geq 10$ †

22. $-2x < 2$ †

In Exercises 23–42, solve each inequality. Write the solution in interval notation and graph the solution set.

23. $x + 3 < 6$ †

24. $x - 2 < 3$ †

25. $1 - x \leq 4$ †

26. $7 - x > 3$ †

27. $2x + 5 < 9$ †

28. $3x + 2 \geq 7$ †

29. $3 - 3x > 15$ †

30. $8 - 4x \geq 12$ †

31. $3(x + 2) < 2x + 5$ †

32. $4(x - 1) \geq 3x - 1$ †

33. $3(x - 3) \leq 3 - x$ †

34. $-x - 2 \geq x - 10$ †

35. $6x + 4 > 3x + 10$ †

36. $4(x - 4) > 3(x - 5)$ †

37. $8(x - 1) \leq 7x - 12$ †

38. $3(x + 2) \geq 5x + 18$ †

39. $5(x + 2) \leq 3(x + 1) + 10$ †

40. $x - 4 > 2(x + 8)$ †

41. $9 - 5x \leq 6 - 8x$ †

42. $3 - 2x \leq -7 + 3x$ †

In Exercises 43–54, solve each combined inequality.

43. $3 < x + 5 < 4$ $(-2, -1)$ 44. $9 \leq x + 7 \leq 12$ $[2, 5]$

45. $-4 \leq x - 2 < 2$ $[-2, 4)$ 46. $-3 < x + 5 < 4$ $(-8, -1)$

47. $-9 \leq 2x + 3 \leq 5$ $[-6, 1]$ 48. $-2 \leq 3x + 1 \leq 7$ $[-1, 2]$

49. $0 \leq 1 - \dfrac{x}{3} < 2$ $(-3, 3]$ 50. $0 < 5 - \dfrac{x}{2} \leq 3$ $[4, 10)$

51. $-1 < \dfrac{2x - 3}{5} \leq 0$ $\left(-1, \dfrac{3}{2}\right]$

52. $-4 \leq \dfrac{5x - 2}{3} < 0$ $\left[-2, \dfrac{2}{5}\right)$

53. $5x \leq 3x + 1 < 4x + 2$ $\left(-1, \dfrac{1}{2}\right]$

54. $3x + 2 < 2x + 3 < 4x - 1$ $\varnothing$

In Exercises 55–58, solve each reciprocal inequality.

55. $(3x + 6)^{-1} < 0$ $(-\infty, -2)$ 56. $(2x - 8)^{-1} < 0$ $(-\infty, 4)$

57. $(2 - 4x)^{-1} > 0$ $\left(-\infty, \dfrac{1}{2}\right)$ 58. $(10 - 5x)^{-1} > 0$ $(-\infty, 2)$

In Exercises 59–74, solve each inequality.

59. $|3x| < 12$ $(-4, 4)$ 60. $|2x| \leq 6$ $[-3, 3]$

61. $|4x| > 16$ 62. $|3x| > 15$

63. $|x + 1| < 3$ $(-4, 2)$ 64. $|x - 4| < 1$ $(3, 5)$

65. $|x| - 2 \geq 1$ 66. $|x| + 2 > 7$ †

67. $|2x - 3| < 4$ † 68. $|4x - 6| \leq 6$ $[0, 3]$

69. $|2x - 5| > 3$ † 70. $|3x - 3| \geq 15$ †

71. $|3x + 4| \leq 19$ † 72. $|9 - 7x| < 23$ †

73. $|2x - 15| < 0$ $\varnothing$ 74. $|x + 5| \leq -3$ $\varnothing$

In Exercises 75–84, solve each inequality and graph the solution set.

75. $\left|\dfrac{x}{2} - 1\right| < 4$ † 76. $\left|2 - \dfrac{x}{2}\right| > 1$ †

77. $\left|\dfrac{x - 8}{2}\right| \leq 4$ † 78. $\left|\dfrac{2x + 15}{8}\right| > 4$ †

79. $\left|\dfrac{3}{8}x - 105\right| \geq 0$ † 80. $|27x - 43| \geq 0$ †

81. $\left|3x - \dfrac{7}{3}\right| - 2 < 3$ † 82. $\left|\dfrac{x}{2} - 1\right| - 1 \leq \dfrac{1}{2}$ †

83. $\left|\dfrac{3}{5}x - 2\right| \leq 4$ † 84. $\left|\dfrac{x}{2} - \dfrac{1}{3}\right| > \dfrac{2}{3}$ †

†Due to space constrictions, answers to these exercises may be found in the Answers beginning on page A–1 in the back of the book.
Answers:
61. $(-\infty, -4) \cup (4, \infty)$ **62.** $(-\infty, -5) \cup (5, \infty)$ **65.** $(-\infty, -3] \cup [3, \infty)$

B EXERCISES Applying the Concepts

85. Poster sales. A student club wants to produce a poster as a fund-raising device. The production cost is $1.80 per poster. There are additional expenses of $1200, independent of production or sales. How many posters must be sold at $2 each for the club to make a profit?
more than 6000 posters

86. Telemarketing. A telemarketer is paid $25 plus 35% of the difference between the item's selling price and the item's cost. All items sell for at least $50 above cost and at most $125 above cost. For each sale made, over what range can a telemarketer's commission vary? $42.50 to $68.75

87. Appliance markup. The markup over the dealer's cost on a new refrigerator ranges from 15% to 20%. If the dealer's cost is $1750, over what range will the selling price vary? $2012.50 to $2100.00

88. Return on investment. An investor has $5000 to invest for a period of 1 year. Find the range of per annum simple interest rates required to generate interest between $200 and $275 inclusive. 4% to 5.5%

89. Hybrid car trip. Sometime after passing a truck stop 300 miles from the start of her trip, Cora's hybrid car ran out of gas. If the tank could hold 12 gallons of gasoline and the hybrid car averaged 40 miles per gallon, find the range of gasoline (in gallons) that could have been in the tank at the start of the trip. 7.5 gal to 12 gal

90. Average grade. Sean has taken three exams and earned scores of 85, 72, and 77 out of a possible 100 points. He needs an average of at least 80 to earn a B in the course. What range of scores on the fourth (and last) 100-point test will guarantee a B in the course? 86 to 100

91. Butterfat content. How much cream that is 30% butterfat must be added to milk that is 3% butterfat to have 270 quarts that are at least 4.5% butterfat? at least 15 qt

92. Amplifier cost. The cost of an amplifier to a retailer is $340. At what price can the amplifier be sold if the retailer wants to make a profit of at least 20% of the selling price? at least $408

93. Pedometer cost. A company produces a pedometer at a cost of $3 each and sells the pedometer for $5 each. If the company has to recover an initial expense of $4000 before any profit is earned, how many pedometers must be sold to earn a profit in excess of $3000? more than 3500 pedometers

94. Car sales. A car dealer has three times as many SUVs and twice as many convertibles as four-door sedans. How many four-door sedans could the dealer have if she has at least 48 cars of these three types? at least 8

95. Temperature conversion. The formula for converting Fahrenheit temperature F to Celsius temperature C is $C = \dfrac{5}{9}(F - 32)$. What range in Celsius degrees corresponds to a range of 68° to 86° Fahrenheit?
20°C to 30°C

96. Parking expense. The parking cost at the local airport (in dollars) is $C = 2 + 1.75(h - 1)$, where h is the number of hours the car is parked. For what range of hours is the parking cost between $37 and $51? 21 to 29 hr

97. Car rental. Suppose a rental car costs $18.00 per day plus $0.25 per mile for each mile driven. What range of miles can be driven to stay within a $30-per-day car rental allowance? 0 to 48 mi

98. Plumbing charges. A plumber charges $42 per hour. A repair estimate includes a fixed cost of $147 for parts and a labor cost that is determined by how long it takes to complete the job. If the total estimate is for at least $210 and at most $294, what interval was estimated for the job? 1.5 to 3.5 hr

Polynomial and Rational Inequalities

Polynomials have a property that allows you to test *any point* in any interval of numbers that does *not* contain a root of the polynomial, and if the value of the polynomial at that point is positive, the value at *every* point in the interval is also positive. Similarly, if the value at the point tested is negative, the value of the polynomial at *every* point in the interval is also negative. We will learn how to use this property to solve inequalities involving polynomials.

◆ **WARNING**

Before attempting to solve any polynomial or rational inequality, rearrange the inequality so that a polynomial or rational expression is on the left-hand side of the inequality symbol and 0 is on the right-hand side.

1 Solve quadratic inequalities.

Using Test Points to Solve Inequalities

The **test-point** method described next works for all quadratic polynomials. We consider quadratic polynomials that have two distinct real roots. In this case, we find the two roots (using the quadratic formula if necessary) and then make sign graphs of the quadratic polynomial. The two roots divide the number line into three intervals, and the sign of the quadratic polynomial is determined in each interval by testing *any* number in the interval.

EXAMPLE 1   **Using the Test-Point Method to Solve a Quadratic Inequality**

Solve $x^2 > 2x + 7$. Write the solution in interval notation and graph the solution set.

SOLUTION

We first rearrange the inequality so that 0 is on the right side.

$x^2 > 2x + 7$	Original inequality
$x^2 - 2x - 7 > 2x + 7 - 2x - 7$	Subtract $2x + 7$ from both sides; remember to change the sign of both terms.
$x^2 - 2x - 7 > 0$	Simplify.

To determine where $x^2 - 2x - 7$ is positive (> 0), we first solve the associated equation.

$$1 \cdot x^2 - 2x - 7 = 0$$

Replace $>$ with $=$ in $x^2 - 2x - 7 > 0$.

$$x = \frac{-b \pm \sqrt{b^2 - 4ac}}{2a}$$

Use the quadratic formula to solve $x^2 - 2x - 7 = 0$.

$$= \frac{-(-2) \pm \sqrt{(-2)^2 - 4(1)(-7)}}{2(1)}$$

Replace a with 1, b with -2, and c with -7 in the quadratic formula.

$$= \frac{2 \pm \sqrt{32}}{2} = 1 \pm 2\sqrt{2}$$

Simplify.

The two roots for $x^2 - 2x - 7 = 0$ are $1 - 2\sqrt{2} \approx -1.8$ and $1 + 2\sqrt{2} \approx 3.8$. These roots divide the number line into three intervals, as shown in Figure A.23. We select a convenient "test point" in each of the three intervals $(-\infty, 1 - 2\sqrt{2})$, $(1 - 2\sqrt{2}, 1 + 2\sqrt{2})$, and $(1 + 2\sqrt{2}, \infty)$. The points we selected, namely, $-3, 0$, and 4, are shown on the number line, but any other three points from each respective interval would work as well. Now evaluate $x^2 - 2x - 7$ for each test point.

Test Interval	Test Point	Value of $x^2 - 2x - 7$	Result
$(-\infty, 1 - 2\sqrt{2})$	-3	$(-3)^2 - 2(-3) - 7 = 8$	Positive
$(1 - 2\sqrt{2}, 1 + 2\sqrt{2})$	0	$(0)^2 - 2(0) - 7 = -7$	Negative
$(1 + 2\sqrt{2}, \infty)$	4	$(4)^2 - 2(4) - 7 = 1$	Positive

The sign of $x^2 - 2x - 7$ at each test point gives the sign of $x^2 - 2x - 7$ for every point in the interval containing that test point. These results are shown in the sign graph in Figure A.23.

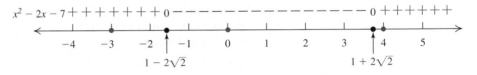

FIGURE A.23

To find where $x^2 - 2x - 7 > 0$, we locate the "+" signs on the sign graph. Because the "+" signs are to the left of $1 - 2\sqrt{2}$, we include the interval $(-\infty, 1 - 2\sqrt{2})$ in the solution set. Because the "+" signs are to the right of $1 - 2\sqrt{2}$ as well, we also include the interval $(1 + 2\sqrt{2}, \infty)$ in the solution set. The solution set consists of the numbers that are in $(-\infty, 1 - 2\sqrt{2})$ or $(1 + 2\sqrt{2}, \infty)$. This can also be written as $(-\infty, 1 - 2\sqrt{2}) \cup (1 + 2\sqrt{2}, \infty)$. The graph of the inequality is shown in Figure A.24.   ■ ■ ■

FIGURE A.24

Practice Problem 1   Solve $x^2 > 2x + 2$. Write the solution in interval notation and graph the solution set.   ■

Once we rearrange a quadratic inequality so that 0 is on the right side, we set the left side equal to 0 and solve the resulting equation. What if the resulting equation has only one real solution or no real solution?

If the resulting equation has only one real solution, this root divides the number line into two sections, and the sign graph will consist of "0" at the one solution and either all "+" signs or all "−" signs everywhere else. This sign graph can then be used to solve the inequality.

If the resulting equation has no real solution, we use the following result, which is true for all polynomial equations (not just quadratic equations).

ONE-SIGN THEOREM

If a polynomial equation has no real solution, then the polynomial is always positive or always negative.

EXAMPLE 2   **Using the One-Sign Theorem to Solve a Quadratic Inequality**

Solve $x^2 - 2x + 2 > 0$.

SOLUTION

This quadratic inequality already has 0 on the right side, so we set the left side equal to 0 to obtain the associated equation. Because there are no obvious factors, we evaluate the discriminant to see if there are any real roots.

$$x^2 - 2x + 2 = 1 \cdot x^2 - 2x + 2 = 0 \qquad a = 1, b = -2, c = 2$$
$$b^2 - 4ac = (-2)^2 - 4(1)(2) \qquad \text{Evaluate the discriminant.}$$
$$= -4 \qquad \text{The discriminant is negative.}$$

Because the discriminant, -4, is negative, $x^2 - 2x + 2 = 0$ has no real roots and $x^2 - 2x + 2$ is always positive (> 0) or always negative (< 0). To find out which, we can use any value of x as a test point. Calculations with $x = 0$ are simple, so we pick 0 as a test point.

Let $x = 0$. Then $(0)^2 - 2(0) + 2 = 2$, which is positive; so $x^2 - 2x + 2$ is *always* positive. The solution set is the set of all real numbers, which is written in interval notation as $(-\infty, \infty)$.   ■ ■ ■

Practice Problem 2   Solve $x^2 + 2x - 2 > 0$.   ■

2 Solve polynomial inequalities.

Other Polynomial Inequalities

Regardless of the degree of the polynomial in an inequality, if we can factor the polynomial, we can use test points and a sign graph to solve it.

EXAMPLE 3   **Solving a Polynomial Inequality**

Solve $x^4 \le 1$. Write the solution in interval notation and graph the solution set.

SOLUTION

We arrange the inequality so that 0 is on the right-hand side.

$$x^4 \le 1 \qquad \text{Original inequality}$$
$$x^4 - 1 \le 0 \qquad \text{Subtract 1 from both sides.}$$

Now we factor the left side.

$$x^4 - 1 \le 0$$
$$(x^2)^2 - 1 \le 0 \qquad \text{Rewrite } x^4 \text{ as } (x^2)^2.$$
$$(x^2 - 1)(x^2 + 1) \le 0 \qquad \text{Factor the difference of squares.}$$
$$(x - 1)(x + 1)(x^2 + 1) \le 0 \qquad \text{Factor } x^2 - 1 = (x - 1)(x + 1).$$

Next, we find the real solutions of the associated equation.

$$(x - 1)(x + 1)(x^2 + 1) = 0 \qquad \begin{array}{l} \text{Replace } \le \text{ with } = \text{ in} \\ (x - 1)(x + 1)(x^2 + 1) \le 0. \end{array}$$

$$x - 1 = 0, \quad \text{or} \quad x + 1 = 0 \quad \text{or} \quad x^2 + 1 = 0 \qquad \text{Zero-product property}$$

Thus, $x = 1$ or $x = -1$. (The equation $x^2 + 1 = 0$ has no real solution.)

The solutions -1 and 1 of $(x - 1)(x + 1)(x^2 + 1) = 0$ divide the number line into three intervals: $(-\infty, -1), (-1, 1),$ and $(1, \infty)$. See Figure A.25.

We select convenient test points in each of these intervals. We pick $-2, 0,$ and 2. Next, we compute the value of $(x - 1)(x + 1)(x^2 + 1)$ at each test point to determine the sign at each point.

Interval	Test Point	Value of $(x + 1)(x + 1)(x^2 + 1)$	Sign
$(-\infty, -1)$	-2	$(-2 - 1)(-2 + 1)((-2)^2 + 1) = 15$	$+$
$(-1, 1)$	0	$(0 - 1)(0 + 1)((0)^2 + 1) = -1$	$-$
$(1, \infty)$	2	$(2 - 1)(2 + 1)((2)^2 + 1) = 15$	$+$

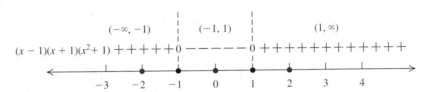

FIGURE A.25

We see from Figure A.25 that the solution set consists of all x between -1 and 1, including both -1 and 1. The solution set is shown in Figure A.26. ■ ■ ■

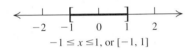

$-1 \le x \le 1$, or $[-1, 1]$

FIGURE A.26

Practice Problem 3 Solve $x^4 \le 3x^2 + 4$. Write the solution in interval notation and graph the solution set. ■

3 Solve rational inequalities.

Rational Inequalities

An inequality involving a rational expression rather than a polynomial is called a **rational inequality**. The additional consideration in a rational inequality is the polynomial in the denominator. As with polynomial inequalities, our first step is to rearrange the inequality (if necessary) so that the right side is 0.

EXAMPLE 4 **Solving a Rational Inequality**

Solve $\dfrac{3}{x - 1} \ge 1$. Write the solution in interval notation and graph the solution set.

SOLUTION

We first rewrite $\dfrac{3}{x - 1} \ge 1$ to get 0 on the right-hand side.

$$\frac{3}{x - 1} \ge 1 \qquad \text{Original inequality}$$

$$\frac{3}{x - 1} - 1 \ge 0 \qquad \text{Subtract 1 from both sides.}$$

$$\frac{3 - (x - 1)}{x - 1} \ge 0 \qquad \text{Use } x - 1 \text{ as a common denominator.}$$

$$\frac{4 - x}{x - 1} \ge 0 \qquad \text{Simplify the numerator.}$$

STUDY TIP

Values of the variable that make the denominator in a rational expression in an inequality equal to 0 cannot be solutions of the given inequality because division by 0 is undefined.

Solve the following, which can often be done mentally.

$$\text{numerator} = 0 \qquad \text{and} \qquad \text{denominator} = 0$$
$$4 - x = 0 \qquad\qquad\qquad x - 1 = 0$$
$$4 = x \qquad\qquad\qquad\qquad x = 1$$

The numbers 1 and 4 divide the number line into three intervals: $(-\infty, 1)$, $(1, 4)$, and $(4, \infty)$. We choose 0, 2, and 5 as convenient test points in these intervals. The test points are shown on the sign graph in Figure A.27.

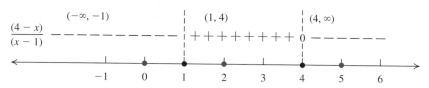

FIGURE A.27

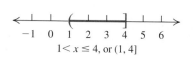

FIGURE A.28

From the sign graph in Figure A.27, we determine the real numbers for which $\dfrac{4 - x}{x - 1}$ is positive (> 0); these are the numbers in the interval $(1, 4)$, as the "+"signs indicate. However, we want to know where $\dfrac{4 - x}{x - 1}$ is *either* positive *or* 0. The expression $\dfrac{4 - x}{x - 1}$ is undefined for $x = 1$ and is 0 only if $x = 4$. The solution set is then $\{x \,|\, 1 < x \le 4\}$, or in interval notation, $(1, 4]$. The graph is shown in Figure A.28. ■ ■ ■

Practice Problem 4   Solve $\dfrac{2}{1 - 3x} > 1$. Write the solution in interval notation and graph the solution set. ■

◆ **WARNING**   You cannot solve a rational inequality by multiplying both sides by the LCD, as you would with a rational equation. Remember that you reverse the sense of an inequality when multiplying by a negative expression. However, when an expression contains a variable, you do not know whether the expression is positive or negative.

SECTION A.11 ■ Exercises

A EXERCISES Basic Skills and Concepts

1. A test point determines the ___sign___ of the polynomial at each point in the interval that contains the test point.

2. The statement that a polynomial with no real solution is always positive or always negative is called the ___one-sign___ theorem.

3. The inequality $3x^2 \le 4 - 2x$ is equivalent to the inequality ___$3x^3 + 2x - 4$___ ≤ 0.

4. You cannot solve a rational inequality by multiplying both sides by the ___LCD___.

In Exercises 5–46, solve each inequality by the test-point method. Write the solution in interval notation.

5. $(x - 3)(x + 2) < 0$ $(-2,3)$ **6.** $(x - 2)(x + 3) < 0$ $(-3,2)$

7. $(x + 5)(x - 2) < 0$ $(-5,2)$ **8.** $(x - 5)(x + 2) > 0$ $(-\infty,-2) \cup (5, \infty)$

9. $(x + 4)(x - 1) \le 0$ $[-4,1]$ **10.** $(x - 4)(x + 1) \le 0$ $[-1,4]$

11. $(x - 6)(x + 1) \ge 0$ $(-\infty,-1] \cup [6, \infty)$ **12.** $(x + 6)(x - 1) \ge 0$

13. $x(x + 3) < 0$ $(-3,0)$ **14.** $x(x - 3) > 0$ $(-\infty,0) \cup (3, \infty)$

15. $x^2 \le 9$ $[-3,3]$ **16.** $x^2 \ge 1$ $(-\infty,-1] \cup [1, \infty)$

17. $25 - x^2 < 0$ **18.** $4 - x^2 > 0$ $(-2,2)$

19. $x^2 + 4x - 12 \le 0$ $[-6,2]$ **20.** $x^2 - 8x + 7 > 0$ †

†Due to space constrictions, answers to these exercises may be found in the Answers beginning on page A–1 in the back of the book.

Answers:

12. $(-\infty, -6] \cup [1, \infty)$ **17.** $(-\infty, -5) \cup (5, \infty)$

21. $6x^2 + 7x - 3 \geq 0$ **22.** $4x^2 - 2x - 2 < 0$

23. $25x^2 + 5x - 6 \leq 0$ **24.** $12x^2 - 5x - 7 < 0$

25. $(x - 2)^2 \leq 1$ $[1, 3]$ **26.** $(x + 3)^2 > 9$

27. $2x^2 + 5x < 12$ **28.** $6x^2 + 11x \geq 7$

29. $2x(x + 3)(x - 3) \leq 0$ **30.** $3x(x + 2)(x - 2) > 0$

31. $(x + 3)(x + 1)(x - 1) \geq 0$ $[-3, -1] \cup [1, \infty)$

32. $(x + 4)(x - 1)(x + 2) < 0$ $[-\infty, -4) \cup (-2, 1)$

33. $x^2 - 2x - 2 < 0$ **34.** $x^2 - 4x - 1 \leq 0$

35. $x^2 \geq -1$ $(-\infty, \infty)$ **36.** $2x^2 \geq -5$ $(-\infty, \infty)$

37. $x^2 + 2x < -1$ $\varnothing$ **38.** $4x^2 + 12x < -9$ $\varnothing$

39. $x^3 - x^2 \geq 0$ $\{0\} \cup [1, \infty)$ **40.** $x^3 - 9x^2 > 0$ $(9, \infty)$

41. $x^4 \leq 2x^2$ $[-\sqrt{2}, \sqrt{2}]$ **42.** $x^4 \geq 3x^2$

43. $x^2 \geq 1$ $(-\infty, -1] \cup [1, \infty)$ **44.** $x^3 < -8$ $(-\infty, -2)$

45. $x^4 \leq 16$ $[-2, 2]$ **46.** $x^4 > 9$ $(-\infty, -\sqrt{3}) \cup (\sqrt{3}, \infty)$

In Exercises 47–66, solve each rational inequality. Write the solution in interval notation.

47. $\dfrac{x + 2}{x - 5} < 0$ $(-2, 5)$ **48.** $\dfrac{x - 3}{x + 1} > 0$ $(-\infty, -1) \cup (3, \infty)$

49. $\dfrac{3x + 2}{x - 3} \leq 0$ $\left[-\dfrac{2}{3}, 3\right)$ **50.** $\dfrac{3 - 2x}{4x + 5} \geq 0$ $\left(-\dfrac{5}{4}, \dfrac{3}{2}\right]$

51. $\dfrac{x + 4}{x} < 0$ $(-4, 0)$ **52.** $\dfrac{x}{x - 2} > 0$ $(-\infty, 0) \cup (2, \infty)$

53. $\dfrac{x + 1}{x + 2} \leq 3$ **54.** $\dfrac{x - 1}{x - 2} > 3$ $\left(2, \dfrac{5}{2}\right)$

55. $\dfrac{(x - 2)(x + 2)}{x} > 0$ **56.** $\dfrac{(x - 1)(x + 3)}{x - 2} < 0$

57. $x + \dfrac{8}{x} \leq 6$ $(-\infty, 0) \cup [2, 4]$ **58.** $x - \dfrac{12}{x} > 1$

59. $2 - \dfrac{2x}{3x - 4} > 0$ **60.** $1 + \dfrac{1}{2 - x} \geq 0$

61. $\dfrac{x + 4}{3x - 2} \geq 1$ $\left(\dfrac{2}{3}, 3\right]$ **62.** $\dfrac{2x - 3}{x + 3} \leq 1$ $(-3, 6]$

63. $3 \leq \dfrac{2x + 6}{2x + 1}$ $\left(-\dfrac{1}{2}, \dfrac{3}{4}\right]$ **64.** $\dfrac{x - 2}{2x + 1} < -1$ $\left(-\dfrac{1}{2}, \dfrac{1}{3}\right)$

65. $\dfrac{x + 5}{x - 2} \geq \dfrac{x}{x + 2}$ **66.** $\dfrac{x - 1}{x + 1} \geq \dfrac{x}{x - 1}$

B EXERCISES Applying the Concepts

67. Free-falling object. The distance d traveled in t seconds by an object dropped from a height h is given by $d = 16t^2$. If a water bottle is dropped from a hot air balloon when the balloon is between 64 and 100 feet above the ground, how long (in seconds) does it take for the bottle to hit the ground? 2 sec to 2.5 sec

68. Interest rate. The amount of money A received at the end of two years when P dollars is invested at a compound rate r is $A = P(1 + r)^2$. A firm is investing \$100,000 for a two-year period and expects to get a return of no less than \$110,000 and no more than \$115,000. What range of interest rates (to the nearest percent) will provide the expected return? 5% to 7%

69. Temperature. The number N of water mites in a water sample depends on the temperature t in degrees Fahrenheit and is given by $N = 110t - t^2$. At what temperature will the number of mites exceed 1000? 10°F to 100°F

70. Falling object. The height h of an object thrown from the top of a ski lift 1584 feet high after t seconds is $h = -16t^2 + 32t + 1584$. For what times is the height of the object at least 1200 feet? 0 to 6 sec

71. Company profit. The profit P (in millions of dollars) of a company for next year is estimated to satisfy the inequality $P(P - 3) < 4(P - 3)$. What is the estimated profit for this company for next year? Between \$3 million and \$4 million

72. Area of a rectangle. The length of a rectangle is 5 meters larger than the width. Find the range of widths that result in such a rectangle whose area is at least 204 square meters. At least 12 m

73. Blackjack. A gambler playing blackjack in a casino using four decks noticed that 20% of the cards that had been dealt were jacks, queens, kings, or aces. After x cards had been dealt, he knew that the likelihood that the next card dealt would be a jack, a queen, a king, or an ace was $\dfrac{64 - 0.2x}{208 - x}$. For what values of x is this likelihood greater than 50%? More than 133 cards

74. Area of a triangle. The base of a triangle is 3 centimeters greater than the height. Find the possible heights h so that the area of such a triangle will be at least 5 square centimeters. At least 2 cm

Answers:

21. $\left(-\infty, -\dfrac{3}{2}\right] \cup \left[\dfrac{1}{3}, \infty\right)$ **22.** $\left(-\dfrac{1}{2}, 1\right)$ **23.** $\left[-\dfrac{3}{5}, \dfrac{2}{5}\right]$ **24.** $\left(-\dfrac{7}{12}, 1\right)$

26. $(-\infty, -6) \cup (0, \infty)$ **27.** $\left(-4, \dfrac{3}{2}\right)$ **28.** $\left(-\infty, -\dfrac{7}{3}\right] \cup \left[\dfrac{1}{2}, \infty\right)$

29. $(-\infty, -3] \cup [0, 3]$ **30.** $(-2, 0) \cup (2, \infty)$ **33.** $(1 - \sqrt{3}, 1 + \sqrt{3})$

34. $[2 - \sqrt{5}, 2 + \sqrt{5}]$ **42.** $(-\infty, -\sqrt{3}] \cup \{0\} \cup [\sqrt{3}, \infty)$

53. $\left(-\infty, -\dfrac{5}{2}\right] \cup (-2, \infty)$ **55.** $(-2, 0) \cup (2, \infty)$

56. $(-\infty, -3) \cup (1, 2)$ **58.** $(-3, 0) \cup (4, \infty)$

59. $\left(-\infty, \dfrac{4}{3}\right) \cup (2, \infty)$ **60.** $(-\infty, 2) \cup [3, \infty)$

65. $\left(-2, -\dfrac{10}{9}\right] \cup (2, \infty)$ **66.** $(-\infty, -1) \cup \left[\dfrac{1}{3}, 1\right)$

Linear Regression

Objectives

1 Define linear regression
2 Find the regression line

---

1 Define linear regression.

Linear Regression

Suppose that in an experiment, we measure two presumably related quantities x and y and obtain n observed ordered pairs $(x_1, y_1), (x_2, y_2), \ldots, (x_n, y_n)$ of measurements. For example, an experiment may involve finding a relationship between the heights and weights of people. A person collects a carefully selected sample of n people and measures the height (y) and the weight (x) of each individual. The ordered pair (x_1, y_1) represents the weight and height of the first person, (x_2, y_2) represents the weight and height of second person, and so on. These observed ordered pairs (x, y) are called **data points**. The interval between the smallest and largest x values of the data points is called the **interval of the data values**. We plot the observed data points on a coordinate plane. The resulting graph is called a **scatter diagram** or **scatterplot**. If the data points do not lie on a straight line (and yet they seem to be scattered close to *some* line), the relationship between the variables is considered linear. We can use a graphing utility to find the line $y = ax + b$ that fits the data points as closely as possible. The line so obtained is called the **regression line** or the **best-fit line**.

Comments on the Regression Line

1. If you use the regression line to predict the values of y corresponding to the values of x *within* the interval of the data values (called **interpolating**), the results are usually quite reasonable.

2. If you use the regression line to predict the values of y corresponding to the values of x well *outside* the interval of the data values (called **extrapolating**), the results become extremely questionable.

3. You can take any set of ordered pairs of measurements and construct a regression line based on these data. However, the results are completely meaningless if the two variables are totally unrelated. For example, you could collect data on students' telephone numbers and their driver's license numbers, construct a scatterplot, and compute the regression line. But because the two variables are unrelated, the result of predicting students' driver's license numbers from their telephone numbers is utterly useless.

So we need a way to determine whether there is indeed a linear relationship between the variables x and y. Statisticians use a number called the **linear correlation coefficient** (or the **correlation coefficient, r**) to measure the strength of a linear relationship (if there is one) between the variables.

The interpretation of the value of the correlation coefficient, r, depends upon the field of application. In the physical sciences using controlled experiments, a value of r in the range $0.9 \leq |r| \leq 1.0$ is considered a high correlation. However, in the social sciences, a value of r in the range $0.7 \leq |r| \leq 1.0$ indicates a high to a very high correlation.

2 Find the regression line.

Finding the Regression Line

Most graphing calculators have built-in programs to find r and the regression line. We illustrate this in the next example by using a TI-83 Plus. (Consult your manual if you are using a different graphing calculator.)

EXAMPLE   Finding Regression Line and *r*

The following table shows the actual temperature and the wind chill temperature, with a wind speed of 30 mph.

Actual Temperature of the Air (°F)	Air Temperature with the Wind Chill Factor (°F)
−40	−80
−30	−67
−20	−53
−10	−39
0	−26
10	−12
20	1
30	15
40	25

Source: www.weather.gov/om/wind chill

Find and graph the regression line relating the variables *x* (actual temperature) and *y* (wind chill temperature). Also find the coefficient of correlation *r*.

SOLUTION

Steps	Calculator Screen
Step 1. Enter the data. Press **STAT 1** for **STAT Edit**. Enter the values of the independent variable *x* in L1 and the corresponding values of the dependent variable *y* in L2.	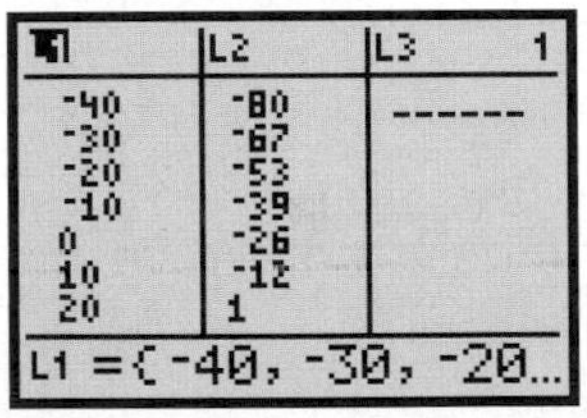
Step 2. Find the equation of the regression line. Press **STAT** and highlight **CALC**. Press **4** for **LinReg** (*ax* + *b*). Press **ENTER**.	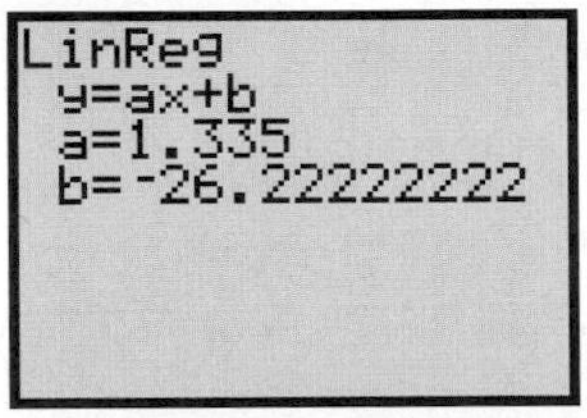
Step 3. Finding *r*. If *r* (and r^2) do not already appear in the figure, press **CATALOG** and select **Diagnostic ON** from the list. Press **ENTER** twice. Repeat Step 2.	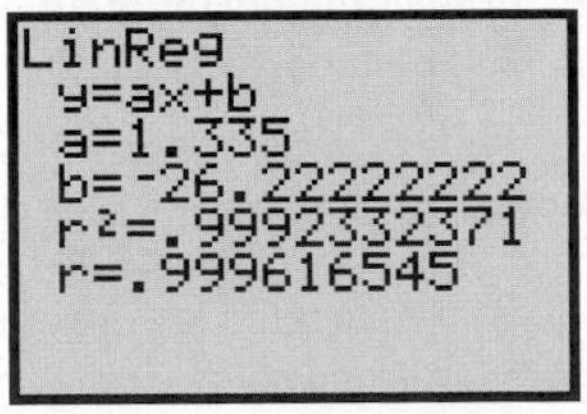
Step 4. Graphing the regression line with the scatterplot. Press *Y* = . Press **VARS 5** for **statistics**. Highlight **EQ** and press **ENTER**. (You should see the regression equation in the *Y* = window.) Press **ZOOM 9** for **ZoomStat**. (You should see the data graphed along with the regression line.)	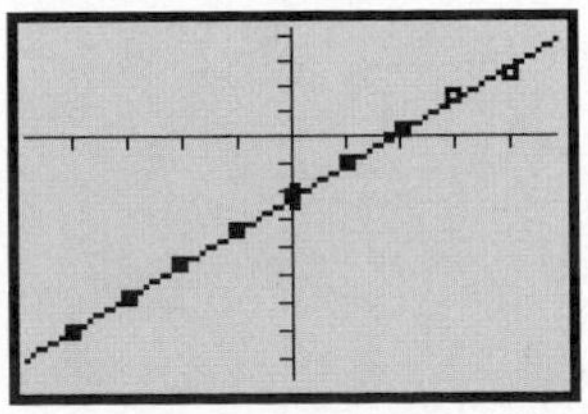

SECTION A.12 ■ Exercises

For each given data set in Exercises 1–7

a. use your graphing calculator to see the scatterplot and the regression line. †

b. write the equation of the regression line and write the coefficient of correlation. †

1. **Weight and height of basketball players.**

Cleveland Cavaliers Professional Basketball Players Height and Weight

Name of Player	Height (in inches)	Weight (in pounds)
Daniel Gibson	74	200
J.J. Hickson	81	242
Zydrunas Ilgauskas	87	260
Darnell Jackson	81	253
LeBron James	80	250
Tarence Kinsey	78	189
Aleksandar Pavlovic	79	235
Joe Smith	82	225
Wally Szczerbiak	79	240
Anderson Varejao	83	260
Ben Wallace	81	240
Delonte West	75	180
Jawad Williams	81	218
Mo Williams	73	190
Lorensen Wright	83	255

Source: http://www.nba.com/cavaliers/roster/

c. Predict the weight of an NBA player whose height is 77 inches. 212 lbs

d. Predict the height of an NBA player who weighs 200 pounds. 75 in.

2. **Consumer price index (CPI).** Think of CPI as an index number for the cost of everything that American consumers buy. The CPI for the year 2000 was 168.8, which means that we consumers spent $168.80 in 2000 to buy goods and services that cost $100 in the 1982–1984 base periods. The following table shows the average annual CPI since 1975.

Year (x)	1975	1977	1979	1981	1983	1985	1987
CPI (y)	53.8	60.6	72.6	90.9	99.6	107.6	113.6

1989	1991	1993	1995	1997	1999	2001	2003
124.0	136.2	144.5	152.4	160.5	166.6	171.1	181.3

Source: Bureau of Labor Statistics

c. Predict the CPI for the year 2016. 247.06

d. When will the CPI be 240? 2014

3. **Temperature and wind chill.** The table shows the actual temperature and wind chill temperature when the wind is blowing at 25 mph. [By the way, the wind chill temperature is calculated by using the formula *wind chill* $(\degree F) = 35.74 + .6215T - 35.75V^{0.16} + .4275TV^{0.16}$, where T = air temperature ($\degree F$) and V = wind speed (mph).]

Actual Temperature of the Air (°F)	Air Temperature with the Wind Chill Factor (°F)
−45	−84
−40	−78
−35	−71
−30	−64
−25	−58
−20	−51
−15	−44
−10	−37
−5	−31
0	−24
5	−17
10	−11
15	−4
20	3
25	9
30	16
35	23

Source: www.nws.gov/om/windchill

c. Estimate the wind chill at a temperature of −60°F.
−104.4°F

4. **"It's not the heat; it's the humidity!"** There is a lot of truth to that statement. As the humidity increases, it is harder for our bodies to cool themselves; so it feels much hotter than it actually is. The *heat index* calculates the apparent temperature (App. Temp.) is based on the actual temperature and relative humidity. The data in the following table represents the apparent temperature versus the percent of relative humidity in a room whose actual temperature is 80°F.

Rel. Humidity (*x*)	0	10	20	30	40	50	60	70	80	90	100
App. Temp. (*y*)	73	75	77	78	79	81	82	85	86	88	91

c. If the relative humidity is 85%, estimate the apparent room temperature. [Note: For a complicated formula relating heat index, relative humidity, and actual temperature, see www.bom.gov.an/olympic/wbgt.htm]. 87.3°F

5. **Inferring height of a female from the humerus bone.** Anthropologists can estimate the height of an adult bases only on the humerus bone (popularly called the funny bone), located between the elbow and the shoulder. The following table shows the approximate relationship between the height *y* (in inches) of females and the length *x* (in inches) of their humerus bone.

x	8	9	10	11	12	13	14
y	50.4	53.5	56.6	59.7	62.8	65.9	69.0

c. Estimate the height of a female whose humerus bone measures 11.3 inches. 60.63 in.

6. **Temperature and cricket chirps.** Biologists have studied the relationship between the temperature and the number of chirps by a cricket. The following table gives the temperature $T(°F)$ and the number *n* of seconds to count 50 chirps.

$T(°F)$	45	55	60	65	70	75	80	84	89	93
n	94	42	32	26	21	20	15	17	20	20

Source: G.W. Pierce, *The Songs of Insects.* Harvard University Press, 1948

c. Use the linear model to estimate *n* if the temperature is 5°F. 110.6 seconds
d. Use the linear model to estimate the temperature if $n = 1$. Is your estimate reasonable?

7. **Inferring height of a male from the humerus bone.** The approximate relationship between the height *y* of a male and the length *x* of his humerus bone is shown in the table. All measurements are in inches.

x	8	9	10	11	12	13	14
y	53	56	59	62	66	68	71

c. Estimate the height of a male whose humerus bone measures 13.7 inches. 69.9

8. a. Suppose from part of a broken humerus bone (the gender is not known) that its length is estimated to be between 10.3 and 10.7 inches. Use the models from Exercises 5 and 6 to approximate the range for the height of an individual of each gender.
   b. Measure your humerus bone and compare your actual height with the estimated height from the appropriate formula.

Answers:
8. a. Females: between 57.53 in. and 58.77 in.; males: between 59.90 in. and 61.10 in.
   b. Answers will vary.

Answers

CHAPTER 1

Section 1.1

Practice Problems: 1.

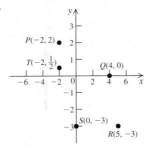

2. $(1998, 21.1), (1999, 20.5), (2000, 20.2), (2001, 19.7), (2002, 19.4),$
$(2003, 18.6), (2004, 17.9)$ **3.** $\sqrt{2}$ **4.** $\left(\dfrac{11}{2}, -\dfrac{3}{2}\right)$

5.
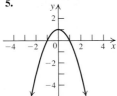
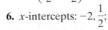

6. x-intercepts: $-2, \dfrac{1}{2}$;
y-intercept: -2

7. Symmetric

8.
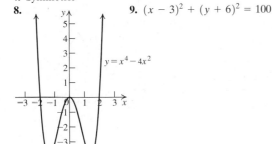

9. $(x - 3)^2 + (y + 6)^2 = 100$

10.
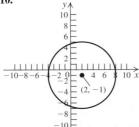

11. Center: $(-2, 3)$; radius: 5

A Exercises: Basic Skills and Concepts:

7.
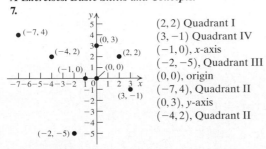

$(2, 2)$ Quadrant I
$(3, -1)$ Quadrant IV
$(-1, 0)$, x-axis
$(-2, -5)$, Quadrant III
$(0, 0)$, origin
$(-7, 4)$, Quadrant II
$(0, 3)$, y-axis
$(-4, 2)$, Quadrant II

8. a. $(1, 0), (2, 0), (3, 0), (4, 0), (5, 0)$. The y-coordinate is 0.
b. Horizontal line intersecting the y-axis at 1

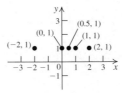

9. b. Vertical line intersecting the x-axis at -1

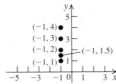

27. On the graph: $(-3, -4), (1, 0), (4, 3)$ Not on the graph: $(2, 3)$
28. On the graph: $(3, 2), (0, 1), (8, 3)$ Not on the graph: $(8, -3)$
29. On the graph: $(1, 1), \left(2, \dfrac{1}{2}\right)$ Not on the graph: $\left(-3, \dfrac{1}{3}\right), (0, 0)$

31.
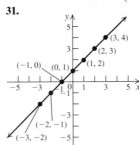

32.
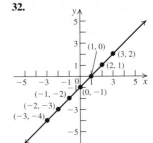

33.
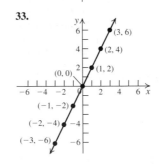

34.
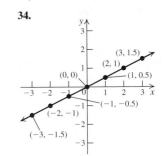

35.
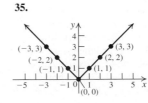

36.
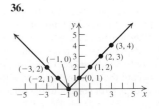

37.

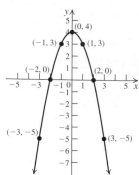

38.

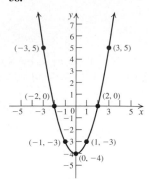

69. $(x - 5)^2 + (y - 1)^2 = 25$
70. $(x - 2)^2 + (y - 5)^2 = 26$

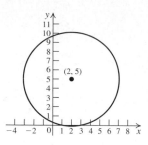

39.

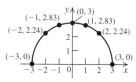

40.

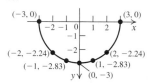

B Exercises: Applying the Concepts:

75.

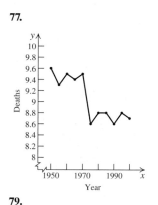

76.

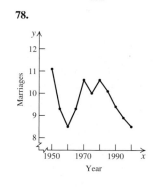

41.

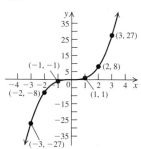

42.

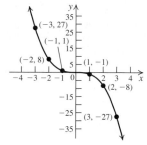

45. *x*-intercept: 4; *y*-intercept: 3   **46.** *x*-intercept: 5; *y*-intercept: 3
47. *x*-intercept: $\frac{5}{2}$; *y*-intercept: $\frac{5}{3}$ **48.** *x*-intercept: 2; *y*-intercept: -3
49. *x*-intercepts: 2, 4; *y*-intercept: 8   **50.** *x*-intercept: 6; *y*-intercepts:
2, 3   **51.** *x*-intercepts: -2, 2; *y*-intercepts: -2, 2   **52.** *x*-intercepts:
-3, 3; *y*-intercept: 3   **53.** *x*-intercepts: -1, 1; no *y*-intercept
55. Not symmetric with respect to the *x*-axis, symmetric with respect
to the *y*-axis, not symmetric with respect to the origin
56. Symmetric with respect to the *x*-axis, not symmetric with respect
to the *y*-axis, not symmetric with respect to the origin   **57.** Not
symmetric with respect to the *x*-axis, not symmetric with respect to
the *y*-axis, symmetric with respect to the origin   **58.** Not symmetric
with respect to the *x*-axis, not symmetric with respect to the *y*-axis,
symmetric with respect to the origin   **59.** Not symmetric with
respect to the *x*-axis, symmetric with respect to the *y*-axis, not
symmetric with respect to the origin   **60.** Not symmetric with
respect to the *x*-axis, symmetric with respect to the *y*-axis, not
symmetric with respect to the origin   **61.** Not symmetric with
respect to the *x*-axis, not symmetric with respect to the *y*-axis,
symmetric with respect to the origin   **62.** Not symmetric with
respect to the *x*-axis, symmetric with respect to the *y*-axis, not
symmetric with respect to the origin   **67.** $(x + 3)^2 + (y + 2)^2 = 9$
68. $(x - 3)^2 + (y + 4)^2 = 97$

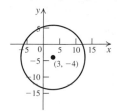

77.

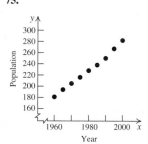

78.

79.

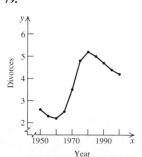

82. c.

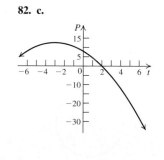

d. -8 and 2. They represent the months when there is neither
profit nor loss.

83. a.

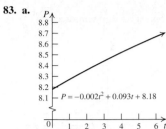

84. a. After 0 second: 320 ft; after 1 second: 432 ft; after 2 seconds: 512 ft; after 3 seconds: 560 ft; after 4 seconds: 576 ft; after 5 seconds: 560 ft; after 6 seconds: 512 ft

b.

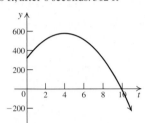

85. a.

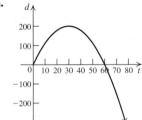

C Exercises: Beyond the Basics:

91.

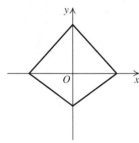

92.

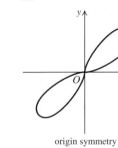

origin symmetry

93.

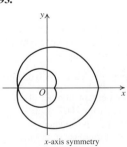

x-axis symmetry

94.

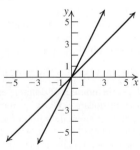

95.

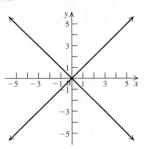

96.

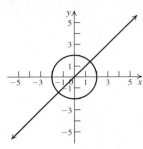

97.

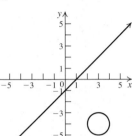

98. Area: 11π

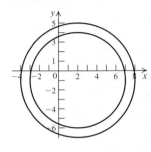

Critical Thinking: 102. The graph is the union of the graphs of $y = \sqrt{2x}$ and $y = -\sqrt{2x}$.

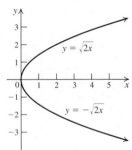

Section 1.2

Practice Problems: **1.** $-\dfrac{8}{13}$ **2.** $y = -\dfrac{2}{3}x - \dfrac{13}{3}$

3. $y = 5x + 11$ **4.** $y = 2x - 3$

5.

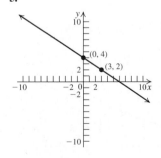

6.

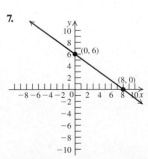

7.

8. Between 176.8 and 179.4 centimeters

9. a. $4x - 3y + 23 = 0$
   b. $5x - 4y - 31 = 0$

A Exercises: Basic Skills and Concepts:

21. $y = \dfrac{1}{2}x + 4$

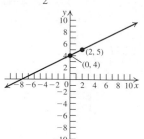

22. $y = -\dfrac{1}{2}x + 4$

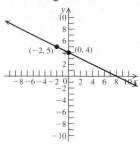

23. $y = -\dfrac{3}{2}x + 4$

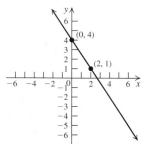

24. $y = \dfrac{2}{5}x + \dfrac{2}{5}$

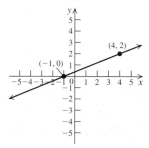

25. $y = -\dfrac{3}{5}x - 1$

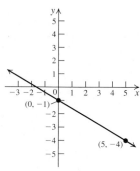

26. $x = 5$

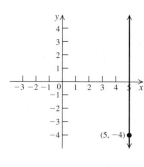

30. $y = \dfrac{6}{7}x + \dfrac{37}{7}$ **31.** $y = \dfrac{2}{3}x + \dfrac{1}{3}$ **32.** $y = -\dfrac{6}{7}x - \dfrac{27}{7}$

43. $y = \dfrac{4}{3}x + 4$ **44.** $y = -\dfrac{2}{5}x - 2$

51. $m = -\dfrac{1}{2}$; y-intercept $= 2$ **52.** $m = \dfrac{1}{3}$; y-intercept $= 3$

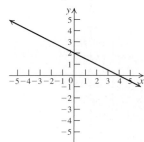

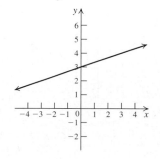

53. $m = \dfrac{3}{2}$; y-intercept $= 3$

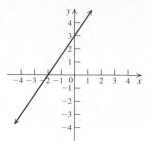

54. $m = -\dfrac{1}{2}$; y-intercept $= \dfrac{15}{4}$

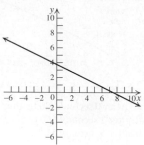

55. m is undefined; no y-intercept

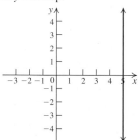

56. $m = 0$; y-intercept $= -\dfrac{5}{2}$

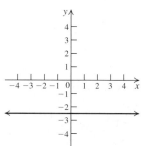

57. m is undefined; y-intercepts $=$ y-axis

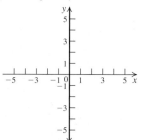

58. $m = 0$; y-intercept $= 0$

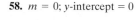
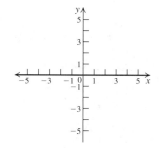

B Exercises: Applying the Concepts: **84. a.** x: number of golf sessions; y: yearly payment to the golf club; $y = 35x + 1000$
b. Slope: cost per golf session; y-intercept: annual membership fee
85. a. x: number of hours worked; y: amount of money earned per week;

$$y = \begin{cases} 11x & x \le 40 \\ 16.5x - 220 & x > 40 \end{cases}$$

b. Slope: hourly wage; y-intercept: salary for 0 hours of work
86. a. x: number of months one owes on the refrigerator; y: amount owed after x months; $y = -15x + 600$ **b.** Slope: monthly payment; y-intercept: initial charge   **87. a.** $x =$ amount of rupees;

$y =$ amount of dollars equal to x rupees; $y = \dfrac{1}{40.5}x$, or $y = 0.02469x$

b. Slope: the number of dollars per rupee; y-intercept: number of rupees for 0 dollars   **88. a.** x: number of years after 2007; y: life expectancy of a female born in the United States in the year $x + 2007$; $y = 0.17x + 80.9$ **b.** Slope: rate of increase in the life expectancy of a female born in the United States; y-intercept: current life expectancy of a female born in the United States;
89. a. x: number of TVs produced; y: cost of production for x TVs; $y = 150x + 10{,}000$ **b.** Slope: marginal cost of a TV; y-intercept: fixed cost independent of the number of TVs produced
90. a. x: demand for the product; y: price per unit when the demand is x; $y = -0.1x + 100$ **b.** Slope: rate of decrease in the price; y-intercept: price per unit when the demand is zero
91. a. $v = \$11{,}200$, when $t = 2$ **b.** $v = \$5600$, when $t = 6$
The tractor's value is $0 at $t = 10$.

92. c.

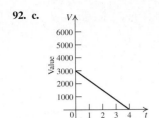

Years

94. b.

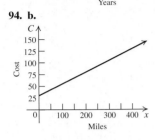

Miles

95. b.

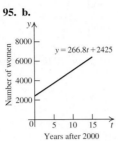

$y = 266.8t + 2425$

Years after 2000

96. b. One degree Celsius change in the temperature is equal to $\frac{9}{5}$ degrees change in Fahrenheit. **c.** 104°F, 77°F, 23°F, 14°F

d. 37.78°C, 32.22°C, 23.89°C, −23.33°C, −28.89°C
100. a. $R = 50x, C = 25x + 75,000, P = 25x − 75,000$
b.

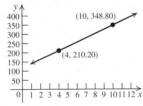

c. Slope = 25; rate of increase in the profit per pair of shoes sold
101. a. $y = 23.1x + 117.8$

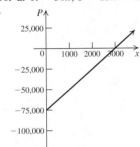

b. Slope: cost of producing one modem; y-intercept: fixed overhead cost
102. b. Slope: monthly increase in the number of viewers;
y-intercept: number of viewers at the beginning of the show

Critical Thinking: **129. a.** This is a family of lines parallel to the line $y = -2x$, and they all have slope equal to −2.

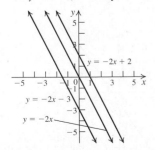

b. This is a family of lines that passes through the point $(0, -4)$, and they all have y-intercept to −4.

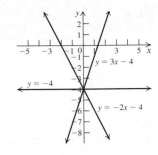

Section 1.3

Practice Problems: **1. a.** No **b.** Yes **2. a.** 0 **b.** −7
c. $-2x^2 - 4hx + 5x - 2h^2 + 5h$ **3.** $(-\infty, 1)$ **4.** No
5. a. Yes **b.** −3, −1 **c.** −5 **d.** −5, 1 **6.** −6 **7.** $-2c - h$
8. $-2x - h + 1$ **9.** Domain; $[1, \infty)$; range: $[6, 12)$ The
function value at 6 is 11.813. **10. a.** $C(x) = 1200x + 100,000$
b. $R(x) = 2500x$ **c.** $P(x) = 1300x - 100,000$ **d.** 77

A Exercises: Basic Skills and Concepts:

30. $f(1) = -1, g(1) = 2, h(1) = 1, g(a) = \dfrac{2}{\sqrt{a}}, g(x^2) = \dfrac{2}{|x|}$

31. $f(-1) = 5, g(-1)$ is not defined, $h(-1) = \sqrt{3}, h(c) = \sqrt{2 - c}$,
$h(-x) = \sqrt{2 + x}$ **32.** $f(4) = 5, g(4) = 1, h(4)$ is not defined,

$g(2 + k) = \dfrac{2}{\sqrt{2 + k}}, f(a + k) = a^2 + 2ak - 3a + k^2 - 3k + 1$

61. b. $x = -1 \pm \sqrt{3}$ **d.** $x = -1 \pm \dfrac{1}{2}\sqrt{14}$ **c.** 1 **75. a.** $x + h$

b. h **c.** 1 **76. a.** $3x + 3h + 2$ **b.** $3h$ **c.** 3
77. a. $x^2 + 2hx + h^2$ **b.** $2hx + h^2$ **c.** $2x + h$
78. a. $x^2 + 2hx - x + h^2 - h$ **b.** $2hx + h^2 - h$ **c.** $2x + h - 1$
79. a. $2x^2 + 4hx + 3x + 2h^2 + 3h$ **b.** $4hx + 2h^2 + 3h$
c. $4x + 2h + 3$ **80. a.** $3x^2 + 6hx - 2x + 3h^2 - 2h + 5$
b. $6hx + 3h^2 - 2h$ **c.** $6x + 3h - 2$ **81. a.** 4 **b.** 0 **c.** 0
82. a. −3 **b.** 0 **c.** 0 **83. a.** $\dfrac{1}{x + h}$ **b.** $-\dfrac{h}{x(x + h)}$

c. $-\dfrac{1}{x(x + h)}$ **84. a.** $-\dfrac{1}{x + h}$ **b.** $\dfrac{h}{x(x + h)}$ **c.** $\dfrac{1}{x(x + h)}$

B Exercise: Applying the Concepts:
97. a. $p(5) = 1150$. If 5000 TVs can be sold, the price per TV is $1150.
$\qquad p(15) = 900$. If 15,000 TVs can be sold, the price per TV is $900.
$\qquad p(30) = 525$. If 30,000 TVs can be sold, the price per TV is $525.
b.

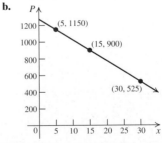

98. b. $R(1) = 1250, R(5) = 5750, R(10) = 10,250, R(15) = 13,500$,
$R(20) = 15,500, R(25) = 16,250, R(30) = 15,750$. They show the
amount of revenue for the given number of TVs sold.
c.

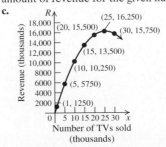

Number of TVs sold
(thousands)

101. d.

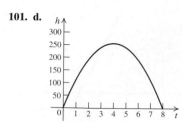

103. After 4 hours: 12 ml; after 8 hours: 21 ml; after 12 hours: 27. 75 ml; after 16 hours: 32. 81 ml; after 20 hours: 36. 61 ml

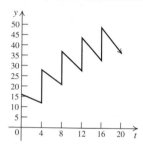

104. 1st day: 400 mg; 2nd day: 500 mg; 3rd day: 525 mg; 4th day: 531.25 mg; 5th day: 532.81 mg; 6th day: 533.2 mg; 7th day: 533.3 mg; 8th day: 533.33 mg; 9th day: 533.33 mg; 10th day: 533.33 mg

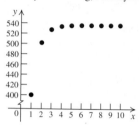

Section 1.4

Practice Problems: **1.** $g(x) = 2x + 6$ **2.** 15.524 m
3. $f(-2) = 4; f(3) = 6$
4.

$f(x) = x^2 \quad x \le -1$
$f(x) = 2x \quad x > -1$

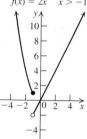

5. $f(-3.4) = -4; f(4.7) = 4$
6. Decreasing on $(-\infty, -3)$, constant on $(-3, 2)$, increasing on $(2, \infty)$

7. $f(-x) = -(-x)^2 = -x^2 = f(x)$; therefore, f is even.

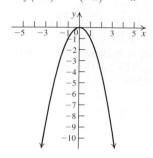

8. $f(-x) = -(-x^3) = x^3 = -f(x)$; therefore, f is odd.

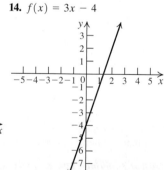

A Exercises: Basic Skills and Concepts:
9. $f(x) = x + 1$ **10.** $f(x) = x - 1$

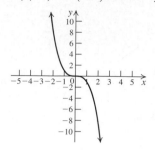

11. $f(x) = 2x + 3$ **12.** $f(x) = 3x - 2$

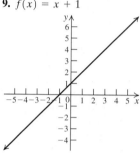 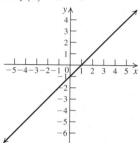

13. $f(x) = -3x + 4$ **14.** $f(x) = 3x - 4$

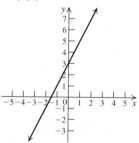 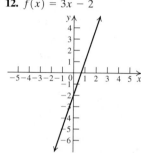

15. $f(x) = \frac{1}{2}x + 3$ **16.** $f(x) = \frac{3}{2}x - 1$

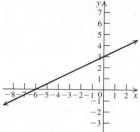 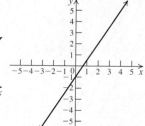

17. $f(x) = -\dfrac{2}{3}x - 1$

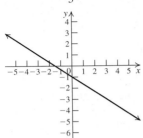

18. $f(x) = -\dfrac{3}{4}x + 1$

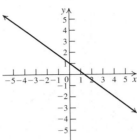

27.

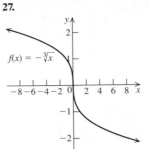

$f(x) = -\sqrt[3]{x}$

28.

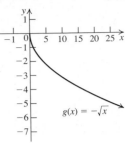

$g(x) = -\sqrt{x}$

19. Constant on $(-\infty, \infty)$

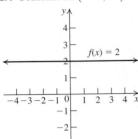

$f(x) = 2$

20.

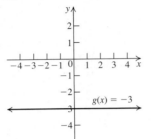

$g(x) = -3$

43. a. Domain: $[-2, 6]$; range: $[-3, 2]$ **b.** x-intercept: $\dfrac{2}{5}$;

y-intercept: $-\dfrac{1}{2}$ **c.** Increasing on $(-2, 2)$, constant on $(2, 6)$

d. Neither even nor odd   **44. a.** Domain: $[-3, 0) \cup (0, 3]$;
range: $\{-1, 1\}$ **b.** No x- or y-intercept   **c.** Constant on $(-3, 0)$ and
$(0, 3)$ **d.** Odd   **45. a.** Domain: $[-2, 4]$; range: $[-1, 2]$
b. x-intercepts: $-2, 0$; y-intercept: 0 **c.** Decreasing on $(-2, -1)$ and
$(2, 4)$, increasing on $(-1, 2)$ **d.** Neither even nor odd
46. a. Domain: $[-3, 5]$; range: $[-2, 3]$ **b.** x-intercept: 0; y-intercept: 0
c. Increasing on $(-3, 5)$ **d.** Neither even nor odd
47. a. Domain: $(0, \infty)$; range: $(0, \infty)$ **b.** No x- or y-intercept
c. Decreasing on $(0, \infty)$ **d.** Neither even nor odd
48. a. Domain: $(-\infty, 3) \cup (3, \infty)$; range: $(-\infty, 3) \cup (3, \infty)$
b. x-intercept: 2; y-intercept: 2 **c.** Decreasing on $(-\infty, 3)$, and $(3, \infty)$
d. Neither even nor odd   **49. a.** Domain: $(-\infty, \infty)$; range: $(0, \infty)$
b. No x-intercept; y-intercept: 1 **c.** Increasing on $(-\infty, \infty)$
d. Neither even nor odd   **50. a.** Domain: $(0, \infty)$; range: $(-\infty, \infty)$
b. x-intercept: 1.5; no y-intercept   **c.** Increasing on $(0, \infty)$
d. Neither even nor odd

21. Increasing on $(-\infty, \infty)$

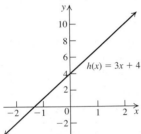

$h(x) = 3x + 4$

22.

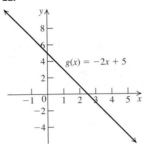

$g(x) = -2x + 5$

51. b.

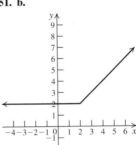

52. b.

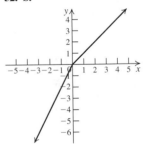

23. Decreasing on $(-\infty, 0)$;
increasing on $(0, \infty)$

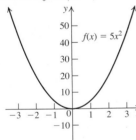

$f(x) = 5x^2$

24.

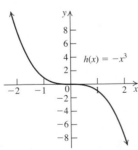

$h(x) = -x^3$

53. b.

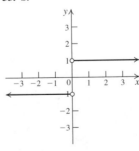

54. b.

$f(x) = x + 2 \quad x \le 1$
$f(x) = 2x + 4 \quad x > 1$

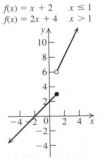

25.

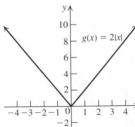

$g(x) = 2|x|$

26.

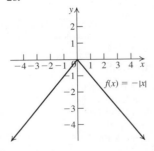

$f(x) = -|x|$

55. Range: $(-\infty, \infty)$

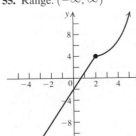

56. Range: $[0, \infty)$

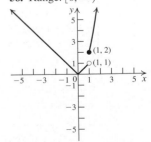

$(1, 2)$
$(1, 1)$

57. Range: $(-\infty, 0] \cup [2, \infty)$

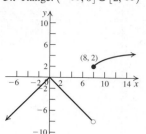

58. Range: $(-\infty, 2)$

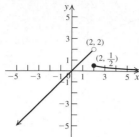

59. Range: $(-\infty, 0] \cup \{1, 2\}$

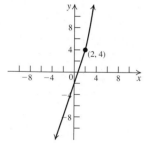

60. Range: $(-\infty, -1] \cup [0, 2]$

B Exercises: Applying the Concepts: **61. a.** $f(x) = \dfrac{1}{33.81}x$;

domain: $[0, \infty)$; range: $[0, \infty)$ **b.** 0.0887. It means that
3 oz = 0.08871.   **62. a.** *h*-intercept: 117.78. It means that water boils
at 0°F at 117.80 feet above sea level. *y*- intercept: 212. If means that
water boils at 212°F at sea level.   **b.** Closed interval from 0 to the
end of the atmosphere, in feet   **63. a.** No *d*-intercept since $d \geq 0$;
y-intercept: 1. It means that the pressure at sea level ($d = 0$) is 1 atm.
b. $P(0) = 1, P(10) \approx 1.3, P(33) = 2, P(100) \approx 4.03$
65. b. The *y*-intercept is the fixed overhead cost.

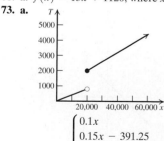

68. a. $f(x) = 15x + 1120$, where *x* is the number of years since 2002
73. a.

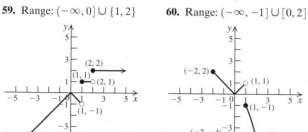

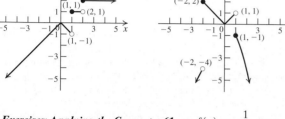

74. a. $f(x) = \begin{cases} 0.1x & \text{if } 0 < x \leq 7825 \\ 0.15x - 391.25 & \text{if } 7825 < x \leq 31{,}850 \\ 0.25x - 3576.25 & \text{if } 31{,}850 < x \leq 77{,}100 \\ 0.28x - 5889.25 & \text{if } 77{,}100 < x \leq 160{,}850 \\ 0.33x - 13{,}931.75 & \text{if } 160{,}850 < x \leq 349{,}700 \\ 0.35x - 20{,}925.75 & \text{if } x > 349{,}700 \end{cases}$

C Exercises: Beyond the Basics:
75. c.

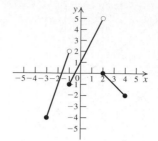

76. b. i. $x = -\dfrac{1}{2}$ or $x = 1$

ii. $x = -\dfrac{1}{4}$ or $x = \dfrac{1}{2}$

c.

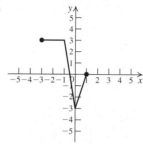

77. a. Domain: $(-\infty, \infty)$; range $[0, 1)$ **b.** Increasing on $(n, n + 1)$
for every integer *n*   **c.** neither

78. a. Domain: $(-\infty, 0) \cup [1, \infty)$; range $\left\{ \dfrac{1}{n} : n \neq 0, n \text{ an integer} \right\}$

b. Constant on $(n, n + 1)$ for every nonzero integer *n*   **c.** Neither

Critical Thinking:
81. b.

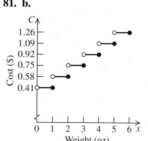

c. Domain: $(0, \infty)$; range:
$\{17n + 41 : n \text{ a nonnegative integer}\}$

83. a. $C(x) = \begin{cases} 150, & \text{if } x \leq 100 \\ 0.2\lfloor x - 100 \rfloor + 150, & \text{if } x > 100 \end{cases}$

b.

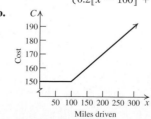

Section 1.5

Practice Problems: **1.**
The graph of *g* is the
graph of *f* shifted
one unit up; the
graph of *h* is the
graph of *f* shifted
two units down.

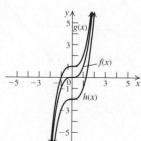

2.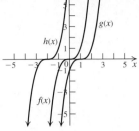

The graph of g is the graph of f shifted one unit to the right; the graph of h is the graph of f shifted two units to the left.

3.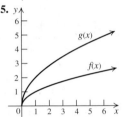

4. Shift the graph of $y = x^2$ one unit to the right, reflect the resulting graph in the x-axis, and shift it two units up.

5.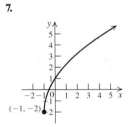

The graph of g is the graph of f vertically stretched by multiplying each of its y-coordinates by 2.

6. a. 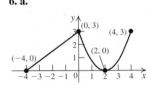 **b.**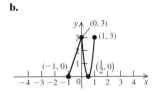

7. 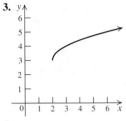 **8.**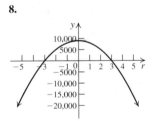

A Exercises: Basic Skills and Concepts:

33. 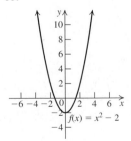 **34.**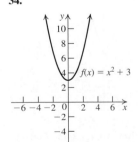

35. 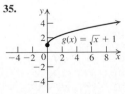 **36.**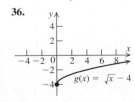

37. 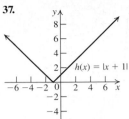 **38.**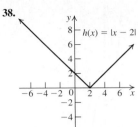

39. 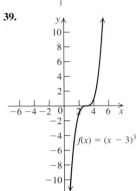 **40.**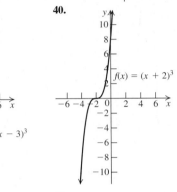

41. $g(x) = (x - 2)^2 + 1$ 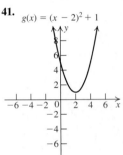 **42.** $g(x) = (x + 3)^2 - 5$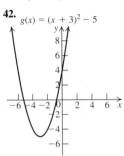

43. 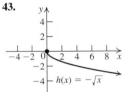 **44.**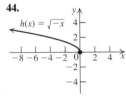

45. 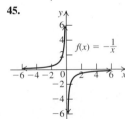 **46.**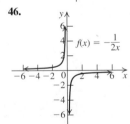

47. 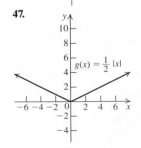 **48.**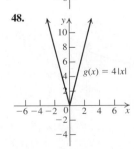

49.

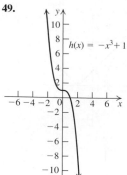

$h(x) = -x^3 + 1$

50.

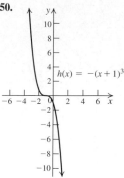

$h(x) = -(x+1)^3$

61. $h(x) = 2[[x+1]]$

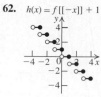

62. $h(x) = f[[-x]] + 1$

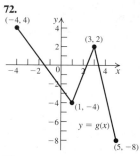

51.

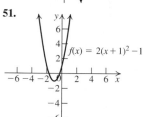

$f(x) = 2(x+1)^2 - 1$

52.

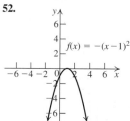

$f(x) = -(x-1)^2$

71.

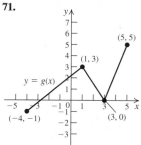

$y = g(x)$, (5, 5), (1, 3), (−4, −1), (3, 0)

72.

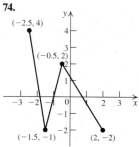

(−4, 4), (3, 2), (1, −4), (5, −8), $y = g(x)$

53.

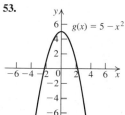

$g(x) = 5 - x^2$

54.

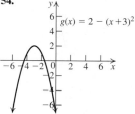

$g(x) = 2 - (x+3)^2$

73.

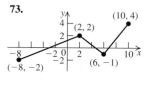

(10, 4), (2, 2), (−8, −2), (6, −1)

74.

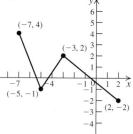

(−2.5, 4), (−0.5, 2), (−1.5, −1), (2, −2)

55.

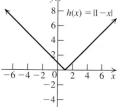

$h(x) = |1 - x|$

56.

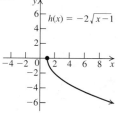

$h(x) = -2\sqrt{x-1}$

75.

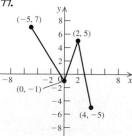

(6, 4), (2, 2), (−3, −2), (4, −1)

76.

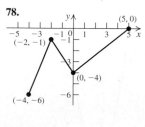

(−7, 4), (−3, 2), (−5, −1), (2, −2)

57.

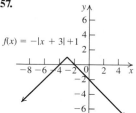

$f(x) = -|x+3| + 1$

58.

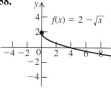

$f(x) = 2 - \sqrt{x}$

77.

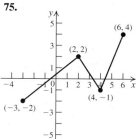

(−5, 7), (2, 5), (0, −1), (4, −5)

78.

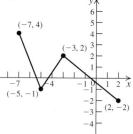

(5, 0), (−2, −1), (0, −4), (−4, −6)

59.

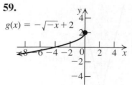

$g(x) = -\sqrt{-x} + 2$

60.

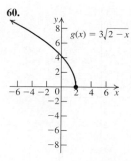

$g(x) = 3\sqrt{2 - x}$

B Exercises: Applying the Concepts

83. a.

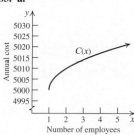

Annual cost; $C(x)$; Number of employees

85. a.

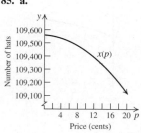

Number of hats; $x(p)$; Price (cents)

87. The first coordinate gives the month. The second coordinate gives the hours of daylight.

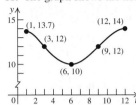

88. The graph shows the number of hours of darkness.

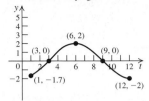

C Exercises: Beyond the Basics: **89.** Shift one unit to the right, stretch by a factor of 2, reflect in the *x*-axis, and shift three units up.

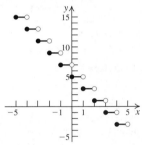

90. Shift two units to the left, stretch by a factor of 3, and shift one unit down.

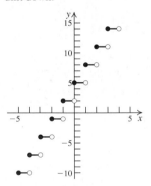

91. Shift two units to the left, and four units down.

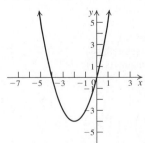

92. $x^2 - 6x = (x^2 - 6x + 9) - 9 = (x - 3)^2 - 9$
Shift three units to the right and nine units down.

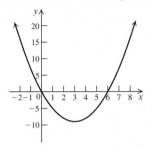

93. $-x^2 + 2x = -(x^2 - 2x + 1) + 1 = -(x - 1)^2 + 1$
Shift one unit to the right, reflect in the *x*-axis, and shift one unit up.

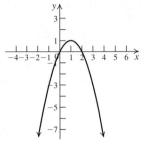

94. $-x^2 - 2x = -(x^2 + 2x + 1) + 1 = -(x + 1)^2 + 1$
Shift one unit to the left, reflect in the *x*-axis, and shift one unit up.

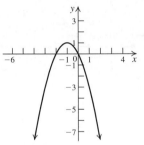

95. $2x^2 - 4x = 2(x^2 - 2x + 1) - 2 = 2(x - 1)^2 - 2$
Shift one unit to the right, stretch by a factor of two, and shift two units down.

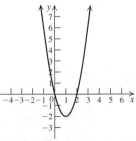

96. $2x^2 + 6x + 3.5 = 2(x^2 + 3x + 2.25) - 1 = 2(x + 1.5)^2 - 1$
Shift 1.5 units to the left, stretch by a factor of 2, and shift one unit down.

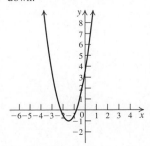

97. $-2x^2 - 8x + 3 = -2(x^2 + 4x + 4) + 11 = -2(x + 2)^2 + 11$
Shift 2 units to the left, stretch by a factor of 2, reflect in the x-axis, and shift 11 units up.

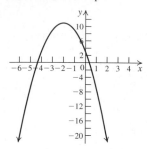

98. $-2x^2 - 2x - 1 = -2(x^2 + x + 0.25) - 0.5 =$
$-2(x - 0.5)^2 - 0.5$ Shift 0.5 unit to the right, stretch by a factor of 2, reflect in the x-axis, and shift by 0.5 unit down.

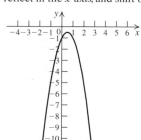

99.   **100.**

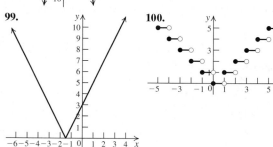

101.   **102.**

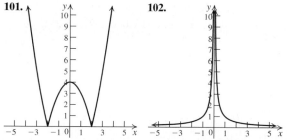

103.   **104.**

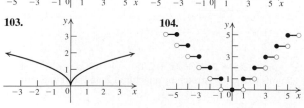

Section 1.6

Practice Problems:

1. $(f + g)(x) = x^2 + 3x + 1$
$(f - g)(x) = -x^2 + 3x - 3$
$(fg)(x) = 3x^3 - x^2 + 6x - 2$
$\left(\dfrac{f}{g}\right)(x) = \dfrac{3x - 1}{x^2 + 2}$

2. a. $(f \circ g)(0) = -5$ **b.** $(g \circ f)(0) = 1$

3. $(g \circ f)(x) = 2x^2 - 8x + 9$
$(f \circ g)(x) = 1 - 2x^2$

4. $(g \circ f)(x) = 2\sqrt{x} + 5$; domain: $[0, \infty)$ **5.** $f(x) = \dfrac{1}{x}$

6. a. $A = 9\pi t^2$ **b.** $A = 324\pi$ square miles

7. a. $r(x) = x - 4500$ **b.** $d(x) = 0.94x$
c. (i) $(r \circ d)(x) = 0.94x - 4500$
   (ii) $(d \circ r)(x) = 0.94x - 4230$
d. $(d \circ r)(x) - (r \circ d)(x) = 270$

A Exercises: Basic Skills and Concepts: **7. a.** $(f + g)(-1) = -1$
b. $(f - g)(0) = 0$ **c.** $(f \cdot g)(2) = -8$ **d.** $\left(\dfrac{f}{g}\right)(1) = -2$

8. a. $(f + g)(-1) = 0$ **b.** $(f - g)(0) = 0$ **c.** $(f \cdot g)(2) = -9$
d. $\left(\dfrac{f}{g}\right)(1) = 0$ **9. a.** $(f + g)(-1) = 0$

b. $(f - g)(0) = \dfrac{1}{2}(\sqrt{2} - 2)$ **c.** $(f \cdot g)(2) = \dfrac{5}{2}$ **d.** $\left(\dfrac{f}{g}\right)(1) = \dfrac{1}{3\sqrt{3}}$

10. a. $(f + g)(-1) = \dfrac{59}{15}$ **b.** $(f - g)(0) = -3$ **c.** Does not exist

d. $\left(\dfrac{f}{g}\right)(1) = \dfrac{1}{6}$

11. a. $(f + g)(x) = x^2 + x - 3$; domain: $(-\infty, \infty)$
b. $(f - g)(x) = -x^2 + x - 3$; domain: $(-\infty, \infty)$
c. $(f \cdot g)(x) = x^3 - 3x^2$; domain: $(-\infty, \infty)$
d. $\left(\dfrac{f}{g}\right)(x) = \dfrac{x - 3}{x^2}$; domain: $(-\infty, 0) \cup (0, \infty)$
e. $\left(\dfrac{g}{f}\right)(x) = \dfrac{x^2}{x - 3}$; domain: $(-\infty, 3) \cup (3, \infty)$

12. a. $(f + g)(x) = x^2 + 2x - 1$; domain: $(-\infty, \infty)$
b. $(f - g)(x) = -x^2 + 2x - 1$; domain: $(-\infty, \infty)$
c. $(f \cdot g)(x) = 2x^3 - x^2$; domain: $(-\infty, \infty)$
d. $\left(\dfrac{f}{g}\right)(x) = \dfrac{2x - 1}{x^2}$; domain: $(-\infty, 0) \cup (0, \infty)$
e. $\left(\dfrac{g}{f}\right)(x) = \dfrac{x^2}{2x - 1}$; domain: $\left(-\infty, \dfrac{1}{2}\right) \cup \left(\dfrac{1}{2}, \infty\right)$

13. a. $(f + g)(x) = x^3 + 2x^2 + 4$; domain: $(-\infty, \infty)$
b. $(f - g)(x) = x^3 - 2x^2 - 6$; domain: $(-\infty, \infty)$
c. $(f \cdot g)(x) = 2x^5 + 5x^3 - 2x^2 - 5$; domain: $(-\infty, \infty)$
d. $\left(\dfrac{f}{g}\right)(x) = \dfrac{x^3 - 1}{2x^2 + 5}$; domain: $(-\infty, \infty)$
e. $\left(\dfrac{g}{f}\right)(x) = \dfrac{2x^2 + 5}{x^3 - 1}$; domain: $(-\infty, 1) \cup (1, \infty)$

14. a. $(f + g)(x) = 2x^2 - 6x + 4$; domain: $(-\infty, \infty)$
b. $(f - g)(x) = 6x - 12$; domain: $(-\infty, \infty)$
c. $(f \cdot g)(x) = x^4 - 6x^3 + 4x^2 + 24x - 32$; domain: $(-\infty, \infty)$
d. $\left(\dfrac{f}{g}\right)(x) = \dfrac{x^2 - 4}{x^2 - 6x + 8}$; domain: $(-\infty, 2) \cup (2, 4) \cup (4, \infty)$
e. $\left(\dfrac{g}{f}\right)(x) = \dfrac{x^2 - 6x + 8}{x^2 - 4}$;
domain: $(-\infty, -2) \cup (-2, 2) \cup (2, \infty)$

15. a. $(f + g)(x) = 2x + \sqrt{x} - 1$; domain: $[0, \infty)$
b. $(f - g)(x) = 2x - \sqrt{x} - 1$; domain: $[0, \infty)$
c. $(f \cdot g)(x) = 2x\sqrt{x} - \sqrt{x}$; domain: $[0, \infty)$
d. $\left(\dfrac{f}{g}\right)(x) = \dfrac{2x - 1}{\sqrt{x}}$; domain: $(0, \infty)$
e. $\left(\dfrac{g}{f}\right)(x) = \dfrac{\sqrt{x}}{2x - 1}$; domain: $\left[0, \dfrac{1}{2}\right) \cup \left(\dfrac{1}{2}, \infty\right)$

16. a. $(f + g)(x) = 1$; domain: $(-\infty, 0) \cup (0, \infty)$
b. $(f - g)(x) = 1 - \dfrac{2}{x}$; domain: $(-\infty, 0) \cup (0, \infty)$
c. $(f \cdot g)(x) = \dfrac{x - 1}{x^2}$; domain: $(-\infty, 0) \cup (0, \infty)$
d. $\left(\dfrac{f}{g}\right)(x) = x - 1$; domain: $(-\infty, 0) \cup (0, \infty)$
e. $\left(\dfrac{g}{f}\right)(x) = \dfrac{1}{x - 1}$; domain: $(-\infty, 0) \cup (0, 1) \cup (1, \infty)$

39. a. $(f \circ g)(x) = 2x + 5$; domain: $(-\infty, \infty)$
  b. $(g \circ f)(x) = 2x + 1$; domain: $(-\infty, \infty)$
  c. $(f \circ f)(x) = 4x - 9$; domain: $(-\infty, \infty)$
  d. $(g \circ g)(x) = x + 8$; domain: $(-\infty, \infty)$
40. a. $(f \circ g)(x) = 3x - 8$; domain: $(-\infty, \infty)$
  b. $(g \circ f)(x) = 3x - 14$; domain: $(-\infty, \infty)$
  c. $(f \circ f)(x) = x - 6$; domain: $(-\infty, \infty)$
  d. $(g \circ g)(x) = 9x - 20$; domain: $(-\infty, \infty)$
41. a. $(f \circ g)(x) = -2x^2 - 1$; domain: $(-\infty, \infty)$
  b. $(g \circ f)(x) = 4x^2 - 4x + 2$; domain: $(-\infty, \infty)$
  c. $(f \circ f)(x) = 4x - 1$; domain: $(-\infty, \infty)$
  d. $(g \circ g)(x) = x^4 + 2x^2 + 2$; domain: $(-\infty, \infty)$
42. a. $(f \circ g)(x) = 4x^2 - 3$; domain: $(-\infty, \infty)$
  b. $(g \circ f)(x) = 2(2x - 3)^2$; domain: $(-\infty, \infty)$
  c. $(f \circ f)(x) = 4x - 9$; domain: $(-\infty, \infty)$
  d. $(g \circ g)(x) = 8x^4$; domain: $(-\infty, \infty)$
43. a. $(f \circ g)(x) = 8x^2 - 2x - 1$; domain: $(-\infty, \infty)$
  b. $(g \circ f)(x) = 4x^2 + 6x - 1$; domain: $(-\infty, \infty)$
  c. $(f \circ f)(x) = 8x^4 + 24x^3 + 24x^2 + 9x$; domain: $(-\infty, \infty)$
  d. $(g \circ g)(x) = 4x - 3$; domain: $(-\infty, \infty)$
44. a. $(f \circ g)(x) = 4x^2 + 6x$; domain: $(-\infty, \infty)$
  b. $(g \circ f)(x) = 2x^2 + 6x$; domain: $(-\infty, \infty)$
  c. $(f \circ f)(x) = x^4 + 6x^3 + 12x^2 + 9x$; domain: $(-\infty, \infty)$
  d. $(g \circ g)(x) = 4x$; domain: $(-\infty, \infty)$
45. a. $(f \circ g)(x) = x$; domain: $[0, \infty)$
  b. $(g \circ f)(x) = |x|$; domain: $(-\infty, \infty)$
  c. $(f \circ f)(x) = x^4$; domain: $(-\infty, \infty)$
  d. $(g \circ g)(x) = \sqrt[4]{x}$; domain: $[0, \infty)$
46. a. $(f \circ g)(x) = x + 2 + 2\sqrt{x + 2}$; domain: $[-2, \infty)$
  b. $(g \circ f)(x) = \sqrt{x^2 + 2x + 2}$; domain: $(-\infty, \infty)$
  c. $(f \circ f)(x) = x^4 + 4x^3 + 6x^2 + 4x$; domain: $(-\infty, \infty)$
  d. $(g \circ g)(x) = \sqrt{\sqrt{x + 2} + 2}$; domain: $[-2, \infty)$
47. a. $(f \circ g)(x) = -\dfrac{x^2}{x^2 - 2}$;
    domain: $(-\infty, -\sqrt{2}) \cup (-\sqrt{2}, 0) \cup (0, \sqrt{2}) \cup (\sqrt{2}, \infty)$
  b. $(g \circ f)(x) = (2x - 1)^2$; domain: $\left(-\infty, \dfrac{1}{2}\right) \cup \left(\dfrac{1}{2}, \infty\right)$
  c. $(f \circ f)(x) = -\dfrac{2x - 1}{2x - 3}$;
    domain: $\left(-\infty, \dfrac{1}{2}\right) \cup \left(\dfrac{1}{2}, \dfrac{3}{2}\right) \cup \left(\dfrac{3}{2}, \infty\right)$
  d. $(g \circ g)(x) = x^4$; domain: $(-\infty, 0) \cup (0, \infty)$
48. a. $(f \circ g)(x) = -\dfrac{1}{x + 1}$; domain: $(-\infty, -1) \cup (-1, \infty)$
  b. $(g \circ f)(x) = \dfrac{x - 1}{x}$; domain: $(-\infty, 0) \cup (0, \infty)$
  c. $(f \circ f)(x) = x - 2$; domain: $(-\infty, \infty)$
  d. $(g \circ g)(x) = \dfrac{x}{2x + 1}$;
    domain: $(-\infty, -1) \cup \left(-1, -\dfrac{1}{2}\right) \cup \left(-\dfrac{1}{2}, \infty\right)$
49. a. $(f \circ g)(x) = 2$; domain: $(-\infty, \infty)$
  b. $(g \circ f)(x) = -2$; domain: $(-\infty, \infty)$
  c. $(f \circ f)(x) = |x|$; domain: $(-\infty, \infty)$
  d. $(g \circ g)(x) = -2$; domain: $(-\infty, \infty)$
50. a. $(f \circ g)(x) = 3$; domain: $(-\infty, \infty)$
  b. $(g \circ f)(x) = 5$; domain: $(-\infty, \infty)$
  c. $(f \circ f)(x) = 3$; domain: $(-\infty, \infty)$
  d. $(g \circ g)(x) = 5$; domain: $(-\infty, \infty)$
51. a. $(f \circ g)(x) = \dfrac{2}{1 + x}$; domain: $(-\infty, -1) \cup (-1, 1) \cup (1, \infty)$
  b. $(g \circ f)(x) = -2x - 1$; domain: $(-\infty, 0) \cup (0, \infty)$
  c. $(f \circ f)(x) = \dfrac{2x + 1}{x + 1}$; domain: $(-\infty, -1) \cup (-1, 0) \cup (0, \infty)$
  d. $(g \circ g)(x) = -\dfrac{1}{x}$; domain: $(-\infty, 0) \cup (0, 1) \cup (1, \infty)$

52. a. $(f \circ g)(x) = \sqrt[3]{x^3 + 2}$; domain: $(-\infty, \infty)$
  b. $(g \circ f)(x) = x + 2$; domain: $(-\infty, \infty)$
  c. $(f \circ f)(x) = \sqrt[3]{\sqrt[3]{x + 1} + 1}$; domain: $(-\infty, \infty)$
  d. $(g \circ g)(x) = (x^3 + 1)^3 + 1$; domain: $(-\infty, \infty)$

B Exercises: Applying the Concepts: **68. c.** $g(f(x)) = 0.9(0.8x) = 0.72x$ **d.** $f(g(x)) = 0.8(0.9x) = 0.72x$ **69. b.** $(f \circ g)(x) = 1.1x + 8.8$; a final test score, which originally was x, after adding 8 points and then increasing the score by 10% **c.** $(g \circ f)(x) = 1.1x + 8$; a final test score, which originally was x, after increasing the score by 10% and then adding 8 points

Section 1.7

Practice Problems: **1.** Not one-to-one. The horizontal line $y = 1$ intersects the graph at two different points. **2. a.** -3 **b.** 4

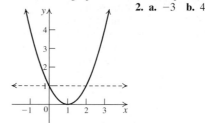

3. $(f \circ g)(x) = 3\dfrac{x + 1}{3} - 1 = x + 1 - 1 = x$ and
  $(g \circ f)(x) = \dfrac{3x - 1 + 1}{3} = \dfrac{3x}{3} = x$; therefore, f and g are inverses of each other.

4.

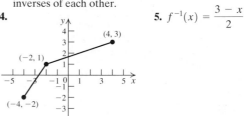

5. $f^{-1}(x) = \dfrac{3 - x}{2}$

6. $f^{-1}(x) = \dfrac{3x}{1 - x}$, $x \neq 1$ **7.** Domain: $(-\infty, -3) \cup (-3, \infty)$; range: $(-\infty, 1) \cup (1, \infty)$ **8.** $g^{-1}(x) = -\sqrt{x + 1}$ **9.** 3630 ft

A Exercises : Basic Skills and Concepts:

35.

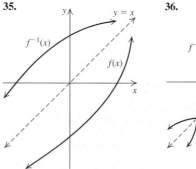

36.

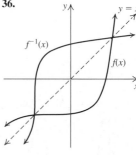

37.

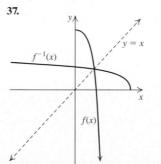

38.

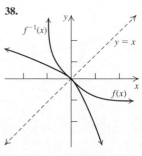

39.

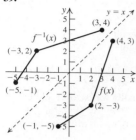

40.

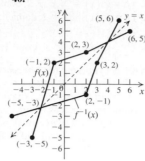

41. a. One-to-one **b.** $f^{-1}(x) = 5 - \dfrac{1}{3}x$

c.

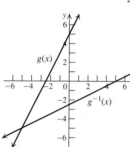

d. Domain f: $(-\infty, \infty)$
x-intercept: 5
y-intercept: 15
Domain g: $(-\infty, \infty)$
x-intercept: 15
y-intercept: 5

42. a. One-to-one **b.** $g^{-1}(x) = \dfrac{1}{2}x - \dfrac{5}{2}$

c.

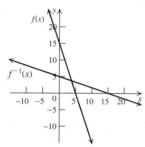

d. Domain g: $(-\infty, \infty)$
x-intercept: $-\dfrac{5}{2}$
y-intercept: 5
Domain g: $(-\infty, \infty)$
x-intercept: 5
y-intercept: $-\dfrac{5}{2}$

45. a. One-to-one **b.** $f^{-1}(x) = (x - 3)^2$

c.

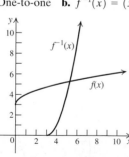

d. Domain f: $[0, \infty)$
x-intercept: 16
y-intercept: 3
Domain f^{-1}: $[3, \infty)$
x-intercept: 3
y-intercept: 16

46. a. One-to-one **b.** $f^{-1}(x) = (x - 4)^2$

c.

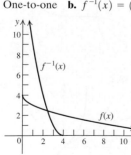

d. Domain f: $[0, \infty)$
No x-intercept
y-intercept: 4
Domain f^{-1}: $(-\infty, 4]$
x-intercept: 4
No y-intercept

47. a. One-to-one **b.** $g^{-1}(x) = (x^3 - 1)$

c.

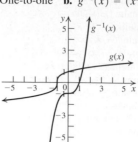

d. Domain g: $(-\infty, \infty)$
x-intercept: -1
y-intercept: 1
Domain g^{-1}: $(-\infty, \infty)$
x-intercept: 1
y-intercept: -1

48. a. One-to-one **b.** $h^{-1}(x) = 1 - x^3$

c.

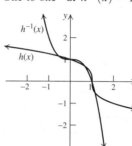

d. Domain h: $(-\infty, \infty)$
x-intercept: 1
y-intercept: 1
Domain h^{-1}: $(-\infty, \infty)$
x-intercept: 1
y-intercept: 1

49. a. One-to-one **b.** $f^{-1}(x) = \dfrac{x + 1}{x}$

c.

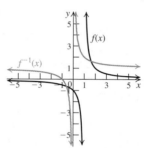

d. Domain f: $(-\infty, 1) \cup (1, \infty)$
No x-intercept
y-intercept: -1
Domain f^{-1}: $(-\infty, 0) \cup (0, \infty)$
x-intercept: -1
No y-intercept

50. a. One-to-one **b.** $g^{-1}(x) = \dfrac{1}{1 - x}$

c.

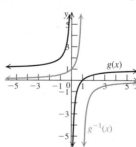

d. Domain g: $(-\infty, 0) \cup (0, \infty)$
x-intercept: 1
No y-intercept
Domain g^{-1}: $(-\infty, 1) \cup (1, \infty)$
No x-intercept
y-intercept: 1

51. a. One-to-one **b.** $f^{-1}(x) = x^2 - 4x + 3$

c.

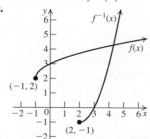

d. Domain f: $[-1, \infty)$.
y-intercept: 3
No x-intercept
Domain f^{-1}: $(2, \infty)$
x-intercept is 3.
No y-intercept

52. a. One-to-one   **b.** $f^{-1}(x) = x^2 + 2x - 1$

c.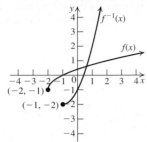

d. Domain $f: [-2, \infty)$
x-intercept: -1
y-intercept: $-1 + \sqrt{2}$.
Domain: $f^{-1}: [-1, \infty)$
x-intercept: $-1 + \sqrt{2}$
y-intercept: -1

55. $f^{-1}(x) = \dfrac{2x + 1}{x - 1}$

Domain: $(-\infty, 2) \cup (2, \infty)$; range: $(-\infty, 1) \cup (1, \infty)$

56. $g^{-1}(x) = \dfrac{x - 2}{1 - x}$

Domain: $(-\infty, -1) \cup (-1, \infty)$; range: $(-\infty, 1) \cup (1, \infty)$

57. $f^{-1}(x) = \dfrac{1 - x}{x + 2}$

Domain: $(-\infty, -1) \cup (-1, \infty)$; range: $(-\infty, -2) \cup (-2, \infty)$

58. $h^{-1}(x) = \dfrac{3x - 1}{x - 1}$

Domain: $(-\infty, 3) \cup (3, \infty)$; range: $(-\infty, 1) \cup (1, \infty)$

59. $y = \sqrt{-x}, x \le 0$ **60.** $y = -\sqrt{-x}, x \le 0$

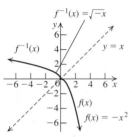   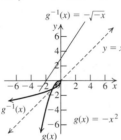

61. $y = x, x \ge 0$ **62.** $y = -x, x \ge 0$

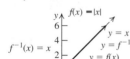    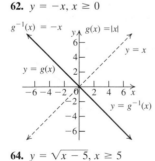

63. $y = -\sqrt{x - 1}, x \ge 1$ **64.** $y = \sqrt{x - 5}, x \ge 5$

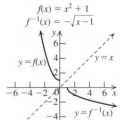   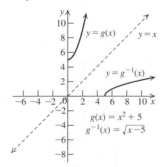

65. $y = -\sqrt{2 - x}, x \le 2$ **66.** $y = \sqrt{-1 - x} \; x \le -1$

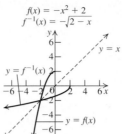   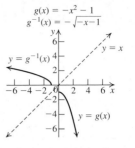

B Exercises Applying the Concepts: 67. a. $C(K) = K - 273$. It represents the Celsius temperature corresponding to a given Kelvin temperature.   **68. a.** $K(F) = \dfrac{5}{9} F + \dfrac{2297}{9}$ **b.** $F(K) = \dfrac{9}{5} K - \dfrac{2297}{5}$.

It represents the Fahrenheit temperature corresponding to a given Kelvin temperature.   **71. a.** Dollars to euros: $f(x) = 0.75x$ (x is the number of dollars; $f(x)$ is the number of euros). Euros to dollars: $g(x) = 1.25x$ (x is the number of euros; $g(x)$ is the number of dollars)   **b.** $g(f(x)) = 0.9375x \ne x$; therefore, g and f are not inverse functions.   **72. a.** $x = 20w - 80$. It gives the food sales in terms of his hourly wage.   **73. a.** $w = \begin{cases} 4 + 0.05x, & \text{if } x > 60 \\ 7, & \text{if } x \le 60 \end{cases}$

b. It does not have an inverse because it is constant on $(0, 60)$ and hence is not one-to-one.

74. a. $l = \dfrac{1}{1.11} T$. It shows the length as the function of the period.

75. a. $x = \dfrac{1}{64} V^2$. It gives the height of the water in terms of the velocity.

76. a. $x = \dfrac{64 - \sqrt{4096 - 8y}}{4}, 0 \le y \le 512$

77. b. $f^{-1}(x) = 60 - \dfrac{1}{600}x$. It shows the number of month that have passed from the first payment until the balance due is $\$x$.

78. a. $p = 2 - \dfrac{1}{4}\sqrt{2x - 2336}, 1168 \le x < 1200$. It shows the price of the computer chips in terms of the demand x.

C Exercises: Beyond the Basics:
81. a.

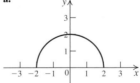

83. a. 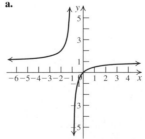   **b.** $f^{-1}(x) = \dfrac{x}{1 - x}$

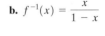

f satisfies the horizontal-line test.

84. a.

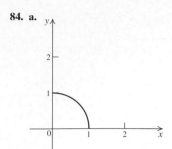

g satisfies the horizontal-line test

85. a. The midpoint is $M(5,5)$. Since its coordinates satisfy the equation $y = x$, it lies on the line.

   b. The slope of the line segment $\overline{PQ}$ is -1, and the slope of the line $y = x$ is 1. Since $(-1)(1) = -1$, the two lines are perpendicular.

87. a. The graph of g is the graph of f shifted one unit to the right and two units up.

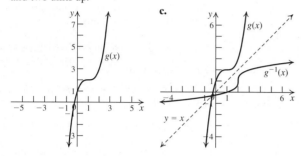

c.

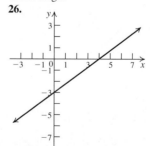

88. a. **(i)** $f^{-1}(x) = \dfrac{1}{2}x + \dfrac{1}{2}$ **(ii)** $g^{-1}(x) = \dfrac{1}{3}x - \dfrac{4}{3}$

   (iii) $(f \circ g)(x) = 6x + 7$ **(iv)** $(g \circ f)(x) = 6x + 1$

   (v) $(f \circ g)^{-1}(x) = \dfrac{1}{6}x - \dfrac{7}{6}$ **(vi)** $(g \circ f)^{-1}(x) = \dfrac{1}{6}x - \dfrac{1}{6}$

   (vii) $(f^{-1} \circ g^{-1})(x) = \dfrac{1}{6}x - \dfrac{1}{6}$

   (viii) $(g^{-1} \circ f^{-1})(x) = \dfrac{1}{6}x - \dfrac{7}{6}$

   b. **(i)** $(f \circ g)^{-1}(x) = \dfrac{1}{6}x - \dfrac{7}{6} = (g^{-1} \circ f^{-1})(x)$

   (ii) $(g \circ f)^{-1}(x) = \dfrac{1}{6}x - \dfrac{1}{6} = (f^{-1} \circ g^{-1})(x)$

89. a. **(i)** $f^{-1}(x) = \dfrac{1}{2}x - \dfrac{3}{2}$

   (ii) $g^{-1}(x) = \sqrt[3]{x + 1}$

   (iii) $(f \circ g)(x) = 2x^3 + 1$

   (iv) $(g \circ f)(x) = 8x^3 + 36x^2 + 54x + 26$

   (v) $(f \circ g)^{-1}(x) = \sqrt[3]{\dfrac{x - 1}{2}}$

   (vi) $(g \circ f)^{-1}(x) = \dfrac{1}{2}\sqrt[3]{x + 1} - \dfrac{3}{2}$

   (vii) $(f^{-1} \circ g^{-1})(x) = \dfrac{1}{2}\sqrt[3]{x + 1} - \dfrac{3}{2}$

   (viii) $(f \circ g)^{-1}(x) = \sqrt[3]{\dfrac{x - 1}{2}}$

   b. **(i)** $(f \circ g)^{-1}(x) = \sqrt[3]{\dfrac{x - 1}{2}} = (g^{-1} \circ f^{-1})(x)$

   (ii) $(g \circ f)^{-1}(x) = \dfrac{1}{2}\sqrt[3]{x + 1} - \dfrac{3}{2} = (f^{-1} \circ g^{-1})(x)$

Critical Thinking: **90.** No. $f(x) = x^3 - x$ is odd, but it does not have an inverse because $f(0) = f(1)$. So it is not one-to-one.
91. Yes. The function $f = \{(0,1)\}$ is even, and it has an inverse: $f^{-1} = \{(1,0)\}$.

Review Exercises

9. a. $2\sqrt{5}$ **b.** $(1,4)$ **c.** $\dfrac{1}{2}$ **10. a.** $6\sqrt{2}$

b. $(0,2)$ **c.** -1 **11. a.** $5\sqrt{2}$ **b.** $\left(\dfrac{13}{2}, -\dfrac{11}{2}\right)$ **c.** -1

12. a. $\sqrt{202}$ **b.** $\left(-\dfrac{5}{2}, -\dfrac{5}{2}\right)$ **c.** $\dfrac{11}{9}$ **13. a.** $\sqrt{34}$

b. $\left(\dfrac{7}{2}, -\dfrac{9}{2}\right)$ **c.** $\dfrac{5}{3}$ **14. a.** $\sqrt{274}$ **b.** $\left(\dfrac{5}{2}, \dfrac{1}{2}\right)$ **c.** $-\dfrac{7}{15}$

16. $d(A, B) = d(B, D) = d(D, C) = d(C, A) = 3\sqrt{5}$, so the four sides have equal length. Therefore, $ABDC$ is a rhombus.   **21.** Not symmetric with respect to the x-axis, symmetric with respect to the y-axis, not symmetric with respect to the origin   **22.** Not symmetric with respect to the x-axis, not symmetric with respect to the y-axis, symmetric with respect to the origin   **23.** Symmetric with respect to the x-axis, not symmetric with respect to the y-axis, not symmetric with respect to the origin   **24.** Symmetric with respect to the x-axis, symmetric with respect to the y-axis, symmetric with respect to the origin

25.

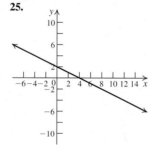

x-intercept: 4; y-intercept: 2; not symmetric with respect to the x-axis, not symmetric with respect to the y-axis, not symmetric with respect to the origin

26.

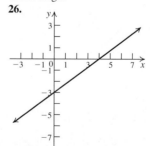

x-intercept: 4; y-intercept: -3; not symmetric with respect to the x-axis, not symmetric with respect to the y-axis, not symmetric with respect to the origin

27.

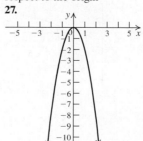

x-intercept: 0; y-intercept: 0; not symmetric with respect to the x-axis, not symmetric with respect to the y-axis, not symmetric with respect to the origin

28.

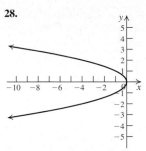

x-intercept: 0; *y*-intercept: 0; symmetric with respect to the *x*-axis, not symmetric with respect to the *y*-axis, not symmetric with respect to the origin

29.

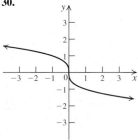

x-intercept: 0; *y*-intercept: 0; not symmetric with respect to the *x*-axis, not symmetric with respect to the *y*-axis, symmetric with respect to the origin

30.

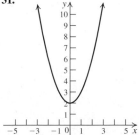

x-intercept: 0; *y*-intercept: 0; not symmetric with respect to the *x*-axis, not symmetric with respect to the *y*-axis, symmetric with respect to the origin

31.

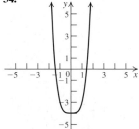

No *x*-intercept; *y*-intercept: 2; not symmetric with respect to the *x*-axis, symmetric with respect to the *y*-axis, not symmetric with respect to the origin

32.

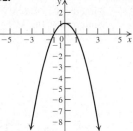

x-intercepts: −1 and 1; *y*-intercepts: 1; not symmetric with respect to the *x*-axis, symmetric with respect to the *y*-axis, symmetric with respect to the origin

33.

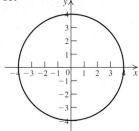

x-intercepts: −4 and 4; *y*-intercepts: −4 and 4; symmetric with respect to the *x*-axis, symmetric with respect to the *y*-axis, symmetric with respect to the origin

34.

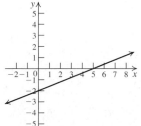

x-intercepts: −2 and 2; *y*-intercept: −4; not symmetric with respect to the *x*-axis, symmetric with respect to the *y*-axis, not symmetric with respect to the origin

38. $2x - 5y = 10 \Rightarrow y = \frac{2}{5}x - 2$; line with slope $\frac{2}{5}$, *y*-intercept: −2, *x*-intercept: 5

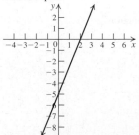

39. $\frac{x}{2} - \frac{y}{5} = 1 \Rightarrow y = \frac{5}{2}x - 5$; line with slope $\frac{5}{2}$; *y*-intercept: −5, *x*-intercept: 2

40. Circle centered at $(-1, 3)$ and with radius 4

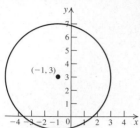

x-intercepts: $-1 \pm \sqrt{7}$
y-intercepts: $3 \pm \sqrt{15}$

41. $x^2 + y^2 - 2x + 4y - 4 = 0 \Rightarrow (x - 1)^2 + (y + 2)^2 = 9$; circle centered at $(1, -2)$ and with radius 3

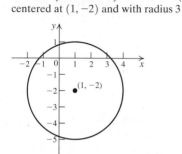

x-intercepts: $1 \pm \sqrt{5}$
y-intercepts: $-2 \pm 2\sqrt{2}$

42. $3x^2 + 3y^2 - 6x - 6 = 0 \Rightarrow x^2 + y^2 - 2x - 2 = 0 \Rightarrow$
$(x - 1)^2 + y^2 = 3$; circle centered at $(1, 0)$ and with radius $\sqrt{3}$

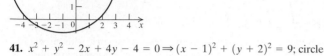

x-intercepts: $1 \pm \sqrt{3}$
y-intercepts: $\pm\sqrt{2}$

49.

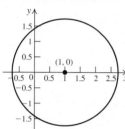

50.

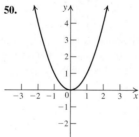

51.

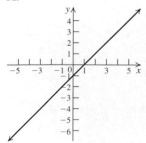

Domain: $(-\infty, \infty)$; range: $(-\infty, \infty)$; function

52.

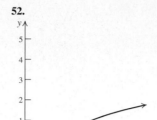

Domain: $[2, \infty)$; range: $[0, \infty)$; function

53.

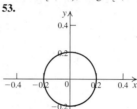

Domain: $[-0.2, 0.2]$; range: $[-0.2, 0.2]$; not a function

54.

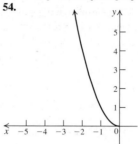

Domain: $(-\infty, 0]$; range: $[0, \infty)$; function

55.

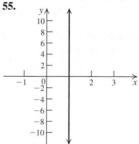

Domain: $\{1\}$; range: $(-\infty, \infty)$; not a function

56.

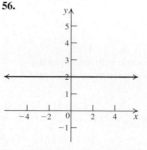

Domain: $(-\infty, \infty)$; range: $\{2\}$; function

57.

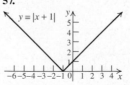

$y = |x + 1|$

Domain: $(-\infty, \infty)$; range: $[0, \infty)$; function

58.

Domain: $[1, \infty)$; range: $(-\infty, \infty)$; not a function

77.

Domain: $(-\infty, \infty)$; range: $\{-3\}$; constant on $(-\infty, \infty)$

78.

Domain: $(-\infty, \infty)$; range: $[-2, \infty)$; decreasing on $(-\infty, 0)$, increasing on $(0, \infty)$

79.

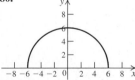

Domain: $\left[\dfrac{2}{3}, \infty\right)$; range: $[0, \infty)$; increasing on $\left(\dfrac{2}{3}, \infty\right)$

80.

Domain: $[-6, 6]$; range: $[0, 6]$; increasing on $(-6, 0)$, decreasing on $(0, 6)$

81.

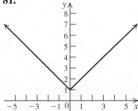

Domain: $(-\infty, \infty)$; range: $[1, \infty)$; decreasing on $(-\infty, 0)$, increasing on $(0, \infty)$

82.

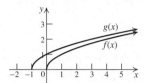

Domain: $(-\infty, \infty)$; range: $[0, \infty)$, decreasing on $(-\infty, 0)$, increasing on $(0, \infty)$

83. The graph of g is the graph of f shifted one unit to the left.

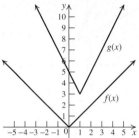

84. The graph of g is the graph of f shifted one unit to the right, stretched by a factor of 2, and shifted three units up.

85. The graph of g is the graph of f shifted two units to the right and reflected in the x-axis.

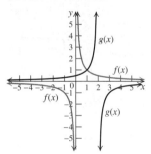

86. The graph of g is the graph of f shifted one unit to the left and two units down.

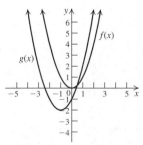

87. Even; not symmetric with respect to the x-axis, symmetric with respect to the y-axis, not symmetric with respect to the origin

88. Odd; not symmetric with respect to the *x*-axis, not symmetric with respect to the *y*-axis, symmetric with respect to the origin
89. Even; not symmetric with respect to the *x*-axis, symmetric with respect to the *y*-axis, not symmetric with respect to the origin
90. Neither even nor odd; not symmetric with respect to the *x*-axis, not symmetric with respect to the *y*-axis, not symmetric with respect to the origin **91.** Neither even nor odd; not symmetric with respect to the *x*-axis, not symmetric with respect to the *y*-axis, not symmetric with respect to the origin **92.** Odd; not symmetric with respect to the *x*-axis, not symmetric with respect to the *y*-axis, symmetric with respect to the origin **93.** $f(x) = (g \circ h)(x)$, where $g(x) = \sqrt{x}$ and $h(x) = x^2 - 4$ **94.** $g(x) = (f \circ h)(x)$, where $f(x) = x^{50}$ and $h(x) = x^2 - x + 2$ **95.** $h(x) = (f \circ g)(x)$, where $f(x) = \sqrt{x}$ and $g(x) = \dfrac{x - 3}{2x + 5}$ **96.** $H(x) = (f \circ g)(x)$, where $f(x) = x^3 + 5$ and $g(x) = 2x - 1$
97. One-to-one; $f^{-1}(x) = x - 2$ **98.** One-to-one;
$$f^{-1}(x) = -\frac{1}{2}x + \frac{3}{2}$$

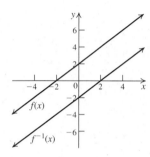

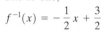
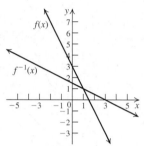

99. One-to-one; $f^{-1}(x) = x^3 + 2$ **100.** One-to-one;
$$f^{-1}(x) = \frac{1}{2}\sqrt[3]{x + 1}$$

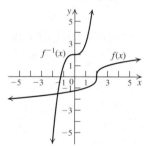

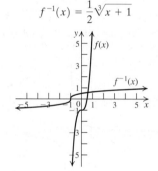

101. $f^{-1}(x) = x^2 - 8x + 17; x \geq 4$
Domain f: $[1, \infty)$; Range f: $[1, \infty)$
102. $f^{-1}(x) = x^2 + 6x + 7; x \geq -2$
Domain f: $[-2, \infty)$; Range f: $[-3, \infty)$
103. a. $f(x) = \begin{cases} 3x + 6, & \text{if } -3 \leq x \leq -2 \\ \dfrac{1}{2}x + 1, & \text{if } -2 < x < 0 \\ x + 1, & \text{if } 0 \leq x \leq 3 \end{cases}$

d.

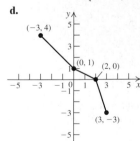

e.

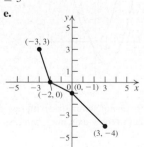

f.

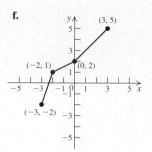

g.

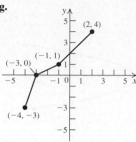

h.

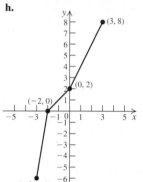

i.

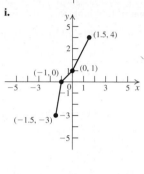

j.

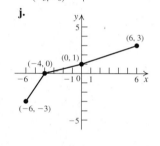

l.

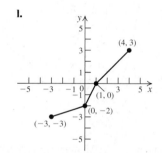

104. b. Slope: the amount of increase in pressure, in pounds per square inch, if the diver descends 1 ft deeper; *y*-intercept: the pressure at the surface of the sea.
105. b. Slope: the cost of disposing 1 pound of waste; *x*-intercept: the amount of waste that can be disposed with no cost; *y*-intercept: the fixed cost

Practice Test A

3.

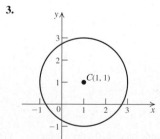

x-intercepts: $1 \pm \sqrt{3}$;
y-intercepts: $1 \pm \sqrt{3}$

CHAPTER 2

Section 2.1

***Practice Problems:* 1.** $y = 3(x - 1)^2 - 5$ **2.** The graph is a parabola. $a = -2, h = -1, k = 3$. It opens downward. Vertex: $(-1, 3)$.

It is a maximum. x-intercepts: $\pm\sqrt{\dfrac{3}{2}} - 1$; y-intercept: 1

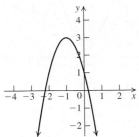

3. The graph is a parabola. $a = 3, b = -3, c = -6$. It opens upward.

Vertex: $\left(\dfrac{1}{2}, \dfrac{-27}{4}\right)$; x-intercepts: $-1, 2$; y-intercept: -6

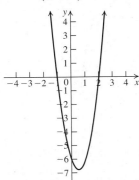

4.   **5.** 192

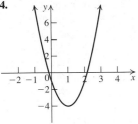

$$\left[\dfrac{3 - 2\sqrt{3}}{3}, \dfrac{3 + 2\sqrt{3}}{3}\right]$$

A Exercises: Basic Skills and Concepts:

26. $y = -\dfrac{4}{9}(x + 3)^2 + 4$ **27.** $y = \dfrac{11}{49}(x - 2)^2 - 3$

28. $y = -\dfrac{2}{3}(x + 3)^2 - 2$ **29.** $y = -12\left(x - \dfrac{1}{2}\right)^2 + \dfrac{1}{2}$

30. $y = \dfrac{3}{2}\left(x + \dfrac{3}{2}\right)^2 - \dfrac{5}{2}$

35. $y = (x + 2)^2 - 4$. The graph is the graph of $y = x^2$ shifted two units to the left and four units down. Vertex: $(-2, -4)$; axis: $x = -2$; x-intercepts: -4 and 0; y-intercept: 0

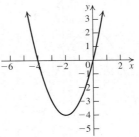

36. $y = (x - 1)^2 + 1$. The graph is the graph of $y = x^2$ shifted one unit to the right and one unit up. Vertex: $(1, 1)$; axis: $x = 1$; no x-intercept; y-intercept: 2

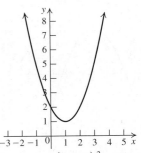

37. $y = -(x - 3)^2 - 1$. The graph is the graph of $y = x^2$ shifted three units to the right, reflected in the x-axis, and shifted one unit down. Vertex: $(3, -1)$; axis: $x = 3$; no x-intercept; y-intercept: -10

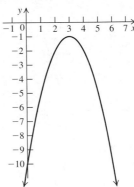

38. $y = -\left(x - \dfrac{3}{2}\right)^2 + \dfrac{41}{4}$.

The graph is the graph of $y = x^2$ shifted $\dfrac{3}{2}$ units to the right, reflected in the x-axis, and shifted $\dfrac{41}{4}$ units up.

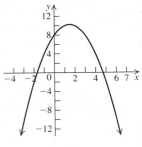

Vertex: $\left(\dfrac{3}{2}, \dfrac{41}{4}\right)$; axis: $x = \dfrac{3}{2}$;

x-intercepts: $\dfrac{3}{2} \pm \dfrac{1}{2}\sqrt{41}$;

y-intercept: 8

39. $y = 2(x - 2)^2 + 1$. The graph is the graph of $y = x^2$ shifted two units to the right, stretched vertically by a factor of 2, and shifted one unit up.

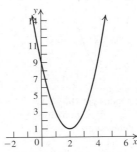

Vertex: $(2, 1)$; axis: $x = 2$; no x-intercept; y-intercept: 9

40. $y = 3(x + 2)^2 - 19$. The graph is the graph of $y = x^2$ shifted 2 units to the left, stretched vertically by a factor of 3, and shifted 19 units down.

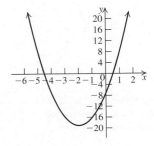

Vertex: $(-2, -19)$; axis: $x = -2$; x-intercepts: $-2 \pm \dfrac{1}{3}\sqrt{57}$; y-intercept: -7

41. $y = -3(x - 3)^2 + 16$. The graph is the graph of $y = x^2$ shifted 3 units to the right, stretched vertically by a factor of 3, reflected in the x-axis, and shifted 16 units up.

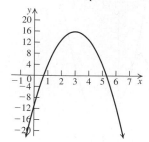

Vertex: $(3, 16)$; axis: $x = 3$; x-intercepts: $3 \pm \frac{4}{3}\sqrt{3}$; y-intercept: -11

42. $y = -5(x + 2)^2 + 33$. The graph is the graph of $y = x^2$ shifted 2 units to the left, stretched vertically by a factor of 5, reflected in the x-axis, and shifted 33 units up.

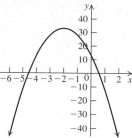

Vertex: $(-2, 33)$; axis: $x = -2$; x-intercepts: $-2 \pm \frac{1}{5}\sqrt{165}$; y-intercept: 13

43. a. Opens up   **b.** $(4, -1)$ **c.** $x = 4$ **d.** x-intercepts: 3 and 5; y-intercept: 15

e.

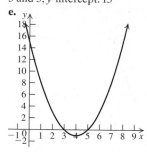

44. a. Opens up   **b.** $(-4, -3)$ **c.** $x = -4$ **d.** x-intercepts: $-4 \pm \sqrt{3}$; y-intercept: 13

e.

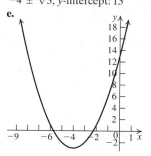

45. a. Opens up   **b.** $\left(\frac{1}{2}, \frac{-25}{4}\right)$ **c.** $x = \frac{1}{2}$ **d.** x-intercepts: -2 and 3; y-intercept: -6

e.

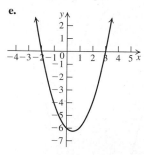

46. a. Opens up   **b.** $\left(-\frac{1}{2}, -\frac{9}{4}\right)$ **c.** $x = -\frac{1}{2}$ **d.** x-intercepts: -2 and 1; y-intercept: -2

e.

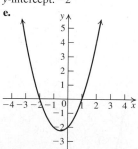

47. a. Opens up   **b.** $(1, 3)$ **c.** $x = 1$ **d.** x-intercepts: none 2; y-intercept: 4

48. a. Opens up   **b.** $(2, 1)$ **c.** $x = 2$ **d.** x-intercepts: none; y-intercept: 5

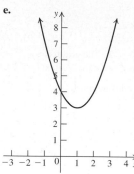

49. a. Opens down   **b.** $(-1, 7)$ **c.** $x = -1$ **d.** x-intercepts: $-1 \pm \sqrt{7}$; y-intercept: 6

e.

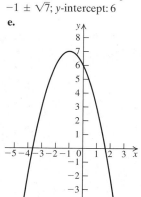

50. a. Opens down   **b.** $\left(\frac{5}{6}, \frac{49}{12}\right)$ **c.** $x = \frac{5}{6}$ **d.** x-intercepts: $-\frac{1}{3}$ and 2; y-intercept: 2

e.

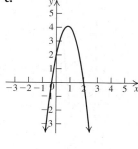

51. a. Minimum: -1 **b.** $[-1, \infty)$ **52. a.** Maximum: 1 **b.** $(-\infty, 1]$ **53. a.** Maximum: 0   **b.** $(-\infty, 0]$ **54. a.** Minimum: 0   **b.** $[0, \infty)$ **55. a.** Minimum: -5 **b.** $[-5, \infty)$ **56. a.** Minimum: -17 **b.** $[-17, \infty)$ **57. a.** Maximum: 16 **b.** $(-\infty, 16]$ **58. a.** Maximum: 3   **b.** $(-\infty, 3]$

59.

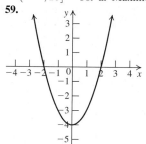

Solution: $[-2, 2]$

60.

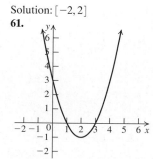

Solution: $\varnothing$

61.

Solution: $(-\infty, 1) \cup (3, \infty)$

62.

Solution: $(-\infty, -2) \cup (1, \infty)$

63.

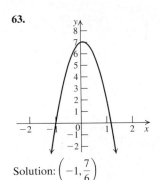

Solution: $\left(-1, \dfrac{7}{6}\right)$

64.

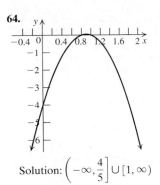

Solution: $\left(-\infty, \dfrac{4}{5}\right] \cup [1, \infty)$

B Exercises: Applying the Concepts: 79. Dimensions of the rectangle: width $= \dfrac{36}{\pi + 4}$ ft; height $= \dfrac{18}{\pi + 4}$ ft

80. a.

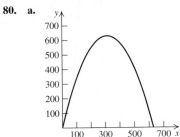

Section 2.2

Practice Problems: 1. a. Not a polynomial function
b. A polynomial function; the degree is 7, the leading term is $2x^7$, and the leading coefficient is 2.

2. $P(x) = 4x^3 + 2x^2 + 5x - 17 = x^3\left(4 + \dfrac{2}{x} + \dfrac{5}{x^2} - \dfrac{17}{x^3}\right)$

When $|x|$ is large, the terms $\dfrac{2}{x}, \dfrac{5}{x^2}$, and $-\dfrac{17}{x^3}$ are close to 0.

Therefore, $P(x) \approx x^3(4 + 0 + 0 - 0) = 4x^3$.

3. $y \to -\infty$ as $x \to -\infty$ and $y \to -\infty$ as $x \to \infty$ **4.** $\dfrac{3}{2}$

5. $f(1) = -7, f(2) = 4$. Since $f(1)$ and $f(2)$ have opposite signs, by the Intermediate Value Theorem, f has a real zero between 1 and 2.
6. $-5, -1$, and 3   **7.** -5 and -3 with multiplicity 1 and 1 with multiplicity 2   **8.** 3
9.

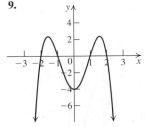

10. 207.378

A Exercises: Basic Skills and Concepts: 9. Polynomial function; degree: 5; leading term: $2x^5$; leading coefficient: 2   **10.** Polynomial function; degree: 4; leading term: $-7x^4$; leading coefficient: -7

11. Polynomial function; degree: 3; leading term: $\dfrac{2}{3}x^3$; leading coefficient: $\dfrac{2}{3}$ **12.** Polynomial function; degree: 3; leading term: $\sqrt{2}x^3$; leading coefficient: $\sqrt{2}$ **13.** Polynomial function; degree: 4; leading term: πx^4; leading coefficient: π **14.** Polynomial function; degree: 0; leading term: 5; leading coefficient: 5   **39.** Zeros: $x = -5$, multiplicity: 1, the graph crosses the x-axis; $x = 1$, multiplicity: 1, the graph crosses the x-axis; maximum number of turning points: 1

40. Zeros: $x = -2$, multiplicity: 2, the graph touches but does not cross the x-axis; $x = 3$, multiplicity: 1, the graph crosses the x-axis; maximum number of turning points: 2   **41.** Zeros: $x = -1$, multiplicity: 2, the graph touches but does not cross the x-axis; $x = 1$, multiplicity: 3, the graph crosses the x-axis; maximum number of turning points: 4   **42.** Zeros: $x = -3$, multiplicity: 2, the graph touches but does not cross the x-axis; $x = -1$, multiplicity: 1, the graph crosses the x-axis; $x = 6$, multiplicity: 1, the graph crosses the x-axis; maximum number of turning points: 3   **43.** Zeros: $x = -\dfrac{2}{3}$, multiplicity: 1, the graph crosses the x-axis; $x = \dfrac{1}{2}$, multiplicity: 2, the graph touches but does not cross the x-axis; maximum number of turning points: 2   **44.** Zeros: $x = 0$, multiplicity: 1, the graph crosses the x-axis; $x = 2$, multiplicity: 1, the graph crosses the x-axis; $x = 4$, multiplicity: 1, the graph crosses the x-axis; maximum number of turning points: 2   **45.** Zeros: $x = 0$, multiplicity: 2, the graph touches but does not cross the x-axis; $x = 3$, multiplicity: 2, the graph touches but does not cross the x-axis; maximum number of turning points: 3   **46.** Zeros: $x = -3$, multiplicity: 1, the graph crosses the x-axis; $x = 3$, multiplicity: 1, the graph crosses the x-axis; maximum number of turning points: 3
51. a. $y \to \infty$ as $x \to -\infty$ and $y \to \infty$ as $x \to \infty$ **b.** No zeros
c. The graph is above the x-axis on $(-\infty, \infty)$.   **d.** 3   **e.** y-axis symmetry.   **f.** 1   **g.**

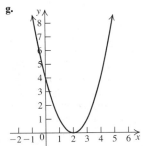

52. a. $y \to \infty$ as $x \to -\infty$ and $y \to \infty$ as $x \to \infty$ **b.** Zero: $x = 2$, the graph touches but does not cross the x-axis.
c. The graph is above the x-axis on $(-\infty, 2) \cup (2, \infty)$.
d. 4   **e.** no symmetries   **f.** 1   **g.**

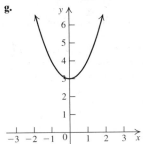

53. a. $y \to \infty$ as $x \to -\infty$ and $y \to \infty$ as $x \to \infty$
b. Zeros: $x = -7$, the graph crosses the x-axis; $x = 3$, the graph crosses the x-axis.
c. The graph is above the x-axis on $(-\infty, -7)$ and $(3, \infty)$ and below the x-axis on $(-7, 3)$.
d. -21 **e.** no symmetries   **f.** 1   **g.**

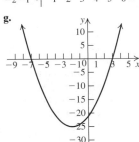

54. a. $y \to -\infty$ as $x \to -\infty$ and $y \to -\infty$ as $x \to \infty$ **b.** Zeros: $x = -2$, the graph crosses the x-axis; $x = 6$, the graph crosses the x-axis.   **c.** The graph is below the x-axis on $(-\infty, -2)$ and $(6, \infty)$ and above the x-axis on $(-2, 6)$ **d.** 12   **e.** no symmetries.   **f.** 1   **g.**

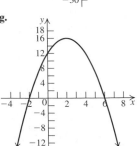

55. a. $y \to \infty$ as $x \to -\infty$ and
$y \to -\infty$ as $x \to \infty$ **b.** Zeros:
$x = -1$, the graph crosses the
x-axis; $x = 0$, the graph touches
but does not cross the x-axis.
c. The graph is above the x-axis
on $(-\infty, -1)$ and below the
x-axis on $(-1, 0) \cup (0, \infty)$.
d. 0 **e.** no symmetries. **f.** 2

g.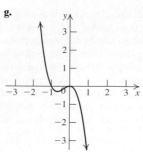

56. a. $y \to \infty$ as $x \to -\infty$ and
$y \to -\infty$ as $x \to \infty$ **b.** Zeros:
$x = -1$, the graph crosses the
x-axis; $x = 0$, the graph crosses
the x-axis; $x = 1$, the graph crosses
the x-axis. **c.** The graph is above
the x-axis on $(-\infty, -1)$ and $(0, 1)$
and below the x-axis on $(-1, 0)$
and $(1, \infty)$. **d.** 0 **e.** origin
symmetry. **f.** 2

g.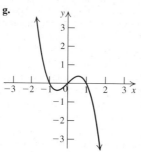

57. a. $y \to \infty$ as $x \to -\infty$ and
$y \to \infty$ as $x \to \infty$ **b.** Zeros:
$x = 0$, the graph touches but
does not cross the x-axis; $x = 1$,
the graph touches but does not
cross the x-axis. **c.** The graph
is above the x-axis
on $(-\infty, 0) \cup (0, 1) \cup (1, \infty)$.
d. 0 **e.** no symmetries **f.** 3

g.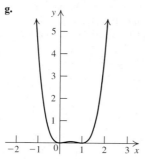

58. a. $y \to \infty$ as $x \to -\infty$
and $y \to \infty$ as $x \to \infty$
b. Zeros: $x = -1$, the graph
crosses the x-axis; $x = 0$, the
graph touches but does not
cross the x-axis; $x = 2$, the
graph crosses the x-axis.
c. The graph is above the
x-axis on $(-\infty, -1)$ and
$(2, \infty)$ and below the x-axis
on $(-1, 0) \cup (0, 2)$. **d.** 0
e. no symmetries **f.** 3

g.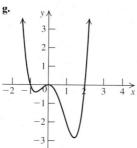

59. a. $y \to \infty$ as $x \to -\infty$
and $y \to \infty$ as $x \to \infty$
b. Zeros: $x = -3$, the graph
crosses the x-axis; $x = 1$, the
graph touches but does not
cross the x-axis; $x = 4$, the
graph crosses the x-axis.
c. The graph is above the
x-axis on $(-\infty, -3)$ and
$(4, \infty)$ and below the x-axis
on $(-3, 1) \cup (1, 4)$. **d.** -12
e. no symmetries **f.** 3

g.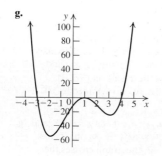

60. a. $y \to \infty$ as $x \to -\infty$
and $y \to \infty$ as $x \to \infty$ **b.** Zero:
$x = -1$, the graph touches but
does not cross the x-axis.
c. The graph is above the x-axis
on $(-\infty, -1) \cup (-1, \infty)$. **d.** 1
e. no symmetries. **f.** 1

g.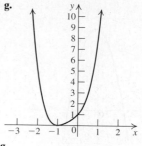

61. a. $y \to \infty$ as $x \to -\infty$
and $y \to -\infty$ as $x \to \infty$
b. Zeros: $x = -1$, the graph
touches but does not cross
the x-axis; $x = 0$, the graph
touches but does not cross the
x-axis; $x = 1$, the graph crosses
the x-axis. **c.** The graph is
above the x-axis on
$(-\infty, -1) \cup (-1, 0) \cup (0, 1)$
and below the x-axis on $(1, \infty)$.
d. 0 **e.** no symmetries. **f.** 4

g.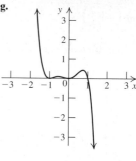

62. a. $y \to \infty$ as $x \to -\infty$ and
$y \to -\infty$ as $x \to \infty$ **b.** Zeros:
$x = -2$, the graph touches but
does not cross the x-axis; $x = 0$,
the graph touches but does not
cross the x-axis; $x = 2$, the graph
crosses the x-axis. **c.** The graph
is above the x-axis on
$(-\infty, -2) \cup (-2, 0) \cup (0, 2)$ and
below the x-axis on $(2, \infty)$.
d. 0 **e.** no symmetries **f.** 4

g.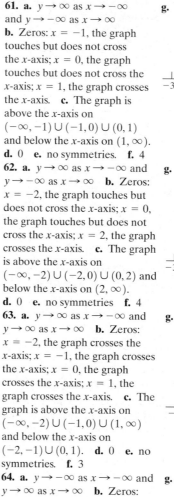

63. a. $y \to \infty$ as $x \to -\infty$ and
$y \to \infty$ as $x \to \infty$ **b.** Zeros:
$x = -2$, the graph crosses the
x-axis; $x = -1$, the graph crosses
the x-axis; $x = 0$, the graph
crosses the x-axis; $x = 1$, the
graph crosses the x-axis. **c.** The
graph is above the x-axis on
$(-\infty, -2) \cup (-1, 0) \cup (1, \infty)$
and below the x-axis on
$(-2, -1) \cup (0, 1)$. **d.** 0 **e.** no
symmetries. **f.** 3

g.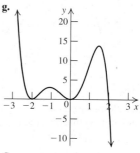

64. a. $y \to -\infty$ as $x \to -\infty$ and
$y \to \infty$ as $x \to \infty$ **b.** Zeros:
$x = 0$, the graph touches but does
not cross the x-axis; $x = 2$, the
graph crosses the x-axis.
c. The graph is below the x-axis
on $(-\infty, 0) \cup (0, 2)$ and above
the x-axis on $(2, \infty)$. **d.** 0
e. no symmetries. **f.** 2

g.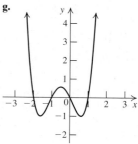

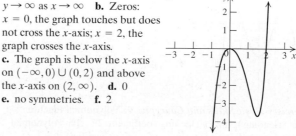

B Exercises: Applying the Concepts: 65. a. Zeros: $x = 0$, multiplicity 2; $x = 4$, multiplicity 1

b.

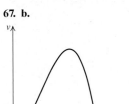

d. Domain: $[0, 4]$. The portion between the x-intercepts constitutes the graph of $R(x)$.

67. b.

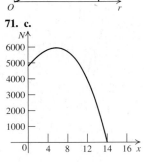

69. c.

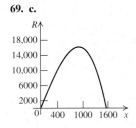

71. c.

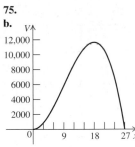

73. b.

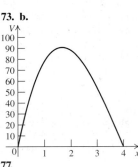

75.
b.

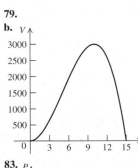

77.
b.

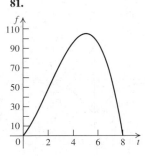

79.
b.

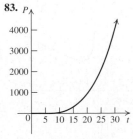

81.

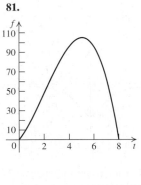

83.

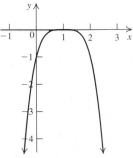

C Exercises: Beyond the Basics:
85. The graph of f is the graph of $y = x^4$ shifted one unit to the right.

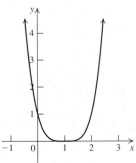

Zero: $x = 1$, multiplicity: 4

86. The graph of f is the graph of $y = x^4$ shifted one unit to the left and reflected in the x-axis.

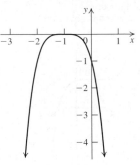

Zero: $x = -1$, multiplicity: 4

87. The graph of f is the graph of $y = x^4$ shifted two units up.

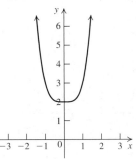

No zeros

88. The graph of f is the graph of $y = x^4$ shifted 81 units up.

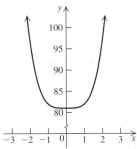

No zeros

89. The graph of f is the graph of $y = x^4$ shifted one unit to the right and reflected in the x-axis.

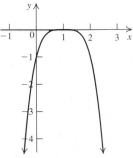

Zero: $x = 1$, multiplicity: 4

90. The graph of f is the graph of $y = x^4$ shifted one unit to the right, compressed by a factor of 2, and shifted eight units down.

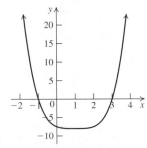

Zero: $x = -1$, multiplicity: 1; $x = 3$, multiplicity: 1

91. The graph of f is the graph of $y = x^5$ shifted one unit up.

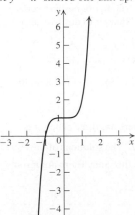

Zero: $x = -1$, multiplicity: 1

92. The graph of f is the graph of $y = x^5$ shifted one unit to the right.

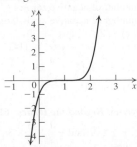

Zero: $x = 1$, multiplicity: 5

93. The graph of f is the graph of $y = x^5$ shifted one unit to the left, compressed by a factor of 4, reflected in the x-axis, and shifted eight units up.

94. The graph of f is the graph of $y = x^5$ shifted 1 unit to the left, compressed by a factor of 3, and shifted 81 units up.

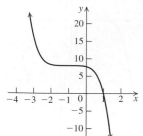 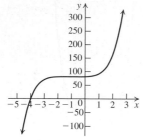

Zero: $x = 1$, multiplicity: 1

Zero: $x = -4$, multiplicity: 1

97. Maximum vertical distance: 0.25. It occurs at $x \approx 0.71$.

98. Maximum vertical distance: 0.19. It occurs at $x \approx 0.77$.

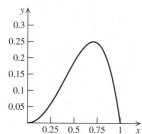 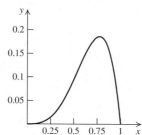

Section 2.3

Practice Problems: **1.** The quotient is $2x^2 + 7x + 3$, and the remainder is 2. **2.** The remainder is 4. **3.** 32 **4.** $\left\{-2, -\dfrac{2}{3}, 3\right\}$ **5.** 18 **6.** $\{-2\}$

Section 2.4

Practice Problems: **1.** Number of positive zeros: 1; number of negative zeros: 0 or 2 **2.** Upper bound: 1; lower bound: -2 **3. a.** $P(x) = 3(x + 2)(x - 1)(x - 1 - i)(x - 1 + i)$ **b.** $P(x) = 3x^4 - 3x^3 - 6x^2 + 18x - 12$ **4.** $-3, -3, 2 + 3i,$ $2 + 3i, 2 - 3i, 2 - 3i, i, -i$ **5.** The zeros are $1, 2, 2i,$ and $-2i$. **6.** The zeros are $2, 4, 1 + i,$ and $1 - i$.

B Exercises: Applying the Concepts: **7.** Number of positive zeros: 0 or 2; number of negative zeros: 1 **8.** Number of positive zeros: 1; number of negative zeros: 0 or 2 **9.** Number of positive zeros: 0 or 2; number of negative zeros: 1 **10.** Number of positive zeros: 1 or 3; number of negative zeros: 1 **11.** Number of positive zeros: 1 or 3; number of negative zeros: 0 or 2 **12.** Number of positive zeros: 1 or 3; number of negative zeros: 1 **13.** Number of positive zeros: 1 or 3; number of negative zeros: 1 **14.** Number of positive zeros: 1 or 3; number of negative zeros: 0 or 2 **39.** $P(x) = 2x^4 - 20x^3 + 70x^2 - 180x + 468$ **40.** $P(x) = -3x^4 + 18x^3 - 114x^2 + 282x - 663$ **53.** $-3, 1, 3 + i, 3 - i$ **55.** $\dfrac{1}{2}, 2, 3, i, -i$ **56.** $-2, 1$ (with multiplicity 2), $1 + i, 1 - i$

C Exercises: Beyond the Basics: **61.** There are three cube roots: $1, -\dfrac{1}{2} + \dfrac{1}{2}\sqrt{3}i,$ and $-\dfrac{1}{2} - \dfrac{1}{2}\sqrt{3}i$. **62.** The solutions of the equation and the zeros of the polynomial are the same. There are n roots. **67.** $f(x) = -4(x - 2)(x^2 - 2x + 5)$
$y \to \infty$ as $x \to -\infty, y \to -\infty$ as $x \to \infty$.

68. $f(x) = 2(x - 1)(x^2 - 4x + 13)$
$y \to \infty$ as $x \to \infty, y \to -\infty$ as $x \to -\infty$.
69. $f(x) = -2(x^2 - 1)(x^2 - 6x + 10)$
$y \to -\infty$ as $x \to -\infty, y \to -\infty$ as $x \to \infty$.
70. $f(x) = 2(x^2 - 2x + 5)(x^2 - 6x + 13)$
$y \to \infty$ as $x \to -\infty, y \to \infty$ as $x \to \infty$.
72. a. According to Descartes's Rule of Signs, the polynomial $x^3 + 6x - 20$ has one positive zero and no negative zero. Therefore, there is exactly one real solution.
b. Substituting $v - u$ for x, we obtain
$$(v - u)^3 + 6(v - u) = v^3 - 3v^2u + 3vu^2 - u^3 + 6(v - u)$$
$$= (v^3 - u^3) + (6 - 3uv)(v - u)$$
$$= 20 - (6 - 3(2))(v - u) = 20.$$
Therefore, $x = v - u$ is the solution.

Section 2.5

Practice Problems: **1.** Domain: $(-\infty, -1) \cup (-1, 5) \cup (5, \infty)$

2. $x = -5$ and $x = 2$ **3.** $x = -3$ **4. a.** $y = \dfrac{2}{3}$ **b.** No horizontal asymptote

5. x-intercept: 0; y-intercept: 0; vertical asymptotes: $x = -1$ and $x = 1$ Horizontal asymptote: x-axis. The graph is symmetric with respect to the origin. The graph is above the x-axis on $(-1, 0) \cup (1, \infty)$ and below the x-axis on $(-\infty, -1) \cup (0, 1)$.

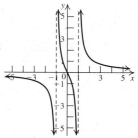

6. x-intercepts: $\pm\dfrac{\sqrt{2}}{2}$; y-intercept: $\dfrac{1}{3}$;

vertical asymptotes: $x = -\dfrac{3}{2}$ and $x = 1$; horizontal asymptote: $y = 1$; symmetry: none. The graph is above the x-axis on $\left(-\infty, -\dfrac{3}{2}\right) \cup \left(-\dfrac{\sqrt{2}}{2}, \dfrac{\sqrt{2}}{2}\right) \cup (1, \infty)$ and below the x-axis on $\left(-\dfrac{3}{2}, -\dfrac{\sqrt{2}}{2}\right) \cup \left(\dfrac{\sqrt{2}}{2}, 1\right)$.

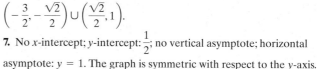

7. No x-intercept; y-intercept: $\dfrac{1}{2}$; no vertical asymptote; horizontal asymptote: $y = 1$. The graph is symmetric with respect to the y-axis. The graph is above the x-axis on $(-\infty, \infty)$.

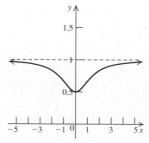

8. No x-intercept; y-intercept: -2; vertical asymptote: $x = 1$; no horizontal asymptote; $y = x + 1$ is an oblique asymptote. Symmetry: none. The graph is above the x-axis on $(1, \infty)$ and below the x-axis on $(-\infty, 1)$.

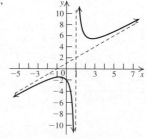

9. a. $R(10) = \$30$ billion
$R(20) = \$40$ billion
$R(30) = \$42$ billion
$R(40) = \$40$ billion
$R(50) = \$35.7$ billion
$R(60) = \$30$ billion

b.

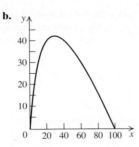

c. 29%

A Exercises: Basic Skills and Concepts:

49. x-intercept: 0; y-intercept: 0; vertical asymptote: $x = 3$; horizontal asymptote: $y = 2$; symmetry: none. The graph is above the x-axis on $(-\infty, 0) \cup (3, \infty)$ and below the x-axis on $(0, 3)$.

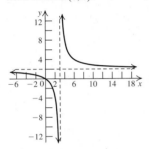

50. x-intercept: 0; y-intercept: 0; vertical asymptote: $x = 1$; horizontal asymptote: $y = -1$; symmetry: none. The graph is above the x-axis on $(0, 1)$ and below the x-axis on $(-\infty, 0) \cup (1, \infty)$.

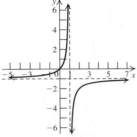

51. x-intercept: 0; y-intercept: 0; vertical asymptotes: $x = -2$ and $x = 2$; horizontal asymptote: x-axis. The graph is symmetric with respect to the origin. The graph is above the x-axis on $(-2, 0) \cup (2, \infty)$ and below the x-axis on $(-\infty, -2) \cup (0, 2)$.

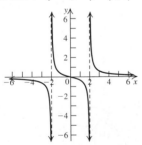

52. x-intercept: 0; y-intercept: 0; vertical asymptotes: $x = -1$ and $x = 1$; horizontal asymptote: x-axis. The graph is symmetric with respect to the origin. The graph is above the x-axis on $(-\infty, -1) \cup (0, 1)$ and below the x-axis on $(-1, 0) \cup (1, \infty)$.

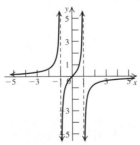

53. x-intercept: 0; y-intercept: 0; vertical asymptotes: $x = -3$ and $x = 3$; horizontal asymptote: $y = -2$. The graph is symmetric in the y-axis. The graph is above the x-axis on $(-3, 0) \cup (0, 3)$ and below the x-axis on $(-\infty, -3) \cup (3, \infty)$.

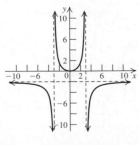

54. x-intercepts: $x = \pm 2$; no y-intercept; vertical asymptote: $x = 0$; horizontal asymptote: $y = -1$. The graph is symmetric with respect to the y-axis. The graph is above the x-axis on $(-2, 0) \cup (0, 2)$ and below the x-axis on $(-\infty, -2) \cup (2, \infty)$.

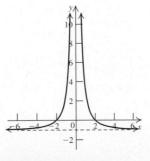

55. No x-intercept; y-intercept: -1; vertical asymptotes: $x = -\sqrt{2}$ and $x = \sqrt{2}$; horizontal asymptote: x-axis. The graph is symmetric with respect to the y-axis. The graph is above the x-axis on $(-\infty, -\sqrt{2}) \cup (\sqrt{2}, \infty)$ and below the x-axis on $(-\sqrt{2}, \sqrt{2})$.

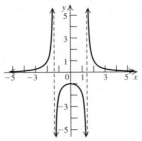

56. No x-intercept; y-intercept: $\dfrac{2}{3}$; vertical asymptotes: $x = -\sqrt{3}$ and $x = \sqrt{3}$; horizontal asymptote: x-axis. The graph is symmetric with respect to the y-axis. The graph is above the x-axis on $(-\sqrt{3}, \sqrt{3})$ and below the x-axis on $(-\infty, -\sqrt{3}) \cup (\sqrt{3}, \infty)$.

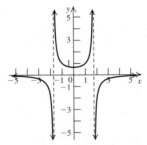

57. x-intercept: -1; y-intercept: $-\dfrac{1}{6}$; vertical asymptotes: $x = -3$ and $x = 2$; horizontal asymptote: x-axis; symmetry: none. The graph is above the x-axis on $(-3, -1) \cup (2, \infty)$ and below the x-axis on $(-\infty, -3) \cup (-1, 2)$.

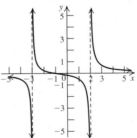

58. x-intercept: 1; y-intercept: $\dfrac{1}{2}$; vertical asymptotes: $x = -1$ and $x = 2$; horizontal asymptote: x-axis; symmetry: none. The graph is above the x-axis on $(-1, 1) \cup (2, \infty)$ and below the x-axis on $(-\infty, -1) \cup (1, 2)$.

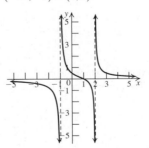

59. *x*-intercept: 0; *y*-intercept: 0; no vertical asymptote; horizontal asymptote: $y = 1$. The graph is symmetric with respect to the *y*-axis. The graph is above the *x*-axis on $(-\infty, 0) \cup (0, \infty)$.

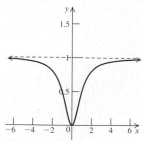

60. *x*-intercept: 0; *y*-intercept: 0; no vertical asymptote; horizontal asymptote: $y = 2$. The graph is symmetric with respect to the *y*-axis. The graph is above the *x*-axis on $(-\infty, 0) \cup (0, \infty)$.

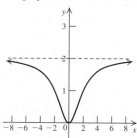

61. *x*-intercept: ± 2; *y*-intercept: $\dfrac{4}{9}$; vertical asymptotes: $x = -3$ and $x = 3$; horizontal asymptote: $y = 1$. The graph is symmetric with respect to the *y*-axis. The graph is above the *x*-axis on $(-\infty, -3) \cup (-2, 2) \cup (3, \infty)$ and below the *x*-axis on $(-3, -2) \cup (2, 3)$.

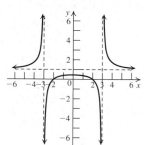

62. No *x*-intercept; *y*-intercept: 4; no vertical asymptote; horizontal asymptote: $y = 1$. The graph is symmetric with respect to the *y*-axis. The graph is above the *x*-axis on $(-\infty, \infty)$.

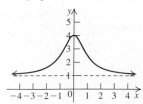

63. No *x*-intercept; *y*-intercept: -2; no vertical asymptote; no horizontal asymptote; symmetry: none. The graph is above the *x*-axis on $(2, \infty)$ and below the *x*-axis on $(-\infty, 2)$.

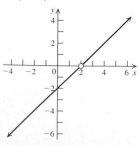

64. No *x*-intercept; *y*-intercept: -1; no vertical asymptote; no horizontal asymptote; symmetry: none. The graph is above the *x*-axis on $(1, \infty)$ and below the *x*-axis on $(-\infty, 1)$.

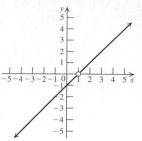

69. Oblique asymptote: $y = 2x$ **70.** Oblique asymptote: $y = x$

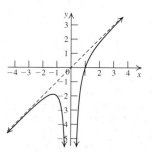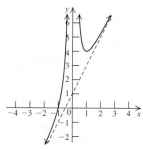

71. Oblique asymptote: $y = x$ **72.** Oblique asymptote: $y = 2x + 1$

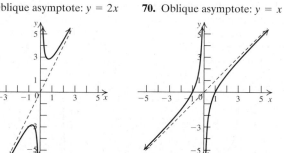

73. Oblique asymptote: $y = x - 2$ **74.** Oblique asymptote: $y = 2x - 1$

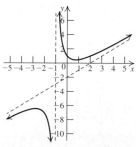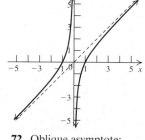

75. Oblique asymptote: $y = x - 2$ **76.** Oblique asymptote: $y = x$

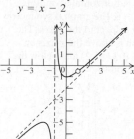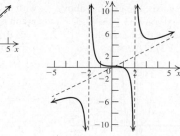

B Exercises: Applying the Concepts: **77. c.** $\overline{C}(100) = 20.5$, $\overline{C}(500) = 4.5$, $\overline{C}(1000) = 2.5$. These show the average cost of producing 100, 500, and 1000 trinkets, respectively.   **d.** Horizontal asymptote: $y = 0.5$. It means that the average cost approaches the fixed daily cost of producing each trinket if the number of trinkets produced approaches ∞.

78. a. $\overline{C}(x) = -0.002x + 6 + \dfrac{7000}{x}$

b. $\overline{C}(100) = 75.8, \overline{C}(500) = 19, \overline{C}(1000) = 11$. These show the average cost of producing 100, 500, and 1000 CD players, respectively.
c. Oblique asymptote: $y = -0.002x + 6$. For large values of x, this is a good approximation of the average cost of producing x CD players.

79. a.

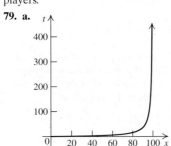

c. ii) Not applicable; the domain is $x < 100$.

80. a. $C(50) = 20, C(75) = 40, C(90) = 100, C(99) = 1000$. These show the cost, in millions of dollars, of catching and convicting 50%, 75%, 90%, and 99% of the criminals, respectively.

b.

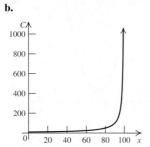

81. b.

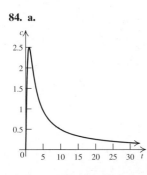

84. a.

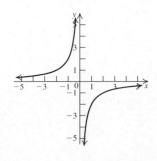

86. b.

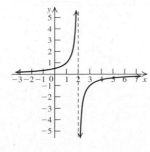

C Exercises: Beyond the Basics:

87. Stretch the graph of $y = \dfrac{1}{x}$ vertically by a factor of 2 and reflect the graph in the x-axis.

88. Shift the graph of $y = \dfrac{1}{x}$ two units to the right and reflect the graph in the x-axis.

89. Shift the graph of $y = \dfrac{1}{x^2}$ two units to the right.

90. Shift the graph of $y = \dfrac{1}{x^2}$ one unit to the left.

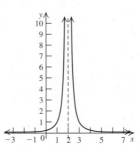

91. Shift the graph of $y = \dfrac{1}{x^2}$ one unit to the right and two units down.

92. Shift the graph of $y = \dfrac{1}{x^2}$ two units to the left and three units up.

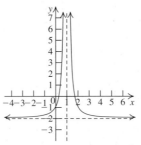

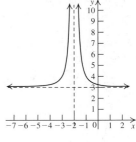

93. Shift the graph of $y = \dfrac{1}{x^2}$ six unit to the left.

94. Shift the graph of $y = \dfrac{1}{x^2}$ three units to the left and three units up.

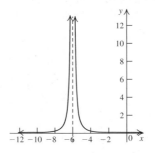

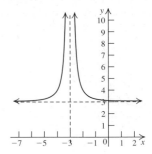

95. Shift the graph of $y = \dfrac{1}{x^2}$ one unit to the right and one unit up.

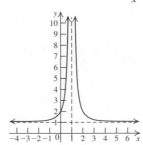

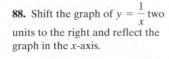

96. Shift the graph of $y = \dfrac{1}{x^2}$ one unit to the left, stretch the graph vertically by a factor of 5, reflect it in the x-axis, and shift it two units up.

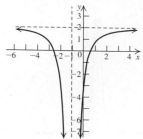

97. a. $f(x) \to 0$ as $x \to -\infty$, $f(x) \to 0$ as $x \to \infty$, $f(x) \to -\infty$ as $x \to 0^-$, and $f(x) \to \infty$ as $x \to 0^+$; no x-intercept; no y-intercept; vertical asymptote: y-axis; horizontal asymptote: x-axis. The graph is symmetric with respect to the origin. The graph is above the x-axis on $(0, \infty)$ and below the x-axis on $(-\infty, 0)$.

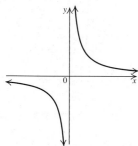

b. $f(x) \to 0$ as $x \to -\infty$, $f(x) \to 0$ as $x \to \infty$, $f(x) \to \infty$ as $x \to 0^-$, and $f(x) \to -\infty$ as $x \to 0^+$; no x-intercept; no y-intercept; vertical asymptote: y-axis; horizontal asymptote: x-axis. The graph is symmetric with respect to the origin. The graph is above the x-axis on $(-\infty, 0)$ and below the x-axis on $(0, \infty)$.

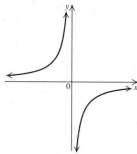

c. $f(x) \to 0$ as $x \to \infty$, $f(x) \to 0$ as $x \to -\infty$, $f(x) \to \infty$ as $x \to 0^-$, and $f(x) \to \infty$ as $x \to 0^+$; no x-intercept; no y-intercept; vertical asymptote: y-axis; horizontal asymptote: x-axis. The graph is symmetric with respect to the y-axis. The graph is above the x-axis on $(-\infty, 0) \cup (0, \infty)$.

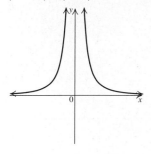

d. $f(x) \to 0$ as $x \to -\infty$, $f(x) \to 0$ as $x \to \infty$, $f(x) \to -\infty$ as $x \to 0^-$, and $f(x) \to -\infty$ as $x \to 0^+$; no x-intercept; no y-intercept; vertical asymptote: y-axis; horizontal asymptote: x-axis. The graph is symmetric with respect to the y-axis. The graph is below the x-axis on $(-\infty, 0) \cup (0, \infty)$.

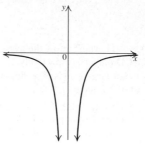

98. a. If c is a zero of $f(x)$, then because c is not a factor of the numerator, $x = c$ is a vertical asymptote. **b.** Because the numerator is positive, the sign of $f(x)$ and $g(x)$ is the same. **c.** If the graphs intersect for some value x, then $f(x) = g(x)$. It implies that $f(x) = \dfrac{1}{f(x)}$, from which it follows that $f(x) = \pm 1$. **d.** If $f(x)$ increases (decreases, remains constant), the denominator of $g(x)$ increases (decreases, remains constant). Therefore, $g(x)$ decreases (increases, remains constant).

99.

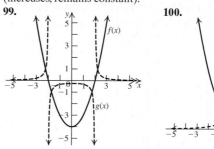

100.

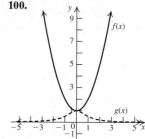

101. $[f(x)]^{-1} = \dfrac{1}{2x + 3}$ and $f^{-1}(x) = \dfrac{1}{2}x - \dfrac{3}{2}$. The two functions are different.

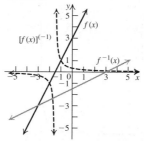

102. $[f(x)]^{-1} = \dfrac{x + 2}{x - 1}$ and $f^{-1}(x) = \dfrac{2x + 1}{1 - x}$.

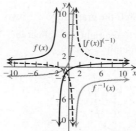

103. $g(x)$ has the oblique asymptote
$y = 2x + 3$; $y \to -\infty$ as $x \to -\infty$, and $y \to \infty$ as $x \to \infty$.

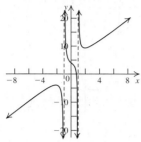

104. $f(x) \to \infty$ as $x \to -\infty$, and $f(x) \to \infty$ as $x \to \infty$. $f(x)$ has no oblique asymptote.

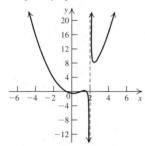

105. Point of intersection: $\left(-\dfrac{1}{3}, 1\right)$; horizontal asymptote $y = 1$

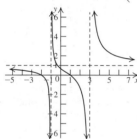

106. Point of intersection: $\left(1, -\dfrac{1}{2}\right)$; horizontal asymptote $y = -\dfrac{1}{2}$

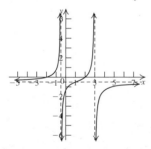

Section 2.6

Practice Problems: **1.** $y = 24$ **2.** 275 **3.** $y = 300$ **4.** $A = 20$
5. $x = \dfrac{9}{4}$ **6.** ≈ 3.7 m/sec²

A Exercises: Basic Skills and Concepts: **21.** $k = \dfrac{1}{17}, P = \dfrac{324}{17}$
22. $k = \dfrac{1}{13}, a = \dfrac{15}{13}$

B Exercises: Applying the Concepts: **31. a.** $y = 30.5x$, where y is the length measured in centimeters and x is the length measured in feet. **b. (i)** 244 cm **(ii)** 162.67 cm **c. (i)** 1.87 ft **(ii)** 4.07 ft
32. a. $y = 2.2x$, where y is the weight measured in pounds and x is the weight measured in kilograms. **43. a.** 1280 candlepower
44. b. (i) 37.95 mph, **(ii)** 60 mph, **(iii)** 69.3 mph
47. a. $H = kR^2N$, where k is a constant.

C Exercises: Beyond the Basics: **49. b.** $k = 0.0375$ **d.** E is multiplied by 8. **e.** E is multiplied by 4. **51. c.** The metabolic rate is multiplied by 2.83. **52.** The gravitational attraction between the sun and Earth is 168.92 times as strong as the gravitational attraction between Earth and the moon.
53. a. $T^2 = \left(\dfrac{4\pi^2}{G}\right)\left(\dfrac{r^3}{M_1 + M_2}\right)$
55. a. $R = kN(P - N)$, where k is the constant of variation.
b. $k = \dfrac{5}{10^6}$ **c.** 125 people per day

Review Exercises

1. (i) Opens up
(ii) Vertex: $(1, 2)$
(iii) Axis: $x = 1$
(iv) No x-intercept
(v) y-intercept: 3
(vi) Decreasing on $(-\infty, 1)$, increasing on $(1, \infty)$

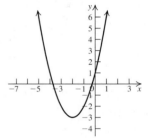

2. (i) Opens up
(ii) Vertex: $(-2, -3)$
(iii) Axis: $x = -2$
(iv) x-intercepts: $-2 \pm \sqrt{3}$
(v) y-intercept: 1
(vi) Decreasing on $(-\infty, -2)$, increasing on $(-2, \infty)$

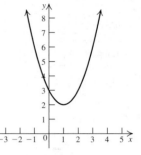

3. (i) Opens down
(ii) Vertex: $(3, 4)$
(iii) Axis: $x = 3$
(iv) x-intercepts: $3 \pm \sqrt{2}$
(v) y-intercept: -14
(vi) Increasing on $(-\infty, 3)$, decreasing on $(3, \infty)$

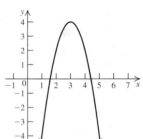

4. (i) Opens down
(ii) Vertex: $(-1, 2)$
(iii) Axis: $x = -1$
(iv) x-intercepts: $-3, 1$
(v) y-intercept: $\dfrac{3}{2}$
(vi) Increasing on $(-\infty, -1)$, decreasing on $(-1, \infty)$

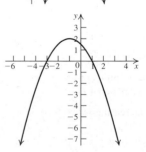

5. (i) Opens down
(ii) Vertex: $(0, 3)$
(iii) Axis: y-axis
(iv) x-intercepts: $\pm\dfrac{\sqrt{6}}{2}$
(v) y-intercept: 3
(vi) Increasing on $(-\infty, 0)$, decreasing on $(0, \infty)$

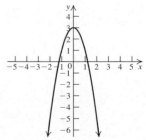

6. (i) Opens up
(ii) Vertex: $(-1, -3)$
(iii) Axis: $x = -1$
(iv) x-intercepts: $-1 \pm \dfrac{\sqrt{6}}{2}$

(v) y-intercept: -1
(vi) Decreasing on $(-\infty, -1)$, increasing on $(-1, \infty)$

7. (i) Opens up
(ii) Vertex: $(1, 1)$
(iii) Axis: $x = 1$
(iv) No x-intercept
(v) y-intercept: 3
(vi) Decreasing on $(-\infty, 1)$, increasing on $(1, \infty)$

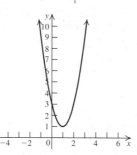

8. (i) Opens down
(ii) Vertex: $\left(-\dfrac{1}{4}, \dfrac{25}{8}\right)$

(iii) Axis: $x = -\dfrac{1}{4}$

(iv) x-intercept: $-\dfrac{3}{2}, 1$

(v) y-intercept: 3

(vi) Increasing on $\left(-\infty, -\dfrac{1}{4}\right)$,

decreasing on $\left(-\dfrac{1}{4}, \infty\right)$

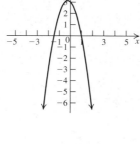

9. (i) Opens up
(ii) Vertex: $\left(\dfrac{1}{3}, \dfrac{2}{3}\right)$

(iii) Axis: $x = \dfrac{1}{3}$

(iv) No x-intercept
(v) y-intercept: 1

(vi) Decreasing on $\left(-\infty, \dfrac{1}{3}\right)$,

increasing on $\left(\dfrac{1}{3}, \infty\right)$

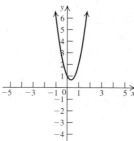

10. (i) Opens up
(ii) Vertex: $\left(\dfrac{5}{6}, \dfrac{23}{12}\right)$

(iii) Axis: $x = \dfrac{5}{6}$

(iv) No x-intercept
(v) y-intercept: 4

(vi) Decreasing on $\left(-\infty, \dfrac{5}{6}\right)$,

increasing on $\left(\dfrac{5}{6}, \infty\right)$

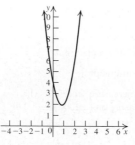

15. Shift the graph of $y = x^3$ one unit to the left and two units down.

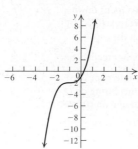

16. Shift the graph of $y = x^4$ one unit to the left and two units up.

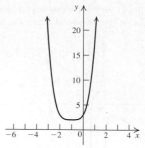

17. Shift the graph of $y = x^3$ one unit right, reflect the resulting graph in the x-axis, and shift it one unit up.

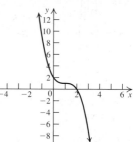

18. Shift the graph of $y = x^4$ three units up.

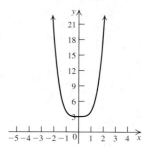

19. (i) $f(x) \to -\infty$ as $x \to -\infty$ and $f(x) \to \infty$ as $x \to \infty$.
(ii) Zeros: $x = -2$, multiplicity: 1, the graph crosses the x-axis; $x = 0$, multiplicity: 1, the graph crosses the x-axis; $x = 1$, multiplicity: 1, the graph crosses the x-axis.
(iii) x-intercepts: $-2, 0, 1$; y-intercept: 0
(iv) The graph is above the x-axis on $(-2, 0) \cup (1, \infty)$ and below the x-axis on $(-\infty, -2) \cup (0, 1)$.
(v) Symmetry: none
(vi)

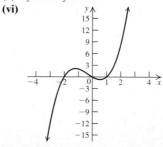

20. (i) $f(x) \to -\infty$ as $x \to -\infty$ and $f(x) \to \infty$ as $x \to \infty$.
(ii) Zeros: $x = -1$, multiplicity: 1, the graph crosses the x-axis; $x = 0$, multiplicity: 1, the graph crosses the x-axis; $x = 1$, multiplicity: 1, the graph crosses the x-axis. **(iii)** x-intercepts: $-1, 0, 1$; y-intercept: 0 **(iv)** The graph is above the x-axis on $(-1, 0) \cup (1, \infty)$ and below the x-axis on $(-\infty, -1) \cup (0, 1)$.
(v) The graph is symmetric with respect to the origin.
(vi)

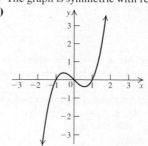

21. (i) $f(x) \to -\infty$ as $x \to -\infty$ and $f(x) \to -\infty$ as $x \to \infty$.
(ii) Zeros: $x = 0$, multiplicity: 2, the graph touches but does not cross the x-axis; $x = 1$, multiplicity: 2, the graph touches but does not cross the x-axis.   (iii) x-intercepts: 0, 1; y-intercept: 0   (iv) The graph is below the x-axis on $(-\infty, 0) \cup (0, 1) \cup (1, \infty)$.   (v) Symmetry: none
(vi)

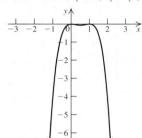

22. (i) $f(x) \to \infty$ as $x \to -\infty$ and $f(x) \to -\infty$ as $x \to \infty$.
(ii) Zeros: $x = 0$, multiplicity: 3, the graph crosses the x-axis; $x = 2$, multiplicity: 2, the graph touches but does not cross the x-axis.
(iii) x-intercepts: 0, 2; y-intercept: 0   (iv) The graph is above the x-axis on $(-\infty, 0)$ and below the x-axis on $(0, 2) \cup (2, \infty)$.
(v) Symmetry: none
(vi)

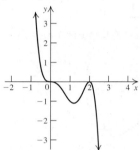

23. (i) $f(x) \to -\infty$ as $x \to -\infty$ and $f(x) \to -\infty$ as $x \to \infty$.
(ii) Zeros: $x = -1$, multiplicity: 1, the graph crosses the x-axis; $x = 0$; multiplicity: 2, the graph touches but does not cross the x-axis; $x = 1$ multiplicity: 1, the graph crosses the x-axis.
(iii) x-intercepts: $-1, 0, 1$; y-intercept: 0   (iv) The graph is above the x-axis on $(-1, 0) \cup (0, 1)$ and below the x-axis on $(-\infty, -1) \cup (1, \infty)$.
(v) The graph is symmetric with respect to the y-axis.
(vi)

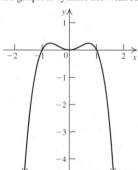

24. (i) $f(x) \to -\infty$ as $x \to -\infty$ and $f(x) \to -\infty$ as $x \to \infty$.
(ii) Zero: $x = 1$, multiplicity: 2, the graph touches but does not cross the x-axis.   (iii) x-intercept: 1; y-intercept: -1 (iv) The graph is below the x-axis on $(-\infty, 1) \cup (1, \infty)$.   (v) Symmetry: none
(vi)

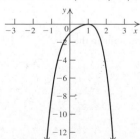

25. Quotient: $2x + 3$; remainder: -7 **29.** Quotient: $x^2 + 3x - 3$; remainder: -6 **30.** Quotient: $-4x^2 - 21x - 131$; remainder: -786

38. $-2, \dfrac{7}{4} - \dfrac{\sqrt{33}}{4}, \dfrac{7}{4} + \dfrac{\sqrt{33}}{4}$ **40.** $\dfrac{1}{4}, -\dfrac{5}{2} + \dfrac{\sqrt{33}}{2}, -\dfrac{5}{2} - \dfrac{\sqrt{33}}{2}$

42. $\pm\dfrac{1}{9}, \pm\dfrac{2}{9}, \pm\dfrac{1}{3}, \pm\dfrac{4}{9}, \pm\dfrac{2}{3}, \pm\dfrac{8}{9}, \pm1, \pm\dfrac{4}{3}, \pm\dfrac{16}{9}, \pm2, \pm\dfrac{8}{3}, \pm4, \pm\dfrac{16}{3}, \pm8,$ ±16 **59.** The only possible rational roots are $-2, -1, 1$, and 2. None of them satisfies the equation.   **60.** The only possible rational roots are $-5, -\dfrac{5}{3}, -1, -\dfrac{1}{3}, \dfrac{1}{3}, 1, \dfrac{5}{3}, 5$. None of them satisfies the equation.

63. x-intercept: -1; no y-intercept; vertical asymptote: y-axis; horizontal asymptote: $y = 1$; symmetry: none. The graph is above the x-axis on $(-\infty, -1) \cup (0, \infty)$ and below the x-axis on $(-1, 0)$.

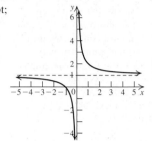

64. x-intercept: 2; no y-intercept; vertical asymptote: y-axis; horizontal asymptote: $y = -1$; symmetry: none. The graph is above the x-axis on $(0, 2)$ and below the x-axis on $(-\infty, 0) \cup (2, \infty)$.

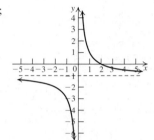

65. x-intercept: 0; y-intercept: 0; vertical asymptotes: $x = -1$ and $x = 1$; horizontal asymptote: x-axis; symmetry: none. The graph is above the x-axis on $(-1, 0) \cup (1, \infty)$ and below the x-axis on $(-\infty, -1) \cup (0, 1)$.

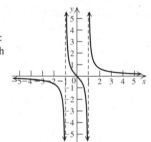

66. x-intercept: ±3; y-intercept: $\dfrac{9}{4}$; vertical asymptotes: $x = -2$ and $x = 2$; horizontal asymptote: $y = 1$. The graph is symmetric with respect to the y-axis. The graph is above the x-axis on $(-\infty, -3) \cup (-2, 2) \cup (3, \infty)$ and below the x-axis on $(-3, -2) \cup (2, 3)$.

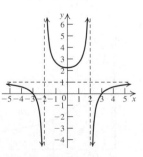

67. x-intercept: 0; y-intercept: 0; vertical asymptotes: $x = -3$ and $x = 3$; no horizontal asymptote; symmetry: none. The graph is above the x-axis on $(-3, 0) \cup (3, \infty)$ and below the x-axis on $(-\infty, -3) \cup (0, 3)$.

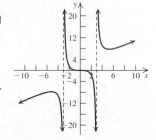

68. x-intercept: -1; y-intercept: $-\dfrac{1}{8}$; vertical asymptotes: $x = -2$ and $x = 4$; horizontal asymptote: x-axis; symmetry: none. The graph is above the x-axis on $(-2, -1) \cup (4, \infty)$ and below the y-axis on $(-\infty, -2) \cup (-1, 4)$.

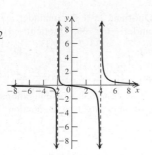

69. x-intercept: 0; y-intercept: 0; vertical asymptotes: $x = -2$ and $x = 2$; no horizontal asymptote. The graph is symmetric with respect to the y-axis. The graph is above the x-axis on $(-\infty, -2) \cup (2, \infty)$ and below the x-axis on $(-2, 0) \cup (0, 2)$.

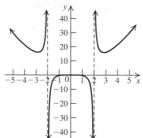

70. x-intercept: 2; y-intercept: $\dfrac{1}{2}$; vertical asymptote: $x = 4$; horizontal asymptote: $y = 1$; symmetry: none. The graph is above the x-axis on $(-\infty, -3) \cup (-3, 2) \cup (4, \infty)$ and below the x-axis on $(2, 4)$.

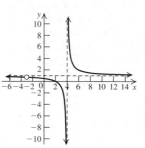

75. Maximum height: 1000. The missile hits the ground at $x = 200$.

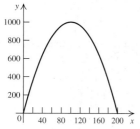

79. a.

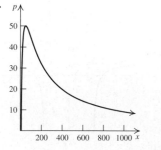

81. c. $\overline{C}(x) = \dfrac{3}{1000}x + 3.9 + \dfrac{150}{x}$

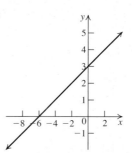

Practice Test A: 2.

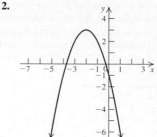

6.

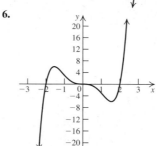

20. $V(x) = x(8 - 2x)(17 - 2x)$

Cumulative Review Exercises (Chapters 1 and 2)

3. x-intercepts: 4, -2; y-intercept: -8

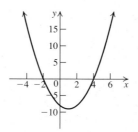

4. Slope is $-\dfrac{1}{3}$; x-intercept: 6; y-intercept: 2

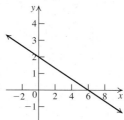

5. Slope is $\dfrac{1}{2}$; x-intercept: -6; y-intercept: 3

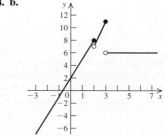

14. b.

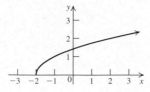

16. a. Shift the graph of $y = \sqrt{x}$ two units to the left.

b. Shift the graph of $y = \sqrt{x}$ one unit to the left, stretch the resulting graph horizontally by a factor of 2, reflect it in the x-axis, and shift it three units up.

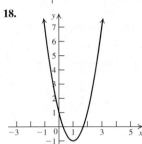

4. $a = \dfrac{1}{5}$ **5.** $x = 2$ **6.** $-8, 16$

7. 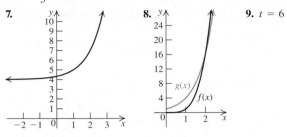   **8.** 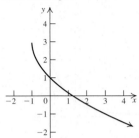   **9.** $t = 6$

18. 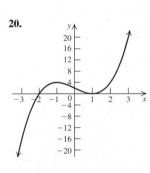   **19.**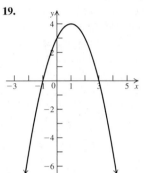

20. 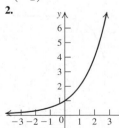   **21.**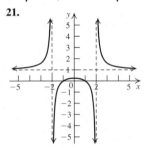

CHAPTER 3

Section 3.1

Practice Problems:

1. $f(2) = \dfrac{1}{16} = 0.06$
 $f(0) = 1$
 $f(-1) = 4$
 $f\left(\dfrac{5}{2}\right) = \dfrac{1}{32} = 0.03125$
 $f\left(-\dfrac{3}{2}\right) = 8$

2. 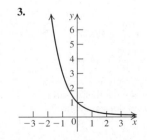   **3.** (graph)

A Exercises: Basic Skills and Concepts:

33. 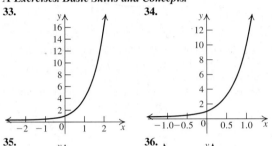   **34.** (graph)

35. 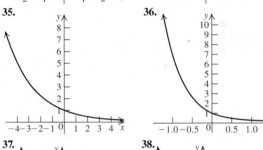   **36.** (graph)

37. 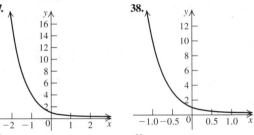   **38.** (graph)

39. 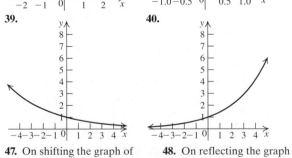   **40.** (graph)

47. On shifting the graph of $f(x) = 4^x$ (Exercise 33) to the right one unit

48. On reflecting the graph of $g(x) = 10^x$ (Exercise 34) in the x-axis

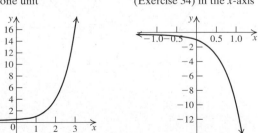

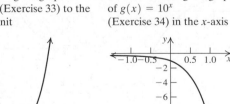

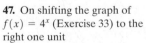

49. Same as the graph of $h(x) = \left(\frac{1}{4}\right)^x$. (See the graph for Exercise 37).

50. On reflecting the graph of $h(x) = \left(\frac{1}{4}\right)^x$ (Exercise 37) across the x-axis

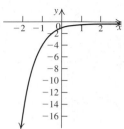

51. On reflecting the graph of $h(x) = 7^{-x}$ (Exercise 36) in the y-axis, followed by shifting two units up

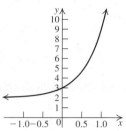

52. On reflecting the graph of $h(x) = 7^{-x}$ (Exercise 36) in the y-axis followed by a reflection across the x-axis and then a shift four units up

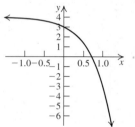

53. On reflecting the graph of $g(x) = \left(\frac{3}{2}\right)^{-x}$ (Exercise 35) across the y-axis and then shifting four units up

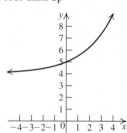

54. On shifting the graph of $f(x) = \left(\frac{1}{10}\right)^x$ (Exercise 38) one unit down

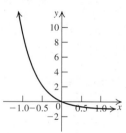

55. On reflecting the graph of $f(x) = (1.3)^{-x}$ (Exercise 39) across the x-axis

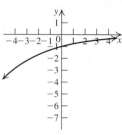

56. On reflecting the graph of $g(x) = (0.7)^{-x}$ (Exercise 40) across the x-axis and then shifting two units up

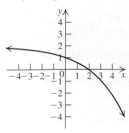

57. Domain $= (-\infty, \infty)$
Range $= (0, \infty)$
Horizontal asymptote: $y = 0$

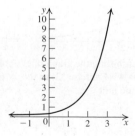

58. Domain $= (-\infty, \infty)$
Range $= (-2, \infty)$
Horizontal asymptote: $y = -2$

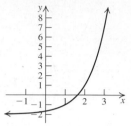

59. Domain $= (-\infty, \infty)$
Range $= (0, \infty)$
Horizontal asymptote: $y = 0$

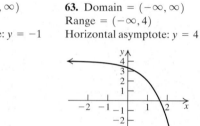

60. Domain $= (-\infty, \infty)$
Range $= (-2, \infty)$
Horizontal asymptote: $y = -2$

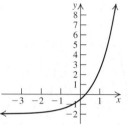

61. Domain $= (-\infty, \infty)$
Range $= (-\infty, 1)$
Horizontal asymptote: $y = 1$

62. Domain $= (-\infty, \infty)$
Range $= (-1, \infty)$
Horizontal asymptote: $y = -1$

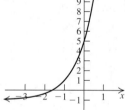

63. Domain $= (-\infty, \infty)$
Range $= (-\infty, 4)$
Horizontal asymptote: $y = 4$

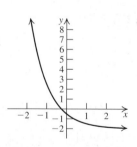

64. Domain $= (-\infty, \infty)$
Range $= (-2, \infty)$
Horizontal asymptote: $y = -2$

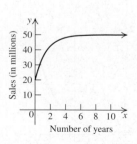

B Exercises: Applying the Concepts:
86. $s(0) = 20$
$s(1) = 35$
$s(5) = \dfrac{1570}{32} = 49.0625$
$s(10) = \dfrac{51170}{1024} = 49.9707$

Horizontal asymptote: $y = 50$. In the long run, the number of sales approaches but does not exceed 50 million.

C Exercises: Beyond the Basics:
93. Range $= [1, \infty)$ symmetric in the *y*-axis.

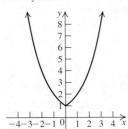

94.

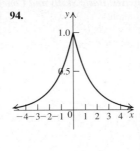

95. Domain $= (-\infty, \infty)$
Range $= [1, \infty)$ symmetric in the *y*-axis.

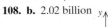
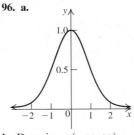

96. a.

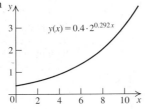

b. Domain $= (-\infty, \infty)$
Range $= (0, 1]$

108. b. 2.02 billion

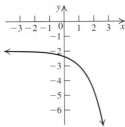

$y(x) = 0.4 \cdot 2^{0.292x}$

Section 3.2

Practice Problems: **1.** $\$11,500$ **2. a.** $A = \$11,485.03$
b. $I = \$3,485.03$ **3. (i)** $\$5325.00$ **(ii)** $\$5330.28$ **(iii)** $\$5333.01$
(iv) $\$5334.86$ **(v)** $\$5335.76$ **4.** $\$14,764.48$ **5. a.** $\$4,284,000.00$
b. $\$1,911,233,655.42$ **c.** $\$2,162,832,947$ **d.** $\$2,256,324,191$
6. $y = -e^{x-1} - 2$

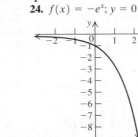

7. a. 11.96229   **b.** 4.76720

A Exercises: Basic Skills and Concepts:
23. $f(x) = e^{-x}$; $y = 0$

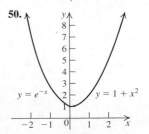

24. $f(x) = -e^{x}$; $y = 0$

25. $f(x) = e^{x-2}$; $y = 0$

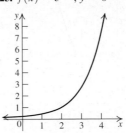

26. $f(x) = e^{2-x}$; $y = 0$

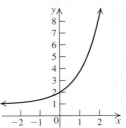

27. $f(x) = 1 + e^{x}$; $y = 1$

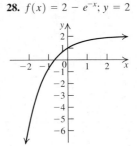

28. $f(x) = 2 - e^{-x}$; $y = 2$

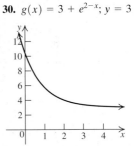

29. $f(x) = -e^{x-2} + 3$; $y = 3$

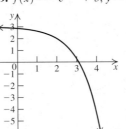

30. $g(x) = 3 + e^{2-x}$; $y = 3$

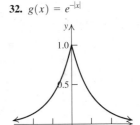

31. $f(x) = e^{|x|}$

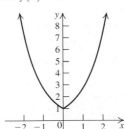

32. $g(x) = e^{-|x|}$

C Exercises: Beyond the Basics:
47.

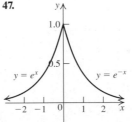

$y = e^{x}$ $y = e^{-x}$

48.

$y = e^{x}$
$y = -e^{x}$

49.

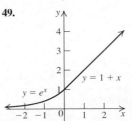

$y = e^{x}$ $y = 1 + x$

50.

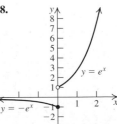

$y = e^{-x}$ $y = 1 + x^2$

51.

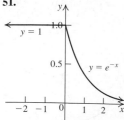

52.

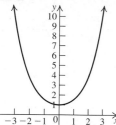

54. $S_5 = 2.716666, S_{10} = 2.7182818, S_{15} = 2.718281828$

55. c.

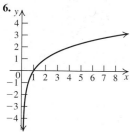

56. c.

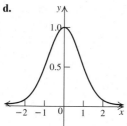

Critical Thinking: **65. d.**

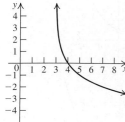

Section 3.3

Practice Problems: **1. a.** $\log_2 1024 = 10$ **b.** $\log_9\left(\dfrac{1}{3}\right) = -\dfrac{1}{2}$

c. $\log_a p = q$ **2. a.** $2^6 = 64$ **b.** $v^w = u$ **3. a.** 2   **b.** $-1/2$
c. -5 **4. a.** 8   **b.** 5   **c.** $x = -2$ or $x = 3$ **5.** $(-\infty, 1)$
6.

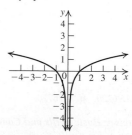

(image 6 — graph for problem 6)

7. Shift the graph of $y = \log_2 x$ three units right and reflect the graph in the *x*-axis.

(graph for problem 7)

8.

(image 8 — graph) $y = \log(x-3) - 2$

9. a. -1 **b.** 0.693
10. a. 17 years   **b.** 21.97%
11. 5.423 min
12. **a.** 30,805.03 m^3
b. 13.82 yr

A Exercises: Basic Skills and Concepts:
63. Domain $= (-3, \infty)$
Range $= (-\infty, \infty)$
Asymptote: $x = -3$

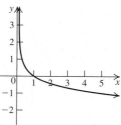

64. Domain $= (1, \infty)$
Range $= (-\infty, \infty)$
Asymptote: $x = 1$

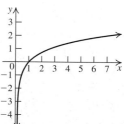

65. Domain $= (0, \infty)$
Range $= (-\infty, \infty)$
Asymptote: $x = 0$

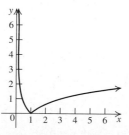

66. Domain $= (0, \infty)$
Range $= (-\infty, \infty)$
Asymptote: $x = 0$

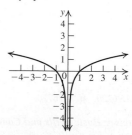

67. Domain $= (-\infty, 0)$
Range $= (-\infty, \infty)$
Asymptote: $x = 0$

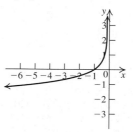

68. Domain $= (-\infty, 0)$
Range $= (-\infty, \infty)$
Asymptote: $x = 0$

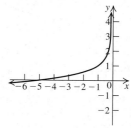

69. Domain $= (0, \infty)$
Range $= [0, \infty)$
Asymptote: $x = 0$

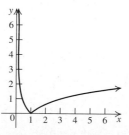

70. Domain $= (-\infty, 0) \cup (0, \infty)$
Range $= (-\infty, \infty)$
Asymptote: $x = 0$

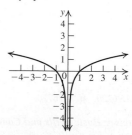

71. Domain $= (1, \infty)$
Range $= (-\infty, \infty)$
Asymptote: $x = 1$

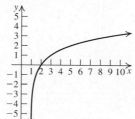

72. Domain $= (-\infty, 0)$
Range $= (-\infty, \infty)$
Asymptote: $x = 0$

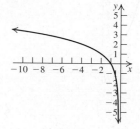

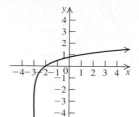
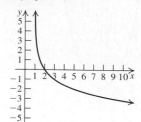

73. Domain = $(-\infty, 3)$
Range = $(-\infty, \infty)$
Asymptote: $x = 3$

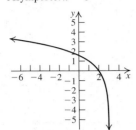

74. Domain = $(0, \infty)$
Range = $(-\infty, \infty)$
Asymptote: $x = 0$

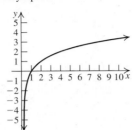

75. Domain = $(-\infty, 3)$
Range = $(-\infty, \infty)$
Asymptote: $x = 3$

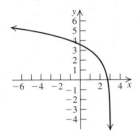

76. Domain = $(-\infty, 3)$
Range = $(-\infty, \infty)$
Asymptote: $x = 3$

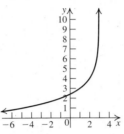

77. Domain = $(-\infty, 0) \cup (0, \infty)$
Range = $(-\infty, \infty)$
Asymptote: $x = 0$

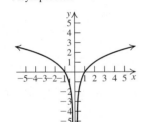

78. Domain = $(-\infty, 0) \cup (0, \infty)$
Range = $(-\infty, \infty)$
Asymptote: $x = 0$

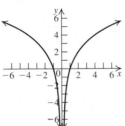

89.

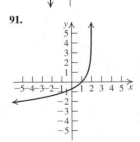

90.

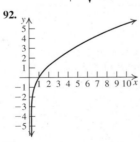

91.

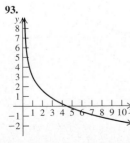

92.

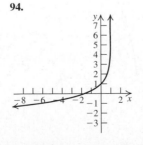

93.

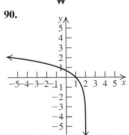

94.

C Exercises: Beyond the Basics:

107.

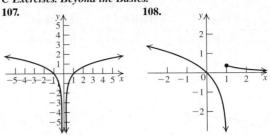

108.

Section 3.4

Practice Problems: **1. a.** -1 **b.** 13
2. a. $\ln (2x - 1) - \ln (x + 4)$

b. $\log 2 + \dfrac{1}{2}\log x + \dfrac{1}{2}\log y - \dfrac{1}{2}\log z$ **3.** $\log \sqrt{x^2 - 1}$

4. $\dfrac{\log 15}{\log 3} \approx 2.46497$ **5.** $f(x) = 5e^{x\ln 4}$ **6.** $\approx 65.34\%$

A Exercises: Skills and Concepts: **19.** $\ln (x) + \ln (x - 1)$

20. $\log_2 (x - 1) - \log_2 (x + 3)$ **21.** $\dfrac{1}{2}\log (x^2 + 1) - \log (x + 3)$

22. $\ln (x) + \ln (x + 1) - 2\ln (x - 1)$
23. $2\log_3 (x - 1) - 5\log_3 (x + 1)$

24. $\dfrac{2}{3}\log_4 (x - 3) + \dfrac{2}{3}\log_4 (x + 3) - \dfrac{2}{3}\log_4 (x - 2)$

$- \dfrac{2}{3}\log_4 (x - 4)$ **25.** $\log_b x + \log_b y + \log_b z$

26. $\dfrac{1}{2}[\log_b x + \log_b y + \log_b z]$ **27.** $2\log_b x + 3\log_b y + \log_b z$

28. $-\log_b x + 2\log_b y - 3\log_b z$ **29.** $\dfrac{1}{2}\log x + \dfrac{1}{4}(\log y + \log z)$

30. $\log_b x + \dfrac{1}{2}(\log_b y + \log_b z)$

C Exercises: Beyond the Basics:

85.

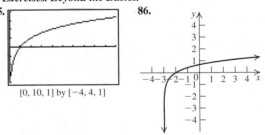

$[0, 10, 1]$ by $[-4, 4, 1]$

86.

Section 3.5

Practice Problems: **1. a.** $x = 5$ **b.** $x = \dfrac{2}{3}$

2. $x = \dfrac{\ln\left(\dfrac{11}{7}\right)}{\ln 3} - 1 (\approx -0.59)$ **3.** $x = 3.82$ **4.** $x = 1.609$

5. a. United States: 329.10 million; Pakistan: 224.14 million
b. Some time in 2020 (after 15.63 yr) **c.** In 27.84 yr (some time
in 2032) **6.** $x = e^{3/2}$ **7.** $\varnothing$ **8.** $x = 9$ **9.** The average growth
rate was approximately 1.74%. **10.** $I_A = \sqrt{10}I_T$

B Exercises: Applying the Concepts:
71. a.

Item	Canada	Mexico	United Kingdom
Population	36.76 million	125.54 million	63.03 million
Milk price	$2.50	$1.38	$1.00
Bread price	$2.61	$2.23	$1.72

b.

Item	Canada	Mexico	United Kingdom
Population	38.80 million	134.91 million	64.17 million
Milk price	$2.82	$1.70	$1.20
Bread price	$2.94	$2.73	$2.06

79. c.

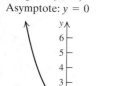

81. a. $a = 4999$ and $k = 0.852$

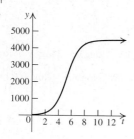

82. c.

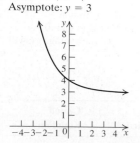

Review Exercises

Basic Skills:

19. Domain $= (-\infty, \infty)$
Range $= (0, \infty)$
Asymptote: $y = 0$

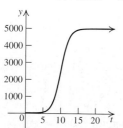

20. Domain $= (-\infty, \infty)$
Range $= (0, \infty)$
Asymptote: $y = 0$

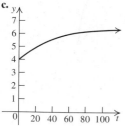

21. Domain $= (-\infty, \infty)$
Range $= (3, \infty)$
Asymptote: $y = 3$

22. Domain $= (-\infty, \infty)$
Range $= [1, \infty)$

23. Domain $= (-\infty, \infty)$
Range $= (0, 1]$
Asymptote: $y = 0$

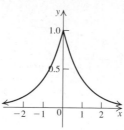

24. Domain $= (-\infty, \infty)$
Range $= (0, \infty)$
Asymptote: $y = 0$

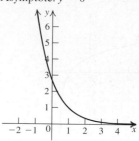

25. Domain $= (-\infty, 0)$
Range $= (-\infty, \infty)$
Asymptote: $x = 0$

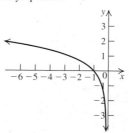

26. Domain $= (-\infty, 0) \cup (0, \infty)$
Range $= (-\infty, \infty)$
Asymptote: $x = 0$

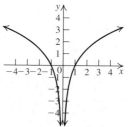

27. Domain $= (1, \infty)$
Range $= (-\infty, \infty)$
Asymptote: $x = 1$

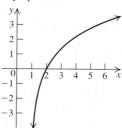

28. Domain $= (-\infty, \infty)$
Range $= (-\infty, 2)$
Asymptote: $y = 2$

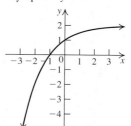

29. Domain $= (-\infty, 0)$
Range $= (-\infty, \infty)$
Asymptote: $x = 0$

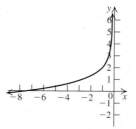

30. Domain $= (-5, \infty)$
Range $= (-\infty, \infty)$
Asymptote: $x = -5$

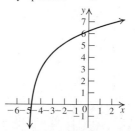

31. a. y-intercept $= 1$
b. $\lim\limits_{x \to \infty} f(x) = 3$
$\lim\limits_{x \to -\infty} f(x) = -\infty$

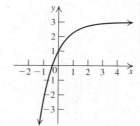

32. a. y-intercept $= 1$
b. $\lim\limits_{x \to \infty} f(x) = 2.5$
$\lim\limits_{x \to -\infty} f(x) = 0$

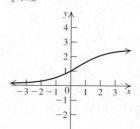

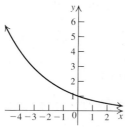
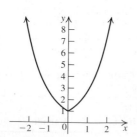

33. a. *y*-intercept = 1
b. $\lim\limits_{x\to\infty} f(x) = 0$
$\lim\limits_{x\to-\infty} f(x) = 0$

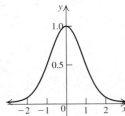

34. a. *y*-intercept = 1
b. $\lim\limits_{x\to\infty} f(x) = -3,$
$\lim\limits_{x\to-\infty} f(x) = 3$

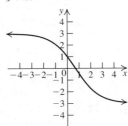

7. Start with $y = \sqrt{x}$, shift one unit left, stretch vertically by a factor of 3, and shift two units down.

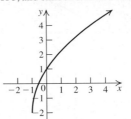

Applying the Concepts: **67.** At 5% compounded yearly, $A = \$9849.70$. At 4.75% compounded monthly, $A = \$9754.74$. The 5% investment will provide the greater return.

70. a.

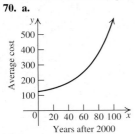

71. a.

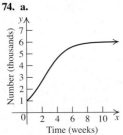

17. d.

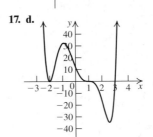

18. d.

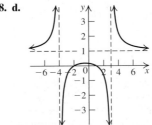

74. a.

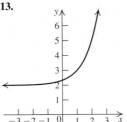

CHAPTER 4

Section 4.1

Practice Problems: **1.**

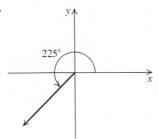

2. a. 13.16°   **b.** 41°16′30″   **3.** $-\dfrac{\pi}{4}$ radians   **4.** 270°
5. Complement = 23°; supplement = 113°   **6.** 7.85 meters
7. ≈ 790 miles   **8.** 36π radians per minute ≈ 1131 feet per minute
9. ≈ 52.36 square inches

A Exercises: Basic Skills and Concepts:
7.

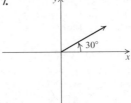

8.

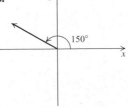

Practice Test A

13.

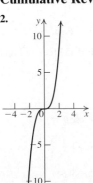

Cumulative Review Exercises (Chapters 1–3)

2.

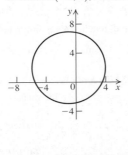

3. Center: $(-1, 2)$; radius: 5

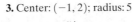

9.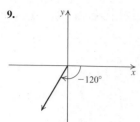

10.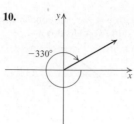

11.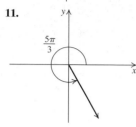

12.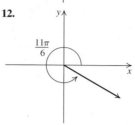

13.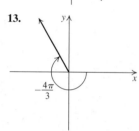

14.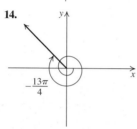

58. Complement: none because the measure of the angle is greater than 90°; supplement: 20°   **59.** Complement: none because the measure of the angle is greater than 90°; supplement: none because the measure of the angle is greater than 180°   **60.** Complement: none because the measure of the angle is negative; supplement: none because the measure of the angle is negative

Section 4.2

Practice Problems: 1. $\sin\dfrac{5\pi}{2} = 1, \cos\dfrac{5\pi}{2} = 0, \tan\dfrac{5\pi}{2}$ is undefined, $\csc\dfrac{5\pi}{2} = 1, \sec\dfrac{5\pi}{2}$ is undefined, $\cot\dfrac{5\pi}{2} = 0$ **2.** $\sin(-270°) = 1$, $\cos(-270°) = 0, \tan(-270°)$ is undefined, $\csc(-270°) = 1$, $\sec(-270°)$ is undefined, $\cot(-270°) = 0$

3. $\sin\theta = -\dfrac{5\sqrt{29}}{29}$ $\csc\theta = -\dfrac{\sqrt{29}}{5}$ **4.** $\sin\dfrac{\pi}{3} = \dfrac{\sqrt{3}}{2}$ $\csc\dfrac{\pi}{3} = \dfrac{2\sqrt{3}}{3}$

$\cos\theta = \dfrac{2\sqrt{29}}{29}$ $\sec\theta = \dfrac{\sqrt{29}}{2}$ $\cos\dfrac{\pi}{3} = \dfrac{1}{2}$ $\sec\dfrac{\pi}{3} = 2$

$\tan\theta = -\dfrac{5}{2}$ $\cot\theta = -\dfrac{2}{5}$ $\tan\dfrac{\pi}{3} = \sqrt{3}$ $\cot\dfrac{\pi}{3} = \dfrac{\sqrt{3}}{3}$

5. 1   **6. a.** $\dfrac{\sqrt{3}}{2}$ **b.** $\dfrac{\sqrt{3}}{2}$ **7. a.** −0.41   **b.** 2.03

A Exercises: Basic Skills and Concepts:

7. $\sin t = \dfrac{1}{3}$ $\csc t = 3$ **8.** $\sin t = \dfrac{\sqrt{3}}{2}$ $\csc t = \dfrac{2\sqrt{3}}{3}$

$\cos t = \dfrac{2\sqrt{2}}{3}$ $\sec t = \dfrac{3\sqrt{2}}{4}$ $\cos t = \dfrac{1}{2}$ $\sec t = 2$

$\tan t = \dfrac{\sqrt{2}}{4}$ $\cot t = 2\sqrt{2}$ $\tan t = \sqrt{3}$ $\cot t = \dfrac{\sqrt{3}}{3}$

9. $\sin t = \dfrac{2\sqrt{2}}{3}$ $\csc t = \dfrac{3\sqrt{2}}{4}$ **10.** $\sin t = \dfrac{1}{2}$ $\csc t = 2$

$\cos t = -\dfrac{1}{3}$ $\sec t = -3$ $\cos t = -\dfrac{\sqrt{3}}{2}$ $\sec t = -\dfrac{2\sqrt{3}}{3}$

$\tan t = -2\sqrt{2}$ $\cot t = -\dfrac{\sqrt{2}}{4}$ $\tan t = -\dfrac{\sqrt{3}}{3}$ $\cot t = -\sqrt{3}$

11. $\sin t = -\dfrac{2\sqrt{6}}{5}$ $\csc t = -\dfrac{5\sqrt{6}}{12}$ **12.** $\sin t = -\dfrac{\sqrt{3}}{2}$ $\csc t = -\dfrac{2\sqrt{3}}{3}$

$\cos t = \dfrac{1}{5}$ $\sec t = 5$ $\cos t = \dfrac{1}{2}$ $\sec t = 2$

$\tan t = -2\sqrt{6}$ $\cot t = -\dfrac{\sqrt{6}}{12}$ $\tan t = -\sqrt{3}$ $\cot t = -\dfrac{\sqrt{3}}{3}$

13. $\sin t = -\dfrac{1}{3}$ $\csc t = -3$ **14.** $\sin t = -\dfrac{1}{5}$ $\csc t = -5$

$\cos t = -\dfrac{2\sqrt{2}}{3}$ $\sec t = -\dfrac{3\sqrt{2}}{4}$ $\cos t = -\dfrac{2\sqrt{6}}{5}$ $\sec t = -\dfrac{5\sqrt{6}}{12}$

$\tan t = \dfrac{\sqrt{2}}{4}$ $\cot t = 2\sqrt{2}$ $\tan t = \dfrac{\sqrt{6}}{12}$ $\cot t = 2\sqrt{6}$

25. $\sin\theta = \dfrac{3}{5}$ $\csc\theta = \dfrac{5}{3}$ **26.** $\sin\theta = \dfrac{5\sqrt{34}}{34}$ $\csc\theta = \dfrac{\sqrt{34}}{5}$

$\cos\theta = -\dfrac{4}{5}$ $\sec\theta = -\dfrac{5}{4}$ $\cos\theta = -\dfrac{3\sqrt{34}}{34}$ $\sec\theta = -\dfrac{\sqrt{34}}{3}$

$\tan\theta = -\dfrac{3}{4}$ $\cot\theta = -\dfrac{4}{3}$ $\tan\theta = -\dfrac{5}{3}$ $\cot\theta = -\dfrac{3}{5}$

27. $\sin\theta = -\dfrac{1}{2}$ $\csc\theta = -2$ **28.** $\sin\theta = -\dfrac{2\sqrt{5}}{5}$ $\csc\theta = -\dfrac{\sqrt{5}}{2}$

$\cos\theta = -\dfrac{\sqrt{3}}{2}$ $\sec\theta = -\dfrac{2\sqrt{3}}{3}$ $\cos\theta = -\dfrac{\sqrt{5}}{5}$ $\sec\theta = -\sqrt{5}$

$\tan\theta = \dfrac{\sqrt{3}}{3}$ $\cot\theta = \sqrt{3}$ $\tan\theta = 2$ $\cot\theta = \dfrac{1}{2}$

29. $\sin\theta = \dfrac{\sqrt{2}}{2}$ $\csc\theta = \sqrt{2}$ **30.** $\sin\theta = -\dfrac{\sqrt{2}}{2}$ $\csc\theta = -\sqrt{2}$

$\cos\theta = \dfrac{\sqrt{2}}{2}$ $\sec\theta = \sqrt{2}$ $\cos\theta = -\dfrac{\sqrt{2}}{2}$ $\sec\theta = -\sqrt{2}$

$\tan\theta = 1$ $\cot\theta = 1$ $\tan\theta = 1$ $\cot\theta = 1$

31. $\sin\theta = -\dfrac{5}{13}$ $\csc\theta = -\dfrac{13}{5}$ **32.** $\sin\theta = -\dfrac{2\sqrt{53}}{53}$ $\csc\theta = -\dfrac{\sqrt{53}}{2}$

$\cos\theta = \dfrac{12}{13}$ $\sec\theta = \dfrac{13}{12}$ $\cos\theta = \dfrac{7\sqrt{53}}{53}$ $\sec\theta = \dfrac{\sqrt{53}}{7}$

$\tan\theta = -\dfrac{5}{12}$ $\cot\theta = -\dfrac{12}{5}$ $\tan\theta = -\dfrac{2}{7}$ $\cot\theta = -\dfrac{7}{2}$

49. $\dfrac{\sqrt{2} + \sqrt{3}}{2}$

C Exercises: Beyond the Basics: **91.** $\pi + 2n\pi$ **92.** $\dfrac{3\pi}{2} + 2n\pi$

93. $\dfrac{7\pi}{4} + 2n\pi$ **94.** $\dfrac{5\pi}{4} + 2n\pi$ **95.** $\dfrac{2\pi}{3} + 2n\pi$ **96.** $\dfrac{11\pi}{6} + 2n\pi$

Critical Thinking: **97.** 1   **98.** −1   **99.** $\sec^2\theta$

Section 4.3

Practice Problems: **1.** Quadrant II   **2.** $\sin\theta = -\dfrac{4\sqrt{41}}{41}$;

$\sec\theta = \dfrac{\sqrt{41}}{5}$ **3. a.** 5°   **b.** $\dfrac{\pi}{3}$ **c.** 1.20   **4.** $\dfrac{\sqrt{2}}{2}$ **5. a.** $-\sqrt{2}$

b. $-\sqrt{3}$ **6.** The ball reaches a maximum height of 153 feet and has a range of 612 feet.   **7.** $\sin t = \dfrac{\sqrt{5}}{3}$; $\tan t = -\dfrac{\sqrt{5}}{2}$

A Exercises: Basic Skills and Concepts:

17. $\sin\theta = -\dfrac{12}{13}$ $\csc\theta = -\dfrac{13}{12}$ **18.** $\sin\theta = -\dfrac{3}{5}$ $\csc\theta = -\dfrac{5}{3}$

$\cos\theta = -\dfrac{5}{13}$ $\sec\theta = -\dfrac{13}{5}$ $\cos\theta = \dfrac{4}{5}$ $\sec\theta = \dfrac{5}{4}$

$\tan\theta = \dfrac{12}{5}$ $\cot\theta = \dfrac{5}{12}$ $\tan\theta = -\dfrac{3}{4}$ $\cot\theta = -\dfrac{4}{3}$

19. $\sin\theta = \dfrac{4}{5}$ $\csc\theta = \dfrac{5}{4}$

   $\cos\theta = -\dfrac{3}{5}$ $\sec\theta = -\dfrac{5}{3}$

   $\tan\theta = -\dfrac{4}{3}$ $\cot\theta = -\dfrac{3}{4}$

20. $\sin\theta = -\dfrac{3}{4}$ $\csc\theta = -\dfrac{4}{3}$

   $\cos\theta = \dfrac{\sqrt{7}}{4}$ $\sec\theta = \dfrac{4\sqrt{7}}{7}$

   $\tan\theta = -\dfrac{3\sqrt{7}}{7}$ $\cot\theta = -\dfrac{\sqrt{7}}{3}$

21. $\sin\theta = \dfrac{3}{5}$ $\csc\theta = \dfrac{5}{3}$

   $\cos\theta = -\dfrac{4}{5}$ $\sec\theta = -\dfrac{5}{4}$

   $\tan\theta = -\dfrac{3}{4}$ $\cot\theta = -\dfrac{4}{3}$

22. $\sin\theta = \dfrac{2\sqrt{13}}{13}$ $\csc\theta = \dfrac{\sqrt{13}}{2}$

   $\cos\theta = \dfrac{3\sqrt{13}}{13}$ $\sec\theta = \dfrac{\sqrt{13}}{3}$

   $\tan\theta = \dfrac{2}{3}$ $\cot\theta = \dfrac{3}{2}$

23. $\sin\theta = -\dfrac{2\sqrt{2}}{3}$ $\csc\theta = -\dfrac{3\sqrt{2}}{4}$

   $\cos\theta = \dfrac{1}{3}$ $\sec\theta = 3$

   $\tan\theta = -2\sqrt{2}$ $\cot\theta = -\dfrac{\sqrt{2}}{4}$

24. $\sin\theta = \dfrac{2\sqrt{5}}{5}$ $\csc\theta = \dfrac{\sqrt{5}}{2}$

   $\cos\theta = -\dfrac{\sqrt{5}}{5}$ $\sec\theta = -\sqrt{5}$

   $\tan\theta = -2$ $\cot\theta = -\dfrac{1}{2}$

Section 4.4

Practice Problems:

1.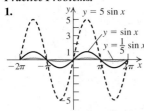

2.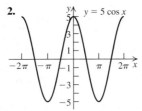

Amplitude $= 5$; range $= [-5, 5]$

3.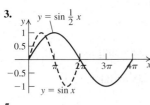

4.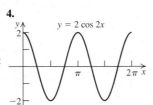

5.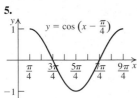

6. Amplitude $= \dfrac{1}{3}$; period $= 5\pi$; phase shift $= -\dfrac{\pi}{6}$

$y = \dfrac{1}{3}\cos\left(\dfrac{2}{5}x + \dfrac{\pi}{6}\right)$

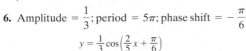

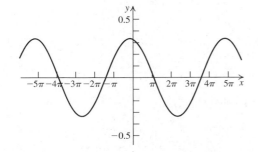

7.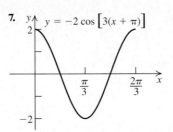

8. a. Period $= \pi$; phase shift $= -\dfrac{\pi}{8}$

   b. Period $= 2$; phase shift $= \dfrac{1}{\pi}$

9.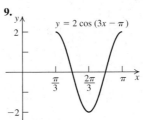

10.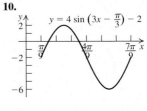

11.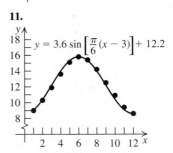

$y = 3.6\sin\left[\dfrac{\pi}{6}(x-3)\right] + 12.2$

12. $y = -4\cos\dfrac{2\pi}{3}t$

A Exercises: Basics Skills and Concepts:

9.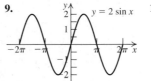

10.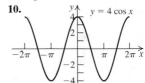

11.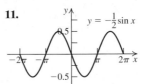

12.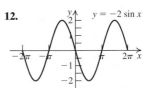

13.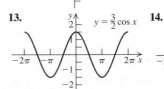

14.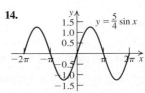

15.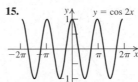

16.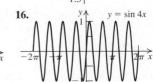

17. 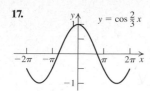 $y = \cos \frac{2}{3} x$

18. 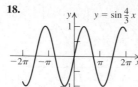 $y = \sin \frac{4}{3} x$

19. 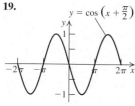 $y = \cos \left(x + \frac{\pi}{2}\right)$

20. 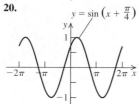 $y = \sin \left(x + \frac{\pi}{4}\right)$

21. 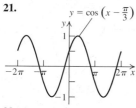 $y = \cos \left(x - \frac{\pi}{3}\right)$

22. 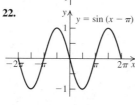 $y = \sin (x - \pi)$

23. 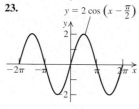 $y = 2 \cos \left(x - \frac{\pi}{2}\right)$

24. 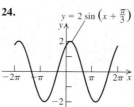 $y = 2 \sin \left(x + \frac{\pi}{3}\right)$

25. 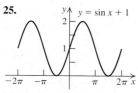 $y = \sin x + 1$

26. 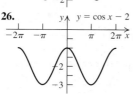 $y = \cos x - 2$

27. 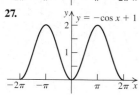 $y = -\cos x + 1$

28. 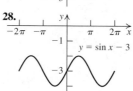 $y = \sin x - 3$

29. Amplitude = 5; period = 2π; phase shift = π

30. Amplitude = 3; period = 2π; phase shift = $\frac{\pi}{8}$

31. Amplitude = 7; period = $\frac{2\pi}{9}$; phase shift = $-\frac{\pi}{6}$

32. Amplitude = 11; period = $\frac{\pi}{4}$; phase shift = $-\frac{\pi}{3}$

33. Amplitude = 6; period = 4π; phase shift = -2

34. Amplitude = 8; period = 10π; phase shift = -9

35. Amplitude = 0.9; period = 8π; phase shift = $\frac{\pi}{4}$

36. Amplitude = $\sqrt{5}$; period = 2; phase shift = -1

37. 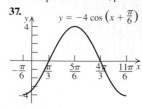 $y = -4 \cos \left(x + \frac{\pi}{6}\right)$

38. 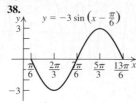 $y = -3 \sin \left(x - \frac{\pi}{6}\right)$

39. 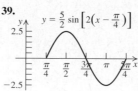 $y = \frac{5}{2} \sin \left[2\left(x - \frac{\pi}{4}\right)\right]$

40. 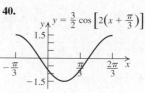 $y = \frac{3}{2} \cos \left[2\left(x + \frac{\pi}{3}\right)\right]$

41. 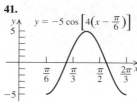 $y = -5 \cos \left[4\left(x - \frac{\pi}{6}\right)\right]$

42. 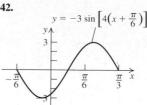 $y = -3 \sin \left[4\left(x + \frac{\pi}{6}\right)\right]$

43. 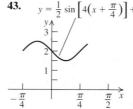 $y = \frac{1}{2} \sin \left[4\left(x + \frac{\pi}{4}\right)\right] + 2$

44. 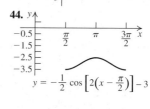 $y = -\frac{1}{2} \cos \left[2\left(x - \frac{\pi}{2}\right)\right] - 3$

45. $y = 4 \cos \left[2\left(x + \frac{\pi}{6}\right)\right]$; period = π; phase shift = $-\frac{\pi}{6}$

46. $y = 5 \sin \left[3\left(x + \frac{\pi}{6}\right)\right]$; period = $\frac{2\pi}{3}$; phase shift = $-\frac{\pi}{6}$

47. $y = -\frac{3}{2} \sin \left[2\left(x - \frac{\pi}{2}\right)\right]$; period = π; phase shift = $\frac{\pi}{2}$

48. $y = -2 \cos \left[5\left(x - \frac{\pi}{20}\right)\right]$; period = $\frac{2\pi}{5}$; phase shift = $\frac{\pi}{20}$

49. $y = 3 \cos \pi x$; period = 2; phase shift = 0

50. $y = \sin \frac{\pi}{3}(x)$; period = 6; phase shift = 0

51. $y = \frac{1}{2} \cos \left[\frac{\pi}{4}(x + 1)\right]$; period = 8; phase shift = -1

52. $y = -\sin \left[\frac{\pi}{6}(x + 1)\right]$; period = 12; phase shift = -1

53. $y = 2 \sin \left[\pi\left(x + \frac{3}{\pi}\right)\right]$; period = 2; phase shift = $-\frac{3}{\pi}$

54. $y = -\cos \left[\pi\left(x - \frac{1}{4\pi}\right)\right]$; period = 2; phase shift = $\frac{1}{4\pi}$

B Exercises: Applying the Concepts: 55. a. Period = $\frac{1}{70}$. The pulse is the frequency of the function; it says how many times the heart beats in one minute.

b. 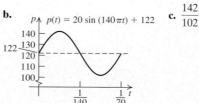 $p(t) = 20 \sin (140\pi t) + 122$ **c.** $\frac{142}{102}$

56. a. Amplitude = 4.5; period = $\frac{1}{80}$

c. 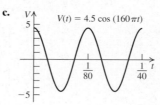 $V(t) = 4.5 \cos (160\pi t)$

57. a. Amplitude = 156; period = $\dfrac{1}{60}$

c.

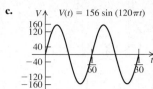

$V(t) = 156 \sin(120\pi t)$

60. a.

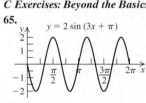

$D(t) = 450 \sin\left(\dfrac{3t}{5}\right) + 1200$

$L(t) = 225 \sin\left[\dfrac{2}{5}(t-2)\right] + 500$

b. As the lion population increases, the deer population decreases. As the lion population decreases, the deer population increases.

63. b.

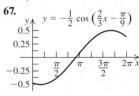

$y = 25.5 \sin\left[\dfrac{\pi}{6}(x-4)\right] + 44.5$

c. January = $f(1) = 19$; April = $f(4) = 44.5$; July = $f(7) = 70$; October = $f(10) = 44.5$. The computed values are very close to the measured values.

64. b.

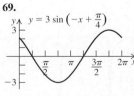

$y = 21.5 \sin\left[\dfrac{\pi}{6}(x-4)\right] + 61$

C Exercises: Beyond the Basics:

65.

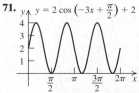

$y = 2 \sin(3x + \pi)$

66.

$y = 4 \cos\left(2x + \dfrac{\pi}{2}\right)$

67.

$y = -\dfrac{1}{2}\cos\left(\dfrac{2}{3}x - \dfrac{\pi}{9}\right)$

68.

$y = -2 \sin(\pi x - 2)$

69.

$y = 3 \sin\left(-x + \dfrac{\pi}{4}\right)$

70.

$y = -2 \sin\left(-2x + \dfrac{\pi}{3}\right)$

71.

$y = 2 \cos\left(-3x + \dfrac{\pi}{2}\right) + 2$

72.

$y = -3 \cos\left(-\dfrac{x}{\pi} + \dfrac{1}{2}\right) + 3$

73.

$y = |\sin x|$

74.

$y = |\cos x|$

75.

$y = x \sin x$

76.

$y = x \cos x$

77.

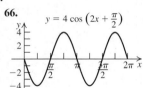

$y = \sqrt{x}\,\sin x$

78.

$y = |x \sin x|$

79.

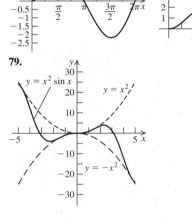

$y = x^2 \sin x$ $y = x^2$ $y = -x^2$

Section 4.5

Practice Problems:

1.

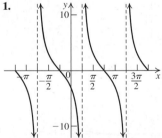

2.

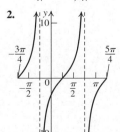

3.

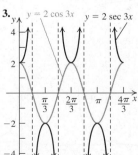

$y = 2 \cos 3x$ $y = 2 \sec 3x$

4. $[2.6, 5.1]$

A Exercises: Basic Skills and Concepts:

9.

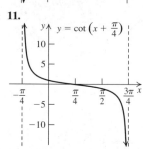

$y = \tan\left(x - \frac{\pi}{4}\right)$

10.

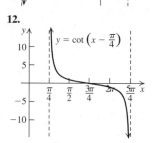

$y = \tan\left(x + \frac{\pi}{4}\right)$

11.

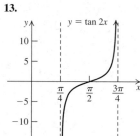

$y = \cot\left(x + \frac{\pi}{4}\right)$

12.

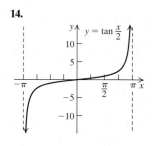

$y = \cot\left(x - \frac{\pi}{4}\right)$

13.

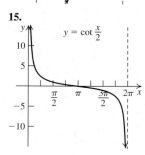

$y = \tan 2x$

14.

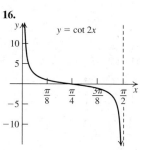

$y = \tan\frac{x}{2}$

15.

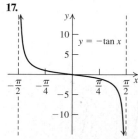

$y = \cot\frac{x}{2}$

16.

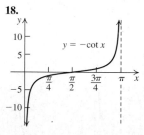

$y = \cot 2x$

17.

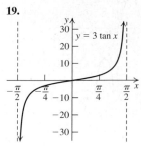

$y = -\tan x$

18.

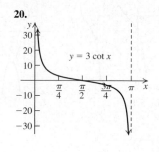

$y = -\cot x$

19.

$y = 3\tan x$

20.

$y = 3\cot x$

21.

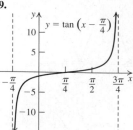

$y = \sec 2x$

22.

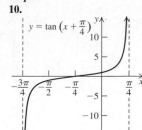

$y = \sec\frac{x}{2}$

23.

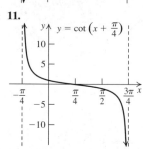

$y = \csc 3x$

24.

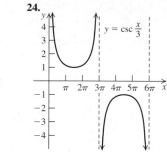

$y = \csc\frac{x}{3}$

25.

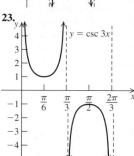

$y = \sec(x - \pi)$

26.

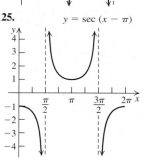

$y = \csc(x - \pi)$

27.

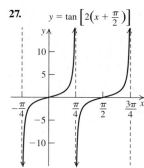

$y = \tan\left[2\left(x + \frac{\pi}{2}\right)\right]$

28.

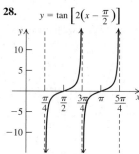

$y = \tan\left[2\left(x - \frac{\pi}{2}\right)\right]$

29.

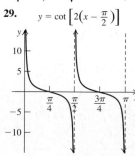

$y = \cot\left[2\left(x - \frac{\pi}{2}\right)\right]$

30.

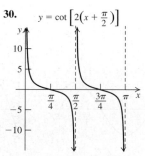

$y = \cot\left[2\left(x + \frac{\pi}{2}\right)\right]$

31.

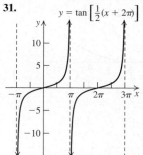

$y = \tan\left[\frac{1}{2}(x + 2\pi)\right]$

32.

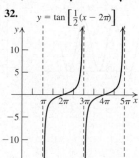

$y = \tan\left[\frac{1}{2}(x - 2\pi)\right]$

33. $y = \cot\left[\frac{1}{2}(x - 2\pi)\right]$

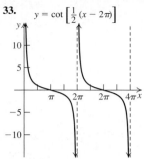

34. $y = \cot\left[\frac{1}{2}(x + 2\pi)\right]$

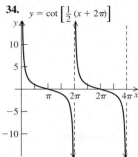

43. $y = \tan\left[\frac{2}{3}\left(x - \frac{\pi}{2}\right)\right]$

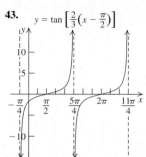

44. $y = 2\cot\left[2\left(x - \frac{\pi}{6}\right)\right]$

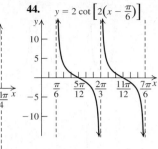

35. $y = -\frac{1}{2}\tan\frac{x}{2}$

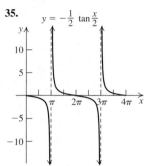

36. $y = -\frac{1}{2}\cot\frac{x}{2}$

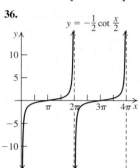

45. $y = -5\tan\left[2\left(x + \frac{\pi}{3}\right)\right]$

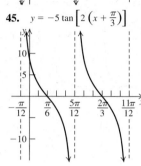

46. $y = -3\cot\left[\frac{1}{2}\left(x - \frac{\pi}{3}\right)\right]$

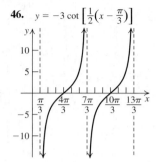

47. $y = \frac{1}{3}\cot\left[2(x - \pi)\right]$

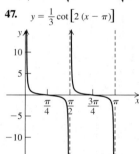

48. $y = \frac{1}{2}\tan\left[4\left(x - \frac{\pi}{6}\right)\right]$

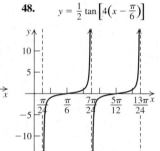

37. $y = \sec\left[4\left(x - \frac{\pi}{4}\right)\right]$

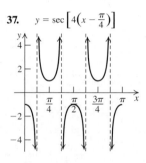

38. $y = \sec\left[\frac{1}{2}\left(x - \frac{\pi}{2}\right)\right]$

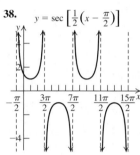

39. $y = -2\sec 3x$

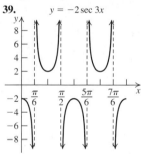

40. $y = -3\csc 4x$

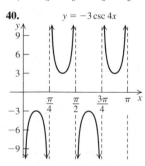

B Exercises: Applying the Concepts:

49. a. $d(t) = 20\tan\frac{\pi t}{5}$

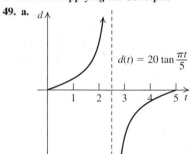

41. $y = 3\csc\left(x + \frac{\pi}{2}\right)$

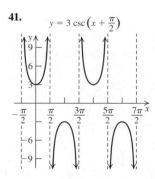

42. $y = 3\sec\left[2\left(x - \frac{\pi}{6}\right)\right]$

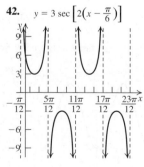

C Exercises: Beyond the Basics:

54. $y = 3\tan(-2x)$

55. $y = 2\cot(-3x)$

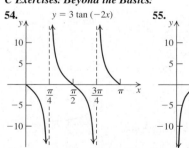
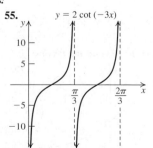

56. $y = -2 \sec\left(-\frac{x}{2}\right)$

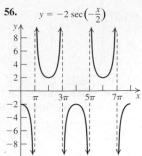

57. $y = -\frac{1}{2} \csc\left(-\frac{x}{5}\right)$

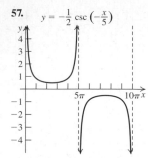

58. $y = \cot(\pi - x)$

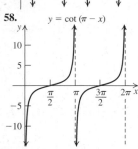

59. $y = -\tan\left(\frac{\pi}{2} - x\right)$

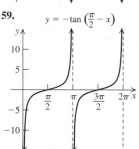

60. $y = -\sec(\pi - 4x)$

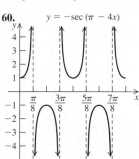

61. $y = 2\sec(\pi - 2x)$

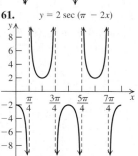

62. $y = 2\tan\left(\frac{\pi}{2} - 2x\right) + 1$

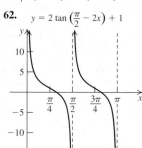

63. $y = 4\cot(\pi - 4x) + 3$

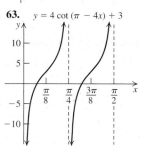

Section 4.6

Practice Problems: **1. a.** $-\frac{\pi}{3}$ **b.** $-\frac{\pi}{2}$ **2. a.** $\frac{3\pi}{4}$ **b.** $\frac{\pi}{3}$ **3.** $\frac{\pi}{6}$
4. $\frac{\pi}{3}$ **5. a.** 1.3490 **b.** 0.2898 **c.** 2.932 **6. a.** 53.1301°
b. 4.4117° **c.** −85.2364° **7.** $-\frac{\pi}{2}$ **8.** $\frac{2\sqrt{2}}{3}$ **9.** 80°

C Exercises: Beyond the Basics:

75.

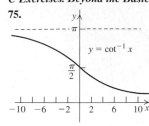
$y = \cot^{-1} x$

76.

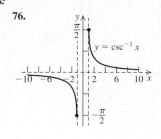
$y = \csc^{-1} x$

77.

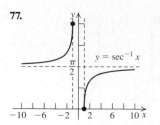
$y = \sec^{-1} x$

Review Exercises

1.

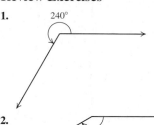
240°

2.

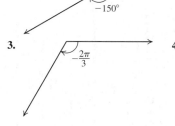
−150°

3.

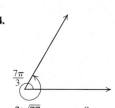
$-\frac{2\pi}{3}$

4.

$\frac{7\pi}{3}$

17. $\sin\theta = \frac{\sqrt{77}}{9}$, $\tan\theta = \frac{\sqrt{77}}{2}$, $\cot\theta = \frac{2\sqrt{77}}{77}$, $\sec\theta = \frac{9}{2}$,
$\csc\theta = \frac{9\sqrt{77}}{77}$ **18.** $\cos\theta = \frac{2\sqrt{6}}{5}$, $\tan\theta = \frac{\sqrt{6}}{12}$, $\cot\theta = 2\sqrt{6}$,
$\sec\theta = \frac{5\sqrt{6}}{12}$, $\csc\theta = 5$ **19.** $\sin\theta = \frac{5\sqrt{34}}{34}$, $\cos\theta = \frac{3\sqrt{34}}{34}$,
$\cot\theta = \frac{3}{5}$, $\sec\theta = \frac{\sqrt{34}}{3}$, $\csc\theta = \frac{\sqrt{34}}{5}$ **20.** $\sin\theta = \frac{4\sqrt{41}}{41}$,
$\cos\theta = \frac{5\sqrt{41}}{41}$, $\tan\theta = \frac{4}{5}$, $\sec\theta = \frac{\sqrt{41}}{5}$, $\csc\theta = \frac{\sqrt{41}}{4}$
21. $\sin\theta = \frac{4\sqrt{17}}{17}$, $\cos\theta = \frac{\sqrt{17}}{17}$, $\tan\theta = 4$, $\cot\theta = \frac{1}{4}$,
$\sec\theta = \sqrt{17}$, $\csc\theta = \frac{\sqrt{17}}{4}$ **22.** $\sin\theta = \frac{7\sqrt{58}}{58}$, $\cos\theta = -\frac{3\sqrt{58}}{58}$,
$\tan\theta = -\frac{7}{3}$, $\cot\theta = -\frac{3}{7}$, $\sec\theta = -\frac{\sqrt{58}}{3}$, $\csc\theta = \frac{\sqrt{58}}{7}$
23. $\sin\theta = \frac{2}{3}$, $\cos\theta = -\frac{\sqrt{5}}{3}$, $\tan\theta = -\frac{2\sqrt{5}}{5}$, $\cot\theta = -\frac{\sqrt{5}}{2}$,
$\sec\theta = -\frac{3\sqrt{5}}{5}$, $\csc\theta = \frac{3}{2}$ **24.** $\sin\theta = -\frac{\sqrt{6}}{3}$,
$\cos\theta = -\frac{\sqrt{3}}{3}$, $\tan\theta = \sqrt{2}$, $\cot\theta = \frac{\sqrt{2}}{2}$, $\sec\theta = -\sqrt{3}$, $\csc\theta = -\frac{\sqrt{6}}{2}$

37. $y = -\frac{3}{2}\cos x$

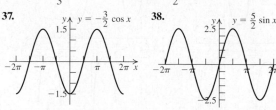

38. $y = \frac{5}{2}\sin x$

39. $y = 3\sin\left(x - \frac{\pi}{3}\right)$

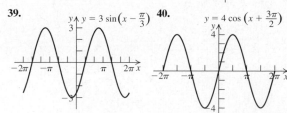

40. $y = 4\cos\left(x + \frac{3\pi}{2}\right)$

41. Amplitude $= 14$; period $= \pi$; phase shift $= -\dfrac{\pi}{14}$

42. Amplitude $= 21$; period $= \dfrac{\pi}{4}$; phase shift $= -\dfrac{\pi}{72}$

43. Vertical stretch $= 6$; period $= \dfrac{\pi}{2}$; phase shift $= -\dfrac{\pi}{10}$

44. Vertical stretch $= 11$; period $= \dfrac{\pi}{6}$; phase shift $= -\dfrac{\pi}{72}$

45.
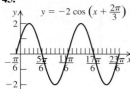
$y = -2 \cos\left(x + \dfrac{2\pi}{3}\right)$

46.
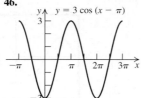
$y = 3 \cos(x - \pi)$

47.
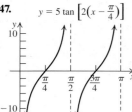
$y = 5 \tan\left[2\left(x - \dfrac{\pi}{4}\right)\right]$

48.
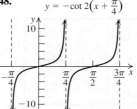
$y = -\cot 2\left(x + \dfrac{\pi}{4}\right)$

49.
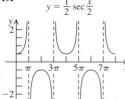
$y = \dfrac{1}{2} \sec \dfrac{x}{2}$

50.
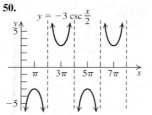
$y = -3 \csc \dfrac{x}{2}$

Practice Test A

15.
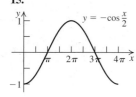
$y = -\cos \dfrac{x}{2}$

16.
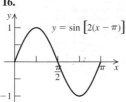
$y = \sin\left[2(x - \pi)\right]$

17.
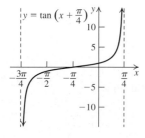
$y = \tan\left(x + \dfrac{\pi}{4}\right)$

Cumulative Review Exercises (Chapters 1–4)

6. $f(x) = x^2$

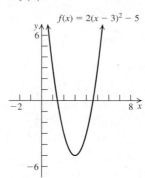
$f(x) = 2(x - 3)^2 - 5$

7.

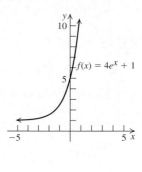
$f(x) = 4e^x + 1$

9. a. $f^{-1}(x) = \dfrac{x}{3} - \dfrac{7}{3}$ **b.** The function does not have an inverse.

12. a. $\dfrac{2}{3}\log x + \dfrac{1}{3}\log y - \dfrac{1}{3}\log z$ **b.** $\ln 1.4 + 4\ln x - \dfrac{1}{2}\ln y$

13.
$$f(x) = \begin{cases} -\ln x & \text{if } x > 0 \\ \ln |x| & \text{if } x < 0 \end{cases}$$

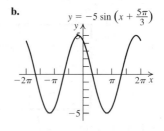

15. $\pm 1, \pm 2, \pm 3, \pm 6, \pm \dfrac{1}{2}, \pm \dfrac{3}{2}$

16. a. $\sin \theta = \dfrac{2\sqrt{10}}{7}$, $\tan \theta = \dfrac{2\sqrt{10}}{3}$,
$\cot \theta = \dfrac{3\sqrt{10}}{20}$, $\sec \theta = \dfrac{7}{3}$,
$\csc \theta = \dfrac{7\sqrt{10}}{20}$

b. $\cos \theta = \dfrac{3\sqrt{13}}{11}$, $\tan \theta = \dfrac{2\sqrt{13}}{39}$,
$\cot \theta = \dfrac{3\sqrt{13}}{2}$, $\sec \theta = \dfrac{11\sqrt{13}}{39}$,
$\csc \theta = \dfrac{11}{2}$

17. a. $\sin \theta = \dfrac{3}{5}$, $\tan \theta = -\dfrac{3}{4}$, $\cot \theta = -\dfrac{4}{3}$,
$\sec \theta = -\dfrac{5}{4}$, $\csc \theta = \dfrac{5}{3}$ **b.** $\sin \theta = -\dfrac{12}{13}$, $\cos \theta = -\dfrac{5}{13}$, $\tan \theta = \dfrac{12}{5}$,
$\sec \theta = -\dfrac{13}{5}$, $\csc \theta = -\dfrac{13}{12}$

18. a.
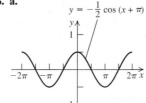
$y = -\dfrac{1}{2}\cos(x + \pi)$

b.

$y = -5\sin\left(x + \dfrac{5\pi}{3}\right)$

CHAPTER 5

Section 5.1

Practice Problems: **1.** $-\dfrac{3}{4}$ **2.** $\sin \theta = \dfrac{\sqrt{2}}{2}$, $\cos \theta = -\dfrac{\sqrt{2}}{2}$,
$\cot \theta = -1$, $\sec \theta = -\sqrt{2}$, $\csc \theta = \sqrt{2}$ **3.** $\dfrac{2}{\sin x}$

A Exercises: Basic Skills and Concepts:

13. $\tan x = -\dfrac{4}{3}$, $\cot x = -\dfrac{3}{4}$, $\sec x = -\dfrac{5}{3}$, $\csc x = \dfrac{5}{4}$

14. $\tan x = -\dfrac{4}{3}$, $\cot x = -\dfrac{3}{4}$, $\sec x = \dfrac{5}{3}$, $\csc x = -\dfrac{5}{4}$

15. $\tan x = \dfrac{\sqrt{2}}{2}, \cot x = \sqrt{2}, \sec x = \dfrac{\sqrt{6}}{2}, \csc x = \sqrt{3}$

16. $\tan x = -\dfrac{\sqrt{2}}{2}, \cot x = -\sqrt{2}, \sec x = -\dfrac{\sqrt{6}}{2}, \csc = \sqrt{3}$

17. $\sin x = -\dfrac{\sqrt{5}}{5}, \cos x = -\dfrac{2\sqrt{5}}{5}, \cot x = 2, \csc = -\sqrt{5}$

18. $\sin x = \dfrac{\sqrt{5}}{5}, \cos x = \dfrac{2\sqrt{5}}{5}, \cot x = 2, \csc x = \sqrt{5}$

19. $\sin x = \dfrac{1}{3}, \cos x = \dfrac{2\sqrt{2}}{3}, \tan x = \dfrac{\sqrt{2}}{4}, \sec x = \dfrac{3\sqrt{2}}{4}$

20. $\sin x = \dfrac{1}{3}, \cos x = -\dfrac{2\sqrt{2}}{3}, \tan x = -\dfrac{\sqrt{2}}{4}, \sec x = -\dfrac{3\sqrt{2}}{4}$

21. $\cos x = -\dfrac{12}{13}, \tan x = \dfrac{5}{12}, \sec x = -\dfrac{13}{12}, \csc x = -\dfrac{13}{5}$

22. $\cos x = \dfrac{12}{13}, \tan x = -\dfrac{5}{12}, \sec x = \dfrac{13}{12}, \csc x = -\dfrac{13}{5}$

23. $\sin x = -\dfrac{2\sqrt{2}}{3}, \cos x = \dfrac{1}{3}, \tan x = -2\sqrt{2},$
$\cot x = -\dfrac{\sqrt{2}}{4}, \csc x = -\dfrac{3\sqrt{2}}{4}$ **24.** $\sin x = \dfrac{2\sqrt{5}}{5}, \cos x = -\dfrac{\sqrt{5}}{5},$
$\cot x = -\dfrac{1}{2}, \sec x = -\sqrt{5}, \csc x = \dfrac{\sqrt{5}}{2}$

35.   An identity

$y = \cos^2 x - \sin^2 x$
$y = 2\cos^2 x - 1$

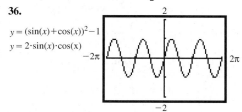

36.   An identity

$y = (\sin(x) + \cos(x))^2 - 1$
$y = 2 \cdot \sin(x) \cdot \cos(x)$

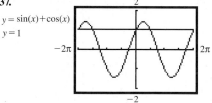

37.

$y = \sin(x) + \cos(x)$
$y = 1$

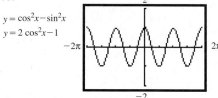

Not an identity

38.

$y = \dfrac{\sin 2x}{2}$
$y = \sin(x)$

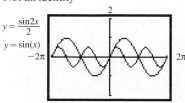

Not an identity

39.

$y = \dfrac{\sin^2(x)}{1 + \cos(x)}$
$y = 1 - \cos(x)$

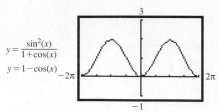

An identity

40.

$y = \sin(x)$
$y = \cos(x) \cdot \tan(x)$

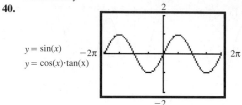

An identity

41.   An identity

X	Y₁	Y₂
-2	-.4161	-.4161
-1.5	.07074	.07074
-1	.5403	.5403
-.5	.87758	.87758
0	ERROR	1
.5	.87758	.87758
1	.5403	.5403

X= -2

42.   An identity

X	Y₁	Y₂
-2	-1.1	-1.1
-1.5	-1.003	-1.003
-1	-1.188	-1.188
-.5	-2.086	-2.086
0	ERROR	ERROR
.5	2.0858	2.0858
1	1.1884	1.1884

X= -2

43.   An identity

X	Y₁	Y₂
-2	.37206	.37206
-1.5	1.1527	1.1527
-1	4.588	4.588
-.5	-3.408	-3.408
0	-1	ERROR
.5	-.2934	-.2934
1	.21796	.21796

X= -2

44.   An identity

X	Y₁	Y₂
-2	5.7744	5.7744
-1.5	199.85	199.85
-1	3.4255	3.4255
-.5	1.2984	1.2984
0	1	1
.5	1.2984	1.2984
1	3.4255	3.4255

X= -2

45.   Not an identity

X	Y₁	Y₂
-2	-1.158	4.3701
-1.5	.14255	-28.2
-1	2.185	-3.115
-.5	-1.557	-1.093
0	0	0
.5	1.5574	1.0926
1	-2.185	3.1148

X= -2

46.   Not an identity

X	Y₁	Y₂
-2	3.6454	-.4161
-1.5	3.99	.07074
-1	3.391	.5403
-.5	2.1887	.87758
0	1	1
.5	.271	.87758
1	.02513	.5403

X= -2

Section 5.2

Practice Problems: **1. a.** $x = \dfrac{\pi}{2} + 2n\pi$ **b.** $x = 0 + 2n\pi$

c. $x = \dfrac{\pi}{4} + n\pi$ **2. a.** $x \approx 48.2°, x \approx 311.8°$

b. $x \approx \pi - 1.1071, x \approx 2\pi - 1.1071$ **3.** $x \approx 105.5°, x \approx 314.5°$

4. $\left\{ \dfrac{\pi}{12}, \dfrac{5\pi}{12}, \dfrac{13\pi}{12}, \dfrac{17\pi}{12} \right\}$ **5.** $x = \dfrac{\pi}{3}$ **6.** $x = \dfrac{\pi}{2}, x = \dfrac{5\pi}{6}, x = \dfrac{11\pi}{6}$

7. $\theta = 2n\pi, \theta = \dfrac{2\pi}{3} + 2n\pi, \theta = \dfrac{4\pi}{3} + 2n\pi$

8. $\theta = \dfrac{\pi}{6}, \theta = \dfrac{\pi}{2}, \theta = \dfrac{5\pi}{6}$ **9.** $\theta = \dfrac{\pi}{3}$ **10.** About 5 days and 24 days

A Exercises: Basic Skills and Concepts:
23. $60° + 360°n, 120° + 360°n$ **24.** $150° + 360°n, 210° + 360°n$
25. $60° + 360°n, 300° + 360°n$ **26.** $210° + 360°n, 330° + 360°n$

41. $\left\{ \dfrac{19\pi}{24}, \dfrac{35\pi}{24} \right\}$ **42.** $\left\{ \dfrac{\pi}{24}, \dfrac{17\pi}{24} \right\}$ **47.** $\left\{ \dfrac{\pi}{6}, \dfrac{5\pi}{6}, \dfrac{7\pi}{6}, \dfrac{11\pi}{6} \right\}$

49. $\left\{ \dfrac{\pi}{12}, \dfrac{7\pi}{12}, \dfrac{13\pi}{12}, \dfrac{19\pi}{12} \right\}$ **50.** $\left\{ \dfrac{\pi}{6}, \dfrac{2\pi}{3}, \dfrac{7\pi}{6}, \dfrac{5\pi}{3} \right\}$

51. $\left\{ \dfrac{\pi}{18}, \dfrac{5\pi}{18}, \dfrac{13\pi}{18}, \dfrac{17\pi}{18}, \dfrac{25\pi}{18}, \dfrac{29\pi}{18} \right\}$

52. $\left\{ \dfrac{\pi}{18}, \dfrac{11\pi}{18}, \dfrac{13\pi}{18}, \dfrac{23\pi}{18}, \dfrac{25\pi}{18}, \dfrac{35\pi}{18} \right\}$ **58.** $\left\{ \dfrac{\pi}{6}, \dfrac{2\pi}{3}, \dfrac{7\pi}{6}, \dfrac{4\pi}{3} \right\}$

59. $\left\{ \dfrac{\pi}{6}, \dfrac{3\pi}{4}, \dfrac{5\pi}{6}, \dfrac{7\pi}{4} \right\}$ **60.** $\left\{ \dfrac{\pi}{6}, \dfrac{2\pi}{3}, \dfrac{5\pi}{3}, \dfrac{11\pi}{6} \right\}$

61. $\left\{ \dfrac{\pi}{6}, \dfrac{3\pi}{4}, \dfrac{5\pi}{6}, \dfrac{7\pi}{4} \right\}$ **62.** $\left\{ \dfrac{\pi}{3}, \dfrac{2\pi}{3}, \dfrac{5\pi}{3} \right\}$ **63.** $\left\{ \dfrac{\pi}{4}, \dfrac{7\pi}{6}, \dfrac{7\pi}{4}, \dfrac{11\pi}{6} \right\}$

64. $\left\{ \dfrac{\pi}{4}, \dfrac{5\pi}{4}, \dfrac{7\pi}{4} \right\}$ **65.** $\left\{ \dfrac{\pi}{6}, \dfrac{5\pi}{6}, \dfrac{7\pi}{6}, \dfrac{11\pi}{6} \right\}$ **66.** $\left\{ \dfrac{\pi}{3}, \dfrac{2\pi}{3}, \dfrac{4\pi}{3}, \dfrac{5\pi}{3} \right\}$

67. $\left\{ \dfrac{\pi}{4}, \dfrac{3\pi}{4}, \dfrac{5\pi}{4}, \dfrac{7\pi}{4} \right\}$ **68.** $\left\{ \dfrac{\pi}{4}, \dfrac{3\pi}{4}, \dfrac{5\pi}{4}, \dfrac{7\pi}{4} \right\}$

69. $\left\{ \dfrac{\pi}{3}, \dfrac{2\pi}{3}, \dfrac{4\pi}{3}, \dfrac{5\pi}{3} \right\}$ **70.** $\left\{ \dfrac{\pi}{3}, \dfrac{2\pi}{3}, \dfrac{4\pi}{3}, \dfrac{5\pi}{3} \right\}$ **71.** $\left\{ \dfrac{\pi}{2}, \dfrac{7\pi}{6}, \dfrac{11\pi}{6} \right\}$

72. $\left\{ \dfrac{\pi}{3}, \dfrac{5\pi}{3} \right\}$ **73.** $\left\{ \dfrac{\pi}{4}, \dfrac{5\pi}{4} \right\}$ **74.** $\left\{ \dfrac{5\pi}{6}, \dfrac{11\pi}{6} \right\}$

75. $\left\{ \dfrac{\pi}{6}, \dfrac{5\pi}{6}, \dfrac{7\pi}{6}, \dfrac{11\pi}{6} \right\}$ **76.** $\left\{ \dfrac{\pi}{3}, \dfrac{2\pi}{3}, \dfrac{4\pi}{3}, \dfrac{5\pi}{3} \right\}$ **78.** $\left\{ 0, \dfrac{2\pi}{3}, \dfrac{4\pi}{3} \right\}$

79. $\left\{ \dfrac{7\pi}{6}, \dfrac{3\pi}{2}, \dfrac{11\pi}{6} \right\}$ **80.** $\left\{ \dfrac{\pi}{3}, \pi, \dfrac{5\pi}{3} \right\}$ **81.** $\left\{ \dfrac{\pi}{3}, \dfrac{5\pi}{6}, \dfrac{4\pi}{3}, \dfrac{11\pi}{6} \right\}$

82. $\left\{ \dfrac{\pi}{6}, \dfrac{\pi}{4}, \dfrac{7\pi}{6}, \dfrac{5\pi}{4} \right\}$ **83.** $\left\{ \dfrac{\pi}{3}, \pi \right\}$

Section 5.3

Practice Problems: **1.** $\dfrac{\sqrt{6} + \sqrt{2}}{4}$ **2.** $\dfrac{\sqrt{2} - \sqrt{6}}{4}$ **5.** $\dfrac{1}{2}$ **6.** $-\dfrac{63}{65}$

8. $y = 2 \sin\left(x + \dfrac{\pi}{3} \right)$ **9.**

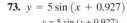

10. Amplitude $= 0.1\sqrt{5}$; period $= \dfrac{\pi}{200}$; frequency $= \dfrac{200}{\pi}$;
phase shift ≈ -0.0028

A Exercises: Basic Skills and Concepts: **7.** $\dfrac{\sqrt{6} + \sqrt{2}}{4}$ **8.** $\dfrac{\sqrt{6} - \sqrt{2}}{4}$

9. $\dfrac{\sqrt{6} - \sqrt{2}}{4}$ **10.** $\dfrac{\sqrt{6} + \sqrt{2}}{4}$ **11.** $-\dfrac{\sqrt{6} + \sqrt{2}}{4}$ **12.** $\dfrac{\sqrt{6} - \sqrt{2}}{4}$

57. $\dfrac{2\sqrt{210} - 6}{35}$ **58.** $\dfrac{-4\sqrt{10} - 3\sqrt{21}}{35}$ **59.** $\dfrac{35\sqrt{210} - 105}{402}$

60. $\dfrac{35(3\sqrt{21} + 4\sqrt{10})}{29}$ **61.** $\dfrac{-4\sqrt{10} - 3\sqrt{21}}{2\sqrt{210} - 6}$ **62.** $\dfrac{3\sqrt{21} - 4\sqrt{10}}{2\sqrt{210} + 6}$

71. $y = \sqrt{2} \sin\left(x + \dfrac{\pi}{4} \right)$ **72.** $y = \sqrt{2} \sin\left(x - \dfrac{\pi}{4} \right)$

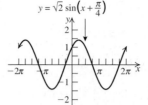

73. $y = 5 \sin (x + 0.927)$ **74.** $y = 5 \sin (x - 0.644)$

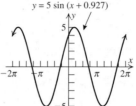

75. $y = 13 \sin (x - 1.176)$ **76.** $y = 13 \sin (x + 0.395)$

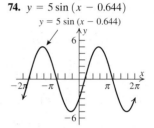

77. $y = \sqrt{10} \sin (x + 2.820)$ **78.** $y = 5 \sin (x + 2.498)$

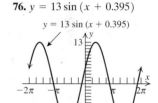

Section 5.4

Practice Problems: **1. a.** $-\dfrac{120}{169}$ **b.** $-\dfrac{119}{169}$ **c.** $\dfrac{120}{119}$ **2. a.** $\dfrac{\sqrt{3}}{2}$

b. $\dfrac{\sqrt{2}}{2}$ **4.** $\dfrac{3}{8} - \dfrac{1}{2} \cos 2x + \dfrac{1}{8} \cos 4x$ **5.** ≈ 99.8 watts

6. $\dfrac{\sqrt{2} + \sqrt{2}}{2}$ **7.** $-\dfrac{\sqrt{26}}{26}$ **9.** $\left\{ \dfrac{\pi}{6}, \dfrac{3\pi}{2}, \dfrac{5\pi}{6} \right\}$ **10.** $\{\pi\}$

A Exercises: Basic Skills and Concepts: **2.** $\dfrac{1 - \cos 2x}{2}$ **3.** $\dfrac{1 + \cos 2x}{2}$

57. a. $\sqrt{\dfrac{13 + 3\sqrt{13}}{26}}$ **b.** $\sqrt{\dfrac{13 - 3\sqrt{13}}{26}}$ **c.** $\sqrt{\dfrac{13 + 3\sqrt{13}}{13 - 3\sqrt{13}}}$

58. a. $\dfrac{2\sqrt{5}}{5}$ **b.** $-\dfrac{\sqrt{5}}{5}$ **c.** -2 **59. a.** $\sqrt{\dfrac{5 + 2\sqrt{6}}{10}}$ **b.** $\sqrt{\dfrac{5 - 2\sqrt{6}}{10}}$

c. $\sqrt{\dfrac{5 + 2\sqrt{6}}{5 - 2\sqrt{6}}}$ **60. a.** $\dfrac{\sqrt{6}}{6}$ **b.** $-\dfrac{\sqrt{30}}{6}$ **c.** $-\dfrac{\sqrt{5}}{5}$

61. a. $\sqrt{\dfrac{5 - \sqrt{5}}{10}}$ **b.** $\sqrt{\dfrac{5 + \sqrt{5}}{10}}$ **c.** $\sqrt{\dfrac{5 - \sqrt{5}}{5 + \sqrt{5}}}$ **62. a.** $\sqrt{\dfrac{7 + \sqrt{42}}{14}}$

b. $\sqrt{\dfrac{7 - \sqrt{42}}{14}}$ **c.** $\sqrt{\dfrac{7 + \sqrt{42}}{7 - \sqrt{42}}}$ **75.** $\left\{ \dfrac{\pi}{6}, \dfrac{5\pi}{6}, \dfrac{7\pi}{6}, \dfrac{11\pi}{6} \right\}$

76. $\left\{ \dfrac{\pi}{3}, \dfrac{2\pi}{3}, \dfrac{4\pi}{3}, \dfrac{5\pi}{3} \right\}$

Section 5.5

Practice Problems: **1.** $\frac{1}{2}\cos 2x + \frac{1}{2}\cos 4x$ **2.** $\frac{2-\sqrt{3}}{4}$

3. a. $2\sin 3x \sin x$ **b.** $\cos 13°$ **4.** $\sqrt{2}\sin\left(x - \frac{\pi}{4}\right)$

6. a. $y = \sin(2\pi \cdot 697t) + \sin(2\pi \cdot 1209t)$
b. $y = 2\cos(512\pi t)\sin(1906\pi t)$ **c.** $f = 953$ Hz, $A = 2\cos(512\pi t)$

A Exercises: Basic Skills and Concepts: **9.** $\frac{1}{2} - \frac{1}{2}\cos 2x$ **10.** $\frac{1}{2}\sin 2x$

11. $\frac{1}{4} + \frac{1}{2}\sin 20°$ **12.** $\frac{1}{2}\cos 20° - \frac{1}{4}$ **13.** $-\frac{1}{4} - \frac{1}{2}\cos 20°$

14. $\frac{1}{2} - \frac{1}{2}\sin 50°$ **19.** $\frac{1}{2}\sin 6\theta + \frac{1}{2}\sin 4\theta$ **20.** $\frac{1}{2}\sin 5\theta - \frac{1}{2}\sin \theta$

21. $\frac{1}{2}\cos x + \frac{1}{2}\cos 7x$ **22.** $\frac{1}{2}\cos 3x - \frac{1}{2}\cos 7x$ **23.** $\frac{\sqrt{3}}{4} - \frac{\sqrt{2}}{4}$

24. $\frac{\sqrt{2}}{4} + \frac{1}{4}$ **25.** $\frac{1}{2} + \frac{\sqrt{2}}{4}$ **26.** $-\frac{1}{4}$ **37.** $2\cos\frac{5}{12}\cos\frac{1}{12}$

43. $\sqrt{2}\cos\left(x - \frac{\pi}{4}\right)$ **44.** $-\sqrt{2}\sin\left(x - \frac{\pi}{4}\right)$ **45.** $\sqrt{2}\sin\left(2x - \frac{\pi}{4}\right)$

46. $\sqrt{2}\cos\left(3x - \frac{\pi}{4}\right)$ **47.** $2\sin\left(\frac{\pi}{4} - x\right)\cos\left(4x - \frac{\pi}{4}\right)$

48. $2\sin\left(3x - \frac{\pi}{4}\right)\cos\left(2x + \frac{\pi}{4}\right)$ **49.** $a\sqrt{2}\cos\left(x - \frac{\pi}{4}\right)$

50. $a\sqrt{2}\cos\left(bx - \frac{\pi}{4}\right)$

B Exercises: Applying the Concepts:
62. a. $y = \sin[2\pi(941)t] + \sin[2\pi(1477)t]$
b. $y = 2\cos(536\pi t)\sin(2\pi(1209)t)$ **c.** $f = 1209$ Hz,
$A = 2\cos(536\pi t)$

C Exercises: Beyond the Basics: **65.** The amplitude is
$2\cos 1 \approx 1.081$. The period is π.

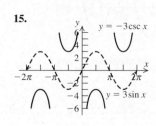
$y = \cos(2x + 1) + \cos(2x - 1)$

66. The amplitude is $2\cos 1 \approx 1.081$. The period is $\frac{2\pi}{3}$.
$y = \sin(3x + 1) + \sin(3x - 1)$

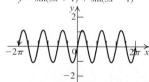

Review Exercises

1. $\cos\theta = -\frac{\sqrt{5}}{3}, \tan\theta = \frac{2\sqrt{5}}{5}, \cot\theta = \frac{\sqrt{5}}{2},$
$\sec\theta = -\frac{3\sqrt{5}}{5}, \csc\theta = -\frac{3}{2}$

2. $\sin\theta = \frac{\sqrt{5}}{5}, \cos\theta = -\frac{2\sqrt{5}}{5}, \cot\theta = -2,$
$\sec\theta = -\frac{\sqrt{5}}{2}, \csc\theta = \sqrt{5}$

3. $\sin\theta = -\frac{2\sqrt{2}}{3}, \cos\theta = \frac{1}{3}, \tan\theta = -2\sqrt{2},$
$\cot\theta = -\frac{\sqrt{2}}{4}, \csc\theta = -\frac{3\sqrt{2}}{4}$

4. $\sin\theta = \frac{1}{5}, \cos\theta = -\frac{2\sqrt{6}}{5}, \tan\theta = -\frac{\sqrt{6}}{12},$
$\cot\theta = -2\sqrt{6}, \sec\theta = -\frac{5\sqrt{6}}{12}$ **23.** $\left\{0, \frac{2\pi}{3}, \frac{4\pi}{3}\right\}$ **24.** $\left\{\frac{7\pi}{6}, \frac{11\pi}{6}\right\}$

25. $\left\{\frac{\pi}{18}, \frac{5\pi}{18}, \frac{13\pi}{18}, \frac{17\pi}{18}, \frac{25\pi}{18}, \frac{29\pi}{18}\right\}$

26. $\left\{\frac{\pi + 3}{6}, \frac{5\pi + 3}{6}, \frac{7\pi + 3}{6}, \frac{11\pi + 3}{6}\right\}$ **33.** $\frac{\sqrt{2 + \sqrt{3}}}{2}$

34. $\frac{\sqrt{6} + \sqrt{2}}{4}$ **36.** $\frac{3 + \sqrt{3}}{3 - \sqrt{3}}$

63.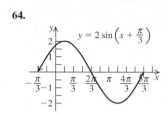

64.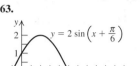

Cumulative Review Exercises (Chapters 1–5)

9.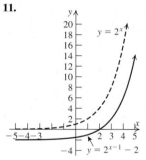

10.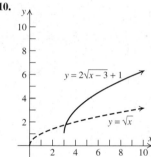

11.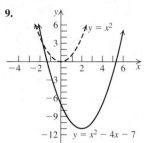

12.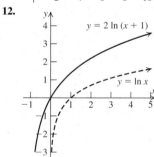

13.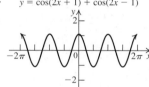

14.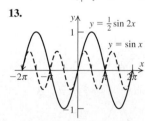

15.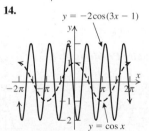

16.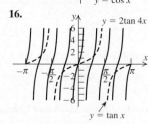

CHAPTER 6

Section 6.1

Practice Problems:

1. $\sin\theta = \dfrac{4}{5}$ $\csc\theta = \dfrac{5}{4}$

$\cos\theta = \dfrac{3}{5}$ $\sec\theta = \dfrac{5}{3}$

$\tan\theta = \dfrac{4}{3}$ $\cot\theta = \dfrac{3}{4}$

2. $\sin\theta = \dfrac{2\sqrt{2}}{3}$ $\csc\theta = \dfrac{3\sqrt{2}}{4}$

$\cos\theta = \dfrac{1}{3}$ $\sec\theta = 3$

$\tan\theta = 2\sqrt{2}$ $\cot\theta = \dfrac{\sqrt{2}}{4}$

3. a. $\sec 69° \approx 2.7904$ **b.** $\cot 15° \approx 3.7321$ **4.** $B = 34°, a \approx 21.4,$
$b \approx 14.4$ **5.** $A \approx 48.4°, B \approx 41.6°, a \approx 27.6$ **6.** ≈ 1.096 miles
7. $\approx 20{,}320$ feet   **8.** ≈ 43 feet   **9.** $80°$

A Exercises: Basic Skills and Concepts:

13. $\sin\theta = \dfrac{\sqrt{5}}{3}$ $\csc\theta = \dfrac{3\sqrt{5}}{5}$

$\cos\theta = \dfrac{2}{3}$ $\sec\theta = \dfrac{3}{2}$

$\tan\theta = \dfrac{\sqrt{5}}{2}$ $\cot\theta = \dfrac{2\sqrt{5}}{5}$

14. $\sin\theta = \dfrac{3}{4}$ $\csc\theta = \dfrac{4}{3}$

$\cos\theta = \dfrac{\sqrt{7}}{4}$ $\sec\theta = \dfrac{4\sqrt{7}}{7}$

$\tan\theta = \dfrac{3\sqrt{7}}{7}$ $\cot\theta = \dfrac{\sqrt{7}}{3}$

15. $\sin\theta = \dfrac{5\sqrt{34}}{34}$ $\csc\theta = \dfrac{\sqrt{34}}{5}$

$\cos\theta = \dfrac{3\sqrt{34}}{34}$ $\sec\theta = \dfrac{\sqrt{34}}{3}$

$\tan\theta = \dfrac{5}{3}$ $\cot\theta = \dfrac{3}{5}$

16. $\sin\theta = \dfrac{11\sqrt{157}}{157}$ $\csc\theta = \dfrac{\sqrt{157}}{11}$

$\cos\theta = \dfrac{6\sqrt{157}}{157}$ $\sec\theta = \dfrac{\sqrt{157}}{6}$

$\tan\theta = \dfrac{11}{6}$ $\cot\theta = \dfrac{6}{11}$

17. $\sin\theta = \dfrac{5}{13}$ $\csc\theta = \dfrac{13}{5}$

$\cos\theta = \dfrac{12}{13}$ $\sec\theta = \dfrac{13}{12}$

$\tan\theta = \dfrac{5}{12}$ $\cot\theta = \dfrac{12}{5}$

18. $\sin\theta = \dfrac{1}{4}$ $\csc\theta = 4$

$\cos\theta = \dfrac{\sqrt{15}}{4}$ $\sec\theta = \dfrac{4\sqrt{15}}{15}$

$\tan\theta = \dfrac{\sqrt{15}}{15}$ $\cot\theta = \sqrt{15}$

52. The Empire State Building is approximately 150 meters away and is approximately 381 meters tall.

Section 6.2

Practice Problems:
1. $C = 45°, b \approx 15.4$ in., $c \approx 12.0$ in.
2. 10,576 ft   **3.** Solution 1: $A \approx 99.2°, B \approx 45.8°, a \approx 20.7$ ft
Solution 2: $A \approx 10.8°, B \approx 134.2°, a \approx 3.9$ ft   **4.** No triangle exists.
5. $A \approx 31.3°, B \approx 88.7°, b \approx 57.7$ ft   **6. a.** ≈ 31.5 mi
b. ≈ 15.6 mi   **7.** 375.2 square feet

Section 6.3

Practice Problems:
1. $b \approx 21.8, A \approx 36.6°, C \approx 83.4°$
2. 5219.5 miles   **3.** $A \approx 41.9°, B \approx 86.0°, C \approx 52.1°$ **4.** No triangle exists.   **5.** 93.5 square meters   **6.** 14,801 gallons

Section 6.4

Practice Problems:
1. a.   **b.**

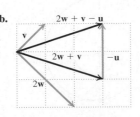

---

2. $\langle 3, -10 \rangle$ **3. a.** $\langle 1, -1 \rangle$ **b.** $\langle 6, -9 \rangle$ **c.** $\langle 8, -13 \rangle$ **d.** $\sqrt{233}$
4. $\left\langle -\dfrac{12}{13}, \dfrac{5}{13} \right\rangle$ **5. a.** $-7\mathbf{i} + 14\mathbf{j}$ **b.** $7\sqrt{5}$ **6.** $\sqrt{3}\mathbf{i} - \mathbf{j}$
7. $303.69°$ **8.** **R** is a force of 50 pounds in the direction N 36.9° W.
9. The ground speed is approximately 801.0 miles per hour, and the bearing is approximately N 62.9° W.   **10.** The magnitude of the resultant is approximately 121.2 lb, and the resultant is in the direction S 7.8° W.

A Exercises: Basic Skills and Concepts:
7.   **8.**

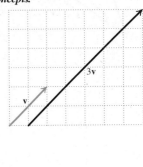

9.   **10.**

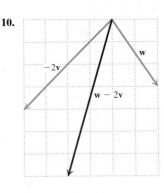

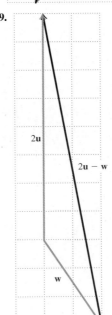

11.

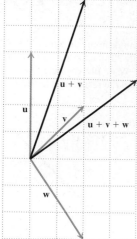

12.

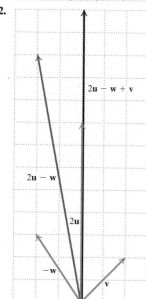

13.

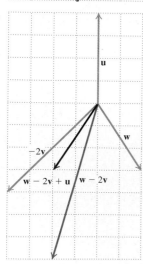

14.

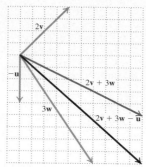

17. $\langle 2, -2 \rangle$ **20.** $\langle -5, -1.4 \rangle$ **21.** $\left\langle -1, -\frac{5}{2} \right\rangle$ **22.** $\left\langle 1, -\frac{10}{9} \right\rangle$

35. $\left\langle \frac{\sqrt{2}}{2}, -\frac{\sqrt{2}}{2} \right\rangle$ **36.** $\left\langle \frac{\sqrt{10}}{10}, \frac{3\sqrt{10}}{10} \right\rangle$ **37.** $\left\langle -\frac{4}{5}, \frac{3}{5} \right\rangle$ **38.** $\left\langle \frac{5}{13}, -\frac{12}{13} \right\rangle$

39. $\left\langle \frac{\sqrt{2}}{2}, \frac{\sqrt{2}}{2} \right\rangle$ **40.** $\left\langle \frac{\sqrt{5}}{3}, -\frac{2}{3} \right\rangle$ **48.** $\frac{5\sqrt{2}}{2}\mathbf{i} + \frac{5\sqrt{2}}{2}\mathbf{j}$

49. $-2\mathbf{i} + 2\sqrt{3}\mathbf{j}$ **50.** $-\frac{3\sqrt{3}}{2}\mathbf{i} + \frac{3}{2}\mathbf{j}$ **51.** $\frac{3}{2}\mathbf{i} - \frac{3\sqrt{3}}{2}\mathbf{j}$

52. $2\sqrt{3}\mathbf{i} - 2\mathbf{j}$ **53.** $\frac{7}{2}\mathbf{i} - \frac{7\sqrt{3}}{2}\mathbf{j}$ **54.** $-4\sqrt{2}\mathbf{i} + 4\sqrt{2}\mathbf{j}$

B Exercises: Applying the Concepts: 69. The resultant is a force of 676.64 pounds making an angle of 47.19° with the positive *x*-axis. **70.** The resultant is a force of 608.28 pounds making an angle of 94.72° with the positive *x*-axis.

Section 6.5

Practice Problems: **1. a.** 8 **b.** 0 **2.** 17.5 **3.** Not parallel
4. 115.3° **5.** ≈ 39 **6. v** and **w** are orthogonal. **7.** $k = 3$
8. a. $\left\langle -\frac{16}{29}, -\frac{40}{29} \right\rangle$ **b.** $-\frac{8\sqrt{29}}{29}$ **9.** $\langle 3, 3 \rangle + \langle 2, -2 \rangle$
10. 3000 foot-pounds

Section 6.6

Practice Problems: **1. a.**

b. $(-2, 30°)$

$P(2, -150°)$

c. $(2, 210°)$ **d.** $(-2, -330°)$ **2. a.** $\left(-\frac{3}{2}, -\frac{3\sqrt{3}}{2} \right)$ **b.** $(\sqrt{2}, -\sqrt{2})$

3. $\left(\sqrt{2}, \frac{5\pi}{4} \right)$ **4.** $(24.8, 36.0°)$ **5.** $r^2 - 2r\sin\theta + 3 = 0$

6. a. $x^2 + y^2 = 25$; a circle centered at the origin with radius 5
b. $y = -x\sqrt{3}$; a line through the origin with slope $-\sqrt{3}$ **c.** $x = 1$; a vertical line through $x = 1$ **d.** $x^2 + (y - 1)^2 = 1$; a circle with center $(0, 1)$ and radius 1

7.

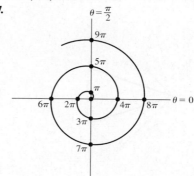

8.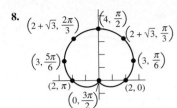

$\left(2+\sqrt{3},\frac{2\pi}{3}\right)$ $\left(4,\frac{\pi}{2}\right)$ $\left(2+\sqrt{3},\frac{\pi}{3}\right)$

$\left(3,\frac{5\pi}{6}\right)$ $\left(3,\frac{\pi}{6}\right)$

$(2,\pi)$ $(2,0)$

$\left(0,\frac{3\pi}{2}\right)$

9.

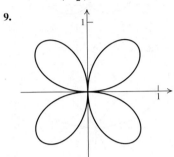

10.

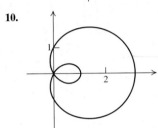

11.

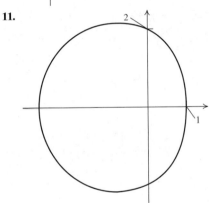

A Exercises: Basic Skills and Concepts:

7.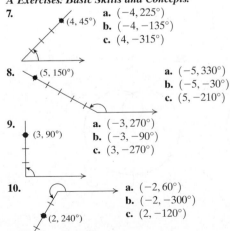
  a. $(-4, 225°)$
  b. $(-4, -135°)$
  c. $(4, -315°)$
  $(4, 45°)$

8. $(5, 150°)$
  a. $(-5, 330°)$
  b. $(-5, -30°)$
  c. $(5, -210°)$

9. $(3, 90°)$
  a. $(-3, 270°)$
  b. $(-3, -90°)$
  c. $(3, -270°)$

10.
  a. $(-2, 60°)$
  b. $(-2, -300°)$
  c. $(2, -120°)$
  $(2, 240°)$

11.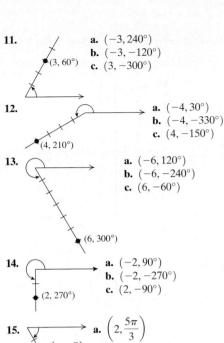
  $(3, 60°)$
  a. $(-3, 240°)$
  b. $(-3, -120°)$
  c. $(3, -300°)$

12.
  $(4, 210°)$
  a. $(-4, 30°)$
  b. $(-4, -330°)$
  c. $(4, -150°)$

13.
  a. $(-6, 120°)$
  b. $(-6, -240°)$
  c. $(6, -60°)$
  $(6, 300°)$

14.
  $(2, 270°)$
  a. $(-2, 90°)$
  b. $(-2, -270°)$
  c. $(2, -90°)$

15.
  $\left(2, -\frac{\pi}{3}\right)$
  a. $\left(2, \frac{5\pi}{3}\right)$
  b. $\left(-2, \frac{2\pi}{3}\right)$

16.
  $\left(\sqrt{2}, -\frac{\pi}{4}\right)$
  a. $\left(\sqrt{2}, \frac{7\pi}{4}\right)$
  b. $\left(-\sqrt{2}, \frac{3\pi}{4}\right)$

17.
  a. $\left(-3, -\frac{11\pi}{6}\right)$
  b. $\left(3, -\frac{5\pi}{6}\right)$
  $\left(-3, \frac{\pi}{6}\right)$

18.
  a. $\left(-2, -\frac{5\pi}{4}\right)$
  b. $\left(2, -\frac{\pi}{4}\right)$
  $\left(-2, \frac{3\pi}{4}\right)$

19.
  $\left(-2, -\frac{\pi}{6}\right)$
  a. $\left(-2, \frac{11\pi}{6}\right)$
  b. $\left(2, \frac{5\pi}{6}\right)$

20.
  a. $(-2, \pi)$
  b. $(2, 0)$
  $(-2, -\pi)$

21.
  a. $\left(4, -\frac{5\pi}{6}\right)$
  b. $\left(-4, \frac{\pi}{6}\right)$
  $\left(4, \frac{7\pi}{6}\right)$

22. $(2, 3)$
  a. $(2, -3.28)$
  b. $(-2, -0.14)$

47. $x^2 + y^2 = 4$; circle centered at $(0, 0)$ with radius 2
48. $x^2 + y^2 = 9$; circle centered at $(0, 0)$ with radius 3
49. $y = -x$; a line with slope -1 and y-intercept 0
50. $y = -\frac{\sqrt{3}}{3}x$; a line with slope $-\frac{\sqrt{3}}{3}$ and y-intercept 0
51. $(x - 2)^2 + y^2 = 4$; a circle centered at $(2, 0)$ with radius 2
52. $x^2 + (y - 2)^2 = 4$; a circle centered at $(0, 2)$ with radius 2

53. $x^2 + (y + 1)^2 = 1$; a circle centered at $(0, -1)$ with radius 1

54. $\left(x + \dfrac{3}{2}\right)^2 + y^2 = \dfrac{9}{4}$; a circle centered at $\left(-\dfrac{3}{2}, 0\right)$ with radius $\dfrac{3}{2}$

55.

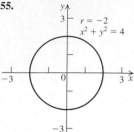

$r = -2$
$x^2 + y^2 = 4$

56.

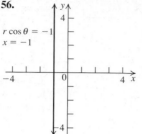

$r \cos \theta = -1$
$x = -1$

57.

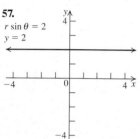

$r \sin \theta = 2$
$y = 2$

58.

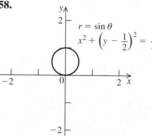

$r = \sin \theta$
$x^2 + \left(y - \dfrac{1}{2}\right)^2 = \dfrac{1}{4}$

59.

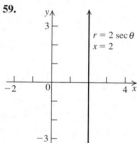

$r = 2 \sec \theta$
$x = 2$

60. $r + 3\cos \theta = 0$
$\left(x + \dfrac{3}{2}\right)^2 + y^2 = \dfrac{9}{4}$

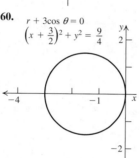

61.

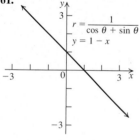

$r = \dfrac{1}{\cos \theta + \sin \theta}$
$y = 1 - x$

62.

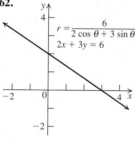

$r = \dfrac{6}{2 \cos \theta + 3 \sin \theta}$
$2x + 3y = 6$

63.

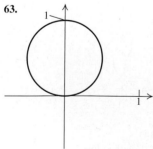

64.

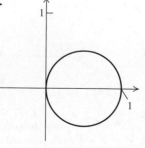

65.

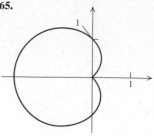

66.

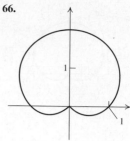

67.

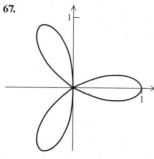

68.

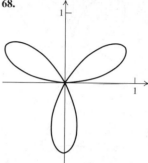

69.

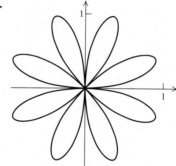

70.

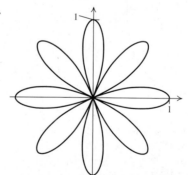

71.

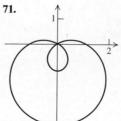

72.

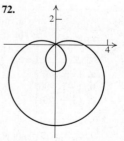

73.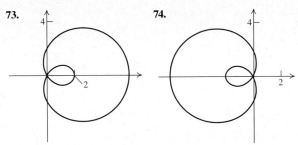

74.

75.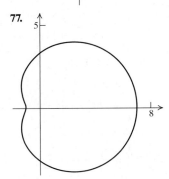

76.

77.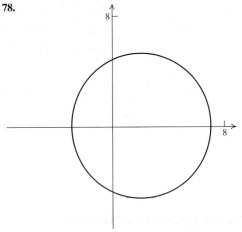

78.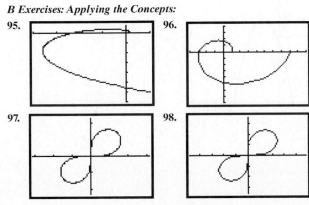

B Exercises: Applying the Concepts:

95.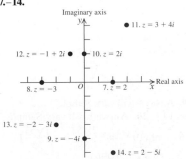

96.

97.

98.

99.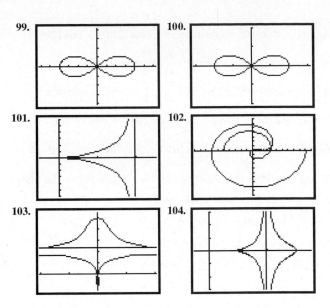

100.

101.

102.

103.

104.

Section 6.7

Practice Problems: 1.

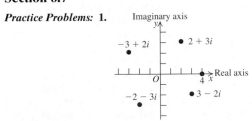

2. a. 13 **b.** 7 **c.** 1 **d.** $\sqrt{a^2 + b^2}$ **3.** $z = \sqrt{2}\left(\cos\dfrac{5\pi}{4} + i\sin\dfrac{5\pi}{4}\right)$

4. $2 - 2\sqrt{3}i$ **5.** $z_1 z_2 = 10(\cos 135° + i\sin 135°); \dfrac{z_1}{z_2} =$

$\dfrac{5}{2}(\cos 15° + i\sin 15°)$ **6.** $2.6(\cos 27.2° + i\sin 27.2°)$ **7. a.** 16

b. $-\dfrac{1}{64}$ **8.** $2^{\frac{1}{6}}(\cos 45° + i\sin 45°), 2^{\frac{1}{6}}(\cos 165° + i\sin 165°),$

$2^{\frac{1}{6}}(\cos 285° + i\sin 285°)$ **9.** $\cos 0° + i\sin 0°, \cos 90° + i\sin 90°,$
$\cos 180° + i\sin 180°, \cos 270° + i\sin 270°$

A Exercises: Basic Skills and Concepts:

7.–14.

23. $3\sqrt{2}(\cos 315° + i\sin 315°)$

39. $z_1 z_2 = 8(\cos 90° + i\sin 90°); \dfrac{z_1}{z_2} = 2(\cos 60° + i\sin 60°)$

40. $z_1 z_2 = 12(\cos 135° + i\sin 135°); \dfrac{z_1}{z_2} = 3(\cos 45° + i\sin 45°)$

41. $z_1 z_2 = 10(\cos 300° + i\sin 300°); \dfrac{z_1}{z_2} = \dfrac{5}{2}(\cos 180° + i\sin 180°)$

42. $z_1z_2 = 40(\cos 0° + i \sin 0°); \dfrac{z_1}{z_2} = \dfrac{5}{2}(\cos(270°) + i \sin(270°))$

43. $z_1z_2 = 15(\cos 60° + i \sin 60°); \dfrac{z_1}{z_2} = \dfrac{3}{5}(\cos 20° + i \sin 20°)$

44. $z_1z_2 = 10(\cos 90° + i \sin 90°); \dfrac{z_1}{z_2} = \dfrac{5}{2}(\cos 40° + i \sin 40°)$

45. $z_1z_2 = 2(\cos 0° + i \sin 0°); \dfrac{z_1}{z_2} = \cos(90°) + i \sin(90°)$

46. $z_1z_2 = 2\sqrt{2}(\cos 105° + i \sin 105°); \dfrac{z_1}{z_2} = \sqrt{2}(\cos 15° + i \sin 15°)$

47. $z_1z_2 = 4\sqrt{2}(\cos 195° + i \sin 195°); \dfrac{z_1}{z_2} = \dfrac{\sqrt{2}}{2}(\cos 105° + i \sin 105°)$

48. $z_1z_2 = 12\sqrt{2}(\cos 345° + i \sin 345°);$
$\dfrac{z_1}{z_2} = \dfrac{3\sqrt{2}}{4}(\cos(105°) + i \sin(105°))$

49. $z_1z_2 = 4\sqrt{3}(\cos 5.3° + i \sin 5.3°);$
$\dfrac{z_1}{z_2} = \sqrt{3}(\cos(-65.3°) + i \sin(-65.3°))$

50. $z_1z_2 = 24\sqrt{2}(\cos 345° + i \sin 345°);$
$\dfrac{z_1}{z_2} = \dfrac{4\sqrt{2}}{3}(\cos(75°) + i \sin(75°))$

65. $4(\cos 0° + i \sin 0°), 4(\cos 120° + i \sin 120°), 4(\cos 240° + i \sin 240°)$
66. $4(\cos 60° + i \sin 60°), 4(\cos 180° + i \sin 180°),$
$4(\cos 300° + i \sin 300°)$ **67.** $\cos 45° + i \sin 45°, \cos 225° + i \sin 225°$
68. $\cos 67.5° + i \sin 67.5°, \cos 157.5° + i \sin 157.5°,$
$\cos 247.5° + i \sin 247.5°, \cos 337.5° + i \sin 337.5°$
69. $\cos 30° + i \sin 30°, \cos 90° + i \sin 90°, \cos 150° + i \sin 150°,$
$\cos 210° + i \sin 210°, \cos 270° + i \sin 270°, \cos 330° + i \sin 330°$
70. $\cos 0° + i \sin 0°, \cos 45° + i \sin 45°, \cos 90° + i \sin 90°,$
$\cos 135° + i \sin 135°, \cos 180° + i \sin 180°, \cos 225° + i \sin 225°,$
$\cos 270° + i \sin 270°, \cos 315° + i \sin 315°$
71. $\sqrt{2}(\cos 150° + i \sin 150°), \sqrt{2}(\cos 330° + i \sin 330°)$
72. $2(\cos 20° + i \sin 20°), 2(\cos 140° + i \sin 140°),$
$2(\cos 260° + i \sin 260°)$

C Exercises: Beyond the Basics:
85. $\cos 0° + i \sin 0°, \cos 90° + i \sin 90°, \cos 180° + i \sin 180°,$
$\cos 270° + i \sin 270°$
86. $\cos 22.5° + i \sin 22.5°, \cos 67.5° + i \sin 67.5°,$
$\cos 112.5° + i \sin 112.5°, \cos 157.5° + i \sin 157.5°,$
$\cos 202.5° + i \sin 202.5°, \cos 247.5° + i \sin 247.5°,$
$\cos 292.5° + i \sin 292.5°, \cos 337.5° + i \sin 337.5°$
87. $2^{\frac{1}{6}}(\cos 15° + i \sin 15°), 2^{\frac{1}{6}}(\cos 135° + i \sin 135°),$
$2^{\frac{1}{6}}(\cos 255° + i \sin 255°)$ **88.** $2^{\frac{1}{7}}(\cos 38.6° + i \sin 38.6°),$
$2^{\frac{1}{7}}(\cos 90° + i \sin 90°), 2^{\frac{1}{7}}(\cos 141.4° + i \sin 141.4°),$
$2^{\frac{1}{7}}(\cos 192.9° + i \sin 192.9°), 2^{\frac{1}{7}}(\cos 244.3° + i \sin 244.3°),$
$2^{\frac{1}{7}}(\cos 295.7° + i \sin 295.7°), 2^{\frac{1}{7}}(\cos 347.1° + i \sin 347.1°)$

Review Exercises

Basic Skills and Concepts: **19.** Answers will vary. Sample
answers: **i.** 18,   **ii.** $10\sqrt{3}$,   **iii.** 10   **20.** Answers will vary. Sample
answers: **i.** not possible,   **ii.** 60,   **iii.** 10   **40.** $\dfrac{2\sqrt{53}}{53}\mathbf{i} - \dfrac{7\sqrt{53}}{53}\mathbf{j}$

41. $\left\langle \dfrac{3\sqrt{34}}{34}, -\dfrac{5\sqrt{34}}{34} \right\rangle$ **42.** $\left\langle -\dfrac{5\sqrt{29}}{29}, -\dfrac{2\sqrt{29}}{29} \right\rangle$

55.

$(24, 30°)$ $(12\sqrt{3}, 12)$

56.
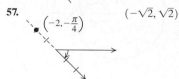
$(6, -6\sqrt{3})$ $\left(12, -\dfrac{\pi}{3}\right)$

57.
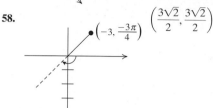
$\left(-2, -\dfrac{\pi}{4}\right)$ $(-\sqrt{2}, \sqrt{2})$

58.
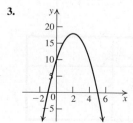
$\left(-3, \dfrac{-3\pi}{4}\right)$ $\left(\dfrac{3\sqrt{2}}{2}, \dfrac{3\sqrt{2}}{2}\right)$

73. $3\left(\cos \dfrac{3\pi}{2} + i \sin \dfrac{3\pi}{2}\right)$ **74.** $\sqrt{2}\left(\cos \dfrac{3\pi}{4} + i \sin \dfrac{3\pi}{4}\right)$
75. $10\left(\cos \dfrac{11\pi}{6} + i \sin \dfrac{11\pi}{6}\right)$ **76.** $4\left(\cos \dfrac{4\pi}{3} + i \sin \dfrac{4\pi}{3}\right)$
77. $\sqrt{2} + \sqrt{2}i$ **78.** $-\dfrac{3}{2} - \dfrac{3\sqrt{3}}{2}i$ **79.** $-3\sqrt{2} + 3\sqrt{2}i$
80. $-2\sqrt{3} - 2i$ **81.** $z_1z_2 = 6(\cos 35° + i \sin 35°);$
$\dfrac{z_1}{z_2} = \dfrac{3}{2}(\cos 15° + i \sin 15°)$ **82.** $z_1z_2 = 8(\cos 320° + i \sin 320°);$
$\dfrac{z_1}{z_2} = 2(\cos 280° + i \sin 280°)$ **83.** $z_1z_2 = 6\left(\cos \dfrac{7\pi}{6} + i \sin \dfrac{7\pi}{6}\right);$
$\dfrac{z_1}{z_2} = \dfrac{2}{3}\left(\cos \dfrac{\pi}{2} + i \sin \dfrac{\pi}{2}\right)$ **84.** $z_1z_2 = 75\left(\cos \dfrac{5\pi}{3} + i \sin \dfrac{5\pi}{3}\right);$
$\dfrac{z_1}{z_2} = \dfrac{1}{3}(\cos \pi + i \sin \pi)$ **85.** $27(\cos 120° + i \sin 120°)$
86. $4096(\cos \pi + i \sin \pi)$ **87.** $4096(\cos 0 + i \sin 0)$
88. $2^{\frac{21}{2}}\left(\cos \dfrac{\pi}{4} + i \sin \dfrac{\pi}{4}\right)$ **89.** $5(\cos 60° + i \sin 60°),$
$5(\cos 180° + i \sin 180°), 5(\cos 300° + i \sin 300°)$
90. $2(\cos 67.5° + i \sin 67.5°), 2(\cos 157.5° + i \sin 157.5°),$
$2(\cos 247.5° + i \sin 247.5°), 2(\cos 337.5° + i \sin 337.5°)$
91. $2^{\frac{1}{5}}(\cos 24° + i \sin 24°), 2^{\frac{1}{5}}(\cos 96° + i \sin 96°),$
$2^{\frac{1}{5}}(\cos 168° + i \sin 168°), 2^{\frac{1}{5}}(\cos 240° + i \sin 240°),$
$2^{\frac{1}{5}}(\cos 312° + i \sin 312°)$ **92.** $2^{\frac{1}{12}}(\cos 52.5° + i \sin 52.5°),$
$2^{\frac{1}{12}}(\cos 112.5° + i \sin 112.5°), 2^{\frac{1}{12}}(\cos 172.5° + i \sin 172.5°),$
$2^{\frac{1}{12}}(\cos 232.5° + i \sin 232.5°), 2^{\frac{1}{12}}(\cos 292.5° + i \sin 292.5°),$
$2^{\frac{1}{12}}(\cos 352.5° + i \sin 352.5°)$

Cumulative Review Exercises (Chapters 1–6)

3.

11.

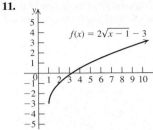

$f(x) = 2\sqrt{x-1} - 3$

12.

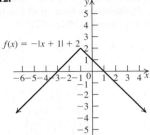

$f(x) = -|x + 1| + 2$

13.

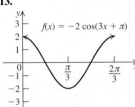

$f(x) = -2\cos(3x + \pi)$

14.

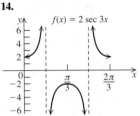

$f(x) = 2\sec 3x$

CHAPTER 7

Section 7.1

Practice Problems: **1.** Equation (1): $(1) + (3) = 4$; equation (2): $3(1) - (3) = 0$ **2.** $\{(-1, 3)\}$ **3.** $\{(4, -1)\}$ **4.** $\varnothing$

5. $\{(x, 2x - 3)\}$ **6.** $\left\{\left(\dfrac{2}{3}, \dfrac{1}{2}\right)\right\}$ **7.** $\{(2, -3)\}$

8. $\{(5700, 31.4)\}$

A Exercises: Basic Skills and Concepts:
13. $(2, 1)$

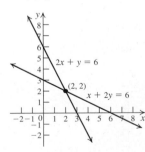

$x - y = 1$ $(2, 1)$ $x + y = 3$

14. $(6, 4)$

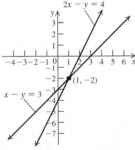

$x + y = 10$ $x - y = 2$ $(6, 4)$

15. $(2, 2)$

$2x + y = 6$ $(2, 2)$ $x + 2y = 6$

16. $(1, -2)$

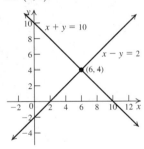

$2x - y = 4$ $(1, -2)$ $x - y = 3$

17. Inconsistent

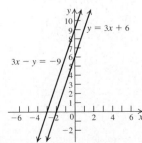

$y = 3x + 6$ $3x - y = -9$

18. Inconsistent

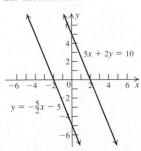

$5x + 2y = 10$ $y = -\dfrac{5}{2}x - 5$

19. $\left(\dfrac{7}{3}, \dfrac{14}{3}\right)$

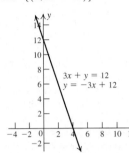

$\left(\dfrac{7}{3}, \dfrac{14}{3}\right)$ $x + y = 7$ $y = 2x$

20. $\left(\dfrac{7}{2}, \dfrac{11}{2}\right)$

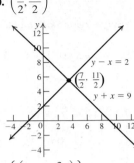

$y - x = 2$ $\left(\dfrac{7}{2}, \dfrac{11}{2}\right)$ $y + x = 9$

21. $\{(x, 12 - 3x)\}$

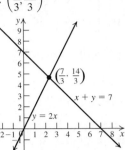

$3x + y = 12$ $y = -3x + 12$

22. $\left\{\left(x, 2 - \dfrac{2}{3}x\right)\right\}$

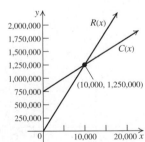

$2x + 3y = 6$ $6y = -4x + 12$

B Exercises: Applying the Concepts: **81.** The diameter of the largest pizza is 21 inches, and the diameter of the smallest pizza is 8 inches.   **82.** The number of calories in a hamburger from Boston Burger is 585, and the number of calories in a hamburger from Carmen's Broiler is 545.   **84.** The average monthly cost of food is $800, and the average monthly cost of clothes is $200.

97. b.

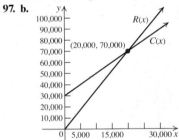

$R(x)$ $(20,000, 70,000)$ $C(x)$

98. b.

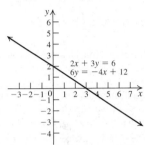

$R(x)$ $C(x)$ $(10,000, 1,250,000)$

Section 7.2

Practice Problems:
1. $(2) + (-3) + (2) = 1$
$3(2) + 4(-3) + (2) = -4$
$2(2) + (-3) + 2(2) = 5$

2. $\{(-2, 1, 3)\}$ **3.** $\varnothing$ **4.** $\left\{\left(8 - \dfrac{7}{3}z, \dfrac{4}{3}z - 3, z\right)\right\}$

5. $[(13 - 17z, -3 + 5z, z)]$ **6.** Cell A contains bone (because $x = 0.4$), cell B contains healthy tissue (because $y = 0.25$), and cell C contains tumorous tissue (because $z = 0.3$).

A Exercises: Basic Skills and Concepts: **15.**
$$\begin{cases} x - y - \dfrac{3}{2}z = \dfrac{1}{2} \\ \quad\ y + \ z = 0 \\ \qquad\qquad z = 1 \end{cases}$$

16.
$$\begin{cases} x \qquad\ + 3z = 8 \\ \quad\ y + \ z = 5 \\ \qquad\qquad z = 2 \end{cases}$$
17.
$$\begin{cases} x + 3y - 2z = 0 \\ \quad\ y - \dfrac{8}{7}z = \dfrac{5}{7} \\ \qquad\quad 0 = k, k \neq 0 \end{cases}$$

18.
$$\begin{cases} x - 4y + \ 7z = 14 \\ \quad\ y - \dfrac{23}{20}z = \dfrac{-29}{20} \\ \qquad\qquad 0 = k, k \neq 0 \end{cases}$$
25. $\left\{\left(\dfrac{35}{18}, \dfrac{29}{18}, \dfrac{5}{18}\right)\right\}$

35. $\left(3 + \dfrac{4}{5}z, -\dfrac{1}{5}z, z\right)$ **36.** $\left(3 - \dfrac{3}{2}z, -1 + \dfrac{13}{2}z, z\right)$

43. Dependent; $\left\{\left(-5 - 9z, \dfrac{-1}{6} + 3z, z\right)\right\}$

***B Exercises: Applying the Concepts:* 47.** Alex worked 2 hours, Becky worked 2.5 hours, and Courtney worked 1.5 hours. **48.** There are twenty-three 18-year-olds, eleven 19-year-olds, and four 20-year-olds. **53.** Cell A contains healthy tissue (because $x = 0.21$), cell B contains tumorous tissue (because $y = 0.33$), and cell C contains healthy tissue (because $z = 0.19$). **54.** Cell A contains tumorous tissue (because $x = 0.35$), cell B contains tumorous tissue (because $y = 0.3$), and cell C contains bone (because $z = 0.45$). **55.** Cells A, B, and C contain healthy tissue (because $x = 0.28$, $y = 0.23$, and $z = 0.21$). **56.** Both cells A and B contain healthy tissue (because $x = 0.21$, $y = 0.23$) and cell C contains metal (because $z = 2$).

Section 7.3

Practice Problems: **1.**

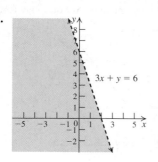

$3x + y = 6$

2.

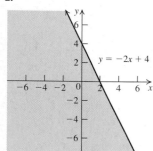

$y = -2x + 4$

3.

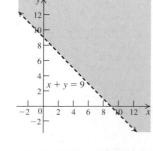

$x + y = 9$

4.

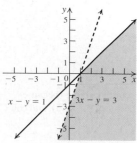

$x - y = 1$ $3x - y = 3$

5. $(0, 4), (5, 2),$ and $(2, 4)$

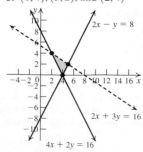

$2x - y = 8$ $2x + 3y = 16$ $4x + 2y = 16$

6. The objective function f has a maximum of 28 at the point $(7, 0)$.
7. The minimum number of calories is 300. The lunch then consists of 10 oz of soup and 0 oz of salad.

A Exercises: Basic Skills and Concepts:

7.

8.

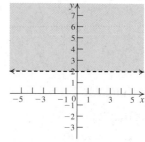

9.

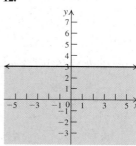

10.

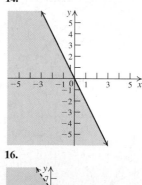

11.

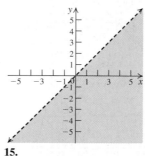

12.

13.

14.

15.

16.

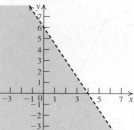

17.
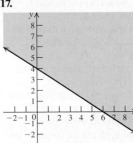

18.
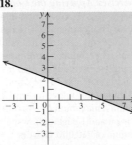

33. There are no vertices of the solution set.
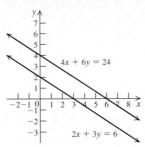

19.
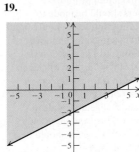

20.
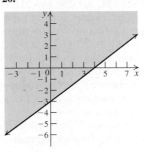

34. There are no vertices of the solution set
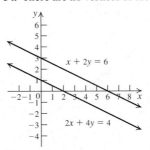

21.
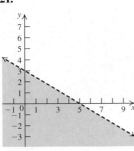

22.
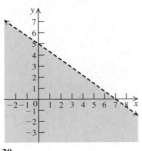

35.
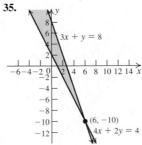

36. There are no vertices of the solution set
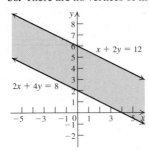

29.
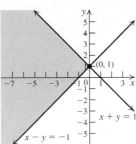

30.
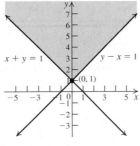

31.
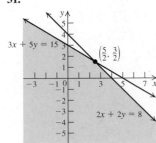

32.
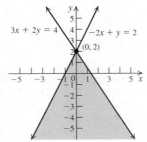

37.
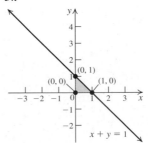

38. There are no vertices of the solution set

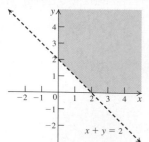

39.

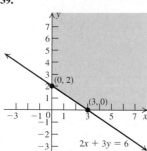

40.

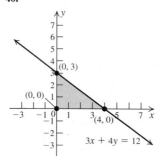

41.

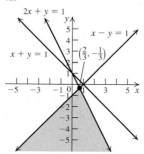

42.

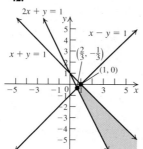

43.

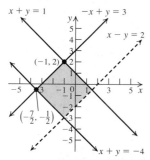

44. There are no vertices of the solution set

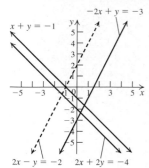

57. Maximum value is $\dfrac{1284}{19}$ at $\left(\dfrac{56}{19}, \dfrac{60}{19}\right)$.

B Exercises: Applying the Concepts: **65.** 80 acres of corn and 160 acres of soybeans to obtain a maximum profit of $10,400 **66.** 0 acres of corn and 240 acres of soybeans to obtain a maximum profit of $9600 **67.** 40 hours for machine I and 20 hours for machine II to obtain a minimum cost of $3600 **68.** Machine I should be operated for zero hours and machine II should be operated for 46.7 hours to obtain a minimum cost of $4200. **69.** 5 minutes of television and 2 pages of newspaper advertisement to obtain a maximum of 340,000 viewers **70.** 5.25 minutes of television and 1 page of newspaper advertisement to obtain a maximum of 710,000 viewers **73.** 400 of both the rectangular and circular table should be made to have a maximum profit of $2800. **74.** 50 GPSs and 20 DVD players should be made to obtain a maximum profit of $380. **75.** The contractor should build 33 terraced houses and 66 cottages to obtain a maximum amount of money: $4,290,000. **76.** 125 female guests and 175 male guests would provide a maximum profit of $4902.50. **78.** The Michigan factory should operate 150 days and the North Carolina factory should operate 200 days for a minimum cost of 17,000,000. **79.** Elisa should buy 10 enchilada meals and 5 vegetable loafs to obtain a minimum cost of $46.25. **80.** 4 workers from the Ready Aid agency and 5 workers from the Super Temps agency should be hired to keep the daily wages at a minimum of $824.

Critical Thinking:
89.

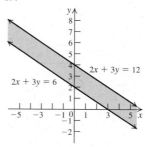

90.

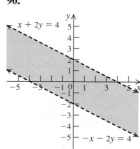

Section 7.4

Practice Problems: **1.** $\{(1,1)\}$ **2.** $\{(4,3),(-4,3),(4,-3),(-4,-3)\}$ **3.** Danielle received $x = 30$ shares as dividends and sold her stock at $p = \$65$ per share.

4.

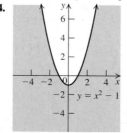

5.

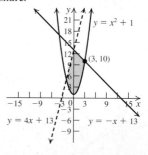

A Exercises: Basic Skills and Concepts:
31. $\{(4,2),(-4,2),(4,-2),(-4,-2)\}$
32. $\{(1,1),(-1,1),(1,-1),(-1,-1)\}$
37. $\{(1,2),(-1,2),(1,-2),(-1,-2)\}$
38. $\{(1,1),(-1,1),(1,-1),(-1,-1)\}$
39. $\left\{(2,1),(2,-1),\left(\dfrac{-11}{3},\dfrac{\sqrt{26}}{3}\right),\left(\dfrac{-11}{3},\dfrac{-\sqrt{26}}{3}\right)\right\}$
40. $\left\{\left(\dfrac{1}{2},\dfrac{\sqrt{35}}{2}\right),\left(\dfrac{1}{2},\dfrac{-\sqrt{35}}{2}\right)\right\}$
44. $\{(-2,1),(2,1),(-2,-1),(2,-1)\}$ **47.** $\left\{\left(-4,\dfrac{-3}{2}\right),(3,2)\right\}$

49. $\{(-2,1),(2,1),(-2,-1),(2,-1)\}$
50. $\{(-2,2),(2,2),(-2,-2),(2,-2)\}$
53. $\left\{(2,-2),(2,2),\left(\dfrac{16}{5},\dfrac{2\sqrt{46}}{5}\right),\left(\dfrac{16}{5},\dfrac{-2\sqrt{46}}{5}\right)\right\}$
54. $\left\{(1,1),(-1,1),\left(\dfrac{\sqrt{10}}{2},2\right),\left(-\dfrac{\sqrt{10}}{2},2\right)\right\}$

55.
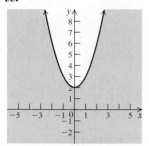

56.
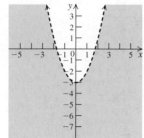

57.
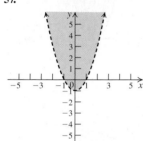

58.
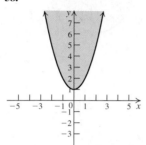

59.
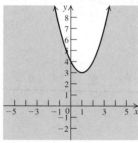

60.
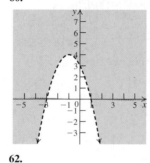

61.
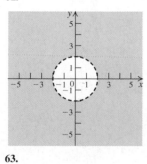

62.
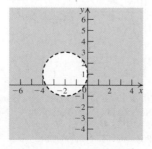

63.
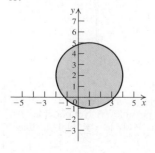

64.

65.
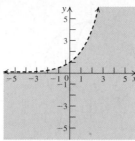

66.
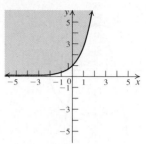

67.
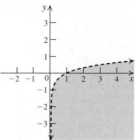

68.
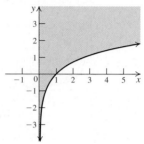

75.
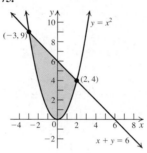

76.
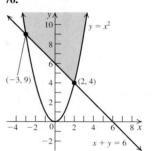

77.
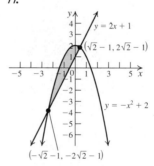

78.
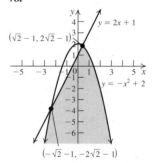

79.
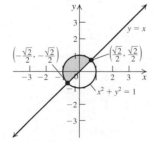

80.
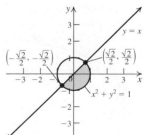

81. **82.**

C Exercises: Applying the Concepts:

93. **94.**

95. **96.**

97. **98.**

99. **100.**

101. **102.**

103. **104.**

105. $\{(1, 7)\}$ **106.** $\varnothing$

107. $\{(1, 5)\}$ **108.** $\{(1, 5)\}$

109. $\left\{(1, 0), \left(2, \dfrac{\ln 2}{\ln 3}\right)\right\}$ **110.** $\left\{(1, 0), \left(2, \dfrac{\ln 2}{\ln 3}\right)\right\}$

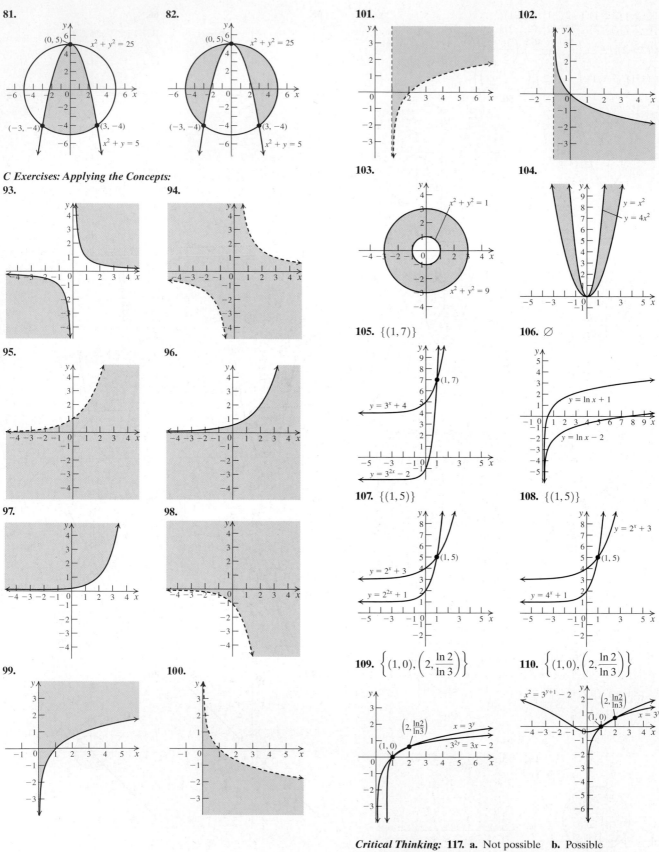

Critical Thinking: 117. a. Not possible   **b.** Possible
c. Not possible   **d.** Possible   **e.** Not possible   **f.** Not possible
118. a. Possible   **b.** Not possible   **c.** Possible   **d.** Not possible
e. Possible   **f.** Not possible

Section 7.5

Practice Problems: **1.** $\dfrac{3}{x+1} - \dfrac{1}{x-2}$ **2.** $\dfrac{5}{x-1} - \dfrac{3}{x} + \dfrac{1}{x+1}$

3. $\dfrac{5}{x} + \dfrac{6}{(x-1)^2} - \dfrac{5}{x-1}$ **4.** $\dfrac{-1}{x} + \dfrac{4x+5}{x^2+2}$

5. $\dfrac{1}{x^2+1} + \dfrac{3x}{(x^2+1)^2}$ **6.** $\dfrac{777}{3112}$

7. $\dfrac{1}{R} = \dfrac{3}{x+1} + \dfrac{1}{x+2} = \dfrac{1}{\dfrac{x+1}{3}} + \dfrac{1}{x+2}$

A Exercises: Basic Skills and Concepts:

12. $\dfrac{A}{(x+2)^2} + \dfrac{B}{x-3} + \dfrac{C}{x+2}$ **13.** $\dfrac{A}{x+1} + \dfrac{Bx+C}{x^2-x+1}$

14. $\dfrac{A}{(x-1)^2} + \dfrac{Bx+C}{x^2+x+1} + \dfrac{D}{x-1}$ **15.** $\dfrac{Ax+B}{x^2+1} + \dfrac{Cx+D}{(x^2+1)^2}$

16. $\dfrac{A}{(2x+3)^2} + \dfrac{Bx+C}{(x^2+3)^2} + \dfrac{D}{2x+3} + \dfrac{Ex+F}{x^2+3}$

19. $\dfrac{-1}{2(x+3)} + \dfrac{1}{2(x+1)}$ **23.** $\dfrac{2}{x+2} - \dfrac{3}{2(x+3)} - \dfrac{1}{2(x+1)}$

24. $\dfrac{16}{15(x+4)} + \dfrac{1}{10(x-1)} - \dfrac{1}{6(x-1)}$

27. $\dfrac{2}{x+1} - \dfrac{3}{(x+1)^2} + \dfrac{1}{(x+1)^3}$ **28.** $\dfrac{1}{2(x-1)} + \dfrac{1}{2(x+1)}$

29. $\dfrac{1}{x} + \dfrac{3}{(x+1)^2} - \dfrac{2}{x+1}$ **30.** $\dfrac{3}{x-1} + \dfrac{2}{x} - \dfrac{1}{(x-1)^2}$

31. $\dfrac{-2}{x} + \dfrac{1}{x^2} + \dfrac{1}{(x+1)^2} + \dfrac{2}{x+1}$

32. $\dfrac{1}{8(x+3)^2} + \dfrac{1}{8(x-1)^2} + \dfrac{1}{16(x+3)} - \dfrac{1}{16(x-1)}$

33. $\dfrac{1}{4(x-1)^2} + \dfrac{1}{4(x+1)^2} + \dfrac{1}{4(x+1)} - \dfrac{1}{4(x-1)}$

34. $\dfrac{-3}{32(x-2)} + \dfrac{3}{16(x+2)^2} + \dfrac{3}{32(x+2)} + \dfrac{3}{16(x-2)^2}$

35. $\dfrac{1}{2(2x-3)} + \dfrac{1}{2(2x-3)^2}$ **36.** $\dfrac{2}{3x+1} - \dfrac{1}{(3x+1)^2}$

37. $\dfrac{-2}{(2x+3)^2} + \dfrac{3}{2x+3}$ **38.** $\dfrac{-1}{3(3x+5)^2} + \dfrac{2}{3(3x+5)}$

42. $\dfrac{2}{x^2+1} + \dfrac{1}{x-1}$ **43.** $\dfrac{-x}{2(x^2+1)} + \dfrac{1}{4(x+1)} + \dfrac{1}{4(x-1)}$

44. $\dfrac{-3}{x^2+4} + \dfrac{1}{x-2} + \dfrac{2}{x+2}$ **45.** $\dfrac{-x}{x^2+1} + \dfrac{1}{x} - \dfrac{x}{(x^2+1)^2}$

46. $\dfrac{-2}{(x^2+2)^2} + \dfrac{1}{x^2+2}$

Review Exercises

7. $\{(1,-1,2)\}$ **8.** $\{(5,1,-2)\}$ **9.** $\{(-1,2,-3)\}$

10. $\left\{\left(2,\dfrac{1}{3},-1\right)\right\}$ **13.** $\{x, 2x-8, 11-3x\}$

14. $\{(x, 23-7x, 16-5x)\}$ **17.** $\left(\dfrac{7}{3} - \dfrac{2}{3}z, \dfrac{2}{3} - \dfrac{1}{3}z, z\right)$

18. $\left(\dfrac{10}{3} - \dfrac{5}{3}z, \dfrac{-2}{3} - \dfrac{2}{3}z, z\right)$

19.
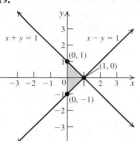

20.
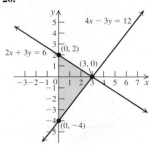

21.
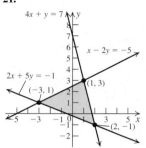

22.
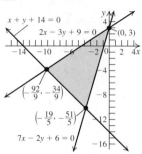

23.
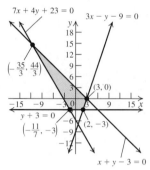

24.
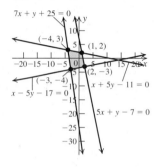

29. $\left\{(-2,1), \left(\dfrac{8}{3}, \dfrac{-5}{9}\right)\right\}$ **30.** $\left\{(1,1), \left(\dfrac{17}{16}, \dfrac{5}{4}\right)\right\}$

31. $\left\{(2,-2), \left(\dfrac{-2}{3}, \dfrac{-14}{3}\right)\right\}$ **32.** $\left\{(1,-3), \left(\dfrac{31}{11}, \dfrac{-23}{11}\right)\right\}$

33. $\left\{(1,2), (-1,-2), \left(2\sqrt{2}, \dfrac{\sqrt{2}}{2}\right), \left(-2\sqrt{2}, \dfrac{-\sqrt{2}}{2}\right)\right\}$

34. $\left\{(3,1), (-3,-1), \left(\dfrac{\sqrt{6}}{2}, \sqrt{6}\right), \left(\dfrac{-\sqrt{6}}{2}, -\sqrt{6}\right)\right\}$

35.
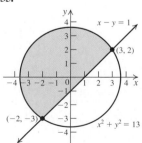

36.
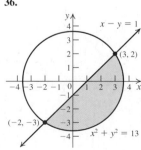

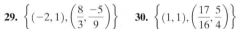

37.

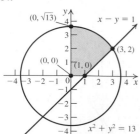

38.

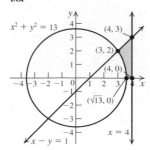

39.

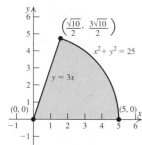

40.

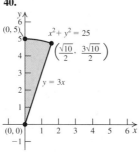

41. $\dfrac{2}{x+2} - \dfrac{1}{x+3}$ **42.** $\dfrac{3}{x-1} - \dfrac{2}{x+4}$

43. $\dfrac{2}{x-1} + \dfrac{1}{x} + \dfrac{5}{(x-1)^2}$ **44.** $\dfrac{2}{x-1} - \dfrac{2}{(x+1)^2} + \dfrac{1}{x+1}$

45. $\dfrac{-1+2x}{(x^2+4)^2} + \dfrac{1}{x^2+4}$ **46.** $\dfrac{1}{2(x-1)} - \dfrac{x}{x^2+1} + \dfrac{1}{2(x+1)}$

54. a. For Budget Rentals, the cost function is $C(x) = 23 + 0.17x$. For Dollar Rentals, the cost function is $C(x) = 24 + 0.22x$.

b.

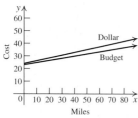

55. a. $C(x) = 60{,}000 + 12x, R(x) = 20x$

b.

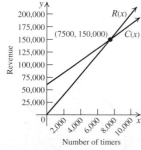

56. c.

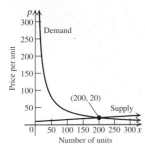

Practice Test A

18.

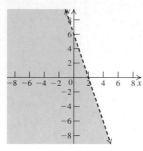

19.

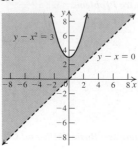

Cumulative Review Exercises (Chapters 1–7)

9.

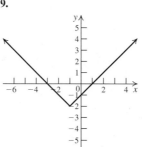

10.

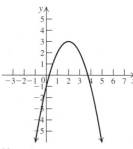

11.

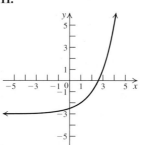

12.

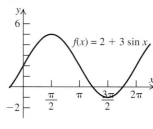

13. a. $f^{-1}(x) = \dfrac{x+2}{2}$ **b.**

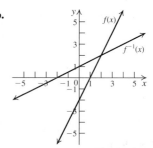

CHAPTER 8

Section 8.1

Practice Problems: **1. a.** 3×2 **b.** 1×2

2. $\begin{bmatrix} 0 & 3 & -1 & | & 8 \\ 1 & 4 & 0 & | & 14 \\ 0 & -2 & 9 & | & 0 \end{bmatrix}$ **3.** $\begin{bmatrix} 1 & 2 & 3 \\ 0 & -2 & -4 \end{bmatrix}$ **4.** $\left\{ \left(-1, \dfrac{-1}{3}, -1 \right) \right\}$

5. $\{(5,2)\}$ **6.** $\{(1,2,-3)\}$ **7.** $\varnothing$ **8.** $\{(1,2,-2)\}$

9. $\left\{ \left(-z-1, \dfrac{5}{3} - \dfrac{2}{3}z, z \right) \right\}$ **10. a.** $\begin{cases} s = 0.01s + 0.1c + 0.1t + 12.46 \\ c = 0.2s + 0.02c + 0.01t + 3 \\ t = 0.25s + 0.3c + 2.7 \end{cases}$

b. $(0.99)(14) - (0.1)(6) - (0.1)(8) = 12.46$ ✓
$(-0.20)(14) + (0.98)(6) - (0.01)(8) = 3$ ✓
$(-0.25)(14) - (0.30)(6) + 8 = 2.7$ ✓

A Exercises: Basic Skills and Concepts:

18. $A = \begin{bmatrix} 6 & -5 & 5 \\ 7 & 2 & 4 \end{bmatrix}$ **19.** $\left[\begin{array}{cc|c} 2 & 4 & 2 \\ 1 & -3 & 1 \end{array}\right]$ **20.** $\left[\begin{array}{cc|c} 1 & 2 & 7 \\ 3 & 5 & 11 \end{array}\right]$

21. $\left[\begin{array}{cc|c} 5 & -7 & 11 \\ -13 & 17 & 19 \end{array}\right]$ **22.** $\left[\begin{array}{cc|c} -3 & 2 & 10 \\ 5 & -1 & -7 \end{array}\right]$

23. $\left[\begin{array}{ccc|c} -1 & 2 & 3 & 8 \\ 2 & -3 & 9 & 16 \\ 4 & -5 & -6 & 32 \end{array}\right]$ **24.** $\left[\begin{array}{ccc|c} 1 & -1 & 0 & 2 \\ 2 & 0 & 3 & -5 \\ 0 & 1 & -2 & 7 \end{array}\right]$

25. $\begin{cases} x + 2y - 3z = 4 \\ -2x - 3y + z = 5 \\ 3x - 3y + 2z = 7 \end{cases}$ **26.** $\begin{cases} -x + 2y + 3z = 6 \\ 2x + 3y + z = 2 \\ 4x + 3y + 2z = 1 \end{cases}$

27. $\begin{cases} x - y + z = 2 \\ 2x + y - 3z = 6 \end{cases}$ **28.** $\begin{cases} x + y + z = 2 \\ x - y - z = 4 \\ 2x + 3y + z = 6 \\ -x + y - z = 8 \end{cases}$ **29.** $\begin{bmatrix} 1 & 2 & 3 \\ 0 & 1 & 1 \end{bmatrix}$

30. $\begin{bmatrix} 1 & 2 & 1 \\ 0 & 1 & 2 \end{bmatrix}$ **31.** $\begin{bmatrix} 1 & 2 & 3 & 4 \\ 0 & 1 & 5 & -3 \\ 0 & 0 & 1 & -1 \end{bmatrix}$ **32.** $\begin{bmatrix} 1 & 1.5 & -2 & .5 \\ 0 & 2 & 3 & 4 \\ 0 & 3 & 4.5 & 6 \end{bmatrix}$

33. $\begin{bmatrix} 4 & 5 & -7 \\ 5 & 4 & -2 \end{bmatrix} \xrightarrow{\frac{1}{4}R_1} \begin{bmatrix} 1 & \frac{5}{4} & -\frac{7}{4} \\ 5 & 4 & -2 \end{bmatrix} \xrightarrow{-5R_1 + R_2 \to R_2}$
$\begin{bmatrix} 1 & \frac{5}{4} & -\frac{7}{4} \\ 0 & -\frac{9}{4} & \frac{27}{4} \end{bmatrix} \xrightarrow{-\frac{4}{9}R_2} \begin{bmatrix} 1 & \frac{5}{4} & -\frac{7}{4} \\ 0 & 1 & -3 \end{bmatrix}$

34. $\begin{bmatrix} 2 & 6 & -8 \\ 3 & -1 & 2 \end{bmatrix} \xrightarrow{\frac{1}{2}R_1} \begin{bmatrix} 1 & 3 & -4 \\ 3 & -1 & 2 \end{bmatrix} \xrightarrow{-3R_1 + R_2 \to R_2}$
$\begin{bmatrix} 1 & 3 & -4 \\ 0 & -10 & 14 \end{bmatrix} \xrightarrow{-\frac{1}{10}R_2} \begin{bmatrix} 1 & 3 & -4 \\ 0 & 1 & -\frac{7}{5} \end{bmatrix}$

35. $\begin{bmatrix} 1 & 4 & 3 & 1 \\ 0 & -3 & -2 & 0 \\ 0 & 7 & 5 & -3 \end{bmatrix} \xrightarrow{-\frac{1}{3}R_2} \begin{bmatrix} 1 & 4 & 3 & 1 \\ 0 & 1 & \frac{2}{3} & 0 \\ 0 & 7 & 5 & -3 \end{bmatrix}$
$\xrightarrow{-7R_2 + R_3 \to R_3} \begin{bmatrix} 1 & 4 & 3 & 1 \\ 0 & 1 & \frac{2}{3} & 0 \\ 0 & 0 & \frac{1}{3} & -3 \end{bmatrix} \xrightarrow{3R_3} \begin{bmatrix} 1 & 4 & 3 & 1 \\ 0 & 1 & \frac{2}{3} & 0 \\ 0 & 0 & 1 & -9 \end{bmatrix}$

36. $\begin{bmatrix} 1 & 1 & 1 & 3 \\ 2 & 3 & 3 & 8 \\ 1 & -3 & -2 & 5 \end{bmatrix} \xrightarrow{-2R_1 + R_2 \to R_2} \begin{bmatrix} 1 & 1 & 1 & 3 \\ 0 & 1 & 1 & 2 \\ 1 & -3 & -2 & 5 \end{bmatrix}$
$\xrightarrow{-R_1 + R_3 \to R_3} \begin{bmatrix} 1 & 1 & 1 & 3 \\ 0 & 1 & 1 & 2 \\ 0 & -4 & -3 & 2 \end{bmatrix}$
$\xrightarrow{4R_2 + R_3 \to R_3} \begin{bmatrix} 1 & 1 & 1 & 3 \\ 0 & 1 & 1 & 2 \\ 0 & 0 & 1 & 10 \end{bmatrix}$

45. $(5, -2)$ $\begin{cases} x + 2y = 1 \\ y = -2 \end{cases}$ **46.** $(2, 3)$ $\begin{cases} x = 2 \\ y = 3 \end{cases}$

47. $(-4 - 4y, y, 3)$ $\begin{cases} x + 4y + 2z = 2 \\ z = 3 \end{cases}$

48. Inconsistent $\begin{cases} x + 2y + 3z = 4 \\ 0z = 1 \end{cases}$

49. $(1, 2, -1)$ $\begin{cases} x + 2y + 3z = 2 \\ y - 2z = 4 \\ z = -1 \end{cases}$ **50.** $(2, -3, 5)$ $\begin{cases} x + 2z = 12 \\ y + 3z = 12 \\ z = 5 \end{cases}$

51. $(2, -5, -2w + 3, w)$ $\begin{cases} x = 2 \\ y = -5 \\ z + 2w = 3 \end{cases}$

52. $(4, 3, z, 2)$ $\begin{cases} x = 4 \\ y = 3 \\ w = 2 \end{cases}$ **53.** $(-5, 4, 3, 0)$ $\begin{cases} x = -5 \\ y = 4 \\ z + 2w = 3 \\ w = 0 \end{cases}$

54. Inconsistent $\begin{cases} x = 3 \\ y = 2 \\ z = 0 \\ 0w = 1 \end{cases}$ **63.** $\{(1, 2, 3)\}$ **64.** $\{(2, -1, 5)\}$

65. $\{(2, 3, 4)\}$ **66.** $\{(1, -1, 2)\}$ **67.** $\{(1, 2, 3)\}$ **68.** $\{(1, 2, -2)\}$

69. $\left\{\left(\frac{5}{2}, \frac{3}{2}, \frac{7}{2}\right)\right\}$ **70.** $\left\{\left(\frac{283}{81}, -\frac{10}{27}, -\frac{113}{81}\right)\right\}$

71. $\{(1 + 4z, 3 - 3z, z)\}$ **72.** $\{(-3, -2 - z, z)\}$ **73.** $\{(1, 2, -1)\}$

74. $\{(1, 2, 1)\}$ **78.** False. For example, $\begin{bmatrix} 1 & 0 & 0 \\ 0 & 1 & 0 \end{bmatrix}$ is in reduced row-echelon form.

B Exercises: Applying the Concepts:

79. a. $\begin{cases} a = 0.1b + 1000 \\ b = 0.2a + 780 \end{cases}$

b. $\left[\begin{array}{cc|c} 1 & -0.1 & 1000 \\ -0.2 & 1 & 780 \end{array}\right]$ **c.** $a = 1100, b = 1000$

80. a. $\begin{cases} a = 0.5a + 0.2b + 50,000 \\ b = 0.3b + 0.6a + 40,000 \end{cases}$ **b.** $\left[\begin{array}{cc|c} 0.5 & -0.2 & 50,000 \\ -0.6 & 0.7 & 40,000 \end{array}\right]$

c. $a = 186,957$ $b = 217,391$

81. a. $l = 0.4t + 0.2f + 10,000$
$t = 0.5l + 0.3t + 20,000$
$f = 0.5l + 0.05t + 0.35f + 10,000$

b. $\left[\begin{array}{ccc|c} 1 & -0.4 & -0.2 & 10,000 \\ -0.5 & 0.7 & 0 & 20,000 \\ -0.5 & -0.05 & 0.65 & 10,000 \end{array}\right]$

c. food, $55,000; transportation, $61,000; labor, $45,400

82. a. $\begin{cases} a = 0.3c + 0.2b + 320 \\ b = 0.1a + 0.4c + 90 \\ c = 0.5b + 0.2a + 150 \end{cases}$ **b.** $\left[\begin{array}{ccc|c} 1 & -0.2 & -0.3 & 320 \\ -0.1 & 1 & -0.4 & 90 \\ -0.2 & -0.5 & 1 & 150 \end{array}\right]$

c. $a = 500, b = 300, c = 400$ **85. a.** $\begin{cases} x - y = 70 \\ -x + z = -120 \\ y - z = 50 \end{cases}$

b. $\{(120 + z, 50 + z, z)\}: 0 \le z \le 130$

86. a. $\begin{cases} x - y = 100 \\ y - w = -200 \\ -x + z = 20 \\ -z + w = 80 \end{cases}$

b. $\{(w - 100, w - 200, w - 80, w)\}: 200 \le w \le 437$

C Exercises: Beyond the Basics:

91. a. $B = \begin{bmatrix} 1 & \frac{3}{2} & 0 & -1 \\ 0 & 1 & 0 & \frac{1}{2} \\ 0 & 0 & 1 & \frac{1}{2} \\ 0 & 0 & 0 & 1 \end{bmatrix}, C = \begin{bmatrix} 1 & 2 & -3 & 1 \\ 0 & 1 & 0 & -\frac{1}{2} \\ 0 & 0 & 1 & \frac{1}{2} \\ 0 & 0 & 0 & 1 \end{bmatrix}$

b. The reduced row-echelon form of the matrices B and C is $\begin{bmatrix} 1 & 0 & 0 & 0 \\ 0 & 1 & 0 & 0 \\ 0 & 0 & 1 & 0 \\ 0 & 0 & 0 & 1 \end{bmatrix}$.   **92. a.** Both $B = \begin{bmatrix} 1 & 0.5 & -0.5 & 0.5 \\ 0 & 1 & -1 & -1 \\ 0 & 0 & 1 & 3 \end{bmatrix}$ and

$C = \begin{bmatrix} 1 & 1 & 1 & 6 \\ 0 & 1 & 0 & 2 \\ 0 & 0 & 1 & 3 \end{bmatrix}$ yield the solution set $\{(1, 2, 3)\}$.

b. The reduced row-echelon form of the matrices B and C is
$\begin{bmatrix} 1 & 0 & 0 & 1 \\ 0 & 1 & 0 & 2 \\ 0 & 0 & 1 & 3 \end{bmatrix}$. **93. a.** $\left\{ \left(\dfrac{dm - nb}{ad - bc}, \dfrac{cm - an}{cb - ad} \right) \right\}$

b. **(i)** $cb \neq ad$ **(ii)** $cb = ad$ and $\dfrac{m}{b} \neq \dfrac{n}{d}$ **(iii)** $cb = ad$ and $\dfrac{m}{b} = \dfrac{n}{d}$

94. $\begin{bmatrix} 1 & 0 & 0 & 1 \\ 0 & 1 & 0 & 0 \\ 0 & 0 & 1 & -2 \end{bmatrix}, \begin{bmatrix} 2 & 1 & 0 & 2 \\ 0 & 1 & 0 & 0 \\ 0 & 0 & 4 & -8 \end{bmatrix}, \begin{bmatrix} 1 & 1 & 1 & -1 \\ 0 & 1 & 0 & 0 \\ 0 & 1 & 1 & -2 \end{bmatrix}$

Section 8.2

Practice Problems: **1.** $x = 1, y = 1$ **2.** $\begin{bmatrix} -6 & 1 & 13 \\ 12 & 3 & 15 \end{bmatrix}$

3. $\begin{bmatrix} 11 & 1 \\ 0 & 6 \\ -15 & -28 \end{bmatrix}$ **4.** $X = \begin{bmatrix} -\dfrac{1}{3} & \dfrac{19}{3} \\ -7 & -\dfrac{35}{3} \end{bmatrix}$

5. Yes, the product AB is defined with order 2×1. **6.** \$2045 thousand **7.** $[31]$ **8.** The product AB is not defined.

$BA = \begin{bmatrix} 42 & -1 \\ 2 & -6 \\ 8 & -4 \end{bmatrix}$ **9.** $AB = \begin{bmatrix} 18 & -3 \\ 12 & 12 \end{bmatrix}, BA = \begin{bmatrix} 14 & -1 \\ 28 & 16 \end{bmatrix}$

10.

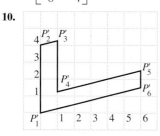

A Exercises: Basic Skills and Concepts:

15. a. $\begin{bmatrix} 0 & 2 \\ 5 & 1 \end{bmatrix}$ **b.** $\begin{bmatrix} 2 & 2 \\ 1 & 7 \end{bmatrix}$ **c.** $\begin{bmatrix} -3 & -6 \\ -9 & -12 \end{bmatrix}$ **d.** $\begin{bmatrix} 5 & 6 \\ 5 & 18 \end{bmatrix}$

e. $\begin{bmatrix} 10 & 2 \\ 5 & 11 \end{bmatrix}$ **f.** $\begin{bmatrix} 6 & 10 \\ 23 & 13 \end{bmatrix}$ **16. a.** $\begin{bmatrix} \dfrac{4}{3} & 1 \\ 1 & \dfrac{3}{2} \end{bmatrix}$ **b.** $\begin{bmatrix} -\dfrac{2}{3} & 1 \\ -3 & \dfrac{5}{2} \end{bmatrix}$

c. $\begin{bmatrix} -1 & -3 \\ 3 & -6 \end{bmatrix}$ **d.** $\begin{bmatrix} -1 & 3 \\ -7 & 7 \end{bmatrix}$ **e.** $\begin{bmatrix} \dfrac{25}{9} & \dfrac{17}{6} \\ \dfrac{17}{6} & \dfrac{13}{4} \end{bmatrix}$ **f.** $\begin{bmatrix} -\dfrac{17}{9} & \dfrac{7}{3} \\ -\dfrac{10}{3} & \dfrac{11}{4} \end{bmatrix}$

17. a, b, d, e, and f are not defined. **c.** $\begin{bmatrix} -6 & -9 \\ 12 & -15 \end{bmatrix}$

18. a, b, d, e, and f are not defined. **c.** $\begin{bmatrix} -3 & -6 & -9 \\ 3 & 9 & -12 \end{bmatrix}$

19. a. $\begin{bmatrix} 7 & 1 & -1 \\ -1 & 1 & 4 \\ 2 & 1 & 4 \end{bmatrix}$ **b.** $\begin{bmatrix} 1 & -1 & -1 \\ -3 & 9 & 0 \\ -2 & -1 & -2 \end{bmatrix}$ **c.** $\begin{bmatrix} -12 & 0 & 3 \\ 6 & -15 & -6 \\ 0 & 0 & -3 \end{bmatrix}$

d. $\begin{bmatrix} 6 & -2 & -3 \\ -8 & 23 & 2 \\ -4 & -2 & -3 \end{bmatrix}$ **e.** $\begin{bmatrix} 46 & 7 & -7 \\ 0 & 4 & 21 \\ 21 & 7 & 18 \end{bmatrix}$ **f.** $\begin{bmatrix} 6 & 1 & -7 \\ -21 & 6 & 16 \\ -13 & -1 & -10 \end{bmatrix}$

20. a. $\begin{bmatrix} 4 & -1 & 4 \\ 2 & 3 & 1 \\ 1 & 5 & 2 \end{bmatrix}$ **b.** $\begin{bmatrix} -2 & 1 & 0 \\ 2 & -1 & -1 \\ -1 & -3 & 4 \end{bmatrix}$ **c.** $\begin{bmatrix} -3 & 0 & -6 \\ -6 & -3 & 0 \\ 0 & -3 & -9 \end{bmatrix}$

d. $\begin{bmatrix} -3 & 2 & 2 \\ 6 & -1 & -2 \\ -2 & -5 & 11 \end{bmatrix}$ **e.** $\begin{bmatrix} 18 & 13 & 23 \\ 15 & 12 & 13 \\ 16 & 24 & 13 \end{bmatrix}$ **f.** $\begin{bmatrix} -10 & -1 & 5 \\ 3 & -7 & 3 \\ 0 & 1 & 2 \end{bmatrix}$

21. a. $\begin{bmatrix} 4 & 3 & -3 \\ 4 & 0 & 7 \\ 4 & 0 & 3 \end{bmatrix}$ **b.** $\begin{bmatrix} -2 & 1 & -3 \\ 2 & 8 & 3 \\ 0 & -2 & -3 \end{bmatrix}$ **c.** $\begin{bmatrix} -3 & -6 & 9 \\ -9 & -12 & -15 \\ -6 & 3 & 0 \end{bmatrix}$

d. $\begin{bmatrix} -3 & 4 & -9 \\ 7 & 20 & 11 \\ 2 & -5 & -6 \end{bmatrix}$ **e.** $\begin{bmatrix} 16 & 12 & 0 \\ 44 & 12 & 9 \\ 28 & 12 & -3 \end{bmatrix}$ **f.** $\begin{bmatrix} -9 & 14 & 5 \\ 22 & -2 & 13 \\ -14 & -1 & -22 \end{bmatrix}$

22. a. $\begin{bmatrix} 4 & -1 & 4 \\ 2 & 3 & 1 \\ 1 & 5 & 2 \end{bmatrix}$ **b.** $\begin{bmatrix} -2 & 1 & 0 \\ 2 & -1 & -1 \\ -1 & -3 & 4 \end{bmatrix}$ **c.** $\begin{bmatrix} -3 & 0 & -6 \\ -6 & -3 & 0 \\ 0 & -3 & -9 \end{bmatrix}$

d. $\begin{bmatrix} -3 & 2 & 2 \\ 6 & -1 & -2 \\ -2 & -5 & 11 \end{bmatrix}$ **e.** $\begin{bmatrix} 18 & 13 & 23 \\ 15 & 12 & 13 \\ 16 & 24 & 13 \end{bmatrix}$ **f.** $\begin{bmatrix} -10 & -1 & 5 \\ 3 & -7 & 3 \\ 0 & 1 & 2 \end{bmatrix}$

23. $X = \begin{bmatrix} -4 & -2 & 1 \\ 1 & 5 & 0 \end{bmatrix}$ **24.** $X = \begin{bmatrix} 4 & 2 & -1 \\ -1 & -5 & 0 \end{bmatrix}$

25. $X = \begin{bmatrix} 0 & 2 & -\dfrac{1}{2} \\ \dfrac{3}{2} & 1 & 4 \end{bmatrix}$ **26.** $X = \begin{bmatrix} 0 & 2 & -\dfrac{1}{2} \\ \dfrac{3}{2} & 1 & 4 \end{bmatrix}$

27. $X = \begin{bmatrix} -4 & -4 & \dfrac{3}{2} \\ -\dfrac{1}{2} & \dfrac{9}{2} & -4 \end{bmatrix}$ **28.** $X = \begin{bmatrix} -2 & -\dfrac{5}{3} & \dfrac{2}{3} \\ 0 & \dfrac{7}{3} & -\dfrac{4}{3} \end{bmatrix}$

29. $X = \begin{bmatrix} \dfrac{1}{2} & \dfrac{9}{4} & \dfrac{1}{2} \\ -2 & -\dfrac{5}{4} & -5 \end{bmatrix}$ **30.** $X = \begin{bmatrix} \dfrac{14}{3} & \dfrac{1}{3} & -\dfrac{2}{3} \\ \dfrac{8}{3} & -\dfrac{19}{3} & -4 \end{bmatrix}$

31. a. $AB = \begin{bmatrix} 4 & 11 \\ 6 & 23 \end{bmatrix}$ **b.** $BA = \begin{bmatrix} 1 & 0 \\ 18 & 26 \end{bmatrix}$

32. a. $AB = \begin{bmatrix} 7 & 19 & -6 \\ 11 & 28 & -2 \\ 2 & 5 & 0 \end{bmatrix}$ **b.** $BA = \begin{bmatrix} 6 & 15 \\ 11 & 29 \end{bmatrix}$

33. a. $AB = \begin{bmatrix} 4 & 7 \\ 3 & -12 \end{bmatrix}$ **b.** $BA = \begin{bmatrix} -13 & 4 & 10 \\ -13 & 5 & 6 \\ 8 & -4 & 0 \end{bmatrix}$

35. a. $AB = [16]$

b. $BA = \begin{bmatrix} 2 & 3 & 5 \\ -4 & -6 & -10 \\ 8 & 12 & 20 \end{bmatrix}$ **36. a.** $AB = \begin{bmatrix} -3 & 0 & 2 \\ -6 & 0 & 4 \\ 3 & 0 & -2 \end{bmatrix}$

b. $BA = [-5]$ **37. a.** $AB = [7 \quad 8 \quad -8]$ **b.** The product BA is not defined. **38. a.** The product AB is not defined.

b. $BA = [-7 \quad 14 \quad -7]$ **39. a.** $AB = \begin{bmatrix} 10 & 7 & 2 \\ 7 & 19 & 4 \\ 10 & 1 & 0 \end{bmatrix}$

b. $BA = \begin{bmatrix} 7 & 4 & 5 \\ 0 & 8 & 3 \\ 19 & 18 & 14 \end{bmatrix}$ **40. a.** $AB = \begin{bmatrix} 13 & -17 & 5 \\ -13 & 8 & -10 \\ 10 & -9 & 6 \end{bmatrix}$

b. $BA = \begin{bmatrix} 6 & -22 & 19 \\ -4 & 11 & -10 \\ 11 & -7 & 10 \end{bmatrix}$

41. $AB = \begin{bmatrix} 13 & 17 & 3 \\ 13 & 8 & 2 \\ 6 & 1 & 6 \end{bmatrix} \neq \begin{bmatrix} 6 & 22 & 19 \\ 2 & 11 & 6 \\ 11 & -1 & 10 \end{bmatrix} = BA$

42. $AB = \begin{bmatrix} 10 & 17 & 36 \\ -25 & 24 & 13 \\ -31 & 13 & 46 \end{bmatrix} = BA$

B Exercises: Applying the Concepts:

47.

	Steel	Glass	Wood	
$C =$	13	5	38	Cost of material
	7	2	7	Transportation cost

48.

	Steel	Glass	Wood	
$C =$	33	13	94	Cost of material
	18	5	17	Transportation cost

51. a.

	Chairman	President	Vice president
Salary	2,500,000	1,250,000	100,000
Bonus	1,500,000	750,000	150,000
Stock	50,000	25,000	5000

b. $\begin{bmatrix} 1 \\ 1 \\ 4 \end{bmatrix}$ Chairman President Vice president **c.** $\begin{bmatrix} 4{,}150{,}000 \\ 2{,}850{,}000 \\ 95{,}000 \end{bmatrix}$ Total salary Total bonuses Total stocks

52.

	Male	Female	Children		Calorie	Protein			
Brown	2	3	1		2400	55	$=$	12,300	278
Newgard	1	1	2		1900	45		7900	166
					1800	33			

For the Brown family, the total calorie requirement is 12,300 and the total protein requirement is 278.
For the Newgard family, the total calorie requirement is 7900 and the total protein requirement is 166.

53. $AD = \begin{bmatrix} 0 & 4 & 4 & 1 & 1 & 0 \\ 0 & 0 & -1 & -1 & -6 & -6 \end{bmatrix}$

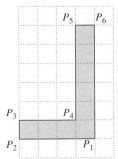

54. $AD = \begin{bmatrix} 0 & -4 & -4 & -1 & -1 & 0 \\ 0 & 0 & 1 & 1 & 6 & 6 \end{bmatrix}$

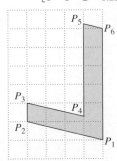

55. $AD = \begin{bmatrix} 0 & 4 & 4 & 1 & 1 & 0 \\ 0 & 1 & 2 & 1.25 & 6.25 & 6 \end{bmatrix}$

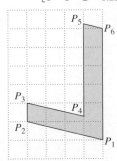

56. $AD = \begin{bmatrix} 0 & 4 & 3.75 & 0.75 & -0.5 & -1.5 \\ 0 & 0 & 1 & 1 & 6 & 6 \end{bmatrix}$

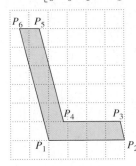

C Exercises: Beyond the Basics:

59. Let $A = \begin{bmatrix} 1 & 2 \\ 3 & 4 \end{bmatrix}$ and $B = \begin{bmatrix} -1 & 0 \\ 2 & -3 \end{bmatrix}$. Then

$(A + B)^2 = \begin{bmatrix} 10 & 2 \\ 5 & 11 \end{bmatrix} \neq \begin{bmatrix} 14 & -2 \\ 17 & 7 \end{bmatrix} = A^2 + 2AB + B^2.$

60. Let $A = \begin{bmatrix} 1 & 2 \\ 3 & 4 \end{bmatrix}$ and $B = \begin{bmatrix} -1 & 0 \\ 2 & -3 \end{bmatrix}$. Then

$A^2 - B^2 = \begin{bmatrix} 6 & 10 \\ 23 & 13 \end{bmatrix} \neq \begin{bmatrix} 10 & 6 \\ 35 & 9 \end{bmatrix} = (A - B)(A + B).$

Section 8.3

Practice Problems:

1. $\begin{bmatrix} 3 & 2 \\ 2 & 1 \end{bmatrix}\begin{bmatrix} -1 & 2 \\ 2 & -3 \end{bmatrix} = \begin{bmatrix} 1 & 0 \\ 0 & 1 \end{bmatrix}$ **2.** $\begin{bmatrix} 3x + z & 3y + w \\ 3x + z & 3y + w \end{bmatrix} \neq \begin{bmatrix} 1 & 0 \\ 0 & 1 \end{bmatrix}$ because this implies that $1 = 3x + z = 0$, which is false. **3.** The inverse of matrix A does not exist.

4. $A^{-1} = \begin{bmatrix} \dfrac{1}{7} & -\dfrac{19}{77} & \dfrac{1}{11} \\ 0 & \dfrac{2}{11} & \dfrac{1}{11} \\ \dfrac{2}{7} & -\dfrac{3}{77} & -\dfrac{1}{11} \end{bmatrix}$

5. The inverse of matrix A does not exist; $B^{-1} = \begin{bmatrix} \dfrac{1}{14} & \dfrac{1}{7} \\ -\dfrac{3}{14} & \dfrac{4}{7} \end{bmatrix}$.

6. $\left\{ \left(\dfrac{11}{2}, -\dfrac{9}{2}, \dfrac{1}{2} \right) \right\}$ **7.** $X = \begin{bmatrix} \dfrac{83{,}200}{19} \\ \dfrac{141{,}600}{19} \end{bmatrix}$

8. $AM = \begin{bmatrix} 1 & 2 & 3 \\ 1 & 3 & 3 \\ 1 & 2 & 4 \end{bmatrix}\begin{bmatrix} 10 & 11 & 19 & 15 & 19 & 5 \\ 1 & 0 & 0 & 23 & 1 & 0 \\ 3 & 9 & 14 & 0 & 6 & 0 \end{bmatrix}$

$= \begin{bmatrix} 21 & 38 & 61 & 61 & 39 & 5 \\ 22 & 38 & 61 & 84 & 40 & 5 \\ 24 & 47 & 75 & 61 & 45 & 5 \end{bmatrix}$

A Exercises: Basic Skills and Concepts:

17. $A^{-1} = \begin{bmatrix} \dfrac{1}{2} & 0 \\ -\dfrac{1}{6} & \dfrac{1}{3} \end{bmatrix}$ **18.** $A^{-1} = \begin{bmatrix} 0 & 1 \\ \dfrac{1}{3} & -\dfrac{4}{3} \end{bmatrix}$

21. $A^{-1} = \begin{bmatrix} 1 & -8 & 10 \\ 0 & 2 & -3 \\ 0 & -1 & 2 \end{bmatrix}$

22. $A^{-1} = \begin{bmatrix} -\dfrac{1}{4} & \dfrac{3}{4} & \dfrac{1}{2} \\ \dfrac{1}{8} & -\dfrac{3}{8} & \dfrac{1}{4} \\ \dfrac{1}{8} & \dfrac{5}{8} & \dfrac{1}{4} \end{bmatrix}$

23. $A^{-1} = \begin{bmatrix} \dfrac{3}{4} & -\dfrac{5}{2} & -\dfrac{1}{4} \\ -\dfrac{1}{2} & 2 & \dfrac{1}{2} \\ \dfrac{1}{2} & -1 & -\dfrac{1}{2} \end{bmatrix}$

24. $A^{-1} = \begin{bmatrix} 1 & 1 & -4 \\ -1 & -1 & 5 \\ \dfrac{1}{5} & 0 & -\dfrac{1}{5} \end{bmatrix}$

25. $A^{-1} = \begin{bmatrix} \dfrac{3}{14} & -\dfrac{1}{14} & \dfrac{5}{14} \\ \dfrac{5}{14} & \dfrac{3}{14} & -\dfrac{1}{14} \\ -\dfrac{1}{14} & \dfrac{5}{14} & \dfrac{3}{14} \end{bmatrix}$

26. $A^{-1} = \begin{bmatrix} \dfrac{3}{2} & -\dfrac{1}{4} & -\dfrac{9}{4} \\ -1 & \dfrac{1}{2} & \dfrac{3}{2} \\ \dfrac{1}{2} & \dfrac{1}{4} & \dfrac{1}{4} \end{bmatrix}$

27. $A^{-1} = \begin{bmatrix} 1 & 0 \\ -\dfrac{3}{2} & \dfrac{1}{2} \end{bmatrix}$

28. $A^{-1} = \begin{bmatrix} -3 & 2 \\ \dfrac{5}{2} & -\dfrac{3}{2} \end{bmatrix}$

29. $A^{-1} = \begin{bmatrix} 5 & 3 \\ 3 & 2 \end{bmatrix}$

30. $A^{-1} = \begin{bmatrix} -1 & 2 \\ -1 & \dfrac{3}{2} \end{bmatrix}$

31. $A^{-1} = \dfrac{1}{-a^2 + b^2} \begin{bmatrix} -a & b \\ -b & a \end{bmatrix}$, where $a^2 \neq b^2$

32. $A^{-1} = \begin{bmatrix} 1 & -1 \\ 1 & -2 \end{bmatrix}$

33. $\begin{bmatrix} 2 & 3 \\ 1 & -3 \end{bmatrix} \begin{bmatrix} x \\ y \end{bmatrix} = \begin{bmatrix} -9 \\ 13 \end{bmatrix}$

34. $\begin{bmatrix} 5 & -4 \\ 4 & -3 \end{bmatrix} \begin{bmatrix} x \\ y \end{bmatrix} = \begin{bmatrix} 7 \\ 5 \end{bmatrix}$

35. $\begin{bmatrix} 3 & 2 & 1 \\ 2 & 1 & 3 \\ 1 & 3 & 2 \end{bmatrix} \begin{bmatrix} x \\ y \\ z \end{bmatrix} = \begin{bmatrix} 8 \\ 7 \\ 9 \end{bmatrix}$

36. $\begin{bmatrix} 1 & 3 & 1 \\ 1 & -5 & 2 \\ 3 & 1 & -4 \end{bmatrix} \begin{bmatrix} x \\ y \\ z \end{bmatrix} = \begin{bmatrix} 4 \\ 7 \\ -9 \end{bmatrix}$

37. $\begin{cases} x - 2y = 0 \\ 2x + y = 5 \end{cases}$

38. $\begin{cases} 2x_1 + 3x_2 = 0 \\ 3x_1 - x_2 = 11 \end{cases}$

39. $\begin{cases} 2x_1 + 3x_2 + x_3 = -1 \\ 5x_1 + 7x_2 - x_3 = 5 \\ 4x_1 + 3x_2 \quad\;\; = 5 \end{cases}$

40. $\begin{cases} 3r - 2s + 3t = 4 \\ 5r + \quad 4t = 3 \\ 2r + 7s \quad\;\; = -8 \end{cases}$

41. $\begin{bmatrix} 1 & 2 & 5 \\ 2 & 3 & 8 \\ -1 & 1 & 2 \end{bmatrix} \begin{bmatrix} 2 & -1 & -1 \\ 12 & -7 & -2 \\ -5 & 3 & 1 \end{bmatrix} = \begin{bmatrix} 1 & 0 & 0 \\ 0 & 1 & 0 \\ 0 & 0 & 1 \end{bmatrix}$

42. $\{(-1, 0, 1)\}$

45. a. $A^{-1} = \begin{bmatrix} 3 & -\dfrac{5}{2} & \dfrac{1}{2} \\ -3 & 4 & -1 \\ 1 & -\dfrac{3}{2} & \dfrac{1}{2} \end{bmatrix}$

46. a. $A^{-1} = \begin{bmatrix} -\dfrac{1}{2} & \dfrac{1}{2} & \dfrac{1}{2} \\ \dfrac{11}{18} & -\dfrac{5}{18} & -\dfrac{7}{18} \\ \dfrac{4}{9} & \dfrac{1}{9} & -\dfrac{5}{9} \end{bmatrix}$

B Exercises: Applying the Concepts:

55. $X = \begin{bmatrix} \dfrac{216{,}000}{277} \\ \dfrac{310{,}000}{277} \\ \dfrac{304{,}000}{277} \end{bmatrix}$

C Exercises: Beyond the Basics:

72. c. $A^{-1} = \begin{bmatrix} \dfrac{3}{4} & \dfrac{1}{4} & -\dfrac{1}{4} \\ \dfrac{1}{4} & \dfrac{3}{4} & \dfrac{1}{4} \\ -\dfrac{1}{4} & \dfrac{1}{4} & \dfrac{3}{4} \end{bmatrix}$

Section 8.4

Practice Problems: 1. a. -38 **b.** 0 **2. a.** $M_{11} = 4, M_{23} = 10,$ $M_{32} = 10$ **b.** $A_{11} = 4, A_{23} = -10, A_{32} = -10$ **3.** 8 **4.** $\{(17, -9)\}$ **5.** $\{(2, -1, 0)\}$

Review Exercises

6. $\begin{bmatrix} -1 & 2 & 5 \\ 4 & 3 & 1 \end{bmatrix}$

7. $\left[\begin{array}{cc|c} 2 & -3 & 7 \\ 3 & 1 & 6 \end{array}\right]$

8. $\left[\begin{array}{ccc|c} 1 & 1 & -1 & 6 \\ 2 & -3 & -2 & 2 \\ 5 & -3 & 1 & 8 \end{array}\right]$

9. $\begin{bmatrix} 1 & -2 & 1 & 7 \\ 0 & 1 & 2 & 1 \\ 0 & 0 & 1 & 2 \end{bmatrix}$

10. $\begin{bmatrix} 1 & 2 & 1 & 7 \\ 0 & 1 & 2 & 5 \\ 0 & 0 & 1 & 2 \end{bmatrix}$

11. $\begin{bmatrix} 1 & 0 & 0 & \dfrac{1}{3} \\ 0 & 1 & 0 & \dfrac{1}{2} \\ 0 & 0 & 1 & -\dfrac{1}{6} \end{bmatrix}$

12. $\begin{bmatrix} 1 & 0 & 0 & \dfrac{1}{2} \\ 0 & 1 & 0 & \dfrac{1}{4} \\ 0 & 0 & 1 & \dfrac{1}{8} \end{bmatrix}$

13. $\{(2, -3, -1)\}$

14. $\{(1, 2, 1)\}$ **16.** $\left\{\left(\dfrac{8}{3}, \dfrac{-4}{3}, 3\right)\right\}$ **19.** $\left\{\left(2, -1, \dfrac{-1}{3}\right)\right\}$

23. a. $\begin{bmatrix} 3 & -1 \\ -8 & 10 \end{bmatrix}$ **b.** $\begin{bmatrix} -1 & 5 \\ 2 & -2 \end{bmatrix}$ **c.** $\begin{bmatrix} 2 & 4 \\ -6 & 8 \end{bmatrix}$ **d.** $\begin{bmatrix} -6 & 9 \\ 15 & -18 \end{bmatrix}$

e. $\begin{bmatrix} -4 & 13 \\ 9 & -10 \end{bmatrix}$

24. a. $\begin{bmatrix} 2 & 1 & -2 \\ 2 & 1 & 2 \\ -3 & 0 & 4 \end{bmatrix}$ **b.** $\begin{bmatrix} 2 & -1 & 0 \\ 0 & 3 & 2 \\ -1 & 0 & 2 \end{bmatrix}$

c. $\begin{bmatrix} 4 & 3 & -5 \\ 5 & 1 & 4 \\ -7 & 0 & 9 \end{bmatrix}$

25. $X = \begin{bmatrix} \dfrac{7}{3} & 0 \\ -\dfrac{19}{3} & 8 \end{bmatrix}$

26. $X = \begin{bmatrix} 1 & 1 & -\dfrac{3}{2} \\ \dfrac{3}{2} & 0 & 1 \\ -2 & 0 & \dfrac{5}{2} \end{bmatrix}$

27. $AB = \begin{bmatrix} -3 & 4 \\ -11 & 10 \end{bmatrix}, BA = \begin{bmatrix} -2 & -4 \\ 8 & 9 \end{bmatrix}$

28. $AB = \begin{bmatrix} 3 & 7 \\ -1 & 2 \end{bmatrix}, BA = \begin{bmatrix} -2 & 1 & 4 \\ 1 & 3 & 5 \\ -4 & 0 & 4 \end{bmatrix}$

29. $AB = [7], BA = \begin{bmatrix} 2 & 4 & -2 \\ 3 & 6 & -3 \\ 1 & 2 & -1 \end{bmatrix}$

30. $AB = \begin{bmatrix} 1 & 2 & 3 & 4 \\ 0 & 5 & 4 & 5 \\ -1 & 8 & 5 & 6 \end{bmatrix}, BA$ is not defined.

35. a. $\begin{bmatrix} 3 & -1 \\ 1 & 0 \end{bmatrix}$ **b.** $\begin{bmatrix} 8 & -3 \\ 3 & -1 \end{bmatrix}$ **c.** $\begin{bmatrix} 8 & -3 \\ 3 & -1 \end{bmatrix}$ **36. a.** $\begin{bmatrix} 3 & -4 \\ -2 & 3 \end{bmatrix}$

b. $\begin{bmatrix} 17 & -24 \\ -12 & 17 \end{bmatrix}$ **c.** $\begin{bmatrix} 17 & -24 \\ -12 & 17 \end{bmatrix}$ **41.** $\begin{bmatrix} \dfrac{2}{5} & -\dfrac{1}{10} \\ -\dfrac{1}{5} & \dfrac{3}{10} \end{bmatrix}$

42. $\begin{bmatrix} \dfrac{1}{5} & \dfrac{2}{5} \\ -\dfrac{1}{10} & \dfrac{3}{10} \end{bmatrix}$ **43.** $\begin{bmatrix} 3 & 2 & 6 \\ 1 & 1 & 2 \\ 2 & 2 & 5 \end{bmatrix}$ **44.** $\begin{bmatrix} -1 & 1 & -1 \\ 1 & -\dfrac{1}{2} & \dfrac{1}{3} \\ 0 & 0 & \dfrac{1}{3} \end{bmatrix}$

47. $\{(-12, 2, 3)\}$ **48.** $\{(2, -4, 4)\}$ **49.** $\left\{\left(\dfrac{7}{6}, \dfrac{1}{6}, 2\right)\right\}$

55. a. $M_{12} = 8, M_{23} = 8, M_{22} = -16$
b. $A_{12} = -8, A_{23} = -8, A_{22} = -16$ **56. a.** $M_{12} = -6,$
$M_{23} = -6, M_{22} = -12$ **b.** $A_{12} = 6, A_{23} = 6, A_{22} = -12$

62. $\{(3, -1)\}$ **63.** $\{(5, 6, 7)\}$ **64.** $\left\{\left(-2, 1, -\dfrac{1}{2}\right)\right\}$

66. $\dfrac{7}{2} \pm \dfrac{5\sqrt{5}}{2}$ **67.** They both satisfy the equation $y = 7x - 11.$

70.

		Male	Female	Children		Calories	Protein		Calories	Protein
A	$\begin{bmatrix} 2 \\ 1 \end{bmatrix}$	$\begin{matrix} 3 \\ 1 \end{matrix}$	$\begin{matrix} 1 \\ 2 \end{matrix}$		$\begin{bmatrix} 2200 \\ 1700 \\ 1500 \end{bmatrix}$	$\begin{matrix} 50 \\ 40 \\ 30 \end{matrix}$	$=$	$\begin{bmatrix} 11{,}000 \\ 6900 \end{bmatrix}$	$\begin{matrix} 250 \\ 150 \end{matrix}$	$\begin{matrix} A \\ B \end{matrix}$

76. The sales of medicine, nonmedical items, and beer and cigarettes were \$11,000, \$4000, and \$3500, respectively.

Practice Test A

2. $\begin{bmatrix} 7 & -3 & 9 & | & 5 \\ -2 & 4 & 3 & | & -12 \\ 8 & -5 & 1 & | & -9 \end{bmatrix}$ **3.** $\begin{cases} 4x - \quad\ \ z = -3 \\ x + 3y \quad\ \ = 9 \\ 2x + 7y + 5z = 8 \end{cases}$

7. $A - B = \begin{bmatrix} 8 & -1 \\ 4 & 8 \\ 11 & 0 \end{bmatrix}$

10. $\begin{bmatrix} -6 & 2 & 0 \\ 10 & 14 & 4 \end{bmatrix}$ **11.** $\begin{bmatrix} 20 & 28 & 8 \\ -9 & 29 & 6 \end{bmatrix}$ **12.** $\begin{bmatrix} 1 & 25 \\ 0 & 16 \end{bmatrix}$

13. $\begin{bmatrix} 1 & -\dfrac{5}{4} \\ 0 & \dfrac{1}{4} \end{bmatrix}$ **14.** $\begin{bmatrix} -11 & 2 & 7 \\ 8 & -1 & -5 \\ 5 & -1 & -3 \end{bmatrix}$ **15.** $\begin{bmatrix} 5 & 2 \\ 3 & 1 \end{bmatrix}\begin{bmatrix} x \\ y \end{bmatrix} = \begin{bmatrix} 32 \\ 18 \end{bmatrix}$

20. $x = \dfrac{\begin{vmatrix} 3 & -1 & 1 \\ 6 & 1 & 1 \\ 4 & 3 & -2 \end{vmatrix}}{\begin{vmatrix} 2 & -1 & 1 \\ 1 & 1 & 1 \\ 4 & 3 & -2 \end{vmatrix}},\ y = \dfrac{\begin{vmatrix} 2 & 3 & 1 \\ 1 & 6 & 1 \\ 4 & 4 & -2 \end{vmatrix}}{\begin{vmatrix} 2 & -1 & 1 \\ 1 & 1 & 1 \\ 4 & 3 & -2 \end{vmatrix}},\ z = \dfrac{\begin{vmatrix} 2 & -1 & 3 \\ 1 & 1 & 6 \\ 4 & 3 & 4 \end{vmatrix}}{\begin{vmatrix} 2 & -1 & 1 \\ 1 & 1 & 1 \\ 4 & 3 & -2 \end{vmatrix}}$

Cumulative Review Exercises (Chapters 1–8)

15.

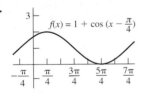

CHAPTER 9

Section 9.2

Practice Problems 1. a.

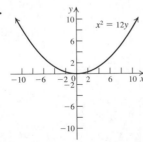

Vertex: $(0, 0)$; focus: $(0, 3)$; directrix: $y = -3$; axis: y-axis

b.

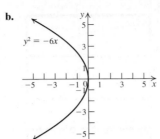

Vertex: $(0, 0)$; focus: $\left(-\dfrac{3}{2}, 0\right)$; directrix: $x = \dfrac{3}{2}$; axis: x-axis

2. a. $x^2 = 8y$ **b.** $y^2 = 4x$

3. Vertex: $(2, -1)$; focus: $\left(2, -\dfrac{7}{8}\right)$; directrix: $y = -\dfrac{9}{8}$

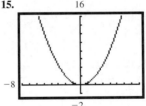

4. Equation: $x^2 = 29.2y$. The thickness of the mirror at the edges is 0.0771 in.

A Exercises: Basic Skills and Concepts:

14. Focus: $\left(-\dfrac{1}{2}, 0\right)$; directrix: $x = \dfrac{1}{2}$; graph: b

15.

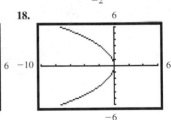

16.

17.

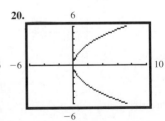

18.

19.

20.

21. $x^2 = -4(y - 3); 4$ **22.** $x^2 = 12(y - 1); 12$

23. $y^2 = -10\left(x - \dfrac{1}{2}\right); 10$ **24.** $y^2 = 2\left(x + \dfrac{3}{2}\right); 2$

25. $(y - 1)^2 = -8(x - 1); 8$ **26.** $(x - 1)^2 = -4(y - 1); 4$
27. $(x - 1)^2 = 16(y - 1); 16$ **28.** $(y - 1)^2 = 12(x - 1); 12$
29. $y^2 = 8(x - 1); 8$ **30.** $x^2 = 4(y - 1); 4$
31. $x^2 = -12(y - 1); 12$ **32.** $y^2 = -8(x + 1); 8$
33. $(y - 3)^2 = -8(x - 2); 8$ **34.** $(x - 2)^2 = -4(y - 3); 4$
35. $(x - 2)^2 = 8(y - 3); 8$ **36.** $(y - 3)^2 = 4(x - 2); 4$

37.

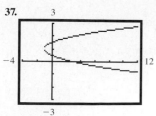

38.

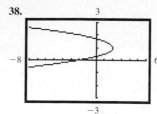

39.

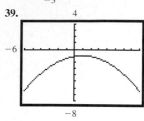

40.

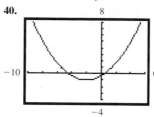

41. Vertex: $(-1, 1)$;

focus: $\left(-\dfrac{1}{2}, 1\right)$;

directrix: $x = -\dfrac{3}{2}$

42. Vertex: $(3, -2)$;

focus: $(5, -2)$;

directrix: $x = 1$

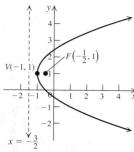

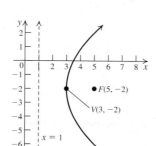

43. Vertex: $(-2, 2)$;

focus: $\left(-2, \dfrac{11}{4}\right)$;

directrix: $y = \dfrac{5}{4}$

44. Vertex: $(3, -1)$;

focus: $(3, 0)$;

directrix: $y = -2$

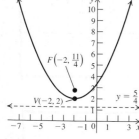

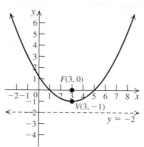

45. Vertex: $(2, -1)$;

focus: $\left(\dfrac{1}{2}, -1\right)$;

directrix: $x = \dfrac{7}{2}$

46. Vertex: $(-3, 2)$;

focus: $(-6, 2)$;

directrix: y-axis

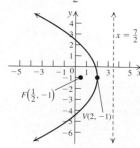

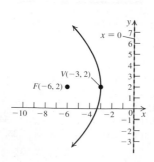

47. Vertex: $(1, 3)$;

focus: $\left(1, \dfrac{1}{2}\right)$;

directrix: $y = \dfrac{11}{2}$

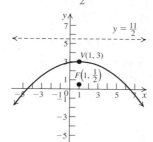

48. Vertex: $(-2, -3)$;

focus: $(-2, -5)$;

directrix: $y = -1$

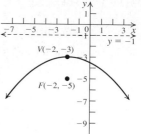

49. Vertex: $(-1, 1)$;

focus: $\left(-1, \dfrac{5}{4}\right)$;

directrix: $y = \dfrac{3}{4}$

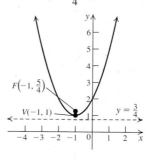

50. Vertex: $(-1, -1)$;

focus: $\left(-1, -\dfrac{11}{12}\right)$;

directrix: $y = -\dfrac{13}{12}$

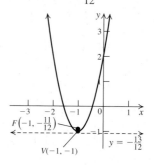

51. Vertex: $\left(-\dfrac{1}{2}, -1\right)$;

focus: $\left(-\dfrac{1}{4}, -1\right)$;

directrix: $x = -\dfrac{3}{4}$

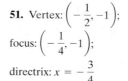

52. Vertex: $(4, 1)$;

focus: $\left(\dfrac{47}{12}, 1\right)$;

directrix: $x = \dfrac{49}{12}$

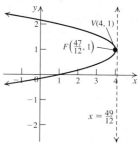

53. Vertex: $\left(-\dfrac{1}{2}, -1\right)$;

focus: $\left(-\dfrac{1}{2}, -\dfrac{5}{4}\right)$;

directrix: $y = -\dfrac{3}{4}$

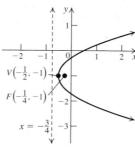

54. Vertex: $\left(-2, -\dfrac{1}{2}\right)$;

focus: $\left(-\dfrac{63}{32}, -\dfrac{1}{2}\right)$;

directrix: $x = -\dfrac{65}{32}$

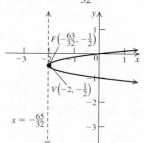

55. Vertex: $(46, 4)$;

focus: $\left(\dfrac{551}{12}, 4\right)$;

directrix: $x = \dfrac{553}{12}$

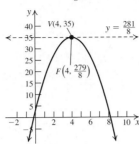

57. $y^2 = 4x$ and $x^2 = \dfrac{1}{2}y$ **58.** $y^2 = -\dfrac{4}{3}x$ and $x^2 = \dfrac{9}{2}y$

59. $(y - 1)^2 = 2x$ and $x^2 = 2(y - 1)$

60. $(y - 2)^2 = x - 1$ and $(x - 1)^2 = -(y - 2)$

61. $(y - 1)^2 = -(x + 2)$ and $(x + 2)^2 = -(y - 1)$

62. $(y + 1)^2 = -(x - 1)$ and $(x - 1)^2 = y + 1$

63. $(y - 1)^2 = x + 1$ and $(x + 1)^2 = y - 1$

64. $(y - 3)^2 = 16(x - 2)$ and $(x - 2)^2 = -\dfrac{1}{4}(y - 3)$

C Exercises: Beyond the Basics: 85. $\left(x - \dfrac{3}{4}\right)^2 = \dfrac{1}{2}\left(y - \dfrac{31}{8}\right)$

86. $\left(y - \dfrac{1}{3}\right)^2 = -\dfrac{1}{3}\left(x - \dfrac{13}{3}\right)$

87. Vertex: $(4, 6)$; focus: $\left(4, \dfrac{11}{2}\right)$; directrix: $y = \dfrac{13}{2}$; axis: $x = 4$

88. Vertex: $\left(-\dfrac{D}{2A}, \dfrac{D^2}{4AE} - \dfrac{F}{E}\right)$; focus: $\left(-\dfrac{D}{2A}, \dfrac{D^2}{4AE} - \dfrac{F}{E} - \dfrac{E}{4A}\right)$;

directrix: $y = \dfrac{D^2}{4AE} - \dfrac{F}{E} + \dfrac{E}{4A}$; axis: $x = -\dfrac{D}{2A}$

89. Vertex: $(-2, 3)$; focus: $\left(-\dfrac{11}{4}, 3\right)$; directrix: $x = -\dfrac{5}{4}$; axis: $y = 3$

90. Vertex: $\left(\dfrac{E^2}{4CD} - \dfrac{F}{D}, -\dfrac{E}{2C}\right)$; focus: $\left(\dfrac{E^2}{4CD} - \dfrac{F}{D} - \dfrac{D}{4C}, -\dfrac{E}{2C}\right)$;

directrix: $x = \dfrac{E^2}{4CD} - \dfrac{F}{D} + \dfrac{D}{4C}$; axis: $y = -\dfrac{E}{2C}$

Critical Thinking: 95. a. No, because if the axis is horizontal, it does not pass the vertical line test.

96. Parabola widens as $|x|$ increases.

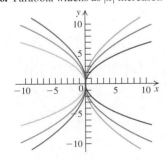

Section 9.3

Practice Problems: 1. $\dfrac{x^2}{36} + \dfrac{y^2}{100} = 1$

2.

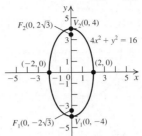

3. $\dfrac{(x - 2)^2}{9} + \dfrac{(y - 1)^2}{25} = 1$ **4.** Center: $(3, -1)$; vertices:

$(3 + \sqrt{42}, -1)$ and $(3 - \sqrt{42}, -1)$; foci: $\left(3 + \dfrac{\sqrt{126}}{2}, -1\right)$ and

$\left(3 - \dfrac{\sqrt{126}}{2}, -1\right)$ **5.** 6.9282 ft

A Exercises: Basic Skills and Concepts:

7. Vertices: $(4, 0)$ and $(-4, 0)$ **8.** Vertices: $(0, 4)$ and $(0, -4)$

foci: $(2\sqrt{3}, 0)$ and $(-2\sqrt{3}, 0)$ foci: $(0, 2\sqrt{3})$ and $(0, -2\sqrt{3})$

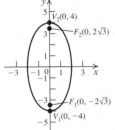

9. Vertices: $(3, 0)$ and $(-3, 0)$ **10.** Vertices: $(0, 2)$ and $(0, -2)$

foci: $(2\sqrt{2}, 0)$ and $(-2\sqrt{2}, 0)$ foci: $(0, \sqrt{3})$ and $(0, -\sqrt{3})$

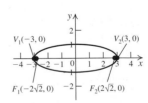

11. Vertices: $(5, 0)$ and $(-5, 0)$ **12.** Vertices: $(5, 0)$ and $(-5, 0)$

foci: $(3, 0)$ and $(-3, 0)$ foci: $(4, 0)$ and $(-4, 0)$

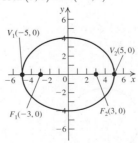

13. Vertices: $(0, 6)$ and $(0, -6)$ foci: $(0, 2\sqrt{5})$ and $(0, -2\sqrt{5})$

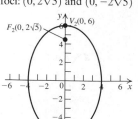

14. Vertices: $(0, 4)$ and $(0, -4)$ foci: $(0, \sqrt{7})$ and $(0, -\sqrt{7})$

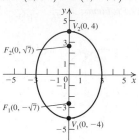

23. Vertices: $(0, \sqrt{5})$ and $(0, -\sqrt{5})$ foci: $(0, \sqrt{3})$ and $(0, -\sqrt{3})$

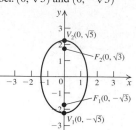

24. Vertices: $(0, \sqrt{7})$ and $(0, -\sqrt{7})$ foci: $(0, 2)$ and $(0, -2)$

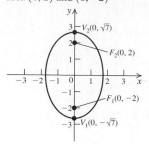

15. Circle centered at the origin with radius 2

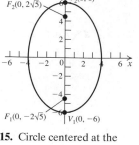

16. Circle centered at the origin with radius 4

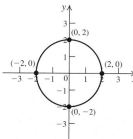

25. Vertices: $\left(\dfrac{\sqrt{14}}{2}, 0\right)$ and $\left(-\dfrac{\sqrt{14}}{2}, 0\right)$; foci: $\left(\dfrac{\sqrt{42}}{6}, 0\right)$ and $\left(-\dfrac{\sqrt{42}}{6}, 0\right)$

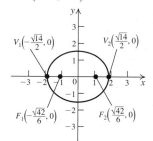

26. Vertices: $\left(\dfrac{\sqrt{33}}{3}, 0\right)$ and $\left(-\dfrac{\sqrt{33}}{3}, 0\right)$; foci: $\left(\dfrac{\sqrt{33}}{6}, 0\right)$ and $\left(-\dfrac{\sqrt{33}}{6}, 0\right)$

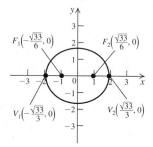

17. Vertices: $(2, 0)$ and $(-2, 0)$ foci: $(\sqrt{3}, 0)$ and $(-\sqrt{3}, 0)$

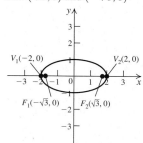

18. Vertices: $(0, 3)$ and $(0, -3)$ foci: $(0, 2\sqrt{2})$ and $(0, -2\sqrt{2})$

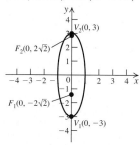

27. $\dfrac{x^2}{9} + \dfrac{y^2}{8} = 1$

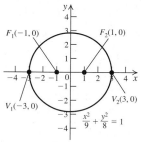

28. $\dfrac{x^2}{25} + \dfrac{y^2}{16} = 1$

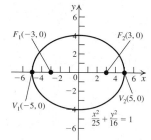

19. Vertices: $(0, 3)$ and $(0, -3)$ foci: $(0, \sqrt{5})$ and $(0, -\sqrt{5})$

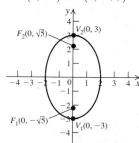

20. Vertices: $(5, 0)$ and $(-5, 0)$ foci: $(\sqrt{21}, 0)$ and $(-\sqrt{21}, 0)$

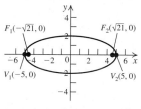

29. $\dfrac{x^2}{12} + \dfrac{y^2}{16} = 1$

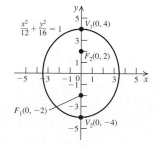

30. $\dfrac{x^2}{27} + \dfrac{y^2}{36} = 1$

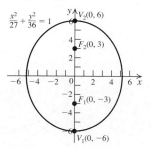

21. Vertices: $(2, 0)$ and $(-2, 0)$ foci: $(1, 0)$ and $(-1, 0)$

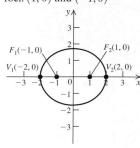

22. Vertices: $(\sqrt{5}, 0)$ and $(-\sqrt{5}, 0)$ foci: $(1, 0)$ and $(-1, 0)$

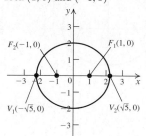

31. $\dfrac{x^2}{25} + \dfrac{y^2}{9} = 1$

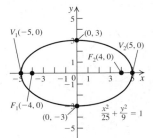

32. $\dfrac{x^2}{25} + \dfrac{y^2}{16} = 1$

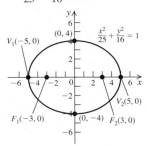

41. Center: $(1, 1)$; foci: $(1, 1 + \sqrt{5})$ and $(1, 1 - \sqrt{5})$; vertices: $(1, 4)$ and $(1, -2)$

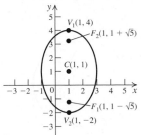

42. Center: $(1, 1)$; foci: $(1, 1 + 2\sqrt{3})$ and $(1, 1 - 2\sqrt{3})$; vertices: $(1, 5)$ and $(1, -3)$

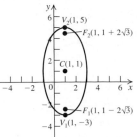

33. $\dfrac{x^2}{16} + \dfrac{y^2}{20} = 1$

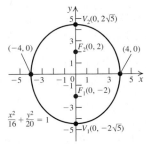

34. $\dfrac{x^2}{25} + \dfrac{y^2}{34} = 1$

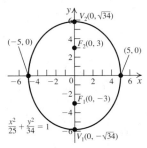

43. Center: $(0, -3)$; foci: $(2\sqrt{3}, -3)$ and $(-2\sqrt{3}, -3)$; vertices: $(4, -3)$ and $(-4, -3)$

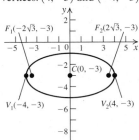

44. Center: $(-2, 0)$; foci: $(-2, \sqrt{5})$ and $(-2, -\sqrt{5})$; vertices: $(-2, 3)$ and $(-2, -3)$

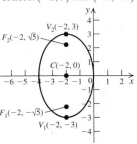

35. $\dfrac{x^2}{9} + \dfrac{y^2}{25} = 1$

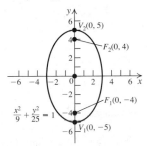

36. $\dfrac{x^2}{4} + \dfrac{y^2}{16} = 1$

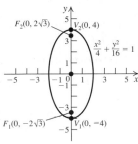

45. Center: $(1, -2)$; foci: $(2, -2)$ and $(0, -2)$; vertices: $(3, -2)$ and $(-1, -2)$

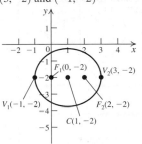

46. Center: $(-1, 2)$; foci: $(-1, 2 + \sqrt{5})$ and $(-1, 2 - \sqrt{5})$; vertices: $(-1, 5)$ and $(-1, -1)$

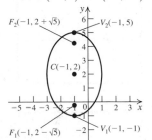

37. $\dfrac{x^2}{36} + \dfrac{y^2}{27} = 1$

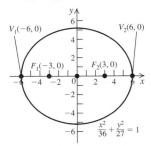

38. $\dfrac{x^2}{16} + \dfrac{y^2}{25} = 1$

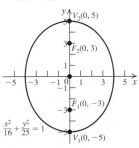

47. Center: $(-3, 1)$; foci: $(-2, 1)$ and $(-4, 1)$; vertices: $(-3 + \sqrt{5}, 1)$ and $(-3 - \sqrt{5}, 1)$

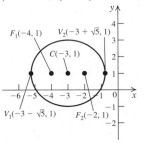

48. Center: $(2, -5)$; foci: $(2, -5 + \sqrt{21})$ and $(2, -5 - \sqrt{21})$; vertices: $(2, 0)$ and $(2, -10)$

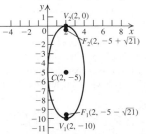

39. $\dfrac{x^2}{9} + \dfrac{y^2}{13} = 1$

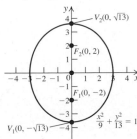

40. $\dfrac{x^2}{13} + \dfrac{y^2}{4} = 1$

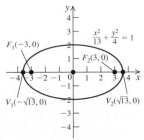

49. Center: $(-1, 2)$; foci: $(1, 2)$ and $(-3, 2)$; vertices: $(2, 2)$ and $(-4, 2)$

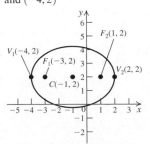

50. Center: $(-4, 3)$; foci: $(-4, 3 + \sqrt{5})$ and $(-4, 3 - \sqrt{5})$; vertices: $(-4, 6)$ and $(-4, 0)$

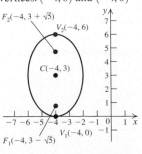

51. Center: $(-2, 4)$; foci: $(-2, 6)$ and $(-2, 2)$; vertices: $(-2, 7)$ and $(-2, 1)$

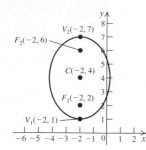

52. Center: $(-2, 2)$; foci: $(-2 + \sqrt{5}, 2)$ and $(-2 - \sqrt{5}, 2)$; vertices: $(-2 + 2\sqrt{5}, 2)$ and $(-2 - 2\sqrt{5}, 2)$

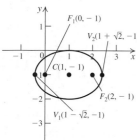

53. Center: $(1, -1)$; foci: $(2, -1)$ and $(0, -1)$; vertices: $(1 + \sqrt{2}, -1)$ and $(1 - \sqrt{2}, -1)$

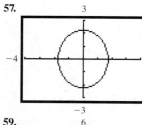

54. Circle centered at $(3, 2)$ with radius $\dfrac{\sqrt{46}}{2}$

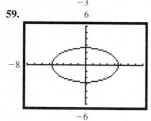

57.

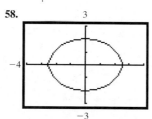

58.

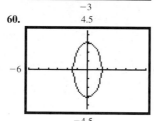

59.

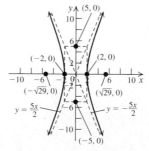

60.

91. Points of intersection: $\left(-\dfrac{17}{13}, -\dfrac{3}{13}\right)$ and $(1, -1)$

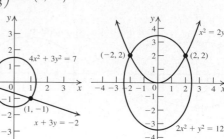

92. Points of intersection: $(-2, 2)$ and $(2, 2)$

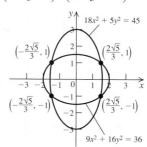

93. Points of intersection: $(\sqrt{2}, 3\sqrt{2}), (\sqrt{2}, -3\sqrt{2}),$ $(-\sqrt{2}, 3\sqrt{2}), (-\sqrt{2}, -3\sqrt{2})$

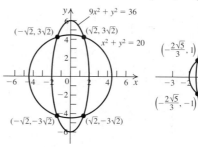

94. Points of intersection: $\left(\dfrac{2\sqrt{5}}{3}, 1\right), \left(\dfrac{2\sqrt{5}}{3}, -1\right),$ $\left(-\dfrac{2\sqrt{5}}{3}, 1\right), \left(-\dfrac{2\sqrt{5}}{3}, -1\right)$

Section 9.4

Practice Problems: 1. y-axis   **2.** Vertices: $(2\sqrt{2}, 0)$ and $(-2\sqrt{2}, 0)$; foci: $(\sqrt{10}, 0)$ and $(-\sqrt{10}, 0)$ **3.** $\dfrac{y^2}{16} - \dfrac{x^2}{20} = 1$ **4.** $y = \dfrac{2}{3}x$ and $y = -\dfrac{2}{3}x$ **5. a.** The standard form of the equation is $\dfrac{x^2}{4} - \dfrac{y^2}{25} = 1$. Vertices: $(2, 0)$ and $(-2, 0)$; endpoints of the conjugate axis: $(0, 5)$ and $(0, -5)$; asymptotes: $y = \dfrac{5}{2}x$ and $y = -\dfrac{5}{2}x$

B Exercises: Applying the Concepts: 65. The bet should not be accepted. The pool shark can hit the ball from any point straight into the pocket, or he can shoot it through the other focus and it will fall into the pocket because of the reflecting property.

68. $\dfrac{x^2}{20{,}067.5556} + \dfrac{y^2}{19{,}894.1067} = 1$

69. $\dfrac{x^2}{8{,}641.5616} + \dfrac{y^2}{8{,}639.0652} = 1$

70. $\dfrac{x^2}{1{,}311.8884} + \dfrac{y^2}{1{,}253.2128} = 1$

71. $\dfrac{x^2}{786{,}254.6241} + \dfrac{y^2}{783{,}788.5085} = 1$

C Exercises: Beyond the Basics: 75. $\dfrac{x^2}{25} + \dfrac{y^2}{16} = 1$ or $\dfrac{x^2}{\dfrac{625}{41}} + \dfrac{y^2}{25} = 1$

b. The standard form of the equation is $\dfrac{y^2}{\frac{1}{9}} - \dfrac{x^2}{1} = 1$. Vertices:

$\left(0, \dfrac{1}{3}\right)$ and $\left(0, -\dfrac{1}{3}\right)$; endpoints of the conjugate axis: $(1, 0)$ and

$(-1, 0)$; asymptotes: $y = \dfrac{1}{3}x$ and $y = -\dfrac{1}{3}x$

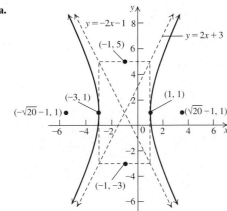

6. a.

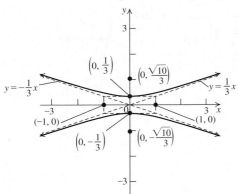

b.

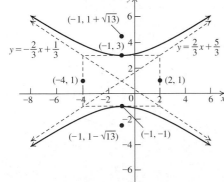

7. $\dfrac{(x-1)^2}{5} - \dfrac{(y-2)^2}{\frac{5}{4}} = 1$. Center: $(1, 2)$; vertices: $(1 + \sqrt{5}, 2)$ and

$(1 - \sqrt{5}, 2)$; endpoints of the conjugate axis: $\left(1, 2 + \dfrac{\sqrt{5}}{2}\right)$ and

$\left(1, 2 - \dfrac{\sqrt{5}}{2}\right)$; asymptotes: $y - 2 = \pm\dfrac{1}{2}(x - 1)$; foci: $\left(\dfrac{7}{2}, 2\right)$

and $\left(-\dfrac{3}{2}, 2\right)$

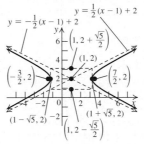

8. $\dfrac{x^2}{16,900} - \dfrac{y^2}{5,600} = 1$

A Exercises: Basic Skills and Concepts:

15. Vertices: $(1, 0)$ and $(-1, 0)$;
foci: $(\sqrt{5}, 0)$ and $1(-\sqrt{5}, 0)$;
transverse axis: x-axis; the
hyperbola opens left and right;
vertices of the fundamental
rectangle: $(1, 2), (-1, 2),$
$(-1, -2), (1, -2)$;
asymptotes: $y = \pm 2x$

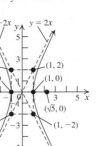

16. Vertices: $(0, 1)$ and $(0, -1)$;
foci: $(0, \sqrt{5})$ and $(0, -\sqrt{5})$;
transverse axis: y-axis; the
hyperbola opens up and down;
vertices of the fundamental
rectangle: $(2, 1), (-2, 1),$
$(-2, -1), (2, -1)$;
asymptotes: $y = \pm\dfrac{1}{2}x$

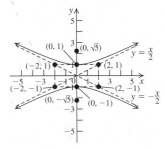

17. Vertices: $(0, 1)$ and $(0, -1)$;
foci: $(0, \sqrt{2})$ and $(0, -\sqrt{2})$;
transverse axis: y-axis; the
hyperbola opens up and down;
vertices of the fundamental
rectangle: $(1, 1), (-1, 1),$
$(-1, -1), (1, -1)$;
asymptotes: $y = \pm x$

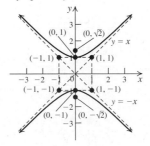

18. Vertices: $(1, 0)$ and $(-1, 0)$;
foci: $(\sqrt{2}, 0)$ and $(-\sqrt{2}, 0)$;
transverse axis: x-axis; the
hyperbola opens left and right;
vertices of the fundamental
rectangle: $(1, 1), (-1, 1),$
$(-1, -1), (1, -1)$; asymptotes:
$y = \pm x$

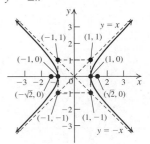

19. Vertices: $(0, 2)$ and $(0, -2)$; foci: $(0, 2\sqrt{10})$ and $(0, -2\sqrt{10})$; transverse axis: y-axis; the hyperbola opens up and down; vertices of the fundamental rectangle: $(6, 2)$, $(-6, 2)$, $(-6, -2)$, $(6, -2)$; asymptotes: $y = \pm\dfrac{1}{3}x$

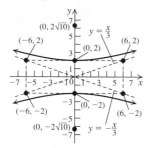

20. Vertices: $(2, 0)$ and $(-2, 0)$; foci: $(2\sqrt{10}, 0)$ and $(-2\sqrt{10}, 0)$; transverse axis: x-axis; the hyperbola opens left and right; vertices of the fundamental rectangle: $(2, 6)$, $(-2, 6)$, $(-2, -6)$, $(2, -6)$; asymptotes: $y = \pm 3x$

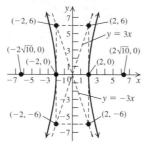

25. Vertices: $\left(\dfrac{1}{3}, 0\right)$ and $\left(-\dfrac{1}{3}, 0\right)$; foci: $\left(\dfrac{\sqrt{10}}{3}, 0\right)$ and $\left(-\dfrac{\sqrt{10}}{3}, 0\right)$; transverse axis: x-axis; the hyperbola opens left and right; vertices of the fundamental rectangle: $\left(\dfrac{1}{3}, 1\right), \left(-\dfrac{1}{3}, 1\right), \left(-\dfrac{1}{3}, -1\right), \left(\dfrac{1}{3}, -1\right)$; asymptotes: $y = \pm 3x$

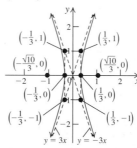

26. Vertices: $(2, 0)$ and $(-2, 0)$; foci: $(2\sqrt{10}, 0)$ and $(-2\sqrt{10}, 0)$; transverse axis: x-axis; the hyperbola opens left and right; vertices of the fundamental rectangle: $(2, 6)$, $(-2, 6)$, $(-2, -6)$, $(2, -6)$; asymptotes: $y = \pm 3x$

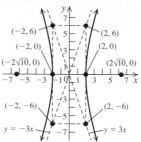

21. Vertices: $(3, 0)$ and $(-3, 0)$; foci: $(\sqrt{13}, 0)$ and $(-\sqrt{13}, 0)$; transverse axis: x-axis; the hyperbola opens left and right; vertices of the fundamental rectangle: $(3, 2)$, $(-3, 2)$, $(-3, -2)$, $(3, -2)$; asymptotes: $y = \pm\dfrac{2}{3}x$

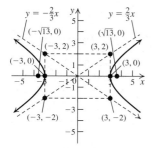

22. Vertices: $(0, 2)$ and $(0, -2)$; foci: $(0, \sqrt{13})$ and $(0, -\sqrt{13})$; transverse axis: y-axis; the hyperbola opens up and down; vertices of the fundamental rectangle: $(3, 2)$, $(-3, 2)$, $(-3, -2)$, $(3, -2)$; asymptotes: $y = \pm\dfrac{2}{3}x$

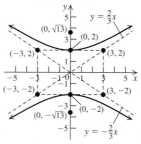

27. $\dfrac{x^2}{4} - \dfrac{y^2}{5} = 1$

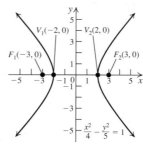

28. $\dfrac{x^2}{9} - \dfrac{y^2}{16} = 1$

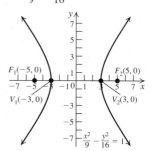

29. $\dfrac{y^2}{16} - \dfrac{x^2}{20} = 1$

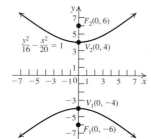

30. $\dfrac{y^2}{25} - \dfrac{x^2}{39} = 1$

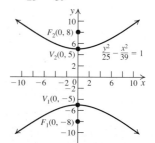

23. Vertices: $(0, 1)$ and $(0, -1)$; foci: $\left(0, \dfrac{\sqrt{5}}{2}\right)$ and $\left(0, -\dfrac{\sqrt{5}}{2}\right)$; transverse axis: y-axis; the hyperbola opens up and down; vertices of the fundamental rectangle: $\left(\dfrac{1}{2}, 1\right), \left(-\dfrac{1}{2}, 1\right), \left(-\dfrac{1}{2}, -1\right), \left(\dfrac{1}{2}, -1\right)$; asymptotes: $y = \pm 2x$

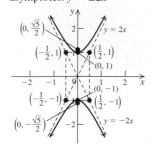

24. Vertices: $(0, 2)$ and $(0, -2)$; foci: $(0, \sqrt{5})$ and $(0, -\sqrt{5})$; transverse axis: y-axis; the hyperbola opens up and down; vertices of the fundamental rectangle: $(1, 2)$, $(-1, 2)$, $(-1, -2)$, $(1, -2)$; asymptotes: $y = \pm 2x$

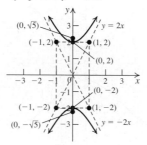

31. $\dfrac{y^2}{4} - \dfrac{x^2}{21} = 1$

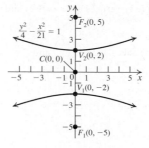

32. $y^2 - \dfrac{x^2}{15} = 1$

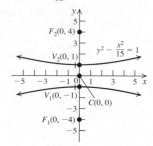

33. $x^2 - \dfrac{y^2}{24} = 1$

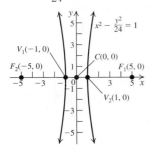

34. $\dfrac{x^2}{9} - \dfrac{y^2}{27} = 1$

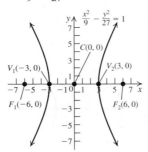

35. $\dfrac{y^2}{9} - \dfrac{x^2}{16} = 1$

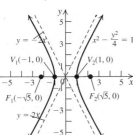

36. $x^2 - \dfrac{y^2}{3} = 1$

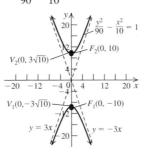

37. $x^2 - \dfrac{y^2}{4} = 1$

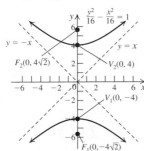

38. $\dfrac{y^2}{90} - \dfrac{x^2}{10} = 1$

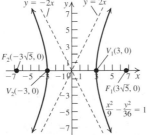

39. $\dfrac{y^2}{16} - \dfrac{x^2}{16} = 1$

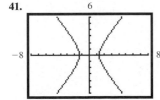

40. $\dfrac{x^2}{9} - \dfrac{y^2}{36} = 1$

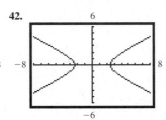

41.

42.

43.

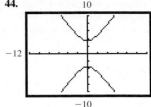

44.

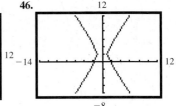

45.

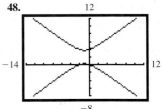

46.

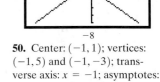

47.

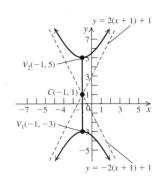

48.

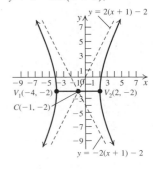

49. Center: $(1, -1)$; vertices: $(4, -1)$ and $(-2, -1)$; transverse axis: $y = -1$; asymptotes:
$$y + 1 = \pm\frac{4}{3}(x - 1)$$

50. Center: $(-1, 1)$; vertices: $(-1, 5)$ and $(-1, -3)$; transverse axis: $x = -1$; asymptotes:
$$y - 1 = \pm 2(x + 1)$$

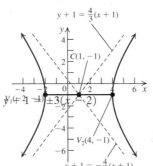

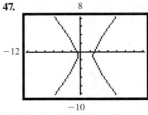

51. Center: $(-2, 0)$; vertices: $(3, 0)$ and $(-7, 0)$; transverse axis: x-axis; asymptotes:
$$y = \pm\frac{7}{5}(x + 2)$$

52. Center: $(-1, -2)$; vertices: $(2, -2)$ and $(-4, -2)$; transverse axis: $y = -2$; asymptotes:
$$y + 2 = \pm 2(x + 1);$$

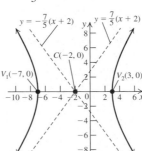

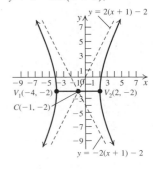

53. Center: $(-4, -3)$; vertices: $(1, -3)$ and $(-9, -3)$; transverse axis: $y = -3$; asymptotes: $y + 3 = \pm\dfrac{7}{5}(x + 4)$

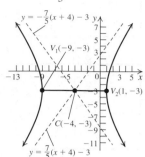

54. Center: $(3, -1)$; vertices: $(6, -1)$ and $(0, -1)$; transverse axis: $y = -1$; asymptotes: $y + 1 = \pm(x - 3)$.

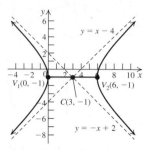

59. Center: $(-3, 0)$; vertices: $(-3 + 3\sqrt{5}, 0)$ and $(-3 - 3\sqrt{5}, 0)$; transverse axis: x-axis; asymptotes: $y = \pm(x + 3)$

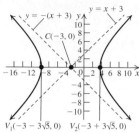

60. Center: $(-2, 0)$; vertices: $(2, 0)$ and $(-6, 0)$; transverse axis: x-axis; asymptotes: $y = \pm\dfrac{1}{2}(x + 2)$

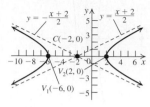

55. Center: $(0, -1)$; vertices: $\left(\dfrac{5}{2}, -1\right)$ and $\left(-\dfrac{5}{2}, -1\right)$; transverse axis: $y = -1$; asymptotes: $y + 1 = \pm 2x$

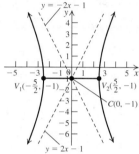

56. Center: $(1, 0)$; vertices: $(5, 0)$ and $(-3, 0)$; transverse axis: x-axis; asymptotes: $y = \pm 3(x - 1)$

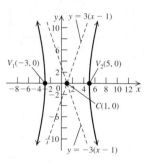

61. Center: $(2, 0)$; vertices: $(4, 0)$ and $(0, 0)$; transverse axis: x-axis; asymptotes: $y = \pm\dfrac{1}{2}(x - 2)$

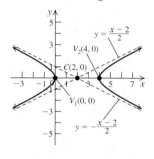

62. Center: $(0, -2)$; vertices: $(2, -2)$ and $(-2, -2)$; transverse axis: $y = -2$; asymptotes: $y + 2 = \pm\dfrac{\sqrt{2}}{2}x$

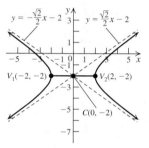

57. Center: $(2, -1)$; vertices: $(2, 4)$ and $(2, -6)$; transverse axis: $x = 2$; asymptotes:

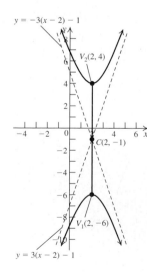

58. Center: $(4, -3)$; vertices: $\left(4 + \dfrac{\sqrt{6}}{3}, -3\right)$ and $\left(4 - \dfrac{\sqrt{6}}{3}, -3\right)$; transverse axis: $y = -3$; asymptotes: $y + 3 = \pm\sqrt{2}(x - 4)$

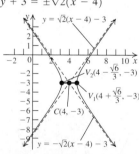

63. Center: $(-3, -4)$; vertices: $(-3, -3)$ and $(-3, -5)$; transverse axis: $x = -3$; asymptotes: $y + 4 = \pm\sqrt{2}(x + 3)$

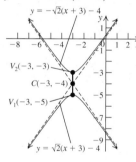

64. Center: $\left(-\dfrac{5}{3}, 2\right)$; vertices: $\left(-\dfrac{5}{3}, 2 + 2\sqrt{3}\right)$ and $\left(-\dfrac{5}{3}, 2 - 2\sqrt{3}\right)$; transverse axis: $x = -\dfrac{5}{3}$; asymptotes: $y - 2 = \pm 3\left(x + \dfrac{5}{3}\right)$

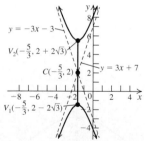

65. Center: $(3, -2)$; vertices: $(3 + \sqrt{6}, -2)$ and $(3 - \sqrt{6}, -2)$; transverse axis: $y = -2$; asymptotes: $y + 2 = \pm \dfrac{\sqrt{6}}{2}(x - 3)$

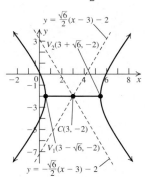

66. Center: $(-2, 1)$; vertices: $(-2, 4)$ and $(-2, -2)$; transverse axis: $x = -2$; asymptotes: $y - 1 = \pm \dfrac{3}{2}(x + 2)$

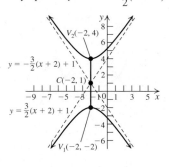

73. Circle

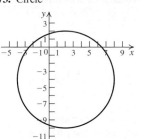

74. Parabola

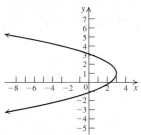

75. Parabola

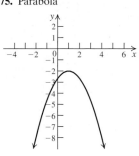

76. Hyperbola

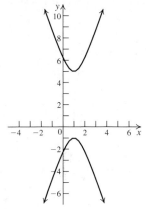

67. Center: $(\sqrt{2}, 0)$; vertices: $(\sqrt{2} + \sqrt{2\sqrt{2} + 1}, 0)$ and $(\sqrt{2} - \sqrt{2\sqrt{2} + 1}, 0)$; transverse axis: x-axis; asymptotes: $y = \pm(x - \sqrt{2})$

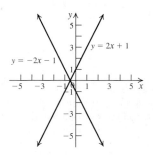

68. This is not a hyperbola, but a pair of lines: $y = \pm 2\left(x + \dfrac{1}{2}\right).$

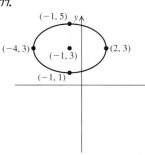

77.

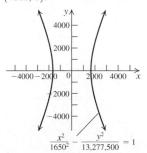

78.

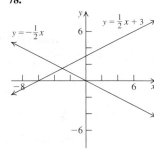

69. Parabola

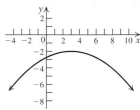

70. Hyperbola

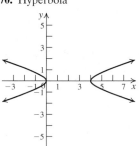

71. Hyperbola

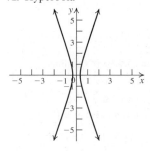

72. Ellipse

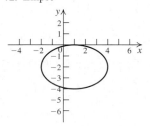

B Exercises: Applying the Concepts:

79. The difference in the distances of the location of the explosion from A and B is given: 600 m. Therefore, the location of the explosion is restricted to points on a hyperbola, the length of whose transverse axis is 600 m. Equation: $\dfrac{x^2}{90,000} - \dfrac{y^2}{160,000} = 1$

80. $(3 \text{ sec}) \times (1100 \text{ ft/sec}) = 3300$ ft; so the explosion is 3300 ft closer to A than to B. Therefore, the location of the explosion is restricted to points on a hyperbola, the length of whose transverse axis is 3300 feet. Equation: $\dfrac{x^2}{2,722,500} - \dfrac{y^2}{25,155,900} = 1$

81. Let the coordinates of Nicole be $(-4000, 0)$ and those of Juan be $(4000, 0)$.

$$\dfrac{x^2}{1650^2} - \dfrac{y^2}{13,277,500} = 1$$

83. Let the coordinates of A be $(-150, 0)$ and the coordinates of B be $(150, 0)$. Equation: $\dfrac{x^2}{5625} - \dfrac{y^2}{16{,}875} = 1$

85. $\dfrac{x^2}{193{,}600} - \dfrac{y^2}{446{,}400} = 1$

86. Let the coordinates of A be $(-125, 0)$ and the coordinates of B be $(125, 0)$. The plane is flying on the line $y = 50$. At the time of receiving the signals, the plane is at $(50.5238, 50)$.

88. Possible locations of the boat: $(-1397.427, 1225.2945)$, $(-281.3655, 209.9383)$, $(359.5894\ 288.219)$, $(1059.7444, 925.19691)$

C Exercises: Beyond the Basics: **89. a.** $x^2 - \dfrac{y^2}{24} = 1$

b. $\dfrac{x^2}{4} - \dfrac{y^2}{21} = 1$ **c.** $\dfrac{x^2}{16} - \dfrac{y^2}{9} = 1$

90.

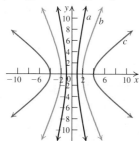

91. a. $y^2 - \dfrac{x^2}{35} = 1$ **b.** $y^2 - \dfrac{x^2}{15} = 1$ **c.** $y^2 - \dfrac{x^2}{8} = 1$

92.

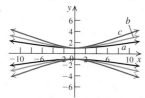

98. $e = \dfrac{5}{4}$; length of the latus rectum: $\dfrac{9}{2}$

99. $e = \dfrac{\sqrt{61}}{5}$; length of the latus rectum: $\dfrac{72}{5}$

100. $e = \sqrt{2}$; length of the latus rectum: 14

101. $e = 3$; length of the latus rectum: 8

102. $e = \dfrac{3}{2}$; length of the latus rectum: 5 **104.** $\dfrac{x^2}{16} - \dfrac{y^2}{20} = 1$

107. Points of intersection: $\left(-\dfrac{91}{20}, \dfrac{109}{10}\right)$

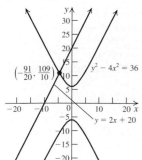

108. Points of intersection: $\left(5, \dfrac{1}{2}\sqrt{10}\right), \left(5, -\dfrac{1}{2}\sqrt{10}\right)$

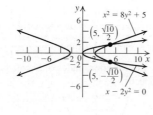

109. Points of intersection: $(-2\sqrt{2}, -\sqrt{7}), (-2\sqrt{2}, \sqrt{7}), (2\sqrt{2}, -\sqrt{7}), (2\sqrt{2}, \sqrt{7})$

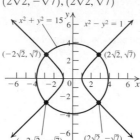

110. Points of intersection: $(-3, 0)$ and $(3, 0)$

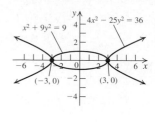

111. Points of intersection: $(-2, -3), (-2, 3), (2, -3), (2, 3)$

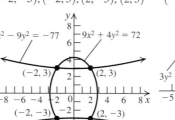

112. Points of intersection: $(-1, -2), (-1, 2), (1, -2), (1, 2)$

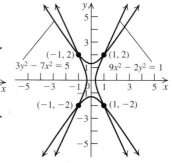

Critical Thinking: **115. a.** Two lines, $y = \pm\dfrac{2}{3}x$ **b.** Because all of the terms on the left-hand side are nonnegative and 6 is positive, the left-hand side is always positive. Therefore, there are no x and y values that satisfy the equation. **c.** A point, $(0, 0)$ **d.** A point, $(2, -4)$ **e.** Two lines, $y = \pm\sqrt{2}\,x$

Group Project:
The rewritten form of the equation is equivalent to

$$A\left(x + \frac{D}{2A}\right)^2 + C\left(y + \frac{E}{2C}\right)^2 = \frac{D^2}{4A} + \frac{E^2}{4C} - F.$$ If the term on the right-hand side is not zero, then the equation can be written as follows:

$$\frac{\left(x + \dfrac{D}{2A}\right)^2}{\dfrac{D^2}{4A^2} + \dfrac{E^2}{4AC} - AF} + \frac{\left(y + \dfrac{E}{2C}\right)^2}{\dfrac{D^2}{4AC} + \dfrac{E^2}{4C^2} - CF} = 1$$

i. If $\dfrac{D^2}{4A} + \dfrac{E^2}{4C} - F = 0$, then the graph consists of one point:

$\left(-\dfrac{D}{2A}, -\dfrac{E}{2C}\right)$. If $\dfrac{D^2}{4A} + \dfrac{E^2}{4C} - F > 0$, then the graph is an ellipse.

If $\dfrac{D^2}{4A} + \dfrac{E^2}{4C} - F > 0$, then there is no solution and the graph is empty.

ii. If $\dfrac{D^2}{4A} + \dfrac{E^2}{4C} - F = 0$, then the graph consists of two lines:

$\sqrt{A}\left(x + \dfrac{D}{2A}\right) = \pm\sqrt{-C}\left(y + \dfrac{E}{2C}\right)$. If $\dfrac{D^2}{4A} + \dfrac{E^2}{4C} - F > 0$, then the graph is a hyperbola with a horizontal transverse axis. If $\dfrac{D^2}{4A} + \dfrac{E^2}{4C} - F > 0$, then the graph is a hyperbola with a vertical transverse axis.

iii. If $\dfrac{D^2}{4A} + \dfrac{E^2}{4C} - F = 0$, then the graph consists of two lines:

$\sqrt{-A}\left(x + \dfrac{D}{2A}\right) = \pm\sqrt{C}\left(y + \dfrac{E}{2C}\right)$. If $\dfrac{D^2}{4A} + \dfrac{E^2}{4C} - F > 0$, then the graph is a hyperbola with a vertical transverse axis. If $\dfrac{D^2}{4A} + \dfrac{E^2}{4C} - F < 0$, then the graph is a hyperbola with a horizontal transverse axis.

iv. If $\dfrac{D^2}{4A} + \dfrac{E^2}{4C} - F = 0$, then the graph consists of one point: $\left(-\dfrac{D}{2A}, -\dfrac{E}{2C}\right)$. If $\dfrac{D^2}{4A} + \dfrac{E^2}{4C} - F > 0$, then there is no solution and the graph is empty. If $\dfrac{D^2}{4A} + \dfrac{E^2}{4C} - F < 0$, then the graph is an ellipse.

Section 9.5

Practice Problems: **1. a.** ellipse **b.** parabola **2.** $\dfrac{x'^2}{4} - \dfrac{y'^2}{4} = 1$

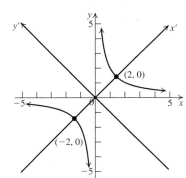

3. $x' = -10.39$, $y' = -6$; the wind speed along the launch path is approximately 10.39 knots (opposite the launch direction), and the wind speed perpendicular to the path is 6 knots in the direction 30° south of due east.

4. $\dfrac{x'^2}{9} + \dfrac{y'^2}{1} = 1$; an ellipse

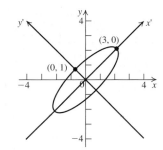

5. $x' = 2y'^2$; a parabola

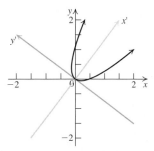

6. an ellipse

A Exercises: Basic Skills and Concepts:

18. $\left(-\dfrac{3}{2} + 2\sqrt{3}, \dfrac{3\sqrt{3}}{2} + 2\right)$

32. $\cos\theta = \dfrac{4}{5}, \sin\theta = \dfrac{3}{5}$ **33.** $\cos\theta = \dfrac{\sqrt{2}}{2}, \sin\theta = \dfrac{\sqrt{2}}{2}$

34. $\cos\theta = \dfrac{\sqrt{2}}{2}, \sin\theta = \dfrac{\sqrt{2}}{2}$ **35.** $\cos\theta = \dfrac{1}{2}, \sin\theta = \dfrac{\sqrt{3}}{2}$

36. $\cos\theta = \dfrac{\sqrt{3}}{2}, \sin\theta = \dfrac{1}{2}$ **40.** $\dfrac{x'^2}{2} + \dfrac{y'^2}{14} = 1$ **41.** $\dfrac{x'^2}{1} - \dfrac{y'^2}{3} = 1$

42. $\dfrac{y'^2}{2/3} - \dfrac{x'^2}{1} = 1$ **43.** $x' = -\dfrac{1}{2}(y' + 1)^2$

46. $\dfrac{x'^2}{2} + \dfrac{y'^2}{4} = 1$ **48.** $4y'^2 - 2x'^2 = 1$ or $\dfrac{y'^2}{1/4} - \dfrac{x'^2}{1/2} = 1$

49.
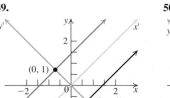

50.
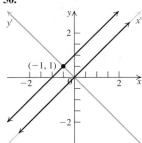

52.
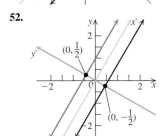

59.
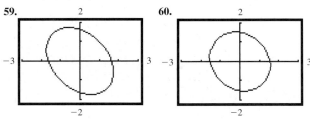

60.

61.
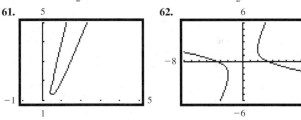

62.

63.
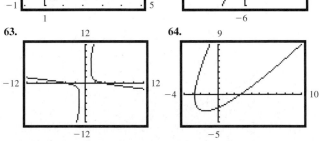

64.

B Exercises: Applying the Concepts: **65.** 7.5 knots opposite the shuttle's direction and 12.99 knots at 30° north of west
66. 9 knots in the direction of the shuttle and 15.59 knots at 30° south of east

Section 9.6

Practice Problems: **1.** $x^2 - 8y^2 - 2x - 68y - 139 = 0$

2. $r = \dfrac{9}{4 + 3\sin\theta}$

3.

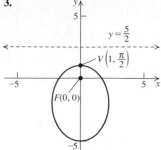

4.

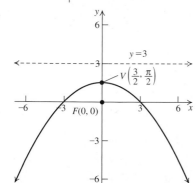

5.

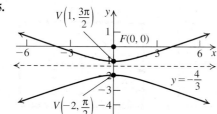

6. $r = \dfrac{20}{3 - 7\cos\theta}$ **7.** $r = \dfrac{3436.96}{1 - 0.25\cos\theta}$

A Exercises: Basic Skills and Concepts:

21. parabola   **22.** parabola

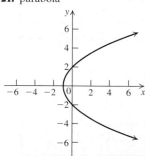

23. hyperbola   **24.** hyperbola

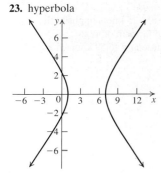   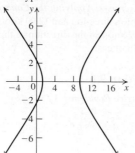

25. ellipse   **26.** ellipse

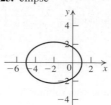   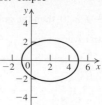

27. ellipse   **28.** ellipse

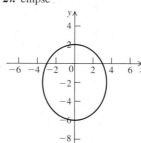   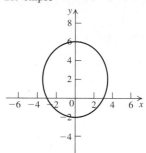

29. hyperbola

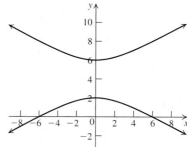

30. hyperbola

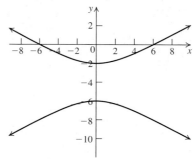

31. ellipse   **32.** ellipse

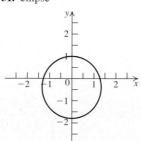   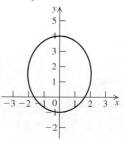

C Exercises: Beyond the Basics:

51. Rotated ellipse

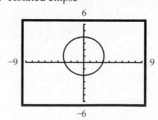

Section 9.7

Practice Problems

1.
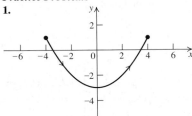

2. The graph is $y = (x - 1)^2 + 3$ starting at the vertex $(1, 3)$ and continuing to the left for $x < 1$.   **3. a.** A semicircle centered at the origin, starting at $(1, 0)$ and continuing counterclockwise to $(-1, 0)$
b. Circles of radius 1 around the origin, starting at $(1, 0)$ and continuing counterclockwise for infinitely many rotations
4. a. $x = \cos t, y = \sin t$; the particle moves counterclockwise starting at $t = 0$ and finishing at $t = \pi$.
b. $x = \cos 2t, y = \sin 2t$; the particle moves counterclockwise starting at $y = 0$ and finishing at $t = \dfrac{\pi}{2}$.
c. $x = -\cos t, y = \sin t$; the particle moves clockwise starting at $t = 0$ and finishing at $t = \pi$.
5. $x = at - b \sin t, y = a - b \cos t$

A Exercises: Basic Skills and Concepts:

9.
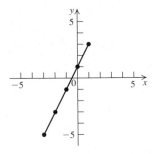

10.
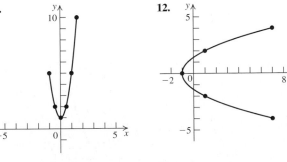

11.
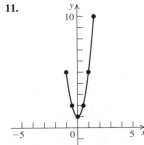

12.
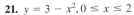

13.
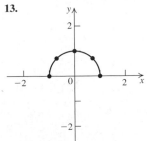

14.
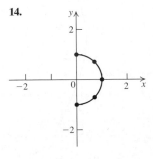

15.
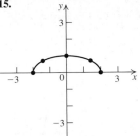

16.
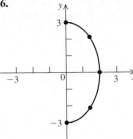

17. $y = 2x$
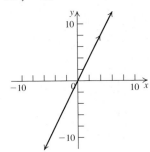

18. $y = \dfrac{x}{3}$
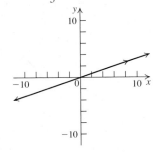

19. $y = 3x, 0 \le x \le 8$
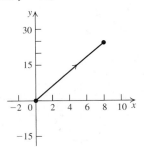

20. $y = -\dfrac{3}{2}x + 3, -1 \le x \le 2$
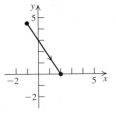

21. $y = 3 - x^2, 0 \le x \le 2$
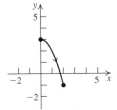

22. $y = \sqrt{x + 2}, -2 \le x \le 4$
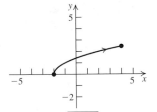

23. $y = \sqrt{4 - x^2}, 0 \le x \le 2$
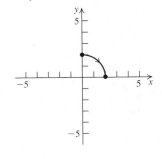

24. $y = -\sqrt{9 - x^2}, 0 \le x \le 3$
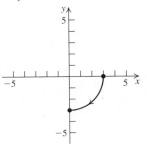

25. $y = 3\sqrt{1 - \dfrac{x^2}{4}}, 0 \le x \le 2$

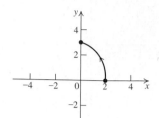

26. $y = 4\sqrt{1 - \dfrac{x^2}{9}}, 0 \le x \le 3$

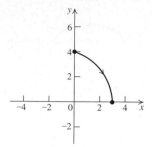

35. $y = (2x^2 - 1)^2, -1 \le x \le 1$, traversed infinitely often in both directions

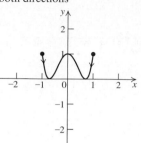

36. $y = (1 - 2x^2)^2, -1 \le x \le 1$, traversed infinitely often in both directions

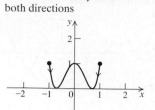

27. $y = \sqrt{x^2 - 4}, x \ge 2$

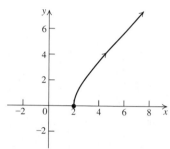

28. $y = 2\sqrt{1 + \dfrac{x^2}{9}}, x \ge 0$

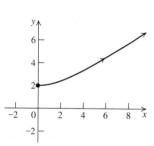

37. $(x - 1)^2 + (y - 2)^2 = 1$, traversed counterclockwise

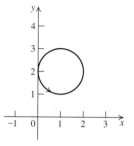

38. $\dfrac{(x - 3)^2}{4} + \dfrac{(y + 1)^2}{25} = 1$, traversed clockwise

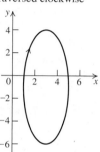

29. $x^2 + y^2 = 1$

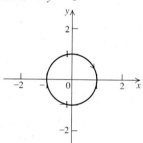

30. $x^2 + y^2 = 1$

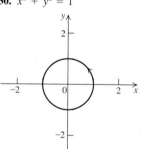

39. $\dfrac{(y - 2)^2}{9} - \dfrac{(x + 1)^2}{4} = 1, x \ge -1, y \ge 5$

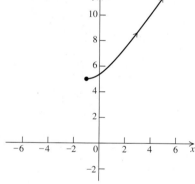

31. $y = 1 - x, 0 \le x \le 1$

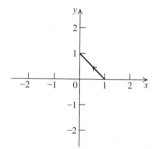

32. $y = 1 - x, 0 \le x \le 1$, traversed twice

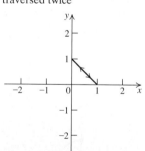

40. $\dfrac{(x - 2)^2}{9} - \dfrac{(y + 1)^2}{25} = 1, x \ge 2$

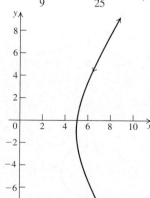

41. $y = x^2, x \ge 1$

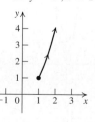

33. $\dfrac{x^2}{9} + \dfrac{y^2}{4} = 1$, traversed counterclockwise

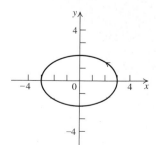

34. $\dfrac{x^2}{9} + \dfrac{y^2}{4} = 1$, traversed twice clockwise

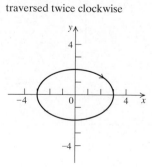

42. $x = 4y^2, x \geq 4, y \leq -1$

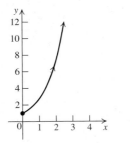

43. $y = \ln x, x \geq 1$

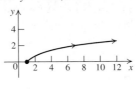

44. $y = e^x, x \geq 0$

45. $x = t, y = 3t + 2$; the particle moves along the entire line from left to right. **46.** $x = \sin t, y = 3 \sin t + 2$; the particle moves back and forth along the segment joining $(-1, -1)$ and $(1, 5)$.
47. $x = e^t, y = 3e^t + 2$; the particle moves to the right on the x-interval $0 < x < \infty$. **48.** $x = e^{-t}, y = 3e^{-t} + 2$; the particle moves to the left on the interval $0 < x < \infty$.
49. $x = -t, y = t^2 - 4$; the particle moves along the entire parabola from right to left. **50.** $x = t^2, y = t^4 - 4$; the particle traverses the right half of the parabola twice, moving from infinity to $x = 0$ and then back out to infinity. **51.** $x = \cos t, y = \cos^2 t - 4$; the particle oscillates on the parabola between the points $(-1, -3)$ and $(1, -3)$.
52. $x = e^t, y = e^{2t} - 4$; the particle moves to the right along the parabola over the x-interval $0 < x < \infty$.

53. $x = \sin t, y = \cos t, -\dfrac{\pi}{2} \leq t \leq \dfrac{\pi}{2}$; the motion is counterclockwise from $(0, -1)$ to $(0, 1)$.

54. $x = \sin 2t, y = \cos 2t, -\dfrac{\pi}{4} \leq t \leq \dfrac{\pi}{4}$; the motion is counterclockwise from $(0, -1)$ to $(0, 1)$.

55. $x = -\sin t, y = \cos t, -\dfrac{\pi}{2} \leq t \leq \dfrac{\pi}{2}$; the motion is clockwise from $(0, 1)$ to $(0, -1)$.

56. $x = -\sin 4t, y = \cos 4t, -\dfrac{\pi}{8} \leq t \leq \dfrac{\pi}{8}$; the motion is clockwise from $(0, 1)$ to $(0, -1)$.

C Exercises: Beyond the Basics:

73. The upper half of the parabola $x = \dfrac{1}{4}(y - 1)^2 + 2$

74. The right branch of the hyperbola $x^2 - y^2 = 1$
75. The upper half of the circle $x^2 + y^2 = 1$, traversed counterclockwise starting at $(1, 0)$ **76.** The lower half of the left branch and the upper half of the right branch of the hyperbola $x^2 - y^2 = 1$

77. The ellipse $\dfrac{x^2}{128} + \dfrac{y^2}{450} = 1$

Group Project: 80. a. The equation is $y = x \tan \theta - \dfrac{g}{2}\left(\dfrac{x}{v_0 \cos \theta}\right)^2$.

b. The vertex is $\left(\dfrac{v_0^2}{g} \sin \theta \cos \theta, \dfrac{v_0^2}{2g} \sin^2 \theta\right)$. **c.** The flight time is $\dfrac{2v_0}{g} \sin \theta$.

d. The range is $\dfrac{2v_0^2}{g} \sin \theta \cos \theta$. **e.** The range is a maximum when $2 \sin \theta \cos \theta = \sin 2\theta$ is a maximum, which happens for $\theta = 45°$.
The maximum range is $\dfrac{v_0^2}{g}$. **f.** Both the range and height at the vertex depend on the square of the speed.

Review Exercises

1. Vertex: $(0, 0)$; focus: $\left(-\dfrac{3}{2}, 0\right)$;
axis: x-axis; directrix: $x = \dfrac{3}{2}$

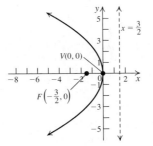

2. Vertex: $(0, 0)$; focus: $(3, 0)$;
axis: x-axis; directrix: $x = -3$

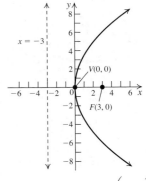

3. Vertex: $(0, 0)$; focus: $\left(0, \dfrac{7}{4}\right)$;
axis: y-axis; directrix: $y = -\dfrac{7}{4}$

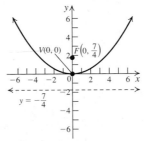

4. Vertex: $(0, 0)$; focus: $\left(0, -\dfrac{3}{4}\right)$;
axis: y-axis; directrix: $y = \dfrac{3}{4}$

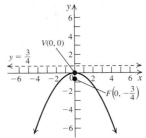

5. Vertex: $(2, -3)$ Focus:
$\left(2, -\dfrac{13}{4}\right)$ Axis : $x = 2$
Directrix: $y = -\dfrac{11}{4}$

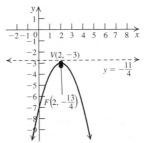

6. Vertex: $(-2, -1)$ Focus:
$\left(-\dfrac{3}{4}, -1\right)$ Axis: $y = -1$
Directrix: $x = -\dfrac{13}{4}$

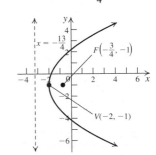

7. Vertex: $\left(-\dfrac{5}{2}, -2\right)$ Focus:
$(-2, -2)$ Axis: $y = -2$
Directrix: $x = -3$

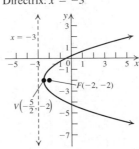

8. Vertex: $(1, -1)$ Focus:
$\left(1, -\dfrac{3}{4}\right)$ Axis: $x = 1$
Directrix: $y = -\dfrac{5}{4}$

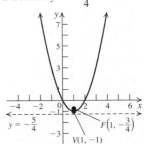

13. Foci: $(\sqrt{21}, 0)$ and $(-\sqrt{21}, 0)$
Vertices: $(5, 0)$ and $(-5, 0)$
Endpoints of the minor axis:
$(0, 2)$ and $(0, -2)$

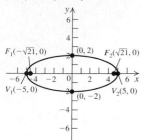

14. Foci: $(0, 3\sqrt{3})$ and $(0, -3\sqrt{3})$
Vertices: $(0, 6)$ and $(0, -6)$
Endpoints of the minor axis:
$(3, 0)$ and $(-3, 0)$

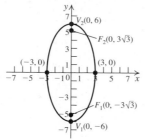

21. $\dfrac{x^2}{16} + \dfrac{y^2}{4} = 1$ **22.** $\dfrac{x^2}{4} + \dfrac{y^2}{36} = 1$

25. Vertices: $(0, 4)$ and $(0, -4)$
Foci: $(0, 2\sqrt{5})$ and $(0, -2\sqrt{5})$
Asymptotes: $y = \pm 2x$

26. Vertices: $(4, 0)$ and $(-4, 0)$
Foci: $(5, 0)$ and $(-5, 0)$
Asymptotes: $y = \pm\dfrac{3}{4}x$

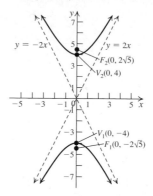

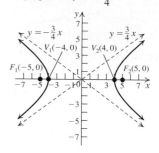

15. Foci: $(0, \sqrt{3})$ and $(0, -\sqrt{3})$
Vertices: $(0, 2)$ and $(0, -2)$
Endpoints of the minor axis:
$(1, 0)$ and $(-1, 0)$

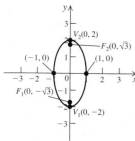

16. Foci: $(0, 2\sqrt{15})$ and $(0, -2\sqrt{15})$;
vertices: $(0, 8)$ and $(0, -8)$;
endpoints of the minor axis:
$(2, 0)$ and $(-2, 0)$

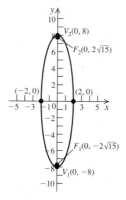

27. Vertices: $(1, 0)$ and $(-1, 0)$
Foci: $(3, 0)$ and $(-3, 0)$
Asymptotes: $y = \pm 2\sqrt{2}x$

28. Vertices: $\left(0, \dfrac{1}{2}\right)$ and
$\left(0, -\dfrac{1}{2}\right)$; foci: $\left(0, \dfrac{\sqrt{2}}{2}\right)$ and
$\left(0, -\dfrac{\sqrt{2}}{2}\right)$; Asymptotes: $y = \pm x$

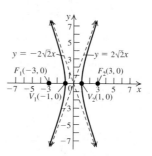

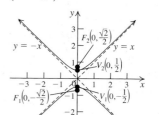

17. Foci: $(-1, -4 + \sqrt{7})$ and
$(-1, -4 - \sqrt{7})$; vertices: $(-1, 0)$
and $(-1, -8)$; endpoints of the
minor axis: $(2, -4)$ and $(-4, -4)$

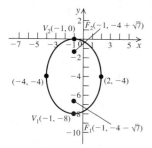

18. Foci: $(1, -1)$ and $(1, -3)$;
vertices: $(1, 0)$ and $(1, -4)$;
endpoints of the minor axis:
$(1 + \sqrt{3}, -2)$ and $(1 - \sqrt{3}, -2)$

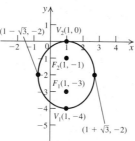

29. Vertices: $(1, 3)$ and $(-5, 3)$;
foci: $(-2 + \sqrt{13}, 3)$ and
$(-2 - \sqrt{13}, 3)$; asymptotes:
$y - 3 = \pm\dfrac{2}{3}(x + 2)$

30. Vertices: $(2, -1 + \sqrt{6})$
and $(2, -1 - \sqrt{6})$; foci:
$(2, -1 + \sqrt{14})$ and $(2, -1 - \sqrt{14})$;
asymptotes: $y + 1 = \pm\dfrac{\sqrt{3}}{2}(x - 2)$

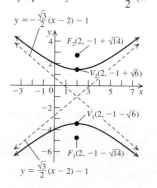

19. Foci: $(-1 + 2\sqrt{2}, 1)$ and
$(-1 - 2\sqrt{2}, 1)$; vertices: $(2, 1)$ and
$(-4, 1)$; endpoints of the minor
axis: $(-1, 2)$ and $(-1, 0)$

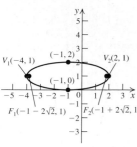

20. Foci: $(-1, 5 + 2\sqrt{3})$ and
$(-1, 5 + 2\sqrt{3})$; vertices: $(-1, 9)$
and $(-1, 1)$; endpoints of the
minor axis: $(1, 5)$ and $(-3, 5)$

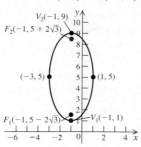

31. Vertices: $(-2, -2)$ and $(-2, -8)$; foci: $(-2, -5 + 3\sqrt{5})$ and $(-2, -5 - 3\sqrt{5})$; asymptotes: $y + 5 = \pm\frac{1}{2}(x + 2)$

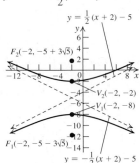

32. Vertices: $(-2, -1)$ and $(-2, -5)$; foci: $(-2, -3 + \sqrt{13})$ and $(-2, -3 - \sqrt{13})$; asymptotes: $y + 3 = \pm\frac{2}{3}(x + 2)$

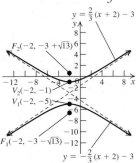

37. Hyperbola

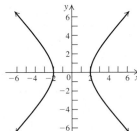

38. Parabola

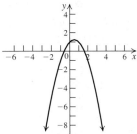

39. Ellipse

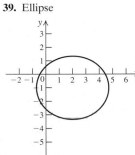

40. Parabola

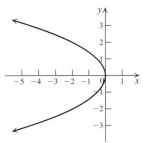

41. Circle

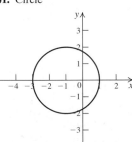

42. Parabola

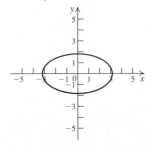

43. Hyperbola

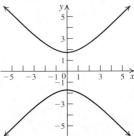

44. Ellipse

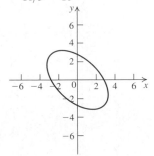

45. Circle

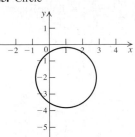

46. Hyperbola

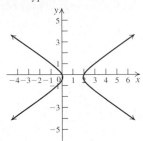

47. Ellipse

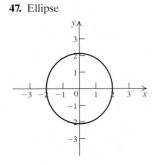

48. Hyperbola

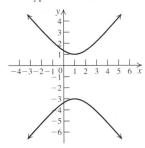

51. Point of intersection: $\left(\dfrac{100}{10}, -\dfrac{91}{20}\right)$

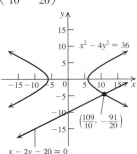

52. Points of intersection: $\left(\dfrac{1}{2}\sqrt{10}, 5\right), \left(-\dfrac{1}{2}\sqrt{10}, 5\right)$

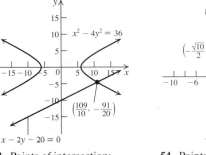

53. Points of intersection: $(2, 1), (2, -1), (-2, 1), (-2, -1),$

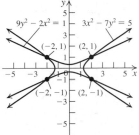

54. Points of intersection: $(2, \sqrt{3}), (2, -\sqrt{3}), (-2, \sqrt{3}), (-2, -\sqrt{3})$

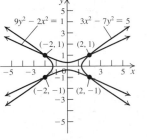

55. $\dfrac{x'^2}{16} - \dfrac{y'^2}{16} = 1$; hyperbola

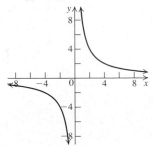

56. $\dfrac{x'^2}{16/3} + \dfrac{y'^2}{16} = 1$; ellipse

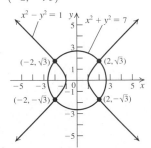

57. $y' = \sqrt{2}x'^2 - 2\sqrt{2}$; parabola   **58.** $\dfrac{y'^2}{4} - \dfrac{x'^2}{9/4} = 1$; hyperbola

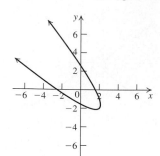

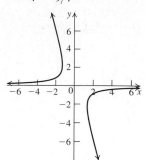

66.

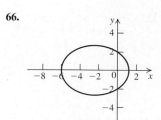

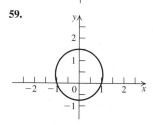

59.

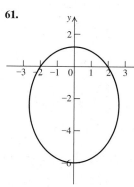

60.

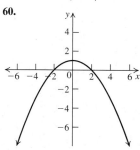

67. $y = 1 - x, 0 \le x \le 1$ **68.** $x^2 + y^2 = 9$

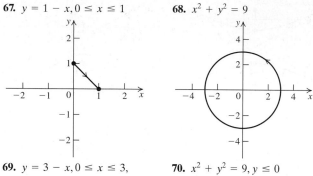

69. $y = 3 - x, 0 \le x \le 3$, traversed twice in each direction   **70.** $x^2 + y^2 = 9, y \le 0$

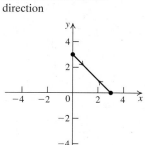

61.

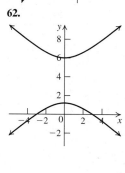

62.

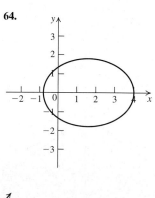

71. $y = -2x^2 + 1, -1 \le x \le 1$, traversed once in each direction   **72.** $\dfrac{y^2}{9} - \dfrac{x^2}{9} = 1, x \ge 3$

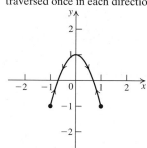

63.

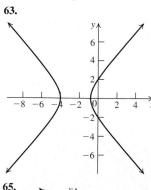

64.

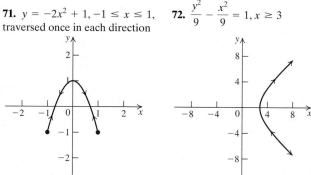

73. $y = x^{2/3}, -\infty < x < \infty$ **74.** $x = y^2 - 2y$

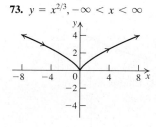

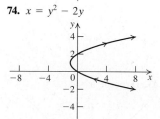

65.

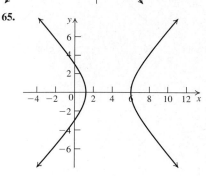

Practice Test A

8.

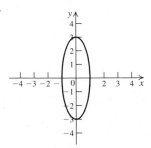

12. Hyperbola

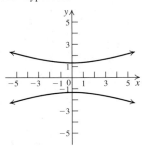

13. Parabola

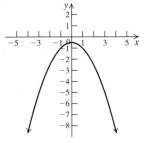

14. Circle

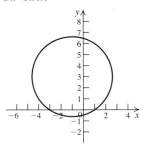

15. Ellipse

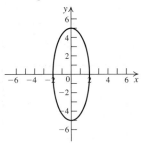

16. Hyperbola

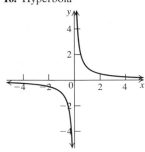

17. Ellipse

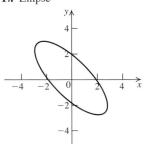

18. Ellipse

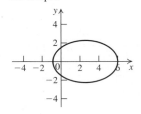

19. Hyperbola

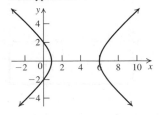

20. Parabola

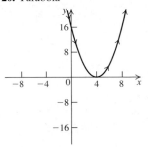

Cumulative Review Exercises (Chapters 1–9)

2.

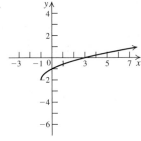

3. $f^{-1}(x) = \dfrac{1}{2}x + \dfrac{3}{2}$

$$f(f^{-1}(x)) = f\left(\dfrac{1}{2}x + \dfrac{3}{2}\right)$$

$$= 2\left(\dfrac{1}{2}x + \dfrac{3}{2}\right) - 3$$

$$= x + 3 - 3 = x$$

4.

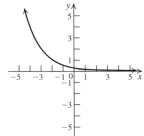

18. Hyperbola

19. Circle

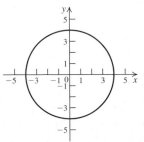

20.

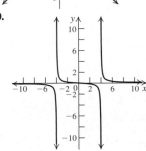

CHAPTER 10

Section 10.1

Practice Problems: 1. $a_1 = -2, a_2 = -4, a_3 = -8, a_4 = -16$

2. $a_1 = -1, a_2 = \frac{1}{4}, a_3 = -\frac{1}{9}, a_4 = \frac{1}{16}, a_5 = -\frac{1}{25}, a_6 = \frac{1}{36}$

3. $a_n = (-1)^{n+1}\left(1 - \frac{1}{n}\right)$

4. $a_1 = -3, a_2 = -1, a_3 = 3, a_4 = 11, a_5 = 27$
5. $a_0 = 1, a_1 = 2, a_2 = 3$, and $a_k = a_{k-2} + a_{k-1}$, for all $k \geq 3$
6. a. 13 **b.** $n(n-1)(n-2)$

7. $a_1 = -2, a_2 = 2, a_3 = -\frac{4}{3}, a_4 = \frac{2}{3}, a_5 = -\frac{4}{15}$

8. -4 **9.** $\displaystyle\sum_{k=1}^{7} (-1)^{k+1}(2k)$

A Exercises: Basic Skills and Concepts:

9. $a_1 = 0, a_2 = \frac{1}{2}, a_3 = \frac{2}{3}, a_4 = \frac{3}{4}$

10. $a_1 = 2, a_2 = \frac{3}{2}, a_3 = \frac{4}{3}, a_4 = \frac{5}{4}$

11. $a_1 = -1, a_2 = -4, a_3 = -9, a_4 = -16$
12. $a_1 = 1, a_2 = 8, a_3 = 27, a_4 = 64$

13. $a_1 = 1, a_2 = \frac{4}{3}, a_3 = \frac{3}{2}, a_4 = \frac{8}{5}$

14. $a_1 = \frac{3}{2}, a_2 = \frac{6}{5}, a_3 = \frac{9}{10}, a_4 = \frac{12}{17}$

15. $a_1 = 1, a_2 = -1, a_3 = 1, a_4 = -1$
16. $a_1 = 1, a_2 = -3, a_3 = 9, a_4 = -27$

17. $a_1 = \frac{5}{2}, a_2 = \frac{11}{4}, a_3 = \frac{23}{8}, a_4 = \frac{47}{16}$

18. $a_1 = \frac{3}{2}, a_2 = \frac{9}{4}, a_3 = \frac{27}{8}, a_4 = \frac{81}{16}$

19. $a_1 = 0.6, a_2 = 0.6, a_3 = 0.6, a_4 = 0.6$
20. $a_1 = -0.4, a_2 = -0.4, a_3 = -0.4, a_4 = -0.4$

21. $a_1 = -1, a_2 = \frac{1}{2}, a_3 = -\frac{1}{6}, a_4 = \frac{1}{24}$

22. $a_1 = 1, a_2 = 1, a_3 = \frac{1}{2}, a_4 = \frac{1}{6}$

23. $a_1 = -\frac{1}{3}, a_2 = \frac{1}{9}, a_3 = -\frac{1}{27}, a_4 = \frac{1}{81}$

24. $a_1 = -1, a_2 = 3, a_3 = -9, a_4 = 27$

25. $a_1 = \frac{e}{2}, a_2 = \frac{e^2}{4}, a_3 = \frac{e^3}{6}, a_4 = \frac{e^4}{8}$

26. $a_1 = \frac{2}{e}, a_2 = \frac{4}{e^2}, a_3 = \frac{8}{e^3}, a_4 = \frac{16}{e^4}$

41. $a_1 = 2, a_2 = 5, a_3 = 8, a_4 = 11, a_5 = 14$
42. $a_1 = 5, a_2 = 4, a_3 = 3, a_4 = 2, a_5 = 1$
43. $a_1 = 3, a_2 = 6, a_3 = 12, a_4 = 24, a_5 = 48$

44. $a_1 = 1, a_2 = \frac{1}{2}, a_3 = \frac{1}{4}, a_4 = \frac{1}{8}, a_5 = \frac{1}{16}$

45. $a_1 = 7, a_2 = -11, a_3 = 25, a_4 = -47, a_5 = 97$
46. $a_1 = -4, a_2 = 7, a_3 = -26, a_4 = 73, a_5 = -224$

47. $a_1 = 2, a_2 = \frac{1}{2}, a_3 = 2, a_4 = \frac{1}{2}, a_5 = 2$

48. $a_1 = -1, a_2 = 1, a_3 = -1, a_4 = 1, a_5 = -1$

49. $a_1 = 25, a_2 = -\frac{1}{125}, a_3 = -25, a_4 = \frac{1}{125}, a_5 = 25$

50. $a_1 = 12, a_2 = -\frac{1}{36}, a_3 = -12, a_4 = \frac{1}{36}, a_5 = 12$

51. a. 2, 11, 26, 47, 74, 107, 146, 191, 242, 299

b.
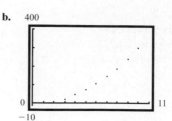

52. a. 1, 2.5, 3, 3.25, 3.4, 3.5, 3.571, 3.625, 3.667, 3.7 (rounded to three decimal places)

b.
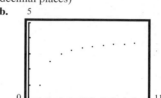

53. a. 0, 1, 2, 3, 4, 5, 6, 7, 8, 9

b.
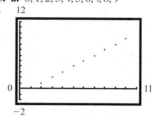

54. a. 0, 4, 18, 48, 100, 180, 294, 448, 648, 900

b.
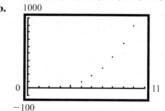

55. a. 0.5, −1, 2, 0.5, −1, 2, 0.5, −1, 2, 0.5

b.
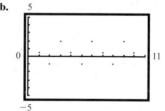

56. a. 1, 1, 1, 1, 1, 1, 1, 1, 1, 1

b.
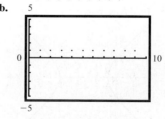

B Exercises: Applying the Concepts:

97. a. $A_1 = 10{,}300, A_2 = 10{,}609, A_3 = 10{,}927.27, A_4 = 11{,}255.09,$ $A_5 = 11{,}592.74, A_6 = 11{,}940.52$

98. a. $A_1 = 102, A_2 = 104.04, A_3 = 106.12, A_4 = 108.24,$ $A_5 = 110.41, A_6 = 112.62$ **99.** 1st year: \$105,000; 2nd year: \$110,250; 3rd year: \$115,762.50; 4th year: \$121,550.63; 5th year: \$127,628.16; 6th year: \$134,009.56; 7th year: \$140,710.04; formula:

$a_n = (100,000)(1.05^n)$ **100.** 1st year: \$45,000; 2nd year: \$40,500; 3rd year: \$36,450; 4th year: \$32,805; 5th year: \$29,524.50; 6th year: \$26,572.05; 7th year: \$23,914.85; 8th year: \$21,523.36; 9th year: \$19,371.02; 10th year: \$17,433.92; formula: $a_n = (50,000)(0.9^n)$

Section 10.2

Practice Problems: **1.** -5 **2.** $a_n = 4n - 7$ **3.** $d = -3$, $a_n = 53 - 3n$ **4.** $\dfrac{85}{6}$ **5.** 2000 ft

A Exercises: Basic Skills and Concepts:
7. Arithmetic; $a_1 = 1, d = 1$ **8.** Arithmetic; $a_1 = 1, d = 2$
9. Arithmetic; $a_1 = 2, d = 3$ **10.** Arithmetic; $a_1 = 10, d = -3$
11. Not Arithmetic   **12.** Not Arithmetic   **13.** Not Arithmetic
14. Arithmetic; $a_1 = -\dfrac{1}{4}, d = \dfrac{1}{2}$ **17.** Arithmetic; $a_1 = 8, d = 2$
18. Arithmetic; $a_1 = -4, d = -5$ **19.** Not arithmetic
20. Not arithmetic   **25.** $a_n = \dfrac{3 - n}{4}$

Section 10.3

Practice Problems: **1.** 3   **2.** It is geometric; $a_1 = \dfrac{3}{2}, r = \dfrac{3}{2}$.

3. a. $a_1 = 2$ **b.** $r = \dfrac{3}{5}$ **c.** $a_n = 2\left(\dfrac{3}{5}\right)^{n-1}$ **4.** 6896.8288

5. 2   **6.** \$91,510.60   **7.** 9

8. $\displaystyle\sum_{n=0}^{\infty} (10,000,000)(0.85)^n = 66,666,666.67$

A Exercises: Basic Skills and Concepts: **7.** Geometric; $a_1 = 3, r = 2$
8. Geometric; $a_1 = 2, r = 2$ **11.** Geometric; $a_1 = 1, r = -3$
12. Geometric; $a_1 = -1, r = -2$ **13.** Geometric; $a_1 = 7, r = -1$
14. Geometric; $a_1 = 1, r = -2$ **15.** Geometric; $a_1 = 9, r = \dfrac{1}{3}$
16. Geometric; $a_1 = 5, r = \dfrac{2}{5}$ **17.** Geometric; $a_1 = -\dfrac{1}{2}, r = -\dfrac{1}{2}$
18. Geometric; $a_1 = \dfrac{10}{3}, r = \dfrac{2}{3}$ **19.** Geometric; $a_1 = 1, r = 2$
20. Geometric; $a_1 = -1, r = 1.06$ **23.** Geometric; $a_1 = \dfrac{1}{3}, r = \dfrac{1}{3}$
24. Geometric; $a_1 = 500, r = 10$ **25.** Geometric; $a_1 = \sqrt{5}, r = \sqrt{5}$
26. Not geometric   **29.** $a_1 = 5, r = \dfrac{2}{3}, a_n = (5)\left(\dfrac{2}{3}\right)^{n-1}$
30. $a_1 = 1, r = \sqrt{3}, a_n = (3)^{\frac{n-1}{2}}$
34. $a_1 = e^2, r = \dfrac{1}{e^2}, a_n = \dfrac{1}{e^{2n-4}}$

Section 10.4

Practice Problems: **1.** $P_{k+1}: (k + 4)^2 > (k + 1)^2 + 9$

Section 10.5

Practice Problems:
1. $(3y - x)^6 = 729y^6 - 1458y^5x + 1215y^4x^2 - 540y^3x^3 + 135y^2x^4 - 18yx^5 + x^6$

2. a. $\dbinom{6}{2} = 15$ **b.** $\dbinom{12}{9} = 220$

3. $(3x - y)^4 = 81x^4 - 108x^3y + 54x^2y^2 - 12xy^3 + y^4$
4. 220   **5.** $1,863,680a^{12}x^3$

A Exercises: Basic Skills and Concepts:
17. $(x + 2)^4 = x^4 + 8x^3 + 24x^2 + 32x + 16$
18. $(x + 3)^4 = x^4 + 12x^3 + 54x^2 + 108x + 81$
19. $(x - 2)^5 = x^5 - 10x^4 + 40x^3 - 80x^2 + 80x - 32$
20. $(3 - x)^5 = -x^5 + 15x^4 - 90x^3 + 270x^2 - 405x + 243$
21. $(2 - 3x)^3 = -27x^3 + 54x^2 - 36x + 8$
22. $(3 - 2x)^3 = -8x^3 + 36x^2 - 54x + 27$
23. $(2x + 3y)^4 = 16x^4 + 96x^3y + 216x^2y^2 + 216xy^3 + 81y^4$
24. $(2x + 5y)^4 = 16x^4 + 160x^3y + 600x^2y^2 + 1000xy^3 + 625y^4$
25. $(x + 1)^4 = x^4 + 4x^3 + 6x^2 + 4x + 1$
26. $(x + 2)^4 = x^4 + 8x^3 + 24x^2 + 32x + 16$
27. $(x - 1)^5 = x^5 - 5x^4 + 10x^3 - 10x^2 + 5x - 1$
28. $(1 - x)^5 = -x^5 + 5x^4 - 10x^3 + 10x^2 - 5x + 1$
29. $(y - 3)^3 = y^3 - 9y^2 + 27y - 27$
30. $(2 - y)^5 = -y^5 + 10y^4 - 40y^3 + 80y^2 - 80y + 32$
31. $(x + y)^6 = x^6 + 6x^5y + 15x^4y^2 + 20x^3y^3 + 15x^2y^4 + 6xy^5 + y^6$
32. $(x - y)^6 = x^6 - 6x^5y + 15x^4y^2 - 20x^3y^3 + 15x^2y^4 - 6xy^5 + y^6$
33. $(1 + 3y)^5 = 243y^5 + 405y^4 + 270y^3 + 90y^2 + 15y + 1$
34. $(2x + 1)^5 = 32x^5 + 80x^4 + 80x^3 + 40x^2 + 10x + 1$
35. $(2x + 1)^4 = 16x^4 + 32x^3 + 24x^2 + 8x + 1$
36. $(3x - 1)^4 = 81x^4 - 108x^3 + 54x^2 - 12x + 1$
37. $(x - 2y)^3 = x^3 - 6x^2y + 12xy^2 - 8y^3$
38. $(2x - y)^3 = 8x^3 - 12x^2y + 6xy^2 - y^3$
39. $(2x + y)^4 = 16x^4 + 32x^3y + 24x^2y^2 + 8xy^3 + y^4$
40. $(3x - 2y)^4 = 81x^4 - 216x^3y + 216x^2y^2 - 96xy^3 + 16y^4$
41. $\left(\dfrac{x}{2} + 2\right)^7 = \dfrac{1}{128}x^7 + \dfrac{7}{32}x^6 + \dfrac{21}{8}x^5 +$ $\dfrac{35}{2}x^4 + 70x^3 + 168x^2 + 224x + 128$
42. $\left(2 - \dfrac{x}{2}\right)^7 = -\dfrac{1}{128}x^7 + \dfrac{7}{32}x^6 - \dfrac{21}{8}x^5 +$ $\dfrac{35}{2}x^4 - 70x^3 + 168x^2 - 224x + 128$
43. $\left(a^2 - \dfrac{1}{3}\right)^4 = a^8 - \dfrac{4}{3}a^6 + \dfrac{2}{3}a^4 - \dfrac{4}{27}a^2 + \dfrac{1}{81}$
44. $\left(\dfrac{1}{2} - a^2\right)^4 = a^8 - 2a^6 + \dfrac{3}{2}a^4 - \dfrac{1}{2}a^2 + \dfrac{1}{16}$
45. $\left(\dfrac{1}{x} + y\right)^3 = y^3 + 3\dfrac{y^2}{x} + 3\dfrac{y}{x^2} + \dfrac{1}{x^3}$
46. $\left(x + \dfrac{2}{y}\right)^3 = x^3 + 6\dfrac{x^2}{y} + 12\dfrac{x}{y^2} + \dfrac{8}{y^3}$

Section 10.6

Practice Problems:
1.

There are $(3)(3) = 9$ ways to choose a course.
2. 35   **3.** 1,000,000   **4.** 5040   **5.** 3024   **6. a.** 72   **b.** 1
7. {bears, bulls, lions}, {bears, bulls, tigers}, {bears, lions, tigers}, {bulls, lions, tigers}; $C(4, 3) = 4$ **8. a.** 36   **b.** 1   **c.** 1
9. 220   **10.** 90

Section 10.7

Practice Problems: 1. a. $E_1 = \{2, 4, 6\}$, $P(E_1) = \dfrac{1}{2}$

b. $E_2 = \{5, 6\}$, $P(E_2) = \dfrac{1}{3}$ **2.** $\dfrac{1}{2}$ **3.** $\dfrac{1}{6}$ **4.** $\dfrac{1}{15{,}890{,}700}$

5. $\dfrac{2}{13}$ **6.** $\dfrac{3}{10}$ **7.** 0.69

A Exercises: Basic Skills and Concepts:

7. $S = \{$(Wendy's, McDonald's, Burger King), (Wendy's, Burger King, McDonald's), (McDonald's, Wendy's, Burger King), (McDonald's, Burger King, Wendy's), (Burger King, Wendy's, McDonald's), (Burger King, McDonald's, Wendy's)$\}$
8. $S = \{$(hamburger, Wendy's), (hamburger, McDonald's), (hamburger, Burger King), (cheeseburger, Wendy's), (cheeseburger, McDonald's), (cheeseburger, Burger King)$\}$
9. $S = \{\{Thriller, The\ Wall\}, \{Thriller, Eagles: Their\ Greatest\ Hits\}, \{Thriller, Led\ Zeppelin\ IV\}, \{The\ Wall, Eagles:\ Their\ Greatest\ Hits\}, \{The\ Wall, Led\ Zeppelin\ IV\}, \{Eagles: Their\ Greatest\ Hits, Led\ Zeppelin\ IV\}\}$
10. $S = \{\{$regular salted, regular unsalted, barbecue$\}$, $\{$regular salted, regular unsalted, cheddar cheese$\}$, $\{$regular salted, regular unsalted, sour cream and onion$\}$, $\{$regular salted, barbecue, cheddar cheese$\}$, $\{$regular salted, barbecue, sour cream and onion$\}$, $\{$regular salted, cheddar cheese, sour cream and onion$\}$, $\{$regular unsalted, barbecue, cheddar cheese$\}$, $\{$regular unsalted, barbecue, sour cream and onion$\}$, $\{$regular unsalted, cheddar cheese, sour cream, and onion$\}$, $\{$barbecue, cheddar cheese, sour cream and onion$\}\}$
11. $S = \{$(white, male), (white, female), (African American, male), (African American, female), (Native American, male), (Native American, female), (Asian, male), (Asian, female), (other, male), (other, female), (multiracial, male), (multiracial, female)$\}$
12. $S = \{$(white, 1), (white, 2), (white, 3), (white, 4), (white, 5), (white, 6), (red, 1), (red, 2), (red, 3), (red, 4), (red, 5), (red, 6), (green, 1), (green, 2), (green, 3), (green, 4), (green, 5), (green, 6)$\}$
27. An event that is very likely to happen has probability 0.999. An event that will surely happen has probability 1. An event that is a rare event has probability 0.001. An event that may or may not happen has probability 0.5. An event that will never happen has probability 0.
28. {the sure event}, {at least one head}, {at least two heads}, {exactly two heads}
63.

Number of people who are	Depressed according to the test	Normal according to the test
actually depressed	90	10
actually normal	135	765

64. Sample points: $\{M_1M_2N, M_1NM_2, M_2M_1N, M_2NM_1, NM_1M_2, NM_2M_1\}$. The probability that Deshawn wins is $\dfrac{2}{3}$.

Review Exercises

1. $a_1 = -1, a_2 = 1, a_3 = 3, a_4 = 5, a_5 = 7$
2. $a_1 = -\dfrac{1}{2}, a_2 = 0, a_3 = \dfrac{3}{2}, a_4 = 4, a_5 = \dfrac{15}{2}$
3. $a_1 = \dfrac{1}{3}, a_2 = \dfrac{2}{5}, a_3 = \dfrac{3}{7}, a_4 = \dfrac{4}{9}, a_5 = \dfrac{5}{11}$
4. $a_1 = 1, a_2 = -2, a_3 = 4, a_4 = -8, a_5 = 16$

Practice Test A

1. $a_1 = 3, a_2 = -9, a_3 = -21, a_4 = -33, a_5 = -45$; arithmetic sequence **2.** $a_1 = -6, a_2 = -12, a_3 = -24, a_4 = -48, a_5 = -96$; geometric sequence **3.** $a_1 = -2, a_2 = -1, a_3 = 2, a_4 = 11, a_5 = 38$

Cumulative Review Exercises (Chapters 1–10)

12.

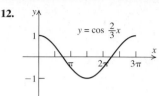

15. Shift the graph of $g(x) = \sqrt{x}$ three units to the right.

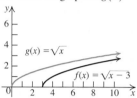

APPENDIX A

Section A.1

Practice Problems: 1. $A \cap B = \{0\}$
$A \cup B = \{-4, -3, -2, -1, 0, 1, 2, 3, 4\}$
2. $A' = \{1, 2, 4, 5, 7, 8\}$ **3. a.** 10 **b.** 1 **c.** 1 **4.** 9
5. a. $\dfrac{10}{3}$ **b.** $\dfrac{22}{3}$ **6.** $(-\infty, -2) \cup (-2, \infty)$ **7. a.** $\dfrac{1}{2}$ **b.** 1 **c.** $\dfrac{4}{9}$
8. a. $\dfrac{1}{4x^8}$ **b.** $-\dfrac{1}{xy^3}$

A Exercises: Basic Skills and Concepts:

39. $|3 - 8| = 5$ **40.** $|2 - 14| = 12$ **41.** $|-6 - 9| = 15$
42. $|-12 - 3| = 15$ **43.** $|-20 - (-6)| = 14$
44. $|-14 - (-1)| = 13$ **45.** $\left|\dfrac{22}{7} - \left(-\dfrac{4}{7}\right)\right| = \dfrac{26}{7}$
46. $\left|\dfrac{16}{5} - \left(-\dfrac{3}{5}\right)\right| = \dfrac{19}{5}$
47. $1 \le x \le 4$

48. $-2 \le x \le 2$

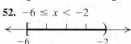

49. $14 < x < 28$

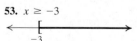

50. $\dfrac{1}{2} < x < \dfrac{9}{2}$

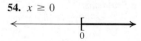

51. $-3 < x \le 1$

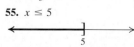

52. $-6 \le x < -2$

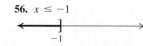

53. $x \ge -3$

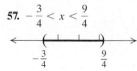

54. $x \ge 0$

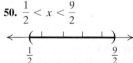

55. $x \le 5$

56. $x \le -1$

57. $-\dfrac{3}{4} < x < \dfrac{9}{4}$

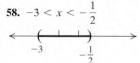

58. $-3 < x < -\dfrac{1}{2}$

73. $(-\infty, -1) \cup (-1, 0) \cup (0, \infty)$ **74.** $(-\infty, 0) \cup (0, 5) \cup (5, \infty)$
75. $(-\infty, -2) \cup (-2, 1) \cup (1, \infty)$ **76.** $(-\infty, -3) \cup (-3, 0) \cup (0, \infty)$
77. $(-\infty, 0) \cup (0, 2) \cup (2, \infty)$ **78.** $(-\infty, 3) \cup (3, \infty)$

B Exercises: Applying the Concepts: **151. a.** people who own
either MP3 or DVD players   **b.** people who own both MP3 and
DVD players   **c.** people who do not own a MP3 player but do own a
DVD player
152. c. {Lexus SC 430, Lincoln Town Car, Infiniti G37}
  e. {Lexus SC 430, Lincoln Town Car}
  f. {Lexus SC 430, Lincoln Town Car}
  g. {Lexus SC 430, Lincoln Town Car, Infiniti G37}
153. $119.5 \le x \le 134.5$ **154.** $30 \le x \le 107$

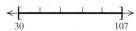

Section A.2

Practice Problems: **1. a.** $4x^2 + 27x - 7$ **b.** $6x^2 - 19x + 10$
2. The quotient is $x^2 + x + 3$, and the remainder is $2x - 3$.
3. The quotient is $2x^2 + 7x + 3$, and the remainder is 2.
4. a. $(x + 2)^2$ **b.** $(2x + 5)(2x - 5)$ **c.** $(x^2 + 9)(x + 3)(x - 3)$
d. $(x - 5)(x^2 + 5x + 25)$ **5. a.** $(5x + 1)(x + 2)$
b. $(x + 3)(x^2 + 1)$

A Exercises: Basic Skills and Concepts: **25.** $x^3 + 3x^2 + 4x + 2$
26. $2x^3 - 13x^2 + 16x - 5$ **27.** $3x^3 - 5x^2 + 5x - 2$
28. $2x^3 - 5x^2 + 5x + 4$ **29.** $x^2 + 3x + 2$
30. $x^2 + 5x + 6$ **31.** $9x^2 + 9x + 2$ **32.** $2x^2 + 11x + 15$
33. $-4x^2 - 7x + 15$ **34.** $-2x^2 + 11x - 5$ **35.** $6x^2 - 7x + 2$
36. $5x^2 - 8x + 3$ **37.** $4x^2 + 4ax - 15a^2$ **38.** $5x^2 + 23ax - 10a^2$
41. $9x^2 + 27x + 27$ **45.** $27x^3 + 27x^2 + 9x + 1$
46. $8x^3 + 36x^2 + 54x + 27$ **47.** $-4x^2 + 25$
57. $3x^2 + 11xy + 10y^2$ **58.** $14x^2 + 11xy + 2y^2$
59. $6x^2 + 11xy - 7y^2$ **60.** $2x^2 - xy - 15y^2$ **61.** $x^4 - 2x^2y^2 + y^4$
62. $16x^4 - 8x^2y^2 + y^4$ **63.** $x^3 - 3x^2y + 4y^3$ **64.** $x^3 + 3x^2y - 4y^3$
65. $x^4 - 4x^3y + 16xy^3 - 16y^4$ **66.** $16x^4 + 16x^3y - 4xy^3 - y^4$
76. Quotient: $3x^5 - \dfrac{9}{2}x^4 + \dfrac{27}{4}x^3 - \dfrac{81}{8}x^2 + \dfrac{267}{16}x - \dfrac{801}{32}$;

remainder: $\dfrac{2723}{32}$ **82.** Quotient: $2x^2 - 5x + 1$ remainder: $-32x$

Section A.3

Practice Problems: **1. a.** $\dfrac{2x}{3}$ **b.** $\dfrac{x - 2}{x + 2}$ **2.** $\dfrac{x(x - 3)}{14}$
3. a. $\dfrac{7}{x - 6}$ **b.** $\dfrac{1}{x + 4}$ **4. a.** $\dfrac{x(5x - 4)}{(x + 2)(x - 5)}$
b. $\dfrac{x(3x + 23)}{(x + 3)(x - 4)}$ **5. a.** $3x^2(x + 2)^2(x - 2)^2$
b. $(x - 5)(x + 5)(x - 1)$ **6. a.** $\dfrac{x^2 + 2x + 8}{(x + 2)(x - 2)^2}$ **b.** $\dfrac{45 - 7x}{3(x - 5)^2}$
7. $\dfrac{1}{x - 5}$ **8.** $\dfrac{25x}{15x - 12}$

A Exercises: Basic Skills and Concepts:
3. $(x^2 + 2x)(x^2 - x - 2)$ or $(x)(x + 2)(x - 2)(x + 1)$
17. $\dfrac{x - 10}{x + 7}, x \ne -7, x \ne 1$ **18.** $\dfrac{x + 5}{x - 4}, x \ne 3, x \ne 4$
19. $\dfrac{x + 2}{x - 2}, x \ne -\dfrac{1}{3}, x \ne 0, x \ne 2$
20. $\dfrac{1}{x - 4}, x \ne -\dfrac{1}{3}, x \ne 0, x \ne 4$
23. $\dfrac{x + 3}{2(x - 3)}$ **24.** $\dfrac{(5x + 3)(x + 2)}{4}$ **25.** $\dfrac{x - 1}{x}$
26. 5   **27.** $\dfrac{x - 1}{x + 3}$ **28.** $\dfrac{(x - 2)(x + 4)}{(x + 5)(x + 1)}$ **33.** $\dfrac{5x(x - 3)}{2}$

48. $\dfrac{2x}{(x - 1)^2}$ **49.** $\dfrac{4x}{x - 3}$ **50.** $\dfrac{3x}{(x + 5)^2}$ **53.** $\dfrac{2(1 - 3x)}{(2x - 1)(2x + 1)}$
54. $\dfrac{1 - 8x}{(4x - 1)(4x + 1)}$ **58.** $21(1 + 3x)$ **59.** $(2x - 1)(2x + 1)^2$
60. $(3x + 1)(3x - 1)^2$ **61.** $(x - 1)(x + 1)(x + 2)$
62. $(x - 3)(x + 3)(x + 2)$ **65.** $\dfrac{7x + 15}{(x - 3)(x + 3)}$
66. $\dfrac{x(3x + 4)}{(x - 1)(x + 1)}$ **67.** $\dfrac{x(4 - x)}{(x - 2)(x + 2)}$ **68.** $-\dfrac{2x^2 + 6x + 5}{(x - 4)(x + 4)}$
71. $\dfrac{2(9x^2 - 5x + 1)}{(3x + 1)(3x - 1)^2}$ **72.** $\dfrac{2(4x^2 + 3x + 1)}{(2x - 1)(2x + 1)^2}$
77. $\dfrac{16a^2}{(x - 5a)(x - 3a)}$ **78.** $\dfrac{5x^2}{(2x - a)(3x + a)}$ **80.** $-\dfrac{h(2x + h)}{x^2(x + h)^2}$

B Exercises: Applying the Concepts:
97. a. $\dfrac{3}{100 + x}$ **98. a.** $\dfrac{1500}{200,000 + x}$ **99. a.** $\dfrac{10\pi x^3 + 240}{x}$

Section A.4

Practice Problems: **1. a.** $\dfrac{7\sqrt{2}}{4}$ **b.** $-\dfrac{1 + \sqrt{7}}{2}$ **c.** $\dfrac{x\sqrt{x} + 3x}{x - 9}$
2. a. -2 **b.** 2   **c.** 3   **d.** Not a real number   **3. a.** $2\sqrt[3]{9}$
b. $2\sqrt[4]{3a^2}$ **4. a.** $13\sqrt{3}$ **b.** $4\sqrt[3]{5x}$ **5. a.** $12x^{7/10}$ **b.** $\dfrac{5}{x^{7/12}}$
c. $\dfrac{1}{x^{2/15}}$ **6.** $\dfrac{(2x + 3)(x + 3)^{1/2}}{x + 3}$ **7.** $\dfrac{1}{3}x^{1/3}(7x + 20)$

A Exercises: Basic Skills and Concepts:
47. $\dfrac{2x + h - 2\sqrt{x(x + h)}}{h}$

Section A.5

Practice Problems: **1.** 10   **2.** 10 ft   **3.** No

Section A.6

Practice Problems: **1.** $\{-1\}$ **2.** \$1500   **3.** 395 m   **4.** 2.5 hr
5. 18 min   **6.** 12.5 gal

Section A.7

Practice Problems: **1. a.** $4 - 2i$ **b.** $-1 + 4i$ **2. a.** $26 + 2i$
b. $-15 - 21i$ **3.** 37   **4. a.** $1 + i$ **b.** $-\dfrac{15}{41} - \dfrac{12}{41}i$

A Exercises: Basic Skills and Concepts:
33. $\bar{z} = 2 + 3i; z\bar{z} = 13$ **34.** $\bar{z} = 4 - 5i; z\bar{z} = 41$
35. $\bar{z} = \dfrac{1}{2} + 2i; z\bar{z} = \dfrac{17}{4}$ **36.** $\bar{z} = \dfrac{2}{3} - \dfrac{1}{2}i; z\bar{z} = \dfrac{25}{36}$
37. $\bar{z} = \sqrt{2} + 3i; z\bar{z} = 11$ **38.** $\bar{z} = \sqrt{5} - \sqrt{3}i; z\bar{z} = 8$

Section A.8

Practice Problems: **1.** $\{-21, -4\}$ **2.** $\{3\}$
3. $\{-2 - \sqrt{5}, -2 + \sqrt{5}\}$ **4.** $\left\{3 - \dfrac{\sqrt{11}}{2}, 3 + \dfrac{\sqrt{11}}{2}\right\}$
5. $\left\{-\dfrac{1}{2}, \dfrac{2}{3}\right\}$ **6.** $\{2 - 3i, 2 + 3i\}$ **7. a.** one real root
b. two unequal real roots   **c.** two nonreal complex roots

A Exercises: Basic Skills and Concepts:

11. $\left\{-3, \dfrac{1}{2}\right\}$ **12.** $\left\{2, \dfrac{5}{2}\right\}$ **13.** $\left\{-1, -\dfrac{2}{3}\right\}$ **14.** $\left\{-\dfrac{4}{3}, -\dfrac{1}{2}\right\}$

15. $\left\{-2, -\dfrac{2}{5}\right\}$ **16.** $\left\{-1, \dfrac{5}{3}\right\}$ **17.** $\left\{-3, \dfrac{5}{2}\right\}$ **18.** $\left\{-\dfrac{1}{2}, \dfrac{1}{3}\right\}$

19. $\left\{-\dfrac{2}{3}, \dfrac{3}{2}\right\}$ **20.** $\{-30, 40\}$ **21.** $\left\{\dfrac{1}{6}, \dfrac{7}{3}\right\}$ **22.** $\left\{-\dfrac{5}{3}, -\dfrac{3}{2}\right\}$

23. $\left\{-\dfrac{25}{2}, 15\right\}$ **24.** $\left\{-\dfrac{9}{4}, -\dfrac{4}{3}\right\}$ **30.** $\left\{-\dfrac{3}{2}i, \dfrac{3}{2}i\right\}$

33. $\left\{\dfrac{2}{3} - \dfrac{4}{3}i, \dfrac{2}{3} + \dfrac{4}{3}i\right\}$ **34.** $\left\{-\dfrac{3}{2} - \dfrac{5}{2}i, -\dfrac{3}{2} + \dfrac{5}{2}i\right\}$

45. $\{-1 - \sqrt{6}, -1 + \sqrt{6}\}$ **46.** $\{-3 - \sqrt{2}, -3 + \sqrt{2}\}$

47. $\left\{\dfrac{3 - \sqrt{13}}{2}, \dfrac{3 + \sqrt{13}}{2}\right\}$ **48.** $\left\{\dfrac{1 - \sqrt{13}}{2}, \dfrac{1 + \sqrt{13}}{2}\right\}$

49. $\left\{-3, \dfrac{3}{2}\right\}$ **50.** $\left\{\dfrac{5}{6} - \dfrac{\sqrt{13}}{6}, \dfrac{5}{6} + \dfrac{\sqrt{13}}{6}\right\}$ **51.** $\{1 - i, 1 + i\}$

52. $\{3 - i\sqrt{2}, 3 + i\sqrt{2}\}$ **53.** $\{5 - i\sqrt{3}, 5 + i\sqrt{3}\}$

54. $\left\{-\dfrac{1}{2} - i, -\dfrac{1}{2} + i\right\}$ **55.** $\{6 - \sqrt{33}, 6 + \sqrt{33}\}$

56. $\left\{-\dfrac{3 + \sqrt{19}}{2}, -\dfrac{3 - \sqrt{19}}{2}\right\}$ **57.** $\left\{-\dfrac{7 + \sqrt{13}}{6}, -\dfrac{7 - \sqrt{13}}{6}\right\}$

58. $\left\{\dfrac{3 - \sqrt{41}}{4}, \dfrac{3 + \sqrt{41}}{4}\right\}$ **59.** $\{-1 - \sqrt{5}, -1 + \sqrt{5}\}$

60. $\{-2, -1\}$ **61.** $\left\{-\dfrac{1}{2}, \dfrac{5}{3}\right\}$ **62.** $\{2 - i\sqrt{3}, 2 + i\sqrt{3}\}$

76. $D = -20$; complex roots **77.** $D = 49$; real unequal roots
78. $D = 0$; real equal roots **79.** $D = 1$; real unequal roots
80. $D = -11$; complex roots **81.** $D = 252$; real unequal roots
82. $D = 2$; real unequal roots **83.** $-2\sqrt{3}$ or $2\sqrt{3}$

84. $-\dfrac{4\sqrt{2}}{3}$ or $\dfrac{4\sqrt{2}}{3}$ **85.** 0 or 8 **86.** $\dfrac{1}{6}$ **87.** $-\dfrac{1}{2}$ **88.** 1 or 9

90. a. $r + s = -\dfrac{5}{3}, rs = -\dfrac{5}{3}$ **b.** $r + s = \dfrac{7}{3}, rs = -\dfrac{1}{3}$

c. $r + s = \sqrt{3}, rs = -\dfrac{4\sqrt{3}}{3}$ **d.** $r + s = 2 - \sqrt{2}, rs = -5 + 5\sqrt{2}$

Section A.9

Practice Problems: **1.** $\{-1, 2\}$ **2. a.** $\{-2, 0, 2\}$ **b.** $\{-2, 2, 5\}$
3. $\{-10, 1\}$ **4.** $\{10\}$ **5.** $\{13\}$ **6.** $\{-6, 7\}$

A Exercises: Basic Skills and Concepts: **13.** $\left\{-\dfrac{7}{6}, \dfrac{11}{6}\right\}$

41. $\{1 - \sqrt{2}, 1 + \sqrt{2}\}$ **42.** $\left\{-\dfrac{1}{2} - \dfrac{\sqrt{5}}{2}, -\dfrac{1}{2} + \dfrac{\sqrt{5}}{2}\right\}$

43. $\left\{\dfrac{7}{2} - \dfrac{\sqrt{53}}{2}, \dfrac{7}{2} + \dfrac{\sqrt{53}}{2}\right\}$ **44.** $\left\{-\dfrac{5}{6} - \dfrac{\sqrt{13}}{6}, -\dfrac{5}{6} + \dfrac{\sqrt{13}}{6}\right\}$
59. $\{-1, 343\}$ **60.** $\{1, 8\}$

Section A.10

Practice Problems:
1. $(2, \infty)$

2. $\left[-1, \dfrac{3}{2}\right)$

3. $(4, \infty)$

4. $[-3, 1]$

5. $\left(-\infty, -\dfrac{9}{2}\right] \cup \left[\dfrac{3}{2}, \infty\right)$

A Exercises: Basic Skills and Concepts:

15. $(-2, 5)$ **16.** $[-5, 0]$

17. $(0, 4]$ **18.** $[1, 7)$

19. $[-1, \infty)$ **20.** $(2, \infty)$

21. $(-\infty, -2]$ **22.** $(-1, \infty)$

23. $(-\infty, 3)$ **24.** $(-\infty, 5)$

25. $[-3, \infty)$ **26.** $(-\infty, 4)$

27. $(-\infty, 2)$ **28.** $\left[\dfrac{5}{3}, \infty\right)$

29. $(-\infty, -4)$ **30.** $(-\infty, -1)$

31. $(-\infty, -1)$ **32.** $[3, \infty)$

33. $(-\infty, 3]$ **34.** $(-\infty, 4]$

35. $(2, \infty)$ **36.** $(1, \infty)$

37. $(-\infty, -4]$ **38.** $(-\infty, -6]$

39. $\left(-\infty, \dfrac{3}{2}\right]$ **40.** $(-\infty, -20)$

41. $(-\infty, -1]$ **42.** $[2, \infty)$

66. $(-\infty, -5) \cup (5, \infty)$ **67.** $\left(-\dfrac{1}{2}, \dfrac{7}{2}\right)$
69. $(-\infty, 1) \cup (4, \infty)$ **70.** $(-\infty, -4] \cup [6, \infty)$
71. $\left[-\dfrac{23}{3}, 5\right]$ **72.** $\left[-2, \dfrac{32}{7}\right]$
75. $(-6, 10)$ **76.** $(-\infty, 2) \cup (6, \infty)$

77. $[0, 16]$ **78.** $\left(-\infty, -\dfrac{47}{2}\right) \cup \left(\dfrac{17}{2}, \infty\right)$

79. $(-\infty, \infty)$

80. $(-\infty, \infty)$

81. $\left(-\dfrac{8}{9}, \dfrac{22}{9}\right)$

82. $[-1, 5]$

83. $\left[-\dfrac{10}{3}, 10\right]$

84. $\left(-\infty, -\dfrac{2}{3}\right) \cup (2, \infty)$

Section A.11

Practice Problems:

1. $(-\infty, 1 - \sqrt{3}) \cup (1 + \sqrt{3}, \infty)$

2. $(-\infty, -1 - \sqrt{3}) \cup (-1 + \sqrt{3}, \infty)$

3. $[-2, 2]$

4. $\left(-\dfrac{1}{3}, \dfrac{1}{3}\right)$

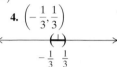

A Exercises: Basic Skills and Concepts:
20. $(-\infty, 1) \cup (7, \infty)$

Section A.12

Group Project:

1. a.

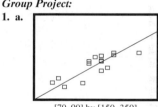

$[70, 90]$ by $[150, 350]$

b. $f(x) = 6.158x - 262.240; r = 0.82675$

2. a.

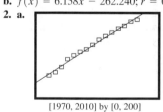

$[1970, 2010]$ by $[0, 200]$

b. $f(x) = 4.633x - 9093.065; r = 0.99555$

3. a.

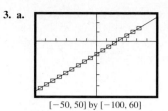

$[-50, 50]$ by $[-100, 60]$

b. $f(x) = 1.339x - 24.069; r = 0.99996$

4. a.

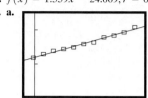

$[-10, 110]$ by $[50, 100]$

b. $f(x) = 0.169x + 72.909; r = 0.99364$

5. a.

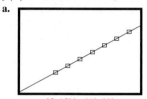

$[5, 15]$ by $[40, 80]$

b. $f(x) = 3.1x + 25.6; r = 1.00000$

6. a.

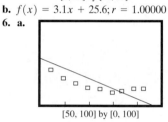

$[50, 100]$ by $[0, 100]$

b. $f(x) = -1.20x + 116.644; r = -0.787$ **d.** 96.4°F; because the scatter appears to level off at high temperatures, this estimate is not reasonable; note that the correlation is low.

7. a.

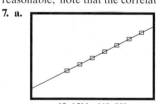

$[5, 15]$ by $[40, 80]$

b. $f(x) = 3.0x + 28.8; r = 1.00000$

Credits

p. 604, Hubble Telescope; European Space Agency (PAL).

p. 605, Appolonius of Perga; Shutterstock.

p. 607, Hubble telescope; European Space Agency (PAL).

p. 607, Edwin Hubble; American Institute of Physics/Emilio Segre Visual Archives (PAL).

p. 616, Parabolic satellite dish; PhotoDisc.

p. 618, Lithotripter; Photo Edit (PAL).

p. 625, Edmund Halley; public domain.

p. 626, Elliptical pool or billiards table; Jill Britton.

p. 629, Alfred Lee Loomis; SAU.

p. 644, Space vehicle launch; NASA.

p. 654, Sputnik; Shutterstock (and Wikipedia public domain).

p. 677, Football; PhotoDisc.

Chapter 10

p. 680, Ancient Babylonia; Bettmann/Corbis.

p. 680, Roulette wheel; PhotoDisc.

p. 680, Biologist; PhotoDisc.

p. 681, Honeybee; Photographer's Choice RF.

p. 684, Leonardo of Pisa; public domain.

p. 692, Space junk, Photo Researchers, Inc. (PAL).

p. 695, Gauss; Library of Congress.

p. 699, ©Ezra Shaw/Getty Images (Sport).

p. 710, Two jars with fleas; Beth Anderson.

p. 711, Falling dominoes; PhotoDisc Blue.

p. 712, Francesco Maurolico; public domain.

p. 715, Towers of Hanoi; Beth Anderson.

p. 716, Blaise Pascal; PAL/Library of Congress.

p. 716, Pascaline, first digital calculator; IBM Corporate Archives (PAL).

p. 723, Social Security card; Silver Burdett Ginn (PAL).

p. 734, Lottery ticket; David Gould/Photographer's Choice.

p. 738, Deck of playing cards; Beth Anderson.

Index of Applications

Index